# History of Shock Waves, Explosions and Impact

A Chronological
and Biographical Reference

*They that know the entire course of the development of science, will, as a matter of course, judge more freely and more correctly of the significance of any present scientific movement than they, who, limited in their views to the age in which their own lives have been spent, contemplate merely the momentary trend that the course of intellectual events takes at the present moment.*

Ernst MACH
German Charles University
Prague 1883

**Ernst (Waldfried Josef Wenzel) Mach**
**(Feb. 18, 1838 – Feb. 19, 1916)**

Austrian philosopher of science, physicist and the father of supersonics, in his study at Vaterstetten, a little town in Upper Bavaria located about 20 km southeast of Munich, where he spent the last three years of his life in the lonely country house of his eldest son Ludwig. He was buried in the cemetery of Haar, a town near Vaterstetten. [Archives of the Ernst-Mach-Institut, Freiburg im Breisgau, Germany]

The quotation above is from his famous book *Die Mechanik in ihrer Entwicklung, historisch-kritisch dargestellt.* Brockhaus, Leipzig (1883). It was translated into English by the American editor Thomas Joseph MCCORMACK, as *The Science of Mechanics: A Critical and Historical Exposition of its Principles.* Open Court Publ. Co., Chicago (1893).

Peter O.K. Krehl

# History of Shock Waves, Explosions and Impact

## A Chronological and Biographical Reference

Springer

Peter O.K. Krehl
Im Lehle 34
79331 Teningen-Nimburg

*Cover figure:* Image of the flow field established by the simultaneous explosion of two small silver azide charges, visualized by a direction-indicating color schlieren technique. A Mach reflection develops along the centerline between the charges. The top part of the blast wave generated by the upper charge reflects from a rough wall, which leads to the 'ripples' seen in this portion of the flow and which delays the onset of Mach reflection at this surface. Image taken by Dr. Harald Kleine in the laboratory of Prof. Kazuyoshi Takayama.

ISBN 978-3-540-20678-1           e-ISBN 978-3-540-30421-0

DOI 10.1007/978-3-540-30421-0

Library of Congress Control Number: 2008934042

© 2009 Springer-Verlag Berlin Heidelberg

This work is subject to copyright. All rights reserved, whether the whole or part of the material is concerned, specifically the rights of translation, reprinting, reuse of illustrations, recitation, broadcasting, reproduction on microfilm or in any other way, and storage in data banks. Duplication of this publication or parts thereof is permitted only under the provisions of the German Copyright Law of September 9, 1965, in its current version, and permission for use must always be obtained from Springer. Violations are liable for prosecution under the German Copyright Law.

The use of general descriptive names, registered names, trademarks, etc. in this publication does not imply, even in the absence of a specific statement, that such names are exempt from the relevant protective laws and regulations and therefore free for general use.

*Typesetting and production:* le-tex publishing services oHG, Leipzig, Germany
*Cover design:* Erich Kirchner, Heidelberg

Printed on acid-free paper

9 8 7 6 5 4 3 2 1

springer.com

# PREFACE

> *The aim of an encyclopaedia is to gather together the knowledge scattered over the face of the Earth, to set forth its general plan to the men with whom we live and to transmit it to the men who will come after us, in order that the labors of past centuries may not have been in vain...*[1]
>
> Denis DIDEROT
> Paris 1755

THIS BOOK is an outgrowth of a previous handbook article[2] on the history of shock waves. While preparing the manuscript for this handbook, it quickly became apparent that this field of enormous breadth could not be adequately treated without also including the evolution of research into percussion and explosion. Since the limited space available in the handbook confined the historical perspectives to a rather limited period of time (1759–1945), the author conceived the idea of providing readers interested in history with a more detailed description of this complex field of physics and engineering, in particular by (i) covering the period beginning from prehistoric to present times; (ii) including biographical sketches of eminent researchers; and (iii) illustrating the milestones using a picture gallery of authentic illustrations. A nine-year study of a large body of original literature brought to light many interesting historical facts which might be unknown in the shock wave physics and detonics community, and which have apparently never been reviewed in a wider context before.

*History*, in general, is understood as being our knowledge and criticisms of the past experiences of mankind. In particular, the *history of science* may be defined as an analytical approach to uncovering the roots of science and providing an overall explanation of its evolution. Traditionally, the history of science faces the problem of judging the significance of past achievements in terms of the experimental methods that are in current use and by the theories that are currently accepted. Since progress in natural science and technology is generally advancing with increasing speed – the rapid progress in computer science and software development is only one striking example of our time – it has become more and more difficult to clearly distinguish between past and present achievements as well as to analyze their impact on other branches of science from the distant view of a historian. Recent achievements may be rated by future generations as important historical milestones, or may be forgotten completely.

The scientific and technical achievements of percussion, explosions and shock wave research made throughout human history are too immense even to be cataloged. Up to now they have been determined by estimating industrial and military applicability, and increasingly also by immediate commercialization. Since these branches of science and technology have grown explosively since the end of World War II, it is obviously difficult to give credit to all recent developments, particularly those from the last 25 years. Nevertheless, it appeared useful to treat the subject historically in an *encyclopedic* approach by illustrating how current developments in various branches of modern science are based upon foundations in classical percussion and shock wave physics, and emphasizing their *phenomenological aspects*.

To provide the reader with quickly accessible and extensive information, the book is divided into six chapters:

**Chapter 1** – INTRODUCTION – determines the position of the subject of this book in the general framework of natural science.

**Chapter 2** – GENERAL SURVEY – provides some general remarks on the historical background of shock and detonation, illustrates the interrelations between the different disciplines, and discloses their genealogy and genesis from the principle of percussion.

**Chapter 3** – CHRONOLOGY – illuminates the historical evolution of percussion, shock wave physics and detonics in terms of milestones. It attempts to specify each contributor's affiliation and the motivation of their research in tabular form. It also contains numerous cross-references to similar studies performed elsewhere, referring the reader to other milestones listed in the chapter and to corresponding illustra-

---

[1] See his article "Encyclopédie" in the famous French *Encyclopédie, ou dictionnaire raisonné des sciences, des arts et des métiers*. S. Faulche, Neufchatel (1751–1772), ed. by Denis DIDEROT and Jean LE ROND D'ALEMBERT.

[2] P. KREHL: *History of shock waves*. In: (G. BEN-DOR ET AL.) *Handbook of shock waves*. Academic Press, New York (2001); vol. 1, pp. 1-142.

tions shown in the next chapter. Many milestones have been provided with comments made from the present point of view and with more recent literature on the subject.

**Chapter 4** – PICTURE GALLERY – illustrates milestone achievements via facsimiles of figures as they were published in the original works.

**Chapter 5** – BIOGRAPHIES INDEX – contains a collection of biographical sketches of eminent researchers in the fields of percussion, explosions, detonations and shock waves. Each biography is supplemented by a list of their most important publications relating to the fields addressed in this book, by secondary literature (such as memoirs, obituaries, Festschriften, *etc.*), and – as far as available – by a portrait picture.

**Chapter 6** – SUPPLEMENTARY REFERENCES – comprises a catalogue of general references that provides the reader with additional biographical and bibliographical sources. It also contains numerous historical review articles and a list of some Internet home pages provided by institutes engaged in shock wave and detonation physics, as well as museums.

The encyclopedic key to the comprehensive body of references, milestones and illustrations given in this book is its detailed NAME INDEX and SUBJECT INDEX.

Hopefully, this encyclopedic approach may provide the reader with a better historical survey than a lengthy narrative description. Since this book contains more than 900 figures and almost 7,000 references, it may also help the modern scientist to illuminate the historical roots of his or her own field of investigation.

Last but not least, the book also tries to pass to the reader the author's joy of learning from early researchers about their motivations and how they proceeded to eventually arrive at significant contributions and ingenious discoveries. For the modern researcher that has quick and easy access to sophisticated instrumentation techniques and vast computer power, it might be amazing to learn about the modest circumstances under which our scientific ancestors found the solution to a problem: often reducing the essentials to a simple model, trying to work out an analytical solution, and checking the theory by rather crude experimental means.

Teningen-Nimburg
March, 2008

*Peter O.K. Krehl*

# Acknowledgements

It took the cooperation of many people to bring a work of this size to reality.

I am especially indebted to the assistance given me by many helpful colleagues around the world, and by the personnel from a large number of archives, libraries and museums: the Académie Royale des Sciences, des Lettres et des Beaux Arts de Belgique at Brussels; the Amalienbibliothek at Weimar; the Archiv zur Geschichte der Max-Planck-Gesellschaft at Berlin-Dahlem; the Bayerische Staatsbibliothek at Munich; the Bibliothèque nationale de France at Paris; the British Library at London; the Bundesarchiv at Koblenz and its Militärarchive at Freiburg and Potsdam; the National Library of Congress at Washington, DC; the Deutsches Museum and the Deutsches Patentamt at Munich; the Ecole des Mines at Paris; the Ecole Polytechnique at Paris; the ETHZ at Zurich; the Historisches Archiv der Deutschen Forschungsanstalt für Luft- und Raumfahrt (DFLR) at Göttingen; the Massachusetts Institute of Technology (MIT) at Cambridge, MA; the Max-Planck-Institut (MPI) für Strömungsforschung at Göttingen; the Minnesota Historical Society at St. Paul, MN; the National Library of Russia at St. Petersburg; the National Library of Scotland at Edinburgh; the Österreichische Nationalbibliothek at Vienna; the Queens University at Belfast; the Raumfahrtmuseum at Peenemünde; the Science Museum Library at London; the Smithsonian National Air & Space Museum at Washington, DC; and the Technisches Museum at Vienna.

The university libraries of the following cities provided valuable assistance with obtaining special literature: Aachen, Adelaide, Berlin, Bratislava, Braunschweig, Budapest, Hamburg, Cambridge, Dresden, Düsseldorf, Freiberg, Freiburg, Göttingen, Halle, Hannover, Harvard, Heidelberg, Karlsruhe, Leipzig, Manchester, Munich, Prague, Regensburg, Riga, Rome, Sheffield, Stockholm, Stuttgart, Tübingen, Turin, Ulm, Vienna and Zurich. Numerous research institutions supplied bibliographical and biographical material such as the David Taylor Model Basin of NSWCCD at Carderick, MD; the Institut für Fluiddynamik (IfA) of the ETH Zürich; the Lick Observatory at Mt. Hamilton, CA; the National Physics Laboratory (NPL) at Teddington, UK; the NASA History Office at Marshall Space Flight Center, Huntsville, AL; the Osteuropa-Institut in Munich; the Poulter Laboratory of Stanford Research International (SRI) at Menlo Park, CA; and the University of Toronto Institute of Aerospace Studies (UTIAS).

In particular, the author wishes to acknowledge the continuous support of librarians and archivists from the University of Freiburg, who also kindly provided access to their old original literature for scanning purposes, thus providing the reader with facsimiles of the best possible quality. Many authors kindly gave their permission for the use of their published figures; individual credits appear in the figure captions.

The author is also grateful to Prof. Eckhart SCHÄFER (Classical Philology, Albert-Ludwigs-Universität, Freiburg) for help with translating Latin literature. Prof. Gary SETTLES (Penn State Gas Dynamics Laboratory, Pennsylvania State University, PA) and Dr. Donald CURRAN (Poulter Laboratory, Stanford International, Menlo Park, CA) kindly read parts of the manuscript, and Prof. emer. William JOHNSON, F.R.S. (Dept. of Mechanical Engineering, Cambridge University, UK), founder editor-in-chief of the *Journal of Impact Engineering*, made a number of worthwhile suggestions relating to early mechanics and ballistics.

I would like to thank Prof. Klaus THOMA, director of the Ernst-Mach-Institute (EMI), for permission to use the services and facilities of the Institute to prepare the draft manuscript. Special thanks are due to some of my former colleagues at EMI who supported me in various ways; particularly to Mrs. Susanne DESCHOUX for her help in translating old English and French literature, to Mr. Bernd GRÜNEWALD and his team at EMI-Medienlabor, and to Dipl.-Ing. Stephan ENGEMANN for computer assistance. All of the drawings shown in the GENERAL SURVEY chapter and some of those in the PICTURE GALLERY chapter were created by Mrs. Elisabeth GÖTZMANN. The cartoons in Fig. 2.10 were kindly provided by my late colleague Dr. Peter NEUWALD.

# TABLE OF CONTENTS

| | | | |
|---|---|---|---|
| **1** | **INTRODUCTION** | | 1 |
| **2** | **GENERAL SURVEY** | | 9 |
| | 2.1 | TERMINOLOGY AND SCOPE | 11 |
| | | 2.1.1 Percussion, Concussion, Impact, and Collision | 12 |
| | | 2.1.2 Explosion and Implosion | 18 |
| | | 2.1.3 Conflagration, Deflagration, Detonation, and Detonics | 22 |
| | | 2.1.4 Hydraulic Jump, Bore, Surge, Tsunami, Seiche, Sea Shock, and Rogue Wave | 25 |
| | | 2.1.5 Shock and Shock Wave | 28 |
| | | 2.1.6 Collisionless Shock Waves | 33 |
| | | 2.1.7 Shock and Vibration | 34 |
| | | 2.1.8 Blast Wave, Blast, and Blasting | 34 |
| | | 2.1.9 Gas Dynamics, Rarefied Gas Dynamics, Magnetogasdynamics, and Cosmical Gas Dynamics | 36 |
| | 2.2 | INITIATION OF PERCUSSION RESEARCH | 38 |
| | | 2.2.1 Natura Non Facit Saltum | 38 |
| | | 2.2.2 Foundation of Dynamics | 41 |
| | | *Classical Percussion Research* • *Center of Percussion* • *Vis Viva Controversy* • *Corpuscular Models* • *Newtonian Demonstrator* | |
| | | 2.2.3 Further Investigations | 46 |
| | | *Hertzian Cone* • *Bulb of Percussion* • *Conchoidal Fracture* • *Percussion Figures* • *Percussion Marks* • *Percussion Force and Contact Time* • *Billiards* • *Ballistic Pendulum* | |
| | | 2.2.4 Applications of Percussion | 49 |
| | | *Pile Driving, Steam Hammer* • *Percussion Drilling* • *Crushing, Fragmentation* • *Fluid Jet Impact* • *Firearms* • *Intense Sound Generation* • *Medical Diagnostics* • *Biomechanics* | |
| | 2.3 | EARLY SPECULATIONS ON SUPERSONIC PHENOMENA | 52 |
| | | 2.3.1 Observations in Nature: Stimulating Riddles | 54 |
| | | *Thunder* • *Hydrometeors* • *Bores* • *Tsunamis* • *Surges* • *Earthquakes and Seaquakes* • *Explosive Volcanic Eruptions* • *Meteorite Impact* • *Cosmic Shock Wave Phenomena* • *Cosmic Explosion Phenomena* • *Cosmogony* | |
| | | 2.3.2 Early Man-Made Shock Generators: Tools and Toys | 69 |
| | | *Clapping of Hands* • *Whip-Cracking* • *Snapping Belts and Snapping Towel* • *Electric Sparks* • *Musical Instruments* | |
| | | 2.3.3 Ballistic Studies: Birth of Supersonics | 71 |

| | | | |
|---|---|---|---|
| 2.4 | EVOLUTION OF SHOCK WAVE PHYSICS | | 72 |
| | 2.4.1 | Nonlinear Acoustics | 72 |
| | 2.4.2 | Main Periods of Evolution | 73 |
| | | *From 1746 to 1808 ▪ From 1808 to 1869 ▪ From 1822 to 1893 ▪ From 1888 to 1930 ▪ From 1930 to 1939 ▪ From 1939 to 1949 ▪ From 1950 to the Present ▪ Documentation and Dissemination* | |
| | 2.4.3 | Aerial Waves of Finite Amplitude: a Challenge for Mathematicians | 76 |
| | | *Water Waves ▪ Approach to Shock Waves ▪ Motivations ▪ Superposition of Shock Waves* | |
| | 2.4.4 | Shock Waves in Gases: First Experimental Proofs of Their Existence | 80 |
| | | *The Roots of Gas Dynamics ▪ First Studies of Intense Air Waves ▪ Mach Reflection ▪ First Laboratory-Scale Supersonic Experiments ▪ Studies of Nozzle Outflow ▪ Wind Tunnels ▪ Shock Tubes* | |
| | 2.4.5 | Shock Waves in a Liquid: the Peculiar Fluid | 86 |
| | | *Shock-Induced Freezing ▪ Liquefaction ▪ Water Hammer, Water Ram, Hydraulic Ram and Hydrodynamic Ram ▪ Underwater Explosions ▪ Water Ricochets ▪ Cavitation ▪ Supercavitation ▪ Photodisruptive Effect ▪ Sonoluminescence ▪ Electrohydraulic Effect* | |
| | 2.4.6 | Solid-State Shock Wave Physics: Initiation by Nuclear Weaponeers | 93 |
| | | *Roots ▪ Establishment and Motivations ▪ Materials Dynamics ▪ Dynamic Fracture ▪ Equations of State ▪ Off-Hugoniot States ▪ Stimulation of Secondary Effects* | |
| 2.5 | PIERCING THE SOUND BARRIER: MYTH AND REALITY | | 100 |
| | 2.5.1 | Unmanned Vehicles: First Demonstrations of Practicability | 101 |
| | 2.5.2 | Manned Vehicles: from Venture to Routine | 103 |
| | 2.5.3 | New Challenges, New Threats | 104 |
| 2.6 | EVOLUTION OF DETONATION PHYSICS | | 105 |
| | 2.6.1 | Black Powder: the Maid of All Work | 106 |
| | 2.6.2 | The Riddle of Detonation: Steps Toward an Understanding | 108 |
| | | *High Explosives ▪ Firing Devices ▪ Firedamp Explosions ▪ Coal Dust Explosions ▪ Detonation Wave ▪ Chapman-Jouguet (CJ) Theory ▪ Zel'dovich-von Neumann-Döring (ZND) Theory ▪ Complex Detonation Processes ▪ Evolution of Chemical Kinetics* | |
| | 2.6.3 | Detonics: the Key to Ultrahigh Shock Pressures, and New Applications | 114 |
| | 2.6.4 | Nuclear and Thermonuclear Explosions: the Ultimate Man-Made Shock Phenomena | 115 |
| | | *Milestones in Nuclear Physics ▪ The First Types of Nuclear Bomb ▪ U.S. Plowshare Program ▪ Soviet Plowshare Program ▪ New Generations of Nuclear Weapons ▪ 'Dirty Bomb' Explosion* | |
| 2.7 | EVOLUTION OF SEISMOLOGY | | 119 |
| | 2.7.1 | Explosion Seismology & Vibroseis | 119 |
| | 2.7.2 | Seismoscopes, Seismographs, and Seismometers | 120 |
| | 2.7.3 | Seismic Prospecting and Research | 121 |
| 2.8 | HIGH-SPEED DIAGNOSTICS | | 122 |
| | 2.8.1 | Precise Time Measurement: the Crucial Condition | 123 |
| | | *Chronoscopes and Chronographs ▪ Electronic Timing Devices ▪ Triggering ▪ Prerigger Framing Photography* | |
| | 2.8.2 | Optical Methods of Visualization: the Key to a Better Understanding | 127 |
| | | *Schlieren Methods ▪ Shadowgraphy ▪ Interferometry ▪ Other Methods* | |
| | 2.8.3 | The Soot Technique: Ingenious 'Black Magic' | 130 |
| | 2.8.4 | High-Speed Photography and Photonics: Freezing the Instant | 130 |
| | | *Single-Shot Photography ▪ High-Speed Cinematography* | |

| | | | |
|---|---|---|---|
| | 2.8.5 | Flash X-Ray Techniques: Visualizing the Hidden................................................................ | 132 |
| | | *Flash Radiography ▪ Flash X-Ray Diffraction Analysis* | |
| | 2.8.6 | The Correct Measurement of Shock Pressure: an Evergreen Problem ........................... | 133 |
| | | *Mechanical Gauges ▪ Piezoelectric Gauges ▪ Piezoresistive Gauges ▪ Examples of Other Methods* | |
| 2.9 | EVOLUTION OF COMPUTATIONAL ANALYSIS ................................................................ | | 137 |
| | 2.9.1 | The Pre-Computer Era: Triumph of Mechanical and Graphical Methods.................... | 138 |
| | | *Digital Mechanical Machines ▪ Analog Mechanical Machines ▪ Graphical Concepts* | |
| | 2.9.2 | Revolution in Calculation: the Automatic Digital Computer ......................................... | 141 |
| | | *Digital Electromechanical Computers ▪ Digital Electronic Computers* | |
| | 2.9.3 | The Tricky Problem: Treating Flow Discontinuities Numerically ................................. | 144 |
| 2.10 | CONCLUDING REMARKS........................................................................................................ | | 146 |

## 3  CHRONOLOGY .............................................................................................................................. 169

*Natural Shock, Explosion & Impact Phenomena: Terrestrial and Extraterrestrial Chronology: ▪ Prehistoric Times ▪ Antiquity ▪ Middle Ages ▪ Modern Times*

## 4  PICTURE GALLERY .................................................................................................................... 845

### 4.1  SHOCK AND PERCUSSION IN NATURE ............................................................................ 847

*Lunar Surface, a Result of Meteorite Impacts ▪ Meteor Crater and Shock Metamorphism ▪ Asteroid Impact: Ries Basin and Steinheim Basin ▪ Two Famous Meteorites: Ensisheim and Murchison ▪ Great Earthquakes: Lisbon (1755) ▪ Great Earthquakes: San Francisco (1906) ▪ Explosive Volcanic Eruption: Krakatau (1883) ▪ Explosive Volcanic Eruption: Mount St. Helens (1980) ▪ Plate Tectonics ▪ Lightning and Thunder ▪ The Riddle of Ball Lightning ▪ Tidal Bores ▪ Hydraulic Jumps ▪ Tsunami Caused by Submarine Volcanic Eruption ▪ Tsunami Caused by Subduction ▪ Examples of Early Tsunami Research ▪ The Tsunami in Indian Mythology ▪ Destructive Tsunami Effects ▪ Rogue or Freak Waves ▪ Sunspots, Solar Flares, and Prominences ▪ Solar Flares and Solar Quakes ▪ Big Bang Portrait ▪ Earth's Bow Shock ▪ Cosmic Jets, Shock Waves, and Mach Cones ▪ Animal World*

### 4.2  PERCUSSION IN THE EVOLUTION OF TECHNOLOGY ................................................ 876

*Basic Tool of Civilization ▪ Early War Machines ▪ Devices Based on Rapid Expansion ▪ Devices Based on Rapid Compression ▪ Periodically Operating Devices ▪ Pile Driver ▪ Forge Steam Ram and Explosion Ram ▪ Explosion Tamper and Wrecking Ball ▪ Percussion Boring ▪ Ricocheting ▪ Big Guns of the 19th Century ▪ Superguns of the 20th Century*

### 4.3  PERCUSSION STUDIES .......................................................................................................... 888

*17th Century: The Pioneering Era of Percussion Research ▪ 18th Century: Percussion Machine for Demonstrating Central Percussion ▪ 18th Century: Percussion Machine for Demonstrating Oblique Percussion ▪ 18th Century: First Ballistic Pendulum ▪ 19th Century: Measurement of Shock Duration ▪ 19th Century: CORIOLIS' Mathematical Studies on Billiards ▪ Oldest Known Seismoscope and 19th-Century Seismography ▪ 19th Century: Percussion Figures ▪ 20th Century: Measurement of Deformation and Force ▪ 20th Century: Pressure-Bar Devices ▪ 20th Century: Taylor Test ▪ 20th Century: Examples of Ball Percussion Studies ▪ 20th Century: Liquids under Impact ▪ 20th Century: Sweet Spots of Sports Equipment ▪ 20th Century: Seismology of Nuclear Explosions*

| | | |
|---|---|---|
| **4.4** | **PERCUSSION AND SHOCK WAVE MODELS** | 906 |

Corpuscular Models and NEWTON's 'Cradle' ▪ Shock Wave Demonstration Apparatus ▪ One-Dimensional Shock Wave Models ▪ Apparatus for Demonstrating Hydraulic Jumps ▪ Traffic Shocks ▪ Amusing Cartoons

| | | |
|---|---|---|
| **4.5** | **SHOCK WAVE VISUALIZATION** | 910 |

TOEPLER's Stroboscopy of Propagating Shock Waves ▪ ANTOLIK's Soot Method ▪ Examples of Shock Wave Photography in Gases ▪ Shock Tube Studies ▪ Whip Cracking ▪ Muzzle Blast and Head Wave ▪ Color Schlieren Photography

| | | |
|---|---|---|
| **4.6** | **HEAD WAVE STUDIES** | 918 |

HUYGENS' Principle of Wave-Front Construction ▪ DOPPLER, Father of the Head Wave Phenomenon ▪ Surface Wave Pattern Produced by a Moving Body in Water ▪ MACH and SALCHER: Prelude to a Pioneering Ballistic Experiment ▪ First Experimental Evidence ▪ Other Optical Methods ▪ Model Sonic Boom Studies ▪ Pressure Measurements Around a Flying Projectile ▪ Blunt Body Concept ▪ Phenomena at Hypersonic Velocities

| | | |
|---|---|---|
| **4.7** | **NOZZLE STUDIES AND APPLICATIONS** | 926 |

Early Safety Valve Constructions ▪ Predecessors of Steam Turbines ▪ Forerunners of the Laval Nozzle ▪ Use of the Venturi Nozzle in the First Wind Tunnel ▪ First Use of Laval Nozzle in a Steam Turbine ▪ Steam Flow in a Divergent Nozzle ▪ Laval Nozzles as Power Generators in Aeronautics

| | | |
|---|---|---|
| **4.8** | **SUPERSONIC JET PHENOMENA** | 931 |

SALCHER and MACH's First Free Air Jet Studies ▪ EMDEN's First Steam Jet Studies ▪ PRANDTL and MEYER's Nozzle and Jet Studies ▪ THOMER's First Radiographs of Detonating Shaped Charges ▪ Formation and Structure of Liquid Jets ▪ Generation of Microjets ▪ Astrophysical Jets

| | | |
|---|---|---|
| **4.9** | **WIND TUNNELS** | 938 |

Pioneering Supersonic Devices in France and England ▪ Prewar Supersonic Facility at TH Aachen ▪ Continuous-Flow Closed-Circuit Supersonic Facility at ETH Zurich ▪ Supersonic Intermittent Indraft Facility at the German Rocket Center, Peenemünde ▪ Slotted Throat of Supersonic Facility at NACA, Hampton, VA ▪ Ludwieg Tube Facility at AVA, Göttingen ▪ First Hypervelocity Facilities

| | | |
|---|---|---|
| **4.10** | **SHOCK TUBES** | 944 |

VIEILLE's Pioneering Setup ▪ BLEAKNEY's Triggerable Shock Tube ▪ Special Types

| | | |
|---|---|---|
| **4.11** | **SHOCK WAVE GENERATION** | 950 |

Snapping Belts and Whip Cracking ▪ Plane-Wave Generators ▪ Gun-Type High-Velocity Accelerators ▪ Other Methods ▪ Laser-Induced Spark ▪ Laser-Supported Detonation (LSD) ▪ Laser-Propelled 'Lightcraft'

| | | |
|---|---|---|
| **4.12** | **SHOCK FRONT ANALYSIS** | 954 |

In Gaseous Matter ▪ In Space ▪ In Solid Matter ▪ Hydraulic Jumps in Water

| | | |
|---|---|---|
| **4.13** | **MACH EFFECT** | 958 |

Interactions of Hydrodynamic Jumps ▪ Shock Interactions in Gases ▪ Shock Interactions in Liquids ▪ Shock Interactions in Solids

| | | |
|---|---|---|
| **4.14** | **SHOCK WAVE EFFECTS** | 966 |

Shock-Induced Creation of Prebiotic Substances ▪ Cavitation ▪ Supercavitation ▪ Condensation ▪ Aerodynamic Shock Heating ▪ $\lambda$-Shock Configuration ▪ Shock Focusing ▪ Transonic Shock Phenomena ▪ Aerodynamic Drag ▪ WHITCOMB's Area Rule ▪ Pseudo Supersonic Wave Effects ▪ Gasdynamic Laser ▪ Shock-On-Shock Problem ▪ Shock Wave Interactions in Metals ▪ Shock-Induced Solidification ▪ Other Phenomena

| | | |
|---|---|---|
| **4.15** | **SHOCK WAVE APPLICATIONS** | 978 |

*Miscellaneous • Determination of Seismic Wave Velocities • Explosion Seismology: the Mintrop Wave • Medical Therapy • Materials Research and Metal Working Industry • Oil Production Industry*

| | | |
|---|---|---|
| **4.16** | **EXPLOSION, IMPLOSION, AND DETONATION** | 984 |

*Early Use of Black Powder • Hollow Charge Effect • Underwater Explosions • Implosion and Explosion in a Gas • Nuclear Implosion Device: The 'Gadget' • Examples of Nuclear Explosions • Electric Guns • Exploding Wires • Large Yield Surface Detonations • Precursor Detonation Phenomenon • One-Dimensional Detonation Front Models • Spinning Detonation • Periodic Cell Structure • Other Explosion Phenomena • Explosive Ablation of Biological Tissue*

| | | |
|---|---|---|
| **4.17** | **EXPLOSION AND DETONATION DIAGNOSTICS** | 998 |

*Maximum Pressure of Fired Gunpowder • Maximum Pressure and Temperature • Chamber Pressure of a Detonating Explosive • Brisance Test of Gunpowder • Brisance Test of a High Explosive • Test of Explosives Used in Mining • Dust Explosion Tester • Detonation Velocity of a High Explosive • Detonation Velocity of a Gaseous Explosion • Interior Ballistic Studies • Blast Wave Recording • Streak Photography in Detonics • Reflected-Light Photography in Detonics*

| | | |
|---|---|---|
| **4.18** | **OPTICAL METHODS FOR FLOW VISUALIZING** | 1008 |

*Schlieren Photography • Shadowgraphy • Interferometry • Holography • Surface Thermography*

| | | |
|---|---|---|
| **4.19** | **HIGH-SPEED DIAGNOSTICS** | 1012 |

*Chronoscopes and Chronographs • Cathode-Ray Oscilloscopes • Time-Delay Generators • Triggered Snapshot Photography • Rotating Mirror • Spark Chronography of a Flying Projectile • Rotating Mirror Streak Cameras • High-Speed Cinematography*

| | | |
|---|---|---|
| **4.20** | **HIGH-SPEED VEHICLES** | 1020 |

*Mythologies • First Supersonic Rocket Flight • First Hypersonic Rocket Flight • First Transonic Rocket Plane • First Supersonic Rocket Plane • Supersonic Propeller • Supersonic Transport (SST) • Hypersonic Aircraft • Manned Spacecraft: Reentry Capsules • Manned Spacecraft: Shuttles • Rocket Sleds • Supersonic Car*

| | | |
|---|---|---|
| **4.21** | **MAN-MADE DISASTERS** | 1028 |

*Steam-Boiler Explosions • Firedamp Explosions • Gun Barrel Bursts • Grain Dust Explosions • Nuclear Reactor Explosion • World Trade Center: Terrorist Bomb Attack • World Trade Center: Terrorist Aircraft Attack*

| | | |
|---|---|---|
| **5** | **BIOGRAPHIES INDEX** | 1035 |

*F.A. ABEL • J. ACKERET • G.B. AIRY • L.V. AL'TSHULER • K. ANTOLIK • R.A. BECKER • D. BERNOULLI • P.E.M. BERTHELOT • H.A. BETHE • W. BLEAKNEY • C.V. BOYS • P.W. BRIDGMAN • R.W.E. BUNSEN • A. BUSEMANN • L. CARRÉ • H. CAVENDISH • J. CHALLIS • D.L. CHAPMAN • G.G. DE CORIOLIS • R. COURANT • C.J. CRANZ • L.M. CROCCO • H. DAVY • H.B. DIXON • C.A. DOPPLER • W.S. DÖRING • H.L. DRYDEN • P.M.M. DUHEM • G.E. DUVALL • S. EARNSHAW • L. EULER • A. FERRI • A.A. FRIEDMANN • K.O. FRIEDRICHS • Y.A. GAGARIN • G.A. GAMOW • I.I. GLASS • H. GLAUERT • V. GOLDBERG • J.S. HADAMARD • C. HAEUSSERMANN • R. HERMANN • H.R. HERTZ • A. HERTZBERG • B. HOPKINSON • E.P. HUBBLE • P.H. HUGONIOT • C. HUTTON • C. HUYGENS • W. JOHNSON • J.C.E. JOUGUET • I. KANT • T. VON KÁRMÁN • G.B. KISTIAKOWSKY • K. KOBES • H. LAMB • O. LAPORTE • C.G.P. DE LAVAL • P.E. LE BOULENGÉ • H.L. LE CHATELIER • H.J.E. LEMAITRE • N. LÉMERY • M.J. LIGHTHILL • E.W.J.W. MACH • L. MACH • N.V. MAIYEVSKII • F.E. MALLARD • R. MALLET • J.M. MARCI VON KRONLAND • E. MARIOTTE • L.H.F. MELSENS • M. MERSENNE • V.A. MIKHEL'SON • L. MINTROP • C.E. MUNROE • J.L. VON NEUMANN • A.B. NOBEL • A. NOBLE • A.K. OPPENHEIM • K. OSWATITSCH • D. PAPIN • W.H. PAYMAN • S.D. POISSON • T.C. POULTER • L. PRANDTL • W.J.M. RANKINE • Lord RAYLEIGH ⇒ J.W. STRUTT • H.V. REGNAULT • G.F.B. RIEMANN • B. ROBINS • T.J. RODMAN • J.S. RUSSELL*

- *A.J.C. DE SAINT-VENANT* ▪ *P. SALCHER* ▪ *R.J. SCHALL* ▪ *H. SCHARDIN* ▪ *O. VON SCHMIDT* ▪ *P. SCHMIDT*
- *F.C. SCHULTZ-GRUNOW* ▪ *L.I. SEDOV* ▪ *N.N. SEMENOV* ▪ *E.M. SHOEMAKER* ▪ *R.I. SOLOUKHIN* ▪ *T.E. STANTON*
- *K.P. STANYUKOVICH* ▪ *A.B. STODOLA* ▪ *G.G. STOKES* ▪ *J.W. STRUTT* ▪ *B. STURTEVANT* ▪ *P.G. TAIT* ▪ *A.H. TAUB*
- *G.I. TAYLOR* ▪ *A.J.I. TOEPLER* ▪ *M.A. TOEPLER* ▪ *A.N. TUPOLEV* ▪ *P.M.E. VIEILLE* ▪ *J. WALLIS* ▪ *H.M. WEBER*
- *C. WIESELSBERGER* ▪ *R.W. WOOD* ▪ *C. WREN* ▪ *C.E. YEAGER* ▪ *Y.B. ZEL'DOVICH* ▪ *G.V. ZEMPLÉN*
- *N.E. ZHUKOVSKY*

## 6 SUPPLEMENTARY REFERENCES ............................................................................................. 1201

### 6.1 GENERAL ENCYCLOPEDIAS ............................................................................................. 1203

### 6.2 SPECIAL ENCYCLOPEDIAS, DICTIONARIES & GLOSSARIES .................................... 1203

### 6.3 HANDBOOKS ............................................................................................................................ 1205

### 6.4 BIOGRAPHICAL CATALOGUES, DICTIONARIES, ENCYCLOPEDIAS & PERIODICALS ............ 1205

### 6.5 BIBLIOGRAPHICAL CATALOGUES ................................................................................... 1207

### 6.6 BOOKS & REVIEW ARTICLES WITH HISTORICAL PERSPECTIVE ............................ 1207

*Acoustics & Nonlinear Acoustics* ▪ *Aerodynamics, Aeronautics & Astronautics* ▪ *Astronomy & Astrophysics* ▪ *Ballistics* ▪ *Bores, Hydraulic Jumps, Solitary Waves & Tsunamis* ▪ *Chemistry* ▪ *Collision, Percussion & Impact* ▪ *Combustion Research* ▪ *Computer, Early Developments & Applications* ▪ *Diagnostics, High-Speed Photography & Scientific Instruments* ▪ *Earthquakes & Seaquakes* ▪ *Explosions & Detonations* ▪ *Fluid Dynamics & Gas Dynamics* ▪ *General Reviews* ▪ *Geology & Geophysics* ▪ *Hydraulics & Hydrodynamics* ▪ *Mathematics* ▪ *Mechanics* ▪ *Mining Industry* ▪ *Nuclear Weapons Development* ▪ *Physics* ▪ *Seismology & Geophysical Prospecting* ▪ *Shock Waves & Blast Waves* ▪ *Shock Waves in Biology & Medicine* ▪ *Steam Boiler Explosions* ▪ *Thermodynamics* ▪ *Thunder* ▪ *Volcanoes*

### 6.7 INTERNET HOMEPAGES ....................................................................................................... 1214
*Research Institutions* ▪ *Technical Museums*

## NAME INDEX ......................................................................................................................................... 1217

## SUBJECT INDEX ................................................................................................................................... 1257

# Abbreviations

AAL – Ames Aeronautical Laboratory
AAS – American Astronomical Society
ACS – American Chemical Society
AEC – Atomic Energy Commission
AEES – Australian Earthquake Engineering Society
AERB – Aerospace and Engineering Research Building
AERL – Aerospace and Energetics Research Laboratory
AFB – Air Force Base
AFWL – Air Force Weapons Laboratory
AGARD – Advisory Group for Aerospace Research & Development
AGI – American Geological Institute
AGN – Active Galactic Nucleus
AGU – American Geophysical Union
AHSPP – Association for High Speed Photography and Photonics
AIA – Aerodynamisches Institut Aachen
AIAA – American Institute of Aeronautics and Astronautics
AIC – Advanced Imaging Conference
AIME – American Institute of Mining, Metallurgical, and Petroleum Engineers
AIRAPT – Association Internationale pour L'Avancement de la Recherche et de la Technologie aux Hautes Pressions
AMIE – Advanced Moon Imaging Experiment
ANL – Argonne National Laboratory
ANRT – Association Nationale de la Recherche Technique
ANSMET – Antarctic Search for Meteorites
ANU – Australian National University
ANU – American Ordnance Association
APS – American Physical Society
ARA – Aeroballistic Range Association
ARC – Aeronautical Research Committee
ARC – Ames Research Center
ARD – Atmospheric Reentry Demonstrator

ARDC – Air Research and Development Command
ARDE – Armament Research and Development Establishment
ARE – Armament Research Establishment
ARL – Admiralty Research Laboratory
ARL – Aerospace Research Laboratory
ARL – Army Research Laboratory
ARS – American Rocket Society
ASA – Acoustical Society of America
ASCE – American Society of Civil Engineers
ASCI – Accelerated Strategic Computing Initiative
ASCI – Advanced Simulation and Computing Initiative
ASCII – American Standard Code for Information Interchange
ASI – Advanced Study Institute
ASM – American Society for Metals
ASME – American Society of Mechanical Engineers
ASNT – American Society for Nondestructive Testing
ASP – Astronomical Society of the Pacific
ASTM – American Society for Testing Materials
ATC – Aberdeen Test Center
AU – International Astronomical Unit
AVA – Aerodynamische Versuchsanstalt
AVCO – Aviation Corporation
AWRE – Atomic Weapons Research Establishment
BAAS – British Association for the Advancement of Science
BAC – British Aircraft Corporation
BALSAD – Ballistic System for Anti-Asteroid Defense
BAM – Bundesanstalt für Materialprüfung
BBC – Brown Boveri & Company
BISRU – Blast Impact and Survivability Research Unit
BMFT – Bundesministerium für Forschung und Technologie

BRL – Ballistic Research Laboratory
BuAer – Bureau of Aeronautics
BuOrd – Bureau of Ordnance
CAGI [ЦАГИ] – Central Aerohydrodynamic Institute
CAL – Cornell Aeronautical Laboratory
CalTech – California Institute of Technology
CANGAROO – Collaboration of Australia and Nippon for Gamma Ray Observatory in the Outback
CARDE – Canadian Armaments and Research Development Establishment
CCAS – Cape Canaveral Air Station
CCD – Charge Coupled Device
CCG – Carl-Cranz-Gesellschaft
CDRC – Civil Defense Research Committee
CDRE – Canadian Defence Research Establishment
CDT – Combustion to Detonation Transition
CEA – Commissariat à l'Energie Atomique
CFD – Computational Fluid Dynamics
CGRO – Compton Gamma-Ray Observatory
CHABA – Committee on Hearing, Bioacoustics and Biomechanics
CIMS – Courant Institute of Mathematical Sciences
CIOS – Combined Intelligence Objectives Subcommittee
CIR – Corotating Interaction Region
CME – Coronal Mass Ejection
CNA – Canadian Nuclear Association
CNES – Centre National d'Etudes Spatiales
CNRS – Centre National de la Recherche Scientifique
COBE – Cosmic Background Explorer
CPU – Central Processing Unit
CREST – Consolidated Reporting of Earthquakes and Tsunamis
CROCCO – Cluster for Research on Complex Computations

CRREL – Cold Regions Research and Engineering Laboratory
CRV – Crew Return Vehicle
CSIRO – Commonwealth Scientific and Industrial Research Organization
CSS – Catalina Sky Survey
CTBT – Comprehensive Test Ban Treaty
DARHT – Dual Axis Radiographic Hydrodynamic Test
DARPA – Defense Advanced Research Projects Agency
DART – Deep-Ocean Assessment and Reporting of Tsunamis
DASA – Defense Atomic Support Agency
DCRC – Disaster Control Research Center
DFVLR – Deutsche Forschungs- und Versuchsanstalt für Luft- und Raumfahrt
DGST – Deutsche Gesellschaft für Extrakorporale Stoßwellentherapie
DIGEST – Deutsche und Internationale Gesellschaft für extracorporale Stoßwellentherapie
DLR – Deutsches Zentrum für Luft- und Raumfahrt e.V.
DMC – Dynamic Materials Corporation
DMR – Double Mach Reflection
DNA – Defense Nuclear Agency
DOE – Department of Energy
DPA – Deutsches Patentamt
DPG – Deutsche Physikalische Gesellschaft
DPMA – Deutsches Patent- und Markenamt
DRDC – Defence Research & Development Canada
DSTO – Defence Science & Technology Organisation
DTMB – David Taylor Model Basin
DVL – Deutsche Versuchsanstalt für Luftfahrt
DWM – Deutsche Waffen- und Munitionsfabrik
DWWSSN – Digital World Wide Standard Seismograph Network
EBW – Exploding Bridge-Wire
EDL – Explosion Dynamics Laboratory
EFP – Explosively Formed Projectile
EHL – Electrohydraulic Lithotripsy
ELSI – Erosion by Liquid and Solid Impact
EMI – Ernst-Mach-Institut
EMP – Electromagnetic Pulse
ENIAC – Electronic Numerical Integrator And Computer

ENSMA – Ecole Nationale Supérieure de Mécanique et d'Aérotechnique
EOS – Equation of State
EPNG – El Paso Natural Gas
EPRI – Electric Power Research Institute
ERA – Explosive Reactive Armor
ERAM – Explicitly Restarted Arnoldi Method
EREA – European Research Establishments in Aerospace
ERW – Enhanced Radiation Warhead
ESA – European Space Agency
ESHPS – Explosion, Shockwave and Hypervelocity Phenomena Symposium
ESWL – Extracorporeal Shock Wave Lithotripsy
ESWT – Extracorporeal Shock Wave Therapy
ETH – Eidgenössische Technische Hochschule
FCT – Flux-Corrected Transport
FDA – Finite Difference Approximation
FEMA – Federal Emergency Management Agency
FhG – Fraunhofer-Gesellschaft
FPST – Free Piston Shock Tunnel
FWHM – Full-Width Half-Maximum
GALCIT – Graduate Aeronautical Laboratories, California Institute of Technology, *in 1961 renamed* Guggenheim Aeronautical Laboratory, California Institute of Technology
GASL – General Applied Science Laboratory
GCVS – General Catalogue of Variable Stars
GD – General Dynamics
GDL – Gas Dynamics Laboratory
GEC – General Electric Company
GEOS – Geodetic Earth-Orbiting Satellite
GISS – Goddard Institute for Space Studies
GPO – Government Printing Office
GRB – Gamma-Ray Burst
GRC – Glenn Research Center
GSA – Geological Society of America
GSFC – Goddard Space Flight Center
HARP – High Altitude Research Program
HEL – Hugoniot Elastic Limit
HERTF – High Energy Research and Technology Facility
HESS – High Energy Stereoscopic System
HETE – High Energy Transient Explorer
HHSTT – Holloman High-Speed Test Track

HIEST – High Enthalpy Shock Tunnel
HMSO – Her Majesty's Stationery Office
HOB – Height of burst
HOPE – H2 Orbiting Plane
HSCT – High-Speed Civil Transport
HMS – Her Majesty's Ship
HST – Hubble Space Telescope
HVA – Heeresversuchsanstalt
HVIS – Hypervelocity Impact Symposium
HVP – Heeresversuchsstelle Peenemünde
HWA – Heereswaffenamt
IAA – International Academy of Astronautics
IAF – International Astronautical Federation
IAPO – International Association of Physical Oceanography
IAS – Institute for Advanced Study
IAS – Institute of Aeronautical Sciences
IAS – International Association of Seismology
IASPEI – International Association of Seismology and Physics of the Earth's Interior
IAT – Institute for Advanced Technology (IAT)
IAU – International Astronomical Union
ICAO – International Civil Aviation Organization
ICCD – Intensified Charge Coupled Device
ICDE – International Colloquium on Dust Explosions
ICDERS – International Colloquium on Dynamics of Explosions and Reactive Systems
ICE – International Cometary Explorer
ICGERS – International Colloquium on Gasdynamics of Explosions and Reactive Systems
ICL – Impact and Crashworthiness Laboratory
ICLRT – International Center for Lightning Research and Testing
ICOGERS – International Colloquia on the Gasdynamics of Explosions and Reactive Systems
ICT – Institut für Chemie der Treibstoffe, *in 1988 renamed* Institut für Chemische Technologie
IDERS – Institute for Dynamics of Explosions and Reactive Systems
IFD – Institute of Fluid Dynamics
IGESTO – Internationale Gesellschaft für Extrakorporale Stoßwellentherapie
IISc – Indian Institute of Science

# ABBREVIATIONS

IMP – Interplanetary Monitoring Platform
IMS – International Monitoring System
IOC – Intergovernmental Oceanographic Commission
IPCR – Institute of Physical and Chemical Research
IPPNW – International Physicians for the Prevention of Nuclear War
IRC – Impact Research Centre
ISAS – Institute of Space and Astronautical Science
ISEE – International Society of Explosives Engineers
ISEE – International Sun-Earth Explorer
ISFV – International Symposium on Flow Visualization
ISIE – International Symposium on Impact Engineering
ISL – Institut Saint-Louis
ISM – Interstellar Medium
ISMST – International Society for Musculoskeletal Shockwave Therapy
ISNA – International Symposium on Nonlinear Acoustics
ISP – Institute for Shock Physics
ISSI – International Space Science Institute
ISSTW – International Symposium on Shock Tubes and Waves
ISSW – International Symposium on Shock Waves
ISWI – International Shock Wave Institute
ITIC – International Tsunami Information Center
ITS – International Tsunami Symposium
ITWS – International Tsunami Warning System
IUEM – Institut Universitaire Européen de la Mer
IUGG – International Union of Geodesy and Geophysics
IUTAM – International Union of Theoretical and Applied Mechanics
IWM – Institut für Werkstoffmechanik
JATO – Jet Assisted Take-Off
JAXA – Japan Aerospace Exploration Agency
JPL – Jet Propulsion Laboratory
JSC – Johnson Space Center
KALIF – Karlsruhe Light Ion Facility
KSC – Kennedy Space Center
KTH – Kungliga Tekniska Högskolan
KTWTL – Kármán Tódor Wind Tunnel Laboratory
KWG – Kaiser-Wilhelm-Gesellschaft
KWI – Kaiser-Wilhelm-Institut

LAL – Langley Aeronautical Laboratory, in 1958 renamed ⇨ LaRC
LANL – Los Alamos National Laboratory
LaRC – Langley Research Center
LASA – Large Aperture Seismic Array
LASCO – Large Angle and Spectrometric Coronagraph
LASL – Los Alamos Scientific Laboratory, in 1981 renamed ⇨ LANL
LCD – Laboratoire de Combustion et de Détonique
LCSE – Laboratory for Computational Science and Engineering
LDA – Laser-Doppler Anemometry
LeRC – Lewis Research Center
LESIA – Laboratoire d'Etudes Spatiales et d'Instrumentation en Astrophysique
LINEAR – Lincoln Near-Earth Asteroid Research
LLL – Lawrence Livermore Laboratory, in 1981 renamed ⇨ LLNL
LLNL – Lawrence Livermore National Laboratory
LMAL – Langley Memorial Aeronautical Laboratory, in 1948 renamed ⇨ LAL
LRL – Lawrence Radiation Laboratory, in 1971 renamed ⇨ LLL
LRSL – Laboratoire de Recherches Techniques de Saint-Louis
LSD – Laser-supported detonation
LTBT – Limited Test Ban Treaty
MABS – Military Applications of Blast Simulation, renamed Military Aspects of Blast and Shock
MCP – MicroChannel Plate
MEMS – Microelectromechanical System
MHD – Magnetohydrodynamics
MIT – Massachusetts Institute of Technology
MMSA – Mining and Metallurgical Society of America
MPE – MPI für Extraterrestrische Physik
MPG – Max-Planck-Gesellschaft
MPI – Max-Planck-Institut
MSFC – Marshall Space Flight Center
MSRC – Major Shared-Resource Center
MST – Microsystems Technology
MSU – Moscow State University
NAA – National Aeronautic Association
NAA – North American Aviation
NA – National Archives
NACA – National Advisory Committee on Aeronautics
NAS – National Academy of Sciences

NASA – National Aeronautics Space Administration
NASM – National Air and Space Museum
NASP – National Aerospace Plane
NATO – North Atlantic Treaty Organization
NBO – Navy Bureau of Ordnance
NBS – National Bureau of Standards, in 1988 renamed ⇨ NIST
NCI – Nuclear Control Institute
NCSA – National Center for Supercomputing Applications
NDBC – National Data Buoy Center
NDC – National Defense Council
NDIA – National Defense Industrial Association
NDRC – National Defense Research Committee
NEA – Near-Earth Asteroid
NEAR – Near-Earth Asteroid Rendezvous
NEIC – National Earthquake Information Center
NEO – Near-Earth Object
NGC – New Galaxy Catalog
NGDC – National Geophysical Data Center
NIAIST – National Institute of Advanced Industrial Science and Technology
NIST – National Institute of Standards and Technology
NMAB – National Materials Advisory Board
NNPT – Nuclear Non-Proliferation Treaty
NOAA – National Oceanic and Atmospheric Administration
NOAO – National Optical Astronomy Observatory
NOL – Naval Ordnance Laboratory
NORSAR – Norwegian Seismic Array
NOTS – Naval Ordnance Test Station
NPL – National Physics Laboratory
NPP – Nuclear Power Plant
NRAO – National Radio Astronomy Observatory
NRC – National Research Council
NRL – Naval Research Laboratory
NSF – National Science Foundation
NSSDC – National Space Science Data Center
NSWC – Naval Surface Warfare Center
NTBT – Nuclear Test Ban Treaty
NWC – Naval Weapons Center
NWL – Naval Weapons Laboratory
OAR – Office of Aerospace Research

OEMU – Oxford Extra-Mural Unit
ONC – Octonitrocubane
ONERA – Office Nationale d'Études
et Recherches Aéronautique
ONR – Office of Naval Research
ORNL – Oak Ridge National Laboratory
OSHA – Occupational Safety and Health
Administration
OSRD – Office of Scientific Research
and Development
PAH – Polycyclic Aromatic Hydrocarbon
PDE – Pulse Detonation Engine
PDF – Planar Deformation Feature
PHERMEX – Pulsed High Energy
Radiographic Machine
Emitting X-rays
PML – Prins Maurits Laboratory
PMMA – Polymethyl methacrylate
PSGDL – Penn State
Gas Dynamics Laboratory
PTBT – Partial Test Ban Treaty
PTWC – Pacific Tsunami Warning Center
PVDF – Polyvinylidene Fluoride
QSO – Quasi-Stellar Object
RAE – Royal Aircraft Establishment
RAL – Rutherford Appleton Laboratory
RAM – Random Access Memory
RAMAC – Ram Accelerator
RARDE – Royal Armament Research
and Development Establishment
RAS – Royal Astronomical Society
R&D – Research and Development
RDD – Radiological Dispersal Device
RLM – Reichsluftfahrtministerium
R&M – Reports and Memoranda
ROF – Royal Ordnance Factory
ROSAT – Roentgen Satellite
ROTSE – Robotic Optical Transient
Search Experiment
RSC – Royal Society of Chemistry
RTO – Research and Technology
Organization
RWTH – Rheinisch-Westfälische
Technische Hochschule
SAF – Schlieren for Aircraft in Flight
SAX – Satellite per Astronomia X
SCEDC – Southern California Earthquake
Data Center
SDC – Shock Dynamics Center
SDI – Strategic Defense Initiative
SDL – Shock Dynamics Laboratory
SECED – Society for Earthquake and
Civil Engineering Dynamics
SGST – Schweizerische Gesellschaft für
Stoßwellentherapie
SI – Système International

SIAM – Society for Industrial
and Applied Mathematics
SIPRI – Stockholm International Peace
Research Institute
SMART – Small Missions for Advanced
Research in Technology
SMIT – Society for Minimally
Invasive Therapy
SMIT – Society for Medical Innovation
and Technology
SMPTE – Society of Motion Picture
and Television Engineers
SMR – Single Mach Reflection
SMRB – Safety in Mines Research Board
SMRE – Safety in Mines
Research Establishment
SMRS – Safety in Mines Research Station
SMS – Seiner Majestät Schiff
SN – Supernova
SNL – Sandia National Laboratories
SNORT – Supersonic Naval Ordnance
Research Track
SNPE – Société Nationale des Poudres
et Explosifs
SNR – Supernova Remnant
SOHO – Solar and Heliospheric
Observatory
SORCE – Solar Radiation and
Climate Experiment
SPE – Space Protection of the Earth
(International conferences)
SPH – Smoothed Particle Hydrodynamics
SPIE – Society of Photo Instrumentation
Engineers
SRI – Stanford Research Institute,
in 1977 renamed SRI International
SSA – Seismological Society of America
SSC – Supersonic Car
SSP – Stockpile Stewardship Program
SST – Supersonic Transport
STAI – Supersonic Tunnel Association,
International
STScI – Space Telescope Science Institute
SUSI – Structures under Shock
and Impact (conference)
SWAN – Simulating Waves Nearshore
SWCMRC – Shock Wave and Condensed
Matter Research Center
SWRC – Shock Wave Research Center
SwRI – Southwest Research Institute
TAL – Technische Akademie
der Luftwaffe
TBRS – Tidal Bore Research Society
TIT – Tokyo Institute of Technology
TN – Technical Note
TSI – Terascale Supernova Initiative

UCRL – University of California
Radiation Laboratory
UDRI – University of Dayton
Research Institute
UMIST – University of Manchester
Institute of Science and Technology
UNAM – Universidad Nacional
Autónoma de México
UNDEX – Underwater Explosions
USACE – U.S. Army Corps of Engineers
USAF – United States Air Force
USBM – United States Bureau of Mines
USGS – United States Geological Survey
USPTO – United States Patent
and Trademark Office
UTIA – University of Toronto Institute
of Aerophysics, in 1964 renamed
⇨ UTIAS
UTIAS – University of Toronto
Institute for Aerospace Studies
UWAL – University of Washington
Aeronautical Laboratories
VDI – Verein Deutscher Ingenieure
VEI – Volcanic Explosivity Index
VIRCATOR – Virtual Cathode Oscillator
VISAR – Velocity Interferometer System
for Any Reflector
VKF – Von Kármán Gas Dynamics Facility
VKI – Von Kármán Institute
for Fluid Dynamics
VLBA – Very Long Baseline Array
VLBI – Very Long Baseline
Interferometry
VNIIEF – All-Union Research Institute
of Experimental Physics
VNIITF – All-Union Research Institute
of Technical Physics
VPI – Virginia Polytechnic Institute
WAC – Women's Army Corps
WADC – Wright Air Development Center
WC/ATWC – West Coast and Alaska
Tsunami Warning Center
WHOI – Woods Hole
Oceanographic Institution
WIS – Weizmann Institute of Science
WiSTL – Wisconsin Shock Tube
Laboratory
WMAP – Wilkinson Microwave
Anisotropy Probe
WMRD – Weapons and Materials
Research Directorate
WSPG – White Sands Proving Ground
WTC – World Trade Center
WWSSN – World-Wide Standardized
Seismograph Network
ZWB – Zentrale für wissenschaftliches
Berichtswesen

# FOREWORD

Modern scientific research, irrespective of the discipline, continues to be dominated by two themes: a growing emphasis on specialization, and an ever-increasing rate of information generation. These trends are not likely to abate soon. Even experienced scientists, let alone new entrants to the field, have difficulty synthesizing seemingly disparate but related information contained in journal articles and conference presentations into a cohesive construct that illustrates the underlying fundamental principles. Apart from the need for synthesis, gaining a proper historical perspective on scientific developments is not easy in an era of information explosion. To fully comprehend the nature of scientific discoveries and developments, whether they are evolutionary or revolutionary, a chronological account of scientific research is an important educational need. Sir Isaac NEWTON's famous statement made in a letter to Robert HOOKE (1676) about seeing further by standing on the shoulders of giants exemplifies how scientific achievements in any era build on previous developments. Because of the sheer volume of information available in the internet age, extracting knowledge from this information represents a significant challenge for the twenty-first century.

The history of impacts, explosions, and shock waves constitutes an interesting window into mankind's use of rapid energy release for both constructive and destructive purposes. Throughout the ages, human beings have sought ever more powerful sources of rapid energy release. This year marks the sixtieth anniversary of the first atomic explosion, and many differing viewpoints have been expressed about the merits and morality of the Manhattan Project. Irrespective of the position one may have on this matter, there can be no argument that the explosive release of nuclear fission energy in 1945 was a profound consequence of twentieth century physics and constituted an epochal event in the history of impact phenomena and explosions. The merit and morality of producing the shock waves that caused and resulted from the 1945 explosions have been and will continue to be debated for a long time. Although it is not surprising that human beings often focus on more recent events, this volume is not about any one event. Instead, it provides valuable insights into the history of shock wave developments over a very broad time period. I know of no other volume that has provided such a history.

The *History of Shock Waves, Explosions and Impact* prepared by Dr. Peter KREHL constitutes a monumental effort on his part and, in my opinion, is an unqualified success in addressing the research and educational needs indicated above. As an active researcher and educator in the field of shock wave science, I was quite curious as to how one synthesizes the history and science of impact phenomena, explosions, and shock waves into a cohesive volume. A careful perusal of this encyclopedic work (since reading it in detail would be a full-time undertaking over an extended period) has been most educational and enjoyable for me. This work is clearly a labor of love and the author has successfully combined history, scientific developments, technology, and social commentary (when appropriate) in an admirable manner. His attention to detail is exemplary.

An important aspect of this encyclopedic review is that, in addition to the historical information, it contains extensive summaries of important scientific developments and papers. The breadth of Dr. KREHL's scientific expertise and knowledge is quite evident. Thus, both experienced and new researchers can use this encyclopedic history as a point of entry to virtually all research areas related to explosions and shock waves. The classical physics concepts and related mathematical developments required to describe the production and propagation of large-amplitude or shock waves in a continuum are presented in a comprehensive manner. Understanding condensed matter response at the microscopic level under shock wave compression, a subject of more recent interest, has provided a natural coupling of classical physics and quantum physics, and these recent developments are also well-referenced. The collection of brief write-ups about individuals who have made significant contributions to this field adds a valuable human dimension to the volume.

Looking into the future, I see shock wave science playing an important and unique role in examining and understanding matter under extreme conditions. With exciting developments in high-intensity lasers, recent breakthroughs in pulsed power capabilities, high time-resolution measurements at different length scales, and advances in computational sciences, shock-wave-related research activities will make valuable contributions to such diverse fields as condensed matter physics and chemistry, applied mathematics, chemical dynamics, materials science, plasma physics, nonlinear wave propagation, planetary science, and astrophysics. The culmination of NASA's "Deep Impact" mission this year represents an exciting, first-of-a-kind shock wave experiment to probe the interior of a comet and, potentially, other objects external to our planet at a future date. The success of this space mission is a testimony to human creativity and curiosity, and such undertakings will surely grow in the future. These new activities will have their origins in mankind's ability to rapidly release energy, which, in turn, is related to developments in impacts, explosions and shock waves. Although the methods for producing shock waves will continue to evolve, the scientific foundations presented in this volume will remain relevant to future discoveries and developments.

This encyclopedic work will occupy a prominent position in my bookshelf. From time-to-time, this unique volume will permit me to reflect on the developments in this fascinating field and to think about the special individuals who made it happen.

Yogendra M. GUPTA<br>
Regents Professor<br>
Dept. of Physics and Institute for Shock Physics<br>
Washington State University<br>
Pullman, Washington 99164, U.S.A.

Progress in understanding the nature of shock wave, detonation and impact phenomena has been particularly governed and greatly assisted by the progress made in high-speed diagnostics of all kinds, including high-speed photography and computer development. Another very important factor has been the invention and steady improvement in ways to produce devices that enable the controllable launch of projectiles at extremely high velocities for impact experiments. Many of the principles and inventions produced, regarded now with the benefit of hindsight, appear obvious and in some instances relatively simple; but as in all great steps in science they required an enormous amount of toil to achieve practical working systems from sometimes just a flash of inspiration.

The nucleus of many of these achievements can be traced back to the impetus for scientific progress provided by the developments in all fields required to satisfy the needs of World War II, and later the desire to begin the exploration of outer space. Over the last sixty years there has been enormous progress in shock wave and detonation physics in many fields, including the behavior and properties of materials under high strain rates.

The study of shock waves obviously requires the ability to produce shock waves in solids, liquids, or gases. This may be done by many methods, but one of the most versatile and important is impacting target materials at high velocities.

After World War II and the onset of the Cold War there was great interest in improving the ballistic performance of guns. Technology from German work on rockets during the war was being extended, in conjunction with nuclear weapons development, to produce ballistic missile systems which would later form the basis for space exploration studies. This work required the laboratory to have the ability to launch models and test pieces at velocities of up to 6.0 km/s or higher. At that time, the only systems available were specialized single-stage solid propellant guns, limited to launch velocities of around 2.7 km/s. The problem lay in the properties of the driver gases formed by propellant combustion. The gas energy is not only used to drive the projectile along the barrel but it also has to accelerate the driving gas itself to projectile velocity. As muzzle velocities approach 2.7 km/s, virtually all of the gas energy is used up in accelerating the driver gas, leaving little to further accelerate the projectile.

The answer lay in the invention of the two-stage light-gas gun in 1948, in which higher launch velocities become possible by using low molecular weight driver gases such as helium. In the first stage barrel, this gas is compressed by a piston driven by conventional propellants, and passes at very high temperature and pressure via a bursting diaphragm into the launch barrel. Unfortunately this technology was not developed until the 1950s, but it has continued to thrive and improve so that velocities of 6 km/s and above are now common, and may be increased to 10 km/s and above using impacted third stage flyer plates.

One of the most widely used configurations for high strain rate material measurements is the split Hopkinson pressure bar or Kolsky apparatus. It was developed by Herbert KOLSKY in 1948 from the original experimental bar system invented by Bertram HOPKINSON in 1914. The Hopkinson bar was originally used to measure the shapes of pressure pulses produced by the impact of bullets or the explosion of detonators, and was used extensively during the First World War. The later split-bar system is very versatile and has been used for a very wide range of experiments.

Another very important experimental area is in the study of supersonic flow around objects. This is crucial in studies for the development of aircraft, spacecraft, ballistic missiles and armaments. These experiments are made possible through the use of well-instrumented high-speed wind tunnels, shock tubes, shock tunnels, plasma jets and aeroballistic range facilities of various kinds.

Shock wave, detonation and impact phenomena require high-speed diagnostics to allow measurements to be made and experimental images to be recorded. Since the beginning of shockwave and detonation research, diagnostic instrumentation has been vital. At first fairly sparse and technically simple, it advanced steadily in the range of types of measurement possible and also achieved steadily increased frequency response, extremely important in shock studies. The invention of the electronic valve was a major step forward and led to the production of vital instruments such as the cathode ray oscilloscope and high input impedance amplifiers.

A further major advance came with the invention of the transistor, leading in time to the production of integrated circuits. It then became possible to downsize equipment and expand the range of useable power sources, often making the burden of field experiments much easier when out of reach of mains power supplies.

Concomitant with the advances in instrumentation and electronics came the development of the computer. Looking back from the twenty-first century it is difficult to imagine how scientific research ever progressed before the digital computer; but of course it did so by using the computing tools available at the time, achieved through arduous effort by dedicated mathematicians. The progress of computation from the earliest mechanical engines, through geared desktop calculators, electrical calculators, early valve computers up to the enormously powerful digital computers of today, is a fascinating story. It is somewhat sad that the impetus to produce early digital computers was provided by war-related requirements such as the production of items like ballistics tables. Now regular advances allow calculations of such complexity and size that earlier scientists could hardly have imagined an apparatus that could possibly perform such miracles.

Today, nearly all areas of measurement are used in shock wave and detonation physics, covering stress and strain, temperature, pressure, shock strength, velocity, shock arrival time, stagnation temperature, stagnation pressure, time intervals, detectors of many kinds, radiation, acceleration, frequency, aerodynamic drag, spin, nutation, precession, yaw, *etc*. In all of these areas, the measurement methods and devices have steadily increased in capabilities and complexity and, since the introduction of the transistor and integrated circuit, they have reduced in size and increased in reliability as well.

Probably the most wide-ranging of measurement and observation techniques has been that of photography. Photography plays a unique role and has the distinction of being able to provide information on many parameters simultaneously. Many measuring devices can provide perhaps two parameters at the most, *e.g.*, a pressure transducer can give a measurement of pressure and frequency. However, a simple photograph may give positional information, subject status, velocity, shock wave information and temperature, and all without physical connection or interference with the experimental procedure involved. With the advantages of modern high-speed photography, it is possible to observe and measure phenomena of extremely short durations with a capability unmatched by any other techniques.

Photography began to be a viable technique in the mid-1850s but did not advance enough to be taken up by scientific experimenters until some twenty years later. Important problems limited its use in science; lenses were small with low light gathering ability, and film emulsions were slow, requiring much light for short duration exposures. The electric spark became an important tool for research, giving a high brightness and a short duration exposure using an open shutter. Ernst MACH used sparks to study shock waves, as did Charles V. BOYS in England. As the 1800s came to a close and research moved into the 1900s, incandescent lamps of many kinds began to appear, together with arc lamps. These were fine for still and cine photography but relied on short-duration shutters to freeze fast motion. Sparks were still used, and then Harold E. EDGERTON invented the flash tube and added another very versatile short-duration light source. The coming of the laser was another landmark: today, extremely short light pulses can be produced by lasers of many kinds, giving exposures down to nanoseconds.

Throughout the 1900s cameras developed rapidly in both still and cine applications. By the late 1940s they had achieved a million frames per second, developed in particular to photograph early atomic bomb research. Eventually film cameras reached the limits of their development and electronic cameras, using image converter techniques, advanced imaging capabilities to 100 million frames per second. The invention of digital cameras marked another leap forward in photographic techniques and offered many advantages over film in terms of instant results and an ability to manipulate images. Digital cameras have produced a revolution in usage in scientific work for both still and cine techniques. It is now possible for the scientist to become his own photographer with relatively little training. At the same time, unfortunately, the professionally trained cameraman is becoming rarer as scientific staff take over his role, perhaps to the detriment of the imaging arts.

What next? Film is still the leader in terms of resolution, but digital developments are already catching up in still photography and will doubtless also reach equality in cine studies in time.

Measurement techniques have evolved steadily in all fields, enabling more and more insight into the areas of shock waves and detonation studies. There will surely be new problems to be studied as space travel advances and the ability to land measuring devices safely on distant worlds increases.

Dr KREHL has produced an extraordinary book, listing the achievements and developments in shockwave and detonics studies over several centuries and the people who made them possible. Its study will inform and entertain us and prepare us for the wonders to come.

Peter W.W. FULLER
M.Sc., M.A., MIET, SMIEEE, C.Eng, FRPS
Former Principal Scientific Officer
Ministry of Defence, Ballistics & Instrumentation
Now, Instrumentation Consultant
United Kingdom BR1 4AR

# 1

## INTRODUCTION

> *Thinking about these shocks reminds me of a problem or, put differently, of a question in the field of mechanics, the answer to which I have not found in any author, not even a hint which would have, at least partly, satisfied me. I wonder what causes the energy and the immense force that one feels occur during the shock and what they might depend on, when, with a single hammer blow of only 8 or 10 pounds, we are able to overcome such resistances which otherwise would not yield to a merely pressing weight without any shock, even if it weighed several hundred pounds.*[1]
>
> Galileo GALILEI
> Arcetri/Florence 1638

THROUGHOUT the evolution of man, the application of the *principle of mechanical percussion* was one of the basic means to produce tools and weapons, which, while steadily promoting his intelligence, also improved his living conditions in the daily struggle for existence. Modern evolutionists assume that the first primitive worked stone tools were applied by the *australopithecines*, the earliest known hominids in human evolution, which lived in Africa about 2.6 million years ago.[2] The positive feedback that occurred from applying tools such as using sharp stone flakes to cut through tough hides and to break bones for marrow – thus obtaining access to high-grade food such as proteins and fats – led to an increase in brain volume and promoted intelligence. This process, when performed steadily over a sufficiently long period of time, was obviously very effective in human evolution.[3]

In today's animal world, the principle of percussion and impact is only rarely used. Some species, provided with horns, antlers or tusks, can focus the force of percussion to a "point," but their rudimentary weapons are only used occasionally for defense purposes and in rival struggles. An unusual exception is the woodpecker which, in order to get access to food, uses his chisel-pointed beak many thousands of times a day by rapidly hammering powerful blows against a tree. Its wonderful skull anatomy, which prevents brain concussion, has been the subject of much research {BECHER ⇨1953; ⇨Fig. 4.1−Z}. The application of external percussion tools is an exception in the animal world and is only used by a few species, *e.g.*, by the Egyptian vulture and the Australian black-breasted buzzard kite {VAN LAWICK-GOODALL ⇨1966; ⇨Fig. 4.1−Z}. A few other animals make use of stones and rocks as "anvils" against which to break open food objects, such as the California sea otter and some primates, while the application of shock waves is apparently limited to a single species only, the snapping shrimp {LOHSE ET AL. ⇨2001; ⇨Fig. 4.1−Z}.

There is no general agreement in the literature as to what is, and what is not, *tool-use* in animals. Tool-use has been defined by the British ethologist Baroness Jane VAN LAWICK-GOODALL,[4] an authority on wild chimpanzees, as "the use of an external object as a functional extension of mouth or beak, hand or claw, in the attainment of an immediate goal. This goal may be related to obtaining food, care of the body, or repulsion of a predator, intruder, *etc*." According to this definition, the behavior of some birds and monkeys, smashing eggs or bones against rocks or dropping them in order to get access to the soft interior, is not tool-using and is termed by various behaviorists and psychologists as *quasi tool-use*. Nevertheless, the physical principle involved in all these examples is the generation of percussion by impact.

The emergence of a flaked-stone technology during the course of hominid evolution marks a radical behavioral departure from the rest of the animal world, constituting the first definite evidence in the prehistoric record of a simple cultural tradition, and laying an important foundation for early civilization. Primitive weapons based on percussion are also more effective when made from stone rather than from wood or bone, because the momentum of the thrown or swung tool is proportional to its mass.

Stone tool-making began in the Early Paleolithic period about 2.6 million years ago, when early hominids began purposefully selecting specific raw materials, and making their own sharp-edged stone tools {Great Rift Valley ⇨*c*.2.6 Ma ago}. In the "Pebble Tool Tradition," the earliest manifestation of this behavior, pebbles were sharpened through the bifacial removal of sharp-edged flakes. In the "Flake Tool Tradition," flakes were used for general-purpose cutting, while the cores were also useful as heavy chopping tools and percussion weapons. The "Chopper-Chopping Tool Tradition," an outgrowth of the two tool traditions, was a technology characterized by broad heavy scrapers or cleavers and implements with an adz-like cutting edge, and practiced throughout a period of about 1.5 million years. The next major technological advance in the production of stone tools

---

[1] G. GALILEI: *Unterredungen und mathematische Demonstrationen über zwei neue Wissenszweige, die Mechanik und die Fallgesetze betreffend (1.−6. Tag, Arcetri 6. März 1638)*. Ed. by A. VON OETTINGEN, Wissenschaftliche Buchgesellschaft, Darmstadt (1964); see *Der 4. Tag*, p. 239.
[2] K. MILTON: *Diet and primate evolution*. Scient. Am. **269**, 70-77 (Aug. 1993).
[3] S. JONES (ed.): *Cambridge encyclopedia of human evolution*. Cambridge University Press, Cambridge (1992), pp. 325-334.
[4] J. VAN LAWICK-GOODALL: *Tool-using in primates and other vertebrates*. In: (D.S. LEHMAN, R.A. HINDE, and E. SHAW, eds.) Advances in the Study of Behavior **3**, 195-246 (1970).

came with the "Acheulean Handaxe Tradition," an era of sophisticated lithic tool production using the principle of percussion with progressive refinement in craftsmanship and practical know-how. The Acheulean Tradition had a great longevity, also on the order of 1.5 million years. In the following Middle and Upper Paleolithic periods, emphasis on tool technologies shifted from core tools to carefully shaped flake tools provided with very efficient blades.

Most weapons invented throughout man's evolution are based upon percussion. Generally, historians differentiate between two types of weapons: *shock weapons* and *missile weapons*. Shock weapons such as the club, halberd, mace, or sword are held in the hands, while missile weapons are thrown with either muscle power or by a delivery system of some sort, such as a catapult. When striking a target, they generate destructive effects that are also based upon percussion. The heavy weaponry of antiquity was also based on percussion, such as the rams used on the bows of warships {⇨Fig. 4.2–F} or operated manually by teams {⇨Fig. 4.2–G}. The most important percussion tool was the chisel used in combination with a wooden mallet {⇨Fig. 4.2–E} or a metal hammer, which played a central role in most early stonemason technologies – and still does. However, compared to the incredibly long era of practical applications of percussion, which is characterized by long periods of technological stagnation, a scientific approach to the phenomenon of percussion began very late: the basic physical laws of mechanical percussion were not discovered until the 17th century.

Since antiquity, the elastic nature of air, whether compressed slowly or rapidly, aroused the curiosity of many natural philosophers and was even used by some primitive races in everyday life, such as in the blow-gun {⇨Fig. 4.2–I} and the pneumatic lighter {⇨Fig. 4.2–K}. The elastic percussion of gases – popularized by the expression "the spring of the air" {BOYLE ⇨1660} – also turned scientific curiosity toward the elasticity of solids, which resulted in the general law of elasticity {HOOKE ⇨1679}. The important observation that each percussion is accompanied by a loss of kinetic energy which is transformed into heat, sound or permanent deformation {Sir NEWTON ⇨1687} stimulated the characterization of so-called "inelastic" percussion processes by a *restitution coefficient*, a term coined by Lord KELVIN and Peter G. TAIT {⇨1879}. In the second half of the 19th century, "elastic" percussion, which also took wave propagation into account, initiated the first theories of shock propagation in solids.

Air consists of a large number of oxygen and nitrogen molecules that are constantly colliding with one another without any loss of total energy due to the perfect elasticity of air. In practice, gases of neutral molecules at pressures of up to hundreds of atmospheres may be considered to be perfect. Up to such pressures, collisions between two molecules dominate (so-called "binary" collisions); the probability of "ternary" (or three-body) collisions is comparatively low.

Elastic percussion also had an enormous impact on the evolution of thermodynamics in the 19th century: the hypothesis that heat could be interpreted as being a collision process between air molecules, and particularly that the mean square velocity of motion is proportional to the (absolute) temperature {HOOKE ⇨1674; BERNOULLI ⇨1738}, was a very successful concept that led to the first kinetic theory of gases {KRÖNIG ⇨1856; CLAUSIUS ⇨1857; MAXWELL ⇨1867}. The motions of the molecules and their interactions may be described by classical Newtonian mechanics; *i.e.*, the collision process must satisfy the laws of conservation of mass, momentum, and energy. Relativistic effects are important only at very high temperatures; *i.e.*, at very large molecular velocities.

Modern particle physics is also based on percussion. The current understanding of subatomic particles, of which atoms are composed, is derived entirely from studying the results of collisions between them. Thus, in modern particle physics, the description of collisions is fundamental to our understanding of matter.

Cosmic rays are a unique example of a complex interaction involving explosion, collision and shock wave phenomena in nature {HESS ⇨1911; BAADE & ZWICKY ⇨1934; ČERENKOV ⇨1934}. Cosmic rays are actually a flux of particles (atomic nuclei, mostly of protons) of ultrahigh energies (up to some $10^{20}$ eV) traveling at nearly the speed of light, rather than rays. Being about $10^8$ times more energetic than the high-energy particles achievable by the most powerful man-made accelerators, they are still a real riddle to astrophysicists, and various hypotheses have been put forward to explain their origin. It is likely that cosmic particles are accelerated up to very high energies in the strong shock wave emerging from a supernova explosion or from the thermonuclear detonation that occurs when a white dwarf star undergoes complete disintegration. When the cosmic particles enter the Earth's atmosphere, complex collision processes occur which produce a number of atomic and subatomic particles creating an "air shower."

Descriptions of particle collisions are of a quantum mechanical rather than of a classical (*i.e.*, nonrelativistic) nature, but they are nevertheless closely based on principles that arise out of fundamental mechanics. While during the inelastic collisions of macroscopic bodies the kinetic energy of the incident body is transformed entirely into heat, collision on a microscopic scale, say between two identical atoms, may cause one or both to be excited into a state of higher internal energy than it started with. This generation of ionized particles and their possible interactions with outer magnetic and/or electric fields is of fundamental concern in plasma physics.[5]

*Stationary* (*i.e.*, nonpropagating) *discontinuities* occur at interfaces between materials and free surfaces. These so-called "contact discontinuities" are very common phenomena in nature: *e.g.*, the abrupt change of density at the interface of water and the atmosphere, or at the interface of two different geological strata. In the Earth's crust this jump in density and elastic properties results in the reflection and refraction of elastic waves, causing *seismic discontinuities*. The Mohorovičić discontinuity is the most prominent seismic discontinuity; it represents the interface between the Earth's crust and the mantle (*see* Sect. 2.2.1). The largest discontinuity in the Solar System is the narrow transition zone between the chromosphere and corona – when compared to the Sun's diameter of about 1,400,000 km, this is a thin shell with a thickness of only about 100 km, in which the temperature and density change sharply.

However, in high-rate dynamic events, *transient discontinuities* in density and other thermodynamic quantities are generated, which travel with supersonic speeds throughout compressed matter as *shock waves* – the modern and most versatile representation of the mighty principle of percussion. Shock waves occupy an interesting position within the more general field of dynamics: they have been observed in the laboratory and in nature, in microscopic as well as in macroscopic dimensions, and even in the infinite, almost "empty" depths of space. Thus shock waves with dimensions covering a range of more than 20 orders of magnitude have been observed.[6]

Shock waves and percussion processes are discontinuous phenomena and are felt by our senses as sudden and violent events. However, no absolute definition exists of what *discontinuous* means in terms of time duration, because any such concept would be of a merely subjective nature. For example, in the intellectual world of a shock physicist, a typical discontinuity would be the shock front, where the shock pressure can rise in tens of nanoseconds, while in the geologic realm, meteorite impacts, explosive volcanic eruptions and earthquakes can provoke drastic and irreversible changes within seconds. Curiously, these two examples – although they take place at greatly different time scales and dimensions – deal with almost the same orders of magnitude when related to life spans as characteristic quantities.[7]

Shock waves were initially regarded as a very specific branch of physics, attracting only minor attention in the scientific community. However, historical studies have shown that shock wave physics transformed from an initially small and specific branch of physics into an important and complex interdisciplinary science, which makes shock physics such a fascinating subject. Furthermore, it was discovered that all violent phenomena in nature, ranging from terrestrial to cosmic dimensions, and often manifesting in dramatic disasters, are governed by shock wave processes. Prominent examples of natural discontinuities on Earth are thunder, meteorite impacts, volcanic explosions, earth- and seaquakes, bores and tsunamis, while in the Universe they encompass stellar explosion and implosion phenomena, bow shocks and planetary shocks, comet and asteroid impacts, and galactic collisions involving tremendous amounts of energy.

The evolution of shock wave physics – mainly an outgrowth of acoustics based upon 17th-century classical percussion and 18th-century aeroballistics – began in the early 19th century with Siméon Denis POISSON's analytical treatment of flow discontinuities {POISSON ⇨1808}, and was initially strictly limited to the gaseous state. In the 19th century, shock wave physics underwent continuous development, predominantly by a small number of British, French and German mathematicians who laid the foundations for modern shock wave theory. Their interest in treating such flow discontinuities mathematically was inspired primarily by academic curiosity rather than by any practical needs. On the other hand, approaching the discontinuity problem experimentally was a tricky task and a real challenge for inventors because of its instantaneous and supersonic motion. Early percussion and shock wave pioneers experimented with crude instrumentation and had only modest financial and personnel means at their disposal. However, this situation changed when experiments on metals performed in the 1870s revealed that dynamic strength under sudden loading is higher than that under static conditions {J. HOPKINSON ⇨1872}. Furthermore, shortly after a demonstration of

---

[5] In the period 1930–1950, physicists significantly revised their views on the elementary constituents of matter, which during the 1920s had been widely assumed to be only the electron and the proton. This revision and the problems it posed gave birth to the field of modern elementary particle physics; *see* L.M. BROWN and L. HODDESON (eds.): *The birth of particle physics*. Cambridge University Press, Cambridge (1986).

[6] Micro cavitation phenomena producing shock waves happen on a micrometer scale (1 μm = $10^{-6}$ m). On the other hand, the tail of a cosmic bow shock {⇨Fig. 4.1−Y} can reach a length of some light-years (1 light-year = $0.946 \times 10^{16}$ m).

[7] For a human being that lives to an age of 75 years, a period of 10 ns corresponds to about $4.2 \times 10^{-18}$ of his life span. On the other hand, a span of 1 s in geological terms corresponds to about $6.9 \times 10^{-18}$ of the Earth's age (ca. 4.6 billion years).

the great potential of high-speed photography {TALBOT ⇨1851} and the first examples of the photographic recording of shock wave phenomena – such as the ballistic head wave {E. MACH & SALCHER ⇨1887}, the cylindrical shock waves surrounding an electric spark, and the irregular reflection of shock waves {E. & L. MACH ⇨1889} – research in supersonics was recognized as a powerful tool for minimizing aerodynamic drag in high-velocity ballistics and optimizing Laval nozzle design for use in steam turbines. At the end of the 19th century, mechanical engineering and ballistics, which had been previously affiliated more with craftsmanship than with the world of research, increasingly entered into the world of experimental and theoretical science.

The next quantum leap in the evolution of supersonics happened in 1935 at the 5th Volta Conference, when aerodynamicists of the leading nations in supersonics discussed problems of high-velocity flight. They met in Guidonia, a research community near Rome noted for its advanced supersonic wind tunnel, which was constructed by the Swiss aeronautical engineer Jakob ACKERET and had just been put into operation. The results of these discussions set in motion an enormous amount of governmentally supported research on a much larger scale around the world: all-round philosophers were now followed by "institutionalized" professionals and managers at many newly established research laboratories and wind tunnel facilities. Pursuing almost exclusively military goals, this led in the 1940s to the foundation of large aerodynamic testing facilities, and to an enormous output of mostly classified scientific and technical literature on high-speed aerodynamics and gas dynamics.

In parallel with this impetuous research into supersonics in gases, underwater shock waves were also studied in more detail. Besides these shock studies in gases (and sometimes also in liquids, mostly in water), interest increasingly turned to the solid state, thus creating an entirely new branch of physics which until the beginning of World War II barely existed, but was regarded as a logical continuation of static high-pressure research established by the U.S. physicist Percy W. BRIDGMAN in the early 1900s.

Since the earliest times, the destructive and disintegrating features of percussion and explosions had found wide applications in the military and in the mining industry, but had gained only little attention and appreciation in pure science. With a deeper knowledge of the physics involved and a better temporal resolution available from visual diagnostics, however, it became more and more evident that high-rate phenomena play an important role not only in terrestrial dynamics, but also in the Universe. With the advent of telescopes in the 17th century, the shape of the Moon's face and the origin of craters aroused the curiosity of scientists, giving ample evidence that the Earth's and other planetary surfaces were shaped by impacting bodies from outer space – and still are, as observed by professional and hobby astronomers when a train of comets impacted on the planet Jupiter {Comet SHOEMAKER–LEVY 9 ⇨1994}. Extraterrestrial examples of supersonic plasma flow which provide good illustrations of the analogy to terrestrial supersonic phenomena include:

▸ the emission of huge high-velocity *flares* from the Sun's surface;
▸ the existence of the *solar wind*, a supersonic plasma flowing radially outward from the Sun;
▸ the *bow shock* around a planet, a collisionless shock wave produced by the interaction of the supersonic solar wind with an obstacle such as the magnetic field of a planet; and
▸ *astrophysical jets*, relativistic beams of stellar matter which emerge from some astrophysical objects.

The progress in nuclear physics that occurred in the first half of the 20th century also stimulated theoretical physicists and astrophysicists to uncover the mechanism of energy production in stars and explosive/implosive peculiarities that occur during stellar evolution. Modern cosmology also assumes that the Universe originated in a huge explosion and expansion of matter, the "Hot Big Bang" which, according to the view of some cosmologists, might end up in a huge implosion – the catastrophic collapse upon itself – thereby fusing once more into the primeval atom, the "Big Crunch" {DAVIES ⇨1994}.

The many studies devoted to shock wave physics resulted in spectacular developments in supersonic aerodynamics, detonation physics, space rocketry and the fabrication of new nuclear weapons with previously unknown destructive effects, thus also vigorously changing worldwide political concepts and military strategies. On the other hand, almost unnoticed by the public and even by many colleagues from other branches of physics, the enormous investments in basic research that have been made since the 1940s have also resulted in much progress in our understanding of the nature of shock waves and their effects – on both microscopic and cosmic levels – which have stimulated other branches of science and technology tremendously. Spectacular examples include high-speed diagnostics, laser fusion, cosmology, astrogeology, chemical kinetics, medical therapy, and, more recently, even biology. In addition, beginning in World War II in the United States in connection with the Manhattan Project, the enormous computer power required to numerically simulate phenomena in nonstationary fluid dynamics, particularly ultrahigh-rate nonlinear phenomena such as those in strong shock waves, constantly challenged the computer industry to increase the computer power available.

Shock waves also play dual roles in the microcosmos of atoms and molecules; roles that are not yet fully understood: laboratory studies have shown that shock waves can create intermediate, short-living states that initiate subsequent self-propagating reactions ("chain reactions") in which even the number of chain carriers can increase in each propagation cycle ("branching chain reactions"). Shock waves can also transform simple molecules into complex ones with promising, unprecedented features ("shock synthesis"). Presumably, this creative effect of shock waves even contributed to the origin of terrestrial life {⇨Figs. 4.14–A, B}.

On the other hand, shock waves can also decompose molecules. It is interesting here to note that this dual tendency for destruction and creation is also reflected in early myths which believe that several worlds had been successively created and destroyed before the present Universe came into being – conceiving the life of the Universe as an endless struggle between benevolent and malevolent forces, or between life and death.[8]

To get a realistic picture of the complex evolution process of percussion and shock wave research – *i.e.*, to reflect on previous states of knowledge, motivations, speculations and achievements – it appeared indispensable to cover the fields of *Explosion* and *Detonation* too, since they are also high-speed phenomena and are related to shock waves in a variety of ways. However, since the greatest progress and the widest applications of the principle of percussion have occurred in the field of shock waves, the present work emphasizes the evolution of this branch of physics. In addition, some of the most important milestones in the advancement of *High-Speed Diagnostics* and *Computer Technology* have been included, since these are key technologies that heavily determined the progress of shock wave physics in the past, and still does so in the present.

---

[8] For example, in Hindu cosmology the Universe is ruled by the three deities *Brahma* (creator of all things), *Vishnu* (protector and preserver of the world) and *Shiva* (destroyer and restorer), the so-called "Hindu Trinity" – the One or Whole with three forms. The U.S. physicist Robert OPPENHEIMER (1904–1967), who was director of the Los Alamos bomb laboratory during World War II and a fan of ancient Hindu meditative poetry, alluding to the Hindu Trinity and the pluralistic tendency of the Universe, named the site of the first atomic bomb explosion "Trinity." This site is located in the *Jornada del Muerto* ("Journey of Death"), in the Alamogordo Desert of New Mexico. When OPPENHEIMER saw the atomic bomb's power at Trinity Test Site in 1945, he began to doubt his work and the West. • Beyond the two atomic bombs dropped in 1945 on Hiroshima and Nagasaki, no further nuclear weapons have been used in times of war. Instead, their existence has stabilized the military balance between the superpowers and prevented global wars ("Peace by Nuclear Deterrence"), thus confirming in a peculiar way this dualism of creation and destruction – or of peace and war.

# 2

## GENERAL SURVEY

## 2.1 TERMINOLOGY AND SCOPE

*The great book of Nature lies ever open before our eyes and the true philosophy is written in it... But we cannot read it unless we have first learned the language and the characters in which it is written... It is written in mathematical language and the characters are triangles, circles and other geometric figures.*[1]

Galileo GALILEI
Florence 1623

*In physics, experiments have a larger power to persuade than reasoning.*[2]

Blaise PASCAL
Paris 1663

PERCUSSION, concussion, collision, impact, explosion, implosion, detonation and shock waves are rapid mechanical phenomena that are related to each other. Since they cannot be resolved by the naked eye, for a long time their sudden and discontinuous nature was hidden behind a veil of vague hypotheses and suppositions. First experimental evidence on physical quantities such as motion (velocity, acceleration), force and thermodynamic state (density, pressure and temperature) were not obtained until significant progress was achieved in the late 19th century in high-speed diagnostics, visualization techniques and photographic recording. Increasing individual knowledge of high-speed events led observers to recognize connections between different phenomena, which stimulated analyses of interrelations. This initiated a purposeful, more systematic research characterized by experimental testing of qualitative hypotheses, which, beginning in the mid-1940s, eventually led to computer-supported quantitative modeling of new concepts.

Mechanics developed in the 17th century from mere contemplations of accounts of observations and experiences into one of the main pillars of the physical sciences. In the late 16th century, the Italian physicist and mathematician Galileo GALILEI reestablished mathematical rationalism over ARISTOTLE's logico-verbal approach and suggested that experimentation should be used primarily to check theoretical deductions, while the French philosopher Blaise PASCAL favored the principle of empiricism and proposed the experimental approach. Both conceptions, supplementing each other in a unique manner, established the principle that was to become the guiding maxim of modern science.

Max PLANCK,[3] an eminent German physicist, appropriately said, "Theory and experiment, they belong together; the one without the other remains unfruitful. We are fully justified in applying KANT's well-known words on the unity of concept and intuition and saying: theories without experiments are empty, experiments without theory are blind. Therefore both – theory and experiment – call for proper respect with the same emphasis."

Mechanics became the most important and advanced branch of *natural philosophy* – a term used by Sir Isaac NEWTON to denote investigations of laws that hold in the material world and the deduction of results that are not directly observable, and which is today covered by the field of *physics*. According to the German philosopher Immanuel KANT, science is characterized by an ordered arrangement of gained knowledge, based on data and observed phenomena as well as on similar cases and their critical testing through the application of "creative inspiration." Phenomena that appeared similar were particularly puzzling to early naturalists, such as

▸ the instantaneous, discontinuous character of the velocities observed during the percussion of tangible bodies and of air molecules at a steep shock front;

▸ the wall-like, crested front of a tidal bore in a river and the steep front of a shock wave in air;

▸ the bow wave generated by a body moving through water (*Kelvin envelope* or *Kelvin wake*) and the head wave (*Mach cone*) generated by a projectile flying supersonically, or by a planet moving through the solar wind;

▸ the irregular reflection properties of hydraulic jumps and shock waves; and

▸ the propagation behavior of a shock wave and a detonation wave.

Since the first scientific investigations of the nature of percussion of tangible bodies, an increasing number of new discontinuous high-rate phenomena have been observed, calling for new definitions, explanations and classifications. To some extent the technical terms used by early natural philosophers to describe high-speed phenomena reflect their early knowledge and understanding. However, as we have looked more deeply at the physical processes over decades of increasing research activity, the meanings of some of the terms have changed, and our insights into their complexity

---

[1] G. GALILEI: *Il saggiatore* ("The Assayer"). Mascardi, Roma (1623).
[2] "Dans la physique, les expériences ont bien plus de force pour persuader que les raisonnements." See B. PASCAL: *Traité de la pesanteur de la masse de l'air*. Desprez, Paris (1663), chap. I.
[3] M. PLANCK: *Zum 25jährigen Jubiläum der von W. FRIEDRICH, P. KNIPPING and M. VON LAUE gemachten Entdeckung* [Lecture held during the meeting of the DPG in Berlin on 9th June 1937]. In: (hrsg. vom Verband dt. Physikalischer Gesellschaften) *Max PLANCK. Physikalische Abhandlungen und Vorträge*. Vieweg & Sohn, Braunschweig (1958), Bd. III, pp. 364-367.

has created a wealth of new, related terms. The main technical terms used in the fields of percussion, collision, impact, explosions, detonations and shock waves are discussed below in more detail.

## 2.1.1 PERCUSSION, CONCUSSION, IMPACT, AND COLLISION

Many early terms describing high-speed phenomena were derived from Latin, then the language used in most learned works: examples include the terms *collision, percussion, explosion*, and *detonation*, which are still in use today. However, reflecting the state of knowledge of each time period, they only gradually evolved into their present-day meanings. This slow process of arriving at clear definitions, starting in the 17th century, was not caused by poor communication. Surprisingly, many early naturalists used to exchange knowledge and ideas with their colleagues abroad at an intense level via correspondence or by traveling. Actually, in order to comprehensively characterize the essentials of high-speed phenomena it is necessary to obtain a deep understanding of how the phenomenon in question evolves over time, both qualitatively and quantitatively. This learning process, which only evolved slowly because of the insufficient temporal resolution of early diagnostics, will be discussed in more detail in Sect. 2.8.

**Percussion.** The term *percussion*[4] designates the action of striking of one moving object against another with significant force. Since the birth of classical percussion in the 17th century {HUYGENS ⇨ 1668/1669; WREN, WALLIS ⇨ 1669; Sir NEWTON ⇨ 1687; ⇨ Fig. 2.16}, the term has been conventionally applied in reference to solid bodies, and in this sense it has also been used throughout the evolution of (terminal) ballistics. The fundamental theory of percussion was based on two spheres of the same material but different masses moving in a straight line and impacting either head-on (*central percussion*) or at an angle (*oblique percussion*).

Real percussion phenomena depend on the shapes of the impacting bodies, their masses, their elastic properties (rigid, perfectly elastic, elastic or inelastic), and their initial velocities. In purely elastic percussion no permanent deformation takes place, and both momentum and kinetic energy are conserved. In the early development of the kinetic theory of gases, percussion models generally assumed that the gas molecules collide with one another or with the wall perfectly elastically, like hard elastic spheres (*e.g.*, glass marbles, billiard balls). In contrast, inelastic percussion between moving bodies produces permanent deformation. While momentum is conserved here as well, kinetic energy is not. Very old examples of inelastic percussion are the soft-hammer percussion method used in flint knapping {⇨Fig. 4.2–D(d)} and the wooden mallet used in combination with a metal chisel in stone masonry {⇨Fig. 4.2–E} and wood carving. Modern vehicle design attempts to largely absorb the kinetic energy in the case of collision accidents by using materials which deform plastically.

The word *percussion* [Germ. *der Stoß*] – used almost exclusively by French and English natural philosophers throughout the 17th and 18th centuries, and also partly in the 19th century – relates to the analysis of the physical process involved when bodies striking each other with some degree of force, while the resulting effect is described by words such as shock, blow, impact, knock *etc*. Percussion refers to solid bodies, more rarely to liquids, and to air.[5] It is interesting to note here that the word "percussion" will prompt most English-speakers to initially think of music rather than physics, ballistics and engineering. On the other hand, some encyclopedias (such as the 1974 edition of the *Encyclopedia Americana*) only refer to the medical meaning of the term *percussion* {VON AUENBRUGGER ⇨ 1754}.

In modern textbooks on mechanics, the classical theory of percussion is often referred to as the "theory of impact," while in physics textbooks it is usually called the "theory of collision." Indeed, Edward J. ROUTH {⇨ 1860} used the term *impact* rather than *percussion* in his widely used textbook *Dynamics of a System of Rigid Bodies*. Thomas J. MCCORMACK, who translated Ernst MACH's book *Die Mechanik in ihrer Entwicklung, historisch-kritisch dargestellt* {E. MACH ⇨ 1883} into English, followed ROUTH's terminology and translated *Die Theorie des Stoßes* into "The Theory of Impact." However, the term *center of percussion* – coined in England in the late 17th century {WALLIS ⇨ 1670/1671} – has long been used by both mechanical engineers and physicists.

The principle of percussion has been widely applied in military technology, but also in civil engineerinmg and medical diagnostics which created a number of percussion-related terms (*see* Sect. 2.2.4).

**Concussion.** The term *concussion*[6] describes the action of violently shaking or agitating, particularly in relation to the shock of impact. In the past it was also used to describe the

---

[4] From Lat: *percutĕre*, meaning to strike or thrust through.

[5] J.A. SIMPSON and E.S.C. WEINER: *The Oxford English dictionary*. Clarendon Press, Oxford (1989), vol. 11, p. 528.

[6] From Lat. *concussio*, meaning a "shock" or "blow."

sudden shaking actions of violent seismic waves,[7] or in gunnery.[8]

Today the term *concussion* is primarily used in medicine to designate a period of paralysis of nervous function resulting from an injury to the brain which, produced by a violent blow to the head, causes temporal unconsciousness. Concussions of the brain can affect memory, judgment, reflexes, speech, balance and coordination. Many encyclopedias, particularly the older editions, refer to this meaning only. In modern geology, a *concussion fracture*[9] designates one of a system of fractures in individual grains of a shock-metamorphosed rock that is apparently formed by violent grain-to-grain contacts in the initial stages of the passage of a shock wave.

The term concussion is also used in weapons technology such as the *concussion grenade* (an antipersonnel device which uses a brilliant flash and loud bang to render an enemy in its vicinity blind, deaf and immobile for a brief period of time) and the *concussion fuse* (a bomb fuse used in ordnance that is set off in the air by the explosion of a previous bomb).

**Impact.** The term *impact*,[10] when used in mechanics, is a single forceful collision and designates the process of momentum transfer between two moving bodies by violent percussion. Impact occurs when two bodies come together with a normal component of relative velocity at an initial point of contact. An impact frequently results in a sudden drop of velocity accompanied by a decrease in dynamic energy. When the two bodies first make contact at a single point, deformation will take place at that point and will progressively spread until the two bodies eventually make contact over the maximum area. In science, the term *impact* was apparently first used by the British chemist Richard WATSON (1781) and the Irish naturalist John TYNDALL (1863) to denote collision effects on an atomic level.[11] In his hypothesis on the formation of the lunar surface by meteorite impact, the U.S. geologist Grove K. GILBERT {⇨1893} first used this term to characterize a typical lunar ring mountain ("impact crater").

(i) *Impact mechanics* is concerned with the reaction forces that develop during a collision and the dynamic response of structures to these reaction forces {STRONGE ⇨2000}.

(ii) *Impact engineering* is a rather new branch of engineering which is concerned with the response of structures and bodies to dynamic loads arising from exposure to blast, collision ("crash tests") or other impact events, and the use of numerical codes in examining impact problems {Int. J. Impact Engng. ⇨1983}. The term *impact engineering* was apparently coined in the early 1980s. The first conference on this subject was held in 1992 {Int. Symp. on Impact Engng. ⇨1992}.[12]

▸ The term *impulsive loading* describes high-velocity processes which, for example, occur when an explosive charge is detonated in intimate contact with a body {RINEHART & PEARSON ⇨1954} or when one body impacts against another at high velocity.

▸ *Structural impact* is concerned with the behavior of structures and components subjected to large dynamic loads produced by explosions, impacts, tornadoes, tsunamis *etc*.

▸ *Structural impact engineering*, which explores the responses and energy-absorbing properties (including damage and failure) of various structural systems, requires a profound knowledge of material characteristics in the dynamic range. This is of particular interest for the crashworthiness design of aircraft, automobiles, trains, ships and marine structures {BARÉNYI ⇨1951; JOHNSON ⇨1978; JONES ⇨1989}.

▸ *Impact welding* is a technique of joining dissimilar materials that cannot be joined by other means (such as aluminium alloys to steel). There are two principle methods of propelling one part into another to form a weld: first by fast burning explosives and second by electro-magnetic induction/repulsion.

Impact engineering now deals with a wide variety of impact situations: it attempts to define the forces that occur during an impact event and the techniques that can be used to absorb these forces; for example, it addresses special but very important problems such as testing and modeling the resistance of aircraft components and jet engines to being struck by birds.

(iii) *Impact physics* is a more recent branch of solid-state shock wave physics that treats the wide spectrum of impact phenomena experimentally and theoretically from the viewpoint of dynamic materials science and thermodynamics, and addresses action as well as reaction effects. Impact physics is mainly derived from terminal ballistics, which

---

[7] J.A. SIMPSON and E.S.C. WEINER: *The Oxford English dictionary.* Clarendon Press, Oxford (1989), vol. II, p. 677. • For example, Don George JUAN and Don Antonio DE ULLOA, referring to the terrible earthquake that happened in 1772 in Quito, wrote in their book *A voyage to South America* [Davis & Reymers, London, 1760], "The terrible concussion was general all over the province of Quito…"

[8] William GREENER wrote in his book *Gunnery in 1858*, a treatise on rifles, cannons and sporting arms [Smith & Elder, London, 1858], "The proper shape and form of cannon to resist concussion…"

[9] J.A. JACKSON (ed.): *Glossary of geology.* Am. Geol. Inst., Alexandria, VA (1997), p. 133.

[10] From Lat. *impactus*, the action of striking or pushing at or against.

[11] J.A. SIMPSON and E.S.C. WEINER: *The Oxford English dictionary.* Clarendon Press, Oxford (1989), vol. VII, p. 695.

[12] M. MACAULAY: *Introduction to impact engineering.* Chapman & Hall, London (1986, 1987).

investigates the various events that occur when a projectile hits a target. Although this field is centuries old, the theory of terminal ballistics is a relatively modern development; it was first developed in the 1960s to satisfy military needs such as improving armor on military vehicles. However, studies of impact and penetration phenomena in terminal ballistics have also been motivated by interests in subjects as diverse as meteorite craters on the Moon and planets, meteoroid protection of space vehicles, and solid state physics of materials at extremely high pressures.[13]

Almost 50 years ago, Robert GRAHAM,[14] a Sandia shock physicist, worked out a 105-page bibliography on *Impact Physics* consisting of a rather complete collection of over 5,000 references and abstracts on the subjects of (1) plastic wave propagation in bounded solids; (2) behavior of metals under explosive conditions; (3) dynamic photoelasticity; (4) penetration phenomena; (5) behavior of material at high strain rates; (6) lateral impact; and (7) impact measurement devices. All these subjects are now as before relevant in modern impact physics. The first book on the physical behavior of impacting solids was published by Werner GOLDSMITH {⇨1960}.

In solid-state shock wave physics, the term *symmetric impact* relates to a laboratory method of generating planar shock waves by shooting a plate, the so-called "flyer plate" {McQUEEN & MARSH ⇨1960}, with a velocity $W$ against a stationary plate of the same material as the flyer plate. In this special case, a pair of shock waves moving out with $\pm U$ from the interface at particle velocities $\pm u = \pm\frac{1}{2}W$ are generated. The symmetric impact allows the generation of a "standard Hugoniot" from a chosen "standard material" such as 24ST aluminum (an alloy of 4.5% Cu, 0.6% Mn, and 1.5% Mg), which facilitates the determination of Hugoniot curves for any solid material {⇨Fig. 2.13}.

Some research laboratories are particularly dedicated to the physical aspects of impact. For example, the U.S. Army Research Laboratory (ARL) at Aberdeen Proving Ground, MD has an *Impact Physics Branch* which is associated with the Terminal Effects Division of the Weapons and Materials Research Directorate (WMRD); and the Ernst-Mach-Institut (EMI) at Freiburg, Germany has an *Impact Physics Division*. Various American universities have *Impact Physics Groups*, such as the University of Dayton Research Institute (UDRI) at Dayton, OH and the Institute for Advanced Technology (IAT) of The University of Texas at Austin. Traditionally, physical aspects of high-velocity impact phenomena are reviewed and discussed at the Hypervelocity Impact Symposia {HVIS ⇨1986}.

(iv) *Hypervelocity impact physics* is an entirely new field of research which is derived from impact physics and was established in the mid-1950s. It should be noted here that *hypervelocity* is a term that is used in different branches of science to classify quite different velocity regimes:

▸ Hypervelocity impact physics deals with the impact of matter at extremely high velocities, resulting in extreme pressure and loading rates. In particular, it encompasses events where, in contrast to the impact of common ordnance, the impact-generated pressures are in excess of the projectile and target strength. For most metals, the lower limit for hypervelocity impact is on the order of (only) 3 km/s.[15]

▸ For aerodynamicists, hypervelocity wind tunnels are testing facilities operated at airflow velocities greater than $M > 12$ (> 4 km/s).

▸ In line with the discovery of shatter cones found in the Kentland quarry in Indiana {BUCHER ⇨1933; BOON & ALBRITTON ⇨1938; DIETZ ⇨1960; SHOEMAKER ET AL. ⇨1961}, geologists have defined the term *hypervelocity* as velocities greater than the speed with which sound travels through average rock (> 5 km/s).

▸ For astronomers and astrophysicists, the term *hypervelocity* covers a very wide range. Meteoroids entering the Earth's atmosphere have velocities of up to 40 km/s. The speeds of extraterrestrial hypervelocity phenomena can range from several hundreds of km/s (*e.g.*, the solar wind) up to 2,000 km/s (*e.g.*, coronal mass ejections), and even to well in excess of 30,000 km/s (*e.g.*, shock waves emitted from hypernovae).

▸ Nuclear physicists use electromagnetic accelerators to accelerate light particles (such as electrons, proton or α-particles) to hypervelocities that are asymptotic to the speed of light.

A hypervelocity impact is accompanied by an *impact flash*. Measurements in several optical, infrared, ultraviolet and even X-ray wavelengths allow astrophysicists to draw conclusions about the kinetic energy of the impact event and possibly about the physical state and nature of the impacted surface {Comet SHOEMAKER-LEVY 9 ⇨1994; Deep Impact Mission ⇨1999; SMART-1 ⇨2003}. Impact-flash phenomenology has been known for many years, and

---

[13] J.L. SUMMERS and B.P. DENARDO: *Terminal ballistics*. In: (T.N. CANNING, A. SEIFF, and C.S. JAMES, eds.) *Ballistic range technology*. AGARDograph No. 138 (1970), pp. 157-165.

[14] R. GRAHAM: *Impact physics*. Sandia Rept. SCR-59 (Dec. 1958); http://stinet.dtic.mil/cgi-bin/GetTRDoc?AD=ADA395534&Location=U2&doc=GetTRDoc.pdf.

[15] H. FAIR: *Hypervelocity then and now*. 1st High Velocity Impact Symposium (HVIS), [San Antonio, TX, 1986]. Publ. in Int. J. Impact Engng. **5**, 1-11 (1987).

is now also being considered for missile-defense applications, in particular, remote diagnostics for kill assessment and target typing.[16]

Hypervelocity impact physics mainly includes studies of the basic phenomenology of cratering, of the impacts of complex structures in space, and of properties of materials at very high pressures. In particular, it is devoted to investigating and assessing the threat of hypervelocity impacts between any spacecraft and man-made debris or naturally occurring meteoroids in the near-Earth space environment, known as "Near-Earth Objects" (NEOs) {Asteroid 2002 MN ⇨ 2002}, as well as protecting spacecraft though passive shielding techniques {WHIPPLE ⇨ 1947}. Hypervelocity impact physics has also gained increasing attention in other branches of science, for example in *astrogeology*, a new branch of geology created in the early 1960s and dedicated to the study of extraterrestrial solid objects {LESEVICH ⇨ 1877}.

(v) *Impact chemistry* is a more recent interdisciplinary branch of science which investigates the shock- and heat-induced chemistry of impacted matter (mostly gases and solids). Examples encompass

▸ detection of unique chemical "signatures" of meteorite impacts in sediment deposits;

▸ possible high-temperature- and shock-wave-induced production of large amounts of nitride oxide in the atmosphere (acid rain) and the stratosphere (ozone depletion);

▸ devolatilization (or pyrolysis) in serpentine [(Mg, Fe)$_3$Si$_2$O$_5$(OH)$_4$], Murchison meteorite, serpentine-iron, and serpentine-pyrrhotite mixtures and their possible role during planetary accretion;[17]

▸ interaction of Jupiter's atmosphere with incandescent fireballs at high altitude produced by the collision of comet SHOEMAKER-LEVY 9 where they formed plumes that subsequently collapsed over large areas;

▸ element anomalies (such as iridium which occurs in meteorites in much greater concentrations than in rocks on Earth); and

▸ the fate of organic matter during planetary accretion.[18]

(vi) *Impact geology* is a branch of planetary geology which treats the complex stages of cratering mechanics (*e.g.*, compression, excavation, emplacement of ejecta, crater collapse), the nature of impact structures (*e.g.*, rock fragmentation, shock metamorphosis, astroblemes), environmental effects on crater formation, the morphology of impact basins, numerical 3-D modeling of large-scale asteroid impact events based on field-structural data, and investigates the role of impact processes in the evolution of the Earth and other planets.

▸ Geologic *impact structures* are generally circular and formed by hypervelocity impact of an interplanetary body (*impactor*) on a planetary surface (target). Craters formed by very oblique impacts may be elliptical.

▸ The impact produces an *impact plume*, a hot cloud of debris. Plume modeling has been proposed by planetary physicists to calculate synthetic plume views, atmospheric infall fluxes, and debris patterns raised by impacting fragments of comets.[19]

▸ Large impacts can cause an *impact metamorphism* of rocks or minerals due to the passage of a shock wave. Strong shock waves can form an *impact melt* due to shock melting of rocks in impact craters.

▸ *Impact erosion* can wear away large rock fragments due to the emission of a strong blast wave during impact. Impact erosion may also account for an early episode of atmosphere loss from Mars.[20]

(vii) In fluid dynamics the term *impact* generally refers to the high-speed interaction of gaseous or liquid matter with a (mostly) solid boundary.

▸ The term *impact loading* was apparently first used by Richard COURANT and Kurt Otto FRIEDRICHS in their textbook *Supersonic Flow and Shock Waves* {COURANT & FRIEDRICHS ⇨ 1948} to describe the rapid loading of a structure by a blast wave {BLEAKNEY ⇨ 1952}.

▸ In high-speed aerodynamics, the *impact pressure* at any point of a fluid is defined as the net balance between the total pressure (or stagnation pressure, also known as the "Pitot pressure") and the static pressure at the point where the impact pressure is to be found.[21]

▸ In underwater explosions near solid structures, the bubble collapse may be accompanied by the formation of a high-speed re-entrant liquid jet directed towards the structure. This phenomenon is known as "jet impact loading."[22]

---

[16] R.J. LAWRENCE ET AL.: *Hypervelocity impact flash at 6, 11, and 25 km/s.* In: *14th APS Topical Conference on Shock Compression of Condensed Matter* [Baltimore, MD, July/Aug. 2005]. APS, College Park, MD (2006), pp. 1349-1352.

[17] J.L. FAUST ET AL.: *Impact chemistry of serpentine-iron planetesimals.* 28th Annu. Lunar and Planetary Science Conf. [Houston, TX, March 1997]. Conf. Proc. CD-ROM (Jan. 1997), Abstract #1711.

[18] T.N. TINGLE ET AL.: *The fate of organic matter during planetary accretion: preliminary studies of the organic chemistry of experimentally shocked Murchison meteorite.* Origins of Life and Evolution of Biospheres **21**, No. 5/6, 385-397 (Sept 1991).

[19] J. HARRINGTON and D. DEMING: *Models of the Shoemaker-Levy 9 impacts.* Astrophys. J. **561**, Part 1, 455-467 (2001).

[20] H.J. MELOSH and A.M. VICKERY: *Impact erosion of the primordial atmosphere of Mars.* Nature **338**, 487-489 (April 6, 1989).

[21] H.W. SIBERT: *High-speed aerodynamics.* Prentice-Hall, New York (1948), pp. 37-38.

[22] G.L. CHAHINE and K.M. KALUMUCK: *The influence of structural deformation on water jet impact loading.* J. Fluids Struct. **12**, 103-121 (1998).

▸ The term *liquid impact* encompasses the dynamics of liquids. Since the 1960s, this has become an area of interdisciplinary research stretching far beyond classical fluid dynamics and materials research {LESSER & FIELD ⇨ 1983}.

▸ The high-velocity impact of liquid droplets against a solid metal surface, which causes serious damage due to erosion, has long been a problem for steam turbine designers and operators {THORNYCROFT & BARNABY ⇨ 1895; S.S. COOK ⇨ 1928; ⇨Fig. 4.14−C}.

▸ With the advent of supersonic and even hypersonic flight velocities, the same phenomenon became of increasing concern in the aircraft and (later) in the aerospace industries due to the rain erosion problems sustained by aircraft {U.S. Air Force ⇨ 1945; The Royal Society ⇨ 1965}, missiles and reentry vehicles {HEYMANN ⇨ 1969}.

▸ The phenomenon of liquid impact is also of interest in the production of high-speed liquid jets for effective cutting and cleaning operations in industry, and even in the design of lithotripters in medicine.

▸ The impact of moving structures on still liquid surfaces gained particular attention in the 1930s with the boom in seaplane construction {WAGNER ⇨ 1932}. On the other hand, the impact of moving waters on resting structures is of great importance when designing effective protection for dams against tsunamis and storm surges.

▸ *Liquid impact* also encompasses the problems of a body impacting a smooth planar surface[23] (the so-called "water-entry problem") and of impacting a bulk volume of water. The latter case, known today as a "hydrodynamic ram" (*see* Sect. 2.4.5), is the oldest example of a liquid impact, and was first demonstrated in France by shooting a bullet into a box filled with water {CARRÉ ⇨ 1705}.

(viii) In climatology and anthropology, the term *impact winter* refers to the immediate result of a major asteroid impact similar to the one that occurred around the Cretaceous-Tertiary (or K-T) boundary {Chicxulub Crater ⇨c.65 Ma ago}. This asteroid is believed to have hit off the Yucatán Peninsula with a force of almost one trillion megatons of TNT equivalent. Similar to effects resulting from a *nuclear winter* and a *volcanic winter*, the ultraviolet rays would be blocked by the global dust cloud, choking the planet in icy winter-like conditions for months (or years) and causing an impact winter with widespread death of plants and the large terrestrial animals that most directly depend on those plants for food. Although the so-called "impact-winter hypothesis" may be popular among the media, little evidence supports it, and other possible extinction mechanisms abound, including global wildfires, greenhouse warming, acid rain, volcanic eruptions, tsunamis and every combination in between.[24]

(ix) Besides all of the scientific and technical aspects discussed above, impact phenomena also have serious human and economic consequences: since the use of automobiles is steadily growing worldwide, and speeds of transportation are rising, the number of traffic accidents and with it the number of people injured or killed by collisions or impacts has increased dramatically.[25] These dangers from modern transportation, recently augmented by the omnipresent risk of terrorist attacks, illustrate the immediate importance of protection measures against shock and impact {⇨Figs. 4.21−G, H}. They also call for a better understanding and modeling of the behavior of materials under shock loading conditions.

**Collisions.** The term *collision*,[26] as used today in science and even in law and insurance, describes a wide range of processes and phenomena ranging from very high, relativistic velocities observed at the molecular, atomic and subatomic levels to very slow velocities seen in earth sciences, as illustrated in the following examples:

(i) In general physics the term *collision* does not necessarily imply actual contact as in classical mechanics: it rather describes any interaction between particles, aggregates of particles, or rigid bodies in which they come near enough to exert a mutual influence, generally with exchange of energy.[27]

(ii) In particle physics, a collision describes an interaction between particles rather than between tangible bodies, and one in which momentum is conserved. *Coulomb collisions* are interactions between two moving electric charges at close range {COULOMB ⇨ 1785}. The term *collision* has been used throughout the evolution of modern particle physics, for more than 120 years (P. SPENCE 1880; E. RUTHERFORD 1904).[28] Elementary (subatomic) particles accelerated to high energies and interacting with matter can either scatter or produce other particles.

---

[23] R. SKALAK and D. FEIT: *Impact on the surface of a compressible fluid.* Trans. ASME, Ser. B (J. Engng. Ind.) **88**, 325-331 (1966).

[24] J. CHAPMAN: *Evidence for impact winter at K/T boundary.* Geotimes (June 24, 2004); http://www.geotimes.org/june04/WebExtra062404.html.

[25] According to the U.S. Statistics the number of deaths resulting from traffic accidents (road, rail & air) in the United States amounted, for example, to more than 45,000 in 1996, and the number of people injured in the United States every year is several times this.

[26] From Lat. *collisio*, meaning "the action of colliding."

[27] S.P. PARKER (ed.): *McGraw-Hill dictionary of scientific and technical terms.* McGraw-Hill, New York (1994).

[28] J.A. SIMPSON and E.S.C. WEINER: *The Oxford English dictionary.* Clarendon Press, Oxford (1989), vol. III, p. 487.

The most violent collisions happen when a particle of matter interacts with a particle of *antimatter* {DIRAC ⇨ 1928; CHAMBERLAIN & SEGRÉ ⇨ 1955; SDI ⇨ 1983}. These antiparticles could potentially combine to form antiatoms, and the antiatoms could form antimatter counterparts to every object in the Universe – antistars, antigalaxies. What is more, if a particle of matter collided with a particle of antimatter, they would both be annihilated in an energetic burst of gamma rays, a process called "annihilation." If a human and an antihuman shook hands, the resulting explosion would be equivalent to 1,000 one-megaton nuclear blasts, each capable of destroying a small city.[29] The annihilation of matter with antimatter has also been discussed as a process that could explain the explosive creation of the Universe in a Big Bang, but it has also stimulated science fiction writers.

(iii) In a chemical detonation, explosive molecules are compressed very rapidly and heated by the shock wave, leading to a chain reaction which results in a propagating detonation wave. There is a wide variety of collision events involving large molecules. These can be roughly categorized according to their duration and the amount of energy transferred, ranging from impulsive single collisions and multiple (or complex) collisions, to *supercollision* events.

(iv) In thermodynamics the classical kinetic theory of gases is based on the perfectly elastic collision of molecules or atoms. Its roots date back to first speculations made in the 17th and 18th centuries to explain the nature of heat from the basic laws of motion of "very small, invisible particles" {HOOKE ⇨ 1665; Sir NEWTON ⇨ 1687} and the kinetic energy of their motion {D. BERNOULLI ⇨ 1738}; important improvements to the kinetic theory of gases were made in the 19th century in terms of the statistical motion of molecules by including the theory of probability {MAXWELL ⇨ 1867; BOLTZMANN ⇨ 1872}.

(v) In chemical kinetics, a branch of physical chemistry, collisions are believed to generate chemical changes, but only if the reaction species – the atoms and molecules – have sufficient internal energy; *i.e.*, a *reactive collision* must be of sufficient energy to break the necessary bonds and must occur in a particular orientation. The rate at which a chemical reaction proceeds is equal to the frequency of these effective collisions. In collision theory models of chemical reactions have been developed to explain the rate laws observed for both one-step and multi-step reactions.

(vi) In mechanical engineering the term *collision* is used to describe a sudden, forceful coming together into direct contact of two moving bodies, resulting in an abrupt change in motion of at least one of the bodies – *i.e.*, it is used synonymously with the terms *percussion* and *impact*.

▸ In a *direct central collision*, the bodies move along the same straight line, and in the special case of a *head-on collision* they move in opposite directions. Everyday examples of head-on collisions with one body at rest and the other in motion include the collision of a golf club and a ball,[30] a hammer and a nail head, and pile-driving. The most prominent examples of direct central collision and elastic head-on collision and chain percussion are provided by an apparatus known as "NEWTON's cradle" {⇨Fig. 4.4–B}.

▸ A *side collision* is a collision between two bodies that are not moving along the same straight line, causing their velocity vectors to intersect at the *point of collision*.

▸ An *oblique central collision* is a special case of a collision which occurs when the colliding bodies are still confined to the same plane.

(vii) In traffic accidents a *head-on collision* is one where either the front ends of two road vehicles, ships, trains or planes hit each other, or a vehicle hits an object from the front. a *rear-end collision* occurs when a vehicle is hit from behind by another vehicle. It is usually caused by the inattentiveness of the driver of the vehicle, which hits the other vehicle from behind, or by the sudden occurrence of a *traffic shock* {LIGHTHILL & WHITHAM ⇨ 1955; ⇨Figs. 4.4–G, H}.

(viii) In stellar dynamics the term *collision* denotes actual physical contact between stars, while the term *encounter* denotes the gravitational perturbation of the orbit of one star by another. In large collisionless stellar systems, such as galaxies, stellar encounters are entirely unimportant when considered over the galaxy's lifetime; however, in smaller stellar systems such as globular clusters they may play a major role.[31]

▸ *Collisional accretion* by gravitational forces is considered to be an important mechanism during the formation of a solar system, particularly during the accretion of small masses into larger ones and eventually the molding of accreting bodies into roughly spherical planets and moons.

▸ *Solar wind collision.* The collision of the solar wind with the interstellar plasma slows the supersonic particles emerging from the Sun to subsonic speeds and results in a real astronomical shock wave effect – known as the "termination shock" (*see* Sect. 2.1.5).

---

[29] G. TARLÉ and S.P. SWORDY: *Cosmic antimatter.* Scient. Am. **278**, 36-41 (April 1998).

[30] For a high-speed video showing the impact of a metal-driver clubhead on a golf ball *see Gallery of high-speed video applictions.* Photron USA Inc., San Diego, CA; http://www.photron.com/gallery/gallery.cfm.

[31] J. BINNEY and S. TREMAINE: *Galactic dynamics.* Princeton University Press, Princeton, NJ (1987).

▸ *Stellar collisions* are common in dense star clusters (*e.g.*, globular clusters and the dense cores of young clusters), where they can result in blue stragglers (stars which are hotter and bluer than other cluster stars having the same luminosity) and very massive stars (> 100 $M_\odot$, where $M_\odot = 1.98 \times 10^{30}$ kg is the mass of the Sun). Direct physical collisions between single stars are rare except in the cores of the densest clusters. Stellar collision researchers distinguish between *parabolic collisions* and *hyperbolic collisions*.[32] A third category are collisions occurring between bound partners in a binary star, either because of the perturbation of the pair by a third star (*elliptic collision*), or as a result of "normal" binary evolution (*circular collision*). Numerous computer simulations using Smoothed Particle Hydrodynamics (SPH) have been carried out in order to explore the evolution of collisionally merged stars.[33]

▸ *Galaxy-galaxy collisions.* Collisions of huge dimensions occur in the Universe when galaxies – *i.e.*, vast systems of stars that also contain gas and dust in various concentrations – interact with each other, thus creating shock waves {Stephan's Quintet ⇨ *c*. 300 Ma ago; NGC 6240 ⇨ 2001; NGC 6745 ⇨ 2004}. Galaxy collisions involve a tremendous amount of energy: two objects with masses of the order of $10^{12}$ solar masses or about $2 \times 10^{42}$ kg meet with typical relative velocities of about 300 km/s, so the collision energy is of order $10^{53}$ J. Galaxy collisions are extremely slow by terrestrial standards, with typical time scales of order $3 \times 10^8$ years, or $10^{16}$ seconds. There is little hope of observing any of the dynamics directly. Since in a galaxy collision the tidal gravity forces are responsible for the most significant effects, the term *tidal interaction* is more commonly used in the field.[34]

When galaxies slam into each other, their individual stars almost never collide. Since the sizes of the stars are very small compared to the average distance between them, the probability of two stars colliding is almost zero. Rather the structure of one or both galaxies gets slowly disrupted, while interior gas condenses to new star forming regions. It is now thought that most galaxies experience several collisions or tidal interactions over the course of their lifetime which are strong enough to profoundly alter their structure and accelerate evolutionary processes. There is also friction between the gas in the colliding galaxies, causing shock waves that can trigger some star formation in the galaxies.

(ix) In Earth sciences the term *collision* is used to describe the motions and interactions of individual plates of varying size on the surface of the Earth. Although the motions involved are very slow compared to all of the collision events mentioned above (in the range of cm/year), tectonic activity can manifest itself as violent earth/seaquakes and volcanism. {LESSER & FIELD ⇨ 1983; Sumatra-Andaman Islands Earthquake ⇨ 2004}.

▸ *Plate convergence* – a major type of plate boundary behavior, and called "a very slow collision" by geologists – can occur between an oceanic and a largely continental plate, between two largely oceanic plates, or between two largely continental plates (*continental collision*). For example, the collision of India into Asia 50 million years ago caused the Eurasian plate to crumple up and override the Indian plate.[35]

▸ *Collisional orogeny* – the formation of mountains due to the thickening of the continental crust – is caused by the convergence of two plates of the continental lithosphere when the intervening oceanic lithosphere is destroyed by subduction.[36]

### 2.1.2 EXPLOSION AND IMPLOSION

Explosions and implosions are extremely rapid phenomena that cause a rapid increase in heat and pressure. In both cases the resulting overpressure, propagating as a wave of condensation, steepens its front, thus turning into a shock wave; *i.e.*, it travels at a supersonic velocity. However, they differ in their geometrical and temporal characteristics: in an explosion, a divergent process, the pressure decays with increasing distance from the explosion source; in an implosion, a convergent process, the pressure increases rapidly with time which can lead to unstable fluid dynamics towards the center of the implosion. Explosions and implosions play an important role during the various stages in stellar evolution.

**Explosion.** There are three main types of explosions: mechanical, chemical and nuclear explosions. Exploding wires, which belong to a particular class of explosions, are impressive examples of extremely fast changes of physical state: the wire material is rapidly transformed from the solid state into

---

[32] M. FREITAG: *Stellar collisions: quantities and definitions.* Modeling Dense Stellar Systems (MODEST) Working Group 4: *Stellar collisions*; http://www.ast.cam.ac.uk/~freitag/MODEST_WG4/Quantities.html.

[33] Examples of computer animations of "Stellar collisions and cosmic catastrophes" can be watched in the Internet {ASP Conference ⇨ 2000}.

[34] C.J. STRUCK: *Galaxy collisions.* Physics Reports **321**, 1-137 (1999).

[35] U.S. Geology Survey (USGS): *The dynamic Earth: understanding plate motions*; http://pubs.usgs.gov/publications/text/understanding.html.

[36] P.L. HANCOCK and B.J. SKINNER (eds.): *The Oxford companion to the Earth.* Oxford University Press, Oxford (2000), pp. 148-151.

the gaseous one, generating heat and an expanding plasma that carries a shock wave. With the advent of high-power pulsed lasers, a new type of explosion source was discovered: the breakdown of a gas caused by the attenuation of an intense pulse of focused laser light produces a kind of spark, the "laser spark" {LEE & KNYSTAUTAS ⇨1969; ⇨Fig. 4.11–G}. However, much higher temperatures and shock wave velocities are achievable than in electric sparks or exploding wires.

The oldest example of an explosion is the explosive rupture of some moist rocks or firewood when exposed to fire, which is actually a steam explosion that may have been noticed by prehistoric man.[37] In the Roman Empire, the word "explosion" [Lat. *explosio*] designated a custom used to express displeasure by clapping one's hands in order to drive (a play) off the stage.[38] Beginning with the advent of black powder in Europe in the 14th century, the term *explosion* was mostly applied to the explosion of gunpowder, but later also to the first high explosives. However, an explosion is not necessarily connected with the exothermic reaction of a chemical explosive. With the advent of the *air gun* (or *wind gun*) in the 15th century, imperfectly constructed high-pressure reservoirs attached to the gun provoked mechanical explosions which could seriously endanger the shooter. On the other hand, the rapid disintegration of Prince RUPERT's drops, which suddenly releases the compressive stress stored internally in the glass, is a comparatively mild explosion and was used as a joke in public demonstrations {HOOKE ⇨1665; ⇨Fig. 4.16–W}. Mechanical explosions caused by overloading pressurized containers happened quite frequently throughout the Industrial Revolution. Prominent examples include the sudden rupture of boiler walls in steam-boiler explosions, which often resulted in many casualties {SS *Le Rhône* ⇨1827; ARAGO ⇨1830; AIRY ⇨1863; HMS *Thunderer* ⇨1876; ⇨Fig. 4.21–A}.

In particular, explosions in air have been defined as when the release of energy is rapid and concentrated enough to produce a pressure wave that one can hear {STREHLOW & BAKER ⇨1976}. Air explosions can be characterized and classified by the generated "blast signature;" *i.e.*, the shock pressure-time profile measured at a certain distance. This depends on the energy rate, the total energy released, and the source geometry. In chemical explosions, some of the total energy released is also transformed into endothermic reactions of the explosion products, and into radiation. Furthermore, the fission energy that is converted into the shock and the blast in a nuclear aerial explosion is reduced due to the additional production of initial and residual nuclear radiation.[39]

The geometry of the explosion source strongly determines how quickly the blast pressure decreases from the center of the explosion, and the following basic cases can be differentiated:

▸ *1-D explosions:* Classic examples are the bursting membrane in a shock tube {VIEILLE ⇨1899; W. BLEAKNEY 1943; ⇨Figs. 4.10–A, B}, and gaseous detonations generated in straight tubes, channels and tunnels. Experimental studies of 1-D explosions have been of particular interest during the history of shock wave and detonation physics, because they first allowed theories to be tested using rather simple one-dimensional fluid dynamic models.

▸ *2-D explosions:* Exploding wires of sufficient length may approach the ideal of *cylindrical explosions*. Exploding thin metal foils, such as those used to produce planar shock waves in solids in contact with such foils {KELLER & PENNING ⇨1962; MÜLLER & SCHULTE ⇨1970}, produce complicated 2-D/3-D wave patterns because of "edge effects." The irradiation of planar solid targets by ultrashort high-power laser pulses {ASKAR'YAN & MOROZ ⇨1962; ⇨Figs. 4.11–H, I} or by gamma-ray pulses emitted from nuclear explosions {TRUNIN ET AL. ⇨1992; ⇨Fig. 4.11–F} are special cases and can produce 2-D blow-offs of target material, thus resulting in an almost planar shock wave propagating into the target.

▸ *3-D explosions:* All natural and most man-made explosions are three-dimensional (3-D) and at large distance roughly approach the ideal case of a true spherical shock. Their analytical treatment is a particularly difficult task and has occupied generations of theoretical gas dynamicists. Compared to 1-D and 2-D explosions, 3-D explosions are characterized by a faster decay of the pressure with distance.

▸ *Point explosions:* Interest in the gas dynamics of spherical shocks (or blasts) initially focused on finding analytical similarity solutions to the hypothetical *point source* in order to describe the propagation of a strong spherical shock wave, which resulted in the famous Taylor-Sedov similarity solution {G.I. TAYLOR ⇨1941}.

---

[37] First evidence for the use of fire was found in Chesowanja, East Africa, dating back to 1.5 million years ago. Fire was also used by early man to heat pebbles, which facilitated stone tool production: when put into a fire, a stone can fracture explosively due to any captured moisture turning rapidly into steam.

[38] P.G.W. GLARE (ed.): *Oxford Latin dictionary*. Clarendon Press, Oxford (1968), p. 651.

[39] For example, for a fission weapon detonating in air at an altitude below 100,000 ft (30.48 km), the fission energy is approximately distributed as follows: about 50% blast and shock, 35% thermal radiation, 5% initial nuclear radiation and 10% residual nuclear radiation. See S. GLASSTONE: *The effects of nuclear weapons*. U.S. Atomic Energy Commission, Washington, DC (1962), pp. 8-9.

Any 3-D explosion approaches a *spherical explosion* at a distance that is much larger than the geometrical dimensions of the source. Classic examples of spherical explosions include detonations of spherical charges of chemical explosives or nuclear devices, and electric discharges of capacitor banks over short spark gaps. In the near field, however, these explosive sources may show considerable deviations from an ideal point source. Even in laser sparks, the driving hot plasma in the breakdown phase is not spherical, but rather deformed in the direction of the incident laser pulse {⇨Fig. 4.11−G}. Stimulated by the need for a better insight into the behavior and effects of nuclear explosions, particular attention was paid to the practical flow phenomena accompanying a *finite source explosion*. Considerable theoretical work was done by Harold L. BRODE {⇨1954, 1959 & 1968; see GLASSTONE ⇨1957} of the Rand Corporation in Santa Monica, CA, on the sudden expansion of a sphere of gas, initially pressurized with air or helium. Experimental studies of the spherical explosions of pressurized thin-walled glass spheres, performed by Canadian fluid dynamicists at the University of Toronto Institute of Aerospace, allowed a comparison to be made with BRODE's numerical solution {GLASS & HALL ⇨1957; ⇨Fig. 4.16−F}. They carried out their experiments in a 3-ft-diameter spherical steel tank, which they called a "shock sphere."

*Steam-boiler explosions* – the most common examples of mechanical explosions in the past – were often caused by unwittingly exceeding the nominal steam pressure, for example by increasing the weight of the safety valve or through poor maintenance {HMS *Thunderer* ⇨1876}. In industry, such actions also had social causes: in the early era of the industrial revolution most workers were paid by the piece, so the steady availability of steam at full power was of vital interest not only to the factory owner but also to the workman's livelihood. This often resulted in boilers overheating and poor inspections of the technical equipment. Another frequent reason for steam-boiler explosions was the sudden release of a large amount of steam, for example through the operation of the boiler at very low water level, or by starting the engine at full steam. The latter case was vividly described in Mark TWAIN's classical autobiography *Life on the Mississippi* (1883) for the example of the side-wheeler SS *Pennsylvania*, which exploded near Ship Island in the State of Mississippi (resulting in 150 casualties). To reduce the increasing number of disastrous boiler explosions – particularly when operated in close contact with the public, like the ones installed on steam boats {Norwich & Yarmouth steamer ⇨1817; SS *Le Rhône* ⇨1827; SS *Princess* (1859) ⇨Fig. 4.21−A; HMS *Thunderer* ⇨1876} or steam locomotives {Locomotion No. 1 ⇨1828; *Best Friend of Charleston* ⇨1831} – private boiler inspection unions were founded in several European countries. These bodies supervised the materials used, the construction and the operation of steam-boilers. They were the seeds for modern governmentally operated inspection authorities that act on modern boiler and pressure-vessel standards and on those for other fields of engineering.[40]

*Thermohydraulic explosions* are violent explosions that can be observed during the contact of a hot liquid with a cool liquid if the temperature of the hot liquid exceeds the homogeneous nucleation temperature of the coolant. This phenomenon, resulting in a violent steam (or vapor) explosion, was named the *fuel-coolant interaction* {Heimaey Eruption ⇨1973}. In the case of water and a hot melt, the term *molten fuel-coolant interaction* has been introduced. As well as being the reason for some severe accidents in industrial plants, molten fuel-coolant interactions were found to play an important role in explosive water-magma interactions (such as in the course of phreatomagmatic volcanic eruptions[41]) and in core disruptive accidents of nuclear reactors {Chernobyl ⇨1986}.

*Micro steam explosions* occur when, for example, matter is exposed to ultraviolet laser radiation of ultrashort duration, and they are used to remove biological tissue from the cornea or skin (for example). Ablation is achieved through a combination of photothermal and photomechanical effects. Thermal denaturation weakens the structural matrix of the tissue, while the explosive transition of water to high-pressure vapor then ruptures the structural matrix, propelling the ablated material from the site of irradiation. Where there is a tissue-air boundary, the second mechanism is clearly seen as an *ablation plume* – a fine mist which contains most of the material ejected during the ablation process {PULIAFITO ET AL. ⇨1987; ⇨Fig. 4.16−Z}. This technique has become an important medical therapy and is widely used in ophthalmology and dermatology.

*Chemical explosions* may be defined as the sudden or extremely rapid conversion of the solid or liquid bulk of an explosive into gas or vapor which is highly expanded by the heat generated during the transformation and so occupies many times the volume of the original substance. The most frequent chemical explosions are gas explosions.

*Gas explosions* are defined as a process where combustion of a premixed gas-air cloud causes a detonation, a rapid

---

[40] For example, in Germany the privately operated Steam-Boiler Inspection Union [Germ. *Dampfkessel-Überwachungsverein*] which successfully reduced the number of boiler explosions during the period 1870–1900, also stimulated the creation of various official Industrial Inspection Boards [Germ. *Gewerbeaufsichtsämter*] and the privately operated, multipurpose Technical Inspection Union [Germ. *Technischer Überwachungsverein*].

[41] R. BÜTTNER and B. ZIMANOWSKI: *Physics of thermohydraulic explosions*. Phys. Rev. **E57** [III], No. 5, 1-4 (1998).

increase of pressure. In fuel-air mixture at atmospheric pressure, the detonation peak pressure is 15–20 bar. A detonation is the most devastating form of a gas explosion. However, a gas explosion might be caused also by the bursting out of gases or vapors from a vessel under high pressure.

*Dust explosion* are the oldest man-made chemical explosions, particularly flour dust explosions that occurred in bakeries {MOROZZO ⇨1785}. Dust explosions occur in modern food processing industries when finely divided combustible matter (*e.g.*, dusts of cereal grain, soy beans, *etc.*) dispersed into an atmosphere containing sufficient oxygen to permit combustion is accidentally ignited – often by friction-generated flames or electrostatic sparks {American Grain Industry ⇨1977; First Int. Colloquium on Explosivity of Industrial Dusts ⇨1978; ⇨Figs. 4.21–D, E}. There is a direct correlation between particle size and its risk of exploding. The smaller the particle, the more reactive the dust. With the beginning of the industrial revolution in the 19th century, dust explosions reached new heights when *coal dust explosions* became a considerable hazard in the coal mining industry {FARADAY & LYELL ⇨1845; ABEL ⇨1881}. *Metal dust explosions* {DORSETT ET AL. ⇨1960}, generally more disastrous than flour and coal dust explosions, became a problem in the 19th century with the advent of new production technologies in the metal industry {Bethlehem Zinc Works ⇨1854}. More recently, it was even discovered that textile dusts produced from processed and treated nylon fiber may be combustible, ignitable and explosive under certain conditions.

*Thermal explosions* are complex events that can involve many chemical and physical processes.[42] For example, thermal explosions occur when matter is evaporated rapidly, such as by focusing a high-power laser on the surface of a solid, which results in a so-called "laser-supported shock wave" {READY ⇨1963; RAMSDEN & SAVIC ⇨1964}. In detonics, a thermal explosion is understood as being a rapid chemical reaction that occurs when the temperature in a high explosive exceeds a certain threshold value.

*Microexplosions* are explosions emitted from very small explosive sources with high energy densities. Since they closely approach the ideal of a point source (previously a mathematical fiction rather than a physical reality), they allow shock and thermal energy to be deposited more economically and precisely at any desired location, thus minimizing the detrimental absorption that occurs upon propagation for focusing shock waves emitted from extended sources {TAKAYAMA ⇨1983}. The oldest example is the detonation of grains of gold fulminate {CROLL ⇨1608},

believed to be the first man-made high explosive, which were ignited by the percussion of a hammer in spectacular demonstrations. It was followed by pulsed electric sparks from capacitor discharges, a very comfortable method of generating shock waves in tight spots made possible by the invention of the Leiden jar {VON KLEIST & CUNAEUS ⇨1745}. In addition, gliding spark discharges {ANTOLIK ⇨1874; E. MACH & WOSYKA ⇨1875} and exploding wires {VAN MARUM ⇨1799; SINGER & CROSSE ⇨1815} allowed the production of shock waves of any desired geometry. With the advent of high-power pulsed lasers of very short duration in the 1960s, it became possible to focus radiation energy in very small volumes of matter.

*Nuclear explosions* results from an instantaneous fusion or fission process that can be many thousands (or millions) of times more powerful than the largest chemical detonations. Nuclear explosions provide access to a realm of high-temperature, high-pressure physics not otherwise available on a macroscopic scale on Earth. Such events are accompanied by the emission of electromagnetic radiation over a wide spectral range, such as light, heat, radio waves and gamma rays, generally referred to as "thermal radiation."

In the Universe, extremely violent explosions occur on a very large scale. The huge and sudden energy release associated with such events exhibits even more complex phenomena than in classic chemical explosions: they emit a broad spectrum of electromagnetic radiation ranging from kilometer-wavelength radio waves to gamma rays with energies of tens of MeV, and the shock waves accelerate energetic particles. Examples include

▸ the *Big Bang*, the largest conceivable explosion of an extremely small but incredibly dense and hot fireball occupying a single point in space, which eventually resulted in the enormous but cold and diffuse present-day Universe {⇨c.14 Ga ago; NASA-GSFC ⇨2003};

▸ cosmic *gamma-ray bursts*, the most violent explosions in the Universe {Vela Satellites ⇨1960s; COLGATE ⇨1967; Gamma-Ray Bursts ⇨1997};

▸ *supernovae* {China ⇨393; China & Switzerland ⇨1006; Near & Far East ⇨1054; BRAHE ⇨1572; KEPLER ⇨1604; SHELTON ⇨1987} and *hypernovae*, very energetic supernovae {Gamma-Ray Bursts ⇨1997; ROSAT ⇨1999};

▸ *solar flares*, enormous explosions of hydrogen and helium above the Sun's surface {CARRINGTON & HODGSON ⇨1859; PARKER ⇨1961; ⇨Fig. 4.1–V}; and

▸ *coronal mass ejections*, huge bubbles of gas that are ejected from the Sun over the course of several hours that can produce interplanetary shocks {Sun & SOHO ⇨2002}.

---

[42] *Thermal explosion: characterization and modeling.* LLNL, Livermore, CA; http://www-cms.llnl.gov/s-t/thermal_explosion.html.

**Implosion.** An *implosion*[43] designates another rapid process which is initially directed opposite to an explosion: the wavefront and the subsequent mass flow first move towards the center of implosion, where they collide with each other and then move outwards; *i.e.*, the latter stages of an implosion also involve explosive behavior. The mathematical treatment of implosions has remained a challenging task. Physically realistic solutions for the center of implosion ($r \rightarrow 0$) – an area of great interest in terms of converging shock waves, but one that is also critical to stability – are particularly difficult to find.

The term *implosion* was apparently coined in the 1880s by the British scientist Sir William THOMSON (Lord KELVIN).[44] Performing implosion experiments on sealed glass tubes, he observed a fine powder of shattered glass, almost like snow, upon implosion. However, inward-bursting phenomena resulting from the violent collapse of vessels due to external pressure would also have been familiar to early vacuum pioneers (VON GUERICKE 1650; HOOKE 1658).

A very simple implosion system consists of a light fluid surrounded by a spherical shell of dense fluid. Very high pressures compress the system. Converging shock waves passing through the interface cause Richtmyer-Meshkov instability {RICHTMYER ⇒ 1960; MESHKOV ⇒ 1969}. Rayleigh-Taylor instability {G.I. TAYLOR ⇒ 1944} occurs at the end of the implosion, when the dense shell is decelerated by the lighter fluid. Today's implosion devices, such as those applied in laser fusion and nuclear weapons, use highly sophisticated technologies that apply converging shock waves in order to achieve ultrahigh compression of matter.

In a supernova explosion, the *forward shock* sweeps up the interstellar medium, while an inward-propagating shock wave forms when the supersonically expanding neutrino-driven wind collides with the slower supernova ejecta thrown off earlier. This *reverse shock* – a term coined in 1974 by the U.S. astronomer Christopher F. MCKEE[45] – that moves within the ejecta heats the material ejected in the explosion itself {NASA's Chandra ⇒ 2000}. When the reverse shock has begun to reverse the direction of the flow it encounters, it essentially sets up an implosion. The hot ball of gas generated in this manner eventually reaches a state from which it proceeds to expand supersonically and generates an outward-propagating *secondary blast-wave shock*.

The supernova explosion of a massive star leaves behind an imploded stellar core which may form a rapidly rotating neutron star that can be observed many years later as a radio pulsar, releasing a huge amount of energy.

Some cosmological models assume that, when the present expansion of the Universe has come to an end, it may reverse and contract under its own gravitation, thus causing the largest implosion imaginable – the so-called "Big Crunch" {DAVIES ⇒ 1994}. Oscillating cosmological models even assume that the Big Crunch is followed by a new Big Bang, thus producing another expanding Universe {FRIEDMANN ⇒ 1922}.[46]

### 2.1.3 CONFLAGRATION, DEFLAGRATION, DETONATION, AND DETONICS

**Conflagration.** Since the late 17th century, the terms *explosion* and *detonation* were used interchangeably. In 18th-century Germany, the term *detonation* [Germ. *Detonation*] was used to designate a progressive combustion [*Verpuffung*], now known as a "deflagration." In those times, a progressive combustion was termed a *conflagration* [*Brunst* or *Verloderung*], a term which today is understood to be a very intense, uncontrolled fire (or an inferno).[47] However, violent volcanic eruptions were also known as "conflagrations" in England.[48]

With the advent of more sensitive and (particularly) faster diagnostics in the 19th century, it became easier to resolve the complex physical processes of combustion and explosion, and new, more rigid definitions of the terms *deflagration* and *detonation* spread among researchers. Detonation waves and deflagration waves are two distinct types of *combustion waves* which proceed nearly homogeneously in premixed systems in a transient process, although at quite different rates.

**Deflagration.** A *deflagration*[49] is a slow combustion process that gives off heat and light but is unable to produce sufficient overpressure to create a shock wave which is strong

---

[43] *Implosion* is an artificial word derived from *explosion* to denote the inwardly directed motion.

[44] J.A. SIMPSON and E.S.C. WEINER: *The Oxford English dictionary.* Clarendon Press, Oxford (1989), vol. VII, p. 725.

[45] C.F. MCKEE: *X-ray emission from an inward-propagating shock in young supernova remnants.* Astrophys. J. **188**, 335-340 (1974).

[46] Ancient Greek philosophers advocated at least two important alternatives: PLATO (428–348 B.C.) accepted the common Greek notion of the Universe as a series of world cycles, each having a beginning and ending, while ARISTOTLE (384–322 B.C.), one of his students, postulated an eternal Universe, a steady-state system without beginning or end.

[47] J.H. ZEDLER (ed.): *Großes vollständiges Universal-Lexicon aller Wissenschaften und Künste.* Zedler, Halle & Leipzig (1732–1754).

[48] In L. SPALLANZANI's book *Travels in the two Sicilies* [Robinson, London, 1798], vol. I, p. 195], it reads: "But the circumference of the Vesuvian crater is never more than half a mile, even when widest distended, and in its most destructive conflagrations."

[49] From Lat. *deflagrare*, meaning "to burn down."

enough to ignite the fuel. Since the mechanism of inflammation occurs via heat transfer, the speed of the flame front propagating through a flammable gas or vapor is less than the speed of sound in that gas or vapor, often far below the velocity of sound in the burnt gases, and on the order of only some 10 cm/s – *i.e.*, very subsonic. Behind the deflagration front the pressure and density decrease. In the simplest theory deflagrations are described as flow discontinuities, which propagate subsonically {COURANT & FRIEDRICHS ⇒1948}. However, real deflagration processes are subjected to a variety of instabilities which can significantly influence their flame shape and propagation speed.

The first man-made deflagrations, which took the form of dust explosions, were probably observed in areas where flour had to be milled, stored and processed in large quantities {MOROZZO ⇒1785}. The term *deflagration* was used as early as 1666 by the British physicist Robert BOYLE.[50] Black powder – the oldest known explosive {M. GRAECUS ⇒c.1250; BACON ⇒1267; SCHWARZ ⇒c.1380} and now classified as a "low" explosive – is a "deflagrating" explosive. In a *confined deflagration*, the expanding combustion products are confined; for example when the flame is traveling within a pipe.

Deflagration is a common decomposition phenomenon in coal mines and it can occur when the ignition of a mixture of air and firedamp (essentially methane) cannot fully develop into an explosion. Deflagration also plays an important role in *pulsejet engines*, such as the Schmidt tube {P. SCHMIDT ⇒1930}, in which the fuel is deflagrated rather than detonated. Deflagrations produce peak pressure rises of only a few bars and propagate with subsonic speeds.

**Detonation.** When a deflagration becomes sufficiently strong, a sudden transition from deflagration to detonation can occur. This transition phenomenon is characterized by very high local pressures, and sometimes very strong damage can be observed at the point of transition to detonation. The term *detonation* designates a violent, supersonic combustion process related to explosive gaseous mixtures and so-called "high explosives," which may take either liquid or solid form.[51] In England, the term was apparently first used by Roger BACON,[52] an eminent English 13th-century natural philosopher. Attacking ARISTOTLE's methodology, he wrote that, for example, the bang generated by hitting a rod (against a hard object for instance) is not an intrinsic property of the rod but is instead caused by the impact induced by the motion. His new method for achieving knowledge, based exclusively on careful observation and cautious eliminative induction, stimulated the evolution of modern science.

The detonation process in explosives, which is characterized by a high rate of heat generation, is initiated by a shock wave which provokes a strong adiabatic compression, and is also sustained by a rapidly progressing wave. This so-called "detonation wave" {BERTHELOT ⇒1881; DIXON ⇒1893}, which appears similar to a strong shock wave, is followed by a combustion zone, in which chemical reactions proceed rapidly to completion. The energy of this reaction maintains constant conditions at the front of the detonation wave, thus leading to a constant detonation rate at or above the velocity of sound in the burnt gases; *i.e.*, on the order of about 2–5 km/s for gases and 6–9.5 km/s for liquid and solid explosives.

In their widely used textbook *Supersonic Flow and Shock Waves*, COURANT and FRIEDRICHS {⇒1948} gave a more specific, mathematical interpretation of the terms *deflagration* and *detonation* based on a simple model proposed previously by CHAPMAN {⇒1899} and JOUGUET {⇒1905}: they demonstrate the characteristic differences of these two processes in terms of the velocities and pressures of the unburnt and burnt gases, noting that "in a *detonation* – in which the pressure increases and the velocity of the gas decreases when the reaction front sweeps over it – the burnt gas is retarded relative to the front," while "in a *deflagration* – in which the pressure decreases – the gas is accelerated away from the reaction front when the reaction front sweeps over the unburnt gas."

The term *detonation* was apparently first used in its modern sense by the British chemist Sir Frederick ABEL {⇒1869}, who studied detonation effects in guncotton. The French chemist P.E. Marcellin BERTHELOT {⇒1870} first termed the rapid motion of the flame front – *i.e.*, the front of the detonation wave – a "shock" [French *un choc*], and in 1881 "explosion wave" [*l'onde explosive*], which appears to have been first taken up in England {DIXON ⇒1893; CHAPMAN ⇒1899}. The term *detonation*, which DIXON initially coined for the expression "rate of explosion," was later also adopted in France {VIEILLE ⇒1900}.

The term *condensed detonation* refers to the detonation of liquids and solids. Condensed substances that are detonable are more commonly referred to as explosives. The most apparent difference between condensed and gaseous detonations are the detonation pressures, which are a factor of $10^3$ to $10^4$ higher for condensed explosives than for gases, due to

---

[50] J.A. SIMPSON and E.S.C. WEINER: *The Oxford English dictionary*. Clarendon Press, Oxford (1989), vol. IV, p. 386.
[51] From Lat. *detonare*, meaning "to thunder out."
[52] R. BACON: *Opera hactenus inedita* (ed. by R. STEELE ET AL.). 16 vols., Clarendon Press, Oxford (1905–1940); vol. VII (1926): *Questiones supra undecimum prime philosophiae Aristotelis*, p. 19.

the large difference in the densities of the media.[53] Unlike gaseous detonations, condensed detonations require a wider use of hydrodynamic theory and a fairly precise knowledge of the equations of state of all materials involved in the detonation and sample compression process.

The term *overdriven detonation* refers to an unstable flame front that propagates through a flammable gas or vapor at a speed in excess of the stable detonation velocity. An overdriven explosion may exist during the transition of a combustion process from a deflagration into a stable detonation, which can produce extremely high pressures in a relatively short time frame.

The term superdetonation (or super detonation) designates a detonation that occurs in an explosive precompressed by an initial shock. This type of detonation has a higher pressure and velocity than a detonation in an uncompressed material would have. A superdetonation wave propagates and eventually overtakes the initial shock. After traveling a certain distance, a superdetonation decays to a steady detonation {CAMPBELL ⇨ 1961; SHEFFIELD ET AL. ⇨ 1989}.

A *retonation* is a wave that moves through the burnt medium from the point of detonation back through burned or burning explosion gases to the ignition source, usually from an area of higher pressure to an area of lower pressure. The term *retonation* was coined in England by DIXON {⇨ 1903} who studied gaseous explosions. Los Alamos physicists have instead used the term *reverse detonation*. Retonation in small sticks of high explosive has been reported by Melvin A. COOK[54] and others.

**Deflagration to Detonation Transition (DDT).** In reactive gases this transition is an extremely complex physical process involving deflagrations, shocks and shock reflections, boundary layers, and all of their interactions with each other, and depending on initial and material conditions. Most of the gasdynamic phenomena associated with the development of detonation in an explosive gas – the process commonly referred to as DDT – have been revealed by high-speed laser schlieren cinematography. This very useful diagnostic method helped to elucidate the role played by exothermic centers in the propagation mechanism of detonation and the cause of cellular structures that had been shown to occur inside the soot-coated wall of a cylindrical tube upon the propagation of a self-sustained detonation wave down the tube {⇨Fig. 4.16–V}.

Since the 1950s, it is known that detonation waves in gaseous mixtures show unstable behavior under near-limit detonations, characterized by very large fluctuations of the detonation velocity that can range from 0.4 to 1.8 times the normal Chapman-Jouguet (CJ) value. Based upon measured velocity profiles two propagation modes can be distinguished and have been used to establish a criterion for detonation limits:

▸ in a *galloping detonation* the reaction front propagates in an overdriven state, but the velocity decreases until attaining a steady state close to the CJ velocity;

▸ in a *stuttering detonation*, however, there exist large velocity fluctuations of the reaction front and a steady state is not reached.

This classification, proposed by detonation researchers at McGill University,[55] allows a qualitative description of the wide range of velocity fluctuations occurring near the detonation limit.

Flame acceleration and DDT are important phenomena in severe accidents, but DDTs also play an important role in the aerothermodynamics of so-called "pulse detonation engines" (PDEs) {IUTAM Symposium on Combustion in Supersonic Flows ⇨1997}, in which the fuel is detonated periodically within an engine tube to produce high forward thrust. Pulse detonation engines, in which propagating detonation waves produce peak pressure rises of 30 bar or more and propagate at Mach 5 or faster, have no moving parts in the power production section and are a promising low-cost alternative to turbojet and liquid propellant rocket engines.[56]

In astrophysics, current *Supernova Type Ia* models {BAADE ⇨1931} invoke a transition from a deflagration wave to a detonation wave.[57]

**Detonics.** The term *detonics* is of more recent origin and was coined by the Swedish explosive specialists Carl H. JOHANSSON and Per A. PERSSON. In their book *Detonics of High Explosives*, they wrote, "The word *detonics* in the title is chosen in preference to the word detonation to indicate the physics of detonating high explosives and their mechanical effects" {JOHANSSON & PERSSON ⇨1970}. Their definition has since been widely accepted.[58]

---

[53] S.J. JACOBS: *Recent advances in condensed media detonations*. ARS J. **30**, 151-158 (1960).
[54] M.A. COOK: *The science of high explosives*. Reinhold, New York (1958), p. 187.
[55] J.J. LEE ET AL: *Doppler interferometry study of unstable detonations*. Shock Waves **5**, 175-181 (1995).
[56] P. PROCTOR: *Pulse detonation technologies advance*. Aviation Week & Space Technology **148**, 48-50 (May 4, 1998).
[57] A.M. KOHKHLOV ET AL: *Deflagration-to-detonation transition in thermonuclear supernovae*. Astrophys. J. **478**, 678-688 (1997).
[58] For example, the *Dictionary of science and technology* (C. MORRIS, ed.), Academic Press, San Diego *etc.* (1992), states on p. 620: "Detonics – the field of study concerned with detonating and the performance of explosives."

**Microdetonics.** Condensed phase explosives used in conventional explosive systems have a charge size on the order of a meter or a sizable fraction of a meter. Microdetonics refers to detonation initiation, detonation acceleration (buildup), and detonation curvature effects in small explosive systems. The miniaturization of exploding systems using high explosives in overall dimension not exceeding 10 cm is a more recent development in detonics which promises many new applications, both in military and industry {STEWART ⇨2002} and in biology and medicine {JAGADEESH & TAKAYAMA ⇨2002}. Since the explosive's reaction zone length apparently scales down linearly with the reduction in overall device dimension, scientist must select the main charge from one of the short reaction zone explosive materials that lie in the class of primary explosives. Unfortunately, primary explosions are often sensitive to low stimulus and accidental initiation, and alternative concepts of realizing miniature exploding systems are required.

## 2.1.4 HYDRAULIC JUMP, BORE, SURGE, TSUNAMI, SEICHE, SEA SHOCK, AND ROGUE WAVE

Surface water waves are fascinating natural phenomena that are, in many cases, easily observable with the naked eye, and so they attracted the curiosities of early natural philosophers, who pondered on the complex propagation properties of waves in general. In particular, "water table experiments" were performed to demonstrate and study the fundamental wave effects of propagation and reflection at obstacles that had a variety of shapes.

Water waves belong to a peculiar family of mechanical waves that have wavelengths ranging from a few millimeters (*ripples*) to hundreds of kilometers (*tsunamis* in deep water). They can have a periodic structure (such as that for ordinary surface waves caused by wind action), and can occur recurrently (*breakers*) or at highly predictable times (*tidal waves and bores*). Surface waves of large amplitude, propagating with steep wavefronts – such as *tsunamis* (or *seismic sea waves*) in shallow water or *rogue* (or *freak*) *waves* in deep water – are episodic, relatively rare phenomena which can however cause tremendous damage and loss of life.

**Hydraulic Jump.** The term *hydraulic jump* designates an abrupt change of depth in a shallow, steady-state flow of water with a free surface, accompanied by turbulent regions in which water depth and flow velocity suddenly change {⇨Fig. 4.1−M}.[59] For example, a hydraulic jump [Germ. *Wassersprung*] is created at the base of a dam when water at a uniformly shallow depth and moving at high velocity in a channel suddenly enters a region of uniformly high depth and low velocity. Another curious example is the ring-shaped hydraulic jump with a radius of a few centimeters which occurs in a kitchen sink when the tap is left running {⇨Fig. 4.1−M}. In both examples, the hydraulic jumps are stationary (*i.e.*, nonpropagating) phenomena.

However, hydraulic jumps can also move in shallow water as propagating discontinuities with almost step-like increases in water depth at their fronts. The flow changes its character upon passage through such discontinuities: it partially supercritical and partially subcritical, which in hydraulics is also called a "tranquil-shooting transition."[60] Hydraulic jumps have a mathematical analog in gas dynamics (steadily progressing one-dimensional compressible supersonic flows in air – *i.e.*, steady 1-D shock waves); small-amplitude gravity waves in water correspond to the acoustic waves in air.

**Bore.** A *bore* is a high tidal wave that rushes up a narrow estuary or tidal river.[61] The *Severn bore* is produced by a tide that rises to about 18 feet (5.5 m) in an hour and a half. This body of water becomes compressed in the narrowing funnel-shaped Severn estuary, and it is piled up into an advancing wave that extends from bank to bank.

From a physical point of view, a *bore* is a nonstationary hydraulic jump. In nature, it represents a great tidal wave or flood wave with a crested front that forms in rivers and estuaries near the coast, and moves swiftly upstream – popularly described as a traveling "wall of water." The largest bores usually occur around high equinoctial tides. There are three main categories of tidal bores:

▸ a *positive tidal bore* of depth $H + h$ running into an adverse current of depth $H$, the most common form, which results in a tongue of water that moves up a river;

▸ a *negative tidal bore* of depth $H - h$ running into still water of depth $H$, which can form (for example) when the water in a canal is suddenly released by the collapse of a lock gate; and

---

[59] An excellent and concise review of shallow-water waves has been given by R.A.R. TRICKER in his article *Water waves*; published in *Encyclopaedia Britannica, Macropaedia*. Benton & Hemingway, Chicago etc. (1974), vol. 19, pp. 654-660.

[60] J.J. STOKER: *Water waves*. Interscience Publ., New York (1957), p. 318.

[61] The origin of the word *bore* is dubious, but it is usually taken to derive from a Scandinavian word, *bára*, meaning "wave" or "billow." The other name by which the phenomenon is known, *eagre*, is also of unknown origin; *see* Encyclopedia Britannica (1911).

▸ a *solitary wave* {RUSSELL ⇨ 1834}, a particular type of wave that only has a crest: its surface profile lies entirely above the still-water level. A solitary wave is formed when a negative bore is followed by an equal positive bore; a positive bore always travels faster than a negative bore of the same height.[62] Note that a solitary wave is an example of a *wave of translation; i.e.,* the water particles advance with the wave and do not return to their original position.

A German encyclopedia[63] states that the term *bore* is of Indian origin, and it means "high tide," since large bores are also observable in the estuary of the Ganges river. The oldest definition of a bore in scientific terms was probably given by the English clergyman and mathematician Samuel EARNSHAW {⇨1858}, "I have defined a bore to be a tendency to discontinuity of pressure." In the early 1900s, the term *bore* was sometimes applied to pressure discontinuities in gases as well, such as the *Riemann wave* {Lord RAYLEIGH ⇨ 1910}.

Bores are also caused by the collision of two tides, and the term *bore* was reportedly in use as far back as the 17th century,[64] when it was described as "a boar, as the seamen term it, and violent encounter of two tides coming in."

The *Severn bore* is is one of the most famous among 60 bores around the world and occurs 12 times a year. Its body of water becomes compressed in the narrowing funnel-shaped Severn estuary {⇨Fig. 4.4–E}, and is piled up into a single advancing wave of up to 2 meters that extends from bank to bank.

**Surge.** *Surges* are unsteady fluid dynamic phenomena in gases and liquids, characterized by a sudden increase in pressure to an extreme or abnormal value and then a drop from this increased value. They have been observed in nature, on the Earth, and on the Sun as well, but they also exist in man-made fluid-related devices: for example, the shutdown of turbines in hydropower systems can cause the dreaded water hammer in adjacent tunnels and shafts {KAREJJSKICH ET AL. ⇨1898}: this is a shock wave that is followed by a slower moving mass of water, the so-called "mass surge." In hydrodynamics, a *surge wave* is a wave propagating at the free surface of a liquid, and it is characterized by a sudden increase in the depth of flow across the wavefront, and severe eddy motion at the wavefront. Hydraulic jumps, bores and tsunamis are sometimes referred to as surge waves. In astrophysics, surges are transient cool plasma jets ejected from small flare-like chromospheric bright points, such as *subflares* (or microflares, small-scale and short-lived flares) and *Ellerman bombs* (tiny fairly bright transient points of light, most often found in emerging flux regions or on the edges of sunspots where the magnetic field is breaking the surface).

*Base surges* are turbulent mixtures of water vapor or condensed droplets and solid particles at or below a temperature of 100 °C; they travel outward along the ground at hurricane velocities. Base surges were first identified during underwater nuclear bomb detonations at Bikini Atoll in the South Pacific {Operation CROSSROADS ⇨1946}. A volcanically produced base surge which resulted from water/magma interactions was noted during the explosive eruption of Taal Volcano {⇨1965}.

*Pyroclastic surges* are "hot" base surges consisting of turbulent, low-density clouds of rock debris and air or other gases that have temperatures appreciably above 100 °C and that move over the ground at high speeds {Mt. Pelée ⇨1902; Mt. St. Helens ⇨1980}.[65]

**Tsunami.** The Japanese name *tsunami* – a term that literally means "harbor wave" [*tsu nami*] and has been used in colloquial Japanese since at least the 1890s[66] – refers to a gravity wave system which forms in the sea following any large-scale, short-duration disturbance of the free surface. It was first proposed for general use by the German-born American geophysicist Beno GUTENBERG {⇨1939} instead of the previous misnomer *tidal wave*. The term *tsunami* was eventually adopted by an international scientific committee at the Tsunami Meeting {⇨1961} in preference to the term *seismic sea wave*, and it is now defined as "a train of progressive long waves generated in the ocean or a small connected body of water by an impulsive disturbance."

Tsunamis are very complex, curious natural phenomena, and because of their enormous destructive power they are popularly known as "killer waves." All large tsunamis can cause severe damage at remote coasts thousands of kilometers from their origin. Prominent examples of so-called

---

[62] A quantum or quasi-particle propagating in the manner of a solitary wave is called a "soliton." This term was coined by the two applied mathematicians Norman J. ZABUSKY and Martin D. KRUSKAL while studying nonlinear interactions among solitary wave pulses propagating in nonlinear dispersive media; *see* Phys. Rev. Lett. **15**, 240-243 (1965).

[63] *Brockhaus Enzyklopädie.* Brockhaus, Wiesbaden, vol. 3 (1967), p. 125.

[64] J.A. SIMPSON and E.S.C. WEINER: *The Oxford English dictionary.* Clarendon Press, Oxford, vol. II (1989), p. 413.

[65] *Description: pyroclastic flows and pyroclastic surges.* USGS; http://vulcan.wr.usgs.gov/Glossary/PyroFlows/description_pyro_flows.html.

[66] Lafcadio HEARN, an Irish-American author who lived in Japan for 14 years, wrote in his book *Gleanings in Buddha-Fields, Studies of Hand and Soul in the Far East* [Houghton & Mifflin, Boston, 1898], "'Tsunami!' shrieked the people, and then all shrieks and all sounds and all power to hear sounds were annihilated by a nameless shock … as the colossal swell smote the shore with a weight that sent a shudder through the hills."

"teletsunamis" are the Alaskan Earthquakes {⇨1946 & 1964}, the Chilean Earthquake {⇨1960}, and the recent seaquake in the Indian Ocean {Sumatra-Andaman Islands Earthquake ⇨2004}.

*Volcanic tsunamis* are generated when some of the tremendous amounts of energy produced during volcanic eruptions are transferred to ocean waters or nearby large lakes. James E. BEGÉT,[67] a volcanologist at Alaska Volcano Observatory, specifies at least nine different mechanisms by which volcanoes produce volcanic tsunamis:

- volcanic earthquakes;
- eruptions of undersea volcanoes;
- movement of pyroclastic flows into the sea, caldera collapse;
- debris avalanches and landslides;
- large lahars entering the sea;
- explosive water-magma interactions caused by the interaction of rising magma and surface water (phreatomagmatic eruptions);
- coupling between water and turbulent air waves traveling from an explosive eruption; and
- collapse of lava benches during effusive (non-explosive) lava eruptions.

*Mega-tsunamis* are defined in the literature as waves that are more than 300 ft (100 m) high; indeed, some tsunami researchers even consider mega-tsunamis to be waves more than a thousand feet (> 300 m) high. The primary sources of such giant tsunamis are the impacts of asteroids with oceans, giant explosive volcanic eruptions, and massive landslides. In 2001, some geophysicists predicted that a future eruption of the Cumbre Vieja volcano on the island of La Palma (one of the Canary Islands, which belong to Spain) may shake loose a gigantic chunk of the mountain's western flank which would slide into the Atlantic and trigger a 100-m mega-tsunami, devastating the coast of northwest Africa and even the far more remote east coast of North America. However, mega-tsunamis are very rare events and, according to a note disseminated in 2003 by The Tsunami Society,[68] "no such event – a mega-tsunami – has occurred in either the Atlantic or Pacific oceans in recorded history."

*Tsunami magnitude* is defined in terms of the highest wave height at the coast. There are six grades (–1, 0, 1, 2, 3, 4) that depend on the maximum wave height. The first grade (–1) includes minor tsunamis with wave heights of less than 0.5 m; the highest grade (4) refers to maximum wave heights in excess of 30 meters {IMAMURA ⇨1949}. However, it is very difficult (and in most cases impossible) to measure the height of a tsunami wave directly. The most common method used to determine tsunami wave height is to measure the *tsunami runup height*, the highest vertical point above sea level onshore reached by the wave. Runup heights are characterized by the distance and extent of vegetation killed by salt, and the debris left once the wave has receded. The U.S. National Geophysical Data Center (NGDC) worked out a Tsunami Runup Database, which contains information on the places where tsunami effects have occurred, and it covers both historic and recent tsunami events, ranging from 1500 B.C. to A.D. 2005.[69]

**Seiche**. The term *seiche* is probably of Swiss origin and denotes rhythmic oscillations of a large water body that depend on the horizontal dimensions and depth of the water.[70] This phenomenon of standing waves – most commonly caused by a change in atmospheric pressure but also caused by seismic waves – can best be observed in long basins such as in harbors, lakes and bays, when the length of the water body corresponds to one of the natural periods of the basin. In Germany, the pendulum-like water movement is appropriately called "rocking waves" [Germ. *Schaukelwellen*]. It was first studied in the 1890s by the Swiss limnologist François Alphonse FOREL in Lake Geneva (length about ca. 72 km, average width ca. 9 km).

Although the phenomena observed are strictly not discontinuous in nature (periods range between a few minutes to an hour or more), seiches may be reinforced by lunar tides or by selective resonance of the waves excited in the extended body of water. For example, the tide in the Bay of Fundy is a very large seiche generated in the embayment by the tides in the adjacent ocean, because the natural period of the bay is the same as that of the semidiurnal tide. Like bores and shock waves, which can intensify dangerously after oblique reflection at a wall, large-amplitude seiches can produce damage when they collide with dock structures or anchored ships.

Large earthquakes can produce both tsunamis and seiches at the same time, as was first noticed by the German natural philosopher Immanuel KANT {⇨1756} while analyzing remote effects of the Great Lisbon Earthquake {⇨1755}.

---

[67] J.E. BEGÉT: *Volcanic tsunamis*. In: (H. SIGURDSSON, ed.) *Encyclopedia of volcanoes*. Academic Press, San Diego etc. (2000), pp. 1005-1013.

[68] *Mega tsunami hazards*. The Tsunami Society, Honolulu, HI (Jan. 15, 2003); http://www.sthjournal.org/media.htm.

[69] *The NGDC tsunami runup database*. National Geophysical Data Center (NGDC); http://www.ngdc.noaa.gov/seg/hazard/tsrnsrch_idb.shtml.

[70] The term *seiche* was apparently first introduced into science by George ROBERTS, who listed it in his *Etymological and explanatory dictionary of the terms and language of geology* [Longman *etc.*, London, 1839].

**Sea Shock.** Akitune IMAMURA,[71] a Japanese seismology professor at Tokyo Imperial University, discussed the effects of earthquakes on water, such as tsunamis and seiches and another rather more rarely reported phenomenon, *sea shocks*, in his textbook *Theoretical and Applied Seismology*. He wrote, "Earthquake motion upon emerging from the sea-bed passes into the water where it takes the form of elastic waves, transmitting its motion to vessels on the surface. These are 'sea shocks.' A ship may receive such shocks without noticing any surface disturbance, although rumbling may at times be heard … In the great Tôkaidô Earthquake of December 23, 1854, a number of vessels were lost off the coast of Idu, but as tsunamis are impotent at great distances offshore, the only conclusion possible is that the wrecks were caused by sea shocks." His interpretation was indeed confirmed by later observations. For example, the Lompoc Earthquake that occurred in southern California on November 4, 1927, also produced a seaquake. This resulted in not only in a tsunami but also a violent compressional shock that was transmitted through the water, which stunned fish near Point Arguello and shook at least two ships in the area.[72]

**Rogue wave.** A *rogue wave* – sometimes referred to as a *"freak" wave* or *monster wave* – is a rare event and generally defined as an unexpectedly high sea wave which may come from a direction other than that of the other sea waves. It is a single, massive wall of water that rises up from an apparently calm sea {⇨Fig. 4.1−T}, or it can develop an "enormous hole" when the troughs of several wave trains coincide. Prof. Robert G. DEAN,[73] a renowned U.S. coastal engineer and wave hydrodynamicist at the University of Florida, defined rogue waves as waves with heights that exceed the significant wave height of a measured wave train by a factor of 2.2. The significant wave height is defined as the average of the one-third highest waves.

There is now growing evidence, based upon various data sources including satellite imagery, that rogue waves are not produced by landslides or earthquakes: the prevailing theory holds that they can result when strong, high storm waves slam headlong into a powerful current traveling in the opposite direction. The interaction can push the storm swells together, so that their frequencies superimpose, creating one tremendously powerful wave that can reach a height of 30 meters or more, which suggests that this mysterious phenomenon is responsible for the loss of many ships. Rogue waves may also result from the focusing of wave energy by ocean currents. More recently, an international forum was set up to discuss rogue wave phenomena and theories on their origin {Rogue Waves 2000 Workshop ⇨2000}.

### 2.1.5 SHOCK AND SHOCK WAVE

In everyday usage, the term *shock* is generally used in medical rather than in a physical or technical sense. With the advent of the first high-voltage generators in the 17th century and the invention of the Leiden jar in the 18th century, the effects of *electric shocks* on humans were widely demonstrated to the public {NOLLET 1746, *see* VON KLEIST & CUNEUS ⇨1745} or accidentally experienced by early electricians when experimenting with dangerously high voltages {BENNET ⇨1789}. Up to the mid-20th century, the term *shock* was generally used in relation to medical[74] and electric phenomena. For example, the 1961 edition of the *Encyclopaedia Britannica* still defined the word "shock" in just two senses: medical and electrical. Over the last 100 years, however, the term *shock* has increasingly been used to refer to a complex set of phenomena related to percussion, shock wave physics, explosion seismology and impact engineering. This wide spectrum of shock-related terms will be discussed in the following.

**Shock.** The term *shock*, derived from the French word *choc*, is frequently used in mechanical engineering in a more general context which is not strictly related to the definition of a shock wave as given below: it instead implies a degree of suddenness and severity – meaning that the excitation is nonperiodic; *i.e.*, applied in the form of an impulse, a step, or a transient vibration. Coupled into an engineering structure, this results in a mechanical response, the so-called "shock response," which is an important criterion of a

---

[71] A. IMAMURA: *Theoretical and applied seismology.* Maruzen, Tokyo (1937), p. 121.
[72] *Lompoc earthquake.* Southern California Earthquake Data Center (SCEDC), Pasadena, CA;
http://www.data.scec.org/chrono_index/lompoc.html.
[73] R.G. DEAN: *Freak waves: a possible explanation.* NATO Advanced Research Workshop on Water Wave Kinematics [Molde, Norway, May 1989]. In: (A. TØRUM and O.T. GUDMESTAD, eds.) *Water wave kinematics.* Kluwer, Amsterdam (1990), pp. 609-612.

[74] In medicine, the word "shock" was originally used as a *physiological* term (designating a state of circulatory collapse caused by an acute loss of blood or fluids from the body), and later also as a *psychological* term (designating a sudden and violent disturbance of emotional or mental equilibrium).

mechanical system related to its ability to resist *impulsive loading*. According to the U.S. researchers John S. RINEHART and John PEARSON, impulsive loading is characterized "by an almost instantaneous, less than a small fraction of a microsecond, rise in load to a quite high but finite value which is followed immediately by a rapid decrease in load; the duration of an impulsive load is usually of the order of microseconds" {RINEHART & PEARSON ⇨1965}. Impulsive loading may develop when an explosive charge is detonated in intimate contact with a body or when one body impacts against another. Its intensity is usually of a sufficient magnitude to produce extensive fracturing and large permanent distortions in the body upon which it acts.

A *shock wave* is sometimes called a "shock."

**Shock Wave.** In the natural sciences, the most common shock-related term is the *shock wave* (also *shock-wave* or *shockwave*).[75] It describes a mechanical wave characterized by a surface or sheet of discontinuity in which, within a narrow region, the pressure, density, particle velocity and temperature change abruptly. Because a shock wave moves faster than the speed of sound, the medium ahead of the shock cannot respond until the shock strikes, and so the shock wave falling upon the initially quiescent particles of matter is a supersonic "hydrodynamic surprise" – similar to a person on the ground being overrun by the thunder-like noise of the sonic boom cone originating from a supersonic aircraft or some other type of aerospace vehicle.[76] In air, a shock wave produced by an explosion and radiating outward from its center is termed a *blast wave*, because it causes a strong wind,[77] while the term *shock* is often used for such waves occurring in water or the ground, because the effect is like that of a sudden impact.

From the point of view of a fixed observer, shock waves can be divided up into *nonstationary shock waves* and *stationary shock waves*. Examples of nonstationary shock waves are the blast wave originating from an explosion, the muzzle blast from a gun, or the head wave emerging from a body flying at supersonic speed. Stationary shock waves – *i.e.*, motionless shock waves with respect to the observer – are generated in supersonic wind tunnels around a test body at rest {LANGEVIN & CHILOWSKY ⇨1918}, or behind nozzles when the outflow velocity exceeds the speed of sound {SALCHER & WHITEHEAD ⇨1889; L. MACH ⇨1897}. Stationary shock waves are also created when the flow on the suction side of a transonic wing is accelerated to a supersonic velocity. Under certain conditions of illumination passengers of transonic airliners can directly watch this curious phenomenon {⇨Fig. 4.14–L}.

*Shock diamonds*, a special kind of stationary shock wave, are disk-shaped shock waves formed behind nozzles due to the presence of reflected shock waves in the exhaust stream and shock heating. They are seen frequently in the exhaust jets of aircraft and rocket engines when viewed from the side {⇨Fig. 4.20–F}. Astronomers have observed knots of bright radio emission in jets of ionized gases – astrophysical jets – which emanate from quasars and other galaxies and range from thousands to millions of light-years in length; these may also be caused by shock heating, as in shock diamonds. "Radio jets" are astrophysical jets that can only be observed with radio telescopes. A radio jet typically ends in a "hot spot," a small region of intense radio emission.

Compared to acoustic waves, which are waves of very small (almost infinitesimal) amplitude, shock waves are "waves of finite amplitude" that can be characterized by six unusual features:

---

[75] EULER {⇨1759}, without yet coining a term, addressed the "size of disturbance" of a sound wave, meaning its amplitude. POISSON {⇨1808} described intense sound as the case "where the molecule velocities can no longer be regarded as very small." STOKES {⇨1848} used the term *surface of discontinuity*, and AIRY {⇨1848, 1849} described the wave as an "interruption of continuity of particles of air." RIEMANN {⇨1859} used the modern terms *condensation shock* [Germ. *Verdichtungsstoß*] and *condensation wave* [*Verdichtungswelle*] to illustrate the jump-like steepening of the wavefront. EARNSHAW {⇨1860} used the terms *positive wave*, to illustrate that the motion of particles is in the direction of wave transmission, and *wave of condensation*, to characterize the increase in density. August TOEPLER {⇨1864} was the first to use the term *shock wave* [Germ. *Stoßwelle*] in the present sense; he originated a shock wave from a spark discharge and first visualized it subjectively using a stroboscopic method. He also used the terms *spark wave* [Germ. *Funkenwelle*] and *air percussion wave* [*Lufterschütterungswelle*] interchangeably, but incorrectly used the term *sound wave* [*Schallwelle*]. RANKINE {⇨1870} used the terms *abrupt disturbance* and *wave of finite longitudinal disturbance*, and HUGONIOT {⇨1885} used the term *discontinuité* [French *discontinuité de la vitesse du gaz et de sa pression*]. Ernst MACH and coworkers (1875–1885) used the terms *shock wave*, *Riemann wave* [Germ. *Riemann'sche Welle*], *bang wave* [*Knallwelle*], and *explosion wave* [*Explosionswelle*]. In the specific case of a supersonic projectile, MACH and SALCHER {⇨1886} used the terms *head wave* or *bow wave* [Germ. *Kopfwelle*] and *tail wave* [*Achterwelle*]. Ernst and Ludwig MACH also designated a shock wave as being a *Schallwelle großer Excursion* {⇨1889}, meaning "a sound wave of large amplitude." VON OETTINGEN and VON GERNET {⇨1888}, when studying oxyhydrogen explosions, called the detonation wave "Stoßwelle." In France, the term *shock wave* [French *onde de choc*] was first used by VIEILLE and HADAMARD {⇨1898}, and later by DUHEM {⇨1901} and JOUGUET {⇨1904}. DUHEM also used the terms *partition wave* [French *onde-cloison*], *true Hugoniot wave*, *surface slope* [French *surface de glissement*], and *quasi shock wave* to characterize special types. Discussing the characteristics of shock waves in air, Lord RAYLEIGH {⇨1910} used the term *aerial waves of finite amplitude*.

[76] D.J. HOLLENBACH and C.F. MCKEE: *Astrophysical shock waves*. In: S.P. MARAN: *The astronomy and astrophysics encyclopedia*. Van Nostrand & Reinhold, New York (1992), pp. 623-627.

[77] For example, at an overpressure of 10 psi (0.69 bar), the maximum wind velocity is about 290 mph (129 m/s); *see* GLASSTONE {⇨1962}, p. 107.

- when the motion is analyzed based on a coordinate system traveling with the shock, then the flow is always supersonic (Mach number $M > 1$) ahead and subsonic ($M < 1$) behind a shock;[78]
- a pressure-dependent, supersonic velocity of propagation;
- the formation of a steep wavefront with a dramatic change in thermodynamic quantities such as density, pressure, temperature, and flow speed;
- for nonplanar shock waves such as spherical blast waves, a strong decrease in the propagation velocity with increasing distance from the center of origin, because some of the energy of the shock wave is expended to heat the medium in which it travels;
- the entropy {CLAUSIUS ⇨ 1865} of the shock-compressed fluid increases and that of the expansion wave decreases compared to the undisturbed fluid. (these changes, however, are small for "weak shock waves"); and
- nonlinear superposition properties are observed during the reflections and interactions of shock waves.

Furthermore, shock waves have the unique property that they accelerate (quiescent) particles at the shock front. The sudden acceleration of charged particles up to relativistic velocities due to repeated scattering across a shock wave (e.g., generated in supernovae or by the collision of black holes and galaxies) remains the most attractive model for the production of high-energy cosmic rays {HESS ⇨ 1912; KRYMSKY ⇨ 1977/1978}.

Shock wave intensity – which can be classified into waves of small but finite amplitude (weak shock regime) and waves of large amplitude (strong shock regime) – can be defined in terms of the pressure ratio across the shock front, $\xi = p/p_0$, a quantity which is also called the "shock strength." Weak shocks for which $\xi$ is barely greater than 1 move approximately with the speed of sound, while strong shocks which are defined by $\xi \gg 1$ always propagate supersonically.[79] There is no clear definition as to what comprises the weak shock regime in air. Some researchers consider that the overpressure $\Delta p = p - p_0$ ranges from about 0.1 to 5 Pa (1–50 μbar) in weak shock waves.

It should be mentioned here that in Germany the term *Stoßwelle* ("shock wave") initially had a different meaning to that invoked today. Throughout the 19th century it designated a seismic sea wave resulting from an earth- or seaquake {KRÜMMEL ⇨ 1887}. The modern meaning of the term *shock wave* was not immediately taken up by encyclopedias. For example, in the German *Meyers Konversationslexikon* (1929), a shock wave was still defined as a "tidal wave originated by an earthquake." Even in the 1960s various prestigious encyclopaedias such as the *Encyclopedia Americana* (1961) and the *Encyclopaedia Britannica* (1962) did not list the term *shock wave*, which may also illustrate the delay between the emergence of a new, rapidly expanding field of science and its acknowledgment and subsequent review by encyclopaedists. Today *shock wave* [French *onde de choc*, Germ. *Stoßwelle* or *Schockwelle*, Span. *onda de choque*, Russ. *ударная волна*] is a clearly defined and well-established term used in science and technology. The huge field of shock and blast waves has now been treated in numerous special encyclopedic articles from different viewpoints for over a decade now, providing the interested layman with an understanding of the peculiar nature of shock waves and their various applications in research and industry.[80]

There are a couple of special shock- and shockwave-related terms in current use that should be mentioned here.

A *reactive shock* is a shock wave supported by an accompanying chemical reaction; the most common example is the

---

[78] It appears that the term *supersonic* was introduced into the English literature by Theodore VON KÁRMÁN in a paper published with N.B. MOORE in 1932. The Oxford English Dictionary [Clarendon Press, Oxford (1989), vol. 17, p. 241] refers to a paper published in the *Journal of the Royal Aeronautical Society* in 1934 in which this term was first used. • Note that the term *supersonics* was previously used to designate periodic sound waves with frequencies greater than those audible to the human ear (> 20 kHz).

[79] From the Rankine-Hugoniot theory, it follows that the propagation velocity of the shock wave, $D$, can be calculated from the shock strength, $\xi$, by the formula $D = c_0 [(\xi + \alpha)/(1 + \alpha)]^{1/2}$, where $\gamma$ is the ratio of the specific heats, $c_0$ the velocity of sound for the gas at rest, and $\alpha = (\gamma - 1)/(\gamma + 1)$. For air with $\gamma = 1.4$, this yields the simple expression $D = c_0 [(6 p/p_0 + 1)/7]^{1/2}$.

[80] The following more recent review articles cover shock wave and detonation phenomena:
(i) K.E. GUBKIN: *Explosion*. In: (A.M. PROKHOROV and J. PARADISE, eds.) *Great Soviet encyclopedia*. MacMillan, New York and Collier MacMillan, London, vol. 5 (1979), pp. 124-126.
(ii) S. LIN: *Shock waves in gases*. In: (S.P. PARKER, ed.) *McGraw-Hill encyclopedia of science & technology*. McGraw-Hill, New York etc., vol. 16 (2002), pp. 435-438.
(iii) J.D. ANDERSON JR.: *Flight (aerodynamics)*. In: (R.A. MEYER, ed.) *Encyclopedia of physical science and technology*. Academic Press, New York, vol. 5 (1991), pp. 389-410.
(iv) M. ROSS: *Shock waves and detonations*. In: (R.G. LERNER and G.L. TRIGG, eds.) *Encyclopedia of physics*. VCH, New York etc. (1991), pp. 1121-1124.
(v) R. RASPET: *Shock waves, blast waves, and sonic booms*. In: (M.J. CROCKER, ed.) *Encyclopedia of acoustics*. Wiley, New York etc., vol. 1 (1997), pp. 329-339.
(vi) R.T. BEYER: *Nonlinear acoustics*. In: (G.L. TRIGG, ed.) *Encyclopedia of applied physics*. VCH, New York etc., vol. 1 (1991), pp. 188-194; J.E. Shepherd: *Compressible flows*. Ibid., vol. 4 (1992), pp. 43-69; and Y. HORIE and A.B. SAWAOKA: *Shock waves*. Ibid., vol. 18 (1997), pp. 29-43.
(vii) S.P. SONETT: *Shock waves*. In: (J.H. SHIRLEY and R.W. FAIRBRIDGE, eds.) *Encyclopedia of planetary sciences*. Chapman & Hall, London (1997), pp. 734-736.

detonation wave generated by the detonation of a high explosive {SCHUSTER ⇨1893; VIEILLE ⇨1900; CRUSSARD ⇨1907}.

In a *rarefaction shock* – i.e., a decompressive shock – the pressure behind the front is smaller than the pressure ahead of it. Initially believed to be physically impossible {RIEMANN ⇨1859; RANKINE ⇨1869; JOUGUET ⇨1904; ZEMPLÉN ⇨1905}, the existence of rarefaction shocks was first predicted by the Soviet theoretical physicist Yakov B. ZEL'DOVICH {⇨1946} in substances that have thermodynamic parameters that are close to critical values. This was later also experimentally confirmed in Freon-13 by his countryman Samson S. KUTATELADZE {⇨1978}.

A *pseudo-shock*, a term coined by the Italian-born U.S. aerodynamicist Luigi CROCCO {⇨1958}, occurs when a supersonic flow is decelerated to subsonic velocities in a duct surrounded by walls. The shape of this *λ-type pseudo-shock* {⇨Fig. 4.14–H} depends on the Mach number of the flow and the conditions of the boundary layer. The term *pseudo-shock* is barely used nowadays, and has been replaced by the more accurate term *shock wave/boundary layer interaction*.

*Micro-shock waves* are spherical shock waves with typical radii of only a few millimeters, both in ambient air as well as in water with peak pressures in the range 1–100 MPa (10–1,000 bar). Micro-shock waves in the cm-range, generated by small amounts of high explosives or electric sparks, played an important role in the discovery of the nature of shock waves {E. MACH & WOSYKA ⇨1875, E. MACH & SOMMER ⇨1877, E. MACH, TUMLIRZ & KÖGLER ⇨1878}. With the advent of pulse lasers, underwater micro-shock waves were produced by focused laser beams {BELL & LANDT ⇨1967}, a method which was taken up in medical therapy by guiding the laser beam in an optical fiber {NAKAHARA & NAGAYAMA ⇨1999; JAGADEESH & TAKAYAMA ⇨2002}. However, there exist some limitations of miniaturization. For example, a wave of finite amplitude needs a certain distance to travel until it has steepened up to become a shock wave, and the structure of a shock wave itself which, only a few mean-free-paths thick, is in the microscopic realm. Furthermore, boundary layer effects can no longer be ignored at increasing miniaturization. The miniaturization of shock waves was also inspired by the arrival of the "new sciences" of microtechnology in the 1960s and nanotechnology in the 1980s with features on a scale near one micrometer ($10^{-6}$ μm) and below hundred nanometers ($10^{-7}$ m), respectively.

A *nanoshock* is a miniature shock wave generated by an ultrashort laser pulse, which can suddenly drive the irradiated material to extreme pressures and temperatures. Since the resulting mechanical transient has a duration of only a few nanoseconds, this shock pulse was termed a *nanoshock* by Dana D. DLOTT, a U.S. chemistry professor at the University of Illinois. The combination of the nanoshock compression technique with time-resolved molecular spectroscopy enables dynamic effects to be studied at the molecular level in chemistry, biology and medicine {DLOTT ⇨2000}.

In magnetohydrodynamics, two basic categories of shock waves – *MHD shocks* – are possible:
- *fast shock waves* with a jump in magnetic pressure acting on the front in the same direction as the jump in gasdynamic pressure, thus resulting in a wave speed greater than the speed of sound in the medium; and
- *slow shock waves* with drops in magnetic pressure and gasdynamic pressure in opposite directions at the wavefront, leading to a slow wave with a speed below the speed of sound.

It has proven helpful to categorize MHD shocks into two further classes of intermediate shocks.[81] In addition, in terms of the direction of material flow, two types of magnetic shock waves can be differentiated: *longitudinal shocks* and *transverse shocks* {DE HOFFMANN & TELLER ⇨1950}.

When a body is slowly subjected to a transient temperature gradient, transient thermal stresses are produced that can be predicted by the methods of *thermoelastostatics* {DUHAMEL ⇨1837}. These stress pulses are not discontinuities according to the definitions of mathematical physics. However, when the change in temperature occurs so rapidly that inertia causes stresses, the resulting dynamic effects can only be predicted by the methods of *thermoelastodynamics* {DANILOVSKAYA 1950, see DUHAMEL ⇨1837}. This, in fact, generates a *thermal shock* with a propagating jump in stress and strain.[82] Such intense shock-like stress pulses arise when, for example, the surface of a solid is irradiated by high-intensity ultrashort pulses of laser light {ASKAR'YAN & MOROZ ⇨1962; READY ⇨1963; WHITE ⇨1963} or X-rays, or by the impact of a burst of particles, for example by pulsed electron beams {WHITE ⇨1963}, pulsed ion beams {BLUHM ET AL. ⇨1985}, or the neutron flux from a nuclear explosion {TRUNIN ET AL. ⇨1992; ⇨Fig. 4.11–F}. The effects of the thermoelastic stress produced by pulsed uniform energy deposition can be described by one-dimensional models using thermoelastic theory {OSWALD ET AL. ⇨1971}.

*Thermal shocking* is a more general term used in engineering. It designates any equipment or system failure caused by

---

[81] P.H. ROBERTS: *Magnetohydrodynamics*. In: (G.L. TRIGG, ed.) *Encyclopedia of physics*. VCH, New York etc. (1991), pp. 680-686.

[82] H.W. BARGMANN: *On the thermal shock parameter in nuclear engineering*. In: *Proc. Conference on the Structural Analysis, Design & Construction in Nuclear Power Plants* [Universidade Federale do Rio Grande do Sul (UFRGS), Porto Alegre, Brasil, April 1978]. Pós-Graduação em Engenharia Civil, UFRGS, Porto Alegre (1978), pp. 681-688.

a sudden change in temperature. Fracture of rocks by thermal shocking has been used in stone tool fabrication and mining since prehistoric times.

In very strong explosions, such as in nuclear explosions, a particular shock has been observed which precedes the original shock wave along the surface and is termed the *thermal precursor shock* {⇨Fig. 4.16–Q}. This unique shock wave phenomenon is generated by preheating the surface via the thermal radiation prior to the arrival of the shock wave, and was first observed in nuclear explosions {SHELTON ⇨1953; BRYANT, ETHRIDGE & KEEFER ⇨1955}; however, it was also later observed in large-yield chemical explosions {CDRE Suffield ⇨1964}. This effect can also be simulated and demonstrated in laboratory shock tube experiments: a shock wave propagating over a heated layer of gas will refract, and its velocity will increase {⇨Fig. 4.13–E}.

A *radiative shock* is characterized by an intense radiation flux that precedes it. The region compressed by the shock heats up and produces photons that ionize the cold gas in which the shock propagates and thus creates a *radiative-precursor shock wave*. Radiative shock waves play a major role in several astrophysical phenomena, such as star formation, supernovae explosion, and stellar winds. Radiative shocks can also be created in the laboratory, for example, by using energetic pulsed lasers: the shock wave is produced on a millimeter scale in a miniaturized shock tube by a piston which is pushed by a high-power laser pulse. The conversion of laser energy into mechanical energy is achieved by using the ablation of a micrometer-thick plastic layer at the top of the piston. This allows to accelerate the piston up to a high speed in the tube and to launch the shock in the gas (Xe) filling the tube.

Since the 17th century, the term *shock* has also been used in the English-speaking world in connection with earthquakes,[83] and it is very common to speak more specifically of a *ground shock* or a *seismic shock* (from the Greek *seismic*, meaning "earthquake"). Today the latter term is used as a synonym for the term *earthquake*,[84] designating "a sudden motion or trembling in the Earth caused by the abrupt release of slowly accumulated strain."

In modern seismic exploration, *seismic waves* are also termed *shock waves*, although they are not characterized by the typical features of a shock wave mentioned above. They rather describe a complex mixture of body waves {POISSON ⇨1831} and surface waves {Lord RAYLEIGH ⇨1885; LAMB ⇨1904} which are modified during propagation by absorption, diffraction and reflection features of the various geologic strata. The sudden force that generates them is not an actual blow, but a wrenching snap, as billions of tons of bedrock, twisted and strained out of shape by the accumulated forces exerted over centuries, rupture along a fault plane and lurch back toward an alignment that relieves the stress {REID ⇨1906}. Consequently, seismic shocks are not characterized by high pressures and steep wavefronts as in true shock waves, but rather by very small earth displacements which happen in the elastic regime; *i.e.*, seismic waves can essentially be treated acoustically. With the advent of the generation of artificial earthquakes by strong infrasound sources (*e.g.*, by detonations or the falling of heavy bodies), the terms *shock* and *seismic shock* are now used in a more specific manner, designating the physical effects which originate from the sudden motion or trembling of the Earth.

A variety of shock-related terms were invented during the period marking the emergence of seismology as a scientific field. Many of these are still in use.[85]

▸ The terms *principal shock* or *main shock* were suggested for the strongest member of a series of earthquakes (J.F. Julius SCHMIDT 1874).

▸ The slighter shocks in a series of seismic waves were termed *accessory shocks*, with those before the principal shock being *preparatory shocks* and those after it *consecutive shocks* (François A. FOREL 1881). The latter two rather historical terms have been replaced by the terms *foreshocks*[86] and *aftershocks*, which designate a series of small seismic shocks preceding and following a large earthquake within minutes, hours, days or even months, respectively.[87] In large earthquakes, geological fault slippage may be announced by foreshocks, preparatory processes for the main rupture; these can occur days to months before the main shock. However, it is still the subject of debate among seismologists as to whether these precursor processes will allow the prediction of earthquakes. Underground nuclear explosions can be followed by aftershocks, and some believe that they may trigger impending earthquakes in the vicinity.

---

[83] An early example is given in Daniel DEFOE's famous novel *The life and strange surprizing adventures of Robinson CRUSOE* [Taylor, London 1719]: "After … I found still no more shocks of the earthquake follow I began to be more compos'd."

[84] R.L. BATES (ed.): *Glossary of geology.* Am. Geol. Inst., Falls Church, VA (1987), p. 198.

[85] C. DAVISON: *The founders of seismology.* Arno Press, New York (1978), p. 121.

[86] In astrophysics, the term *foreshock* designates the region upstream of a collisionless shock (such as the Earth's bow shock), which contains the accelerated ions and associated MHD waves.

[87] *Encyclopædia Britannica, Micropædia.* Benton & Hemingway, Chicago etc. (1974), vol. I, p. 127 and vol. IV, p. 228.

- *Microseismic shocks* are very small geological disturbances which are normally only detectable with sensitive instruments and are unlikely to cause damage, but are still clearly distinguishable from background noise.
- The term *shock line* [Germ. *Stoßlinie*] was coined in Austria {HÖFER ⇨1880}. It was used to designate the vertical projection of the trajectory of the seismic discontinuity propagating in the Earth's crust onto its surface.

In the near-field of large chemical and nuclear explosions initiated underground, at the surface or just above ground, the underground compression is no longer elastic, and true *seismic shock waves* are generated. However, they quickly decrease in amplitude with increasing distance from the explosion source.

In astrophysics, the term *termination shock* [Germ. *Endstoßwelle*, French *terminaison choc*] designates the boundary that precedes the heliosphere, where particles from the Sun (the "solar wind") drop abruptly from supersonic to subsonic speeds and clash with atomic matter from deep space (the "interstellar wind"). The origin of the term *termination shock* is unknown, but may have arisen from the application of the Laval nozzle in steam turbine work in the late 19th century.[88] This heliospheric shock phenomenon, which is believed to occur about 13 billion kilometers from Earth (or more than two times farther out than Pluto, the most distant planet) is currently being studied "live" via the two Voyager space probes which, after traveling through the Solar System for more than 25 years, have now reached the region where a termination shock is expected by astrophysicists {Voyager 1 & 2 ⇨2003; Voyager 1 ⇨2004}.

The term *terminal jet shock*, another astrophysical term, refers to a galactic jet phenomenon. At the jet terminus two shocks are formed: the *jet shock* (or *terminal Mach disk*), which effectively stops the incoming jet, and the *standoff shock* (or *bow shock*), which acts to accelerate and heat the ambient interstellar medium {M.D. SMITH ET AL. ⇨1985; ⇨Fig. 4.8–L}.

## 2.1.6 COLLISIONLESS SHOCK WAVES

In the 1950s, theoretical studies carried out by plasma physicists indicated that shock waves could form even in the near-vacuum of outer space, where particle collisions are extremely rare. They proposed that the collective electrical and magnetic properties of plasmas could produce interactions that take the place of collisions and permit shocks to form. a magnetic field endows collisionless plasmas with elastic properties analogous to those of a dense gas, and so a plasma wave crossing a magnetic field behaves somewhat like an ordinary sound wave. The theoretical analysis of so-called "collisionless shock waves" (or "collisionfree shock waves") therefore initially followed the ideas developed from earlier research on aerodynamic shocks. The first experimental confirmation was given by Norman F. NESS and his colleagues at NASA's Goddard Space Flight Center. Using data collected from the IMP 1 spacecraft, they detected clear signs that a collisionless shock exists where the solar wind encounters the Earth's magnetic field {NESS ET AL. ⇨1964}.

Shock waves in hot and low-density plasmas are anomalous in classical shock physics in that they cannot be interpreted on the basis of interparticle binary collisions. In hot plasmas, particle-wave interactions generated by fluctuating fields in the plasma dominate over the particle-particle interactions (*collisional shocks*) that dominate in classical fluid dynamics. For example, collisions of charged particles in the solar wind are so rare that they do not affect the formation of the shock or the dissipation of the solar wind's kinetic energy. Shock waves in plasmas have been observed and studied in both laboratory plasmas and space, and they have been simulated in computer-based "experiments." There are some fundamental differences between ordinary, collision-dominated shocks and collisionless shocks:[89]

- The plasma is generally not in thermodynamic equilibrium behind the shock.
- Jump conditions do not completely determine the downstream state.
- Collisionless shock fronts have widths that are less than – sometimes much less than – typical collisional mean free paths. A good example of such a shock is the Earth's bow shock, where the scale length of the transition region between the upstream and downstream states is several hundreds of kilometers or less, while the mean free path of the solar wind ions is on the order of $10^7$ km.[90]
- The thickness of the shock front also depends upon the direction of the magnetic field.

---

[88] Private communication by J. Randy JOKIPII, Regent's Professor at the Dept. of Planetary Sciences, University of Arizona (March 16, 2006).

[89] J.L. THIFFEAULT: *Collisionless shocks and the Earth's bow shock.* Presented in the Fall of 1994 for a Gas Dynamics astronomy class at Imperial College London; http://www.ma.imperial.ac.uk/~jeanluc/talks/bowshock.pdf.

[90] E.W. GREENSTADT and R.W. FREDRICKS: *Shock systems in collisionless plasmas.* In: (E.N. PARKER ET AL., eds.) *Solar System plasma physics.* North-Holland, Amsterdam (1979); vol. III: *Solar System plasma processes*, p. 5.

Depending on the wind velocity, magnetic field and ion density of the preshock material, shock waves associated with molecular outflows may be of the jump (J) type or the continuous (C) type.

- If the magnetic field is weak or nonexistent, all components (*e.g.*, atoms, ions, and electrons) have the same velocity. One may then observe a *J-shock*, because thermodynamic quantities such as temperature and density undergo a discontinuity.
- If the magnetic field is strong enough, it can provoke a partial decoupling of the flows of the different components of the medium through the shock: the charged particles gyrate around the magnetic field lines and are consequently coupled to this field, while the neutral particles are affected only indirectly by the magnetic field, through collisions with the positive ions (and electrons). The magnetic field accelerates the preshock material without producing a sudden jump in the temperature or velocity of the gas. Such a continuous, magnetically dominated shock is called a "C-shock," as first discussed in 1980 by the U.S. astrophysicist Bruce T. DRAINE.[91]

Exploding stars – supernovae – create very strong shocks that speed into the interstellar medium {HUGGINS ⇨1864} at tens of thousands of kilometers per second, generating interstellar shock waves {DRAINE & MCKEE ⇨1993} propagating at up to hundreds of kilometers per second in the hot components of the interstellar medium. Using the emission spectrum produced by interstellar shock waves, it is possible to differentiate between C-shocks and J-shocks.

Interstellar shock waves are likely to be responsible for the acceleration of cosmic rays {HESS ⇨1912; KRYMSKY ⇨1977/1978}. Elements beyond iron are formed during a supernova explosion and transported by the blast into outer space. They can later be captured into other clouds and become part of new stars and new planets. Future generations of stars formed from this material will therefore start life with a richer supply of heavier elements than the earlier generations of stars.

## 2.1.7 SHOCK AND VIBRATION

The phrase *shock and vibration* describes a special field of mechanical engineering that is related to a variety of vibration phenomena and their practical countermeasures, such as damping by shock absorbers {1st Symposium on Shock & Vibration ⇨1947; Handbook of Shock & Vibration ⇨1961}. In many practical applications, classical shock and vibration engineering and advanced shock physics converge. For example, in order to better protect personnel carrying large-caliber guns from extreme acoustic and mechanical shock loading, the issues that this problem imposes on gun construction require joint efforts from mechanical engineers to ensure that the enormous recoil momentum and vibrations are efficiently absorbed and from gas dynamicists to reduce the volume of noise produced by the strong muzzle blast.[92] Since the human body can be regarded as a biological as well as a mechanical system, shock and vibration effects on man are very complex, and so tolerance criteria for shock and vibration exposure are difficult to derive.[93]

## 2.1.8 BLAST WAVE, BLAST, AND BLASTING

Finite-amplitude waves in gases – particularly in the atmosphere – are the most common representatives of propagating discontinuities. Although relatively simply structured in comparison to shock waves in solids, it took decades of improvements in diagnostic techniques and theoretical refinements to fully uncover their true properties and to understand the various phenomena associated with the reflections from and interactions with structures and flows that had been previously observed in both laboratory and full-scale experiments.

**Blast Wave.** A shock wave in air – such as that emitted from an atmospheric explosion, a fired gun or some such phenomenon – is generally referred to as a *blast wave*, because it is accompanied by a strong wind, as felt by a fixed observer when the wave passes by.

While the term *shock wave* is used in a more general manner – to depict a pressure wave with a steep front but where the wave profile is not specified in more detail – the pressure-time profile of an ideal *blast wave* can be characterized

---

[91] B.T. DRAINE: *Interstellar shock waves with magnetic precursors.* Astrophys. J. **241**, 1021-1038 (1980); *Erratum.* Ibid. **246**, 1045 (1981).

[92] The ear is the part of the human body most sensitive to shock and blast injury. For pure tones, the maximum sound level which the ear can tolerate with pain but without immediately being damaged amounts to about 140 dB ref $2 \times 10^{-4}$ μbar, corresponding to a pressure of 2 mbar (1 mbar corresponds to 134 dB, 1 bar to 194 dB). The damage response to pulses, however, depends in a complex manner on the rise time, amplitude and pulse duration of the pressure, and there are wide variations in individual susceptibility to ear injuries. The ear is particularly sensitive to short-duration blast waves. Peak pressures of only a few 10 psi (1 psi = 68.9 mbar) can rupture the eardrum, and still smaller pressures can damage the conducting mechanism and the inner ear {HIRSCH ⇨1968}.

[93] H.E. VON GIERKE and D.E. GOLDMAN: *Effects of shock and vibration on man.* In: (C.M. HARRIS, ed.) *Shock and vibration handbook.* McGraw-Hill, New York etc. (1988), pp. **44**.1-58.

by its rise time, peak overpressure, positive phase duration, and total wave duration {FRIEDLÄNDER ⇨1946}. In the most common case of spherical explosions in air originating from chemical explosives, these quantities can be scaled precisely in terms of the released energy {B. HOPKINSON ⇨1915; CRANZ 1926; KENNEDY ⇨1946}, and they are related to the masses of standard high explosives, mostly to TNT {HAEUSSERMANN ⇨1891; DEWEY ⇨1964; BAKER, WESTINE & DODGE ⇨1973}. In nuclear explosions, the positive phase duration is longer than that arising from a chemical explosion with the same energy yield.

In astrophysics, a term coined in Germany {ZÖLLNER ⇨1865}, the shock wave that emerges from a supernova explosion and ejects the star's envelope into interstellar space is also called a "blast wave." A *blast wave shock* results from the interaction of ambient gas with the stellar material ejected by a supernova and this is the shock wave which precedes the ejecta.

The *blast wave accelerator* is a concept that describes the propulsion of a projectile through a gun tube via the sequential detonation of charges of high explosives, which exert a force on the base of the projectile. This concept allows hypersonic velocities (on the order of 10 km) to be obtained, and so it has been discussed as a way to launch materials into space inexpensively {WILSON ⇨1993}.

**Blast.** The term *blast* is used as a shorthand for the term *blast wave*, and/or to designate the ignition of gunpowder or other explosive. A blast wave always propagates with supersonic velocity. The steep-fronted pressure wave causes a sudden shock in the surrounding air or ground, and at a sufficiently large distance from the source this approaches spherical geometry. Blast waves are the most common form of shock waves and they are generated in most man-made explosions, ranging from nuclear blasts to the blasts from powerful electric spark discharges, and by other means where energy is suddenly released in a small space, such as in chemical microexplosions and focused laser beams of high intensity and short duration.

The *muzzle blast* [Germ. *Mündungsknall*, French *l'onde de bouche*] is the shock wave produced in the air by the violent eruption of propellant gases, as generated when a projectile exits the muzzle of a gun (which is appropriately described by ballisticians as "uncorking" the barrel). The muzzle blast actually involves a definite sequence of events: it is a system of normal and oblique shock waves that form the boundaries of a central supersonic region in front of the muzzle where the principal expansion and cooling of the gases occurs. This typical barrel shock pattern is called a "shock bottle" {SLADE ⇨1946}. The first pictures of this complex shock interaction phenomenon occurred in CRANZ's textbook on ballistics {CRANZ ⇨1925}.

The muzzle blast is preceded by a less intense precursor blast caused by the piston-like motion of the projectile when it is still moving in the barrel and compressing the air ahead of it. The German ballistician Carl CRANZ was the first to study the evolution of the muzzle blast from a fired rifle cinematographically. His series of pictures clearly show that this precursor is established before the bullet has emerged from the barrel. In the case of supersonic shots, the head wave is generated at the moment when the projectile outruns the front of the blast wave {⇨Fig. 4.5–L}.

In a volcanic eruption, pyroclastic flows can be generated during the climactic phase when, for example, slope failure unroofs a magma conduit or hydrothermal system, thus generating surges that have been known by volcanologists and geologists as *volcanic blasts* or *hot hurricanes* since 1980 {Mt. St. Helens ⇨1980; STURTEVANT ⇨1991}. Accounts of such blast effects arising from explosive volcanic eruptions and observed at different distances were first collected and analyzed from the Krakatau event {⇨1883}. Eruption-induced atmospheric shock waves have been discussed in a USGS note which also reviewed violent explosive volcanic eruptions in history and addressed the magnitudes of the blast pressures and energies released.[94]

According to the USGS geologists Dwight R. CRANDELL[95] and Rick P. HOBLITT, "a volcanic explosion that has a significant low-angle component and is principally directed toward a sector of no more than 180 degrees is referred to as a *lateral blast*. Such a blast may produce a mixture of rock debris and gases hundreds of meters thick that moves at high speed along the ground surface as a pyroclastic flow, pyroclastic surge, or both … Lateral blasts may affect only narrow sectors or spread out from a volcano to cover a sector as broad as 180 degrees, and they can reach distances tens of kilometers from a vent." Lateral blasts produced by volcanic eruptions may propagate supersonically; however, experimental evidence is difficult to obtain because of the extremely dangerous environmental conditions involved. It is believed that most lateral blasts propagate subsonically (see Sect. 2.3.1).

When a strong blast wave – for example that produced by a large-yield chemical explosion, a nuclear explosion or an explosive volcanic eruption – arrives at its target, it may

---

[94] *Hazards – Atmospheric shock waves.* USGS (1987); http://vulcan.wr.usgs.gov/Hazards/NRC_Definitions/shock_waves.html.

[95] D.R. CRANDELL and R.P. HOBLITT: *Lateral blasts at Mount St. Helens and hazard zonation.* Bull. Volcanol. **48**, 27-37 (1986).

creat destructive effects known as "blast damage." The air immediately behind the shock front is accelerated to high velocities and creates a powerful wind. The wind in turn, creates a dynamic pressure against the sides of objects facing the blast. The combination of the pressure jump (called the "overpressure") and the dynamic pressure causes blast damage; both immediately jump to their peak values when the shock wave arrives. They then decay over a period ranging from a few tenths of a second to several seconds, depending on the strength and yield of the blast and the geometry and dimensions of the object.

There is a definite relationship between the overpressure and the dynamic pressure. The overpressure and dynamic pressure are both the same at 70 psi (4.81 bar), and the wind speed is 1.5 times the speed of sound (about 502 m/s). Below an overpressure of 70 psi, the dynamic pressure is less than the overpressure; above 70 psi it exceeds the overpressure. Since the relationship is fixed, it is convenient to use the overpressure alone as a yardstick for measuring blast effects. At 20 psi (1.38 bar) overpressure, the wind speed is 500 mph (223 m/s), higher than any tornado wind.

The danger from overpressure comes from the collapse of buildings that are not as resilient as most. The violent implosion of windows and walls creates a hail of deadly missiles, and the collapse of the structure above can crush or suffocate those caught inside. City areas are completely destroyed (with massive loss of life) by overpressures of 5 psi (0.34 bar), which produced wind speed of 162 mph (72 m/s) – *i.e.*, close to the peak wind speeds of the most intense hurricanes.[96] For comparison purposes: the category-five Hurricane Katrina, which devastated New Orleans in August 2005, had an estimated maximum wind speed of about 175 mph (78 m/s) {FUJITA ⇨ 1971}.

**Blasting.** The term *blasting* has been used in mining engineering since the beginning of the 19th century at least, in order to describe the operation of breaking up coal, ore, rock, or other minerals using explosives – also called "shot firing." However, the use of black powder, now considered a "low" explosive, is much older, dating back to the 16th century {Venetian Mining Industry ⇨ 1572}.

*Blasting* also describes the operation of breaking up ice using chemical explosives {BARNES ⇨ 1927}.

### 2.1.9 GAS DYNAMICS, RAREFIED GAS DYNAMICS, MAGNETOGASDYNAMICS, AND COSMICAL GAS DYNAMICS

**Gas Dynamics.** This particular branch of fluid dynamics evolved at the end of the 19th century from attempts to understand the fundamentals of high-speed compressible flow through nozzles and passages. Gas dynamics has grown with the development of high-speed flight and has became an area of research for physicists, chemists, applied mathematicians and astrophysicists. Historically, one of the most important areas of application of gas dynamics was also the theoretical treatment of detonation in gases, in particular the description of the fundamental properties of the detonation wave – a reactive shock wave – and the unsteady motion of the detonation products.

The term *gas dynamics* suggests the idea that the field is exclusively related to the state of gaseous matter. However, Kirill P. STANYUKOVICH, an international authority on gas dynamics, proposed to include "the flow of all compressible media, including liquids and solids" (under conditions of high pressures) {STANYUKOVICH ⇨ 1955}. Today gas dynamics is understood as "the study of the motion of gases and of the nature and effect of such motion."[97] Klaus OSWATITSCH,[98] a renowned German physics professor who taught gas dynamic at KTH Stockholm and the University of Freiburg, considered gas dynamics to be a generalization rather than a specialization of hydrodynamics, since it was linked to technical thermodynamics through the inclusion of thermal processes, to mathematics by the development of methods for solving hyperbolic differential equations, and to traditional mechanical acoustics by the inclusion of shock waves. Because of these broad links to other branches of science, in the 1950s he recommended that more chairs dedicated to gas dynamics should be established at German universities.

The term *gas dynamics* (or *gasdynamics*) was apparently coined by Jakob ACKERET {⇨ 1927} in his handbook article entitled *Gasdynamik*. However, modern gas dynamics takes a wider, more general view than in the pioneering days and is nowadays concerned with the causes and effects arising from high-speed flows – steady and unsteady, viscous and nonviscous, conducting and nonconducting – for which compressibility is physically significant. Gas dynamics brings together concepts and principles from several branches of science, and

---

[96] *Blast damage and injury.* Federation of American Scientists (FAS), Washington, DC; http://www.cartage.org.lb/en/themes/Sciences/Chemistry/NuclearChemistry/NuclearWeapons/FirstChainReaction/EffectsNucl/Mechanisms.htm.

[97] C. MORRIS (ed.): *Dictionary of science and technology.* Academic Press, San Diego *etc.* (1992), p. 905.

[98] K. OSWATITSCH: *Die Gasdynamik als Hochschulfach.* Physik. Blätter **9**, 271-272 (1953).

includes mechanics, thermodynamics, aerodynamics, and chemical kinetics.

Besides the term *gas dynamics*, other names have been considered, such as *compressible flow, compressible aerodynamics, supersonic flow, thermofluid dynamics* and *aerothermodynamics*. However, aside from the more general term *gas dynamics*, only *compressible flow* (beginning at $M > 0.3$; $\Delta P/P > 5\%$) and *supersonic flow* ($M > 1$) are currently used routinely.

In the case of very high temperature flow fields, a complete analysis should be based upon the simultaneous study of both the gasdynamic field and the thermal radiation field. This particular branch of fluid mechanics, created in the 1960s and termed *radiation gas dynamics* {PAI ⇨1966}, combines ordinary gas dynamics with the physics of radiation. Gas temperatures may be very high and gas densities very low, such as for the hypersonic flow associated with aerospace vehicles (particularly during the reentry of a space vehicle), and for flows associated with nuclear reactions (particularly in the blast wave of a nuclear bomb).

**Rarefied Gas Dynamics.** This rather new offshoot of gas dynamics, which quickly developed into a large field of its own, covers phenomena in a gas or at a surface in contact with a gas when the gas density becomes sufficiently low that the mean free path is no longer negligibly small compared to the characteristic dimension of the flow geometry. Rarefied gas dynamics touch upon the subject of this book in relation to processes of momentum and energy exchange in high-speed gas-surface interactions during hypersonic flight at high altitudes (*e.g.*, during interplanetary rocket flight or for orbiting satellites), the formation and structure of shock waves and associated boundary layer effects, and the reflection of shock waves at bounding surfaces.

The theory of rarefied gas dynamics is based to a large extent on the familiar kinetic gas theory {MAXWELL ⇨1878}. A program for the study of these problems was defined by the Chinese-born U.S. engineer Hsue-Shen TSIEN {⇨1946}, a well-known aerodynamics specialist who suggested the standard classification of rarefied gas flows based on the Knudsen number $K_n = l/L$, where $l$ is the mean-free path of a molecule, and $L$ is the characteristic length of the object stationed in the flow {KNUDSEN ⇨1934}. $K_n$ may, in general, take any value. Conveniently, rarefied gas dynamics is subdivided into four different flow regimes:[99]

- free-molecular flow (extremely rarefied, $K_n \gg 1$);
- near-free-molecular flow (highly rarefied);
- transition flow (moderately rarefied); and
- slip flow (only slightly rarefied, $K_n \ll 1$).

Flows of highly rarefied gas are studied via the kinetic theory, while flows of slightly rarefied gas are treated from the standpoint of the gas dynamic theory of a continuous medium.

With the advent of the first artificial satellites in the late 1950s, the Soviet Sputnik 1 {⇨1957} and the U.S. Explorer 1 (1958), rarefied gas dynamics quickly attracted increasing attention in the study of hypersonic flow and drag problems, and the study of cosmic gases {Int. Symposium on Rarefied Gas Dynamics ⇨1958}.

**Magnetogasdynamics.** Studies of the interaction between magnetic fields and moving, electrically conducting gases shed light on a great variety of new phenomena, ranging from laboratory to cosmic dimensions. To unite the three disciplines of gas dynamics, electrodynamics and plasma physics, the term *magnetogasdynamics* was created, apparently in the early 1950s. It embraces most gaseous and compressible media. However, the term *magnetohydrodynamics*, which implies that the subject pertains to applications in water or at least in incompressible fluids, is also frequently used in the literature when referred to the flow of ionized gases in the presence of magnetic fields. Theodore VON KÁRMÁN {⇨1959} suggested the more general term *magnetofluidmechanics*, thus embracing both magnetogasdynamics and magnetohydrodynamics. The first monograph dedicated to plasma physics and magnetofluidmechanics appeared in 1963 and was written by Ali Bulent CAMBEL,[100] an Italian-born applied scientist and engineer at George Washington University.

**Cosmical Gas Dynamics.** This particular branch of astrophysics is mainly an outgrowth of classical gas dynamics, rarefied gas dynamics and magnetogasdynamics. Essentially pioneered by American, Soviet and British researchers in the late 1940s, cosmical gas dynamics was initially termed *cosmical aerodynamics*. However, before the establishment of cosmic gas dynamics, flow problems of cosmical dimensions were also treated in *cosmical electrodynamics*.

*Cosmical gas dynamics* appeared to enter general usage by both astrophysicists and fluid dynamicists in the 1950s. It was used as the title for the 2nd Symposium of Gaseous

---

[99] S.A. SCHAAF: *Rarefied gas dynamics*. In: (W.T.H. LOH, ed.) *Modern developments in gas dynamics*. Plenum Press, New York & London (1969), pp. 235-254.

[100] A.B. CAMBEL: *Plasma physics and magnetofluidmechanics*. McGraw-Hill series in missile and space technology, McGraw-Hill, New York (1963).

Masses of Cosmical Dimensions {⇨1949}, which was renamed the Symposium on Cosmical Gas Dynamics, and was held in 1953 at Cambridge, MA. Gas dynamical effects govern the physics of many objects in the Universe such as
- the dynamics of interstellar gas and effects of gravitational fields, a branch of astrophysics termed *interstellar gas dynamics*;
- shock waves caused by supernova and solar flare explosions;
- shock wave interactions with magnetic fields and high-energy particles;
- planetary bow waves;
- extragalactic and stellar jets;
- red supergiants in their final stage, their explosion into a supernova and subsequent collapse into a neutron star or a black hole; and
- spinning pulsars, which are remnants of exploded supernovae.

## 2.2 INITIATION OF PERCUSSION RESEARCH

*If a ball strikes another equal stationary ball, it comes to rest when that has been driven out.*[101]

Marcus MARCI VON KRONLAND
Prague 1639

IN physics, the course of history usually proceeds from simple to more complicated problems. The mechanical speculations of the ancient Greeks, as principally evoked by the works of ARCHIMEDES, related wholly to statics or to the doctrine of equilibrium. However, their thinking only extended into dynamics along the most unsuccessful paths.[102] Since the earliest times, however, dynamic processes rather than static ones have been of primary concern to man in his daily struggle of life, and they have been applied throughout his evolution in order to improve his tools and weapons, and to make their application more efficient. Dynamics is an entirely modern science that began with Galileo GALILEI's questioning of "why" and "how" the many motions that can be observed take place.

### 2.2.1 NATURA NON FACIT SALTUM

*In fact, such a principle of hardness [as assumed by advocates of the Atom in their corpuscular models] could not exist. It is something impossible, contradicting this general law, that nature constantly observes in all its operations. I am talking about this immutable and perpetual order established since the creation of the Universe that can be called the Law of Continuity, according to which everything that takes place does so in infinitely small steps. It seems that common sense dictates that no change can be made through fault; Natura non operatur per saltum; nothing can pass from one extremity to the other without passing through all the degrees in between.*[103]

Johann BERNOULLI
University of Basel 1727

*The third of the abovementioned theorems relates to the continuity of all mechanical effects – in former times a controversial supposition of all physical theories which, freely borrowed from ARISTOTLE, proclaimed to the well-known dogma: natura non facit saltum. However, modern research has also broken seriously through this hitherto always respected stronghold of physical science. This time, these are the principles of thermodynamics which, based upon more recent facts derived from experience, clashed with that theorem, and if all the signs are to be believed, its days are numbered. Indeed, nature seems to make jumps of a rather odd kind… In all cases the quantum hypothesis resulted in the conception that there are changes in nature which occur not steadily but rather explosively. I hardly need to remind you that the discovery and more detailed research of radioactive phenomena have gained considerably in clearness…*[104]

Max PLANCK
Friedrich-Wilhelms-Universität, Berlin 1913

The natural sciences – today understood to be a broad spectrum of disciplines concerned with objects or processes that occur in nature, and including fields such as physics, chemistry, biology, geology, astronomy, *etc.* – partly emerged from physics and mechanics of the 17th and 18th century, and grew from the soil of the "harmony of continuity." Mechanics and phenomenological thermodynamics regarded all processes as continuous, at least as a first approximation, and attempted to express them analytically.

---

[101] See MARCI {⇨1639}, *Propositio* XXXVII, *Porisma* I.
[102] For more on the development of the principles of statics and dynamics, *see* Ernst MACH's book *The science of mechanics*. Open Court, La Salle, IL (1960).
[103] See J. BERNOULLI {⇨1727}. The Swiss mathematician and physicist Johann BERNOULLI (1667–1748) was a brother of the mathematician Jakob BERNOULLI (1654–1705) and father of the mathematician and hydrodynamicist Daniel BERNOULLI (1700–1782); for the latter *see* the Biographies Index.
[104] M. PLANCK: *Neue Bahnen der physikalischen Erkenntnis* [Rede gehalten beim Antritt des Rektorats der Friedrich-Wilhelms-Universität Berlin, am 15. Oktober 1913]. *Max PLANCK. Physikalische Abhandlungen und Vorträge.* Vieweg & Sohn, Braunschweig (1958), vol. III, pp. 65-76.

The conception of the *Principle of Continuity* [Lat. *lex continui*], based upon the thoughts of some Greek philosophers and resumed in the Renaissance {TISSOT ⇨1613}, was taken up by numerous natural philosophers. For example, in his *Theory of Monads*, the German philosopher, mathematician and physicist Gottfried W. LEIBNIZ[105] developed the idea that each individual substance is subject to a perpetual change of state, leading to the metaphor that "the present is pregnant with the future." Such changes are without jumps, and the transition of a substance from one state to another is always continuous and orderly. He wrote in his *Nouveaux essays* (1704), "Tout va par degrés dans la nature et rien par sauts." The Principle of Continuity was also assumed by the German philosophers Johann Gottfried VON HERDER and Friedrich VON SCHLEGEL throughout their sketches of literary history.

This concept certainly affected subsequent generations of natural philosophers in a disadvantageous manner, resulting in their aversion to permit sudden changes to their analysis and mathematical-physical models. For subsequent naturalists it took a further 150 years to accept discontinuities because it involved the abandonment of the continuity principle *Natura non facit saltus* ("Nature does not make leaps")[106] – *i.e.*, the denial of the discontinuity of dynamic effects. This may be illustrated in the following examples:

- In classical mechanics, discussions on discontinuous processes and their actual existence were initiated by studying percussion in more detail. The eminent Swiss mathematician Johann BERNOULLI considered it absurd to apply the laws of percussion to hard and perfectly elastic bodies, because upon collision this would result in sudden velocity changes which – as he argued – cannot happen in nature {J. BERNOULLI ⇨1727; DIDEROT & D'ALEMBERT ⇨1751; ⇨Fig. 2.1}.
- The discussions were later extended to discontinuous wave motion, then termed *sound waves of finite amplitude*, and today known as "shock waves." For early physicists, shock waves were difficult to accept, because they were characterized by a stepped wavefront. In mathematical nomenclature, shock waves are unsteady planes of the first order; *i.e.*, discontinuities where fluid dynamic quantities – such as the density and velocity at both sides of the shock front – differ by finite amounts.[107] The assumption of discontinuities in analysis, the most powerful instrument of mathematics, did not evolve until the 18th century {EULER ⇨1748}. It reached its first milestone with the work of Jean B.J. FOURIER (1807), who showed that any arbitrary function – including functions with discontinuities, such as steps – can be expressed by a trigonometric series.
- Other, more prominent examples in the history of science of permitting discontinuities include so-called "energy leaps" or "energy jumps" in quantum mechanics, based upon the discovery of the energy quantum in 1900 by the German physicist Max PLANCK, which eventually forced scientists to give up on the Principle of Continuity. The British physical chemist David CHAPMAN,[108] cofounder of the first theory of detonation {CHAPMAN ⇨1899; JOUGUET ⇨1905}, wrote in 1914, "PLANCK's quantum law, in a simple form, is this: particles of matter emit and absorb energy not slowly and continuously, but in 'jerks.' In other words, the process of emission of energy is assumed in all cases to be analogous to the process which occurs when a molecule changes into an isomeric form. The latter process has always been assumed by chemists to be a sudden one, and therefore accompanied by a sudden evolution of energy. PLANCK's hypothesis, therefore, is equivalent to the assertion that all energy changes in matter are of the same character as those which occur in chemical change. The discontinuous character of all chemical change has become so familiar to chemists that it has ceased to be regarded as strange, or as needing explanation. Yet PLANCK's generalization is considered by some physicists to involve the abandonment of the principle – *Natura non facit saltum* – and the denial of the continuity of dynamical effect."
- Stimulated by PLANCK's discovery of the structure of electromagnetic radiation, the Danish physicist Niels H.D. BOHR realized in 1913 that atomic stability is also related to the notion of discontinuity. He postulated that an atom is capable of subsisting in a series of discrete stationary states without radiating energy, and that the radiation of energy occurs only when it makes a complete transition from one stationary state to another by emitting one quantum of energy in the form of electromagnetic radiation.

---

[105] G.W. LEIBNIZ: *Nouveaux essais sur l'entendement humain* (1704). This essay, which presented a detailed criticism of John LOCKE's position, was not published during his lifetime; it first appeared in print in *Œuvres philosophiques latines et françaises de feu Mr. DE LEIBNITZ...* (R.E. RASPE, ed.). J. Schreuder, Amsterdam/Leipzig (1765), vol. 3, chap. IV, p. 16.

[106] Instead of the plural *saltus* ("leaps"), the singular *saltum* ("leap") is often found in the literature.

[107] Discontinuities of the second, third, *etc.*, order are those where only the second, third, *etc.*, local or temporal derivatives of the fluid quantities are unsteady.

[108] D. CHAPMAN: *General and physical chemistry.* Annu. Repts Chem. Soc. **11**, 1-33 (1914).

- In physics, another important example of discontinuous action relating to atomic stability is *radioactivity* – the spontaneous disintegration of an atomic nucleus by the emission of some form of matter and/or energy. Decay rates of such nuclear processes are characterized by so-called "half-lives." Measured half-lives range from $3 \times 10^{-7}$ seconds to $10^{15}$ years for alpha decay, and from $10^{-3}$ seconds to $10^{16}$ years for beta decay. Electromagnetic decay, characterized by the emission of gamma rays, is much faster and usually occurs after on the order of only $10^{-15}$ seconds – *i.e.*, the emission of gamma rays from nuclei carries away energy and angular momentum in an extremely discontinuous manner.

- The most strongly discontinuous processes may have happened in the very initial phase of the creation of the Universe which, according to the standard Big Bang model, consisted of a mixture of radiation (photons) and particles at an extremely high temperature. In accordance with the equivalence of mass and energy inherent in Albert EINSTEIN's Special Theory of Relativity, collisions between high-energy photons would have transformed radiation into particles of matter. Collision processes between sufficiently energetic photons created particle-antiparticle pairs which immediately underwent mutual annihilation.[109]

- In seismology, stationary discontinuities created by changes in chemical composition or physical properties (*e.g.*, phase changes) are marked by a sudden or rapid increase in the speed of seismic waves with depth; *i.e.*, nonpropagating discontinuities can provoke dynamic discontinuities in terms of velocity changes. Classical examples include the *Conrad discontinuity* between the Earth's upper and lower crust (at a depth of 7.5–8.6 km) and the *Mohorovičić discontinuity*, the boundary between the crust and upper mantle (about 35–50 km below the continents and about 10 km below the oceans).

- Fractures in solid media represent mechanical discontinuities that strongly affect the propagation of elastic waves either across or along the fracture plane.[110] One of the physical models used to analyze the seismic properties of fracture is the *displacement discontinuity model*, which assumes that the stresses across a fracture are continuous, but that the displacements are not.[111]

- In contrast to early physicists, most early chemists took the discontinuous, instantaneous character of chemical changes to be self-evident and did not question it. Studies performed in the first half of the 19th century by renowned chemists – *e.g.*, by J. Jakob BERZELIUS, Jean-Baptiste A. DUMAS, Justus VON LIEBIG and Friedrich WÖHLER – revealed that chemical reactions are indeed rather complex processes and might involve a chain of metastable atomic arrangements, so-called "free radicals" {LIEBIG & WÖHLER ⇨1832}. These phenomena gained great importance when attempting to understand very rapid self-supporting chemical reactions. Prominent examples include chain reactions in detonations {BODENSTEIN ⇨1913; SEMENOV ⇨1934} and supersonic combustion {BILLIG ⇨1959; IUTAM Symposium on Combustion in Supersonic Flows ⇨1997}.

- Giving up the rigid Principle of Continuity also opened up new ways of interpreting mutation leaps in genetics. The Swedish botanist Carl VON LINNÉ, attempting to describe and name plants correctly and to group them systematically into categories in his *Philosophia botanica* (1751), considered the species – the basic unit of botany – to be fixed and unchangeable, and concluded that "we can count as many species now as were created at the beginning."[112] Addressing appropriate principles of botanical nomenclature, he wrote: "The fragments of the Natural Method are to be sought out studiously. This is the beginning and the end of what is needed in botany. *Natura non facit saltus*. All plants exhibit their contiguities on either side, like territories on a geographical map."[113] But the British naturalist Charles Robert DARWIN encountered the motto in a sharp and interesting form posing an alternative meaning of terrible import,[114] "Nature makes no leaps, but God does." Hence, if one wants to know whether something of interest is of natural origin or

---

[109] *Big Bang theory*. In: (P. MURDIN, ed.) *Encyclopedia of astronomy and astrophysics*. Institute of Physics, Nature Publ. Group, London *etc.*, vol. 1 (2001), pp. 173-174.

[110] L.J. PYRAK-NOLTE and N.G.W. COOK: *Elastic interface waves along a fracture*. Geophys. Res. Lett. **14**, No. 11, 1107-1110 (1987).

[111] M. SCHOENBERG: *Elastic wave behavior across linear slip interfaces*. JASA **68**, 1516-1521 (1980).

[112] S. LINDROTH: *Carl LINNAEUS (or VON LINNÉ)*. In: (C.C. GILLESPIE, ed.) *Dictionary of scientific biography*. Ch. Scribner's Sons, New York, vol. 8 (1973), pp. 374-381.

[113] C. VON LINNÉ: *Philosophia botanica, in qua explicantur fundamenta botanica*. G. Kiesewetter, Stockholm (1751); Engl. translation by S. FREER: *LINNAEUS' philosophia botanica*. Oxford University Press, New York (2003); § 77, p. 40.

[114] H.E. GRUBER: *DARWIN on man: a psychological study of scientific creativity. Together with DARWIN's early and unpublished notebooks*. Wildwood House, London (1974), p. 125.

supernatural one must ask, "Did it arise gradually out of that which came before, or suddenly without any evident natural cause?"

## 2.2.2 FOUNDATION OF DYNAMICS

> *Dynamics* [French *La Dynamique*] *is the science of accelerating and decelerating forces and the variable motions which they must produce. This science is entirely due to modern time, and* GALILEI *is the one who has laid the first fundaments.*[115]
>
> Joseph-Louis LAGRANGE
> Paris 1788

The evolution of mechanics began in antiquity, with general reflections on motion, but it then stagnated for a long period until it was resumed during the Renaissance. The birth of "modern times" – characterized by the rediscovery of certain ancient philosophers and, according to Leonardo DA VINCI (1452–1519), an era that was "ruled by numbers" – provoked considerable progress in mathematics, thus enriching the ways in which scientists conceived of phenomena. His countryman, the astronomer, physicist and mathematician Galileo GALILEI (1564–1642), one of the most outstanding representatives of this era and the "founder of modern experimental science," became widely known for his contributions to the Law of Gravity, which he ingeniously demonstrated using the examples of free fall and inclined throw. He also pondered on the enormous forces of percussion {GALILEI ⇨1638; ⇨Fig. 4.3−C}. Everyday percussion phenomena, such as a strike with a hammer, allow one to generate tremendous dynamic effects very easily, which can be reproduced by static means only with much greater effort.

The foundations for *dynamics*, laid in the 16th century by GALILEI, quickly grew into a new branch of mechanics. He derived the motion of a ball rolling down an inclined plane, a falling body, a pendulum, and a ball thrown into the air. His important result that applying a force to a body causes its velocity to change were later formulated in Newton's Second Law of Motion {Sir NEWTON ⇨1687}. Nearly all subsequent dynamics has been based upon his conclusions.

The German all-round natural philosopher Gottfried Wilhelm LEIBNIZ (1690–1730) substituted kinetic energy for the conservation of movement. In his work *La Monadologie* ("Theory of Monads"), he postulated in 1714 that the states of substances – which he assumed to be individual, indestructible and indivisible, so-called "monads" – can be ordered continuously into smooth transitional sequences; *i.e.*, all changes occur without jumps (*Lex continui*, 1704), and that the world is a harmonious cosmos, not a chaos. Consequently, his view of dynamic processes excluded discontinuities altogether. Percussion and shock waves, however, are discontinuous natural phenomena that occur widely in the Universe and are related to each other. Figure 2.1 schematically illustrates the generation of such discontinuities. Obviously, an ambitious analytical approach is required to describe the steep gradients of velocities and forces that occur in real percussion processes and the jumps in pressure, density and temperature that occur across the shock front for shock waves.

**Classical Percussion Research.** Scientific research in mechanical percussion – now generally termed *classical percussion* – did not start until the beginning of the 17th century. Many prominent naturalists of that century contributed to the understanding of percussion. The first significant progress made in this field in the 17th century is illustrated in more detail in the CHRONOLOGY {GALILEI ⇨1638; MARCI ⇨1639; DESCARTES ⇨1644; WALLIS ⇨1669 & 1670/1671; WREN ⇨1669; HUYGENS ⇨1652 & 1668/1669; MARIOTTE ⇨1671; Sir Isaac NEWTON ⇨1687; HUYGENS ⇨1703}. Percussion studies initially used tangible bodies like billiard balls or cannonballs and were mainly based on observations. These early investigations on the nature of percussion revealed that

- in a closed system, assuming that there is no friction, energy is conserved as kinetic, potential and elastic energy;
- percussion phenomena are material-dependent, and depend particularly strongly on the hardness of the bodies involved in the collision, which means that percussion can be classified into *elastic percussion* and *inelastic percussion*;
- in the case of *elastic percussion*, the velocities of the percussion partners can be determined from their masses and their initial velocities by applying the two Laws of Conservation of Momentum and Energy {⇨Fig. 2.16};
- in the case of *inelastic percussion* the kinetic energy is partly transformed into heat, but momentum is conserved;
- in the case of two colliding bodies, the ratio $\Delta c/\Delta v$ of the relative velocities after and before the collision ($\Delta c$ and $\Delta v$, respectively) is constant {Sir NEWTON ⇨1687}. This quantity depends on the material and geometry of the collision partners and significantly determines the reflection behavior of colliding bodies

---

[115] J.L. DE LAGRANGE: *Mécanique analytique*. Veuve Desaint, Paris (1788), vol. I (1811), p. 221.

{⇨Fig. 2.5}. Two hundred years later, this ratio was termed the *coefficient of restitution e* (or $\varepsilon$) {⇨Fig. 4.3–F} by Lord KELVIN {⇨1879}; and
▸ the phenomena associated with the percussion are determined by the body's geometry; in particular where it is struck in regard to its center of gravity, leading to the differentiation between so-called "central percussion" and "eccentric percussion" {⇨Fig. 2.2}.

**Center of Percussion.** It has been known for a long time that a hammer blow can be transmitted up the arm and an uncomfortable shock is felt when a hammer is held too far from its head. The English mathematician John WALLIS {⇨1670/1671} noticed that an impacted body begins to rotate upon experiencing an impulsive force, and concluded that the percussion or striking of a moving body can be greatest at a particular point – the so-called "center of percussion" [Lat. *centrum percussionis*] – in which the whole percussive force of the body can be assumed to be concentrated.

Figure 2.9 illustrates the characteristic features of the center of percussion in a body of mass $m$ freely rotating around a fixed axis and struck at a distance $X$ from the pivot. The distance of the center of percussion to the axis is given by $L_{CP} = \Theta/mL_{CG}$, where $\Theta$ is the *moment of inertia* of the body with respect to the axis, and $L_{CG}$ is the distance between the pivot and the center of gravity. Generally, the center of percussion will be away from the center of gravity ($L_{CP} > L_{CG}$) and positioned on a line that connects the pivot center with the center of gravity. In the simple case of a straight bar of uniform cross-section and length $L$ shown here, the center of gravity is $\frac{1}{2}L$ and the center of percussion $\frac{2}{3}L$ away from the pivot – a result which was provided by HUYGENS in his *Horologium oscillatorium* (Paris 1673).

There are numerous practical and scientific applications where the center of percussion plays an important role, for example:
▸ *In ballistics.* The ballistic pendulum, which is used to measure projectile velocities, is perhaps the oldest example of the scientific application of the concept of the center of percussion. An optimized ballistic pendulum {CASSINI JR. ⇨1707; ROBINS ⇨1740} is constructed such that the bullet hits the pendulum at its center of percussion. The projectile's kinetic energy is then optimally transferred to the pendulum mass, and the impulsive force generated by the projectile impact causes no reaction force at the pivot.[116]

▸ *In hand-held tools.* Percussive tools such as hammers, sledges, axes and adzes are best designed when the center of percussion is placed as close as possible to the tool head, which can be achieved for example by choosing a heavy tool head and a light handle.[117] A hammer can only be stopped from jiggling upon use when it is properly held at a certain distance from the tool head.[118]
▸ *In hand-held sports kits.* (i) When a baseball bat (or a cricket bat) strikes the ball at the center of percussion, it both maximizes the kinetic energy of a blow transmitted to the ball and minimizes the sting from the handle in the batsman's hand; *i.e.*, no shock will be felt in his hands {BRODY ⇨1979}. In the case of an aluminum softball bat of length 0.81 m, the center of percussion is about 0.17 m away from the fat or distal end of the bat {⇨Fig. 4.3–Y} – somewhat less than one-third of the length of along a cylindrical bar of length $L$ {⇨Fig. 2.9}, because the center of gravity is further away from the batman's hand than $\frac{1}{2}L$. (ii) A tennis player wants to hit the ball in such a way as to achieve the greatest momentum transfer to the ball with the least reaction force on his wrist and elbow. When the ball hits the racket at the center of percussion, or the "sweet spot" {BRODY ⇨1979; HATZE ⇨1998} of the racket, the rotational reaction on his wrist and elbow is minimized, and therefore the risk of getting "tennis elbow" [*lateral epicondylitis*] is reduced. On the other hand, shock actions produced by shock waves are also routinely used to heal a "tennis elbow" in modern medical therapies {DGST & IGESTO ⇨1995}.
▸ *In hand-held weapons.* A sword will handle effectively (or feel "alive") if it is balanced properly and the "sweet spot" exhibits the lowest tendency to vibrate. Therefore, the sweet spot is the most effective portion of the blade to strike the target with. A well-designed sword should structurally reinforce the sweet spot in order to reduce "wobbliness."[119] George L. TURNER,[120] a member of the American Association of Renaissance Martial Arts, has written a long, readable article on the dynamics of hand-held weapons.

---

[116] I. SZABÓ: *Einführung in die technische Mechanik*. Springer, Berlin *etc.* (1959), pp. 324-326.

[117] R.S. HARTENBERG: *Hand tools*. In: *The new Encyclopædia Britannica, Macropædia*. Benton & Hemingway, Chicago *etc.* (1974), vol. 8, pp. 605-624.

[118] I. SZABÓ: *Repertorium und Übungsbuch der Technischen Mechanik*. Springer, Berlin *etc.* (1960), p. 216.

[119] *Discerning a well-made sword*. Sword Forum International (Mesa, Arizona); http://swordforum.com/sfu/primer/wellmade.html.

[120] G.L. TURNER: *Dynamics of hand-held impact weapons*. Association of Renaissance Martial Arts; http://armor.typepad.com/bastardsword/sword_dynamics.pdf.

▸ *In mechanical engineering.* In the automotive industry, rear axles of the crank and swing type should be constructed such that the center of percussion coincides with the pivot of the stub axle.[121] In a hammer mill, the shock action between the lever and the rotating cam can be minimized by making use of the concept of the center of percussion.[122, 123]

In the case of hand-held implements, however, more recent studies performed in the United States and Austria have shown that the problem of minimizing shock action in the hand is more complex than the case of a fixed axis of rotation: the point on a hand-held implement that feels "sweetest" is *not* the center of percussion, because the hand, which extends over a finite length of the handle, exerts an opposing force on the handle {HATZE ⇨1998}.

**Vis Viva Controversy.** The percussion studies performed in the 17th century resulted in a challenging question on the nature of force. Obviously, the enormous force involved plays a major role in all percussion processes. However, early endeavors to measure the "force of percussion" by comparing it with the pressure of a weight at rest failed {GALILEI ⇨1638; ⇨Fig. 4.3−C}. Since the force of percussion typically acts during a very brief period of time − an "infinite" force acts for an "infinitesimal" time producing an instantaneous change in the velocity of the impacting body in the limiting case {⇨Fig. 2.1} − its measurement was not accessible with early diagnostic means (neither were the very short contact times of percussion and temporal deformations). Therefore, early percussion theories such as those proposed by HUYGENS, MARIOTTE, WALLIS, and WREN, relinquished the difficult task of evaluating the enormous instantaneous force from the beginning.

An impressive and characteristic feature of all percussion processes is the *impetus* (= $mv$) of a mass $m$ moving with velocity $v$, a term coined in the late Middle Ages by the French philosopher Jean BURIDAN in his *Theory of Impetus* to describe the enormous action which a moving mass can provoke {BURIDAN ⇨c.1350}. He correctly theorized that the mover imparts a power proportional to both $v$ and $m$ to the moved object, which keeps it moving. In order to characterize this action of force, 17th-century natural philosophers proposed different physical quantities:

▸ The French philosopher and mathematician René DESCARTES used the momentum (or impulse) given by $mv$, a value which he called the "quantity of motion" {DESCARTES ⇨1644}.

▸ On the other hand, the German scientist and philosopher Gottfried W. LEIBNIZ {⇨1686} favored the quantity $mv^2$ as a "true measure of force" [Germ. *wahres Kraftmaß*], and he labeled this quantity the "living force" [Lat. *vis viva*; Germ. *lebendige Kraft*] − in contrast to the "dead force" [Lat. *vis mortua*, Germ. *tote Kraft*], which produces no active work (such as forces from weight or static pressure).

▸ In his Second Law of Motion, Sir Isaac NEWTON recognized force as being associated with the acceleration of a mass {Sir NEWTON ⇨1687}.

▸ In his book *Essai d'une nouvelle théorie de la manœuvre des vaisseaux* ("Essay on a New Theory of the Handling of Ships," Basel 1714), the Swiss mathematician Johann BERNOULLI the Elder exposed the confusion in Cartesian mechanics between force and living force.

▸ The Scottish mathematician Colin MACLAURIN {⇨1724} argued against the mensuration of the forces of bodies by the square of the velocities.

A reconciliation could be attained by comparing the "efficiency of action" [Germ. *Wirkungsfähigkeit*] of a moving body against a force such as gravity. This can be illustrated using the following example: a body thrown up vertically with a velocity $v$ of 2 climbs for a period of time $t$ of 2, but it travels a distance $h$ of 4 against gravity, because $t = (2h/g)^{1/2}$ and $h = v^2/2g$. Hence, when the efficiency of action of a moving body of mass $m$ is measured in relation to the time over which the force acts against gravity, then $mv$ (the momentum or impulse) is the correct quantity. On the other hand, when the efficiency of action is related to the distance traveled $h$, then $2mgh = mv^2$ (or or *vis viva*) is the correct quantity {HUYGENS ⇨1668/1669}. Both concepts − based upon the two Laws of the Conservation of Momentum and Energy − are the basis for all theories on percussion and shock waves. Kinetic energy is now taken to represent one-half of the *vis viva*.

The "question of *vis viva*" was used to make claims about what the point of physics should be, what techniques should be used to gain physical knowledge, and what metaphysics made sense.[124] The *vis viva* dispute dragged on for over half a century. Eventually, it was recognized by the French physicist and mathematician Jean LE ROND D'ALEMBERT {⇨1743} who returned to previous debates on this subject in the foreword of

---

[121] H. FRANKE (ed.): *Lueger Lexikon der Technik*. Dt. Verlagsanstalt, Stuttgart, Bd. 12 (1967): *Lexikon der Fahrzeugtechnik*, p. 653.

[122] I. SZABÓ: *Einführung in die technische Mechanik*. Springer, Berlin *etc.* (1959), pp. 334-336.

[123] *Hütte* (ed. by Akademischer Verein Hütte, e.V., Berlin). W. Ernst & Sohn, Berlin, vol. I (1931), p. 297.

[124] M. TERRALL (Dept. of History, UCLA): *Vis viva revisited*. Hist. Sci. **42**, 189-209 (2004).

his treatise *Traité de dynamique* ("Treatise on Dynamics") that both concepts were two different ways of looking at the same problem: on the one hand, force can be defined in terms of the velocity of an objects, on the other hand, in terms of resistance that has to be overcome to stop a moving body – probably having here in mind impact experiments with objects stopped by springs as carried out previously by the Dutch mathematician and experimental physicist Willem Jacob's GRAVESANDE {⇨1720; ⇨Figs. 2.7, 4.3–G & 4.3–O}. However, in spite of D'ALEMBERT's clarifying comments, the interest in the *vis viva* controversy among natural philosophers persisted for a long time {WOLLASTON ⇨1805}.

In military applications, measures to increase the *vis viva* and, therefore, the destructive power of a bombardment, were achieved by intuition and practicability rather then as a result of mathematical studies. Throughout the 15th century, individually carved stone cannonballs were increasingly replaced by heavier wrought iron and later cast iron balls, thus leading to the first ballistic revolution. It is interesting to note here that this principle has recently been resumed and refined through the invention of so-called "kinetic energy (KE) projectiles" [Germ. *Wuchtgeschosse*]. Today high-velocity armor-piercing rounds with heavy cores are considered to be the most effective anti-tank weapons. Heavy projectiles are fabricated from high-density materials such as tungsten carbide or depleted uranium (DU). Indeed, DU projectiles are not only very hard, which provides a high penetration efficiency, but they are also pyrophoric – *i.e.*, they react chemically upon impact, causing an explosion inside whatever they penetrate. Furthermore, the depleted uranium used in the Balkans War in the 1990s is being blamed for a number of deaths from leukemia ("Balkans Syndrome").

**Corpuscular Models.** A major question that troubled most early natural philosophers was that of the nature of matter. Theories on the constitution of bodies suppose them either to be continuous and homogeneous, or to be composed of a finite number of distinct particles – such as atoms or molecules. In certain applications of mathematics to physical problems, it is convenient to suppose bodies to be homogeneous in order to make the quantity of matter in each differential element a function of the coordinates. On the other hand, molecular theories can be divided into *static theories*, which assume that molecules are at rest in the body, and *dynamic theories*, which suppose the molecules to be in motion, even while the body is apparently at rest {MAXWELL ⇨1867}. In order to explain the properties of matter, any dynamic molecular (or atomic) theory will use a moving particle model, also known as a "corpuscular model."

The main impetus to develop such a dynamic theory came from Greek philosophers, particularly from DEMOCRITUS (460–370 B.C.) of Abdera and EPICURUS (341–270 B.C.) of Samos, who attempted to explain matter as being composed of invisible atoms. LUCRETIUS (94–55 B.C.), a Roman poet and philosopher, modified and refined these Greek theories {⇨1st Century B.C.}. He described these invisible and impenetrably hard atoms as all moving downwards in infinite space with equal velocities, which suffer an imperceptible change at random times and positions, just enough to allow occasional collisions to take place between the atoms. Curiously enough, this rather modern concept based on infinitely small particles was proposed almost 1700 years before the scientific study of collisions between tangible bodies was initiated during the Renaissance.

The corpuscular hypothesis was widely discussed among 17th-century philosophers in England and France, resulting in an updated version of the atomic philosophies of antiquity:

▸ René DESCARTES and Pierre GASSENDI explained natural phenomena in terms of small, invisible particles of matter: DESCARTES proposed in his *Principia philosophiae* (1644) the existence of relatively hard but divisible corpuscles [Lat. *corpusculae*] that filled all space. In contrast, GASSENDI, drawing in his *Syntagma philosophicum* (1658) upon EPICURUS and LUCRETIUS, theorized about indivisible atoms in motion in an extended void.

▸ Robert BOYLE said that matter was made up of small corpuscles and explained all natural phenomena through the motion and organization of "primary particles" which move freely in fluids, less freely in solids and which produce corpuscles by coalition. In his treatise *Origin of Forms and Qualities According to the Corpuscular Philosophy* (1666), he advanced a view that, following DESCARTES, avoided taking a stand upon contemporary disputed issues.

▸ Robert HOOKE {⇨1665} suggested a corpuscular model of percussion on an atomic level in his *Micrographia*, in order to explain the properties of gases in terms of the motion and collision of atoms.

▸ The ideas of Edmé MARIOTTE {⇨1673} regarding the constitution of air and the role of corpuscles in the propagation of sound resulted in a hypothetical corpuscular model.

▸ Sir Isaac NEWTON {⇨1687} evolved a corpuscular theory of matter in his *Principia*. Conceiving of the ideal atom as being perfectly hard, he also used a corpuscular model {⇨Fig. 4.4–A} to explain the nature of light and to illustrate that the propagation of sound occurs via percussion from one particle to another.

In the following two centuries, the crude corpuscular model was refined by researchers in France, Switzerland, England and Germany:

- Daniel BERNOULLI {⇨1738} used NEWTON's corpuscular model in his *Hydrodynamica*, in which he expressed the phenomenon of heat via the velocity of colliding atoms {JOULE ⇨1850}, thus producing the first thermodynamic theory of heat.
- Jean LE ROND D'ALEMBERT {⇨1743} essentially accepted NEWTON's model in his *Traité de dynamique*; however, to explain elasticity he evolved a model of the atom as a hard particle connected to its neighbors by springs. His model, which in some ways represents the archetype of many subsequent shock wave models, stimulated other naturalists to also explain the propagations of other types of mechanical waves, such as those of seismic shocks {DESMAREST ⇨1756}, in the same manner. The mass/spring model was resumed in the 1960s by the Russian shock physicist Lev. V. AL'TSHULER in his shock wave model {⇨Fig. 4.4–D (d)}.
- NEWTON's corpuscular model stimulated the assumption of "dark stars" (or "black stars") {MICHELL ⇨1783; DE LAPLACE ⇨1796}, which anticipated the modern concept of "black holes" {WHEELER ⇨1968; Supermassive Black Hole ⇨1994}.
- Claude-Louis NAVIER {⇨1822} used a corpuscular model to derive the Laws of Motion for continuous media.
- August Karl KRÖNIG {⇨1856} proposed a corpuscular model consisting of discrete particles that have only translatory motion.
- Rudolf J.E. CLAUSIUS {⇨1857}, assuming that translational motion alone could not account for all the heat present in a gas, proposed a corpuscular model of translational, rotary and vibratory energy in which collisions can cause transformations of one form of motion into another.
- James Clerk MAXWELL {⇨1867}, who introduced statistics into thermodynamics, worked out the distribution of velocities among the molecules of a gas and the mean free path between molecular collisions. For gases at near-atmospheric densities, all of the particles spend most of their time moving with a constant speed, called the "thermal speed" {JOULE ⇨1850}, and any particle behaves as if it were alone in its container. At any time instant, however, about one particle in 100,000 will be colliding with another particle or with the container walls, and, over a time interval of one second, each particle of the gas will experience about $10^9$ collisions.[125]

The first experimental evidence for the concept that heat is related to the movement of particles was provided by the American-British Count VON RUMFORD {⇨1798}, following his heat and friction experiments. The French chemist Claude-Louis BERTHOLLET {⇨1809} described a percussion experiment that showed that heat could no longer be produced in a metal once hammering produced no further decrease in volume (a decrease that would cause caloric to be expelled like water from a sponge). Therefore, he associated the rise in temperature with a decrease in the volume of a solid body. However, the French engineer Sadi CARNOT rejected BERTHOLLET's explanation of heating by percussion, and in particular the supposed association between a decrease in volume and a rise in temperature {BERTHOLLET ⇨1809}.

The Scottish botanist Robert BROWN {⇨1827} gave the first visual demonstration of the random movement of microscopic particles suspended in a liquid or gas – so-called "Brownian motion." Experiments performed in 1909 by the French physicist Jean-Baptiste PERRIN confirmed the physical theory of Brownian motion {⇨Fig. 4.4–A}. He was honored with the 1926 Nobel Prize for Physics "for his work on the discontinuous structure of matter, and especially for his discovery of sedimentation equilibrium."

**Newtonian Demonstrator.** Edmé MARIOTTE {⇨1676}, an early French natural philosopher and mathematician, came up with the idea of studying the percussion of two balls suspended on long threads {⇨Figs. 4.3–D, G}. This arrangement was based on a famous experiment devised by Marcus MARCI {⇨1639}, an early Bohemian naturalist and physician, where a cannonball would be fired towards a row of other cannonballs which would stop the flying ball fully, but in doing so the last ball of the row would be expelled with a velocity equal to the velocity of the impacting ball {⇨Fig. 4.3–B}. MARIOTTE's set-up, which eliminated detrimental friction effects between ball and table that occurred in MARCI's set-up, opened the door to the construction of spectacular apparatuses demonstrating chain percussion. Sir Isaac NEWTON used a two-ball pendulum with balls of equal as well as different diameters (*i.e.*, masses) to demonstrate his rules of percussion. The eminent Dutch experimental physicist Willem Jacob's GRAVESANDE {⇨1720} devised various percussion machines suitable for demonstrating and quantitatively studying the straight central and oblique central percussion of elastic and inelastic bodies {⇨Figs. 4.3–G, H}.

---

[125] M. MCCHESNEY: *Gaseous state*. In: *Encyclopaedia Britannica, Macropaedia*. Benton & Hemingway, Chicago *etc.* (1974), vol. 7, pp. 914-922.

The multiple percussion pendulum {⇨Fig. 4.4–B} – also known in England as the "Newtonian demonstrator" (or more popularly as "NEWTON's cradle" or "balance balls") and in Germany as the "Klick-Klack" (due to its sound) – has been in use since at least the early 18th century in order to demonstrate the phenomenon of *chain percussion*, and exists in almost every modern collection of physical instruments.[126] Applying a number of balls of equal size, mass and composition, each bifilarly suspended in order to ensure that oscillations occurred in only one dimension, this ingenious set-up quickly became a spectacular system for demonstrating the Laws of Conservation of Momentum and Energy: the only way that momentum and energy can both be conserved is if the number of impacting balls is the same as the number of ejected balls, and if the ejected balls reach the same height as were reached by the impactors. For common table devices the ball impact velocities are in the order of some 100 cm/s.

An extended version of the multiple-ball pendulum, with additional springs arranged between each ball – thus forming a mechanical transmission line, now frequently used for deformable object simulation in engineering and even in medicine – was also proposed in order to demonstrate the propagation of a shock wave {BURTON ⇨1893; AL'TSHULER ⇨1965}. Modern studies have revealed that chain percussion in the Newtonian demonstrator is actually a rather complex process involving nonlinear dispersion and resulting in a unique solitary wave {HERRMANN & SCHMÄLZLE ⇨1981; NESTERENKO ⇨2001}.

## 2.2.3 FURTHER INVESTIGATIONS

Percussion research reached its next peak in the second half of the 19th century. The German physicist Franz NEUMANN {⇨1885}, the French mathematician Adhémar J.C. DE SAINT-VENANT {⇨1867} and the German physicist Heinrich HERTZ {⇨1882} developed (partly contradictory) percussion theories in which they included HOOKE's Law of Linear Elasticity {HOOKE ⇨1679}.[127] This also allowed theoretical determinations of the instantaneous stress distribution or percussion force. However, experimental evidence was difficult to obtain because of the crude high-speed diagnostics available at that time and, with a few exceptions, this topic remained open for investigation by 20th-century percussion researchers.

**Hertzian Cone.** The German physicist Heinrich HERTZ {⇨1882} theoretically demonstrated that the stress distribution in a plate that is normally impacted by a hard sphere has a conical geometry that extends symmetrically from the point of impact at the surface into the impacted body – the "Hertzian cone" [Germ. *Hertz'scher Kegel*], known among prehistorians as the "bulb of percussion" {⇨Fig. 4.2–A}. The important result obtained by HERTZ confirmed various hypotheses from archaeologists about how handaxes, arrowheads, knives, and other objects made from flint or similarly hard minerals were produced by primitive man via percussion {KERKHOF & MÜLLER-BECK ⇨1969}.

In the Stone Age, the essential elements of stone-tool making were *flakes and cores*: flakes are the relatively thin pieces that are detached under this Hertzian cone angle, and cores are their sources. Flaking can be carried out

▸ by percussion, with a pointed hammerstone;
▸ by means of a cylindrical hammer – for example, the shaft of a long bone;
▸ by pressure; or
▸ in the crudest manner, by battering the piece of stone to be flaked against another stone serving as an anvil.

The early stone industries are distinguished from one another by the different methods of tool-making, by the size of the flakes removed, and by the types and variety of tools produced.

**Bulb of Percussion.** Percussion applied in a small area on the surface of a brittle stone (such as the crypto-crystalline silica rocks chert and flint) typically produces a swelling on the flake at the point where it has been struck, detaching it from the core. This percussion mark is known as the "bulb of percussion" [Germ. *Schlagzwiebel* or *Schlagkegel*; French *bulbe de percussion* or *cône de percussion*]. In particular, among prehistorians and flint-knappers the semi-cones and bulbs are usually termed *positive bulbs of percussion*, and the hollows are termed *negative bulbs of percussion*. In the interest of strict accuracy the words *of percussion* should be altered to *of applied force*, because there are other ways of producing them besides percussion, such as by static pressures {LEAKEY ⇨1934}.

Distinctive kinds of flakes which show direct superposition of positive and negative bilbs of percussion on the interior and exterior flake surfaces are generally attributed to human manufacture. They may be produced in two ways: by being struck from a core from which a primary flake has previously been struck, or by being produced simultaneously

---

[126] E.R. JONES and R.L. CHILDERS: *Contemporary college physics*. Addison-Wesley, Reading, MA (1993). • For a demonstration experiment of the multiple-ball pendulum which allows one to select the number of striking balls *see* http://www.walter-fendt.de/ph11e/ncradle.htm.

[127] R. HOOKE: *De potentia restitutiva* [Lectures]. Martyn, London (1678).

with the primary flake.[128] Since violent point impacts are extremely rare events in nature, modern archeologists consider stones showing these typical bulb features to be important evidence of man-made stone tools, particularly when these strokes were applied to the stone repeatedly and in a controlled manner, generally in a row in order to provide a cutting edge, as seen on the oldest chopper stones made in East Africa {LEAKEY ⇨ c.2.6 Ma}.

Results obtained in modern anthropology have not only illustrated the enormous time span during which percussion played an important role in man's evolution, but also underline the great value of interdisciplinary cooperation in prehistoric research. The French archeologist Jacques BOUCHER DE PERTHES {⇨1837}, upon finding a variety of flint tools in the Somme Valley which had been worked on by man, was the first to relate man's antiquity to periods of geological time. His discovery also certified that man had existed far earlier than the widely accepted date of 4004 B.C. – the year of creation according to the Book of Genesis and a chronology written in the 1650s by the Irish Anglican Bishop James USSHER.

The very important phenomenon of a "bulb of percussion" was apparently first correctly described by the Englishman Sir John EVANS,[129] the cofounder of prehistoric archaeology, who collected and classified flint implements. He wrote, "The character of fracture is at first at the point of impact… in all cases where a splinter of flint is struck off by a blow, there will be a bulb or projection, of a more or less conical form, at the end where the blow was administered, and a corresponding hollow in the block from which it was dislodged. This projection is usually known as the 'bulb of percussion' – a term, I believe, first applied to it by the late Dr. Hughes FALCONER, F.R.S." FALCONER was an English palaeontologist and botanist who supervised the organization of Indian fossils for the British Museum and pursued palaeontological research while traveling in southern Europe.

Sir John EVANS, who searched for traces of early man in Britain, stated the following important conclusion: "If on a splinter of flint such a bulb occurs, it proves that it must have resulted from a blow, in all probability, but not of necessity, given by human agency; but where the bulb is on the principal face, and analogous depressions, or portions of them, are visible on the several other faces, and at the same end of a flake, all of them presenting the same character, and in a definite arrangement, it is in the highest degree probable that such a combination of blows must be the result of design, and the features presented are almost as good as a warrant for the *human origin of the flake* as would be the maker's name upon it."

**Conchoidal Fracture.** Various fine-grained rocks show typical *conchoidal fracture* which is a smooth but curved fracture surface resembling the interior surface of a shell – a curious phenomenon explained by the interaction and reflection of shock waves resulting from percussion. Such waves run through the stone along a curved route, detaching a curved flake. Rocks of this type include many lavas, as well as obsidian, flint and chert, and they were used in tool-making from the earliest times in the evolution of man.[130] To gain further insights into the past, the basics of stone tool technology are now being studied by performing "experimental archeology." Today many prehistoric museums demonstrate the art of flint knapping to their visitors.

**Percussion Figures.** When a blow from a sharp point is applied to the surface of a crystal, a fracture pattern is produced which is closely related to the internal structure of the crystalline lattice {REUSCH ⇨1867}. This "percussion figure" [Germ. *Schlagfigur*] is characterized by lines that are parallel to the plane of symmetry of the crystal. For example, a sharp blow applied to a cleavage flake of mica produces a six-rayed star of cracks, while a point-loaded cubic crystal such as rock salt produces a four-rayed fracture pattern {⇨Fig. 4.3–N}. These cracks coincide with planes of easy separation or of gliding in the crystal. Percussion figures are especially useful to mineralogists since they allow them to quickly determine the crystallographic orientations of some minerals.

**Percussion Marks.** These are crescent-shaped scars produced on hard, dense pebbles (especially ones of chert or quartzite) by a sharp blow, as by the violent impact of one pebble on another.

▸ Percussion marks are also created by rock to rock collisions and may be indicative of high-velocity flows.[131] In 1997, NASA's Mars Pathfinder Lander detected near its landing site several rocks who show percussion marks. Based upon other surface features on Mars, researchers assume that Mars was subjected to multiple flood episodes. Observations made during a

---

[128] A.J. JELINEK, B. BRADLEY, and B. HUCKELL: *The production of secondary multiple flakes.* Am. Antiquity **36**, No. 2, 198-200 (1971).

[129] J. EVANS: *The ancient stone implements, weapons and ornaments of Great Britain.* Longmans et al., London (1872), pp. 247-248.

[130] J.A.J. GOWLETT: *Tools – the palaeolithic record.* In: (S. JONES, S. BUNNEY, and R. DAWKINS, eds.) *The Cambridge encyclopedia of human evolution.* Cambridge University Press, Cambridge (1992), p. 350.

[131] K.K.E. NEUENDORF ET AL. (eds.): *Glossary of geology.* American Geological Institute, Alexandria, VA (2005), p. 481.

flood in southern Iceland revealed that collision of rocks is audible and sounds like distant thunder.[132]

▸ Percussion marks play a prominent role in interpreting early archaeological site formation and hominid behavior. Percussion marks are closely associated with hammerstone impact notches and show consistent micromorphological features which distinguish them from tooth marks and other classes of bone surface modification. Given indications of prehistoric hammer-stone breakage of marrow bones, an awareness of percussion marks is critical for accurately identifying the biological agents of bone modification at archaeological sites and provides a new diagnostic of carcass processing by hominids.[133]

**Percussion Force and Contact Time.** Percussion forces of enormous magnitude can arise when "hard" bodies strike one another violently. However, they are only effective during the short duration when the striking bodies are in direct contact with each other. This *contact time* is typically composed of a *compression period* and a *restitution period* {⇨Fig. 2.8}. Contact times during percussion were first measured electrically by the British engineer Robert SABINE {⇨1876}, who revealed that they are of very short duration – somewhere in the microsecond regime – and dependent on the mass and initial velocities of the percussion partners, their hardness, and the angle of impact. Peter G. TAIT {⇨1892}, a Scottish professor of physics and mathematics, devised a percussion machine that was very similar in construction to a guillotine. This machine permitted the first continuous recordings of contact times during percussion events. Using a graphical method, he also evaluated the duration of percussion for various practical examples, such as contact times between a golf ball and a club, one billiard ball with another, and a hammer with a nail.

It is interesting here to note that in the 18th century the concepts of percussion and impact were also applied on a merely theoretical basis to illuminate and prove the nature of motion and equilibrium. In the first part of his *Traité de dynamique* (Paris 1743), the French mathematician and encyclopedist Jean LE ROND D'ALEMBERT was inclined to reduce every mechanical situation to one of impact rather than to resort to the effects of continual forces. The U.S. historian J. Morton BRIGGS,[134] addressing this unique procedure in his biographical note, wrote, "D'ALEMBERT's Third Law dealt with equilibrium, and amounted to the principle of the conservation of momentum in impact situations… His proof rested on the clear and simple case of two equal masses approaching each other with equal but opposite speeds. They will clearly balance one another, he declared, for there is no reason why one should overcome the other. Other impact situations were reduced to this one; in cases where the masses or velocities were unequal, the object with the greater quantity of motion (defined as $m\,v$) would prevail. In fact, D'ALEMBERT's mathematical definition of mass was introduced implicitly here; he actually assumed the conservation of momentum and defined mass accordingly."

**Billiards.** This game, which fascinates game players and laymen as well as physicists and mathematicians, became an important stimulus for percussion research {MARCI ⇨1639; HUYGENS[135] 1669}, because it allowed elastic collisions to be easily studied under almost ideal conditions. Its country of origin is unknown.[136] In the early days of the game, billiards players used a rather simple crooked stick, as shown in MARCI's book on percussion entitled *De proportione motus…* ("On Proportions of Motion…"), which enabled the player to push rather than to strike the ball precisely {MARCI ⇨1639; ⇨Figs. 4.3–A, B}. However, from 1735 onwards a straight tapered stick of polished wood, which was well balanced and fitted with an ivory reinforced tip – the "cue" –

---

[132] J.W. RICE JR.: *Flooding and ponding on Mars: field observations and insights from the polar realms of the Earth.* In: Lunar & Planetary Science Conference XXXI, Lunar & Planetary Institute, Houston, TX (2000), CD-ROM, Abstract No. 2067 (2000).

[133] R.J. BLUMENSCHINE and M.M. SELVAGGIO: *Percussion marks on bone surfaces as a new diagnostic of hominid behavior.* Nature **333**, 763-765 (June 23, 1988).

[134] J. MORTON-BRIGGS: *Jean LE ROND D'ALEMBERT.* In: (C.C. GILLESPIE, ed.) *Dictionary of scientific biography.* Ch. Scribner's Sons, New York, vol. 1 (1970), pp. 110-117.

[135] In a letter to Henry OLDENBURG, HUYGENS – then secretary of the Royal Society of London – solved the famous "carom" [French *carambolage*] problem of finding the point of reflection between two eccentrically colliding billiard balls by applying a hyperbola rather than a circle [*See: Œuvres complètes de Christiaan HUYGENS*, publiées par la Société Hollandaises des Sciences. Swets & Zeitlinger, Amsterdam; vol. 6 (1976): *Correspondence 1666–1669*, No. 1745: and HUYGENS' letter to OLDENBURG (June 1669) on the *Problema Alhaseni*, p. 462].

[136] Some notes about the historical background of billiards might be useful to the reader. Its origin is obscure; its name is perhaps derived from the French word *billiart*, meaning a "stick." There is reference to the game in SHAKESPEARE's *Antony and Cleopatra* II, 5 (Cleopatra: "Let it alone, let's to billiards," written in 1606/1607), but the existence of billiards in pre-Christian times has not been substantiated. Encyclopedias such as the German *Meyer* and *Brockhaus* and the *Encyclopedia Americana* favor Italy as the country of origin, but the *Encyclopædia Britannica* also states that France, England, Spain and China have been credited with the invention of the game, and that the earliest references to the game in Europe occurred in the 15th century. The first concrete evidence of the existence of billiards traces back to France during the reign (1461–1483) of LOUIS XI, but its spread to other European countries did not start until 200 years later, during the unusually long reign of LOUIS XIV, covering the period 1643–1715 {⇨Fig. 4.3–A}.

came into use. In a more advanced form of the game, the cue was turned into a precise, high-tech percussion tool. In the 18th century, Captain François MINGAUD, a French billiards champion, was the first to provide the cue with a leather tip.

John CARR, an English billiards teacher, invented the "side stroke" (1818), or "side" as it has come to be known; however, he had trouble with his glossy leather tip. In order to increase the adhesion between cue and ball, he was the first to use chalk, which he applied uniformly to the cue tip. This allowed him to to impart a spinning motion to the cue ball by striking it off-center; striking the cue ball (a cue-tip width) above-center to impart overspin will cause the ball to roll forward, or to follow the object ball – the so-called "follow shot;" while striking it below-center will cause it to roll backward – known as a "draw shot."[137] Striking the ball laterally off-center, which is called "side" in Great Britain, "English" in the United States, and *effet* in France, introduced new and amazing phenomena: striking the ball left and right of center causes the ball to spin clockwise and counterclockwise, respectively. On the other hand, striking the ball above and below the center increases and decreases the speed of the cue ball compared to normal speed, respectively.

These eccentric twists first illustrated the complex nature of percussion, particularly the influence of *spin*
- on the collision and reflection behavior of the cue ball after impact with an object ball or with the cushion;
- on the friction between a spinning ball and the cloth; and
- on the resulting velocity and direction of the struck ball.

These new phenomena stimulated the French engineer and mathematician Gustave-Gaspard DE CORIOLIS {⇨1835} to reflect on collision in the presence of friction and to develop the first mathematical theory of billiards, which he himself considered to be his best work. CORIOLIS, a professor at the Ecole Polytechnique who held the prestigious position of a *répétiteur*, was supported by General Henri A. DE THOLOZÉ, then director of this distinguished school and an ardent fan of billiards. He demonstrated various billiards *effet* phenomena to CORIOLIS, who analyzed them mathematically – resulting in a unique cooperation between a superior and a subordinate. His studies, laid down in his book *Théorie mathématique des effets du jeu de billard* (Paris 1835), also resulted in an important by-product – the discovery of the "Coriolis force."

**Ballistic Pendulum.** The application of the Laws of the Conservation of Momentum and Energy, one of the basic findings of 17th-century percussion research, resulted in a fundamental theory of impact and proved most useful in the *ballistic pendulum*. This simple but most ingenious instrument, first suggested by a French astronomer {CASSINI JR. ⇨1707}, was introduced into practical ballistics by the English mathematician and military engineer Benjamin ROBINS {⇨1740} in order to quantitatively determine the velocity of a moving bullet. The impulsive penetration of the bullet is so close to being instantaneous, and the inertia of the pendulum's block is so large compared with the momentum of the shot, that the ball and the pendulum move as one mass before the pendulum has been sensibly deflected from the vertical. From $\theta$, the angle of deflection, it is possible to calculate the impact velocity $v_p$ of a musket or even of a cannon shot ($v_p \propto \sin \theta/2$). ROBINS also used the ballistic pendulum to study projectile drag as a function of its velocity, thus creating *aeroballistics* – an important branch of ballistics that deals with the aerodynamics of projectiles moving through the atmosphere; since World War II this field has also involved research on rockets and guided missiles, and on bombs dropped from aircraft.

ROBINS' remarkable supersonic ballistic experiments, which he performed with spherical projectiles traveling at velocities of up to about 1,670 Engl. ft/s (509 m/s, $M \approx 1.5$ at 20 °C), revealed that aerodynamic drag increases considerably when approaching the velocity of sound. These experiments were repeated in 1958 and analyzed with modern instrumentation by the U.S. aerodynamicist Sighard F. HOERNER, who proved that ROBINS must have reached supersonic velocities with his gun shots {ROBINS ⇨1746}. However, the ballistic pendulum – albeit based upon an impressively simple principle – produces a number of errors in measurement, and thus the data obtained from it must be carefully analyzed {CRANZ 1927, see ROBINS ⇨1740}.

Percussion has not become a branch of science of its own, but has instead acted as an important multidisciplinary tool (and it still does), thereby stimulating science and technology in a variety of ways and initiating new fields in applied physics. Examples include *shock wave physics* in general, and *solid-state shock wave physics* and *impact physics* in particular. In addition, applications of percussion phenomena and methods have created new disciplines of great economic significance, such as *impact engineering, dynamic materials testing, seismic surveying,* and *biomechanics*.

### 2.2.4 APPLICATIONS OF PERCUSSION

Applications of the principle of percussion in classical terms – *i.e.*, the direct impact of one body against another – are as

---

[137] A *follow shot* is when the cue ball stops momentarily, then follows the object ball's direction upon contact with the object ball. The *draw shot* makes the cue ball spin back after it impacts the object ball.

old as mankind itself and encompass most types of weapons and tools that have been invented. Percussion is used for special purposes in industry and science, and even in medicine. However, compared to the huge number of applications of shock waves, pure percussion techniques are limited to rather few applications. In modern science and engineering, the term *percussion* has been replaced almost entirely by the term *impact*, an important branch of dynamics covering a wide range of loading rates which is now treated in special textbooks {*e.g.*, GOLDSMITH ⇨1960; KINSLOW ⇨1970}, journals {Int. J. Impact Engng. ⇨1983} and technical reports, and at symposia, conferences and workshops {*e.g.*, Int. Symp. on Structural Crashworthiness ⇨1983; DYMAT ⇨1985; HVIS ⇨1986; SUSI ⇨1989; Int. Conf. on Large Meteorite Impacts and Planetary Evolution ⇨1992; ISIE ⇨1992; crashMAT ⇨2001}.

**Pile Driving, Steam Hammer.** The principle of percussion has been used since the earliest times for *pile driving* {Old World ⇨*c.*10,000 B.C.}. Numerous tools and weapons based upon percussion were developed in antiquity, for example war galleys with rams on their bows, wall hammers, battering rams [Lat. *aries*], and catapults [Lat. *ballistae*] {⇨Figs. 4.2–F, G}. With the advent of pile engines {DA VINCI ⇨1490s; BESSON ⇨1578; GROLLIER DE SERVIÈRE ⇨1719; ⇨Fig. 4.2–N}, the question of how to optimize their efficiency was discussed, and the influence of friction between the pile and the soil was recognized {HUTTON ⇨1812}.[138] The steam hammer, a British invention {NASMYTH ⇨1838} based on the principle of percussion, soon became an indispensable tool in most pile driving and drop forge techniques. The first huge drop forging presses were built and operated in the early 1890s, *e.g.*, at the Bethlehem Iron Company in Pennsylvania (weight of the ram block: 113.4 tons, dropping height: 6 m), at Schneider & Co in Le Creusot, France (80 tons, 5 m), and at the Kruppwerke in Essen, Germany (50 tons, 3 m). These were important requirements for the evolution of the heavy industry, as well as proud national symbols of progress and unlimited technical feasibility. *Drop forging*, where the metal is hammered between two dies, also became an important mass production technique, although it is only suitable for producing small and medium size objects. The hammer is dropped from its maximum height, to which it is usually raised by steam or air pressure.

**Percussion Drilling.** This technique of sinking a borehole was invented by the German technician Karl G. KIND {⇨1834; ⇨Fig. 4.2–Q}, who used a tool that repeatedly dropped onto the same spot. This technique allowed the first deep-drilling of holes in rocks. The method stimulated the invention of the pneumatic drilling hammer, which proved its great utility during the construction of the 15-km-long St. Gotthard Tunnel (1872–1881) connecting Switzerland with Italy. Drilling hammers became indispensable tools in the construction industry, particularly when working heavy concrete. Today electric and pneumatic drilling hammers have become widespread tools for the professional as well as for the do-it-yourselfer.

A modern outgrowth of percussion drilling is *laser percussion drilling*, a more recent development. The term *percussion* refers to the repeated operation of the laser in short pulses ($10^{-3}$ s), which are separated by longer time periods ($10^{-2}$ s). The energy supplied by the laser is bounded, and pulse-wise behavior allows for large bursts of energy. The actual drilling predominantly consists of removal by melt ejection, a process initiated by the sudden expansion of the vapor evaporating from the surface heated by absorption of laser energy. The gas dynamics for this complex process is similar to the well-known model of a shock tube.[139]

**Crushing, Fragmentation.** The understanding of percussion phenomena also provided new insights into traditional engineering applications, for example the mechanical processes of crushing and fragmentation of brittle materials, as used in the industrial treatment of stones and ores.[140] The *percussion grinder* is a machine for crushing quartz or other hard material by a combined rubbing and pounding process. The *wrecking ball*, a medieval technique, is still widely used to demolish concrete and masonry structures {⇨Fig. 4.2–P}. Suspended from a cable attached to a crane, the wrecking ball, which is usually a pear-shaped iron body that can weigh of up to 10 tons, is either dropped onto or swung into the structure that is to be demolished.

**Fluid Jet Impact.** The interaction between a moving solid body and a liquid jet emerging from a nozzle at high pressure can provoke the plastic deformation of the liquid jet or the brittle fracture of the jet, depending on the viscosity of the liquid and the impact velocity {KORNFELD ⇨1951; ⇨Fig. 4.3–V}. This curious phenomenon has been explained by a relaxation mechanism {KOZYREV & SHAL'NEV

---

[138] Kinetic (or dynamic) friction occurs when two objects are moving relative to each other and rub together. The *coefficient of kinetic friction* depends on the pair of surfaces in contact (roughness) and on the sliding velocity; it is usually less than the *coefficient of static friction*.

[139] K. VERHOEVEN: *Modelling laser percussion drilling*. Proefschrift Technische Universiteit Eindhoven, The Netherlands (2004).
[140] H.H. GILDEMEISTER: *Spannungszustand und Bruchphänomene in prallbeanspruchten Kugeln*. Ph.D. thesis, Fakultät für Chemie-Ingenieurwesen der TH Karlsruhe (June 1976).

⇒1970; ⇒Fig. 4.3–W}. On the other hand, a characteristic peculiarity of the impacts of miniature volumes of fluid (drops, jets) onto a solid is the generation of a pressure peak in the fluid pressure-time profile acting on the solid. This pressure pulse can be regulated as desired by utilizing fluids with different relaxation times, changing their temperature, or by adding polymers.

Studying and understanding impact phenomena of a high-velocity liquid (water) jet on a solid surface {HEYMANN ⇒1969} are also important tasks when attempting to raise the efficiency of hydraulic equipment operated by liquid jet impact, such as in water jet cutting technology[141] and hydraulic mining.[142] The latter technique was one of the dominant processes used by the California gold mining industry and was in use from the mid-1850s until 1884. It involved dropping water almost vertically hundreds of feet down to the mining site and channeling it through heavy iron pipes, where it exploded from nozzles in so-called "water cannons," thus disintegrating gold-laden banks of soft gravel and leaving huge craters.

**Firearms.** The principle of percussion has also been widely used in the construction of reliant firearms. The disadvantages of flintlocks became particularly obvious during the Napoleonic Wars (1803–1815). The inventions of the *percussion lock* {FORSYTH ⇒1805}, a system that used a small pill of detonating explosive (fulminating powder) placed below a plunger at the entrance to the touchhole, which overcame the disadvantages of the flintlock ignition, and the invention of the *percussion cap* (or *perciussion primer*) {SHAW ⇒1815}, an outgrowth of FORSYTH's idea {⇒1805}, created a wealth of new percussion-related terms, such as

- the *percussion arm* – usually a small-arms type weapon based on the detonation of an explosive when struck sharply;
- the *percussion bullet* (or *explosive bullet*) – a bullet containing a substance which is exploded by percussion;
- the *percussion fuse* – a fuse used in shells and case shots which ignite by the impact of the projectile striking the target;
- the *percussion gun* – the name given to firing a gun by means of a percussion cap placed over the flash hole;
- the *percussion lock* – the lock of a gun that is fired by percussion upon fulminating powder;
- the *percussion match* – a match which ignites by percussion; and
- the *percussion powder*, the powder used in percussion caps (since about 1832 consisting of mercury fulminate).

Percussion systems combining propellant, bullet and primer {DREYSE ⇒1827} eventually led to the modern cartridge, which revolutionized the science of war. To some extent the percussion cap can also be regarded as the archetype of the blasting cap, the so-called "detonator" {A. NOBEL ⇒1863/1864}.

**Intense Sound Generation.** In musical acoustics, the most intense percussion instruments are huge metal bells which are struck near the rim by a metal striker inside the bell or an exterior hammer. The acoustical structure of the sound of the bell is a complex mixture of partials that depends not only on the bell's geometry but also on the metals used for the two percussion partners (bell and striker), and the locus of percussion.

Artificial seismic waves are also generated via the principle of percussion, such as by weight-dropping units, high explosives ("explosion seismology," *see* Sect. 2.7.1) or by "vibrator trucks" where hydraulic pistons oscillate a mass in a vertical manner imparting a force through a base plate attached to the ground, thus inducing vibrations directed down into 'the earth. Pneumatic air-guns are also used, particularly in marine environments, which generate strong sound waves by releasing a burst of extremely high-pressured air into the water. So-called "seismic air-guns" usually operate at about 2,000 psi and typical volumes of air expelled vary from 30 in.$^3$ (492 cm$^3$) to about 800 in.$^3$ (13,104 cm$^3$).[143] Large air-gun arrays normally employed produce sound levels of 260–270 dB (re 1 µPa) at 1 m.

**Medical Diagnostics.** In medicine, percussion is used in physical diagnostics to obtain information about what lies below the skin (in physics, a similar procedure is used to determine the natural frequencies of a body by striking it with a hammer). Although it is a rather ancient method {VON AUENBRUGGER ⇒1754}, it is still routinely used in chest and abdomen diagnostics: the clinician strikes the skin with a finger or a hammer, setting up vibrations; the resulting sounds are then interpreted. If the stroke is made upon the body, it is called "immediate percussion;" if upon something placed against the body (*e.g.*, a finger of the other hand, or a small instrument made for this purpose), "mediate percussion."

---

[141] F. HAMMELMANN (1997), *see* LEACH & WALKER {⇒1965}.
[142] D.A. SUMMERS: *Hydraulic mining: jet assisted cutting.* In: (H.L. HARTMAN, ed.) *SME (Society of Mining Engineers) handbook.* Society of Mining, Metallurgy and Exploration, Littleton, CO (1992), chap. 2.2.3.

[143] W. DRAGOSET: *A brief overview of seismic air-gun arrays.* The Leading Edge **19**, No. 8, 898-902 (Aug. 2000).

**Biomechanics.** By definition, *biomechanics* is the analysis of the mechanics of living organisms and their parts, which makes use of the laws of physics and engineering concepts to describe the motion of body segments, and the effects of internal and external forces which act upon them during movement and rest. A particular branch of this discipline is the study of the response of muscles, joints, bones and other parts of the human body to the sudden application of forces resulting from a single blow or repetitive blows, and from interactions with shock waves. The subject has gained increasing attention from not only the military, industry and those working in sport, but also by insurance companies and courts. Examples of shock-type biodynamics addressed in this book include:

- skull injuries caused by percussion {MARCI ⇨ 1639};
- explosion-like injuries in soft tissue resulting from projectile impact {MELSENS ⇨ 1872};
- injuries to the cerebral spine and its tissues caused by sudden forceful flexion or extension to the neck – so-called "whiplash" {CROW ⇨ 1928};
- severe injuries like swelling of the brain, hemorrhaging, and neck injuries that result when a baby (or child) is shaken – so-called "shaken baby syndrome" {GOLDSMITH ⇨ 1960};
- injuries to the muscles and tendons of the outside of the elbow that result from overuse or repetitive stress in the application of hand-held percussion tools and certain sports equipment – such as "tennis elbow" and "golfers elbow" {BRODY ⇨ 1979; HATZE ⇨ 1998};
- the intense hammering shock that should be experienced by the brain of a woodpecker but which is, in fact, heavily reduced by internal protection mechanisms {BECHER ⇨ 1953};
- human injuries caused by strong blast waves resulting from guns and explosions (*see* Sect. 2.1.5); and
- injuries to marine animals caused by the impact of underwater shock waves resulting from underwater detonations {LEWIS ⇨ 1996}.

Biomechanics even touches upon such daily activities as walking and running: during each step, a rapid deceleration occurs at the foot/ground interface, resulting in a shock being imparted to the musculoskeletal system, and insufficient shock attenuation may result in severe overuse injuries. This has brought up the important question of how footwear can be designed to attenuate the *foot/ground impact shock*, and how such properties can be tested.[144]

## 2.3 EARLY SPECULATIONS ON SUPERSONIC PHENOMENA

> *The unprecedented identification of the spectrum of an apparently stellar object [quasar 3C 273] in terms of a large red-shift suggests either of the two following explanations. (i) The stellar object is a star with a large gravitational red-shift. Its radius would then be of the order of 10 km... (ii) The stellar object is the nuclear region of a galaxy with a cosmological red-shift of 0.158, corresponding to an apparent velocity of 47,400 km/sec.*[145]
>
> Maarten SCHMIDT
> Mt. Wilson and Palomar Observatories
> Southern California 1963

HIGH-SPEED natural phenomena have aroused the curiosity as well as the anxiety of man from the earliest times. With the evolution of fluid dynamics and thermodynamics beginning in the 19th century, the causes of most terrestrial high-rate phenomena have been understood using scientific methods and improved instrumentation. Furthermore, with the progress in radiation and nuclear physics in the 20th century many solar phenomena have been understood as well.

With the rapid development of astronomical diagnostics in the last 50 years and the increasing use of space probes, it became possible to observe celestial objects and cosmical nonstationary phenomena up to remote distances of the Universe. However, numerous phenomena, having enormous velocities of recession ranging from a few percent up to nearly the velocity of light and associated with shock waves, are still a mystery to modern science – such as the nature of quasars, the fluid dynamics of astrophysical jets and supermassive black holes, and the origin of cosmic rays and gamma-ray bursts. Science has now reached a higher level of understanding cosmic phenomena but at the same time also produced more complex questions and new riddles.

Compared to terrestrial natural violent events – perceived by a remote observer after delay times ranging from milli-

---

[144] This problem has been discussed and studied by numerous biomechanics researchers; *see*, for example, the review paper by Joseph HAMILL, a biomechanics professor at the University of Massachusetts, Amherst, MA, entitled *Evaluation of Shock Attenuation*, which he presented at the 4th Symposium of the Technical Group on Footwear Biomechanics [Canmore, Canada, Aug. 1999].

[145] M. SCHMIDT: *3C 273: a starlike object with large red-shift.* Nature **197**, 1040 (1963). He reached the conclusion that the quasar 3C 273 was not a star, but the enormously bright nucleus of a distant galaxy. • Quasars are extremely distant celestial objects whose power output is several 1,000 times that of our entire galaxy. His discovery that their spectra have a large red shift revolutionized quasar research. Subsequently, the radio source 3C 48 was found to have a redshift corresponding to a velocity of recession of 37% the speed of light. Strong evidence now exists that a quasar is produced by gas falling into a supermassive black hole in the center of a galaxy. • Estimations of these enormous velocities are based on the assumption that the red shifts are *cosmological*, which implies that the objects have to be billions of light-years away and therefore extremely luminous to look as bright as they do in our night sky.

seconds/seconds (thunder), to seconds/minutes (seismic shocks, blasts from explosive volcanic eruptions) up to minutes/hours (tsunamis) – the electromagnetic waves, originating from remote violent cosmic events and received by optical, radio or X-ray telescopes, have traveled millions to billions of light-years until arriving at a terrestrial observer. This implies that astrophysical diagnostics can only provide a picture of the cosmical past: current events at remote distances in the Universe are not accessible by any diagnostics, only by computational simulation. Furthermore, it is well-known from the study of terrestrial chemical and nuclear detonations that their durations as well as their effects at remote distances increase with increasing yield. Cosmic detonations have enormous yields and consequently their effects can last hundreds to thousands of years. For example, the famous Crab Nebula is a remnant of the supernova SN 1054 which astronomers had witnessed in A.D. 1054. The "detonation products" of this event, a cloud of hot ionized gases creating the 11-light-years-diameter Crab Nebula, are still expanding into the cold gases (now at a rate of 1,500 km/s), thereby producing shock waves.

The most common high-rate terrestrial phenomenon is thunder and attempts to understand this phenomenon promoted the evolution of supersonics.[146] However, before we begin to discuss this fascinating pioneering phase which took place in the 19th century, it first seems worthwhile to look back to the very beginning of supersonics. Its roots reach back as far as to the 1630s, when percussion research was still in its infancy and ballistics was limited to empirical testing only. Some unique highlights on the road towards supersonics are illustrated as cartoons in Fig. 2.10.

In the 1630s, the French clergyman, natural philosopher and mathematician Marin MERSENNE first determined the velocity of sound in air by firing a cannon and observing the delay between the arrivals of the muzzle flash and the muzzle blast a large distance away {MERSENNE ⇨1636; ⇨Fig. 2.10–A}. He correctly assumed that the velocity of light is extremely large compared to the velocity of sound, which we know to be true today but was then a much more controversial assumption then, and one that resulted from much debate. Using his so-called "flash-to-bang rule," he also proposed to evaluate the distance of a lightning strike.[147] Later, this method was also used to determine the velocity of sound in water {COLLADON ⇨1826}.

In the early 1640s, MERSENNE resumed his ballistic studies and considered how he might determine the velocity of a musket ball. When positioning himself close to the target, he noticed that the impact of the ball was heard at almost the same time as the muzzle flash {MERSENNE ⇨1644; ⇨Fig. 2.10–B}. From this he concluded that the ball must propagate with a velocity close to that of sound. Since the muzzle blast of a musket initially propagates at supersonic velocities, and even the speed of the ball fired from a musket, as first measured one hundred years later by Benjamin ROBINS using the ballistic pendulum {ROBINS ⇨1740; ⇨Fig. 2.10–C}, can be supersonic, it is quite possible that he had already experimented with supersonics – but without knowing it.

Based on this very simple though ingenious method of comparison, we can place MERSENNE at the front line of early sonic – perhaps even supersonic – pioneers. His idea of comparing a rapid phenomenon of unknown velocity with another one of known velocity reduces the problem, such that the ear only has to evaluate whether there is a time delay between two events.[148] In the Great Lisbon Earthquake {⇨1755}, it was observed that the velocity with which the quake and its aftershocks were propagated "was the same, being at least equal to that of sound; for all followed immediately after the noise that preceded them, or rather the noise and the earthquake came together" {MICHELL ⇨1760}, thus providing a first quantitative estimate of the velocity of propagation of seismic shocks.

In the 19th century, MERSENNE's method of comparison played an important role in the genesis of shock wave physics, serving as an experimental proof of the first mathematical theory of shock waves {EARNSHAW ⇨1860}: the British naval officers and artic explorers Sir William E. PARRY, Henry FOSTER and Sir John ROSS {⇨1824/1825}, who were searching for the Northwest Passage, observed that at a fairly large distance from a gun the officer's word of command "Fire" was distinctly heard *after* the report of the gun {⇨Fig. 2.10–D}. This proved in a unique way that the muzzle flash, a weak shock wave, propagates faster than sound.

---

[146] In the 1950s, the term *supersonics* was also applied in physical acoustics to the study of high-frequency sounds, above-audio frequencies – a branch of acoustics now termed *ultrasonics*.

[147] MERSENNE's method for determining the sound velocity $c$ is the oldest one known and is now termed the *direct method* ($c = \Delta s/\Delta t$). To determine the velocity of sound in a solid in this way on a laboratory scale requires advanced instrumentation, which was not available to 19th-century physicists. The easier *indirect method* ($c = \lambda \nu$), which only requires a Kundt tube (A. KUNDT 1866) and a tuning fork, was instead used to obtain the first data on the velocities of sound in solids. However, since this method only works with standing acoustic waves, it is not applicable to shock waves.

[148] The temporal resolution of the ear is about 2 ms, while the just-noticeable interaural delay ranges between 30 and 200 µs. See E. ZWICKER and H. FASTL: *Psychoacoustics: facts and models*. Springer, Berlin *etc.* (1990), p. 262.

## 2.3.1 OBSERVATIONS IN NATURE: STIMULATING RIDDLES

> *I strongly suspect it will soon be established as a law of nature, that the sound of a thunder-clap is propagated with far greater rapidity than ordinary sounds.*[149]
>
> The Rev. Samuel EARNSHAW
> Church and Parish of Sheffield 1860

> *ABBE's theory reminds one of the explanations of detonation given by JOURNÉE, SEBERT, DE LABOURET, in a certain way, regarding the concentration of noise, also of BOSSCHA's hypothesis. However, nowhere does ABBÉ speak of a head wave which the meteorite, moving at large speed, is carrying with it, and which, according to our explanation, is the primary cause of the main detonation heard during a meteorite fall. Since this view has been expressed, further observations have been made which further confirm the correctness of this widely supported theory.*[150]
>
> Ernst MACH
> Karl-Ferdinand-Universität
> Prague 1893

Shock waves are a common natural phenomenon on Earth. They are produced in the form of blast waves in the near-field of a thunderclap and during the fall of meteorites. Under certain conditions they can also arise during phreatic (or explosive) eruptions of volcanoes, for example when water and heated volcanic rocks interact to produce a violent expulsion of steam and pulverized rocks. Earthquakes and volcanic eruptions can generate seismic shocks and tsunamis, capable of producing disastrous effects also in remote areas. Asteroids, rocky "mini planets" moving in orbit around the Sun and ranging in size from about 1 to 1,000 km, can produce the most massive terrestrial shock effects when they impact the Earth, resulting in global catastrophes with biological mass extinctions. In the following, these natural supersonic phenomena will be illuminated in more detail.

**Thunder.** In prehistoric times, thunder was regarded as a deity and in antiquity it was worshipped by various nations {⇨Fig. 4.1−J}. Thunder is probably the most common of all loud natural sounds. Calculations performed in the 1920s estimated that there were several thousand lightning flashes per hour around the world. However, in spite of improved diagnostics, exact data on this are still difficult to obtain, because many lightning discharges occur between clouds, which may not be detectable, even with modern satellite surveillance. Since thunder also plays an important role in the evolution of shock physics, it appears useful to address this striking phenomenon more specifically.

Early naturalists were fascinated by thunder and proposed somewhat curious hypotheses about its origin {ARISTOTLE ⇨333 B.C.; SENECA ⇨A.D. 63; DESCARTES ⇨1637; MERSENNE ⇨1644; MARIOTTE ⇨1673; LÉMERY ⇨1700; WALL ⇨1708; FRANKLIN ⇨1749; ARAGO ⇨1838}. Upon applying the flash-to-bang method {MERSENNE ⇨1636}, some French scholars {MONTIGNY ET AL. ⇨1860; LAURENT 1860; HIRN ⇨1860} estimated unrealistically high propagation velocities of thunder, ranging from the velocity of sound in air to extreme values of above 6,000 m/s. The Rev. Samuel EARNSHAW {⇨1851}, an English mathematician, performed key work due to his experiences with lightning and thunder {⇨Fig. 2.10−E}. His conclusions − though partly incorrect from a modern perspective − stimulated him to such a degree that he worked out the first mathematical solution to POISSON's crude concept {POISSON ⇨1808} of treating sounds of finite amplitude {EARNSHAW ⇨1860}.

Today thunder is not only of interest to atmospheric acousticians due to its relevance to atmospheric properties, but it has also prompted modern shock and plasma physicists to develop numerical models that describe the complex generation and propagation of the shock wave from a rapidly heated and expanding discharge channel. Since the 1970s, a considerable number of experimental and theoretical research papers dealing with the generation of thunder have been published which, partly based on reproducible laboratory conditions, have been able to explain some of the many puzzling phenomena. More recently, various review articles have been published on this complex subject.[151, 152, 153]

In spite of the many theories on the cause of thunder proposed in recent years and the progress made in understanding this complex phenomenon, it is still a scientific riddle and remains a challenge to theoreticians as well as experimentalists. For example, Peter GRANEAU, a U.S. physicist at the Northeastern University in Boston, speculated that thunder may not be caused by the thermal expansion of the lightning channel, but may instead be driven by the sudden liberation of chemical bond energy from the nitrogen and oxygen molecules of the air {GRANEAU ⇨1989}.

---

[149] See EARNSHAW {⇨1860}.
[150] See E. MACH & DOSS {⇨1893}.
[151] A.A. FEW JR.: *Acoustic radiation from lightning.* In: (E.P. KRIDER, ed.) *The Earth's electrical environment.* National Academy, Washington, DC (1986), pp. 46-60.
[152] M.A. UMAN: *All about lightning.* Dover Publ., New York (1986); *Lightning.* In: (R.G. LERNER and G.L. TRIGG, eds.) *Encyclopedia of physics.* VCH, New York etc. (1991), pp. 637-639.
[153] H.E. BASS: *Atmospheric acoustics.* In: (G.L. TRIGG, ed.) *Encyclopedia of applied physics.* VCH, New York etc.; vol. 2 (1991), pp. 145-179.

Modern theories of chemical evolution suggest that the early Earth was covered largely with a warm, slightly alkaline ocean. In 1953, Stanley MILLER, a graduate student working with Nobel Prize winner Dr. Harold UREY, published experiments on the synthesis of amino acids in a simulated primordial Earth environment using electric sparks. The *Miller-Urey experiment*, an experiment that produced an "organic soup," became a landmark in chemical evolution {S.L. MILLER ⇨1953; ⇨Fig. 4.14–A}. Similar experiments were later carried out using high-temperature shock waves generated in a shock tube {⇨Fig. 4.14–B}. George T. JAVOR,[154] an associate professor of microbiology at the Loma Linda University in California, who discussed modern views on the origin of life in 1987, wrote: "Though rich in carbon monoxide, carbon dioxide, ammonia, methane, hydrogen, and nitrogen, the atmosphere [of the early Earth] definitely did not contain atomic or molecular oxygen. Ultraviolet light from the Sun, geothermal energy from volcanoes, shock waves from thunder, and cosmic radiation acted upon gases of the primitive atmosphere causing the formation of biomonomers such as amino acids, sugars, purines, pyrimidines, and fatty acids. These substances polymerized to form the prototypes of more recent proteins, nucleic acids and cell membranes. In time they coalesced to form the first proto-cell, a collection of polymers enclosed in a membrane. Eventually these protocells became increasingly complex, until the first true living cell was born."

Thunder has a very complicated acoustic structure, because it is the result of lightning, itself a complex phenomenon. It is difficult to realistically simulate thunder in a laboratory. For example, Guy G. GOYER {⇨1964/1965}, a research physicist at the National Center for Atmospheric Research (NCAR) in Boulder, Colorado, used a long detonation cord ("Primacord") with a specific charge weight of 12 grams of PETN per meter, a value which he obtained based on theoretical calculations of the shock wave generated by lightning. However, the detonation of the Primacord does not occur at all points simultaneously, but rather represents a point source propagating with the velocity of the detonating PETN cord. A line source closely approaching the characteristics of lightning can only be realized by using long spark channels, which require the use of very high voltages. Laboratory experiments on lightning and measurements of the shock pressures generated were conducted by Martin A. UMAN {⇨1970} at Westinghouse Research Laboratories, while Henry E. BASS {⇨1980}, a physics professor at the University of Mississippi, performed various numerical simulations. These studies have indeed shown that the disturbance starts out as a shock wave in the close vicinity of the stroke. However, just a few yards from the channel, the shock wave quickly turns into a weak shock wave – *i.e.*, it propagates at almost the velocity of sound, thus confirming the long-known rule of thumb {MERSENNE ⇨1636} that the distance of the nearest flash of lightning from the observer can be estimated by measuring the time lag between the flash and thunderclap.

The source of most lightning is the electric charge separation that occurs in a thunderstorm, although other forms of lightning can occur, *e.g.*, due to charge separation in snow- and sandstorms, in the clouds generated by some volcanoes, and near thermonuclear explosions. Most lightning flashes produced by thunderstorms are intracloud discharges, but cloud-to-ground lightning – also known as "streaked lightning" – has been studied more than any other form of lightning because of its practical interest and (relative) ease of observation. The discharge channel, which rapidly ionizes, heats and expands the gas along its path, can reach a length of a few kilometers and a diameter of up to a few centimeters, thereby typically carrying a current of some 10 kA and dissipating an energy of about 3 kJ per meter.

The development of the shock from this rapidly expanding channel can be calculated using *weak shock theory* {WHITHAM ⇨1956 & 1974}. Since the strike has a finite length, the shock wave is primarily cylindrical close to the spark, but becomes increasingly spherical at long distances from the spark. However, in real lightning the discharge channel has a tortuous geometry, giving rise to a "string of pearls," which means that thunder must be modeled as a long line of spherically spreading sources – *i.e.*, the solution represents a complex combination of the classical examples of a "point source" explosion {G.I. TAYLOR ⇨1941} and a "line source" explosion {LIN ⇨1954}.

The sharp report that can be heard from a thunderclap – which strongly resembles the muzzle blast of a rifle, a shock wave – is only heard when lightning strikes nearby. At larger distances, however, the sound of thunder is modified during its propagation by nonlinear effects in the atmosphere and ground effects that shift the spectrum to lower frequencies, giving rise to the low-frequency rumbling typically heard. In addition, the cloud-to-ground discharges, which are best investigated using high-speed diagnostics, show a rapid sequence involving a moving stepped leader which initially descends from the cloud to the ground and then returns to the cloud; 10 to 20 strikes following this first one then follow. The first return strike of the lightning flash produces a much louder sound than the leader, because it transmits a more powerful electric current. In his work on thunder, the

---

[154] G.T. JAVOR: *Origin of life: a look at late 20th-century thinking.* Origins **14**, No. 1, 7-20 (1987).

Rev. Samuel EARNSHAW {⇨1851 & 1858} observed that a lightning stroke – obviously belonging to this particular type of cloud-to-ground discharge – was heard in the village *almost simultaneously* with the flash although the lightning struck about a mile outside of the village. From this observation, he used the flash-to-bang method to estimate the propagation velocity of thunder, which he found to be supersonic. However, he may have erroneously taken the sequence of rapid flashes and strikes to be a single event.

The loudest natural sounds in the ocean are lightning strikes, with a source level of about 260 dB (re 1 μPa). Strikes occur at a rate of about two per km$^2$ per year in most of the world's coastal regions, where coincidentally marine mammals abound.[155]

Lightning is not unique to Earth. In 1979, cameras on NASA's Voyager 1 planetary explorer spacecraft found lightning on Jupiter.[156] Both Voyager 1 and 2 detected electrical signals from Jupiter characteristic of lightning. This discovery was the first hard evidence that such violent electrical discharges take place on other planets. In 1997, the Galileo spacecraft also photographed what appear to be visible lightning flashes in Jupiter's turbulent atmosphere {RINNERT ⇨1985}. Detection of electrostatic discharges on Saturn and Uranus by Voyager 2, along with radio signals associated with lightning picked up by the Pioneer Venus orbiter and the Russian Venera probe, may indicate that lightning – most probably also accompanied by violent thunderclaps; i.e., by weak atmospheric shock waves – are commonplace on some planets in our Solar System.

**Hydrometeors.** The generic term *hydrometeor* is used to designate all products of the condensation or sublimation of atmospheric water – *e.g.*, rain drops, ice crystals, fog, hail, and clouds. The physical constitutions of hydrometeors can actually be modified by acoustic waves of finite amplitude. In the Middle Ages, church bells were often rung in the hope of reducing the damage due to heavy storms. In the 1890s, experiments were carried out in Texas aimed at producing artificial rain by firing explosive charges on the ground or at an altitude of several thousand feet {DYRENFORTH ⇨1891}. During the First and Second World Wars, several reports indicated the occurrence of heavy rain immediately after ferocious artillery battles. Explosive charges have been used since the 1950s in various European countries and in South Africa to generate shock waves in clouds in an effort to reduce crop damage from hail. Roland IVES {⇨1941}, a U.S. meteorologist, noticed that the firing of a revolver triggers super-cooled fog droplets to become frozen {MAURIN & MÉDARD ⇨1947}. Guy GOYER {⇨1964/1965}, a U.S. research physicist, detonated a long detonation cord (a "Primacord") 300 ft above the cone of the Old Faithful Geyser at Yellowstone Park and observed that hail fell immediately following the detonation. He also reported on previous speculations that lightning, which acts as a shockwave generator, can affect hailstones, supercooled water droplets, and large cloud droplets.

As the velocities at which aircraft travel have increased, the problem of rain erosion has became more and more obvious. In 1945, U.S. Air Force personnel reported on observations of rain erosion phenomena in plastic materials {U.S. Air Force ⇨1945}. With the advent of supersonic flight {YEAGER ⇨1947} and of hypersonic flight {X-15 ⇨1961}, this serious problem became an urgent research topic, focusing particularly on the interaction of the head wave – a weak shock wave – with rain droplets, and the disintegration process itself {HANSON ET AL. ⇨1963; The Royal Society ⇨1965; BOWDEN & MCONIE ⇨1965; BRUNTON & BOWDEN ⇨1965}.

**Bores.** Since bores are primarily caused by tides – *i.e.*, by the gravitational forces exerted by the Moon and the Sun – they are also appropriately termed *tidal bores*. Bores are surface waves that propagate in shallow water with a steep, almost discontinuous front similar to a shock wave in air. In particular, when the Earth is aligned with the Sun and the Moon (a phenomenon known as "syzygy"), tides of maximum range, so-called "spring tides," result {WANG CH'UNG ⇨A.D. 85}. In his account on the invasion of Britain, the famous Roman dictator Julius CAESAR[157] alluded to the nature of spring tides [Lat. *maritime aestus maximi*] as being perfectly well understood in connection with the phases of the Moon. Bores of significant height – described as a "wall of water" by observers – are generated either by the rushing of the tide up a narrowing estuary or by the collision and oblique interaction of two tides. The latter form, representing the oldest known natural phenomenon of nonlinear wave superposition {⇨Fig. 4.1–M}, can be followed by the naked eye and must have been noted by coastal dwellers long ago.

Bores can be observed in the rivers of many countries, such as the *barre* of the Seine, the *mascaret* (or *raz de marée*) of the Gironde, and the *pororoca* of the Amazon. a large amount of literature exists on some of the major bores,

---

[155] *Where lightning strikes.* Science@NASA (Dec. 5, 2001); http://science.nasa.gov/headlines/y2001/ast05dec%5F1.htm.
[156] D.A. GURNETT, R.R. SHAW, R.R. ANDERSON, and W.S. KURTH: *Whistlers observed by Voyager 1 – detection of lightning on Jupiter.* Geophys. Res. Lett. **6**, 511-514 (June 1979).

[157] G.I. CAESAR: *De bello gallico.* Liber **IV**, 29 (55 B.C.).

while information on others is lacking, and evidence rests purely on observation. The Tidal Bore Research Society compiled a catalog of bores around the world, which (as of May 2006) encompassed 55 bores.[158]

Large bores have been recorded in the Ganges and Amazon with heights of up to 5–6 m. However, the highest waves have been observed in China in the estuary of the Qiantang River, with heights of up to 8 m {⇨Fig. 4.1–L}. Tide-watching at the Qiantang River near Hangzhou, a town about 170 km southwest of Shanghai, has been performed for over 2,000 years: each year, on the 18th day of the 8th month of the Chinese Lunar Calendar (*i.e.,* near the time of the Autumnal Equinox), thousands of tourists make their way to the China International Qiantang River Tidal Bore Festival to watch this spectacular event. In England, the bore on the River Severn is certainly the best known example of this phenomenon, and the one that has been studied the most and for the longest as well {⇨Fig. 4.4–E}.[159] The *Severn bore* is produced by a tide that rises about 18 feet (5.5 m) in an hour and a half. This body of water becomes compressed in the narrowing funnel-shaped estuary, and heaped up into an advancing wave extending from bank to bank. Measurements carried out in 1849 by Captain F.W. BEECHEY had revealed that the Severn bore can reach heights up to 1.5 m and propagate with a velocity of about 12.5 mph (20 km/h); *i.e.,* it moves about four times faster than walking speed.[160]

In Germany, tidal bores are appropriately called "jump waves" [Germ. *Sprungwellen*]. Since the height of the water behind the wave front is greater than that ahead of the front and the propagation velocity $V$ increases with the water depth $d$ according to the relation $V \propto d^{1/2}$ {DE LAGRANGE ⇨1781; BÉLANGER ⇨1828}, the front must propagate "supersonically" compared to the water at rest ahead. An approaching bore does not produce a report like the sonic boom of a supersonically moving body; it instead announces itself to the observer through an increasingly loud roaring sound. However, it is this discontinuous nature of the wave itself and its close mathematical analogy to a gaseous shock wave which attracted the curiosity of many early naturalists and stimulated the interest of modern physicists and mathematicians in nonlinear wave phenomena {JOUGUET ⇨1920; PREISWERK ⇨1938; H.A. EINSTEIN ET AL. ⇨1946/1947; GILMORE ET AL. ⇨1950}.

A bore can be intensified considerably by reflection; for example, by forcing it to striking a rigid obstacle at an angle which produces an enormous impulsive force on it. From a mathematical point of view, bores are like nonlinear shock wave phenomena. Similar to the oblique reflection of a shock wave, which more than doubles the pressure behind the reflected wave, the height of a bore can be more than doubled when it strikes a rigid wall obliquely.[161] Depending on the angle of incidence, this can result in irregular reflection, creating a Mach stem {⇨Fig. 4.13–A}.

**Tsunamis.** Tsunamis are catastrophic seismic sea waves and are mainly a Pacific phenomenon {Komaishi Seaquake ⇨1896; Myojin-sho Reef Eruption ⇨1952}: they are rare in the Atlantic Oceans. In the Indian Ocean, the denser oceanic Indian plate subducts beneath the overriding continental Burma plate, which resulted very recently in the most destructive tsunami in recorded history. It devastated the coastal areas of several countries and killed almost 230,000 people from many nations {Sumatra-Andaman Islands Earthquake ⇨2004; ⇨Fig. 4.1–S}.

While tidal bores are recurrent events – *i.e.,* they are predictable and confined to estuaries and so are of only limited danger – tsunamis are (like earth- and seaquakes) unpredictable events, in both time and magnitude. Because of their suddenness and enormous destructive power, even at remote coasts, tsunamis were the subject of much discussion among early natural philosophers {MARCELLINUS ⇨A.D. 365; KANT ⇨1756; MICHELL ⇨1760} as well as among early seismologists {MALLET ⇨1846; RUDOLPH ⇨1887; ROTTOK ⇨1890; MILNE ⇨1898; Count BERNARD ⇨1907}. Modern research has revealed that the origins of tsunamis are quite complex.[162] Tsunamis can be generated

▸ by impulsive large *vertical* tectonic displacements of the sea floor, which may reach five meters or more, resulting in (shallow) earthquakes in *subduction zones* with epicenters close to the coast, particularly when the earthquakes have large amplitudes and shallow foci ranging from 0 to 40 km – such as those resulting from the slippage of denser oceanic plates beneath continental plates {Aegean Earthquake ⇨365; Lisbon

---

[158] *Documenting the worldwide tidal bore phenomena.* Tidal Bore Research Society (TBRS); http://tidal-bore.tripod.com/catalogue.html.

[159] F. ROWBOTHAM: *The Severn bore.* Dawlish MacDonald, London (1964).

[160] R.A.R. TRICKER: *Bores, breakers, waves and wakes.* Am. Elsevier, New York (1965), p. 38.

[161] Reflection intensification effects for the Qiantang bore {⇨Fig. 4.1–L} are illustrated in the book by John E. SIMPSON: *Gravity currents in the environment and the laboratory* [Cambridge University Press, Cambridge (1997), pp. 98-99].

[162] K. IIDA: *Magnitude, energy, and generation mechanisms of tsunamis and a catalogue of earthquakes associated with tsunamis.* In: (D.C. COX, ed.) *Proc. Tsunami Meeting* [University of Hawaii, Honolulu, HI, 1961]. Union Géodésique et Géophysique International, Monographie No. 24, L'Institut Géographique National, Paris (1963), pp. 7-18.

Earthquake ⇨1755; Chilean Earthquake ⇨1960; Sumatra-Andaman Islands Earthquake ⇨2004}. On the contrary, *horizontal* tectonic displacements may reach twenty meters or more, but cannot produce tsunamis;

- by rapid injection of large masses of material into the ocean by violent volcanic eruptions {Santorin Volcano ⇨1645 B.C.; Krakatau ⇨1883; Myojin-sho Reef Eruption ⇨1952};
- by earthquake-induced submarine landslides {Coast of Yorkshire ⇨1856};
- by earthquake-induced coastal landslides {Lituya Bay ⇨1958; ⇨Fig. 4.1–Q};
- through the decomposition of a gas hydrate (a solid compound containing natural gas and water) at the shelf;[163]
- by impact of large objects from outer space such as meteorites, asteroids, *etc.*, into the sea {Chicxulub Crater ⇨c.65 Ma ago}; and
- by underwater explosions of nuclear devices {Test BAKER ⇨1946}.

*Subduction* is the commonest mechanism of tsunami generation. It represents a "line" source that produces rapid displacement of a long but comparatively narrow vertical water column which extends from the surface to the seafloor, can extend over hundreds of kilometers in length, and can therefore result in very destructive tsunamis. Shores located in the vicinity of and facing such line sources are particularly endangered by tsunamis, as has been demonstrated recently by the Sumatra-Andaman Islands Earthquake {⇨2004}, which devastated the west coast of Sumatra in particular.

The other mechanisms listed above are limited to comparably small dimensions, and thus approach "point" sources which result in tsunamis of a smaller magnitude. Tsunamis are less frequent along Atlantic coastlines than in Asia-Pacific ones, but are still a danger; the only subduction zones around the Atlantic are the Puerto Rico Trench and the Antilles subduction zone around the eastern Caribbean, and the South Sandwich Trench south of South America.

The *characteristics of tsunamis* differ distinctively from wind-generated surface waves. In contrast to surface waves, the entire water column, from the surface to the seafloor, begins to move and transfer energy. In deep water, the wavelengths (the distance from crest to crest) are enormous, about 100–200 km, but the wave heights (or amplitudes) are small, only 0.3–0.6 m; thus, in the open ocean on board a ship, a tsunami is barely noticeable. The resulting wave steepness, or ratio of height to length, is on the order of only $10^{-6}$.

Since the waves have long periods of an hour or more, the corresponding rise times are on the order of many minutes – *i.e.*, extremely long compared to rise times in shock waves, which are on the order of nano- to microseconds. Since tsunami wavelengths are much longer than even the greatest ocean depths, they behave as *shallow water waves*, regardless of the depth $H$, and their velocity $v$ can be described by the simple relation $v = (gH)^{\frac{1}{2}}$, where $g$ is the acceleration due to gravity {DE LAGRANGE ⇨1781}.[164] This formula was used in the 19th century by geologists to determine the average depths of the oceans, long before deep-sea soundings were taken {Arica Earthquake ⇨1868; Iquique Earthquake ⇨1877}. In addition, the relationship also has enormous practical value, because it enables seismologists to issue warnings to remote endangered coasts immediately after a seaquake and several hours before it arrives there. For example, an earthquake that occurred in Alaska generated a tsunami which, after traveling more than 3,800 miles (6,114 km), destroyed various costal towns in Hawaii {Aleutian Islands 1946; ⇨Fig. 4.1–R}.

Tsunamis that are generated in deep water radiate out at high speed from the triggering event. For example, at a water depth of 5,000 m, they propagate with a speed of about 800 km/h (220 m/s) – the speed of a jet airplane! Since the rate at which a wave loses its energy is inversely related to its wavelength, a tsunami will lose little energy as it propagates and can travel great distances. However, when it reaches the shallow water of a continental shelf, the wavelength of the tsunami shortens and its velocity drops tremendously due to friction with the sea bottom. For example, at a depth of 10 m, its velocity reduces to only about 36 km/h (10 m/s), which causes its front to compress into a towering wall of water. Shortly after crashing ashore and impacting buildings, an event which is often accompanied by a loud bang, the enormous masses of water flow back out to sea, thereby creating dangerous reverse currents. The impact forces from some tsunamis are enormous. Large rocks weighing several tons, along with boats and other debris, can be moved inland hundreds of feet by tsunami wave activity {MARCELLINUS ⇨365}. Tsunamis can also travel quickly up rivers and streams.

Coastal areas located close to the epicenter of a seaquake cannot be warned effectively because of the short arrival time of the resulting tsunami {Okushiri Earthquake ⇨1993}.

---

[163] E.D. SLOAN ET AL.: *Future of gas hydrate research* [Abstract]. EOS, Trans. Am. Geophys. Union **80**, No. 22, 247 (1999).

[164] According to this simple relationship, tsunamis in very deep water ($H > 11,400$ m) would propagate with supersonic speeds in relation to the velocity of sound in air at normal conditions. The Mariana trench, however, which is the deepest part of all of the Earth's oceans, only has a depth of about 10,910 m.

However, the much longer arrival times of tsunamis at remote coasts, up to several hours, allow the use of efficient early warning systems {Unimak Island ⇨1946}. Local and historical knowledge of earth- and seaquakes around the world has made it possible to create an International Tsunami Information Center (ITIC). Initiated in 1965 by the United States, this Center has proven to be of great benefit to all countries with Pacific coasts {ITIC ⇨1965}. A similar system does not yet exist for nations bordering the Atlantic and Indian Oceans, although the famous Lisbon Earthquake {⇨1755}, and very recently the Sumatra-Andaman Islands Earthquake {⇨2004}, demonstrated the enormous potential for destruction posed by tsunamis produced in coastal regions bordering these oceans in dramatic fashion.

A tsunami consists of a series of waves. Often the first wave may not be the largest. The danger from a tsunami can last for several hours after the arrival of the first wave. Usually the wave of greatest height is not the first one, but it most commonly occurs among the first ten waves or so. For example, during the Sumatra-Andaman Islands Earthquake {⇨2004} the third wave of the tsunami was the most powerful and reached highest, occurring about an hour and a half after the first wave.

Tsunamis are not typically characterized by a steep, single front like a hydraulic jump: the arrival of a tsunami is sometimes indicated by a withdrawal of water which may be preceded by short-period, low-amplitude oscillations, known as "forerunners."[165] In many cases it has been observed that a tsunami causes the water near the shore to first recede, similar to an extremely low tide, thus exposing the ocean floor {MARCELLINUS ⇨365}. Shortly after the Lisbon Earthquake {⇨1755}, this strange phenomenon attracted curious people to the bay floor, where many of them were drowned by the subsequent wave crest that arrived only minutes later. A pronounced withdrawal of water from the shore has also been observed during the recent South East Asia tsunami disaster {Sumatra-Andaman Islands Earthquake ⇨2004}.

*Tsunami animation* is increasingly used to illustrate the generation and propagation of tsunamis produced by landslides, projectile impacts, asteroid impacts, and large chemical and nuclear explosions. In recent years, a collection of Tsunami Computer Movies was generated at Los Alamos National Laboratory using various numerical codes.[166] Some animations of tsunami propagation prepared by USGS and Japanese researchers can be watched on the Internet; for example, a hypothetical tsunami generated in the U.S. Pacific Northwest,[167] and the tsunamis generated by the 1960 Chilean Earthquake[168] and the 2004 Sumatra-Andaman Islands Earthquake.[169]

The effects of the force from a tsunami wave on structures have been modeled both numerically and in the laboratory. A review of previous work on this important subject was given by Nicolas HARITOS and colleagues at the 2005 Conference of the Australian Earthquake Engineering Society (AEES) held at Albury, New South Wales.

**Surges.** Sudden or violent changes of pressure in a gas, a liquid or a plasma are termed *surges*. They have been observed on Earth and also on the Sun. In the atmosphere, they are caused by sudden increases in the wind speed of a large wind stream, especially in the tropics (*e.g.,* the so-called "surge of the trades" in the trade-wind belts or the "surge of the monsoon" in the monsoon currents).[170] A *storm surge* is simply water that is pushed towards the shore by the force of the high-speed winds swirling around the storm, resulting in exceptionally high water levels. It is generated by an extreme meteorological event, such as a winter storm or cyclone. When combined with *storm tides*; *i.e.,* high-water levels above normal (astronomical) tide levels, such events create *hurricane storm tides* which can increase the mean water level by 15 feet (4.6 m) or more, causing severe flooding in coastal areas {FUJITA ⇨1981}. Storm surge flooding, a threat to many of the world's coastlines, and one that presents a great potential for loss of life, can occur due to cyclones in the Bay of Bengal {G.H. DARWIN ⇨1898}, typhoons on the east coast of Japan, and hurricanes in the Gulf of Mexico and on the Atlantic coast of the United States.

In astronomy, surges are sudden, violent events that send jets of cool plasma into the Sun's corona at speeds of up to 200 km/s along a straight or slightly curved path, thereby reaching heights up to 100,000 km.[171]

---

[165] T.S. MURTY: *Seismic sea waves. Tsunamis.* Bulletin 198, Dept. of Fisheries and the Environment. Fisheries & Marine Service, Ottawa, Canada (1977), p. x and 2.

[166] *Tsunami computer movies.* LANL, Los Alamos, NM (Oct. 2002); http://t14web.lanl.gov/Staff/clm/tsunami.mve/tsunami.htm.

[167] *An animation of a hypothetical tsunami generated by oblique slip.* USGS; http://walrus.wr.usgs.gov/tsunami/OTanimation.html.

[168] N. SHUTO: *Propagation of the earthquake-generated 1960 Chilean Tsunami across the Pacific.* Disaster Control Research Center, Tohoku University, Sendai, Japan; http://www.geophys.washington.edu/tsunami/general/historic/models_60.html.

[169] K. SATAKE: *2004 Indian Ocean Earthquake.* National Institute of Advanced Industrial Science & Technology (NIAIST), Tsukuba, Japan; http://www.stormcenter.com/media/envirocast/tsunami/animation.gif.

[170] *Encyclopædia Britannica, Micropædia.* Benton & Hemingway, Chicago etc. (1974), vol. IX, p. 690.

[171] K. SHIBATA: *Surges.* In: (P. MURDIN, ed.) *Encyclopedia of astronomy and astrophysics.* Institute of Physics Publ., London *etc.*, vol. 4 (2001), pp. 3258-3261.

**Earthquakes and Seaquakes.** *Earthquakes* are rather common natural phenomena in many countries and can occur at any time. Seaquakes occur in all oceans and also in some inland seas like the Mediterranean Sea. The term *seaquake* [*Seebeben*] was coined in the 1880s by Emil RUDOLPH {⇨1887}, a German geology professor at the University of Strassburg.

Surprisingly, the main causes of earth- and seaquakes were not discovered until the *plate tectonics revolution* of the 1960s. Based upon the theory of continental drift (Alfred WEGENER 1915), the theory of mantle thermal convection (Arthur HOLMES 1929), and the theory of "sea-floor spreading" (Harry HESS & Robert DEITZ 1961), it was recognized that the nature of the Earth's surface is largely determined by the motions of rigid plates that slowly drift over geological timescales, and that the relative tectonic motions between adjacent plates give rise to large earthquakes along the plate boundaries. Large, devastating earthquakes – so-called "megathrust earthquakes" – are mostly of tectonic origin; they occur when one plate slips beneath another, a process termed *subduction*, which can generate significant tsunamis {Sumatra-Andaman Islands Earthquake ⇨2004}.

For example, the San Andreas Fault, a major fracture of the Earth's crust close to the west coast of North America, and well known for producing occasional large earthquakes {San Francisco Earthquake ⇨1906}, is a result of shear between the Pacific and North American Plates.

Until the middle of the 18th century, naturalists drew their ideas on the mechanisms of earthquakes from contemporary records of the writings of some ancient philosophers {ARISTOTLE ⇨333 B.C.; SENECA ⇨A.D. 63; PLINY the Younger ⇨A.D. 79}. However, in England, in the "Year of Earthquakes" {⇨1750}, a remarkable series of earthquakes occurred which stimulated the imagination of British naturalists to explain their origin, and before the end of the year nearly 50 articles were communicated to the Royal Society in London. Only five years later, Lisbon {⇨1755} was destroyed by one of the largest earthquakes in recorded history. This disaster, which prompted investigations into its causes and proposals for possible countermeasures, happened towards the end of the Baroque period (ca. 1600–1780) – an era of strong contrasts, with much splendor and enjoyment on the one hand, and much escapism and *Memento mori* on the other. Many renowned naturalists, philosophers and even some poets commented on the disaster and started intellectual debates in attempts to find plausible reasons for it {DESMAREST, JACOBI, KANT, KRÜGER, LEBRUN, MAYER, ROUSSEAU, VOLTAIRE, all ⇨1756}. The Marquis de POMBAL, then Prime Minister of Portugal, survived the earthquake and ordered an inquiry into the Lisbon Earthquake {⇨1755} and its effects to be established in all parishes of the country {⇨Fig. 4.1−E}; this inquiry may represent the birth of seismology, the scientific study of earthquakes.

The British geologist and astronomer John MICHELL {⇨1760}, today considered to be one of the founders of seismology, and the Dutch theoretician Antonius G. DRYFHOUT {⇨1766} tried to explain the Lisbon Earthquake by invoking some sudden explosions in the internal parts of the Earth. However, instead of assuming underground explosions of oxyhydrogen and quoting the sensational oxyhydrogen explosion experiment previously demonstrated publicly by the French alchemist Nicolas LÉMERY {⇨1700}, MICHELL postulated that earthquakes arose from subterranean fires which produced violent steam explosions upon contact with large quantities of water. He also explained the origins of seaquakes likewise, assuming that the sudden collapse of the roof over an subterranean fire produced the quake. He regarded volcanoes as useful "safety valves" which tended to prevent earthquakes by providing passage to high-pressure vapors.

Shortly after the Great Lisbon Earthquake, the German philosopher Immanuel KANT {⇨1756} attempted to deduce its cause. Although he initially erroneously followed LÉMERY's volcanic hypothesis – then widely supported by other contemporary naturalists – he correctly recognized that the resulting great seismic sea waves (tsunamis) along the English, African, Caribbean and south European coasts were caused by a seaquake rather than by an earthquake. Very recently, KANT's concept was partially confirmed by Charles L. MADER, a retired U.S. shock physicist from Los Alamos National Laboratory {Lisbon Earthquake ⇨1755}. Using tsunami wave characteristics to numerically simulate the possible source of the historical Lisbon Earthquake, MADER found that its epicenter may have been close to the 1969 earthquake epicenter located south of the Gorringe Bank, a region southwest of Portugal at the eastern end of the Azores-Gibraltar plate boundary.

Beginning in the early 1800s, the theory of elastic wave propagation was developed by CAUCHY, POISSON, STOKES, Lord RAYLEIGH and others who first described the main wave types to be expected in solid materials. These include compressional waves and shear waves, which are termed, respectively, *body waves* (because they travel through solid volumes) and *surface waves* (because they travel along free surfaces). Since compressional waves travel faster than shear waves and therefore arrive early on in a seismic event, they are termed *primary waves* (or *P-waves*), whereas the shear waves that arrive after are termed *secondary waves* (or

*S-waves*). The shearing ground motion caused by the S-wave produces no change in volume.

This complicated mixture of wave types, which is particularly pronounced in extended volumes like the Earth's crust, becomes much more apparent after they have propagated for sufficiently large distances. Seismic studies performed at large distances require either strong seismic shock sources (such as high explosives) or very sensitive seismic detectors. However, neither of these were available before the 1860s. On the other hand, laboratory studies, involving much smaller dimensions, were ruled out, because wave transit times would be only very short (on the order of hundreds of nanoseconds to tens of microseconds), which were not measurable until some 80 years later.

The scientific investigation of earthquakes did not begin until the middle of the 19th century, with the detailed study of the propagation, reflection and refraction of seismic shocks made by the Irish engineer Robert MALLET {⇨1846} who also termed this new branch of geophysics *seismology*. Progress in this new discipline was significantly determined by *seismometry*, a new diagnostic technique concerned with the detection and measurement of seismic ground motions. It comprises the design of *seismographs*, their calibration and installation, and the quantitative interpretation of seismograms in terms of ground motion. The early history of seismometry up to 1900 has been reviewed in an excellent article written by James DEWEY[172] and Perry BYERLY, two seismologists at the USGS National Earthquake Information Center in Denver, Colorado.

The seismic shocks generated by earthquakes produce not only violent vibrations and shakings of the ground but also audible sounds: near the source the sound sometimes includes sharp snaps; further away the sounds have been described as low and booming, like a distant clap of thunder or the boom of a distant cannon or explosion.

In seismology, the size of an earthquake is given by its magnitude $M$, a logarithmic measure. There are a number of different magnitude scales, such as those based on earthquake duration ($M_d$), locality ($M_l$), surface waves ($M_s$), body waves ($M_b$), and moment ($M_w$).[173] The moment magnitude $M_w$, which is calculated using the seismic moment, is now the one most commonly used to measure large-earthquake magnitudes {HANK & KANAMORI ⇨1979}.

Historically, seismologists assumed that a fault wouldn't break faster than the shear (or S) wave that radiates from an earthquake's epicenter. Limits on the speed at which a rupture can propagate down a fault stem from past observations as well as assumptions about classical fracture mechanics and the material strength of rocks. Recent analyses of an historic earthquake {San Francisco Earthquake ⇨1906} and observations of earthquakes in California, Alaska, Turkey, and China {Kunlun Shan Earthquake ⇨2001} have given an inkling that a fault could break faster than the velocity of the shear wave, with rock motions outrunning the shear wave along a fault – a curious phenomenon called "supershear." Researchers at CalTech demonstrated in laboratory experiments that *supershear fault rupture* is a real possibility rather than a mere theoretical construct {ROSAKIS ET AL. ⇨1999}. Using high-speed photography they recorded the rupture and stress waves as they propagated through their model fault, a set consisting of two plates of polymer material forced under pressure and a tiny wire inserted into the interface. In order to trigger the "earthquake" the wire was suddenly turned into an expanding plasma (exploding wire) by discharging a capacitor. The supersonic rupture produced a Mach cone similar to a projectile flying at supersonic speed {E. MACH & SALCHER ⇨1886}. In aerodynamics the Mach cone is characterized by a shock wave, a sudden increase in pressure. Similarly, in large earthquakes supershear may produce a sudden "super shake," thus amplifying the destructive power of ground motion.

The discovery of supershear has complicated a relatively simple picture of how earthquakes unfold. Obviously, it requires more advanced theoretical models to better understand the conditions for supershear in terms of fault properties and earthquake magnitude, and to predict the length of travel before the transition to supershear.

**Explosive Volcanic Eruptions.** These are the most spectacular and most dangerous types of volcanic eruptions. *Explosive volcanic eruptions* are classified into four eruption types and, placed in order of increasing violence, they are termed:
▸ Hawaiian-type eruptions {Mt. Kilauea ⇨1959};
▸ Strombolian-type eruptions {Stromboli ⇨1930};
▸ Vulcanian-type eruptions {Vulcano ⇨1888}; and
▸ Peléan-type eruptions {Mt. Pelée ⇨1902}.[174]

The lavas of volcanoes that erupt explosively are distinguished by their large viscosity and small mass diffusivity, which retard the release of dissolved vapors during eruption until the magma arrives at or near the Earth's surface, where

---

[172] J. DEWEY and P. BYERLY: *The early history of seismometry (to 1900)*. Bull. Seismol. Soc. Am. **59**, No. 1, 183-227 (1969).
[173] *Recent earthquakes, glossary*. USGS, Earthquake Hazards Program; http://earthquake.usgs.gov/recenteqsww/glossary.htm#magnitude.

[174] H. SIGURDSSON (ed.): *Encyclopedia of volcanoes*. Academic Press, San Diego, CA (2000), pp. 249-269.

degassing can then be explosive. A volcano may also violently explode when

- the internal pressure of the gas and lava increases such that lava breaks through the flanks of the volcano. When the whole vent and the upper part of the magma chamber are rapidly emptied, the central, unsupported part may collapse, thus forming a "caldera;"
- ground water placed in contact with the magma is superheated, resulting in a violent steam-blast eruption; and/or
- the vent, which is initially blocked by a plug of solidified lava, is suddenly reopened by an increase in internal pressure, the dreadful "uncorking effect." This prompts volcanic ejecta to exit at supersonic speeds.

Depending on the viscosity of the magma, the pressure and the gas content, extreme violent explosive eruptions may also induce atmospheric blast waves – known as "volcanic blasts" {Santorini ⇨1645 B.C.; Mt. Vesuvius ⇨ A.D. 79; Krakatau ⇨1883; Mt. Bandai-san ⇨1888; Mt. Pelée ⇨1902; Mt. St. Helens ⇨1980; Mt. Pinatubo ⇨1991}.

Furthermore, mass outflow occurring at supersonic speeds can provoke an energetic fragmentation of liquid magma into fine ash particles and their ejection into the upper atmosphere. Atmospheric shock waves (blasts) can damage structures far from their source: for example, directly by breaking windows, or indirectly by inducing ground shocks, which can damage buildings.[175]

A *lateral blast* is produced by an explosive volcanic eruption in which the resultant cloud of hot ash and other material moves horizontally rather than upward. Lateral blasts can generate erosional phenomena in the surroundings of the crater – known as "furrows" {KIEFFER & STURTEVANT ⇨1988; STURTEVANT ET AL. ⇨1991} – which appear similar to high-speed ablation effects due to aerodynamic heating {JOULE & THOMSON ⇨1856; ⇨Fig. 4.14-W} such as those observed at the nose of a body flying at supersonic speed. In Japan, a country which has more than 80 active volcanoes, many of them located very close to residential areas – shock wave dynamics is applied extensively to understand volcanic eruptions. Current studies encompass analog experiments into magma fragmentation, magma water vapor explosions, numerical simulations of explosive eruptions, and in situ blast-wave measurements. Very recently, Tsutomu SAITO[176] and Kazuyosjhi TAKAYAMA, two Japanese gas dynamicists, reported on the application of a new 3-D computer code to an imaginary eruption of Mount Fuji (a 3,776-m-high stratovolcano about 100 km west of Tokyo), which provided useful information on how interactions of blast waves depend on the geometry of the ground.

It should be noted here that the term *blast* as used by geologists in the designations "volcanic blast" and "lateral blast" is not used in the strict sense defined by gas dynamicists for a point explosion resulting in a well-defined, steep-fronted, wave profile, but rather as a strong compressive atmospheric wave with amplitudes in the lower weak shock wave region. Just like in the case of thunder, the source of the blast is rather extended, and nonlinear propagation and refraction behavior of the ejecta-loaded and heated atmosphere as well as reflection effects at ground level mitigate the steepness of the emitted pressure waves. The nature of eruption-induced blasts was studied in more detail during the violent explosion of Mount St. Helens on May 18, 1980. Witnesses nearby did not observe any atmospheric shocks or sonic booms, which indicates that the furious blast, also called a "superheated hurricane" by some reporters, was subsonic. However, USGS geologists cautiously noted that "in some areas near the blast front the velocity may have approached, or even exceeded, the supersonic rate for a few moments" {Mt. St. Helens ⇨1980}. The late Bradford STURTEVANT {⇨1991}, a U.S. fluid dynamicist at CalTech, was a proponent of the idea that volcanoes could erupt supersonically.

One curious ejection phenomenon associated with explosive volcanic eruptions is known as a "volcanic bomb:" here, large individual fragments of molten lava ejected from the vent at high (supersonic) velocities in a viscous state are shaped during flight and assume an aerodynamically rounded form as they solidify before striking the earth. Various types of volcanic bomb can form, depending on the kind of volcanic material ejected, its temperature and viscosity, its gas content and its ejection velocity, *e.g.*, spheroidal bombs, ribbon bombs, spindle bombs (with twisted ends), and breadcrust bombs (with a compact outer crust or a cracked surface). Irregular flattish bombs that do not get hard before hitting the ground are called "pancake bombs" or "cow-dung bombs."

Similar to the *Richter scale* {RICHTER ⇨1935} used to classify the magnitude (energy) of an earthquake, the *Volcanic Explosivity Index* (*VEI*) was introduced into volcanology in order to standardize the assignment of the size of an explosive eruption {NEWHALL ⇨1982}.

---

[175] R.P. HOBLITT, C.D. MILLER, and W.E. SCOTT: *Volcanic hazards with regard to siting nuclear-power plants in the Pacific Northwest.* USGS, Open-File Rept. No. 87-287 (1987).

[176] T. SAITO and K. TAKAYAMA: *Applying shock-wave research to volcanology.* Computing Sci. Engng. **7**, No. 1, 30-35 (2005).

**Meteorite Impact.** Depending on the size and mechanical strength of the celestial body involved (comet, asteroid or meteoroid[177]), a terrestrial impact can dramatically affect not only the close environment of the impact, but also remote areas as well. According to a widely accepted theory, the biggest terrestrial impact happened when the proto-Earth and a Mars-sized protoplanet collided with each other, eventually resulting in the creation of the Moon {⇨c.4.5 Ga ago}.

Meteorite impacts are sudden and violent natural phenomena. Many of these huge catastrophic impact events have occurred since the Earth was formed. Perhaps the most spectacular was one that happened long before the evolution of man; this impact created the *Chicxulub Crater*, some 170 km in diameter, and shaped the Yucatan peninsula. This event possibly led to the "Cretaceous-Tertiary mass extinction" {Chicxulub ⇨c.65 Ma; ALVAREZ ⇨1978}, marking the end of the dinosaur era. In Germany, about 15 million years ago, a large meteorite, smaller than the one which created the Chicxulub Crater but certainly still large enough to cause massive effects, shaped the *Ries Basin* {⇨c.15 Ma ago; ⇨Fig. 4.1–C}, a crater about 20 km in diameter that is located in northern Bavaria. This impact event was probably accompanied by a second, simultaneous impact nearby, which created the smaller *Steinheim Basin*, a unique crater structure with a central hill in the inner plain (a structure typically found in small lunar craters) {Ries Basin ⇨c.15 Ma ago; ⇨Fig. 4.1–C}.

Other huge terrestrial impact craters have also been identified elsewhere, *e.g.*, the *Manicouagan Crater* (70 km in diameter, about 200 Ma) in northern Canada, one of the oldest and certainly largest impact events in Earth's history. However, the total number of huge impact events that occurred throughout the evolution of man (< 3.5 million years ago) is still unknown, because only a small number of impact sites created within this period have been identified, such as the 10.5-km-diameter *Bosumtwi Crater* in central Ghana, which was created about 1.07 million years ago when primitive man had already fabricated stone tools via percussion in East Africa, only some 3,500 km away from the impact site.[178]

Impact events caused by celestial objects that resulted in big crater structures killed most life in the surrounding area, and also must have affected the evolutionary course of man in more remote areas too. By the year 2000, almost 160 impact sites had been identified on Earth,[179] and this number is still increasing.

The most recent impact event of a significant magnitude happened on June 30, 1908 in Russia {Stony Tunguska ⇨1908}. Since this impact site is located in Siberia, in an isolated, almost uninhabited area, the occurrence of the event did not come to the attention of the international scientific community immediately. However, at 5.30 P.M. on the day of the collision, a series of unusual pressure waves in the atmosphere were recorded by microbarographs at Kew Observatory in England, and European seismographs also recorded a strong ground wave. The exact impact site was not investigated until 1928: the Russian scientist Leonid KULIK, who was selected by the Soviet Academy of Sciences to investigate what happened, led an expedition to the Tunguska region. He discovered a zone approximately 50 km in diameter that exhibited the effects of the intense heat produced by the meteorite, which was surrounded by another area 30 km in diameter where trees had been flattened by the air wave; these trees all pointed away from the central zone of the impact – a feature also observed more recently during the explosive eruption of Mount St. Helena {⇨1980; ⇨Fig. 4.1–G}. KULIK found a total of at least ten craters, ranging from 30 to 175 ft (9–53 m) in diameter, but searches to find the body of the meteorite were fruitless. An unusual meteorite fall was also observed in 1947 in the Sikhote-Alin Mountains, about 430 km northeast of Vladivostok {Sikhote Alin Meteorite ⇨1947}. The meteorite, which broke up during its passage through the atmosphere, causing violent explosions, produced a number of craters with diameters ranging from about 0.5 to 26 meters.

Fortunately, however, most meteoroids were small, and when entering the atmosphere they burned up by aerodynamic heating or fragmented into small pieces. This spectacular phenomenon of disintegration is often accompanied by a barrage of air shocks. The fall of a single meteorite, or sometimes the fall of a meteor swarm (a "meteoric shower"), is a spectacular event, and one which has long fueled superstitions and mystified man until very recently. In the late 18th century, the German physicist Ernst F.F. CHLADNI {⇨1794}, today considered to be the father of acoustics,

---

[177] A *meteoroid* is a mass that orbits the Sun (or another star) or any other object in interplanetary space that is too small to be called an "asteroid" or a "comet." A *meteorite* is a meteoroid that strikes the Earth (or another large body) and reaches the surface without being completely vaporized. One speaks of a *meteor* or *shooting star* when the mass is small enough to be totally destroyed in the atmosphere. Until two or three centuries ago, the term *meteor* included other phenomena such as halos, auroras and rainbows [after *Encyclopedia Americana*. Americana Corp., New York, vol. 18 (1974), pp. 713-716].

[178] In East Africa – the cradle of mankind – no meteorite impacts before this have been discovered. The only significant impact event that happened during the existence of prehistoric man was the Bosumtwi event in Ghana.

[179] A catalog and an atlas of impact craters around the world was prepared by the Geophysical Data Centre (GDC), Geological Survey of Canada. See *Impact cratering on Earth*;
http://gdcinfo.agg.nrcan.gc.ca/crater/paper/index_e.html.

also studied meteorites. Calling them "fireballs" [Germ. *Feuerkugeln*], he was apparently the first to maintain that meteorites really fell from the sky – as many observers claimed, but scientists denied. Eventually, investigations of the L'Aigle Fall {BIOT ⇨1803} carried out by a French commission firmly established that the stones did actually fall from outer space. However, this result was not immediately accepted by learned men. For example, the U.S. President Thomas JEFFERSON refused to believe that a meteorite that fell in Weston, Connecticut (in 1807) was extraterrestrial in origin. Accounts of the famous Washington Meteor (1873) that produced "short, hard reports like heavy cannon" provoked disputes among contemporary scientists about the possible causes of the shock phenomena observed {ABBE ⇨1877}. The Austrian philosopher Ernst MACH, who, together with the Austrian physicist Peter SALCHER discovered the ballistic head wave {E. MACH & SALCHER ⇨1886}, correctly explained that this "sonic boom" phenomenon was caused by the supersonic motion of the meteorite in the atmosphere {E. MACH & DOSS ⇨1893}.

But how were these meteorites produced in space? Some early natural philosophers proposed the idea that meteorites were ejecta from lunar volcanoes. Modern hypotheses on the formation of meteorites assume that they come from small masses of gas and dust which condensed in space due to gravity, or that they are the debris from larger, disintegrated bodies. After GALILEI's first telescopic observations of the face of the Moon {⇨Fig. 4.1–A}, published in 1610,[180] it was generally supposed that lunar craters[181] were artifacts of intelligent beings (Johannes KEPLER 1634), or were the result of volcanic action as proposed by Robert HOOKE {⇨1665}, William HERSCHEL[182] (1787), Johann Hieronymus SCHRÖTER {⇨1802}, and James Dwight DANA[183] (1846). The idea that craters could be formed by the impact of "comets" was first conceived in the 1840s by the Bavarian astronomer and physician Franz VON PAULA GRUITHUISEN[184] [Germ. "*kometarischer Weltkörper*"], although the credit is generally given to the British astronomer Richard A. PROCTOR {⇨1873} in the literature. In the early 1890s, the U.S. geologist Grove K. GILBERT {⇨1893} was the first to provide support for this idea through sound arguments, stating that the lunar craters, which he called "impact craters," were generated by meteoritic impacts in the geologic past {⇨Fig. 4.1–A}. His impact hypothesis became the basis for all subsequent thinking along these lines; however, until the 1930s most astronomers still believed that the Moon's craters were giant extinct volcanoes.

The first systematic studies of meteoritic material collected from various craters around the world {SPENCER ⇨1933} supported the impact theory {BUCHER ⇨1933; BOON & ALBRITTON ⇨1938}, particularly the discovery of the following geological structures:

▸ *Shatter cones* [Germ. *Strahlenkegel*], curiously striated conical geologic structures {⇨Fig. 4.1–C}, are formed when an intense shock wave travels through rock. Since they point toward the center of the impact, they provide some of the best macroscopic evidence for an impact.

▸ *Impactites* are composed of fused meteoric material and slag, but mostly of material from the impact site – so-called "target rock." An *impactite* is a vesicular, glassy to finely crystalline geological material formed by the impact of a meteorite, asteroid or comet. This material often suffers shock metamorphosis, which happens when a high-pressure shock wave passes through the target rock, generating pressures of over 40 GPa (400 kbar) at the point of impact.

▸ The sensational discovery of *coesite* {COES ⇨1953} and *stishovite* {STISHOV & POPOVA ⇨1961}, two high-pressure shock-induced polymorphs of quartz, in Arizona's Meteor Crater[185] (now officially called

---

[180] G. GALILEI: *Siderius nuncius* ("The Starry Messenger"). Venice, Italy (March 4, 1610).

[181] The word *crater* was first mentioned by the Ionian poet HOMER (*c.*8th century B.C.); it derives from the Greek word *krater* designating a vessel used for diluting wine with water. It was first used in 1791 to classify lunar surface phenomena by the German astronomer Johann Hieronymus SCHRÖTER (1745–1816). See § 507-516 in his *Selenotopographische Fragmente*. 2 vols, Lilienthal & Helmst, Göttingen (1791–1802).

[182] W. HERSCHEL: *An account of three volcanos in the Moon*. Phil. Trans. Roy. Soc. Lond. **77**, 229-232 (1787).

[183] J.D. DANA: *On the volcanoes of the Moon*. Am. J. Science & Arts **11**, 335-355 (1846).

[184] F. VON PAULA GRUITHUISEN: *Der Mond und seine Natur* [Besonderer Abdruck aus seinem astronomischen Jahrbuche für 1848]. München (1848), p. 35.

[185] The famous *Meteor Crater* in Arizona {⇨*c.*50,000 years ago; ⇨Fig. 4.1–B} is the most recently produced large terrestrial crater and certainly the most studied one. It was first made famous in 1891 due to the discovery of many masses of meteoritic iron scattered around it and the further discovery of diamonds inside these iron masses. Harvey H. NININGER, a famous U.S. meteorite hunter, wrote in his book *Arizona's Meteor Crater:* "Formerly it was supposed that the early inhabitants had actually witnessed the event which produced the crater, but archaeologists seem now to be in agreement that man has been in this part of the continent only during the last 20,000 or 25,000 years." Since modern geologists assume that Meteor Crater has an age of about 50,000 years, it is not very likely that the impact event was witnessed by prehistoric Arizonian hunters. The numerous arrow points picked up in the 1950s by NININGER were estimated by the Museum of Northern Arizona to be only about 800–900 years old. Some archeological evidence indicates that aborigines definitely lived in this area at least as early as 11,000 years ago. On the other hand, Robert S. DIETZ, a U.S. geologist, noted in his article *Astroblemes* [Scient. Am. **205**, 51-58 (Aug. 1961)] that "the Hopi Indians are said to retain the legend that one of their gods descended here from the sky in fiery grandeur."

"Barringer Crater"), constituted further evidence for meteoritic impact scars {CHAO, SHOEMAKER & MADSEN ⇨1960}. Shortly after, a very similar discovery was also made in the Ries Basin in Bavaria {⇨c.15 Ma ago; SHOEMAKER & CHAO ⇨1961; ⇨Fig. 4.1–C}.

The term *astrobleme* was originally a Greek word meaning "star wound," but it was later revived by the U.S. geologist Robert S. DIETZ to refer to scars on the Earth caused by meteoritic impacts. Astrobleme identification is based on finding the presence of the shock-induced structures mentioned above. Among the first terrestrial impact structures to have been identified as (probable) astroblemes were the *Bosumtwi Crater* in Ghana and the *Vredefort Ring*, the world's largest impact crater structure, located in the Republic of South Africa {Vredefort Basin ⇨c.2 Ga}. The astrobleme concept of developing criteria that could be used to recognize impact structures significantly promoted our knowledge of the geological history of the Earth, Moon, and other planets. An important milestone in the understanding of shock-induced geologic effects was reached at the First Conference on Shock Metamorphism of Natural Materials {⇨1966}. Laboratory studies of shock metamorphism have long provided the basis for recognizing the diagnostic features of ancient terrestrial impact craters. However, Paul DECARLI,[186] a shock physicist at SRI, recently pointed out that there are significant differences between the range of parameters accessible in small-scale laboratory impact experiments and the conditions of large-scale natural impact events – in regard to both the peak pressures involved and the duration of the shock pressure.

Today we know that there are a huge number of small bodies in the Solar System, termed *near-Earth objects* (*NEO*s), with orbits that regularly bring them close to Earth and which, therefore, may someday strike our planet. Astronomers estimate that there are approximately one million bodies larger than 50 meters in diameter, a size which is generally considered to be the threshold for the body to be able to penetrate through the Earth's atmosphere without fracturing. Several teams of international astronomers are currently surveying the sky with electronic cameras to find such NEOs and to quantify the danger of future Earth impacts. For example, the Catalina Sky Survey (CSS), based at the Lunar and Planetary Laboratory of the University of Arizona, is a consortium of three operating surveys aiming at discovering comets and asteroids and identifying potentially hazardous NEOs. The omnipresent danger of collisions with NEOs became starkly apparent in 2002 when a celestial body about 100 m in diameter passed the Earth at a distance of only 120,000 km {Asteroid 2002 MN ⇨2002}.

In 1994, it became possible to observe the effects of a cosmic impact from a safe distance for the first time when a series of icy and stony fragments of comets impacted on Jupiter, spawning spectacular fireballs and producing impact scars that were visible to even amateur astronomers {Comet SHOEMAKER-LEVY 9 ⇨1994}.

**Cosmic Shock Wave Phenomena.** In ancient times, people thought of the night sky as being permanent, and the few changes that were observable with the unaided eye were usually repetitive, such as the changing phases of the Moon and the slow drift of the constellations. The positions of the stars were also regarded as being fixed and eternal. Besides the motion of a few "wandering stars," later determined to be the planets, not much appeared to happen in the cosmos. Therefore, the occasional burst from an exploding star or the visit of a comet was a terrifying and portentous event.

In the past 50 years, however, astrophysics research using optical microwave and X-ray diagnostics has shown that dynamic phenomena are much more common in the cosmos than previously anticipated. Compared to terrestrial shock and detonation phenomena, cosmic dynamic processes reach enormous dimensions and relativistic velocities. For example, the *solar wind*, a stream of ionized gas particles emitted from the corona of the Sun, is our closest dynamic flow phenomenon of cosmic dimensions. It was first predicted by Prof. Eugene N. PARKER {⇨1958} at the University of Chicago, a distinguished expert in the theory of cosmic magnetic fields. Only four years later, his hypothesis on the solar wind was confirmed experimentally {NEUGEBAUER & SNYDER ⇨1962}.

The solar wind, which is accelerated in the magnetic field of the Earth, produces a *bow shock* around the Earth {AXFORD ⇨1962; KELLOG ⇨1962; ⇨Fig. 4.1–X}, which is analogous to the bow wave of a boat or the head wave of an object moving supersonically in the atmosphere.[187] The bow wave is a jump in plasma density, temperature and magnetic field associated with the transition from supersonic to subsonic flow. The *foreshock* is the region upstream of the bow shock, where energetic protons reflected back toward the

---

[186] P.S. DECARLI ET AL.: *Laboratory impact experiments versus natural impact events.* In: (C. KOEBERL and K.G. MACLEOD, eds.) *Catastrophic events and mass extinctions: impacts and beyond.* Geological Society of America, Special Paper No. 356 (2002), pp. 595-605.

[187] D. BURGESS: *Magnetosphere of Earth: bow shock.* In: (P. MURDIN, ed.) *Encyclopedia of astronomy and astrophysics.* Institute of Physics, Bristol etc. and Nature Publ. Group, London etc., vol. 2 (2001), pp. 1564-1568.

Sun from the shock may help to heat, decelerate and deflect the solar wind.

A similar boundary layer separates the heliosphere from interstellar space: at about 80–100 AU,[188] a *termination shock* is formed as the solar wind slows down from supersonic to subsonic speeds. The solar wind finally stops at the heliopause, the "stagnation" surface between the solar wind and the ions of the interstellar medium, which occurs about 130–150 AU from the Sun, and which forms the boundary of the heliosphere, the immense magnetic bubble containing our Solar System {Voyager 1 & 2 ⇨ 2003; Voyager 1 ⇨ 2004}.[189]

Bow shocks also form around other planets of the Solar System {Mariner 4 ⇨ 1965; Voyager 2 ⇨ 1989; Cassini ⇨ 2004} as well as around hot young stars, where vigorous stellar winds slam into the surrounding interstellar medium {⇨ Fig. 4.1–Y}.

Colliding and fusing galaxy clusters should produce *cosmic shock waves*. The outlines of these giant shock waves have now been seen as radioemitting structures.[190] Cosmic shock waves around distant clusters of galaxies could be generating some of the mysterious cosmic rays that strike Earth.

*Accretion shock waves* are spherical shock waves that arise when material (usually gas) spirals inward to a gravitational source. Accretion shocks arise in core-collapse supernovae, star formation, and accreting white dwarfs and neutron stars. In 2003, U.S. astrophysicists showed that small perturbations to a spherical shock front can lead to rapid growth of turbulence behind the shock, driven by the injection of vorticity from the now nonspherical shock. In 2007, they proposed a new explanation for the generation of neutron star spin which, for the first time, matches astronomical observations. Their results are based on simulations run on the Leadership Computing Facility Cray X1E at ORNL as part of the SciDAC TeraScale Supernova Initiative {⇨ 2001}.[191]

**Cosmic Explosion Phenomena.** Shocks of much larger dimensions are generated during stellar explosions of dying stars, termed *supernovae* {BAADE & ZWICKY ⇨ 1931}. The earliest accounts of observations of a supernova (SN 1006), called a "guest star" at the time, can be found in Chinese and Swiss annals a thousand years old {Guest Star ⇨ 1006}. The name *nova* – a shorthand for *nova stella* (from the Latin, meaning "new star") – was coined by the German astronomer and mathematician Johannes KEPLER {⇨ 1604}. He actually discovered an unstable star that was more than 100 times brighter than the modern definition of a nova; this was later recognized as being a supernova (SN) and was cataloged as SN 1604. Prior to KEPLER, the Danish astronomer Tycho BRAHE had already observed a supernova; today this is classified as SN 1572 {BRAHE ⇨ 1572}.

The most recent supernova that could be seen with the naked eye, named SN 1987A {SHELTON ⇨ 1987}, was discovered on February 23, 1987, in the Large Magellanic Cloud. Because of its relative proximity to Earth (only 168,000 light-years), SN 1987A became easily the best-studied supernova of all time. Supernovae are among the most energetic explosions in the Universe and can temporarily rival the energy release of an entire galaxy. Most of the energy, however, is not emitted as electromagnetic radiation, but rather in the form of kinetic energy imparted to stellar gases, which are accelerated into space, reaching relativistic velocities of up to one tenth of the velocity of light and causing discontinuities in pressure in the interstellar medium.

Supernova remnants, the remains of exploded stars, are actually hot gases that have been hurtled into space by the force of a supernova explosion. Some of these remnants are thousands of years old and many hundreds of light-years wide. For example, the famous Crab Nebula M1 (or NGC 1952) is the remnant of the supernova SN 1054. Many supernova remnants (SNR) have been found and cataloged. Supernova remnants are the major source of energy, heavy elements and cosmic rays in our Galaxy. In 1995, the National Radio Astronomy Observatory (NRAO) in Socorro, NM published the first "movie" showing the development of the remnant of SN 1993J in the galaxy M81 (or NGC 3031) over a period of one year from September 1993 to September 1994, as it expands with near-circular symmetry {DÍAZ ⇨ 1993}.[192] An analysis of these "radio" images by NRAO researchers revealed that the debris shell of SN 1993J showed no signs of slowing due to interactions with material surrounding it yet. The material from the star's explosion is moving at nearly 16,000 km/s. At that speed, the material would travel the distance from the Earth to Saturn in one day.

---

[188] The astronomical unit (AU) is the mean distance of the Earth from the Sun (ca. 150 million km).

[189] P. FRISCH: *The galactic environment of the Sun*. Am. Sci. **88**, No. 1, 52-59 (2000).

[190] T.A. ENßLIN: *Radio traces of cosmic shock waves*. Science **314**, No. 5800, 772-773 (Nov. 2006).

[191] J.M. BLONDIN and A. MEZZACAPPA: *Pulsar spins from an instability in the accretion shock of supernovae*. Nature **445**, 58-60 (Jan. 4, 2007).

[192] *Astronomers make "movie" of radio images showing supernova explosion.* National Radio Astronomy Observatory (NRAO), Socorro, NM (Nov. 30, 1995); http://www.nrao.edu/pr/1995/supermovie/.

One possible cause of the supernova phenomenon is that the core undergoes gravitational collapse, which generates a *pulsar* {HEWISH & BELL ⇨1967}, a rapidly spinning *neutron star* {LANDAU ⇨1932; BAADE & ZWICKY ⇨1934}. Experimental evidence for this was first provided by huge Čerenkov detectors located in the United States and Japan, which both recorded neutrino bursts from SN 1987A {SHELTON ⇨1987}. The name *pulsar* is an abbreviation of "pulsating radio star." The detectable radiation from pulsars occurs entirely in the radio wavelength region of the electromagnetic spectrum, ranging from about 40 MHz to 2 GHz, with periods of about 33 milliseconds to 3.7 seconds. However, the pulsar in the Crab Nebula is so far the only remnant found to simultaneously emit radio, optical and X-ray pulses.

Stellar explosions that are even more powerful than supernovae are termed *hypernovae* {Beppo-SAX & CGRO ⇨1997; ROSAT ⇨1999}, and these are possibly the most powerful type of cosmic explosion to occur in the Universe since the Big Bang. A hypernova is an exceptionally large star which has collapsed because nuclear fusion is no longer taking place within its core. Its total collapse produces two highly collimated relativistic jets and is believed to creating a black hole rather than a neutron star. The concept of a hypernova was introduced by Bohdan PACZYŃSKI {⇨1997}, an astrophysics professor at Princeton University, NJ, in order to explain gamma-ray bursts (GRBs).[193] The *collapsar* {GRB 030329 ⇨2003} is an extremely attractive model that fits a wide range of observed gamma-ray bursts. The *collapsar model* predicts highly beamed energy deposition responsible for the GRB along the symmetry axis.

These mysterious γ-ray burst phenomena were first observed by space-based military detectors {Vela Satellites ⇨1960}, and were later examined in more detail by three satellite observatories:

- NASA/ESA's Hubble Space Telescope (HST), launched in 1990 and able to detect ultraviolet, visible and near-infrared wavelengths;
- NASA's Compton Gamma-Ray Observatory (CGRO), launched in 1991 and dedicated to observing the high-energy Universe; and
- the ASI/NIVR BeppoSAX, an X-ray astronomy satellite, allowing observations in the spectral range 0.1–300 keV (launched in 1996).

Hypotheses on the origin of the tremendous energies observed in *quasars* are associated with stellar explosions, the gravitational collapse of massive stars, supernova explosions, the conversion of gravitational energy into particle energy by magnetic fields, matter-antimatter annihilation, and the rotational energy of a very compact mass (as proposed for pulsars). The term *quasar*, a contraction of "quasi-stellar radio source," was originally applied only to the star-like counterparts of certain strong radio sources whose optical spectra exhibit red shifts that are much larger than those of galaxies {M. SCHMIDT ⇨1963}. Subsequently, however, a class of quasi-stellar objects was discovered with large red shifts that exhibit little or no emission at radio wavelengths. The term *quasi-stellar object* (QSO) is now commonly applied to star-like objects with large red shifts regardless of their radio emissivity. Based on the hypothesis that the quasar red shifts are cosmological and somehow related to the Big Bang – *i.e.*, that they are a consequence of the expansion of the Universe and thus directly related to the distance of the object {HUBBLE ⇨1929} – quasars are considered to be the remotest objects located at the edge of the visible Universe that are moving with very high velocities, approaching up to 80% the velocity of light.

It is generally agreed that the enormous energy emitted by quasars is gravitational and not thermonuclear in origin, and perhaps generated

- by multiple supernova outbursts, each supernova collapsing and releasing a large amount of gravitational energy;
- by collisions between stars; or
- by the gravitational collapse of a single supermassive star.

The energy source of a quasar is widely believed to be a supermassive black hole of several billion solar masses that is accreting matter from its surroundings; the black hole is surrounded by hot gas clouds revolving at speeds of up to several 1,000 km/s.

**Cosmogony.** The questions of the origin, the age and the evolution of the Universe have occupied man since the earliest times, but was not approached scientifically until the Age of Enlightenment, in the works of such eminent scientists as Thomas WRIGHT (1711–1786), Pierre-Simon DE LAPLACE (1749–1827) and Immanuel KANT (1724–1804). However, the foundations for such studies lie much further back in time: early contributions were made by Tycho BRAHE (1564–1601), Galileo GALILEI (1564–1642), Johannes KEPLER (1571–1630), René DESCARTES (1596–1650), and Sir Isaac NEWTON (1643–1727).

Based on his General Theory of Relativity, the German-born theoretical physicist Albert EINSTEIN (1879–1955) laid the mathematical foundations for the structure of the Universe as a whole {A. EINSTEIN ⇨1917}. He constructed a

---

[193] R. IRION: *Gamma beams from a collapsing star.* Science **283**, 1993 (March 26, 1999).

static model that was finite but unbounded and had a spherical geometry. The Russian physicist and fluid dynamicist Alexander FRIEDMANN {⇨1922} noticed that EINSTEIN's field equations allowed two kinds of nonstatic solutions that are consistent with EINSTEIN's Cosmological Principle: these two models describing the dynamics of the Universe assumed either a negative spatial curvature, resulting in the continuous expansion of the Universe, or a positive spatial curvature, resulting in cycles of expansion and contraction – a Big Bang followed by a Big Crunch {DAVIES ⇨1994}.

Independently of FRIEDMANN's pioneering work, the Belgian mathematician, physicist and priest Georges LEMAÎTRE {⇨1927} also published a paper on the cosmology of an expanding Universe resulting from the catastrophic explosion of an extremely high-condensed state containing all the matter of the Universe. Although expanding models of the Universe had previously been considered by other researchers, LEMAÎTRE's model, which assumed that the expansion would accelerate, has become the leading theory of modern cosmology. The discovery that galaxies were, in general, receding from us {HUBBLE ⇨1929} provided the first clue that the Big Bang Theory might be correct, and today FRIEDMANN and LEMAÎTRE are both considered to be the main founders of the Big Bang Theory.

Albert EINSTEIN, who welcomed FRIEDMANN's and LEMAÎTRE's results, then revised his cosmological model in the early 1930s, and, in cooperation with the Dutch astronomer Willem DE SITTER, constructed the simplest form of an expanding world model – the so-called "Einstein-de Sitter Universe" – a simple solution of the field equations of general relativity for an expanding Universe with zero cosmological constant and zero pressure.[194] Their homogeneous and isotropic model assumes that the Universe (i) expands from an infinitely condensed state at time $t = 0$ at such a rate that the density varies as $1/t^2$, and (ii) contains large amounts of matter that do not emit light and, therefore, had not been detected. This matter – now called "dark matter" – has since been shown to exist by observing its gravitational effects.

A further milestone in cosmogony was the development of the "Hot Big Bang Model" by the Russian-born U.S. physics professor George GAMOW and his collaborators Ralph A. ALPHER and Robert C. HERMAN, two young physicists {GAMOW, ALPHER & HERMAN ⇨1948}. Stating that the Universe began in a gigantic explosion, they predicted that there should be a relic radiation field from the Big Bang, which resulted from the primordial fireball that LEMAÎTRE called *L'atome primitif* ("The Primeval Atom") or *L'œuf cosmique* ("The Cosmic Egg"). Based upon EINSTEIN's General Theory of Relativity and his Cosmological Principle, the Hot Big Bang Model is supported mainly by two important observations:

▸ it predicts that the light elements (such as H, He and Li) should have been created from protons and neutrons just a few minutes after the Big Bang. The observed abundance of light elements, especially that of helium, is hard to explain without invoking this theory.

▸ The predicted background radiation was indeed discovered; it took the form of a residual black-body radiation at about 3 K {PENZIAS & WILSON ⇨1965}.

In February 2003, NASA released a "baby" picture of the Universe {NASA-GSFC ⇨2003; ⇨Fig. 4.1–W}, the earliest ever taken, which captured the afterglow of the Big Bang. NASA scientists claimed that, based on an analysis of the data obtained for the cosmic microwave background, the Universe has an age of 13.7 ± 0.2 billion years. Previous theories on the age of the Universe were based upon the reciprocal of the Hubble constant (which is close to 10 billion years), and the ages of stars in globular clusters (which are among the oldest in the Universe), resulting in an uncertainty in the order of several billion years.

The British astronomer Fred HOYLE, who, like his colleagues and countrymen Hermann BONDI and Thomas GOLD, was an eager advocate of a rival steady-state relativistic cosmological model, was the first to (dismissively) call the expansion model the "Big Bang" {Sir Fred HOYLE 1940s, *see* BONDI ⇨1948}. The discovery of the excess microwave radiation, however, suggests that the steady-state model is incorrect and that the Universe as a whole changes over time.

Some modern cosmogonists assume that space, time and matter originated together, and that in the very initial phase of the explosion energetic photons created particle-antiparticle pairs which collided with each other, resulting in annihilation and their conversion into photons.

In the very early Universe, the temperature was so great that all of the matter was fully ionized and dissociated. Roughly three minutes after the Big Bang itself, the temperature of the Universe had rapidly cooled down from its initial phenomenal $10^{32}$ Kelvin to approximately $10^9$ Kelvin. At this temperature, the production of light elements (namely deuterium, helium, and lithium) was able to take place – a process known as "Big Bang nucleosynthesis." Elements heavier than helium are thought to have originated

---

[194] A. EINSTEIN and W. DE SITTER: *On the relation between the expansion and the mean density of the Universe.* Proc. Natl. Acad. Sci. U.S.A. **18**, 213 (1932).

in the interiors of stars that formed much later in the history of the Universe {Sir HOYLE ⇨1946}. David TYTLER[195] (an astrophysics professor at the Center for Astrophysics and Space Sciences, UC San Diego) and his collaborators reviewed the historical development of and recent improvements in the theory of Big Bang nucleosynthesis.

Big Bang cosmologists have devised a chronology for the history of our Universe {WEINBERG ⇨1977}, covering the period from $1.7 \times 10^{-43}$ seconds after creation – the *quantum of time* or *Planck time* (PLANCK 1899),[196] the earliest known time that can be described by modern physics – up to the present time, and illustrating the main events on a diagram.[197] During this enormous time span, the initial temperature of about $10^{32}$ K decreased to the current background radiation temperature of only 2.725 K {PENZIAS & WILSON ⇨1965}.

## 2.3.2 EARLY MAN-MADE SHOCK GENERATORS: TOOLS AND TOYS

> *I feel that E. MACH and B. DOSS give a correct interpretation of the cracking sound of a falling meteorite which, moving with great speed, is generated by the head wave. As a new example I would like to present the whip which, when skillfully used, also cracks; therefore we may conclude that the outermost end of the whip lash moves with a velocity which exceeds that of sound; i.e., it moves faster than 335 m/s. An experimental test confirmed my speculation.*[198]
>
> Otto LUMMER
> Schlesische Friedrich-Wilhelms-Universität
> Breslau 1905

Until the advent of gunpowder, the only means available for man to produce very loud sounds were the clapping of hands, whip-cracking, and belt-snapping.

**Clapping of Hands.** In his Ramsden Memorial Lecture on *Shock Waves*, Sir James Michael LIGHTHILL,[199] who was Beyer professor of applied mathematics at Manchester University and one of the leading British fluid dynamicists, mentioned that "*weak shock waves* are produced by the clap of a hand… how fascinating at all ages from three months onwards is the capacity of a gentle movement of the hands to produce a pressure wave whose duration when it passes your ears is the thousandth part of a second, although echoes in a hall like this make it appear longer." A strong hand clap is perceived by the ear as a sharp tone with a surprisingly high peak level;[200] indeed, an *explosio* in its antique sense. However, the rise time of the pressure pulse is much less steep than that typical of a shock wave.

**Whip-Cracking.** Whip-cracking has probably been used since antiquity as an aid for drovers, tamers and coachmen, and as a children's toy. Since the Middle Ages, whip-cracking has traditionally been practiced in southern Germany at Shrovetide to generate noise in contrast to Lent, a period of silence and contemplation, and it has been used for centuries in Switzerland for communication purposes. However, it was only rarely investigated by early scientists because the mechanism of shock generation in whip-cracking and its analysis are rather complicated. The German physicist Otto LUMMER {⇨1905} was the first to speculate that the shock might be caused by supersonic motion of the whip tip. The solution to this riddle requires ambitious diagnostics and so it was not resolved until the advent of more advanced high-speed photography {CARRIÈRE ⇨1927; BERNSTEIN, HALL & TRENT 1958}. More recent experiments carried out at the Ernst-Mach-Institut (EMI) in Freiburg, using large-field-of-view shadowgraphy combined with laser stroboscopy and high-speed videography, have clearly shown that a shock wave is emitted from the whip tip (which does indeed move with a supersonic velocity of about 700 m/s), but that the decisive mechanism for generating strong reports is the abrupt flapping of the tuft at the turning point {KREHL ET AL. ⇨1995; ⇨Fig. 4.5−K}. Although the shock wave quickly reduces in strength because of its spherical expansion, it is still perceived at a distance of many meters away as a sharp report, very similar to the one from a starter's pistol.

Recently, it has been proposed that the tails of some dinosaurs were also capable of whip-cracking {ALEXANDER ⇨1989}. The giant sauropod dinosaurs of the family *Diplodocidae* were known to have enormous and graceful tails

---

[195] D. TYTLER ET AL.: *Review of Big Bang nucleosynthesis and primordial abundances* [with 194 refs.]. Phys. Scripta **85**, 12-31 (2000).

[196] All matter, energy, space and time are presumed to have exploded outward from the original singularity at $t = 0$. Nothing is known of the very earliest period of the history of the Universe: 0 to $1.7 \times 10^{-43}$ seconds.

[197] P. SHELLARD: *A brief history of the Universe*. Cambridge Cosmology Hot Big Bang, Relativity & Gravitation Group, Cambridge University, U.K.; http://www.damtp.cam.ac.uk/user/gr/public/bb_history.html.

[198] See LUMMER {⇨1905}.

[199] M. LIGHTHILL: *Shock waves*. Mem. Manch. Lit. Phil. Soc. **101**, 1-6 (1959).

[200] According to Dr. Joachim FELDMANN at the Institut für Technische Akustik, TU Berlin, who kindly performed some measurements for the author, the peak pressures of a hand clap and a cap gun, recorded at a distance of 0.4 m using a Bruel & Kjaer ¼-in.(6.35-mm)-diameter capacitor microphone, amount to about 125 dB (0.36 mbar) and 138 dB (1.59 mbar), respectively. At these small overpressures the shock wave is still quite weak, propagating at practically the velocity of sound.

that tapered to thin tips. Nathan MYRHVOLD, chief technology officer and senior vice president of Advanced Technology at Microsoft, showed that computer modeling indicated that diplodocid tails could have reached supersonic velocities, and argued that this was physically plausible {MYRHVOLD & CURIE ⇨1997}. He was supported in his studies by Philip CURIE, a Canadian palaeontologist at the Royal Tyrrell Museum of Palaeontology in Drumheller, Alberta, Canada. Support for their hypothesis comes from the shape and mass distribution of the tail, which seem optimized for supersonic cracking. In addition, they speculated that a popper just a centimeter or two in length, made of skin and tendon, would improve shock wave generation and protect other tissues from the stress of cracking. They rather spectacularly concluded that, "Finally, we must confess that it is pleasing to think that the first residents of Earth to exceed the sound barrier were not humans, but rather the diplodocid sauropods. Following their demise, a hiatus in supersonic motion of over a hundred million years ensued until this capability was rediscovered by our species."

**Snapping Belts and Snapping Towel.** The "snapping of belts" {⇨Fig. 4.11–A} is, much like the whip, another very primitive method of generating waves of finite amplitude in air. Its origin is unknown, but it might have been discovered by chance by primitive man when he learned to tan skins and fabricate smooth leather belts. When performed in an appropriate manner (*i.e.*, the ends of the two spread belts are pulled rapidly apart), the air enclosed between the belts is pushed to the sides, thus producing a loud sound. This simple device, more a toy than a useful tool, produces an impressive cracking sound in a more comfortable way than whip cracking does, although it is not as loud, and to some extent it even permits the emitted pulse to be tailored by varying the widths of the belts and their flexibility. Recently, this belt-snap phenomenon was investigated in more detail using a high-speed digital camera and a schlieren optical system of large aperture.[201] Results show that compression of the air between the two rapidly-approaching leather bands first causes a spherical shock wave to form near one hand. The compression then runs along the belt length toward the other hand at supersonic speed, producing a stronger oblique shock wave that is believed responsible for the audible crack.

The flicking of a wet towel produces not only a painful sting when applied to the skin, but when it is cast forth and withdrawn sharply in the right way it also produces a cracking sound in the air akin to shooting a gun. This cracking noise, a shock wave, is the result of the towel tip reaching supersonic velocities {LEE ET AL. ⇨1993}.

**Electric Sparks.** Compared with the purely mechanical means mentioned above, the application of electric sparks to generate shock waves is a rather new approach. After the invention of the electrostatic generator {VON GUERICKE ⇨1663} and the Leiden jar {VON KLEIST & CUNEUS ⇨1745}, it became possible for the first time to store considerable amounts of electric charge and to discharge them in a very short time. The discharge is typically accompanied by a spectacular flash and a sharp report, an impressive demonstration that, particularly in the era when the application of electricity was being pioneered, was often shown in university lectures and private circles, thus also stimulating discussions on the nature of lightning and thunder {WALL ⇨1708}. The electric spark proved to be not only very useful for generating shock waves at any time and in any space with any desired geometry, but it was also an alternative method to chemical explosives for generating shock waves, thus allowing one to differentiate between electrical and chemical secondary effects of observed shock wave interaction phenomena – an important advantage that, for example, considerably facilitated the interpretation of Mach reflection {E. MACH & WOSYKA ⇨1875}.

**Musical Instruments.** Although the sounds of many musical instruments are, in terms of intensity, still within the realms of acoustics, there are unique exceptions, such as some percussion instruments (*e.g.*, drums, bells, gongs) and some wind instruments (*e.g.*, whistles, sirens, horns). The latter can produce sounds with very large amplitudes, particularly when they are operated with pressurized air. The oldest account of the effects of intense sound waves is given in the Bible for the example of the "trumpets of Jericho"[202] – in modern terminology fanfares rather than trumpets. Prof. Gary SETTLES and collaborators at Penn State University used their unique large-scale schlieren system to visualize the weak shock wave emitted from a modern trumpet {⇨Fig. 4.5–N}.

---

[201] G. SETTLES, M. HARGATHER, M. LAWSON, and R. BIGGER: *Belt-snap shock wave*. 26th Int. Symp. on Shock Waves, Göttingen, Germany (July 2007), *see* abstract.

[202] (Probably around 1550 B.C.), on the seventh day, God told JOSHUA, MOSES' successor, to have the priests march around the walls of Jericho seven times, blowing their trumpets. The priests gave one long blast on their trumpets. The strong, thick walls of Jericho crumbled to the ground! JOSHUA and his armed men rushed across the rubble and took the city of Jericho (*see* the Holy Bible, Joshua **6**:1-20).

### 2.3.3 BALLISTIC STUDIES: BIRTH OF SUPERSONICS

> *The last particular I shall here take notice of, is a most extraordinary, and astonishing encrease of the resistance, and which seems in a manner to take place all at once, and this when the velocity comes to be that, of between eleven and twelve hundred feet in one second of time. This encrease however only concerns the absolute quantity of the resistance, the law of it continuing in other respects nearly the same as before: and it is remarkable farther, that the case wherein this encrease of resistance becomes observable, is that, wherein the velocity of the shot, is at least equal to that velocity with which sounds are propagated: whence Mr. Robins has with great fagacity offered his reasons to believe, that in this case the air does not make its vibrations sufficiently fast, to return instantaneously into the place the bullet has left; but that the bullet then leaves a vacuum behind it; whereby it becomes exposed to the whole resistance, the body of air before it is capable of giving…*[203]
>
> Martin FOLKES
> London 1747

In the early 16th century, ballistics reached an important milestone due to the introduction and general use of spherical iron shot fired from iron or bronze cannons, instead of massive stone shot. This "ballistic revolution" also started the so-called "terminal ballistics cycle," which initiated improvements in weapon design and, in turn, new designs for protecting from such improved weapons:[204] an effective self-supporting process which continues even now. For example, with the advent of modern high-quality concrete – based on the invention of Portland-cement made by the English mason Joseph ASPDIN (1824) and on the idea of reinforcing concrete by metal wires ("ferroconcrete"), an invention of the French gardener Joseph MONIER (1867) – a low-cost and effective material for protecting against heavy impact became available. It was first used to strengthen military positions against artillery attacks on a large scale between the two World Wars to construct the French Maginot Line in the 1930s and the German West Wall in the period 1938–1939. On the other hand, the increasing use of concrete in the construction of fortifications and shelters stimulated military engineers to develop "superguns" in both World Wars {⇒Fig. 4.2−T}. Recent developments in bunker- and tunnel-busting weapons include "thermobaric" bombs which use the combined effects of heat and explosive pressure against certain types of tunnel targets to maximize the kill rate {NAGLE ⇒2002}.

The steady progress made in gun technology over more than six centuries has improved gunshot reliability and reproducibility, both of which are important preconditions for deriving empirical laws. Beginning in the 17th century, gunnery evolved into a science of its own and the gunner into a lauded expert in applied mechanics, then the most advanced discipline of physics. High-ranked artillerists were often graduates from prestigious polytechnic schools or military academies, well trained in physics and higher mathematics. Early ballisticians had already noticed the importance of aerodynamic drag and its dependence on projectile geometry and velocity, which is essential for precise aiming. Sir Isaac NEWTON was the first to study the problem of fluid resistance on a scientific basis. He stated that the resistance depends on three factors: the density of the fluid, the velocity, and the shape of the body in motion. In his *Principia* {Sir NEWTON ⇒1687}, he specified that the resistance of a body moving through a fluid consists of three parts: a first part which is constant, a second part which is proportional to the velocity, and a third part which is proportional to the square of the same, the latter part being the most important.[205]

At low speeds, the air behaves like an incompressible fluid. The classical theory of hydrodynamics, which involves no viscosity and is concerned only with irrotational motion (*i.e.*, motion where the vorticity is zero everywhere), predicts that a body moving steadily will experience no resistance or lift. At higher speeds, however, energy is increasingly dissipated, so that bodies moving at speeds faster than that of sound encounter considerable resistance. Up to the 18th century, the aerodynamic drag of bodies was measured by timing their free fall, mounting the body on a pendulum, or suspending the body in a flow. The English military engineer Benjamin ROBINS devised a rotating arm machine that permitted the rotation of the test object in a reproducible manner by means of a falling weight. ROBINS {⇒1746} also performed systematic aeroballistic studies at substantial velocities and first measured the supersonic velocity of a musket ball – thus giving birth to a new discipline of fluid dynamics: *supersonics*.

The British mathematician and gun expert Charles HUTTON {⇒1786} followed in ROBINS' footsteps; he first extended the ballistic pendulum technique to cannon shots, and he measured supersonic velocities as well. In 1932, the famous Hungarian-born U.S. aerodynamicist and applied

---

[203] From the *laudatio* given by Martin FOLKES, then President of the Royal Society of London, on the occasion of awarding the Copley Medal to Benjamin ROBINS. See J. WILSON: *Mathematical tracts of the late Benjamin ROBINS*. Nourse, London, vol. 1 (1761), Preface, p. xxix.

[204] L.W. LONGDON: *Terminal ballistics*. In: (R. VINCENT, ed.) *Textbook of ballistics and gunnery*. H.M.S.O., London, vol. 1 (1987), pp. 641-644.

[205] See his *Principia, Lib. II, Scholium* at the end of Sections I and III, respectively.

mathematician Theodore VON KÁRMÁN,[206] who coined the term *wave drag* for a new type of drag encountered at supersonic velocities, appropriately called these pioneering studies of early ballisticians "the theoretical-empirical preschool of supersonic aerodynamics."

In the following period, drag research proceeded along three main lines:
- measurement of the drag force as a function of the body's geometry and velocity – a difficult and cumbersome enterprise, particularly in the early days of high-speed instrumentation;
- derivation of general (mostly empirical) rules from these data for practical applications; and
- the development of a general dynamic theory of drag.

In the literature, the history of drag research has been illuminated from both a historical[207, 208, 209] and a ballistic[210] viewpoint. In his famous review article *Ballistik* ("Ballistics") written for the Encyclopedia of Mathematical Sciences (1901), the German physicist and patriarch of ballistic research Carl CRANZ[211] addressed both historic and ballistic aspects of previous drag research. In a single diagram he compared data for drag *vs.* velocity obtained from measurements on various projectile geometries {⇨Fig. 4.14−M}. He thus illustrated that drag initially increases strongly in the transonic regime, but after passing through the "sound barrier" {HILTON ⇨1935}, at which the projectile equals the velocity of sound in the surrounding air, it decreases. His diagram also nicely demonstrated the high standard of ballistic drag research at a time when high-speed drag research in aeronautics[212] was still some way off, and even the feasibility of motor flight had not yet been proven {WRIGHT Bros. ⇨1903}. In addition, it clearly showed that the transonic phase ($M = 0.7–1.3$) is of great practical importance in both ballistics and high-speed aeronautics.

## 2.4 EVOLUTION OF SHOCK WAVE PHYSICS

*A shock wave is a surface of discontinuity propagating in a gas at which density and velocity experience abrupt changes. One can imagine two types of shock waves: (positive) compression shocks which propagate in the direction where the density of the gas is a minimum, and (negative) rarefaction waves which propagate in the direction of maximum density.*[213]

Györy ZEMPLÉN
Műegyetem (Royal Josephs Polytechnic)
Budapest 1905

SHOCK WAVES are constant companions in most high-speed events and arise when matter is subjected to rapid compression – for example, by the violent expansion of the gaseous products from a high explosive or by an object moving faster than the speed of sound in the surrounding fluid. The large number of disciplines that now fall into the category of shock wave research did not evolve along a straight path to their present state. Rather, they emerged from complex interactions among shockwave-related disciplines, or independently from other branches of science {⇨Fig. 2.12}. One practical means of obtaining a useful survey of the development of shock wave physics is to classify the large number of milestone achievements and observed phenomena in terms of states of matter {⇨Fig. 2.11} – *i.e.*, shock waves in gases, liquids, solids, and plasmas. The following paragraphs will refer to the first three states of matter only.

### 2.4.1 NONLINEAR ACOUSTICS

*Gas dynamics*, a field of shock wave physics relating to the gaseous state, emerged from *nonlinear acoustics* – a particular branch of acoustics covering all such acoustical phenomena which are amplitude-dependent due to the nonlinear response of the medium in which the sound propagates; *i.e.*, phenomena which are beyond classical acoustics and can no longer be described by the infinitesimal theory.

It is quite possible that early acousticians reflected on some of the unusual phenomena associated with intense sound. Sound waves propagate linearly only when both their

---

[206] Th. VON KÁRMÁN: *Höher, schneller und heißer.* Interavia **11**, 407 (1956).
[207] O. FLACHSBART: *Geschichte der experimentellen Hydro- und Aeromechanik, insbesondere der Widerstandsforschung.* In: (L. SCHILLER, ed.) *Handbuch der Experimentalphysik.* Akad. Verlagsgesell., Leipzig, vol. IV, 2 (1932), pp. 1-61.
[208] A.R. HALL: *Ballistics in the seventeenth century.* Cambridge University Press, Cambridge (1952).
[209] R. GIACOMELLI and E. PISTOLESI: *Historical sketch.* In: (W.F. DURAND, ed.) *Aerodynamic theory.* Dover Publ., New York, vol. 1 (1963), pp. 305-394.
[210] C. CRANZ: *Äußere Ballistik.* Springer, Berlin, vol. 1 (1925), chap. 2.
[211] C. CRANZ: *Ballistik.* In: (F. KLEIN and C. MÜLLER, eds.) *Enzyklopädie der mathematischen Wissenschaften.* Teubner, Leipzig (1901–1908), vol. IV: *Mechanik.* Part 18, pp. 185-279.
[212] According to NASA-GSFC, the term *aeronautics* [French *aéronautique*, derived from the Greek words for "air" and "to sail"], being concerned with flight within the Earth's atmosphere, originated in France.

[213] This modern and concise definition of a shock wave was first given by the young Hungarian physicist G. ZEMPLÉN {⇨1905}. Visiting Germany and France on a research fellowship (1904–1906), his interest in shock waves was obviously stimulated by the mathematicians Felix KLEIN, Pierre DUHEM and Jacques HADAMARD.

amplitudes are very small *and* the times and distances over which they are observed are not too great.[214] If either of these conditions is violated, one may have to account for nonlinear effects, which result in severe waveform distortion. The first condition of large disturbance amplitudes is quite familiar to shock physicists. The second condition requires comment. Curiously, in nonlinear acoustics the steepening of initially small disturbances is a cumulative, long-duration evolutionary process that might be very small after a single wavelength but may grow into a serious distortion after the wave has propagated a distance of thousands of wavelengths. In shock wave physics, however, the waveform is confined to a single pulse − for example, in the case of a blast wave of the Friedlander type {FRIEDLÄNDER ⇒1946}, and in sonic booms of the N-wave type {DUMOND ⇒1946} − and the steepening process, favored by the larger pressure amplitudes involved, is confined to shorter distances, unless the shock wave is heavily damped by dissipation and geometrical expansion.[215]

The British professor Sir Richard SOUTHWELL,[216] while commemorating at the University of Glasgow the centennial of RANKINE's appointment to the Queen Victoria Chair of Civil Engineering and Mechanics, addressed the peculiarities of nonlinearities in acoustics, making the interesting comment that words spoken at a civilized volume will pass through a speaking tube unaltered, but they become increasingly distorted when the volume is raised. Prof. Robert T. BEYER[217] at Brown University, who discussed the early history of nonlinear acoustics and reviewed modern achievements, appropriately wrote, "Only a few decades ago, nonlinear acoustics was little more than the analysis of shock waves and large-amplitude mechanical vibrations. Gradually, however, more and more of acoustics has been examined for its nonlinear aspects until today one can write a nonlinear supplement to virtually every chapter of a text on acoustics and vibrations…"

Modern nonlinear acoustics is said to comprise the fields of
▸ aeroacoustics (the study of noise generation);
▸ finite-amplitude waves, understood by acousticians as being the branch of acoustics lying between linear acoustics and weak shock waves;
▸ shock waves;
▸ phenomena associated with the passage of intense sound beams, such as radiation pressure, streaming and cavitation; and
▸ nonlinear acoustic phenomena resulting from the presence of cracks in solids, which can be used in crack diagnostics.[218]

Obviously, for shock physicists treating gas dynamics, the close relationship to nonlinear acoustics is striking, and modern literature on acoustic research − such as papers published in the journals *JASA* (since 1929) and *Acustica* (since 1951), and in the proceedings of the *International Congress on Acoustics* (since 1953) − continues to be a rich source of information.

### 2.4.2 MAIN PERIODS OF EVOLUTION

From a historical point of view, the evolution of modern shock wave physics can be roughly divided into seven partly overlapping periods:

**From 1746 to 1808.** The birth of supersonic aeroballistics, beginning with drag studies of musket shots {ROBINS ⇒1746} and later of gun shots {HUTTON ⇒1783}. Scientific discussions are initiated regarding whether sounds of finite amplitude propagate differently to sound waves of infinitesimal amplitude {EULER ⇒1759}. The method of characteristics is developed which, much later, will prove most useful in gas dynamics and high-speed aerodynamics for solving first-order partial differential equations, particularly of the hyperbolic type {MONGE ⇒1770s}.

**From 1808 to 1869.** Measurements of the velocity of sound in air are performed using a gun; it is observed that at a large distance from the gun the command "Fire" is heard *after* the report of the gun, which proves that loud sounds must propagate supersonically {PARRY ⇒1824/1825}. A first mathematical theory for *waves of finite amplitude*, later termed *shock waves*, is developed {POISSON ⇒1808; AIRY ⇒1848 & 1849; CHALLIS ⇒1848; EARNSHAW ⇒1858 & 1860; RIEMANN ⇒1859; RANKINE ⇒1869}.

---

[214] D.G. CRIGHTON: *Propagation of finite-amplitude waves in fluids.* In: (M.J. CROCKER, ed.) *Encyclopedia of acoustics.* Wiley, New York etc., vol. 1 (1997), pp. 203-218.

[215] In the acoustic case, the pressure $p$ of a spherical and cylindrical wave decays with distance $r$ from the source according to $p \sim 1/r$ and $p \sim 1/r^{0.5}$, respectively, while for a point and a cylindrical explosion the shock wave pressure decays according to $p \sim 1/r^3$ and $p \sim 1/r^2$, respectively {TAYLOR ⇒1941; LIN ⇒1954}.

[216] R. SOUTHWELL: *W.J.M. RANKINE: a commemorative lecture delivered on 12 December, 1955, in Glasgow.* Proc. Inst. Civ. Eng. (Lond.) **5**, 178-193 (1956).

[217] R.T. BEYER: *Nonlinear acoustics.* Acoustical Society of America, Woodbury, NY (1997).

[218] A.M. SUTIN and V.E. NAZAROV: *Nonlinear acoustic methods of crack diagnostics.* Radiophys. Quant. Electr. **38**, No. 3-4, 109-120 (1995).

**From 1822 to 1893.** This is the classical era of pioneering experimental studies, ranging from more precise measurements of the velocity of sound in the free atmosphere performed using the muzzle blasts from firearms {Bureau des Longitudes ⇨1822} to the first measurements of the velocities of shock waves. First experimental evidence is provided that a shock wave does indeed propagate supersonically {REGNAULT ⇨1863}, as later confirmed by laboratory-scale experiments using electric sparks and small amounts of high explosives {E. MACH & SOMMER ⇨1877}. Schlieren observations made using a stroboscopic method reveal that a spark discharge in air is surrounded by a wave with a sharp front, the so-called "shock wave" {A. TOEPLER ⇨1864}. This period also sees various discoveries of shock wave effects that had been unknown previously in linear acoustics, such as the dependency of velocity of the shock wave on the shock strength {E. MACH & SOMMER ⇨1877}, and the discovery of *irregular reflection* {E. MACH & WOSYKA ⇨1875}, later to be called the "Mach effect" {E. MACH & WOSYKA ⇨1875}. A dimensionless parameter is proposed in order to distinguish among laminar and turbulent flow regions, later called the "Reynolds number" {REYNOLDS ⇨1883}. The predicted head wave phenomenon {DOPPLER ⇨1847} is first experimentally proved using schlieren photography {E. MACH & SALCHER ⇨1886} and shadowgraphy {BOYS ⇨1890}. The Rankine-Hugoniot shock theory is established {RANKINE ⇨1869; HUGONIOT ⇨1887}. The first photographs of a shock wave, generated by the discharge of a Leiden jar in air, are obtained using gelatin dry plates {E. & L. MACH ⇨1889}. The Mach-Zehnder interferometer is invented {L. MACH & ZEHNDER ⇨1891} and five years later first used to measure the density distribution around a supersonic bullet {L. MACH ⇨1896}. It is recognized that the sharp report accompanying the fall of a meteorite is caused by the head wave phenomenon {E. MACH & DOSS ⇨1893}.

**From 1888 to 1930.** This is an era of refined, more systematic experimental and mathematical studies on the nature of shock waves, particularly on shock waves in air. The invention of the Laval nozzle {DE LAVAL ⇨1888} tremendously stimulates the construction of more efficient steam turbines and initiates supersonic flow studies inside and outside of nozzles of various geometry. First investigations of supersonic free air jets using high-speed photography reveal a "lyre" pattern of reflected shock waves, later called "shock diamonds" {SALCHER & WHITEHEAD ⇨1889}; these studies prompt the idea of using a supersonic blow-down wind tunnel {SALCHER ⇨1889}. It is recognized that a supersonic flow passing around a sharp corner expands through a "fan" of Mach lines centered at the corner, later called the "Meyer-Prandtl expansion fan" {MEYER & PRANDTL ⇨1908}. It is first demonstrated theoretically that a shock wave may propagate as a condensation or as an expansion {DUHEM ⇨1909}. A dimensionless parameter is introduced that characterizes a fluid property in the regime of convection, later called the "Prandtl number" {PRANDTL ⇨1910}. In the late 1920s, investigations originating from the practical needs of aeroballistics (minimization of wave drag), aeronautics (high-speed propellers) and steam turbine development (optimal Laval nozzle geometry) eventually lead to the establishment of *gas dynamics*, a new branch of fluid dynamics. The first supersonic wind tunnel is set in operation {STANTON ⇨1921}. A dimensionless parameter characterizing the flow velocity with respect to the sound velocity of the surrounding (quiescent) medium is introduced; this is later called the "Mach number" {ACKERET ⇨1928}. A nonstationary cosmological model is also first proposed in this period {FRIEDMANN ⇨1922; LEMAÎTRE ⇨1927}, which later resulted in the famous *Big Bang* theory on the origin of the Universe.

**From 1930 to 1939.** This period is the early era of high-speed aviation and supersonic wind tunnel testing, which reaches its first culmination at the 5th International Volta Congress {Rome ⇨1935}, where numerous researchers from prestigious research institutes around the world present their pioneering contributions on this subject. In the following years, a number of unique transonic and supersonic wind tunnels will be set in operation {ACKERET ⇨1933; BUSEMANN & WALCHNER ⇨1933; WIESELSBERGER ⇨1934; HVA Peenemünde ⇨1939}.

**From 1939 to 1949.** This era is characterized by an enormous growth in shock wave research (when then also included the solid state) – research of the highest priority that is almost exclusively initiated to satisfy military needs. Large groups or teams of research workers are formed to concentrate effort on individual problems; such team research will become a common feature of postwar science. Though relatively brief and suffering from a shortage of trained personal, time and money, this era revives interest in the shock tube technique {BLEAKNEY ⇨1946}, air blast {G.I. TAYLOR ⇨1941; KENNEDY ⇨1946; SLADE JR. ⇨1946; Helgoland Blast ⇨1946; Texas City Explosion Disaster ⇨1947}, sonic booms {DUMOND ⇨1946}, underwater explosions {DTMB & Kriegsmarine ⇨1941; KIRKWOOD & BETHE ⇨1942; Operation CROSSROAD ⇨1946}, and shock wave interaction phenomena such as the Mach effect {VON NEUMANN ⇨1943; SPITZER & PRICE ⇨1943; WOOD ⇨1943; CHARTERS ⇨1943; L.G. SMITH ⇨1945}. The first hypersonic flow studies performed close to Mach 9 are carried out in a modified supersonic wind tunnel {ERDMANN ⇨1943/1944}.

This era is also characterized by the design and testing of atomic bombs {Trinity Test ⇨1945; Semipalatinsk-21 ⇨1949} and their first military use to destroy whole cities {Hiroshima & Nagasaki Bombing ⇨1945}. In Germany, the United States and England, the new types of engines for use in aircraft, missiles and rockets are developed, such as the pulsejet {P. SCHMIDT ⇨1930}, transonic aircraft {WARSITZ (Heinkel 176) ⇨1939; Messerschmitt AG (Me 163 ⇨1941 & Me 262 ⇨1942); GILKE (Lockheed P-38) ⇨1941 & 1945}, anti-aircraft supersonic missiles {HVA Peenemünde ⇨1939} and large supersonic rockets {HVA Peenemünde ⇨1942}, two-stage rocket {1949, ⇨Fig. 4.20-D}, and supersonic aircraft {YEAGER (Bell XS-1) ⇨1947}.

Shock wave physics proves to be an indispensable tool in nuclear weapons technology, thus initiating modern solid-state shock wave physics {GORANSON ⇨1944} and detonation physics {ZEL'DOVICH ⇨1940; DÖRING (Secret Workshop "Probleme der Detonation") ⇨1941; KISTIAKOWSKY ⇨1941; VON NEUMANN ⇨1942; DÖRING ⇨1943; JOHNSTON ⇨1944; LANDAU & STANYUKOVICH ⇨1945}. This, in turn, also stimulates significant advances in submicrosecond high-speed diagnostics {REYNOLDS ⇨1943; "pin method" ⇨1944; LIBESSART ⇨1944}, and ultrahigh-speed photography {STEENBECK ⇨1938; Trinity Test (BRIXNER) ⇨1945; C.D. MILLER ⇨1946}. Last but not least, various electronic computing machines are developed in the U.S.A., U.K. and Germany in the 1940s in order (amongst other reasons) to solve the *hydrodynamic equations* (or *shock wave equations*) used in nuclear bomb physics (*see* Sect. 2.9.2).

**From 1950 to the Present.** The postwar era, the most complex one, is characterized by the establishment of a large number of governmentally and privately operated research institutes. Shock wave research becomes intimately connected with industry, defense, and politics, thus rendering the classical ideal of pure science obsolete. Almost all research dedicated to all branches of shock and detonation physics is now done by highly trained experts, employed wholly or mainly for this work within such special institutions. Due to competition, there is also a general tendency for research workers to become very specialized. On the other hand, independent individual research and invention – still one of the main pillars of progress in science and technology in the 19th century – has degraded to a curiosity.

The tremendous progress made in shock wave physics, which has already provided an enormous body of worthwhile fundamental theoretical and practical knowledge, further promotes interdisciplinary research in a unique manner. It stimulates new branches within the classical sciences such as in astronomy and astrophysics, cosmology and cosmogony, geology and geophysics, and even in medicine and biology.

This era sees many spectacular milestone achievements in practical aeronautics and astronautics built upon the foundations laid in previous eras. The spectacular first hypersonic manned space flight {GAGARIN ⇨1961} demonstrates that reentry at hypersonic speed can be successfully controlled. In the same year, the U.S. hypersonic research aircraft makes its first successful hypersonic flight {X-15 ⇨1961}. With the development of supersonic airliners {Tupolev Tu-144 ⇨1968; Concorde ⇨1969}, the experience of flying supersonically becomes routine: regular civil supersonic flights from British Airways and Air France begin in 1976 but end in 2003 because of the age of the aircraft, possibly terminating the spectacular era of supersonic civil transportation forever.

In the 1950s, cosmical gas dynamics develops into an important branch of astrophysics. Numerous space research programs as well as military interests stimulate studies in hypersonic aerodynamics, reentry and hypervelocity impact. In the 1960s – perhaps the Golden Age of funding – research in shock physics enters new dimensions, reaching its culmination with the first manned flight to the Moon {Apollo 11 ⇨1969}. The discovery of shock polymorphism in meteorite craters {CHAO, SHOEMAKER & MADSEN ⇨1960}, which advances our understanding of the Earth's past, also initiates numerous astrogeological research programs. The invention of the pulsed laser {MAIMAN ⇨1960} enables researchers to deposit enormous power densities in very small volumes, thus allowing the generation of very strong shock waves at microscopic dimensions. On the other hand, the discovery of the Earth's bow shock {AXFORD & KELLOGG ⇨1962; ⇨Fig. 4.12-D} first illustrates the enormous dimensions shock waves can assume in nature. This discovery also stimulates interest in the generation of the solar wind and in the physics of the corona of the Sun, our closest star. Furthermore, it prompts the question of whether other planets of the Solar System are also surrounded by bow shocks. The question of the existence of a *heliospheric termination shock* is the subject of present investigations {Voyager 1 & 2 ⇨2003}. The discoveries of quasars {MATTHEWS & SANDAGE ⇨1960; M. SCHMIDT ⇨1963} and pulsars {BELL & HEWISH ⇨1967} using optical, radio and X-ray astronomy initiate discussions on the role of shock phenomena of galactic dimensions, such as the origin of cosmic rays and possible mechanisms of their production in supernovae by high Mach number shock waves {BAADE & ZWICKY ⇨1934; GINZBURG & SYROVATSKII ⇨1961; ISEE Program ⇨1977; HESS ⇨2004}.

**Documentation and Dissemination.** Over the past five decades, shock wave physics and detonation physics have grown into huge fields of their own: new results in these fields are published not only in a large number of textbooks, technical reports and patents, but also in an increasing number of specific journals addressing almost every offshoot of these disciplines. For example, journals covering nonlinear acoustics, high-speed aerodynamics, gas dynamics, supersonic combustion, detonation, high-velocity impact, *etc.*, are listed in the CHRONOLOGY and include the following:
- *Journal of the Acoustical Society of America (JASA)* {⇨1929};
- *Combustion and Flame* {⇨1954};
- *Astronautica acta* {⇨1955};
- *AIAA Journal* {⇨1963};
- *Combustion, Explosion, and Shock Waves* {⇨1965};
- *Shock and Vibration Digest* {⇨1969};
- *Propellants, Explosives, Pyrotechnics* {⇨1976};
- *International Journal of Impact Engineering* {⇨1983};
- *Experiments in Fluids* {⇨1983};
- *Physics of Fluids A* {⇨1989}; and
- *Shock Waves* {⇨1990}.

In addition, a considerable number of conferences dedicated to shock wave physics and shock effects have been organized, *e.g.*, the
- Symposia on Shock & Vibration {1947};
- Symposia on Cosmical Gas Dynamics {⇨1949};
- Biennial Gas Dynamics Symposium {VON KÁRMÁN ⇨1955};
- Int. Symposia on Shock Tubes {⇨1957};
- Int. Symposia on Rarefied Gas Dynamics {⇨1958};
- Meetings of the Aeroballistic Range Association, ARA {⇨1961};
- Symposia Transsonica {⇨1962};
- AIRAPT International High Pressure Conferences {⇨1965};
- Conferences on Shock Metamorphism of Natural Materials {⇨1966};
- Int. Symposia on Military Applications of Blast Simulation, MABS {⇨1967};
- Int. Colloquia on Gas Dynamics of Explosions {⇨1967};
- Oxford Conferences on Mechanical Properties of Materials at High Rates of Strain {⇨1974};
- APS Conferences on Shock Waves in Condensed Matter {⇨1979};
- Int. Mach Reflection Symposia {⇨1981};
- Int. Conferences on Mechanical and Physical Behavior of Materials under Dynamic Loading {DYMAT ⇨1985; EURODYMAT 1994};
- High Velocity Impact Symposia, HVIS {⇨1986}; and
- crashMat Conferences {⇨2001}.

Other conferences, particularly those devoted to high-speed photography and diagnostics, traditionally serve as an important vehicle for exchanging new ideas and results among researchers working worldwide in shock physics, detonics and supersonic combustion research, such as the International Congresses on High-Speed Photography (& Photonics) {⇨1952} and the International Symposia on Flow Visualization {⇨1977}.

## 2.4.3 AERIAL WAVES OF FINITE AMPLITUDE: A CHALLENGE FOR MATHEMATICIANS

*The particles on the crest are themselves moving in the direction of the wave motion, and with a velocity which becomes greater and greater (for the particles which happen to be on the crest) as the wave approaches the shore. It is evident that the limit to these circumstances is, that the front of the* wave becomes as steep as a wall, while the uppermost particles are moving towards the shore and the lowermost from the shore; that the former, therefore, will tumble over the latter; and this is the motion of a surf.[219]

Sir George B. AIRY
University of Cambridge
Cambridge 1845

Surprisingly, the discontinuity problem closely associated with a shock wave was successfully tackled by neither experimentalists nor philosophers, but rather by mathematical physicists. The French engineer and mathematician Emile JOUGUET[220] wrote, "The shock wave represents a phenomenon of rare peculiarity such that it has been discovered by the pen of mathematicians, first by RIEMANN, then by HUGONIOT. The experiments did not happen until afterwards." RIEMANN {⇨1859} and HUGONIOT {⇨1887}, how-

---
[219] G.B. AIRY: *Tides and waves.* In: (E. SMEDLEY, Hugh J. ROSE, and Henry J. ROSE, eds.) *Encyclopaedia Metropolitana.* B. Fellowes, Oxford & Cambridge, vol. 5 (1845), §298, p. 301, and §249, p. 314.
[220] E. JOUGUET: *Résumé des théories sur la propagation des explosions.* La Science Aérienne **3**, No. 2, 138-155 (1934).

ever, were not the only pioneers. As shown in Fig. 2.15 and in the CHRONOLOGY, they had a surprisingly large number of predecessors who substantially contributed to this new field, thus paving the way for a gradual increase in our understanding of discontinuous wave propagation.

**Water Waves.** Wave phenomena on the surface of water – ranging from waves of small amplitude (*ripples*) to those of large amplitude (*tidal bores, tsunamis*) – have fascinated man since the earliest times. The unsteady nature of a tidal bore, with its well-defined front, is a spectacular event that occurs periodically in many estuaries of large rivers around the world. It is certainly the most illustrative demonstration of a propagating discontinuity, considering its enormous kinetic energy and huge destructive power, and that it advances with a roaring sound several times faster than walking speed and moves up the river for many miles.

The first steps toward the development of a theory of tides and waves can be ascribed to Sir Isaac NEWTON's theory[221] of the equilibrium tide, in which he investigated the forces that raise tides, and to Daniel BERNOULLI's theory[222] of ocean tides (1740). These studies were considerably extended by numerous French contributions, such as from Pierre-Simon DE LAPLACE[223] (1790), Joseph L. DE LAGRANGE[224] (1811–1815), Augustin L. CAUCHY[225] (1815), and Siméon D. POISSON[226] (1815–1816). Originally, the mathematicians CAUCHY and POISSON were dissatisfied with the fact that their predecessors had only dealt with the problem of preformed waves. Independently, they discussed the problem of the "generation of a wave," produced when the surface of deep water is disturbed momentarily. The solution of their theory is the so-called "Cauchy-Poisson wave" {CAUCHY ⇒1815; POISSON ⇒1815}, a wave in which the wavelength increases with the distance from the disturbance. Wave gauge records obtained in 1952 from tsunamis originating from a series of violent seaquakes in Japan showed close agreement with the theory of the Cauchy-Poisson wave, thus confirming that this wave type actually does exist in nature {UNOKI & NAKANO ⇒1952}.

AIRY, CHALLIS, STOKES, Lord RAYLEIGH, JOUGUET, and LAMB – to mention only a few of the early contributors to hydrodynamics and shock wave physics – started from investigations of water wave phenomena and then turned to shock waves. Obviously, for mathematicians the analogy between the equations of the shallow water theory and the fundamental differential equations describing a one-dimensional compressible flow in air was particularly striking. Since discontinuous wave propagation of water waves – contrary to shock waves in gases – can easily be followed by the naked eye, even on a laboratory scale, no expensive high-speed cameras and special high-intensity short-duration light sources were necessary. In so-called "water table experiments," the benefits of this technique were studied in great detail {PREISWERK ⇒1938; H.A. EINSTEIN ⇒1946; CROSSLEY JR. 1949}.

However, these classical studies also revealed some limitations of this method. The approximate theory of waves of finite amplitude propagating in shallow water is not a linear one, because the wave crests are higher above the mean water line than the troughs are below the mean water line. James J. STOKER,[227] a U.S. applied mathematician at the Courant Institute in New York who specialized in using mathematical analysis to determine water flow and flood waves of rivers and large reservoirs, wrote in his textbook *Water Waves*: "The theory is often attributed to STOKES [1845, ⇒1849] and AIRY [⇒1845], but was really known to DE LAGRANGE [⇒1781]. If linearized by making the additional assumption that the wave amplitudes are small, the theory becomes the same as that employed as the mathematical basis for the theory of the tides in the oceans. In the lowest order of approximation, the nonlinear shallow water theory results in a system of hyperbolic partial differential equations, which in important special cases can be treated in a most illuminating way with the aid of the method of characteristics."

STOKER attributed this analogy to Dmitri RIABOUCHINSKY {⇒1932}, a Soviet aerodynamicist renowned for constructing Russia's first important wind tunnel in 1904. But prior to him, the French engineer and mathematician Emile JOUGUET

---

[221] See Sir Isaac NEWTON {⇒1687}; *Principia*, Book I, Prop. 66, and Book III, Prop. 24.

[222] In the year 1740, the *Académie Royale des Sciences de Paris* asked the following question "Quelle est la cause du flux et du reflux de la mer?" The prize was awarded for four memoirs, submitted by R.P.A. CAVALLERI, D. BERNOULLI, L. EULER, and C. MACLAURIN. BERNOULLI's contribution [see *Prix* 1740, pp. 55-191] is the longest in the volume, and after his *Hydrodynamica* it is the longest of all his works. The problem of tides must have occupied him for a long time.

[223] P.S. DE LAPLACE: *Mémoire sur le flux et le reflux de la mer*. Mém. Acad. Roy. Sci. (1790–1797), pp. 45-181.

[224] J.L. DE LAGRANGE: *Mécanique analytique*. Courcier, Paris (1811–1815), vol. II.

[225] L. CAUCHY: *Mémoire sur la théorie de la propagation des ondes à la surface d'un fluide pesant d'une profondeur indéfinie* [1815]. Mém. Sav. Étrang. I, 3-312 (1827). • In 1816, CAUCHY won a contest held by the French Academy on the propagation of waves at the surface of a liquid. His solution became a milestone in the evolution of hydrodynamics.

[226] D.S. POISSON: *Mémoire sur la théorie des ondes* [1815]. Mém. Acad. Sci. Inst. France I, 71-186 (1816).

[227] See STOKER {⇒1957}, pp. x-xi, 25-32.

{⇨1920}, then one of the leading shock wave pioneers and today better known as one of the intellectual fathers of the first detonation theory, had already discussed the similarity between shooting channel flow and supersonic compressible flow. On the other hand, JOUGUET was stimulated by his countryman Jean Baptiste BÉLANGER {⇨1828}, a professor of mechanical engineering who first theoretically studied the propagation of hydraulic jumps in open water channels. In the 1840s, the English mathematician and astronomer George B. AIRY was the first to work out a theory of river tides {AIRY ⇨1845}. He also first predicted that the propagation velocity of a tidal wave is amplitude-dependent, and concluded that "a wave of this type cannot be propagated entirely without change of profile, since the speed varies with height" {AIRY ⇨1845}.

**Approach to Shock Waves.** The theoretical approach to treating shock waves can be traced back as far as to the *Principia* {Sir NEWTON ⇨1687}, in which sound propagation in a fluid is explained by the transport of impulses between individual particles – an illustrative model which stimulated not only the genesis of thermodynamics, but was also resumed by some early shock pioneers and even by modern researchers {⇨Fig. 4.4–A}. Assuming, erroneously, that sound is an isothermal process, Sir Isaac NEWTON made a crude calculation of the sound velocity in air. The great French mathematician, astronomer, and physicist Pierre-Simon DE LAPLACE {⇨1816}, noticing a discrepancy of almost 20% between NEWTON's theoretical result and preexisting measured data {MERSENNE ⇨1636 & 1644}, improved the theory by assuming that sound is an *adiabatic* process – a term coined by RANKINE {⇨1859}. Prior to this, POISSON {⇨1808}, stimulated into working on this subject by DE LAPLACE, had mathematically tackled the sound velocity problem in a paper published in the Journal de l'Ecole Polytechnique (Paris). Under the heading *Mouvement d'une ligne d'air dans le cas où les vitesses des molécules ne sont pas supposées très-petites* ("One-dimensional Movement of Air in the Case that the Velocities of the Molecules are No Longer Very Small"), he also approached the basic question of how to solve the wave equation in the case of finite amplitude, thus laying the foundations for the first shock wave theory {⇨Fig. 2.17}. It is worth noting that this happened at a time when an experimental verification of such discontinuities, propagating invisibly through the air as a wave of condensation, was still pending.

POISSON's early approach, first resumed by the English astronomer James CHALLIS {⇨1848}, was extended in the following decades by numerous researchers in England, France and Germany {AIRY ⇨1848; STOKES ⇨1848 & 1849; RANKINE ⇨1858 & 1870; EARNSHAW ⇨1858, 1859 & 1860; RIEMANN ⇨1859; CHRISTOFFEL ⇨1877; HUGONIOT ⇨1885–1887; TUMLIRZ ⇨1887; BURTON ⇨1893; HADAMARD ⇨1903; WEBER ⇨1901; DUHEM 1901–1909; JOUGUET 1901–1910; LUMMER ⇨1905; ZEMPLÉN ⇨1905; Lord RAYLEIGH ⇨1910; G.I. TAYLOR ⇨1910; *etc.*}. However, the transition to present-day shock wave theory, largely a result of many more advanced contributions made by international researchers, was not straightforward. We often take this knowledge for granted, ignoring the partially impetuous disputes[228] and cumbersome struggles that it took to achieve our current understanding of the shock wave riddle. Details of this gradual process of understanding may be found in the CHRONOLOGY.

**Motivations.** In this context, some remarks concerning the motivation for tackling the problem of shock waves seem to be worth mentioning here:

▸ AIRY, CHALLIS, and JOUGUET first studied tidal bores out of mathematical curiosity.

▸ In 1834, John S. RUSSELL {⇨1834}, an English naval architect best known for his research into ship design, chanced upon the *great solitary wave* (or *wave of translation*) – a wave phenomenon consisting of a single crest that arises when a single negative bore immediately follows an equal positive bore. In a wave of translation, the water particles advance with the wave and do not return to their original position.

▸ RUSSELL was also the first to observe a curious wave reflection phenomenon {E. MACH & WOSYKA ⇨1875; ⇨Fig. 4.13–A} which was later termed the *Mach effect* (or *Mach reflection effect*) {VON NEUMANN ⇨1943}.

▸ In 1845, the Rev. Samuel EARNSHAW, an English mathematician, began to treat bores in more depth theoretically. A few years later, after he had a crucial experience with thunder {EARNSHAW ⇨1851}, he turned his attentions to this subject, before finally investigating airborne waves of finite amplitude.

---

[228] For example, *see* the disputes between EULER and DE LAGRANGE {EULER ⇨1759}, PARRY and GALBRAITH {PARRY ⇨1824/1825}, CHALLIS and STOKES {STOKES ⇨1848}, CHALLIS and AIRY {CHALLIS ⇨1848}, EARNSHAW and LE CONTE {LE CONTE ⇨1864}, and Lord RAYLEIGH and STOKES {Lord RAYLEIGH ⇨1877}.

- Bernhard RIEMANN, the famous German mathematician, took a keen interest in the nature of *Luftwellen von endlicher Schwingungsweite* ("Aerial Waves of Finite Amplitude"). This interest did not arise from pure mathematical curiosity, but rather was stimulated by the German physicist Hermann VON HELMHOLTZ. In 1856, while studying combination tones, VON HELMHOLTZ had discovered the unusual phenomenon that the sounding of two musical tones of high intensity results in the appearance of a sum frequency as well as a difference frequency. VON HELMHOLTZ attributed the presence of these combination tones to the nonlinearity of the ear. However, the British physics professor Arthur W. RÜCKER[229] and his collaborator Edwin EDSER were the first to prove that the interaction actually occurs in the medium, by exciting a tuning fork at the sum frequency.
- William J.M. RANKINE in England and Pierre-Henri HUGONIOT in France approached the shock wave phenomenon from a thermodynamics viewpoint.
- The British physicist Lord RAYLEIGH, who wrote the first textbook on acoustics, became interested in aerial shock waves due to his curiosity about the physical conditions at the shock front.
- The German mathematician Heinrich M. WEBER treated shock waves as a mathematical problem in order to find solutions for various types of partial differential equations. He later edited RIEMANN's lectures on mathematical physics and extended his own theoretical studies of shock waves through numerous examples and comments.

**Superposition of Shock Waves.** From a mathematical physics perspective, shock waves are nonlinear problems because the equations of state, $p(\rho)$, are usually nonlinear. For all normal fluids, the equation of state produces a plot that curves concave upwards in the $(p,\rho)$-plane, so disturbances in pressure $p > p_0$ and consequently in density $\rho > \rho_0$ propagate with supersonic velocity because of $c = (dp/d\rho)^{1/2} > c_0$ {⇨Fig. 2.1}. Here $p_0$ and $\rho_0$ are the pressure and density of the undisturbed fluid, respectively.

The German theoretical physicist Werner HEISENBERG[230] also showed a recurrent interest in nonlinear problems related to fluid dynamics and shock waves. In two papers – one published in 1924 on vortex motion entitled *Nichtlineare Lösungen der Differentialgleichungen für reibende Flüssigkeiten* ("Nonlinear Solutions of Differential Equations for Frictional Liquids"), and the other published in 1953 on meson production entitled *Theorie der Explosionsschauer* ("Theory of Explosion Showers"), in which he treated mesons (a group of subatomic particles) as a shock wave problem – he pointed out the general difficulties of obtaining physically meaningful solutions for nonlinear equations. Obviously, in the case of shock waves the behavior of the substance at very small dimensions becomes important. Fortunately, as he pointed out, in gas dynamics it is often not necessary to go into the microscopic details: in order to determine the course of the shock wave, it is instead enough to describe the irreversible processes in the shock through an increase in entropy.

The superposition of shock waves is itself another nonlinear phenomenon. This "nonlinear interaction" of "nonlinear processes" in the shock wave is, therefore, a very complex phenomenon. Consequently, the resulting shock pressure in the region of interaction is not the arithmetic sum of the two incident pressures (as in the acoustic case): it is always larger, and depends on the shock strength and the angle at which the two shock fronts interact with each other. In the case of the *Mach effect* {E. MACH & WOSYKA ⇨1875}, a third shock wave – the *Mach shock* – is generated besides the *incident shock* and the *reflected shock*. Depending on the interaction geometry, this Mach shock also assumes different names {⇨Fig. 2.14}: an oblique reflection of a (planar or curved) shock front at a solid boundary produces the so-called "Mach stem," while an oblique interaction of two propagating shock waves of equal strengths and geometries (either planar or curved) produces a symmetric phenomenon – termed a *Mach disk* (or *Mach bridge*).

Since the establishment of the four basic types of oblique shock wave reflection {WHITE ⇨1951}, a number of special cases of single and double Mach reflection have been observed and named {⇨Fig. 2.14}. In 1985, a common nomenclature was proposed in order to avoid confusion and to improve communications between investigators[231] – a difficult goal which has so far been only partially realized.

---

[229] See their paper *On the objective reality of combination tones.* Proc. Phys. Soc. (Lond.) **13**, 412-430 (1895).

[230] W. HEISENBERG: *Nonlinear problems in physics.* Phys. Today **20**, 27-33 (May 1967).

[231] G. BEN-DOR and J.M. DEWEY: *The Mach reflection phenomenon: a suggestion for an international nomenclature.* AIAA J. **23**, 1650-1652 (1985).

## 2.4.4 SHOCK WAVES IN GASES: FIRST EXPERIMENTAL PROOFS OF THEIR EXISTENCE

> *It seems as if the locus of interference between the two original [shock] waves has become the source of a third wave which, during propagation, interferes with the two original ones. This V-propagation can be explained as an area of superposition of the two waves which is attributed to a larger propagation velocity – this locus can be understood as a wave in the wave; i.e., as a second wave which propagates with an excessive velocity within the first wave. This interpretation fully agrees with the quantitative results of my third and fourth work.*[232]
> 
> Ernst MACH
> Karl-Ferdinand-Universität
> Prague 1878

A gas is a collection of molecules separated by distances so large that most of the time the molecules interact only weakly with each other. The brief periods during which the molecules interact strongly are considered to be *collisions*.

Shock waves can be regarded in one sense as the modern version of percussion. However, compared to classical percussion dealing with tangible bodies, shock transmission occurs on a microscopic level in a gas: the momentum is transmitted via innumerous individual collision processes from one molecule (or atom) to another. For example, in a monatomic gas it takes only a few collisions to adjust to the different equilibrium states between the upstream and downstream sides of a shock wave – i.e., a shock front is only a few mean free paths thick. The unique behavior of a traveling shock can best be illustrated to the novice using the various chain percussion models that have been proposed in the literature. Some of them are shown in Fig. 4.4–D.

The rapid development of experimental shock wave physics started with studies of gases usually performed for the following reasons:

▸ In the 17th century, the elastic nature of air – popularized by Robert BOYLE as "the spring of the air" – was studied experimentally and was also used in practice. Prominent examples include various constructions of the so-called "air gun" or "wind gun" {GUTER ⇨1430; BOYLE ⇨1647; ⇨Fig. 4.2−I} in which the force employed to propel the bullet is the elasticity of compressed atmospheric air, and pneumatic lighters {⇨Fig. 4.2−K}, which impressively demonstrate the adiabatic properties of quickly-compressed air. The first step towards a scientific theory of shock waves was the determination of the isothermal equation of state of a gas {BOYLE & TOWNLEY ⇨1660}.

▸ The relatively low velocity of sound in air in comparison to that in a liquid or solid (it is, for example, smaller by factors of about 5 and 20 in water and iron, respectively) was advantageous for 19th-century experimentalists, when high-speed diagnostics were still in their infancy, and the smallest time resolution achievable was mostly limited to the medium microsecond regime.

▸ All three optical standard methods – schlieren, shadowgraphy and interferometry – are light transmission techniques; *i.e.* they require a transparent medium and are therefore ideally suited for studies in gases.

▸ In practice, the majority of shock wave applications still occur in air.

**The Roots of Gas Dynamics.** Shock waves in gases fall under the remit of the wide field of *gas dynamics* or *compressible flow*. The first memoir on this subject was apparently published by the Polish physicist Marian SMOLUCHOWSKI {⇨1903}. He termed this branch of fluid dynamics *aerodynamics*. In the same year, the Hungarian-Swiss engineer Aurel STODOLA {⇨1903} first treated the thermodynamics of high-speed flows in his classical textbook *Die Dampfturbinen und die Aussichten der Wärmekraftmaschinen* ("Steam Turbines, With an Appendix on Gas Turbines and the Future of Heat Engines"). The first handbook articles on gas dynamics, which also emphasized shock wave phenomena, were published by Felix AUERBACH (1908), Ludwig PRANDTL (1913), Jakob ACKERET (1927), Adolf BUSEMANN (1931), and Albert BETZ (1931) {AUERBACH ⇨1908}. In the following years, important contributions to the fundamental treatment of gas dynamics were made in England by Geoffrey I. TAYLOR[233] and John W. MACCOLL, and in Germany by Robert SAUER.[234]

The roots of gas dynamics can be traced back to early attempts to calculate the velocity of sound in air {Sir NEWTON ⇨1687; BIOT ⇨1802; DE LAPLACE ⇨1816; Bureau des Longitudes ⇨1822; POISSON ⇨1823}. It was recognized early on that a long baseline is needed to measure the sound velocity in order to compensate for the limited accuracy of the clocks available at the time {MERSENNE ⇨1636; CASSINI

---

[232] This is Ernst MACH's ingenious interpretation of "Mach reflection." See his paper *Über den Verlauf der Funkenwellen in der Ebene und im Raum.* Sitzungsber. Akad. Wiss. Wien **77** [Abth. II], p. 819 (1878).

[233] G.I. TAYLOR and J.W. MACCOLL: *The mechanics of compressible fluids.* In: (W.F. DURAND, ed.) *Aerodynamic theory.* Springer, Berlin, vol. 3 (1935), chap. H, pp. 209-250.

[234] R. SAUER: *Theoretische Einführung in die Gasdynamik.* Springer, Berlin (1943).

DE THURY ET AL. ⇨1738}. This method, however, was not directly transferable to the crucial test of whether waves of intense sound propagate faster than the velocity of sound: since the pressure rapidly decreases with the distance from the source, the region of supersonic velocity is limited to the near-field of the explosion source.

**First Studies of Intense Air Waves.** The first attempt to mathematically predict the velocity of sound {Sir NEWTON ⇨1687} had shown a considerable discrepancy between theory and measurement, and subsequent generations of naturalists attempted to develop a general mathematical theory of sound that was capable of correctly predicting the velocity of sound as a function of the density, temperature and humidity of the air, as well as the pitch and intensity of the sound.

Experiments made by Jean-Baptiste BIOT {⇨1809} in France and William Henry BESANT {⇨1859} in England provided evidence that whatever the pitch and loudness, all musical sounds are transmitted with precisely the same velocity. Some researchers, extrapolating these results without hesitation to violent sounds in general, denied any effects of sound intensity on the velocity of sound at all {LE CONTE ⇨1864}. On the other hand, there was also experimental evidence that the velocity of sound does indeed depend on intensity. In order to eliminate all possible influences of humidity on sound propagation, the English arctic explorers William E. PARRY, Henry FOSTER and James C. ROSS {⇨1824/1825} performed measurements of the velocity of sound at very low temperatures in the North Polar region; *i.e.*, in perfectly dry air. They observed by chance the curious phenomenon that the report of a gun was heard at their furthest station *before* the command to fire {⇨Fig. 2.10–D}. This suggested that intense air waves (here the muzzle blast) travel more quickly than weak air waves (here the commander's voice). The Rev. Samuel EARNSHAW {⇨1858 & 1864} considered their observation to be important proof of his mathematical theory of sound.

Evidence that the velocity of sound depends on its intensity was also observed in France. The skillful experimentalist Henri REGNAULT {⇨1863}, then widely known for his sophisticated test methods and careful measurements, originally intended to measure sound velocities in various gases and liquids. In order to get a long baseline, he performed his experiments in the public sewage channels and gas pipelines of Paris, which advantageously confined the sound to two dimensions. To secure sufficient sound intensity at the receiver station, he generated the sound at the tube entrance with small amounts of explosives {⇨Fig. 2.10–F}. He was initially unaware that in doing so he introduced shock (blast) waves rather than sound waves into his measurement method. His remarkable results, published in various international journals but now almost forgotten, obviously quantitatively proved the existence of supersonic velocities for the first time and certainly must have encouraged contemporaries from other countries to tackle this subject further.

**Mach Reflection.** Ernst MACH, who in the period 1867–1895 held the chair of experimental physics at the German *Karl-Ferdinand-Universität* ("Charles University") in Prague, was interested in physical and physiological acoustics. He was supported by a team of coworkers, which later also included his oldest son Ludwig MACH, and he had the opportunity to systematically continue his research in this particular field through a period of almost 28 years, which was most unusual in the research scene of the 19th century. Curiously enough, E. MACH began his gas dynamic studies with research into one of the most difficult subjects in shock wave physics, the oblique interaction of shock waves {E. MACH & WOSYKA ⇨1875; ⇨Figs. 4.5–D & 4.13–C}. This subject presented a particularly difficult challenge to researchers during the evolution of gas dynamics. Later termed the *Mach effect* by the Hungarian-born U.S. mathematician John VON NEUMANN {⇨1943}, this interaction is a complex nonlinear superposition phenomenon and still remains a fascinating subject for thorough experimental and theoretical research {Mach Reflection Symposium ⇨1981}.

The Mach effect is characterized by a triple point at which three branches of shock waves intersect: the incident shock wave, the reflected shock wave, and a third shock wave – the *Mach stem* – behind which the pressure is larger than that behind the incident and reflected shock wave {⇨Fig. 2.14}. This curious so-called "three-shock problem" {VON NEUMANN ⇨1943} was first studied systematically in shock tube experiments at Princeton University under the guidance of the physics professor Walker BLEAKNEY {BLEAKNEY & TAUB ⇨1949}. When he applied stronger incident shock waves, Donald R. WHITE {⇨1951}, one of BLEAKNEY's coworkers, observed a new, very complex shock interaction geometry with two triple points {⇨Figs. 2.14(d) & 4.13–D}. Later, this type of shock wave interaction was termed *double Mach reflection* in order to differentiate it from the more common one-triple-point geometry, which was termed *single Mach reflection*.

Richard COURANT and Kurt O. FRIEDRICHS, when discussing the reflection of shock waves in their book *Supersonic Flow and Shock Waves* {⇨1948}, only differentiated between three types of Mach reflection (MR): (i) *direct MR* when the triple point (TP) moves away from the reflecting surface; (ii) *stationary MR* when TP moves parallel to the reflecting surface; and (iii) *inverted MR* when TP moves towards the reflecting surface. They also proposed the term *λ-configuration* instead of *Mach reflection*, which, however, was not adopted by others.

Particularly in gas dynamics, Mach reflection is an almost omnipresent phenomenon and on that can manifest itself in a number of different wave configurations. This created a wealth of new special terms, some of which were placed in a more recent classification scheme. A brief historical review of this particular branch of shock wave physics was given by Gabi BEN-DOR,[235] a shock physics professor at Ben-Gurion University of the Negev.

The Mach effect exists in all states of matter. In 1842, the Scottish naval engineer John S. RUSSELL {⇨1842} studied the reflection of hydraulic jumps at a solid boundary, during which he observed incidentally Mach reflection[236] (*i.e.*, before even Ernst MACH in 1875); this achievement is barely known in the shock physics community. RUSSELL documented his observation using a pen-and-ink drawing {⇨Fig. 4.13–A} because high-speed photography {TALBOT ⇨1851; MADDOX ⇨1871} had not yet been established. Later, this unusual wave interaction phenomenon was also photographed in nature by the Englishman Vaughan CORNISH {⇨1910} and published in his book *Waves of the Sea and Other Water Waves* {⇨Fig. 4.1–M}.

After the discovery of Mach reflection in air {E. MACH & WOSYKA ⇨1875}, there was speculation about whether this effect might also exist in other states of matter. In shock wave studies performed in the United States during World War II, it was shown that Mach reflection also exists in a liquid (water) {SPITZER & PRICE ⇨1943}. Later, it was shown in the Soviet Union that Mach reflection also occurs when detonation waves generated in a triangular prism constructed from a solid high explosive interact obliquely {E.A. FEOKTISTOVA ⇨1960}. Soviet physicists also first experimentally demonstrated the existence of Mach reflection in solids such as aluminum and iron {AL'TSHULER ET AL. ⇨1962}.

**First Laboratory-Scale Supersonic Studies.** Although supersonic aerodynamics was not a completely new branch of fluid dynamics by the 1870s – its validity had already been proven experimentally {ROBINS ⇨1746; HUTTON ⇨1783; PARRY ⇨1824/1825; REGNAULT ⇨1863} – Ernst MACH initiated a scientific investigation of shock waves performed at the laboratory scale that introduced the use of high-speed visualization with microsecond resolution. Based on his numerous fundamental contributions to shock waves, he is now generally considered to be the father of supersonics. After his discovery of "Mach reflection," he proved experimentally, together with his student Jan SOMMER, that a shock wave does indeed propagate at supersonic velocities, but it rapidly approaches the velocity of sound as the distance from the source increases {E. MACH & SOMMER ⇨1877}, thus confirming on a laboratory scale REGNAULT's previous results from full-scale propagation studies.

MACH's most famous experiments, however, were certainly his ballistic studies. Together with Peter SALCHER, an Austrian physicist and professor at the Austrian-Hungarian Marine Academy in Rijeka [now Fiume, Croatia], he first showed that a projectile flying supersonically produces a shock wave with a roughly hyperbolic shape which is fixed to the projectile – the "head wave" or "bow wave" [German *Kopfwelle* or *Bugwelle*], known in France as "l'onde balistique" {E. MACH & SALCHER ⇨1886; ⇨Figs. 4.6–E to G}. This curious wave phenomenon had already been predicted by the Austrian physicist Christian A. DOPPLER {⇨1847} using a simple graphical method {⇨Figs. 4.6–B, C} based upon the application of the Huygens principle {HUYGENS ⇨1678; ⇨Fig. 4.6–A}. However, there were no experimental diagnostic means at DOPPLER's disposal to prove his ingenious concept. The discovery of the supersonic head wave immediately prompted great interest in military circles, since fundamental research in this area could result in a better understanding of aerodynamic drag and ultimately projectile geometries with minimum drag – thus significantly increasing the effective ballistic ranges of fire arms and improving the theoretical prediction of trajectories.

These pioneering experimental investigations performed by Ernst MACH and his team – together with theoretical studies performed in England, France and Germany – resulted in a basic knowledge of supersonic flows in the 1880s. Practical aerodynamics, however, was still in its infancy, and the first flight of man {VON LILIENTHAL ⇨1891} had not yet been achieved.

The ballistic head wave strongly reminds one of the bow wave generated by an object moving through water. However, there is a significant difference: while in gas dynamics the conical angle decreases as the speed of the object increases, this is not the case with the wake generated by a ship moving through water {⇨Fig. 4.6–D}. All objects moving

---

[235] G. BEN-DOR: *Oblique shock wave reflections*. In: (G. BEN-DOR ET AL., eds.) *Handbook of shock waves*. Academic Press, New York, vol. 2 (2001), pp. 67-179.

[236] The author owes this reference to Prof. Manuel G. VELARDE, a fluid physicist at the Instituto Pluridisciplinar of the Universidad Complutense de Madrid, Spain.

through water have a pseudo-effective Mach number of 3, so long as the water is deep compared to the wavelength of the waves produced, the course of the ship is straight, and its speed is constant.[237] Curiously enough, none of the famous European marine painters of previous centuries had ever reproduced this striking bow wave phenomenon, although it can be clearly observed with the naked eye.

**Studies of Nozzle Outflow.** It appears that studies of the exhaust of compressed gas from an orifice originated from malfunctions or imperfect construction of the safety valve. This simple device was invented by the Frenchman Denis PAPIN and used by him in his *steam digester*, the first pressure cooker {PAPIN ⇨1679; ⇨Fig. 4.7–A}. With the advent of the first steam engines in the early 18th century, such valves – which initially only consisted of a hole in the boiler wall covered by a plate loaded with a weight – were often applied. However, to avoid the need to use large weights, these valves were often made too narrow to be able to vent dangerous overpressures quickly, resulting in steam-boiler explosions that often produced many casualties and a great deal of material damage {⇨Fig. 4.21–A}. Boiler explosions were believed to be a complex chain of different causes {ARAGO ⇨1830; AIRY ⇨1863}. Eager attempts to avoid them prompted not only engineers but also scientists to undertake detailed studies that became the roots of early supersonic research {VENTURI ⇨1797; DE SAINT-VENANT & WANTZEL ⇨1839; FLIEGNER ⇨1863; NAPIER ⇨1866; EMDEN ⇨1903; LORENZ ⇨1903; FLIEGNER ⇨1903; PRANDTL ⇨1908}.

The *Laval nozzle* – a nozzle with a convergent-divergent geometry that was invented by the prominent Swedish industrialist and engineer Carl Gustaf Patrick DE LAVAL and used by him to deliver steam to turbine blades – was the first to permit supersonic flow velocities at its exit {DE LAVAL ⇨1888}. Thus, the Laval nozzle became an important device in steam turbine engineering {⇨Figs. 4.7–F, G & 4.8–C}, because it increased the efficiency of steam turbines. In aeronautical engineering, it not only had an enormous impact on rocket engine design {⇨Fig. 4.7–H}, but it also triggered much supersonic flow research and the development of supersonic wind tunnels {⇨Figs. 4.8–E & 4.9–A to I }.

These successful studies of outflow phenomena that occur at the exit of a nozzle also stimulated the development of new diagnostic techniques for high-speed visualization, which facilitated the interpretation of supersonic flow phenomena {SALCHER & WHITEHEAD ⇨1889; L. MACH ⇨1897; Figs. 4.8–A, B}. Further experimental and theoretical flow studies of the inside and outside of a Laval nozzle provided a basic knowledge of supersonic flow as a function of nozzle geometry and operational parameters, which stimulated the conception of improved supersonic wind tunnel designs {REYNOLDS ⇨1883; PRANDTL ⇨1908; MEYER & PRANDTL ⇨1908}.

**Wind Tunnels.** In order to study aerodynamic drag quantitatively, Sir Isaac NEWTON dropped spheres from the dome of St. Paul's Cathedral in London, and from these observations he developed the first theory of aerodynamic drag. Benjamin ROBINS used a whirling-arm device: the object under study was placed at the end of a rotating rod in order to observe the object's rapid passage through the air. Another method was to tow models through still water, since air is a fluid whose behavior is in many ways comparable to that of water[238]. A big step forward in drag research was the direction of a homogeneous high-speed jet of air at scale models in tunnel-like passages – so-called "wind tunnels" – a diagnostic principle that had already been proposed by Leonardo DA VINCI {⇨1490s}, but does not appear to have actually been applied by him. Later, the Austrian physicist Peter SALCHER {⇨1889}, while visualizing the flow around supersonic projectiles, hit upon the same idea of likewise investigating the inverse case of the flow of air against a body at rest in order to confirm the results already obtained.

*Subsonic Wind Tunnels ($M < 0.8$).* Subsonic tunnels are the simplest types of wind tunnels. Making use of the Bernoulli theorem {D. BERNOULLI ⇨1738}, air velocities in this range can use a pipe with a constriction which decreases the pressure and increases the flow velocity {VENTURI ⇨1797; ⇨Fig. 4.7–D}. The scale model under study is mounted in the contraction in the throat of the tube. The first facilities built according to this concept were rather crude, low-velocity devices used in pioneering British aerofoil studies {WENHAM ⇨1871; PHILLIPS ⇨1885; ⇨Fig. 4.7–E}. In the early 1900s, the WRIGHT Brothers {⇨1903} also built and used two wind tunnels: the second one, which was more advanced and was driven by a 2-hp gasoline engine, produced a maximum wind velocity of 27 mph (about 12 m/s). Ludwig PRANDTL {⇨1907} constructed the first large subsonic wind tunnel with a test cross-section of $2 \times 2$ m$^2$ using a closed-circuit air flow, an operational principle which also proved very useful in the supersonic regime {ACKERET ⇨1933}.

*Transonic Wind Tunnels ($0.8 < M < 1.2$) and Supersonic Wind Tunnels ($1.2 < M < 5$).* In addition to low-velocity

---

[237] R.A.R. TRICKER: *Bores, breakers, waves and wakes*. Am. Elsevier, New York (1965).

[238] The smooth free-surface flow of a liquid (water) over a horizontal bed obeys equations of motion similar to those for the isentropic 2-D flow of a hypothetical gas with specific heat ratio $\gamma = 0.2$. For air this value is 1.4.

aerofoil studies, ballisticians and mathematicians – who had known of the supersonic velocities of small- and large-caliber shots {ROBINS ⇨1742; EULER ⇨1745; HUTTON ⇨1783; MAYEVSKI ⇨1881} since the 18th century – asked for experimental data on aerodynamic drag and stability, ranging from supersonic muzzle velocities ("$v_0$") down to almost zero, in order to allow better predictions of projectile trajectories. Of particular interest was the transonic regime, in which aerodynamic drag changes tremendously {⇨Fig. 4.14–M}.

Pioneers of supersonic wind tunnel design were confronted with a number of complex problems in this case, for example:
- to design new nozzle that could produce transonic or supersonic flow in the test section;
- to develop an appropriate method of supporting the model in the test section in such a way that the flow around the model remains essentially equivalent to that around the model in free flight;
- to adapt standard methods of optical visualization (*e.g.*, shadow, schlieren, interferometer techniques) to the special requirements of a wind tunnel set-up;
- to measure the forces on the test model; and
- to determine the power requirements at various air speeds.

This resulted in some basic design concepts for high-speed wind tunnels, such as continuous closed-circuit tunnels, intermittent indraft tunnels, blow-down tunnels, and pressure-vacuum tunnels.

The first test facilities designed to investigate the supersonic regime were not constructed until the 1920s by the French aeroballisticians Paul LANGEVIN and Constantin CHILOWSKY {⇨1918} and Eugène HUGUENARD and Jean André SAINTE-LAGUË {⇨Fig. 4.9–A}. The British engineer Sir Thomas E. STANTON adopted the Laval nozzle in the world's first supersonic wind tunnel, which he called "wind channel." It was constructed and operated at the National Physics Laboratory (NPL) in Teddington {STANTON ⇨1921; ⇨Fig. 4.9–B}. In this mini blow-down facility, which had a useful diameter of only 0.8 in. (20.3 mm), supersonic flow velocities of up to a Mach number of M = 2 could be achieved.

Supersonic wind tunnels using atmospheric air need drying devices to avoid condensation, a detrimental phenomenon which, producing a dense fog in the tunnel, hinders any optical diagnostics and potentially results in local changes in the Mach number {5th Volta Conf. (PRANDTL) ⇨1935}. By the late 1930s, a small number of useful supersonic wind tunnels had been built in Europe {STANTON ⇨1928; ACKERET ⇨1933; BUSEMANN & WALCHER ⇨1933; WIESELSBERGER ⇨1934; CROCCO ⇨1935; 5th Volta Conf. (ACKERET, CROCCO) ⇨1935; HVA Peenemünde ⇨1939; BETZ ⇨1939}.

In the United States, the Langley Memorial Aeronautical Laboratory was home to the few transonic and supersonic tunnels owned by the Government before World War II:
- the 11-in. (27.9-cm)-dia. high-speed tunnel (1928);
- the 24-in. (61-cm)-dia high-speed tunnel (1934);
- the 8-ft (244-cm)-dia. high-speed tunnel (Mach 0.5 until 1945, enhanced to Mach 1 in early 1945);
- the 9-in. (22.9-cm)-dia. supersonic tunnel (designed in 1939, operational in 1942);
- the 11-in. (27.9-cm)-dia. hypersonic tunnel (designed in 1945, operational in 1947); and
- the 4 × 4-ft$^2$ (122 × 122-cm$^2$) supersonic pressure tunnel (designed in 1945, operational in 1948).

A 1 × 3-ft$^2$ (30.4 × 91.4-cm$^2$) supersonic tunnel (completed before 1945) also existed at the NACA Ames Laboratory, Moffet Field, CA. Furthermore, three universities (MIT, Cornell and CalTech) owned high-speed wind tunnels.[239] U.S. wind tunnels, ranging from the NACA Wind Tunnel No. 1 (operational in 1920) to advanced wind tunnel technology (up to 1981), were discussed by Donald D. BAALS[240] and William R. CORLISS at NASA's Scientific & Technical Information Branch. Their report contains constructional details and photographs of all of them.

***Hypersonic Wind Tunnels (5 < M < 12).*** Hypersonic aerodynamics became important when designing military aircraft, missiles, the space shuttle and other high-speed objects. In order to study and simulate hypersonic flight, new methods of aerodynamic testing were developed. Besides high-speed instrumentation for measuring pressures, densities, forces and temperatures, conventional shadowgraph, schlieren and interferometer techniques were supplemented by new airflow visualization techniques, such as *thermography* {⇨Fig. 4.18–H}, and *particle tracing,* a method in which the test model is coated with a mixture of oil and fluorescent powder and then exposed to ultraviolet light.

Hypersonic flow velocities of up to almost $M = 9$ {ERDMANN ⇨1943/1944} were first reached by using a special nozzle design in the large supersonic wind tunnel facility at the *Heeresversuchsanstalt Peenemünde* ("Army Rocket Testing Center Peenemünde"), the main research center for German rocketry during World War II. These studies at Peenemünde had already revealed that realistic simulations

---

[239] Priv. comm. by Erik M. CONWAY, a visiting historian at NASA Langley Research Center (March 1, 2001).
[240] D.D. BAALS and W.R. CORLISS: *Wind tunnels of NASA.* NASA Rept. SP-440, NASA History Office, Washington, DC (1981); http://www.hq.nasa.gov/office/pao/History/SP-440/cover.htm.

of hypersonic flows (*i.e.*, of Mach numbers above 5) would require a more sophisticated wind tunnel design. Hypersonic research was not resumed until the late 1940s in the United States {BECKER ⇨1947; WEGENER ⇨1951}.

In the following years, a number of hypersonic facilities were constructed using different methods of generating hypersonic flows with large Reynolds numbers, *e.g.*, by modifying conventional hypersonic wind tunnels using a "shroud" technique {FERRI & LIBBY ⇨1957}, by inventing new testing facilities such as hotshot tunnels, plasma jets and shock tunnels {WITTLIFF, WILSON & HERTZBERG ⇨1959}, or by using high-velocity ballistic accelerators such as light-gas guns {CROZIER & HUME ⇨1946; ⇨Fig. 4.11–D}.[241] A historical perspective of early hypersonic wind tunnel design was given by Julius LUKASIEWICZ,[242] a Canadian professor of aerospace engineering at Carleton University in Ottawa, Ontario.

***Hypervelocity Wind Tunnels (M > 12).*** Hypervelocity aerodynamics involves the physics of atmospheric reentry. However, there is a fundamental problem with operating wind tunnels at hypervelocities: the ambient temperature in the test section decreases as the Mach number increases until the air liquefies. This would result in inaccurate experimental results. In order to prevent condensation, the stagnation temperature of the air must be very high, *e.g.*, the stagnation temperature in the stilling section must be above 4,700 °C for the ambient temperature in the test section to remain above 0 °C. This requires special heating techniques and vast electrical energy. Examples of hypervelocity tunnels, referred to in the CHRONOLOGY, are the *gun tunnel* or *Stalker tube* {STALKER ⇨1965} and the *shock tunnel* {WITTLIFF ET AL. ⇨1959}.

In 1991, Hans GRÖNIG,[243] a German professor of mechanical engineering at the RWTH Aachen, reviewed the development of European hypersonic and hypervelocity wind tunnels used to test models of new space vehicles.

**Shock Tubes.** The shock tube, invented in France by the physicist and explosives specialist Paul VIEILLE {⇨1899} as a by-product of his detonation studies, became the most important measuring and testing device in gas dynamic research. VIEILLE applied the shock tube in order to demonstrate that shock waves generated by the detonation of explosives propagate in essentially the same manner as shock waves generated by the bursting diaphragm of the high-pressure section that formed one end of his tube. The basic theory for the shock tube was laid down by the German researchers Karl KOBES {⇨1910}, Frederick HILDEBRAND {⇨1927}, and Hubert SCHARDIN {⇨1932}. KOBES and HILDEBRAND had a rather curious approach to gas dynamics: they investigated whether it would be possible to improve the performance of air suction brakes on long railway trains by using shock waves. Their experiments essentially confirmed this hypothesis.

The shock tube – rediscovered during World War II by the U.S. physicist Walker BLEAKNEY and associates at Princeton University {BLEAKNEY ⇨1946 & 1949}, and further improved by his group through the introduction of a trigger pin to pierce the diaphragm {BLEAKNEY & TAUB ⇨1949} – soon proved its excellent applicability for quantitatively investigating 1-D propagation and interaction phenomena of shock waves within a wide range of gas dynamic parameters. Furthermore, the shock tube was quickly introduced into other laboratories around the world for studying the interactions of shock waves with scaled architectural structures such as model houses, plants, shelters, and vehicles {BLEAKNEY ⇨1952; ⇨Fig. 4.5–J}. During the long postwar period of Cold War, such model interactions were of great practical concern because of the constant threat of the use of nuclear blasts against a wide range of civil and military installations.

Special shock tube constructions were produced in the 1960s to simulate the blast waves from nuclear explosions {CULBERTSON ⇨1970} – a particularly challenging task because higher peak overpressures had to be provided than for chemical explosions, combined with longer pressure durations. In addition, such blast simulators were supplemented with thermal radiation simulators to get a realistic picture of the overall effects of nuclear damage – so-called "synergetic effects" {MORRIS ⇨1971}. Since the shock tube is also a perfect, low-cost tool for studying supersonic flow patterns over an enormous range of Mach numbers, its use even replaced traditional wind tunnel work to some extent {MAUTZ, GEIGER & EPSTEIN ⇨1948; BLEAKNEY & TAUB ⇨1949}. However, the electronic diagnostics used in shock tube facilities must be fast, and were therefore initially more expensive than those in wind tunnels. This delayed wide application of the shock tube for such purposes at first, but the costs of such equipment have since considerably decreased.

The shock tube also proved to be most useful for studying the ignition and detonation behavior of explosive gaseous mixtures {PAYMAN & SHEPHERD ⇨1937 & 1940; SHEPHERD ⇨1948}. That the shock tube was also useful for generating

---

[241] A. POPE and K.L. GOIN: *High-speed wind tunnel testing.* Wiley, New York *etc.* (1965), pp. 442-460.

[242] J. LUKASIEWICZ: *Experimental methods of hypersonics.* M. Dekker, New York (1973), pp. 25-51.

[243] H. GRÖNIG: *Shock tube application: high enthalpy European wind tunnels.* In: (K. TAKAYAMA, ed.) *Proc. 18th Int. Symposium on Shock Waves* [Sendai, Japan, July 1991]. Springer, Berlin *etc.* (1992), vol. 1, pp. 3-16.

high temperatures in gases was first recognized and exploited in high-speed spectroscopic studies by Otto LAPORTE {⇨1953}, a German-born U.S. physicist who first applied reflected shock waves to produce high temperatures in gases. The shock tube was also successfully modified for use in studies of chemical kinetics {GLICK, SQUIRE & HERTZBERG ⇨1957}. The magnetically driven shock tube {KOLB ⇨1957} first allowed Mach numbers in excess of 100 and extended the highest shock temperatures that could be attained up to one million degrees. However, with the advent of high-power pulsed lasers in the 1960s, expectations that this technique would eventually be used to help produce controlled nuclear fusion dropped.

### 2.4.5 SHOCK WAVES IN A LIQUID: THE PECULIAR FLUID

> *The famous Florentine experiment, which so many Philosophical writers have mentioned as a proof of the incompressibility of water, will not, when carefully considered, appear sufficient for that purpose… But it was impossible for the gentlemen of the Academy del Cimento to determine, that the water which was forced into the pores and through the gold, was exactly equal to the diminution of the internal space by the pressure.*[244]
> 
> John CANTON
> London 1761

> *Shock waves in water differ greatly from those in air. If the pressure difference is the same, the shock velocity and the mass velocity are considerably less, and the temperature rise is enormously less. If the piston velocity (equals mass velocity) is the same in air and in water, then the shock velocity is considerably higher, the pressure difference very much higher and the temperature rise lower in the water case.*[245]
> 
> George B. KISTIAKOWSKY
> Harvard University
> Cambridge, MA 1941

Liquids occupy an interesting intermediate position between the gaseous and solid state, and were regarded for a long time as being incompressible matter {Florence Academy ⇨1660s}. Edmé MARIOTTE, the eminent French natural philosopher, asserted in his treatise *Mouvement des eaux et des autres corps fluides* ("Motion of Water and Other Fluidic Objects," published posthumously in 1686) that water is incompressible and hence has no elastic force. He also studied the force of impact of water in the form of its speed of efflux through a small hole at the base of a reservoir. It took further 76 years until the English naturalist John CANTON {⇨1762} first demonstrated its – albeit very small – compressibility. In any liquid, the compressibility increases and the density decreases as the temperature rises. Therefore, unlike gases, the velocity of sound in liquids decreases (approximately linearly) as the temperature rises. However, water occupies a special position amongst liquids: the compressibility drops initially as the temperature rises to a minimum of about 60 °C, and only then does it increase. Therefore, initially the velocity of sound in water has a positive temperature coefficient and reaches a maximum value of 1,557 m/s at 74 °C. Above this temperature the velocity of sound in water decreases.

Real liquids, which have small but finite viscosities, difference widely in terms of their compressibility: in the low-pressure regime, in which the molecules are being pushed into effective contact, the viscosities of organic liquids increase under pressure at a rate that increases rapidly with increasing pressure; however, for water or monatomic mercury this rise in viscosity is comparatively small. At higher pressures, when the compressibility arises from a decrease in molecular volume, the volumes of ordinary liquids become surprisingly similar. Obviously, it is very difficult to model the complex behavior of liquids by defining a "perfect liquid" in analogy to the "perfect gas" that has played such an important role in the kinetic theory of gases. Eric D. CHISOLM[246] and Duane C. WALLACE, two LANL scientists who recently proposed a theory for the dynamics of monatomic liquids, appropriately wrote, "Despite a long history of physical studies of the liquid state, no single theory of liquid dynamics has achieved the nearly universal acceptance of BOLTZMANN's theory of gases or BORN's theory of lattice dynamics of crystals. This shows the extraordinary theoretical challenge that liquids pose; they enjoy none of the properties that make either crystals or gases relatively tractable. A great deal of effort has been devoted to understanding liquids as hard-sphere systems, which do model the core repulsion present in real liquids, but omit the important potential energy effects…"

In shock wave physics, liquids and gases are often treated as *compressible fluids*. At pressures of 1 kbar or more, the densities of gases become of the same order of magnitude as those of their liquid phase, and there ceases to be any significant difference between the gas and the liquid. On the other hand, liquids and solids are also considered to be phases of *condensed matter*, and at sufficiently high pres-

---

[244] J. CANTON: *Experiments to prove that water is not incompressible*. Phil. Trans. Roy. Soc. Lond. **52**, 640-643 (1761/1762).

[245] See KISTIAKOWSKY ET AL. {⇨1941}.

[246] E.D. CHISOLM and D.C. WALLACE: *Dynamics of monatomic liquids*. J. Phys. Condens. Matter **13**, R739-R769 (2001).

sures polymorphism phase transitions can occur. At first, the existence of dynamic polymorphism was never seriously considered by static high-pressure workers, since transformation rates observed in static experiments are classified as "sluggish," ranging from minutes to days. However, the discovery of polymorphism in dynamic experiments demonstrated that

- the transformation rate between two polymorphs is large enough to be detectable in shock wave experiments; and
- the pressure at which transformation occurs dynamically agrees with an established static transition pressure.

"Shock wave splitting" provides clear evidence of first-order phase transitions, which are characterized by discontinuities in volume and entropy. In contrast, shock-induced second-order transitions involve phases for which volume and entropy are continuous, but higher derivatives of energy, specific heat, compressibility and thermal expansion are discontinuous {DUVALL & GRAHAM ⇨1977}. Second-order transitions may play an important role in shock-compressed solids such as ferromagnetic-to-paramagnetic iron alloys {CURRAN ⇨1961}, but second-order transitions and some effects that may be associated with them have not yet been studied in shock-compressed liquids.

Generally, a fluid is defined as a nonsolid state of matter in which the atoms or molecules are free to move past each other; *i.e.*, they offer no permanent resistance to change of shape. Liquids, most of which approach an ideal fluid in the sense that they cannot support shear stress, are much more difficult to compress than gases, since liquid molecules are already arranged such that they are close together. Liquids do not have unit cells; they are, to a first approximation, irregularly packed volumes of spheres; the arrangement of their atoms can be visualized as billiard balls jumbled in a bag. As a consequence, typical shock wave properties such as wave-steepening effects and supersonic propagation are clearly observable only at much higher shock pressures than for shock-compressed gases. Additionally, at the leading edge of the shock, dense fluids ($\rho > 1$ g/cm$^3$) show thermodynamic characteristics that are markedly different from those of a viscous, heat-conducting gas, and above the viscous fluid limit they show a nonplastic solid response. For example, in the case of shock-compressed water, this limit is reached just below 13 kbar {WALLACE ⇨1982}.

There is also a remarkable difference if we compare gases with liquids in terms of their compressibility $\xi = \rho/\rho_0$: while gases can highly be compressed by static means up to $\xi > 1,000$, the maximum compression across a shock wave in a gas cannot exceed a surprisingly small value which, for example for air ($\gamma = 1.4$) amounts to just six – a value which can easily be obtained by using a modern bicycle tire pump.[247] On the other hand, liquids and solids can be shock-compressed to a higher degree than is possible with any static means presently available.

Depending on the pressure applied, a compressed liquid can solidify (or freeze), while a compressed solid can liquefy (or melt). Shock waves propagating in a gas-liquid mixture, such as in moist air, can generate shock-induced condensation effects, a phenomenon that caused much puzzlement in the pioneering era of supersonic wind tunnels {5th Volta Conf. ⇨1935}, and one that occurs when an aircraft is flying under particular conditions at supersonic speed: in this case it reveals itself as clouds partly that surround the fuselage or wings {⇨Fig. 4.14–F}.

Shock-compressed fluids in the liquid state, when compared to gaseous fluids, may show unusual properties (*e.g.*, high viscosity, low compressibility, phase transformations), and may generate complicated side effects (*e.g.*, cavitation). Since the density of a liquid is much higher than that of a gas – for instance, water is some 770 times more dense than air – hydrostatic buoyancy effects resulting from differences in hydrostatic pressure at different depths cannot be neglected. Compared to a gaseous environment, such effects occur in gases at much smaller dimensions, leading to typical phenomena such as upward movement, deformation and jet formation {⇨Fig. 4.16–E} of the gas bubble in an underwater explosion {BLOCHMANN ⇨1898; RAMSAUER ⇨1923; LAMB ⇨1923; KIRKWOOD & BETHE ⇨1942; COLE ⇨1948; Undex Reports ⇨1950}.

Shock waves in liquids, particularly those in water, were barely studied until the beginning of World War II. However, a few remarkable contributions, described in more detail in the CHRONOLOGY, should be emphasized here.

**Shock-Induced Freezing.** Water – the most abundant liquid on Earth – is a highly unusual substance at atmospheric pressure, and so physicists expected it to be similar under high static and/or shock pressures.

The U.S. physicist Percy W. BRIDGMAN {⇨1931}, who performed high-pressure studies on compressed liquids, demonstrated that water solidifies under static pressure in the

---

[247] According to elementary shock theory, for an ideal gas, the *maximum compression ratio* at the shock front is given by $\xi = (\gamma+1)/(\gamma-1)$, where $\gamma$ is the ratio of the specific heats at constant pressure and constant temperature, respectively. The limiting compression for a monatomic gas (*e.g.*, He, Ne, Ar, Kr, Xe, Rn) is $\xi = 4$, and for a diatomic gas (*e.g.*, H$_2$, O$_2$, N$_2$) it is $\xi = 6$.

pressure range 0–25 kbar, thereby producing seven different crystalline modifications of ice: *Ice I* to *Ice VII*, with *Ice II* being the most dense structure. During World War II, the German physicist and ballistician Hubert SCHARDIN {⇨1940/1941}, while studying ballistic phenomena in water and other liquids, was apparently the first to suggest *shock-induced freezing*. However, his ballistic impact experiments, in which he measured the temporal optical transparency of shock-compressed liquids, showed that water remains transparent behind the shock front. After World War II, with the advent of more advanced diagnostic techniques, this line of research was resumed both in the United States and the Soviet Union, but it gave ambiguous results. The Los Alamos scientists shocked water to 100 kbar, but found no sign of opacity due to freezing {WALSH & RICE ⇨1957}. Soviet researchers noticed a kink in the measured Hugoniot curve of water, from which they concluded that a shock-induced phase transition beginning at 115 kbar had occurred {AL'TSHULER ET AL. ⇨1958}. However, this result could not be confirmed by another Soviet group {ZEL'DOVICH ET AL. 1961, *see* AL'TSHULER ET AL. ⇨1958}. Very recently, this fascinating subject in shock physics was resumed in the United States: a multiple shock compression technique was used to prove that shock-compressed water does indeed freeze, even on a nanosecond time scale {DOLAN & GUPTA ⇨2004}.

Flash X-ray diffraction of shock-solidified liquids may provide valuable information on whether a new polycrystalline solid state has indeed been produced or whether this is a pseudo-effect (*e.g.*, due to a shock-induced large increase of viscosity). However, initial experiments carried out in this area in the early 1970s by the author on shock-compressed $CCl_4$ {SCHAAFFS & TRENDELENBURG ⇨1948} using characteristic radiation (such as Mo-K$\alpha$ and Cu-K$\alpha$ from standard vacuum discharge flash X-ray tubes), were just too ambitious for their time and so they failed. The application of soft flash X-rays from laser-produced plasmas {WARK ET AL. ⇨1991} or further progress in the development of high-intensity pulsed soft X-ray lasers, combined with high-sensitivity MCP detectors {FARNSWORTH ⇨1930}, may enable shock researchers to follow the transition of a liquid from its amorphous state into a shock-induced polycrystalline or polycrystalline-like state in the future.

Corresponding static X-ray diffraction studies of the structures of liquids at high pressures and temperatures have been a long-standing goal in static high-pressure research. With the maturation of third-generation synchrotron sources, this goal was eventually attained a few years ago in Japan by Yoshinori KATAYAMA[248] and collaborators at the Japan Atomic Energy Research Institute, who observed a first-order liquid-liquid phase transition in phosphorus in a large-volume press. In addition to finding a known form of liquid phosphorus at low pressure – a molecular liquid comprising tetrahedral $P_4$ molecules – they found a high-pressure polymeric form at pressures above 1 GPa (10 kbar).

**Liquefaction.** The curious and unusual phenomenon of liquefaction can be achieved by starting from the gaseous as well as from the solid state: a *liquefaction shock wave* is a compression shock that converts vapor (*i.e.*, a gas) into a liquid state. The existence of a liquefaction shock wave was first predicted on physical grounds and was later also proved experimentally {DETTLEFF ET AL. ⇨1979}. The liquefaction shock is a new phenomenon and one that is quite distinct from the well-known condensation of vapor in an expanding flow {5th Volta Conf. (PRANDTL) ⇨1935; HVA Peenemünde ⇨1939}.

On the other hand, the term *shock liquefaction* is used in solid-state shock wave physics to describe a first-order phase transition from the solid to the liquid state, also known as "shock melting" {KORMER ET AL. ⇨1965; ASAY ET AL. ⇨1976; DUVALL ET AL. ⇨1977}. Solids that are shock-compressed far beyond their yield stress can be treated as fluids to a first approximation.

Furthermore, the term *shock liquefaction* is also used in geology and seismology, but here it takes a quite different meaning: it designates a sudden, large decrease in the shearing resistance of a water-saturated cohesionless soil (such that as caused by seismic shocks or some other strain) which involves a temporary transformation of the material into a viscous fluid mass. Shock liquefaction is among the most catastrophic of all ground failures, and it has been observed to occur during almost all large earthquakes; it was extremely dramatic during the Kobe Earthquake {⇨1995} in Japan for example. The term *liquefaction* was apparently first used in a paper by A. HAZEN[249] in which he described the failure of hydraulic fill sands in Calaveras Dam, located near San Francisco, CA: on March 24, 1918, the upstream toe of the under construction Calaveras dam suddenly flowed, the water gate tower collapsed and approximately 800,000 cubic yards (731,520 $m^3$) of material moved around 300 ft (91.4 m). Apparently at the time of the failure none special disturbance was noticed.

---

[248] Y. KATAYAMA, T. MIZUTANI, W. UTSUMI, O. SHIMOMURA, M. YAMAKATA, and K. FUNAKOSHI: *A first-order liquid-liquid phase transition in phosphorus.* Nature **403**, 170-173 (2000).

[249] A. HAZEN: *Hydraulic-fill dams.* ASCE Trans. **83**, 1713-1745 (1920).

**Water Hammer, Water Ram, Hydraulic Ram and Hydrodynamic Ram.** These four terms relate to devices and/or effects that occur when a liquid mass is suddenly accelerated or decelerated, causing pressure pulses (weak shock waves) in the liquid.

The term *water hammer* or *water ram* originally referred to an instrument that has been used since at least the late 18th century to illustrate the fact that liquids and solids fall at the same rate in a vacuum. It consists of a hermetically sealed glass tube devoid of air and partially filled with water. When the tube is quickly reversed, the water falls to the other end with a sharp and loud noise like that of a hammer, thus impressively demonstrating that a liquid in a tube without air behaves as a compact mass. The instrument was made and sold by glass-blowers and barometer-makers.[250] In the 19th century, this phenomenon was also known by the name *Kryophor* in Germany,[251] because it could be demonstrated with WOLLASTON's kryophor,[252] a curved glass tube provided with two spherical balloons at the end and partially filled with a liquid.

In hydraulics, the term *water hammer* [Germ. *Wasserschlag*, French *coup de bélier*] also describes the concussion or sound of water in a long pipe when its flow is suddenly stopped or when a live stream is suddenly admitted, such as by closing or opening a valve inserted in the line, respectively. This phenomenon, termed the *water hammer effect*, is detrimental in pipe systems, because the pressure pulse produced can propagate to remote areas and destroy tubes, valves, and other installations. This effect would most probably have been observed as far back as the time of the Roman Empire. For example, in Rome's highly impressive water supply system, which was constructed in the first century A.D., the water was brought overland in conduits and then fed into a vast network of small bronze, lead or ceramic pipes. Beginning in the 19th century, the problem of water hammer or hydraulic shocks in water supply lines became important in countries where large water supply systems had to be built to satisfy the rapidly increasing demands of urbanization, industry, and agriculture. The water hammer problem was first treated scientifically in Russia at the turn of the 19th century {KARELJKICH & ZHUKOVSKY ⇨1898}.

Drops of liquid impacting onto a solid surface are also termed a *water hammer*. This water hammer effect can cause serious material damage it, and long proved to be a problem to steam turbine {⇨Fig. 4.14−C} and high-speed naval propeller {Sir PARSONS ⇨1897; Brit. Committee of Invention & Research ⇨1915; S.S. COOK ⇨1928} designers and operators. In the 1960s, the same phenomenon became of increasing concern in the aerospace industry due to both serious rain erosion problems sustained by supersonic aircraft and missiles and anticipated blade erosion problems in the turbines of space-based power plants (for example, in megawatt or even gigawatt turbines earmarked for asteroid tug propulsion purposes), which would use liquid metals as the working fluid {HEYMANN ⇨1969}.

The term *hydraulic ram* originally referred to a device for pumping water. The French hot-air balloon pioneer Joseph-Michel MONTGOLFIER {⇨1796}, supported by the Swiss physicist Aimé ARGAND, successfully used the water hammer phenomenon to pump water {⇨Fig. 4.2−M}. They called their pump a "hydraulic ram" [French *bélier hydraulique*]. In their water pump, the natural downward flow of running water was intermittently halted by a valve, so that the flow was forced upward through an open pipe into a reservoir.

The term *hydraulic ram* also designates an effect that can be generated by an object impacting and penetrating into a liquid − probably the oldest known shock wave effect in a liquid. The French natural philosopher Louis CARRÉ {⇨1705} observed a curious phenomenon where a bullet that is shot into a wooden box filled with water causes the box to blow up. The impacting bullet transfers a large amount of momentum to the water, generating a shock wave that ruptures the walls. In the 19th century, as the projectile velocities achieved by small firearms increased, a new type of injury from penetrating bullets was diagnosed: a high-velocity bullet not only cuts through tissues directly in its path but it also imparts sufficient kinetic energy to adjacent tissues to cause an explosion-like effect that results in injury to an area many times the diameter of the bullet. This dreaded phenomenon, based on the hydraulic ram effect, was first observed in the Prussian-French War (1870−1871), and both nations erroneously accused each other of having used explosive bullets, which were banned {St. Petersburg Declaration ⇨1868}.

---

[250] J.A. SIMPSON and E.S.C. WEINER: *The Oxford English dictionary.* Clarendon Press, Oxford (1989), vol. 19, p. 996.
[251] *Meyers Konversations-Lexikon.* Bibl. Inst., Leipzig & Berlin (1897), vol. 15, p. 1003 "Sieden."
[252] W.H. WOLLASTON: *On a method of freezing at a distance.* Phil. Trans. Roy. Soc. Lond. **103**, 71-74 (1813).

Ever since the first air battles were fought in World War I, the susceptibility of aircraft fuel tanks to ballistic impact has been a serious issue for military aircraft, whose fuel tanks cannot be fully armored against gun shots. When sufficiently energetic debris impact and penetrate a tank below the fuel level, the result of the energy transfer is an increase in pressure that can tear the tank apart. The subsequent fuel release can have catastrophic consequences. This hazard has also been termed a *hydraulic ram* {ANKENEY ⇨1977}.

In 1977, a Hydraulic Ram Seminar was held on this particular subject at the U.S. Air Force Flight Dynamics Laboratory (Wright-Patterson Air Force Base, OH).[253] In the 1980s, empirical codes such as the Explicitly Restarted Arnoldi Method (ERAM) and the ABAQUS software for finite element analysis (FEA) were coupled to provide a complete hydraulic ram analysis: ERAM was used to determine hydrodynamic ram loads, and ABAQUS was used as a nonlinear quasi-static analysis tool that took the peak pressures output from ERAM.

The study of hydraulic rams in commercial aviation was largely ignored until the Concorde disaster of July 25, 2000 in the commune of Gonesse after takeoff from Paris (113 casualties). This disaster resulted from the high-speed impact of an exploded tire fragment on a fuel tank on a wing, causing "hydraulic ram" shock waves to be set up in the fuel that led to tank rupture and then a catastrophic fire.[254] Since then, such studies have received a lot of attention, particularly in Europe.

The term *hydraulic ram* is somewhat misleading, since it suggests a quasi-static compression phenomenon,[255] whereas in reality the effect comprises at least four distinct sequential dynamic phenomena (or phases).[256] Each of the four phases of a hydraulic ram contributes to structural damage and so they must all be accounted for in ballistic analysis.

- The *shock wave phase* occurs during the penetration of the tank by a projectile. Note that this occurs even though the projectile may be traveling at a subsonic speed with respect to the fuel tank.
- The *pressure field phase* (or *drag phase*) is characterized by relatively low pressures and long durations. Since the pressure field lies ahead of the projectile, pressure field damage loads are greatest on the exit wall and lowest on the entrance wall.
- The *cavity collapse phase* occurs due to the fact that a high-velocity projectile imparts a large radial velocity to the fuel it displaces along its path.
- The *free surface phase* depends on the depth to which the tank is filled, and may occur either during or after the cavity collapse phase. The upward motion of the fuel reaching the top of the tank results in the generation of a large impulsive load on the top tank wall.

Instead of the term *hydraulic ram*, modern hydrodynamicists are increasingly tending to use the term *hydrodynamic ram* to denote the overall effect of the four phases listed above.

**Underwater Explosions.** Compared to gaseous explosions, underwater explosions encompass a wealth of complex and different phenomena. Their investigation is more challenging to the shock physicist and requires cameras with higher frame rates and shorter exposure times, because the velocities involved are higher (by a factor of almost five) than for explosions in air. In addition, proper illumination is more difficult to achieve because of the higher attenuation and greater scattering of light in water.

A distinctive feature of explosions in water is that the gas bubble pulsates between a series of maximum and minimum values for the bubble radius. In the early stages of the explosion, the bubble expands until the mean radius inside it falls below the pressure in the compressed water surrounding it. It then stops expanding and instead starts to contract. It contracts until its mean pressure exceeds that in the surrounding water, at which point it again expands outwards. During the contracting phase, compressive waves are sent inwards, giving rise to a converging spherical shock wave that is reflected at the center as an outgoing shock. A new shock wave is formed during each contraction, so multiple shocks are one characteristic of underwater explosions. However, beginning with the first contraction phase, the buoyancy effect distorts the spherical form of the bubble significantly, leading to a kidney-shaped bubble geometry and the generation of a vortex ring {COLE ⇨1948; SNAY ⇨1958; HOLT ⇨1977}. Novel effects observed for nuclear explosions generated underwater, at the surface or above it stimulated numerous investigations into their mechanisms and energy release {COLLINS & HOLT ⇨1968}.

---

[253] S.J. BLESS and A.J. HOLTEN (eds.): *Hydraulic Ram Seminar*. Tech. Rept. AFFDL-TR-77-32. Wright-Patterson AFB, OH (May 1977).

[254] J. MORROCCO and P. SPARACO: *Concorde team activates return-to-flight plan.* Aviation Week & Space Technology **154**, 38-40 (Jan. 22, 2001).

[255] The large output piston (plunger) of a hydraulic press, an invention of the Englishman Joseph BRAMAH {⇨1795}, is also called a "hydraulic ram."

[256] R.F. WILLIAMS: *Shock effects in fuel cells.* Final Rept. prepared for McDonnell-Douglas Corporation, Long Beach, CA. Poulter Laboratory, SRI, Menlo Park, CA (1969).

Shock wave effects in water resulting from underwater explosions were first observed in military applications. For example, Henry L. ABBOT {⇨1881} in the United States and Rudolf BLOCHMANN {⇨1898} in Germany studied underwater explosion phenomena associated with submarine mines, a subject that had become of increasing interest to all navies since the invention of the torpedo in the 1860s. During World War II, research on underwater explosions was driven forward on a large scale by the United States and England. Their UNDEX Reports {⇨1946}, published after the end of the war, include a wealth of data on underwater explosion phenomena and their analytical treatment, and even today are a rich source of information. Robert COLE's book *Underwater Explosions* {COLE ⇨1948}, which largely built on the results from research performed by the Allies during World War II, became a bestseller in this field, and is probably still the textbook consulted most often on the propagation and effects of shock waves in liquids.

**Water Ricochets.** Rebounding or "skipping" is a phenomenon that can easily be demonstrated by throwing flat stones along the surface of a pond at a low angle of elevation so as to make them rebound repeatedly from the surface of the water, raising a succession of jets – an action often called "playing ducks and drakes." It was first studied scientifically by the Bohemian naturalist Marcus MARCI {⇨1639}. By observing their trajectory, he explained the rebound effect using the Law of Reflection {⇨Fig. 4.3–B}, which was already known in optics. *Ricochet firing*, which is also based upon rebound, was a French discovery made in 1697 by the military engineer Sébastien LE PRESTRE DE VAUBAN: a shot could be aimed so as to skip over the ground (or water) many times, which allowed areas protected from direct fire to be targeted. It also increased ballistic ranges considerably {DOUGLAS ⇨1851}. For example, test firings aboard HMS *Excellent*, used as a floating gunnery school at Portsmouth from 1859 onwards, revealed that at an elevation of one degree, the 800-yard (732-m) range of a 32-pound (14.5-kg) gun could be increased to 2,900 yards (2,652 m) after 15 grazes.[257] Notes written at the time recommended that, "Ricochet firing requires a perfectly smooth sea. The closer the gun is placed to the water, the farther it ranges the shot. It might be advisable to heel the ship over by running in the opposite guns."

This peculiar percussion phenomenon gained new interest with the advent of seaplanes and the need for them to land at high speed or on rough seas. Investigations performed in various countries, including the United States (KÁRMÁN & WATTENDORF 1929), Germany {WAGNER ⇨1932}, and the former Soviet Union (SEDOV & WLADIMIROW 1942), revealed that this skipping effect is a complicated combination of gliding and periodic bouncing which also generates finite-amplitude waves in the water. To extend the rough-water capabilities of seaplanes, important scientific progress was made in the theory of water impact and submerged lifting elements (hydrofoils). In the 1930s, the largest and fastest airplanes in the world were seaplanes, but after the outbreak of World War II their commercial and military significance diminished.

It is interesting to note here that "skipping" was also proposed in 1944 in Germany by the rocket engineer Eugen SÄNGER[258] and by the mathematician Irene BREDT for long-range hypervelocity vehicles, where the vehicle would "bounce" along the denser air layers surrounding the Earth. Using such a "skip trajectory," a combination of a ballistic trajectory with a glide trajectory, ballistic flight would go hand-in-hand with lifting flight.

**Cavitation.** Historically, Osborne REYNOLDS,[259] a British engineer and physicist, was the first to observe cavitational phenomena in water flowing through a tube with a local constriction. He described these phenomena in a paper dated 1894. In accordance with the Bernoulli law, a region of lowered pressure is formed in the narrow part of the tube. If the flow rate is sufficiently great, the pressure falls to a value corresponding to the vapor pressure, which at room temperature is only about 20 mbar, causing the water to boil in the narrow cross-section of the tube. These empty *bubbles* (or *cavities*) carried along by the current collapse and vanish upon entering a region of increased pressure. The collapse of each bubble, which proceeds at a very high rate, is accompanied by a kind of a *hydraulic blow* that causes a sound (a click). The clicks from a large number of bubbles collapsing unite into one continuous sound, and this what is heard when water flows through a tube with a local constriction {KORNFELD & SUVOROV ⇨1944}.

---

[257] *Early life and royal naval service. Muzzle-loading naval guns – some data*. Maritime history; http://www.cronab.demon.co.uk/info.htm.

[258] E. SÄNGER and I. BREDT: *Über einen Raketenantrieb für Fernbomber*. Bericht U93538, Dt. Luftfahrtforsch. (1944). Engl. translation in: *A rocket drive for long-range bombers*. Rept. No. CGD-32, Tech. Info. Branch, Bureau of Aeronautics, Navy Dept., Washington, DC (1946).

[259] O. REYNOLDS: *Experiments showing the boiling of water in an open tube at ordinary temperatures*. In: *Papers on mechanical and physical subjects*. Cambridge University Press, Cambridge (1901), vol. 2 (1881–1900), pp. 578-587.

*Cavitation* is understood as being the violent agitation of a liquid caused by the rapid formation and collapse of bubbles, transforming a liquid into a complex two-phase, liquid/vapor system. The term *cavitation* was coined in England {THORNYCROFT & BARNABY ⇨1895}. Cavitation became a subject of intense research in science and engineering shortly after the first use of steam turbines. In the 1880s, at the beginning of the age of steam turbines, erosion effects caused by cavitation were observed not only on the blade tips of turbine wheels but also on marine propellers that were initially driven at very high speeds in order to avoid losses involving high gear reduction between turbine and propeller. Studies of cavitation phenomena were prompted by both engineering requirements (to avoid cavitation erosion) {THORNYCROFT & BARNABY ⇨1895; S.S. COOK ⇨1928} and scientific curiosity {Lord RAYLEIGH ⇨1917; PRANDTL 1925; JOUGUET 1927; ACKERET 1938, LAUTERBORN ET AL. ⇨1972}. The central implosion of cavitation bubbles is accompanied by the emission of shock waves, which can have destructive effects. Russian researchers were the first suggest that the cavities can collapse asymmetrically and produce a liquid jet {KORNFELD & SUVOROV ⇨1944}. John P. DEAR and John E. FIELD, two British physicists at Cavendish Laboratory, performed detailed investigations of this phenomenon: using high-speed photography they observed that asymmetric bubble collapse is indeed accompanied by jet formation, which they suggested as being the major mechanism for cavitation damage {DEAR & FIELD ⇨1988}. In addition, jet produced by the shock collapse of cavities might play a role in the ignition and propagation of explosive reactions {BOWDEN & MCONIE ⇨1965; COLEY & FIELD ⇨1970}.

**Supercavitation.** An extreme version of cavitation is *supercavitation*. In order to reduce the hydrodynamic drag of an object moving at high speed through a liquid (water), a single bubble is artificially formed that envelops the object almost completely. This technique has been pursued since the 1960s in naval warfare by both Soviet and western researchers. At velocities over about 50 m/s (180 km/h), blunt-nosed projectiles – so called "cavitators" – or prow-mounted gas-injection systems are used to produce these low-density gas pockets – so-called "supercavities." With slender, axisymmetric bodies, such supercavities take the shape of elongated ellipsoids, ideally beginning at the forebody and trailing behind, with the length dependent on the speed of the body.[260] Supercavitation allows objects to achieve supersonic velocities in water {Naval Undersea Warfare Center ⇨1997; ⇨Fig. 4.14–E}. However, it is apparently still difficult to maintain a complete artificial bubble over a longer period of time.

**Photodisruptive Effect.** Bubble collapse in a liquid and its associated shock pressure effects can now be generated over a very wide spatial/temporal range, covering meters/milliseconds down to nanometers/femtoseconds. An example of the upper limit of bubble dynamics is the gas sphere of an underwater explosion.[261] On the other hand, bubble dynamics at the lower limit encompass all microcavitation phenomena: for example, the irradiation of biological tissue with femtosecond laser pulses can result in ultrashort mechanical shock pulses causing a mini explosion – a phenomenon which has been termed the *photodisruptive effect*. It has been used in femtosecond laser nanosurgery to create a "nanoscalpel" capable of cutting nanometer-sized particles, such as chromosomes in a living cell {KÖNIG ET AL. ⇨1999}.

Cavitation is a very effective way of generating shock waves, and it is even used in the animal world. For perhaps millions of years, the species *Alpheus heterochaelis* of the family *Alpheidae* – the largest snapping shrimp to inhabit tropical and shallow seawaters, and one that reaches a body length of up to 55 mm – can clamp its claw so rapidly that a water jet is generated which causes a vapor bubble in the jet to swell and collapse with a distinct "snap." Recent studies performed at the University of Twente by the Dutch scientist Michael VERSLUIS and German collaborators at the Universities of Munich and Marburg revealed that the very short shock pulse emitted has a peak pressure of about 80 bar measured with a needle hydrophone at a distance of 4 cm from the claw {LOHSE ET AL. ⇨2000; ⇨Fig. 4.1–Z}. The shock wave is apparently used to stun or kill small prey. In Germany, the snapping shrimp is appropriately called "pistol shrimp" [Germ. *Pistolenkrebs*].[262]

**Sonoluminescence.** The conversion of acoustical or shock wave energy into optical energy is a mysterious phenomenon. The light emission produced in this way from gaseous bubbles in liquids is termed *sonoluminescence*. In the early 1930s, it was discovered in France that a photographic plate

---

[260] S. ASHLEY: *Warp drive underwater*. Scient. Am. **284**, 62-71 (May 2001).

[261] In an underwater explosion the maximum bubble radius $R_{max}$ is given by $R_{max} = J(W/Z)^{1/3}$, where $W$ = charge weight in lb, $Z$ = total hydrostatic head (including that of the atmosphere) in ft, $J$ = radius constant (= 12.6 for TNT). For instance, the maximum bubble radius of a 300-lb charge of TNT which explodes in 30-ft depth is about 21 ft; {SNAY ⇨1958}, p. 267.

[262] R. ALTEVOGT: *Höhere Krebse*. In: (B. GRZIMEK, ed.) *Grzimeks Tierleben. Enzyklopädie des Tierreiches*. Kindler-Verlag, Zürich, vol. I (1971), p. 487.

could be fogged in the presence of a cavitation field generated by ultrasound in water.²⁶³ It had previously been assumed that this effect arose from the nucleation, growth and collapse of gas-filled bubbles in a liquid. Sonoluminescence produced from single bubbles allowed much better insights into this effect and its characteristics.²⁶⁴ However, theories on the origin of sonoluminescence are still somewhat controversial.²⁶⁵ The most plausible explanation for the origin of the extremely short bursts of light emitted from the bubble is that an imploding shock wave is generated within the bubble during the final stages of collapse which heats up gas inside the bubble and generates light. Experimental results obtained by three groups at American Universities (Yale, WA and Los Angeles, CA) support this explanation for sonoluminescence – namely that a collapsing bubble creates an imploding shock wave. These three groups all recorded sharp acoustical pops during the sonoluminescence process.²⁶⁶

Very recently, it was observed that the snapping shrimp also creates a flash of light during shock wave emission, which the discoverers have appropriately called "shrimpoluminescence" {LOHSE ET AL. 2000}.

**Electrohydraulic Effect.** First observed in England {SINGER & CROSSE ⇨1815} and later rediscovered in the United States {RIEBER (1947), ⇨1951} and the former Soviet Union {JUTKIN ⇨1950}, the *electrohydraulic effect* uses a powerful high-voltage discharge to generate strong shock waves, with the resulting strong current pulse being fed into either a thin metallic wire submersed in water or a spark gap submerged in an electrically nonconducting (organic) liquid. This effect was first highlighted by the Latvian urologist Viktor GOLDBERG {⇨1959}, who first successfully applied it to the disintegration of bladder stones in man. His method has been called "electrohydraulic shock lithotripsy." Later, this effect was also used for the disintegration of rocks, in well-drilling, and for grinding materials. Applications in production technology also include high-speed deformation of metals and intensification of processes in chemical and chemical-metallurgical engineering.

## 2.4.6 SOLID-STATE SHOCK WAVE PHYSICS: INITIATION BY NUCLEAR WEAPONEERS

> *Shock wave physics as we know it today owes its genesis and entire formation period to a minor industrial revolution of precision casting, pressing and machining of explosives. This revolution and in addition the long-term support for equation-of-state research were entirely the result of the World War II crash program to develop a nuclear weapon.*²⁶⁷
>
> John W. TAYLOR
> Los Alamos National Laboratory
> Los Alamos, NM 1983

> *In early 1948, the scientific leadership of the All-Union Research Institute of Experimental Physics (VNIIEF) set experimenters a fundamental task of determining the equations of state and the shock compressibility of fissioning materials at megabar pressures. Because of the lack of accurate equations of state, it was impossible to predict unambiguously the power of the first A-bomb variant then in preparation for a test, nor could alternative bomb models be readily compared and assessed.*²⁶⁸
>
> Lev V. AL'TSHULER
> Institute of High Temperatures
> Moscow 1996

In solid amorphous or noncrystalline materials, the component atoms or molecules are usually hold together by strong forces, which therefore occupy definite positions relative to one another. In crystalline solids, however, the atoms (and molecules) are not only held in place, they are also arranged in a definite order that is constantly repeated throughout the sample; this periodic arrangement of atoms in a crystal is termed a *lattice*. In a crystal, an atom never strays far from a single, fixed position; the thermal vibrations associated with the atom are centered about this position. Shock propagation through such a lattice diminishes the *unit cell* that is representative of the entire lattice. Obviously, shock wave propagation through solids is far more complex than through a gas.

The Rankine-Hugoniot relationships {RANKINE ⇨1869; HUGONIOT ⇨1887}, which relate the thermodynamic state of the medium behind the shock front to its initial state, the shock wave velocity and the particle velocity, also proved most useful in the further evolution of solid-state shock wave physics. detailed experimental investigations of shock waves and the

---

[263] N. MARINESCO and J.J. TRILLAT: *Actions des ultrasons sur les plaques photographiques.* C. R. Acad. Sci. Paris **196**, 858-860 (1933).
[264] D.F. GAITAN: *An experimental investigation of acoustic cavitation in gaseous liquids.* Ph.D. thesis, University of Mississippi (1990).
[265] C.M. SEHGAL: *Sonoluminescence.* In: (G.L. TRIGG, ed.) *Encyclopedia of applied physics.* VCH, New York etc., vol. 19 (1997), pp. 1-23.
[266] *Sonoluminescence research vibrates with activity.* AIP Bulletin of Physics News No. 299 (Dec. 13, 1996).

[267] J.W. TAYLOR: *Thunder in the mountains.* In: (J.R. ASAY, R.A. GRAHAM, and G.K. STRAUB, eds.) *Proc. 3rd Conference on Shock Waves in Condensed Matter* [Santa Fe, NM, July 1983]. North-Holland, Amsterdam etc. (1984), pp. 3-15.
[268] L.V. AL'TSHULER: *Experiment in the Soviet atomic project.* In: (I.M. DREMIN and A.M. SEMIKHATOV, eds.) *Proc. 2nd Int. A.D. Sakharov Conference on Physics* [Moscow, USSR, May 1996]. World Scientific, Singapore (1997), pp. 649-655.

shock-induced properties of materials have essentially been made possible by devising ways to measure some of these parameters, from which the others could be derived by the Rankine-Hugoniot relationships. Initially, however, experimentalists lacked the diagnostic means to temporally resolve the shock wave parameters, and for many years progress in understanding shock waves in solids was closely dependent on progress in fast electronics and high-speed photography.

The most important (largely theoretical) works on shock waves in condensed matter dating from the period 1808–1949 have been brought together into a single collection.[269] Important papers from 1948 to the present were later added to this original collection, allowing research workers and historians to trace the major developments leading to the establishment of the unique field of shock compression in solids.[270]

**Roots.** Historically, *solid-state shock wave physics* is an outgrowth of classical percussion, which was established in the 17th century, but the main push into this new area of research came from 19th century pioneers of classical shock wave theory, who did not limit their analyses to fluids, but also reflected on the peculiarities of shock waves in solids. In his famous treatise *On the Thermodynamic Theory of Waves of Finite Longitudinal Disturbance*, the Scottish civil engineer William J.M. RANKINE {⇨1869} clearly states that his derived relations are valid "for any substance, gaseous, liquid or solid." Subsequent works on this subject addressed the shock-compressed solid state in more detail {CHRISTOFFEL ⇨1877; HUGONIOT ⇨1889; DUHEM ⇨1903; HADAMARD ⇨1903; JOUGUET ⇨1920}. In addition, important contributions from theoretical physicists also stimulated the evolution of shock wave physics in solids. Prominent examples include:

- various laws and theories of elastic percussion {HUYGENS ⇨1668/1669; EULER ⇨1737; D. BERNOULLI ⇨1770; HODGKINSON ⇨1833; Lord KELVIN ⇨1879; HERTZ ⇨1882};
- the first theory of elasticity {MAXWELL ⇨1850};
- the equation of state for solid matter based on lattice vibration theory, developed by the German physicists Gustav MIE (1903) and Eduard GRÜNEISEN {1912 & ⇨1926}; and
- theories on the dynamic plasticity of metals proposed by Geoffrey I. TAYLOR (1942), Theodore VON KÁRMÁN (1942) and Khalil A. RAKHMATULIN (1945) {B. HOPKINSON ⇨1905}.

In contrast to the rapid and steady progress made in shock wave physics in gaseous matter since the 1870s, research on shock-compressed solids evolved only very slowly. The main reason for this was certainly the need for high-speed diagnostics with submicrosecond resolution, which were not − with just a few exceptions − available until after World War II, and were even then initially confined to only a few laboratories.[271] Early researchers studying the dynamic properties of solids, particularly those interested in their rate-dependent strength, had to rely on simple experimental techniques. For example, the British engineer John HOPKINSON {⇨1872} measured the strength of a steel wire when the wire was suddenly stretched by a falling weight. He made the important observation that the strength is much greater under rapid loading than in the static case, a puzzling phenomenon that was later investigated in more detail by his son Bertram HOPKINSON {⇨1905}, who occupied the chair of applied mechanics at the University of Cambridge. He also discovered *spallation*, a curious fragmentation phenomenon that occurs at the back side of a metal plate loaded on its front side by a high explosive {B. HOPKINSON ⇨1912}, or by the hypervelocity impact of a solid body {⇨Fig. 4.3–U}. HOPKINSON's results sowed the seeds of modern materials dynamics, a field that has grown steadily ever since, and which has become an indispensable tool in the computer modeling of structural behavior under high loading rates.

In the 19th century, the British engineer Charles A. PARSONS {⇨1892}, inventor of the multistage steam turbine, and the French chemist Henri MOISSAN, a professor at the University of Paris, attempted to use shock waves to induce polymorphism in solids (particularly in carbon to produce artificial diamonds).[272] However, their efforts did not provide any clear evidence, and were just too ambitious for their time. An important step toward this goal was the later work on the effects of static high pressure on a large number of liquid and solid substances carried out by the U.S. experimental physicist Percy W. BRIDGMAN at Harvard College, Cambridge, MA. Throughout a long-lasting campaign (1903–1961), BRIDGMAN laid the foundations for understanding the behav-

---

[269] J.N. JOHNSON and R. CHÉRET (eds.): *Classic papers in shock compression science.* Springer, New York (1998).

[270] J.N. JOHNSON and R. CHÉRET: *Shock waves in solids: an evolutionary perspective.* Shock Waves **9**, No. 3, 193-200 (1999).

[271] At this point, the diagnostic instruments with the highest time resolution were mechanically driven streak cameras, which provided records of distance as function of time. At the end of World War II, the fastest camera available was the U.S. Bowen RC-3 rotating mirror camera, which had a writing speed of 3.1 mm/μs − *i.e.*, 1 mm on the film plane corresponded to a time span of about 320 ns.

[272] A medium undergoing shock-induced polymorphism − *i.e.*, a first-order phase transition ($\Delta V < 0$) − exhibits discontinuities in slope of the Hugoniot curve at the boundaries of the mixed phase region. • In solid matter, the Hugoniot curve can also show a discontinuity (known as a "cusp"), where elastic failure occurs.

ior of matter under static high pressures, which earned him the 1946 Nobel Prize for Physics.

In this early era, when even static high-pressure research was still in its infancy and had to battle against many technical problems with achieving such high pressures, shock waves generated by high explosives were already considered to be an easy way of reaching high pressures, although appropriate high-speed instrumentation for detecting and recording physical parameters of interest was not yet available. Compared to static methods, the use of detonation to achieve very high pressures is indeed surprisingly simple, particularly when using Mach reflection of cylindrically convergent shock waves. For example, the U.S. physicists George R. FOWLES and William M. ISBELL at SRI's Poulter Laboratory devised a coaxial set-up comprising a hollow cylinder of high explosive that generated shock pressures of up to 1.9 Mbar in the center of an enclosed test sample – a cylindrical copper rod in their experiment. Their device was only about the size of a Coca-Cola bottle {FOWLES & ISBELL ⇨1965; ⇨Fig. 4.13–L}.

**Establishment and Motivations.** Contrary to the foundations of gas dynamics, which were mainly built upon European contributions, the foundations of modern solid-state shock wave physics were laid mainly in the United States and the former Soviet Union. Initiated during World War II, such investigations were exclusively motivated by military requirements.

On the one hand, research in the United States followed classical paths aimed at better understanding the dynamic responses of materials, particularly those of metals, under rapid loading conditions such as the penetration of a high-speed projectile into armor or its attack by high explosives, as well as appropriate protection measures – *i.e.*, research was carried out mostly according to the classic "terminal ballistics cycle" (*see* Sect. 2.3.3). During the war, U.S. activities in this subject of vital interest were coordinated by the National Defense Research Committee (NDRC), and the results were summarized in a confidential report[273] distributed in 1946, which covered the topics of terminal ballistics, dynamic properties of matter, and protection. On the other hand, solid-state shock wave physics was an outgrowth of nuclear weapons research related to the Manhattan Project {⇨1942}. The results from these secret studies were initially classified for years.

A historic review of this new branch of dynamic high-pressure physics was apparently first published in 1983 by the physicist John W. TAYLOR,[274] a former participant of the Manhattan Project. He remembered the beginning of shock wave physics at Los Alamos, "The history begins with a few hastily constructed demonstration experiments and ends with a new field of scientific inquiry." Another historic review was given in 1991 by the U.S. solid-state physicist Charles E. MORRIS,[275] which emphasized the pioneering shock wave research and equation-of-state studies at Los Alamos in the 1950s.

The first paper on the experimental study of shock wave propagation in solids, which was entitled *The Propagation of Shock Waves in Steel and Lead* and published in a British journal, was written by Donald C. PACK and coworkers {PACK ET AL. ⇨1948}, three physicists at the Armament Research Department (ARD) of the British Ministry of Supply. They investigated the stress system set up by an explosive detonating in contact with a metal surface, then the most basic and commonly used arrangement for generating high-pressure shock waves in any solid specimen.

The first review article on the progress of shock wave research in solids was not given until 1955: the U.S. physicist Roy W. GORANSON, who in 1945 initiated a program at Los Alamos Scientific Laboratory to systematically study the equations of state of various materials, presented the first data on the dynamic compressibility of metals. This allowed the first quantitative comparisons with BRIDGMAN's static measurements {GORANSON ET AL. ⇨1945 & 1955}. The classic, most widely cited, work in this new field of science was published in 1958 by three Los Alamos scientists {RICE, WALSH & MCQUEEN ⇨1958}. Their pioneering paper *Compression of Solids by Strong Shock Waves*, a landmark in the evolution of solid-state shock wave physics, was later appropriately commented on by the Sandia shock scientist Robert GRAHAM {⇨1993}, "Almost overnight, ordnance laboratories throughout the world were able to convert technology developed for weapons into technology for visionary studies of matter in a new regime. Swords forged for nuclear weapon development were beaten into high pressure science plowshares." Since then, an ever-increasing number of investigations have led to a considerable improvement in methods of generating dynamic pressures and their diagnostics as well as to refinements of numerical thermodynamic models.

One of the main goals for postwar solid-state shock wave research was the development of the hydrogen bomb – a complex and difficult task that required basic knowledge of how solid matter behaves under ultrahigh dynamic pressures that was previously not available. Progress in solid-state

---

[273] E.P. WHITE (ed.): *Effects of impact and explosions*. Summery Tech. Rept. AD 221 586, Div. 2 of NDRC, Washington, DC (1946).

[274] *See* reference given in footnote 267.

[275] C.E. MORRIS: *Shock-wave equation-of-state studies at Los Alamos*. Shock Waves **1**, 213-222 (1991).

shock wave physics is rather well documented in the open literature. In the Western World, the stepwise progress made and developmental highlights have been reviewed over the years in numerous articles contributed by researchers working at the forefront of this field, such as by

- Melvin H. RICE and associates {⇨1958} at Los Alamos Scientific Laboratory (LASL), New Mexico;
- George E. DUVALL[276] and George R. FOWLES, and William J. MURRI and associates {⇨1974} at Poulter Laboratory of SRI, Menlo Park, CA;
- Sefton D. HAMANN {⇨1966} at the Commonwealth Scientific and Industrial Research Organization (CSIRO) in Melbourne, Australia;
- Robert A. GRAHAM[277] and Lee DAVISON at Sandia Laboratories in Albuquerque, New Mexico;
- Budh K. GODWAL[278] and associates at Bhabha Atomic Research in Bombay, India; and
- George DUVALL[279] and Yogendra M. GUPTA[280] at the Shock Dynamics Laboratory of Washington State University in Pullman, Washington.

Independent of nuclear weapons research performed in the framework of the Manhattan Project, similar work was also initiated in the former Soviet Union's Nuclear Center located in Arzamas-16 [now Sarov]. This institution [now the Russian Federal Nuclear Center of the All-Russian Scientific Research Institute of Experimental Physics (RFNC-VNIIEF)], then headed by Academician Yulii B. KHARITON, was, from its very conception up to 1996, in many senses a "hidden world," where a multi-disciplinary team of physicists, mathematicians, designers and high-speed instrumentation engineers was provided with highly favorable working conditions, and where fundamental science and defense mutually benefited from one another.[281] Motivation and development of solid-state shock wave physics went along similar lines. A considerable number of eminent physicists, in particular Yakov B. ZEL'DOVICH and Lev V. AL'TSHULER, significantly contributed to the fields of detonation physics and dynamic high-pressure physics.[282] The first Soviet results on dynamic equations of state for a number of metals were not published in the open literature until 1958.[283] In 1965, AL'TSHULER,[284] the patriarch of Russian solid-state shock wave physics, reviewed the use of shock waves in high-pressure physics, and, more recently, the development of dynamic high-pressure techniques in the former Soviet Union, some of which were more advanced than those developed in the Western World.[285]

Beginning in the early 1950s, dynamic high-pressure physics was also increasingly applied beyond military circles, thereby pushing the upper limit of achievable pressures far beyond the static pressures achievable at that time. Dr. Thomas POULTER,[286] the first director of the Poulter Laboratory in Menlo Park, CA, remembered anecdotally, "We found that with the instrumentation available to us in the Poulter Laboratory we could conduct most of the experiments that I had done earlier with static pressures, but all of which were under one million psi [about 68 kbar]. After attending the Gordon Conference on High Pressure Physics [at Kimball Union Academy in Meriden, NH] where pressures were plotted up to 3 feet high and on a 3' × 4' chart and pressures obtained and studied by P.W. BRIDGMAN (who had probably done more high pressure work than all of us combined) were on a 3" × 4" lower left corner of the same chart, he [BRIDGMAN] commented, 'one hates to spend a lifetime in a field and see it pushed into such a small corner as that.'"

In particular, ambitious static high-pressure researchers targeted the "magic" limit of about 3.65 Mbar, a value considered by geophysicists to be the gravitationally generated pressure at the center of the Earth. Modern static compression techniques have even surpassed this limit. In 1990, Arthur L. RUOFF,[287] a professor of materials science at Cornell University, and his collaborators produced static pressures of

---

[276] G.E. DUVALL and G.R. FOWLES: *Shock waves.* In: (R.S. BRADLEY, ed.) *High-pressure physics and chemistry.* Academic Press, New York, vol. 2 (1963), pp. 209-291.

[277] R.A. GRAHAM and L. DAVISON: *Shock compression of solids.* Phys. Repts. **55**, 255-379 (1979).

[278] B.K. GODWAL, S.K. SIKKA, and R. CHIDAMBARAM: *Equation of state theories of condensed matter up to about 10 TPa.* Phys. Repts. **102**, 121-197 (1983).

[279] G.E. DUVALL: *Shock wave research: yesterday, today and tomorrow.* In: (Y.M. GUPTA, ed.) *Proc. 4th APS Topical Conference on Shock Waves in Condensed Matter* [Spokane, WA, July 1985]. Plenum Press, New York (1986), pp. 1-12.

[280] Y.M. GUPTA: *Shock waves in condensed materials.* In: (S.P. PARKER, ed.) *The encyclopedia of science and technology.* McGraw-Hill, New York, vol. 16 (1987), pp. 382-386.

[281] V.A. TSUKERMAN and Z.M. AZARKH: *Arzamas-16: Soviet scientists in the nuclear age: a memoir.* Bramcote, Nottingham (1999).

[282] Y. KHARITON, K. ADAMSKII, and Y. SMIRNOV: *The way it was.* Bull. Atom. Sci. **52**, No. 6, 53-59 (1996).

[283] L.V. AL'TSHULER ET AL.: *Dynamic compressibility and equation of state of iron under high pressure.* Sov. Phys. JETP **7**, 606-614 (1958); *Dynamic compressibility of metals under pressures from 400,000 to 4,000,000 atmospheres.* Ibid. **7**, 614-619 (1958).

[284] L.V. ALT'SHULER: *Use of shock waves in high pressure physics.* Sov. Phys. Uspekhi **8**, 52-91 (1965).

[285] L.V. AL'TSHULER ET AL.: *Development of dynamic high-pressure techniques in Russia.* Physics Uspekhi **42**, 261-280 (1999).

[286] T.C. POULTER: *Over the years.* Library of Stanford Research International (SRI), Menlo Park, CA (1978), p. 94.

[287] A.L. RUOFF, H. XIA, H. LUO, and Y.K. VOHRA: *Miniaturization techniques for obtaining static pressures comparable to the pressure at the center of the Earth: X-ray diffraction at 416 GPa.* Rev. Scient. Instrum. **61**, 3830-3833 (1990).

up to 4.16 Mbar in the laboratory via special diamond-anvil cells. Two years later they pushed this limit even further, up to 5.6 Mbar.[288]

BRIDGMAN's results {⇨1911, 1925 & 1931} gave modern shock physicists their first clues about the static compressibility of solids at high pressures and the stress-dependent plasticity of metals, thus arousing their curiosity about how substances would behave under dynamic pressures. This also promoted various other spectacular investigations, *e.g.*, on shock-induced polymorphic transitions in iron {BANCROFT ET AL. ⇨1956}, on possible ice modifications of shock-compressed water {SCHARDIN 1940/1941; RICE & WALSH ⇨1957; AL'TSHULER ET AL. ⇨1958}, on shock-induced transformations of quartz into high-pressure polymorphs {CHAO ET AL. ⇨1960; SHOEMAKER & CHAO ⇨1961}, and of shocked graphite into diamond {DECARLI & JAMIESON ⇨1961}. Today dynamic and static high-pressure methods of generation, calibration and measurement do not compete with one another, but are used instead to complement each other, as shown for the example of the extension of the "ruby scale" {BARNETT EL AL. ⇨1973}.[289]

**Materials Dynamics.** Modern testing methods for studying the responses of materials under shock loading conditions are closely related to percussion. The *Charpy test* {CHARPY ⇨1901} and the *Izod test* {IZOD ⇨1905}, two fracture toughness tests used at relatively low strain rates ($c. 10^1$–$10^2 \, s^{-1}$), use the impact energy of a single blow of a swinging pendulum. The energy absorbed, as determined by the subsequent rise in the pendulum, is a measure of the impact strength or notch toughness.

To a large extent, many high-rate testing methods are based on the planar impact of two bars, a basic 1-D arrangement which had already been treated by numerous researchers, both theoretically {EULER ⇨1737 & 1745; W. RICHARDSON 1769, see GALILEI ⇨1638; POISSON ⇨1811; CAUCHY ⇨1826; F.E. NEUMANN 1857/1858, see NEUMANN ⇨1885; DE SAINT-VENANT ⇨1867} and experimentally {RAMSAUER ⇨1909; DONNELL ⇨1930}. Today numerical analyses of 2-D and 3-D impact phenomena, ranging from stiff to flexible structures and also including vibration and wave effects, are of great practical concern in engineering {STRONGE ⇨2000}.

Today a variety of planar impact tests are widely used in high-rate materials testing; prominent examples include

▸ the *Hopkinson pressure bar* {B. HOPKINSON ⇨1914}, which was further developed into the *split Hopkinson pressure bar* or *Kolsky bar* {KOLSKY ⇨1949};
▸ the *Taylor test* {G.I. TAYLOR & WHIFFIN ⇨1948};
▸ the *flyer-plate method* {MCQUEEN & MARSH ⇨1960}; and
▸ the *planar impact test* using a high-velocity projectile {HUGHES & GOURLEY 1961, see CROZIER & HUME ⇨1946}.

Modern methods of performing high-rate materials testing are based on the generation and careful diagnostics of uniaxial shock waves in the test specimen as well as on analyses supported by complex numerical modeling on a microscopic level, thereby averaging micro-processes over a "relevant volume element" in a so-called "mesomodel" which contains a statistical number of micro-units in order to get a continuum model {CURRAN ET AL. ⇨1987}.[290] The response of materials to high-velocity impact loading, which is governed by both thermomechanical and physical processes, spans a wide region of material behavior, ranging from high impact pressures and temperatures where thermodynamic effects prevail, to low pressures where mechanical properties are important.[291]

**Dynamic Fracture.** *Fracture mechanics* is a branch of solid mechanics that is concerned almost entirely with fracture-dominant failure and deals with the behavior of cracked bodies subjected to static and/or dynamic stresses and strains. While such research was stimulated during the Industrial Revolution by the many accidents resulting from the failure of mechanical structures (such as steam-boilers, railway equipment, *etc.*), fracture mechanics as a science originated in the 1920s from a study of crack propagation in brittle materials, predominantly in glass.[292] *Fracturing* is a natural dynamic process in which material bonds are broken and voids are created in a previously intact material.

---

[288] A.L. RUOFF, H. XIA, and Q. XIA: *The effect of a tapered aperture on X-ray diffraction from a sample with a pressure gradient: studies on three samples with a maximum pressure of 560 GPa.* Rev. Scient. Instrum. **63**, 4342-4348 (1992).

[289] W.B. DANIELS: *High-pressure techniques.* In: (G.L. TRIGG, ed.) Encyclopedia of applied physics. VCH, New York *etc.*, vol. 7 (1993), pp. 495-509.

[290] The word *mesomodel* is a contraction of *meso-scale model* (meso means "medium"). Mesomodels are used, for example, for numerical simulations of the multiphase behavior of materials under dynamic load, and for the localization of and damage computation in laminates. A *laminate* is considered, at the mesoscopic scale, to be a stack of homogenized plies and interfaces.

[291] J.R. ASAY and G.I. KERLEY: *The response of materials to dynamic loading.* Int. J. Impact Engng. **5**, 69-99 (1987).

[292] A historical review of fracture mechanics was given by M. JANSSEN, J. ZUIDEMA, and R.J.H. WANHILL in their book *Fracture mechanics.* Dup Blue Print, Delft (2002) & Spon Press, London (2004), chap. 1.

The most striking fracture phenomena are the initiation, propagation and branching of cracks, which were first investigated in transparent brittle materials such as glass and Plexiglas using high-speed cinematography {SCHARDIN ⇨1954}. These studies showed that cracks propagate slower than the longitudinal speed but faster than the shear wave speed in the material. Based upon predictions from classic continuum mechanics, it was widely believed that a brittle crack could not propagate faster than the longitudinal wave speed. However, later studies performed at high energy rates using a ruby laser showed that cracks can also propagate along weak crystallographic planes supersonically in regard to the longitudinal wave speed {WINKLER ET AL. ⇨1970}. More recent studies in brittle polyester resin have shown that, when cracks propagate faster than the shear wave speed, a traveling Mach cone is created which consists of two shear shock waves. Apparently, these results have similarities to shallow earthquake events {ROSAKIS ET AL. ⇨1999}.

The term *dynamic fracture* denotes the effects of inertia resulting from the rapid propagation of a crack, and the label "dynamic loading" is attached to the effects of inertia on fracture resulting from rapidly applied loads.[293] In the past, various mathematical attempts were made to understand dynamic fracture: both the threshold conditions that trigger this process, and the kinetics by which it proceeds. A landmark in the evolution of modern fracture mechanics was the development of mesomechanical[294] failure models by the U.S. metallurgist Troy W. BARBEE[295] and coworkers. They realized that the nucleation, growth, and coalescence of microscopic voids, cracks, or shear bands in the failing material could be treated as the development of a new phase, and that quantitative post test analysis of recovered specimens could be used to develop nucleation and growth functions for the process. This era of dynamic fracture research was documented in a 1987 review paper {CURRAN ET AL. ⇨1987}. The latest development in this field are atomistic models for simulating materials failure that involve up to one billion atoms and the use of supercomputers; this area of research is known as "molecular dynamics" {ABRAHAM & GAO ⇨2000}. Such simulation studies have demonstrated that the limiting speed of fracture is the longitudinal wave speed in harmonic (linear) solids. Also, in anharmonic solids the crack velocity can even exceed the longitudinal wave speed, thus becoming truly supersonic.

**Equations of State.** The large number of investigations on shock waves and their physical effects that have been carried out so far in solids have revealed that solids exhibit rather complex behavior in comparison to gases and liquids. Most solids exhibit an elastic-plastic behavior – *i.e.*, they will obey the Law of Linear Elasticity {HOOKE ⇨1679} until a yield condition is reached above which the solid will deform plastically. For stresses much higher than the so-called "yield stress," a solid behaves like a fluid to a first approximation, since the percentage deviations from the isotropic stress distribution are small. Therefore, in order to describe the shock compression of a solid material in terms of an equation of state, a form similar to the equation of state for a gas or a liquid was assumed.

In the early period of shock wave research, the simple *Murnaghan equation* {MURNAGHAN ⇨1937} was proposed, which is similar to the *Tait equation* {TAIT ⇨1888} for shock-compressed liquids. However, neither of these equations consider any shock heating. One major advance was the application of the *Mie-Grüneisen equation* {GRÜNEISEN ⇨1926} to shock-compressed solids {WALSH, RICE, MCQUEEN & YARGER ⇨1957}, which became probably the most commonly used equation of state in solid-state shock wave physics. However, theories describing the thermodynamic state under shock loading – which therefore also take into account rate-dependent and structural properties such as porosity and viscosity – have proved to be very complex and are still in development. In addition, the shock characterization and thermodynamic behavior of composite materials, which have been increasingly used since the 1970s in military and space applications to provide effective shields against high-velocity impact {WHIPPLE ⇨1947}, are a particular challenge to theoreticians.

The shock-compressed state of liquids and solids is not usually described and characterized by its "rapidity," as in gas dynamics (with the Mach number), but rather by the shock pressure – an historic relic of static high-pressure research which, as it steadily pushed the pressure limit higher and higher, used to announce its progress in terms of the maximum hydrostatic pressure achieved. In the 1950s and 1960s, many solid-state shock physicists followed this trend. However, materials response to shock loading is better described by the concept of stresses and strain rates, particularly when shear behavior cannot be neglected. The stress

---

[293] J.D. ACHENBACH: *Dynamic effects in brittle fracture.* In: (S. NEMAT-NASSER, ed.) *Mechanics today.* Pergamon Press, New York etc., vol. 1 (1972), pp. 1-57.

[294] According to ANTOUN ET AL. {⇨2003}, *mesomechanics* is a new approach used in fracture mechanics "to average the behavior of the individual voids or crack over a 'relevant volume object' representing a continuum point in space."

[295] T.W. BARBEE, L. SEAMAN, and R.C. CREWDSON: *Dynamic failure criteria of homogeneous materials.* Tech. Rept. No. AWFL-TR-70-99, Air Force Weapons Laboratory, Kirtland Air Force Base, NM (1970).

generated by a planar normal shock wave, which results in a uniaxial strain, can be decomposed into a hydrostatic component and a deviator.[296]

For over 30 years the shock community has grappled with how to measure the complete stress in and behind a shock wave. The most accurate shock wave measurements of pressure or stress have always been made in uniaxial stress, for example by planar impact. Such measurements typically provide stress, shock velocity, and particle velocity histories in the direction of shock propagation. The use of Lagrangian analysis yields stress/strain paths under uniaxial strain loading. In solid-state shock wave physics, correct measurements of stress histories are essential when testing numerical codes describing dynamic materials behavior. On the other hand, more refined numerical modeling has disclosed problems with measuring lateral shock stresses accurately (*see* Sect. 2.8.6).

A useful quantity to use to characterize a shock-compressed solid is the strain rate, since the response of a solid is dependent on the rate at which strain occurs {EXPLOMET 80 ⇨ 1980}. The global 1-D strain rate is defined as the rate of change of strain $\varepsilon$ with time $t$, given by

$$d\varepsilon/dt = 1/L \times dL/dt \approx 1/L \times \Delta L/\Delta t = v/L,$$

where $L$ is the initial length, $\Delta L$ is the elongation, and $v$ is the speed of deformation. At high rates, many materials deform by a different mechanism to that which occurs at low rates. High strain rates enter into most fracture, impact, erosion and shock loading situations. Depending on the method applied, strain rates obtainable in practice range from about $10^1 - 10^2$ s$^{-1}$ (Charpy test and Izod test) to about $10^3$ s$^{-1}$ (split-Hopkinson pressure bar) up to $10^7$ s$^{-1}$ (high explosives, hypervelocity impact). Extreme strain rates of up to some $10^9$ s$^{-1}$ are possible in thin targets subjected to irradiation from ultrahigh-power laser pulses of picosecond duration.

With continuous technical improvements in the generation and recording of flash X-ray diffraction patterns at submicrosecond timescales, this promising diagnostic method was first successfully applied in the 1970s to shock-compressed materials. This allowed the interpretation of the physical quantity "strain rate" – originally introduced to characterize dynamic loading in the macroscopic world – to be extended to the microscopic regime too: in a crystal lattice the uniaxial compression of the unit cell with interplanar spacing $d$, caused by a shock wave with rise time $\Delta \tau$, is given by $\Delta d/d$,

and the corresponding strain rate $d\varepsilon/dt$ is approximately given by

$$d\varepsilon/dt \approx (\Delta d/d)/\Delta \tau.$$

A crystalline powder sample with a cubic lattice such as potassium chloride [KCl], when isotropically compressed by a planar 12-kbar shock wave with a rise time of some 10 ns {JAMET & THOMER ⇨ 1972}, would produce a microscopic strain rate on the order of $10^6$/s.

**Off-Hugoniot States.** *Principal Hugoniots* – *i.e.*, the loci of single shock states for shocks of varying strength, starting from materials at normal density, atmospheric pressure, and room temperature (~ 300 K) – do not cover all states of interest in dynamic high-pressure physics. In order to obtain a more complete description of achievable states, it is necessary to access ($p,v$)-data of shock pressure $p$ and density $\rho$ (or specific volume $v = 1/\rho$) off the principal Hugoniot, which requires more sophisticated techniques and analytical methods. Traditionally, multiple or reflected shock waves have been used to achieve this goal. Such *off-Hugoniot* states are of interest, for example, when modeling

▸ materials under extreme conditions of pressure and temperature, such as those used in inertial confinement fusion studies;
▸ detonation products of solid high explosives; and
▸ planetary and stellar interiors at extreme conditions.

The impressive advancement achieved in experimental solid-state shock wave physics and improvements made in numerical modeling of dynamic materials response were mainly based on

▸ the availability of principal Hugoniot data for a large number of elements and compounds;
▸ the availability of more complete ($p, v, e, T$) equation of state data by extending principal Hugoniot data using the Mie-Grüneisen theory;
▸ the generation of well-defined and reproducible shock wave profiles;
▸ advances in submicrosecond measurements and visualization techniques;
▸ more refined analyses of thermodynamic states off the principal Hugoniot; and
▸ increased computer power, which permitted more detailed numerical material modeling.

Today, solid-state shock wave physics, now a well-established branch of high-pressure physics, provides a rich source of principal Hugoniot data for all kinds of solids: so-called "Hugoniot Data Banks" cover almost all solid

---

[296] R.J. WASLEY: *Stress wave propagation in solids.* M. Dekker, New York (1973).

elements, a large number of technically relevant metal alloys, as well as many common minerals {WALSH, RICE, MCQUEEN & YARGER ⇨1957; VAN THIEL, SHANER & SALINAS ⇨1977}. In addition, solid-state shock wave physics has also placed a wealth of unique shock compression and diagnostic techniques at the disposal of scientists of other disciplines. This has significantly contributed not only to progress in traditional fields such as impact physics, ballistics and crash engineering, but also to new applications in geophysics, planetary sciences, seismology, fracture mechanics, laser fusion, materials science, and more recently also in medicine and biology.

## 2.5 PIERCING THE SOUND BARRIER: MYTH AND REALITY

> *"That is correct,"* answered NICHOLL, *"it is now 11 P.M. Thirteen minutes since we left America."*
>
> *"One question, however, still remains unsolved"* said BARBICANE, *"why have we not heard the bang when the Columbiad was fired?"*
>
> *Less than a quarter of an hour later, BARBICANE stood up and shouted in a piercing voice: "I have it. Because our projectile flew faster than sound!"*[297]
>
> Jules VERNE<br>Paris 1869

COVERING large distances in a short time – *i.e.*, traveling at high speed, much faster than the fastest animals can move along – has been a dream of man since the earliest times, and has been the subject of numerous myths and legends {⇨Figs. 4.20−A, B}. Since until very recently man was incapable of achieving this goal through his own technology, he projected his ambition onto the supernatural powers of deities. For example, in early Greek mythology, the god HELIOS, who represented the Sun, was believed to traverse the heavens by day and to sail around the Earth (then viewed as a flat disc afloat on the river of Ocean) at night. His high-speed vehicles, which would have needed to be capable of supersonic velocities to perform this enormous task, were a four-horse chariot and a mystical boat in the form of a golden bowl. Even now, only a few decades have passed since man first realized this dream, although the velocity required by HELIOS' to sail around the world in a night has been exceeded by more than one order of magnitude {GAGARIN ⇨1961}.

Flying supersonically – let alone from one planet to another – was regarded as the wildest fiction at a time when the first flight of man with an "apparatus heavier than air" {VON LILIENTHAL ⇨1891} was still the pipe dream of a few outsiders. However, when the French author Jules VERNE wrote his fictional works *De la terre à la lune* (1865) and *Autour de la lune* (1872), substantial insights into supersonics had already been gained:

▸ the English military engineer Benjamin ROBINS {⇨1746} had proven with his ballistic pendulum that a musket ball can fly supersonically and that drag increases dramatically when approaching the velocity of sound, a finding which the British gun expert and mathematician Charles HUTTON {⇨1786} extended to cannon shots forty years later;

▸ the Austrian physicist Christian A. DOPPLER {⇨1847} had pondered on wave generation by a supersonic object, thereby predicting that the object is surrounded by a conical wave, the "head wave" {E. MACH & SALCHER ⇨1886};

▸ the English physicist James Prescott JOULE and the Scottish engineer and physicist William THOMSON {⇨1856} had predicted "aerodynamic heating" at high flight velocities;

▸ the U.S. ballistician General Thomas J. RODMAN {⇨1864} had demonstrated with his 15- and 20-in. monster guns – so-called "columbiads"[298] – that with his method of hollow casting guns of an even larger caliber should be realizable in principle; and

▸ the inventions of "mammoth powder" {RODMAN ⇨1857}, a new, progressive-burning propellant, and of a number of new high explosives with hitherto unprecedented detonation capabilities, such as guncotton {SCHÖNBEIN ⇨1845}, nitroglycerin {SOBRERO ⇨1846} and dynamite {A. NOBEL ⇨1867}, stimulated the fantasies of laymen as well as the expectations of experts that they could be applied for new, unprecedented propulsion purposes.

---

[297] J. VERNE: *Autour de la lune*. J. Hetzel, Paris (1872). • In his work of fiction, three men and two dogs were shot to the Moon in a huge bullet by a monster cannon which had to achieve a minimum velocity of 11.2 km/s (when neglecting aerodynamic drag) in order to escape the Earth's gravitational attraction. This so-called "escape velocity" became a much discussed quantity, and one that is still unattainable by single-stage powder guns. Jules VERNE was scientifically advised by his cousin Henri GARCET, who was a mathematics professor at the Ecole Henri IV in Paris.

[298] The origin of the term "columbiad" is obscure. Some believe it was named after Joel BARLOW's popular poem *The Columbiad* (1807), while others suggest that the term was applied to any cannon manufactured at Henry FOXALL's Columbian Foundry near Washington, DC, which produced cannons and munitions in the period 1803−1854. • The term was widely used in the late 19th century and was also taken up early by the German encyclopedia *Meyers Konversations-Lexikon* (1894).

### 2.5.1 UNMANNED VEHICLES: FIRST DEMONSTRATIONS OF PRACTICABILITY

> *We kept hoping that this rocket [the A-5, predecessor of the V2] would exceed the velocity of sound. The big question was: would there be stronger oscillations around the trajectory tangent due to the accompanying increase in aerodynamic drag and the variability of the center of gravity, thus causing the rocket to explode?*[299]
>
> Walter DORNBERGER
> Heeresversuchsanstalt Peenemünde
> Peenemünde-Ost 1938

When aerodynamicists and aircraft designers began to ponder on the problems of supersonic flight, supersonics as a scientific discipline was already well advanced, and the following examples highlight illustrate the progress made in this field:

- German ballisticians had found the optimum shape for supersonic projectiles used in firearms, the famous S-bullet [Germ. *Spitzgeschoss*]. This pointed projectile, which was the fastest infantry projectile at that time ($v_0$ = 893 m/s), was adopted by many armies around the world {DWM ⇨1903};
- the German professor Ludwig PRANDTL at Göttingen University, together with his Ph.D. student Theodor MEYER, had studied the expansion of a supersonic flow around a corner, which resulted in the "Prandtl-Meyer theory of expansion" {MEYER ⇨1908};
- British studies on various airscrew geometries revealed a significant loss of thrust when the airscrew tips approached the velocity of sound {LYNAM ⇨1919; REED ⇨1922}. However, both theoretical and experimental requirements for the development of a supersonic airscrew {XF-88B ⇨1953; ⇨Fig. 4.20–G} were still many years from being fulfilled;
- the British engineer Sir Thomas STANTON, who built the world's first supersonic wind tunnel at NPL in Teddington {STANTON ⇨1921}, demonstrated that such a device is not only useful for aerodynamic drag studies of model projectiles, but that it is also useful for pressure measurements on scale models of aerofoils {STANTON ⇨1928};
- the Swiss aerodynamicist Jakob ACKERET devised a theory for thin sharp-edged supersonic aerofoils {ACKERET ⇨1925};
- the Hungarian-born U.S. aeronautical engineer Theodore VON KÁRMÁN and his Ph.D. student Norton B. MOORE studied the aerodynamic drag of slender, spindle-like bodies at supersonic speed at CalTech's Guggenheim Aeronautical Laboratory {VON KÁRMÁN & MOORE ⇨1932};
- the German aeronautical engineer Adolf BUSEMANN investigated the flow around aerofoils at supersonic speeds and observed that thin profiles are superior to thick ones {BUSEMANN & WALCHNER ⇨1933};
- the U.S. aeronautical engineers John STACK and Eastman N. JACOBS at NACA's Langley Aeronautical Laboratory visualized the generation of a shock wave above an aerofoil in a transonic flow {STACK & JACOBS ⇨1933; ⇨Fig. 4.14–L};
- Albert BETZ {⇨1939}, a German professor of applied mechanics at AVA in Göttingen, gave the first experimental evidence of the superiority of the sweptback wing design, a concept previously proposed by BUSEMANN, who was originally derided for his idea {5th Volta Conf. ⇨1935}. However, the results of his studies, which were summarized in various top secret German patents, remained unknown outside Germany until the end of Word War II; and
- in 1945, still before the first manned supersonic flight, the British applied mathematician M. James LIGHTHILL[300] published a theory on the aerodynamic drag on finely pointed bodies of revolution. Obviously, the results from a new area of research, *gas dynamics of thin bodies*, had suggested to aircraft designers that a thin and slender design would be best for realizing supersonic velocities.

In the mid-1930s, issues relating to high-speed aviation were still of academic rather than of military relevance, and were freely discussed in the international aeronautical community {5th Volta Conf. ⇨1935}. However, shortly after 1935, a veil of secrecy was drawn over this subject. The vital importance of speed was impressively demonstrated in the numerous air battles during World War II. Although manned flight through the sound barrier didn't appear to be close at hand in the early 1940s, various detrimental supersonic phenomena resulting from shock waves had already been observed in transonic aviation: in the United States, England and Germany, test pilots had experienced dramatic and dangerous increases in drag when their fighter planes had approached transonic speeds in dives. These increases in drag were accompanied by a loss of control which were obviously caused by aerodynamic compressibility effects and disturbed airflow {VON KÁRMÁN & DRYDEN ⇨1940; DITTMAR, Me-163

---
[299] W. DORNBERGER: *V2 – der Schuß ins Weltall*. Bechtle, Esslingen (1952), p. 66.

[300] M.J. LIGHTHILL: *Supersonic flow past bodies of revolution*. ARC Repts. & Mem. No. 2003. Ministry of Supply, H.M.S.O., London (1945).

⇨1941; GILKE, P-38 ⇨1941; MUTKE, Me-262 ⇨1945}. Some pilots of transonic planes lost control of their aircraft due to these shock waves, and barely escaped or were killed in a crash. Subsequent wind tunnel studies of model aircraft revealed the existence of shock waves when approaching the speed of sound. These shock waves indirectly caused tremendous forces capable of tearing the fuselage and wings apart.

Long before the birth of supersonic aeronautics, ballisticians and physicists had pondered on the reasons for the increased aerodynamic drag in the transonic regime {Sir NEWTON ⇨1687; ROBINS ⇨1746; HUTTON ⇨1786; BASHFORTH ⇨1864; GREENHILL ⇨1879; MAIYEVSKII ⇨1881; VIEILLE ⇨1900; BENSBERG & CRANZ ⇨1910}, some of them also attempting to find practical ways of diminishing wave drag. Obviously, these mysterious drag phenomena in the transonic regime impaired the ability to aim accurately and reduced the effective range and terminal velocity of shots when impacting the target. Their studies gave birth to *aeroballistics* – a new branch of ballistics dealing with the motion and behavior of projectiles, missiles and high-speed aircraft in the Earth's atmosphere.

Ballistic studies using flash photography performed in Austria {E. MACH & SALCHER ⇨1886} and England {BOYS ⇨1890} had already proved that bullets traveling at supersonic speed do not burst into pieces upon passing the sound barrier. However, piercing the "magic sound barrier" with a complex, lightweight vehicle structure, such as a full-scale manned airplane, appeared very risky to pilots and aerodynamicists, and many aircraft designers thought that it was an impossible task to manage the pressure forces generated by strong shock waves.

Flying at $M > 1$ should easiest to realize when the velocity of sound in the atmosphere is a minimum. Since the Mach number is defined as the ratio of the velocity of an object (such as a vehicle) to the velocity of sound, which decreases with increasing altitude, supersonic flight is most likely to be achieved at large altitudes.[301] Last but not least, the reduction in aerodynamic drag at such altitudes would also reduce the power required to break the "sonic wall."

The development of an appropriate propulsion system played a central role in supersonic aviation. In the early 1940s, the most promising low-weight high-power propulsion was either the turbojet engine {He-178 ⇨1939; Me-262 ⇨1941} or the rocket motor {He-176 ⇨1939; Me-163 ⇨1944; ⇨Fig. 4.20–E}.

The proof of technical feasibility came from military rocketry. In the 1930s, General Walter DORNBERGER and Wernher VON BRAUN, two eminent rocket jet experts, had experimented with liquid-fuel-propelled rockets. With generous support provided by the German Army, they were directed to develop a guided rocket capable of transporting a 1-ton payload over a distance of 300 km at high supersonic velocities ($M > 4$). The aerodynamics of the 14-m-long A-4, which was reminiscent of the shape of an S-bullet but with four fins at its rear for control and stabilization at high speeds, was carefully studied using a number of rocket models in supersonic wind tunnels at the Heeresversuchsanstalt (HVA) Peenemünde, the German Rocket Center at the Baltic Sea. The results proved that stable flight could be achieved up to Mach 4 {HVA Peenemünde ⇨1939}. On October 3, 1942, the A-4, later named the V2, reached a velocity of $M > 4$, thereby covering a distance of almost 320 km and attaining a record height of 84.5 km – thus opening the door to the "realm" of space[302] {HVA Peenemünde ⇨1942; ⇨Fig. 4.20–C}. This test proved that a complex and delicate structure such as a large guided rocket could survive not only the passage through the dreadful transonic regime, but also the much-discussed aerodynamic heating that occurs during the supersonic phase. For the A-4, this phase lasted about six minutes,[303] and the heating was estimated by VON BRAUN to be around 800 °C. During the experimental stage of the A-4 he had ordered the aluminum skin to be replaced with a 0.64-mm-thick steel sheet. This measure was also useful due to the general shortage of aluminum in Germany in those years. The nose of the rocket, initially intended to carry a 1-ton charge of TNT, was fabricated from 6-mm-thick steel sheet. However, in order to reach the required range of 250 km, the charge weight had to be reduced.[304] Last but not least, the successful test at Peenemünde also first demonstrated in a spectacular manner an environmental problem connected with supersonic flight – the *sonic*

---

[301] The sound velocity $u$ is a rather complex function of altitude $h$ which resembles a zigzag line when plotted in the $(h, u)$-plane. In the range 0–10 km, $u$ diminishes with increasing $h$, dropping from initially 340 m/s at sea level to a minimum value of 296 m/s at about 10 km. Then $u$ remains almost constant over 10–20 km. At altitudes of 20–50 km, $u$ increases steadily up to about 330 m/s at 50 km, before dropping steadily from 55–85 km to a value of about 275 m/s. See *Standard atmosphere* (ed. by NOAA). U.S. Govt. Printing Office, Washington, DC (1976).

[302] There is no definite boundary between the Earth's atmosphere and space. Above 500 km, in the exosphere, the Earth's atmosphere merges with the gases of interplanetary space, which some scientists consider to be the beginning of space. Others consider space to begin at an altitude of 160 km above the Earth's surface. According to the latter definition, the V2/WAC-Corporal test {⇨1949; ⇨Fig. 4.20–B} was the first successful project to launch a man-made vehicle into space.

[303] In the test performed on October 3, 1942, the rocket engine operated for 58 seconds. Its radio transmitter worked for another five minutes until the rocket hit the Baltic Sea. See E. STUHLINGER: *Wernher VON BRAUN*. Bechtle, Esslingen & München (1992), p. 68.

[304] W. DORNBERGER: *Peenemünde. Die Geschichte der V-Waffen*. Bechtle, Esslingen (1981), p. 246.

*boom effect.* Although this phenomenon was already known to ballisticians as the *ballistic crack,* the enormous size of the shock cone produced by a large object flying at high Mach number, which affects much larger areas on the ground than a projectile shot from a gun. The extent of the boom's "carpet" (the distance that it can be heard on each side of the aircraft's path along the ground, as well as its intensity), varies with the aircraft's height, speed, flight conditions, and with the prevailing atmospheric conditions. Typical maximum figures for SST aircraft ± 20 km from the ground are positive peak overpressures of between 50 and 100 Pa.[305]

The first vehicle to reach hypersonic velocities ($M > 5$) was the U.S. WAC (Women's Army Corps)-Corporal rocket {⇨1949}, developed by the Jet Propulsion Laboratory for the U.S. Army. Combined with a captured German V2 as a booster, it reached a velocity of $M > 7$ {⇨Fig. 4.20–D}. Beginning with the launch of the first Earth satellite {Sputnik 1 ⇨1957}, space vehicles and probes were routinely accelerated to hypersonic velocities which ranged between 7.9 and 11.2 km/s (decreasing with altitude) in order to enter Earth's orbit.

## 2.5.2 MANNED VEHICLES: FROM VENTURE TO ROUTINE

> *I do not imagine for one moment that man will be happy until he has conquered the "sonic barrier;" any experiments that we can perform to help him over the difficult threshold to supersonic flight will be of great value and may well save the lives of intrepid pioneers in this new and hazardous venture.*[306]
>
> Earnest F. RELF
> London 1946

> *For me personally, however, I prefer a slower pace. I travel constantly by jet, but I like nothing better than to think of myself riding through the Paris boulevards as my parents did in old Budapest in a fiacre with a coachman and two horses.*[307]
>
> Theodore von KÁRMÁN
> CalTech
> Pasadena, CA 1962

The first official demonstration of manned supersonic flight was provided on October 14, 1947 over Rogers Dry Lake in southern California by the USAF test pilot Charles YEAGER {⇨1947}. His test rocket plane, the Bell X-1, was attached to a B-29 mother ship and carried to an altitude of about 7,600 m. It then rocketed separately to about 12 km, reaching a velocity of $M = 1.06$ in level flight for 18 seconds. YEAGER's first supersonic flight reconfirmed previous observations that shock waves can vigorously shake a plane when approaching the sound barrier, but that beyond $M = 1$ the bumpy road turns into "a perfectly paved speedway" (YEAGER). His little plane, currently displayed at the Smithsonian National Air & Space Museum, looks rather clumsy according to our present understanding of supersonic aircraft design. It has neither sweptback wings {BUSEMANN ⇨1935} nor the typical "thinness" of fuselage exhibited by later supersonic aircraft, the key to achieving low supersonic drag.

With his legendary flight YEAGER not only pierced the sound barrier physically, but he also cleared away the doubts and hesitations that were present in the minds of numerous contemporaries that human supersonic flight was not possible. The next achievements in high-speed flight occurred in rapid succession. On December 1953, in an improved version of his rocket-powered research aircraft, YEAGER became the first man to exceed Mach 2.[308] On April 12, 1961, the Soviet cosmonaut Yuri A. GAGARIN {⇨1961} became the first human to fly hypersonically during the reentry of his Vostok 1 craft {⇨Fig. 4.20–J}, which began at $M > 25$. In the same year, the U.S. test pilot Robert WHITE {X-15 ⇨1959 & 1961} became the first pilot to reach hypersonic velocities ($M = 6$) with the X-15, the first hypersonic rocket plane {⇨Fig. 4.20–I}. In the years following, this remarkable aircraft reached an altitude of 67 miles (108 km) and attained hypersonic speeds of up to $M = 6.72$.

Beginning in the 1960s, various nations started to design a supersonic civil passenger plane (also known as "supersonic transport" or "SST" {⇨Fig. 4.20–H}): the American Boeing 2707-300 SST {⇨1968}; the Soviet Tupolev Tu-144 {⇨1968}; and the British/French Concorde {⇨1969}. By the early 1960s, the construction of passenger aircraft capable of speeds of 2,000 mph (939 m/s) appeared feasible. The chief issues were then to make such flights economically viable and to solve the special structural design problems involved. However, the environmental problems imposed on the population in the vicinity of airports and along

---

[305] J.R. HASSALL and K. ZAVERI: *Acoustic noise measurements.* Brüel & Kjaer, Naerum, Denmark (1979), p. 220.

[306] E.F. RELF: *Recent aerodynamic development* [34th Wilbur Wright Memorial Lecture]. J. Roy. Aeronaut. Soc. **50**, 421-449 (1946), p. 445. • Ernest F. RELF, F.R.S., was a staff scientist at the Aerodynamic Dept. of the National Physical Laboratory (NPL). In 1946, he became Principal of the new College of Aeronautics at Cranfield, U.K.

[307] Th. VON KÁRMÁN and L. EDSON: *The wind and beyond. Theodore VON KÁRMÁN, pioneer and pathfinder in space.* Little, Brown & Co., Boston etc. (1967), p. 234.

[308] Flying twice as fast as the velocity of sound was a technical challenge rather than a venture. Many laymen predicted further barrier effects at Mach 2. The question of what happens upon passing Mach 2 was also frequently asked by passengers who were using the Concorde for the first time.

the flight path were not seriously considered.³⁰⁹ Only the Concorde, a delta-winged jetliner designed for $M = 2$ and 144 passengers, achieved the goal of being used routinely for long-distance civil flights over a long period of time. Flights continued up to October 2003, when regular service was shut down. Although it had faced numerous environmental and economic problems since being put into service in 1976, the Concorde was a landmark in the design of supersonic aircraft, and its elegant futuristic appearance compared to other airliners always made it a spectacle.

The Tupolev TU-144 {⇨1968} is another great example of the successful design of a supersonic liner; it even made its maiden flight prior to the Concorde. The TU-144 first flew passenger services in 1977, but unfortunately was not an economic success. One of the seventeen TU-144 built was modified to act as a supersonic flight laboratory {Tu-144LL ⇨1996}: NASA teamed with American and Russian aerospace industries over a five-year period in a joint international research program to develop technologies for a proposed future second-generation supersonic airliner to be developed in the 21st century.

The huge progress in supersonic flight, from YEAGER's first bumpy supersonic flight in a small research aircraft through to the smooth, almost unnoticeable passage of the huge Concorde through Mach 1, is perhaps best illustrated in an editorial published in 1971 in the American magazine *Aviation Week & Space Technology*.³¹⁰ Robert HOTZ, editor-in-chief of this journal, who participated in a demonstration flight at the 1971 Paris Air Show, wrote: "The most sensational aspect of flying as a passenger at Mach 2 in a supersonic transport is that there are no sensations whatsoever that differ from those in the current generation of subsonic jets... At Mach 1 there was a slight tremor that felt much the way an automobile coughs with a fouled spark plug... During the climb from 20,000 to 50,000 feet [6,096 to 15,240 m] and acceleration from Mach 1 to Mach 2, the flight was smooth as silk. When Aerospatiale test pilot Jean FRANCHI leveled off at 50,500 feet [15,392 m] and the Machmeter needle flickered just past two on the dial and steadied for normal Concorde cruise, one French journalist exploded in disbelief: 'I don't believe it,' he said. 'You must have a mouse inside that instrument that winds it up to Mach 2'" Later British Airways provided the following information for world travelers who had not yet experienced the event of passing the sound barrier in a Concorde, "There will also be a new Machmeter [⇨Fig. 4.20–H], which displays the aircraft speed to passengers, and to celebrate the breaking of the sound barrier, a subtle wash of blue light will pass through the cabin as Concorde passes through Mach 1."

All the high-speed record flights mentioned above were performed at large altitudes; *i.e.*, under conditions of considerably reduced aerodynamic drag. However, for decades breaking the sound barrier with a land vehicle seemed to be an insurmountable task. On October 13, 1997 – one day before the fiftieth anniversary of YEAGER breaking the sound barrier in an airplane – the sound barrier was finally broken by a land vehicle. In northwestern Nevada, on the imperfectly smooth pavement of the salt flats of Black Rock Desert, the British jet test pilot Andy GREEN {⇨1997} reached a speed of $M = 1.007$ with his jet car *ThrustSSC* {⇨Fig. 4.20–N}, a four-wheel vehicle provided with special high-speed rubber tires. Two days later, GREEN set the supersonic world land speed record of 766.609 mph ($M = 1.02$). The appearance of a bow shock wave in a photo of the car also proved its supersonic status {⇨Fig. 4.6–J}.

Land vehicles without wheels can potentially reach substantially higher velocities than this. Indeed, by the mid-1950s, rail-guided rocket sleds were pushing medium-scale test models weighing up to 91 kg up to speeds of Mach 3.4 {SNORT ⇨1955; ⇨Fig. 4.20–M}. More recent developments of such test facilities have extended the range of test velocities available into the hypersonic regime. Presently, the Holloman High Speed Test Track (HHSTT) at Holloman Air Force Base in New Mexico is the largest facility capable of studying the effects of traveling at hypersonic speeds on full-scale models. A world speed record was achieved for a sled test recently when a 87-kg payload was accelerated up to a velocity of 2,885 m/s (about Mach 8.6) {Holloman Air Force Base ⇨2004}.

## 2.5.3 NEW CHALLENGES, NEW THREATS

The ultimate goal of high-speed flight would be the achievement of cruising velocities approaching the velocity of light (299,776 km/s). The term *astronautics*, used throughout the Western World, literally means navigation among the stars.³¹¹ Human voyages to other stars, however, are still science fiction and will not be attained in the foreseeable future.³¹² The

---

[309] R.L. BISPLINGHOFF: *The supersonic transport*. Scient. Am. **210**, 25-35 (June 1964).
[310] R. HOTZ (ed.): *Martinis at Mach 2*. Aviation Week & Space Technology **94**, 7 (June 7, 1971).
[311] In his book *The Wind and Beyond* (1967), Theodore VON KÁRMÁN appropriately wrote on page 238: "My friend Dr. DRYDEN has proposed that we substitute for astronautics the word cosmonautics, the science of travel in the cosmos, which is the Universe bounded by one star, our Sun. But the United States never did anything about this suggestion, perhaps because the Russians had already taken over the word."
[312] *Non est ad astra mollis e terries via* ("There is no easy way from the Earth to the stars"), Lucius Annaeus SENECA (c.4 B.C. – c.A.D. 65).

light from the nearest star, *Proxima Centauri* – a member of *Alpha Centauri*, a triple star system in the southern constellation of Centaurus – takes a little longer than four years to reach the Solar System. Space vehicles propelled by nuclear fusion (fusing hydrogen into helium), a concept much discussed in the 1960s, would require a huge takeoff mass in order to attain velocities approaching the velocity of light. Similar problems also face laser-driven vehicles, where a train of high-energy laser pulses provided by huge terrestrial laser systems would be employed. These so-called "lightcraft" are still at the experimental stage and have only reached very modest velocities so far {MEAD & MYRABO ⇨1997}.

A terrestrial challenge also remains. The ultimate demonstration of human flight piercing the sound barrier is still pending: skydiving from the stratosphere in free fall, protected only by a pressurized "shockwave-proof" spacesuit. The U.S. Air Force pilot Joseph W. KITTINGER was said to be the first human to exceed the sound barrier in free fall. On August 16, 1960, he jumped from an open gondola suspended from a giant helium-filled balloon at 31.3 km. However, his jump record was unofficial and not sanctioned by the Fédération Aéronautique Internationale (FAI). Moreover, KITTINGER used a stabilizing parachute early in his jump, so technically his jump wasn't considered free fall.

Another attempt was made by Michel FOURNIER, a retired French army officer and an experienced parachutist. He hoped to reach a maximum speed of 1,600 km/h (about Mach 1.5). His "Super Jump," a free fall from a helium-filled balloon at an altitude of more than 40 km, was predicted to last about six minutes and 25 seconds. FOURNIER suggested that preparations for the jump have proven useful for several fields, such as aerospace medicine, and especially the technology associated with high-altitude rescue jumps for the crews of endangered space shuttles.[313, 314] His attempt, planned for September 9, 2002 above Saskatchewan, Canada, had to be cancelled because of strong jet streams at 7,000 meters. Unfortunately, another attempt scheduled for August 25, 2003 at the North Battleford municipal airport, Saskatchewan, also failed because the helium-filled balloon (which was 25 stories high) burst during launch. In February 2006, he announced publicly to make another attempt to jump above the large Canadian plains, with the ambition to become the first man to cross the sonic wall in free fall. In September 2006, it was reported that his "Big Jump" will be postponed until the next years because of budget problems.

Achieving high cruising velocities under water is also another great challenge in terms of both research and technology. Supercavitation would allow naval weapons and vessels to travel submerged at hundreds of miles per hour. Laboratory studies performed in 1997 at the Naval Undersea Warfare Center (Newport, RI) demonstrated that a blunt-nose projectile launched underwater acts as a "cavitator" {NUWC ⇨1997; ⇨Fig. 4.14–E}. Generating a large cavitation sheet that surrounds the whole projectile, it can move even faster than the speed of sound in water, thus producing a bow wave in water that is very similar to that of a supersonic shot in air {E. MACH & SALCHER ⇨1886}. Unknown to the Western World, the Soviet Union began to consider the idea of high-speed torpedoes decades ago.[315] In the late 1970s, the Russian Navy developed a rocket-powered supercavitating torpedo, the revolutionary *Shkval* [Russ. шквал meaning "squall"], being capable of 230 mph (about 102 m/s) – *i.e.*, it was 4 to 5 times faster than other conventional torpedoes.[316]

Future research into high-speed torpedoes would aim at speeds approaching or even surpassing the sound velocity. If such torpedoes were to be realized, they would potentially change the face of naval warfare.

## 2.6 EVOLUTION OF DETONATION PHYSICS

> *The kinetic theory of gases has for us artillerists a special charm, because it indicates that the velocity communicated to a projectile in the bore of a gun is due to the bombardment of that projectile by myriads of small projectiles moving at enormous speeds, and parting with the energy they possess by impact to the projectile... But in the particular gun under discussion, when the charge was exploded there were no less than 20,500 cubic centimetres of gas, and each centimetre at the density of explosion contained 580 times the quantity of gas, that is, 580 times the number in the exploded charge is $8\frac{1}{3}$ quadrillions, or let us say approximately for the total number eight-followed by twenty-four ciphers...*[317]
>
> Sir Andrew NOBLE
> London 1900

HISTORICALLY, investigations into the nature of shock waves have been closely related to the riddle of detona-

---

[313] *Skydiver to break sound barrier.* BBC News (Jan. 22, 2002); http://news.bbc.co.uk/2/hi/uk_news/england/1774130.stm.
[314] *Historic balloon jump deflates fast.* Saskatchewan News Network (Aug. 25, 2003); http://freerepublic.com/focus/f-news/971578/posts?page=14.

[315] S. ASHLEY: *Warp drive underwater.* Scient. Am. **284**, 62-71 (May 2001).
[316] *VA-111, Shkval (Squall).* Military Scope, U.S.A. (Dec. 1, 2001); http://www.periscope.ucg.com/mdb-smpl/weapons/minetorp/torpedo/w0004768.shtml.
[317] From an evening lecture presented at the Royal Institution; *see* A. NOBLE: *Some modern explosives.* Not. Proc. Meet. Memb. Roy. Inst. Lond. **16**, 329-345 (1902).

tion, a highly transient thermochemical wave phenomenon. The correct interpretation of detonation was not achieved until the period 1880–1905 – almost 300 years after the invention of *gold fulminate*, the first "high explosive." Detonation physics, a branch of fluid mechanics that has since grown into a huge field of its own, is closely related to thermodynamics, shock wave physics and chemical kinetics. Like shock wave physics, it relies strongly on high-speed – in many cases ultrahigh-speed – diagnostics. The following short sketches focus on some of the circumstances related to the step-by-step process of investigating the thermodynamic nature of chemical explosions and detonations.

## 2.6.1 BLACK POWDER: THE MAID OF ALL WORK

> *Gunpowder, that old mixture, possesses a truly admirable elasticity which permits its adaptation to purposes of the most varied nature. Thus, in a mine it is wanted to blast without propelling; in a gun to propel without blasting; in a shell it serves both purposes combined; in a fuse, as in fireworks, it burns quite slowly without exploding. But like a servant for all work, it lacks perfection in each department, and modern science armed with better tools, is gradually encroaching on its old domain.*[318]
>
> Alfred NOBEL
> London 1876

An *explosive* has been defined as a material – either a pure single substance or a mixture of substances – which is capable of producing an explosion through its own energy. Black powder or gunpowder is the most important "low" explosive. It is a combustible material that inherently contains all of the oxygen needed for its combustion; it burns but does not explode; and it produces a voluminous quantity of gas in a very short time which itself produces an explosion.

The origin of black powder has been disputed for centuries. It may have been imported from China, where it has long been used in fireworks and propelling rockets {Pienching ⇨1232; ⇨Fig. 4.20−B}. Curiously, however, Marco POLO (1254–1324), the famous Italian traveler, never mentioned such a curiosity in his writings. It may have been discovered in northern Europe {GRAECUS ⇨c.1250; BACON ⇨1267} or in Germany {SCHWARZ ⇨c.1380}. Black powder was apparently first applied for ballistic purposes in the early 14th century both in Europe {GHENT ⇨1313/1314} and Arabia {SHEM ED DIN MOHAMED ⇨1320}. It may have been invented in Italy, because chemicals of the proper quality were mined there. Whatever the source, gunpowder became the "fuel" most commonly used to power firearms of all calibers over the centuries, and it partly replaced muscle in the rapidly growing mining industry. On the other hand, the rapidity of its transformation accompanied by the release of enormous dynamic forces challenged natural philosophers to understand the chemical reactions involved and to develop theoretical tools to optimize its use in ballistics.

Black powder consists of a mixture of potassium nitrate or ordinary saltpeter [$KNO_3$], sulfur and charcoal, and was first described in Europe by the English scholar Roger BACON {⇨1267} for incendiary and explosive applications. The fuel in black powder is charcoal and sulfur, which are burned by the oxygen contained in potassium nitrate or sodium nitrate {DU PONT DE NEMOURS ⇨1858}. Black powder will not produce high pressures unless it is confined (*e.g.*, in a hole for blasting purposes). Since it will only burn rapidly – it cannot detonate[319] – it cannot be used to generate shock waves. However, when used in firearms the hot gases of the reacting gunpowder are initially confined in the barrel but they are suddenly released at the moment when the projectile leaves the muzzle; this "uncorking effect" produces an impressive *muzzle blast*, which is actually a shock wave. By the 18th century, projectiles propelled by charges of gunpowder reached supersonic velocities {ROBINS ⇨1746; HUTTON ⇨1783}. In addition to the muzzle blast, supersonic shots always generate a *head wave*: a second shock wave which is carried along with the flying projectile.

The first recorded nonmilitary use of black powder was for mining {Upper Leogra Valley ⇨1572; Schemnitz ⇨1627}, thus clearly illustrating that laborious methods of breaking ore or coal by hand could be replaced by the more economical and faster method of blasting. Black powder was also used for various spectacular construction jobs, such as for the first canal tunneling in France {Malpas Tunnel ⇨1679}. The application of black powder for blasting purposes in the mining industry was facilitated by the invention of the *Bickford fuse* or *safety fuse* {BICKFORD ⇨1831}. However, the need to produce potassium nitrate from certain soils imported from Spain, India, Italy and Iran limited its widespread use as an industrial explosive. With the invention of *B Blasting Powder* by the U.S. industrialist Lammot DU PONT {⇨1858}, a modified black powder in which the expensive potassium nitrate

---

[318] A. NOBEL: *On modern blasting agents*. Am. Chemist **6**, 60-68, 139-145 (1876).

[319] Depending on the grain size and the confinement, black powder burns at a relatively slow rate ranging between 560 and 2,070 ft/s (171–631 m/s). *See Blasters' Handbook*, E.I. du Pont, Wilmington, DE (1967), p. 26. Black powder, according to modern classification, is a "low" explosive. The arbitrary cut-off speed between "high" and "low" explosives is 3,300 ft/s (about 1,000 m/s).

was replaced by the cheaper sodium nitrate[320] or Chile saltpeter [$NaNO_3$] imported from Peru and Chile, black powder was widely accepted for industrial use. It was also used in one of the most difficult construction jobs of the 19th century, that of the first large railroad tunnel between France and Italy {Mont Cenis Tunnel ⇨1857}.

The nature of the reaction that takes place when gunpowder is fired has long fascinated the minds of chemists. Until around the year 1856, the metamorphosis of gunpowder was assumed to take place according to the equation found by the French chemist Michel E. CHEVREUL {⇨1825}

$$2\,KNO_3 + 3\,C + S \rightarrow K_2S + N_2 + 3\,CO_2.$$

One gram of burning black powder produces about 718 cal, and over 40% of its original weight is transformed into hot gases at a temperature of almost 3,900 °C.[321]

However, the accuracy of this simple equation was subsequently found to be doubtful. The first exact investigation of the compositions of both the gases and the solid products in the explosion was carried out at the University of Heidelberg by the German chemistry professor Robert W. BUNSEN {⇨1857} and his former student Léon SCHISCHKOFF, a Russian chemistry professor at the Artillery School of St. Petersburg. Starting from the concept of developing a "chemical theory of gunpowder," they eventually discovered that a large number of salts are produced whose presence had not been detected before, besides the known constituents of smoke and solid residues. A comprehensive theory that explained the presence of the solid and gaseous residues after the explosion of gunpowder proved to be very difficult to devise. In the 1870s, Andrew NOBLE and Frederick A. ABEL {⇨1875}, two British military chemists, pursued the same idea of developing a theory for the firing of gunpowder. Although further progress was made, they arrived at essentially the same conclusion as BUNSEN and SCHISCHKOFF before, namely that the explosion products are not only very numerous, but also vary considerably in terms of their proportions according to the initial conditions (e.g., the levels of moisture and impurities, and the grain size), especially the pressure and therefore the temperature under which the explosion takes place. The German chemist Heinrich DEBUS {⇨1882}, one of BUNSEN's oldest pupils and friends, resumed this subject in the early 1880s. He was the first to make the important observation that, depending on the mixing ratio of the ingredients, fired gunpowder can also produce endothermic reactions that decrease the temperature of the explosion and therefore have a detrimental effect on the explosion. This finding further illustrated the complex nature of fired gunpowder.

As well as the chemical aspects of fired gunpowder, the physical ones were also discussed. The main questions for practical applications in artillery were:

▸ What is the maximum temperature of the powder burnt under ordinary pressure?
▸ What is the maximum pressure in the bore of a gun, and how long does it need to reach this pressure?
▸ How long does it take to convert the powder into gas?
▸ What is the maximum volume of the gaseous products generated during the explosion in relation to the initial volume?
▸ What is the "theoretical work" of gunpowder?

The answers to this catalog of complex problems, which were tackled both experimentally and theoretically by early chemists, ballisticians and physicists, led to somewhat contradictory results. They were discussed in the introduction of NOBLE's and ABEL's classical memoir on fired gunpowder {⇨1875}. For 1 kg of fired gunpowder, NOBLE and ABEL estimated that

▸ the temperature of the explosion was about 2,200 °C;
▸ the maximum pressure is 6,400 bar when the powder fills the space entirely;
▸ the permanent gases generated by the explosion occupy about 280 times the volume of the original powder at 0 °C and 1 bar; and
▸ the total work is 332,128 m kg* – a value which is about ten times smaller than the total work of 1 kg of coal.[322]

Black powder remained the sole explosive and propellant material until the mid-19th century, when nitroglycerin {SOBRERO ⇨1846} and cellulose nitrate {SCHÖNBEIN ⇨1845} were invented. However, it completely vanished from coal mining applications. Since exploding black powder could ignite firedamp, it was replaced by "permissible" explosives {Gelsenkirchen ⇨1880; MALLARD & LE CHÂTELIER ⇨1883; PENNIMAN ⇨1885} which produce a cooler flame and therefore a less intense flash of light {SIERSCH ⇨1896; ⇨Fig. 4.17–G}. Today black powder is still used as an igniter for smokeless cannon propellants in large-caliber guns, and in fireworks, saluting charges and safety fuses.

The investigations of fired gunpowder were a tremendous challenge for early diagnostics. The biggest problems were (i) how to measure how long the combustion takes in the

---

[320] Sodium nitrate can be synthesized from ammonia, which can be produced in large quantities by a process invented in 1908 by the German chemist Fritz HABER. This enabled Germany to manufacture explosives during World War I, when foreign supplies of nitrates were cut off.
[321] T.L. DAVIS: The chemistry of powder and explosives. Angriff Press, Las Vegas, NV (1972), pp. 42-43.
[322] For the meaning of kg* see ROUX & SARRAU {⇨1873}.

bore of a gun, and (ii) how to investigate how the combustion process is related to the motion of the projectile in the barrel – a classical problem in interior ballistics {DE LAGRANGE ⇨1793; A. NOBLE ⇨1894; ⇨Fig. 4.17–L}. The principal task of interior ballistics is, given a particular gun, a particular shot, and a particular kind of powder, and a given load, to find

- the position and velocity of the shot;
- the mean pressure (and incidentally the temperature) of the gases in the gun; and
- the fraction of the powder burned,

all as functions of the time or of each other until the exit of the shot from the muzzle of the gun.[323] Since the chemical and physical processes during the explosion all occur within the micro- to millisecond range, depending on the size of the reaction vessel and the grain size of the powder, their temporal resolution requires high-speed recording techniques, and so research in this field greatly stimulated advances in mechanical/electrical chronoscope technology. These technical challenges imposed on early gun propellant researchers – which were often combined with requirements to record the blast wave generated by the explosion as well – were an important "preschool" for tackling the even more difficult problem of how to record the much faster phenomena generated by the numerous new high explosives invented in the second half of the 19th century.

It is amusing to note here that black powder was also used in the past for another, very particular, civil purpose. Hundreds of years ago, black powder was used to test the alcohol content of whiskey: saturation with a 50% alcohol/water mixture will still allow black powder to catch fire; it burns with a clear blue flame. Both were mixed and ignited. If the black powder flashed, then it provided "proof" that it was good whiskey; *i.e.*, that it contained at least 50% alcohol. However, if there was too much water in the whiskey, the powder would be too wet to ignite: proof that it was *not* good whiskey. Thus, the 50% alcohol test became 100% proof that it was good whiskey, which is why it is now called "100-proof whiskey."[324] This unique application of black powder – though not related to shocks and explosions – provides further evidence of Alfred NOBEL's view that black powder has long been "the maid of all work."

## 2.6.2 THE RIDDLE OF DETONATION: STEPS TOWARD AN UNDERSTANDING

> *Recent experiment has shown that the rapidity with which gun-cotton detonates is altogether unprecedented, the swiftness of the action being truly marvelous. Indeed, with the exception of light and electricity, the detonation of gun-cotton travels faster than anything else we are cognizant of...*[325]
>
> Sir Frederick A. ABEL
> Royal Artillery Institution
> Woolwich 1873

Since the discovery of gold fulminate, man's first high explosive, by the German alchemist Oswald CROLL {⇨1608}, an impressive number of other explosive materials have been discovered by chance, or invented and modified by systematic investigations of industrial chemists to match particular requirements.

**High Explosives.** A high explosive (HE) is a compound or mixture which, when initiated, is capable of sustaining a detonation wave, producing a powerful blast effect. Examples of the most important HEs include silver fulminate {BERTHOLLET ⇨1786/1787}, picric acid {WOULFE ⇨1775}, mercury fulminate {HOWARD ⇨1800}, nitrogen trichloride {DULONG ⇨1812}; nitroglycerin {SOBRERO ⇨1846}, gun-cotton {SCHÖNBEIN ⇨1846}, dynamite {A. NOBEL ⇨1867}, tetryl (Wilhelm MICHLER & Carl MEYER 1879), ammonium nitrate (AN) {GLAUBER ⇨1659; PENNIMAN ⇨1885}, TNT {HAEUSSERMANN ⇨1891}, PETN {TOLLENS & WIGAND ⇨1891}, lead azide {CURTIUS ⇨1891}, and RDX {HENNING ⇨1899}.[326]

Shortly before World War II, the high explosive HMX (**h**igh **m**elting **ex**plosive) or cyclotetramethylenetetranitramine – also known as "octogen" or "homocyclonite," as well as by other names – was discovered. HMX is one of the densest and most heat-stable high explosives known, and together with RDX it is the basis for almost all modern high explosives. HMX [$C_4H_8N_8O_8$] is similar to RDX [$C_3H_6N_6O_6$] but it has a higher molecular weight (296.2) and a much higher melting point (256 to 281 °C). Moldable plastic explosives were also developed during World War II. In the early 1950s, the first polymer-bonded explosives were developed in the United States at LASL, (Los Alamos, NM).

---

[323] A.G. WEBSTER: *Some new methods in interior ballistics*. Proc. Natl. Acad. Sci. U.S.A. **6**, No. 11, 648-659 (Nov. 15, 1920).

[324] D.J. HANSON: *Alcohol glossary*, see definition of the term *proof* given in http://www2.potsdam.edu/alcohol-info/AlcoholGlossary/GlossaryP.html.

[325] From an editorial note on Sir Frederick A. ABEL's discovery made at Woolwich; see his article *The rapidity of detonation*. Nature **8**, 534 (1873).

[326] For a review of the discoveries of various high explosives up to the late 1930s; see G. BUGGE: *Schieß- und Sprengstoffe und die Männer, die sie schufen*. Franckh, Stuttgart (1942).

Based upon HMX, and using Teflon as the binder, they were developed for projectiles and lunar active seismic experiments {Apollo 14 ⇒1971; Apollo 16/17 ⇒1972}.

The super high explosive CL-20 was first synthesized in the United States {NIELSEN ⇒1987}. Its extreme detonation velocity of 9,380 m/s (TNT: 6,930 m/s; HMX: 9,110 m/s) is of particular interest for achieving extremely high dynamic pressures, but this new highly energetic material is also currently being investigated for use as a rocket propellant. More recently, U.S. chemists synthesized another new type of very powerful high explosive, which they called "cubane," a high-density hydrocarbon {ZHANG ET AL. ⇒2000}.

Generally, explosives are classified in terms of their explosivity:
- *low explosives* (such as black powder, the first known explosive) only deflagrate;
- *high explosives* detonate. They are usually subdivided into two categories: *primary high explosives* (such as some fulminates and lead azide), which detonate promptly upon ignition by a spark, flame, or heat generated by a light impact, and *secondary high explosives* (such as TNT, RDX, PETN), which require a detonator and, in some cases, a supplementary booster charge.

Explosives can also be classified in terms of their inner structure, which is useful for understanding the decomposition processes at the detonation front and for developing appropriate theoretical detonation models:
- *homogeneous explosives* include gases, liquids without bubbles or suspended solids, and perfect crystals of solid explosives. In these materials, planar shock waves uniformly compress and heat the explosive material. The release of chemical heat is controlled by the bulk temperature and pressure;
- *heterogeneous explosives* include liquids with bubbles or suspended particles and pressed or cast solids with voids, binders, metal particles, *etc*. The release of chemical heat by the initiating shock wave is controlled by local high-temperature and high-pressure regions, such as "hot spots" due to collapsing voids.

Detailed reviews of the large number of explosives and propellants available are given in special encyclopedias such as in the *Ullmann's Encyclopedia*[327] and the *Encyclopedia of Chemical Technology*[328] (these also include tabular comparisons of their physical properties), or in handbooks such as Dupont's *Blaster's Handbook* (1967) and the *LLNL Explosives Handbook* (1985).

**Firing Devices.** The greatest advance in the science of explosives after the discovery of black powder was certainly the invention of the *blasting cap* {A. NOBEL ⇒1863/1864}. Combined with the *safety fuse*, invented by the Englishman William BICKFORD {⇒1831}, NOBEL's blasting cap provided a dependable means for detonating dynamite and the many other high explosives that followed it. His "detonators" contained mercury fulminate in a copper capsule which initiated the explosive reaction in a column of explosive by percussion rather than by the local heat from an electric spark or an electrically heated wire. From the 1880s onwards, a modification of this detonator (that still, however, worked according to the same principle of mechanical percussion) became widely used as an explosive signaling device. This so-called "railway torpedo" [Germ. *Knallkapsel*], which was provided with clips to fix it onto the railhead, was detonated when a vehicle passed by in order to attract the attention of railwaymen.

The blasting caps that followed were metal capsules containing a secondary explosive (such as TNT, tetryl or PETN) and a primary explosive (mostly mercury fulminate or lead azide). They are still used today to initiate secondary explosives. In the mid-1930s, the British chemist William PAYMAN and his collaborators visualized the detonation process for detonators using schlieren photography {⇒Fig. 4.16–X}. Depending on the detonator geometry and the casing material, they observed jetting of gases and particles which are sent out ahead of the main shock wave, and are crucial for initiating detonation in the adjacent secondary explosive. The *safety detonator*, an "exploding bridgewire" version of the detonator, invented during World War II by the U.S. physicist Lawrence H. JOHNSTON {⇒1944}, was one of the technical prerequisites for realizing the first nuclear implosion bomb {VON NEUMANN & NEDDERMEYER 1943; Trinity Test ⇒1945}. Since the safety detonator contains no primary explosive at all, it is extremely safe. The secondary explosive inside is initiated with submicrosecond accuracy via a shock wave generated by an exploding wire. This safety detonator has become an indispensable high-precision ignition tool in modern detonics.

The high detonation velocities of modern high explosives are an easy and very effective means of generating ultrahigh dynamic pressures {FOWLES & ISBELL ⇒1965}. However, the detonation velocity is not a constant quantity; *i.e.*, it is dependent not only on the chemical composition and ambi-

---

[327] J. BOILEAU, C. FAUQUIGNON, and C. NAPOLY: *Explosives*. In: (L. KAUDY, J.F. ROUNSAVILLE, and G. SCHULZ, eds.) *Ullmann's encyclopedia of industrial chemistry*. VCH, Weinheim/Bergstr., vol. A10 (1987), pp. 143-172.

[328] V. LINDNER: *Explosives and propellants*. In: (M. HOWE-GRANT, ed.) *Encyclopedia of chemical technology*. Wiley, New York *etc.* (1991–1998), vol. 10 (1993), pp. 1-125.

ent physical conditions, but unfortunately also on the geometry of the high explosive bulk. For example, the detonation front in a stick diverges spherically as it progresses through the explosive, but when it reaches the stick's surface, the energy leaving the stick decreases the velocity locally. These so-called "edge effects" {G.I. TAYLOR ⇨ 1949} are particularly detrimental in all precision detonation configurations and were first noticed in the implosion assembly of the first atomic bomb (S. NEDDERMEYER 1943). Here, this phenomenon played an important role in the design of three-dimensional high explosive lenses made of fast- and slow-detonating high explosives that were used to shape the detonation wave in order to make it converge inward on a central point.

In the open literature, the use of a composite high-explosive charge for detonation wave shaping was apparently first discussed by the Englishman J.H. COOK {⇨ 1948}, who used this method to engrave small designs on metal plates. Today detonation wave shaping is also important in explosive-formed projectiles (EFPs), a unique method in which a high explosive is used to simultaneously form a projectile and propel it up to hypersonic velocities. In the more sophisticated Multi-EFP concept, these munitions are set to explode ahead of the incoming threat, thus forming a dense "curtain" of high-velocity fragments within one millisecond.

By the year 1880, some of the explosives mentioned above were already being used for military and civil purposes. For example, mercury fulminate was used extensively. In 1835, France alone produced 800 million percussion caps that used mercury fulminate. However, the physico-chemical processes of explosion and detonation were not yet known. The increasing number of new explosive substances discovered in the following decades stimulated physicists and chemists to uncover the riddle of detonation, which was felt to be somehow closely related to the rapidly propagating mechanical (shock) wave created by the detonation.

In the period 1869–1874, the first detonation velocity measurements were carried out by the British chemist Sir Frederick A. ABEL. These studies, performed in long, confined charges of various high explosives, revealed unusually high velocities in the range of some thousands of meters per second. However, the rationale for detonation was not yet understood. The experiments of Henri J.B. PELLET[329] and P. CHAMPION in France and Charles L. BLOXAM in England appeared to indicate that the detonating agent exerted some kind of vibratory action upon the particles of the substance to be exploded. Studying the behavior of unconfined and confined charges, ABEL speculated that detonation in a high explosive might be transmitted by means of some "synchronous vibrations" {ABEL ⇨ 1869}, which BLOXAM, a professor of chemistry at King's College in London, called "sympathetic explosions." Shortly thereafter, the French chemist P.E. Marcellin BERTHELOT {⇨ 1870} was the first to correctly assume that detonation might be caused by a strong moving mechanical shock – une onde de choc (a "shock wave") – but experimental proof of this was yet to appear {WENDLANDT ⇨ 1924}.

BERTHELOT's hypothesis, however, was not immediately accepted by the explosion research community, and the 1878 edition of the prestigious *Encyclopaedia Britannica*[330] reads, "An explosive molecule is most unstable, certain very delicately balanced forces preserving the chemical and physical equilibrium of the compound. If these forces be rapidly overthrown in succession, we have explosion; but when, by a blow of a certain kind, they are instantaneously destroyed, the result is detonation. Just as a glass globe may withstand a strong blow, but be shattered by the vibration of a particular note, so it is considered by some authorities that, in the instance cited, the fulminate of mercury communicates a vibration to which the gun-cotton molecule is sensitive, and which overthrows its equilibrium; it is not sensitive to the vibration caused by the nitroglycerin, which only tears and scatters it mechanically." During this period, Hermann HELMHOLTZ's "theory of resonance" (1857) and the demonstration of resonance performed using his "Helmholtz resonator" were very popular among acousticians, and Heinrich HERTZ first demonstrated the detection of weak electromagnetic waves with a resonance circuit (1887), certainly one of the greatest triumphs for the mighty principle of resonance. However, the riddle of detonation in liquid and solid explosives couldn't be solved using this concept of resonance.

At this point, it is useful to look back on previous attempts to understand detonation in gases. The discovery of oxyhydrogen and its violent explosive properties {TURQUET DE MAYERNE ⇨ 1620; CAVENDISH ⇨ 1760s} stimulated not only a crude theory on the origin of earthquakes {LÉMERY ⇨ 1700; KANT ⇨ 1756}, but it also turned the interest of naturalists to other explosive gaseous mixtures, particularly to firedamp, which had been a hazard to miners since the beginning of hard coal mining in the 12th century. In the second half of the 19th century, coal mining increased tremendously in most European countries due to the rapid

---

[329] H.J.B. PELLET and P. CHAMPION: *Sur la théorie de l'explosion des composes détonants*. C. R. Acad. Sci. Paris **75**, 210-214 (1872); *Explosions produced by high tones*. Phil. Mag. **46** [IV], 256 (1873).

[330] W.H. WARDELL: *Explosives*. In: *Encyclopaedia Britannica*. Black, Edinburgh (1878), vol. 8, pp. 806-813.

adoption of steam engines for use in ironworks, the production industry, and advances in railroad and shipping traffic. In addition, hard coal was increasingly used as a fuel in homes and for producing city gas. For example, in England the productivity of hard coal mining increased from ten million tons at the beginning of the 19th century to 184 million tons in 1890.[331] Black powder, first used in the 16th century as a blasting agent in mines, was also applied in underground coal mining due to its relatively gentle, heaving action, leaving the coal in a good position for rapid loading. However, black powder has a dangerous tendency to ignite coal dust and coal gas, which mostly consists of methane. As a consequence, many disastrous mine explosions occurred with a large loss of life {ABEL ⇒1881}.

**Firedamp Explosions.** The scientific investigation of firedamp explosions was initiated by Sir Ralph MILBANKE {⇒1813}, founder and first chairman of the Society for Preventing Accidents in Coal Mines. He asked the English chemist Humphry DAVY to develop a safe lamp for working in coal mines. DAVY {⇒1816}, partly supported by his assistant Michael FARADAY, analyzed the explosivity of firedamp and discovered that the critical mixture for explosion is 9% methane and 91% air. The result of DAVY's studies was the "Davy lamp" {⇒Fig. 4.21−B}, certainly his most famous invention. Unfortunately, however, this lamp could only partly mitigate the risk of explosion accidents. For example, William GALLOWAY {⇒1874}, a British mine inspector, demonstrated that the blast wave generated in a firedamp explosion forces the flame through the mesh of a Davy-type safety lamp. There were also other sources of open fire, such as those resulting from explosives used for blasting purposes, one of the oldest and most important civil applications of explosives since the 16th century.

DAVY's combustion studies, though worthwhile from a practical point of view, could not solve the riddle of the nature of an explosion. Similarly, studies carried out almost 50 years later by Robert BUNSEN {⇒1857 & 1867}, a German chemist at the University of Heidelberg, were only partly successful. BUNSEN, cofounder of chemical spectroscopy, and an international authority on analytical chemistry, who turned the analysis of gases into an exact science, measured the strength and rate of combustion of various explosive gaseous mixtures, including oxyhydrogen. However, he used an experimental set-up that could not provoke detonation, only deflagration, which is a much slower chemical reaction than detonation by several orders of magnitude.

**Coal Dust Explosions.** Dust explosions, certainly the oldest type of man-made explosion, frequently occurred in flour mills and bakeries {MOROZZO ⇒1785}, and later also in metal powder works {Bethlehem Zinc Works ⇒1854}. Such explosions stimulated various hypotheses that coal-dust-laden air might cause explosions in coal mines in an analogous manner {FARADAY & LYELL ⇒1845; RANKINE & MACADAM[332] 1872}. Eventually, a series of tragic firedamp explosions in the French coal mining industry {Jabin de Saint-Etienne ⇒1876} led to the foundation of the French Firedamp Commission {⇒1877} with the objective to scientifically investigate possible causes of these explosions and to suggest efficient and economic countermeasures. Additional mining accidents due to firedamp explosions in France, England and the United States soon afterward (some of them probably also produced by the presence of coal dust) initiated the foundation of similar national institutions in these countries.

**Detonation Wave.** In Paris at the Collège de France, the chemist P.E. Marcellin BERTHELOT initiated comprehensive studies on the mechanism of detonation in gases. These investigations were carried out together with Paul VIEILLE, a skilful chemist and explosives specialist at the Central Laboratory of the Service des Poudres et Salpêtres in Paris, and they revealed

▸ that an explosive wave − later generally termed a *detonation wave* − exists in explosive gaseous mixtures and propagates at a tremendous speed of up to 2,500 m/s; and

▸ that the propagation velocity only depends on the mixture composition, not on the tube diameter, as long as that diameter is not too small {BERTHELOT ⇒1881; BERTHELOT & VIEILLE ⇒1882}. This explained BUNSEN's failure to provoke detonation in narrow tubes {BUNSEN ⇒1867}.

Another pair of prominent researchers were also enlisted by the French Firedamp Commission: François E. MALLARD, a professor of mining engineering, and Henry L. LE CHÂTELIER, a chemistry professor who both taught at the distinguished Ecole des Mines in Paris. They were asked by the Commission to examine the best means of guarding against firedamp explosions in mines. This led to a series of investigations on the specific heat of gases at high tempera-

---

[331] *Meyers Conversationslexikon.* Bibliographisches Institut, Leipzig & Wien, vol. 16 (1897), pp. 374-375.

[332] S. MACADAM: *On flour-mill fire-explosions.* Proc. Phil. Soc. Glasgow **8**, 280-288 (1873).

tures, the ignition temperatures and the flame propagation velocities of gaseous mixtures {MALLARD & LE CHÂTELIER ⇨1881}. They first recorded the propagation of an explosive wave in a long tube with a rotating drum camera – a milestone in detonation diagnostics. Most surprisingly, they observed that the transition from combustion to detonation occurs suddenly, and that the detonation velocity is comparable to the velocity of sound of the burnt detonation products {MALLARD & LE CHÂTELIER ⇨1883}.

**Chapman-Jouguet (CJ) Theory.** The results from these two French teams prompted a new era in understanding explosions, and promoted the first theory of detonation. A simple and convincing explanation for the "explosive wave" was given by the English chemist David L. CHAPMAN {⇨1899}, who – stimulated by previous English studies on detonation {SCHUSTER ⇨1893; DIXON ⇨1893} – assumed that the chemical reaction takes place due to a sharply defined front sweeping over the unburnt gas and changing it instantaneously into burnt gas {⇨Fig. 4.16–S}. The term *explosive wave* was understood by CHAPMAN "to limit the space within which chemical change is taking place. This space is bounded by two infinite planes. On either side of the wave are the exploded and unexploded gases, which are assumed to have uniform densities and velocities." In comparison to a normal shock wave with its discontinuous transition from uncompressed to compressed gas across the shock front, the detonation front additionally separates two chemically different states of unburnt and burnt gases. A detonation wave is therefore also called a "reactive shock wave." When CHAPMAN put forward his pioneering hypothesis on detonation, little was known about the atomic processes of initiating detonation. Avoiding a discussion of this problem, and instead focusing on the macroscopic behavior of explosive waves, he wrote, "How the true explosive wave is actually generated in practice is a question without the scope of the present investigation. In order to avoid the discussion of this point, I shall substitute for it a physical conception, which, although unrealizable in practice, will render aid in illustrating the views here advanced."

CHAPMAN's assumption was also made independently by the French engineer and mathematician Emile JOUGUET {⇨1904/1906 & 1917}, a professor at the Ecole des Mines in Saint-Étienne. The so-called "Chapman-Jouguet (CJ) theory" – a term apparently later coined at the U.S. Bureau of Mines {LEWIS & FRIAUF ⇨1930} – independently evolved in Russia {MIKHEL'SON ⇨1890}, France {JOUGUET ⇨1904/1906; CRUSSARD ⇨1907; JOUGUET ⇨1917}, and England {SCHUSTER ⇨1893; DIXON ⇨1893; CHAPMAN ⇨1899}. The CJ theory, which treats the detonation wave as a discontinuity with an infinite reaction rate, assumes that the hot products of the combustion wave act as an expanding hot-gas piston that accelerates the unburnt mixture ahead, thereby forming a sharply defined front called the "explosive wave" (or, in modern terms, the "detonation wave"), which sweeps over the unburnt gas {⇨Fig. 4.16–S}. Furthermore, the understanding of detonation phenomena was significantly advanced owing to the classical work of the French physicist Louis CRUSSARD {⇨1907}.

The CJ theory does not require any information about the chemical reaction rate; *i.e.*, about the chemical kinetics. More details about the evolution of this first detonation model can be found in the CHRONOLOGY. The conservation equations for mass, momentum and energy across the one-dimensional wave give a unique solution for the detonation velocity (or CJ velocity) which correspond to the *Mikhel'son line* {MIKHEL'SON ⇨1893} or *Rayleigh line* {Lord RAYLEIGH ⇨1878}. In the pressure-volume diagram the Mikhel'son-Rayleigh-line corresponds to the CJ solution which is tangent of the Hugoniot curve for the burnt gas: the tangent line inclination to the Hugoniot curve is the local sound speed squared in the burnt gas. A higher detonation velocity than the CJ detonation velocity is possible by constructing an explosive in a two-layer arrangement and then using the so-called "channel effect" {BAKIROV & MITROFANOV ⇨1976}.

**Zel'dovich-von Neumann-Döring (ZND) Theory.** The assumption of a sharp detonation front is an idealization, and the Chapman-Jouguet theory, which treats the detonation wave as a discontinuity with an infinite reaction rate, was later refined by introducing a three-layer model of the detonation front {⇨Fig. 4.16–S}. This improved model, which also takes the reaction rate into account, was advanced independently during World War II in the former Soviet Union by the chemist Yakov B. ZEL'DOVICH (1940), in the United States by the mathematician John VON NEUMANN (1942), and in Germany by Werner DÖRING, a theoretical physics professor at the Reichsuniversität Posen {Secret Workshop "Probleme der Detonation" ⇨1941; DÖRING ⇨1943}. Today known as the "Zel'dovich-von Neumann-Döring (ZND) theory," it provides the same detonation velocities and pressures as the CJ theory; the only difference between the two models is the thickness of the wave {ZEL'DOVICH ⇨1940}. In this so-called "ZND model," the detonation process consists of a shock wave that takes the material from its initial state to a "von Neumann" spike point on the unreacted

Hugoniot. The ZND reaction zone is traversed by proceeding down the detonation Rayleigh line from the spike point to the CJ condition; *i.e.*, the fully reacted state.

The ZND model was originally derived for a one-dimensional steady state detonation wave in a gaseous explosive, and modifications of this model have since been proposed by numerous detonation researchers in order to get a better match to experimental conditions. For example, a more refined *one-dimensional nonequilibrium ZND model* was developed {TARVER ⇨1982} that also includes the thermal relaxation processes which precede and follow the exothermic chemical reconstitution reactions that take place in condensed explosives. Since the 1950s, numerous theoretical and experimental studies have been performed in order to precisely determine the CJ pressure and the length of the reaction zone in a detonating high explosive {DUFF & HOUSTON ⇨1953; DEAL ⇨1957}. For solid high explosives, the reaction zone length − *i.e.*, the distance from the spike point to the CJ state − is very short; for the high explosive HMX, this amounts to only some hundreds of microns, which corresponds to a reaction zone time of only a few tens of nanoseconds {GUSTAVSEN ET AL. ⇨1997}.

The CJ and the ZND theories provide a macroscopic picture of the mechanism of detonation, but they cannot describe the chemical processes of detonation at the molecular level. Obviously, in order to "wake up" or "trigger" an explosive molecule using an impacting shock wave, the energy that must be deposited by the shock wave to initiate exothermic chemical decomposition must be distributed to the vibrational modes of the explosive molecule within the appropriate time window to form a detonation wave.[333]

**Complex Detonation Processes.** Numerous experimental studies have revealed that many self-sustaining detonations − *i.e.*, detonations furnished by the driving energy of the chemical reaction − are not strictly one-dimensional, but that they contain transverse waves which may be quite strong, leading to a *periodic cell structure* {BONE ET AL. ⇨1936}. The Los Alamos detonation physicists Wildon FICKETT[334] and William C. DAVIS gave an intuitive picture of how cellular structures might arise from a grid of regularly spaced perturbations interacting with each other in terms of individual Mach reflections {VON NEUMANN ⇨1943}. They wrote, "The key feature of the structure is the transverse wave, an interior shock joined to the leading shock in the conventional three-shock configuration. The Mach stem and the incident shock are part of the leading shock, and the transverse wave is the reflected shock. The transverse waves move back and forth across the front. Groups of them moving in the same direction take up a preferred spacing on the order of 100 reaction-zone lengths. They are not steady waves, but are continually decaying, and stay alive only by periodic rejuvenation through collision with other transverse waves moving in the opposite direction."

Transverse waves are particularly strong in *spinning detonation*, which is often observed in near-limit mixtures in round tubes {CAMPBELL & WOODHEAD ⇨1926; BONE & FRASER ⇨1929}.[335] Based upon images obtained by different high-speed recording techniques, the two-dimensional cell structure is now fairly well understood, but the processes that occur in the three-dimensional reaction zone and the wave structure behind the front are difficult to resolve optically. With the advent of powerful computers, so-called "supercomputers," the riddle of three-dimensional structures of detonation waves has also been tackled through numerical simulation {FUJIWARA & REDDY ⇨1993}.

**Evolution of Chemical Kinetics.** The classical chlorine-hydrogen explosion − a puzzling photochemical-induced reaction discovered by the French chemists Joseph-Louis GAY-LUSSAC and Louis J. THÉNARD {⇨1809} − was investigated in more detail by the English chemist David L. CHAPMAN in the period 1909–1933. The German chemist Max E.A. BODENSTEIN {⇨1913} had first advanced the concept of reactive intermediates as part of a "chain reaction" [Germ. *Kettenreaktion*], a term which he coined. In 1918, Walther H. NERNST, another German chemist studying photochemistry, postulated his "atom chain reaction theory." This assumed that once the energy of a quantum has initiated a reaction in which free atoms are formed, these formed atoms can themselves decompose other molecules, resulting in the liberation of more free atoms, and so on… Both of their concepts explained detonation not as an instantaneous, single-stage chemical reaction, but rather as branched chain reactions that pass through various short-living intermediate states.

Their findings stimulated the evolution of *chemical kinetics*, a new and exciting branch of physical chemistry that deals with the rates and mechanisms of chemical reactions {SEMENOV & HINSHELWOOD 1928; Nobel Prize for Chemistry ⇨1956}.

---

[333] C.M. TARVER: *What is a shock wave to an explosive molecule?* In: (M.D. FURNISH, N.N. THADHANI, and Y. HORIE, eds.): *Shock compression of condensed matter − 2001.* AIP Conference Proceedings, Melville, NY (2002), pp. 42-49.

[334] See FICKETT & DAVIS {⇨1979}, pp. 291-300.

[335] An *explosion limit* usually refers to the range of pressure and temperature for which an explosive reaction at a fixed composition mixture is possible.

*Combustion theory,* a more general discipline which grew steadily in importance in the 20th century, also tackled the difficult problem of understanding explosion processes in order to provide an insight into the fundamentals of chemical kinetics. It turned out that two qualitatively different mechanisms can produce explosions in homogeneous combustion systems:

▸ one mechanism is that of a *thermal explosion,* in which heat released by the reaction raises the temperature. This, in turn, accelerates the rate of heat release;

▸ the other mechanism is that of a *branched-chain explosion,* in which large numbers of highly reactive intermediate chemical species – so-called "free radicals" – are produced in the combustion reactions, and these radicals accelerate the reaction rate.

The increasing number of studies aimed at improving our fundamental understanding of all aspects of combustion from a theoretical and a mathematical (numerical) modeling perspective led to the establishment of the journal *Combustion Theory and Modelling* {⇨1997}.

Since the invention of the four-stroke internal combustion engine by the German mechanical engineer Nikolaus A. OTTO in 1876, and its widespread use in automobiles (which were first invented in 1885–1886 by Carl BENZ and Gottlieb DAIMLER), the study of *knocking* (or *pinging*) has occupied generations of chemists {NERNST ⇨1905}. This detrimental detonation phenomenon, which occurs during combustion in an instantaneous, uncontrolled manner, reduces both the power output and the lifetime of the engine. Detailed studies that made use of the basic knowledge of detonation obtained previously from investigations of firedamp explosions showed that an explosive, premature self-ignition of the fuel takes place in Otto-engines before the flame front, thus leading to a detonation wave. Knocking can be prevented in internal combustion engines through the use of high-octane gasoline, the addition of lead (which increases the octane rating) or isooctane additives to the gasoline, or by retarding spark plug ignition.

## 2.6.3 DETONICS: THE KEY TO ULTRAHIGH SHOCK PRESSURES, AND NEW APPLICATIONS

High explosives played an important role in the evolution of shock wave physics, particularly in the study of solid matter under high dynamic pressures. At first, simple arrangements with an explosive in contact with the test target were used. This straightforward technique was later refined by employing an explosive lens {J.H. COOK ⇨1948}, which allowed the controlled generation of planar shock waves for the first time, and the measurement of Hugoniot data from one-dimensionally shocked materials {GORANSON ET AL. ⇨1955; DICK ⇨1970}. The method was further improved by using an explosively driven plate arrangement, the "flyer plate method" {MCQUEEN & MARSH ⇨1960}, which allowed the generation of shock pressures of up to 2 Mbar. In the 1950s, Soviet researchers developed cascade detonation devices incorporating spherical flyer plates which allowed the production of ultrahigh pressures {AL'TSHULER ⇨1996}. Another method of generating high dynamic pressures using high explosives is the application of the Mach effect, a nonlinear superposition of two obliquely interacting shock waves {AL'TSHULER ET AL. ⇨1962; FOWLES & ISBELL ⇨1965; NEAL ⇨1975; ⇨Figs. 4.13–H, L}.

As solid-state shock wave physics has evolved further, other, more appropriate, methods of generating very high dynamic pressures in the laboratory that do not require high explosives have been developed. Prominent examples include:

▸ hypervelocity planar impact methods that use the light-gas gun {CROZIER & HUME ⇨1946}, the railgun {RASHLEIGH & MARSHALL ⇨1978}, or the electric gun {STEINBERG ET AL. ⇨1980};

▸ pulsed radiation methods that use giant laser pulses {ASKAR'YAN & MOROZ ⇨1962}, or high-intensity pulsed soft X-ray pulses;

▸ pulsed beams of electrons, neutrons or ions {BLUHM ET AL. ⇨1985}; and

▸ a strong flux of neutrons from an underground nuclear explosion contained within rocks {TRUNIN ET AL. ⇨1992}, which has been the only type of nuclear test permitted since the Moscow International Nuclear Test Ban Treaty came into effect {⇨1965}.

However, high explosives are still an inexpensive and indispensable way of producing special high-pressure effects or other physical effects in a simple way. For example, they proved very useful for generating very high magnetic fields by *magnetic field compression* {TERLETSKII ⇨1957; FOWLER ET AL. ⇨1960}, a promising method of generating strong current pulses when operating pulsed radiation sources such as high-intensity pulsed laser radiation, electromagnetic microwaves and bursts of γ-rays, which (amongst other applications) can also be used in special experimental arrangements to generate shock waves under extreme conditions.

The *shaped charge cavity effect*, a curiosity in the history of explosives and a staple in detonics, has proven its wide applicability in both military and civil circles. This unique effect was originally discovered in charges without liner, first in a charge of black powder {VON BAADER ⇨1792} and then also in high explosives {VON FÖRSTER ⇨1883; MUNROE ⇨1888}, and it is particularly striking when a metal liner is used. The resulting hypervelocity jet of molten liner material is capable of penetrating even thick steel plates. This *shaped charge lined cavity effect* was invented independently by various detonation researchers in the late 1930s {THOMANEK ⇨1938}, and it immediately gained the greatest importance during World War II as an inexpensive but highly effective armor-piercing weapon. The "bazooka" [Germ. *Panzerfaust*], an anti-tank device, was based on the shaped-charge concept. Used as the "poor man's high-velocity gun," the shaped charge has become a standard weapon in many military arsenals. Shaped charges are also used to perforate oil-well casings {MCLEMORE ⇨1946; ⇨Fig. 4.15−L} – this has long been their main civilian application – as well as for tapping open-hearth steel furnaces. This important cavity effect was studied in great detail (particularly the jet formation) both analytically {BIRKHOFF & G.I. TAYLOR ⇨1948} and experimentally using flash radiography {STEENBECK ⇨1938; ⇨Fig. 4.8−F}. In addition, flash X-ray diffraction provided the first insights into the fine structure of the high-velocity jet of the liner material {JAMET & THOMER ⇨1974; GREEN ⇨1974}.

*Blasting* describes the process of reducing a solid body, such as rock and ice, to fragments. Besides its various applications in the military, this method is also used for civil purposes, such as to move large masses of earth when building canals {⇨Fig. 4.16−K} and dams, for tunneling {Musconetcong Tunnel ⇨1872} and other construction works, and to demolish natural obstacles {Hell Gate ⇨1885; Ripple Rock ⇨1958} or man-made structures {Helgoland Blast ⇨1947}. Historically, black powder was used as early as the 17th century for mining purposes. However, beginning with the invention of dynamite {NOBEL ⇨1867}, black powder was increasingly replaced by more efficient high explosives, and since about 1900 so-called "permissible explosives" have been employed, which considerably reduce the chances of triggering firedamp explosions. In the oil industry, explosives are also used for *oil well shooting* {ROBERTS ⇨1864} and to "snuff out" oil well fires {KINLEY ⇨1913; ⇨Fig. 4.15−K}. Other large fields in which high explosives are applied include *explosive working* {RINEHART & PEARSON ⇨1963; ⇨Figs. 4.15−I, J}, the *shock synthesis* of new materials {PRÜMMER ⇨1987}, and *explosion seismology* (*see* Sect. 2.7.1).

## 2.6.4 NUCLEAR AND THERMONUCLEAR EXPLOSIONS: THE ULTIMATE MAN-MADE SHOCK PHENOMENA

> *As the Director of the Theoretical Division at Los Alamos, I participated at the most senior level in the Manhattan Project that produced the first atomic weapons. Now, at the age of 90, I am one of the few remaining senior project participants… In my judgment, the time has come to cease all physical experiments, no matter how small their yield, whose primary purpose is to design new types of nuclear weapons, as opposed to developing peaceful uses of nuclear energy. Indeed, if I were President, I would not fund computational experiments, or even creative thought designed to produce new categories of nuclear weapons…*[336]
> 
> Hans A. BETHE
> Cornell University
> Ithaca, NY 1997

A proper historical account of the Manhattan Project {⇨1942} and the development of the first atomic bombs – even one confined solely to the shock and detonation research that took place during the Project – is beyond the scope of this book. However, some of the most important achievements are listed in the CHRONOLOGY, and the most important steps that led to the first atomic bombs based upon uranium and plutonium will be summarized in this chapter. Their construction presented many difficulties, and such a task was actually considered to be impossible by many renowned physicists of that time. A few examples of problems related to nuclear blast wave effects {G.I. TAYLOR ⇨1941 & 1944; VON NEUMANN ⇨1943} and precise triggering of multiple detonations {JOHNSTON ⇨1944} are also given in the CHRONOLOGY.

The development, testing and use of the first nuclear weapons are subjects that have been treated extensively in the literature and widely illuminated from scientific, technical,

---

[336] From an open letter by Prof. em. Hans A. BETHE – recipient of the 1961 Enrico Fermi Award, winner of the 1971 Nobel Prize for Physics, and previously a scientific adviser at the nuclear test-ban talks in Geneva – sent on April 25, 1997 to U.S. President Bill CLINTON; *see* Federation of American Scientists (FAS), http://www.fas.org/bethepr.htm. • President CLINTON, answering BETHE's letter on June 2, 1997, diplomatically pointed out that he had directed "that the United States maintain the basic capability to resume nuclear test activities prohibited by the Comprehensive Test Ban Treaty in the unlikely event that the United States should need to withdraw from this treaty."

biological, medical, political and logistical viewpoints.³³⁷ It is a tale that is both fascinating and repulsive: on the one hand, it was an incredibly challenging task that involved physicists, chemists, engineers and technicians exploring a completely new area of research, and the pressures of war demanded that their work had to be both swift and successful. On the other hand, the nuclear weapons that resulted from the work provided their possessors with massive potential for destruction and terror, which could threaten even the existence of life on Earth. Furthermore, the progress made in nuclear weaponry, together with the significant advances obtained in supersonic and hypersonic aerodynamics, rocketry and digital computers since the 1940s, completely changed the international political scene, military strategies, the global economy, and even our own personal lives, right up to the present day. All of these dramatic changes created a new world of previously unknown threats and anxieties.

---

[337] The classic of this genre, which describes American efforts, is the book by Richard RHODES, *The Making of the Atomic Bomb* [Simon & Schuster, New York, 1986]. The book by Lillian HODDESON ET AL.: *Critical Assembly. A Technical History of Los Alamos during the Oppenheimer Years, 1943–1945* [Cambridge University Press, Cambridge, 1993], treats in detail (i) the research and development that led to implosion and gun weapons; (ii) the chemistry and metallurgy that enabled scientists to design these weapons; and (iii) the conception of the thermonuclear bomb, the so-called "Super." An excellent survey of U.S. nuclear weapons development, covering the pioneering period 1939–1963 and including many unique illustrations, is given by Frank H. SHELTON in his book *Reflections of a Nuclear Weaponeer* [Shelton Enterprise, Inc., Colorado Springs, CO, 1988]. The Los Alamos Museum edited the informative brochure *Los Alamos 1943-1945; The Beginning of an Era* [Rept. LASL-79-78, July 1986]. The more recent book *Picturing the Bomb* [Abrams, New York, 1995] by Rachel FERMI (granddaughter of Enrico FERMI) and Ester SAMRA contains a unique gallery of photographs from the secret world of the Manhattan Project.

Similar activities in this field that were carried out in the Soviet Union during the period 1939–1956 are comprehensively described in David HOLLOWAY's book *STALIN and the Bomb* [Yale University Press, New Haven, CT & London, 1994]. In addition, several renowned Soviet pioneers of shock wave and detonation physics have more recently given most interesting accounts of their roles and tasks in the development of nuclear weapons, such as Lev. V. AL'TSHULER, Yulii B. KHARITON, Igor V. KURCHATOV, Andrei D. SAKHAROV, and Yakov B. ZEL'DOVICH. Models of the first American and Soviet atomic bombs are shown in the Bradbury Science Museum at Los Alamos, New Mexico, and in the Museum of Nuclear Weapons at Arzamas-16, Sarov, Russia, respectively.

Nuclear research carried out in Germany during World War II was not directed towards building atomic bombs; *see* H.A. BETHE: *The German uranium project.* Phys. Today **53**, 34-36 (July 2000). However, speculations by numerous German scientists who fled Nazi Germany in the 1930s stimulated the governments of the United States and Great Britain to begin their own efforts in order to anticipate possible German atomic weaponry. Recent investigations carried out by some German science journalists have brought to light the fact that a few scientists tried to develop a simple fusion bomb in Germany during World War II {HAJEK ⇨ 1955}. However, these scientists were not seriously supported by the German Ministry of War.

---

The development of nuclear weapons encompassed unprecedented and complex shock- and detonation-related problems that required comprehensive theoretical and experimental studies for their solutions. When the Manhattan Project was set up in June 1942, considerable research into fission had already been performed in previous years, particularly in some European countries. There was also a solid foundation of basic knowledge of shock wave and detonation physics, although this was only understood and applied by a minority of scientists and engineers. European aerodynamicists and ballisticians had also developed a basic understanding of high-speed diagnostics and microsecond high-speed photography, but the application of their methods as well as the duplication of their apparatus was an art rather than a consolidated technique that was commercially available. The challenging goal of building an atomic bomb within the short period of three years only became possible because a significant number of top scientists from American universities were hired and incorporated into the Manhattan Project. In addition, a considerable number of renowned scientists from Europe, many of whom had escaped Nazi Germany in the 1930s and emigrated to the United States or England, brought their competence in nuclear physics, fluid dynamics, chemistry, metallurgy and high-speed diagnostics to the project. Their substantial knowledge could be rapidly focused on the problems involved and effectively exchanged between the Allied research institutions.

**Milestones in Nuclear Physics.** Numerous important discoveries, technological achievements and speculations were used by early nuclear weaponeers to develop a nuclear bomb small enough to be dropped from an aircraft. These main milestones include:

▸ the discovery of radioactivity (A.H. BECQUEREL 1896);
▸ the first concept of an "atomic bomb," which originated in science fiction {WELLS ⇨ 1914} but was first patented by L. SZILARD in 1934;
▸ the invention of the cyclotron (E. LAWRENCE 1930);
▸ the discovery of the hydrogen isotope "deuterium" (H.C. UREY 1931);
▸ the discoveries of the "neutron" {CHADWICK ⇨ 1932} and the "positron" (C.D. ANDERSON 1932);
▸ the splitting of the lithium nucleus into two alpha particles (J.D. COCKROFT & E.T.S. WALTON 1932);
▸ the discovery of the uranium isotope U-235 {DEMPSTER ⇨ 1935};
▸ the first demonstration of nuclear "fission" in uranium {HAHN & STRASSMANN ⇨ 1938; FRISCH & MEITNER ⇨ 1939};

- the recognition of the enormous energy released in fission processes in Germany {HAHN & STRASSMANN ⇒1938; S. FLÜGGE 1939; HEISENBERG ⇒1939}, Denmark {FRISCH & MEITNER ⇒1939; N. BOHR 1939}, and the Soviet Union (Y.B. ZEL'DOVICH & Y.B. KHARITON 1939);
- the assumption (based on theoretical analysis) that the component most likely to undergo fission when bombarded with neutrons, thus yielding more neutrons in sufficient numbers to possibly sustain an explosive chain reaction, was not U-238 but instead the rather less common isotope U-235 (N. BOHR & J.A. WHEELER 1939); the first experimental evidence of such a "chain reaction" was provided by I. JOLIOT-CURIE in 1939;
- the initiation of the American program to build an atomic bomb by the Einstein-Szilard letter addressed to the U.S. President Theodore ROOSEVELT {A. EINSTEIN ⇒1939};
- the discovery of spontaneous fission in uranium – i.e., fission that occurs without the need for neutron bombardment (K.A. PETRZHAK & G.N FLEROV 1940);
- the successful isotope separation of natural uranium by gaseous diffusion (O.R. FRISCH & R.E. PEIERLS 1940; J.R. DUNNING & E.T. BOOTH 1941);
- the (secret) discovery of plutonium Pu-239, the first transuranic element {SEABORG, KENNEDY, WAHL & SEGRÉ ⇒1941}; and
- the first evidence of a self-sustaining chain reaction in an uranium "reactor" (E. FERMI 1942).

**The First Types of Nuclear Bombs.** Nuclear weaponeers elaborated on two bomb concepts:
- the *gun-assembly device* proposed by Los Alamos scientists in the early 1940s and, independently, by Soviet scientists (Y.B. KHARITON, Y.B. ZEL'DOVICH & G.N. FLEROV 1941); and
- the *implosion-assembly device* proposed by U.S. scientists (J. VON NEUMANN & S. NEDDERMEYER 1943).

The applicability and effects of nuclear weapons based upon fission were demonstrated impressively by the successful test of the first fission bomb {Trinity Test ⇒1945}, and its first use for military purposes {Hiroshima & Nagasaki Bombing ⇒1945}. However, at the end of World War II the idea of a weapon based upon fusion, a so-called "thermonuclear bomb" or "H-bomb," was still the subject of much discussion {FERMI & TELLER ⇒1941}. Edward TELLER's "Super," which used liquid deuterium as a fuel kept at about –250 °C, was successfully tested in 1952 {MIKE Test ⇒1952}. In the former Soviet Union, the first successful thermonuclear test was performed only a few months after the American MIKE test {Semipalatinsk ⇒1953}. However, the Soviet device was a more advanced design and had already been constructed as a compact, aircraft-deliverable weapon.

Compared to fission bombs, once the explosion is started (by a fission bomb) in a fusion bomb, the production of additional explosive energy is relatively inexpensive. Therefore, large-yield bombs are most efficiently based upon fusion. The shock and heat effects of a fusion bomb are similar to the effects of fission bombs, although they are usually more powerful and, therefore, all effects must be scaled up.

Although over 50 years have passed since the MIKE Test and a number of other countries have developed fusion bombs since then, the construction details for TELLER's Super are still secret. TELLER,[338] who is generally considered to be the father of the hydrogen bomb, wrote an encyclopedia article in 1974 on this subject, concluding that: "The development of the technical ideas that led to the first man-made thermonuclear explosion is secret, as is the history of construction of the hydrogen bomb itself … After the American and Russian tests, thermonuclear explosions were produced by the British (1957), the Chinese (1967), and the French (1968). It became evident that secrecy does not prevent the proliferation of thermonuclear weapons. It is doubtful whether an international nonproliferation treaty will be effective. Since secret tests of thermonuclear bombs are not easy, such a treaty has a somewhat better chance to limit the capability for hydrogen bomb warfare." TELLER only gave a very general schematic of the operation of a fusion bomb. However, in the following edition of *The Encyclopedia Americana*, Mark CARSON,[339] a competent Manhattan Project scientist who led the team of physicists that developed the hydrogen bomb at Los Alamos, became more specific in his article *Atomic Bomb*, in which he also provided a more detailed schematic on the design of nuclear weapons, including the hydrogen bomb.

**U.S. Plowshare Program.** Compared to the use of nuclear explosions for military applications, attempts to use them for peaceful purposes have only been partially successful. The Plowshare Program {⇒1958} was established in 1958 by the U.S. Atomic Energy Commission (AEC) – now the Dept. of

---
[338] E. TELLER: *Hydrogen bomb*. In: *The Encyclopedia Americana*. Americana Corp., New York, vol. 14 (1974), pp. 654-656.
[339] M. CARSON. *Atomic bomb*. In: *The Encyclopedia Americana*. Americana Corp., New York, vol. 2 (1997), pp. 641-642.

Energy (DOE) – in order to explore the technical and economic feasibility of using nuclear explosives for industrial applications, such as for excavation {Test SEDAN ⇨1962, ⇨Fig. 4.16–J; Test BUGGY ⇨1968, ⇨Fig. 4.16–K}: including the creation of canals and harbors, the carving of highways and railroads through mountains, open pit mining, dam construction, and quarry projects. In addition, underground nuclear explosions were believed to be applicable for stimulating natural gas production {Test GASBUGGY ⇨1967}, the creation of underground natural gas and petroleum storage reservoirs, *etc.* The most spectacular project proposed was certainly the construction of a larger Panama Canal {Atlantic-Pacific Interoceanic Canal Study Commission ⇨1970} using nuclear explosions. The Plowshare Program comprised 27 Plowshare nuclear tests with a total of 35 individual detonations. It was discontinued in 1975.

Nuclear explosions were also used for some scientific purposes. Examples given in the CHRONOLOGY include:

▸ the generation of an artificial radiation belt to test the confinement of charged particles in magnetic fields on a very large scale {Project STARFISH ⇨1962}; and
▸ the generation of ultrahigh shock pressures using nuclear explosions {RAGAN ET AL. ⇨1977; TRUNIN ET AL. ⇨1992; ⇨Fig. 4.11–F}.

**Soviet Plowshare Program.** The Soviet program "Nuclear Explosions for the National Economy" was the equivalent of the U.S. program Operation Plowshare {⇨1958}. The best known of these nuclear tests in the West was the Chagan Test {1965} as radioactivity from this underground test was detected over Japan by both the U.S.A and Japan in apparent violation of the 1963 Limited Test Ban Treaty {Nuclear Test Ban Treaty ⇨1963}.

**New Generations of Nuclear Weapons.** Today's thermonuclear weapons consist of two separate stages. The first stage is a nuclear fission weapon that acts as a trigger for the second stage. The explosion of the fission trigger produces a temperature and pressure that is high enough to fuse together the hydrogen nuclei contained in the second stage. The energy from this nuclear fusion produces a large explosion and a large amount of radioactivity.

Tomorrow's thermonuclear weapons will probably not rely on a nuclear-fission trigger to provide the conditions needed for nuclear fusion. Instead, they may use new types of very powerful but conventional high explosives, arranged, for example, in a spherical shell around a capsule containing the hydrogen gases tritium and deuterium. When the explosives are detonated, the capsule will be crushed inwards and the gases rapidly heated to a temperature high enough to allow the fusion of hydrogen nuclei to take place. Such a pure nuclear fusion weapon based upon inertial confinement would be relatively simple to design and construct. Since no fissible materials are required, this would be the ideal terrorist weapon for the 21st century. In order to achieve levels of nuclear fusion that are militarily useful, new explosives are being developed that can produce energy concentrations that are much greater than those produced by today's conventional high explosives.[340]

Similar concepts, combined with a hollow-charge compression technique, were already envisaged during World War II in the United States (J. VON NEUMANN 1943) and studied in Germany (later continued in France) {HAJEK ⇨1955}, but obviously without succeeding in initiating by this simple way a fusion reaction.

**'Dirty Bomb' Explosion.** Over the past few years a new type of nuclear explosion has been discussed in the press: so-called "dirty nuclear explosions." A "dirty bomb"[341] has been defined by the U.S. Department of Defense (DOD) as a "Radiological Dispersal Device" (RDD). The bomb produces a conventional chemical explosion, but it contains radioactive material, and so it is used to spread radiation over a wide area. Although a dirty bomb is not a true nuclear bomb and does not produce the heavy damage of a nuclear blast, it could have a significant psychological impact when applied in a city, causing fear, panic and disruption, and forcing costly cleanup operations.[342] In addition, exposure to radioactive contamination could increase the long-term risk of cancer. Such a bomb could therefore provide a simple but effective weapon for terrorist warfare when used in cities and other densely populated areas.

---

[340] From a lecture Frank BARNABY presented at the 13th World Congress of International Physicians for the Prevention of Nuclear War (IPPNW), held at Melbourne, Australia, in 1998. BARNABY, a British nuclear physicist who worked at the Atomic Weapons Research Establishment (AWRE), Aldermaston and was director of the Stockholm International Peace Research Institute (SIPRI), wrote the book *How to Build a Nuclear Bomb: and Other Weapons of Mass Destruction* [Granta, London, 2003].

[341] The term *dirty bomb* was coined by the media to describe a radiological dispersion device that combines conventional explosives, such as dynamite, with radioactive materials in the form of powder or tiny pellets packed around the explosive material. The effects of such a bomb are limited to the conventional blast damage at the site of the explosion and the contamination from radioactive materials spread by the blast (from WHO/RAD Information Sheet, Feb. 2003).

[342] M.A. LEVI and H.C. KELLY: *Weapons of new mass destruction*. Scient. Am. **287**, 58-63 (Nov. 2002).

## 2.7 EVOLUTION OF SEISMOLOGY

> *Now the recent improvements in the art of exploding, at a given instant, large masses of gunpowder, give us the power of producing an artificial earthquake at pleasure; we can command with facility a sufficient impulse to set in motion an earth wave that shall be rendered evident by suitable instruments at the distance.*[343]
>
> Robert MALLET
> Victoria Works
> Dublin 1846

SEISMOLOGY – now widely recognized as being the Earth science that is concerned with the scientific study of natural and man-made earthquakes, and the movement of waves through the Earth – has its roots in percussion rather than in classical shock wave physics. The term *seismology*, derived from the Greek word *seismos*, meaning "earthquake," was coined by Robert MALLET {⇨1857}, an Irish engineer and a cofounder of seismology. Originally, one of the main aims of investigations in this field was to study the seismic (elastic) waves generated by an earthquake, a complex mixture of various types of body waves and surface waves. These "seismic shocks" are not shock waves in the conventional sense; *i.e.*, they do not show the typical steep shock front, the high pressure, or the extremely short shock durations (on the order of micro- or even nanoseconds) of shock waves.

Classical seismology primarily involves the study of earthquakes as a geophysical phenomenon and the study of the internal structure of the Earth; important practical applications include seismic prospecting {⇨Fig. 4.15–E} and the seismic monitoring of nuclear explosions {⇨Fig. 4.3–Z}.[344] However, more recently, the field of seismology has expanded to encompass the study of elastic waves in other celestial bodies too.

- The first extraterrestrial *passive seismic experiments* were carried out on the Moon during the four Apollo (12/14/15/16) missions, and they have also been performed on Mars {Viking 2 Mission ⇨1975}.
- *Active seismic experiments* were among the first to be deployed on the Moon. During the Apollo 14/16/17 missions, active seismic experiments using explosives were conducted on the Moon to determine the structure of and the velocity of sound through its upper crust {Apollo 14 ⇨1971; Apollo 17 ⇨1972}.
- Asteroid seismology can help us to understand their internal structure and to get information on their material properties, which are currently unknown {NEAR ⇨1996}.
- Modern seismology has even been extended to the Sun, thus creating in the 1960s *helioseismology*, a branch of solar physics that investigates pressure wave oscillations in the Sun caused by processes in the larger convective region. Since the Sun's "surface" is not directly accessible via landers, and so it is not possible to install seismometers on the Sun as done on the Moon, helioseismology makes use of astronomical (optical) observations of seismic oscillatory waves in order to determine the inner structure of the Sun.[345] Acoustic, gravity and surface gravity waves generate different resonant modes in the Sun which appear as up and down oscillations of the gases: the oscillation modes are observed as Doppler shifts of spectrum lines and they can be used to sample different parts of the solar interior {WOLFF ⇨1972; KOSOVICHEV & ZHARKOVA ⇨1998; ⇨Fig. 4.1–V}.

The novel research domain of *asteroseismology* refers to studies of the internal structures of pulsating stars, which involve interpretating their frequency spectra. Asteroseismologists make great use of such oscillations to probe stellar interiors that are not observable directly. The basic principles of asteroseismology are very similar to those developed to study the seismology of the Earth.[346]

### 2.7.1 EXPLOSION SEISMOLOGY & VIBROSEIS

The new field of seismology was significantly boosted when it became possible to produce seismic shocks artificially by a concussive force, for example by using explosives {MALLET ⇨1846, 1849 & 1860; ABBOTH ⇨1878; MILNE & GRAY ⇨1883; DUCAN ⇨1927}. Use of the explosion method, which is the oldest one, created the new discipline of *explosion seismology*, which is of great economic and scientific value. Explosions are useful in this sense because the source mechanism is less complex than for most earthquakes, and the locations and detonation times are generally known precisely. Almost half of the seismic data collected on land have been acquired using explosive sources such as charges of dynamite. This enabled not only rapid testing and

---

[343] R. MALLET: *On the dynamics of earthquakes*. Trans. Roy. Irish Acad. **21**, 50-106 (1846).
[344] C. KISSLINGER: *Seismology*. In: (G.L. TRIGG, ed.) *Encyclopedia of applied physics*. VCH, New York etc., vol. 17 (1996), pp. 155-174, and *Update 1* (1999), pp. 415-422.
[345] D.O. GOUGH ET AL.: *Perspectives in helioseismology*. Science **272**, 1281-1283 (May 31, 1996).
[346] *Asteroseismology*. Institute of Astronomy, Katholieke Universiteit, Leuven, Belgium; http://www.ster.kuleuven.be/research/asteroseism/.

improvements in seismic diagnostics, but it also stimulated subsurface exploration. In particular, the application of refraction and reflection techniques {KNOTT ⇨ 1899; MINTROP ⇨ 1919 & 1924; DUNCAN 1927; ⇨Fig. 4.15–E} in conjunction with strong explosions has provided detailed knowledge of the structure of the Earth's crust and even of its inner core {GUTENBERG ET AL. ⇨1912}. Furthermore, evidence that earthquakes can be triggered by detonating chemical explosives or nuclear devices underground has enhanced man's understanding of earthquake mechanisms and has led to more accurate predictions of seismic shock arrival times.

Historically, explosion seismology started with black powder. However, since black powder cannot be detonated, the seismic shocks generated using it were not strong enough to be recorded at great distances, which was initially an important prerequisite for better resolving and differentiating between P- and S-waves using the seismographs available at that time. With the advent of high explosives, these seismic studies could be successfully extended to measure seismic velocities in all kinds of geological materials, and to investigate focal mechanisms and effects of large disturbances at greater depths.

Modern seismography allows one to differentiate between natural disturbances (*e.g.*, earthquakes, volcanic eruptions, meteorite impacts) and those generated artificially by man (*e.g.*, by large chemical or nuclear explosions), and to determine their locations and strengths. The discovery of the "boundary wave" [Germ. *Grenzwelle*] by the German geologist and seismologist Ludger MINTROP {⇨1919}, a pseudo-supersonic wave phenomenon, allowed the depths and thicknesses of subterranean layers to be ascertained for the first time {⇨Fig. 4.15–E}. This so-called "Mintrop wave" [Germ. *MINTROPsche Welle*], which was initially kept secret because of its commercial applicability, was later rediscovered and visualized on a laboratory scale by the German physicist Oswald VON SCHMIDT {⇨1938}. Among fluid dynamicists and acousticians, the Mintrop wave is better known as the "Schmidt head wave" [Germ. *VON SCHMIDTsche Kopfwelle*] – also a pseudo-supersonic wave propagation phenomenon {⇨Fig. 4.14–O}, which allows one to optically determine the various wave speeds in optically transparent as well as opaque solid materials.

The expensive nature of the drilling required to deploy explosive sources can be circumvented by using nonexplosive sources: such approaches involve mechanically impacting the Earth's surface, achieved by applying a large dropping weight for example {MILNE 1880s; MINTROP ⇨1908}, or by shaking the surface with mechanical devices of various construction using vibrator trucks. Today, most onshore seismic data are acquired using the latter method, known as "Vibroseis®." This important seismic reflection method that was invented and developed by John CRAWFORD, Bill DOTY, and Milford LEE at the Continental Oil Company (Conoco) in the early 1950s, imparts coded seismic energy into the ground. The energy is recorded with geophones and then processed using the known (coded) input signal. The resulting time-domain representation of vibroseis data is an impulsive wavetrain with wavelet properties consistent with the coded input signal convolved with the Earth's reflectivity series. Historically, vibratory seismic surveys collect data from one source location at a time, summing one or more sources at each location.

In *structural geology* – the branch of geology concerned with the deformation of rock bodies and with interpreting the natural forces that caused the deformations – artificial seismic waves became an important tool for the discovery of undersea mountain ranges with central rifts and massive transform faults. They are also applied to determine large-scale structures, for example to map sea floor stratigraphy and sedimentary structures. Geophones laid out in lines measure how long it takes the waves to leave the seismic source, reflect off a rock boundary, and return to the geophone. The resulting two-dimensional image, which is called a "seismic line," is essentially a cross-sectional view of the Earth oriented parallel to the line of geophones. Seismology, which was originally developed by the oil industry to perform seismic surveys, also became an important tool for the discovery of undersea mountain ranges with central rifts and massive transform faults.

## 2.7.2 SEISMOSCOPES, SEISMOGRAPHS, AND SEISMOMETERS

Proper measurement techniques were critical to progress in seismology, and the pioneers of seismology contributed considerably to the development of new diagnostic methods and instrumentation.

Early *seismoscopes* were rather crude instruments and were only capable of detecting the occurrence of earthquakes; some constructions were also capable of detecting the azimuths of their origin from the observer's location. The oldest known construction dates back to about A.D. 130 and was built in China {⇨Fig. 4.3–L}. *Seismometers* are sensors that quantitatively detect ground motions. Early seismometers were pendulums that were not, however, capable of recording high-frequency ground shaking. Modern seismometers produce an electric signal that can be recorded. At the beginning of the 20th century, electrodynamic systems were introduced that indicated ground velocity instead of displacement. In the early 1980s, force-balanced systems became available, which cover a high dynamic range (> 140 dB).

Great advances in seismology were achieved in the late 19th century with the invention of *seismographs*, which provided the first records (*seismograms*) of ground motion as a continuous function of time. Seismographs are ingenious instruments for measuring and recording seismic shock waves; their contribution to geophysical and geological knowledge is comparable to the contribution of the telescope to astronomy, or high-speed photography to shock wave physics. In contrast to modern shock pressure gauges such as those used in modern solid-state shock wave physics, which have rise times on the order of nanoseconds, seismographs occupy the very low frequency regime (down to $10^{-4}$ Hz), and are capable of recording ground motions in three directions over a large dynamic range. Examples of famous seismograph constructions include:

▸ the *Palmieri seismograph* (1856), which is thought to have been the first to record the times of seismic shocks. The more advanced *electromagnetic Palmieri seismograph* (1877) served as an effective detector on Mt. Vesuvius in Italy for many years;

▸ the rolling-sphere *Gray seismograph* (1881), which was the first to allow ground tremors in any horizontal direction to be recorded {⇨Fig. 4.3−M};

▸ the pendulum-based *Ewing seismograph* (1883), which was the first seismograph to be installed in the United States; it recorded the Great San Francisco Earthquake in 1906 {⇨Figs. 4.1−E & 4.3−M}; and

▸ the photographically recording *Wiechert seismograph* {WIECHERT ⇨1898}, which used a viscously-damped pendulum as a sensor to lessen the effects of pendulum eigen-oscillations.

Generally, seismographic records can detect four separate groups of waves:[347]

▸ *compressional waves* or *longitudinal waves*, also known as "P-waves" (or "primary waves"), which have the highest propagation velocities of all seismic waves, ranging from 1.5 to 8 km/s in the Earth's crust;

▸ *shear waves*, also known as "S-waves" (or "secondary waves"), which involve mainly transverse motion, usually at 60–70% of the speed of P-waves. S-waves are felt in an earthquake as the "second" wave, they do not travel through liquids;

▸ *Rayleigh surface waves* {Lord RAYLEIGH ⇨1885} have a rotating, up-and-down motion like that of breakers at sea. They have the smallest propagation velocities, but often exhibit much larger amplitudes than the other waves, and are then very destructive. Most of the shaking felt during an earthquake is due to Rayleigh waves; and

▸ *Love surface waves*, the fastest type of surface wave, discovered by the British geophysicist August E.H. LOVE (1911). These move transversely, whipping back and forth horizontally without agitating the surface vertically or longitudinally.

It is possible to determine the distance and origin of an earthquake from the arrival times of the different waves. However, the complex nature of seismic waves made it very difficult for early earthquake researchers to correctly read and interpret seismograms. Just to add to their troubles, these seismograms were also obtained by crude and imperfect instrumentation with low temporal resolution, low sensitivity, and a small dynamic range. Today sophisticated short-, intermediate-, and/or long-period multi-component seismometers are operated in networks, and the seismic data, which is continuously digitally recorded at many locations around the world, are exchanged among seismic observatories via the Internet or other interconnected computer networks and they are analyzed using sophisticated computer programs.

### 2.7.3 SEISMIC PROSPECTING AND RESEARCH

The most valuable economic spin-off resulting from seismology was *seismic prospecting*. Initially this method of exploration used chemical explosives to induce a percussion force in the ground {MALLET ⇨1849 & 1860; ABBOT ⇨1876}, but other artificial sources were subsequently developed, such as heavy-duty thumper trucks (*vibroseis*), which create vibrations by hammering the ground. These trucks produce a repeatable and reliable range of frequencies, and are a preferred source of vibrations compared to dynamite. In offshore locations, specially designed vessels are deployed. They are equipped with arrays of air guns which shoot out highly pressurized air into the water, creating a concussion that hits and vibrates the sea floor. Shock guns {Europe ⇨1957} have also been used for marine prospecting.

During the 1920s and 1930s, two seismic techniques – the *reflection method* and the *refraction method* {⇨Fig. 4.15−E} – were developed and immediately applied for prospecting purposes in the oil-producing regions of the United States {MINTROP ⇨1919; Orchard Dome ⇨1924} and Mexico. The U.S. general scientist Dr. Thomas POULTER,[348] who participated in Admiral Richard BYRD's Antarctic expedition to the South Pole (1933–1935), was the first to measure the ice

---

[347] Illustrious schematics of the wave motion of these four wave types were given by Bryce WALKER in his book *Earthquake*. Time-Life Books, Alexandria, VA (1982), p. 79.

[348] T. POULTER: *Seismic measurements on the Ross Shelf Ice.* Trans. Am. Geogr. Union **28**, 162-170, 367-384 (1947).

thickness and the contour of the bottom of the Ross Shelf using seismology. Today seismology plays an important role in both land and marine seismic surveys.

The two different and competing methods are based on the reflection or refraction of seismic waves and their proper detection by seismographs. Initiated by some concussive force, these seismic or elastic earth waves travel down to a dense or high-velocity bed; they are then carried along that bed until they are re-refracted or reflected upwards, respectively, to seismic detectors located on the surface some distance from the shot point. It is the time required for the seismic wave to reach each detector from the shot point that is recorded. The speed of transmission of the waves through different geological structures is proportional to the density or the compactness of the formation. Unconsolidated formations such as sands transmit waves at a low velocity, and massive crystalline rocks such as rock salt allow high propagation speeds. Generally, the density of the rock near the surface of the Earth increases with depth. When a seismic wave is refracted at the boundary of a deeper layer, the pulse travels at the velocity of sound in the lower layer; *i.e.*, supersonically with respect to the velocity of sound in the upper layer {⇨Fig. 4.15–E}. The wave propagation is therefore analogous to the Schmidt head wave {⇨Fig. 4.14–O}. During its propagation in the boundary layer it sends secondary waves into the upper medium, which arrive at the surface where they are recorded as function of time in a seismogram. On the other hand, seismic waves reflected at the boundary of two layers return immediately to the surface. However, the refracted wave arrives prior to the reflected wave at greater distances between the shot point and the receiving station.

Seismic surveying proved very useful, not only when exploring for oil and gas, coal, minerals and groundwater, but also in other fields. In geophysical research, the achievement of more accurate mapping is of broad interest in scientific studies of the Earth and of earthquake physics. For example, the systematic use of seismographs on an international scale allowed the exchange of global seismic data, which enormously enhanced not only our understanding of earthquakes, but also our knowledge of the Earth's interior. The German geophysicist Beno GUTENBERG {⇨1912} from Göttingen University made the first correct determination of the radius of the Earth's fluid core while carefully analyzing seismic data from a remote earthquake. In 1936, the Danish seismologist Inge LEHMANN interpreted waves in the shadow zone as P-waves that were reflected on a 5,000-km-deep discontinuity, thus indicating a region with different properties inside the fluid core. The solidity of this inner core was first suggested in the 1940s by the Australian geophysicist and mathematician Keith Edward BULLEN, and it was proven in 1971 by the Polish-born U.S. geophysicist Adam M. DZIEWONSKI and the U.S. earth scientist Freeman GILBERT using observations of the Earth's free oscillations.

High-resolution marine seismic techniques are currently being applied at the University of Southampton to archaeological remains and for mapping purposes, and the theoretical and practical advantages of this technique are being investigated for geophysical surveys of wooden artifacts and shallow intertidal sites.

Seismic event detection and location are also considered to be the single most important research issues involved in monitoring the Comprehensive Test-Ban Treaty {CTBT ⇨1996}. However, nuclear explosion and nonproliferation monitoring {⇨Fig. 4.3–Z} requires the processing of huge amounts of seismic sensor data – a difficult and challenging task which can only be achieved by using complex automated algorithms to characterize seismic events.

## 2.8   HIGH-SPEED DIAGNOSTICS

*Today's science strives to create its world view based not on speculations but – if possible – rather on observable facts: now it examines its constructions by observation. Each newly observed fact supplements the world view, and each deviation of a construction from the observation draws attention to an imperfection, a gap. The visualized is examined and supplemented by the imagined, which itself is the result of the previously visualized. Therefore, there is a special charm to verifying things by observation – i.e., by perceptions that have only been theoretically developed or assumed.*[349]

Ernst MACH
Universität Wien
Vienna 1897

WHEN the English monk, philosopher and statesman Sir Francis BACON suggested his experimental methodology, which involved taking things to pieces [Lat. *dissecare naturam*], it led to a preferred way of dividing the world into object and observing systems. In particular, it stimulated the evolution of classical mechanics tremendously. The application of this methodology led to the dissection of complex mechanical systems into their components in order to calculate the forces and momentums involved in the systems. This method was rendered most useful with the advent of Sir Isaac NEWTON's three Laws of Motion. In addition, the dissection of time, the "fourth dimension," also proved to be most useful. The analysis of

---

[349] E. MACH: *Populär-wissenschaftliche Vorlesungen*. Barth, Leipzig (1903), pp. 351-352.

physical and chemical processes performed by splitting them into a series of single time steps, one following the other, was one of the most successful ideas in the evolution of science and technology – and it still is today. High-speed cinematography is the most effective technique to use to reach this goal, and it is the ideal way to understand the nature of rapid events that are otherwise inaccessible to the naked eye. In particular, military research has prompted much development in high-speed diagnostics.

The basic questions that are usually asked in the study of any high-speed event are (i) how rapid is it and (ii) how long does it go on for. A classic example is the temporal measurement of pressure in a gun barrel. This was not only of vital interest to early ballisticians attempting to improve firearms, but such an analysis was also crucial to obtaining a basic understanding of wave propagation in gaseous matter, and it is no accident that fast chronoscopes originated in this area of research. Another important condition is triggering – *i.e.*, the appropriate timing of the diagnostics in order to catch the high-speed event within the selected time window of recording. Since supersonic phenomena cannot be resolved with the naked eye, *high-speed visualization* is an indispensable tool for shock wave and detonation research, which therefore relies on three essential "ingredients:"

- an appropriate optical method of visualization;
- a method of recording the data for later temporal and spatial analysis; and
- a reliable method of triggering.

These three basic conditions will be discussed in the following paragraphs.

## 2.8.1 PRECISE TIME MEASUREMENT: THE CRUCIAL CONDITION

> *It was my intention to visualize the compression wave which, carrying the sound originating from an electric spark, spreads into the atmosphere in all directions. Eventually I let the sound-producing spark discharge closely in front of the schlieren head and – very shortly thereafter – there was a second spark at the position of the illuminator, illuminating the field of view over a very short time interval so that the sound wave should still be visible in the field of view.*[350]
>
> August TOEPLER
> Königl. Landwirthschaftliche Akademie
> Poppelsdorf/Bonn 1864

The ambitious development of precision measurement devices that took place in the 19th century was almost entirely due to the requirements of basic ballistics. For example, in interior ballistics the measurement of the actual position of the projectile in the barrel is an important condition for calculating its acceleration, and for estimating the burning behavior and efficiency of a new propellant. In exterior ballistics, measurements of the time of flight allowed the velocity of the projectile as it left the muzzle (the so-called "muzzle velocity $v_0$") and the decrease in the projectile velocity along its trajectory to be studied.[351] These studies provided a deeper look into various fundamental problems related to aerodynamic drag at supersonic velocities, and they were first performed by Benjamin ROBINS {⇒1746} for musket shot and by Charles HUTTON {⇒1783} for cannon shot. In addition, the mass production of firearms required reliable testing and quality control, which also included the determination of $v_0$. However, the application of the classical ballistic pendulum {CASSINI JR. ⇒1707; ROBINS ⇒1740 & 1746} made such measurements rather cumbersome, not very accurate and difficult to apply to large caliber guns.

**Chronoscopes and Chronographs.** The evolution of fast chronoscopes – instruments for precisely measuring the duration of a single high-speed event or of several events, and their temporal correlation – began in Germany (Preußische Artillerie-Prüfungskommission 1838) and England {WHEATSTONE ⇒1839; ⇒Fig. 4.19–A} with the incorporation of an electromagnetic trigger into a mechanical clock in order to start it and stop it. The so-called "Wheatstone chronoscope" allowed one to measure the time of flight of a projectile by positioning a breaking contact in front of the muzzle and a closing one at the target. In the early 1840s, the Wheatstone chronoscope was improved considerable by the German-born Swiss watchmaker Matthias HIPP. This "Hipp-Wheatstone clock" {⇒Fig. 4.19–A} was widely used in ballistics, astronomy and physiology.

Shortly after, special mechanical-electrical devices – *chronographs* – were devised for measuring, indicating and permanently recording the duration of an event. Prominent examples include a falling vertical pendulum invented in 1848 by Captain A.J.A. NAVEZ[352] of the Belgian Army, the "Navez chronograph" (1853), and a unique dropping-weight timing system constructed in the early 1860s by the Belgian military engineer Major Paul-Emile LE BOULENGÉ {⇒1860s & 1882}

---

[350] See A. TOEPLER {⇒1864}, pp. 30-31.

[351] In order to characterize the drag of a projectile, 19th-century ballisticians also began to use quantities other than $v_0$ – for example $v_{50}$, the projectile velocity measured 50 meters away from the muzzle.

[352] A.J.A. NAVEZ: *Application de l'électricité à la mesure de la vitesse des projectiles.* J. Corréard, Paris (1853); *Instruction sur l'appareil électro-balistique.* J. Corréard, Paris (1859).

of the Belgian Artillery. This "Le Boulengé chronograph" {⇨Fig. 4.19−B} recorded the time elapsed through knife marks made on the surface of a falling cylindrical rod. This instrument was used for around a hundred years, and proved to be versatile and surprisingly accurate. It was widely used to measure the flight times in open ballistic ranges as well as the detonation velocities of explosive gaseous mixtures in a laboratory environment {BERTHELOT & VIEILLE ⇨1882}.

The idea of electrical triggering − a high-tech revolution of tremendous potential at that time − resulted in a variety of other self-recording instruments which used electrical-mechanical and photographic methods of recording. For example, in the 1840s Werner VON SIEMENS,[353] then a Prussian artillery officer and not yet ennobled, constructed an electroballistic chronograph, the "Siemens chronograph." Using a spark gap as the writing element, it consisted of a fixed needle electrode and a polished steel surface of a rotating drum, which was used as the counter-electrode. At the moment of departure from the muzzle, the projectile prompted the discharge of a Leiden jar across this needle-drum gap via a two-wire-gauge positioned at the muzzle of the test gun, which generated a small, sharply defined mark on the drum surface. After a specified flight distance, a second Leiden-jar circuit was discharged via a second two-wire-gauge with a second needle electrode in series, thus creating a second mark on the rotating drum. The elapsed time was determined as the distance between the two spark marks on the drum.

The enormous progress made at this initial stage of electrical chronometry also benefited other branches of science; for example it was used in astronomy to determine the longitudes and culminations of stars with greater precision. Starting in the 1870s, the development of short-timescale chronometry in England, France and Germany was pushed forward by the rapidly increasing number of new ballistic propellants and high explosives, as well their increased use in civil and military applications. In addition, the increasing pressure that was placed on the mining industry to prevent firedamp explosions required more precise research instruments in order to better resolve the nature of these rapid explosion processes.

In the early 1860s, the Rev. Francis A. BASHFORTH, a British professor of mathematics at Woolwich, devised an electric chronograph that measured down to fractions of a second by interpolation. This "Bashforth chronograph" was used by him to measure aerodynamic drag in the supersonic range {BASHFORTH ⇨1864}. In his instrument, a platform was arranged to descend slowly alongside a vertical rotating cylinder. The platform carried two markers controlled by electromagnets, which produced a double spiral on the surface of the cylinder. One electromagnet was linked by a circuit to a clock, and the marker actuated by it marked seconds on the cylinder. The circuit of the other electromagnet was completed through a series of contact pieces attached to the screens through which the shot passed in succession. When the shot reached the first screen, it broke a weighted cotton thread which kept a flexible wire in contact with a conductor. When the thread was broken by the shot, the wire left the conductor and almost immediately established the circuit through the next screen by engaging with a second contact. The time of the occurrence of the rupture was recorded on the cylinder by the second marker.[354]

The British ballistician Sir Andrew NOBLE {⇨1873} constructed a chronograph that used spark traces generated by inductance coils, which were recorded on a rotating drum covered with paper: the "Noble chronograph" {⇨Fig. 4.17−L}. Later extended to a system of multiple-spark traces − the first multichannel chronograph − it was applied by NOBLE to interior ballistics in order to record the arrival times of a projectile at various points within a gun barrel. Each recording channel was provided with little wire gauges which protruded into the barrel and were connected in series with each inductance coil; these were cut upon the passage of the projectile, thus interrupting the flow of current in the inductance coil and stopping the spark trace. By graphically differentiating the distance-time profile obtained twice, NOBLE obtained the acceleration of a fired projectile and, using Sir NEWTON's Second Law of Motion, the corresponding instantaneous force that acted on its base.

The French engineer Marcel DEPREZ constructed a chronograph similar to the Siemens chronograph. However, he used a soot layer on a revolving drum for recording purposes. Furthermore, unlike the Noble chronograph, he included an electrically activated tuning fork that produced a second trace for accurate time calibration purposes. Later DEPREZ improved his construction and extended it so that it could record 20 individual channels. The "Deprez chronograph" {DEPREZ ⇨1874} proved to be very useful for determining detonation velocities in gaseous mixtures {MALLARD & LE CHÂTELIER ⇨1881; BERTHELOT & VIEILLE ⇨1882}. In the years following, Hippolyte SÉBERT[355] (1881), a French lieutenant colonel, improved the Deprez chronograph.

---

[353] W. VON SIEMENS: *Über die Anwendung des elektrischen Funkens zu Geschwindigkeitsmessungen.* Ann. Phys. **66** [II], 435-444 (1845).

[354] *The Encyclopaedia Britannica.* Cambridge University Press, Cambridge, vol. 6 (1911), p. 305.

[355] H. SEBERT: *Notice sur de nouveaux apparats balistiques employés par la service de l'artillerie de la marine.* L. Baudoin, Paris (1881).

SÉBERT's instrument – the "Sébert chronograph" – was capable of precisely measuring times to an accuracy of less than $1/_{50,000}$ second (20 μs). It was widely applied in ballistics and detonation studies as a *velocimeter*; for example, it was used to measure projectile motion inside gun barrels and in free flight, detonation velocities of high explosives, and recoil motions of guns.

The "electric-tram chronograph," which was invented by Frederick J. JERVIS-SMITH {⇨1889} at Oxford, greatly differed from all previous time-measuring instruments. It used a moving plate which was carried on wheels and ran on rails, on which traces were made by means of electromagnetic styli. Capable of recording a large number of events separated by small periods of time, it was used by him to measure the acceleration periods of explosions, the velocities of bullets, and it was also used in many physiological time measurements. His tram chronograph was commercialized by the Elliot Brothers, two London instrument-makers.

All of the chronoscopes and chronographs discussed above were constructions that used a combination of mechanical and electrical elements. However, it is interesting to note here that fully electrical time measurement devices were also devised during this same pioneering period. The French physicist Claude S.M. POUILLET {⇨1844} invented his famous "ballistic galvanometer," a chronoscope for measuring projectile velocities. The galvanometer was connected to a battery via two grids which were connected in series and were activated as switches by the flying projectile. The deflection of the galvanometer was proportional to the quantity of electricity in the short current pulses and, therefore, to the projectile velocity. In England, the engineer Robert SABINE {⇨1876} invented another chronoscope that was based on discharging a capacitor for a short period of time. The time elapsed was determined by the decrease in the voltage at the capacitor. The "Sabine chronoscope" was used in exterior ballistics; in this case the projectile traveled through two contact grids which, acting as switches, briefly discharged the capacitor by a small amount over a resistor with a constant resistance.

In the early 1910s, Carl CRANZ and Karl BECKER, two eminent ballisticians at the Berlin Military Academy, invented an electric spark photochronograph, the "Cranz-Becker chronograph" {⇨Fig. 4.19–J}, which they used in high-precision aerodynamic drag studies. During the following decades, many more constructions and derivatives of these classical instruments were invented to serve the needs of laboratory physicists and testing ground technicians who required them due to the fast progress being made in high explosives, shock waves and ballistics. A detailed review of early chronoscope and chronograph developments was given in the 1910 Encyclopaedia Britannica[356] and in CRANZ's famous textbook on ballistics.[357]

**Electronic Timing Devices.** The first electronic device for measuring time spans based upon an electronic binary digital circuit was constructed by Charles E. WYNN-WILLIAMS, a British physics professor at Cambridge University (1931). It employed thyratron tubes and was used to count radioactive rays registered by a Geiger counter. This electronic counter became increasingly popular before World War II. Since it was the first practical digital electronic counting device, it played a crucial role in the development of nuclear physics and paved the way to digital electronic computers. Based upon the *multivibrator* or *flip-flop*, which was also an English invention {ECCLES & JORDAN ⇨1919}, it had the great advantage that the elapsed time – which was later preselectable in units of micro- and even nanoseconds – could be read directly from the face of the instrument, which considerably facilitated the measurement of shock front velocities and improved its accuracy.

The first real electronic *oscillographs* were built prior to electronic counter development. They were based upon the cathode ray tube (c.r.t.), a significant instrument invented in 1897 by the German physicist Ferdinand BRAUN. Oscillographs are electronic instruments that (similar to chronographs) measure, indicate and permanently record quantities that vary of time using a c.r.t. or another electronic display. While early *oscilloscopes* also used a c.r.t., they were only capable of displaying fluctuating electrical signals on a fluorescent screen and, like chronoscopes, they could not record the signal permanently.

Unlike all mechanical chronographs, cathode-ray oscillographs are equally sensitive at all frequencies, from zero to the highest frequency of oscillation, and are therefore perfectly suited to recording impulsive or "transient" phenomena. However, these unique instruments were not initially available commercially, and shock and explosion pioneers had to build their own rudimentary oscillographs. Various recording methods were applied during this exciting pioneering period of cathode-ray oscillograph development.

The *Dufour-type oscillograph*,[358] developed in France by Alexandre E. DUFOUR during and after World War I, was the

---

[356] *The Encyclopaedia Britannica*. University Press, Cambridge, vol. VI (1910), pp. 301-305.
[357] C. CRANZ: *Lehrbuch der Ballistik*. Springer, Berlin, vol. III (1927) *Experimentelle Ballistik*, pp. 35-133.
[358] A.E. DUFOUR: *Oscillographe cathodique pour l'étude des basses, moyenne & hautes fréquences*. L'onde électrique **1**, 638-663, 699-715 (1922); Ibid. **2**, 19-42 (1923).

most advanced instrument of the time for recording impulsive phenomena electronically. It used a classical high-voltage (60-kV) Braun tube with a cold cathode. The electrons, generated via field emission, were accelerated in a soft vacuum to a high velocity in order to achieve an intense photographic effect on film. The film plate was positioned in the vacuum chamber and it was directly exposed to the cathode ray. Time sweeping was achieved through the use of magnetic coils, and a maximum cathode-ray spot velocity of 4,000 km/s (4 mm/ns) was realized. However, because of the high velocity of the cathode ray, the input voltage sensitivity was quite low.

The *Wood-type oscillograph*, the archetype of modern oscillographs, was devised by Joseph J. THOMSON, a British professor who received the Nobel Prize for Physics in 1906 for his discovery of the electron in 1897. The oscillograph was constructed and tested by his countryman Albert B. WOOD {⇨1923}. In contrast to the Dufour-type oscillograph, it used a hot cathode, which allowed a lower anode/cathode voltage (3 kV), and was constructed in a compact and robust manner. Very similar to the Dufour-type oscillograph, the cathode ray was recorded directly onto photographic film. Since the time sweeping of the c.r.t. was achieved by applying a high-frequency periodic voltage at the horizontal deflection plates rather than through a sawtooth voltage, the signal-time displays recorded had to be converted into a scale that was linear with time – a cumbersome task which the user had to perform stepwise and by hand. David A. KEYS, a physicist at McGill University, was the first to succeed in recording pressure-time profiles of underwater explosions with high temporal resolution {KEYS ⇨1921; ⇨Fig. 4.19–C}; he used a Wood-type oscillograph and a tourmaline gauge. KEYS performed these studies on the orders of the British Admiralty.

In parallel with the development of fully electronic oscillographs, *hybrid-type oscillographs* were also constructed. They only included a c.r.t. for vertical deflection; the horizontal deflection was realized mechanically by a rotating drum covered with film paper, onto which the screen of the c.r.t. was imaged. During 1919–1920, DUFOUR[359] applied this technique to record low- and medium-frequency waveforms. This recording method was resumed and perfected in Germany. The dynamic gas pressure in fired rifles was successfully recorded in this way using a piezoelectric gauge {JOACHIM & ILLGEN ⇨1932; ⇨Fig. 4.19–D}.

The recording of very fast transient signals using single-shot triggering oscilloscopes was not possible until the advent of phosphors with improved persistence. In addition, a quantum leap in recording high-speed c.r.t. traces occurred upon the invention of high-sensitivity photographic materials, such as the well-known *Polaroid film* {LAND ⇨1948}. Furthermore, electronic storage mesh devices arranged behind the tube's phosphor were developed in the 1960s which allowed one to capture and store transient pictures that could be recorded with conventional still cameras.

A significant milestone was reached with the invention of *digital storage oscilloscopes*, which first became commercially available in the 1970s. These instruments digitize the input signals and store them into digital memories. The key advantages of this approach are that the stored waveform can easily be reviewed on the display at any time, long after the actual original analog signal has disappeared, and that the digitized data can be used directly in subsequent computational analyses.[360]

Modern *transient recorders* (or *transient digitizers*) that store the data in memories are capable of recording transient wave forms up to very high sampling rates. Combined with general-purpose PC-based instrumentation for data presentation and analysis they form very powerful and versatile data acquisition systems. Due to technological advances, the distinction between transient digitizers and digital oscilloscopes is blurring, and either are being used for recording transient signals in shock and detonation physics.

**Triggering.** Correct triggering always was and still is one of the most crucial factors in shock wave and explosion research: it is required, for example, to precisely activate pressure gauge amplifiers, cameras and light sources in order to capture the motion of the shock wave in the field of view. During the pioneering period of shock wave research, triggering was a real art. The "Knochenhauer circuit" {KNOCHENHAUER ⇨1858; ⇨Fig. 4.19–E}, a coupling of two high-voltage capacitor discharge circuits, was used by August TOEPLER {⇨1864} to generate a shock wave at a first spark gap and to illuminate the shock wave in the given field of view using a second spark gap fired after an adjustable delay time, which could be as short as tens of microseconds {⇨Fig. 4.18–A}. This first fully electric delay circuit was improved by Ernst MACH and Gustav GRUSS, and they used it in their optical studies of Mach reflection {E. MACH &

---

[359] A.E. DUFOUR: *Oscillographe cathodique.* J. Phys. & Rad. **1** [VI], No. 5, 147-160 (1920).

[360] N. BROCK: *Oscilloscopes, analog and digital.* In: (G.L. TRIGG, ed.) *Encyclopedia of applied physics.* VCH, Weinheim/Bergstr., vol. 13 (1995), pp. 37-57.

GRUSS ⇨1878}. Since then their circuitry has been known as the "Mach circuit" {⇨Fig. 4.19–E}.

The appropriate triggering of a mechanically generated shock wave – such as the head wave generated by a supersonic bullet fired from a rifle {E. MACH & SALCHER ⇨1886} – in the highly sensitive environment of an optical interferometer was even more tricky. Eventually, it was satisfactorily resolved by Ludwig MACH {⇨1896}, Ernst MACH's oldest son, who combined the fluid dynamic delay with an electric switching method {⇨Fig. 4.19–G}. This unique triggering technique was later reused to generate a series of light flashes in a multiple-spark camera {CRANZ & SCHARDIN ⇨1929; ⇨Fig. 4.19–M}.

In detonics, individual delay times between the initiations of multiple charges can easily be achieved by using a detonation cord. This consists of a core made of a high explosive, usually PETN, surrounded by a waterproof covering. The type most frequently used is Primacord®, a flexible linear detonating cord manufactured by Dyno Nobel Inc. (Salt Lake City, UT). It detonates with a velocity of about 20,350 ft/s (6.2 mm/μs). Much longer delay times can be achieved with the *Bickford fuse* or *safety fuse*, which consists of a central hemp cord surrounded by a core of black powder enclosed in a PE water-resistant cover. Since it is intended for the ignition of black powder charges at rather low rates, it burns at a velocity of about 1 ft/min (5.1 mm/s).[361]

It was soon recognized by shock experimentalists that a rectangular current pulse of any desired duration is very useful for solving many trigger problems. Beginning in the 1940s, mechanical time delay devices like the *Helmholtz pendulum* {⇨Fig. 4.19–F}, which were initially used to generate such pulses, were quickly superseded by electronic circuits. For the generation of rectangular current pulses, a simple transmission line consisting either of lumped L,C elements or a coaxial cable was used. Charged up to a DC potential and discharged via a mercury-wetted contact, this provided very steep rectangular current pulses in the microsecond or even the nanosecond regimes. In the analog era, the *monostable multivibrator* (or *monoflop*), a derivative of the *flip-flop* {ECCLES & JORDAN ⇨1919}, became the most widely used electronic delay generator for "one shot" triggering.[362] To obtain a delayed pulse for trigger purposes, an electric pulse derived from the high-speed event was steepened using a *Schmitt trigger circuit* (a useful circuit invented in 1938 by the German engineer Otto H. SCHMITT[363]) which started a monoflop. Its rectangular output wave profile, differentiated, rectified and inverted, was applied in order to control a high-speed camera or other diagnostic equipment.

With the arrival of the digital era, however, these analog delay generators were increasingly superceded by IC circuitry, encompassing numerous flip-flop-based digital counters. Today delay generators are already integrated into most electronic multiple-frame cameras for triggering purposes, which makes it much easier to record within the desired time window.

**Prerigger Framing Photography.** This is a special, extended recording method of real-time cinematography in which the camera continuously records images until the trigger signal is received and the camera is stopped. Depending on the chosen trigger position within the total series of recorded images, the system saves in memory a preselected number of pretrigger frames. The high-speed video camera Kodak model 4540 is provided with such a pretrigger capability {⇨Fig. 4.19–O}. It allows to study unpredictable events, such as the sudden and uncontrolled failure of a structure under dynamic mechanical or thermal loading.

Unpredicted failure mechanisms up to almost 1,000 h can be captured also with *Timelapse Video Recording*, a digital technique which is readily provided with a time/date generator (Victor Company, Japan).

## 2.8.2 OPTICAL METHODS OF VISUALIZATION: THE KEY TO A BETTER UNDERSTANDING

*Although the schlieren method is only one of the many methods I had to use, it is a very important one, and I believe that you will enjoy the results as much as I do.*[364]

Ernst MACH
Karl-Ferdinand-Universität
Prague 1887

Photographs are usually taken in reflected light (a method known as "reflected-light photography") in everyday life, as well as in science and engineering. However, this approach is not directly applicable for shock wave visualization because generally the density jump at the shock front in a

---

[361] Primacord and the Bickford fuze were manufactured by the Ensign-Bickford Co. (Simsbury, CT). In 2003, the Ensign-Bickford Co. merged with Dyno Nobel ASA, with the new entity to be called "Dyno Nobel Inc." (Salt Lake City, UT).
[362] B. CHANCE: *Waveforms*. McGraw-Hill, New York (1949), pp. 166-171, 179-182.

[363] O.H. SCHMITT: *A thermionic trigger*. J. Scient. Instrum. **15**, No. 1, 24-26 (1938).
[364] Taken from a letter of Ernst MACH to August TOEPLER (dated July 11, 1887) which is now kept in the Archives of the TU Dresden.

transparent fluid cannot be resolved in this way under normal conditions. This also holds for shocks in a solid: most solids are opaque anyway, and the shock wave is hidden somewhere inside. The key to investigating fluid phenomena is to select and apply the optical method that delivers the most appropriate and accurate picture of the most important physical quantities, such as the density and the pressure distributions in the desired direction and at the desired time instant.

Important scientific advances often happen when complementary investigational techniques are brought together. The three basic optical methods – schlieren, shadowgraphy and interferometry – gave early shock researchers their first insights into an abundance of completely new supersonic flow phenomena. These three principal optical techniques of modern flow visualization, which fulfill all of the requirements described above, were invented within the short period 1864 to 1891. However, it is impossible to determine the velocity, pressure and density using only optical experiments; some gas dynamics equations must be applied. If the flow structure (such as the configurations and locations of the shock waves) and the density field (obtained by optical methods) are known, the process of integrating the fluid dynamic equations is greatly simplified.[365]

**Schlieren Methods.** The schlieren method was invented by the English natural philosopher Robert HOOKE {⇨1672} and it was used by the French physicist Léon FOUCAULT (1858) to test the optical surfaces of mirrors for use in telescopes. However, August TOEPLER {⇨1864}, a lecturer in physics and chemistry at the Royal Agricultural College in Bonn-Poppelsdorf, Germany, devised a modification of the schlieren method, the "Toepler schlieren method" {⇨Fig. 4.18–A}, which proved most useful in fluid dynamics. From an historical point of view it is remarkable that one of the first applications that TOEPLER used his method for was the visualization of a propagating spark wave – a weak shock wave. TOEPLER's schlieren method was also used in the famous ballistic experiments performed to visualize the head wave generated by a supersonic projectile {E. MACH & SALCHER ⇨1886}, and in the study of high-pressure free air jets {SALCHER & WHITEHEAD ⇨1889}.

TOEPLER's classical schlieren method is a two-dimensional technique; *i.e.*, it is only capable of measuring the total deviation of a light beam that passes through a test section containing the gas under study. Consequently, there is no way to separate out the effects of density gradients at different positions in the test section. This disadvantage was overcome by the invention of a "three-dimensional schlieren system" that uses pulsed laser holography {BUZZARD ⇨1968}.

*Color schlieren methods* {⇨Fig. 4.18–B} are increasingly used in shock tube and wind tunnel facilities to visualize supersonic flows. Color schlieren photography is not only more appealing to the eye, but it also facilitates analysis. Its principle, devised in 1896 by the London amateur microscopist Julius H. RHEINBERG, was first applied to gas dynamics by the German physicist Hubert SCHARDIN {⇨1942; ⇨Fig. 4.6–I}. The introduction of the "constant deviation dispersion prism," which was positioned between the white light source and the first schlieren lens, stimulated the application of the color schlieren method to fluid dynamics {HOLDER & NORTH ⇨1952; ⇨Fig. 4.18–B}. This resulted in a variety of new modifications being proposed and applied by others. Further examples of color schlieren pictures taken by different methods are shown in Figs. 4.5–M, N.

**Shadowgraphy.** The shadowgraph method was invented at the University of Agram, Austro-Hungarian Empire (now Zagreb, Kroatia) by the physics professor Vincenz DVOŘÁK, who was one of Ernst MACH's assistants at the Charles University in Prague during the period 1871–1875 {DVOŘÁK ⇨1880; ⇨Fig. 4.18–C}. Widely applied by the English physicist and inventor Charles V. BOYS, this technique considerably simplified the visualization of supersonic flows in ballistic testing ranges, where it has since become a standard technique {BOYS ⇨1890; ⇨Fig. 4.6–H}. Using an intense pulsed point light source (*e.g.*, a Libessart spark) and a retroreflective background (*e.g.*, a Scotchlite screen) positioned at a distance several meters away from the spark, it is even possible to obtain shadowgraphs from large objects moving at high speed with sufficient film exposure {EDGERTON ⇨1958; ⇨Fig. 4.18–D}.

A detailed historical review of schlieren and shadowgraph techniques was given recently {SETTLES ⇨2001}.

**Interferometry.** Historically, interference phenomena were used to establish the nature of light.[366] The first interference phenomena to be noticed, the colors exhibited on soap bubbles and thin films on glass surfaces, were studied on a scientific basis by Robert BOYLE (1664), Robert HOOKE (1672)

---

[365] S.M. BELOTSERKOVSKY: *Anwendungsmöglichkeiten und Perspektiven optischer Methoden in der Gasdynamik.* In: (N.R. NILSSON and L. HÖGBERG, eds.) *Proc. 8th Int. Congress on High-Speed Photography.* Almqvist & Wiksell, Stockholm, and Wiley, New York *etc.* (1968), pp. 410-414.

[366] M. BORN and E. WOLF: *Principles of optics.* Pergamon Press, London *etc.* (1959).

and Sir Isaac NEWTON (1672 & 1675), who studied "Newton rings." Subsequent studies of the superposition of two beams of light resulted in numerous interferometer constructions, most of them invented and applied in the second half of the 19th century {JAMIN ⇨1856; Armand H.L. FIZEAU 1862; Albert A. MICHELSON 1881; ZEHNDER ⇨1891; L. MACH ⇨1891; Lord RAYLEIGH 1896; Charles FABRY & Alfred PÉROT 1899}. In particular, the *Jamin interferometer* {JAMIN ⇨1856}, the prototype for many subsequent interferometer techniques, paved the way for interferometry to be used as a diagnostic tool in fluid dynamics. First used to measure the amplitude of acoustic waves at the threshold of hearing {A. TOEPLER & BOLTZMANN ⇨1870; ⇨Fig. 4.18−E} − a masterpiece of experimental physics which demonstrated the enormous sensitivity of this method − it must have also stimulated Ernst MACH to apply this method to scan the (previously unknown) density profile at the shock front of a "spark" wave, a weak aerial shock wave generated by discharging a Leiden jar {E. MACH & von WELTRUBSKY ⇨1878; ⇨Fig. 4.18−E}.

Independently, Ludwig MACH at Charles University in Prague and Ludwig ZEHNDER at Würzburg University improved the Jamin interferometer and came up with a new configuration, now known as the "Mach-Zehnder interferometer" {⇨Fig. 4.18−F}. Advantageously, it allows the object beam and the reference beam to be separated by a large distance. Ludwig MACH {⇨1896} subsequently also demonstrated the great potential of interferometry by visualizing the flow around a supersonically flying bullet and obtaining quantitative data for the region behind the shock wave in a subsequent analysis, which was an important milestone in fluid dynamics and high-speed diagnostics. His next impressive application was the interferometric recording of free air jets at high speed {L. MACH ⇨1897}. Since the Mach-Zehnder interferometer measures variations in refractive index, and hence in density, it is particularly appropriate for flow visualization studies in ballistic tunnels, shock tubes and wind tunnels.[367]

**Other Methods.** Other ingenious − albeit rather exotic − methods for optically visualizing the propagation of shock waves have been applied for special applications. Examples include:

▸ *pulsed-laser holography*, a technique for recording high-speed events during the duration of the laser pulse. For example, in fluid dynamics it can be applied to measure the size, position, displacement and velocity of particles in a flow field and allows a shock wave to be recorded in three dimensions {GABOR ⇨1947; BROOKS ET AL. ⇨1966; LAUTERBORN ET AL. ⇨1972; ⇨Fig. 4.18−G};

▸ various surface-supported *optical reflection techniques*, which allow the instant of arrival of a shock wave at the surface of an (opaque) solid to be visualized {FEOKTISTOVA ⇨1960; FOWLES & ISBELL ⇨1965; ⇨Fig. 4.13−L};

▸ *moiré techniques*, which, for example, are useful for visualizing the displacement fields of impacted bodies, the movements, deformations and vibrations of a model in a wind tunnel, or the propagation of plastic waves in shock-compressed solids {KORBEE ET AL. ⇨1970};

▸ *laser speckle photography*, a noncontact technique that relies on the speckle effect produced when laser light is scattered at a diffusing surface. The method is suitable for measuring displacement components of specimens with rough surfaces as well as the density fields of compressible fluid flows over a wide dynamic range, such as those generated by the Mach reflection of shock waves and by thermal convection;[368]

▸ *dynamic photoelasticity*, a method for visualizing stress and fracture in dynamically loaded model structures {MAXWELL ⇨1850};

▸ *surface thermography*, a global and nonintrusive technique that uses color paints, liquid crystals or infrared cameras and is particularly suited for hypersonic flow and reentry studies {KLEIN ⇨1968; ⇨Fig. 4.18−H};

▸ *smoke flow visualization*, a unique method for visualizing supersonic flow {GODDARD ⇨1959}; and

▸ *particle tracer analysis*, an outgrowth of Vincent P. GODDARD's method which allows the reconstruction of physical properties of large spherical explosions in free air when combined with high-speed photography of smoke trail tracers introduced into the ambient gas immediately before the arrival of the shock wave {CDRE Suffield ⇨1964; ⇨Fig. 4.16−Q}.

---

[367] R. LADENBURG and D. BERSHADER: *Interferometry.* In: *High speed aerodynamics and jet propulsion.* Oxford University Press, London; vol. IX (1955): (R.W. LADENBURG, B. LEWIS, R.N. PEASE, and H.S. TAYLOR, eds.) *Physical measurements in gas dynamics and combustion*, pp. 47-78.

[368] M. KAWAHASHI and H. HIRAHARA: *Velocity and density field measurements by digital speckle method.* Opt. Laser Technol. **32**, 575-582 (2000).

## 2.8.3 THE SOOT TECHNIQUE: INGENIOUS 'BLACK MAGIC'

> *If the sound [shock] waves are the originators of the [Antolik] soot figures, then it should be possible to use them to study shock reflection. This is indeed the case.*[369]
>
> Ernst MACH
> Karl-Ferdinand-Universität
> Prague 1875

The soot technique, discovered by chance in the early 1870s by the Hungarian schoolmaster Károly ANTOLIK, is both a visualization and recording method {ANTOLIK ⇨1874; ⇨Fig. 4.5–C}. However, although it is amazingly simple to use, it is not so easy to understand the mechanism by which soot is removed behind shock and detonation wave interactions. This aroused the curiosity of various researchers to investigate this phenomenon in more detail such as by applying high-speed cinematography {CRANZ & SCHARDIN ⇨1929; SCHULTZ-GRUNOW ⇨1969}.

In contrast to the visualization methods discussed above, the soot method is a time-integrating recording method; *i.e.*, it is not capable of taking snapshots of propagating and interacting shock waves but is instead limited to reproducing irregular shock reflection phenomena as irreversible traces of the triple points in the soot, known as "triple point trajectories." However, by ingeniously using an anti V-arrangement (*i.e.*, two V-arrangements facing each other {⇨Fig. 4.5–D}), Ernst MACH and Jaromir WOSYKA {⇨1875} were able to show for the first time that it is possible to capture the motion of two Mach disks colliding head-on. The disks were stopped and reproduced as a thin line of piled-up soot, thus making the width of the Mach disk visible in the plane of symmetry: an unusual and simple but ingenious way to visualize a supersonic phenomenon without using expensive high-speed photographic instrumentation!

The soot method has also been used successfully to record the cellular nature of the detonation front in gaseous explosive mixtures, for example by simply blackening the inner wall of the reaction vessel.[370] Recently, the soot technique even proved useful for recording detonation phenomena in dust explosions {⇨Fig. 4.16–U}. Soot-covered foils have also been used to characterize the regular cellular structure of a detonation. This cellular structure – a pattern of multiple V-arrangements, like that observed by MACH and ANTOLIK in the case of interacting shock waves – is inscribed by the triple points of transverse waves interacting in the detonation wavefront.

ANTOLIK's soot technique can also be adapted to record shock interactions at higher pressure levels by increasing the adhesion of the soot particles, *e.g.*, by sandblasting the glass plates prior to applying the soot layer. This procedure even allows one to record double Mach reflection, which results in two concentric soot funnels {⇨Fig. 4.5–F}: at transition point $P_T$ at which *Double Mach Reflection* (*DMR*) turns into *Single Mach Reflection* (*SMR*) they merge into a single soot funnel.

Recently, Japanese researchers explored an explanation of detonation soot track formation and compared it with previous hypothesis of formation mechanism. Investigating soot track formation numerically, they assumed that the soot tracks were due to variations in the direction and magnitude of the shear stress created by the boundary layer over the soot foil.[371]

## 2.8.4 HIGH-SPEED PHOTOGRAPHY AND PHOTONICS: FREEZING THE INSTANT

> *In the month of June last a successful experiment was tried at the Royal Institution, in which the photographic image was obtained of a printed paper fastened upon a wheel, the wheel being made to revolve as rapidly as possible during the operation. From this experiment the conclusion is inevitable, that it is in our power to obtain the pictures of all moving objects, no matter in how rapid motion they may be, provided we have the means of sufficiently illuminating them with a sudden flash.*[372]
>
> Henry Fox TALBOT
> Lacock Abbey 1851

The advancement of high-speed photography has always been crucial to detailed analyses of shock and detonation effects and applications of them. In this regard, Ernst MACH's scientific way of experimenting was very successful and instructive to his contemporaries. He was not only an eminent philosopher of science, but he can also be regarded as being the first gas dynamicist and high-speed photographer, who carried out his shock and explosion research according to the motto *Sehen heißt verstehen* ("seeing is understanding").

---

[369] E. MACH and J. WOSYKA: *Über einige mechanische Wirkungen des elektrischen Funkens*. Sitzungsber. Akad. Wiss. Wien **72** (Abth. II), 44-52 (1875).

[370] R.I. SOLOUKHIN: *Shock waves and detonations in gases*. Mono Book Co., Baltimore (1966).

[371] K. INABA, M. YAMAMOTO, J.E. SHEPHERD, and A. MATSUO: *Soot track formation by shock wave propagation*. ICCES **4**, No. 1, 41-46 (2007).

[372] W.H.F. TALBOT: *On the production of instantaneous images*. Phil. Mag. **3** [IV], 73-77 (1852).

The challenge of attempting to record dynamic events in compressible gas flows encouraged the development of new high-speed photographic equipment which, in turn, enabled the discovery of new shock phenomena. During the pioneering period of such research, gas dynamicists were often also high-speed photographers who invented, developed, built, applied and improved their own equipment.

**Single-Shot Photography.** Snapshot photography of a dynamic event was first demonstrated by the English chemist and pioneer photographer W. Henry Fox TALBOT {⇨1851}. However, the complicated process he used to prepare and develop the film did not permit its immediate application to ambitious plans to freeze the motions of events occurring at supersonic speeds. Since high-speed films were not yet available to August TOEPLER {⇨1864}, he visualized shock waves subjectively using a stroboscopic set-up – a curiosity in the early history of fluid dynamics. With the advent of highly sensitive gelatin dry plates {MADDOX ⇨1871} and electric spark light sources of a high intensity but a short duration (about 1 μs or less), it became possible to both capture the motions of propagating aerial shock waves with practically no motion blur, and to obtain a sufficient exposure density on photographic film. The first photographed shock wave was generated by the discharge of a Leiden jar, visualized with the schlieren method, and photographed on a gelatin dry plate {E. MACH & WENTZEL 1884; E. & L. MACH ⇨1889; ⇨Fig. 4.5–G}.

**High-Speed Cinematography.** The evolution of high-speed photography, ranging from single-shot photography to high-speed cinematography, is a story of its own {*e.g.*, RAY ⇨1997}, but a few milestones can be mentioned here. The *rotating mirror* {⇨Fig. 4.19–I}, a mechanical device for resolving the motion of an object in one dimension, was apparently first used in England by Sir Charles WHEATSTONE {⇨1834}, the legendary inventor of the Wheatstone bridge.[373] The rotating mirror was first used in Germany for resolving the propagation and reflection of detonation waves in gaseous combustible mixtures {⇨Fig. 4.19–K}. This principle of streak recording was later adopted in England and became known under the name *wave speed camera* {PAYMAN & WOODHEAD ⇨1931; ⇨Fig. 4.19–L}.[374] It is in-teresting here to note that this task was not set by military needs – unlike most high-speed diagnostic techniques developed over the following decades – but was rather initiated by safety considerations, namely attempts to better understand the mechanism of firedamp explosions in coal mines and thus to initiate more effective preventive measures. In the United States, the streak recording technique that used rotating mirror cameras was primarily applied in basic detonation research to better understand the ultrafast processes of initiation and propagation in detonating condensed high explosives as well as for the routine testing of high explosives and detonators {⇨Figs. 4.17–P, Q, R}.

High-speed cinematography reached its first peak with the invention of the *Cranz-Schardin multiple-spark camera* {CRANZ & SCHARDIN ⇨1929; ⇨Fig. 4.19–M}. This was a big step forward in the evolution of high-speed photography, because the new device immediately allowed one to resolve high-speed events at 300,000 frames per second for a total of eight frames, which was soon extended to the present-day standard of 24 frames. Based on a unique recording principle that permits almost any frame rate desired to be realized and almost any type of pulsed light source of small dimension to be used, it was later also modified into cineradiography using a number of flash X-ray tubes {TSUKERMAN & MANAKOVA ⇨1957; CHANG ⇨1983}.

The ambitious U.S. program of atomic weapons development and testing during and after World War II resulted in the further development of *mechanical framing cameras* with high and ultrahigh frame rates (up to some $10^6$ frames/s) {⇨Figs. 4.17–P, R}. In addition, the various requirements of dynamic plasma diagnosis in numerous fusion research programs stimulated new developments in *ultrafast image tube camera technology*, particularly in the United States, England, France, and the former Soviet Union. With the advent of the *microchannel plate* (*MCP*) in the 1980s – a U.S. invention based on the electron multiplier {FARNSWORTH ⇨1930} – a new optoelectronic device with excellent gating capability in the low nanosecond regime and high light intensification became available that could be combined very successfully with the preexisting *charge-coupled device* (*CCD*), a U.S. invention {BOYLE & G.E. SMITH ⇨1969}. This resulted in the *Intensified CCD (ICCD)* which, when applied in a multiple arrangement with optical image splitting, created a new generation of *ultrafast multiple digital framing cameras* that are useful for recording all kinds of shock wave, detonation and impact phenomena {Hadland Photonics Ltd. ⇨1993}.

---

[373] The Wheatstone bridge, operated in a pulsed mode and modified to eliminate large D.C. offset voltages, has become an important device for supplying a rectangular current pulse for all kinds of piezoresistive shock pressure gauges.

[374] W. PAYMAN: *The wave-speed camera*. Safety of Mines Research Board (SMRB), U.K., Papers Nos. 18 and 29 (1926).

## 2.8.5 FLASH X-RAY TECHNIQUES: VISUALIZING THE HIDDEN

> *It was a lucky chance that even before the beginning of war it occurred to STEENBECK, while studying pulsed gas discharges at low pressure at the Siemens Company, to investigate the emitted X-radiation in more detail. It showed that it was sufficiently intense to expose photographic plates within a fraction of a microsecond. STEENBECK visited me on July 25, 1938, asking whether such short and intense flash X-rays would not be of great importance in ballistics, and proposing cooperation…Of course, the most important of these applications became the revelation of the hollow charge effect.*[375]
>
> Hubert SCHARDIN
> Technische Akademie der Luftwaffe
> Berlin-Gatow 1938

Radiography (in optical terms a shadowgraph method) has been around since Wilhelm C. RÖNTGEN's spectacular X-ray experiments at Würzburg University in 1895.[376] However, capturing the motions of high-rate phenomena – particularly those of shock and detonation waves with velocities of up to almost 10,000 m/s and jets from shaped charges with velocities of up to 12,000 m/s – requires X-ray pulses of submicrosecond duration, which also need to be of a very high intensity in order to provide a sufficient exposure density on the film. Unfortunately, however, pulsed X-ray sources that fulfilled these two requirements were not available to shock and explosion researchers until the 1960s. Curiously enough, this urgently required diagnostic method, which was later called "flash radiography," was then invented almost simultaneously by the German physicist Max STEENBECK {⇨1938} at Siemens AG in Berlin and the U.S. physicists Kenneth H. KINGDON and H.E. TANIS {⇨1938} at General Electric Co. in Schenectady, NY.

**Flash Radiography.** During World War II, flash radiography rapidly grew into an important diagnostic tool for detonics and the first flash X-ray systems using vacuum discharge tubes became commercially available: in Germany by Siemens AG (Berlin) and in the U.S.A. by Westinghouse Electric Corporation (Bloomfield, NJ). Flash X-ray systems using pure field emission tubes were produced not until the late 1950s by Field Emission Corporation (McMinnville, OR). During the war flash radiography was particularly applied in Germany to secret military studies of shaped charges. Later it became the preferred (and often the only applicable) method of visualizing nuclear implosion bomb assemblies and laser fusion experiments. Applications in detonation physics considerably stimulated advances in precision detonation techniques and the development of other high-intensity pulsed radiation sources too. Modern commercially available flash X-ray systems cover the range from 75 kV to 2 MV. Exotic examples include pulsed electron beam machines, flash neutron radiography using pulsed nuclear reactors, and high-voltage linear electron accelerators such as the PHERMEX facility at the Los Alamos Scientific Laboratories {VENABLE & BOYD ⇨1965}.[377]

Unlike optical methods, flash radiography is insensitive to the self-luminous events that accompany all detonation processes, and the smoke resulting from detonation products cannot obscure the test object when using this technique. In addition, X-rays provide insight into the interiors of shock-loaded solids and make it possible to measure temporal shock front positions. These particular properties of flash X-rays allowed phenomena connected with the emission of very bright light that had previously not been accessible to optical methods to be visualized for the first time, for example:

▸ the creation of shock waves emerging from exploding wires during the very initial stage of wire explosion;
▸ the formation of jets in shaped charges;
▸ the propagation of detonation fronts in liquid and solid high explosives, and even in gases, and
▸ the ontact areas of objects that impact at high velocities.

Furthermore, it also became possible to visualize shock wave propagation and interaction phenomena in optically opaque media (*i.e.*, the majority of solids), and to quantitatively determine the density jump across the shock front using photodensitometry. This was first demonstrated not only in high-density matter, such as in liquids and solids {SCHALL ⇨1950}, but also in various gases using high X-ray-absorbing additives {SCHALL & THOMER ⇨1951}. In the 1950s, *flash soft radiography*, a technique which is also particularly well-suited to visualizing density variations in low-density matter, was developed. This technqiue uses special X-ray tubes that preferably emit an intense soft X-ray spectrum, and it combines highly sensitive X-ray films with

---

[375] H. SCHARDIN: *Über die Entwicklung der Hohlladung.* Wehrtech. Monatshefte **51**, Nr. 4, 97-120 (1954).
[376] It is interesting to note here that only one year after his famous discovery of X-rays, RÖNTGEN made a radiograph of a firearm – his personal rifle – which (together with numerous handwritten comments) he sent to his friend Franz S. EXNER, a professor of physics at the University of Vienna and pioneer of modern physics. However, ROENTGEN's technique only allowed him to take radiographs of objects at rest. *See* O. GLASSER: *Wilhelm Conrad Röntgen und die Geschichte der Röntgenstrahlen.* Springer, Berlin *etc.* (1995), pp. 272-273.
[377] L.E. BRYANT (ed.): *Proceedings of the Flash Radiography Symposium* [Houston, TX, Sept. 1976]. The American Society for Nondestructive Testing, Columbus, OH (1977).

high-conversion-efficiency intensifying screens. This allowed shock waves in gases of low atomic numbers to be visualized, even without using X-ray-absorbing additives {HERRMANN ⇨ 1958}.

Numerous examples of various applications of flash X-ray diagnostics are provided in the PICTURE GALLERY {⇨Figs. 4.5–H; 4.8–F, G, H, I; 4.13–H; 4.14–T; 4.15–B; 4.16–O}.

**Flash X-Ray Diffraction Analysis.** X-ray diffraction, a technique first established to analyze the fine structure of unloaded crystalline materials (M. VON LAUE 1912; W.H. BRAGG 1912; P.J.W. DEBYE and P. SCHERRER 1916), and later extended to statically loaded targets, was also extended to measure the lattice compression of crystals *during* shock compression, by applying high-intensity flash soft X-ray pulses of submicrosecond duration – preferably characteristic (or line) radiation such as Cu-Kα and Mo-Kα. This new and important method, known as "flash X-ray diffraction," was combined with a precisely triggerable planar shock compression technique, thus providing the first insights into the microscopic regime of a shock-loaded crystal lattice. However, flash X-ray diffraction is presently limited to single- and polycrystalline substances of low atomic number only, and it focuses on the region immediately behind the shock front. Spectacular demonstrations using flash X-ray diffraction have provided evidence that phase transformations can occur on nanosecond timescales in shock-compressed solids {JOHNSON ET AL. ⇨ 1970 & 1972}. Furthermore, the first flash X-ray diffraction patterns obtained from an aluminum shaped charge indicated that the high-velocity jet (6.4 km/s) consists of a particulate solid – *i.e.*, it has not fully transformed into a molten state {JAMET & THOMER ⇨ 1975; GREEN ⇨ 1975}.

In the early 1990s, an even more advanced experimental technique was applied in England, which used a high-power pulsed laser to simultaneously shock-compress the target and generate an intense soft X-ray pulse for diffraction purposes {WARK ET AL. ⇨ 1991}. In the late 1990s, some classic flash X-ray diffraction experiments of the 1970s were performed on essentially the same crystal types at Washington State University {RIGG & GUPTA ⇨ 1998}, although a number of worthwhile technical improvements were made and the loading conditions of the target under planar impact were carefully reevaluated. These studies resulted in a better understanding of the compression of the unit cell under planar shock waves.

### 2.8.6 THE CORRECT MEASUREMENT OF SHOCK PRESSURE: AN EVERGREEN PROBLEM

> *There are perhaps few questions upon which, till within quite a recent date, such discordant opinions have been entertained as upon the phenomena and results which attend the combustion of gunpowder. As regards the question alone of the pressure developed, the estimates are most discordant, varying from the 1000 atmospheres of* ROBINS *to the 100,000 atmospheres of* RUMFORD…[378]
>
> Sir Andrew NOBLE
> W.G. Armstrong & Co., Elswick
> Sir Frederick A. ABEL
> British War Dept., London
> 1874

Pressure, temperature and density (or specific volume, the reciprocal value) are all basic quantities used to describe the thermodynamic state of matter. A knowledge of the pressure is also of particular importance when characterizing dynamic processes.

In ballistics, it allows one to estimate the time-dependent force exerted on the base of the projectile by the propellant gases during its passage through the barrel. Using the Second Law of Motion {Sir NEWTON ⇨ 1687}, the acceleration of the projectile and its velocity at the moment that it leaves the muzzle ("$v_0$;" an important ballistic quantity for gunners) can be determined.

In shock wave physics, the pressure-time profile $p(t,R)$ allows one to determine the mechanical impulse $I$ in the blast field of an explosion at a distance $R$ using a simple integration procedure given by $I(t,R) = \int p(t,R)\,dt$. The knowledge of the impulse that acts on a structure positioned at $R$ is an important quantity that is used to estimate the damage caused to structures exposed to blast waves.

In classical percussion, the percussion pressure – defined in terms of the normal percussion force per contact area – can be very large, because the short-acting force of percussion is typically already quite large and the contact area is small in "hard" percussion partners, such as those for a metal plate hit by a hammer via a lathe center or a sharp edged chisel.

In aerodynamics, the *total pressure* or *stagnation overpressure p* that is recorded by a gauge placed face-on to a blast wave is equal to $\rho v^2 f_p(M)$, where $v$ is the particle flow velocity, $\rho$ the gas density and $f_p$ is a pressure coefficient. It is difficult to determine $p$ by measuring $\rho$ and $v$ because $f_p$

---

[378] A. NOBLE and F.A. ABEL: *Researches on explosives. Fired gunpowder. Part II.* Phil. Trans. Roy. Soc. Lond. **171**, 50 (1880).

depends on the Mach number $M = v/a$, where $a$ is the velocity of sound.

Pressure gauges should cause as little disturbance as possible to the shock wave flow, and miniaturization was a logical step toward this goal. Since the increase in pressure in a shock wave is accompanied by an increase in temperature, pressure gauges must be shielded or designed in such a way that the sensor element is only pressure-dependent (not temperature-dependent). To a certain extent this is achieved, for example, in the case of piezoresistive gauges by choosing manganin, a metal alloy with a very low temperature coefficient. Furthermore, an ideal pressure gauge should be free of any effects due to stress history (hysteresis) and should have a "bulk-intrinsic" gauge coefficient – *i.e.*, it should be dependent only on the pressure-sensitive material of the gauge and not on the particular gauge geometry. Even piezoelectric and piezoresistive pressure gauges, with their material-dependent piezo coefficients, approach this ideal response only partially and thus require careful analysis of the stress loading conditions applied.

It is small wonder that the measurement of pressure is associated with the earliest efforts of ballisticians as well as percussion, explosion and shock wave researchers who were searching for realistic records of pressure-time profiles under various experimental conditions and within a wide temporal range. The difficult task of constructing appropriate pressure gauges that are inexpensive, small-sized and have high temporal resolution – rise times in the low nanosecond regime are required in shock wave physics and rise times of as little as picoseconds are used in micro shock wave applications – has occupied generations of inventors and researchers, and still does. Although pressure sensors are now a billion-dollar industry (and one that is still growing), only a few companies have specialized in the development of sensors appropriate for use in shock wave diagnostics. The need to record fast pressure-time profiles stimulated the development of not only the gauge technique itself, but also recording devices. In the 1920s, mechanical chronographs were superseded by c.r.t. oscilloscopes {WOOD ⇨1923}, and today those analog oscilloscopes have been replaced by either digital oscilloscopes or digitizers in most laboratories (*see* Sect. 2.8.1).

The use of pressure gauges capable of recording shock pressures in all states of aggregation has grown into a special branch of diagnostics and is almost as complex and bizarre as shock wave physics itself. Apparently, it has never been reviewed in context. The true time-resolved measurement of shock wave profiles in shock-loaded matter is a difficult task, particularly in solids, because transit times are often very small, which requires techniques with ultrashort response times. The use of piezoelectric and piezoresistive gauges in solid-state shock wave physics was reviewed up to the early 1970s by William MURRI and associates {⇨1974}. Robert A. GRAHAM[379] and James R. ASAY, two Sandia shock physicists, reviewed diagnostic methods (optical, capacitive, piezoelectric and piezoresistive) developed up to 1978 to measure shock wave profiles in solids. Various designs of pressure gauges are now used routinely in large numbers. However, with the steady refinement of numerical models developed in dynamics materials science over the last few decades, measured pressure data have increasingly become the subject of critical analyses {GUPTA ET AL. ⇨1980}. Experimental techniques used to measure pressures in blast waves were reviewed in 2001 by John M. DEWEY, a noted Canadian shock physicist {DEWEY ⇨1964}. Some developments, which are also given in the CHRONOLOGY and in the PICTURE GALLERY, will be summarized in the following.

**Mechanical Gauges.** Diaphragm sensors that measure pressure through the elastic deformation of a diaphragm – such as that of a membrane, a plate or a capsule – are particularly well suited to mechanically measuring dynamic pressures. When directly coupled with a mechanical chronograph, they were used almost exclusively in the 19th century to record pressure-time profiles of rapid events. Membranes tend to eigen vibrations and hence tend to modify the pressure signal transmitted. These natural oscillations can be damped, but this measure would also reduce the displacements of the membrane – compensation by electronic amplifiers was not possible at that time.

Examples of early pressure recording using membranes that are described in this book include:
- blast-wave profiles from large explosions of gunpowder in the open air {WOLFF ⇨1899; ⇨Fig. 4.17–M};
- primary and secondary pressure pulses emitted from underwate explosions {BLOCHMANN ⇨1899; ⇨Fig. 4.16–D}; and
- detonation pressures for firedamp or other combustible gases {MALLARD & LE CHÂTELIER ⇨1883; ⇨Fig. 4.17–C}, or for solid explosives detonating in a closed reaction vessel {BICHEL ⇨1898; ⇨Fig. 4.17–D}.

Although mechanical gauges have rather large rise times (in the millisecond regime), they can provide fairly good results for large explosions which produce long-duration pressure pulses.[380]

---

[379] R.A. GRAHAM and J.R. ASAY: *Measurement of wave profiles in shock-loaded solids.* High Temperatures, High Pressures **10**, 355-390 (1978).

[380] According to *Hopkinson scaling* {B. HOPKINSON ⇨1915}, an observer, stationed at a distance $\lambda R$ from the center of an explosive source with a charac-

In the mid-19th century, mechanical gauges for recording peak pressures of high-rate events were developed mostly for ballistic purposes. The famous *Rodman gauge* {RODMAN ⇨1857; ⇨Fig. 4.17−K} consists of an indentation tool that is placed into a hole bored into the powder-chamber of a gun or fastened in the base of the shot. The *crusher gauge* {⇨NOBLE 1872; ⇨Fig. 4.17−K}, an improvement of the Rodman gauge, is based on the deformation of a metal (copper) cylinder. Both gauges were first calibrated with static means and then used to measure the peak pressure in the bore of a fired gun.[381] Crusher gauges based on the principle of the deformation of a ball by a pressure-driven piston saw a renaissance in World War II; for example, they were used extensively in Operation CROSSROADS {⇨1946}, which comprised two nuclear events that were designed primarily to provide information on the effects of atomic bombs on naval vessels. These gauges were deployed at various water depths near ships and on hulls.

Weight- or spring-loaded devices, similar in construction to the safety valves used in steam-boilers {PAPIN ⇨1679; ⇨Figs. 4.7−A, B}, were used by early ballisticians and chemists, for example to measure the peak pressure of fired gunpowder {Count VON RUMFORD ⇨1797; ⇨Fig. 4.17−A}; to compare different qualities of gunpowder in regard to their "brisance" {⇨Fig. 4.17−E}; and to evaluate peak detonation pressures in various combustible gaseous mixtures {BUNSEN ⇨1857; ⇨Fig. 4.17−B}.

**Piezoelectric Gauges.** The *direct piezoelectric effect* {CURIE Bros. ⇨1880} – *i.e.*, the generation of an electric charge in a crystal or a ceramic by subjecting it to a mechanical stress – has been used since the early 1920s to measure pressure-time profiles of dynamic events. *Piezoelectric pressure gauges* typically feature high-frequency response and are particularly useful for recording shock waves. Examples of recorded pressure-time profiles described in this book include:

▸ pressure pulses emerging from an electrolytic gas detonation {KEYS ⇨1921};
▸ shock waves resulting from an underwater explosion {KEYS 1923, *see* KEYS ⇨1921; WOOD ⇨1923; ⇨Fig. 4.19−C};
▸ gas pressure histories in fired rifles {JOACHIM & ILLGEN ⇨1932; ⇨Fig. 4.19−D};

▸ blast pressure histories in the areas around detonating solid explosives {CROW & GRIMSHAW ⇨1932};
▸ gauge calibration in shock tube studies {REYNOLDS ⇨1943}; and
▸ studies of interactions of high-energy laser pulses with solid targets {JONES ⇨1962; KREHL ET AL. ⇨1975}.

Early researchers favored tourmaline crystals because of their high piezoelectric coefficient. Today piezoelectric gauges based on quartz and various piezoelectric ceramics dominate the market and are mainly used in acoustics and gas dynamics. In addition, polyvinylidene fluoride (PVDF) foils {KAWAI ⇨1969} are becoming increasingly popular in shock wave diagnostics, because this inexpensive material has a very short response time in the nanosecond regime. Advantageously, it can be formed into a variety of shapes to suit special applications. Modern needle hydrophones use very thin PVDF foils. Because of their high upper frequency limit of several MHz, they are also appropriate for diagnosing weak shock waves in water {⇨Fig. 4.1−Z}.

The *indirect* or *inverse piezoelectric effect* (CURIE Bros. 1881) – *i.e.*, the generation of a mechanical stress in a crystal or a ceramic by subjecting it to an applied voltage – is utilized in piezoelectric shock wave generators that are used in extracorporeal shock wave lithotripsy. Typically, hundreds of small oscillatory piezoceramic crystals (that are arranged hemispherically and pulsed via an electric discharge) generate pressure pulses which are focused onto a small zone of about one square centimeter.

**Piezoresistive Gauges.** The *piezoresistive effect* is the change in the resistivity of certain materials due to the application of mechanical strain and was discovered in 1856 by Lord KELVIN. The first *piezoresistive pressure gauge* – a manganin pressure gauge – was apparently constructed by the U.S. physicist Percy W. BRIDGMAN {⇨1911}, who made piezoresistive measurements on several polycrystalline metals under static high pressures {BRIDGMAN ⇨1925}. He later also outlined the formal nature of the piezoresistive effect in single crystals. George HAUVER and coworkers at BRL, Aberdeen Proving Grounds seem to have been the first to exploit the piezoresistive nature of materials to measure pressure in a shock-loaded sample, using a thin disc of sulfur {HAUVER ⇨1960}. Researchers in England {FULLER & PRICE ⇨1962} and independently in the United States {BERNSTEIN & KEOUGH ⇨1964} first used manganin gauges to measure the longitudinal stress in a shock-loaded sample under uniaxial stress. Standard carbon composition resistors

---

teristic dimension $\lambda d$ will feel a blast wave of the same amplitude $P$ from an explosive charge of dimension $d$ detonating at distance $R$, but an increased positive pulse duration $\lambda T$ {BAKER, WESTINE & DODGE ⇨1973}.

[381] A historical review of gauges for explosive pressure measurement and recording up to 1906 was given by J.E. PETAVEL: *The pressure of explosions – experiments on solid and gaseous explosives. Parts I and II*. Phil. Trans. Roy. Soc. Lond. **A205**, 357-398 (1906).

like those used in electronics have also been used as inexpensive piezoresistive gauges to measure shock pressures {WATSON ⇨1967}. In 1967, the first attempts were made to measure the lateral stress in a shock-loaded sample {BERNSTEIN ET AL. ⇨1967} – a difficult task which has been the subject of much debate among shock researchers since then {GUPTA ET AL. ⇨1980}.

Whereas piezoelectric gauges are primarily used to measure shock pressures in gases and liquids, piezoresistive gauges dominate pressure diagnostics in solid-state shock wave physics. Because of their very short rise time, their small size, low impedance and moderate price, they are ideal sensors for measuring shock pressures, particularly in a noisy environment and in destructive "one-shot" shock experiments. Comparative measurements of the piezoresistive coefficients of various semiconductors have indicated that the gauge factors of C, Ge and Si could be 10–20 times larger than those based on metal films. However, semiconductors have a nonlinear pressure-resistivity characteristic and are more temperature-dependent than standard metal foil gauges. Therefore, they are only used for special applications, such as in "carbon gauges," which use a thin graphite film encapsulated between two Kapton foils. They also proved useful for measuring pressures in weak shock waves in gas dynamics {⇨Fig. 4.12–C}.

**Examples of Other Methods.** One unusual *direct method* of pressure measurement is the ruby fluorescence pressure gauge. It makes use of the fact that the fluorescence of some refractory phosphor materials – such as ruby and alexandrite – is pressure-dependent. The wavelength of the fluorescent emission is linearly dependent upon the pressure from about 1 MPa to 43 GPa (0.01–430 kbar). This linearity up to very high pressures makes the ruby gauge a particularly attractive measurement technique in shock wave physics. Unfortunately, however, the wavelength shift also depends upon the temperature {HORN & GUPTA 1986, *see* BARNETT ⇨1973}. The ruby pressure gauge may be very useful for measurements performed at extremely high pressures or within harsh chemical environments, where traditional elastic-member deformation techniques cannot be applied.

There are many *indirect methods* of pressure measurement; some examples are given in this book:

▸ Ernst MACH used a Jamin interferometer to measure the density jump at the front of an aerial shock wave generated by a spark discharge {E. MACH & VON WELTRUBSKY ⇨1878; ⇨Fig. 4.12–A}. His son Ludwig MACH applied this method to measure the jump in air density generated by firing a supersonic projectile from a rifle {L. MACH ⇨1896; ⇨Fig. 4.6–L}. By converting density data into pressure data using equation-of-state data, he found that the peak pressure jumps at the head wave and tail wave are about +0.2 bar and –0.085 bar, respectively.

▸ It is possible to determine the shock front velocity $U$ from a series of flash radiographs and the compression ratio $\rho_1/\rho_0$ at the shock front using photodensitometry. a combination of the Rankine-Hugoniot equations allows one to determine the shock pressure $p_1$ {HUGONIOT ⇨1887, *see* Eq. (14)} at selected moments in time. Examples of flash radiographs are given in the PICTURE GALLERY {⇨Figs. 4.5–H, 4.8–F to I, 4.13–H, 4.14–T, 4.15–B, 4.16–O}.

▸ Some of the authors have also included photodensitometer curves of shock-compressed matter in their publications. On the other hand, the measurement of density-time profiles in a shocked target using pulsed radiography is a difficult task, because it requires a high-intensity X-ray pulse of uniform amplitude, which cannot be realized with conventional methods of flash X-ray generation.

▸ Modern high-speed diagnostics also offer numerous indirect pressure measurement techniques. For example, British ballisticians used a calibrated time fuse in order to measure the pressure acting on the head of an actual 3.3-in. caliber shell shot by an 18-pdr. field gun that traveled at between Mach 0.9 and 1.2 – definitely an ingenious although unusual method of pressure measurement {BAIRSTOW ET AL. ⇨1920; ⇨Fig. 4.6–M}.

▸ The U.S. shock researcher William C. HOLTON used water as a pressure gauge {M.A. COOK ⇨1962}. In order to measure detonation pressures in solid explosives, he coupled the detonation wave into a water tank with transparent windows and photographed the propagation of the shock wave at different moments in time – the so-called "aquarium technique." Using the known Hugoniot data of water, he converted measured shock front propagation velocities into shock pressures.

▸ Physicists at the University of Stuttgart developed a needle hydrophone based on an optical fiber with an ultrashort rise time in order to measure compression pulses in water of up to +1,000 bar and rarefaction pulses of up to –100 bar {STAUDENRAUS & EISENMENGER ⇨1988}.

## 2.9 EVOLUTION OF COMPUTATIONAL ANALYSIS

> *The question as to whether a solution which one has found by mathematical reason really occurs in nature and whether the existence of several solutions with certain good or bad features can be excluded beforehand, is a quite difficult and ambiguous one... Mathematically, one is in a continuous state of uncertainty, because the usual theorems of existence and uniqueness of a solution, that one would like to have, have never been demonstrated and are probably not true in their obvious forms... Thus there exists a wide variety of mathematical possibilities in fluid mechanics, with respect to permitting discontinuities, demanding a reasonable thermodynamic behavior etc., etc... It is difficult to say about any solution which has been derived, with any degree of assurance, that it is the one which must exist in nature.*[382]
>
> John von NEUMANN
> Institute for Advanced Study
> Princeton, NJ 1949

FOR over half a century, the use of computer methods to study problems in almost all branches of science and engineering has grown in popularity. Such methods are used as both an aid to planning and interpreting experiments, and as a way to provide theoreticians with insights into the analysis of dynamic phenomena. The use of computers has caused numerical analysts to reconsider the classical methods associated with their research area. For example, the widespread application of computers has stimulated investigations into the basic computational problems of numerical stability, the relationship between the number of iterations required and the precision that must be carried through the calculation, as well as the relationship between the selected mesh size and the minimum time step required when simulating dynamic phenomena. Typically, shock waves and detonation waves in space are governed by nonlinear partial differential equations. One of the methods most frequently used to obtain approximate solutions to these is the *method of finite differences*, which essentially consists of replacing each partial derivative with a difference quotient. The digital computer has proved to be a very efficient tool for solving the resulting set of difference equations, and appropriate numerical codes such as the FDA (Finite Difference Approximation) method were developed to solve partial differential equations. Today a wide spectrum of numerical codes for flow simulation purposes is readily available to the fluid dynamicist {LANEY ⇨ 1998}.

During World War II, a quantum leap in the evolution of computational techniques was achieved in a few countries, including the United States, England and Germany; these were intended for use in military applications. In the United States, this development resulted in the first large-scale mechanical calculating machines, which were immediately applied to nuclear bomb physics. After the war, the development of high-speed electronic digital computers was mainly the result of urgent numerical analyses necessitated by the development of thermonuclear weapons, as well as numerical simulations of their new destructive effects – a consequence of the perceived Soviet threat to U.S. (and *vice versa*) during the long period of the Cold War (1945–1989). This placed heavy emphasis on research and development in ballistics, nuclear physics, detonics, aerodynamics, astronautics and related technical fields, such as high-speed photography and diagnostics, image processing, automatic control systems, lasers and pulsed power technology.[383] In turn, this led to

- mounting interest in fast computation and in stored-program computers from universities and governmental and industrial research laboratories;
- the formation of professional groups concerned with computers;
- the staging of influential technical conferences;
- the foundation of computer science periodicals; and
- the creation of the Internet, which connects governmental institutions, companies, universities, corporate networks and hosts, thus providing quick and easy access to a huge number of different databases, which has tremendously enhanced communication among individual researchers.

According to a modern definition,[384] *computers* are "electronic devices that are capable of accepting data and instructions, executing the instructions to process the data, and presenting the results." The majority of historians also do not consider any machine to be a "computer" unless it embodies the stored-program concept. It may be interesting to note at

---

[382] J. VON NEUMANN: *Discussion of the existence and uniqueness or multiplicity of solutions of the aerodynamical equations.* Proc. of the Symposium on the Motion of Gaseous Masses of Cosmical Dimensions [Paris, France, Aug. 1949]. In: *Problems of cosmical aerodynamics*. Central Air Documents Office, Dayton, OH (1951), pp. 75-84.

[383] F. HARLOW and N. METROPOLIS: *Computing and computers. Weapons simulation leads to the computer era.* Los Alamos Science **4**, 132-141 (1983).

[384] C. MORRIS (ed.): *Dictionary of science and technology*. Academic Press, San Diego etc. (1992), p. 489.

this point that the term *computer* originally designated "a person employed to make calculations in an observatory, in surveying, *etc.*" – not a machine.[385] It is not known with any certainty who first used the term *computer* to denote a calculating machine. Obviously, however, the modern term was not immediately accepted. For example, the 1961 edition of the *Britannica Encyclopaedia* still used the old-fashioned terms *calculating machine* and *computing machine* rather than the modern term *computer*.

The history of computing, which ranges from the first crude mechanical instruments to modern electronic computers, has been documented and discussed in a large number of books from technical, scientific, economic and human/social viewpoints;[386] some of them, especially those written by distinguished computer pioneers, are particularly informative.[387] In addition, a collection of essays[388] on the history of computer technology, written by some of the foremost contributors in the field, provides first-hand information on a subject which for decades was covered by a veil of secrecy. Prominent examples include Arthur W. BURKS, J. Presper ECKERT, John W. MAUCHLY, Nicholas C. METROPOLIS, George R. STIBITZ, Stanislaus ULAM, Maurice V. WILKES and Konrad ZUSE – to name but a few of the 39 contributors to this collection of essays.

This chapter can only briefly review some of the most important milestones in the early era of the development of computers, and thus it attempts only to illustrate their close interrelations with the subject matter of this book. Obviously, during the pioneering era up to the end of World War II, the main impetus for developing faster and smaller electronic computers in Europe and the United States was almost exclusively the immediate solution of vital military problems such as those imposed by ballistics, fluid dynamics, shock wave physics and detonics – a stimulating process of complex interrelations which continued throughout the period of Cold War and right up to the present day.

## 2.9.1 THE PRE-COMPUTER ERA: TRIUMPH OF MECHANICAL AND GRAPHICAL METHODS

*The same thing which you have done by hand calculation, I have just recently tried to do in a mechanical way. I have constructed a machine which automatically reckons together the given numbers in a moment, adding, subtracting, multiplying and dividing.*[389]

Wilhelm SCHICKARD
Eberhard Karls Universität
Tübingen 1623

The two domains of *geometry* and *numbers* are the basis for and the essence of all mathematics. The development of calculating machines for providing such numbers was the main aim of generations of mathematicians, engineers and instrument-makers. The invention of infinitesimal calculus in the 17th century initiated rapid progress in formulating complex problems in fluid dynamics and ballistics via differential equations; however, their solution by analytical methods imposed great difficulties, and numerical approaches were beyond the abilities of the calculating machines that existed at that point. Geometry embraces far more than purely spatial concepts. In the 18th century, the eminent mathematicians Leonard EULER and Gaspard MONGE applied differential calculus to geometry in order to find solutions of differential equations by applying graphical procedures, thus creating differential geometry, the "modern language of physics." Numerical and geometrical approaches to computation, which complement each other in a unique manner, are briefly discussed in the following.

**Digital Mechanical Machines.** The bead-and-wire *abacus*, invented in Egypt in around 500 B.C., is the oldest known example of arithmetic computational aid. The first significant steps toward the construction of mechanical computers were made as early as the 17th century by some distinguished natural philosophers. Wilhelm SCHICKARD, a German mathematician and orientalist, invented (before even PASCAL) the first *digital computing machine* (1623) that (according to his words): "…immediately computes the given numbers automatically; adds, subtracts, multiplies, and divides." Unfortunately, his machine was destroyed in a fire shortly before his death. However, a few of his drawings survived and were

---

[385] J.A. SIMPSON and E.S.C. WEINER: *The Oxford English dictionary*. Clarendon Press, Oxford, vol. III (1989), pp. 640-641.

[386] See, for example M.R. WILLIAMS: *A history of computing technology*. IEEE Computer Society Press, Los Alamitos, CA (1997), pp 66-83.

[387] M.V. WILKES: *Memoirs of a computer pioneer*. MIT Press, Cambridge, MA (1985); H.H. GOLDSTINE: *The computer. From PASCAL to VON NEUMANN*. Princeton University Press, NJ (1993); W.J. ECKERT and R. JONES: *Faster, faster: a description of a giant electronic calculator and the problems it solves*. McGraw-Hill Book Co., New York (1955); K. ZUSE: *The computer – my life*. Springer, Berlin (1993); and J. VON NEUMANN: *The computer and the brain*. Oxford University Press, London (1958).

[388] N. METROPOLIS, J. HOWLETT, and G.C. ROTA (eds.): *A history of computing in the twentieth century*. Academic Press, New York etc. (1980).

[389] From a letter to the German astronomer Johannes KEPLER (dated Sept. 20, 1623). In a later letter to KEPLER (dated Feb. 25, 1624), SCHICKARD enclosed a sketch of his calculating machine. Both letters have survived. *See Wilhelm SCHICKARD: Briefwechsel* (F. SECK, ed.). 2 vols., Frommann-Holzboog Verlag, Stuttgart (2002).

discovered in the 1950s in papers left by the German astronomer Johannes KEPLER, with whom he had corresponded. a working model reconstructed from his notes proved the soundness of his construction.[390] In the early 1640s, Blaise PASCAL, a French philosopher, built a calculator to assist his father (a mathematician and judge who presided over the tax court) with local administration. Several arithmetic machines manufactured by him have survived. Gottfried W. LEIBNIZ, a German mathematician and philosopher, also constructed a calculating machine, a perfected version of an earlier machine developed by PASCAL, and presented it to the Royal Society in London (1673). His machine could add, subtract, multiply, divide, and extract roots.

Charles BABBAGE, an English mathematician and inventor, was the first to conceive of an automatic computer for mechanically calculating mathematical tables – the forerunner of the modern stored-program computer. However, his *analytical engine* (1833), which was designed to produce mathematical tables by accepting data and performing a sequence of different mathematical operations according to instructions from punched cards, was not completed. The U.S. statistician and inventor Herman HOLLERITH perfected the idea of punched cards from a technical standpoint and developed the *automatic tabulating machine* (1890) for use in statistics.

Eventually, these step-by-step improvements and inventions led to various kinds of mechanical and/or electromechanical *hand calculators* and *punched-card machines* for adding, subtracting, multiplying and dividing two numbers, which found wide application in science, commerce and statistics.

In 1928, the first important scientific application of HOLLERITH's punched-card technique was demonstrated by Leslie John COMRIE, an astronomer and pioneer in mechanical computation from New Zealand. He succeeded in numerically solving the classical "three-body problem" of Newtonian celestial mechanics (the gravitational interaction of the masses of Sun, Earth and Moon), a surprisingly difficult mathematical task which had previously frustrated not only such eminent mathematicians as D'ALEMBERT, EULER, DE LAGRANGE and DE LAPLACE, but also subsequent generations of astronomers.[391] Using the *Hollerith tabulator*, an 80-column punched card machine from IBM, he calculated lunar orbits. COMRIE's method of applying punched-card computation to astronomy was picked up in the United States by Wallace J. ECKERT, an astronomer by training and a computer engineer at Columbia University, who became one of America's leading proponents of punched-card computation in the 1930s.

During World War II, theoretical physicists at Los Alamos, who were facing problems that were insoluble by analytical means and were also far too complicated to be solved by desk computing machines, reviewed ECKERT's successful application and adapted his computing method to their own needs, decisively stimulating the further development of such machines in the United States (*see below*).

**Analog Mechanical Machines.** The idea of the *slide rule* (1614) rests upon the idea of logarithms formulated by the Scottish mathematician John NAPIER, who also built a primitive calculating machine that used calculating rods to quickly multiply, divide and extract roots. His mechanical numbering device was made of horn, bone or ivory, later known as "Napier bones." Around 1620 the English mathematician William OUGHTRED came up with the idea of using a circular slide rule to improve accuracy. The standard rectangular slide rule, based upon a construction devised by the French military officer Amédée MANNHEIM (1859), was in common use until the early 1970s. The *planimeter* (1814), which was also an analog instrument, was devised by the Bavarian engineer Johann H. HERMANN to measure the value of a definite integral provided graphically as a planar surface, by tracing its boundary lines.

William THOMSON (later Lord KELVIN) and his brother James THOMSON, an engineering professor at the University of Glasgow, devised the *differential analyzer* (1876), one of the earliest mechanical analog computers capable of solving differential equations, which they used to calculate tide tables. In the late 1920s, the U.S. engineer Vannevar BUSH and his colleagues at MIT developed the *network analyzer*, a system for setting up miniature versions of large and important electrical networks. At the same time, they developed a prototype of the differential analyzer, the first analog computer capable of solving integral equations. BUSH suggested its application to solve numerical problems of linear and nonlinear electrical networks. During World War II, he worked on radar antenna profiles and calculated artillery firing tables.

The differential analyzer was later also extended to solve ordinary second-order differential equations. In the early 1930s, the Mexican-born U.S. scientist Manuel Sandoval VALLARTA and the Belgian mathematician and abbot Georges LEMAÎTRE {⇨1927}, cocreator of the Big Bang Model

---

[390] B. VON FREYTAG LÖRINGHOFF: *Über die erste Rechenmaschine*. Physik. Blätter **14**, 361-365 (1958); *Wiederentdeckung und Rekonstruktion der ältesten neuzeitlichen Rechenmaschine*. VDI-Nachrichten **14**, Nr. 39, 6 (1960).

[391] E.O. LOVETT: *The problem of several bodies: recent progress in its solution*. Science **29**, 81-91 (Jan. 15, 1909).

for the evolution of the Universe, used BUSH's machine in their complex calculations on the distribution of primary cosmic radiation.

The ability to solve ordinary differential equations is also highly relevant to exterior ballistics. Here the main task is to describe, for a given fuse/shell/gun combination, the position and velocity of the projectile when the gun is fired at various elevations and initial velocities. The problem becomes even more complex when aiming at a moving target from a mobile position. In 1935, the German Siemens AG built C35, the "Artillerie-Schußwert-Rechner" which was installed on the battleships *Gneisenau* and *Scharnhorst* for guiding surface-to-air missiles (SAMs). The roughly 1-ton mechanical device contained about 2,000 gear-wheels.[392] An analog computer was also used in the German V2 rocket for guidance purposes.

In World War II, the requirements of anti-aircraft artillery "predictors," a particularly challenging task, stimulated activities aimed at improving existing analog computers. In these, the variables were the latitude, longitude, and height of the target and those of the projectile, respectively, all of which varied rapidly with time. The input data were the muzzle velocity, the ballistic characteristics of the projectile and atmospheric conditions like the wind velocity and air density. The analog computer had to solve two simultaneous equations so that the target and the projectile, each of which was moving along its own course, would arrive at the point of intersection at the same time. In 1940, Clarence A. LOVELL and David B. PARKINSON at Bell Telephone Laboratories developed an analog computer that could be used to control aircraft guns. Two years later, they built the M-9 GUN PREDICTOR, another "ballistic computer" for predicting the future position of an aircraft (based on its precise current position, its course and its speed) and firing data for an anti-aircraft gun. This more advanced analog device, which used a set of wheels to set potentiometer resistances, accepted electrical inputs from the XT-1, a truck-mounted microwave radar, and sent electrical outputs to the anti-aircraft guns to direct them accordingly. One radar and one M-9 could control a battery of four guns. In 1941, at the Peenemünde Rocket Center the German electrical engineer Helmut HOELZER[393] developed an electronic circuit with phase-shifting elements that could integrate differential equations – the first fully electronic general-purpose analog computer – which was applied to the guidance system of the A-4, the world's first long-range ballistic missile {HVA Peenemünde ⇨1942}.

Another application of analog computers was to aerodynamics, where they were particularly useful for solving equations relating lift to air flow characteristics, yaw angle and airfoil geometry. The British physicist and mathematician Douglas R. HARTREE[394] developed the *differential analyzer*, an analog computer. He first proposed its use to solve partial differential equations – particularly of the hyperbolic type, which are of the greatest importance in fluid dynamics – and in 1955 he eventually fabricated such a mechanical device. He succeeded in solving parabolic partial differential equations, but after also attempting to treat the hyperbolic type, he had to admit that "a few attempts have been made to apply it to equations of hyperbolic type, with only partial success." His attempts were too ambitious for his time and this problem was not successfully solved until the advent of digital calculating machines.

Some stages in the unique history of mechanical analog computers used to control the firing of large guns, ranging from early developments up to their golden age in World War II and their subsequent obsolescence in the 1950s, have been reviewed fairly recently.[395]

**Graphical Concepts.** In addition to numerical analysis, various graphical methods were used. High-speed flows are termed *hyperbolic* and can be solved graphically. This *method of characteristics* {MONGE ⇨1770; PRANDTL & BUSEMANN ⇨1929} was used in unsteady gas dynamics to solve 1-D and 2-D fluid dynamic problems, *e.g.*:

▸ to solve the "Lagrange problem" in ballistics {DE LAGRANGE ⇨1793};
▸ to shape the geometries of Laval nozzles in supersonic wind tunnels {PRANDTL 1906; WIESELSBERGER & HERMANN ⇨1934};
▸ to analyze the operating cycle of the pulse-jet engine invented by the German engineer Paul SCHMIDT {⇨1930}, the "Schmidt tube" (P. SCHMIDT ⇨1930; BUSEMANN 1936 and SCHULTZ-GRUNOW 1943, *see* Sect. 5);
▸ to describe the propagation of spherical blast waves (SCHULTZ-GRUNOW 1943);
▸ to determine the lateral expansion of the gases behind a detonating slab of explosive {HILL & PACK ⇨1947};

---

[392] H. PETZOLD: *Moderne Rechenkünstler.* Beck, München (1992).
[393] H. HOELZER: *Anwendung elektrischer Netzwerke zur Lösung von Differentialgleichungen und zur Stabilisierung von Regelvorgängen (gezeigt an der Stabilisierung des Fluges einer selbst- bzw. ferngesteuerten Großrakete).* Ph.D. thesis, TH Darmstadt (1946).
[394] D.R. HARTREE: *Calculating instruments and machines.* University of Illinois Press, Urbana, IL (1950), pp. 33-34.
[395] A. BEN CLYMER: *The mechanical analog computers of Hannibal* FORD *and William* NEWELL. IEEE Ann. Hist. Comput. **15**, No. 2, 19-34 (1993).

## 2 GENERAL SURVEY

- to solve various problems in hydrodynamics {OSWATITSCH ⇨ 1947}
- to determine the flow in a steady supersonic jet of air issuing from a slightly supersonic circular orifice into a vacuum {OWEN & THORNHILL ⇨ 1948}; and
- to treat steady 2-D and 3-D flows with shock waves (FERRI 1954).

In hydrodynamics, the *LÖWY-SCHNYDER method* {BERGERON ⇨ 1937} was used to calculate the propagation and reflection of hydrodynamic shocks in water supply lines. Such shocks (the *water hammer effect* {CARRÉ ⇨ 1705; ZHUKOVSKY ⇨ 1898}) can cause damage to pipes in hydraulic installations.

### 2.9.2 REVOLUTION IN CALCULATION: THE AUTOMATIC DIGITAL COMPUTER

> *The ENIAC project was funded to facilitate the preparation of firing tables. The calculation of trajectories was a good problem for computing machines, because the calculation depends critically on the drag function $G(v^2)$, which gives the resistance of the air to the movement of the shell as a function of the velocity squared. This function is an ill-behaved function of the shell's velocity, especially as the shell passes through the sound barrier. In hand calculation the resistance was read from printed tables, and on the differential analyzer the resistance was fed in from an input table… All of these table-input methods were too slow for ENIAC. Hence, we used variable resistor matrices set by hand switches…*[396]
>
> Arthur W. BURKS
> University of Michigan
> Ann Arbor, MI 1980

**Digital Electromechanical Computers.** While studying magnetomechanics of telephone relays, the U.S. applied mathematician George R. STIBITZ turned his attention in 1937 to relay-controlled binary circuits. In the following year, with the support of Bell Laboratories in New Jersey, he developed a *two-digit binary adder*, which was operational late in 1939 and was demonstrated in 1940 by remote control between several U.S. cities. During the war he designed various circuits and supervised the design of a *relay interpolator* and a *relay ballistic computer* (named the "Stibitz Computer") for the National Defense Council (NDC).

Today STIBITZ is internationally recognized as being the father of the modern digital computer. However, in Germany, the engineer Konrad ZUSE also independently created a binary calculator that he named ZUSE 1 (or Z1). This was the first *binary digital computer*, and it became operational as a test model in 1938, one year prior to STIBITZ's calculator. During World War II, financially supported by the Reichsluftfahrtministerium (German Ministry of Aerial Warfare), he built several special relay-based machines for the Henschel Flugzeugwerke AG Berlin, which were immediately applied to aerodynamic calculations of the Hs 293, a radio-guided winged bomb provided with a rocket engine.[397] In the period 1942–1945, he designed his Z4, a fully programmable computer which became operational in the beginning of 1945, for the Aerodynamische Versuchsanstalt (AVA) at Göttingen. His instrument, which was renovated and supplemented after the war, was initially operated at the Institute for Applied Mathematics of the ETH Zurich (1950–1955). It was programmed by the Swiss applied mathematician Heinz RUTISHAUSER, one of the founders of ALGOL (1958–1960). The calculating machine was then moved to France, to the Laboratoire de Recherches Techniques de Saint-Louis {⇨1945}, where it was used for defense work until 1960. a reconstruction of ZUSE's computer Z3 (operational in 1941, destroyed during the war) and his original Z4 (operational at the end of war) are now on permanent display at the exhibition "Informatik" of the Deutsches Museum in Munich.

The first *automatic digital computer* was conceived by Howard H. AIKEN of Harvard University (Cambridge, MA). Planned long before the war, and built in cooperation with the IBM Corporation,[398] this large-scale, relay-based punched-card machine was not completed until 1944, when it was installed at Harvard University. Named ASCC (Automatic Sequence Controlled Calculator), or MARK I, it was mostly an electromechanical device, a "relay calculator" that was capable of performing only about three additions per second, and it was primarily used for computing ballistic data. a second machine, MARK II, was built in 1944 for the Naval Proving Ground Group (Dahlgren, VA) in order to provide vital ballistic calculations. Both the MARK I and MARK II were decimal machines, and operations – not yet programs in the modern sense – could be tailored to a specific application by inserting "plugwires" into a "plugboard."

---

[396] BURKS was one of the leading pioneers in his field. He developed and built ENIAC, the world's first electronic, digital, general-purpose scientific computer, at the Moore School of Electrical Engineering, University of Pennsylvania. See his article *From ENIAC to the stored-program computer: two revolutions in computers*. In: (N. METROPOLIS ET AL, eds.): *A history of computing in the 20th century*. Academic Press, New York etc. (1980), pp 311-344.

[397] K. ZUSE: *Der Computer – mein Lebenswerk*. Springer, Berlin etc. (1984), pp. 62-64.
[398] C.J. BASHE, L.R. JOHNSON, J.H. PALMER, and E.W. PUGH: *IBM's early computers*. MIT Press, Cambridge, MA (1986), pp. 25-33.

The new computing machines were mostly used to solve problems encountered in the development of the first atomic bomb, for example:

- to calculate the neutron diffusion for a sphere of fissionable material surrounded by a spherical shell of inert, scattering material;
- in the development of an equation of state (EOS) for uranium and plutonium by interpolating between high-pressure data obtained by numerically solving the Thomas-Fermi differential equation and low-pressure data obtained experimentally;
- in hydrodynamic calculations, particularly when calculating the propagation of shock waves and compression in the spherically symmetric implosion of the Trinity atomic bomb (VON NEUMANN & NEDDERMEYER 1943); and
- to solve the three-shock problem or Mach reflection {VON NEUMANN ⇒ 1943}, which was of immediate interest for determining the "optimum" height of burst (VON NEUMANN, Sept. 1944) in the planned bombing of Hiroshima and Nagasaki.

In 1944, the Los Alamos Scientific Laboratory (LASL) ordered ten IBM machines in order to help calculate the critical masses of odd-shaped bodies and to solve hydrodynamic equations for implosion. These IBM machines could not be programmed, and a considerable number of intermediate steps had to be done by hand. Hans A. BETHE,[399] head of the Theoretical Division at Los Alamos during World War II, remembered, "In our wartime calculations, the Hugoniot conditions at the shock front were fitted by hand, at each time step in the calculation. The position of the shock front was also calculated by hand, using the shock velocity deduced from the Hugoniot equation at the previous time step. This was a somewhat laborious procedure; but each time step on the computer took about one hour, so that the hand calculations could easily keep in step with the machine."

The U.S. mathematician and computer pioneer Herman H. GOLDSTINE,[400] who became a major contributor to the logical design of the EDVAC (*see below*), and who cooperated with John VON NEUMANN on various numerical analysis projects, remembered that "the main motivation for the first electronic computer was the automation of the process of producing firing and bombing tables." GOLDSTINE, who headed a section of the Ballistic Research Laboratory (BRL) at the Moore School of Electrical Engineering of the University of Pennsylvania in 1942, was himself engaged in generating new calculating instruments for the faster production of such shooting tables.

**Digital Electronic Computers.** The first electronic digital computer was named the ENIAC (Electronic Numerical Integrator and Calculator), a large programmable digital machine initially designed to recompute artillery firing and bombing tables for the Ordnance Corps of the U.S. Army. It was built primarily to integrate the equations of external ballistics in a step-by-step process, but it was flexible enough to be applied to a wide range of large-scale computations other than the numerical integration of differential equations.[401]

The ENIAC was developed by John W. MAUCHLY, a professor of electrical engineering, and J. Presper ECKERT, a young electronic engineer. Both worked at the Moore School of Electrical Engineering at the University of Pennsylvania, which was a center for calculating firing tables and trajectories at that time. In August 1942, MAUCHLY wrote a memo entitled *The Use of High Speed Vacuum Tube Devices for Calculating* in which he proposed the idea that the speed of calculation "can be made very much higher than of any mechanical device." VON NEUMANN and ECKERT proposed to achieve a high computing speed by operating vacuum-tube circuits at 100,000 pulses per second – *i.e.*, using pulses at 10 μs intervals – which was indeed realized with the final version of the ENIAC. This computing machine became operational in 1946 at the Moore School and in 1947 at the Ballistic Research Laboratory (BRL) in Aberdeen Proving Ground, MD.[402] The 30-ton ENIAC contained about 18,000 electron tubes and 70,000 resistors, and used punched cards for input and output data. It was more than 1,000 times faster than its electromechanical predecessors and could execute up to 5,000 additions per second.

The ENIAC was designed primarily for use in exterior ballistics, particularly for the step-by-step integration of differential equations in order to calculate ballistic trajectories of bombs and shells. However, from the beginning, ECKERT's goal was to make it generally useful for other military problems too, such as those in interior ballistics and for all kinds of data reduction. In the period from December 1945 to January 1946, VON NEUMANN used the ENIAC at Los Alamos for a pre-

---

[399] H.A. BETHE: *Introduction*. In: (S. FERNBACH and A. TAUB, eds.): *Computers and their role in the physical sciences*. Gordon & Breach, New York *etc.* (1970), pp. 1-9.

[400] H.H. GOLDSTINE: *The computer. From PASCAL to VON NEUMANN*. Princeton University Press, NJ (1993), p. 135.

[401] K. KEMPF: *Electronic computers within the ordnance corps*. Historical monograph covering the pioneer efforts and subsequent contributions of the U.S. Army Ordnance Corps in the field of automatic electronic computing systems during the period 1942−1961; http://ftp.arl.mil/~mike/comphist/61ordnance/.

[402] W.T. MOYE: *ENIAC: The army-sponsored revolution*. ARL (Jan. 1996); http://ftp.arl.mil/~mike/comphist/96summary/index.html.

liminary study on the hydrogen bomb, Edward TELLER's planned thermonuclear "super bomb." This challenging task amounted to solving a system of partial differential equations. However, it turned out that the "super problem" was too complicated for the ENIAC with its 1,000 bits of memory, and only a highly simplified version of the calculation was run, which revealed very little about how such a weapon might work.[403] In September 1946, Abraham H. TAUB, a mathematician and theoretical physicist at Princeton University, used the ENIAC to calculate reflection and refraction phenomena of shock waves {BLEAKNEY ⇨ 1946}. In the same year, the machine was moved to BRL, which was in the process of becoming one of the great wartime and postwar computing centers in the United States. An informative collection of review papers on the development of ENIAC and early computer applications at BRL was published in 1996.[404, 405]

By March 1945, VON NEUMANN had begun to consider a computer that was not assigned to definite, often very specialized (military) purposes, but was instead allowed to run quite freely and be governed by scientific considerations. In particular, he had in mind a new computing device powerful enough to solve nonlinear partial differential equations with two or three independent variables, such as those used for weather prediction and in the study of the properties of numbers. Indeed, VON NEUMANN proposed that this planned computer, dubbed EDVAC (Electronic Discrete Variable Automatic Computer), should be a *stored-program serial computer*; this architectural design was used in several subsequent generations of computers.[406] He described his concept in a paper entitled *First Draft of a Report on the EDVAC*, written in the spring of 1945 for the U.S. Army Ordnance. A "program" is a sequence of instructions on how to perform particular operations on data contained in a memory. His stored-program concept, the so-called "von Neumann architecture," translated mathematical procedures into a machine language of instructions. The "von Neumann machine" was characterized by a large Random Access Memory (RAM) that was used to address memory locations directly, and a Central Processing Unit (CPU) that possessed a special working memory.

VON NEUMANN also devised a method for converting the relay-based ENIAC concept of an externally programmed machine into that used by EDVAC. The plugboards and programming switches of the ENIAC were replaced in the EDVAC by an electrically alterable memory that could store both the instructions and numbers to be used in a calculation at electronic speeds. The EDVAC was built by the U.S. National Bureau of Standards and operated by the Ballistic Research Laboratory at Aberdeen from 1950 onwards. EDVAC was the first American stored-program computer. Prior to this, however, the first computer to meet the criterion of VON NEUMANN's stored-program concept was the EDSAC (Electronic Delay Storage Automatic Calculator), which performed its first calculations in May 1949. This binary, serial-type computer, which was the first to use supersonic delay lines for memory,[407] was built in Great Britain by Maurice V. WILKES[408] and his team at the Mathematical Laboratory of Cambridge University, later known as the "Computer Laboratory." In 1945, a number of British visitors came to the Moore School, and these visits prompted the computerization of Great Britain, which also resulted in the MADM (Manchester Automatic Digital Machine) at the University of Manchester and the ACE (Automatic Computing Engine) at NPL in Teddington.

The first generation of *stored-program computers* also included the IAS, a computer built at the Institute for Advanced Study (IAS) at Princeton University; the early real-time WHIRLWIND designed under the leadership of Jay FORRESTER at MIT's Digital Computer Laboratory; the UNIVAC I (Universal Automatic Computer) at the Moore School of the University of Pennsylvania; and many others. The UNIVAC I, designed by ECKERT and MAUCHLY, was the first general-purpose computer to be made commercially available – before this, computers had only been rented and had to be constantly serviced and often required repairs. Nicholas C. METROPOLIS and associates at Los Alamos built the MANIAC (Mathematical Analyzer, Numerical Integrator and Computer), which began operation in March 1952. The MANIAC,

---

[403] A. FITZPATRICK: *TELLER's technical nemeses: the American hydrogen bomb and its development within a technological infrastructure.* Phil Tech **3**, No. 3, 10-17 (1998).

[404] M.H. WEIK: *The ENIAC story.* Ordnance [A journal of the American Ordnance Assoc., Washington, DC] (Jan./Feb. 1961); http://ftp.arl.mil/~mike/comphist/eniac-story.html.

[405] See (T.J. BERGIN, ed.) *50 years of army computing. From ENIAC to MSRC.* Record of a symposium and celebration, Aberdeen Proving Ground, MD (Nov. 13-14, 1996). Rept. ARL-SR-93, Army Research Laboratory (ARL), Aberdeen Proving Ground, MD (Sept. 2000); http://stinet.dtic.mil/oai/oai?verb=getRecord&metadataPrefix=html&identifier=ADA431730.

[406] In the mid-1930s, the German engineer Konrad ZUSE developed a concept which closely approached VON NEUMANN's idea of a program stored in the memory of a calculating machine. In 1936, ZUSE applied for a German patent, which, however, he did not get; *see* K. ZUSE: *Verfahren zur selbsttätigen Durchführung von Rechnungen mit Hilfe von Rechenmaschinen.* Dt. Patentanmeldung Z 23139/GMD Nr. 005/021 (1936).

[407] The EDVAC used supersonic (meaning here ultrasonic) delay lines – acoustic high-speed internal memories consisting of mercury-filled tubes about 5' long and 1" in diameter in contact with a quartz crystal at each end.

[408] M.V. WILKES: *Memoirs of a computer pioneer.* MIT Press, Cambridge, MA (1985).

like the EDVAC and the IAS, was also used for calculations related to the development of the first hydrogen bomb {MIKE Test ⇨ 1952}.

In the Soviet Union, work on the development of computers had begun in 1947 in order to use this new technology in the fields of nuclear physics, reactor technology, ballistics, electronics, gas dynamics, *etc.* Several different computers were developed in the 1950s. The most important of these was the BESM (*Bolshaja Elektronno-Schetnaja Mashina*, meaning "Large Electronic-Computing Machine"), which was designed under the leadership of Sergei Alexeevich LEBEDEV and set in operation in 1952. Employing about 5,000 vacuum tubes and a William-tube memory, it was then Europe's fastest computer and it closely matched the American IBM 701 (introduced in 1954) in terms of performance. One of the first uses for Russian computers was to perform calculations related to the development of the first Soviet thermonuclear bomb {Semipalatinsk Test Site ⇨ 1955}.[409] In the 1950s, further Russian digital computers were designed and built, for example STRELA (1953), a large William-tube machine; URAL (1955), a magnetic drum computer; and others. Herman H. GOLDSTINE[410] provided a general survey of worldwide computer development, also including the Soviet Union.

### 2.9.3 THE TRICKY PROBLEM: TREATING FLOW DISCONTINUITIES NUMERICALLY

> *In the investigation of phenomena arising in the flow of a compressible fluid, it is frequently desirable to solve the equations of fluid motion by stepwise numerical procedures, but the work is usually severely complicated by the presence of shocks. The shocks manifest themselves mathematically as surfaces on which density, fluid velocity, temperature, entropy and the like have discontinuities.*[411]
>
> John VON NEUMANN
> Robert D. RICHTMYER
> Institute for Advanced Study
> Princeton, NJ 1950

In the pioneering era of shock wave research, digital computers proved very useful for solving problems involving continuous quantities; *i.e.*, those that change smoothly rather than "jump," like the thermodynamic quantities at a shock front. Many problems relevant to the military, however, were related to supersonic flow and detonation, like the propagation and reflection of a blast wave originating from a strong (nuclear) explosion. This immediately raised the problem of how to treat wave discontinuities numerically. In particular, the problem of simulating implosions became crucial to the planned design for the implosion bomb during 1943 and 1944, and this required the integration of hyperbolic partial differential equations and the use of realistic high-pressure equations of state for the high explosive, the uranium tamper and the plutonium core. This was a very difficult and complex task that could not be done by hand. In addition, plutonium was only available in very small quantities at that time, and experimentally obtained equation-of-state data were not yet available. Other important applications of digital computers included those to computational aerodynamics, particularly in the design of so-called "shock-free" configurations for transonic flight − a speed regime that all high-speed tactical aircraft were still confined to during World War II.

The first hydrodynamic initial value problems with prescribed boundary conditions that were fed into a computer were one-dimensional, and so the *Lagrangian* coordinates − *i.e.*, fixed-in-the-material coordinates − were most appropriate. The first two-dimensional codes were also in Lagrangian coordinates, but instability problems were encountered when treating discontinuities like shock waves. These were not understood until VON NEUMANN's rediscovery of a pre-World War II paper on approximation mathematics {COURANT, FRIEDRICHS & LÉWY ⇨ 1928}, in which a condition for stability in the solution of partial differential equations was worked out; this was later called by VON NEUMANN the "Courant criterion." However, in the early days of computers, there were two limiting factors which rendered numerical fluid dynamics difficult: the limited memory available to store the data for the cells, and the computer time required to process the solution to completion.

VON NEUMANN was the first to attempt to apply digital computers to solve true shock wave problems, at the Institute for Advanced Study at Princeton University. According to William ASPRAY,[412] a notable computer historian, VON NEUMANN developed three of the earliest numerical approaches for treating shock waves. Following ASPRAY's view, his methods were these:

---

[409] D. HOLLOWAY: *STALIN and the bomb.* Yale University Press, New Haven, CT & London (1994), p. 314.

[410] H.H. GOLDSTINE: *The computer from PASCAL to VON NEUMANN.* Princeton University Press, Princeton (1972), pp. 349-362.

[411] J. VON NEUMANN and R.D. RICHTMYER: *A method for the numerical calculation of hydrodynamic shocks.* J. Appl. Phys. **21**, 232-237 (1950).

[412] W. ASPRAY: *John VON NEUMANN and the origins of modern computing.* MIT Press, Cambridge, MA (1992), pp. 108-110, 112, 288.

His oldest attempt to treat shock discontinuities goes back to a report published in 1944.[413] He proposed replacing the full hydrodynamic phenomena with a simple kinetic model that simulated the fluid using a line of beads connected by springs, a simple model for simulating continuum mechanics that DE LAGRANGE {⇨1788} had already proposed in his *Mécanique analytique*. In this one-dimensional shock wave model {⇨Fig. 4.4–D} – previously also proposed by Charles V. BURTON (1893) in England and later picked up by Lev V. AL'TSHULER (1965) in the former Soviet Union – the beads represent the idealized molecules and the springs the intermolecular forces. ASPRAY wrote, "VON NEUMANN was convinced that 14 bead molecules would be satisfactory for shocks in one dimension but that as many as 3,000 beads may be required in two- and three-dimensional problems which would probably have placed the problem beyond the range of calculating equipment then available." In order to calculate a similar collision problem imposed by the development of nuclear weapons using the ENIAC, VON NEUMANN suggested his Monte Carlo method (1946–1947), which he worked out together with the Polish-born U.S. mathematician Stanislaus ULAM. He proposed to calculate the actions of 100 neutrons throughout the course of 100 collisions each in order to trace the isotropic generation of neutrons from a variable composition of active material along the sphere radius.

His second approach was developed in collaboration with the Los Alamos physicist Robert D. RICHTMYER. They introduced an artificial viscosity into numerical calculations to "smear out the discontinuity" such that the shock thickness becomes larger than the spacing on the grid points. The difference equations could then be solved directly, as if there were no shocks {VON NEUMANN & RICHTMYER ⇨1950}.

His third approach to the shock problem, undertaken in collaboration with the U.S. computer expert Hermann H. GOLDSTINE, was a direct numerical assault, one of the first attempts to solve a complex problem involving shocks directly by numerical means.[414] Considering the delay and propagation of a shock wave from a strong point-source explosion in an ideal gas, they applied an iterative method to calculate the Rankine-Hugoniot conditions {HUGONIOT ⇨1887} across the shock.

Retrospectively, the second method of simulating shock waves, where an artificial viscosity was added to the hydrodynamic equations, became the most important method of treating flow discontinuities numerically. Early computers, however, barely had enough capacity to solve 2-D shock problems with sufficient resolution in time and space, and the numerical treatment of true three-dimensional shock problems was still far beyond them.

The enormous computational resources at Los Alamos, Livermore and Albuquerque (Sandia), which were provided for programs of national interest, particularly the development of nuclear weapons, were most useful for analyzing materials under high shock compression, which stimulated the evolution of *Computational Fluid Dynamics (CFD)*, an entirely new discipline creating a wealth of new numerical simulation techniques.[415] Early codes for solving shock wave problems in solids were called "hydrocodes," because the deviatoric stresses – stresses that cause the volume to deviate from its original proportions – were neglected, an assumption typically made for fluids. Originally, hydrocodes were developed for modeling fluid flow at all speeds, but they also proved useful to study natural high-dynamic events (such as the hypervelocity impact of an asteroid on a planet) which are complex enough that an analytical solution is not possible.[416]

Ideally, a computer simulation should apply an appropriate algorithm, generally called a "numerical method," to produce sharp approximations to discontinuous solutions automatically; *i.e.*, without explicit tracking and the use of jump conditions. Methods that attempt to do this are called "shock-capturing methods." a number of numerical methods that use conservation laws to solve Euler and Navier-Stokes equations have been available for at least ten years now.[417] A recent review of numerical methods was given by Philip L. ROE,[418] a renowned numerical fluid dynamicist at the Dept. of Aerospace Engineering at the University of Michigan.

An increasing number of commercial numerical codes for treating shock discontinuities have been made available. However, comparisons of the codes with each other and with experimental results provide the crucial test in this case, and VON NEUMANN's previous reflections on this problem, on whether an obtained solution really occurs in nature – see the citation provided at the start of this section – is still the fundamental question. In this way, Kazuyoshi

---

[413] J. VON NEUMANN: *Proposal and analysis of a numerical method for the treatment of hydrodynamical shock problems*. Rept. 108.1R AMG-IAS No. 1, Applied Mathematics Group, Institute for Advanced Study, Princeton, NJ (1944).

[414] H. GOLDSTINE and J. VON NEUMANN: *Blast wave calculation*. Commun. Pure Appl. Math. **8**, 327-353 (1955).

[415] N.L. JOHNSON: *The legacy and future of CFD at Los Alamos*. Rept. LA-UR-96-1426, LANL, Los Alamos, NM (1996).

[416] W.E. JOHNSON and C.E. ANDERSON: *History and application of hydrocodes in hypervelocity impact*. Int. J. Impact Engng. **5**, 423-439 (1987).

[417] R.J. LEVEQUE: *Numerical methods for conservation laws*. Birkhäuser, Basel *etc.* (1992).

[418] P.L. ROE: *Numerical methods*. In: (G. BEN-DOR ET AL., eds.) *Handbook of shock waves*. Academic Press, New York, vol. 1 (2001), pp. 787-876.

TAKAYAMA,[419] a shock physics professor at the Shock Wave Research Center of Tohoku University in Sendai, initiated an interesting comparison test in 1997 by calculating the Mach reflection of a shock wave that propagates at Mach 2 in air and strikes a 46°- and a 49°-wedge on a computer. This unique international benchmark test, so far supported by eighteen numerical contributions (using different numerical methods) and three experimental contributions (using contact shadowgraphy and holographic interferometry), has shown that the structures of shock waves passing over wedges – a rather difficult problem that has occupied shock physicists for almost fifty years – can indeed successfully be captured by present numerical techniques.

In astrophysics, numerical hydrodynamic simulations have become an indispensable tool for linking observations with theory, thus promoting the understanding of many astrophysical flow phenomena such as shock wave formation in core collapse supernovae and astrophysical jets. Such objects are not accessible to any kind of experimental manipulation, so astrophysicists must rely solely on information they can receive from astrophysical phenomena via electromagnetic radiation (*e.g.*, from radio, Roentgen or optical telescopes), measured particle radiation (*e.g.*, cosmic rays), and magnetic fields. The situation is further complicated by three facts:[420]

- the physical processes that give rise to the astrophysical phenomena may occur deep inside the observed object and are inaccessible to observation;
- these processes may involve extreme conditions which are experimentally inaccessible in the laboratory; and
- such processes often occur on timescales that are much longer than a human life span; *i.e.*, one obtains only a snapshot of the phenomena.

Since the mid-1990s, supercomputers have increasingly been used worldwide to numerically simulate complex 3-D physical phenomena in climate research, materials science and nuclear weapons development. One particular challenge is to investigate events in astrophysics and cosmology using numerical simulation methods, such as the collisions of asteroids and planets, supernova explosions and gamma-ray bursts, collisions of two neutron stars, galaxy and quasar formation, galaxy-galaxy collisions, and convection and magnetic-field generation in the fluid interiors of planets and stars.

In 2003, on the occasion of the 50-year anniversary of coining the term *computer experiment*, Steven STROGATZ,[421] a professor in the Department of Theoretical and Applied Mechanics at Cornell University, wrote in *The New York Times*: "In 1953, Enrico FERMI and two of his colleagues at Los Alamos Scientific Laboratory, Jon PASTA and Stanislaw M. ULAM, invented the concept of a 'computer experiment.' Suddenly the computer became a telescope for the mind, a way of exploring inaccessible processes like the collision of black holes or the frenzied dance of subatomic particles – phenomena that are too large or to fast to be visualized by traditional experiments, and too complex to be handled by pencil-and-paper mathematics. The computer experiment offered a third way of doing science. Over the past 50 years, it has helped scientists to see the invisible and imagine the inconceivable... But perhaps the most important lesson of FERMI's study is how feeble even the best minds are at grasping the dynamics of large, nonlinear systems. Faced with a thicket of interlocking feedback loops, where everything affects everything else, our familiar way of thinking fall apart. To solve the most important problems of our time, we're going to have to change the way we do science."

## 2.10 CONCLUDING REMARKS

*Scientists in the shock wave field are becoming too closely programmed. There is a feeling among productive scientists that a minute in which progress is not made toward some programmed objective is a minute wasted... This philosophy produces a lot of data and a lot of papers, but it has some negative aspects. It deprives the scientist of the perspective required to evaluate what he's doing. It encourages the scientist to go on doing what he has done in the past for too long because study and thought take time. The net result may be a net loss to science... It is dangerous to proceed far without tapping the knowledge of the scientists in the laboratory.*[422]

George E. DUVALL
Washington State University
Pullman, WA 1985

RETROSPECTIVELY, the evolution of percussion, explosion and shock wave research as illustrated here in a rather condensed form has taken such a bizarre and complex

---

[419] K. TAKAYAMA: *Shock wave reflection over wedges: a benchmark test for CFD and experiments*. Shock Waves **7**, 191-203 (1997).

[420] E. MÜLLER: *Simulation of astrophysical fluid flow*. In: (R.J. LEVEQUE and O. STEINER, eds.) *Computational methods for astrophysical fluid flow*. Springer, Berlin (1998).

[421] S. STROGATZ: *The real scientific hero of 1953*. The New York Times (March 4, 2003).

[422] G.E. DUVALL: *Shock wave research: yesterday, today and tomorrow*. In: (Y.M. GUPTA, ed.) *Proc. 4th APS Topical Conference on Shock Waves in Condensed Matter* [Spokane, WA, July 1985]. Plenum, New York (1986), pp. 1-12.

course that predictions of future developments and applications in this field run obvious risks. However, some trends are recognizable. The enormous progress in shock wave and detonation physics achieved within the last 50 years has created a solid foundation of basic knowledge – including shock generation and diagnostic techniques, computational methods of shock analysis, and Hugoniot data on diverse materials and their combinations – which has increasingly also stimulated other disciplines in science and technology. It has created a multitude of interrelations which extend well beyond the historical roots of shock wave physics.[423] Unfortunately, however, this enormous diversity in shockwave-related disciplines has promoted a specialization of knowledge which renders discussions among shock scientists increasingly difficult. For example, gas dynamicists and aeronautical engineers – who are accustomed to working with gases and thinking in terms of mean-free path lengths, viscosity effects, boundary layers, Mach and Reynolds numbers, vortices, *etc.* – can now barely communicate with solid-state shock physicists, who treat shock waves in terms of Hugoniot elastic limits, plastic waves, shear bands, spallation, grain deformation, lattice compression, shock polymorphism, *etc.* Moreover, astrophysicists who studying stellar structures, stellar atmospheres, interstellar material and galaxies in terms of relativity and gravitation, have developed complex mathematical models that describe the physical and nuclear processes in cosmic objects, which are far more complex than the models used in classical gas dynamics and detonation physics.

The ultrarapid nature associated with shock and detonation phenomena have initiated a wide range of new methods for high-speed diagnostics, particularly for high-speed visualization and recording. Obviously, the era of high-speed analog diagnostics is now definitely coming to an end. One of the few exceptions is the pressure gauge technique, which is still analog. Modern shock physics laboratories use digital, mostly menu-controlled measurement equipment, including not only typical electronic measuring instruments (such as pulsers, delay generators, counters, storage oscilloscopes, pressure-gauge amplifiers, *etc.*), but also digital high-speed cameras. This also implies that photographic film is increasingly being replaced by electronic storage, and that in the near future, high-speed and ultrahigh-speed mechanical cameras – ingenious, optomechanical precision instruments and true servants throughout pioneering decades of shock wave research – will only be seen in technical museums. Surprisingly however, the old-fashioned spark light source – which has been used since August TOEPLER's first shock wave studies in the 1860s and widely modified since then in regard to light intensity, spectral emission, pulse duration and repetition frequency – remains an indispensable and reliable piece of equipment that is applied in many shock tube facilities and indoor ballistic test ranges.

Future trends in shock wave physics are definitely heading towards new frontiers, increasingly encompassing both cosmic and microscopic dimensions. The recently taken "baby" picture of the Universe {⇨Fig. 4.1–W} confirmed the Big Bang Theory and the previously estimated age of the Universe. Extraterrestrial shock wave diagnostics incorporated in a 720-kg space probe are heading out of the Solar System: as in April 2007, Voyager 1 (launched in September 1977) was over 101 AU (15 billion km) from the Sun, and had thus entered the heliosheath, the so-called "heliospheric termination shock", also known as the "solar wind termination shock" {Voyager 1 & 2 ⇨2003}, where the supersonic plasma of the solar wind begins to slow down as it encounters the interstellar medium. Astrophysicists expect that the termination shock is responsible for the acceleration of interstellar particles which are ionized in the heliosphere and become charged with energies of the order of 20–300 MeV, a fascinating hypothesis known as the "anomalous cosmic-ray component." At a distance of about 230 AU (34.5 billion km) from the Sun, the Voyager space probes may eventually reach a "bow shock," caused by the heliosphere itself moving supersonically through the interstellar medium. However, the Voyagers will have exhausted their power supply long before this, by around 2020.[424] Indeed, the probing of cosmic shock waves with spacecraft will become a very long-term research program not comparable with any one previously performed in the history of science and – if ever started – it will occupy several generations of astrophysicists.

---

[423] This suggests the idea of introducing a new term for this interdisciplinary field of research (*e.g., superdynamics*) in order to better differentiate such research from the standard (low- and medium-rate) dynamical methods used in classical mechanical engineering. • In mathematical physics, *superdynamics* denotes a method in which certain terms in the equations of motion are replaced by arbitrary functions [Phys. Rev. Lett. **80**, 972-975 (1998)]. A German and a U.S. company have also adopted this term. When the late Karl VOLLRATH and Gustav THOMER (two German scientists that used to work at ISL) edited in 1967 a compendium of articles reviewing the state of high-speed diagnostics used for the study of shock and detonation phenomena, they chose the book title *Kurzzeitphysik* ("Short Time Physics") to emphasize the fact that both shock wave and detonation phenomena are high-speed in nature and, therefore, can both be studied using very similar experimental techniques and mathematical methods. The Ernst-Mach-Institut (EMI) in Freiburg, which investigates a wide range of high-rate phenomena, adopted the subname *Institut für Kurzzeitdynamik* in 1979.

[424] L.A. FISK: *Over the edge?* Nature **426**, 21-22 (2003).

On the other hand, the ultimate goal of materials research is to understand dynamical processes on a microscopic level and to develop mathematical models. More advanced high-speed diagnostics with higher spatial and temporal resolutions may eventually unveil how highly compressed matter with a complex structure (solids and liquids) behaves microscopically directly at and far behind the shock front. Shock compression of solid or porous matter – appropriately called "a physical-chemical-mechanical process" by the Sandia shock physicist Robert GRAHAM[425] – is very complex and is not yet fully understood. Over timescales from hundreds of picoseconds to hundreds of nanoseconds, solids are converted to thermodynamic states in which the deformation is fluid-like to a certain approximation. Shock-induced defects at atomic and microstructural levels might lead to local concentrations of mechanical and kinetic energy, resulting in "hot spots." These defects may play a leading role during the transition from the solid state into fluid-like flow. In addition, changes in chemical composition that occur over ultrashort time durations may introduce substantial complications when analyzing and interpreting the shock-compression process.

In the inanimate realm of simply structured molecules, efforts to understand *high-rate chemical processes* have been pursued for more than a century. Research into shock-induced chemical changes, such as the thermal decomposition of explosive gaseous mixtures, led to the discovery of chain reactions, an important phenomenon in technology, which resulted in the new field of *chemical reaction kinetics*. On the other hand, methods of studying *high-rate physical processes* in the micro world are a challenge to modern dynamic materials diagnostics – such as the fine-structural behavior and rearrangement under the passage of shock waves of crystal lattices, mixtures of fine polycrystalline or even amorphous materials, matrix structures like ceramics, liquids and solids undergoing different types of phase transitions; and porous or multiphase materials. Flash X-ray diffraction experiments have already contributed to a better understanding of the shock loading effects of the unit cell. These exciting studies are presently limited to crystals and microcrystalline substances consisting of elements with low atomic numbers. However, it is quite possible that the "storehouse of creation" (Lord KELVIN) will furnish physicists with more appropriate techniques that will allow this method to be extended to technically relevant high atomic number materials (elements and alloys).

Modern shock wave physics and detonics have also resulted in a broad spectrum of diagnostic methods and instruments for *high-speed visualization* – a rich mine for other branches of science; for example, some of them are now also used in microactuator technology.[426] A microactuator is a device a few micrometers to a few centimeters in size which transforms electrical or laser energy into motion. MEMS (Micro-Electro-Mechanical System) technology has been applied in fluid dynamics for active boundary layer control and drag reduction, and has been proposed for use in ballistics for missile and guidance control {LIPELES & BROSCH ⇨2002}. Microbiology studies that use high-speed visualization to disclose shock-induced rupture effects in laser-irradiated biological microstructures (such as chromosomes) led to the application of this effect as a nano cutting tool (a "nanoscalpel") in biological cell research {KÖNIG ET AL. ⇨1999}. It is quite possible that in the future similar feedback effects will also result from the combination of *microsystems technology* with *gene technology* and *microbiology*, where shock waves generated as a single pulse or repetitively in a limited space at a well-defined strength would play the role of a *microactuator*. Present ideas for applying shock waves in the bioworld include:

▸ the destruction of fatal bacterium and virus strains;
▸ the elimination of cancer cells or the prevention of further cancer cell growth;
▸ gene manipulation in order to eliminate previously incurable hereditary diseases; and
▸ the destruction of white blood cells when treating patients for leukemia, HIV and other diseases, rather than using conventional blood irradiation therapy.

The miniaturization of shock and detonation systems is increasingly discussed in the shock physics community because of numerous promising applications in science, engineering, biology and medical therapy. However, detonation waves as well as shock waves produced by conventional methods in the cm- and m-range cannot directly be scaled down to a microscopic scale, both in regard to their methods of generation and their behavior of propagation and stability. Obviously, the scaling of shock waves {BROUILLETTE ⇨2003} and detonation waves {STEWART ⇨2002} requires new knowledge to be acquired through improved modeling and thoughtful experimentation.

Miniature nanometric shock waves {DLOTT ⇨2000} on a pico-/nanosecond time scale and in a very small shocked volume (a few ng) – so-called "nanoshocks" – can be gener-

---

[425] R.A. GRAHAM: *Shock compression of solids as a physical-chemical-mechanical process.* In: (S.C. SCHMIDT and N.C. HOLMES, eds.) *Shock waves in condensed matter 1987.* North-Holland, Amsterdam *etc.* (1988), pp. 11-18.

[426] P. KREHL, S. ENGEMANN, C. REMBE, and E. HOFER: *High-speed visualization, a powerful diagnostic tool for microactuators – retrospect and prospect.* Microsyst. Technol. **5**, 113-132 (1999).

ated in polymeric and polycrystalline solids by irradiation with ultrashort pulses from a Nd:YLF laser. Coherent Raman spectroscopy during nanoshock propagation in a very thin sample can be used to determine the nanoshock wave form which is characterized by the shock front rise and decay times, shock duration, peak pressure, and velocity. The study of microscopic shock effects on a nanosecond time and nanometer size scale, pointing into a new branch of *Nano-Shock Wave Physics*, may be very useful for a number of important applications in macroscopic shock and detonation physics as well as in nano- and microsystems technology. Examples include the understanding of shock-induced (1) chemical reactions; (2) material transformation and mechanical deformation processes; (3) structural compression and relaxation dynamics in organic polymers and biomolecular materials, *e.g.*, proteins; and (4) initiation of detonation in energetic materials, *e.g.*, chemical high explosives. Such experimental studies involve new theoretical model considerations at nanometer size scale.

A much discussed hypothesis on the possible origin of life is how life became organized by spontaneous generation from nonliving matter – a fascinating subject of research which brings together shock physics on a molecular scale (chemical reaction kinetics) with shock physics on a cosmic scale (physical cosmology). The famous Miller-Urey experiment {S.L. MILLER ⇨1953; ⇨Fig. 4.14–A} gave evidence that amino acids, out of which all life's proteins are made (with few exceptions), can be created by strictly physical-chemical processes, without the help of living organisms. This suggests the idea that on the early Earth the building blocks of life were created from a prebiotic primordial atmosphere under the action of shock and heat. The mix of amino acids found in the well-studied stony Murchison Meteorite {⇨1969} was similar to those produced in Miller-Urey-type experiments. It is generally assumed that the structures of most meteorites were modified on their parent asteroids by a variety of processes including thermal metamorphism and shock metamorphism.

There are still many open questions that laboratory-scale shock wave studies may answer one day. How were these chemical building blocks, essential for life, originally produced? Could shock-induced synthesis have delivered them, for example by meteorite bombardment, in Earth's early history? Future studies in shock wave and impacts physics may play an important role to solve these fundamental questions of biology.

Polycyclic aromatic hydrocarbons (PAHs) are the most complex organic molecules, to that date, found in space {WITT, VIJH & GORDON ⇨2004}. PAH molecules are thought to be widely present in many interstellar and circumstellar environments in our Galaxy as well as in other galaxies. From the interstellar medium (ISM), a dilute gas and dust that pervades interstellar space, stars are born and when they die stars eject gas and dust back into the ISM. Due to this recycling of material, from stars to the ISM and back into stars, the ISM is enriched with more complex materials and molecules {HOYLE ⇨1946}; one such group of molecules is the PAHs which is considered of being important for early life.

The discovery of complex organic molecules in space brings the evolution of organic matter into a context with the evolution of stars which, governed by destructive explosion and implosion processes, transmit their "detonation products" via shock waves to remote distances in the galaxy. The concept of modern science that stellar explosions can create complex organic molecules – and possibly proteins, the building blocs of terrestrial life – would confirm in an amazing manner the dualistic principle perceived by some old religions: the opposing ambiguity of creation and destruction; *i.e.*, of life and death.

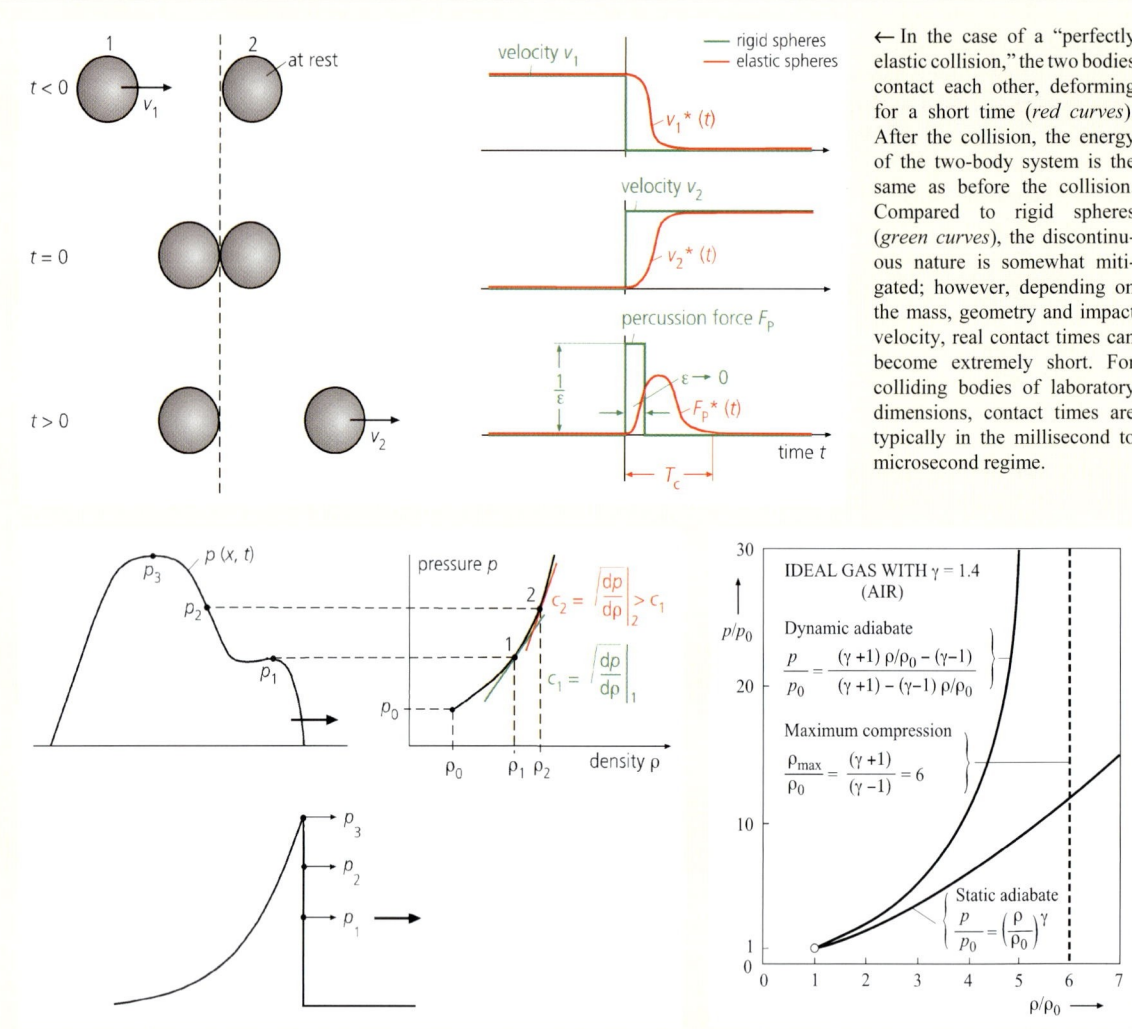

**Fig. 2.1** Illustration of the discontinuity problem in classical percussion and shock wave physics. *Top:* When a rigid sphere *1*, moving with velocity $v_1$ along a straight line, strikes a second sphere *2* of the same mass initially at rest head-on, sphere *1* is immediately halted, thereby transferring all of its kinetic energy to sphere *2*, which begins to move with velocity $v_2 = v_1$. The velocity-time profiles $v_1$ and $v_2$ (*green curves*) are discontinuous and are, in mathematical terms, called "step functions." The percussion force $F_P$ becomes extremely large, and acts during only an extremely short time interval – the so-called "contact time" $T_c$ – which can be described by a delta impulse function. [By the author] *Bottom, left:* In shock waves, the discontinuity in most thermodynamic quantities at the shock front is not immediately present, but instead builds up during its propagation and so needs a certain amount of time to develop. For all normal fluids, the plot of adiabatic compression, $p = p(\rho)$, curves upward; *i.e.*, the velocity of any wave disturbance at pressure $p$ and density $\rho$, measured with respect to coordinates moving with the fluid, is given by $c = (dp/d\rho)^{1/2}$. Therefore regions of higher pressure in the wave-time profile move with a higher velocity, and the ultimate result of this overtaking effect will be to make the front of the shock wave very steep. Typically, pressure rise times at the shock front are very short, on the order of nanoseconds. [After R.H. COLE: *Underwater explosions*. Dover Publ., New York (1965), p. 24] *Bottom, right:* Schematics of equations of state for an ideal gas with $\gamma = 1.4$ (air). Comparison between the isentrope or "static adiabat" and the "dynamic adiabat" which the French engineer Pierre-Henri HUGONIOT derived in the mid-1880s – the so-called "Hugoniot curve" (or the "Hugoniot" for short). Note that the dynamic adiabat lies above the static adiabat – *i.e.*, it requires a higher pressure to compress a gas to a given density ratio $\rho/\rho_0$ using a shock wave than in the static case. [By the author]

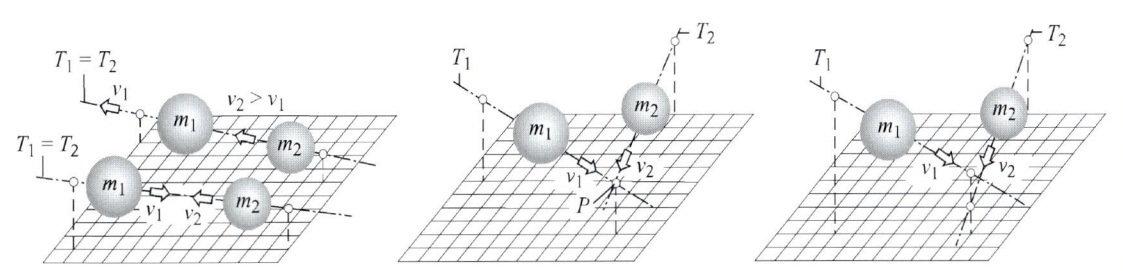

**Fig. 2.2** Collisions can take place in one, two and three dimensions. *Left:* Two examples of a *direct central collision* (or *collinear collision*): bodies *1* and *2* move along the same straight line of impact $T_1 = T_2$, the trajectories of the centers of gravity of masses *1* and *2*. In this case, they can meet traveling either in opposite directions (a "head-on" collision), or in the same direction with velocity $v_2 > v_1$ (a "front-end" collision). *Center: Oblique central collision:* bodies *1* and *2* meet in the same plane, and trajectories $T_1$ and $T_2$ intersect at point *P*. *Right: Oblique eccentric collision:* bodies *1* and *2* move along the trajectories $T_1$ and $T_2$, respectively, which don't intersect. In the special case of the oblique collision of two smooth spheres, there is no tangential force. On the other hand, if friction is present, the translational kinetic energies of the colliding bodies may partly be transformed into rotational kinetic energy and heat, depending on the roughness of the bodies. [By the author]

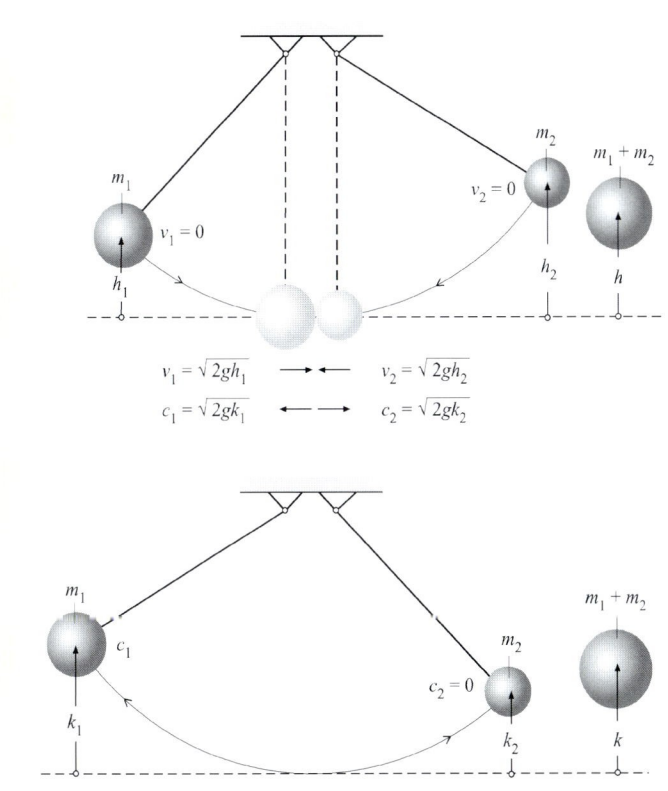

**Fig. 2.3** In 1644, the French natural philosopher René DESCARTES defined force $F$ as the product of mass $m$ and velocity $v$ ($F = mv$) which he called the "quantity of motion." Galileo GALILEI found that a body dropped at height $h$ reaches a velocity $v = (2gh)^{1/2}$. Likewise, a body thrown up vertically with a velocity $v$ reaches a height $h = v^2/2g$. In 1669, the Dutch physicist Christiaan HUYGENS followed GALILEI's concept and recognized that force can also be defined by its ability to surpass resistance such as gravity, hence $F \propto mh$ or $F \propto mv^2$. He postulated by intuition that "the sum of the products of the magnitudes and the squares of the velocities of the bodies before and after impact are always equal," thus anticipating the Law of the Conservation of Kinetic Energy. *Top:* In the case of a twin pendulum with masses $m_1$ at $h_1$ and $m_2$ at $h_2$, the center of mass is given by $h = (m_1 h_1 + m_2 h_2)/(m_1 + m_2)$. *Bottom:* After releasing the balls simultaneously, they collide with velocities $v_1$ and $v_2$. Rebounding with velocities $c_1$ and $c_2$, they reach the heights $k_1$ and $k_2$, respectively. Then the center of the two masses $m_1$ and $m_2$ is given by $k = (m_1 k_1 + m_2 k_2)/(m_1 + m_2) \leq h$. In the ideal case of perfectly elastic impact and no aerodynamic drag, one has $h = k$, which leads to the Law of the Conservation of *Vis Viva* or "living force"): $m_1 v_1^2 + m_2 v_2^2 = m_1 c_1^2 + m_2 c_2^2$. This relation is identical to the Law of the Conservation of Kinetic Energy, given by $½ m_1 v_1^2 + ½ m_2 v_2^2 = ½ m_1 c_1^2 + ½ m_2 c_2^2$. [By the author]

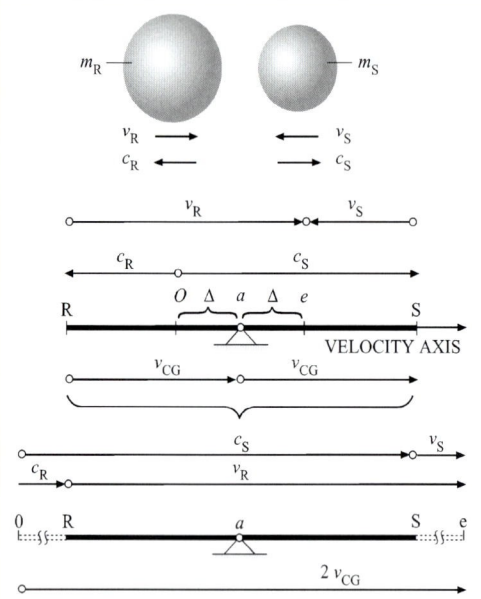

**Fig. 2.4** In 1668, the English mathematician and architect Sir Christopher WREN communicated his memoir "Law of Nature in the Collision of Bodies of Motion" to the Royal Society of London. Considering the case of two perfectly elastic bodies $R$ and $S$ meeting each other along a straight line with given velocities $v_R$ and $v_S$, he proposed an ingenious geometrical solution to find the velocities $c_R$ and $c_S$ after impact. He discovered that "the collision of bodies is equivalent to a balance reciprocating upon two centers equidistant either side of the center of gravity: for the balance may be extended into a yoke when the need arises." WREN started his geometrical method by plotting the (given) velocity vectors $v_R$ and $v_S$ along the velocity axis. The velocity of the center of gravity of the two masses $m_R$ and $m_S$ is given by $v_{CG} = (m_R v_R + m_S v_S)/(m_R + m_S)$. By applying the known length $\Delta = v_R - v_{CG}$ left from the point "$a$," the fulcrum of his balance model, he obtains the point "$0$." The velocities after impact, $c_R$ and $c_S$, are represented by the two distances $0 \to R$ and $0 \to S$ and are given by $c_R = 2v_{CG} - v_R$ and $c_S = 2v_{CG} - v_S$, respectively. Note that since no external forces act upon the system during impact, the velocity of the center of mass is the same before and after impact, hence $v_{CG} = (m_R v_R + m_S v_S)/(m_R + m_S) = (m_R c_R + m_S c_S)/(m_R + m_S)$.

This equation leads to the important Law of the Conservation of Momentum: $m_R v_R + m_S v_S = m_R c_R + m_S c_S$. [After Phil Trans. Roy. Soc. Lond. **3**, 867 (1669); schematic by the author]

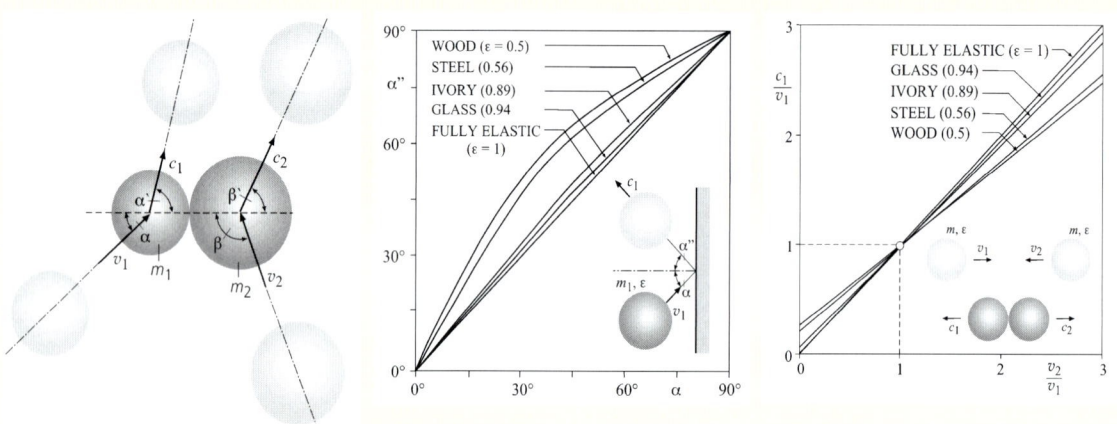

**Fig. 2.5** *Left:* The treatment of the *oblique central collision* of two smooth spheres of masses $m_1$ and $m_2$ moving in the same plane is one of the fundamentals of percussion theory. Because of no friction, the tangential forces at the point of contact are infinitesimal, and the tangential component of the momentum of each ball is conserved; i.e., $v_1 \sin\alpha = c_1 \sin\alpha'$ and $v_2 \sin\beta = c_2 \sin\beta'$. By applying the Conservation Laws of Momentum and Energy for the normal velocity components, and a given coefficient of restitution $\varepsilon$, one obtains $c_1 \cos\alpha' = v_1 \cos\alpha - (v_1 \cos\alpha - v_2 \cos\beta) \times (1+\varepsilon)/(1 + m_1/m_2)$, and $c_2 \cos\beta' = v_2 \cos\beta - (v_2 \cos\beta - v_1 \cos\alpha) \times (1+\varepsilon)/(1 + m_2/m_1)$. [I. SZABÓ: *Einführung in die Technische Mechanik*. Springer, Berlin (1966), p. 372] *Center & right:* There are two special cases of practical interest. For the reflection of a ball of mass $m_1$ against a solid plane wall (*center*) with $m_2 = \infty$ and $v_2 = c_2 = 0$, one obtains $\alpha'' = \arctan(1/\varepsilon \tan\alpha)$ and $c_1 = -v_1 \cos\alpha (\varepsilon^2 + \tan^2\alpha)^{1/2}$. Note that in the special case of a perfect elastic percussion ($\varepsilon = 1$) the mass is reflected at the angle of incidence ($\alpha = \alpha''$) and $c_1 = -v_1$. For central collision (*right*), one obtains $c_1 = v_1 - (v_1 - v_2) \times (1+\varepsilon)/(1 + m_1/m_2)$ and $c_2 = v_2 - (v_2 - v_1) \times (1+\varepsilon)/(1 + m_2/m_1)$. [By the author]

**Fig. 2.6** In 1656, the Dutch natural philosopher Christiaan HUYGENS sent a letter to the French scholar Claude MYLON discussing an interesting three-body percussion problem which has been of enormous importance in practical operations involving hammer/chisel percussion tools since the earliest times. He stated that "a small body striking a larger one, gives it a higher velocity than by direct percussion when a third body of medium size is inserted between both." *Top:* The problem is illustrated in more detail. The small sphere *1* of mass $m_1$ strikes a large sphere *3* of mass $m_3$ directly, which results in a velocity $c_3$. However, when sphere *1* strikes sphere *3* via an intermediate sphere *2* of medium size and mass $m_2 > m_1$, sphere *3* moves with a velocity $c_3^* > c_3$. *Bottom:* As shown in the diagram, the ratio of velocities, $\xi = c_3^*/c_3$, which represents a kind of "gain in percussion," increases as the ratios $m_2/m_1$ and $m_3/m_2$ increase; *i.e.*, the momentum of sphere *3*, given by $m_3 c_3^*$, can be increased by incorporating an additional body *2*, in a manner analogous to placing a chisel between a hammer and work-piece. [By the author]

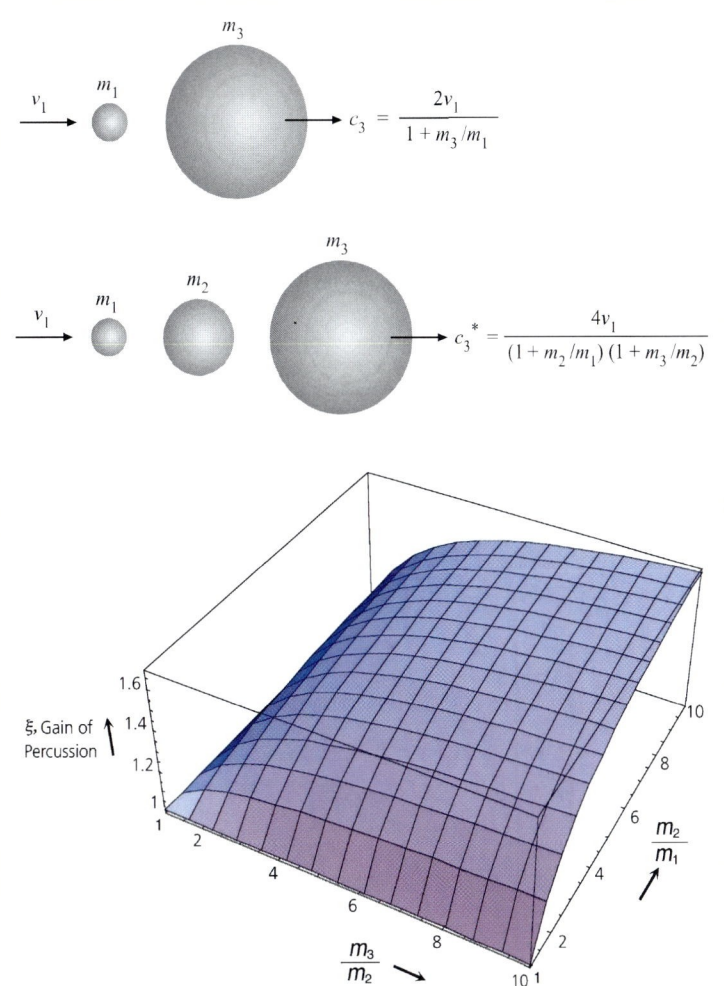

**Fig. 2.7** In the pioneering era of percussion research, the *force of percussion* was a rather obscure quantity and was erroneously interpreted by leading naturalists as either the momentum ($mv$) or the "living force" ($mv^2$). The maximum force of percussion $F_{max}$ can be estimated by providing the impacted body with a short spring and recording its maximum deflection $\Delta x$ during impact. This is demonstrated here for the example of a ball pendulum impacting an ideal short helical spring of stiffness $c$. *Left:* The mass $m$, which is initially at rest ($v = 0$), is released at height $h$. *Center:* At the moment when it touches the spring, the ball has the maximum velocity $v_{max} = (2gh)^{1/2}$, where $g$ is the gravitational acceleration. *Right:* At the moment of maximum spring deflection, the total kinetic energy of the ball, given by $mgh = \frac{1}{2}mv^2$, is completely transformed into the spring energy $\frac{1}{2}c\Delta x^2$; hence $\Delta x = v(m/c)^{1/2}$ and $F_{max} = v(mc)^{1/2}$. Note that for a rigid impact ($c \to \infty$), the maximum force of percussion becomes infinity ($F_{max} \to \infty$). [By the author]

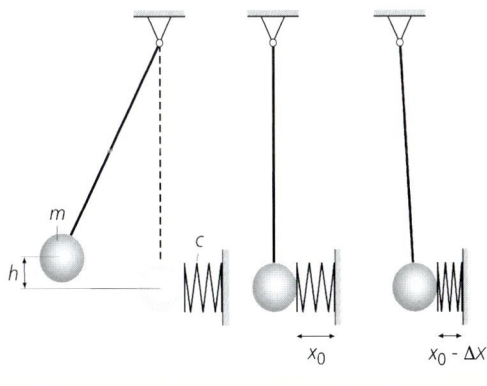

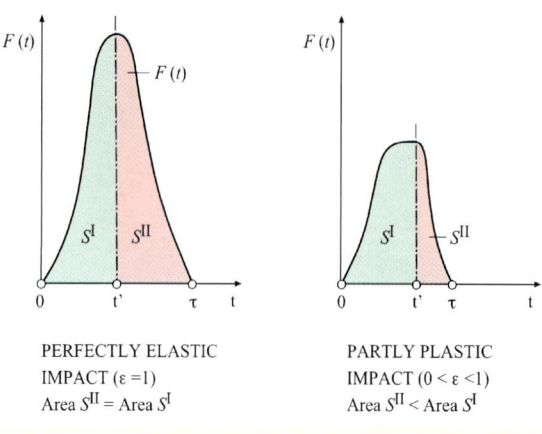

**Fig. 2.8** The force of percussion, particularly the way it changes with time $F(t)$, was the source of much speculation among early researchers, because it was not accessible to measurement using the diagnostics available at the time. In his *Traité de mécanique* (Paris 1833), the French mathematician and physicist Siméon-Denis POISSON distinguished for the case of *elastic percussion* (left) between two phases: the first one begins with the contact of the bodies and ends at the highest compression; the second phase begins at that instant and ends when the bodies begin to separate. There is no loss in kinetic energy, and the areas under $S^I$ and $S^{II}$ for both phases, which represent the partial impulses $\int F(t)dt$, are equal. However, in the case of a *plastic impact* (right), the kinetic energy is partly transformed into heat and, therefore, lost in the restitution process; hence $S^{II} < S^I$. [After P. GUMMERT and K.A. RECKLING: *Mechanik*. Vieweg & Sohn, Braunschweig *etc.* (1986), p. 565]

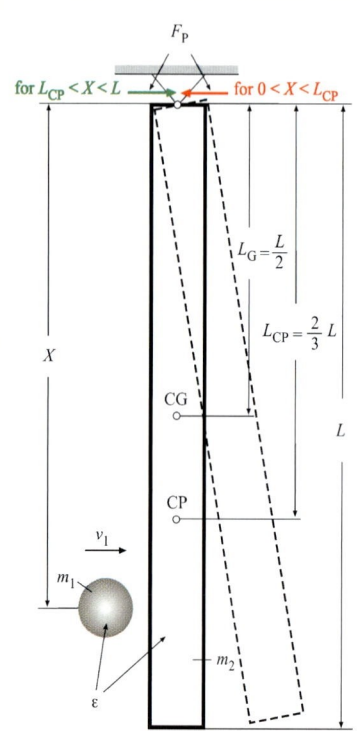

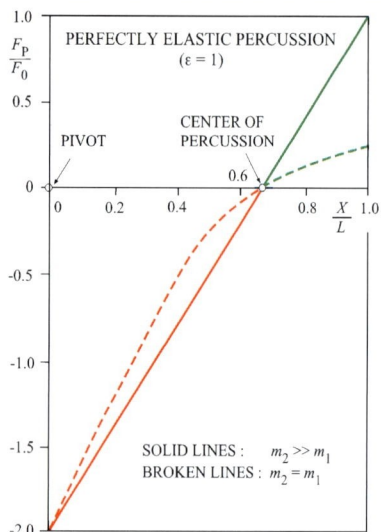

*Left:* A vertical compound pendulum consisting of a cylindrical bar of mass $m_2$ and length $L$ is impacted at distance $X$ from the pivot by a ball of mass $m_1$ moving at a velocity $v_1$. *Right:* Depending on the distance $X$ from the pivot and the coefficient of restitution $\varepsilon$, the impact produces either a positive or a negative force $F_P$ in the pivot, which is given by
$F_P/F_0 = (1 + \varepsilon) \times (\frac{1}{2}X/L - \frac{1}{3})/[\frac{1}{3} + (m_1/m_2)(X/L)^2]$,
where $F_0 = m_1 v_1/\Delta t$ is the impulsive force. Note that at $X = \frac{2}{3}L$ the force $F_P$ becomes zero; *i.e.*, there is no reaction in the pivot.

**Fig. 2.9** In his treatise *Horologium oscillatorium* ("The Pendulum Clock," Muguet, Paris 1673), the Dutch physicist Christiaan HUYGENS described a method that could be used to determine the "center of oscillation" of a compound pendulum – *i.e.*, the point that is vertically below the point of suspension when the pendulum is at rest, at a distance equal to the length of the equivalent simple pendulum. Shortly after, the English mathematician John WALLIS was the first to notice that an impulsive force can be fully transmitted to a freely rotating body when it is struck at its center of oscillation. This will exert no reaction force in the body's pivot, and so the center of oscillation (*CO*) is also called the "center of percussion" (*CP*). Note that the center of gravity (*CG*) is located above *CP*. [By the author]

## 2 GENERAL SURVEY

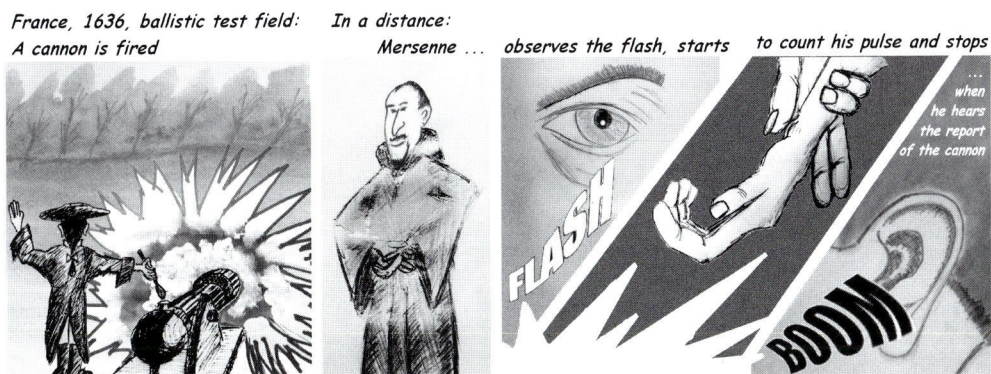

***A:*** In 1636, Marin MERSENNE, a French natural philosopher and priest, determined the velocity of sound in air by firing a cannon and noting the delay between the flash from the muzzle and the arrival of the sound a large distance away – the "flash-to-bang method." Like Galileo GALILEI in his pendulum experiments, he used his own pulse as a clock. Assuming that one pulse beat takes a second, he obtained a velocity for sound of 450 m/s, which was too large. Today we know that the pulse rate of a healthy person ranges between 60 and 80 beats per minute. In 1737, a French commission was set up to determine the velocity of sound precisely, and they obtained a value of 337 m/s using a chronometer. In his *Principia* (1687), Sir Isaac NEWTON calculated a velocity for sound of only 298 m/s, assuming an isothermal equation of state. The considerable discrepancy between theory and experiment stimulated subsequent generations of natural philosophers, until Pierre-Simon DE LAPLACE (1816) solved this puzzle by showing that sound is an adiabatic process.

***B:*** In 1644, MERSENNE showed that the velocity of a musket ball must be of the same order as the velocity of sound, which he believed to be about 450 m/s (*see above*). He observed that a person positioned near a solid target hears the impact of a musket ball at almost the same instant as the report of the musket. Using this simple method of comparison, he placed the speed of larger missiles at about 180–275 m/s. In order to gauge the possible effect of aerodynamic drag, he also calculated that a musket ball with a velocity of 600 ft/s (209 m/s) must displace 14,400 times its own volume of air during each second of its flight. He concluded that the aerodynamic drag increases with the velocity of the projectile, and inversely with its radius and density, thus explaining the relatively long ranges of large cannon shots [A.R. HALLER: *Ballistics in the seventeenth century*. University Press, Cambridge (1952), p. 107]. Note that in MERSENNE's era it was not yet technically possible to determine the velocity of a flying projectile. However, his observation that the sound of the impact with the target was heard at almost the same time as the arrival of the muzzle blast was the first indication that the projectile velocity must be on the order of the velocity of sound.

**Fig. 2.10** Illustration of six pioneering milestones, *A* to *F*, which initiated scientific research into supersonic phenomena. [Cartoons courtesy of Dr. Peter NEUWALD, EMI, Freiburg]

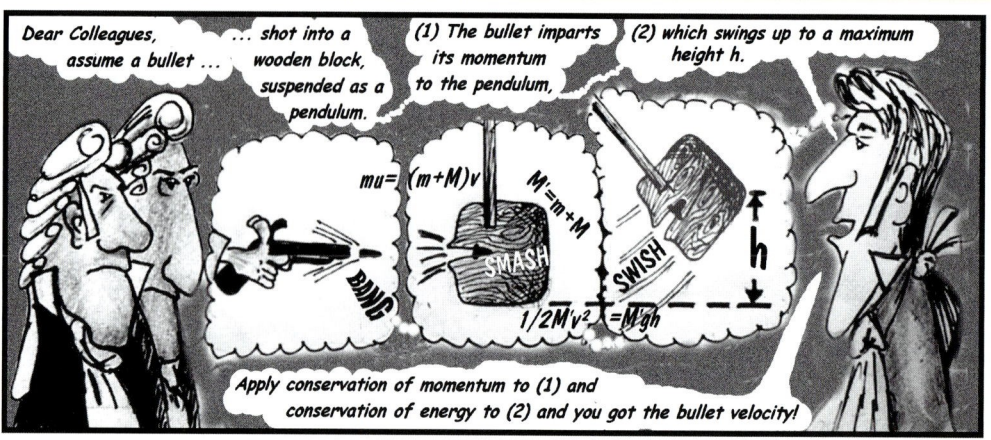

***C:*** In 1707, the Frenchman Jacques CASSINI JR. invented the ballistic pendulum in order to transfer the high velocity of a moving body to a large mass, thus facilitating observation by reducing its velocity. The ballistic pendulum was used in the 1740s in England by the military engineer Benjamin ROBINS, who first measured supersonic velocities (Mach number $M \approx 1.5$) for a musket ball. In the 1780s, his countryman Charles HUTTON, a mathematician and ballistician at Woolwich Royal Military Academy, continued ROBINS' studies. Using a 2-in. (5.08-cm)-caliber cannon and iron balls, he measured supersonic velocities of up to $M \approx 1.87$. It is worth noting that both ballisticians also speculated on aerodynamic drag at high projectile velocities and performed the first quantitative measurements.

***D:*** During the period 1824–1825, the English polar explorers William E. PARRY, Henry FOSTER and James C. ROSS, whilst searching for the Northwest Passage in Northern Canada, performed sound velocity measurements in dry air at Port Bowen. They used a six-pounder brass gun to produce the sound, and measured the time interval between the muzzle flash and the arrival of the report using a pocket chronometer at a distance of 3.9 km. PARRY, who made the measurement, noticed to his great surprise that the officer's command, "Fire," was distinctly heard to occur about one beat of the chronometer *after* the report of the gun on several occasions. From this observation, he concluded that "the velocity of sound depends in some measure upon its intensity." When Samuel EARNSHAW, an English mathematician and Chaplain of the Queen Mary Foundation in the church and parish of Sheffield, presented the first *Theory of Sound of Finite Amplitude* in 1858 at the meeting of the British Association for the Advancement of Science (BAAS), he (correctly) cited PARRY's observation as experimental proof of his theory that intense sound waves must propagate faster than ordinary sound. However, some contemporary physicists such as the Englishman William GALBRAITH sharply refused this interpretation, arguing that PARRY's observation was influenced by the wind and humidity, neither of which, according to Captain PARRY's journal, were recorded properly.

**Fig. 2.10** (cont'd)

2 GENERAL SURVEY

***E:*** In 1851, Samuel EARNSHAW, an English clergyman and natural philosopher, observed that "a thunder-storm which lasted about half an hour was terminated by a flash of lightning of great vividness, which was instantly – *i.e.*, without any appreciable interval between – followed by an awful crash, that seemed as if by atmospheric concussion alone it would crush the cottages to ruins. Every one in the village had felt at the moment of the crash that the electric fluid had certainly fallen somewhere in the village … But, to the surprise of everybody, it turned out that no damage had been done in the village, but that that flash of lightning had killed three sheep, knocked down a cow, and injured the milkmaid at a distance of more than a mile from the village."

In another case illustrated here, EARNSHAW (erroneously) observed that the thunderclap originating from lightning striking a house 5 km away was heard only two seconds after seeing the lightning. From these curious observations, he concluded that intense sound, such as that emitted from a bolt of lightning, must propagate at an enormous supersonic velocity. A few years later, similar observations were also reported by French natural philosophers. His observations, which he communicated to the British Association for the Advancement of Science, increased his interest in intense acoustic waves, eventually resulting in the first mathematical Theory of Shock Waves (1858).

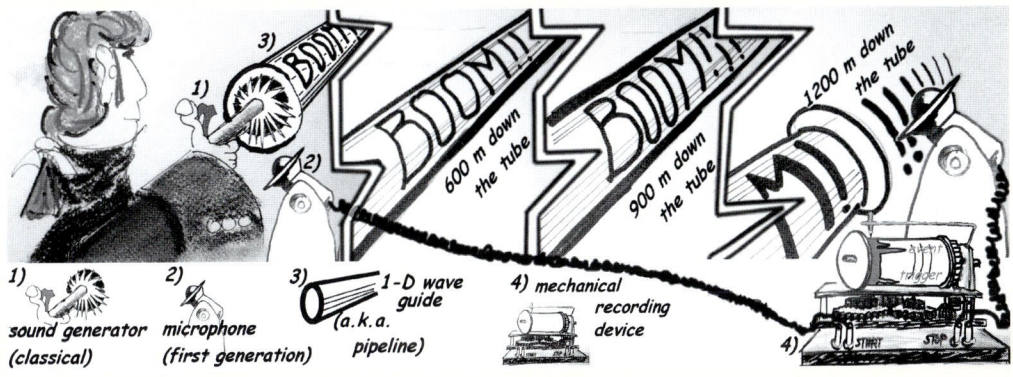

***F:*** In the mid-1860s, the famous French experimental chemist and physicist Henri V. REGNAULT made careful measurements of the velocity of sound in various gases. He noticed that sound propagating in a tube is less attenuated than sound propagating freely in the atmosphere. In order to obtain precise data on the velocity of sound and to compensate for the limited accuracy of his mechanical chronograph, he used long tubes such as the gas pipeline at Ivry-sur-Seine (a town south of Paris), which was 10.8 cm in diameter and 1,150 m long. By simultaneously triggering his drum chronograph and firing a pistol of one gram of powder at the entrance to the a long pipe, he was able to measure the arrival time of the blast wave at the other end of the tube, using a membrane connected to a mechanical contact as a microphone. This allowed him to determine the average propagation velocity of a blast wave in a tube. In a second experiment, he measured the arrival time of a blast wave propagating freely in air, which quickly loses energy through expansion and turns into an ordinary sound wave. Thus, by simply comparing both of the measured velocities for the propagation of sound, he was able to provide the first quantitative experimental proof that the velocity of sound does indeed depend on its intensity, and that intense sound travels faster than weak sound.

**Fig. 2.10** *(cont'd)*

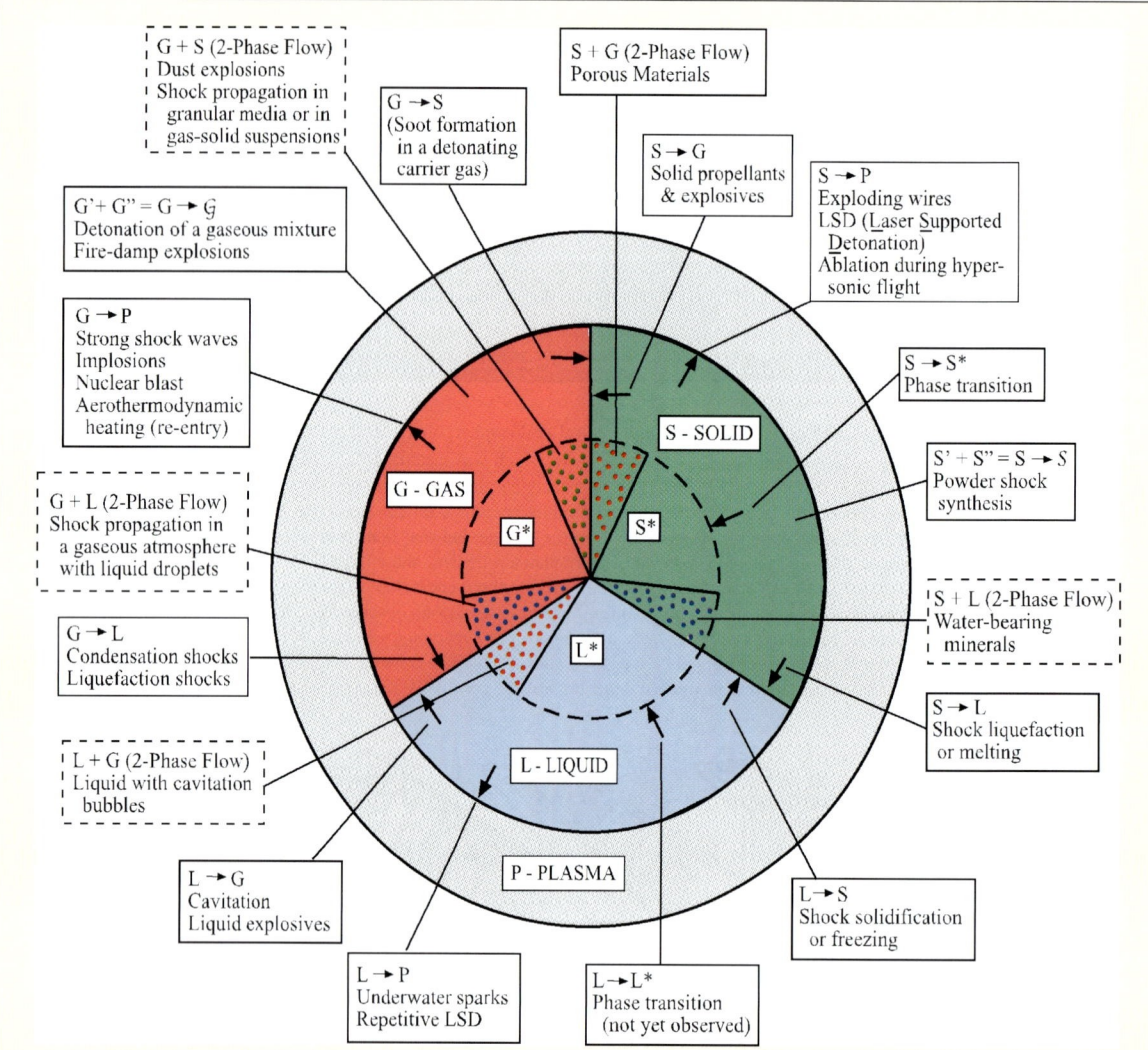

**Fig. 2.11** Classification of shock wave physics in terms of phenomena observed in different states of shock-compressed matter (*green:* S - solid, *blue:* L - liquid, *red:* G - gas, *gray:* P - plasma). Starting from the initial states of matter at rest – G, L and S – shock-induced physical changes are designated by G*, L* and S*, and shock-induced chemical changes by $\mathcal{G}$, $\mathcal{L}$ and $\mathcal{S}$. Historically, equation-of-state studies were initially confined to just a single phase, with no chemical or physical changes involved. In many practical cases, however, shock waves propagate in a multiphase medium, which either already exists when the shock is initially generated (*e.g.,* a blast wave propagating in a coal-dust-loaded atmosphere) or is due to the interaction of the shock with the medium itself (*e.g.,* the generation of a cavitation zone by shock reflection at a solid boundary in an underwater explosion). Over the last few decades, shock propagation and interaction phenomena in two-phase or multiphase flows – here indicated by two-color zones: *red/blue* (gas ↔ liquid), *red/green* (gas ↔ solid) and *green/blue* (solid ↔ liquid) – have become of increasing interest to researchers.
[By the author]

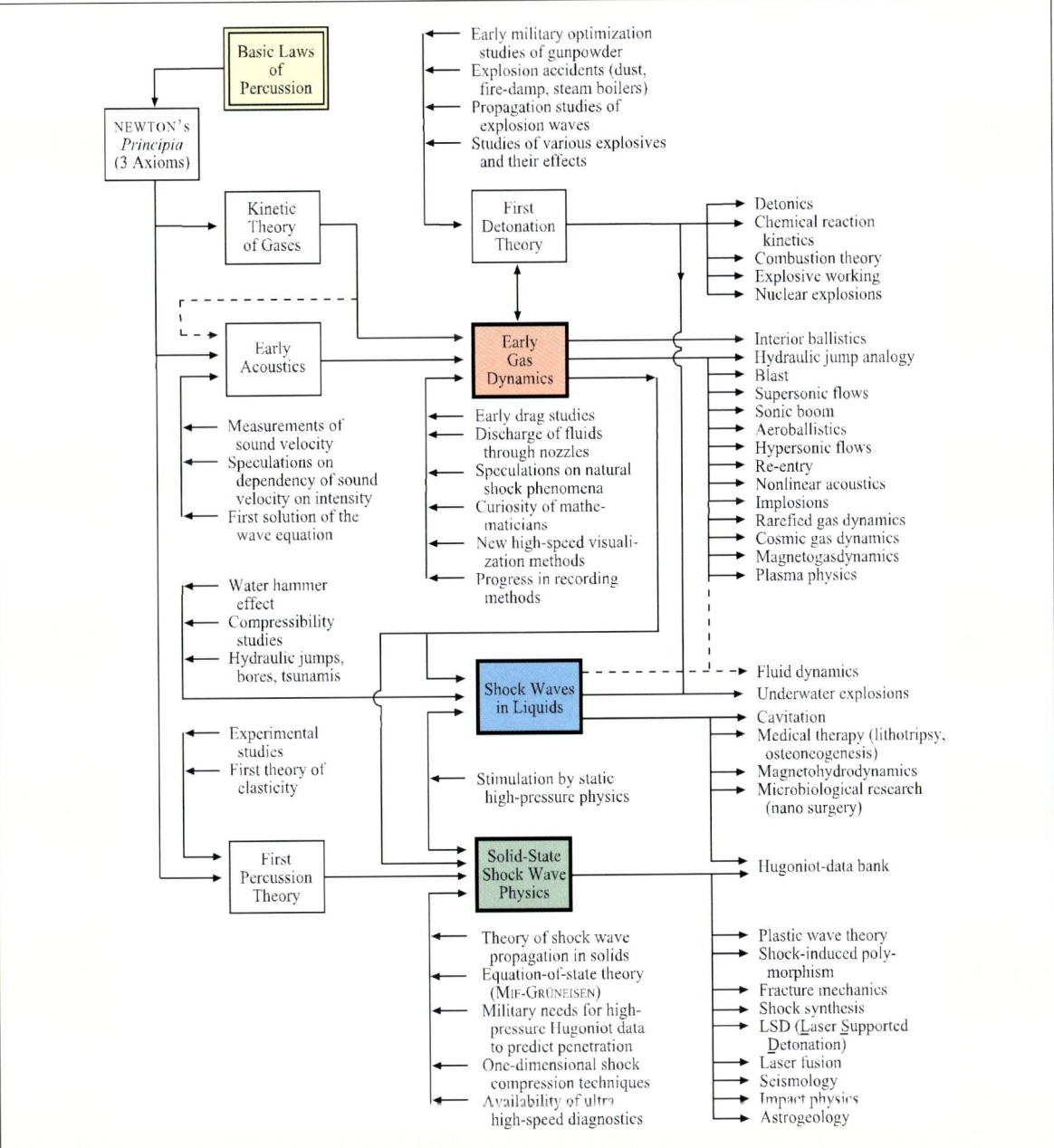

**Fig. 2.12** This is an attempt to, through the use of a flow chart, illuminate the complex evolution of the current wealth of shock-wave-related branches of science and technology, which arose through interactions between the various disciplines. This evolution began in the 17th century with studies of classical percussion. *Shock waves* would now appear to be an inadequate term to use to link the many new branches that have evolved over the last 50 years in this field, perhaps a new and more general term would be more appropriate. [By the author]

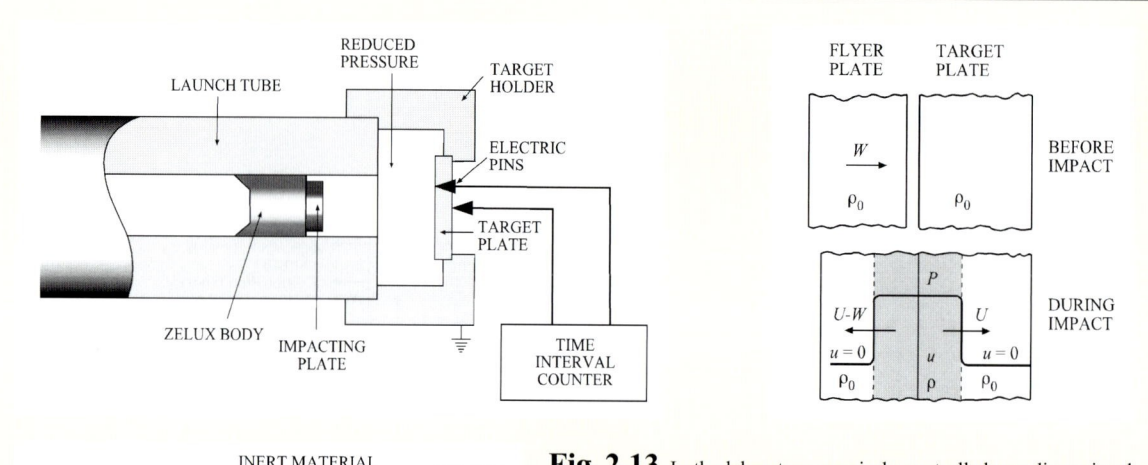

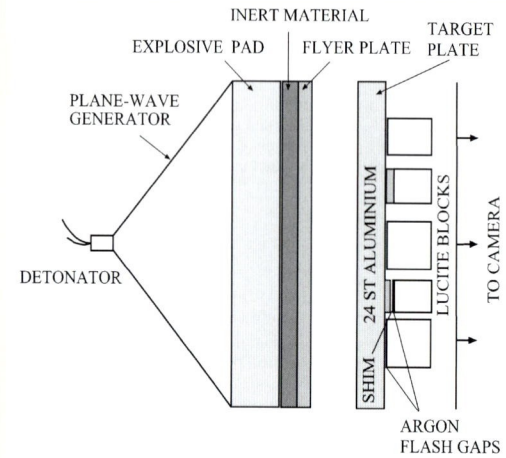

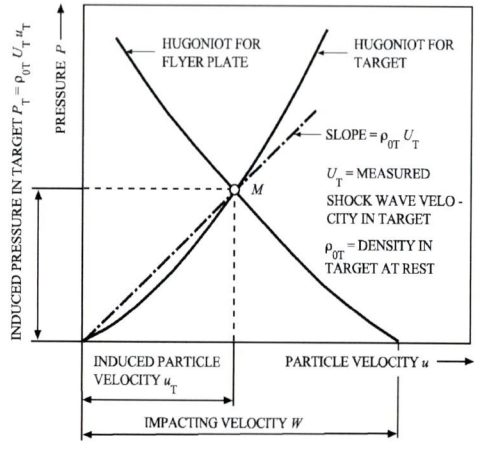

**Fig. 2.13** In the laboratory, precisely controlled one-dimensional shock waves are generally produced through the planar impact of a high-density flat plate moving at high velocity into a stationary plate, the so-called "target plate." *Top, left:* Example of a planar impact using a light-gas gun; here the shock front velocity $U$ in the target plate is measured by electric pins. [After A.H. JONES ET AL.: J. Appl. Phys. **37**, 3493 (1966)] *Center:* Another example of a controlled planar impact is the "flyer plate" method, which uses an explosively driven plate arrangement; $U$ is measured by the flash-gap technique in this case. The schematic shown here relates to the special case that the target material consists of 24 ST aluminum (94% Al), an alloy that is commonly used as a standard target material for determining Hugoniot curves of other solid materials. [After G.E. DUVALL and G.R. FOWLES: *Shock waves*. In: *High Pressure Physics and Chemistry*. Academic Press, New York (1963), vol. 2, p. 209] *Top, right:* Schematic of shock propagation in target and flyer plate shortly after impact, illustrated here for the special case where the flyer and target are made of the same material. Before impact (*top*) the flyer is traveling at a velocity $W$, and the target is at rest. Just after impact (*bottom*), the material on either side of the impact interface has been compressed to a density $\rho$, raised to a pressure $P$, and accelerated to a particle velocity $u$. The pressure as a function of the distance is also shown, superimposed on the schematic. The material at the contact surface is shown propagating into the flyer and the specimen at a velocity $U$ relative to the material, the "shock front velocity." The motion of the contact surface, which creates shock waves in both the flyer plate and the target plate, can be compared to the piston surface of the piston model used in gas dynamics {⇒Fig. 4.4–D}. The standard material most widely used for the flyer plate is the alloy 24 ST aluminum; its Hugoniot has been determined very precisely. [After W.J. MURRI and D.R. CURRAN, Tech. Rept. 001-71, Poulter Lab., SRI, Menlo Park, CA (1971)] *Bottom, left:* Schematic of plane shock waves induced by planar impact. For a "symmetric" impact (*i.e.*, when the flyer and the target are made of the same material) the particle velocity $u$ imparted to the target is exactly one-half of the projectile velocity $W$. The impact produces a wave (traveling to the right) in the target, which is initially in the state $P = 0$, $u = 0$. The stopping shock produced in the flyer plate lies on the reflected Hugoniot through $u = W$, and the direct shock induced in the target lies on the direct Hugoniot curve of the target. The common state produced by impact lies at the intersection $M$ of the two curves. [By the author]

## 2 GENERAL SURVEY

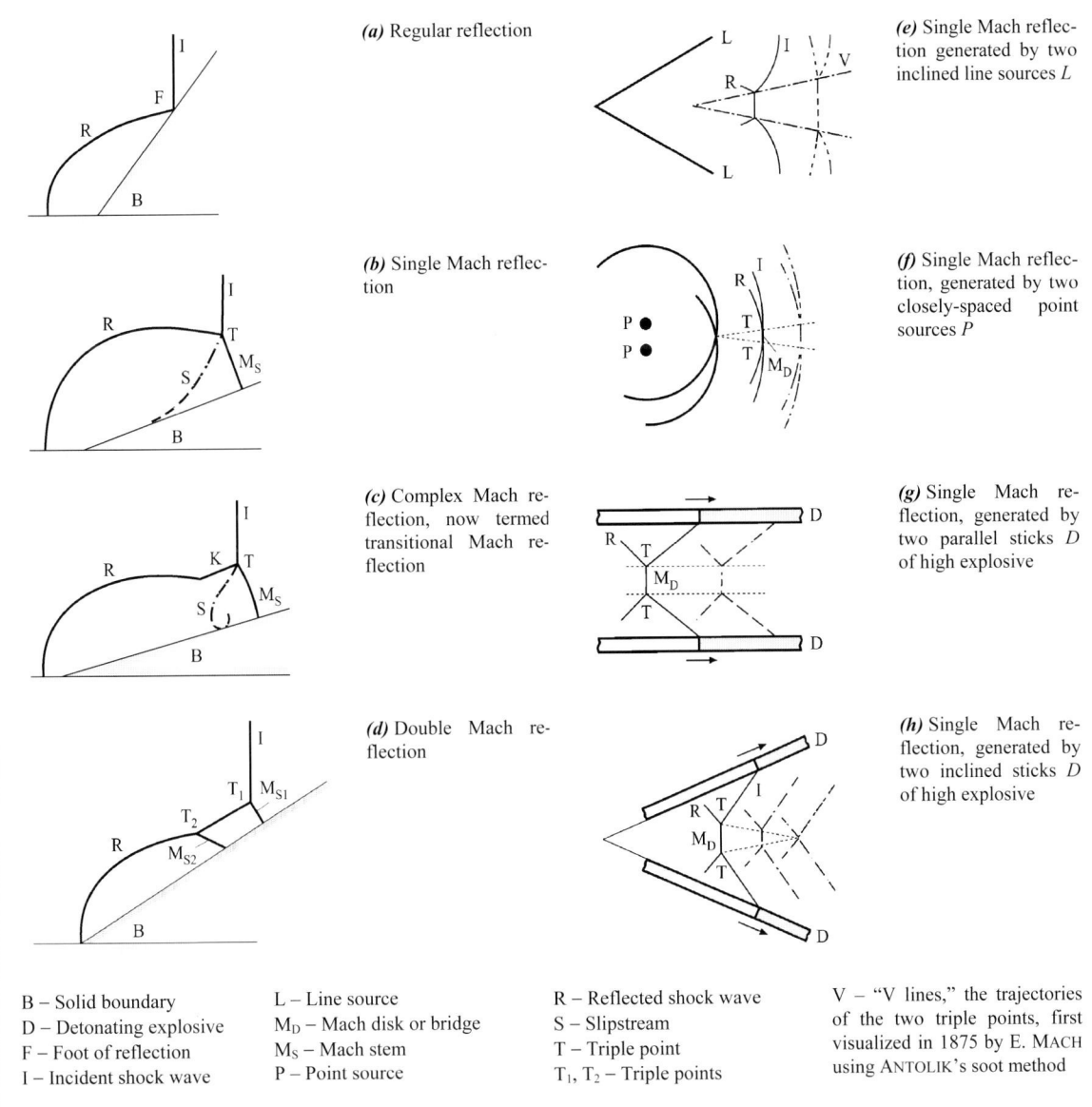

**Fig. 2.14** There exist four basic types of nonstationary oblique shock-wave reflection: *(a) regular reflection* and *(b) single Mach reflection* {E. MACH & WOSYKA ⇨1875}; *(c) complex Mach reflection* or *transitional Mach reflection* {SMITH ⇨1945}, and *(d) double Mach reflection* {WHITE ⇨1951} – the last three types being termed *irregular reflection*. Mach reflection, an important area of shock wave research, has been studied in great detail since the 1940s using either a plane shock wave generated in a shock tube and impinging on a plane wedge, or by the interaction of spherical or cylindrical shock waves emerging from electric sparks, exploding wires, or by the detonation of high explosives. Depending on the selected geometry, one can differentiate between asymmetric Mach reflection phenomena *(a–d)* and symmetric Mach reflection phenomena *(e–h)*. Advantageously, symmetric arrangements don't require a solid boundary and are particularly useful for producing extremely high dynamic pressures. In the cases *(b–d)* and *(e–f)*, the Mach stems and Mach disks, respectively, increase with time ("progressive Mach reflection"). In case *(g)*, the size of the Mach disk remains the same over time ("steady Mach reflection"). Case *(h)*, which depicts a configuration in which the Mach disk decreases in time, is called "inverse Mach reflection" (or "regressive Mach reflection"). [By the author]

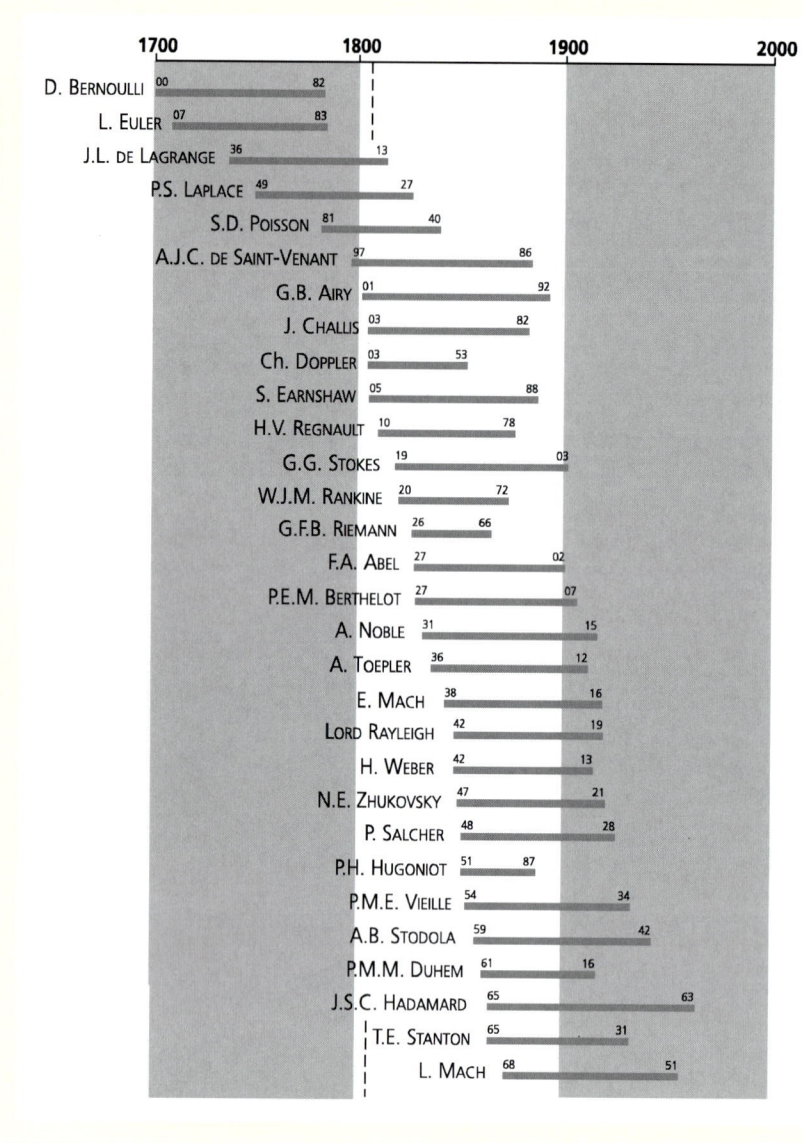

**Fig. 2.15** Life spans of some renowned percussion, explosion and shock wave researchers. The beginning of the *Shock Wave Era*, marked by the vertical broken line, can be attributed to the French mathematician and physicist Siméon-Denis POISSON, who in 1808 was the first to analytically treat "waves in which the velocities of the molecules are not supposed to be very small" {⇨ Fig. 2.17}. [By the author]

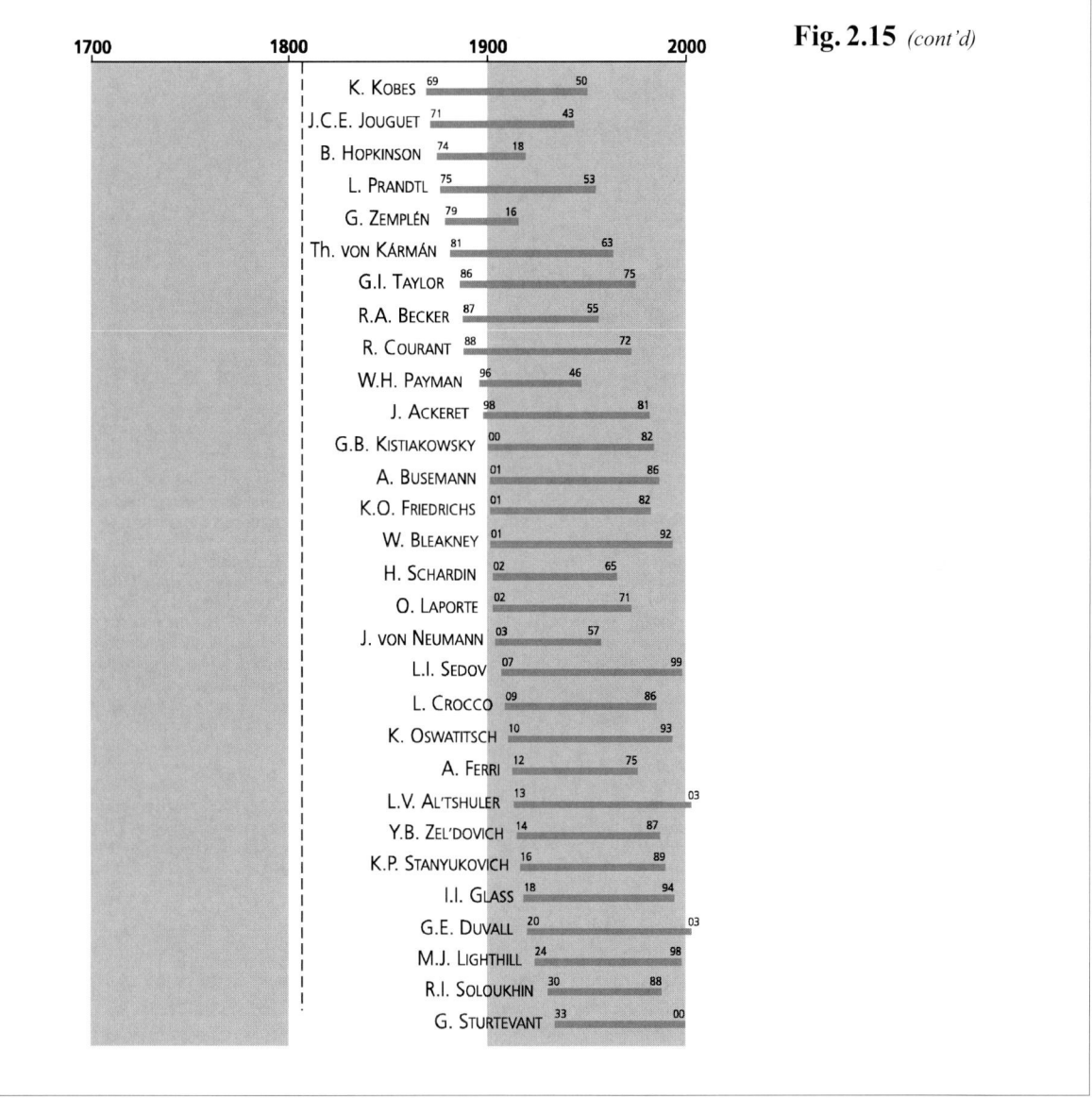

**Fig. 2.15** (cont'd)

## 22  LE IOVRNAL

**EXTRAIT D'VNE LETTRE DE M. HVGENS**
*à l'Auteur du Journal.*

IE vous enuoye, comme i'auois promis, mes propositions touchant le mouuement de percuſſion, c'eſt à dire le mouuement qui eſt produit par la rencontre des corps. Cette matiere a déja été examinée par pluſieurs excellens Hommes de ce ſiecle, comme Galilée, Deſcartes, le P. Fabri, & depuis peu par M. Borelli, deſquels ie ne rapporteray pas maintenant les diuers ſentimens : Mais ie vous diray ſeulement que ma Theorie s'accorde parfaitement auec l'experience, & que ie la crois fondée en bonne demonſtration, comme i'eſpere de faire voir bientoſt en la donnant au public.

*Regles du mouuement dans la rencontre des Corps.*

1. Quand vn corps dur rencontre directement vn autre corps dur, qui luy eſt égal & qui eſt en repos, il luy transporte tout ſon mouuement, & demeure immobile apres la rencontre.

2. Mais ſi cet autre corps égal eſt auſſi en mouuement, & qu'il ſoit porté dans la meſme ligne droite ; ils font vn échange reciproque de leurs mouuemens.

3. Vn corps, quelque petit qu'il ſoit, & quelque peu de viteſſe qu'il ait, en rencontrant vn autre plus grand qui ſoit en repos, luy donnera quelque mouuement.

4. La regle generale pour determiner le mouuement qu'acquierent les corps durs par leur rencontre directe, eſt telle.
Soient les corps A & B, deſquels A ſoit meu auec la viteſſe AD, & que B aille à ſa rencontre ou bien vers le meſme coſté auec la viteſſe BD, ou que meſmes il ſoit en repos, le point D en ce cas étant le meſme que B : Ayant trouué dans la ligne AB le point C 'centre de grauité des corps AB, il faut prendre CE égale à CD, & l'on aura EA pour la viteſſe du corps A apres la rencontre, & EB pour celle du corps B, & l'vne & l'au-

## DES SCAVANS.  23

tre vers le coſté que montre l'ordre des points EA, EB: Que s'il arriue que le point E tombe en A ou en B, les corps A ou B feront reduits au Repos.

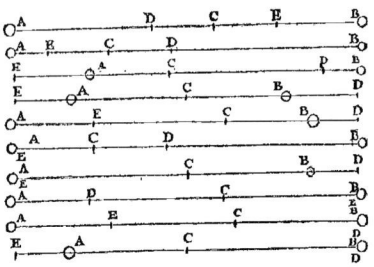

5. La quantité du mouuement qu'ont deux corps, ſe peut augmenter ou diminuer par leur rencontre ; mais il y reſte toujours la meſme quantité vers le meſme coſté, en ſouſtrayant la quantité du mouuement contraire.

6. La ſomme des produits faits de la grandeur de chaque corps dur, multiplié par le quarré de ſa viteſſe, eſt toujours la meſme deuant & apres leur rencontre.

7. Vn corps dur qui eſt en repos, receura plus de mouuement d'vn autre corps dur plus grand ou moindre que luy, par l'interpoſition d'vn tiers de grandeur moyenne, que s'il en eſtoit frappé immediatement : Et ſi ce corps interpoſé eſt moyen proportionel entre les deux autres, il fera le plus d'impreſſion ſur celuy qui eſt en repos.

Ie conſidere en tout cecy des corps d'vne meſme matiere, ou bien i'entends que leur grandeur ſoit eſtimée par le poids.

Au reſte i'ay remarqué vne loy admirable de la Nature, laquelle ie puis démontrer en ce qui eſt des corps Spheriques, & qui ſemble eſtre generale en tous les autres

**Fig. 2.16** In 1669, the Dutch physicist Christiaan HUYGENS published his *Règles du mouvement dans la rencontre des corps* ("Laws of Motion on the Impact of Bodies") for the first time in the French *Journal des Sçavans*. An English translation of his rules can be found in the CHRONOLOGY {HUYGENS ⇨1669}. His most important result is rule 6 (*right*), which says that "*the sum of the products of the magnitudes and the squares of the velocities of the bodies before and after impact are always equal*" – this represents the birth of the Law of the Conservation of Kinetic Energy. In the same year, HUYGENS also published the same article in Latin (which was the international language among scholars at that time) in the English journal *Philosophical Transactions*. His important contributions to the laws governing impact made his work a cornerstone of classical mechanics. [Journal des Sçavans (Paris) **5**, 22-24 (March 18, 1669)]
• The *Journal des Sçavans*, the earliest scientific journal, was established in January 1665 in Paris. The figure shows facsimiles of the first two pages of HUYGENS' paper at a somewhat reduced scale. Note the small format of this journal. The first issues of this journal were reprinted in 1679 in Amsterdam at an even smaller scale: they had a page size of only $13 \times 7$ cm$^2$.

364　　　　　　　　ANALYSE.

que *Newton* a remarquée, le premier, entre la vitesse du son donnée par le calcul, et celle qui résulte de l'observation.

*Mouvement d'une Ligne d'air dans le cas où les Vitesses des molécules ne sont pas supposées très-petites.*

[23.] EN supposant que la masse d'air se réduise à une simple ligne, et en prenant cette ligne pour l'axe des abscisses $x$, les quantités $p$ et $\varphi$ ne seront plus fonctions que de $x$ et $t$; et si l'on fait abstraction de la pesanteur, les équations (a) et (b) du n.° 1 deviendront

$$\int \cdot \frac{dp}{p} + \frac{d\varphi}{dt} + \frac{1}{2} \cdot \frac{d\varphi^2}{dx^2} = 0,$$

$$\frac{d\rho}{dt} + \rho \cdot \frac{d^2\varphi}{dx^2} + \frac{d\rho}{dx} \cdot \frac{d\varphi}{dx} = 0.$$

Si l'on suppose de plus la température constante dans toute la ligne d'air, la pression $p$ sera proportionnelle à la densité $\rho$, et l'on aura $p = a^2 \cdot \rho$, $a$ étant un coefficient constant qui représente la racine carrée du rapport de l'élasticité de l'air à la densité, avant que le mouvement ait commencé. La première de ces équations deviendra donc

$$a^2 \cdot \log \frac{\rho}{D} + \frac{d\varphi}{dt} + \frac{1}{2} \cdot \frac{d\varphi^2}{dx^2} = 0, \quad (1)$$

$D$ étant la densité de l'air avant le mouvement; éliminant $\rho$ entre celle-ci et l'équation précédente, on trouve

$$\frac{d^2\varphi}{dt^2} + 2\frac{d\varphi}{dx} \cdot \frac{d^2\varphi}{dx\,dt} + \frac{d\varphi^2}{dx^2} \cdot \frac{d^2\varphi}{dx^2} = a^2 \frac{d^2\varphi}{dx^2}. \quad (2)$$

L'intégrale complète de cette équation est impossible sous forme finie. Il est facile de s'en assurer, en lui appliquant la méthode que M. *Laplace* a donnée pour intégrer les équations aux différences partielles du second ordre, après l'avoir ramenée à la forme linéaire, que cette méthode suppose, au moyen d'une transformation connue que l'on doit à M. *Legendre*. Cette transformée linéaire est comprise parmi les équations que M. *Laplace* a intégrées, par le moyen d'une intégrale

**Fig. 2.17** Facsimile of the page of Siméon-Denis POISSON's *Mémoire sur la théorie du son* ("Memoir on the Theory of Sound") in which he first addresses the special case that "the velocities of the molecules are *not* very small." His result – essentially a simple wave solution of the differential equation of flow in an isothermal gas – stimulated subsequent researchers to work in this area. The first was James CHALLIS {⇨1848}, who observed that POISSON's solution cannot always be solved uniquely for a given velocity, followed by Sir George G. STOKES {⇨1848}, who was the first to apply both the Law of Conservation of Mass and the Law of Conservation of Momentum to the problem, thus deducing two discontinuity conditions for an isothermal gas. [J. Ecole Polytech. 7, 319 (1808)]

# JOURNAL
DE
# L'ÉCOLE POLYTECHNIQUE.

MÉMOIRE

SUR LA PROPAGATION DU MOUVEMENT DANS LES CORPS

ET SPÉCIALEMENT DANS LES GAZ PARFAITS ($^1$);

Par H. HUGONIOT,

Répétiteur de Mécanique à l'École Polytechnique.

## PREMIÈRE PARTIE.

INTRODUCTION.

**1.** Les questions qui font l'objet de ce Mémoire intéressent aussi bien l'Analyse que la Physique mathématique. Les mouvements des corps naturels sont régis par des équations aux dérivées partielles dont la théorie est encore aujourd'hui bien obscure. Tout progrès fait dans l'étude de ces mouvements fournit un renseignement nouveau sur la nature de ces

---

($^1$) Ce Mémoire est textuellement celui qui a été présenté à l'Académie des Sciences par M. Hugoniot et déposé au Secrétariat de l'Institut le 26 octobre 1885.

L'Auteur, enlevé par une mort prématurée, n'a pu apporter à sa rédaction primitive les modifications et les compléments qu'il avait, paraît-il, l'intention d'y introduire encore; mais, tel qu'il est, ce Mémoire suffira pour faire apprécier au lecteur le talent élevé que possédait Hugoniot et l'étendue de la perte que la Science a faite en sa personne.

Voulant respecter la pensée de l'Auteur, nous n'avons fait subir aucune retouche à son travail; nous nous sommes bornés à le diviser en deux Parties, en vue des nécessités de l'impression : la seconde Partie sera insérée dans un prochain Cahier.

Dans celle que nous publions aujourd'hui, le lecteur reconnaîtra plusieurs résultats qui ne sont pas nouveaux; c'est la preuve, non d'un défaut d'érudition, mais du souci constant qu'avait l'Auteur de présenter, dans chacun de ses Mémoires, tout ce qui était nécessaire pour le comprendre en son entier. (*Note de la Rédaction.*)

**Fig. 2.18** Facsimile of the first page of Pierre-Henri HUGONIOT's first part of his famous memoir *Sur la propagation du mouvement dans les corps et plus spécialement dans les gaz parfaits* ("On the Propagation of Motion in Bodies, and Especially in Perfect Gases"), which was published posthumously in 1887, the year of his death. Note the footnote, the first two paragraphs of which read as follows: "This memoir is textually identical to the one that was presented to the Academy of Science by Mr. HUGONIOT and was given in to the Secretariat of the Institute on October 26, 1885. The author, who died prematurely, was not able to bring to his primitive work the modifications and complements that he had intended, it appears, to bring; but such as it is, this essay will be enough to make the reader appreciate the high capacity that HUGONIOT possessed and the immensity of his loss to science. Out of respect for the author's thoughts, we have not made any revisions to his work…" [J. Ecole Polytech. 57, 3-97 (1887)]

88      H. HUGONIOT.

qui ne sont autres que celles des paragraphes précédents, où l'on a changé le signe de $\frac{dx}{dt}$.

La deuxième de ces relations montre que $v - v_1$ et $z_1 - z$ sont finies en même temps. Ainsi toute variation brusque de vitesse entraîne pour la tranche une variation brusque de dilatation.

On déduit d'ailleurs de ces relations

$$p_1 = p\, \frac{2(1+z) - (m-1)(z_1-z)}{2(1+z) + (m-1)(z_1-z)},$$

formule d'une haute importance qui permet de calculer la pression finale au moyen des données initiales et de la variation survenue dans la dilatation.

La formule peut être mise encore sous une autre forme. Soient $\rho$ et $\rho_1$ les densités correspondant aux dilatations $z$ et $z_1$; il est visible que

$$p_1 = p\, \frac{(m+1)\rho_1 - (m-1)\rho}{(m+1)\rho - (m-1)\rho_1}$$

ou

$$\boxed{p_1 = p\, \frac{(m+1)\frac{\rho_1}{\rho} - (m-1)}{(m+1) - (m-1)\frac{\rho_1}{\rho}}.}$$

Le rapport des pressions est ainsi une fonction rationnelle du rapport des densités.

La pression $p_1$ est essentiellement positive, ce qui fixe des limites que ne peut dépasser le rapport $\frac{\rho_1}{\rho}$.

Supposant d'abord ce rapport supérieur à 1, ce qui arrive quand la variation brusque a pour effet de comprimer la tranche, le numérateur est toujours positif; il faut qu'il en soit de même du dénominateur.

Donc

$$\frac{\rho_1}{\rho} < \frac{m+1}{m-1}.$$

Si, au contraire, la variation brusque a pour effet de dilater la tranche,

**Fig. 2.18** *(cont'd)* Facsimile of page 88 from HUGONIOT's second part of his great memoir in which he derived his famous "dynamic adiabat" $p = p(\rho_1/\rho)$ for an ideal gas – today known as the "Hugoniot" (short for "Hugoniot curve"), which is given by the highlighted equation. He used the letter $m$ (instead of $\gamma$) for the ratio of specific heats, $c_P/c_V$. Note that in the $(p,\rho)$-plane the dynamic adiabat is steeper than the static adiabat, as illustrated in Fig. 2.1. [J. Ecole Polytech. **58**, 88 (1889)]

# 3

## CHRONOLOGY

# 3 CHRONOLOGY

**GUIDE TO USING THE CHRONOLOGY**

- This table covers all kinds of high-rate phenomena, such as percussion, blast, shock waves, explosions, implosions and impact as well as milestones in the development of basic high-speed diagnostic techniques.
- Years marked with the symbol "⇨", *e.g.* {⇨1803}, refer to other milestones in the CHRONOLOGY. Figures marked with "⇨", *e.g.* {⇨Fig. 2.6} and {⇨Fig. 4.5–A}, refer to the GENERAL SURVEY (Chap. 2) and the PICTURE GALLERY (Chap. 4), respectively.
- Notes following the symbol "•" refer to later achievements and/or comments on the milestone from today's point of view.
- Notes in the citations given in "[…]" are amendments made by the author.
- Ga [Lat. *giga-annum*] and Ma [Lat. *mega-annum*] are units of time equal to one billion ($10^9$) years and one million ($10^6$) years, respectively. They are commonly used in scientific disciplines to signify very long time periods in the past.

# NATURAL SHOCK, EXPLOSION & IMPACT PHENOMENA

## TERRESTRIAL:

▸ **Animal world: percussion tool-using Egyptian vulture and California sea otter** {Sect. 1; ⇨Fig. 4.1–Z}, **"pulse-jet" bombardier beetle** {SCHNEPF & WENNEIS ⇨1969}, **supersonic botfly?** {TOWNSEND ⇨1927}; **supersonic tail cracking of giant sauropod dinosaurs?** {ALEXANDER ⇨1989}, **underwater shock wave producing snapping shrimp** {LOHSE ET AL. ⇨2000; ⇨Fig. 4.1–Z}.

▸ **Asteroid and meteorite impacts** {Vredefort Basin ⇨c.2 Ga ago; Sudbury Basin ⇨c.1.85 Ga ago; Chicxulub Crater ⇨c.65 Ma ago; Popigai Crater ⇨35.7 Ma ago; Ries Basin & Steinheim Basin ⇨c.15 Ma ago; Meteor Crater ⇨c.50,000 years ago; Ensisheim Meteorite ⇨1492; L'Aigle Fall ⇨1803; Krähenberg Meteorite ⇨1869; Sikhote-Alin Meteorite ⇨1947; Allende Meteorite ⇨1969; Murchison Meteorite ⇨1969; Allan Hills Meteorite 1977, *see* CARTER & KENNEDY ⇨1964, CLARKE JR. ET AL. ⇨1981, and Allan Hills ⇨1984}.

▸ **Ball lightning?** {ARAGO ⇨1838; FARADAY ⇨1841; Lord KELVIN ⇨1888; RAÑADA ET AL. ⇨1998}.

▸ **Barisal guns (or "fog" guns)?** {Barisal ⇨1870; G.H. DARWIN ⇨1895}.

▸ **Bores** {CH'UNG ⇨c.85; NENNIUS ⇨c.800; RUSSELL ⇨1834; G.H. DARWIN ⇨1898; WHITHAM ⇨1958}.

▸ **Collision (interaction) of two bores** {NENNIUS ⇨c.800; CORNISH ⇨1910; ⇨Fig. 4.1–M}.

▸ **Collision of two lithospheric plates** {Sumatra-Andaman Islands ⇨2004}.

▸ **Collisional orogeny** {*see* Sect. 2.1.1}.

▸ **Earthquakes** {MARCELLINUS ⇨A.D. 365; Genroku Kanto ⇨1703; Kamchatka Peninsula ⇨1737; Comte DE BUFFON ⇨1749; England ⇨1750; Lisbon ⇨1755; Boston ⇨1755; Messina 1783, *see* SPALLANZANI ⇨1798 and MALLET ⇨1846; Coast of Yorkshire ⇨1856; Naples ⇨1857; Arica ⇨1868; Iquique ⇨1877; Mt. Bandai-san ⇨1888; Tokyo 1889, *see* VON REBEUR-PASCHWITZ ⇨1889; Assam 1897, *see* Count BERNARD ⇨1907; San Francisco ⇨1906; Atacama 1922, *see* GUTENBERG ⇨1939; Taisho Kanto 1923, *see* Genroku Kanto ⇨1703; Unimak Island ⇨1946; Kern County, CA 1952, *see* GUTENBERG ⇨1955; Lituya Bay ⇨1958; Chile ⇨1960; Denver ⇨1962; Alaska ⇨1964; Kunlun Shan ⇨2001; Basel 2006, *see* Denver ⇨1962}.

▸ **Exploding comet or asteroid?** {Stony Tunguska ⇨1908}.

▸ **Exploding meteors** {CHLADNI ⇨1794; L'Aigle Fall ⇨1803; Los Alamos ⇨1998; Northern Germany ⇨1999}.

▸ **Fulgurites** {WITHERING ⇨1790; BEUDANT ET AL. ⇨1828; DARWIN ⇨1832; SPENCER ⇨1932; MCCOLLUM ⇨1997}.

▸ **Impact metamorphism** ➡ **Shock metamorphism**.

▸ **Rogue waves (or "freak waves")** {Workshop on Rogue Waves ⇨2000}.

▸ **Seaquakes** {Lisbon ⇨1755; MALLET ⇨1846; KRÜMMEL ⇨1887; RUDOLPH ⇨1887; ROTTOK ⇨1890; Kamaishi ⇨1896; Okushiri ⇨1993; Sumatra-Andaman Islands ⇨2004}.

▸ **Seiches** {KANT ⇨1756; FOREL 1901, *see* KANT ⇨1756}.

▸ **Shatter cones** {Sudbury Basin ⇨c.1.85 Ga ago; Popigai Crater ⇨c.35.7 Ma ago; BUCHER ⇨1933; BOON & ALBRITTON ⇨1938; DIETZ ⇨1946 & 1960; SHOEMAKER ET AL. ⇨1961; SAGY ET AL. ⇨2004}.

▸ **Shock metamorphism** {FOOTE ⇨1891; FRIEDEL ⇨1892; COES ⇨1953; CHAO ET AL. ⇨1960; STISHOV & POPOVA ⇨1961; CARTER & KENNEDY ⇨1964; 1st Conf. on Shock Metamorphism of Natural Materials ⇨1966; FRENCH ⇨1998}.

- ▸ **Supershear (or intersonic) rupture earthquakes?** {San Francisco Earthquake ⇨1906; ROSAKIS ET AL. ⇨1999; Kunlun Shan Earthquake ⇨2001}.
- ▸ **Supersonic tornadoes?** {FUJITA ⇨1981}.
- ▸ **Tektites of terrestrial origin?** {C.R. DARWIN ⇨1844; O'KEEFE ⇨1967 & 1976; CHAPMAN ⇨1971}.
- ▸ **Thunder** {ARISTOTLE ⇨303 B.C.; SENECA ⇨A.D. 63; DESCARTES ⇨1637; MERSENNE ⇨1644; MARIOTTE ⇨1673; LÉMERY ⇨1700; WALL ⇨1708; COLLINSON ⇨1749; KRÜGER ⇨1756; BENNET ⇨1789; VON HELVIG ⇨1815; ARAGO ⇨1838; EARNSHAW ⇨1851; MONTIGNY ⇨1860; HIRN ⇨1860; RAILLARD ⇨1860; UMAN ⇨1970; BASS ⇨1980; GRANEAU ⇨1989}.
- ▸ **Tsunamis** {MARCELLINUS ⇨365; Genoku Kanto ⇨1703; Kamchatka Peninsula ⇨1737; Lisbon ⇨1755; KANT ⇨1756; MICHELL ⇨1760; Volcano Unzen-dake ⇨1792; Mt. Tambora ⇨1815; Coast of Yorkshire ⇨1856; Arica ⇨1868; Iquique ⇨1877; Krakatau ⇨1883; RUDOLPH ⇨1887; Kamaishi ⇨1896; MILNE ⇨1898; Count BERNARD ⇨1907; GUTENBERG ⇨1939; Unimak Island ⇨1946; IMAMURA ⇨1949; Myojin-sho Reef ⇨1952; Lituya Bay ⇨1958; Valdivia ⇨1960; Prince William Sound ⇨1964; Mt. St. Helens ⇨1980; Okushiri ⇨1993; Sumatra-Andaman Islands ⇨2004}.
- ▸ **Volcanic explosive eruptions** {Fish Canyon Tuff ⇨$c$.28 Ma ago; Yellowstone Caldera ⇨$c$.2 Ma ago; Mt. Toba ⇨$c$.71,000 years ago; Santorini Island ⇨$c$.1645 B.C.; Mt. Vesuvius ⇨A.D. 79 & 1944; Mt. Tambora ⇨1815; Krakatau ⇨1883; Mt. Bandai-san ⇨1888; Mt. Vulcano ⇨1888; Mt. Pelée ⇨1902; Stromboli ⇨1930; Mt. Kilauea-Iki ⇨1959; Taal Volcano ⇨1965; Mt. St. Helens ⇨1980; Mt. Pinatubo ⇨1991}.

**EXTRATERRESTRIAL:**

- ▸ **Accretion shock waves** {⇨Sect. 2.3.1}.
- ▸ **Astrophysical jets** {CURTIS ⇨1918; SEYFERT ⇨1943/1944; HERBIG & HARO ⇨1950s; COURTÈS & CRUVELLIER ⇨1961; SPIEGEL ⇨1978; Supermassive Black Hole ⇨1994} **and terminal jet shock** {M.D. SMITH ET AL. ⇨1985}.
- ▸ **Big Bang** {The Universe ⇨$c$.14 Ga ago; EINSTEIN ⇨1917; FRIEDMANN ⇨1922; LEMAÎTRE ⇨1927; HUBBLE ⇨1929 **and Hot Big Bang (or "Universal Fireball")** {ALPHER, HERMAN & GAMOW ⇨1948; WEINBERG ⇨1977; WMAP ⇨2001; NASA-GSFC ⇨2003}.
- ▸ **Big Crunch (or Antibang)?** {DAVIES ⇨1994}.
- ▸ **Big Whack (or Big Splash)?** {Earth-Moon Formation ⇨$c$.4.5 Ga ago}.
- ▸ **Black holes** {MICHELL ⇨1783; DE LAPLACE ⇨1796; SEYFERT ⇨1943/1944; WHEELER ⇨1968; Supermassive Black Hole ⇨1994}.
- ▸ **Collapsar-type supernova** {GRB 030329 ⇨2003}.
- ▸ **Collapsed stars:** ⇨**Black holes,** ⇨**Neutron stars and pulsars, and** ⇨**White dwarf stars.**
- ▸ **Collision of a planet with a comet, meteoroid or asteroid** {Callisto ⇨$c$.4 Ga ago; G.D. CASSINI ⇨1690; DE MAUPERTUIS ⇨1742; DE LAPLACE ⇨1796; Comet SHOEMAKER-LEVY 9 ⇨1994}.
- ▸ **Collision of a star (Sun) with a comet** {Comte DE BUFFON ⇨1778; Comet 1979 XI ⇨1979}.
- ▸ **Collision of a stellar wind with a nebula flow** {LL Orionis ⇨1995}.
- ▸ **Collision of two galaxies** {Cartwheel Galaxy ⇨$c$.500 Ma ago; NGC 2207 & IC 2163 ⇨$c$.140 Ma ago; Antennae Galaxies ⇨$c$.63 Ma ago; NGC 6240 ⇨2001; NGC 6745 ⇨2004}, **collision of two neutron stars** {SWESTY ET AL. ⇨1998}, **collision of two planets** {Pluto-Charon Formation ⇨$c$.4.5 Ga ago; Earth-Moon Formation ⇨$c$.4.5 Ga ago}, **collision of two stars** {Sir JEANS ⇨1929; HILLS & DAY ⇨1976; ASP Conference on Stellar Collisions, Mergers and their Consequences ⇨2000}, **collision of two supermassive black holes** {NGC 6240 ⇨2001}, **and collision of two supersonic stellar winds** {Binary System (WR+OB) ⇨1976}.
- ▸ **Cosmic bow shocks** {ASCHENBACH ET AL. ⇨1995; LL Orionis ⇨1995}.
- ▸ **Cosmic implosion** ⇨**Big Crunch.**
- ▸ **Cosmic Mach cones** ⇨**Cosmic bow shocks.**
- ▸ **Cosmic particle-accelerator shock** {GINZBURG & SYROVATSKII ⇨1961}.
- ▸ **Cosmic-ray-induced spallation** {B. HOPKINSON ⇨1912}.
- ▸ **Cosmic rays** {HESS ⇨1912; BAADE & ZWICKY ⇨1934; GINZBURG & SYROVATSKII ⇨1961; COLGATE ⇨1967; KRYMSKY ⇨1977/1978; HESS Telescope Array ⇨2004}.

- **Cosmic shock waves** {⇨Sect. 2.3.1}.
- **Cosmic "tornadoes"** {HH 49/50, *see* HERBIG & HARO ⇨1950s}.
- **Eruptive variable stars** {General Catalogue of Variable Stars ⇨1948; Wolf-Rayet Variables ⇨1976; Dual Hour-Glass Shaped Nebula ⇨1999}.
- **Exploded planet?** {Solar System ⇨3.2 Ma ago; OLBERS ⇨1804; DE LAGRANGE ⇨1814; OVENDEN ⇨1972; VAN FLANDERN ⇨1993}.
- **Explosive nucleosynthesis** {HOYLE ⇨1946}.
- **Explosive (or catastrophic) variable stars: novae** {ANDERSON ⇨1901; General Catalogue of Variable Stars ⇨1948; Nova CYGNI 1992 ⇨1992; Bz Cam ⇨Fig. 4.1–Y}, **supernovae** {Cygnus Loop ⇨$c.$15,000 years ago; Guest Star ⇨393; Guest Star ⇨1006; Guest Star ⇨1054; BRAHE ⇨1572; KEPLER ⇨1604; BAADE & ZWICKY ⇨1931; I. SHELTON ⇨1987; DÍAZ ⇨1993}, **hypernovae** {Beppo-SAX & CGRO ⇨1997; ROSAT ⇨1999; HESS Telescope Array ⇨2004}, **and dwarf novae** {HIND ⇨1855}.
- **Extragalactic jets** ⇨Astrophysical Jets.
- **Gamma-ray bursts** {Vela Satellites ⇨1960s; COLGATE ⇨1967; Beppo-SAX & CGRO ⇨1997; ROSAT ⇨1999}.
- **Intergalactic shock wave** {STEPHAN's Quintet ⇨$c.$300 Ma ago}.
- **Lightning (and thunder?) on Jupiter and other planets of the Solar System** {PLINY the Elder ⇨A.D. 77; RINNERT ⇨1985}.
- **Mach cones in Saturn's dusty rings?** {SAMSONOV ET AL. ⇨1999}.
- **Neutron stars and pulsars** {LANDAU ⇨1932; BAADE & ZWICKY ⇨1934; BELL & HEWISH ⇨1967; 1st Pulsar Conf. ⇨1968; I. SHELTON ⇨1987; ARDAVAN ⇨1989}.
- **Planetary impact phenomena** {Callisto ⇨$c.$4 Ga ago; Moon's *Mare Imbrium*, *see* GILBERT ⇨1893; Comet SHOEMAKER-LEVY 9 ⇨1994; Mima's *Herschel Crater*, *see* Voyager 1 ⇨1980}.
- **Planetary standing bow shocks: (i) natural as on Earth** {GOLD ⇨1953; DUNGEY ⇨1958; AXFORD & KELLOGG ⇨1962; NESS ⇨1964; GREENSTADT 1971, *see* NESS ⇨1964; ISEE Program ⇨1977}, **Jupiter and Neptune** {Voyager 2 ⇨1989}, **and Saturn** {Cassini Spacecraft ⇨2004}, **and (ii) induced as on Mars** {Mariner 4 ⇨1965} **and Venus** {RUSSELL ET AL. ⇨1979}.
- **Quasars** {MATTHEWS & SANDAGE ⇨1960; M. SCHMIDT ⇨1963}.
- **Red giant stars** {FABRICIUS ⇨1596; RUSSELL ⇨1913; PORTER ET AL. ⇨1997} **and red supergiant stars** {CHANDRASEKHAR ⇨1931; General Catalogue of Variable Stars ⇨1948; SHELTON ⇨1987; Nova CYGNI 1992 ⇨1992; PORTER ET AL. ⇨1997}.
- **Seismic phenomena: Marsquakes** {Viking 2 Lander ⇨1976; FERRILL ET AL. ⇨2004}; **Moonquakes** {Apollo 11 ⇨1969}; **and Sunquakes** {KOSOVICHEV & ZHARKOVA ⇨1998}.
- **Shock metamorphism** {Apollo 11 mission ⇨1969; CLARKE JR. ET AL. ⇨1981; LEWIS ET AL. ⇨1987}.
- **Solar phenomena: coronal mass ejections** {TEMPEL ET AL. ⇨1860; LASCO ⇨2002}, **flares** {CARRINGTON ⇨1859; HODGSON ⇨1859; CHAPMAN & FERRARO ⇨1931; PARKER ⇨1961; ISEE Program ⇨1977; KOSOVICHEV & ZHARKOVA ⇨1998}, **heliospheric termination shock** {PARKER ⇨1961; Voyager 1 & 2 ⇨2003; Voyager 1 ⇨2004}, **solar wind** {BIERMANN ⇨1951; GOLD ⇨1953; PARKER ⇨1958; NEUGEBAUER & SNYDER ⇨1962; ISEE Program ⇨1977; Voyager 1 & 2 ⇨2003}, **and sunspots** {J. FABRICIUS ⇨1611; ALFVEN 1950, *see* DUNGEY ⇨1958}.
- **Supersonic hurricanes on Neptune?** {Voyager 2 ⇨1989}.
- **Stellar implosions** ⇨Collapsar-type supernova.
- **Tektites of lunar origin?** {C.R. DARWIN ⇨1844; O'KEEFE ⇨1967 & 1976; CHAPMAN ⇨1971}.
- **Volcanic (explosive) phenomena: on Io** {MORABITO ⇨1979}, **on Mars** {Mars Global Surveyor ⇨2002}, **and on the Moon** {HOOKE ⇨1665; SPURR ⇨1949; KOZYREV ⇨1958}.
- **White dwarf stars** {CLARK ⇨1862; RUSSELL ⇨1913; LUYTEN ⇨1922; CHANDRASEKHAR ⇨1931; Nova CYGNI 1992 ⇨1992}.

# PREHISTORIC TIMES

| | | |
|---|---|---|
| c.14 Ga ago | The Universe | **Creation of the Universe in a gigantic explosion – the Hot Big Bang or "Universal Fireball."** ▪ Its subsequent evolution from one hundredth of a second up to the present day can be described reliably by the Big-Bang model {LEMAÎTRE ⇨1927; ALPHER, HERMAN & GAMOW ⇨1948; WEINBERG ⇨1977}. This includes the expansion of the Universe {HUBBLE ⇨1929}, the origin of light elements and the relic radiation from the initial fireball {PENZIAS & WILSON ⇨1965}, as well as a framework for understanding the formation of galaxies and other large-scale structures. In fact, the Big-Bang model is now so well-regarded that it is known as the "standard cosmology."[1]<br><br>In February 2003, **NASA released a "baby" picture of the Universe showing the cosmic microwave background radiation of the Big Bang that had traveled 13 billion light-years and was generated 380,000 years after the Big Bang** {⇨Fig. 4.1–W}. Based upon this, it has been concluded that the Universe was created 13.7 ± 0.2 billion years ago {NASA-GSFC ⇨2003}. |
| 4.57 ± 0.11 Ga ago | Milky Way | **Creation of the Sun** (age determined using computer models of stellar evolution and cosmochronology).[2] ▪ The Sun is kept "puffed up" to its apparent size by the heat which nuclear reactions create in its core. Once its nuclear fuel is used up, the pull of gravity overcomes all other forces and makes the star contract to a very small size. The star's atoms or even its nuclei are then crushed, and the process may turn it into a pulsar {BELL & HEWISH ⇨1967} or black hole {MICHELL ⇨1783; DE LAPLACE ⇨1796}. An enormous amount of energy is released in this last collapse, blowing off the star's outer layers as a rapidly expanding cloud of gas. **It is widely believed that powerful shock fronts form ahead of this cloud's advance,** and through them some ions get accelerated to the very high energies of cosmic rays.[3] |
| c.4.5 Ga ago | Solar System | **(I) Formation of the Earth-Moon system by the oblique collision between the proto-Earth and Theia, a Mars-sized** (*i.e.*, a third or half the size of Earth) **protoplanet acting as the impactor.** ▪ In 1975, William K. HARTMANN[4] and Donald R. DAVIS, two U.S. space scientists, were the first to speculate that the formation of the Moon resulted from a giant impact into the early Earth from an object (named Theia), which was at least the size of Mars. Among the first to advocate this now widely accepted theory was Alastair G.W. CAMERON,[5] a noted Canadian-born U.S. astrophysicist at Harvard-Smithsonian Center for Astrophysics.<br><br>The leading theory today for the Moon's formation postulates that some material from that collision was ejected into orbit and condensed into the Earth's Moon. Only the so-called **"Giant Impact Theory"** (or "Giant Impact Hypothesis") – also known as **"Big Splash"** (or "Big Whack") – provides a simple explanation for the Moon's chemistry,[6] as revealed by more than 800 pounds of lunar rock and soil returned by the Apollo lunar landings for analysis on Earth {Apollo 11 ⇨1969; Apollo 14 ⇨1971; Apollo 16 & 17 ⇨1972}.[7] |

---

[1] P. SHELLARD: *The hot Big Bang model.* In: *Cambridge cosmology.* Cambridge University Press, Cambridge (1996); www.damtp.cam.ac.uk/user/gr/public/bb_home.html.
[2] A. BONANNO, H. SCHLATTL, and L. PATERN: *The age of the Sun and relativistic corrections in the EOS.* Astron. & Astrophys. **390**, 1115-1118 (2002).
[3] D.P. STERN: *The exploration of the Earth's magnetosphere.* Glossary, NASA-GSFC (Nov. 2001); http://www.phy6.org/Education/gloss.html.
[4] W.K. HARTMANN and D.R. DAVIS: *Satellite-sized planetesimals and lunar origin.* Icarus **24**, No. 4, 504-515 (1975).
[5] A.G.W. CAMERON and W.R. WARD: *The origin of the Moon.* Proc. Lunar Sci. Conf. **7**, 120-122 (1976).
[6] W.K. HARTMANN and D.R. DAVIS: *Satellite-sized planetesimals and lunar origin.* Icarus **24**, No. 4, 504-515 (1975); W.K. HARTMANN: *A brief history of the Moon.* Planet. Rept. **17**, 4-11 (1997).
[7] H.J. MELOSH: *Impact cratering.* In: (J.H. SHIRLEY and R.W. FAIRBRIDGE, eds.) *Encyclopedia of planetary sciences.* Chapman & Hall, London *etc.* (1997), pp. 326-335.

(i) Calculations made in the 1970s by Sir Harold JEFFREYS,[8] a British astronomer and geophysicist, supplanted the three classical theories of lunar origin (capture, fission and co-accretion). He calculated an age for the Earth of about 4 Ga. Today a somewhat larger value of 4.54 Ga is generally accepted, based on measurements of isotopes in the Earth's oldest lead ores.

(ii) From numerical simulations using Smoothed Particle Hydrodynamics (SPH), it has been concluded that the Moon was formed almost exclusively by material from the mantle of the impactor.[9, 10] ▪ In the SPH method the domain of the computation is not divided into spatial cells (as in ordinary hydrodynamics), but rather the matter in the domain of computation is divided into smooth overlapping spheres. The density of a sphere has a radial distribution that is bell-shaped, with highest density at the center and a sharp outer edge. The particles carry individual internal energies, but such properties as density and pressure are collectively determined by the overlap between the density distributions of the particles; these are meaningful only when typically a few tens of particles contribute to the overlap.[11]

**(II) Formation of the the Pluto-Charon binary by a giant impact** (according to an impact hypothesis), in much the same way as the Earth-Moon binary. ▪ The Pluto-Charon binary was first suggested by William B. McKINNON,[12] a U.S. planetary scientist who drew analogies from theories being developed at that time for possible formation scenarios of binary asteroids.

Most astronomers assume that all of the planets were formed 4.57 billion years ago at the same time from the large body of gas and dust that once surrounded the Sun. Charon is the largest satellite relative to its parent planet in the Solar System; its diameter is 1,172 km, just over half the size of Pluto. The ages of Pluto (discovered in 1930/1931) and Charon (discovered in 1978) are unknown. Pluto may represent a remnant planetesimal, a rare survivor from the earliest Solar System. Evidence for the Pluto-Charon impact hypothesis comes from the system's high specific angular momentum, its high axial obliquity, and the large mass ratio of the binary. Very recently, a numerical hydrodynamic simulation was published that demonstrated that the formation of Pluto-Charon via a large collision is plausible.[13]

Planet-planet collisions are also expected during the early stages of the formation of extrasolar planets and are possible in mature planetary systems through secular planet-planet perturbations. Since planetary collisions are accompanied with electromagnetic signals, there exists the possibility of directly detecting such events.[14]

| c.4 Ga ago | Callisto [the farthest satellite of Jupiter] | **An asteroid, hitting Callisto and exploding on impact, heats the icy surface to above the melting point and produces a huge impact crater called "Valhalla."** ▪ Valhalla seems to be surrounded by frozen shock waves – a curious ripple pattern away from the impact site that, freezing into place and giving it a bull's-eye appearance, extends about 2,000 kilometers from its center.[15] |
|---|---|---|

Callisto is the second largest of the Galilean satellites (Ganymede, Callisto, Io, and Europa; G. GALILEI & S. MARIUS 1610) and the farthest from Jupiter. This heavily-cratered moon has a

---

[8] H. JEFFREYS: *The Earth: its origin, history and physical constitution.* Cambridge University Press, Cambridge, U.K. (1976).
[9] W. BENZ, W.L. SLATTERY, H.J. MELOSH, and A.G.W. CAMERON: *The origin of the Moon and the single impact hypothesis.* Icarus **66**, No. 6, 515-535 (1986); Ibid. **71**, No. 7, 30-45 (1987); Ibid. **81**, No. 2, 113-131 (1989).
[10] R.M. CANUP: *Simulations of a late lunar forming impact.* Icarus **168**, No. 2, 433-456 (2004).
[11] A.G.W. CAMERON: *The origin of the Moon and the single impact hypothesis.* Harvard-Smithsonian Center for Astrophysics, Cambridge, MA; http://www.xtec.es/recursos/astronom/moon/camerone.htm.
[12] W.B. MCKINNON: *On the origin of Triton and Pluto.* Nature **311**, 355-358 (1984).
[13] R.M. CANUP: *A giant impact origin of Pluto-Charon.* Science **307**, 546-550 (Jan. 28, 2005).
[14] B. ZHANG and S. SIGURDSSON: *Electromagnetic signals from planetary collisions.* Astrophys. J. **596**, Part 2, L95-L98 (2003).
[15] C.J. HAMILTON: *Views of the Solar System: Callisto's Valhalla region.* Solarviews.com (Feb. 10, 1995); http://www.solarviews.com/cap/jup/callist2.htm.

density of 1.8 times that of water and consists of a rocky core surrounded by frozen water. Callisto is possibly the oldest landscape in the Solar System; its surface does not appear to have undergone any sort of geological activity. All the craters on Callisto are flattened because the surface, largely composed of ice, tends to settle and flow.

| | | |
|---|---|---|
| *c.*2 Ga ago | *Vredefort Basin [ca. 120 km southwest of Johannesburg], Northwest and Free State Provinces, South Africa* | **Creation of the 300-km-dia. Vredefort Basin by the impact of a meteorite.**[16] ▪ This basin is the oldest and largest known impact structure on Earth. The ring of hills now present in the center of Vredefort Basin are the eroded remains of a dome created by the rebound of the rock below the impact site after the asteroid hit. The central uplift of this impact structure – the so-called "Vredefort Dome" – has been added to UNESCO's World Heritage List.[17]<br><br>Vredefort Basin, Sudbury Basin {⇨*c.*1.85 Ga}, and Chicxulub Crater {⇨*c.*65 Ma ago} are the largest known terrestrial impact structures. Each has some multiple-ring attributes. The three impact events are also the reason for the existence of world-class mineral and hydrocarbon deposits.[18] |
| *c.*1.85 Ga ago | *Sudbury Basin [ca. 340 km north of Toronto], southeastern Ontario, Canada* | **Creation of the Sudbury Basin by meteoritic or cometary impact.**[19] ▪ Evidence for the Basin's impact origin are shatter cones ranging in size from inches to tens of feet across. Today the Sudbury impact basin is about 60 km long and 30 km wide. Some geologists believe that the strongly deformed crater, a multi-ringed impact structure, was originally about 250 km in diameter, making the Sudbury Basin the second largest terrestrial impact structure.<br><br>In 1883, rich nickel and copper ore deposits were accidentally discovered during excavations for the main line of the Canadian Pacific Railroad. The Sudbury region produces more than three quarters of the world's nickel. |
| *c.*500 Ma ago | *Sculptor ["Sculptor's Studio," a constellation in the Southern Hemisphere]* | **A head-on collision of two galaxies creates the Cartwheel Galaxy,** a group of galaxies located about 500 million light-years away from Earth in *Sculptor*. This immense ring-like structure, about 150,000 light-years in diameter, is composed of newly formed, extremely bright, massive stars. ▪ When galaxies collide, they pass through each other – their individual stars rarely come into contact. It has become accepted that many of the "classical" ring galaxies are formed from a head-on collision between a small intruder galaxy and a larger disk system.[20, 21] The ring forms as gas and stars are crowded into an expanding wave that moves radially through the disk.<br><br>Pictures taken in October 1994 with the Wide-Field & Planetary Camera of NASA/ESA's Hubble Space Telescope allowed to obtain high-resolution images of the structure of the Cartwheel Galaxy over a broad range of wavelengths (1,150 to 11,000 Å).[22] The new details of star birth resolved by Hubble provide an opportunity to study how extremely massive stars are born in large fragmented gas clouds. The ring-like shape is the result of the gravitational disruption caused by a small intruder galaxy passing through a large one, compressing the interstellar gas and dust, and causing a wave of star formation to move out from the impact point like a ripple across the surface of a pond.[23] |

---

[16] W.U. REIMOLD and R.L. GIBSON: *Meteorite impact! The Vredefort structure.* Van Rensburg Publ., Melville, South Africa (2005).
[17] *Vredefort Dom.* World Heritage Center, UNESCO; http://whc.unesco.org/en/list/1162.
[18] R. GRIEVE and A. THERRIAULT: *Vredefort, Sudbury, Chicxulub: three of a kind?* Annu. Rev. Earth Planet. Sci. **28**, 305-338 (2000).
[19] F.W. BEALES and G.P. LOZEJ: *Sudbury basin sediments and the meteoritic impact theory of origin for the Sudbury structure.* Can. J. Earth Sci. **12**, 629-635 (1975); Ibid. **13**, 179-181 (1976).
[20] R. LYNDS and A. TOOMRE: *On the interpretation of ring galaxies: the binary ring system II Hz4.* Astrophys. J. **209**, 382-388 (1976).
[21] J.C. THEYS and E.A. SPIEGEL: *Ring galaxies. II.* Astrophys. J. **212**, 616-619 (1977).
[22] *A fiery galactic collision.* Dept. of Physics & Astronomy, University of Tennessee; http://csep10.phys.utk.edu/astr162/lect/galaxies/Cartwheel.html.
[23] *Astronomy picture of the day: the Cartwheel Galaxy.* NASA-GSFC (June 12, 2001); http://antwrp.gsfc.nasa.gov/apod/ap010612.html.

| | | |
|---|---|---|
| c.300 Ma ago | Pegasus ["The Winged Horse," a constellation in the Northern Hemisphere] | **Creation of a huge *intergalactic shock wave* in the middle of STEPHAN's Quintet,** an unusual group of five galaxies located about 300 million light-years away from Earth in *Pegasus*. • Based upon infrared observations made with NASA's Spitzer Space Telescope, this unique shock phenomenon was discovered by an international research team including scientists from the Max-Planck-Institute for Nuclear Physics (MPIK) in Heidelberg.[24, 25]<br><br>STEPHAN's Quintet has often been used as proof that the redshift is not truly a distance indicator, which would completely overturn current cosmology, because although four of the galaxies (NGC 7317, 7318A, 7318B, 7319) have similar, large redshifts, the fifth (NGC 7320) – although apparently a member of the group – shows a much smaller redshift. |
| c.140 Ma ago | Canis Major ["Larger Dog," a constellation in the Southern Hemisphere] | **Two spiral galaxies** – located about 140 million light-years away from Earth in *Pegasus* and consisting of the large galaxy NGC 2207 and the smaller galaxy IC 2163 – **are in the process of colliding and merging.**[26] Strong gravitational forces from the large galaxy will increasingly distort the shape of the smaller one. In about a billion years time the two galaxies will merge and become an elliptical galaxy.<br><br>Pictures taken in 1999 with the NASA/ESA Hubble Space Telescope revealed that the two galaxies are about 100,000 light-years apart. A merger between two spiral galaxies condenses the gas and dust clouds, igniting star birth. In galaxy collisions, the stars in them usually do not collide, because the space between stars is so vast. Infrared pictures, taken in 2005 with NASA's Spitzer Space Telescope revealed that the white balls on the arms of both galaxies, called "beads on a string" by astronomers, are clusters of newborn stars created by condensation of dust and gas.[27] |
| c.65 Ma ago | Chicxulub Crater, Yucatan Peninsula, Mexico | **Collision of the Earth with an about 15-km-dia. comet or asteroid.**[28] • The 170-km-dia. buried *Chicxulub Crater* (or *Yucatan Crater*), which was not discovered until the 1990s in the neighborhood of the village Puerto Chicxulub [*Chicxulub* means "Fire of the Beast" in the Mayan language] is one of the largest impact structures on Earth, and if the size estimate is proven correct, perhaps one of the largest produced in the inner Solar System in the last four billion years.<br><br>(i) Experimental evidence for this popular impact hypothesis was first obtained through the detection of an "iridium signature" – a concentration of iridium hundreds of times greater than normal – which can be found at about one hundred locations worldwide {ALVAREZ ⇒ 1978}.<br><br>(ii) **Using Ar-40/Ar-39 isotope dating methods, the Yucatan impact crater has been dated with high precision, and was found to be 64.98 million years old.**[29]<br><br>(iii) The time of the impact is called the **"K/T boundary,"** marking the end of the Cretaceous (K) Period (146–65 Ma ago), and the beginning of the Tertiary (T) Period (65–2.5 Ma ago). However, whether the impact event that occurs at approximately the level of the K/T boundary was the cause of the enhanced iridium concentrations as well as the wave of biological extinctions is still the subject of numerous disputes.<br><br>(iv) In the Hell Creek Formation – a distinct layer of rock that is recognizable over a large geographic area including Montana, North Dakota and South Dakota – this famous K/T bound- |

---

[24] *Gigantic cosmic cataclysm in Stephan's Quintet of galaxies.* PHYSORG.com (March 3, 2006); http://www.physorg.com/news11392.html.
[25] P.N. APPLETON ET AL.: *Powerful high-velocity dispersion molecular hydrogen associated with an intergalactic shock wave in Stephan's Quintet.* Astrophys. J. **639**, L51-L54 (March 10, 2006).
[26] *A grazing encounter between two spiral galaxies.* HUBBLESITE, NewsCenter (Nov. 1999); http://hubblesite.org/newscenter/archive/releases/1999/41/.
[27] *Ready for the cosmic ball.* Spitzer Space Telescope, NASA-JPL (2006); http://sscws1.ipac.caltech.edu/Imagegallery/image.php?image_name=ssc2006-11a.
[28] A.R. HILDEBRAND ET AL.: *Chicxulub Crater: a possible Cretaceous/Tertiary boundary impact crater on the Yucatan peninsula, Mexico.* Geology **19**, 867-871 (1991).
[29] C.C. SWISHER III ET AL.: *Coeval Ar-Ar ages of 65 millions years ago from Chicxulub crater melt-rock and Cretaceous-Tertiary boundary tektites.* Science **257**, 954-958 (1992).

ary consists of a 2-cm-thick highly acid-leached *boundary bed* and a 1-cm-thick *impact bed* with shocked quartz and an iridium anomaly.[30]

(v) A review of the Chicxulub Crater deposits was given by Philippe CLAEYS, a research geologist at UC Berkeley.[31] More recent research using seismic reflection data from the offshore portion of the crater shows that the overall crater had three concentric rings: a central 80-km-dia. peak, a 130-km-dia. inner ring, and a 195-km-dia. outer ring.[32]

(vi) This most recent major impact event on Earth may have resulted in
- an "impact winter," caused by the extra dust thrown into the atmosphere;
- a "greenhouse crisis," caused by the release of large amounts of $CO_2$ from the carbonate-rich rocks of the Gulf coast during the impact;
- an "atmospheric pollution," caused by acid rain as the result of the loading of the atmosphere with sulfur aerosols;
- "global wildfires," causing the consumption of most of the standing biomass of the planet in huge fires; and
- "tsunami effects," causing the devastation of coastal areas and the coverage of previously fertile soils with salt.[33]

**These proposed catastrophic mechanisms would have horrific ecological effects, such as a mass biological extinctions, killing about 85% of all species** – including not only the formidable dinosaurs[34, 35] and other reptiles on land, but also numerous marine animals and organisms such as *calcareous nanoplankton*, an important key element at the bottom of marine food chains during the Cretaceous.[36] The palaeontologists William A.S. SARJEANT,[37] a geology professor at the University of Saskatchewan, and Philip J. CURRIE, a curator of the Royal Tyrrell Museum, discussed the pros and cons of the meteoritic impact theory and the concept of the "Great Extinction" in detail.

**In addition to the Chicxulub impact event at the end of the Cretaceous there was possibly another huge impact event caused by an asteroid which created the about 600×450-km² underwater *Shiva Crater* in the Indian Ocean west of India.**[38] This event, perhaps triggering extensive volcanic eruptions in hot-spot areas of the Indian Ocean, may have contributed to the worldwide extinction of the dinosaurs.

| | | |
|---|---|---|
| c.63 Ma ago | *Corvus ["The Crow," a small constellation of the Southern Hemisphere]* | **Formation of the Antennae Galaxies by collision of two spiral galaxies NGC 4038 and NGC 4039.** ▪ Discovered in 1785 by the British astronomer Sir Frederick William HERSCHEL, Antennae is located some 63 million light-years away from Earth in *Corvus*. Conventional observation from the Earth's surface had only been able to detect the two long antenna-like arms extending from the galaxy nucleus. From its orbit of Earth, however, the NASA/ESA Hubble Space Telescope has been clearly able to identify the detailed nature of this fascinating galaxy |

---

[30] G.J. RETALLACK: *Acid trauma at the Cretaceous-Tertiary boundary in eastern Montana*. G.S.A. Today **6**, No. 5, 1-7 (1996).
[31] P. CLAEYS: *When the sky fell on our heads: identification and interpretation of impact products in the sedimentary record*. Rev. Geophys. Suppl. **33**, 95-100 (1995).
[32] J.V. MORGAN ET AL.: *Size and morphology of the Chicxulub impact crater*. Nature **390**, 472-476 (1997).
[33] J. BOURGEOIS, T.A. HANSEN, P.L. WIBERT, and E.G. KAUFFMAN: *A tsunami deposit at the Cretaceous-Tertiary boundary in Texas*. Science **241**, 567-570 (July 29, 1988).
[34] A.R. HILDEBRAND: *The Cretaceous/Tertiary boundary impact (or the dinosaurs didn't have a chance)*. J. Roy. Astron. Soc. Can. **87**, 77-118 (1993).
[35] K.O. POPE, A.C. OCAMPO, and C.E. DULLER: *Surficial geology of the Chicxulub impact crater, Yucatan, Mexico*. Earth Moon Planets **63**, 93-104 (1993); *Meteorite impact and the mass extinction of species at the Cretaceous Tertiary boundary*. Proc. Natl. Acad. Sci. (Washington) **95**, 11028-11029 (Sept. 15, 1998).
[36] J.J. POSPICHAL: *Calcareous nanoplankton mass extinction at the Cretaceous/Tertiary boundary: an update*. In: (G: RYDER ET AL., eds.) *The Cretaceous-Tertiary event and other catastrophes in Earth history*. Special paper – Geological Society of America (GSA), vol. 307. GSA, Boulder (1996), pp. 335-360.
[37] W.A.S. SARJEANT and P.J. CURRIE: *The "Great Extinction" that never happened: the demise of the dinosaurs considered*. Can. J. Earth Sci. **38**, 239-247 (2001).
[38] L. MULLEN: *Shiva: another K-T impact?* NASA Astrobiology Mag. (Nov. 3, 2004).

collision.[39] During the slow, hundred-million-year collision, one galaxy can rip the other apart gravitationally, and dust and gas common to both galaxies does collide. A close-up of the galaxy collision showed more details of the collision scenario: dark dust pillars mark massive molecular clouds, which are being compressed during the galactic encounter, causing the rapid birth of millions of stars.[40]

The two galaxies are known as the "Antennae" because the two long tails of stars, gas, and dust thrown out of the galaxies as a result of the collision resemble the antennae of an insect. Computer simulations,[41] using intermediate masses of dark matter in the galaxy model,
- provided a chronological sequence (movie) of how colliding galaxies evolve; and
- showed that long tails only arise when the mass of dark matter is less than 10 times the mass of the stars.

c.35.7 Ma ago

*Popigai Crater [ca. 100 km from the coast of the Laptev Sea], northern Siberia, Russia*

**A large asteroid impacts the Siberian platform to form an about 100-km-dia. crater filled with melted and shocked material including shock-generated impact diamonds.**[42] ▪ The well-preserved crater, named ***Popigai Crater***, is the fourth largest impact crater on Earth.[43] The impact created shatter cones in Archean gneisses, PDFs, lechatelierite (amorphous $SiO_2$) and diaplectic glasses, coesite and stishovite. Shock pressures instantaneously transformed graphite to diamond within 13.6 km of ground zero.[44] Individual crystals are about one micrometer in diameter and consist, about half-and-half, of cubic diamond and hexagonal diamond (named "lonsdaleite").

General criteria for recognizing terrestrial impact structures encompass surface form and geological structure, geophysical characteristics, rock type characteristics, and microscopic deformation and melting features.[45] Unique features which provide definite evidence for meteorite impact origin are[46, 47]

- shatter cones – striated conical fracture surfaces produced by meteorite impact into fine-grained brittle rocks, such as limestone;
- planar deformation features (PDFs) – microscopic features in grains of (for example) quartz or feldspar consisting of very narrow planes of glassy material arranged in parallel sets that have distinct orientations with respect to the grain's crystal structure;
- high-pressure mineral phases – such as coesite and stishovite (dense, high-pressure phases of quartz that have so far been identified in nature only in shock-metamorphosed quartz-bearing rocks from meteorite impact craters), and diamond (cubic and hexagonal); and
- diaplectic glasses – amorphous forms of crystals resulting from shock wave compression and subsequent pressure release.

---

[39] *Hubble reveals stellar fireworks accompanying galaxy collisions.* HubbleSite, Space Telescope Science Institute (STSI), Office of Public Outreach (Oct. 21, 1997); http://hubblesite.org/newscenter/archive/releases/1997/34.
[40] *Astronomy picture of the day: close-up of Antennae Galaxy collision.* NASA-GSFC (Oct. 27, 1997); http://antwrp.gsfc.nasa.gov/apod/ap971027.html.
[41] J. DUBINSKI: *NGC 4038/4039 – the Antennae.* Lick Observatory, UC Santa Cruz, CA (Oct. 28, 2001); http://www.cita.utoronto.ca/~dubinski/antennae/antennae.html.
[42] V.L. MASAITIS, M.V. MIKHAILOV, and T.V. SELIVANOVSKAYA: *Popigai Basin – an explosion meteorite crater.* Meteoritics **7**, No. 1, 39-46 (1972).
[43] The three largest impact craters on Earth are: Vredefort (c.2 billion years, 300 km dia.), Sudbury (c.1.85 billion years, 250 km dia.), and Chicxulub (c.65 million years, 170 km dia.).
[44] V.L. MASAITIS: *Popigai crater: origin and distribution of diamond-bearing impactites.* Meteoritics & Planetary Science **33**, 349-359 (1998).
[45] B.M. FRENCH: *Traces of catastrophe.* Lunar and Planetary Institute, Houston, TX (1998); *see Appendix*, pp. 107-110.
[46] *Terrestrial impact craters. Glossary.* Lunar and Planetary Institute, Houston, TX (2007); http://www.lpi.usra.edu/publications/slidesets/craters/glossary.shtml.
[47] D. STÖFFLER and R.A.F. GRIEVE: *Classification of the term "impactite."* A proposal on behalf of the International Union of Geological Sciences (IUGS) Subcommission on the Systematics of Metamorphic Rocks (April 2003); http://www.bgs.ac.uk/scmr/docs/paper_12/scmr_paper_12_1.pdf.

| | | |
|---|---|---|
| c.28 Ma ago | La Garita Caldera, San Juan Mountains, south-western Colorado | **The largest known explosive volcanic eruption on Earth occurs.** ▪ The so-called "Fish Canyon Tuff Eruption" ejected about 5,000 km$^3$ of igneous fragments, thereby producing the huge La Garita Caldera (about $35 \times 75$ km$^2$). For comparison purposes: the recent eruption of Mount St. Helens {⇨1980} produced only 1.2 km$^3$ of fragments.[48]<br><br>The Fish Canyon Tuff, La Garita Caldera, and newly identified precaldera lava-like rocks were the focus of an informal four-day field workshop in early September 1997, involving 18 participants from five countries. |
| c.15 Ma ago | Nördlinger Ries, northern Bavaria, Germany | **Creation of the Ries Basin** [Germ. *Nördlinger Ries*], a shallow, nearly circular depression about 20 km in diameter {⇨Fig. 4.1–C}, **by an asteroid impact.** ▪ Modern geologists assume that the **3.5-km-dia. Steinheim Basin** [Germ. *Steinheimer Becken*] **30 km away** was formed at the same time by the same asteroid, which broke apart during its passage through the atmosphere.[49] This basin represents a complex impact crater with a well-developed central uplift, a hill about 50 m high called "Klosterberg" {⇨Fig. 4.1–C}. |
| c.3.2 Ma ago | Solar System | (According to the exploded planet hypothesis) **the explosion of a modest, probably moon-sized, planetary body produces the asteroid belt** (astronomically dated at 3.2 Ma ago) **and all comets.** ▪ The idea of a "lost planet" has not only a mythological origin {PLATO ⇨4th century B.C.}, but has also occupied generations of astronomers {MAUPERTUIS ⇨1742; OLBERS 1804 and DE LAGRANGE 1814, see PIAZZI ⇨1801; OVENDEN ⇨1972; NAPIER[50] & DODD 1973; VAN FLANDERN ⇨1993}.<br><br>The problem of how asteroids were formed is far from being settled.<br>▸ Three mechanisms have been proposed as possible causes of planetary explosions: phase changes, natural fission reactors, and gravitational heat energy.[51]<br>▸ The breakup of a planet might also be possible by rapid rotation, tidal disruption and (oblique) collision of two small bodies.[52] |
| c.2.6 Ma ago | Paleontological site of Koobi Fora [near Lake Rudolf, now Lake Turkana], northern Kenya | ▪ In 1959, **Louis S.B. LEAKEY**, a Kenyan archaeologist and anthropologist, **and his wife Mary D. LEAKEY discovered the oldest known stone tools in the Great Rift Valley of East Africa.** The primitive tools, consisting of five choppers, a number of flakes, and a couple of battered stones {⇨Fig. 4.2–C}, were found partly on the surface, partly below in a tuff sediment dated to about 2.6 Ma ago. These artifacts are crude but can be distinguished from naturally broken rocks by regular patterns of scars over a rock from which slivers of stones – so-called **"flakes"** – were detached by consistent fracture {⇨Fig. 4.2–A}.<br><br>Typically, their tools were of a crude, single type; a general-purpose implement that hardly changed in form during the next two million years. It is variously known as a "pebble tool," a "pebble chopper," a "chopping tool," or simply as a "chopper."<br>▸ The pebble, which was waterworn and hence rounded and could be as large as a fist, and was preferably flattish rather than spherical, was given a few violent but skillfully applied blows by a hammer-stone or pounder, using the ***direct percussion technique (I)*** {⇨Fig. 4.2–D(a)}. Several large flakes or chips were knocked off the tool stone to create a sharp and roughly serrated crest or ridge on it, yielding an implement that was edged at one end and that could be gripped at the other end. |

---

[48] M. ORT: *Largest explosive eruptions: new results for the 27.8 Ma Fish Canyon Tuff and the La Garita Caldera, San Juan Volcanic Field, Colorado.* Northern Arizona University, Flagstaff, AZ; http://staff.aist.go.jp/s-takarada/CEV/newsletter/lagarita.html.
[49] C.R. MATTMÜLLER: *Ries und Steinheimer Becken.* F. Enke Verlag, Stuttgart (1994).
[50] W.Mcd. NAPIER and R.J. DODD: *The missing planet.* Nature **242**, 250-251 (March 23, 1973).
[51] T. VAN FLANDERN: *Planetary explosion mechanisms.* Meta Res. Bull. **11**, No. 3, 33-38 (2002); http://www.metaresearch.org/solar%20system/eph/PlanetExplosions.asp.
[52] C. OLIVIER: *Asteroids.* In: *Encyclopedia Americana* (Int. edn.). Americana Corporation, New York, Vol. 2 (1974), pp. 548-551.

|  |  |  |
|---|---|---|
|  |  | ▸ These so-called **"Olduwan** (or Oldowan) **pebble tools,"** named after Olduvai Gorge, could be used to hack, mash, cut, grub roots, and to scrape and break bones and skulls for their marrow and brain.[53] ▪ A glossary of stone tools of the Olduwan industry can be found in Mary D. LEAKEY's book *Olduvai Gorge: My Search for Early Men*, (Collins, London, 1979). |
|  |  | **The discovery has been taken as a key indicator of a human species.** Similar stone tools of earliest hominoids dated from 2.5 to 2 million years old were also discovered in Kenya, and later in Malawi, Tanzania and Zaire. |
| *c.*2 Ma ago | *Yellowstone Caldera, Wyoming, Montana, Idaho* | A great explosive eruption ($VEI = 8$) produces the Huckleberry Ridge Tuff (> 2,450 km$^3$) and a composite caldera more than 75 km long. ▪ Further catastrophic eruptions occurred *c.*1.3 million years ago, and then *c.*630,000 years ago. This cycle of approx. 600,000–700,000 years suggests another eruption could be brewing. These giant eruptions were fueled by large volumes of magma. However, recent research by the USGS and others suggests there may not be a deep plume beneath Yellowstone. |
|  |  | The latest eruption spewed out nearly 1,000 km$^3$ of debris. What is now the Yellowstone Park's central portion then collapsed, forming an about $45 \times 76$-km$^2$ caldera (or basin). **The hotspot Yellowstone Caldera is one of the largest and most active calderas in the world.**[54] |
| *c.*1.75 Ma ago | *Olduvai Gorge, eastern Serengeti Plains [near Lake Victoria], Tanzania* | ▪ In 1960, **Louis S.B. LEAKEY and Mary LEAKEY found stone tools as well as the jaw and teeth of a man,** fortuitously lying under volcanic rock (tuff) and having a potassium-argon date of about 1.75 million years, dating them to the Lower Pleistocene.[55] The stone tools were probably manufactured by early hominoids, the first human species, *Homo habilis* (*c.*2.2–1.6 Ma ago), which was distinguished from preceding ape men or *Australopithecus* (*c.*5–2.2 Ma ago) by larger brains. |
| *c.*1.4– 0.3 Ma ago | *Olduvai Gorge, Tanzania; Middle East & Western and Central Europe* | More advanced hominoids of the species *Homo erectus* (*c.*2–0.4 Ma ago) slowly develop the primitive so-called **"Olduwan-type choppers"** into large, almond-shaped bifacial cutting tools (*Acheulean handaxes*). **Dietary increases in protein and fat from large animals result in an increase in their brain size from about a third to a half that of modern humans** (*Homo sapiens sapiens*). ▪ Throughout this era, the interest of primitive man shifted from core tools like the Olduwan and Acheulean industries to flake tools by creating new technologies: the ***anvil technique (II)*** {⇨ Fig. 4.2–D(b)} employs a large stationary stone against which the workpiece was swung to batter off some large flakes, while the ***soft hammer technique (III)*** {⇨ Fig. 4.2–D(d)} is based on a discovery that hard stone could be chipped by striking it with a softer material such as a piece of bone, antler, or even wood. While the older technique using a stone or cobble as a pounding or pecking tool – the so-called "hammerstone technique" – produced short and deep flake scars, this new method gave more uniform and hence sharper cutting edges. |
| *c.*0.25 Ma ago | *Clacton-on-Sea, Essex, England* | *Homo sapiens* (*c.*0.3–0.2 Ma ago), the earliest form of our own species, **invents the spear – man's first projectile weapon.** Functioning as an extension of the arm, it can deliver con- |

---

[53] R.S. HARTENBERG: *Hand tools*. In: *The new Encyclopædia Britannica, Macropædia*. Benton & Hemingway, Chicago *etc.*, vol. 8 (1974), pp. 605-624. A glossary of stone tools of Olduwan industry can be found in the book by M.D. LEAKEY: *Olduvai Gorge: my search for early men*. Collins, London (1979).

[54] *Description: Yellowstone Caldera, Wyoming*. USGS (Jan. 22, 2003); http://vulcan.wr.usgs.gov/Volcanoes/Yellowstone/description_yellowstone.html.

[55] L.S.B. LEAKEY: *Finding the world's earliest man*. Natl. Geogr. Mag. **118**, 420-435 (Sept. 1960); *Exploring 1,750,000 years into man's past*. Ibid. **120**, 564-589 (Oct. 1961); *Adventures in the search of man*. Ibid. **123**, 132-152 (Jan. 1963). ▪ At first LEAKEY estimated that *Zinjanthropus* (now termed *Australopithecus boisei*), was 600,000 years old. He was wrong. Using a new method of dating, the carbon-14 technique, geophysicists from UC Berkeley concluded that the site was 1.75 Ma (million years) old.

siderable kinetic energy, focused onto a flint or bone tip, to a remote target. • **An early example of a spear**[56] **dating back some 250,000 years was discovered in England in 1911.**[57]

However, in the period 1995–1998, eight wooden spears were found together with tenthousands of animal bones in the southern part of the brown coal open mining Schöningen (Lower Saxony, Germany). A very recent investigation has shown that the spears are approximately 400,000 years old – so far **the oldest known hunting weapons of mankind!** They indicate that early man (*Homo erectus*) even in this evolutionary stage was a specialized and experienced hunter.

*c.*71,000 years ago — *Mt. Toba, Central Sumatra, Indonesia*

**Giant explosive eruption of volcano Mt. Toba.** • The super eruption ($VEI = 8$) – also designated as a mega-eruption – causes a great collapse to occur, forming a caldera, which fills with water creating Lake Toba. • Later, the floor of the caldera was uplifted to form Samosir ("Island of the Dead") in the lake.

Volcanic ash from Mount Toba has been traced northwest across India, where a widespread terrestrial marker bed of primary and reworked airfall ash exists. According to Michael R. RAMPINO[58] and Stephen SELF, two Earth scientists and volcanologists at New York University, **the super-eruption,** displacing about 800 km$^3$ of ash and burying most of India under ash, **may have caused a six-year global *volcanic winter*,** caused by light-reflecting sulfur particles lingering in the atmosphere. • For comparison purposes: the explosion of Mount Tambora {⇨1815}, the largest known historic eruption, displaced 20 km$^3$ of ash, and the recent eruption of Mount St. Helens {⇨1980} only produced a tiny 0.2 km$^3$ of ash.

Stanley H. AMBROSE,[59] a professor of anthropology at the University of Illinois in Urbana, has suggested that this horrific volcanic winter 70,000 years ago, followed by the coldest thousand years of the last Ice Age, resulted in widespread famine and death in modern human populations around the world, such as the species *Homo sapiens neanderthalensis* (*c.*200,000–30,000 years ago) and *Homo sapiens sapiens* (*c.*130,000 years ago to present). The abrupt "bottleneck," or decrease, in our ancestors' populations, in turn, might have brought about the rapid "differentiation" – or genetic divergence – of the surviving populations. AMBROSE arrived at the conclusion that "Toba may have caused modern races to differentiate abruptly only 70,000 years ago, rather than gradually over one million years."

*c.*50,000 years ago — *Meteor Crater [ca. 50 km east of Flagstaff], northern Central Arizona*

**Creation of Meteor Crater by the impact of an asteroid known as the "Canyon Diablo meteorite"** (or "Arizona meteorite"). • Today Meteor Crater {⇨Fig. 4.1–B} is officially named **"Barringer Crater"** {BARRINGER & TILGHMAN ⇨1905}. The Meteor Crater Visitor Center[60] located at the crater rim has a 1,406-lb (638-kg) meteorite fragment, the largest ever found in the area, on display "for visitors to view and touch."

Based on modern knowledge gained from laboratory hypervelocity impact experiments, more recent simulation studies have shown that the 1.2-km-dia. Meteor Crater was generated by an asteroid of (only) about 60 meters in diameter weighing several hundred thousand tons which approached the Earth with a velocity of some 10 km/s. • As illustrated in Fig. 4.3–U, a hypervelocity impact can produce a crater much larger than the size of the impacting body.

---

[56] The spear, in which a "handle" is fixed to a pointed tool, is also the first evidence of a "compound tool," even preceding stone axes and adzes. A large number of uni- and bifacially flaked pointed stone artifact forms dating from the Middle Palaeolithic have been discovered in the Old World, which are usually assumed to be projectile points for spears.
[57] I. TATTERSALL ET AL. (eds.): *Encyclopedia of human evolution and prehistory*. Garland, New York & London (1988), p. 535.
[58] M.R. RAMPINO and S. SELF: *Volcanic winter and accelerated glaciation following the Toba super-eruption*. Nature **359**, 50-52 (1992).
[59] S.H. AMBROSE: *Late Pleistocene human population bottlenecks, volcanic winter, and the differentiation of modern humans*. J. Hum. Evol. **34**, 623-651 (1998); *Did the super-eruption of Toba cause a human population bottleneck? Reply to* GATHORNE-HARDY *and* HARCOURT-SMITH. Ibid. **45**, 231-237 (2003).
[60] *Meteor Crater Visitor Center.* Winslow, AZ 86047; http://www.meteorcrater.com/visitorcenter/overview.htm.

| | | |
|---|---|---|
| c.45,000 years ago | *Near East* | Start of the Upper Palaeolithic in the Near East (it began in Europe about 10,000 years later). Toolmakers gradually begin to use special techniques to produce blade tools from a carefully prepared core:

(i) The **indirect percussion technique (IV)** {⇨Fig. 4.2–D(e)}, also called the **"punch technique,"** allows one to detach blades by exerting pressure on the core with a pointed tool, such as a sharpened piece of antler or bone.

(ii) The **pressure flaking technique (V)** {⇨Fig. 4.2–D(f)} uses a short, pointed instrument of bone, antler or wood to pry off – rather than to strike off by a blow – many parallel wafer-thin flakes to leave the smallest scars. Pressure flaking was the least violent and most sophisticated of all methods of working stone in a controlled manner by chipping or knapping the edge. Until the advent of the percussion lock {FORSYTH ⇨1805; SHAW ⇨1814}, it was one of the main techniques used to sharpen the flint used in flint locks.

Technique (I) {Koobi Fora ⇨c.2.6 Ma ago}, techniques (II) and (III) {Olduvai Gorge ⇨c.1.4–0.3 Ma ago} and technique (IV) are dynamic methods, while technique (V) is a purely static one. In all cases, however, the generation of **Hertzian cracks**, resulting in a **Hertzian cone** fracture {HERTZ ⇨1882; ⇨Fig. 4.2–B}, is the essential process used to provoke fragmentation. Note that the fracture, governed by elastic wave propagation in the work-piece, is itself a dynamic process, regardless of whether it is triggered by dynamic or static pressures. Typically, in hard, fine-grained materials – such as flint, quartzite, quartz, volcanic rocks (including obsidian and volcanic glass) – breaking occurs along long curved surfaces that look like a shell, so-called **"conchoidal fractures"** {KERKHOF & MÜLLER-BECK ⇨1969}. |
| c.15,000 years ago | *Cygnus ["The Swan," a constellation of the Northern Hemisphere]* | **A supernova explosion occurs in the constellation *Cygnus*.** ▪ **The stellar explosion forms a huge expanding bubble of hot gas which produces the *Cygnus Loop*, a group of bright nebulae in Cygnus.**

The Cygnus Loop is one of the largest supernova remnants; it is (only) about 2,500 light-years away, in Cygnus.

▸ **The shock wave from this old supernova explosion is still expanding past nearby stars** {PIKELNER ⇨1954}. The collision of this gaseous shock wave with a stationary gas cloud heats the gas, causing it to glow in a spectacular array of colors. Unique astronomical pictures of supernova remnants have been taken with a special wide-field camera on board of the NASA/ESA Hubble Space Telescope.[61]

▸ The shock wave is also a thermal bremsstrahlung source of soft X-rays, with a spectral temperature of about $2 \times 10^6$ K.[62] X-ray images of the Cygnus Loop obtained in the 1990s during the ROSAT (Roentgen Satellite) All-Sky Survey revealed gas that was heated to millions of degrees by faster shocks earlier in its history. |
| c.10,000 B.C. | *Old World* | **Pile driving becomes important in Neolithic times.** Fortified villages and lake dwellings are built in large numbers using piles as essential constructional elements. ▪ The origin of pile driving – using a succession of hammer blows to drive (wooden) piles into the ground – is unknown. More recent examples of this technique can be seen in Amsterdam, Venice and other cities that were constructed upon piles. |
| c.9000 B.C. | *Vela ["The Sail," a constellation of the Southern Hemisphere]* | **A supernova occurs in the constellation *Vela*.** ▪ The supernova remnant (SNR), which results from the interaction of ambient gas with stellar material ejected by the supernova, forms the Gum Nebula (named after the Australian astronomer Colin GUM).[63] Although it is the closest supernova remnant, it is so large and faint that it is actually hard to see. High-resolution X-ray |

---

[61] R. NEMIROFF and J. BONNELL: *Cygnus Loop supernova shockwave*. NASA-GSFC (July 18, 1995); http://antwrp.gsfc.nasa.gov/apod/ap950718.html.
[62] *ROSAT PSPC Cygnus Loop soft band*. Max-Planck-Institut für Extraterrestrische Physik (1992); http://wave.xray.mpe.mpg.de/rosat/calendar/1992/aug.
[63] *The Gum Nebula supernova remnant*. Astronomy picture of the day. NASA-GSFC (Nov. 7, 2000); http://antwrp.gsfc.nasa.gov/apod/ap001107.html.

c.6000–2500 B.C. — Middle East, Near East, Asia & Europe

**Transition from the Stone Age to the Bronze Age** for much of humanity living in North Africa, Asia, and Europe.

(i) With the advent of metals (beginning with smelted copper, then bronze and eventually iron) it becomes possible to manufacture metallic hand percussive tools such as hammers, adzes and axes that are far more efficient than those obtained earlier with stone tools.

(ii) Since the metals applied are about three times heavier than rocks, it is possible to shift the center of gravity closer to the tool-head, which better approximates the ideal configuration of a hand-held tool – *i.e.*, a light handle combined with a heavy tool head {WALLIS ⇨ 1670/1671}. Furthermore, **with heavier, more compact metallic tool heads, stronger blows can easily be made, thus initiating a technical revolution in the design of percussive hand tools.**

(iii) **Because copper is comparatively soft, it can be worked by *cold hammering* with stone tools,** a technique which can more than double the hardness of copper to the point where it equals pure iron, although the copper is then brittle and cracks very easily. However, some gentle heating will anneal the copper, which can then be cold-worked again to give a fairly heavy tool with a comparatively hard surface that can, if desired, be shaped to form a sharp edge. ▪ The earliest objects produced by cold hammering techniques were found in Anatolia, in western Iran, and date back to the ninth millennium B.C.; *i.e.*, long before the Bronze Age. In around 6500 B.C., centers of metallurgy appeared in Anatolia and Mesopotamia. In around 4500 B.C., copper objects were produced in the Balkans, the technique later spreading west through the Mediterranean and along the Danube valley.[64]

# ANTIQUITY

c.1645 B.C. — Santorini Island [now Thera], eastern Mediterranean Sea

**Destruction of the *Santorini volcano* by a giant explosive eruption** ($VEI = 6$), probably the greatest volcanic catastrophe of the Bronze Age (number of fatalities unknown). ▪ The sudden collapse and sinking of the island's center beneath the sea formed a submerged caldera, and may have generated a huge 10- to 50-meter-high tsunami, inundating the coastlines of the eastern Mediterranean.

In 1939, **Spyridon MARINATOS**,[65] a Greek professor of archeology, **speculated that the Thera event destroyed the Minoan civilization, the Bronze Age (Aegean) civilization of Crete,** and was probably the source of PLATO's legend of Atlantis. He proposed a date of 1500 B.C. for the eruption. More recently, Danish scientists,[66] studying cores from the Greenland icecap, reported evidence that Thera exploded in 1645 B.C. (with an estimated standard deviation of +7 years and an estimated error limit of +20 years) – *i.e.*, some 150 years before the date usually assumed. That put so much time between the natural disaster and the decline of the Minoan civilization that the link began to be doubted, seeming far-fetched at best. ▪ The date of the eruption of Thera is still a matter of debate. By the time of the 3rd Int. Congress on Thera, Greece (1989), it was clear from archeological evidence that the Minoan civilization on Crete was *not* destroyed by the eruption of Thera.

---

[64] *The secrets of alloys*. Standards, Measurements and Testing. European Commission, Brussels, Belgium. RTD Info Nr. 19 (June/July 1998); http://europa.eu.int/comm/research/rtdinf19/19e07.html.

[65] S. MARINATOS and O. IMBODEN: *Thera, key to the riddle of Minoan*. Natl. Geogr. Mag. **141**, 702-726 (1972).

[66] C.U. HAMMER, H.B. CLAUSEN, W.L. FRIEDRICH, and H. TAUBER: *The Minoan eruption of Santorini in Greece dated to 1645 B.C.?* Nature **328**, 517-519 (Aug. 6, 1987).

In 2006, an international team of scientists from the University of Rhode Island (URI) and the Hellenic Center for Marine Research found that this historic volcanic eruption was much larger and more widespread than previously believed (now estimated at $VEI = 7$). They found deposits of volcanic pumice and ash 10 to 80 m thick extending out 20 to 30 km in all directions from Santorini. The new evidence of the marine deposits resulted in an upward adjustment from a previous estimate of 39 km$^3$ to about 60 km$^3$ of material ejected. An eruption of this size likely had far-reaching impacts on the environment and civilizations in the region (such as the disappearance of the Minoan culture on nearby Crete).[67]

| | | |
|---|---|---|
| c.1400 B.C. | Greece and Aegean Sea | From the beginning of the Mycenaean period, **the early Greeks attributed earthquakes to the angry outbursts of the sea god POSEIDON,** with his three-pronged spear controlling the waves and the sea, shattering rocks, and shaking the shores. ▪ According to the legendary Greek poet HOMER (8th century B.C.) who provided the oldest known literary source of Greek myths and legends, POSEIDON was worshipped in regions affected by earthquakes. |
| c.1200 B.C. | Near East and southeast Europe | Beginning of the Iron Age, in which iron largely replaces bronze in implements and weapons. With the invention of steel (a mixture of iron and carbon), **the elementary principle of percussion is used in fire-making:** the steel is smacked across the edge of a piece of flint to make sparks which are caught in dry tinder. ▪ **Before the introduction of steel, sparks were obtained in the Stone Age using a piece of iron pyrite [$FeS_2$] for the striker.**[68] This technique was also used by Eskimos and certain North America Indians up to modern times. It is still used by some small Indian tribes on the Island of Tierra del Fuego, southern South America. |
| 4th century B.C. | Classical Greece | (According to a widespread old Greek myth) PHAETON, the son of the Sun god HELIOS, took the sun-chariot and drove it too close to the Earth, scorching the surface, until ZEUS cast a thunderbolt and caused Phaeton to fall to his death. The Phaeton myth stimulated both antique authors as well as modern mythologists, catastrophists and astronomers: |

(i) **The Greek philosopher PLATO ($c.429–347$ B.C.) wrote in his treatise *Timaeus*,** one of his later dialogs, **that SOLON ($c.638–558$ B.C.) had been told by an Egyptian priest that the Phaeton myth owed its origin to one of a series of cosmic disturbances which produced periodic catastrophes on Earth.** The priest claimed,[69] "That story, as it is old, is in the style of a legend, but the truth of it lies in the shifting of the heavenly bodies which move around the Earth, and a destruction of many things on the Earth by fierce fire, which recurs after long intervals."

(ii) In the 1920s, the German Jesuit priest Franz Xaver KUGLER,[70] an authority on Babylonian and Hebrew astronomy and mythology, reached the conclusion that Phaeton had been a very bright celestial object which fell to Earth as a shower of large meteorites, causing catastrophic fires and floods in Africa and elsewhere. He wrote: "Repeatedly meteorites have been observed in both past and present times which, resembling the Sun in regard to size and brightness, passed across the sky in different directions at high velocity, then often exploding with lightning and thunder, sometimes setting terrestrial homes and fields on fire."

(iii) Some modern astronomers have speculated that a former major planet of the Solar System, now missing between Mars and Jupiter, was destroyed in the comparatively recent past

---

[67] *Santorini eruption much larger than originally believed.* Dept. of Communications, University of Rhode Island, Kingston, RI (Aug. 23, 2006); http://www.uri.edu/news/releases/?id=3654.

[68] D. STAPERT and L. JOHANSEN: *Flint and pyrite: making fire in the stone age.* Antiquity **73**, No. 282, 765-777 (1999).

[69] PLATO: *Timaeus, Critias, Cleitophon, Menexenus, Epistles* (translated by R.G. BURY). Plato with an English translation, vol. 7. Loeb Classical Library, Harvard University Press, Cambridge, MA (1929), chap. 22, c & d.

[70] F.X. KUGLER: *Sybillinischer Sternkampf und Phaëthon in naturgeschichtlicher Beleuchtung* ("The Sybilline battle of the stars and Paeton seen as natural history"). Aschendorff, Münster (1927), pp. 36-39; http://abob.libs.uga.edu/bobk/kugler/.

333 B.C. *Classical Greece*

**ARISTOTLE (384–322 B.C.),**[71] one of the most eminent Greek philosophers and naturalists, **reflects in a surprisingly detailed manner on the nature of sound, intense sound (thunder) and the possible origins of earthquakes and volcanism.**

(i) Explaining the generation of sound waves in water and air by percussion, he states, "All sounds are produced by the meeting of bodies or of the air with bodies, because the air is set in motion in the same way in which, in other cases, bodies are moved, whether by contraction or expansion or compression, or again when it clashes together by an impact from the breath or from the strings of musical instruments" [*On Things Heard*; *see* the introduction, p. 122]. "Sound is heard both in air and in water. **What is required for the production of sound is an impact of two solids against one another and against the air.** The latter condition is satisfied when the air impinged upon does not retreat before the blow; *i.e.*, is not dissipated by it. That is why it must be struck with a sudden sharp blow, if it is to sound – the movement of the whip must outrun the dispersion of the air, just as one might get in a stroke at a heap or whirl of sand as it was traveling rapidly past" [*On the Soul*; Book II, Chap. 8, p. 668].

(ii) **He explains an echo by a rebound of aerial masses:** "An echo occurs when the air rebounds from a mass of air, having been unified, bounded and prevented from dissipation by the containing walls of a vessel, rebounding like a ball from a wall" [*On the Soul*; Book II, Chap. 8, p. 668].

(iii) **Addressing the action of intense sound waves** (*i.e.*, weak shock waves), **he says, "If the moving bodies are so great, and the sound which penetrates to us is proportionate to their size, that sound must reach us with an intensity many times that of thunder,** and the force of its action must be immense" [*On the Heavens*; Book II, Chap. 9, p. 479].

(iv) **He explains thunder by the collision of two clouds:** "There are two kinds of exhalation, moist and dry, and their combination contains them both potentially. Now the heat that escapes disperses to the upper region. But if any of the dry exhalation is caught in the process as the air cools, it is squeezed out as the clouds contract, and it forcibly carries on and collides with the neighboring clouds, and the sound of this collision is what we call 'thunder' [*Meteorology*; Book II, Chap. 9; p. 596] ... A wind which is whirled along in a dense watery cloud and being driven forth through it violently breaks up the continuous masses of the cloud, causing a roar and crash, which we call 'thunder' " [*On the Universe*; p. 631].

(v) Contrary to contemporary mythical and astrological theories, **he considers earthquakes to be analogous to atmospheric events and explains volcanism likewise.** His concept of winds produced by evaporation ascribes great power to "exhalations" generated by the Earth's own internal fire.[72] "The earth is essentially dry, but rain fills it with moisture. Then the Sun and its own fire warm it and give rise to a quantity of wind both outside and inside it. But not water nor earth is the cause of earthquakes but wind – that is, the inrush of the external exhalation. The severest earthquakes take place where the earth is spongy and cavernous. A violent wind is thrust back into the earth by the onrush of the sea in a great mass [*Meteorology*; Book II, Chap. 8, p. 592.] ... As the earth contains many sources of water, so also it contains many sources of wind and fire. Of these, some are subterranean and invisible, but many have vents and blow-holes. Often, a strong current becomes caught in the hollows of the earth, and, being cut off, it shakes the earth violently, seeking an exit, and sets up the condition which we commonly call an 'earthquake.' We find analogous phenomena occurring in the sea. Chasms form in it and its waters often recede or the waves rush in" [*On the Universe*; pp. 632-633].

---

[71] J. BARNES (ed.): *The complete works of ARISTOTLE*. Bollingen Series LXXI, Princeton University Press, Guildford, Surrey (1984).

[72] E. OESER: *Historical earthquake theories from ARISTOTLE to KANT*. Abhandl. Geol. Bundesanstalt Wien **48**, 11-31 (1992).

(vi) **In his theory of motion, ARISTOTLE assumes that a body in constant motion** – such as a flying projectile – **requires a constant cause;** in other words, as long as a body remains in motion, a force must be acting on that body [*Physica*; Books III-VII]. ▪ This incorrect assumption was not recognized until the late Middle Ages {BURIDAN ⇨ c.1350}.

(vii) Like other ancient philosophers, he takes the view that the starry sky is eternally invariable. ▪ In this way, the discovery of variable stars[73] contributed to the "Astronomical Revolution" of the 16th and early 17th centuries {BRAHE ⇨ 1572; D. FABRICIUS ⇨ 1596; KEPLER ⇨ 1604}.

| | | |
|---|---|---|
| 246–146 B.C. | Mediterranean | **Application of the catapult or *ballista*** {⇨Fig. 4.2–F} **on a large scale in the Punic Wars** – a long-lasting conflict between the Roman Republic and the Carthaginian (or Punic) Empire that resulted in the final destruction of Carthage. ▪ The heaviest engines were capable of throwing balls weighting from 100 to 300 kg at a range of 750 m [1000 *Schritte*].[74] The enormous force was generated by the torsion of two strong bundles of tendons. The projectile, a ball of stone or lead, was accelerated along a groove with an inclination of about 45°. The considerable kinetic energy of the projectile impacting the target was transformed via percussion into destructive power. |

**The word "ballistic" was derived from the Latin word *ballista*.** *Ballistics* is a term that encompasses a wide variety of disciplines and can be subdivided into three main categories:
▸ *aeroballistics,* dealing with ballistics (especially of missiles) in the atmosphere;
▸ *aerospace ballistics,* dealing with the flight path trajectory of a spacecraft through a gaseous atmosphere (air) into space and sometimes back through the same or another atmosphere, and;
▸ *hydroballistics,* dealing with the flight path trajectory of powered/unpowered wetted, cavitating and supercavitating projectiles fired, launched or propelled in liquids (mostly water).

Each of these categories has three main aspects:
▸ *internal ballistics,* the study of a projectile's behavior from the time it ignites the propellant to the time it exits from the barrel of the firearm;
▸ *external ballistics,* the study of what happens during the bullet's flight after leaving the muzzle and before reaching the target, and;
▸ *terminal ballistics,* the study of how a projectile penetrates the (solid) target.

| | | |
|---|---|---|
| 1st century B.C. | Roman Empire | **Titus LUCRETIUS CARUS** (*c*.99–*c*.55 B.C.),[75] a Latin poet and philosopher, **writes the monumental poem *De rerum natura*** ("On the Nature of Things"). It consists of six books in which he strongly propounds the atomic theory of EPICURUS (341–270 B.C.), a Greek philosopher. In Book I and II, **LUCRETIUS states that all matter is composed of very small, invisible entities** – the indivisible (hence, structureless) atoms – **which move rapidly with equal velocities** and, at random times and positions, suffer an imperceptible change, **just enough to allow occasional collisions to take place between the atoms.** The only forces are those of the collisions between atoms. Apart from atoms, the only reality is infinite empty space, through which the atoms move. ▪ **His postulated principles,** which were surprisingly ahead of their time and partly mirror the basic tenets of modern atomism, **also stimulated the genesis of the kinetic theory of gases** {D. BERNOULLI ⇨ 1738; LESAGE ⇨ 1784; JOULE ⇨ 1850; CLAUSIUS ⇨ 1857; MAXWELL ⇨ 1867}. |

---

[73] For the classification of variable stars *see* General Catalogue of Variable Stars ⇨ 1948.
[74] *Meyers Konversations-Lexikon*. Bibliographisches Institut, Leipzig, vol. II (1894), p. 391.
[75] C. BAILEY: *TITI LUCRETI CARI De rerum natura* [in Lat.]. 3 vols., Clarendon Press, Oxford (1947).

| | | |
|---|---|---|
| c.60 | Alexandreia, Roman Empire [now Alexandria, northern Egypt] | **HERO [or HERON] of Alexandreia** (c.10–c.70), a Greek mathematician and inventor, **describes in his treatise *Pneumatica* ("Pneumatics") an *aeolipile*** [Lat. *aeoli pila*] – **a little steam turbine** which consists of a hollow ball pivoted so that it can turn on a pair of hollow tubes through which steam is supplied from a cauldron {⇨Fig. 4.7–C}. The steam, jetting out of the ball from one or more projecting bent tubes, causes the ball to revolve. ▪ His invention can be regarded as the forerunner of the reaction steam turbine {Sir PARSONS ⇨1884}. |
| A.D. 63 | Roma, Imperium Romanum | **Lucius Annaeus SENECA** (c.4 B.C.–A.D. 65),[76] a Roman philosopher and former tutor of Emperor NERO, **publishes his *Naturales quaestiones* ("Treatises on Natural Phenomena")** – a popular sketch of astronomy and meteorology that uses physics as the basis for ethical reflections. He also summarizes opinions of contemporary naturalists on the origin of lightning and thunder, and writes, "Fire is produced by the percussion of flint and steel, and by the friction of two pieces of wood; it may happen, therefore, that the clouds, hurried away by the wind, are likewise inflamed by means of percussion and friction" [Book II, § 22]. **Concerning the noise of thunder, he has the idea that, analogous to a hand clap, it is generated by the collision of two enormous clouds** [Book II, § 27]. |
| 77 | Misenum [now Miseno], Gulf of Naples, Roman Empire | **Gaius PLINIUS SECUNDUS** [known as "PLINY the Elder" (23–79)],[77] an ancient Roman natural philosopher, administrator and author who knew the origin of lightning in the friction of clouds, **publishes his *Historia naturalis* ("Natural History"), an epoch-making encyclopedic work in 37 books.** He writes: "It is not generally known what has been discovered by men who are the most eminent for their learning, in consequence of their assiduous observations of the heavens, that the fires which fall upon the Earth, and receive the name of thunderbolts [*fulminum nomen habeant*] proceed from the three superior stars [*siderum*], but principally from the one which is situated in the middle … and hence it is commonly said, **the thunderbolts are darted by Jupiter.**" ▪ Modern space research confirmed that lightning (and most probably also thunder) exist on other planets, including Jupiter {RINNERT ⇨1985}. |
| 79 | Mt. Vesuvius, Gulf of Naples, central Italy | On August 24, there is a **catastrophic eruption of the volcano *Mount Vesuvius*** (*VEI* = 5), number of casualties unknown. Great quantities of ash-laden gas are violently discharged to form a cauliflower-shaped eruption column that rises to heights estimated at up to 30 km. This causes an intense rain of pumice and volcanic ash, which destroys the nearby towns of Pompeii [now Pompei] and Herculaneum [now Ercolano]. Gaius PLINIUS CAECILIUS SECUNDUS [known as **"PLINY the Younger"** (c.63–113)],[78] a Roman author, administrator and nephew of PLINY the Elder {⇨A.D. 77}, **witnesses the eruption from about 30 km west of the volcano.** ▪ Later he recorded his observations in two letters (Book VI, letters 16 & 17):<br>(i) Describing the huge eruption cloud towering over the area, he wrote: "Its general appearance can best be expressed as being like a pine rather than any other tree, for it rose to a great height on a sort of trunk and then split off into branches, I imagine because it was thrust upwards by the first blast and left unsupported as the pressure subsided, or else it was borne down by its own weight so that it spread out and gradually dispersed. Sometimes it looked white, sometimes blotched and dirty, according to the amount of soil and ashes it carried with it." ▪ About 4 km$^3$ of ash was erupted in about 19 hours.<br>(ii) He described the effects of the eruption on buildings and the fall of ashes and pumice lapilli: "They debated whether to stay indoors or take their chance in the open, for **the buildings were now shaking with violent shocks, and seemed to be swaying to and fro as if they** |

---

[76] L.A. SENECA: *Naturales quaestiones* [Books I-III]; Engl. translation by T.H. CORCORAN. The Loeb classical library, vol. 7. Harvard University Press, Cambridge, MA (1999).

[77] PLINY the Elder: *Historia naturalis* ("Natural History"). Engl. translation by J. BOSTOCK and H.T. RILEY. London, Book. II (1865): *An account of the world and the elements*, chap. 18: *Why thunder is ascribed to Jupiter.*

[78] B. RADICE (ed.): *PLINY: letters and Panegyricus.* Harvard University Press, London (1969), pp. 427-445.

were torn from their foundations. Outside on the other hand, there was the danger of falling pumice-stones, even though these were light and porous…"

(iii) He even described the tsunami generated by the earthquake: "**We** [PLINY the Younger and his mother who tried to escape from the approaching volcanic conflagration] **also saw the sea sucked away and apparently forced back by the earthquake:** at any rate it receded from the shore so that quantities of sea creatures were left stranded on dry sand. On the landward side a fearful black cloud was rent by forked and quivering bursts of flame, and parted to reveal great tongues of fire, like flashes of lightning magnified in size…"

In the following centuries, the most violent and devastating explosive eruptions of Vesuvius happened in 1631 (emission of an approx. 50-km-high eruption cloud, pyroclastic flows and lahars; collapse of the crater accompanied by earthquakes and a tsunami; ca. 4,000 casualties) and in 1944, hitherto the last eruption {Mt. Vesuvius ⇨ 1944}.

In so-called **"Plinian eruptions,"** named after PLINY the Younger, the dissolution of magmatic volatiles in the volcano's conduit leads to

▸ **disruption, the explosive ejection of pyroclastic material;** and
▸ the formation of a high-altitude eruption column – the "umbrella pine" described by PLINY the Younger – which is sustained for hours or days above the volcano.

The velocities of ejecta can approach or even exceed the sound velocity in air, thus producing blast waves and seismic shocks, particularly when the vent is suddenly uncorked in the initial phase of eruption. Experimental evidence for the emission of atmospheric shock waves (blast waves) during volcanic eruptions has been found by various researchers.[79]

*c.*85    *K'uei-chi [now Cheki-ang], eastern China*

WANG CH'UNG (27–97), a Chinese philosopher and writer, **correctly recognizes a connection between tides and tidal bores, and the lunar cycle.**

(i) In his essay *Lun-hêng* ("Disquisitions") he states,[80] "When the rivers fall into the ocean, they merely accelerate their course, but, upon entering the three rivers [*i.e.*, the Ch'ien-tang, the Shan-yin and the Shang-yü], they begin to roar and foam in their channels, which is usually shallow and narrow, and thus rise as great waves … The Ch'ü river of Kwang-ling [a place in Kiangsu] has such great waves. They are caused by the narrow passage … In deep channels the water flows quietly, but where there are shallows, sands, or stones, it rushes through, swells, and forms rapids … Billows and rapids are identical … When the billows enter the three rivers, they boil and wallop against the banks, while in the middle no sound is produced … The rising of the waves follows the growing and waning, the bigness and smallness, the fullness and extinction of the Moon."

(ii) WANG CH'UNG also referred to the **Hang-Chau bore**, the famous spring tide occurring at regular intervals and entering the Ch'ien-tang [or Qiantang, Quiantang] river {⇨Fig. 4.1−L}.

In the late 1890s, William U. MOORE,[81] a navy captain of the British Admiralty, observed that the bore originates not at the mouth of the river, as was hitherto expected, but 12 or 15 miles (19.2 or 24 km) outside it.

▸ It is divided into two branches, which join together 4 miles (6.4 km) from Haining, making a continuous white line two miles (3.2 km) in length.
▸ It shortly afterwards contracts in width, and increases in speed and height, rising 8–11 ft (2.4–3.3 m) high, and traveling between 12 and 13 knots an hour (6–6.7 m/s), which is pretty close to modern estimations: in a more recent note, it is reported that

---
[79] R.P. HOBLITT, C.D. MILLER, and W.E. SCOTT: *Atmospheric shock waves induced by eruptions*. In: *Volcanic hazards with regards to sitting nuclear-power plants in the Pacific northwest*. Rept. No. 87-297, USGS (1987).
[80] A. FORKE: *Miscellaneous essays of WANG CH'UNG*; Engl. translation by LUN-HÊNG. Paragon, New York, vol. II (1962), pp. 250-251.
[81] W.U. MOORE: *The bore of the Tsien-Tang Kiang (Hang-Chau Bay)*. J. Chin. Roy. Asiat. Soc. **23**, 185-247 (1889); Nature **40**, 163 (1889).

| | | the Qiantang tidal bore speeds along at approximately 15 mph (6.7 m/s), and its crest can attain heights of 15 ft (4.6 m).[82] |
|---|---|---|
| c.130 | *Imperial Chancellery for Astronomical and Calendrical Sciences, Sian [now Xi'an], central China* | **CHANG HÊNG** [or ZHANG HENG] (78–139), a Chinese philosopher, astronomer and mathematician, **invents the first seismoscope** {⇨Fig. 4.3–L}. ▪ The ingenious instrument, which unfortunately has not survived, was a large bronze urn, consisting of an 8-*shaku* (about 2-m)-dia. copper vessel with eight dragon heads gazing in eight directions, and apparently provided with an inner pendulum. Bryce WALKER,[83] showing a marvelous model in his book *Earthquake*, described its function as follows, "Eight dragons held in their mouths bronze balls; an internal mechanism, activated by even a slight tremor, opened the mouth of one dragon, releasing the ball to sound an alarm as it clanked into the open mouth of a metal toad below. Imperial watchmen determined the direction of the quake from the orientation of the empty-mouthed dragon."<br><br>In about A.D. 132, a large earthquake struck a city about 600 km northwest of the Chinese capital city of Sian. It was not felt in Sian at the location of the seismoscope, but was reportedly announced by the loud ringing of a bronze ball falling from the northwest dragon's mouth into that of a toad.[84, 85]<br><br>Seismoscopes of limited effectiveness were used in the 18th century. For example, in 1751 the Italian earthquake researcher Andrea BINA used a simple pendulum above a tray of sand. |
| 365 | *Aegean Sea and Alexandria, Roman Empire* | On July 21, **a great seismic sea wave, shortly following an earthquake, overflows the Aegean coasts and islands.** Thousands of people die and innumerable buildings are washed away.<br><br>Later, **Ammianus MARCELLINUS** (330–395),[86] a Greek-Roman historian, **gave in his** ***Rerum gestarum libri qui supersunt*** **("On Currently Existing Books on Deeds"), an account of this event,**[87] which was probably the first detailed description in history on the characteristics and effects of seismic sea waves (or tsunamis). **He particularly addresses the receding water level,** which, as we know today, is caused by the wave propagating landward with a leading trough – a curious phenomenon which typically precedes the arrival of the first tsunami wave {Lisbon Earthquake ⇨1755; Arica Earthquake ⇨1868; Sumatra-Andaman Islands Earthquake ⇨2004}. He wrote:<br><br>(i) "On the 21st of July, in the first consulship of [Flavius] VALENTINIAN with his brother [Valens], horrible phenomena suddenly spread through the entire extent of the world, such as are related to us neither in fable nor in truthful history. For a little after daybreak, preceded by heavy and repeated thunder and lightning, **the whole of the firm and solid Earth was shaken and trembled, the sea with its rolling waves was driven back and withdrew from the land,** so that in the abyss of the deep thus revealed, men saw many kinds of sea-creatures stuck fast in the slime; and vast mountains and deep valleys, which Nature, the creator, had hidden in the unplumbed depths, and then, as one might well believe, first saw the beams of the Sun.<br><br>(ii) Hence, many ships were stranded as if on dry land, and since many men roamed about without fear in the little that remained of the waters, to gather fish and similar things with their hands, **the roaring sea, resenting, as it were, this forced retreat, rose in its turn; and over the boiling shoals it dashed mightily upon islands and broad stretches of the mainland,** and leveled innumerable buildings in the cities and wherever else they were found; so that amid the mad discord of the elements the altered face of the Earth revealed marvelous sights. |

---

[82] *Science question of the week.* NASA-GSFC (Dec. 21, 2001); http://www.gsfc.nasa.gov/scienceques2001/20011221.htm.
[83] B. WALKER: *Earthquake*. Time-Life Books Inc., Alexandria, VA (1982), p. 81.
[84] J. MILNE: *Seismology in Japan*. Nature **26**, 627-631 (1882).
[85] J. DEWEY and P. BYERLY: *The early history of seismometry (to 1900)*. Bull. Seism. Soc. Am. **59**, No. 1, 183-227 (1969).
[86] J.C. ROLFE: *Ammianus MARCELLINUS*. Harvard University Press, Cambridge, MA (2000), pp. 649-651.
[87] See his *Book XXVI*, pp. 10, 15-19.

(iii) **For the great mass of waters, returning when it was least expected, killed many thousands of men by drowning;** and by the swift recoil of the eddying tides a number of ships, after the swelling of the wet element subsided, were seen to have foundered, and the lifeless bodies of shipwrecked persons lay floating on their backs or on their faces.

(iv) Other great ships, driven by the mad blasts, landed on the tops of buildings (as happened at Alexandria), and some were driven almost two miles inland, like a Laconian ship which I myself in passing that way saw near the town of Mothone, yawning apart through long decay."

This unprecedented Aegean Earthquake was probably responsible for a 9-m uplift of western Crete, resulting from the reactivation of a major thrust fault along the Hellenic (or southern Aegean) volcanic arc.[88]

| 393 | China | During March, **a "guest star" appears in *Scorpius*** ("The Scorpion," a prominent constellation of the Southern Hemisphere) and is visible for eight months (according to the records of the Chinese CHIN-SHU, A.D. 635).[89] • It is likely to have been a supernova. In 1997, Chinese astronomers have proposed that the shell-type supernova remnant RX J1713.7-3946 is a possible relic of the A.D. 393 guest star {HESS Telescope Array ⇨2004}.

The impact of Chinese historical astronomical records is important in the study of astronomy today. In particular, the impact of the Chinese records related to historical supernovae have made important contributions to modern astronomy, contributing to the rapid progress of space sciences and high-energy astrophysics made in the recent two decades.[90]

# MIDDLE AGES

| c.800 | Sabrina, Britannia [now River Severn, England] | NENNIUS, a Welsh Christian monk, compiles or revises the *Historia Britonum* ("History of the Britons"), a collection of historical and topographical information.

(i) In the attached section *De Mirabilibus Britanniae* ("On the Wonders of Britain") he records thirteen wonders. The fifth wonder is that of the bubbles of froth at the *Fluminis Sabrinae* – i.e., the River Severn [Sabrina was the Roman transmogrification of "Habren," the name of both the river and a Celtic princess]. **The "bubbles of froth" refer to the Severn's tidal bore, where water flowing downstream meets the upcoming tide to create an amazing water form.**[91]

(ii) In one reference to the Severn Bore, he writes: "Another wonder is Dan Ri Hafren, that is, the Two Kings of the Severn. When the sea floods into the Severn estuary, two heaped-up wave crests are built up separately, and fight each other like rams. One goes against the other, and they clash in turn, and then one withdraws from the other and they go forth together again at each tide. This they have done from the beginning of the world to the present day…"[92] • There are only two locations on the Severn where such an event can be witnessed. Before the weir was built at Maisemore, the west channel tide would wrap around the Upper Parting into the eastern channel and collide with the flood tide running up that channel. But more famous is the collision that occurs in the eastern channel of the Noose, as the leading west-channel tide rebounds off Hock Cliff and flows straight back into collision with the advancing flood tide in the eastern channel.[93]

---

[88] S.C. STIROS and S. PAPAGEORGIOU: *Seismicity of western Crete and the destruction of the town of Kisamos at A.D. 365: archaeological evidence.* J. Seismol. **5**, 381-397 (2001).

[89] R. STOTHERS: *Is the supernova of A.D. 185 recorded in ancient Roman literature?* Isis **68**, No. 3, 443-447 (Sept. 1977).

[90] Z.R. WANG: *The impact of historical Chinese astronomical records.* Astrophys. & Space Sci. **305**, 207-210 (2006).

[91] S.B. PALMER: *De mirabilibus Britanniae.* What Planet is this? (2005); http://inamidst.com/notes/mirabilibus.

[92] NENNIUS: *British history and the Welsh annals* (edited and translated by J. MORRIS). Phillimore, London (1980).

[93] This note was taken from *Severn Bore, the wonder of Britain.* Tidal Bore Research Society (TBRS), U.K.; http://www.tidalbore.info/england/severn_wonder.html.

| 1006 | China & Switzerland | **Chinese and Swiss annals report on the observation of a "guest star"** which, previously not present, makes a transient visit to the heavens. ▪ Its exact position in the southern constellation near the star Beta Lupi, on the border of *Centaurus* ("The Centaur," a constellation of the Southern Hemisphere), was only recently figured out by locating its nebulous remnant.

(i) In China, the new star [Lat. *nova stella*] has an extraordinary brilliance and is visible for over a year.[94] ▪ It was probably seen first on April 30, 1006, according to Chinese and Japanese records. The term *guest star* was used in Chinese astronomical records to designate various novae, comets and even meteors.[95]

(ii) In Switzerland, the event is described in the *Annales Sangallenses Maiores*, the annals of the monastery St. Gallen: "A new star of unusual size appeared, glittering in aspect, and dazzling the eyes, causing alarm. In a wonderful manner this was sometimes contracted, sometimes diffused, and moreover sometimes extinguished. It was seen likewise for three months in the innermost limits of the south, beyond all the constellations which are seen in the sky."

Such spontaneous galactic events, accompanied by a spectacular and rapid increase in brightness, were later recognized as stellar explosions {BAADE & ZWICKY ⇨1934}, indicating the violent end to the evolution of a certain type of star. Today we know that the first **nova** (an exploding star) observed in 1006, and named as such, was actually a **supernova** (later termed SN 1006), a violently exploding star which is more than 100 times brighter than an ordinary nova.[96] In 1998, Japanese astrophysicists reported the discovery of TeV gamma ray emission from the remnant of SN 1006 made with the Australian-based CANGAROO 3.8-m telescope.[97]

Further historical supernovae explosions, following the event of A.D. 1006, happened in the constellations of *Taurus* {Guest Star ⇨1054}, *Cassiopeia* {BRAHE ⇨1572}, and *Serpens* {KEPLER ⇨1604}, and, more recently, in the Tarantula Nebula of the Large Magellanic Cloud {L. SHELTON ⇨1987}.[98] Supernovae are the source of the heavier elements from which plants and animals are made. Since supernovae emit these elements into interstellar space and transport them over distances of millions of light-years to remote galaxies, supernovae are therefore fundamental to our own life as well as to any other life forms that may exist in the Universe. |
| 1054 | Far East (China, Korea, and Japan) & Near East | Around July, **astronomers of various countries witness a stellar explosion, the famous supernova – later to be termed SN 1054 – which flares in *Taurus*** ("The Bull," a winter constellation of the Northern Hemisphere). The supernova, first noted in China on July 4 and reported by Chinese astronomers, can be seen in daylight for 23 days, but is not nearly so bright as its predecessor SN 1006 {Guest Star ⇨1006}. ▪ The "guest star" remained visible to the naked eye for about two years. There is a strong suspicion that Native Americans recorded the event in rock paintings and perhaps on pottery. However, there is no confirmed record of observations in Europe.

**The explosion of SN 1054 produced the rapidly expanding shell of gas that modern astronomers identified as the *Crab Nebula***[99] in M1 (or NGC 1952) – the most well known supernova remnant, which was discovered in 1731 by the English physician and amateur astronomer John BEVIS. The first suggestion of a connection between the Crab Nebula and |

---

[94] Actually a *nova* is not a new star, but an already existing star which due to the eruption has become very luminous and thus visible.

[95] H. PENG-YOKE, F.W. PAAR, and P.W. PARSONS: *The Chinese guest star of A.D. 1054 and the Crab Nebula*. Vistas Astron. **13**, 1-13 (1972).

[96] L. QIBIN: *A recent study on the historical novae and supernovae*. In: (Max-Planck-Gesellschaft & Academia Sinica, eds.) *High energy astrophysics* [2nd Workshop on High Energy Astrophysics [Schloss Ringberg, Germany, July 1987]. Springer, Berlin *etc.* (1988), pp. 2-25.

[97] T. TANIMORI ET AL.: *Discovery of TeV gamma rays from SN 1006: further evidence for the supernova remnant origin of cosmic rays*. Astrophys. J. **497**, L25-L28 (April 10, 1998).

[98] D.H. CLARK and F.R. STEPHENSON: *The historical supernovae*. Pergamon Press, Oxford (1977); D.H. CLARK: *Supernovae, historical*. In: (J.H. SHIRLEY and R.W. FAIRBRIDGE, eds.) *Encyclopedia of planetary sciences*. Chapman & Hall, London *etc.* (1997), pp. 886-887.

[99] The Crab Nebula was given its name in about 1850 by the Irish astronomer William PARSON, Third Earl of Rosse at Birr in Ireland, who observed the appearance of the nebula through his 72-in. (183-cm) reflecting telescope.

supernova SN 1054 seems to have been made by Knut Emil LUNDMARK,[100] a Swedish astronomer. ▪ The hot gas is still expanding so rapidly, at approximately 1,500 km/s, that actual changes can be seen by comparing new photographs with those taken early in the last century. **The nebula expands like a hot bubble of gas into the cold, surrounding interstellar gas, resulting in an expanding shock wave.** At the center of the Crab Nebula is a pulsar {BELL & HEWISH ⇨1967}, a pulsating neutron star {LANDAU ⇨1932; BAADE & ZWICKY ⇨1934} formed from the remains of the star, which produces pulses at radio wavelengths and, unusually for a pulsar, also flashes in the visible and X-ray parts of the spectrum.

| | | |
|---|---|---|
| 12th century | *Germany* | With the beginning of the hard coal-mining industry, it is quite possible that **the first accidents occur due to firedamp (methane/air) explosions.** ▪ Studies in the 19th century showed that to trigger an explosion with an open flame, *e.g.*, with a miner's lamp, the critical mixture is about 9% methane and 91% air {DAVY ⇨1815}. Methane is an odorless, colorless, hydrocarbon gas which results from the decay of organic matter.<br><br>Later it was also realized that coal dust could ignite and explode, even in the absence of firedamp {FARADAY & LYELL ⇨1845}. |
| 1232 | *Pien-ching [now Kaifeng], East China* | A Chinese chronicler gives **the oldest known account of the production and effects of man-made chemical explosions.** He reports that during the siege of the town Pien-ching by Mongolian troops, the citizens used explosive projectiles (assumed to be rockets) for defense purposes. Known as "sky-shaking thunder" [Chin. *zhentianlei*], they produced a loud report, audible up to a distance of 100 *li* (about 50 km), and provoking "shattering penetrations" among the enemies. These explosive projectiles contained *huoyao* ("black powder"), packed in tube-like iron vessels which were ignited via a slow-match and then catapulted.[101] |
| c.1250 | *Northern Europe* | **Marcus GRAECUS** [Mark "The Greek"], an otherwise unheard-of author, and possibly a fictitious one, **describes black powder, and its use in rockets and fireworks, in a six-page tract in Latin entitled *Liber ignium ad comburendos hostes*** ("Book of Fires for Burning Enemies").[102] For use in rockets, he recommends a 6:2:1 formula (by weight) for the three ingredients saltpeter ("Chinese snow"), charcoal and sulfur, respectively, which will become in the 17th century a standard recipe for English gunpowder {Governmental studies ⇨1635}. He even recommends the use of willow charcoal, which modern experience has shown to be one of the best ingredients of use in gunpowder. ▪ His tract was probably translated from an original Greek text and was certainly written no later than the beginning of the 9th century. |
| 1267 | *Oxford, England* | **Roger [Rogerus] BACON** (*c.*1214–*c.*1294),[103] an eminent English monk and scholar, **publishes his *Opus majus* ("Great Work"), his largest treatise.** In the Sixth Part, entitled *La scienza sperimentale* ("On Experimental Science"), he mentions saltpeter and its use in firecrackers, which are obviously already being used as childrens' toys. Describing the explosive properties of black powder, the early form of gunpowder, he writes: "Certain inventions disturb the hearing to such a degree that, if they are set off suddenly at night with sufficient skill, neither city nor army can endure them. No clap of thunder could compare with such noises. Certain of these strike such terror to the sight that the coruscations of the clouds disturb it incomparably less … We have an example of this in that toy of children which is made in many parts of the |

---

[100] K.E. LUNDMARK. *Suspected new stars recorded in old chronicles and among recent meridian observations.* Publ. Astron. Soc. Pac. **33**, No. 192, 225-238 (1921).

[101] H.U. VOGEL: *Das Schießpulver und dessen militärische Nutzung.* In: (A. EGGEBRECHT, ed.) *China, eine Wiege der Weltkultur: 5000 Jahre Erfindungen und Entdeckungen.* Hildesheimer Ausstellungskatalog, Römer- und Pelizaeus-Museum. Von Zabern, Mainz (1994), pp. 148-151.

[102] *Liber ignium ad comburendos hostes auctore Marco GRÆCO.* See F.J.G. DE LAPORTE DU THEIL: *Traité des feux propres à détruire les ennemis, composé par Marcus LE GREC.* Publié d'après deux manuscrits de la Bibliothèque Nationale, Paris (1804).

[103] R. BACON: *Fratris Rogeri BACON, ordinis Minorum, opus majus ad Clementem IV, pontificem Romanum.* Typis Gulielmi Bouyer, Londini (1733); R.B. BURKE (editor and translator): *The opus maius of Roger BACON.* University of Pennsylvania Press, Philadelphia, PA (1928); vol. 2, pp. 629-630.

world, namely, an instrument as large as the human thumb. **From the force of the salt called 'saltpeter,' so horrible a sound is produced at the bursting of so small a thing, namely, a small piece of parchment, that we perceive it exceeds the roar of sharp thunder,** and the flash exceeds the greatest brilliancy of the lightning accompanying the thunder." ▪ BACON read Arabic, and it is possible that he got his knowledge from Arabic sources.

| | | |
|---|---|---|
| 1313/ 1314 | *Ghent, capital of Flanders [now northwestern Belgium]* | **Authentic town records first indicate the existence of firearms.** ▪ Today it is difficult to say exactly when cannons were invented; many records of particular events or battles mentioning firearms were written some time after the incident. Some of these also suggest the existence of firearms at the siege of Metz (France) in 1324 and its use by English troops against the Scots in 1327. According to Father Angelo BELLANI (1776–1852), an Italian priest, the English army used cannons – short-barreled columbiads – at the Battle of Crécy (France) {⇨1346}. However, English and French writers do not mention the use of cannon in their description of the battle. |
| 1320 | *Russia* | **A manuscript illustration by SHEM ED DIN MOHAMED shows tubes for shooting arrows and balls by means of gunpowder.** ▪ Today this oldest known manuscript showing a firearm is kept at the Asiatic Museum in St. Petersburg, Russia. |
| 1326 | *Library of Christ Church, Oxford, England* | **First known illustration of a cannon shot provided by the British clergyman Walter DE MILIMETE** in his manuscript *De nobilitatibus, sapientiis et prudentiis regum* ("On the Dignity, Wisdom and Prudence of Kings"), which he presented to king Edward III of England. **It shows an arrow projectile at the moment of leaving the muzzle** {⇨Fig. 4.2–H}. ▪ This early type of cannon, which was called "iron pot" [French *pots de fer*] or "vase" [French *vase*], was manufactured at Ghent. One was sent to England in 1314.<br><br>In the same year, a Florentine document mentions that the Council of Florence directs the manufacture of metal cannons and bullets [Lat. *pilas seu pallectas ferreas et canones de metallo*]. |
| 1346 | *Crécy-en-Ponthieu Dépt. Somme, northern France* | In the Battle of Crécy, the first major land battle of the Hundred Years' War (1337–1453), French troops and their allies under king Philip VI are defeated by English troops under king Edward III.<br><br>(i) **This battle,** in which the English longbow triumphed over Genoese crossbowmen and armored knights, **established the longbow as one of the most feared weapons of the medieval period.** In addition, a longbowman could shoot two to five times more frequently in a given time than a crossbowman.<br>▸ William Forbes PATERSON,[104] a British crossbow expert, published data about an experiment comparing a longbow and a crossbow that was spanned with a cranequin. A longbow using a draw weight of about 31 kg accelerated a 70.9-g bolt to a velocity of about 40 m/s. This velocity could also be obtained by a crossbow, although it could only be imparted to a 35.5-g bolt and only by applying a much larger draw weight of 336 kg.<br>▸ Modern hunting crossbows are engineered to launch 25-g bolts at initial velocities in excess of 60 m/s, with draw weights of about 68 kg.<br><br>(ii) **The English army also successfully used a type of short-barreled cannon, called a "bombard,"** which frightened the horses, thus adding to the carnage. |
| 1350s | *Milan, northern Italy* | **Francesco PETRARCA,** an Italian scholar and poet, **refers to the recently used cannons at the siege of Metz** (1324) introduced via the Arabs or Mongols. In his treatise *De remedies utriusque fortunae* ("Remedies for Fortune Fair and Foul") he writes, "These instruments which discharge balls of metal with most tremendous noise and flashes of fire … were a few |

---

[104] W.F. PATERSON: *A guide to the crossbow.* Society of Archer-Antiquaries, Bridlington, U.K. (1990).

years ago very rare and were viewed with greatest astonishment and admiration, but now they are become as common and familiar as any other kinds of arms."

c.1350 — Faculty of Arts, University of Paris, France

**Jean BURIDAN** (c.1300–1358),[105] a little-known French philosopher of the later Middle Ages, **names the motion-maintaining property *impetus*.**

(i) Revising ARISTOTLE's theory of motion {ARISTOTLE ⇨ 333 B.C.}, he promotes his "Theory of Impetus," arguing that a projectile continues in motion not – as ARISTOTLE had held – because it is supported by the surrounding air, but because of the force transmitted to it by the object that launched it. He writes: "… after leaving the arm of the thrower, the projectile would be moved by an impetus given to it by the thrower and would continue to be moved as long as the impetus remained stronger than the resistance, and would be of infinite duration were it not diminished and corrupted by a contrary force resisting it or by something inclining it to a contrary motion." [*Physica*, Book VIII]

(ii) He further holds that the impetus increases with the speed with which it was set in motion, and with its quantity of matter.

**BURIDAN anticipated Sir Isaac NEWTON's First Law of Motion** {Sir NEWTON ⇨ 1687}. Clearly, his term *impetus* or *impressed force* of a body of mass $m$ moving with velocity $v$ is closely related to the modern concept of momentum given by $m \times v$. The idea of impetus was not completely new on the 14th-century scene; the term had been used in biblical and Roman literature in the general sense of a thrust toward some goal, and John PHILOPONUS, a Greek commentator on ARISTOTLE, had written in the 6th century of an "incorporeal kinetic force" impressed on a projectile as the cause of its motion.[106] More than 400 years later, the English physicist WOLLASTON {⇨ 1805} had still thought it convenient to give the *vis viva* ($mv^2$) of a body in motion the name of impetus.

c.1380 — Freiburg, im Breisgau, southwestern Germany

**Legendary invention of *Schwarzpulver* ("black powder") or gunpowder by Berthold SCHWARZ**, a Franciscan friar and alchemist. ▪ There has been much difference of opinion regarding his share in the discovery attributed to him. Many later writers placed him vaguely in the 14th century, and while some, being more specific, give 1354 – the date inscribed upon his monument at Freiburg – as the time of his discovery; others simply give him credit for the invention of firearms, and notably of brass cannon.[107, 108]

A more recent historical study performed by W. Gerd KRAMER,[109] a German physicochemist and historian, brought to light that in the 1370s the nigromant (meaning "alchemist") Berthold ANKLITZEN – in an old chronicle also named "Bertholdus niger" [Germ. *der schwarze Berthold*] and erroneously retranslated into Berthold SCHWARZ – invented a granular, more efficient gunpowder for use in the *Steinbüchse* ("stone gun"), a short brass cannon with a stepped caliber for shooting stone balls, that he had also invented. SCHWARZ (correctly) recognized that not the fire but rather the pressure of the fired gunpowder propels the projectile. In order to prevent leakage of the hot propellant gases, he applied a wooden plug in the narrow part of the barrel carrying the gunpowder. The ball was placed in the wide part of the barrel just in front of this plug.

---

[105] J. BURIDAN: *Quaestioners super octo libros Physicorum ARISTOTELIS* ("Questions about the eight books of ARISTOTLE"). Paris (c. 1350); *Subtilisimae quaestiones super octo Physicorum libros ARISTOTELIS*. Edn. Paris (1509).

[106] W.A. WALLACE: *Experimental science and mechanics in the Middle Ages*. In: (P.P. WIENER, ed.) *The dictionary of the history of ideas: studies of selected pivotal ideas*. 4 vols., Ch. Scribner's Sons, New York (1973/1974), vol. 2, pp. 197-205.

[107] H. HANSJAKOB: *Der schwarze Berthold: der Erfinder des Schießpulvers und der Feuerwaffen*. Herder, Freiburg (1891).

[108] F.M. FELDHAUS: *Berthold der Schwarze, anno 1380*. Z. ges. Schieß- u. Sprengstoffwesen **1**, 413-415 (1906).

[109] W.G. KRAMER: *Der Fall Berthold SCHWARZ*. Weber, Freiburg (1993).

| | | |
|---|---|---|
| c.1382 | Ghent, East-Flanders, Belgium | **First construction of a monster gun, the *Dulle Griete* ("Mad Margaret")**, a 25-in. (635-mm)-caliber, 197-in. (5-m)-long bombard which is capable of firing 700-lb (317-kg) granite balls. ▪ According to the *Guide des Voyageurs dans la Ville de Gand* (1826), the people of Ghent used it in 1411. |

In the following centuries, the increasing interest in big guns led to numerous spectacular and prestigious developments, for example

- the 19.5-in. (495-mm) *Mons Meg* siege gun (1460) of Edinburgh Castle;
- the 887-mm Dardanelles Guns of the Ottoman Sultan MEHMED II {⇨1453};
- the 34.63-in. (880-mm) Tsar Cannon (1586) of Moscow;
- the Great Cannon of Jaiwaan (18th century) at Jaigarh Fort in Rajasthan, then renowned as the largest in all Asia;
- Robert MALLET's 30-in. (762-mm) Mammoth Mortar (1858) and Lord PALMERSTON's 930-mm mortar *Folly* (1858), two British pieces of ordnance; and
- the American 25-in. (635-mm) gun *Beelzebub* {RODMAN ⇨1864} and the 35.1-in. (892-mm) mortar *The Dictator* (1864–1865).

In England, scientific studies of big gun design began in the 1880s at Woolwich Military Academy {GREENHILL ⇨1890}. In Germany, the idea of building monster guns {⇨Fig. 4.2–T} was revived in the 20th century, resulting in four unusual super cannons:

- the WWI 420-mm howitzer *Dicke Bertha* ("Big Bertha") {RAUSENBERGER ⇨1908};
- the WWI 210-mm long-range gun ("Paris gun") {RAUSENBERGER ⇨1916}.
- the WWII 800-mm 1,350-ton *Dora,* the world's biggest artillery gun; and
- the WWII "V3" (an abbreviation used for *Vergeltungswaffe 3*, meaning "Vengeance Weapon 3") which consisted of a battery of five by five 150-mm supersonic guns. Codenamed *Hochdruckpumpe* ("High Pressure Pump") and known among German ballisticians by the code-name *Tausendfüßler* ("Millipede"), such secret batteries were located in France at the English Channel and were aimed at London. Each gun had a barrel length of about 130 m, with side chambers carrying propellant charges; each charge was fired just as the shell passed by, thus imparting more energy and speed to the shell.

Generally, superguns proved to be too expensive, too bulky for quick transportation and too ambitious in service; none proved to be crucial to the course of war. A historical perspective on the development of monster guns and other unconventional guns up to the end of World War I – the first "War of Engineers" – was given by William JOHNSON,[110] a retired professor of mechanical engineering at the University of Cambridge.

| | | |
|---|---|---|
| 1430 | Nuremberg, Bavaria, Germany | **Invention of the *Windrohr* or *Windbüchse* ("wind-gun" or "air-gun")** by the German gunsmith GUTER.[111] Advantageously, it produces practically no noise, and there are no powder fumes. ▪ The distinguished English naturalist Robert BOYLE {⇨1647} named the air-gun a "wind-gun." This curious invention, which can be attributed to numerous nations, has a rather complex history, and over the centuries numerous European gunsmiths have attempted to improve the performance of this somewhat mysterious weapon. However, in practice the wind-gun has barely been used and has only been applied for special purposes. The high air pressure of a loaded air-gun, in more advanced constructions amounting to almost 200 bar, has provoked numerous accidents. The firing of a wind-gun is accompanied by a curious luminous phenomenon emerging from the muzzle {VON HELVIG ⇨1815}. |

---

[110] W. JOHNSON: *Some monster guns and unconventional variations.* Int. J. Impact Engng. **11**, 401-439 (1991).
[111] *Brockhaus' Konversations-Lexikon.* F.A. Brockhaus, Leipzig (1908), vol. 16, p. 755.

Pieter van Musschenbroek,[112] a Dutch professor of physics at the University of Leiden, reported that in 1474 the German gentleman von Schmettau had a rather crude air-gun in his gun cabinet. The air-gun was improved in 1566 by Hans Lobsinger, a Nuremberg mechanic, and mentioned by David de Flurance Rivault (1605) in his work *Les éléments d'artillerie* ("Elements of Artillery").[113]

In 1788, around a thousand repeater wind-guns were used by an Austrian shooting battalion.[114] Recently, the air-gun saw a renaissance and was used

▸ **as a seismic source in marine seismic prospecting,** thus offering an environmentally friendly alternative to explosive sources[115] (From the early 1970s to the present, air-gun tuned arrays have been the leading source for marine seismic operations.[116]); and

▸ to accelerate frozen deuterium pellets for refueling fusion plasmas in a tokamak experimental reactor {⇨Fig. 4.2–J}.

**The wind-gun stimulated the curiosity of generations of naturalists to experiment on the "elastic nature" of air,** to evaluate the maximum outflow velocity of a gas through a small opening (such as a valve or a nozzle), and to reflect on the thermodynamic behavior of slow and fast compression of matter.[117] The wind-gun was also extended to large calibers. Its bizarre early history up to the mid-18th century was reviewed by Johann Heinrich Zedler,[118] a German bookseller and publisher, in his famous 68-volume universal encyclopedia.

# MODERN TIMES

| | | |
|---|---|---|
| 1453 | *Constantinople Byzantine/East Roman Empire [now Istanbul Turkey]* | Turkish troops under the command of Mehmed II, Sultan of the Ottoman Empire, blockade the Bosporus by erecting a strong fortress at its narrowest point. At this fortress, Rumeli Hisari, **they use a clumsy bronze cannon charged with gunpowder and stone balls to batter the ancient walls of the Byzantine city.** The city is captured on May 29, 1453, marking the final destruction of the Eastern Roman Empire. ▪ The monster gun had a caliber of 88.7 cm, was 17 ft (5.18 m) long and weighed 19 tons. The spherical solid projectile was just less than 600 lb (272 kg) weight.[119]<br><br>It was soon discovered that smaller iron bullets, of much less weight than large stone ones, would be more efficacious if impelled at higher velocity by using greater quantities of stronger powder, thus impacting the target at a higher kinetic energy. This occasioned an alteration in the matter and form of the cannon, which became lighter and more manageable; at the same time they became stronger in proportion to their bore. This change took place at about the close of the 15th century. |
| 1490s | *Florence, Tuscany, central Italy* | **Leonardo da Vinci** [a nickname meaning "Leonard from the town of Vinci"], the most famous Italian painter, draftsman, sculptor, architect, and engineer of the Renaissance, **increasingly shifts his interest from the elementary theory of mechanics to applied mechanics,** becoming particularly concerned with problems of friction and resistance in percussion and fluid dynamics. |

---

[112] P. van Musschenbroek: *Introductio ad philosophiam naturalem.* Luchtmans, Lugduni Batavorum (1762), Part II, § 2111, seq.
[113] J.G. Doppelmayr: *Historische Nachricht von den Nürnbergischen Mathematicis u. Künstlern.* Monath, Nürnberg (1730); Olms, Hildesheim (1972).
[114] H. Müller: *Gewehre, Pistolen, Revolver.* Kohlhammer, Stuttgart (1979), see chap. *Windbüchsen und elektrische Zündung,* pp. 154-156.
[115] B.F. Giles: *Pneumatic acoustic energy source.* Geophys. Prospecting **16**, 2-53 (1968).
[116] B.F. Giles and R.C. Johnston: *System approach to air-gun array design.* Geophys. Prospecting **21**, 77-101 (1973).
[117] A. Hoff: *Airguns and other pneumatic arms.* Barrie & Jenkins, London (1972).
[118] J.H. Zedler: *Großes vollständiges Universal-Lexikon aller Wissenschaften und Künste.* Zedler, Leipzig/Halle (1732–1754). Akad. Druck- und Verlagsanstalt, Graz (1961–1964); see vol. 57 (1748, 1962), pp. 660-668.
[119] W. Johnson: *Some monster guns and unconventional variations.* Int. J. Impact Engng. **11**, 401-439 (1991).

(i) **He outlines an engine for driving piles** {⇨Fig. 4.2–N}, a technique which is extensively used in his lifetime to pile hydraulic locks in order to alter the course of the Arno River.[120]
• According to his drawing, his percussion machine consisted of a vertical frame with a hoist for lifting weights. The head of the hoist was equipped with a gripping device made of two flexed crossbows which were released when the pile reached the maximum height, thus transmitting to it all of the power available. This action could be repeated several time over, until the pile was driven into the ground at the required depth. A practical problem with any efficient pile driving technique is the increase in friction between the pile and the earth as the pile is driven {HUTTON ⇨1812}.

(ii) Regarding the movement of a body through a resisting medium, **he presents the modern view**[121] that "the air becomes condensed before bodies that penetrate it swiftly, acquiring more or less density as the speed is more violent or less."

(iii) **He states the cardinal principle of wind-tunnel testing:**[122] "For since the action of the medium upon the body is the same whether the body moves in a quiescent medium, or whether the particles of the medium impinge with the same velocity upon the quiescent body; let us consider the body as if it were quiescent and see with what force it would be impelled by the moving medium…" He also develops a model wind tunnel to test the effects of waves on submerged objects. • Almost two hundred years later, Sir Isaac NEWTON independently asserted this same "reciprocity principle."

(iv) **He also uses markers of all shapes and sizes to trace the movements of water and wind in order to detect the geometrical shapes of their lines of movement.**[123] • The tracer method was rediscovered in the 1890s: Ludwig MACH,[124] Ernst MACH's oldest son, used the tracer method in his wind tunnel to photograph the flow of air, and simultaneously the British engineer Osborne REYNOLDS[125] used colored bands to visualize fluid motion. Much later, the method of using markers was also introduced into shock wave diagnostics using smoke filaments {GODDARD ⇨1959; CDRE Suffield ⇨1964}.

(v) Leonardo[126] also proposes a wedge-type *shock absorber* (for a flying machine to provide a smooth landing) **and a** *telescope shock absorber* ("to allow a man to fall from great heights").

| 1492 | *Ensißhein, seat of the Austrian regency [now Ensisheim, Alsace, France]* | On November 7 at 11.30 a.m., a boy observes the fall of a single meteorite which, accompanied by a very loud explosion, lands in a wheat field close to the town of Ensisheim. The stone meteorite, which weighs about 127 kg, is kept in the parish church for a long time.[127, 128] • **The so-called "Ensisheim meteorite"** {⇨Fig. 4.1–D} **is the oldest preserved meteorite of which appreciable remains are extant and available for research.**[129] Today its residual mass of about 55 kg is kept at the Regency Museum in the town hall of Ensisheim, but small |

---

[120] The drawing, together with plans for the Arno project, were printed in the facsimile edition *Il codice atlantico di Leonardo DI VINCI*. Barbera, Firenze (1975); vol. IX, p. 785. The original drawing of his pile driver machine {⇨Fig. 4.2–N} is kept in the Biblioteca Ambrosiana in Milan.
[121] A.R. HALL: *Ballistics in the seventeenth century*. Cambridge University Press, Cambridge (1952), p. 109.
[122] R. GIACOMELLI and E. PISTOLESI: *Historical sketch*. In: (W.F. DURAND, ed.) *Aerodynamic theory*. Springer, Berlin, vol. I (1934), p. 313. *The aerodynamics of Leonardo DA VINCI*. J. Roy. Aeronaut. Soc. **34**, 1016-1038 (1930).
[123] K.D. KEELE: *Leonardo DA VINCI's elements of the science of man*. Academic Press, New York etc. (1983), pp. 137-140.
[124] L. MACH: *Sichtbarmachung von Luftstromlinien*. Z. Luftschiffahrt & Physik der Atmosphäre **15**, Heft 6, 129-139 (1896).
[125] O. REYNOLDS: *Study of fluid motion by means of colored bands*. Proc. Roy. Inst. **14**, 129-138 (1896); *Flow of water shown by colored bands*. Nature **58**, 467-468 (1898).
[126] Manuscript B (IFP), fol. 89r and Madrid Manuscript I (BNM), fol. 62v. *See* online exhibit: *Leonardo and the engineers of the Renaissance*. Istituto e Museo di Storia della Scienza, Firenze, Italy (2004);
http://brunelleschi.imss.fi.it/genscheda.asp?appl=LIR&xsl=paginamanoscritto&lingua=ENG&chiave=100931 and
http://brunelleschi.imss.fi.it/genscheda.asp?appl=LIR&xsl=paginamanoscritto&lingua=ENG&chiave=100930.
[127] Confrérie St. Georges des Gardiens de la Météorite d'Ensisheim: *L'histoire d'une météorite*. Brochure du Musée de la Régence, Ensisheim, France.
[128] I.D. ROWLAND: *A contemporary account of the Ensisheim meteorite, 1492*. Meteoritics **25**, 19-22 (1990).
[129] U.B. MARVIN: *The meteorite of Ensisheim: 1492 to 1992*. Meteoritics **27**, 28-72 (1992).

| | | |
|---|---|---|
| | | specimens of it are kept in many museums. The Ensisheim meteorite, a stony meteorite classified as LL6, contains $SiO_2$ (40.65%), $MgO$ (25.54%), $FeO$ (17.26%), $FeS$ (5.35%), $Al_2O_3$ (2.25%), $CaO$ (2.24%), $Na_2O$ (0.98%), $Ni$ (1.05%), *etc.*[130] |
| c.1500 | China | According to an ancient legend, **a Chinese official named WAN-HOO attempts a flight to the Moon using a large wicker chair to which were fastened 47 large rockets** {⇨Fig. 4.20–B}. Forty seven assistants, each armed with torches, rushed forward to simultaneously light the fuses. In a moment there was a tremendous roar accompanied by billowing clouds of smoke. When the smoke cleared, the flying chair and WAN-HOO were gone.[131] |
| 1572 | Herrevadsklos-ter [near Ljung-byhed], Skåne, Denmark | In the evening of November 11, the Danish nobleman and astronomer **Tycho BRAHE sees a new star in** *Cassiopeia* ("The Lady in Chair," an autumnal constellation of the Northern Hemisphere), **where none had been seen before.** It is visible for about 18 months before fading from view.<br><br>In the following year, he published a brief tract entitled *De nova stella* ("On a New Star"), which made him famous all over Europe ("TYCHO's nova"). ▪ The extreme brilliance of the "new star" was actually due to the fact that it was a *supernova*, later termed SN 1572. |
| 1572 | Upper Leogra Valley [above Schio], Province of Vicenza, northern Italy | **First use of black powder for blasting purposes in the Venetian mining industry.** Giovanni Battista MARTINENGO addresses a petition to the Council of Ten "with skill and my new method to extract from the said places greater benefit than the mind of any man has ever imagined." ▪ MARTINENGO gave no hint as to what his new method was, but in 1595 Filippo DE ZORZI, Inspector-General of the Vicentino mining area, gave the following unfavorable report on MARTINENGO's work, "This man did not work his mines, timbering the galleries, tunneling little by little and dumping the spoil outside as in the ordinary way of working, but in an eccentric fashion making a little hole in the mountain rock with artillery powder, wished to open up and shatter the mountain by force, and thus discover whatever was within that had hitherto remained undiscovered."[132]<br><br>However, **MARTINENGO's method became one of the main and most important applications of pyrotechnics for peaceful purposes.** Partly replacing the very cumbersome manual work with hammer and chisel, it became increasingly established in the 17th century in numerous European countries. Early blasting powder had the same ingredients as black powder, but contained a smaller amount of potassium nitrate [$KNO_3$] which favorably resulted in a slower reaction and a lower price. The shattering action of the blast wave – which depends on the number, diameter and length of the bore holes, the charge weight and the type of rock – was optimized empirically. |
| 1578 | Lyon, Dépt. Rhône, east central France | Jacques BESSON,[133] a French engineer and consultant of king Francis II, **publishes the first book on mechanical engineering of modern times, entitled** *Théâtre des instruments mathématiques et méchaniques.* Richly illustrated, it also shows machinery for obliquely driving piles into the ground (at an angle of about 55° in the example shown in his book). His pile-driving hammer, a lever device mounted on existing piles, is operated by two men. |
| 1596 | Osteel, East Frisia, northwest Germany | On August 3, **David FABRICIUS**, a Dutch Lutheran pastor and astronomer, **discovers Mira** ("The Miraculous"), a faint star located in *Cetus* ("The Whale," a constellation of the Southern |

---

[130] E. JAROSEWICH: *Chemical analyses of meteorites: a compilation of stony and iron meteorites analyses*. Meteoritics **25**, 323-337 (1990).
[131] *Rocketry through the ages: a timeline of rocket history*. NASA History Office, Marshall Space Flight Center (MSFC), Huntsville, AL; http://history.msfc.nasa.gov/rocketry/index.html.
[132] G.J. HOLLISTER-SHORT: *Gunpowder and mining in sixteenth- and seventeenth-century Europe*. Hist. Technol. (London & New York) **10**, 31-66 (1985).
[133] J. BESSON: *Théâtre des instruments mathématiques & méchaniques* [avec l'interprétation des figures d'iceluy, par François BEROALD]. B. Vincent, Lyon (1578).

Hemisphere) that within a few weeks disappears. ▪ When he saw Mira brighten again in 1609, it became obvious that he had discovered a new kind of object in the sky.

*Mira Ceti* **is the first known** *pulsating variable star.* Mira stars are a special type of variable red giant star which pulsate.[134] Over the course of 80 to 1,000 days, a Mira star can vary in brightness by a factor of ten times or more during the cycle. **The change in observed brightness is believed to be the result of a pulsation, and the maximum light is associated with the emergence of a shock wave into the atmosphere of the star.** ▪ An international team of astronomers led by Guy PERRIN[135] from the Paris Observatory/LESIA (Meudon, France) and Stephen T. RIDGWAY from the National Optical Astronomy Observatory (Tucson, Arizona) used interferometric techniques to observe the close environments of five Mira stars. They found that they are surrounded by a shell of water vapor and carbon monoxide; this makes them seem larger than they actually are. These new observations bring the size of Mira stars in line with mathematical models that predict their size and composition.

| | | |
|---|---|---|
| 1604 | *Prague, Bohemia, Habsburg Monarchy* | On October 17, **Johannes KEPLER,**[136] a German astronomer who was previously assistant to the Danish astronomer Tycho BRAHE and is now Court Mathematician to Emperor RUDOLF II, **witnesses the explosion of a supernova in** *Serpens* ("The Serpent," a summer constellation of the Northern Hemisphere) while watching a rare conjunction of Mars, Jupiter and Saturn. The exploding star remains visible for 17 months. |

In the same year, Galileo GALILEI at Padua observes the new star for the first time on December 24. In January of the following year, he will deliver three lectures on the new star at the University of Padua. ▪ The unusual event – known as "KEPLER's nova" and later termed SN 1604 – was evidence that the realm of the fixed stars, considered since ancient times to be pure and changeless, could indeed undergo change. It was the second supernova to be observed in a generation {BRAHE ⇒ 1572}.

| | | |
|---|---|---|
| 1608 | *Prague, Bohemia, Habsburg Monarchy* | **Oswald CROLL,**[137] a German physician and iatrochemist (one who explains or treats diseases following chemical principles), **first describes the preparation of** *aurum fulminans* ("fulminating gold").[138] ▪ More important than the function of *aurum fulminans* as a putative medicament was its usefulness when investigating the nature of combustion, a question which exercised a long series of natural philosophers. Although *aurum fulminans* was derived from a nitrous agent, it was not clear what part the *aqua fortis* ("nitric acid") played in its formation. It provided, therefore, an ideal control for experiments to test theories which purported to explain the explosion of gunpowder. Robert BOYLE found that its action differed significantly from that of gunpowder, in that it could be ignited in a vacuum by both a burning-glass and a hot iron.[139] |

*Fulminating gold* [$(ClAuNH_2)_2NH$] is prepared from gold chloride and aqueous ammonia. Note that fulminating gold is distinct from "gold fulminate." *Gold fulminate* [gold hydrazide, $AuN_2H_3$] is a water-soluble olive-green powder and more explosive. Unlike conventional gunpowder, it explodes violently when dry and exposed to friction, shock and heat, and so is obviously the first very sensitive man-made high explosive ("exploding gold"). Gold fulminate is a salt of fulminic acid [$C_2H_2N_2O_2$] – in Germany appropriately called *Knallsäure* ("bang acid")

---

[134] For the classification of variable stars *see* General Catalogue of Variable Stars ⇒ 1948.
[135] G. PERRIN, S.T. RIDGWAY ET AL.: *Confirmation of the molecular layer model with narrow band near-infrared interferometry.* Astron. Astrophys. **426**, 279-296 (2004).
[136] I. KEPPLERI: *De stella nova in pede serpentarii* ("The new star in the foot of the Serpent-Bearer"). P. Sessius, Prague (1606).
[137] O. CROLL: *Osvaldi CROLLII Basilica chymica.* P. Chovet, Geneva (1609).
[138] In his book *Geschichte der Explosivstoffe* (Oppenheim, Berlin, 1895/1896), S.J. VON ROMOCKI credited Basilius VALENTINUS, an alchemist and Benedictine monk, as the first person to produce *aurum fulminans* at the end of the 16th century.
[139] R.G. FRANK JR.: *John AUBREY, F.R.S, John LYDALL, and science at Commonwealth Oxford.* Not. Rec. Roy. Soc. **27**, 193-217 (1973), *Aurum fulminans*, pp. 196-198.

– which was discovered in 1824 by Justus VON LIEBIG.[140] In 1659, Thomas WILLIS, Robert HOOKE's first Oxford employer, also experimented with gold fulminate to show that it can explode on a mere concussion, without the need for air or sparks (which were once thought to be required for any kind of ignition). He placed a little piece of it onto a spoon, and covered the spoon with a heavy coin. When gently tapping the spoon onto a table top, the chemical exploded violently, blowing the coin up to the ceiling.

**Other, very similar compounds of high brisance were discovered subsequently, such as silver fulminate [$Ag_2(CNO)_2$] by Comte Claude-Louis BERTHOLLET {⇒1788} and mercury fulminate [$Hg(CNO)_2$] by Edward C. HOWARD {⇒1800}**. Initially, these fulminates were only useful as explosives in toys and tricks. However, in particular mercury fulminate, which is less sensitive to shock and vibration than silver fulminate, quickly gained increasing import-ance in the fabrication of percussion caps {FORSYTH ⇒1805; SHAW ⇒1814} and blasting caps {NOBEL ⇒1865}. Besides, they also stimulated the curiosity of early naturalists about explosion phenomena, and represented a useful alternative to electric sparks in model shock interaction experiments {E. & L. MACH ⇒1889}.

| | | |
|---|---|---|
| 1610 | Chair of Mathematics, University of Padua, Veneto, north Italy | Galileo GALILEI, an eminent Italian natural philosopher, **publishes his treatise *Sidereus Nuncius* ("The Starry Messenger"), which contains his observations and research on lunar craters** {⇒Fig. 4.1–A}.[141] He correctly recognizes that the circular "spots" are depressions. However, he records no opinion about the origin of these structures. ▪ Three centuries passed before the process of impact cratering was properly understood, and it took another half century before most of the doubters were convinced that the Moon's craters were caused by the impact of large meteorites {GIFFORD ⇒1924; DIETZ ⇒1946 & 1960; BALDWIN ⇒1949}.<br><br>On September 3, 2006, ESA's SMART-1 spacecraft impacted into the Moon, creating a man-made impact crater {SMART-1 ⇒2003}. |
| 1611 | Osteel, East Frisia, northwest Germany | On March 9, **Johann FABRICIUS**, a 24-year old university student returning from the Netherlands with one or more telescopes, **first observes sunspots**.[142]<br><br>(i) Together with his father David FABRICIUS {⇒1596}, a Lutheran pastor and astronomer, he continues his studies. Observing sunspots using the projection method by means of a camera obscura, they correctly interpret the day-to-day motions of sunspots as an indication of the Sun's axial rotation.<br><br>(ii) In June of the same year, J. FABRICIUS publishes their observations and interpretations in a small pamphlet entitled *De Maculis in Sole observatis et apparente earum cum Sole conversione, Narratio* ("Narration on Spots Observed on the Sun and their Apparent Rotation with the Sun"), which is printed in Wittenberg, eastern Germany.<br><br>It appears that the English astronomer and naturalist Thomas HARRIOT first observed sunspots – even prior to J. FABRICIUS – as evidenced by entries in his notebook dated December 8, 1610. The text does not mention the spots explicitly, even though they are clearly indicated on his drawing.[143] ▪ The study of sunspots and their possible origin stimulated subsequent generations of astronomers, leading to the discovery of solar flares {CARRINGTON ⇒1859} and stimulating the evolution of solar physics, an important branch of astrophysics dedicated to the study of the physical activities associated with the Sun. |

---

[140] F. KURZER: *Fulminic acid in the history of organic chemistry.* J. Chem. Educ. **77**, 851-857 (2000).
[141] E.A. WHITAKER: *Galileo's lunar observations and the dating of the composition of Sidereus Nuncius.* J. Hist. Astron. **9**, 155-169 (1978).
[142] *David (1564–1617) and Johannes (1587–1616) FABRICIUS.* The Galileo Project. Rice University, Houston, TX (1995); http://galileo.rice.edu/sci/fabricius.html.
[143] W.M. MITCHELL: *The history of the discovery of solar spots.* Popular Astron. **24**, 22, 82, 149, 206, 290, 341, 428, 488, 562 (1916).

| | | |
|---|---|---|
| 1613 | France | Jacques TISSOT,[144] a little-known French philosopher and writer, **assumes that all natural phenomena occur continuously and unbroken – expressed in the motto** *Natura non facit saltus* ("Nature does not make leaps"); TISSOT rather used the scholastic form *Natura in operationibus suis non facit saltum* ("Nature in her operations does not proceed by leaps").[145] ▪ This idea, however, was not new and already anticipated by the Greek philosopher ARISTOTLE (384–322 B.C.) in his *Scala naturae* ("Chain of Being"). He regarded the classes of nature as unchangeable, just as, for him, the Universe was unchangeable; therefore, the idea of evolution of living beings – that is, the idea that one species of living beings gradually changed into another – did not come to his mind at all. |
| 1620–1640 | Flanders, The Netherlands | Johan B. VAN HELMONT,[146] a Belgian physiologist, physician, and a leading chemist of his time, **studies "aeriform bodies"** (*i.e.*, gases).<br><br>(i) **He recognizes the existence of discrete gases, thereby coining the term** *gas*. ▪ This term, possibly derived from the Greek word *chaos* ("disorder"), was used by Philippus A. PARACELSIUS to specify "common air." Although today it denotes the third state of matter, it was not used in this general sense until after the discovery of "artificial aeriforms" by Antoine L. LAVOISIER in the middle of the 18th century.<br><br>(ii) He notices that gases can be condensed into liquids, and writes, "Gas is composed of invisible atoms which can come together by intense cold and condense into minute liquid drops."<br><br>(iii) Referring to the fact that gas can explode glass containers, **he correctly explains the explosion of gunpowder based on the theory of the rapid expansion of gases.**<br><br>(iv) Addressing the possible cause of earthquakes, his explanation is based on the popular belief that a wrathful God unleashes them on a sinful world. He suggests that an avenging angel strikes a huge celestial bell, thus inducing violent atmospheric vibrations that shake the ground upon their arrival on Earth. ▪ Albeit prejudiced by contemporary religious belief, he thus correctly anticipated that **there was a wave character to the seismic shocks** radiating out from the earthquake's focus. |
| 1620 | London, England | Théodore TURQUET DE MAYERNE,[147] a physician and chemist, **describes the explosion of a "certain kind of air" – meaning an explosion of oxyhydrogen.** This explosive gaseous mixture of hydrogen [$H_2$] and oxygen [$O_2$] at the volume ratio 2:1 results, for example, from mixing iron filings and sulfuric acid [$H_2SO_4 + Fe \rightarrow H_2 + FeSO_4$] and the molecular oxygen [$O_2$] of the atmosphere. |
| 1621 | Coal Mines of Gateshead, England | On October 14, **an accident takes place in a pit in Gateshead [near Newcastle upon Tyne].** In the register of St Mary's Church, under this date, is recorded the interment of "Richard BACKAS burn'd in a pit" – probably the earliest recorded accident caused by a firedamp explosion.[148] |
| 1623 | Rome, Italy | Galileo GALILEI, the famous Italian astronomer and mathematician who made fundamental contributions to mechanics and fluid dynamics, **assumes that heat is caused "by a multitude of fire-corpuscles having certain shapes and moving with certain velocities."** In his work *Il saggiatore* ("The Assayer") he wishes to stress that something moving rapidly through the air does not necessarily grow hot, as reported in an old story that the Babylonians cooked their eggs by whirling them in slings.[149] He is rather of the opinion that the eggs would not cook but |

---

[144] J. TISSOT: *Discours véritable de la vie, mort, et des os du géant* THEUTOBOCUS *[...]*. Poyet, Lyon (1613). In: (E. FOURNIER, ed.) *Variétés histoires et littéraires*, I-IX. Jannet, Paris (1855–1863).
[145] *Lessico Universale Italiano di lingua, lettere, arti, scienze e tecnica*. Istituto della Enciclopedia Italiana, Roma, vol. XIV (1974), p. 517.
[146] J.B. VAN HELMONT: *Ortus medicinae*. Ludovicum Elzevirium, Amsterdam (1648).
[147] J. BROWNE: *Sir Théodore* TURQUET DE MAYERNII *[1573–1655]. Opera medica*. Typis R.E., London (1703).
[148] *A history of mine safety research in Great Britain*; http://www.users.zetnet.co.uk/mmartin/fifepits/starter/safe-2.htm.
[149] M. BOAS: *The scientific Renaissance 1450–1630*. Collins, London (1962), pp. 261-264.

| | | merely cool down. ▪ **Aerodynamic heating,** a subject of debate among early naturalists, **was first proved experimentally using a whirling thermometer** {JOULE & THOMSON ⇨1856}. |
|---|---|---|
| 1627 | *Oberbieberstollen, Schemnitz, Komitat Hont, Hungary* | **February 8 marks the first successful demonstration of black powder by the Tirol mining engineer Caspar WEINDL in the Royal mines of Schemnitz,** which are famous for their large ore deposits containing gold, silver, copper and lead {⇨Fig. 4.16–A}.[150] ▪ Soon afterwards, black powder was also at work in other parts of Germany (Graslitz/Böhmen 1627, Clausthal/Harz 1632, Freiberg/Sachsen 1643, and Rheinland-Westfalen 1650), in Austria (Radmer/Steiermark 1635), in Sweden (Nasafjäll/Lapland 1635), in Norway (Rövas 1644), and in England (North Derbyshire 1670).[151] |
| 1635 | *England* | **Governmental studies on finding the *optimum formula*[152] of black powder for use in guns result in a 6:1:1 formula** (*i.e.*, 75% saltpeter, 12.5% charcoal, and 12.5% sulfur), the quickest and most vigorous composition.[153] ▪ Similar attempts to find the optimum formula when using black powder as a propellant in military firearms were also attempted in other countries {French Commission ⇨1794}. The 6:1:1 formula closely approaches the formula used in the 20th century for military and sporting purposes.<br><br>Until about 1450, gunpowder was simply a pulverized 2:1:1 mixture (50% saltpeter, 25% charcoal, and 25% sulfur). From the middle of the 16th century, other ways of augmenting the violence of explosion were tried, especially the addition of chemicals prominent in the pharmacopoeia such as antimony, mercury, vinegar, and alcohol to the ordinary powder. Prince RUPERT OF THE PALATINATE discovered that ordinary gunpowder – prepared from standard ingredients but with more particular attention paid to their purity, fineness and incorporation – can significantly increase its efficiency.[154] |
| 1636 | *Minim Convent de l'Annonciade, Paris, France* | **Marin MERSENNE,**[155] a Minim friar and natural philosopher, **performs the first measurement of the velocity of sound in air** by firing a cannon, and observing the delay between the muzzle flash and arrival of sound {⇨Fig. 2.10–A}. He finds a value of "230 toises par seconde" (about 450 m/s), to which he will adhere in a later work {MERSENNE ⇨1644}. **He also discovers that the intensity of sound, like that of light, is inversely proportional to the distance from its source.** ▪ It is commonly believed – on the authority of many textbooks and some encyclopedias[156] – that Pierre GASSENDI, a friend of MERSENNE, was the first to measure the velocity of sound in air. However, GASSENDI, knowing MERSENNE's results, did not perform any experiments himself, and his collected works[157] do not contain any numerical measurements of the sound velocity.[158] |

---

[150] Schemnitzer Berggerichtsbuch, Jahrgang 1627, p. 37. *See* the monograph by O. GUTTMANN: *Handbuch der Sprengarbeit*. Vieweg, Braunschweig (1892), pp. 2-3.
[151] M. FUNK and H. KREBS: *Die Entwicklung der Sprengtechnik im Bergbau, dargestellt am Beispiel des Freiberger Bergbaus*. Sprengstoffe, Pyrotechnik [Schönebeck/Elbe] **24**, Nr. 2, 3-10; Nr. 3, 3-15; Nr. 4, 3-16 (1987).
[152] In a truly "optimal" mixture only exothermic reactions occur, and any endothermic reactions reducing the temperature of fired gunpowder are avoided {BUNSEN & SCHISCHKOFF ⇨1857; NOBLE & ABEL ⇨1875; DEBUS ⇨1882}. This would result in the maximum muzzle velocity for a given amount of gunpowder.
[153] T.L. DAVIES: *The chemistry of powder and explosives*. Angriff Press, Las Vegas (1972), p. 39.
[154] A.R. HALL: *Ballistics in the seventeenth century*. Cambridge University Press (1952), p. 61.
[155] M. MERSENNE: *De l'utilité de l'harmonie*. S. Cramoisy, Paris (1636), p. 44.
[156] According to the *Encyclopaedia Britannica, Macropaedia* (1974, vol. 17, p. 19), GASSENDI, who made in 1635 one of the first recorded measurements of the sound velocity in air, determined a value of 478.4 m/s.
[157] P. GASSENDI: *Opera omnia*. 6 vols., Montmort & Sorbière, Lyon (1658).
[158] J.M.A. LENIHAN: *MERSENNE and GASSENDI, an early chapter in the history of sound*. Acustica **1**, 96-99 (1951).

| | | |
|---|---|---|
| 1637 | Leiden, The Netherlands | René DESCARTES,[159] a French philosopher and mathematician, **addresses in his philosophical work** *Discours de la méthode* ("Discourse on Method") **the possible causes of lightning and thunder,** thereby essentially following the ideas of the philosophers of antiquity {ARISTOTLE ⇨ 333 B.C.; SENECA ⇨ A.D. 63} that the noise of a thunder clap is the consequence of the violent collision of two large air masses (clouds). |

In a treatise entitled *Les météores* ("The Meteors"), DESCARTES, attempting to strengthen this idea by comparison, writes, "As to violent storms which are accompanied with whirlwinds, thunder and lightning, those which I have been able to examine leave no doubt on my mind that they are caused in the following manner: when it happens that a number of clouds collect one above another, it is no uncommon circumstance for the more elevated to descend violently upon the lower ones, in the same manner, as I remember formerly to have seen in the Alps, about the month of May, that the snows being warmed and moistened, and so made heavier by the Sun, the slightest commotion in the air was sufficient to cause the sudden descent of a great mass, known under the name of an avalanche, which resounding in the valleys, was not unlike the sound of thunder" (Discours VII).

| | | |
|---|---|---|
| 1638 | Arcetri [a town near Florence], Tuscany, central Italy | **Publication of GALILEI's treatise** *Discorsi e dimostrazioni matematiche intorno à due nuove scienze* ("Discourses and Mathematical Demonstrations Relating to Two New Sciences"), in which he sums up his life's work.[160] • His "First New Science," discussed in the **"First and Second Day"** of the *Discorsi*, deals with the strength of materials. His "Second New Science," presented in the **"Third and Fourth Day,"** deals mostly with accelerated motion of objects, but also briefly addresses the **phenomena of percussion.** The **"Fifth and Sixth Day"** were later added. His contributions to the subject of this book may be summarized as follows: |

(i) In his "Fourth Day," he demonstrates the parabolic path of projectiles – known as **"GALILEI's vacuum theory"** or **"parabolic trajectory theory"** – and even gives a table of firing ranges. • This led to more mathematicians and scientists working in the area of projectile trajectory. Sir Isaac NEWTON {⇨ 1687} expanded on GALILEI's work by taking into account the effects of aerodynamic drag on a spherical projectile; for military reasons the early experimenters were mainly concerned with spheres. Benjamin ROBINS {⇨ 1742}, resuming NEWTON's studies on aerodynamic drag, expanded them to supersonic velocities. The German ballistician Carl CRANZ gave an impressive example of the enormous influence of aerodynamic drag on the flight of the widely used infantry S-bullet {Prussian Army ⇨ 1903}.

(ii) In his "Fourth Day," **GALILEI also briefly addresses the puzzling phenomenon of percussion:** "My difficulty and surprise consists in not being able to see whence and upon what principle is derived the energy and immense force [Ital. *energia e forza immensa*] which makes its appearance in a blow; for instance we see the simple blow of a hammer, weighing not more than 8 or 10 pounds [3.63 or 4.53 kg], overcoming resistances which, without a blow, would not yield to the weight of a body producing impetus by pressure alone, even if that body weighed many hundreds of pounds. **I would like to discover a method of measuring the force of such a percussion.** I can hardly think it infinite, but incline rather to the view that it has its limit and can be counterbalanced and measured by other forces, such as weights, or by levers or screws or other mechanical instruments which are used to multiply forces in a manner which I satisfactorily understand."

---

[159] C. ADAM: *Oeuvres de DESCARTES*. Nouvelle édition, Vrin, Paris (1996), vol. 6: *Discours de la méthode & essais*, pp. 315-316.

[160] G. GALILEI: *Discorsi e dimostrazioni matematiche intorno à due nuove scienze*. Appresso gli Elseviril, Leida (1638); Germ. translation by A. VON OETTINGEN: *Unterredungen und mathematische Demonstrationen über zwei neue Wissenszweige*. In: *Ostwald's Klassiker der exakten Wissenschaften*. Nr. 24 & 25, Engelmann, Leipzig (1908); Engl. translation by H. CREW and A. DE SALVIO: *Dialogues concerning two new sciences* [with an introduction by A. FAVARO]. Dover Publ., New York (1954).

(iii) In his "Sixth Day,"[161] **he first treats the problem of percussion scientifically.** He obtains the important result that – contrary to a "dead" weight [Ital. *Peso morto*] – each collision accomplishes work, and that the collision efficiency depends on the product of mass and velocity of both collision partners. This allows one to determine the ratios of velocities during collision, but not the impulsive force. **In order to measure the force of percussion, he proposes a clever experiment** by using a jet of water emerging from a bucket hanging at one side of a balance and impinging a scale arranged below the bucket, while the opposite scale is loaded by a weight such that the balance is in equilibrium prior to opening the valve in the bottom of the bucket {⇨Fig. 4.3–C}. However, his experiment gives a negative result, because force cannot directly be compared with the force-time integral – in modern notation named **"impulse."**
▪ GALILEI's classic dynamic puzzle about how to describe time and the force of percussion (or the force of body's impact) remained unsolved, a problem which occupied subsequent generations of natual philosopher and initiated the *vis viva* controversy {DESCARTES ⇨1644; LEIBNIZ ⇨1686; J. BERNOULLI 1714, *see* Sect. 2.2.2; MACLAURIN ⇨1724; EULER ⇨1737 & 1745; D'ALEMBERT ⇨1743; William RICHARDSON[162] 1769; WOLLASTON ⇨1805}. GALILEI could not, throughout his life, find systematic relations among specific gravities, height of fall and percussive forces.

(iv) In his "Sixth Day," also addressing the example of pile driving, **he correctly recognizes that, compared to the force of gravity, the force of percussion must be enormously large or even infinite** [Ital. *"La forza della percossa essere immense, o infinita."*]. ▪ Today GALILEI is generally considered to be the initiator of scientific percussion research.

In 1633, the Roman Inquisition had banned publication of any future work by GALILEI. His treatise was printed in 1638 by Lowys ELSEVIER in Leiden, The Netherlands, where the writ of the Inquisition was of little account. However, in the following year, its distribution was prohibited by an ecclesiastical decree. In 1635, a translation into Latin was published by Matthias BERNEGGER in Straßburg, Germany.

| | | |
|---|---|---|
| 1639 | *Charles University [Karl-Ferdinand-Universität, founded in 1366 as Collegium Carolinum], Prague, Bohemia, Habsburg Monarchy* | Marcus MARCI DE KRONLAND,[163] an eminent Bohemian physicist, mathematician and physician, **publishes his book *De proportione motus: seu Regula sphygmica…*** ("Of Proportions in Motion: the Sphygmic[164] Rule…"), which contains many illustrated examples. |

(i) He discusses various phenomena of **perfectly elastic central and lateral percussion** in a rather qualitative manner. Although he does not attempt to bring his results into an analytical form, they represent an important milestone towards the Laws of Percussion. ▪ The modern definition of "perfectly elastic percussion" is that the kinetic energy is conserved. By MARCI's time, however, this important law of conservation had not yet been discovered {HUYGENS ⇨1668/1669}.

(ii) Experimenting mainly with wooden balls, but also with billiard balls and cannonballs, he classifies collisions into those between hard, soft and fragile bodies, thereby already recognizing that the velocity is an important quantity for characterizing the intensity of the impulse.

(iii) MARCI is the first to make the important observation that in the case of elastic percussion, a moving body colliding with a body at rest of the same mass abruptly ceases its motion

---

[161] The manuscript of his "Sixth Day," containing his latest theory of impact, was not finished by 1638 and is therefore not included in his *Discorsi*. It was later added; *see* T. BUONAVENTURI, G. GRANDI, and B. BRESCIANI (eds.): *Collected works of G. GALILEI*. 3 vols., Tartini & Franchi, Firenze (1718). ▪ GALILEI (1564–1642) who became blind in 1638, dictated the manuscript at around 1640 to his disciples Vincenzo VIVIANI (1622–1703) and Evangelista TORRICELLI (1608–1647).
[162] W. RICHARDSON: *An essay on the force of percussion*. Phil. Trans. Roy. Soc. Lond. **58**, 17-23 (1769).
[163] I.M. MARCI: *De proportione motus: seu regula sphygmica ad celeritatem et tarditatem pulsuum ex illius motu ponderibus geometricis librato absque errore metiendam*. Ioannis Bilinae, Pragae (1639); *see* facsimile.
http://archimedes.mpiwg-berlin.mpg.de/cgi-bin/toc/toc.cgi?dir=marci_regul_062_la_1639;step=thumb.
[164] The word *sphygmic* is probably of Greek origin, meaning "of or relating to the pulse;" *see* the *Academic Press Dictionary of Science and Technology*. Academic Press, San Diego, CA *etc.* (1992).

and transfers it fully to the other body, thus giving a unique and surprising example of the conservation of momentum, which he put down in his Propositio[165] XXXVII:

- Porisma I: **"If a ball strikes another equal stationary ball, it comes to rest when that ball has been driven out"** [Lat. *"Si globus alium globum percutiat quiescentem & aequalem, illo expulso quiescit."*]. • In order to better illustrate his Porisma I, MARCI shows a picture in his booklet of the collision of a cannonball with a line of stationary cannonballs lying in contact on a table and concludes that the striking ball will come to rest while the last ball of the row will be driven out {⇨Fig. 4.3–B}.

(iv) Other important conclusions on the percussion of elastic bodies are given by him in the following porisms:

- Porisma II: "If a larger ball strikes a smaller stationary ball, the larger continues to move in the same direction when the smaller has been driven out."
- Porisma III: "If two balls of the same mass or weight collide in motion, each rebounds."
- Porisma IV: "If a small ball strikes a larger stationary ball, the small ball rebounds and the larger remains at rest."
- Porisma V: "If two balls of the same mass or weight collide in motion, each rebounds."
- Porisms VI–VIII concern the percussion of unequal balls, both in motion, and involve the comparison of impulse and weight.

(v) MARCI discovers that a ball, striking a plane obliquely, will be reflected at an angle which equals the angle of incidence, and applies the Law of Reflection to the game of billiards.

In a later work, MARCI,[166] **also applying mechanics to a problem of medical interest, attempts to explain why the human skull, upon receiving a blow, fractures on the opposite side.**

| | | |
|---|---|---|
| 1640s | *France* | Blaise PASCAL, a French mathematician, physicist and religious philosopher, when showing that the phenomenon attributed in TORRICELLI's barometer (1643) to the *"horror vacui"* is caused by the weight of the air, **advances the idea that air pressure decreases with altitude.** • On September 19, 1648, his hypothesis was confirmed by measurements of his brother-in-law Florin PÉRIER in Paris (alt. 34 m above sea level) and at the top of the Puy de Dôme (alt. 1,570 m), a mountain overlooking Clermont-Ferrand (Dépt. Auvergne). Around 1650, PASCAL discovered that pressure, when applied to a confined liquid at rest, is transmitted undiminished through the liquid in all directions, regardless of the area to which the pressure is applied – the so-called **"Pascal law."**[167] |

The unit *Pascal (Pa)* **has since been adopted as the derived SI unit for pressure** {Gen. Conf. on Weights and Measures ⇨1960}. Pressure, defined as the perpendicular force to unit area [*e.g.*, in CGS units: $dyn/cm^2$] can also be related to specific energy [$dyn/cm^2 = dyn\ cm/cm^3 = erg/cm^3$]:

- In detonics, **the specific inner chemical energy $e$ of an explosive can be characterized by an "intrinsic" pressure.** For example, pure TNT {HAEUSSERMANN ⇨1891} with $\rho_0 = 1.654$ g/cm$^3$ and $e = 4.184$ kJ/g $= 6.92$ kJ/cm$^3$ has an intrinsic pressure of 69.2 kbar. Modern high explosives have intrinsic pressures < 100 kbar. Note that these intrinsic pressures are lower than the Chapman-Jouguet (CJ) pressures {DEAL ⇨1957}.

---

[165] A "proposition" [Lat. *propositio*] is a formal statement of a truth to be demonstrated (*i.e.*, a theorem), while a "porism" [Lat. *porisma*, a term of Greek origin] or a "corollary" [Lat. *corollarium*] is a deduction from a previous demonstration or problem; *see The Oxford English dictionary*. Clarendon Press, Oxford (1989).

[166] I.M. MARCI: *De proportione motus figurarum rectilinearum et circuli quadratura ex motu*. Ex typographia academia, Pragae (1648), p. N1 recto.

[167] B. PASCAL: *Traité de l'équilibre des liqueurs et de la pesanteur de la masse de l'air*. Guillaume Desprez, Paris (1663).

▸ In magnetohydrodynamics, the electrons carried in a current through a straight conductor experience a magnetic pressure $P_M$ which acts perpendicular to field lines and is given by

$$P_M = \tfrac{1}{2} B^2 / \mu_0,$$

where $B$ is the magnetic field and $\mu_0 = 4\pi \times 10^{-7}$ H/m. Using this formula, **one can characterize magnetic fields by magnetic pressures:** for example, a magnetic field of 250 T produces a pressure of 25 GPa (250 kbar).[168]

| 1644 | The Netherlands |
|---|---|

René DESCARTES [or Renati DES-CARTES],[169] an eminent French mathematician and philosopher, **publishes his two-book treatise *Principia philosophiae* ("Principles of Philosophy"),** an account of his philosophical and scientific views, in which he attempts to give a logical account of all natural phenomena in one single system of mechanical principles rather than to solve specific technical problems:

(i) In his mechanistic cosmology – today known as "mechanical philosophy" – DESCARTES (like GALILEI) explains all of the phenomena of nature (including light) in terms of matter and motion. Here his "Theory of Percussion" plays a central role insofar as he tries to view mechanical phenomena in a global manner and to explain them in terms of the direct interaction of the bodies that constituted it. For example, **he interprets the force of a moving body by the expression $mv$ (mass × velocity), which he calls the "quantity of motion,"** and conceives the idea that in the Universe this quantity remains constant, so that when one body loses momentum the loss is immediately compensated for by an increase of momentum in other bodies – hence, **the only true measure of a force is the change in momentum it produces in a given time.** ▪ The German natural philosopher Gottfried Wilhelm LEIBNIZ {⇨1686} later attacked this view and argued that the true measure of a force is the change it produces in *vis visa* ("living force"), a quantity taken to be twice what we now call "kinetic energy."

(ii) In Book II, §§ 46–52 of his treatise **he propounds seven laws of impact of perfectly elastic bodies meeting in the same straight line.** However, only his first law of percussion concerning the impact of two equal bodies is correct and intuitively obvious: if these have equal velocities, they rebound; if their velocities are unequal, they will move on together after percussion.[170] ▪ DESCARTES' conception of motion had analyzed impact in terms of the force of the moving body impinging on others. Sir NEWTON {⇨1687} treated the moving body as the passive subject of external forces acting upon it, and this new approach to impact dynamics remains the basis for the analysis of impact today. DESCARTES' view of treating the subject of percussion incompletely and incorrectly stimulated the Dutch mathematician and physicist Christiaan HUYGENS to work out a more general and advanced theory of impact {HUYGENS ⇨1652, 1668/1669 & 1703}.

(iii) **DESCARTES gives an interesting new hypothetical explanation of the movement of a body through a resisting medium,** stating that all bodies consist of innumerable infinitely divisible particles [Lat. *corpuscula*] – thus anticipating Isaac NEWTON's corpuscular model {Sir NEWTON ⇨1687}. According to his idea, each particle moves at a great velocity in a perfect medium so that an immersed body is subjected to an equal pressure on all sides, and in whatever direction it moves, the impact of the particles remains the same because their movement is

---

[168] C.M. FOWLER, R.S. CAIRD, R.S. HAWKE, and T.J. BURGESS: *Future pulsed magnetic field applications in dynamic high pressure research.* In: (K.D. TIMMERHAUS and M.S. BARBER, eds.) *Proc. 6th AIRAPT Int. High Pressure Conference* [Boulder, CO, July 1977]. Plenum Press, New York etc. (1979), vol. 2, pp. 981-992.

[169] R. DESCARTES: *Principia philosophiae.* Elzevir, Amstelodami (1644); (V.R. MILLER and R.P. MILLER, eds.) *Principles of philosophy.* Kluwer, Dordrecht (1991).

[170] For a discussion of DESCARTES' rules of percussion *see*, for example, (i) I. SZABÓ: *Geschichte der mechanischen Prinzipien.* Birkhäuser Verlag, Basel & Stuttgart (1977), pp. 436-439; and (ii) F. CHAREIX: *Les règles du choc dans les Principia II de DESCARTES.* Origine des Rationalités à l'Age Classique (ORACL); http://www.age-classique.fr/article.php3?id_article=14.

so much more rapid than that of any solid body. Since all physical fluids are not ideal – *i.e.*, they contain much larger corpuscles which impact against a moving body – they cause a reduction of its velocity. ▪ **Since DESCARTES believed that the same quantity of matter must always occupy the same space, his fluid model didn't allow any compression effects.** Therefore, he explained drag as a process of stirring up the corpuscles, thus opposing Leonardo DA VINCI's previous view of a compressible medium {DA VINCI ⇨ 1490s}.

| | | |
|---|---|---|
| 1644 | *Minim Convent de l'Annonciade, Paris, France* | **Marin MERSENNE,**[171] a French theologian, natural philosopher and mathematician, **publishes a physical-mathematical treatise** entitled *Cogitata physico-mathematica* ("Physical-Mathematical") which contains a chapter on ***Phenomena ballistica*** ("Ballistic Phenomena"):<br><br>(i) Under the heading *Soni velocitas maior est globorum explosorum velocitate* ("The Sound Velocity is Always Larger than the Velocity of Fired Balls"), he discusses phenomena of sound propagation. **He** (incorrectly) **states that the velocity of intense sound is not reduced when its level is reduced** [Lat. *nec enim soni velocitas ex illius debilitate minuitur, cum soni auditu perceptibilis pars vltima primæ velocitatem æmuletur*].<br><br>(ii) **However, he correctly states that, from observing the delay between muzzle flash and report, it is easily possible to deduce the distance of the cannon fired against besiegers or besieged.** Similarly, MERSENNE concludes that from the delay between lightning and thunder the observer can determine the distance at which the stroke of lightning happened – the so-called **"flash-to-bang rule."**<br><br>(iii) From these observations he concludes that an attentive soldier who sees the muzzle flash from a cannon fired at a distance of 600 feet (183 m) has at least one second of time to withdraw from the impact by doing three or four steps to the side.[172] But he cautiously recommends, "I don't want to give anybody reason for testing it, unless he supplies himself with armor, helmet and other kinds of equipment, being thus fully prepared when moving into position. However, using a protective screen positioned in front, everybody can find out which reaches them earlier: the sound or the ball [Lat. *Sed & pariete interiecto quispiam id explorare potest, ad quem priùs fragor, quàm globus perveniet*]." ▪ **His ingenious method allowed a crude estimation to be made of the velocity of a cannonball,** although a quantitative measurement was not possible until the advent of the ballistic pendulum {J. CASSINI JR. ⇨ 1707; ROBINS ⇨ 1740; HUTTON ⇨ 1783; ⇨ Fig. 2.10−C}. |
| 1647 | *Private laboratory at Stalbridge Manor, Dorset, southwest England* | **Robert BOYLE,** an English natural philosopher and theological writer, **corresponds with Samuel HARTLIB,** a learned man of Polish origin, **on new mechanical inventions such as the *wind gun*.**[173] BOYLE writes to him, "As for the pneumatic engine, that I use to call a 'wind-gun,' which you mention in your letter as presented to the King, and forbidden by him to have any companions, sure the artist, that received the command, was more ingenious than obedient; for I remember very well to have seen one of them not exceeding in bigness, nor differing much in shape from an ordinary carabine, which being charged by the sole impression of the air, would by violence of the contracted Boreas, send forth a leaden bullet, just the caliber, with force to kill a man at twenty-five or thirty paces from him. This wind-gun I saw both charged and discharged; and now it comes into my mind, I read, not long since, in a late mechanical treatise of the excellent MERSENNUS, both the construction and use of this engine; and amongst the uses one, whose stratagem obliged me to take of it particular notice; and it was, how by the help of |

---

[171] M. MERSENNE: *Cogitata physico-mathematica.* A. Bertier, Paris (1644); *see Phenomena ballistica*, Propositio XXXV [*Soni velocitas maior est globorum explosorum velocitate, 230 sexpedas spatio vnius secundi minuti conficit*], pp. 138-140. ▪ A facsimile of his *Cogitata physico-mathematica* (1644) can be found in the catalogue *Gallica, la bibliothèque numérique*, Bibliothèque nationale de France; http://gallica.bnf.fr/.

[172] MERSENNE found a sound velocity of about 448 m/s {MERSENNE ⇨ 1636}. By applying his flash-to-bang rule he concluded that a cannonball needs at least one second to cover a distance of 600 ft, hence he assumed an average velocity of ≤ 183 m/s.

[173] M. BOAS-HALL: *L'élasticité de l'air avant MARIOTTE et en Angleterre.* In: (P. COSTABEL, ed.): *MARIOTTE, savant et philosophe* (1684). *Analyse d'une renommée.* Librairie Philosophique, Paris (1986), pp. 55-63.

| | | |
|---|---|---|
| | | this instrument, to discover the weight of the air; which, for all the prattling of our book-philosophers, we must believe to be both heavy and ponderable, if we will not refuse belief to our sense…" |
| 1652 | *Private study at his family home in Voorburg, [a town near The Hague], The Netherlands* | **Christiaan HUYGENS,** an eminent Dutch natural philosopher, **formulates the rules of elastic percussion.**[174] ▪ One of the motivations for his study was his lack of belief in DESCARTES' laws of percussion {DESCARTES ⇒1644}.<br><br>On July 6, 1656, referring to DESCARTES' work on percussion, he wrote in a letter[175] to Claude MYLON, a lawyer and amateur mathematician: "Before long I shall have finished a treatise which deals with the mutual interaction of the motion of hard bodies when they meet each other. For this case, I found rules which are completely different from the ones stated by Mr. DE CARTES [DESCARTES], with the exception of the first rule. And I demonstrate that a body at rest, independently of its size, can be set into motion when meeting with another body – however small it may be. **Another remarkable proposition says that when a small body strikes a larger one, it gives it a higher velocity than that achieved by direct percussion when a third body of medium size is inserted between both bodies.** The same happens when the large body hits the small one." ▪ In 1656, HUYGENS collected the results from his percussion studies, supported by proofs, in a treatise *De motu corporum ex percussione* ("On the Motion of Bodies after Impact"), which was published posthumously {HUYGENS ⇒1703}. A brief summary of his results was published in the *Philosophical Transactions of the Royal Society* and in the French *Iournal des Sçavans* {HUYGENS ⇒1668/1669}. |
| 1659 | *Private laboratory at Amsterdam, The Netherlands* | **Johann Rudolf GLAUBER,** an eminent German chemist ("the German BOYLE") and apothecary, **first synthesizes ammonium nitrate** [$NH_4NO_3$] **by combining ammonium carbonate** [$(NH_4)_2CO_3$] **and nitric acid** [$HNO_3$]. He calls it "nitrum flammans" because of the difference of its yellow flame from that of potassium nitrate.[176]<br><br>Ammonium nitrate (AN), also known as "ammonium salt," is a strong oxidizer. When heated in a confined space or brought in contact with other materials, it can undergo explosive decomposition. Its decomposition under shock and heat may take place according to<br>$NH_4NO_3 \rightarrow N_2O + 2 H_2O$ when heated to 200–300 °C, or<br>$2 NH_4NO_3 \rightarrow 2 N_2 + 4 H_2O + O_2$ when detonated in a high explosive.<br>The full power of ammonium nitrate as an explosive was not discovered until the end of World War I.<br><br>(i) In September 1921, **a violent explosion occurred at Oppau, Germany, when a congealed pile of 4,500 tons of an ammonium sulfate-nitrate double salt** [$(NH_4)_2SO_4/NH_4NO_3$] **detonated with tremendous force.** It was the largest man-made explosion prior to the advent of the atomic bomb, which left a crater 60 m deep and 129 m in diameter, and caused severe damage to a radius of 6 km.[177] Charles E. MUNROE,[178] a U.S. explosives expert who, stimulated by the disaster, thoroughly studied the explosivity of ammonium nitrate, and found it altogether safe for handling when properly stored. He concluded that "ammonium nitrate when stored by itself … and apart from explosive substances is, for transportation and storage, not an explosive." However, a subsequent disaster which took place in the United States {Texas City ⇒1947}, clearly demonstrated again the potential dangers of its storage and transportation. |

---

[174] H.J.M. BOS: *Ch. HUYGENS.* In: (C.C. GILLESPIE, ed.) *Dictionary of scientific biography.* Ch. Scribner's Sons, New York, vol. 6 (1972), pp. 597-613.

[175] *Œuvres complètes de Christiaan HUYGENS* (ed. by the Société Hollandaise des Sciences). Nijhoff, The Hague, vol. 1 (1888): *Correspondance 1638–1656*, p. 448.

[176] R.E. KIRK and D.F. OTHMER (eds.): *Encyclopedia of chemical technology.* The Interscience Encyclopedia, Inc., New York, vol. 1 (1947), pp. 817-823.

[177] (Ed.): *Bauliche Lehren aus dem Oppauer Unglück.* Z. ges. Schieß- u. Sprengstoffwesen **20**, No. 11, 109 (1925).

[178] C.E. MUNROE: *The explosivity of ammonium nitrate.* Chem. Met. Engng. **26**, 535-542 (1922).

(ii) Ammonium nitrate, which has about 70% the strength of nitroglycerin has been mixed with TNT to make an explosive nearly as powerful as TNT alone. It has also been used to modify the detonation rate of other explosives, such as nitroglycerin in so-called **"ammonia dynamites"** {NORRBIN & OHLSSON ⇨1867}.

(iii) **Ammonium nitrate is widely used as an artificial nitrogen fertilizer.** In combination with fuel oil or other liquid hydrocarbons – known as "ANFO" (ammonium nitrate/fuel oil) – it became an important industrial explosive which proved useful in open pit mining, quarrying, and civil construction. However, it is well-known among terrorists that ANFO can also be used to easily make bombs {Oklahoma City ⇨1995}. In the U.S. no permit is required to sell or buy it, and there is no register of buyers or sellers.[179]

| | | |
|---|---|---|
| 1660s | *Accademia del Cimento, Florence, Tuscany, central Italy* | At the Florence Academy, **an experiment is performed to find out whether water is compressible:** water is sealed into a sphere of lead which is then flattened between the jaws of a press until the water exudes through the lead walls. A measurement of the distortion of the lead shows no measurable loss of volume.[180] ▪ The first experimental evidence of the compressibility of liquid was not given until 100 years later {CANTON ⇨1762}. |
| 1660 | *Gresham College, London, England* | November 28 marks the **official foundation of the *Royal Society of London*.**[181] The founder-members are Christopher WREN, Robert BOYLE, John WILKINS, Sir Robert MORAY, and William BROUNCKER, 2nd Viscount BROUNCKER. The first Curator of Experiments is Robert HOOKE. Viscount BROUNCKER, an English mathematician, will become the first president (1662–1677). ▪ The origins of the Royal Society lie in an "invisible college" of natural philosophers, essentially formed from the founder members, who began meeting in the mid-1640s to discuss the ideas of Francis BACON (1561–1626). His concept of contemplating "things as they are, without superstition or imposture, error or confusion" gave a powerful impetus to the development of science and technological process in the 17th century. |
| 1660 | *Private laboratory at Oxford, England* | Robert BOYLE[182] **discovers that at constant temperature the volume *V* of a given mass of gas is inversely proportional to the pressure *p*.**<br><br>In the same year, BOYLE finds that enclosed columns of air exhibit a high degree of elasticity, thus popularizing the expression "the spring of the air." ▪ Two years later, his disciple Richard TOWNLEY first put it into an analytical form, $pV$ = const. This so-called **"Townley-Boyle law"** – often referred to as the **"Boyle law"** – describes a gas under *isothermal* conditions. The problem of finding the corresponding law for *adiabatic* conditions, an important step towards the analytical treatment of shock waves, was not tackled successfully until 163 years later {POISSON ⇨1823}.<br><br>BOYLE studied not only the behavior of slow (isothermal) gas compression, but also showed a keen interest in rapid gas compression effects, such as those applied in the **wind-gun** {BOYLE ⇨1647} and the **blow-gun** {⇨Fig. 4.2–I}.[183] Besides, another pneumatic device, the **pneumatic lighter** (or tachopyrion) {MOLLET ⇨1804; ⇨Fig. 4.2–K}, was well-suited to illustrating the adiabatic nature of the "dynamic elasticity" of air. |
| 1663 | *Magdeburg, Saxony-Anhalt, east Germany* | Otto VON GUERICKE, the celebrated natural philosopher and consul of Magdeburg, **builds the first high-voltage generator based upon friction electricity.** It consists of a globe of sulfur that sparks and crackles when rotated and rubbed by hand. ▪ His machine did not have a metallic conductor and rotating glass disc rubbing against leather cushions, which were typical ele- |

---

[179] F.E. FOLDVARY: *Ammonium nitrate: regulate it*. The Progress Report (Editorial); http://www.progress.org/fold226.htm.
[180] Cited in P.W. BRIDGMAN: *The physics of high pressure*. Bell & Sons, London (1958), p. 1.
[181] In the Royal Charter of 1663, the Society is referred to as "The Royal Society of London for Improving Natural Knowledge."
[182] R. BOYLE: *A continuation of new experiments, physico-mechanical: touching the spring and weight of the air, and their effects*. Hall, Oxford (1669–1682).
[183] E.G. WOLFF: *Air guns*. Milwaukee Public Museum, Milwaukee, WI (1958).

ments of improved friction machines added by subsequent inventors. With the discovery of electrostatic induction, electric friction machines were quickly replaced by the more efficient electric influence machines. They were invented independently in 1865 by the German physicists August TOEPLER[184] and Wilhelm HOLTZ,[185] and later improved by the British engineer James WIMSHURST.[186]

After the invention of dynamite {NOBEL ⇨1867}, **high-voltage electric friction machines were used for a brief period in blasting operations to fire detonation caps by an electric spark,** but soon became superseded with the invention of magneto-electric and dynamo-electric blasting machines {H.J. SMITH ⇨1878}.

| 1665 | Paris, France | On January 5, **publication of the first issue of the French *Le Journal des Sçavans* (1665–1792) – the earliest scholarly journal.** The first editor is Denis DE SALLO, a member of the French parliament who belongs to a group of intellectuals forming the predecessors of the Académie des Sciences. Since "news ages quickly," as DE SALLO states, the aim of this journal, as itemized in the premier issue, is to publish weekly in order |

▸ to provide a catalogue and short description of books, and obituaries of famous men;
▸ to publish experiments in physics and chemistry, observations of astrological phenomena, new anatomical findings, useful machines, *etc.*;
▸ to publish current events in academia; and
▸ to print decisions of tribunals and universities.

In part as a response, and within only two months of establishment of the *Journal des Sçavans*, **the English journal *Philosophical Transactions of the Royal Society of London* is published on March 6, 1665** as a monthly serial.[187] ▪ In 1887, the Journal was split into *Series A* (Mathematical, physical and engineering sciences) and in 1888 into *Series B* (Biological sciences).

**Soon after the establishment of these two journals, Christiaan HUYGENS {⇨1668/1669} first published his "Laws of Percussion"** – a landmark in classical mechanics – in both journals.

| 1665 | Gresham College, London, England | **Robert HOOKE,** an eminent English physicist and inventor, and at that time BOYLE's assistant, **publishes his *Micrographia*** ("Small Drawings"), a series of observations made with the aid of magnifying lenses; some of these on very small things, some on astronomical bodies. His book also contains numerous new and revolutionary ideas: |

(i) **He speculates on a *corpuscular model*,**[188] stating in general that all matter expands when heated and that air is made up of particles separated from each other by relatively large distances. Prior to NEWTON's *Principia* {Sir NEWTON ⇨1687}, **HOOKE propounds a theory where the properties of matter, particularly of gases, are to be understood in terms of the motion and collision of atoms** – a concept later termed the *kinetic theory of gases* {D. BERNOULLI ⇨1738; KRÖNIG ⇨1856; CLAUSIUS ⇨1857; MAXWELL ⇨1867}. Referring to the nature of a "fluid body," he states in the Preface of his *Micrographia*, "What is the *cause of fluidness*; And this, *I conceive*, to be nothing else but a certain *pulse* or *shake* of *heat*; for Heat being nothing else but a very *brisk* and *vehement agitation* of the parts of a body (as I have elsewhere made *probable*) the parts of a body are thereby made so *loose* from one another, that they easily *move any way*, and become *fluid*."

---

[184] A. TOEPLER: *Über die Erzeugung einer eigenthümlichen Art von intensiven electrischen Strömen vermittels eines Influenz-Electrometers.* Ann. Phys. **125** [II], 469-496 (1865).

[185] W. HOLTZ: *Über eine neue Elektrisiermaschine.* Ann. Phys. **127** [II], 320-327 (1865).

[186] J. WIMSHURST: *Electrical influence machines.* Proc. Roy. Inst. **12**, 300-305 (1889); *A new form of influence machine.* Phil. Mag. **36** [V], 264-267 (1893).

[187] Facsimiles of the early rare volumes of *Le Journal des Sçavans* as well as of the *Philosophical Transactions of the Royal Society of London* can be found in the catalog *Gallica, la bibliothèque numérique,* Bibliothèque nationale de France; http://gallica.bnf.fr/.

[188] R. HOOKE: *Micrographia.* Martyn & Allestry, London (1665). Facsimile reproduction published by Dover Publ., New York (1961); *see Preface* and pp. 13-14.

(ii) **He experimentally modeled the formation of lunar craters and describes experiments in which he dropped bullets into wet clay.** He notices that the muddy splash crater so produced looked like features on the Moon, but he doubts that lunar craters formed in this manner because, "… it would be difficult to imagine whence those bodies [the projectiles] should come; and next, how the substance of the Moon should be so soft." He also made "volcanic" craters by cooling boiled plaster of Paris, which preserved the bubbles, and proposes the idea that the lunar craters were formed by the collapse of huge bubbles of gas – the so-called **"Hooke gas-bubble theory."** • Meteoric craters, however, remained virtually unknown in contemporary terrestrial experience, and the impact hypothesis did not prevail until the early 1890s {GILBERT ⇨ 1893; WEGENER ⇨ 1920; SPENCER ⇨ 1932}.[189]

(iii) **HOOKE first investigates scientifically the explosion-like disintegration of so-called "Prince Rupert drops"** (or "Prince RUPERT's drops") and visualizes the fracture behavior using a microscope. • Prince Rupert drops {⇨Fig. 4.16–W} are blobs of molten glass dropped into cold water. Note that soda-lime glass is best suited for demonstration purposes. When Prince Rupert drops are broken in air, the noise produced by the fragmentation is not sharp like the explosion of a small crystal of silver azide or lead azide, but instead is like the sound produced when a 6- to 7-mm-dia. glass rod is broken by bending it.

▶ These curious drops were introduced to England in the 1640s by Prince RUPERT OF THE PALATINATE (1619–1682), grandson of king JAMES I of England. Prince RUPERT who, acting as a Royalist commander of the English Civil War, brought these to the attention of CHARLES II, king of Great Britain and Ireland. Prince Rupert drops were not only demonstrated in public lectures {FARADAY ⇨ 1859/1860}, but also used as a joke: the king would have a subject hold the bulb end in the palm of the hand, and then break off the tip, giving the startled person a small explosion right there in a closed hand. It was harmless fun, though, as the glass shatters into powder, not into jagged shards.[190]

▶ More recent studies at Cavendish Laboratory (Cambridge, U.K.), using high-speed photography, revealed that **fracture at the tail end releases a considerable amount of stored compressed stress, thereby sending shock waves through the interior and causing a sudden fragmentation of the entire drop in all directions.**[191]

(iv) **HOOKE makes the first schlieren experiments.** • *Schlieren* [a German word meaning "streaks" or "striae"] designates regions in a transparent substance in which changes in density are revealed by changes in refractive index, under a special type of illumination typically causing light and dark areas in a "schlieren picture" {A. TOEPLER ⇨ 1864}.

| | | |
|---|---|---|
| 1666 | *Arundel House, The Royal Society of London, England* | On October 17, at a meeting of the Royal Society **a demonstration of a "percussion machine" is given:**[192] "An experiment was tried of the propagation of motion by a contrivance, whereby two balls of the same wood, and of equal bigness, were so suspended, that one of them being let fall from a certain height against the other, the other was impelled upwards to near the same height, from which the first was let fall, the first becoming then almost quiescent, and the other returning, impelled the first upwards again to almost the same height it had fallen from before, itself becoming then in a manner motionless, till after some return they both vibrated together. It was ordered that this experiment be prosecuted, and others of that kind thought upon…" |

---

[189] E.T. DRAKE and P.D. KOMAR: *Origin of impact craters; ideas and experiments of HOOKE, GILBERT and WEGENER*. Geology (GSA) **12**, 408-411 (1984).
[190] *Prince RUPERT's drop and glass stress*, Corning Museum of Glass, Corning, NY; http://www.cmog.org/index.asp?pageId=735.
[191] S. CHANDRASEKHAR and M.M. CHAUDRI: *The explosive disintegration of Prince RUPERT's drops*. Phil. Mag. **70B** [VII], 1195-1218 (1994).
[192] T. BIRCH: *The history of the Royal Society of London for improving of natural knowledge from its first rise*. Culture et Civilisation, Brussels *etc.* (1756/1757), vol. II, pp. 116-117.

In the minutes of the following meeting on October 24, it reads, **"The experiment about propagating of motion was prosecuted with three balls, of which the middle remained almost quiescent, though struck by either of the lateral ones, which impelled each other upwards."** ▪ This percussion machine, which had been used by the Dutch natural philosopher Christiaan HUYGENS and British savants since 1661, is shown schematically in MARIOTTE's work {⇨1671}, and in more detail in the *Principia* {Sir NEWTON ⇨1687}. However, the idea of using a pendulum apparatus rather than spheres which collide with each other on a planar table {MARCI ⇨1639} or a billiard table already goes back to GALILEI, who used it in his *Discorsi* to illustrate a thought experiment {GALILEI ⇨1638}. Advantageously, a pendulum arrangement eliminates friction effects between ball and table, and the essential source of friction is due to deformation losses only.

For centuries, the multiple-ball pendulum {⇨Fig. 4.4−B} – better known as **"NEWTON's cradle"** (sometimes referred to as **"NEWTON's demonstrator"**) and in Germany as the **"Klick-Klack"**[193] – has become an indispensable instrument for demonstrating chain percussion to generations of ballisticians, physicists and mechanical engineers:

(i) In 1914, the German ballistician Carl CRANZ[194] and collaborators used their ballistic cinematograph optical shadowgraphy to visualize the motions of the balls in a three-steel-ball pendulum at a frame rate of 2,500 $s^{-1}$.

(ii) In 1976, William JOHNSON,[195] a British mechanical engineering professor at the University of Cambridge, discussed the assumptions and shortcomings in conventional treatments of NEWTON's cradle and reconsidered the elementary theory of the impact of a line of particles usually taught. He showed that the "simple" impact of impinging spheres is actually a strength of materials problem which requires the consideration of a constitutive law (Hooke law) and the loss of kinetic energy in real impact situations.

(iii) Various analyses on the dynamics of NEWTON's cradle, which have recently been reviewed {NESTERENKO ⇨2001}, have shown that – contrary to most previous assumptions – momentum is not concentrated in a single ball moving through the chain, as commonly supposed; *i.e.*, momentum and energy are not conserved through all cycles {HERRMANN & SCHMÄLZLE ⇨1981}. For example, the impact of one ball is followed mainly by the breaking of one ball from the other end; however, a small amount of momentum is also carried by the second ball. **Thus, the chain of contacting balls is not "dispersion-free,"– particularly for a long chain, but rather characterized by a strongly nonlinear dispersion resulting in the generation of a solitary wave** {RUSSELL ⇨1834}.

| | | |
|---|---|---|
| 1666 | *Private Library of king Louis XIV, Rue Vivienne, Paris, France* | On December 22, **foundation of the *Académie Royale des Sciences* ("Royal Academy of Sciences")** at the instigation of Jean-Baptiste COLBERT, then French minister of finance, with the objective to study a broad range of scientific activities, in particular to advance the science of navigation with the help of mathematics and astronomy. |

In the same year, the Dutch physicist Christiaan HUYGENS, one of the ten founding members, draws up the research agenda of the Académie. He proposes to advance knowledge of natural history,[196] particularly

---

[193] W. BÜRGER: *Klick-Klack. Das Geheimnis der schwingenden Stahlkugeln.* Bild der Wissenschaft **34**, 100-1012 (Nov. 1997).

[194] C. CRANZ, P.A. GÜNTHER, and F. KÜLP: *Photographische Aufnahme von sehr rasch verlaufenden Vorgängen, insbesondere von Schussvorgängen, mittels Vorderbeleuchtung durch das Licht elektrischer Funken.* Z. ges. Schieß- u. Sprengstoffwesen **9**, 1-4 (1914).

[195] W. JOHNSON: *Simple linear impact.* IMechE & UMIST (University of Manchester) **4**, 167-181 (1976).

[196] C. HUYGENS: *Pour l'Assemblée de Physique.* Académie Royale des Sciences (1666). Collection Boulliau, Bibliothèque Nationale, Paris. Translated by R.A. HATCH; http://web.clas.ufl.edu/users/rhatch/pages/03-Sci-Rev/SCI-REV-Home/resource-ref-read/major-individuals/huygens/08sr-hygn-CB.htm.

- "to perform experiments on the vacuum with the instrument and otherwise, and determine the weight of air;
- **to examine the explosive force of gunpowder enclosed, in small quantities, in an iron or very thick copper box;**
- to examine in the same manner the force of water rarified by fire;
- to examine the force and speed of wind, and the uses drawn from it for navigation and in machines; and
- **to examine the force of percussion or the communication of motion in the impact of bodies,** concerning which I believe I have been the first to provide true principles."

| | | |
|---|---|---|
| 1668 | *Arundel House, The Royal Society of London, England* | At a meeting of the Royal Society, **a resolution is passed that the phenomenon of percussion should be treated in more detail and its results passed to a committee of the Society.**[197] • The coordinator was the German-born English scholar Henry OLDENBURG, first secretary of the Society and first editor of its journal *Philosophical Transactions* {⇨1665}.<br><br>In the same year, OLDENBURG[198] writes to HUYGENS in a letter dated November 18: "When the contents of your letter indicating your willingness to impart to the Royal Society the rules and theorems you have found with respect to all kinds of motion was communicated to the Society, your offer was accepted with much satisfaction and I was ordered to let you know of this as soon as possible, and to request you at the same time to be so good as to send them first that part which (according to the method you have adopted in this matter) enlighten that which comes next, and so in succession. Meanwhile the Society think that if you were disposed to let them see the whole scheme, and to indicate together with that the principal experiments that you have made to establish those rules, they could the better judge of the whole … I will speak to Mr. WREN, when I see him, of your desire to see his work on this topic, since he is one of the chief of those who have investigated the nature of motion…" • OLDENBURG got responses from WALLIS {⇨1669}, WREN {⇨1669}, and HUYGENS {⇨1668/1669}. These three eminent geometricians had already performed experimental and/or theoretical studies on percussion in previous years. |
| 1668 | *Gresham College, London, England* | **Robert HOOKE,**[199] the famous English natural philosopher, **writes his *Discourse on Earthquakes,*** which at a later date will be read before the Royal Society of London. Speculating on the presence of fossil shells on mountains and in inland regions, he writes: "It seems not improbable, that the tops of the highest and most considerable Mountains in the World have been under Water, and that they themselves most probably seem to have been the Effects of some very great Earthquake." • His discourse on the effects of earthquakes, though containing many passages of considerable merit, generally tended to incorrectly interpret the phenomenon in question. |
| 1668/ 1669 | *L'Académie Royale des Sciences, Paris, France* | In March, **Christiaan HUYGENS,** a renowned Dutch mathematician, physicist and astronomer, **announces his** *Regulae de motu corporum ex mutuo impulsu* ("Laws of Motion on the Impact of Bodies") – **a summary of previous results of his investigations on the percussion of "hard bodies"** [Lat. *corpora dura*]; *i.e.*, **of *totally elastic bodies*** recovering their form after impact {HUYGENS ⇨1652}. He publishes his results first in French in the *Journal des Sçavans*[200] {⇨Fig. 2.16} and shortly after in Latin in the *Philosophical Transactions*.[201] |

---

[197] R. HALL: *Mechanics and the Royal Society.* Brit. J. Hist. Sci. **3**, 24-38 (1966).
[198] A.R. HALL and M.B. HALL (eds.): *The correspondence of Henry OLDENBURG.* University of Wisconsin Press, Madison, WI, vol. 5 (1968): *1668–1669,* pp. 177-178.
[199] R. WALLER (ed.): *The posthumous works of Robert HOOKE.* Smith & Walford, London (1705), p. 288.
[200] C. HUYGENS: *Règles du mouvement dans la rencontre des corps [Extrait d'une lettre de M. HVGENS a l'Auteur du Iournal].* Journal des Sçavans (Paris) **5**, 22-24 (March 18, 1669).
[201] (Foreword by the editor) *A summary account of the laws of motion, communicated by Mr. Christian HUYGENS in a letter to the R. Society, and since printed in French in the Iournal des Sçavans of March 18, 1669.* Phil. Trans. Roy. Soc. Lond. **4**, No. 46, 925-927 (1669).

- HUYGENS was displeased by the appearance of WALLIS' and WREN's papers on percussion {WALLIS ⇒ 1669; WREN ⇒ 1669} without mention of his own work of the same kind, and dispatched a summary of his own studies to the *Journal des Sçavans*.

(i) Introductorily, HUYGENS states in a letter to the editor: "I send you, as promised, my propositions concerning the motion of percussion, that is the motion which is produced by the impact of bodies. This subject has already been studied by several excellent men of this century such as GALILEI [⇒ 1638], DESCARTES [⇒ 1644], FABRI [[202]] and more recently by Mr. BORELLI [[203]]; their contributions I shall not review here. But I tell you only that my theory perfectly agrees with the experiment and, as I believe, is built upon demonstration which, as I hope, will soon be given publicly."

(ii) **HUYGENS[204] postulates seven tenets [Lat. *regulae*] of impact,** the most important of which are

- Tenet 1: "A body striking another body at rest loses its own motion and communicates an equal quantity to the other." • This observation had already been made by MARCI {⇒ 1639}.

- Tenet 2: "Two equal bodies, colliding with equal and opposite velocities, separate after impact with the same velocities." • This means that for two perfectly elastic bodies with masses $m_1 = m_2 = m$ impacting with velocities $v_1 = v_2 = v$, the "quantity" (*see* Tenet 5 below) of the motion is conserved; *i.e.*, $mv + mv = 2\,mv = $ const. In the general case with $m_1 \neq m_2$ and $v_1 \neq v_2$ the quantity of motion is also conserved, hence $m_1v_1 + m_2v_2 = $ const.

- Tenet 3: "A body at rest, however great, is set in motion by a body which strikes it, however small {GALILEI ⇒ 1638}." • HUYGENS probably refers to the fourth tenet of DESCARTES' percussion theory, where he had (incorrectly) denied the possibility that a large body can be set in motion by the impact of a smaller one {DESCARTES ⇒ 1644}.

- Tenet 4: This refers to a geometrical method for determining the velocities after impact which is very similar to the one proposed by WREN {⇒ 1669}.

- Tenet 5: "The quantity of motion [*la quantité du mouvement*] which two bodies have can be increased or decreased by their impact: however, counted in the same direction it remains constant at any time when the oppositely directed value is subtracted from it." • Probably referring to DESCARTES, the term *quantity of motion* of a body is defined as the product of mass and absolute velocity, given by $m \times |v|$. The velocity $v$ is a vector quantity which, depending on its direction, can be positive or negative. Thus, HUYGENS rule is indeed correct.

- Tenet 6: "The sum of the products of the magnitudes and the squares of the velocities of the bodies before and after impact are always equal." • **He first extends existing percussion theories – until then exclusively governed by the *Law of Conservation of Momentum* – by a concept which, in modern terms, corresponds to the *Law of Conservation of Kinetic Energy*.** In a later memoir published posthumously he gave a proof of this most important rule {HUYGENS ⇒ 1703}. Unfortunately, HUYGENS didn't say how he found this rule. However, he apparently followed GALILEI's concept, who had proposed to characterize force by its ability to overcome resistance, such as the

---

[202] H. FABRI: *Philosophia 2: Tractatus physicus de motu localis: in qvo effectvs omnes, qvi ad impetvm, motum naturalem, violentum, & mixtum pertinent, explicantur, & ex principiis physicis demonstantur*. I. Champion, Lugduni (1646). • The French Jesuit Honoré FABRI (1607–1688) was a natural philosopher, mathematician and astronomer.

[203] G.A. BORELLI: *De vi percussionis*. Giacopo Monte, Bologna (1667). • In this book, Giovanni Alfonso BORELLI (1608–1679), an Italian mathematician and physicist, discusses percussion in detail, some general problems of motion, gravity, magnetism, the motion of fluids, the vibrations of bodies, and pendular motion, to cite just a few items.

[204] C. HUYGENS: *Regulae de motu corporum ex mutuo impulsu* [submitted on Jan. 4, 1669]. Phil. Trans. Roy. Soc. Lond. **4**, No. 46, 927-928 (1669). *See also* *Œuvres complètes de HUYGENS*. M. Nijhoff, The Hague (1895), vol. VI, 383-385.

ability of a body moving at velocity $v$ to climb to a height $h$ under the action of gravity ($v^2 = 2gh$). Based upon this concept, a schematic presentation leading to HUYGENS' sixth rule is given in Fig. 2.3. ▪ The more general Law of Conservation of Energy (the First Law of Thermodynamics) was not discovered in the early 1840s by the German physicists Hermann VON HELMHOLTZ and Julius Robert MAYER.[205]

▸ Tenet 7: "A hard [meaning a perfectly elastic] body at rest will receive more motion from another, second hard body, being larger or smaller, by the interposition of a third body of medium size [meaning mass]. And when this interposed body is averaged proportional to the two others, he provokes the largest action on the quiescent body."
▪ This remarkable effect is illustrated in Fig. 2.6.

(iii) Finally, he states that his rules are fully in conformity with WREN's rules {WREN ⇨ 1669}, and annotates that he was much surprised to see that **the velocity of the center of mass of two bodies is not changed by their collision.** ▪ The collision does not change the total momentum of the system of two bodies, only the distribution of momentum between the two bodies. David HALLIDAY[206] and Robert RESNICK, two U.S. physics professors, illustrated in their textbook *Physics* by a series of "snapshots" taken at equal time intervals that, in the case of two bodies elastically colliding head-on, the center of mass moves in the laboratory reference frame with steady velocity along a straight line.

HUYGENS rules were communicated to the Royal Society on November 17, 1668. Two days later, John WALLIS, who attended the meeting, wrote to Henry OLDENBURG, secretary of the Society: "… What you say of M. HUYGENS his account he gave of some experiments of motion, is true inough; (I was then present:) but it is true allso that he gave us no account of any principles by which he did calculate but onely of ye Result; (and I remember, yt when hee had so done it, he was carefull to blott out all his writing, that it might not from thence appear how he did calculate:) But hee is to remember allso; that the thing had before been done by Mr. [Lawrence] ROOK & Dr. [Christopher] WREN; & that his calculation at that time was but to try whether his & theirs did agree: & it was found so to do: His answers proving to bee the same which theirs before had done…"

In the following year on January 12, WALLIS[207] wrote to OLDENBURG: "… Mr HUYGENS Hypotheses & Propositions of Motion; I have scarce had time more then to read over. But they seem to mee very ingenious, & his way of demonstration also. Hee proceeds from other principles; but, if I mistake not, his & mine may be well inough accommodated. Taking his *duram* ["hard"] in Hyp. 2 as opposed to *molle* ["soft"] in such a sense as this admits a transposition of parts (as in Lead & other soft bodies which upon percussion change their figure) and supposing his *dura* there to be also *aequaliter Elastica* ["equally elastic"], from which I am to **derive** that Resilition which he **postulates**…"

In 1703, his rules of percussion, with proofs added, were published posthumously {HUYGENS ⇨ 1703} and called by Ernst MACH {⇨1883} in his book *Die Mechanik in ihrer Entwicklung, historisch-kritisch dargestellt* the "Laws of Impact" [Germ. *Die Gesetze des Stoßes*]. They form an important cornerstone of classical mechanics.

---

[205] This law states that energy can be transformed from one form to another, but that the total amount is always conserved. ▪ Conservation of energy is a basic principle of modern physics that is used when analyzing the very smallest (subatomic) domains and the largest known structure (the Universe), and just about everything in-between. *See also* the footnote on BODENSTEIN {⇨1913}.
[206] D. HALLIDAY and R. RESNICK: *Physics. Parts I and II*. Wiley, New York *etc.* (1966), pp. 220-222.
[207] This letter and the above mentioned letter were reprinted in the book by A.R. HALL and M.B. HALL (eds.): *The correspondence of Henry OLDENBURG*. University of Wisconsin Press, Madison, WI, vol. 5 (1968): *1668–1669*, pp. 192-194, 336-337.

| 1669 | Chair of Geometry, University of Oxford, England | John WALLIS,[208] an English mathematician, **presents to the readers of the *Philosophical Transactions* a summary of the results of his percussion studies** ("General Laws of Motion").

(i) Very similar to DESCARTES {⇨1644}, **he considers the product of weight** [Lat. *pondus*] **and velocity** [Lat. *celeritas*] – in modern terms the product of mass and velocity ($m \times v$), which is the momentum of a body in motion – **to be an important feature of characterizing an impact, and he calls it the "impetus."** ▪ In modern science the term *impetus* [from Lat. *in-petere* "to go to, to attack"] has two meaning: a force that moves something along (as understood by WALLIS), and the act of applying force suddenly.

(ii) The tenets 9-11 relate to the impact of ***totally inelastic bodies*** – *i.e.*, to bodies which, subjected to pressure, alter their form without recovering it. ▪ Although WALLIS, like many of his contemporaries, attempted to explain kinetics via the concept of "balance," a concept borrowed from statics, he nevertheless obtained some important results. They will be discussed by him in connection with his more detailed treatise on percussion published two years later, which also addressed elastic percussion perfectly {WALLIS ⇨1670/1671}.

(iii) **WALLIS is the first to address the peculiarities of oblique percussion.** In his tenet 12, he says: "If the heavy bodies are not proceeding along the same straight line and do not meet directly, but collide against each other obliquely, the preceding computation must be adjusted according to the degree of the obliquity. For the impetus of a body colliding obliquely is to the impetus that there would be if the collision were direct (other things being equal) as the radius to the secant of the angle of obliquity … When this consideration is duly introduced into the former calculus it will yield the velocity, impetus, and direction: that is to say, with what velocity, impetus, and in what direction the two bodies will severally rebound, which thus collide together." ▪ The oblique impact is dependent on the friction between the two bodies. When the bodies have a smooth surface (*i.e.*, when there is no friction), then the tangential forces of impact are conserved {⇨Fig. 2.5}.

**The analysis of the dynamics occurring in the contact zone during oblique percussion between rough bodies has become a subject of modern research.** In the case of rigid-body percussion in noncollinear configurations, a slip reversal occurs at the contact point if the angle of incidence is small and there is significant friction. However, at increasing angles of incidence, the contact process may change to slip-stick. For a noncollinear collision between a pair of hard, inelastic bodies, a discrete-element model of local contact compliance has been proposed in order to analyze effects due to tangential compliance.[209] |
| 1669 | Chair of Astronomy, University of Oxford, England | Christopher WREN,[210] an eminent English architect, astronomer and geometrician, **treats the problem of percussion in a 1½-page treatise entitled *Lex naturae de collisione corporum*** ("The Law of Nature in the Collision of Bodies"). Like HUYGENS {⇨1668/1669} he limits his study to the percussion of ***totally elastic bodies*** – *i.e.*, to bodies which completely recover their form after collision:

(i) Introductorily, the publisher annotates: "Theory concerning the same subject, imparted to the R. Society Dec. 17 last, though entertain'd by the Author divers years ago, and verified by many experiments, made by Himself and That other excellent Mathematicians Mr [Lawrence] ROOK before the said Society, as is attested by many Worthy Members of that Illustrious Body." |

---

[208] J. WALLIS: *A summary account of the general laws of motion, by way of letter written by him to the Publisher, and communicated to the Royal Society* [in Latin, submitted on Nov. 26, 1668]. Phil. Trans. Roy. Soc. Lond. **3**, No. 43, 864-866 (Jan. 11, 1669). An English translation can be found in the book by A.R. HALL and M.B. HALL (eds.): *The correspondence of Henry OLDENBURG*. University of Wisconsin Press, Madison, WI, vol. 5 (1968): *1668–1669*, pp. 167-170.

[209] W.J. STRONGE, R. JAMES, and B. RAVANI: *Oblique impact with friction and tangential compliance*. Phil. Trans. Roy. Soc. Lond. **A359**, 2447-2465 (2001).

[210] C. WREN: *Lex Naturae de collisione corporum* [in Latin, submitted on Dec. 17, 1668]. Phil. Trans. Roy. Soc. Lond. **3**, No. 43, 867-868 (1669). An Engl. translation can be found in the book by A.R. HALL and M.B. HALL (eds.): *The correspondence of Henry OLDENBURG*. University of Wisconsin Press, Madison, WI, vol. 5 (1968): *1668–1669*, pp. 320-322.

(ii) **WREN introduces the term *proper velocities of bodies*** [Lat. *velocitates corporum propriae*], **which are reciprocally proportional to their masses.** For example, for two elastic bodies $A$ and $B$ of masses $m_A$ and $m_B$ with "proper velocities" $v_A$ and $v_B$, respectively, the following rule applies:

$$v_A : v_B = m_B : m_A \text{ or } v_A m_A = v_B m_B.$$

WREN compares the dynamic process of collision with the balance beam in statics: equilibrium [meaning "balance"] corresponds to a collision situation in which masses approach one another at proper velocities inversely proportional to their masses. He postulates that when two bodies collide head-on with their proper velocities (*i.e.*, $v_A m_A = -v_B m_B$), they also retain their proper velocities, even after collision. **The collision of bodies when "proper velocities" are considered is equivalent to a balance swinging about its center of gravity.**

(iii) **The collision of bodies when "improper velocities"** [*i.e.*, velocities other than proper velocities $vp_A$ and $vp_B$] **are considered is equivalent to a balance beam reciprocating upon two centers equidistant either side of the center of gravity:** the balance may be extended into a yoke when the need arises. **His ingenious balance model** {⇨Fig. 2.4} **is based upon the *Law of the Center of Mass*,** whose velocity is given by

$$v_G = (m_A v_A + m_B v_B)/(m_A + m_B) = (m_A c_A + m_B c_B)/(m_A + m_B)$$

which (in modern notation) **leads to the *Law of Conservation of Momentum*:**

$$m_A v_A + m_B v_B = m_A c_A + m_B c_B. \tag{1}$$

Here $v_A$ and $v_B$ are the velocities before collision and $c_A$ and $c_B$ the ones after collision. His geometrical method permits the determination of the velocities after collision, which (again in modern notation) are given by:

$$c_A = 2(m_A v_A + m_B v_B)/(m_A + m_B) - v_A = 2v_G - v_A, \tag{2}$$

and

$$c_B = 2(m_A v_A + m_B v_B)/(m_A + m_B) - v_B = 2v_G - v_B. \tag{3}$$

Note that when for example $m_A$, $m_B$, $v_A$ and $v_B$ are given, a second equation (generally the Law of Conservation of Kinetic Energy) is necessary besides Eq. (1) to determine $c_A$ and $c_B$. Apparently, WREN found the correct solution merely by his geometrical method.[211] ▪ Unfortunately, even today we don't know how WREN proceeded in detail to find this unusual geometrical solution. Contrary to HUYGENS and WALLIS he didn't resume the subject of percussion in later years. After the Great Fire of London (1666) he was appointed by king Charles II to take charge of rebuilding London, including the enormous task of designing and building over 50 new churches in London, such as St. Paul's Cathedral.

(iv) **WREN states that** – prior to publication – **he tested his theorems "by experiments made by myself and M. [Mister] ROOK**[212] **before the Royal Society"** (at the Society's meeting on December 17, 1668). ▪ Sir Isaac NEWTON referred in his *Principia* to WREN's and HUYGENS' experiments which, together with his own ones, supported his proposition of the "equality of action and reaction" – later named the *Third Law of Motion* {Sir NEWTON ⇨1687}.

---

[211] WREN's unusual method was discussed by (i) A.R. HALL: *Mechanics and the Royal Society.* Brit. J. Hist. Sci. **3**, 24-38 (1966/1967), pp. 30-32; (ii) R.S. WESTFALL: *Force in NEWTON's physics: the science of dynamics in the seventeenth century.* Macdonald, London (1971), pp. 203-206; and (iii) I. SZABÓ: *Geschichte der mechanischen Prinzipien.* Birkhäuser, Basel & Stuttgart (1977), pp. 443-445. ▪ István SZABÓ (1906–1980), a notable Hungarian-born German professor of mechanical engineering at the Technical University Berlin, demonstrated that WREN's geometrical rule, based upon his balance beam model, indeed confirms the conservation law of *vis visa*; *i.e.*, $m_A v_A^2 + m_B v_B^2 = m_A c_A^2 + m_B c_B^2$.

[212] WREN apparently referred to Lawrence ROOK (1623–1662), who was a professor (first of astronomy, later of geometry) at Gresham College in Holborn, London, where WREN taught as a professor of astronomy in the period 1657–1661.

| 1670/ 1671 | Chair of Geometry, University of Oxford, England | **John WALLIS,** an English mathematics professor, **publishes his book *Mechanica, sive de moti tractatus geometricus*** ("Mechanics, or Geometrical Tracts on Motion"), **which also includes a treatise on percussion entitled *De percussione* ("On Percussion");**[213] a summary of his main findings, which he had already published {WALLIS ⇒1669}. |
|---|---|---|

(i) His treatise contains 13 tenets on the central percussion of two bodies: tenets 1–8 relate to straight translatory motion of a body ("General Laws of Motion"), tenets 9–12 to **perfectly inelastic percussion,** and tenet 13 to **perfectly elastic percussion.**

(ii) **If two (perfectly) inelastic bodies which have equal momenta** [$m_A v_A = m_B v_B$] strike each other head-on [$v_A = -v_B$], **rest will ensue after impact.** However, if their momenta are unequal [$m_A v_A \neq m_B v_B$], the two masses $m_A$ and $m_B$ stick together after collision, so that there will be a final common velocity $u$. The sum of the momenta [$m_A v_A + m_B v_B$] will be the momentum after impact. If one divides this momentum by the sum of the masses, one will obtain the velocity of the motion after impact:

$$u = (m_A v_A + m_B v_B)/(m_A + m_B).$$

(iii) **His simple equation is a special case of the general Law of Conservation of Momentum which holds for both elastic and inelastic impact.** ▪ For example, for two masses $m_A$ and $m_B$ with velocities $v_A$ and $v_B$ before impact and $c_A$ and $c_B$ after impact, respectively, the Law of Conservation of Momentum yields:

$$m_A v_A + m_B v_B = m_A c_A + m_B c_B.$$

When $m_A$, $m_B$, $v_A$ and $v_B$ are known, a second equation is required in order to determine $c_A$ and $c_B$. This is the Law of Conservation of (Kinetic) Energy, which was first applied by HUYGENS {⇒1668/1669}, and apparently found by intuition rather than recognizing it as an important principle of mechanics {⇒Fig. 2.3}.

(iv) WALLIS notices that if a body freely rotates about a fixed axis and its motion is suddenly checked by the retention of one of its points, the force of percussion will vary with the position (the distance from the axis) of the point arrested. **He calls the point at which the intensity of the impact is greatest the "center of percussion"** [Lat. *centrum percussionis*]. ▪ The center of percussion is the ideal point in a body freely rotating around a fixed axis at which the struck body will experience no jarring action at the pivot {⇒Fig. 2.9}. The center of percussion is important in engineering, science and sports kits {*see* Sect. 2.2.2; ⇒Figs. 4.3–X, Y}. In 1976, George W. FICKEN JR.[214] described the construction of an apparatus for demonstrating the center of percussion and pivot point of a ball being struck.

In 1695, WALLIS[215] also included his treatise *De percussione* ("On Percussion") in his three-volume *Opera mathematica*. Slightly revised, it contains a final note in which he points out that **his center of percussion is identical to HUYGENS' *center of oscillation*.** ▪ The Dutch mathematician, astronomer and physicist Christiaan HUYGENS,[216] who in 1673 treated in his work *Horologium oscillatorium sive de motu pendulorum* ("The Oscillating Clock, or On the Motion of Pendulums") the oscillation of a body about a stationary axis, had discovered the so-called **"center of oscillation"** of a compound pendulum; *i.e.*, the point at which the whole pendulum mass can be supposed to be concentrated. HUYGENS mentioned in his memoir how he became involved in tackling this difficult problem: "Some time ago, when I was still a boy, the most learned MERSENNE proposed to me and to many others the investigation of centers of oscillation or agitation, a very famous problem among mathematicians of that time, as far as I can gather from the letters he sent me and

---

[213] J. WALLIS: *Mechanica: sive de motu tractatus geometricus*. G. Godbid, London (1670/1671); *see De percussione*, pp. 660-682.
[214] G.W. FICKEN: *Center of percussion demonstration*. Am. J. Phys. **44**, No. 8, 789 (1976).
[215] J. WALLIS: *Opera mathematica* [in Latin]. Sheldon, Oxoniae (1695); *see De percussione*, pp. 1002-1015.
[216] Part 4 of HUYGENS' treatise "The Pendulum Clock" was entitled "On the Center of Oscillation." It was translated into English by Michael S. MAHONEY, a history professor at Princeton University; http://www.princeton.edu/~hos/mike/texts/huygens/centosc/huyosc.htm.

| 1671 | L'Académie Royale des Sciences, Paris, France | Edmé MARIOTTE,[218] a French physicist and plant physiologist, **presents to the French Academy a comprehensive treatise on the laws of *elastic* and *inelastic* percussion of bodies** – obviously partly based on previous works {MARCI ⇨ 1639; WREN ⇨ 1669; WALLIS ⇨ 1669; HUYGENS ⇨ 1652 & 1668/1669} – but which approaches the problem by linking theory with many new experiments and applications. For example, by letting ivory balls fall from varying heights onto a steel anvil coated lightly with dust, MARIOTTE first shows that "hard" (*i.e.*, elastic) bodies do indeed deform upon impact. Circles of varying widths show that the degree of flattening is dependent on the speed of the ball at impact. |

from the recently edited letters of DESCARTES[217] containing his response to MERSENNE on these matters."

1671    L'Académie Royale des Sciences, Paris, France

Edmé MARIOTTE,[218] a French physicist and plant physiologist, **presents to the French Academy a comprehensive treatise on the laws of *elastic* and *inelastic* percussion of bodies** – obviously partly based on previous works {MARCI ⇨ 1639; WREN ⇨ 1669; WALLIS ⇨ 1669; HUYGENS ⇨ 1652 & 1668/1669} – but which approaches the problem by linking theory with many new experiments and applications. For example, by letting ivory balls fall from varying heights onto a steel anvil coated lightly with dust, MARIOTTE first shows that "hard" (*i.e.*, elastic) bodies do indeed deform upon impact. Circles of varying widths show that the degree of flattening is dependent on the speed of the ball at impact.

His treatise, which was published two years later, is the first comprehensive treatment of the laws of inelastic and elastic impact and of their application to various physical problems; it long served as the standard work on the subject and went through three editions in MARIOTTE's lifetime.

▸ In the first section MARIOTTE gives definitions of elastic and inelastic bodies, laws of elastic collision, and transmission of impulse through a chain of elastic bodies – a problem proposed by DESCARTES.

▸ In the second part, MARIOTTE tests the problem of oblique collision by employing HUYGENS' method of impact on a moving boat without crediting HUYGENS.

In the third edition of his *Principia* (1713), Sir Isaac NEWTON[219] praised MARIOTTE's contribution on the Laws of Percussion, calling him *"Mariottus Clarissimus"* ["The Most Illustrious Mariotte"]. In 1690, however, HUYGENS[220] accused MARIOTTE of plagiarism and wrote, **"MARIOTTE took everything from me** as can attest those of the Academy, M. DU HAMEL, M. GALLOIS, and the registers. [He took] the machine, the experiment on the rebound of glass balls, **the experiment of one or more balls pushed together against a line of equal balls** [this phenomenon, however, was first described by MARCI ⇨ 1648], the theorems that I had published [*Phil. Trans.* (1668); *J. des Sçavans* (1669)]. He should have mentioned me. I told him that one day, and he could not respond…" The principle of his percussion machine was indeed demonstrated previously {The Royal Society of London ⇨ 1666}.

1672    Arundel House, The Royal Society of London, England

At a meeting of the Royal Society, **Robert HOOKE,** then a professor of geometry at Gresham College in London, **gives a first demonstration of a new optical method for visualizing density variations** which he had described seven years before in his *Micrographia* ("Small Drawings") {HOOKE ⇨ 1665}. In his demonstration, he uses a candle as a light source, which is positioned in the focus of a lens, and a second candle positioned near the lens, "to shew, that, besides the flame and smoke of a candle there is a continual stream rising up from it, distinct from the air…" These observations not only first prove the suitability of the schlieren method for visualizing flow phenomena, but also stimulate him to develop a theory of combustion.[221]

HOOKE's new method was later modified by August TOEPLER {⇨1864} who, terming it the *Schlierenmethode* ("schlieren method"), demonstrated its excellent suitability for visualizing shock waves. The term *schlieren* is a German word meaning "streaks" or "striae."

---

[217] *Renati DESCARTES epistolae omnes.* Ex Typographia Blaviana, Amstelodami (1683), *Pars III*, p. 317.
[218] E. MARIOTTE: *Traité de la percussion ou choc des corps*. E. Michallet, Paris (1673).
[219] I. NEWTON: *Philosophiae naturalis principia mathematica*. Crownfield, Cantabrigae (1713), p. 47 (*see Scholium* after the paragraph *Laws of Motion*).
[220] *Œuvres complètes de Christiaan HUYGENS* (ed. by the Société Hollandaises des Sciences). Swets & Zeitlinger, Amsterdam (1929), vol. 16: *Percussion*, p. 209.
[221] R. RIENITZ: *Schlieren experiment 300 years ago*. Nature **254**, 293-295 (1975).

| | | |
|---|---|---|
| 1673 | L'Académie Royale des Sciences, Paris, France | **Christiaan HUYGENS,** a Dutch mathematician, astronomer and physicist, **and Denis PAPIN,** a French-born British physicist, **demonstrate their *pompe balistique* ("ballistic pump") to Jean-Baptiste COLBERT,** then French Secretary of State and Minister of Finance. They prove the feasibility of gunpowder to operate a *moteur à explosion* ("explosion engine") for pumping water or lifting weights {⇨Fig. 4.2–L}.[222] Their device doesn't use the explosive action directly, but rather the vacuum which arises when the hot explosion products have left the combustion chamber via a valve and the cylinder has cooled down. Work is then performed by atmospheric pressure, which resets the piston to its initial position. ▪ HUYGENS proposed the principle in order to pump water from the Seine to the gardens of Versailles.<br><br>Five years later, **Jean DE HAUTE-FEUILLE,** a French physicist and inventor, **proposed a similar concept of using gunpowder to fuel the explosion engine.** In 1926, Carl J. CRANZ,[223] a German ballistician, annotated that combustion engines operated with explosives would be less efficient than those with conventional fuels: the energy content of black powder is only about 1.3 kcal/g, that of nitroglycerin is 2.9, while for hard coal it amounts 14.5, for city gas 11, and for gasoline 18. |
| 1673 | Private laboratory at Paris, France | **Edmé MARIOTTE**[224] **discovers by quantitative measurements that the drag of a body is proportional to the second power of the flow velocity.** He performed his experiments in a river by exposing a square plate to a flow of water. ▪ In the 1673 edition of his *Traité de la percussion ou chocq des corps* ("Treatise on Percussion or Impact of Bodies"), MARIOTTE annotated that his derived laws of collision might also be applicable when treating natural intense sound phenomena, such as thunder. It appears that he had already anticipated a corpuscular model with air particles for the transmission of sound, even prior to a concept published fourteen years later in the *Principia* {Sir NEWTON ⇨1687}. |
| 1678 | Arundel House, The Royal Society of London, England | At a meeting of the Society, **Robert HOOKE,**[225] an English all-round natural philosopher ("England's Leonardo"), **announces his Law of Elasticity, also known as the "Law of Deformation."** In his *Lectiones cutlerianae*[226] "Cutlerian Lectures") he states that *"ut tensio sic vis"* ("as the extension, so the force") – *i.e.,* the stretching of a solid elastic body (*e.g.*, a helical spring) is proportional to the force applied to it. ▪ In 1660, he already discovered an instance of his law while working on designs for the balance springs of clocks.<br><br>HOOKE did not clearly differentiate between what we today hold to be quantitatively distinct physical relations: BOYLE's law for gases {BOYLE ⇨1660} and his own law for elastic solids.[227] His one-dimensional approach laid the foundation for scientific studies on stress and strain in the 1820s; the French mathematician Augustin-Louis CAUCHY,[228] stimulated by the French engineer Claude Louis M.H. NAVIER[229] developed the concept of stress and strain, thus creating the fundamental mathematical apparatus for the modern theory of elasticity.[230] CAUCHY also formulated the linear stress-strain relationship that is now called **"HOOKE's law"** (or the "Hooke law"). |

---

[222] E. GERLAND: *LEIBNIZens und HUYGENS' Briefwechsel mit PAPIN, nebst der Biographie PAPIN's.* Abhandl. Königl. Akad. Wiss. Berlin (1881), pp. 42-45.
[223] C. CRANZ: *Lehrbuch der Ballistik: II. Innere Ballistik.* Springer, Berlin (1926), p. 429.
[224] E. MARIOTTE: *Harmonie universelle, contenant la théorie et la pratique de la musique.* Livre VIII: *Traité du mouvement des eaux et des autres corps fluides.* E. Michallet, Paris (1686).
[225] R. HOOKE: *De potentia restitutiva, or of spring: explaining the power of springing bodies.* J. Martyn, London (1678).
[226] In 1664, Sir John CUTLER (1607–1693), F.R.S., an English merchant and financier, founded a lectureship in mechanics for HOOKE which carried an annual salary of £50. In total, HOOKE gave six so-called "Cutlerian Lectures." See *The Cutler lectures of Robert HOOKE.* Oxford University Press, Oxford (1931).
[227] A.E. MOYER: *Robert HOOKE's ambiguous presentation of "HOOKE's law."* Isis **68**, No. 2, 266-275 (June 1977).
[228] A.L. CAUCHY: *Recherches sur l'équilibre et le mouvement intérieur des corps solides ou fluides, élastiques ou non élastiques.* Bull. Soc. Philom. Paris (1823), pp. 9-13.
[229] C.L.M.H. NAVIER: *Mémoire sur la flexion des plans élastique* [read 1820]. Bull. Soc. Philom. Paris (1823), pp. 92-102.
[230] S.P. TIMOSHENKO: *History of strength of materials.* McGraw-Hill, New York *etc.* (1953), chap. 5: *The beginning of the mathematical theory of elasticity,* pp. 104-122.

| 1678 | L'Académie Royale des Sciences, Paris, France | **Christiaan HUYGENS writes his *Traité de la lumière* ("Treatise on Light")**[231] which, however, will not be published until 1690.
| | | (i) Proposing a new way of thinking about how waves propagate, he concludes that the critical structures for waves are surfaces of equal amplitude: *wavefronts*. **HUYGENS suggests that each point on the surface of a wavefront acts as a point source for outgoing spherical waves ("wavelets"), and that the sum of the wavelets produces a new wavefront** {⇨Fig. 4.6–A}. ▪ The so-called **"Huygens principle"** is not only very useful for explaining a number of optical and acoustical wave phenomena – *e.g.*, the property of diffraction, the capacity of sound waves to bend around corners and to spread out after passing through a small hole or slit – but was also successfully applied by him to construct the wavefront of a source of sound moving supersonically {DOPPLER ⇨1847}. The Huygens principle was slightly modified by Augustin FRESNEL to explain why no back wave was formed, and Gustav R. KIRCHHOFF demonstrated that the principle could be derived from the wave equation.[232] Attempts to extend his principle to shock wave propagation remained unfinished {LUMMER ⇨1905; VON NEUMANN ⇨1942}.
| | | (ii) From his simple principle of wavelet construction, he also derives the Laws of Reflection and Refraction.
| | | (iii) At the end of his treatise, he gives a mathematical demonstration that DE FERMAT's principle (1658) – a principle stating that a ray of light follows the path that requires the least time to travel from one point to another – follows from his principle of wavelet construction.
| 1679 | Malpas Tunnel, southwest France | **First use of black powder in civil engineering in the 165-m-long and 8-m-wide Malpas Tunnel of the Canal du Midi,** the oldest canal tunnel in the world. ▪ The tunnel was planned and constructed under the supervision of Pierre-Paul RIQUET, a rich tax-farmer and engineer. The Canal du Midi was officially opened to shipping in 1681.
| 1679 | London, England | **Denis PAPIN,**[233] a French experimental physicist and assistant of Robert BOYLE, **invents a pressure cooker which he calls a "digester"** {⇨Fig. 4.7–A}, a forerunner of the autoclave. It consists of a closed vessel with a tightly fitting lid that confines the steam until a high pressure is generated, raising the boiling point of the water considerably and even allowing to bones to be softened. In order to prevent explosions, the top of the vessel incorporates a *safety valve* of his own invention. ▪ This safety valve, later modified in a manifold manner, became a technologically important device in the development of steam power {⇨Figs. 4.7–B & 4.21–A}.
| 1686 | Princely House of Brunswick [Braunschweig], Hannover, northern Germany | **Gottfried Wilhelm LEIBNIZ,**[234] one of Germany's greatest mathematicians and philosophers, working as a historian for ERNEST AUGUSTUS I, Duke of Lüneburg, **publishes in *Acta eruditorum*** – then one of the most important international scientific journals – **a short treatise in Latin on the conservation of motion.** Criticizing the Cartesian Principle, he favors the method which Galileo GALILEI {⇨1638} and Christiaan HUYGENS {⇨1668/1669} had used – *i.e.*, **characterizing the force of a body in motion by the product of the mass *m* and the height *h* to which the body could rise against resistance (gravity), hence ~*mh* or *mv*²;** *v* being the initial velocity to which the body must attain to reach the height *h*.
| | | Nine years later, in an essay on dynamics, LEIBNIZ[235] disclosed his view in more detail: **for the quantity *mv*² he coined the expression *vis visa* ("living force"),** which he regarded – in

---

[231] Ch. HUYGENS: *Treatise on light* [translated by S.P. THOMPSON]. Macmillan, London (1912).
[232] M. BELGER, R. SCHIMMING, and V. WÜNSCH: *A survey on HUYGENS' principle*. J. Anal. Appl. **16**, No. 1, 9-36 (1997).
[233] D. PAPIN: *A new digester or engine for softening bones, containing the description of its make and use in these particulars*. Bonwicke, London (1681).
[234] G.W. LEIBNIZ: *Brevis demonstratio erroris memorabilis cartesii et aliorum circa legem naturae...* ("A Short Demonstration of a Remarkable Error of DESCARTES and Others, Concerning the Natural Law by which they think that the Creator Always Preserves the Same Quantity of Motion; by Which, However, the Science of Mechanics is Totally Perverted"). Acta Erud. **5**, 161-163 (1686).
[235] G.W. LEIBNIZ: *Specimen dynamicum*. Acta Erud. **14**, 145-157 (1695).

opposition to René DESCARTES {⇨1644} – as the true measure of force of a body in motion. On the other hand, he termed the gravitational force acting on a body at rest *vis mortua* ("dead force"). ▪ **This started a debate on the true measure of force – the famous *vis visa* controversy.** The need to augment Newtonian mechanics to encompass systems more complex than collections of point masses engendered a century-long dispute about conservation.[236] **The problem was eventually clarified by Jean LE ROND D'ALEMBERT** {⇨1743}. Nevertheless, the *vis visa* controversy remained a subject of discussions among natural philosophers, and even in 1805 the English scientist William Hyde WOLLASTON[237] addressed this issue in his *Bakerian Lecture*[238] on the Force of Percussion, which he delivered at the Royal Society of London.

In his *Mathematical and Philosophical Dictionary* (1745), the English mathematician and ballistician Charles HUTTON,[239] upon reviewing the dispute on moving bodies, wrote: "… the controversy became more general, and was carried on for several years by LEIBNIZ, Johann and Daniel BERNOULLI, POLENI, WOLFIUS, GRAVESANDE, CAMUS, MUSSCHENBROEK, & on one side; and PEMBERTON, EAMES, DESAGULIERS, Dr. S. CLARK, M. DE MAIRAN, JURIN, MACLAURIN, ROBINS, & on the other side … The defenders of LEIBNIZ's principle, beside the arguments above-mentioned, refer to the spaces that bodies ascend to, when thrown upwards, or the penetrations of bodies let fall into soft wax, tallow, clay, snow, and other soft substances, which spaces are always as the squares of the velocities of the bodies. On the other hand, their opponents retort, that such spaces are not the measures of the force in question, which is rather percussive and momentary, as those above are passed over in unequal times, and are indeed the joint effect of the forces and times."

| | | |
|---|---|---|
| 1686 | Arundel House, The Royal Society of London, England | **Denis PAPIN,**[240] a French experimental physicist and inventor, **gives a demonstration before the Royal Society of a new sort of gun that "shoots by rarefaction of the air"** – in contrast to ordinary wind-guns which use compressed air {GUTER ⇨1430; BOYLE ⇨1647}. The gun is operated by first evacuating a pipe by a "pneumatic engine" [probably meaning a vacuum pump of the von-Guericke type], then a valve attached to this pipe is suddenly opened and the atmospheric pressure, rushing into the barrel and acting on the projectile base, drives it swiftly along the pipe. He reports that the projectile, in his example a 2-ounce (62.2-g) lead disc fitted to the pipe, was accelerated to a speed a "little less than of a wind-gun by compression." |
| 1687 | Trinity College, Cambridge, England | In July, **the Royal Society of London publishes Isaac NEWTON's *Philosophiae naturalis principia mathematica*** ("The Mathematical Principles of Natural Philosophy"),[241] an exact quantitative description of the motions of visible bodies, resting on his three Laws of Motion, which will make him to the foremost figure of the scientific revolution of the 17th century. |

---

[236] G.E. SMITH: *The 'vis viva' dispute: a controversy at the dawn of dynamics.* Phys. Today **59**, 31-36 (Oct. 2006).
[237] W.H. WOLLASTON: *The Bakerian Lecture on the force of percussion* [1805]. Phil. Trans. Roy. Soc. Lond. **96**, 13-22 (1806).
[238] The *Bakerian Lecture* is the Royal Society's premier lecture in the physical sciences. It originated in 1775 through a bequest by Mr. Henry BAKER F.R.S. (1698–1774), of £100 for an oration or discourse on natural history or experimental philosophy. The first Bakerian Lecture was delivered in the same year by the Irish chemist and mineralogist Peter WOULFE.
[239] C. HUTTON: *A philosophical and mathematical dictionary.* Johnson & Robinson, London (1745). G. Olms, Hildesheim *etc.* (1975), vol. I, pp. 494-498.
[240] D. PAPIN: *An account of an experiment shown before the Royal Society, of shooting by the rarefaction of the air.* Phil. Trans. Roy. Soc. Lond. **16**, 21-22 (1686).
[241] I. NEWTON: *Philosophiae naturalis principia mathematica* [in Latin, imprimatur Julii 5.1686]. S. Pepys, Londini (1687, 1713, 1726) – Translated into English by A. MOTTE in 1729, revised by F. CAJORI: *Sir Isaac NEWTON's mathematical principles of natural philosophy and his system of the world.* University of California Press, Berkeley, CA (1960); *see Corollary VI,* pp. 21-28. All citations given below were taken from this English translation. ▪ A facsimile of the 1687 edition of his *Principia* can be found in the catalogue *Gallica, la bibliothèque numérique,* Bibliothèque nationale de France; http://gallica.bnf.fr/.

Significant milestones contributing to the evolution of mechanics and fluid dynamics are as follows:[242]

(i) In his *Preface*, NEWTON (from 1705 Sir NEWTON) explains how he derived the Law of Gravitation from astronomical phenomena and how he deduced from it the motions of the planets, comets, and the seas. Then he expresses his wish that the rest of nature could be derived from the attracting and repelling forces of particles and the results cast in the deductive mode of his *Principia*.

(ii) In the following chapter *Definitiones*, NEWTON[243] postulates a giant cannon that could launch a projectile into orbit; his thesis is that if a cannon is placed horizontally on a mountain top and the charge increased with each succeeding round, the range would increase until, eventually, the shot would not fall back to Earth at all but remain in orbit, pulled back only by atmospheric drag. He writes, "If a leaden ball, projected from the top of a mountain by the force of gunpowder, with a given velocity, and in a direction parallel to the horizon, is carried in a curved line to the distance of two miles before it falls to the ground; the same, if the resistance of the air were taken away, with a double or decuple velocity, would fly twice or ten times as far. **And by increasing the velocity … [the projectile] even might go quite round the whole Earth before it falls; or lastly, so that it might never fall to the Earth, but go forwards into the celestial spaces, and proceed in its motion *in infinitum*.**" ▪ His idea was resumed in the 20th century by some ballistic researchers who proposed to put payloads into a low-Earth orbit using superguns such as the Babylon Gun (or Iraq Gun), which, however, was never completed.[244, 245]

(iii) In an introductory chapter, entitled *Axiomata sive leges motus* ("Axioms, or Laws of Motion"), **NEWTON postulates his three Laws of Motion:**

▸ **First Law:** "Every body continues in its state of rest or of motion in a straight line unless it is compelled to change that state by force impressed on it."

▸ **Second Law:** "The change of motion is proportional to the motive force impressed and is made in the direction of the straight line in which that force is impressed." ▪ By measuring the acceleration of a body (or by measuring the deformation-time profile and integrating it twice) it is possible to determine the "force of percussion" during the impact of bodies {DUNN ⇨1897; HÖNIGER ⇨1912; RAMSAUER ⇨1909; BERGER ⇨1924; ⇨Figs. 2.7 & 4.3-O}.

▸ **Third Law:** "For every action (force) in nature there is an equal and opposite reaction." NEWTON gives credit to WREN, WALLIS and HUYGENS for establishing the truth of the Third Law.[246] ▪ The Third Law is important not only in the impact of bodies {HUYGENS ⇨1703}, but also for all jet engines, which produce thrust through action and reaction propulsion, allowing the realization of supersonic and hypersonic flight velocities. According to modern view the Newtonian symmetrical model of impact involves two equal and oppositely directed forces which arise in the impact, and act equally and oppositely to change the motion of the two bodies. However, it was not familiar to his contemporaries and immediate successors who still thought in terms of the transfer of forces of motion which – as they assumed – are directly related to the speed or to the mass of the body (or to both) and not to the nature of any interaction in

---

[242] For additional comments on NEWTON's propositions, *see* (i) W.D. HAYES: *The Newtonian pressure law*. In: (A. MIELE, ed.) *Theory of optimum aerodynamic shapes*. Academic Press, New York (1965), pp. 189-193; and (ii) W.D. HAYES and R.F. PROBSTEIN: *Hypersonic flow theory*. Academic Press, New York (1959), sect. 3.1.

[243] F. CAJORI: *Sir Isaac NEWTON's mathematical principles of natural philosophy and his system of the world*. University of California Press, Berkeley, CA (1960); *see Definition* V, pp. 3-4.

[244] *Babylon gun*. Encyclopedia Astronautica; http://www.astronautix.com/lvs/babongun.htm.

[245] W. LOWTHER: *Arms and the man: Dr. Gerald BULL, Iraq, and the supergun*. Macmillan, London (1991).

[246] C. GAULD: *The historical context of NEWTON's Third Law and the teaching of mechanics*. Res. Sci. Education **23**, 95-103 (1993).

which that body might be involved. Roger Joseph [Ruggero Giuseppe] BOSCOVICH (1711–1787), an eminent Croatian astronomer and mathematician, sought to explain in his book *Theoria philosophiae naturalis redacta ad unicam legem virium in natura existentium* ("Theory of Natural Philosophy Derived to the Single Law of Forces, Which Exist in Nature"), published at Vienna in 1758, all the chemical and physical properties of matter in terms of the interactions of indivisible point-like structures (atoms), each being surrounded by a *force field* that is alternately attractive and repulsive at varying distances from the center. According to BOSCOVICH's view, this field prevents point-particles from actually coming into contact. On the contrary, Sir NEWTON's contemporaries still thought of impact in terms of contact.[247]

(iv) **NEWTON also addresses the problem of percussion** [Scholium]:

▸ He starts from the rules given previously for hard and perfectly elastic bodies {HUYGENS ⇨1668/1669; WREN ⇨1669; WALLIS ⇨1669 & 1670/1671; MARIOTTE ⇨1671}, and describes experiments with two equal spheres suspended on 10-ft (3.05-m)-long threads. He writes, "… together with Law III, Sir Christopher WREN, Dr. WALLIS and Mr. HUYGENS, the greatest geometers of our times, did severally determine the rules of the impact and reflection of hard bodies, and about the same time communicated their discoveries to the Royal Society, exactly agreeing among themselves as to those rules. Dr. WALLIS, indeed, was somewhat earlier in the publication; then followed Sir Christopher WREN, and, lastly, Mr. HUYGENS. But Sir Christopher WREN confirmed the truth of the thing before the Royal Society by the experiments on pendulums, which M. MARIOTTE soon after thought it to explain in a treatise entirely upon that subject." ▪ In the 1713 edition, Sir Isaac NEWTON modified this passage and called MARIOTTE *"Clarissimus Mariottus"* ["The Most Illustrious Mariotte"]. The question of priority was discussed by the Scottish physicist Peter G. TAIT.[248] He came to the conclusion that "Clarissimus Mariottus" is meant in a sarcastic manner, because MARIOTTE had not mentioned WREN's contribution in his treatise *Traité de la percussion ou choc des corps* {MARIOTTE ⇨1671}.

▸ He makes the following remarkable observation: "I must add, that the experiments we have been describing, by no means depending upon that quality of hardness, do succeed as well in soft as in hard bodies. For if the rule is to be tried in bodies not perfectly hard, we are only to diminish the relation in such a certain proportion as the quantity of the elastic force requires. By the theory of WREN and HUYGENS, bodies absolutely hard return one from another with the same velocity with which they meet … **In *bodies imperfectly elastic* the velocity of the return is to be diminished together with the elastic force; because that force is certain and determined, and makes the bodies to return one from the other with a relative velocity, which is in a given ratio to that relative velocity with which they met.**"

▸ He recognizes that the ratio of the relative velocities of two bodies $A$ and $B$ – colliding with velocities $v_A$ and $v_B$ and repelling with velocities $c_A$ and $c_B$ – is constant; *i.e.*, in modern terms:

$$(v_A - v_B) : (c_A - c_B) = \text{const.}$$

The Scottish physicists William THOMSON and Peter G. TAIT {⇨1879}, resuming NEWTON's idea, will call this ratio the "coefficient of restitution" {⇨Fig. 4.3–F}.

---

[247] D. PAPINEAU: *BOSCOVICH and the Newtonian analysis of impact*. In. (P. BURSILL-HALL, ed.) *R.J. BOSCOVICH: vita e attivita scientifica: his life and scientific work*. Atti del Convegno Roma (1988). Istituto della Enciclopedia Italiana, Roma (1993), pp. 183-194.

[248] P.G. TAIT: *Note on a singular passage in the 'Principia'*. Proc. Roy. Soc. Edinb. **13**, 72-78 (1886).

(v) In Book II, entitled *De motu corporum* ("The Motion of Bodies"), NEWTON discusses the following subjects:

- **He tackles the difficult problem of aerodynamic drag for a body moving at high speed through a gas obeying BOYLE's law.** Based upon a molecular hypothesis, he shows that drag increases with the square of the velocity [Prop. XXXIII], thus resuming a view previously proposed by MARIOTTE {⇨1673}. He establishes a relation equivalent to our familiar Mach number relation, with the pressure proportional to the density times the velocity squared [Props. XXXII & XXXIII].[249] ▪ Later Émile JOUGUET {⇨1920} showed that NEWTON's result for aerodynamic drag at high velocity can also be derived from the perfect gas law.

- He states that projectiles excite a motion in fluids as they pass through them: "… bodies moving in elastic fluids, if they are obtuse behind and before, condense the fluid a little more at their fore parts, and relax the same at their hinder parts; and therefore meet also with a little more resistance than if they were acute at the head and tail" [Prop. XXXVII].[250] ▪ Experimental evidence of the existence of a condensation at the head and a rarefaction at the tail of a flying projectile was first reported exactly 200 years later {E. MACH & SALCHER ⇨1886}.[251]

- Based on his sine-squared pressure law, **NEWTON shows that the resistance of a sphere is half that of a circular cylinder of equal diameter with the axis in the flow direction** [Prop. XLV] – the so-called **"Newtonian pressure law."** ▪ Today the Newtonian pressure law is commonly used to estimate the pressures on bodies of various shapes exposed to high-speed gas flows. The German aerodynamicist Adolf BUSEMANN[252] introduced a centrifugal correction term equal to the momentum flow in the shock layer times the curvature of the layer – the so-called **"Newton-Busemann pressure law."**[253]

- **He presents his *corpuscular fluid model*** {⇨Fig. 4.4–A} showing how a body vibrating in an elastic medium will propagate pulses through the medium [Prop. XLI]. He states that "shocks are propagated in a fluid in the manner of successive shocks" [Prop. L, *Scholium*]. In an attempt to investigate the exact nature of the motion of vibrating particles, he concludes that this motion follows the oscillating pendulum. ▪ It should be noted here that the ideas of Edmé MARIOTTE {⇨1673} and Robert HOOKE {⇨1674} on the constitution of air and the role of corpuscles resulted in similar models. From the continuum viewpoint,[254] NEWTON's corpuscular model was a setback for the evolution of fluid dynamics. However, it proved true in superaerodynamics {TSIEN ⇨1946}, where the average distance traveled between collisions of molecules is comparable to a significant body dimension.

- **Treating fluid flow analytically, NEWTON makes the first attempt at an *isothermal theory of sound*.** By a devious argument, he obtains for the sound velocity *a* the formula

---

[249] T. VON KÁRMÁN: *Isaac NEWTON and aerodynamics*. J. Aeronaut. Sci. **9**, 521-522, 548 (1942).

[250] F. CAJORI: *Sir Isaac NEWTON's mathematical principles of natural philosophy and his system of the world* [Engl. translation by A. MOTTE in 1729]. University of California Press, Berkeley (1960); pp. 351 and 366.

[251] E. MACH and P. SALCHER: *Photographische Fixierung der durch Projectile in der Luft eingeleiteten Vorgänge*. Sitzungsber. Akad. Wiss. Wien **95** (Abth. II), 764-780 (1887).

[252] A. BUSEMANN: *Flüssigkeits- und Gasbewegung*. In: (E. KORSCHELT, ed.) *Handwörterbuch der Naturwissenschaften*. Fischer, Jena (1933), vol. 4, pp. 244-279.

[253] W.D. HAYES: *The Newton-Busemann pressure law*. In: (A. MIELE, ed.) *Theory of optimum aerodynamic shapes*. Academic Press, New York (1965), pp. 343-349.

[254] In classical gas dynamics the medium is considered to be a *continuum* in which each elementary volume of space contains a large number of particles (molecules); *i.e.*, the distance between molecules is small and parameters of the medium vary continuously in space and time. However, in the case of a very highly rarefied medium – for example the medium filling interstellar space – the medium cannot be idealized as a continuum: it must be treated on a discontinuous basis; *i.e.*, it must be assumed to consist of discrete particles.

$a = (p_0/\rho_0)^{1/2} = 979$ ft/s (298 m/s),

thus obtaining the same result as if he had assumed the motion to be isothermal using BOYLE's law, $p/\rho$ = const. Correcting this value by assuming some detrimental influences on the propagation of sound in the air, such as by foreign bodies and vapors, he then obtains a value of 1,142 ft/s (348 m/s) [Prop. L]. • Although NEWTON's treatment of the sound problem was incomplete, it contained true principles, and Pierre-Simon DE LAPLACE[255] appreciatively wrote, "Sa théorie, quoique imparfaite, est un monument de son génie." In 1863 the Austrian physics professor Josef STEFAN[256] – who, together with Ludwig BOLTZMANN, will later derive the famous Stefan-Boltzmann law (1879) – showed that NEWTON's formula for the sound velocity can be derived from the kinetic theory of gases {KRÖNIG ⇨ 1856} and that the molecular velocity is $3^{1/2}$ times larger that NEWTON's (isothermal) sound velocity. STEFAN's paper stimulated Ernst MACH to derive a simple percussion model for understanding sound propagation on a molecular level and the influence of sound intensity on the sound velocity {E. MACH & SOMMER ⇨1877; ⇨Fig. 4.4–D(a)}.

(vi) In Book III, entitled *De mundi systemate* ("The System of the World"), **NEWTON explains the regularities of the planets and their motions around the Sun,** as observed by the world in its present state of evolution. However, he does not explain – on the basis of his own mechanical principles – how the Solar System originated and reached its present state of uniformity.

The mechanics of NEWTON's Principia seemed to manage quite well without appealing to hypothesis. However, despite its enormous achievements in shaping subsequent work in mechanics, it was too idiosyncratic from an epistemic standpoint to serve as general model for the natural sciences.[257]

| 1687 | Parthenon, Acropolis, Athens, Greece | On September 26, during a siege of the Acropolis by Francisco MOROSINI, Commander in chief of the Venetian fleet, **a cannonball pierces the roof of the (then intact) structure of the Parthenon** [finished in 432 B.C.], **causing the explosion of the gun powder stored in its interior by the Turks.** • The explosion destroyed a great part of the monument which was constructed of Pentelicon, the world's most purest and most precious white marble. Modern archeologists are interested in this explosion, because they want to know how far and in what size distributions the resulting fragments have flown. They believe that this will help them in the reconstruction of this three-dimensional puzzle.

At GALCIT (Pasadena, CA) a research project has been initiated involving the complete constitutive and fracture characterization of both ancient and newly quarried marble pieces subjected to a variety of loading rates.[258] Eventually this information will form the basis for the construction of numerical models simulating the explosion and fragmentation of the monument. Initial activities included the experimental and numerical study of damage created due to individual and multiple cannonball impacts on column drums in order to estimate their residual strength and load carrying ability. To achieve this the dynamic constitutive behavior of Pentelicon marble has been studied in detail. |
|---|---|---|
| 1690 | Paris Observatory, France | In December, **Jean-Dominique** [Giovanni Domenico] **CASSINI,** a French astronomer, **reports an unusual feature on the disk of Jupiter.** He records the time variation of a dark spot, which |

---

[255] P.S. DE LAPLACE: *Traité de mécanique céleste.* Duprat, Paris, vol. V (1825), Book 12, p. 95.
[256] J. STEFAN: *Über die Fortpflanzungsgeschwindigkeit des Schalles in gasförmigen Körpern.* Ann. Phys. **118** [II], 494-496 (1863).
[257] E. MCMULLIN, R.M. SARGENT, J.R. VOELKEL, and G.E. SMITH: *The impact of NEWTON's Principia on the philosophy of science. Commentary. Authors' reply.* Phil. Sci. **68**, No. 3, 279-345 (2001).
[258] A. ROSAKIS: *Dynamic deformation and fracture behavior of "Pentelicon" marble.* A joint project of GALCIT, United States, and NTUA, Greece. GALCIT, Pasadena, CA (April 2005); http://www.solids.caltech.edu/~rosakis/interests.html#ddf and http://www.solids.caltech.edu/~rosakis/research/marble.pdf.

appeared in the period December 5–23, 1690. ▪ A detailed description of this spot appeared in his note *Nouvelles déscouvertes dans le globe de Jupiter* ("New discoveries on the globe of Jupiter"), which is kept in the Library of the Paris Observatory. His drawings of a possible impact spot on Jupiter show a remarkable similarity to the phenomenon observed after the collision of comet SHOEMAKER-LEVY 9 {⇨1994} with Jupiter; such collision events are very rare.[259]

| 1690 | *University of Marburg, Hesse, central Germany* | Denis PAPIN,[260] a French physicist and inventor, **proposes to replace gunpowder in his pompe balistique ("ballistic pump") by steam,** which would considerably aid the speed of the refueling procedure after each working cycle {HUYGENS & PAPIN ⇨1673; ⇨Fig. 4.2−L}. ▪ His idea initiated the **birth of the atmospheric steam engine,** which was realized shortly after in England by Thomas SAVERY (1698) and Thomas NEWCOMEN (1705). |
|---|---|---|
| 1692 | *Port Royal, Jamaica, Caribbean Sea* | **On June 7, a violent earthquake strikes the city of Port Royal,** then England's richest New World possession. Subsequently, a tsunami generated by the earthquake destroys the city, sweeping it into the sea (resulting in about 2,000 casualties). ▪ In an account of the so-called "Jamaica Earthquake,"[261] a unique wave phenomenon of **"visible waves"** is described by an observer who "saw the ground rise like the sea in a wave as the earthquake passed along," and "could distinguish the effects of it to some miles distance by the motion of the tops of the trees on the hills." A similar phenomenon of traveling seismic "visible waves," albeit on much smaller scale, was observed during the Boston Earthquake {WINTHROP ⇨1755}.

In 1959, American archeologists started marine excavation of old Port Royal, probing down into 6–10 feet (1.8–3 m) of silt. The studies revealed surprising results: though many buildings had been toppled by the tremors, entire blocks of shops and homes had been carried virtually intact beneath the sea by the gliding sections of land.[262] |
| 1700 | *L'Académie Royale des Sciences, Paris, France* | Nicolas LÉMERY,[263] a French alchemist and physician, **publishes his earthquake model experiment,** which will become known as the **"Volcan de Lémery"** – a self-triggered oxyhydrogen explosion. He mixed 50 pounds (22.7 kg) of iron filings with the same amount of flowers of sulfur, made a moist, well-kneaded dough out of it, put it in a pot and buried it 1 ft (0.3 m) deep in the ground. The dough, heated up by the chemical reaction, caused the soil to be cracked and to belch fire. He also explains other violent natural phenomena by the spontaneous reaction of iron and sulfur likewise, such as lightning, thunder, subterranean fires and hurricanes. ▪ LÉMERY's "iron-sulfur hypothesis," a historical curiosity from our present point of view, was taken up by some prominent natural philosophers of his time to explain the cause of earthquakes {KRÜGER ⇨1756; KANT ⇨1756}.

Humphry DAVY,[264] after watching explosive eruptions of Vesuvius in Italy in 1819/1820, critically wrote, "If the idea of LÉMERY were correct, that the action of sulphur on iron may be a cause of volcanic fires, sulphate of iron ought to be the great product of the volcano; which is known not to be the case; and the heat produced by the action of sulphur on the common metals, is quit inadequate to account for the appearances. When it is considered that volcanic fires occur and intermit with all the phenomena that indicate intense chemical action, it seems not unreasonable to refer them to chemical causes. But for phenomena upon such a scale, an im- |

---

[259] I. TABE, J. WATANABE, and M. JIMBO: *Discovery of a possible impact spot on Jupiter recorded in 1690.* Publ. Astron. Soc. Japan (Letters) **49**, No.1, L1-L5 (1997).
[260] D. PAPIN: *Nova methodus ad vires motrices validissimas levi pretio comparandas.* Actis Eruditorum Lipsiae (Aug. 1690), p. 410.
[261] Cited in MICHELL's memoir {MICHELL ⇨1760}.
[262] B. WALKER: *Earthquake.* Time-Life, Alexandria, VA (1982), p. 51.
[263] N. LEMERY: *Chimique des feux souterrains, des tremblements de terre, des ouraganes, des éclairs et du tonnerre.* Mém. Acad. Paris (1700), pp. 140-152; *Physische und chymische Erklärung der unterirdischen Feuer, der Erdbeben, Stürme, des Blitzes und Donners.* Phys. Abhandl. Erster Theil, Königl. Akademie der Wissenschaften in Paris, Breslau (1748), pp. 418-420.
[264] H. DAVY: *On the phenomena of volcanoes.* Phil. Trans. Roy. Soc. Lond. **118**, 241-250 (1828).

mense mass of matter must be in activity, and the products of volcano ought to give an idea of the nature of the substances primarily active."

| 1703 | Genroku Kanto, Sagami Bay, Honshu, central Japan | **On December 31, a great earthquake (magnitude 8.2) produces strong ground motions and large tsunamis with maximum height up to 5–6 meters on the north to west coast of Sagami Bay.** The earthquake occurs along the convergent plate boundary of the Sagami Trough, where the Philippine Sea Plate subducts beneath the North American Plate, so-called "interplate earthquake." ▪ It is listed as having killed 100,000 people, but its death toll was probably much smaller (5,233 dead).[265]

Other disastrous tsunamis with estimated maximum water heights ($h_{max}$) include:
▸ Greece (1410 B.C.): $h_{max}$ unknown, 100,000 dead;
▸ Chile (1570): $h_{max}$ = 4 m, 5,000 dead;
▸ Peru (1687): $h_{max}$ = 8 m, 5,000 dead;
▸ Jamaica (1692): $h_{max}$ = 1.8 m, 2000 dead;
▸ Peru (1746): $h_{max}$ = 24 m, 4,800 dead;
▸ Indonesia (1883): $h_{max}$ = 35 m, 36,500 dead; and
▸ Indonesia (2004): $h_{max}$ = 50 m, 297,248 dead.

On September 1, 1923, the Taisho Kanto Earthquake (magnitude 7.9) occurred, another historical earthquake along the convergent plate boundary of the Sagami Trough. It generated a tsunami approximately 30–40 ft (9–12 m) high which crashed unto shore about 5 minutes later. |
| 1703 | France | Jean DE HAUTE-FEUILLE,[266] a French abbot and physicist, **proposes a simple fluid seismoscope to predict earthquakes.** It consists of a bowl filled to the brim with mercury so that seismic foreshocks would cause some of the mercury to spill out. ▪ It is not known whether he ever realized his idea. In 1784, the Italian poet and Carmelite Atanagio CAVALLI reinvented the **mercury seismoscope** and also first built such an instrument.

In 1855, the Italian physicist and meteorologist Luigi PALMIERI designed a mercury seismometer which used mercury-filled glass tubes and was provided with an electrical contact that stopped a clock. It was the first device that recorded the time of an earthquake. ▪ The mercury seismometer proved to be a useful instrument and was frequently used in the 19th century to observe the arrival of seismic shocks which produced ripples on the surface of the mercury. To resolve weak seismic shocks, such as those resulting from artificial earthquakes {MALLET ⇨1846}, the surface was observed under glancing incidence or strewn with dust. |
| 1703 | Leiden, The Netherlands | Publication of HUYGENS' treatise *De motu corporum ex percussione* ("On the Motion of Bodies after Impact") which, according to HUYGENS will, is published posthumously by the Dutch editors Burcherus DE VOLDER and Bernhardus FULLENIUS.[267] Refuting DESCARTES' laws of percussion {DESCARTES ⇨1644}, **HUYGENS gives the first thorough treatment of perfectly elastic percussion.** He does not treat the difficult problem of how to evaluate the large forces that act very briefly during impact, but rather limits his study to determining the velocities after impact.

(i) He starts from the four assumptions that
▸ the Principle of Inertia also prevails in all percussion processes, thus restating the First Law of Motion {Sir NEWTON ⇨1687};
▸ percussion phenomena as considered in his treatise are elastic in nature; |

---

[265] *U.S. National Geophysical Data Center tsunami event database*; http://www.ngdc.noaa.gov/seg/hazard/tsevsrch_idb.shtml.

[266] J. DE HAUTE-FEUILLE: *Microscope micrométrique ... gnômon horizontal, et instrument astronomique pour prendre la hauteur des astres ... avec un moyen de faire des observations sur les tremblements de terre et de les pouvoir prédire.* Paris (1703), pp. 25-28.

[267] HUYGENS' manuscript originating from 1656 was first published by B. DE VOLDER and B. FULLENIUS (eds.): *Christiani HUGENII opuscula posthuma*. Lugduni Batavorum (1703). ▪ A German translation appeared in: (F. HAUSDORFF, ed.): *Ch. HUYGENS' nachgelassene Abhandlungen: Über die Bewegung der Körper durch den Stoss. Über die Centrifugalkraft.* In: Ostwald's Klassiker der exakten Wissenschaften. Nr. 138, Engelmann, Leipzig (1903), pp. 3-34.

▸ the relativity of impact motions should be considered (meaning the relative velocities of impacted bodies); and
▸ the quantity of "motion" is proportional to $m|v|$; *i.e.*, to the "magnitude" (meaning "the mass") of the body multiplied by the absolute value of its velocity.

(ii) **He derives 13 tenets;** his most important results are as follows:
▸ Tenet 1: He investigates the influence of relativity on the motion of colliding bodies. Illustrated via his famous boat experiment {⇒Fig. 4.3–E}, he considers the impact of two equal bodies colliding head-on with the same but opposite velocities $v$: for a spectator in the boat moving with velocity $+v$ they collide with velocities $+v$ and $-v$ and rebound with velocities $-v$ and $+v$, while a spectator on the shore has the impression that they impact with velocities $v(+v) = 2v$ and $-v(+v) = 0$, and rebound with $-v(+v) = 0$ and $+v(+v) = 2v$. From this he concluded that, independently of the velocity of the coordinate system in which the impact takes place, equal elastic masses exchange their velocities during impact. For example, an elastic body impacting on another of equal mass at rest transfers to the latter its entire velocity while itself remaining at rest after the impact {MARCI ⇒ 1639} – *i.e.*, no perceptible *vis visa* is lost.

▸ Tenet 3: He assumes that a body at rest, however great, is set in motion by a body which strikes it, however small {GALILEI ⇒ 1638}.

▸ Tenet 4: He postulates that the relative velocities of two equal bodies (*i.e.*, of equal masses $m_A = m_B$) before and after percussion are equal and oppositely directed, and are given by

$$v_A - v_B = c_B - c_A,$$

where $v_A, v_B$ and $c_A, c_B$ are the velocities before and after impact, respectively.

▸ Tenet 8: Building on previous results {GALILEI ⇒ 1638}, he considers the velocities $v_A$ and $v_B$ acquired through freefall from heights $h_A$ and $h_B$, and supposes that the bodies are directed upward after collision, rising to heights $h_A'$ and $h_B'$. Because of the axiom that the center of gravity of a mechanical system cannot rise under the influence of gravity alone, the centers of the two systems $(m_A, m_B, h_A, h_B)$ and $(m_A, m_B, h_A', h_B')$ must be at the same height, from which it can be calculated that the height of the center of gravity $h_G$ is

$$h_G = (m_A h_A + m_B h_B)/(m_A + m_B) = (m_A h_A' + m_B h_B')/(m_A + m_B) \qquad (1)$$

or

$$m_A h_A + m_B h_B = m_A h_A' + m_B h_B'. \qquad (2)$$

HUYGENS considers the velocities $v_A$ and $v_B$ before impact as acquired through freefall from heights $h_A = v_A^2/2g$ and $h_B = v_B^2/2g$. Here $g$ is the acceleration due to gravity. The masses rise after collision to the heights $h_A' = c_A^2/2g$ and $h_B' = c_B^2/2g$. Since the collision is reversible, one has $c_A = -v_A$ and $c_B = -v_B$. Replacing the heights $h_A, h_A', h_B$ and $h_B'$ in Eq. (2) yields

$$m_A v_A^2 + m_B v_B^2 = m_A c_A^2 + m_B c_B^2. \qquad (3)$$

▸ Tenet 11: Based upon Eq. (3), **HUYGENS states that, in the case of two elastic bodies, the sum of the products of the magnitudes and the squares of the velocities of the bodies before and after impact are equal;** *i.e.*, **the conservation of living force is given by $\Sigma m v^2$**. Generally in two-body percussion problems, the masses, $m_A$ and $m_B$, and the initial velocities, $v_A$ and $v_B$, of the bodies are known. For two elastic bodies impacting head-on, the Law of the Conservation of Momentum (as shown by him in his Tenet 9) gives

$$m_A v_A + m_B v_B = m_A c_A + m c_B. \qquad (4)$$

Thus, for the two unknown quantities $c_A$ and $c_B$, one has the two equations (3) and (4). **At this point he recognizes a fundamental law of physics – the Law of Conservation of Kinetic Energy – thus anticipating LEIBNIZ's concept of *vis visa*** {LEIBNIZ ⇨ 1686}. ▪ In 1669, when HUYGENS' results were first published by the Royal Society {HUYGENS ⇨ 1668/1669}, LEIBNIZ was 22 years old and had just finished his education as a lawyer. HUYGENS first met LEIBNIZ in 1672 in Paris, and the two enjoyed a lifelong friendship.

▸ Tenet 12: HUYGENS also addresses the scientifically interesting and practically important three-body percussion problem {⇨Fig. 2.6}, which he had already discussed in 1652 in a letter to Claude MYLON {HUYGENS ⇨ 1652}.[268]

| | | |
|---|---|---|
| 1704 | Ducal Library, Wolfenbüttel, Lower Saxony, Germany | Gottfried W. LEIBNIZ,[269] a German natural philosopher and general genius of the Baroque period who made lasting contributions to philosophy, mathematics, physics and history, **speculates on the inner explosive force of steam, which he calls the "fulmination of water."** In a letter to Denis PAPIN, with whom he corresponds frequently, **LEIBNIZ reports that water droplets sprayed on a glowing iron block and hit by a hammer produce sharp reports.** ▪ In the early years of steam engines (the late 17th century), steam-boiler explosions also stimulated the idea that earthquakes might be caused in a similar way {*e.g.,* JACOBI ⇨ 1756}. |
| 1705 | Paris, France | Louis CARRÉ,[270] a French mathematician and secretary of the French philosopher Nicole MALEBRANCHE, **describes the firing of a bullet from a shotgun into a wooden box filled with water.** He observes that the sudden stroke compresses the water, bursting the box – this phenomenon is later termed the *hydraulic ram* or *hydrodynamic ram*. ▪ Wolfgang GENTNER, an experimental physics professor at the University of Freiburg, used to demonstrate CARRÉ's spectacular experiment in his lectures. Using a cigar box and an air-gun, he demonstrated that a bullet impacting an empty box enters and leaves the box structure essentially undestroyed, while a water-filled cigar box explodes upon impact.[271]

When a fast moving object (such as a bullet) strikes a water tank, it transfers a large amount of momentum to the water, thus forming a shock wave traveling out from the point of impact. Since the central portion is the most strongly compressed and therefore travels faster than the portion near the surface, the wave front is not hemispherical but rather ellipsoidal {⇨Fig. 4.11−E}.[272] This hydraulic ram phenomenon is of particular importance to the durability of military aircraft and spacecraft {ANKENEY ⇨ 1977}. However, the hydraulic ram effect is also of great interest to military surgeons, because bullets penetrating soft tissue produce a peculiar type of gunshot wound. In 1914 in Berlin, the eminent German ballistician Carl CRANZ[273] studied the explosive effects of S-projectiles {Prussian Army ⇨ 1903} in model ballistic experiments using high-speed photography. |
| 1707 | L'Académie Royale des Sciences, Paris | Jacques CASSINI,[274] a French geodesist and astronomer, and son of the famous astronomer Jean-Dominique CASSINI {⇨1690}, **proposes a method to transfer the high velocity of a moving body to a large mass, thus facilitating its observation by reducing its velocity.** In a |

---

[268] See also the editor's discussion note on the three-body percussion problem in Ch. HUYGENS' *nachgelassene Abhandlungen: Über die Bewegung der Körper durch den Stoss. Über die Centrifugalkraft* (edited by F. HAUSDORFF). In: *OSTWALD's Klassiker der exakten Wissenschaften*. Nr. 138, Engelmann, Leipzig (1903), pp. 70-71.
[269] E. GERLAND: *LEIBNIZens und HUYGENS' Briefwechsel mit PAPIN, nebst der Biographie PAPIN's*. Königl. Akad. Wiss., Berlin (1881), pp. 306-307.
[270] L. CARRE: *Expériences physiques sur la réfraction des balles de mousquet dans l'eau et sur la résistance de ce fluide*. Mém. Acad. Roy. Sci. Paris (1705), Ibid. Correard, Paris (1846).
[271] Dr. Heinz REICHENBACH, EMI, Freiburg; private communication (June 2003).
[272] J.H. MCMILLEN and E.N. HARVEY: *A spark shadowgraphic study of body waves in water*. J. Appl. Phys. **17**, 541-555 (1946).
[273] C. CRANZ, P.A. GÜNTHER, and F. KÜLP: *Photographische Aufnahme von sehr rasch verlaufenden Vorgängen, insbesondere von Schussvorgängen, mittels Vorderbeleuchtung durch das Licht elektrischer Funken*. Z. ges. Schieß- u. Sprengstoffwesen **9**, 1-4 (1914).
[274] J. CASSINI: *Sur les armes à feu. Différemment chargées*. Histoire de l'Académie Royale des Sciences (Paris), pp. 4-5 (1707).

notice to the Paris Academy, it says, "Mr. [Louis] CARRÉ having reported to the Academy some experiments that one of his friends had made on firearms charged in different manners, it was desirable to verify them and CASSINI le fils was charged to do this. He [J. CASSINI] made a sort of machine which consists of a piece of wood reinforced at one end by a metal sheath of sufficient thickness to withstand all the impacts from the same gun fired always from the same distance. This piece was mobile and depending on the force should give more or less way to the impact, and at the same time should mark, by the construction of the machine, how much it gave way."

**J. CASSINI's important concept was later resumed in the ballistic pendulum** {ROBINS ⇨1740; HUTTON ⇨1783; ⇨Fig. 4.3–I} and, to some extent, in the Hopkinson pressure bar {B. HOPKINSON ⇨1914; ⇨Fig. 4.3–P}.[275]

| 1708 | England | **Doctor WALL**,[276] an English physician and scientific writer, **notices the crackling sound of sparks.** Studying frictional electricity, which he generates by drawing a piece of amber swiftly through a woolen cloth, he makes the following important observation: "The crackling is full as loud as that of charcoal on fire; nay, five or six cracklings, or more, according to the quickness of placing the finger, have been produced from one single friction, light always succeeding each of them. Now I doubt not, but on using a longer and larger piece of amber, both the crackling and light would be much greater, because I never yet found any crackling from the head of my cane, though it is a pretty large one; and it seems, in some degree, to represent thunder and lightning." ▪ Twelve years later, **the English electrician Stephen GRAY**,[277] resuming WALL's remarkable discussion on the possible identity of lightning and the "electrical fluid," stated that **"electricity seems to be of the same nature as thunder and lightning, if we may compare great things with small."**

Various early observations indicated that strong thunderclaps must propagate with supersonic velocity {EARNSHAW ⇨1851; MONTIGNY ⇨1860; HIRN ⇨1860; RAILLARD ⇨1860}. This supported the idea that mechanical waves of finite amplitude might propagate faster than ordinary sound such as that predicted by theory, which was first proved on a laboratory scale by studying spark discharges {E. MACH & SOMMER ⇨1877; E. MACH, TUMLIRZ & KÖGLER ⇨1878}. On the other hand, analogical extrapolations from small spark discharges to large ones also introduced errors in concepts; *i.e.*, of polarity of the thundercloud charge, and the nature of the discharge, whether unidirectional or oscillating. |
| 1719 | Abbaye de Savigny [near Saint-Hilaire-du-Harcouët], Normandy, northern France | **Gaspard GROLLIER DE SERVIÈRE**, Grand-Prieur de l'Abbaye de Savigny, **publishes a book on a collection of models of machinery created by his grandfather Nicolas GROLLIER DE SERVIÈRE,**[278] a French inventor and descendant of the famous French bibliophile Jean GROLIER. This book contains unique descriptions and illustrations of numerous models of lathe work, clocks, bridges, sluices, locks, various water-raising machines, water and wind mills, pontoon bridges, an ingenious cyclometer mounted on a wheelbarrow, a revolving machine for reference books, a wheelchair, and drawing apparatus. **He also constructed a full-sized pile-driver in his gardens, powered by a water-wheel held steady by two boats.** ▪ It is not clear whether all the mechanisms described were actual models, or whether some drawings and descriptions were included. |

---

[275] W. JOHNSON: *The origin of the ballistic pendulum: the claims of Jacques CASSINI and Benjamin ROBINS.* Int. J. Mech. Sci. **32**, 345-374 (1990).
[276] Dr. WALL: *Experiments on the luminous qualities of amber, diamonds, and gum lac.* Phil. Trans. Roy. Soc. Lond. **26**, 69-76 (1708/1709).
[277] S. GRAY: *An account of some new electrical experiments.* Phil. Trans. Roy. Soc. Lond. **31**, 104-107 (1720/1721).
[278] N. GROLLIER DE SERVIÈRE: *Recueil d'ouvrages curieux de Mathématique et de Mécanique ou description du Cabinet de Mr. GROLLIER DE SERVIERE.* Forey, Lyon (1719, 1733); C.A. Jombert, Paris (1751).

| | | |
|---|---|---|
| 1720 | *University of Leiden, The Netherlands* | **Willem-Jacob's** [or Gulielmo Jacobo] **GRAVESANDE**,[279] a Dutch natural philosopher and professor of mathematics and astronomy, **dedicates two chapters in his textbook** *Physices elementa mathematica, experimentis confirmata. Sive introductio ad philosophiam Newtonianam* ("Mathematical Elements of Natural Philosophy, Confirmed by Experiments; or, An Introduction to Newtonian Philosophy") **to the percussion of bodies.** He shows how to experimentally treat various cases of head-on and oblique elastic collision of two and three bodies, and how to determine the center of percussion {WALLIS ⇨ 1669 & 1670/1671}. ▪ His book, which passed through six editions and was also translated into English and French, is richly illustrated by numerous apparatus appropriate for demonstrating to an audience {⇨Figs. 4.3–G, H}.<br><br>Two years later, GRAVESANDE[280, 281] presented some new results on the impact of bodies in his *Essai d'une nouvelle théorie du choc des corps fondée sur l'expérience* ("Essay on a New Theory of Percussion of Bodies Based upon Experiments"):<br><br>(i) In Proposition VIII, entitled *Dans les corps égaux les forces sont en raison des quarrez de leurs vitesses*, GRAVESANDE postulates that the forces of identical bodies correspond to their velocities. He demonstrates this rule by using copper balls of the same diameter but different masses which he drops from different heights onto a plane consisting of soft clay. The impacts of the balls produce craters of different diameters depending on the force of impact. For example, a full sphere *1* of diameter $D$ and mass $m_1$, dropped from a height $h_1$ and impacting with velocity $v_1$, produces the same crater diameter in the clay as a hollow sphere *2* of diameter $D$ and mass $m_2 = m_1/3$ dropped from a height $h_2 = 3h_1$ and impacting with $v_2$. ▪ The same impact energy that GRAVESANDE called the "force of percussion" is given when<br>$$m_1/m_2 = h_2/h_1 = v_2^2/v_1^2.$$<br>(ii) In Proposition X, entitled *La force d'un corps est proportionnelle à sa masse multipliée par le quarré de sa vitesse*, GRAVESANDE **affirms that the force of a body is proportional to its mass multiplied by the square of its velocity** – thus, departing from his customary attachment to the Newtonian school and adopting the Huygens-Leibniz concept of *vis visa* {HUYGENS ⇨ 1668/1669 & 1703; LEIBNIZ ⇨ 1686}. |
| 1724 | *L'Académie Royale des Sciences, Paris, France* | **As their Prize question, the French Academy of Sciences proposes that the Laws of Motion should be demonstrated in bodies impinging on one another.** ▪ Among the prominent candidates for this task were Colin MACLAURIN and Johann BERNOULLI; the former actually won the prestigious 1724 Prize {MACLAURIN ⇨1724}.<br><br>However, three years later, J. BERNOULLI {⇨1727} also published a treatise on this subject which eventually earned him the 1727 Prize from the French Royal Academy. |
| 1724 | *Lorraine, northeastern France* | **Colin MACLAURIN,** a Scottish professor of mathematics at Marischal College in Aberdeen, **writes an essay entitled** *Démonstration des loix du choc des corps* ("Demonstration of the Laws of Percussion of Bodies") **during a visit to France, which will win him a prize offered by the Royal Academy of Sciences at Paris.** The substance of this tract will later be inserted into his two-volume book *A Treatise of Fluxions*.[282] ▪ An anonymous referee later wrote in the *Philosophical Transactions*,[283] "The author illustrates an argument which he had proposed in a |

---

[279] W.J.'s GRAVESANDE: *Physices elementa mathematica, experimentis confirmata. Sive introductio ad philosophiam Newtonianam.* Vander Aa, Lugduni Batavorum (1720); Langerak & Verbeeck, Leiden (3rd edn. 1742). *See De corporum collisione simplici, directa et obliqua*, pp. 255-331 and *De collisione composita*, pp. 331-365. ▪ His book was translated into English by John T. DESAGULIERS, a chaplain and fellow of The Royal Society of London: *Mathematical elements of natural philosophy confirmed by experiments, or, an introduction to Sir Isaac NEWTON's philosophy.* Senex & Taylor, London (1721–1726). Note that the English edition contains many new illustrations of experimental setups, such as a new machine for demonstrating the effects of the oblique percussion of elastic spherical bodies {⇨Fig. 4.3–H}.

[280] W.J.'s GRAVESANDE: *Essai d'une nouvelle théorie du choc des corps fondée sur l'expérience.* J. littéraire de la Haye **12**, 1-54, 190-197 (1722).

[281] J.N.S. ALLAMAND (ed.): *Œuvres philosophiques et mathématiques de Mr. G.J.'s GRAVESANDE.* Rey, Amsterdam (1774), vol. 1, pp. 217-252.

[282] C. MACLAURIN: *A treatise of fluxions.* Ruddimans, Edinburgh (1742), vol. I, chap. 12.

[283] (Anonymous): *An account of a book intituled "A Treatise of Fluxions," in two books.* Phil. Trans. Roy. Soc. Lond. **42**, No. 468, 325-363 (1742/1743).

piece that obtained the Prize proposed by the Royal Academy of Science at Paris in 1724 against the mensuration of the forces of bodies by the square of the velocities, shewing that if this doctrine was admitted, the same power or agent, exerting the same effort, would produce more force in the same body when in a space carried uniformly forwards, than if the space was at rest; or that springs acting equally on two equal bodies in such a space, would produce unequal changes in the forces of those bodies."

In 1740, MACLAURIN submitted another essay to the Royal Academy for a prize, entitled *De causa physica fluxus et refluxus maris* ("On the Physical Cause of Tides"), in which he discussed Sir Isaac NEWTON's theory of the tides. He shared the prize with Daniel BERNOULLI and Leonard EULER.

| | | |
|---|---|---|
| 1727 | *University of Basel, northern Switzerland* | **Johann BERNOULLI,**[284] an eminent Swiss mathematician and a brother of the mathematician Jakob BERNOULLI, **publishes a *Discours sur les loix de la communication du mouvement* ("Discourse on the Laws of the Communication of Motion").** ▪ For this treatise J. BERNOULLI won the 1727 Prize of the French Academy.[285] In two further prize-winning papers he treated the motion of the planets at aphelion (1730) and the cause of the inclination of the planetary orbits relative to the solar equator (1735).<br><br>(i) **He considers it absurd to apply the Laws of Percussion to hard and perfectly elastic bodies** {*e.g.*, DESCARTES ⇨1644; HUYGENS ⇨1652, 1668/1669 & 1703}, because collision of these would result in sudden velocity changes which – as he argues – cannot happen in nature (J. BERNOULLI: *"Natura non operatur per saltum"*). ▪ In his contribution to the *Encyclopédie* {DIDEROT & D'ALEMBERT ⇨1751}, the French mathematician D'ALEMBERT didn't agree with BERNOULLI's argument.<br><br>(ii) **He attempts to defend LEIBNIZ's doctrine of *vis visa* ("living force")** {LEIBNIZ ⇨1686}. ▪ This was criticized by some contemporary natural philosophers. For example, James JURIN,[286] an English physician and mathematician who in 1721 acted as secretary of the Royal Society, had great disputes about moving bodies with the philosophers of the school of LEIBNIZ. He proposed an *experimentum crucis* to decide the matter. This famous *vis visa* controversy was eventually clarified by D'ALEMBERT {⇨1743}.<br><br>In the same year, Sir Isaac NEWTON dies, and Johann BERNOULLI is generally regarded as the leading mathematician in Europe. |
| 1732 | *Paris, France* | **Henri PITOT,**[287] a French hydraulic engineer, **publishes a paper in which he describes an instrument which measures the flow velocity in a water stream.**<br><br>(i) His probe essentially consists of an open-end tube facing the oncoming flow ("impact tube"), where the other end of the tube is connected to a pressure-sensitive element.<br><br>(ii) Excited about its simple construction, he enthusiastically writes, "The idea of this machine, known as a "Pitot," is so simple and so natural that as soon as it had come to me, I ran at once to the [Seine] river to carry out the first test with a simple tube of glass, and the effect met my expectation perfectly. After this first test, I could not imagine that such a simple, and at the same time very useful thing, had escaped so many skilled people who wrote and worked on the movement of water." |

---

[284] J. BERNOULLI: *Discours sur les lois de la communication du mouvement.* In: (J.E. HOFMANN, ed.) *Johannis BERNOULLI: opera omnia.* Olms, Hildesheim (1968), vol. 3, chap. 1, pp. 7-14.

[285] See *Recueil des pièces qui ont remporté les prix fondés dans l'Académie Royale des Sciences par M. ROUILLÉ de Meslay ... depuis l'année 1720 jusqu'en 1728.* Académie des Sciences, Paris (1728).

[286] J. EAMES: *A short account of Dr. JURIN's ninth and last dissertation "De Vi Motrice."* Phil. Trans. Roy. Soc. Lond. **41**, 607-610 (1740/1741).

[287] H. PITOT: *Description d'une machine pour mesurer la vitesse des eaux courantes et le sillage des vaisseaux.* Mém. Math. & Phys. Acad. Roy. (Paris) **XXXIV**, 504-519 (1732).

The so-called **"Pitot tube"** became a standard instrument for measuring the total pressure of a moving gas stream, which can be related to the free stream velocity and the static pressure. The Pitot-static combination allows the measurement of both subsonic and supersonic Mach numbers, therefore constituting a Machmeter {Concorde ⇒1975; ⇒Fig. 4.20-H}.[288]

- In a subsonic flow the Pitot tube measures the local total pressure $p_L$. Together with a measurement of the static pressure $p_S$, the Mach number $M$ can be computed from the $p_L/p_S$ ratio by the relation[289]

$$M^2 = [2/(\gamma-1)] \times (p_L/p_S)^{(\gamma-1)/\gamma} - 1].$$

- In a supersonic flow a detached bow shock is formed ahead of the probe which complicates the flow measurement, because the bow shock will cause a drop in the total pressure from the local total pressure $p_L$ to $p_L^*$, the latter being sensed by the Pitot port. The difference $p_L - p_L^*$ becomes substantial at large Mach numbers, and the function is available in table form.

| | | |
|---|---|---|
| 1737 | Kamchatka Peninsula, Siberia, Russia | On October 16, **an earthquake generates a very large tsunami with an estimated height of 210 ft (64 m).** It hits Cape Lopatka at the southern tip of Kamchatka Peninsula (number of casualties unknown). • The 1737 tsunami is known among seismologists as the largest and earliest known historical seismic event for Kamchatka. |
| 1737 | University of Basel, northern Switzerland | Daniel BERNOULLI,[290] an eminent Swiss mathematician and all-round physicist, **extends the classical problem of collision** – hitherto only derived for spherical bodies striking each other head-on {MARCI ⇒1639; WALLIS ⇒1669 & 1670/1671; WREN ⇒1669; HUYGENS ⇒1652, 1668/1669 & 1703; MARIOTTE ⇒1671} – **to the general case of the asymmetric collision of two bodies of arbitrary shapes** (*e.g.*, of eggs). To illustrate his method, he considers the case of the collision of a bar with a spherical body and assumes that the rod is initially at rest and then hit orthogonally. Using his Principle of Minimal Inertia, he determines<br>▸ the point on the bar which, at the instant of the collision, remains fixed (the "instant point of suspension");<br>▸ the "effective mass of the bar;" and<br>▸ its final velocity at the point of impact.<br>Hence, **his method allows one to calculate the impact of a bar with a sphere in the same way as the impact between two spheres.**<br>In his later years, BERNOULLI {⇒1770} returned to the subject of impact and studied the influence of elastic vibrations on the living force (*vis viva*). |
| 1737 | Academy of Sciences, St. Petersburg, Russia | Leonard EULER,[291] the great Swiss mathematician and one of the founders of pure mathematics, **treats the central and eccentric (oblique) percussion of two bodies, and specifies in more detail his previous hypothesis on the nature of elastic and inelastic percussion.**<br>In the early 1770s, EULER resumed his study of percussion problems. He treated the general case of the impact of two spheres under consideration of both translatoric and rotatoric motion,[292] and the oblique impact of two spherical pendulums.[293] |

---

[288] P.L. CHAMBRÉ and S.A. SCHAAF: *The impact tube*. In: *High speed aerodynamics and jet propulsion*. Princeton University Press, Princeton, NJ; vol. IX (1954): (R.W. LADENBURG ET AL., eds.) *Physical measurements in gas dynamics and combustion*, pp. 111-123.

[289] *Fluids – Lecture 16 Notes*. MIT Open Course Ware; http://ocw.mit.edu/NR/rdonlyres/Aeronautics-and-Astronautics/16-01Fall-2005-Spring-2006/BFB1C79B-7498-4F9C-9E6D-CC24FE5A7CFD/0/f16_sp.pdf.

[290] D. BERNOULLI: *De variatione motuum a percussione excentrica*. Comm. Acad. Sci. Imp. Petropol. **IX**, 189-206 (1737).

[291] L. EULER: *De communicatione motus in collisione corporum sese non directe percutientium* [1737]. Comm. Acad. Scient. Imp. Petropol. **IX**, 50-76 (1744).

[292] L. EULER: *De collisione corporum gyrantium* [1772]. Comm. Acad. Scient. Imp. Petropol. **XVII**, 272-314 (1773).

[293] L. EULER: *De collisione corporum pendulorum* [1772]. Comm. Acad. Scient. Imp. Petropol. **XVII**, 315-332 (1773).

| 1738 | L'Académie Royale des Sciences, Paris, France | **A French commission,** consisting of the French astronomer and geodesist César François CASSINI DE THURY (son of Jacques CASSINI), the Italian-born French astronomer Jean Dominique MARALDI, and the French astronomer and abbot Nicolas Louis DE LA CAILLE, **is set up in order to precisely measure the velocity of sound in air.**[294] Cannons are fired at half-hour intervals, alternately at Montmartre and Montlhéry, two suburbs in the north and south of Paris, respectively, about 27 km apart. The times are measured by pendulum clocks. **The result at 6 °C yields 173** *toises/s* (337.18 m/s) which, compared to previous measurements {MERSENNE ⇨ 1636}, represents a considerable improvement.[295] ▪ Precisely measured data on sound velocities were of great importance for comparing with existing acoustic theories. Moreover, such experiments stimulated not only the evolution of gas dynamics, but also challenged high-speed diagnostics. |
|---|---|---|
| 1738 | University of Basel, northern Switzerland | **Daniel BERNOULLI**[296] **publishes (in Latin) his most important treatise** *Hydrodynamica* ("Hydrodynamics"), mainly the result of his previous studies performed in St. Petersburg: |

(i) **He coins the word "hydrodynamics"** in order to analytically cover hydrostatic as well as hydraulic (*i.e.*, dynamic) phenomena by a single method.

(ii) **He first applies differential equations to the motion of (ideal) fluids,** and also derives a formula for the flow velocity of air streaming from a small opening.

(iii) In chapter X, BERNOULLI defines the characteristics of elastic fluids, such as air: "they are heavy, expand in all directions, unless they are confined, and allow themselves to be compressed continuously more and more as the compressing forces increase." Using the example of a cylinder/piston model, he regards gas as consisting of "extremely small bodies … infinite in number … [which] agitate in a very rapid motion; thus the small bodies, while they impinge on the lid *EF* [the piston surface] and also support the same by their continually repeated impacts compose an elastic fluid which expands if the weight [on the piston] is removed or diminished…"

(iv) He shows that the pressure $p$ will be proportional to the kinetic energy of the particles. The energy $e$ of a single particle of mass $m$ and velocity $v$ is given by $e = \frac{1}{2} mv^2$. The frequency of impacts is proportional to the speed $v$, and the force of each impact is proportional to the momentum, given by $mv$. ▪ This allows one to express the thermodynamic state by a mean squared velocity $v_m^2$ of the particles rather than by an empirical temperature {JOULE ⇨ 1850}.

(v) **He anticipates the formulation of the first kinetic theory of gases,** a subject which will be resumed much later {KRÖNIG ⇨ 1856}.

(vi) **He applies the laws of conservation of mass and energy to a fluid** and states that for nonturbulent, nonviscous flow (streamlines), the sum of gravitational potential energy ($gz$), kinetic energy ($\frac{1}{2}v^2$) and another potential energy[297] ($p/\rho$) does not change; *i.e.*,

$$gz + \tfrac{1}{2}v^2 + p/\rho = \text{const.}$$

Here $g$ is the acceleration due to gravity, $v$ is the velocity of the fluid particles, $p$ is the fluid pressure, and $z$ is a coordinate in the vertical direction. ▪ **This so-called "Bernoulli principle" or "Bernoulli theorem" predicts that fluid pressure is reduced wherever the speed of flow is increased.** It proved very useful when designing nozzles and wings.

(vii) **He also theoretically analyzes the muzzle velocity of a gun fired by compressed air.**

Alternative treatments on the motion of fluids – albeit less advanced – were published in 1743 in the *Hydraulica* ("Hydraulics") by Johann BERNOULLI (who backdated it to 1734 and

---

[294] C.F. CASSINI DE THURY: *Sur la propagation du son.* Hist. Mém. Acad. Roy. Sci. Paris (1738), pp. 183-208.
[295] *L'acoustique du XVIe au XVIIIe siècle.* In: (R. TATON, ed.): *La science moderne.* Presses Universitaires de France, Paris (1969), vol. II, pp. 528-532.
[296] D. BERNOULLI: *Hydrodynamica, sive de viribus et motibus fluidorum commentarii.* Dulsecker, Strassburg (1738). Engl. translation by T. CARMODY and H. KOBUS: *Hydrodynamics.* Dover Publ., New York (1968).
[297] In Germany, the term $p/\rho$, resulting from the work done by unit mass of the fluid against the pressure difference along the flow, is appropriately called the *Druckenergie* ("pressure energy").

partly plagiarized his son Daniel BERNOULLI),²⁹⁸ and in 1744 by Jean LE ROND D'ALEMBERT in his *Traité de l'équilibre et du mouvement des fluides* ("Treatise on Equilibrium and on the Movement of Fluids").

1740   London, England   **Benjamin ROBINS,**²⁹⁹ a British mathematician and military engineer, **first uses the ballistic pendulum for small shot** {J. CASSINI ⇨1707}:

(i) His iron pendulum is faced with wood to catch the bullet, and in order to determine the deflection of the pendulum he uses a narrow measuring tape passing through two steel edges pressed lightly together {⇨Fig. 4.3–I}. He writes: "This instrument thus fitted, if the weight of the pendulum be known, and likewise the respective distances of its center of gravity, and of its center of oscillation, from its axis of suspension, it will thence be known, what motion will be communicated to this pendulum by the percussion of a body of a known weight moving with a known degree of celerity, and striking it in a given point." ▪ Note that in the collision of a simple ballistic pendulum of mass $M$, impacted by a projectile of mass $m$ and velocity $v$, the momentum of the projectile $mv$ is conserved, hence

$$mv = (m + M)V.$$

Under the assumption that during impact the total kinetic energy of the projectile given by $\frac{1}{2}m v^2$ is conserved (*i.e.*, is not partly transformed into heat), then this amount of kinetic energy is fully transferred into potential energy to swing the combined masses of pendulum and projectile up to a height $h$, hence

$$\tfrac{1}{2}mv^2 = (m + M)gh,$$

where $g$ is the acceleration due to gravity. Thus, the simple measurement of $h$ allows the determination of $v$ {⇨Fig. 2.10–C}. ROBINS' crude ballistic pendulum and its theory were subsequently refined so that it became possible to also determine projectile velocities with sufficient accuracy by using compound pendulums.³⁰⁰ Beginning in the mid-19th century, however, ballistic pendulums became increasingly superseded by the development of electromechanical chronoscopes {⇨Fig. 4.19–A} and electric-spark photochronographs {BECKER & CRANZ ⇨1912; ⇨Fig. 4.19–J}.

(ii) **From these experiments, as well as from his theory on the force of exploded gunpowder, ROBINS deduces many important conclusions** which are useful to practical artillery use and the theory of projectiles. For example, the velocity of expansion of the flame of gunpowder, when fired in a piece of artillery, without either bullet or other body before it, is prodigiously great, *viz.* 7,000 ft/s (2,134 m/s) or upwards, as appears from his experiments.

The ballistic pendulum proved to be a very versatile tool for research:
- ▸ **Charles HUTTON,** following an idea initiated by ROBINS, **further developed the recoil pendulum into a handy instrument − the so-called "éprouvette,"** a French word meaning "test tube" − to ascertain the quality or strength of gunpowder. Éprouvettes were always fired without any ball {⇨Fig. 4.17–E}.³⁰¹ Similarly, ***ballistic mortars*** were also used to determine the strength or "power" of a high explosive with a

---

²⁹⁸ C.A. TRUESDELL, III: *The hydrodynamics of Daniel and John BERNOULLI (1727–1740)*. In: (A. SPEISER ET AL., eds.) *Leonhardi EULERI opera omnia.* Teubner, Lausanne (1954). See vol. XII [2nd series]: (C.A. TRUESDELL, III, ed.) *Leonhardi EULERI commentationes mechanicae. Editor's introduction to EULERI opera omnia II 12*, pp. xxiii-xxxviii.

²⁹⁹ B. ROBINS: *New principles of gunnery*. Nourse, London (1742).

³⁰⁰ C. CRANZ: *Das ballistische Pendel und seine Theorie*. In: (C. CRANZ, ed.) *Lehrbuch der Ballistik: III. Experimentelle Ballistik*, Springer, Berlin (1927), pp. 39-44.

³⁰¹ Prof. emer. William JOHNSON, F.R.S., drew the author's attention to a book by G.C. STONE entitled *A glossary of the construction, decoration and use of arms and armor...* [Southworth Press, Portland, Maine (1934); copied by J. Brussel, New York (1961), p. 514] in which the author refers to gunpowder testers prior to HUTTON, such as those invented by (i) William BOURNE (1578) [*Inventions or devices very necessary for all generalles and captaines or leaders of men, as well by sea as by land*. Th. Woodcock, London (1578)], (ii) Joseph FURTTENBACH (1627) [*Halinitro-Pyrobolia: Beschreibung einer newen Büchsenmeisterey, nemlichen: Gründlicher Bericht, wie der Salpeter, Schwefel, Kohlen unnd das Pulfer zu præpariren...* Saur, Ulm (1627)], and (iii) Nathanael NYE (1647) [*The art of gunnery*. W. Leak, London (1647)]; the latter invented the mortar éprouvette.

precision of < 1% by measuring the angle of recoil produced by a 10-g charge of the explosive.[302]

▸ In 1915, **the U.S. rocket pioneer Robert GODDARD began to experiment on his ballistic pendulum with various nozzle designs,** using a small metal combustion chamber filled with a type of gunpowder, ignited by electricity. The end of the chamber was threaded, so that nozzles of various designs could be screwed onto it and tested. Using a Laval nozzle {DE LAVAL ⇨1888}, he obtained jet velocities of between 7,000 and 8,000 ft/s (2,134–2,438 m/s) and efficiencies of up to 63%.[303]

▸ During World War II, **a *double ballistic pendulum* for normal impact was developed in the United States** which allowed the measurement of both striking and residual velocities: it consisted of a transmission pendulum supporting the target, and a terminal pendulum which stopped the projectile after perforation.[304]

▸ In the mid-1960s, U.S. physicists at Lawrence Radiation Laboratory (Livermore, CA) measured the total momentum delivered by a focused giant laser pulse to a solid metallic target using a simple ballistic pendulum {GREGG & THOMAS ⇨1966}.

▸ In the early 1980s, Dutch ballisticians at TNO Prins Maurits Laboratory (Rijswijk, The Netherlands) used the ballistic pendulum to study armour/anti-armour interactions, in particular the oblique impact and perforation of a kinetic energy (KE) penetrator onto/through an armor.[305]

| | | |
|---|---|---|
| 1742 | *Royal Arsenal, Woolwich, southeast London, England* | In the old records[306] of Woolwich Arsenal of the year 1742, **an accident due to a bursting gun is described:** "On Friday there was a proof of iron ordnance at Woolwich, when a 24-pdr. burst, and a piece of metal about 4 cwt. [203 kg] flew near 300 yards [274 m] over the heads of the people, and fell upon the top of a chimney of a house adjoining the founder's, broke through the roof and through three stories down to the ground floor, and providentially did no other mischief. The gun, in its agitation before bursting, turned that which lay next to it so as to point its muzzle towards the spectators and the storekeeper's house, and had it not struck out the portfire, which was lighted, great mischief would probably have ensued." ▪ Accidents due to bursting guns {⇨Fig. 4.21−C} happened quite frequently, both during proof-testing and in the midst of battle, but reports of this phenomenon and their causes were mostly kept classified. The reasons why gun barrels occasionally burst are very complex and depend on the rate of fire, incorrect selection of propellant, low fabrication quality of the shells used and their low resistance to shock, the condition of the barrel, its wear, soiling, the amount of deposits originating from the shell mantle, *etc.* Cast iron cannons that burst catastrophically during firing were particularly dangerous, because this delivered high-velocity fragments in all directions, thereby often killing or badly wounding the gun crew.[307] In contrast, cast bronze cannons can only split or tear open. |

The earliest effective cannons were rather crude tubes fabricated from wrought iron bars and reinforced by iron rings or hoops, and the projectiles were made of stone, individually shaped by masons. Subsequent cast bronze cannons, which were first used in the 15th century along with wrought iron and cast iron balls, had a much better performance, but were too expensive and would have required an enormous amount of tin and copper to supply the armies with a sufficient number of such guns, which no European state could afford. The complete re-

---

[302] J. TAYLOR ET AL.: *Improved operation of the ballistic mortar for determining the "power" of high explosives*. J. Sci. Instrum. **26**, 266-268 (1949).
[303] D.P. STERN: *Robert GODDARD and his rockets*. NASA-GSFC, Greenbelt, MD (Jan. 2005); http://www-istp.gsfc.nasa.gov/stargaze/Sgoddard.htm#q24.
[304] G.T. REYNOLDS and R.L. KRAMER: *A double pendulum for use in studies of the ballistic behavior of armor*. Rept. NDRC A-52 (July 1942).
[305] W.J. KOLKERT and R.J.M. VAN AMELSFORT: *Study of armour/anti-armour interactions at oblique impact with a ballistic pendulum methology*. In: *Proc. 7th Int. Symp. on Ballistics* [The Hague, The Netherlands, April 1983]. American Defence Preparedness Association (ADPA), Arlington, VA (1983).
[306] W.T. VINCENT: *The records of the Woolwich district* (ed. by J.P. JACKSON). J.S. Virtue & Co, London (1888), vol. 1, p. 339.
[307] W. ROSTOKER: *Troubles with cast iron cannon*. Archeomaterials **1**, No. 1, 69-90 (1986).

| | | placement of bronze cannons with cast iron cannons, however, was never achieved – in spite of technical improvements like the system of hollow-casting {RODMAN ⇨ 1864}. Both cannon types were eventually superseded by cast steel and forged steel guns in the second half of the 19th century. |
|---|---|---|
| 1742 | Académie des Sciences, Paris, France | **Pierre-Louis M. DE MAUPERTUIS,**[308] a famous French mathematician and astronomer who will first formulate the Principle of Least Action (1750), **outlines in a letter the devastating effects which would result from the collision of a comet with the Earth.**[309] Referring to Edmund HALLEY, an English astronomer who in 1705 calculated the orbit of a comet named after him, he writes, "But the most severe accident of all would be that a comet came to shake the Earth, surging it and smashing it into a thousand pieces; the two bodies would without a doubt be destroyed, but the gravity would immediately rebuild one or several other planets. A less severe shock that would not shatter our planet entirely would still cause great changes concerning the continents and the seas: due to such a shock, the waters would rise to great heights in some places and would flood large regions on the surface of the Earth which they would abandon afterwards. It is to such a shock that M. HALLEY attributes the cause of the Flood" (pp. 243-244).<br><br>More than 50 years later, a similar scenario of devastating effects was described by DE LAPLACE {⇨1796}, however, without any knowledge of DE MAUPERTUIS' letter. |
| 1742 | London, England | **Benjamin ROBINS,**[310] a British military engineer and ballistician, **publishes his book *New Principles of Gunnery* which, for the first time, brings science into the field of practical gunnery.** ▪ The careful application of the ballistic pendulum {J. CASSINI ⇨1707; ROBINS ⇨1740} enabled the velocity of projectiles to be judged accurately at any stage of their flight. **ROBINS' work revolutionized gunnery and founded the science of *aeroballistics*.** It also gave increased impetus to the work of gun designers. Although the average gunner may not have understood this work, it showed the increasing relationship between warfare and mathematics that continues to this day.<br><br>Charles HUTTON,[311] who was the first to considerably extend ROBINS' ballistic experiments and analyze them thoroughly {HUTTON ⇨1783}, later wrote, "Before the discoveries and inventions of that gentleman [Mr. ROBINS], very little progress had been made in the true theory of military projectiles. His book however contained such important discoveries, that it was soon translated into several languages on the continent, and the late famous Mr. L. EULER honored it with a very learned and extensive commentary, in his translation of it into the German language [EULER ⇨1745]. That part of Mr. ROBINS' book has always been much admired, which relates to the experimental method of ascertaining the actual velocities of shot, and in imitation of which, but on a large scale, those experiments were made which were described in my paper…" |
| 1743 | Paris, France | **Jean LE ROND D'ALEMBERT,**[312] a 26-year French mathematician, **publishes his important book *Traité de dynamique* ("Treatise on Dynamics"):** |

---

[308] P.L.M. DE MAUPERTUIS: *Lettre sur la comète qui paroissoit en 1742* (Paris, 1742). In: *Oeuvres de Mr. DE MAUPERTUIS*. J. M. Bruyset, Lyon (1756), tome 3, pp. 207-256.

[309] This letter was written on the occasion of a comet – first observed in 1742 by Jean-Dominique MARALDI and now designated as C/1742 C1 – which, compared to the famous Halley comet, was a rather weak object [DE MAUPERTUIS: "Cette comète qui fait tant de bruit, est une des plus chétives qui ait jamais paru"].

[310] B. ROBINS: *New principles of gunnery*. J. Nourse, London (1742); republished by The Richmond Publ. Co., Richmond, Surrey, U.K. (1972).

[311] C. HUTTON: *New experiments in gunnery, for determining the force of fired gunpowder, the initial velocity of cannonball, the ranges of projectiles at different elevations, the resistance of the air to projectiles, the effect of different lengths of guns, and of different quantities of powder, &,&.* In: *Tracts on mathematical and philosophical subjects*. T. Davison, London (1812), vol. 2, Tract XXXIV, pp. 306-384.

[312] J. LE ROND D'ALEMBERT: *Traité de dynamique: dans lequel les loix de l'équilibre & du mouvement des corps sont réduites au plus petit nombre possible, & démontrées d'une manière nouvelle…* David, Paris (1743, 1758); Fuchs, Paris (1796).

(i) He writes, "The second part, in which I set myself the task of treating the interactive motion of bodies, constitutes the main part of my work. For this reason I have given this book the title *Treatise on Dynamics*. This name, which actually means the science of forces or causes of motion, might not appear correct for this work at a first glance, because I understand mechanics as the science of effects rather than of causes. However, since today the term *dynamics* is very common among scholars to designate the science of motion of bodies which act somehow upon each other, I wish to adhere to this term in order to indicate to geometricians by this title of the treatise that I set myself the task of essentially contributing to perfection and enrichment of this branch of mechanics." ▪ According to modern definition,[313] dynamics is understood "as the field of mechanics that deals with the study of motion and the forces that bring about this motion."

(ii) **He states that the Third Law of Motion** {Sir NEWTON ⇨1687} **is true for bodies that are free to move as well as for bodies that are rigidly fixed** – the so-called **"d'Alembert principle."**

(iii) Referring to discussions among contemporary scientists about whether kinetic energy or momentum is the true measure of the effect of a force on a body – the famous *vis visa* controversy (*see* Sect. 2.2.2) – he dismisses the argument as being pointless and arising from a confusion of terminology.

(iv) **He proves the Law of the Conservation of Living Forces** which was first proposed by Christiaan HUYGENS {⇨1652, 1668/1669 & 1703} in his treatise on the Laws of Percussion. ▪ The "living force" [Lat. *vis visa*] of a mass $m$ moving with velocity $v$ is given by $mv^2$ {LEIBNIZ ⇨1686}. **Today the term *vis visa* is understood as twice the kinetic energy; it does not include that part of the kinetic energy of the body which is due to the vibrations and rotations of its molecules.**

In his *Traité de mécanique* (1829), the French mathematical physicist and engineer Gaspard G. DE CORIOLIS found it more appropriate to give the term $½ mv^2$ the name *vis visa* – *i.e.*, the modern definition of kinetic energy – rather than the quantity $mv^2$.

| | | |
|---|---|---|
| 1743 | *Paris, France* | Jean Antoine NOLLET,[314] a French abbot and amateur natural philosopher, **gives the first experimental evidence that sound also propagates through water.** Diving into the Seine river, he notices that a sound signal is transmitted with the same frequency as in air, but is somewhat modified in intensity. ▪ NOLLET's finding was important insofar as it confirmed that water, like gases, is also an "elastic fluid." |
| | | It was not until 19 years later that it also became possible to demonstrate the compressibility of liquids in the laboratory {CANTON ⇨1762}. Hitherto, liquids were generally regarded as being incompressible, even by such distinguished natural philosophers as Galileo GALILEI (*c*.1500) and Edmé MARIOTTE (1686). |
| 1744 | *Royal Academy of Sciences, Berlin, Kingdom of Prussia, Germany* | At the inauguration of the Royal Academy of Sciences and in the presence of FREDERICK II the Great, king of Prussia, **Christian Friedrich LUDOLFF JR.**,[315] a physician and natural philosopher, **gives an experimental lecture on electricity to the General Assembly.**[316] Apparently using an electric friction machine {VON GUERICKE ⇨1663}, he demonstrates the use of electricity to ignite a mixture of explosive gases (*e.g.*, a spirit/air mixture) by an electric spark. ▪ In particular, he noticed that iron electrodes produce the most intense sparks – then an important requirement for the success of the experiment, because high-energy spark discharges were not |

---

[313] C. MORRIS (ed.): *Dictionary of science and technology*. Academic Press, San Diego, CA *etc.* (1992), p. 698.
[314] J.A. NOLLET: *Sur l'ouïe des poissons et sur la transmission des sons dans l'eau*. Histoire de l'Académie Royale des Sciences (Paris), pp. 29-36 (1743).
[315] Editor's note: *Physique générale: Sur l'électricité*. Histoire de l'Académie Royale des Sciences et des Belles Lettres de Berlin **1** [Classe Physique], 10-12 (1745).
[316] J. HEILBRON: *B. FRANKLIN: Briefe von der Elektrizität*. Vieweg, Braunschweig & Wiesbaden (1983), p. xlvii.

yet available to him; the Leiden jar {VON KLEIST & CUNEUS ⇨1745}, which allows one to generate strong sparks, was also not yet invented.

| 1745 | Royal Academy of Sciences, Berlin, Kingdom of Prussia, Germany |

Leonard EULER,[317] an eminent Swiss mathematician and physicist, **publishes the translation of ROBINS' book *New Principles of Gunnery*** {ROBINS ⇨1742}, into which he included many lengthy and most worthwhile comments, thus making ROBINS' work a standard source for subsequent generations of ballisticians. **He also discovers the similarity law for the resistance of air to the motion of bodies at high speeds, and establishes a parameter which will be known 184 years later as the Mach number** {ACKERET ⇨1929}.[318]

In the same year, EULER[319] **submits to the Academy a study on the calculation of the force of percussion, and its time duration.** Using the example of two elastic ("hard") parallelepipeds, one impacting head-on and the other initially at rest – *i.e.*, using the basic configuration of many modern planar shock impact methods {HOPKINSON ⇨1914; CROZIER & HUME ⇨1957; MCQUEEN & MARSH ⇨1960} – he calculates the period of the excited harmonic oscillation, $T$, and the maximum percussion force, which, as he notes, occurs at $T/4$. For two such bodies weighing 1 pound (0.453 kg) and impacting at 100 ft/s (30.5 m/s), he obtains a percussion force of 4,000 pounds (1814 kg, corresponding to 18,495 N) and a duration of 1.2 ms.

| 1745 | Kammin Cathedral, Pomerania & Leiden, The Netherlands |

**Georg Ewald VON KLEIST,** dean of Kammin Cathedral and an amateur scientist, **invents the capacitor,** which becomes popularly known in Germany by the name *Kleistsche Flasche* ("Kleist's jar"). Independently, the Dutch private gentleman Andreas CUNAEUS gets the same idea. **His invention is communicated by Pieter VAN MUSSCHENBROEK,** a Dutch physics professor at the University of Leiden, who also coins the term *Leidener Flasche* ("Leiden jar").[320]
• In England, the Leiden jar (or Leyden jar) was also humorously called the **"shock-bottle."** In the United States, the invention was highly appreciated for its great versatility in experimenting with electricity. Benjamin FRANKLIN was enthusiastic about "MUSSCHENBROEK's wonderful bottle."[321]

Perhaps the most spectacular demonstration of the Leiden jar was given in 1746 by a Frenchman, Abbé Jean-Antoine NOLLET, before members of the court of king Louis XV at a monastery in Paris: 700 monks joined hands, and one end of this long line of men was connected to the outer contact of a charged Leiden jar. When the man at the other end of the line was connected to the other terminal of the Leiden jar, all 700 monks simultaneously leaped into the air, thus convincingly demonstrating the effects of an *electric shock.*[322]

The easy availability of powerful high-voltage discharges resulted in great popular interest in electrical experiments, such as for electric gun demonstrations {VOLTA ⇨1777; ⇨Fig. 4.16–L}, model lightning/thunder experiments, and the electric ignition of gunpowder and fulminates {LUDOLFF JR. ⇨1744; WATSON ⇨1745; FRANKLIN ⇨1751}. **The loud report produced by the discharge of a battery of Leiden jars became an impressive demonstration of air shocks. It stimulated curiosity about sounds of finite amplitude and short duration, which played an important role in the discovery of shock waves** {A. TOEPLER ⇨1864; ANTOLIK ⇨1874; E. MACH & WOSYKA ⇨1875; E. MACH, TUMLIRZ & KÖGLER ⇨1878; WOOD ⇨1899; M. TOEPLER ⇨1904}.

---

[317] L. EULER: *Neue Grundsätze der Artillerie.* A. Haude, Berlin (1745).
[318] F.I. FRANKL: *On the priority of EULER in the discovery of the similarity law for the resistance of air to the motion of bodies at high speeds* [in Russ.]. Dokl. AN (SSSR) **70**, No. 1, 39-42 (1950).
[319] L. EULER: *De la force de percussion et de sa véritable mesure* [read 1745]. Mém. Acad. Berlin **1**, 21-53 (1746).
[320] *Biographisch-Literarisches Handwörterbuch* (ed. by J.C. POGGENDORFF). Barth, Leipzig (1863), vol. 1, p. 1271.
[321] B. FRANKLIN in a letter to Peter COLLINSON, dated Philadelphia, July 28, 1747.
[322] A.F. KIP: *Electricity.* In: (B.S. CAYNE, ed.) *The encyclopedia Americana.* Americana Corp., New York, vol. 10 (1974), p. 136.

The ability of mild electric shocks to temporarily stun a person but not do any permanent damage has stimulated numerous inventors: after World War II various so-called "electroshock guns" or "stun guns" were developed. Electroshock guns are generally used for self defense. However, the use of these devices, particularly by law enforcement officers, is sometimes the subject of debate. For example, concerns have been raised about the risks that stun guns pose to people with heart disease, and there have been reports of deaths of such individuals after stun gun use.[323]

| 1745 | Crane Court, The Royal Society of London, England | William WATSON,[324] a British physician and apothecary, **reports on his electrical experiments to the Royal Society.**

(i) Stimulated by the Berlin experiments {LUDOLFF JR. ⇨1744}, **he succeeds in igniting inflammable gas – a hydrogen/air mixture – by an electric spark** which "gets off with an explosion equal to the firing of a large pistol." **However, similar experiments with ordinary gunpowder fail.**

(ii) He writes, "I have made several trials in order to fire gunpowder alone, which I tried both warm and cold, whole and powdered, but never could succeed … **But I can, at pleasure, fire gunpowder, and even discharge a musket, by power of electricity, when the gunpowder has been ground with a little camphor, or with a few drops of some inflammable chemical oil.** The oil somewhat moistens the powder, and prevents its flying away: The gunpowder then being warmed in a spoon, the electrical flashes fire the inflammable vapor, which fires the gunpowder: But the time between the vapor firing the powder is so short, that frequently they appear as the same, and not successive operations, whereas the gunpowder itself seems fired by the electricity…" ▪ Apparently, WATSON used an electrostatic machine, but not Leiden jars {VON KLEIST & CUNEUS ⇨1745}, which were invented that same year. This must have resulted in rather weak sparks incapable of igniting gunpowder {FRANKLIN ⇨1751}. |

| 1746 | London, England | Benjamin ROBINS,[325] using the ballistic pendulum {J. CASSINI ⇨1707; ROBINS ⇨1740}, observes that the drag of a projectile increases with the square of the velocity. However, upon performing experiments at firing speeds of up to 1,670 ft/s (509 m/s; $M \approx 1.5$), **ROBINS makes the startling discovery that sphere drag values increase with about the third power at high speeds,** and states that "the velocity at which the moving body shifts resistance is nearly the same with which sound is propagated through the air." ▪ ROBINS' experiments, which mark the **birth of supersonic ballistics,** were first extended by Charles HUTTON {⇨1783} to large shot. In the 20th century, the German-born U.S. fluid dynamicist Sighard F. HOERNER[326] repeated ROBINS' gun shots and essentially confirmed his measured supersonic drag data.[327] Eric William E. ROGERS,[328] a British aerodynamicist and deputy director at the Royal Aircraft Establishment (RAE), Farnborough, compared ROBINS' results for sphere drag values with HOERNER's data in a single diagram ranging from Mach 0.6 to 1.6.

ROBINS' results were incredible to scientists of that time. It had been thought that the influence of drag was about the same as that of gravity, but later experiments have shown it to be many times greater. **Ballistic drag effect can be broken up into three influences: those due** |

---

[323] *Electroshock guns.* Wikipedia, the free-content encyclopedia. Wikimedia Foundation, Inc., St. Petersburg, FL; http://en.wikipedia.org/wiki/Taser.
[324] W. WATSON: *Experiments and observations tending to illustrate the nature and properties of electricity.* Phil. Trans. Roy. Soc. Lond. **43**, 481-501 (1745).
[325] B. ROBINS: *An account of the experiments relating to the resistance of the air, exhibited at different times before the Royal Society in the year 1746.* In: (J. WILSON, ed.) *Mathematical tracts of the late Benjamin ROBINS.* Nourse, London (1761); vol. 1.
[326] S.F. HOERNER: *Aerodynamic drag.* Dayton, OH (1951). Extended edition: *Fluid-dynamic drag.* Midland Park, NJ (1958). ▪ Both monographs were published privately by S.F. HOERNER.
[327] H.M. BARKLA: *Benjamin ROBINS and the resistance of air.* Ann. Sci. **30**, 107-122 (1973).
[328] E.W.E. ROGERS: *Aerodynamics, retrospect and prospect* [Lanchester Memorial Lecture]. Aeronaut. J. **86**, 43-67 (1982), *see* Fig. 9.

- ▸ **to the pressure of the atmosphere** on the bow of the projectile;
- ▸ **to the skin friction** from the atmosphere along the body of the projectile, which depends on its surface roughness and minor convolutions; and
- ▸ **to base drag,** which results from the turbulence induced in the atmosphere at the rear of the projectile.

BECKER & CRANZ {⇨1912} were the first to compile and illustrate drag coefficient data, obtained by various researchers and for various cylindrical projectile geometries, as a function of Mach number in a single diagram {⇨Figs. 4.14–M & 4.19–J}. The subject of ballistic drag was discussed in more detail by Carl CRANZ in his textbook *Äußere Ballistik* {CRANZ ⇨1925} and resumed by Hubert SCHARDIN in his lecture *Die Bedeutung der Gasdynamik in der Ballistik* ("The Importance of Gas Dynamics in Ballistics"), which he presented in October 1946 at the Laboratoire de Recherches Techniques de Sain-Louis, France.[329] **The optimum ballistic shape has the so-called "Haack-Sears body"** {HAACK ⇨1941}.

| | | |
|---|---|---|
| 1746 | *Birmingham, central England* | **John ROEBUCK,** a British physician, inventor and chemist, **devises the lead-chamber process, which permits the relatively inexpensive production of large quantities of sulfuric acid** – an important substance which was previously expensive and available only in limited quantities. ▪ The rapidly increasing use of sulfuric acid, which became the most important of all industrial chemicals, also increased the number of accidents provoked by improper handling, particularly when mixing with water: **if water is added to sulfuric acid, explosive spattering may occur because of the large quantity of heat released.** At atmospheric pressure a mixture of sulfuric acid and water boils at a temperature of 338 °C, and mixing produces 20.4 kcal/Mol (0.21 kcal/g),[330] – a considerable amount of energy which is about one fifth of the energy of TNT {HAEUSSERMANN ⇨1891}. Therefore, it is safer to add acid to the water rather than water to the acid, because the dense acid sinks through the bulk of the water, thus helping to rapidly disperse the heat evenly. |
| 1747 | *L'Académie Royale des Sciences, Paris, France* | **Jean LE ROND D'ALEMBERT,**[331] a French mathematician, scientist and writer, **gives the solution to the one-dimensional wave equation for sound propagation in air.** ▪ Subsequently, Leonard EULER {⇨1759} and Joseph L. DE LAGRANGE[332] attempted to calculate the sound velocity. Their approach was based on D'ALEMBERT's solution of the one-dimensional wave equation with an equation of state $p = K\rho$, thus achieving an important milestone in the theoretical treatment of sound and shock waves.[333] The modern explanation, based upon the concept of an adiabatic process, did not become possible until after the creation of the mechanical theory of heat in the 19th century {KRÖNIG ⇨1856}. |
| 1748 | *Royal Academy of Sciences, Berlin, Kingdom of Prussia, Germany* | **Leonard EULER,**[334] a Swiss mathematician and physicist and one of the founders of pure mathematics, **makes a distinction between a *continuous* and a *discontinuous* curve or function** while studying the analytic nature of functions. He considers a *discontinuous function* "as being composed of the algebraic expressions appropriate to their 'continuous' portions." ▪ EULER did not consider functions which are discontinuous in a modern sense, such as step discontinuities. His understanding of *discontinuous* rather corresponded to curves with a corner, which today are considered to be *continuous*. |

---

[329] W. HAACK: *The calculation of stability and damping of spin-stabilized projectiles.* In: (W.C. NELSON, ed.) *Selected topics on ballistics* [Cranz Centenary Colloquium, Freiburg, Germany, April 1958], Pergamon Press, London etc. (1959), pp. 125-139.

[330] E. WIBERG: *Lehrbuch der anorganischen Chemie.* De Gruyter, Berlin (1960), p. 209.

[331] J. LE ROND D'ALEMBERT: *Recherche sur la courbe que forme une onde tendue mise en vibration.* Hist. Acad. Roy. Sci. & Belles Lettres (Berlin) **3**, 214-249 (1747).

[332] J.L. DE LAGRANGE: *Mémoire sur la théorie du mouvement des fluides.* Nouv. Mém. Acad. Roy. Sci. & Belles-Lettres Berlin (1781), pp. 151-198.

[333] C.A. TRUESDELL, III: *The slow emergence of the one-dimensional wave equation for aerial propagation (1687–1759).* In: (A. SPEISER ET AL., eds.) *Leonardi EULERI opera omnia* **XIII** [2nd series]. Teubner, Lausanne (1954); see *Editor's introduction*, pp. xxxvi-xxxvii.

[334] L. EULER: *Introductio in analysis infinitorum.* 2 vols., Bousquet, Lausanne (1748).

In EULER's era, the problem of discontinuity was pressing since it was applicable to the analytical treatment of the string problem. In 1791, the French mathematician Louis F.A. ARBOGAST,[335] regarding the nature of arbitrary functions to be used when solving partial differential equations, made a distinction between **continuity, discontinuity** and **discontiguity** of a function. His use of the term *discontinuity* corresponds to EULER's view, but his new term *discontiguity* addressed mathematical discontinuity in a modern sense, since it covered functions that were discontinuous at isolated points.

In 1807, Jean-B.-J. FOURIER,[336] while trying to find the solutions to a partial differential equation describing heat diffusion into a solid, became the first to treat real discontinuities in a physical context. He discovered that any arbitrary function, including mixed and step functions, can be represented by a trigonometric series.[337, 338] **The adoption and treatment of discontinuities [Lat. *saltus*] were important requirements for mathematically tackling the difficult problem of "sounds of finite amplitude" – *i.e.*, of mechanical waves which can develop into shock waves.**

| | | |
|---|---|---|
| 1749 | *Crane Court, The Royal Society of London, England & Philadelphia, Pennsylvania, Colonial America* | On November 16, **Peter COLLINSON,** a British naturalist and antiquary in London who is a fellow scientist and friend of Benjamin FRANKLIN, **reads FRANKLIN's famous paper**[339] **on his studies of electricity,** entitled *Opinions and Conjectures Concerning the Properties and Effects of the Electrical Matter, Arising from Experiments and Observations, Made at Philadelphia, 1749.* His results are as follows: |

(i) Treating the possible causes of thunder, lightning, meteors and earthquakes, he tries to attribute them to electricity. ▪ Studies on electricity were made popular by the recent invention of the Leiden jar {VON KLEIST & CUNEUS ⇨1745}, and FRANKLIN's own sensational electrical experiments {FRANKLIN ⇨1751}.

(ii) He explains lightning by the interactions of two different types of clouds: nonelectric clouds raised from exhalations of the sea, and electrified clouds raised from the land.

(iii) Similar to ARISTOTLE {⇨333 B.C.} **he assumes that thunder is the result of the touching of clouds, which is re-echoed from cloud to cloud and across the extent of the firmament.** ▪ FRANKLIN's concept of "electrical atmospheres" – speculating that clouds are electrified and that the lightning discharge is a rapid release of electric fluid from clouds – resulted shortly after his famous invention of the lightning conductor and his use of the lightning kite to conduct the electricity from lightning. His hypothesis also stimulated other researchers in France (ISNARD 1757; P. BERTHOLON DE ST. LAZARE 1779), Italy (A. BINA 1751; G.B. BECCARIA 1758) and England (W. STUKELEY 1756), who attempted to prove the identities of electrical shocks and earthquakes.[340]

Benjamin FRANKLIN was an American publisher, author, inventor, scientist and diplomat, and one of the most admired men of his time, who reached the height of his fame in the following years {FRANKLIN ⇨1751; WILCKE ⇨1758} and then turned more fully to a career in politics.

---

[335] L.F.A. ARBOGAST: *Mémoire sur la nature des fonctions arbitraires qui entrent dans les intégrales des équations aux différences partielles.* Imprimerie Académie Impériale des Sciences, St. Pétersbourg (1791).

[336] J. FOURIER: *Mémoire sur la propagation de la chaleur dans les corps solides.* Nouv. Bull. Sci. Soc. Philomat. Paris **1**, 112-116 (1807); *Théorie analytique de la chaleur.* Didot, Paris (1822).

[337] A.P. YOUSCHKEVITCH: *The concept of function up to the middle of the 19th century.* Arch. Hist. Ex. Sci. **16**, 37-85 (1976/1977).

[338] I. GRATTAN-GUINNESS: *The development of the foundations of mathematical analysis from EULER to RIEMANN.* MIT Press, Cambridge (1970).

[339] B. FRANKLIN: *Tracts on electricity* (ed. by P. COLLINSON). E. Cave, London (1751).

[340] E. OESER: *Historical earthquake theories from ARISTOTLE to KANT.* Abhandl. Geol. Bundesanstalt Wien **48**, 11-31 (1992).

| 1749 | Montbard, Dépt. Bourgogne & Paris, France | **Georges-Louis LECLERC, Comte DE BUFFON,**[341] a French naturalist, **publishes the first three volumes of his encyclopedic work *Histoire naturelle* ("Natural History").** In vol. 1 he explains earthquakes in the framework of a general theory of the Earth, linking their origin to volcanism. He states that
▸ earthquakes fall into two classes: those caused by subterranean fires and explosions of volcanoes, and those with no clear connections to volcanoes or subterranean fires;
▸ nothing is comparable to the violence originating from subterranean fires which produce huge depressions (caving-ins); and that
▸ even big earthquakes only play a secondary role in the evolution of the Earth. |
|---|---|---|
| 1750 | United Kingdom | **Five strong earthquakes occur in England in a single year:** on February 19 and March 19 in London and the Home Counties, on March 29 in Portsmouth and the Isle of Wight, on April 13 in the northwest of England and the northeast of Wales, and on October 11 in Northamptonshire and the surrounding counties. ▪ The year 1750 therefore became known as the **"Year of Earthquakes."**
This activity excited much interest in earthquakes {HOLLMANN ⇒1750}, and before the end of 1750, nearly 50 articles were communicated on the subject to the Royal Society of London.[342] These reports described the quake events themselves, speculated about their causes, and collected and reviewed previous quake accounts. Although valuable from a historical point of view, they did not offer new scientific concepts but instead stimulated contemporary naturalists {WINTHROP ⇒1755; MICHELL ⇒1760}. |
| 1750 | The Royal Society of Sciences, Göttingen, Lower Saxony, Germany | At a meeting of the Society, **Samuel C. HOLLMANN,**[343] a professor of natural philosophy at the University of Göttingen, **takes the view that earthquakes are caused by the buildup of pressurized vapors in the interior of the Earth,** and not by electric phenomena as assumed by some of his contemporaries. **In order to prevent earthquakes, he recommends releasing these internal pressures by drilling the Earth's crust.** ▪ The idea of releasing the built-up of internal stress by triggering small earthquakes in order to prevent large earthquakes in densely populated areas has been discussed at length {Denver Earthquakes ⇒1962}. |
| 1751 | Crane Court, The Royal Society of London, England & Philadelphia, Pennsylvania, Colonial America | In his fifth letter, dated July 27, to Peter COLLINSON, a member of the Royal Society, **Benjamin FRANKLIN**[344, 345] **reports on his experiments and observations on electricity performed in Philadelphia. He also addresses the possibility of firing gunpowder by an electric spark:** "I have not heard that any of your *European* electricians have ever been able to fire gunpowder by the electric flame. We do it here in this manner. A small cartridge is filled with dry powder, hard rammed, so as to bruise some of the grains; two pointed wires are then thrust in, one at each end, the points approaching each other in the middle of the cartridge till within the distance of half an inch; then, the cartridge being placed in the circle, when the four jars are discharged, the electric flame leaping from the point of one wire to the point of the other, within the cartridge amongst the powder, *fires it*, and the explosion of the powder is at the same instant with the crack of the discharge." ▪ FRANKLIN's memoir was translated into German and supplemented by comments referring to recent results obtained in Europe {WILCKE ⇒1758}. |

---

[341] G.L. LECLERC, Comte DE BUFFON: *Histoire naturelle: générale et particulière* [with the assistance of L.J.M. DAUBENTON]. 15 vols., Imprimerie Royale, Paris (1749–1767); vol. 1 (1749), pp. 109-110, 502-535. *See also* Internet edition: http://www.buffon.cnrs.fr/index.php?lang=fr#hn.

[342] Phil. Trans. Roy. Soc. Lond. **46**, 601-750 (1749/1750).

[343] Göttingische Anzeigen von gelehrten Sachen. 20. Stück (14. Feb. 1756), pp. 161-165.

[344] FRANKLIN's correspondence with P. COLLINSON was published in England under the title *New experiments and observations on electricity, made at Philadelphia in America, by Mr. Benjamin FRANKLIN, and communicated to Mr. COLLINSON, F.R.S. at London*. E. Cave, St. John's Gate (1751).

[345] *See also* A.H. SMYTH: *The writings of Benjamin FRANKLIN*. Haskell House Publ., New York (1970), vol. II, pp. 423-426.

Two years later, FRANKLIN (using the pseudonym Richard SAUNDERS or "Poor Richard") first published his idea of the lightning rod in a short note entitled *How to Secure Houses, etc. from Lightning* in Poor Richard's Almanack for the Year of Christ 1753.

| | | |
|---|---|---|
| 1751 | Paris, France | **The publication of the 28-volume *Encyclopédie*,**[346] later popularly known as the "Bible of Enlightenment," begins. The science editors of this first modern encyclopedia are the philosopher Denis DIDEROT and the mathematician Jean LE ROND D'ALEMBERT, two of the most dynamic leaders of the Enlightenment. **The *Encyclopédie* also addresses the topics *explosion*, *percussion* and *choc*, which are reviewed by D'ALEMBERT:** |

(i) Differentiating between *explosions* arising from firing a cannon, from the inflammation of gases during the fermentation process of liqueurs, *etc.*, and *fulminations* ("chemical explosions"), he asks the following questions: "But how does it come that the inflammation of gunpowder and the mixture of two liqueurs produces this sudden and thundering dilatation? How and why should the air have been compressed before? That's what cannot be explained and, to tell the truth, has hitherto been fully ignored."

(ii) His review article on *percussion* refers to previous works {J. BERNOULLI ⇒1727; DESCARTES ⇒1644; WREN ⇒1669; WALLIS ⇒1669 & 1670/1671; HUYGENS ⇒1652, 1668/1669 & 1703}, but does not mention the important contributions made by MARCI {⇒1639} and MARIOTTE {⇒1671}. Addressing J. BERNOULLI's previous critique that the Laws of Percussion cannot be applied to perfectly hard bodies, he argues that perfectly hard bodies don't really exist in nature and, even if they did, the contact surface would only be limited to a point-like area in which the percussion partners lose their velocity almost instantaneously; *i.e.*, the contact time would approach zero. From this, **D'ALEMBERT concludes that the general laws of percussion might be applicable to elastic, perfectly hard bodies as well, thus even furnishing an example where the common doctrine of *Natura non operatur per saltum* may no longer be maintained.**

(iii) Referring to the action which a moving body produces when striking against another, he considers the term *choc* to be another expression for the term *percussion*.

(iv) Giving a definition for the term *dynamics*, he states: "**Actually, dynamics [French *dynamique*] signifies the science of forces [*la science des puissance*] or causes of motion [*causes motrices*]; *i.e.*, the forces which set bodies in motion** ... Mr. Leibniz first used this term [in his treatise *Specimen dynamicum* (1695); LEIBNIZ ⇒1686] to designate the most transcendental part of mechanics which treats the motion of bodies, such as caused by motive forces acting momentarily and continuously. In this sense, the general principle of dynamics is that the product of accelerating or decelerating force and time is equal to the element of velocity; the reason for this is that the velocity increases or decreases in each instant, by virtue of the sum of small repetitive shocks which the motive force imparts to the body during this instant."

DIDEROT's and D'ALEMBERT's *Encyclopédie*, a deep and extensive fount of knowledge in its time, is still used by historians today as a valuable source.

| | | |
|---|---|---|
| 1753 | Berlin, Kingdom of Prussia, Germany | Leonard EULER,[347] one of the greatest Swiss mathematicians and a member of the Berlin Royal Academy of Sciences, **publishes an analysis of projectile trajectories which incorporates empirical aerodynamic drag values.** Assuming a square-dependency of aerodynamic drag with velocity {⇒Sir NEWTON 1687}, he solves the equations of subsonic ballistic motion |

---

[346] D. DIDEROT and J. LE ROND D'ALEMBERT: *Encyclopédie ou dictionnaire raisonné des sciences, des arts et des métiers*. 35 vols., S. Faulche, Neufchatel (1751–1780).

[347] L. EULER: *Recherches sur la véritable courbe que décrivent les corps jetés dans l'air ou dans un autre fluide quelconque*. Hist. Acad. Roy. Sci. & Belles Lettres (Berlin) **9**, 321-352 (1753).

and summarizes some of the results into convenient numerical tables which allow one to determine the resulting ballistic range as a function of muzzle velocity and launch angle.

| 1754 | Vienna, Habsburg Monarchy | Leopold VON AUENBRUGGER, an Austrian physician, **invents a medical diagnostic procedure based on percussion** which he will not publish until seven years later.[348] His method entails striking the (human) body directly or indirectly with short, sharp taps of a hammer or finger in order to determine the condition of the parts beneath by the pitch or character of the sound. • His procedure has become a routine diagnostic method for investigating diseases of the chest or abdomen: <br> ▸ if the stroke is made directly upon the body, it is called **"immediate percussion;"** <br> ▸ if made upon something placed against the body – *e.g.*, a finger of the other hand, or a small instrument made for the purpose, like the *percussion hammer* {WINTRICH ⇨1841} – it is called **"mediate percussion."** <br> VON AUENBRUGGER's epoch making discovery of percussion as a medical diagnostic tool was extended to the examination of tendon and muscular reflexes by the invention of the *reflexion hammer* {WINTRICH ⇨1841}. |
|---|---|---|
| 1755 | Lisbon and Algarve, Portugal & Epicenter in Atlantic Ocean [ca. 200 km west-southwest of Cape St. Vincent] | On November 1 (All Saints' Day) at 9.30 A.M., **Lisbon is largely destroyed by a heavy earthquake of magnitude 8.7 with its epicenter in the Atlantic Ocean, about 200 km WSW of Cape St. Vincent.** <br> (i) A series of violent seismic shocks with an estimated magnitude of 8.5–8.6, which knock over candles and thus cause many fires, level the Portuguese capital. About 10,000 people are killed in Lisbon and about 40,000–50,000 are injured {⇨Fig. 4.1–E}. <br> (ii) **About half an hour after the earthquake, a total of three seismic sea waves swamp the town;** the maximum height of the waves is estimated at six meters. Curiously, at the arrival of the first seismic sea wave, the waters of the Tejo River recede, thus exposing the shallow sea floor of Lisbon's Oeiras Bay to more than a mile out. This unique phenomenon attracts many curious people to the bay floor, who are drowned when the wave crest follows only minutes later. <br> (iii) The destruction is greatest in Algarve, southern Portugal, where the tsunami dismantles some coastal fortresses and, in the lower areas, levels houses. In some places the waves crest at more than 30 m. • Seismic sea waves, later called "tsunamis" {Tsunami Meeting ⇨1961}, were also observed as far away as England, Norway and the West Indies. <br> (iv) In order to facilitate a hazard analysis, in 1758 Sebastião José DE CARVALHO E MELLO (later Count of Oeiras and Marquês DE POMBAL), then Minister of Foreign Affairs and virtual ruler of Portugal, instructs the priest Luis CARDOSO, a member of the Royal Academy of History, to hold enquiries in all parishes. The first such enquiries, concerning information about lands, rivers and mountains, took place in 1721 and 1747, but the 1758 enquiry was supplemented by more questions concerning the 1755 Earthquake {⇨Fig. 4.1–E}.[349] • When CARDOSO died in 1769, some of the parishes were yet to send their answers. Finally, in 1832, the answers kept safely at the Bibliotheca do Convento das Necssidades were united in the 43-volume *Dicionário Geográfico* or *Memórias Paroquiais*.[350] <br> (v) The Lisbon disaster frightens the Baroque world,[351] which had previously been regarded as perfect and harmonic, a world best expressed in such mottos as |

---

[348] L. VON AUENBRUGGER: *Inventum novum ex percussione thoracis humani, ut signo, abstrusos interni pectoris morbos detegendi* [*Neue Erfindung, mittelst des Anschlagens an den Brustkorb, als eines Zeichens, verborgene Brustkrankheiten zu entdecken*]. Trattner, Vindobonae (1761).
[349] G. PEREIRA: *Interrogatorios para a organização do «Dictionario Geographico» do P.ᵉ Luis CARDOSO*. Archeólogo Português (Lisboa) **1**, 268-271 (1895).
[350] At the Instituto dos Arquivos Nacionais/Torre do Tombo (IAN/TT) in Lisbon, only the enquiry of 1756 (some parishes) and the enquiry of 1758 still exist. The enquiries of 1721 and 1747 have been lost. Miguel Bandeira VELOSO at IAN/TT; private communication (June 2005).
[351] The impact of the Lisbon Earthquake on contemporary philosophy and European earthquake research were discussed by W. BREIDERT: *Die Erschütterung der vollkommenen Welt*. Wiss. Buchgesell., Darmstadt (1994).

- "Le tout est bien!" (Jean-Jacques ROUSSEAU, 1749);
- "Whatever is, is right!" (Alexander POPE, 1733); and
- "... the best of all possible worlds!" (Gottfried W. LEIBNIZ, c.1694).

(vi) However, the disaster also stimulates the international scientific community {DESMAREST, JACOBI, KANT, KRÜGER, MAYER ⇨1756; MICHELL ⇨1760} to explain its causes.[352]
• In the following year, almost 50 accounts on the Lisbon Earthquake were published by the Royal Society of London.[353] They were used by several pioneer seismologists to challenge existing ideas about earthquake generation.

The famous 1755 Lisbon Earthquake also stimulated modern researchers: through careful archival research it has proved possible to obtain relatively value-free accounts of most aspects of the earthquake and to use these to not only model the physical characteristics of and the damage caused by the earthquake, but also to consider the implications for present day hazard assessment and urban planning. Only a few examples can be given here:

- In 1996, Arch C. JOHNSTON,[354] a U.S. seismology research professor at the Mid-America Earthquake Center of the University of Memphis, TN, estimated the seismic moments of historic earthquakes in stable continental regions, including the 1755 Lisbon Earthquake.
- In 2001, Reverend David K. CHESTER,[355] a British geography professor at the University of Liverpool, reviewed the progress that has been made in identifying the source and faulting mechanisms, the processes involved in the generation and impact of tsunamis, the damage caused to different types of building, and the use being made of historical earthquakes of different sizes in defining future hazard scenarios for Lisbon and other areas of Iberia.
- In 2001, Charles L. MADER,[356] a former Los Alamos staff scientist, numerically modeled the generation and propagation of the Lisbon tsunami. By comparing his calculated wave amplitude data with those data observed in 1755 along the European, African and American coasts, he evaluated some of the characteristics of the earthquake source, particularly that the epicenter may have been close to the earthquake epicenter located on February 28, 1969, south of the Gorringe Bank – a ridge in the Atlantic Ocean off the coast of Portugal which was uplifted by the northward movement of the African Plate against the Eurasian Plate.
- In 2004, Marc-André GUTSCHER,[357] an American-born French geophysicist at the Institut Universitaire Européen de la Mer (IUEM) in Plouzané, France, discussed current research on the tectonic source of the Lisbon Earthquake, which may have been caused by an eastward-dipping subduction zone beneath Gibraltar.[358]

| 1755 | Boston, Massachusetts, New England & London, England | On November 18, only 17 days after the Lisbon Earthquake, **Boston is hit by a series of violent seismic shocks.** Shortly after, **John WINTHROP,** an American mathematician and astronomer, **gives a lecture in the Chapel of Harvard College "… in occasion of the Earthquake which stroke New-England the week before."** He reports on curious effects of an *aftershock* while sitting by his fireplace, with his feet on the hearth. Startled to feel and see the bricks jiggling, he observed that they seemed to rise up in sequence, one after another, then |

---

[352] E. OESER: *Historical earthquake theories.* University of Vienna, Austria; http://www.univie.ac.at/Wissenschaftstheorie/heat/heat.htm.
[353] Published in Phil. Trans. Roy. Soc. Lond. **49**, 351-444 (1756).
[354] A. JOHNSTON: *Seismic moment assessment of earthquakes in stable continental regions – III. New Madrid 1811–1812, Charleston 1886, and Lisbon 1755.* Geophys. J. Int. **126**, 314-344 (1996).
[355] D.K. CHESTER: *The 1755 Lisbon Earthquake.* Progr. Phys. Geography **25**, 363-383 (2001).
[356] C.L. MADER: *Modeling the 1755 Lisbon Tsunami.* Sci. Tsunami Haz. **19**, 93-98 (2001).
[357] M.-A. GUTSCHER: *What caused the great Lisbon Earthquake?* Science **305**, No. 5688, 1247-1248 (2004).
[358] J.F.B.D. FONSECA and M.-A. GUTSCHER: *The source of the Lisbon Earthquake.* Science **305**, No. 5718, 50 (2005).

quickly drop back into place, as though jarred successively from underneath. **"It was not a motion of the whole hearth together,"** he explained, **"either from side to side; or up and down; but of each brick separately by itself."**

Two years later, WINTHROP presented a paper[359] on various accounts given from the Boston Earthquake, thereby **first describing the earthquake in terms of a sequence of waves:** "The earthquake began with a roaring noise in the N.W. like thunder at a distance; and this grew fiercer, as the earthquake drew nearer; which was almost a minute in coming to this place … My neighbour tells me, that, as soon as he heard the noise, he stopt, knowing, that it was an earthquake, and waiting for it; and he reckoned he had stood still about 2', when the noise seemed to overtake him, and the Earth began to tremble under him … By his account, as well as that of others, the first motion of the Earth was what may be called a 'pulse,' or rather an *undulation*; and resembled that of a long rolling, swelling sea, and the swell was so great, that he was obliged to run and catch hold of something, to prevent being thrown down … there were two of these great wavings, succeeded by one, which was smaller … I suppose, most others, imagined, that the height of the shock was past. But instantly, without a moment's intermission, the shock came on with redoubled noise and violence; though the species of it was altered to a tremor, or quick horizontal vibratory motion, with sudden jerks and wrenches."

▪ **His detailed report, which anticipated the modern concept of "propagating seismic waves," is a milestone in the evolution of seismology.**

Fifty years later, this concept was independently resumed by Thomas YOUNG, a British physicist and important contributor to the theory of elasticity. In his *Lectures on Natural Philosophy* (1807), **he remarked that an earthquake "is probably propagated through the earth nearly in the same manner as a noise is conveyed through the air"** – *i.e.*, as a *propagating wave*.

| | | |
|---|---|---|
| 1755 | *The Royal Academy of Sciences, Berlin, Kingdom of Prussia, Germany* | Leonard EULER, director of the Mathematical Class of the Academy (founded in 1744) and a member of the board, **presents two papers on the motion of fluids on the basis of differential calculus:**<br><br>(i) In his first paper,[360] he works out a set of equations describing the motion of an inviscous incompressible liquid, gas, *etc.*, which will later become known as the **"general Euler equations of hydrodynamics."**<br><br>(ii) In his second paper,[361] **EULER first treats compressible fluids analytically.** Neglecting viscosity effects he shows that a moving compressible fluid can be described in terms of pressure, density and a velocity vector by a set of coupled differential equations which consists of a time-dependent continuity equation for conservation of mass and three time-dependent momentum equations.[362] In particular, he treats the flow of a compressible fluid in a straight horizontal tube. Introductorily he writes,[363] "Since in my preceding memoirs I reduced all the theory of fluids to two analytic equations, the consideration of these formulae appears to be of the greatest importance, for they include not only all that has been discovered by methods very different and for the most part slightly convincing, but also all that one could desire further in this science. However, sublime are the researches on fluids which we owe to the Messrs. [Daniel and Johann] BERNOULLI, [Alexis Claude] CLAIRAUT and [Jean LE ROND] D'ALEMBERT; |

---

[359] J. WINTHROP: *An account of the earthquake felt in New England and the neighboring parts of America, on the 18 November 1755*. Phil. Trans. Roy. Soc. Lond. **50**, 1-18 (1757).

[360] L. EULER: *Principes generaux du mouvement des fluides*. Mém. Acad. Sci. Berlin **11**, 274-315 (1757).

[361] L. EULER: *Continuitation des recherches sur la théorie du mouvement des fluides*. Mém. Acad. Sci. Berlin **11**, 316-361 (1757).

[362] *Euler equations*. NASA-GRC, Cleveland, OH; http://www.grc.nasa.gov/WWW/K-12/airplane/eulereqs.html.

[363] C.A. TRUESDELL: *Rational fluid mechanics, 1687–1765*. In: (C.A. TRUESDELL, ed.) *Euleri opera omnia*. Orell, Füssli & Turici, Lausanne (1954); *see Series Secunda*, vol. XII, pp. LXXXIV-C.

they flow so naturally from my two general formulae that one cannot sufficiently admire this accord of their profound meditations with the simplicity of the principles from which I have drawn my two equations, and to which I was led immediately by the first axioms of mechanics."

Later, the British John CHALLIS claimed that he was first to derive the equations for compressible flow.

| | | |
|---|---|---|
| 1755 | *University of Königsberg, East Prussia, Germany* | Immanuel KANT, a 31-year old German philosopher, **publishes – shortly before the event of the Great Lisbon Earthquake {⇨1755} – his cosmology** *Allgemeine Naturgeschichte und Theorie des Himmels* ("Universal Natural History and the Theory of the Heavens"), in which "he tried to combine Sir NEWTON's mechanics with the advantages of the Cartesian cosmology" (Friedrich VON WEIZSÄCKER[364]).

(i) Provided with the subtitle *Versuch von der Verfassung und dem mechanischen Ursprunge des ganzen Weltgebäudes nach Newtonischen Grundsätzen* ("An Essay on the Constitution and Mechanical Origin of the Whole Universe Treated According to Newtonian Principles,") he goes far beyond anything that is to be found in NEWTON's own writings {Sir NEWTON ⇨1687}.

(ii) His concept of the Universe is based on a dynamic mechanism in which various regions undergo cyclical changes. He claims that the Universe, which owes its genesis to a Mosaic deity (as also assumed by NEWTON and LEIBNIZ), had a beginning in time and is infinite in spatial extent.

(iii) **He claims that the Sun, the planets of the Solar System, and their moons all arose from the condensation of primordial, widely diffused matter.** ▪ His nebular hypothesis, partly anticipated by the Swedish scientist and theologian Emanuel SWEDENBORG,[365] was later resumed in France {DE LAPLACE ⇨1796}.

James W. ELLINGTON[366] at the University of Connecticut, one of KANT's numerous biographers and translators, summarized his contributions to astronomy as follows: "He anticipated astronomical facts that were later confirmed by very powerful observational techniques and with the help of relativistic cosmological theory. He conjectured that the Solar System is a part of a vast system of stars making up a single galaxy, that the so-called 'nebular stars' are galactic systems external to but similar to our own galaxy (a fact that was not confirmed until the 20th century), and that there are many such galaxies that make up the Universe as a whole…" |
| 1756 | *France* | Nicolas DESMAREST,[367] a French geologist, **using the model for the elastic collision of billiard balls, attempts to explain the propagation of seismic shocks as a chain of percussions,** passing from one mountain range to the other. ▪ His crude but modern wave model, which assumes the transport of energy and impulse, was an important step towards understanding the riddle of wave propagation without mass transport. |
| 1756 | *Hannover, Lower Saxony, Germany* | Johann F. JACOBI,[368] a German philosopher with mathematical background, **considers the Great Lisbon Earthquake to be a result of volcanism:**

(i) He assumes that seismic shocks originate from the sudden mixing of molten rock with water: "When cold water comes in contact with a considerable amount of molten material, a large bursting and cracking sound will arise. If a distinct stream of cold water flows into a place where a large quantity of molten minerals is contained, the flow will explode with the greatest violence, |

---

[364] C.F. VON WEIZSÄCKER: *Die Tragweite der Wissenschaft*. Hirzel, Stuttgart (1966), vol. 1: *Schöpfung und Weltentstehung: Die Geschichte zweier Begriffe*, p. 131.
[365] E. SWEDENBORG: *Opera philosophica et mineralia* ("Philosophical and mineralogical works"). 3 vols., Hekel, Dresden & Leipzig (1734); vol. 1: *Principia*.
[366] J.W. ELLINGTON: KANT, *Immanuel*. In: (C.C. GILLESPIE, ed.) *Dictionary of scientific biography*. Ch. Scribner's Sons, New York, vol. 7 (1973), pp. 224-235.
[367] N. DESMAREST: *Conjectures physico-mécaniques sur la propagation des secousses dans les tremblements de terre, et sur la disposition des lieux qui en ont ressenti les effets*. Paris (1756).
[368] J.F. JACOBI: *Sammlung einiger Erfahrungen und Mutmaßungen vom Erdbeben*. In: *Nützliche Sammlungen*, 15.-18. Stück, pp. 225-288, Hannover (1756).

whereby the water itself will immediately boil and roar and transforms into hot steam which expands with the greatest force." ▪ His earthquake model approaches the modern concept of a steam explosion.

(ii) Referring to Denis PAPIN's digester {⇨1679}, which shows how much hot steam may contract when cooling down, he (erroneously) explained the rise and fall of the sea level observed during the Lisbon Earthquake likewise.

(iii) He also reports on a digester explosion: "It once happened to the late Privy Councilor HAMBURGER at Jena that the force of the expanding vapors and air have driven the lid of this pot with its five screws such that all screw threads were peeled off and the screws were so smooth as if turned off on a turning-lathe. I myself have seen them… [*see* § 10]."

| | | |
|---|---|---|
| 1756 | University of Königsberg, East Prussia, Germany | Immanuel KANT, one of Germany's most eminent philosophers, who published his essay on the mechanical origin of the Universe the previous year {KANT ⇨1755}, **writes two essays on the possible causes of the Lisbon Earthquake.** Contrary to most of his contemporaries, KANT – then still in his precritical stage – strictly avoids any emotions and religious explanations of the possible causes. In his first treatise[369] on earthquakes, he rather focus his investigations on rational facts: |

(i) Trying to explain the active causes of earthquakes, he gives a chemical explanation by quoting an experiment for setting up an oxyhydrogen explosion {LÉMERY ⇨1700}. ▪ In this point, KANT erroneously followed concepts proposed by contemporary scientists {KRÜGER ⇨1756}.

(ii) He explains the transport mechanism of seismic energy by subterranean caves – a result of the Earth's evolution as supposed since antiquity. These caves and galleries, running parallel to mountains and rivers, can carry the effects of wind, blast or fire over longer distances. Thus, he explains the violence of the Lisbon Earthquake by the fact that the city was built alongside the river and recommends its rebuilding at a different location. He points out that mountainous countries such as Peru and Chile are particularly endangered by earthquakes. However, where the Earth's crust is even (such as in his homeland Prussia), earthquakes are rare phenomena.

(iii) In order to explain the effects of tidal and compressional waves caused by earthquakes, he quotes Louis CARRÉ's hydrodynamic shock experiment {CARRÉ ⇨1705}. Repeating his experiment by shooting a bullet into a $12 \times 6 \times 6$ in.$^3$ ($30 \times 15 \times 15$ cm$^3$) wooden box filled with water, KANT observes that water can react like a solid body when suddenly compressed. He concludes that **water can almost transmit shocks without softening them** – thus approaching DESMAREST's concept of wave propagation {DESMAREST ⇨1756}.

(iv) **He correctly attributes the movement of those inland lakes which are not connected with the sea not to varying atmospheric pressure, but rather to the slightly varying tilt of the Earth's surface.** This may occur during big earthquakes at great distances. ▪ This concept was later resumed by François Alphonse FOREL,[370] a Swiss physician and scientist. Using the example of Lake Geneva, FOREL described such movements by oscillations of surface water in lakes and landlocked seas, which he termed ***seiches***.

(v) **He correctly recognizes that the observed *great sea waves* (which we now call "tsunamis") are related to earthquake-induced vertical displacements of the sea floor.**

---

[369] I. KANT: *Von den Ursachen der Erschütterungen bei Gelegenheit des Unglücks, welches die westlichen Länder von Europa gegen Ende des vorigen Jahres getroffen hat*. Königsbergische wöchentliche Frag- und Anzeigungs-Nachrichten, Nr. 4 & Nr. 5 (Jan. 1756); *Geschichte und Naturbeschreibung der merkwürdigen Vorfälle des Erdbebens, welches an dem Ende des 1755sten Jahres einen großen Teil der Erde erschüttert hat*. J.H. Hartung, Königsberg (März 1756); *Fortgesetzte Betrachtung der seit einiger Zeit wahrgenommenen Erderschütterungen*. Königsbergische wöchentliche Frag- und Anzeigungs-Nachrichten, Nr. 15 & Nr. 16 (April 1756).

[370] F.A. FOREL: *Handbuch der Seenkunde: allgemeine Limnologie*. Engelhorn, Stuttgart (1901).

| | | |
|---|---|---|
| 1756 | *University of Helmstedt, Lower Saxony, Germany* | **Johann Gottlob KRÜGER,**[371] a German professor of medicine and philosophy and a member of the Prussian Academy of Sciences, **reflects on the causes and beneficial aspects of the Lisbon disaster for the progress of science.** Similar to Immanuel KANT {⇨1756}, he explains the causes of the Great Lisbon Earthquake by the fermentation and self-ignition of a mixture of iron and sulfur {LÉMERY ⇨1700}, and the causes of thunder, hurricanes and subterranean fires likewise. His model assumes that enormous masses of air continuously penetrate via crevices and galleries into large depths, and that God only requires a spark to ignite the "winds of sulfur" accumulated in subterranean vaults. He supplements his reflections with a moral observation and interprets this catastrophe as God's punishment. ▪ Eleven years prior to the Lisbon Earthquake, KRÜGER[372] had already written about the causes of earthquakes, thereby quoting LÉMERY's experiment in his *History of the Earth* (1746). |
| 1756 | *University of Göttingen, Lower Saxony, Germany* | **Tobias MAYER,**[373] a German professor of mathematics and physics and director of the Observatory, **explains the Lisbon Earthquake as the result of a sudden, short-duration local change in the direction of gravity,** with the ground itself remaining unmoved and the objects on it moving around. ▪ MAYER refused an answer to the question of where such a change in the direction of gravity came from, referring instead to NEWTON's *Principia*, which had explained the movements of celestial bodies based on gravity without being able to explain the causes of gravity itself {Sir NEWTON ⇨1687}. |
| 1756 | *Paris, France* | **VOLTAIRE** (alias François-Marie AROUET),[374] one of the greatest of all French writers, **publishes a poem on the Lisbon disaster** entitled *Poème sur le désastre de Lisbonne, ou Examen de cet axiome "Tout est bien,"* thereby criticizing the philosophy of his contemporaries. This immediately provokes objections from the French philosopher and writer Jean-Jacques ROUSSEAU, who defends his thesis of optimism in a letter to VOLTAIRE.[375]<br><br>Shortly after the Lisbon Earthquake, the French poet Ponce-Denis E. LEBRUN published a poem entitled *Ode sur les désastres de Lisbonne*. |
| 1758 | *Wismar, Swedish Dominion, Baltic Sea [now northern Germany]* | **Johann Carl WILCKE,**[376] a German "electrician," **publishes a German translation of** FRANKLIN's memoir *New Experiments and Observations on Electricity, Made at Philadelphia in America* {FRANKLIN ⇨1751}.<br><br>(i) In a footnote WILCKE annotates that the German professor BROSE first succeeded in **igniting vapors of heated gunpowder by an electric spark.**<br><br>(ii) WILCKE also describes his own, hitherto unpublished experiments. For example – very similar to FRANKLIN – **he reports that he succeeded in igniting common gunpowder by an electric spark discharge:** he pressed gunpowder in a glass tube to form a 1-in. (25.4-mm)-long column, covered both ends with metal foils, connected one end to the conductor of his electrostatic machine, and inserted a thin wire into the other end of the glass tube, which he connected to a switching gap leading to the other pole of his machine. The electrically triggered explosion of the gunpowder ruptured the glass tube. |

---

[371] J.G. KRÜGER: *Gedanken von den Ursachen des Erdbebens, nebst einer moralischen Betrachtung*. C.H. Hemmerde, Halle *etc.* (1756).
[372] J.G. KRÜGER: *Geschichte der Erde in den alleraltesten Zeiten*. Lüderwald, Halle (1746).
[373] T. MAYER: *Versuch einer Erklärung des Erdbebens*. Nützliche Sammlungen [Hannoverische gelehrte Anzeigen], 19. Stück, pp. 289-296 (5. Martius 1756).
[374] VOLTAIRE [or F.M. AROUET]: *Poème sur la destruction de Lisbonne; ou examen de cet axiome, Tout est bien*. Œuvres complètes, **XII**, 107-130. Imprim. Soc. Litt. typogr., Kehl (1785).
[375] Jean-Jacques ROUSSEAU (1712–1778), one of the greatest French philosophers, whose philosophy marked the end of the Age of Reason, later resumed his thesis in his book *Émile ou De l'éducation* (Paris 1762). It begins with the revolutionary statement: *"Tout est bien sortant des mains de l'Auteur des choses, tout dégénère dans les mains de l'homme."* ("Everything is good as it leaves the hands of the Author of things, everything degenerates in the hands of man.").
[376] J.C. WILCKE: *Des Herrn Benjamin FRANKLIN's Esq. Briefe von der Electricität*. Kiesewetter, Leipzig (1758).

| | | |
|---|---|---|
| 1759 | The Royal Academy of Sciences, Berlin, Kingdom of Prussia, Germany | Leonard EULER,[377] a Swiss mathematician, **addresses in a letter to Joseph L. DE LAGRANGE, an Italian-born French mathematician, the possibility that the propagation of sound might depend on the "size of the disturbances," which would mean the "size of the displacements" or the "intensity of sound" when expressed in modern terms.** He writes to DE LAGRANGE: "It is very remarkable that the propagation of sound actually takes place more rapidly than the theory indicates, and at present I renounce the opinion I had formerly, that the following disturbances could accelerate the propagation of the preceding ones, in such a way that the higher is the sound the greater is its speed, as possibly you have seen in our latest memoirs. It has also come into my mind that the size of the disturbances might cause some acceleration, since in the calculation they have been supposed infinitely small, and it is plain that [finite] size would change the calculation and render it intractable. But, in so far as I can discern, it seems to me that this circumstance would rather diminish the speed." ▪ EULER's hypothesis was indeed correct insofar as the amplitude of sound, which he called the "size of disturbance," can influence the speed of sound. However, he (incorrectly) assumed that the velocity would diminish with increasing amplitude. |
| 1760s | Private laboratory at 13 Great Marlborough Street, London, England | Henry CAVENDISH,[378] a British physicist and chemist, **carries out a series of studies on the chemistry of "airs" and waters** that show that there are an assortment of chemical airs or gases of different specific gravities.<br>(i) He investigates "fixed air" [$CO_2$], generated in fermentation and putrefaction.<br>(ii) **He also experiments with various mixtures of "dephlogisticated air" [$O_2$] and "inflammable air" [$H_2$], and their ignition by an electric spark from a Leiden jar** {VON KLEIST & CUNEUS ⇨1745}. In order to quantify the mechanical energy released from a confined volume of that gas (oxyhydrogen), CAVENDISH[379] constructs a **"measurer of explosions of inflammable air."** |
| 1760 | University of Cambridge, England | Reverend John MICHELL,[380] a British natural philosopher and geologist, **publishes a great memoir on the causes of earthquakes** which shows a remarkable knowledge of the strata in various parts of England and abroad:<br>(i) Starting with a discussion of previous reviews given by Élie BERTRAND[381] and John BEVIS,[382] **he argues that earthquakes are caused by waves originating in the Earth.** He addresses the various motion phenomena of the Great Lisbon Earthquake {⇨1755}, the position of the seismic focus, and the propagation of seismic waves and sea waves.<br>(ii) MICHELL already distinguishes between primary causes and secondary effects of seismic shocks. According to his hypothesis, one primary cause of earthquakes is subterranean fires which produce high-pressure steam when they are exposed to water, resulting in sudden and violent effects (explosions).<br>(iii) Similar to the German natural philosopher Johann F. JACOBI {⇨1756}, **MICHELL anticipates the modern concept of *hydroexplosions*.** He writes, "These fires, if a large quantity of water should be let out upon them suddenly, may produce a vapour whose quantity and elastic force may be fully sufficient for the purpose." |

---

[377] C.A. TRUESDELL, III: *EULER's two letters to LAGRANGE in October, 1759*. In: (C.A. TRUESDELL, ed.) *Leonardi EULERI opera omnia* **XIII** [II]. Teubner, Leipzig *etc.* (1926); *see* Editor's *Introduction*, pp. xxxvii-xli.

[378] H. CAVENDISH: *Three papers, containing experiments on factitious air* [1766]. Phil. Trans. Roy. Soc. Lond. **56**, 141-184 (1767).

[379] H. CAVENDISH: *A measurer of explosions of inflammable air* [Laboratory note]. In: (E. THORPE, ed.) *The scientific papers on the honorable Henry CAVENDISH, F.R.S.* In: *Chemical & Dynamical*, Cambridge University Press, vol. II (1921), pp. 318-320.

[380] J. MICHELL: *Conjectures concerning the cause and observations upon the phenomena of earthquakes: particularly of that great earthquake of the first of November, 1755, which proved so fatal to the city of Lisbon, and whose effects were felt as far as Africa, and more or less throughout all Europe* [read Feb. and March 1760]. Phil. Trans. Roy. Soc. Lond. **51**, 566-634 (1760); Phil. Mag. **52** [I], 186-195, 254-270, 323-340 (1818).

[381] E. BERTRAND: *Mémoires historiques et physiques sur les tremblemens de terre*. P. Gosse, La Haye (1757).

[382] J. BEVIS: *The history and philosophy of earthquakes*. Nourse, London (1757).

(iv) **He regards volcanic eruptions as useful safety valves.**

(v) From records of arrival times of seismic shock waves resulting from the Lisbon Earthquake and measured at more distant places, **he first estimates the velocity of the seismic wave at more than 20 miles a minute (536 m/s).**

(vi) He also states that **the great waves of the sea** – the term *tsunami* was not yet coined – which frequently follow earthquakes and roll to shore at an enormous height, **are due to an undulation transferred to the ocean water at a point directly above that at which the primary undulation takes place.**

Compared to contemporary hypotheses on the causes of earth- and seaquakes {KANT, JACOBI, MEYER & DESMAREST ⇒1756}, MICHELL's theory was certainly the most advanced, and today he is justly considered to be one of the main founders of *seismology*, a term which was invented in the 19th century by the Irish civil engineer Robert MALLET {⇒1846, 1849, 1857 & 1860}.

| | | |
|---|---|---|
| 1762 | *Crane Court, The Royal Society of London, England* | John CANTON,[383] a British physicist, **scientifically demonstrates the limited compressibility of water, which has previously been regarded as being incompressible.** He places the test liquid in a thermometer-like arrangement and, compressing the bulb, obtains an observable increase in the change of volume in the capillary. **This represents the birth of static high-pressure research!** |

Two years later, CANTON[384] reported on further compression experiments, which proved that sea- and rainwater, mercury, olive oil and spirit of wine are compressible when subjected to static pressures too. He observed that compressed liquids behave elastically, and that, for example, at 34 °F (1.1 °C) spirit of wine has the highest compressibility ($66 \times 10^{-6}$/bar), water has medium compressibility ($49 \times 10^{-6}$/bar), and mercury has the lowest compressibility ($3 \times 10^{-6}$/bar).[385] ▪ CANTON's observations were not verified until sixty years later by the U.S. inventor Jakob PERKINS,[386] who constructed a piezometer and high-pressure apparatus that generated a pressure of some two kilobars in water.

In 1822, Hans Christian ÖRSTED,[387] a Danish natural philosopher in Copenhagen, designed an apparatus to measure the compressibility of liquids. He observed that the compressibility of water diminishes as the temperature increases, while those of the other liquids tested (such as mercury, alcohol, methyl alcohol, ether, chloroform, carbon disulfide) increases. He also found that the compressibility of seawater is slightly less than that of fresh water, while the addition of salt diminishes it further in proportion to the quantity of salt in solution.[388]

| | | |
|---|---|---|
| 1766 | *The Netherlands* | Johan François DRYFHOUT,[389] a Dutch lawyer and amateur naturalist, **attributes the Lisbon Earthquake to a subterranean explosion.** Speculating that mines of gunpowder-like materials are running through the Earth, he explains the quake by the explosion of such a mine beneath Lisbon, lifting it and settling it back "like a marble bounced on stone."[390] |
| 1769 | *Brescia, Lombardy, northern Italy* | On August 18, **the city is devastated when the tower of the church of San Nazaro is strucked by lightning.** The discharge current, passing through vaults where 200,000 lb (90,000 kg) of gunpowder have been stored for safe-keeping, causes a massive explosion |

---

[383] J. CANTON: *Experiments to prove that water is not incompressible*. Phil. Trans. Roy. Soc. Lond. **52**, 640-643 (1762).
[384] J. CANTON: *Experiments and observations on the compressibility of water and some other fluids*. Phil. Trans. Roy. Soc. Lond. **54**, 261-262 (1764).
[385] CANTON's data closely approach modern compressibility data: water $45 \times 10^{-6}$/bar and mercury $3.8 \times 10^{-6}$/bar at 20 °C; *see*, for example, E.U. CONDON and H. ODISHAW (eds.): *Handbook of physics*. McGraw-Hill, New York *etc.* (1967), pp. 3-13.
[386] J. PERKINS: *On the compressibility of water*. Phil. Trans. Roy. Soc. Lond. **110**, 324-330 (1820).
[387] H.C. ÖRSTED: *Experiment, die Compression des Wassers zu zeigen*. Journal für Chemie und Physik **36**, 332-339 (1823).
[388] *Apparatus of Örsted*. The Manufacturer & Builder **1**, 272 (1869).
[389] J.F. DRYFHOUT: *Natuurkundige Aanmerkingen on Onderstellingen ... de zeldzame Waterberoerngen op den 1° November 1755*. Ottho van Thiel, 'sGravenhage (1755); *Nodere Aanmerkingen over de Oorzaak en Werkinge der Waterberoeringen van den eersten Nov. 1755*. Ibid. (1762).
[390] B. WALKER: *Earthquake*. Time-Life, Alexandria, VA (1982), p. 54.

which destroys one sixth of the city and killes about 3,000 people. ▪ Previously, it was first demonstrated in laboratory experiments in Colonial America {FRANKLIN ⇨1751} and later in a more detailed study in Germany {WILCKE ⇨1758} that common gunpowder can be ignited by an electric spark discharge.

In response to this disaster, the British parliament passed two acts establishing standards for the manufacture and storage of gunpowder in private hands. In 1782, Benjamin FRANKLIN made recommendations for protecting the new British powder magazines at Purfleet (near London) from the effects of lightning. The magazines had been equipped with pointed rods as recommended by FRANKLIN. After the American Revolution began, the Purfleet powder magazines were struck by lightning. However, it was later determined that the buildings at Purfleet were ruined simply because the conductor was defective in certain parts and not laid deep enough into the ground.

| | | |
|---|---|---|
| 1770s | Ecole Royale du Génie, Mézières, Dépt. Ardennes, northeastern France | Gaspard MONGE,[391] a French mathematician, physicist and public official, enters a field of study that will hold his interest for many years: DE LAGRANGE's theory of general partial differential equations and their solution by means of MONGE's theory of surfaces. |

(i) MONGE calls his method of geometric construction of particular solutions the "method of characteristics."

(ii) Starting from a first-order partial differential equation, he gives a geometric interpretation of the method of the variation of parameters and writes, "**In the following I will use, as I have always done, different characteristics for the different ways of differentiating; this method is more practical, as it is not necessary to find a fractional form in order to present a partial differential … This memoir contains the constructions of integrals of partial differential equations** that are more general than the ones I had constructed until now, and there I demonstrate that the geometrical places of these integrals generally satisfy their partial differential equations, which is what I had thought to myself."

(iii) His geometric construction of a particular solution to the equations under consideration allows him to determine the general nature of the arbitrary function involved in the solution of a partial differential equation. ▪ **The key idea of MONGE's work was that each step in an analytic calculation corresponds to some geometrical construction.** His method of characteristics was widely applied and further developed by subsequent mathematicians and physicists who introduced such basic notations as the *characteristic curve* (sometimes referred to as *MONGE's curve*), the *characteristic cone* (or *MONGE's cone*), the *trajectory of characteristics*, the *characteristic variable*, etc. A detailed discussion of MONGE's work was given by René A. TATON,[392] a French science historian and scientific director of the Centre National de la Recherche Scientifique (CNRS). Together with Carl Friedrich GAUSS and Leonhard EULER, MONGE is considered a founder of *differential geometry*, the study of curves and surfaces in space.

The Englishman Samuel EARNSHAW {⇨1858} and, independently, the German G.F. Bernhard RIEMANN {⇨1859} first applied MONGE's classical work on characteristics to the propagation of waves of finite amplitude; *i.e.*, shock waves. Their studies were continued and greatly extended by various other prominent shock wave pioneers {HUGONIOT ⇨1887; HADAMARD ⇨1903; PRANDTL & BUSEMANN ⇨1929}. Reviews on the progress of further developments of this important mathematical tool and its application to gas dynamics and hydrodynamics were

---

[391] G. MONGE: *Mémoire sur la construction des fonctions arbitraires dans les intégrales des équations aux différences partielles*. Mémoires des mathématiques et de physique présentés à l'Académie . ..par divers sçavans … **7** [2e partie], 267-300, 305-327 (1773). Imprimerie Royale, Paris (1776); *Sur la détermination du fonctions arbitraires dans les intégrales de quelques équations aux différences partielles*. Miscellanea taurinensia (Torino) **5**, 16-78 (1770–1773).

[392] R. TATON: *L'œuvre scientifique de Gaspard MONGE*. Presse Universitaire de France, Paris (1951); *Gaspard MONGE*. In: (C.C. GILLESPIE, ed.) *Dictionary of scientific biography*. Ch. Scribner's Sons, New York, vol. 9 (1973), pp. 469-478.

given by Robert SAUER,[393] Klaus OSWATITSCH,[394] COURANT & FRIEDRICHS {⇨1948}, and Michael B. ABBOTT.[395] Antonio FERRI[396] wrote on applications to steady 2-D and 3-D flows, and flow patterns characterized by the presence of shock waves. Richard MADDEN[397] applied the method of characteristics to axisymmetric hypervelocity impact problems.

| | | |
|---|---|---|
| 1770 | *University of Basel, northern Switzerland* | **Daniel BERNOULLI,**[398] the famous Swiss mathematician and physicist, treats the problem of collision by applying the theory of elasticity, and **develops the first "wave theory" of elastic percussion.** Assuming a freely suspended straight elastic rod hit at its center of gravity perpendicular to its axis by a force of percussion, he calculates the deformation of the rod and, from this, the loss of kinetic energy as a result of harmonic elastic vibrations. **His result shows that elastic vibrations diminish the living force** (*vis viva*) by a factor of five ninths; *i.e.*, the elastic oscillations consume more than half the total kinetic energy.[399] |
| 1774 | *London, England* | **Edward NAIRNE,**[400] a London instrument supplier and experimentalist, **studies electrical explosions of thin wires.** Upon discharging a battery of Leiden jars {VON KLEIST & CUNAEUS ⇨1745} initially charged up to a high voltage through a 1-m-long iron wire 0.15 mm in diameter, he observes that "it flew about the room in innumerable red hot balls, on examining these balls, they were in general hollow, and seemed to be nothing but scoria." |
| 1775 | *London, England* | **Peter WOULFE,**[401] a British chemist and mineralogist, **first prepares** *picric acid* [2,4,6-trinitrophenol, $C_6H_3N_3O_7$] by treating blue indigo dye [$C_{16}H_{10}N_2O_2$] with nitric acid [$HNO_3$]. ▪ The substance was subsequently used principally as a yellow dye until in 1885 the French organic chemist Jean-Baptiste André DUMAS realized that the substance was highly explosive. In 1886, **the French chemist Eugène TURPIN showed that the black powder in artillery shells could be replaced with picric acid,** and in the Russian-Japanese War (1904/1905) picric acid was the most widely used military explosive.<br><br>Picric acid was apparently first used as an explosive in 1867 by the Italian engineer Luigi BORLINETTO.[402] It was used in warheads under the name **melinite** in France, **lyddite** in Britain and **shimose** in Japan. However, there were a number of practical problems, *e.g.*, corrosion, self-ignition, and toxicity. Just before World War I, when an economical process was developed for nitrating toluene, TNT {HAEUSSERMANN ⇨1891} began to replace picric acid in artillery ammunition. |
| 1777 | *Royal School, Como, Lombardy, northern Italy* | **Count Alessandro VOLTA,** an Italian high-school professor of physics, **invents the** *spark eudiometer* **for analyzing swamp gases and testing the "goodness" of air.** It consists of a stout glass tube of constant inner diameter closed at the top, where two electrodes pass through the glass and form a spark gap. The lower end of the tube is placed in a dish of water, and the air to be tested is introduced into the tube and its volume noted. A known volume of hydrogen is then let into the tube, and the mixture ignites by static electricity. The experiment could then determine the "goodness of the air" (*i.e.*, its oxygen content) from measurements of the volume of the re- |

---

[393] R. SAUER: *Charakteristikenverfahren für Kugel- und Zylinderwellen reibungsloser Gase*. ZAMM **23**, 29-32 (1943).
[394] K. OSWATITSCH: *Über die Charakteristikenverfahren der Hydrodynamik*. ZAMM **25/27**, 195-208, 264-270 (1947).
[395] M.B. ABBOTT: *An introduction to the method of characteristics*. American Elsevier, New York (1966), pp. 35-44, 128-163.
[396] A. FERRI: *The method of characteristics*. In: (W.R. SEARS, ed.) *General theory of high-speed aerodynamics*. Princeton University Press, Princeton, NJ (1954).
[397] R. MADDEN: *The application of the method of characteristics in three independent variables to the hypervelocity impact problem*. Ph.D. thesis, Virginia Polytechnic Institute (June 1967).
[398] D. BERNOULLI: *Examen physico-mechanicum de motu mixto qui laminis elasticis a percussione simul imprimitur*. Novi Commentarii Academiae Scientiarum Imperialis Petropolitanae **XV**, 361-380 (1770).
[399] I. SZABÓ: *Geschichte der mechanischen Prinzipien*. Birkhäuser, Basel *etc.* (1977), pp. 465-470.
[400] E. NAIRNE: *Electrical experiments by Mr. Edward NAIRNE, of London, mathematical instrument-maker, made with a machine of his own workmanship, a description of which is prefixed*. Phil. Trans. Roy. Soc. Lond. **64**, No. 1, 79-89 (1774).
[401] P. WOULFE: *Experiments on a new colouring substance from the island of Amsterdam in the South Sea*. Phil. Trans. Roy. Soc. Lond. **64**, 91-93 (1775).
[402] M. GIUA: *Chimica delle sostanze esplosive*. U. Hoepli, Milano (1919), pp. 287, 296.

maining gas.⁴⁰³ ▪ Later VOLTA suggested the construction of a gun using inflammable air – the so-called "**Volta pistol**" – in which a spark from a Leiden jar {VON KLEIST & CUNEUS ⇨ 1745} fired a mixture of hydrogen and oxygen enclosed in the barrel by a cork.⁴⁰⁴ VOLTA's early pistols were capable of sending a lead ball 20 ft (6 m) and denting a board. VOLTA's pistol, a ballistic curiosity, remained a favorite lecture hall demonstration for a long time. Another device that produced loud bangs was the ***Volta cannon***, which was based on the same principle. His ***electric mortar*** used an electric spark to explode gunpowder.

Similarly, **Benjamin FRANKLIN's thunderhouse** – a three-dimensional wooden model house, equipped with a lightning rod, hinged walls and a detachable roof and a small charge of gunpowder – **had demonstrated the benefit of his lightning rod** and broken the barriers to its acceptance. When an electric charge was applied to the lightning rod in his model house, electricity would pass through with no effect. However, when a wooden block was positioned to break the grounding circuit, the spark would ignite the gunpowder with convincing and entertaining effects: the roof and walls literally blew off.⁴⁰⁵

| | | |
|---|---|---|
| 1778 | *Montbard Dépt. Bourgogne & Paris, France* | **Georges-Louis LECLERC, Comte DE BUFFON**,⁴⁰⁶ a French naturalist, **proposes the first version of the cataclysmic hypothesis for the origin of the Solar System in his memoir *Les époques de la nature*** ("Epochs of Nature"). |

(i) He assumes that the impact of a massive "comet" tore some material out of the Sun, which then cooled and condensed into planets. The impact was off-center, so that not only were the planets thrown out into their orbits, but they (together with the Sun) were set spinning on their axes, some so rapidly that they threw off additional material from which their moons were made. ▪ Comets are icy chunks of water and dust. They originate from two regions in the outer Solar System: the Kuiper Belt (a belt of icy bodies that extends from Neptune out to past Pluto), and the Oort cloud (a spherical shell that surrounds the Solar System). Today it is known that comets lack sufficient mass to accomplish such a task, because they are simply too small. It has been estimated that there are billions of comets in the Solar System, with a total mass of only one tenth that of the Earth's mass. Furthermore, when comets come near the Sun they vaporize, developing a bright tail. The first known case of a comet actually colliding with the Sun happened in August 1979 {Comet HOWARD-COOMEN-MICHELS (1979 XI) ⇨ 1979}.

(ii) Based upon the cooling rate of iron, he calculates that the age of the Earth is 75,000 years; *i.e.*, much older than the 6,000 years proclaimed by the Church.

In the following centuries, his (incorrect) cataclysmic theory – later known as the **"Buffon cometary collision hypothesis"** – was resumed and modified by a number of astronomers that the Sun and another star collided to form the planets. However, it has been decided that the hot gases produced by such a collision would have escaped the Sun's gravity and would not have condensed into bodies as small as planets. Catastrophic theories therefore have been abandoned and generally nebular theories {KANT ⇨ 175; DE LAPLACE ⇨ 1796} are favored to explain the formation of the Solar System.

| | | |
|---|---|---|
| 1780 | *Diocese Chiemsee, Bavaria, southern Germany* | In Salzburg, **Franz Adam WISSHOFER**, a German priest and amateur naturalist, **publishes a pamphlet on the design of an *electric gun*** {VOLTA ⇨ 1777} that uses the discharge of a Leiden jar {VON KLEIST & CUNEUS ⇨ 1745} to ignite an inflammable gas that propels the projectile.⁴⁰⁷ ▪ Much later, the idea of an "electric gun" was resumed and developed further in the |

---

⁴⁰³ T.B. GREENSLAD JR.: *Instruments for natural philosophy*; http://www2.kenyon.edu/depts/physics/EarlyApparatus/index.html.
⁴⁰⁴ T.B. GREENSLADE JR.: *VOLTA's pistol*. Rittenhouse [The journal of the American Scientific Instrument Enterprise] **1**, 55-57 (1987); http://www2.kenyon.edu/depts/physics/EarlyApparatus/Titlepage/Static_Electricity.html.
⁴⁰⁵ Thunderhouse (2007); http://thunderhouse.org/legacy.cfm.
⁴⁰⁶ G.L. LECLERC, Comte DE BUFFON: *Les époques de la nature*. In: *Histoire naturelle* [with the assistance of L.J.M. DAUBENTON]. Supplément, tome cinquième. Imprimerie Royale, Paris (1778); 2 vols., Imprimerie Royale, Paris (1780).
⁴⁰⁷ H. MÜLLER: *Gewehre, Pistolen, Revolver*. Kohlhammer, Stuttgart (1979), chap. *Windbüchsen und elektrische Zündung*, pp. 154-156.

| 1781 | Private laboratory at 13 Great Marlborough Street, London, England | **Henry CAVENDISH, an English physicist and chemist, resumes his previous investigations on oxyhydrogen explosions** {CAVENDISH ⇨ 1760s}. **He observes that mixtures of "common air"** [$N_2/O_2$] **and "inflammable air"** [$H_2$], **enclosed in a vessel and electrically fired, are converted into a deposit of dew that is pure water inside the vessel. All of the inflammable air is converted but only about four-fifths of the common air is converted.** ▪ CAVENDISH was stimulated by an experiment which the English scientist John WARLTIRE had carried out in the same year. WARLTIRE, who exploded a mixture of air and hydrogen, found that the weight of the residual gases was less than that of the original mixture. He attributed the loss to the weight of the heat emitted in the reaction. CAVENDISH concluded that some substantial error was involved, since he did not believe that heat possesses enough weight to impact on the experiment. CAVENDISH found that the weights involved did balance if he included the weight of a thin film of liquid produced in the reaction – a film that had been ignored in all previous work on the combustion of hydrogen and oxygen.[408]

CAVENDISH[409] did not publish his important results until three years later. In this famous paper on the synthesis of water he concluded: "From the fourth experiment it appears, that 423 measures of inflammable air are nearly sufficient to completely phlogisticate 1,000 of common air; and that the bulk of the air remaining after the explosion is then very little more than four-fifths of the common air employed; so that, as common air cannot be reduced to a much less bulk than that, by any method of phlogistication [[410]], we may safely conclude, that when they are mixed in this proportion, and exploded, almost all the inflammable air, and about one-fifth part of the common air, lose their elasticity, and are condensed into the dew which lines the glass." Fortunately, archival records supported his priority when many individuals, including the French chemist Antoine-Laurent LAVOISIER, claimed to be the discoverer.

Inflammable air was discovered by CAVENDISH in 1766. In 1788, LAVOISIER named "inflammable air" *hydrogen* [from the Greek *hydrogenium*, meaning "water-former"], because as hydrogen burns, it produces water. Hydrogen is the most abundant element in the Universe. |
| 1781 | The Royal Academy of Sciences, Berlin, Kingdom of Prussia, Germany | **Joseph L. DE LAGRANGE,**[411] an eminent Italian-born French mathematician and astronomer, **publishes a fundamental treatise on the motion of fluids.** In the last chapter, entitled *Du mouvement d'un fluide contenue dans un canal peu profond et presque horizontal, et en particulier du mouvement des ondes* ("On the Motion of a Fluid Confined in a Shallow and Almost Horizontal Duct, and Particularly on Wave Motion"), **he considers a surface wave of infinitesimal height in shallow water in a duct of finite length:**<br>(i) He derives the famous formula for the propagation velocity as<br>$$v = (gh)^{1/2}, \qquad (1)$$<br>where $g$ is the acceleration due to gravity and $h$ the water depth of the liquid at rest. This wave speed is independent of wavelength, provided the latter is long compared with $h$. ▪ His shallow-water approximation employed the *method of parallel sections*, whereby all liquid at each value of the horizontal coordinate $x$ is assumed to have the same horizontal velocity, but this velocity and the disturbed depth are allowed to vary slowly in $x$ and time $t$. |

---

[408] F. SEITZ: *Henry CAVENDISH: the catalyst for the chemical revolution.* Proc. Am. Phil. Soc. **148**, No. 2, 151-179 (June 2004).

[409] H. CAVENDISH: *Experiments on air.* Phil. Trans. Roy. Soc. Lond. **74**, 119-153, 170-177 (1784).

[410] "Phlogistication" is the act or process of combining with *phlogiston* – a hypothetical substance once believed to be present in all combustible materials and to be released during burning.

[411] J.L. DE LAGRANGE: *Mémoire sur la théorie du mouvement des fluides.* Nouv. Mém. Acad. Roy. Sci. & Belles-Lettres Berlin (1781), pp. 151-198.

(ii) Referring to the known formula describing the velocity $v$ of a free-falling body, given by
$$v = (2gh)^{1/2},$$
and drawing a comparison between sound waves in air and gravity waves, he states, "Thus, as the velocity of propagation of sound is found equal to that which a weight would require in falling from the height of the atmosphere (assumed homogeneous), the velocity of propagation of waves will be the same as that which a weight would acquire in falling from a height equal to half the depth of water in the canal."

(iii) He refers to measurements previously performed by the French engineer Philippe DE LA HIRE who had observed a velocity of 1.412 *pied par seconde* (0.46 m/s) in a water depth of eight-tenths of a *pouce* (about 2.2 cm), thus essentially confirming DE LAGRANGE's theoretical result.

(iv) He (erroneously) claims that his shallow-water results should remain a good approximation for deep-water waves because most of the motion is confined near the surface.

**In the case of a *hydraulic jump*** (also called a "shooting flow"), **the velocity $v$ is calculated by**
$$v \approx [g(h_1 + h_2)h_2/2h_1]^{1/2}, \tag{2}$$
where $h_1$ and $h_2$ are the water depths ahead and behind the jump, respectively.[412] For increasingly weaker jumps (so-called "streaming flow"), $h_2$ approaches $h_1$ – thus eventually converging with DE LAGRANGE's solution for the velocity of infinitesimal surface waves. Frictional effects such as those caused by bottom roughness ("boundary friction") reduce the propagation velocity of shallow water waves below the frictionless values given in Eqs. (1) and (2).

| | | |
|---|---|---|
| 1783 | *Dijon, Dépt. Côte d'Or, east central France* | **General Lazare N.M. CARNOT,**[413] a French mathematician and father of the famous physicist Sadi CARNOT {Comte BERTHOLLET ⇨1809}, **publishes his book *Essai sur les machines en général*** ("Essay on Engines in General") which contains all of the elements of his engineering mechanics and wins a prize in a competition. |

(i) In order to postulate his Principle of Continuity in relation to the transmission of power, he considers a machine (*e.g.*, the lever, the wedge, the pulley, the screw, the inclined plane, and so forth) to be an intermediary body serving to transmit motion between two or more primary bodies that do not act directly on one another. **Wishing to attribute no properties except those common to all forms of matter to such machines, he envisages them as intrinsically nothing more than systems composed of corpuscles.** Beginning his analysis with the action of one corpuscle upon another in machine motion, he obtains an equation stating that, in a system of "hard" (*i.e.*, perfectly elastic) bodies, the net effect of mutual interaction among the corpuscles that constitute the system is zero.[414]

(ii) **He treats the fully inelastic percussion of two bodies,** defined such that after percussion both masses $m$ and $M$ – impacting each other head-on with velocities $v_0$ and $V_0$ – move with a common velocity
$$v = V = (mv_0 + MV_0)/(m + M).$$
He finds that the kinetic energy of both masses after percussion is equivalent to the kinetic energy $E_{kin}$ of a reduced mass
$$\mu = mM/(m + M)$$

---

[412] F.R. GILMORE, M.S. PLESSET, and H.E. CROSLEY JR.: *The analogy between hydraulic jumps in liquids and shock waves in gases.* J. Appl. Phys. **21**, 243-249 (1950).

[413] L.M.N. CARNOT: *Essai sur les machines en général, par un officier du Corps Royal du Génie.* A.M. Defay, Dijon (1783).

[414] C.C. GILLESPIE: CARNOT, Lazare-Nicolas-Marguerite. In: (C.C. GILLESPIE, ed.) *Dictionary of scientific biography.* Ch. Scribner's Sons, New York, vol. **3** (1971), pp. 70-79.

moving with a relative velocity $\Delta v = v_0 - V_0$; *i.e.*, $E_{kin} = \frac{1}{2}\mu\Delta v^2$. ▪ In 1803, **CARNOT**[415] resumed the subject of percussion and arrived at important conclusions. Today his result is known as the **"Carnot theorem."**[416] It states that the energy dissipated during impact will be equal to the kinetic energy corresponding to the lost velocity.

The British mathematician Edward J. ROUTH[417] differentiated between three parts of CARNOT's theorem:

- Part 1: **In the impact of inelastic bodies, *vis visa* or "living force"** {LEIBNIZ ⇨1686} **is always lost.**
- Part 2: **The amount of lost *vis visa* equals the living force of the relative motion of the two bodies;** *see* (ii) above. In cases of explosion [meaning here the restitution period in which the impacting bodies or particles separate], *vis visa* is always gained.
- Part 3: **In the case of perfectly elastic percussion, the whole process consists of two phases: first a** *compression force* (as if the bodies were inelastic), **and then a** *restitution force* (being of the nature of an explosion). The circumstances of these two forces are equal and opposite to each other. Hence the *vis visa* lost in compression is exactly balanced by the *vis visa* gained during the period of restitution {POISSON ⇨1833; ⇨Fig. 2.8}.

| | | |
|---|---|---|
| 1783 | City of York, North Yorkshire, England | John GOODRICKE,[418] a young British amateur astronomer, **first calculates the period of the variable star Algol** (Beta Persei) – an eclipsing multi-star system located in *Perseus* ("The Champion," a constellation of the Northern Hemisphere) which he discovered the previous year – **to be 68 hours and 50 minutes.** The star changes its brightness by more than a magnitude as seen from Earth.[419] **Attempting to explain this puzzling phenomenon of a "variable star," he (correctly) suggests that the distant sun is periodically occulted by a dark body.** ▪ The Royal Society of London awarded him the 1783 Copley Medal "for his discovery of the Period of the Variation of Light in the Star Algol." <br><br> This particular type of variable star is now known as an **"eclipsing binary."** There are a number of other reasons why variable stars can change their brightness.[420] |
| 1783 | Chair of Mathematics, Royal Military Academy at Woolwich southeast London, England | Charles HUTTON,[421] a British mathematician and ballistician, **extends the ballistic pendulum to cannon shot and measures supersonic muzzle velocities of up to 2,030 ft/s** ($M \approx 1.8$). He uses charges of up to 16 oz (454 g) of gunpowder; and his projectiles are 1.96-in. (5-cm)-dia. iron cannonballs, weighing up to 16 oz, 13 dr. (477 g). ▪ His studies, which he continued with in the following years, first confirmed that supersonic vel-ocities are not only obtainable for small firearms {ROBINS ⇨1746}; they can also be attained by large-caliber guns. |

---

[415] L.M.N. CARNOT: *Principes fondamentaux de l'équilibre et du mouvement*. Detereille, Paris (1803).

[416] A. SOMMERFELD: *Vorlesungen über theoretische Physik*. Dieterich'sche Verlagsbuchhandl., Wiesbaden (1949), vol. 1: *Mechanik*, p. 27.

[417] E.J. ROUTH: *The elementary part of a treatise on the dynamics of a system of rigid bodies*. Macmillan, London (1897, 1905); Dover Publ., New York (1960), pp. 302-304. German translation: *Die Dynamik der Systeme starrer Körper*. Teubner, Leipzig (1898), vol. I, pp. 335-337.

[418] J. GOODRICKE: *A series of observations on, and a discovery of the period of the variation of the light of the bright star in the head of Medusa, called Algol*. Phil. Trans. Roc. Soc. Lond. **73**, 474-482 (1783).

[419] The Greek defined the very brightest stars as being of the "first magnitude" and the faintest observable stars as of the sixth magnitude, with gradations of brightness in between. In 1856, the English astronomer Norman Robert POGSON proposed the system now in use: the difference in magnitude between two stars is given by 2.5 times the logarithm (to base 10) of the brightness ratio. For example, a star of magnitude 5.0 is 2.5 as bright as one of magnitude 6.0; *i.e.*, the brighter the object, the lower the number assigned as a magnitude. A difference of five magnitudes in brightness, or $(2.5)^5$, would be very nearly to 100, so that a first magnitude star is about 100 times as bright as a sixth magnitude star.

[420] For the classification of variable stars *see* General Catalogue of Variable Stars ⇨1948.

[421] C. HUTTON: *New experiments in gunnery, for determining the force of fired gunpowder, the initial velocity of cannonball, the ranges of projectiles at different elevations, the resistance of the air to projectiles, the effect of different lengths of guns, and of different quantities of powder, &, &*. In: *Tracts on mathematical and philosophical subjects*. T. Davison, London (1812), vol. 2, Tract XXXIV, pp. 306-384.

In another treatise published in the same year, HUTTON[422] **investigates the drag of projectiles within a wide range of speeds, up to supersonic** (20–2,000 ft/s or 6.1–610 m/s), "to show according to what power of the velocity, at every point, the resistance increases."

(i) He observes that "commencing with the second power or square of the velocity, at the very beginning or slow motion [5 ft/s], the exponent of the power gradually increases, till at the velocity of 1,500 or 1,600 ft/s [457–488 m/s], it arrives at the 2.153 power of the same … After the 1,600 feet [488 m/s] velocity, where the exponent (2.153) is greatest, it gradually decreases again to the end [towards 2,000 ft/s or 610 m/s]."

(ii) He explains the velocity-dependent drag of a projectile by a vacuum generated at its rear, "The circumstance of the variable and increasing exponent in the ratio of the resistance is owing chiefly to the increasing degree of vacuity left behind the ball, in its flight through the air, and to the condensation of the air before it. It is well known, that air can only rush into a vacuum with a certain degree of velocity, *viz.*, about 1,200 or 1,400 feet [366 or 427 m] in a second of time; therefore, as the ball moves through the air, there is always left behind a kind of vacuum, either partial or complete; that as the velocity is greater, the degree of vacuity behind goes on increasing, till at length, when the ball moves as rapidly as the air can rush in and follow it, the vacuum behind the ball is complete, and to complete ever after, as the ball continues to move with all greater degrees of velocity. Now the resistance, which the ball suffers in its flight, is of a triple nature; one part of it being in consequence of the *vis inertia* of the particles of air, which the ball strikes in its course; another part from the accumulation of the elastic air before the ball; and the third part arises from the continued pressure of the air on the forepart of the ball, when the velocity of this is such as to leave a vacuum behind it in its flight, either wholly or in part … As soon as the motion of the ball becomes equal to that of the air, and always when greater [*i.e.*, at supersonic speeds], then the ball has to sustain the whole pressure of the atmosphere on its forepart, without having any aid from a counter-pressure behind…"
• HUTTON's explanation illustrates well the attempts of early supersonic pioneers to find a plausible reason for the puzzling phenomenon of the strong increase in drag in the transonic regime.

| | | |
|---|---|---|
| 1783 | *University of Cambridge, England* | **The British Revd John MICHELL,**[423] an astronomer, geologist and the "father of seismology" {MICHELL ⇨1760}, **develops the idea of invisible (black) stars.**[424] He writes: "… if the semi-diameter of a sphere of the same density with the Sun were to exceed that of the Sun on the proportion of 500 to 1 … all light emitted from such a body would be made to return towards it, by its own proper gravity." • Thirteen years later in France, Pierre-Simon DE LAPLACE {⇨1796} – apparently independently of MICHELL – propounds the same concept of **"black holes,"** a term coined in 1967 by the U.S. theoretical physicist John A. WHEELER {⇨1968}. |
| 1784 | *City of York, North Yorkshire, England* | **John GOODRICKE,**[425] a young British astronomer, **discovers the variability of Beta Lyrae,** an *eclipsing variable star* {GOODRICKE ⇨1783}, **and the periodic variability of Delta Cephei,** the second known Cepheid variable and the source of the name of this class of *pulsating variable stars.*[426] • Two years later, GOODRICKE died at the early age of 22 from pneumonia, which he probably caught when observing Delta Cephei. |

---

[422] C. HUTTON: *Theory and practice of gunnery, as dependent on the resistance of the air.* In: *Tracts on mathematical and philosophical subjects.* T. Davison, London (1812), vol. 3, Tract XXXVII, pp. 209-315.

[423] J. MICHELL: *On the means of discovering the distance, magnitude, &. of the fixed stars, in consequence of the diminution of the velocity of their light, in case such a diminution should be found to take place in any of them, and such other data should be procured from observations, as would be farther necessary for that purpose* [read Nov. 27, 1783]. Phil. Trans. Roy. Soc. Lond. **74**, 35-57 (1784).

[424] S. SCHAFFER: *John MICHELL and black holes.* J. Hist. Astron. **10**, 42-43 (1979).

[425] J. GOODRICKE: *A series of observations on, and a discovery of the period of the variations of light of the star marked δ by Bayer, near the head of Cepheus.* Phil. Trans. Roy. Soc. Lond. **76**, 48-61 (1786).

[426] For the classification of variable stars *see* General Catalogue of Variable Stars ⇨1948.

| 1784 | Geneva, western Switzerland | **Georges-Louis LeSage,**[427] a Swiss mathematician, physicist and inventor, **explains gravity in terms of the impact of "ultramundane corpuscles" on bodies** – later to be known as the "**kinetic theory of gravity.**" He speculates that these particles, coming from outside the natural world and moving at very high speeds in all directions, also set in motion the particles of light and various æthereal media which, in their turn, act on the molecules of gases and maintain their motions. ▪ James Clerk MAXWELL {⇨1867} who read LeSage's hypothesis with interest, later wrote, "His theory of impact is faulty, but his explanation of the expansive force of gases is essentially the same as in the dynamical theory [the kinetic theory of gases] as it now stands." |
|---|---|---|
| 1784 | Royal Artillery, England | **Henry Shrapnel,** a British officer, **invents an antipersonnel projectile timed to burst in the air toward the end of its flight and to discharge a large number of small bullet-like balls over a wide area,** each capable of killing or wounding a man – so-called "**shrapnel.**" ▪ Adopted by the British artillery in 1803, the shrapnel shell remained a major artillery missile until World War I. It has since been replaced by high-bursting-charge projectiles, in which the steel balls are replaced by projectile fragments.<br><br>When for example a pipe bomb or a steam boiler explodes, the casing breaks up, causing shrapnel fragments to scatter off with high velocity – the so-called "shrapnel problem." Shrapnel motion, which has been studied using gasdynamic methods, is important to the understanding and modeling of the effects of violent mechanics in fragmentation warhead technology.[428] |
| 1784/ 1785 | The Academy of Sciences, Turin, Piemonte, northwestern Italy | **Joseph L. de Lagrange,**[429] an Italian-born French mathematician, **studies the percussion force of a water jet impinging perpendicularly or obliquely on a plane.** Measurements of the percussion force performed by Georg W. Krafft (1736) gave a much smaller value than predicted theoretically by Daniel Bernoulli (1736), Jean Le Rond d'Alembert (1769), and Charles Bossut (1772). Assuming a simple model of fluid flow with a central core of stagnated liquid surrounded by a shell of streaming and laterally deviated liquid, de Lagrange derives a simple formula for the percussion force that better matched experimental results.<br><br>This basic problem of *hydrodynamics* – a term coined by Daniel Bernoulli {⇨1738} to analytically cover hydrostatic as well as hydraulic (*i.e.,* dynamic) phenomena by a single method – anticipated the difficult task of evaluating the flow of water impacting the blade of a water turbine. |
| 1785 | Académie des Sciences, Paris, France | **Charles Augustin Coulomb,**[430] a French physicist and member of the mechanics section of the Académie des Sciences, **presents his three reports on electricity and magnetism,** in which he explains the laws of attraction and repulsion between electric charges and magnetic poles, **and treats the basic interaction between two charged particles in relative motion.** ▪ So-called "Coulomb collisions" play an important role in many areas of plasma physics, accelerator physics and astrophysics. |
| 1785 | University of Turin, Piemonte, northwestern Italy | **Count Morozzo,**[431] a Turin university administrator and naturalist who studied the constitution of air and phosphorescence phenomena, **reports on a flour-dust explosion** in a Turin flour warehouse – probably the earliest account of such a phenomenon.<br><br>(i) He describes the circumstances as follows: "On the 14th of December, 1785, about six o'clock in the evening, there took place in the house of Mr. Giacomelli, baker in this city, an explosion which threw down the windows and window-frames of his shop, which looked into |

---

[427] G.L. LeSage: *Physique mécanique.* In: P. Prévost: *Deux traités de physique mécanique.* Paschoud, Genève (1818).

[428] C.J. Poole: *Shrapnel motion.* Masters thesis, University of Oxford (2001).

[429] J.L. de Lagrange: *Sur la percussion des fluides.* Mém. Acad. Roy. Sci. Turin **I**, 95-108 (1784/1785).

[430] C.A. Coulomb: *Premier mémoire sur l'électricité et le magnétisme.* Hist. Acad. Roy. Sci. (Paris), pp. 569-577 (1785); *Second mémoire...* Ibid. pp. 578-611 (1785); *Troisième Mémoire...* Ibid. pp. 612-638 (1785).

[431] Count Morozzo: *Account of a violent explosion which happened in a flour-warehouse, at Turin, Dec. 14th, 1785; to which are added some observations on spontaneous inflammations.* The Repertory of Arts and Manufactures **2**, 416-432 (1795).

the street; the noise was as loud as that of a large cracker, and was heard at a considerable distance. At the moment of the explosion, a very bright flame, which lasted only a few seconds, was seen in the shop; and it was immediately observed that the inflammation proceeded from the flour-warehouse, which was situated over the back shop, and where a boy was employed in stirring some flour by the light of a lamp. The boy had his face and arms scorched by the explosion; his hair was burnt, and it was more than a fortnight before his burns were healed…"

(ii) He speculates that the flour might have produced "inflammable air" by fermentation (such as is supposed to exist in dampened hay), which, mixed with air and dispersed flour dust, was ignited by the light of the lamp, thus leading to this violent inflammation. However, he critically remarks that model experiments did not prove this hypothesis, although upon examination the flour used was found to be extraordinarily dry and originated from the Piedmont area, which had no rain for five or six months.

| | | |
|---|---|---|
| 1786/ 1787 | Paris, France | Comte Claude-Louis BERTHOLLET,[432] an eminent French chemist who as one of the band of scientific men will later accompany Napoleon to Egypt, **prepares "fulminating silver" – a black powder –** by acting with ammonia on precipitated silver oxide. ▪ BERTHOLLET's detonating silver was not a fulminate. |

▸ Fulminating compounds are reaction products of Hg, Ag, Au, or Pt salts with ammonia, with a complex, ill-defined composition. According to modern terminology, classified as "primary explosives," they are too sensitive for any practical use. Fulminating gold was the first known high explosive {CROLL ⇨ 1608}.

▸ In the late 1790s, the Italian chemist Luigi Valentino BRUGNATELLI[433] discovered that if silver be dissolved in nitric acid and the solution added to spirits of wine, *silver fulminate* [AgONC] – **a highly explosive white powder** – was obtained which explodes more easily than mercury fulminate {HOWARD ⇨ 1800}.

▸ Later BRUGNATELLI worked out a more satisfactory method for preparing silver fulminate.

In the same year, Comte **BERTHOLLET obtains potassium chlorate** [KClO$_3$], later also named *Berthollet salt*, by the action of chlorine on caustic potash. He observes that a mixture of potassium chlorate and carbon explodes energetically. **This stimulates him to try to replace saltpeter [KNO$_3$] in classical gunpowder by potassium chlorate,** a powerful oxidizing agent. In a public experiment carried out with this new type of gunpowder, the director of the plant and four other people are killed on the spot.[434] ▪ Unfortunately, the mixture of sulfur, charcoal and potassium chlorate is unstable, dangerous to mix or handle in large quantities, and cannot safely be utilized as a substitute for gunpowder as a propellant. However, his work on potassium chlorate stimulated the invention of the percussion lock {FORSYTH ⇨ 1805}. Today gunpowder containing potassium chlorate instead of nitrate is commonly used in fireworks and matches.

| | | |
|---|---|---|
| 1789 | Wirksworth, Derbyshire, England | Reverend Abraham BENNET,[435] an English hobby electrician and Curate of Wirksworth, **publishes the book** *New Experiments on Electricity.* |

---

[432] C.L. BERTHOLLET: *Observations sur quelques combinaisons de l'acide marin dephlogistiqué ou de l'acide muriatique oxygéné*. Mém. Acad. Roy. Sci. Turin **III**, 385-396 (1786/1787).

[433] L.V. BRUGNATELLI: *Sur les fulminations*. Ann. Chim. **27**, 331-332 (1798).

[434] S.K. KAPOOR: *BERTHOLLET, Claude-Louis*. In: (C.C. GILLESPIE, ed.) *Dictionary of scientific biography*. Scribner's Sons, New York, vol. 2 (1970), pp. 73-82.

[435] A. BENNET: *New experiments on electricity, wherein the causes of thunder and lightning as well as the constant state of positive or negative electricity in the air or clouds, are explained; with experiments on clouds of powders and vapours artificially diffused in the air. Also a description of a doubler of electricity, and of the most sensible electrometer yet constructed with other new experiments and discoveries in the science*. J. Drewry, Derby (1789); http://www.wirksworth.org.uk/NEWEXP-4.htm#8-45.

(i) In order to investigate the nature of electricity – "an extremely subtle fluid which pervades metals with astonishing facility" – he performed numerous laboratory experiments using a gold leaf electrometer that he had invented.

(ii) He also determined the polarity of thunder clouds with the help of this instrument. Similar to previous experiments carried out in Colonial America {FRANKLIN ⇒1749}, he used a kite to direct the electricity of a thunder cloud. In his book, he gives the following account of an experiment made on July 5, 1788: "Large dark clouds frequently passed over; therefore about 2 o'clock in the afternoon a kite was raised, with a soften'd brass wire in the string about 200 yards long. When the kite had been flying about an hour, a dark cloud appeared at a great distance, and changed the electricity from positive to negative, which increased till the cloud came nearly over, and some large drops of rain fell, and to secure the string from being wet I endeavoured to tie it on the opposite side of a post to which it was before fastened; but when my hand came near the string I received so severe a shock that my arm was deprived of sensation during a few seconds, and I was obliged to let the string go, first terrified at the supposed loss of my left arm, and then gratefully rejoicing to feel the returning sensibility after rubbing with the other hand. The explosion was heard at the distance of about 40 yards like the loud crack of a whip. The kite was raised often before and since this time, but without any remarkable appearance."

| | | |
|---|---|---|
| 1790 | Birmingham, West Midlands, central England | William WITHERING,[436] an English physician and naturalist, **describes some imperfect siliceous tubes and a melted pebble of quartz which he found while in digging into the ground under a tree where a man had been killed by lightning.** He writes,<br><br>(i) "The specimens which accompany this Paper … will demonstrate the intense heat which must have existed to bring such materials into fusion.<br><br>(ii) Sand, unmixed with calcareous matter, agglutinated by the heat. Within the hollow part of this mass, the fusion has been so perfect, that the melted quartzose matter has run down the hollow, and assumed nearly a globular figure.<br><br>(iii) Mr. [James] WATT suggested to me, that the hollows had been occasioned by the expansion of moisture whilst the fusion existed."<br><br>**These curious siliceous tubes – so-called fulgurites – were first studied in more detail in the early 1820s by D. François J. ARAGO,**[437] a French physicist at the Bureau des Longitudes in Paris. By carefully analyzing their appearance and color and analyzing various recorded accounts of eyewitnesses he came to the conclusion that fulgurites must be the result of a lightning strike penetrating into the moist soil, which creates a cavity due to the expansion of steam and fuses the loose sand grains together. |
| 1792 | Volcano Unzen-dake [near Nagasaki], central Japan | On May 21, **after explosive activity of the volcano Unzen-dake, the collapse of one of its several lava domes triggers a great landslide** which discharges an enormous quantity of debris and a series of mud streams (lahars) into the bay, **resulting in a gigantic tsunami** (about 10,450 casualties).[438] ▪ This eruption of Unzen-dake was Japan's worst-ever volcanic disaster.<br><br>The conclusion that landslides as well as submarine slides must produce tsunamis, and are therefore one particular cause of the large waves, evolved only slowly {MILNE ⇒1898; Count BERNARD ⇒1907; GUTENBERG ⇒1939}. |
| 1792 | Munich, Bavaria, southern Germany | Franz Xaver VON BAADER,[439] a German natural philosopher and *Bergrat* ("mining engineer") interested in the application of explosives in the mining industry, **observes that the energy of a blast can be focused on a small area by forming a hollow in the charge** that increases the |

---

[436] W. WITHERING: *An account of some extraordinary effects of lightning*. Phil. Trans. Roy. Soc. Lond. **80**, 293-295 (1790).
[437] D.F.J. ARAGO: *Sur les tubes vitreux qui paraissent produits par des coups de foudre*. Ann. Chim. Phys. **19** [II], 290-303 (1822).
[438] A. IMAMURA: *Theoretical and applied seismology*. Maruzen Co., Tokyo (1937), p. 126.
[439] F. VON BAADER: *Versuch einer Theorie der Sprengarbeit*. Bergmännisches Journal **1**, No. 3, 193-212 (1792).

explosive effect and saves powder – the so-called **"cavity effect."** ▪ VON BAADER's publication was apparently read and put into practice in Norway, and for a short time also in Germany in the Harz Mines.[440] However, since he used black powder, which is not capable of detonation, his arrangement was not a shaped charge device {VON FÖRSTER ⇨1883} in the modern sense.

Seven years later VON BAADER makes the observation that the surface relief of an explosive is reproduced on a closely facing steel plate by the focusing of explosion products – so-called **"explosive engraving"** {MUNROE ⇨1888}.

| | | |
|---|---|---|
| 1793 | Paris, France | During the French Revolution (1789–1815), **the mathematician Joseph-Louis DE LAGRANGE attempts to analytically solve two important ballistic problems:** |

▸ **what is the motion of the fired gunpowder in a firearm during the acceleration of the projectile, and**

▸ **what is the temporal and local distribution of the pressure in the barrel?** ▪ In the time of DE LAGRANGE, quick-burning black powder was used, as opposed to slow-burning smokeless powders of today. As a result, the powder was completely burned, developing its maximum pressure before the projectile moved very far inside the gun.[441]

His results were published posthumously in 1832 by Siméon-Denis POISSON,[442] who annotated in the introduction that DE LAGRANGE apparently performed the study by orders of the French government.

DE LAGRANGE's approach to the problem – in the literature widely known as the **"Lagrange problem"** – was resumed by succeeding generations of interior ballisticians:

(i) The most important writings in which his problem is dealt with are HUGONIOT's famous memoir *Mémoire sur la propagation du mouvement dans les corps…* {HUGONIOT ⇨1887}, and HADAMARD's *Leçon sur la propagation des ondes* {HADAMARD ⇨1903}.

(ii) The end conditions of the Lagrange problem of interior ballistics are the initial conditions of the problem of emptying a gun. The motion of the powder gas tends to a simplified analytical expression which has been used as the basis for a system of interior ballistics.[443]

(iii) In 1948, Pierre DE HALLER,[444] chief engineer at the Swiss company Sulzer Frères SA in Winterthur, proposed a graphical solution of the Lagrange problem using the method of characteristics.

(iv) The Soviet gas dynamicist Kirill P. STANYUKOVICH[445] solved the Lagrange problem for the two cases where, over the entire volume of the gas, the energy is liberated both instantaneously and non-instantaneously.

| | | |
|---|---|---|
| 1794 | Paris, France | On September 28, **creation of the *Ecole Centrale des Travaux Publics*** ("Central School for Public Works") by General Lazare N.M. CARNOT and Gaspard MONGE and its establishment by the National Convention. ▪ In 1795, the school was renamed ***Ecole Polytechnique*** – its present name – and in 1804 it was transformed into a military school by Napoléon BONAPARTE who gave its motto "Pour la patrie, les sciences et la gloire." In 1976, the Ecole moved to Pa- |

---

[440] D.R. KENNEDY: *History of the shaped charge effect. The first 100 years*. Company brochure prepared by D.R. Kennedy & Associates, Inc., Mountain View, CA (1983).

[441] R.H. KENT: *Some hydrodynamical problems related to ballistics*. Am. Math. Monthly **48**, No. 1, 8-14 (Jan. 1941).

[442] S.D. POISSON: *Relatives au mouvement du boulet dans l'intérieur du canon, extraites des manuscrits de LAGRANGE*. J. Ecole Polytech. **13**, No. 21, 187-204 (1832).

[443] A.E.H. LOVE and F.B. PIDDUCK: *LAGRANGE's ballistic problem*. Phil. Trans. Roy. Soc. Lond. **A222**, 167-226 (1922).

[444] P. DE HALLER: *Une solution graphique du problème de Lagrange, en balistique intérieure*. Bulletin technique de la Suisse romande (Lausanne) **74**, No. 1, 1-5 (Jan. 3, 1948).

[445] K.P. STANYUKOVICH: *Unsteady motion of continuous media*. Gostekhizdat, Moscow (1955). Pergamon Press, London *etc.* (1960), pp. 587-605.

laiseau (a town 15 miles out of Paris).[446] A historical perspective on this famous institution was provided by Gaston PINET.[447]

Many prominent French pioneers of fluid mechanics, percussion, explosion, and shock wave physics studied and/or taught here, *e.g.*, D. François J. ARAGO (1776−1853), Jean B. BÉLANGER (1790−1874), Jean-Baptiste BIOT (1774−1862), Sadi CARNOT (1796−1832), Augustin L. CAUCHY (1789−1857), Gaspard G. DE CORIOLIS (1792−1843), Jean Marie C. DUHAMEL (1797−1872), Joseph-Louis GAY-LUSSAC (1778−1850), Jacques S. HADAMARD (1865−1963), Pierre-Henri HUGONIOT (1851−1887), Joseph L. DE LAGRANGE (1736−1813), Pierre-Simon DE LAPLACE (1749−1827), Henry LE CHÂTELIER (1850−1936), Gaspard MONGE (1746−1818), Arthur J. MORIN (1795−1880), C. Louis M.H. NAVIER (1785−1836), Siméon D. POISSON (1781−1840), Jean-Victor PONCELET (1788−1867), Claude S.M. POUILLET (1791−1868), Adhémar DE SAINT-VENANT (1797−1886), Henri Victor REGNAULT (1810− 1878), Emile SARRAU (1837−1904), Paul VIEILLE (1854−1934), and Pierre L. WANTZEL (1814−1848).

| 1794 | France | **A French commission consisting of chemists and artillerists makes more than 200 test shots in order to determine the optimum formula of gunpowder for use in firearms.** Performing numerous ballistic tests, they find that saltpeter/charcoal/sulfur mixtures on a percentage basis (by weight) of 76/14/10 and 76/15/9 give the largest firing ranges, but the large amount of hygroscopic charcoal required is prohibitive. ▪ After a few years, the recommended mixture was changed to 77/12.5/10.5.[448] In Germany, the optimum composition of gunpowder for use in the infantry rifle M/71.84 (introduced to the army in 1871) was determined to be 76/15/9.

Similar optimization studies on gunpowder were also carried out systematically in other European countries. They can be considered to be the first step towards scientifically uncovering the puzzling chemical reactions that take place in fired gunpowder {BUNSEN & SCHISCHKOFF ⇨1857; NOBLE & ABEL ⇨1875; DEBUS ⇨1882}. |
|---|---|---|
| 1794 | Wittenberg, Saxony-Anhalt, eastern Germany | Ernst F.F. CHLADNI,[449] a German physicist and an authority on acoustics, **starts his 30-year research campaign into meteorites, which he calls *Feuerkugeln* ("fireballs").** Based on observations by reliable witnesses and data on samples of meteorites, **he first concludes that meteorites do not originate on the Earth, but indeed fall from the sky.** ▪ CHLADNI's interest in meteorites was inspired the year before by a conversation with Georg LICHTENBERG, a German professor of physics at the University of Göttingen, who had witnessed a fireball and encouraged his assistant to study the puzzling phenomenon in more detail. CHLADNI's hypoth-esis that meteorites were extraterrestrial in origin, which he later supported with new results,[450] was not widely accepted until the L'Aigle Fall {⇨1803}, a shower of stony meteorites.

Today CHLADNI's term *fireball* is used to describe a streak of light produced by a meteoroid entering the Earth's atmosphere that is particularly brilliant and spectacular: according to the definition from the International Astronomical Union (IAU), brighter than magnitude −4 (about the same as Venus). **Exploding fireballs are called "bolides"** [Greek βολις, meaning a "missile" (thrown spear)]. When fireballs explode near the end of their flight, the meteoric debris from the fall can scatter over a wide area in hundreds of fragments. Freshly fallen fragments, called "meteorites," are of great scientific value. |

---

[446] *A brief history of the Ecole Polytechnique*; http://www.polytechnique.edu/page.php?MID=28.
[447] G. PINET: *Histoire de l'Ecole Polytechnique*. Baudry, Paris (1887).
[448] A. URE: *Über Schießpulver und Knallpulver*. Dingler's Polytech. J. **39**, 269-287 (1831).
[449] E.F.F. CHLADNI: *Über den Ursprung der von Pallas gefundenen und anderer ähnlicher Eisenmassen, und über einige damit in Verbindung stehende Naturerscheinungen*. J.F. Hartknoch, Riga (1794).
[450] E.F.F. CHLADNI: *Neue Beiträge zur Kenntnis der Feuermeteore und der herabfallenden Massen*. Gilbert Ann. **68**, 329-370 (1821); Ibid. **71**, 359-386 (1822); Ibid. **75**, 229-257 (1823).

| | | |
|---|---|---|
| 1794 | *Louvain [or Leuven], Flanders, central Belgium* | Jean Baptiste VON MONS,[451] a Belgian professor of chemistry and agronomy, reports in a letter to Friedrich A.C. GREN (editor of the German *Journal der Physik*) on unusual explosion and implosion studies: **in a dark room, Antoine DE PARCIEUX** [or DEPARCIEUX],[452] a French mathematician and physicist, **observed "a vivid flame similar to an electric spark"** [Germ. *eine lebhafte Flamme gleich einem electrischen Funken*] **during the instant that thin-walled glass spheres explode or implode.** VON MONS produced |

- an explosion, by placing a sealed glass sphere filled with air at atmospheric pressure in a receptacle, and then evacuating the receptacle; and
- an implosion upon evacuating a glass receptacle. Since it was not capable of withstanding the atmospheric pressure, it imploded during evacuation.

Similar experiments into imploding and exploding glass spheres were performed in Sweden by Carl G. VON HELVIG {⇒1815} and more recently in Canada at UTIA, Toronto {GLASS & HALL 1957; ⇒Fig. 4.16–F}.

| | | |
|---|---|---|
| 1795 | *London, England* | **Joseph BRAMAH,** a British inventor in London, **obtains a patent on his *hydraulic press*** – the so-called "Bramah press." His idea, based upon PASCAL's law {PASCAL ⇒1640s} and quickly taken up by the manufacturing industry {GILBERT ⇒1819}, later became the basic tool used in static high-pressure research {BRIDGMAN ⇒1904}, which, in turn, also stimulated dynamic high-pressure research into liquids and solids using shock waves in the 1940s. |
| 1796 | *Paris, France* | **Pierre Simon Marquis DE LAPLACE,**[453] a French mathematician, physicist and astronomer, **publishes his semipopular book** *Exposition du système du monde* **("The System of the World").** |

(i) Based upon a theory previously developed in Germany {KANT ⇒1756}, he speculates that a rotating gas cloud contracted, and the outer region of the cloud developed into a cool gaseous disk from which the planets condensed, while the inner region contracted to form the Sun. ▪ Based upon this so-called "Kant-Laplace nebular hypothesis" – which assumes a cooling and contraction of a gaseous nebula for the origin of the Solar System – Alastair G.W. CAMERON,[454] a noted Canadian-born U.S. astrophysicist and space scientist at Harvard-Smithsonian Center for Astrophysics, developed in the early 1960s the *theory of nucleosynthesis,* the production of the chemical elements in stars.

(ii) Similar to DE MAUPERTUIS {⇒1742}, DE LAPLACE suggests that an impact or even a glancing encounter with a comet might change the axis of rotation of the Earth and cause cataclysmic floods. He writes, "The seas would abandon their ancient positions, precipitating themselves towards a new equator; a great portion of the human race and the animals would be drowned in the universal deluge, or destroyed by the violent shock imparted to the terrestrial globe; entire species would be annihilated; all the monuments of human history destroyed. Such are the disasters which the collision of a comet could produce should its mass becomes comparable to that of the Earth … Our species would be reduced to a small number of individuals and to most miserable living conditions, solely concerned with staying alive, and thereby completely losing our memory of the sciences and arts (Book IV, Chap. 4)."

(iii) **Similar to MICHELL {⇒1783}, DE LAPLACE anticipates the concept of a *black hole.*** Agreeing with Sir Isaac NEWTON that light is made up of particles, he theorizes that if enough mass were added to a star, the gravitational force would become so great that its escape velocity (for the Earth 11 km/s, for the Sun 617 km/s) would equal the speed of light. Then, light

---

[451] Editor's note. J. der Physik **8**, 20-21 (1794).
[452] A. DE PARCIEUX: *Observations sur un nouveau phénomène de lumière.* Bull. Sci. Soc. Philomat. Paris **1**, 58 (1791).
[453] P.S. DE LAPLACE: *Exposition du système du monde.* Imprimerie de Cercle-Social, Paris (1796, 1800, 1808, 1813, 1827, 1835). Reprint of the 1835 edition (M. SERRES, ed.): *Pierre-Simon LAPLACE: exposition du système du monde.* Corpus des oeuvres de philosophie en langue française. Fayard, Paris (1984).
[454] A.G.W. CAMERON: *Interstellar communication.* W.A. Benjamin, New York (1963); D. TER HAAR and A.G.W. CAMERON: *Origin of the Solar System.* Academic Press, London and New York (1963), p. 1.

particles wouldn't be able to leave the star – it becomes a "black" star. DE LAPLACE writes, "A luminous star, of the same density as the Earth, and whose diameter should be two hundred and fifty times larger than that of the Sun, would not in consequence of its attraction, allow any of its rays to arrive at us; it is therefore possible that the largest luminous bodies in the Universe may, through this cause, be invisible (Book V, Chap. 6; first and second editions only)."
▪ When Albert EINSTEIN published his Special Theory of Relativity (1905), he argued that nothing can move faster than light. This means that DE LAPLACE's black stars are also black holes, because if light can't escape, all other matter would also be trapped. The concept of a black hole was expanded upon by the German astronomer Karl SCHWARZSCHILD[455] just months after EINSTEIN had published his Theory of Gravitation (1915). **The term *black hole* was coined by the U.S. theoretical physicist John A. WHEELER {⇨1968}.**

| 1796 | *Paris, France* | Thirteen years after his sensational hot-air balloon ascents, **Joseph-Michel MONTGOLFIER,**[456] a French manufacturer, **invents** – with the assistance of the Swiss technician Aimé ARGAND – **the *bélier hydraulique* ("hydraulic ram"),** a water pump that uses the kinetic energy from a copious flow of running water under a small head to force a small portion of that water to a higher level {⇨Fig. 4.2–M}. ▪ In Germany, this curious device is appropriately called "shock lifter" [Germ. *Stoßheber* or *hydraulischer Widder*].
     Hydraulic shocks are detrimental in common water pipelines {KARELJSKICH & ZHUKOVSKY ⇨1898}; however, when used in this type of pump they should be as strong as possible to provoke efficient pumping. The efficiency of this principle is quite low, because about 70% of the water that passes through the system exits via the escape valve. Since the hydraulic ram does not require any additional source of energy and is very simple in construction, it is still in current use in mountainous regions. For example, modern ram pumps can deliver 700 liters/min up to a height of 300 m.[457] |
|---|---|---|
| 1797 | *Artillery Arsenal, Munich, Bavaria, southern Germany & England* | Sir Benjamin THOMPSON, **Count VON RUMFORD** [Germ. *Reichsgraf von Rumford*],[458] an American-British physicist temporarily acting as a government administrator in Bavaria, **publishes the description of a device for proving gunpowder** {⇨Fig. 4.17–A}. Contrary to Benjamin ROBINS, he correctly supposes that "though the inflammation of gunpowder is very rapid, the progress of the combustion is by no means so instantaneous as has been imagined." He also describes a number of remarkable experiments on the discharge of firearms, showing that a considerable amount of unconsumed grains of gunpowder are blown out of them: not only out of their muzzles, but also out of their vents. Of course, these unconsumed grains cannot contribute to the acceleration of the projectile. ▪ His method of proving gunpowder, a simple device which balances the dynamic pressure with a static pressure (which was later resumed by BUNSEN {⇨Fig. 4.17–B} in his studies on gas explosions) was generally accepted as standard by both British and Bavarian armies. |
| 1797 | *University of Modena, northern Italy* | **Giovanni Battista VENTURI,**[459] an Italian physics professor working on hydraulics and acoustics, **studies the effects of constricted channels on water flow in order to reduce the turbulence and losses caused by such velocity changes.** Demonstrating his hydraulic experiments publicly in the physics auditorium of the university, he shows that the outflow of water from a |

---

[455] K. SCHWARZSCHILD: *Über das Gravitationsfeld eines Massenpunktes nach der EINSTEINschen Theorie.* Sitzungsber. Königl. Preuss. Akad. Wiss. **1**, 189-196 (1916).
[456] J.M. MONTGOLFIER: *Note sur le bélier hydraulique, et sur la manière d'en calculer les effets.* J. des Mines **XIII**, 42-51 (1802/1803).
[457] *Weinmann Sondermaschinen- und Steuerungsbau GmbH,* D–91217 Hersbruck, Germany.
[458] B. THOMPSON: *Experiments to determine the force of fired gunpowder.* Phil. Trans. Roy. Soc. Lond. **87**, 222-292 (1797).
[459] G.B. VENTURI: *Mémoire sur la communication latérale du mouvement dans les fluides, appliqué à l'explication de différents phénomènes hydraulique.* Bull. Sci. Soc. Philomat. Paris **II**, 60-61 (1797); *Untersuchungen und Erfahrungen über die Seitenmittheilung der Bewegung in flüssigen Körpern, angewandt auf die Erklärung verschiedener hydraulischer Erscheinungen.* Ann. Phys. **2** [I], 418-465 (1799); **3** [I], 35-47, 129-166 (1800).

vessel can be accelerated by using a conically divergent rather than a straight tube. He explains this effect by the "lateral communication" of the fluid particles produced by the atmospheric pressure [French *principe de communication latérale dans les fluides*]. Using a convergent-divergent tube, he also shows that the water passing through the tube speeds up as it enters the throat, and the pressure drops {⇨Fig. 4.7−D}. ▪ VENTURI's invention, the so-called "Venturi tube" − a convergent-divergent nozzle, and in a sense the forerunner of the Laval nozzle {DE LAVAL ⇨1888} − has been used for many purposes, *e.g.*, for flow-rate measurement (*Venturi meter*) of both gases and liquids using the pressure differential based on Daniel BERNOULLI's principle {D. BERNOULLI ⇨1738}, in early aerodynamic drag studies {PHILLIPS ⇨1885; ⇨Fig. 4.7−E}, in blasting nozzles,[460] and in automobile carburetors.

| 1798 | *Artillery Arsenal, Munich, Bavaria, southern Germany* | Count VON RUMFORD {⇨1797},[461] an American-born British physicist and government administrator, **stimulated by his interest in gunpowder and weaponry begins his studies on the nature of heat and friction.** He writes, "Being engaged, lately, in superintending the boring of cannon, in the workshops of the military arsenal at Munich, I was struck with the very considerable degree of heat which a brass gun acquires, in a short time, in being bored; and with the still more intense heat (much greater than that of boiling water, as I found by experiment) of the metallic chips separated from it by the borer … What is heat? Is there any such thing as an *igneous fluid?* Is there anything that can with propriety be called *caloric?* We have seen that a very considerable quantity of heat may be excited in the friction of two metallic surfaces and given off in a constant stream or flux, *in all directions,* without iteration or intermission, and without any signs of diminution or exhaustion … It is hardly necessary to add that anything which any *insulated* body, or system of bodies, can continue to furnish *without limitation* cannot possibly be a *material substance*, and it appears to me to be extremely difficult, if not quite impossible, to form any distinct idea of anything capable of being excited and communicated in the manner **the heat was excited and communicated in these, except if it be MOTION.**" ▪ At the time, it was considered that the heat fluid, *caloric*, was squeezed out of the metal during boring like water out of a sponge {Comte BERTHOLLET ⇨1809}, but RUMFORD's experiment seemed to show that it could be created in unlimited amounts, raising serious doubts as to its material nature. Count VON RUMFORD's experiments revealed evidence for the concept of heat as a movement of particles {HOOKE ⇨1674; D. BERNOULLI ⇨1738}. He also made one of the earliest measurements of the equivalence of heat and mechanical energy. |
|---|---|---|
| 1798 | *Chair of Natural History, University of Pavia, Lombardy, northern Italy* | Lazzaro SPALLANZANI,[462] an Italian abbot and distinguished scientist, **addresses various earthquake and volcanic phenomena in his book *Viaggi alle due Sicilie…*** ("Travels in the Two Sicilies…"):<br><br>(i) **Giving an account of the 1783 Calabrian Earthquake at Messina, he observed a *progressive motion* of the earthquake wave or shock,** particularly that each portion of the country disturbed is shaken *in succession* by the progress of the earthquake wave, not the whole country at once, as very commonly believed, "A very violent noise, resembling that of a number of carriages rattling over a stone-bridge, was the first symptom, while at the same time a thick cloud arose from Calabra, which was the center of the earthquake, the propagation of which was *successively* apparent by the fall of buildings from the point of the Faro, to the city of Messina, as if at that point a mine had been fired which extended along the shore and con- |

---

[460] G.S. SETTLES and S.T. GEPPERT: *Redesigning blasting nozzles to improve productivity.* J. Protect. Coat. Linings **13**, No. 10, 64-72 (1996).
[461] B. THOMPSON: *Untersuchung über den Ursprung der durch Friction bewirkten Wärme.* Allgem. J. Chem. **1**, 9-37 (1798); *(IV) An inquiry concerning the source of the heat which is excited by friction.* Phil. Trans. Roy. Soc. Lond. **88**, 80-102 (1798).
[462] L. SPALLANZANI: *Viaggi alle due Sicilie e in alcune parti dell'Appennino.* Comini, Pavia (1792−1797); Engl. translation: *Travels in the two Sicilies, and some parts of the Appennines.* 4 vols., Robinson, London (1798). *See* vol. III, pp. 356-357, and vol. IV, pp. 156-159.

tinued into the city. The shock was most violent, and the motion irregular." ▪ Independently of SPALLANZANI, a similar observation of "successive propagation" had already been made during the Boston Earthquake {WINTHROP ⇨ 1755}.

(ii) He recognizes that several gases are important in lavas and volcanic regions, including "hydrogenous gas, sulfurated hydrogenous gas, carbonic acid gas, sulfurous acid gas, azotic gas." **But another, more powerful agent, he notices, is "water, principally that of the sea," which communicates by passages under volcanoes. Upon reaching the subterranean fires, the water suddenly turns to vapor, and the elastic gas expands rapidly, causing explosive volcanic eruptions.**

(iii) Supporting his hypothesis on a more practical level, SPALLANZANI cites accidents in glass-making factories, in which molten glass was poured into molds that were not completely dry, causing dreadful steam explosions (*see* Memoirs of the Academy of Bologna, vol. 4).

| 1799 | *Museum of the Teyler Foundation, Haarlem, The Netherlands* | **Martin VAN MARUM,**[463] a Dutch physician-scientist, **first reports on exploding-wire experiments which he began in the early 1790s.**

(i) He constructed a large battery of 100 Leyden jars which, connected in parallel, are charged up to a high voltage using a Teyler-type friction electrical machine. Voltage monitoring occurs via a Brook-type electrometer.

(ii) In order to test the potentials of his battery he discharged it into a 100-ft (30.48-m)-long iron wire of $1/240$ in. (0,105 mm) in dia. He observes a total disintegration of the wire into small molten glowing droplets (a phenomenon which requires an energy in the order of 500 to 1,000 J). **Birth of exploding wire research!**

(iii) He also observes that after wire explosion the battery is not completely discharged: the residual charge suffices to disintegrate a 6-ft-long $1/240$-in.-dia. iron wire in the same manner. |

| 1800 | *Somerset House, The Royal Society of London, England* | In a paper presented at a meeting of the Royal Society, **Edward Charles HOWARD,**[464] a notable English chemist, **first describes the preparation and properties of mercury fulminate** [$Hg(ONC)_2$]. The new substance, resulting from a reaction of mercury with nitric acide and alcohol, is a white powder which is extremely sensitive to friction and percussion and violently explodes upon impact. ▪ The exact date of his discovery is unknown.

Mercury fulminate – in Germany appropriately called "detonating mercury" [Germ. *Knallquecksilber*] – is a high explosive with a large brisance (the measure of how rapid an explosive develops its maximum pressure) and has a heat of detonation of 104.8 Cal. Depending on the density $\rho$, its detonation velocity ranges from 2,760 m/s (for $\rho = 1.66$ g/cm$^3$) to 5,400 m/s (for $\rho = 4.2$ g/cm$^3$).[465] Used to ignite gunpowder {FORSYTH ⇨ 1805; SHAW ⇨ 1814}, it was also incorporated by Alfred NOBEL in his invention of the blasting cap {NOBEL ⇨ 1865}. However, since the middle of the 20th century, it has been largely replaced by the high explosives PETN {TOLLENS ⇨ 1871} and RDX {HENNING ⇨ 1899}.

The unusual explosive properties of fulminates {CROLL ⇨ 1606; BERTHELOT & VIEILLE ⇨ 1880}, the first man-made high explosives, were frequently shown in public demonstrations and private circles. In July 1798, when Napoléon BONAPARTE arrived with a 34,000-man army in Egypt, he was accompanied by nearly 1,000 civilians; as well as administrators, these were mostly scientists, artists, engineers and handcrafters, some of them who gave public demonstrations of their abilities. For example, Sheikh Abdarrahman AL-ĞABARTĪ,[466] a chronicler of the French occupation in Cairo (1798–1801), gave an account of a French demonstration of an ex- |

---

[463] M. VAN MARUM: *Beschreibung einer großen Batterie von 550 Quadratfuß Belegung und einiger damit angestellter Versuche*. Ann. Phys. **1** [I], 68-87 (1799).

[464] E.C. HOWARD: *On a new fulminating mercury*. Phil. Trans. Roy. Soc. Lond. **90**, 204-238 (1800).

[465] H. KAST and A. HAID: *Über die sprengtechnischen Eigenschaften der wichtigsten Initialsprengstoffe*. Z. ges. Schieß- u. Sprengstoffwesen **20**, No. 11, 44-45 (1925).

[466] A. AL-ĞABARTĪ: *BONAPARTE in Ägypten*. Artemis, Zürich & München (1983), p. 168.

plosive white powder – most probably mercury fulminate. He wrote, "In the house of Hasan KĀŠIF ČERKES they [the Frenchmen] installed a locality serving for medical sciences and chemical pharmacy ... Once he [one of the Frenchmen] took a small quantity like a white dust particle [BERTHOLLET ⇨ 1786/1787] and put it on an anvil; he hit it slightly with a hammer, and a horrible report resounded like from a rifle. We were startled by it, and they laughed at us ... They have there objects, conditions and apparatus of strange kinds incomprehensible with our intelligence."

1801 — *Palermo Observatory, Sicily, southern Italy*

On the night of January 1, **Giuseppe PIAZZI,** an Italian astronomer, **observes a faint 7th magnitude star where none had been a few days before.** Initially terming it a *star-like comet*, without nebulosity, and naming it "Ceres" (after the patron goddess of Sicily), he soon realizes that it must be a new *minor planet* (also termed a *planetoid* or *asteroid*). It is now designated 1 CERES and with a diameter of 940 km is the largest known asteroid. ▪ The thousandth minor planet to be discovered was named 1000 PIAZZI in his honor.

▸ **The term *asteroid* was coined in 1802 by the famous German-born British astronomer Sir Frederick William HERSCHEL,**[467] who wrote, "From this, their asteroidical appearance, if I may use that expression ... I shall take my name and call them 'Asteroids.'" The asteroid 2000 HERSCHEL is named for him. ▪ Although the term *asteroid* (meaning "star-like") is a misnomer because these objects are not stars, it became widely use.

▸ In the following years, further asteroids were discovered: 2 PALLAS (580 km) in 1802 and 4 VESTA (578 km) in 1807, both by Heinrich W.M. OLBERS, a German physician and hobby astronomer, and 3 JUNO (240 km) in 1804 by Karl HARDING, a German astronomer. All these first four planetoids had similar orbits.

▸ In 1804, OLBERS proposed in a letter to Sir F. William HERSCHEL the idea that a larger planet which had formerly orbited at the same distance had been broken into fragments: this was the so-called **"planetary explosion theory"** of the origin of asteroids. He wrote: "Could it be that CERES and PALLAS are just a pair of fragments ... of a once great planet which at one time occupied its proper place between Mars and Jupiter?"

▸ In 1814, the French astronomer Joseph L. DE LAGRANGE extended OLBERS' theory to explain comets, pointing out that their extremely elongated orbits would also be a natural by-product of an explosion.

▸ In 1972, a Canadian astronomer proposed a missing planet theory. The planet, which is missing from the gap between Mars and Jupiter where the main belt of asteroids is found, was perhaps Saturn-sized {OVENDEN ⇨ 1972}.

▸ Although some important arguments support the exploded planet hypothesis {VAN FLANDERN ⇨ 1993}, it is now generally believed that no complete planet ever formed in this region of the Solar System due to the gravitational influence of Jupiter.

The subject of asteroids is receiving increasing attention, but the problem of how they were formed is far from being settled. Most of the asteroids travel around the Sun in elliptical paths, but some have highly eccentric orbits. A collision of a sufficiently large asteroid with the Earth would result in a global catastrophe {⇨c.65 Ma ago; PROCTOR ⇨1873; Stony Tunguska ⇨1908; ALVAREZ ⇨1978; NEAR Shoemaker ⇨1996; Asteroid 2002 MN ⇨2002}.

1802 — *Collège de France, Paris, France*

**Jean-Baptiste BIOT,**[468] a French professor of mathematical physics, **publishes the first (adiabatic) theory of sound** and acknowledges the assistance of Pierre-Simon DE LAPLACE. He advances physical arguments in favor of

$$p = K\rho^{\gamma},$$

---

[467] W. HERSCHEL: *Observations on the two lately discovered celestial bodies.* Phil. Trans. Roy. Soc. Lond. **92**, 213-232 (1802), *see* p. 228.

[468] J.B. BIOT: *Théorie mathématique de la propagation du son.* J. Phys. Théor. Appl. **55** [I], 173-182 (1802).

later to be called the "Laplace-Poisson law" or "adiabatic (gas) law" {DE LAPLACE ⇒1816; POISSON ⇒1823} – which results for the sound velocity a at sea level

$$a = (\gamma p_0/\rho_0)^{1/2},$$

where $\gamma = c_p/c_v$ is the ratio of specific heats at constant pressure and constant volume, and $p_0$ and $\rho_0$ are the pressure and density at the initial state "0," respectively. • Compared to the isothermal theory of sound {Sir NEWTON ⇒1687}, BIOT's theory was a tremendous achievement.

Substituting the ratio $p_0/\rho_0$ by using the equation of state for an ideal gas ($p_0/\rho_0 = RT/M$), the sound velocity $a$ can be expressed by

$$a = (\gamma p_0/\rho_0)^{1/2} = (\gamma RT/M)^{1/2},$$

where $R$ is the universal gas constant, $M$ is the molar mass, and $T$ is the absolute temperature of the gas at state "0." At terrestrial temperatures and pressures, the speed of sound tends to be a fraction of a kilometer per second. However, **in space the temperatures can be very high and the density very low. Thus, the interstellar speed of sound is typically on the order of 10 km/s.**

| | | |
|---|---|---|
| 1802 | *Private observatory at Lilienthal, Duchy of Bremen, northern Germany* | **Johann Hieronymus SCHRÖTER,**[469] a German astronomer, pondering on the crater-forming processes of lunar craters [Germ. *Ringcrater*], **postulates that the rim materials of lunar craters would just fill the pit,** "For each crater, the part of the material above the surface is approximately equal to the volume of the interior depression below the surface" – the so-called **"Schröter rule."** • A large number of craters do fulfil the requirements of his rule, which strongly supported the hypothesis that the displaced material contained in the raised rims is actually shattered rock resulting from an explosion not far beneath the surface. The origin of such an explosion was later explained by the high-velocity impact of a meteorite {GIFFORD ⇒1924; BALDWIN ⇒1949}. |
| 1803 | *L'Aigle, Dépt. Orne, Normandie, France* | On April 26, **an unusual meteorite fall is eye-witnessed by a number of persons: the appearance of a fireball in broad daylight which explodes with a loud sound,** scattering thousands of stony fragments over the surface of a territory some miles in extent. The young French physicist Jean-Baptiste BIOT is asked by Jean-Antoine DE CHAPTAL, a professor of chemistry and then French Minister of the Interior, to investigate whether such a form of lightning as that reported – appearing in a clear sky and literally hurling thunderbolts – really existed.<br><br>(i) After having questioned the people in the locality where the meteorite had fallen, BIOT[470] **gives his famous account of the so-called "L'Aigle Fall:"** "On Tuesday, April 26, 1802, about one in the afternoon, the weather being serene, there was observed from Caen, Pont-Audenen and the environs of Alençon, Falaise and Verneuil, a fiery globe of a very brilliant splendor, which moved in the atmosphere with great rapidity. Some moments after, there was heard at L'Aigle, and in the environs of that city to the extent of more than thirty leagues [145 km] in every direction, a violent explosion, which lasted five or six minutes. At first there were three or four reports like those of a cannon, followed by a kind of discharge which resembled a firing of musketry; after which there was heard a dreadful rumbling like the beating of a drum, a multitude of mineral masses were seen to fall…"<br><br>(ii) From the number of shock waves or explosions felt by the observer, BIOT correctly concluded that the meteorite must have broken into a large number of pieces upon striking the dense atmosphere at low levels. • The L'Aigle meteorite – a stony meteorite classified as L6 – contains $SiO_2$ (39.20%), MgO (24.31%), FeO (13.46%), FeS (6.58%), Al2O3 (2.19%), CaO |

---

[469] J.H. SCHRÖTER: *Selenotopographische Fragmente*. 2 vols, Lilienthal & Helmst, Göttingen (1791–1802).

[470] J.B. BIOT: *Account of a fire-ball which fell in the neighborhood of L'Aigle: In a letter to the French minister of the Interior.* Phil. Mag. **16** [I], 224-228 (1803).

(1.82%), Na2O (0.96%), Ni (1.29%), *etc.*[471] Its chemical composition is very similar to the Ensisheim meteorite {⇨1492}.

In the following years, about 3,000 meteorite fragments were collected, ranging in mass from 9 g to 8 kg.[472] Although CHLADNI {⇨1794} had already proposed an extraterrestrial origin for meteorites, his hypothesis was questioned by most members of the French Academy. BIOT's objective report, however, marked the beginning of a general recognition of the reality of meteorites in France.

| 1804 | Prague, Bohemia, Austro-Hungarian Empire | **Franz Joseph Ritter VON GERSTNER,**[473] an Austrian professor of mathematics and mechanics at the Institute of Engineering Education, **gives the first exact nonlinear solution for waves of finite amplitude on deep water.** ▪ The elevation profile of the so-called "Gerstner wave" takes the form of a trochoidal curve. GERSTNER's wave theory, rediscovered by William J.M. RANKINE,[474] who also revived interest in it, has been used in many engineering applications. |

| 1804 | Académie de Lyon, Dépt. Rhône, east central France | **Joseph MOLLET,**[475] a professor of natural philosophy and a member of the Lyon Academy, **reinvents the pneumatic lighter** [French *briquet pneumatique*].[476] ▪ The so-called "Mollet pump" can sometimes be found in technical and ethnological museums. The device is often used to demonstrate the effect of adiabatic gas compression {BIOT ⇨1802; DE LAPLACE ⇨1816; POISSON ⇨1823; RANKINE ⇨1859} in physics lectures.[477] The adaptation of air compression for fire-making was almost simultaneously discovered by William NICHOLSON in England, and a device was patented in 1807 by Richard LORENTZ.[478] The principle of adiabatic compression used in the pneumatic lighter inspired Rudolf DIESEL {⇨1893} to invent his self-igniting "diesel engine." |

The origin of the pneumatic lighter or "fire-pump" {⇨Fig. 4.2–K} is unknown, but in several Asian countries – such as China, Indonesia, the Philippines and Burma – they have apparently been used for a long time.[479] In the mid-19th century, Frederick BOYLE,[480] a British world traveler and ethnologist, mentioned that he had actually found pneumatic tinder-boxes made of bamboo in use among the Dyaks, a tribe in Borneo notorious for head-hunting and their dreadful blowguns.[481] He saw a Dyak place some tinder on a broken piece of earthenware, holding it steady with his thumb while he struck it a sharp blow with a piece of bamboo. The tinder caught fire. In Europe, however, the pneumatic lighter was never widely used in households and was quickly superseded by the invention of the friction match in the 1820s.

---

[471] E. JAROSEWICH: *Chemical analyses of meteorites: a compilation of stony and iron meteorites analyses*. Meteoritics **25**, 323-337 (1990).

[472] *Brockhaus' Konversationslexikon*. Brockhaus, Leipzig (1908), vol. 11, p. 815.

[473] F.J. VON GERSTNER: *Theorie der Wellen*. Abhandl. Königl. Böhm. Gesell. Wiss. (Prag) **1** [III], 176-195 (1804); *Theorie der Wellen samt der daraus abgeleiteten Deichprofile*. Ann. Phys. **32** [I], 412-440 (1809).

[474] W.J.M. RANKINE: *On the exact form of waves near the surface of deep water*. Phil. Trans. Roy. Soc. Lond. **153**, 127-138 (1863).

[475] J. MOLLET: *Mémoire sur deux faits nouveaux: l'inflammation des matières combustibles et l'apparition d'une vive lumière, obtenues par la seule compression de l'air*. [lu dans la séance publique de l'Académie de Lyon, mars 27, 1804]. Ballanche, Lyon (1811).

[476] R. FOX: *The fire piston and its origin in Europe*. Technol. Culture **10**, 355-370 (1969).

[477] For an ideal gas, originally at volume $V_1$ and temperature $T_1 = 293$ K and then adiabatically compressed to a volume $V_2$, the temperature $T_2$ is given by $T_2 = T_1 (V_1/V_2)^{\kappa-1}$. For example, for air ($\kappa = 1.4$) and compression to 10% of the original volume, the temperature of the compressed gas increases to $T_2 = 736$ K.

[478] M.D. LAW (ed.): *Chambers's encyclopaedia*. International Learning Systems Corporation Limited, London (1973), vol. 5, p. 642. ▪ The Meyer encyclopedia in Germany and the Larousse encyclopedia in France each credit the Frenchman DU MONTIER and the French Colonel (Jacques François ?) GROBERT (1806) as the inventor of the pneumatic lighter, respectively.

[479] H. BALFOUR: *The "fire-piston."* In: (H. BALFOUR ET AL., eds.): *Anthropological essays presented to Edward Burnett TYLOR in honor of his 75th birthday Oct. 2, 1907*. Clarendon Press, Oxford (1907).

[480] F. BOYLE: *Adventures among the Dyaks of Borneo*. Hurst & Blackett, London (1865).

[481] Blowguns, used primarily against small game but also in warfare, have a very long history, but their origin is unknown. Blowguns were used in tropical South America, Southeast Asia, Indonesia and the Philippines, as well as by certain North American indians and even in Europe in the Middle Ages to shoot incendiary arrows. Blowguns are still used today for hunting by Amazonian Indians, who use poison-dipped darts {⇨Fig. 4.2–I}, and by Pygmies in Africa. Modern veterinarians also use the blowgun with a tranquilizer dart to narcotize wild animals before medical treatment.

Blowguns, used primarily against small game and in warfare as well, have a very long history, but their origin is unknown. Blowguns were not only known in tropical South America, Southeast Asia, Indonesia and the Philippines, but also used by certain North American indians and in the Middle Ages even in Europe for shooting incendiary arrows. Blowguns are still used today for hunting by Amazonian Indians using poison-dipped darts {⇨Fig. 4.2−I} and by Pygmies in Africa. Modern veterinarians also use the blowgun with tranquilizer darts to narcotize wild animals before medical treatment.

| | | |
|---|---|---|
| 1804 | Private observatory, Bremen, northern Germany | **Heinich W.M. OLBERS,** a German physician and hobby astronomer, **proposes in a letter to Sir Frederick William HERSCHEL,** a German-born British astronomer, **the idea that a larger planet which had formerly circulated at the same distance as the recently discovered two minor planets CERES and PALLAS** [PIAZZI ⇨1801] **had been broken into fragments.** He writes, "Could it be that CERES and PALLAS are just a pair of fragments … of a once great planet which at one time occupied its proper place between Mars and Jupiter?" ▪ OLBERS' hypothesis on the origin of asteroids – the so-called **"planetary explosion theory"** – was resumed by others {DE LAGRANGE ⇨1814; OVENDEN ⇨1972; VAN FLANDERN ⇨1993}. |
| 1805 | Somerset House, The Royal Society of London, England | On November 14 at a meeting of the Society, the English chemist and physicist **William Hyde WOLLASTON,**[482] an associate of Humphry DAVY at the Royal Institute who will become famous in the optics community for inventing a polarizing beam splitter (so-called "Wollaston prism"), **addresses the principle of *vis viva* – an early formulation of the conservation of energy.**

(i) Introductorily he states, "When different bodies move with the same velocity, it is universally agreed that the forces, which they can exert against any obstacle opposed to them, are in proportion to the quantities of matter contained in the bodies respectively. But, when equal bodies move with unequal velocities, the estimates of their forces has been a subject of dispute between different classes of philosophers. LEIBNIZ [⇨1686] **and his followers have maintained that the forces of bodies are as the masses multiplied into the *squares* of their velocities, (a multiple to which I shall for conciseness give the name of *impetus*);** while those, who are considered as 'Newtonians,' conceive that the forces are in the *simple ratio* of the velocities, and consequently as the *momentum* or *quantitas motus*, a name given by NEWTON to the multiple of the velocity of a body simply taken into its quantity of matter.

(ii) The objective of his Lecture is "to consider which of these opinions respecting the force exerted by moving bodies is most conformable to the usual meaning of that word, and to shew that the explanation given by NEWTON of the Third Law of Motion [Sir NEWTON ⇨1687] is in no respect favorable to those who in their view of this questions have been called 'Newtonians.'"

(iii) Defending LEIBNIZ's view and giving numerous examples, he concludes: "… whether we are considering sources of extended exertion or of accumulated energy, whether we compare the accumulated forces themselves by their gradual or by their sudden effects, the idea of mechanic force in practice is always the same, and is proportional to the space through which any moving force is exerted or overcome, or to the square of the velocity of a body in which such force is accumulated" – thus closely following D'ALEMBERT's view {⇨1743}. |
| 1805 | Belhelvie, Aberdeenshire, Scotland, U.K. | **Alexander John FORSYTH,** a Scottish clergyman fond of game shooting, **discovers the principle of percussion ignition,** for which he will obtain a patent in 1807. Realizing the major problem with the flint-lock gun – in particular, its unreliability in damp conditions – he employs in his "percussion lock" a small charge of mercury fulminate {HOWARD ⇨1800} for ignition which, encapsulated for protection from the elements in paper, foil or thin metal, deto- |

---

[482] W.H. WOLLASTON: *On the force of percussion* [Bakerian Lecture]. Phil. Trans. Roy. Soc. **96**, 13-22 (1806).

nates upon being struck by the hammer of the gun. ▪ **His invention, superseding the flint lock, completely revolutionized firearms development and, consequently, the science of war.** Today FORSYTH's own models of the percussion lock which he invented after years of experiments in the Tower of London are now on display in the Tower Museum.

In the following year, FORSYTH was called upon by the government to continue the development of his system in the Tower of London. In 1807, he patented an ignition mechanism which relied on a "detonating powder" consisting mainly of potassium chlorate {Comte BERTHOLLET ⇨ 1788}. ▪ After the invention of the percussion cap {SHAW ⇨ 1814}, mercury fulminate was widely used to ignite gunpowder. For example, in 1835 France alone produced 800 million pieces.[483] The mass fabrication of percussion caps, however, resulted in previously unknown explosive hazards. Compared to the operation of powder mills, which was prohibited by law in densely populated areas, the production of percussion caps was not initially restricted, which caused numerous serious accidents.[484]

1807  Paris, France  **Joseph Nicéphore NIÉPCE and his brother Claude NIÉPCE,** two French inventors, **obtain from Napoléon BONAPARTE a ten-year patent for their "pyréolophore," the world's first explosion-driven piston engine.** This ancestor to the modern internal combustion engine is supposed to be driven by lycopodium dust explosions.[485] ▪ The combustion of lycopodium dust is a useful experiment to demonstrate to students the relationship between the rate of a reaction and the surface area of the reactants: the exposed surface of a pile of lycopodium is relatively small and the rate so slow as to be non-existent; however, when the powder is blown into the air, the surface area is huge and the combustion reaction rate is so fast that it becomes explosive.

In the previous year (on December 15, 1806), General Lazare CARNOT,[486] a statesman and military engineer, and Comte Claude L. BERTHOLLET, a notable chemist, presented a report to the Paris Institute stating that: "**The fuel ordinarily used by M.M. NIÉPCE is made of lycopodium spores, the combustion of which is the most intense and the easiest; however, since this material is costly, they replaced it with pulverized coal and mixed it if necessary with a small portion of resin, which works very well, as was proved by many experiments.** In M.M. NIÉPCE's machine, no portion of heat is dispersed in advance; the moving force is an instantaneous result, and all of the effect of the fuel is used to produce the dilatation that causes the moving force. In another experiment, a machine installed on a boat with a prow about two feet wide by three feet high, reduced in the underwater part and weighing about 2,000 pounds, went up the Saône [a river in eastern France] on just engine power, with a speed greater than the river's in the opposite direction; the amount of fuel burnt was around one hundred and twenty-five grains per minute [8.1 g/min], and the number of pulsations was twelve to thirteen in the same amount of time [12–13 rpm]. The Commissioners then conclude that the machine proposed under the name *Pyréolophore* by M.M. NIÉPCE is ingenious, that it could become very interesting due to its physical and economical results, and it deserves the approbation of the Commission." ▪ In 1813, the Niépce Brothers turned to the problem of photography {NIÉPCE ⇨ 1826}.

In the same year, **J.N. NIÉPCE invents a sort of hydraulic ram** {MONTGOLFIER ⇨ 1796}, devised in response to a government competition to replace the apparatus used to supply Versailles with water from the Seine.[487]

---

[483] I.S. ERSCH and J.G. GRUBER (eds.): *Allgemeine Encyklopädie der Wissenschaften und Künste*. Brockhaus, Leipzig (1885); 2. Sect. (H-N), 37. Teil, p. 265.
[484] (Ed.): *Schreckliche Explosion des Knallpulvers*. Dingler's Polytech. J. **10**, 513 (1823).
[485] F.L. NEHER: *Die Erfindung der Photographie*. Kosmos, Stuttgart (1938), p. 36.
[486] L.N.M. CARNOT and C.L. BERTHOLLET: *Rapport sur une nouvelle machine inventée par MM. NIEPCE et nommée par eux: "Pyréolophore."* Mém. de l'Institut (Paris 1807), pp. 146-153. ▪ For an Engl. translation of their report *see The pyreolophore*; http://www.nicephore-niepce.com/pagus/pireus1.html.
[487] R. SOULARD: *Joseph N. NIÉPCE*. In: (C.C. GILLESPIE, ed.) *Dictionary of scientific biography*. Ch. Scribner's Sons, New York, vol. **10** (1974), pp. 118-119.

| 1808 | Ecole Polytechnique, Paris, France | **Siméon-Denis POISSON**,[488] a young French mathematician and physicist and previous student of Pierre-Simon DE LAPLACE and Joseph L. DE LAGRANGE, **presents his theory of sound for perfect fluids** {⇨Fig. 2.17}. At the end of his treatise, **he extends his theory to the special case that "the velocities of the molecules in an air column are not supposed to be very small."** |
|---|---|---|

(i) He assumes that disturbances with a finite amplitude propagate in an ideal gas in one (positive) direction $x$, and applies the law
$$a^2 = dp_0/d\rho_0,$$
with $a$ being the sound velocity; *i.e.*, the limit to which the velocity of propagation of the wave approximates when the particle velocity becomes indefinitely small.

(ii) He arrives at the following general equation:
$$d\varphi/dt + a\,d\varphi/dx + \tfrac{1}{2}d\varphi^2/dx^2 = 0.$$
Here $\varphi$ is the velocity function, and $d\varphi/dx$ is the velocity of disturbance (or particle velocity) at time $t$ of a particle whose distance from the origin is $x$.

(iii) His exact solution for a wave traveling in one (positive) direction reveals that the particle velocity $d\varphi/dx$ behind a pressure disturbance can be expressed as
$$d\varphi/dx = F[x - a\,t - d\varphi/dx\, t],$$
where $F$ denotes an arbitrary function. His solution differs from the equation given previously by D'ALEMBERT {⇨1747} only in that $d\varphi/dx$ also appears in the argument of the function $F$. POISSON notes that DE LAGRANGE[489] had already obtained a very similar result; *i.e.*,
$$d\varphi/dx = f[x - a\,t - t \times f(x - a\,t)].$$

(iv) POISSON himself does not pursue the implications of his calculation, which obviously indicate that parts of the wave where the particle velocity is large must propagate more rapidly than parts where the disturbances and therefore the particle velocity are small. • This must lead to increasing distortion of the wave profile as it travels through the medium – a very important feature that was first recognized by James CHALLIS {⇨1848}, and qualitatively worked out and illustrated by George G. STOKES {⇨1848}.

Retrospectively, **POISSON's treatise may be regarded as the foundation for shock wave theory as well as theoretical nonlinear acoustics.** He also coined the symbol $\gamma$ for the ratio of the specific heats, which, later resumed by William J.M. RANKINE {⇨1869}, is used even to this day.

| 1809 | Private residence and chemical laboratory at Arcueil [a suburb south of Paris], France | **Comte Claude-Louis BERTHOLLET**,[490] an eminent French chemistry professor who had accompanied Napoléon on an expedition to Egypt, **discusses the heating effects of percussion.**[491] |
|---|---|---|

(i) Like most contemporary calorists, he associates the rise in temperature with a decrease in the volume of the solid body being struck, a decrease that would cause *caloric* (an imponderable "heat fluid") to be expelled like water from a sponge and so present an obvious analogy to the "adiabatic" heating of a gas.

(ii) In support of this explanation, he describes experiments performed around 1803 with Jean-Baptiste BIOT and Marc-Auguste PICTET. They showed that heat could no longer be produced once the hammering produced no further decrease in volume. By measuring the density of his samples of various metals before and after they had been struck powerfully in a coining

---

[488] S.D. POISSON: *Mémoire sur la théorie du son*. J. Ecole Polytech. (Paris) **7**, 319-392 (1808).
[489] J.L. DE LAGRANGE: *Sur une nouvelle méthode de calcul intégral pour les différentielles affectées d'un radical carré, sous lequel la variable ne passe pas le 4e degré*. Mem. Acad. Roy. Sci. Turin **II**, 218-290 (1784/1785).
[490] C.L. BERTHOLLET: *Sur la chaleur produite par le choc et la compression*. Mém. Phys. Chim. Soc. Arcueil **2**, 441-448 (1809).
[491] R. FOX: *S. CARNOT: reflexions on the motive power of fire – a critical edition with the surviving scientific manuscripts*. Manchester University Press, Manchester (1986).

press at the Mint, BERTHOLLET is able to give some indication of the relationship between the compression and the consequent rise in temperature.

**In the 1820s, the famous French army engineer and physicist Sadi CARNOT,** son of General Lazaré N.M. CARNOT {⇨1783}, **resumed the matter:**

(I) In regard to BERTHOLLET's hypothesis, S. CARNOT[492] arrived at a different explanation: "Hitherto there have been very few studies of the changes in the temperature of bodies resulting from the effects of motion. But this type of phenomenon deserves the attention of experimenters. When bodies are in motion, especially when motive power is consumed or produced, marked changes occur in the distribution, and perhaps also in the quantity, of heat. I shall summarize a few observations which display the phenomenon most clearly … The impact between bodies. We know that whenever bodies strike one another, motive power is consumed. The only exception is for perfectly elastic bodies, and these do not exist in nature. We also know that when bodies strike one another, there is a change, a rise, in temperature. It should be hard to account for the release of heat in this case, as M. BERTHOLLET has done, by a reduction in the volume of the body. For when the reduction in volume has been taken to its limit, no more heat should be released. This does not happen. **For heat to be released, it is sufficient that the body should be capable of being deformed by percussion.**"

(II) S. CARNOT[493] speculated that "heat is nothing but motive power, or rather motion, which has changed its form. It is motion of the particles of bodies. Wherever motive power is destroyed, there is a simultaneous production of an amount of heat exactly proportional to the motive power that is destroyed. Conversely, wherever there is destruction of heat, motive power is produced. **Hence we may state, as a general proposition, that the quantity of motive power in nature is fixed and that, strictly speaking, motive power is neither produced nor destroyed. It is true that it changes its form; that is, it sometimes produces one kind of motion, sometimes another. But it is never annihilated.**"

(III) Stimulated by Count VON RUMFORD's experiments on the heat produced by friction {Count VON RUMFORD ⇨1798}, S. CARNOT[494] also pondered the heating of gases by friction and discussed the sudden expansion of compressed gases through small orifices. ▪ Modern fluid dynamicists have resumed this subject, which is important when estimating losses in thermal turbines.[495] So-called "Carnot flow" is defined as the quasi-steady, detached flow through the sudden enlargement in area between two straight cylindrical pipes of different diameters.[496] For a supersonic flow entering a very large volume, a so-called **"Carnot shock"** can develop, in which the flow loses most of its kinetic energy.

| | | |
|---|---|---|
| 1809 | Collège de France, Paris, France | Jean-Baptiste BIOT,[497] a French professor of mathematical physics, **and collaborators give evidence that musical sounds of every pitch and violence are transmitted with precisely the same velocity.** They prove that the harmony of musical notes is not distorted by transmission through the air contained in a 951-m-long pipe to the slightest degree. ▪ His experiment, together with similar results obtained by William H. BESANT {⇨1859}, was cited by others |

---

[492] This manuscript – a single-folded sheet enumerated *IV-F°8* by Camille RAVEAU in *Sadi CARNOT. Biographie et manuscrit* – is one of nearly 40 pages of rather disjointed notes which remained unknown until his brother Hippolyte CARNOT presented them to the Académie des Sciences in 1878. They were translated into English and annotated by Robert FOX, a British historian at the University of Manchester, in his book mentioned in the previous footnote; *see* Annex III: *Notes on mathematics, physics, and other subjects*. ▪ Sadi CARNOT, who died prematurely of cholera in 1832 at the age of 36, only published the book *Réflexions sur la puissance motrice du feu et sur les machines propres à développer cette puissance* (Paris 1824).
[493] S. CARNOT's single-sheet manuscript *III-F° recto*. See previous footnotes.
[494] S. CARNOT's single-sheet manuscripts *VI-F°11 recto* and *VI-F°11 recto*. See previous footnotes.
[495] *See*, for example, K. OSWATITSCH: *Gasdynamik*. Springer, Wien (1952), pp. 162-165.
[496] K. FÖRSTER: *All about Carnot flow*. ZAMM **82**, No. 3, 177-190 (2002).
[497] J.B. BIOT: *Sur la propagation du son à travers les corps, et à travers l'air dans les tuyaux cylindriques très-allongés*. Mém. Phys. Chim. Soc. Arcueil **2**, 405-423 (1809).

{LE CONTE ⇒ 1864} and (incorrectly) generalized as an experimental proof that aerial waves of finite amplitude – *i.e.*, shock waves – can also only propagate at the velocity of sound.

| 1809 | Ecole Polytechnique, Paris, France | **Joseph-Louis GAY-LUSSAC and Louis J. THENARD,**[498] two French chemists, **study various mixtures of hydrogen and chlorine.** By exposing a 1:1 mixture [$H_2/Cl_2$] to diffuse daylight, they gradually obtain hydrogen chloride [HCl, French *acide muriatique*]. On the other hand, **when being exposed to direct sunlight, the gaseous mixture explodes violently, destroying the glass balloon.** Therefore, they call hydrogen chloride "chlorine detonating gas." ▪ The heat released by a chlorine-hydrogen explosion is only around two-thirds that released in an oxy-hydrogen explosion {TURQUET DE MAYERNE ⇒ 1620; CAVENDISH ⇒ 1760s & 1784}. The chlorine-hydrogen reaction

$$H_2 + Cl_2 \rightarrow 2\ HCl + 43.8\ kcal$$

is a classical lecture demonstration of a photochemical reaction initiated by blue light but not red light, and this remained a riddle to chemists and physicists for a long time. **Work on this phenomenon eventually led to the discovery of "chain reactions"** {BODENSTEIN ⇒ 1913}. |
|---|---|---|
| 1811 | Faculty of Sciences, University of Paris, France | **Siméon-Denis POISSON,**[499] a French professor of mechanics, **publishes his two-volume textbook *Traité de mécanique* ("Treatise of Mechanics").** In volume II, he treats elastic and inelastic percussion of two homogeneous bodies of spherical or arbitrary geometry, and the simultaneous percussion of an arbitrary number of spherical homogeneous bodies.

In the second edition, which was published in 1833 and became the standard work on mechanics for many years, he considerably extended the subject of percussion {ROUTH ⇒ 1860}:

(i) He considers the problem of the impact of two smooth inelastic bodies: under the assumption that the motion of each body just before impact is given, he derives six equations of motion for each body in order to determine the motion just after impact.

(ii) In the case of the elastic percussion of two bodies colliding with each other, he distinguishes between a series of three stages:
  ▸ the *contact stage* when the bodies begin to touch;
  ▸ the *compression stage* during which normal velocities are equal and are pointing into the same direction, and the maximum compression is attained; and
  ▸ the *final phase* in which the bodies begin to separate {⇒Fig. 2.8}.

(iii) He treats the influence of friction and rotation on two colliding balls, such as occur in the game of billiards {MORIN ⇒ 1833; DE CORIOLIS ⇒ 1835}.

(iv) He modifies the general percussion formula for the case when two bodies colliding with each other are not entirely free, and extends the general percussion formula for the case when three or more bodies collide with each other. |
| 1812 | Private chemical laboratory at Arcueil [a suburb south of Paris], France | **Pierre Louis DULONG,** a French chemist and assistant to Comte Claude-Louis BERTHOLLET, a renowned chemistry professor, **discovers nitrogen trichloride** [$NCl_3$], a yellow oily liquid which forms when chlorine reacts with ammonium chloride. **The new chemical – called "DULONG's explosive oil" – detonates violently upon shock or heat.** His investigation cost him an eye and several fingers. ▪ Other nitrogen halides include the lower nitrogen fluorides [$NF_2$ and $N_2F_2$], and nitrogen iodine [$NI_3 \cdot 2\ NH_3$]. The latter is a complex compound and stable only when wet; after drying it explodes at the slightest contact with a moving solid object, and can be set off by the touch of a feather. So far, attempts to prepare nitrogen bromide [$NBr_3$] have been unsuccessful. |

---

[498] J.L. GAY-LUSSAC and L.J. THENARD: *De la nature et des propriétés de l'acide muriatique et de l'acide muriatique oxigéné.* Mém. Phys. Chim. Soc. Arcueil **2**, 339-358 (1809).

[499] S.D. POISSON: *Traité de mécanique.* 2 vols., Bachelier, Paris (1811, 2nd edn. 1833). Germ. translation by I.E.E. SCHMIDT: *Lehrbuch der Mechanik.* 2 vols., Cotta'sche Buchhandl., Stuttgart *etc.* (1825, 1826); Reimer, Berlin (1835, 1836). Engl. translation by H.H. HARTE: *A treatise of mechanics.* Longman, London (1842).

In the following year, the famous British chemist Humphry DAVY[500] reports on his own studies on the explosive properties of the new high explosive $NCl_3$, which he immediately initiated after hearing about DULONG's discovery. In a letter to Sir Joseph BANKS, then president of the Royal Society, DAVY wrote:

(i) "I think it right to communicate to you and through you to the Royal Society, such circumstances as have come to my knowledge respecting a new and very extraordinary detonating compound. I am anxious that those circumstances should be made public as speedily as possible, because experiments upon the substance may be connected with very dangerous results … I attempted to collect the products of the explosion of the new substance, by applying the heat of a spirit lamp to a globule of it, confined in a curved glass tube over water: a little gas was at first extricated, but long before the water had attained the temperature of ebullition, a violent flash of light was perceived, with a sharp report; the tube and glass were broken into small fragments, and I received a severe wound in the transparent cornea of the eye, which has produced a considerable inflammation of the eye. This experiment proves what *extreme* caution is necessary in operating on this substance, for the quantity I used was scarcely as large as a grain of mustard seed [about 1 mm in diameter]…"

(ii) "The mechanical force of this compound in detonation, seems to be superior to that of any other known, not even excepting the ammoniacal fulminating silver [E.C. HOWARD ⇨1800]. The velocity of its action appears to be likewise greater."

| | | |
|---|---|---|
| 1812 | Royal Military Academy, Woolwich, southeast London, England | **Charles HUTTON**,[501] a British mathematician and ballistician, **determines the efficiency of pile-engines.**<br>(i) In order to optimize pile-engines {⇨Fig. 4.2–N}, he writes that the following basic question must be solved: by which means, in the shortest time, and with the fewest men, or the least force, can the most piles be driven to the greatest depth?<br>(ii) Contrary to previous theories[502] deduced from the laws of the percussion of bodies only, he also considers the friction of the surface penetrated, and calculates the depth sunk by the pile at each stroke of the ram. • The effect of the blow is proportional to the weight of the hammer or ram and to the height that the head of the pile has dropped. However, the resisting force between the pile and soil increases with increasing depth of the pile, and successive penetrations gradually diminish until they become so small as to be almost imperceptible. |
| 1813 | Tyne and Wear County, North East England | Numerous explosion hazards in the Wear Valley coal mines near Sunderland [now part of the City of Sunderland in Tyne and Wear] draw attention to flame propagation. **Sir Ralph MILBANKE**, chairman of the Society for Preventing Accidents in Coal Mines (founded in 1813), **asks the British chemist Humphry DAVY**, a professor at the Royal Institution and one of the most famous chemists of his time, **to develop a safe lamp for use in coal mines** {DAVY ⇨1815}. |
| 1814 | Paris, France | **Joseph L. DE LAGRANGE**, an Italian-born French mathematician and astronomer, **extends OLBERS' exploded planet hypothesis** {OLBERS ⇨1804} **to explain the comets,** suggesting that the highly elongated orbits of comets could be explained by a planetary explosion. |
| 1814 | London, England | **Joshua SHAW**, a British artist, **invents the *percussion cap.*** It consists of a small truncated cone of metal, preferably of copper, which contains a small amount of fulminate of mercury inside its crown, protected by foil and shellac. This cap is shaped to fit tightly over a steel nipple set in the gun barrel or an extension thereto. When the cap is detonated, a jet of flame passes |

---

[500] H. DAVY: *On a new detonating compound, in a letter from Sir Humphrey DAVY, L.L.D. F.R.S. to the Right Honourable Sir Joseph BANKS, Bart. K.B. P.R.S.* Phil. Trans. Roy. Soc. Lond. **103**, 1-7 (1813); *Some further observations on a new detonating substance.* Ibid. **103**, 242-251 (1813).

[501] C. HUTTON: *Tracts on mathematical and philosophical subjects.* T. Davison, London (1812), vol. 3, pp. 317-321.

[502] T. BUGGE: *Tentamen continens theoriam machinae sublicarum.* Phil. Trans. Roy. Soc. Lond. **69**, 120-129 (1779).

down an open channel in the nipple into the powder charge. ▪ He kept his discovery secret until his arrival in Philadelphia, PA, where he established himself. In 1822, he obtained a patent for the percussion cap and percussion lock for small arms. The validity of his claim, however, was later disputed, the invention being attributed in part to Alexander J. FORSYTH {⇨1805}.[503]

The advantages of SHAW's new system over the flintlocks related to its speed and sureness of firing. **The invention of the percussion cap was an important milestone towards the modern cartridge, which unifies bullet, powder charge, and percussion primer** {DREYSE ⇨1827}. The percussion cap can also be regarded as the archetype of the blasting cap, the so-called "detonator" {A. NOBEL ⇨1863/1864}.

| | | |
|---|---|---|
| 1815 | *Mt. Tambora Volcano, Sumbawa Island, Java Sea* | On April 5, **the eruption of the volcano Mount Tambora begins with a moderately large explosion** from which ash falls in east Java and thunder-like sounds are heard up to 1,400 km away. Shortly after, from April 10 to 11, a much larger eruption occurs, ejecting about 50 km$^3$ of magma and leaving a deep summit caldera (and about 92,000 casualties). ▪ The Tambora explosive eruption, the largest volcanic eruption in recorded history ($VEI$ = 7), produced tsunami waves with a height exceeding ten meters and blast waves (atmospheric shock waves) that reverberated around the world several times, breaking windows at a distance of up to about 400 km.[504]<br><br>The super-eruption of 1815 which sent a plume of volcanic ash high into the Earth's atmosphere was followed by "the year without summer" in 1816 – the coldest year in the last few centuries. This "nuclear winter" devastated crops in the Northern Hemisphere, yielding famine and plague.[505] |
| 1815 | *Hebburn & Wallsend Collieries [near Newcastle-on-Tyne], northeast England* | On August 24, **Humphry DAVY**, an eminent English chemist, **visits Wallsend Colliery** (North Tyneside) **and Hebburn Colliery** (South Tyneside). Firedamp gas samples from Hebburn Colliery, taken in wine bottles, are sent to London to the Royal Institute for investigation.<br><br>In the same year, on November 1, **DAVY describes the basic principle of his safety lamp in a letter to the Royal Society: "This invention consists in covering or surrounding the flame of a lamp or a candle by a wire sieve…"** On November 9, DAVY[506] **lectures at the Royal Society on his recent investigations on the nature of firedamp:**<br>(i) Based upon previous studies performed in 1814 in the Grand Duke's Laboratory at the Florentine Academy on test samples of firedamp gas which he collected with his assistant Michael FARADAY[507] during a journey to the Apennines and Hebburn Colliery, DAVY first concludes that<br>    ▸ "1 part of gas inflamed with 6 parts of air in a similar bottle, produced a slight whistling sound;<br>    ▸ 1 part of gas with 8 parts of air, rather a louder sound;<br>    ▸ 1 part with 10, 11, 12, 13 and 14 parts, still inflamed, but the violence of combustion diminished;<br>    ▸ in 1 part of gas and 15 parts of air, the candle burnt without explosion with a greatly enlarged flame…" |

---

[503] Anonymous: *Joshua SHAW, artist and inventor. The early history of the copper percussion cap.* Scient. Am. **21**, 90 (Aug. 1869).
[504] G.A. STEWART: *Description of a volcanic eruption in the island of Sumbawa.* Trans. Literary Soc. Bombay **2**, 109-114 (1820).
[505] J.Z. DE BOER and D.T. SANDERS: *Volcanoes in human history: the far-reaching effects of major eruptions.* Princeton University Press, Princeton, NJ (2001).
[506] H. DAVY: *(I) On the fire-damp of coal mines, and on methods of lighting the mines so as to prevent its explosion.* Phil. Trans. Roy. Soc. Lond. **106**, 1-22 (1816); *(II) An account of an invention for giving light in explosive mixtures of Fire-damp in coal mines, by consuming the Fire-damp.* Ibid. **106**, 23-24 (1816).
[507] M. FARADAY's *Journal of Oct. 1814.* In: (B. BOWERS and L. SYMONS, eds.) *Curiosity perfectly satisfied. FARADAY's travels in Europe 1813–1815.* P. Peregrinus Ltd., London (1991), pp. 126-128.

(ii) Firedamp was often supposed to be hydrogen [$H_2$]; however, DAVY first proves that it is methane [$CH_4$] which, due to DAVY, consists of "4 proportions of hydrogen in weight 4, and 1 proportion of charcoal in weight 11.5."

(iii) Using gas from the distillation of coal mixed with eight times its volume of air, he also determines the rate at which the gas explosion propagates in a tube and makes the first rough estimate of the temperature reached in an explosion. The flame, fired in a 1-ft (305-mm) long tube with ¼ in. (6.35 mm) in diameter, takes more than a second to traverse the tube.

(iv) He also observes that the same mixture that burns in a wide tube may not support flame propagation in a narrow tube with a diameter less than a certain "critical diameter," and concludes: **"Without doubt to prevent explosions it is necessary, to use lamps, where the surrounding air is allowed to access only through small tubes or wire gauzes to the lamp flame."**

(v) His studies on flame propagation in tubes will result in him constructing a safe mining lamp – the so-called **"Davy lamp"** {⇨ Fig. 4.21−B} − in which a copper mesh with small openings prevents flame propagation from the inside of the lamp to the atmosphere of the mine.[508] ▪ Later it was demonstrated that the blast wave generated in a firedamp explosion forces the flame through the meshes of a Davy-type safety lamp {GALLOWAY ⇨ 1874}.

In the same year,
- George STEPHENSON,[509] an English self-educated colliery engineer and famous inventor of the railroad locomotive (1814), tests his flame safety lamp at Killingworth Colliery (Durham County). ▪ In 1813, William Reid CLANNY,[510] an English physician, had actually invented the first practical safety lamp for use in coal mines, for which he was awarded the Silver Medal by the Royal Society of Arts in 1816. But since Sir Humphry DAVY refused to patent his lamp, priority tests were never performed in court. The three men refused to patent their individual inventions. When asked if he was going to cover his invention with a patent, Sir Humphry DAVY replied: "No, my good friend, I never thought of such a thing, my sole object was to serve the cause of humanity, and, if I have succeeded, I am amply rewarded in the gratifying reflection of having done so."
- **Archduke JOHN of Austria** [*Johann, Erzherzog von Österreich* (1782–1859)] pays a visit to DAVY and **becomes one of the first promoters of the Davy lamp.** ▪ Shortly after, he introduced the Davy lamp to the Styrian mining industry.[511]

In the course of his work, DAVY found that firedamp and air will combine without explosion in the presence of a coil of platinum wire – the first example of what came to be known in kinetics as *heterogeneous catalysis*, which is of great importance to the chemical industry and is used in living organisms. ▪ The principle of the Davy lamp was later also successfully used in Germany by Jaaks & Behrns Co. at Lübeck for the construction of explosion-proof flour-dust filters.[512]

**After World War II, the first electrical instruments to measure the percentage of methane in the air in underground coal mines − so-called "methanometers" − were designed to alert miners to the presence of potentially dangerous concentrations of this gas.**[513] Modern hand-held instruments are capable of giving a flashing signal or sounding an alarm

---

[508] H. DAVY: *(II) An account of an invention for giving light in explosive mixtures of fire-damp in coal mines by consuming the fire-damp*. Phil. Trans. Roy. Soc. Lond. **106**, 23-24 (1816).

[509] G. STEPHENSON: *On safe-lamps for coal mines*. Phil. Mag. **46** [I], 458-460 (1815); *Letter with a few remarks on his claim to priority in the invention of the safety-lamp*. Ibid. **49** [I], 204-206 (1817).

[510] W.R. CLANNY: *On the means of producing a steady light in coal mines, without danger of explosion*. Phil. Trans. Roy. Soc. Lond. **103**, 200-205 (1813).

[511] Communicated by the *Technisches Museum Wien*, A-1140 Vienna, Austria.

[512] R. WEBER: *Über die Entstehung von Bränden in Mahlmühlen*. Z. für Tech. Hochschulen **3**, 51-53 (1878).

[513] *The methanometer*. App-Tek International Pty Ltd, Brendale, Australia (Aug. 2002); http://www.methanometer.com/#Q5.

| | | |
|---|---|---|
| 1815 | Ecole Polytechnique, Paris, France | **Augustin-Louis CAUCHY,**[515] a young French mathematician, **analytically investigates the problem of water waves generated by sudden disturbances of a free surface** – such as the ring-shaped waves on the water surface of a pond excited by a stone falling in it. In the same year he submits his memoir to the French Academy of Sciences which had announced a mathematical prize competition on surface wave propagation on a liquid of indefinite depth two years earlier. ▪ In the following year, CAUCHY won the *Grand Prix* of the Institut de France for his paper on wave propagation.<br><br>In the same year, the French mathematician **Siméon-Denis POISSON,**[516] an examiner at the Ecole Polytechnique and one of the judges, independently **relates a memoir of his own to record his work on this subject.** ▪ The so-called "Cauchy-Poisson wave problem," a classic in the evolution of water wave theories,[517] has also been used to model tsunami generation by assuming that the initial surface elevation of the water waves is similar to the displacement of the seafloor and by deriving the generation and propagation of the surface water waves through the solution of an initial-value problem. |
| 1815 | Swedish Artillery, Stockholm, Sweden | **Major-General Carl Gottfried VON HELVIG,**[518] a German military engineer, chief of the Swedish Artillery and member of the Swedish Royal Academy of Sciences, **reports on the observations of luminous phenomena from imploding glass bombs** [Germ. *gläserne Knallbomben*] – evacuated thin-wall glass bulbs with an outer diameter ranging between 52 and 78 mm.<br><br>(i) When a glass bomb is smashed on the floor in a dark room, the breaking of the glass bulb causes an implosion which is always accompanied by a sharp report and – sometimes, but not always – by a flash of light. Since the strongest flashes are observed on clear and warm days, it appears that the luminous implosion phenomenon is dependent on the humidity of the air. ▪ The French natural philosopher Antoine DE PARCIEUX {⇨1794} was the first to notice that the implosion of a glass bulb is accompanied by a flash of light.<br><br>(ii) The light from an imploding glass bomb has a pale white color – similar to the one observable at the muzzle of a wind-gun {GUTER ⇨1430; BOYLE ⇨1647} – and appears to implode into the vacuum from all sides; *i.e.*, it is undirected, with the exception of a dark region in the center. In contrast, light emerging from the muzzle of a fired wind-gun is directed and appears to emerge from the muzzle.<br><br>In the same year, VON HELVIG[519] also reflects on the cause of lightning and thunder. He differentiates between a short blast [Germ. *kurzer Schall*] originating from a small-sized source (such as the report produced by an imploding glass bomb) and a long blast [*gedehnter Schall*] produced from an extended source (such as the jet of burnt gases emerging from the muzzle of a fired cannon or the long channel of a flash of lightning). |

---

[514] D.S. BÀRRIE: *The wand of science – story of the flame safety lamp*. D. Barrie Risk Management, Sutton Coldfield, U.K. (May 2006).

[515] A.L. CAUCHY: *Mémoire sur la théorie de la propagation des ondes à la surface d'un fluide pesant d'une profondeur indéfinie* [1815]. Mém. Sav. Étrang. (Prix Acad. Roy. Sci., concours de 1815 et de 1816) **I**, 3-312 (1827).

[516] S.D. POISSON: *Mémoire sur la théorie des ondes* [1815]. Mém. Acad. Sci. Inst. France **1**, 71-186 (1816).

[517] For more information on the Cauchy-Poisson wave problem, *see* STOKES {⇨1849} and H. LAMB: *Hydrodynamics*. Cambridge University Press, Cambridge (1932), pp. 384-387.

[518] C.G. VON HELVIG: *Einige Versuche mit gläsernen sogenannten Knallbomben*. Ann. Phys. **51** [I], 112-115 (1815).

[519] C.G. VON HELVIG: *Bemerkungen über Blitze und Donner, nebst Vermuthungen über das Entstehen der Luft-Erscheinungen*. Ann. Phys. **51** [I], 117-148 (1815).

| | | |
|---|---|---|
| 1815 | Somerset House, The Royal Society of London, England | At a meeting of the Royal Society, **George John SINGER and Andrew CROSSE**,[520] two English electricians, **report on the effects of electric wire explosions** {VAN MARUM ⇨ 1799}. In these experiments, an exploding lead wire was placed in the axial direction of a thin-walled metallic cylinder filled with water and pulsed from a large battery of charged Leiden jars {VON KLEIST & CUNEUS ⇨ 1745; ⇨Fig. 4.16−N}. The cylinder itself expanded more or less in proportion to its resistance, usually becoming undulated on the surface or bursting open. Upon generating more violent wire explosions using a larger battery, they observe that even iron cylinders with thicknesses greater that those of the strongest muskets are heavily damaged by cracks. They state that "the expansive power of electricity acting in this way is therefore vastly superior to the most potent gunpowder." ▪ Their remarkable results anticipated the electrohydraulic effect that Lev A. JUTKIN {⇨ 1950} rediscovered in the former Soviet Union. |
| 1816 | L'Académie des Sciences, Paris, France | **Pierre-Simon DE LAPLACE**,[521] a French astronomer and mathematician, **publishes his previous hypothesis** {BIOT ⇨ 1802} **that the production of a sound wave is an adiabatic process;** *i.e.*, no heat is added or removed from the medium. He states, without demonstration, a correction of NEWTON's formula that was published in his *Principia* {Sir NEWTON ⇨ 1687}.[522] |
| 1816 | Saint-Loup de Varennes, Dépt. Saône-et-Loire, east central France | **Nicéphore NIÉPCE**, a French inventor, **positions sheets of paper covered with silver salts, known to blacken in daylight, at the back of a *camera obscura*. He thus produces the first (non-fixed) photographic image of nature**, a view from his workshop window. It is a negative, and the image vanishes because in broad daylight the coated paper becomes completely black. He calls these images "retinas." ▪ Until then, the camera obscura – a box with a lens fitted onto a hole, that projects an inverted image of the outside view onto the back of the box – had only been used as a drawing aid for artists and for observing solar eclipses, even though it had been known about since antiquity.<br><br>In 1824, NIÉPCE put lithographic stones coated with bitumen at the back of a camera obscura and obtained a fixed image of a landscape for the first time. **In 1828, he found a method that led to images of superior quality with half-tones.**[523] |
| 1817 | River Yare at Norwich, East Anglia, England | On April 4, Good Friday morning, **a boiler explosion occurs on board the Norwich and Yarmouth steamer,** just as she is leaving the Foundry Bridge, Norwich (ten people are killed and five injured). ▪ This latter event led to the establishment of a Select Committee in the House of Commons to investigate the best way to prevent such explosions on steam boats.[524] |
| 1819 | University of Leipzig, Saxony, east-central Germany | **Ludwig Wilhelm GILBERT**,[525] a physics professor and editor of the eminent German journal *Annalen der Physik*, **reviews the first practical constructions and applications of the hydraulic press** {PASCAL ⇨ 1640s; BRAMAH ⇨ 1795} which is first shown this year at the Paris Industrial Exhibition. He annotates that, according to the Berlin newspaper *Spenersche Zeitung* (dated May 19, 1818), the German mechanic Erwin NEUBAUER constructed in 1818 at the Nathusius Machine Factory in Hundisburg (a town near Magdeburg, Germany) a hydraulic press capable of producing a "pressure of 300,000 pounds," which was used primarily for pressing out sugarbeets and olives.[526] |

---

[520] G.J. SINGER and A. CROSSE: *An account of some electrical experiments by M. DE NELIS.* Phil. Mag. **46** [I], 161-166 (1815).
[521] P.S. DE LAPLACE: *Sur la vitesse du son dans l'air et dans l'eau.* Ann. Chim. Phys. **3** [II], 238-241 (1816).
[522] B.S. FINN: *LAPLACE and the speed of sound.* Isis **55**, 7-19 (1964).
[523] Nicéphore NIÉPCE. A biography provided by Spéos, Paris Photographic Institute; http://www.nicephore-niepce.com/home-us.html.
[524] W. FINCH-CRISP: *Chronology retrospect 1800–1877*; http://www.ean.co.uk/Data/Bygones/History/Local/Norfolk/Great_Yarmouth/Crisp/html/body_crisp2.htm.
[525] L.W. GILBERT: *Beschreibung und Beurtheilung der von dem Mechaniker Joseph BRAMAH erfundenen Wasserpresse, in welcher das Wasser nach dem Prinzip des hydrostatischen Paradoxes wirkt.* Ann. Phys. **60** [I], 1-13 (1819).
[526] It is not possible to estimate the pressure actually achieved from these data. NEUBAUER's new press was capable of producing a gravitational force corresponding to the weight of almost 150 tons which, however, was distributed over a greater area than just a few square centimeters in his juice press.

| 1822 | *Bureau des Longitudes, Paris, France* | **A commission of the Bureau des Longitudes makes a series of sound velocity measurements** between Montlhéry and Villejuif, two communes in the suburbs of Paris that are 11 miles (about 18 km) apart. Cannons are fired at two stations every five minutes, and chronometers are used for timing. The result at 15.9 °C is $u = 340.9$ m/s, which improves upon previous results {French Commission ⇨1738}.[527] ▪ Outdoor experiments to determine the sound velocity are also dependent on humidity and wind. Modern handbook data,[528] often excluding the influence of humidity and wind, and treating air as a perfect gas, show that the sound velocity $u$ depends on the temperature $\vartheta$ (in °C) according to the formula
$$u(\vartheta) = 331.45\,[1+\vartheta/273.16]^{1/2} \text{ m/s}$$
which, for $\vartheta = 15.9$ °C, would result in $u = 340.96$ m/s – a value incredibly close to the one obtained in 1822. |

| 1822 | *Ecole des Ponts et Chaussées, Paris, France* | **Claude Louis Marie Henri NAVIER,**[529] a French civil engineer, **presents a paper on the law of motion of continuous media.** He considers fluids and solids to be made up of particles that are close to each other and act upon each other by attraction or repulsion, resulting from caloric heat. He derives partial differential equations of motion for a fluid particle in terms of the shear and normal forces exerted on it, to which he applies FOURIER's method in order to find particular solutions. ▪ Though NAVIER did not comprehend the essential mechanism of viscous action, his results were mathematically correct.

In the following years, the same equations were developed with greater comprehension by Augustin L. CAUCHY,[530] Siméon-Denis POISSON,[531] and George G. STOKES.[532] The latter, in particular, made the final derivation of the equations of motion in terms of a coefficient of viscosity and applied the equations to the resistance of small spheres. The so-called **"Navier-Stokes equations"** are an extension of the Euler equations {EULER ⇨1755}.

A more general form of the Navier-Stokes equations was developed by Adhémar DE SAINT-VENANT[533] and later found to be applicable not only to the laminar phase of viscous flow but also to that known as "fluid turbulence." James C. MAXWELL {⇨1867}, supplementing his own dynamical theory of gases, derived the Navier-Stokes equations by assuming a distribution function of gas molecules. **A first attempt to solve the Navier-Stokes equations for shock waves was first successfully achieved by Richard A. BECKER** {⇨1921}. |

| 1823 | *Ecole Polytechnique, Paris, France* | **Louis Joseph GAY-LUSSAC,** a professor of chemistry who pioneered investigations into the behavior of gases, **communicates a report of his experiments on the decomposition products of gunpowder to the *Comité des Poudres et Salpêtres* at Paris.** These products are obtained by allowing small quantities of gunpowder to fall into a tube arranged to receive the gases, and heated to redness. He finds that the volume of these gases, at 0 °C and 1 bar, is 450 times the volume of the gunpowder. ▪ He later corrected his result and estimated the permanent gases to be about 250 times as voluminous as the gunpowder – a value pretty close to the one later obtained in British studies, which yielded a factor of 280 {NOBLE & ABEL ⇨1875}.[534] |

---

[527] F.J.D. ARAGO: *Résultats des expériences faites par ordre du Bureau des Longitudes, pour la détermination de la vitesse du son dans l'atmosphère.* Connaissance des temps ou des mouvements célestes… (Paris, 1822), pp. 361-371.

[528] D.E. GRAY (ed.): *American Institute of Physics handbook.* McGraw-Hill, New York *etc.* (1972), pp. 3-74.

[529] C.L.M.H. NAVIER: *Mémoire sur les lois du mouvement des fluides.* Mém. Acad. Roy. Sci. Paris **6**, 389-440 (1823).

[530] A.L. CAUCHY: *Exercices de mathématiques.* De Bure, Paris (1826–1829).

[531] S.D. POISSON: *Mémoire sur les équations générales de l'équilibre et du mouvement des corps solides élastiques et des fluides.* J. Ecole Polytech. (Paris) **13**, Cahier 20, 1-174 (1831).

[532] G.G. STOKES: *On the theories of the internal friction of fluids in motion, and of the equilibrium and motion of elastic solids.* Trans. Cambr. Phil. Soc. **8**, 287-319 (1849).

[533] A.J.C. DE SAINT-VENANT: *Note à joindre au Mémoire sur la dynamique des fluides.* C. R. Acad. Sci. Paris **17**, 1240-1243 (1843).

[534] This work of GAY-LUSSAC is not listed in the British *Catalogue of scientific papers.* NOBLE & ABEL {⇨1875} who discussed the discrepancy in his result, tried in vain to obtain a copy of his report.

| 1823 | Ecole des Ponts et Chaussées, Paris, France | Claude Louis M.H. NAVIER,[535] a French mathematician, **first performs systematic studies of the dynamic strengths of materials.** He investigates how large the maximum stress and strain at each point along an elastic thread become when a weight attached to the lower end of the thread is suddenly accelerated downwards, for example by dropping it from a height. |
|---|---|---|
| 1823 | Faculté des Sciences, Paris, France | **Siméon D. POISSON,**[536] a French mathematician, **reviews the present state of investigations into the velocity of sound.** He writes, "The sound velocity in air, derived from a formula given by NEWTON [*Principia*, Book II, Scholium, after Proposition], differs from the observed velocity and surpasses the calculated velocity by a fifth; *i.e.*, by 20%. When LAGRANGE [Misc. Taur. **1**, I-X, 1-112 (1759)] in his first studies on the theory of sound arrived at the same formula, he tried of course to explain this discrepancy between calculation and observation. His analysis was based on two suppositions: the minuteness of the air vibrations, and the proportionality of the elastic force with density. He [DE LAGRANGE] proved first, against EULER's opinion [Mém. Acad. Sci. Berlin **15** (1759); EULER ⇨ 1759], that the amplitude of vibration does not influence the magnitude of the sound velocity; in addition he mentioned that one could coincide this velocity with the result from observation by supposing that the elastic force increases to a larger extent than the density; but he could not give any particular reason for this increase of elasticity, which cannot be described by the general law of compression for air. It is not less true that this increase is really due to the air motion: **It is Mr. LAPLACE** [Ann. Chim. Phys. **3**, 238-241 (1816)] **who pointed out the true reason and fully explained and eliminated the difference between NEWTON's formula and the measurement.** This reason is the release of heat, which always occurs during the compression of air or the production of coldness which goes along with dilatation, likewise..." |

In the same year, POISSON,[537] stimulated by DE LAPLACE's approach, **derives the two gas laws for adiabatic compression.**

(i) A gas with a ratio of specific heats $k$ that is initially at pressure $p$, has a density $\rho$ and a temperature $\Theta$, and is then adiabatically compressed to a density $\rho'$, reaches a new state ($p'$, $\rho'$, $\Theta'$), which is given by

$$p' = p (\rho'/\rho)^k$$

and

$$\Theta' = (266.67 + \Theta)(\rho'/\rho)^{k-1}.$$

(ii) He writes, "These equations contain the elasticity and temperature laws of gases which are compressed or expanded without a variation in their heat quantity; which will happen when the gases are in a heat-proof glass container, or when the compression, as with the sound phenomenon, is so fast that one can assume that the heat loss is negligible..." ▪ POISSON assumed that absolute zero was at a temperature of –266.67 °C, which we have now replaced with the modern value of –273.15 °C. His equation of state for adiabatic compression – later to be termed the ***Poisson isentrope***, the ***Poisson adiabatic law*** or the ***static adiabat*** – is only applicable to sound waves of infinitesimal amplitude. It was a major step towards deriving the ***dynamic adiabat*** or ***Hugoniot curve*** of a shock-compressed gas {HUGONIOT ⇨ 1887}.

| 1824 | London, England | Jacob PERKINS, a U.S. inventor who had been working in England since 1818, **exhibits his steam gun.**[538] ▪ The idea of using steam to drive projectiles was first proposed by Leonardo DA VINCI.[539] In the early 19th century, and prior to PERKINS, the concept of using steam instead of gunpowder to accelerate projectiles had already been proposed in England by the engineers |
|---|---|---|

---

[535] C.L.M.H. NAVIER: *Sur les effets des secousses imprimées aux poids suspendus à des fils ou à des verges élastiques*. Nouv. Bull. Sci. Soc. Philomat. Paris (1823), pp. 73-76.
[536] S.D. POISSON: *Sur la vitesse du son*. Ann. Chim. Phys. **23** [II], 5-16 (1823).
[537] S.D. POISSON: *Sur la chaleur des gaz et des vapeurs*. Ann. Chim. Phys. **23** [II], 337-353 (1823).
[538] *The Encyclopædia Britannica*. Adam & Black, Edinburgh (1885), p. 549.
[539] *Leonardo DA VINCI*. Published by Leisure Arts, Istituto Geografico De Agostini, Novara, Italy (1964), vol. II, Ms. B. fol. 33r.

and inventors James WATT and Henry BESSEMER, and in France by the generals Francois DE CHASSELOUP-LAUBAT and Philippe Henri DE GIRARD.

In the following years, PERKINS designed a steam rifle and a steam gun capable of firing spherical projectiles at a rate of 420 and 80 shots per minute, respectively.[540] In the early 1860s, PERKINS' concept of using steam in guns was resumed in the United States and it was proposed that steam guns could be used to defend harbors and to arm fortifications and war vessels.[541] In a letter to U.S. President Abraham LINCOLN, it says: "… we beg leave to ask your attention to Messrs. PERKINS' invention of the Steam Gun, as presented in the brief description brought from London by Pliny MILES [an American postal journalist]. The statement of the results of the trial of their last gun of moderate size with a rifled barrel and superheated steam, together with the illustrative facts and arguments adduced of its vast power and great utility, if constructed on a large scale, seem to us to justify the very moderate expenditure required ($5,000) for the construction of a twelve-pounder gun with every modern improvement…"
▪ However, although the idea attracted much attention among contemporary ballisticians, the requirement of highly compressed steam prevented its practical use.

| 1824/ 1825 | Port Bowen, Prince Regent's Inlet, northern Canada |

**William Edward PARRY,**[542] a captain of the Royal Navy and an Arctic explorer, **attempts to find a northwest passage out of Hudson Bay,** and spends the winter in Port Bowen.

(i) While waiting for the ice to break through, he studies the Eskimos and gathers scientific data. Together with Lt. Henry FOSTER, participating in the expedition as an assistant surveyor, **he performs experiments on the velocity of sound in order "to determine the rate at which sound travels at various temperatures and pressures of the atmosphere."**

(ii) Measurements at very low temperatures (*i.e.*, in air of perfect dryness) are considered to be of particular interest because they avoid the need to make any corrections of sound velocity data due to the humidity of the atmosphere. Using a six-pounder brass gun placed on the beach at the head of Port Bowen and fired on signal from the HMS *Hecla*, PARRY and FOSTER, by carefully noting the interval between the flash and report at a distance of about 3.9 km via the beats of a pocket chronometer[543] held at the ear of each observer, measure an anomalously high velocity of sound {⇨Fig. 2.10–D}.

PARRY,[544] who had already made similar observations during his second polar voyage (1821–1823), also included FOSTER's curious observation on sound velocity measurements in his report to the British Admiralty: "The experiments on the 9th February 1822 were attended with a singular circumstance, which was – **the officers' word of command 'Fire,' was several times distinctly heard both by Captain PARRY and myself, about one beat of the chronometer <u>after</u> the report of the gun; from which it would appear, that the velocity of sound depended in some measure upon its intensity**…" ▪ FOSTER used an Arnold chronometer with a 19,200 train; *i.e.*, it made 16 swings in three seconds, producing eight beats in that

---

[540] *Meyers Konversations-Lexikon.* Bibliographisches Institut, Leipzig & Wien (1894); see *Dampfgeschütz*, vol. 4, p. 514.

[541] *To His Excellency, Abraham LINCOLN, President of the United States, and Hon. Simon CAMERON, Secretary of War: copy of a letter from eminent citizens of New York, to the President and Secretary of war, requesting a trial of PERKINS' steam gun for defending harbors, and arming fortifications and war vessels.* New York (Oct. 15, 1861). Library of Congress, American Memory Historical Collections; http://memory.loc.gov/ammem/rbpehtml/rbpebibTitles87.html.

[542] W.E. PARRY: *Journal of the third voyage for the discovery of a North-West Passage from the Atlantic to the Pacific; performed in the years 1824–1825, in His Majesty's ships "Hecla" and "Fury,"* under the orders of W.E. PARRY. J. Murray, London (1826); see *Experiments to determine the rate at which sound travels at various temperatures and pressures of the atmosphere*; see Appendix, p. 86.

[543] By 1800, the pocket chronometer – a portable and accurate time keeping device which allowed mariners to determine their longitudinal position – was already commercially available.

[544] W.E. PARRY: *Journal of a second voyage for the discovery of a North-West Passage from the Atlantic to the Pacific; performed in the years 1821–22–23, in His Majesty's ships "Fury" and "Hecla,"* under the orders of W.E. PARRY. J. Murray, London (1824/1825), p. 140, and Appendix *Abstract of experiments to determine the velocity of sound at low temperature*, pp. 237-239.

time.[545] Thus, a time delay of one beat as observed by FOSTER and PARRY would correspond to a time span of 0.375 s.

Contemporary naturalists attributed their unusual findings to possible influences of humidity and wind, arguing that they had not precisely recorded these parameters.[546] But PARRY[547] and FOSTER replied, "It was certainly far from our intention to oppose our opinions on these points to those of NEWTON and DE LAPLACE. We considered our remark at the time, as a fair deduction from our own experiments, without at all considering with what theory it might be at variance: our only wish being, to furnish data for philosophers to arrive at such laws as will make the computed and observed velocities of sound agree more exactly with each other, than appears to be the case, in the present state of our information of all the modifying circumstances to which the motion of sound is subjected." **PARRY's and FOSTER's unusual observations** – supported also by those of British James C. ROSS (who participated in the expeditions and later became a famous arctic explorer) – **were later cited by the Rev. Samuel EARNSHAW {⇨1858 & 1864} as an important experimental proof of his mathematical theory of sound that intense air waves do indeed travel more quickly than weak air waves.**

| 1825 | Musée d'Histoire Naturelle, Paris, France | Michel Eugène CHEVREUL,[548] a professor of chemistry and director of dying at the Manufacture des Gobelins in Paris, **contributes an article on nitrates to the *Dictionnaire des Sciences Naturelles* ("Dictionary of Natural Sciences"), in which he also discusses the decomposition of gunpowder.** He indicates the difference between when the decomposition occurs explosively (as in the bore of a gun) and when it takes place slowly (such as by ignition in open air). In the former case, he supposes that the decomposition can be represented by the equation $$2\,KNO_3 + S + C_3 \rightarrow K_2S + N_2 + 3\,CO_2.$$ This equation was taken up by numerous chemical textbooks, but later studies showed that many more compounds are produced upon the ignition of gunpowder {BUNSEN & SCHISCHKOFF ⇨1857; NOBLE & ABEL ⇨1875; DEBUS ⇨1882}. |
|---|---|---|
| 1825 | University of Leipzig, Saxony & University of Halle, Saxony-Anhalt, east-central Germany | Ernst Heinrich WEBER, a professor of human anatomy and physiology at Leipzig, **and his younger brother Wilhelm WEBER,**[549] a physicist at Halle, **publish their textbook *Wellenlehre auf Experimente gegründet...* ("Wave Theory Based upon Experiments...") – a rich source of all kinds of wave phenomena.** For example, the WEBER brothers (i) observe and describe the wave pattern generated on the surface of a liquid when the liquid is contained in a vessel which is struck at its border by a percussive force; and (ii) treat the problem where under certain circumstances the wave front can form a crest or a trough.<br><br>In the case of shock waves propagating in an elastic medium, the pressure always increases at the shock front; *i.e.*, the incident shock wave starts with a crest. However, in the case of seismic sea waves or tsunamis, it was observed that in many cases the sea first recedes before the main wave arrives {MARCELLINUS ⇨A.D. 365; Lisbon Earthquake ⇨1755; Sumatra-Andaman Islands Earthquake ⇨2004}. This curious phenomenon was also confirmed by analyzing water gauge records taken during the Arica Earthquake {⇨1868}, the Iquique Earthquake {⇨1877}, and the explosive eruption of Krakatau {⇨1883}. |

---

[545] From Jonathan BETTS, Curator of Horology at the Royal Observatory Greenwich; private communication (May 2002).

[546] See, for example, W. GALBRAITH: *On the velocity of sound*. Phil. Mag. **68** [I], 214-219 (1826).

[547] W.E. PARRY and H. FOSTER: *Reply to Mr. GALBRAITH's remarks on the experiments for ascertaining the velocity of sound at Port Bowen*. Phil. Mag. **1** [II], 12-13 (1827).

[548] M.E. CHEVREUL: *Nitrates*. In: (F.G. CUVIER, ed.) *Dictionnaire des sciences naturelles*. Levrault, Strasbourg (1816–1826), vol. 35 (1825): *Produits de la combustion de la poudre*, p. 58; *Produits de la poudre lente et de la combustion rapide*, pp. 58-62.

[549] E.H. WEBER and W. WEBER: *Wellenlehre auf Experimente gegründet oder über die Wellen tropfbarer Flüssigkeiten mit Anwendung auf die Schall- und Lichtwellen*. Fleischer, Leipzig (1825).

| 1826 | Lake of Geneva, western Switzerland | In November, **Daniel Colladon**, a Swiss physicist and apothecary, **measures the velocity of sound in water quite accurately.**

(i) For strong source of sound, he uses a church bell placed under water and triggered simultaneously with a cannon; for sound detection he uses a long ear-trumpet submerged 5 m below the surface. To obtain high accuracy he chooses a long baseline. He performs the measurements at night so that he can spot the firing of the cannon more easily.

(ii) Carrying out his experiments at a water temperature of 8 °C, he measures an average time of 9.4 s for a sound wave to travel 13,487 m, from which he calculates a sound velocity of 1,434.79 m/s {Colladon & Sturm ⇨1828} – a value closely approximating modern data.[550] |
|---|---|---|
| 1826 | Ecole Polytechnique, Paris, France | **Augustin-Louis Cauchy,**[551] a French mathematical physicist, **treats the impact of two cylindrical rods which meet each other head-on with equal or different velocities analytically:**

(i) Applying the differential equations derived by him in 1822 for the vibrations of an elastic or inelastic solid body means that motion in this one-dimensional percussion problem is governed by just two variables: the time $t$ and the abscissa $x$. Thus, the partial differential equations can immediately be integrated provided that the principal variables satisfy the conditions given by the problem.

(ii) He notices that there are differences between the results derived from ordinary percussion theory of elastic bodies and results obtained from experiments or from a more advanced theory better conforming with nature, as had already been pointed out previously by de Coriolis.
▪ Cauchy, who had developed a strong interest in elasticity since the publication of Navier's paper on equilibrium and vibration of elastic solids (1821), worked out the fundamental mathematical apparatus of the theory of elasticity in the 1830s.

A theory of impacting cylinders which also provides information on the contact time and the velocities after impact was eventually established by Barré de Saint-Venant {⇨1867} and, independently, by Franz E. Neumann {⇨1885}. |
| 1827 | Pont de la Guillotière [near Lyon], Dépt. Rhône, east central France | On March 3, **the steamboat *Le Rhône***, assigned to tow ships between Arlet and Lyon, **violently explodes in the vicinity of Lyon harbor. The accident,** which largely destroys the boat, **is accompanied by the puzzling but frequently observed phenomenon of "synchronous" boiler explosions.**

In the years following the accident, the French physicist D. François J. Arago began to carefully analyze the possible causes of steam-boiler explosions {⇨Fig. 4.21−A}. He described the accident as follows {Arago ⇨1830}: "The steamboat was provided with an unusually large engine of good workmanship made by a Paris machine shop and fed with steam by four boilers made from iron sheet. Each boiler had a diameter of 1.3 m. After the accident, it was shown that at many locations the iron had a thickness of only 5 mm. The explosion occurred on March 4, 1827 … many persons became victims of this sad event. Even on the quay of the Rhône, some spectators were killed by flying woodwork. The deck was completely blown off; the chimneys, which weighed more than 30 quintals [3,000 kg], were shot up almost vertically to a considerable height; the vaulted dome of one boiler was thrown out 250 m even though it weighed at least 20 quintals. This dreadful event was an unavoidable consequence of the carelessness of the operator. Annoyed at not being able to overcome the velocity of the current as fully as anticipated, **Mr. Steel** locked the safety valves of all four boilers, thus re- |

---

[550] For a given temperature $\theta$, ranging from 0 to 30 °C, the velocity of sound, $u$ in m/s, can be approximated by the empirical relationship
$u = 1,449.2 + 4.6\theta - 0.055\theta^2 + 0.00029\theta^3 + 35 (0.01\theta - 1.34)$. See C.S. Clay and H. Medwin: *Acoustical oceanography: principles and applications*. Wiley, New York (1977). Using this relation, and assuming a temperature of 8 °C, this yields a value of 1,438.53 m/s, which is in excellent agreement with the data of Colladon, who used a rather simple experimental method. The error is less than 0.3%.

[551] A.L. Cauchy: *Mémoire sur le choc des corps élastiques*. Nouv. Bull. Sci. Soc. Philomat. Paris (1826), pp. 180-182.

moving all of their mobility. This fact – though appearing incredible – has been authentically recorded. We have already mentioned that the boat was equipped with four boilers; two of them undoubtedly exploded simultaneously. If I have been informed correctly, the third boiler, which was also found to have burst, was later taken out of the Rhône. The bursting of two or three boilers within the same second is a strange fact to which we shall return when speaking about the various explanations given of these phenomena…" ▪ **In the example of simultaneous boiler explosions, other neighboring boilers, stressed by excessive steam pressure, can be triggered to burst too by either the blast wave or by boiler debris emitted from the first boiler explosion.** A very similar phenomenon of simultaneous explosions was also observed in arrangements of multiple explosive charges {ABEL ⇨1869}. It was correctly interpreted as being due to the blast wave from one explosion, which can trigger a neighboring charge {E. MACH & WENTZEL ⇨1885}.

| 1827 | Soho Square [formerly Sir Joseph BANK's private botanic garden], London, England | Robert BROWN,[552] a Scottish botanist studying microscopic life, **notices that little particles of plant pollen** [*clarkia pulchella*], about $1/4{,}000$ to $1/5{,}000$ of an inch (6.4–5.1 µm) in length and suspended in water, **constantly and rapidly jiggle about in random directions – so-called "Brownian motion"** {⇨Fig. 4.4–A}.<br><br>(i) He makes the important statement that the motion of the particles, which he terms *active molecules*, "arose neither from currents in the fluid, nor from its gradual evaporation, but belonged to the particles itself."<br><br>(ii) Further experiments carried out with inanimate powdered specimens (*e.g.*, pit coal, metals, glass, rocks "of all ages," and even a fragment of the Egyptian Sphinx statue), such as pit coal, metals, glass, rocks "of all ages," and even a fragment of the Egyptian Sphinx statue, revealed the same phenomenon. ▪ BROWN's discovery – a result of statistical fluctuations in number of fluid molecules, which are in constant thermal motion, bombarding each side of each particle – first illustrated the important role that collisions plays in nature at the molecular level.<br><br>**The phenomenon of Brownian motion confirmed Daniel BERNOULLI's crude hypothesis that temperature can be expressed by the motion of molecules** {⇨1738}, thus stimulating the evolution of the kinetic theory of gases {KRÖNIG ⇨1856}. The analysis of Brownian motion by Albert EINSTEIN (1905) led to the formulation of the Boltzmann constant. The French physicist Jean-Baptiste PERRIN,[553] resuming BROWN's study in the late 1900s, investigated the successive motions of small particles experimentally by using an automatic camera to record the positions of a particle at 30-second intervals {⇨Fig. 4.4–A}. His analysis confirmed the atomic nature of matter.<br><br>More recently, BROWN's classical experiment was repeated by Brian J. FORD,[554] a scientific writer and lecturer. Using BROWN's original microscope, he made a movie of "Brownian" pollen motion, which he posted on the Internet {⇨Fig. 4.4–A}. |
| 1827 | Dreyse & Collenbusch Company, Sömmerda, Prussia, Germany | **Nikolaus VON DREYSE,** a German gunsmith, stimulated by contemporary experiments to convert the flint lock into the more reliable percussion lock {FORSYTH ⇨1805}, **invents his famous *Zündnadelgewehr*** ("needle gun").[555] It uses a new kind of cartridge which combines propellant, bullet and primer. The percussion cap is located in the base of the bullet and is triggered by a long, slender firing pin which pierces the paper cartridge from the rear. ▪ The disadvantages of flintlocks became particularly obvious during the Napoleonic Wars (1803–1815). |

---

[552] R. BROWN: *A brief account of microscopical observations made in the months of June, July and August 1827, on the particles contained in the pollen of plants; and on the general existence of active molecules in organic and inorganic bodies.* Phil. Mag. **4** [II], 161-173 (1828); Supplement Ibid. **6** [II], 161-166 (1829).

[553] J. PERRIN: *Mouvement Brownien et réalité moléculaire.* Ann. Chim. Phys. **18** [VIII], 5-114 (1909).

[554] B.J. FORD: *Brownian movement in clarkia pollen: a reprise of the first observations.* The Microscope **40**, 235-241 (1992). For a reproduction of this paper and a movie of Brownian movement *see* http://www.sciences.demon.co.uk/wbbrowna.htm.

[555] F. PFLUG: *Nikolaus VON DREYSE und die Geschichte des preußischen Zündnadelgewehrs.* Haude & Spener, Berlin (1866).

| | | |
|---|---|---|
| 1828 | Aycliffe Level, Stockton & Darlington Railway, England | **On July 1, a steam boiler explosion of the *Locomotion No. 1* kills the driver and maims the water pumper.** Wreckage from the engine covers several fields. ▪ This famous locomotive was constructed by Robert STEPHENSON, the world's first locomotive builder, and built in his factory at Forth Banks, Newcastle. ▪ It hauled the first train on the opening day of the Stockton & Darlington Railway on September 27, 1825. After the explosion, the locomotive was completely overhauled and remained in service until 1841. *Locomotion No. 1* is now on display at the Darlington Railway Centre and Museum.[556] |
| 1828 | Compagnie du Canal des Ardennes, France | In a pioneering study, **Jean Baptiste C. BÉLANGER,**[557] a French professor of mechanical engineering, **investigates the behavior of water flow in an open channel and high-speed shooting with sudden changes in depth — known as a "hydraulic jump"** — the oldest type of discontinuous wave motion and easily resolvable with the naked eye in nature. For a hydraulic jump of water height $h_2$, he derives a formula for the step height $\Delta h = h_2 - h_1$ in terms of the initial water depth $h_1$ and the velocity $v$ of the jump, given by $$\Delta h = 0.5\varepsilon - h_1 + (0.25\varepsilon^2 + h_1)^{1/2}$$ with $\varepsilon = v^2/2g$, where $g$ is the acceleration due to gravity. ▪ His remarkable study is an early attempt to characterize the propagation speed of a discontinuous wavefront by its strength, in his case a hydraulic jump propagating in (incompressible) shallow water by its step height $\Delta h$. Analogously, for the case of a shock wave advancing in a compressible fluid in a layer of invariant thickness, this would correspond to a step increase in density at the shock front given by $\Delta \rho = \rho - \rho_0$.<br><br>Using his classical theory, Émile JOUGUET {⇨1920} showed that the loss of internal energy [French *perte de charge*] of a hydraulic jump can be described in terms of the difference in water heights, which is a particular case of HUGONIOT's law of the dynamic adiabat or "Hugoniot curve" {HUGONIOT ⇨1887} when the water is considered to be an adiabatically moving "hydraulic gas" with $\gamma = 2$. The analogy between a hydraulic jump and a shock wave has long fascinated shock wave researchers {PREISWERK ⇨1938; H.A. EINSTEIN & BAIRD ⇨1946/1947}. |
| 1828 | Faculté des Sciences, Paris, France | **François Sulpice BEUDANT,**[558] a French professor of mineralogy, **reports on the first artificial production of *fulgurites* or *lightning tubes*.** Natural fulgurites — a word derived from the Latin *fulgur*, meaning "lightning" — are glass-like substances, the result of lightning striking the Earth's surface and melting soil or rock {DARWIN ⇨1832; WITHERING ⇨1879; ⇨Fig. 4.1–J}.<br><br>(i) BEUDANT performed the experiments together with Jean N.P. HACHETTE, a French mathematician, and Felix SAVART, a French physicist and surgeon. They used glass or quartz powder and the largest battery of Leiden jars then available in Paris, which was kept at the Charles Cabinet of the Conservatoire des Arts et Métiers. Storing the testing powder in a hole drilled into a brick, they initiated a concentric electric discharge through the powder column.<br><br>(ii) BEUDANT reports that a void was created by the expansion and escape of gas and moisture resulting from the explosive fusion during the strike. In the case of glass powder, the high |

---

[556] *Locomotion*. Darlington Railway Centre and Museum, Darlington, Tees Valley, U.K.; http://www.drcm.org.uk/Content/Collections/Locomotion.htm.
[557] J.B.C. BELANGER: *Essai sur la solution numérique de quelques problèmes relatifs au mouvement permanent des eaux courantes*. Carilian-Gœury, Paris (1828).
[558] F.S. BEUDANT: *Expérience sur la formation des tubes fulminaires*. Ann. Chim. Phys. **37** [II], 319-321 (1828); F.S. BEUDANT: *Versuche über die Bildung von Blitzröhren*. Ann. Phys. **13** [II], 117-118 (1828).

temperature of the arc, which melted the powder around the arc, produced 25-mm long fulgurites with inner diameters of 0.5 mm and outer diameters ranging from 1.5 to 3 mm and appearances very similar to natural fulgurites. However, experiments performed with quartz powder failed.[559]

The production of quartz fulgurites requires higher pulse energies than were available to BEUDANT, and was first demonstrated in the United States {MCCOLLUM ⇨ 1997}.

| | | |
|---|---|---|
| 1828 | Geneva, western Switzerland | Daniel COLLADON,[560] together with Charles-François STURM, a Swiss mathematician and private tutor, **publishes experimental data on the compressibility of various substances and on the velocity of sound in water** {COLLADON ⇨ 1826}. They show that putting this data into POISSON's formula for the speed of sound yields a value of 1,437.8 m/s, which closely agrees with their measured value of 1,435 m/s (at 8 °C). They also report on the measurement of heat emitted by liquids following the application of strong and sudden pressures. Their results earn them a prize from the Paris Academy. |
| 1830 | Ecole Polytechnique, Paris, France | D. François J. ARAGO,[561] a French physicist and astronomer, **discusses the possible causes of frequent boiler explosions of steam engines,** which typically result in many casualties and heavy damage to adjacent facilities {SS *Le Rhône* ⇨ 1827}. Addressing the dangers emanating from the use of PAPIN's safety valve {⇨ 1679}, he points out that many valve constructions are too narrow to allow quick release when the internal boiler pressure increases suddenly {AIRY ⇨ 1863}, a dangerous phenomenon which can be ascribed to various causes.<br><br>The limitation of the outflow of fluid through small openings became a subject of much discussion among engineers and scientists {D. BERNOULLI ⇨ 1738; DE SAINT-VENANT & WANTZEL ⇨ 1839; NAPIER ⇨ 1866; REYNOLDS ⇨ 1885; SALCHER & WHITEHEAD ⇨ 1889; L. MACH ⇨ 1897; P. EMDEN ⇨ 1903; STODOLA ⇨ 1903; PRANDTL ⇨ 1904; *etc.*}. The topic stimulated not only the evolution of supersonic flows, but also promoted the effective operation of steam turbines {DE LAVAL ⇨ 1888}. |
| 1831 | Charleston, South Carolina | On **June 17, the steam-boiler of America's first passenger railroad locomotive, the *Best Friend of Charleston*, explodes and kills one person.** The first accident of this kind was caused by an incorrect operation: the steam was allowed to accumulate by the negligence of the fireman, who pressed on the safety valve and therefore prevented the surplus steam flow escaping, causing the boiler to burst at the bottom. ▪ The *Best Friend of Charleston* was the first steam locomotive in the U.S.A. to establish a regular scheduled passenger service. Operated by the South Carolina Canal & Railroad Company, it made its premier trip on Christmas Day, 1830. The locomotive is now housed in the Engine House at Ann Street, Charleston, SC.[562]<br><br>In America, **the first head-on collision of two trains happened on May 6, 1853** at Secaucus, NJ (46 people died). |
| 1831 | Cornwall, England | William BICKFORD,[563] a British leather merchant, **conceives a safe device for igniting gunpowder** – the so-called **"Bickford fuse,"** often referred to as the **"safety fuse"** – a slow burning fuse used to either prepare internal shell delays or to time sequential firing. It consists of a core of black powder wrapped in textile (*e.g.*, jute yarn) coated with a waterproofing agent (*e.g.*, asphalt or natural rubber).[564] ▪ Prior to BICKFORD's invention, blasting was done with gunpowder, ignited by a fuse made of wheat straws or goose quills nested together and filled |

---

[559] Quartz glass (SiO$_2$) has a higher melting temperature of about 1,600 °C compared to modern soda-lime glass which has a transition point at about 500 °C.
[560] D. COLLADON and C.F. STURM: *Über die Zusammendrückbarkeit der Flüssigkeiten*. Ann. Phys. **12** [II], 161-197 (1828).
[561] D.F.J. ARAGO: *Über die Explosionen der Dampfmaschinen*. Ann. Phys. **18** [II], 287-314, 415-436 (1830).
[562] *Best Friend of Charleston*. Charleston Chapter of the National Railway Historical Society; http://bestfriendofcharleston.org/.
[563] W. BICKFORD: *Safety fuze for igniting gunpowder used in blasting rocks*, &. Brit. Patent No. 6,159 (1831).
[564] N.G. JOHNSON: *Explosives*. In: *The new Encyclopædia Britannica, Macropædia*. Benton & Hemingway, Chicago *etc.*, vol. 7 (1974), p. 84.

with powder. One day, while visiting a rope maker, the idea came to BICKFORD that if a funnel filled with powder could be arranged to pour a stream into the center of the twisted strands, and if the rope could be securely fastened and waterproofed, he would have a slow-burning fuse.[565]

BICKFORD established a factory in Cornwall to serve the Cornish tin mines. The safety fuse provided a dependable means for conveying flame to the charge and was very useful in underground coal mining for igniting black powder. Its timing – *i.e.*, the time required for a given length to burn – is slow, about 2–4 ft/min (1–2 cm/s), but amazingly accurate. Bickford fuses remained in general use until the 1920s.

| | | |
|---|---|---|
| 1831 | *Ecole de Guerre, Metz, Dépt. Moselle, northeastern France* | **Général Guillaume PIOBERT,**[566] a professor of gunnery and member of the French Academy of Sciences, **begins a series of experiments on military gunpowder.** He advances the hypothesis |

  (i) that **in a closed chamber the combustion of grains of black powder proceeds in parallel layers** [French *par couches parallèles*] in the direction normal to the surface – the so-called "Piobert law;" and

  (ii) to operate fire arms effectively, **the barrel length should be matched to the charge weight,** such that the powder should be burnt completely at the moment that the projectile leaves the barrel. [French *«La poudre la plus convenable, pour une arme déterminée, est celle qui, brûlant d'une manière complète dans le temps que met le projectile à parcourir l'âme de la pièce, lui imprime non instantanément mais graduellement, toute la force de projection dont elle est susceptible.»*]

Studies performed almost sixty years later by his countryman Paul VIEILLE[567] revealed that PIOBERT's hypothesis could not be confirmed for black powder of ordinary density ($\rho = 1.65$ g/cm$^3$); however, it could be proved when heavily compressed ($\rho \geq 1.85$ g/cm$^3$).[568]

| | | |
|---|---|---|
| 1831 | *Faculté des Sciences, Paris, France* | **In a continuation of his previous theoretical studies on the equilibrium and motion of elastic bodies,**[569, 570] **Siméon-Denis POISSON,**[571] an eminent French mathematician and one of the main founders of modern mathematical physics, **shows that a disturbance produced in a small portion of a solid body results in two kinds of waves:** |

  ▸ in the faster wave, the motion of each particle is normal to the wavefront and accompanied by volume changes (compression or dilatation); and

  ▸ in the other (slower) wave the motion of each particle is tangential to the wavefront, and there is only distortion without volume change during the motion.

The discovery of the existence of longitudinal and transverse waves in a solid body was a significant milestone in fundamental mechanics. In seismology, these two types of waves are termed

  ▸ *longitudinal waves* or *primary (P-) waves* with velocity $v_P = [(\lambda + 2\mu)/\rho]^{1/2}$, and
  ▸ *transverse waves* or *secondary (S-) waves* with velocity $v_S = [\mu/\rho]^{1/2} < v_P$.

Here $\rho$ is the density, and $\lambda$ and $\mu$ are elastic constants. **Seismic shock waves comprise *body waves*** (P- and S-waves), which can travel through the Earth's inner layers, **and *surface waves***

---

[565] William BICKFORD, *the inventor of the safety fuse*; http://www.cornwall-calling.co.uk/famous-cornish-people/bickford.htm.
[566] G. PIOBERT: *Traité d'artillerie théorique et pratique: précis de la partie élémentaire et pratique.* Levrault, Metz (1836); *Mémoires sur les pouvoirs de guerre des différents procédés de fabrication, avec résumés des épreuves comparatives faites sur ces poudres à Esquerdes en 1831 et 1832 et à Metz en 1836 et 1837.* Bachelier, Paris (1844).
[567] P. VIEILLE: *Etude sur le mode de combustion des matières explosives.* Mém. Poudres & Salpêtres **6**, 256-391 (1894).
[568] C. CRANZ: *Lehrbuch der Ballistik: II. Innere Ballistik.* Springer, Berlin (1926), p. 121.
[569] S.D. POISSON: *Mémoire sur l'équilibre et le mouvement des corps élastiques* [read April 14, 1828]. Mém. Acad. Sci. Inst. France **8**, 357-570, 623-627 (1829).
[570] S.D. POISSON: *Mémoire sur les équations générales de l'équilibre et du mouvement des corps élastiques et des fluides* [1829]. J. Ecole Polytech. (Paris) **13**, No. 20, 1-174 (1831).
[571] S.D. POISSON: *Mémoire sur la propagation du mouvement dans les milieux élastiques* [read Oct. 11, 1830]. Mém. Acad. Sci. Inst. France **10**, 549-606 (1831).

(Rayleigh waves and Love waves), which can only move along the Earth's surface like ripples on water. Rayleigh waves {Lord RAYLEIGH ⇨1885}, a combination of longitudinal and transverse motions, are particularly destructive in earthquakes.

| 1832 | *Maldonado [a small town on the northern bank of the Rio de la Plata, 140 km from Montevideo], Uruguay, South America* | On July 26, during his voyage on the HMS *Beagle*, **Charles R. DARWIN,**[572] the famous British naturalist, **makes an entry in his journal noting that he has found a group of vitrified, siliceous tubes formed by lightning entering loose sand near the sand-hillocks of Maldonado** {WITHERING ⇨1790}. He writes, "The sand-hillocks of Maldonado, not being protected by vegetation, are constantly changing their position. From this cause the tubes projected above the surface; and numerous fragments lying near, showed that they had formerly been buried to a greater depth. Four sets entered the sand perpendicularly: by working with my hands I traced one of them two feet deep; and some fragments which evidently had belonged to the same tube, when added to the other part, measured five feet three inches. The diameter of the whole tube was nearly equal, and therefore we must suppose that originally it extended to a much greater depth … Such easy fusibility as glass was to form tubes so diminutive, **we must feel greatly astonished at the force of a shock of lightning, which, striking the sand in several places, has formed cylinders, in one instance of at least thirty feet long, and having an internal bore, where not compressed, of full an inch and a half; and this in a material so extraordinarily refractory as quartz!** … The tubes, as I have already remarked, enter the sand nearly in a vertical direction. One, however, which was less regular than the others, deviated from a right line, at the most considerable bend, to the amount of thirty-three degrees. From this same tube, two small branches, about a foot apart, were sent off; one pointed downwards, and the other upwards. This latter case is remarkable, as the electric fluid must have turned back at the acute angle of 26°, to the line of its main course … **The internal surface is completely vitrified, glossy, and smooth.** A small fragment examined under the microscope appeared, from the number of minute entangled air or perhaps steam bubbles, like an assay fused before the blowpipe. The sand is entirely, or in greater part, siliceous; but some points are of a black color, and from their glossy surface possess a metallic luster. The thickness of the wall of the tube varies from a thirtieth to a twentieth of an inch, and occasionally even equals a tenth. On the outside the grains of sand are rounded, and have a slightly glazed appearance … **The tubes are generally compressed, and have deep longitudinal furrows,** so as closely to resemble a shrivelled vegetable stalk, or the bark of the elm or cork tree. Their circumference is about two inches, but in some fragments, which are cylindrical and without any furrows, it is as much as four inches. The compression from the surrounding loose sand, acting while the tube was still softened from the effects of the intense heat, has evidently caused the creases or furrows … Judging from the uncompressed fragments, the measure or bore of the lightning (if such a term may be used) must have been about one inch and a quarter … **At Paris, M. HACHETTE and M. BEUDANT** [BEUDANT ⇨1828] **succeeded in making tubes, in most respects similar to these fulgurites, by passing very strong shocks of galvanism through finely-powdered glass.** When salt was added, so as to increase its fusibility, the tubes were larger in every dimension. They failed both with powdered feldspar and quartz." |
| 1832 | *University of Leipzig, Saxony, east-central Germany* | **Gustav Theodor FECHNER,**[573] a German editor and translator of physics and chemistry literature, publishes his book *Repertorium der Experimentalphysik* ("Repertory of Experimental Physics"), in which he reviews recent advances made in this science. On page 102 he **addresses different kinds of surface water wave patterns produced by waves of short** |

---

[572] C.R. DARWIN: *A naturalist's voyage around the world. Journal of researches into the geology and natural history, of HMS Beagle*. J. Murray, London (1913), chap. III, p. 53. The University of Adelaide Library; http://etext.library.adelaide.edu.au/d/darwin/charles/beagle/.

[573] G.T. FECHNER: *Repertorium der Experimentalphysik. Enthaltend eine vollständige Zusammenstellung der neueren Fortschritte dieser Wissenschaft*. Voß, Leipzig (1832).

wavelengths – so-called "ripples." Ripples form around objects moving slowly across a water surface or when a slow current flows past an obstacle; in both cases the ripples are reminiscent of the wake generated by a moving ship. ▪ FECHNER, who was also a professor of experimental physics, failed to derive a satisfactory solution to the problem. However, he urged his readers to approach this curious wave phenomenon analytically.

It is a difficult mathematical task to describe the pattern of ripples in the neighborhood of an object moving across water.

(i) Jean-Victor PONCELET,[574] a French professor of mechanics at the Ecole d'Application in Metz and previously a pupil of the eminent geometrician Gaspard MONGE, tried in vain to find a geometrical solution that would enable the wave pattern to be determined from the speed of the object.

(ii) DOPPLER {⇨1847}, stimulated by FECHNER's book and PONCELET's studies, **tackled the problem from a different point of view, which eventually led to his geometric construction of the Mach head wave** {E. MACH & SALCHER ⇨1886}.

(iii) In 1871, William THOMSON[575] (from 1892 Lord KELVIN) outlined the cognate theory of the ripples and waves produced when a local disturbance travels over a horizontal sheet of water at a speed exceeding a certain minimum. Using the theory, he determined a minimum velocity of 23 cm/s, which was in excellent agreement with the experimental value (mean value 23.22 cm/s).[576]

The wake pattern of a body (*e.g.*, a ship) moving on the surface of a liquid is rather complex and best visualized by optical aerial photography: it consists of a bow wave, a stern wave, a transverse wave, a turbulent (or vortex) wake trailing the body, and a turbulent region adjacent to the body's surface. The so-called **"Kelvin wake"** is formed by cusp waves and consists of two arms – the **"Kelvin arms"** – which trail the body in the form of a V-shaped pattern, forming the **"Kelvin envelope"** {⇨Fig. 4.6–D}. The turbulent wake is also often visible on images taken by the Synthetic Aperture Radar (SAR) aboard the European Remote Sensing (ERS) satellites ERS-1 and ERS-2, but the visibility of the Kelvin arms strongly depends on the azimuth angle of the radar relative to the ship's heading.[577]

| | |
|---|---|
| 1832 *University of Gießen & Technical School, Kassel, Hesse, Germany* | **Justus VON LIEBIG**[578] **and Friedrich WÖHLER,** two eminent German chemists, **study the properties of bitter almond oil** [benzaldehyde, $C_6H_5C-OH$] **and find that the same group, benzoyl** [$C_6H_5CO-$], **passes unchanged through a great variety of reactions – it is a so-called "radical."** ▪ VON LIEBIG and WÖHLER were stimulated by French studies: in 1830, the chemists Pierre ROBIQUET[579] and Antoine BOUTRON-CHARLAND converted the oil of bitter almonds to benzoic acid by oxidation. They also created a neutral compound which decomposed to benzoic acid and another crystalline substance which formed the same acid from the oil. They named the crystalline material "amygdalin," and inferred that the oil was a *benzoyl radical*. |

**Attempting to systematize organic chemistry, VON LIEBIG later developed his radical theory,** which presumes that organic compounds are the result of interactions and transformations of radicals rather than of atoms. Later the term *radical* was used to designate an atom or group of atoms containing an unpaired electron, and thus became synonymous with *free radical*, meaning "uncombined radical." Free radicals are formed as transient intermediates in a large

---

[574] J.V. PONCELET: *Notice sur quelques phénomènes produits à la surface libre des fluides, en repos ou en mouvement, par la présence des corps solides qui y sont plus ou moins plongés, et spécialement sur les ondulations et les rides permanents qui en résultent*. Ann. Chim. Phys. **46** [II], 5-25 (1831).

[575] W. THOMSON: *The influence of wind on waves in water supposed frictionless*. Phil. Mag. **42** [IV], 368-374 (1871); *Waves under motive power of gravity and cohesion jointly, without wind*. Ibid. **42** [IV], 374-377 (1871).

[576] W. THOMSON: *Ripples and waves*. Nature **5**, 1-3 (Nov. 2, 1871).

[577] W. ALPERS ET AL.: *The tropical and subtropical ocean viewed by ERS SAR*. ESA; http://earth.esa.int/applications/ERS-SARtropical/.

[578] J. VON LIEBIG and F. WÖHLER: *Untersuchungen über das Radikal der Benzoesäure*. Ann. Pharm. **3**, 249-282 (1832).

[579] P. ROBIQUET and A. BOUTRON-CHARLAND: *Nouvelle expériences sur les amandes amères, et sur l'huile volatile qu'elles fournissent*. Ann. Chim. **44** [II], 352-382 (1830).

number of organic reactions. The first free radical was produced in 1900 by the Russian-born U.S. chemist Moses GOMBERG. It is generally considered that short-lived free radicals play an important role in self-propagating reactions. In *branched-chain explosions*, which are generally very rapid reactions, large numbers of such highly reactive intermediate chemical species are produced which further accelerate the rates of these reactions {BODENSTEIN ⇨1913}.

| 1833 | Manchester Mechanics Institute, Manchester, England | Eaton HODGKINSON,[580] a British engineer and experimentalist, **studies the collision of a beam with a ball.** His experimental set-up consists of a 1.3-m long steel beam supported at both ends which is impacted transversally by a 22-kg cast iron ball suspended on a 5-m-long wire. He records the maximum deflections using a stylus fixed along the beam, which writes into a layer of soft clay. • HODGKINSON showed that permanent deformation takes place under the smallest load – thus no material can be regarded as perfectly elastic.

In the following two years, HODGKINSON continued his experiments and, by comparing static with dynamic deflections, tried to derive general laws for this particular case of impact loading, which was then of practical concern in the rapidly growing construction and railroad industry. • Although he was not successfully from the present point of view, his studies stimulated subsequent researchers like the British mathematician Homersham COX,[581] who tackled the problem more on a scientific basis and worked out an approximation method. |

| 1833 | Ecole d'Application de Metz, Dépt. Moselle, northeastern France | At the behest of Siméon-D. POISSON, Arthur Jules MORIN,[582] a French professor of mechanics who will later follow Gaspar G. DE CORIOLIS to the Chair of Mechanics at the Ecole Polytechnique, **begins a two-year campaign of experiments to study the influence of friction on percussion phenomena – "shock friction," sometimes referred to as "impulsive friction" –** presumably a third type of friction between impacting bodies beyond the two ordinary types of static and dynamic (or kinetic) friction. **For example, he finds that the same ratio of shock friction to the normal force of percussion exists as observed in the case of ordinary friction, independent of velocity.**

It had already been noticed in early pile-driving techniques that friction plays an important role in practical percussion {GALILEI ⇨1638; HUTTON ⇨1812; ⇨Fig. 4.2−N}. The fundamentals of friction were first investigated by Charles Augustin COULOMB, a French physicist, in his prize-winning memoir *Théorie des machines simples* (1781). Studying both static and dynamic friction of sliding surfaces and friction in rolling, he found that friction can be described by a two-term equation: the first term is a constant and the second term varies with time, velocity, velocity rate (acceleration), or other parameters. Today COULOMB and MORIN are considered to be the founders of the new science of tribology. Impulsive friction was first treated analytically by POISSON {⇨1811} and DE CORIOLIS {⇨1835}, and in more detail by ROUTH {⇨1860}. |

| 1834 | Stotternheim [a town near Erfurt], Thuringia, east central Germany | Karl G. KIND,[583] a German mining engineer, **invents *Schlagbohren* ("percussion boring").** His new free-fall method uses a heavy tool which, when repeatedly dropped on the same spot from a suitable height, pulverizes the rock and gradually penetrates it {⇨Fig. 4.2−Q}. Together with E.H. Karl VON OEYNHAUSEN's *Rutschschere* ("gliding jar"), a device for coupling and decoupling the drill bit with the bore rods, it provides the first effective technique for drilling deep holes. • In 1846, at Mondorf (a little town at the French-Luxembourg border), he |

---

[580] E. HODGKINSON: *On the effect of impact on beams.* Rept. Meet. Brit. Assoc. **3**, 421-422 (1833); *On the collision of imperfectly elastic bodies.* Ibid. **4**, 534-543 (1834); *On impact upon beams.* Ibid. **5**, 93-116 (1835). • In 1841, Eaton HODGKINSON (1789–1861) became professor of mechanics and engineering at University College in London. In later years, he was regarded as a European authority upon the properties of iron (cast or wrought) used in construction.

[581] H. COX: *On impacts on elastic beams.* Trans. Cambr. Phil. Soc. **9**, 73-78 (1849).

[582] A. MORIN: *Lettre de M. MORIN, capitaine d'artillerie, à M. ARAGO sur diverses expériences relatives au frottement et au choc des corps.* Ann. Chim. Phys. **56** [II], 194-198 (1834); *Nouvelles expériences faites à Metz en 1833 sur le frottement, sur la transmission du mouvement par le choc, sur les résistances des milieux imparfaits à la pénétration des projectiles, et sur le frottement pendant le choc.* Mém. Sav. Acad. Sci. Inst. France **VI** [II], 641-785 (1835).

[583] K.G. KIND: *Anleitung zum Abteufen der Bohrlöcher.* G. Michaelis, Luxemburg (1842).

proved the soundness of his method by drilling a 736-m-deep hole: a world record drilling depth which was not surpassed until 1871.

**KIND's drilling method stimulated others to construct effective pneumatic drilling hammers.** Their great practicability was first proved in France during the construction of the Mont Cenis Tunnel {⇨1857} – the largest construction job performed with black powder up to that time – and in Switzerland during the construction of the St. Gotthard Tunnel (1872–1881).

| | | |
|---|---|---|
| 1834 | *Union Canal at Hermiston [near Riccarton Campus of Heriot-Watt University], Edinburgh, Scotland, U.K.* | While conducting experiments to determine the most efficient design for canal boats, **the Scottish engineer and naval architect John S. RUSSELL chances across a *solitary wave*** – a single positive wave that continues to travel down the canal, thereby maintaining its shape for a great distance.<br><br>In the following years, RUSSELL[584] made field observations of its height and speed, supplemented by small-scale experiments. He then proposed the well-known equation for the velocity of propagation of water waves in shallow water {DE LAGRANGE ⇨1781}, which he modified into<br>$$v = (g h')^{1/2},$$<br>where $g$ is the acceleration due to gravity, and $h'$ denotes the mean depth of the water at the wave crest (rather than at rest) for any cross-sectional shape of channel.<br><br>Ten years after his discovery, RUSSELL {⇨1844} described his important discovery as follows, "I was observing the motion of a boat which was rapidly drawn along a narrow channel by a pair of horses, when the boat suddenly stopped – not so **the mass of water in the channel** which it had put in motion; it accumulated round the prow of the vessel in a state of violent agitation, then suddenly leaving it behind, rolled forward with great velocity, **assuming the form of a large solitary elevation, a rounded, smooth and well-defined heap of water, which continued its course along the channel apparently without change of form or diminution of speed.** I followed it on horseback, and overtook it still rolling on at a rate of some eight or nine miles an hour [3.6 or 4 m/s; 12.9 or 14.5 km/s], preserving its original figure some thirty feet [9.1 m] long and a foot to a foot and a half [0.3 to 0.45 m] in height. Its height gradually diminished, and after a chase of one or two miles [1.6 or 3.2 km] I lost it in the windings of the channel. Such, in the month of August 1834, was my first chance interview with that singular and beautiful phenomenon which I have called the **'Wave of Translation.'**"<br><br>Almost 25 years after his discovery, RUSSELL[585] wrote: "This is the most beautiful and extraordinary phenomenon: the first day I saw it was the happiest day of my life. Nobody had ever had the good fortune to see it before or, at all events, to know what it meant. It is now known as the **'solitary wave of translation'**…"<br><br>A solitary wave has gradually leading and trailing edges; *i.e.*, it is not a shock wave. The water is still before and after the passage of the wave, but it moves forward under the wave as it passes. In mathematical physics, **the solitary wave – also known as the "soliton" – is a wave-like solution to a nonlinear differential equation describing a wave which maintains a constant shape as it propagates.**[586]<br><br>▸ In 1895, the Dutch mathematicians Diederik Johannes KORTEWEG[587] and Gustave DE VRIES described the behavior of certain types of waves that occurred in a shallow canal using a nonlinear differential equation, known as the "cnoidal theory." The so-called **"Korteweg-de Vries equation"** describes the spatial changes in the amplitude of the wave with time. Its solutions are *cnoidal waves* – periodic waves with sharp crests separated by wide flat troughs, where the wave crests are higher above the still- |

---

[584] J.S. RUSSELL and J. ROBISON: *Report of the Committee on Waves*. Rept. Meet. Brit. Assoc. **7**, 417-496 (1838).
[585] J.S. RUSSELL: *The modern system of naval architecture*. Day & Son, London (1865), p. 208.
[586] A.C. NEWELL: *Solitons in mathematics and physics*. Society for Industrial and Applied Mathematics (SIAM), Philadelphia, PA (1985).
[587] D.J. KORTEWEG and G. DE VRIES: *On the change of form of long waves advancing in a rectangular canal, and on a new type of long stationary waves*. Phil. Mag. **39** [V], 422-443 (1895).

water level (SWL) than the troughs are below the SWL.[588, 589] Long finite-amplitude permanent waves propagating in shallow water may be described by the cnoidal theory. Note that the solitary wave, a special form of the cnoidal wave, has *all* of its profile above the SWL.

- Modern numerical calculations of solitary waves in shallow water have been reviewed by John W. MILES[590] (Institute of Geophysics and Planetary Physics, UC La Jolla, CA), Leonard W. SCHWARTZ[591] (School of Mathematics, University of New South Wales), together with J.D. FENTON (Dept. of Civil Engineering, University of Auckland), and Mykola YASKO[592] (Applied Mathematics Faculty, Dnipropetrovsk National University, Ukraine).

- Solitary waves appear in many fields of science, ranging from classical fluid dynamics, solid state and elementary particle physics, to astro- and biophysics. For example, **solitons play an important role in the chain percussion of tangible bodies.** Prominent examples relating to the subject of this book encompass NEWTON's cradle {The Royal Society of London ⇨1666} and the impact of granular materials {NESTERENKO ⇨2001}.

The history of solitons or solitary waves was reviewed by John E. ALLEN,[593] a British engineering professor at the University of Oxford who in the 1950s first studied solitary waves in a collisionless plasma containing a magnetic field.

| | | |
|---|---|---|
| 1834 | *Chair of Experimental Philosophy, Kings College, London, England* | **Charles WHEATSTONE,**[594] a British professor of physics and an inventor, **first uses the *revolving mirror* as a diagnostic tool to resolve high-speed phenomena;** in particular, to measure the speed of electricity in a conductor. ▪ The rotating mirror {⇨Fig. 4.19–I}, which was subsequently used by the French experimental physicist J.B. Léon FOUCAULT[595] (1850) and the German physics professor B. Wilhelm FEDDERSEN[596] (1858) in sensational experiments, later became an important element in high-speed rotating-mirror framing and streak cameras. Stimulated in the early 1940s by the urgent need to study shock wave effects associated with the development of nuclear weapons, work with these cameras eventually resulted in very sophisticated ultrahigh-speed cameras incorporating helium- (instead of air-) driven turbines.[597, 598] |

Thirty-four years later, Sir Edward SABINE,[599] then president of the Royal Society of London, stated at his presentation of the Copley Medal to WHEATSTONE, "But no series of his researches have shown more originality and ingenuity than those by which he succeeded in measuring the velocity of the electric current and the duration of the spark. The principle of the rotating mirror employed in these experiments, and by which he was enabled to measure time to the millionth part of a second, admits of application in ways so varied and important that it may be regarded as having placed a new instrument of research in the hands of those employed in delicate physical inquiries of this order."

---

[588] The Korteweg-de Vries equation had already been found earlier by the French physicist and mathematician Joseph V. BOUSSINESQ: *Essai sur la théorie des eaux courantes*. Mém. Acad. Sci. Inst. France **23** [II], 1-680 (1877).
[589] J.W. MILES: *The Korteweg-de Vries equation: a historical essay*. J. Fluid Mech. **106**, 131-147 (1981).
[590] J.W. MILES: *Solitary waves*. Annu. Rev. Fluid Mech. **12**, 11-43 (1980).
[591] L.W. SCHWARTZ and J.D. FENTON: *Strong nonlinear waves*. Annu. Rev. Fluid Mech. **14**, 39-60 (1982).
[592] M. YASKO: *Numerical method for calculation of the solitary and other nonlinear waves on water of arbitrary depth*. ZAMM **76**, Suppl. 2, 707-708 (1995).
[593] J.E. ALLEN: *The early history of solitons (solitary waves)*. Physica Scripta **57**, 436-441 (1998).
[594] C. WHEATSTONE: *An account of some experiments to measure the velocity of electric light*. Proc. Roy. Soc. Lond. **3**, 299-300 (1834).
[595] L. FOUCAULT: *Méthode général pour mesurer la vitesse de la lumière dans l'air et dans les milieux transparents*. C. R. Acad. Sci. Paris **30**, 551-560 (1850).
[596] B.W. FEDDERSEN: *Beiträge zur Kenntnis des elektrischen Funkens*. Ann. Phys. **103** [II], 69-88 (1858).
[597] P.W.W. FULLER and J.T. RENDELL: *The development of high speed photography*. In: (S.F. RAY, ed.) *High speed photography and photonics*. Focal Press, Oxford *etc.* (1997), pp. 21-23.
[598] V. PARKER and C. ROBERTS: *Rotating mirror and drum cameras*. In: (S.F. RAY, ed.) *High speed photography and photonics*. Focal Press, Oxford *etc.* (1997), pp. 167-180.
[599] E. SABINE: *President's address*. Proc. Roy. Soc. Lond. **18**, 145-147 (1869).

| 1835 | Ecole Centrale des Arts et Manufactures, Paris, France | Gaspar G. DE CORIOLIS,[600] a French mathematical physicist, **publishes his book *Théorie mathématique des effets du jeu de billard* ("Mathematical Theory of the Effects of Billiards"), in which he analytically treats the basic collision problems of billiards, thereby also considering the influence of friction** {POISSON ⇒ 1811; MORIN ⇒ 1833}: |
|---|---|---|

(i) He writes, "The game of billiards as it presents today by the use of appropriate cues for providing the ball with sufficient rotational force imposes diverse dynamic problems for which the solutions can be found in this work. I suppose that persons who have knowledge of rational mechanics like the students of the Ecole Polytechnique will look forward with interest to all the particular effects which can be observed in ball motion. After having seen the production of these effects by the famous player MINGAUD a long time ago, I have attempted to subject them to calculation. My results are the subject of the first and eighth chapter of this work. Furthermore, I have completed what in particular refers to the percussion of balls in consideration of friction. I am obliged to Mr. MINGAUD[601] in that, when watching his play, I ascertained the conformation of the derived formula and constructions as a result of the experience … Monsieur POISSON has treated, in the new edition of his *Traité de Mécanique*, friction effects of a sphere moving along a straight line: this problem is a special case which has to be solved in the game of billiards. The son [Johann Albrecht EULER[602]] of the famous [Leonard] EULER has been occupied with the movement of a sphere on a plane, thereby considering only friction by gliding. His memoir, which I did not find out about until I finished my work, is published in the *Proceedings of the Berlin Academy 1758*. The only propositions which my work have in common are in chapter one and are of a simple nature: it consists of assuming that the curve described by the ball center is a parabola when friction by gliding is not considered. This surveyor has not provided the theorem that yields the motion of the ball until it comes to rest, even when considering the two types of friction. **Concerning the effect of friction on the percussion of balls among one another and with the cushion, and all that refers to inclined strokes by the cue, I don't believe that it has been treated previously.**"

(ii) He considers the impact of two rough spheres sliding on each other during the whole of the impact, and shows that if two rough spheres impinge on each other, the direction of sliding is the same throughout the impact.

(iii) For the basic billiard "shots" [French *coups*] – obtained when the cue hits the ball at dead center (exact middle), above its center ("follow"), below its center ("draw"), and right or left of center, so-called "English" [French *effet*] – he treats central and oblique (sometimes referred to as eccentric) collisions between balls of equal sizes and collisions of a ball with the cushion in an elegant manner. **His theory shows that, after reflection at the cushion or collision with another resting ball, a billiard ball with spin moves along a *curved* trajectory** {⇒ Fig. 4.3–K}.

DE CORIOLIS' theory on billiards was resumed more recently by Arthur SOMMERFELD,[603] a German professor who treated the problem in his lectures on theoretical physics given at Munich University.

| 1835 | Dublin, Ireland | Robert MALLET, an Irish civil engineer, **bores a deep well for Guinness's brewery in Dublin and obtains a good supply of water by setting off an explosive charge at the well bottom.** ▪ His new explosive technique – today known as **"fracturing"** in the oil business – was later described in an article published in the *Practical Mechanic's Journal*[604] in relation to |
|---|---|---|

---

[600] G. CORIOLIS: *Théorie mathématique des effets du jeu de billard*. Carilian-Goeury, Paris (1835).
[601] Captain François MINGAUD (1774–1847) of the French infantry, a billiard enthusiast, practiced the game while being a political prisoner. When liberated, he went on a tour demonstrating billiard shots unlike any seen before.
[602] J.A. EULER: *Recherches des mouvemens d'un globe sur un plan horizontal*. Hist. Acad. Roy. Sci. Belles Lettres (Berlin) **14**, 284-353 (1758).
[603] A. SOMMERFELD: *Vorlesungen über theoretische Physik*. DTB, Wiesbaden (1949), vol. 1: *Mechanik*, pp. 151-155, 259-261.
[604] *Oil springs in Canada*. The Practical Mechanic's Journal **2** [III], 323 (Feb. 1867).

"a well in argillaceous limestone at the Constabulary Barracks, Phoenix Park, Dublin, in 1851." a waterproof canvas cartridge containing 18 lbs (8.1kg) of powder was lowered into the hole to about two-thirds the full depth and ignited by a piece of Davy-lamp fuse. "The shock of the explosion under 10 or 12 feet [3 or 3.6 m] head of water above the top of the bore was felt like that of a small earthquake. The water rose up immediately to within 4 feet [1.2 m] of the surface and has continued to afford an abundant supply."

In about 1866, Canadian oil well borers used a similar technique, which is described in the same article. They used a torpedo "ignited from the surface of the ground that the strata were dislodged and fissured so much that if oil exists then it is found to flow on until the well is spent ... The notion of thus producing fissures extending for considerable distances from a deep bore-hole into the surrounding masses is not new, though its application to oil wells may be so. Mr. MALLET employed this method about sixteen years ago in the course of his practice." Present techniques don't use torpedoes {ROBERTS ⇨1864}, but rather shaped charges {MCLEMORE ⇨1946; ⇨Fig. 4.15–L}.

1837 — *Moulin-Quignon [near Abbeville], Dépt. Somme, northern France*

Jacques BOUCHER DE PERTHES,[605] a French writer and archeologist, **discovers flint tools which had been worked on by primitive man, such as hand axes and other stone tools,** some embedded with the bones of extinct mammals, in gravel deposited during the last ice age (which concluded $c.10,000$ years ago). His discovery of flint implements certify that man existed far earlier than 4004 B.C. – then the widely accepted date of creation.[606] ▪ BOUCHER DE PERTHES was **one of the first prehistorians to develop the idea of measuring man's antiquity by geological time,** and today he is regarded as the founder of prehistoric archaeology. His three-volume monograph on primitive tool-making, *Antiquités celtiques et antédiluviennes* ("Celtic and Antediluvian Antiquities"), stimulated subsequent researchers to investigate techniques that would have allowed stone tools to have been produced using the principle of percussion. It eventually resulted in the milestone discovery that primitive man created crude percussion tools much earlier than previously anticipated {Great Rift Valley ⇨$c.2.6$ Ma ago}.

Today, experimental archeology has become an important scientific method for understanding ancient techniques by re-establishing and practicing them. In relation to lithic analysis, there are three main areas:

(i) **Knapping experiments performed to understand the percussion process of the technology involved in producing particular types of stone tools.** ▪ Today many museums of prehistory also demonstrate basic percussion techniques for making and using stone tools.

(ii) Functional experiments using replicas of tools to test a particular hypothesis concerning tool use and to provide reference material for a use-wear analysis.

(iii) Taphonomic studies performed by recreating sites; this may involve burying stone tool replicas and debitage – the flakes and chips from stone tool production, the most abundant artifact type in prehistoric archaeological sites – along with bone fragments in order to simulate a site and to test environmental influences.

1837 — *Ecole Polytechnique, Paris, France*

Jean Marie C. DUHAMEL,[607] a French mathematician, **lays down the foundations of thermoelastic theory** by deriving the coupled heat conduction equation using the corresponding equations of motion. With respect to thermally-induced waves and vibrations, he states, "It is therefore allowable, particularly with regard to the slowness with which the cooling always takes place, to completely neglect these small movements of the molecules around their equi-

---

[605] J. BOUCHER DE PERTHES: *Antiquités celtiques et antédiluviennes. Mémoire sur l'industrie primitive et les arts à leur origine*. 3 vols., Treuttel & Wurtz, Paris (1847–1864); *De l'homme antédiluvian et de ses œuvres*. Jung & Treuttel, Paris (1860); *Des outils de pierre*. Jung & Treuttel, Paris (1865).

[606] In his *Chronologies of the Old and New Testaments* (published 1650–1654) the Irish Anglican Bishop James USSHER (1580–1656) stated that the Earth was created by God on Sunday, 23 October 4004 B.C. He arrived at his conclusion by carefully considering historical biblical events and tracing the lineages in the King James Bible.

[607] J.M.C. DUHAMEL: *Second mémoire sur les phénomènes thermo-mécaniques*. J. Ecole Polytech. (Paris) **15**, 1-57 (1837), p. 31.

librium positions, and to consider the equilibrium as strictly taking place at any time and varying with the inner propagation of the heat: one can point out, by the way, that these oscillatory movements would produce alternating rarefactions and compressions at each point, the effects of which would tend to compensate each other." Thus, the time rate of the change in temperature is considered slow enough that inertia effects can be disregarded in the equations of thermoelasticity.[608] • The corresponding high-rate dynamic problem was first solved by the Soviet scientist V.Y. DANILOVSKAYA {⇨1950}.

| | | |
|---|---|---|
| 1838 | *L'Académie des Sciences, Paris, France* | **D. François J. ARAGO,**[609] a French physicist, **prepares an essay on thunder** for the Annuaire du Bureau des Longitudes, and gives a masterly historical sketch of the real facts about thunder that have been accumulated thus far. He deduces the scientific and practical inferences, that can legitimately be drawn from these. **He also discusses *ball lightning* and analyzes a number of evidently reliable observations,** pointing out that an observer viewing the descent of the ball at an angle from the side is not subject to an optical illusion, as had often been assumed {FARADAY ⇨1841; Lord KELVIN ⇨1888}. |
| 1838 | *ROF Bridgewater Foundry, Patricroft, Greater Manchester, England* | **James NASMYTH,** a British engineer at the Royal Ordnance Factory (ROF), **invents the *steam hammer*.** In his construction, the cylinder is fixed and the hammer is attached to the piston rod. • Anchors and similar large forgings could be made in one piece with this new machine, which greatly increased the quality of the work and reduced the price.[610] The invention of the steam hammer {⇨Fig. 4.2−O} gave significant impetus to the Industrial Revolution. In 1861, the Scottish engineer Robert WILSON of Dunbar perfected NASMYTH's steam hammer and constructed a ***double-acting hammer*** at Woolwich Arsenal which falls under the action of both gravity and power.<br><br>In 1843, NASMYTH **also applied the principle of his steam hammer to pile-driving − the *steam ram* was born!** |
| 1839 | *Paris, France* | **Louis-Jacques-Mandé DAGUERRE,** a French painter, inventor and physicist, **and Isidore NIÉPCE,** son of J. Nicéphore NIÉPCE {⇨1816}, **transfer all copyrights of their photographic method to the French government,** which in turn grants a life-annuity and releases the patent for free use by all nations.<br><br>In the same year, the French term *photographie* ("photography") is coined. |
| 1839 | *Faculté des Sciences, Paris, France* | **Jean-Victor PONCELET,**[611] a French professor of mathematics and mechanics, **makes an important contribution to the theory of percussion.** In his textbook on mechanics, he treats the following interesting cases in the section entitled *Examen des principales circonstances du mouvement oscillatoire des prismes sous l'influence de charges constants et de chocs vifs* ("Investigation on the Principal Circumstances of Oscillatory Motion of Rods Under the Influence of Static and Shock Loading"):<br>▸ A mass-free rod is suddenly loaded at one end by a weight; the initial velocity of the rod is zero. He shows that a two-fold extension is produced in this case compared to static loading by the same weight.<br>▸ He studies the same configuration, but where the weight already has an initial velocity.<br>▸ The rod has a mass and can oscillate around its rest position.<br>▸ The rod is permanently loaded with a weight which suddenly receives a percussion. |

---

[608] H.W. BARGMANN: *Dynamic thermal shock resistance.* In: (J.L. ZEMAN and F. ZIEGLER, eds.) *Topics in applied continuum mechanics.* Springer, Vienna *etc.* (1974), pp. 174-181.

[609] D.F.J. ARAGO: *On thunder and lightning.* Edinb. New Phil. J. **26**, 81-144, 275-291 (1839); *Meteorological essays, with an introduction by A. VON HUMBOLDT* [translated by R.A. SABINE]. Longman & Co, London (1855).

[610] W. JOHNSON: *Indentation and forging and the action of NASMYTH's anvil.* The Engineer (Lond.) **205**, 348-350 (1958).

[611] J.V. PONCELET: *Introduction à la mécanique industrielle, physique et expérimentale.* Thiel, Metz (1839).

| 1839 | Ecole des Ponts et Chaussées, Paris, France | Adhémar-Jean-Claude DE SAINT-VENANT and Pierre Laurent WANTZEL,[612] two French mathematicians and engineers, **study compressible flow in a duct with varying cross-sectional area and the exhaust of compressed air from a small opening.** Using the Laplace-Poisson law of adiabatic compression {DE LAPLACE ⇨1816; POISSON ⇨1823} and a one-dimensional form of BERNOULLI's energy equation, they assume compressible flow with a pressure-dependent density $\rho = \rho(p)$. They express the difference in enthalpy via the pressure integral, which leads to their famous fundamental formula relating the outflow velocity $v$ to a given pressure ratio $p/p_0$ across the nozzle
$$v^2 = 2p_0 [1 - (p/p_0)^m]/m\rho_0,$$
where $m = (\kappa - 1)/\kappa$ and $\kappa$ is the ratio of specific heats. • The *critical speed* of a nozzle is reached when the gas flow precisely equals the local speed of sound. The Laval nozzle {DE LAVAL ⇨1888} allows the flow velocity to exceed the speed of sound.

The outflow into a vacuum is of particular interest, because the jet experiences no resistance, and a maximum jet velocity should be expected.

(i) For an outflow of air under atmospheric pressure into a vacuum ($p = 0$), the maximum outflow velocity is given by
$$v_{max} = (2p_0/m\rho_0)^{1/2}$$
which in the case of air ($\kappa = 1.405$) amounts to 757 m/s – *i.e.*, **the outflow into a vacuum is supersonic** ($M = 2.22$).

(ii) The British mathematician and ballistician Charles HUTTON,[613] who had previously tackled the same problem, (incorrectly) calculated the outflow velocity $v$ of common air into vacuum to be 1,335 ft/s (406.9 m/s) according to the simple formula
$$v = 2(Ag)^{1/2},$$
where $A$ is the height of a uniform air column reaching up to the edge of the atmosphere (assumed value: $A = 27{,}720$ ft or 8,449 m), and $g = g/2 = 16^{1}/_{12}$ ft/s$^2$ (4.902 m/s$^2$), as derived by him by a devious argument ($g =$ acceleration due to gravity). • The flow in a supersonic jet emerging from a near-sonic orifice into vacuum was calculated by British researchers {OWEN & THORNHILL ⇨1948}.

When a high-pressure gas is exhausted through a nozzle to an ambient pressure much below the nozzle exit pressure, the free jet expands to the ambient pressure very quickly, thereby increasing the cross-sectional area of the jet and creating a "normal" shock and the typical "shock bottle" configuration of a highly underexpanded jet.[614] Underexpanded free jets are found in a number of applications, for example:

> • in the muzzle blasts of guns, which are jets of short-duration formed by the hot, high-pressure powder gases released at shot ejection {SLADE JR. ⇨1946}; and
> • in rocket exhausts at high altitude, which exhibit highly underexpanded plumes. |
| 1839 | Kings College, London, England | Charles WHEATSTONE,[615] an eminent English experimental physicist, **invents the first electric chronoscope** – the so-called **"Wheatstone chronoscope"** – an electromagnetically controlled mechanical stopwatch driven by a weight; the dial pointer is started and stopped by the action of an electromagnet which moves a pawl that engages with a toothed wheel fixed onto the axle to which the dial pointer is attached {⇨Fig. 4.19–A}. The main advantage of the chronoscope is that it has a visible dial from which a time value can be read, thus eliminating the tedious and potentially unreliable counting of cycles on a chronograph. However, numerous technical |

---

[612] A.J.C. DE SAINT-VENANT and P.L. WANTZEL: *Mémoire et expériences sur l'écoulement de l'air.* J. Ecole Polytech. (Paris) **16**, 85-122 (1839).
[613] C. HUTTON: *Tracts on mathematical and philosophical subjects.* T. Davison, London (1812), vol. 3, p. 353.
[614] S. CRIST, P.M. SHERMAN, and D.R. GLASS: *Study of the highly underexpanded sonic jet.* AIAA J. **4**, 68-71 (1966).
[615] C. WHEATSTONE: *Description of the electromagnetic clock.* Proc. Roy. Soc. Lond. **4**, 249-278 (1840); *Note sur le chronoscope électro-magnétique.* C. R. Acad. Sci. Paris **20**, 1554-1561 (1845).

problems remain, such as the difficulty in obtaining precise onsets and offsets with the primitive relays of the time, and wet-battery instabilities. ▪ Besides other applications, **WHEATSTONE used his chronoscope at an English gunnery to measure projectile velocities.**

In 1843, the German-born Swiss watchmaker Matthias HIPP − popularly known as the "Swiss EDISON" − improved WHEATSTONE's design of the electric chronoscope, making the movement of the clockwork more uniform.[616] Astronomers were the first to appreciate the unparalleled accuracy of this so-called **"Hipp-Wheatstone clock"** {⇨Fig. 4.19−A}. The instrument was also widely used by physiologists to measure reaction times.

| | | |
|---|---|---|
| 1840 | *London, England* | **The British Association for the Advancement of Science (BAAS) appoints a Committee for the Study of Earthquakes** − the first of its kind worldwide. Practically, it is confined to the analysis of earthquakes that have occurred in the United Kingdom. ▪ Subsequently, further societies were founded in a number of other earthquake-endangered countries, *e.g.*, in Switzerland (1878), Japan (1880), Italy (1895), Austria (1895), the United States (1906), France (1921), and Germany (1922).[617] |
| 1841 | *Royal Military Academy, Woolwich, southeast London, England* | **Michael FARADAY,**[618] a professor of chemistry, **addressing the phenomenon of ball lightning** gives essentially the same explanation of as proposed previously in France {ARAGO ⇨1838}. However, he states that the optical illusion is caused by an afterimage perceived by eyes that just have seen the dazzling flash of an ordinary bolt. ▪ The subject was later resumed by Lord KELVIN {⇨1888}. |
| 1841 | *Juliusspital, Würzburg, Bavaria, Germany* | **Max Anton WINTRICH,** a German clinician, **creates the first popular *percussion hammer*** consisting of a metal handle and a hard rubber head.[619] ▪ It was used in neurology to examine muscle stretch reflexes such as the patellar reflex (or knee-jerk) which was first recognized in 1870 by Wilhelm H. ERB, a German neurologist.<br><br>In the following years, a wide variety of percussion hammers were developed, each one known by the name of its developer. The first *reflex hammer*, based upon the percussion hammer and provided with a triangular rubber head, was conceived by the U.S. neurologist John M. TAYLOR,[620] so-called "Taylor hammer" (or "tomahawk reflex hammer"). It was first demonstrated to the Philadelphia Neurological Society in 1888.[621] |
| 1842 | *Polytechnic Institute [now TU Prague], Prague, Bohemia, Austro-Hungarian Empire* | On May 25, **Christian Andreas DOPPLER,**[622] an Austrian physicist, **presents a paper to the Academy on the possible origin of the colored light from double stars, which contains the first statement of his "Doppler effect."** ▪ The Doppler effect was first proven experimentally<br>(i) in **acoustics:** In 1844, the Dutch meteorologist Christophorus H.D. BUYS-BALLOT[623] carried out his famous railway experiments at Utrecht using an open car with several trumpeters. In 1860, the Austrian physicist Ernst MACH[624] at Vienna University reconfirmed these results at the laboratory scale using a rapidly revolving whistle. |

---

[616] B. EDGELL and W.L. SYMES: *The Wheatstone-Hipp chronoscope. Its adjustments, accuracy, and control.* Brit. J. Psychol. **2**, 58-88 (1906).
[617] G. DAVISON: *The founders of seismology.* Arno, New York (1978).
[618] M. FARADAY: *On some supposed forms of lightning.* Phil. Mag. **19** [III], 104-106 (1841).
[619] M.A. WINTRICH: *Beschreibung einer neuen Percussions-Methode.* Berliner Medicinische Central-Zeitung **10**, Sp. 1-4 (1841).
[620] J.M. TAYLOR: *New form of percussion hammer.* J. Nerv. & Ment. Disease **15**, 253 (1888).
[621] F. PINTO: *A short history of the reflex hammer.* Practical Neurology **3**, 366-371 (2003).
[622] C.A. DOPPLER: *Über das farbige Lichtes der Doppelsterne und einiger anderer Gestirne des Himmels.* Abhandl. Königl. Böhm. Gesell. Wiss. Prag **2** [V], 465-482 (1841/1842).
[623] C.H.D. BUYS-BALLOT: *Versuche auf der Niederländischen Eisenbahn, nebst gelegentlichen Bemerkungen zur Theorie des Hrn. Prof. DOPPLER.* Ann. Phys. **66** [II], 321-351 (1845).
[624] E. MACH: *Über die Änderung des Tones und der Farbe durch Bewegung.* Sitzungsber. Akad. Wiss. Wien **41** (Abth. II), 543-560 (1860).

(ii) in **optical astronomy:** In 1868, the London astronomer William HUGGINS[625] first tried to measure the radial velocities of stars via the Doppler shifts $\Delta\lambda$ of their spectral lines. The relative change in the wavelength, $\Delta\lambda/\lambda_0$, of the light from a star moving with velocity $v$ is given by the simple relation

$$\Delta\lambda/\lambda_0 = [(1 + v/c)/(1 - v/c)]^{1/2} - 1,$$

where $c$ is the velocity of light. ▪ The Hubble law {HUBBLE ⇨ 1929} is based on the redshifts observed in the spectra of stars. From the large redshifts of quasars, it was concluded that they recede with extremely high velocity; for example the quasar 3C-9 recedes with $v = 0.8c$ {M. SCHMIDT ⇨ 1963}; and

(iii) in **high-speed diagnostics:** In the 1950s, the ***Doppler radar technique*** was found to be particularly useful in hypervelocity range facilities for providing accurate velocity and therefore aerodynamic drag data {KOCH ⇨ 1952}. However, this method was superceded by the invention of the laser. Applications of the Doppler effect also proved most useful in fluid dynamics and gas dynamics using ***laser Doppler anemometry***,[626] and in terminal ballistics and solid-state shock wave physics using ***laser Doppler velocimetry*** {BARKER & HOLLENBACH ⇨ 1965}.

| | | |
|---|---|---|
| 1842 | *Chair of Chemistry, University of Marburg, Hesse, west-central Germany* | **Robert W. BUNSEN,**[627] an associate professor of chemistry, **succeeds in synthesizing the arsenic compound *Kakodyl* [(CH$_3$)$_2$As–As(CH$_3$)$_2$].**[628] **Kakodyl, mixed with oxygen gas and inflamed by an electrical spark, explodes more powerfully than fulminating gas,** generally breaking the vessel holding it. (The name *Kakodyl*, which was coined by BERZELIUS, was later changed to *cacodyl*.) ▪ BUNSEN's studies on cacodyl derivatives, performed in the period 1836–1842, showed that the cacodyl radical [(CH$_3$)$_2$As–] is preserved as an "unchangeable member" through numerous reactions – thus supporting the radical theory of organic compounds advocated by the Swedish chemist Jöns Jakob BERZELIUS, who in 1818 had announced his "dualistic theory," and by the German chemists Justus VON LIEBIG and Friedrich WÖHLER who discovered the benzoyl radical in 1832.<br><br>In the following year, **BUNSEN experimented with various cacodyl compounds such as cacodyl cyanide [(CH$_3$)$_2$AsCN],** a white explosive powder which gives off an extremely poisonous vapor when exposed to air. When working on this extremely poisonous cacodyl cyanide, he wore a mask with a long tube linked to fresh air. In 1843, an explosion shattered the mask, destroying his sight in his right eye, and nearly ended his life. This put an end to his organic research, and he turned to inorganic chemistry. However, in later years he resumed his interest in the chemical nature of exploding gunpowder {BUNSEN ⇨ 1857} and in the physical properties of explosive gaseous mixtures {BUNSEN ⇨ 1867}. |
| 1842 | *France* | **Chevalier TRÉVILLE DE BEAULIEU,** a French colonel, **invents the first muzzle brake by drilling holes into the muzzle region of the barrel of a rifle.** The holes are sloped backward to divert the expanding gases in that direction. ▪ 21 years later, the French military conducted tests with a 106-mm gun with 36 60-mm-dia. holes inclined rearward at 45°. Data published by TRÉVILLE DE BEAULIEU disclosed the great success of this brake, since it doubles the accuracy and reduces the recoil distance to 25% of its normal distance with a loss of only 6% in muzzle velocity.[629] |

---

[625] W. HUGGINS: *Further observations on the spectra of some of the stars and nebulae, with an attempt to determine therefrom whether these bodies are moving towards or from the Earth.* Phil. Trans. Roy. Soc. Lond. **158**, 529-564 (1868).

[626] Y. YEH and H.Z. CUMMINS: *Localized fluid flow measurements with a He-Ne laser spectrometer.* Appl. Phys. Lett. **4**, 176-178 (1964).

[627] R.W. BUNSEN: *Untersuchungen über die Kakodylreihe.* Ann. Chem. Pharm. **37**, 1-57 (1841); Ibid. **42**, 14-46 (1842); and Ibid. **46**, 1-48 (1843); *On the radical of the Kakodyl series.* Phil. Mag. **20** [III], 382-393 (1842).

[628] Jöns Jacob BERZELIUS, a Swedish chemist, suggested to BUNSEN the name "cacodyl" [Germ. *Kakodyl*] – a name of Greek origin meaning "stinking" – due to the nauseous smell of its compounds.

[629] C.J. LANG (ed.): *Engineering design handbook.* Guns series, muzzle devices. AMC Pamphlet No. 706-251. Headquarters, U.S. Army Materiel Command, Washington, DC (May 1968), p. 1-1.

The addition of a muzzle brake reduces the forward blast, but at the same time also increases the blast pressure to the rear – thus increasing the effects on those handling the rifles. Since then, the muzzle brake has been developed for a wide range of guns. A limited series of two-dimensional model tests on muzzle brakes were performed during World War II in Göttingen under the supervision of Klaus OSWATITSCH.[630]

| 1843 | University of New York, New York City, New York | **John William DRAPER,**[631] an English-born U.S. professor of medicine and chemist, **presents an investigation on the chemical action of light on chlorine gas to the British Association.** He shows that this gas undergoes a definite chemical modification, due to its absorption of the rays from sunlight. • He also invented a reaction vessel which (although somewhat altered) was used in photochemical research involving the light-induced chemical combination of equal volumes of hydrogen and chlorine – a gaseous mixture previously called "chlorine detonating gas" {GAY-LUSSAC & THÉNARD ⇨1809; BUNSEN ⇨1853; BUNSEN & ROSCOE ⇨1857}. |
|---|---|---|
| 1843 | Stuttgart, Baden-Württemberg, southern Germany | **Ferdinand KRAUSS,**[632] a German physician, **first observes the way snapping shrimps** [*Alpheus heterochelis*] **produce a characteristic sharp acoustic pulse with their large claw** [*chela*], but he cannot prove the exact origin of this noise. • Recent studies have revealed that the snapping claw generates weak shock waves in water by cavitation {LOHSE ET AL. ⇨2001; ⇨Fig. 4.1–Z}. |
| 1844 | Private study at Down House, Downe, Kent, England | **Charles R. DARWIN,**[633] the eminent British naturalist, **first reports on the discovery of curious lens-like "volcanic bombs"** which he collected in New South Wales, Australia, on a large sandy plane between the Darling and Murray rivers during the voyage of the HMS *Beagle* (Dec. 27, 1831 – Oct. 2, 1836).<br><br>(i) The bombs have a glassy, bottle-green appearance, looking like strongly-flattened oval spheres of obsidian with a striped surface structure. • In the last two centuries, the origin of volcanic bombs – later to be termed ***australites*** – was a subject of intense debate {NININGER ⇨1956}: they were believed to come from terrestrial volcanoes (such as supposed by DARWIN), from lunar volcanoes, from the Sun, from lightning fusing atmospheric dust, from comets, from the region where the asteroids are now, from the fixed stars, and from Jupiter and the other planets.[634]<br><br>(ii) He gives a short account of the origin and nature of volcanic bombs: "The geology of this island is in many respects interesting. In several places I noticed volcanic bombs, that is, masses of lava which have been shot through the air whilst fluid, and have consequently assumed a spherical or pear-shape. Not only their external form, but, in several cases, their internal structure shows in a very curious manner that they have revolved in their aerial course … The central part is coarsely cellular, the cells decreasing in size towards the exterior; where there is a shell-like case about the third of an inch in thickness, of compact stone, which again is overlaid by the outside crust of finely cellular lava. I think there can be little doubt, first, that the external crust cooled rapidly in the state in which we now see it; secondly, that the still fluid lava within, was packed by the centrifugal force, generated by the revolving of the bomb, against the external cooled crust, and so produced the solid shell of stone; and lastly, that the |

---

[630] *See* OSWATITSCH in the BIOGRAPHIES INDEX.
[631] J.W. DRAPER: *On a change produced by exposure to the beams of the Sun in the properties of an elementary substance.* Rept. Meet. Brit. Assoc. **13**, 9 (1943).
[632] F. KRAUSS: *Die südafrikanischen Crustaceen. Eine Zusammenstellung aller bekannten Malacostraca, Bemerkungen über deren Lebensweise und geographische Verbreitung, nebst Beschreibung und Abbildung mehrerer neuer Arten*. Schweizerbart, Stuttgart (1843), pp. 1-68. Cited in R.G. BUSNEL (ed.): *Acoustic behavior of animals*. Elsevier, Amsterdam etc. (1963), p. 306.
[633] Ch. DARWIN: *Geological observations: on the volcanic islands and parts of South America visited during the voyage of HMS "Beagle."* Smith & Elder, London (1844). Republished by Appleton & Co., New York (1891), pp. 44-45.
[634] *See* J.A. O'KEEFE (ed.): *Tektites*. The University of Chicago Press, Chicago, IL (1963).

centrifugal force, by relieving the pressure in the more central parts of the bomb, allowed the heated vapours to expand their cells, thus forming the coarsely cellular mass of the centre."

**The particular shape of australites was successfully reproduced by aerodynamic ablation in a hypervelocity air stream:** laboratory experiments carried out by Dean R. CHAPMAN[635] and Howard K. LARSON, two U.S. aeronautical engineers at NASA's Ames Research Center (Moffet Field, CA), suggested the idea that australites have been sculptured by aerodynamic heating of rigid tektite glass during the process of a descent into the Earth's atmosphere at hypervelocities.

*Australites* belong to the class of objects known as **"tektites"** [from Greek *tektos*, "melted"] – a term proposed in 1900 by the British-born Austrian geologist Franz Eduard SUESS.[636] Tektites (or tectites) are small glassy bodies of natural but unknown origin found in various "strewn fields" on Earth, ranging from a few tens of microns up to about 10 cm in diameter. In the 1950s and early 1960s, the leading theories were that tektites were of lunar origin, ejected by either impacts or lunar volcanism {O'KEEFE ⇒1967 & 1976}, or they were formed by large asteroid impacts on Earth which melted the rocks at the impact site splashing liquid droplets to large distances.

**Tektites are probably formed by the impacts of large meteorite or comets, which melt the rocks at the impact site, splashing liquid droplets over large distances.** They are mostly high in silica and very low in water content. One can group tektites in terms of their origin and age or form:
- *North American* (34 Ma), found in Texas, Georgia, MA;
- *Czechoslovakian* (15 Ma), found in Bohemia and Moravia (*moldavites*);
- *African* (1 Ma), found in the Ivory Coast and in the Libyan Desert; and
- *Australian* (0.7 Ma), found in Tasmania, Australia, Indonesia, Southeast Asia, the Philippines, and in various marine sediments of the Indian and Pacific oceans.

One can also group tektites in terms of their sculpture:
- *microtektites* (mostly spheroids, diameters < 2 mm);
- *splash-form tektites* (mostly round shapes of some cm in diameter, with typical corrosion marks such as grooves and pits);
- *Muong-Nong tektites* (mostly elliptical, ranging in size from cm to dm), found in Laos; and
- *australites* (typical lens-like forms < 25 mm, with a flange around the edge).

1844 *Conservatoire des Arts-et-Métiers, Paris, France*

Claude S.M. POUILLET,[637] a French physicist and administrator, **describes an electric circuit used to measure the duration of short current pulses** by studying its action of the pulses on the magnetic needle of a galvanometer. His method is based on the principle of the ballistic pendulum {J. CASSINI ⇒1707; ROBINS ⇒1740} and is later termed the ***ballistic galvanometer*** {⇒Fig. 4.3–J}. This method even allows the measurement of time spans in the minor microsecond regime. He considers the precise measurement of time to be crucial to a better understanding of high-speed events, such as the ignition process of gunpowder and the contact duration of impacting bodies.

His so-called **"Pouillet chronoscope"** was later refined in England by Robert SABINE {⇒1876}, and in Germany by Carl W. RAMSAUER {⇒1909}.

---

[635] D.R. CHAPMAN and H.K. LARSON: *The lunar origin of tektites*. NASA Tech. Note D-1556 (Feb. 1963).

[636] F.E. SUESS: *Die Herkunft der Moldavite*. Jb. K.u.K. Geol. Reichsanstalt Wien **50**, 192-382 (1900).

[637] C.S.M. POUILLET: *Note sur un moyen de mesurer des intervalles de temps extrêmement courts, comme la durée du choc des corps élastiques, celle du débandement des ressorts, de l'inflammation de la poudre etc.; et sur un moyen nouveau de comparer les intensités des courants électriques, soit permanents, soit instantanés*. C. R. Acad. Sci. Paris **19**, 1384-1389 (1844).

| 1844 | City of Manchester, England | At the meeting of the British Association for the Advancement of Science (BAAS), **John S. Russell,**[638] a Scottish engineer and naval architect, **reports on his studies into the behavior of the *wave of translation* or *great solitary wave*** {Russell ⇒1834}:
(i) This particular wave type, which has a symmetrical wave form with a single hump and moves at constant speed on the surface of an inviscid incompressible fluid, is capable of traveling a considerable distance in a uniform channel with almost no change.
(ii) Referring to his former studies performed in the period 1833–1840, he reports on the reflection of this type of wave at a solid boundary, "When the angle of the ridge with the surface is small, not greater than 30°, the reflexion is complete in angle and in quantity. When the ridge of the wave makes an angle greater than 30°, the angle of reflexion is still equal to the angle of incidence, but the reflected wave is less in quantity than the incident wave … when the angle of the ridge of the wave is within 15° or 20° of being perpendicular to the plane [*i.e.*, at an angle of incidence of less than 75° or 70°], reflexion ceases, the size of the wave near the point of incidence and its velocity rapidly increases, and it moves forward rapidly with a high crest at right angles to the resisting surface. Thus, at different angles we have the phenomena of total reflexion, partial reflexion, and, non-reflexion and lateral accumulation; phenomena analogous in name, but dissimilar in condition from the reflexion of heights…" ▪ His observation of a "lateral accumulation" of the reflected wave front is most remarkable, in particular the fact that the reflected wave merges with the front of the incident wave, thus creating a new wavefront that extends at right angles to the boundary {⇒Fig. 4.13–A}.
**The phenomenon of irregular wave reflection,** found by him accidentally during the study of interactions of hydraulic jumps, **was later rediscovered for the case of interacting shock waves in air** {E. Mach & Wosyka ⇒1875}. **It was brought to light again by John von Neumann** {⇒1943}, **who termed it *Mach reflection*.** |
|---|---|---|
| 1845 | Royal Observatory, Greenwich, Kent, England | George B. Airy,[639] an English mathematician and Royal Astronomer, **publishes a long memoir on the properties of tides and water waves** in the British *Encyclopaedia Metropolitana*:
(i) He reviews ordinary tidal phenomena and the classical equilibrium theory of tides given by Sir Isaac Newton {⇒1687} in his *Principia* (Lib. I), and extended in the early 1820s by Pierre-Simon de Laplace in his *Mécanique céleste* ("Celestial Mechanics").
(ii) He presents an extended theory of waves on water. In particular, he analytically treats the motion of waves of finite amplitude in a uniform water canal with a rectangular section and a small breadth. Via the method of successive approximation he finds that different parts of a progressive wave travel with different velocities. **Airy makes the important statement that the form of the wave changes during its propagation – such that the crests tend to gain upon the hollows so that the anterior slopes become steeper and steeper.**
(iii) He also investigates the motion of waves in extended surfaces of water:
▸ In a tabular form (on p. 291 of his memoir) he treats the relations that exist between the wavelength, the propagation velocity and the water depth. For large wavelengths $> 10^6$ ft (305 km) the propagation velocity $v$ is almost independent of the water depth.
▸ **For an infinite wavelength, the velocity approaches the result for a *shallow water wave*** given by the simple relation
$$v = (gh)^{1/2},$$
where $g$ is the acceleration due to gravity {de Lagrange ⇒1781}. ▪ Since the wavelength of a tsunami is in excess of 100 km – *i.e.*, very large compared to the water depth, even in very deep waters – the velocities of tsunami waves can be estimated by |

---

[638] J.S. Russell: *Report on waves. Made to the Meeting of the British Association for the Advancement of Science (1834–1843).* Rept. Meet. Brit. Assoc. **14**, 311-390 (1844); later published in Russell's book *The wave of translation in the oceans of water, air and ether.* Trübner & Co, London (1885), p. 315.

[639] G.B. Airy: *Tides and waves* [1842]. In: *Encyclopaedia Metropolitana.* Fellowes, London (1845), vol. 5, Art. 208, pp. 241-396.

this relationship, or **the average depth of the ocean can be determined from the measured propagation velocity of a tsunami.** The latter method was applied by VON HOCHSTETTER {Arica Earthquake ⇨1868} before deep-sea soundings were taken, and by Franz E. GEINITZ and Alexander G. SUPAN {Iquique Earthquake ⇨1877} to check measured deep-sea data {HMS *Challenger* Expedition ⇨1872}.

In the same year, **John Scott RUSSELL,**[640] a London naval architect and civil engineer, **states independently that the tidal (lunar) wave is a wave of the first order or "wave of translation"** {RUSSELL ⇨1834} which – when the amplitude is small compared to the water depth $h$ – propagates with a velocity $v = (gh)^{1/2}$.

| 1845 | *School of Geographical Arts, Potsdam, Kingdom of Prussia, Germany* | **Heinrich BERGHAUS,**[641] a German geographer and professor of mathematics, **publishes a chart of the volcanic phenomena of Europe, South America, most of Central America and the eastern half of North America,** in which the regions disturbed by earthquakes are indicated by shading, where the thickness of the shading indicates the intensity and frequency of the shocks. Curves are also drawn bounding the areas affected by three remarkable earthquakes – the Lisbon Earthquake (1755), the Caracas Earthquake (1812), and the Southeast Europe Earthquake (1838). His map is mostly based on data collected by the German geologist Karl E.A. VON HOFF.[642]

In the 1850s, BERGHAUS' map stimulated the Irish engineer and earthquake researcher Robert MALLET[643] to incorporate his own data and to produce the first seismic map of the distribution of earthquakes over the globe {⇨Fig. 4.1–H}. This illustrates that areas frequently disturbed by earthquakes – **"seismic bands"** – follow the lines of elevation that divide the great oceanic basins as well as mountain chains and volcanic vents,[644] thus approaching modern knowledge, based upon plate tectonics {⇨Fig. 4.1–I}. |

| 1845 | *Haswell Collieries, Durham District, England* | **Michael FARADAY,** a chemistry professor at the Royal Military Academy in Woolwich, **and Charles LYELL,** a Scottish geologist, **investigate possible causes of a serious explosion at Haswell Collieries.**[645] They observe many signs of the coal dust being partly burned and partly subjected to a charring or choking action. **They conclude that coal dust adds considerably to the disastrous effects of firedamp explosions and that *proper ventilation* is an effective means of preventing similar accidents.** ▪ The first experimental evidence that coal dust plays an important role in the intensity of firedamp explosions was found in France (C. VERPILLEUX 1867; P. VITAL 1875).[646] Their results were later confirmed in British studies {GALLOWAY ⇨1875; ABEL ⇨1881} and by French mining engineers {CHARPY & LE CHÂTELIER ⇨1890}.

In 1832, the Englishman George STEPHENSON, principal inventor of the railroad locomotive, had called attention to the need for proper ventilation – both from the viewpoint of preventing firedamp explosions and to improve the health of miners. Referring to the prevention of firedamp explosions, he wrote in a letter to the South Shields Committee: "Generally speaking, there has been some fault in the ventilation of the mines when accidents have occurred … and the same opinion is held by many of the most experienced authorities at the present day. In this matter, **the one cry, whether we look to security against explosion, or to the affording to miners an atmosphere which they can breathe without injury to health, is 'More Air!'**" ▪ One of the most efficient measures for preventing firedamp explosions in |

---

[640] J.S. RUSSELL: *On the terrestrial mechanism of the tides.* Proc. Roy. Soc. Edinb. **1**, 179-182 (1845).
[641] H. BERGHAUS: *Physikalischer Atlas.* J. Perthes, Gotha (1845, 1848); Sect. 3: *Geologie,* No. 7.
[642] K.E.A. VON HOFF: *Geschichte der natürlichen Veränderungen der Erdoberfläche.* 5 vols. (1822–1841), vols. 4 and 5: *Chronik der Erdbeben und Vulkan-Ausbrüche.*
[643] R. MALLET: *Fourth report upon the facts and theory of earthquake phenomena.* Rept. Meet. Brit. Assoc. **28**, 1-136 (1858), plate XI.
[644] C. DAVISON: *The founders of seismology.* Arno Press, New York (1978), pp. 73-75.
[645] M. FARADAY and C. LYELL: *On explosions in coal mines.* Phil. Mag. **26** [III], 16-35 (1845).
[646] F.A. ABEL: *Résultat des expériences faites avec des poussières provenant de la houillère de Seaham.* Ann. Chim. Phys. **24** [V], 384-432 (1881).

coal mines, which the French Firedamp Commision later recommended, was indeed the provision of an effective ventilation system {CHARPY & LE CHÂTELIER ⇨ 1890}. Natural ventilation was created by level drainage tunnels connecting the sloping surface to the shaft; surface stacks above the shaft increased the efficiency of ventilation. The most positive method, before the introduction of fans in the 18th century, was the use of a furnace on the surface.

| | | |
|---|---|---|
| 1845 | *Chair of Chemistry, University of Basel, northern Switzerland* | **Christian F. SCHÖNBEIN,** a German-Swiss chemist, **invents *Schießbaumwolle* ("guncotton") – cellulose nitrate with a high nitrogen content.** Experimenting with the nitration of cellulose, he hit upon this new explosive substance when dipping cotton in a mixture of nitric and sulfuric acid, and washing to remove the excess acid.[647] SCHÖNBEIN communicates his process to John TAYLOR,[648] who is granted an English patent in the following year. ▪ In *The Annual Register* (London, Dec. 1846), it reads: "The interest of scientific men and of the public has been strongly excited by the discovery, by Professor SCHÖNBEIN, of an explosive compound, appearing to possess many advantages over gunpowder, called 'Gun Cotton' … On the application of a spark, the solid mass is at once converted to a gaseous state, producing scarcely any smoke, and, if carefully prepared, leaving no residuum behind." <br><br> Guncotton (also known by the chemically incorrect name "nitrocellulose") violently explodes when struck or strongly heated. However, the detonation was initially difficult to control because acid residues in guncotton originating from the production process provoked dangerous self-ignitions, and it took further research efforts to convert guncotton into a reliable gun propellant: <br> ▸ In 1862, the Austrian General and Baron Wilhelm LENK VON WOLFSBERG, trying to use guncotton as an explosive, improved its storage life. <br> ▸ Eventually, in 1866 the British chemist Sir Frederick A. ABEL[649] showed how guncotton could be rendered stable and safe. His method was to remove all traces of the sulfuric and nitric acids used in its manufacture by mincing, washing in soda until all the acid had been removed, and drying. <br> ▸ Around 1885, guncotton was first used for military purposes by the French. Since it is more powerful than black powder, rifle projectiles could be given higher muzzle velocities of up to 850 m/s. |
| 1846 | *Royal Irish Academy, Dublin, Ireland* | **Robert MALLET,**[650] an Irish civil engineer and a co-founder of seismology, **publishes a milestone memoir entitled *On the Dynamics of Earthquakes*:** <br><br> (i) **He first explains the structural damage produced by earthquakes due to the passage of "waves of elastic compression" through the ground.** Likewise he explains curious rotations of a pair of obelisks at the Convent of St. Bruno in the small town of Stefano del Bosco, the blocks of which had suffered differential rotation during the Great Calabrian Earthquake in 1783 {SPALLANZANI ⇨ 1798}.[651] <br><br> (ii) **He correctly speculates that seismic sea waves are generated by seaquakes** when the water is displaced vertically {⇨Fig. 4.1–N}. <br><br> (iii) **In order to study the velocity of seismic waves, he first proposes the use of explosives (gunpowder) to generate artificial quakes and to detect their arrival time at remote distances using a mercury seismoscope.** In more detail, he suggests, "For, knowing the time of transit of the [earth] wave, we can find the modulus of elasticity which corresponds to it, and |

---

[647] C. HAEUSSERMANN: *Gedächtnisrede auf Christian Friedrich SCHÖNBEIN* [Referat]. Z. ges. Schieß- u. Sprengstoffwesen **4**, 433-434 (1909).
[648] J. TAYLOR: *Improvements in the manufacture of explosive compounds, communicated to me from a certain foreigner residing abroad.* Brit. Patent No. 11,407 (Oct. 8, 1846).
[649] F.A. ABEL: *On the manufacture and composition of gun-cotton.* Phil. Trans. Roy. Soc. Lond. **156**, 269-308 (1866).
[650] R. MALLET: *On the dynamics of earthquakes; being an attempt to reduce their observed phenomena to the known laws of wave motion in solids and fluids.* Trans. Roy. Irish Acad. **21**, 50-106 (1846).
[651] C. LYELL: *Principles of geology.* J. Murray, London (1830), vol. 1, p. 418.

finding this, discover the particular species of rock formation to which this specific elasticity belongs … **Now the recent improvements in the art of exploding, at a given instant, large masses of gunpowder, at great depths under water, give us the power of producing, in fact, an *artificial earthquake* at pleasure;** we can command with facility a sufficient impulse to set in motion an earth wave that shall be rendered evident by suitable instruments at the distance, probably, of many miles, and there is no difficulty in arranging such experiments, so that the explosion shall be produced by the observer of the time of transit himself, though at the distance of twenty or thirty miles [32 or 48 km], or that the moment of explosion shall be fixed, and the wave period registered by chronometers, at *both* extremities of the line of transit."

MALLET {⇨1860} later performed such experiments, but he had to battle against various technical problems. However, his sound idea was proven thirty years later with more technically advanced equipment {ABBOT ⇨1876}. Today, the study of the properties of the materials and structure of the Earth by artificial means – such as by explosives – is referred to as ***explosion seismology*** {MALLET ⇨1849 & 1860; ABBOT ⇨1876; MINTROP ⇨1919 & 1924}, a branch of geophysics of considerable scientific as well as commercial significance.

| | | |
|---|---|---|
| 1846 | *Chair of Applied Chemistry, University of Turin, Piemonte, northwestern Italy* | Ascanio SOBRERO,[652] an Italian professor of chemistry and a former student of Théophile J. PELOUZE and Justus VON LIEBIG, **discovers nitroglycerin** [glycerol trinitrate, $C_3H_5O_3(NO_2)_3$] by treating glycerol with a mixture of nitric and sulfuric acid. In a letter to PELOUZE, he first reports on the amazing properties of the new liquid explosive, which he calls "pyroglycerin." ▪ SOBRERO himself considered nitroglycerin – a molecular explosive which contains the fuel and oxidizer in a single molecule – to be far too dangerous to be of any practical use. He is quoted to have said,[653] "When I think of all the victims killed during nitroglycerin explosions, and the terrible havoc that has been wreaked, which in all probability will continue to occur in the future, I am almost ashamed to admit to being its discoverer."<br><br>**The production of nitroglycerin** [Engl. also *nitroglycerine*, French *nitroglycérine*] **on an industrial scale did not begin until Immanuel NOBEL,** a Swedish inventor of considerable merit, **and his son Alfred NOBEL initiated practicability studies in the 1850s** {A. NOBEL ⇨1862 & 1867}. In parallel with these studies, the U.S. chemist George MOWBRAY, following the work of SOBRERO and others in Europe, began to ponder the safe production of nitroglycerin. Nitroglycerin first replaced black powder on a large scale during the construction of the four-mile (6.4-km) Hoosac Tunnel at North Adams, MA, an ambitious American railway project (1855–1866). In order to minimize the dangers of transportation and storage, the nitroglycerin was supplied by MOWBRAY's plant, which he had erected near North Adams.[654] |
| 1847 | *Technical State Academy, Prague, Bohemia, Austro-Hungarian Empire* | On April 23, **Christian Andreas DOPPLER,**[655] an Austrian physicist, **presents a paper on the acoustical consequences of the *Doppler effect* to the Academy** {DOPPLER ⇨1842}:<br><br>(i) **He speculates that the propagation velocity of sound should increase with intensity.** He discusses what might happen to a disturbance that, propagating with a velocity $u$ = const, moves faster than the sound velocity $a$ of the surrounding medium. Assuming a sequence of "wavelets" {HUYGENS ⇨1678} propagating with the velocity of sound, he graphically constructs a conical geometry for four cases of disturbances propagating along a straight line with<br>▸ a subsonic velocity stationary in time {⇨Fig. 4.6–B};<br>▸ a supersonic velocity stationary in time {⇨Fig. 4.6–B}; |

---

[652] A. SOBRERO: *Sur plusieurs composés détonants produits avec l'acide nitrique et le sucre, la dextrine, la lactine, la marnite, et la glycérine.* C. R. Acad. Sci. Paris **25**, 247-248 (1847).
[653] *Nobel e-Museum.* The Official Web Site of The Nobel Foundation; http://www.nobel.se/index.html.
[654] N.G. JOHNSON: *Explosives.* In: *Encyclopedia Britannica. Macropaedia.* Benton & Hemingway, Chicago *etc.* (1974), vol. 7, p. 85.
[655] C.A. DOPPLER: *Über den Einfluß der Bewegung des Fortpflanzungsmittels auf die Erscheinungen der Äther-, Luft- und Wasserwellen* [vorgetragen 1847]. Abhandl. Königl. Böhm. Gesell. Wiss. Prag **5** [V], 293-306 (1848).

▸ a supersonic velocity increasing in time {⇨Fig. 4.6–C}; and
▸ a supersonic velocity decreasing in time {⇨Fig. 4.6–C}.

(ii) **For the supersonic case ($u > a$) with $u$ = const, he derives the famous formula for the half-angle, $\alpha$, between the newly created wavefront and the line at which the disturbance propagates, as given by**

$$\alpha = \arcsin a/u.$$

- In 1887, Ernst MACH termed the envelope of this cone *Kopfwelle* ("head wave"); this cone has also been termed *MACHscher Kegel* ("Mach cone") to honor MACH's experimental work, or *MACHsche Kopfwelle* ("Mach head wave") to better distinguish from the *von SCHMIDTsche Kopfwelle* ("Schmidt head wave") {VON SCHMIDT ⇨1938}. PRANDTL {⇨1913} named the half-angle $\alpha$, associated with the Mach cone, the *MACHscher Winkel* ("Mach angle").

(iii) **He also applies his construction of wavefronts to curved trajectories of disturbances** – again for subsonic, sonic and supersonic speeds. He even considered moving sources in dispersive media, albeit without conclusive results. ▪ These illustrations clearly show how deeply DOPPLER explored the problem of moving sound sources. His purely theoretical results were confirmed forty years later by supersonic ballistic experiments {E. MACH & SALCHER ⇨1886}. A curved bow wave – such as that produced by an aircraft flying a curve supersonically – may result in shock-focusing effects.[656]

| | | |
|---|---|---|
| 1848 | *Royal Observatory, Greenwich, Kent, England* | **George B. AIRY,**[657] an eminent English natural philosopher and seventh Astronomer Royal, **shows that "the [sound] velocity $u$ does not depend on the absolute pressure of the air in its normal state of density, but rather upon the proportion of the change of pressure to the change of density;** *i.e.*, $u \propto \Delta p / \Delta \rho$. This ratio is increased by the suddenness of condensation in one part, which, when the elastic force is great, makes it still greater, and by the suddenness of rarefaction in another part, which, when the elastic force is small, makes it still smaller – thus in both ways increasing the change of pressure." |
| 1848 | *Chair of Astronomy and Experimental Philosophy, Cambridge Observatory, England* | **James CHALLIS,**[658] a British clergyman, physicist and astronomer, **resumes the classical analytical problem of the velocity of disturbances** – he is obviously not aware of POISSON's previous work {⇨1808}. He concludes, "The velocity of transmission of a vibration in a medium, for which the relation between the pressure and the density is given by the equation $p = a^2 \rho$, is not simply $a$, but a greater quantity $a(1 + k)^{1/2}$ [with $k$ being a small numerical quantity], which is the same for vibrations of different magnitude … I cannot avoid adverting here a difficulty which has long presented itself to me, with respect to the explanation usually given of the excess of the velocity of sound above the value $a$. Admitting that a sudden condensation by developing heat produces a higher degree of temperature, and therefore of elastic force, than would exist in the same state of density without such development, does it not thence follow, that a sudden rarefaction, by absorbing heat, produces a lower temperature and less elastic force than would exist in the same state of density without such absorption?" |

In the same year, CHALLIS,[659] replying to AIRY's comments on his own theory of sound {AIRY ⇨1848}, **finds that, for waves of finite amplitude propagating in a perfect gas, the velocity of propagation alters as it advances and tends ultimately to become a series of sudden compressions followed by gradual dilatations. The velocity of propagation is greater than the sound velocity, and certain faster parts in the wave profile will overtake the slower ones,** thus leading to ambiguous mathematical solutions – the so-called **"Challis paradox."**

---

[656] H. GOBRECHT (ed.): *Lehrbuch der Experimentalphysik.* De Gruyter, Berlin (1970), vol. 1, p. 496.
[657] G.B. AIRY: *The Astronomer Royal's remarks on Prof. CHALLIS' theoretical determination of the velocity of sound.* Phil. Mag. **32** [III], 339-343 (1848).
[658] J. CHALLIS: *Theoretical determination of the velocity of sound.* Phil. Mag. **32** [III], 276-284 (1848).
[659] J. CHALLIS: *On the velocity of sound, in reply to the remarks of the Astronomer Royal.* Phil. Mag. **32** [III], 494-499 (1848).

| | | |
|---|---|---|
| 1848 | Heilbronn, Baden-Württemberg, southern Germany | **Julius Robert MAYER,**[660] a German physician and natural philosopher, **speculates that the enormous energy emitted by the Sun is caused by the daily impact of meteoric showers.** Upon striking the Sun, the kinetic energy of the matter is converted to light and heat. ▪ In the early 1840s, MAYER (and independently afterwards the Scottish engineer John James WATERSTON, an ignored pioneer of the kinetic theory of gases who explained gravity by the impact of particles) had suggested that solar radiation results from the conversion of gravitational energy into heat. However, MAYER's paper was rejected by the Academy in Paris, and WATERSTON's by the Royal Society in London.<br><br>Since chemical action could only generate heat for a few thousand years, their so-called "meteoric hypothesis of the Sun's heat" was much discussed among contemporaries. It was resumed in Germany by Hermann VON HELMHOLTZ and in England by William THOMSON, the later Lord KELVIN {VON HELMHOLTZ ⇨1854}. |
| 1848 | Pembroke College, Cambridge, England | **George G. STOKES,**[661] a British physicist, fluid dynamicist and mathematician, **replies to CHALLIS' claim of a contradiction in the commonly accepted theory of sound** {CHALLIS ⇨1848}:<br><br>(i) Assuming an isothermal gas, **he introduces surfaces of discontinuity in the velocity and density of the medium, thereby circumventing the Challis paradox.**<br><br>(ii) **He indicates that small pressure disturbances might create compression waves with discontinuous fronts, because each subsequent sound wave will propagate in a medium with a slightly higher sound velocity.** He writes, "Of course, after the instant at which the expression (*A*) becomes infinite, some motion or other will go on, and we might wish to know what the nature of the motion was. **Perhaps the most natural supposition to make for trial is that a surface of discontinuity is formed, in passing across which there is an abrupt change of density and velocity. The existence of such a surface will presently be shown to be possible** … The strange results at which I have arrived appear to be fairly deducible from the two hypotheses already mentioned. It does not follow that the discontinuous motion considered can ever take place in nature, for we have all along been reasoning on an ideal elastic fluid which does not exist in nature. In the first place, it is not true that the pressure varies as the density, in consequence of the heat and cold produced by condensation and rarefaction respectively. But it will be easily seen that the discontinuous motion remains possible when we take account of the variation of temperature due to condensation and rarefaction, neglecting, however, the communication of heat from one part of the fluid to another. Indeed, so far as the possibility of discontinuity is concerned, it is immaterial according to what law the pressure may increase with the density."<br><br>(iii) **He also first applies overall the conservation relations of mass and momentum across the shock front that are now usually attributed to William J.M. RANKINE** {⇨1869} **and Pierre H. HUGONIOT** {⇨1887}. |
| 1849 | Royal Observatory, Greenwich, Kent, England | **George B. AIRY,**[662] an English natural philosopher and seventh Astronomer Royal, **first points out the analogy between the changes in velocity of sound waves of finite amplitude and those that take place in sea waves when they roll into shallow water.** |
| 1849 | Killiney Beach, County Dublin, Ireland | **Robert MALLET,**[663] an Irish engineer, resumes his previous seismologic studies {MALLET ⇨1846} and **conducts seismic experiments to determine the speed of "earth waves of elastic compression,"** particularly in wet quartz sand. The experiments consist of exploding charges of 25 pounds (11.33 kg) of gunpowder buried in casks 6 feet 6 inches (1.98 m) down |

---

[660] J.R. MAYER: *Beiträge zur Dynamik des Himmels in populärer Darstellung*. Published privately, Heilbronn (1848).
[661] G.G. STOKES: *On a difficulty in the theory of sound*. Phil. Mag. **33** [III], 349-356 (1848).
[662] G.B. AIRY: *On a difficulty in the problem of sound*. Phil. Mag. **34** [III], 401-405 (1849).
[663] R. MALLET: *Second report on the facts of earthquake phenomena*. Rept. Meet. Brit. Assoc. **21**, 272-320 (1851).

into the sand, and then measuring the transit time of the resultant shock wave over a range of half a mile (804.5 m). Using a mercury seismometer, he determines a speed of only 824 ft/s (251 m/s), admitting that the speed is far lower than he expected.

One year later, he repeated his experiments on Dalkey Island, County Dublin, where he measured a speed of 1,664 ft/s (507 m/s) in massive granite instead of the 8,000 ft/s (2,438 m/s) he had expected. Later, he speculated that "owing to discontinuity of rocky masses as found in nature, nearly $^7/_8$ of the full velocity is lost." • In rocks, the elastic constants and the wave velocities usually increase with increasing pressure and decrease with increasing temperature. For example, for granite the longitudinal wave velocity ranges between 5,000–6,250 m/s, and the transversal wave velocity between 2,000–3,500 m/s.[664]

| | | |
|---|---|---|
| 1849 | *Ecole des Mines, Saint Étienne, Dépt. Loire, southeastern France* | Edouard PHILLIPS,[665] a French mathematician and engineer, **considers the problem of the impact of two rough inelastic bodies of any form when the direction of friction is not necessarily the same throughout the impact** {ROUTH ⇒ 1860}:<br><br>(i) Assuming that the sliding does not vanish during the impact, he divides the period of impact into elementary portions and applies POISSON's rule for the magnitude and direction of the friction to each elementary period. He discusses the case in which the two bodies have their principal axes parallel to each other at the point of contact, and each body has its center of gravity on the common normal at the point of contact.<br><br>(ii) However, he does not examine the impact of elastic bodies in detail, though he remarks that the period of impact must be divided into two portions which must be considered separately. |
| 1849 | *Pembroke College, Cambridge, England* | George G. STOKES,[666] an English physicist and mathematician, **reinvestigates the theory of oscillatory waves such as those that occur on the surface of the sea** – so-called "Cauchy-Poisson waves" {CAUCHY ⇒ 1815; POISSON ⇒ 1815}. He pursues various approximations so as to cover **the case where the height of the wave is *not* very small in comparison with the wavelength.** Drawn to this subject for pure mathematical reasons, he obtains several important results:<br><br>(i) He finds that the expression for the velocity of propagation is independent of the height of the waves to a second approximation.<br><br>(ii) **With respect to the form of the waves, he notices that the elevations are no longer similar to the depression,** as is the case to a first approximation, but that the elevations are narrower than the hollows, and the height of the former exceeds the depth of the latter.<br><br>(iii) He also illustrates possible consequences for practical navigation, and states, "There is one result of a second approximation which may possibly be of practical importance. It appears that the forward motion of the particles is not altogether compensated by their backward motion; so that, in addition to their motion of oscillation, the particles have a progressive motion in the direction of propagation of the waves. In this case in which the depth of the fluid is very great, this progressive motion decreases rapidly as the depth of the particle considered increases. Now when a ship at sea is overtaken by a storm, and the sky remains overcast, so as to prevent astronomical observations, there is nothing to trust to for finding the ship's place but the dead reckoning. But the estimated velocity and direction of motion of the ship are her velocity and direction of motion relatively to the water. If then the whole of the water near the surface be moving in the direction of the waves, it is evident that the ship's estimated place will be erroneous. If, however, the velocity of the water can be expressed in terms of the length and |

---

[664] D.E. GRAY (ed.): *American Institute of Physics handbook.* McGraw-Hill, New York *etc.* (1972), p. 2-105.
[665] E. PHILLIPS: *Sur les changements instantanés de vitesse qui ont lieu dans un système de points matériels.* Math. Pures Appl. **14** [I], 300-336 (1849).
[666] G.G. STOKES: *On the theory of oscillatory waves.* Trans. Cambr. Phil. Soc. **8**, 441-455 (1849).

height of the waves, both which can be observed approximately from the ship, the motion of the water can be allowed for in the dead reckoning."

(iv) **He also proposes a method for calculating the shape of the waves.** As an illustration of his results he gives a diagram showing the shape of a deep-sea wave in which the difference in level between crest and hollow is $^7/_{40}$ of the wavelength. ▪ Later, valuable appendices pushing the approximation further using a new method, and showing that the slopes meeting at the crest of the highest possible oscillatory wave that could be propagated without changing their form enclose an angle of 120° were added to the reprint of this memoir in his *Mathematical and Physical Papers* (1880–1905).

STOKES' results, demonstrated for the example of water waves, first revealed that the amplitude of a wave exerts significant influence on the form of the wave profile and its wavefront, particularly that **wave steepening occurs at large levels** and that **with increasing amplitude the waves flatten in troughs, but sharpen at the crests.** His surprising findings stimulated analogous studies on sound waves of finite amplitude, thus paving the way to shock waves. His solution for periodic waves progressing over deep water without changing type was extended in 1915 by Lord RAYLEIGH.[667] He derived general equations that can be applied to progressive waves as a particular case, or to stationary (*i.e.*, nonpropagating) waves in which the principal motion is proportional to a simple circular function of time, and also to ascertain what occurs when the conditions required for the particular cases are not satisfied.

| 1850 | Faculty of Science, University of Paris, France | **P.E. Marcellin BERTHELOT,**[668] a French organic and physical chemist, **studies the tensile strengths of a number of liquids, including water.** He uses a glass ampoule which, with the exception of a small air bubble, is completely filled with pure water. After slowly heating up the ampoule, the bubble disappears. Then while slowly lowering the temperature, he observes that the slightest shock or vibration instantaneously separates the gas from the water, accompanied by **"a kind of fizz, a gentle sound and a more or less noticeable concussion"** [French *une sorte d'ébullition, un léger bruit et une secousse plus ou moins notable*]. From the recorded temperature difference he estimated a (negative) pressure of –50 bars at the moment that the water filament is broken. ▪ The precise measurement of negative pressures in water under various conditions has actually been a challenge to scientists for more than a century {STAUDENRAUS & EISENMENGER ⇨1988}. Modern measurements on negative pressures in water vary greatly, ranging up to –1.4 kbar.[669] |
|---|---|---|
| 1850 | Royal Artillery and Engineering School, Berlin, Kingdom of Prussia, Germany | **Rudolf J.E. CLAUSIUS,**[670] a German mathematical physicist, **presents his great memoir on the theory of heat.** He first recognizes that two independent principles are involved in thermodynamics: the equivalence of heat and energy (the First Law of Thermodynamics, then known as the "Law of Conservation of Heat"), and the inability of heat to pass spontaneously from a colder to a warmer body (Second Law of Thermodynamics).<br><br>It took a further 15 years before he introduced his concept of *entropy* {CLAUSIUS ⇨1865}, **which is hugely important to the theoretical treatment of irreversible processes such as shock waves.** Like the other thermodynamic quantities in normal shock waves, the entropy is discontinuous at the shock front and always increases in the medium through which it passes. This places important restrictions on the manner in which quantities can vary across a shock. The increase in entropy across the shock front is produced by either collisions between |

---

[667] Lord RAYLEIGH (J.W. STRUTT): *Deep water waves, progressive or stationary, to the third order of approximation.* Proc. Roy. Soc. Lond. **A91**, 345-353 (1915).
[668] P.E.M. BERTHELOT: *Sur quelques phénomènes de dilatation forcée des liquides.* Ann. Chim. Phys. **30** [III], 232-237 (1850).
[669] H. MARIS and S. BALIBAR: *Negative pressures and cavitation in liquid helium.* Phys. Today **53**, 29-34 (Feb. 2000).
[670] R.J.E. CLAUSIUS: *Über die bewegende Kraft der Wärme und die Gesetze, welche sich daraus für die Wärmelehre selbst ableiten lassen.* Ann. Phys. **79** [II], 368-397, 500-524 (1850).

particles in the shocked and unshocked fluids, or (as in astrophysics) by the generation and dissipation of plasma turbulence.

| | | |
|---|---|---|
| 1850 | Private laboratory at Salford, Lancashire, England | **James Prescott JOULE,**[671] a British physicist, **explains the pressure of gases by molecular impacts.** Using some simplifying assumptions regarding the size and velocities of molecules in a gas, he shows that the exact relation between the pressure $p$ exerted by a gas and the mean translational kinetic energy of its molecules is given by $$p = \tfrac{1}{3}\rho v_m^2,$$ where $v_m$ is the root-mean-square velocity of the molecules, and $\rho$ is the density of the gas. ▪ For example, for air at 1 bar and 0 °C, the density is $\rho = 0.00129$ g/cm$^3$. The velocity that air molecules must have in order to produce a pressure of 1 bar at 0 °C is calculated to be 482 m/s.<br><br>According to the classical kinetic theory of gases, the mean-square-root velocity of the molecules of the gas is given by $$V_m = (3RT/M)^{1/2},$$ where $M$ is the molar mass, $R$ is the universal gas constant (= 8.3143 J/mol K), and $T$ is the temperature. For oxygen ($M \approx 32$ g/mol) and nitrogen ($M \approx 28$ g/mol) at $T = 0$ °C = 273.15 K, the root-mean-square molecular velocities are then 461.4 m/s and 493.3 m/s, respectively. This approaches JOULE's value. Although the gas molecules are moving supersonically, their diffusion is slow, because they travel only short distances in straight lines before they are deflected in a new directions by collisions with other gas molecules. |
| 1850 | University of Edinburgh, Scotland, U.K. | **James Clerk MAXWELL,**[672] an eminent Scottish theoretical physicist, **publishes a paper on the theory of elasticity.** He shows that two elastic constants are necessary to describe the elastic behavior of an isotropic solid, and completely develops the technique of *photoelastic stress analysis*. ▪ MAXWELL's photoelastic method permits the entire stress field to be determined, which proved very useful in the study of impact-induced shock wave propagation in two-dimensional birefringent solid specimens.[673] |
| 1851 | John Murray Publisher, London, England | **Sir Howard DOUGLAS,**[674] a lieutenant-general of the British Navy, **publishes his book *A Treatise on Naval Gunnery*, in which he also reports on supersonic ricochet studies previously performed at Portsmouth.** Firing a 32-pounder cannonball at a relatively low velocity of 455 m/s almost horizontally such that it bounces off the water surface, he observes that the ball can cover a distance of 2,470 meters, thereby rebounding an average of 22 times and only reducing its velocity by about 150–170 m/s, of which, according to British firing tables, 124 m/s are due to aerodynamic drag alone. **DOUGLAS' experimental studies prove that the velocity decrease during recurrent rebound is only small, and that the cannonball, which barely penetrates into the water like a flat stone thrown obliquely, repeatedly rebounds on the surface of the water.** ▪ In 1883, Ernest DE JONQUIÈRES,[675] a French mathematician and vice admiral, was stimulated by DOUGLAS' results to work out a simple theoretical model of ricocheting based on an energetic approach. It showed that, in order to match the velocity loss observed by Sir Howard DOUGLAS, for each reflection cycle the angle of reflection must always be a little greater than the angle of incidence (by about 0° 8'). <br><br>Ricochet firing has long been an important method in ballistics, because it allows the effective range to be increased and ground otherwise hidden by direct fire to be targeted: |

---

[671] J.P. JOULE: *Some remarks on heat, and the constitution of elastic fluids*. Mem. Manchester Phil. Soc. **9**, 107-114 (1851).
[672] J.C. MAXWELL: *On the equilibrium of elastic solids*. Trans. Roy. Soc. Edinb. **20**, 87-120 (1850).
[673] W. GOLDSMITH: *Dynamic photoelasticity*. In: (K. VOLLRATH and G. THOMER, eds.) *Kurzzeitphysik*. Springer, Wien etc. (1967), pp. 579-619.
[674] H. DOUGLAS: *A treatise on naval gunnery: dedicated by special permission to the Lords Commissioners of the Admiralty*. J. Murray, London (1851).
[675] E. DE JONQUIERES: *Sur le ricochet des projectiles sphériques à la surface de l'eau*. C. R. Acad. Sci. Paris **97**, 1278-1281 (1883).

▸ In 1692, this method was first practiced by the French military engineer Sébastien LE PRESTRE DE VAUBAN at the siege of Ath (southwest Belgium).

▸ In 1757, a practice of this kind was successfully used by Frederick II, the Great, king of Prussia, at the battle of Roßbach against French, Austrian and Russian troops. He had several 6-inch (152.4-mm) mortars, made with trunnions, and mounted on traveling carriages, which were fired obliquely upon the enemy's lines, and among their horses.

▸ In World War II, the ricochet method was also used successfully by the British Air Force to attack and break the German Möhne and Eder dams in the Ruhr using back-spinning spherical bombs.[676]

| 1851 | Parish of Sheffield, South Yorkshire, northern England | **The Rev. Samuel EARNSHAW,** a British mathematician, **observes unusual sound phenomena during thunder,** that he will later describe as follows:[677] "A thunder-storm which lasted about half an hour was terminated by a flash of lightning of great vividness, which was instantly (*i.e.*, without any appreciable interval between) followed by an awful crash, that seemed as if by atmospheric concussion alone it would crush the cottages to ruins. Every one in the village had felt at the moment of the crash that the electric fluid had certainly fallen somewhere in the village … But, to the surprise of everybody, it turned out that no damage had been done in the village, but that that flash of lightning had killed three sheep, knocked down a cow, and injured the milkmaid at a distance of more than a mile from the village." ▪ Since sound needs about 4½ seconds to cover an English mile (1,609 m), and EARNSHAW felt that lightning flash and thunder happened almost simultaneously, even though the lightning struck more than a mile away, he concluded that intense sound, such as that originating from a thunderclap, must propagate with supersonic velocity. More recent studies on lightning and thunder, however, revealed that thunder only propagates supersonically in the very near vicinity of the stroke {UMAN ⇨ 1970; BASS ⇨ 1980; GRANEAU ⇨ 1989; ⇨ Fig. 2.10−E}.

Seven years later, he presented his *Theory of Sound of Finite Amplitude* {EARNSHAW ⇨ 1858}, a milestone in early shock wave physics. |
| --- | --- | --- |
| 1851 | Consultant Engineering, Edinburgh, Scotland, U.K. | **William J.M. RANKINE,**[678] a Scottish engineer and physicist, **addresses the sound problem and previous arguments given by the French mathematician Pierre-Simon DE LAPLACE** {⇨1816}. He states, "Now the velocity with which a disturbance of density is propagated is proportional to the square root, not of the total pressure divided by the total density, but of the variation of pressure divided by the variation of density … It is therefore greater than the result of NEWTON's calculation, and this, whether the disturbance is a condensation or a dilatation, or compounded of both." ▪ He closely followed a correct concept proposed by the English physicist George B. AIRY {⇨1848}, that any disturbance propagates with the velocity

$$c = (\partial p/\partial \rho)^{1/2},$$

an important discovery for tackling a difficult problem with waves of finite amplitude (*i.e.*, of shock waves): since the curve of adiabatic compression for most fluids, $p = p(\rho)$, curves upward, $c$ increases with compression. **This "overtaking effect" means that regions of higher pressure will approach those of lower pressure ahead of it, thus eventually leading to a steep wavefront** {⇨Fig. 2.1}. |

---

[676] W. JOHNSON: *Ricochet of non-spinning projectiles, mainly from water. Part I: Some historical contributions.* Int. J. Impact Engng. **21**, 15-24 (1998); *The ricochet of spinning and non-spinning spherical projectiles, mainly from water. Part II: An outline of theory and applications.* Ibid. **21**, 25-34 (1998).

[677] S. EARNSHAW: *On the velocity of the sound of thunder.* Phil. Mag. **20** [IV], 37-41 (1860).

[678] W.J.M. RANKINE: *On LAPLACE's theory of sound.* Phil. Mag. **1** [IV], 225-227 (1851).

| | | |
|---|---|---|
| 1851 | *Pembroke College, Cambridge, England* | **George G. STOKES,**[679] an English physicist and mathematician, **supports DE LAPLACE's theory that sound is an adiabatic process** {DE LAPLACE ⇨1816}. He submits the view that during the propagation of pulses in an elastic fluid, compressions and expansions of the particles take place so rapidly that there is no time for any appreciable transmission of heat between different particles, thus showing that CHALLIS' supposition[680] that the heat developed is lost by radiation is untenable, but that DE LAPLACE's view has a real physical foundation. |
| 1851 | *Private laboratory at Lacock Abbey, Wiltshire, England* | **William Henry Fox TALBOT,**[681] a British physicist, **takes the first microsecond snapshot photo** of a page of the *London Times* rotating at high speed on a revolving disk using an electric spark from a Leiden jar {VON KLEIST & CUNEUS ⇨1745} as a flash light source. He states, "It is in our power to obtain the pictures of all moving objects, no matter in how rapid motion they may be, provided we have the means of sufficiently illuminating them with a sudden electric flash." To obtain the high sensitivity required, he uses plates covered with albumin (a protein), which he exposes immediately after sensitization.[682] ▪ This experiment was an important step toward single-shot high-speed photography, because it first proved the excellent ability of photographic film to freeze rapid events for later detailed analysis as well as the suitability of electric sparks for short-time illumination purposes. |
| 1853 | *Chemistry Laboratory, University of Heidelberg, Germany* | **Robert W. BUNSEN,**[683] a German chemist, **performs experiments on explosions of mixtures of electrolytic gas** – so-called "detonating gas" [$2H_2/O_2$, Germ. *Knallgas*] – **with increasing amounts of carbonic oxide.** In order to measure the peak pressure of the explosion, he uses a "lever device" [Germ. *Hebelvorrichtung*], a mechanical apparatus very similar to a weight-loaded safety valve {⇨Fig. 4.17−B}. He claims that his results are inconsistent with the principle of "mass action." ▪ Four years later, BUNSEN reported on quantitative pressure measurements of gaseous explosive mixtures using his lever device {BUNSEN ⇨1957}.<br><br>In 1805, Comte Claude-Louis BERTHOLLET, an eminent French chemist, had already proposed the (correct) idea that the rate of a chemical reaction is dependent upon the concentrations of reactants. This fact was later called the "Law of Mass Action." The British chemist Harold B. DIXON {⇨1877}, however, showed that BUNSEN's conclusion was incorrect. |
| 1853 | *Erckmann Telegraph Wire Factory, La Villette, Dépt. Calvados, northern France* | **Gregorio VERDÚ,**[684] an engineer at the Spanish Army Corps of Engineers, **performs experiments to fire gunpowder by an electric spark.** He is assisted by the mechanic Heinrich D. RÜHMKORFF, the famous inventor of the Rühmkorff high-voltage induction coil (1851) who was born in Hannover, Germany but spent most of his life in Paris. VERDÚ applies his generator to produce sparks in gap-type detonators. They consist of two isolated parallel copper wires with bare ends, 1.5 mm apart, both projecting into the powder. He succeeds in provoking ignition, even when using cable lengths of up to 26 km.<br><br>In the same year, Moritz F. GÄTZSCHMANN,[685] a German professor of mining at Freiberg in Saxony, uses a highly explosive powder – a mixture of antimony sulfide [$Sb_2S_3$] and potassium chlorate [$KClO_3$] – as a booster to amplify the heat effects of a spark discharge, thus improving the firing quality of gunpowder. Using a 78.5-m long cable and eight spark-gap-type detona- |

---

[679] G.G. STOKES: *An examination of the possible effect of the radiation of heat on the propagation of sound.* Phil. Mag. **1** [IV], 305-317 (1851).
[680] J. CHALLIS: *Theoretical determination of the velocity of sound.* Phil. Mag. **32** [III], 276-284 (1848); *On the theory of the velocity of sound.* Ibid. **1** [IV], 405-408 (1851).
[681] W.H.F. TALBOT: *On the production of instantaneous images.* Phil. Mag. **3** [IV], 73-77 (1852).
[682] G. BUCKLAND: *Fox TALBOT and the invention of photography.* Scolar, London (1980).
[683] R.W. BUNSEN: *Untersuchungen über die chemische Verwandtschaft.* Ann. Chem. Pharm. **85**, 137-155 (1853); *Gasometrische Methoden.* Vieweg & Sohn, Braunschweig (1857).
[684] G. VERDÚ: *Memoria sobre nuevos esperimentos para dar fuego à las minas por medio de la electricidad.* Revista de Madrid **III**, 159-164 (1853); *Versuche über das Zünden von Sprengminen mittels Electricität.* Dingler's Polytech. J. **128**, 421-423 (1853).
[685] M.F. GÄTZSCHMANN: *Die Zündung von Sprengschüssen durch den elektrischen Funken.* Dingler's Polytech. J. **128**, 424-428 (1853).

| | |
|---|---|
| 1854 *Physiological Institute, University of Königsberg, East Prussia & University of Glasgow, Scotland, U.K.* | tors, he first demonstrates that simultaneous electrical firing of gunpowder is possible at a safe distance.

On February 7, in a popular lecture delivered at Königsberg commemorating the 50th anniversary of Immanuel KANT's death, **Hermann von HELMHOLTZ,**[686] a German assistant professor of physiology and physics, **discusses the *Mayer-Waterston solar meteoric hypothesis* on the origin of the Sun's heat** {MAYER & WATERSTON ⇨1848} **based on true thermodynamic principles.**

(i) His more advanced meteoric theory arise from supposing the Sun and its heat to have originated in a coalition of smaller bodies, falling together by mutual gravitation, and generating, as they must do according to JOULE's principle (1843), an equivalent amount of heat to the motion lost during the collision.

(ii) His ***gravitational implosion model*** proposes that the huge weight of the Sun's outer layers should cause the Sun to gradually contract − thus compressing the gases in its interior and converting gravitational energy into heat. This does in fact happen in the protostar phase of stellar formation, as has been discovered by modern astrophysicists.

In the same year, William THOMSON[687] (from 1892 Lord KELVIN), a Scottish physics professor, suggests the idea that the Sun's heat might be produced continually by the impact of meteors falling onto its surface. • In 1862, THOMSON,[688] forced by astronomical evidence to modify his hypothesis, then argued that the primary source of the energy available to the Sun was the gravitational energy of the primordial meteors from which it was formed. However, due to the additional mass, the period of the Earth (length of the year) would have decreased by about 1.5 months since the beginning of the Christian era. Therefore, he later abandoned his meteoric theory in favor of VON HELMHOLTZ's gravitational contraction theory.

According to this so-called "Helmholtz-Kelvin contraction theory," the Sun could not be more than about 20 million years old. The British naturalist Charles DARWIN was so shaken by the power of THOMSON's analysis and by the authority of his theoretical expertise that in the last editions of *On The Origin of the Species* he eliminated all mention of specific time scales. Many 19th-century geologists and evolutionary biologists had already concluded that the Sun must have been shining for at least several hundred million years in order to account for geological changes and the evolution of living things, both of which depend critically upon energy from the Sun. This contradiction initiated the famous ***Age of the Earth Debate*** which took place in 1872 at the Royal Institution in London.[689, 690]

The source of solar energy was not explained until the discovery of atomic transmutation processes. In the late 1930s, the sequence of nuclear reactions within the Sun in which hydrogen nuclei were burned using carbon as a catalyst was worked out independently by two German theoretical physicists, Hans A. BETHE[691] and Carl F. VON WEIZSÄCKER.[692] This process is known as the **"carbon-nitrogen-oxygen (CNO) cycle"** or the **"Bethe-Weizsäcker cycle."** |

---

[686] H. VON HELMHOLTZ: *Über die Wechselwirkung der Naturkräfte* [Vortrag gehalten in Königsberg am 7. Feb. 1854]. In: *Vorträge und Reden von H. VON HELMHOLTZ.* Vieweg & Sohn, Braunschweig (1884), vol. I, pp. 25-77; Engl. translation: *On the interaction of natural forces.* Phil. Mag. **11** [IV], 489-518 (1856).

[687] W. THOMSON: *On the mechanical energies of the Solar System.* Phil. Mag. **8** [IV], 409-430 (1854).

[688] W. THOMSON: (i) *On the age of the Sun's heat.* Macmillan's Mag. **5**, 288-293 (1862); http://zapatopi.net/kelvin/papers/on_the_age_of_the_suns_heat.html. (ii) *The secular cooling of the Earth* [read April 28, 1862]. Trans. Roy. Soc. Edinburgh **23**, 157-159 (1864).

[689] L. BADASH: *The Age-of-the-Earth debate.* Scient. Am. **261**, No. 2, 90-96 (1989).

[690] A. STINNER and J. TEICHMANN: *Lord KELVIN and the Age-of-the-Earth debate: a dramatization.* Sci. Educ. **12**, 213-228 (2003); http://www.sci-ed.ca/Publications/Lord%20Kelvin%20Drama.pdf.

[691] H.A. BETHE: *Energy production in stars.* Phys. Rev. **55** [II], 103, 434-456 (1939).

[692] C.F. VON WEIZSÄCKER: *Über Elementumwandlungen im Innern der Sterne.* Physik. Z. **38**, 176-191 (1937); Ibid. **39**, 633-646 (1938).

| | | |
|---|---|---|
| 1854 | *Bethlehem, Pennsylvania* | **An unusually violent accidental explosion of finely powdered zinc happens at the Bethlehem Zinc Works,**[693] a classical example of a metal-dust explosion. ▪ Explosive dusts may be classified according to their explosivity into four dust explosion classes:<br>▸ *ST 0:* no explosions;<br>▸ *St 1:* weak explosions (grain dust and coal dust explosions),<br>▸ *St 2:* strong explosions (explosions of organic pigments); and<br>▸ *St 3:* very strong explosions (explosions of fine metal dusts).<br>For example, for aluminum dust the rate of pressure rise can be 200 times higher than for coal dust.[694] U.S. researchers compared the explosion hazards of various metal dusts {DORSETT ET AL. ⇨1960}. |
| 1854 | *Academy of Sciences, St. Petersburg, Russia* | **Mikhail Vasilievich OSTROGRADSKY,**[695] a Russian applied mathematician and member of the Academy, **presents a memoir on a general theory of percussion in which he investigates the impact of systems,** thereby assuming that the constraints arising at the moment of impact are preserved after impact. He extends the principle of virtual displacements to the phenomenon of inelastic impact, and derives the basic formula for the analytical theory of impact. ▪ In the period 1822–1827, OSTROGRADSKY, then in his early twenties, attended courses by outstanding French mathematicians such as CAUCHY, FOURIER, DE LAPLACE, LEGENDRE, and POISSON, who taught at the Ecole Polytechnique {⇨1794/1795} or other educational institutions. |
| 1855 | *Private Observatory, Regent's Park, London, England* | On December 15, **John Russell HIND,**[696] a British astronomer and director of George BISHOP's private observatory, **discovers the variable star U Gem (Geminorum),** a repeating nova which is located in *Gemini* ("The Twins," a spring constellation of the Northern Hemisphere).<br>(i) A few days later, the brightness of this previously unlisted star drops from nearly 9th magnitude to below the 13th magnitude. ▪ It became again bright 100 days later, on March 24, 1856.<br>(ii) This sudden rise and decrease which had never before been observed in a star causes a great deal of interest, and HIND concludes, "It is evidently a variable star of a very interesting description, inasmuch as the minimum brightness appears to extend over a great part of the whole period, contrary to what happens with Algol and S Cancri." ▪ Besides his early discovery of U Gem, HIND discovered further 21 variable stars.<br>U Gem is a dwarf-nova type cataclysmic variable that brightens abruptly and unpredictably and belongs to the class of *explosive* (or *catastrophic*) *variable stars.*[697] It is a compact, interacting binary system made up of a primary star which is a dense, small, white (hot) dwarf, and a secondary red (cool) Sun-like star. The two stars orbit around each other at very high speeds. It takes only 251 minutes for the stars to spin around one another. The secondary star is so close to the primary that it loses material from its surface towards the primary. This material collects in a disk, called an "accretion disk," around the primary. **Time-resolved spectroscopic observations of U Gem during its 2000 March outburst showed strong spiral shocks in the accretion disk.**[698] |
| 1855 | *Newcastle upon Tyne, northeast England* | **George ROBINSON,**[699] a British physician to the Newcastle and Gateshead Dispensaries and a scientific writer, stimulated by previous exploding wire experiments carried out in his country {SINGER & CROSSE ⇨1815}, **performs pioneering experiments to disintegrate various urinary calculi by repeatedly discharging a Leiden jar** {VON KLEIST & CUNEUS ⇨1745}. |

---

[693] C.E. MUNROE: *Modern explosives.* Scribner's Mag. **3**, 563-576 (1888).
[694] J. CROSS and D. FARRER: *Dust explosions.* Plenum Press, New York & London (1982), p. 151.
[695] M.V. OSTROGRADSKY: *Mémoire sur la théorie générale de la percussion* [read 1854]. Mém. Acad. Sci. St. Pétersbourg **VII**, 267-304 (1857).
[696] J.R. HIND: *On a new variable star.* Month. Not. Roy. Astron. Soc. **16**, 56 (1855/1856).
[697] For the classification of variable stars *see* General Catalogue of Variable Stars ⇨1948.
[698] P.J. GROOT: *Evolution of spiral shocks in U Geminorum during outburst.* Astrophys. J. **551**, Part 2, L89-L92 (2001).
[699] G. ROBINSON: *On the disintegration of urinary calculi by the lateral disruptive force of the electrical discharge.* Proc. Roy. Soc. Lond. **7**, 99-102 (1854/1855).

(i) He explains his experimental set-up as follows, "Two copper wires, one-twentieth of an inch [1.27 mm] in diameter, were connected ... with a Leyden jar ... These copper wires were soldered to platinum wires half an inch [12.7 mm] long and one-thirtieth of an inch [0.85 mm] diameter. Each wire was drawn through a fine gutta percha tube, and the tubes, having first been placed perfectly parallel, were warmed and gently pressed together so as to assume somewhat of the appearance of a flexible bougie; the platinum wires projecting beyond the gutta percha to the extent of one-eighth of an inch [3.18 mm], and their free extremities being slightly everted and separated from each other by an interval of one tenth of an inch [2.54 mm]. In experimenting, the united gutta percha tubes were grasped and the projecting platinum points pressed against the surface of the calculus; the jar was then discharged by another person, and a series of such discharges thus passed between the free extremities of the parallel platinum wires while resting upon the surface of the stone."

(ii) With this simple arrangement, lithic, acid and phosphatic calculi could be fragmented as well as an oxalate of lime or mulberry calculus.

(iii) **He concludes, "These experiments appear to demonstrate the practicability of applying the lateral disruptive force of the electrical discharge to the disintegration of calculi in the bladder."** ▪ His experimental work with shock waves, however, was not accepted by contemporary surgeons. It took more than 100 years until the Latvian urologist Viktor GOLDBERG {⇨1959}, resuming ROBINSON's idea but obviously having no knowledge of it, first successfully fragmented a calculus in the bladder *in vivo*.

| | | |
|---|---|---|
| 1855 | *Private study at Villeporcher-Saint-Ouen/Vendôme, Dépt. Loir-et-Cher, central France* | While performing extensive research into the theory of elasticity, **Adhémar-Jean-Claude Barré DE SAINT-VENANT**,[700] a French mathematician and engineer, **observes that local effects of loading do not affect global strain – later to be known as the "Saint-Venant principle."** He writes,[701] "If the forces acting on a small portion of the surface of an elastic body are replaced by another statically equivalent system of forces acting on the same portion of the surface, this redistribution of loading produces substantial changes in the stresses locally but has a negligible effect on the stresses at distances which are large in comparison with the linear dimensions of the surface on which the forces are changed." ▪ The Saint-Venant principle is important in the application of elasticity to many practical situations in which boundary conditions are satisfied in terms of stress and moment resultants rather than by point-wise specifications. |
| | | Research in the area of the Saint-Venant principle has been reviewed by Cornelius O. HORGAN,[702] a professor of applied mathematics and mechanics at the University of Virginia in Charlottesville, VA. Theoretical and experimental work carried out to check whether this principle could also be extended to dynamic "nonequilibrium" problems was discussed by Bruno A. BOLEY,[703] a professor of civil engineering at Columbia University, New York City. |
| 1856 | *Coast of Yorkshire, England* | On March 2, **the sea rises and falls a considerable distance at many places along the coast of Yorkshire**. At Whitby, the tide ebbs and flows six times per hour, covering such a distance that a vessel entering the harbor is alternately afloat and aground. The strange phenomenon is not accompanied by earthquakes. |
| | | Two years later, the Irish seismologist Robert MALLET[704] suggested the idea that sometimes waves like the above might be produced by an underwater slippage of the material forming the face of a submarine bank. By degradation and deposition caused by currents, the slope |

---

[700] A.J.C. DE SAINT-VENANT: *De la torsion des prismes, avec des considérations sur leur flexion, etc.* Mém. Recueil Savants Étrangers **XIV**, 233-560 (1855).
[701] S. TIMOSHENKO and J.N. GOODIER: *Theory of elasticity*. McGraw-Hill, New York (1934), p. 33.
[702] C.O. HORGAN: *Recent developments concerning SAINT VENANT's principle: an update*. Appl. Mech. Rev. **42**, 295-303 (1989).
[703] B.A. BOLEY: *Application of SAINT-VENANT's principle in dynamical problems*. J. Appl. Mech. **22**, 204-206 (1955); *On a dynamical Saint-Venant principle*. Trans. ASME: J. Appl. Mech. **27**, 74-78 (1960).
[704] R. MALLET: *Fourth report upon the facts and theory of earthquake phenomena*. Rept. Meet. Brit. Assoc. **28**, 1-136 (1858).

| | | |
|---|---|---|
| | | of this bank reaches an angle exceeding the limit of response of the material from which it was formed. **MALLET thus anticipated the modern view that great sea waves (tsunamis) can be caused by suboceanic landslides** {MILNE ⇨1898; GUTENBERG ⇨1939}. |
| 1856 | *Ecole Polytechnique, Paris, France* | **Jules Célestin JAMIN,**[705] a French physicist, **invents the first optical interferometer** {⇨Fig. 4.18−E}, which he uses to measure the refractive index of gases. ▪ His instrument is the archetype of many subsequent interferometer constructions.<br><br>Already 14 years later, August TOEPLER[706] and Ludwig E. BOLTZMANN, two noted physics professors at the University of Graz, introduced optical interferometry into fluid dynamics in order to determine the amplitude at the threshold of hearing − a masterpiece of experimental acoustics {⇨Fig. 4.18−E}. They find that at the threshold of hearing, the displacements of the air molecules amount to only one-tenth of the wavelength of green light; *i.e.*, to about 52 nm. Modern measurements, however, give a value of only 0.01 nm, which is only approximately one-tenth of the diameter of a hydrogen molecule.[707] |
| 1856 | *Royal Secondary School, Berlin, Prussia, Germany* | **August Karl KRÖNIG,**[708] a German high-school professor, **publishes the first kinetic theory of gases.** Following a model proposed by Daniel BERNOULLI {⇨1738}, he assumes that a gas consists of discrete particles (molecules), each of which behaves according to universal mechanical laws, and that the molecules have only translational motion. ▪ Rudolf J.E. CLAUSIUS {⇨1857}, James C. MAXWELL {⇨1867}, and Ludwig E. BOLTZMANN {⇨1872} subsequently made important improvements to this theory, and today they are regarded as the main founders of the kinetic theory of gases.[709] |
| 1856/<br>1857 | *Private laboratory in Joule's brewery at Salford, Lancashire, England & Chair of Physics, University of Glasgow, Scotland, U.K.* | **James P. JOULE,**[710] a British physicist, **and William THOMSON,** an engineer from Glasgow (from 1892 Lord KELVIN, in recognition of his work in engineering and physics), **experimentally study thermal effects of bodies moving rapidly through air:**<br><br>(i) They conclude that "a body round which air is flowing rapidly acquires a higher temperature than the average temperature of the air close to it all round." ▪ This phenomenon was later dubbed the **"Joule-Thomson effect."**<br><br>(ii) They annotate that "the same phenomenon must take place universally whenever air flows against a solid or a solid is carried through air. If the velocity of 1,780 feet per second [542.5 m/s] in the foregoing experiment gave 137° cent. difference of temperature between the air and the solid, how probable is it that meteors moving at from six to thirty miles per second [ca. 9.6 to 48.3 km/s], even through a rarefied atmosphere, really acquire, in accordance with the same law, all the heat which they manifest! On the other hand, it seemed worth while to look for the same kind of effect on a much smaller scale in bodies moving at moderate velocities through the ordinary atmosphere … **we have tried and found, with thermometers of different sizes and variously shaped bulbs, whirled through the air at the end of a string,** with velocities of from 80 to 120 feet per second [24.4 to 36.6 m/s], **temperatures always higher than when the same thermometers are whirled in exactly the same circumstances at smaller velocities.**" |

---

[705] J. JAMIN: *Description d'un nouvel appareil de recherches, fondé sur les interférences.* C. R. Acad. Sci. Paris **42**, 482-485 (1856); *Neuer Interferential-Refractor.* Ann. Phys. **98** [II], 345-349 (1856).

[706] A. TOEPLER and L. BOLTZMANN: *Über eine neue optische Methode, die Schwingungen tönender Luftsäulen zu analysieren.* Ann. Phys. **141** [II], 321-352 (1870).

[707] H. GOBRECHT: BERGMANN-SCHAEFER: *Lehrbuch der Experimentalphysik.* De Gruyter, Berlin (1970), vol. 1, p. 561.

[708] A. KRÖNIG: *Grundzüge einer Theorie der Gase.* Ann. Phys. **99** [II], 315-322 (1856).

[709] T. KOGA: *Introduction to kinetic theory stochastic processes in gaseous systems.* Pergamon Press, Oxford etc. (1970); see Appendix G: *A historical sketch of kinetic theory,* pp. 261-277.

[710] J.P. JOULE and W. THOMSON: *On the thermal effects of fluids in motion. On the temperature of solids exposed to currents of air.* Proc. Roy. Soc. Lond. **8**, 178-185 (1856/1857); *On the thermal effects of fluids in motion. Temperature of a body moving through air.* Ibid. **8**, 556-567 (1856/1857).

(iii) In the case where the velocity of translation of the body $v$ is a small fraction of the velocity of sound (for example, $a$ = 1,115 ft/s or 339.85 m/s at 17 °C), they estimate for the "hot spots" at the body's surface – *i.e.*, at those points where the flow velocity is slowed down to zero – a temperature increase of

$$\Delta\Theta \, [°C] = 58.8 \, (v/a)^2.$$

For example, for a bulb thermometer moving at $v$ = 183 ft/s (55.78 m/s), they measure a temperature rise of $\Delta\Theta$ = 1 °C. ▪ The theoretical increase in temperature given by their formula above would be 1.58 °C, which is surprisingly close.

In JOULE's era, the phenomenon of aerodynamic heating – first discussed in the Renaissance using the example of the **"Babylonian eggs"** {GALILEI ⇨1623} – was a subject of pure academic curiosity. The determination of the temperature taken up by a solid surface in an airstream is a complex problem and depends on the type of boundary layer: a turbulent boundary layer will generate more frictional heat, and thus be hotter, than its laminar counterpart {POHLHAUSEN ⇨1921}. With the advent of high flight velocities in the late 1930s, the importance of aerodynamic heating was rediscovered by the Italian General G. Arturo CROCCO {⇨1931}, who proposed the term *aerothermodynamics*. It then became a practical problem when supersonic flight was realized, not only in the military realm of rocketry and aircraft {HVA Peenemünde ⇨1942; YEAGER ⇨1947; X-15 ⇨1959 & 1961}, but later also in civil aviation {Tupolev Tu-144 ⇨1968; Concorde ⇨1969}.

| | | |
|---|---|---|
| 1857 | *Gulf of Naples, central Italy* | On December 16, **a highly destructive earthquake near Naples kills more than 12,000 people, and leaves numerous villages and small towns in ruins.** ▪ The principal shock had been preceded by minor disturbances and was followed by numerous after-shocks which continued for many months. |

Shortly after, the Irish civil engineer **Robert MALLET travels to Italy to study the fractured walls and overthrown pillars.** Having been interested in the mechanism of earthquakes since the early 1840s {MALLET ⇨1846}, and supported by a grant from the Royal Society of London, he hopes to determine the velocities and directions of the seismic shocks that produced them.

(i) He starts from the crude concept that earthquakes are explosive in origin and primarily generate compressional waves originating from a seismic point source, followed by dilatation waves.

(ii) From cracks found in buildings extending perpendicular to the propagation of seismic waves, he constructs a map of "isoseismal lines," and calculates the depth of the hypocenter to be around 10 km.

(iii) He determines the velocity of the particles during wave propagation. ▪ Later he arrived at the conclusion that "the velocity of the particles in even the greatest shocks is extremely small, not exceeding 20 ft/s [6 m/s] in very great earthquakes, and probably never having reached 80 ft/s [2.4 m/s] in any shock that has occurred in history."

(iv) He also constructs the trajectory of a body thrown down by a seismic shock wave.[711]

**Although some of MALLET's results are incorrect from the present viewpoint, they nevertheless mark** – together with his subsequent determinations of the velocity of earth waves {MALLET ⇨1860} – **a milestone in *seismology*, a term which he coined** besides other important parameters, such as the terms

▸ ***seismic focus*** (the place of origin of the earthquake, practically a locus or space of three dimensions);

▸ ***angle of emergence*** (= arctan $h/a$, where $h$ is the depth of the center of earthquake, and $a$ is the axial distance);

---

[711] *See* his book on the Great Neapolitan Earthquake (*see* next footnote), vol. 1, pp. 67-68; *See also* http://www.univie.ac.at/Wissenschaftstheorie/heat/gallery/fig3-91f.htm.

> *seismic vertical* (a vertical line joining the epicentre and the focus);
> *meizoseismic line* (the line of maximum destruction); and
> *meizoseismal area* (the most disturbed area within the innermost isoseismal line).

Five years later, MALLET[712] published his results in a report which was reviewed by the *Saturday Review*[713] as follows: "Any one who looked for the first time at the ruins of a town overthrown by a violent shock of earthquake would laugh at the idea of extracting from that mass of confused piled-up rubbish any substantial information as to the precise subterranean centre of disturbance, or the character of the earthquake oscillations. Yet this was the task which Mr. MALLET, with the support of the Royal Society undertook to perform … The result of the expedition is given in two large volumes, magnificently illustrated with photographs … if we are to take his own estimate of the success of his investigations, we must admit that he has ascertained with precision the exact position, depth, and in addition, measures of the amplitude and velocity of the seismic perturbation, which he evidently relies on with the most absolute confidence … There is no doubt that he has proved that a science of seismology is capable of being built up … to an extent which would scarcely have been anticipated." ▪ Because of his epochal contributions to the understanding of earthquakes, the 1857 Naples Earthquake later became renowned among seismologists as "MALLET's Earthquake."

| | | |
|---|---|---|
| 1857 | Mont Cenis, Dépt. Savoie, southeastern France | Begin of the construction of a 13.7-km-long Mont Cenis Tunnel, a railroad tunnel, which will be driven through the Alps between France and Italy. Technical innovations such as pneumatic drilling machines and electrical ignition of explosive charges are used. ▪ **The Mont Cenis Tunnel, finished in 1871, became known as the largest construction job with black powder up to that time.** It runs under the Col de Fréjus from Modane (France) to Bardonecchia (Italy). |
| 1857 | Chemistry Laboratory, University of Heidelberg, Germany | Robert W. BUNSEN,[714] a German professor and an international authority on chemistry, **attempts to determine the explosion pressure produced by the gases reacting inside a closed vessel.**<br><br>(i) He describes the explosion of mixtures of hydrogen and of carbonic oxide with oxygen in a strong vessel sealed with a loaded lid {BUNSEN ⇨1853; ⇨Fig. 4.17–B}. When the pressure produced in the vessel exceeded the pressure on the lid, the latter was raised in the manner of a safety valve {⇨Fig. 4.7–A, B}, and some of the heated gas escaped; when the pressure was less than that on the lid, no gas escaped and little noise was heard.<br><br>(ii) He finds the pressures produced by the explosions of gaseous mixtures of $2CO/O_2$ and $2H_2/O_2$ to be about 10 and 9.5 atmospheres, respectively.<br><br>In the same year, BUNSEN,[715] **in collaboration with the Russian Léon SCHISCHKOFF** [or SHISHKOFF], a chemistry professor at the Artillery School in St. Petersburg, **investigates the products of the firing of gunpowder – both gaseous and solid products of the explosion – for the first time.**<br><br>(I) For the firing of powder whose gases are formed in a closed space and cannot freely expand, they determine the maximum temperature (3,340 °C), the maximum pressure (4,374 atm), and the maximum mechanical effect of powder or "theoretical work" (67,410 m kg* for 1 kg of powder).[716] |

---

[712] R. MALLET: *Great Neapolitan Earthquake of 1857: the first principles of observational seismology.* Chapman & Hall, London (1862).
[713] Saturday Review (London) **15**, 215 (1863).
[714] R.W. BUNSEN: *Gasometrische Methoden.* Vieweg & Sohn, Braunschweig (1857); *Gasometry: comprising the leading physical and chemical properties of gases* (translated by H.E. ROSCOE). Walton & Maberly, London (1857).
[715] R.W. BUNSEN and L. SCHICHKOFF: *Chemische Theorie des Schießpulvers.* Ann. Phys. **102** [II], 321-353 (1857); *On the chemical theory of gunpowder.* Phil. Mag. **15** [IV], 489-512 (1858).
[716] For the meaning of the unit kg*, see ROUX & SARREAU {⇨1873}.

(II) They also ascertain that a large number of salts whose presence had not previously been detected are normal constituents of the smoke and solid residue, and that many other gaseous products besides carbon dioxide and nitrogen are formed. ▪ Other investigators {BERTHELOT[717] 1871; NOBLE & ABEL ⇨1875; DEBUS ⇨1882} that took up this complex subject arrived at BUNSEN's and SCHISCHKOFF's conclusion; namely, that it is not possible to give any simple expression for the reaction, because the products are not only very numerous, but also vary considerably in their proportions depending on the conditions (especially the pressure, and therefore the temperature) under which the explosion in a closed space takes place.

In the same year, BUNSEN[718] **describes in a letter to Sir Henry Enfield ROSCOE the demonstration of a light-initiated explosion of hydrogen and chlorine gas.** He writes, "The amount of gas produced in our hydrogen chlorine generator is led in the dark through the system of glass bulbs B, which are about the size of a dove's egg and blown so *thin* that they can be crushed *with the finger*, and prior to passage of the gas have been thoroughly wetted inside with water. Once the apparatus is full, the rubber ligatures are tied together at *a* and cut cleanly at *b*, and the rubber ends are dipped into molten yellow wax. These bulbs can be kept several days without spoiling, and if one of them is held near an open window it explodes instantaneously. The bulbs can be held in the hand quite safely by their stems provided one puts on a glove and holds a small glass plate between the bulb and the face. The explosion occurs on the greyest of days, even with rather heavy fog, and it is scarcely more dangerous than the ignition of a soap bubble filled with O and H. **I have now had a red and a blue pane of glass installed in the auditorium window: if a bulb is held in front of the former it remains unchanged, whereas in front of the latter it explodes instantaneously.** There is scarcely another lecture demonstration that could be more beautiful than this." ▪ According to another source,[719] this experiment was frequently carried out by "filling a bottle with a mixture of chlorine and hydrogen, and throwing it from an open window into the bright sunshine."

In the period 1856–1859, BUNSEN[720] and ROSCOE, his former student and now professor of chemistry at Owens College in Manchester, jointly published a five-part memoir on their photochemical research.

| | | |
|---|---|---|
| 1857 | *Polytechnikum Zürich, Switzerland* | Rudolf J.E. CLAUSIUS,[721] a German theoretical physicist, **initiates a *Dynamical Theory of Gases*** which will quickly evolve into an important field of physical research:<br><br>(i) **His concept of molecular motion goes far beyond the billiard model of previous writers such as the German August K. KRÖNIG** {⇨1856}. He shows that translational motion alone cannot account for all of the heat present in a gas. If rotatory, vibratory, and translational motion is ascribed to the molecules, the conservation of translational kinetic energy can no longer be assumed, because collisions might cause transformations of one form of motion into another. Consequently, the idea of a constant equal velocity for all molecules – an assumption of many early corpuscular models – becomes untenable.<br><br>(ii) He shows how one can estimate the fraction of the total energy that is translational using heat data. Based upon this qualitative model, CLAUSIUS develops a theory of changes of state. |

---

[717] P.E.M. BERTHELOT: *Sur la force de la poudre et des matières explosives*. Gauthier-Villars, Paris (1871).
[718] A facsimile of BUNSEN's letter (dated Nov. 10, 1857) has been published by O. KRÄTZ: *Historische Experimente (1857). R. BUNSEN: Versuche zur Explosion von Wasserstoff/Chlor-Gemischen am Licht.* Chemie, Experimente und Didaktik (Thieme, Stuttgart) **1**, 63-66 (1975). Engl. translation of BUNSEN's letter in: H.W. ROESKY and K. MÖCKEL: *Chemical curiosities*. VCH, New York (1996), p. 279.
[719] G.S. NEWTH: *Chemical lecture experiments, non-metallic elements*. Longmans, Green & Co., London *etc.* (1899), p. 89.
[720] R.W. BUNSEN and H.E. ROSCOE: *Photo-chemical researches*. Phil. Trans. Roy. Soc. Lond. **147**, 355-380, 381-402, 601-620 (1857); Ibid. **149**, 879-926 (1859), Ibid. **153**, 139-160 (1863).
[721] R. CLAUSIUS: *Über die Art der Bewegung, welche wir Wärme nennen*. Ann. Phys. **100** [II], 353-380 (1857); *On the kind of motion which we call heat*. Phil. Mag. **14** [IV], 108-127 (1857).

(iii) He explains the evaporation of a liquid by assuming that even though the average motion of its molecules may not be sufficient to carry them beyond the range of the attractive forces of their neighbors, "we must assume that the velocities of several molecules deviate within wide limits on both sides of this average value," and therefore a few molecules will be moving fast enough to escape from the surface of a liquid even at temperatures below the boiling point.

In the following year, CLAUSIUS improved his kinetic theory by introducing a **mean free path** representing the average distance a molecule may travel before interacting with another molecule. Later, the Scottish physicist James Clerk MAXWELL {⇒1867} incorporated the mean free path into his own kinetic theory and showed that it could be related to macroscopic gas properties such as viscosity.

| | | |
|---|---|---|
| 1857 | Paris, France | **Louis POINSOT,**[722] a member of the French Academy and senator, **publishes a paper on the properties of the center of percussion of a body set into sudden rotation.** He also shows that other, new, centers may play an important role in the dynamics of impacted bodies, referring to the generation of singular motions such as those for richocets. |
| 1857 | Allegheny Arsenal, U.S. Army Ordnance Dept., Pennsylvania | **Thomas J. RODMAN,**[723, 724] a U.S. military officer, **invents the *indentation gauge*.** It consists of a piston working in a hole bored into the wall of a gun and acting on an indenting tool, a hardened steel knife edge penetrating into a disc of soft copper. The calibration is performed mechanically, and the pressure needed to make a similar cut in the copper can be determined {⇒Fig. 4.17–K}. |

(i) The chief objectives of his experiments are:
- ▸ to ascertain the pressure exerted on the bores of service guns;
- ▸ to determine the pressure in guns of different calibers, with the charges and projectiles in each caliber being arranged such that an equal column or weight of powder is behind an equal column or weight of shot;
- ▸ to investigate the effect produced on the gaseous tension in the bore of a gun by an increase in the size of the grains of the powder; and
- ▸ to determine the ratio of the tension of fired gunpowder to its density.

(ii) With his new gauge – so-called **"Rodman gauge"** – he determines the maximum internal pressure in the bore of a gun at different points. • British investigations, however, later showed that RODMAN had greatly overestimated the maximum pressure that arises in gun barrels {NOBLE ⇒1894}.

(iii) **He also discovers that the maximum pressure in a gun decreases with increasing gunpowder grain size,** leading to the invention of ***mammoth powder*** {RODMAN ⇒1860}.

| | | |
|---|---|---|
| 1858 | Leeds, West Yorkshire, central England | On November 20, at the meeting of the British Association for the Advancement of Science (BAAS), **the Rev. Samuel EARNSHAW,**[725] a British mathematician, **presents his famous *Mathematical Theory of Sound*** which is the most complete of his time. |

(i) Describing the aim of his work, he writes, "I consider this article as tending to account for the discrepancy between the calculated and observed velocities (which most experimentalists have remarked and wondered at), when allowance is made (as will be done in a future part of this paper) for change of temperature…" His achievements are as follows:

(ii) He improves POISSON's one-dimensional theory of finite amplitude disturbances {POISSON ⇒1808}, converting the equations into *Lagrangian* coordinates. This has the advantage of following fluid particles, so that interfaces can be immediately identified at each time.

---

[722] L. POINSOT: *La percussion des corps*. J. Math. Pures & Appl. **2** [II], 281-352 (1857); Ibid. **4** [II], 161-170, 421-426 (1859).

[723] T.J. RODMAN: *Reports of experiments on the properties of metals for cannon, and the qualities of cannon powder; with an account of the fabrication and trial of a 15-inch gun*. Crosby, Boston, MA (1861); Mitteilungen über Gegenstände des Artillerie- und Geniewesens, Wien (1881), see *Notizen*, p. 57.

[724] C. CRANZ: *Lehrbuch der Ballistik: III. Experimentelle Ballistik*. Springer, Berlin (1927).

[725] S. EARNSHAW: *On the mathematical theory of sound* [communicated in Nov. 1858]. Phil. Trans. Roy. Soc. Lond. **150**, 133-148 (1860).

(iii) Using the adiabatic law
$$p = p_0 (\rho/\rho_0)^k,$$
where $k = c_p/c_v$ is the ratio of the specific heat of the gas at constant pressure to its specific heat at constant temperature, and $p_0$ and $\rho_0$ are pressure and density at rest, respectively, he obtains the complete solution for a wave progressing in one direction in a medium in which the pressure is a function of the density in the wave, and observes that the differential equations of motion need not necessarily possess a unique solution for the velocity.

(iv) Assuming that heat conduction and viscosity might prevent the true formation of a discontinuity in a real fluid, he speculates, "**I have defined a bore to be a tendency to discontinuity of pressure; and it has been shown that as a wave progresses such a tendency necessarily arises. As, however, discontinuity of pressure is a physical impossibility, it is certain Nature has a way of avoiding its actual occurrence.** To examine in what way she does this, let us suppose a discontinuity to have actually occurred at the point $A$, in a wave which is moving forwards. Imagine a film of fluid at $A$ forming a section at right angles to the tube. Then on the back of this film there is a certain pressure which is discontinuous with respect to the pressure on its front. To restore continuity of pressure, the film at $A$ will rush forward with a sudden increase of velocity, the pressure in the front of the film not being sufficient to preserve continuity of velocity. In so doing the film will play the part of a piston generating a bit of wave in front, and a small regressive wave behind. The result will be a prolongation of the wave's front, thereby increasing the original length of the wave, and producing simultaneously a feeble regressive wave of a negative character…"

(v) He draws the important conclusion that "the velocity with which a sound is transmitted through the atmosphere depends on the degree of violence with which it was produced … **The report of fire-arms will travel sensibly faster than a gentle sound, such as the human voice.**"

The transactions of this meeting[726] later read, "Fortunately, it transpired at the Meeting, that in Captain PARRY's Expedition to the North [PARRY ⇨ 1824/1825; ⇨ Fig. 2.10–D], whilst making experiments on sound, during which it was necessary to fire a cannon at the word of command given by an officer, **it was found that the persons stationed at the distance of three miles [4.8 km] to mark the arrival of the report of the gun, always heard the report of the gun before they heard the command to fire;** thus proving that the sound of the gun's report had outstripped the sound of the officer's voice; and confirming in a remarkable manner the result of the author's mathematical investigation, that the velocity of sound depends in some degree on its intensity." It was James C. ROSS, who later become a famous South Pole explorer carrying out important Arctic and Antarctic magnetic surveys, that was in command of the cannon during PARRY's expedition.

It is interesting to note that EARNSHAW's solution of POISSON's formulation – albeit exact – was only provided in an implicit form by him until it was solved explicitly in 1935 by the Italian engineer Eugenio FUBINI-GHIRON.[727] A review of this important theory and its frequent rediscovery was given by David T. BLACKSTOCK, a researcher at General Dynamics, New York.[728]

| | | |
|---|---|---|
| 1858 | Herzogliche Realschule, Meiningen, Saxony, Germany | Karl Wilhelm KNOCHENHAUER,[729] a German high-school professor, studies an electric discharge circuit that consists of two Leiden jars coupled to each other. • **The so-called "Knochenhauer circuit" was modified by August TOEPLER and first applied by him to visualize** |

---

[726] S. EARNSHAW: *On the mathematical theory of sound*. Rept. Meet. Brit. Assoc. **28**, 34-35 (1858).
[727] E. FUBINI-GHIRON: *Anomalie nella propagazione di onde acustiche di grande ampiezza*. Alta Frequenza **4**, 530-581 (1935).
[728] D.T. BLACKSTOCK: *Propagation of plane sound waves of finite amplitude in nondissipative fluids*. JASA **34**, 9-30 (1962).
[729] K.W. KNOCHENHAUER: *Über den elektrischen Zustand der Nebenbatterie während ihres Stromes*. Sitzungsber. Akad. Wiss. Wien **33** (Abth. II), 163-204 (1858).

**a propagating shock wave** by stroboscopic illumination using a repetitively operated electric spark discharge {A. TOEPLER ⇨1864; ⇨Fig. 4.19–E}.

In the 1870s, the Knochenhauer circuit was further improved by Ernst MACH {⇨1878} to reliably control a delay pulse in the microsecond regime – the so-called **"Mach circuit"** {⇨Fig. 4.19–E} – which was an important requirement in single-shot photography for precisely stopping the motion of a shock wave within a given, limited field of view. The potential and further improvements in the Knochenhauer circuit were discussed by Bruno GLATZEL.[730]

| | | |
|---|---|---|
| 1858 | *E.I. DuPont de Nemours & Company, Wilmington, Delaware* | **Lammot DU PONT DE NEMOURS**, a U.S. industrialist, **modifies the formula of black powder by substituting the costly potassium nitrate** [$KNO_3$] **for the much cheaper sodium nitrate or Chile saltpeter** [$NaNO_3$].[731] ▪ Although it did not produce an explosive that is as high-quality as potassium nitrate, it was suitable for most mining and construction applications. His new powder came to be known as **"B blasting powder"** to distinguish it from conventional black powder, named **"A blasting powder"** by his company. |
| 1859 | *Private observatory at Redhill, Reigate, Surrey, southern England* | On September 1, three important observations relating to the Sun are made which will stimulate increasing research and theoretical studies on our closest star, eventually leading to the establishment of *Solar Physics*, now a branch of geophysics:<br><br>(i) **Richard Christopher CARRINGTON**,[732] a British amateur astronomer at Redhill, **observes a violent and rapid eruption moving along the Sun's surface with an average velocity of about 186 km/s near a large sunspot at a latitude of about 10°** {⇨Fig. 4.1–U}. Using a telescope which projects the image of the Sun's disk onto a painted plate of glass, he gives the following account of his observation, "While engaged in the forenoon of Thursday, September 1, in taking my customary observation of the forms and positions of the solar spots, an appearance was witnessed which I believe to be exceedingly rare. The image of the Sun's disk was, as usual with me, projected on to a plate of glass coated with distemper of a pale straw color, and at a distance and under a power which presented a picture of about 11 inches [28 cm] diameter. I had secured diagrams of all the groups and detached spots, and was engaged at the time in counting from the chronometer and recording the contacts of the spots with the cross-wires used in the observation, when within the area of the great north group (the size of which had previously excited general remark), two patches of intensely bright and white light broke out, in the positions indicated in Fig. 1 … My first impression was that by some chance a ray of light had penetrated a hole in the screen attached to the object-glass, for the brilliancy was fully equal to that of direct sun-light; but, by at once interrupting the current observation, and causing the image to move … I saw I was an unprepared witness of a very different affair. I therefore noted down the time by the chronometer, and seeing the outburst to be very rapidly on the increase, and being somewhat flurried by the surprise, I hastily ran to call some one to witness the exhibition with me, and on returning within 60 seconds, was mortified to find that it was already much changed and enfeebled. Very shortly afterwards the last trace was gone. In this lapse of 5 minutes, the two patches of light traversed a space of about 35,000 miles [56,315 km]." ▪ CARRINGTON had seen a flare of the rare variety that is visible in white light, apparently propagating across the Sun's surface with an average velocity of about 188 km/s. The flare became known as **"CARRINGTON's solar flare"** among astronomers. Solar flares emit in a wide spectral range, from radio to hard X-rays. |

---

[730] B. GLATZEL: *Elektrische Methoden der Momentphotographie*. Vieweg & Sohn, Braunschweig (1915), chap. II: *Methoden zur Herstellung kleiner Zeitdifferenzen*, pp. 18-70.
[731] *Blasters' handbook*. E.I. du Pont de Nemours & Co., Wilmington, DE (1989), p. 4.
[732] R. CARRINGTON: *Description of a singular appearance seen in the Sun on Sept 1, 1859*. Month. Not. Roy. Astron. Soc. **20**, 14-16 (1860).

(ii) **At the same time Richard HODGSON,**[733] another British amateur astronomer, **makes the same observation:** "While observing a group of solar spots on the 1st September, I was suddenly surprised at the appearance of a very brilliant star of light, much brighter than the sun's surface, most dazzling to the protected eye, illuminating the upper edges of the adjacent spots and streaks, not unlike in effect the edging of the clouds at sunset; the rays extended in all directions; and the centre might be compared to the dazzling brilliancy of a bright star. It lasted for some five minutes, and disappeared instantaneously about 11.25 A.M."

(iii) **At the same moment that CARRINGTON and HODGSON made their observations, modest but very marked disturbances of three magnetic elements are observed at Kew Observatory** (Richmond, Surrey). All of the elements are affected simultaneously and quite abruptly. CARRINGTON and HODGSON notice that a magnetic storm on Earth intensifies soon afterwards, but they refrained from connecting the observation of a solar flare with this occurrence of a magnetic storm. • **Their unique observations made highlighted the enormous dimensions and dynamics of solar flare explosions for the first time** {CHAPMAN & FERRARO ⇒1931; GOLD ⇒1949; PARKER ⇒1961}, which prompted much discussion on the relationship between solar eruptions and magnetic disturbances.

(iv) About 17 hours later, other researchers also observe considerable magnetic disturbances. • In the previous year, the German natural philosopher Alexander VON HUMBOLDT coined the term *magnetisches Ungewitter* ("geomagnetic storm") for sudden magnetic disturbances.[734]

Besides solar coronal mass ejections {LASCO ⇒2002}, large solar flares are among the most dramatic events that occur on the Sun. They occur near sunspots {J. FABRICIUS ⇒1611}, usually along the dividing line (neutral line) between areas of oppositely directed magnetic fields.[735] In a matter of just a few minutes, they heat material to many millions of degrees and release as much energy as a billion megatons of TNT, ejecting masses of gases exceeding the escape velocity of the Sun (about 610 km/s), and causing intense radio and X-ray emission. **Impulsive flares caused by chromospheric eruptions can produce expanding magnetohydrodynamic waves** (similar to the blast wave originating from a chemical explosion) **which, propagating through the corona, spread over the surface with extremely high velocities, on the order of 1,000 km/s.**

With the advent of steadily improving spectral diagnostics, more complex flare phenomena have been observed, rendering their classification a difficult task, and one which has occupied solar physicists since the mid-1940s. Today there are two widely used systems for the classification of a flare: the first one is based on an optical observation of the brightness of white light (like those first made by CARRINGTON and HODGSON), but usually based on the hydrogen-alpha line, while the second one is based on X-ray observations of the Sun.[736] Using hard X-ray and soft X-ray images taken in space with special X-ray photon detectors, it is possible to estimate the mean energy deposited into the flare plasma. Hard X-rays (> 20 keV), radio waves and gamma-rays are mainly emitted in the impulsive phase of a flare, which can last from milliseconds to seconds.[737]

---

[733] R. HODGSON: *On a curious appearance seen in the Sun*. Month. Not. Roy. Astron. Soc. **20**, 16-20 (1860).

[734] A. VON HUMBOLDT: *Kosmos. Entwurf einer physischen Weltbeschreibung*. Cotta, Stuttgart *etc.* (1858), vol. III, p. 127.

[735] D.H. HATHAWAY: *Solar Flares: flare characteristics, flare observations, flares and magnetic shear*. Science Directorate. NASA-MSFC, Huntsville, AL (Jan. 2003); http://science.nasa.gov/ssl/pad/solar/flares.htm.

[736] E. CLIVER: *Solar flare classification*. In: (P. MURDIN, ed.) *Encyclopedia of astronomy and astrophysics*. 4 vols., Institute of Physics, London *etc.* (2001), vol. 3, pp. 2515-2518.

[737] G.D. HOLMAN: *Solar flare theory*. Laboratory for Astronomy and Solar Physics, NASA-GSFC, Greenbelt, MD (Oct. 2003); http://hesperia.gsfc.nasa.gov/sftheory/.

| 1859 | *The Royal Society of Science, Göttingen, Lower Saxony, Germany* | On November 22, **G.F. Bernhard RIEMANN,**[738] one of Germany's greatest mathematicians, **presents his famous *Theorie der Wellen endlicher Schwingungsweite* ("Theory of Waves of Finite Amplitude")** which is not limited to a single progressive wave as is EARNSHAW's solution {⇨1858}, and is therefore more general and suited to calculating the propagation of planar waves of finite amplitude proceeding in both directions: |
|---|---|---|

(i) Speaking on his motivation for the work, RIEMANN[739] writes, "This investigation does not claim to provide useful results to experimental research; rather the author merely wishes to consider it to be a contribution to the theory of nonlinear partial differential equations. In the same way as the most fruitful methods for the integration of linear partial differential equations were not uncovered by the development of the general notion of this task, but rather evolved from the treatment of special physical problems, it also appears that the theory of nonlinear partial differential equations can best be promoted by the treatment of special physical problems taking into account all secondary conditions. In fact, the solution to this very special problem, the subject of this article, has required new methods and interpretation, thus leading to results which probably will play a role in more general problems…"

(ii) Limiting his study to steady one-dimensional (1-D) flow, and considering motions occurring at a fixed point in the gas (*Eulerian* coordinates), **he assumes a pressure-density relation $p = p(\rho)$ that depends only on density and holds for all particles and all times, even across shocks** – *i.e.,* it reduces to adiabatic motion in the case of weak shocks.

(iii) To find the essential propagation properties of waves of finite amplitude described by nonlinear hydrodynamic equations, **he integrates the partial differential equations** (of the hyperbolic type) **using the method of characteristics** {MONGE ⇨1770s}, which simplifies under the assumption that the speed of sound is a function of density alone (*Riemann invariants*). ▪ The 1-D motion of an isentropic gas can be characterized by the presence of Riemann invariants, their choice as independent variables results in an essential simplification of the system of Euler equations which becomes linear. Another case which also presents Riemann invariants was discovered in 1954 in which the LMS-gas[740] and the isentropic gas have been shown to be mathematically equivalent.[741]

(iv) **He shows that an original disturbance splits into two opposite waves:** the rarefaction wave grows thicker while the condensation wave (a shock wave) grows thinner; he calls this *Verdichtungsstoß* ("condensation shock"). The gas passed over by the shock wave will be compressed and heated while that passed over by the rarefaction will be expanded and cooled.

His results are an important step toward a mathematical treatment of shock wave steepening and formation. However, **using the static adiabat, he erroneously assumes that the entropy** {CLAUSIUS ⇨1865} **remains unchanged through the shock wave** – *i.e.,* **he believes shock waves to be an *isentropic process*,** which assumes no heat transfer and neglects frictional effects. ▪ The total energy content (enthalpy) remains unchanged, whereas **the entropy always increases with a shock wave, because it is a irreversible process.** This was first recognized by the Scottish engineer William J.M. RANKINE {⇨1869} and later, independently, by the French military engineer Pierre-Henri HUGONIOT {⇨1887}.

RIEMANN's one-dimensional shock problem, defined as an initial value problem for a system based upon the laws of conservation of mass, momentum and energy, is known as the

---

[738] B. RIEMANN: *Über die Fortpflanzung ebener Luftwellen von endlicher Schwingungsweite*. Abhandl. Königl. Gesell. Wiss. Gött. **8** [Math. Physik. Kl.], 243-265 (1860).

[739] B. RIEMANN: Abstract of preceding paper in: Nachr. Georg-August-University & Königl. Gesell. Wiss. Gött. Nr. 19 (5. Dez. 1859), pp. 192-196.

[740] A *Ludford-Martin-Stanyukovich (LMS) gas* is an ideal gas with constant specific heats but with a special entropy distribution, so that generalized Riemann invariants exist; *see* K.P. STANYUKOVICH: *General solutions of the gasdynamic equations for one-dimensional motions for a given equation of state or process* [in Russ.]. Dokl. AN (SSSR) **96**, 441-444 (1954), and M.H. MARTIN and G.S.S. LUDFORD: *One dimensional anisentropic flows*. Comm. Pure & Appl. Math. **7**, 45-63 (1954).

[741] B. GAFFET: *The nonisentropic generalization of the classical theory of Riemann invariants*. J. Phys. A: Math. Gen. **20** [II], 2721-2731 (1987).

"**Riemann problem.**" His treatise is now accepted as a classic in gas dynamics, and his method of integrating linear partial differential equations – although limited to the 1-D case only – has since become a useful method in applied mathematics.

| 1859 | *St. John's College, Cambridge, England* | **William Henry BESANT,**[742] an English lecturer of mathematics, **mentions an experiment showing that musical sounds of every pitch and violence are transmitted with precisely the same velocity:** "On a fine and still evening of June 1858, the *Messiah* was performed in a tent, and the Hallelujah Chorus was distinctly heard, *without loss of harmony*, at a distance of two English miles [3.046 km]." His observation fully confirms previous experiments {BIOT ⇨1809}. ▪ Since the human ear can pick up even slight differences in musical intervals, this fact was (incorrectly) considered to provide proof that violent sound in general, regardless of its amplitude, is transmitted with the same velocity {LE CONTE ⇨1864}. |
|---|---|---|
| 1859 | *University of Glasgow, Scotland, U.K.* | **William J.M. RANKINE,** a Scottish professor of engineering, **publishes his *Manual of the Steam Engine,*** in which he reduces the theory of thermodynamics to rules that are easy to understand and that are adapted for use by engineers. **While doing so he coins the term *adiabatic*** [from the Greek, meaning "not to be passed through"]. |
| 1859/ 1860 | *Royal Institution of Great Britain, London, England* | During the Christmas holidays of 1859/1860, **Michael FARADAY,**[743] the renowned English physicist and chemist, delivers a lecture entitled *Gravitation – Cohesion* before a young audience,[744] and **gives an illustrious example of the properties of *Prince Rupert drops*** {HOOKE ⇨1665; ⇨Fig. 4.16–W}. He says, "Here is a piece of glass (*he produces a piece of plate glass about two inches square* [$5 \times 5$ $cm^2$]). I shall want this afterward to look to and examine its internal condition, and here is some of the same sort of glass differing only in its power of cohesion, because while yet melted it had been dropped into cold water (*he exhibits a Prince Rupert drop*), and if I take one of these little tear-like pieces and break off ever so little from the point, the whole will at once burst and fall to pieces. I will now break off a piece of this. (*The lecturer nips off a small piece from the end of one of the drops, whereupon the whole drop immediately falls to pieces.*) There! you see **the solid glass has suddenly become powder, and more than that, it has knocked a hole in the glass vessel in which it was held.** I can show the effect better in this bottle of water, and it is very likely the whole bottle will go. (*A 6-oz [187-ml] vial is filled with water, and a Prince Rupert drop is placed in it with the point of the tail just projecting out; upon breaking the tip off, the drop bursts, and* **the shock, which is transmitted through the water to the sides of the bottle, shatters the latter to pieces**)." |
| 1860s | *Ecole Pyrotechnique, Brussels, Belgium* | **Paul Emile LE BOULENGÉ,** a Belgian lieutenant-general and experimental ballistician, **invents an electrically triggerable dropping weight timing system** – the so-called **"Le-Boulengé chronograph"** {⇨Fig. 4.19–B}. ▪ After making some technical improvements it became a robust and accurate instrument with a temporal resolution of less than one millisecond, and was even applicable in harsh environments such as in "open" ballistic ranges.[745] It was widely used until the end of World War I. |
| 1860 | *Europe & Asia* | On July 18, **various European astronomers** – such as E. Wilhelm TEMPEL, Fabian C.O. VON FEILITZSCH and Carl F. VON WALLENBERG in Germany, Frederico A. OOM in Portugal, Francis GALTON in England, and E.W. MURRAY (during a Himalayan expedition) – **make** |

---

[742] W.H. BESANT: *A treatise on hydrostatics and hydrodynamics.* Deighton & Bell, Cambridge (1859), p. 187
[743] C.W. ELIOT (ed.): *Scientific papers: physics, chemistry, astronomy, geology, with introductions, notes and illustrations.* Harvard Classics, No. XXX, Collier & Son, New York (1910), pp. 29-30.
[744] The *Royal Institution Christmas Lectures* for young people were founded by FARADAY in 1826; his goal was to communicate the excitement of scientific discovery to children. The Christmas Lectures have continued ever since.
[745] P.E. LE BOULENGÉ: *Mémoire sur un chronographe électro-balistique* [présenté 5 Déc. 1863]. Mém. Cour. Mém. Sav. Etrang. Acad. Roy. (Bruxelles) **32**, 1-39 (1864/1865). ▪ For example, an original Le-Boulengé chronograph is kept at the museum of the Etablissement Technique de Bourges (ETBS), France.

thorough observations of the total solar eclipse which crosses Spain. Independently, they draw peculiar features of the corona.[746] ▪ Based on comparison with modern coronal observations, it is quite likely that these drawings represent the first records of a coronal mass ejection in progress {LASCO ⇨2002}.

**Coronal mass ejections typically drive shock waves in space that produce energetic particles.** For example, on February 17, 2000 a full halo coronal mass ejection erupted from the Sun, and three days later the shock passage of an interplanetary wave of ionized gas and magnetic fields was recorded on Earth.[747]

| | | |
|---|---|---|
| 1860 | *Parish of Sheffield, South Yorkshire, England* | The Rev. Samuel EARNSHAW,[748] resuming his mathematical theory of sound {EARNSHAW ⇨1858}, **discusses the issue of whether violent sounds propagate more rapidly than gentle sounds.** He distinguishes three kinds of waves, all propagating with different velocities $v$ compared to the observed sound velocity $a$, thus giving rise to what he designates the ***triplicity of sound:*** |

- ▸ *minute* waves: their velocity varies from 0 to $0.5a$;
- ▸ *ordinary* waves: their velocity varies from $0.5a$ to $a$; and
- ▸ *violent* waves: their velocity varies from $a$ to infinity.

Although this arbitrary classification is rather hypothetical, he draws a very important conclusion: **"If the theory here advanced be true, the report of fire-arms should travel faster than the human voice, and the crash of thunder faster than the report of a cannon."**

In the same year, EARNSHAW[749] publishes another paper on this subject. He shows that planar sound waves of finite amplitude and of a stationary form are mathematically impossible in a gas which vibrates adiabatically – the so-called **"Earnshaw paradox."** Such wave motion is possible only if the fluid satisfies an equation of state which has the form

$$p = p_0 - C^2/\rho,$$

where $C$ is a constant. However, there is no known gas whose adiabatic equation of state takes this special form. If this relation held, one would have $d^2p/d\rho^2 < 0$, which would contradict the Second Law of Thermodynamics.[750]

| | | |
|---|---|---|
| 1860 | *Logelbach [near Colmar], Alsace, France* | Gustav Adolph HIRN,[751] a French autodidact and independent scholar, **speculates on possible reasons why the velocity of sound depends on intensity** by assuming a pressure-dependent ratio of the specific heats at constant pressure and volume, respectively – *i.e.*, $\gamma(p) = c_p/c_v$. Using an unusual phenomenon associated with thunder to support his hypothesis, he writes, "… whatever the distance from our village where I saw the lightning struck, *provided that the thunder-cloud extended to the zenith*, I noticed that the noise – not always, this would be too absolute, but very often – arrived much more rapidly than it should correspond to the distance between me and the strike." ▪ HIRN communicated his observation on thunder to MONTIGNY {⇨1860}, who forwarded it to the Belgian Academy. |

28 years later, HIRN,[752] readdressing the riddle of thunder, took the view that "the beginning of the thunder clap gives us the minimum distance of the lightning [MERSENNE ⇨1644], and the length of the thunder clap gives us the length of the column."

---

[746] C.A. RANYARD: *Observations made during total solar eclipse.* Mem. Roy. Astron. Soc. **41** (1879), chap. *Eclipse of 1860, July 18th*, pp. 520-579; *Note with respect to the rate of motion of gaseous matter projected from the Sun.* Month. Not. Roy. Astron. Soc. **41**, 77-79 (1881).
[747] *Interplanetary shock wave passes Earth.* NASA Space Science News (Feb. 21, 2000); http://www.spacescience.com/headlines/y2000/ast21feb_1.htm.
[748] S. EARNSHAW: *On the velocity of the sound of thunder.* Phil. Mag. **20** [IV], 37-41 (1860); *On the triplicity of sound.* Ibid. **20** [IV], 186-192 (1860).
[749] S. EARNSHAW: *On the mathematical theory of sound.* Phil. Trans. Roy. Soc. Lond. **150**, 133-148 (1860).
[750] G. BIRKHOFF: *Hydrodynamics.* Princeton University Press, Princeton, NJ (1960), pp. 20-21.
[751] G.A. HIRN: *Sur le bruit du tonnerre.* Cosmos (Paris) **16** [I], 651-655 (1860).
[752] Editor's note: *The sound of thunder.* Scient. Am. **59**, 201 (1888).

| 1860 | Royal Athenaeum, Antwerp, Belgium | **Charles M.V. Montigny,**[753] a Belgian astronomer and royal council, **reports on his observation that thunder sometimes propagates with a velocity far greater than ordinary sound:**
(i) He had noticed that the time lag between lightning striking a dairy farm 5.2 km distant (producing the typical dry sound) and its associated thunderclap amounted to just two seconds at the most – a curious phenomenon that the English mathematician Rev. Samuel Earnshaw {⇨1851} had already noticed, and one that was much discussed among contemporary natural philosophers {Hirn ⇨1860; Raillard ⇨1860; Laurent[754] 1860}.
(ii) He also reports on a number of similar observations on thunder phenomena made by him and others, and concludes, "I feel obliged to announce to the Academy this formal assertion which supports the facts I had the honor to report previously. This confirmation can only attract anew the attention of observers from various countries on the anomaly which the sound of thunder seems to present in regard to his propagation velocity."
Montigny's observation that thunder would propagate with about 2.6 km/s (Mach > 7) was totally unrealistic. More recent studies have revealed that thunder only propagates supersonically in the very close vicinity of the lightning stroke {Uman ⇨1970; Bass ⇨1980; Graneau ⇨1989}. |
|---|---|---|
| 1860 | Courchamp [near Gray], Dépt. Haute-Saône, France | **F. Raillard,**[755] a little known French abbot and amateur naturalist, **speculates on the propagation velocity of thunder.** Estimating abnormally high velocities ranging from 5,000 to 6,600 m/s, he writes, "I heard the first outbursts of thunder three or four seconds after the lightning had appeared; however, according to the delay of the reinforcement of the noise originating from the stem of the lightning, and to its orientation, I estimated that the fire was lit in the vicinity of Gray, about 20 km from Courchamp where I was." Raillard refers to Jean-Baptiste Biot, who had a discussion with the French mathematician and physicist Siméon Denis Poisson on irregular propagation phenomena of thunder many years before. Poisson, however, although essentially supporting this idea, did not resume it in his *Mémoires sur la théorie du son* ("Memoir on the Theory of Sound"), which he published in 1808 at Paris. |
| 1860 | Allegheny Arsenal, U.S. Army Ordnance Dept., Pennsylvania | **Thomas J. Rodman,** a U.S. military engineer, **invents *mammoth powder.*** It consists of large pellets of powder pressed into hexagonal grains perforated with several longitudinal holes to allow the charge to burn smoothly while the projectile travels down the barrel. Compared to normal powder, it maintains a higher bore pressure and increases the muzzle velocity, thus giving the cannon more range. Rodman also tested powder in perforated cakes. ▪ This finding, supported by measurements made using his crusher gauge {Rodman ⇨1857}, became important for large-caliber guns {Rodman ⇨1864} since it reduced the chance of damaging the barrel.
In the following year, his mammoth powder – known as **"Rodman powder"** – was adopted for heavy ordnance. The perforated cake powder for rifled cannon of large caliber was immediately adopted by the Russian government, which obtained specimens from Fortress Monroe in 1860, and it also came into use in Prussia soon afterward. Military authorities in England decided to use the mammoth powder – known as **"pebble powder"** in England – in their big rifled guns.[756] |

---

[753] C.M.V. Montigny: *Note sur la vitesse du bruit du tonnerre.* Bull. Acad. Sci. Brux. **IX** [II], 36-46 (1860); *Observations sur l'accélération de la vitesse du bruit du tonnerre.* Ibid. **X** [II], 62-63 (1860).
[754] A. Laurent: *Intensité et vitesse du bruit du tonnerre.* Cosmos (Paris) **17** [I], 7-10 (1860).
[755] F. Raillard: *Sur le bruit du tonnerre.* Cosmos (Paris) **16** [I], 373-374 (1860); *Du bruit du tonnerre, de ses variations ou de ses roulements, de sa vitesse &.* Ibid. **17** [I], 166-172, 675-677 (1860).
[756] *Thomas Jefferson Rodman.* Virtual American Biographies; http://www.famousamericans.net/thomasjeffersonrodman/.

| | | |
|---|---|---|
| 1860 | Peterhouse College, University of Cambridge, England | Edward John ROUTH,[757] a Canadian-born English mathematician, **publishes the two-volume textbook** *An Elementary Treatise on the Dynamics of a System of Rigid Bodies*, **in which he also analytically treats impulsive forces in great detail.**

(i) He gives a historical review of previous attempts to consider the influence of friction on percussion events – so-called **"shock friction" or "impulsive friction"** {POISSON ⇨1833; MORIN ⇨1833; DE CORIOLIS ⇨1835; PHILLIPS ⇨1849}.

(ii) Addressing the problem of impulsive friction in more detail, he states, "If the bodies are rough an impulsive friction will clearly be called into play. Since an impulse is only the integral of a very great force acting for a very short time, we might suppose that impulsive friction obeys the laws of ordinary friction. But these laws are founded on experiment, and we cannot be sure that they are correct in the extreme case in which the forces are very great…" **He presents a method for including friction in the analysis of impact events,** which he illustrates via numerous practical examples for two- and three-dimensional motions of single and composed bodies, both elastic and inelastic. For example, he treats the general problem of the motion of a spherical ball projected without initial rotation against any rough elastic plane.

(iii) Although his book focuses on rigid bodies in which elastic and plastic deformation is ignored, **he also discusses the important general problem of motion just after impact when two bodies of any form – smooth or rough, elastic or inelastic – impinge on each other in a given manner (such as eccentrically and obliquely).** ▪ For the important special case of the oblique central collision of two smooth, nonrotating elastic spheres, the result can be found in many textbooks on classical mechanics {⇨Fig. 2.5}.

(iv) Addressing MALLET's column seismometer used in the Great Neapolitan Earthquake {MALLET ⇨1857}, he also discusses the problem of determining the intensity and direction of seismic shocks from the geometry of the overturned columns.

In the following years, his book became a bestseller on the continent and went through seven editions. For the German edition (1898) the famous mathematician Felix KLEIN at Göttingen University wrote a foreword: "In fact, ROUTH's work also occupies a very specific position within the English textbook literature. It is – one could say – the result of a teaching method developed at the University of Cambridge, of which the author was a recognized master for many years. By strongly recommending a thorough study based on individual examples, this method unquestionably advances the student's ability in a certain direction leading to extraordinary development." |
| 1861 | Somerset House, The Royal Society of London & Holyhead Mountain, Anglesey, England | Robert MALLET,[758] an Irish civil engineer and scientific investigator, **reports the results from his seismic experiments to the Society** {MALLET ⇨1849}. These experiments had been performed over the previous ten years with the support of the famous physicist Sir Charles WHEATSTONE and the astronomer Thomas Romney ROBINSON.

In the same year, resuming his seismic studies, **MALLET employs large-scale explosions** in the governmental quarries on Holyhead Mountain {⇨Fig. 4.15–D}. Using single charges of gunpowder of up to 5,000 kg – high explosives such as dynamite {NOBEL ⇨1867} were not yet available to him – and horizontal mercury seismoscopes {DE HAUTE-FEUILLE ⇨1703}, he hopes to observe the arrival of the elastic wave. Unfortunately, however, the induced seismic shock is not strong enough to detect the precursor at a distance beyond 2 km from the point of explosion.
▪ **MALLET's pioneering experiments, albeit not immediately successful, mark the birth of** |

---

[757] E.J. ROUTH: *An elementary treatise on the dynamics of a system of rigid bodies*. London (1860); 7th enlarged edition: Macmillan, London (1905). Reprinted by Dover Publ., New York (1960); for sections on *Impulsive friction* see §§ 181, 187-198, 315-331, 389.

[758] R. MALLET: *Account of experiments made at Holyhead (North Wales) to ascertain the transit-velocity of waves, analogous to earthquake waves, through the local rock formations*. Phil. Trans. Roy. Soc. Lond. **151**, 655-679 (1861); *Appendix*. Ibid. **152**, 663-676 (1862).

| | | |
|---|---|---|
| | | *explosion seismology* – a geophysical method of generating artificial earthquakes in order to measure the velocity of elastic waves in a variety of geological materials {MALLET ⇨1846}. |
| 1862 | Dearborn Observatory of Northwestern University, Chicago, Illinois | On the night of January 31, **Alvan CLARK,** a renowned U.S. telescope maker, **discovers a faint companion to Sirius** while testing the lens of the Dearborn Telescope.[759] ▪ In the period 1834–1844, the German astronomer Friedrich Wilhelm BESSEL had noticed some irregularities in the movement of Sirius. He predicted this companion from the wobbling motion of Sirius, the brightest star in the northern sky – now called "Sirius A" (or "The Dog Star"), a normal star more than twice as massive as the Sun.<br><br>Subsequent studies of the companion, which was called "Sirius B" (or "The Pup"), showed that its blackbody spectrum peaks at 110 nm, corresponding to a temperature of 26,000 K. This led to the conclusion that it was hotter than the Sun. The radius was calculated to be just 4,200 km. Although smaller than the Earth, Sirius B is almost as massive as the Sun. **Sirius B is a *white dwarf*.** The theory of white dwarf stars was developed at Trinity College by the Indian-born astrophysicist Subrahmanyan CHANDRASEKHAR {⇨1931}. |
| 1862 | Marathon, New York & Harvard College Observatory, Cambridge, Massachusetts | On July 16, the two U.S. astronomers **Lewis A. SWIFT** (New York) **and Horace P. TUTTLE** (Cambridge, MA) **independently discover the comet 109P** in *Camelopardalis* ("The Giraffe," a constellation in the Northern Hemisphere). ▪ Later rediscovered in 1992 by the Japanese amateur astronomer Tsuruhiko KIUCHI, **this comet, also known as "Comet SWIFT-TUTTLE," is the largest and therefore the most dangerous object known to make repeated passes near the Earth.**[760]<br><br>Comet 109P/SWIFT-TUTTLE is thought to be about six miles (9.7 km) across, about the same size of the asteroid that killed the dinosaurs {Chicxulub ⇨c.65 Ma ago}. In the early 1990s, Brian G. MARSDEN,[761] a British astronomer at the Harvard-Smithsonian Center for Astrophysics, predicted that the comet and the Earth might be in the same place in space at the same time on August 14, 2126. The comet was last seen in March 1995 by observers at Siding Spring Observatory (New South Wales, Australia). |
| 1862 | Off the west end of Santo Domingo (Dominican Republic), Caribbean Sea | On August 22, **the British brig *Alice*,** according to his captain ATWOOD, **experienced "a severe shock of an earthquake which shook the vessel like a leaf, lasting about ten or twelve seconds."**[762]<br><br>A so-called **"sea shock"** is understood as an acoustic wave that a ship senses in the sea, originating from an earthquake, reaching the ship through sea water, and giving it a short vertical motion {*see also* Sect. 2.1.4}. Recently, Japanese mariners of a training ship encountered at Nagasaki harbor Mie open sea wharf reported on a sea shock caused by the Fukuoka Open Sea Earthquake that occurred on March 20, 2005.[763] The epicenter of the earthquake of magnitude 7 was in shallow water in the Sea of Japan and produced strong ground motions. A tsunami warning was issued, but curiously in this earthquake no tidal level changes were observed. |
| 1862 | Nitroglycerine Ltd. Stockholm, Heleneborg [a town near Stockholm], Sweden | Alfred B. NOBEL,[764] a Swedish chemist, engineer, industrialist and inventor, **applies for a patent on improvements made to the process of manufacturing nitroglycerin** {SOBRERO ⇨1846}. He erects works at Heleneborg, an isolated area outside Stockholm, where nitroglycerin is manufactured for the first time on a commercial scale. ▪ Although he initially called it "pyroglycerin," it soon became known as "nitroglycerin" or "NOBEL's blasting oil." |

---

[759] *The Dearborn Telescope, circa 1864, Adler Planetarium & Astronomy Museum.* The Deller Conservation Group, Ltd., Paper presented at the American Institute for Conservation of Historic and Artistic Works (AIC) Annual Meeting, St. Louis (1999); http://deller.com/newpage6.htm.
[760] D.L. CHANDLER: *Comet put on list of potential Earth impactors.* NewScientist.com (June 2005); http://www.newscientist.com/article.ns?id=dn7449.
[761] B.G. MARSDEN ET AL.: *Periodic comet SWIFT-TUTTLE (1992t).* IAU Circ. **5670**, 1 (Dec. 1992); *Update on Comet SWIFT-TUTTLE.* Icarus **105**, No. 2, 420-426 (1993).
[762] *Earthquake at sea. Serious shock experienced by the brig Alice.* The New York Times (Sept. 2, 1862).
[763] Y. HIROSHI, G. MASAJI, A. TAKASHI, and M. AKIRA: *On sea shock with earthquake (Mj7.0) in west-off Fukuoka Prefecture.* Navigation **163**, 92-94 (2005).
[764] A. NOBEL: *Nitroglycerine* [on manufacturing and fining]. Brit. Patent No. 1,813 (July 20, 1864).

Two years later, his works were entirely wrecked by an explosion which cost the lives of NOBEL's youngest brother Emil Oskar and his chemist Carl Eric HERTZMANN.

| 1863 | Newcastle upon Tyne, north-east England |

At a meeting of the British Association for the Advancement of Science (BAAS), **George B. AIRY**,[765] an English natural philosopher and Astronomer Royal, **reviews previous theoretical and experimental attempts**[766, 767] **to calculate the destructive energy of steam boiler explosions.**

(i) He writes, "Very little of the destructive effect of an explosion is due to the steam contained in the steam-chamber at the moment of explosion. The rupture of the boiler is effected by the expansive power common at the moment to the steam and the water, both at a temperature higher than the boiling-point; but as soon as steam escapes, and thereby diminishes the compressive force upon the water, a new issue of steam takes place from the water, reducing its temperature; when this escapes, and further diminishes the compressive force, another issue of steam of lower elastic force from the water takes place, again reducing its temperature; and so on; till at length the temperature of the water is reduced to the atmospheric boiling-point, and the pressure of the steam is reduced to zero. It is this enormous quantity of steam, of gradually diminishing power, which is thus produced from water during the course of the explosion, that causes the disastrous effects of the explosion."

(ii) Based upon a numerical estimate, **he concludes that one cubic foot of water at 60 psi (4.12 bar) is equal to the destructive energy of one pound of gunpowder;** *i.e.*, 1 $cm^3$ of steam at about 4.1 bar equals the energy of 16 mg of gunpowder.

In the following years, **various hypotheses for possible causes of steam boiler explosions were discussed,**[768] encompassing

▸ the generation and ignition of oxyhydrogen which occurs when water chemically reacts with the overheated iron walls when the water level is low in the boiler;

▸ the sudden destruction of the initial isolation of water from the boiler walls, thus nullifying the protecting *Leidenfrost layer*;

▸ the sudden generation of large quantities of steam by the *delayed boiling* phenomenon;

▸ the reduction in mechanical strength of the boiler material at high temperature; and

▸ the increasing instability of the boiler walls when firing sulfurous coal.

The accidents prompted engineers and metallurgists to study dynamic materials behavior under thermal and mechanical stress and to improve production technology. They also prompted the foundation of the first official safety inspection authorities.

Steam boiler explosions – puzzling events that accompanied the evolution of steam engines in the 17th century and quickly increased in number – were pondered by generations of mechanical engineers and physicists. For example, in 1879 Alfred NOBEL[769] patented a system of nonexplodable boilers.

| 1863 | Eidgenössische Technische Hochschule (ETH), Zurich, Switzerland |

**Albert FLIEGNER**,[770] a Swiss professor of mechanical engineering, **studies the air flow through well-rounded mouthpieces and first observes the flow phenomenon of "choking."**

• If the initial tank pressure is sufficiently large compared to the backpressure, a nozzle will be

---

[765] G.B. AIRY. *Report on steam-boiler explosions*. Rept. Meet. Brit. Assoc. **33**, 686-688 (1863); *On the numerical expression of the destructive energy in the explosions of steam-boilers, and on its comparison with the destructive energy of gunpowder*. Phil. Mag. **26** [IV], 329-336 (1863).

[766] F.J.D. ARAGO: *Résultats des expériences faites par ordre du Bureau des Longitudes, pour la détermination de la vitesse du son dans l'atmosphère*. Connaissance des Temps (Paris), pp. 361-371 (1822).

[767] R. ARMSTRONG and J. BOURNE: *The modern practice of boiler engineering, containing observations on the construction of steam-boilers*. Spon, Leipzig (1856), chap. III: *Explosions: an investigation into some of the causes producing them, and the deterioration of boilers generally*.

[768] F. FISCHER: *Zur Geschichte der Dampfkesselexplosionen*. Dingler's Polytech. J. **213**, 296-308 (1874).

[769] A. NOBEL: *Pour des brûleurs perfectionnés le gaz et les huiles d'éclairage*. French Patent No. 129,316 (Feb. 27, 1879).

[770] A. FLIEGNER: *Über das Ausströmen der Luft durch gut abgerundete Mündungen*. Vierteljahrsschrift der Naturforschenden Gesellschaft in Zürich **VIII**, 282-284 (1863).

choked during the initial portion of the discharge process. According to the so-called "Fliegner formula," the choked mass flow rate d$m$/d$t$ [kg/s] for air is given by

$$\mathrm{d}m/\mathrm{d}t = 0.04042\, A\, P_0\, (T_0)^{-\frac{1}{2}},$$

where $P_0$ [Pascal] is the stagnation pressure, $T_0$ [K] is the stagnation temperature, and A [m$^2$] is the throat area.

Ten years later, FLIEGNER[771] showed that, for air flowing from well-rounded mouthpieces, there is no discontinuity of the law of flow as NAPIER's hypothesis implies, but the curve of flow bends so sharply that NAPIER's rule {NAPIER ⇒1866} may be taken to be a good approximation to the true law. With the advent of the Laval steam turbine {DE LAVAL ⇒1888}, FLIEGNER[772] also studied the outflow of air and steam from conically divergent nozzles. ▪ His incorrect view that the flow velocity in a divergent nozzle cannot surmount the sound velocity – to which he adhered for many years – was criticized by Ludwig PRANDTL {⇒1908}.

| 1863 | *Collège de France, Paris, France* | **Henri V. REGNAULT,** a French chemist and physicist, **starts a five-year campaign to precisely measure the velocity of sound in air and other gases:**

(i) To exclude negative side-effects (such as the wind), he performs his experiments in long pipes with lengths up to 20 km and diameters ranging from 0.108 to 1.1 m, thereby using the gas pipeline and sewage channel system of Paris. This allows him to use long baselines to compensate for the limited accuracy of his mechanical chronoscope.

(ii) Discharging a small quantity of gunpowder at the pipe entrance, he determines the average blast velocity by mechanically recording the arriving pressure signal at the pipe end using a membrane microphone combined with a rotating-drum chronograph {⇒Fig. 2.10−F}. ▪ With his ingenious method of placing an explosive energy source in front of a long tube and using a cylindrical section of the blast from an open-air explosion for blast propagation studies, **REGNAULT can be considered as the inventor of the *blast tube*.**[773]

(iii) He creates waves of equal intensity by discharging one gram of gunpowder from the same pistol at the orifices of conducting tubes with very different cross-sectional areas. According to Hermann VON HELMHOLTZ and Gustav R. KIRCHHOFF, the velocity in a tube should be less than it is in free air by a factor that depends on the diameter of the tube, the frequency of sound, and the viscosity of the applied gas.[774] By reducing the tube diameter $D$ in his experiments, he also diminishes the intensity of the propagating sound wave. For example, for air he observes that for $D = 1.1$ m the sound velocity amounts to 330.6 m/s, but reduces to 324.25 m/s for $D = 0.108$ m – **thus first showing quantitatively that the velocity of sound decreases with decreasing intensity.**

(iv) He speculates that "the emission of a strong wave always causes "a true displacement [French *un véritable transport*] of the first gaseous layers, which considerably increases the propagation velocity of the wave, especially through the first portion of its course" – thereby mentioning the steepening process at the wave front, an essential feature of a shock wave.

Accounts of his experiments, published in 1868 in French and British journals, illustrate the extraordinary care and ingenuity of REGNAULT's work.[775] Since his remarkable achievements have barely been acknowledged by the modern shock physics community, he is cited here in more detail: "The theoretical calculation assumes that the excess of compression which

---

[771] A. FLIEGNER: *Ergebnisse einiger Versuche über das Ausströmen der atmosphärischen Luft* [1873]. Der Civilingenieur **XX**, col. 13-47 (1874); Ibid. **XXIII**, col. 443-510 (1877); Ibid. **XXIV**, col. 39-48 (1878).

[772] A. FLIEGNER: *Versuche über das Ausströmen von Luft durch konisch divergente Rohre.* Schweiz. Bauz. **XXXI**, 68-70, 78-80, 84-86 (1898); *Die größte Ausströmungsgeschwindigkeit elastischer Flüssigkeiten.* Ibid. **XLIII**, 104-108, 140-145 (1904).

[773] R. ROBEY: *Blast tubes.* In: (G. BEN-DOR ET AL., eds.) *Handbook of shock waves.* Academic Press, New York (2001), vol. 1, pp. 623-650.

[774] Lord RAYLEIGH (J.W. STRUTT): *The theory of sound.* Macmillan, London (1878); vol. II, §§ 347-348.

[775] H.V. REGNAULT: *Sur la vitesse de propagation des ondes dans les milieux gazeux.* Mém. Acad. Sci. Inst. France **37**, 3-575 (1868); *On the velocity of the propagation of waves in gaseous media.* Phil. Mag. **35** [IV], 161-171 (1868).

exists in the wave is infinitely small compared with the barometric pressure supported by the gas. But the experiments done to determine the rate of sound in free air have been previously performed using a cannon, and the wave has been reckoned from its source, namely the cannon's mouth. Now, as it leaves the cannon, this wave is under enormous compression – a compression, it is true, that diminishes very rapidly as the wave spreads spherically through space; but during the first part of its course it cannot be supposed that its compression is infinitely small … **When the excess of compression in the wave is a sensible fraction of the compression of the gaseous medium at rest, we can no longer employ DE LAPLACE's formula, but must have recourse to a more complex formula embracing the true elements of the problem.** Even the formula which I have given in my Memoir [Mém. Acad. Roy. Paris **37**, 3-575 (1868)] is only an approximation; for it implicitly admits MARIOTTE's law and all its consequences … In short, the mathematical theory has as yet only touched upon the propagation of waves in a perfect gas – that is to say, in an ideal fluid possessing all the properties which have been introduced hypothetically into the calculation. It is therefore not surprising that the results of my experiments often disagree from theory." ▪ His remarkable results were not immediately accepted.[776]

Fourteen years later, Ernst MACH and Jan SOMMER {⇨1877} first confirmed REGNAULT's observations also on a laboratory scale.

| | | |
|---|---|---|
| 1863/ 1864 | *Nitroglycerine Ltd. Stockholm, Heleneborg, Sweden* | **Alfred B. NOBEL,**[777] a Swedish chemist and industrialist, **realizes that "nitroglycerin and other analogous substances," unlike gunpowder, cannot be set off by a flame, but instead require a shock to initiate an explosion.** ▪ His important discovery ushered in a new era in the history of explosives. The introduction of the "initial ignition principle" – the use of a strong blast wave rather than heating – was a significant achievement in the practical technique of blasting.<br><br>(i) His first "igniter" is a small wooden plug filled with gunpowder, which could be detonated by lighting a fuse. This, in turn, causes an explosion of the surrounding nitroglycerin.<br><br>(ii) In his patent application he writes, "… the detonation of explosive bodies, either non-inflammable, or which may be ignited without explosion by spontaneous decomposition of a small portion of their mass, thus producing an initiative explosion, which spreads of itself."<br>Based upon these experiments, NOBEL begins to experiment with small metal receptacles loaded with mercury fulminate {HOWARD ⇨1800} mixed with gunpowder or nitrate of potash, known as **"NOBEL's igniter."** It is made of a tin plate rolled into the shape of a sugarloaf bag, with the free edges soldered.<br><br>Three years later, NOBEL stopped using the igniter and started to use a copper detonator {NOBEL ⇨1867}. The *safety detonator,* a "bridgewire" version of the detonator, was invented in the United States during World War II {JOHNSTON ⇨1944}.<br><br>Ten years after having perfected his invention of the igniter, NOBEL bluntly stated:[778] "But the real era of nitroglycerin opened with the year 1864, when a charge of pure nitroglycerin was first set off by means of a minute charge of gunpowder." |
| 1864 | *Fort Hamilton, New York Harbor, New York* | On October 25, **Thomas J. RODMAN,** a U.S. military engineer, **performs the first tests of his 20-in. (50.8-cm)-caliber gun** {⇨Fig. 4.2–S}.[779] Made from cast iron, it has a total length of 20 feet and 3 in. (7.036 m), and the barrel weighs 116,497 pounds (52,815 kg).[780] ▪ RODMAN's guns belong to the type of so-called **"columbiads"** – *i.e.,* pieces of ordnance that were very thick in the breech and tapered off gradually from the trunnions to the muzzle. |

---

[776] H.J. RINK: *Über die Geschwindigkeit des Schalls nach Hrn. REGNAULT's Versuchen.* Ann. Phys. **149** [II], 533-546 (1873).
[777] A. NOBEL: *Nitroglycerine.* Brit. Patent No. 1,813 (July 20, 1864).
[778] E. BERGENGREN: *Alfred NOBEL. The man and his work.* Nelson & Sons, London *etc.* (1962).
[779] Anonymous: *The biggest gun in the world.* Harper's Weekly: J. Civiliz. (New York) **5**, No. 222 (March 30, 1861).
[780] D.B. WEBSTER JR.: *RODMAN's great guns.* Ordnance **47**, 60-62 (July/Aug. 1962).

In the same year, vivid descriptions of the huge dimensions of RODMAN's new cannon as well as of the first problems encountered when firing it are given in the press:

(i) On July 23, 1864, the Pittsburgh *Gazette* writes: "Juveniles, aged from ten to fifteen years, were amusing themselves today in crawling into the bore on their hands and knees. a good sized family including ma and pa, could find shelter in the gun, and it would be a capital place to hide in case of a bombardment…"

(ii) In two short notes, the *Scientific American*[781] reports, "*Beelzebub* is the name given to the new twenty-inch gun just completed and prepared for proving at the Fort Pitt Works. It is a Dahlgreen navy gun, weighs as finished 98,915 pounds [44,844 kg], and will throw a solid shot weighing 1,080 pounds [489.9 kg] to a distance, it is calculated, of over six miles [9.7 km]. The gun was cast on the 5th of May last, was taken from the pit on the 22nd of June, and has since been brought to its present perfection of finish, the equal of any gun in the world … This immense 'shooting-iron' is now in readiness for removal to the proving ground, when it is to be tested by nine discharges, firing a solid shot with each, three of which will be with a charge of 60 pounds [27.2 kg] of powder, three with 80 pounds [36.3 kg], two with 100 pounds [45.4 kg], and one, in all probability with 125 pounds [56.7 kg]. No notice will be given publicly of the day of proving as it is desirous to avoid the presence of a crowd on the occasion." ▪ The firing worked well enough with charges of 60 and 80 pounds (27.2 and 36.3 kg), but when the second charge of 100 pounds (45.3 kg) was fired, "immediately after the recoil, a wrought-iron bolt, three inches in thickness, suddenly snapped and the breachband broke, letting the enormous mass to the ground, crushing as if they were made of timber, the T-rails beneath. This sudden accident brought the test to a conclusion…"

Unfortunately, these two accounts didn't address the projectile velocities achieved. In the following year, however, Jules VERNE {⇨1865}, also referring in his novel *From the Earth to the Moon* to the *Scientific American* account, mentioned that the velocity of RODMAN's super shots was 800 yards per second (732 m/s; $M \approx 2.2$) – *i.e.*, faster than twice the velocity of sound, but he gave no reference. With his 15-in. (38.1-cm) caliber gun, built and tested in previous years, RODMAN measured a muzzle velocity of 1,735 ft/s (529 m/s; $M \approx 1.6$) for a 330-pound (149.7-kg) shell.

| | | |
|---|---|---|
| 1864 | *Royal Military Academy, Woolwich, southeast London, England* | **Rvd. Francis A. BASHFORTH**, first British professor of mathematics to the Advanced Class of Artillery Officers, **finds that solutions to real gunnery problems require an increase in the quality of experimental data on aerodynamic drag.** Despite military opposition, he designs an electric high-precision rotating-drum chronograph – the so-called **"Bashforth chronograph"** – which records a circuit break each time the shot passes through one of ten "screens" 150 ft (45.7 m) apart.[782] A separate pen records circuit-break times from a pendulum clock. These time-distance data yield the aerodynamic drag as well as space and time ballistic tables. This first multiple-screen arrangement is the precursor to modern aeroballistic ranges. |

In the following years, BASHFORTH extended his aerodynamic drag measurements:
- in the period 1867–1868:[783] for spherical projectiles in the range 850–2,150 ft/s (259.1–655.3 m/s) and for projectiles with ogival heads ($R = 1.5$ calibers) in the velocity range 900–1,700 ft/s (274.3–518.2 m/s); and
- in the period 1878–1880:[784] for projectiles with ogival heads only in the ranges 100–900 ft/s (30.5–274.3 m/s) and 1,700–2,800 ft/s (518.2–853.4 m/s).

---

[781] Anonymous: *The 20-inch navy gun.* Scient. Am. **12**, 280, 288 (1865).

[782] F.A. BASHFORTH: *A revised account of the experiments made with the Bashforth chronograph to find the resistance of the air to the motion of projectiles, with the application of the results to the calculation of trajectories according to J. BERNOULLI's method.* Cambridge University Press, Cambridge (1890).

[783] F.A. BASHFORTH: *Reports on experiments made with the Bashforth chronograph to determine the resistance of the air to the motion of projectiles, 1865–1870.* W. Clowes & Son, London (1870).

BASHFORTH's results verified the hypotheses that aerodynamic drag is a function of velocity and air density {Sir NEWTON ⇒ 1687; ROBINS ⇒ 1746; HUTTON ⇒ 1783} as well as the "sectional density" $m/d^2$, as used in the modern definition of the "ballistic coefficient" ($m$ = mass and $d$ = diameter of projectile). ▪ Sphere drag comparisons by Donald G. MILLER[785] (LLNL, CA) and Allan B. BAILEY (Arnold Air Force Station, TN), two scientists involved in the analysis of drag measurements, have shown that BASHFORTH's data scatter no more than modern spark range data.

| | | |
|---|---|---|
| 1864 | *Private Observatory at Tulse Hill, London, England* | **Sir William HUGGINS,**[786] a British astronomer, **studies the spectra of various nebulae by applying Robert Wilhelm BUNSEN's and Gustav Robert KIRCHHOFF's discoveries in spectral analysis to astronomy.** The spectra he obtains of various nebulae establish that they are composed mainly of incandescent gas. He also shows that the Orion Nebula is a cloud of gas.<br><br>The space between the stars – the so-called *interstellar medium* – which is emptier than the best vacuums created on the Earth, is composed of about 99% gas (consisting of about 90% hydrogen in atomic or molecular form and 10% helium), while the rest is dust (made of thin, highly flattened flakes or needles of graphite and silicates coated with water ice). The dust is probably formed in the cool outer layers of red giant stars. **Shock waves produced by supernova explosions inject vast quantities of mechanical energy into the interstellar medium,** which can sweep the surrounding medium into expanding shells filled with hot gas.[787]<br><br>▸ Interstellar shock waves cause the interstellar gas to radiate, providing an invaluable diagnostic for violent events in the interstellar medium. ▪ In 1930, the Swiss-born U.S. astronomer Robert J. TRUMPLER[788] discovered interstellar dust absorption by comparing the angular sizes and brightnesses of globular clusters.<br><br>▸ Interstellar shocks are likely to be responsible for the acceleration of cosmic rays {HESS ⇒ 1912; KRYMSKY ⇒ 1977/1978}. |
| 1864 | *Chair of Natural Philosophy, South Carolina College, Columbia, South Carolina* | **Joseph LE CONTE,**[789] a U.S. physician and lecturer in physics and mathematics, **reviews the large body of international literature relating to the obvious discrepancy between the velocity of sound given by physical theory and that obtained through direct experiment:**<br><br>(i) He addresses previous theories of violent sound {AIRY ⇒ 1849; EARNSHAW ⇒ 1858 & 1860; CHALLIS ⇒ 1848 & 1851; STOKES ⇒ 1848 & 1851; *etc.*}, and previous observations {BIOT ⇒ 1809; PARRY & FOSTER ⇒ 1824/1825; BESANT ⇒ 1859; MONTIGNY ⇒ 1860; *etc.*}.<br><br>(ii) He correctly considers observations previously made on thunder {EARNSHAW ⇒ 1851 & 1860; RAILLARD ⇒ 1860; MONTIGNY ⇒ 1860} to be a psychological illusion {UMAN ET AL ⇒ 1970}.<br><br>(iii) **However, he rejects all hypotheses of wave propagation attributed to the peculiarities of large amplitudes,** and states: "It is true there may be nothing à priori improbable in the assumption that the velocity of sound might be related to the violence of the disturbance; but the fact that the analytical investigations conduct to such extreme results as to set at nought all our physical conceptions, originate a strong presumption that they belong to that class of mathematical fictions which have frequently sharpened the ingenuity and brightened the imagination of some of the most eminent geometers." |

---

[784] F.A. BASHFORTH: *Final reports on experiments made with the Bashforth chronograph to determine the resistance of the air to the motion of elongated projectiles, 1878–1880.* W. Clowes & Son, London (1880).

[785] D.G. MILLER and A.B. BAILEY: *Sphere drag at Mach numbers from 0.3 to 2.0 at Reynolds numbers approaching $10^7$.* J. Fluid Mech. **93**, 449-464 (1979).

[786] W. HUGGINS: *On the spectra of some of the nebulae.* Phil. Trans. Roy. Soc. Lond. **154**, 437-444 (1864).

[787] *Multi-wavelength images help astronomers study star birth, death.* Science Daily, University of Illinois at Urbana-Champaign, IL (Jan. 17, 2006); http://www.sciencedaily.com/releases/2006/01/060115173654.htm.

[788] R.J. TRUMPLER: *Absorption of light in the galactic system.* Publ. Astron. Soc. Pacific **42**, No. 248, 214-227 (1930); http://articles.adsabs.harvard.edu/cgi-bin/nph-iarticle_query?1930PASP...42..214T&amp;data_type=PDF_HIGH&amp;type=PRINTER&amp;filetype=.pdf.

[789] J. LE CONTE: *On the adequacy of LAPLACE's explanation to account for the discrepancy between the computed and the observed velocity of sound in air and gases.* Phil. Mag. **27** [IV], 1-32 (1864).

(iv) Supporting DE LAPLACE's view {⇨1816}, he concludes that "… the accuracy of the physical reasoning upon which DE LAPLACE's formula is based has not been invalidated by the recent discussions on the mathematical theory of sound." ▪ LE CONTE's paper is very interesting from a historical point of view, because the numerous reasons discussed illustrate not only the keen interest of contemporary naturalists in this subject, but they also reveal how difficult it was for them to accept the notion that hitherto unknown mechanisms generate large-amplitude sound waves that propagate in air at supersonic velocities.

In the same year, **the Rev. Samuel EARNSHAW,**[790] an English mathematician and clergyman who in previous years had made important contributions to the mathematical theory of sound {EARNSHAW ⇨1858 & 1860}, **raises some serious objections to LE CONTE's treatise on sound:**

(I) EARNSHAW defends the observations of sounds of finite amplitude made by Prof. Charles M.V. MONTIGNY {⇨1860} and particularly by Captain William E. PARRY {⇨1824/1825}, which he had already cited previously as important experimental proof of his own theory of sound.

(II) EARNSHAW writes, "Dr. LE CONTE asserts that I have *'failed to produce a single unexceptionable fact, or a single satisfactory observation, in verification of my theoretical deductions.'* Such assertion as this are easily made, even in the presence of plain evidence to the contrary; but they should never be introduced into inquiries after scientific truth … And again, with regard to Captain PARRY's record of the word *'fire'* having been heard after the report of the gun in his experiments on sound. How does Dr. LE CONTE dispose of that fact? There are persons now living who were present at the experiment and noticed the fact. There was no room for doubtful opinions. The fact is of such a nature that they could not be mistaken. They testify that they distinctly heard the word *'fire'* after the sound of the gun's report had passed them. But Dr. LE CONTE gets rid of this well-authenticated fact in a very simple and summary way, a way by which any other unpleasant fact may also be got rid of. He tells us that *'the records of the most trustworthy experimentalists concur in establishing the general fact that all sounds travel at the same rate.'* Captain PARRY's officers we see, then, were not trustworthy experimentalists … For let it be remembered that **the question at issue is not whether all gentle sounds (*i.e.*, sounds of musical type) travel at the same rate, but whether a violent sound travels at the same rate as a musical sound;** and I deny that experimentalists concur in establishing this to be a fact, as Dr. LE CONTE asserts they do … Captain PARRY's fact still therefore stands an obstinate witness with Prof. MONTIGNY's facts in favor of the truth of my theory; and Dr. LE CONTE's assertion that I have not produced a single fact confirmatory of my theory is disposed of…"

| | | |
|---|---|---|
| 1864 | *Titusville & Ladies Well, Oil Creek, northwestern Pennsylvania* | Edward A.L. ROBERTS, a U.S. lieutenant colonel of the 29th New Jersey Volunteers, **makes the first successful oil well shots on the Ladies Well.** He will obtain a patent for his technique of "shooting" the wells to increase oil flow round a bore hole – the so-called **"Roberts torpedo."** ▪ At first, gunpowder, eight pounds (3.63 kg) per charge, was used to torpedo the wells by cracking strata and loosening accumulated paraffin. With the introduction of nitroglycerin {SOBRERO ⇨1846} in the late 1860s, oil well shooting became more successful but also more dangerous.[791] For deep wells, a more recent technique uses shaped charges to increase the flow of oil {McLEMORE ⇨1946; ⇨Fig. 4.15−L}. |

In 1865, ROBERTS founded the Roberts Petroleum Torpedo Company in New York. His exorbitant charges caused many "wild cat" shooters to make money out of torpedoing the wells under cover of darkness. Many fights between the licensees and the "wild catters" occurred, and thousands of lawsuits arose over the illegal shooting of wells. In these suits, due to his in-

---

[790] S. EARNSHAW: *Reply to some remarks of Dr. LE CONTE in his paper on the problem of the velocity of sound.* Ibid. **27** [IV], 98-104 (1864).

[791] A brief introduction and history of oil well shooting was provided by Analog Service Inc., Fordsville, KY (2000); http://www.logwell.com.

fluence and wealth, ROBERTS always won. From *The Titusville Morning Herald* of June 17, 1866, "Our attention has been called to a series of experiments that have been made in the wells of various localities by Col. ROBERTS, with his newly patented torpedo. The results have in many cases been astonishing. The torpedo, which is an iron case, containing an amount of powder varying from fifteen to twenty pounds, is lowered into the well, down to the spot, as near as can be ascertained, where it is necessary to explode it. It is then exploded by means of a cap on the torpedo, connected with the top of the shell by a wire. **The object of the torpedo is to clean out all the deposits at the bottom of the well such as gravel, pieces of seed-bag, *etc.*, as well as to open the fissures, where the oil comes through.** These frequently become perfectly clogged with paraffin, and other matter that effectually prevents the production of oil from the well. It is claimed that the explosion of the torpedo at so great a depth, cannot result in any serious damage to the well … A large number of wells have been experimented on, and the most signal success has attended the same. Several wells on the Tarr farm, on Oil creek, have proved the most productive as yet. One of these was increased in production by the use of the torpedo from ten to over one hundred barrels [1,160 to 11,600 liters] per day."

| | | |
|---|---|---|
| 1864 | *Royal Agricultural Academy, Poppelsdorf [a little town near Bonn], Germany* | **August TOEPLER,**[792] a German experimental physicist of great skill, **first publishes his *schlieren* method** {⇨Figs. 4.5–A, B & 4.18–A}, which gives ground-breaking results: |

(i) Although the principle of the method was previously discovered in England by Robert HOOKE[793] (1665) and was rediscovered in France by J.B. Léon FOUCAULT[794] (1859), TOEPLER uses a simple arrangement consisting of a point light source, an adjustable knife-edge cutoff, and a telescope, which proves extremely useful in the visualization of compressible flow phenomena.

(ii) **TOEPLER directly visualizes the propagation and reflection of shock waves in air, thereby noticing their sharp wavefronts.** Since high-sensitivity photographic films {MADDOX ⇨1871} are not yet available to him, **he studies the shocks subjectively using a sophisticated stroboscopic arrangement and a modification of the Knochenhauer circuit** {KNOCHENHAUER ⇨1858; ⇨Fig. 4.19–E}, which delays the illumination spark relative to the spark generating the shock wave.[795]

(iii) At first confused by the appearance of several shock fronts, he annotates: "Apart from the envelope and little clouds, the spark seems to be surrounded by concentric spheroids *a b c* with rather sharp boundaries. They are never disrupted or bulged; with increasing size they approach a spherical geometry. Close to the spark they resemble a cylinder which is bounded by two hemispheres. Operating the induction coil at a high repetition rate they give the impression of soap bubbles which, forming around the spark, immediately disappear again. It makes one believe that always several, usually three or four, are visible simultaneously in the field of view. However, in the case that the coil is working at the lowest possible rate so that the ear is capable of clearly differentiating between each stroke, it is obvious that each discharge corresponds to only a single one of the spheroids described above, but that from spark to spark, the phenomenon strongly varies in size and formation."

(iv) **To illustrate this discontinuous wave phenomenon, he first uses the correct terms *Stoßwelle* ("shock wave") and *Lufterschütterungswelle* ("air percussion wave"), but likewise also the incorrect term *Schallwelle* ("sound wave").**

(v) **TOEPLER notices that the spark channel is not a homogeneous cylindrical plasma column but is instead pinched,** thereby showing constrictions in the axial direction. • With

---

[792] A. TOEPLER: *Beobachtungen nach einer neuen optischen Methode.* M. Cohen & Sohn, Bonn (1864), p. 43.
[793] H.W. ROBINSON and W. ADAMS (eds.): *The diary of R. HOOKE.* Taylor & Francis, London (1953).
[794] L. FOUCAULT: *Mémoire sur la construction des télescopes en verre argenté.* Ann. Observ. Impér. Paris **5**, 197-237 (1859).
[795] U. NEUBERT: *Beitrag zur TOEPLERschen Schaltung der Funkenkinematographie.* Z. Tech. Phys. **24**, 179-187 (1943).

this important observation, he anticipated the ***pinch effect*** discovered in 1907 by Edwin NORTHRUP, who named it the "pinch phenomenon" {NORTHRUP ⇨1907; TONKS ⇨1937}.

TOEPLER's schlieren method has been widely applied and modified in fluid dynamics, particularly in high-speed aerodynamics and nonstationary gas dynamics. Very recently, Prof. Gary S. SETTLES {⇨2001} gave a comprehensive review of the history and applications of the schlieren method, which is still widely used for visualizing compressible flow and shock waves.

***Microschlieren*** is a technique used to study compressible flows in great detail. It was developed in the 1950s by Walter S. BRADFIELD[796] and Joseph J. SHEPPARD JR., two aerodynamicists at the University of Minnesota, in order to study small flow fields in detail, such as flows near leading edges, those about probes immersed in high-speed thin boundary layers, shock wave boundary-layer interactions, *etc.*

| | | |
|---|---|---|
| 1865 | *Polytechnikum, Zurich, Switzerland* | Rudolf J.E. CLAUSIUS,[797] a German mathematical physicist, **introduces his concept of entropy** – calling his new theorem the "principle of equivalence of transformations."[798] According to his concept, the entropy change per unit mass $\Delta s$ is equal to or greater than the heat change $\Delta q_H$ divided by the absolute temperature T (*i.e.*, $\Delta s \geq \Delta q_H/T$). He states, "*Die Energie der Welt ist constant; die Entropie strebt einem Maximum zu.*" ("The energy in the world is constant; the entropy tends towards a maximum.") |

**All spontaneous processes, such as shock waves, are irreversible processes;** *i.e.*, **entropy increases during such processes.**[799] Due to dissipative processes in the shock front, the entropy always increases with a shock wave {JOUGUET ⇨1904; DUHEM ⇨1909; BECKER ⇨1922; BETHE ⇨1942}, which was not immediately recognized in the early era of shock wave physics {RIEMANN ⇨1859}.

▸ In particular, viscosity produces an increase in entropy, while heat conduction produces an increase or decrease in entropy, depending on the increase in the entropy of fluid particles resulting from heat conduction in the colder and hotter layers, respectively {MORDUCHOW & LIBBY ⇨1949}.

▸ The entropy jump at a shock front is especially high when a medium contains a large number of pores (small empty holes randomly distributed). Shock compression of a porous medium leads to intense heating.[800]

| | | |
|---|---|---|
| 1865 | *Logelbach [a little town near Colmar], Alsace, France* | Gustave Adolphe HIRN,[801] an Alsatian physicist-engineer, philosopher and independent scholar, **describes a simple impact experiment for determining the mechanical heat equivalent.** His setup consists of two cylindrical rods which are suspended horizontally on threads, one moving rod (acting as an impactor) strikes the other resting rod (acting as an anvil) head-on. His calorimeter, a hollow leaden cylinder filled with water, is inserted between anvil and impactor. He obtains for the mechanical heat equivalent a value of 425 mkg. • In the mid- |

---

[796] W.S. BRADFIELD and J.J. SHEPPARD JR.: *Microschlieren.* Aero/Space Engng. **18**, 37-40, 56 (May 1959).

[797] R. CLAUSIUS: *Über verschiedene für die Anwendung bequeme Formen der Hauptgleichungen der mechanischen Wärmetheorie.* Ann. Phys. **125** [II], 353-400 (1865).

[798] Concerning the meaning of the new term *entropy*, CLAUSIUS wrote, "Sucht man für S einen bezeichnenden Namen, so könnte man, ähnlich wie von der Größe U gesagt ist, sie sey der *Wärme- und Werkinhalt* des Körpers, von der Größe S sagen, sie sey der *Verwandlungsinhalt* des Körpers. Da ich es aber für besser halte, die Namen derartiger für die Wissenschaft wichtiger Größen aus den alten Sprachen zu entnehmen, damit sie unverändert in allen neuen Sprachen angewandt werden können, so schlage ich vor, die Größe S nach dem griechischen Worte ἡ τροπή, die Verwandlung, die *Entropie* des Körpers zu nennen. Das Wort *Entropie* habe ich absichtlich dem Worte *Energie* möglichst ähnlich gebildet, denn die beiden Größen, welche durch diese Worte benannt werden sollen, sind ihren physikalischen Bedeutungen nach einander so nahe verwandt, dass eine gewisse Gleichartigkeit in der Benennung mir zweckmäßig zu seyn scheint."

[799] J.B. RUSSELL: *A note on the entropy change across a normal shock wave.* Aeronaut. J. **78**, 431-432 (1974).

[800] L.G. BOLKHOVITINOV and Y.B. KHVOSTOV: *The Rankine-Hugoniot relation for shock waves in porous media.* Nature **274**, 882-883 (Aug. 31, 1978).

[801] G.A. HIRN: *Théorie mécanique de la chaleur.* Gauthier-Villars, Paris (1865), Part I: *Exposition analytique et expérimentale*, pp. 58-62.

1840s, James Prescott JOULE found that a mechanical energy of 430 mkg is required to increase the temperature of 1 kg water by one degree.

Already in the 1850s, Charles LABOULAYE,[802] a French engineer and industrialist, tried to determine the mechanical equivalent of heat by an impact experiment: he dropped a 440-kg weight from a height of 1.045 meter on a 5-kg cup-like body of lead which, filled with 2 kg water and provided with a thermometer, served as a calorimeter. He observed that the temperature of the water increased by 0.8 °C after impact and estimated a mechanical equivalent of (only) 247 mkg.

| 1865 | *Nitroglycerine Ltd. Stockholm, Heleneborg, Sweden & Krümmel [near Hamburg], northern Germany* | **Alfred B. NOBEL,**[803] a Swedish chemist and industrialist, **addresses the advantages and multi-purpose applications of nitroglycerin in the mining industry.** He writes, "The greatest advantage of nitroglycerin consists in the fact that when it is used a force can be introduced into the blast-hole of a mine ten times as great as when powder is used. Hence arises a great economy in manual labor, the importance of which is understood when it is remembered that **the labor of the miner represents, according to the hardness of the rock, from five to twenty times the value of the powder required,** a saving therefore which will often amount to 50 percent. The use of this substance is very simple. If the blast-hole of the mine is fissured, it must be lined with clay in order to render it tight. Nitroglycerin is then poured in, and the upper part of the hole is filled with water; in the nitroglycerin is then introduced a safety-match of suitable length, at the end of which is pressed a strong percussion-cap. The operation is finished, and it is only necessary to put fire to the match."

In the same year, NOBEL, upon receiving financial support in Hamburg, erects a new works at Krümmel on the Elbe, which quickly developed into the largest explosive factory in Europe. |
| 1865 | *Athenaeum, Deventer, The Netherlands* | **Heinrich Wilhelm SCHRÖDER VAN DER KOLK,**[804] a Dutch professor of mathematics and physics, **correctly assumes that intense sound propagates faster than weak sound.** He tries to derive a formula for the sound velocity in terms of the ratio of the specific heats, the velocity of sound at infinitely small amplitude, and the specific volume reduction caused by the intense sound. However, since he assumes compression along the static adiabat, often referred to as the Poisson law, and not along the "dynamic adiabat" or "Hugoniot curve," his equation underestimates the velocity increase. ▪ The problem was first solved in a general manner in France by Pierre-Henri HUGONIOT {⇨1887} and was later addressed in a practicable equation by Paul VIEILLE {⇨1900}. |
| 1865 | *Paris, France* | **Jules VERNE,** the celebrated French science fiction author, **describes the journey of two Americans and one Frenchman (and two dogs) in a space vehicle in his two famous novels** *De la terre à la lune* ("From the Earth to the Moon") and *Autour de la lune* ("Around the Moon"). **The vehicle is launched from Florida, flies around the Moon, and lands safely in the Pacific.** In the books, he addresses various scientific and technological problems:

(i) Obviously stimulated by contemporary ballisticians and the widespread enthusiasm for big guns {RODMAN ⇨1864}, VERNE's Moon travelers use a huge, 270-m-long vertical gun {⇨Fig. 4.2−S} to accelerate a projectile to the "escape velocity" of 11 km/s. This term for the velocity needed to escape the Earth's gravitational pull was commonly used in subsequent discussions of how manned space-flights could be realized {Sputnik 1 ⇨1957}.

(ii) Aerodynamic drag at this hypersonic projectile velocity is believed to be negligible by the Moon travelers, who argue that the passage through the atmosphere would only last a few |

---

[802] C. LABOULAYE: *Note sur des expériences à l'aide desquelles on détermine la valeur de l'équivalent mécanique de la chaleur.* C. R. Acad. Sci. Paris **46**, 773-775 (1858).

[803] A.B. NOBEL: *Results of blasting experiments made with nitroglycerine at Vieille-Montagne mine.* Phil. Mag. **30** [IV], 236-238 (1865).

[804] H.W. SCHRÖDER VAN DER KOLK: *On the velocity of sound.* Phil. Mag. **30** [IV], 34-49 (1865); *Note on the velocity of sound, and on the mechanical energy of chemical actions.* Ibid. **30** [IV], 391-392 (1865).

seconds anyway. However, the crew is wary of the tremendous aerodynamic heating involved {GALILEI ⇨1623; JOULE & THOMSON ⇨1856}, and the omnipresent danger of colliding with meteoroids.

(iii) VERNE's crew also speculates on the origin of lunar craters {HOOKE ⇨1665; SCHRÖTER ⇨1802; PROCTOR ⇨1873; GILBERT ⇨1893; GIFFORD ⇨1924}. Although correctly considering the possibility that they originated from the impacts of celestial bodies (such as asteroids or comets), VERNE favors the popular (at that time) volcanic hypothesis.

VERNE's dream provided tremendous stimulation for subsequent generations of scientists and engineers "of landing a man on the Moon and returning him safely to Earth" {U.S. President KENNEDY ⇨1961}. Indeed, the dream became reality 104 years later when Apollo 11 {⇨1969} landed on the Moon.

| 1865 | University of Leipzig, Saxony, east-central Germany | **Johann Friedrich Karl ZÖLLNER**, a German astronomer, **coins the term *Astrophysik* ("astrophysics") to unite the distinct disciplines of physics and chemistry with astronomy.** ▪ Today astrophysics is understood as being the branch of astronomy that deals with the physical and chemical nature of celestial objects and events.<br><br>ZÖLLNER who worked on comets, the constitution of the Sun, and the thermal conditions of planets, attempted to explain sunspots as huge areas of cinders floating on the Sun's surface.[805] This principle was extended to other stars, some of which are thought to have much more active "starspot" activity. As a star with starspots rotates, its brightness changes slightly. Stars exhibiting such behavior are called "rotating variables."[806] |
|---|---|---|
| 1866 | The Napier Brothers, Glasgow, Scotland, U.K. | **Robert D. NAPIER**,[807] a Scottish mechanical engineer, **studies the flow characteristics of a gas flowing from the vessel in which it is compressed through an orifice and into the atmosphere.** He observes that the rate of discharge increases as the ratio of the receiver pressure to the initial pressure decreases from unity to about 0.5, but that when the latter stage is reached, a further reduction in the receiver pressure has no effect on the rate of discharge, which remains constant – the so-called **"choking effect."** ▪ An important step forward in the theory of orifice discharge was achieved by the British physicist and engineer Osborne REYNOLDS {⇨1885}, who assumed a continuous fall of pressure along the axis of the jet. |
| 1867 | Chair of Chemistry, University of Heidelberg, Germany | **Robert W. BUNSEN**,[808] a German professor and authority on chemistry, makes the first careful measurement of the rate at which an explosion is propagated in gases.<br><br>(i) He describes a method of measuring the rapidity of the flame in gaseous explosions by passing a mixture of explosive gas through an orifice at the end of a tube and igniting the gases as they issue into the air. He determines the rate at which the gases must be driven through the tube to prevent the flame passing back through the opening. **Using this method, he finds that the rate of propagation of the ignition of pure electrolytic gas [$2H_2/O_2$] is 34 m/s, while the rate of ignition of carbonic oxide and oxygen [$2CO/O_2$] is less than 1 m/s.** ▪ BUNSEN (incorrectly) applied these results to the rate of explosion of gases in closed vessels. His results were accepted as being authoritative until the discovery of high detonation velocities {HARCOURT ⇨1880; BERTHELOT ⇨1881; BERTHELOT & VIEILLE ⇨1882; MALLARD & LE CHÂTELIER ⇨1883; VON OETTINGEN & VON GERNET ⇨1888; DIXON ⇨1893}. These studies demonstrated that BUNSEN's results only apply to the mild and usually short initial phases of such explosions. In his detonation studies, **BUNSEN did not provoke a detonation but rather a deflagration, which explained his obtained low velocities.** |

---

[805] J.F.K. ZÖLLNER: *Über die Periodizität und heliographische Verbreitung der Sonnenflecken*. Ann. Phys. **142** [II], 524-539 (1871).
[806] For the classification of variable stars, *see* General Catalogue of Variable Stars ⇨1948.
[807] R.D. NAPIER: *On the velocity of steam and other gases, and the true principles of the discharge of fluids*. Spon, London (1866).
[808] R.W. BUNSEN: *Über die Temperatur der Flammen des Kohlenoxyds und Wasserstoffs*. Ann. Phys. **131** [II], 161-179 (1867); *On the temperature of a flame of carbonic oxide and hydrogen*. Phil. Mag. **34** [IV], 489-502 (1867).

(ii) He also performs the first systematic research into the pressures and temperatures produced by the explosion of gases in closed vessels.

(iii) Based upon his results, he concludes that there is discontinuous combustion during explosions. BUNSEN speculates that in a gaseous explosion "the total gaseous mass does not explode at once, but rather successively in discontinuous partial explosions that propagate stepwise through the gas" [Germ. *"eine discontinuierliche, gleichsam stufenweise erfolgende Verbrennung"*]. ▪ These deductions were criticized by the French chemist P.E. Marcellin BERTHELOT {⇨1881}, who pointed out that they assumed that the specific heats of steam and of carbonic acid at high temperatures were constant. German researchers attempted to visualize these "successive explosions," but came to erroneous conclusions {VON OETTINGEN & VON GERNET ⇨1888}.

| | | |
|---|---|---|
| 1867 | *Friedrich-Wilhelms-Universität [from 1945 Humboldt-Universität], Berlin, Prussia, Germany* | **Heinrich MAGNUS,** a German physicist and chemist, **receives August TOEPLER's paper on schlieren observations of spark waves** {A. TOEPLER ⇨1864}. **He criticizes TOEPLER's use of the term *Schallwelle* ("sound wave") for the visualized spark wave, a weak shock wave.** In a letter[809] to TOEPLER, he states, "I was never in doubt about the correctness of your observations … however, I have declared myself against the expression 'sound wave' as, I suppose, I already did previously. Now you state in your kind letter that the air is expanded by the spark which causes a compression propagating with the speed of sound: this is clear and nobody will contest it, just as little as this compression was reflected. However, it is not the sound which is visible, but rather the air which, heated and perhaps colored by the spark, expands from the position of the spark and is reflected, because the compressed air itself is visible with your apparatus … A designation can easily give reason for a misinterpretation. Who will not imagine waves emitted by a sounding body when hearing about 'sound waves'?" ▪ This stimulated TOEPLER[810] to give a more detailed definition in his next paper, "The electric spark is a very favorable source of sound; it can be used to provide single shocks which [at increasing repetition rate] can be driven up to the generation of a tone. The expression 'sound' has been used for any perceptible impression to the sense of hearing, likewise the word 'sound wave,' also in case that the air particles do not experience a full oscillation…" |
| 1867 | *Private residence at Glenlair, Kirkcudbrightshire, Scotland, U.K.* | **James Clerk MAXWELL,**[811] a Scottish physicist, **publishes his memoir *On the Dynamical Theory of Gases,* his most important single paper.** <br><br>(i) He discusses previous attempts to establish a general theory of gases {D. BERNOULLI ⇨1738; LESAGE ⇨1784; J. HERAPATH 1847; JOULE ⇨1850; CLAUSIUS ⇨1857; O.E. MEYER 1865}. He works out the distribution of velocities among the molecules of a gas, and the mean free path of a molecule between molecular collisions. ▪ In the following years, MAXWELL extended his own "dynamical theory."[812] <br><br>(ii) During the experiments on gases, which he made together with his wife in the year 1865, his attention was directed to viscoelastic phenomena in fluids. **He introduces the concept of viscoelasticity in order to explain the behavior of fluids.** In his model, the elastic behavior (HOOKE's law) is represented by an ideal spring, and the viscous behavior by an ideal dashpot. An analysis of this model gives the *relaxation time* – a measure of the time required for the energy stored in the spring to shift to the dashpot and dissipate. ▪ For stresses of sufficiently short duration – less than the shortest relaxation time – liquids can behave in the same way as solids. |

---

[809] This letter (dated April 29, 1867) is now kept in the archives of the TU Dresden.
[810] A. TOEPLER: *Die vom elektrischen Funken in Luft erzeugte Welle.* Ann. Phys. **131** [II], 180-215 (1867).
[811] J.C. MAXWELL: *On the dynamical theory of gases.* Phil. Trans. Roy. Soc. Lond. **157**, 49-88 (1867).
[812] J.C. MAXWELL: *Illustrations of the dynamical theory of gases. Part I: On the motion and collisions of perfectly elastic spheres.* Phil. Mag. **19** [IV], 19-32 (1860); *Part II: On the process of diffusion of two or more moving particles among one another.* Ibid. **20** [IV], 21-37 (1860); *On the dynamical theory of gases.* Phil. Trans. Roy. Soc. Lond. **157**, 49-88 (1867).

| 1867 | | For example, viscoelastic liquids, when impacted by a solid, show either elastic or brittle fracture behavior {KORNFELD ⇨1951; ⇨Fig. 4.3–V}. |
|---|---|---|
| 1867 | Nitroglycerine Ltd. Stockholm, Heleneborg, Sweden | **Alfred B. NOBEL**,[813] a Swedish industrialist, realizing the hazards involved in the handling of liquid nitroglycerin, **invents *dynamite*, which he also terms *Guhr dynamite*.**

(i) The new high explosive – named after the Greek word *dynamis* (meaning "power") – is a mixture of three parts of nitroglycerin and one part of *Kieselguhr*, a German word designating a diatomaceous (siliceous) earth, a nonexplosive porous absorbent. ▪ Two types of dynamite were manufactured: No. 1 with 75% nitroglycerin, and No. 2 with 64%.

(ii) NOBEL states that **"the principle for its action lay in the sudden development of a very intense pressure or shock."** ▪ Dynamite is sensitive to the action of a blasting cap {NOBEL ⇨1864}, but relatively insensitive to ordinary shock. For dynamite with a density loading of 1.5 the detonation velocity varies from 6,650 to 6,800 m/s. Dynamite releases energy at a very high rate. For example, an eight-ounces (225-g) stick of an average-energy dynamite may have a working capacity of about 40,000 horsepower.

(iii) His invention of dynamite will become a great commercial success and fully established nitroglycerin as the leading blasting agent. ▪ In the period 1867–1874, NOBEL founded 15 factories worldwide, which increased dynamite production from 11 tons in 1867 to 3,120 tons in 1874.[814]

In the same year, **NOBEL** stops using his *igniter* {NOBEL ⇨1863/1864} and **develops his *fulminating cap*,** as it is then commonly termed. Containing 80% mercury fulminate and 20% potassium chlorate in a copper cap, the detonator – also known as the **"copper detonator"** – is used to safely, and under all conditions, initiate the explosive reaction in a column of dynamite by percussion rather than by the local heat from an electric spark or an electrically heated wire. |
| 1867 | Sweden | **Johan Henrik NORRBIN**[815] and **Johan OHLSSON**, two Swedish chemists, **observe that the explosive properties of dynamites are enhanced by the addition of ammonium nitrate** [$NH_4NO_3$] {GLAUBER ⇨1659}. ▪ The Swedish industrialist Alfred NOBEL subsequently acquired their patent and used it in his explosive compositions.

Dynamite, which is high in ammonium nitrate and contains a cooling agent such as sodium nitrate (*i.e.*, common salt), is preferably used in underground coal-mining, because it produces a flame that is sufficiently cool to avoid ignition of underground gases such as methane and/or coal dust, thus considerably reducing the danger of disastrous mine explosions {DAVY ⇨1815; FARADAY & LYELL ⇨1845; Jabin de Saint-Etienne ⇨1876; MALLARD & LE CHÂTELIER ⇨1881 & 1883; CHARPY & LE CHÂTELIER ⇨1890; SIERSCH ⇨1896; U.S. Bureau of Mines ⇨1908}.

Explosives that have been approved for safe use in coal mines by governmental authorities are known in the United States as ***permissible explosives***; in England they are referred to as ***permitted explosives***, in France as ***explosives antigrisouteux***, in Belgium as ***explosifs S.G.P.*** (*sécurité, grisou, poussière*), and in Germany and Austria as ***schlagwettersichere Sprengstoffe*** or ***Wetter-Sprengstoffe***. *Ammonia dynamites* contain 65–82% ammonium nitrate and 4% nitroglycerin, or 27–56% ammonium nitrate and 12–30% nitroglycerin. Also, almost all permitted explosives contain admixtures, such as 1–4% wood meal, 10–40% alkali chloride and 2–8% DNT or TNT. |

---

[813] A. NOBEL: *Explosive compounds*. Brit. Patent No. 1,345 (May 7, 1867).

[814] H. DE MOSENTHAL: *The life-work of Alfred NOBEL*. J. Soc. Chem. Ind. **18**, 443-451 (1899).

[815] J.H. NORRBIN and J. OHLSSON: *Sätt och medel för tillblandning af sprängämnen samt för deras förvaring och antändning* ("Methods and Means for Mixing Explosives and Their Storage and Ignition"). Swed. Patent No. 188 (1867).

| 1867 | University of Tübingen, southern Germany | **Friedrich Eduard REUSCH,**[816] a professor of physics and technology, **discovers that a star-like pattern of fine cracks is formed around the point of impact when a crystal is loaded at its surface by a sharp blow from a pointed instrument** such as a center punch {⇨Fig. 4.3–N}. For rock salt and mica, he obtains crack patterns resembling four-rayed and six-rayed stars, respectively – so-called **"percussion figures"** [Germ. *Schlagfiguren*]. ▪ His **"center punch test"** [Germ. *Körnerprobe*] has since been adopted by mineralogists to quickly identify the crystal symmetry in the absence of faces.

When the crystal surface is gradually loaded with a blunt punch, a similar pattern – a so-called **"pressure figure"** [Germ. *Druckfigur*] – is obtained. The resulting rays are similar in character but not necessary in the same position as the ones seen in a percussion figure. |
|---|---|---|
| 1867 | Private study at Villeporcher-Saint-Ouen/ Vendôme, Dépt. Loir-et-Cher, central France | **Adhémar-Jean-Claude Barré DE SAINT-VENANT,**[817] a French engineer and mathematician, **analytically treats the longitudinal impact of elongated bars.** For simplicity, the impacted bar is assumed to be made of the same material and to be of the same thickness as the impacting bar, but to have a different length {CAUCHY ⇨1826}.

(i) He shows that, except when the lengths are equal, a considerable fraction of the original energy takes the form of vibrations in the longer bar, so that the translational velocities after impact are less than those calculated by Sir Isaac NEWTON for "perfectly elastic" bodies. His solution reveals that after impact the short bar will take the initial velocity of the longer bar and become free of tension. ▪ This particular feature was later applied to the concept of the Hopkinson pressure bar {B. HOPKINSON ⇨1914}.

(ii) He attempts to calculate the stresses in the bars, but the convergence of his solution, an infinite trigonometric series, does not allow a definite result to be obtained. ▪ The problem was finally solved in terms of discontinuous functions by Hippolyte SÉBERT[818] and Pierre-Henri HUGONIOT, two French artillery officers. |
| 1867 | Polytechnikum Riga, Latvia | **August TOEPLER,**[819] a professor of chemistry and chemical technology, resumes his previous schlieren experiments {A. TOEPLER ⇨1864}. He displays his improved schlieren apparatus at the 4th International World Fair in Paris and **gives a public demonstration of the propagation of spark waves – *i.e.*, of weak shock waves.** |
| 1868 | Arica, Peru [now Arica, Chile] | On August 13, **a series of heavy seismic shocks begins which lasts almost six days, creating strong seismic sea waves (tsunamis).** The natural disaster – which later became known as the **"Arica Earthquake"** – kills about 70,000 people, mostly due to the destructive tsunamis.

(i) At Arica, a town at the Pacific coast, English naval officers observe that the precursor wave first provokes an increase in the sea level of about 2–3 meters. Then, after the water level drops again, an enormous tsunami occurs, with a height of up to 17 meters.

(ii) In Sydney harbor,[820] the self-registering tide gauge records the seismic sea waves as small perturbations – in total about 150 within three days – which are superimposed as sharp spikes on the ordinary smooth tidal oscillations {⇨Fig. 4.1–P}. |

---

[816] F.E. REUSCH: *Über eine besondere Gattung von Durchgängen in Steinsalz und Kalkspath.* Ann. Phys. **132** [II], 441-451 (1867); *Über die Körnerprobe am zweiachsigen Glimmer.* Ibid. **136** [II], 130-135 (1869).

[817] A.J.C. DE SAINT-VENANT: *Sur le choc longitudinal de deux barres élastiques de grosseurs et de matières semblables ou différentes et sur la proportion de leur force vive, qui est perdue pour la translation ultérieure. Et généralement sur le mouvement longitudinal d'un système de deux ou plusieurs prismes élastiques.* J. Math. Pures Appl. **12** [II], 237-376 (1867).

[818] H. SÉBERT and P.H. HUGONIOT: *Sur les vibrations longitudinales des barres élastiques dont les extrémités sont soumises à des efforts quelconques.* C. R. Acad. Sci. Paris **95**, 213-215, 278-281, 338-340 (1882); *Sur le choc longitudinal d'une tige élastique fixée par l'une de ses extrémités.* Ibid. **95**, 381-384 (1882).

[819] F.J. PISKO: *Optische, astronomische und geodätische Instrumente.* In: (K.u.K. Österreichisches Central-Comité, Hrsg.) *Officieller Ausstellungs-Bericht. Instrumente für Kunst und Wissenschaft auf der Welt-Ausstellung zu Paris im Jahre 1867.* Braumüller, Wien (1867), pp. 118-119.

[820] F. VON HOCHSTETTER: *Die Erdbebenfluth im Pazifischen Ocean vom 13. bis 16. August 1868, nach Beobachtungen an der Küste von Australien.* Sitzungsber. Akad. Wiss. Wien **60** (Abth. II), 818-823 (1869).

In the same year, **Ferdinand VON HOCHSTETTER,**[821] a German professor of geology, **begins to analyze measured arrival times of the seismic sea wave at remote coasts in the Pacific Ocean.**

(I) Addressing previous theories on the propagation of a "wave of translation" in deep water of uniform depth {AIRY ⇨1845; RUSSELL ⇨1845}, he finds that **the seismic sea wave propagates with the velocity of the tidal wave,** and confirms that its wave speed depends on water depth. For example, a wave propagated along the Chilean coast from Arica to Valdivia (distance 2,640 km) at about 527 km/h, but when crossing the deep ocean from Arica to Honolulu (10,350 km) it propagated with about 820 km/h.

(II) **He proposes to calculate the average depth of the ocean,** $h_{AV}$**, from the average velocity of the seismic sea wave,** $c_{av}$**,** by the simple relation

$$h_{av} = c_{av}^2 / g,$$

where $g$ is the acceleration due to gravity {DE LAGRANGE ⇨1781; AIRY ⇨1845}.

Nine years later, another great earthquake happened on the west coast of South America, only about 175 km south of Arica {Iquique Earthquake ⇨1877}. A comparison and analysis of the data collected from both events promoted tsunami research.

| | | |
|---|---|---|
| 1868 | *St. Petersburg, Russia* | On December 11, **the *St. Petersburg Declaration* is ratified,** which bans the use of explosive bullets below 400 grams in weight for purposes of war. It is signed by all European countries and North America. ▪ The Declaration of Saint Petersburg is the first formal agreement prohibiting the use of certain weapons in war. It was prompted by the invention in 1863 by Russian military authorities of a bullet which exploded on contact with a hard substance and whose primary object was to blow up ammunition wagons. In 1867, the projectile was modified so as to explode on contact with a soft substance. As such, the bullet would have been an inhuman instrument of war, and so the Russian Government, unwilling to use the bullet itself or to allow another country to take advantage of it, suggested that the use of the bullet be prohibited by international agreement. |
| 1868 | *University of Heidelberg, Germany* | **Gustav R. KIRCHHOFF,**[822] a German physicist and authority on natural sciences, **comments upon August TOEPLER's paper on the schlieren visualization of spark waves** {A. TOEPLER ⇨1867}. He writes to TOEPLER, "… without doubt the expression 'sound wave' as you use it is justified, and an aerial percussion [Germ. *Luftschütterung*] makes an impression on the ear even if it is of only very short duration but of sufficiently high intensity." |
| 1869 | *Krähenberg [a village near Zweibrücken], southwestern Palatinate, Germany* | On May 5, **a 16.5-kg stony meteorite, lenticular in shape, hits the area close to the village of Krähenberg.** The fall of the so-called **"Krähenberg meteorite,"** a chondrite, is accompanied by a cannon-like report, followed by a rumbling noise like distant thunder. ▪ The structure and appearance of this meteorite resembles the Ensisheim meteorite {⇨1492}.<br><br>In the same year, Christian L.O. BUCHNER,[823] a German high-school professor and author of the catalog *Die Meteoriten in Sammlungen* ("The Meteorites in Collections," Leipzig 1863), reports that the meteorite penetrated about 1 ft (0.3 m) deep into the ground and was immediately dug out, whereupon it was still found to be hot – an obvious example of aerodynamic heating. |

---

[821] F. VON HOCHSTETTER: *Über das Erdbebeben in Peru. Am 13. August 1868 und die dadurch veranlassten Fluthwellen im Pacifischen Ocean, namentlich an den Küsten von Chili und von Neu-Seeland.* Sitzungsber. Akad. Wiss. Wien **58** (Abth. II), 837-860 (1868); *Die Erdbebenfluth im Pazifischen Ocean vom 13. bis 16. August 1868 und die mittleren Tiefen dieses Oceans.* Ibid. **59** (Abth. II), 109-132 (1869).
[822] KIRCHHOFF' letter (dated April 7, 1868) is now kept in the Archives of the TU Dresden.
[823] C.L.O. BUCHNER: *Der Meteorstein von Krähenberg.* Ann. Phys. **137** [II], 176 (1869).

| 1869 | Royal Military Academy at Woolwich Arsenal, southeast London, England | Frederick A. ABEL,[824] a British professor of chemistry and ordnance chemist, **shows that unconfined high-explosive charges** (*e.g.*, guncotton, nitroglycerin, dynamite, mercury fulminate) **only burn if ignited by a flame or a hot wire, but they detonate if subjected to an impulsive force** such as that applied by a hammer blow, a detonator cap, or the impact of a projectile. ▪ **In this paper, ABEL was the first to use the term** *detonation* in the modern sense, since he describes "the detonation of guncotton…" Hitherto, the terms *explosion* and *detonation* – a term apparently used from the late 17th century onwards[825] – were applied interchangeably. |
|---|---|---|

In the same year, ABEL[826] **reports on his observation that one detonating dynamite cartridge can trigger another that is positioned in its vicinity.** He speculates that detonation is transmitted by means of some "synchronous vibrations" and states, "The vibrations produced by a particular explosion, if synchronous with those which would result from the explosion of a neighboring substance which is in a state of high chemical tension, will, by their tendency to develop those vibrations, either determine the explosion of that substance, or at any rate greatly aid the disturbing effect of mechanical force suddenly applied, while, in the case of another explosion which produces vibrations of different character, the mechanical force applied by its agency has to operate with little or no aid." ▪ However, Ernst MACH, refuting ABEL's "queer" hypothesis, correctly attributed this phenomenon to the mechanical effect of the shock wave {E. MACH & WENTZEL ⇨1885}.

| 1869 | Chair of Civil Engineering & Mechanics, University of Glasgow, Scotland, U.K. | William J.M. RANKINE,[827] a Scottish engineer and physicist, and one of the founders of the field of thermodynamics, **submits a paper to the Royal Society of London "on adiabatic waves,** that is waves of longitudinal disturbance in which there is no transfer of heat" **and on the problem of "how to determine the relations that must exist between the laws of the elasticity and heat of any substance, gaseous, liquid or solid, and those of the wave-like propagation of a finite longitudinal disturbance in that substance."** RANKINE's significant achievements can be summarized as follows: |
|---|---|---|

(i) Treating the shock wave as a one-dimensional discontinuity and assuming a dissipative fluid – *i.e.*, it is conductive but nonviscous – **he applies the laws of conservation of mass, momentum and energy to states far up- and downstream of the shock front,** thus obtaining three equations. ▪ They are equivalent to those of HUGONIOT in the case of a perfect gas, later referred to as the ***Rankine-Hugoniot equations*** {HUGONIOT ⇨1887}.

(ii) **He introduces the expression "adiabatic"** (from Greek, meaning "to pass through") to characterize a change in the volume and pressure of the contents of an enclosure without any exchange between the enclosure and its surroundings {RANKINE ⇨1859}.

(iii) **He introduces the term *adiabatic curve*** for a $p(v)$ diagram, which he obtained by plotting the pressure $p$ against the specific volume $v$ in the adiabatic equation. ▪ Shortly after, J. Willard GIBBS,[828] a U.S. theoretical physicist and chemist, proposed the expression "isentropic curve" or "isentrope," since the entropy remains constant in a reversible adiabatic process.

▸ For compressible flows with little or small flow turning (such as the ideal flow through a nozzle), the flow process is *reversible* and the entropy is constant.

▸ But across a shock wave, the changes in the flow properties are *irreversible* and the entropy of the entire system increases.

---

[824] F.A. ABEL: *(II) Contributions to the history of explosive agents* [abstract]. Proc. Roy. Soc. Lond. **18**, 395-397 (1869).
[825] *Etymologisches Wörterbuch des Deutschen* (ed. by W. PFEIFER). Verlag dtv, München (1999); *Duden* (ed. by G. DROSDOWSKI). *(VII) Etymologie*. Duden-Verlag, Mannheim *etc.* (1997).
[826] F.A. ABEL: *(XIV) Contribution to the history of explosive agents*. Phil. Trans. Roy. Soc. Lond. **159**, 489-516 (1869).
[827] W.J.M. RANKINE: *On the thermodynamic theory of waves of finite longitudinal disturbance* [read Dec. 16, 1869]. Phil. Trans. Roy. Soc. Lond. **160**, 277-286 (1870).
[828] J.W. GIBBS: *Graphical methods in the thermodynamics of fluids*. Trans. Connecticut Acad. **II**, 309-342 (1871–1873).

(iv) **Also addressing the rarefaction wave phenomenon, which he calls "sudden rarefaction," he refers to a discussion with Sir William THOMSON** and annotates in a footnote, "Sir William THOMSON has pointed out to the author, that a wave of sudden rarefaction, though mathematically possible, is an unstable condition of motion; any deviation from absolute suddenness tending to make the disturbance become more and more gradual. Hence the only wave of sudden disturbance whose permanency of type is physically possible, is one of sudden compression; and this is to be taken into account in connexion with all that is stated in the paper respecting such waves." ▪ Discussions on the possible existence of rarefaction shocks were later resumed by numerous researchers {JOUGUET ⇨1904; ZEMPLÉN ⇨1905; DUHEM ⇨1909; ZEL'DOVICH ⇨1946; DRUMMOND ⇨1957; KUTATELADZE ET AL. ⇨1978}.

(v) He also measures the ratio of specific heats $\gamma$ and finds that "$\gamma$ is nearly 1.41 for air, oxygen, nitrogen and hydrogen, and for steam-gas nearly 1.3."

(vi) Asked by the editor of the *Proceedings of the Royal Society* to also give credit to previous investigators of waves of finite longitudinal disturbance and to point out to what extent the results arrived at his paper are identical with previous research, RANKINE,[829] citing the works of POISSON, STOKES, AIRY, and EARNSHAW, claims that "the new results, then, obtained in the present paper may be considered to be the following:

▸ the conditions as to transformation and transfer of heat which must be fulfilled, in order that permanence of type may be realized, exactly or approximately;

▸ the types of wave which enable such conditions to be fulfilled, with a given law of the conduction of heat; and

▸ the velocity of advance of such waves.

The method of investigation in the present paper, by the aid of mass-velocity to express the speed of advance of a wave, is new, so far as I know; and it seems to me to have great advantages in point of simplicity."

| | | |
|---|---|---|
| 1870s | United States, France, Italy & Germany | **Underwater explosion studies are initiated by the naval forces from a number of maritime countries** – a consequence of the American Civil War (1861–1865), which had clearly demonstrated the importance of mines in naval warfare. ▪ Most technical reports on this subject were prepared by companies and governmental agencies and classified, although some were published, such as those containing first results |

▸ **on sea mine explosions and associated surface phenomena,** performed for the American Navy by Henry L. ABBOT {⇨1881}, for the French Navy by Arthur MOISSON[830] and Joseph-Marie-Suzanne AUDIC,[831] and for the Italian Navy by Timoteo BERTELLI;[832]

▸ **on torpedo development and application,** performed for the British Navy by Sir George E. ARMSTRONG,[833] and for the German Navy by Franz PLACH;[834] and

▸ **on gas globe oscillation,** performed for the German Navy by G.F. Rudolph BLOCHMANN {⇨1898}.

---

[829] W.J.M. RANKINE: *Supplement to a paper "On the thermodynamic theory of waves of finite longitudinal disturbance."* Phil. Trans. Roy. Soc. Lond. **160**, 287-288 (1870).

[830] A. MOISSON: *Des explosions au sein de l'eau*. Rev. Maritime et Coloniale **52**, 744-770 (1877); Ibid. **53**, 86-120 (1877).

[831] J.M.S. AUDIC: *Étude sur les effets des explosions sous-marines*. Rev. Maritime et Coloniale **54**, 561-601 (1877).

[832] T. BERTELLI: *Studi comparativi fra alcune vibrazioni meccaniche artificiali e le vibrazioni sismiche*. Annali dell'Ufficio Centrale Meteorologico e Geodinamico Italiano **X** [II], Parte IV, 5-44 (1892).

[833] G.E. ARMSTRONG: *Torpedoes and torpedo-vessels*. Bell, London (1896).

[834] F. PLACH: *Die gepresste Schiesswolle. Eine Abhandlung über die Beurtheilung, Verwendung und Handhabung dieses Explosiv-Preparates für Torpedo- und Seeminen-Lehrcurse…* E. Schaff, Pola (1891).

| | | |
|---|---|---|
| 1870 | *Barisal [a small port at the mouth of a river of the same name], northern shore of the Bay of Bengal, India* | **In the Proceedings of the Asiatic Society of Bengal,**[835] **two letters about remarkable sounds much like gun reports which British officials at Barisal had previously called "Barisal Guns" are published.** One writer, Mr. F.H. PELLEW, explains the sound phenomenon as being the result of the oblique superposition of two peaks. He writes, "In regard to the *Barisal Guns*, my notion was that waves of a length of a mile or two each, advancing obliquely from the S.S.W. would break successively on the coast from W. to E. To a person close by, the sound of each wave would be somewhat continuous; but to a person 40 or 60 miles [64 or 97 km] off, if the wave broke simultaneously, the sound would be a boom like that of a gun, because both extremities of the wave would be nearly at the same distance from the hearer as the center … This phenomenon often occurred, the wave rising to an immense height and breaking over a mile or two of beach at one moment." ▪ The curious phenomenon is called **"Barisal guns"** (or "fog guns") in England, **"mist pouffers"** in Belgium and The Netherlands, **"brontidi"** (or "baturlio marinas") in Italy, and **"Seeschießen"** in Germany. Obviously, it was first heard near the town of Barisal and in the channels of the Brahmaputra Delta when the tide rose, but later also along the Belgian coast, in northern France, and in some parts of Scotland and Germany.<br><br>In 1895, the discussion of this mysterious phenomenon of thundering noises was resumed by one of Charles DARWIN's sons {G.H. DARWIN ⇨1895}. Some observers believed it to originate inside the Earth, comparing the noise "to the shock which the internal fluid mass might give to the Earth's crust." However, it is highly improbable that such movements are the cause. No correlations between the noise from Barisal guns and earthquakes or thunderstorms have been reported; thunder is not usually heard more than a few kilometers away. |
| 1870 | *Chair of Organic Chemistry, Collège de France, Paris, France* | **P.E. Marcellin BERTHELOT,**[836] a French chemist, **defines the "strength" of condensed and gaseous explosives** which primarily depends on the volume of the gaseous detonation products and their temperature.<br><br>(i) **He also emphasizes the role of a mechanical shock which propagates from "layer to layer"** [French "*couche par couche*"] **during detonation,** thus arriving independently at the same conclusion as BUNSEN {⇨1867} and anticipating an important assumption of the later Chapman-Jouguet (CJ) theory {CHAPMAN ⇨1899; JOUGUET ⇨1904/1906}.<br><br>(ii) He writes, "In order to transmit the transformation of a detonating bulk which is not subjected in all parts to the same action, it is necessary that the same conditions of temperature, pressure, *etc.* which have provoked the phenomenon in one point propagate successively, layer to layer, through all parts of the bulk…" |
| 1870 | *Union College, Schenectady, New York* | Edward Everett HALE,[837] an American author and clergyman, **first describes an artificial Earth satellite in his short story *The Brick Moon*,** telling of a sphere 200 ft (60 m) in diameter built of bricks to "stand fire, very, very well," that is due to be launched into an orbit 4,000 miles (6,400 km) high.<br><br>(i) He writes: "If from the surface of the Earth, by a gigantic peashooter, you could shoot a pea upward from Greenwich, aimed northward as well as upward; if you drove it so fast and far that when its power of ascent was exhausted, and it began to fall, it should clear the Earth, and pass outside the North Pole; if you had given it sufficient power to get it half round the Earth without touching, that pea would clear the Earth forever. It would continue to rotate above the North Pole, above the Feejee Island place, above the South Pole and Greenwich, forever, with the impulse with which it had first cleared our atmosphere and attraction. If only we could see that pea as it revolved in that convenient orbit, then we could measure the longitude from that, as soon as we knew how high the orbit was…" |

---

[835] Proc. Asiatic Soc. Bengal (Calcutta), pp. 289-291 (Nov. 1870).
[836] P.E.M. BERTHELOT: *Sur la force des poudres et des matières explosives*. C. R. Acad. Sci. Paris **71**, 619-625, 667-677, 709-728 (1870).
[837] E.E. HALE: *The brick Moon, and other stories*. Published in 1870 and 1871 in the magazine *Atlantic Monthly* (Boston, MA).

(ii) Specifying the necessary properties of this "pea" in more detail, he says, "Large – that it might be seen far away by storm-tossed navigators; and light – that it might be the easier blown four thousand and odd miles into the air; lest it should fall on the heads of the Greenlanders or the Patagonians; lest they should be injured and the world lose its new moon. But, of course, all this lath-and-plaster had to be given up. For the motion through the air would set fire to this moon just as it does to other aerolites, and all your lath-and-plaster would gather into a few white drops, which no Rosse telescope even could discern … It must stand fire well, very well. Iron will not answer. It must be brick; we must have a Brick Moon."

| 1871 | *Private laboratory at London, England* | Richard Leach MADDOX,[838] a British physician and experimental photographer, **prepares emulsions of silver bromide in essentially the same manner as that used for making colloidal emulsions, but replaces collodion by gelatin coated on a glass plate** – the so-called **"gelatin dry plate."**[839] ▪ His invention initiated the manufacture of sensitized materials. The gelatin dry plate, which was commonly used in the period 1880–1920, could be factory-produced and stored for months (and eventually even years) before use. |
|---|---|---|

Improvements to this new photographic process were made by the British manufacturer and experimental photographer Charles BENNETT (1878) in particular, who discovered that the sensitivity of the gelatin-silver bromide emulsion can be increased by a prolonged heat treatment, known as "ripening."[840] Further contributions were made by the German photochemist Hermann Wilhelm VOGEL (1873), the British photochemist Sir William DE WIVELESLIE ABNEY (1874), the Belgian chemist Désiré VAN MONKHOVEN (1879), and the Austrian photochemist Josef Maria EDER (1880). Their combined effort eventually resulted in the highly sensitive **"photo-gelatin dry plate"** which had improved upon the speed of previous wet plates by a factor of two by 1880, a factor of ten by 1885 and a factor of fifty by 1900. Coated plates were on sale by 1876; the first orthochromatic plates were marketed in 1884.

High-speed gelatin dry plates became one of the basic requirements for obtaining the first photographs of ballistic head waves {E. MACH & SALCHER ⇒1886} and shock waves emerging from electric sparks {E. & L. MACH ⇒1889}.

| 1871 | *Private laboratory at London, England* | Hermann SPRENGEL,[841] a retired German-born British chemist and inventor of the mercury vacuum pump (1865), **introduces new types of mining explosives which are prepared by mixing an oxidizing substance with a combustible one.** |
|---|---|---|

(i) About his motivation he writes, "Led by the idea that (as a rule), an explosion is a sudden combustion, I have submitted a variety of mixtures of oxidizing and combustible agents to the violent shock of a detonating cap. These mixtures were made in such proportions, that their mutual oxidation and deoxidization should be theoretically complete."

(ii) He concludes, "Lastly, though not least, to avoid the danger of a spontaneous explosion of these compounds during their manufacture, storage, and transportation, **we can keep the oxidizing agent apart from the combustible agent, until their chemical combination is to take place in obedience to the will of man.** I am aware that this way has been pointed out before, and has been abandoned as impracticable; but formerly both the existing and the combustible agents were solids. Now, however, we have two liquids, or a liquid and a solid, rendering their admixture easier."

These so-called **"Sprengel explosives,"** which consist of an oxidizing agent (such as chlorates, nitrates and nitric acid) and a fuel (such as benzene, nitronaphthalene and nitrobenzene),

---

[838] R.L. MADDOX: *An experiment with gelantino-bromide*. Brit. J. Photogr. **18**, 422-423 (1871).
[839] W. BAIER: *Quellendarstellungen zur Geschichte der Photographie*. VEB Fotokinoverlag, Halle (1964), pp. 262-264, 410.
[840] *History of photography in Brighton: 'Dry plate' photography*; http://www.spartacus.schoolnet.co.uk/DSphotodry5E1.htm.
[841] H. SPRENGEL: *Safety explosive compounds*. Brit. Patent No. 921 (April 6, 1871), and *Explosive compounds*. Brit. Patent No. 2,642 (Oct. 5, 1871); *On a new class of explosives which are non-explosive during their manufacture, storage, and transport*. J. Chem. Soc. (Lond.) **11**, 796-808 (1873).

were mixed together at the blasting site just before use, which eliminated the dangers associated with the transportation and storage of live explosives for a prolonged time. They were mostly used in Europe but also in the United States, such as in the spectacular blast in New York harbor {Hell Gate ⇨ 1885}.

| | | |
|---|---|---|
| 1871 | *Aeronautical Society [now Royal Aeronautical Society], London, England* | **Francis Herbert** W<small>ENHAM</small>, a British marine engineer and a pioneer of aeronautics who is interested in the flight of birds, proposes to test wings of various shapes in a wind tunnel. In collaboration with John B<small>ROWNING</small>, he **designs and builds the world's first wind tunnel.** A fan-blower upstream of the model, driven by a steam engine, propels air down a tube 12 ft (3.65 m) long and 18 in.$^2$ (116 cm$^2$) in cross-section towards the model. **He conducts extensive studies on cambered wings and aspect ratios, and discovers the advantages of high-aspect-ratio airfoils.** ▪ W<small>ENHAM</small>, a founding member of the Aeronautical Society of Great Britain, had already delivered the first lecture to the Society in London on June 27, 1866, entitled *On Aerial Locomotion*[842] – a milestone in early aeronautical research, and today an aviation classic. One of his suggestions was that aerial research should begin with a thorough study of "kites" (meaning small winged structures or model flying machines). |
| 1872 | *British expedition of HMS Challenger* | On December 7, **beginning of a four-year oceanographic exploration**[843] **carried out by the British Admiralty and supported by the Royal Society of London.** The British Naval research vessel HMS *Challenger* sails the Atlantic and Pacific Oceans performing scientific research such as measuring currents, depths {Iquique Earthquake ⇨ 1877}, and ocean basin contours {⇨ Fig. 4.1–O}. The Scottish physicist **Peter G.** T<small>AIT</small> {⇨ 1879} **participates and measures deep-sea temperatures.** ▪ These readings required corrections because of the great pressures to which the thermometers had been subjected. This task aroused T<small>AIT</small>'s interest in high pressure physics, particularly in the compressibilities of water, glass and mercury. |
| 1872 | *Musconetcong Tunnel, New Jersey* | **Construction is begun on the Musconetcong Tunnel,** a 1.491-km-long railroad tunnel located 24 km east of Phillipsburg in the state of New Jersey. ▪ It was the first major project in the United States in which dynamite {N<small>OBEL</small> ⇨ 1867} was successfully used. It enabled driving to proceed at a pace of 41 meters per month.[844]<br><br>Henry Sturgis D<small>RINKER</small>,[845] then a young U.S. mining engineer, was supervisor. He recorded his experience in his book *Tunneling, Explosive Compounds and Rock Drills* (1878). |
| 1872 | *Chair of Theoretical Physics, University of Graz, Austro-Hungarian Empire* | Ludwig E. B<small>OLTZMANN</small>,[846] an eminent Austrian physicist, **publishes his famous paper (entitled in translation)** *Further Studies on the Thermal Equilibrium Among Gas Molecules.* Assuming that the gas has a low density (so only binary collisions need be considered), and that the molecules are completely uncorrelated before the collision ("molecular chaos"), he derives an equation which describes the macroscopic quantities (such as mass, velocity and energy) of a fluid system in terms of a distribution function – the so-called **"Boltzmann equation"** – which satisfies the fluid conservation equations. ▪ The solution to the classical Boltzmann equation in terms of this distribution function (an integro-differential equation) is a very complex problem, and approaches to solving it in relation to applications in physical gas |

---

[842] F.H. W<small>ENHAM</small>: *On aerial locomotion and the laws by which heavy bodies impelled through air are sustained.* Aeronaut. Soc. Report **I**, 10-40 (1866); http://invention.psychology.msstate.edu/library/Wenham/WenhamLocomotion.html.

[843] P.G. T<small>AIT</small>: *Voyage of HMS »Challenger« Report.* H.M.S.O., London (1888).

[844] D.R. T<small>ONE</small>: *Blasting.* In: (B.S. C<small>AYNE</small>, ed.) *The encyclopedia Americana.* Americana Corp., New York (1974), vol. 4, p. 66.

[845] H.S. D<small>RINKER</small>: *Tunneling, explosive compounds, and rock drills* [Giving the details of practical tunnel work; the constituents and properties of modern explosive compounds; the principles of blasting; the history of … machine rock-drilling; and detailed descriptions of the various rock-drills and air-compressors in use … with a history of tunneling from the reign of R<small>AMSES</small> II. to the present time]. Wiley, New York (1878).

[846] L. B<small>OLTZMANN</small>: *Weitere Studien über das Wärmegleichgewicht unter Gasmolekülen.* Sitzungsber. Akad. Wiss. Wien **66** (Abth. II), 275-370 (1872).

dynamics and rarefied gas dynamics have been addressed by David HILBERT, Sydney CHAPMAN, David ENSKOG, and David BURNETT.[847]

Richard BECKER {⇨1921} was apparently the first who calculated the thickness of a shock wave using the Boltzmann equation. In 1951, Harold M. MOTT-SMITH,[848] a staff scientist at the U.S. Atomic Energy Commission (Washington, DC), solved the Boltzmann equation for a strong shock wave by assuming the distribution function to consist of a sum of two Maxwellian terms with temperatures and mean velocities corresponding to the subsonic and supersonic streams. He reported that the shock thickness found is considerably greater than that given by previous theories. The problem of finding an exact shock wave solution of the Boltzmann equations is one of the most challenging problems in kinetic theory.[849] Discrete velocity models achieved great success in describing many aspects of rarefied gas dynamics, a subject which gained increasing importance with the advent of satellites {Sputnik ⇨1957} and manned space flights {GAGARIN ⇨1961}.

| | | |
|---|---|---|
| 1872 | Trinity College, Cambridge, England | **John HOPKINSON,**[850] a British engineer and physicist, **measures the strength of steel wires when they are suddenly stretched by a falling weight.** He observes that the minimum height from which a weight has to be dropped to break the wire is independent of the size of the weight and depends only on the velocity. He explains this surprising result in terms of the propagation of elastic waves up and down the wire. ▪ His experiments were later repeated by his son {B. HOPKINSON ⇨1905}. |
| 1872 | Royal Academy of Sciences, Brussels, Belgium | **Louis H.F. MELSENS,**[851] a Belgian chemist and physician, reports on the severe crater-like nature of wounds inflicted by the French Army's new high-speed bullets, as used in the recent Prussian-French War (1870–1871), and **speculates that they were caused not by explosive bullets banned by the St. Petersburg Declaration** {⇨1868}, **but rather by air condensation phenomena in front of the projectile.**<br><br>Nine years later, MELSENS returned to the same subject in a lecture presented at Paris. His hypothesis that the existence of compressed air may cause mechanical, explosion-like effects upon striking the human body inspired Ernst MACH, who attended his lecture to investigate these possible ballistic phenomena experimentally.[852] However, MACH had to wait almost six years before his photographic technique was mature enough to catch the motion of a high-speed bullet in supersonic flight. **MELSENS' speculation was indeed confirmed by the discovery of the head wave, a thin shell of compressed air in front of a supersonic projectile** {E. MACH & SALCHER ⇨1886}. |
| 1872 | Elswick Ordnance Co., Elswick, Newcastle-upon-Tyne, England | **Andrew NOBLE,**[853] a Scottish physicist and gunnery expert, **invents the *crusher gauge*** – a modification and improvement of RODMAN's indentation gauge {RODMAN ⇨1857}, and a byproduct of his investigation into the behavior of explosives and artillery. The new gauge will make his name famous. ▪ In the early period of shock and explosion research, the crusher gauge was the instrument most commonly used to evaluate the maximum pressure of high-rate thermodynamic phenomena such as explosions and detonations. Since the gauge is simple in construction, inexpensive and insensitive to electromagnetic radiation, it was subsequently modified in many ways: for example, in order to measure maximum pressures generated in un- |

---

[847] S. CHAPMAN and T.G. COWLING: *The mathematical theory of nonuniform gases.* Cambridge University Press, London (1952), chap. 4.
[848] H.M. MOTT-SMITH: *The solution of the Boltzmann equation for a shock wave.* Phys. Rev. **82**, 885-892 (1951).
[849] G. SPIGA and S. OGGIONI: *On shock wave solutions in extended discrete kinetic theory.* SIAM J. Appl. Math. **52**, No. 1, 105-111 (Feb. 1992).
[850] J. HOPKINSON: *On the rupture of iron wire by a blow.* Proc. Manch. Lit. Phil. Soc. **11**, 40-45, 119-121 (1872).
[851] L.H.F. MELSENS: *Note sur les plaies des armes à feu.* J. Méd. Chir. Pharm. (Soc. Roy. Sci. Méd. Natur. Bruxelles) **54**, 421- 437, 513-527 (1872); Ibid. **55**, 21-29, 121-126, 217-222, 293-303 (1872).
[852] E. MACH: *Über Erscheinungen an fliegenden Projektilen.* In: E. MACH: *Populär-wissenschaftliche Vorlesungen.* J. Barth, Leipzig (1903), p. 352.
[853] A. NOBLE: *On the tension of fired gunpowder.* Proc. Roy. Inst. **6**, 274-283 (1872); *Sur la force explosive de la poudre à canon.* Rev. Scient. France & Etrang. **48** [II], 1125-1141 (1872).

derwater explosions, Lt.-Col. Henry L. ABBOT {⇒1881}, a U.S. engineer, used crusher gauges in which a steel piston acted on a small lead cylinder fixed on a massive support.[854]

The idea saw a renaissance in World War II: **the physicist Sir William G. PENNEY**, a member of the joint research team at Los Alamos and known as the "British Oppenheimer," **used his "five-gallon-can blast pressure gauges" to map the overpressure in the Mach stem region of a nuclear explosion.** Blast pressures were computed from how much the cans were crushed.[855]

| | | |
|---|---|---|
| 1872 | Somerset House, The Royal Society of London, England | At a meeting of the Society, **Robert H. SCOTT**,[856] a British meteorologist, **and Sir William GALLOWAY**, a British mining engineer, **discuss the connection between colliery explosions and weather.** They compare actual dates of explosions for the years 1868–1870 with meteorological records and show that, "out of 550 explosions, 48% might be attributed to the state of the barometer, 22% to the state of the thermometer, and 30% remained unaccounted for on meteorological grounds." Since the natural temperature of a mine increases with depth, the warm weather might stop natural ventilation. On the other hand, **the largest group of accidents were reported when a serious break in the barometer records occurred after a period of calm weather.** |
| 1873 | Woolwich Arsenal, southeast London, England | **Frederick A. ABEL**,[857] a British ordnance chemist, **and his assistant Edwin O. BROWN use the Noble chronograph** {NOBLE & ABEL ⇒1875} **to measure the detonation velocity of guncotton at around 20,000 ft/s** (6,096 m/s). They state, "Recent experiment has shown that the rapidity with which gun-cotton detonates is altogether unprecedented, the swiftness of the action being truly marvelous. Indeed, with the exception of light and electricity, the detonation of gun-cotton travels faster than anything else we are cognizant of…" |
| 1873 | Royal Astronomical Society (RAS), London, England | **Richard PROCTOR**,[858] a British astronomer and honorary secretary of the RAS, **speculates that lunar craters were produced by the impact of meteorite-like bodies with the Moon.** He proposes that "we should look for the existence of a zone of small bodies like the asteroids … I have based my opinion that the Solar System had its birth, and long maintained its fires, under the impact and collisions of bodies gathered in from outer space." ▪ **The dominant geological process that affects asteroids is collision with other asteroids, with most of these collisions producing craters.** Modern biologists assume that a massive biological extinction occurred in the Cretaceous-Tertiary era due to the impact of an asteroid with the Earth {⇒c.65 Ma ago; ALVAREZ ⇒1978}.<br><br>Several hundred thousand asteroids – also called "minor planets" – have been discovered so far {PIAZZI ⇒1801}, most of which orbit the Sun in a belt between the orbits of Mars and Jupiter, although some asteroids are located in the orbit of Jupiter and some other asteroids have been detected as far away as the orbit of Saturn. |
| 1873 | Dépôt Central des Poudres et Salpêtres des Manufactures de l'État, Paris, France | **Louis ROUX**[859] **and Emile SARRAU,** two French military engineers, **differentiate between an "explosion of the first kind"** (*detonation*) **and an "explosion of the second kind"** (*deflagration*), thus confirming previous observations {ABEL ⇒1869}.<br><br>In the same year, **ROUX**[860] **and SARRAU carefully determine the energy contents of gunpowder and various high explosives** by carrying out explosions in a water bath and measuring the temperature increases due to the explosions. **They determine a value of 310,590 mkg\*** |

---

[854] R.H. COLE: *Underwater explosions*. Dover Publ., New York (1948), pp. 147-148, 427.
[855] F.H. SHELTON: *Reflections of a nuclear weaponeer*. Shelton Enterprise Inc., Colorado Springs (1988), pp. **2**:38-39.
[856] R.H. SCOTT and W. GALLOWAY: *The connection between colliery explosions and weather*. Nature **5**, 504 (1872).
[857] F.A. ABEL and E.O. BROWN: *The rapidity of detonation*. Nature **8**, 534 (1873).
[858] R. PROCTOR: *Condition of the Moon's surface*. Quart. J. Sci. **3**, 29-55 (1873).
[859] L. ROUX and E. SARRAU: *Expériences sur les effets de la dynamite*. C. R. Acad. Sci. Paris **76**, 1089-1092 (1873).
[860] L. ROUX and E. SARRAU: *Sur la chaleur de combustion des matières explosives*. C. R. Acad. Sci. Paris **77**, 138-142 (1873); *Recherches expérimentales sur les matières explosives*. Ibid. **77**, 478-481 (1873).

for the theoretical work done by 1 kg of gunpowder, which is much greater than previously reported data, *e.g.*,
- 67,410 mkg* {BUNSEN & SCHISCHKOFF ⇒1857};
- 88,157 mkg* found by Joseph STADLER[861] (1866); and
- 161,500 mkg* {BERTHELOT ⇒1870}.

For 1 kg of dynamite (75/25 nitroglycerin/kieselguhr), they determine a theoretical value for the work done of 548,250 mkg*.[862] • According to Franz VON RZIHA,[863] a professor of railroad and tunnel construction at the Technische Hochschule Wien, the results obtained by ROUX and SARRAU were essentially confirmed in practical blasting operations.

| | | |
|---|---|---|
| 1873 | University of Leiden, The Netherlands | Johannes Diederik VAN DER WAALS, a self-educated Dutchman, **obtains his doctorate for a thesis entitled *Over de Continuiteit van den Gas- en Vloeistoftoestand* ("On the Continuity of the Gaseous and Liquid State"),** which immediately marks him out as one of the foremost physicists. • He introduced two parameters representing size and attraction into the ideal gas law, and worked out a more exact formula for the equation of state of real fluids by decreasing the specific volume by a certain constant amount referred to as the covolume. He received the 1910 Nobel Prize for Physics for this outstanding contribution to the physics of fluids, particularly for his work on the equation of state for gases and liquids.[864]<br><br>The so-called "van der Waals equation" which takes the effect of intermolecular attractions at high densities into account as well as the reduction in effective volume due to the actual volume of the molecules **has been used widely to interpret experiments in which the pressure is measured after an explosion in a closed bomb.**[865] |
| 1874 | Realschule Kaschau, Austro-Hungarian Empire | Károly ANTOLIK,[866] a Hungarian schoolmaster, **publishes his "soot method" and uses this unique method to record strange interference patterns in the vicinity of crooked gliding spark discharges** {⇒Fig. 4.5–C}. He observes that "conically shaped branches" [Germ. *kegelartige Ausläufer*] originate from the concave parts of a spark path which, however, disappear when the discharge occurs in a vacuum. • ANTOLIK explained this phenomenon by the behavior of the gliding spark itself, which prefers to follow a straight rather than a crooked path in vacuum. In reality, however, **ANTOLIK's soot figures are the very first records of irregular interactions of shock waves in air** {E. MACH & WOSYKA ⇒1875}. |
| 1874 | Paris, France | Marcel DEPREZ,[867] a French engineer, **reports on his new chronograph construction:** a cylinder to receive records is driven at high speed, obtaining a surface velocity of 4–5 m/s. The velocity is determined by means of an electrically driven tuning fork, with the traces read by means of a vernier gauge. A mercury speed indicator of the Ramsbottom type enables the rotation to be continuously controlled.[868] • The so-called **"Deprez chronograph"** was used in early studies measuring detonation velocities in gaseous mixtures {MALLARD & LE CHÂTELIER ⇒1881; BERTHELOT & VIEILLE ⇒1882}. |

---

[861] J. STADLER: *Über die Wirkung des Pulvers beim Sprengen.* Österr. Z. Bergwesen **14**, 33-38 (1866).
[862] In the old Technical System of Measurement the unit of force was 1 kg* (now 1 kp); and an energy of 1 kcal corresponded to 427 mkg*.
[863] F. VON RZIHA: *Über die mechanische Arbeit der Sprengstoffe.* Dingler's Polytech. J. **262**, 128-134 (1886).
[864] J.D. VAN DER WAALS: *The equation of state for gases and liquids.* In: *Nobel lectures in physics, 1901–1921.* Elsevier, Amsterdam (1967).
[865] C. CRANZ: *Lehrbuch der Ballistik: II. Innere Ballistik.* Springer, Berlin (1926), pp. 40, 107.
[866] K. ANTOLIK: *Das Gleiten elektrischer Funken.* Ann. Phys. **151** [II], 127-130 (1874).
[867] M. DEPREZ: *Études sur les chronographes électriques et recherches sur l'étincelle d'induction et les électro-aimants.* C. R. Acad. Sci. Paris **78**, 1427-1430 (1874); *Perfectionnement aux chronographes électriques et recherches sur les électro-aimants.* Ibid. **78**, 1562-1565 (1874).
[868] A. FAVARGER: *L'électricité et ses applications à la chronométrie.* L.-É Privat, Genève (1886).

| 1874 | London, England | **William GALLOWAY,**[869] an inspector of mines in Scotland, **clearly establishes that a blast wave forces the flame of the safety-lamp** {DAVY ⇨1815} **through the mesh of its gauze, thus leading to the ignition of the explosive gas mixture surrounding the latter.** • His studies, carried out both in the laboratory and in coal-pits, were the first to reveal that blast waves produced either by the firing of a shot or especially by a sharp explosion where tamping is shot out of a hole are very effective, even at considerable distances.
| | | In the following years, GALLOWAY[870] investigated between 60 and 80 cases of firedamp explosions in Scotland. He thought that coal dust, both alone and with firedamp, was explosive {FARADAY & LYELL ⇨1845}. His theory that coal dust played an important part in explosions was later confirmed by his experiments in South Wales.
| 1875 | Free Church Training College for Teachers, Glasgow, Scotland, U.K. | **John KERR,**[871] a Scottish physicist, **observes that certain transparent dielectric media, particularly liquids** (*e.g.*, nitrobenzene), **become birefringent when placed in strong electric fields.** • This **"Kerr electro-optic effect"** which rotates the plane of polarization was successfully used as an optical high-speed shutter in early film and TV equipment (KAROLUS 1923 & 1925), as well as in high-speed ballistics {CRANZ, SCHARDIN & KUTTERER 1932–1933; FÜNFER ⇨1940; MÜLLER ⇨1957}, in high-speed diagnostics[872] (C.J. CRANZ, F.W. DEUTSCH & R.E. KUTTERER 1940), and in early high-power pulsed laser systems for Q-switching {MCCLUNG & HELLWARTH ⇨1963}.
| | | In 1967, applications of the Kerr cell to high-speed diagnostics were reviewed by Werner MÜLLER,[873] a German applied physicist. The use of the high-voltage Kerr cell as a rapid shutter in high-speed photography has since been superseded by ultrafast controllable image converter tubes and microchannel plates. The *Pockels cell*, which is based upon the change of birefringence in some crystals, is a shutter that works at a lower voltage than a Kerr cell {POCKELS ⇨1893}, and it is widely used for Q-switching in pulsed laser systems.
| 1875 | Chair of Experimental Physics, German Charles University, Prague, Bohemia, Austro-Hungarian Empire | **Ernst MACH,**[874] an Austrian professor of natural philosophy, assisted by his student **Jaromir WOSYKA, repeats ANTOLIK's soot pattern experiments** {ANTOLIK ⇨1874}. Curiously, MACH dives into a completely new area of research related to puzzling supersonic phenomena, starting with the nonlinear reflection and interaction properties of shock waves – one of the most challenging problems of shock wave physics:
| | | (i) MACH begins his investigations by (correctly) assuming that the soot pictures are indeed of acoustic and not of electric origin, as hitherto speculated on by ANTOLIK.
| | | (ii) By reducing ANTOLIK's complex zigzag gliding spark geometry to a single V-shaped gliding spark, but otherwise leaving his soot method unchanged, MACH records the funnel-like trajectories of the triple points of two obliquely interacting shock waves – later termed ***Mach funnels*** {⇨Fig. 4.5–D}.
| | | (iii) Using an ingenious arrangement of two oppositely facing V-shaped gliding sparks, he generates two Mach funnels that simultaneously interact with each other, each producing a new shock wave – phenomena later termed ***Mach disks***. These two Mach disks collide head-on, producing along the axis of symmetry a fine line of piled-up soot {⇨Fig. 4.5–D}.[875]

---

[869] W. GALLOWAY: *Experiments with safety-lamps.* Proc. Roy. Soc. Lond. **22**, 441-451 (1874).
[870] W. GALLOWAY: *On the influence of coal dust in colliery explosions.* Proc. Roy. Soc. Lond. **24**, 354-372 (1876); Ibid. **28**, 410-421 (1879); Ibid **32**, 454-455 (1881); Ibid. **33**, 437-445, 490-495 (1882).
[871] J. KERR: *On a new relation between electricity and light; dielectrified media.* Phil. Mag. **50** [IV], 337-348, 446-458 (1875).
[872] H. SCHARDIN: *Carl CRANZ als Mitbegründer der Kurzzeitphotographie.* Proc. 4th Int. Congress on High-Speed Photography [Cologne, Germany, Sept. 1958]. In: (O. HELWICH and H. SCHARDIN, eds.) *Kurzzeitphotographie.* Helwich, Darmstadt (1959), pp. 1-8.
[873] W. MÜLLER: *Elektro-optische Verschlüsse.* In: (K. VOLLRATH and G. THOMER, eds.) *Kurzzeitphysik.* Springer, Wien etc. (1967), chap. *Der Kerr-Zellenverschluß,* pp. 209-258.
[874] E. MACH and J. WOSYKA: *Über einige mechanische Wirkungen des elektrischen Funkens.* Sitzungsber. Akad. Wiss. Wien **72** (Abth. II), 44-52 (1875).
[875] F. SCHULTZ-GRUNOW: *Über die MACH'sche V-Ausbreitung.* ZAMM **28**, 30-31 (1948).

(iv) Based on these observations, MACH arrives at the following important conclusion: "It should be pointed out that ANTOLIK's simple and ingenious method of preliminary tracing of the spark enables various applications in the field of acoustics, because it can be used to create intense sound waves with an arbitrary initial shape." ▪ **MACH and WOSYKA were the first to systematically study the irregular reflection of shock waves** – later be termed *Mach reflection* or the *Mach effect* {VON NEUMANN ⇨ 1943}. Alois DE WAHA,[876] a professor of mathematics and physics at the Athénée de Luxembourg, and Mathias SCHULLER,[877] a professor at the Polytechnicum Budapest, performed similar soot experiments. Both had duplicated some of ANTOLIK's and MACH's results using explosives instead of electric sparks. Unfortunately, SCHULLER did not consider his results to be important enough for publication.

**The superposition of Mach funnels that periodically increase and decrease in width plays an important role in the propagation of gaseous detonations, both in two dimensions** (*e.g.*, in channels) **and in three dimensions** (free-field). For example, the two families of incident and reflected (or transverse) detonation waves form a typical cell pattern on a soot-coated channel wall {DENISOV & TROSHIN ⇨ 1959; SHCHELKIN & TROSHIN ⇨ 1963; SCHULTZ-GRUNOW ⇨ 1969; ⇨ Fig. 4.16–U}.

The soot method, although very simple in principle, is somewhat tricky to use in practice, particularly when attempting to provide a homogeneous and well-adhering soot layer reliable enough to obtain a high spatial resolution and a wide dynamic range for pressure recording. However, by increasing the adhesion of the soot layer on the glass plate, it is even possible to record *double Mach reflection* {WHITE ⇨ 1951}, which results in two concentric Mach funnels {⇨ Fig. 4.5–F}.[878]

| | | |
|---|---|---|
| 1875 | Private laboratory at Paris, France | Alfred B. NOBEL,[879] a Swedish chemist and inventor, **makes a far-reaching discovery when dissolving collodion cotton in nitroglycerin.** This act results in a gelatinous (jelly-like) mass which produces an even more powerful blast than that from dynamite {A. NOBEL ⇨ 1867}. ▪ NOBEL exploited his invention commercially and produced a number of very effective blasting explosives which were used in mining, hard-rock tunneling and underwater blasting, such as *blasting gelatin* (nitroglycerin gelatinized with nitrocellulose), *gelatin dynamite* (blasting gelatin doped with active ingredients) and *ballistite* {A. NOBEL ⇨ 1887}. |
| 1875 | Somerset House, London, England | At a meeting of the Royal Society for the Encouragement of Arts, Manufacture and Commerce, Alfred NOBEL[880] **reads his paper entitled** *Modern Blasting Agents*.<br><br>(i) He gives information regarding his invention of dynamite {A. NOBEL ⇨ 1867} and the difficulties involved in its practical use, stating, "The concentration of power, velocity of explosion, and immunity from danger, are the three points on which mainly depend the success or non-success of a new explosive substance."<br><br>(ii) Speaking of gunpowder he says, "That old mixture possesses a truly admirable elasticity which permits its adaptation to purposes of the most varied nature. Thus, in a mine it is wanted to blast without propelling; in a gun to propel without blasting; in a shell it serves both purposes combined; in a fuse, as in fireworks, it burns quite slowly without exploding. Its pressure exercised in those numerous operations, varies between 1 oz (more or less) to the square inch [0.9 bar] in a fuse, and 85,000 pounds to the square inch [5,862 bar] in a shell. But like a servant for all work, it lacks perfection in each department, and modern science armed with better tools, is gradually encroaching on its old domain." |

---

[876] A. DE WAHA: *Sur l'interférence des explosions électriques.* Publ. Inst. Roy. Luxemb. **16** (1878); Ann. Phys. Chem. [Beiblätter] **2**, 158-160 (1878).
[877] Mentioned in the paper by E. MACH & SIMONIDIS {⇨ 1879}.
[878] P. KREHL: *Single and double Mach reflection – its representation in Ernst MACH's historical soot recording method.* In: (K. TAKAYAMA, ed.) *Proc. 18th Int. Symposium on Shock Waves* [Sendai, Japan, July 1991]. Springer, Berlin *etc.* (1992), vol. 1, pp. 221-226.
[879] A. NOBEL: *Exploding compounds* [Blasting gelatin]. Brit. Patent No. 4,179 (1875).
[880] A. NOBEL: *On modern blasting agents.* Am. Chemist **6**, 60-68, 139-145 (1876).

| | | |
|---|---|---|
| 1875 | *Elswick Ordnance Co., Newcastle-upon-Tyne & Woolwich Arsenal, southeast London, England* | **Captain Andrew NOBLE**[881] **and Prof. Frederick A. ABEL,** two military chemists, **report first results from a research program on gunpowder and its explosive effects in guns.** Their ambitious program encompasses the following goals: |

- "to ascertain the products of combustion of gunpowder fired under circumstances similar to those which exist when it is exploded in guns or mines;
- to ascertain the 'tension' of the products of combustion at the moment of explosion, and to determine the law according to which the tension varies with the gravimetric density of the powder;
- to ascertain whether any, and if so what, well defined variation in the nature or proportions of the products accompanies a change in the density or size of grains of the powder;
- to determine whether any, and if so what, influence is exerted on the nature of the metamorphosis by the pressure under which the gunpowder is fired;
- to determine the volume of permanent gases liberated by the explosion;
- to compare the explosion of gunpowder fired in a close vessel with that of similar gunpowder when fired in the bore of a gun;
- to determine the heat generated by the combustion of gunpowder, and thence to deduce the temperature at the instant of explosion; and
- to determine the work which the gunpowder is capable of performing on a shot in the bore of a gun, and thence to ascertain the total theoretical work if the bore be supposed of indefinite length."

**In their memoir, they give a masterly review of recent investigations into the firing of gunpowder covering the period 1702–1875.**[882] Some of their most important results can be summarized as follows:

(i) They find that a powder composed of *pure* carbon, saltpeter, and sulfur furnishes potassium carbonate, potassium sulfate, potassium disulfide, carbonic acid, carbonic oxide and nitrogen as chief products upon its complete combustion, according to the equation

$$16\,KNO_3 + 21\,C + 5\,S \rightarrow 5\,K_2CO_3 + K_2SO_4 + 2\,K_2S_2 + 13\,CO_2 + 3\,CO + 8\,N_2.$$

(ii) **They compare previous estimations from various historic studies and contemporary "reliable" handbooks for the elastic force of fired gunpowder in cannons,** which range between 100 and 100,000 atmospheres from the studies: *e.g.*, 100 by John BERNOULLI;[883] 1,000 by Benjamin ROBINS (1743); 2,000 by Charles HUTTON; 10,000 by Daniel BERNOULLI; 12,400 by Thomas J. RODMAN (1857–1859); 100,000 by Count VON RUMFORD {⇨1797}; and between 2,200 by Col. Charles L. BLOXAM (1867), and 29,000 by Gen. Guillaume PIOBERT[884] (1859) from the handbooks.

(iii) With the help of the crusher gauge {NOBLE ⇨1872}, they determine the maximum pressure produced when a charge of gunpowder is exploded in a confined space (such as in a cannon). Based on the Boyle-Mariotte law and the Gay-Lussac law (*see* pp. 102–104 of their memoirs), **they derive a simple relation for the maximum pressure for the case that the powder is fired in a closed space** which Carl CRANZ[885] will later call the **"Abel equation"** or **"Abel law"** [Germ. *Abelsche Gleichung*]. NOBLE and ABEL find that the tension achieved by the products of combustion when the powder entirely fills the space in which it is fired is about

---

[881] A. NOBLE and F.A. ABEL: *Researches on explosives – fired gunpowder. (Part I.)* Phil. Trans. Roy. Soc. Lond. **165**, 49-155 (1875); *Recherches sur les substances explosives. Combustion de la poudre.* C. R. Acad. Sci. Paris **89**, 155-164 (1879).

[882] Their introduction was also translated into German; *see* B. PLEUS: *Die ältesten Untersuchungen über die Verbrennung des Schwarzpulvers. Geschichtliche Einleitung zu den "Researches on Explosives" von Andrew NOBLE und Frederick ABEL*. Z. ges. Schieß- u. Sprengstoffwesen **5**, Nr. 5, 84-87 (1910); Ibid. Nr. 6, 104-107 (1910).

[883] The estimations of John and Daniel BERNOULLI were mentioned by C. HUTTON in a note to the new edition of ROBIN's book *Gunnery* (1805), p. 57.

[884] G. PIOBERT: *Traité d'artillerie: théorique et expérimentale*. Mallet-Bachelier, Paris (1859), pp. 354-360.

[885] C. CRANZ: *Lehrbuch der Ballistik: II. Innere Ballistik*. Springer, Berlin (1926), pp. 38-42.

42 tons per square inch (6.4 kbar). This pressure correlates well with theoretical estimates based on measurements of in-barrel projectile velocities.

(iv) They find that the permanent gases produced by the explosion occupy about 280 times the volume of the original powder (at 0 °C and 1 bar).

(v) **They determine a value of 332,000 mkg\* for the total theoretical work done by the exploded gunpowder when it is allowed to expand indefinitely,** which closely approaches a value found two years before by French researchers {ROUX & SARREAU ⇨1873}.[886]

(vi) They employ a a rotating-drum chronograph previously developed by NOBLE – the **"Noble chronograph"** {⇨Fig. 4.17–L} – to ascertain the velocity acquired by shot at different parts of the bore when moving from a state of rest inside the gun by recording the signals from electric "cutting-wire" gauges. This instrument is particularly well-suited to the measurement of very small intervals of time and could potentially measure intervals of less than 1 μs; it was however only accurate to around 10 μs when used for practical purposes (*i.e.*, when used in the field).

Initially, the results from their first memoir were well-received abroad. The French Academy of Sciences appointed a commission (*Comité consultatif des Poudres et Salpêtres*) to investigate the results of the British memoir in more detail. Based on some critical comments made by the renowned French chemist P.E. Marcellin BERTHELOT,[887] NOBLE and ABEL reinvestigated some points and published their revised results in a second memoir five years later.[888] Their results on fired gunpowder became the basis for modern internal ballistics.

| 1876 | *Jabin de Saint-Etienne, Graissessac, France* | On February 4, **a serious firedamp explosion occurs in the French hard coal mining industry** (186 casualties). • This severe accident which was only one of many others which already happened in this colliery prompted the foundation of the French Firedamp Commission {⇨1877}, a research commission supported by the government, in the following year. |
|---|---|---|
| 1876 | *Stokes Bay, Hampshire, England* | On July 14, between Portsmouth Harbor and Stokes Bay, **the most disastrous accidental boiler explosion in the history of the Royal Navy occurs during a full-power trial of the battleship HMS *Thunderer*,** killing 45 men and seriously injuring 40 others. • The ship was fitted with eight rectangular box boilers, built for a working pressure of 30 psi (2.06 bar). The upper part of the front of the front starboard boiler was blown out and the stokeholds and engine room were immediately filled with steam. It was believed that **the real cause of this great disaster was the failure of one of the safety valves,** probably due to corrosion. The safety valves were of the dead-weight type and $5^7/_8$ in. (14.9 cm) in bore diameter.<br><br>The accident resulted in the introduction of the spring-loaded safety valve, which had an alarm whistle fitted which blew when the pressure reached a few psi above the set pressure of the main valve.[889] |
| 1876 | *Hallet's Point, East River, New York* | On September 24, **General John NEWTON,** an American military engineer, **with the aid of ex-confederate General Mansfield LOVELL, extends the gneiss riverbed of the East River, New York. He removes obstructions using 50,000 pounds** (22,679 kg) **of dynamite.** Lt. Col. Henry Larcom ABBOT[890] from the U.S. Corps of Engineers, stimulated by previous British seismologic studies {MALLET ⇨1861}, takes advantage of the seismic shock produced by this work, and measures the longitudinal velocity with which the seismic shock is transmitted through the ground. ABBOT has more sensitive seismometers than were previously available to MALLET and makes use of the fact that the detonation of dynamite induces stronger seismic |

---

[886] For the meaning of the unit kg* used in this book *see* ROUX & SARREAU {⇨1873}.
[887] P.E.M. BERTHELOT: *Observation sur le mémoire de M.M. NOBLE et ABEL relative aux matières explosives.* C. R. Acad. Sci. Paris **89**, 192-196 (1879).
[888] A. NOBLE and F.A. ABEL: *Researches on explosives – fired gunpowder (Part II).* Phil. Trans. Roy. Soc. Lond. **171**, 203-279 (1880).
[889] P. RUSSELL: *Navies in transition: HMS Thunderer boiler explosion* (1999); http://www.btinternet.com/~philipr/thexp.html.
[890] H.L. ABBOT: *On the velocity of transmission of earth waves.* Am. J. Sci. **15**, 178-184 (1878).

shocks than the explosion of gunpowder, as used by MALLET. **ABBOT is able to detect the arrival of the longitudinal wave up to a distance of 280 km away.** It takes 45 seconds to cover this distance, corresponding to an average seismic wave velocity of 6,222 m/s. ▪ **ABBOT's work, which was a landmark in seismology, first proved the applicability of high explosives to geophysical investigations.**

The successful operation at Hallet's Point also stimulated John NEWTON to remove other obstacles in the East River using high explosives {Hell Gate ⇒1885}.

| | | |
|---|---|---|
| 1876 | *German Charles University, Prague, Bohemia, Austro-Hungarian Empire* | Wenzel ROSICKÝ,[891] a coworker of Ernst MACH, **uses the soot method** {ANTOLIK ⇒1874} **to visualize unique shock-focusing phenomena in an elliptic reflector** {⇒Fig. 4.5–E}. The pictures of soot reveal that the focal point is completely free of soot, and that it forms the center of a sharply pronounced star-like soot triangle. ▪ Today ellipsoidal reflectors are also used to focus shock waves in extracorporeal shock wave lithotripsy {CHAUSSY, EISENBERGER ET AL. ⇒1970; FORSSMANN ⇒1976}.[892] |
| 1876 | *British Telegraph Manufactory, London, England* | Robert SABINE,[893] chief engineer of the company and a scientific writer, **measures the shock contact time of colliding elastic bodies** using the ballistic galvanometer, an ingenious electrical method {POUILLET ⇒1844; ⇒Fig. 4.3–J}. His instrument – the **"Sabine chronoscope"** – allows the time between two successive mechanical movements to be measured with a considerable degree of accuracy. The method is based on the fact that a charged capacitor can only be discharged at a certain definite rate through a given circuit. For the duration of a blow of a light hammer weighing about 1 ounce (28.35 g) against a steel anvil, he obtains contact times of around 50 µs. Further experiments reveal that the contact time decreases with increasing impact velocity. ▪ **His important results reliably proved for the first time that contact times of impacting elastic bodies are indeed very short.**<br><br>In the following years, Heinrich HERTZ {⇒1882} was the first to confirm theoretically and Peter G. TAIT {⇒1892} then reconfirmed experimentally the existence of very short contact times during the collisions of solid bodies. |
| 1877 | *Paris, France* | On March 26, **foundation of the French Firedamp Commission** (*Commission d'étude des moyens propres à prévenir les explosions du grisou*) **aimed at uncovering the mysterious phenomenon of detonation, and investigating how firedamp explosions can be prevented.**[894] Prominent members of this commission throughout its existence include P.E. Marcellin BERTHELOT, François Earnest MALLARD, Henry LE CHÂTELIER, Hippolyte SÉBERT, and Paul VIEILLE. ▪ Shortly after, similar commissions were also formed in Austria, Belgium, England (with Sir Frederick ABEL as president), Prussia, and Saxony. The numerous published results[895, 896, 897, 898, 899] from these investigations illuminated not only the complex causes of firedamp explosions and suggested possible countermeasures, but also became an important step towards an understanding of the nature of detonation. |

---

[891] W. ROSICKÝ: *Über mechanische Wirkungen des elektrischen Funkens.* Sitzungsber. Akad. Wiss. Wien **73** (Abth. II), 629-650 (1876).

[892] F. RIEBER: *Shock wave generator.* U.S. Patent No. 2,559,227 (July 3, 1951).

[893] R. SABINE: *On a method of measuring very small intervals of time.* Phil. Mag. **1** [V], 337-346 (1876); *Dauer eines Schlages.* Dingler's Polytech. J. **222**, 499-500 (1876).

[894] *Etude des questions relatives à l'emploi des explosifs en présence du grisou* [Premier Rapport présenté à la Commission des Substances Explosives, Sous-Commission spéciale]. Ann. Mines **14** [VIII], 197-318 (1888).

[895] J.N. HATON DE LA GOUPILLIERE: *Rapport présenté au nom de la commission d'étude des moyens propres à prévenir les explosions de grisou.* Ann. Mines **18**, [VII], 193-411 (1880); Germ. translation in: Z. für Berg-, Hütten- u. Salinenwesen im Preußischen Staate **29**, 281-394 (1881).

[896] *Bestimmungen über die Vorsichtsmaßregeln gegen schlagende Wetter* (herausgegeben von der Preußischen Schlagwetter-Kommission). Marcus, Bonn (1884).

[897] *Final Report. Royal Commission to inquire into accidents in mines, and means of preventing their occurrence, or limiting their disastrous consequences.* Command Paper 4699 in *Parliamentary Papers*, London (1886), vol. 16, pp. 411-811.

[898] A. HASSLACHER: *Haupt-Bericht der preußischen Schlagwetter-Commission.* Ernst & Korn, Berlin (1886/1887).

[899] *Verhandlungen des Centralcomites der österreichischen Commission zur Ermittlung der zweckmäßigsten Sicherheitsmaßregeln gegen die Explosion schlagender Wetter in Bergwerken.* 4 vols., Hof- und Staatsdruckerei, Wien (1888/1889).

| | | |
|---|---|---|
| | | Statistical data on mining accidents caused by firedamp explosions revealed that the average number of miners killed in the period 1820–1830 amounted to 27, and the production rate from mining was 4 million tons per year. In the period 1860–1870, however, the number of miners killed was successfully reduced to an average number of 17, even though production increased to 17 million tons per year.[900] |
| 1877 | *Iquique, northern Chile* | On May 9, **a great earthquake of unusually long duration, lasting about 4.5 minutes, nearly destroys the coastal town of Iquique as well as other coastal towns, and generates Pacific-wide destructive tsunamis.** This disaster, later known among seismologists as the "**Iquique Earthquake,**" provokes similar tsunami effects along coasts across the Pacific, just like the Arica Earthquake {⇨1868}. Wave heights reach over 20 m in Chile and up to 5 m in Hawaii.<br>• An ocean-crossing tsunami that, after traveling some hours, can cause damage far away from its source (> 1,000 km) has been termed a *teletsunami* (or *far-field tsunami*). On the contrary, a *local tsunami* (or *near-field tsunami*) has a very short travel time (< 30 min).<br>One year later, two German scientists used Ferdinand VON HOCHSTETTER's crude model of tsunami propagation {Arica Earthquake ⇨1868} to analyze data on the arrival times of seismic sea waves recorded at a number of harbors along the Pacific coast and at different islands in the Pacific Ocean:<br>(i) **Franz Eugen GEINITZ**,[901] a German professor of mineralogy and geology at the University of Rostock, **finds good agreement with VON HOCHSTETTER's results.** In particular, he finds that the primary seismic wave propagated across the Pacific with about the same velocity as the one in 1868, and that it was followed by secondary waves. However, he observes a discrepancy between the depths measured by the HMS *Challenger* Expedition {⇨1872} along the coast of Iquique and the average depths calculated by applying VON HOCHSTETTER's method.<br>(ii) **Alexander Georg SUPAN**,[902] a German geography lecturer at Czernowitz in Bukovina (a province of the Austro-Hungarian Empire), **resumed VON HOCHSTETTER's idea of calculating the average depths of parts of the Pacific from seismic sea wave arrival times.** He subdivided the surface of the Pacific into 10°-longitude × 10°-latitude wide stripes, resulting in a total of 151 stripes. Based upon August PETERMANN's depth chart of the Pacific Ocean {⇨Fig. 4.1−O}, he calculated average depths for 83 stripes; measured depth data were not yet available for the remaining stripes. Fortunately, however, the Iquique seismic sea waves passed over a number of stripes for which measured arrival times and average depth data were available. His method of comparison essentially confirms VON HOCHSTETTER's theory. However, its application to the complex sea-bottom structure of the Pacific provoked critical comments by some contemporary geologists.[903, 904] |
| 1877 | *Washington, DC* | A committee formed by the Philosophical Society of Washington publishes the observations of 48 people who witnessed **the fall of the "Washington Meteorite"** on December 24, 1873.[905] Descriptions of the acoustic phenomena observed range between "no sound" and a "very violent sound" (*e.g.*, "short, hard reports like heavy cannon and continued resounding"). The committee concludes that the sound could have been generated by sound focusing and the |

---

[900] *Analyse synoptique des rapports officiels sur les accidents de grisou en France de 1817 à 1881.* Annales des Mines (Mémoires) **1** [VIII], 293-365 (1882); **2** [VIII], 393-476 (1882); **4** [VIII], 67-127 (1883); **6** [VIII], 73-189 (1884); **8** [VIII], 195-299 (1885); **9** [VIII], 31- 109 (1886); **10** [VIII], 11- 91 (1886).

[901] E. GEINITZ: *Das Erdbeben von Iquique am 9. Mai 1877 und die durch dasselbe verursachte Erdbebenfluth im großen Ocean.* Nova Acta Acad. Caes. Leop.-Carol. Germ. Nat. Curios. (Halle/Saale) **40**, 383-444 (1878).

[902] A.G. SUPAN: *Die mittlere Tiefe des Großen Oceans.* Dr. A. Petermann's Mitteilungen aus Justus Perthes' Geographischer Anstalt (Gotha) **24**, 213-215 (1878).

[903] G.G. VON BOGUSLAWSKI: *Handbuch der Ozeanographie. Band I.* Von Engelhorn, Stuttgart (1884), pp. 109-110.

[904] G.G. VON BOGUSLAWSKI and O. KRÜMMEL: *Handbuch der Ozeanographie. Band II.* Von Engelhorn, Stuttgart (1887), pp. 131-134.

[905] *Report of the committee to collect information relative to the meteor of December 24, 1873 Washington, DC.* Bull. Phil. Soc. Washington **2**, 139-161 (1874–1877).

Doppler effect {DOPPLER ⇒1842}. ▪ Ernst MACH, later referring to this "queer theory" {E. MACH & DOSS ⇒1893}, correctly explained that the loud report was not the result of any explosion or focusing effects, but was due to the head wave phenomenon {E. MACH & SALCHER ⇒1886} alone.

The mean velocity of the "Christmas Meteor" was later estimated to 38.5 miles per second (61.6 km/s).[906] A meteorite traveling at such a high velocity through the Earth's atmosphere is heated up at the surface by colliding with the air molecules and forms a cylindrical plasma tail.

---

**1877** — *Lehrstuhl für Mathematik, Kaiser-Wilhelms-Universität, Straßburg, Germany*

Elvin Bruno CHRISTOFFEL,[907] a German professor of mathematics renowned for his work in mathematical analysis, **studies theoretically the propagation of a planar shock wave,** and extends RIEMANN's one-dimensional shock theory {RIEMANN ⇒1859}. He derives the mechanical conditions for percussion between two fluidic particles, discusses discontinuities compatible with the continued existence of partial differential equations, and studies the general case of the reflection of a shock wave at a rigid boundary of any geometry.

In the same year, CHRISTOFFEL[908] also treats **the propagation of shocks through an elastic solid medium,** and possible modifications at the boundary between two elastic solid media. ▪ Corresponding experimental studies, however, were still out of reach due to the lack of appropriate high-speed diagnostics.

---

**1877** — *Trinity College, Oxford, England*

Harold B. DIXON, a young British lecturer of chemistry, **hits upon the idea of using the rate of explosion as a way to determine the course of a chemical reaction.**

(i) While investigating the influence of steam on the union of carbonic oxide and oxygen, DIXON makes the epochal discovery that **the prolonged drying of a mixture of carbonic oxide and oxygen in combining proportions renders it nonexplosive when subjected to electric sparks of an intensity sufficient to readily ignite an undried medium.** ▪ In a later paper, DIXON {⇒1893} remembered, "If steam acts as a carrier of oxygen to the carbonic oxide by a series of alternate reductions and oxidations, an increase in the amount of steam present, beyond that required to initiate the reaction, should be accompanied by an increase in the rate of combination up to a certain limit. Attempts were therefore made to detect such an increase by measuring the velocity of the flame in a tube … In the spring of 1881 I attempted to measure the rate of explosion of carbonic oxide and oxygen with varying quantities of steam by photographing on a moving plate the flashes at the beginning and end of a closed tube 20 feet [6.1 m] long. The two flashes appeared to be simultaneous to the eye, but no record of the rate was obtained, for the apparatus was broken to pieces by the violence of the explosion."

(ii) He advances the correct idea that in the presence of water the oxidation of carbonic oxide occurs according to the equations

$$CO + H_2O = CO_2 + H_2$$

and

$$2\,H_2 + O_2 = 2\,H_2O.$$

(iii) Testing BUNSEN's conclusion about the inconsistency of his experiments with BERTHOLLET's Law of Mass Action {BUNSEN ⇒1853}, he finds that BUNSEN's conclusion was erroneous and that the Law of Mass Action applies to gaseous explosions.

---

[906] H.A. PECK: *The Christmas meteor of 1873 at Washington, DC.* Month. Weather Rev. **35**, 447-448 (Oct. 1907).
[907] E.B. CHRISTOFFEL: *Untersuchungen über die mit dem Fortbestehen linearer partieller Differentialgleichungen verträglichen Unstetigkeiten.* Ann. di Mat. Pura Appl. **8** [II], 81-112 (1877).
[908] E.B. CHRISTOFFEL: *Fortpflanzung von Stößen durch elastische feste Körper.* Ann. di Mat. Pura Appl. **8** [II], 193-243 (1877).

| | | |
|---|---|---|
| | | DIXON[909] first announced his discovery in a talk presented at the British Association for the Advancement of Science (BAAS), held at Swansea on August 28, 1880, where he also demonstrated it experimentally. |
| 1877 | *Russia* | **Vladimir V. LESEVICH**,[910] a Russian natural philosopher who elaborated a "scientific philosophy," **is credited with inventing** a branch of astronomy based primarily on the study of the geology of solid extraterrestrial objects (such as meteorites), and secondarily on telescopic spectroscopy. ▪ This idea was later resumed by the U.S. geologist Eugene M. SHOEMAKER who stimulated the creation of the world's first **Center of Astrogeology** at the USGS in Flagstaff, Arizona, in the early 1960s.<br><br>Astrogeology is now defined[911] as "a science that applies the principles and techniques of geology, geochemistry, and geophysics to the study of the nature, origin, and history of the condensed matter and gases in the Solar System (usually excluding the Earth)." It encompasses:<br>▸ remote-sensing observations and *in situ* manned exploration of other planetary bodies (such as the Moon and Mars);<br>▸ the study of the chemistry, mineralogy, and history of objects that occur on the Earth but are of known or possible extraterrestrial origin (such as meteorites and tektites), or that are returned to the Earth (such as lunar samples); and<br>▸ the study of the effects of extraterrestrial processes (such as meteorite impacts, solar energy changes, and tides) on the Earth in the present and past. |
| 1877 | *German Charles University, Prague, Bohemia, Austro-Hungarian Empire* | Ernst MACH[912] and his collaborator Jan SOMMER **study the propagation of blast waves on the laboratory scale.**<br><br>(i) **They give experimental evidence that a shock wave travels faster than a sound wave and that the velocity of the shock wave increases with shock strength** – thus reconfirming previous large-scale measurements made by Henri V. REGNAULT {⇨1863} and achieving an important milestone in the evolution of shock wave physics.<br><br>(ii) Using a linear percussion model – a row of gas molecules arranged two by two along a straight line {⇨Fig. 4.4–D} – they illustrate that the velocity of percussion must also increase when the velocity of sound increases, such as in the case of violent sound.<br><br>(iii) They conclude, "It does not contradict the theory to assume that the velocity of sound increases with the intensity of the impulse. **Only for very small vibrations does the velocity of sound not depend on the amplitude. But this is not valid for vibrations of finite amplitude** as has been proved by RIEMANN in 1860 in his paper *Über die Fortpflanzung ebener Luftwellen von endlicher Schwingungsweite* ('On the Propagation of Plane Aerial Waves of Finite Amplitude'). 'Velocity of sound' hereby receives a quite different meaning; it is different at every point of the wave and alters during the wave motion. It appears that we deal in our experiments with such waves as described by RIEMANN." |
| 1877 | *Private study at Terling Place, Witham, Essex, England* | Lord RAYLEIGH (born John William STRUTT), an eminent British physicist and acoustician who is in the final phase of writing his book *The Theory of Sound*, **has a disagreement with Prof. George G. STOKES**, his former teacher at Cambridge University **who had published a paper on sounds of finite amplitude almost 30 years previously** {STOKES ⇨1848}.[913] |

---

[909] H.B. DIXON: *On the influence of water on the union of carbonic oxide with oxygen at high temperature*. Rept. Meet. Brit. Assoc. **50**, 503-504 (1880).

[910] V.V. LESEVICH: *Opyt kriticheskogo izsledovaniia osnovonachal pozitivnoi filosofii*. M. Stasiulevich, St. Petersburg (1877); V.V. LESEVICH: *Sobranie sochinenii*. IU.V. Leontovich, Moscow (1915), vol. 1, pp. 189-454.

[911] R.L. BATES (ed.): *Glossary of geology*. American Geological Institute (AGI), Falls Church, VA (1987), pp. 40, 751.

[912] E. MACH and J. SOMMER: *Über die Fortpflanzungsgeschwindigkeit von Explosionsschallwellen*. Sitzungsber. Akad. Wiss. Wien **75** (Abth. II), 101-130 (1877).

[913] The following correspondence between STOKES and Lord RAYLEIGH has been taken from J. LARMOR (ed.): *Memoir and scientific correspondence of the late Sir George Gabriel STOKES*. Cambridge University Press (1907), pp. 102-104.

(i) On June 2, 1877, Lord RAYLEIGH writes to him: "In consequence of our conversation the other evening I have been looking at your paper 'On a difficulty in the theory of sound,' *Phil. Mag.* Nov. 1848. The latter half of the paper appears to me to be liable to an objection, as to which (if you have time to look at the matter) I should be glad to hear your opinion." After deriving a relation between the velocities and densities at the shock front imposed by the energy condition, RAYLEIGH concludes, "It would appear therefore that on the hypotheses made, no discontinuous change is possible. I have put the matter very shortly, but I dare say what I have said will be intelligible to you."

(ii) STOKES admits that William THOMSON (later Lord KELVIN) had already made similar objections that the proposed "queer kind of motion" would violate the Law of Conservation of Energy. Avoiding a confrontation with his former student, STOKES gently replies to Lord RAYLEIGH on June 5, 1877, "It seemed, however, hardly worth while to write a criticism on a passage in a paper which was buried among other scientific antiquities. PS: You will observe I wrote somewhat doubtfully about the possibility of the queer motion."

In the second (revised and enlarged) edition of his two-volume book *The Theory of Sound* (1894 & 1896), **Lord RAYLEIGH returned to the source of his previous disagreement with STOKES.** He writes (vol. 2, p. 40), "… but it would be improper to pass over in silence an error on the subject of discontinuous motion into which RIEMANN [⇨1859] and other writers have fallen. It has been held that a state of motion is possible in which the fluid is divided into two parts by a surface of discontinuity propagating itself with constant velocity, all the fluid on one side of the surface of discontinuity being in one uniform condition as to density and velocity, and on the other side in a second uniform condition in the same respects…" There follows a proof of the impossibility of such motions based on a "violation" of the conservation of energy, although the conditions for the conservation of mass and momentum are satisfied. ▪ **It was not yet recognized that dissipative effects within the shock front play an important role, thus leading to an increase in entropy in a shock wave** {ZEMPLÉN ⇨1905}. Furthermore, Lord RAYLEIGH failed to notice that contemporary experimentalists had already proven the existence of a sharply pronounced discontinuity at the shock front via both the schlieren method {A. TOEPLER ⇨1865; E. MACH & WENTZEL ⇨1884; E. MACH & SALCHER ⇨1886; E. & L. MACH ⇨1889} and the interferometer technique {E. MACH & VON WELTRUSBY ⇨1878}.

| 1878 | *German Charles University, Prague, Bohemia, Austro-Hungarian Empire* | **Ernst MACH**[914] **and J. VON WELTRUBSKY use the Jamin interferometer** {JAMIN ⇨1856} **to visualize the local density profile through the shock front for the first time** {⇨Figs. 4.12–A & 4.18–E}. The shock wave is generated by the electric discharge of a capacitor (Leiden jar) in air. ▪ They confirmed previous theoretical ideas {RIEMANN ⇨1859} that there was a density jump at the shock front. |

In the same year, MACH and collaborators also obtain the following results:

(i) **Ernst MACH,**[915] **Ottokar TUMLIRZ, and Carl KÖGLER measure the velocity-distance profile of a blast wave** generated by an electric spark discharge. Their thorough study confirms previous observations {E. MACH & SOMMER ⇨1877} that the blast wave velocity approaches the velocity of sound as the distance from the origin of the blast increases. Using an ingenious simple experimental arrangement {⇨Fig. 4.17–N}, they construct a curve for a spark discharge in air showing how the front speed of the emitted shock wave decreases with distance. Their diagram reveals that the shock velocity of 756 m/s measured at a distance of 80 mm decreases to only 373 m/s at 977 mm, indicating that a shock wave must approach the speed of sound (about 340 m/s) as the

---

[914] E. MACH and J. VON WELTRUBSKY: *Über die Formen der Funkenwellen.* Sitzungsber. Akad. Wiss. Wien **78** (Abth. II), 551-560 (1878).

[915] E. MACH, O. TUMLIRZ, and C. KÖGLER: *Über die Fortpflanzungsgeschwindigkeit der Funkenwellen.* Sitzungsber. Akad. Wiss. Wien **77** (Abth. II), 7-32 (1878).

distance increases. ▪ In 1920, Arthur L. FOLEY,[916] a U.S. researcher at the Waterman Institute for Research at Indiana University, resumed the subject: he reported on a new and direct photographic method for finding the instantaneous spark wave velocity – in contrast to MACH's method, which only allowed average velocities to be determined. He concluded: "Since the velocity very close to the source is very high, average velocities run high. However, previous results have been entirely too high due to errors inherent in the experimental methods used."

(ii) **Ernst MACH**[917] **and Gustav GRUSS study *Mach reflection* in a V-shaped gliding spark arrangement** {E. MACH & WOSYKA ⇨ 1875} in more detail. Their publication contains correct drawings of the complex shock interaction phenomenon {⇨Fig. 4.13–C}, but not photographs. They also establish the fact that the speed of the shock fronts increases with the spark intensity.

(iii) **Ernst MACH**[918] **gives an ingenious interpretation of the generation of the Mach disk and its increasing width over time.** He writes, "It seems as if the locus of interference between the two original waves has become the source of a third wave which during propagation interferes with the two original ones. **The V-propagation can be explained as an area of superposition of the two waves which is attributed to a larger propagation velocity – this locus can be understood as** *a wave in the wave* **– i.e., as a second wave which propagates with an excessive velocity in the first wave.** This interpretation fully agrees with the quantitative results of my third and fourth work." ▪ MACH's term of a *third wave* is used among modern shock physicists to denote the *Mach disk* and *Mach stem* in a symmetric and asymmetric shock interaction, respectively.

| | | |
|---|---|---|
| 1878 | *Private residence at Glenlair, Kirkcudbrightshire, Scotland, U.K.* | James Clerk MAXWELL,[919] an eminent Scottish physicist, **theorizes on the interaction of gas molecules with a surface and calculates a distribution function for the reflected molecules.** His collision model is based on the assumptions that a fraction $(1-\alpha)$ of the molecules is reflected in a specular fashion from the surface while another fraction, $\alpha$, is re-emitted diffusely. This allows one to calculate the tangential momentum transferred to the wall by the incident molecules. ▪ MAXWELL thus created the science of *rarefied gas dynamics,* a new field of research based on the molecular theory of gases and statistical mechanics. |
| 1878 | *Private study at Terling Place, Witham, Essex, England* | Lord RAYLEIGH (born John William STRUTT), an eminent English physicist, **publishes his two-volume monograph** *The Theory of Sound.* In volume 2, §250, he addresses previous studies on waves of finite amplitude in air {POISSON ⇨ 1808; STOKES ⇨ 1848; EARNSHAW ⇨ 1859 & 1860; RIEMANN ⇨ 1859}. He derives a linear relationship between the pressure $p$ and specific volume $v\ (=1/\rho)$ of a gas, given by $$p = \text{const} - u_0^2 \rho_0^2/\rho = \text{const} - u_0^2 \rho_0^2 v.$$ This equation represents a straight line in the $(p, v)$-plane which is the locus for all jump discontinuities $\Delta p$ and $\Delta \rho$ starting from an initial state $(p_0, v_0 = 1/\rho_0)$ and satisfying the equations of conservation of mass and momentum. ▪ **Note that this so-called "Rayleigh line" is identical to the "Mikhel'son line"** {MIKHEL'SON ⇨ 1893}. |
| 1878 | *U.S.A.* | H. Julius SMITH, a U.S. inventor, **constructs the first satisfactory *magneto-induction exploder* for the ignition of electric detonator caps.** His blasting machine comprises a gear-type arrangement of rack bar and pinion that operates a small magnetoelectric machine. When the rack bar is pushed down rapidly, it revolves the pinion and generates an electrical current that sets off blasting caps {A. NOBEL ⇨ 1864}, which in turn ignite dynamite or other high explo- |

---

[916] A.L. FOLEY: *A photographic method of finding the instantaneous velocity of spark waves.* Phys. Rev. **16** [II], 449-463 (1920).
[917] E. MACH and G. GRUSS: *Optische Untersuchungen der Funkenwellen.* Sitzungsber. Akad. Wiss. Wien **78** (Abth. II), 467-480 (1878).
[918] E. MACH: *Über den Verlauf der Funkenwellen in der Ebene und im Raum.* Sitzungsber. Akad. Wiss. Wien **77** (Abth. II), 819 (1878).
[919] J.C. MAXWELL: *On stresses in rarefied gases arising from inequalities of temperature.* Proc. Roy. Soc. Lond. **27**, 304-308 (1878); Phil. Trans. Roy. Soc. Lond. **170**, 231-256 (1880).

sives.⁹²⁰ • The machine, then known as the "Magneto No. 3," was produced and marketed by the Laflin & Rand Powder Company of New York. In 1905, the Du Pont Company in Wilmington, DE, purchased the Laflin & Rand Powder Company and assumed control over production of the exploder. However, a serious competitor appeared in the form of the more efficient and more reliable dynamoelectric blasting machine based on Werner VON SIEMENS' dynamo principle.⁹²¹ These so-called **"dynamo exploders"** became the devices most widely used in blasting operations. A review of early blasting machines up to 1904 was given by Prof. Fritz HEISE,⁹²² director of the Royal Mining College in Berlin.

**Prior to the magnetoelectric machine, blasting engineers used electrostatic generators based upon friction** {VON GUERICKE ⇨1663}. Operated by a crank via a gear, it charged two Leiden jars up to 30–40 kV which, together with the generator, were housed in a hermetically sealed box to avoid the penetration of moisture.⁹²³ Ignition of the high explosive was provoked by discharging the Leiden jars across a gap-type blasting cap or a number of them connected in series. However, since the lead wires carried a high voltage, this often resulted in ignition problems, particularly when the operation took place in the moist environment of a mine.

| 1878 | *Polytechnische Hochschule, Berlin-Charlottenburg, Germany* | **Rudolph WEBER,**⁹²⁴ a professor of chemistry, **reports on the possible causes of flour dust explosions** {⇨Fig. 4.17–H}. He observes that they can be caused by strong electric sparks which provoke self-triggering of flour dust, and finds that air with a flour-dust concentration ranging between 20 and 30 mg in one liter of air can be caused to explode by a flame. • His work won him a prize offered by the *Verein für Gewerbfleiß in Preussen* ("Prussian Trade Union").⁹²⁵ |
|---|---|---|
| 1879 | *Sea of Marmora* | In January, **a severe accident on board HMS *Thunderer* occurs when a 12-in. 38-ton muzzle-loading gun burst during firing practice,** killing 11 men and injuring 35 others. • A commission was established to look into the disaster at Malta. At first it was suggested that the gun burst due to some defect in the hydraulic breech-loading machinery supplied from Elswick ordnance works (W.G. Armstrong & Company, Ltd.). The ballistician Andrew NOBLE, partner of Sir William G. ARMSTRONG, maintained that the gun had been double-loaded. The intention had been to fire both guns in the turret together, and it was not noticed that one had misfired. This gun was then reloaded, and a second charge rammed in on the top of the unexploded charge. Subsequent experiments confirmed NOBLE's view, showing that – whatever the drawbacks of breech-loading – the muzzle-loading system was not without certain dangers of its own.⁹²⁶<br><br>Barrel bursts {⇨Fig. 4.21–C} – unfortunately not a rare phenomenon with the frequent use of firearms, and often fatal to the attendant artillery crew – arise from various causes, *e.g.*,<br>▸ flaws in the original material;<br>▸ material fatigue by overheating;<br>▸ increased projectile friction by soiling; and<br>▸ use of inappropriate propellants that burn too violently. |

---

⁹²⁰ *Dynamic blasting machine.* Western Museum of Mining & Industry, Colorado Springs, CO; http://www.wmmi.org/html/collections/index.htm.

⁹²¹ W. VON SIEMENS: *Über die Umwandlung von Arbeitskraft in electrischen Strom ohne Anwendung permanenter Magnete.* Ann. Phys. **130** [II], 332-335 (1867); *Die dynamoelectrische Maschine.* Ibid. **14** [III], 469-482 (1881).

⁹²² F. HEISE: *Sprengstoffe und Zündung der Sprengstoffe mit besonderer Berücksichtigung der Schlagwetter- und Kohlenstaubgefahr auf Steinkohlengruben.* Springer, Berlin (1904), pp. 168-197.

⁹²³ An electrostatic blasting machine, manufactured by A. BORNHARDT, a mechanic to the court at Braunschweig, is illustrated in Isidor TRAUZL's booklet: *Die Dynamite, ihre Eigenschaften und Gebrauchsweise sowie ihre Anwendung in der Landwirthschaft und im Forstwesen.* Wiegandt, Hempel & Parey, Berlin (1876), pp. 25-28.

⁹²⁴ Anonymous: *Über die Entstehung von Bränden in Mahlmühlen.* Z. für Tech. Hochschulen **3**, 51-53 (1878).

⁹²⁵ R. WEBER: *Preisgekrönte Abhandlung über die Ursachen von Explosionen und Bränden in Mühlen, sowie über die Sicherheitsmaßregeln zur Verhütung derselben.* Verhandlungen des Vereins zur Beförderung des Gewerbfleißes. **57**, 83-103 (1878).

⁹²⁶ P.W. (anonymous): *Sir Andrew NOBLE, 1831–1915.* Proc. Roy. Soc. Lond. **A94**, i-xvi (1918), pp. v-vi.

Only a few burst barrels from historic guns have survived, particularly from large cannons. Most of them, however, were immediately melted down because metals were often in short supply in times of war.

The HMS *Thunderer*, launched in 1872, was the fifth ship of the Royal Navy with this name and one of the first ironclad British warships. She was also an unlucky ship: three years before this gun burst accident, one of her steam boilers exploded, killed many men {HMS *Thunderer* ⇒1876}. Later, she was refitted with Whitehead torpedoes, used as a training ship, and eventually broken up in 1909.

| 1879 | Royal Military Academy, Woolwich, southeast London, England | **Sir Alfred George GREENHILL,**[927] a British professor of mathematics to the Advanced Class of Artillery Officers, **observes that mortars are being wrecked by rifling twist rates that are too high, and so he attempts to find the theoretically optimum rifling twist rate (that just sufficient to stabilize a projectile).** He works out a simple rule of thumb for estimating optimum twist rates in artillery and small arms, the **"Greenhill rule,"** based on a prolate spheroid (a surface of revolution obtained by rotating an ellipse about its major axis) in laminar, subsonic flow. While empirically successful, it ignores the (as yet unknown) transonic and supersonic velocity dependence of the aerodynamic coefficients. ▪ Very recently, Donald G. MILLER,[928] a retired LLNL physical chemist at Livermore, CA, reviewed GREENHILL's original contributions to this important practical artillery problem, as well as further improvements made by modern ballisticians.

*Rifling* – a term applied to a system of helical grooves cut into the bore of a gun in order to impart spin to the projectile, which improves the accuracy of a projectile – was probably invented by the German gunsmith Augustin KUTTER (who died in 1630).[929] With the advent of elongated rifled projectiles, it was soon recognized that missile wobble produces a great deal of drag which makes it very difficult to compute the trajectory of the projectile. |
| 1879 | U.S. Arsenal, Watertown, Massachusetts | **William W. JACQUES,**[930] a U.S. physicist and fellow at Johns Hopkins University, **experimentally studies the (supersonic) velocity phenomenon associated with very loud sound.**

(i) He refers to an interesting observation, "It is very well known that the velocity of a musical sound is, within very wide limits, sensibly independent of its intensity and of its pitch. The experimental proof of this is that a piece of music, played by a military band at a considerable distance, comes to the ear of the observer with its harmony entirely undisturbed … When, however, we come to the consideration of a loud and sharp shock or explosion, in which the disturbances are very violent and abrupt, we cannot be at all sure that the changes of density are negligibly small, and hence that the velocity of sound for such cases would be a constant." ▪ A similar observation had already been made in England {BESANT ⇒1859}.

(ii) **Investigating the transmission of a report from a cannon in different directions, he finds that at a position 70–90 ft** (21.3–27.4 m) **to the rear it rises to a maximum of 1,267 ft/s** (386 m/s) – *i.e.*, **to a supersonic velocity – and then falls off.** ▪ It appears that JACQUES' work is the earliest American contribution to the investigation of shock (blast) waves. His work in this area was obviously stimulated by his previous studies at various German universities (Göttingen, Berlin, and Leipzig). |

---

[927] A.G. GREENHILL: *On the rotation required for the stability of an elongated projectile.* Minutes of the Proceedings of the Royal Artillery Institution **X**, No. 7, 577-593 (1879).
[928] D.G. MILLER: *Where did GREENHILL's twist rule come from?* Int. J. Impact Engng. **32**, No. 11, 1786-1799 (2006).
[929] *Meyers Konversations-Lexikon.* Bibliographisches Institut, Leipzig & Wien, vol. V (1894), p. 162.
[930] W.W. JACQUES: *On the velocity of very loud sounds.* Am. J. Sci. **17** [III], 116-119 (1879).

| | | |
|---|---|---|
| 1879 | *Chemisch-analytisches Laboratorium, Polytechnikum, Zurich, Switzerland* | **Wilhelm MICHLER,**[931] a German instructor [Germ. *Privatdozent*] in chemistry and assistant to Prof. Viktor MEYER, **and Carl MEYER discover the high explosive *tetryl*** [2,4,6-trinitrophenylmethylnitramine, $C_7H_5N_5O_8$]. ▪ Tetryl, like RDX and HMX, belongs to the nitramine class of organic explosives. Its structure was established in 1886 by the Dutch chemist **Karel Hendrik MERTENS.**[932] In 1889, the Dutch chemist Pieter VAN ROMBURGH[933] proved its structure by synthesizing it from picryl chloride and potassium methylnitramine. In both World Wars, tetryl was used in torpedoes and artillery shells as well as in blasting caps or booster charges. Detonating by friction, shock, or spark, tetryl is more powerful then TNT, although its instability compared to TNT made it less attractive to the military.<br><br>***Tetrytol*** is a cast mixture of tetryl and TNT (70/30 or 75/25) and is used in burster tubes for chemical bombs, demolition blocks, and cast shaped charges. |
| 1879 | *Chair of Natural Philosophy, University of Edinburgh, Scotland, U.K.* | **Peter G. TAIT,** a Scottish professor of natural philosophy, **begins his research into the corrections that need to be made to the deep-sea temperatures** that he had collected during the HMS *Challenger* Expedition {⇨1872} because of the great pressures to which the thermometers had been subjected. ▪ **His studies,** performed in the period 1879–1888, **led to important experimental results on the static compressibility of seawater** {TAIT ⇨1888}, glass, and mercury, and stimulated his interest in the compression behavior of solids under impact {TAIT ⇨1892}. |
| 1879 | *University of Glasgow, Scotland, U.K.* | **William THOMSON**[934] (later Lord KELVIN) and Peter G. TAIT, two Scottish professors of natural philosophy, **discuss the collision of spherical bodies in their *Treatise on Natural Philosophy.***<br><br>(i) Using NEWTON's general rule of percussion {Sir NEWTON ⇨1687}, they assume that two spheres with masses $m_A$ and $m_B$ move along a straight line with velocities $v_A$ and $v_B$, respectively. After colliding head-on, they recede from one another with velocities $c_A$ and $c_B$. Using the conservation of momentum<br>$$m_A v_A + m_B v_B = m_A c_A + m_B c_B,$$<br>and the conservation of energy<br>$$m_A v_A^2 + m_B v_B^2 = m_A c_A^2 + m_B c_B^2,$$<br>they show that $(c_B - c_A) < (v_A - v_B)$ always holds.<br><br>(ii) **They introduce a *coefficient of restitution e*, a numerical factor which they define as the quotient of the relative speed of separation to the relative speed of approach:**<br>$$e = \Delta c/\Delta v = (c_B - c_A):(v_A - v_B) = -(c_B - c_A):(v_B - v_A),$$<br>where $0 < e < 1$. For the ideal case of perfectly elastic bodies ($e = 1$), one has<br>$$(v_B - v_A) = -(c_B - c_A).$$<br>(iii) In a footnote, they say, "In most modern treatises this is called a "coefficient of elasticity," which is clearly a mistake; suggested, it may be, by NEWTON's words, but inconsistent with his facts, and utterly at variance with modern language and modern knowledge regarding elasticity."<br><br>(iv) They (incorrectly) state that *e* is a constant and only dependent on the material. ▪ Indeed, Samuel HAUGHTON,[935] an Irish professor of geology at the University of Dublin, had already shown experimentally in 1864 that *e* is not a constant. Experimenting with spherical bodies, he observed that *e* decreases with increasing impact velocity. More recent studies have confirmed |

---

[931] W. MICHLER and C. MEYER: *Verhalten von Sulfochloriden zu Aminen. III. Mittheilung.* Ber. Dt. Chem. Ges. **12**, 1791-1793 (1879).
[932] K.H. MERTENS: *Die Nitrierung von Di- und Monomethylanilin mit verdünnter Salpetersäure.* Ber. Dt. Chem. Ges. **19**, 2123-2127 (1886).
[933] P. VAN ROMBURGH: *Synthèse de la trinitrophénylméthylnitramine.* Rec. Trav. Chim. Pays-Bas **8**, 215-216 (1889).
[934] W. THOMSON and P.G. TAIT: *Treatise on natural philosophy.* Clarendon Press, Oxford (1879); Cambridge University Press (1923), vol. I, p. 278. Germ. translation by H. HELMHOLTZ and G. WERTHEIM: *Handbuch der theoretischen Physik.* 2 vols., Vieweg, Braunschweig (1871, 1874).
[935] S. HAUGHTON: *On the dynamical coefficients of elasticity of steel, iron, brass, oak and teak.* Proc. Roy. Irish Acad. **8**, 86-87 (1864).

this behavior. Furthermore, **in an oblique impact $e$ can even exceed unity,** because rebound processes depend on the impact angle.[936]

The coefficient of restitution $e$ – also designated by the Greek letter $\varepsilon$ in some textbooks on mechanics – is a measure of the elasticity in a one-dimensional collision. The following four cases can be distinguished:
- $e = 1$ for perfectly elastic collisions. Kinetic energy is conserved;
- $0 < e < 1$ for partially elastic collisions. In the low-velocity regime between 2 and 3 m/s, $e$ is 0.5 for wood, 0.56 for steel, 0.89 for ivory, and 0.94 for glass {⇨Fig. 2.5}.[937] Kinetic energy is not conserved;
- $e = 0$ for totally inelastic collisions between bodies made of soft materials (such as lead and putty). This gives the maximum possible loss of kinetic energy; and
- $e > 1$ for hyperelastic collisions. Here kinetic energy is gained as a result of some internal change in one or both bodies (*e.g.*, due to a chemical explosion).

The coefficient of restitution can be determined by dropping a small ball on the horizontal surface of a large mass of the same material and comparing the height $h$ to which the ball bounces to the height $H$ from which it is dropped {⇨Fig. 4.3–F}; *i.e.*,

$$e = (h/H)^{1/2}.$$

The total dropping time $T$, which can involve numerous rebounds before the mass eventually comes to rest, is described by a geometric series in terms of $e$:

$$T = (2H/g)^{1/2} + 2(2H/g)^{1/2}(e + e^2 + e^3 + \ldots) = (2H/g)^{1/2}(1 + e)/(1 - e),$$

where $g$ is the acceleration due to gravity.[938]

| | | |
|---|---|---|
| 1880 | *Gelsenkirchen, North-Rhine Westphalia, Germany* | **The first testing gallery for studying the suitability of newly developed explosives as "permitted explosives" in coal mines is established.** ▪ The testing of permitted explosives is complicated by the fact that any of the explosives available may ignite firedamp if favorable conditions are chosen. Official tests can therefore only serve to distinguish between<br>▸ the explosives which will presumably not cause ignition under normal conditions of use underground, and will only do so under unusually severe and exceptional conditions; and<br>▸ those which are more likely to cause ignition and must therefore be rejected.<br><br>In subsequent years, similar facilities were also constructed in other European countries in order to simulate the gas and dust conditions in coal mines as closely as possible. For example, a station for testing explosives for use in coal mines was erected at Woolwich in England and opened for testing in June 1897. In the early 1940s, the ***Buxton test*** was developed at the Safety in Mines Research Station (Buxton, England) to determine the likelihood or limits at which an explosive will ignite gas or coal dust, before it can be placed on the official permitted list. This testing gallery uses a steel cannon combined with a gas explosion chamber to standardize the conditions when testing permitted mining explosives.[939] |
| 1880 | *Collège de France & Service des Poudres et Salpêtres, Paris, France* | P.E. Marcellin BERTHELOT[940] and Paul VIEILLE, two French chemists, **study the explosive properties of mercury fulminate** {HOWARD ⇨1800}. They measure its density ($\rho_0 = 4.42$ g/cm³) and determine a maximum pressure of 4,272 atm when the detonation takes place in a closed chamber. However, when detonating in contact with a target, mercury fulminate |

---

[936] H. HAYAKAWA and H. KUNINAKA: *Theory of the inelastic impact of elastic materials.* Phase Transitions: A Multinational Journal **77**, 889-909 (2004); http://arxiv.org/PS_cache/cond-mat/pdf/0312/0312005.pdf.
[937] *Hütte I* (herausgegeben vom Akadischen Verein Hütte, e.V. Berlin). W. Ernst & Sohn, Berlin (1931), p. 296.
[938] I. SZABÓ: *Repertorium und Übungsbuch der Technischen Mechanik.* Springer, Berlin *etc.* (1960), pp. 215-216.
[939] W. PAYMAN: *The testing of permitted explosives. Basis of the official test.* The Iron and Coal Trades Rev. (Lond.) **142**, 157-158 (1941).
[940] P.E.M. BERTHELOT and P. VIEILLE: *Etude des propriétés explosives du fulminate de mercure.* C. R. Acad. Sci. Paris **90**, 946-952 (1880).

generates a much higher pressure of 48,000 atm. • In comparison, guncotton generates a pressure of only 24,000 atm due to its lower density ($\rho_0 = 1.1$ g/cm$^3$).

| 1880 | *Laboratory of Mineralogy, Sorbonne University, Paris, France* | **Pierre CURIE**, a French physicist, **and his brother Jacques discover the phenomenon of piezoelectricity.** They notice that specially prepared crystals (*e.g.*, tourmaline, quartz, topaz, cane sugar, and Rochelle salt) produce surface charges when subjected to mechanical stress. • This so-called **"direct piezoelectric effect"** is very useful for measuring dynamic pressures of high-rate phenomena.<br><br>Initially, large arrays of crystals with a high piezoelectric constant (such as tourmaline) had to be used to obtain enough electric charge that small variations in charge could be recorded with a cathode-ray tube by directly connecting the gauge to the deflection electrodes {KEYS ⇨1921}. However, when electronic amplifiers became available in the 1920s, small piezoelectric pressure gauges were increasingly applied to the diagnostics of shock waves and detonation waves {CROW & GRIMSAW ⇨1932; JOACHIM & ILLGEN ⇨1932; REYNOLDS ⇨1943; GORANSON ET AL. ⇨1955; JONES ET AL. ⇨1962; LAUTERBORN ⇨1972}. |
|---|---|---|
| 1880 | *University of Agram, Kingdom of Croatia-Slovenia [now Zagreb, Croatia]* | **Čeněk** (or Vincent) **DVOŘÁK**,[941] a Bohemian physicist and previously assistant to Ernst MACH, **describes "a simple kind of schlieren observation," later termed** *shadowgraphy*. The method does not require any lenses or concave mirrors, thus allowing a large field of view {⇨Fig. 4.18–C}. • Shadowgraphy was invented as a flow visualization method in England by Robert HOOKE around 1672, using only the Sun and a white surface upon which to cast the shadow {SETTLES ⇨2001}. The change in illumination is roughly proportional to the rate of change in the density gradient. For this reason, DVOŘÁK 's method for observing certain flow phenomena may be superior to the schlieren method {A. TOEPLER ⇨1864}, in which the change of illumination is roughly proportional to the density gradient only.[942]<br><br>Large-field shadowgraphy, which is used to visualize propagating shock waves for example, can be realized using a retroreflective screen {EDGERTON ⇨1958; SETTLES ⇨1983; KREHL ET AL. ⇨1995}. |
| 1880 | *Tottenham, Middlesex, England* | **A. Vernon HARCOURT**, a British professor of chemistry, **investigates the explosion of a mixture of coal gas and air that happened in a large gas main near Tottenham Court Road.** In a report to the Board of Trade, he arrives at the conclusion that the flame traveled at a rate exceeding 100 yards per second (91 m/s) – a result which seems to be inconsistent with BUNSEN's conclusions, that the rate of propagation for the ignition of carbonic oxide and oxygen was less than 1 m/s {BUNSEN ⇨1857}. • On the other hand, Harold B. DIXON,[943] referring to his combustion studies carried out in the period 1876–1880, later annotated that he had observed that the flame produced by igniting a mixture of moist carbonic oxide and oxygen traveled through a long eudiometer too rapidly to be followed by the eye on several occasions. For example, in the winter of 1880/1881, DIXON was startled by the rapid increase in velocity and violence as a flame of carbon disulfide with nitric oxide traveled down a long glass vessel. The first measurements of detonation velocities were performed in France {MALLARD & LE CHÂTELIER ⇨1881; BERTHELOT ⇨1881}. |
| 1880 | *Bergakademie Příbram, Austro-Hungarian Empire* | **Hans HÖFER VON HEIMHALT**,[944] an Austrian professor of geology, **discusses how seismology can benefit mining and geologic exploration.** |

---

[941] C.(V.) DVOŘÁK: *Über eine neue einfache Art der Schlierenbeobachtung*. Ann. Phys. **9** [III], 502-511 (1880).
[942] D.W. HOLDER and R.J. NORTH: *Schlieren methods*. National Physical Laboratory (NPL), Notes on Applied Science No. 31. H.M.S.O., London (1963), p. 36.
[943] H. DIXON: *The rate of explosion in gases* [Bakerian Lecture]. Phil. Trans. Roy. Soc. Lond. **A184**, 97-188 (1893), see p. 98.
[944] H. HÖFER VON HEIMHALT: *Die Seismologie (Erdbebenkunde) im Dienste des Bergbaus*. Österr. Z. Berg- & Hüttenwesen **28**, 277-278, 290-292 (1880).

(i) He points out that most seismic shocks are produced along faults and are mechanical waves which do not originate from a "point" in the Earth's crust (as some contemporaries still assumed), but rather emerge from an area. **He coins the term *shock line*** [Germ. *Stoßlinie*], which is, according to his definition, a vertical projection of the loci of seismic excitation onto the surface. In order to improve prospecting, he recommends that the locations of shock lines determined from seismograph records or seismoscopes should be compared with corresponding geologic maps.

(ii) He illustrates the benefits of his method by referring to his previous studies of various earthquakes in Kärnten (1848–1876), which always resulted in the same shock line. Furthermore, shock lines determined from earthquakes that happened in the Herzogenrath/ Aachen region (1873, 1877), an important area of hard coal mining, reveal a system of faults which have so far only partially been used for mining.

In the same year, HÖFER VON HEIMHALT[945] also publishes a contribution to the theory of blasting.

| 1881 | *U.S. Army Corp of Engineers, Washington, DC* | **Lt.-Col. Henry L. ABBOT,**[946] a U.S. military engineer and inspector, **reports on various puzzling underwater explosion phenomena** observed over the period 1869–1879 associated with 697 underwater explosions performed with different charges, at different distances, and under different conditions of depth. |

(i) He attempts to measure the intensity of the pressure pulse at different points in the water.
• ABBOT had no means for determining the duration of the pressure or the characteristics of its rise and fall, even though these factors may be of equal importance in terms of their effects on a ship. Pressure measurements carried out during World War II revealed that ABBOT's gauges were too sluggish to give accurate results.

(ii) **He reports on strange surface phenomena associated with underwater explosions** that he observed about 0.1 seconds after the explosion: "The surface of the water around the torpedo over a distance of 200 feet [61 m] is covered by a misty spray resembling rain, which has been thrown upward from the surface by the shock. Over the torpedo appears a dome of water of which the diameter is about 100 feet [30.5 m] and the extreme height about 20 feet [6.1 m]. The surface of this dome is a fleecy texture; and through the top are bursting upward many spear-like jets, which cover a space about 50 feet [15.2 m] in diameter and attain in the middle an extreme height of 105 feet [32 m]." • Similar surface phenomena from underwater explosions had also been described by a number of military researchers in other countries {⇨1870s}. Accounts of observations of surface phenomena resulting from sea mine explosions were later collected and discussed by RUDOLPH {⇨1887}, who attempted to correlate them with similar sea phenomena that had already been observed in seaquakes.

Further studies on surface phenomena from underwater explosions, mostly performed by a number of the leading maritime nations during World War II, eventually revealed that three main phases can be clearly distinguished, particularly when the charge is fired near the surface:[947]
> ▸ the arrival of the primary shock at the surface, which produces a rounded dome of whitish water above the charge and a rapidly advancing ring of darkened water – known as the "slick," which spreads out from the charge;

---

[945] H. HÖFER VON HEIMHALT: *Beiträge zur Spreng- oder Minentheorie.* Österr. Z. Berg- & Hüttenwesen **28**, 135-138, 152-155, 161-163, 178-182, 189-191, 203-206, 213-215 (1880).
[946] H.L. ABBOT: *Report upon experiments and investigations to develop a system of submarine mines for defending the harbors of the United States.* Professional Papers of the Corps of Engineers of the U.S. Army, Rept. No. 23, Washington, DC (1881).
[947] R.H. COLE: *Underwater explosions.* Dover Publ., New York (1965), pp. 12, 210-211, 392-399.

- the reflection of the incident shock wave at the surface, causing an impressive vertical plume of spray – the so-called "dome." Another interesting observation is the occurrence of other domes caused by "secondary pulses" (or "bubble pulses") {BLOCHMANN ⇨1898} if the charge is fired at sufficient depth for the gas bubble to contract one or more times before venting; and
- the breakthrough of gases to the atmosphere, causing further plumes of spray mixed with explosion products.

| 1881 | *Royal Commission on Accidents in Mines, England* | Frederick A. ABEL,[948] an English chemist and explosives specialist, **carries out experimental explosions with coal dust in large mine galleries.** These studies, stimulated by the calamitous accident in the Seaham Colliery in Durham in the fall of 1880, lead to the following conclusions:<br>(i) Finely divided particles suspended in air provide a similar source of danger to that occasionally experienced in flour mills.<br>(ii) Some perfectly noncombustible powders, which are also not susceptible to any chemical change when exposed to a flame, are barely inferior to the most inflammable or sensitive of the Seaham dust samples in their power to bring about the ignition of an otherwise nonflammable mixture of firedamp and air.<br>(iii) Mixtures of firedamp and air, present in proportions bordering on those that will ignite upon the approach of a flame, combust instantly if they contain just a few particles of such noncombustible dusts in suspension, or Seaham or other dusts from coal pits.<br>(iv) When very highly inflammable dust is suspended in air in which no trace of hydrocarbon gas (firedamp) is present, a blown-out shot can produce ignitions that extend as far as the mixture of air with sufficient dust to maintain the flame.<br>(v) However, ABEL cautiously limits the applicability of the results of his model studies to real firedamp/coal-dust explosions: "In conclusion, it may be admitted as possible that, with the large volume of flame, and the great disturbing effect, of a blown-out shot, as the initiatory cause of the ignition of dust, and its suspension in the surrounding air, such inflammation may, in the complete absence of firedamp, be propagated to a greater distance than the results of small experiments would warrant one in assuming. But it can scarcely be maintained that the air of a mine in which the coal gives off gas at all can be, at any time, *free* from firedamp; and as the existence of very small and unsuspected quantities of that gas in the air of a mine may suffice to bring about the ready propagation of flame by coal-dust, and thus to develop violent explosive effects, it would appear needless to assume that coal-dust may, in the entire absence of firedamp, give rise to explosions, even of only limited character, in coal-mines, in order to account for casualties which cannot be ascribed to the existence of accumulations or sudden outbursts of firedamp."<br>Some years later, ABEL's experimental observations were essentially confirmed by the Prussian Firedamp Commission at Neunkirchen in the Saarbrücken District, which considerably extended his studies by conducting large-scale experiments. ▪ Discussions on the role of dust in firedamp explosions in the Seaham Colliery were again resumed in the early 1890s. **Since the only portions of the mine untouched by the explosion were those which were damp, and therefore free of dust, it was impossible to explain the propagation of this explosion in any other way than by the dust theory.** |
| 1881 | *Collège de France, Paris* | P.E. Marcellin BERTHELOT,[949] a French professor of organic chemistry and member of the French Firedamp Commission {⇨1877}, studies the propagation speeds of the flames obtained from essentially the same gaseous explosive mixtures as those investigated by MALLARD & LE CHÂTELIER {⇨1881} in the same year. However, unlike them, he **announces the discovery** |

---

[948] F.A. ABEL: *On colliery explosions* [Report on experiments with dust from Seaham Colliery]. Chemical News **44**, 16-18, 27-31, 39-42 (1881).
[949] P.E.M. BERTHELOT: *Sur la vitesse de propagation des phénomènes explosifs dans les gaz.* C. R. Acad. Sci. Paris **93**, 18-22 (1881).

of flame propagation speeds from explosions of gaseous mixtures that are much greater than had been measured previously. His important results are as follows:

(i) Using long tubes up to 5 m in length and 8 mm in diameter, he makes the important observation that the initially slow flame speed approaches a characteristic high limiting value after propagating a sufficiently long distance in the tube. That value is independent of the pressure of the gases, the material of the tube, and the inner diameter of the tube (above a small limit), but is constant for each gaseous mixture.

(ii) **He measures flame velocities of up to 2,500 m/s, calling this thermochemical phenomenon "explosion wave"** [French *l'onde explosive*], later generally called a "detonation wave." He does not explain this supersonic velocity of combustion as being due to any thermal conductivity and diffusion process that governs the propagation of a slow flame, but instead proposes that it is due to **the impact of the products of combustion "from layer to layer"** {BERTHELOT ⇨ 1870}, thus achieving an important step forward in the theory of detonation.

(iii) He also states that the velocity of the explosion wave could be predicted if the heats of combustion, the densities and the specific heats of the products are known. ▪ His discovery that each mixture of inflammable gases has a definite maximum velocity of explosion – *i.e.*, that the rate of explosion in terms of the mean translational velocity of the gaseous molecules is a new physicochemical constant – kickstarted a new era in the theory of explosions.

In the same year, BERTHELOT[950] **is the first to demonstrate that a shock wave can chemically decompose a gas into its elements.** Generating the shock wave with a high explosive (mercury fulminate), he uses acetylene [$C_2H_2$] as an example to show that the violent explosion (which is accompanied by a flash of light) transforms the initial gas into fine carbon particles dispersed in a hydrogen atmosphere. He reaches the important conclusion that **"these phenomena give evidence that direct thermodynamic relations exist between chemical and mechanical actions."** ▪ More recent studies performed by Carl Faust ATEN[951] and Edward F. GREENE at the Metcalf Chemical Laboratories of Brown University, who used infrared analysis of the quenched products, have shown that diacetylene [$C_4H_2$] is the most important product of acetylene pyrolysis upon shock heating: it can form over 5% of all reaction products, and it decomposes into hydrogen and carbon [$C_4H_2 \rightarrow C_4 + H_2$].

| | |
|---|---|
| 1881 *Chair of Experimental Physics, University of Graz, Austro-Hungarian Empire* | Ludwig E. BOLTZMANN,[952] an Austrian professor of mathematics and physics, **checks the theories previously derived for the longitudinal planar impact of two cylindrical bars experimentally** {CAUCHY ⇨ 1826; DE SAINT-VENANT ⇨ 1867}.<br><br>(i) According to these theories, the intensity of two impacting bars – even under the assumption of perfect elasticity – depends not only on the masses of the bars but also on their lengths and their longitudinal wave speeds: the impact produces longitudinal waves in each bar which are reflected at the free end and return to the point of impact, interfering with each other and thus altering the intensity of impact.<br><br>(ii) To approach perfectly elastic impact conditions, he performs experiments with glass and natural rubber bars of equal mass but different lengths and diameters, and measures their velocities before and after impact. In the case where both bars have the same length, he measures velocities in agreement with the Cauchy-Saint-Venant theory. However, for the impact of bars of different lengths, the velocity loss is less than predicted by theory, which he attributes to errors in measurement. |

---

[950] P.E.M. BERTHELOT: *Détonation de l'acétylène, du cyanogène et des combinaisons endothermiques en général*. C. R. Acad. Sci. Paris **93**, 613-619 (1881).
[951] C.F. ATEN and E.F. GREENE: *The rate of formation of carbon from the pyrolysis of acetylene in shock waves*. Disc. Faraday Soc. **22**, 162-166 (1956).
[952] L. BOLTZMANN: *Einige Experimente über den Stoß von Cylindern*. Ann. Phys. **17** [III], 343-347 (1882).

| 1881 | Russian Artillery Academy, St. Petersburg, Russia | **Nikolai V. MAIYEVSKII**,[953] a Russian physicist interested in the ballistics of cannon shells, **publishes a work on the aerodynamic drag of projectiles at high speeds $v$ that exceed the velocity of sound $a$.** For velocity ratios of up to $v/a < 1.1$, he finds that the "ballistic resistance" of spherical projectiles is proportional to the quantity
$$c_x = 1 + 3.34\,(v/a)^2.$$
Very similar to his predecessor Leonard EULER {⇨1745}, he considers the ratio $v/a$ – later termed *Mach number* {ACKERET ⇨1929} – to be an important quantity that governs aerodynamic drag at high velocities.[954] ▪ Today the term *ballistic resistance* is used in modern armor research to characterize the ability of a protective material to stop projectile penetration. |
|---|---|---|
| 1881 | Ecole des Mines, Paris, France | **François Earnest MALLARD**,[955] a professor of geology, mineralogy and physics, **and Henry L. LE CHÂTELIER**, a professor of chemistry, **measure the propagation of the flame front in gaseous explosive mixtures of $H_2/O_2$, $CO/O_2$ and $CH_4/O_2$.**

(i) They trace the progress of the flame photographically on sensitized paper moving vertically on a revolving cylinder using the Deprez chronograph {DEPREZ ⇨1874}; *i.e.*, **they produce a streak record of the propagating detonation front.** ▪ Seven years later, this method was improved by using a rotating mirror {VON OETTINGEN & VON GERNER ⇨1888}.

(ii) In short tubes with a length of 1.35 m, they observe a speed of 570 m/s in oxyhydrogen which, however, reduces to only 70 m/s when the tube length is reduced to about a quarter of this. Very similar to DAVY {⇨1816}, they notice that the flame is not propagated in narrow tubes.

(iii) They observe that the detonation wave is
▸ not progressive, it is always instantaneous; and
▸ characterized not only by its great speed, but also by its intense luminosity and the very high pressures associated with it.

In the same year, the French chemist P.E. Marcellin BERTHELOT {⇨1881} independently discovers flames from gaseous explosions that travel even faster. ▪ MALLARD and LE CHÂTELIER as well as BERTHELOT and VIEILLE, who were all members of the French Firedamp Commission {⇨1877}, formed two very effective teams investigating explosion phenomena in gaseous mixtures {BERTHELOT & VIEILLE ⇨1882; MALLARD & LE CHÂTELIER ⇨1883} – then a new area of combustion research. |
| 1881 | Sevran [a town near Paris], France | **Alfred NOBEL**, the renowned Swedish inventor of numerous high explosives and gun propellants, **transfers his chemical laboratory from Paris to Sevran.**
Two years later, NOBEL obtained permission from the French Government to place a cannon and establish a small shooting range in an abandoned fort near his laboratory, which allowed him to test newly developed gun propellants on his own.[956] |
| 1882 | Service des Poudres et Salpêtres, Paris, France | **Foundation of the French journal *Mémorial des Poudres et Salpêtres* by the order of a ministerial decree.**[957] The editorial board consists of Emile SARRAU, C. ARNOULD, and E. DÉSORTAUX. ▪ It was the first professional journal exclusively dedicated to the communication of results from research into explosives, and their application for civil and military purposes. |

---

[953] N. MAIYEVSKII: *Sur les résultats des expériences concernant la résistance de l'air et leur application à la solution des problèmes du tir*. Bull. Acad. Imp. Sci. St. Pétersbourg **27**, 1-14 (1881).

[954] F.I. FRANKL: *On the priority of EULER in the discovery of the similarity law for the resistance of air to the motion of bodies at high speeds* [in Russ.]. Dokl. AN (SSSR) **70**, No. 1, 39-42 (1950).

[955] E. MALLARD and H.L. LE CHATELIER: *Sur les vitesses de propagation de l'inflammation dans les mélanges gazeux explosifs*. C. R. Acad. Sci. Paris **93**, 145-148 (1881).

[956] H. DE MOSENTHAL: *The life-work of Alfred NOBEL*. J. Soc. Chem. Ind. **18**, 443-451 (1899).

[957] The journal was published under this name from 1882/1883 (vol. 1) to 1913/1914 (vol. 17). It continued as *Mémorial des Poudres* from 1921 (vol. 18) to 1939 (vol. 29), and from 1948 (vol. 30) to 1964/1965 (vol. 46/47).

Initially created with the aim of improving communications between French researchers regarding explosion technology, it soon grew into an international scientific forum.

| | | |
|---|---|---|
| 1882 | *Collège de France & Laboratoire Central, Service des Poudres et Salpêtres, Paris, France* | **P.E. Marcellin BERTHELOT**[958, 959] **and Paul VIEILLE,** two French chemists, **measure the detonation velocities in around 50 mixtures of fuels and oxidizers diluted by different amounts of nitrogen:**

(i) They use the Deprez chronograph {DEPREZ ⇒1874} and later the Le Boulengé chronograph {LE BOULENGÉ ⇒1860s}. Both are electromechanical instruments that are triggered and stopped by electrical pulses. The start pulse is provided by the same electric spark used to ignite the gaseous mixture, and the stop pulse is generated by suspending a thin foil strip covered with a small amount of fulminate perpendicularly to the tube axis, which explodes when the detonation front arrives, thus breaking the holding current in the chronograph.

(ii) **They observe a quasi-stationary detonation velocity that only depends on the composition of the mixture, but not on the tube material and diameter, so long as the latter is not too small.**

(iii) They conclude that the velocity of the "explosion wave" {BERTHELOT ⇒1881} is quite independent of the material and diameter of the tube employed, provided that a certain small limiting diameter is exceeded {BUNSEN ⇒1867}, and it is also independent of the pressure.
▪ The conclusion about the pressure was subsequently found to be erroneous by a researcher in England {DIXON ⇒1893}.

(iv) They state that "the velocity of the explosion wave is determined by the same general laws which apply to the sound velocity" [*"La vitesse de propagation de l'onde explosive est réglé par les mêmes lois générales que la vitesse du son."*]. ▪ Their "sound-wave theory" was not superseded until the advent of fundamental works on shock waves and detonation {HUGONIOT ⇒1887; CHAPMAN ⇒1899; VIEILLE ⇒1899 & 1900; JOUGUET ⇒1904–1906; CRUSSARD ⇒1907}, which found that "detonation" is essentially a "shock wave" propagating through a medium that is discontinuous in the vicinity of the wave front.

It is interesting to note here that August TOEPLER {⇒1864}, who first observed the discontinuous front of a propagating spark wave (*i.e.*, a weak shock wave) with his schlieren method, also erroneously used the conventional term *sound wave* to designate the new wave type, although this immediately provoked criticism by the reviewers of his submitted manuscript {MAGNUS ⇒1867; KIRCHHOFF ⇒1868}. This illustrates how slowly the new phenomenon of a "shock wave" was adopted by the scientific community of that time. |
| 1882 | *University of London, England* | **Heinrich DEBUS,**[960] a German professor of chemistry, stimulated by previous investigations into the chemical reactions of gunpowder {BUNSEN & SCHISCHKOFF ⇒1857; NOBLE & ABEL ⇒1875}, **proposes a *chemical theory of gunpowder*** which – as he emphasizes – "explains in a satisfactory manner the chemical reactions which occur during and after the explosion, not only of a powder of normal composition, but, generally, of a mixture of $x$ molecules of saltpeter, $y$ atoms of carbon and $z$ atoms of sulfur." ▪ However, ordinary gunpowder contains more carbon and sulfur than is required for complete combustion {NOBLE & ABEL ⇒1875}. This leads to two endothermic processes which do not produce heat, but rather consume it. |

---

[958] P.E.M. BERTHELOT: *Sur l'onde explosive*. C. R. Acad. Sci. Paris **94**, 149-152 (1882); Ann. Chim. Phys. **28** [V], 289-332 (1883).
[959] P.E.M. BERTHELOT and P. VIEILLE: *Nouvelles recherches sur la propagation des phénomènes explosifs dans les gaz*. C. R. Acad. Sci. Paris **95**, 151-157 (1882); *Sur la période d'état variable qui précède le régime de détonation et sur les conditions d'établissement de l'onde explosive*. Ibid. **95**, 199-205 (1882).
[960] H. DEBUS: *Chemical theory of gunpowder* [Bakerian Lecture]. Phil. Trans. Roy. Soc. Lond. **173**, 523-594 (1882); *Chemische Theorie des Schießpulvers. Erste Abhandlung*. Justus Liebig's Ann. Chem. **212**, 257-315, Ibid. **213**, 15-65 (1882); *Zweite Abhandlung*. Ibid. **265**, 257-315 (1891).

| 1882 | Physikalisches Institut, Friedrich-Wilhelms-Universität, Berlin, Germany | **Heinrich R. HERTZ,**[961] the German physicist more famous for his pioneering experiments in broadcasting and receiving radio waves than for his contributions to mechanics, **treats the collision of solid bodies analytically.**

(i) He views the contact of two bodies as an equivalent problem to that encountered in electrostatics, and **by applying the potential theory he calculates the stress and strain resulting from the force acting in the contact area.** He first makes the important assumption that the duration of the percussion is much longer than the time that the elastic wave needs to travel through the colliding body. For straight, elastic percussion of two spheres colliding head-on with relative velocity $v$, he obtains a solution in the form of a potential – the **"Hertz law of contact,"** sometime referred to as the **"Hertz theory of impact"** – which describes the stresses and deformation near the contact point as a function of the geometric and elastic properties of the bodies ("Hertz contact model"). From this he derives a simple formula for two steel spheres of radius $R$ [mm] which contact each other with a relative velocity $v$ [mm/s]. Using the force unit kg* (now kp), he obtains for

▸ the maximum pressure: $P = 29.1\, v^{2/5}$ [kg/mm$^2$],
▸ the duration of contact: $T = 0.000024 R\, v^{-1/5}$ [s],
▸ and the diameter of the contact area: $a_m = 0.0020 R v^{2/5}$ [mm].

(ii) He gives an excellent example of the dependency of the contact time on the diameters of the spherical bodies: two 50-mm-dia. steel balls impacting head-on with a relative velocities of 10 mm/s contact each other with a maximum pressure of 73 kp/mm$^2$ in a tiny circular contact area 0.26 mm in dia. for a duration of 380 μs; however, if the two steel spheres had the same diameters as the Earth, the contact time would be almost 27 hours.

(iii) He also measures the size of the contact surface by covering one of the colliding bodies with soot, thus providing experimental proof of his theory.

HERTZ's theory applies to normal contact between two elastic solids that are smooth and can be described locally with orthogonal radii of curvature, but does not account for tangential forces that may develop in applications where the surfaces slide or carry traction. His theory of impact became an essential component of classical percussion theory and generations of physicists and engineers have applied it to quite different cases of percussion:

▸ Pioneering experimental studies revealed that the duration of impact, $T$, decreases with increasing percussion velocity $v$ {HAMBURGER ⇨1885; TAIT ⇨1892; BERGER ⇨1924}.
▸ August FÖPPL,[962] a famous German professor of mechanical engineering at TH Munich, tried to apply HERTZ's method to two circular cylinders of equal diameter and made from the same ductile material, where the axes of the cylinders were perpendicular to each other. However, it is very difficult to find the load at which permanent setting begins in ductile materials.
▸ In 1906, the Russian mechanical engineer Aleksandr Nikolaevich DINNIK[963] performed impact experiments that involved measuring contact times using the ballistic galvanometer {POUILLET ⇨1844}. He showed the validity of HERTZ's formula for the impact of steel spheres; however, for zinc and (particularly) for lead spheres he observed considerable discrepancies from the formula.
▸ HERTZ's theory also holds well for the interaction of the spheres in NEWTON's cradle {The Royal Society of London ⇨1666}, regardless of whether balls of the same size or mass are used. However, as the collision perturbation passes through the chain, ad- |

---

[961] H. HERTZ: *Über die Berührung fester elastischer Körper.* J. Reine & Angew. Math. **92**, 156-171 (1882).
[962] A. FÖPPL: *Dauerversuche von BAUSCHINGER, ausgeführt in den Jahren 1886–1893.* Mittheilungen aus dem Mechanisch-Technischen Laboratorium der K. Technischen Hochschule München. 25. Heft (Neue Folge), 51 pages (1897).
[963] A.N. DINNIK: *HERTZ's formula and its experimental verification* [in Russ.]. Zurnal Russkago Fiziko-Chimiceskago Obscestva pri Imperatorskom Sankt-Peterburgskom Universitete **38**, 242-249 (1906).

ditional collision times for subsequent collisions are only half the collision time for the first collision.[964] This shows that **chain percussion cannot be broken down into components as if each ball is separated by a small distance, but rather that the interaction times for each pair of neighboring balls overlap.**

▸ Heinrich GOBRECHT,[965] an experimental physics professor at TU Berlin, **extended HERTZ's theory for the important case of two colliding balls with masses $m_1$ and $m_2$.** The relationship between contact time $T$ and the impact velocities $v_1$ and $v_2$ is given by

$$T = 2.94 \{25C^3(m_1 m_2)^2/[(16\, v_1 - v_2) \times (m_1 + m_2)^2]\}^{1/5},$$

where

$$C = \{[(9R_1 + R_2)/(64 R_1 R_2)] \times [(1 - \mu_1)G_1 + (1 - \mu_2)G_2]\}^{1/3}.$$

Here $C$ is a constant that depends on the radii $R_1$ and $R_2$, $\mu_1$ and $\mu_2$ are the Poisson ratios, and $G_1$ and $G_2$ are the shear moduli of balls "1" and "2," respectively.

▸ Yoh-Han PAO,[966] a Chinese-born U.S. physicist at E.I. du Pont de Nemours (Wilmington, DE) **extended the Hertz theory of contact to certain classes of viscoelastic bodies.** For example, he carried out numerical calculations for the case of plastic objects dropped onto a planar steel surface from a height of 4 ft (1.22 m). His results show that very high stresses can be obtained in impacts, even when the impact velocity is quite low (about 4.9 m/s).

▸ HERTZ's theory created **Contact Mechanics,** a new branch of engineering mechanics.[967] Contact problems are an area of great technical importance in industrial applications in mechanical and civil engineering, however only very few problems involving contact can be solved analytically. For most industrial applications, numerical methods have to be applied since the contacting bodies have complex geometries, undergo large deformations (including time dependent responses) or are affected by other factors.[968]

▸ William J. STRONGE,[969] an engineering professor at Cambridge University, **extended HERTZ's theory for elastic-plastic impact** which results in permanent (*i.e.*, irreversible) indentation of the contact surfaces of the colliding bodies.

In the same year, HERTZ,[970] continuing his mechanical studies, **calculates the elastic stress distribution for the contact of a hard sphere on a plate.** He finds that in the half-space compressive stress is limited to a conical region which extends from the point of contact into the half-space, while tensile stresses only exist outside this conical region – the so-called **"Hertzian cone"** {⇨Fig. 4.2–B} – and a paraboloidal area near the axis of symmetry. ▪ The apex of the Hertzian cone extends from the point of contact under a cone angle of 130 ± 10°; this is independent of the radius of the tip, load, and the type of material.[971] It is a well-known experimental fact that surface cracks systematically form outside the edge of the circle of contact, where the maximum tensile stress occurs. The fracture of a brittle solid under a spherical indenter – **"Hertzian fracture"** – is the best-studied case for fracture in a strongly inhomogeneous, well-defined, stress field.[972]

---

[964] D.R. LOVETT, K.M. MOULDING, and S. ANKETELL-JONES: *Collisions between elastic bodies: NEWTON's cradle.* Eur. J. Phys. **9**, 323-328 (1988).
[965] H. GOBRECHT: *Lehrbuch der Experimentalphysik.* W. de Gruyter & Co, Berlin (1970), vol. I, p. 264.
[966] Y.H. PAO: *Extension of the Hertz theory of impact to the visco-elastic case.* J. Appl. Phys. **26**, 1083-1088 (1955).
[967] K.L. JOHNSON: *Contact mechanics.* Cambridge University Press, Cambridge (1985, 2003).
[968] P. WRIGGERS: *Computational contact mechanics.* Springer, Berlin (2006).
[969] W.J. STRONGE: *Contact problems for elastic-plastic impact in multi-body systems.* In: (B. BROGLIATO, ed.) *Impacts in mechanical systems: analysis and modelling.* Lecture Notes in Physics, Springer, Berlin *etc.* (2000), pp. 189-234.
[970] H. HERTZ: *Über die Berührung fester elastischer Körper und über die Härte.* Verhandlungen des Vereins zur Beförderung des Gewerbefleißes (Berlin) **61**, 449-463 (Nov. 1882).
[971] G.P. CHEREPANOV: *Mechanics of brittle fracture.* McGraw-Hill Int. Book Co., New York *etc.* (1979), pp. 550-556.
[972] F.C. FRANK and B.R. LAWN: *On the theory of Hertzian fracture.* Proc. Roy. Soc. Lond. **A299**, 291-306 (1967).

"**Hertzian cracks**" are best observed in transparent brittle and hard media such as glass; they can be generated either statically by pressing a small hard sphere on a planar surface or dynamically by a low-velocity impact {KERKHOF & MÜLLER-BECK ⇨1969; ⇨Fig. 4.2–B}. The Hertzian cone also exists in flint stone, obsidian, and other hard (and rather isotropic) materials. In prehistoric times, the splitting of fragments from stones by Hertzian cone fracture – "**flakes**" – was used to create handaxes, arrowheads, scrapers, knifes, *etc.* {⇨ *c.*45,000 years ago; ⇨Fig. 4.2–A}.

| | | |
|---|---|---|
| 1882 | Stockholm, Sweden | **Carl Gustav P. DE LAVAL,**[973] a Swedish engineer, **invents the first steam turbine.** He uses steam jets issued from nozzles, which release their energy by pushing a vane; such a system is called the "**action turbine.**" His steam turbine requires steam jets with velocities that are as high as possible, which will eventually lead to the invention of the Laval nozzle, which in turn will stimulate studies in supersonic aerodynamics. ▪ Earlier, when working as a metallurgical engineer at the iron works in Klosterbruck (Moravia), DE LAVAL had begun to consider a rotating prime mover that did not have pistons and crankshafts. His diary of 1876 shows a sketch of an action turbine, although not his expansion nozzle {DE LAVAL ⇨1888}.[974] |
| 1882 | Imperial College of Engineering, Tokyo, Japan | **John MILNE,**[975] a British professor of geology who has been resident in Japan since 1876, and will become one of the main founders of seismology, **reviews the present status of seismology in Japan. He states that "the greater the initial disturbance the greater the velocity of propagation."** ▪ MILNE's observation that the seismic shock propagates more quickly as the strength of the disturbance increases – a striking wave propagation effect that is typical of shock waves – may have been prompted by the compression of previously loose and inhomogeneous strata material.<br><br>In shock wave physics, a similar effect is observed when a porous substance is compressed by the passage of a shock wave. Plotting the equation of state of a porous material in a diagram of $p = p(\rho)$, porosity shows up as a pronounced plateau in the curve concave upwards. As soon as all of the voids have been closed by the increase in pressure, the curve again begins to rise {MURRI ET AL. ⇨1974}. |
| 1882 | Laboratoire Central, Service des Poudres et Salpêtres, Paris, France | **Paul M.E. VIEILLE,**[976] a French physicist and explosives chemist, **first uses a dynamic method to determine the accelerating force of an explosion.** Arranging a small piston in the wall of his explosion test chamber, he records its displacement-time profile under the action of the expanding gases on a soot-covered rotating drum, and obtains the acceleration by twice differentiating the measured profile. |
| 1883 | Perbuatan Volcano, Krakatau Island (Indonesia) in the Sunda Strait, Indian Ocean | Between August 26 at about 6 P.M. and August 27 at about 10 A.M., three violent underwater volcanic explosions of Perbuatan Volcano – better known as "Krakatau (or Krakatao, Krakatoa) Volcano" – destroy a large part of the volcano, thereby partly exposing its inner structure {⇨Fig. 4.1–F}. The last eruption, which is the most violent (*VEI* = 6), produces a partial collapse of the crater. **The enormous mass of lava spilled into the ocean produces the greatest steam explosion in historic times.** A series of huge seismic sea waves (tsunamis) as high as 40 meters inundate the adjacent coasts up to 5 km inland, killing more than 36,000 people. ▪ Before the 1883 eruption, the island of Krakatau was made up of three cones: Perbuatan (altitude 122 m), Danan (445 m) and Rakata (823 m). During the last phase of eruption nearly the whole of the island disappeared except for a horseshoe-shaped remnant of the cone of Rakata |

---

[973] C.G.P. DE LAVAL: *Turbine.* Swed. Patent No. 325 (1883).
[974] I. JUNG: *Dr. DE LAVAL and his early work with the steam turbine* [De Laval Memorial Lecture]. Aktiebolaget de Lavals Ångturbin, Stockholm (1957).
[975] J. MILNE: *Seismology in Japan.* Nature **26**, 627-631 (1882).
[976] P. VIEILLE: *Sur la mesure des pressions développées en vase close par les mélanges gazeux explosifs.* C. R. Acad. Sci. Paris **95**, 1280-1282 (1882); *Etude sur le rôle des discontinuités dans les phénomènes de propagation.* Mém. Poudres & Salpêtres **10**, 177-260 (1900).

{⇒Fig. 4.1–F}. New eruptions at the volcano since 1927 have built *Anak Krakatau* ("Child of Krakatau"), a new island.

Later investigations brought to light the following unusual phenomena:

(i) At Merak, a community only 35 miles (56.3 km) from Krakatau, the wave's waterline along several hills located two miles (3.2 km) inland was over 100 ft (30.5 m) high.[977]

(ii) At Moltke Harbor in South Georgia (southern Atlantic Ocean), the water gauge recorded a small number of precursors followed by a large number of strong seismic sea waves, with both series of waves superimposed as spikes on the daily tidal fluctuations.[978]

(iii) The mighty, eruption-induced atmospheric blast wave, encountered as a "roar" at Batavia [now Djakarta] that caused windows to break, was still audible at a distance of 3,500 km in Australia and 4,750 km on Rodriguez Island {⇒Fig. 4.1–F}.[979]

(iv) **A barograph deflection of about 7 millibar is recorded 150 km from Krakatau by Sir Richard STRACHEY,**[980] an Indian administrator and chairman of the meteorological council of the Royal Society of London. Residues of the blast wave are also recorded as barometric fluctuations around the globe.[981]

| | | |
|---|---|---|
| 1883 | Somerset House, The Royal Society of London, England | At a meeting of the Society, first details of the violent Krakatau eruption are communicated, particularly of the "great air-wave," its nature and extent.<br><br>In the following year, a Krakatau Committee was appointed by the British Royal Society with the goal "to collect the various accounts of the volcanic eruption at Krakatau, and attendant phenomena, in such a form as shall best provide for their preservation, and promote their usefulness." ▪ **The report of the Krakatau Committee became the first comprehensive scientific documentation of an explosive volcanic eruption.** It included:<br>▸ accounts on the great air-wave and sounds, the seismic sea wave and on unusual optical atmospheric phenomena;<br>▸ studies on the nature and distribution of the materials ejected from Krakatau; and<br>▸ reports of unusual optical, magnetic and electrical phenomena which accompanied the Krakatau eruption, based on a series of observations made in various parts of the world.<br><br>Based upon hundreds of accounts, it also brought to light new knowledge on the propagation behavior of blast waves and tsunamis over large distances, and even around the globe.[982] Since Krakatau sent material into the stratosphere to a height of about 30 km near the equator, **Sir Harold JEFFREYS,**[983] a British geophysicist and astronomer, **suggested that the average vertical velocity of ejected material was at least 0.8 km/s (*i.e.*, supersonic),** but that the actual initial velocity would even be greater, since energy would be lost through aerodynamic drag. A detailed summary of the effects of the Krakatau eruption was given more recently.[984] |

---

[977] J.D. TRUBY: *Krakatoa – the killer wave.* Sea Frontiers **17**, No. 3, 130-139 (1971).

[978] G.B. NEUMAYER: *Die Katastrophe in der Sunda-Straße.* Ann. Hydrogr. **12**, 201-207, 254-259, 359-369 (1884).

[979] R.D.M. VERBEEK: *Krakatau 1883, the volcanic eruption and its effects* (T. SIMKIN and S. FISKE, eds.). Landsdrukkerij, Batavia (1885). Smithsonian Institution Press, Washington, DC (1983).

[980] R. STRACHEY: *On the air waves and sounds caused by the eruption of Krakatoa in August, 1883.* In: (G.J. SYMONS, ed.): *The eruption of Krakatoa and subsequent phenomena: report of the Krakatoa Committee of the Royal Society.* Trübner, London (1888), pp. 57-88.

[981] G.B. NEUMAYER: *Über die durch den Ausbruch des Vulkans Krakatau am 26.–27. August hervorgerufenen atmosphärischen Erscheinungen.* Mitth. Geogr. Gesell. Hamburg (1884), pp. 309-312.

[982] G.J. SYMONS (ed.): *The eruption of Krakatoa and subsequent phenomena: report of the Krakatoa Committee of the Royal Society.* Trübner, London (1888).

[983] H. JEFFREYS: *The Earth. Its origin, history and physical constitution.* Cambridge University Press, Cambridge (1976), p. 500.

[984] T. SIMKIN and R.S. FISKE: *Krakatau 1883 – the volcanic eruption and its effects.* Smithsonian Institution Press, Washington, DC (1983).

| | | |
|---|---|---|
| 1883 | *Schießbaumwoll-fabrik Wolff & Co., Walsrode, Lower Saxony, Germany* | **Max VON FÖRSTER,**[985] chief company engineer, experiments with guncotton and **rediscovers the *hollow cavity effect* for high explosive charges without inlet** {VON BAADER ⇨1792} – also called the **"von Foerster (Förster) effect"** in Europe.

(i) Based on his working hypothesis "to give the detonating gases of gun-cotton a certain direction aiming towards the target," he performs a series of experiments with hollow cartridges, which he positioned on top of a thick wrought-iron plate {⇨Fig. 4.16–B}.

(ii) In conclusion, he writes, "Estimated on the whole, it appears that the effect of the hollow cartridge of the same size and less weight is superior to the full one of more weight…"

The important discovery and rediscovery of the *hollow charge effect* (or *shaped charge unlined cavity effect*) can be ascribed to other inventors of different nationalities {MUNROE ⇨1888}. Historical perspectives on this unique phenomenon of great practical (military) importance were published in Germany during World War II by Heinz FREIWALD,[986] and in 1954 by Hubert SCHARDIN,[987] two research physicists. A more recent review was given by Donald R. KENNEDY,[988] a U.S. consulting engineer, to commemorate the 100th anniversary of VON FÖRSTER's discovery of the shaped charge effect.

Compared to the shaped charge *unlined* cavity effect, the shaped charge *lined* cavity effect is a phenomenon of greater importance for both military and civil applications, as discovered much later in SCHARDIN's institute by Franz Rudolf THOMANEK {⇨1938}. The tip velocities of shaped charges with lined cavities can reach values – the highest velocity is 90 km/s for beryllium, 70 km/s for copper, and 46 km/s for lead[989] – and ultrashort X-ray pulses are required to stop their motion on film {⇨Figs. 4.8–F, G}. |
| 1883 | *Chair of Experimental Physics, German Charles University, Prague, Bohemia, Austro-Hungarian Empire* | **Ernst MACH,**[990] an Austrian philosopher of science and pioneer of supersonics and gas dynamics {E. MACH & WOSYKA ⇨1875; E. MACH & SOMMER ⇨1877; E. MACH & GRUSS ⇨1878; E. MACH & VON WELTRUSBY ⇨1878; E. MACH & WENTZEL ⇨1884 & 1885; E. MACH & SALCHER ⇨1886; E. & L. MACH ⇨1889; E. MACH & SALCHER ⇨1889; E. MACH & DOSS ⇨1893}, **publishes his book *Die Mechanik in ihrer Entwicklung, historisch-kritisch dargestellt.*** ▪ His book went through 9 German and 6 English editions. The second German edition was first published in 1893 in English translation as "The Science of Mechanics: A Critical and Historical Exposition of its Principles."

(i) In the preface to the first German edition, he specifies the objective of his book: "The present volume is not a treatise upon the application of the principles of mechanics. Its aim is to clear up ideas, expose the real significance of the matter, and get rid of metaphysical obscurities."

(ii) In the second chapter, entitled "The Development of the Principles of Dynamics," he discusses the fundamental principles of dynamics. Emphasizing the crucial importance of a critical analysis of physical principles, he also addresses the particular importance of studying the Laws of Percussion (Germ. *Stoßgesetze*, translated into "Laws of Impact") in the evolution of dynamics. In a section entitled "The Laws of Impact," he states: **"The laws of impact were the** |

---

[985] M. VON FÖRSTER: *Versuche mit komprimirter Schießbaumwolle in der Schießbaumwollfabrik Wolff & Co., Walsrode.* Mittler & Sohn, Berlin (1883). See Van Nostrand's Engng. Mag. **31**, 13-119 (July/Dec. 1984).

[986] H. FREIWALD: *Zur Geschichte der Hohlraumwirkung bei Sprengladungen* [mit einem Vorwort von Hubert SCHARDIN]. Schriften Dt. Akad. Luftfahrtforsch. Nr. 1046 (1941). ▪ This rare and then secret report is now kept at the Deutsches Museum München; *see* Doc. No. DAL Schr. 1046/41g.

[987] H. SCHARDIN: *Über die Entwicklung der Hohlladung.* Wehrtech. Monatshefte **51**, Nr. 4, 97-120 (1954).

[988] D.R. KENNEDY: *History of the shaped charge effect. The first 100 years.* Company brochure prepared by D.R. Kennedy & Associates, Inc., Mountain View, CA (1983).

[989] P.V. PIPICH: *Experimental investigation of fast shaped-charge jets.* J. Appl. Mech. Tech. Phys. **41**, 818-823 (2006).

[990] E. MACH: *Die Mechanik in ihrer Entwicklung, historisch-kritisch dargestellt.* Brockhaus, Leipzig (1883). Engl. translation by Thomas J. MCCORMACK: *The science of mechanics: a critical and historical account of its development.* Open Court, Chicago, IL (1893). ▪ Further editions of MACH's book were published in Germany (1888, 1897, 1901, 1904, 1908, 1912, 1921, 1933, 1988) and in the United States (1893, 1902, 1915, 1919, 1942, 1960, 1974).

occasion of the enunciation of the most important principles of mechanics, and furnished also the first examples of the application of such principles."

(iii) He discusses previous contributions to impact {GALILEI ⇨1638; DESCARTES ⇨1644; MARCI ⇨1639; HUYGENS ⇨1652, 1668/1669 & 1703; WALLIS ⇨1669 & 1670/1671; WREN ⇨1669; MARIOTTE ⇨1671; ROBINS ⇨1742}, but does not address shock waves – then still commonly designated as "waves of finite amplitude" – although by that time he had already made a number of pioneering discoveries on their nature of propagation and interaction.

Ernst MACH's critical reflections on mechanics gave rise to a spiritual discussion of the scientific, historical, and philosophical foundations of classical physics.[991] After MACH's death in 1916, Albert EINSTEIN, referring to MACH's book *Die Mechanik*, wrote in his obituary on him,[992] "There you will find set forth brilliantly ideas which by no means as yet have become the common property of physicists," and in his autobiography EINSTEIN[993] remembered, "This book exercised a profound influence upon me … while I was a student."

| | | |
|---|---|---|
| 1883 | *Ecole des Mines, Paris, France* | **François Earnest MALLARD**[994] **and Henry LE CHÂTELIER,** two French chemists, **study the explosivity of mixtures of air and methane.** |

(i) They observe that there is a certain delay or "period of induction" before the gaseous mixture actually explodes. At 650 °C this delay amounts to about 190 seconds, at 1,000 °C to about 1 second, and at 2,200 °C the explosion occurs momentarily without any appreciable delay. ▪ The French Firedamp Commission {⇨1877}, which initiated these tests, later published a temperature of 1,500 °C as the maximum permissible for explosives in coal seams, and 1,900 °C for those intended to be used in the accompanying rock.

(ii) Using a rapidly rotating drum covered with film, upon which the image of the flame is projected {⇨Fig. 4.17–C}, they make streak records of flames from explosions in glass tubes ranging from 10 to 20 mm in dia. and from 1 to 3 m in length. ▪ Their technique of photographically recording accelerating flames and detonations was used and improved in England, France and later also in Germany. These studies were mostly concerned with fully developed detonation. However, photographic recording also proved very useful when investigating the process of transition from deflagration to detonation. For example, in the early 1960s, combustion researchers at UC Berkeley studied how flame propagation and induction distances depend on the tube diameter.[995]

(iii) Depending on the tube length and boundary conditions at the tube exit, they observe the following propagation phenomena:
- a phase of constant low propagation velocity ("deflagration");
- a phase of vibrations ("intermediate state"); and
- a phase of rapid wave propagation ("detonation wave").

They conclude that the detonation wave heats up the gaseous explosive mixture by adiabatic compression, thus reaching ignition temperature.

| | | |
|---|---|---|
| 1883 | *Akabane Engineering Works, Tokyo, Japan* | **John MILNE,**[996] an English seismologist and geologist, **resumes the method of generating artificial earthquakes** {MALLET ⇨1846 & 1861} while continuing his previous seismological studies {MILNE ⇨1882}. |

---

[991] E.N. HIEBERT: *Ernst MACH.* In: (C.C. GILLESPIE, ed.) *Dictionary of scientific biography.* Ch. Scribner's Sons, New York, vol. 8 (1973), pp. 595-607.
[992] A. EINSTEIN: *Ernst MACH.* Physik. Z. **17**, 101-104 (1916).
[993] *Albert EINSTEIN: philosopher-scientist.* In: (P.A. SCHILPP, ed.) *Library of living philosophers.* Evanston, IL (1949), pp. 20-21.
[994] F.E. MALLARD and H.L. LE CHATELIER: *Recherches sur la combustion des mélanges gazeux explosifs.* Annales des Mines **4** [VIII], 274-568 (1883).
[995] W. BAUMANN, P.A. URTIEW, and A.K. OPPENHEIM: *On the influence of tube diameter on the development of gaseous detonation.* Z. Elektrochemie **65**, 898-902 (1961) ▪ In this paper, the authors briefly reviewed previous studies of the development of detonation based on photographic records, as first used by MALLARD & LE CHÂTELIER.
[996] J. MILNE and T. GRAY: *On seismic experiments.* Phil. Trans. Roy. Soc. Lond. **173**, 863-883 (1883).

(i) In order to investigate a number of phenomena connected with earthquake motion, he attempts to determine in particular
- the difference between the magnitudes and characteristics of the motions produced at stations situated at various positions with regard to the point at which the blow was struck;
- the relation between normal and transverse vibration, as simultaneously exhibited at the various stations; and
- the transmission velocities of normal and transverse vibrations.

(ii) With the assistance of his colleague Thomas GRAY, a professor of telegraphic engineering, **MILNE produces seismic disturbances by percussion by dropping a 1,800-lb** (816.5-kg) **weight from a 35-ft** (10.7-m)**-high tower.** By varying the dropping height, the impact velocities could be increased up to 14.5 m/s. • MILNE's method of generating artificial earthquakes by dropping a large weight from a tower – a simple but unwieldy method compared to the use of explosives {MALLET ⇨1846} – was later resumed by MINTROP {⇨1911} at Göttingen University.

(iii) MILNE uses the **horizontal *Ewing seismometer*** (1880), an invention by James A. EWING, a professor of mechanical engineering and physics at Tokyo's Imperial College, and the **vertical *Gray-Ewing seismograph*** (1882) {⇨Fig. 4.3−M}. **Using both of these seismometers,** set into motion by an auxiliary seismometer shortly before the arrival of the seismic shock wave, **all three of ground motion components are recorded for the first time** (*i.e.*, up and down, back and forth, and side to side) on moving glass plates. MILNE and GRAY find that
- the amplitude of motion is nearly inversely related to the distance;
- the direct vibrations, although dominant at first, die out more rapidly than the transverse motions; and
- **the normal motions have a quicker rate than the transverse movements.** • At that time, the occurrence of seismic surface shocks was yet to be recognized {Lord RAYLEIGH ⇨1885}.

(iv) They arrive at the conclusion that the slow rate of propagation seems to account for an observation often made when earthquakes take place in Tokyo – that rumbling sounds precede them. This is because the rumbling sounds are caused by the cracking and creaking of buildings. So these sounds travel through the air and thus reach the ear before the tremors are felt through through the ground.

| | | |
|---|---|---|
| 1883 | France | Arthur MOISSON,[997] a little-known French naval captain and ballistician, **theoretically investigates the aerodynamic drag of cylindrically, spherically, and ogivally shaped projectiles traveling at speed $u$ as a function of the ratio $u/a$ in the range 0.2 to 2, where $a$ is the sound velocity.** This ratio will later be termed the *Mach number* {ACKERET ⇨1929}. |
| 1883 | Owens College, Manchester, England | Osborne REYNOLDS,[998] a British engineer and physicist, studies viscous effects in a moving airstream at low speeds and **demonstrates that if the nondimensional number $vl/\nu$ – later known as the "Reynolds number" – is the same in two geometrically similar experiments, then (and only then) the flow patterns are geometrically similar in both experiments** ($v$ = mean flow velocity, $l$ = characteristic length, in circular pipes the pipe dia., and $\nu$ = kinematic fluid viscosity). • The Reynolds number ($Re$) distinguishes between flow regions such as laminar or turbulent flow in pipes, in the boundary layer or around immersed objects. REYNOLDS furnished the first definite experimentally verified information on the nature of the variations in the force resulting from an airstream as a function of its speed.[999] |

---

[997] A. MOISSON: *Evaluation de la résistance de l'air*. Extraits Mém. Artill. Marine **11**, 421-457 (1883).
[998] O. REYNOLDS: *An experimental investigation of the circumstances which determine whether the motion of water shall be direct or sinuous, and of the law of resistance in parallel channels*. Phil. Trans. Roy. Soc. Lond. **174**, 935-982 (1883).
[999] N. ROTT: *Note on the history of the Reynolds number*. Annu. Rev. Fluid Mech. **22**, 1-11 (1990).

In addition to REYNOLDS' original definition, the Reynolds number can be expressed physically as the ratio of the inertial forces to the frictional forces. **Thus, viscous effects are important at small Reynolds numbers, while inertial effects predominate at high Reynolds numbers.**[1000] For geometrically similar flow patterns to exist, it is not sufficient for the bodies to be geometrically similar; the Reynolds number must also be the same. This is an important requirement for correctly correlating full-scale tests with tests performed in a wind tunnel.

1884 · *Collège de France & Laboratoire Central, Service des Poudres et Salpêtres, Paris, France*

**P.E. Marcellin BERTHELOT**[1001] **and Paul VIEILLE,** two French detonation researchers, **invent the "bomb calorimeter."** Based on a new thermochemical method, they apply this new instrument to measure the specific heats of various gases up to 2,000 °C with an accuracy hitherto unattainable.

In the same year, **VIEILLE invents a smokeless powder which he calls** *poudre B* – an abbreviation for *poudre blanche* ("white powder"), as opposed to *poudre N* or *poudre noir* ("black powder"). Test shots performed with a 65-mm-caliber cannon showed that **the new powder permits the ballistic effects of black powder to be achieved with the same pressure but with only a third of the charge, thus allowing a significant increase in the efficiency of fire arms.** ▪ Details of his invention were not published in the literature until six years later.[1002]

1884 · *Ecole d'Artillerie de la Marine, Lorient, France*

**Pierre-Henri HUGONIOT and Hippolyte SÉBERT,**[1003] two French physicists, **analytically examine a one dimensional discontinuous gas flow.** They assume that the flow parameters before and after the discontinuity behave adiabatically (Poisson law). ▪ These studies – later significantly improved and extended by HUGONIOT by assuming a steeper, dynamic equation of state – eventually led to the first general shock theory {HUGONIOT ⇨ 1887}.

1884 · *Vienna Academy, Vienna, Austro-Hungarian Empire*

The *Anzeiger der Wiener Akademie* ("Gazette of the Vienna Academy") **publishes a note reporting that Ernst MACH,** a professor and senior administrative officer at Prague University, **and his assistant Josef WENTZEL have successfully taken photographs of a flying bullet and the "sound wave" emitted from an electric spark for the first time.**[1004] For optical visualization, they used the sensitive schlieren method {A. TOEPLER ⇨ 1864}, and for permanent recording they selected the most sensitive silver bromide gelatin dry plates {MADDOX ⇨ 1871} commercially available at that time. ▪ The short note, however, doesn't contain any photographs, because the exposure density was too low for reproduction.

In the following years, they improved the method and demonstrated the first schlieren photos of a supersonic bullet {E. MACH & SALCHER ⇨ 1886}, as well as the propagation and interaction of shock waves {E. & L. MACH ⇨ 1889}. ▪ A large number of Ernst MACH's original photo plates have survived. They were donated to the Ernst-Mach-Institut by Karma MACH, Ernst MACH's daughter-in-law.[1005] Together with MACH's notebooks and correspondence, they are now kept in the Archives of the Deutsches Museum at Munich.[1006]

---

[1000] W.F. HILTON: *High-speed aerodynamics*. Longmans, Green & Co., London *etc.* (1952), pp. 488-489, 584-585.
[1001] P.E.M. BERTHELOT and P. VIEILLE: *Sur la chaleur spécifique des éléments gazeux, à très hautes températures*. C. R. Acad. Sci. Paris **98**, 770-775 (1884).
[1002] P. VIEILLE: *Etude des pressions ondulatoires produites en vase clos par les explosifs*. Mém. Poudres & Salpêtres **3**, 177-236 (1890).
[1003] P.H. HUGONIOT and H. SEBERT: *Sur la propagation d'un ébranlement uniforme dans un gaz renfermé dans un tuyau cylindrique*. C. R. Acad. Sci. Paris **98**, 507-509 (1884).
[1004] Anzeiger der Kaiserlichen Akademie der Wissenschaften zu Wien, Nr. XV (1884), pp. 121-122.
[1005] Dr. Heinz REICHENBACH, former director of EMI, wrote a short internal note on the history of the Ernst-Mach-Archives entitled *Zur Geschichte des Ernst-Mach-Archivs.* EMI, Freiburg (July 10, 1997).
[1006] The online presentation of Ernst MACH's photographs is still under construction. It will encompass 480 pictures relating to ballistics and gas dynamics in total. Some of his original photo plates can already be found on the Internet; *see* http://www.deutsches-museum.de/bib/archiv/mach/.

| 1884 | Clark, Chapman, Parsons & Company, Gateshead, Northumberland, England | **Sir Charles Algernon PARSONS,** a British mechanical engineer and industrialist, **invents the first multistage steam turbine** in which the steam is discharged from moving nozzles against stationary blades – the so-called **"reaction turbine."** His machine utilizes several stages in series. In each stage, the expansion of the steam is restricted to allow the greatest extraction of kinetic energy without causing the turbine blades to turn too quickly. Initial applications of the turbine for marine propulsion purposes, however, will uncover serious limitations caused by cavitation problems at the propeller. This problem will stimulate systematic studies on cavitation phenomena {Sir PARSONS ⇨1897 & 1915}.[1007]

DE LAVAL's turbine {⇨1882} was based on the mechanical principle of action, and was therefore known as the "action turbine." In modern steam turbines, however, the principles of action and reaction are both frequently employed in different sections of the same turbine. |
|---|---|---|
| 1885 | Hell Gate, New York City, New York | On October 10, **a huge charge consisting of 75,000 pounds** (34 tons) **of No. 1 dynamite** (with 75% nitroglycerin) **and 240,000 pounds** (108.9 tons) **of nitrobenzene/potassium chlorate** [$C_6H_5NO_2/KClO_3$] **is fired to remove "Flood Rock"** (sometimes referred to as the "Middle Reef") **in the Hell Gate section of the East River** in order to form the East River Channel, thus connecting Long Island Sound with Upper New York Bay.[1008] ▪ Hitherto, the channel of the East River between Wards Island and Astoria, Long Island – named "Hell Gate" by the Dutch mariner Adriaen BLOCK who first passed the Hell Gate Section in 1614 – was a serious menace to navigation. Starting in 1851, the reefs were blasted and the channel was dredged, but work continued for almost 50 years before the passage was safe for large vessels.

(i) The operation is planned and carried out by General John NEWTON, who had already successfully mined a reef at Hallet's Point in Hell Gate {J. NEWTON ⇨1876}. ▪ Reports claimed that the shock of the blast was felt as far away as Princeton, NJ, well beyond the Hudson. The *New York Times* devoted its entire cover page to the event and interpreted it smugly as "another triumph of human will over nature." This explosion was the young nation's earliest and most spectacular earth-moving project, and was financially supported by Congress.

(ii) **Lt.-Col. Henry L. ABBOT** {⇨1876 & 1881}, performing measurements on the propagation of the seismic wave, **observes that the speed of a seismic disturbance can undergo strange variations, even appearing to accelerate as its intensity diminishes** {KNOTT ⇨1899}. |
| 1885 | Faculty of Philosophy, University of Breslau, Lower Silesia, Germany | Max HAMBURGER,[1009] a doctoral student under Prof. Oskar E. MEYER, **provides the first experimental evidence of HERTZ's theory of impact** {HERTZ ⇨1882}.

(i) For example, when using 300-mm-long brass cylinders with diameters of 10.3 mm, he finds that the percussion for an impact velocity of 404 mm/s lasts for 587 µs. Since sound waves in brass propagate with a velocity of 3,200 m/s (3.2 mm/µs), the longitudinal wave needs 187.5 µs to return to the impacted side of the cylinder – *i.e.*, during percussion, the wave makes more than three round trips.

(ii) **He also finds that in rods the duration of percussion**
  ▸ **decreases with increasing impact velocity,** thus confirming HERTZ's theory of impact derived for impacting spheres {HERTZ ⇨1882};
  ▸ **but increases almost linearly with the length and diameter of the impacting rod.** |

---

[1007] L.C. BURRILL: *Sir Charles PARSONS and cavitation* [Parsons Memorial Lecture]. Trans. Inst. Marine Eng. **63**, 149-167 (1951).
[1008] *Blasters' handbook*. E.I. du Pont de Nemours & Co., Wilmington, DE (1989), p. 6.
[1009] M. HAMBURGER: *Untersuchungen über die Zeitdauer des Stoßes elastischer zylindrischer Stäbe*. Ph.D. thesis, University of Breslau (1885).

| 1885 | Chair of Experimental Physics, German Charles University, Prague, Bohemia, Austro-Hungarian Empire | Ernst MACH,[1010] a professor of natural philosophy, **and his student Josef WENTZEL publish a study on blast waves originated from chemical explosions.** Setting up a pair of parallel line charges of silver fulminate, they record, using the soot method {ANTOLIK ⇨ 1874}, the interference of the two head waves produced by the detonation fronts, from which they find detonation velocities ranging from 1,700 to 2,000 m/s {⇨Fig. 4.17−I}. ▪ It is interesting to note that MACH and WENTZEL correctly use the term *Stoßwelle* ("shock wave") in their work to appropriately describe the observed abrupt pressure increase. They write, "The propagation of the shock wave can be felt by the hand, and optically (using the schlieren method) it can be shown that this wave consists of a single shock (without periodicity)."
Previously, MACH and coworkers had experimented with spark discharges and chemical explosives in order to generate shock waves, thus using the terms *Funkenwelle* ("spark wave") and *Explosionswelle* ("explosion wave"), respectively, as well as the terms *Knallwelle* ("blast wave") and *Verdichtungsstoß* ("condensation shock"), the latter term being adopted from RIEMANN {⇨1859}. |
|---|---|---|
| 1885 | University of Königsberg, East Prussia, Germany | Franz Ernst NEUMANN,[1011] a German professor of physics and crystallography and director of the Mathematical-Physical Seminar, **publishes a treatise on the longitudinal impact of two thin cylindrical rods** – a subject on which he had lectured previously (1857–1858). His approach, based on D'ALEMBERT's solution of the one-dimensional wave equation {D'ALEMBERT ⇨ 1747}, allows the normal velocity and axial stress in the rods to be evaluated as a function of time. |
| 1885 | Jenkintown, Pennsylvania | Russell Sylvanus PENNIMAN,[1012] a U.S. chemist, **obtains a patent on a method of using ammonium nitrate** {NORRBIN & OHLSSON ⇨ 1867} **as a reliable explosive.** Since this substance is rather hygroscopic, he proposes to coat it with a small percentage of paraffin prior to use.[1013] ▪ This development soon led to a series of ammonia dynamites becoming popular. |
| 1885 | Brighton, Sussex, southeast England | Horatio Frederick PHILLIPS,[1014] a British pioneer of aviation who in 1907 will perform the first powered flight in England, demonstrates the improved lifting qualities of mildly cambered surfaces in wind tunnel studies:
(i) To avoid wind fluctuations, he creates his tunnel using a steady stream of vapor emerging from a system of fine nozzles {⇨Fig. 4.7−E}.
(ii) Remarkably, **his 43×43-cm² wind tunnel is followed by an "expanding delivery tube of sheet iron"** of length of 1.83 m, which flares from a width of about 20.3 cm to about 61 cm, thus acting as a diffuser {VENTURI ⇨ 1797}. ▪ The incorporation of a diffuser into a wind tunnel was not resumed until 1911 by the French engineer Gustave EIFFEL in his design of a 32-m/s wind tunnel which he operated at Auteuil, a western suburb of Paris.[1015]
(iii) Using a mechanical balance, PHILLIPS measures the resistance of curved air foils up to a velocity of 18 m/s.
(iv) He also tests first designs of thick wing sections with curved upper and lower surfaces (now used on all airplanes), for which he will obtain several patents. |

---

[1010] E. MACH and J. WENTZEL: *Ein Beitrag zur Mechanik der Explosionen*. Sitzungsber. Akad. Wiss. Wien **92** (Abth. II), 625-638 (1885).
[1011] F. NEUMANN: *Vorlesungen über die Theorie der Elastizität der festen Körper und des Lichtäthers*. Teubner, Leipzig (1885), chap. 20: *Theorie des geraden Stoßes cylindrischer Körper*, pp. 332-350.
[1012] R.S. PENNIMAN: *Protected nitrate of ammonia for use in explosive compounds*. U.S. Patent No. 312,010 (Feb. 1885).
[1013] N.G. JOHNSON: *Explosives*. In: *The new Encyclopædia Britannica, Macropædia*. Benton & Hemingway, Chicago *etc.*; *see* vol. 7 (1974), p. 85.
[1014] A first note on PHILLIPS' *Experiments with currents of air* (without mentioning his name) was published in: Engineering (Lond.) **40**, 160-161 (Aug. 14, 1885).
[1015] W. TOLLMIEN, H. SCHLICHTING, and H. GÖRTLER (eds.): *L. PRANDTL: gesammelte Abhandlungen*. Springer, Berlin *etc.* (1961), vol. 3, pp. 1333-1334.

| 1885 | *Private laboratory at Terling Place, Witham, Essex, England* | Lord RAYLEIGH,[1016] an English physicist and acoustician, **shows theoretically that waves passing along the free boundary of an elastic body may play an important part in earthquakes,** since their intensity will dominate over waves spreading through the interior of the Earth at great distances because they spread over just two dimensions. He shows that their influence decreases rapidly with depth, and that their velocity of propagation is smaller than that of body waves. ▪ His surmise was fully confirmed and such a wave has since been called a **"Rayleigh wave."** For example, in the Great Messina Earthquake (1908), Prince Boris B. GALITZIN[1017] [or GOLITSYN], a Russian seimologist at Petrograd, used his electromagnetic seismograph to trace seismic surface shocks from the earthquake that had traveled around in the Earth in opposite directions. His measured data on the surface wave velocity showed good agreement with RAYLEIGH's theory.

Rayleigh surface waves are generated not only by earthquakes but also by chemical and nuclear explosions, and by meteorite impact. U.S. seismologists observed that even a bolide sonic boom (an aerial shock wave) induced seismic ground motions, including Rayleigh waves.[1018] They are also generated by the high-velocity impact of a projectile with a solid target; in this case, however, the disturbance only extends a small distance from the free surface.[1019]

In 1995, Japanese researchers at Tohoku University discovered that the **Rayleigh surface wave can induce a conical shock in air.**[1020] |
|---|---|---|
| 1885 | *Manchester, North West England* | In a paper read before the Manchester Literary and Philosophical Society in November 1885, Osborne REYNOLDS,[1021] a British engineer and physicist, **considers the thermodynamics of fluid flow in the case of a gas or vapor discharging from one vessel into another through an orifice or nozzle.**

(i) He calls attention to the fallacy of the assumption that the pressure in the receiving vessel is the same as that at the orifice, and goes on to consider the observation that the rate of flow is only affected by the pressure in the receiver if it is greater than about half of the pressure in the upstream vessel.

(ii) Moreover, he establishes that the reason for this is that a limit to the flow is reached when the velocity at the orifice becomes equal to the velocity of sound at that point.[1022] ▪ Later, Ludwig PRANDTL, a physics professor at Göttingen University, showed that this view was incorrect {PRANDTL ⇒ 1908}. |
| 1886 | *Fortress Monroe of the U.S. Army, Virginia* | Colonel James Monroe INGALLS,[1023] a renowned U.S. ballistician who established the Dept. of Ballistics at Fortress Monroe in 1882 and also teaches there, **publishes his book *Exterior Ballistics in the Plane of Fire*,** in which he studies the motion of a projectile in three different scenarios. **He coins the following terms:**<br>▸ ***Interior ballistics*** (later also termed ***internal ballistics***): "Interior Ballistics treats of the motion of a projectile within the bore of the gun while it is acted upon by the highly elastic gases into which the powder is converted by combustion…" |

---

[1016] Lord RAYLEIGH (J.W. STRUTT): *On waves propagated along the plane surface of an elastic solid.* Proc. Lond. Math. Soc. **17**, 4-11 (1885/1886).
[1017] Fürst B. GALITZIN: *Vorlesungen über Seismometrie* (given in 1911 and edited by O. HECKER). Teubner, Leipzig *etc.* (1914), p. 78.
[1018] C.A. LANGSTON: *Seismic ground motions from a bolide shock wave.* J. Geophys. Res. **109**, Paper B12309 (2004).
[1019] N.L. HICKERSON: *Stress wave propagation in solids.* In: (R. KINSLOW, ed.) *High-velocity impact phenomena.* Academic Press, New York & London (1970), pp. 23-43.
[1020] H.H. SHI, K. TAKAYAMA, and N. NAGAYASU: *The measurement of impact pressure and solid surface response in liquid-solid impact up to hypersonic range.* Wear **186/187**, 352-359 (Aug. 1995).
[1021] O. REYNOLDS: *On the flow of gases* [read 1885]. Phil. Mag. **21** [V], 185-199 (March 1886).
[1022] J.D. JACKSON: *Osborne REYNOLDS, scientist, engineer, and pioneer.* Proc. Math. Phys. Sci. **451**, No. 1941 (Osborne Reynolds Centenary Volume, 1995), pp. 49-86.
[1023] J.M. INGALLS: *Exterior ballistics in the plane of fire.* Van Nostrand, New York (1886).

- ▸ ***Exterior ballistics:*** "Exterior Ballistics considers the circumstances of motion of a projectile from the time it emerges from the gun until it strikes the object aimed at."
- ▸ ***Ballistics of penetration*** (now called **"terminal ballistics,"** which also encompasses ***wound ballistics***).

In 1887, the British author James Atkinson LONGRIDGE published the book *Internal Ballistics*, the first monograph dedicated exclusively to this particular branch of ballistics, and in 1894 INGALLS published the book *Interior Ballistics*. Three editions of both books were published. The two traditional branches of interior and exterior ballistics were treated in detail by the German ballistician Carl CRANZ in his classic textbooks {CRANZ ⇨1925}.

In the period between the two World Wars, a fourth branch of ballistics dedicated to the transition between interior and exterior ballistics – ***intermediate ballistics*** or ***transitional ballistics*** – evolved. With the advent of high-speed cinematography in the 1920s, a variety of unusual flow phenomena in the muzzle environment were visualized {⇨Fig. 4.5–L}. *Intermediate ballistics* [Germ. *Zwischenballistik*] – a term apparently coined by the German ballistician Theodor ROSSMANN[1024] – is marked by the emission of gas from the barrel in front of the projectile and by the discharge of propellant gases behind the projectile. The effect on the delivery of the projectile and the utilization of the momentum of the propellant gas are of special interest in muzzle brakes.[1025] ROSSMANN also proposed the branch of ballistics called "gun carriage ballistics" [Germ. *Lafettenballistik*] to investigate the motion of the gun, the forces involved in recoil and their effective reduction via muzzle brakes. However, this term was not adopted by the ballistics community.

With the advent of satellites and other hypervelocity space vehicles in the late 1950s, another branch was added to these four branches of ballistics – ***geoballistics*** – which is related to very long orbital ranges and times of flight, where the Coriolis force and local variations in the magnitude and direction of gravitational forces can no longer be neglected.

| | |
|---|---|
| 1886<br><br>German Charles University, Prague, Bohemia & Royal Imperial Austro-Hungarian Naval Academy, Fiume [now Rijeka, Croatia], Austro-Hungarian Empire | **Ernst MACH and Peter SALCHER begin a long series of unique ballistic experiments.** SALCHER, a professor of physics at the Royal Austrian Naval Academy, performs the experiments at the Adriatic Naval Test Station in Fiume. Via correspondence they discuss the step-by-step progress in their work: **MACH**, a physics professor at Prague, sketching the experimental set-up in a letter[1026] to SALCHER {⇨Fig. 4.6–E} states, **"I expect that the projectile will carry an envelope of compressed air of an approximate geometry as shown opposite. The apex of the truncated cone will doubtless be dependent on the ratio of the sound velocity to the projectile velocity."** ▪ Unfortunately, MACH and SALCHER did not consider themselves obliged to cite Christian DOPPLER {⇨1847} in their paper as the spiritual originator of the head wave phenomenon. It is interesting here to note that MACH himself instead referred to the Huygens principle {HUYGENS ⇨1678}. In MACH's letter to SALCHER mentioned above, MACH made a small sketch of the head wave and the cone angle, and annotated: "According to HUYGHENS' principle it should be: sound velocity/projectile velocity = sin$\alpha$, so that the projectile velocity can be directly read from the picture."<br><br>About three months later, in a letter to MACH, SALCHER[1027] reports on his successful attempt to photograph the head wave:<br>(i) Using two supersonic infantry rifles, the Austrian *Werndl* (muzzle velocity $v_0$ = 438 m/s; projectile dimensions 11 mm dia. × 27 mm) and later the Portuguese *Guedes* (530 m/s; 8 mm |

---

[1024] T. ROSSMANN (ROßMANN): *Ballistik auf dem Schießplatz*. In: (H. SCHARDIN, ed.) *Beiträge zur Ballistik und technischen Physik: verfasst von Schülern des Herrn Geheimrat Professor Dr. phil. Dr.-Ing. E.h. Carl CRANZ anlässlich seines 80. Geburtstages am 2. Januar 1938*. J.A. Barth, Leipzig (1938), pp. 91-103.
[1025] K. OSWATITSCH: *Zwischenballistik*. DLR-Forschungsbericht Nr. 64-37 (1964).
[1026] This letter (dated Feb. 16, 1886), hitherto unknown to the scientific community, was sent to the author in 1994 by Dr. Günter SALCHER (Hermagor, Austria), who keeps a collection of letters from E. MACH in his private archives.
[1027] This letter (dated May 21, 1886) is now kept in the Archives of the Deutsches Museum, München.

dia. × 33 mm), SALCHER takes the first photographs of supersonic projectiles in free flight. His first photos {⇨Fig. 4.6−F} reveal that **a supersonic projectile produces a hyperbolic-like "head wave"** [Germ. *Kopfwelle*] **followed by a "tail wave"** [*Achterwelle*], similar to a "bow wave" [*Bugwelle*] of a ship. Since they are both weak shock waves, they approach a conic geometry which is primarily dependent on the projectile velocity, but also to some minor extent on the projectile geometry.

(ii) SALCHER notices a series of intermediate waves,[1028] later called **"Mach lines"** [Germ. *Machlinien*] which arise from the rough surface of the projectile. ▪ In 1907, Ludwig PRANDTL at Göttingen University noticed that Mach lines also arise in a supersonic flow propagating above a rough surface, thereby producing a Scotch plaid pattern of diagonal Mach lines {⇨Fig. 4.8−E}. Each Mach line is a small shock wave that commences at the site of the imperfection and projects at an angle dependent on the speed of the air in the tunnel. At Mach 1, the Mach lines are perpendicular to the flow; at Mach 2, the angle is 30° (the angle whose sine is 0.5); and in general, at Mach $M$, the angle is arcsine $1/M$. For example, small pieces of Scotch tape just 0.003 in. (0.076 mm) thick placed on the tunnel wall give rise to strong artificial Mach lines on schlieren photos.[1029]

In the same year, the *Anzeiger der Kaiserlichen Akademie der Wissenschaften* [Wien, vol. 23, p. 136] reports in a short note on their first successful experiments of visualizing the head wave phenomenon.

In the following year, MACH[1030] and SALCHER published their celebrated results in a sensational paper entitled *Photographische Fixierung der durch Projectile in der Luft eingeleiteten Vorgänge* ("Photographic Fixation of Processes Initiated in Air by Projectiles"):

(I) **They show that the velocity $v$ of a supersonic projectile can be determined from the head wave cone geometry by the simple relation**

$$\sin \alpha = c_0/v = 1/M,$$

where $c_0$ is the sound velocity in the ambient gas at rest, $\alpha$ is the half-cone angle, and $M$ is the Mach number. ▪ Later, this relation was called **"Mach equation"** [Germ. *MACHsche Gleichung*], although it was first derived by Christian DOPPLER {⇨1847}. **Ludwig PRANDTL** {⇨1913} **termed the cone geometry *Mach cone*** [Germ. *MACHscher Kegel*], **and the cone angle *Mach angle*** [*MACHscher Winkel*].

(II) **They realize that the nose waves in the photographs are, in fact, shock waves and not sound waves,** and that the propagation speed of these nose waves exceeds that of sound. They write, "The compression in front of a projectile at supersonic speed has to become so great … until its propagation velocity equals that of the projectile … Following the nose wave backwards, the angle $\alpha$ of the element relative to the line of flight must fall off slowly and approach the limiting value arcsin $c_0/v$ … the curve is thus comparable to a hyperbola … The vortex of the wave, therefore, lies nearer the nose for higher velocities … Pointing the projectiles also brings the wave vortex nearer the nose." ▪ Since the Mach cone is never a straight cone in practice, but rather approaches a hyperboloid, it may be difficult to determine the appropriate cone angle $\alpha$ {PRANDTL ⇨1913}.

(III) **Addressing the similarity of the head wave to the motion of a body in water,** they write, "It is possible to reproduce this phenomenon [the head wave] if we take a rod of cross-section *AB* in a large water tank and move it at a velocity which exceeds the velocity of wave propagation." ▪ E. MACH and SALCHER erroneously assumed that in the hydraulic analogy the

---

[1028] H. REICHENBACH: *Contributions of Ernst MACH to fluid dynamics.* Annu. Rev. Fluid Mech. **15**, 1-28 (1983).

[1029] D.D. BAALS and W.R. CORLISS: *Wind tunnels of NASA.* NASA Rept. SP-440, NASA History Office, Washington, DC (1981), chap. 5: *The era of high-speed flight. The Scotch tape enigma.* See also http://www.hq.nasa.gov/office/pao/History/SP-440/ch5-4.htm.

[1030] E. MACH and P. SALCHER: *Photographische Fixierung der durch Projectile in der Luft eingeleiteten Vorgänge.* Sitzungsber. Akad. Wiss. Wien **95** (Abth. IIa), 764-780 (1887).

Mach angle would increase when the body moves more quickly through the water. However, the half-cone angle of the head wave drawn in water is independent of the body's velocity and is always 39° – *i.e.*, **the Mach number for motion on water is always 3** {⇨Fig. 4.6–D}.

(IV) To find the density distribution around a supersonic projectile, they propose an electrolytic method first demonstrated in the 1820s by the Italian physicist Leopoldo NOBILI[1031] (so-called **"Nobili rings"**) and later studied in more detail by the Frenchman Adrien GUÉBHARD,[1032] that involves the use of "a silver-coated copper sheet on the bottom of a container filled with an electrolyte, placing a non-conducting model projectile on the sheet and dipping metal probes connected to a battery to find the equipotentials."[1033] ▪ This method was later successfully used to investigate the field of flow of a compressible fluid past a cylinder {G.I. TAYLOR & SHARMAN ⇨1928}.

(V) From the photographs, they conclude that the air condensation in front of the projectile must be considerable. This prompts the idea that the explosion-like crater-shaped injuries previously observed in the Prussian-French War (1870–1871) were not caused by explosive projectiles fragmenting explosively upon impact with a target {MELSENS ⇨1872}, but rather by pressure effects originated by the supersonic speed of the projectile. ▪ The experimental evidence of a condensation in front of the projectile and a rarefaction at its tail confirmed a previous hypothesis {Sir NEWTON ⇨1687}.

| 1887 | Ecole Polytechnique, Paris, France | **Pierre Henri HUGONIOT**,[1034] an eminent French physicist, obviously not aware of RANKINE's previous work {⇨1869} **formulates a general theory of discontinuous one-dimensional flow using Lagrangian coordinates;** *i.e.*, coordinates that follow fluid particles in space and time – hence also called **"material coordinates"** – so that interfaces (such as shocks) can be immediately identified at each point in time. His most remarkable results can be summarized as follows: |

(i) Assuming an ideal *polytropic gas*,[1035] he applies the principle of the conservation of energy to a region bounded by two fixed planes, including the region with the discontinuity. Noting that the increase in the kinetic energy of the fluid must balance the change in the internal energy in the initial and final states, **for a perfect gas with a constant ratio of specific heats, $m$ ($= \gamma$) he obtains the following relation between the pressures and volumes in the two states** (*see* p. 88, second part of his memoir; ⇨Fig. 2.18):

$$p_1 = p_0 [(m + 1) \rho_1/\rho_0 - (m - 1)] / [(m + 1) - (m - 1) \rho_1/\rho_0] \tag{1}$$

which can also be written as

$$p_1 = p_0 [(m + 1) v_0/v_1 - (m - 1)] / [(m + 1) - (m - 1) v_0/v_1] \tag{2}$$

Here $p_0$ and $p_1$, $\rho_0$, and $\rho_1$, and $v_0$ and $v_1$ are pairs of pressures, densities and specific volumes at the initial state ("0") and the final state ("1"), respectively. This equation implies that the $(p, \rho)$-data are not positioned along the **isentrope** or **static adiabat** as given by the Poisson law {POISSON ⇨1823}

$$p_1 = p_0 (\rho_1/\rho_0)^\gamma = p_0 (v_0/v_1)^\gamma, \tag{3}$$

---

[1031] L. NOBILI: *Sur une nouvelle classe de phénomènes électro-chimiques*. Bibliothèque Universelle de Genève **33**, 302-314 (1825) & **34**, 194-213 (1826); *Sur les apparences et les mouvements électro-chimiques de mercure*. Ibid. **35**, 261-284 (1827).

[1032] A. GUÉBHARD: *Figuration électrochimique des lignes équipotentielles sur les portions quelconques du plan*. J. Phys. Thèor. Appl. **1** [II], 205-222, 483-492 (1882).

[1033] E. MACH: *Über Herrn A. GUÉBHARD's Darstellung der Äquipotentialcurven*. Sitzungsber. Akad. Wiss. Wien **86** (Abth. IIa), 8-14 (1882).

[1034] P.H. HUGONIOT: *Mémoire sur la propagation du mouvement dans les corps et plus spécialement dans les gaz parfaits. 1e Partie*. J. Ecole Polytech. (Paris) **57**, 3-97 (1887); *2e Partie*. Ibid. **58**, 1-125 (1889). ▪ HUGONIOT's manuscript, entirely ready for print before his premature death in 1887, was edited by R. LIOUVILLE, a retired chief engineer at the Service des Poudres in Paris.

[1035] In a *polytropic gas* – a special case of an *ideal gas* – the internal energy $e$ is simply proportional to the temperature $T$ and is given by $e = c_v T$, where $c_v$ is the specific heat at constant volume. In this particular case, $e$ can be expressed in terms of the specific volume $v$ and pressure $p$ by the equation $e = p v / (\gamma - 1)$. *See*, for example, the book by COURANT & FRIEDRICHS {⇨1948}, pp. 6-7.

but rather along a ***dynamic adiabat*** or ***shock adiabat*** – the so-called **"Hugoniot curve"** – which lies above the static adiabat centered at the same point $(p_0, v_0 = 1/\rho_0)$ {⇨Fig. 2.1}.[1036] ▪ Lord RAYLEIGH {⇨1910} discussed RANKINE's and HUGONIOT's different approaches to the shock wave problem. The investigations carried out by William J.M. RANKINE were based upon conduction of heat in the gas {RANKINE ⇨1869}, while HUGONIOT considered his gas to be nonconducting. RAYLEIGH later concluded, "But a little examination reveals that this [Hugoniot's] law *is precisely the same* as given 15 years earlier by RANKINE, a fact which is the more surprising inasmuch as the two authors start from quite different points of view."

(ii) **He shows that for a perfect gas the maximum shock compression ratio is given by**
$$\rho_1/\rho_0 = (m+1)/(m-1)$$
which for air ($m = 1.4$) is equal to (just) 6 {⇨Fig. 2.1}. ▪ HUGONIOT used the letter $m$ instead of $\gamma$, (as used by POISSON {⇨1808} and by RANKINE {⇨1869}), which will later be adopted in general.

(iii) **He also addresses the propagation of shock waves in solids.** Considering the conditions at the contact area of two colliding bodies, he states, "It is doubtful whether the discontinuities which are described by the theory of wave propagation are only a simplified analytical fiction or whether they correspond to the physical reality. This is an open question which is difficult to answer at the present state of science."

It is useful to illustrate HUGONIOT's approach here using a modern representation with different coordinates:

(A) Using ***Lagrangian*** **coordinates** {*e.g.*, COURANT & FRIEDRICHS ⇨1948}, which allow one to follow the fates of individual particles, it follows from the Laws of Conservation of Mass and Momentum that

$$\rho_0 v_0 = \rho_1 v_1; \quad \text{conservation of mass.} \tag{4}$$

$$p_0 + \rho_0 v_0^2 = p_1 + \rho_1 v_1^2; \quad \text{conservation of momentum.} \tag{5}$$

Here the flow to one side of the shock is specified by the velocity $v_0$, the density $\rho_0$, and the pressure $p_0$, with corresponding values of $v_1, \rho_1, p_1$, on the other side. Noting that the kinetic energy is transformed into internal energy at the discontinuity – *i.e.*, energy is conserved – he derives the relation

$$p_0 v_0 + \rho_0 v_0 (e_0 + \tfrac{1}{2} v_0^2) = p_1 v_1 + \rho_1 v_1 (e_1 + \tfrac{1}{2} v_1^2); \quad \text{conservation of energy.} \tag{6}$$

Eliminating $v_1$ and $v_2$ between Eqs. (1), (2) and (3) yields, for a polytropic gas,

$$e_1 - e_0 = \tfrac{1}{2}(p_0 + p_1)(v_0 - v_1), \tag{7a}$$

where $e_0$ and $e_1$ are the internal energies per unit mass in regions "0" and "1." In the special case of an ideal polytropic gas, the energy increase $\Delta e$ at the shock front is given by

$$\Delta e = e_1 - e_0 = (p_1 v_1 - p_2 v_2)/(\gamma - 1). \tag{7b}$$

Combining Eq. (7a) with Eq. (7b) and noting that $\gamma = m$ yields HUGONIOT's solution Eq. (2).

HUGONIOT considered a nonreactive shock wave in an ideal gas of constant $\gamma$, and $e_0$ and $e_1$ are the heat energies per unit mass on the two sides before and after passing the shock wave, respectively. However, CRUSSARD {⇨1907} first demonstrated that Eq. (7) also holds for reactive shock waves such as detonation waves. In this important case, $e_0$ and $e_1$ are the chemical plus heat energies per unit mass.

The following terms are commonly in use:

---

[1036] The curve through $(p_0, v_0)$ is said to be centered at $p_0, v_0$ and is also known variously as the *Rankine-Hugoniot p,v curve*, *R-H curve* or the *Hugoniot*. The shock process is adiabatic but not isentropic {COURANT & FRIEDRICHS ⇨1948; DUVALL & GRAHAM ⇨1977; ⇨Fig. 2.1}. If the initial state is the standard laboratory state – *i.e.*, at 25 °C and 1 bar – the Hugoniot curve is called the "principal Hugoniot." ▪ Some researchers considered an Hugoniot to be a plot of shock pressure *vs.* a *condensation factor* defined as $(1 - \rho_0/\rho)$, for example see GOLDSMITH {⇨1960}. However, this presentation was not generally adopted.

▸ The Eqs. (4), (5) and (7) are termed the **Rankine-Hugoniot equations** (or *Rankine-Hugoniot shock relations*).
▸ More specifically, Eq. (7a) is termed the **Hugoniot equation** (or *Hugoniot relation*).
▸ The thermodynamic nature of the fluid is introduced from the $(e, p, v)$-equation of state solely through Eq. (7). When $e$ is eliminated between Eq. (7) and the $(e, p, v)$-equation of state for the fluid, the resulting $p = p(v)$ relationship is known as the **"Hugoniot."** The curve $p(v)$, beginning at the initial state through $(p_0, v_0)$, is also known variously as the **Rankine-Hugoniot p,v curve** or the **R-H curve.** If the initial state refers to the standard laboratory conditions – *i.e.*, to room temperature (25 °C) and atmospheric pressure (1 bar) – the Hugoniot curve is termed the ***principal Hugoniot.***
▸ Note that in the shock physics literature the term **Hugoniot relation** also frequently takes different meanings; for example, it connects the pressure jump across a shock wave with the corresponding velocity jump; *i.e.*, it refers to the Rankine-Hugoniot relation Eq. (5). The dependency between the shock front velocity $U$ and the particle velocity $u$ which is approximately linear and given by

$U = A + Bu$,

where $A$ and $B$ are constants, is also termed *Hugoniot relation* among solid-state shock physicists.

(B) ***Eulerian* coordinates** {*e.g.*, COLE ⇨1948} – which determine quantities characterizing the motion and the state of the medium at a given point in space and at a given instant of time – are preferable from both mathematical and physical points of view; the Rankine-Hugoniot equations are then:

$$(U - u_1)/v_1 = (U - u_0)/v_0. \tag{8}$$

$$p_1 + (U - u_1)^2/v_1 = p_0 + (U - u_0)^2/v_0. \tag{9}$$

$$p_1 v_1 + \tfrac{1}{2}(U - u_1)^2 - e_1 = p_0 v_0 + \tfrac{1}{2}(U - u_0)^2 - e_0. \tag{10}$$

Here $e_0, e_1$; $p_0, p_1$; and $v_0, v_1$ are the specific internal energies, pressures and specific volumes for the undisturbed and disturbed states, "0" and "1," respectively; $U$ is the shock front velocity and $u_1$ is the particle velocity to which the shock-compressed material is accelerated. Assuming that the shock moves into undisturbed gas ($u_0 = 0$) and eliminating $U$ and $u_1$ between Eqs. (8), (9) and (10) yields the Hugoniot relation:

$$e_1 - e_0 = \tfrac{1}{2}(p_1 + p_0) \times (v_0 - v_1). \tag{11}$$

Note that the Hugoniot relation, Eq. (7) or Eq. (11), contains no velocity terms, only thermodynamic quantities. Therefore, this equation is independent of the coordinates chosen. The following expressions for $U$ and $u_1$ can be derived from Eqs. (8) and (9):

$$U = v_0 [(p_1 - p_0)/(v_0 - v_1)]^{1/2} \tag{12}$$

and

$$u_1 = [(p_1 - p_0) \times (v_0 - v_1)]^{1/2}. \tag{13}$$

Since generally $p_1 \gg p_0$, Eq. (12) can be written as

$$p_1 = \rho_0 U^2 (1 - \rho_0/\rho_1) = \rho_0 U^2 (1 - v_1/v_0). \tag{14}$$

Equation (14) represents a straight line in the $(p, v)$-plane, known as the **"Rayleigh line"** {Lord RAYLEIGH ⇨1878} or **"Mikhel'son line"** {MIKHEL'SON ⇨1893}. Hence, by measuring the shock front velocity $U$ and the compression ratio $\rho_1/\rho_0$ – for example from a series of flash X-ray photographs using flash radiography {STEENBECK ⇨1938; KINGDON & TANIS ⇨1938} – and then using Eq. (14), it is possible to determine the shock pressure $p_1$ from the measured results and the Hugoniot curve of shock-compressed matter {SCHALL ⇨1950; SCHALL & THOMER ⇨1951}. The same method has been also proposed for measuring the detonation pressure in a high explosive {SCHALL ⇨1955}.

Depending on the particular application, Lagrangian coordinates[1037] might be more convenient than Eulerian ones, particularly in the case of hydrodynamic initial value problems in one dimension,[1038] or when numerically calculating the blast wave that emerges from the detonation of a spherical charge of TNT {BRODE ⇒ 1959}.

(C) *Hybrid numerical methods* have been proposed in order to accelerate the convergence and/or to improve the accuracy of the solutions to some scientific computing problems. Examples of such methods include:

- The **Particle-In-Cell (PIC) method**,[1039] developed in 1955 at Los Alamos. This combines some of the best features of both the Lagrangian and Eulerian representations. It has been used successfully for a number of problems which other methods have not been able to handle, such as for calculations in fluid dynamics involving large distortions of the fluid, large slippages, colliding interfaces, shock propagation in discontinuously enlarging channels, shock refraction at an oblique gaseous interface, and shock interactions with deformable objects.
- In the 1960s, the first reactive Eulerian hydrodynamic **EIC (Explosive-In-Cell) code**, based upon PIC, was used in detonation physics to calculate the interaction of a shock wave from a homogeneous explosive with a cylindrical void in order to study chemical decomposition by a "hot spot" and the resulting initiation of a propagating detonation.
- **Coupled Euler-Lagrange (CEL) or Arbitrary Lagrangian Eulerian (ALE) techniques** are traditionally used to solve high-velocity impact problems and to perform hydrodynamic ram and fuselage decompression analyses.
- **Smoothed Particle Hydrodynamics (SPH)** is a computational technique for the numerical simulation of the equations of fluid dynamics without the use of an underlying numerical mesh. Although originally developed for use in astrophysical gas dynamics it has also been applied to tsunami modeling, planetary and ballistic impact phenomena, and cosmological simulations. In a hybrid approach, the best features of a Lagrangian approach are combined with the SPH method.

| | | |
|---|---|---|
| 1887 | *Naval Academy & University of Kiel, northern Germany* | Otto KRÜMMEL,[1040] a German professor of geography, **publishes the second part of his *Handbuch der Ozeanographie*** ("Handbook of Oceanography"). Besides discussing tides and contemporary theories on ocean currents and circulations, he reviews the classical theories of water waves in shallow and deep water. In the chapter *Seebeben- oder Stoßwellen* ("Seaquake or Shock Waves") he discusses historic accounts of seismic sea wave phenomena and more recent results obtained by the German geologists Ferdinand VON HOCHSTETTER and Franz E. GEINITZ from analyses of the Arica Earthquake {⇒1868}, the Iquique Earthquake {⇒1877}, and the Krakatau Eruption {⇒1883}. |
| 1887 | *Saint Sevran, Dépt. Seine-et Oise & Paris, France* | Alfred NOBEL,[1041] a Swedish chemist and industrialist, **introduces the first smokeless powder, which he calls "ballistite"** (49% nitroglycerin, 49% nitrocellulose, and 2% aniline/diphenylamine). Cut into flakes, it makes an excellent gun propellant. ▪ It was subsequently used for over 75 years. |

---

[1037] In a private communication to the German mathematician Heinrich M. WEBER, Hermann HANKEL also considered EULER to be the originator of the hydrodynamic equations in *Lagrangian* representation; see H.M. WEBER: *Über eine Transformation der hydrodynamischen Gleichungen*. J. Math. **68**, 286-292 (1868). The English mathematician William H. BESANT showed that the Euler equations can be deduced from the Lagrangian equations; see Quart. J. Math. **11**, 203-205 (1871).
[1038] R. COURANT and K.O. FRIEDRICHS: *Supersonic flow and shock waves*. Interscience, New York (1948), chap. 1.
[1039] F.H. HARLOW: *The particle-in-cell computing method for fluid dynamics*. In: (B. ALDER, S. FERNBACH, and M. ROTENBERG, eds.) *Methods in computational physics*. Academic Press, New York & London (1964), pp. 319-343.
[1040] O. KRÜMMEL: *Handbuch der Ozeanographie. Band II*. Engelhorn, Stuttgart (1887).
[1041] A. NOBEL: *Improvements in the manufacture of explosives*. Brit. Patent No. 1,471 (1888).

Two years later, **the two British chemists James DEWAR and Frederick ABEL invented a similar product to ballistite which they called "cordite"** (58% nitroglycerin, 37% nitrocellulose, and 5% petroleum jelly). Alfred NOBEL claimed that the patent for ballistite also covered cordite, but the British courts ruled against him (1895). ▪ During World War II, *cordite N*, a triple-base smokeless explosive, was developed. This explosive contains nitrocellulose, nitroglycerin and nitroguanidine, and has a higher detonation velocity than ballistite. Since it is very cool-burning with little smoke and no flash, it is used as a propellant in aircraft gun ammunition.

| | | |
|---|---|---|
| 1887 | Kaiser-Wilhelms-Universität, Straßburg, Germany | Emil RUDOLPH,[1042] a German professor of geophysics, **publishes the first part of his detailed work on *Seebeben* ("seaquakes"), a term which he coined in the early 1880s.**<br><br>(i) Particularly interested in the causes of earthquakes, and having reviewed previous hypotheses on the origin of seaquakes and associated tidal waves, **he speculates that submarine explosions cause seaquakes** – in particular that submarine volcanic eruptions at the sea-bottom cause *Erdbebenflutwellen* (meaning "earthquake-induced tidal waves"); the term *tsunami* had not yet been adopted {10th Pacific Science Congress ⇨1961}.<br><br>(ii) Stimulated by accounts given by sea captains of strange encounters in which their vessels suffered severe damage from vibrations, RUDOLPH reviewed a large number of ship's logs and newspaper accounts. Many of these events were accompanied by explosive sounds like "distant thunder" or "cannon fire." Other witnesses reported that, for example, "the sea was thrown up to a great height, possibly 80 feet [24.4 m] or more, in a column" and that "the ship was violently shocked from stem to stern, accompanied by a rumbling noise like distant thunder, but seemed close to us." ▪ RUDOLPH correctly correlated these accounts to seaquakes. His hypothesis on their origin, however, was not immediately accepted by contemporary geologists.<br><br>In the following years, RUDOLPH[1043] **contributed two further memoirs to the phenomenon of seaquakes.** He gave a comprehensive collection of eyewitness accounts of seaquakes, and reviewed previous studies on the explosion of sea mines {Underwater Explosion Studies ⇨1870s; ABBOT ⇨1881} in the context of surface phenomena observed during seaquakes. |
| 1887 | German Charles University, Prague, Bohemia, Austro-Hungarian Empire | **Ottokar TUMLIRZ,**[1044] a German instructor of physics who obtained his Ph.D. under the supervision of Ernst MACH, **presents his shock wave theory.** It is based on RIEMANN's mathematical model; *i.e.*, it also assumes an adiabatic law {RIEMANN ⇨1859}. However, to avoid RIEMANN's error, he explicitly uses the principle of energy conservation applicable to continuous motion, in place of the principle of momentum conservation. He concludes that as soon as a discontinuity is formed, it will immediately disappear again, and it will be accompanied by a lengthening of the wave and the disturbance speeds up. He takes this process to be an explanation for the increased velocity of the wave, thus explaining the increased velocity of very intense sounds, such as those generated by electric spark discharges and investigated by Ernst MACH and coworkers in the late 1870s. ▪ TUMLIRZ's work was first critically reviewed by the Englishman Charles V. BURTON {⇨1893}. |
| 1888 | Mt. Bandai-san, Fukushima Prefecture, Honshu Island, central Japan | On July 15, **explosive eruption of volcano Mount Bandal-san.** The activity begins with a violent earthquake, followed by a major debris avalanche and many large explosions (461 casualties). One of the volcano's four peaks is blown off (the volume of the ejected material was later estimated to be 1.2 km$^3$). ▪ This episode was well documented by Seiki SEKIYA,[1045] a professor occupying the newly established Chair of Seismology at the Imperial University of |

---

[1042] E. RUDOLPH: *Über submarine Erdbeben und Eruptionen*. Ph.D. thesis, Kaiser-Wilhelms-Universität Straßburg. Schweizerbart, Stuttgart (1887). Also published in: Beiträge zur Geophysik **1**, 133-373 (1887).

[1043] E. RUDOLPH: *Über submarine Erdbeben und Eruptionen* [Fortsetzung]. Beiträge zur Geophysik **2**, 537-666 (1895), Ibid. **3**, 273-336 (1898).

[1044] O. TUMLIRZ: *Über die Fortpflanzung ebener Wellen endlicher Schwingungsweite*. Sitzungsber. Akad. Wiss. Wien **95** (Abth. IIa), 367-387 (1887).

[1045] S. SEKIYA and Y. KIKUCHI: *The eruption of Bandai-san*. Trans. Seism. Soc. Jap. **13**, 139-222 (1890); *see also* STURTEVANT ET AL. {⇨1991}.

Tokyo, and Yasushi KIKUCHI who arrived at the mountain a few days later and collected many eyewitness reports.

(i) Based on these accounts, they write, "**The explosions were accompanied by terrible wind blasts or coup de vent.** In the parts most exposed to the fury of these blasts, houses were leveled to the ground and trees torn up by their roots … It would appear that the tremendous explosions of steam at quick intervals, lasting for about a minute, produced violent disturbances of the air, consequent upon the sudden radial expansion of the liberated volumes of steam … The eruptions of Mt. Bandai-san may be aptly compared to the firing of a tremendous gun – such as one, however, as can only be forged by nature … The destroying tempests … were something more than atmospheric, consisting besides of heated blasts of steam and air, thickly mixed with dust and rock-fragments fierce enough to crush the trees and to strip them not only of branches but even of their bark, and withering scoring, and scorching everything in their course."

(ii) Also addressing the observed phenomenon of a *lateral blast* {KIEFFER & STURTEVANT ⇒ 1984}, they notice, "On the west side of Biwa-sawa, the effects of the storm were especially striking; trees with a diameter of more than a meter had been lain prostrate on the ground in thousands; and a forest was thickly encumbered with fallen trees … **Here as everywhere else, the trees fell with their heads pointing away from the crater.**" ▪ A similar eruption phenomenon was observed more recently in the United States {Mt. St. Helens ⇒ 1980}.

| | | |
|---|---|---|
| 1888 | *Mount Vulcano, Vulcano Island, Aeolian Islands, southern Italy* | On August 2, **Mount Vulcano** [Roman *Vulcania*, Greek *Hiera*], a volcano on an island in the Tyrrhenian Sea, **starts to erupt; the eruption will last for two years** (until March 22, 1890). ▪ Vulcano, which has been active throughout recorded history, typically erupts explosively, shattering its plug with tremendous violence and shooting out clouds black with particles of solid lava.<br><br>So-called **"Vulcanian-type eruptions"** – which are more explosive than Strombolian-type eruptions {Stromboli Volcano ⇒ 1930} – eject large proportions of volcanic ash and new lava fragments that do not take on a rounded shape during their flight through the air. This may be because the lava is too viscous or because it has already solidified. These eruptions commonly also eject breadcrust bombs and blocks. |
| 1888 | *Bath, Somerset, South West England* | In September, at the 58th meeting of the British Association for the Advancement of Science (BAAS), **William THOMSON,**[1046] a Scottish physics professor (since 1892 Lord KELVIN), **comments that ball lightning** {ARAGO ⇒ 1838} **is an optical illusion from a bright light associated with the blind spot of the eye,** "When people had been looking in some direction or other when the flash came; at the instant that the flash came there was an intense action on the center of the retina, especially if they chanced to see the flash in the sky; naturally after such a startling incident the eyes are moved and the person after seeing the flash looks about to see what has happened – looks on the floor, looks along the wall, looks up at the window, and a spot of light follows, so that he believed this marvelous ball of lightning could be seen by every person present going out of any window that he happened to look out of" – thus stating the same theory as Michael FARADAY {⇒ 1841} did previously.<br><br>**Based on the large number of recent observations, however, ball lightning appears to be a real phenomenon.** The ball has a diameter somewhere between those of a golf ball and a large beach ball; it moves horizontally at low speed, and can decay silently or explode violently. Although rather well-documented since the Middle Ages as a natural but rare phenomenon associated with thunder, it is still an enigma to modern science. The phenomenon of ball lightning {⇒Fig. 4.1–K}, which has occupied generations of naturalists, is still the object of |

---

[1046] From a comment made by Sir William THOMSON in a session on lightning conductors; *see* Rept. Meet. Brit. Assoc. **58**, 603-605 (1888).

speculation and has, more recently, undergone numerical computer modeling {RAÑADA ET AL. ⇨1998}. A comprehensive review of ball lightning, building on a monumental catalog of about 2,400 references prior to 1999, was given recently by Mark STENHOFF,[1047] a U.S. ball lightning researcher and scientific writer.

1888 *Ecole Normale, Paris & Tir du Champ de Châlon, Dépt. Saône-et-Loire, France*

**Félix Albert JOURNEE,**[1048] **Charles Louis E. DE LABOURET,**[1049] and **Hippolyte SEBERT,**[1050] three French military researchers, **perform systematic supersonic ballistic experiments.** In these studies, carried out on the orders of the French Navy Artillery, they do not use high-speed photography to visualize the head wave, but rather measure the front velocity of the head (shock) wave along a line perpendicular to the periphery of the Mach cone and compare the data with Christian DOPPLER's cone model of wave propagation {DOPPLER ⇨1847}.

In the same year, Ernst MACH[1051] criticizes their method and claims priority in the discovery of the head wave phenomenon {E. MACH & SALCHER ⇨1886}. He writes, "Based on experiments of Captain JOURNÉE and deductions of Mr. LABOURET, Colonel SÉBERT has recently derived his view on the velocity of propagation of shots with live ammunitions. Although it is far from me to doubt that the said gentlemen worked completely independently of me, one circumstance appears conspicuous to me. Mr. SÉBERT who mentioned my name in passing, has not perceived that most of his explanations are contained *implicite* in my work published together with Professor SALCHER, while another set of his views are not compatible with our experiments and developments."

1888 *A.B. Separator Company, Stockholm, Sweden*

**Carl Gustav DE LAVAL,**[1052] a Swedish engineer and industrialist, **receives a Swedish patent for his *Laval nozzle*** {⇨Fig. 4.7–F}.

(i) Apparently aware that previous studies on straight nozzles had shown that gas expands best at the velocity of sound – the so-called **"critical speed"** {DE SAINT-VENANT & WANTZEL ⇨1839} – and following his own intuition, **he uses a convergent-divergent nozzle geometry.** This expands the gas isentropically from subsonic to supersonic speeds, thus converting heat energy into motion very efficiently. ▪ He immediately commercialized his new nozzle geometry in his *impulse steam turbine*, a few turbines have survived {DE LAVAL ⇨1893}.

(ii) In his patent he describes his nozzle as follows:[1053] "The present design allows the steam to expand completely before it reaches the rotating, working part of the steam engine, in order to completely use the inherent capacity for work and thus give the steam the most conceivable amount of power. This intention is achieved by making the steam inlet channel in the vicinity of the rotating part of the steam engine shaped in that way that the cross-sections of the channel increase towards the rotating part. This enlargement of the cross-section is controlled in such a way that the ratio between the smallest and the largest section as well as the distance between them shall be such that a 'permanent' fluid flow in an isentropic expansion of the steam will be achieved between those sections."

(iii) In his patent he claims, "In rotating steam engines, a steam inlet channel that has a cross-section in the vicinity of the rotating part of the steam engine that increases in the direction of the said rotating part **in order to expand the steam in such a way that the steam will achieve its highest possible speed before its contact with the rotating, working part of the**

---

[1047] M. STENHOFF: *Ball lightning: an unsolved problem in atmospheric physics.* Kluwer Academic Press & Plenum Press, New York (1999).
[1048] F.A. JOURNEE: *Sur la vitesse de propagation du son produit par les armes à feu.* C. R. Acad. Sci. Paris **106**, 244-246 (1888).
[1049] C.L.E. DE LABOURET: *Sur la propagation du son produit par les armes à feu.* C. R. Acad. Sci. Paris **106**, 934-936 (1888); Ibid. **107**, 85-88 (1888).
[1050] H. SEBERT: *Sur le mode de propagation du son des détonations, d'après les expériences faites au camp de Châlons par M. le capitaine* JOURNÉE. Séances Soc. Franc. Phys. (1888), pp. 35-61.
[1051] E. MACH: *Über die Fortpflanzungsgeschwindigkeit des durch scharfe Schüsse erregten Schalles.* Sitzungsber. Akad. Wiss. Wien **97** (Abth. IIa), 1045-1052 (1888).
[1052] G.P. DE LAVAL: *Steam inlet channel for rotating steam engines.* Swed. Patent No. 1,902 (Nov. 24, 1888).
[1053] This part and the following patent claim were translated by Prof. Olof SÖDERBERG, Finspång, Sweden; private communication (March 14, 1999).

steam engine." ▪ DE LAVAL had already experimented with nozzles 12 years previously. As recorded in his personal notes from 1876, he observed that the severe shock behind a nozzle consumes a lot of energy.[1054] It appears that he resumed this subject when he experimented with an S-shaped rotating channel – the classical reaction turbine {HERO ⇨c.60; Sir PARSONS ⇨1884}. In his notebook from 1886, there are entries on a turbine 10 in. (25.4 cm) in dia. "with conically widened nozzles at the orifice." A dimensioned sketch of an original Laval nozzle was reproduced in the Ph.D. thesis of Paul EMDEN {⇨1903; ⇨Fig. 4.8–C}.

DE LAVAL's idea of using a convergent-divergent nozzle geometry did not appear to be completely new:

▸ Walter TRAUPEL,[1055] a German professor of mechanical engineering, annotated in his textbook on steam turbines that the German KÖRTING Bros. used conical nozzles for steam injectors in 1878 {⇨Fig. 4.7–D}.

▸ Ludwig PRANDTL,[1056] a professor of fluid dynamics at Göttingen University, commented in his textbook on fluid mechanics that, "Hans Johann KÖRTING communicated to me that his grand-uncle Ernst KÖRTING, the widely-known inventor of numerous jet apparatus, already used conically-divergent steam nozzles in 1878. The Laval steam turbine became known in 1883."

▸ Ernst KÖRTING (a mechanical engineer and previously a student of Prof. August RITTER at the Hannover Polytechnic) and his brother Berthold KÖRTING (a businessman) founded the Gebrüder Körting OHG (now Körting Hannover AG), a factory for steam jet apparatus at Hannover. According to his descendant Johannes KÖRTING,[1057] Ernst KÖRTING had already applied in 1869 for a patent on a divergent nozzle geometry which was, however, rejected. Thereafter, the special nozzle geometry used in his steam jet pump was considered to be a company secret.[1058]

In 1903, Ludwig KLEIN,[1059] an assistant at the Königl. Technische Hochschule München, first determined the steam exit velocity at each turbine wheel for a 23-stage steam turbine using a graphical method.

| | |
|---|---|
| 1888 | U.S. Naval Torpedo Station, Newport, Rhode Island |

Charles E. MUNROE,[1060] a U.S. chemist and high explosives specialist, accidentally **discovers how to shape explosives in order to concentrate energy.**

(i) He observes that increasing the depth of the cavity in an explosive generates greater and greater effects on a metal plate facing the explosive. He explains this by an oblique collision of explosive waves. His discovery is partly a rediscovery of the effect of an unlined shaped charge {VON BAADER ⇨1792; VON FÖRSTER ⇨1883}. ▪ In the American literature, MUNROE is usually considered to be the discoverer of the hollow cavity effect, and so it is called the **"Munroe effect."** Similarly, the high-velocity jet of molten metal emerging from a shaped charge with a metal liner – which MUNROE never studied {THOMANEK ⇨1938} – is called a **"Munroe jet."**

(ii) The hollow cavity effect is used by him to imprint designs on iron plates by interposing a stencil between the explosive and the iron plates – a unique technique which he terms *explosive*

---

[1054] Prof. Carl-Göran NILSON, Djursholm, Sweden; private communication (May 1999).
[1055] W. TRAUPEL: *Thermische Turbomaschinen*. Springer, Berlin *etc.* (1988), vol. 1, p. 100.
[1056] L. PRANDTL: *Strömungslehre*. Vieweg & Sohn, Braunschweig (1957), p. 253.
[1057] J. KÖRTING: *Ernst KÖRTING 1842–1921. Ein Ingenieur und Unternehmer im kaiserlichen Deutschland.* Technikgeschichte in Einzeldarstellungen, Nr. 34. Verein Deutscher Ingenieure (VDI), Düsseldorf (1975), pp. 37-38, 54-55.
[1058] Dr. York FUSCH, a great-grandson of Ernst KÖRTING and member of the board of Körting Hannover AG, sent two copies of drawings on convergent-divergent nozzles such as those used in E. KÖRTING's steam injector pumps to the author. Made in October 27, 1876 and August 4, 1878, they prove that Ernst KÖRTING had already used this particular nozzle geometry many years prior to DE LAVAL.
[1059] L. KLEIN: *Theorie, Konstruktion und Nutzeffekt der Dampfturbinen*. VDI-Z. **39**, 1189-1195 (1895).
[1060] C.E. MUNROE: *On certain phenomena produced by the detonation of gun-cotton*. Proc. Newport Natl. Hist. Soc. Rept. No. 6 (1883–1888); *Wave-like effects produced by the detonation of gun-cotton*. Am. J. Sci. **36**, 48-50 (1888).

| | | |
|---|---|---|
| | | ***engraving***.[1061] He writes, "What is surprising is to find that the impression produced by the exploding mass is an almost exact copy of the face of the explosive which was in contact with the metal. This is best observed with gun-cotton, for, from the nature of the material, it can be shaped according to fancy, and such figures and designs as one wishes can be stamped upon its surface…" ▪ The explanation given for the Munroe phenomenon – that waves of hot compressed gases increase in force through a hollow charge of explosive – led to much discussion.[1062] |
| 1888 | *German University of Dorpat, Russia [now Tarpu, Estonia]* | Arthur J. VON OETTINGEN and Arnold VON GERNET,[1063] two Estonian-German chemists, **set out to demonstrate that Robert W. BUNSEN's principle of successive partial explosions is valid** {BUNSEN ⇨1867}.<br><br>(i) To resolve the hypothetical discontinuous nature of one-dimensional oxyhydrogen explosions, **they use a high-speed rotating mirror and a still camera and produce time-resolved records for the propagation of the flame front** in a straight 40-cm-long eudiometer tube {⇨Fig. 4.19–K}. Since the spectrum emitted by the flame is beyond the spectral range of even the most sensitive photographic plate, they add a small quantity of a salt to the system [cuprous chloride, CuCl].<br><br>(ii) They obtain streak records showing the passage of compression waves moving through the products of combustion, which are sharply reflected backwards and forwards from the end of the tube and gradually diminish in intensity and velocity.<br><br>(iii) They correctly state that a true detonation wave proceeds from the spark which ignited the electrolytic gas, but incorrectly conclude that its compression wave traverses the tube several times before it becomes visible because it raises the salt to incandescence – *i.e.*, that the explosion itself is quite invisible. ▪ Detonation studies carried out at Owens College in Manchester showed that the detonation does not occur immediately; it only occurs after the flame has moved some distance, which varies with the nature of the mixture and the position of the spark {DIXON ⇨1903}.<br><br>(iv) **Using their unique streak diagnostics, they determine an initial explosion velocity of 2,560 m/s**, thus confirming previous results obtained in France {BERTHELOT ⇨1881}. After several reflections in the tube, the shock wave has diminished to a velocity of 600 m/s.<br><br>(v) They call the explosion wave "main wave" [Germ. *Hauptwelle*] or "Berthelot wave" [*Berthelot'sche Welle*] and its reflection at the tube end "shock wave" [*Stoßwelle*], but also observe various secondary waves which they call "Bunsen waves" [BUNSEN*'sche Wellen*]. |
| 1888 | *Chair of Natural Philosophy, University of Edinburgh, Scotland, U.K.* | Peter G. TAIT,[1064] a Scottish physicist, **proposes an empirical isothermal equation of state to fit data for the compressibility of seawater up to 500 bar,** which is given by<br>$$(p + B)/(p_0 + B) = (\rho/\rho_0)^A,$$<br>where $A$ and $B$ are two empirical functions of temperature, and $p$ and $\rho$ are the pressure and density, respectively. The initial state is given by $p_0$ and $\rho_0$. For $B \gg p_0$, this equation can be approximated by the **"Tait equation"**<br>$$p = C\,[(\rho/\rho_0)^A - 1],$$<br>which resembles an isentrope in a perfect gas. ▪ At 20 °C, one obtains $C = 3{,}047$ bar and $a = 7.15$ {COLE ⇨1948}. The quantity $C$ is actually a weak function of entropy, but for many applications can be assumed to be a constant. |

---

[1061] C.E. MUNROE: *Modern explosives*. Scribner's Mag. **3**, 563-576 (1888).
[1062] A. MARSHALL: *The detonation of hollow charges*. J. Soc. Chem. Ind. **29**, 35-T (1920).
[1063] A. VON OETTINGEN and A. VON GERNET: *Über Knallgasexplosionen*. Ann. Phys. **33** [III], 586-609 (1888).
[1064] P.G. TAIT: *Scientific papers*. Cambridge University Press, vol. II (1900); see *Report on some of the physical properties of fresh water and of sea-water. From the "Physics and Chemistry" of the voyage of HMS Challenger*. Paper No. LXI, pp. 1-68, and *On the compressibility of liquids in connection with their molecular pressure*. Paper No. CVII, pp. 334-338.

▸ During World War II, the U.S. physicists John G. KIRKWOOD[1065] and John M. RICHARDSON at Cornell University used the Tait equation to describe seawater from initial conditions ($p_0, \rho_0$) up to a pressure of 25 kbar.

▸ Yuan-Hui LI,[1066] a researcher at Lamont Geological Observatory (Columbia University, New York) concluded that the relationship given by the Tait equation represents the thermodynamic relationship of water very well.

▸ A modified version of the Tait equation has also been used to describe the $p(\rho)$-relations of some organic liquids.[1067, 1068]

| | | |
|---|---|---|
| 1889 | W. & J. Mackay & Co., Chatham, Kent, England | **Publication of the first *Dictionary of Explosives*, compiled by the British Mayor John P. CUNDILL.**[1069] ▪ It was also translated and published in France (1893).<br><br>The increasing number of new explosives invented in the second half of the 19th century meant that a detailed classification and description of their physical and chemical properties was required. In 1890, the Englishman Manuel EISSLER[1070] published his ***Handbook on Modern Explosives***, and in the period 1886–1893 the renowned U.S. chemist Charles MUNROE[1071] prepared an *Index to the Literature of Explosives*. |
| 1889 | *II. Physikalisch-Chemisches Laboratorium, Universität Leipzig, Saxony, east-central Germany* | **Svante August ARRHENIUS,**[1072] a Swedish chemist who had worked in the previous year as a visiting scientist at Wilhelm OSTWALD's laboratory in Leipzig and subsequently at Jacobus H. VAN'T HOFF's institute in Amsterdam, **works out a chemical reaction rate law based upon his concept that molecules must collide with sufficient energy if they are to react;** *i.e.*, where the temperature is a decisive parameter in the initiation of a chemical reaction.<br><br>(i) He studies the influence of an increase in temperature on the velocity of a chemical reaction. Starting from VAN'T HOFF's equilibrium equation,[1073] he provides an analytical relation between the velocity coefficient and the temperature.<br><br>(ii) **He states that the specific rate of a chemical reaction $K$ is proportional to the concentration of activated molecules.** His so-called **"Arrhenius equation"** (or "Arrhenius rate law") is given by<br>$$K = A \exp(-E/RT),$$<br>where $A$ is a constant, $E$ is the activation energy, $T$ is the temperature in degrees Kelvin, and $R$ is the universal gas constant. ▪ This important equation not only explains ordinary chemical reaction rates but also many biological reaction rates.<br><br>The Arrhenius equation even proved useful for describing strong exothermic reactions such as detonations {DOVE & TRIBBECK ⇨1970}.[1074] |
| 1889 | *Trinity College, Oxford, England* | **Frederick John JERVIS-SMITH,**[1075] a British vicar and a Millard lecturer in mechanics, **determines the rate of change of the velocity of propagation of the explosion for various gaseous mixtures.** |

---

[1065] J.G. KIRKWOOD and J.M. RICHARDSON: *The pressure wave produced by an underwater explosion. Part II*: Rept. OSRD-670 (June 1942); *Ibid., Part III.* Rept. OSRD-813 (Aug. 1942).

[1066] Y.H. LI: *Equation of state of water and sea water*. J. Geophys. Res. **72**, 2665-2678 (1967).

[1067] J.O. HIRSCHFELDER, C. CURTISS, and R. BIRD: *Molecular theory of gases and liquids*. Wiley, New York (1964), p. 261.

[1068] Y.A. ATANOV: *An approximate equation for the liquid state at high pressures*. Russ. J. Phys. Chem. **40**, 655-656 (1966).

[1069] J.P. CUNDILL: *A dictionary of explosives*. W. & J. Mackay & Co., Chatham (1889).

[1070] M. EISSLER: *A handbook on modern explosives, being a practical treatise on the manufacture and application of dynamite, gun-cotton, nitro-glycerine, and other explosive compounds, including the manufacture of collodion-cotton*. C. Lockwood, London (1890).

[1071] C.E. MUNROE: *Index to the literature of explosives. Part I*. Friedenwald, Baltimore (1886); *Part II*. Deutsch Lithographic Printing Company, Baltimore (1893).

[1072] S.A. ARRHENIUS: *Über die Dissociationswärme und den Einfluß der Temperatur auf den Dissociationsgrad der Elektrolyte*. Z. Physik. Chem. **4**, 96-116 (1889); *Über die Reaktionsgeschwindigkeit bei der Inversion von Rohrzucker durch Säuren*. Ibid. **4**, 226-248 (1889).

[1073] J.H. VAN'T HOFF *Études de dynamique chimique*. Muller, Amsterdam (1884).

[1074] Y.B. ZEL'DOVICH and Y.P. RAIZER: *Physics of shock waves and high-temperature hydrodynamics phenomena*. Dover Publ., Mineola, NY (2002), pp. 368-374.

(i) He was stimulated by previous French studies which showed that explosive gaseous mixtures do not reach their maximum velocity of propagation immediately after their ignition, but that a certain maximum velocity is attained soon after initial ignition {BERTHELOT & VIEILLE ⇨ 1881; MALLARD & LE CHÂTELIER ⇨ 1881}.

(ii) Reasoning that few advances could be made in this branch of research into explosions unless very small periods of time could be measured with certainty, he invents a chronograph which records signals on a soot-covered glass plate propelled by means of a falling weight – the so-called **"electric tram chronograph."** It is capable of measuring $1/20,000$th second (a time span of 50 μs) with ease, and periods of time differing from $1/10$th second to $1/20,000$th second on the same moving surface.[1076]

JERVIS-SMITH also served on a committee appointed by the Home Secretary to inquire into the causes of explosions due to compressed gases.

| | |
|---|---|
| 1889 *Ballistic Test Site, Pola, Austro-Hungarian Empire* | **Ernst MACH**[1077] **and Peter SALCHER supervise free-field supersonic ballistic experiments on large shots using single-shot high-speed photography.** The study is financially supported by the German Krupp Company. |

(i) **They use a *gliding spark* arrangement as a flash light source.** It consists of two magnesium wires used as electrodes which are embedded into a groove in a hard rubber plate and are confined by a glass plate. Simple to construct and providing a high light output, this linear geometry is able to match the schlieren knife in TOEPLER's famous schlieren method {A. TOEPLER ⇨ 1864}. ▪ **The gliding spark or "guided spark,"** which was perfectly suited to generating shock waves of different geometries and strengths, **played an important role in the evolution of shock wave physics** {ANTOLIK ⇨ 1874; E. MACH & WOSYKA ⇨ 1875}. Gliding sparks have been applied and modified in various ways for use in schlieren and reflected-light photography.[1078] At EMI, Freiburg, a multiple gliding-spark light source of ultrashort duration (180 ns FWHM) and of possessing an extremely luminous peak intensity (200 Mcd) was developed for reflected-light color photography of EFPs (explosively formed projectiles), using a nine-stage 35-kV Marx-surge generator for current pulsing.[1079]

(ii) In order to photograph the head wave geometry of a 9-cm-caliber projectile flying at 438 m/s, they use a darkened wooden hut (floor size $14 \times 2.5$ m$^2$) provided with a small entrance and exit portholes for the projectile trajectory. It houses the schlieren arrangement and sophisticated electric trigger circuitry {⇨Fig. 4.19–H}, previously developed by Ernst and Ludwig MACH and first used by them at the Meppen Ballistic Test Site (northern Germany) to record small shots photographically. Their diagnostic method even permitted schlieren photos of sufficient exposure density to be obtained in a harsh environment – a masterpiece of early ballistic photography {⇨Fig. 4.6–G}.

---

[1075] F.J. JERVIS-SMITH and M.A. MILLARD: *An experimental investigation of the circumstances under which a change of the velocity in the propagation of the ignition of an explosive gaseous mixture takes place in closed and open vessels. Part I: Chronographic measurements.* Proc. Roy. Soc. Lond. **45**, 451-452 (1889).

[1076] F.J. JERVIS-SMITH: *A new form of electric chronograph.* Phil. Mag. **29** [V], 377-383 (1890).

[1077] E. MACH and P. SALCHER: *Über die in Pola und Meppen angestellten ballistisch-photographischen Versuche.* Sitzungsber. Akad. Wiss. Wien **98** (Abth. IIa), 41-50 (1889).

[1078] E. FÜNFER: *Gleitfunken als Lichtquelle für Funkenkinematographie.* Rept. No. 9/46, D.E.F.A., L.R.S.T.A., Saint-Louis, France (April 1946).

[1079] P. KREHL and S. ENGEMANN: *A high-intensity diffuse light source of ultrashort duration for reflected-light color photography.* Rev. Scient. Instrum. **64**, 1785-1793 (1993); *600 million candles in a flash.* Physics World **7**, No. 4, 33 (April 1994).

| 1889 | German Charles University, Prague, Bohemia, Austro-Hungarian Empire | **Ernst MACH**[1080] **and his son Ludwig MACH,** a physician and experimental physicist, **make the first schlieren photo** {⇨Fig. 4.5−G} **of a shock wave in air, which they call** *Schallwelle großer Excursion* ("sound wave of large excursion"). Also studying the interaction phenomena of two shock waves emerging from two closely spaced point sparks, **they obtain the first schlieren photos of the Mach disk** {⇨Fig. 4.13−C}. Their experiments fully confirm the triple-point model of shock wave reflection that they had previously assumed to be true based on soot records alone {E. MACH & WOSYKA ⇨1875; E. MACH & GRUSS ⇨1878}.

54 years later, U.S. researchers using two detonator caps in a very similar geometry were the first to prove that Mach reflection also exists in water {SPITZER & PRICE ⇨1943}. |
|---|---|---|
| 1889 | Potsdam Observatory & Wilhelmshaven Observatory, northern Germany | **Ernst VON REBEUR-PASCHWITZ,**[1081] a German astronomer, geophysicist and seismologist, **uses delicate photographically recording seismographs to obtain the world's first records of a teleseismic event − an earthquake which happens in the same year on April 18 near Tokyo, Japan.** The waveform is recorded simultaneously on two identical sensitive horizontal pendulums, one located at the Telegrafenberg of the Potsdam Observatory (8,211 km from Tokyo) and the other at Wilhelmshaven (8,307 km from Tokyo). ▪ The realization that strong earthquakes could be recorded at great distances helped to usher in the modern era in the fields of seismology and the physics of the Earth's interior. |
| 1889 | Royal Imperial Austro-Hungarian Naval Academy, Fiume [now Rijeka], Croatia, Austro-Hungarian Empire | **Peter SALCHER,**[1082] an Austrian physics professor, **and Robert WHITEHEAD,** a British engineer and factory owner, **study the discharge parameters of a "free air jet"** exhausting from a pressure reservoir through a small opening, and compare their experimental data with various existing theories. SALCHER performs the experiments in WHITEHEAD's torpedo factory (Whitehead & Co.) at Fiume. WHITEHEAD, who invented the first self-propelling torpedo (1866), gained much experience in the generation and storage of high-pressure gases because his torpedoes are propelled by engines that use compressed air at pressures of up to 2,000 psi [137 bar]. Their experimental results are as follows:

(i) By illuminating the free air jet with a light source of very short duration (such as an electric spark) or a light source of long duration (such as the light from a Geisler discharge tube or even sunlight), it is possible to visualize nonstationary or stationary flow characteristics, respectively.

(ii) They are the first to make the startling observation that **the jet emerging from a pressurized nozzle contains a crossed wave pattern.** Since this interference pattern reminds SALCHER of an ancient Greek harp, he calls it a "lyre" [Germ. *Lyra*] in a letter to Ernst MACH.[1083] They correctly interpret this as a superposition of reflected waves. ▪ In subsequent years, the free air jet experiments described above were resumed at Charles University, Prague by Ludwig MACH, who was the first to obtain interferograms of excellent quality from this important phenomenon {L. MACH ⇨1897}.

In the same year, **SALCHER,** inspired by his previous study on free air jets, **suggests a supersonic blow-down wind tunnel where the air flows through the tunnel and the test body is at rest inside it.** In a paper written together with Ernst MACH,[1084] it says, "On the occasion of the experiments on projectiles SALCHER hit upon the idea of likewise investigating the inverse case of the flow of air against a body at rest in order to confirm the results already obtained." They confirm that the inverse case is indeed possible, but head wave studies on model projectiles were not practicable with existing equipment due to the small jet diameter available. |

---

[1080] E. MACH and L. MACH: *Über die Interferenz von Schallwellen von großer Excursion.* Sitzungsber. Akad. Wiss. Wien **98** (Abth. IIa), 1333-1336 (1889).
[1081] E. VON REBEUR-PASCHWITZ: *The earthquake of Tokyo, April 18, 1889.* Nature **40**, 294-295 (1889).
[1082] P. SALCHER and J. WHITEHEAD: *Über den Ausfluß stark verdichteter Luft.* Sitzungsber. Akad. Wiss. Wien **98** (Abth. IIa), 267-287 (1889).
[1083] SALCHER's letter to E. MACH (dated April 19, 1888) is now in the Archives of the Deutsches Museum, München.
[1084] E. MACH and P. SALCHER: *Optische Untersuchungen der Luftstrahlen.* Sitzungsber. Akad. Wiss. Wien **98** (Abth. IIa), 1303-1309 (1889).

- SALCHER's idea was apparently first realized in France by Eugène HUGUENARD and Jean André SAINTE-LAGUË for measurements of aerodynamic drag on projectiles at supersonic speeds {LANGEVIN & CHILOWSKY ⇒1918; ⇒Fig. 4.9−A}. In the United States, MAUTZ ET AL. {⇒1948} were apparently the first to extend this idea by striking a test body at rest with a shock wave.

| | | |
|---|---|---|
| 1889 | Würzburg, Bavaria, Germany | Max ZWERGER,[1085] a German geometrician and high-school professor, **publishes a historical-critical study on the center of oscillation for a compound pendulum.** He gives a brief historical perspective of the subject and discusses how to find the center of gravity of a rigid body. For flexible bodies, he refers to the fundamental works of Daniel BERNOULLI and Leonard EULER, who treated this difficult problem in a general manner but were not able to apply it to practical examples because of the enormous computational effort involved. ▪ Since the center of oscillation of a compound pendulum is identical to the center of percussion of a body rotating around a fixed axis and struck eccentrically by an impulsive force {WALLIS ⇒1670/1671}, ZWERGER's study is also useful for applications in mechanical engineering. |
| 1890 | Royal College of Science, South Kensington, Middlesex, England | Charles V. BOYS,[1086] a British physicist and inventor of numerous sensitive instruments, **studies the air flow around bullets and the interactions of multiple shock waves using shadowgraphy** {DVOŘÁK ⇒1880; ⇒Fig. 4.6−H}. **He also succeeds in measuring the spin rate of a shot.** Since a professional ballistic range is not available to him, he performs his shot experiments in a long hallway in his institute. ▪ BOYS,[1087] who repeated some of Ernst MACH's ballistic experiments {E. MACH & SALCHER ⇒1886}, promoted the spread of their methodology in England. Today, however, shadowgraphy is used more frequently than the schlieren method in most outdoor ballistic facilities due to its simplicity and minor sensitivity toward temperature fluctuations. |

In 1892, BOYS[1088] wrote to the Austrian philosopher Ernst MACH, "I am much obliged to you for your kindly sending me copies of your papers and the two photographs. I have when speaking on bullet photography thoroughly recognized that the whole credit of bullet photography is yours, as you were the first, to carry it out successfully and that your apparatus answers perfectly *i.e.*, so far as I can judge from your account of them. In the English papers were inaccurate reports and in one case of a scientific paper I corrected it, as stated, what you had done. The daily papers are always so untrustworthy that it is absurd to credit them. I do not think I have failed to appreciate or to recognize what you have done … If you should think I have not properly recognized your work, I am exceedingly sorry, that it should be so, but I am sure if you had heard what I have said at the Royal Society and elsewhere, that you would not think so."

In the same year, Ernst MACH[1089] writes in a paper: "BOYS' method is certainly a simplification when it is used merely for demonstration purposes in a lecture. However, I suppose that everybody who wants to study this matter in more detail will prefer an optical image which allows one to estimate the condensation by its shading, rather than a mere silhouette which only reveals the contours of the air waves … Nevertheless, I am grateful to Mr. BOYS that he has taken over this assignment, previously not touched by others, and I hope that he intends to continue it in future."

---

[1085] M. ZWERGER: *Der Schwingungsmittelpunkt zusammengesetzter Pendel: historisch-kritische Untersuchung nach den Quellen bearbeitet.* Lidauscher, München (1889).
[1086] C.V. BOYS: *Notes on photographs of rapidly moving objects, and on the oscillating electric spark.* Phil. Mag. **30** [V], 248-260 (1890).
[1087] C.V. BOYS: *Photography of flying bullets, etc.* [with discussion]. Photogr. J. **16**, 199-209 (1892).
[1088] This letter (dated July 29, 1892) is now kept in the Archives of the Deutsches Museum, München.
[1089] E. MACH: *Ergänzungen zu den Mitteilungen über Projektile.* Sitzungsber. Akad. Wiss. Wien **101** (Abth. IIa), 977-983 (1892), p. 983.

| 1890 | *French Firedamp Commission, Paris, France* | Georges CHARPY,[1090] a French chemist and metallurgist, **and Henry LE CHÂTELIER,**[1091] a French chemist and mining engineer, **review the first results from experiments performed by the French Firedamp Commission to investigate possible causes of firedamp explosions** [French *le grisou*] **in coal mines.** To avoid such explosions, the commission recommends |

- the provision of an effective ventilation system to prevent sudden outbursts of firedamp, which will reduce the concentration of methane in the air below 5%, particularly in all higher gallery sections;
- the use of safe explosives;
- the avoidance of open fire, sparks, *etc.*; and
- the exclusive use of miner's lamps that remain safe even at higher air speeds.

At that time, pure coal dust explosions – *i.e.*, those that occur without the presence of firedamp, which had been frequently observed by British but only rarely by French mining engineers – were not yet considered a real hazard.

| 1890 | *Chemical Laboratory, University of Kiel, northern Germany* | Theodor CURTIUS,[1092] a German professor of chemistry, **is the first to characterize azide [$-N_3$] salts and to discover silver azide** [$AgN_3$], which he calls *"Stickstoffsilber."* He reports that "the detonation of only a few milligrams of silver azide, produced by the action of heat or percussion, is unprecedented and only comparable to the sharp report of the discharge of a Leiden jar." **He also discovers mercury azide** [$Hg_2(N_3)_2$], which he calls *"Stickstoffcalomel."*

In the following year, CURTIUS[1093] reported on the discovery of *Bleiazid* ("lead azide") [$Pb(N_3)_2$]. This chemical detonates with great force under the action of mild heating or when subjected to sudden impact, and it took many years of research to control the manufacture of lead azide. Because of its great stability to moisture, it was introduced into detonator caps in the late 1920s, replacing mercury fulminate {HOWARD ⇨1800} and thus making blasting operations safer. |

| 1890 | *Royal Military Academy, Woolwich, southeast London, England* | Alfred G. GREENHILL,[1094] a mathematics professor at Woolwich stimulated by the design and use (at sea or in forts) of monster guns such as the British 110-ton and the Krupp 135-ton guns, **investigates the scientific principles involved in making big guns:**

(i) A simple rule implies that the size of the gun depends on the thickness of armor it is required to attack: the caliber practically increasing with the thickness to be pierced, but the weight of the gun increases as the cube of the caliber. This imposes a number of difficulties on big guns, such as metallurgical limitations, handling issues and transportation problems.

(ii) Assuming elastic material behavior and hydrostatic loading by internal and external pressures, he calculates the internal stresses and strains in thick cylinders of compound guns. ▪ The analysis of dynamic material phenomena that arise from the propagation of radial vibrations was not yet accessible.

(iii) He discusses particular problems related to shrinkage when making compound gun barrels consisting of a tube, a jacket and a series of hoops, and theoretical estimations for optimizing their design.

(iv) Also, addressing the much-discussed possibility of strengthening guns with wire, **he proposes a theory for estimating the circumferential strength of the so-called "wire gun."**
▪ The wire gun was apparently first proposed in the 1850s by the American gun expert William E. WOODBRIDGE. Wire guns were cheaper to produce and allowed the weight of artillery guns to be reduced. |

---

[1090] G. CHARPY: *Les travaux de la Commission du grisou.* Rev. Gén. Sci. Pures Appl. **1**, 540-546 (1890).
[1091] H.L. LE CHATELIER: *Le grisou et ses accidents.* Rev. Gén. Sci. Pures Appl. **1**, 630-635 (1890).
[1092] T. CURTIUS: *Über Stickstoffwasserstoffsäure (Azoimid) N₃H.* Ber. Dt. Chem. Gesell. **23**, 3023-3033 (1890).
[1093] T. CURTIUS: *Neues vom Stickstoffwasserstoff.* Ber. Dt. Chem. Gesell. **24**, 3341-3349 (1891).
[1094] A.G. GREENHILL: *The scientific principles involved in making big guns.* Nature **42**, 304-309, 331-334, 378-381 (1890).

The enthusiasm for monster guns lasted up to World War II {⇨Fig. 4.2–T}. However, difficulties with transportation, handling and maintenance have meant that they have been of limited utility in practical warfare.

| | | |
|---|---|---|
| 1890 | *Imperial Navy, Germany* | **Ernst C.R. ROTTOK,**[1095] a German captain lieutenant and admiralty councilor, **discusses the various sound, sea surface and seismic shock phenomena that have been observed during seaquakes.** He summarizes contemporary hypotheses developed by the German geologists Emil RUDOLPH {⇨1887}, and Ferdinand VON HOCHSTETTER and Franz E. GEINITZ {Arica Earthquake ⇨1868}. |
| 1890/ 1891 | *Midland, Texas* | **The U.S. Congress gives the Dept. of Agriculture $9,000 to conduct experiments to produce rain artificially by disrupting the "air currents" overhead using explosions,** a method previously proposed by the U.S. civil engineer Edward POWERS.[1096] The German-born government official and Union veteran Gen. Robert DYRENFORTH is appointed to take charge. |

In the following year, DYRENFORTH led an expedition to Texas and began performing his main set of blast experiments using large quantities of gunpowder on the ground, sticks of dynamite attached to high-flying kites, and oxyhydrogen balloons detonating at an altitude of several thousand feet. One observer wrote, "At the touch of the electric discharge, the balloon suddenly is transformed into a brilliant globe of fire, which instantly swells to monstrous size, casting a flash of light over every object within several miles, and then, after a few moments of darkness … the tremendous crash of the explosion comes rolling on and shakes the very ground by its concussion." Since rain did indeed fall in the area during the experimental period, POWERS' concept was initially considered to be a useful method of producing rain at will. However, later experiments revealed that this method cannot produce rain.[1097]

The use of explosions to influence weather phenomena was resumed in 1896 by the Honorable Albert STIGER, mayor of the town of Windisch-Feistritz in Styria. He attempted to reduce damage from hailstorms by firing cannonballs into thunderclouds. However, his method proved ineffective. ▪ Indeed, detonation experiments carried out in supercooled clouds from the Old Faithful Geyser at Yellowstone National Park, WY, in the 1960s showed that a shock wave can actually trigger (rather than prevent) the generation of hail {GOYER ⇨1964/1965}.

The modern approach to this problem is based on the principle of seeding the thunderstorm with particles of silver iodide using timed artillery shells or rockets.

| | | |
|---|---|---|
| 1891 | *Philadelphia, Pennsylvania* | **Albert Edward FOOTE,**[1098] a physician, mineralogist and private dealer in minerals, **first describes meteoritic diamonds in the Arizona meteorite,** also known as the "Canyon Diablo meteorite," the impactor that produced Meteor Crater {⇨c.50,000 years ago}. ▪ It was later proposed that the meteoritic diamonds arose due to the conversion of graphite to diamond by shock metamorphism upon the meteorite's impact with Earth. Analysts studying the Canyon Diablo diamonds found that up to a third of them had a hexagonal atomic structure never before seen in diamond; ordinary Earth-grown diamonds have a cubic structure. **Mineralogists named the new hexagonal variant of diamond *lonsdaleite* after Dame Kathleen LONSDALE,** an Irish-born British mineralogist at the Royal Institution and University College, London. She developed an X-ray technique for accurate measuring the distance between carbon atoms in diamond.[1099] |

---

[1095] E.C.R. ROTTOK: *Unterseeische vulkanische Eruptionen und Erdbeben.* Himmel und Erde (Leipzig & Berlin) **2**, 509-518 (1890).
[1096] E. POWERS: *War and the weather.* E. Powers, Delavan, WI (1890).
[1097] W.B. MEYER: *The life and times of U.S. weather (Part III). Rain made to order: preliminary experiments in Texas prove successful (August 1891).* American Heritage Magazine, New York (June/July 1986); http://www.americanheritage.com/articles/magazine/ah/1986/4/1986_4_38.shtml.
[1098] A.E. FOOTE: *A new locality for meteoric iron with a preliminary notice of the discovery of diamonds in the iron.* Am. J. Sci. **42**, 413-417 (1891).
[1099] D.K. LONSDALE: *Diamonds, natural and artificial.* Nature **153**, 669 (1944).

| 1891 | *French Navy Artillery, Lorient, Dépt. Morbihan, France* | **Fernand-François Gossot**,[1100] a French ballistician and captain of the French Navy who had resumed previous French supersonic ballistic-range studies {Journée, Labouret & Sébert ⇒1888}, **reports that the pressure in the ballistic head wave is not a vibration but rather consists of a single shock.** He recorded the pressure signal using a mechanical chronograph and a contact pressure gauge which, upon receiving the head wave, interrupted the electric current of an electromagnetic stylus. ▪ His result fully confirmed previous optical studies which had shown that the head wave is indeed a single pressure discontinuity {E. Mach & Salcher ⇒1886}. |
|---|---|---|
| 1891 | *Chemische Fabrik Griesheim, Hesse, Germany* | **Carl Haeussermann**,[1101] a German chemist, **discovers the explosive properties of trinitrotoluene** [2,4,6-trinitrotoluene, $C_7H_5N_3O_6$], **later known as "TNT."** The substance was first crudely synthesized by the German chemist Julius Wilbrand[1102] via the reaction of toluene with nitric acid in the presence of sulfuric acid; pure TNT was first prepared by the German chemist Paul Hepp[1103] in 1880. ▪ Haeussermann, who undertook the manufacture of TNT on an industrial scale, was also the first to suggest its military use: pressed into shells, it can withstand the shock developed by the propellant when a shell is fired from a gun barrel. In 1902, TNT was adopted for use by the German Army. |

(i) According to the modern classification of explosives, TNT is a "secondary explosive," which is still very important in the field of high explosives. Its detonation velocity was later determined to be 6,380 m/s (for a density of 1.51 g/cm$^3$).

(ii) TNT gained great military importance in both World Wars and became an important primary product for a number of other explosives, such as the military explosives *amatol* (50/50 AN/TNT), *cyclotol* (60/40 RDX/TNT), *nitrolite* (76/12/5 AN/TNT/nitroglycerin), *pentolite* (50/50 PETN/TNT), and *tritonal* (80/20 TNT/aluminum).

(iii) TNT has also been chosen for use as a standard: the equivalent energy yield of an explosive is usually related to that of pure TNT (loading density $\rho_0 = 1.654$ g/cm$^3$), which, by definition, has a specific chemical energy of 1 kcal/g or 4.184 kJ/g.[1104]

(iv) Since the first nuclear explosion {Trinity Test ⇒1945}, the yield of each nuclear weapon has been related to the so-called **"TNT equivalent energy."** In addition, the huge energy released by large natural catastrophes (such as earthquakes, explosive volcanic eruptions, and asteroid impacts) is frequently computed in megatons of TNT equivalent, or in terms of the energy released by the detonation of the atomic bomb over Hiroshima {Hiroshima & Nagasaki Bombing ⇒1945}, which released an energy of 15 kilotons (kT) of TNT equivalent or $15 \times 10^{12}$ cal.

(v) One of the first calculations of the equation of state for the detonation gas from a solid high explosive was also related to TNT: British detonation researchers calculated the changes in the data for the equation of state (*i.e.*, $p, \rho$ values) during an isentropic expansion after detonation {Jones & A.R. Miller ⇒1948}.

DNT [2,4-dinitrotoluene or 2,4-DNT; $C_7H_6N_2O_4$], a by-product of the production of TNT, is a gelatinizing and waterproofing agent. It is less reactive than TNT and is therefore used in small amounts in "permitted" (or "permissible") explosives {Norrbin & Ohlsson ⇒1867}.

---

[1100] F.F. Gossot: *Détermination des vitesses des projectiles.* Mémorial de l'Artillerie de la Marine **19**, 181-234 (1891).
[1101] C. Haeussermann: *Über die explosiven Eigenschaften des Trinitrotoluols.* Z. Angew. Chem. **4**, 508-511 (1891).
[1102] J. Wilbrand: *Notiz über Trinitrotoluol.* Ann. Chem. Pharm. **128**, 178-179 (1863).
[1103] P. Hepp: *Über Trinitroderivate des Benzols und Toluols: über Additionsprodukte von Nitroderivaten mit Kohlenwasserstoffen.* Ph.D. Thesis, Universität Straßburg (1881).
[1104] S. Glasstone (ed.): *The effects of nuclear weapons.* U.S. Atomic Energy Commission, Washington, DC (1962). ▪ Referring to the *TNT equivalent energy*, he wrote on p. 14, "This value is based on an accepted, although somewhat arbitrary, figure of $10^{12}$ calories (= $4.184 \times 10^{12}$ J) as the energy released in the explosion of 1 kiloton of TNT." However, the released energy of a detonating TNT charge also depends on the experimental parameters (*e.g.*, the charge density and its geometry). Besides the standardized value given above, higher values (ranging up to $5.1 \times 10^{12}$ J/kT) can be found in the literature. One ton of TNT releases approximately 1.2 billion calories (that is, 5.1 kJ/g). Nuclear explosions are usually measured in kilotons (kT) or megatons (MT).

| 1891 | Derwitz/Krilow [near Potsdam], northern Germany | Otto VON LILIENTHAL, a German aeronautical pioneer, **achieves the longest controlled flight of a human (> 25 m). His achievement marks the birth of flight with heavier-than-air machines.** ▪ Over a period of only five years, VON LILIENTHAL developed 18 different models of gliders and made nearly 2,000 successful glider flights.<br><br>By this time, the basics of supersonics were already understood reasonably well, but practical aviation was still very much in its infancy. |
|---|---|---|
| 1891 | University of Würzburg, Bavaria, Germany & German Charles University, Prague, Bohemia, Austro-Hungarian Empire | **Invention of the *Mach-Zehnder interferometer* by Ludwig MACH**[1105] **and, independently, by Ludwig ZEHNDER.**[1106] Ludwig MACH first uses his interferometer in nonstationary gas dynamics. ZEHNDER[1107] then applies this new type of interferometer in work performed for his doctorate at the University of Würzburg (under the guidance of Wilhelm C. RÖNTGEN) in order to investigate the pressure dependency of the refractive index of water. ▪ Consisting of two beam splitters and two mirrors, the Mach-Zehnder interferometer {⇨Fig. 4.18−F} divides the source beam into two different parallel light paths of arbitrary distance: an object beam and a reference beam, which proved to be most worthwhile for measuring variations in refractive index in a compressible gas flow.<br><br>In the late 1890s, Ludwig MACH was employed as a scientific consultant by the Carl Zeiss Company at Jena, which commercialized his invention. In 1928, Zeiss developed a special Mach-Zehnder interferometer to study flows in close proximity to heated bodies for a Soviet research institute;[1108] this experimental task proved a particular challenge due to the high sensitivity of this type of interferometer. In the 1930s, the Mach-Zehnder interferometer was increasingly adopted by German ballistic and wind-tunnel testing facilities. From the late 1940s onwards, the Mach-Zehnder interferometer was also used in the United States to study shock waves[1109] and supersonic air jets,[1110] and to evaluate the density field behind obliquely reflecting shock waves − such as those associated with single and double Mach reflection {BLEAKNEY & TAUB ⇨1949; WHITE ⇨1951}. |
| 1891 | Newport Torpedo Station, Rhode Island | Charles E. MUNROE,[1111] a U.S. military chemist and explosives expert, **invents a smokeless powder**, a mixture of nitrocellulose [$(C_6H_7O_2(OH)_3)_x$] and nitrobenzene [$C_6H_5NO_2$], which he makes from guncotton freed from lower nitrates by washing with methyl alcohol and colloided with nitrobenzene. The colloid is rolled to the desired thickness and cut into strips which are hardened (or indurated) by the action of hot water or steam. Most of the nitrobenzene is distilled out by this treatment and the colloid is left as a very hard and tough mass which he calls **"indurite."** ▪ Although the powder was remarkably stable and effective when properly made, it was not adopted for military service {U.S. President HARRISON ⇨1892}. |
| 1891 | Agricultur-chemisches Laboratorium, Universität Göttingen, Germany | Bernhard C.G. TOLLENS[1112] and Peter WIEGAND, two research German chemists, **discover pentaerythritol** [$C_5H_{12}O_4$], an odorless, white, crystalline, solid compound.<br><br>In 1894, the new substance was first used in Germany by the Rheinisch-Westphälische Sprengstoff AG at Troisdorf to produce **pentaerythritol tetranitrate** [$C_5H_8N_4O_{12}$], better known as **"PETN"** or **"nitropenta."** It is one of the strongest known high explosives and has a detonation velocity of 8,300 m/s. Classified as a "secondary explosive," it is more sensitive to |

---

[1105] L. MACH: *Über ein Interferenzrefraktor*. Sitzungsber. Akad. Wiss. Wien **101** (Abth. IIa), 5-10 (1892).

[1106] L. ZEHNDER: *Ein neuer Interferenzrefraktor*. Z. Instrumentenkunde **11**, 275-285 (1891).

[1107] L. ZEHNDER: *Über den Einfluß des Druckes auf den Brechungsexponenten des Wassers*. Ann. Phys. **34** [III], 91-121 (1888).

[1108] G. HANSEN: *Über ein Interferometer nach Zehnder-Mach*. Z. Tech. Phys. **12**, 436-440 (1931).

[1109] R. LADENBURG, J. WINCKLER, and C.C. VAN VOORHIS: *Interferometric studies of faster than sound phenomena. Part I. The gas flow around various objects in a free, homogeneous, supersonic air stream*. Phys. Rev. **73** [II], 1359-1377 (1948); *Part II. Analysis of supersonic air jets*. Ibid. **76** [II], 662-677 (1949).

[1110] J. WINCKLER: *The Mach interferometer applied to studying an axially symmetric supersonic air jet*. Rev. Scient. Instrum. **19**, 307-322 (1948).

[1111] C.E. MUNROE: *Indurite*. U.S. Patent No. 489,684 (1891).

[1112] B.C.G. TOLLENS and P. WIEGAND: *Über den Penta-Erythrit, einen aus Formaldehyd und Acetaldehyd synthetisch hergestellten 4-wertigen Alkohol*. Justus Liebig's Ann. Chem. **265**, 316-340 (1891).

shock than TNT. For decades PETN was not used on a practical basis, but during the period between the two World Wars, the development of formaldehyde and acetaldehyde as cheap commercial raw materials changed PETN from being an expensive material produced in small quantities to one produced on a large scale at a reasonable cost.[1113] The composition known as **"pentolite"** (50/50 TNT/PETN) is used in ***Primacord*** ™**,** a detonation fuse manufactured by the Ensign-Bickford Company, Connecticut [now part of Dyno Nobel], which reliably detonates at nearly four miles per second (6.4 mm/μs).

| | | |
|---|---|---|
| 1891/ 1892 | *Chair of Natural Philosophy, University of Edinburgh, Scotland, U.K.* | **Peter G. TAIT,**[1114] a Scottish physics professor, **makes inelastic percussion experiments on materials with a coefficient of restitution between 0.3 and 0.8.** ▪ For a sample of vulcanized India rubber ($e \approx 0.78$), he found a compressive force proportional to $z^{3/2}$ (with $z$ as the compression at contact) which is in agreement with the Hertz theory of impact {HERTZ ⇨1882} |

In 1893, TAIT[1115] wrote in a letter to Heinrich R. HERTZ, a physics professor at the University of Bonn: "Some months ago, I was told by Lord KELVIN that you had brilliantly attacked the problem of the impact of elastic spheres. Being very busy at the time, I glanced over your paper in Crellos **92** [J. Reine & Angew. Math. **92**, 156-171 (1881)], but did not attempt to read it. I had been working for some years at direct experiments on impact, but I used a mass of 2; 4; and 8 kg falling through 1 m or so, and the elastic body on which it fell was a cylinder whose upper surface was very slightly convex. The amount of longitudinal distortion was, in some cases, as much as 30 mm. I found, by a graphical method, that the force called into play was at the power $^{3/2}$ of the distortion thus measured. On lately reading your paper with same care, I found to my great surprise that this is the same law which you have theoretically deduced for spherical bodies…" ▪ Unfortunately, we do not know today HERTZ 's answer. TAIT's records, now kept at the Archives of the University of Edinburgh, do not contain any letters from HERTZ. It is possible that HERTZ who prematurely died on January 1, 1894, was unable to answer TAIT's letter, because he was already ill with a series of infections.

| | | |
|---|---|---|
| 1892 | *Sorbonne & Ecole Supérieure de Pharmacie, Paris, France* | **Charles FRIEDEL,**[1116] a French chemistry professor and mineralogist at the Sorbonne, **reports that careful analysis has proved beyond doubt the existence of diamond in a portion of the Arizona meteorite** {Meteor Crater ⇨c.50,000 years ago; FOOTE ⇨1891}. ▪ In the following year, Charles FRIEDEL[1117] and F.F. Henry MOISSAN,[1118] a French professor of chemistry at the Ecole Supérieure de Pharmacie, found that the Arizona meteorite contained the three varieties of carbon: diamond (transparent and black), graphite, and amorphous carbon. |

The nature, origin and artificial production of diamonds stimulated the curiosity of numerous researchers:

- ▸ In 1796, the English chemist Smithson TENNANT proved that diamond consists solely of carbon by burning a diamond in an atmosphere of oxygen.
- ▸ In the early 1900s, the French chemist F.F. Henri MOISSAN[1119] **was the first to attempt to produce artificial diamonds.** He used a solution of carbon in a suitable molten metal at high temperature, which he quenched rapidly in water, and provided a demonstration of this to the French Academy of Sciences. However, **there was no clear evidence of any incipient transformation of carbon into diamond.**

---

[1113] R.E. KIRK and D.F. OTHMER (eds.): *Encyclopedia of chemical technology*. Interscience Encyclopedia, New York, (1951), vol. 6, pp. 32-34.
[1114] P.G. TAIT: *On impact* [1891/1892]. Trans. Roy. Soc. Edinb. **36**, 225-252 (1892); Ibid. **37**, 381-397 (1895).
[1115] This letter (dated Jan. 4, 1893) is now kept in the Archives of the Deutsches Museum, München.
[1116] C. FRIEDEL: *Sur l'existence du diamant dans le fer météorique de Cañon Diablo*. C. R. Acad. Sci. Paris **115**, 1037-1041 (1892).
[1117] C. FRIEDEL: *Sur la reproduction du diamant*. C. R. Acad. Sci. Paris **116**, 224-226 (1893).
[1118] F.F.H. MOISSAN: *Etude de la météorite de Cañon Diablo*. C. R. Acad. Sci. Paris **116**, 288-290 (1893); *Sur quelques propriétés nouvelles du diamant*. Ibid. **116**, 460-463 (1893).
[1119] F.F.H. MOISSAN: *Sur quelques expériences nouvelles relatives à la préparation du diamant*. C. R. Acad. Sci. Paris **140**, 277-283 (1905).

▸ In the late 1910s, Sir Charles A. PARSONS,[1120] a British engineer and inventor of the steam turbine, used a 0.303-in. (6.9-mm)-caliber rifle to fire steel bullets at 1,500 m/s into an armored pressed steel house filled with graphite powder. All of his attempts, however, also gave negative results. ▪ MOISSAN and his contemporaries believed that diamonds could be synthesized successfully by this method, but later investigations rejected this conclusion.[1121]

▸ In the 1940s, **Percy W. BRIDGMAN**,[1122] a U.S. physics professor, **showed that diamond rather than graphite is the stable form of carbon in the presence of high temperatures and high pressures.** ▪ The idea that this process could be reproduced in a laboratory captured the imagination of chemists and physicists. This remarkable feat was first achieved through the use of static high pressures {BUNDY ⇨ 1955}, and later of shock waves {DECARLI & JAMIESON ⇨ 1961}.

▸ In the 1960s, interest was revived in the origin of nanometer-sized diamonds in meteorites {CARTER & KENNEDY ⇨ 1964}. Nanometer sized diamond has been found in meteorites, proto-planetary nebulae and interstellar dusts, as well as in residues of detonation and in diamond films. Remarkably, the size distribution of diamond nanoparticles appears to be peaked around 2–5 nm.

| | | |
|---|---|---|
| 1892 | London, England | Oscar GUTTMANN,[1123] a German blasting engineer working as a consultant in London, **publishes his *Handbuch der Sprengarbeit* ("Handbook of Blasting Operation")**. ▪ Addressing the blasting community in the mining industry in particular, it discusses properties and applications of various high explosives, their safe handling and proper ignition, and the production of blastholes by pneumatic drilling machines. |
| 1892 | Washington, DC | U.S. President Benjamin HARRISON – referring to *indurite* {MUNROE ⇨ 1891}, the first smokeless powder used by the U.S. Navy for large guns – states in his farewell message to U.S. Congress,[1124] "I consider one of the great achievements of my administration the invention of smokeless powder [indurite] by Charles E. MUNROE." ▪ However, **the chief obstacle that prevented the widespread use of indurite by the U.S. military was a lack of consistency of composition** that resulted from the use of improperly nitrated guncotton and to difficulties with removing the residual solvent. |
| 1892 | Chair of Natural Philosophy, University of Edinburgh, Scotland, U.K. | Peter G. TAIT,[1125] a Scottish physics professor, **describes a simple but very effective apparatus to study the laws of impact of various materials** {⇨Fig. 4.3–J}.<br>(i) In order to measure the shock duration between an impinging block and the material to be studied he uses an apparatus that he humorously calls "my guillotine." The impactor – a block sliding freely between vertical guide rails precisely like the axe of a guillotine – is attached with a pointer to continuously record the block movement on a revolving plate-glass wheel that is coated with soot. For time measurement he uses a tuning fork that simultaneously produces a second trace on the revolving plate.<br>(ii) The curve traced out in this manner contains a complete record of the whole motion of the impinging block; and from this record all the relevant numerical data of the experiment can be obtained, such as the successive heights of rebound, the time of duration of impact, and the amount of compression of the substance on which the block fell. |

---

[1120] C.A. PARSONS: *Experiments on the artificial production of diamond* [Bakerian Lecture, 1918]. Phil. Trans. Roy. Soc. Lond. **A220**, 67-107 (1919).
[1121] M. SEAL: *Comments on some early attempts to synthesize diamond: the experiments of HANNAY and MOISSAN*. In: *Proc. First Int. Congress on Diamonds in Industry* [Paris, France, 1962]. Industrial Diamond Information Bureau, London (1963), pp. 313-327.
[1122] P.W. BRIDGMAN: *An experimental contribution to the problem of diamond synthesis*. J. Chem. Phys. **15**, 92-98 (1947).
[1123] O. GUTTMANN: *Handbuch der Sprengarbeit*. F. Vieweg & Sohn, Braunschweig (1892).
[1124] C.A. BROWNE: *Charles Edward MUNROE 1849–1938*. J. Am. Chem. Soc. **61**, 1301-1316 (1939), see p. 1306.
[1125] P.G. TAIT: *Sur la durée du choc*. Rev. Gén. Sci. Pures & Appl. **3**, 777-781 (1892).

(iii) In the first series of experiments he studied the impact properties of plane tree, cork, vulcanite, and vulcanized India rubber, and in the second series lead, steel, glass, new native India rubber, and various kinds of golf balls.

(iv) **Based upon his percussion studies he estimates the duration of impact or contact time between hammer and nail** (200 µs), **and the associated time-average force** (300 lb-wt).

(v) TAIT, a keen devotee of the game of golf, made a very interesting series of experiments on the velocities and spin of golf balls when driven from the tee {⇨Fig. 4.3–T} and the resistance they experience when moving through the air. Thereby he first recognizes that the rough surface, in combination with backspin (underspin), creates lift (Magnus force).[1126] He also estimates that the time-average of the force during the collision (which may have lasted < 1 ms) of the golf club and ball must be reckoned in tons' weight (> 5 tons). ▪ However, TAIT's results were not directly applicable to the golf impact problem, because his apparatus was restricted to much smaller speeds of approach. The problem was resumed more recently at the World Scientific Congresses of Golf.[1127]

| | | |
|---|---|---|
| 1893 | Linde's Eismaschinen AG, Technisches Büro Berlin, Germany | On February 23, **Rudolf DIESEL**,[1128] a German mechanical engineer and inventor, **obtains a patent on an ignition-compression engine – the "Diesel engine."** It relies on heat generated by adiabatically compressing the air in the cylinder to ignite the fuel, rather than on an electric spark. ▪ The first model was successfully operated on February 17, 1897 at the Maschinenfabrik Augsburg AG.[1129] |

When DIESEL was still a student at the Industrieschule (Polytechnic) Augsburg (1874–1876), he discovered there a pneumatic lighter {MOLLET ⇨1804} in a collection of physical instruments, which stimulated his lifelong interest in adiabatic compression of gases, eventually resulting in his invention of the "Diesel engine." DIESEL later demonstrated the pneumatic lighter to his children. His younger son Eugen DIESEL[1130] remembered, "Eventually, after trying the third time, we observed that the tinder exposed to the compressed hot air began to glow. Seeing this tinder turning red and smouldering through the glass walls touched me in a magic way. 'Now you imagine', said my father, 'if there had been some gasoline or petroleum or coal dust in there: such a fuel would ignite, and the increased gas pressure due to the combustion – heat expands objects and, of course, also the air – would push up the piston. The *Diesel engine* is nothing more than just such a pneumatic lighter, with the difference that the fuel, which is finely dispersed, is injected into the compressed glowing air. There it ignites by itself and performs work since the hot and high-tension gas pushes the piston, which, by means of the crank, turns the flywheel.'"

| | | |
|---|---|---|
| 1893 | London, England | At the meeting of the Royal Physical Society, **Charles V. BURTON**,[1131] a British researcher, **reports on the results from his treatment of the difficult problem of whether the motion of spherical waves of finite amplitude can become discontinuous in the absence of viscosity, as in the case of planar waves of finite amplitude:** |

(i) Resuming Lord RAYLEIGH's critiques on RIEMANN's theory of planar waves of finite amplitude (*see also* RAYLEIGH's *Theory of Sound*, vol. II, p. 41; Lord RAYLEIGH ⇨1878) and addressing a previously proposed theory of the propagation of waves of finite amplitude {TUMLIRZ ⇨1887}, he writes, "The assumed motion here criticized is one in which density and velocity are constant for all points on the same side of the surface of discontinuity, while this surface itself is

---

[1126] P.G. TAIT: *Some points in the physics of golf.* Nature **42**, 420-423 (1890); Ibid. **44**, 487-498 (1891); Ibid. **48**, 202-205 (1893).
[1127] W. GOBUSH: *Impact force measurement on golf balls.* In: (A.J. COCHRAN, ed.) *Proc. 1st World Scientific Congress of Golf* (St. Andrews, Scotland, July 1990). Spon, Cambridge (1990), pp. 219-224; C.E. SCHEIE: *The golf club - ball collision 5,000 g?*, Ibid., pp. 237-240.
[1128] R. DIESEL: *Arbeitsverfahren und Ausführungsart für Verbrennungskraftmaschinen.* Germ. Patent No. 67,207 (eingereicht am 28. Feb. 1892).
[1129] E. RÖDL: *Ein Selbstzünder erobert die Welt.* Deutsches Museum, München; http://www.deutsches-museum.de/ausstell/meister/diesel.htm.
[1130] Eugen DIESEL: *DIESEL: Der Mensch, das Werk, das Schicksal.* Hanseatische Verlagsanstalt, Hamburg (1937), pp. 86-88.
[1131] C.V. BURTON: *On plane and spherical waves of finite amplitude* [read Feb. 24, 1893]. Phil. Mag. **35** [V], 317-333 (1893).

propagated through the fluid with constant velocity. He [RIEMANN ⇨ 1859] shows that the same objection applies when, on either side of the surface, velocity and density vary continuously in the direction of propagation, while the velocity of propagation of the surface is also allowed to vary … The real source of error lies in RIEMANN's fundamental hypothesis. At the outset he supposes the expansion and contraction of the air to be either purely isothermal or purely adiabatic, and thenceforward he treats the air as a frictionless and mathematically continuous fluid, in which pressure and density are connected by an invariable law. But in general the existence of such a fluid is contrary to the conservation of energy; for as soon as discontinuity arises, energy will be destroyed."

(ii) Referring to (divergent) spherical waves of finite amplitude, he states that "… if viscosity be neglected, we must conclude that under any practically possible law of pressure the motion in spherical sound-waves always becomes discontinuous; and *a fortiori* the same will be true of cylindrical waves."

| 1893 | *Chair of Chemistry, Owens College, University of Manchester, England* | **Harold B. DIXON**,[1132] a British professor of chemistry, resuming his previous idea of using the rate of explosion as a means for determining the course of a chemical reaction {DIXON ⇨ 1877}, **reports on his observations of the high velocities of explosions in gases and puts forth the view that the detonation wave travels at the velocity of sound through the burning gases,** thus essentially supporting Prof. Arthur SCHUSTER's view of the steady motion of the detonation front. • DIXON's concept of detonation was indeed correct and stimulated his student David L. CHAPMAN {⇨1899} to construct the first theory of detonation.

(i) Using a 55-m-long, coiled-up lead pipe with an inner dia. of 8 mm {⇨Fig. 4.17−J} and an Elliot chronograph (a heavy pendulum carrying a smoked glass plate), he measures in oxyhydrogen a detonation velocity of 2,821 m/s oxyhydrogen, thus essentially confirming previous measurements made by P.E. Marcellin BERTHELOT {⇨1881}. However, he finds that the "rate of explosion" (meaning the detonation velocity) increases slightly with pressure in the case of hydrogen and oxygen, at least up to two atmospheres.

(ii) His experiments on the propagation of explosion showed that the low velocity of a few m/s previously measured by Robert BUNSEN {⇨1867} refers to only the initial period of combination before the explosion wave attains its maximum velocity.

(iii) In an appendix to DIXON's memoir, **Arthur SCHUSTER**,[1133] a British professor of applied mathematics and director of the Physics Laboratory at Owens College, **proposes to apply previous theories of shock waves to the detonation process,** such as those worked out by RIEMANN {⇨1859} and HUGONIOT {⇨1887}.

▸ SCHUSTER derives a simple formula to calculate the velocity $V$ of the detonation front as given by
$$V = [(\rho/\rho_0) \times (p - p_0)/(\rho - \rho_0)]^{1/2},$$
where $p$, $p_0$ and $\rho$, $\rho_0$ are pressures and densities, respectively, and $p_0$, $\rho_0$ denote the initial state of the fluid.

▸ SCHUSTER reports that he observed good agreement with experimentally determined rates of explosion in various explosive gaseous mixtures, and writes, "Lord RAYLEIGH [see his *Theory of Sound*, vol. II, p. 4] criticizing his [RIEMANN's] investigation, draws attention to the fact that a steady wave is only possible for a particular relation between the pressure and density of the gas, which is different from the one that actually holds. In the case of the explosion waves, it seems possible, however, that the temperature, pressure, and density of the gas should adjust themselves so as to make RIEMANN's equations applicable. In fact, they must do so if the front of the wave keeps its type, which it |

---

[1132] H. DIXON: *The rate of explosion in gases* [Bakerian Lecture]. Phil. Trans. Roy. Soc. Lond. **A184**, 97-152, 154-188 (1893).
[1133] A. SCHUSTER: *Note by professor SCHUSTER, F.R.S.* Phil. Trans. Roy. Soc. Lond. **A184**, 152-154 (1893).

probably does when the velocity has become constant ... In the strict sense of the word I do not think the explosion wave can be steady, because if the motion is, as assumed, linear, compression must precede the explosion, and Lord RAYLEIGH's objection would hold for the front part of the wave in which no combination takes place. But it seems possible to me that the motion may not be strictly linear, and that if the average velocities over a cross-section of the tube were taken the ordinary equations would apply. **It seems probable that jets of hot gases are projected bodily forward from that part of the wave in which the combination takes place, and that these jets, which correspond to the spray of a breaking wave, really fire the mixture.**"

A similar idea had already been developed in France {BERTHELOT ⇨1870; BERTHELOT & VIEILLE ⇨1882}. SCHUSTER's correct supposition of a steadily moving detonation wave eventually led to the first theory of detonation {CHAPMAN ⇨1899}. The nonuniformity of the detonation front can result in a curious cell structure {DENISOV & TROSHIN ⇨1959; ⇨Fig. 4.16–U}.

---

1893 — *U.S. Geological Survey (USGS), Washington, DC*

During his retirement address, the President of the Philosophical Society of Washington, **Grove Karl GILBERT,**[1134] chief geologist of the USGS, **presents the first comprehensive hypothesis of meteorite impact:**

(i) He compares lunar craters with those made by dropping pebbles into mud paste, or by firing projectiles into plastic materials. Investigating the origin of the lunar surface and applying geological insight and knowledge, he refutes the common assertion that the lunar craters are volcanoes {⇨Fig. 4.1–A}. Rather, he concludes that they were formed by giant impacts.

(ii) **He is the first to recognize the radiating structure around *Mare Imbrium* ("The Sea of Rains"), and he attributes this "sculpture" to a giant impact** which resulted "in the violent dispersion in all directions of a deluge of material, solid, pasty, liquid." ▪ Recently, a team of scientists have precisely dated Mare Imbrium, the youngest of the large meteorite craters on the Moon, to 3.909 ± 0.013 billion years, when an asteroid colliding with the Moon formed the 1,160-km-dia. Imbrium impact basin.[1135]

(iii) He estimates that the impact of a body striking the Moon's surface with a parabolic velocity would raise the temperature to 3,500 °F (1,927 °C) or higher. This would melt much of the rock and account for the level floors of some of the craters.

GILBERT's lunar scientific contributions rank with those of Galileo GALILEI, Johann H. MADLER and Ralph B. BALDWIN. His work was little known nor appreciated until it was rediscovered in the early 1960s. **His impact hypothesis became the basis for all subsequent thought along these lines.** In the period 1920–1933, Alfred WEGENER {⇨1920} and Leonard J. SPENCER {⇨1932} published more detailed comparative studies that lent increased support to the impact thesis. On the other hand, however, GILBERT ironically failed to explain the origin of Meteor Crater {⇨c.50,000 years ago}, which he attributed to volcanic origin.

Indeed, lunar impacts continue to occur. For example, on November 7, 2005, researchers from NASA's Marshall Space Flight Center (Huntsville, AL) were testing a new 10-in. telescope and video camera assembled to monitor the Moon for meteor strikes. They actually managed to record a meteoroid impact near the edge of Mare Imbrium in real time. The 12-cm-wide meteoroid, which was traveling at 27 km/s upon impact, produced a crater approximately 3 m in dia. and 0.4 m deep. The blast was equal in energy terms to about 70 kg of TNT.[1136]

---

1893 — *World Colombian Exposition, Chicago, Illinois*

**Gustav DE LAVAL**, a Swedish engineer and inventor, **displays his single-stage impulse steam turbine.** The engine (15 hp at 16,000 rpm), which is reversible and designed for marine use,

---

[1134] G.K. GILBERT: *The Moon's face*. Bull. Phil. Soc. Washington **12**, 241-292 (1893).
[1135] E. GNOS ET AL.: *Pinpointing the source of lunar meteorite: implications for the evolution of the Moon*. Science **305**, 657-659 (July 30, 2004).
[1136] *Explosion on the Moon*. NASA press release (Dec. 23, 2005); http://science.nasa.gov/headlines/y2005/22dec_lunartaurid.htm.

has been tested in the vicinity of Stockholm on Lake Mölaren, where it was used to drive a launch. Its novelty is that
- the turbine blades are driven by a stream of hot, high-pressure steam emerging from **a series of unique convergent-divergent nozzles** {DE LAVAL ⇨1888}; and
- the turbine has a flexible shaft; and double helical gears.

Initially, it could be produced in different-sized units ranging from 5 to 200 horsepower. In 1896, in what was possibly the first commercial application of a steam turbine for electric power generation in the United States, the New York Edison Company imported and installed two 300-hp De Laval turbines, establishing the first electric generating stations in New York City. Today specimens of his first turbine are included in the collections of the Smithsonian Institution (History of Technology Building, Washington, DC) and the Deutsches Museum (Munich, Germany) {⇨Fig. 4.7−F}, where they are on permanent display.

| 1893 | German Charles University, Prague, Bohemia, Austro-Hungarian Empire & Polytechnikum Riga, Latvia | Ernst MACH,[1137] the distinguished Austrian philosopher and physicist, **and C. Bruno DOSS**, a German mineralogist and geologist, **assume that the sharp bang of a meteorite approaching the Earth is a supersonic phenomenon, and thus is due to the creation of a ballistic head wave:**

(i) MACH's motivation for treaing the phenomenon of meteoric showers was a letter by Cleveland ABBE, a senior professor of meteorology at the U.S. Weather Bureau, and a member committee of the Philosophical Society of Washington, which analyzed the fall of the Washington Meteorite {⇨1877}. ABBE, who in 1877 claimed to have already provided a "true theory of thunder and meteorite explosions," states: "We are disposed to consider the so-called 'explosion', and subsequent 'rumbling' not as due to a definite explosion of the meteor, but as a result of the concentration at the observer's ear of the vast volume of sound emanating, almost simultaneously, from a large part of the meteor's path, being, in that respect, not dissimilar to ordinary thunder." ABBE then tries to explain the violent sound by the Doppler effect {DOPPLER ⇨1842}, and concludes, "We may remark that it requires only comparatively feeble noises distributed along the entire path of the meteor to produce, by their concentration at the observer's station, a sound equal to that of loud thunder."

(ii) MACH, rejecting ABBE's theory and his prior claim, replies that **only the head wave phenomenon is the true cause of the explosion-like sound effects.** ▪ MACH's interpretation was indeed correct; however, although he pioneered supersonics and was ahead of his time, he did not realize that the head waves of meteorites, which can enter the Earth's atmosphere at ultra-high speeds of up to 70 km/s, are closely wrapped around the meteorites, forming so-called **"hypersonic boundary layers"** {⇨Fig. 4.6−O}, and that air passing through such a head wave (a shock wave) is heated to very high temperatures − thus creating previously unknown surface heating and erosion effects. |

| 1893 | Moscow Agricultural Institute [now K.A. Timiryazev Agricultural Academy], Soviet Union | Vladimir A. MIKHEL'SON,[1138] a Russian physicist, **develops the first theory of detonation based on the theory of shock waves:**

(i) **He assumes that the detonation wave consists of a shock wave** which compresses the combustible mixture {⇨Fig. 4.16−S}.

(ii) **In his theory of detonation, he first proposes a linear law and then assumes steady propagation of the reaction products;** *i.e.*, the intermediate states propagate at similar velocities. He writes, "As regards detonation, here we encounter an extraordinarily interesting case in which, owing to the specific chemical and thermal processes involved, the conditions for con- |

---

[1137] E. MACH and B. DOSS: *Bemerkungen zu den Theorien der Schallphänomene bei Meteoritenfällen*. Sitzungsber. Akad. Wiss. Wien **102** (Abt. IIa), 248-252 (1893).

[1138] V.A. MIKHEL'SON: *On the normal ignition velocity of explosive gaseous mixtures* [in Russ.]. Ph.D. thesis. Moscow University Printing Service, Moscow (1890); republished in: Scientific Papers of the Moscow Imperial University on Mathematics & Physics, vol. **10**, pp. 1-93 (1893).

stant velocity propagation are indeed satisfied." Starting from the equations of mass and momentum, he derives the elementary relation

$$p = p_0 + (U/v_0)^2 \times (v_0 - v),$$

which is also derived in the same year in England {SCHUSTER ⇨1893} and represents a straight line in the $(p, v)$-plane. Here $U$ denotes the shock front velocity, and $v$ and $v_0$ are the specific volumes at pressures $p$ and $p_0$, respectively. ▪ Yakov B. ZEL'DOVICH[1139] termed this line the *Mikhel'son line* in honor of MIKHEL'SON's early contribution to the theory of detonation, which was unknown for a long time to contemporary scientists outside Russia. In the Western World, this line is called the **"Rayleigh line,"** referring to the work of Lord RAYLEIGH {⇨1878}. Horace LAMB,[1140] England's famous applied mathematician, however, attributed this result to William J.M. RANKINE {⇨1869}. Referring to his work and the "mass-velocity" $m$ of the wave, LAMB showed in a different manner that

$$p_1 + m^2 v_1 = p_0 + m^2 v_0,$$

and concluded, "Hence a wave of finite amplitude could not be propagated unchanged except in a medium such that $p + m^2 v = $ const … It may be noticed that the relation is represented on WATT's diagram [*i.e.*, a plot $p = p(v)$] by a *straight line*."

1893 — Imperial College of Engineering, Tokyo, Japan

John MILNE,[1141] an English seismologist and geologist, **adapts his seismograph so that it records photographically on light-illuminating film,** thus eliminating the problem of friction between pen and (smoked) paper. ▪ The so-called **"Milne seismograph"** became the standard instrument for seismologists around the world, since it clearly distinguished between *longitudinal waves* (back-and-forth waves which he called "condensational waves") and *transversal waves* (up-and-down waves or side-to-side waves, which he called "distortional waves"). The different types of wave travel at different velocities, as first predicted analytically by Siméon D. POISSON {⇨1831}.

1893 — Physical Institute, University of Göttingen, Germany

Friedrich C.A. POCKELS,[1142] an Italian-born German physicist, **publishes a long memoir about the influence of electrostatic fields on the optical behavior of piezoelectric crystals,** which wins him a prize set by the Göttingen Royal Academy.

(i) He defines the goal of his studies: "In order to decide the important question whether electro-optic phenomena in piezoelectric crystals are only the result of deformations caused by the electric field or whether also a direct effect of electrostatic forces on the motion of light occurs, quantitative investigations are necessary."

(ii) He reports on his discovery that the application of a steady electric field to certain birefringent materials causes the refractive index to vary, approximately in proportion to the strength of the field. ▪ This electro-optical phenomenon is now called the **"Pockels effect."** The "Pockels coefficient" is on the order of $10^{-10}$ to $10^{-12}$ m/V.

Advantageously, a Pockels cell typically uses voltages 5–10 times lower than those used in an equivalent Kerr cell {KERR ⇨1875}. Both Pockels and Kerr effects can be used to construct ultrafast (nanosecond) optical shutters by placing them between crossed polarizers. Prior to the advent of microchannel plates {FARNSWORTH ⇨1930} controllable optical shutters like these were extremely useful devices for photographing strongly self-luminous phenomena such as those related to detonations. Pockels cells also proved to be very useful for Q-switching solid-state lasers, which are widely applied in high-speed diagnostics for both shock and detonation waves, and to generate shock waves in small areas.

---

[1139] Y.B. ZEL'DOVICH and A.S. KOMPANEETS: *Theory of detonation*. Academic Press, New York (1960).
[1140] H. LAMB: *Hydrodynamics*. Cambridge University Press (1932), p. 485.
[1141] J. MILNE: *A note on horizontal pendulums*. Seismol. J. Japan **3**, 55-60 (1894).
[1142] F. POCKELS: *Über den Einfluß des elektrostatischen Feldes auf das optische Verhalten piezoelektrischer Krystalle*. Abhandl. Königl. Gesell. Wiss. Gött. **39** [Math.-Physik. Cl.], 1-204 (1893).

| 1894 | *Naval Royal Engineering College, Keyham, Devonport, England* | On May 18, in a discourse delivered at the Royal Institution of Great Britain, **Arthur Mason Worthington**,[1143] a British professor of practical physics, **reports on the physical behavior of falling drops** – a subject which attracted his attention since the mid-1870s.[1144] He observed that when a drop hits a liquid surface, a crown-like splash ("Worthington crown") is produced, followed by the formation of a central jet ("Worthington jet") which shoots upwards from the impact site. ▪ The splashing phenomena take place over a very short time-scale, and in his early experiments Worthington had to rely on eyesight only. However, in the following years he refined his drop impact studies[1145] using a still camera and a flash light source (an electric spark "in absolute darkness") – a method called today "single-shot photography." |
| | | In the early 1950s, the matter was resumed by the MIT professor Harold E. Edgerton[1146] using flash stroboscopy. He produced a fine series of a drop of milk falling into water which showed that the impact creates a crown-like splash of thorns. His ***Milk-Drop Coronet*** became the logo of the Congresses on High-Speed Photography {1st Int. Symposium on High-Speed Photography ⇨1952}. More recently, theoretical fluid dynamicists have simulated numerically Worthington's curious crown-and-jet phenomenon. |
| 1894 | *Elswick Ordnance Co., Elswick, Newcastle-upon-Tyne, England* | **Andrew Noble**,[1147] a British physicist and ballistician, **critically compares previous pressure data obtained by the indentation gauge** {Rodman ⇨1857} **with that obtained with his own crusher gauge** {Noble ⇨ 1872}. He first correlates measured pressure data in the bore with theory using measured kinematic data for the projectile {⇨Fig. 4.17–L}. He states, "It is curious that so distinguished an artillerist as Major Rodman should never have taken the trouble to calculate what energies the pressure which his instrument gave would have generated in a projectile; had he done so he would have found that many of the results indicated by his instrument were not only improbable but were absolutely impossible." |
| 1895 | *University of Cambridge, England* | **Sir George Howard Darwin**,[1148] a British astronomer and second son of the famous naturalist Charles R. Darwin, **again draws the attention of the readers of the British journal *Nature* to "Barisal Guns"** {Barisal ⇨1870}. Addressing this unusual sound phenomenon, he writes, "The detonations are dull, more or less resembling distant artillery and are repeated a dozen times in the day-time when the sky is clear, and especially towards evening after a very hot day." |
| | | One year later, referring to Prof. Darwin's letter to *Nature*, George B. Scott[1149] gave an account of the Barisal gun phenomena, "At Chilmari [the landing place for Tura and distant about 300 miles from the mouths of the Brahmaputra and Ganges], near the river-bank, about 10 A.M. in the day, weather clear and calm, we heard the booming distinctly, about as loud as heavy cannon would sound on a quiet day about ten miles off, down the river. Shortly after, we heard a heavy boom very much nearer, still south. Suddenly we heard two quick successive reports, more like horse-pistol or musket (not rifle) shots close by…" |
| | | Nine years after Darwin's note, the German Friedrich Krauss[1150] published a paper on observations of the same phenomenon made at Lake Constance. ▪ No satisfactory explanation has ever been given for this strange sound phenomenon, which resembles distant thunder or |

---

[1143] A.M. Worthington: *The splash of a drop and allied phenomena* [read 1894]. Proc. Roy. Inst. **14**, 289-303 (1896); *A study of splashes.* Society for Promoting Christian Knowledge, London (1895); Longmans, Green & Co, London (1908); Macmillan, New York (1963).
[1144] A.M. Worthington: *On the form assumed by drops of liquids falling vertically on a horizontal plate.* Proc. Roy. Soc. Lond. **25**, 261-271 (1876).
[1145] A.M. Worthington: *Impact with a liquid surface, studied by the aid of instantaneous photography* [1896–1899]. Phil. Trans. Roy. Soc. **A189**, 137-148 (1897); Ibid. **A194**, 175-199 (1900).
[1146] H.E. Edgerton and J.R. Killian Jr.: *Flash, seeing the unseen.* C.T. Branford Co., Newton Centre, MA (1954).
[1147] A. Noble: *On methods that have been adopted for measuring pressures in the bores of guns.* Rept. Meet. Brit. Assoc. **64**, 523-540 (1894).
[1148] G.H. Darwin: *"Barisàl guns" and "mist pouffers."* Nature **52**, 650 (1895).
[1149] G.B. Scott: *Barisal guns.* Nature **53**, 197 (1896).
[1150] F. Krauss: *Über das Seeschießen.* Blätter des Schwäbischen Albvereins **16**, 75-78, 144-146, 170-174 (1904).

cannot heard from the direction of the sea. **In KRAUSS' era, some natural philosophers believed the Barisal gun phenomenon to have a seismic origin.**

| | | |
|---|---|---|
| 1895 | *University of Würzburg, Bavaria, Germany* | **Wilhelm Conrad RÖNTGEN,**[1151] a German physics professor, **discovers a strange invisible radiation which he calls "X-rays" to indicate their unknown nature** while investigating the effects of cathode rays. ▪ In the laboratory, X-rays can be produced by accelerating electrons through a high voltage in a vacuum and stopping them abruptly in a dense target material. According to MAXWELL's classical electromagnetic theory, the deceleration of electrons will produce electromagnetic radiation with wavelengths ranging from about 5 nm to some 10,000 nm. The conversion efficiency of the electron energy into radiation, however, is very small (*e.g.*, 0.74% for a tungsten anode at 100 kV).[1152] In pulsed high-current X-ray tubes, the high-velocity impact of electrons on the target surface and their partial penetration into the target can produce unusual subsurface explosion and melting phenomena {⇨Fig. 4.16−Y}.<br><br>In nature, X-rays can also be generated when particles moving at relativistic velocities impact matter, for example by a fractured comet impacting the gaseous atmosphere of a massive planet {Comet SHOEMAKER-LEVY 9 ⇨ 1994}. |
| 1895 | *John Thornycroft & Co., Ltd., Southampton, England* | **John Isaac THORNYCROFT,**[1153] a British naval architect of high-speed torpedo boats, and inventor of the air-cushion vehicle, **and Sidney Walter BARNABY,**[1154] a lecturer at the Royal Naval College and member of the council of the Institution of Naval Architects, **investigate why a British destroyer has failed to meet its design speed. They observe that a marine screw propeller – if turned too fast – could waste its effort by creating voids in the water which then suddenly collapse. They coin this phenomenon** *cavitation* {Brit. Admiralty ⇨ 1915}. ▪ In several cases it was observed that only a few hours of running were sufficient to make the propeller completely unfit for work. The remedy was to operate at lower rpm with more turbines and to use propellers with very wide blades in order to enlarge the blade surface in contact with the water. An analogous phenomenon for water and steam turbines was discovered at almost the same time as the discovery of cavitation for propeller screws.<br><br>In 1973, at the Symposium on Finite-Amplitude Wave Effects in Fluids, held at Copenhagen, **Werner LAUTERBORN,**[1155] an experimental physics professor at Göttingen University, **gave the following general definition of the term** *cavitation*: "*Cavitation* **is a group of phenomena** which are associated with the occurrence of *cavities in liquids*, especially their formation, motion, and the physical, chemical, and biological effects thereby produced. a *cavity in a liquid* (often called a 'cavitation bubble') is any bounded volume in space being empty or containing vapor and/or gas with a defined physical boundary at least part of which must be made up of liquid … Cavitation is introduced in this definition as a group of phenomena. The name is also used for the area of research investigating these phenomena … The definition also contains implicitly that cavitation is not simply a nonlinear phenomenon of some kind but a phenomenon of its own. The notice of this may greatly influence the way of attacking problems and may be used as a guideline where the usual nonlinear approaches may fail in describing the phenomena observed." |
| 1896 | *Kamaishi, Honshu, Japan* | During the evening of June 15, **a huge tsunami resulting from a seaquake in the Pacific destroys the coastal town of Kamaishi,** drowning about 27,000 people and wounding about |

---

[1151] W.C. RÖNTGEN: *Über eine neue Art von Strahlen*. Sitzungsber. Phys.-Med. Gesell. Würzburg (1895), pp. 132-141; Ibid. (1896), pp. 11-19.
[1152] F. JAMET and G. THOMER: *Flash radiography*. Elsevier, Amsterdam *etc.* (1976), p. 4.
[1153] J.I. THORNYCROFT and S.W. BARNABY: *Torpedo boat destroyers*. Minutes Proc. Inst. Civil Eng. **122**, 51-103 (1895).
[1154] S.W. BARNABY: *On the formation of cavities in water by screw propellers at high speeds*. Trans. Inst. Nav. Archit. **39**, 139-143 (1898).
[1155] W. LAUTERBORN: *General and basic aspects of cavitation*. In: (L. BJOERNOE, ed.) *Proc. 1973 Symposium on finite-amplitude wave effects in fluids* [Copenhagen, Denmark, Aug. 1973]. IPC Science and Technology Press, Ltd., Guildford, Surrey, U.K. (1974), pp. 195-202.

5,000 people.[1156] The approach of the tsunami is described by the few survivors as "a dull, increasing sound, like from a distant storm" or "a roaring surf, eventually sounding like heavy artillery fire." The origin of this seismic disturbance will later be pinpointed to the bottom of the Tuscarora Deep (max. depth 8,513 m), which runs from the Bering Islands to the Isle of Hokkaido. ▪ In 1928, a large tsunami again devastated the coastal region of Kamaishi. Therefore, the deepest breakwaters (63 m) in the world were then constructed in Kamaishi Bay in order to reduce the opening to the bay in order to lessen tsunami run-up height and to protect the port area from tsunamis as well as storm waves.

| 1896 | German Charles University, Prague, Bohemia, Austro-Hungarian Empire | **Ludwig MACH**[1157] first uses the Mach-Zehnder interferometer {L. MACH & ZEHNDER ⇨1891} **to visualize the flow field around a supersonic bullet** ($v$ = 620 m/s).<br>(i) In order to trigger his point light source (an electric spark) at the precise moment when the projectile is passing the entrance of his interferometer, he uses an ingenious fluid-mechanical method of producing an electric delay: prior to entering the interferometer, the head wave partly enters into a pipe, and triggers the electric spark upon its arrival at the end of this pipe {⇨Fig. 4.19–G}.<br>(ii) Surprisingly, his interferometric results show that the bell-shaped head wave is only a thin shell, and that the compressions are quite modest, barely more than a fifth of an atmosphere {⇨Fig. 4.6–L}. |
|---|---|---|
| 1896 | England | **Percy Gerald SANFORD,**[1158] a British chemist, public analyst and consultant, publishes his book *Nitro-Explosives*. His work which begins with a historical review describes a large number of explosives based upon the basic types of nitro explosives, such as nitroglycerin, nitrocelluloses, dynamite, nitro-benzol, the fulminates, and smokeless powders. ▪ For more than 100 years, his classic volume has served as the most important single textbook, operating handbook, laboratory manual, and technical reference on the subject of explosives chemistry, engineering, manufacture, testing, handling, and storage. |
| 1896 | Pressburg [now Bratislava], Austro-Hungarian Empire | **Alfred SIERSCH,**[1159] director of Dynamit AG in Vienna, is concerned about the safe use of explosives in coal mines, and so he **uses photography to classify the nature and intensity of the flash emitted by an exploding charge** {⇨Fig. 4.17–G}. Using a still camera and photographing the flash with an open shutter at night, he observes that the flash intensity from an explosive depends on the geometry, density, mode of stemming, and admixtures. **He concludes that the shape and dimensions of the flash afford a clue to the eventual security of the explosive, since the smaller the flash, the greater its relative security for use in the coal mining industry.** ▪ On the whole, this straightforward method proved to be useful. However, he was not yet aware that shock wave reflection and interaction phenomena can also contribute considerably to the geometry and intensity of the flash {MICHEL-LÉVY & MURAOUR ⇨1934}.<br>SIERSCH's experiments are only described here as an example; around 1880, several European governments that were seeking to develop safer substitutes for black powder set up testing stations {Gelsenkirchen ⇨1880}. Black powder was formerly used in large quantities in the underground coal mining industry, and from a performance standpoint it is probably the best explosive for that purpose. However, it has a dangerous tendency to ignite firedamp (mostly methane). Today, the use of black powder in underground coal mines is no longer allowed in |

---

[1156] J. REIN: *Das Seebeben von Kamaishi am 15. Juni 1896*. Dr. A. Petermann's Mitteilungen aus Justus Perthes' Geographischer Anstalt (Gotha) **43**, 34-37 (1897).

[1157] L. MACH: *Weitere Versuche über Projektile*. Sitzungsber. Akad. Wiss. Wien **105** (Abth. IIa), 605-633 (1896).

[1158] P.G. SANFORD: *Nitro-explosives: a practical treatise concerning the properties, manufacture, and analysis of nitrated substances, including the fulminates, smokeless powders, and celluloid*. Crosby Lockwood & Son: London (1896); http://library.case.edu/ksl/ecoll/books/sannit00/sannit00.pdf.

[1159] A. SIERSCH: *Photography in the technology of explosives*. Trans. Am. Inst. Min. Eng. **11**, 2-8 (1896).

1897    Ordnance Dept. of the U.S. Army & Pennsylvania Military Academy, Chester, Pennsylvania

most countries, and it has been replaced by special explosives approved for use in gassy and dusty coal mines {PENNIMAN ⇨ 1885}.

**Lieut. Beverly W. DUNN,**[1160] a professor of military science and tactics, **is the first to use a photographic method to measure the *impulsive force* during the planar impact of two elastic bodies.**

(i) He states, "The usual classification of variable forces as impulsive and nonimpulsive depends upon their rates of change of intensity. As ordinarily understood, an impulsive force is one whose intensity changes too rapidly to permit successive measures of it. The main object of all engineering and structural work may be said to be the supplying of adequate resistances to neutralize dangerous forces … There are two methods for such measurement, the static and the dynamic methods. The former, which is quite limited in its application, consists essentially in balancing the unknown force by a known one [GALILEI ⇨ 1638; ⇨ Fig. 4.3–C]. The latter, which is the true scientific method, requires the accurate measurement of extremely minute intervals of time and space."

(ii) In order to measure dynamic forces during impact, he records the temporal displacement $s(t)$ of a falling anvil of mass $m$ using a lever connected to a mirror to magnify the tiny displacement. A beam of light reflected from this mirror falls on a ***photo-chronograph***, a rotating drum covered with photographic paper. For time calibration he simultaneously records the vibrations of a 500-Hz tuning fork. By twice differentiating the measured displacement-time curve $s(t)$ and using the Second Law of Motion {Sir NEWTON ⇨ 1687}, he finds the dynamic force given by $F = m d^2 s/dt^2$.

**DUNN's work – apparently the first significant American contribution to impact diagnostics – stimulated subsequent dynamic materials research:**

▸ Beginning in 1894, George Owen SQUIER, a general in the U.S. Army, and Albert Cushing CREHORE,[1161] his civilian chief assistant, invented a ***photo-chronograph*** that used a new method of securing light signals in order to plot the movements of projectiles in gun bores.

▸ 1904, the U.S. engineer W. Kendrich HATT[1162] investigated samples under impulsive tension using a stylus directly connected to the impacting anvil, thus avoiding any pivot looseness incurred in DUNN's lever method.

▸ In 1909, the German physicist Carl RAMSAUER {⇨ 1909}, studying the head-on impact of cylindrical rods, used a photographic method which incorporated a lever-mirror arrangement very similar to that used in DUNN's optical method to magnify the small deformations during impact.

▸ In 1912, the German engineer Walter HÖNIGER[1163] also adapted DUNN's method to a Martens-type pendulum machine in order to measure the impulsive force and contact time during impact.

▸ In 1924, the Austrian engineer Franz BERGER {⇨ 1924} used single-shot reflected-light photography of the impacting bodies to record the temporal displacement during impact.

---

[1160] B.W. DUNN: *A photographic impact testing machine for measuring the varying intensity of an impulsive force.* J. Franklin Institute **144**, No. 5, 321-348 (Nov. 1897).
[1161] A.C. CREHORE and G.O. SQUIER: *Experiments with a new polarizing photo-chronograph as applied to the measurement of the velocity of projectiles.* Phys. Rev. **2** [I], 122-137 (1894); Ibid. **3** [I], 63-70 (1895).
[1162] W.K. HATT: *Tensile impact tests of metals.* Proc. Ann. Meet. Am. Soc. Testing Materials (ASTM) **IV**, 282-315 (1904).
[1163] W. HÖNIGER: *Anwendung der Kinematographie zur Ermittlung der Stoßkraft bei Schlagversuchen.* VDI-Z. **56**, 1501-1505 (1912).

| 1897 | German Charles University, Prague, Bohemia, Austro-Hungarian Empire | Ludwig MACH,[1164] resuming previous outflow experiments {SALCHER & WHITEHEAD ⇨1889}, **visualizes free air jets emerging from nozzles with various exit geometries.** Applying not only the schlieren technique but also the interferometer technique {⇨Figs. 4.8–A, B}, he makes the important observation that as the driving pressure increases
▸ the jet diameter surmounts the nozzle diameter; and
▸ the reflected wavefronts no longer intersect in a point (as in the case of regular reflection), but rather form a sequence of new waves which result from the Mach effect {E. MACH & WOSYKA ⇨1875}. ▪ This structure was later termed ***shock diamonds*** – a sequence of pairs of oblique shock fronts caused by repeated reflections and re-reflections, each interacting irregularly and creating a sequence of Mach disks until the disturbances are damped out by viscous effects.

Since shock diamonds are a stationary (*i.e.*, non-propagating) wave phenomenon, they can be observed with the naked eye and are seen for example in the exhausts from rockets and jet aircraft {⇨Fig. 4.20–F}. Because the shock disks raise the temperature of the gas moving through them, excess fuel is ignited, causing the glow recognized as the shock diamond. |
|---|---|---|
| 1897 | C.A. Parsons & Co., Newcastle-upon-Tyne, England | Sir Charles A. PARSONS,[1165] a British engineer and inventor, **begins a three-decade study into marine propulsion.** High propeller speeds are generally advantageous for the steam turbine; but if they are too high, they lead to much cavitation. With the help of flow visualization diagnostics, he minimizes cavitation effects, thus also improving the propulsive efficiency. **He ascribes cavitation to the "water-hammer of collapsing vortices,"** and compares this phenomenon to whip-cracking, "whereby nearly all the energy of the arm that swings the whip is finally concentrated in the tag." ▪ The first mathematical treatment of collapsing cavitation bubbles was performed by Lord RAYLEIGH {⇨1917}. |
| 1898 | Sprengstoff-AG Carbonit, Schlebusch/Köln, Germany | Christian Emil BICHEL,[1166] director of Sprengstoff-AG Carbonit, **determines the maximum pressures of various explosives** (*e.g.*, black powder, blasting gelatin, carbonite, and guhr dynamite) **when they detonate in a closed chamber 20 liters or less in volume.**

(i) He records the pressure-time profile in the chamber, $p(t)$, using a mechanical "brisance" gauge – which is an ordinary *Dampfmaschinen-Indikator* ("steam engine indicator") combined with a rotating drum which has a writing speed of 2–3 m/s in his study {⇨Fig. 4.17–D}. Deviations of the estimated pressures from those found for the tested explosives are less than 4%.

(ii) From this $p(t)$ record, he also determines the mechanical work performed by the detonating explosive, which he compares with data obtained from the Trauzl Lead Block Test {BRUNSWICK ⇨1903; ⇨Fig. 4.17–F}. |
| 1898 | Torpedo Inspection Organization, German Imperial Navy, Kiel, northern Germany | G.F. Rudolph BLOCHMANN,[1167] a German military scientist, **is the first to correlates the local, temporal, and causal aspects of numerous underwater explosion phenomena:**

(i) In order to measure the shock wave pressures emitted during underwater explosions, he uses a calibrated "dynamometer" {⇨Fig. 4.16–D}, a spring-loaded pressure gauge connected with a stylus, the motion of which is recorded on a revolving drum covered with a wax-coated paper.

(ii) **He develops the most advanced theory of underwater explosion that, along with gas bubble oscillation, allows the shock pressure in the water to be predicted any distance from the explosive.** |

---

[1164] L. MACH: *Optische Untersuchung der Luftstrahlen*. Sitzungsber. Akad. Wiss. Wien **106** (Abth. IIa), 1025-1074 (1897).
[1165] J.A.E. (anonymous): *Sir Charles PARSONS*. Proc. Roy. Soc. Lond. **A131**, v-xxv (1931).
[1166] C.E. BICHEL: *Experimentelle Untersuchung von Sprengstoffen mit Hülfe eines neuen Brisanzmessers*. Mittler, Berlin (1898). Translated and edited by A. LARSEN: *New methods of testing explosives*. C. Griffin & Co., London (1905).
[1167] R. BLOCHMANN: *Die Explosion unter Wasser*. Marine-Rundschau **9** (1. Teil), 197-227 (1898).

(iii) He compares the results from his theory with his measurements and obtains fairly good agreement.

(iv) Using his analytical results for gas bubble oscillation, he attempts to explain strange surface phenomena associated with explosions {ABBOT ⇒1881}, such as the dome of spray thrown up from the surface, and the subsequent formation of spear-like plumes of spray.

| | | |
|---|---|---|
| 1898 | *Chair of Experimental Philosophy, University of Cambridge, England* | George Howard DARWIN,[1168] a British astronomy professor and second son of the famous naturalist Charles DARWIN, **reports on a 5- to 6-m-high tidal bore on the Ganges (at Bangladesh).** • There are two separate classes of bore that occur in Asia: those that are determined by large spring tides, and those that are a by-product of cyclonic activity during the monsoon season (May to October). The Ganges has a mean spring mesotidal range of only about two meters, so it seems that the bore DARWIN had observed resulted from cyclonic activity (*see also* Sect. 2.3.1: "storm surges"). |
| 1898 | *Aleksejew Water Line Station, Moscow, Russia* | K. KARELJSKICH, W. OLDENBURGER, and N. BERJOSOWSKY, three hydraulic engineers, **experimentally study the propagation of hydraulic shocks in water pipes.** They use pipes with diameters of up to 6 in. (15.2 cm) and lengths of up to 2,494 ft (868 m) which they connected to Moscow's 24-in. (61-cm)-dia. main water line via a fast-closing valve. For pipes ranging in diameter between 2 to 6 in. (5.1 to 15.2 cm), they record pressure jumps of between three and four bar. With the aid of an electrical chronograph, they measure speeds ranging from 3,290 to 4,200 ft/s (1,003–1,280 m/s), regardless of whether the pressure jump is generated by a sudden opening or closing of the valve. • Note that the velocity of sound in free water amounts to 1,485 m/s (at 20 °C), but – similar to the speed of sound of air waves in pipes (G.R. KIRCHHOFF) – decreases with decreasing pipe diameter. |

Nikolai E. ZHUKOVSKY,[1169] a Russian professor of theoretical mechanics who supervises and theoretically analyzes these experiments, **notices that *water-hammer waves* in plumbing systems** {MONTGOLFIER ⇒1796} **are related to shock discontinuities** that propagate with constant speed; they are dependent on the wall material and thickness of the pipe but independent of the shock intensity.

(i) He shows that the simple formula for the propagation velocity of sound in water pipes, as previously derived by the Dutch mathematician Diederik Johannes KORTEWEG,[1170] can also be applied to the propagation of hydraulic shocks. KORTEWEG, who considered the pipe to be an elastic membrane, obtained for the velocity of sound $v$ in an elastic pipe the following simple formula:

$$v = v_1 v_2 / (v_1^2 + v_2^2)^{1/2} \text{ with } v_1 = (k/\rho_0)^{1/2} \text{ and } v_2 = [eE/(2R_0\rho_0)]^{1/2}.$$

Here $R_0$ is the inner dia. of the pipe, $\rho_0$ is the density of the water at rest, $k$ is the modulus of elasticity of water, $E$ is the modulus of elasticity of the pipe material, and $e$ is the wall thickness of the pipe.

(ii) **For a rigid solid surface, the pressure jump or *water-hammer pressure* $p$ can be estimated via the simple relation**

$$p = \rho_0 c_0 U,$$

where $\rho_0$ and $c_0$ are the density and acoustic velocity of the undisturbed liquid, and $U$ is the velocity of the discontinuity carried by the wave, which, he assumes, is approximately the same as the acoustic velocity; *i.e.*,

$$p = \rho_0 c_0^2.$$

---

[1168] G.H. DARWIN: *The tides and kindred phenomena in the Solar System* [based upon lectures delivered in 1897 at the Lowell Institute, Boston, MA]. J. Murray, London (1898).

[1169] N.E. ZHUKOVSKY: *Über den hydraulischen Stoß in Wasserleitungsröhren.* Mém. Acad. Sci. Imp. St. Pétersbourg **9** [VIII], No. 5, 1-72 (1900).

[1170] D.J. KORTEWEG: *Over voortplating-snelheid van golven in elastische buizen.* Van Doesburgh, Leiden (1878); Ph.D. thesis, University of Amsterdam (1878).

This simple equation, derived from momentum considerations, will also been widely applied in elastic liquid-solid impact theory to evaluate the "one-dimensional water hammer" {HEYMANN ⇨1969}.

(iii) At transitions from large to small pipe diameters, the shock intensity can double and, under unfavorable reflection conditions, even increase further up to a fatal level of shock loading.

(iv) Reflected shocks can generate detrimental periodic oscillations in the pipe system.

(v) Hydraulic shocks can be prevented by using slowly closing valves, with a closing time proportional to the length of the water pipe, and by installing wind tanks in the close vicinity of the valves.

In the 1890s, the problem of "water hammering" – an impulsive force created by the rapid deceleration of water resulting from closing a valve or nozzle too quickly – was a constant companions in urban infrastructures. It was also tackled by Irving P. CHURCH[1171] in England and Rolla Clinton CARPENTER[1172] in the United States. ▪ A similar but even more serious problem of water hammering is the generation of hydraulic shocks in penstocks resulting from accelerated or decelerated water masses when the water flow to a hydraulic turbine must be quickly controlled upon sudden load changes in the electric generator. This complex problem was later treated theoretically in great detail by the Italian civil engineer Lorenzo ALLIÉVI.[1173]

| | | |
|---|---|---|
| 1898 | *Bristol, southwest England* | At a meeting of the British Association for the Advancement of Science (BAAS), **John MILNE**,[1174] a British seismologist, delivers his third report on seismological investigations **addressing suboceanic changes in relation to earthquakes and sea waves** ("tsunamis"). |

(i) He states that "**sea waves can be caused by the excessive deposition of sediments,** the suboceanic escape of waters from subterranean sources, the sudden release of waters backed up in bays by gales, changes in the magnitude and direction of ocean currents, and by suboceanic seismic and volcanic action, sudden and extensive yieldings might take place along the face of slopes in a critical condition. That such **suboceanic landslides** had often taken place was proved by an appeal to the experience of cable engineers, who often found that cable interruptions were the result of their burial along lengths of several miles, the materials covering the lost sections having fallen from the faces of slopes along the base of which the cables had been laid … **The fact that earthquakes originating in deep water,** as, for example, at a depth of 4,000 fathoms [7,316 m] off the N.E. coast of Japan, **have been accompanied by a series of sea waves** which may agitate an ocean for 24 hours tells us that there must have been a sudden suboceanic displacement of a very large body of material, accompanying some form of bradyseismical adjustment."

(ii) MILNE[1175] also discussed this subject in his book *Earthquakes and Other Earth Movements* published in the same year. Referring to the example of tsunamis produced by an earthquake on the slope of the Tuscarora Deep {Kamaishi Earthquake ⇨1896}, **he pondered the mechanism for such large disturbances, and hit upon the steepening shock wave model,** "For small waves, the velocity with which they travel depends upon the square root of their lengths; but with larger waves, like earthquake waves, the velocity practically depends upon the square root of the depth of water, and these latter travel more quickly than the former … If, therefore, we have a series of disturbances of unequal magnitude producing sea waves, which, from the series of shocks which have been felt upon shores subsequently invaded by waves, seems in all probably often to have been the case, **it is not unlikely that the waves of an early**

---

[1171] I.P. CHURCH: *"Water ram" in pipes*. Franklin Inst. J. **129**, 328-336, 374-383 (1890).
[1172] R.C. CARPENTER: *Some experiments on the effect of water hammer*. Engng. Rec. **30**, 173-175 (1894).
[1173] L. ALLIÉVI: *Teoria generale del moto perturbato dell' acqua nei tubi in pressione*. Annali della Società degli Ingeneri ed Architetti (1903); L. ALLIÉVI and R. DUBS: *Allgemeine Theorie über die veränderliche Bewegung des Wassers in Leitungen*. Springer, Berlin (1909), pp. 1-155.
[1174] J. MILNE: *Third report of the Committee on Seismological Investigation*. Rept. Meet. Brit. Assoc. **68**, 179-272 (1898), *see* p. 272.
[1175] J. MILNE: *Earthquakes and other earth movements*. Paul, Trench, Trübner & Co., London (1898), pp. 165-175.

disturbance may be overtaken and interfered with a series which followed ... These considerations help us to understand ... the phenomena observed by those who have recorded tidal waves as they swept inwards upon the land. For instance, we understand the reason why sea waves, as observed at places at different distances from the origin of a disturbance, should be of different heights."

At that time, submarine volcanic explosions were thought to be the most general cause of tsunamis {Krakatau Eruption ⇨1883; Myojin-sho Reef Eruption ⇨1952}; however, the Irish civil engineer **Robert MALLET had already suggested that suboceanic landslides were a possible cause of such sea waves** {Coast of Yorkshire ⇨1856}. ▪ Modern studies, correlating earthquake characteristics with tsunami characteristics, have provided evidence of direct relationship between these two sets of phenomena {GUTENBERG ⇨1939}, reinforcing the belief that tectonic displacements of the sea floor along faults are the most common generators of tsunamis.[1176]

| | | |
|---|---|---|
| 1898 | *Laboratoire Central, Service des Poudres et Salpêtres, Paris, France* | **Paul VIEILLE**,[1177] a French physicist and chemist, **measures the shock front velocities originating from small amounts of explosives** by placing and igniting them at one end of a 4-m-long air-filled tube 22 mm in dia. Using chronography, he measures supersonic velocities for both gunpowder (337–1,268 m/s) and mercury fulminate (359–1,138 m/s). ▪ **His measurements of a fast-propagating discontinuity confirm previous theoretical models** {RIEMANN ⇨1859; HUGONIOT ⇨1885 & 1887}, **as well as MACH's previous observations on supersonic waves generated by powerful electric sparks** {E. MACH ET AL. ⇨1875 & 1878} **and explosives** {E. MACH & SOMMER ⇨1877}. |
| 1898 | *Lehrstuhl und Institut für Geophysik, Universität Göttingen, Germany* | **Emil WIECHERT**,[1178] a German physics professor and founder of the Göttingen Institute of Geophysics[1179] – the world's first chair devoted to this new branch of physics – **introduces the first photographically recording seismometer with a viscously damped horizontal pendulum as a sensor.** ▪ This new seismometer was the first instrument capable of producing useful records for the entire duration of an earthquake. Known as the **"Wiechert seismograph"** or "Göttingen seismograph," it is still in operation at the Institute. Horizontal and vertical seismographs using WIECHERT's design were commercialized by the Göttingen company Spindler & Hoyer GmbH (founded in 1898). |
| 1899 | *Owens College, University of Manchester, England* | **David L. CHAPMAN**,[1180] a British chemist and previously a student of Prof. Harold B. DIXON, **treats unsupported detonation analytically.**<br>(i) He recognizes that when detonation occurs in a tube, so that the motion is confined to one dimension, the detonation wave must be followed by a region of forward-moving gas, and that the length of this region must increase continually.<br>(ii) **He assumes that, once the maximum velocity is reached, the detonation front** – *i.e.*, the front of the *explosive wave*, a term which he proposes to limit only to the space within which chemical changes are taking place – **moves steadily, the flow is planar, and the chemical reaction occurs instantaneously.**<br>(iii) **He discovers that a unique solution to the one-dimensional conservation equations across the detonation wave corresponds to the minimum wave velocity solution, where the Rayleigh line is tangent to the equilibrium Hugoniot curve – the so-called "tangency solution."** ▪ CHAPMAN's idea, that the detonation velocity corresponds to the tangency solution, |

---

[1176] D.C. COX: *Status of tsunami knowledge.* In: (D.C. COX, ed.) *Proceedings of the Tsunami Meetings Associated with the 10th Pacific Science Congress* [Honolulu, HI, 1961]. Institut Géographique National, Paris (1963), pp. 1-6.
[1177] P. VIEILLE: *Sur la vitesse de propagation d'un mouvement dans un milieu en repos.* C. R. Acad. Sci. Paris **126**, 31-33 (1898); Ibid. **127**, 41-43 (1898).
[1178] E. WIECHERT: *Seismometrische Beobachtungen im Göttinger Geophysikalischen Institut.* Nachr. Gesell. Wiss. Gött. (1899), pp. 195-208.
[1179] J. RITTER: *History of seismology in Göttingen*; http://www.uni-geophys.gwdg.de/~eifel/Seismo_HTML/history.html.
[1180] D.L. CHAPMAN: *On the rate of explosion in gases.* Phil. Mag. **47** [V], 90-104 (1899), *see* p. 95.

is identical to a detonation theory later put forward independently in France by J.C. Émile JOUGUET {⇨1904–1906}, who demonstrated that the detonation velocity is equal to the local velocity of sound in the reaction products.

(iv) He analyzes the solutions to the shock jump conditions for explosive gaseous mixtures and observes that the minimum wave velocity solution agrees with experimental measurements previously made by Harold B. DIXON {⇨1893}.

(v) He concludes, "**When an explosion starts, its character and velocity are continually changing until it becomes a wave permanent in type and of uniform velocity.** I think it is reasonable to assume that this wave – *i.e.*, the wave of which the velocity has been measured by professor DIXON – is that steady wave which possesses minimum velocity; for, once it has become a permanent wave with uniform velocity, no reason can be discovered for its changing to another permanent wave having a greater uniform velocity and a greater maximum pressure…"

| | | |
|---|---|---|
| 1899 | *Faculty of Sciences, University of Marseille, southern France* | **Charles FABRY**[1181] **and Alfred PÉROT,** two French physicists, **design a powerful new interferometer – the so-called "Fabry-Pérot interferometer,"** which represents a significant improvement over the Michelson interferometer (1881). Unlike the Michelson interferometer, the Fabry-Pérot design contains planar surfaces that are all partially reflecting so that multiple rays of light are responsible for creating the observed interference patterns. ▪ Fabry-Pérot interferometers (FPIs) are widely used in shock wave and detonation physics for precisely measuring the displacements and velocities of materials accelerated by shocks. Multibeam FPIs are useful tools, particularly when the experiments performed are either very expensive or are poorly reproducible.[1182] The Large Angle and Spectrometric Cronograph (LASCO) aboard the SOHO spacecraft, which studies the structure and evolution of the solar corona, employs a tunable FPI {LASCO ⇨2002}. |
| 1899 | *Germany* | **G.F. HENNING,**[1183] a German chemist, **is the first to prepare cyclotrimethylenetrinitramine [$C_3H_6O_6N_6$], and he observes that the substance "when quickly heated, explodes with a bang."** However, instead of using it for blasting purposes, he suggests that it could be used as a base material for drugs. ▪ Cyclotrimethylenetrinitramine or **RDX** – known in Germany as **hexogen,** in Italy as **T4,** and in the United States as **cyclonite** – **is a powerful secondary explosive with a detonation velocity of 8,640 m/s.** The name RDX was coined by the British as a contraction of Research Department Explosive. In order to create more powerful weapons to fill bombs and artillery shells during World War II, both sides used RDX together with TNT on a large scale in castable compounds such as **Composition B** or Comp-B (60/40 RDX/TNT), also known as **"cyclotol,"** which is used for loading shaped-charge bombs, special fragmentation projectiles, and grenades.<br>In 1920, the Austrian chemist Edmund VON HERZ[1184] reinvented cyclotrimethylenetrinitramine and proposed an effective method of preparation. He wrote, "We have a perfectly new, hitherto unknown explosive, which combines in itself in an ideal way the advantages both of the ethereal salts of nitric acid and also those of the aromatic nitro compounds, combined with remarkable stability and non-sensitiveness while it surpasses all hitherto known and practically usable explosives in energy, shattering power and density … The absolute specific gravity is 1.82 and this is a maximum not possessed by any nitro compound hitherto. This fact enables very high loading densities to be obtained which is of great importance for many purposes, *e.g.*, bursting charges for projectiles, detonators, and percussion caps." |

---

[1181] C. FABRY and A. PÉROT: *Théorie et application d'une nouvelle méthode de spectroscopie interférentielle.* Ann. Chim. **16** [VII], 115-144 (1899).
[1182] D.R. GOOSMAN: *The multibeam Fabry-Pérot velocimeter: efficient measurement of high velocities.* Sci. Technol. Rev. **2**, 85-93 (July 1996).
[1183] G.F. HENNING: *Verfahren zur Darstellung eines Nitrokörpers aus Hexamethylentetramin.* Germ. Patent No. 104,280 (Juni 1899).
[1184] E. VON HERZ: *Improvements relating to explosives.* Brit. Patent No. 145,791 (March 1920).

| | | |
|---|---|---|
| 1899 | University of Edinburgh, Scotland, U.K. | **Cargill G. KNOTT,**[1185] a British lecturer in applied physics who had long been interested in the physics of earthquake phenomena, **presents his theory of reflection and refraction of seismic waves for application in seismology:**

(i) KNOTT, who had previously worked as a physics professor at Tokyo University, writes, "At Lord KELVIN's suggestion I reproduce, with additions and extensions, a paper I published eleven years ago in the *Transactions of the Seismological Society of Japan* [Feb. 23, 1888] … in which the problem of the behavior of an elastic wave incident on the interface of rock and water was for the first time fully worked out. In that paper also, I believe, the sound method of treating the general problem when the two media are elastic solids was first explicitly stated."

(ii) To the previously treated cases of the reflection and refraction of seismic disturbances through rock and water, KNOTT adds the important cases of those through rock and rock, rock and air, and solid rock and fluid rock, thereby assuming certain relations among the densities.

(iii) **He distinguishes between** *purely elastic wave* **and** *quasi-elastic wave* **propagation,** saying "Purely elastic vibrations are propagated with high speed, while quasi-elastic disturbances, straining the material distinctly beyond the *limit of elasticity*, travel much more slowly." KNOTT speculates that the latter type causes the destructive effects of an earthquake.

The two most important techniques of explosion prospecting – the *reflection method* and the *refraction method* – are based upon the analysis of reflected and refracted seismic waves, respectively, measured at the surface by geophones {MINTROP ⇨ 1919 & 1924}. |
| 1899 | Laboratoire Central, Service des Poudre et Salpêtres, Paris, France | **Paul VIEILLE,**[1186] a French physicist, chemist and inventor, **constructs the first bursting diaphragm shock tube** {⇨Fig. 4.10–A} to demonstrate that a shock wave propagates with a speed greater than the speed of sound. However, he does not propose a theory to account for the shock speeds observed.

(i) His device consists of a 6-m-long steel tube with a constant cross-section 22 mm in dia. which is divided by a diaphragm into two parts: a 2-m-long driver (or compression) chamber followed by a 4-m-long expansion chamber.

(ii) He uses collodion, paper, glass, or steel as the diaphragm material. The diaphragm ruptures automatically upon reaching a certain overpressure, *e.g.*, 35 bar for a 1.5-mm-thick glass plate. **When a pressure difference is developed and the diaphragm breaks, a compressive disturbance will propagate into the end that was initially at the lower pressure, and a rarefaction is sent into the other end. The compression rapidly becomes a shock wave.**

(iii) In air he achieves shocks with Mach numbers of up to $M = 2$. He concludes that "explosives do not play any essential role in phenomena of propagation at great speeds," meaning that the phenomenon of supersonics is not limited to the use of explosives; supersonic waves can for example also be generated by a bursting membrane.

His *shock tube* – a term coined much later by U.S. gas dynamicists {BLEAKNEY, WEIMER & FLETCHER ⇨1949} – **became an important diagnostic tool for a number of scientific disciplines:**
   ▸ for aerodynamic purposes;
   ▸ in high-temperature chemical physics;[1187]
   ▸ in pulsed gas dynamics laser systems, where the shock tube is used as a driver unit;
   ▸ in the kinetics of chemical reactions for studying vibrational and rotational energy transfer;
   ▸ in plasma spectroscopy; |

---

[1185] C.G. KNOTT: *On reflection and refraction of elastic waves, with seismological application.* Phil. Mag. **48** [V], 64-97 (1899).
[1186] P. VIEILLE: *Sur les discontinuités produites par la détente brusque des gaz comprimés.* C. R. Acad. Sci. Paris **129**, 1228-1230 (1899).
[1187] A.G. GAYDON and I.R. HURLE: *The shock tube in high-temperature chemical physics.* Reinhold, New York (1963).

- in investigations of vapor bubble dynamics in two-phase flow and Richtmyer-Meshkov instabilities {MESHKOV ⇨1969}; and
- even for fertilizer production as a possible alternative to the Haber process.[1188]

A critical survey of shock tube research up to 1970 was provided by Abraham HERTZBERG,[1189] a professor of aero- and astronautics at the University of Washington.

A modern modification of the classical shock tube is the "electrothermal plasma gun" {⇨Fig. 4.16–M}. Instead of applying a high static pressure in the driver section, it generates the high pressure required by impulsively heating the driver gas via a powerful capacitor discharge.

---

**1899**

*Army Artillery Test Range, Cummersdorf [near Berlin], Germany*

Walther WOLFF,[1190] a professor of military engineering, **is the first to investigate large-scale explosions in air** by order of the Prussian Ministry of War. His important results can be summarized as follows:

(i) Using an electrical contact microphone which triggers a Le-Boulengé chronograph {LE BOULENGÉ ⇨1860s; ⇨Fig. 4.19–B}, he measures the velocity of the spherical blast wave emerging from the explosion of large quantities of black powder or trinitrophenol. His *Luftstoßanzeiger* ("blast indicator"), a mechanical microphone, uses a thin rubber membrane which is directly coupled to a rotating-drum chronograph {⇨Fig. 4.17–M}. **His observations fully confirm Ernst MACH's previous results that the blast wave is supersonic close to the charge but rapidly decreases with increasing distance** {E. MACH, TUMLIRZ & KÖGLER ⇨1878}. For trinitrophenol with a charge weight of 1,500 kg, WOLFF measures a velocity of 858 m/s at a distance of ten meters.

(ii) **His mechanically recorded pressure-time profiles already show all the typical characteristics of a blast wave, such as the steep rise, the rapid decay, and the phase of negative pressure (or the "suction" phase).** Addressing such negative blast pressures, he notes that, "The shards from the windows positioned close (at a distance of 25 m) to the origin of the explosion were mostly thrown away from the explosion, while a small number also fell towards it. In the adjacent zones, the shards (from the same windows) were thrown about equally towards the origin of the explosion and away from it. With increasing distance from the explosion the percentage of debris thrown towards the source increased until eventually the broken windows were exclusively thrown towards the source, and no debris could be detected in opposite direction. This striking distribution of glass fragments – already frequently observed previously though incompletely in accidental explosions – appears to provide evidence that both positively directed forces pointing away from the explosion, as well as negatively directed ones moving bodies towards the explosion, can occur at the same location in the surroundings of an explosion."

(iii) **He also studies the responses of structures to blasts.**

In the following year, WOLFF[1191] **reported on the use of his microphone to measure supersonic projectile velocities.** He positioned two microphones provided with electrical contacts along the trajectory – one at the left-hand side and the other on the right hand side to improve the accuracy of the measurement – which, when successively activated by the head wave, triggered two mechanical chronographs synchronously. • Although such instruments have rise times on the order of 100 ms and therefore cannot truly record pressure-time profiles, they nevertheless provided worthwhile information on the propagation of blast waves gener-

---

[1188] A. HERTZBERG: *Nitrogen fixation for fertilizers by gasdynamic techniques*. In: (G. KAMIMOTO, ed.) *Proc. 10th Int. Shock Tube Symposium* [Kyoto, Japan, 1975]. Shock Tube Research Society, Japan (1975), pp. 17-28.

[1189] A. HERTZBERG: *Shock tube research, past, present and future*. In: (I.I. GLASS, ed.) *Proc. 7th Int. Shock Tube Symposium* [Toronto, Canada, 1969]. University of Toronto Press (1970), pp. 3-5.

[1190] W. WOLFF: *Über die bei Explosionen in der Luft eingeleiteten Vorgänge*. Ann. Phys. **69** [III], 329-371 (1899).

[1191] W. WOLFF: *Die Messung von Geschossgeschwindigkeiten*. Mutter Erde (Berlin & Stuttgart) **2**, Nr. 8, 145-148 (1900).

ated by explosions and muzzle blasts. WOLFF's method of measuring blast pressures by using mechanical barographs was resumed by various other military investigators.[1192]

| | | |
|---|---|---|
| 1899 | Physical Laboratory, University of Wisconsin, Madison, Wisconsin | Robert W. WOOD,[1193] a U.S. experimental physicist, **repeats August TOEPLER's spark wave propagation and reflection experiments** {A. TOEPLER ⇨1864}, but instead of observing the phenomena through a schlieren telescope **he uses photographic film to record it.** WOOD states, "I have always felt that the very beautiful method derived in 1867 by TOEPLER for the study of *schlieren* or *striae*, is not as well known outside of Germany as it deserves to be, and trust that the photographs illustrating this paper are sufficient excuse for bringing it before the readers of the *Philosophical Magazine*. Sound waves in air were observed by TOEPLER, but they have never to my knowledge been photographed. When seen subjectively, the wave-fronts, if at all complicated, cannot be very carefully studied, as they are only illuminated for an instant, and appear in rapid succession in different parts of the fields of the viewing-telescope." ▪ WOOD failed to notice that the difficult task of photographing a shock wave emerging from a spark discharge had already been tackled by German researchers {E. MACH & WENTZEL ⇨1884; E. MACH & L. MACH ⇨1889}. Nevertheless, **WOOD obtained interesting results on the reflection, refraction, and diffraction of spark waves (*i.e.*, of weak shock waves).**<br><br>In the following year, WOOD reported on further experiments with spark waves which made the great potential of shock wave photography widely known to the Anglo-American public {WOOD ⇨1900}. |
| 1900 | K.u.k. Kriegsschule Wien, Vienna, Austro-Hungarian Empire | **Artillery-General Philipp HESS,** a military engineer and head of the Austrian Military & Aeronautics Organization [Germ. *Militär- & Luftschifffahrtwesen*], **reports** in the Austrian bulletin *Mitteilungen des technischen Militär-Comités* **that he has photographed a luminous band at the point of collision of two compression waves,** produced by two simultaneously detonating 100-g cartridges of an ammonium safety explosive suspended in the air a short distance apart. In the same year, his results, which are highly relevant to the safe use of explosives in a mining atmosphere, are published and discussed by the German mining assessor Fritz HEISE.[1194]<br><br>Harold B. DIXON and collaborators at Owens College in Manchester resumed the subject: they studied luminous phenomena during head-on collisions of two detonation waves using streak photography {DIXON ⇨1903}. |
| 1900 | Laboratoire Central, Service des Poudres et Salpêtres [from 1971 Société Nationale des Poudres et Explosifs, SNPE], Paris, France | **Paul VIEILLE,** a French physicist and explosives expert, **publishes three papers with important results:**<br><br>(i) **Starting from HUGONIOT's shock wave theory** {HUGONIOT ⇨1887}, VIEILLE[1195] **first derives for a gas with a constant ratio of specific heats** $m$ $(=\gamma)$ **a relation between the shock front velocity** $U$ **and the overpressure at the shock front** $\Delta p = p - p_0$. This relation is given by<br>$$U = c_0 [1 + \Delta p/p_0 \times (m+1)/2m]^{1/2},$$<br>where $p_0$ and $c_0$ denote the pressure and sound velocity at rest, respectively. He also confirms this relationship experimentally.<br><br>(ii) VIEILLE[1196] **is the first to speculate on hypersonic flight, thereby predicting stagnation pressures and temperatures for flight in ideal air at speeds of up to** $M \approx 30$, **as well as associated surface phenomena** (such as incandescence and erosion, leading to thermal rup- |

---

[1192] D.C. MILLER: *Sound waves: their shape and speed; a description of the phonodeik and its application, and a report on a series of investigations made at Sandy Hook Proving Ground.* Macmillan, New York (1937).
[1193] R.W. WOOD: *Photography of sound-waves by the "Schlierenmethode."* Phil. Mag. **48** [V], 218-227 (1899).
[1194] F. HEISE: *Zur Theorie der Sicherheitssprengstoffe.* Glückauf **36**, Nr. 13, 265-267 (1900).
[1195] P. VIEILLE: *Etude sur le rôle des discontinuités dans les phénomènes de propagation.* Mém. Poudres Salpêtres **10**, 177-260 (1900).
[1196] P. VIEILLE: *Sur la loi de résistance de l'air au mouvement des projectiles.* C. R. Acad. Sci. Paris **130**, 235-238 (1900).

tures in meteorite falls). He concludes, "Without admitting an absolute value for these numbers, one can imagine that the incandescence of meteorites, the erosion of the surface and the rupture which accompanies their passage through our atmosphere are explicable by pressures and temperatures that can be predicted using the law of the propagation of discontinuities, even when taking the rarefaction of the medium it passes through into account."

(iii) Resuming his previous hypothesis of detonation, and stimulated by his colleague Pierre M.M. DUHEM,[1197] a physicist who in 1896 had reflected on how chemical phenomena might modify the elasticity of a medium, VIEILLE[1198] **states that "the detonation wave is a discontinuity maintained in this state by a chemical reaction"** [French *une discontinuité à l'état de régime par la réaction chimique...*] induced by the increase in temperature due to adiabatic compression. This reaction, however, may occur somewhat after the passage of the shock wave. He writes, "A first step in the interpretation consists of dividing – so to speak – the *detonation* phenomenon into two parts: the reaction is regarded as the product of an elevated temperature primarily due to the physical phenomenon of adiabatic compression in an inert environment. The chemical phenomenon, according to this hypothesis, is a consequence of [this] compression and may itself involve a certain delay with respect to the passage of the mechanical wave; in all cases, it is not necessary to modify the properties of the environment in which the mechanical wave propagates, and its role is simply to maintain the elevated value of condensation propagating with the wave..." ▪ VIEILLE's view of detonation is similar to previous hypotheses proposed in England by Arthur SCHUSTER {DIXON ⇨1893} and David L. CHAPMAN {⇨1899}, which apparently he didn't know about.

| | | |
|---|---|---|
| 1900 | *University of Wisconsin, Madison, Wisconsin* | Robert W. WOOD,[1199] a U.S. experimental physicist, **photographs focused spark waves (weak shock waves)** using spherical, parabolic and elliptical mirrors. His snapshots of wave reflection and refraction phenomena show interesting new focusing effects {⇨Fig. 4.14–I}. |
| 1901 | *Edinburgh, Scotland, U.K.* | On February 21/22, **Thomas D. ANDERSON**, a Scottish clergyman, **discovers the first new star (nova) of the 20th century.** It is located in *Perseus* ("Perseus," a constellation of the Northern Hemisphere) and of magnitude 2.7. |

(i) Over the next two days, it reaches zero magnitude, thus becoming the brightest star in the sky, but after that it decreases rapidly. On March 15, it is a 4th-magnitude star; during the next three months it repeatedly oscillates between magnitudes 4 and 6, and by the end of the year it has faded to a magnitude of 7.[1200]

(ii) There is evidence that the outburst must have been extremely rapid, for the region where the new star appears was photographed repeatedly at Harvard during February 1901, and no trace of the star was found on a plate taken on February 19, 1901, which showed 11th-magnitude stars. Thus, a rise of at least eight magnitudes in two days must have occurred.[1201]

(iii) The new star was later recognized as being an *explosive* (or *catastrophic*) *variable star*,[1202] and has since been classified as **Nova PERSEI 1901**. A nova occurs in a close binary

---

[1197] P.M.M. DUHEM: *Théorie thermodynamique de la viscosité, du frottement et des faux équilibres chimiques*. Mém. Soc. Sci. Phys. Nat. (Bordeaux) **2** [V], 1-208 (1896).
[1198] P. VIEILLE: *Rôle des discontinuités dans la propagation des phénomènes explosifs*. C. R. Acad. Sci. Paris **131**, 413-416 (1900).
[1199] R.W. WOOD: *The photography of sound-waves and the demonstration of the evolutions of reflected wave-fronts with the cinematograph*. Nature **62**, 342-349 (1900); *Photography of sound-waves and the cinematographic demonstration of the evolution of reflected wave fronts*. Proc. Roy. Soc. Lond. **66**, 283-290 (1900); Phil. Mag. **50** [V], 148-156 (1900).
[1200] The Greek defined the very brightest stars as being of the "first magnitude" and the faintest observable stars as of the sixth magnitude, with gradations of brightness in between. In the 1850s, the English astronomer Norman Robert POGSON (1829–1891) proposed the form of logarithmic system now in use: the difference in magnitude between two stars is given by 2.5 times the logarithm (to base 10) of the brightness ratio. For example, a star of magnitude 5.0 is 2.5 times as bright as one of magnitude 6.0; *i.e.*, the brighter the object, the lower the number assigned as a magnitude. A difference of five magnitudes in brightness, or $(2.5)^5$, would be very nearly to 100, so that a first magnitude star is about 100 times as bright as a sixth magnitude star.
[1201] From the *Encyclopædia Britannica* (1911); http://41.1911encyclopedia.org/S/ST/STAR.htm.
[1202] For the classification of variable stars *see* General Catalogue of Variable Stars ⇨1948.

system and is characterized by a rapid and unpredictable rise in brightness of 7–16 magnitudes within a few days, followed by a steady decline back to the pre-nova magnitude over a few months. Modern astronomers assume that the event is caused by **an accreting white dwarf which draws material (hydrogen) from its close binary companion until there is sufficient fuel to trigger a thermonuclear explosion, which blasts a shell of hydrogen-rich matter into space.**

In the same year, on August 21, six months after ANDERSON's discovery, **the two French astronomers Camille FLAMMARION and Eugene M. ANTONIADI discover that a nebula surrounds Nova PERSEI 1901.** Subsequent photographs show that this nebula, which consists mainly of two incomplete rings of nebulosity, is expanding outwards at a rate of 2–3" per day. This expansion continues at the same rate until the following year. Two possible explanations for the phenomena of temporary stars are proposed:

▸ the *collision theory* supposes that the outburst is the result of a collision between two stars or between a star and a swarm of meteoric or nebulous matter; and

▸ the *explosion theory* regards the outburst as similar to the sudden start of activity of a long period variable.

Their puzzling observation implied that the debris ejected from the nova explosion was traveling at the astonishing rate of 11 arc minutes per year – ten times the speed of light! It typically takes *years* before the shell of matter ejected from such an event can be resolved by Earth-based telescopes. This extraordinarily rapid motion caused much excitement among astronomers and in the popular press.[1203] ▪ The unusual phenomenon was later explained by a sheet of dust around PERSEI 1901 (now more commonly called "GK Persei"), which caused super-luminal light echoes.[1204]

---

1901 — *Straßburg, Germany*

In April, at the First International Conference on Seismology, **it is proposed to found the International Association of Seismology (IAS)** in order to promote the collaboration of scientists and engineers studying earthquakes. ▪ IAS became an association of states rather than a society of individuals. The Association was eventually formed on April 1, 1904, and included 18 countries. The history of IAS up to 1951 was reviewed by the French seismology professor Jean-Pierre ROTHÉ,[1205] an honorary secretary-general of the IAS.

In 1951, the name of the IAS was extended to the **International Association of Seismology and Physics of the Earth's Interior (IASPEI)** in order to include the activities of the Association related to the study of physical phenomena of the Earth's interior (*e.g.*, tectonophysics, geothermy, radioactivity, elasticity, and plasticity).

---

1901 — *University of Christiana, Christiana [now Oslo], Norway*

**Kristian BIRKELAND,**[1206] a prominent Norwegian physics professor and founder of magnetospheric physics, **receives the first world patent for an electromagnetic launcher – in today's nomenclature a *coil gun*.** In his gun, a magnetized iron projectile is accelerated by a series of solenoids: as the projectile passes each solenoid, an attached wedge pushes apart contacts, opening the circuit of each solenoid in succession. In order to research and produce electromagnetic cannons, a joint-stock company is formed in the same year, called "Birkeland's Firearms." ▪ **The highest speed achieved was 100 m/s, with a mass of 10 kg, fired from a 4-m-long cannon.** BIRKELAND officially demonstrated his gun at a lecture given at the University of Oslo on March 6, 1903, during which his cannon short-circuited and exploded. The following day, the

---

[1203] K. DAVIS: *Variable star of the month February 2000: Nova Persei 1901.* American Association of Variable Star Observers (AAVSO), Cambridge, MA; http://www.aavso.org/vstar/vsots/1100.shtml.
[1204] J.E. FELTON: *Light echoes of Nova Persei 1901.* Sky & Telescope **81**, 153-157 (Feb. 1991).
[1205] J.P. ROTHÉ: *Fifty years of history of the International Association of Seismology (1901–1951).* Bull. Seismol. Soc. Am. **71**, 905-923 (1981).
[1206] K. BIRKELAND: *Fremgangsmåde til udslyngnin af projektiler ved hjelp af magnetisk kraft.* Norwegian Patent No. 11,201 (Sept. 16, 1901); *Elektromagnetisk kanon.* Ibid. No. 11,228 (April 22, 1902); *Projektil for elektromagnetiske kanoner.* Ibid. No. 11,342 (Dec. 11, 1901).

national daily newspaper *Morgenbladet* wrote: "The shot fired yesterday with a 10 kilo projectile was not altogether successful in that during the firing some wiring was burned and the cannon was rendered useless. According to the professor, this would only take half an hour to repair." The history of electromagnetic guns was reviewed by Alv EGELAND,[1207] a Norwegian plasma physics professor at the University of Oslo.

It has been proposed that lunar soil could become a source of relatively inexpensive oxygen propellant for vehicles moving from low Earth orbit to geosynchronous Earth orbit and beyond. This lunar oxygen could replace the oxygen propellant that, in current plans for these missions, is launched from the Earth's surface and amounts to approximately 75% of the total mass. Electromagnetic launchers could provide a way to get this lunar oxygen off the lunar surface at minimal cost.[1208]

| 1901 | *Forges de Châtillon-Commentry et Neuves-Maisons, France* | Georges CHARPY,[1209] a French chemist and metallurgist, **describes a single-blow technique for determining the toughness of a material – the "Charpy impact test."** It measures the energy absorbed by a small notched specimen held horizontally and supported at its ends when it is struck and broken by the impact of a heavy pendulum. The energy absorption is determined from the change of kinetic energy of the pendulum after it strikes the material. ▪ In 1905, CHARPY devised a machine that is remarkably similar to present designs.<br><br>In the same year, a similar impact test is proposed in England {IZOD ⇨1905}. ▪ In 1933, the American Society for Testing and Materials (ASTM) first published a standard test method for pendulum impact testing that, after numerous revisions and updates, is still in use today and is dominant in Europe. The history of impact testing methods from the CHARPY test to the present was reviewed recently.[1210] |
| --- | --- | --- |
| 1901 | *Ecole des Mines, Saint-Étienne, Dépt. Loire, France* | J.C. Émile JOUGUET,[1211] a French physicist and mathematician and a disciple of professor Pierre M.M. DUHEM, **uses DUHEM's *Méthode de l'énergétique* ("energetic method")[1212] in order to generalize previous results on (nonreactive) shock wave propagation** {RIEMANN ⇨1859; HUGONIOT ⇨1887}. ▪ JOUGUET's paper, entitled (in translation) "On the Propagation of Discontinuities in Fluids," became the starting point for all of his subsequent studies on the propagation of *reactive discontinuities*; *i.e.*, of ***detonation waves*** {JOUGUET ⇨1904–1906 & 1917}. |
| 1901 | *Lehrstuhl für Mathematik, Kaiser-Wilhelms-Universität, Straßburg, Germany* | Heinrich M. WEBER,[1213] a German professor of mathematics, **presents his revised edition of RIEMANN's lectures on mathematics,** which he had delivered in the period 1855–1866 at the University of Göttingen and which were first edited in 1869 by RIEMANN's previous student Karl HATTENDORFF. In addition, **treating shock waves** in two chapters entitled *Fortpflanzung von Stößen in einem Gase* ("Propagation of Shocks in a Gas") and *Luftschwingungen von* |

---

[1207] A. EGELAND: *BIRKELAND's electromagnetic gun: a historical overview*. IEEE Trans. Plasma Sci. **17**, No. 2, 73-82 (April 1989).

[1208] W.R. SNOW and H.H. KOLM: *Electromagnetic launch of lunar material*. In: (M.F. MCKAY, D.S. MCKAY, and M.B. DUKE, eds.) *Space resources: energy, power and transport*. NASA Rept. SP-502, NASA Scientific & Technical Information Program, Washington, DC (1992) vol. 2; http://www.belmont.k12.ca.us/ralston/programs/ltech/SpaceSettlement/spaceresvol2/electromag.html.

[1209] G. CHARPY: *Note sur l'essai des métaux à la flexion par choc de barreaux entaillés*. Annuaire (Société des Ingénieurs Civils de France) **54**, 848 (June 1901).

[1210] D. FRANÇOIS and A. PINEAU: *From CHARPY to present impact testing*. In: *Charpy Centenary Conference* [Poitiers, France, Oct. 2001]. Elsevier, Amsterdam, (2002).

[1211] J.C.E. JOUGUET: *Sur la propagation des discontinuités dans les fluides*. C. R. Acad. Sci. Paris **132**, 673-676 (1901). ▪ This paper, together with the milestone papers of ZEMPLÉN {⇨1905} and DUHEM {⇨1909}, was translated into English by G. GENDRON: *English translation of three milestone papers on the existence of shock waves*. Rept. VPI-E-89-12, Virginia Polytechnic Institute, Blacksburg, VA (1989).

[1212] P.M.M. DUHEM, then a professor of physics at the University of Bordeaux, tried to approach continuum mechanics via generalized thermodynamics, using for example thermodynamic potentials and the principle of virtual work. His attempts culminated in his memoir *Traité d'énergétique ou de thermodynamique générale*. Gauthier-Villars, Paris (1911).

[1213] H. WEBER: *Die partiellen Differential-Gleichungen der mathematischen Physik (nach RIEMANN's Vorlesungen)*. Vieweg & Sohn, Braunschweig (1901), vol. II, pp. 469-521.

*endlicher Amplitude* ("Aerial Vibrations of Finite Amplitude"), **WEBER extends RIEMANN's theory** {⇨1859} **to special cases.** Returning to Lord RAYLEIGH's previous objection {⇨1878} to RIEMANN's theory, he demonstrates that RIEMANN's theory is indeed compatible with the Law of the Conservation of Energy.[1214]

Nine years later, Lord RAYLEIGH {⇨1910} resumed this problem in his classical review paper on the evolution of shock wave theories.

| | | |
|---|---|---|
| 1902 | *Mt. Pelée, Martinique Island, French West Indies* | On May 8, **the volcano Mount Pelée erupts violently, blowing off over 55 meters of the summit.** A glowing cloud bursts out with enormous force and rolls down the side of the mountain. The pyroclastic surges sweep mainly over the port of Saint-Pierre, a town about 6 km south of the volcano (nearly 30,000 casualties). On August 30 of the same year, a second eruption destroys a number of other localities, killing about 2,500 people.

In the same year, Alfred LACROIX,[1215] a professor of mineralogy at the National Museum of Natural History in Paris, travels to Mount Pelée, spending there a year and doing research on the volcano. He discovers an uncommon phenomenon which he calls a "fiery cloud" [French *nuée ardente*] – a destructive avalanche of glowing clouds of black volcanic particles emitted by an explosive eruption. ▪ It has been estimated that a *nuée ardente* may attain speeds as great as 160 km/h (44 m/s; *i.e.*, still subsonic).[1216]

**A so-called "Peléan-type eruption" is the most violent type of explosive volcanic eruption.** Such eruptions are usually associated with stratovolcanoes that have magmas with a relatively high level of silica. They are a formidable combination of dome formation, glowing avalanches of hot gas and other volcanic fragments of all size, and lateral blasts {Mt. St. Helens ⇨1980; KIEFFER & STURTEVANT ⇨1984; Mt. Bandai-san ⇨1988}. |
| 1903 | *Kitty Hawk, North Carolina* | On December 17, **Wilbur WRIGHT**, a U.S. inventor and aviation pioneer, **performs the first controlled motor flight over a distance of 260 meters.** The total duration is 59 seconds; *i.e.*, the average flight velocity is about 4.4 m/s.

**In the early 1900s, the WRIGHT Bros. started their wind tunnel experiments in order to optimize the design of wings and propeller blades.**[1217]

(i) They built and used two wind tunnels: the second one, which was more advanced and was driven by a 2-hp gasoline engine, produced a maximum wind velocity of 27 mph (12 m/s). ▪ Theodore VON KÁRMÁN,[1218] a Hungarian-born aeronautical research engineer who visited the WRIGHT Bros. in 1926, stated in his memoirs, "The peak event of this part of my visit to the U.S.A. was my meeting in Dayton, OH with Orville WRIGHT … To my surprise and enormous interest, I found that Orville WRIGHT was familiar with the fundamentals of aerodynamic theory. He told me that before the historic flight at Kitty Hawk, he and his brother spent almost two thousand hours with their small wind tunnel, studying the relative merits of various wing shapes."

(ii) The WRIGHT Bros. demonstrated that results from the wind tunnel and spinning arm are different, due to the circular motion of the latter. As a result, the spinning arm was no longer used in testing. |

---

[1214] Lord RAYLEIGH (J.W. STRUTT): *Theory of sound*. Macmillan, London (1878), vol. II, p. 41.
[1215] A. LACROIX: *Sur les cendres des éruptions de la Montagne Pelée de 1851 et de 1902*. C. R. Acad. Sci. Paris **134**, 1327-1329 (1902).
[1216] See *Encyclopædia Britannica, Micropaedia*. Benton & Hemingway, Chicago *etc.* (1974), vol. VII, p. 436.
[1217] Replicas of the wind tunnel used by the Wright Bros. in their wind tunnel research in the period 1901–1903 are on display at several U.S. museums; *see*, for example, http://www.centennialofflight.gov/wbh/loc_wb_pdf/pdf_files/wind-tunnel.pdf.
[1218] Th. VON KÁRMÁN: *The wind and beyond. Theodore VON KÁRMÁN: pioneer in aviation and pathfinder in space*. Little, Brown & Co, Boston *etc.* (1967), pp. 128-129.

| 1903 | Germany | **Introduction of the *Spitzgeschoss* ("S-bullet") into the Prussian Army. • The infantry S-bullet, a low-drag supersonic projectile** ($v_0$ = 895 m/s) **with a pointed, ogival-shaped nose equal to seven calibers, was a landmark in the development of infantry cartridges.**

(i) It was adopted in the calibers 6.5, 7, 7.65, and 8 mm by a large number of foreign military administrations – partly directly, partly after being further developed.

(ii) Before World War I, the German ballisticians Karl BECKER and Carl CRANZ {⇨1912} carried out free-flight ballistic studies using the schlieren method and their invention, **electric spark photochronography** {⇨Fig. 4.19–J}, in order to determine the aerodynamic drag up to a velocity of 1,025 m/s. Lt. STRÖDEL, one of CRANZ's students, also performed schlieren studies on the head and tail waves generated by an S-projectile {Prussian Army ⇨1903}.[1219]

(iii) CRANZ {⇨1925} gave an impressive example of the enormous influence of aerodynamic drag on the flight of an S-bullet with an initial velocity ("$v_0$") of 895 m/s: it would have an effective range of 82,000 m in a vacuum. However, when fired in the atmosphere at an elevation of 32°, the maximum would be only 3,400 m.[1220]

In the following year, the Deutsche Waffen- und Munitionsfabrik (DWM) AG at Berlin-Borsigwalde, which was in charge of the development and production of the S-bullet, obtained a German patent[1221] for a "pointed bullet for fire arms, particularly for high flight velocities, characterized in that the pointed head has a length about half the total length of the bullet and is curved by a radius about four to nine times the caliber." |
|---|---|---|
| 1903 | Berlin, Germany | At the 5th International Congress on Applied Chemistry, **H. BRUNSWICK**,[1222] a German explosives expert, **recommends that the *Trauzl Lead Block Test* (or *Trauzl Test*) should be standardized in order to allow better comparison between different high explosives in terms of their brisance:** ten grams of the test explosive, wrapped in tinfoil and stemmed with sand, are fired in a hole 25 mm in dia. and 125 mm deep drilled into a cylindrical lead block that is 200 mm in diameter and 200 mm in height {⇨Fig. 4.17–F}. The resulting increase in the cavity volume is measured and taken as a criterion for the brisance of detonation. • The Austrian blasting engineer Isidor TRAUZL, an early expert on dynamite,[1223] proposed his Lead Block Test in the 1880s. He was also a busy writer of popular technical booklets, such as those on the properties of new high explosives and their practical use in military and civil blasting operations (*e.g.*, in mining, agriculture, and forestry).[1224]

Apparently, the German mining engineer Fritz WINKHAUS[1225] and the tunnel construction engineer Franz VON RZIHA {ROUX & SARRAU ⇨1873} were the first to make TRAUZL's test popular among blasting engineers. TRAUZL's simple method of classifying the *brisance* of a detonating explosive – using BRUNSWICK's proposed geometry – was adopted worldwide as a standard method. It is still in use today. |

---

[1219] C. CRANZ: *Lehrbuch der Ballistik. I. Äußere Ballistik.* Springer, Berlin (1925), vol. 1, p. 40.

[1220] C. CRANZ: *Lehrbuch der Ballistik. Ergänzungen.* Springer, Berlin (1936), p. 1.

[1221] Deutsche Waffen- und Munitionsfabrik (DWM), Berlin: *Spitzgeschoß für Handfeuerwaffen.* Germ. Patent No. 204,660 (Nov. 17, 1908).

[1222] H. BRUNSWICK. *Methoden zur Prüfung von Sprengstoffen mit besonderer Berücksichtigung der TRAUZL'schen Bleiblockprobe.* In: (O.N. WITT und G. PULVERMACHER, eds.) *Bericht über den 5. Int. Kongress für Angewandte Chemie* [Berlin, Germany, June 1903]. Dt. Verlag, Berlin (1904), vol. 2, sect. IIIb, pp. 286-292.

[1223] I. TRAUZL: *Das Dynamit.* Österreich. Z. Berg- u. Hüttenwesen **17**, Nr. 30, 234-237; Nr. 31, 243-245; Nr. 32, 252-254; Nr. 33, 262-263; Nr. 34, 268-270; Nr. 35, 277-279 (1869).

[1224] TRAUZL's books on the practical applications of dynamite are:
(i) *Explosive Nitrilverbindungen: insbesondere das Dynamit, dessen Eigenschaften und Verwendung in der Militär- und Civil-Technik.* Gerold in Comm, Wien (1869); (ii) *Explosive Nitrilverbindungen: Dynamit und Schießwolle, deren Eigenschaften und Verwendung in der Sprengtechnik.* Gerold in Comm, Wien (1870); (iii) *Die Dynamite, ihre Eigenschaften und Gebrauchsweise sowie ihre Anwendung in der Landwirthschaft und im Forstwesen.* Wiegandt, Hempel & Parey, Berlin (1876); (iv) *Dynamite. Ihre ökonomische Bedeutung und ihre Gefährlichkeit.* Lehmann & Wentzel, Wien (1876); (v) *Zur Schlagwetter-Frage.* Spiess & Co, Wien (1885).

[1225] F. WINKHAUS: *Versuche zur Ermittlung der Sprengwirkung von Sprengstoffen, ausgeführt auf der Versuchsstrecke der Westfälischen Berggewerkschaftskasse auf der Zeche Consolidation, Schacht I bei Schalke i.W.* Glückauf **31**, Nr. 50, 875-879 (22. Juni 1895).

| 1903 | Owens College, Manchester, England | **Harold B. Dixon,**[1226] a professor of chemistry, **publishes a brilliant memoir embodying his photographic research into the full course of a gaseous explosion,** from its initial phase of slow uniform flame movement up to its culmination in detonation.

(i) He first reviews the historical background of the field, discussing previous studies on the rate of movement of the flame and the pressures produced during the explosion of gases {Davy ⇨1815; Bunsen ⇨1867; Berthelot ⇨1881; Mallard & Le Châtelier ⇨1881 & 1883; Berthelot & Vieille ⇨1882 & 1883; Liveing[1227] & Dewar 1884}, and photographic records of the moving flame {von Oettingen & von Gernet ⇨1888}.

(ii) He illustrates the influence of compression waves and the collision of detonation waves.

(iii) **He reports on his discovery of "reflection waves,"** which arise when a detonation wave is either arrested by the closed end of a tube or is momentarily retarded upon passing a constriction inside it.

(iv) **Addressing the curious phenomenon of a backward-propagating "retonation wave,"** he writes, **"The strongly luminous wave thrown back from the point where the detonation is started I propose (*nominis egestate*) to call the 'retonation-wave.'** This wave has not the same constant characteristics that mark the detonation, but when generated under certain conditions it resembles detonation most closely … A study of a number of photographs leads to the conclusion that the retonation is faster and more luminous when no other bright waves have been thrown back by the advancing flame before the point of detonation is reached."

Investigation into the subject of retonation was resumed after World War II by numerous detonation researchers {*e.g.*, Oppenheim et al. ⇨1962} and discussed at International Symposia on Detonation, *e.g.*, by Kistiakowsky (1.1951), Clark & Schwartz (3.1960), Marlow (4.1965), Calzia & Carabin (5.1970), Held, Ludwig & Nikowitsch (6.1976), Chick & Hatt (7.1981), and Jing et al. (8.1985). Los Alamos detonation physicists used the term *reverse detonation* rather than *retonation* to describe the process of shocking an explosive up to a higher density and then having a high-pressure reflected shock initiate a propagating detonation back through the higher density explosive.[1228] |
| 1903 | Faculty of Philosophy, University of Basel, Switzerland | **Paul Emden,**[1229] a Swiss physicist, **studies steam jets emerging from orifices in detail using optical shadowgraphy,** a method which his brother Robert Emden[1230] had previously used in his investigations of gas jets.

(i) He uses weakly conical nozzles ranging from 0.3 to 3.63 mm in dia. as well as an original Laval nozzle provided by the German company Maschinenbauanstalt Humboldt AG in Köln-Kalk {⇨Fig. 4.8–C}.

(ii) He makes two important observations. First, that above a pressure of 1.8 at, stationary sound waves are generated in the steam jet that are very similar in appearance to the stationary waves previously observed by his brother in gas jets. Second, the "wavelength" of the periodic wave pattern increases as the pressure and the diameter of the nozzle exit increases. |

---

[1226] H.B. Dixon: *VIII. On the movements of the flame in the explosion of gases.* Phil. Trans. Roy. Soc. Lond. **A200**, 315-352 (1903). Regarding the term "retonation," *see* Part V (written in conjunction with R.H. Jones and J. Bower): *On the initiation of the detonation-wave and on the wave of "retonation,"* pp. 339-342.

[1227] G.D. Liveing and J. Dewar: *Spectroscopic studies on gaseous explosions.* Proc. Roy. Soc. Lond. **36**, 471-478 (1884).

[1228] Dr. Charles L. Mader, Honolulu, HI; private communication (Jan. 2006).

[1229] P. Emden: *Die Ausströmungserscheinungen des Wasserdampfes.* Ph.D. thesis, University of Basel (1903), published by R. Oldenbourg, München (1903).

[1230] R. Emden: *Über die Ausströmungserscheinungen permanenter Gase.* Habilitationsschrift, Königl. Technische Hochschule München (1898), published by Barth, Leipzig (1899); Ann. Phys. **69** [III], 246-289, 426-453 (1899).

| | | |
|---|---|---|
| 1903 | Polytechnikum, Zurich, Switzerland | Albert FLIEGNER,[1231] a Swiss professor of mechanical engineering, **repeats his (incorrect) view that the velocity of a jet of pressurized gas emerging from a nozzle cannot exceed the velocity of sound.** Based upon the periodic jet structure observed and published in the same year by Paul EMDEN in his doctoral thesis {EMDEN ⇨1903}, he argues that changes in density propagate with the velocity of sound, and hence the jet does too. • He continued to adhere to this view, which was criticized by Ludwig PRANDTL {⇨1908}, in subsequent years. |
| 1903 | Collège de France, Paris, France | Jacques S. HADAMARD,[1232] a French mathematician, **treats discontinuities mathematically in a general form:**<br>(i) **For an ideal gas he derives the Hugoniot curve** [French *loi adiabatique dynamique*] {HUGONIOT ⇨1887} as<br>$$\tfrac{1}{2}(p_1 + p_0)(v_0 - v_1) = (p_1 v_1 - p_0 v_0)/(\gamma - 1),$$<br>which, plotted in the $(p,v)$-plane, is steeper than POISSON's adiabatic law [French *loi adiabatique statique*] {POISSON ⇨1923; ⇨Fig. 2.1}.<br>(ii) While studying the works of G.F. Bernhard RIEMANN {⇨1859} and Pierre-Henri HUGONIOT {⇨1887}, he noticed that the shock front can be considered separately, and that it can be mathematically transformed by a particular simple procedure not connected to any specific problem. This procedure can be fully described by the **"identity and kinematic conditions,"** as well as their derivatives. He postulates, "If a function of the coordinates and of time, together with all its derivatives, is defined both outside of and at the surface of discontinuity, then the rule for compound differentiation can be applied to it at the surface of discontinuity."<br>(iii) **He presents a general theory of their characteristics** and distinguishes them as propagation paths of vanishingly small shock waves, since the energy defect across them becomes zero. • Referring to HADAMARD's theorem, Theodore VON KÁRMÁN,[1233] a Hungarian-born U.S professor of aeronautical engineering, later annotated, "According to his theorem, a vortex-free flow ahead of a shock wave can remain vortex-free after passing through the shock only when the wave is straight. If the shock wave is curved, it produces vorticity. This is a fact which makes the analysis of motion behind a shock wave rather complicated."<br>(iv) HADAMARD also uses the terms *onde de choc* ("shock wave") to illustrate the wave-type character of this phenomenon, and *onde d'accélération* ("acceleration wave") – an ordinary wave with a continuous increase in velocity behind the front, but with a discontinuous velocity gradient at the front.[1234] |
| 1903 | Institute of Technical Physics, University of Göttingen, Germany | Hans LORENZ,[1235] a German fluid dynamicist, **is the first to analytically treat stationary flow though a divergent tube.** He shows that<br>▸ a maximum flow velocity is only possible in a tube where the cross-section increases;<br>▸ for the special case of cylindrical divergent tubes, the maximum velocity is attained at the tube exit; and<br>▸ a single cross-section exists where the maximum flow velocity is identical to the sound velocity, and this depends on the local thermodynamic properties of the applied gas. |

---

[1231] A. FLIEGNER: *Noch einmal die Düse der DE LAVAL'schen Dampfturbine*. Schweiz. Bauz. **XLI**, 175-177 (1903).
[1232] J. HADAMARD: *Leçon sur la propagation des ondes et les équations de l'hydrodynamique*. A. Hermann, Paris (1903).
[1233] Th. VON KÁRMÁN: *Aerodynamics: selected topics in the light of their historical development*. Cornell University Press, Ithaca, NJ (1954), p. 119.
[1234] E. JOUGUET: *Mécanique des explosives*. Doin, Paris (1917), pp. 210-239.
[1235] H. LORENZ: *Die stationäre Strömung von Gasen durch Rohre mit veränderlichem Querschnitt*. Physik. Z. **4**, 333-337 (1903).

| | | |
|---|---|---|
| 1903 | Academy of Sciences, Krakow, Poland | **Marian M. SMOLUCHOWSKI**,[1236] a Polish physicist and pioneer in the field of fluctuation theory, **presents a memoir on the present state of theoretical aerodynamics to the Academy.** He also addresses the *aerodynamics of compressible flows* – a branch of physics later termed *gas dynamics* {ACKERET ⇨1927} which will result in a wealth of new nonstationary flow compression and high-temperature phenomena. |
| 1903 | Institut für Thermische Maschinen, Polytechnikum, Zurich, Switzerland | **Aurel B. STODOLA**,[1237] a Hungarian–Swiss mechanical engineer, **publishes his famous book *Die Dampfturbinen und die Aussichten der Wärmekraftmaschinen*** ("The Steam Turbines and the Future of Heat Engines"), in which steam turbines are treated in great detail. ▪ His book became a classic and was translated into English (1905) and French (1927).<br><br>(i) **His book also contains the first studies on the characteristics of supersonic flow through a cylindrical divergent nozzle** {⇨Fig. 4.7–G}. In his nozzle testing set-up, he was able to vary the back-pressure over any range desired by closing a valve downstream of the nozzle exit. ▪ The possibility of supersonic flow in such nozzles, although established theoretically, had not yet been verified experimentally, and was therefore a matter of some controversy.<br><br>(ii) While measuring the pressure distribution along the nozzle axis at different back-pressures, he noticed a sequence of steep pressure increases and concluded, "I see in these extraordinary heavy increases of pressure a realization of the 'compression shock' theoretically derived by RIEMANN, because steam particles of great velocity strike against a slower moving steam mass and are therefore compressed to a higher degree." ▪ **Each zone of maximum pressure is visible in a photograph as a vertical line with respect to the nozzle axis, which was called a "barrier line"** [Germ. *Staulinie*] **by Carl J. CRANZ.**[1238]<br><br>A historical review on the outflow of gases and steam from orifices was given by Ludwig PRANDTL in 1903.[1239] |
| 1903 | Kaluga [a city 190 km southwest of Moscow], central Russia | **Konstantin Eduardovich TSIOLKOVSKY**,[1240] a Russian teacher of mathematics and an early rocket pioneer, **publishes his most important work, entitled** (in translation): **"Exploration of Cosmic Space by Means of Reaction Devices."** It is the world's first scientific publication on this complex subject. He clearly outlines how a reaction thrust motor could use the Third Law of Motion {Sir NEWTON ⇨1687} to allow humans to escape from Earth {⇨Fig. 4.7–H}, which represents the **birth of astronautics!** Apparently inspired by the French science fiction writer Jules VERNE {⇨1865}, TSIOLKOVSKY accurately describes the state of weightlessness, the theoretical function of rockets in a vacuum, and the problem of heating resulting from air friction. He demonstrates why rockets would be needed for space exploration, and also advocates the use of liquid propellants, which are still used today. ▪ In his later writings, he correctly determined that the velocity required for an object to escape from the Earth into orbit is about 8 km/s {Sputnik 1 ⇨1957}, and speculated that this could be achieved by using **a multi-stage rocket fueled by liquid oxygen and liquid hydrogen.** |

---

[1236] M.M. SMOLUCHOWSKI: *Sur les phénomènes aérodynamiques et les effets thermique qui les accompagnent*. Bull. Int. Acad. Sci. de Cracovie No. 3, 143-182 (1903).

[1237] A.B. STODOLA: *Die Dampfturbinen und die Aussichten der Wärmekraftmaschinen*. Springer, Berlin (1903, 1904, 1905, 1910, 1922, 1924); *Steam turbines, with an appendix on gas turbines and the future of heat engines*. Van Nostrand, New York (1905, 1906) and P. Smith, New York (1927); *Turbines à vapeur et à gaz*. Dunod, Paris (1925).

[1238] C. CRANZ: *Lehrbuch der Ballistik: II. Innere Ballistik*. Springer, Berlin (1926), pp. 190-196.

[1239] L. PRANDTL: *Strömende Bewegung der Gase und Dämpfe*. In: *Encyclopädie der mathematischen Wissenschaften*. Teubner, Leipzig (1903–1921); see vol. V, 1 (1903), chap. 18: *Ausströmen aus Öffnungen und Mundstücken*, pp. 287-319.

[1240] K. TSIOLKOVSKY: *Exploration of cosmic space by means of reaction devices* [in Russ.]. Nautchnoye Obozreniye ("Scientific Review," St. Petersburg) **5** (1903). ▪ Due to political problems in Russia this paper was not published until 1911, where it appeared in Vozkukhoplavania, a Russian aircraft magazine.

| | | |
|---|---|---|
| 1904 | Ecole des Mines, Saint-Étienne, Dépt. Loire, France | **J.C. Émile JOUGUET,**[1241] a French professor of mechanics, **derives an expression for the change in entropy in a small-amplitude shock wave in terms of the second derivative of the specific volume** $v$ **with respect to the pressure** $p$, $(\partial^2 v/\partial p^2)_S$. Since the adiabatic curve in the pressure-volume diagram is curved downward for practically all substances – *i.e.*, this expression is always negative – **he concludes that a rarefaction shock is impossible** {ZEMPLÉN ⇨1905; DUHEM ⇨1909}. ▪ Yakov B. ZEL'DOVICH {⇨1946}, however, showed theoretically that rarefaction shocks are actually possible, which was also later proven experimentally {KUTATELADZE ET AL. ⇨1978}. In addition, William E. DRUMMOND {⇨1957} showed that a rarefaction shock should be possible in matter undergoing a phase transition, which produces a kink in the Hugoniot curve. |
| 1904 | Owens College, University of Manchester, England | **Horace LAMB,**[1242] a British physicist and mathematician, **solves the theoretical problem of surface waves excited by an impulsive line or point load acting on an homogeneous linear elastic half-space.** He finds that the surface disturbance can be roughly divided into two parts:<br>(i) **a minor tremor,** consisting of both longitudinal and transverse waves, which starts with some abruptness and may be described as a long undulation leading up to the main shock and decaying gradually after this has passed; and<br>(ii) **the main shock,** which propagates as a solitary wave {RUSSELL ⇨1834} at the velocity found by Lord RAYLEIGH {⇨1887}.<br>LAMB's contribution – the proof of the existence of so-called **"Lamb waves"** – is of fundamental importance to theoretical seismology. |
| 1904 | Institute of Applied Mechanics, University of Göttingen, Germany | **Ludwig PRANDTL,**[1243] a German engineer and fluid dynamicist, **begins a study on wave propagation phenomena inside and outside of nozzles of various geometries when high-pressure air is exhausted through them:**<br>(i) Starting from RIEMANN's shock theory {RIEMANN ⇨1859}, he gives a quantitative explanation for the periodic formation of stationary waves in free air jets {SALCHER ⇨1889}: expansion waves originating at the edge of the outlet are reflected at the boundary of the free jet as compression waves, which in turn are reflected as expansion waves. This process repeats periodically, thus resulting in crossed lines – later termed ***shock diamonds.***<br>(ii) He also deduces the "wavelength" of the crossed wave pattern in the photograph from the ratio $c/w$ ($w$ = supersonic flow velocity along the axis, $c$ = sound velocity at that state), which can be estimated with sufficient accuracy from the inclinations of the characteristic lines with respect to the axis of the jet using DOPPLER's law, $\sin\alpha = c/w$ {DOPPLER ⇨1847}. ▪ PRANDTL {⇨1913} **later termed this angle the** *MACHscher Winkel* **("Mach angle").**<br>In the same year, at the Third International Congress on Mathematics in Heidelberg, PRANDTL[1244] **proposes his idea of a** *Grenzschicht* **("boundary layer") near the surface of a body moving through a fluid.** ▪ His concept proved extraordinarily fruitful for the development of fluid mechanics. During World War II, some high-speed aerodynamicists considered removing some of the boundary layer air by sucking it through a porous surface or a number of slots in order to increase the laminar stability, delay transition, and reduce drag.[1245] The |

---

[1241] E. JOUGUET: *Remarques sur la propagation des percussions dans les gaz.* C. R. Acad. Sci. Paris **138**, 1685-1688 (1904).

[1242] H. LAMB: *On the propagation of tremors over the surface of an elastic solid.* Phil. Trans. Roy. Soc. Lond. **A203**, 1-42 (1904).

[1243] L. PRANDTL: *Über stationäre Wellen in einem Gasstrahl.* Physik. Z. **5**, 599-601 (1904); *Beiträge zur Theorie der Dampfströmung durch Düsen.* VDI-Z. **48**, 348-350 (1904).

[1244] L. PRANDTL: *Über Flüssigkeitsbewegung bei sehr kleiner Reibung.* In: (A. KRAZER, ed.) *Verhandl. des 3. Int. Mathematiker-Kongresses* [Heidelberg, Germany, Aug. 1904]. Teubner, Leipzig (1905), pp. 484-491.

[1245] H.L. DRYDEN: *Mechanics of boundary layer flow.* In: (R. VON MISES and Th. VON KÁRMÁN, eds.) *Advances in applied mechanics.* Academic Press, New York (1948), pp. 1-40.

flow in a boundary layer may likewise be laminar or turbulent, and the flow patterns and locations of shock waves are dependent on the type of flow in the boundary layer.[1246]

1904 — TH Dresden, Dresden, Saxony, Germany

Maximilian TOEPLER,[1247] a German physicist and the eldest son of August TOEPLER, **visualizes and photographs spark waves** (*i.e.*, weak shock waves) **using his father's schlieren method** {A. TOEPLER ⇨1864}. ▪ Although his research interests shifted towards electrical spark phenomena rather than shock waves in later years, he still maintained an interest in the schlieren method. In the 1930s, he wrote a review article on the schlieren method,[1248] and became a consultant to the German Rocket Center at Peenemünde, which had installed a large field-of-view schlieren system at the new supersonic wind tunnel {HVA Peenemünde ⇨1939}.

1904 — Ries Basin, northern Bavaria, Germany

Ernst WERNER,[1249] a German amateur geologist based at Gmünd, **first suggests that the Ries Basin** [Germ. *Nördlinger Ries*] {⇨*c.*15 Ma ago; ⇨Fig. 4.1−C] **was caused by an impact rather than by volcanism.** He writes, "Another 80 km away [from the almost horizontal mesa of the *Schwäbische Alb*] brings us to the puzzling, nearly circular valley of the Ries which, while collapsed into the White Jura, doesn't show any essential features of exterior volcanism. A huge amphitheatre of 25 km in dia. In its oddness, it reminds one of the strange ring mountains of the Moon, and it is easy to think that the origins of these as well as of the Ries Basin are due to the same causes … Surveying the processes uncovered by scientific research in the Ries Alb I could not help feeling that the disaster came suddenly and from outside of the area and that it must have taken an unimaginable long period of evolution on our planet, while it was exposed to innumerous and even more violent catastrophes." ▪ At that time, however, many German geologists, such as Walter KRANZ[1250] and Richard LÖFFLER,[1251] still favored the volcanic explosion hypothesis, which explained the shattered rock fragments with glassy inclusions as volcanic ejecta.

In the case of the Ries Basin, **the impact breccia was called *Suevit* ("suevite") by the German geologist Adolf SAUER,**[1252] although it is popularly known in southern Germany as *Schwabenstein* {⇨Fig. 4.1−C}. There are two kinds of suevite: material which was ejected into the air by the meteorite impact and then fell back into the crater ("fall-back suevite") and suevite that fell outside the rim ("fall-out suevite"). The 15th century parish church (which has a tower 90 m high) of St. Georg at Nördlingen, a town located in the center of the Ries Basin, was built mostly from suevite.

1904– 1906 — Ecole des Mines, Saint-Étienne, Dépt. Loire, France

Émile JOUGUET,[1253] a professor of mechanical engineering and a mathematician, **formulates his detonation theory** after having studied detonation in great detail. **He correctly postulates that a detonation wave comprises a shock wave followed by a combustion wave,** and independently arrives at very similar conclusions to those of CHAPMAN {⇨1899}. Assuming that the detonation products are at thermodynamic equilibrium, and using previously measured data for heat capacities at high temperatures, JOUGUET calculates the velocity of the detonation

---

[1246] W. GRIFFITH: *Shock tube studies of transonic flow over wedge profiles.* J. Aeronaut. Sci. **19**, 249-264 (1952).
[1247] M. TOEPLER: *Objektive Sichtbarmachung von Funkenschallwellen nach der Schlierenmethode mit Hilfe von Gleitfunken.* Ann. Phys. **14** [IV], 838-842 (1904); *Neue, einfache Versuchsanordnung zur bequemen subjektiven Sichtbarmachung von Funkenschallwellen nach der Schlierenmethode.* Ibid. **27** [IV], 1043-1050 (1908).
[1248] M. TOEPLER: *Schlierenmethode.* In: (R. DITTLER and G. JOOS, eds.) *Handwörterbuch der Naturwissenschaft.* G. Fischer, Jena (1931–1935), vol. VIII, pp. 924-929.
[1249] E. WERNER: *Das Ries in der schwäbisch-fränkischen Alb.* Blätter des Schwäbischen Albvereins **16**, 154-167 (1904).
[1250] W. KRANZ: *Zum Nördlinger Ries-Problem.* Centralblatt Mineralogie, Stuttgart (1923), pp. 278-285, 301-309.
[1251] R. LÖFFLER: *Der Eruptionsmechanismus im Ries.* Z. Dt. Geol. Gesell. **78**, 177-178 (1926).
[1252] A. SAUER: *Petrographische Studien an den Lavabrocken aus dem Ries.* Jahreshefte des Vereins für vaterländische Naturkunde in Württemberg **57**, LXXXVIII (1901).
[1253] E. JOUGUET: *Sur l'onde explosive.* C. R. Acad. Sci. Paris **139**, 121-124 (1904); *Sur la propagation des réactions chimiques dans les gaz.* J. Math. Pures Appl. **1** [VI], 347-425 (1905); Ibid. **2** [VI], 5-86 (1906).

wave for various gaseous mixtures and obtains good agreement with previously measured values {BERTHELOT & VIEILLE ⇨1883; DIXON ⇨1893}. He concludes that
- the chemical reaction at the detonation front occurs instantaneously from unburnt into burnt gas {⇨Fig. 4.16−S};
- at the end of the "reaction zone" (which is some distance behind the detonation front), the detonation products approach a state of chemical equilibrium;
- the detonation approaches a steady state − *i.e.*, the detonation products propagate at constant velocity;
- the flow is laminar and one-dimensional; and
- behind the detonation front, the velocity of the detonation products with respect to this front is equal to the local velocity of sound.

The contributions of CHAPMAN and JOUGUET have been commemorated by detonation physicists by the following terms:

(i) CHAPMAN's and JOUGUET's widely applicable theory was first called the **"Chapman-Jouguet (CJ) theory"** (1899–1905) by LEWIS & FRIAUF {⇨1930}. This theory is essentially hydrodynamical and furnishes no satisfactory explanation for the extreme rapidity at which chemical reactions occur in gaseous explosions. The CJ theory is governed by four equations – the three conservation equations {HUGONIOT ⇨1887}, Eqs. (1) − (3), and the *CJ hypothesis,* Eq. (4):[1254]

$$\text{Momentum:} \quad p_0 + \rho_0 D^2 = p_{CJ} + \rho_{CJ}(D - u_{CJ})^2 \tag{1}$$

$$\text{Mass:} \quad \rho_0 D = \rho_{CJ}(D - u_{CJ}) \tag{2}$$

$$\text{Energy:} \quad e_0 + \tfrac{1}{2} D^2 + p_0 v_0 = e_{CJ} + \tfrac{1}{2}(D - u_{CJ})^2 + p_{CJ} v_{CJ} - Q \tag{3}$$

$$\text{CJ hypothesis:} \quad \rho_0 D^2 = \rho_{CJ}^2 a_{CJ}^2 = \rho_{CJ} \gamma p_{CJ} \tag{4}$$

Here $\rho$ is the density, $p$ is the pressure, $a$ is the sound velocity, $D$ is the detonation speed, $u$ is the flow speed, $e$ is the internal energy, $Q$ is the reaction energy, and $\gamma = c_p/c_v$. The indices "0" and "CJ" refer to the initial state and the CJ point, respectively.

(ii) The state of the exploded gas immediately behind the explosion wave is termed the *Chapman-Jouguet state.* It is the point where the Rayleigh line {Lord RAYLEIGH ⇨1878} or Mikhel'son line {MIKHEL'SON ⇨1893}, drawn through the initial state at $p_0$ and $1/\rho_0$, is tangent to the Hugoniot of the reaction products. This point − also termed the *Chapman-Jouguet point* − is the end-state for self-sustaining detonation waves. At this point, the velocities of the reaction zone, reaction products and the shock front in the solid explosive are all the same, and so this is the only point on the detonation Hugoniot where a steady-state detonation can occur.[1255] Many experimental studies in gaseous explosive mixtures have shown that measured detonation velocities are in excellent agreement with the Chapman-Jouguet point.

(iii) The *Chapman-Jouguet condition* is usually expressed by the equation

$$(dp/dv)_H = -(p - p_0)/(v - v_0) \approx -p/(v_0 - v),$$

in which $(dp/dv)_H$ denotes the gradient of the curve which represents the Hugoniot in the $p$-$v$ diagram. Here $p$ and $p_0$ are the pressures, while $v$ and $v_0$ are the specific volumes in the shock-wave front and at rest, respectively. The pressure of the unexploded material, $p_0$, is completely negligible compared with $p$.

(iv) The *Chapman-Jouguet model* assumes a homogeneous reaction layer; however, most surprisingly, later studies on detonation waves in gases instead showed that complicated patterns occurred, such as a **spinning structure** {CAMPBELL & WOODHEAD ⇨1926; BONE & FRASER ⇨1929} or a **periodic cell structure** {DENISOV & TROSHIN ⇨1959; SHCHELKIN & TROSHIN ⇨1963}. The **reaction zone** is an important attribute of an explosive, since it greatly

---

[1254] R. SCHALL: *On the hydrodynamic theory of detonation.* Propellants, Explosives, Pyrotechnics **14**, 133-139 (1989).
[1255] P.W. COOPER: *Explosives engineering.* Wiley-VCH, New York (1997).

influences its detonation velocity $D$. Robert J. FINKELSTEIN[1256] and George GAMOW determined the length of the reaction zone for several explosives. For example, for nitroglycerin ($D$ = 7,600 m/s) and TNT ($D$ = 6,940 m/s) the length of the reaction zone amounts to 0.4 mm and 3 mm, respectively; *i.e.*, **slower reacting explosives have longer reaction zones.**

| | | |
|---|---|---|
| 1905 | *Meteor Crater, northern Central Arizona* | **Daniel M. BARRINGER**, a Philadelphia mining engineer and geologist, **and Benjamin Chew TILGHMAN**,[1257] a Philadelphia mathematician and applied physicist, **suggest that *Meteor Crater* {⇨$c$.50,000 years ago; ⇨Fig. 4.1–B} has an extraterrestrial origin.** ▪ BARRINGER, together with his friend TILGHMAN, formed the Standard Iron Company and drilled for iron at Meteor Crater from 1903 to 1905. Although their drilling proved fruitless, they provided the first evidence for the impact origin of the crater. Meteor Crater – then known as "Coon Mountain Butte" or "Coon Butte" – is today officially named **"Barringer Crater."** |

At that time, many U.S. geologists still assumed that the huge hole in Arizona – about 4,000 ft (1,219 m) in dia. and 600 ft (183 m) deep, and considered to be "the most mysterious geologic feature in the West" – was produced by a violent **underground steam explosion.** The steam was assumed to have accumulated in the pores of the sandstone until the limit of the stone was eventually reached. However, in the official *Guidebook of the Western United States*,[1258] edited by the USGS ten years after BARRINGER's suggestion, a footnote states cautiously, "The cause of this great hole in the ground has not been ascertained. Several geologists believe that it was made by the impact of a great meteor, a view suggested by the occurrence of many small masses of meteoric iron in the vicinity, as well as elsewhere in the surrounding country. But a mining company organized to find and work the large mass supposed to be buried in the hole failed to obtain any evidence of its existence…"

| | | |
|---|---|---|
| 1905 | *Cambridge Engineering School, University of Cambridge, England* | **Bertram HOPKINSON**,[1259] a British engineer and materials dynamicist, **repeats previous experiments on the dynamic strength of steel wires performed by his father John HOPKINSON** {⇨1872}, but instead uses an apparatus which enables him to measure the maximum strain at the top of the wire, and to apply smaller weights so that the rate of exponential decay in the tail of the stress wave is very much more rapid. **He reconfirms that the tensile strength of steel wires is indeed much greater under rapid conditions than when loaded statically.** |

In the early 1940s, this important result stimulated much research into dynamic elastic behavior of solids and shock-loaded materials, which stimulated the development of the **"one-dimensional finite amplitude theory"** on the dynamic plasticity of metals, as derived by Geoffrey I. TAYLOR,[1260] Theodore VON KÁRMÁN,[1261] and Khalil A. RAKHMATULIN.[1262]

| | | |
|---|---|---|
| 1905 | *United Kingdom* | **E.A. IZOD**,[1263] a British physicist, **proposes a single-blow impact test using a swinging hammer to measure the resistance of a metal or plastic specimen to impact.** However, unlike the Charpy impact test {CHARPY ⇨1901}, the specimen stands erect, like a fence post. *Izod impact* **is defined as the kinetic energy needed to initiate and continue fracture until the specimen is broken.** ▪ The "Izod impact test" is used mainly in Anglo-Saxon countries, and has become the most common test of its kind in the United States. The energy absorbed, as |

---

[1256] R. FINKELSTEIN and G. GAMOW: *Theory of the detonation process.* NavOrd Rept. No. 90-46, Navy Dept., Bureau of Ordnance (1947).
[1257] D.M. BARRINGER and B.C. TILGHMAN: *Coon Mountain and its crater.* Proc. Acad. Natl. Sci. (Philadelphia) **57**, 861-886 (1905).
[1258] N.H. DARTON ET AL. (eds.): *Guidebook of the western United States.* USGS Bull. No. 613. Govt. Print. Office, Washington, DC (1915), chap. *The Santa Fe route*, pp. 112-113.
[1259] B. HOPKINSON: *The effects of momentary stresses in metals.* Proc. Roy. Soc. Lond. **A74**, 498-506 (1905).
[1260] G.I. TAYLOR: *The plastic wave in a wire extended by an impact load.* Brit. Minist. Home Sec. Civ. Def., Res. Com., Rept. R.C. 329 (1942); *The testing of materials at high rates of loading.* J. Inst. Civ. Eng. **26**, 486-518 (1946).
[1261] Th. VON KÁRMÁN: *On the propagation of plastic deformation in solids.* Rept. OSRD-365 (1942).
[1262] K.A. RAKHMATULIN: *Propagation of a wave of unloading* [in Russ.]. Prikl. Mat. Mekh. (SSSR) **9**, 91-100 (1945).
[1263] (E.A. IZOD) *Impact testing of notched bars.* The Engineer **99**, 249-250 (March 10, 1905).

measured by the subsequent rise of the pendulum, is a measure of impact strength or notch toughness. There are two different kinds of Izod tests:

- The *notched* Izod test is best applied when determining the impact resistances of parts with many sharp corners (such as ribs, intersecting walls and other stress risers), and it has become the standard test for comparing the impact resistance of plastic materials.
- The *unnotched* Izod test uses the same loading geometry, except that a notch is not cut into the specimen.

CHARPY's and IZOD's impact tests are widely used to characterize the impact resistances of plastics, and are therefore of great engineering importance. The history of impact testing was reviewed recently.[1264]

| | | |
|---|---|---|
| 1905 | Dept. of Physics, University of Breslau, Lower Silesia, Germany | Otto R. LUMMER,[1265] a German physicist, **publishes his *Theorie des Knalls* ("Shock Theory"):**<br><br>(i) As he briefly outlines, he approaches the shock problem using the Huygens principle of wavefront propagation {HUYGENS ⇨ 1678} and by referring to Christian DOPPLER's method of wavefront construction for a rapidly moving sound source {DOPPLER ⇨ 1847}. • It seems that John VON NEUMANN {⇨ 1942} conceived of a similar approach to modeling shock wave propagation as LUMMER, but obviously did not follow it up.<br><br>(ii) **He also speculates that whip cracking is a supersonic phenomenon.** • Throughout the 19th century, it had been assumed that the sharp report from a whip was due to the creation of a vacuum by the rapidly moving whip which was suddenly and violently filled by the surrounding air. This concept is similar to that illustrated by an implosion experiment occasionally shown in public lectures, where a report is produced by destroying an evacuated thin-walled glass sphere [Germ. *Knallkugel*].[1266] The first successful attempts to tackle the riddle of whip cracking were performed until the 1920s {CARRIÈRE ⇨ 1927}. |
| 1905 | New York City, New York | Hiram Percy MAXIM,[1267] a U.S. gunsmith and son of the legendary engineer Sir Hiram Stevens MAXIM, the father of the automatic portable machine gun (1885), **invents the first silencer for small firearms.** His "Maxim silencer" gun attachment {⇨Fig. 4.15–A}, for which he obtained U.S. and German patents in 1910, is based on the modern concept of a multiple baffle arrangement that is screwed onto the barrel end.[1268] In his legendary indoor demonstrations, he proves that the attachment efficiently reduces the muzzle blast, and founds the Maxim Silent Firearms Company. • Silencers were innovatively marketed as the "gentleman's way of target shooting." They turned out to be quite popular, and sold quite well in the United States throughout the 1920s and into the early 1930s, until the advent of high taxation by the Federal Firearms Act in 1934. However, interest from military circles in introducing "silent firearms" into the army was small.<br><br>A list of twenty U.S. patents for firearm silencers granted in the period 1888–1959 is given in a U.S. Army handbook.[1269] |

---

[1264] T.A. SIEWERT ET AL.: *The history and importance of impact testing.* In: (T.A. SIEWERT and M.P. MANAHAN, eds.) *Pendulum impact testing: a century of progress.* ASTM, West Conshohocken, PA (1999).

[1265] O.R. LUMMER: *Über die Theorie des Knalls.* Jb. Schles. Gesell. Vaterl. Cultur **83** [II], 2-11 (1905).

[1266] *Bilder-Conversations-Lexikon für das Deutsche Volk.* Brockhaus, Leipzig; *see* chap. *Knall*, vol. 2 (1838), pp. 620-621.

[1267] H.P. MAXIM: *Silent firearms.* U.S. Patent No. 880,386 (March 1910); *Knalldämpfer.* Germ. Patent No. 220,470 (June 1910).

[1268] K.R. PAWLAS: *Entwicklung und Konstruktion der Schalldämpfer, Teil III.* In: *Waffengeschichte.* Chronica-Reihe W123, Publizistisches Archiv, Nürnberg (1983).

[1269] C.J. LANG (ed.): *Engineering design handbook.* Published in: *Guns series, muzzle devices.* AMC Pamphlet No. 706-251. Headquarters, U.S. Army Materiel Command, Washington, DC (May 1968); *see* Appendix A-6.

| | | |
|---|---|---|
| 1905 | Friedrich-Wilhelms-Universität, Berlin, Germany | **Walther Hermann NERNST,**[1270] director of the Institut für Physikalische Chemie, **studies the shock phenomenon of "knocking"** [Germ. *Klopfen, Schlagzündung*] **in reciprocating internal combustion engines.** ▪ NERNST – an avid automobile fan since the late 1890s, when he became the owner of one of the first automobiles in Göttingen – dedicated much time to this problem and arrived at the conclusion that this phenomenon might be due to the creation of a detonation wave. His correct hypothesis initiated a long period of international research into this important technical problem.<br><br>NERNST, who helped establish modern physical chemistry, also worked on theories related to thermodynamics, chemical equilibria and photochemistry {BODENSTEIN ⇨1913; WENDTLAND ⇨1924}. He received the 1920 Nobel Prize for Chemistry for his formulation of the Third Law of Thermodynamics. |
| 1905 | University of Göttingen, Germany | **Győző ZEMPLÉN,**[1271] a Hungarian physicist and visiting scientist from the Múegyetem (Royal Josephs Polytechnic) in Budapest, **publishes a pioneering paper on the nature of shock waves in gases.**<br><br>(i) Considering an ideal gas with constant specific heat, he shows that the entropy changes in a shock wave: it rises with increasing pressure and falls with decreasing pressure. **From this, he concludes that a rarefaction shock (across which entropy would have to diminish) is not possible – the so-called "Zemplén theorem."**<br><br>(ii) In his paper, **he gives the first concise and modern definition of a shock wave:** "A shock wave is a surface of discontinuity propagating in a gas at which density and velocity experience abrupt changes. One can imagine two types of shock waves: (positive) compression shocks which propagate into the direction where the density of the gas is a minimum, and (negative) rarefaction waves which propagate into the direction of maximum density." |
| 1906 | San Francisco, northern California | Early in the morning of April 18, **the city of San Francisco and a large area around it are hit by a series of heavy earthquake shocks with a maximum magnitude of 7.8.**<br><br>(i) The slippage along the San Andreas Fault, the boundary between the North American and the Pacific tectonic plates, begins offshore just west of San Francisco Bay. From that epicenter, the rupture spreads along the fault to the southeast and the northwest. Slippage occurred along almost 500 km of the fault ▪ Based upon a discrepancy between magnitude data obtained by seismic and geodetic measurements, **U.S. seismologists recently arrived at the conclusion that the slipping part of the fault moved faster than the shear-wave velocity in the crust of the San Andreas Fault** (about 3 km/s) – *i.e.*, supersonically (up to 5 km/s), a phenomenon called "supershear."<br><br>▸ In this case the seismic waves form an intense pressure pulse that is analogous to a sonic boom in the air. Such pressure pulses can do a lot more damage than the seismic waves that travel at lower speeds can.<br><br>▸ If the rupture propagated at much higher velocity than previously assumed, this would suggest an earthquake of shorter duration, which in fact was observed on faraway seismographs.[1272, 1273]<br><br>(ii) Ruptured gas mains quickly catch fire, and the broken water mains make it necessary to fight the fire with dynamite. For three days, uncontrolled fires rage and about a third of the city |

---

[1270] W. NERNST: *Physikalisch-chemische Betrachtungen über den Verbrennungsprozeß in den Gasmotoren*. VDI-Z. **49**, 1426-1431 (1905).
[1271] G. ZEMPLEN: *Sur l'impossibilité d'ondes de choc négatives dans les gaz*. C. R. Acad. Sci. Paris **141**, 710-712 (1905); Ibid. **142**, 142-143 (1906). ▪ Also see the footnote of JOUGUET {⇨1901}.
[1272] S.G. SONG, G.C. BEROZA, and P. SEGALL: *Evidence of supershear rupture during the 1906 San Francisco Earthquake*. Paper presented at the American Geophysical Union conference (San Francisco, CA, Dec. 2005).
[1273] R. PROPPER: *Earthquake: 1906 revisited a new model for an old quake*. Stanford Scient. Mag. **4**, No. 2 (Dec. 2005); http://www.stanford.edu/~dgermain/volume4-2/articles4-2/earthquakeRevisited.html.

is razed {⇨Fig. 4.1–E}. • Of the city's 400,000 inhabitants, about 3,000 were killed and about 225,000 were injured; the damage to property amounted to about $400 million in 1906 US $.[1274]

After this event – later called the "Great 1906 San Francisco Earthquake" – **Harry F. Reid,**[1275] a U.S. professor of dynamic geology at Johns Hopkins University in Baltimore, **developed his *elastic rebound theory.***

(I) It explained the relative displacements that cause an earthquake as being the result of a gradual build-up of strain until the strength of the rock is exceeded, the rock ruptures and the two sides of the rupture suddenly rebound elastically – that is, they snap back to almost the positions that they should have occupied had there been no strain.

(II) He based his conclusions on various geodetic surveys conducted in the San Francisco region over a 55-year period before the 'quake. His last survey, performed just after the 'quake, indicated that the same roads, fences and streams measured previously had been shifted dramatically. For example, in the vicinity of San Francisco, the roads and fences had been shifted horizontally by as much as 20 ft (6 m).

(III) **His model correctly described the generation of seismic shocks via the slippage of masses of rock along earth fractures – so-called "faults" – the most common mechanism of earthquakes.**

An interesting collection of photos of various types of faults and their effects has been compiled by the U.S. National Geophysical Data Center (NGDC), and can be accessed on the Internet.[1276] It encompasses examples of normal faults, lateral strike-slip faults, oblique slip faulting, and reverse and thrust faults.

| | | |
|---|---|---|
| 1906 | *Stuttgart, southern Germany* | In September, at the 78th Meeting of the Gesellschaft Deutscher Naturforscher und Ärzte (GDNÄ), **Ludwig Prandtl,**[1277] a German professor of applied mechanics at the University of Göttingen, **resumes his previous studies on supersonic wave propagation of gases and steam vented from nozzles** {Prandtl ⇨1904}. |

(i) He briefly reviews previous nozzle outflow problems: "The problem of the outflow of vessels was studied much earlier, when the remarkable fact was discovered [Reynolds ⇨1885; Hugoniot[1278] 1886] that the velocity at the narrowest part of the outflow opening cannot become larger than the sound velocity corresponding to the local state of the gas. Thereupon, and based upon the fact that waves have been observed in the exhaust jet, which were interpreted as usual sound waves, **the curious legend has spread that a larger velocity than the sound velocity can never be achieved in the case of gaseous flow through quiescent tubes and vessels.** Even now this view has found an eager advocate in Mr. A. Fliegner in Zurich [Fliegner ⇨1903; Fliegner[1279] 1906], although it has already been disproved by Stodola's measurements of the pressure distribution in flowing steam [Stodola ⇨1903] and by similar experiments."

(ii) Making use of the flow theory of stationary and adiabatic motion, where the sum of the kinetic and the potential energy is constant in each gas particle, he discusses the flow characteristics in a Laval nozzle.

---

[1274] *1906 San Francisco Earthquake and fire.* Zpub San Francisco pages; http://www.zpub.com/sf/history/1906earth.html.

[1275] H.F. Reid: *The elastic rebound theory of earthquakes.* California University Press, Berkeley, CA (1911).

[1276] *Geologic hazards photos, vol. 1: faults.* National Geophysical Data Center (NGDC), Boulder, CO; http://www.ngdc.noaa.gov/seg/cdroms/geohazards_v1/document/647010.htm.

[1277] L. Prandtl: *Neue Untersuchungen über die strömende Bewegung der Gase und Dämpfe.* Physik. Z. **8**, 23-30 (1907).

[1278] P.H. Hugoniot: *Sur le mouvement varié d'un gaz comprimé dans un réservoir qui se vide librement dans l'atmosphère.* C. R. Acad. Sci. Paris **103**, 1002-1004 (1886); *Sur un théorème relatif au mouvement permanent et à l'écoulement des fluides.* Ibid. **103**, 1178-1181 (1886); *Sur la vitesse limitée d'écoulement des gaz.* Séances Soc. Franç. Phys. (Paris), pp. 120-124 (1886).

[1279] A. Fliegner: *Beiträge zur Dynamik der elastischen Flüssigkeit.* Schweiz. Bauz. **XLVII**, Nr. 3, 30-32; Nr. 4, 41-46; Nr. 9, 103-110 (1906).

(iii) In his experiments PRANDTL used planar nozzles confined between two glass plates, which allowed him to perform schlieren photography of the formation of various wave phenomena, such as

> the intersection of shock waves inside a free jet {⇨Fig. 4.8–D}; and
> **the propagation of *Mach waves*** (*i.e.*, weak shock waves produced by small disturbances in the flow which propagate with sound velocity). **They cause *Mach lines*** (expansion waves) **inside the divergent section of a super-sonic nozzle** {⇨Fig. 4.8–E} – thus indicating the need to consider appropriate area ratio distributions to obtain uniform supersonic flow.

| | | |
|---|---|---|
| 1906 | *Lehmanns Verlag, Munich, Bavaria, Germany* | **Foundation of the German journal *Zeitschrift für das gesamte Schieß- und Sprengstoffwesen*,** in order to "improve and promote the communication between science and industry, and to advance the development and application of explosives and propellants." The editor-in-chief is Richard ESCALES. ▪ It was the second international journal that to be exclusively dedicated to the rapidly growing field of explosives, ballistics and shock waves {*Mémorial des Poudres et Salpêtres* ⇨1882}. The journal was published from 1906 (vol. 1) to 1944 (vol. 39). |
| 1906 | *Sprengstoff-AG Carbonit, Schlebusch/Köln, Germany* | **Christian E. BICHEL,**[1280] a German explosive specialist, **obtains a British patent for his invention of a *plastic explosive*.** ▪ His motivation was to develop an explosive which could easily be placed into shells, torpedoes and mines without introducing (detrimental) cavities. His plastic explosive consisted of a mixture of trinitrotoluene (or TNT) and resins. |
| 1906 | *Laboratoire de la Commission des Substances Explosives, Paris, France* | **Henri Joseph DAUTRICHE,**[1281] a French chemist and blasting engineer, **describes a simple method of measuring the detonation velocity of a test explosive:** (i) His "difference method" {⇨Fig. 4.17–I} uses a match [French *cordeau*] of known detonation velocity (6,500 m/s) placed on a lead plate, with its two ends inserted into the test cartridge at a known distance. The ends are ignited by the passage of the detonation wave in the cartridge. When the two waves in the cordeau meet, they make a sharp furrow in the lead plate, and the displacement of this furrow from the midpoint of the cordeau is a measure of the detonation velocity in the cartridge. ▪ His method greatly resembles Ernst MACH's interference method {E. MACH & WENTZEL ⇨1885}, which he used to determine the propagation velocities of explosion waves. (ii) For confined dynamite, he measures detonation velocities ranging from 1,991 to 6,794 m/s, depending on the initial density of the cartridges. ▪ The detonation velocity of an explosive first increases with increasing density, passes a maximum value, and then decreases with increasing density. DAUTRICHE, using his difference method, demonstrated this behavior on the example of cheddite (a French explosive consisting of a high proportion of inorganic chlorates mixed with nitroaromatics plus a little paraffin or castor oil as a moderant for the chlorate). |
| 1906 | *University of Bordeaux, Dépt. Gironde, France* | **Pierre M.M. DUHEM,**[1282] a French physicist and scientific philosopher, **analytically demonstrates that true shock waves** – *i.e.*, waves that have a discontinuous front according to RIEMANN's and HUGONIOT's theory – **are only stable in a perfect fluid with zero viscosity.** In real fluids with small but finite viscosities, only **"quasi shock waves"** are possible. |

---

[1280] C.E. BICHEL: *Process of producing a plastic explosive from trinitrotoluol*. Brit. Patent No. 16,882A (1906).
[1281] H.J DAUTRICHE: *Sur les vitesses de détonation*. C. R. Acad. Sci. Paris **143**, 641-644 (1906); Ibid. **144**, 1030-1032 (1907); *Cheddites au perchlorate d'ammoniaque*. Mém. Poudres Salpêtres **14**, 206-233 (1906/1907), chap. VI: *Vitesse de détonation*, pp. 215-218.
[1282] P.M.M. DUHEM: *Sur les quasi-ondes de choc*. C. R. Acad. Sci. Paris **142**, 324-327, 377-380, 491-493, 612-616, 750-752 (1906).

| | | |
|---|---|---|
| 1906 | University of Göttingen, Germany | **Ludwig PRANDTL,**[1283] a German professor of applied mechanics, **makes a first estimation of shock front thickness.** Starting from heat conduction processes in the transition layer of an ideal gas of constant viscosity and heat conductivity, he calculates a shock front thickness of 0.5 µm for a shock wave in air with a pressure jump of 0.2 atm, and states that "**the thicknesses of shock layers range within the wavelengths of visible light** [about 0.39–0.78 µm]." |
| 1907 | Consolidated Coal Company mine at Monongah, Marion County, West Virginia | On December 6, **an underground firedamp explosion kills 362 miners, the deadliest mining disaster in American history.**[1284] ▪ The explosion was thought to have been caused by the ignition of firedamp. This in turn ignited the highly flammable coal dust, which is found in all West Virginia bituminous coal mines. What ignited the "black damp" (methane) is unknown although two theories emerged: carelessness with an open lamp or a dynamite blast gone wrong. |
| 1907 | Servicio Sismológico de Chile, Santiago, central Chile | **Fernand Marie BERNARD, Comte DE MONTESSUS DE BALLORE,**[1285] a French seismologist who pioneered seismology in Chile, (correctly) **states that tsunamis can originate from phenomena which are indirectly seismic in origin, such as slides,** "Tsunamis can originate only from phenomena which are indirectly seismic; for example, slides [French *éboulements*] … In the Assam Earthquake of 1897, large landslides occurred, and there is nothing hypothetical about the idea that large waves will result from submarine slides. Thus, this is not only a possible but even a sure cause of the production of tsunamis." |
| 1907 | Ecole Nationale des Mines, Saint Etienne, Dépt. Loire, France | **Louis CRUSSARD,**[1286] a French physicist and detonation researcher, **is the first to apply the Rankine-Hugoniot equations** {HUGONIOT ⇨1887} **to a reactive fluid,** using a graphical representation.[1287]<br>(i) He suggests that the explosion wave is composed of a shock wave and a combustion wave [French *"une onde de choc et combustion"*], and that the flame moving in the shock front burns the combustible gas completely [*"une réunion d'une onde de choc et d'une flamme brûlant totalement le gaz combustible"*].<br>(ii) According to his model, the shock and flame front propagate with a velocity equal to the speed of sound in the medium that follows them (*i.e.*, the burnt gas) – thus anticipating the **supplementary Chapman-Jouguet (CJ) relation** {JOUGUET ⇨1917}. |
| 1907 | The Lanchester Motor Company, Birmingham, Werwickshire, England | **Frederick William LANCHESTER,**[1288] a British automotive engineer, scientist and inventor, **anticipates the fundamental principle governing supersonic flight.** In his two-volume book *Aerodynamics*, he states, "In the extreme case when the velocity of flight becomes equal to the velocity of sound, no disturbance can precede the aerofoil in its flight, and the whole reaction will be due to the communication of downward momentum; the cyclic component in the motion round the wing vanishes." |
| 1907 | Leeds & Northrop Company, Philadelphia, Pennsylvania | **Edwin F. NORTHRUP,**[1289] a U.S. electrothermic engineer and manufacturer of electrical instruments, **reports at the New York meeting of the American Physical Society on experiments in which he passed a large current through liquid mercury.** |

---

[1283] L. PRANDTL: *Zur Theorie des Verdichtungsstoßes*. Z. ges. Turbinenwesens **3**, 241-245 (1906).
[1284] *Monongah mine disaster*. West Virginia Archives & History; http://www.wvculture.org/HISTORY/disasters/monongah03.html.
[1285] F.M. BERNARD DE MONTESSUS DE BALLORE: *La science séismologique*. A. Colin, Paris (1907).
[1286] L. CRUSSARD: *Ondes de choc et onde explosive*. Bull. Soc. Industrie Minérale **6** [IV], 257-364 (1907); *Propriété de l'onde explosive*. C. R. Acad. Sci. Paris **144**, 417-420 (1907).
[1287] M. ROY: *Sur la contribution de Louis CRUSSARD à la théorie de la propagation des combustions et à ses applications*. Ecole Polytechnique, Paris; http://www.annales.org/archives/x/cru.html.
[1288] F.W. LANCHESTER: *Aerodynamics, constituting the first volume of a complete work on aerial flight*. Archibald Constable & Co., London (1907); *Aerodynamik: ein Gesamtwerk über das Fliegen*. Teubner, Leipzig & Berlin (1909).
[1289] E.F. NORTHRUP: *Some newly observed manifestations of forces in the interior of an electric conductor*. Phys. Rev. **24** [I], 474-497 (1907).

(i) **He observed that a pressure difference exists between the axis and the outer edge of the column, resulting in a constriction of the liquid metal conductor.**

(ii) He writes, "Some months ago, my friend Carl HERING [a U.S. electrical engineer] described to me a surprising and apparently new phenomenon which he had observed. He found, in passing a relatively large alternating current through a non-electrolytic, liquid conductor contained in a trough, that the liquid contracted in cross-section and flowed up hill lengthwise of the trough, climbing up upon the electrodes ... Mr. HERING suggested the idea that this contraction was probably due to the elastic action of the lines of magnetic force which encircle the conductor, which lines, he said, acted on the conductor like stretched rubber bands, tending to compress it, especially at its weakest point. **As the action of the forces on the conductor is to squeeze or pinch it, he jocosely called it the 'pinch phenomenon.'**" ▪ HERING also observes longitudinal forces, which had previously been neglected by others. A discussion on longitudinal forces in conductors has been conducted over the last decade, although such discussions are not new – they have occurred periodically since the time of Ampère.[1290]

The theory for a constricted gas current was first developed in 1934 by Willard Harrison BENNETT,[1291] a U.S. physicist at Ohio State University. The *pinch effect,* a term coined by TONKS {⇨1937}, has also been used to produce high dynamic pressures of some 100 kbar in small solid cylindrical samples {BLESS ⇨1972}.

| | | |
|---|---|---|
| 1907 | *Institute of Applied Mechanics, University of Göttingen, Germany* | **Ludwig PRANDTL,** a German professor of applied mechanics, **establishes the *Modellversuchsanstalt* ("Model Testing Facility") at Göttingen University.** ▪ The facility was later renamed the *Aerodynamische Versuchsanstalt (AVA) Göttingen* ("Aerodynamic Testing Facility Göttingen").[1292, 1293]<br><br>Two years later, PRANDTL[1294] reported on the construction of the first *closed-circuit wind tunnel.* ▪ The most important characteristic of his tunnel concept is that the air circulates in a closed duct, which is a more energy-efficient approach than using an open tunnel. The tunnel had a large test section of $2 \times 2$ m$^2$, but a rather low speed of 10 m/s. The second tunnel, built in 1917, also had a closed circuit, and reached a maximum speed of 50 m/s. The principle of this so-called **"Göttingen wind tunnel"** (or "Prandtl wind tunnel") was later extended into the supersonic regime by the Swiss aerodynamicist Jakob ACKERET {⇨1933}. |
| 1908 | *Podkamennaja Tunguska [ca. 3,400 km east of Moscow], Siberia, Russia* | On June 30, at about 7 A.M. local time, **an unidentified object enters the Earth's atmosphere above the Stony Tunguska and generates a huge blast wave equivalent to the energy liberated by the explosion of about $10^7$ tons of TNT equivalent.**<br><br>(i) Accounts from the town of Kansk (located about 600 km south from the impact site) and from the village of Kuriski-Popovich (in the Kansk District) state that "a first shock caused the doors, windows and votive lamp to shake, a minute later a second shock followed, accompanied by subterranean rumbling," and "a severe earthquake and two loud bursts, like the firing of a large caliber gun, were observed in the vicinity," respectively.<br><br>(ii) Seismic shocks are recorded at numerous observatories, such as at Jena, Hamburg, Irkutsk, Tashkent and Tiflis, but the fall of the meteor is not immediately brought to the notice of the scientific world, although strange barometric phenomena resulting from the blast wave are recorded in England {SHAW ⇨1908}. ▪ The **"Tunguska meteorite"** devastated an unpopu- |

---

[1290] L. JOHANSSON: *Longitudinal electrodynamic forces – and their possible technological applications.* M.S. thesis, Dept. of Electroscience, Lund University, Sweden (Sept. 1996).
[1291] W.H. BENNETT: *Magnetically self-focusing streams.* Phys. Rev. **45** [II], 890-897 (1934).
[1292] L. PRANDTL, C. WIESELSBERGER, and A. BETZ: *Ergebnisse der Aerodynamischen Versuchsanstalt zu Göttingen.* R. Oldenbourg, München (1923–1927).
[1293] J.C. ROTTA: *Die Aerodynamische Versuchsanstalt in Göttingen – ein Werk Ludwig PRANDTLs.* Vandenhoeck & Ruprecht, Göttingen (1990).
[1294] L. PRANDTL: *Die Bedeutung von Modellversuchen für die Luftschifffahrt und Flugtechnik und die Einrichtungen für solche Versuche in Göttingen.* VDI-Z. **53**, 1711-1719 (1909).

lated, about 720-square-mile (1850-km$^2$) area of Siberian forest, but did not form a crater.[1295] The Tunguska event was probably due to the explosion of a relatively small body in the atmosphere. Christopher F. CHYBA[1296] (a NASA meteorite researcher) and his collaborators have estimated that the Tunguska event may have been caused by the explosion of a stony meteoroid about 30 meters in diameter that was traveling at about 15 km/s; it released an energy of 10–20 megatons TNT equivalent at an altitude of about 10 km.

In the 1970s, **Victor P. KOROBEINIKOV**[1297] (a Soviet applied mathematician and a member of the Soviet Academy of Sciences) **and coworkers investigated the various geophysical effects of the Tunguska explosion.** They numerically simulated the blast of the meteorite in the atmosphere via an explosion of a semi-infinite cylindrical charge with a variable specific energy along its axis, which is inclined at an angle to the Earth's surface. To determine the propagation of the shock waves, they took into account the nonuniformity of the atmosphere. They also solved the inverse problem based on information obtained about the zone of forest flattening – the main consequence of the Tunguska meteorite explosion. ▪ It is still not clear whether the great explosion was due to a comet, an asteroid or something else (made of ice or rock).[1298] Since no impact crater has been found and, more tellingly, patches of trees were left untouched near the epicenter of the blast, a German researcher proposed a geophysical hypothesis (such as a gas explosion, because a large natural gas deposit lies below the site). Andrei Yu. OL'KHOVATOV, a Russian researcher, believes that the Tunguska event was not an impact of a stony asteroid/meteorite or a comet, but a manifestation of geophysical (terrestrial) process: roughly speaking, a result of coupling between tectonic and atmospheric processes in very rare combination of favourable geophysical factors.[1299]

| 1908 | *University of Jena, Thuringia, Germany* | Felix AUERBACH,[1300] a German theoretical physicist, **reviews the present state of the art in physical acoustics, thereby also addressing the enormous progress achieved in supersonics since ANTOLIK's gliding spark soot experiments** {ANTOLIK ⇨1874}. Under the headings *Aeromechanik und Anomalien der Ausbreitungsgeschwindigkeit* ("Aeromechanics and Anomalies of the Propagation Velocity"), he discusses many notable early contributions to instationary gas dynamics, which was a rapidly expanding branch of fluid dynamics at the time.

Subsequently, further handbook articles on compressible flow were prepared by Ludwig PRANDTL[1301, 1302] (1913 & 1921), Jakob ACKERET {⇨1927}, Adolf BUSEMANN[1303] (1931), and Albert BETZ[1304] (1931). |
|---|---|---|
| 1908 | *Compagnie des Omnibus, Paris, France* | René LORIN,[1305] a French engineer, **obtains a French patent on his *propulseur à réaction* ("propulsive duct") – a compressorless jet engine that he then proposes for use in aeronautics.**[1306] Based on the ram effect, this new "ramjet" engine is supposed to derive its thrust by the addition and combustion of fuel with air compressed solely as a result of the forward speed of the engine, using a specially shaped intake (an "air-breathing jet engine"). ▪ In France, |

---

[1295] E.L. KRINOV: *Giant meteorites.* Pergamon Press, New York (1966).
[1296] C.F. CHYBA, P.J. THOMAS, and K.J. ZAHNLE: *The 1908 Tunguska explosion: atmospheric disruption of a stony asteroid.* Nature **361**, 40-44 (1993).
[1297] V.P. KOROBEINIKOV, P.I. CHUSHKIN, and L.V. SHURSHALOV: *Mathematical model and computation of the Tunguska meteorite explosion.* In: *Cinquième Colloque sur la Gazodynamique des Explosions et des Systèmes Réactifs* [Bourges, France, Sept. 1975]. Acta Astronaut. **3**, 615-622 (1976).
[1298] (Note): *Tunguska revisited.* Science **285**, 1205 (Aug. 20, 1999).
[1299] (Note): *More theories on Tunguska.* Science **297**, 1803 (Sept. 13, 2002).
[1300] F. AUERBACH: *Allgemeine Physik* (vol. I), *Akustik* (vol. II). In: (A. WINKELMANN, ed.) *Handbuch der Physik.* Barth, Leipzig (1908/1909).
[1301] L. PRANDTL: *Gasbewegung.* In: (E. KORSCHELT, ed.) *Handwörterbuch der Naturwissenschaften* **4**, 544-560. Fischer, Jena (1913).
[1302] L. PRANDTL: *Strömende Bewegung der Gase und Dämpfe.* Enzyklopädie der mathematischen Wissenschaften **V** (1), 287-319. Teubner, Leipzig (1903–1921).
[1303] A. BUSEMANN: *Gasdynamik.* In: (L. SCHILLER, ed.) *Handbuch der Experimentalphysik.* Akad. Verlagsgesell., Leipzig, vol. IV, 1 (1931), pp. 343-460.
[1304] A. BETZ: *Gasdynamik.* In: *Hütte. Des Ingenieurs Taschenbuch.* Ernst & Sohn, Berlin, vol. I (1931), pp. 412-422.
[1305] R. LORIN: *Système de propulsion.* French Patent No. 390,256 (May 1908).
[1306] R. LORIN: *La propulsion à grande vitesse des véhicules aériens: Etude d'un propulseur à réaction directe.* L'Aérophile **17**, 463-465 (1909).

the ramjet is called *"statoréacteur,"* a word created in 1945 by Maurice ROY (before the terms *trompes thermopropulsives* or *tuyères thermopropulsives* were used). The British adopted *athodyd* (contraction of aerothermodynamic duct) before using ramjet. In Germany, the ramjet was named the *Lorinflugrohr* or *Staustrahltriebwerk*. In the Soviet Union, the ramjet was named the *PVRD (prjamotyènyj vozdušno-reaktyvnyj dvigati)*.[1307]

At that time, only five years after the first motor flight {WRIGHT Bros. ⇨1903}, applications of LORIN's engine were difficult to envisage. However, the ramjet provides a simple and efficient means of propulsion for aircraft moving at relatively high supersonic flight speeds. It is, however, quite inefficient at the transonic flight speeds that it was originally intended for by LORIN. The first successful application of the ramjet to flight was not achieved until 1945, when supersonic flight was maintained by a ramjet developed by the Applied Physics Laboratory at Johns Hopkins University and associated contractors under the sponsorship of the U.S. Navy Bureau of Ordnance. A review of early ramjet development was given by the U.S. physical chemist William H. AVERY.[1308] The principle of the ramjet was later developed into the scramjet {BILLIG ⇨1959; NASA-LaRC ⇨1998; HyShot Program ⇨2002; Hyper-X ⇨2004}.

| 1908 | Somerset House, London, England | At a meeting of the Royal Society of London, **Henry Reginald A. MALLOCK,**[1309] a British consulting engineer and physicist, **first reports on crumpling phenomena observed when thin-walled metal tubes are axially crushed by impact.**[1310] Under end-compression, the surface of the cylinder becomes unstable and folds are generated, which can be either circular in plan and independent of $\theta$ (the angle which the radius makes with a fixed dia. of the tube), square-shaped in plan; or hexagonal in plan. He concludes that "the crushing force requisite therefore undergoes periodic variations, being a maximum at the beginning of the formation of a new fold and a minimum when the fold is nearly completed." |
|---|---|---|

Investigation into this curious crumpling effect was resumed in the following years by a number of applied mathematicians and mechanical engineers, who were pursuing quite different goals, such as:

▸ to obtain a physical understanding of the crumpling process of very thin cylinders, and to estimate the amount of energy absorbed by the deformation of the diamond pattern to the folded rhomboid pattern;[1311]

▸ to determine by an approximate analysis the load under which a thin cylinder of given diameter and shell thickness begins to collapse;[1312]

▸ to apply this effect to car bumper design (especially by using certain kinds of plastic materials in tubular form, which recover their shapes completely after a large amount of compression);[1313] and in the crush analysis of vehicle structures (which also involves considering the axial impact crushing of thin square tubes);[1314]

▸ to compare existing theories on the dynamic and static axial crushing of circular cylinders to empirical results, and to work out the influence of strain rate effects;[1315] and

▸ to determine the energy dissipated by a thin-walled column.[1316]

---

[1307] P. KUENTZMANN and F. FALEMPIN: *Ramjet, scramjet & PDE – an introduction.* ONERA, France; http://www.onera.fr/conferences/ramjet-scramjet-pde/#ramjet.

[1308] W.H. AVERY: *Twenty-five years of RAMJET development.* Jet Propulsion **25**, 604-614 (1955).

[1309] A. MALLOCK: *Note on the instability of tubes subjected to end pressure, and on the folds in a flexible material.* Proc. Roy. Soc. Lond. **A81**, 388-393 (1908).

[1310] For example, a photo of this unusual crumpling phenomenon is shown on the front cover of Norman JONES' book *Structural Impact* {JONES ⇨1989}.

[1311] A. PUGSLEY and M. MACAULAY: *The large-scale crumpling of thin cylindrical columns.* Quart. J. Mech. Appl. Math. **13**, 1-10 (1960).

[1312] J.M. ALEXANDER: *An approximate analysis of the collapse of thin cylindrical shells under axial loading.* Quart. J. Mech. Appl. Math. **13**, 10-15 (1960).

[1313] W. JOHNSON: *Crashworthiness of vehicles.* Mechanical Engineering Publ. Ltd., London (1978), pp. 29-30.

[1314] H.C. WANG and D. MEREDITH: *The crush analysis of vehicle structures.* Int. J. Impact Engng. **1**, 199-225 (1983).

[1315] W. ABRAMOWICZ and N. JONES: *Dynamic axial crushing of circular tubes.* Int. J. Impact Engng. **2**, 263-281 (1984); *Dynamic progressive buckling of circular and square tubes.* Ibid. **4**, 243-270 (1986).

| | | |
|---|---|---|
| 1908 | Institute of Applied Mechanics, University of Göttingen, Germany | Theodor MEYER,[1317] one of PRANDTL's Ph.D. students, resumes PRANDTL's previous nozzle flow studies on stationary waves in gas jets {PRANDTL ⇨ 1904 & 1907}. **PRANDTL and MEYER recognize that a series of expansion waves – the so-called "Prandtl-Meyer expansion fan" – originates at sharp corners. These waves accelerate and rarefy the flow while moving round the corner.**<br><br>(i) **They analytically treat the oblique interaction of shock waves and the deflection of a supersonic flow round a corner.** In their "Prandtl-Meyer theory of expansion" (1907/1908) they show that under certain conditions a supersonic flow can navigate around a sharp corner using this expansion, without causing a shock wave. This is caused by an acceleration of supersonic flow, while at subsonic speeds the flow round a sharp corner is decelerated, causing a reversal of flow in the boundary layer. ▪ The "Prandtl-Meyer expansion" often occurs at the point of maximum thickness of supersonic shapes, such as the double wedge (rhombus) wing {HILTON ⇨ 1952}. The equations of Prandtl-Meyer expansions for both perfect and imperfect gases have been tabulated by the Ames Research Staff {Ames Aeronautical Laboratory ⇨ 1953}.<br><br>(ii) The so-called **"Prandtl-Meyer angle"** is the angle through which a supersonic stream turns to expand from $M = 1$ to $M > 1$.<br><br>(iii) From the laws of the conservation of mass, momentum, and energy for very small (differential) deflections, MEYER derives the so-called **"Prandtl-Meyer function,"** and presents shock wave tables of pressure ratios for various angles of incidence and reflection. ▪ The Prandtl-Meyer function is useful for calculating the Prandtl-Meyer expansion fan. Contrary to a shock wave, across which the Mach number decreases and the static pressure increases, the Mach number increases across an expansion fan, the static pressure decreases, and the total pressure remains constant. |
| 1908 | Institute of Geophysics, University of Göttingen, Germany | **Ludger MINTROP,**[1318] a German seismologist, **begins the development of the first portable, highly sensitive seismograph – the so-called "Mintrop seismograph."**[1319] ▪ His seismograph was used during World War I to locate Allied artillery firing positions: by setting up his seismographs opposite the Allied guns, their firing positions could be triangulated. After the war, MINTROP reversed the process, which resulted in the ***seismic reflection method*** {MINTROP ⇨ 1919} – a new technique that was used in geophysical prospecting, and good example of an efficient conversion of a military application into a civil one.<br><br>Independently of MINTROP, the American John Clarence KARCHER, an employee of the U.S. Bureau of Standards, invented a similar seismograph in 1917. Reginald Aubrey FESSENDEN, a Canadian-born U.S. radio pioneer and professional inventor, proposed a seismic method similar to MINTROP's one, and obtained a U.S. patent for it in 1919.[1320] |
| 1908 | Friedrich Krupp AG, Essen, Ruhr Area, Germany | **Fritz RAUSENBERGER,** a German professor of mechanical engineering and chief artillery designer at Krupp Company, **begins to design the *Kurze Marinekanone*** ("short naval gun") – a 420-mm howitzer with a range of 14 km, capable of firing up to 800-kg shells. ▪ The first gun was ready in August 1914. Used successfully in World War I by the Germans against Belgian |

---

[1316] V. LEIBER: *Modelluntersuchung zum Stauchverhalten von Längsträgern.* Masters thesis, Ecole Nationale Supérieure de Physique de Strasbourg (Sept. 1999). ▪ The work was carried out in 1999 at the Ernst-Mach-Institut (EMI), Freiburg, Germany.

[1317] T. MEYER: *Über zweidimensionale Bewegungsvorgänge in einem Gas, das mit Überschallgeschwindigkeit strömt.* Ph.D. thesis, University of Göttingen (1908), Sonderabdruck aus den *Mitteilungen über Forschungsarbeiten auf dem Gebiete des Ingenieurwesens* des VDI, published by Schade, Berlin (1908).

[1318] L. MINTROP: *Verfahren zur Ermittlung des Ortes künstlicher Erschütterungen.* Germ. Patent No. 304,317 (May 17, 1917); *Erforschung von Gebirgsschichten mittels künstlicher Erdbebenwellen.* Germ. Patent No. 371,963 (Dec. 7, 1919).

[1319] N. DOMENICO: *The Mintrop mechanical seismograph.* The Leading Edge **15**, 1049-1052 (Sept. 1996). ▪ Pictures of MINTROP's seismograph can also be seen in the Virtual Museum of the Society of Exploration Geophysics, Tulsa, OK; http://www.seg.org/museum/VM/Mintrop.html.

[1320] R.A. FESSENDEN: *Method of destroying enemy gun positions.* U.S. Patent No. 343,241 (1919).

| | | forts, it was nicknamed ***Dicke Bertha* ("Big Bertha"),** after Frau Bertha KRUPP VON BOHLEN UND HALBACH, head of the Krupp family {⇨Fig. 4.2–T}. |
|---|---|---|
| 1908 | *Dublin, Ireland* | At the annual meeting of the British Association for the Advancement of Science (BAAS), **William Napier SHAW** (later Sir NAPIER SHAW), a British meteorologist and secretary of the Meteorological Council at London, **discusses wave motion and shows microbarograms of a series of aerial waves**[1321] which were taken over the region from Cambridge to Petersfield on the day of the meteorite fall in Siberia {Stony Tunguska ⇨1908}. These records clearly reveal "the succession of **four undulations,** commencing with a range of about five thousandths of an inch [ca. 1.67 mbar], lasting about a quarter of an hour and then **violently interrupted by a sudden, though slight explosive disturbance,** which set up different and much faster oscillations for a similar interval." He concludes, "It would seem that the disturbance, if not simultaneous at the different places, traveled faster than 100 miles per hour [160 km/h]." |
| 1909 | *University of Bordeaux, Dépt. Gironde, France* | **Pierre M.M. DUHEM,**[1322] a French professor of physics and philosophy, **analytically investigates the thermodynamic state on each side of a shock front in terms of the first three derivatives of the entropy *s* with respect to the density $\rho$** – i.e., of $ds/d\rho$, $d^2s/d\rho^2$ and $d^3s/d\rho^3$. Depending on the ratio of specific heats (either $c_p/c_v < 3$ or $c_p/c_v > 3$) and the curvature of the isentrope in the $(p, v)$-diagram in terms of $dp/d\rho$ and $d^2p/d\rho^2$, **he first demonstrates that a shock wave may propagate as a condensation** if the isentropes in the $(p, v)$-diagram are curved up (as in the case of perfect gases); **or as an expansion** if the isentropes are curved down. Here $v$ is the specific volume (= $1/\rho$).<br><br>His conclusion resumed previous discussions {RANKINE ⇨1869; JOUGUET ⇨1904; ZEMPLÉN ⇨1905} regarding so-called **"rarefaction shocks,"** which, it was said, cannot form in real fluids since they would result in a decrease of entropy. However, they also stimulated further fruitful discussions on this subject in later decades {ZEL'DOVICH ⇨1946; KUTATELADZE ET AL. ⇨1987}. ▪ Recently, the Austrian fluid dynamicist Alfred KLUWICK[1323] reviewed previous investigations on the conditions under which rarefaction shocks may form in single-phase fluids in a handbook article. |
| 1909 | *University of Heidelberg, Germany* | **Carl W. RAMSAUER,**[1324] a German physicist and habilitation candidate at Prof. LENARD's institute, **experimentally and theoretically investigates the phenomenon of elastic percussion in detail:**<br><br>(i) He measures deformation during percussion, its duration, and velocities before and after percussion using an optical method and photographic recording {⇨Fig. 4.3–O}.<br><br>(ii) In order to improve the elasticity of the impacting bodies, he uses helical springs and natural rubber cylinders with ivory plates at their heads to achieve a homogeneous percussion pressure distribution. When this set-up is used, the elastic wave theory {DE SAINT-VENANT ⇨1867} is confirmed, thus showing that discrepancies with theory are due to imperfect elastic behavior.<br><br>(iii) He attempts to approach the ideal of perfectly elastic percussion by mounting springs on the heads of impacting steel cylinders {GRAVESANDE ⇨1722}.<br><br>(iv) He demonstrates that the resulting complex impact process can be divided into a shock due to percussion, and another (impairing) shock due to oscillation. |

---

[1321] F.J.W. WHIPPLE: *The great Siberian meteor and the waves, seismic and aerial, which it produced.* Quart. J. Roy. Meteorol. Soc. **56**, 287-304 (1930).
[1322] P.M.M. DUHEM: *Sur la propagation des ondes de choc au sein des fluides.* Z. Physik. Chem. **69**, 169-186 (1909). *See also* footnote of JOUGUET {⇨1901}.
[1323] A. KLUWICK: *Rarefaction shocks.* In: (G. BEN-DOR ET AL., eds.) *Handbook of shock waves.* Academic Press, New York (2001), vol. 1, pp. 339-411.
[1324] C. RAMSAUER: *Experimentelle und theoretische Grundlagen des elastischen und mechanischen Stoßes.* Ann. Phys. **30** [IV], 417-495 (1909).

| 1910 | U.S. Bureau of Mines, Pittsburgh, Pennsylvania | On May 16, **Congress creates the U.S. Bureau of Mines (USBM)** in order to promote improved safety in mining through research and training – in response to the growing number of fatalities in the coal mining industry which were partly attributed to coal dust explosions. ▪ Focusing on the nature of firedamp and dust explosions, USBM immediately initiated tests on explosives permissible for use in gaseous and dusty coal mines, and supported studies on the explosion properties of hundreds of different powders over a period of 60 years.<br><br>In 1995, USBM, a U.S. Government agency, was closed and certain functions partly transferred to other Federal agencies (such as USGS). |
|---|---|---|
| 1910 | Royal Seismological Observatory, Breslau-Krietern, Lower Silesia, Germany | Georg VON DEM BORNE,[1325] a German professor of geology at the TH Breslau and director of the seismological observatory, **discusses possible causes of anomalous sound propagation effects which have been observed in disasters caused by large explosions and during volcanic explosive eruptions** (e.g., those of Krakatau in Indonesia and Cotopaxi in Ecuador) – such as an inner *zone of normal audibility*, which is then followed by an approximately 100-km-wide *zone of silence* which, at a distance of about 100 to 200 km from the acoustic source, is surrounded by a *zone of abnormal audibility*. From the existence of these three zones he concludes that the sound must propagate along curved rays. He proposes the following model:<br><br>(i) In the lower altitudes of the atmosphere in which the temperature decreases in a vertical direction and in which the molecular weight of air is almost constant, acoustic rays curve upwards. At high altitudes, however, the molecular weight of the air decreases and the velocity of sound increases – thus the acoustic rays curve downwards, and consequently the sound returns to the ground.<br><br>(ii) Estimations show that the influence of wind on this curious audibility effect is comparatively small. |
| 1910 | Royal Colonial Institute, London, England | Vaughan CORNISH,[1326] a British scientific writer who spent half a century collecting data on waves and an associate of Owens College (Manchester, U.K.), **publishes his book *Waves of the Sea and Other Water Waves*.** ▪ For his contemporaries, his book was a rich mine of wave phenomena, and continues to be for modern fluid dynamicists: it contains photographs of all kinds of wave phenomena in nature – such as the **oblique interaction of hydraulic jumps in very shallow water, including Mach reflection** {⇨Fig. 4.1–M}, propagation phenomena of snow and sand waves, as well as water waves in rivers. |
| 1910 | Technische Hochschule Wien, Austro-Hungarian Empire | Karl KOBES,[1327] an Austrian mechanical engineer, **investigates the question of whether the application of shock waves could improve the performance of railway air-suction brakes.** This was an important practical problem at the time, particularly for long trains, which needed to stop the last cars from overrunning. **He publishes the first shock tube theory.** ▪ His "shock tube" – a term not yet coined {BLEAKNEY ET AL. ⇨1949} – was not a laboratory-type, smooth and straight pipe, but rather consisted of a test train with 71 cars (total length 746 m); i.e., it was provided with a common arrangement of brake hoses, conduits, elbows, joints, and valves. Using the shock arrival time measured at the last car, he determined an average shock wave velocity of 370 m/s, thus showing that fast brake activation can indeed be accomplished with supersonic velocities. |

---

[1325] G. VON DEM BORNE: *Über die Schallverbreitung bei Explosionskatastrophen.* Physik. Z. **11**, 483-488 (1910).
[1326] V. CORNISH: *Waves of the sea and other water waves.* T. Fisher Unwin, London (1910), Frontispiece and p. 173.
[1327] K. KOBES: *Die Durchschlagsgeschwindigkeit bei den Luftsauge- und Druckluftbremsen.* Z. österr. Ing.- & Architektenverein **62**, 553-579 (1910).

| 1910 | Chair of Applied Mechanics, University of Göttingen, Germany | Ludwig PRANDTL,[1328] a German professor of applied mechanics, **introduces a dimensionless parameter that characterizes fluid properties in the convection regime.** • This so-called *Prandtlsche Zahl* (**"Prandtl number"**) provides an analytical link between the velocity and the temperature field of a fluid and is defined as
$$P_r = \nu/\kappa.$$
Here $\nu$ is the kinematic viscosity, and $\kappa$ is the thermal diffusivity of a substance. A low Prandtl number – such as the value of 0.7 obtained for air and many other gases – indicates high convection. • In a shock wave, the distribution of the thermodynamic quantities through the shock front are dependent on heat conduction and viscosity {BECKER ⇨1922; MORDUCHOW & LIBBY ⇨1949} – hence they are governed by the Prandtl number. |
|---|---|---|
| 1910 | Private laboratory at Terling Place, Witham, Essex, England | Lord RAYLEIGH,[1329] England's leading acoustician, **publishes a lengthy memoir entitled *Aerial Plane Waves of Finite Amplitude*** – now an antiquated term for "plane shock waves in air." **He thoroughly reviews and comments on previous theories,** thereby also treating in detail the nature of the shock process in a viscous heat-conducting gas:

(i) Starting from the Navier-Stokes equations {NAVIER ⇨1822}, **he investigates possible influences of heat conduction on the shape of the discontinuity.** Resuming his earlier critiques on this subject {Lord RAYLEIGH ⇨1877 & 1878}, he states, "The problem now under discussion is closely related to one which has given rise to a serious difference of opinion. In his paper of 1848, STOKES [⇨1848] considered the sudden transition from one constant velocity to another, and concluded that the necessary conditions for a permanent regime could be satisfied … Similar conclusions were put forward by RIEMANN [⇨1859] in 1860. Commenting on these results in the *Theory of Sound* (1878), I pointed out that, although the conditions of mass and momentum were satisfied, the condition of energy was violated, and that therefore the motion was not possible; and in republishing this paper STOKES [⇨1877] admitted the criticism, which had indeed already been made privately by Lord KELVIN. On the other hand, BURTON [⇨1893] and H. WEBER [⇨1901] maintain, at least to some extent, the original view … Inasmuch as they ignored the question of energy, it was natural that STOKES and RIEMANN made no distinction between the cases where energy is gained or lost. As I understand, WEBER abandons RIEMANN's solution for the discontinuous wave (or a bore, as it is sometimes called for brevity) of rarefaction, but still maintains it for the case of the bore of condensation. No doubt there is an important distinction between the two cases; nevertheless, I fail to understand how a loss of energy can be admitted in a motion which is supposed to be the subject to the isothermal or adiabatic laws, in which no dissipative action is contemplated. In the present paper, the discussion proceeds upon the supposition of a gradual transition between the two velocities or densities. It does not appear how a solution which violates mechanical principles, however rapid the transition, can become valid when the transition is supposed to become absolutely abrupt. All that I am able to admit is that under these circumstances dissipative forces (such as viscosity) that are infinitely small may be competent to produce a finite effect."

(ii) **He derives a simple formula for estimating the shock front thickness $x$, which is on the order of $u \times \mu/\rho$, where $u$ is the velocity of the wave and $\mu/\rho$ the specific gas viscosity.** He writes, "For the present purpose we may take $u$ as equal to the usual velocity of sound; *i.e.*, $3 \times 10^4$ cm per second. For air under ordinary conditions the value of $\mu/\rho$ in c.g.s. measure is 0.13; so that $x$ is of the order of $\frac{1}{3} \times 10^{-5}$ cm. That the transitional layer is in fact extremely thin is proved by such photographs as those of BOYS [⇨1890], of the aerial wave of approximate discontinuity which advances in front of a modern rifle bullet; but that according to calculation this thickness should be well below the microscopic limit may well occasion surprise." |

---

[1328] L. PRANDTL: *Eine Beziehung zwischen Wärmeaustausch und Strömungswiderstand der Flüssigkeiten.* Phys. Z. **11**, 1072-1078 (1910).

[1329] Lord RAYLEIGH (J.W. STRUTT): *Aerial plane waves of finite amplitude.* Proc. Roy. Soc. Lond. **A84**, 247-284 (1910).

(iii) **He examines the conditions under which a compressive shock can propagate as a continuous steady wave.** He shows that when the gas has heat conduction but no viscosity, this is possible only for weak shocks; *i.e.*, with shock compression ≤ 1.4. With viscosity but no heat conduction, the continuous steady wave is always possible, and the same is presumably true with both viscosity and heat conduction.

| | | |
|---|---|---|
| 1910 | *Cavendish Laboratory, Cambridge, England* | Geoffrey I. TAYLOR,[1330] a 25-year-old British physicist, **theoretically investigates the thermodynamic conditions at the shock front:** (i) He extends Lord RAYLEIGH's approach {⇨1910} by including not only heat conduction, but viscosity as well, and establishes theoretically that a propagating sharp transition layer of permanent type is possible only when the pressure increases across the layer, and when diffusion processes operate in its interior. (ii) He demonstrates that in a real gas the discontinuity would be eliminated by dissipative effects (both viscosity and thermal heating). (iii) To obtain an estimate for the thickness of a shock wave, he constructs continuum equations and assumes a perfect gas with constant viscosity $\mu$ and heat conductivity $\kappa$. He shows that they can be solved exactly if either $\mu = 0$ or $\kappa = 0$, and approximately if the velocity jump across the layer is relatively small. His scientific paper (his second one) was awarded the Smith Prize for senior mathematics students at Cambridge University. |
| 1911 | *Harvard University, Cambridge, Massachusetts* | Percy W. BRIDGMAN,[1331] a U.S. physicist dedicated to static high pressure research, **describes a gauge for the measurement of static pressures based on the piezoresistive properties of manganin,** an alloy consisting of Cu (84%), Mn (12%) and Ni (4%). The resistance of manganin is shown to be a linear function of pressure up to 12 kbar and, by extrapolation, enables pressure measurements to be made with some certainty up to 20 kbar. ▪ BRIDGMAN later extended the linear range of the gauge response up to 30 kbar.[1332] After World War II, manganin gauges became an important diagnostic tool for measuring dynamic pressures (or stresses), such as in shock-loaded samples {HAUVER ⇨1960; FULLER & PRICE ⇨1962; BERNSTEIN & KEOUGH ⇨1964; BERNSTEIN ET AL. ⇨1967; GUPTA ET AL. ⇨1980}. In gas dynamics, however, piezoresistive gauges are hardly used, because the small changes of resistivity that occur in the low-pressure regime require high amplification and very careful compensation of the tremendous offset caused by temperature effects – even when carbon, which has a higher piezoresistive coefficient than ytterbium and manganin, is applied {⇨Fig. 4.12–C}. |
| 1911 | *New York City, New York* | Robert J. COLLIER, a prominent aviator and an early president of the Aero Club of America, **establishes the "Robert J. Collier Trophy" to encourage the American aviation community to strive for excellence and achievement in aeronautic development.** ▪ The award, first administered by the Aero Club of America and later by the National Aeronautic Association {NAA ⇨1913}, is presented annually "for the greatest achievement in aeronautics or astronautics in America, with respect to improving the performance, efficiency, and safety of air or space vehicles, the value of which has been thoroughly demonstrated by actual use during the preceding year." The first Collier Trophy was awarded in 1911 to Glenn H. CURTIS for developing the hydro airplane, which was the first airplane to use ailerons for lateral control. **The award was also** |

---

[1330] G.I. TAYLOR: *The conditions necessary for discontinuous motion in gases.* Proc. Roy. Soc. Lond. **A84**, 371-377 (1910).
[1331] P.W. BRIDGMAN: *The measurement of hydrostatic pressures up to 20,000 kilograms per square centimeter.* Proc. Am. Acad. Arts Sci. **47**, 321-343 (1911).
[1332] P.W. BRIDGMAN: *Physics above 20,000 kg/cm² [Bakerian Lecture].* Proc. Roy. Soc. Lond. **A203**, 1-17 (1950).

given to those who had made significant contributions to transonic/supersonic/hypersonic aerodynamic research and the testing of air or space vehicles, such as

- in 1948 to John STACK, Lawrence D. BELL and Chuck YEAGER {⇨1947} for determining the physical laws affecting supersonic flight;
- in 1951 to John STACK {⇨1951} and associates for the development and use of the slotted throat wind tunnel for transonic speed research;
- in 1953 to James H. KINDLEBERGER for development of the F-100 ("Super Sabre"), USAF's first operational aircraft capable of flying faster than the speed of sound (max. speed 864 mph or Mach 1.3 at an altitude of 35,000 ft);
- in 1954 to Richard T. WHITCOMB {⇨1951} for the development of the Whitcomb area rule used to reduce the increase in wing drag associated with transonic flight; and
- in 1961 to Robert M. WHITE, Joseph A. WALKER, A. Scott CROSSFIELD and Forrest PETERSEN, four U.S. test pilots of the X-15 airplane {⇨1959 & 1961}, for the scientific advances resulting from the X-15 test program.

| 1911 | Institute of Geophysics, University of Göttingen, Germany | Ludger MINTROP, a Ph.D. student of Prof. Emil WIECHERT, one of the most famous geophysicists of his time, finishes his thesis entitled *Über die Ausbreitung der von den Massendrucken einer Großgasmaschine erzeugten Bodenschwingungen* ("On the Propagation of Ground Movements Generated by a Large Gas Engine"). In his Ph.D. thesis, **MINTROP reproduces a seismogram of an artificial earthquake that was produced in 1908 by dropping a 4,000-kg iron ball from a 14-m-high-tower, impacting on rocky soil.**[1333] This enabled the first seismogram {⇨Fig. 4.3−S} showing the fine details of body waves (P- and S-waves) from a controlled source to be obtained. ▪ The dropping weight he used − the so-called "Mintrop ball" [Germ. *Mintrop-Kugel*] − and the tower have survived and can be seen at the University Campus Göttingen {⇨Fig. 4.3−S}. |
|---|---|---|
| 1912 | The Royal Institution, London, England | On January 26, in an Evening Discourse with Lord RAYLEIGH in the chair, **Bertram HOPKINSON,**[1334] a British engineer and son of the renowned engineer John HOPKINSON {⇨1872}, **gives a lecture entitled *The Pressure of a Blow*.**<br><br>(i) **He reports on new fracture phenomena** that occur in metal specimens when small quantities of explosive in contact with them are detonated. Using charges of guncotton placed upon steel plates, he observes that for plates with a thickness greater than ½ in. (12.7 mm), a circular disk of metal from the opposite side of the plate is broken away and thrown off; an effect which he calls **"scabbing"** {⇨Fig. 4.14−S}. This curious phenomenon of separation − the so-called **"Hopkinson effect"** (also known as "back-spalling" or "spallation")[1335] − occurs when a strong compressive shock of short duration is reflected from the back surface of a body, thus producing a tensile wave. ▪ Dynamic fracture based on spalling has become an interesting subject in modern ballistic research {ANTOUN ET AL. ⇨2003}.<br><br>(ii) **He provides details on experiments into the penetration of metal by bullets and shells,** including those showing that a hard-nosed missile striking armor plate shatters, while missile heads with soft metal caps covering the hard noses are able to penetrate the armor {CRANZ & SCHEEL ⇨1927}. |

---

[1333] L. MINTROP: *Über künstliche Erdbeben.* In: *Internationaler Kongress für Bergbau, Hüttenwesen, Angewandte Mechanik und Praktische Geologie* [Düsseldorf, Germany, June 1910]; see *Berichte der Abteilung für praktische Geologie.* Arbeitsausschuss des Kongresses, Düsseldorf (1910), pp. 98-112.

[1334] B. HOPKINSON: *The pressure of a blow* [Evening discussion on Jan. 26, 1912]. In: (J.A. EWING and J. LAMOR, eds.) *The scientific papers of Bertram HOPKINSON.* Cambridge University Press (1921), pp. 423-437; *A method of measuring the pressure produced in the detonation of high explosives, or by the impact of bullets.* Proc. Roy. Soc. Lond. **A89**, 411-413 (1914).

[1335] In nuclear physics, the term *spallation* has a differing meaning: it refers to a high-energy nuclear reaction that occurs when a target nucleus is struck by a high-energy particle (>50 MeV), thereby ejecting numerous lighter particles. The product nucleus is correspondingly lighter than the original nucleus. The term *spallation* has als been adopted by astrophysicists: *cosmic-ray-induced spallation* − a form of naturally occurring nuclear fission and nucleosynthesis − refers to the formation of elements from the impact of cosmic rays on an object.

(iii) He first describes his remarkable new method of measuring the effect of the detonation of high explosives as well as that of the impact of bullets – which later became known as the **"Hopkinson pressure bar"** {B. HOPKINSON ⇨ 1914}.

| | | |
|---|---|---|
| 1912 | *Institut für Radiumforschung, Wien, Austro-Hungarian Empire* | On April 17 during a solar eclipse, **Victor Franz Hess,**[1336] an Austrian physicist, **detects Ultragammastrahlung ("ultra γ-radiation") in a balloon-borne experiment with an ionization chamber.** ▪ HESS had already performed numerous balloon ascents during which he had measured this new type of radiation. However, since it was also observed during a solar eclipse, a solar origin for this radiation could be excluded with certainty. He was the first to show that the Earth is constantly being bombarded by ionizing radiation of **extraterrestrial origins** – later termed *cosmic rays*[1337] or, more appropriately, *cosmic radiation.* When cosmic radiation interacts with atoms in the Earth's atmosphere, cascades of secondary particles (such as electrons, protons, neutrons, neutrinos and muons) are produced due to numerous collisions.[1338] For his pioneering discovery, HESS received the 1936 Nobel Prize for Physics.<br><br>*Cosmic radiation* consists of a flux of particles (mostly hydrogen nuclei) with extremely high energies, reaching up to $3 \times 10^{20}$ eV (or 48 J),[1339] which is about $10^{21}$ times the energy of an air molecule at room temperature and about $10^8$ times the energy achievable by the most powerful earthbound accelerators.[1340] The origin of cosmic radiation is not yet known, but **astrophysicists speculate that these ultrahigh-energy particles are generated by supernova explosions, which drive strong,** *super-relativistic blast waves* **through the surrounding medium,** which in turn accelerate nuclei from the material they pass through, resulting in cosmic radiation.[1341, 1342] However, theorists also have proposed other possible sources for cosmic radiation, such as hypernovae, mutually annihilating neutron stars, radio galaxies, X-ray binaries (such as Cygnus X-3), active galactic nuclei (AGNs), and black holes. |
| 1912 | *Militärtechnische Akademie, Berlin, Germany* | **Captain Karl BECKER**[1343] **and Prof. Carl J. CRANZ,** two top German top ballisticians, **publish a series of quantified drag measurements on 8-mm projectiles moving at speeds ranging from 300 to 1,050 m/s.**<br><br>(i) **They use a new type of electric-spark photochronograph – the so-called "Cranz-Becker chronograph"** {⇨Fig. 4.19–J} – which enables precise measurement of the projectile velocity along its trajectory, an important precondition for determining aerodynamic drag data.<br><br>(ii) They publish a quantitative graph showing that the drag coefficient of a *Spitzgeschoss* or *S-Projektil* ("S-Projectile") {Prussian Army ⇨ 1903}<br>▸ is constant below 300 m/s;<br>▸ largely increases in the region between 300 and 400 m/s, thus exhibiting a typical transonic drag rise; and<br>▸ gradually decreases as the velocity increases above 400 m/s. |

---

[1336] V.F. HESS: *Beobachtungen der durchdringenden Strahlung bei sieben Freiballonfahrten.* Sitzungsber. Akad. Wiss. Wien **121** (Abt. IIa), 2001-2032 (1912).

[1337] The term *cosmic ray* is misleading. Rather than being an actual ray – a thin beam of radiant energy or particles – it is a single particle (usually of one the lightest elements, such as hydrogen and helium) which has been accelerated to enormous speeds.

[1338] J.W. CRONIN, T.K. GAISSER, and S.P. SWORDY: *Cosmic rays at the energy frontier.* Scient. Am. **276**, No. 1, 44-49 (Jan. 1997).

[1339] For Example, this enormous energy, packed into the size of a proton (rest mass $1.6726 \times 10^{-24}$ g), would correspondent to the kinetic energy of a cricket ball (160 grams) moving at 24.5 m/s (88 km/h).

[1340] Particles with energies of $>10^{20}$ eV strike the Earth's atmosphere at a rate of only about one per square kilometer per year.

[1341] M.E. DIECKMANN, K.G. MCCLEMENTS, S.C. CHAPMAN, R.O. DENDY, and L.O'C. DRURY: *Electron acceleration due to high frequency instabilities at supernova remnant shocks.* Astron. Astrophys. **356**, 377-388 (2000).

[1342] H. SCHMITZ, S.C. CHAPMAN, and R.O. DENDY: *The influence of electron temperature and magnetic field strength on cosmic ray injection in high Mach number shocks.* Astrophys. J. **570**, 637-646 (2002).

[1343] K. BECKER and C. CRANZ: *Messungen über den Luftwiderstand bei großen Geschwindigkeiten.* Artilleristische Monatshefte **69**, 189-196 (1912); Ibid. **71**, 333-368 (1912).

(iii) **Their graph reconfirmed early British drag measurements, performed for speeds of up to 610 m/s, that aerodynamic drag decreases after the sound barrier is exceeded** {HUTTON ⇨1763; BASHFORTH ⇨1864}, as well as later studies performed in England, The Netherlands and Russia for speeds of up to about 650 m/s and in Germany at Krupp Company at speeds of up to about 900 m/s {⇨Fig. 4.14–M}.

The analytical treatment of aerodynamic drag at high velocities presented a challenge to early fluid dynamicists, and various aerodynamic models were proposed. For example, the German physicist Hans LORENZ {⇨1903} modeled the channel of compressed gas produced by a bullet flying at high velocity by the transient flow in a tube with a varying cross-section. He thus derived a curve for aerodynamic drag *vs.* projectile velocity, which has two branches that asymptotically approach very large drag values at Mach one.

| | | |
|---|---|---|
| 1912 | *Geophysical Institute, University of Göttingen, Germany* | **Beno GUTENBERG,**[1344] **Karl ZOEPPRITZ and Ludwig GEIGER,** three German geophysicists, **make the first correct determination of the radius of the Earth's fluid core** while analyzing seismographic material originating from distant earthquakes. Using the strange phenomena of the late compressional wave (*primary wave* or *P-wave*) and the missing shear wave (*secondary wave* or *S-wave*), they conclude that the Earth has a fluid core. ▪ They obtain a value of 2,900 km for the depth at which the (outer) core starts, which is very close to the modern value of 2,889 km. **The so-called "Gutenberg discontinuity" marks the boundary between the Earth's mantle and the outer core.** However, the two distinct regions of the core – *i.e.*, the solid inner core almost entirely composed of iron (dia. *c.*2,550 km) and the liquid outer core composed mainly of a nickel-iron alloy (*c.*2,200 km thick) – were yet to be recognized.

Prior to GUTENBERG's studies, Prof. Emil WIECHERT, director of the Göttingen Geophysical Institute, had predicted that the Earth must have a central iron core. GUTENBERG and associates had based their research on data obtained with the horizontal seismograph, which WIECHERT {⇨1898} had invented for studying the Earth's structure, and which he had tested at the campus of Göttingen University using artificial earthquakes produced by dropping a 1,000-kg iron ball from a 14-m-high tower {MINTROP ⇨1911; ⇨Fig. 4.3–S}. |
| 1912 | *Kgl. Materialprüfungsamt, Groß-Lichterfelde, Berlin, Germany* | **Walter HÖNIGER,**[1345] a German engineer, **determines the force of percussion in impact tests using a *Martens precision pendulum*.**

(i) The test sample is mounted between the impacting ram and an anvil. Applying high-speed streak photography, he records the distance-time profile *s(t)* of a ram of mass *m* after it is struck.

(ii) By differentiating the curve *s(t)* twice, and using the Second Law of Motion {Sir NEWTON ⇨1687}, he calculates the force of percussion, given by $F_P = m\, d^2s(t)/dt^2$. |
| 1913 | *Elektrochemisches Institut, TH Hannover, Germany* | **Max E.A. BODENSTEIN,**[1346] an eminent German physical chemist, **first advances the concept of reactive intermediates as part of a chain reaction mechanism.**

(i) While studying the photochemically-induced chlorine-hydrogen explosion {GAY-LUSSAC & THÉNARD ⇨1809; BUNSEN ⇨1857}, he finds that the reaction velocity is proportional to the square of the chlorine concentration and inversely proportional to the oxygen concentration.[1347] |

---

[1344] B. GUTENBERG, K. ZOEPPRITZ, and L. GEIGER: *Über Erdbebenwellen. V. Konstitution des Erdinnern, erschlossen aus dem Bodenverrückungsverhältnis der einmal reflektierten zu den direkten Longitudinalwellen, und einige andere Beobachtungen über Erdbebenwellen.* Nachr. Königl. Gesell. Wiss. Gött. [Math.-Phys. Kl.] (1912), pp. 121-206; *Über Erdbebenwellen. VI. Konstitution des Erdinnern, erschlossen aus der Intensität longitudinaler und transversaler Erdbebenwellen, und einige Beobachtungen an den Vorläufern.* Ibid., pp. 623-675.

[1345] W. HÖNIGER: *Anwendung der Kinematographie zur Ermittlung der Stoßkraft bei Schlagversuchen.* VDI-Z. **56**, 1501-1505 (1912).

[1346] M. BODENSTEIN: *Photochemische Kinetik des Chlorknallgases.* Z. Physik. Chem. **85**, 297-328 (1913); *Eine Theorie der photochemischen Reaktionsgeschwindigkeiten.* Ibid. **85**, 329-397 (1913).

[1347] M. BODENSTEIN: *100 Jahre Photochemie des Chlorknallgases.* Ber. Dt. Chem. Gesell. **75A**, 119-125 (1942).

(ii) Through the concept of a *chain reaction* – a series of reactions in which the product of each step is a reagent for the next step – he correctly explains this law and, simultaneously, the fact that the photochemical yield exceeds Albert EINSTEIN's Law of Equivalents[1348] by a factor of $10^4$.

The concept of a chemical "chain reaction" [Germ. *Kettenreaktion*] gained great importance in detonation and combustion physics and occurs in three steps:

- *initiation* in which a reactive intermediate (*e.g.*, an atom, an ion, or a neutral molecular fragment) is formed, usually through the action of an agent such as light, heat, or a catalyst;
- *propagation*, whereby the intermediate reacts with the original reactants, producing stable products and another intermediate, whether of the same or different kind. The new intermediate reacts as before, so a repetitive cycle begins; and
- *termination*, which may be natural, as when all the reactants have been consumed or the containing vessel causes the chain carriers to recombine as fast as they are formed, but more often is induced intentionally by introduction of substances called "inhibitors" (or "antioxidants").

In the example of a detonating chlorine-hydrogen mixture – for a long-time an irritating riddle to physico-chemists and called in 1918 by Walther H. NERNST *"... die boshafte Chlorknallgasexplosion"* ("... the malicious chlorine-hydrogen explosion") – the reactions begin with the dissociation of the chlorine molecule into chlorine atoms, $Cl_2 \rightarrow 2Cl$, which is followed by the two reactions

$$Cl + H_2 \rightarrow HCl + H \text{ and } H + Cl_2 \rightarrow HCl + Cl.$$

In this reaction, new unstable molecules are formed besides stable reaction products and so on, constituting a chain reaction and continuing with explosive rapidity.

| | | |
|---|---|---|
| 1913 | KT & O, Midway Field, Taft, California | Myron M. KINLEY, an oil field worker called to dislodge a stuck valve in a blazing well in the Taft Field, **discovers by accident that the blast wave generated by an explosive can be used to shoot out the flame from an oil well fire** {⇨Fig. 4.15–K}. • He later noticed that gas fires have a mixing chamber, a 3 to 15-ft (0.9 to 4.6-m)-wide space between the bottom of the blaze and the mouth of the well in which gas and air mix. He figured out that an explosion set off close to that chamber breaks the gas flow long enough for the blaze to disappear before the gas can resume its upward direction.[1349] The use of dynamite to blow out the flames of a high-pressure oil or gas well fire is still regarded as the most reliable method of doing so.[1350] |
| 1913 | Point Hawkins [ca. 7 km southeast of Baltimore], Maryland | Charles E. MUNROE,[1351] a U.S. chemist and explosives specialist, **gives the first detailed documentation on the destructive effects of a large-yield chemical explosion.** The British cargo ship SS *Alum Chine*, which is assigned to transport explosives to the Panama Canal for use in blasting operations and has 285 tons of dynamite on board, explodes during freight loading at Point Hawkins. The explosion is felt as a blast and/or seismic shock, depending on the distance and the direction from the origin of the explosion. The disaster leaves 62 people dead and 60 wounded.<br><br>Twelve years later, MUNROE,[1352] upon analyzing a large number of observations made at distances ranging from 6.4 to 160 km, concluded that these differences can be attributed to the |

---

[1348] This law, formulated by A. EINSTEIN in 1905, says that the sum of the energy equivalents of the masses (according to $E = mc^2$) and the kinetic energy of the particles – including the energy of photons, involved in a process of change – remains unchanged. It distinctly differs from the classical conservation law of energy which only considers the sum of kinetic and potential energies to be conserved.

[1349] J.D. KINLEY: *Call KINLEY. Adventures of an oil well firefighter*. Cock-A-Hoop Publ., Tulsa, OK (1995), pp. 16-20.

[1350] Karl KINLEY JR., Kinley Corporation, Houston, TX; private communication (Jan. 2001).

[1351] C.E. MUNROE: *Zones of silence in sound areas from explosion*. Prof. Mem. Corps Eng. (U.S.A.) **10**, 253-260 (1918).

[1352] C.E. MUNROE: *Distribution des ondes de l'explosion de l'Alum Chine*. Mém. Artill. Franç. **4**, 545-552 (1925).

acoustic phenomenon of *zones of silence* or *shadow zones* {VON DEM BORNE ⇨ 1910; GUTENBERG ⇨ 1926}, and the orientation of the ship at the moment of the explosion.

| 1913 | Institute of Applied Mechanics, University of Göttingen, Germany | Ludwig PRANDTL,[1353] a German professor of applied mechanics, **reviews progress in compressible flows and supersonics, and coins the terms *MACHscher Winkel* ("Mach angle") and *MACHscher Kegel* ("Mach cone") while doing so.** ▪ The simple formula for the Mach angle, given by |

$$\sin\alpha = c/v = 1/M,$$

suggests the idea that, when the velocity of sound $c$ is known, an unknown projectile velocity $v$ can be found by sinply measuring the angle $\alpha$. However, Ernst MACH and Peter SALCHER had already pointed out that **a real head wave has a hyperbolic rather than a straight geometry, and it only approaches a straight cone geometry at a large distance from the projectile** {E. MACH & SALCHER ⇨ 1886}. For example, for the German infantry projectile model M.88 ($v_0$ = 642 m/s), Carl J. CRANZ[1354] compared the calculated projectile velocity, $v_C$, obtained using the measured Mach angle according to the formula given above, with the measured velocity, $v_M$, obtained using his electric-spark photochronograph {⇨Fig. 4.19–J}.[1355] He noticed that $v_C < v_M$ always holds. However, the discrepancy is often quite small and amounted to less than 2% in his example.

In the case of the sonic boom phenomenon – *e.g.*, that resulting from an aircraft flying at a supersonic speed {DUMOND ET AL. ⇨ 1946; ACKERET ⇨ 1952; ROY ⇨ 1952} – the regions outside and inside of the Mach cone have been termed the *zone of silence* and the *zone of action*, respectively. The latter zone is a continuous field of flow which may contain further shock waves. Typically, **there is no approach noise from supersonic aircraft,** and pilots or passengers seated ahead of the engine will not hear the engine, except for any noise transmitted through the structure of the aircraft.

The Mach cone – the common envelope of all shock waves emerging from a supersonic body – is a basic phenomenon in physics. First discovered in air, it also exist in other states of agregation, for example

▸ when a projectile moves through water at a velocity greater than the sound velocity in water {⇨Fig. 4.14–E};

▸ when a fracture propagates faster than the shear velocity in the material, so-called "supershear" {ROSAKIS ET AL. ⇨ 1999};

▸ when an object moves in a shallow vibrofluidized granular layer of micro spheres at a velocity greater than a "critical" velocity;[1356]

▸ when an energetic charged particle passes through a transparent nonconductive material {CHERENKOV ⇨ 1934}; or

▸ when a particle moves in a dusty plasma {SAMSONOV ET AL. ⇨ 1999}.

| 1913 | Princeton University Observatory, Princeton, New Jersey | Henry Norris RUSSELL,[1357] a U.S. astronomer and director of the Observatory, **first uses the terms *giant* and *dwarf* in his talk to the Royal Astronomical Society in London.** ▪ Ejnar HERTZSPRUNG, a Danish chemist and astronomer, had earlier used the German term *Giganten* ("giants"). |

---

[1353] L. PRANDTL: *Gasbewegung*. In: (E. KORSCHELT, ed.) *Handwörterbuch der Naturwissenschaften* (Fischer, Jena) **4**, 544-560 (1913).
[1354] C. CRANZ: *Lehrbuch der Ballistik: I. Äußere Ballistik*. Springer, Berlin (1925), pp. 38-42.
[1355] C. CRANZ and K. BECKER: *Messungen über den Luftwiderstand bei großen Geschwindigkeiten*. Artilleristische Monatshefte **69**, 189-196 (1912); Ibid. **71**, 333-368 (1912).
[1356] P. HEIL ET AL.: *Mach cone in shallow granular fluid*. Phys. Rev. E **70** [III], Paper No. 060301(R) (2004).
[1357] H.N. RUSSELL: "*Giant*" *and* "*dwarf*" *stars*. Observatory **36**, 324-329 (1913).

Bright red stars are called "red giants," because they must be very large to be simultaneously cool (thus red) and bright. Likewise, dim blue stars are called "white dwarfs" {LUYTEN ⇨ 1922}, since they must be very small in order to be simultaneously hot (thus white) and dim.

| | | |
|---|---|---|
| 1913 | London, England | Herbert G. WELLS,[1358] a British novelist and today widely considered to be the father of modern science fiction, **predicts in his book *The World Set Free* the development of atomic power and the use of atomic bombs in a global war** ("a war to end all wars"). ▪ In WELLS' book HOLSTEN, an atomic physicist (the main character) is made to discover artificial radioactivity. He "set up atomic disintegration in a minute particle of bismuth; it exploded with great violence into a heavy gas of extreme radioactivity which, in its turn, disintegrated over the course of seven days." WELLS settled for bismuth, which is the heaviest of the stable elements. Its relationship to the radioactive elements (radium, thorium, actinium) was being interpreted at that time.[1359] Today, although bismuth can be atomically disintegrated, the result is not an explosion but radioactivity which persists for seven-and-a-half minutes, and the end product is not gold − as WELLS suggested − but actinium-lead.<br><br>His book made a great impression on the 34-year old Hungarian Leo SZILARD, a physicist and *Privatdozent* ("Instructor") at the University of Berlin, who read WELL's novel in 1932. In the following year, at which point he was living as a refugee in England, he had the idea that it might be possible to achieve a nuclear chain reaction. Coincidentally, the neutron was discovered in the same year {CHADWICK ⇨ 1932}. A year later, **SZILARD filed a patent application for an atomic bomb, which described not only the basic concept of using neutron-induced chain reactions to create a nuclear explosion, but also the key concept of the *critical mass*** {PERRIN ⇨ 1939}. SZILARD[1360] obtained various patents for his unique ideas, making him the legally recognized inventor of the atomic bomb.[1361] |
| 1914 | Cambridge Engineering School, University of Cambridge, England | Bertram HOPKINSON,[1362] a British engineer and materials dynamicist, **describes a novel and ingenious variant of the ballistic pendulum that can be used to analyze the force and time of a blow.**<br><br>(i) He uses the principle of the ballistic pendulum {J. CASSINI ⇨ 1707; ROBINS ⇨ 1740}, which measures the momentum of a blow, but by using an ingenious modified form of this device he manages to separate out its two factors, namely force and time. His pendulum device consists of a uniform long and thin high-strength steel rod, which is divided by a transverse joint (maintained by magnetism) into a long and a short portion − the so-called **"Hopkinson pressure bar"** {⇨Fig. 4.3−P}.<br><br>(ii) The rod takes the blow longitudinally and transmits it as a wave of elastic compression proceeding from the long to the short section. At the extreme end of the short section, the wave of compression is reflected back along the rod as a wave of tension.<br><br>(iii) When the reflected wave reaches the joint, the short piece flies off and carries with it a fraction of the total initial momentum that depends on the length of the short piece. Adjustment of the short piece enables a situation to be created whereby the whole of the momentum of the blow is absorbed by it, leaving the main portion of the rod at rest. Thus it is possible to deter- |

---

[1358] H.G. WELLS: *The world set free. A story of mankind*. Macmillan, London (1914); Collins, London & Glasgow (1956); *see* Introduction by R. CALDER.

[1359] F. SODDY: *The interpretation of radium, being the substance of six free popular experimental lectures delivered at the University of Glasgow, 1908*. J. Murray, London (1909).

[1360] L. SZILARD: *Improvements in or relating to the transmutation of chemical elements*. Brit. Patents No. 440,023 (Appl. March 1934) and No. 630,726 (Appl. June/July 1934); *Neutronic reactor*. U.S. Patent No. 2,708,656 (Appl. Dec. 1944); *Apparatus for nuclear transmutation*. U.S. Patent No. 263,017 (Appl. March 1939). *See* L. SZILARD: *The collected works of Leo SZILARD*. MIT Press, London & Cambridge (1972); *see* vol. 1: *Scientific papers*, pp. 527-530, 622-696.

[1361] B. BERNSTEIN: *The unsung father of the A-bomb*. Discover. The Magazine of Science, pp. 37-42 (Aug. 1985).

[1362] B. HOPKINSON: *A method of measuring the pressure produced in the detonation of high explosives, or by the impact of bullets*. Proc. Roy. Soc. Lond. **A89**, 411-413 (1914).

mine the length of the pressure wave, and from that it is possible to infer the duration of the blow. The maximum pressure can also be measured by using a very short length for the detachable piece.

(iv) He examines the blow imparted by a bullet striking the end of the rod normally, and shows that four-fifths of the impulse from the blow caused by the detonation of guncotton positioned at or close to one end of the rod is delivered in 20 µs.

Although HOPKINSON's method has the advantage of simplicity, it suffers from various serious limitations; for example, it does not provide the shape of the pressure-time curve of a pulse, only its duration and the value of the maximum pressure. ▪ Later developments – known as the **"Davies bar"** {DAVIES ⇨1948} and the **"Kolsky bar"** or **"split Hopkinson pressure bar"** {KOLSKY ⇨1949} – avoided these limitations. In a more recent review, André LICHTENBERGER,[1363] a research scientist at ISL (Saint-Louis, France), discussed the applicability and limitations of the Hopkinson bar test for obtaining data on dynamic materials at strain rates ranging between $5 \times 10^2 \, s^{-1}$ and $5 \times 10^3 \, s^{-1}$.

| 1914 | Earthquake Observatory of the Miner's Union, Bochum, Ruhr Area, Germany | Ludger MINTROP,[1364] a German seismologist and director of the Bochum Earthquake Observatory, **investigates whether there is a connection between earthquakes and firedamp explosions in coal mines,** which was a popular subject for discussion among experts at the time, as well as among the public. Based on statistical data for firedamp explosions in German collieries and earthquakes recorded by German observatories in the period 1909–1913, he proves that there is definitely no connection between these phenomena, and that the tiny number of occasions when both phenomena occur coincidentally fully corresponds with the probability expected for such coincidences. This result also holds for *microseismic shocks* – a curious phenomenon usually observed in the winter months.<br><br>In the vicinity of active volcanoes, microseismicity can also result from microearthquakes which have a magnitude of 2 or less on the Richter magnitude scale {RICHTER ⇨1935}. They can increase in magnitude within hours of a potential eruption. |
|---|---|---|
| 1915 | Washington, DC | **Creation of the National Advisory Committee on Aeronautics (NACA)**[1365] by the U.S. Congress at the request of U.S. President Woodrow WILSON, with the goal "to supervise and direct the scientific study of the problems of flight, with a view to their practical solution, and to determine the problems which should be experimentally attacked, and to discuss their solution and their application to practical questions … the committee may direct and conduct research and experiment in aeronautics." ▪ NACA was modeled on the British Advisory Committee for Aeronautics (ACA) which was established in 1909 at the National Physics Laboratory (NPL).<br><br>In 1920, NACA opened the Langley Memorial Aeronautical Laboratory, its own research facility at Langley Field near Hampton, Virginia.[1366] Studies into high-speed aerodynamics, which were initiated as early as the 1920s at NACA {BRIGGS, HALL & DRYDEN ⇨1922}, gathered pace in the 1940s with the construction and operation of a special airplane for research into the problems of transonic and supersonic flight {YEAGER ⇨1947}.[1367] In 1948, Langley Memorial Aeronautical Laboratory (LMAL) was renamed Langley Aeronautical Laboratory |

---

[1363] A. LICHTENBERGER: *Compression dynamique par barres d'HOPKINSON*. Rept. No. RE/002/87, DYMAT, Etablissement Technique Central de l'Armement, Arcueil, France (1987).

[1364] L. MINTROP: *Erdbeben, Schlagwetterexplosionen und Stein- und Kohlenfall*. Glückauf **50**, 330-339 (1914).

[1365] The National Advisory Committee for Aeronautics (NACA). U.S. Centential of Flight Commission; http://www.centennialofflight.gov/essay/Evolution_of_Technology/NACA/Tech1.htm.

[1366] J.R. HANSEN: *Engineer in charge. A history of the Langley Aeronautical Laboratory, 1917–1958*. Rept. NASA SP-4305, NASA History Office, Washington, DC (1987).

[1367] J.V. BECKER: *The high speed frontier. Case histories of four NACA Programs, 1920–1950*. Rept. NASA SP-445, NASA History Office, Washington, DC (1980).

(LAL). In 1958, when NACA was subsumed by the newly established NASA {⇨1958}, it was renamed Langley Research Center (LaRC).[1368]

Renowned members of the Committee over the years have included a number of famous personalities from the fields of science and engineering, as well as pioneers of aviation, *e.g.*, Orville WRIGHT, Charles A. LINDBERGH, Henry H. ARNOLD, David W. TAYLOR, Joseph S. AMES, Frederick C. CRAWFORD, and Harry F. GUGGENHEIM.

| 1915 | British Admiralty, London, England | **The British Admiralty appoints a "Committee of Invention and Research – Erosion of Propellers,"** which is given the goal of systematically searching for the origin of destructive effects by cavitation bubbles and determining the cause of the severe erosion of the propeller blades of the British ocean liners HMS *Lusitania* and HMS *Mauretania*[1369] {Lord RAYLEIGH ⇨1917; S.S. COOK ⇨1928}.[1370] The chairman of this Committee, a subcommittee of the Board of Invention and Research, is Sir Charles A. PARSONS.

In 1919, at the Spring Meeting of the Institution of Naval Architects, Sir PARSONS and Stanley S. COOK read a paper describing the work of the Committee at considerable length.

(i) They referred to calculations that had been carried out but which were published much later {S.S. COOK ⇨1928}, and gave a description of the experimental apparatus by which they had been confirmed.[1371]

(ii) They observed that **the destruction of the screws takes place as a result of the numerous repeated "hydraulic blows" accompanying the collapse of cavities.** Somewhat later, the German fluid dynamicist Hermann FÖTTINGER[1372] came to the same conclusion in regard to hydrotechnical constructions.

(iii) **They concluded "that the corrosion of propellers is very slight, but that erosion is serious and is caused by the hammer action of the water on the propeller blades, produced by cavities closing up on the surface of the blades.** This action may be caused either by cavitation of the propeller itself, occurring more generally when the propeller is in a varying wake, or by the cavities and vortices formed by the action of other propellers ahead of it, and the erosive action is generally aggravated upon a propeller which works in the wake of another." |
|---|---|---|
| 1915 | Chair of Mechanism and Applied Mechanics, University of Cambridge, England | Bertram HOPKINSON,[1373] a British engineering professor, while reflecting on damage phenomena from explosions, **proposes his cube-root law for scaling the blast field around conventional explosive charges** at sea-level – the **"Hopkinson law"** (or "Hopkinson rule"). He states that when two explosive charges made from the same explosive have similar shapes but different sizes, and are detonated in the same atmosphere, the blast waves produced by the two charges are similar at the same distance from each charge if the distance is scaled by the cube root of the size of the charge. • The Hopkinson rule proved to be most valuable for predicting explosions in air {KENNEDY ⇨1946; DEWEY ⇨1964}. Hopkinson scaling was reviewed by BAKER & WESTINE {⇨1973}. |

---

[1368] R.E. BILSTEIN: *Orders of magnitude. A history of the NACA and NASA, 1915–1990*. The NASA History Series, Rept. NASA SP-4406. Office of Management, Scientific and Technical Information Division, Washington, DC (1989); http://www.hq.nasa.gov/office/pao/History/SP-4406/cover.html.

[1369] Both passenger liners were launched in 1906 and made their maiden voyages in 1907. The HMS *Mauretania* was revolutionary in that it was the very first passenger vessel fitted with the new steam turbine engine developed by Sir Charles A. PARSONS {⇨1884}, holding the record for the fastest westbound crossing of the Atlantic for over 20 years.

[1370] D. SILBERRAD: *Propeller erosion*. Engineering (Lond.) **93**, 33-35 (Jan. 1912). • Other examples of the destruction of screw propellers were given in the same year, see W. RAMSAY: *The erosion of bronze propeller-blades*. Engineering (Lond.) **93**, 687-691 (1912).

[1371] C. PARSONS and S.S. COOK: *Investigations into the causes of corrosion or erosion of propellers*. Trans. Roy. Inst. Nav. Arch. **61**, 223-231 (1919).

[1372] H. FÖTTINGER (ed.): *Hydraulische Probleme*. Wissenschaftlicher Beirat des Vereins Deutscher Ingenieure (VDI), Berlin (1926).

[1373] B. HOPKINSON: Brit. Ordnance Board Minutes No. 13,565 (1915).

During the First World War, H.W. HILLIAR,[1374] a British researcher, first applied the Hopkinson rule to underwater explosions and found good agreement. In 1919, he wrote, "HOPKINSON's rule can be deduced theoretically as an extension of the principle described in Section 9 *Comparison of Large and Small Charges*. Its validity has been proved experimentally for charges differing very widely in magnitude. The rule is obviously of great value in enabling full-scale inferences to be drawn from model experiments." However, subsequent underwater explosion studies performed at the U.S. Naval Ordnance Laboratory {SNAY ⇒1959} revealed that effects of gravity – which changes the hydrostatic pressure governing bubble pulsation and provokes *gravity migration* and *jet formation* inside the bubble {⇒Fig. 4.16–E} – cannot be scaled. This is also the case for the viscosity, which changes the profile of the propagating shock wave.

| 1915 | *Private laboratory at Terling Place, Witham, Essex, England* | Lord Rayleigh,[1375] an eminent English physicist, **discusses the scientific small-scale model based on the principle of similitude.**

(i) Introductorily he indignantly writes, "I have often been impressed by the scanty attention paid even by original workers in the field to the great principle of similitude. It happens not infrequently that results in the form of 'laws' are put forward as novelties on the basis of elaborate experiments, which might have been predicted *a priori* after a few minutes consideration!"

(ii) **He establishes the fundamentals of dimensional analysis based on FOURIER's work** (1822). ▪ Jean Baptiste Joseph FOURIER,[1376] a French physicist and mathematican, was the first to apply the geometrical concept of dimension to physical magnitudes. He recognized the existence of dimensionless groups in his equations, "It should be noted that each physical quantity, known or unknown, possesses a *dimension* proper to itself and that the terms in an equation cannot be compared one with another unless they possess the same *dimensional exponent*." However, he did not see most of the consequences that were drawn out later.

It seems to have been from about this time that the method became standard fare for the physicist and engineer. Similitude – *i.e.*, similarity of behavior of different systems, now called "dimensionless analysis" – is a method for producing dimensionless numbers and requires that the geometric (*e.g.*, angles, length ratios), kinematic (*e.g.*, displacement ratios, velocity ratios) and dynamic (*e.g.*, force ratios, stress ratios, pressure ratios) similarity parameters are equal for the model and the real world. "Dimensional analysis" (or "scaling") is a powerful technique widely employed in science. It has also been successfully applied on compressible flows and associated dimensionless numbers (*e.g.*, Reynolds number, Mach number, Prandtl number, specific heat ratio). Prior to the age of powerful digital computers and numerical methods for complex design problems, dimensional analysis was critical to the development of large scale complex dynamic physical systems. |
|---|---|---|
| 1916 | *Washington, DC* | On September 20, at a meeting held in New York City, **creation of the U.S. National Research Council (NRC),** by the National Academy of Sciences at the request of U.S. President Woodrow WILSON, for the purposes of "promoting research in the natural sciences and encouraging the application and dissemination of scientific knowledge." The chairman is George E. HALE, director of Mt. Wilson Solar Observatory. ▪ The council does not maintain its own scientific laboratories; its chief concern is rather the cooperation and integration of research activities. |

---

[1374] H.W. HILLIAR: *Experiments on the pressure wave thrown out by submarine explosions*. Engl. Dept. of Scientific Research & Experiment (1919). Reprinted in *Underwater explosion research; a compendium of British and American reports*. Office of Naval Research (ONR), Washington, DC (1950); *see* vol. 1: *The shock wave*, pp. 83-158.

[1375] Lord RAYLEIGH (J.W. STRUTT): *On the principle of similitude*. Nature **95**, 66-68, 644 (March 1915).

[1376] J.B.J. FOURIER: *Théorie analytique de la chaleur*. Didot, Paris (1822). *The analytical theory of heat*. Dover Publs., New York (1955).

| | | |
|---|---|---|
| 1916 | *Friedrich Krupp AG, Essen, Ruhr Area, Germany* | **Fritz RAUSENBERGER,** a professor of mechanical engineering and chief artillery designer at Krupp Company, **and his team begin the development of a long-range gun.** |

This 140-ton supergun – later officially called "Kaiser-Wilhelm-Geschütz," or the "Paris gun" in the ballistic literature – had some remarkable features:
- The caliber was 210 mm and the gun had an exceptionally long range (120 km). The muzzle speed was 1,610 m/s.
- The barrel of this most unusual gun consisted of a rifled section followed by a smoothbore section.
- Because the barrel had an enormous length of 33.5 m, it needed a special suspension-bridge-like support to hold it straight {⇨Fig. 4.2–T}.
- Since the Earth's atmosphere grows thinner with increasing altitude, maximum ranges were best achieved at elevations of around 50° to 55°.[1377, 1378]

The Paris gun was first used in France: on March 23, 1918, Paris was hit by a 100-kg shell from a distance of 122 km. In the early 1960s, the concept of the Paris gun was resumed by U.S. scientists at McGill University with the original goal of studying the ballistics of reentry at low cost. This goal was later extended by the project leader to fire a payload into space from a gun {HARP ⇨1961}.

| | | |
|---|---|---|
| 1916/ 1917 | *Ministère de l'Armement et des Fabrications de Guerre, Paris, France* | **André L.A. FAUCHON-VILLEPLÉE,** a French engineer, **invents the first electromagnetic rail gun.** He also builds a working model, which uses shells fitted with "wings" that serve as the armature. • His main motivation for creating an electromagnetic gun was apparently to compete with the **Kaiser-Wilhelm-Geschütz** or **Parisgeschütz** {RAUSENBERGER ⇨1916, ⇨Fig. 4.2–T}. |

In 1920, FAUCHON-VILLEPLÉE[1379] obtained a German patent on his new gun device. After World War II, his idea was resumed in order to accelerate macro particles to hypersonic velocities {RASHLEIGH & MARSHALL ⇨1978} in high-velocity impact studies and in materials research.

| | | |
|---|---|---|
| 1917 | *Halifax, Nova Scotia, southeastern Canada* | On December 6, **the SS *Mont Blanc*,** a French-owned freighter loaded to the gunnels with a large cargo of picric acid, TNT, guncotton, and benzene (in total 2,653 tons), **collides with the Belgian relief ship SS *Imo* and explodes in Halifax Harbor.** |

(i) The explosion – later known as the "Halifax explosion," is the greatest disaster caused by a man-made explosion up to that point. The explosion is so large that it kills more than 2,000 people, injured 9,000 more and completely flattened two square kilometers of northern Halifax.

(ii) **The blast also initiates a tsunami** that reached as high as 18 meters above the harbor's high-water mark on the Halifax side. People blown off their feet by the explosion hung on for their lives as water rushed over the shoreline and through the dockyard. The tsunami lifted the SS *Imo*, which had no cargo on board, onto the shore of Dartmouth, where the ship stayed until spring.[1380, 1381]

| | | |
|---|---|---|
| 1917 | *Friedrich-Wilhelms Universität, Berlin, Germany* | **Albert EINSTEIN,**[1382] a professor of theoretical physics and director of the Institut für Physik of the Kaiser-Wilhelm-Gesellschaft (KWG) zur Förderung der Wissenschaften e.V. [since 1948 the Max-Planck-Gesellschaft (MPG) zur Förderung der Wissenschaften], **proposes the first relativistic cosmological model, envisaging a static, finite spherical Universe.** In his *Allge-* |

---

[1377] R. TODD: *A brief history of the Paris gun*; http://www.landships.freeservers.com/parisgun_history.htm.
[1378] H.W. MILLER: *The Paris Gun: the bombardment of Paris by the German long range guns and the great German offensives of 1918*. Harrap, London (1930).
[1379] A. FAUCHON-VILLEPLÉE: *Elektrische Kanone*. Germ. Patent No. 376,391 (July 1920); *Canons électriques, système Fauchon-Villeplée*. Berger-Levrault, Nancy etc. (1920).
[1380] J.G. ARMSTRONG: *The Halifax explosion and the Royal Canadian Navy: inquiry and intrigue*. UBC, Vancouver, B.C. (2002).
[1381] *The Halifax explosion*. CBC/Radio-Canada; http://www.cbc.ca/halifaxexplosion/he2_ruins/index.html.
[1382] A. EINSTEIN: *Kosmologische Betrachtungen zur allgemeinen Relativitätstheorie*. Sitzungsber. Königl. Preuss. Akad. Wiss. (Berlin) **1**, 142-152 (1917). Engl. translation in: (H.A. LORENTZ ET AL., eds.) *The principle of relativity*. Dover Publ., New York (1952), pp. 177-188.

*meine Relativitätstheorie* ("General Theory of Relativity," 1916) – the modern theory of gravitation – both gravitation and cosmology are geometrized.

(i) He shows how a "scale factor" of the Universe varies with time; the gravitational effects of a given mass are described by 16 coupled hyperbolic-elliptic nonlinear partial differential equations: the so-called **"Einstein field equations."**[1383] They permit the existence of three kinds of space: flat (Euclidean), spherical (closed), and hyperbolic (open).

(ii) In an attempt to obtain a solution that yields a Universe that does not expand – which was believed to be the case by most cosmologists at the time – from the field equations, he introduces two parameters: a "cosmological" constant $A$ and a density parameter $D$. ▪ Thus, **EINSTEIN imagined a spatially closed,** *static* (*i.e.*, a time-invariant) **Universe in which space has a constant curvature.**

The subsequent observation of redshifts in the spectra of galaxies {HUBBLE ⇨ 1929}, which confirmed the idea that the Universe evolves through its expansion, contradicted EINSTEIN's static Universe. However, his solution forecast the existence of masses enormously greater than any known at the time and that the radius of the Universe was hundreds of times greater than the most distant objects hitherto observed. The various cosmological models proposed by others differed in their choice of the value of $A$ and $D$. **Nonstatic models were first studied by Alexander FRIEDMANN** {⇨1922} **and, independently, by Georges Edouard LEMAÎTRE** {⇨1927}.

In the same year, the Dutch astronomer Willem DE SITTER[1384] alternatively proposes a flat Universe with a positive cosmological constant that is empty of matter – the so-called **"de Sitter cosmological model."** It expands at an exponential rate forever, with no initial Big Bang {The Universe ⇨ *c.* 14 Ga ago; NASA-GSFC ⇨2003}, nor with a final Big Crunch {DAVIES ⇨1994}. ▪ DE SITTER's model proved useful for explaining the observed recession velocities of extragalactic nebulae as being a simple consequence of the properties of the gravitational field.

| | | |
|---|---|---|
| 1917 | *Ecole des Mines, Saint-Étienne, Dépt. Loire, France* | **Émile JOUGUET**,[1385] a French professor of mechanical engineering and *Répétiteur* at the Ecole Polytechnique, **publishes his book *Mécanique des explosifs* ("Mechanics of Explosives"), in which he also resumes his previous studies on detonation** {JOUGUET ⇨ 1904–1906}: (i) He give a historical perspective on the development of the theory of detonation, referring to the works of BERTHELOT, CHAPMAN, DIXON, LE CHATELIER, MALLARD, SCHUSTER, and VIEILLE, but also to lesser known contributions made by the French researchers CRUSSARD, DUHEM, DAUTRICHE, and TAFFANEL. However, he doesn't mention the important Russian contribution {MIKHEL'SON ⇨1893}. (ii) Referring to CRUSSARD's graphical method {CRUSSARD ⇨1907}, **JOUGUET assumes that the Rankine-Hugoniot equations are not only valid for discontinuities (such as shock waves) propagating in the same fluid, but can also be used to describe two separate, chemically distinct environments – so-called "reactive waves."** (iii) He explains the mechanism of detonation at constant speed by considering the detonation front to be a special type of shock wave to which the Rankine-Hugoniot equations can be applied by including the part due to the chemical reaction in the energy balance – the **"supplementary Chapman-Jouguet (CJ) condition."** |

---

[1383] A. EINSTEIN: *Die Grundlage der allgemeinen Relativitätstheorie.* Ann. Phys. **49** [IV], 769-822 (1916).

[1384] W. DE SITTER: *On the relativity of inertia. Remarks concerning EINSTEIN's latest hypothesis.* Koninklijke Akademie van Wetenschappen te Amsterdam. Section of Sciences **19**, 1217-1225 (1916/1917); *On EINSTEIN's theory of gravitation, and its astronomical consequences.* Third Paper. Month. Not. Roy. Astron. Soc. **76**, 699-728 (1916); Ibid. **77**, 155-183 (1916); Ibid. **78**, 3-28 (1917).

[1385] E. JOUGUET: *Mécanique des explosifs, étude de dynamique chimique.* O. Doin & Fils, Paris (1917).

| 1917 | Private laboratory at Terling Place, Witham, Essex, England | Lord RAYLEIGH,[1386] an English physicist and associated with the British Admiralty at that time, **analytically treats the important engineering problem of the sudden collapse of an empty spherical cavity in a large mass of an incompressible liquid.** |
|---|---|---|

(i) He writes, "I learned from Sir C. PARSONS [Sir PARSONS ⇨ 1897 & 1915] that he also was interested in the same question in connection with cavitation behind screw-propellers, and that at his instigation Mr. S. COOK [[1387]], on the basis of an investigation by BESANT [[1388]], had calculated the pressure developed when the collapse is suddenly arrested by impact against a rigid concentric obstacle … It appears that before the cavity is closed these pressures may rise very high in the fluid near the inner boundary."

(ii) He assumes "ideal fluid" conditions, (*i.e.*, incompressible, inviscid liquid), neglects surface tension, and assumes that the cavity was empty or contained a fluid (presumably vapor) which remained at constant pressure (less than liquid pressure) throughout the bubble collapse.

(iii) In order to solve the cavitation bubble dynamics problems he uses an energy balance method.

- He calculates the velocity $v$ of contraction of a single cavitation bubble as function of time, given by
$$v(t) = \{\tfrac{2}{3} P_0/\rho \, [(R_0/R)^3 - 1]\}^{1/2},$$
where $P_0$ is the hydrostatic pressure, $\rho$ is the density of the liquid, $R_0$ is the initial (max.) radius of the bubble, and $R(t)$ is the radius of the bubble at time instant $t$.

- To find the pressure $P$ in the interior of the fluid during collapse, he extends a previous calculation made by the English mathematician William Henry BESANT and shows that the final volume is extremely small when the initial pressure of the gas is only a small fraction of $P_0$, the pressure of the surrounding fluid. His analysis shows that
$$P(t) \propto P_0 \, [R_0/R(t)]^3;$$
*i.e.*, when the bubble collapses ($R \to 0$), the pressure near the boundary layer becomes very great. A similar result, he annotates, was obtained by S.S. COOK, who estimated a pressure of 10,300 atm for $R = R_0/20$.

(iv) He also considers the problem where the cavity contains a small amount of gas which is isothermally compressed, and converts the energy of collapse into the pressure of this imprisoned gas {Sir PARSONS ⇨ 1884 & 1897; S.S. COOK ⇨ 1928}. ▪ In reality, however, the bubble undergoes isentropic compression, and a high temperature is attained as well as a high dynamic pressure.

Lord RAYEIGH's famous paper stimulated many subsequent investigations into cavitation. The first experimental evidence of the high-pressure pulse originating from a collapsing bubble – later also called **"transient cavitation"** – was given by Mark HARRISON[1389] at David Taylor Model Basin in Carderock, MD, using acoustic diagnostics, and by Wernfried GÜTH,[1390] a physicist and acoustician at Göttingen University, using an optical schlieren technique. **Bubble jet formation**, a result of the collapse of an unstable, asymmetric bubble wall, **was first suggested by Soviet researchers as a possible mechanism of damage in cavitation erosion** {KORNFELD & SUVOROV ⇨ 1944}. It was first experimentally confirmed by Charles F. NAUDÉ[1391] and Albert T. ELLIS, two fluid dynamicists at CalTech (Pasadena, CA). A detailed study of the collapse of an array of cavities using high-speed photography was performed by British researchers {DEAR & FIELD ⇨ 1988}. An excellent review of cavitation-generated ero-

---

[1386] Lord RAYLEIGH (J.W. STRUTT): *On the pressure developed in a liquid during the collapse of a spherical cavity*. Phil. Mag. **34** [VI], 94-98 (1917).
[1387] *Erosion of propellers*. Rept. of the Propeller Sub-Committee of the Board of Invention and Research (Sept. 17, 1917), sect. III.
[1388] W.H. BESANT: *A treatise on hydrostatics and hydrodynamics*. Deighton & Bell, Cambridge (1859), § 158.
[1389] M. HARRISON: *An experimental study of single bubble cavitation noise*. JASA **24**, 776-782 (1952).
[1390] W. GÜTH: *Zur Entstehung der Stoßwellen bei der Kavitation*. Acustica **6**, 526-531 (1956).
[1391] C.F. NAUDÉ and A.T. ELLIS: *On the mechanism of cavitation damage by non-hemispherical cavities in contact with a solid boundary*. Trans. ASME D: J. Basic Engng. **83**, 648-656 (1961).

| | | |
|---|---|---|
| 1918 | Moscow, Russia | sion phenomena was given more recently by Achim PHILIPP[1392] and Werner LAUTERBORN, two German physicists at Göttingen University.

On December 1, **foundation of the Central Aerohydrodynamic Institute** [CAGI, Russ. *ЦАГИ*] by Andrei N. TUPOLEV and Nikolai E. ZHUKOVSKY, the father of Soviet aviation. ▪ This Soviet aircraft institution became the most prominent State Aerospace Research Center of Russia. Already in 1922 TUPOLEV became head of the Institute's design bureau. |
| 1918 | Engineering Division, U.S. Army, McCook Field, Dayton, Ohio | **Frank Walker CALDWELL**[1393] **and Elisha Noel FALES,** two U.S. aeronautical mechanical engineers, **design and build the first American high-speed wind tunnel** (14 in. in dia., 200 m/s) in order to study compressibility effects on airfoil sections suitable for use as propeller blade sections. **They notice that there is a large decrease in lift coefficient accompanied by a large increase in drag coefficient at a "critical speed" in the transonic region,** so the lift-to-drag ratio is enormously decreased. ▪ Their wind tunnel has survived and can be seen today in the U.S. Air Force Museum at Dayton, OH.

Following the introduction of the Mach number {ACKERET ⇒1928}, the term *critical Mach number* was also coined, which relates the critical speed to the local sound velocity.[1394] |
| 1918 | Lick Observatory, Santa Cruz, California | **Heber D. CURTIS,**[1395] an eminent U.S. astronomer, **discovers the first example of an** *extragalactic jet*. In M87 (or NGC 4486), a giant elliptical galaxy located in the heart of the *Virgo* cluster, he notices a "curious straight ray … apparently connected with the nucleus by a thin line of matter." ▪ Extragalactic jets are relativistic gas flows that are located or originating beyond the Milky Way and generated in the centers of active galaxies. During their lifetimes of up to 100 million years they transport tremendous amounts of energy up to millions of light-years into intergalactic space. Nowadays several hundreds of these extragalactic jets are known, which are detectable by the synchrotron and inverse Compton radiation they emit at radio frequencies.

An extragalactic jet is a particular example of an *astrophysical jet*, a highly collimated supersonic flux of plasma that some galaxies, quasars, microquasars, X-ray binaries, *etc.*, may produce due to the accretion of matter onto massive objects. High-resolution radio interferometry has shown that extragalactic jets are extremely well collimated with opening angles of a few degrees only, extending to distances of up to several $10^5$ parsecs ($3.262 \times 10^5$ light-years). Today, several hundred of these extragalactic jets are known, which are detected by the synchrotron and inverse Compton radiation they emit at radio frequencies. |
| 1918 | Collège de France, Paris & French Governmental Research Laboratory, France | **Paul LANGEVIN and Constantin CHILOWSKY,** two French physicists, **suggest the first supersonic wind tunnel for use in aeroballistics.** In order to test a new type of projectile, they propose to use a high-speed current of air moving at supersonic velocities emerging from a Laval nozzle. ▪ Later, a British ballistic commission that visited them initiated similar studies at the National Physical Laboratory (NPL) in Teddington, U.K. {STANTON ⇒1921}.[1396]

High-speed wind tunnels with a closed test section can be used up to a Mach number of about 0.85. At higher Mach numbers, however, they "choke," meaning that the Mach number in the neighborhood of the model rapidly increases to unity, and a shock wave extends right across the air stream. **The idea of using a free jet rather than a closed test section that avoided the choking problem was first realized in Paris by Eugène HUGUENARD,**[1397] a |

---

[1392] A. PHILIPP and W. LAUTERBORN: *Cavitation erosion by single laser-produced bubbles*. J. Fluid Mech. **361**, 75-116 (1998).
[1393] F.W. CALDWELL and E.N. FALES: *Wind tunnel studies*. In: *Aerodynamic phenomena at high speeds*. Rept. NACA-TR 83 (1920).
[1394] J.D. ANDERSON JR.: *Research in supersonic flight and the breaking of the sound barrier*. In: (P.E. MACK, ed.) *From engineering science to big science*. The NASA History Series, Rept. NASA SP-4219 (2001); http://history.nasa.gov/SP-4219/Cover4219.htm, chap. 3.
[1395] H.D. CURTIS: *The planetary nebulae*. Publ. Lick Observatory **13**, 55-74 (1918).
[1396] T.E. STANTON: *The development of a high-speed wind tunnel for research in external ballistics*. Proc. Roy. Soc. Lond. **A131**, 122-132 (1931).
[1397] E. HUGUENARD: *Les souffleries aérodynamiques à très grande vitesse*. La Technique Aéronautique **15**, 346-355, 378-392 (1924); Rept. NACA-TM 318 (1925).

physicist and inventor at the Conservatoire National des Arts et Métiers, **and Jean André Sainte Laguë,** a mathematics professor at the Lycée Janson-de-Sailly (located in the XVIe arrondissement of Paris). They carried out such experiments on a stationary high-speed current of air moving at velocities greater than the velocity of sound {⇨Fig. 4.9–A}. Initially using an 8-cm-dia. Laval working section, they obtained a Mach number that was barely above one ($M = 1.07$). Later, however, after changing the diffuser angle, they reached Mach numbers of up to 1.4. The drag was measured by a torsion balance.[1398]

In the 1920s, the free jet method was also taken up in the United States by the two aeronautical engineers Lyman J. Briggs[1399] and Hugh L. Dryden in their 2-inch jet at Edgewood Arsenal, Maryland. By employing a convergent-divergent supersonic nozzle, they obtained a maximum Mach number of $M = 1.08$. It was the first American supersonic facility.

| 1919 | National Physics Laboratory, Teddington, England | By order of the Advisory Committee for Aeronautics (ACA) **British researchers begin to study airscrew tip phenomena at high speeds.**[1400] They recognize that a propeller is a wing whose flow characteristics and therefore propulsion efficiency vary along the span. |
|---|---|---|

- E.J.H. Lynam,[1401] a British aeronautical researcher, performs zero-advance tests of a low-pitch propeller model in free air at tip speeds of up to 1,180 ft/s (360 m/s) – apparently the structural limit for "thoroughly well-seasoned American black walnut" test blades. At higher velocities, he proposes to use airscrews constructed of Stringy Bark [*Eucalyptus conglomerata*], a tough fibrous bark which has a higher bonding strength than that of walnut. Using propellers with blunt-edged, thick blades, **Lynam observes a loss in thrust and a large increase in blade drag when the rotational speed of the blade tips approaches the velocity of sound.** However, he doesn't observe any perceptible discontinuity in the sound emitted by the airscrew as the tip velocity approaches and exceeds the velocity of sound in air, as had been supposed by many contemporary engineers.
- Sylvanus Albert Reed,[1402] a U.S. aeronautical engineer and designer, subsequently used thin-bladed metal propellers, and made the remarkable observation that **thick, blunt-edged propellers are not appropriate for high-speed applications.**

Contemporary theories of the airscrew were still limited to subsonic tip speeds, and progress on the challenging task of treating high propeller speeds was only made incrementally:

- In 1926, the distinguished British aeronautical engineer Hermann Glauert[1403] commented, "Little is known on the effect of the compressibility of the air on the characteristics of an aerofoil moving with high velocity and further progress, both in theory and in experiment, is necessary before the theory of the airscrew can be modified to take account of this effect."
- The Austrian physicist Christian Doppler {⇨1847} had already shown that a body moving in a circle at supersonic speed produces a rotating head wave. In the late 1930s, this problem was investigated theoretically by Ludwig Prandtl.[1404]

---

[1398] H. Dryden: *Supersonic travel within the last two hundred years.* Sci. Month. **78**, 289-295 (May 1954).

[1399] L.J. Briggs and H.L. Dryden: *Pressure distribution over airfoils at high speeds.* Rept. NACA-TR 255 (1926).

[1400] L. Bairstow, A. Fage, and H.E. Collins: *The relation between the efficiency of a propeller and its speed at rotation.* Brit. Advisory Committee of Aeronautics *(ACA),* Rept. R&M No. 259 (May 1916); Causton & Sons, London (1919).

[1401] E.J.H. Lynam: *Preliminary report of experiments on a high-tip-speed airscrew at zero advance.* Rept. R&M No. 596, Brit. ACA (March 1919).

[1402] S.A. Reed: *Air reactions to objects moving at rates above the velocity of sound, with application to the air propeller.* Rept. NACA-TM 168 (Nov. 1922); http://naca.central.cranfield.ac.uk/reports/1922/naca-tm-168.pdf. • In 1932, Reed co-founded the Institute of Aeronautical Sciences (IAS) with headquarters at New York City. This institute presents the Sylvanus Albert Reed Award, the highest award which an individual can receive for achievement in the field of aeronautical science and engineering.

[1403] H. Glauert: *The elements of aerofoil and airscrew theory.* Cambridge University Press, Cambridge (1926).

[1404] L. Prandtl: *Über Schallausbreitung bei rasch bewegten Körpern.* Schriften Dt. Akad. Luftfahrtforsch. Nr. 7, 1-14 (1939).

| | | |
|---|---|---|
| | | ▸ The high-speed propeller problem turned out to be rather complex and can be divided into three cases: subsonic tip velocity; subsonic flight and supersonic tip velocity; and supersonic flight velocity. In the early 1940s, a theory of high-speed propellers covering all three cases was first derived in the former Soviet Union by F.I. FRANKL.[1405] Experimental studies on a transonic airscrew in real flight were not performed until the early 1950s {McDonnell XF-88B ⇨1953; ⇨Fig. 4.20–G}. |
| 1919 | *Chair of Applied Physics and Electrical Engineering, City & Guilds College, Finsbury, North London, England* | **William Henry ECCLES,**[1406] a British professor of physics who had previously coined the term *diode*, **and his collaborator F.W. JORDAN invent a two-position vacuum tube electronic circuit that alternates positions with successive pulses – the so-called "bistable multivibrator"** (or "flip-flop"). They use the rectangular wave form to sustain the vibration of a tuning fork, which could thus substitute for the contact-driven fork.[1407] ▪ The bistable multivibrator or "Eccles-Jordan circuit" is an extremely versatile logic device which is used in most digital circuits, including those in digital computers. **It also became the most basic element used in electronic counters, which are widely used in shock wave diagnostics.**<br><br>The **"monostable multivibrator"** – also known as a **"one-shot multivibrator"** (or "Eccles-Jordan circuit") – is a derivative of the flip-flop. However, it has only one stable state and a quasi-stable state into which it can be driven for a limited period of time. In addition to further pulse-forming circuitry, **it can provide an electric pulse with a desired delay time, and has become an important device in shock wave physics since it is used to trigger high-speed instrumentation from a high-speed event,** in many cases from the shock wave itself. |
| 1919 | *Earthquake and Geomagnetic Observatory, Bochum, Ruhr Area, Germany* | **Ludger MINTROP,**[1408] a German seismologist, **discovers a new seismic wave phenomenon:** while analyzing contemporary seismic theories and trying to match them with recent measurements on "artificial earthquakes" obtained by using explosives or by dropping a heavy weight {MINTROP ⇨1911}, he noticed in a seismogram that a precursor wave clearly precedes the direct wave. From this **he correctly concludes that a *Grenzwelle* ("boundary wave") or an *elastische Kopfwelle*** ("elastic head wave") **exists;** these later become known among geophysicists as the ***MINTROPscheWelle*** ("Mintrop wave") and among fluid dynamicists as the *von SCHMIDTsche Kopfwelle* ("Schmidt head wave") {VON SCHMIDT ⇨1938}. ▪ Generated at the boundary between two media with different velocities of sound, the Mintrop wave and the Schmidt head wave always propagate along the boundary layer of geological strata at the highest of the two velocities of sound in the two media; *i.e.*, pseudo-supersonically with respect to the medium with the lowest velocity of sound. The phenomenon of the Mintrop wave is used in two important seismic methods:<br><br>(i) The ***seismic refraction method*** {⇨Fig. 4.15–E} has its origin in the *Mintrop seismograph* {MINTROP ⇨1908}. Based on the arrival times of reflected (directly returned) waves and refracted (bent) waves, as measured by a number of seismographs positioned along the surface, MINTROP was able to estimate the depths of subsurface geological formations for the first time. According to Walter KERTZ,[1409] a German geophysicist, MINTROP hit upon the method of seismic surveying by pondering upon the idea that "it couldn't be that nothing rose back to the surface" [Germ. „*Es konnte doch nicht sein, dass nach oben nichts passiert.*"]. ▪ This important method, which proved useful in many practical tests, induced MINTROP to establish Seismos |

---

[1405] F.I. FRANKL: *Theory of propellers with a finite number of blades at high progressive and circular speeds*. Trudy CAHI [Central Aerohydrodynamic Institute, Moscow] **540** (1942).

[1406] W.H. ECCLES: *The use of the triode valve in maintaining the vibration of a tuning fork*. Proc. Phys. Soc. (Lond.) **31**, 269 (1919); W.H. ECCLES and F.W. JORDAN: *Sustaining the vibration of a tuning fork by a triode valve*. The Electrician **82**, 704-705 (June 20, 1919).

[1407] W.A. MARRISON: *The evolution of the quartz crystal clock*. The Bell System Tech. J. **27**, 510-588 (1948).

[1408] According to his own account, he discovered the seismic head wave phenomenon in August 1919. See L. MINTROP: *100 Jahre physikalische Erdbebenforschung*. Die Naturwissenschaften **9/10**, 258-262, 289-295 (1947).

[1409] W. KERTZ: *Ludger MINTROP, der die angewandte Geophysik zum Erfolg brachte*. Mitteil. Dt. Geophys. Gesell. **3**, 2-16 (1991).

GmbH at Hannover in 1921 – the world's first geophysical company. The aim of this German company was to use artificial earthquakes caused by the detonation of small charges of dynamite in a shallow bore-hole to explore for valuable oil, coal and ore deposits. MINTROP's seismic refraction method was first used by him and his crew in Mexico (1922) for oil exploration. Employed by the Gulf Oil Company, his crew discovered the first salt dome ("Orchard Dome") to produce oil in 1924 in Fort Bend County, Texas. In the following years, this most profitable method was used almost exclusively for the detection of salt domes in the Texas-Louisiana Gulf Coast region, an important source of oil and gas.

(ii) The *seismic reflection method* {⇨Fig. 4.15–E} – a competitive technique – **has dominated seismic prospecting: it requires smaller explosive charges than the refraction method,** because the geophones are smaller distances away from the shot point, and it provides better structural detail. Note that when the distance between the shot point and the receiving point is relatively short, reflected waves usually reach the receiving point prior to refracted waves. • The seismic reflection method was first successfully used in 1927 in Oklahoma for oil exploration by the U.S. geologist Jones E. DUNCAN, who supervised an experimental crew from the U.S. Geophysical Research Corporation (Tulsa, OK).[1410]

| | | |
|---|---|---|
| 1920 | Mt. Wilson Solar Observatory, Pasadena, California | John August ANDERSON,[1411] a U.S. astronomer, **uses exploding wires to produce high temperatures in excess of about 3,000 °C available at that time for high-temperature spectroscopic studies.** He investigates the pressure shifts of spectral lines that occur during explosions and concludes that the brilliant flash has an intrinsic intensity that corresponds to a temperature of about one hundred times the intrinsic brilliancy of the Sun. Using a rotating mirror, he visualizes the dynamics of the size of the flash: the shock wave emitted. In open air he measures speed of propagation of about 3,300 m/s. • ANDERSON, today considered to be the father of scientific exploding wire research {1st Conf. on the Exploding Wire Phenomenon ⇨1959}, inspired many subsequent researchers in this field. However, the clear visualization of the shock and flow field around exploding wires – a low-level phenomenon that requires optical magnification and therefore increases the velocity on the film plane in practice – requires ambitious ultrahigh-rate diagnostics, particularly ultrahigh-speed photographic recording, which was not realized until after World War II {MÜLLER ⇨1957; ⇨Fig. 4.16–P}. |
| 1920 | Munitions Inventions Dept. & Ballistic Range of the Experimental Dept. of HMS Excellent, Portsmouth, Hampshire, England | Leonard BAIRSTOW[1412] (a professor of aviation at Imperial College and England's leading aerodynamicist), **Ralph Howard FOWLER** (a mathematical physicist), **and Douglas Rayner HARTREE** (a physics student), each of whom is involved in an anti-aircraft gunnery project, **measure the dynamic pressures acting on the head of an actual 18-pdr. (8.16-kg) shell between $M = 0.9$ and 1.2** {⇨Fig. 4.6–M}:<br>(i) The idea is to use a service time-fuse as a manometer to determine the pressure under which powder is burning. From laboratory experiments, it is known that the burn rate of a service time-fuse is affected by the total external pressure on the vents.<br>(ii) A series of observations, for shells fired along the same trajectory at short intervals of time, determines a relation between time of burning and length of fuse burnt. By numerical differentiation of the observations, they obtain a relation between the rate of burning and the time, and therefore, by comparison with the laboratory experiments, between the external pressure and the time.<br>(iii) They use shells fitted with caps entirely enclosing the fuse; each cap has a series of holes drilled at the same distance from the nose. |

---

[1410] R. KELLER: *Exploration seismic techniques*. Dept. of Geol. Sciences, University of Texas, El Paso; http://www.geo.utep.edu/pub/ortega/introduction.PDF.
[1411] J.A. ANDERSON: *Spectra of explosions*. Proc. Natl. Acad. Sci. (U.S.A.) **6**, 42-43 (1920); *The spectrum of electrically exploded wires*. Astrophys. J. **51**, 37-43 (1920).
[1412] L. BAIRSTOW, R.H. FOWLER, and D.R. HARTREE: *The pressure distribution on the head of a shell moving at high velocities*. Proc. Roy. Soc. Lond. **A97**, 202-218 (1920).

| | | |
|---|---|---|
| | | (iv) Their experimental results show that the dynamic pressure has a maximum positive value at the nose of the shell of between 0.7 and 0.3 bar, but this falls rapidly as the observation point moves towards the base; for example, at 2 inches (5 cm) behind the projectile head it drops to about 0.1 bar. The pressure even becomes negative (max. –0.15 bar) before the cylindrical part of the shell is reached. ▪ For $M = 1.2$, their measured pressure data along a 18-pdr shell boundary, starting about half a calibre away from the projectile point, coincide fairly well with the results obtained with an approximation method developed some time later {VON KÁRMÁN & MOORE ⇒ 1932}. |
| 1920 | Ecole Nationale Supérieure des Mines, Paris, France | Émile JOUGUET,[1413] a professor of mechanical engineering, **discusses the similarity between shooting channel flow and supersonic compressible flow.** He also suggests that this analogy can be used to study two-dimensional gas flows via experiments with a rectangular water channel. ▪ An extension to three-dimensional motion was provided in the 1930s by Dimitri P. RIABOUCHINSKY {⇒ 1932}. |
| 1920 | Deutsche Seewarte & Universität Hamburg, northern Germany | Alfred WEGENER,[1414] an eminent German meteorologist and geophysicist best known for his development of the continental drift hypothesis,[1415] **performs a series of excellent experiments to verify the idea that lunar craters** {HOOKE ⇒ 1665; GILBERT ⇒ 1893} **and the terrestrial Meteor Crater** {⇒ c.50,000 years ago; BARRINGER & TILGHMAN ⇒ 1905; CHAO ET AL. ⇒ 1960} **originated from meteorite impacts.**<br>(i) He caries out impact experiments into gypsum powder. Subsequently sprayed with water, the hardened plaster crater model was then photographed under low-angle light.<br>(ii) He successfully experimentally reproduces a crater geometry with a central hill, and concludes, "The cone-shaped central hill consists of that part of the loose layer [covering the ground prior to the impact event] which because of his central location experiences no radial acceleration, but only a compression." |
| 1921 | Sheffield, South Yorkshire, England | **Establishment of the Safety in Mines Research Board (SMRB)** by the British Mines Department.<br>In the following years, the organization acquired the Harpur Hill, Buxton Site (1924) for performing mining safety work on a large scale, and opened up central laboratories (1928) in Portobello Street, Sheffield {PAYMAN ⇒ 1928; PAYMAN & WOODHEAD ⇒ 1931; PAYMAN & SHEPHERD ⇒ 1937 & 1940}. In 1947, the Safety in Mines Research Establishment (SMRE) was formed as part of the British Ministry of Fuel and Power, bringing together the work at Sheffield and Buxton. |
| 1921 | Friedrich-Wilhelms-Universität, Berlin, Germany | Richard A. BECKER,[1416] a German theoretical physicist, **presents his thesis entitled *Stoßwelle und Detonation* ("Shock Wave and Detonation") for the certificate of habilitation.** Published one year later, his thesis will become a classic in gas dynamics. His achievements can be summarized as follows:<br>(i) To illustrate how shock waves in gases are formed on a qualitative basis, he proposes his simple ***Becker piston model*** {⇒ Fig. 4.4–D(e)}. Assuming the stepwise motion of a piston in a tube and the coalescence of pressure pulses, he qualitatively explains how shock waves in fluids are formed. ▪ This phenomenon of the "shocking-up" process of a finite pressure pulse has been observed in measurements of waves generated by bursting air-filled pressure vessels.[1417] |

---

[1413] E. JOUGUET: *Quelques problèmes d'hydrodynamique général.* J. Math. Pures Appl. **3** [VIII], 1-63 (1920).

[1414] A. WEGENER: *Die Aufsturzhypothese der Mondkrater.* Sirius (Leipzig) **53**, 189-194 (1920); *Versuche zur Aufsturztheorie der Mondkrater.* Nova Acta (Abhandl. der Kaiserl. Leopold.-Carol. Dt. Akad. Naturforscher) **106**, Nr. 2, 109-117 (1920); *Die Entstehung der Mondkrater.* Vieweg, Braunschweig (1921).

[1415] A. WEGENER: *Die Entstehung der Kontinente und Ozeane.* Vieweg, Braunschweig (1915).

[1416] R. BECKER: *Stoßwelle und Detonation.* Z. Phys. **8**, 321-362 (1922); *Impact, waves and detonation.* Rept. NACA-TM 505 (1929).

[1417] R.J. LARSON and W. OLSON: *Measurements of air blast effects from simulated nuclear reactor core excursions.* Memo No. 1102, BRL, Aberdeen Proving Ground, MD (1957).

(ii) He concludes that weak shocks are many free paths thick and may be treated by ordinary hydrodynamics including the effects of viscosity and thermal conductivity of the medium.

(iii) **Treating the Navier-Stokes equations** {NAVIER ⇨ 1822} **for nonweak shocks, he obtains the first exact solution to the one-dimensional equations for a real fluid** – *i.e.*, involving both viscosity and heat conduction. ▪ This problem was resumed by the two British researchers Geoffrey I. TAYLOR[1418] and John W. MACCOLL, who gave an approximate solution for weak shock waves.

(iv) **He calculates the thickness of a shock front in air** by assuming constant values for the transport coefficient and the specific heats. For air initially at 1 bar and 0 °C, he finds for a shock pressure of 8 bar that the front thickness is already smaller than the mean free path length, which amounts to about 90 nm, and for strong shocks at 2,000 bar even remains below the mean distance between two molecules (about 3.3 nm). Using the Boltzmann equation {BOLTZMANN ⇨ 1872} **he concludes that classical kinetic theory is inapplicable to very intense shock waves in gases,** because the temperature increase in the shock front is the result of only a few collision processes. ▪ In 1944, it was shown that BECKER's conclusion that the Boltzmann equation is no longer applicable rests on an oversight: **the thickness of a shock front is always at least of the order of magnitude of a free path,** and it is to be expected that the Boltzmann equation can be applied even for the most violent shocks {THOMAS ⇨ 1944}.

(v) Starting from the Tamman equation of state,[1419] **BECKER also estimates shock wave data in liquids.** For example, for a strong shock wave propagating in ethyl ether [$C_4H_{10}O$] at 10 kbar, the calculated shock front thickness is 0.65 nm, which is comparable to 0.55 nm, the mean distance between two molecules.

(vi) **He calculates detonation velocities in gases and essentially confirms JOUGUET's theory** {⇨1905}. ▪ However, similar calculations were not yet possible for liquid and solid explosives because the equations of state for the hot burnt gases were unknown.

In 1963, **Jürgen ZIEREP**,[1420] a German applied mathematician, resuming BECKER's piston model **explained the formation of the compression shock graphically using the elegant method of characteristics** {MONGE ⇨ 1770s; HADAMARD ⇨ 1903}.

| | | |
|---|---|---|
| 1921 | *Dept. of Physics, McGill University, Montreal, eastern Canada* | **David A. KEYS,**[1421] a Canadian physicist, **uses the piezoelectric effect to record, "with negligible inertia," very rapid pressures obtained when mixtures of electrolytic gas and air are detonated at constant volume.** He uses tourmaline crystals and records the change in electrical charge with a Wood-type cathode-ray oscillograph {WOOD ⇨ 1923}. ▪ The principle underlying the design of his gauge was first suggested by the British physics professor Joseph John THOMSON, discoverer of the electron and winner of the 1906 Nobel Prize for Physics. He had demonstrated the piezoelectric properties of certain types of crystal and their potential for performing measurements of rapidly fluctuating pressures in the low microsecond regime.<br><br>Two years later, KEYS[1422] applied his diagnostic method in studies for the British Admiralty. He used a tourmaline pressure gauge to measure shock pressures of underwater explosions resulting from the detonation of small charges of guncotton and TNT {⇨Fig. 4.19–C}. |
| 1921 | *Moscow University, Russia* | **Aleksandr Ivanovich NEKRASOV,**[1423] a Soviet mathematician and fluid dynamicist, **first proves the existence of progressing periodic waves of finite amplitude in water of infinite** |

---

[1418] G.I. TAYLOR and J.W. MACCOLLL: *The mechanics of compressible fluids*. In: (W.F. DURAND, ed.) *Aerodynamic theory*. Springer, Berlin (1935), vol. III, pp. 209-250.
[1419] G. TAMMANN: *Über Zustandsgleichungen im Gebiete kleiner Volumen*. Ann. Phys. **37** [V], 975-1013 (1912).
[1420] J. ZIEREP: *Vorlesungen über theoretische Gasdynamik*. G. Braun, Karlsruhe (1963), p. 96.
[1421] D.A. KEYS: *A piezoelectric method of measuring explosion pressures*. Phil. Mag. **42** [VI], 473-488 (1921).
[1422] D.A. KEYS: *The cathode-ray oscillograph and its application to the exact measurement of explosion pressures, potential changes in vacuum tubes and high tension magnetos*. J. Franklin Institute **196**, 576-591 (1923).

depth from a purely mathematical point of view. ▪ In the following year, NEKRASOV was awarded the N.E. Zhukovsky Prize for his work.

In 1925, Tullio LEVI-CIVITA,[1424] an Italian professor of mechanics and mathematics, independently tackled the same problem via a different method which proved useful for analyzing steady surface waves.

| 1921 | *Flugzeugbau Friedrichshafen, Werft Warnemünde, northeastern Germany* | Ernst POHLHAUSEN,[1425] a former Ph.D. student of Prof. Ludwig PRANDTL, **discusses the heat exchange between solid bodies and fluids in the presence of only minor friction and heat conduction.** He first shows that in a flow of velocity $v$, a solid surface with a laminar boundary layer should experience a temperature rise given by $$\Delta T = \tfrac{1}{2} v^2 (1 - C_0)/c_p,$$ which, for an airstream with $C_0 = 0.844$ and $c_p = 0.24 \text{ cal g}^{-1} \text{degree}^{-1}$, reduces to $$\Delta T = 7.766 \times 10^{-5} v^2,$$ where $v$ is in m/s and $\Delta T$ is in degrees. ▪ English researchers experimentally verified POHLHAUSEN's equation at high subsonic velocities on a flat plate, and found it to be reasonably true.[1426] |
|---|---|---|
| 1921 | *Dept. of Engineering, National Physics Laboratory (NPL), Teddington, Middlesex, England* | Sir Thomas E. STANTON,[1427] a British engineer and former student of Osborne REYNOLDS, **sets up a mini supersonic wind tunnel** {⇨Fig. 4.9–B} – obviously the world's first to have a Mach number that is significantly above 1. The blow-down facility, which has a diameter of only 0.8 in. (20.3 mm) but approaches $M = 2$, is first used for Pitot-tube calibrations in supersonic flow. **The tunnel will also be used to quantitatively investigate drag, lift and upsetting moment on scale models of projectiles** with diameters not exceeding 0.09 in. (2.29 mm); drag forces in the order of 10 grams are measured with a little balance.[1428] ▪ STANTON's mini wind tunnel has survived and is now kept at the NPL Museum in Teddington.<br><br>Later, STANTON used a more advanced wind tunnel especially designed for airfoil studies {STANTON ⇨1928}. |
| 1922 | *Friedrich-Wilhelms-Universität, Berlin, Germany* | Richard BECKER,[1429] a German physicist, **develops a hydrodynamic theory for the detonation of condensed explosives in order to compute the detonation velocities of nitroglycerin and mercury fulminate.**<br><br>(i) His theory is based on applying a simple equation of state for the detonation products, according to the form $$p(V - \alpha) = RT,$$ where $p$ is the pressure, $V$ is the volume, $T$ is the temperature, $R$ is the universal gas constant, and $\alpha$ is the covolume representing the incompressible portion of the detonation products – the so-called **"Becker equation of state."**<br><br>(ii) He states that the computed detonation velocities were determined "with an accuracy indicating the order, at least, of the magnitude observed."<br><br>Since the application of his theory led to considerable discrepancies with experimental data for a number of explosives, various improvements were proposed in the following years {ZEL'DOVICH ⇨1940; KISTIAKOWSKY & WILSON ⇨1941; VON NEUMANN ⇨1942; DÖRING ⇨1943; LANDAU & STANYUKOVICH ⇨1945; JONES & A.R. MILLER ⇨1948; MADER ⇨1963}. |

---

[1423] A.I. NEKRASOV: *On steady waves.* Izvestija Ivanovo-Voznesenskogo Politechniceskogo Instituta [Bull. Inst. Politech. Ivanovo-Vosniesensk], No. 3 (1921); *The exact theory of steady waves on the surface of a heavy liquid.* Izdat. Akad. Nauk SSSR, Moscow (1951).

[1424] T. LEVI-CIVITA: *Détermination rigoureuse des ondes permanents d'ampleur finie.* Math. Ann. **93**, 264-314 (1925).

[1425] E. POHLHAUSEN: *Der Wärmeaustausch zwischen festen Körpern und Flüssigkeiten mit kleiner Reibung und kleiner Wärmeleitung.* ZAMM **1**, 115-121 (1921).

[1426] W.F. HILTON: *High-speed aerodynamics.* Longmans, Green & Co., London etc. (1952), p. 528.

[1427] T.E. STANTON: *The development of a high-speed wind tunnel for research in external ballistics.* Proc. Roy. Soc. Lond. **A131**, 122-132 (1931).

[1428] The results were described in a report to the Ordnance Committee in 1922.

[1429] R. BECKER: *Stoßwelle und Detonation.* Z. Phys. **8**, 321-362 (1921/1922).

| | | |
|---|---|---|
| 1922 | General Electric Company (GEC), Lynn, Massachusetts | **Lyman James** BRIGGS,[1430] a U.S. aeronautical engineer, **and collaborators begin measurements of the characteristics of wing sections moving at sonic and supersonic speeds,** with the wing sections corresponding to tip sections of propeller blades. They do not use a wind tunnel but rather open-air jets ranging from 2 to 12 in. (5.1–30.5 cm) in diameter, thus unwittingly following Peter SALCHER's previously suggested method of supersonic drag testing {SALCHER ⇨1889}. |
| 1922 | Faculty of Physics and Mathematics, Polytechnic Institute, St. Petersburg, Russia | Alexander FRIEDMANN,[1431] a Soviet physicist and astronomer, **publishes a paper in a German journal entitled** *Über die Krümmung des Raumes* **("On the Curvature of Space")**. Based upon Albert EINSTEIN's static cosmological model {A. EINSTEIN ⇨1917}, FRIEDMANN **formulates the first theory of a variable type of model Universe.** Assuming that the average mass is constant and that all fundamental parameters are known except the expansion factor, he shows that the radius of curvature of the Universe can either increase or it can be a periodic function of time. ▪ His work, which was published prior to the discovery of redshifts {HUBBLE ⇨1929}, stimulated others to derive nonstationary cosmological models from EINSTEIN's general theory of relativity {LEMAÎTRE ⇨1927; ALPHER, HERMAN & GAMOW ⇨1948}. |
| 1922 | Lick Observatory, Santa Cruz, California | Willem J. LUYTEN,[1432] a Dutch-born U.S. astronomer, **coins the term** *white dwarf.* ▪ It entered general circulation when the British astrophysicist Sir Arthur Eddington picked it up and used it in his famous mass-luminosity relation paper in early 1924.[1433] ▪ A white dwarf is a "collapsed star" that has exhausted most or all of its nuclear fuel. Its outer layers has drifted away into space, while its core has collapsed into a planet-sized ball of very high density (in the order of $10^6$ g/cm$^3$). **A white dwarf that consumes matter from its companion can explode as** *Type Ia Supernovae* {CHANDRASEKHAR ⇨1931}, causing its complete thermonuclear disruption.[1434]<br><br>The term *black dwarf* – a small, very dense, cold, dead star – was coined by John A. WHEELER, a U.S. theoretical physicist at Princeton University, in 1969. The term *brown dwarf* – a sub-stellar object which is not really brown but a very dull red – was coined by Jill C. TARTER, a U.S. astrophysicist at NASA's Ames Research Center in 1975. The upper mass for a brown dwarf is that which is just insufficient for normal hydrogen fusion to be triggered in the core. |
| 1923 | Hercules Powder Co., Wilmington, Delaware | **Foundation of the American journal** *The Explosives Engineer*, with the goal to serve industries that use explosives. The first editors are Harry ROBERTS JR. and Nelson S. GREENSFELDER. ▪ The publication of this journal ceased in 1961 with vol. 39. |
| 1923 | British Aeronautical Research Committee, Farnborough & London, England | G.P. DOUGLAS[1435] **and R.M.** WOOD, two British aeronautical engineers, **resume previous high-speed propeller tests in free air** {LYNAM ⇨1919} by studying the motion of airscrews 2 ft (0.61 m) in dia. in their 7-ft (2.13-m) low-speed wind tunnel.<br><br>(i) They developed a powerful turbine-driven propeller dynamometer suitable for testing propellers at high tip speeds in their wind tunnel. |

---

[1430] L.J. BRIGGS, G.F. HALL, and H.L. DRYDEN: *Aerodynamic characteristics of airfoils at high speeds*. Rept. NACA-TR 207 (1925).

[1431] A.A. FRIEDMANN: *Über die Krümmung des Raumes*. Z. Phys. **10**, 377-386 (1922).

[1432] W.J. LUYTEN: *Third note on faint early-type stars with large proper motions*. Publ. Astron. Soc. Pac. **34**, 356-357 (1922); *White dwarfs and degenerate stars*. Vistas in Astronomy **2**, 1048-1056 (1956).

[1433] Private communication by Dr. Jay HOLBERG, Dept. of Planetary Sciences, Lunar and Planetary Laboratory (LPL), University of Arizona, Tucson, Arizona. See also his recent paper *How degenerate stars came to be known as white dwarfs*. Paper No. 205.1, presented at the AAS 207th Meeting, Washington, DC (Jan. 8-12, 2006).

[1434] D. BRANCH: *When a white dwarf explodes*. Science **299**, No. 5603, 53-54 (2003).

[1435] G.P. DOUGLAS and R.M. WOOD: *The effects of tip speed on airscrew performance. An experimental investigation of an airscrew over a range of speeds of revolution from "model" speeds up to tip speeds in excess of the velocity of sound in air*. Brit. ARC-R&M No. 884, London (June 1923).

| | | |
|---|---|---|
| | | (ii) **They find that whenever airscrews revolve close to the speed of sound, the lift drops sharply and the drag rises.** At their highest tip speed of 1,180 ft/s (359 m/s, $M = 1.08$), the propeller efficiency drops from 0.67 to 0.36. |
| 1923 | *Private residence at Cambridge, England* | Horace LAMB,[1436] a retired professor of mathematics at Owens College of the University of Manchester and 74 years old at the time, **analytically treats the early expansion of the gas bubble of an underwater explosion.** |
| | | In the same year, S. BUTTERWORTH[1437] at the British Admiralty Research Laboratory **extends LAMB's theoretical treatment by including the effect of a constant external hydrostatic pressure.** Obviously not aware of previous analytical results {BLOCHMANN ⇨ 1898}, he confirms that the gas bubble of an underwater explosion must oscillate in size. He finds that the time of oscillation is on the order of one second for a 100-lb (45.359-kg) charge, and correctly concludes that in deep water this may result in a succession of pulses which would, however, rapidly diminish in amplitude. |
| 1923 | *General Electric Company (GEC), Schenectady, New York* | Irving LANGMUIR, a U.S. physical chemist and associate director at GEC who had studied gas discharges in electron tubes with his colleagues since the early 1920s, **coins the term *plasma* for a quasi-neutral system of ionized gas** consisting of neutrals, free positive ions, and free electrons. **To distinguish it from solids, fluids and electrically neutral gases, he considers the plasma to be a fourth state of matter** – in analogy with the theory held by ancient Greek philosophers that the Universe consists of four elements: earth, water, air, and fire. ▪ Harold M. MOTT-SMITH,[1438] a collaborator of LANGMUIR for many years, later remembered: "We noticed the similarity of the discharge structures [between mercury vapor discharges, Geissler tubes and gas-filled thermionic tubes] they revealed. LANGMUIR pointed out the importance and probable wide bearing of this fact. We struggled to find a name for it. For all members of the team realized that the credit for a discovery goes not to the man who makes it, but to the man who names it … We tossed around names … but one day LANGMUIR came in triumphantly and said he had it. He pointed out that the 'equilibrium' part of the discharge acted as a sort of substratum carrying particles of special kinds, like high-velocity electrons from thermionic filaments, molecules and ions of gas impurities. This reminds him of the way blood plasma carries around red and white corpuscles and germs … So he proposed to call our 'uniform discharge' a *plasma*." LANGMUIR[1439] did not use the word *plasma* in a scientific paper until 1929. |
| | | Shock waves propagating in a plasma without an applied external magnetic field behave much like shock waves propagating in a neutral gas, because submicroscopic forces among the electrically charged particles can be neglected to the first order. **In the case of strong shock waves, however, collision processes at the shock front may provoke an ionization of the neutral gas particles** – thus leading to spectacular luminous phenomena {MICHEL-LÉVY ET AL. ⇨ 1941}. |
| 1923 | *Dept. of Physics, University of Danzig, Germany* | Carl W. RAMSAUER,[1440] a German professor of experimental physics, **systematically studies full-scale underwater explosions** with charges of gun-cotton up to 1.91 kg which are fired at depths of up to eight meters: |

---

[1436] H. LAMB: *On the early stages of a submarine explosion.* Phil. Mag. **45** [VI], 257-265 (1923).
[1437] S. BUTTERWORTH: *Report on the theoretical shape of the pressure-time curve and on the growth of the gas-bubble.* In: (G.K. HARTMANN and E.G. HILL, eds.) *Underwater explosions research. A compendium of British and American reports.* Office of Naval Research (ONR), Dept. of the Navy, Washington, DC (1950); *see* vol. 2: *The gas globe*, pp. 1-11.
[1438] H.M. MOTT-SMITH: *History of "plasmas."* Nature **233**, 219 (Sept. 17, 1971).
[1439] I. LANGMUIR: *The interaction of electron and positive ion space charge in cathode sheaths.* Phys. Rev. **33** [II], 954-989 (1929).
[1440] C. RAMSAUER: *Die Massenbewegung des Wassers bei Unterwasserexplosionen.* Ann. Phys. **72** [IV], 265-284 (1923); *The movement of water in a body caused by explosions* [Summary by G.F.S. (anonymous)]. J. Franklin Institute **196**, 850-851 (1923).

(i) **He first determines the position of the gas bubble boundary using an ingenious "electrolytic probe" method** {⇨Fig. 4.16−C}. It consists of an arrangement of electrodes supported at suitable distances from the charge by a rigid frame which, together with a common electrode, forms conducting circuits with the seawater acting as an electrolyte.

(ii) The expansion of the bubble successively interrupts the electrode circuits, and these interruptions are recorded with a mechanical chronograph. He finds the relation

$$r_{max} \sim (m/P)^{1/3}$$

for the maximum radius, where $m$ is the mass of the explosive, and $P$ is the total hydrostatic pressure at the depth of the explosion.

(iii) **He also makes the important observation that the bubble migrates upwards.** With a charge of 1.91 kg, the water is moved outward by 166 cm when the cartridge is situated 3 m below the water surface, and by 146 cm when the depth is 8 m. ▪ His method was limited to the recording of the bubble expansion; it could not detect its oscillation, which was recognized during World War II as providing a further source of underwater shock waves that endanger submarines {United States & Germany ⇨1941}.

| 1923 | *Admiralty Research Laboratory, Teddington, Middlesex, England* | **Albert Beaumont WOOD**,[1441] a British physicist, **reports on a new high-speed single-shot cathode-ray tube (c.r.t.) oscillograph for recording rapid impulsive phenomena such as pressure-time profiles of underwater explosions with high temporal resolution** {KEYS ⇨1921}. The instrument was devised by the British physics professor Joseph J. THOMSON and constructed by WOOD in the early 1920s. It uses a hot cathode and is provided with two pairs of plates for horizontal and vertical electrostatic deflection. The c.r.t. is triggered by a high-frequency periodic voltage at the horizontal deflection plates. Recording proceeds by exposing the photographic film (Schumann plates) in the vacuum directly to the cathode rays.<br><br>In 1951, WOOD (1890−1964) received the prestigious Duddell Medal and Prize,[1442] and the Underwater Acoustics Medal (1961) from the Acoustical Society of America (ASA) "for the development of the cathode-ray oscillograph and its adaptation to the study of underwater explosions." Since 1970, the A.B. Wood Medal and Prize from the Institute of Acoustics (United Kingdom) has been presented to individuals who have made distinguished contributions to the application of acoustics associated with the sea. |
|---|---|---|
| 1924 | *University of Vienna, Austria* | **Franz BERGER**,[1443] an Austrian physicist attempting to study the collision of solid bodies in greater detail than many of his predecessors, **investigates how forces, displacements and velocities conform with the natural laws of elastic collision.**<br><br>(i) He measures the force-time profile of colliding bodies using a twin-pendulum apparatus {⇨Fig. 4.3−O} in which a spring is inserted between test body and impactor {⇨Fig. 2.7} and the deformation of the spring is recorded using single-flash photography.<br><br>(ii) He experimentally proves that the shocked contact surface of the collision moves impulsively and attributes this phenomenon to the rarefaction wave created by reflection of the compression wave at the free surface − a phenomenon today known as **"surface shock unloading."** |

---

[1441] A.B. WOOD: *The cathode ray oscillograph*. Proc. Phys. Soc. (Lond.) **35**, 109-124 (1923); J. Inst. Electr. Eng. **63**, 1046-1055 (1925).
[1442] The *Duddell Medal* was instituted by the Council of The Physical Society in 1923 as a memorial to the English electrical engineer William DU BOIS DUDDELL (1872−1917), the inventor of the electromagnetic oscillograph.
[1443] F. BERGER: *Das Gesetz des Kraftverlaufes beim Stoß. Untersuchungen über die gesetzmäßigen Beziehungen beim Stoß elastischer Körper*. Vieweg & Sohn, Braunschweig (1924).

| | | |
|---|---|---|
| 1924 | *Hector Observatory [now Dominion Observatory], Wellington, New Zealand* | **Algernon Charles** GIFFORD,[1444] a New Zealand astronomer and public lecturer, **points out – and is probably the first to do so – that lunar craters may have been formed by explosions rather than by the simple process of splashing, resulting from the impact of a high-velocity punch:**

(i) Carefully analyzing the effects of meteoric impact on the Moon, he emphasizes the importance of explosion on impact.

(ii) Proposing his **"impact-explosion analogy,"** he states, "The fact which has not been taken into account hitherto in considering the meteoric hypothesis is that **a meteor, on striking the Moon, is converted, in a very small fraction of a second, into an explosive compared with which dynamite and TNT are mild and harmless** ... It is known that the meteors which enter the Earth's atmosphere do so with velocities ranging from 10 to 45 miles a second [16 to 72 km/s]. Those which collide with the Moon should have almost the same speed."

(iii) He points out that the collision energy of meteoritic bodies would be sufficient to form explosion pits and that the resulting craters would be nearly circular regardless of the angle of approach. ▪ High-speed impact experiments carried out in the early 1950s, however, revealed that the circular shape of a crater becomes increasingly elliptically with increasing obliquity of impact {RINEHART & WHITE ⇨1952}.

GIFFORD's hypothesis on the origin of lunar craters supported SCHRÖTER's rule {SCHRÖTER ⇨1802} and extended GILBERT's lunar impact theory {GILBERT ⇨1893}. |
| 1924 | *Faculty of Sciences, University of Marseille, southern France* | **Joseph Jean Camille** PÉRÈS,[1445] a French professor of rational and applied mechanics, **analytically treats the general case of two- and three-dimensional impact of two solid bodies of arbitrary contours.** Using a geometric representation, he investigates the influence of friction for given coefficients of restitution and friction, and determines the trajectory of the point of contact for both impacting bodies. |
| 1924 | *Physikalisch-Chemisches Institut, Friedrich-Wilhelms-Universität, Berlin, Germany* | **Rudolf** WENDLANDT,[1446] previously a Ph.D. student of Prof. Walther H. NERNST, **reports on detonation experiments in gaseous mixtures of $H_2/O_2$ and $CO/O_2$.**

(i) Studying detonation in a 21-mm-dia. tube of variable length (0.8–8.93 m), he uses a ballistic galvanometer {POUILLET ⇨1844} to measure the detonation velocity. The gaseous mixtures were ignited by a strong spark generated from a Ruhmkorff induction coil.

(ii) By plotting the detonation velocity (on the ordinate) as a function of the hydrogen content (on the abscissa) for different hydrogen/air mixtures, he observes that – contrary to detonation velocities calculated according to the classical theory – **the detonation velocity drops sharply at about 20% $H_2$ within a narrow range of the mixing ratio,** while less violent $CO/O_2$ mixtures reveal only a weakly pronounced velocity maximum. He explains this unusual phenomenon by the fact that the reaction does not have time to occur within an "accessible" time, and that the heat released in the wave is less than the total reaction heat.

(iii) He concludes that the ability of an explosive mixture to transmit detonation at constant velocity ends at a certain limit, depending on the mixing ratio and velocity. ▪ Since his measurements for the decrease in the detonation velocity of reactive gaseous mixtures with tube length agreed quantitatively with VIEILLE's previous shock tube measurements in nonreactive gaseous mixtures {VIEILLE ⇨1899}, **WENDLANDT might be considered to be the first to provide experimental evidence that a detonation wave behaves like a shock wave.** |

---

[1444] A.C. GIFFORD: *The mountains of the Moon*. J. Sci. Technol. New Zealand **7**, 129-142 (1924); *The origin of the surface features of the Moon*. Ibid. **11**, 319-327 (1928).

[1445] J.J.C. PERES: *Choc en tenant du frottement*. Nouv. Ann. Math. **2** [V], 98-107 (1924); *Choc de deux solides avec frottement*. Ibid. **2** [V], 216-231 (1924); With E. DELASSUS: *Note sur le choc en tenant compte du frottement de glissement*. Ibid. **2** [V], 383-391 (1924).

[1446] R. WENDLANDT: *Detonationsgrenze und Detonation gasförmiger Gemische*. Ph.D. thesis, Friedrich-Wilhelms-Universität, Berlin (1923); *Experimentelle Untersuchungen zur Detonationsgrenze gasförmiger Gemische*. Z. Physik. Chem. **110** [Nernst-Band], 637-655 (1924); *Die Detonationsgrenzen in explosiven Gasgemischen*. Ibid. **116**, 227-260 (1925).

| | | |
|---|---|---|
| 1925 | *Aerodynamische Versuchsanstalt (AVA), University of Göttingen, Germany* | Jakob ACKERET,[1447] a Swiss aerodynamicist, **publishes his famous** *two-dimensional (2-D) linearized wing theory,* which neglects second-order terms and skin friction.

(i) He considers a thin wing with sharp leading and trailing edges exposed to uniform and parallel supersonic flow.

(ii) According to his theory, the deflection of the stream causes an increase in pressure at a concave corner and a pressure drop at a convex corner. Consequently, in the case of supersonic flow, **a** *shock wave* **emanates from the concave corner and an** *expansion wave* **(or** *rarefaction wave)* **from the convex corner.**

(iii) The 2-D cases for both wave types, first observed by Ernst MACH and Peter SALCHER in their pioneering supersonic ballistic experiments {E. MACH & SALCHER ⇨1886}, had already been solved by Ludwig PRANDTL {⇨1907} and Theodor MEYER {⇨1908} – known as the **"Prandtl-Meyer theory of expansion."** ▪ Thus, ACKERET's concept allowed one to calculate the flow past a 2-D wing at supersonic speeds more readily than that at subsonic speeds, and has been widely applied by succeeding generations of fluid dynamicists to a variety of practical problems in aeronautical engineering. Fritz SCHULTZ-GRUNOW, a former Ph.D. student of professor PRANDTL, compared ACKERET's supersonic wing theory to the development of the Kutta-Zhukovsky equation (1912) used at subsonic speeds.

The experimental testing of ACKERET's 2-D wing theory was delayed, probably due to the worldwide shortage of supersonic tunnels available in the mid-1920s. 2-D balance tests were performed in Italy in the Guidonia supersonic tunnel in 1939 and in England at NPL in 1943–1945. In Germany, first studies were carried out to find the optimum aerodynamic geometry of wings and fins used in guided supersonic missiles {BETZ ⇨1939; HVA Peenemünde ⇨1939}. ACKERET's theory was resumed by the British physicist Geoffrey I. TAYLOR {⇨1932} and the German aeronautical engineer Adolf BUSEMANN {5th Volta Conf. (BUSEMANN) ⇨1935}. The rapid evolution of practical aeronautics in the supersonic and hypersonic regimes demanded new concepts for treating fluid flow problems, which resulted in a new discipline – **Gas Dynamics of Thin Bodies.**[1448] |
| 1925 | *Harvard University, Cambridge, Massachusetts* | Percy W. BRIDGMAN,[1449] a U.S. physics professor, **resuming his previous studies on piezoresistivity** {BRIDGMAN ⇨1911}, **makes first systematic measurements for a variety of metals.** He also recognizes the tensor nature of this effect in crystals. ▪ In 1950, BRIDGMAN[1450] used manganin as a secondary pressure gauge and reported a linear resistance change with static pressure up to 30 kbar.

In the 1960s, this so-called **"piezoresistive effect"** became important in solid-state shock wave physics as a diagnostic tool for measuring the dynamic pressure responses of shock-loaded materials {HAUVER ⇨1960; FULLER & PRICE ⇨1962; BERNSTEIN & KEOUGH ⇨1964; BERNSTEIN ET AL. ⇨1967}. |
| 1925 | *Institut für Technische Physik, TH Berlin-Charlottenburg, Germany* | Carl CRANZ,[1451] an international authority on ballistics, **publishes the monograph** *Äußere Ballistik* ("Exterior Ballistics"), **the first volume of his four-volume textbook** *Lehrbuch der Ballistik* ("Textbook of Ballistics"). ▪ Addressing both theoretical and experimental methods, it became the book that was probably consulted the most often by generations of ballisticians. In a previous contribution to a dictionary of physics, CRANZ[1452] already had given a definition of the two main branches of ballistics: |

---

[1447] J. ACKERET: *Luftkräfte auf Flügel, die mit größerer als Schallgeschwindigkeit bewegt werden.* Z. Flugtech. & Motor-Luftschiffahrt **16**, 72-74 (1925).
[1448] F.I. FRANKL and E.A. KARPOVICH: *Gas dynamics of thin bodies* [translated from Russian]. Interscience Publ., London & New York (1953).
[1449] P.W. BRIDGMAN: *The effect of tensions on the transverse and longitudinal resistance of metals.* Proc. Am. Acad. Arts Sci. **60**, 423-449 (1925).
[1450] P.W. BRIDGMAN: *Physics above 20,000 kg/cm²* [Bakerian Lecture]. Proc. Roy. Soc. Lond. **A203**, 1-17 (1950).
[1451] C. CRANZ: *Lehrbuch der Ballistik.* Springer, Berlin. Vol. I (1925): *Äußere Ballistik;* vol. II (1926): *Innere Ballistik;* vol. III (1927): *Experimentelle Ballistik;* vol. IV (1936): *Supplement.*
[1452] C. CRANZ and O. VON EBERHARD: *Ballistik.* In: (A. BERLINER and K. SCHEEL, eds.) *Physikalisches Handwörterbuch.* Springer, Berlin (1924), pp. 74-75.

(i) *Theoretical Ballistics* deals with all kinematic phenomena associated with a fired gun; *i.e.*, the motions of projectile and gun as well as the treatment of these phenomena via mathematical, physical and chemical methods of analysis. In particular, one must differentiate between two main tasks:

- the *innerballistisches Hauptproblem* ("interior ballistic problem"); *i.e.*, how to determine, for any given instant of time, the gas pressure and projectile velocity inside the barrel as a function of time and distance, as well as the amount of gunpowder burned; and
- the *außenballistisches Hauptproblem* ("exterior ballistic problem"); *i.e.*, how to calculate or graphically determine the projectile trajectory for any given set of initial parameters, such as projectile geometry, muzzle velocity, air density, wind velocity, and the rotation of the Earth.

(ii) *Experimental Ballistics* encompasses all methods of measurement, observation and registration related to the complex motions of the projectile, barrel and powder/smoke.

| | | |
|---|---|---|
| 1925 | *University of Strasbourg, France* | **Ernest ESCLANGON**,[1453] a professor at the Faculty of Sciences and director of the Strasbourg Observatory, **writes a long memoir on gun muzzle blasts and sonic boom effects generated by projectiles.** ▪ The first systematic muzzle blast studies were performed in Germany {CRANZ ⇨1925; ⇨Fig. 4.5–L}. High-speed photography revealed that the rapidly expanding propellant gas acts like a piston to drive a blast wave into the surrounding air. For most of the short period in which the barrel empties, there is an almost stationary inward-facing system of interacting shocks close to the muzzle – the so-called **"shock bottle"** (or "barrel shock pattern") {SLADE JR. ⇨1946}.[1454] |
| 1926 | *Private laboratory, Auburn, Massachusetts* | On March 16, **Robert H. GODDARD**, a U.S. physicist and inventor, **achieves the first flight of a liquid-fueled rocket, lasting 2.5 seconds.** His 3.5-m-long rocket, propelled by a mixture of liquid oxygen and gasoline, reaches a height of 12.5 m. He makes the following note in his diary about this ground-breaking test: "The first flight with a rocket using liquid propellants was made yesterday at Aunt Effie's farm in Auburn … It looked almost magical as it rose, without any appreciably greater noise or flame … Some of the surprising things were the absence of smoke, the lack of very loud roar, and the smallness of the flame." ▪ Financially supported by the Guggenheim Foundation, he continued his development of the liquid-fueled rocket. On March 8, 1935 at Roswell, NM, **GODDARD became the first person to create a liquid-fueled rocket that traveled at faster than the speed of sound**,[1455] and in the following year he published his second Smithsonian Paper *Liquid-Propellant Rocket Development*.[1456] Because of his pioneering experiments, GODDARD is today considered to be the father of American rocketry.<br><br>Unlike engines based upon solid propellants, liquid-fueled rocket engines are easily controlled by varying the rate at which fuel and oxidizer flow into the combustion chamber. The first large vehicles to reach supersonic and hypersonic velocities were liquid-fueled rockets {HVA Peenemünde ⇨1942; White Sands Proving Ground ⇨1949}. |
| 1926 | *Orica Germany GmbH, Troisdorf, North-Rhine Westphalia, Germany* | **Foundation of the journal** ***Nobel-Hefte***, edited by the Sprengtechnischer Dienst ("Blasting Service") of Orica Germany GmbH in cooperation with the Sprengtechnischer Dienst of Dynamit Nobel AG, Troisdorf and the Wasag-Chemie AG, Essen. It is targeted at explosives experts working in the mining, quarrying and construction industries. ▪ The journal ceased with volume 71 (2005). |

---

[1453] E. ESCLANGON: *L'acoustique des canons et des projectiles*. Mémorial de l'Artillerie Française **4**, No. 3, 639-1026 (1925).

[1454] Note that the term *shock-bottle* (also *shock bottle*) was used in the United States for the Leiden jar {VON KLEIST ⇨1745}.

[1455] *The chronological history of the scientific accomplishments of Robert H. GODDARD*. George C. Gordon Library, Worcester Polytechnic Institute, Worcester, MA; http://www.wpi.edu/Academics/Library/Archives/Goddard/chronology2.html.

[1456] Smithsonian Miscellaneous Collections, Smithsonian Institution Press, Washington, DC, vol. 95, No. 3 (1936).

| | | |
|---|---|---|
| 1926 | *Physikalisch-Technische Reichsanstalt (PTR), Berlin, Germany* | Eduard GRÜNEISEN,[1457] a German physicist, **proposes an equation of state for solid matter based on his lattice vibration theory,** which he derived from investigations performed by Gustav MIE[1458] (a German physicist) and himself.[1459] ▪ The so-called **"Mie-Grüneisen equation"** can be written as $$pv + G(v) = \Gamma(v)\,e,$$ where $e$ is the specific internal energy, $v$ is the specific volume, $p$ is the pressure, and $\Gamma(v)$ is the Grüneisen coefficient for the material. The function $G(v)$ is related to the lattice potential.<br><br>The Mie-Grüneisen equation, first applied at Los Alamos to shock-compressed matter {WALSH, RICE, MCQUEEN & YARGER ⇨1957}, proved most useful for calculating off-Hugoniot states. |
| 1926 | *Private study at Darmstadt, Germany* | Beno GUTENBERG,[1460] a German geophysicist and university lecturer, **reports on his analytical studies on the structure of the upper atmosphere in terms of temperature and sound velocity.** Starting from the well-known phenomenon that strong air blasts are surrounded by three concentric zones – an inner zone of normal audibility, then a zone of silence, followed by an outer zone of abnormal audibility {VON DEM BORNE ⇨1910; MUNROE ⇨1913} – he derives the general curves for temperatures in the stratosphere.<br><br>His work on the structure of the atmosphere originated in World War I: working as a meteorologist attached to the chemical warfare engineers at the Russian, French and Belgian fronts, he was also assigned the problem of predicting the likelihood of the backward drift of the poison gases released by the German army onto German soldiers. In addition, he was assigned to **measure the locations of cannons from the travel-times of their reports** – a similar problem to locating earthquake sources. |
| 1926 | *Trinity College, Oxford, England* | Cyril Norman HINSHELWOOD, a British physical chemist and tutor, **publishes his book *The Kinetics of Chemical Change in Gaseous Systems*.** ▪ Later HINSHELWOOD also extended his studies to chemical reactions in living organisms; the results of this line of investigation appeared in his book *Chemical Kinetics of the Bacterial Cell* (1946). |
| 1926 | *Delmag Co., Esslingen, southern Germany* | **Invention of the *explosion ram* by Albert PFLÜGER**[1461] **and Konrad HAAG,** two German mechanical engineers at Delmag Company.[1462] ▪ The new method of generating significant impulsive forces of percussion was mainly used for pile-driving, paving, and soil condensation purposes {⇨Figs. 4.2–O, P}. |
| 1926/ 1927 | *University of Manchester, England* | Colin CAMPBELL[1463] and Donald W. WOODHEAD, two British physical chemists, **investigate gaseous detonation in mixtures of carbon monoxide and oxygen** ($CO/O_2$). Using a rotating drum smear camera, they obtain photographic records that are very unusual. In one of these smear camera photographs, **they notice that "the burning gases behind the wave front show marked horizontal bands"** {⇨Fig. 4.16–T}. Apparently, this is caused by a region of higher luminosity (and, therefore, higher temperature) in the detonation front, which rotates around the tube axis as the detonation advances. ▪ **Their observation of this "spin phenomenon" was the first indication that real detonations might be more complex than postulated by the simple CJ theory** {CHAPMAN ⇨1899; JOUGUET ⇨1904 & 1904–1906}. |

---

[1457] E. GRÜNEISEN: *Zustand des festen Körpers.* In: (H. GEIGER and K. SCHEEL, eds.) *Handbuch der Physik.* Springer, Berlin (1926), vol. X, pp. 1-59.

[1458] G. MIE: *Zur kinetischen Theorie der einatomigen Körper.* Ann. Phys. **11** [IV], 657-697 (1903).

[1459] E. GRÜNEISEN: *Theorie des festen Zustandes einatomiger Elemente.* Ann. Phys. **39** [IV], 257-306 (1912).

[1460] B. GUTENBERG: *Die Geschwindigkeit des Schalles in der Atmosphäre.* Physik. Z. **27**, 84-86 (1926); *Schallgeschwindigkeit und Temperatur in der Stratosphäre.* Gerlands Beiträge zur Geophysik **27**, 217-225 (1930).

[1461] A. PFLÜGER and K. HAAGE: *Durch eine Verpuffungskraftmaschine betriebene Ramme.* Germ. Patent No. 488,179 (July 16, 1926).

[1462] A. BONWETCH: *Neuerungen im Baumaschinenwesen. Straßenbaumaschinen.* VDI-Z. **73**, No. 36, 1269-1274 (1929).

[1463] C. CAMPBELL and D.W. WOODHEAD: *The ignition of gases by an explosion wave. (I) Carbon monoxide and hydrogen mixtures.* J. Chem. Soc. (Lond.) **129**, 3010-3021 (1926).

In the following year, CAMPBELL[1464] and WOODHEAD studied this puzzling phenomenon in more detail.

(i) They wrote: "In a previous paper [CAMPBELL & WOODHEAD ⇨ 1926–1927] attention was directed to the notable appearance of a moving-film record of the explosion-wave passing through the mixture $2\,CO + O_2$. Fig. 11 in that paper is remarkable in that closely striated illumination is recorded behind the trace of the wave-front, and not the uniform illumination which occurs in the great majority of previously published photographic records of explosion-waves. Close examination of some of DIXON's records (*Phil. Trans.* 1903, **A200**, 315; notably Fig. 11, $2\,CO + O_2$ and Fig. 21, $CS_2 + 2\,O_2$) disclose striations similar to those now described, but they are ill-defined and do not appear to have been noticed hitherto. The present paper deals with experiments made with a view to define some of the conditions required for the production of such striated records."

(ii) They state that "it does not indicate the striated records are due to induced vibrations of the apparatus or to a periodic propagation of the explosion-wave." However, they give a (correct) interpretation communicated to them by Mr. E.F. GREIG, "who suggested that **a part of the wave-front,** the light of which affects the photographic film, **traverses a helical path on the walls of the tube.** The extension of this interpretation to the illumination from the outer layers of the burning gases behind the wave-front would also imply a rotation of the source of light affecting the photographic film. It is clear that a wave being propagated along a helical path could account for the dependence of the length of the undulations on the bore of the tube and for the alternate production of undulations in the records from the upper and lower windows…" ▪ Experiments have shown that this phenomenon – called **"spinning detonation"** (or "detonation spin") – is easily reproducible {BONE & FRASER ⇨ 1929}. It always arises in a pipe of a given diameter for a mixture of known composition. **The phenomenon of detonation spin has also been observed in condensed explosives.**[1465]

| | | |
|---|---|---|
| 1927 | *Kaiser-Wilhelm Institut (KWI) für Strömungsforschung, Göttingen, Germany* | Jakob ACKERET,[1466] a Swiss aerodynamicist and director of the famous KWI für Strömungsforschung,[1467] **presents his handbook article *Gasdynamik* ("Gas Dynamics"):**<br><br>(i) Generally understood as being a study on the dynamic properties of gases, he provides a more specific definition of this term: "True gas dynamics is the theory of compressible flow when flow velocities are on the order of magnitude of the sound velocity, and the resulting changes in density, pressure and temperature are no longer negligible." ▪ **Apparently, the term *gas dynamics* – according to Klaus OSWATITSCH[1468] a discipline linking thermodynamics with hydrodynamics – was first used by ACKERET in his handbook article.**<br><br>(ii) Starting from a memoir on aerodynamics written by Marian SMOLUCHOWSKI {⇨1903}, he discusses various two-dimensional high-speed flow phenomena – such as those occurring in pipes, nozzles, around corners and associated with objects flying supersonically. |
| 1927 | *McGill University, Montreal, Canada* | Howard T. BARNES,[1469] an American-born Canadian physics professor, **reports on a new method of destroying an iceberg by a thermite reaction.** A charge consisting of a few hundred pounds of thermite – a mixture of metal oxide (mostly iron oxide) and aluminum powder |

---

[1464] C. CAMPBELL and D.W. WOODHEAD: *Striated photographic records of explosion waves.* J. Chem. Soc. (Lond.) **130**, 1572-1578 (1927).

[1465] A.N. DREMIN: *Detonation waves in condensed media* [in Russ.]. Nauka, Moscow (1970), p. 186.

[1466] J. ACKERET: *Gasdynamik.* In: (H. GEIGER and K. SCHEEL, eds.) *Handbuch der Physik.* Springer, Berlin (1927), vol. VIII, pp. 289-342.

[1467] Its eventful history has been reviewed by (i) C. TOLLMIEN: *Das Kaiser-Wilhelm-Institut (KWI) für Strömungsforschung verbunden mit der Aerodynamischen Versuchsanstalt.* In: (H. BECKER, H.J. DAHMS, and C. WEGELER, eds.) *Die Universität Göttingen unter dem Nationalsozialismus. Das verdrängte Kapitel ihrer 250-jährigen Geschichte.* Saur, München & London (1987), pp. 464-488; and (ii) K. KRAEMER: *Geschichte der Gründung des Max-Planck-Instituts für Strömungsforschung in Göttingen.* In: (G. GRABITZ and H.O. VOGEL, eds.) *Max-Planck-Institut für Strömungsforschung Göttingen 1925–1975.* Festschrift zum 50-jährigen Bestehen des Instituts, Göttingen (1975).

[1468] K. OSWATITSCH: *Was ist Gasdynamik?* Physik. Blätter **14**, 108-116 (1958).

[1469] H. BARNES: *Some physical properties of icebergs and a method for their destruction.* Proc. Roy. Soc. Lond. **A114**, No. 767, 161-168 (1927).

heated to ignition temperature by an explosive charge – was used to produce extremely high temperatures in the ice. An iceberg was cracked this way into smaller pieces through "thermal shocking." ▪ In 1924, BARNES had observed at the Newfoundland Coast the interesting fact that the greatest calving and most numerous cracking took place in the early morning at and immediately after sunrise.

In the period 1959–1960, the Ice Patrol of the U.S. Coast Guard conducted similar tests on two icebergs, thereby also using the violent combustion of thermite. However, these experiments did not corroborate BARNES' original findings of thermal fracturing. Further attempts to determine means for disintegrating icebergs included gunfire, mines, torpedoes, depth charges, and bombing. The Coast Guard concluded that such methods would be economically as well as practically unsound,[1470] "The use of conventional explosives or combustibles proves difficult. In addition to the operational hazards of approaching and boarding an iceberg in a seaway, the theory of explosive demolition shows that a 1,000-lb (453-kg) charge of conventional explosives would be needed to break up approximately 70,000 cubic ft (1,982 m$^3$) of ice and a hundred such charges would be needed for the destruction of an average iceberg. Furthermore, to melt a medium-size iceberg of 100,000 tons would require the complete theoretical heat of combustion of over a quarter of a million gallons (946,325 liters) of gasoline."

| 1927 | Institut Catholique de Toulouse, Dépt. Haute-Garonne, southwest France | Zéphirin CARRIÈRE,[1471] a French clergyman and professor of physics, **studies the phenomenon of whip cracking, the oldest man-made method of generating shock waves.** He simulates whip cracking in the laboratory using a machine-driven whip and high-speed schlieren diagnostics. He estimates an average velocity of about 350 m/s for the cracker velocity at the turning point – the "critical point" at which the cracker reaches its maximum velocity. The wave emitted is barely recognizable in pictures taken of the motion of the whip tip and reproduced in his article. However, he comments that the original photos show a spherical wave rather than an expected head wave geometry. ▪ Almost 30 years later, CARRIÈRE[1472] resumed his studies in this area and published photos of higher contrast which clearly show the circular shape of the emitted shock wave {⇨Fig. 4.5–K}.

The first pictures showing the shock wave emerging from a real whip were obtained by U.S. researchers {BERNSTEIN ET AL. ⇨1958; ⇨Fig. 4.5–K}. In 1965, Harold E. EDGERTON,[1473] a MIT professor and a widely renowned high-speed photographer, recorded the flip of the popper on the end of a whip with a still camera using a high-speed xenon multi-flash system operated at a 2,000 flashes per second. However, he was not able to visualize the emitted shock wave. More recent studies using high-speed videography combined with copper-vapor-laser stroboscopy have provided the first detailed picture of the mechanism of shock wave generation {KREHL ET AL. ⇨1995; ⇨Figs. 4.5–K & 4.11–A}. |
|---|---|---|
| 1927 | Institut für Technische Physik, TH Berlin-Charlottenburg, Germany | Carl CRANZ,[1474] a famous German ballistician, **and Karl Friedrich SCHEEL,** a physics professor and renowned editor of various German scientific journals and handbooks, **discuss various ballistic paradoxes** which up to now have been only partly understood, and have not yet been investigated quantitatively by theoretical physicists. Prominent examples, discussed by CRANZ[1475] in more detail in his four-volume textbook on ballistics {CRANZ ⇨1925}, include curious exterior ballistic and impact phenomena: |

---

[1470] *Is it practical to destroy icebergs before they reach the shipping lanes?* Int. Ice Patrol, U.S. Coast Guard R&D Center, Groton, CT; http://www.uscg.mil/LANTAREA/IIP/FAQ/ReconnOp_5.shtml.
[1471] Z. CARRIERE: *Le claquement du fouet.* J. de Physique et le Radium **8** [VI], 365-384 (1927).
[1472] Z. CARRIERE: *Exploration par le fouet des deux faces du mur du son.* Cahiers de Physique No. 63, pp. 1-17 (Nov. 1955).
[1473] H.E. EDGERTON: *Applications of xenon flash.* Proc. 7th Int. Congr. High-Speed Photography [Zurich, Switzerland, Sept. 1965]. In: (O. HELWICH and H. SCHARDIN, eds.) *Kurzzeitphotographie.* Helwich, Darmstadt (1967), pp. 3-16.
[1474] C. CRANZ and K. SCHEEL: *Ballistische Paradoxa.* Z. Tech. Phys. **8**, 359-362 (1927).
[1475] *See* vol. I (1925), §§ 75A, 75D, 77 and 78; vol. II (1926), §§ 19 and 24; and vol. III (1927), §§ 4 and 63.

- Depending on the position of an observer, **a single supersonic shot may produce at least three distinct reports.** These are caused by the muzzle blast {⇨Fig. 4.5−L}, the head wave {E. MACH & SALCHER ⇨1886}, and the impact with the target. Further reports may be felt by the observer due to reflections from the ground or obstacles.
- A high-speed projectile − such as an S-bullet traveling at $v_0 = 893$ m/s {DWM ⇨1903} − does not penetrate deepest into soil (sand) close to the muzzle, but rather into a target at a considerable distance of some 100 m.
- An armor-piercing shell provided with a cap of soft steel has a higher penetration power than one provided with a pointed steel head {B. HOPKINSON ⇨1912}.
- **The muzzle velocity of a lubricated gun barrel is smaller than that of a dry one.**
- Impact fragments, for example those resulting from the impact of an S-bullet on a steel plate in the normal direction, are mostly distributed in a vertical direction, rather than towards the barrel.
- An S-bullet fired in water in a horizontal direction a few centimeters below the surface travels upwards; *i.e.*, against gravity.
- An S-bullet fired from a short distance into a block of pine wood produces a cylindrical hole which has almost the same diameter as the projectile caliber. However, when a block of plastic clay is used instead of pine wood, the bullet produces a large central hole.

| 1927 | TH Berlin-Charlottenburg, Germany | **Friedrich HILDEBRAND,**[1476] a Ph.D. student, **investigates nonstationary flow in long lines of railway airbrake systems.** Working under the supervision of Georg K.W. HAMEL, a professor of mathematics who had recently tackled a related problem on compressible flow,[1477] HILDEBRAND extends the initial, crude, shock tube theory of Karl KOBES {⇨1910}. ▪ HILDEBRAND was certainly stimulated by his father, Wilhelm HILDEBRAND,[1478] who was then director at the Knorr-Bremse AG in Berlin-Lichtenberg, and who had invented the **"Hildebrand-Knorr pressure brake"** for railroad cars. Later further improved by his son Friedrich, it earned the company the "Grand Prix" at the International Exposition in Paris in 1937. |
|---|---|---|
| 1927 | Institut für Aerodynamik Aachen (IAA), TH Aachen, Germany | **Theodore VON KÁRMÁN,** a Hungarian-born aeronautical research engineer, **constructs the first Japanese wind tunnel similar to the one at Aachen, a closed-circuit subsonic facility** which was completed in 1914. ▪ The new tunnel at Aachen, following a concept developed at Göttingen, used the same air continuously by keeping it circulating in a closed loop.<br><br>In the following year, the new facility was installed in the Research Department of the Kawanishi Airplane Company (Kobe, Honshu, Japan) and made operational by the German mechanical engineer Erich KAYSER, one of VON KÁRMÁN's assistants.[1479] |
| 1927 | University of Louvain, central Belgium | **Georges Edouard LEMAÎTRE,**[1480] a Belgian professor of astrophysics and a priest, **publishes a paper on a dynamic cosmological model of the Universe.**<br><br>(i) He attempts to combine the advantages of two currently discussed cosmological models of the Universe {A. EINSTEIN ⇨1917; DE SITTER ⇨1917}.<br><br>(ii) Obviously not aware of the nonstatic model previously proposed in the Soviet Union by Alexander FRIEDMANN {⇨1922}, he works out an intermediate solution such that |

---

[1476] F. HILDEBRAND: *Über unstationäre Luftbewegungen in langen Rohrleitungen und ihre Beziehungen zu den Steuervorgängen der indirekten Luftbremsen.* Ph.D. thesis, TH Berlin-Charlottenburg (1927), VDI-Verlag, Berlin (1927).
[1477] G. HAMEL: *Das Ausströmen von Gasen durch Düsen.* VDI-Z. **55**, 1895-1898 (1911).
[1478] W. HILDEBRAND: *Die Entwicklung der selbsttätigen Einkammer-Druckluftbremse bei den europäischen Vollbahnen.* Springer, Berlin (1927).
[1479] Th. VON KÁRMÁN: *The wind and beyond.* Little, Brown & Co., Boston *etc.* (1967), pp. 129-134.
[1480] G. LEMAITRE: *Un Univers homogène de masse constante et de rayon croissant, rendant compte de la vitesse radiale des nébuleuses extra-galactiques.* Ann. Soc. Scient. Brux. **47** [Série A], 49-59 (1927); *A homogenous Universe of constant mass and increasing radius accounting for the radial velocity of extragalactic nebulae.* Month. Not. Roy. Astron. Soc. **91**, 483-490 (1931).

- the mass of the Universe is a constant related to the cosmological constant;
- the radius of the Universe increases without limit; and
- the recession velocities of extragalactic nebulae are an effect of the expansion of the Universe.

(iii) He concludes, "It remains to find the cause of the expansion of the Universe. We have seen that the pressure of radiation does work during the expansion. This seems to suggest that the expansion has been set up by the radiation itself. In a static Universe, light emitted by matter travels round space, comes back to its starting-point, and accumulates indefinitely. It seems that this may be the origin of the velocity of expansion $R'/R$ which EINSTEIN assumed to be zero and which in our interpretation is observed as the radial velocity of extragalactic nebulae."

▪ His important paper gave a relativistic explanation for the reddening observed in the spectra of distant nebulae {HUBBLE ⇨1929}. Originally published in a Belgian journal that was virtually unheard of abroad, it was subsequently translated into English and published in 1931 in *Monthly Notices of the Royal Astronomical Society*. **Today FRIEDMANN and LEMAÎTRE are considered to be the main founders of the *Big Bang theory*.**

In the early 1930s, LEMAÎTRE formulated the modern Big Bang theory: he proposed the idea that the expansion of the Universe began with the explosion of a highly-condensed "super-atom" which he called ***L'atome primitif*** ("the primeval atom") or ***L'œuf cosmique*** ("the cosmic egg"). ▪ On the back of this work, the name of LEMAÎTRE became known overnight and his "fireworks theory" – his own expression – captured the imagination of the public.[1481] In 1933, LEMAÎTRE and EINSTEIN traveled together to California for a series of seminars. At Mt. Wilson, LEMAÎTRE detailed his theory that the Universe had been created by the explosion of a "primeval atom" and that it was still expanding. Gleefully, EINSTEIN jumped to his feet, applauding, and said, "This is the most beautiful and satisfactory explanation of creation to which I have ever listened."[1482]

According to an ancient creation myth, the Universe was created from a primordial egg which (in some mythologies) was laid by a bird. Modern cosmologists assume that perhaps the Universe was generated by a high-energy radioactive process at the beginning of the expansion phase, when the portion of the Universe that we can see today was only a few millimeters across. Estimates for the age of the Universe, based upon the Big Bang theory, have ranged from 10 to 20 billion years {⇨c.14 Ga ago}. However, recent studies based upon measured cosmic microwave background radiation data {NASA-GSFC ⇨2003} have determined the age of the Universe at 13.7 ± 0.2 billion years and have confirmed that the geometry of the Universe is Euclidean, and that space is flat resembling a sheet in 2-D. From the Hubble law {HUBBLE ⇨1929}, one would expect that the Universe expands homogeneously, as assumed in the Big Bang theory. Most cosmologists, however, assume that the expansion rate has changed over time: that the rate was faster in the past and that it slowed down to its current rate.

| 1927 | United States | Charles H.T. TOWNSEND,[1483] a U.S. entomologist, **publishes an article devoted to the wing structure of the deer botfly** [*Cephenomyia pratti*], a small, blunt-headed insect which sprays its eggs into the nostrils and throats of deer, scattering them like tiny bombs while on the wing. At the end of his article, he writes: "On 12,000-foot [3,658-m] summits in New Mexico I have seen pass me at an incredible velocity what were quite certainly the males of *Cephenomyia*. I could barely distinguish that something had passed, only a brownish blur in the air of about the right size for these flies and without sense of form. As closely as I can estimate, their speed |

---

[1481] A.V. DOUGLAS: *Georges LEMAÎTRE, 1894–1966*. J. Roy. Astron. Soc. Canada **61**, 77-80 (1967).
[1482] P. MICHELMORE: *EINSTEIN, Albert*. In: *The new Encyclopædia Britannica, Macropædia*. Benton & Hemingway, Chicago *etc.*, vol. 6 (1974), pp. 510-514.
[1483] C.H.T. TOWNSEND: *On the Cephenemyia mechanism and the daylight-day circuit of the Earth by flight*. J. N. Y. Entomol. Soc. **35**, 245-252 (1927).

must have approximated 400 yards per second [about 366 m/s]" – *i.e.*, **estimating a supersonic flight velocity.**

Scientists took his claim seriously for years. But in 1938, **Irving LANGMUIR,**[1484] a U.S. chemist and physicist at General Electric Co. and recipient of the 1932 Nobel Prize in Chemistry, **demolished this astounding claim of supersonic botflies:**

▸ If the botfly flew at 800 mph (1,287 km/h) the wind pressure against its head would be 8 psi (about 0.5 at), "probably enough to crush the fly." The power needed to maintain such a velocity would be 370 watts or about one-half horsepower, "a good deal for a fly!"

▸ According to his calculations, if one of TOWNSEND's botflies were stopped over a distance of one centimeter, it would generate a force of impact of 310 pounds (140 kg). He concluded, "It is obvious that such a projectile would penetrate deeply into human flesh" – whereas, in reality, "it produced a very noticeable impact, far greater than that of any other insect I have met" but bounced off the skin after the collision.

In order to find the speed at what a flying insect turns into a blur, he got a lump of metal the size of a deer botfly, tied it to a thread and swung it around. At 13 mph (20 km/h), it was a blur, "the shape could not be seen, but it could be recognized as a small object of about the correct size." At 26 mph (42 km/h), "the fly was barely visible as a moving object." So TOWNSEND probably saw deer botflies cruising at around 42 km/h.

| | | |
|---|---|---|
| 1928 | *Swampscott, Massachusetts* | In September, **the First Symposium on Combustion,** sponsored by the Gas and Fuel Division of the American Chemical Society (ACS), **is held during the 76th meeting of the ACS,** with the goal "to emphasize the practical significance of combustion research, particularly in the area of high-output combustion in aviation power plants." The general chairman is George Granger BROWN, a professor at the University of Michigan. ▪ The 2nd Symposium was held at Rochester, NY (1937). Beginning with the 3rd Symposium held at Madison, WI (1948), it was renamed the **Symposium on Combustion and Flame and Explosion Phenomena.** Beginning with the 4th Symposium held at Cambridge, MA (1952), all subsequent conferences were named the **Symposium (International) on Combustion** and held at Pittsburgh, PA (1954); New Heavens, CT (1956); London & Oxford, U.K. (1958); Pasadena, CA (1960); Ithaca, NY (1962); Cambridge, U.K. (1964); Berkeley, CA (1966); Poitiers, France (1968); Salt Lake City, UT (1970); University Park, PA (1972); Tokyo, Japan (1974); Cambridge, MA (1976); Leeds, U.K. (1978); Waterloo, Canada (1980); Haifa, Israel (1982); Ann Arbor, MI (1984); Munich, Germany (1986); Seattle, WA (1988); Orléans, France (1990); Sydney, Australia (1992); Irvine, CA (1994); Naples, Italy (1996); Boulder, CO (1998); Edinburgh, Scotland (2000); Sapporo, Japan (2002); Chicago, IL (2004); Heidelberg, Germany (2006); and Montreal, Canada (2008).<br><br>The success of the 3rd Symposium not only prompted a continued series of symposia but it also acted as a catalyst for the formation of **The Combustion Institute** {⇨1954}. The Symposium on Combustion has become an important forum for presenting and discussing papers on classical combustion research into flames and fires, and also on high-rate discontinuous phenomena such as on chemical kinetics, supersonic combustion, gaseous and dust explosions, and detonation. |
| 1928 | *ETH Zurich & Escher-Wyss AG, Zurich, Switzerland* | **Jakob ACKERET,**[1485] a Swiss aeronautical engineer, delivers his inaugural lecture as *Privatdozent* ("Instructor") on aerodynamic drag at very high velocities, during which he **coins the term *MACHsche Zahl* ("Mach number")**[1486] "since the well-known physicist E. MACH clearly recognized the fundamental significance of this ratio in our field and confirmed it by ingenious |

---

[1484] I. LANGMUIR: *The speed of the deer fly.* Science **87**, No. 2254, 233-234 (March 11, 1938).
[1485] J. ACKERET: *Der Luftwiderstand bei sehr großen Geschwindigkeiten.* Schweiz. Bauz. **94**, 179-183 (1929).
[1486] His argument, fully cited, says: "In der Aerodynamik höherer Geschwindigkeiten tritt das Verhältnis $v/a$ dauernd auf ($v$ = Geschwindigkeit des betrachteten Körpers bzw. der Luftströmung, $a$ = Schallgeschwindigkeit). Es empfiehlt sich deshalb, eine abkürzende Bezeichnung einzuführen. Da der bekannte Physiker Ernst MACH auf unserem Gebiete die grundlegende Bedeutung dieses Verhältnisses besonders klar erkannt und durch geniale experimentelle Methoden bestätigt hat, scheint es mir sehr berechtigt, $v/a$ als MACHsche Zahl zu bezeichnen."

experimental methods." ▪ Nowadays MACH's name is used by almost anyone to describe something that is very fast. In fact, MACH is more famous for this than for his numerous other contributions to the philosophy of science.

(i) The Mach number $M = v/a_0$ denotes the ratio of the velocity of motion $v$ of an object (or a high-speed flow) to the velocity of sound $a_0$ in the surrounding (quiescent) fluid.[1487] In addition, special Mach numbers have been introduced into fluid dynamics:

▸ The **cutoff Mach number** $M_{\text{cutoff}}$ is the limiting value of the Mach number at which, because of atmospheric effects, no sonic boom is heard {DUMOND ET AL. ⇨1946}.
▪ Studies have established that the *cutoff Mach number* ranges roughly between about 1.0 and 1.3 depending on atmospheric conditions and the altitude of the object (plane).[1488, 1489] This means that sonic booms produced by objects moving faster than 1.3 times the sound speed should be heard at ground level.

▸ The **stream Mach number** $M_\infty$ is the ratio of the stream velocity to the velocity of sound.

▸ The **shock Mach number** $M_S$ is the ratio of the shock front velocity to the upstream velocity of sound and is conveniently used to characterize the strength of a shock wave. For a perfect gas, the jump conditions across a steady shock can be expressed in terms of the ratio of specific heats $\gamma$ and $M_s$. For example, the pressure ratio at the shock front is given by the simple relation (*see* Sect. 2.1.5, footnote 79)

$$p_1/p_0 = 2\gamma M_s^2/(\gamma+1) - (\gamma-1)/(\gamma+1).$$

Conditions upstream and downstream of the shock front are denoted with the suffixes "0" and "1," respectively.[1490]

▸ If the velocity is variable in the field, the ratio between the velocity at an arbitrary point and the sound velocity corresponding to the temperature at that point is termed the **local Mach number**.[1491]

▸ The **critical Mach number** $M_{\text{crit}}$ is the lowest value of the stream Mach number for which a local Mach number of unity occurs in a flow. As first suggested by the British mathematical physicist Geoffrey I. TAYLOR, the terminology is based on the reasoning that $M_{\text{crit}}$ provides at least a local limit for the onset of local shock waves – in high-speed aerodynamics this usually occurs near the point of maximum suction on the wing {CALDWELL & FALES ⇨1918}.[1492]

▸ In magnetohydrodynamics, the **magnetic Mach number** $M_\alpha$ or **Alfvén Mach number** is the dimensionless ratio of the local flow speed and the local Alfvén speed.[1493]

▸ In astrophysics, *the* **cosmic Mach number** $M_{\text{cos}}$ which characterizes the coldness of the velocity field was introduced in 1990 in order to test cosmological models.[1494] Given a galaxy sample, $M_{\text{cos}}$ is defined as the ratio of the streaming velocity to the random velocity dispersion of galaxies in a given patch of the Universe. $M_{\text{cos}}$ has been calculated using large-scale cold dark matter hydrodynamical simulations or measured via the Sunyaev-Zel'dovich effect.

---

[1487] For example, due to the Earth's rotation, any quiescent object positioned at the equator moves with an absolute velocity of about $M = 1.4$. However, since the atmosphere moves together with the Earth, the effective Mach number is zero. DOPPLER {⇨1846}, whilst studying wavefront geometries resulting from moving sources, also discussed the cases of when the atmosphere (for sound waves) or the ether (for light waves) moves in relation to a quiescent source.

[1488] H.A. WILSON JR.: *Sonic boom*. Scient. Am. **206**, 36-43 (Jan. 1962).

[1489] E.J. KANE: *Some effects of the nonuniform atmosphere on the propagation of sonic booms*. JASA **39**, S26-S30 (Sept. 1966).

[1490] Acoustic waves propagate with Mach 1. Even at the threshold of pain ($\Delta P = P_1 - P_0 \approx 100$ Pa), the Mach number is only $M = 1.0004$.

[1491] T. VON KÁRMÁN: *Aerodynamics: selected topics in the light of their historical development*. Cornell University Press, Ithaca, NJ (1954), pp. 103-123.

[1492] W. SEARS: *Small perturbation theory*. In: *High speed aerodynamics and jet propulsion*. Princeton University Press, Princeton, NJ; vol. VI (1954): (W.R. SEARS, ed.) *General theory of high speed aerodynamics*, pp. 61-121.

[1493] The Alfvén speed is the speed at which Alfvén waves are propagated along the magnetic field and this can be much larger than the speed of sound.

[1494] J.P. OSTRIKER and Y. SUTO: *The Mach number of the cosmic flow: a critical test for current theories*. Astrophys. J. **348**, 378-382 (1990).

(ii) The Mach number also has a physical meaning in addition to being the ratio of the speed of movement to the speed of sound. Since the square of the velocity of sound is given by
$$a_0^2 = \gamma RT,$$
it follows that
$$M^2 = v^2/(\gamma RT);$$
i.e., $M^2$ **represents the ratio of the directed kinetic energy of motion** ($\propto v^2$) **of the molecules of a gas to their random energy of motion** ($\propto T$).[1495] For air, the directed and random (or heat) energies are equal at $M = (3/\gamma)^{1/2} = 1.46$. At higher Mach numbers, the directed kinetic energy of the molecules is greater than their random energy.[1496]

(iii) The Mach number also governs the nature of the mathematical flow equations: a subsonic flow ($M < 1$) has an elliptical solution, while a supersonic flow ($M > 1$) has a hyperbolic solution.

The term *Mach number* was not immediately accepted by the Russians, who at one point preferred **Bairstow number**,[1497] nor by the French, who proposed **Moisson number**[1498] {MOISSON ⇨ 1883}.[1499] The Mach number was introduced into the English literature in the late 1930s. Nicholas ROTT,[1500] an American-born Swiss theoretical fluid dynamicist, published an article on the history of the Mach number.

| 1928 | Committee of Erosion Research, England | Stanley S. COOK,[1501] a British research engineer, **reports on his investigations of the hydrodynamic properties of collapsing cavities** {⇨Fig. 4.14–C}.

(i) Assuming an incompressible fluid, he calculates the pressures that might arise from the collapsing vortices of cavitating propellers.

(ii) He first notes that high dynamic pressures can occur in the liquid as a result of the compressive normal impact of a liquid, and introduces the acoustic pressure limit – i.e., the "water hammer" pressure – into the liquid-impact literature.

(iii) He also discusses the possible effects of the shapes of the impacting surfaces.

His studies, verified by experimental methods, convinced the Propeller Committee of the British Navy {⇨1915} that the deterioration of propeller blades of cruisers and destroyers by erosion was indeed caused by the water hammer effect – i.e., cavitation – thus essentially confirming a previous hypothesis {Sir PARSONS ⇨1897}. |
| 1928 | Mathematical Institute, University of Göttingen, Germany | Richard COURANT,[1502] Kurt O. FRIEDRICHS, and Hans LÉWY, three German mathematicians, **discuss the possibility of catastrophic instabilities that can arise when partial differential equations are approximated by difference equations, and their avoidance.** They first show that there is a maximum limit on the time increment to get a stable solution – the so-called **"Courant condition"** (or "Courant criterion"). • With the advent of high-speed digital computers, this stability criterion became very important in numerical analyses of problems related to fluid dynamic. It indicates that high time resolution is also required to achieve high spatial resolution in any hydrodynamic simulation: the solution to the differential equation is stable if and only if $c\Delta t < \Delta r$, where $c$ is the local velocity of sound, and $\Delta t$ and $\Delta r$ are the increments in time and radius, respectively. |

---

[1495] N.P. BAILEY: *The thermodynamics of air at high velocities.* J. Aeronaut. Sci. **11**, No. 3, 227-238 (July 1944).
[1496] W.F. HILTON: *High-speed aerodynamics.* Longmans, Green & Co., London *etc.* (1952), pp. 488-489.
[1497] Sir Leonard BAIRSTOW (1880–1963) was then Great Britain's leading aerodynamicist. See G. TEMPLE: *Leonard BAIRSTOW.* Biogr. Mem. Fell. Roy. Soc. (Lond.) **11**, 23-40 (1965).
[1498] Arthur MOISSON (1842–?) was a captain of the French Navy. See M.L. GABEAUD: *Le nombre de MOISSON.* Mém. Artill. Franç. **21**, 857-868 (1947).
[1499] J. BLACK: *Ernst MACH. Pioneer of supersonics.* J. Roy. Aeronaut. Soc. **54**, 371-377 (1950).
[1500] N. ROTT: *Jakob ACKERET and the history of the Mach number.* Annu. Rev. Fluid Mech. **17**, 1-9 (1985).
[1501] S.S. COOK: *Erosion by water-hammer.* Proc. Roy. Soc. Lond. **A119**, 481-488 (1928).
[1502] R. COURANT, K. FRIEDRICHS, and H. LÉWY: *Über die partiellen Differentialgleichungen der mathematischen Physik.* Math. Ann. **100**, 32-74 (1928).

| | | |
|---|---|---|
| 1928 | Los Angeles, California | **Harold E. CROW,** an American orthopedist, **first introduces the term** *whiplash* **at a meeting,** describing it as a sudden acceleration-deceleration force on the neck and upper trunk from external forces exerting a "lash-like effect."[1503] • The term *whiplash* suggests the idea that when a whip is cracked the lash moves in an acceleration-deceleration manner. While this is true for the whip handle, the lash itself starts moving near the handle in a sharp loop, which propagates down toward the tip with increasing acceleration until the cracker produces the shock wave at the moment when the turning point is reached {KREHL ET AL. ⇨1995}.<br><br>In the case of a rear impact of a motor vehicle, one differentiates between two phases:<br>▸ During the first phase of whiplash, the head moves rearward relative to the torso, so-called "head lag."<br>▸ In the following phase of whiplash, the head moves forward, so-called "reentry."<br>Medical studies have revealed that about 10–20% of occupants involved in rear-end car collisions suffer from whiplash injuries; most patients recover relatively well within three to six months post-injury. • In the literature, the term *whiplash* is sometimes also used to describe an injury from any type of crash scenario. |
| 1928 | Saint John's College, Cambridge, England | **Paul Adrien Maurice DIRAC,**[1504] a British physicist, **predicts the existence of a positively charged particle with the same mass and charge as the (negatively charged) electron.** He claims that for every particle of ordinary matter there is an antiparticle with the same mass but an opposite charge. • Today, DIRAC is considered to be the "father of antimatter." This subject stimulated the imaginations of science fiction writers, who wrote of antistars, antiuniverses and antiworlds.[1505]<br><br>Four years later, his theory was confirmed when Carl D. ANDERSON, a U.S. physicist at the California Institute of Technology, detected the first antiparticle – the *positron* – which has the same mass and magnitude of charge as the electron. • **Positrons interact with electrons shortly after they are created and thus disappear in a burst of radiation.** |
| 1928 | Budapest, Hungary | **Albert FONO,**[1506] a Hungarian engineer, **obtains a German patent for a propulsive device which he calls "air jet engine"** [Germ. *Luftstrahlmotor*]. The design of the device shows a convergent-divergent inlet, and he describes it as being specifically intended for future supersonic flight at large altitude. • His engine is clearly recognizable as a prototype of a modern ramjet. |
| 1928 | Royal Aircraft Establishment (RAE), Farnborough, England | **Hermann GLAUERT,**[1507] a British aeronautical engineer, **investigates how air compressibility effects the forces, such as drag and lift, acting on an airfoil.** He gives the first interpretation of PRANDTL's "Aerodynamic theory of airfoils" [Germ. *Aerodynamische Tragflügeltheorie*][1508] ("Aerodynamic Theory of Airfoils") in England. The so-called **"Prandtl rule"** [Germ. *PRANDTLsche Regel*] or "Prandtl-Glauert analogy" is the starting point for all similarity laws used in gas dynamics.[1509] It enables designers to calculate the amount of lift needed at high speed – up to the speed of sound, but not beyond. • The method was later extended to the supersonic and hypersonic ranges.[1510] |

---

[1503] J.M.S. PEARCE: *Polemics of chronic whiplash injury.* Neurology **44**, 1993-1997 (1994).

[1504] P.A.M. DIRAC: *The quantum theory of the electron. Part I.* Proc. Roy. Soc. Lond. **A117**, 610-624 (1928); *Part II.* Ibid. **A118**, 351-361 (1928).

[1505] A book by L.M. KRAUSS [*The physics of star trek.* Basic Books, New York (1995)] is a constructive criticism of antimatter science fiction.

[1506] A. FONO: *Luftstrahlmotor für Hochflug.* Germ. Patent No. 554,906 (May 1928).

[1507] H. GLAUERT: *The effect of compressibility on the lift of an airfoil.* Proc. Roy. Soc. Lond. **A118**, 113-119 (1928).

[1508] L. PRANDTL: *Tragflügeltheorie. I. Mitteilung.* Nachr. Akad. Wiss. Göttingen [Math.-Physik. Kl.] (1918), pp. 451-477; *II. Mitteilung.* Ibid. (1919), pp. 107-137. • Engl. translations: *Theory of lifting surfaces. Part I.* Rept. NACA-TN 9 (1920); *Theory of lifting surfaces. Part II.* Rept. NACA-TN 10 (1920).

[1509] L. PRANDTL: *Über Strömungen, deren Geschwindigkeit mit der Schallgeschwindigkeit vergleichbar sind.* J. Aeronaut. Res. Inst. (Tokyo Imperial University) **5**, No. 65, 25-34 (1930).

[1510] J. ZIEREP: *Ludwig PRANDTL, Leben und Wirken.* In: (G.E.A. MEIER, ed.) *Ludwig PRANDTL, ein Führer in der Strömungslehre.* Vieweg & Sohn, Braunschweig (2000), pp. 1-16.

| 1928 | Safety in Mines Research Board (SMRB), Sheffield, England | **William H. PAYMAN**,[1511] a British physical chemist and principal scientific officer at SMRB, **studies the mechanism of shock wave formation from a detonation in air.** He recognizes the vast difference between the initial wave, which can outrun the hot gas cloud, and the succeeding wave, calling them the **"shock wave"** and **"pressure wave,"** respectively. Typically, this pressure wave, which is emitted from the expanding gas cloud, is much broader and more energetic than the shock wave, but has a much lower peak pressure. ▪ This concept was later resumed by Melvin A. COOK,[1512] who worked out a simplified model of the pressure wave and found qualitative agreement with observations made by Richard G. STONER[1513] and Walker BLEAKNEY at the Palmer Physical Laboratory (Princeton University, NJ). |
|---|---|---|
| 1928 | Institute of Chemical Physics, Leningrad, Soviet Union | **Nikolai N. SEMENOV**,[1514] a Soviet physical chemist and head of the Dept. of Combustion and Explosion Research, **formulates his two theories of thermal explosions.** He had previously realized that any exothermic reaction with an overall positive energy of activation – whether or not it proceeds by a chain mechanism – will accelerate and perhaps be explosive if the rate of heat loss cannot keep pace with the rate of heat generation and, therefore, the temperature and rate of reaction increase.<br><br>(i) His theory states that homogeneous gas reactions may become explosive in either of two ways:<br>▸ According to one mechanism, the heat of reaction simply accumulates in the mass of gas faster than it can be removed through conduction to the walls, thus leading to a continual rise in temperature and a consequent acceleration of the rate of the reaction, eventually yielding an explosion.<br>▸ The other mechanism involves the formation of active molecules in the initial reaction step which carry on the reaction; if one step occasionally produces two or more of these active bodies, the rate of reaction may accelerate rapidly under certain conditions until an explosion occurs.<br><br>(ii) He shows that, for any exothermic reaction, there is a lower pressure limit to thermal explosion, $p_1$, given by<br>$$\ln p_1/T = B + A/T,$$<br>where $T$ is the temperature of the wall in a parallel plane-sided vessel, $A$ is proportional to the energy of activation of the reaction, and $B$ is a dimensionless quantity that increases with the coefficient of thermal conductivity and diminishes with increasing heat of reaction and frequency factor. This expression was found to represent the explosion limit for the decompositions of chlorine monoxide and azomethane reasonably well {ALLEN & RICE ⇨1935}. |
| 1928 | Dept. of Engineering, National Physics Laboratory (NPL), Teddington, Middlesex, England | **Sir Thomas STANTON**,[1515] a British mechanical engineer, **studies airfoils of various geometries at supersonic speeds of up to $M = 3.25$.** Using an advanced, continuously driven wind tunnel 3.07 in. (78 mm) in diameter, a high-speed stream is obtained by compressing air to 80 psi (5.5 bar) in a plant capable of delivering 2,800 ft³ (79.3 m³) of free air per minute. The air is discharged through a convergent-divergent conical nozzle into an approximately parallel tunnel, in which the model being tested is supported. ▪ STANTON's three-inch wind tunnel was later converted into an approximately parallel working jet by using inlet nozzles with profiles determined graphically {PRANDTL & BUSEMANN ⇨1929}, and confining the working jet into a square glass- |

---

[1511] W. PAYMAN: *The detonation wave in gaseous mixtures and the pre-detonation period.* Proc. Roy. Soc. Lond. **A120**, 90-109 (1928).
[1512] M.A. COOK: *The science of high explosives.* R.E. Krieger Publ., Huntington, NY (1971), pp. 322-327.
[1513] R.G. STONER and W. BLEAKNEY: *The attenuation of spherical shock waves in air.* J. Appl. Phys. **19**, 670-678 (1948).
[1514] N.N. SEMENOV: *Theorie des Verbrennungsprozesses.* Z. Physik **48**, 571-582 (1928); *Zur Theorie der chemischen Reaktionsgeschwindigkeit.* Z. Physik. Chem. **B2**, 161-168 (1929).
[1515] T.E. STANTON: *A high-speed wind channel for tests on aerofoils.* Brit. ARC-R&M 1130 (1928); *The development of a high-speed wind tunnel for research in external ballistics.* Proc. Roy. Soc. Lond. **A131**, 122-132 (1931).

| | | sided chamber suitable for observation with the schlieren method.[1516] Measurements of the wave systems set-up round conical-headed model cones indicated agreement with a theory published in 1933 by Geoffrey I. TAYLOR and John W. MACCOLL {VON KÁRMÁN & MOORE ⇨1932}. |
|---|---|---|
| 928 | *Cavendish Laboratory, Cambridge University, England* | Geoffrey I. TAYLOR,[1517] a mathematical physicist and research professor at Yarrow, **investigates, in collaboration with C.F. SHARMAN, the flow field of a compressible fluid past a cylinder using an electrical analogy.** They show that their method, based on a technique proposed previously by Leopoldo NOBILI and Adrien GUÉBHARD {E. MACH & SALCHER ⇨1886}, gives results of high enough accuracy in a reasonable time. However, it is limited to only two-dimensional flow problems – a handicap for most aeronautical applications. ▪ This area of research, supported by a grant from the British Air Ministry, originated from the realization by TAYLOR that the equations describing steady two-dimensional irrotational flow of a compressible fluid are identical to those describing the flow of electric current in a planar conducting sheet of variable thickness. In the former problem, the fluid density is related to the fluid velocity at the same point by D. BERNOULLI's equation, whereas the analogous quantity in the latter problem, the sheet thickness, may be chosen arbitrarily. TAYLOR perceived that this analogy enabled the problem of the steady flow of air at high speeds to be solved by performing an iterative set of measurements of electric current flow through a shallow layer of electrolyte in a tank with paraffin wax at the bottom, where the bottom is carved into the depth distribution required by BERNOULLI's equation after each measurement.[1518] |
| 1929 | *Chicago, Illinois* | On May 10, **the Acoustical Society of America (ASA) is founded,** with the goal "to increase and diffuse the knowledge of acoustics and promote its practical applications." The first president is H. FLETCHER (Bell Telephone Laboratories, New York City), and the first honorary member is the famous inventor Thomas A. EDISON (then at the age of 82). ▪ Initially intended to serve the growing practical requirements of architectural acoustics, its official bulletin – the ***Journal of the Acoustical Society of America (JASA),*** established in 1929 – soon began to cover all branches of theoretical and applied acoustics, **including nonlinear acoustics, historically one of the major foundations of shock wave physics.** |
| 1929 | *Chair of Chemical Technology, Imperial College of Science and Technology, London, England* | William Arthur BONE,[1519] a British chemist and fuel specialist, **and Reginald P. FRASER study detonation in a moist gaseous mixture (2 CO + $O_2$).** Similar to previous French researchers in this field {MALLARD & LE CHÂTELIER ⇨1883}, they use a high-speed rotating drum camera to measure the flame speeds of detonation phenomena in straight tubes. **Streak records taken with their "Fraser high-speed photographic machine,"** which has a writing speed of about 200 m/s, **show a periodic band structure; they also observe a helical track** formed by the detonation "head" in the explosion tube {⇨Fig. 4.16−T}.<br><br>In a later paper, BONE[1520] **and collaborators summarized their detailed photographic investigations into the phenomenon of "spin" in gaseous detonations** {CAMPBELL & WOODHEAD ⇨1926}. They stated: "A new view of the detonation-wave in gaseous explosions is advanced. For it can no longer be regarded as simply a homogeneous 'shock wave', in which an abrupt change in pressure in the vicinity of the wave-front is maintained by the adiabatic combustion of the explosive medium through which it is propagated; but it must now be viewed as |

---

[1516] A. BAILEY and S.A. WOOD: *The conversion of the Stanton 3-inch high-speed wind tunnel to the open jet type.* Proc. Inst. Mech. Eng. (Lond.) **135**, 445-466 (1937).

[1517] G.I. TAYLOR and C.F. SHARMAN: *A mechanical method for solving problems of flow in compressible fluids.* Proc. Roy. Soc. Lond. **A121**, 194-217 (1928).

[1518] G.K. BATCHELOR: *G.I. TAYLOR.* Biogr. Mem. Fell. Roy. Soc. (Lond.) **22**, 565-633 (1976), see p. 584.

[1519] W.A. BONE and R.P. FRASER: *V. Photographic investigation of flame movements in carbonic oxide – oxygen explosions.* Phil. Trans. Roy. Soc. Lond. **A228**, 197-234 (1929); *X. Photographic investigation of flame movements in gaseous explosions. Parts IV, V. and VI.* Ibid. **A230**, 363-385 (1932).

[1520] W.A. BONE, R.P. FRASER, and W.H. WHEELER: *II. A photographic investigation of flame movements in gaseous explosions. Part VII: The phenomenon of spin in detonation.* Phil. Trans. Roy. Soc. Lond. **A235**, 29-67 (1935).

a more or less stable association, or coalescence, of two separate and separable components, namely of an intensively radiating flame-front with an invisible shock wave immediately ahead of it; and whether persistent 'spin' is developed or not depends upon the stability or otherwise of their association…" ▪ Their observations stimulated other researchers who, although coming to a different explanation for the origin of the periodic phenomena, essentially confirmed the inhomogeneity of the detonation front.

Yakov B. ZEL'DOVICH[1521] and Kirill I. SHCHELKIN,[1522] two Soviet physicists, developed a *theory of spinning detonation* in which a portion of the detonation front moves circularly and at an oblique angle to the main direction of wave propagation {DENISOV & TROSHIN ⇨1959; SHCHELKIN & TROSHIN ⇨1963}.

## 1929 — Institut für Technische Physik, TH Berlin-Charlottenburg, Germany

Carl CRANZ,[1523] a German professor of applied physics and expert on ballistics, **and Hubert SCHARDIN,** a physicist and his close assistant, **publish a paper with milestone achievements in high-speed photography and gas dynamics:**

(i) They report on a new high-speed framing camera with no moving parts – the so-called **"Cranz-Schardin multiple-spark camera"** [Germ. *CRANZ-SCHARDINsche Mehrfachfunken-Kamera*], which they invented in the period 1927–1928. It uses an optical arrangement where a multiple array of spark gaps, each matched to a field lens, forms successive images on the common film plate of events occurring in the plane of the field lens. The images are then switched at a rate limited only by the electronic circuits relaying the air spark gap discharges. Their prototype provides eight frames of excellent quality at a maximum frame rate of $3 \times 10^5$ frames per second {⇨Fig. 4.19−M}.

(ii) **They first cinematographically record phenomena associated with oblique interactions of shock waves** {⇨Fig. 4.13−C}, which they appropriately call "V-propagation" [Germ. *V-Ausbreitung*]; the term *Mach reflection* {E. MACH & WOSYKA ⇨1875} had not yet been coined {VON NEUMANN ⇨1943}. They are also the first to study the Mach effect quantitatively. For the case of two point sources of equal strength, they derive a formula for the condition under which Mach reflection occurs. This condition is based upon the angle of incidence of the two interacting shock waves and the lateral velocity of the point of interaction.

(iii) Also investigating the well-known conundrum of **why an implosion is always accompanied by a sharp report,** they show that – contrary to widespread opinion – the report is not caused by the rarefaction wave itself but rather by the blast wave created shortly after the implosion. They demonstrate this phenomenon using an evacuated tube 34 m long that they suddenly open at one end; the air rushes in violently and is reflected at the other end of the tube, returning to the open end of the tube after about 0.2 s, yielding a perceptible report. **The rarefaction wave cannot be heard.**

(iv) **Herman H. ZORNIG discusses historic soot recording experiments on irregular shock reflection** {E. MACH & WOSYKA ⇨1875}, **their possible interpretation, and related problems of nonacoustic reflection of air shocks with SCHARDIN.**[1524] ▪ ZORNIG,[1525] a former chief of the Technical Division of the U.S. Army, came to Berlin as an assistant military attaché and studied ballistics under Prof. CRANZ in the period 1927–1929 – *i.e.,* at the time when CRANZ and SCHARDIN first used high-speed cinematography to uncover Mach reflection phenomena.[1526, 1527]

---

[1521] Y.B. ZEL'DOVICH: *On the theory of spinning detonation* [in Russ.]. Dokl. AN (SSSR) **52**, No. 2, 147-150 (1966).

[1522] K.I. SCHELKIN: *On a theory of the phenomenon of spinning detonation* [in Russ.]. Dokl. AN (SSSR) **47**, No. 7, 482-484 (1945).

[1523] C. CRANZ and H. SCHARDIN: *Kinematographie auf ruhendem Film und mit extrem hoher Bildfrequenz*. Z. Phys. **56**, 147-183 (1929).

[1524] R.J. EMRICH: *Early development of the shock tube and its role in current research*. In: (Z.I. SLAWSKY, J.F. MOULTON JR., and W.S. FILLER, eds.) *Proc. 5th Int. Shock Tube Symposium* [White Oak, Silver Spring, MA, April 1965]. APS, Fluid Dynamics Div. (1965), pp. 1-10.

[1525] Herman H. ZORNIG (1888–1973) graduated from Iowa State College (1909) and entered the Army. After his return from Berlin to the United States, he held several leading positions in ammunition and ordnance divisions of the U.S. Army, and was named the first Director (1938–1946) of the Ballistic Research Laboratory (BRL) at Aberdeen.

[1526] H.H. GOLDSTINE: *The computer from PASCAL to VON NEUMANN*. Princeton University Press, Princeton, NJ (1972), p. 127.

This subject later became important for the Manhattan Project {⇨1942}, when the "optimum" Height of Burst (HOB) for the two nuclear explosions in Japan {Hiroshima & Nagasaki Bombing ⇨1945} had to be selected such that the damage at ground level, provoked by Mach reflection {VON NEUMANN ⇨1943 & 1945}, would be maximized.[1528]

In the following years, CRANZ and SCHARDIN **extended the frame rate of their new camera up to $10^6$/s and the number of frames to 24,** which became the standard version. In spite of the enormous advances made in digital high-speed framing photography, this camera type is still the workhorse in some gas dynamics laboratories and is highly regarded due to its excellent resolution {⇨Figs. 4.5–I, J}. In addition, it is possible to transform individual frames from a shadowgraph into a schlieren recording by a simple procedure prior to recording, which later facilitates the analysis of flow phenomena.

| | | |
|---|---|---|
| 1929 | Mt. Wilson Observatory, San Gabriel Mountains [near Pasadena], California | Edwin P. HUBBLE,[1529] an extragalactic astronomer from the U.S., **publishes a plot of the Doppler shift in the light from 22 galaxies *vs.* their distances, and reports that while galaxies are receding from us, more distant galaxies are receding from us more rapidly – the so-called "Hubble law."** |

Two years later, after having compiled more data on the velocities of nebulae together with his assistant Milton L. HUMASON, he stated,[1530] "The relation [between radial velocity and distance] is a linear increase in the velocity amounting to about +500 km/s per million parsecs of distance." • HUBBLE's simple statement had an enormous impact on cosmology: **it promoted the idea that the expanding Universe may have originated in a huge explosion – thus supporting the Big Bang theory** {FRIEDMANN ⇨1922; LEMAÎTRE ⇨1927}. Astronomers have since named the rate at which the Universe is expanding the ***Hubble constant*** in honor of him; it is denoted by

$$H_0 = v/r,$$

where $v$ is the recession velocity of the galaxy in question, obtained from its redshift, and $r$ is its distance away from the Earth.

Since the late 1950s astronomers have been arguing for an $H_0$ value of between 50 and 100 km/s/megaparsec, where one megaparsec is equal to $3.262 \times 10^6$ light-years (= $3.085 \times 10^{18}$ km). Assuming a value of 72 km/s/megaparsec, as recently determined with an uncertainty of only 10% (itself a milestone achievement in cosmology[1531]), this means that a galaxy 1 megaparsec away will be moving away from us at a speed of 72 km/s, while another galaxy 100 megaparsec away will be receding at 100 times this speed; *i.e.*, at 7,200 km/s.

| | | |
|---|---|---|
| 1929 | Mt. Wilson Observatory, Pasadena, California | Sir James JEANS,[1532] a British physicist and astronomer, **publishes the popular book *The Universe Around Us.*** |

(i) **Ignoring the idea of stellar collisions,** he writes, "The Universe consists in the main not of stars but of desolate emptiness – inconceivably vast stretches of desert space in which the presence of a star is a rare and exceptional event … The stars move blindly through space, and the players in the stellar blind-man's-buff are so few and far between that the chance of encountering another star is almost negligible."

---

[1527] H. SCHARDIN: *Über die Entwicklung der Hohlladung*. Wehrtech. Monatshefte **51**, Nr. 4, 97-120 (1954).
[1528] F. REINES and J. VON NEUMANN: *The Mach effect and height of burst*. In: (H.A. BETHE, ed.) *Blast wave*. Rept. LA-2000, LASL, Los Alamos, NM (Aug. 13, 1947), chap. 10. • A declassified version of this chapter has been published in A.H. TAUB (ed.): *John VON NEUMANN. Collected works*. Pergamon Press, Oxford *etc.* (1963), vol. VI, pp. 309-347.
[1529] E.P. HUBBLE: *A relation between distance and radial velocity among extra-galactic nebulae*. Proc. Natl. Acad. Sci. **15**, 168-173 (1929).
[1530] E.P. HUBBLE and M.L. HUMASON: *The velocity-distance relation among extra-galactic nebulae*. Astrophys. J. **74**, 43-80 (1931).
[1531] W. FREEDMAN ET AL.: *Final results from the Hubble Space Telescope Key Project to measure the Hubble constant*. Astrophys. J. **553** (1), 47-72 (May 2001).
[1532] J. JEANS: *The Universe around us*. Cambridge University Press, Cambridge (1929).

(ii) He performs some simple calculations that indicate that a collision is so unlikely that not a single one of the hundred billion stars in the Milky Way galaxy's disk has been involved in a collision in the entire lifetime of our galaxy. ▪ JEANS and others had calculated that a stellar collision should occur in our Galaxy only once in $10^{12}$ to $10^7$ years, whereas several novae – which may originate from collisions of two stars – occur each year.[1533]

American astronomers later showed that the probability of a stellar collision is not simply a function of their cross-sectional area and the average distance between the objects, but that gravitational focusing is an important effect {HILLS & DAY ⇨1976}.

| | | |
|---|---|---|
| 1929 | Kaiser-Wilhelm-Institut (KWI) für Strömungsforschung, Göttingen, Germany | **Ludwig PRANDTL**[1534] **and Adolf BUSEMANN**, two eminent German aeronautical engineers, **develop a graphical solution based on the method of characteristics – the so-called "Prandtl-Busemann method of characteristics"** [Germ. *PRANDTL-BUSEMANNsches Charakteristikenverfahren*]. |

(i) The changes of state occurring in a supersonic flow are manifested in expansion or compression waves. Therefore, when the flow passes through these lines of disturbance, the flow velocity and the flow direction are modified by definite values which can easily be determined from the graphical representation of the characteristics.[1535]

(ii) It allows one to approximately determine smooth supersonic flows at arbitrary initial and boundary conditions in Laval nozzles and other profiles. Their method replaces the stationary two-dimensional supersonic potential flow with a crossing system of stationary sound waves. ▪ BUSEMANN later mentioned that PRANDTL had already used a primitive form of the characteristics method as early as 1906 to shape the exit of his Laval nozzles for parallel supersonic jets.[1536]

The Prandtl-Busemann method of characteristics is very versatile and, for example, was used to approximately calculate compressible flows about profiles with local regions of supersonic velocity.[1537]

| | | |
|---|---|---|
| 1929 | University of Cambridge, England | At the 4th Pacific Science Congress in Java, the great British physicist, mathematician and meteorologist **Geoffrey I. TAYLOR**[1538] **reports on his investigation into long gravity waves in the atmosphere.** |

(i) **These waves can result from extraordinary strong blast waves,** such as the one observed by a number of observatories during the final volcanic eruption of Krakatau volcano {⇨1883}. This pressure wave, which was observed in all of the pressure traces recorded, traveled several times around the Earth (taking about 33 hours each time) at great altitude with an average (supersonic) velocity of about 330 m/s.[1539]

(ii) TAYLOR shows that this is consistent with the propagation of a nondispersive wave with properties similar to those of barotropic waves in a constant-depth fluid. His analytical results show good agreement with Krakatau data recorded in 1883.

---

[1533] F.L. WHIPPLE: *Supernovae and stellar collisions*. Proc. Nat. Acad. Sci. **25**, No. 3, 118-125 (1939).
[1534] L. PRANDTL and A. BUSEMANN: *Näherungsverfahren zur zeichnerischen Ermittlung von ebenen Strömungen mit Überschallgeschwindigkeit*. Festschrift zum 70. Geburtstag von Prof. A. STODOLA. Füßli, Zürich (1929), pp. 499-509.
[1535] L. PRANDTL: *Führer durch die Strömungslehre*. Vieweg & Sohn, Braunschweig (1942), p. 247.
[1536] A. BUSEMANN: *Ludwig PRANDTL 1875–1953*. Mem. Fell. Roy. Soc. **5**, 193-205 (1959).
[1537] B. GÖTHERT and K.H. KAWALKI: *Berechnung kompressibler Strömungen mit örtlichen Überschallfeldern*. Forschungsbericht Nr. 1794, Zentrale für wissenschaftliches Berichtswesen der Luftfahrtforschung des Generalluftzeugmeisters (ZWB), Berlin-Adlershof (Aug. 1943); *The calculation of compressible flows with local regions of supersonic velocity*. Rept. NACA-TM 1114 (1947).
[1538] G.I. TAYLOR: *The air wave from the great explosion at Krakatoa*. In: *Proc. 4th Pacific Science Congress* [Java, Indonesia, May/June 1929], Pacific Science Association, Batavia (1930), vol. IIB: *Physical papers*, pp. 645-655.
[1539] The velocity of sound is a rather complex function of altitude, *see* Sect. 2.5.1.

| | | |
|---|---|---|
| 1930 | Stromboli, Aeolian Islands, southern Italy | On September 30, **two violent eruptions of the volcano Stromboli take place.** ▪ Stromboli has been erupting almost continuously for at least the past 2,400 years, but violent eruptions are relatively rare events. Typically, the eruptions produce an occasional lightshow that give rise to its nickname, the "Lighthouse of the Mediterranean."<br><br>So-called **"Strombolian-type eruptions"** (or "Strombolian bursts"), which are more explosive than Hawaiian-type eruptions {Mt. Kilauea-Iki ⇨1959}, are characterized by the intermittent explosion or fountaining of relatively fluid basaltic lava from a single vent or crater. Each episode is caused by the release of volcanic gases, and they typically occur every few minutes or so, sometimes rhythmically and sometimes irregularly. The lava fragments generally consist of partially molten volcanic bombs that become rounded as they fly through the air. |
| 1930 | Applied Mechanics Division, University of Michigan, Ann Arbor, Michigan | Lloyd Hamilton DONNELL,[1540] an assistant professor of mechanical engineering, **studies longitudinal shock wave transmission in solid bodies** when impacted and when the dimensions are not very small compared to the velocities of such waves. He treats various cases of practical importance theoretically, such as<br>▸ the impact of thin bars with free or fixed ends;<br>▸ effects of a sudden change in the cross-section or material of a bar; and<br>▸ waves due to the application of a force at an intermediate section. |
| 1930 | Television Laboratories Inc., San Francisco, California | Philo T. FARNSWORTH,[1541] a U.S. electrical engineer, **proposes the continuous dynode "electron multiplier."** It consists of a long and narrow hollow tube. When an appropriate bias voltage is applied, secondary electrons generated by intercepted photons or charged particles on the inside surface of the channel tube are multiplied down the tube through additional secondary electron emission. By properly choosing the inner surface material, a gain of $10^4$ can be achieved.<br><br>In the 1960s, this simple but very effective and versatile principle was further developed at the Bendix Research Laboratories (Southfield, MI), leading to a two-dimensional derivative consisting of a multitude of FARNSWORTH's single channel, continuous dynodes – a so-called **"MicroChannel Plate (MCP)."** These proved to be very useful devices in high-speed diagnostics.[1542]<br>▸ In fast oscilloscopes, MCPs multiplied the trace intensities of cathode-ray tubes by a factor of about 1,000, which allowed single-sweep viewing up to the oscilloscope's rise time specification at bandwidths of as much as 1 GHz for the first time. Before fast digitizers became available, these oscilloscopes were indispensable for analyzing ultrafast shock waves, and were used for example to record pressure profiles from piezoelectric and piezoresistive gauges, and shock arrival times using electric-pin methods.<br>▸ In optical applications, the MCP is encased in a glass vacuum envelope between a photocathode and a phosphor screen. The outstanding features of such a system are its high light intensification and fast shuttering, which permitted CCD cameras {BOYLE & G.E. SMITH ⇨1969} to be designed with ultrashort exposure times down to only a few nanoseconds in the 1990s. Such **"Intensified CCDs (ICCDs)" became an important requirement for realizing applications of shock wave recording in solids and high-temperature plasmas** {Hadland Photonics Ltd. ⇨1993}.<br>▸ **MCPs are also sensitive to X-rays in the keV to MeV region.** Since they have a high temporal resolution (nanoseconds) as well as a good geometric resolution (about |

---

[1540] L.H. DONNELL: *Longitudinal wave transmission and impact.* Trans. Am. Soc. Mech. Eng. **52**, 153-167 (1930).
[1541] P.T. FARNSWORTH: *Electron multiplier.* U.S. Patent No. 1,969,399 (March 3, 1930).
[1542] A.W. WOODHEAD and G. ESCHARD: *Microchannel plates and their application.* Acta Electronica **14**, 181-200 (1971).

| | | |
|---|---|---|
| | | 20 µm), they can be used as detectors and amplifiers in flash radiography {CHANG ⇒1983} and flash X-ray diffraction.[1543] |
| 1930 | U.S. Bureau of Mines, Pittsburgh Experimental Station, Pittsburgh, Pennsylvania | **Bernhard LEWIS,**[1544] a British-born U.S. physical chemist, **and James Byron FRIAUF,** a U.S. consultant physicist, **carry out the first critical test of the Chapman-Jouguet (CJ) theory** {CHAPMAN ⇒1899; JOUGUET ⇒1904–1906}. They measure and calculate detonation velocities in mixtures of hydrogen and oxygen ($2H_2 + O_2$) diluted with several gases such as $N_2$, $O_2$, He, and A.<br>(i) On the basis of the CJ theory they computed detonation velocities for each of the two following assumptions:<br>▸ the composition of the burned gases corresponds to complete combustion; and<br>▸ the composition of the burned gases corresponds to chemical equilibrium for the dissociation of water vapor into hydrogen and oxygen, and into hydrogen and hydroxyl, and for the dissociation of molecular into atomic hydrogen.<br>They obtain very good agreement between measured and calculated values.<br>(ii) **They introduce the term *Chapman-Jouguet theory* into detonation physics,** arguing that "in view of the equivalence of the relation developed by CHAPMAN and JOUGUET, it seems preferable to refer to the Chapman-Jouguet theory rather than the Jouguet theory, even though CHAPMAN's investigations are less comprehensive." |
| 1930 | Maschinen- und Apparatebau München, Germany | **Paul SCHMIDT,** a German mechanical engineer and inventor, stimulated by previous works performed in the 1900s by the Frenchmen Victor KARAVODINE and Georges MARCONNET, develops his *Schmidtrohr* ("Schmidt tube"), **a pulsejet engine,** and obtains German and British patents for it.[1545] Based on a 3.6-m-long tube resonator with a valve matrix at the entrance and a Laval nozzle at the exit, it uses the reflected shock wave for periodical reignition (at about 50 Hz).[1546] ▪ Further developed in Berlin by Argus Motoren GmbH, the *Argus-Schmidtrohr* ("Argus Schmidt tube") was applied in World War II by the German Air Force to power the long-range flying bomb (or "buzz-bomb"). This subsonic, pilotless aircraft was the world's first cruise missile. Codenamed *Kirschkern* ("Cherry Pit"), it could carry a 830-kg explosive warhead. Later, it became widely known as the "V1" (an abbreviation used for *"Vergeltungswaffe 1,"* meaning "Vengeance Weapon 1").[1547] Throughout the war, many improvements were made to it; the most advanced construction reached a velocity of up to 800 km/h ($M = 0.66$).[1548] During World War II, the operation cycle of the Schmidt tube was analyzed in detail by the German mechanical engineers Adolf BUSEMANN[1549] and Fritz SCHULTZ-GRUNOW.[1550] |

---

[1543] T. EBDING: *Verwendung von Microchannelplates in der Röntgenkristallographie.* ISL Rept. R 124/84. ISL, Saint Louis, France (1984); *Einsatz von Microchannelplates (Vielkanalplatten) als Röntgenbildverstärker in der Röntgenblitztechnik.* ISL Tech. Rept. RT 505/84, ibid. (1984); *Aufbau eines schnellen Röntgenblitzverstärkers für Anwendung in der Kinematographie.* ISL Tech. Rept. RT 514/87, ibid. (1987).

[1544] B. LEWIS and J.B. FRIAUF: *Explosions in detonating gas mixtures. I. Calculation of rates of explosions in mixtures of hydrogen and oxygen and the influence of rare gases.* J. Am. Chem. Soc. **52**, 3905-3920 (1930).

[1545] P. SCHMIDT: *Verfahren zum Erzeugen von Antriebskräften (Reaktionskräften) an Luftfahrzeugen.* Germ. Patents No. 523,655 (1930) and No. 567,586 (1931), and classified Germ. Patent No. Sch 98,044 I/46g (1932); *Improved method of producing motive forces for the propulsion of vehicles or aircraft.* Brit. Patent No. 368,564 (1931).

[1546] F. SCHULTZ-GRUNOW: *Wirkungsweise des Paul-Schmidt-Verpuffungsstrahlrohres. V1-Propulsion.* Flugabwehr & Flugtech. **19**, 141-143 (1948).

[1547] The "Vengeance Weapon 2" (V2 or A-4) was a German ballistic missile used in World War II, and was the forerunner of modern space rockets and long-range missiles {VON BRAUN ⇒1936; HVA Peenemünde ⇒1940/1941 & 1942}. The "Vengeance Weapon 3" (V3) consisted of a battery of 5×5 supersonic guns; see {RODMAN ⇒1864} and the book by David IRVING: *Die Geheimwaffen des Dritten Reiches.* Arndt-Verlag, Kiel (2000), pp. 238-246, 275-283, 343-344, 359.

[1548] W. HELLMOND: *Die V1. Eine Dokumentation.* Bechtle, Esslingen/München (1991); *see Appendix.*

[1549] A. BUSEMANN: *Bericht über den Paul SCHMIDTschen Strahlrohrantrieb.* Bericht FB530, Deutsche Versuchsanstalt für Luftfahrtforschung (DVL), Berlin-Adlershof (1936).

[1550] F. SCHULTZ-GRUNOW: *Gasdynamische Untersuchungen am Verpuffungsstrahlrohr.* Institut für Mechanik, TH Aachen. (ZWB) FB 2015/1 (1943) und FB 2015/2 (1944); Engl. translation: *Gas dynamic investigations of the pulse jet tube.* Rept. NACA-TM 1131 (1947).

After World War II, the study of the pulsejet engine was resumed in the United States[1551] and used to propel guided missiles to supersonic velocities, *e.g.*, the North American "Hound Dog" (first launch on April 23, 1959), which exceeded Mach two.

| 1931 | *Norman Bridge Laboratory, Cal-Tech, Pasadena, California* | **Wilhelm Heinrich W. BAADE,** a German astronomer, **and Fritz ZWICKY,** a Bulgarian-born U.S. theoretical physicist, **introduce the term *supernova*.** ▪ In 1940, ZWICKY[1552] stated in a paper on the various types of novae, "BAADE and I introduced the term *supernova* in seminars given at the California Institute of Technology in 1931." ▪ The debate about the nature of spiral galaxies already led previously to the realization that there must be "giant novae" (K.E. LUNDMARK, 1920), novae of "impossible great absolute magnitude" (H.D. CURTIS, 1921), "exceptional novae" (E.P. HUBBLE, 1929), and "Hauptnovae" (W.H.W. BAADE, 1929).

ZWICKY distinguished between "common novae" and "supernovae." Astronomers have classified supernovae according to the absorption lines of different chemical elements that appear in their spectra {CHANDRASEKHAR ⇨1931}.

Currently, a U.S. multi-institution, multi-disciplinary collaboration of astrophysicists, nuclear physicists, applied mathematicians, and computer scientists develops models to better understand core collapse supernovae {Terascale Supernova Initiative ⇨2001}. |
|---|---|---|
| 1931 | *Dept. of Physics, Harvard University, Cambridge, Massachusetts* | **Percy W. BRIDGMAN,**[1553] a U.S. physics professor, **publishes his book *The Physics of High Pressure*, in which he describes his measurements of the compressibilities, viscosities, polymorphic transitions**[1554] **and other properties of many elements and compounds subjected to high pressures.** ▪ Although it exclusively dealt with static pressure phenomena in compressed liquids and solids, his book stimulated later generations of shock physicists to study the dynamic properties of materials under pressures not accessible by static compression techniques – and it still does.

In 1946, BRIDGMAN received the Nobel Prize for Physics in recognition of his pioneering research into the effects of high pressures on the behavior of matter. |
| 1931 | *Trinity College, Cambridge, England* | **Subrahmanyan CHANDRASEKHAR,**[1555] an Indian-born U.S. astrophysicist, **studies astrophysical models of white dwarf stars** {CLARK ⇨1862} **and comes to the conclusion that a white dwarf cannot be more massive than about 1.2 solar masses** (solar mass: $1.989 \times 10^{30}$ kg). ▪ This became known as the **"Chandrasekhar limit."** A more accurate modern value for this limit is 1.44 solar masses.[1556]

▹ Solitary stars with masses below this limit are stable white dwarfs. A white dwarf represents the final stage in the life of a low-mass star. The white dwarf takes billions of years to cool off and eventually becomes a ***black dwarf***, a burned-out star which no longer emits visible light.

▹ In a binary star system, gravitational forces may cause a white dwarf to accrete matter from its companion until it reaches the Chandrasekhar limit.

▹ Stars which have masses above this limit have sufficient gravity to collapse further, resulting in a tremendous thermonuclear explosion of the now super-dense white dwarf and the formation of a ***Type Ia Supernova*. The resulting shock wave,** which expands for hundreds of thousands of years and interacts with the interstellar medium, **can create perturbations that cause gravitational collapse and new star formation.** |

---

[1551] J.V. FOA: *Intermittent jets*. In: (O.E. LANCASTER, ed.) *Jet propulsion engines*. Princeton University Press, Princeton, NJ (1959), pp. 377-438.
[1552] F. ZWICKY: *Types of novae*. Rev. Mod. Phys. **12**, 66-85 (1940).
[1553] P.W. BRIDGMAN: *The physics of high pressure*. Dover Publ., New York (1931, 1970); Bell & Sons, London (1949, 1952, 1958).
[1554] *Polymorphism* is a phenomenon in which identical chemical compounds can exhibit different crystal structures.
[1555] S. CHANDRASEKHAR: *The density of white dwarf stars*. Phil. Mag. **11** [VII], 592-596 (1931); *The maximum mass of ideal white dwarfs*. Astrophys. J. **74**, 81-82 (1931).
[1556] F.H. SHU: *The physical Universe: an introduction to astronomy*. University Science Books, Mill Valley, CA (1982), p. 128.

▸ If the original star was so massive that its strong stellar wind had already blown off the hydrogen from its atmosphere by the time it explodes, then its supernova does not show hydrogen spectral lines: it is a *Type Ib Supernova*. It has a high abundance of He lines in its spectra, while a *Type Ic Supernova* has a low abundance of such lines when measured soon after the supernova event. The subclasses Type Ib and Type Ic have provided new insights into the variety of conditions under which stars explode.

▸ Supernova SN 1987A {SHELTON ⇨ 1987} was found to be a new, unusual type of supernova in terms of its relatively compact progenitor, dim light curve, and some special spectral characteristics. It was classified as *Type II-P* (*P* standing for peculiar).[1557]

▸ *Type II Supernova* **occurs at the end of the lifetime of a supergiant star,** when its nuclear fuel is exhausted and its iron core collapses due to gravity. This causes the outer layers to fall inward, bounce off the core, and detonate as a supernova, **sending shock waves outward.**[1558] Type II supernovae show Balmer hydrogen lines in their spectra.

In 1983, CHANDRASEKHAR was awarded the Nobel Prize for Physics "… for his theoretical studies of the physical processes of importance to the structure and evolution of the stars."

| | | |
|---|---|---|
| 1931 | Imperial College, London, England | Sydney CHAPMAN,[1559] a British mathematician and geophysicist, **and Vincenzo C.A. FERRARO,** one of CHAPMAN's students, **tackle the puzzling phenomenon of magnetic storms.**

(i) **They suggest that the "sudden commencement"** – a small step-like jump in the magnetic field observed all over the world {CARRINGTON ⇨ 1869} – **occasionally comes from the Sun and is the front of a huge plasma cloud emitted from the Sun that hits the Earth's magnetic field.** • They proposed that such jumps marked the cloud's arrival. However, they had not yet recognized that the flow of plasma from the Sun is not confined to isolated clouds, but occurs continuously, in the form of the *solar wind* {BIERMANN ⇨ 1951; PARKER ⇨ 1958}.

(ii) Using a schematic, they propose the idea that these plasma clouds envelop the Earth, thereby carving a "cavity" in the cloud in which the Earth and its magnetic field are confined.

Their hypothesis – the so-called **"Chapman-Ferraro cavity"** – was a landmark in astrophysics and led to the prediction of planetary bow shocks {GOLD ⇨ 1953; AXFORD & KELLOGG ⇨ 1962}.[1560] |
| 1931 | School of Aeronautical Engineering, University of Rome, Italy | **General G. Arturo CROCCO,**[1561] a pioneer of Italian aviation and Director of Research for the Italian Air Force, **coins the term *aerothermodynamics*** – a combination of fluid mechanics and thermodynamics – to refer to aerodynamic heating at supersonic speeds that, at the time, could only be observed for the propeller tips of high-speed aircraft. • This particular branch of super/hypersonic flow, later introduced and propagated by Theodore VON KÁRMÁN,[1562] has since become important in the design of high-speed air- and spacecraft, because one must account for the reduction in materials strength that occurs at elevated temperatures. Indeed, this effect starts at rather low Mach numbers. Using the example of the Soviet SST, the Tupolev Tu-144 {⇨ 1968}, it was observed that air friction heated up the airframe to temperatures over 150 °C |

---

[1557] J.C. WHEELER: *Supernovae, new types*. In: (S.P. MARAN, ed.) *The astronomy and astrophysics encyclopedia*. Van Nostrand & Reinhold, New York (1992), pp. 887-890.

[1558] N. STRAUMANN: *Physics of type II supernova explosions*. In: (H. MITTER and F. WIDDER, eds.) *Particle physics and astrophysics*. Springer, Berlin *etc.* (1989).

[1559] S. CHAPMAN and V.C.A. FERRARO: *A new theory of magnetic storms. Part 1: the initial phase*. Terr. Magn. Atmos. Electr. **36**, 171-186 (1931); Ibid. **37**, 147-156 (1932).

[1560] V.C.A. FERRARO: *The birth of a theory* [an account of the beginnings of the Chapman-Ferraro theory]. In: (S.I. AKASOFU, B. FOGLE, and B. HAURWITZ, eds.) *Sidney CHAPMAN, eighty: from his friends*. University of Colorado Press, Boulder, CO (1969), pp. 14-18.

[1561] G.A. CROCCO: *Sui corpi aerotermodinamici portanti*. Rendiconti della Academia Nazionali die Lincei **16** [VI], 161-166 (1931).

[1562] Th. VON KÁRMÁN: *Aerodynamics*. Cornell University Press, Ithaca, NY (1954).

| | | above that of the surrounding air when it flew for hours at Mach 2, with heat concentrated on the nose and leading edges of the wings.[1563] |
|---|---|---|
| 1931 | Safety in Mines Research Board (SMRB), Sheffield, England | **William H. PAYMAN,**[1564] a British physical chemist and principal scientific officer at SMRB, **and Donald W. WOODHEAD first apply the streak technique in order to photograph the initial processes of various explosion and shock wave phenomena in straight tubes.** In order to get the required time resolution, they use the classical rotating mirror {WHEATSTONE ⇨1834; VON OETTINGEN & VON GERNET ⇨1888}, but they incorporated a slit positioned perpendicular to the wave propagation in the optical path. In this way, they obtain a continuous radius-time photographic record of the event on a drum rotating at high speed – a **"streak record."**<br><br>In the following years, they refined their streak camera, which they called a **"wave speed camera"** {⇨Fig. 4.19–L}, and obtained a writing speed of 0.5 mm/µs (500 m/s). |
| 1932 | Wahar Crater, Arabian Desert, Saudi Arabia | In February, **Harry St. John B. PHILBY,**[1565] a British explorer and Arabist, **crosses the Arabian Desert (the "Empty Quarter") in search of the legendary city of Wahar** (or Wabar). He finds the "walls" of the city to be the rims of a series of craters, and the abundant "cinders" of the city that had been "destroyed by fire from heaven" proved upon examination to be nearly pure silica glass. Nearby, rusted pieces of meteoritic iron are also found, the largest remnant weighing 25 pounds (11.3 kg). ▪ **PHILBY's** discovery of *silica glass* stimulated contemporary theories on meteorite impact {SPENCER ⇨1932}. |
| 1932 | Cavendish Laboratory, Cambridge, England | **James CHADWICK,**[1566] a British physicist, exposes beryllium to bombardment from alpha particles and **discovers the *neutron*.** ▪ This discovery, for which he received the 1935 Nobel Prize for Physics, provided a new tool for atomic disintegration and initiation of chain reactions {HAHN & STRASSMANN ⇨1938; FRISCH & MEITNER ⇨1939}. The discovery of the neutron opened up new methods of building nuclear weapons {SZILARD ⇨1913; FRISCH & PEIERLS ⇨1941}, and of producing new fissionable bomb material, like plutonium {SEABORG ET AL. ⇨1941}. Since hydrogenous compounds strongly absorb neutrons, *flash neutron radiography* – a shadowgraphy technique like flash radiography, but one that uses very strong fluxes of fast neutrons instead of X-rays – enables the rapid motion of organic hydrogenous substances contained in a metallic envelope (*e.g.*, a detonating chemical explosive in a bomb shell) to be visualized.[1567]<br><br>In the same year, the Soviet physicist Lev D. LANDAU speculates on the possible existence of stars composed entirely of neutrons – **"neutron stars"** {BAADE & ZWICKY ⇨1934} – a subject which will stimulate generations of astrophysicists and cosmologists. |
| 1932 | Milan, Lombardy, northern Italy | At the 20th meeting of the Italian Association for the Advancement of Science, **G. Arturo CROCCO,**[1568] Director of Research for the Italian Air Force, **reads a paper that discusses the possibility of *superaviation* –** *i.e.*, flight at very high altitudes (above 37,000 ft or 11,278 m) and at speeds of up to about $M = 3$. He addresses not only the particular problems of high-speed flight in the stratosphere (such as lift and propulsion), but also discusses the economic efficiency of such flights. |

---

[1563] H. MOON: *Soviet SST. The technopolitics of the Tupolev-144*. Orion Books, New York (1989), p. 121.
[1564] W. PAYMAN and D.W. WOODHEAD: *Explosion waves and shock waves*. Proc. Roy. Soc. Lond. **A132**, 200-213 (1931).
[1565] H.St.J.B. PHILBY: *The empty quarter: being a description of the Great South Desert of Arabia known as Rub al Khali*. Constable & Co., London (1933).
[1566] J. CHADWICK: *Possible existence of a neutron*. Nature **129**, 312 (1932); *The neutron and its properties*. Brit. J. Radiol. **6**, 24-32 (1933).
[1567] H. BERGER: *Neutron radiography*. Elsevier, Amsterdam (1965).
[1568] G.A. CROCCO: *Flying in the stratosphere*. Aircraft Engng. **4**, 171-175, 204-209 (1932).

In the same year, the subject of superaviation was resumed in the nine-volume encyclopedia *Mezhplanetnye Soobschniya* ("Interplanetary Communications"), written by the Soviet aerospace researcher Nikolai Alekseevich RYNIN.[1569]

1932 — Research Dept., Woolwich, southeast London, England

**Alwyn Douglas CROW,**[1570] a physics professor and ballistician, **and W.E. GRIMSHAW use a piezoelectric gauge to measure the pressures given by various explosives at various loading densities in order to derive the equations of state for propellant gases.**

(i) For the quantity "ballistic force" of the propellant they obtain agreement to within 2% between the values derived from direct experiment and those obtained from thermodynamic considerations.

(ii) They also show that the covolume from the constituents of the gas complex can be evaluated by using the hard kernel value of the molecular radii increased by some 2.5%.

1932 — Zeiss-Ikon AG, Dresden, Saxony, Germany

**Hans JOACHIM**[1571] **and Hans ILLGEN,** two German research physicists, **measure rifle gas pressure using their "piezo-indicator,"** an instrument consisting of a piezo-crystal, an electronic amplifier and an electronic/mechanical oscillograph: vertical displacement of the pressure signal is monitored by a cathode-ray tube (c.r.t.), while time sweeps are monitored mechanically by imaging the screen of the c.r.t. onto a rotating drum covered with photo paper {⇨Fig. 4.19–D}.

1932 — Guggenheim Aeronautical Laboratory, CalTech, Pasadena, California

**Theodore VON KÁRMÁN,**[1572] a Hungarian-born professor of aeronautical engineering, **and Norton B. MOORE,** his Ph.D. student, **calculate the pressure distribution at the head of a flying projectile by numerically integrating the hydrodynamical equations for the steady flow of a compressible fluid with axial symmetry.** Their calculation, covering Mach numbers from 1.0 to 2.5, takes into account a new type of drag – the **"wave drag"** – that occurs when the body approaches the velocity of sound.

(i) They consider the aerodynamic drag of projectiles to be composed of three components:
- the head resistance;
- the skin friction acting along the body; and
- the wave resistance contributed by the underpressure at the base.

(ii) They demonstrate how to find the optimum shape for a body of revolution with a given caliber and nose length. Their calculated drag-velocity profile is lower than that obtained for the S-projectile {DWM ⇨1903} and deduced from ballistic experiments {BAIRSTOW, FOWLER & HARTREE ⇨1920} – probably because in actual flight the drag is increased by yaw and oscillations of the shell.

**In the following year, Geoffrey I. TAYLOR**[1573] **and John W. MACCOLL extended the study to the more general case of a flow around an unyawed circular cone.** They calculated the pressure exerted by a supersonic flow on a conical surface and showed that there is a particular minimum Mach number below which the shock wave will cease to be attached to the nose of the cone. ▪ For narrow cones, the Kármán-Moore method was consistent with the Taylor-Maccoll method. The Taylor-Maccoll method is very useful when dealing with bullets with conical heads, but useless for estimating pressures over a long, slender, pointed body.

---

[1569] N.A. RYNIN: *Interplanetary flight and communications: superaviation and superartillery* [in Russ.]. Leningrad, vol. 2 (1929), No. 6; Engl. technical translation: Rept. NASA-TT F 645, NASA and NSF, Washington, DC (1971).

[1570] A.D. CROW and W.E. GRIMSHAW: *On the equation of state of propellant gases*. Phil. Trans. Roy. Soc. Lond. **A230**, 39-73 (1932).

[1571] H. JOACHIM and H. ILLGEN: *Gasdruckmessungen mit Piezoindikator*. Z. ges. Schieß- u. Sprengstoffwesen **27**, 76-79, 121-125 (1932).

[1572] Th. VON KÁRMÁN and N.B. MOORE: *The resistance of slender bodies moving with supersonic velocities with special reference to projectiles*. Trans. ASME **54**, 303-310 (1932).

[1573] G.I. TAYLOR and J.W. MACCOLL: *The air pressure on a cone moving at high speed*. Proc. Roy. Soc. Lond. **A139**, 278-311 (1933).

| | | |
|---|---|---|
| 1932 | *Fluid Dynamics Laboratory, University of Paris, France* | **Dimitri Pavlovitch RIABOUCHINSKY**,[1574] a Russian-born fluid dynamicist who founded the world's first aerodynamic institute in 1904 at Koutchino (Russia) and emigrated to France in 1919, **investigates the hydrodynamic analogy between the shallow-water equations and gas dynamics** on the orders of the French Ministry of Civil Aviation. ▪ The connection between the shallow-water equations and gas dynamics allows the application of sophisticated numerical techniques developed to capture shocks and to directly identify the effects of rotation and nonlinearity on their formation and evolution in the case of bores.[1575] The quantitative aspects of the analogy are deduced based on the analogous equations of continuity and energy conservation for 2-D isentropic flow of a gas having a ratio of specific heats $\kappa = 2.0$, and quasi-2-D frictionless flow of an incompressible liquid.[1576] Using these relations, the local liquid height can be related to the local gas density, temperature and pressure.<br><br>The application of the hydraulic analogy to transonic and supersonic problems was studied and reviewed by Donald R.F. HARLEMAN[1577] and Arthur T. IPPEN,[1578] two fluid dynamicists at MIT's Hydrodynamics Laboratory. |
| 1932 | *Institut für Technische Physik, TH Berlin-Charlottenburg, Germany* | **Hubert SCHARDIN**,[1579] a German applied physicist and assistant to Prof. Carl CRANZ, **investigates shock propagation in tubes and the conditions required for the ignition of explosive gaseous mixtures ($H_2/O_2$) by incident and reflected shock waves theoretically.** He formulates the so-called **"shock tube equations,"** which relate the selected diaphragm pressure ratio to the shock strength achieved:<br><br>(i) He assumes the use of a straight cylindrical tube extended in the *x*-direction which is provided with a diaphragm at $x = 0$, separating a left section ($x < 0$) of high static pressure $p_0'$ from a right section ($x > 0$) of low static pressure $p_0''$. When the diaphragm is suddenly removed, a shock wave is produced which moves from left to right with velocity<br>$$w_1 = \tfrac{1}{4}(\kappa+1)u_1 + [(\kappa+1)^2 u_1^2/16 + a_0^2]^{1/2}.$$<br>The pressure jump at the front is then given by<br>$$p_1/p_0'' = 1 + \kappa u_1 w_1/a_0^2.$$<br>Here $a_0$ is the sound velocity, $\kappa$ is the ratio of specific heats, and $u_1$ is the flow velocity behind the shock front. The rarefaction wave starts at $x = 0$ and moves from right to left with velocity $-a_0$.<br><br>(ii) For air at given ratios of $u_1/a_0$, he presents the calculated jumps in pressure, density and temperature at the shock front in tabular form.<br><br>(iii) When the same gas is used in both the high- and low-pressure sections of the shock tube, the maximum shock Mach number $M_s$ {ACKERET ⇨1928} amounts to $M_s = 4.24$ for a monatomic gas ($\gamma = 5/3$) and $M_s = 6.16$ for a diatomic gas ($\gamma = 7/5 = 1.4$). ▪ For the standard type of shock tube, and using air, maximum shock velocities of up to about 2,000 m/s are achievable, which are far to low to simulate spacecraft reentry, for example. Much higher velocities can be obtained in special hypervelocity facilities. Extreme shock velocities of up to 67 km have been achieved in the compressor-driven Voitenko shock tube {VOITENKO ⇨1964}.<br><br>SCHARDIN's work, building on previous investigations on this subject {VIEILLE ⇨1899; KOBES ↪1910, HILDEBRAND ➡1927}, was apparently first resumed during World War II by Abraham H. TAUB,[1580] a mathematician at the Institute for Advanced Study in Princeton, NJ, |

---

[1574] D. RIABOUCHINSKY: *Sur l'analogie hydraulique des mouvements d'un fluide compressible.* C. R. Acad. Sci. Paris **195**, 998-999 (1932); *Quelques nouvelles remorques sur l'analogie hydraulique des mouvements d'un fluide compressible.* Ibid. **199**, 632-634 (1934); *Aérodynamique.* Ibid. **202**, 1725-1728 (1936).

[1575] A.C. KUO and L.M. POLVANI: *Time-dependent fully nonlinear geostrophic adjustment.* J. Phys. Oceanography **27**, 1614-1634 (1997).

[1576] A.H. SHAPIRO: *Free surface water table.* In: *High speed aerodynamics and jet propulsion.* Princeton Univ. Press, Princeton, NJ. Princeton Series, vol. IX (1954): (R.W. LADENBURG ET AL., eds.) *Physical measurements in gas dynamics and combustion*, pp. 309-321.

[1577] D.R.F. HARLEMAN and A.T. IPPEN: *The range of application of the hydraulic analogy in transonic and supersonic aerodynamics.* Mémoire sur la Mécanique des Fluides offerts a M. Dimitri P. RIABOUCHINSKY. Publ. Scient. & Tech. Ministère de l'Air, Paris (1954), pp. 91-112.

[1578] A.T. IPPEN and D.R.F. HARLEMAN: *Studies on the validity of the hydraulic analogy to supersonic flow – part III.* Tech. Rept. AF-TR-5985 (Oct. 1950).

[1579] H. SCHARDIN: *Bemerkungen zum Druckausgleichsvorgang in einer Rohrleitung.* Physik. Z. **33**, 60-64 (1932).

[1580] See the Appendix *Theoretical relationships* given in W. BLEAKNEY's paper published in Rev. Scient. Instrum. **20**, 814-815 (1949).

and formed the basis for modern shock tube theory.[1581] The exact time-dependent solution of the shock tube equations is known and can be compared with the 1-D solution computed applying numerical discretizations, the so-called **"shock tube problem"** (or "Sod problem") – a special case of the **Riemann problem** with velocities on both sides of discontinuity set to zero.[1582]

| 1932 | British Museum of Natural History, London, England | Leonard James SPENCER,[1583] a British geologist and mineral collector, **strongly supports** GILBERT's impact theory {GILBERT ⇒1893} **that craters are due to the impact of small bodies moving in orbits around the Earth or the Sun.**

(i) Addressing the remarkable discovery of silica glass in the Wahar (or Wabar) Crater, which was found as cindery and slaggy masses and even as complete "bombs," he wrote, "Silica glass is, in fact, of rare occurrence in nature. It is best known in the form of fulgurites or lightning tubes, which are formed when sand dunes are struck by lightning. Since a temperature of about 1,700 °C is required to melt quartz sand, the development of a large amount of heat is here indicated … Now the close association of silica glass and meteoric iron with a group of craters in a sandy desert can be accounted for in no other way but by the impact of a shower of large meteoric masses."

(ii) SPENCER summarizes the available information on meteorite craters, citing five terrestrial craters or crater clusters with associated meteoritic material: the Meteor Crater in Arizona {⇒c.50,000 years ago}, the Odessa Crater in Texas, the Henbury Crater field in Australia, the Wahar Crater in Arabia, and the Campo del Cielo Crater field in Argentina.

(iii) Silica structures found at Wahar {PHILBY ⇒1932} "suggest that there was a pool of molten and boiling silica and that a rain of molten silica was shot out from the craters through an atmosphere of silica, iron, and nickel produced by the vaporization of the desert sand and part (perhaps a large portion) of the meteorite. The minute pimples on the surface were dewdrops from these vapors formed in the last stages."

(iv) **Meteorite craters "are not merely dents or holes made just by the projectile force of the meteorite as hitherto supposed. They appear, rather, to be explosion craters due to the sudden vaporization of part of the material, both of the meteorite and of the earth, in the intense heat developed by the impact."**

Various authors have suggested that impacts and explosion of meteorites can account for the Ries and Steinheim Basins in Germany {⇒c.15 Ma ago; WERNER ⇒1904; STUTZER ⇒1936; SHOEMAKER & CHAO ⇒1961; ⇒Fig. 4.1–C}, the Lake Bosumtwi Crater in the Ashanti region of Ghana, the Köfels Crater of the Tyrolian Alps, and the Pretoria Salt-Pan of South Africa. However, most geologists at the time still favored some form of cryptovolcanic hypothesis, maintaining that the explosions were due to the expansion of gases associated with ascending magmas. |
| 1932 | Cavendish Laboratory, Cambridge, England | Geoffrey I. TAYLOR,[1584] a British physics professor and eminent fluid dynamicist, uses ACKERET's **airfoil theory** {ACKERET ⇒1925} **to calculate the forces on a thin biconvex airfoil moving at supersonic speed.** He compares his result with STANTON's drag data obtained in the wind tunnel at the National Physics Laboratory in Teddington {STANTON ⇒1928}. |

---

[1581] W. BLEAKNEY and R.J. EMRICH: *The shock tube*. In: *High speed aerodynamics and jet propulsion*. Princeton University Press, Princeton, NJ; vol. VIII (1961): (F.E. GODDARD, ed.) *High speed problems of aircraft and experimental methods*, pp. 596-647.

[1582] G.A. SOD: *A survey of several finite difference methods for systems of non-linear hyperbolic conservation laws*. J. Comput. Phys. **27**, 1-31 (1978).

[1583] L.J. SPENCER: *Meteorite craters*. Nature **129**, 781-784 (1932); *Meteorite craters as topographical features on the Earth's surface*. Ann. Rept. Smithsonian Inst. (1933), pp. 307-325; Geograph. J. **81**, 227-248 (March 1933).

[1584] G.I. TAYLOR: *Applications to aeronautics of ACKERET's theory of aerofoils moving at speeds greater than that of sound*. Brit. Rept. ARC-R&M 1467, WA-4218-5a (1932).

| | | |
|---|---|---|
| 1932 | Flugwissen-schaftliches Institut, TH Berlin-Charlottenburg, Germany | Herbert WAGNER,[1585] a German professor of aircraft construction, **studies the fundamental processes of percussion and gliding when a body at high speed hits the free surface of a fluid.** • The impact of a body onto water under a small angle of incidence leads to it periodically bouncing along the surface, which can easily be demonstrated by throwing a small stone at the surface of a pond at a low angle. This phenomenon – known as **"ricocheting"** (or "skipping") – had already attracted the thoughts of some early natural philosophers. For example, the Bohemian percussion researcher Johann Marcus MARCI VON KRONLAND {MARCI ⇨1639} had explained this effect in his treatise *De proportione motus* [*see* Propositio XXXX] via the Law of Reflection {⇨Fig. 4.3–B}. Since the late 17th century, ballisticians have used ricochet firing to increase the range of fire or to probe areas otherwise immune from direct fire {DOUGLAS ⇨1851; ⇨Fig. 4.2–R}.<br><br>In the period between the two World Wars, the skipping effect to which seaplanes (the largest and fastest airplanes in the world in the 1930s) are subjected when they land on water was found to be of great practical importance when constructing floats, and was tackled in the United States[1586] and the former Soviet Union.[1587] |
| 1933 | Institut für Aerodynamik, ETH Zürich, Switzerland | **The world's first continuous-flow closed-loop supersonic wind tunnel, designed by Jakob ACKERET, begins operating.**[1588] The facility {⇨Fig. 4.9–D} has a test cross-section of $40 \times 40$ cm$^2$ and uses a Laval nozzle ($M = 2$). It is anticipated that it will be used not only to test model aircraft but also for research into ballistics and steam and gas turbine design. • He designed the facility when he was still working at Escher-Wyss AG in Zurich which also built the device, in cooperation with Brown Boveri & Company (BBC) in Baden, Switzerland.<br><br>In 1935, BBC delivered a second supersonic wind tunnel [Ital. *galleria ultrasonica*] closely based on ACKERET's design to the Italian Aeronautical Laboratory at Città aeronautica di Guidonia (inaugurated on April 27, 1935), an aircraft research center near Rome which was named after the Italian General Alessandro GUIDONI (1880–1928), a renowned pioneer of aeronautics. By increasing the power up to 3,000 hp, the maximum Mach number could be increased from 2 to 2.7. |
| 1933 | Kentland Impact Crater, Newton County, northwestern Indiana | Walter Herman BUCHER,[1589] a U.S. geologist at the University of Cincinnati, **discovers structures with curious striations ("horsetail texture") – later known as "shatter cones"** {⇨Fig. 4.1–C} **in a large quarry about two miles** (3.2 km) **east of Kentland.** These cones have apical angles ranging from 75° to nearly 90°, and they can be 2 m long in limestone and 12 m long in shale. Believing that these unusual "cryptovolcanic structures" (BRANCA & FRAAS 1905) are disturbances in deranged Paleozoic beds, he ascribes their origin to a deep-seated explosion of gases derived from an igneous intrusion. • The term *cryptovolcanic structure* (or *cryptoexplosion structure*) – referring to large-scale explosions in the Earth's crust – is now largely obsolete.<br><br>Today shatter cones are recognized as being important "index fossils" for the shock-structure remains of ancient meteoritic impact structures on the Earth's surface – so-called **"astroblemes"** (a word with Greek roots, meaning "star-wound") which formed directly under the point of impact {BOON & ALBRITTON ⇨1938; DIETZ 1947, *see* DIETZ ⇨1946}. Unlike, for example, coesite {COES ⇨1953} and stishovite {STISHOV & POPOVA |

---

[1585] H. WAGNER: *Über Stoß- und Gleitvorgänge an der Oberfläche von Flüssigkeiten.* ZAMM **12**, 193-215 (1932).
[1586] Th. VON KÁRMÁN and F.L. WATTENDORF: *The impact on seaplane floats during landing.* Rept. NACA-TN 321 (1929).
[1587] L.I. SEDOV and A.N. WLADIMIROW: *Das Gleiten einer flach-kielartigen Platte.* Dokl. AN (SSSR) **33**, No. 3, 116-119 (1941); *Water ricochets.* Ibid. **37**, No. 9, 254-257 (1942).
[1588] J. ACKERET: *Der Überschallwindkanal des Instituts für Aerodynamik an der ETH.* Aero-Revue Suisse **10**, 112-114 (1935).
[1589] W.H. BUCHER: *Cryptovolcanic structures in the United States.* In: *Proc. 16th Int. Geological Congress* [Washington, DC, June 1933]. Banta Publ. Co., Menash, WI (1936), vol. 2, pp. 1055-1084.

⇨1961}, shatter cones are the only known diagnostic shock features to be megascopic in scale, ranging from centimeters to meters.

Shatter cones can also be generated artificially by high-velocity impact {SHOEMAKER ET AL. ⇨1961} and large-yield explosions {RODDY & DAVIS ⇨1977}. **It is generally accepted that shatter cones are formed by the shock wave propagating within the outer region of the impact zone** {FRENCH ⇨1998} and not within the immediate area of the impact where rocks are mostly evaporated, melted, or ejected.

| 1933 | *Kaiser-Wilhelm-Institut (KWI) für Strömungsforschung, Göttingen, Germany* | Adolf BUSEMANN[1590] and Otto WALCHNER, two German aerodynamicists, **measure the lift and aerodynamic drag of nine different aerodynamic profiles at supersonic speeds of up to $M = 1.47$.**

(i) **Their high-speed wind tunnel is based upon the famous Göttingen Wind Tunnel devised by Ludwig PRANDTL.** Being of the intermittent indraft type, it consists of a test chamber with a rectangular cross-section of $6 \times 7.3$ cm$^2$, followed by a Laval nozzle carefully designed to get a homogeneous supersonic flow, a diffuser, a valve, and a vacuum reservoir. **This facility is believed to be the first German supersonic wind tunnel ($M = 1.47$).**

(ii) They observe that thin profiles are superior to thick ones, and that a symmetric biconvex profile is superior to a segmental (*i.e.*, like a segment of a circle) profile. Upon comparing the experimental results with his newly derived approximation method for frictionless supersonic flow, BUSEMANN finds that they only agree well for small yaw angles. ▪ Their study stimulated the design of aerofoils of supersonic aircraft and supersonic propellers. **The German rocket A-4b,** a long-range version of the A-4 {HVA Peenemünde ⇨1942} that was provided with wings with a sweep of 52°, **marked the first successful attempt to realize a high-Mach-number supersonic winged vehicle** {⇨Fig. 4.20–C}.

In the same year, the German space flight pioneer and rocket propulsion engineer **Eugen SÄNGER** makes use of the results obtained at Peenemünde in his own research and **develops studies for a winged hypersonic high-altitude rocket plane** that could be boosted to an Earth orbit and then glide back to land. ▪ His work partly stimulated subsequent American studies into rocket-propelled hypersonic aircraft. |
| --- | --- | --- |
| 1933 | *Forschungsinstitut der AEG, Berlin, Germany* | Carl W. RAMSAUER,[1591] a German physics professor, **reports on a new method of rapidly heating a test gas up to extreme temperatures that uses a high-velocity projectile as a piston.** He used an airgun 9 mm in caliber to which he smoothly attached a second barrel containing the test gas and incorporating a quartz window that allows the projectile motion to be recorded photographically. The far end of the second barrel is closed. For example, for an adiabatically compressed ideal gas, the temperature increase can be calculated via the relation

$$T = T_0 \left[1 + mv^2/(102 \, qLC_v)\right],$$

where $T_0$ (degree K) is the initial gas temperature, $m$ (g) is the projectile mass, $v$ (m/s) is the projectile velocity, $q$ (cm$^2$) is the inner cross-sectional area of the barrel, $L$ (cm) is the length of the barrel, and $C_v$ (cal/Mol degree) is the specific heat at constant volume. For example, by performing the experiment in $CO_2$ gas at laboratory conditions ($T_0 = 293$ K, $p_0 = 1$ atm) and choosing $m = 20$ g, $v = 1,000$ m/s, and $L = 100$ cm, the temperature could be increased to $T = 89,000$ K. However, since the escape of compressed gas between the barrel wall and the projectile cannot be avoided, the temperatures attainable in practice are much smaller than those predicted by theory.

In the 1940s, some German nuclear physicists proposed to increase the efficiency of RAMSAUER's method by shooting two projectiles into a gaseous column enclosed in the test barrel from opposite directions to attempt to initiate nuclear fusion in a gaseous mixture. |

---

[1590] A. BUSEMANN and O. WALCHNER: *Profileigenschaften bei Überschallgeschwindigkeiten*. Forsch. Ing. Wes. **4A**, 87-92 (1933).
[1591] C. RAMSAUER: *Über eine neue Methode zur Erzeugung höchster Drucke und Temperaturen*. Physik. Z. **34**, 890-894 (1933).

| | | |
|---|---|---|
| 1933 | *NACA's Langley Memorial Aeronautical Laboratory [now NASA's Langley Research Center], Hampton, Virginia* | **John STACK and Eastman N. JACOBS,** two U.S. aeronautical engineers, **take the first photograph** {⇨Fig. 4.14–L] **of the transonic flow field over airfoils at speeds above the critical Mach number** {ACKERET ⇨1928}. They use the schlieren technique and correlate their flow analysis with detailed pressure measurements. **Shock waves and attendant flow separations are seen for the first time starting at subsonic stream speeds of about Mach 0.6** {5th Volta Conf. (JACOBS) ⇨1935}. ▪ Since shock waves were observed for subsonic stream flow, the phenomenon was first considered by Langley's theorist Theodore THEODORSEN to be an optical illusion.[1592] |
| 1934 | *Dept. of Mechanical Engineering, Stanford University, Palo Alto, CA & Springer-Verlag, Berlin, Germany* | **Publication of the first volume from the six-volume compendium** *Aerodynamic Theory*,[1593] a general review of progress under a grant of the Guggenheim Fund for the promotion of aeronautics; the editor-in-chief is William Frederick DURAND, professor emeritus at Stanford University and an eminent U.S. researcher in aeronautical engineering.[1594] The compendium consists of |

  ▸ vol. 1 (1934): *Mathematical Aids, Fluid Mechanics, Historical Sketch* [W.F. DURAND ET AL.];
  ▸ vol. 2 (1935): *General Aerodynamic Theory, Perfect Fluids* [T. VON KÁRMÁN and J.M. BURGERS];
  ▸ vol. 3 (1935): *Theory of Single Burbling, Mechanics of Viscous Fluids, Mechanics of Compressible Fluids, Experimental Methods – Wind Tunnels* [C. WITOSZYŃSKI ET AL.];
  ▸ vol. 4 (1935): *Applied Airfoil Theory, Airplane Body Non-Lifting System, Drag and Influence on Lifting System, Airplane Propellers, Influence of the Propeller on Other Parts of the Airplane Structure* [A. BETZ ET AL.];
  ▸ vol. 5 (1935): *Dynamics of the Airplane, Airplane Performance* [B.M. JONES ET AL.]; and
  ▸ vol. 6 (1936): *Airplane as a Whole, Aerodynamics of Airships, Performance of Airships, Hydrodynamics of Boats and Floats, Aerodynamics of Cooling* [W.F. DURAND ET AL.].

The work was so well-received that early in 1939 the publishers proposed the preparation of a supplementary volume to cover new developments. Again, working with VON KÁRMÁN, a plan was worked out and potential authors contacted. However, the outbreak of World War II put an end to the project.[1595]

| | | |
|---|---|---|
| 1934 | *Mt. Wilson Observatory & CalTech, Pasadena, California* | Only 18 months after the discovery of the neutron {CHADWICK ⇨1932}, **Wilhelm Heinrich W. BAADE,**[1596] a German astronomer, **and Fritz ZWICKY,** a Bulgarian-born U.S. theoretical physicist, **link supernova explosions** {BAADE & ZWICKY ⇨1931} **with the formation of neutron stars.** |

  (i) They state, "With all reserve, we advance the view that a supernova represents the transition of an ordinary star into a neutron star." ▪ ZWICKY was the first to understand that supernovae resulted from the explosion of massive stars, and he still holds the record for the most supernovae discovered. Today a neutron star is considered to be the crushed ultradense core of an exploded star.

---

[1592] J.V. BECKER: *The high speed frontier.* Rept. NASA SP-445 (1980), chap. II: *The high-speed airfoil program.*
[1593] W.F. DURAND (ed.): *Aerodynamic theory.* 6 vols., Springer, Berlin (1934–1936); Caltech, Pasadena, CA (1943); Dover Publ., New York (1967); P. Smith, Gloucester, MA (1976).
[1594] F.E. TERMAN: *William Frederick DURAND: March 5, 1859 – August 9, 1958.* Biogr. Memoirs (National Academy of Sciences, Washington, DC) **48**, 152-176 (1976); http://darwin.nap.edu/books/0309023491/html/152.html.
[1595] From reference given in the preceding footnote.
[1596] W. BAADE and F. ZWICKY: *Supernovae and cosmic rays.* Phys. Rev. **45** [II], 138 (1934); *On super-novae.* Proc. Natl. Acad. Sci. U.S.A. **20**, No. 5, 254-259 (May 15, 1934).

(ii) They advance the idea that supernovae might be responsible for the Galaxy's supply of high-energy cosmic rays, and that they might leave neutron stars behind as end-products.

**33 years later, the first observational evidence of the Baade-Zwicky hypothesis was obtained with the discovery of a rapidly rotating neutron star – a "pulsar"** {BELL & HEWISH ⇨1967} – in the center of the Crab Nebula, which is a remnant of a stellar explosion observed centuries ago by the Chinese {Guest Star ⇨1054}. Further evidence was provided by the spectacular supernova SN 1987A {L. SHELTON ⇨1987}.

| | | |
|---|---|---|
| 1934 | *Institute of Physics, Moscow, U.S.S.R. Academy of Sciences, Soviet Union* | Pavel A. CHERENKOV [or ČERENKOV],[1597] then a postgraduate student, **observes that light of a bluish color is emitted when an energetic charged particle passes through a transparent nonconductive material** (such as glass or water) **at a velocity greater than the velocity of light within the material.** ▪ This emission effect of an "electromagnetic shock wave" has been used by nuclear physicists to detect rapidly moving charged particles. |

In the years following, the Soviet physicists Igor Y. TAMM and Ilya M. FRANK (1937) attempted to theoretically interpret this **"Cherenkov effect."** They concluded that velocity phenomena similar to a head wave in supersonic aerodynamics {DOPPLER ⇨1847; E. MACH & SALCHER ⇨1886} exist in the microcosmos when an energetic particle moves through a medium at a velocity greater than the phase velocity of light in this medium; *i.e.*, **the Cherenkov radiation is an optical analog of a sonic boom, and manifests itself as very short pulses in a cone around the particle's direction of travel.** ▪ For this unique discovery and interpretation, the three Soviet scientists CHERENKOV, TAMM and FRANK jointly earned the 1958 Nobel Prize for Physics.[1598]

In quantum mechanical terms, the Cherenkov condition corresponds to momentum and energy conservation. Cherenkov radiation has scientific and technological uses, *e.g.*, in determining velocities in particle physics experiments and detecting high-energy charged particles in nuclear reactors, respectively. The Cherenkov effect is also observed when cosmic rays – which are actually cosmic particles (atomic nuclei, mostly of hydrogen) traveling close to the speed of light, rather than rays – enter the atmosphere: these particles prompt cascades of molecular collisions and are traveling so fast that they exceed the speed of light in the tenuous upper atmosphere (which is only a little less than the speed of light in a vacuum), and so emit Cherenkov radiation.[1599] These flashes can be detected by large mirror arrays on the ground.

| | | |
|---|---|---|
| 1934 | *Technical University Copenhagen, Denmark* | Martin KNUDSEN,[1600] a Danish physics professor, **publishes a booklet entitled** *The Kinetic Theory of Gases – Some Modern Aspects,* which is based upon three lectures delivered in the previous year at the University of London. |

(i) The first part outlines the foundations of the kinetic theory of gases, the rudiments of which were postulated some 250 years ago as a very daring hypothesis, and which became a useful theory through the work of Rudolf J.E. CLAUSIUS {⇨1857} and James C. MAXWELL {⇨1867}.

(ii) According to KNUDSEN, the foundations of modern kinetic theory are based upon four important assumptions:
- ▸ "Any gas consists of separate particles called 'molecules.' In a pure gas these are all alike;
- ▸ the molecules move about in all directions;

---

[1597] P.A. CHERENKOV: *Visible faint luminosity of pure liquids under the action of γ-radiation* [in Russ.]. Dokl. AN (SSSR) **2**, No. 8, 451-454 (1934).
[1598] P.A. CHERENKOV: *Strahlung von Teilchen, die sich mit Überlichtgeschwindigkeit bewegen und einige ihrer Anwendungsmöglichkeiten in der Experimentalphysik* [Nobel Lecture]. Physik. Blätter **15**, 385-397 (1959).
[1599] J.W. CRONIN, T.K. GAISSER, and S.P. SWORDY: *Cosmic rays at the energy frontier.* Scient. Am. **276**, No. 1, 44-49 (Jan. 1997).
[1600] M. KNUDSEN: *The kinetic theory of gases.* Methuen, London (1934, 1946, 1950, 1952).

▸ the pressure caused by the movement of the molecules is the only one that exists in a gas when it is in the ideal state; and

▸ the molecules are not infinitely small. Thus they collide with one another."

(iii) Addressing the molecular flow of gases through tubes − studied experimentally by him since 1907 − KNUDSEN discusses the conditions required for equilibrium: it exists when the pressure is so low that the mean free path is much greater than the radius of the tube (*free molecule flow*). However, "if the pressure in the tube is augmented so much that the number of mutual encounters of molecules in the tube cannot be disregarded in comparison with the number of impacts against the wall, then a back flow of gas occurs along the axis of the tube" (*viscous flow* or *Poiseuille flow*). ▪ The ratio of the mean free path to some characteristic dimension of the flow field is termed the **Knudsen number**, $K_n$. The region where the mean free path is long in comparison with the apparatus is known as the **"Knudsen region."**

Note that the flow of a highly rarefied gas ($K_n \gg 1$) is studied via the kinetic theory, while the flow of a slightly rarefied gas ($K_n \ll 1$) is treated via the gas dynamic theory of a continuous medium. After the advent of the first artificial satellites {Sputnik 1 ⇨1957}, the dynamics of rarefied gases developed into **Rarefied Gas Dynamics**, a research field of major importance, especially in the United States and the Soviet Union.

| | | |
|---|---|---|
| 1934 | *East Africa* | **Louis S.B. LEAKEY,**[1601] a notable Kenyan-born British anthropologist who for eight years has led a series of expeditions in East Africa in search of fossils of mankind's ancestors, **publishes his book *Adam's Ancestors*,** in which he tries to bring together all of the latest discoveries about the Stone Age and to present an up-to-date account of what is presently known about our ancient ancestors and cousins. In particular, he tries to answer some of the numerous questions that prehistorians are often asked to explain − for example, **how it is possible to distinguish between a piece of flint which was chipped by a Stone Age man and one which has been chipped by natural agencies.**<br><br>In the 1950s, L.S.B. LEAKEY, together with his wife Mary Douglas LEAKEY née NICOL, collected early man-made tools (mostly made of basalt and quartzite) and fossilized bones of many extinct mammals {Koobi Fora ⇨*c.*2.6 Ma ago; Olduvai Gorge ⇨*c.*1.75 Ma ago; ⇨Fig. 4.2−C}. These sensational discoveries provided the first evidence that percussion had been used by early man more than 2.5 million years ago. |
| 1934 | *Services des Poudres, Paris, France* | **Albert MICHEL-LÉVY**[1602] **and Henri MURAOUR,** two renowned French military engineers, **start pioneering experimental investigations into the luminous and spectroscopic phenomena that appear along with intense shock waves in gases.**<br><br>(i) They notice curious high-intensity luminous phenomena while studying shock wave interactions, particularly during the interaction of a single shock wave generated by a single explosive with a solid boundary, and the interaction of two shock waves generated by two explosives fired simultaneously.<br><br>(ii) They also observe luminous phenomena when a shock wave is reflected from a very light obstacle such as cigarette paper {⇨Fig. 4.14−G}.<br><br>(iii) They correctly surmise that the luminosity is solely attributed to the shock wave itself and not to any phenomena due to the explosion process, such as the emission of burnt particles, thus rejecting a previous hypothesis presented by the Austrian Alfred SIERSCH {⇨1896}. |

---

[1601] L.S.B. LEAKEY: *Adam's ancestors − an up-to-date outline of what is known about the origin of man*. Methuen, London (1934).
[1602] A. MICHEL-LEVY and H. MURAOUR: *Sur la luminosité des ondes de choc*. C. R. Acad. Sci. Paris **198**, 1760-1762 (1934).

| | | |
|---|---|---|
| 1934 | *Institute of Chemical Physics of the U.S.S.R., Leningrad, Soviet Union* | **Nikolai N. SEMENOV,**[1603] a leading Soviet physical chemist, **publishes his monograph *Chain Reactions* in which he develops a theory of *nonbranching chain reactions*** – a special form of chain reaction in which the number of chain carriers increases in each propagation. As a result the reaction accelerates extremely rapidly, sometimes being completed in less than one millisecond.<br><br>(i) It is the result of previous discoveries that he and his team have made by studying critical phenomena such as the ignition limit during the oxidation of vapors of phosphorus, hydrogen, carbon monoxide and other compounds.<br><br>(ii) He writes, "In 1927 and 1928 in Oxford, Leningrad and partly at Princeton, the chain theory was applied to a study of the reactions leading to inflammation and explosion. Most importantly, the theory has advanced here hand-in-hand with new experiments which led to a discovery of new and the explanation of old, long-forgotten, and quite unintelligible phenomena, and they have outlined the field of those reactions which are specific to the new conception. They have aroused a broad interest in this new reaction field and in 1930, 1931, 1932 and 1933 they brought to life a wave of new kinetical investigation … It is hoped that the analysis given here will enable us to make some new generalization and thus to advance the question of the classification of reactions and the detection of new laws common to wide classes of chemical change somewhat further."<br><br>His thorough and continuous investigations which are important for understanding fast chemical combustion processes (*e.g.*, in internal combustion engines, jet engines, and high explosives) earned him the 1956 Nobel Prize for Chemistry {SEMENOV & HINSHELWOOD ⇨1956}. |
| 1934 | *Lehrstuhl für Luftfahrttechnik, TH Aachen, Germany* | **Installation of the second German supersonic wind tunnel** {⇨Fig. 4.9–C} **at Prof. Carl WIESELSBERGER's institute, under the leadership of Rudolf HERMANN,** a German applied physicist.<br><br>(i) The design of their supersonic wind tunnel is essentially based on a facility previously developed at the Göttingen KWI für Strömungsforschung {BUSEMANN & WALCHNER ⇨1933}, but their tunnel has a larger test section of $10 \times 10$ cm$^2$ and can be operated at up to $M = 3$.<br><br>(ii) The Laval nozzle is covered with a layer of plaster of Paris, which ensures sufficient surface smoothness but is easier to form than wood or metal. The ideal nozzle geometry was determined graphically using the method of characteristics. • The modern design of the Aachen Supersonic Wind Tunnel was used as the basis for subsequent constructions and, for example, was also adapted to the requirements of the Peenemünde facility {HVA Peenemünde ⇨1939}.[1604, 1605]<br><br>Two years later, the Aachen facility was used to test models of various liquid propellant rockets such as the A-3 (short for "Aggregat 3"),[1606] which was the predecessor of the A-4, later called the "V2."[1607] |
| 1935 | *Rome, Italy* | In the period from September 30 to October 6, **the Fifth Volta Conference on the topic *Le alte velocità in aviazione* ("High Velocity in Aviation") is held.** The president is General Gaetano Arturo CROCCO, a renowned Italian pioneer of aeronautics. For the first time, leading |

---

[1603] N.N. SEMENOV: *Cepnye reakcii*. Goschimizdat, Leningrad (1934); *Chemical kinetics and chain reactions*. Clarendon Press, Oxford (1935).

[1604] C. WIESELSBERGER: *Die Überschallanlage des Aerodynamischen Instituts der Technischen Hochschule Aachen*. Luftwissen **4**, 301-303 (1937); VDI-Z **82**, 1230 (1938); *Arbeiten des Aerodynamischen Instituts der Technischen Hochschule Aachen*. Forsch. Gebiet Ing. Wes. **10**, 55-56 (1939).

[1605] A. NAUMANN (ed.): *50 Jahre Aerodynamisches Institut der RWTH Aachen. 1913–1963*. Abhandl. Aerodyn. Inst. der RWTH Aachen Heft 17 (1963); See also *The history of the AIA*. Institute of Aerodynamics, RWTH Aachen; http://www.aia.rwth-aachen.de/index.php?id=80&L=3.

[1606] The 6.75-m-long A-3 was a test rocket and designed to study the potential use of liquid fuel to propel missile-like bodies [see *Studiengerät zur Erprobung des Flüssigkeitsantriebes für geschossähnliche Körper gemäß Geheimer Kommandosache vom 29.11.1937*].

[1607] H. KURZWEG: *The aerodynamic development of the V2*. In: (T. BENECKE and A.W. QUICK, eds.) *History of German guided missiles development*. AGARD First Guided Missiles Seminar [Munich, Germany, April 1956]. Appelhaus, Brunswick (1957), pp. 50-69.

supersonic aerodynamic engineers from around the world meet to discuss the possibilities of supersonic flight.[1608] Numerous milestone contributions are presented:

(i) **Adolf Busemann**,[1609] an instructor at the University of Dresden, **presents his famous concept of** *sweepback wings*.

- He predicts that his arrow wings — which have a shape that ensures that they remain within the shock cone at supersonic speeds — would have less drag than straight wings exposed to the head wave, a weak shock wave.

- He reports that he has extended the existing linear airfoil theory by including terms of higher orders. According to his theory, which considers an infinite wing of constant cord yawed at an angle $\gamma$ measured from a position perpendicular to the flight velocity $U$, only the normal component of the velocity along the wing chord

$$U_n = U \cos\gamma$$

and the downwash velocity need to be considered. The tangential velocity

$$U_t = U \sin\gamma$$

produces no effect.

At a dinner following the meeting, Luigi Crocco, the prominent Italian aerodynamicist, sketched an airplane with swept wings "and a swept propeller," labeling it "the airplane of the future."[1610] • Since propeller-driven aircraft of the mid-1930s lacked the ability fly supersonically, Busemann's idea was not immediately realized, but it did influence most subsequent high-speed aircraft designs. German wartime research {Betz ⇨ 1939} showed that the use of sweepback wings did indeed ease the shock wave problem associated with flying at very high speeds {Messerschmitt Me 163 ⇨ 1941; Messerschmitt Me 262 ⇨ 1942}.

(ii) **Adolf Busemann also presents a more accurate calculation of the two-dimensional pressure and force coefficients of a thin wing at supersonic speeds.** Using Ackeret's thin two-dimensional wing theory {Ackeret ⇨ 1925}, but also taking second-order terms into account, he worked out a much better approximation to wing pitching moment.[1611]

(iii) **Jakob Ackeret**,[1612] an aerodynamics professor at ETH Zurich, **discusses the supersonic wind tunnel** {Ackeret ⇨ 1933} that he just completed for the Italians at Guidonia.

(iv) **Theodore von Kármán**,[1613] a Hungarian-born professor of aeronautical engineering and director of the Guggenheim Aeronautical Laboratory, **presents a new theory of supersonic flow from the viewpoint of drag.**

(v) **Eastman N. Jacobs**,[1614] a U.S. research scientist at NACA, **presents first schlieren photos of shock waves generated around model airfoils exposed to a subsonic flow** {Stack & Jacobs ⇨ 1933}.

(vi) **Luigi Crocco**,[1615] an aeronautical engineer at the University of Rome, **reports on the Italian high-speed wind tunnel** {Ackeret ⇨ 1933}.

---

[1608] In 1935, the idea of flying at supersonic speeds was far from the consciousness of the aeronautical community: the fastest aircraft of the time was the British seaplane *Supermarine S6B*, which won the last Schneider Trophy in 1931, setting a speed record at 547.19 km/h (152 m/s). • The Schneider Trophy, a pure speed contest, prompted significant advances in aircraft design, particularly in the fields of aerodynamics and engine design.

[1609] A. Busemann: *Aerodynamischer Auftrieb bei Überschallgeschwindigkeit*. V Convegno Volta su "Le alte velocità in aviazione"[Rome, Italy, Sept./Oct. 1935]. Reale Accademia d'Italia, Roma (1936), pp. 328-360; Luftfahrtforsch. **12**, 210-220 (1935).

[1610] R.T. Jones: *Adolf Busemann: 1901–1986*. Mem. Tributes: Nat. Acad. Engng. **3**, 62-67 (1989).

[1611] W.F. Hilton: *High-speed aerodynamics*. Longman, Green & Co. London etc. (1952); see § 10.5: *Busemann's second-order theory*, pp. 226-241.

[1612] J. Ackeret: *Windkanäle für hohe Geschwindigkeiten*. V Convegno Volta su "Le alte velocità in aviazione" [Rome, Italy, Sept./Oct. 1935]. Reale Accademia d'Italia, Roma (1936), pp. 487-536; L'Aerotecnica **16**, 885-925 (1936).

[1613] Th. von Kármán: *The problem of resistance in compressible fluids*. V Convegno Volta su "Le alte velocità in aviazione" [Rome, Italy, Sept./Oct. 1935]. Reale Accademia d'Italia, Roma (1936), pp. 222-276.

[1614] E. Jacobs: *Methods employed in America for the experimental investigation of aerodynamic phenomena at high speeds*. V Convegno Volta su "Le alte velocità in aviazione" [Rome, Italy, Sept./Oct. 1935]. Reale Accademia d'Italia, Roma (1936), pp. 369-401.

[1615] L. Crocco: *Galleria aerodinamiche per alto velocità*. V Convegno Volta su "Le alte velocità in aviazione" [Rome, Italy, Sept./Oct. 1935]. Reale Accademia d'Italia, Roma (1936), pp. 542-562.

(vii) **Ludwig PRANDTL**,[1616] a professor of applied mechanics at the University of Göttingen, **reports on strange shock-like phenomena that he observed downstream in the nozzle throat, and shows a schlieren picture taken in his supersonic tunnel** at Göttingen University {⇨Fig. 4.14–F}. In the discussion that follows, Carl WIESELSBERGER, a professor of aerodynamics at the University of Aachen, surmises that this phenomenon, which had also been observed in the supersonic wind tunnel at Aachen, might be caused by condensed water vapor when atmospheric (*i.e.*, moist) air is used. This phenomenon, which is particularly detrimentally to the visualization of high-speed flows, was later named **"condensation shock."**
▪ Experimental studies performed in the early 1940s at the Kaiser-Wilhelm-Institut (KWI) für Strömungsforschung in Göttingen by Klaus OSWATITSCH[1617] and at the Heeresversuchsanstalt (HVA) Peenemünde by Rudolf HERMANN[1618] proved that WIESELSBERGER's supposition was indeed correct.

Shortly after the Conference, some nations (starting with HITLER's Germany and MUSSOLINI's Italy) began to classify the subject of high-speed flight due to its military relevance.

| | | |
|---|---|---|
| 1935 | *Aberdeen Proving Ground, Maryland* | **Establishment of the Research Division of the Aberdeen Proving Ground by the U.S. Army.** The first director of the Research Division is Colonel Herman H. ZORNIG. ▪ In 1938, the laboratory was renamed the Ballistic Research Laboratory (BRL), which was reorganized – along with a number of sister laboratories – into the newly established **Army Research Laboratory** (ARL) in 1992. |
| 1935 | *Mallinckrodt Laboratory, Harvard University, Cambridge, Massachusetts* | Augustine O. ALLEN,[1619] a chemist and Ph.D. student, **and Oscar K. RICE,** a physical chemistry instructor, **study the mechanism for the explosion of azomethane** [$CH_3N_2CH_3$]. <br>(i) The experimental procedure consists of introducing a known pressure of azomethane gas into an evacuated bulb kept in an air bath at a known temperature, and then following the changes in pressure with a mercury manometer. At a given temperature, the explosion starts at a "critical pressure" which depends on the temperature of the bulb, the size of the bulb, and the presence of inert gases. Helium, for example, is a relatively good heat conductor, so it conducts the heat to the walls more rapidly than pure azomethane, meaning that its explosion limit is higher. <br>(ii) In order to explain their experimental results, they start from the hypothesis that the explosion of azomethane is caused purely thermally, so an azomethane explosion may offer a favorable case for the study of thermal explosions. **They apply their experimental data to SEMENOV's theory of thermal explosions** {SEMENOV ⇨1928}, which states that the heat of reaction that accumulates in a reacting gas leads to an explosion under certain conditions, **and they find that this theory explains their data satisfactorily.** <br>(iii) As a warning to other chemists who may wish to use azomethane, they note that the gas can be ignited by static electric sparks, "The violence of the resulting detonation was evidenced by the fact that small pieces of glass from the trap containing the liquid were driven through other pieces of apparatus, leaving clean bullet holes." |
| 1935 | *Rostock and Warnemünde, northern Germany* | The German BUTTER Brothers[1620] invent the *Sprengniet* ("explosive rivet") and assign their patent to the German Heinkel Aircraft Company. ▪ The explosive rivet {⇨Fig. 4.15–J} – a particular technique used in explosive metal forming – allows one to rivet aircraft sections while working on only one side – resulting in a so-called **"blind rivet technique."** The technique involves drilling the hole, inserting the rivet and then exploding it, using either a hot iron or a |

---

[1616] L. PRANDTL: *Allgemeine Überlegungen über die Strömung zusammendrückbarer Flüssigkeiten*. V Convegno Volta su "Le alte velocità in aviazione" [Rome, Italy, Sept./Oct. 1935]. Reale Accademia d'Italia, Roma (1936), pp. 168-197, 215-221; ZAMM **16**, 129-142 (1936).
[1617] K. OSWATITSCH: *Die Nebelbildung in Windkanälen und ihr Einfluß auf Modellversuche*. Jb. Dt. Akad. Luftfahrtforsch. (1941), pp. 692-703.
[1618] R. HERMANN: *Der Kondensationsstoß in Überschall-Windkanaldüsen*. Luftfahrtforsch. **19**, 201-209 (1942).
[1619] A.O. ALLEN and O.K. RICE: *The explosion of azomethane*. J. Am. Chem. Soc. **57**, 310-317 (1935).
[1620] K. BUTTER and O. BUTTER: *Niet, dessen Schließkopf durch Verformung mittels Sprengung gebildet wird*. Germ. Patent No. 655,669 (May 1935).

| | | |
|---|---|---|
| 1935 | School of Aeronautical Engineering, University of Rome, Italy | **Luigi CROCCO,**[1622] an Italian aeronautical engineer and the son of General G. Arturo CROCCO, **publishes a fundamental theoretical study on the relative merits of different types of supersonic wind tunnels** {L. CROCCO ⇨1931 & 1932; 5th Volta Conf. (CROCCO) ⇨1935}. ▪ Theodore VON KÁRMÁN later referred to this review article, which was also translated into French, as the "bible of supersonic wind tunnels." |
| 1935 | University of Chicago, Illinois | **Arthur J. DEMPSTER,** an American-Canadian physicist, passes natural uranium through his mass spectrograph and **detects a second, lighter isotope – uranium U-235.** ▪ He later announced in a lecture that "it was found that a few seconds' exposure was sufficient for the main component at 238 reported by Dr. [Francis W.] ASTON, but on long exposures a faint companion of mass number 235 was also present."<br><br>Three years later, **Alfred O.C. NIER,** a U.S. physicist at Minnesota University, **measured the ratio of U-235 to U-238 in natural uranium at 1:139, which meant that only about 0.7% of natural uranium is U-235.**[1623] He then investigated the separation of these two isotopes, an important prerequisite for developing atomic bombs based upon fission of U-235 {Manhattan Project ⇨1942}. |
| 1935 | National Physical Laboratory (NPL), Teddington, Middlesex, England | **Origin of the term *sound barrier* (and the myth surrounding it) by William Frank HILTON,** a British aerodynamicist who conducts high-speed experiments at NPL. ▪ HILTON[1624] later remembered: "That year the annual NPL show day came on a very hot summer day. Just before closing time a deputation of reporters introduced themselves, said that they had been blinded by science all the day, had no 'story' as yet, and in words of one syllable, what did I do in the HST [High-Speed Tunnel]? Dropping the usual pattern I seized a curve of $C_D$ against $V/a$ (as Mach number was then called), and said: **'See how the resistance of a wing shoots up like a barrier against higher speed as we approach the speed of sound.' Next day I was crucified by all the leading dailies for having coined the phrase *sound barrier*.**"<br><br>The idea of an invisible "sound barrier" (or "sonic barrier") – a physical barrier to supersonic airplane flight – became widespread among the public, but proved to be a myth. Thanks to powerful engines and design features that minimize drag, airplanes exceeded the speed of sound and now routinely fly faster than it {YEAGER ⇨1947; X-15 ⇨1959 & 1961; Concorde ⇨1969}. Besides the term *sound barrier*, which was also adopted by the Russians [звуковой барьер], the term *sonic wall*, which suggests a even stronger obstacle to surmount before the speed of sound is achieved, is also commonly used. The term *sonic wall* was adopted in a number of other languages [*e.g.* Germ. *Schallmauer*, French *mur du son*, Ital. *muro del suono*, Span. *muro sónico*]. |
| 1935 | Seismological Laboratory, Cal-Tech, Pasadena, California | **Charles F. RICHTER,**[1625] a U.S. geophysicist and seismologist, **and Beno GUTENBERG,** a German-born U.S. seismologist, **introduce an instrumentation-based magnitude scale for earthquakes** in order to improve earthquake statistics and to eliminate confusion between the terms *intensity* and *magnitude*. ▪ Older arbitrary scales were based on intensity as judged by the |

high-frequency gun. Explosive rivets have since been used for numerous commercial applications other than in the aircraft industry.

In 1939, Heinkel sold its patent to the DuPont Company. At the time, the sole use of explosive rivets was in the manufacture of aircraft. Ironically, many thousands of American planes containing explosive rivets were later used to fight against Germany in World War II.[1621]

---

[1621] Hagley Museum and Library, Wilmington, DE; private communication (May 2002).
[1622] L. CROCCO: *Gallerie aerodinamiche per alte velocità.* L'Aerotecnica **15**, 237-275, 735-778 (1935); also published as: *Tunnels aérodynamiques pour grandes vitesses.* Extrait du Mémorial de L'Artillerie, Imprimerie Nationale, Paris (1938), pp. 358-442.
[1623] R. RHODES: *The making of the atomic bomb.* Simon & Schuster, New York (1986), p. 285.
[1624] W.F. HILTON: *British aeronautical research facilities.* J. Roy. Aeronaut. Soc. **70**, 103-107 (Jan. 1966).
[1625] C.F. RICHTER: *Elementary seismology.* Freeman & Co, San Francisco, CA (1958).

degree of shaking and local destructive effects, for example the Rossi-Forel 10-degree scale and the Mercalli 12-degree scale.

(i) They develop a method of determining the magnitude $M$ of earthquakes in southern California – the **"Richter scale."** Based on recordings made with instruments such as the Wood-Anderson seismometer, RICHTER took the logarithm of the amplitude $A$ of the recorded ground motion and added a distance correction factor $D$ to take into account distance from the station:

$$M = \log_{10} A \text{ (mm)} + D.$$

The arbitrary logarithmic Richter scale begins at 0 and has no upper limit. For example, the famous San Francisco Earthquake of 1906 was 7.8 on the Richter scale, and the Great Alaskan Earthquake of 1964 was 9.2. The largest earthquake ever recorded had a magnitude of 9.5 {Great Chilean Earthquake ⇨1960}.

(ii) To relate the magnitude $M$ of an earthquake to the total amount of radiated energy $E$ released in seismic waves during a (local) earthquake, they propose the equation

$$\log_{10} E \text{ (erg)} = 11.8 + 1.5M,$$

the so-called **"Gutenberg-Richter magnitude-energy relation."** ▪ For example, for a large earthquake with a magnitude of $M = 9$, such as the recent Sumatra-Andaman Islands Earthquake {⇨2004}, the emitted seismic energy amounts to

$$E = 10^{1.5M + 11.8} = 10^{25.3} \text{ erg} = 1.995 \times 10^{15} \text{ kJ},$$

corresponding to 476,817 tons of TNT equivalent, or about 32,000 Hiroshima bombs {HAEUSSERMANN ⇨1891; Hiroshima & Nagasaki Bombing ⇨1945}. Similarly, an earthquake of magnitude $M = 6$ corresponds roughly to the energy of the Hiroshima bomb, which is equivalent to 15,000 tons of TNT, or about $6.28 \times 10^{10}$ kJ.

A more consistent measure of large earthquakes used nowadays is the so-called **"moment magnitude"** ($M_w$), which is calculated basis on the seismic moment and is completely independent of the type of recording instrument used {HANKS & KANAMORI ⇨1979}.

| | | |
|---|---|---|
| 1935 | *French Navy, France* | H. ROGIER,[1626] chief engineer of the French Naval artillery, **publishes a richly illustrated memoir entitled** *La mort des canons sur le champ de bataille* **("The Death of Gun Barrels on the Battlefield").** It which is based upon five technical reports written during World War I by Paul CLEMENCEAU, a French lieutenant and gun expert at the Atelier de Construction de Bourges, France. |

(i) ROGIER discusses various causes of gun barrel damage observed in 75-mm guns, such as
- the enemy's artillery action (*e.g.*, barrel puncture from impacting shell-splinters or barrel deformation from the blast wave produced by a shell detonating nearby);
- heavy wear and erosion (originating from the hot propellant gases and abnormally high friction between the rifling and the driving band); and
- accidents during firing (*e.g.*, from a shell that detonates prematurely, which can result in an enlargement of the inner barrel diameter or in a fatal "banana split"-type barrel rupture).

(ii) He reports that the omnipresent danger from barrel bursts can be reduced considerably by providing the shell's base with a metallic plate which stops the hot propellant gases from entering the explosive charge. ▪ Barrel bursts are a serious problem, particularly for the gun crew {⇨Fig. 4.21–C}. They typically happen in times of war due to the lower quality of munitions available and the higher rate of fire required. For example, one barrel burst happened every (approximately) 500,000 shots in France before World War I, but this rate increased during World War I to less than one burst every 3,000 shots.

---

[1626] H. ROGIER: *La mort des canons sur le champ de bataille.* Mémorial de l'Artillerie Française **XIV**, No. 1, 3-60 (1935).

| | | |
|---|---|---|
| 1935 | Princeton University, Princeton, New Jersey | **Eugene P. WIGNER,**[1627] a Hungarian-born U.S. professor of mathematical physics, **and Hillard Bell HUNTINGTON,** a U.S. physicist, **suggest that diatomic solid molecular hydrogen (an insulator), may transform into a metallic monatomic solid phase when subjected to very high pressures.** They estimate that the transition pressure is at least 250 kbar. ▪ In a hydrogen phase diagram, the area "metallic phase" designates a state at which the electrons behave like a degenerated Fermi gas and the estimated conductivity is high.<br><br>Their hypothesis stimulated numerous static and dynamic high-pressure studies on hydrogen {WEIR, MITCHELL & NELLIS ⇒1996}. Different molecular phases (I, II, and III) exist in solid hydrogen at about 1 Mbar, all of which have been studied intensely using diamond anvil cell experiments.[1628] |
| 1936 | Guggenheim Aeronautical Laboratory of the California Institute of Technology (GALCIT) Pasadena, California | **Students at GALCIT, directed by Dr. Theodore von KÁRMÁN, begin design and experimental work with liquid-propellant rocket engines in the Arroyo Seco just outside of Pasadena.** ▪ After the CalTech group's successful rocket experiments, the Army helped CalTech acquire land in the Arroyo Seco for test pits and temporary workshops. Airplane tests at nearby air bases proved the concept. This resulted in designs for new jet-assisted takeoff (JATO) rockets that involved the use of small solid-fuel rockets as boosters to shorten the runways required by aircraft in short field or overload conditions. The first U.S. airplane to make use of JATOs took off in August 1941.<br><br>During World War II, the GALCIT Rocket Research Project also began work on high-altitude rockets. Reorganized in November 1944 under the name *Jet Propulsion Laboratory (JPL)*, the facility continued postwar research and development on tactical guided missiles, aerodynamics, and broad supporting technology for U.S. Army Ordnance. JPL was transferred from Army jurisdiction to NASA's control shortly after the establishment of the new organization {NASA ⇒1958}. Currently, JPL is a government-owned facility located 32 km northeast of Los Angeles. Many of the JPL's main research activities are related to the exploration of the Earth and the Solar System by automated spacecraft. |
| 1936 | Truppenübungsplatz Berlin-Kummersdorf & Institut für Aerodynamik Aachen (IAA), TH Aachen, Germany | **Werner VON BRAUN,** a young German mechanical engineer working in the design and testing of liquid-propellant rockets at Berlin-Kummersdorf, **visits the Aachen Supersonic Wind Tunnel Facility ($M = 3.3$, working section $10 \times 10$ cm$^2$).** He brings along a drawing of a pointed, slender body with fins – the design of the A-3 rocket – which needs to be tested for drag and stability at the maximum Mach number available. Rudolf HERMANN, an assistant to Prof. Carl WIESELSBERGER at the Aachen wind tunnel test facility, performs aerodynamic model tests. **By increasing the length of the tail unit, HERMANN and collaborators succeed in verifying that the flight of the rocket will be stable at subsonic to supersonic velocities** (up to $M = 3.3$).[1629] ▪ The A-3 was the first large liquid fuel rocket. It had a length of 6.74 m and a weight of 740 kg, and was the forerunner to the A-4 {HVA Peenemünde ⇒1939 & 1942}.<br><br>In the following year, Carl WIESELSBERGER[1630] described his unique and highly advanced supersonic facility in the open literature. The Aachen facility was built in the period 1934-1937 and financed by the German Air Force. Some of the first models tested were anti-aircraft projectiles. |

---

[1627] E. WIGNER and H.B. HUNTINGTON: *On the possibility of a metallic modification of hydrogen.* J. Chem. Phys. **3**, 764-770 (1935).
[1628] I.F. SILVERA and M.G. PRAVICA: *Hydrogen at megabar pressures and the importance of ortho-para concentration.* J. Phys. Cond. Matter **10**, 11169-11177 (1998).
[1629] M.J. NEUFELD: *The rocket and the Reich. Peenemünde and the coming of the ballistic missile era.* The Free Press, New York (1995), chap. 3.
[1630] C. WIESELSBERGER: *Die Überschallanlage des Aerodynamischen Instituts der Technischen Hochschule Aachen.* Luftwissen **4**, 301-303 (1937). *See also* the article *Das Institut unter der Leitung von C. WIESELSBERGER.* In: (A. NAUMANN, ed.) *50 Jahre Aerodynamisches Institut der RWTH Aachen.* Abhandl. Aerodyn. Inst. RWTH Aachen, Heft 17 (1963), pp. 20-24.

| | | |
|---|---|---|
| 1936 | Bergakademie Freiberg, Saxony, Germany | K. Otto STUTZER,[1631] a German geologist, **speculates that the Meteor Crater in Arizona** {⇨$c.$50,000 years ago; ⇨Fig. 4.1–B} **and the Ries Basin in northern Bavaria** {⇨$c.$15 Ma ago; ⇨Fig. 4.1–C} **are the result of asteroid impacts in the geologic past rather than that of huge steam explosions caused by volcanic activity.** ▪ The discovery of coesite {COES ⇨1953}, a high-pressure polymorph of silica, in Meteor Crater provided new evidence for STUTZER's hypothesis {CHAO ET AL. ⇨1960; SHOEMAKER & CHAO ⇨1961}. |
| 1937 | Hereswaffenamt (HWA), Berlin, Germany | **Plans are drawn up at the** *Heereswaffenamt* **("Army Ordnance Office")** **to build an aerodynamics-ballistics research institute,** "capable of furnishing all of the data on aerodynamics, stability, aerodynamic control, and heat transfer needed for the development of numerous military projects, such as supersonic projectiles (fired from guns), rocket-powered flying vehicles without wings (called 'missiles'), that use fins to stabilize their flight, and rocket-powered supersonic vehicles with wings and fin-assemblies, or with delta wings (R. HERMANN)." ▪ **This triggered the construction of the** *Heeresversuchsanstalt (HVA) Peenemünde*, **the German rocket center at Peenemünde-Ost (Baltic Sea), which included a large supersonic wind tunnel facility** {⇨Fig. 4.9–E}.[1632] |
| 1937 | Ecole Centrale des Arts et Manufactures, Paris, France | Louis J.B. BERGERON,[1633] a French professor of electrical engineering, **presents a graphical method of determining current and voltage transients of pulsed electrical transmission lines** – the **"Bergeron method."** His method is based on a theory worked out by the Austrian Richard LÖWY[1634] and the Swiss-born Othmar SCHNYDER[1635] for calculating **hydraulic shocks or "water hammers"** [French *coup de bélier*, German *Wasserhammer*] **in water supply lines** {KARELJSKICH & ZHUKOVSKY ⇨1898} – the **"Löwy-Schnyder method."** ▪ BERGERON's method also proved useful for calculating the transients of pulsed electrical transmission lines when the load resistance is time-dependent, such as in vacuum-discharge flash X-ray tubes.[1636] |
| 1937 | School of Aeronautical Engineering, University of Rome, Italy | Luigi CROCCO,[1637] an Italian aeronautical engineer, **investigates fluids in chemical equilibrium.** Combining the entropy equation with the momentum equation, he obtains a relation between the flow velocity vector $\vec{V}$, the vorticity $\nabla \times \vec{V}$ (= curl $\vec{V}$), and other thermodynamic properties. The so-called **"Crocco equation"** (or **"Crocco vorticity law"**) suggests the important result that a vortex-free flow behaves isentropically across the whole flow field. ▪ His equation was later extended by Andrew VAZSONYI,[1638] a Hungarian-born U.S. mathematician at Harvard University, in order to take into account fluid viscosity – the so-called **"Crocco-Vazsonyi equation."** |

---

[1631] O. STUTZER: *"Meteor Crater" (Arizona) und Nördlinger Ries.* Z. Dt. Geol. Gesell. **88**, 510-523 (1936).

[1632] R. HERMANN: *The supersonic wind tunnel installations at Peenemünde and Kochel, and their contributions to the aerodynamics of rocket-powered vehicles.* In: (L.G. NAPOLITANO, ed.) *Space: mankind's fourth environment* [Selected papers from the 32nd Int. Astronaut. Congress]. Pergamon Press, Oxford *etc.* (1982), pp. 435-446.

[1633] L. BERGERON: *Propagation d'ondes le long des lignes électriques. Méthode graphique.* Bull. Soc. Franç. Électr. **7** [V], 979-1004 (1937); *Du coup de bélier en hydraulique au coup de foudre en électricité.* Dunod, Paris (1949). Engl. translation: *Water hammer in hydraulics and wave surges in electricity.* Wiley, New York (1961).

[1634] R. LÖWY: *Druckschwankungen in Druckrohrleitungen.* Springer, Wien (1928).

[1635] O. SCHNYDER: *Druckstöße in Pumpensteigleitungen.* Schweiz. Bauz. **94**, 271-273, 283-286 (1929); *Druckstöße in Rohrleitungen.* Wasserkraft und Wasserwirtschaft **27**, 49-54, 64-70 (1932); *Druckstöße in Rohrleitungen.* Vogt-Schild, Solothurn (1943).

[1636] P. KREHL: *Analytical study on the maximization of bremsstrahlung and K-series production efficiencies in flash X-ray tubes.* Rev. Scient. Instrum. **57**, 1581-1589 (1986).

[1637] L. CROCCO: *Eine neue Stromfunktion für die Erforschung der Bewegung der Gase mit Rotation.* ZAMM **17**, 1-7 (1937).

[1638] A. VAZSONYI: *On rotational gas flows.* Quart. Appl. Math. **3**, 29-37 (1945).

| CHRONOLOGY | 1937 – 1938 | 509 |

| 1937 | Johns Hopkins University, Baltimore, Maryland & Institute for Advanced Study (IAS), Princeton, New Jersey | Francis D. MURNAGHAN,[1639] an Irish-born U.S. professor of mathematics, **derives a simple "one constant formula" for an isotropic elastic solid under hydrostatic pressure,** given by
$$p = a\,(f + 5f^2),\ \text{with}\ f = \tfrac{1}{2}\,[(v/v_0)^{2/3} - 1]\ \text{and}\ a = 3\lambda + 2\mu.$$
Here $\lambda$ and $\mu$ are the two elastic constants, and $v/v_0$ is the ratio of specific volumes at compression and rest, respectively. In the case of sodium, his calculated data fit BRIDGMAN's static compressibility data to an accuracy of within 1.5% over the range 2,000–20,000 atm. ▪ **The so-called "Murnaghan equation of state" was used extensively in the early era of shock wave research.** However, since it has no temperature dependence, it is unsuitable for cases where shock heating is important. **The Tait equation of state {TAIT ⇨1888} has a similar form and is still used in underwater explosion research.** |

| 1937 | Safety in Mines Research Board, Sheffield, England | William H. PAYMAN[1640] and Wilfred Charles F. SHEPHERD, two British detonation researchers, **rediscover the *shock tube*** – although this term has not yet been coined {BLEAKNEY, WEIMER & FLETCHER ⇨1949} – **as a powerful tool for studying combustion processes in air-methane mixtures** using a camera with a schlieren optical system, and to clarify whether a shock wave alone could start an explosion in a firedamp/air atmosphere. ▪ Today, the shock tube, in its various forms {⇨Figs. 4.10–A to L}, is used extensively in universities and industrial research laboratories in many countries. |

| 1937 | General Electric Company (GEC), Schenectady, New York | Levi TONKS,[1641] a U.S. research physicist, **coins the term *pinch effect*** while studying high-current-density phenomena in low-pressure arcs. ▪ The phenomenon had already been discovered in 1907 by the U.S. electrical engineer Carl HERING, who referred to it as the "pinch phenomenon" {NORTHRUP ⇨1907}.

In the dynamic pinch, the radius of the plasma column decreases with time, and the cylindrical current shell moves inward, thus acting like a magnetic piston and sweeping up all of the charged particles it encounters – the **"snowplow concept."** The pinch effect is commonly used to compress a plasma {LANGMUIR ⇨1923}; *i.e.*, gaseous matter. However, using pinched hollow metal conductors, **the suitability of the pinch method was first demonstrated in a Z-pinch arrangement for shock compression of miniature solid specimens** {BLESS ⇨1972}. The Z-pinch is a constriction of a conductor carrying a large current, caused by the interaction of that current with its own encircling magnetic field.

In the early 1950s, the pinch effect was first applied by James TUCK, an English physicist, at Los Alamos Scientific Laboratory (LASL) to rapidly compress and thereby heat a fusion plasma using a magnetic confinement device (code-name "Project Sherwood"). |

| 1938 | Germany & United States | **Invention and first applications of *flash radiography*** – a method of generating high-intensity X-ray pulses of microsecond/submicrosecond duration.

(i) Max STEENBECK,[1642] a German research physicist at the Siemens-Werke Berlin, **invented the method the year previously,** calling it the *"Röntgenblitztechnik"* ("Roentgen flash technique"). It uses a high-voltage capacitor that discharges through a mercury-vapor-filled capillary discharge tube, thus channeling the discharge and providing a small focus for radiographic purposes. His flash X-rays have a duration of approximately 1 µs. He immediately recognizes that flash radiography could provide an outstanding diagnostic tool for |

---

[1639] F.D. MURNAGHAN: *Finite deformations of an elastic solid*. Am. J. Math. **59**, 235-260 (1937).
[1640] W. PAYMAN and W.C.F. SHEPHERD: *(IV) Quasi-detonation in mixtures of methane and air*. Proc. Roy. Soc. Lond. **A158**, 348-367 (1937).
[1641] L. TONKS: *Theory and phenomena of high current densities in low pressure arcs*. Trans. Electrochem. Soc. **72**, 167-182 (1937).
[1642] M. STEENBECK: *Über ein Verfahren zur Erzeugung intensiver Röntgenblitze*. Wiss. Veröff. aus den Siemens-Werken (Springer, Berlin) **17**, 363-380 (1938).

- capturing the motions of projectiles in flight;[1643]
- visualizing shock waves in optically opaque media; and
- resolving self-luminous events, such as electrical discharges and detonation waves.

His published flash radiographs – showing a small-caliber lead projectile penetrating in soft and hard wood – clearly demonstrated the usefulness of flash radiography.

(ii) **Kenneth Hay KINGDON**[1644] **and H.E. TANIS,** two researchers at General Electric Company (Schenectady, NY), **generate intense flash X-rays in order to study mutation effects in biological samples,** such as wheat seeds and *Drosophila* eggs. Very similar to STEENBECK, they also use a mercury-pool X-ray tube, but with a different diode geometry. Applying a high-voltage capacitor discharge, they obtain a dose output of 3.5 roentgens in about 5 μs.

In the same year, STEENBECK's new diagnostic technique, which is proposed to Hubert SCHARDIN in July, is immediately applied by Franz R. THOMANEK {⇨1938}, who takes the first flash radiographs of detonating hemisphere-shaped charges. ▪ **First reviews on the history of flash radiography were provided by the three flash X-ray pioneers Werner SCHAAFFS**[1645] (a German research physicist at Siemens Company in Berlin), **Lucien BEAUDOUIN**[1646] (a French radiation expert at the Institut National des Sciences et Techniques Nucléaires in Saclay), **and Gustav THOMER**[1647] (a German detonation and radiation physicist at ISL, France).

| | | |
|---|---|---|
| 1938 | *Soviet Union* | **A.F. BELAJEV**[1648] **first uses exploding wires to initiate detonation in nitrogen chloride and nitroglycerin.** ▪ When it was found that exploding wires could produce detonation in less sensitive secondary explosives, such as PETN and RDX, ordnance designers became extremely interested. Not only could the electrical safety of electro-explosive devices be improved, but the handling hazards associated with sensitive primary explosives could be eliminated. From the 1960s onwards, exploding bridge-wire devices were increasingly used in missile technology, for stage igniters, stage separators, explosive bolts, cable cutters, thrust terminators, destruction systems, *etc.* |
| 1938 | *Southern Methodist University, Dallas, Texas* | **John Daniel BOON**[1649] **and Claude C. ALBRITTON,** two U.S. geologists, **show that geologic structures of the Kentland type – "shatter cones"** {BUCHER ⇨1933; ⇨Fig. 4.1–C} **– are the products of meteorite impacts.**<br><br>(i) According to their theory, a high-velocity impact – many times faster than the velocity of a shock wave in any type of rock – compresses the rocks elastically rather than deforming them plastically, after which they are "backfired" into a damped-wave disturbance.<br><br>(ii) **They assume that typical shatter-cones pointing toward the impinging body are formed during the initial or compressional stage of such a meteoroid impact.** ▪ Shatter cones provide a useful criterion for recognizing astroblemes, because this kind of shattering cannot be produced by any other natural means. |

---

[1643] M. STEENBECK: *Röntgenblitzaufnahmen von fliegenden Geschossen.* Die Naturwissenschaften **26**, 476-477 (1938).
[1644] K.H. KINGDON and H.E. TANIS: *Experiments with a condenser discharge.* Phys. Rev. **53** [II], 128-134 (1938).
[1645] W. SCHAAFFS: *Erzeugung und Anwendung von Röntgenblitzen.* Ergebn. exakt. Naturwissenschaften **28**, 1-46 (1955).
[1646] L. BEAUDOUIN: *La radiographie éclair.* Institut National des Sciences et Techniques Nucléaires, Presses Universitaires de France, Paris (1968).
[1647] G. THOMER: *History of flash radiography.* In: (L.E. BRYANT JR. , ed.) *Flash Radiography Symposium. 36th National Fall Conference of the American Society for Nondestructive Testing (ASNT)* [Houston, TX, Sept. 1976]. ASNT, Columbus, OH (1977), pp. 1-14.
[1648] A.F. BELAJEV: *The production of detonation in explosives under the action of a thermal pulse* [in Russ.]. Dokl. AN (SSSR) **18**, No. 4/5, 267-269 (1938).
[1649] J.D. BOON and C.C. ALBRITTON JR.: *Established and supposed examples of meteoritic craters and structures.* Field & Laboratory (Dallas, TX) **6**, 44-56, 57-64 (1938).

| | | |
|---|---|---|
| 1938 | Kaiser-Wilhelm-Institut (KWI) für Chemie, Berlin-Dahlem, Germany | Otto HAHN,[1650] a German radiochemist, **and Fritz STRASSMANN,** a German inorganic chemist, **perform the first artificial nuclear fission of the uranium isotope U-235 using neutrons.** They cautiously comment that their results, published on January 6, 1939, "are in opposition to all the phenomena observed up to the present in nuclear physics."<br><br>In the following year, **Siegfried FLÜGGE,**[1651] a German theoretical physicist at KWI, **estimated the energy released by the uranium fission process:**<br><br>(i) He concluded that one cubic meter of uranium oxide [$UO_2$, $UO_3$] – corresponding to approximately 4.2 tons of the mineral pitchblende, which contains 50–80% uranium – provides sufficient fission energy to supply electricity to central Germany for a period of eleven years.<br><br>(ii) **He also speculated on the huge quantity of explosive energy that could be released within milliseconds by artificial nuclear fission.** In nature, however, this event is quite unlikely because the concentration of uranium, even in highly enriched deposits, is far too low to maintain a chain reaction.<br><br>(iii) Shortly thereafter, FLÜGGE[1652] stated in a Berlin newspaper that the fission energy of 4.2 tons of uranium oxide would be sufficient "to throw the water masses of Lake Wannsee [a well-known lake in the southwest of Berlin with a length of about 3 km] into the stratosphere." ▪ Written in a "popular science" manner, this was the earliest example of a public illustration of the huge amount of energy which could be released explosively by nuclear fission. The first atomic bomb was detonated less than six years later {Trinity Test ⇨ 1945}.<br><br>It is interesting to note here that in the same year that FLÜGGE published his estimations, **Yakov B. ZEL'DOVICH and Yulii B. KHARITON,** two renowned Soviet physicists and nuclear experts, delivered a classified report on this topic at a seminar held at the Leningrad Physico-Technical Institute, in which they **elucidated the conditions required for a nuclear explosion and estimated its destructive force.**[1653] Later, both physicists became the main contributors to the development of the first Soviet atomic bomb {Test Site Semipalatinsk 21 ⇨ 1949}. |
| 1938 | National Physical Laboratory (NPL), Teddington, Middlesex, England | **William F. HILTON,**[1654] a British aerodynamicist, **reports on previous studies performed in 1934 in order to photograph the shock wave patterns arising from propeller blades at high rotational speeds.**<br><br>(i) He uses the direct shadow method, and in order to capture the motion he uses a microsecond-duration spark as a light source.<br><br>(ii) He finds that at subsonic tip speeds each propeller blade (observed at $M = 0.98$) has a single *spiral shock wave* of limited size attached to it, with each turn of the spiral indicating one revolution of the propeller; and that for supersonic tip speeds (at $M = 1.21$) a **second shock wave is formed ahead of the blade,** in the same way as a detached bow wave is formed for supersonic bullets, and the spiral waves become infinitely long.<br><br>(iii) His photographs also show the turbulent wake behind the blade, and in some cases the tip vortex trailing back from the blade tip. |

---

[1650] O. HAHN and F. STRASSMANN: *Über den Nachweis und das Verhalten der bei der Bestrahlung des Urans mittels Neutronen entstehenden Erdalkalimetalle.* Die Naturwissenschaften **27**, 11-15 (1939); *Über die Bruchstücke beim Zerplatzen des Urans.* Ibid. **27**, 89-95 (1939); *Nachweis der Entstehung aktiver Bariumisotope aus Uran und Thorium durch Neutronenbestrahlung. Nachweis weiterer aktiver Bruchstücke bei der Uranspaltung.* Ibid. **27**,163-164 (1939).
[1651] S. FLÜGGE: *Kann der Energieinhalt der Atomkerne technisch nutzbar gemacht werden?* Die Naturwissenschaften **27**, 402-410 (1939).
[1652] S. FLÜGGE: *Die Ausnutzung der Atomenergie.* Deutsche Allgemeine Zeitung (Berlin) **78**, Nr. 387 (Aug. 15, 1939).
[1653] Y. KHARITON and Y. SMIRNOV: *The Khariton version.* Bull. Atomic Scientists **49**, 20-31 (May 1993).
[1654] W.F. HILTON: *The photography of airscrew sound waves.* Proc. Roy. Soc. Lond. **A169**, 174-190 (Dec. 1938).

| 1938 | Institut für Technische Physik, TH Berlin-Charlottenburg, Germany | Heinz LANGWEILER,[1655] a collaborator with Prof. Carl CRANZ, **investigates the question of the maximum velocity that can be reached with firearms using modern propellants:**

(i) Under the ideal assumption that the total specific energy content $e_n$ of the propellant is fully transformed into the kinetic energy of the projectile, he obtains the simple relation

$$v_{max} = (2e_n)^{1/2}$$

for the maximum velocity. For nitrocellulose ($e_n = 950$ cal/g), this yields a maximum velocity of 2,820 m/s.

(ii) Using high-speed cinematography achieved by applying the Cranz-Schardin multiple-spark camera {CRANZ & SCHARDIN ⇒1929}, he measures a maximum velocity of 2,790 ms, which is very close to the theoretical value. In order to increase the initial velocity up to 3,800 m/s for nitrocellulose and up to 7,000 m/s for more energetic explosives, he proposes to incorporate detonation phenomena into the acceleration process, without going into any more detail however.

(iii) He attempts to calculate the length of the region of forward-moving gas following the detonation wave for TNT and assumes that the burnt gases preserve a high particle velocity $u_{BG}$ until the passage of a rarefaction shock wave reduces this velocity to zero. However, he recognizes that a rarefaction shock wave cannot occur, and tries to overcome this difficulty by saying that the region of transition where the velocity $u_{BG}$ is reduced to zero may be assumed to be small compared with the distances traveled by the detonation and rarefaction waves. ▪ LANGWEILER's assumption – albeit not fully correct – partly anticipated the "Taylor wave" phenomenon {G.I. TAYLOR ⇒1949}.

LANGWEILER's motivation for tackling this question was a military one. The famous *Parisgeschütz* ("Paris gun") reached a muzzle velocity of only 1,600 m/s {RAUSENBERGER ⇒1916; ⇒Fig. 4.2–T}. With the development of high-speed aircraft in the late 1930s, anti-aircraft guns needed a higher muzzle velocity to compensate for their higher aerodynamic drag and achieve more destructive action in the target. This objective also initiated German studies into using electromagnetic railguns {HÄNSLER ⇒1944/1945}. |
|---|---|---|
| 1938 | Institut für Aerodynamik, ETH Zürich, Switzerland | Ernst PREISWERK,[1656] a Swiss fluid dynamicist and one of Prof. ACKERET's Ph.D. students, **reports on the applicability and limitations of the analogy between a hydraulic jump and a two-dimensional compressible gas flow in his Ph.D. thesis.** While studying the propagation of a hydraulic jump – *i.e.*, a horizontal water flow at low depth and with a free surface – through a planar Laval nozzle, or its oblique reflection at a solid boundary, he notices that data obtained by water flow measurements increasingly deviate from the gas dynamic solution with increasing strength. ▪ At that time, this analogy was of particular interest because it would have allowed shock waves (and with them expensive high-speed diagnostics) to be replaced by relatively simple water table installations (and conventional movie recording).

Similar studies were also carried out in the late 1940s at CalTech {H.A. EINSTEIN ET AL. ⇒1946/1947; GILMORE ET AL. ⇒1950}. |
| 1938 | Technische Akademie der Luftwaffe (TAL), Berlin-Gatow, Germany | Oswald VON SCHMIDT,[1657] a German physicist, **treats wave propagation at the boundary between two media with different wave speeds.** He observes that any wave that enters a material with a higher wave propagation velocity produces a head wave in the material with the lower propagation velocity {⇒Fig. 4.14–O}. **The new wave phenomenon – known in Germany as the *VON SCHMIDTsche Kopfwelle* ("Schmidt head wave") – appears to be similar** |

---

[1655] H. LANGWEILER: *Zur Frage der mit den heutigen Treibpulvern maximal erreichbaren Geschossgeschwindigkeiten.* Z. Tech. Phys. **19**, 410-421 (1938).

[1656] E. PREISWERK: *Anwendungen gasdynamischer Methoden auf Wasserströmungen mit freier Oberfläche.* Mitteilungen aus dem Institut für Aerodynamik, Nr. 7, ETH Zürich (1938); *Application of the methods of gas dynamics to water flows with free surface. Part I: Flows with no energy dissipation.* Rept. NACA-TM 934 (1940); http://naca.larc.nasa.gov/digidoc/report/tm/34/NACA-TM-934.PDF

[1657] O. VON SCHMIDT: *Über Knallwellenausbreitung in Flüssigkeiten und festen Körpern.* Z. Tech. Phys. **19**, 554-561 (1938).

to the head wave produced by a supersonic bullet {E. MACH & SALCHER ⇨1886}. ▪ VON SCHMIDT[1658] noticed that the head wave he observed was analogous to the "migrating reflection in seismology" he had observed two years earlier; this phenomenon had previously been discovered in Germany by Ludger MINTROP {⇨1919}.

**The Schmidt head wave (SHW) is a pseudo-supersonic phenomenon which is independent of the presence of any shock waves and is also observable with sound waves.** It has some unusual features:

> ▸ In the acoustic case, the wave front of the SHW is a straight line. However, in the case of shock waves where the propagation velocity of the spreading shock wave rapidly diminishes, such as for cylindrical and spherical geometries, it becomes a curved line.
> 
> ▸ The SHW phenomenon is also a very common phenomenon in detonics: since the detonation velocities of most solid high explosives are higher than the shock propagation velocities in solids and liquids, the detonation wave generates a head wave in a solid or liquid body bordering the detonating high explosive.
> 
> ▸ Other wave types in solids can also produce SHWs, but at different angles: Erwin MEYER[1659] at Göttingen University, who applied VON SCHMIDT's method in order to study wave propagation in solids, showed that bending waves and shear waves generated in a solid can also create head waves in a bordering liquid {⇨Fig. 4.14−O}, *e.g.*, when the sound wave is generated at the solid-liquid boundary by a spark discharge.
> 
> ▸ In the case of explosions generated at the bottom of the sea, the SHW is not a straight line, but instead has a concave shape, because the underwater shock decays more rapidly then the ground shock propagating in the soil.
> 
> ▸ In the case of a strong explosion generated at a certain height above the ground, the resulting Mach reflection also produces a SHW which precedes the Mach stem and runs into the triple point {⇨Fig. 4.13−J}.

| | |
|---|---|
| 1938 *Luftfahrtforschungsanstalt Braunschweig & Technische Akademie der Luftwaffe (TAL), Berlin-Gatow, Germany* | **Franz Rudolf THOMANEK,** a young Austrian detonation physicist, while studying the cavity effect for high-explosives {VON BAADER ⇨1792; VON FÖRSTER ⇨1883; MUNROE ⇨1888}, **discovers the importance of the cavity liner and documents the "shaped charge lined cavity effect."**[1660]<br><br>(i) He started his studies due to Hubert SCHARDIN's hypothesis that the cavity effect might be caused by the Mach effect. SCHARDIN closely followed a previous notion {E. MACH & WOSYKA ⇨1875} that acoustic effects depend on the ambient pressure and must disappear when the area of wave interactions is evacuated.<br><br>(ii) THOMANEK, using a glass recipient positioned inside the cavity, observes to his great surprise that the target action of a shaped charge with a lined cavity increases significantly. ▪ In the following year, Gustav THOMER began to visualize the formation of a jet using the recently developed flash radiography technique {STEENBECK ⇨1938; ⇨Fig. 4.8−F},[1661] which allowed the collapse of the liner to be studied without interference from the smoke and flame associated with the detonation.<br><br>In August 1942, the Soviet physicists Yulii B. KHARITON and Veniamin A. TSUKERMAN first recognized the great potentials of flash radiography and discussed the possibility of re- |

---

[1658] O. VON SCHMIDT: *Zur Theorie der Erdbebenwellen. Die "wachsende" Reflexion der Seismik als Analogon zur "Kopfwelle" der Ballistik.* Z. Geophysik **12**, 199-205 (1936).

[1659] E. MEYER: *Über Ultraschallversuche im Physikalischen Praktikum.* Il Nuovo Cimento, Supplemento **7** [IX], No. 2, 1-7 (1950); *Messungen zur Körperschallübertragung an Hand von Modellen.* Acustica **6**, 51-58 (1956).

[1660] H. SCHARDIN: *Über die Entwicklung der Hohlladung.* Wehrtech. Monatshefte **51**, Nr. 4, 97-120 (1954).

[1661] A set of some of the first flash radiographs of shaped charges were donated to Dr. C. FAUGUIGNON, Adjoined Scientific Director of ISL, upon Dr. G. THOMER's retirement from ISL. See D.R. KENNEDY: *History of the shaped charge effect. The first 100 years.* Company brochure prepared by D.R. Kennedy & Assoc., Inc., Mountain View, CA (1983).

cording the explosion of the German *Faustpatrone*, a forerunner of the German *Panzerfaust* ("tank fist").[1662]

After World War II, Henry Hans MOHAUPT,[1663] a Swiss inventor, claimed in an article that he actually discovered the lined cavity effect as far back as 1935.

| 1939 | *Institute of Theoretical Physics, University of Copenhagen, Denmark* | On January 13 – only one week after the publication of the discovery of fission in uranium {HAHN & STRASSMANN ⇒ 1938} – **Otto R. FRISCH,** a German physicist and nephew of Lise MEITNER, **finishes an experiment that conclusively demonstrates that the uranium atom can be split into two almost equal pieces.**

(i) FRISCH discusses the implications of this new type of nuclear reaction and the resulting **enormous energy release** with the Danish physicist Niels BOHR and, together with MEITNER, publishes this epochal result, coining the term *Kernspaltung* (**"nuclear fission"**) for the splitting of a nucleus by neutron bombardment while doing so. ▪ The new name was immediately adopted by the nuclear physics community.[1664]

(ii) FRISCH also provides first experimental evidence that the fission products of the uranium nucleus must each fly apart with kinetic energies on the order of a hundred million electronvolts (eV).[1665]

FRISCH's sensational discovery – made almost simultaneously by the Austrian physicists Willibald JENTSCHKE (II. Physikalisches Institut, Universität Wien) and Friedrich PRANKL (Institut für Radiumforschung, Wien) as well as by the U.S. chemists Robert D. FOWLER and Richard W. DODSON (Johns Hopkins University, Baltimore, MD) – **was an important milestone in the realization of the atomic bomb** {WELLS ⇒ 1914; SZILARD 1934; HAHN & STRASSMANN ⇒ 1938; FLÜGGE 1939; HEISENBERG ⇒ 1939; Trinity Test ⇒ 1945}. |
| 1939 | *Sorbonne University, Paris, France* | In May, **Francis PERRIN,**[1666] a French physicist, **coins the term** *critical mass* [French *masse critique*], the smallest amount of material needed to sustain a nuclear chain reaction. He estimated that a critical mass of forty tons would be necessary in order to achieve a moderated-neutron chain reaction in a solid sphere of uranium oxide [$U_3O_8$]. ▪ The concept of a critical mass in a nuclear chain reaction goes back to patent applications filed by Leo SZILARD in 1934, in which he referred to being able to produce a nuclear explosion in a sufficiently thick mass of material {WELLS ⇒ 1913}.

PERRON's concept was extended by Rudolf E. PEIERLS, a refugee from Nazi Germany after the outbreak of World War II and a physics professor at Birmingham University. The resulting calculations were of considerable importance in the development of the first atomic bomb {Manhattan Project ⇒ 1942; Trinity Test ⇒ 1945}. |
| 1939 | *Private residence at Old Grove Road, Nassau Point, Peconic, Long Island* | On August 2, **Albert EINSTEIN,**[1667] **addressing the dangers of nuclear chain reactions, writes in a letter to U.S. President Franklin D. ROOSEVELT:** "Some recent work by E. FERMI and L. SZILARD, which has been communicated to me in manuscript, leads me to expect that the element uranium may be turned into a new important source of energy in the immediate future … This new phenomenon would also lead to the construction of bombs, and **it is conceivable** – though much less certain – **that extremely powerful bombs of a new type may thus be constructed** … I understand that Germany has actually stopped the sale of ura- |

---

[1662] V.A. TSUKERMAN and Z.A. AZARKH: *People and explosions.* VNIIEF, Arzamas-16 (1994).
[1663] H. MOHAUPT: *Shaped charges and warheads.* In: (F.B. POLLAD and J.A. ARNOLD, eds.) *Aerospace ordnance handbook.* Prentice-Hall, Englewood Cliffs, NJ (1966), chap. 11.
[1664] O. FRISCH and L. MEITNER: *Disintegration of uranium by neutrons: a new type of nuclear reaction.* Nature **143**, 239-240 (1939); *Products of the fission of the uranium nucleus.* Ibid. **143**, 471-472 (1939).
[1665] O. FRISCH: *Physical evidence for the division of heavy nuclei under neutron bombardment.* Nature **143**, 276 (1939).
[1666] F. PERRIN: *Calcul relatif aux conditions éventuelles de transmutation en chaîne de l'uranium.* C. R. Acad. Sci. Paris **208**, 1394-1396, 1573-1575 (1939).
[1667] This letter is now kept in the National Archives at the F.D. Roosevelt Library, Hyde Park, New York, NY; http://www.aip.org/history/einstein/ae43a.htm.

nium from Czechoslovakian mines, which she has taken over. That she should have taken such early action might perhaps be understood on the ground that the son of the German Under-Secretary of State, VON WEIZSÄCKER, is attached to the Kaiser-Wilhelm-Institut in Berlin where some of the American work on uranium is now being repeated." • This famous letter, today known as the "Albert Einstein letter" (although it was actually the first of four letters to ROOSEVELT), was most likely written by Leo SZILARD, the scientist who invented the chain reaction {WELLS ⇨ 1913}. Nevertheless, EINSTEIN took full responsibility for its consequences. In November 1954, five months before his death, he summarized his feelings about his role in the creation of the atomic bomb: "I made one great mistake in my life … when I signed the letter to President [Franklin D.] ROOSEVELT recommending that atomic bombs be made; but there was some justification – the danger that the Germans would make them."[1668]

In October 1939, **ROOSEVELT establishes a governmental Advisory Committee on Uranium in response to this letter.** Based on the results of the MAUD Committee's work {FRISCH & PEIERLS ⇨ 1940}, it will lead to the construction of the first atomic bomb {Manhattan Project ⇨ 1942; Trinity Test ⇨ 1945}. • However, post-war analyses of various historic sources have revealed that the "race for the atomic bomb" between the Allies and Germany was a myth.[1669]

| | | |
|---|---|---|
| 1939 | *Ernst-Heinkel Flugzcugwerke, Warnemünde, northern Germany* | On August 27 – four days before the outbreak of war between Germany and Poland – **the first successful flight of the Heinkel He 178, the world's first true turbojet-propelled aircraft, is achieved.** This single-seater jet research aircraft (max. velocity 700 km/h) was designed by Hans-Joachim PABST VON OHAIN, a graduate from the University of Göttingen who developed the first theory of gas turbine power plants in 1935 and later constructed a working model of the first jet turbine engine – thus leading to a revolution in aviation {Messerschmitt Me 262 ⇨ 1942}. • Jet engines are air-breathing devices. Many jet engine variants have been developed; the main ones are turbojet, fanjet, turboprop {McDonnell XF-88B ⇨ 1953}, ramjet {LORIN ⇨ 1908; FONO ⇨ 1928} and scramjet {BILLIG ⇨ 1959; NASA-LaRC ⇨ 1998; Hyshot Program ⇨ 2002}. Turbojet engines have also been used in supersonic civil transportation {Tupolev Tu-144 ⇨ 1968; Concorde ⇨ 1969}.

**Jet propulsion was first proposed by the 21-year old Englishman Frank WHITTLE.** He published his fundamental thesis *Future Developments in Aircraft Design*, which discusses the use of both gas turbines and jet propulsion for aircraft, while he was a flight cadet at Cranwell, in 1928. In January 1930, he applied for his first patent, which defined a gas turbine with a centrifugal compressor as a high-velocity jet source for propulsion. His design was first tested in 1937. However, the first British flight of a jet-powered aircraft using a Whittle W.1 engine was not performed until on May 15, 1941, with a Gloster E.28/39 ("Gloster Whittle"). |
| 1939 | *Moffet Field, California* | On December 20, **foundation of the Ames Aeronautical Laboratory (ARL) as the second NACA laboratory.** • It was named after Joseph Sweetman AMES, a former president of Johns Hopkins University (1929–1935), and longtime NACA chairman (1919–1939). With the creation of NASA {⇨ 1958}, the Laboratory was renamed Ames Research Center (ARC).

In 1943, serious plans were being made for the construction of a supersonic wind tunnel to study compressibility effects over airfoils. The first supersonic tunnel at Ames was a pressurized closed-loop $1 \times 3$ ft$^2$ ($30.48 \times 91.44$ cm$^2$) system powered by four compressors, each driven by a 2,500-hp electric motor. The pressure in the system was to be variable from 0.3 to |

---

[1668] R.W. CLARK: *EINSTEIN: the life and times*. World Publishing, New York (1971), p. 752.
[1669] W. HEISENBERG: *The Third Reich and the atomic bomb*. Bull. Atom. Sci. **24**, 34-35 (June 1968).

4.0 atmospheres, giving a Reynolds number range of from 0.5 to $10 \times 10^6$ per foot of model length; the Mach number range of the tunnel would be from 1.4 to 2.2.[1670]

| 1939 | *Aerodynamische Versuchsanstalt (AVA), Braunschweig, Germany* | Albert BETZ,[1671] a professor of applied mechanics and director of AVA, **looks into Adolf BUSEMANN's previous suggestion to use swept-back wings for high-speed aviation** {5th Volta Conf. (BUSEMANN) ⇨1935}. Model experiments on sweepback wings are performed secretly at the AVA in a high-speed wind tunnel (cross-section $11 \times 11$ cm$^2$) at velocities close to the velocity of sound. They give first experimental evidence of the superiority of the swept-back wing design. For example, the study shows that **aerodynamic drag increases by about 500% between $M = 0.7$ and 0.9 for straight wings, while the increase amounts to only about 20% for swept-back wings.**[1672] ▪ These milestone results were immediately taken up by the German aircraft industry, leading to the world's first tactical aircraft with swept-back wings, the Me 163 {Messerschmitt AG ⇨1941} and the Me 262 {Messerschmitt AG ⇨1942}.

Over the next two years, BETZ obtained three patents for modified swept-back wing designs, all classified by the German Government as top secret.[1673] These studies were not published until after World War II, and caused considerable excitement among foreign aerodynamic experts.[1674] The concept of a swept wing was resumed at NACA's Langley Field in early 1947. Two transonic jet fighters, the Soviet MiG-15 (operational in 1947) and the North American XP-86 ("Sabre") which became operational in 1948, had swept-back wings. Both set the pattern for most fighters that followed, although some adopted a delta-wing layout. |
|---|---|---|
| 1939 | *Balch Graduate School of the Geological Sciences, Caltech, Pasadena, California* | Beno GUTENBERG,[1675] a German-born U.S. geophysicist, **discusses the question of how tsunamis are linked to earthquakes.**

(i) In particular, addressing triggering of tsunamis by submarine slides, he refers to previous statements given by the seismologists Count Fernand M. BERNARD DE MONTESSUS DE BALLORE {⇨1907} and John MILNE {⇨1898}.

(ii) Based on numerous accounts of earthquakes accompanied by tsunamis, GUTENBERG arrives at the conclusion that "**tsunamis may be produced by submarine volcanic eruptions, submarine slides started by earthquakes, submarine faulting, and atmospheric conditions** [*e.g.*, by strong storms] … The macroseismic as well as the microseismic data of the Atacama Earthquake of November 11, 1922 indicate clearly that the fault movement occurred inland; the tsunamis originated from a submarine slide near a relatively feebly shaken stretch of the coast where the surface slopes steeply to a considerable depth. On gently sloping coasts, such as those of California, large tsunamis are rare and the relatively small tsunamis there are probably produced by faulting at the bottom." ▪ The occasional occurrence of tsunamis associated with earthquakes whose epicenters were located on land has led to the alternate notion that submarine landslides are the generating mechanism. |

---

[1670] E.P. HARTMAN: *Adventures in research. A history of Ames Research Center 1940–1965.* NASA Rept. SP-4302, NASA Center History Center, Washington, DC (1970); http://history.nasa.gov/SP-4302/sp4302.htm.

[1671] A. BETZ: *Flugzeug mit Geschwindigkeiten in der Nähe der Schallgeschwindigkeit.* Secret Germ. Patent No. 732/42 (1939).

[1672] A. BETZ: *Sonderaufgaben der aerodynamischen Forschung.* Schriften Dt. Akad. Luftfahrtforsch. Nr. 1020/40g, 113-118 (1940/1941).

[1673] A. BETZ: *Tragflügel für sehr hohe Geschwindigkeiten (Pfeilstellung).* Germ. Patent No. G732/42; *Tragflügel für sehr große Geschwindigkeiten (mit längs der Spannweite wechselndem Pfeilwinkel).* Germ. Patent No. G790/42; *Flügel mit veränderlicher Pfeilstellung (verstellbare Flügel).* Germ. Patent No. G799/42.

[1674] R. SMELT: *A critical review of German research on high-speed airflow.* J. Roy. Aeronaut. Soc. **50**, 899-934 (1946).

[1675] B. GUTENBERG: *Tsunamis and earthquakes.* Bull. Seismol. Soc. Am. **29**, 517-526 (1939).

| 1939 | Heereswaffenamt (HWA), Berlin & Dept. of Physics, University of Leipzig, Saxony, Germany | Werner HEISENBERG,[1676] one of the most eminent theoretical physicists of his time, **is placed in charge of investigations into the possibility of developing a nuclear bomb by the Heereswaffenamt** ("Army Ordnance Office"). On December 6, he sends his conclusions to Berlin, stating that **enrichment** – i.e., the increase in the proportion of U-235 to U-238 – **is "the only method of producing explosives several orders of magnitude more powerful than the strongest explosives yet known."**

In the following year, HEISENBERG sent a second report (dated Feb. 10, 1940) on this subject to Berlin. In February 1942, he delivered a lecture[1677] on nuclear bomb physics to top German officials at the House of German Research in Berlin, in which he stated, "Obtaining energy from uranium fission is undoubtedly possible if enrichment in the U-235 isotope is successful. **Production of pure U-235 would lead to an explosive of unimaginable force.** Natural uranium can also be used for energy production in a layered arrangement with heavy water. A layered arrangement of these substances can transfer its great energy reserves over a period of time to a thermal power machine. Such a reactor provides a means of liberating very large, usable quantities of energy from relatively small quantities of substance … That such a machine does not burn any oxygen would be a particular advantage if used in submarines … An operational machine can also be used to obtain a hugely powerful explosive; over and above that, it promises a number of other scientifically and technically important applications, which go beyond the scope of this talk."

The Farm Hall tapes later showed however that HEISENBERG did not pursue the idea of developing atomic bombs for Nazi Germany any further during the war.[1678, 1679, 1680] |
| --- | --- | --- |
| 1939 | Heeresversuchsanstalt (HVA) Peenemünde des Heereswaffenamtes, Peenemünde-Ost, Baltic Sea, Germany | **Construction of the world's most advanced supersonic wind tunnel under the leadership of Rudolf HERMANN,**[1681] a former coworker of professor Carl WIESELSBERGER {VON BRAUN ⇨1936}.

(i) The famous HVA Wind Tunnel {⇨Fig. 4.9–E} is a blowdown-to-vacuum complex ($M = 4.4$, later extended to $M = 5.3$) and has a working section of $40 \times 40$ cm².[1682, 1683, 1684] **Its main task will be the aerodynamic optimization of the A-4** { HVA Peenemünde ⇨1939 & 1942} **and A-5 rockets as well as the first guided anti-aircraft supersonic missile, code-name *Wasserfall*.**[1685]

(ii) The aerodynamic characteristics of these models – such as drag, lift and pitching moment – can be measured using an electromagnetic three-component balance.

(iii) **Operation above $M = 5$ reveals that air condensation effects become significant and impair visualization.** This discovery will eventually lead to the installation of the world's first dryer system to remove moisture from the air before it enters the nozzle. |

---

[1676] W. HEISENBERG: *Die Möglichkeiten der technischen Energiegewinnung aus der Uranspaltung*. Bericht an das Heereswaffenamt Berlin vom 6. Dez. 1939 • The report is now kept in the archives of the Deutsches Museum, München [Archiv, Atomdokumente G-39 und G-40].

[1677] W. HEISENBERG: *A lecture on bomb physics: February 1942*. Phys. Today **48**, 27-30 (Aug. 1995).

[1678] J. BERNSTEIN and D. CASSIDY: *Bomb apologetics: Farm Hall, August 1945*. Phys. Today **48**, 32-36 (Aug. 1995).

[1679] J. LOGAN: *The critical mass*. Am. Scient. **84**, No. 3, 263-277 (1996).

[1680] H.A. BETHE: *The German uranium project*. Phys. Today **53**, 34-36 (July 2000).

[1681] R. HERMANN: *The supersonic wind tunnel of the Heereswaffenamt and its application in external ballistics*. Kochel, Germany (June 16, 1945). Microfilm Mi 56-4553, National Union Catalogue, Library of Congress, Washington, DC (1972), vol. 242, p. 259.

[1682] W. KRAUS: *Der Überschall-Windkanal von Peenemünde*. Interavia **6**, 558-561 (1951).

[1683] P.P. WEGENER: *The Peenemünde wind tunnels. A memoir*. Yale University Press, New Haven, CT etc. (1996), pp. 69-70.

[1684] The aerodynamic testing of A-4 models in the Peenemünde wind tunnel and the first successful ignition of the A-4 {HVA Peenemünde ⇨1942} have also been described by Ruth KRAFT in her novel *Insel ohne Leuchtfeuer*. Vision Verlag GmbH, Berlin (1991), pp. 18-24, 193-208.

[1685] The 7.85-m-long surface-to-air missile *Wasserfall* reached a maximum velocity of Mach 2. See J. HERMANN: *Die Peenemünder Flugabwehrrakete Wasserfall*. In: T. BENECKE, K.H. HEDWIG, and J. HERMANN: *Flugkörper und Lenkraketen*. Bernard & Graefe Verlag, Koblenz (1987), pp. 131-136.

| | | Based upon a recommendation of Fritz ZWICKY, an American-Swiss astrophysics professor at CalTech {BAADE & ZWICKY ⇨1934} and a technical representative of a team from the Combined Intelligence Objectives Subcommittee (CIOS), this highly advanced wind tunnel was confiscated after the war by the U.S. Army, and dismantled and shipped to the Naval Ordnance Laboratory at White Oak, Maryland. The supersonic tunnel was later operated by the Naval Surface Warfare Center (now defunct). |
|---|---|---|
| 1939 | Siemens AG, Berlin-Siemens-stadt, Germany | **Justus MÜHLENPFORDT,**[1686] a German research engineer, **improves upon the operation of flash X-ray tubes** {STEENBECK ⇨1938; KINGDON & TANIS ⇨1938} **by developing continuously pumped tubes known as "open-vacuum discharge flash X-ray tubes"** (or "Siemens tubes"). He also introduces the conical anode, which is a good compromise between the contradictory demands for a high current and a small effective focus. ▪ Today the majority of flash X-ray tubes used around the world in ballistics, detonics and shock wave physics apply his tube concept.<br>In 1956, Walter P. DYKE[1687] and co-workers in the United States generated X-ray pulses using pure field emission in sealed-off tubes, later known as "field-emission tubes," which were operated via Marx-surge generators. Commercialized by Field Emission Corporation (McMinnville, OR) they almost completely dominated the market for decades. However, beginning in the 1970s, vacuum discharge flash X-ray tubes saw a renaissance because of their great versatility for diagnostic use in shock and detonation physics (*e.g.*, small focal size, high output intensity, broad range of operation voltages, spectral emission characteristics, long durability). |
| 1939 | Versuchsstelle der Luftwaffe Peenemünde-West des Reichs-luftfahrtminis-teriums (RLM), Peenemünde-West, Baltic Sea, Germany | **Erich WARSITZ,** one of the most experienced German test pilots of high-speed aircraft, **reaches a velocity of 850 km/h (236 m/s) at an altitude of 700 to 800 m in the Heinkel He 176.**[1688] ▪ Further development of this highly promising aircraft, however, was not approved by the Reichsluftfahrtministerium ("German Air Force Ministry"), which instead favored Willi MESSERSCHMIDT's new high-speed jet aircraft {Messerschmitt Me 262 ⇨1942; MUTKE ⇨1945}.<br>The He 176 was conceived in 1937 by Ernst HEINKEL, an eminent German professor of aeronautical engineering, industrialist and successful designer of high-speed aircraft, in order to achieve a maximum velocity of up to 1,000 km/h (278 m/s). It was the first high-speed aircraft provided with a liquid fuel rocket motor, which was designed by the German mechanical engineer and inventor Hellmuth WALTER, who made a number of important contributions in the field of air- and spacecraft propulsion, such as:<br>▸ the invention of the $H_2O_2$ turbine that powered the fuel pumps of the V2 rocket {HVA Peenemünde ⇨1942};<br>▸ the construction of the launching device that was vital to the successful flight of the V1 cruise missile {P. SCHMIDT ⇨1930};<br>▸ the design of the rocket motor for the 1,000-km/h Messerschmitt Me 163 ("Komet") {Messerschmitt AG ⇨1941; ⇨Fig. 4.20–E} – the fastest aircraft used in World War II, and the only rocket-powered aircraft ever to see active military service; and<br>▸ the production of successful rockets used to assist aircraft takeoff and a range of rocket motors for air/ground and ground/air missiles. |

---

[1686] J. MÜHLENPFORDT: *Verfahren zur Erzeugung kurzzeitiger Röntgenblitze*. Germ. Patent No. 748,185 (issued 1939, publ. 1944).
[1687] W.P. DYKE and W.W. DOLAN: *Field emission*. In: (L. MARTON, ed.) *Advances in electronics and electron physics*. Academic Press, New York, vol. 8 (1956), p. 89.
[1688] B. STÜWE: *Peenemünde West*. Bechtle Verlag, Esslingen & München (1998), p. 181.

| | | |
|---|---|---|
| 1940 | School of Physics, University of Birmingham, Werwickshire, England | In the spring, **Otto R. FRISCH and Rudolf E. PEIERLS**, two nuclear physicists as well as refugees from Austria and Germany, respectively, **write a top-secret two-part memorandum on nuclear bomb physics**:[1689]<br><br>(i) The first part, entitled *On the Construction of a "Superbomb," Based on a Nuclear Chain Reaction in Uranium*, shows that a nuclear bomb is much more feasible than people think.<br>   ▸ They conclude that a fast-neutron chain reaction would be possible in 1 kg of metallic U-235, and that the destructive effect of a 5-kg bomb would be equivalent to that of several kilotons of dynamite, "It is necessary that such a sphere should be made in two (or more) parts which are brought together first when the explosion is wanted. Once assembled, the bomb would explode within a second or less, since one neutron is sufficient to start the reaction and there are several neutrons passing through the bomb in every second, from the cosmic radiation ... A sphere with a radius of less than 3 cm could be made up in two hemispheres, which are pulled together by springs and kept separated by a suitable structure which is removed at the desired moment..."<br>   ▸ They also propose that U-235 could be separated from natural uranium by a thermal diffusion method. ▪ Their prediction of the fissionable mass required was quite close to the one used in the first atomic bomb {Trinity Test ⇒1945}. However, instead of using a spring-loaded mechanism to suddenly bring together two subcritical masses, as proposed in their memorandum, the Trinity bomb used high explosives to implode and compress two subcritical hemispheres of plutonium (forming a sphere with a diameter of about 8.1 cm, total weight about 6 kg) in order to reach the desired supercriticality.<br><br>(ii) In the second part of their memorandum, entitled *On the Properties of a Radioactive "Superbomb,"* they correctly conclude that nothing could be used to defend against such a superbomb when applied as a weapon, because there is no material or structure that can withstand the impact of the explosion.<br><br>Their memorandum formed the basis of the later report of the MAUD Committee (a codename for a top-secret committee) on the use of uranium as a source of power, which was presented to the British Government. A copy of this report was passed to the United States, who subsequently started their massive nuclear weapons program {Manhattan Project ⇒1942}. |
| 1940 | U.S.A. | On June 27, **establishment of the National Defense Research Committee (NDRC)** by the U.S. government, which is given the goal of coordinating planning and expenditure on military research. The new organization is headed by Vannevar BUSH, President of the Carnegie Institution. ▪ It originally consisted of five divisions (those related to the subject of this book were the Armor & Ordnance division and the Bombs, Fuels, Gases & Chemical Problems division). After a reorganization in 1942, 19 divisions were established, including those for Ballistic Research, Effects of Impact and Explosion {NDRC ⇒1946}, Explosives, New Missiles, and War Metallurgy.<br><br>NDRC was superseded by the Office of Scientific Research and Development (OSRD) in 1941, and eventually dissolved in 1947. |

---

[1689] The Frisch-Peierl Memorandum has been reproduced in the book by M. GOWING: *Britain and atomic energy*. Macmillan, London (1964); *see Appendix 1*, pp. 389-393.

| | | |
|---|---|---|
| 1940 | *Kaiser-Wilhelm-Institut (KWI) für Strömungsforschung, Göttingen, Germany* | In July, **Heinrich** [or Henry] **GÖRTLER**,[1690, 1691] a Canadian-born German applied mathematician, **first describes the formation of centrifugal instabilities in boundary layers along concave walls in his Habilitationsschrift ("habilitation thesis")**. Typically, the axes of the cellular vortices produced point in a streamwise direction. ▪ The British physicist Geoffrey I. TAYLOR[1692] had previously discovered and analytically explained a similar phenomenon in the flow between two concentric cylinders, with one cylinder rotating with respect to the other. The unstable flow, governed by centrifugal and viscous force fields, turns from a system of alternate annular vortices into a random, turbulent motion – the so-called **"Taylor instability"** {⇨Fig. 4.14–V}. The term ***Taylor-Görtler vortices*** was obviously coined by Hermann SCHLICHTING[1693] in his textbook *Grenzschichttheorie* ("Boundary Layer Theory"). Taylor-Görtler vortices are thought to be the cause of spiral grooves and crosshatching which develop on bodies upon supersonic reentry {⇨Fig. 4.14–W},[1694, 1695] and the so-called **"furrows"** found around the cones of some volcanoes after supersonic explosive eruptions {KIEFFER & STURTEVANT ⇨1984}. |
| 1940 | *Kaiser-Wilhelm-Institut (KWI) für Chemie, Berlin-Dahlem, Germany* | On July 17, **Carl Friedrich VON WEIZSÄCKER**,[1696] a German theoretical physicist, **anticipates the idea of a plutonium bomb by generating a new nuclear fuel via the fission of U-238 using thermal neutrons in a nuclear reactor.** (i) He suggests to Heereswaffenamt ("Army Ordnance Office") that this new material could be used for three purposes: ▸ for the construction of very small "machines" (*i.e.*, reactors) to produce energy; ▸ as a high explosive; and ▸ "to generate other elements in large quantities by a mixing process." (ii) He writes about element 94 – plutonium, "As soon as such a machine is on operation, the question of how to obtain explosive material, according to an idea of VON WEIZSÄCKER, takes a new turn. In the transmutation of the uranium in the machine, a new substance comes into existence, element 94, which very probably – just like U-235 – is an explosive of equally unimaginable force. This substance is much easier to obtain from [the common] uranium [U-238] than U-235, however, since it can be separated from uranium by chemical means. "[1697] According to the British journal *Physics World* (June 2005), two patent applications for these important discoveries under his name were found in Russian archives, *e.g.*, describing how plutonium could be used in a powerful bomb: "a process for the explosive production of energy from the fission of element 94, whereby element 94 … is brought together in such amounts in one place, for example a bomb, so that the overwhelming majority of neutrons produced by fission excite new fissions and do not leave the substance." ▪ VON WEIZSÄCKER participated at the meeting in Berlin on September 17, 1939, when the German *Uranprojekt* was begun. The Americans seized his papers when they captured his laboratory at the Universität Straßburg in December 1944. However, they revealed that Germany had not come close to developing a nuclear weapon. |

---

[1690] H. GÖRTLER: *Über eine dreidimensionale Instabilität laminarer Grenzschichten an konkaven Wänden*. Nachr. Wiss. Gesell. Gött. [Math.-Phys. Kl.: Fachgruppe 1, Nachrichten aus der Mathematik] N. F. **2**,1 (1940). ▪ Engl. translation: *On the three-dimensional instability of laminar boundary layers on concave walls*. Rept. NACA-TM 1375 (1954).

[1691] H. GÖRTLER: *Instabilität laminarer Grenzschichten an konkaven Wänden gegenüber gewissen dreidimensionalen Störungen*. ZAMM **21**, 250-252 (1941).

[1692] G.I. TAYLOR: *Stability of a viscous liquid contained between two rotating cylinders*. Phil. Trans. Roy. Soc. Lond. **A223**, 289-343 (1923).

[1693] H. SCHLICHTING: *Grenzschichttheorie*. G. Braun, Karlsruhe (1951), p. 413. Engl. translation: *Boundary layer theory*. McGraw-Hill, New York (1960).

[1694] T.N. CANNING, M.E. WILKENS, and M.E. TAUBER: *Ablation patterns on cones having laminar and turbulent flows*. AIAA J. **6**, 174-175 (1968).

[1695] M. TOBAK: *Hypothesis for the origin of cross-hatching*. AIAA Paper 69-11, Am. Inst. Aeronaut. & Astron. (1969).

[1696] C.F. VON WEIZSÄCKER: *Eine Möglichkeit der Energiegewinnung aus Uran 238* (17. Juli 1940). Dok. Nr. G-59, Archiv des Deutschen Museums, München.

[1697] J. BERNSTEIN: *HITLER's uranium club*. Copernicus Books, New York (2001), p. 340.

| | | |
|---|---|---|
| 1940 | Harvard University, Cambridge, Massachusetts | Percy W. BRIDGMAN,[1698] a U.S. professor of physics and the father of static high-pressure physics, **performs some orienting experiments on the effect of high mechanical stress on various solid explosives at the request of George B. KISTIAKOWSKY,** a chemistry professor at Harvard University.<br>(i) He applies two types of stress:<br>▸ hydrostatic pressures of up to 50 kbar with a large shearing stress superimposed; and<br>▸ hydrostatic pressures of up to 100 kbar with a low-level shearing stress.<br>From the eleven tested explosives, seven survive stress loadings of the first type, and three survive loadings of the second type without detonation.<br>(ii) He attributes the triggering of detonation to "secondary effects"– such as the sparks resulting from a fractured metal part and local elevations of temperature – and concludes "that mechanical stresses of themselves of the magnitude of those reached in these experiments cannot be counted on to initiate explosion in explosives of the type of those investigated." ▪ His results were not published until after the war.<br>In the 1960s, Edward TELLER,[1699] a Hungarian-born U.S. theoretical physicist at Lawrence Radiation Laboratory (LRL) of UC Berkeley, resumed the subject. Starting from a working hypothesis that "no explosion occurs if the pressure is purely hydrostatic," he postulates that mechanical stresses can result in a lowering of the activation energy. |
| 1940 | Technische Akademie der Luftwaffe (TAL), Berlin-Gatow, Germany | Ewald FÜNFER,[1700] a research physicist at the German Air Force Technical Academy, **describes the construction of a new high-speed stereoscopic camera in a classified institute report.** Using a Kerr cell {KERR ⇨ 1875} as an ultrafast, controllable shutter, it allows the photography of self-luminous phenomena in reflected light at submicrosecond exposure times, for example from an exploding incendiary shrapnel – an extremely difficult task and a masterpiece of high-speed photography. |
| 1940 | U.S.A. | Theodore VON KÁRMÁN and Hugh L. DRYDEN, two aeronautical engineers, **discuss shock phenomena that occur in planes at "transonic" speeds,** while on a business trip to Washington, DC. In his memoirs, VON KÁRMÁN[1701] writes, "We talked about the phenomenon and decided that if we invented a word, it had to be something between subsonic and supersonic to indicate that the body travels 'through' the speed of sound and back. We chose 'trans-sonic'. However, there was an argument as to whether to spell it with one s or with two s's. My choice was one s. Dr. DRYDEN favored two s's … We agreed on the illogical single s and thus it has remained. Incidentally, I used this new expression in a report to Wright Field. Although we just made up the word, nobody asked me what it meant. They just accepted transonic as if it had always belonged to the language." ▪ The transonic regime was then of great practical importance: at high flight speeds, the air moves over certain parts of the wing and tail at speeds greater than the speed of the plane because of the curvature of these sections. This phenomenon, also called **"shock-stall,"** creates shock waves that dance forward and back, causing a sudden partial loss of lift and dangerous vibrations of the skin structure. Pilots of the Lockheed P 38 ("Lightning") reported that their aircraft shook wildly and lost equilibrium at around Mach 0.8. In 1941, a Lockheed test pilot died when it is believed that shock waves from the plane's wings created turbulence that tore away the horizontal stabilizer – thus sending the plane into a fatal plunge. |

---

[1698] P.W. BRIDGMAN: *The effect of high mechanical stress on certain solid explosives.* J. Chem. Phys. **15**, 311-313 (1947).
[1699] E. TELLER: *On the speed of reactions at high pressures.* J. Chem. Phys. **36**, 901-903 (1962).
[1700] E. FÜNFER: *Funkenkinematographische Aufnahmen von Panzerplattenbeschüssen auf einem Schiessplatz bei Tageslicht.* Bericht 7/40, Ballistisches Institut, Technische Akademie der Luftwaffe (TAL), Berlin-Gatow (1940). ▪ This report is now kept at the Bundesarchiv/Militärarchiv Freiburg, Doc. No. RL39/17.
[1701] Th. VON KÁRMÁN: *The wind and beyond. Theodore VON KÁRMÁN pioneer in aviation and pathfinder in space.* Little, Brown & Co., Boston *etc.* (1967), p. 233.

> The term *shock-stall* originated in 1936 during high-speed aerodynamic wing studies carried out in England at the National Physical Laboratory (NPL).[1702] In order to describe the compressibility stalling speed, Ernest F. RELF, a staff scientist at the Aerodynamic Department of NPL, chose the term *shock-stall* as a shorter and more explicit expression than "compressibility stall" or "compressibility burbling speed." The British aerodynamicist William Frank HILTON,[1703] creator of the term *sound barrier* {HILTON ⇒ 1935}, wrote in his textbook *High-Speed Aerodynamics*: "The formation of a 'shock wave' on a wing will normally cause a separation of flow, and this may be called a **'stall.'"**

| | | |
|---|---|---|
| 1940 | *Safety in Mines Research Establishment, Sheffield, England* | William H. PAYMAN[1704] and Wilfred C.F. SHEPHERD, two British detonation researchers, continue their shock tube detonation studies {PAYMAN & SHEPHERD ⇒ 1937}. They use different driver gases and take schlieren photographs of the shock wave. **They notice that the use of hydrogen as a driver gas results in higher shock pressures in the test (expansion) chamber.** ▪ This classic paper, which they did not publish until 1946, provides a modern experimental verification of the shock wave equations obtained by measuring the speed of a shock wave moving into a gas at rest, and the speed of the gas following the shock wave. The relation between the measured speeds is very close to the theoretical relation implied by the shock wave relations. |
| 1940 | *Institute of Chemical Physics, Leningrad, Soviet Union* | Kirill I. SHCHELKIN,[1705] a Soviet combustion researcher, studies the detonation of gaseous mixtures in a shock tube and **discovers the influence of wall roughness on the transition from combustion to detonation. He explains this influence via the turbulization of the gas.** <br><br> In 1947, Yakov B. ZEL'DOVICH[1706] noted the role of the velocity profile – *i.e.*, the nonuniform distribution of velocity over the tube cross-section – on the detonation process. ▪ It has since become clear that both factors are very significant. Wall roughness also has a significant influence on the losses, the detonation speed, and the detonation limits. |
| 1940 | *Institute of Chemical Physics, Leningrad, Soviet Union [now St. Petersburg, Russia]* | Yakov B. ZEL'DOVICH,[1707] an eminent Soviet chemical physicist, **presents his planar "steady detonation" model** {⇒Fig. 4.16–S}. It assumes that a one-dimensional nonreactive shock wave is the leading element in the detonation, followed by a one-dimensional reaction zone in which detonation is initiated and completed, thereafter followed by nonreactive flow. ▪ His theoretical model, which also considered the corresponding energy balance around this zone, served as the starting point for all detonation-hydrodynamic calculations, and even now remains a topic of interest. <br><br> In the following three years, the same idea was worked out independently in the United States by the mathematician John VON NEUMANN {⇒1942} and in Germany by Werner DÖRING, a theoretical physicist at the Reichsuniversität Posen {Secret Workshop on "Probleme der Detonation" ⇒1941; DÖRING ⇒1943}. **Today known as the "Zel'dovich-von Neumann-Döring (ZND) theory," it describes a steady-state planar detonation wave as a shock followed by a reaction zone of decreasing pressure terminating at the Chapman-Jouguet (CJ) plane.** More realistic detonation models, however, include transport processes which re- |

---

[1702] W.F. HILTON: *British aeronautical research facilities*. J. Roy. Aeronaut. Soc. **70**, 103-107 (Jan. 1966).
[1703] W.F. HILTON: *High-speed aerodynamics*. Longman, Green & Co. London *etc.* (1952), p. 15.
[1704] W. PAYMAN and W.C.F. SHEPHERD: *(VI) The disturbance produced by bursting diaphragms with compressed air*. Proc. Roy. Soc. Lond. **A186**, 293-321 (1946).
[1705] K.I. SHCHELKIN: *Influence of tube wall roughness on the origin and propagation of detonation in gases* [in Russ.]. Zh. Eksp. Teor. Fiz. (SSSR) **10**, 823-827 (1940).
[1706] Y.B. ZEL'DOVICH: *Theory of the detonation onset in gases* [in Russ.]. Zh. Tekhn. Fiz. (SSSR) **17**, 3-26 (1947).
[1707] Y.B. ZEL'DOVICH: *On the theory of the propagation of detonation in gaseous systems* [in Russ.]. Zh. Eksp. Teor. Fiz. (Moscow-Leningrad) **10**, 542-568 (1940). ▪ Engl. translation: Rept. NACA-TM 1261 (Nov. 1950).

sult in the shock zone and the reaction zone becoming continuous in space and overlapping somewhat.[1708]

| 1940/ 1941 | Heeresversuchs-anstalt (HVA) Peenemünde, Peenemünde-Ost, Baltic Sea, Germany | **Some German aerodynamicists and rocket engineers at HVA begin to discuss the use of multistage rockets;** calculations are made for a two-stage 25.8-m-long rocket vehicle – called the **"A-9/A-10 concept"** – which uses an A-10 as a launching rocket for a winged missile (the A-9), which would skip and glide to its target.[1709] ▪ In 1944, plans were drawn up for a transatlantic two-stage rocket that would be guided by submarines to the U.S. coast (Germ. codename "Projektil Amerika"). Even more ambitious plans proposed a three-stage rocket – the *A-9/A-10/A-11 concept* – intended for flights into space.[1710]

The Russian rocket pioneer Konstantin TSIOLKOVSKY {⇨1903} was the first to propose space exploration by means of multistage (or multistep) rockets, using propellants such as liquid oxygen and liquid hydrogen. Hypervelocity flights were first achieved with a two-stage rocket using a captured German V2 rocket as a booster {WAC Corporal ⇨1949}. |

| 1940/ 1941 | *Abteilung für Technische Physik und Ballistik, Technische Akademie der Luftwaffe (TAL), Berlin-Gatow, Germany* | Hubert SCHARDIN,[1711] a German physicist and departmental head, **first suggests the possibility that phase transformations in liquids might be induced by shock waves – "shock-induced freezing."** By firing bullets at speeds ranging from 800 to 1,800 m/s into a tank filled with carbon tetrachloride [$CCl_4$] or water, and photographing this process, he finds the region surrounding the bullet to be opaque in $CCl_4$ at 1,200 m/s and in water at 1,800 m/s, whereas the water remains transparent at 800 m/s. He arrives at the conclusion that $CCl_4$ partially freezes under shock compression. ▪ Werner DÖRING,[1712] a theoretical physics professor who later resumed this phenomenon from a theoretical point of view, confirmed SCHARDIN's conclusion of shock-induced partial freezing.

At the time, BRIDGMAN's static high-pressure experiments on water had already been published, and these revealed the existence of seven ice configurations – **"Ice I"** (hexagonal phase) to **"Ice VII"** (cubic phase).[1713] His studies certainly stimulated early shock wave pioneers to investigate the corresponding problem of whether water freezes when it is rapidly compressed – such as by a shock wave {WALSH & RICE ⇨1957; AL'TSHULER ET AL. ⇨1958; HAMANN ⇨1966; DOLAN & GUPTA ⇨2004}. |

| 1941 | Radiation Laboratory, UC Berkeley, California | On March 28, the *fissile* **plutonium isotope Pu-239 is discovered in secret by Glenn T. SEABORG, Joseph W. Kennedy, Arthur C. WAHL,** three U.S. radiochemists, **and Emilio SEGRÉ,** an Italian-born U.S. nuclear physicist.

(i) They produced the new element in a cyclotron by bombarding the uranium isotope U-238 with neutrons.[1714] ▪ Already in 1940 SEABORG and collaborators had first produced artificially the isotope Pu-238 by deuteron bombardment of U-238 in the cyclotron at Berkeley. **Pu-238 is a *non-fissile* isotope** with a half life of 87 years.

(ii) Their important discovery will not be published until 1946. ▪ The atomic bomb dropped on Nagasaki was the first nuclear weapon to use plutonium Pu-239 {Hiroshima & Nagasaki Bombing ⇨1945}. Until recently, the major use of Pu-239 has been for nuclear weaponry.

Pu-239, the 94th element, fissions when bombarded by neutrons. Although the spontaneous fission rate of plutonium was measured to be higher than uranium (E. SEGRÉ 1943) – a detri- |

---

[1708] W. FICKETT and W.C. DAVIS: *Detonation.* University of California Press, Berkeley etc. (1979), pp. 191-199.
[1709] *A9/A10.* Encyclopedia astronautica; http://www.astronautix.com/lvs/a9a10.htm.
[1710] E. STÜHLINGER and F.I. ORDWAY: *Wernher von Braun: Aufbruch in den Weltraum.* Bechtle, Esslingen & München (1992), pp. 74-75.
[1711] H. SCHARDIN: *Experimentelle Arbeiten zum Problem der Detonation.* Jb. Dt. Akad. Luftfahrtforsch. (1940/1941), pp. 314-334.
[1712] W. DÖRING and G. BURKHARDT: *Beiträge zur Theorie der Detonation.* Zentrale für wiss. Berichtswesen (ZWB) der Luftfahrtforschung des Generalluftzeugmeisters, Berlin-Adlershof, Forschungsbericht Nr. 1939 (1944), pp. 55-58.
[1713] P.W. BRIDGMAN: *The phase diagram of water to 45,000 kg/cm$^2$.* J. Chem. Phys. **5**, 964-966 (1937).
[1714] Cited in R.G. HEWLETT and O.E. ANDERSON JR.: *A history of the United States Atomic Energy Commission.* University of California Press, Berkeley (1962); *see* vol. 1: *The new world, 1939–1946,* pp. 41-42.

mental property which might blow the fissile material apart before it has chance to undergo an efficient chain reaction, causing an atomic weapon to predetonate ("nuclear fizzle") – this disadvantage was circumvented by **applying plutonium in an implosion assembly** (J. VON NEUMANN & S. NEDDERMEYER 1943) rather than in a gun assembly device {Manhattan Project ⇨1943; Hiroshima Bombing ⇨1945; ⇨Fig. 4.16–G}. This allows one to bring the subcritical masses together more rapidly until it reaches criticality and detonates. In addition, the critical mass is inversely proportional to the square of the density. For example, if the density of the sphere is increased by a factor of two, the initial one critical mass becomes four critical masses in the compressed state (SERBER 1943). Therefore, the first nuclear test was carried out using a plutonium implosion bomb {Trinity Test ⇨1945}.

In the early stage of nuclear weapons development, **the experimental determination of criticality was an extremely risky enterprise** {FRISCH ⇨1944}. **For a spherical geometry, criticality is reached for U-235 and Pu-239 at 110 pounds** (49.9 kg) **and 35.2 pounds** (16 kg), **respectively.** When an external shield – a **"tamper"** – is used, it reflects neutrons back into the fissionable material that would otherwise leave. For example, a tamper of U-238 one inch thick around a sphere of Pu-239 reduces the mass required to produce criticality from 16 kg to 10 kg.[1715] In addition, it was much easier to create Pu-239 in large quantities (as urgently required for the first atomic bombs) than U-235.

| | | |
|---|---|---|
| 1941 | *University of Kyoto, Honshu, Japan* | In May, **Tokutaro HIGAWARA,** a Japanese physicist, **suggests in a lecture that a thermonuclear reaction among hydrogen nuclei could be triggered by an explosive chain reaction resulting from U-235 fission** {VON WEIZSÄCKER ⇨1940}.[1716] |

In September of the same year, Enrico FERMI presents a similar idea to Edward TELLER, a Hungarian-born U.S. physicist at Columbia University, while in New York.[1717] ▪ **The discussion between FERMI and TELLER resulted in the concept of using a nuclear explosion to initiate thermonuclear reactions in deuterium – producing a "hydrogen bomb"** {MIKE Test ⇨1952}. According to Hans A. BETHE,[1718] theoretical division leader at Los Alamos during the period 1943–1945, the H-bomb was first suggested by TELLER in 1942.

| | | |
|---|---|---|
| 1941 | *U.S. Air Corps* | During late spring, **Major Signa A. GILKE encounters serious trouble during a dive from an altitude of about 9,000 m attempted while flying a Lockheed fighter plane P-38 ("Lightning") at high speed.** When an airspeed of 515 km/h (143 m/s) is reached, he notices that the airplane's tail begins to shake violently, the nose begins to drop until the dive becomes almost vertical and it is impossible to operate the elevators. GILKE manages somehow to recover and to land safely. ▪ In November 1941, Lockheed test pilot Ralph VIRDEN had a fatal accident when transonic effects stopped him from being able to pull his airplane out of a steep dive. 17 months passed before Lockheed engineers began to determine what caused the Lightning's nose to drop. They tested a scale model of a P-38 in the wind tunnel at NACA's Ames Laboratory and found that shock waves formed when airflow over the wing reached transonic speeds and became turbulent. In 1944, Lockheed engineers installed electrically powered dive recovery flaps under each wing to restore lift and smooth the airflow.[1719] Eventually, these corrective dive flaps were installed on all P-38's, starting with the later P-38J versions. |

According to a more recently published interview,[1720] Hans G. MUTKE, a former German World War II test pilot of the Messerschmitt Me 262 {⇨1942}, reported in 2001 that during a test flight near Innsbruck (on April 9, 1945) his plane went into a steep dive of 40° at an alti-

---

[1715] J.S. FOSTER JR.: *Nuclear weapons.* In: (B.S. CAYNE, ed.) *The encyclopedia Americana.* Americana Corp., New York (1974), vol. 20, p. 521.
[1716] G.A. GONCHAROV: *American and Soviet H-bomb development programmes: historical background.* Phys. Uspekhi **39**, 1033-1044 (1996).
[1717] (Ed.) *This month in physics history: November 1, 1952: TELLER and the hydrogen bomb.* APS News **12**, No. 10, 2 (Nov. 2003).
[1718] H.A. BETHE: *Comments on the history of the H-bomb.* Los Alamos Science **3**, No. 3, 43-53 (1982).
[1719] *Lockheed P-38 Lightning.* The Aviation History Online Museum; http://www.aviation-history.com/lockheed/p38.htm.
[1720] M. SCHULZ: *Flammenritt über dem Moor.* Der Spiegel (Hamburg) **55**, Nr. 8, 214-218 (2001).

tude of 11,000 m, thereby passing the sound barrier for a duration of about seven seconds (his airspeed indicator was stuck against its limit of 1,100 km/h). The "breakthrough" was accompanied by strong shocks and vibrations which damaged parts of the fuselage, but he still enabled a happy landing. However, it is unlikely that the Me 262 (max. speed 870 km/h) broke the sound barrier; he rather experienced the typical shock wave effects ("buffeting") which already begin at about Mach 0.8.

| | | |
|---|---|---|
| 1941 | *Messerschmitt AG, Augsburg, Bavaria, Germany* | On October 2, **the rocket-powered Messerschmitt Me 163 ("Komet") interceptor** {⇨Fig. 4.20−E}, one of the first tailless aircraft designs, and also one of the earliest to incorporate substantial wing sweepback, **reaches a record velocity of 1,003 km/h in a secret test flight at an altitude of about 4,000 m.** The speed was measured by a ground team using cine-theodolites.[1721] ▪ The Me 163, designed by the famous German aircraft designer Alexander LIPPISCH, had a critical Mach number of 0.84, while other early World War II fighters − notably Lockheed's P-38 ("Lightning") {GILKE ⇨1941} − began to run into transonic aerodynamic effects at Mach numbers as low as 0.68.[1722]<br><br>Thirty-seven years later, Heini DITTMAR, previously a test pilot for the Me-163, gave the following account of his record flight,[1723] "My speedometer was soon reading 910 km/h and kept on increasing, soon topping the 1,000 km/h mark. Then the needle began to waver, there was a sudden vibration in the elevons and the next moment the aircraft went into an uncontrollable dive, causing high negative 'g' … I immediately cut the rocket and for a few moments I thought that I had really had it at last! Then, just as suddenly, the controls reacted normally again and I eased the aircraft out of its dive." ▪ DITTMAR described the "buffeting" typically experienced in the transonic regime. Pilots on both sides had experienced this when strange buffetings and even reversals of their controls took place during high-speed power dives. DITTMAR, however, was apparently the first pilot to experience it at these speeds. |
| 1941 | *Deutsche Akademie der Luftfahrtforschung (DAL), Berlin, Germany* | On October 25, **leading German shock wave and detonation experts meet at a secret workshop on the "Probleme der Detonation" to discuss urgent theoretical and experimental problems on detonation.**[1724] The chairman is Prof. Richard GRAMMEL. Topics of high priority up for discussion include:<br>▸ properties of blast and detonation waves (presented by Richard BECKER);<br>▸ pressure distributions in burnt detonation gases and surrounding objects (Werner DÖRING);<br>▸ theoretical treatment of spherical blast waves (Fritz SAUER);<br>▸ graphical solutions for nonstationary gaseous flows (Adolf BUSEMANN);<br>▸ experimental studies to clear up physical problems with detonation (Hubert SCHARDIN);<br>▸ recent studies on the shaped charge cavity effect (Gerhard HENSEL);<br>▸ methods used to measure pressure-time profiles in shock waves (Günther TURETSCHEK);<br>▸ measurements of air blast waves using a condenser microphone (Wilhelm SCHNEIDER); and<br>▸ studies on gas flow and blast effects of large-caliber guns (Theodor ROSSMANN). |

---

[1721] B. STÜWE: *Peenemünde-West. Die Erprobungsstelle der Luftwaffe für geheime Fernlenkwaffen und deren Entwicklungsgeschichte.* Bechtermünz, Augsburg (1998), pp. 200-269.

[1722] T. ATWOOD: *Komet 163. Chief test pilot Rudy OPITZ tells it like it was.* Flight Journal; Ridgefield, CT (August 1997).

[1723] B. JOHNSON: *The secret war.* Methuen, New York etc. (1978), pp. 275-283.

[1724] The lectures presented were published in a secret volume of the Schriften der Deutschen Akademie der Luftfahrtforschung Nr. 1023/41G, 31-45 (1941). ▪ Apparently, just a few copies have survived. For example, a copy is kept at the Archives of the Deutsches Museum, München. Prof. DÖRING's private copy, donated to the author by him in June 2003, has been transferred to the library of EMI, Freiburg.

| | | |
|---|---|---|
| 1941 | David Taylor Model Basin (DTMB), Carderock, Maryland & Chemisch-Physikalische Versuchsanstalt der Kriegsmarine, Kiel, northern Germany | **Underwater explosion studies are initiated, particularly into the oscillation of the gas globe – known as the "bubble"** – which generates the initial shock wave and a train of acoustic pulses.[1725] ▪ The experimental investigations on bubble motion and multiple shock wave generation, beginning on a laboratory scale and using Edgerton stroboscopy,[1726] also stimulated theoretical studies at other laboratories in the U.S.A and the U.K.[1727] The results of these studies, leading throughout the war to a wealth of new data on underwater shock wave propagation and interaction phenomena with boundary surfaces, were partly published after the war as *UNDEX (Underwater Explosions) Reports* {U.S.A./U.K. ⇨1950}.<br><br>During World War II, similar experiments were also carried out in Germany, but using high-speed cinematography instead of stroboscopy.[1728] **Early studies on a single-charge underwater explosion had already revealed that the main shock is followed by a second large pulse and further small ones** {BLOCHMANN ⇨1898}. Probably during or before the submarine warfare of World War I, this phenomenon was recognized as being a threat to a submarine's hull. In 1943, the U.S. lieutenant D.C. CAMPBELL[1729] appropriately wrote, "For some time, submarine personnel have noticed that more than one impact results from a single nearby underwater explosion such as a depth charge. Successive shocks were noted, and it was believed that the intensity and the time between blows decreased with each successive blow. Motion pictures of the action of floating models subjected to underwater explosions corroborated this impression." |
| 1941 | Heeresversuchsanstalt (HVA) Peenemünde, Peenemünde-Ost, Baltic Sea, Germany | Wolfgang Siegfried HAACK,[1730] a German mathematician, **first proposes a projectile geometry with minimum wave drag;** *i.e.*, which minimizes the energy radiated through the shock waves. His projectile has a slender geometry with zero nose and base areas, where the wave-drag volume term is proportional to $V^2/L^4$, $V$ being its volume and $L$ its length. Therefore, for a given volume it is convenient to have a very long body. Doubling the length $L$ at constant $V$ will reduce the wave drag by a factor of 16. ▪ The idea was later resumed in the United States by the U.S. aerospace engineer William R. SEARS.[1731] The so-called **"Haack-Sears body"** is superior in terms of wave drag to the **von Kármán ogive**, which is also a slender body of revolution with minimum wave drag, but it is only pointed at one end.[1732] |
| 1941 | University of Colorado, Boulder, Colorado | Roland L. IVES,[1733] a U.S. geophysicist and meteorologist, **demonstrates that sounds of finite amplitude** – such as a blast from an automobile horn or the sound of a revolver firing – **trigger supercooled fog droplets to freeze, causing some of them to "flash" into ice.** |

---

[1725] *A photographic study of small-scale underwater explosions*. David W. Taylor Model Basin, Confidential Test Rept. R-39 (Aug. 1941). ▪ Cited in: D.C. CAMPBELL: *Motions of a pulsating gas globe under water – a photographic study*. David W. Taylor Model Basin Rept. 512 (1943).

[1726] H.E. EDGERTON: *Electronic flash, strobe*. MIT Press, Cambridge, MA (1979).

[1727] M. EWING and A. CRARY: *Multi impulses from underwater explosions* (1941). Published in: Woods Hole Oceanogr. Inst. Ann. Rept.; see R.H. COLE: *Underwater explosions*. Dover Publ., New York (1948).

[1728] WEINERT: *Unterwasserzeitlupenaufnahmen von Gasblasenschwingungen*. Berichte der chemisch-physikalischen Versuchsanstalt der Kriegsmarine (Kiel) **245**, Heft VI (1941). ▪ This report was cited by W. DÖRING and H. SCHARDIN in a post-war review paper on detonations. See (A. BETZ, ed.) *Naturforschung und Medizin in Deutschland 1939–1946*. Verlag Chemie, Weinheim/Bergstr. (1953); Bd. 11: *Hydro- und Aerodynamik*, pp. 97-125.

[1729] D.C. CAMPBELL: *Motions of a pulsating gas globe under water – a photographic study*. David W. Taylor Model Basin Rept. 512 (1943).

[1730] W. HAACK: *Geschoßformen kleinsten Wellenwiderstandes*. Lilienthal-Gesell. Ber. **139**, 14-29 (1941).

[1731] W.R. SEARS: *On projectiles of minimum wave drag*. Quart. Appl. Math. **4**, 361-366 (1946).

[1732] A. MIELE (ed.): *Theory of optimum aerodynamic shapes: extremal problems in the aerodynamics of supersonic, hypersonic, and free-molecular flows*. Academic Press, New York (1965).

[1733] R.L. IVES: *Detection of supercooled fog droplets*. Aeronaut. Sci. **8**, 120-122 (1941).

| | | |
|---|---|---|
| 1941 | Harvard University, Cambridge, Massachusetts | George B. KISTIAKOWSKY,[1734] a Russian-born U.S. chemist, **and Edgar Bright WILSON**, a U.S. physicist, **improve the Becker equation of state** {BECKER ⇨1922} **for the detonating gas of a high explosive by including the temperature** $T$ **as an explicit state variable:** $$pv/RT = 1 + x\exp(\beta x),$$ where $p$ is the pressure, $R$ is the gas constant, $v$ is the specific volume, the variable $x = k/vT^{\alpha}$, and $k$ is the covolume. Experimentally determined detonation velocities for numerous explosives were reproduced by choosing values for $\alpha = 0.25$ and $\beta = 0.3$.<br><br>The so-called **"Becker-Kistiakowsky-Wilson (BKW) equation of state"** – a semi-empirical equation of state based upon a repulsive potential applied to the virial equation of state – has been used in thermochemical calculations with considerable success to predict the Chapman-Jouguet conditions {JOUGUET ⇨1904–1906; WALKER & STERNBERG ⇨1965} for a variety of solid explosives. Charles L. MADER,[1735] an eminent Los Alamos detonation researcher, reviewed the historical background of the BKW equation of state. |
| 1941 | Institut für Mechanik, TH Berlin-Charlottenburg, Germany | Walther KUCHARSKI,[1736] a German professor of mechanical engineering, **analytically treats the kinetics of folded stretch-free ropes.**<br><br>(i) He writes, "Unsteadiness of important physical quantities plays an important role in the kinetics of continua. By applying and further developing methods which – at least in technical mechanics – were previously barely known, one can draw important conclusions on the behavior of such unsteady phenomena; this will be demonstrated in some special papers which will be published in the near future. In the following communication, the problem is, so to speak, the inverse. The stretch-free ideal rope treated here shows types of motions of folding parts which can be described completely by simple mathematical means, and many curious, partly paradoxical phenomena are revealed which, while inherently interesting in their own right, also demonstrate the influence of looping."<br><br>(ii) He treats the example of a rope folded 180°, which is accelerated at one end while the other end is free, thus closely approaching the motion of a whip lash during whip cracking.<br><br>(iii) He finds that the speed of the free end increases without limit when the folding location approaches the end of the rope, and correctly concludes that the "flapping around" of the end of the rope can occur extraordinarily fast.<br><br>**This so-called "Kucharski effect" plays an important role in the mechanism of whip cracking** {GRAMMEL & ZOLLER ⇨1949}. |
| 1941 | Services des Poudres, Paris, France | Albert MICHEL-LÉVY,[1737] a French engineer, **and collaborators study luminous phenomena behind the shock fronts in various gases,** an extension of their previous experiments {MICHEL-LÉVY & MURAOUR ⇨1934}. They generate strong shock waves through the head-on collisions of shock waves emitted by high explosives. In argon, they observe intense light emission that increases towards the ultraviolet. |
| 1941 | Cavendish Laboratory, Cambridge, England | Geoffrey I. TAYLOR,[1738] an eminent British mathematical physicist and fluid dynamicist, **carries out an investigation into the possibility of detonation waves in three dimensions,** as distinct from the familiar one-dimensional detonation of gases in tubes. He theoretically exam- |

---

[1734] G.B. KISTIAKOWSKY and E.B. WILSON JR.: *Report on the prediction of detonation velocities of solid explosives.* Rept. OSRD-69 (1941); *The hydrodynamic theory of detonation and shock waves.* Rept. OSRD-114, NDRC (1941).

[1735] C.L. MADER: *Numerical modeling of detonations.* University of California Press, Berkeley etc. (1979). See Appendix E: *Numerical solution of equilibrium detonation properties using the B-K-W equation of state*, pp. 412-448.

[1736] W. KUCHARSKI: *Zur Kinetik dehnungsloser Seile mit Knickstellen.* Ing.-Archiv **12**, 109-123 (1941).

[1737] A. MICHEL-LEVY, H. MURAOUR, and E. VASSY: *Répartition spectrale énergétique dans la lumière émise lors de la rencontre d'ondes de choc.* Rev. Opt. Théor. Instrum. **20**, 149-160 (1941).

[1738] G.I. TAYLOR: *Detonation waves* [Paper prepared for the Civil Defence Research Committee]. Ministry of Home Security of Great Britain (Jan. 1941); *The dynamics of the combustion products behind plane and spherical detonation fronts in explosives.* Proc. Roy. Soc. Lond. **A200**, 235-247 (1950).

ines the detonation of a spherical charge of TNT ignited at its center, and analyzes the motion of the gases produced by the explosion behind the detonation front. **In contrast to the French engineer Émile JOUGUET,[1739] he arrives at the conclusion that spherical detonation waves *are* possible,** provided that infinitely rapid rates of change of velocity and pressure are admissible exactly at the detonation front. ▪ The Soviet physicist Yakov B. ZEL'DOVICH[1740] had tackled the same problem independently of and at almost the same time as TAYLOR. Similar to TAYLOR, he concluded that spherical detonation waves should be propagated with the same velocities as planar waves, provided that the pressure and velocity gradients behind the waves are infinite.

The effects predicted by TAYLOR's analysis, which assumes an infinitely thin shock wave, have in fact been observed. In the 1920s, Paul Frédéric LAFFITTE,[1741] a French physical chemist at the University of Paris, had demonstrated experimentally that stable spherical detonation waves could be initiated and sustained in perfect (stoichiometric) mixtures of $CS_2$ and $H_2$ with oxygen; however, only under the provision that sufficiently powerful explosives were used as initiators. LAFFITTE used 24-cm-dia. glass spheres filled with an explosive gaseous mixture. He initiated the explosion at the center of the sphere using a detonator containing 1 g of mercury fulminate {HOWARD ⇨1800}. The spherical detonation wave produced by this traveled at the same speed as that found when the same mixture was exploded in a tube.

In the same year, **TAYLOR,**[1742] one of the main British contributors to the Manhattan Project {⇨1942}, **analytically studies a high-intensity point explosion and provides the first exact similarity solution.** ▪ His results, then of the greatest military importance and top secret, were not published in the open literature until 1950.

(i) At the beginning of his paper he states, "This paper was written early in 1941 and circulated to the Civil Defense Research Committee of the Ministry of Home Security in June of that year. The present writer had been told that it might be possible to produce a bomb in which a very large amount of energy would be released by nuclear fission – the name **atomic bomb** had not been used – and the work here described represents his first attempt to form an idea of what mechanical effects might be expected if such an explosion could occur. In the then-common explosive bomb mechanical effects were produced by the sudden generation of a large amount of gas at a high temperature in a confined space. The practical question which required an answer was: would similar effects be produced if energy could be released in a highly concentrated form unaccompanied by the generation of gas? This paper has now been declassified, and though it has been superseded by more complete calculations, it seems appropriate to publish it as it was first written, without alterations."

(ii) He finds that **the shock wave moves with a steady speed** ($D$ = const), analogous to the case of a planar detonation, but only for the ideal case of a point explosion with instantaneous energy release. However, **the pressure decreases very rapidly with increasing distance,** according to the law $p \sim 1/r^3$. (Notice that in the acoustic case $p \sim 1/r^2$.) ▪ Shortly after World War II, an analogous solution for a point explosion with an instant energy source for the shock was also worked out independently in the former Soviet Union by Leonid I. SEDOV[1743] (known as the

---

[1739] E. JOUGUET: *Sur la propagation des réactions chimiques dans les gaz.* J. Math. Pures Appl. **1** [VI], 347-425 (1905); *Sur les ondes de choc et de combustion sphériques.* C. R. Acad. Sci. Paris **144**, 632-633 (1907).

[1740] Y.B. ZEL'DOVICH: *Pressure and velocity distributions in the detonation products of a divergent, spherical wave* [in Russ.]. Zh. Eksp. Teor. Fiz. (SSSR) **12**, 389-406 (1942).

[1741] P.F. LAFFITTE: *Sur la propagation de l'onde explosive.* C. R. Acad. Sci. Paris **177**, 178-180 (1923); *Recherches expérimentale sur l'onde explosive et l'onde de choc.* Ann. Phys. **4** [X], 587-694 (1925); see *L'onde explosive sphérique*, pp. 645-652.

[1742] G.I. TAYLOR: *The formation of a blast wave by a very intense explosion.* Ministry of Home Security, Brit. Rept. RC-210, II-5-153 (1941); Proc. Roy. Soc. Lond. **A201**, 159-186 (1950).

[1743] L.I. SEDOV: *Propagation of strong explosive waves* [in Russ.]. Prikl. Mat. Mekh. (SSSR) **10**, No. 2, 241-250 (1946). See also his book *Similarity and dimensional methods in mechanics.* Infosearch, London (1959), pp. 233-251.

"Sedov-Taylor similarity solution") and in the United States by John VON NEUMANN.[1744] **Later, the corresponding cylindrical problem was also solved** {LIN ⇨ 1954}.

If a density gradient exists in the atmosphere — even in just one direction — the point explosion problem is no longer self-simulating, and an exact solution cannot be obtained. **A point explosion in an inhomogeneous atmosphere was first treated in the Soviet Union by the theoretical physicist Aleksandr S. KOMPANEETS,**[1745] who concluded, "No matter how great the energy of the explosion, a strong shock wave cannot penetrate downwards further than 11 km. For any penetration below this altitude, the shock wave will be rapidly weakened because of the rarefaction waves traveling upwards from it, into the region that is open to the empty space above, where these waves will be lost."

| | | |
|---|---|---|
| 1942 | *New York City, New York* | In June, **work on the secret "Manhattan Project"** (or "Project Y") — a code-name for the efforts of the United States during the period 1942–1945 to produce the first atomic bomb {Trinity Test ⇨ 1945} — **is started.**[1746, 1747, 1748, 1749] ▪ The project was named after the Manhattan Engineer District of the U.S. Army Corps of Engineers, because much of the initial research was done in New York City.<br><br>In April 1943, the physicist Robert SERBER from the University of Wisconsin, one of J. Robert OPPENHEIMER's assistants and Theoretical Group Leader of the bomb project, defined its goal more specifically, "The object of the project is to produce *a practical military weapon* **in the form of a bomb** in which the energy is released by a fast neutron chain reaction in one or more of the materials to show nuclear fission." ▪ **The Manhattan Project had an enormous impact on the subsequent evolution of shock wave and detonation physics, but also strongly stimulated the evolution of digital computers and pushed high-speed photography and electronic diagnostics towards nanosecond time resolution.** These activities led to the installation of a large number of special laboratories and test sites operated by governmental agencies, private research organizations and universities. |
| 1942 | *Messerschmitt AG, Augsburg, Bavaria, Germany* | On July 18, **first test flight of the twin-jet Messerschmitt Me 262** (Pilot Fritz WENDEL). The first jet fighter is powered by the new BMW gas turbine engines, each producing 2,000 pounds of thrust. ▪ **The Me 262 became the world's first sweepback jet fighter to enter operational service.** It was the fastest aircraft of its time (870 km/h at an altitude of 6,100 m). Prior to this, the usefulness of sweepback wings had been proven by wind tunnel tests at the AVA Göttingen {BETZ ⇨ 1939}, and the Messerschmitt AG.[1750] |
| 1942 | *Heeresversuchsanstalt (HVA) Peenemünde, Peenemünde-Ost, Germany* | On October 3, **first test flight of the supersonic A-4 rocket.** This huge, 14-m-long ballistic missile {⇨Fig. 4.20–C}, developed by the German Army under the supervision of General Walter DORNBERGER and Wernher VON BRAUN and designed to transport a payload (a TNT warhead) of almost one ton at a velocity of up to Mach 4, is propelled by a particular fuel mixture (75% ethyl alcohol and 25% water), and liquid oxygen. The two liquids were delivered to the thrust chamber {⇨Fig. 4.7–H} by two rotary pumps, driven by a steam turbine. |

---

[1744] J. VON NEUMANN: *The point source solution*. In: (K. FUCHS, J.O. HIRSCHFELDER, J.L. MAGEE, R. PEIERLS, and J. VON NEUMANN, eds.) *Blast wave*. Rept. LA (1947), Los Alamos, NM (2000), chap. 2, pp. 27-55. ▪ See also A.H. TAUB (ed.): *The collected works of John VON NEUMANN*. Pergamon Press, New York (1963), vol. VI, pp. 219-237.
[1745] A.S. KOMPANEETS: *A point explosion in an inhomogeneous atmosphere*. Sov. Phys. Dokl. **5**, 46-48 (1960).
[1746] *Los Alamos, NM (Project Y)*. The Manhattan Project Heritage Preservation Association, Inc.; http://www.childrenofthemanhattanproject.org/LA/Photo-Pages-2/LAPG_13.htm.
[1747] F.H. SHELTON: *Reflections of a nuclear weaponeer*. Shelton Enterprise Inc., Colorado Springs, CO (1988), chap. 1: *The Manhattan Project*.
[1748] M.B. STOFF (ed.): *The Manhattan project: a documentary introduction to the atomic age*. Temple University Press, Philadelphia, PA (1991).
[1749] L. HODDESON, P.W. HENRIKSEN, R.A. MEADE, and C. WESTFALL: *Critical assembly: a technical history of Los Alamos during the Oppenheimer years, 1943–1945*. Cambridge University Press, Cambridge (1993).
[1750] Th. VON KÁRMÁN and L. EDSON: *The wind and beyond. Theodore VON KÁRMÁN, pioneer and pathfinder in space*. Little, Brown & Co, Boston *etc.* (1967), p. 224.

(i) The first test on June 13 fails, because shortly after the ignition of the engine the fuel supply cut off, but the rocket had already passed Mach 1 by this point. **It is the first time that a man-made vehicle has passed through the dreaded sound barrier.**

(ii) During the second test on August 16, the A-4 reaches a velocity of more than Mach 2, but again problems occur with the fuel supply.

(iii) **Eventually, in a third test on October 3, Wernher VON BRAUN's team succeeds in launching the first man-made vehicle penetrating into the thermosphere,** thus entering the "vestibule" of space. **The missile accelerates to over Mach 4, reaches a record height of 84.5 km, and covers a range of 191 km** {⇨Fig. 4.20–C}.[1751]

In 1943, HITLER decided to use the A-4 as a "vengeance weapon," and towards the end of war, the A-4 became widely known as the "V2" (an abbreviation for *"Vergeltungswaffe 2,"* meaning "Vengeance Weapon 2"). The first combat V2 was launched toward western Europe on September 7, 1944. Approximately 6,500 V2s were manufactured during the period 1944–1945.

| | | |
|---|---|---|
| 1942 | *Los Alamos Scientific Laboratory (LASL), Los Alamos, north central New Mexico* | On November 1942, the U.S Army acquires 54,000 acres of the Pjarito Plateau for use as a "demolition range." <br><br> In the following year, **on January 1, 1943, the University of California was selected to operate the new Los Alamos Scientific Laboratory (LASL) by the U.S. Government in order to centralize nuclear bomb research and the development of the Manhattan Project** {⇨1942}. LASL is managed by the University of California (UC).[1752] The theoretical physicist J. Robert OPPENHEIMER becomes the first director of LASL. The work at Los Alamos, which has hitherto been performed at the Universities of Chicago, Cornell, Minnesota, Purdue, Stanford and Wisconsin, and at the Carnegie Institution in Washington, DC,[1753] receives its own designation as "Project Y." • In the spring of the following year, the staff at Los Alamos began research into largely unknown radioactive materials, such as the rare uranium isotope U-235 {DEMPSTER ⇨1935}, and the newly discovered plutonium Pu-239 {SEABORG ET AL. ⇨1941} which were later used in the first atomic bombs {Trinity Test ⇨1945; Hiroshima & Nagasaki Bombing ⇨1945}. Very recently, a historical perspective on the work performed at Los Alamos was provided by Harris MAYER,[1754] a former staff scientist. <br><br> On January 1, 1981, the Laboratory was renamed LANL (Los Alamos National Laboratory) at the request of the United States Congress. |
| 1942 | *Dept. of Physics, Cornell University, Ithaca, New York* | Hans A. BETHE,[1755] a German-born U.S. theoretical physicist, **calculates the stability of shock waves for an arbitrary equation of state and discusses the shape of the Hugoniot curve at high temperatures and pressures.** He also deals with the case where a phase transition is induced by the shock – a phenomenon that was thought to be possible in strong underwater explosions at the time. <br><br> (i) In his introduction, he circumscribes the goal of his study, "The theory of shock waves thus far has been developed mainly for ideal gases. Even for these, the question of stability of shock waves has received little attention. Recently, the problem of shock waves in water has gained much practical importance. Therefore, it seems worthwhile to investigate the properties of shock waves under conditions as general as possible." <br><br> (ii) He treats the Hugoniot curve $H(v, s)$ in terms of volume per unit mass ($v$) and entropy per unit mass ($s$), and derives the three stability conditions of a shock wave: <br><br> $\partial^2 p(v, s)/\partial v^2 > 0$ {DUHEM ⇨1909}; $v\partial p(v, e)/\partial e > -2$; and $\partial p(v, e)/\partial v < 0$. |

---

[1751] W. DORNBERGER: *V2 – Der Schuß ins Weltall.* Bechtle, Esslingen (1952).
[1752] *Los Alamos National Laboratory – history.* Los Alamos National Laboratory (LANL); http://www.lanl.gov/history/index.shtml.
[1753] *Los Alamos 1943–1945; The beginning of an era.* Brochure LASL-79-79, LASL, Los Alamos, NM (1979).
[1754] H. MAYER: *People of the hill – the early days.* Los Alamos Science No. 28 (2003), pp. 2-29.
[1755] H.A. BETHE: *On the theory of shock waves for an arbitrary equation of state.* Rept. OSRD-545, NDRC Div. B (1942).

(iii) He concludes that **the transition from solid to liquid, from solid to gas and from liquid to gas, as well as the reverse transitions, should not affect the stability of the shock, while in the solid-solid transitions the shock front would split into successive shocks,** the first one raising the medium to a metastable state and the second one transforming it into the new stable phase.

At that time, BETHE believed that the duration of the shock pressure was not sufficient to induce these new thermodynamically stable phases. Later experiments, however, clearly showed that shock waves can induce polymorphic transitions, even for submicrosecond durations {JOHNSON & MITCHELL ⇨ 1972}. The splitting of shock fronts into independent waves moving at different speeds has been observed in materials undergoing a phase transition {BANCROFT ET AL. ⇨ 1956; AL'TSHULER ET AL. ⇨ 1958} and transitions from elastic into inelastic behavior {PACK ET AL. ⇨ 1948; MINSHALL ⇨ 1955; GORANSON ET AL. ⇨ 1955}. This interesting subject of shock-induced phase transitions – certainly one of the most challenging problems in shock wave physics – was later resumed and discussed in more detail by leading U.S. researchers {DUVALL & GRAHAM ⇨ 1977}.

| 1942 | *Institut für Gasdynamik, Braunschweig, Lower Saxony, Germany* | K. Gottfried GUDERLEY,[1756] a German aeronautical engineer, **analytically treats the two basic cases of cylindrical and spherical implosions produced by converging shock waves.** He predicts a power law increase in Mach number as the shock approaches the axis or point of symmetry. At the instant of shock collapse, the solution is singular, meaning that infinite shock strengths exist at the center of the implosion, but the subsequent reflected shock motion again follows a simple power law. He points out that real gas effects might furnish a natural limit to these theoretical singularities. ▪ Implosion phenomena are complicated superposition phenomena and can be accompanied by strong velocity jumps of the shock front caused by Mach reflection {⇨Fig. 4.14–K}.<br><br>GUDERLEY's classical theory of implosion was later resumed in the United Kingdom, and a more accurate similarity solution was derived by D.S. BUTLER.[1757] In 1986, GUDERLEY, then a retired chief mathematician at the Aeronautical Research Laboratories (Wright-Patterson AFB, Dayton, OH) delivered the 11th G.I. Taylor Memorial Lecture entitled *Implosions*. |
|---|---|---|
| 1942 | *Cornell University, Ithaca, New York* | John G. KIRKWOOD,[1758] a U.S. physicist, **and Hans A. BETHE,** a German-born U.S. professor of theoretical physics, **derive a theory on the shock wave produced by an underwater explosion.**<br><br>(i) Their so-called "Kirkwood-Bethe theory" predicts, for longer ranges, an asymptotic peak pressure-distance relationship of the form<br>$$p(r) = K / \{(r/a_0) [\ln(r/a_0)]^{1/2}\},$$<br>where $p$ is the pressure, $r$ is the pressure gauge distance from the explosion site, $K$ is an empirical value, and $a_0$ is the radius of the equivalent spherical charge. ▪ Using the principle of similarity {WHITE ⇨ 1946}, underwater shock wave peak pressures are customarily expressed by<br>$$p(r) = K (W^{1/3}/r)^\alpha,$$<br>where $W$ is the charge weight in pounds. Later measurements determined the empirical values for TNT as $K = 2.16 \times 10^4$ and $\alpha = 1.13$ when $p$ is measured in psi. ▪ This approximation is in good agreement with the Kirkwood-Bethe theory and is applicable in the range 10–10,000 |

---

[1756] G. GUDERLEY: *Starke kugelige und zylindrische Verdichtungsstöße in der Nähe des Kugelmittelpunktes bzw. der Zylinderachse.* Luftfahrtforsch. **19**, 302-312 (1942).

[1757] D.S. BUTLER: *Converging spherical and cylindrical shock waves.* Rept. 54/54, Armament Research Establishment (ARE), Fort Halstead, Kent, U.K. (1954).

[1758] J.G. KIRKWOOD and H.A. BETHE: *The pressure wave produced by an underwater explosion I.* Progr. Rept. OSRD-588, NDRC Div. B (May 1942).

charge radii.[1759] However, at very short ranges, one has $\alpha > 1.13$. At very large ranges the theoretical pressure function must decrease with $1/r$ (*i.e.*, $\alpha = 1$, the acoustic approximation).

(ii) They theoretically demonstrate that in an underwater explosion, after reaching the rebound point in bubble dynamics, a shock wave is emitted into the surrounding liquid, thus essentially confirming previous photographic studies {David Taylor Model Basin ⇨1941; Kriegsmarine Kiel ⇨1941}.

The initial pressure wave measured at modest distances from an underwater explosion is often modeled as a spherical shock wave with an exponential decay. In the mid-1970s, Peter H. ROGERS,[1760] a shock researcher at Naval Research Laboratory (Orlando, FL) worked out a closed-form analytical **"weak-shock" solution** for the subsequent propagation of such a wave. He reported that his results were in good agreement with the Kirkwood-Bethe theory, existing measurements, and the widely used experimentally based semi-empirical similarity formulas.

1942 *Institute for Advanced Study (IAS), Princeton, New Jersey*

**John VON NEUMANN,**[1761] a Hungarian-born U.S. mathematician and a faculty member of the IAS since 1933, **theoretically treats the process of detonation, hereby extending the CJ theory** {CHAPMAN ⇨1899; JOUGUET ⇨1904–1906}.

(i) Similar to Yakov B. ZEL'DOVICH {⇨1940} in the Soviet Union and to Werner DÖRING {⇨1943} in Germany, he assumes a three-layer model for the detonation front – the so-called **"Zel'dovich-von Neumann-Döring (ZND) theory."** ▪ It should be noted here that DÖRING[1762] first reported on his detonation theory at a lecture delivered in 1941 in Berlin {Secret Workshop on the "Probleme der Detonation" ⇨1941}, and so chronologically speaking it should really be the ZDN theory!

(ii) The ZND model assumes that the detonation process consists of a planar shock wave that takes the unreacted material from its initial state ($P_0$, $v_0$) to the post-shock, or ***von Neumann state.*** The chemical reaction is initiated at the ***von Neumann spike*** point ($P_S$, $v_S$) which is given by the intersection of the Rayleigh line {Lord RAYLEIGH ⇨1878} with the unreacted Hugoniot. In traversing the ZND reaction zone, one proceeds down the Rayleigh line from the spike point to the CJ point ($P_{CJ}$, $v_{CJ}$), thereby crossing the full set of partial reaction Hugoniots in the process. The rate at which an element of explosive passes from the point ($P_S$, $v_S$) to the CJ state depends on the kinetics of the reaction and cannot be determined from hydrodynamic considerations. ▪ For details, the reader is referred to the pioneering paper written by DUFF & HOUSTON {⇨1955}, to a review article on explosives,[1763] and to measurements of detonation wave profiles in some HMX-based explosives taken using two VISAR interferometers {GUSTAVSEN ET AL. ⇨1997}. The conditions under which the structure of a von-Neumann spike can be characterized were first studied at Göttingen University by careful optical measurements.[1764, 1765]

In the same year, **VON NEUMANN also speculates on a method to find an equivalent to Huygens principle** {HUYGENS ⇨1678} **for waves of finite amplitude** – *i.e.*, for shock waves.[1766] ▪ Unfortunately, details of his interesting approach have not been passed on to us.

---

[1759] A.B. ARONS: *Underwater explosion shock wave parameters at large distances from the charge.* JASA **26**, 343-346 (1954).

[1760] P.H. ROGERS: *Weak-shock solution for underwater explosive shock waves.* JASA **62**, 1412-1419 (1977).

[1761] J. VON NEUMANN: *Theory of stationary detonation waves.* Progr. Rept. OSRD-549 (April 1942).

[1762] W. DÖRING: *Der Druckverlauf in den Schwaden und im umgebenden Medium bei der Detonation.* In: *Probleme der Detonation.* Schriften Dt. Akad. Luftfahrtforsch. Nr. 1023/41G, 31-45 (1941).

[1763] R. ENGELKE and S.A. SHEFFIELD: *Explosives.* In: (G.L. TRIGG, ed.) *Encyclopedia of applied physics.* VCH, New York, vol. 6 (1997), 327-357; *see* pp. 335-336.

[1764] W. JOST, T. JUST, and H.G. WAGNER: *Investigation of the reaction zone of gaseous detonations.* In: *8th Symp. (Int.) on Combustion* [held at Pasadena, CA in Aug. 1960]. William & Wilkins, Baltimore, MD (1960), pp. 582-588.

[1765] T. JUST and H.G. WAGNER: *Die Reaktionszone in Gasdetonationen.* Z. Phys. Chem. **13**, 241-243 (1957); *Untersuchung der Reaktionszone von Detonationen in Knallgas.* Z. Elektrochem. **64**, 501-513 (1960).

[1766] R.J. SEEGER: *On MACH's curiosity about shock waves.* In: (R.S. COHEN and R.J. SEEGER, eds.) *Ernst MACH, physicist and philosopher.* Boston Studies Phil. Sci. **6**, 42-67 (1970).

| | | |
|---|---|---|
| 1942 | *Technische Akademie der Luftwaffe (TAL), Berlin-Gatow, Germany* | Hubert SCHARDIN,[1767] a German physicist, **presents a comprehensive review article on schlieren methods and their basic theory.** Numerous examples of applications to ballistics and shock waves illustrate the versatility of the classical schlieren method {A. TOEPLER ⇨1864; ⇨Fig. 4.6–I}. He describes twelve schlieren methods, of which four produce colored images:<br>▸ the simple background distortion method;<br>▸ the background grid distortion method;<br>▸ the large colored grid background method; and<br>▸ the lens-and-grid technique.<br>In later decades, new color schlieren methods for use in gas dynamics were invented, which have been reviewed more recently {SETTLES ⇨2001}. |
| 1942 | *Institut für Aerodynamik Aachen (IAA), TH Aachen, North-Rhine Westphalia, Germany* | Carl WIESELSBERGER,[1768] a German professor of aeronautical engineering and head of the famous Aachen Aerodynamic Institute, **discusses the influence of a test body in the high-speed flow of a wind tunnel.** He concludes that this would result in a velocity increase at the location of the model when the walls of the tunnel are closed, and a velocity decrease when the walls are open; *i.e.*, when a free jet is used. This phenomenon can falsify measurement data for the test model. ▪ Therefore, **WIESELSBERGER proposed a semi-open tunnel solution, which in effect is a slotted tunnel with two slots.**<br>The Italians had already succeeded in obtaining airfoil force data during the war in semi-open high-speed wind tunnels up to Mach 0.94, and the Germans up to about 0.92, thus anticipating the ***many-slotted transonic tunnel*** configuration realized in the 1940s at NACA by the U.S. engineers Ray H. WRIGHT, John STACK and associates {⇨Fig. 4.9–F}.[1769] |
| 1943 | *Manhattan Project, Los Alamos Scientific Laboratory (LASL), Los Alamos, New Mexico* | In April, **Robert SERBER**, a physicist at UC Berkeley and a member of the Manhattan District team, **delivers five top-secret lectures on how to build an atomic bomb – summarized in the brochure *Los Alamos Primer* – in order to introduce scientists arriving at Los Alamos to the current state of atomic research.** ▪ These lectures summarized all that was known at the time about designing and building an atomic bomb (such as the energy fission process, fast neutron chain reactions, fission cross-sections, neutron spectrum, number and capture, bomb material properties, the effect of a tamper and its efficiency, probability of predetonation, detonation sources, and problems related to mechanical shooting). The classified lectures, which also contained crude sketches of a possible future bomb design {⇨Fig. 4.16–G}, were published for the first time in 1992.[1770]<br>In the same year, Captain William Sterling ("Deke") PARSONS, an ordnance officer of the U.S. Navy, joins the Manhattan Project as leader of the Ordnance Division. His task is to helping to create "a perfectly functioning atomic bomb that could end the war." ▪ Since the uranium bomb was mostly an ordnance problem, he was particularly involved in the research and development of the gun assembly of LITTLE BOY {Hiroshima ⇨1945} which, because of its unstable design, was armed by himself in-flight in the bomb bay of the *Enola Gay*.[1771] |
| 1943 | *Washington, DC* | In December, at a NACA meeting, **leading American aerodynamic experts discuss how to provide aerospace companies with better information on high-speed flight in order to improve aircraft design.** A full-scale high-speed aircraft is proposed that would help to investi- |

---

[1767] H. SCHARDIN: *Die Schlierenverfahren und ihre Anwendungen*. Ergebn. exakt. Naturwiss. **20**, 303-349 (1942).
[1768] C. WIESELSBERGER: *Über den Einfluß der Windkanalbegrenzung auf den Widerstand insbesondere im Bereiche der kompressiblen Strömung*. Luftfahrtforsch. **19**, 124-128 (1942); published posthumously.
[1769] J.V. BECKER: *The high speed frontier*. Rept. NASA SP-445, Washington, DC (1980), chap. 2.
[1770] R. SERBER: *The Los Alamos primer* (with an introduction by R. RHODES). University of California Press, Berkeley *etc.* (1992). ▪ SERBER died in 1997 at 88, see obituary in: Reflections (LANL) **2**, No. 7, 5 (July 1997).
[1771] William "Deke" PARSONS, Rear Admiral (1901–1953); http://www.atomicarchive.com/Bios/Parsons.shtml.

gate difficult compressibility and control problems, power plant issues and the effects of higher Mach and Reynolds numbers. It is thought that a full-scale airplane with a trained pilot at the controls would yield more accurate data than could be obtained in a wind tunnel. • Discussions continued through 1944 and eventually led to the XS-1 Program {YEAGER ⇨1947}.

1943 *U.S. Ballistic Research Laboratory (BRL), Aberdeen Proving Ground, Maryland*

**Alexander Crane CHARTERS JR.,** a U.S. aeronautical engineer, while analyzing WOOD's soot experiments {WOOD ⇨1943}, **discovers *lines of discontinuity*** in Ernst MACH's "opposite V gliding spark" arrangement {E. MACH & WOSYKA ⇨1875} that separate areas of equal pressure but different densities – *i.e.*, **they are not true shock waves.**[1772] Furthermore, they represent discontinuity lines for entropy and temperature.[1773] • **They were later called "contact discontinuity lines"** {COURANT & FRIEDRICHS ⇨1948} **or "slipstreams"** {BLEAKNEY & TAUB ⇨1949}. Since flash radiography is sensitive to changes in density rather than to changes of pressure, these lines can be recorded by this method {⇨Fig. 4.13–M}.

1943 *Institute of Theoretical Physics, Reichsuniversität Posen, Germany*

**Werner DÖRING,**[1774, 1775] a German professor of theoretical physics, resuming previous work on detonation {Secret Workshop "Problems of Detonation" ⇨1941}, **presents a 1-D model of the detonation process** which, as he cautiously notes, "may be valid for detonations in gases with sufficiently slow reaction velocity." In his model, the detonation wave is identical to a shock wave which, by increasing the pressure and temperature, lowers the barrier to the reaction that exists under the initial conditions, thus provoking reaction. In the *zone of reaction* [Germ. *Umsetzungszone*] behind the shock wave, the laws of continuum physics are applied, and friction and heat conduction are neglected. Based upon this model, he develops his Theory of Detonation and obtains the following results:

(i) Equation-of-state data located along the Hugoniot curve below that point which corresponds to the smallest, normal detonation velocity cannot occur behind a stationary planar detonation wave, because this would require physically impossible stationary rarefaction in or behind the zone of reaction.

(ii) A planar detonation wave propagates after a certain distance with normal detonation velocity, which is also valid however small the reaction velocity is. General shock theory shows that an initially high detonation velocity is reduced by rarefaction waves and that an initially slow detonation wave is accelerated by compression waves.

(iii) He shows qualitatively that wall effects can produce a detonation wave which propagates at a slower than normal detonation velocity and no longer has a planar wavefront.

(iv) He also shows that wall effects can trigger pulsation of the detonation wave.

DÖRING's contribution to the modern theory of detonation, which he worked out independently of Yakov B. ZEL'DOVICH {⇨1940} in the Soviet Union and John VON NEUMANN {⇨1942} in the United States, has been acknowledged by calling their hypothesis the ***Zel'dovich-von Neumann-Döring (ZND) theory***. • It should be noted that DÖRING[1776] first reported on his detonation theory at a lecture delivered in 1941 in Berlin {Secret Workshop "Probleme der Detonation" ⇨1941}, and so this should actually be the ZDN theory, chronologically speaking!

---

[1772] R.J. SEEGER: *On MACH's curiosity about shock waves*. In: (R.S. COHEN and R.J. SEEGER, eds.) *Ernst MACH, physicist and philosopher*. Boston Studies Phil. Sci. **6**, 42-67 (1970), p. 63.

[1773] A.H. TAUB (ed.): *J. VON NEUMANN: Collected works*. Pergamon Press, Oxford *etc.* (1963), vol. VI, pp. 309-347.

[1774] W. DÖRING: Über den Detonationsvorgang in Gasen. Ann. Phys. **43** [V], 421-436 (1943).

[1775] W. DÖRING and G. BURKHARDT: *Beiträge zur Theorie der Detonation*. Zentrale für wissenschaftliches Berichtswesen (ZWB) der Luftfahrtforschung des Generalluftzeugmeisters, Berlin-Adlershof, Forschungsbericht Nr. 1939 (1944). • Engl. translation: *Contributions to the theory of detonation*. Tech. Rept. No F-TS-1226-IA (GDAM A9-T-45), Wright-Patterson Air Force Base, Dayton, OH (1949).

[1776] W. DÖRING: *Der Druckverlauf in den Schwaden und im umgebenden Medium bei der Detonation*. In: *Probleme der Detonation*. Schriften Dt. Akad. Luftfahrtforsch. Nr. 1023/41G, 31-45 (1941).

Theoretical detonation research performed up to 1939 was first summarized in Germany by Wilhelm JOST,[1777] then a professor at the Physical-Chemical Institute of the University of Leipzig. Research carried out in the United States during World War II was summarized by COLE {⇒1948} and COURANT & FRIEDRICHS {⇒1948}. Research on detonation carried out in the Soviet Union up to 1950 was briefly reviewed by ZEL'DOVICH[1778] and coworkers.

| 1943 | Applied Mathematics Group, Institute for Advanced Study (IAS), Princeton, New Jersey |

John VON NEUMANN,[1779] a Hungarian-born U.S. mathematician, **develops a "two-shock theory"** of *regular reflection* **and a "three-shock theory"** of *irregular reflection* **or** *Mach reflection*:

(i) He coins the term **Mach effect**[1780] to denote the three-shock configuration {E. MACH & WOSYKA ⇒1875; E. MACH & L. MACH ⇒1889}. Based on his two-shock theory, he proposes the so-called **"von Neumann criterion"** (or "detachment criterion") for the termination of regular reflection in a two-dimensional pseudo-stationary, inviscid gas flow. ▪ Independently of VON NEUMANN in the United States, Werner DÖRING,[1781] a German professor of theoretical physics at the Reichsuniversität Posen, calculated in the period 1943–1944 the oblique reflection of planar aerial shock waves at a rigid planar boundary. He determined the angle of reflection as a function of the angle of incidence and the incident pressure, and obtained very similar results to VON NEUMANN in his two-shock theory. DÖRING correctly concluded that complex solutions should correspond to the Mach effect, which had already been studied in Germany experimentally by that point using high-speed cinematography {CRANZ & SCHARDIN ⇒1929}.

(ii) The quantitative experimental evidence for VON NEUMANN's theory of oblique reflection of shock waves comes mainly from five sources which were all obtained in 1943:
- ballistic photographs taken at Aberdeen, MD;
- shock tube photographs taken at Princeton University, Princeton, NJ;
- supersonic wind tunnel photographs taken at Teddington, U.K.;
- WOOD's model shock interaction experiments made at Johns Hopkins University, MD {WOOD ⇒1943}; and
- open-air shock wave experiments {REYNOLDS ⇒1943}.

It is not clear whether VON NEUMANN was also aided by the French Colonel Paul LIBESSART {⇒1944}, who studied Mach reflection experimentally in England at that time.

(iii) **VON NEUMANN also predicts the existence of Mach reflection in water.** He writes, "The theory permits us to derive for any substance – that is, for any 'caloric equation of state' in the sense of (5) in paragraph 11 – the necessary criteria for the classifications regular and Mach, increasing and decreasing strength in the regular case, extraordinary and ordinary type in the Mach case … Experiments in water would be particularly important, and it is hoped that they will be undertaken shortly." ▪ In the very same year, shock reflection experiments in water proved that Mach reflection does indeed also exist in water {SPITZER & PRICE ⇒1943}.

The Mach effect was actually used when the first atomic bombs were dropped on two Japanese cities {Hiroshima & Nagasaki Bombing ⇒1945} in order to determine the appropriate altitude for the bomb explosion in terms of achieving optimum (*i.e.*, maximum) damage to

---

[1777] W. JOST: *Explosions- und Verbrennungsvorgänge in Gasen*. J. Springer, Berlin (1939); Engl. translation by H.O. CROFT: *Explosion and combustion processes in gases*. McGraw-Hill, New York & London (1946).
[1778] Y.T. GERSHANIK, Y.B. ZEL'DOVICH, and A.I. ROZLOVSKII: *Adiabatic combustion processes of inflammable gas mixtures* [in Russ.]. Zh. Fiz. Khimii (SSSR) **24**, 85-95 (1950).
[1779] J. VON NEUMANN: *Oblique reflection of shocks*. Navy Dept., Bureau of Ordnance, Explosives Res. Rept. No. 12, Washington, DC (Oct. 1943).
[1780] The term *Mach effet* is also used in physiology to designate the phenomenon of *Mach bands*. In fluid dynamics, the terms *Mach effect* and *Mach reflection effect* are used likewise.
[1781] W. DÖRING and G. BURKHARDT: *Beiträge zur Theorie der Detonation*. Zentrale für wiss. Berichtswesen (ZWB) der Luftfahrtforschung des Generalluftzeugmeisters, Berlin-Adlershof, Forschungsbericht Nr. 1939 (1944), pp. 120-130.

the target on the ground – the **optimum HOB ("Height of Burst")**.[1782] Later, the following definitions were given by the U.S. Dept. of Defense (DOD) and NATO[1783]

> ▸ for the *height of burst:* "The vertical distance from the Earth's surface or target to the point of burst. Also called '**HOB;**'" and
>
> ▸ for the *optimum height of burst:* "For nuclear weapons and for a particular target (or area), the height at which it is estimated a weapon of a specified energy yield will produce a certain desired effect over the maximum possible area."

For targets with a complex structure there is no single HOB for any specified explosion yield in terms of blast effects. Samuel GLASSTONE,[1784] an international authority on nuclear blast effects, commented that, "As a rule, strong (or hard) targets will require the equivalent of a low air burst or a surface burst. For weaker targets, which are destroyed or damaged at relatively low overpressures or dynamic pressures, the height of burst may be raised to increase the damaged areas, since the required pressures will extend to a larger range than for a low air or surface burst."

The large amount of experimental data on Mach reflection obtained since then have revealed several significant discrepancies between experimental data and VON NEUMANN's classical theory of shock reflection.[1785]

> ▸ The U.S. physicist Lincoln G. SMITH {⇨1945} first found that regular reflection exists beyond the limit of VON NEUMANN's detachment criterion, and Mach reflection does not occur immediately when the theoretical regular reflection limit is exceeded – the so-called **"von Neumann paradox,"** also known as the "Birkhoff triple shock paradox" {BIRKHOFF ⇨1950}.
>
> ▸ Another, different criterion for the termination of regular reflection was derived by Le Roy F. HENDERSON[1786] and Andrei LOZZI, two Australian fluid dynamicists at the University of Sydney. Their "mechanical equilibrium criterion" is based on the three-shock theory as well.
>
> ▸ Hans G. HORNUNG[1787] (a U.S. fluid dynamicist at the Australian National University in Canberra) and his collaborators explained discrepancies between theory and experiment on the basis of the viscous boundary layer displacement thickness at the reflection point on the wedge surface.

| 1943 | Princeton University Station, Physics Dept., Princeton University, Princeton, New Jersey | George T. REYNOLDS,[1788] a young physicist in Prof. Walker BLEAKNEY's group, uses a pressurized pot terminated by a diaphragm which, when pierced, **produces a steep pressure pulse in a gas with a rise time of only a few nanoseconds.** His device, a kind of short shock tube {BLEAKNEY ⇨1946}, is very useful **for calibrating piezoelectric (tourmaline) gauges.** • Since then, the piezoelectric gauge has become the standard pressure gauge in most shock tube facilities and is routinely used to measure shock front positions and pressure-time profiles "head-on" and "side-on."<br><br>Asked by John VON NEUMANN, REYNOLDS applied the tourmaline gauge to study regular and Mach reflection of shock waves in air. Using a single-sweep oscilloscope, they observed that gauges positioned above ground at the height of a small explosive charge detected **two** |

---

[1782] F. REINES and J. VON NEUMANN: *The Mach effect and height of burst.* In: (A.H. TAUB, ed.) *J. VON NEUMANN. Collected works.* Pergamon Press, Oxford *etc.* (1963), vol. VI, pp. 309-347.

[1783] *Dept. of Defense dictionary of military and associated terms.* Joint Publication 1-02 (April 2001); http://www.asafm.army.mil/pubs/jp1-02/jp1-02.pdf.

[1784] S. GLASSTONE: *The effects of nuclear weapons.* U.S. Govt. Print. Office, Washington, DC (April 1962), § 3.28, pp. 114-115.

[1785] R.J. VIRGONA, L.F. HENDERSON, H. HONMA, and D.Q. XU: *Discrepancies between theory and experiment for shock reflection.* In: (K. TAKAYAMA, ed.) *Proc. 18th Int. Symposium on Shock Waves* [Sendai, Japan, July 1991]. Springer, Berlin *etc.* (1992), pp. 189-192.

[1786] L.F. HENDERSON and A. LOZZI: *Experiments on transition of Mach reflection.* J. Fluid Mech. **68**, 139-155 (1975).

[1787] H.G. HORNUNG, H. OERTEL, and R.J. SANDEMAN: *Transition to Mach reflection of shock waves in steady and pseudo-steady flow with and without relaxation.* J. Fluid Mech. **90**, Part 3, 541-560 (1979).

[1788] G.T. REYNOLDS: *A preliminary study of plane shock waves formed by bursting diaphragms in a tube.* Rept. OSRD-1519 (1943).

shocks (resulting from the incident shock and reflected shock), but when they were placed near the ground only **one shock** could be detected (resulting from the Mach stem) – thus confirming Mach's observation that at glancing incidence an irregular kind of shock reflection occurs, **"Mach reflection"** {VON NEUMANN ⇨ 1943}.[1789]

1943 — *Institut für Mechanik, TH Aachen, Nort-Rhine Westphalia, Germany*

Fritz SCHULTZ-GRUNOW,[1790] a German professor of mechanical engineering, **publishes a pioneering paper on shock wave propagation in ducts with a section where the cross-sectional area changes** (such as in a conical section). He correctly treats the flow up- and downstream of this section as an unsteady, one-dimensional flow using the theory of characteristics, but approximates the flow in the section as being quasi-one-dimensional and steady. ▪ His method of approximation was an important achievement that had many engineering applications in the pre-computer era. For example, it allowed the first determinations of the exhaust flows from internal combustion engines, the flows in diffusers of shock tunnels, and those in converging/diverging nozzles such as those used in rockets and jet propulsion engines.

Today the flow through ducts with inserted area change segments can easily be handled as a true two- or three-dimensional unsteady flow using numerical methods.[1791]

1943 — *Underwater Explosives Research Laboratory, Woods Hole, Massachusetts*

Ralph W. SPITZER,[1792] a physical chemist, **and Robert S. PRICE,** a chemical engineer, **study interference effects of spherical shock waves resulting from two underwater detonations of small charges. Using single-shot photography, they first prove that Mach reflection also occurs in water** {⇨Fig. 4.13–G}, thus confirming John VON NEUMANN's hypothesis {VON NEUMANN ⇨ 1943}. However, they cannot detect any slipstreams {CHARTERS ⇨ 1943} which is possibly due to the fact that the pressure levels of the intersecting shock waves are too low. In the same year, D.C. CAMPBELL[1793] at David Taylor Model Basin photographs intersecting shock waves in water.

In the following year, **Duncan P. MACDOUGALL,**[1794] a physical chemist, **and his collaborators reconfirmed the existence of Mach reflection in water** using two cylindrical sticks of pentolite (50/50 TNT/PETN) inclined at an angle to form a V-geometry. They were the first to observe that slipstreams are produced behind a Mach disk in water too. ▪ The early history of Mach reflection in water was first reviewed by Robert H. COLE.[1795] Further work performed up to 1977 in this specific area of shock wave physics has been cited elsewhere {KREHL, HORNEMANN & HEILIG ⇨ 1977}.

1943 — *Johns Hopkins University, Baltimore, Maryland*

Robert W. WOOD,[1796] a U.S. professor emeritus of physics, **repeats Ernst MACH's historic soot experiments** {E. MACH & WOSYKA ⇨ 1875} having been asked about Mach reflection by John VON NEUMANN {VON NEUMANN ⇨ 1943}. **WOOD's results fully confirm the existence of Mach disk or Mach bridge formation.** ▪ Raymond J. SEEGER,[1797] a U.S. physicist at the U.S. National Science Foundation who reviewed Ernst MACH's early shock wave experiments in 1970, showed photos of WOOD's soot plate experiments in his paper. In the reference sec-

---

[1789] R.J. EMRICH: *Walker BLEAKNEY and the development of the shock tube at Princeton.* Shock Waves **5**, 327-339 (1996).
[1790] F. SCHULTZ-GRUNOW: *Nichtstationäre, kugelsymmetrische Gasbewegung und nichtstationäre Gasströmung in Düsen und Diffusoren.* Ingenieur-Archiv **14**, 21-29 (1943).
[1791] O. IGRA, L. WANG, and J. FALCOVITZ: *Nonstationary compressible flow in ducts with varying cross-section.* J. Aerospace Engng. **212**, Part G, 225-243 (1998).
[1792] R.W. SPITZER and R.S. PRICE: *Photographs of intersecting shock waves (Mach effect).* Interim Rept. UE No. 16, NDRC Div. 8 (1943).
[1793] D.C. CAMPBELL: *Experiments in the production and photography of intersection underwater shock waves.* Progr. Rept. R-203, David Taylor Model Basin (1943).
[1794] D.P. MACDOUGALL, G.H. MESSERLY, and E.M. BOGGS: *Note on Mach reflection in water.* Interim Rept. UE No. 20, NDRC Div. 8 (1944).
[1795] R.H. COLE: *Underwater explosions.* Dover Publ., New York (1948), pp. 255-261.
[1796] R.W. WOOD: *On the interaction of shock waves.* Rept. OSRD-1996 (1943).
[1797] R.J. SEEGER: *On MACH's curiosity about shock waves.* In: (R.S. COHEN and R.J. SEEGER, eds.) *Ernst MACH, physicist and philosopher.* Boston Studies Phil. Sci. **6**, 42-67 (1970).

tion, he notes that "the photographs used in this article are from the residue of WOOD's scientific materials at the [Henry Augustus] Rowland Physics Laboratory, courtesy of J. WOSYKA MANDANSKY." Jaromir WOSYKA was the name of a student of MACH who participated in the discovery of "Mach reflection" in 1875. Since the name WOSYKA is a pretty rare one, this suggests that WOOD was somehow affiliated with WOSYKA, who may have demonstrated the soot recording technique to him – thus linking MACH's pioneering era in the 19th century to the atomic weapons era of the 1940s.[1798]

| | | |
|---|---|---|
| 1943/ 1944 | *Mount Wilson Observatory, San Gabriel Mountains, southern California* | **Carl Keenan SEYFERT**, a U.S. astronomer and National Research Fellow, **studies a series of about a dozen active spiral galaxies** which have barely perceivable arms and very small, exceptionally bright centers or nuclei – later termed *Seyfert galaxies.*<br>(i) His investigations reveal that the nuclei of these active galaxies<br>▸ contain a central concentration of hot, violent gas moving at high speeds, ranging from several hundred miles per second to several thousand miles per second;<br>▸ show broad emission lines; and<br>▸ are extremely strong sources of radio waves – an example being the galaxy M106 {COURTÈS & CRUVELLIER ⇨1961}, where the radio emission is believed to be synchrotron emission from the jet – and infrared energy.<br>(ii) He classifies the Seyfert galaxies depending upon whether their spectra show both narrow and broad emission lines (Type I), or only narrow lines (Type II).<br>Data obtained from studying Seyfert galaxies, appear to show that each galaxy contains a massive black hole within its nucleus. Surrounding this black hole is a disk of spiraling matter called an "accretion disk;" the material falling into the intense gravitational field of the black hole releases copious amounts of energy, sometimes in the form of powerful, highly collimated beams of matter and energy – "astrophysical jets." Around one percent of all spiral galaxies are thought to be Seyfert galaxies. **Some Seyfert galaxies** (such as the galaxy M87 or NGC 4486) **also emit unusually large quantities of X-rays, which suggests that extremely violent explosions occur in the nuclei of these galaxies** {COURTÈS & CRUVELLIER ⇨1961; Supermassive Black Hole ⇨1994}. |
| 1943/ 1944 | *Heeresversuchsanstalt(HVA) Peenemünde, Peenemünde-Ost, Baltic Sea & Kochel, Upper Bavaria, Germany* | **Siegfried F. ERDMANN**, head of the Basic Research Group, **performs the first hypersonic wind tunnel tests close to Mach 9** {⇨Fig. 4.9–H} by modifying the nozzle of the HVA's supersonic wind tunnel. • Peter WEGENER,[1799] a former colleague of ERDMANN who eye-witnessed some of the preparations for the experiment, later remembered, "ERDMANN prepared a final grand experiment before the last tunnel was to be dismantled. More or less on his own initiative, he wanted to achieve a flow of close to nine times the speed of sound. His design to achieve this goal was actually superior to the design plans of the new hypersonic tunnel at Kochel. He intended to replace the entire test section of the remaining large tunnel with a special box that could take high pressure and house a nozzle designed for $M = 8.8$. The design was based on experience that had served well up to $M = 4.4$. The resulting shape looked unusual. An extremely narrow slit at the throat – the narrowest part of the nozzle, where the speed of sound is attained – was followed by two symmetrical, sharply opening nozzle walls, again enclosed in parallel plate-glass sidewalls. ERDMANN correctly anticipated that a hypersonic nozzle must be fed with dry air at a substantially higher pressure than that of the atmosphere."<br>Shortly afterwards, the evacuation of the Peenemünde Supersonic Laboratory at the Baltic Sea to Kochel (alt. 605 m), a little town in the Bavarian Alps, begins. **Plans are devised to use the hydroelectric plant on the nearby Lake Walchensee** (alt. 802 m) **to directly provide the** |

---

[1798] WOOD's manuscripts (Ms. 96) are kept at the Milton S. Eisenhower Library, The Johns Hopkins University, Baltimore, MD. His soot records are referred to in "*Interactions of shock waves*" (Mimeograph). Glass slides: deposited smoke on glass."

[1799] P.P. WEGENER: *The Peenemünde wind tunnels. A memoir.* Yale University Press, New Haven, CT *etc.* (1996), pp. 69-70.

enormous power (about 60 MW) needed to operate a planned huge hypersonic wind tunnel ($1 \times 1$ m$^2$, $M = 10$). ▪ After the war, hypersonic wind tunnel studies were first resumed in the United States in various laboratories {BECKER ⇨1947; WEGENER ⇨1951}.

1944  Mt. Vesuvius, Gulf of Naples, central Italy

On March 22, **begin of explosive eruptions of the volcano,** accompanied by the formation of pyroclastic flow on the sides of the cone and the emission of two eruption plumes, thus forming a large crater about 300 m deep and 600 m across, damaging Naples and its outlying towns, and reburying the ruins of Pompeii under nearly a foot of ash (28 casualties).

**Some months before the eruption plans were discussed,** mostly by armchair strategists, **for triggering an eruption of Mt. Vesuvius by dropping bombs down the vent** in order to cause discomfort to the Axis troops then in its environs.[1800] Shortly before the volcano erupted the Allied forces had taken over, and were now in charge of disaster relief. The massive eruption on March 22nd caused significant damage to the USAF 340th bombardment group, all 88 B-25s were completely destroyed.[1801]

1944  Manhattan Project, Los Alamos Scientific Laboratory (LASL), Los Alamos, New Mexico

In May, **Geoffrey I. TAYLOR**, a British mathematical physicist and one of the major contributors to the Manhattan Project {⇨1942}, **gets involved in the implosion bomb program by first pointing out implosion instability problems.**

(i) He demonstrates the conditions under which the boundary between two materials is stable and unstable mathematically, thus adding his expertise to a problem that has become central to the implosion bomb design.

(ii) He shows that when a heavy (high-density) material is accelerated against a light (low-density) material, the boundary between the two becomes stable. Conversely, **when a light material impacts a heavy material, the boundary becomes unstable** and the two materials will mix in a manner that is not very amenable to calculations. In the case of the planned implosion bomb, the spherical high-explosive surface material is lighter than the tamper materials inside of the sphere, and the tamper materials are lighter than the fissile material (plutonium) at the center of the sphere.[1802]

(iii) He points out that in many of the applications being considered for the implosion phase of an atomic bomb explosion, there is in fact such instability that the nuclear explosion would start too soon.

G.I. TAYLOR[1803] didn't publish his results in the open literature until 1950. The so-called **"Rayleigh-Taylor (R-T) instability"** can be attributed to Lord RAYLEIGH,[1804] who studied the phenomena of a heavy fluid sitting above a lighter fluid in a gravitational field. The flow can then be described in terms of bubbles of light fluid rising into the heavier medium and spikes of heavy fluid dropping into the lighter medium. **R-T instabilities are ubiquitous, occurring in both man-made and natural phenomena:**

▸ A very large explosion, such as a nuclear explosion, momentarily produces a large mass of hot low-density gas near the ground which, due to an a R-T instability, results in the typical *mushroom cloud.*

▸ R-T instabilities have been observed in many settings, including not only in hydrodynamics but also in plasma physics and laser-driven spherical implosions of fusion pellets.

---

[1800] *Vesuvius wages war its own way.* In: The Washington Post, Washington, DC (April 9, 1944), p. B5.
[1801] *The Mount Vesuvius eruption of March 1944.* War Wings Art.com; http://www.warwingsart.com/12thAirForce/Vesuvius.html.
[1802] F.H. SHELTON: *Reflections of a nuclear weaponeer.* Shelton Enterprise Inc., Colorado Springs (1988), p. **1**:21.
[1803] G.I. TAYLOR: *The instability of liquid surfaces when accelerated in a direction perpendicular to their planes (Part I).* Proc. Roy. Soc. Lond. **A201**, 192-196 (1950); D.J. LEWIS: *Experiments (Part II).* Ibid. **A202**, 81-96 (1950).
[1804] Lord RAYLEIGH (J.W. STRUTT): *Investigation of the character of the equilibrium of an incompressible heavy fluid of variable density.* Proc. Lond. Math. Soc. **14**, 170-177 (1882/1883); *On the dynamics of revolving fluids.* Proc. Roy. Soc. Lond. **A93**, 148-154 (1917).

- In volcanology, R-T instabilities are possible in a solidifying lava lake and may also occur in a magma chamber, where the mushy layer develops from the roof.
- R-T instabilities are also common in supernova remnants in which hot gas from the explosion is slamming into the surrounding interstellar medium. They are especially obvious in the Crab Nebula, and give rise to the familiar clumpy appearance of material in these objects. Observations of the famous supernova SN 1987A {⇒ 1987} have strongly suggested the occurrence of large-scale mixing in the ejecta during the explosion.[1805]

In 1984, David H. SHARP,[1806] a Los Alamos theoretical physicist, provided a survey on R-T instabilities, describing the phenomenology that occurs at a Taylor unstable interface, and reviewing attempts to understand these phenomena quantitatively.

| | | |
|---|---|---|
| 1944 | *Los Alamos Scientific Laboratory (LASL), Los Alamos, New Mexico* | In the fall, **Otto Robert FRISCH,** an Austrian nuclear physicist participating in the design of the first atomic bomb {Manhattan Project ⇒ 1942}, **checks on neutron multiplication and considers how to determine the precise critical mass of uranium and plutonium.** ▪ These extremely dangerous studies involved working with near-critical assemblies of fissionable materials; *i.e.*, with amounts that are just small enough to avoid starting a chain reaction. He had already theoretically investigated such studies previously {FRISCH & MEITNER ⇒ 1939; FRISCH & PEIERLS ⇒ 1940}.<br><br>Rudolf E. PEIERLS,[1807] an Anglo-German nuclear physicist, later remembered: "FRISCH had a narrow escape when he was working with a bare mass of fissible material without a surrounding scatter. As he bent over the specimen, slow neutrons scattered back from his body added sufficiently to the multiplication factor to pass criticality. Only his alertness in observing the behavior of the indicator lights attached to the counters, and a quick reaction in pulling away part of the fissible material, stopped the chain reaction reaching a dangerous radiation level … The most daring experiment of this kind was the 'dragon,' so-called because it was like tickling the tail of a dragon. The idea, proposed by FRISCH, was to have a piece of fissible material which, because of a hole through its middle, was sub-critical. A plug, which fitted loosely in the hole, would make it supercritical. The plug was dropped through the hole from a suitable height, so that it passed through rapidly. The system would thus become supercritical for a brief interval, and during this time a chain reaction would develop but would not have time to build up to a dangerous level. FRISCH said he was somewhat surprised to be given approval for this experiment, which was carried out safely and provided valuable information." |
| 1944 | *U.S. Ballistic Research Laboratory (BRL), Aberdeen Proving Ground, Maryland* | In December, **the first large modern supersonic wind tunnel in the United States is set in operation.** The facility, which has a power of 13,000 hp, has a working section of $15 \times 20$ in.$^2$ ($38.1 \times 50.8$ cm$^2$), and was designed at the Guggenheim Aeronautical Laboratory of California Institute of Technology (GALCIT) by Theodore VON KÁRMÁN with the assistance of Allen E. PUCKETT, an aeronautical engineer.[1808] ▪ The famous astronomer Edwin P. HUBBLE temporarily acted as a director of this facility. |

---

[1805] T. KUMAGAI ET AL.: *Gamma rays, X-rays, and optical light from the cobalt and the neutron star in SN 1987A.* Astrophys. J. **345**, 412-422 (1989).
[1806] D.H. SHARP: *An overview of the Rayleigh-Taylor instability.* Physica **D12**, 3-10 (1984).
[1807] R. PEIERLS: *Otto Robert FRISCH.* Biogr. Mem. Natl. Acad. Sci. (U.S.A.) **27**, 283-306 (1981).
[1808] Th. VON KÁRMÁN: *The wind and beyond.* Little, Brown & Company, Boston & Toronto (1967), pp. 230-231.

| | | |
|---|---|---|
| 1944 | Manhattan Project, Los Alamos Scientific Laboratory (LASL), Los Alamos, New Mexico | Before the end of World War II, NACA had in operation – in addition to a modified version of John STACK's earlier blow-down wind tunnel – a $9 \times 9$-in.$^2$ ($22.9 \times 22.9$-cm$^2$) supersonic wind tunnel and a $1 \times 3$-ft$^2$ ($30.5 \times 91.4$-cm$^2$) supersonic wind tunnel.[1809]

**An electrical diagnostic technique – known as the "electric pin method" – is derived from suggestions made by several researchers,** including Otto FRISCH, Darol FROMAN, Rudolf PEIERLS, Ernest TITTERTON, Philip MOON, and Alvin GRAVES. In this technique, which is of great importance to the success of the nuclear bomb implosion assembly {Trinity Test ⇨1945}, metal pins are erected in the space near the imploding object and connected to electric circuits. As the implosion proceeds, each of these pins is struck, and an oscilloscope displays clearly related current bursts with time. ▪ By erecting a series of pins at the same distance from the imploding shell, it is also possible to determine the symmetry of collapse. **The electric pin method ultimately gave the most accurate implosion timing information and was particularly useful when studying detrimental jet formation in a nuclear bomb implosion** {G.I. TAYLOR ⇨1944}.[1810]

After World War II, the electric pin method was also successfully applied to measure the velocity ("velocity pins") of the plastic wave {PACK, EVANS & JAMES ⇨1948; MINSHALL ⇨1955}, as well as in early equation-of-state (EOS) studies {GORANSON ET AL. ⇨1955}. |
| 1944 | Buffalo, New York | Leroy R. CARL[1811] reports in a brief publication that **two brass discs have been welded together under high velocity impact by loading them by a high explosive.** Since his photomicrographs do not reveal melting at the bond line, he concludes that the weld was not a fusion weld but was instead formed by a solid-state mechanism. In a series of experiments he reproduces the "solid-state weld," as he calls it, which represents the **birth of *explosion welding*** (or *explosive welding*). ▪ Explosion welding, a process where metals are joined metallurgically, is also known as **"explosive bonding"** (or "explosive cladding"), and it can be used to join a wide variety of similar and even dissimilar metals. The controlled detonation of high explosives accelerates one or both of the constituent metals into each other, ejecting a plasma which removes impurities from the two metal surfaces in front of the collision point, leaving behind a clean metal for joining. The transition joint has a high mechanical strength and is ultra vacuum-tight, because the process creates mechanically interlocking surfaces {⇨Fig. 4.15–I}.

The textbook *Explosive Welding of Metals and its Application*, written by Sir Bernard CROSSLAND[1812] (one of the most distinguished university-based engineers in Britain and Ireland and an international authority on metal fatigue, exploding welding and friction welding) became the classic reference on this subject, and also addressed its bizarre history. Explosion welding was not actually used commercially until the 1960s. S.H. CARPENTER[1813] and R.H. WITTMANN reviewed developments in the theory and application of explosion welding up to 1975. Recently, the history of explosion welding was reviewed by George YOUNG,[1814] marketing director of Dynamic Materials Corporation (DMC) Clad Metal Group, Boulder, CO. |

---

[1809] H.L. DRYDEN: *Supersonic travel within the last two hundred years*. Sci. Month. **78**, 289-295 (May 1954). ▪ In the Appendix of his report *The high speed frontier* [Rept. NASA SP-445 (1980)], J.V. BECKER, covering the period 1923–1950, gave a survey of wind tunnels ($M > 0.9$) used in the United States and Europe, which does not conform with DRYDEN's presentation in terms of achievable maximum Mach numbers and cross-sections. However, it should be kept in mind that it was a common practice to experiment with different nozzle geometries, which changed the operational parameters within certain limits.

[1810] L. HODDESON, P.W. HENRIKSEN, R.A. MEADE, and C. WESTFALL: *Critical assembly*. Cambridge University Press, Cambridge (1993), p. 156.

[1811] L.R. CARL: *Brass welds, made by detonation impulse*. Metal Progress **46**, 102-103 (1944).

[1812] B. CROSSLAND: *Explosive welding of metals and its application* [with 266 refs.]. Clarendon Press, Oxford (1982), pp. 7-8.

[1813] S.H. CARPENTER and R.H. WITTMANN: *Explosion welding*. Annu. Rev. Mat. Sci. **5**, 177-199 (1975).

[1814] G. YOUNG: *Explosion welding, technical growth and commercial history*. Conference Paper publ. in: *Stainless Steel World America 2004 Conference* [Houston, TX (2004)]. KCI Publ. BV, Zutphen (2004), Paper P0439; http://www.dynamicmaterials.com/data/brochures/1-%20Young%20Paper%20on%20EXW%20History.pdf.

| 1944 | Los Alamos Scientific Laboratory (LASL), Los Alamos, New Mexico | **Roy W. GORANSON**, a former student of Prof. Percy W. BRIDGMAN, **initiates a program to determine equation-of-state (EOS) data for shock-compressed materials** – a subject of immediate interest to those involved in the proper design of nuclear weapons and their effects. It will also stimulate other laboratories in the United States and abroad to initiate similar research in shock wave physics on a large scale. **This work represents the birth of modern solid-state shock wave physics!** • It should be noted that some theoretical studies on the behavior of shock waves in solids had already been performed by early pioneers such as CHRISTOFFEL {⇒1877}, HUGONIOT[1815] (1889), DUHEM[1816] (1903), HADAMARD {⇒1903} and JOUGUET[1817] (1920), and work on plastic wave theory had already been done {B. HOPKINSON ⇒1905}. The long-planned systematic campaign at Los Alamos, which provided the first important experimental results, was efficiently supported by close cooperation with other national research laboratories and private research organizations.[1818, 1819] GORANSON's results, initially classified, were later published in the open literature {GORANSON ET AL. ⇒1955}. His paper became a classic in solid-state shock wave physics.<br><br>Independent of this work, shock compression methods were also developed in the former Soviet Union in the period 1947–1948.[1820, 1821] Research was undertaken by two groups there, both performing similar studies, one consisting of Lev V. AL'TSHULER, Veniamin A. TSUKERMAN, K.K. KRUPNIKOV and Samuil B. KORMER, and the other involving Filipp A. BAUM, Kirill P. STANYUKOVICH and Boris I. SEKHTER.[1822] |
|---|---|---|
| 1944 | Los Alamos Scientific Laboratory (LASL), Los Alamos, New Mexico | **Lawrence H. JOHNSTON**,[1823] a student of Prof. Luis ALVAREZ at the University of California Radiation Laboratory, **observes that the shock wave generated by an exploding wire can also produce detonation in PETN** {TOLLENS ⇒1871}, one of the most sensitive secondary explosives in which detonation cannot normally be effected by a heated wire. He even succeeds in triggering high explosives with a delay of less than a microsecond, which will be very important when constructing the ambitious implosion assembly of the first atomic bomb {Trinity Test ⇒1945}.[1824] • The ability of an exploding wire to initiate detonation in a secondary explosive had already been discovered previously in the Soviet Union {BELAJEV ⇒1938}.<br><br>The important discovery of this **"Exploding Bridge-Wire (EBW) detonator"** allowed the pill of primary explosive in the conventional detonator {NOBEL ⇒1863} to be replaced with a secondary explosive, thus substantially reducing the handling hazards. However, detonation is not immediately initiated by an EBW device, but rather after a build-up phase. In studies performed in the 1960s at LASL, this time delay was found to be on the order of 1 μs {⇒Fig. 4.17–Q}. The exploding bridge-wire detonator – also called a **"safety detonator"** – is widely used in explosive-operated missile and space vehicle technology. Another great advantage is the reduction in the time delay from milliseconds to microseconds or even below, which |

---

[1815] P.H. HUGONIOT: *Mémoire sur la propagation du mouvement dans les corps et plus spécialement dans les gaz parfaits. 1e Partie.* J. Ecole Polytech. (Paris) **57**, 3-97 (1887).

[1816] P.M.M. DUHEM: *Sur les théorèmes d'HUGONIOT, les lemmes de M. HADAMARD et la propagation des ondes dans les fluides visqueux.* C. R. Acad. Sci. Paris **132**, 1163-1167 (1901); *Sur les ondes longitudinales et transversales dans les fluides parfaits.* Ibid. **132**, 1303-1306 (1901).

[1817] J.C.E. JOUGUET: *Sur les ondes de choc dans les corps solides.* C. R. Acad. Sci. Paris **171**, 461-464 (1920); *Sur la célérité des ondes dans les solides élastiques.* Ibid. **171**, 512-515 (1920); *Sur la variation d'entropie dans les ondes de choc des solides élastiques.* Ibid. **171**, 789-791 (1920).

[1818] J.W. TAYLOR: *Thunder in the mountains.* In: (J.R. ASAY, R.A. GRAHAM, and G.K. STRAUB, eds.) *Proc. 3rd Conference on Shock Waves in Condensed Matter – 1983* [Santa Fe, NM, July 1983]. North-Holland, Amsterdam etc. (1984), pp. 3-15.

[1819] C.E. MORRIS: *Shock-wave equation-of-state studies at Los Alamos.* Shock Waves **1**, 213-222 (1991).

[1820] Y. KHARITON, K. ADAMSKII, and Y. SMIRNOV: *The way it was.* Bull. Atomic Scientists **52**, No. 6, 53-59 (1996).

[1821] L.V. AL'TSHULER, R.F. TRUNIN, V.D. URLIN, V.E. FORTOV, and A.I. FUNTIKOV: *Development of dynamic high-pressure techniques in Russia.* Phys. Uspekhi **42**, 261-280 (1999).

[1822] L.V. AL'TSHULER: *Use of shock waves in high pressure physics.* Sov. Phys. Uspekhi **8**, 52-91 (1965), p. 55.

[1823] L.H. JOHNSTON: *Electric initiator with exploding bridge wire.* U.S. Patent No. 3,040,660 (filed Nov. 8, 1944; applied June 26, 1962).

[1824] L. HODDESON, P.W. HENRIKSEN, R.A. MEADE, and C. WESTFALL: *Critical assembly.* Cambridge University Press, Cambridge (1993), pp. 169-173.

| | | |
|---|---|---|
| | | allows the exact synchronization of individual detonation fronts emerging from multi-explosive devices. |
| 1944 | The Leningrad Physico-Technical Institute, Leningrad, Soviet Union | Mark I. KORNFELD[1825] and L. SUVOROV, two Soviet fluid dynamicists, **first suggest that cavitation bubbles might collapse asymmetrically and produce a liquid jet which, when it hits a solid boundary, could produce damage.** ▪ The jet is formed by the inversion of one side of the cavity (*i.e.*, one side of the bubble is pulled back inside the bubble). The jet passes across the cavity and penetrates the far surface. If this asymmetrical collapse occurs near a solid surface and the jet forms in the direction of the solid surface, then there is a liquid/solid impact with the generation of a water hammer pressure {DEAR & FIELD ⇨ 1988}. |
| 1944 | Oxford Extra-Mural Unit (OEMU), England | Colonel Paul LIBESSART,[1826] a French engineer who joined the OEMU after the fall of France in the war, **studies the Mach effect in air using a particular spark photography technique** which enables successive shadowgraphs of a target to be taken at short intervals of time. ▪ His optical recording method, which apparently applies a Cranz-Schardin arrangement {CRANZ & SCHARDIN ⇨ 1929}, was also used at OEMU in studies on wound ballistics.<br><br>Among high-speed photographers, **LIBESSART became widely known as the inventor of a near-point high-intensity spark light source of coaxial design.**<br>▸ The classical Libessart spark is an ***open Libessart spark*** which uses the bright light emitted from the open end of a spark generated in a narrow channel of a dielectric material.[1827] This unique instrument allows one to make sharp, fully exposed shadowgraphs. Pulsed light sources of submicrosecond duration obtained using his principle are still in use in some modern wind tunnel and shock tube facilities.<br>▸ The principle of the point Libessart spark also proved most useful for generating light flashes of some 100 μs duration, which are required in all schlieren, shadowgraph and interferometer methods when the event has to be illuminated during a certain period of time in order to take a sequence of pictures with a high-speed cine camera. The ***confined Libessart spark*** is a classical Libessart spark confined end-on by a transparent window (such as Plexiglas). Advantageously, this produces a stable high-temperature plasma column which emits end-on a high-intensity light-time pulse of nearly rectangular shape and also avoids a flickering of the source size.[1828] |
| 1944 | Institute for Advanced Study (IAS), Princeton, New Jersey | John VON NEUMANN,[1829] a Hungarian-born U.S. mathematician, **proposes a new approach to the hydrodynamic shock problem.**[1830] Based on a simple pressure-density relationship proposed by RIEMANN {⇨ 1859}, he provides a computational procedure which he applies to the collision of shock waves as well as to rarefaction waves – also called **"negative shocks"** by Theodore VON KÁRMÁN. He finds that the particles acquire small oscillations that are superimposed on their true paths after passing through the location of the shock. He interprets this in terms of internal energy. ▪ His method was resumed in the following years by Hilda GEIRINGER[1831] and Richard VON MISES.[1832] |

---

[1825] M.I KORNFELD and L. SUVOROV: *On the destructive action of cavitation.* J. Appl. Phys. **15**, 495-506 (1944).

[1826] P. LIBESSART: *Spark photographs of the Mach effect.* Rept. RC-417 (May 1944). See Doc. No. HO 195/15/417, Public Record Office, Kew, U.K.

[1827] W.G. HYZER: *Engineering and scientific high-speed photography.* Macmillan, New York (1962), pp. 300-301.

[1828] P. KREHL and J.B. HAGELWEIDE: *Adjustable long duration high-intensity point light source.* Rev. Scient. Instrum. **52**, 863-868 (1981).

[1829] J. VON NEUMANN: *Proposal and analysis of a new numerical method for the treatment of hydrodynamical shock problems.* Repts. OSRD-3617 and NDRC-108, Appl. Math. Group, Institute for Advanced Study (IAS), Princeton, NJ (1944).

[1830] The hydrodynamic shock problem was defined as follows: find a solution, allowing discontinuous solutions, to the three conservation laws, the increasing-entropy law, and the ideal-gas law when given physically acceptable initial and boundary values. See D.L. HICKS: *The convergence of numerical solutions of hydrodynamic shock problems.* Rept. AD0849487, Air Force Weapons Laboratory, Kirtland AFB, NM (1969).

[1831] H. GEIRINGER: *On numerical methods in wave interaction problems.* Adv. Appl. Mech. **1**, 201-248 (1948).

[1832] R. VON MISES: *Mathematical theory of compressible fluid flow.* Academic Press, New York (1958), pp. 224-229, 481-482.

In June of the same year, VON NEUMANN designs an explosive lens implosion system for use in the first atomic bomb {Trinity Test ⇒1945}. James TUCK, a British physicist and director of the British delegation to the Manhattan Project {⇒1942}, came up with the critical idea of explosive lenses for detonation wave shaping.

| | | |
|---|---|---|
| 1944 | University of Göttingen, Germany | Klaus OSWATITSCH,[1833] an Austrian physicist and notable gas dynamicist, **performs the first theoretical and experimental studies to determine the factors influencing muzzle brake (or recoil) efficiency** by order of the Heereswaffenamt ("Army Ordnance Office") in Berlin. These brakes recover momentum from the exhausting propellant gases by deflecting the flow away from the direction of fire. However, they also significantly increase the blast overpressure behind the gun in the vicinity of crew members. ▪ This problem remained a permanent challenge to postwar designers of large-caliber cannons.[1834, 1835] |
| 1944 | Ballistic Research Laboratory (BRL), Aberdeen Proving Ground, Maryland | Llewellyn Hilleth THOMAS,[1836] a physicist and ballistician, **extends Richard A. BECKER's theory of the shock front** {BECKER ⇒1921} **by using variable coefficients of viscosity $\mu$ and heat conduction $k$, assuming that $\mu$ und $k$ are proportional to $T^{\frac{1}{2}}$ ($T$ = absolute temperature). He shows that all shocks in air aside from the very weakest are a few free paths thick, and that the Boltzmann equation** {BOLTZMANN ⇒1872} **can be applied to even for the most violent shocks.**<br><br>In the same year, **Robert Green SACHS,**[1837] a theoretical physicist at BRL, **proposes an extension of the Hopkinson scaling law** {B. HOPKINSON ⇒1915} to account for effects of altitude or other changes in ambient conditions on air blast waves – the so-called **"Sachs scaling law."** |
| 1944 | Institute for Advanced Study (IAS), Princeton, New Jersey | Hermann WEYL,[1838] a German-born U.S. mathematician, **presents a report to the Applied Mathematics Panel on the problem of the stability of shock profiles in an arbitrary fluid** {BETHE ⇒1942}. He investigates the question: what conditions are required for the equation of state of a fluid that allows shocks, with their distinctive qualitative features, to be produced? He shows that these conditions have some differential (local) structure and some global structure. In the second part of his paper, he investigates the structure of a shock layer whose width is of the same small order of magnitude as the heat conductivity and viscosity, and finds that the problem has a unique solution. |
| 1944/ 1945 | Gesellschaft für Gerätebau mbH, Klaist, Upper Bavaria, Germany | Joachim HÄNSLER,[1839] a German electrical engineer, **designs and builds the first electromagnetic railgun.** His 20-mm-caliber cannon has a length of two meters, is powered by car batteries, and is capable of accelerating 2-g aluminum cylinders to a maximum velocity of 1,080 m/s. He obtains a velocity of 1,210 m/s with a two-stage device. The new accelerator is apparently conceived as being a potential new type of weapon. ▪ His method was later resumed in Australia {RASHLEIGH & MARSHALL ⇒1978}. |

---

[1833] K. OSWATITSCH: *Flow research to improve the efficiency of muzzle brakes*. Rept. R1001. German Army Ordnance Office [*Heereswaffenamt*], Berlin (1944).

[1834] E.M. SCHMIDT: *Muzzle devices, a state-of-the-art survey*. Rept. No. 2276, BRL, Aberdeen Proving Ground, MD (Feb. 1973).

[1835] K.C. PHAN and J.L. STOLLERY: *The effect of suppressors and muzzle brakes on shock wave strength*. In: (R.D. ARCHER and B.E. MILTON, eds.) *Shock tubes and waves*. In: *Proc. 14th Int. Symposium on Shock Tubes and Waves* [Sydney, Australia, Aug. 1983]. New South Wales University Press, Kensington, N.S.W. (1983), pp. 519-526.

[1836] L.H. THOMAS: *Note on BECKER's theory of the shock front*. J. Chem. Phys. **12**, 449-453 (1944).

[1837] R.G. SACHS: *The dependence of blast on ambient pressure and temperature*. BRL Rept. 466, Aberdeen Proving Ground, MD (1944); *Some properties of very intense shock waves*. Phys. Rev. **69** [II], 514-515 (1946).

[1838] H. WEYL: *Shock waves in arbitrary fluids*. Comm. Pure Appl. Math. **2**, 103-122 (1949).

[1839] J. HÄNSLER: *Ein Beitrag zum Problem des elektrischen Geschützes; Versuche mit dem elektrischen Geschütz*. Versuchsberichte I-III, Gesellschaft für Gerätebau mbH, Klais, Oberbayern (Sept. 1944 – Jan. 1945). See also *Electric gun and power source*. Armour Research Foundation, Illinois Institute of Technology, Rept. No. 3 on Project No. 15-391-E (1946).

| 1945 | Trinity Site [ca. 60 miles northwest of the town of Alamogordo], Alamogordo Desert, New Mexico | On Monday, July 16, at about 5.30 A.M, the **Trinity Test – the ignition of the first nuclear fission bomb named "The Gadget"** {⇨Fig. 4.16–G} – is carried out successfully at an altitude (height of burst, HOB) of 100 ft (30.5 m).[1840] The bomb is an implosion-type weapon that uses lenses of high explosive to rapidly implode a hollow subcritical sphere of fissionable material (Pu-239) into a solid supercritical sphere {VON NEUMANN ⇨1944}.[1841] |
|---|---|---|

(i) **The nuclear detonation is caught on film by 37 motion picture cameras running at different speeds under the direction of Berlyn BRIXNER,** an expert on high-speed photography based in Los Alamos {⇨Fig. 4.16–H}. Cinematographic records are made by Fastax cameras {⇨Fig. 4.19–O} which can be operated at up to 10,000 frames per second. Pictures of the expansion of the shock wave are taken by cameras placed at stations located every half-mile from the explosion. Upon analysis of the results, a total yield of 19 kilotons of TNT equivalent will be found.

(ii) Measurements of the mass velocity are carried out using suspended Primacord {TOLLENS & WIEGAND ⇨1891} and magnesium flash powder.

(iii) The peak blast pressure is recorded using spring-loaded piston gauges.

(iv) The excess shock velocity in relation to the velocity of sound is measured with a moving-coil loudspeaker pickup.

(v) **The Italian-born U.S. physicist Enrico FERMI**[1842] **devises his own order-of-magnitude method for roughly determining the blast yield:** "About 40 seconds after the explosion the air blast reached me. I tried to estimate its strength by dropping from about six feet small pieces of paper before, during, and after passage of the blast wave. Since, at that time, there was no wind, I could observe very distinctly and actually measure the displacement of the pieces of paper that were in the process of falling while the blast was passing. The shift was about 2½ meters, which, at the time, I estimated to correspond to the blast that would be produced by 10,000 tons of TNT…"▪ Photographs of the bomb explosion released later that year showed the radius $R$ (in meters) of the blast wave as a function of time $t$ (in milliseconds), fitting with $t_0 = 1$ ms the approximation $R \approx 60(t/t_0)^{0.4}$ up to about $t = 100$ ms when the shock Mach number {ACKERET ⇨1928} fell to about unity. Combining this information with the Sedov-Taylor similarity solution {G.I. TAYLOR ⇨1941}, the Soviet physicist Leonid I. SEDOV and others were able to infer the total energy released, which was an official American secret at that time.[1843]

(vi) Kenneth GREISEN,[1844] a young nuclear phyicist, observes the propagation of the shock wave: "When the intensity of the light had diminished, I put away the glass and looked toward the tower directly. At about this time I noticed a blue color surrounding the smoke cloud. Then someone shouted that we should observe the shock wave traveling along the ground. The appearance of this was a brightly lighted circular area, near the ground, slowly spreading out towards us. The color was yellow. At what I presume was about 50 seconds after the shot, the ground shock and sound reached us almost simultaneously."

(vii) The Austrian-born theoretical physicist Victor WEISSKOPF,[1845] who had emigrated to the U.S.A., notices that "the path of the shock wave through the clouds was plainly visible as an expanding circle all over the sky where it was covered by clouds."

(viii) Seismographs and other devices are used to measure ground shock motions within 800 yards and 100 miles (0.732–161 km) from ground zero, because it is feared that the explo-

---

[1840] R. RHODES: *The making of the atomic bomb*. Simon & Schuster, New York (1986), pp. 617-678.
[1841] *Trinity test: historic documents, photos and video*. AJ Software & Multimedia (2005); http://www.trinityremembered.com/maps/index.html.
[1842] E. FERMI: *My observations during the explosion at Trinity on July 16, 1945*; http://www.trinityremembered.com/documents/Fermi.html.
[1843] R.D. BLANDFORD and K.S. THORNE: *Applications of classical physics*. CalTech, Pasadena, CA (Sept. 2004), chap. 16: *Compressible and supersonic flow*; http://www.pma.caltech.edu/Courses/ph136/yr2004/0416.1.K.pdf.
[1844] K. GREISEN: *My observations during the explosion at Trinity on July 16, 1945*; http://www.trinityremembered.com/documents/Greisen.html.
[1845] V. WEISSKOPF: *My observations during the explosion at Trinity on July 16, 1945*; http://www.trinityremembered.com/documents/Weisskopf.html.

sion might be noticed in neighboring towns and prompt lawsuits when the Manhattan Project becomes public knowledge.

Two days later, General Leslie R. GROVES, administrative head of the Manhattan Project, states in a memorandum to the Secretary of War about the Trinity Test:

(I) "The test was successful beyond the most optimistic expectations of anyone. Based on the data which it has been possible to work up to date, I estimate the energy generated to be in excess of the equivalent of 15,000 to 20,000 tons of TNT; and this is a conservative estimate. Data based on measurements which we have not yet been able to reconcile would make the energy release several times the conservative figure.

(II) **There were tremendous blast effects.** For a brief period there was a light intensity within a radius of 20 miles [32.2 km] equal to several midday suns; a huge ball of fire was formed which lasted for several seconds. This ball mushroomed and rose to a height of over ten thousand feet before it dimmed. The light from the explosion was seen clearly at Albuquerque, Santa Fe, Silver City, El Paso and other points generally to about 180 miles [290 km] away. **The sound was heard to the same distance in a few instances but generally to about 100 miles [161 km].** Only a few windows were broken although one was some 125 miles [201 km] away.

(III) A massive cloud was formed which surged and billowed upward with tremendous power, reaching the sub-stratosphere at an elevation of 41,000 feet [12,497 m], 36,000 feet [10,973 m] above the ground, in about five minutes, breaking without interruption through a temperature inversion at 17,000 feet [5,182 m] which most of the scientists thought would stop it. **Two supplementary explosions occurred in the cloud shortly after the main explosion.** The cloud contained several thousand tons of dust picked up from the ground and a considerable amount of iron in the gaseous form. Our present thought is that this iron ignited when it mixed with the oxygen in the air to cause these supplementary explosions. Huge concentrations of highly radioactive materials resulted from the fission and were contained in this cloud."[1846]

Later investigations showed that, depending on the height of burst, about 45–55% of the fission energy appeared as the blast and shock. Since the positive duration of a blast wave from a nuclear explosion is longer than that from a chemical one, damage effects due to air blast loading are more severe than hitherto observed from conventional explosions.

1945　*Hiroshima, southwestern Honshu & Nagasaki, northwestern Kyushu, Japan*

In August, **the U.S. Air Force drops two atomic bombs on Japan, in order to terminate the Japanese war in an unprecedentedly sudden manner.** (The war in Europe had ceased almost three months earlier.)

(i) On August 6 at 8:15 A.M., **the first atomic bomb is dropped on Hiroshima,** then the seventh largest city in Japan. The gun-assembly type uranium bomb {Manhattan Project ⇨1943} – codenamed LITTLE BOY – was not tested before it was dropped on Hiroshima. Its detonation is equivalent to about 15,000 tons of TNT (Height Of Burst, HOB ≈ 1,900 ft or 579 m). Most of the city is destroyed, and about 75,000 people killed or fatally injured.

(ii) On August 9 at 11:02 A.M., **the second atomic bomb is dropped on Nagasaki,** a major shipbuilding center. The detonation of the plutonium implosion-type bomb – codenamed FAT MAN and identical in design to "The Gadget" {Trinity Test ⇨1945} – is equivalent to about 21,000 tons of TNT (HOB ≈ 1,850 ft or 564 m). Most of the city is destroyed, and about 74,000 people are killed. • Later, the crew of the B-29 Superfortress (named *Enola Gay*) that dropped the Nagasaki bomb, having witnessed the explosion, contributed to the following official account[1847] of the first atomic air raid in history: "The flash after the explosion was deep purple,

---

[1846] *Memorandum for the Secretary of War (18 July 1945)*; http://www.trumanlibrary.org/whistlestop/study_collections/bomb/small/mb03.htm.
[1847] *Administrative history, history of 509th Composite Group, 313th Bombardment Wing, 20th U.S. Air Force, Activation to 15 August 1945.* • This information was provided to the author by the Smithsonian National Air & Space Museum, Washington, DC.

then reddish and reached to almost 8,000 feet [2,438 m]; the cloud, shaped like a mushroom, was up to 20,000 feet [6,096 m] in one minute, at which time the top part broke from the 'stem' and eventually reached 30,000 [9,144 m]. The stem of the mushroom-like column of smoke, looking now like a giant grave marker, stood one minute after the explosion upon the whole area of the city, excepting the southern dock area. This column was a thick white smoke, darker at the base, and interspersed with deep red. Though about fifteen miles (slant range) from the target when the explosion occurred, both escort aircraft, as well as the strike plane, reported feeling **two shock waves jar the aircraft.** Approximately 390 statute miles [628 km] away from the target area, the column of smoke still could be seen piercing the morning sky." The second shock felt by the crew of the *Enola Gay* was caused by the reflection of the primary shock from the ground. The crew of the other B-29 (named *Bock's Car*), which dropped the bomb over Nagasaki, noticed not two but **five shock waves,** which were caused by reflections of the primary shock wave from the mountains surrounding Nagasaki.[1848]

The precise yields of these two bomb explosions was difficult to gauge for these early bomb types and remained the subject of later discussions and investigations.[1849] For example, Sir William G. PENNEY[1850] and collaborators found that the Hiroshima explosion was $12 \pm 1$ kilotons and the Nagasaki explosion was $22 \pm 2$ kilotons of TNT equivalent. During the long Cold War that followed, a knowledge of the yields of such bombs was of particular interest when attempting to understand the mechanism of damage caused to a wide spectrum of civilian targets, the effects of bomb-related radiation on man, and when predicting damage scenarios in a the event of a nuclear war.

In November 1947, **Robert OPPENHEIMER**[1851] {Manhattan Project ⇒1942}, referring to Hiroshima and Nagasaki bombing, **said in a lecture presented at MIT: "In some sort of crude sense, which no vulgarity, no humor, no overstatement can quite extinguish, the physicists have known sin; and this is a knowledge which they cannot lose."**

| | | |
|---|---|---|
| 1945 | *Saint Louis, Alsace, Dépt. Haute Rhin, France* | In August, **the Laboratoire de Recherches Balistiques et Aérodynamiques de Saint-Louis is established.** The Laboratory, partly dedicted to military research, is directed by the French Ingénieur-Général Robert CASSAGNOU (administration) and the German professor Hubert SCHARDIN (science and technology). ▪ In 1951, the institute was renamed the **Laboratoire de Recherches Techniques de Saint-Louis (LRSL).** In 1959, the LRSL was transformed into a joint French-German research institute in order to promote the scientific cooperation between France and Germany, and renamed the **ISL** (short for the **Institut franco-allemand de recherches de Saint-Louis,** the **Deutsch-Französisches Forschungsinstitut Saint-Louis**).<br><br>Historical reviews on the research carried out at the ISL in ballistics and gas dynamics have been provided by Günter WEIHRAUCH,[1852] a German research engineer, and Herbert OERTEL,[1853] a German physicist and fluid dynamicist. |
| 1945 | *U.S. Air Force* | **First observation of the phenomenon of rain erosion.** U.S. Air Force personnel reports on the deterioration of a wing-shaped radar antenna installation on a Boeing B-29 aircraft ("Superfortress," max. flight speed 357 mph or 160 m/s; flight duration in rain unknown). This so- |

---

[1848] L. HODDESON, P.W. HENRIKSEN, R.A. MEADE, and C. WESTFALL: *Critical assembly.* Cambridge University Press, Cambridge (1993), p. 396.

[1849] W.G. PENNEY, D.E.J. SAMUELS, and G.C. SCORGIE: *The nuclear explosive yields at Hiroshima and Nagasaki.* Phil. Trans. Roy. Soc. Lond. **A266**, 357-424 (1970).

[1850] J. MALIK: *The yields of the Hiroshima and Nagasaki nuclear explosions.* Rept. LA-8819, LANL, Los Alamos, NM (1985).

[1851] J.R. OPPENHEIMER: *Physics in the contemporary world* [Lecture No. 25]. Technology Rev. **50** (1948). ▪ This remark became famous when it was quoted in *Time* (on Feb. 23, 1948 and on Nov. 8, 1948).

[1852] *Ballistische Forschung im ISL von 1945 bis 1994* (ed. by G. WEIHRAUCH). ISL, Saint-Louis, France (1994).

[1853] H. OERTEL: *33 years of research by means of shock tubes at the French-German Research Institute at Saint-Louis.* Proc. 14th Int. Symposium on Shock Tubes [Sydney, Australia, Aug. 1983]. In: (R.D. ARCHER and B.E. MILTON, eds.) *Sydney Shock Tube Symposium.* University of New South Wales Press, Sydney (1984), pp. 3-13.

called "eagle wing antenna" was made of plastic reinforced with glass fabric. ▪ While the phenomenon had also occurred at much lower speeds, it had been attributed to other causes such as erosion by sand or dust, or the impacts of stones during takeoff or landing.[1854]

The severity of rain erosion damage increased with the advent of higher flight speeds, and the extent of the erosion increased with the widespread use of nonmetallic materials in aircraft construction {Int. Conference on Rain Erosion ⇨1965}.

| | | |
|---|---|---|
| 1945 | Engineering Committee of the Red Army, Institute of Physical Problems, Academy of Sciences of the U.S.S.R., Soviet Union | Lev D. LANDAU[1855] and Kirill P. STANYUKOVICH, two Soviet physicists, **study the equation of state for detonation products.** Starting from the covolume equation of state, $$p(V-\alpha) = RT,$$ they turn to the concept of a barotropic relationship $$pV^k = \text{const}$$ for the adiabatic expansion of detonation products from the Chapman-Jouguet (CJ) state, and suggest a value of $k = 3$ as a first approximation. ▪ Their hypothesis was confirmed experimentally for most other relevant high explosives (*e.g.*, $k = 3.12$ for TNT, $k = 2.9$ for PETN, and $k = 3.1$ for HMX). **A very useful consequence of the so-called "Landau-Stanyukovich polytrope" – given by $pV^k = \text{const}$, with $k = 3$ – is that the coefficient $2/(k-1)$ in the Riemann invariant becomes equal to one,** which provides a series of analytical solutions for one-dimensional flows of detonation products. These solutions are the foundations of modern detonation physics. |

Similar studies on the equation of state of detonation products were also being carried out at the same time at Los Alamos by Dennison BANCROFT[1856] and Roy W. GORANSON.

| | | |
|---|---|---|
| 1945 | Palmer Physical Laboratory, Princeton University, Princeton, New Jersey | Lincoln G. SMITH,[1857] a U.S. experimental physicist, **uses a shock tube and photographs the oblique reflection of planar shocks in air at a rigid planar wall,** thus providing the first quantitative information about the validity of VON NEUMANN's two- and three-shock solutions {VON NEUMANN ⇨1943}:<br><br>(i) **He discovers that, contrary to the reflection of a sound wave, a shock wave reflects at a larger angle than the angle of incidence.** At large shock strengths, the Mach reflection begins at angles close to those at which the theory says regular reflection is not possible. For weak shocks, regular reflection continues to be seen at large angles of incidence that are (at the time) thought to be impossible from theory.[1858] ▪ Later, his observed discrepancy was named the *von Neumann paradox* {BIRKHOFF ⇨1950}.<br><br>(ii) SMITH's shadow and schlieren pictures also show a reverse curvature (kink) of the reflected shock near the triple point – a wave configuration which has been named *complex Mach reflection.* ▪ As the strength of the incident shock wave increases, complex Mach reflection turns into double Mach reflection {WHITE ⇨1951}; *i.e.*, complex Mach reflection can be regarded as a transient state between single and double Mach reflection. This is the reason why complex Mach reflection was later appropriately renamed *transitional Mach reflection.* The different types of Mach reflection are illustrated schematically in Fig. 2.14. |

---

[1854] A.A. FYALL: *Practical aspects of rain erosion of aircraft and missiles.* Phil. Trans. Roy. Soc. Lond. **A260**, 161-167 (1966).

[1855] L.D. LANDAU and K.P. STANYUKOVICH: *On a study of detonations of condensed substances* [in Russ.]. Dokl. AN (SSSR) **46**, No. 9, 362-364 (1945).

[1856] D. BANCROFT and R.W. GORANSON: *A method for determining equations of state and reaction zones in detonation of high explosives, and its application to pentolite, composition-B, baratol, and TNT.* Class. Rept. LA-487, Los Alamos, NM (Jan. 11, 1946).

[1857] L.G. SMITH: *Photographic investigation of the reflection of plane shocks in air.* Rept. OSRD-6271, Palmer Physical Laboratory, Princeton University (1945). ▪ Today, almost 60 years after its preparation, his report is apparently still classified. Only a short abstract of his work had been published in Phys. Rev. **69** [II], 678 (1946).

[1858] P. COLELLA and L.F. HENDERSON: *The VON NEUMANN paradox for the diffraction of weak shock waves.* In: (H. REICHENBACH, ed.) *Proc. 9th Int. Mach Reflection Symposium* [EMI, Freiburg, Germany, June 1990]. Ernst Mach-Institut, Freiburg (1990).

| | | |
|---|---|---|
| | | In the same year, **John VON NEUMANN**[1859] treats various shock wave interaction phenomena analytically. In the case of Mach reflection, **he coins the term for the new shock wave, the *Mach stem*** – a merging of the reflected shock with the incident shock in the vicinity of the reflecting wall. |
| 1946 | *Unimak Island, Aleutian Islands, Alaska* | On April 1, **an earthquake triggers one of the largest trans-Pacific tsunamis known, which produces considerable damage and large numbers of casualties in remote parts of the Pacific Ocean.** For example, the tsunami exceeds 30 meters in height on Unimak Island, and strikes the Hawaiian Islands with a runup height of more than 16 meters, causing there great damage in Hilo and killing 159 persons {⇨Fig. 4.1–R}.[1860] ▪ **The 1946 Aleutian Earthquake was a typical "tsunami earthquake," in that it generated larger tsunamis than expected from its seismic waves.** The puzzling mechanism of the 1946 Aleutian Earthquake is still subject to debate and has been investigated via computer-supported numerical analyses.[1861] |

> In 1949, the U.S. Coast and Geodetic Survey (USCGS) established the Pacific Tsunami Warning Center (PTWC) on Hawaii.[1862] This Center not only warns of local and regional tsunamis in Hawaii, but also provides teletsunami warnings for most countries in the Pacific Basin as well as Hawaii and all other U.S. interests in the Pacific outside of Alaska and the U.S. West Coast. Those areas are served by the West Coast & Alaska Tsunami Warning Center (WC/ATWC).[1863]

> In 1965, due to the efforts of the UNESCO Intergovernmental Oceanographic Commission (UNESCO/IOC), the U.S. National Tsunami Warning Center in Honolulu was expanded and became the International Tsunami Information Center {ITIC ⇨1965}.

> In 1968, the International Coordination Group for the Tsunami Warning System in the Pacific (ICG/ITSU[1864]) was established by UNESCO/IOC to provide information on tsunamis in the Pacific.

> In 1997, CREST (Consolidated Reporting of Earthquakes and Tsunamis),[1865] an improved system for providing tsunami warnings, was implemented on the West Coast of the United States up to Alaska, and on Hawaii by the USGS, NOAA, the Pacific Northwest Seismograph Network, and three other university networks. The goal of this project was to reduce the time needed to issue a tsunami warning by providing the warning centers (such as PTWC and WC/ATWC) with high-dynamic-range, broadband waveforms in near real time.

> The 2004 Indian Ocean Tsunami {Sumatra-Andaman Islands ⇨2004} brought the urgent need to be better prepared for such events to the world's attention. In 2005, it initiated the establishment of the Intergovernmental Coordination Group for the Indian Ocean Tsunami Warning and Mitigation System (ICG/IOTWS).

---

[1859] J. VON NEUMANN: *Refraction, intersection and reflection of shock waves.* NavOrd Rept. No. 203-45, Navy Dept. Bur. Ord., Washington, DC (July 16, 1945); In: (A.H. TAUB, ed.) *John VON NEUMANN. Collected works.* Pergamon Press, Oxford *etc.* (1963), vol. VI, pp. 300-308.

[1860] T.A. JAGGAR: *The great tidal wave of 1946.* Natl. History **55**, No. 6, 263-268 (1946).

[1861] Y. TANIOKA and T. SENO: *Detailed analysis of tsunami waveforms generated by the 1946 Aleutian Tsunami Earthquake.* Natl. Haz. Earth Syst. Sci. **1**, 171-175 (2001).

[1862] *Pacific Tsunami Warning Center (PTWC), Ewa Beach, Hawaii.* National Oceanographic & Atmospheric Administration (NOAA), Washington, DC; http://www.prh.noaa.gov/pr/ptwc/.

[1863] *West Coast & Alaska Tsunami Warning Center (WCATWC), Palmer, Alaska.* NOAA; http://wcatwc.arh.noaa.gov/.

[1864] It appears that ITSU is not a strict acronym but instead stands for "International TSUnami." Chip MCCREERY, Director of PTWC, Ewa Beach, Hawaii, private communication (Aug. 26, 2006). For abbreviations, *see also* the *IOC Tsunami glossary*; http://ioc3.unesco.org/itic/contents.php?id=24.

[1865] D. OPPENHEIMER ET AL.: *The CREST project: consolidated reporting of earthquakes and tsunamis.* Proc. Int. Tsunami Symposium (2001), National Tsunami Hazard Mitigation Program Session, Paper R-5 (2002).

| | | |
|---|---|---|
| 1946 | *Arzamas-16 [ca. 400 km east of Moscow], Soviet Union* | On April 13, **establishment of KB-11 (Design Bureau 11), the Soviet Union's nuclear weapons research and development facility at Arzamas-16.** It is located about 60 km north of the city of Arzamas. The first laboratories are set up in the buildings of the Sarovskaya Pustyn monastery. ▪ Later on, KB-11 became the All-Union Research Institute of Experimental Physics (VNIIEF). In 1990, the Institute was renamed the All-Russian Scientific Research Institute of Experimental Physics (VNIIEF). In November 1992, an agreement was signed to create joint ventures between VNIIEF and LANL, particularly in the areas of pulsed power and high magnetic fields. In 1994, Arzamas-16 was renamed Kremlev, and a Sister Cities relationship was created between Los Alamos and Kremlev. In 1995, the Russian President Boris N. YELTSIN signed a law which officially changed the city's name to Sarov – its original name.<br><br>The first Soviet atomic bomb was developed at Arzamas-16 in the late 1940s, after which it continued as a center for nuclear weapons research.[1866] Most of the pioneering Soviet shock wave studies were also performed at VNIIEF. In 1996, Vladimir A. BELUGIN,[1867] director of VNIIEF, reviewed the 50-year history of the Institute. |
| 1946 | *Bikini Atoll, Marshall Islands [part of UN trusteeship under U.S. jurisdiction], Micronesia, western Pacific* | On July 1, **Operation CROSSROADS – the first series of American underwater nuclear weapons tests – begins.** Both weapons used are FAT MAN-type atomic bombs with a yield of 23 kt of TNT. The operation, consisting of Test ABLE (July 1, 1946) and Test BAKER (July 25, 1946), it is carried out in order to study the effects of nuclear weapons on ships, equipment, and material. In particular, in order to study damage scenarios on warships, a target fleet consisting of nearly 100 vessels of different sizes up to the size of a battleship (the USS *Saratoga*) is positioned close to the center of the explosion.<br><br>(i) In Test ABLE the bomb detonated at an altitude of 520 ft (160 m); history's 4th atomic explosion.<br><br>(ii) In Test BAKER, a plutonium bomb is dropped from a B-29 bomber.[1868] The nuclear explosion is initiated 90 ft (27.5 m) beneath the surface of the lagoon.<br><br>▸ From a distance of 15 nautical miles (27.8 km), BAKER {⇨Fig. 4.16–I} was described as follows: "The flash seemed to spring from all parts of the target fleet at once. A gigantic flash – then it was gone. And where it had been now stood a white chimney of water reaching up and up. Then a huge hemispheric mushroom of vapor appeared like a parachute suddenly opening … By this time, the great geyser had climbed to several thousand feet. It stood there as if solidifying for many seconds, its head enshrouded in a tumult of steam. Then, slowly, the pillar began to fall and break up. At its base a tidal wave of spray and steam rose to smother the fleet and move on toward the islands."<br><br>▸ **The underwater shock wave spread throughout the lagoon at 1.55 km/s as a shock disk that was clearly visible in aerial photographs.** The fireball bubble reached the surface of the lagoon within a few milliseconds. As the steam and fission products hit the atmosphere, a huge dome-shaped condensation cloud flashed into view, shooting upward at 3.2 km/s. The dome was lit from within by the explosion and burst in another hundredth of a millisecond. 400 million kilograms of water were ejected as a hollow column about a kilometer in diameter. The explosion column consisted of water droplets and gases mixed as an aerosol. **When the aerosol column collapsed back to the surface of the lagoon, it created a hurricane-velocity wave called a "base surge."**[1869]<br><br>The two tests of Operation CROSSROADS, which was preceded by a similar nuclear test of the same yield but with the bomb timed to explode above the ocean surface, revived interest in |

---

[1866] V. TSUKERMAN and A. ZINAIDA: *Arzamas-16: Soviet scientists in the nuclear age.* Bramcote, Nottingham (1999).
[1867] V.A. BELUGIN: *50 years serving motherland.* VNIIEF, Sarov, Russia (June 26, 1996); http://www.lanl.gov/orgs/pa/Director/belugin.html.
[1868] F.H. SHELTON: *Reflections of a nuclear weaponeer.* Shelton Enterprise Inc., Colorado Springs, CO (1988), pp. **2**:44-48.
[1869] R.V. FISHER: *Operation CROSSROADS.* Dept. of Geology, UC Santa Barbara; http://volcanology.geol.ucsb.edu/bikini.htm.

World War II underwater explosion research. The various new effects observed demonstrated that a specific knowledge of underwater explosion phenomena is required when applying nuclear weapons underwater.

| 1946 | *Office of Scientific Research and Development (OSRD), Washington, DC* | In September, **publication of the National Defense Research Committee (NDRC) report entitled *Effects of Impact and Explosion,*** edited by Merit P. WHITE,[1870] a U.S. mechanical engineer and bomb damage analyst. The report is mostly concerned with providing information, and not with the development or improvement of devices. It discusses the following major topics:<br>▸ underwater explosives and explosions (with 147 refs.);<br>▸ explosions and explosives in air (230 refs.);<br>▸ explosions in earth (28 refs.);<br>▸ the muzzle blast, its characteristics, effects and control (36 refs.);<br>▸ fundamentals of terminal ballistics (no refs.);<br>▸ terminal ballistics of armor protection (71 refs.), concrete protection (58 refs.), soil protection (10 refs.), and plastic protection (3 refs.);<br>▸ model supersonic wind tunnel design (4 refs.);<br>▸ the behavior of materials under dynamic loads (68 refs.);<br>▸ defense against shaped charges (33 refs.);<br>▸ structural protection (37 refs.); and<br>▸ target analysis and weapon selection (15 refs.). |
|---|---|---|
| 1946 | *Palmer Physics Laboratory, Princeton University, Princeton, New Jersey* | Walker BLEAKNEY,[1871] a U.S. physicist and group leader in military research, **publishes the first results in the open literature on the propagation of shock waves in a tube.** His *shock tube* {⇨Fig. 4.10–B} – a term which he will coin three years later {BLEAKNEY ⇨1949} – is essentially based on Paul VIEILLE's concept {⇨1899}, but the basic concept is improved technically and supplemented with modern high-speed diagnostics. ▪ The development of the shock tube was a significant advance in experimental gas dynamics. BLEAKNEY's numerous collaborators at Princeton – such as George T. REYNOLDS, Lincoln G. SMITH, Charles H. FLETCHER, Wayland C. GRIFFITH, Richard G. STONER, and David K. WEIMER, to name just a few – made valuable contributions, which were mostly published as classified OSRD Reports {REYNOLDS ⇨1943; L.G. SMITH ⇨1945; WHITE ⇨1951}.<br><br>Based on BLEAKNEY's careful shock tube experiments, Abraham H. TAUB,[1872] a U.S. mathematician and theoretical physicist at the Institute for Advanced Study in Princeton, NJ, worked out a theory for the reflection and refraction of shock waves. |
| 1946 | *New Mexico Institute of Mining and Technology, Socorro, New Mexico* | William D. CROZIER,[1873] a U.S. atmospheric physicist and acoustician, **and William HUME,** a U.S. physicist, **develop the first *light gas gun*** – a two-stage piston-compression gun {⇨Fig. 4.11–D}. In the first stage, the piston is accelerated in a conventional powder chamber loaded with nitrocellulose propellant. In the second stage, the traveling piston compresses helium acting as a propellant gas in order to accelerate projectiles to very high velocities. The gun has since been used in exterior and terminal ballistics.<br><br>Eleven years later, they published their results in the open literature, stating, "In connection with some special investigations … it was proposed late in 1946 that a new range of muzzle velocities might be reached if the column of conventional powder gas with an effective mole- |

---

[1870] M.P. WHITE (ed.): *Effects of impact and explosion.* Summary Tech. Rept. of Div. 2. NDRC, OSRD, Washington, DC (Sept. 1946). This report was declassified in 1960.
[1871] W. BLEAKNEY: *Shock waves in a tube.* Phys. Rev. **69** [II], 678a (1946).
[1872] A.H. TAUB and L.G. SMITH: *Theory of reflection of shock waves.* Phys. Rev. **69** [II], 678 (1946); A.H. TAUB: *Refraction of plane shock waves.* Ibid. **72** [II], 51-60 (1947).
[1873] W.D. CROZIER and W. HUME: *High-velocity, light-gas gun.* J. Appl. Phys. **28**, 892-894 (1957).

cular weight of 20 to 25 in the bore of a gun were replaced by a column of gas of low molecular weight, such as hydrogen or helium [John CORNER[1874]]. Several methods for accomplishing this were considered, and a year or so later a 'gun' was designed and built which performed substantially as predicted."

Two-stage piston-compression guns can be grouped into two general classes according to the manner in which the propellant gas is compressed and heated:

▸ in *isentropic-compression guns,* the piston speed is low, and compression occurs almost isentropically; and

▸ In *shock-compression guns,* the piston speed is high, and compression is accomplished by strong shock waves which precede the piston.

There are considerable differences between the pressure and temperature histories associated with each type, as well as differences in gun geometry.[1875]

The light gas gun became a versatile tool in scientific research:

▸ **The achievement of higher projectile speeds with light gas guns than hitherto obtainable with powder guns** {LANGWEILER ⇨1938} **allowed aerodynamic studies to be performed at up to about 10,000 m/s, which considerably promoted research in reentry physics and hypervelocity impacts.**[1876, 1877]

▸ **Subsequently the light gas gun was also used to extend the range of Hugoniot data determined up to the Mbar range.** Researchers at General Motors used the light gas gun to measure dynamic properties of materials (*e.g.*, steel, tungsten and gold) up to about 6 Mbar {JONES ET AL. ⇨1966}. Darrel S. HUGHES,[1878] a professor of physics at the University of Texas in Austin, TX, and his collaborators produced precisely controlled shock waves over an intermediate range of pressures between the highest that could then be produced statically and the lowest that could be produced by explosives.

▸ **The light gas gun was also considered for use as a high-velocity pellet injector for "deep fueling" a high-temperature fusion plasma** in the tokamak facility at JET (Joint European Torus) in Culham, U.K. Solid 6-mm-dia. deuterium pellets, generated in a Grenoble-type cryostat at about 8 K, were successfully accelerated at EMI, Freiburg, to velocities of up to 4,300 m/s – still a world speed record for shooting frozen gas pellets in this very low temperature range.[1879, 1880]

| | | |
|---|---|---|
| 1946 | *U.S. Naval Electronics Laboratory, Point Loma District, San Diego, California* | **Robert S. DIETZ,** a U.S. geologist, applies geological techniques to study the Moon's surface features and **concludes that lunar craters were caused by impacts.**<br><br>In the following year, DIETZ[1881] found peculiar fractures in the rock at Kentland, Indiana, that caused it to break into striated cones – **"shatter cones"** {BUCHER ⇨1933; BOON & ALBRITTON ⇨1938} – the first reliable geologic criterion for the recognition of impact structures in the absence of meteorites {DIETZ ⇨1960}. He noticed that the apices of the cones gen- |

---

[1874] J. CORNER: *Theory of the interior ballistics of guns.* Wiley, New York (1950), p. 96.
[1875] R.E. BERGGREN and R.M. REYNOLDS: *The light-gas-gun model launcher.* In: (T.N. CANNING, A. SEIFF, and C.S. JAMES, eds.) *Ballistic-range technology.* AGARDograph No. 138, NATO-AGARD, Neuilly-sur-Seine, France (1970).
[1876] C.J. MAIDEN: *Meteoroid impact.* In: (D.P. LEGALLEY and J.W. MCKEE, eds.) *Space exploration.* McGraw-Hill, New York (1964), p. 236.
[1877] A.C. CHARTERS, J.W. GEHRING, and C.J. MAIDEN: *Impact physics, meteoroids, and spacecraft structures.* In: (B.H. GOETHERT and H.H. KURZWEG, eds.) *Fluid dynamic aspects of space flight.* In: *AGARD-NATO Specialists' Meeting* [Marseille, France, April 1964]. AGARDograph No. 87, Gordon & Breach, New York (1966), vol. 1, pp. 247-297.
[1878] D.S. HUGHES, L.E. GOURLEY, and M.F. GOURLEY: *Shock-wave compression of iron and bismuth.* J. Appl. Phys. **32**, 624-629 (1961).
[1879] P. KREHL and J. HELM: *Development of a high-speed $D_2$-pellet injector.* EMI Rept. E16/88, Ernst-Mach-Institut, Freiburg, Germany (1988).
[1880] K. SONNENBERG, P. KUPSCHUS, J. HELM, and P. KREHL: *Ein Hochgeschwindigkeits-Pelletinjektor für die Kernfusion.* Physik. Blätter **45**, 121-122 (1989).
[1881] R.S. DIETZ: *Meteorite impact suggested by the orientation of shatter-cones at the Kentland, Indiana, disturbance.* Science **105**, No. 2715, 42-43 (Jan. 10, 1947).

erally pointed toward the center of the crater, indicating the source of the shock that fractured the rock.

| 1946 | CalTech, Pasadena, California | Jesse W.M. DUMOND,[1882] a French-born U.S. physicist, **and collaborators coin the term *N-wave* for the pressure-time profile of the *sonic boom* (or *sonic bang*) phenomenon.** The flow pattern from a supersonic projectile as studied by them is precisely the same as the weak shock wave pattern due to a supersonic flight.

(i) The N-wave signature is generated from steady flight conditions at supersonic velocity (about 750 miles per hour at sea level). Its pressure wave is shaped like the letter "N." So-called "N-waves" have a front shock with a positive peak overpressure which is followed by a linear decrease in the pressure until the rear shock returns to ambient pressure {WHITHAM ⇨1950}. ▪ In 1945, the Soviet physicist Lev D. LANDAU[1883] had analyzed the weak shock waves from a supersonic projectile and predicted an N-wave shape for the pressure signature in the far-field; *i.e.*, at distances large compared with the dimensions of the body.

(ii) They make the interesting remark that **"the width of the N-wave at sufficiently large distances is determined solely by the shock strength and the distance from the profile, independently of the specific shape of the profile."** ▪ The German-born U.S. mathematician Kurt O. FRIEDRICHS[1884] showed that this apparent paradox – that the width of the N-wave should be independent of the shape (length) of the wing profile – is indeed correct.

**A special case of the sonic boom is the so-called "U-wave" (or "focused boom"), which is generated from manoeuvring flight.** Its pressure wave is shaped like the letter "U." U-waves have positive shocks at the front and rear of the boom in which the peak overpressures are increased compared to the N-wave. For an N-wave boom, the positive peak overpressure varies from less than 1 psi to about 10 psi (69 to about 690 mbar). Peak overpressures for U-waves are two to five times greater than those for the N-wave. The strongest sonic boom ever recorded was 144 pounds per square foot (1 psi) and it did not cause injury to the researchers who were exposed to it. The boom was produced by a F-4 ("Phantom II") flying just above the speed of sound at an altitude of 100 ft (30.5 m).[1885]

With the advent of supersonic fighters in the 1950s, the piercing of the "sound barrier" {HILTON ⇨1935} – at first regarded as an appealing novelty by the public[1886] – soon grew to be a serious environmental problem, particularly in Europe with its densely populated areas.

Partly stimulated by various planned SST projects {Tupolev Tu-144 ⇨1968; Boeing 2707-300 SST ⇨1968; Concorde ⇨1969}, numerous special conferences on sonic boom phenomena have been organized, *e.g.*,
- ▸ the NASA Conferences on Sonic Boom Research (1967–1970);
- ▸ the ASA (Acoustic Society of America) Sonic Boom Symposia (1965–1970);
- ▸ the ICAO (International Civil Aviation Organization) Sonic Boom Panel Meetings (1969–1970);
- ▸ the High Speed Research Program Sonic Boom Workshops (1992–1994);
- ▸ the Southern California Forum on Sonic Boom Impact Research (2002–); and
- ▸ the ISNA (International Symposium on Nonlinear Acoustics) International Sonic Boom Forum (2005–). |

---

[1882] J.W.M. DUMOND, E.R. COHEN, W.K.H. PANOFSKY, and E. DEEDS: *A determination of the wave forms and laws of propagation and dissipation of ballistic shock waves*. JASA **18**, 97-118 (1946).

[1883] L.D. LANDAU: *On shock waves at large distances from the place of their origin* [in Russ.]. J. Phys. Acad. Sci. USSR **9**, 496-500 (1945).

[1884] K.O. FRIEDRICHS: *Formation and decay of shock waves*. Comm. Pure Appl. Math. **1**, 211-245 (1948).

[1885] *Sonic boom*. U.S. Air Force Fact Sheet. Armstrong Laboratory, Wright-Patterson A.F.B., OH (2003); http://www.af.mil/factsheets/factsheet.asp?fsID=184. This contains the following interesting note: "The strongest sonic boom ever recorded was 144 pounds per square foot [1 bar], and it did not cause injury to the researchers who were exposed to it. The boom was produced by a F-4 flying just above the speed of sound at an altitude of 100 feet [30.5 m]."

[1886] *Breaking the sound barrier*. An English (fictional) movie directed by David LEAN (produced by Lopert Pictures Corporation and released in Nov. 1952). It won one Oscar in 1953.

An excellent review of the effects of sonic booms, covering the early period 1946–1968 and comprising more than 800 references, was given by John H. WIGGINS JR.,[1887] a U.S. structural dynamicist and expert on sonic boom and earthquake effects.

Various ideas for boomless supersonic flight have been proposed. One idea includes flying at Mach number and altitude combinations for which no sonic boom is observed on the ground, known as "flight below the Mach cutoff." **At *Mach cutoff*,** a term used by sonic boom researchers, **rays from the sonic boom wavefront are reflected back into the atmosphere at some altitude above ground level.**[1888] Recently, Vehicle Research Corporation (Pasadena, CA) in cooperation with the English Universities of Oxford, Cambridge, Southampton and Nottingham, proposed a mechanism for eliminating the sonic boom generated by supersonic flight. It involves the use of a nozzle-shaped wing underside, together with a planar underwing jet of engine compressor air.[1889]

| 1946 | NDRC Explosives Research Laboratories, Bruceton, Pennsylvania | John E. ELDRIDGE,[1890] Paul M. FYE and Ralph W. SPITZER, three U.S. physical chemists, **develop the "explosive flash charge" for snapshot underwater photography.** The light source for the flash consists of an explosive charge surrounded by a concentric layer of argon gas at atmospheric pressure which is compressed by the shock wave ("argon flash bomb"). Depending on the thickness of the argon layer, it provides a very intense light pulse with a short duration of only one microsecond, appropriate for full-scale underwater shock wave photography. |
|---|---|---|
| 1946 | Institut für Aerodynamik, ETH Zürich, Switzerland | Fritz FELDMANN[1891] and Jakob ACKERET, two Swiss aerodynamicists, **study airfoils at high speed using schlieren photography.** A detailed analysis of supersonic regions discloses the important result that **the interaction of the shock wave with the boundary layer produces a new shock configuration** which emerges from the surface of the airfoil. Since the new shock phenomenon resembles the Greek letter λ, they call it **"λ-shock"** {⇨Fig. 4.14–H}. |
| 1946 | University of Manchester, England | Friedrich Gerhart FRIEDLÄNDER [or FRIEDLANDER],[1892] an Austrian-born British applied mathematician, **publishes a four-part memoir on the diffraction effects of planar sound pulses by a semi-infinite plane, an infinite wedge, a semi-infinite screen, as well as on a paradox in the theory of reflection.** He assumes that the pressure of the incident pulse, traveling in such a manner that its wavefront is parallel to the plane of the wall, rises instantaneously and then decays exponentially, according to the function $$p(z) = (1-z)\,e^{-z},$$ where $z = ct/\lambda$. In this so-called "Friedlander function," $c$ denotes the velocity of sound, $t$ is the running time, and $\lambda$ is a parameter representing the "pulse thickness." a remarkable feature occurs in all cases where a pulse with a discontinuous initial pressure rise is diffracted: an initial pressure discontinuity across the boundary of the shadow. ▪ These investigations were suggested by Prof. Geoffrey I. TAYLOR and carried out by FRIEDLÄNDER at the University of Cambridge.<br><br>In freely expanding spherical shock waves (or blast waves) in which no reflective forces are encountered, the time histories of most relevant physical properties – such as the hydrostatic overpressure, density, temperature, dynamic pressure and particle velocity (but |

---

[1887] J.H. WIGGINS JR.: *Effects of sonic boom.* J.H. Wiggins Co., Palos Verdes Estates, CA (1969).
[1888] C.M. DARDEN ET AL.: *Status of boom methodology and understanding.* Proceedings of a conference held in Jan. 1988 at NASA-LRC, Hampton, VA. NASA Conference Publication NASA-CP-3027, NASA Scient. & Tech. Information Div. (1989).
[1889] S. RETHORST: *Shock-free supersonic transport.* Smith Institute, Guildford, Surrey, U.K.; http://www.smithinst.ac.uk/Projects/ESGI40/ESGI40-VRC.
[1890] P.M. FYE, R.W. SPITZER, and J.E. ELDRIDGE: *Photography of underwater explosions.* Rept. OSRD-6246 (1946).
[1891] F. FELDMANN: *Kompressibilitätseffekte in der Flugtechnik.* Interavia **1**, 1-8 (1946).
[1892] F.E. FRIEDLANDER: *The diffraction of sound pulses.* Proc. Roy. Soc. Lond. **186**, 322-367 (1946).

not entropy) – **may be expressed at a specified radius *R* by a *modified Friedlander function*.**
For example, the most important blast wave property, the pressure $P(t)$, can be expressed by

$$P(t, R) = P_S \, e^{-\alpha t} (1 - t/T_+).$$

Here $t$ is the time from the moment that the shock arrives at the gauge, $P_S$ is the peak value of the pressure immediately behind the shock front, $\alpha$ is a decay constant, and $T_+$ is the total time that the pressure is above ambient.[1893] In the case of weak blast waves at large distances from the charge, however, inhomogeneities in the air may cause the measured blast pressure-time profiles – including the rise time and the positive pulse duration $T_+$ (if there is a pressure reversal) – to deviate considerably from the modified Friedlander function, meaning that the impulsive loading of the target structure is different to that predicted from ideal propagation behavior.[1894] ▪ The modified Friedlander equation with exponential decay is not only important in detonics, but also in other fields of physics, engineering and chemistry. For example, typical gun blast pressure traces can be approximated by two Friedlander functions, with the second Friedlander function resulting from the reflected wave, which closely follows the first Friedlander function resulting from the free-air wave.[1895]

| 1946 | Dept. of Mathematics, University of Cambridge, Cambridge, England |
|---|---|

Fred HOYLE,[1896] a distinguished British astronomer, mathematician, and popularizer of science, **first realizes**
- that stars can produce heavy elements and that these can be spread into the surrounding interstellar medium by explosive processes or by stellar winds; and
- that in massive stars which evolve to have very hot dense interiors statistical equilibrium would produce the iron-peak elements. **This, followed by explosive ejection, would enrich the interstellar gas in these elements.**[1897]

By the mid-1940s heavy elements were thought to originate in an initial dense hot phase of the Universe. HOYLE's work focused people's attention on the idea that all heavy elements are made from hydrogen by nucleosynthesis in stars. This is the standard paradigm today – except for D, $^4$He, $^3$He, and $^7$Li, most of which is produced in the Hot Big Bang {ALPHER, HERMAN & GAMOW ⇨1948; WEINBERG ⇨1977; WMAP ⇨2001; NASA-GSFC ⇨2003}. **So-called "explosive nucleosynthesis" is believed to occur in supernovae** {BAADE & ZWICKY ⇨1931; TeraScale Supernova Initiative ⇨2001}.
- ***Explosive carbon burning*** occurs at about $2 \times 10^9$ K and produces the nuclei from neon to silicon. ***Explosive oxygen burning*** occurs near $4 \times 10^9$ K and produces nuclei between silicon and calcium in atomic weight.
- At higher temperatures, still heavier nuclei, up to and beyond iron, are produced.

| 1946 | Tennessee Eastman Company, Kingsport, Tennessee |
|---|---|

WILLIAM D. KENNEDY,[1898] a U.S. physical chemist and explosives expert, **performs blast wave studies** by order of the National Defense Research Committee (NDRC). **He confirms Hopkinson scaling** {B. HOPKINSON ⇨1915} **for blast measurements** obtained for charge weights ranging from a few pounds of explosives up to several thousand pounds.

---

[1893] J.M. DEWEY: *Spherical shock waves.* In: (G. BEN-DOR ET AL., eds.) *Handbook of shock waves.* Academic Press, New York (2001), vol. 2, pp. 441-481.
[1894] F. WECKEN and M. FROBÖSE: *Über die Frontsteilheit von Luftstoßwellen bei Ausbreitung über große Entfernungen.* Tech. Mitteilung T27/62, ISL, Saint-Louis, France (1962).
[1895] M.F. WALTHER: *Gun blast from naval guns.* Tech. Rept. TR-2733, Naval Weapons Laboratory (NWL), Dahlgren, VA (Aug. 1972).
[1896] F. HOYLE: *The synthesis of the elements from hydrogen.* Month. Not. Roy. Astron. Soc. **106**, 343 (1946).
[1897] W.L.W. SARGENT: *Fred HOYLE's major work in the context of astronomy and astrophysics today.* In: (D. GOUGH, ed.) *The scientific legacy of Fred HOYLE.* Cambridge University Press, Cambridge (2005), pp. 1-8.
[1898] W.D. KENNEDY: *Explosions and explosives in air.* In: (W.T. WHITE, ed.) *Effects of impact and explosions.* Summary Tech. Rept. AD 221 586, Div. 2 of NDRC, Washington, DC (1946), vol. 1, part II, chap. 2.

| | | |
|---|---|---|
| 1946 | Well Explosives, Inc., Forth Worth, Texas | **Robert H. McLemore,**[1899] a U.S. petroleum engineer, is the **first to propose the use of shaped charges in the oil well industry to increase the flow of oil.** He writes, "Much has been written about the need for a method of increasing the secondary recovery of oil. It is well known that many fields have produced a small part of the total oil in place. It is believed that, with a process such as this [the shaped explosive charge], much of this oil will eventually be recovered, at an economical rate." ▪ In 1976, McLemore received the DeGolyer Service Medal for his distinguished and outstanding service to the petroleum industry.<br><br>The method of using shaped charges {von Förster ⇨1883; Munroe ⇨1888; Thomanek ⇨1938} was an important and necessary development for deep wells that require the use of steel tubes or cement dressing for well casing. The charges are arranged in a tool called a "gun," which is lowered on a wire-line into the well opposite the producing zone. When the gun is in position, the charges are fired electronically from the surface. After the perforations are made, the tool is retrieved. The new method stimulated research and design into minimizing jet size and maximizing jet efficiency, and into the prevention of shock interference between neighboring charges.[1900] The use of shaped charges is now a standard technique in the oil well industry {⇨Fig. 4.15–L}. |
| 1946 | NACA's Langley Memorial Aeronautical Laboratory [now NASA's Langley Research Center], Hampton, Virginia | **Cearcy D. Miller,**[1901] a U.S. engineer, **obtains a patent on a high-speed camera that uses an optical system based on a rotating mirror and refocused revolving beams, and operates at 40,000 frames per second.** ▪ Miller invented the NACA camera in February 1936 at the Langley Field laboratory because there was a demonstrable need for a faster camera than was anything available commercially for use in the study of spark-ignition engine knock.<br><br>In the same year, Miller[1902] and collaborators successfully apply the "NACA camera" in order to study combustion and knock in the cylinders of internal combustion engines. ▪ In the early 1950s, **Miller's idea for operating cameras was taken up by Berlyn Brixner,**[1903] an engineer at Los Alamos Scientific Laboratory, **in order to record nuclear explosion and shock wave phenomena.** Brixner developed a mechanical ultrahigh-speed camera capable of attaining a photographic speed of 15,000,000 frames per second {⇨Fig. 4.19–N}. The main feature that made this ultrahigh speed possible was a three-sided steel mirror revolving at 23,000 times a second (1,380,000 rpm) powered by a small helium-driven turbine. The Los Alamos camera, a landmark in ultrahigh-speed cinematography, used ordinary 35-mm film and provided up to 96 consecutive pictures from a single event. |
| 1946 | Institute for Advanced Study (IAS), Princeton, New Jersey | **John von Neumann,**[1904] **Arthur W. Burks,** and **Herman H. Goldstine,** three U.S. mathematicians and computer scientists, **review the entire field of automatic computation in detail and present new comprehensive designs for a parallel, *stored-program* computer.** ▪ Their concept of storing sets of instructions or commands in memories – thus giving the computer a repertoire of procedures it can follow automatically whenever necessary – was an important breakthrough in the evolution of digital computers. It strongly influenced the design of all sub- |

---

[1899] R.H. McLemore: *Formation penetrating with shaped explosive charges.* The Oil Weekly (Houston, TX), No. 8, 56-58 (July 1946).

[1900] T. Poulter: *The development of shaped charges for oil well completion.* Petroleum Trans. **210**, 11-18 (1957).

[1901] C.D. Miller: *High-speed motion picture camera.* U.S. Patent No. 2,400,887 (May 28, 1946). *The NACA high-speed motion-picture camera optical compensation at 40,000 photographs per second.* NACA-TR 856 (1946); http://ntrs.nasa.gov/archive/nasa/casi.ntrs.nasa.gov/19930091928_1993091928.pdf.

[1902] C.D. Miller: *Relation between spark-ignition engine knock, detonation waves, and auto-ignition as shown by high-speed photography.* NACA-Rept. No. 855 (1946), see http://ntrs.nasa.gov/archive/nasa/casi.ntrs.nasa.gov/19930091927_1993091927.pdf. C.D. Miller et al.: *Analysis of spark-ignition engine knock as seen in photographs taken at 200,000 frames per second.* NACA-TR 857 (1946); http://ntrs.nasa.gov/archive/nasa/casi.ntrs.nasa.gov/19930091929_1993091929.pdf.

[1903] B. Brixner: *Fifteen million frames per second.* Proc. 2nd Int. Congress on High-Speed Photography [Paris, France, Sept. 1954]. In: (P. Naslin and J. Vivie, eds.) *Photographie et cinématographie ultra-rapides.* Dunod, Paris (1956), pp. 108-113.

[1904] J. von Neumann, A.W. Burks, and H.H. Goldstine: *Preliminary discussion of the logical design of an electronic computing instrument.* Part 1, vol. I. Report prepared for U.S. Army Ordnance Dept. under contract W-36-034-ORD-7481 (1946).

| | | |
|---|---|---|
| | | sequent digital computers. For example, early electronic computers based on the IAS architecture were
▸ EDVAC (Electronic Discrete Variable Automatic Computer), a *stored program, serial computer*; and
▸ ORDVAC (Ordnance Discrete Variable Automatic Computer), a *stored-program, parallel* computer. |
| 1946 | *Laboratoire de Recherches Balistiques et Aérodynamiques, Saint-Louis, Alsace, France* | **Hubert SCHARDIN,** a German physicist and co-director of the institute, **delivers a lecture entitled** *Die Bedeutung der Gasdynamik in der Ballistik* ("The Importance of Gas Dynamics in Ballistics"). He gives an illustrative survey of developments in this field up to the end of World War II by compiling the drag coefficients of numerous projectiles and showing them in a single $c_w$-diagram within the range $M = 0.2$–$10$.[1905] His selected projectiles include munitions used in both World Wars, such as
▸ the Krupp 10-cm cylindrical standard projectile and modifications;
▸ the Cranz-Becker 8-cm cylindrical S-projectile {DWM ⇨ 1903};
▸ bullets of different geometries studied in wind tunnel tests in Germany and France;
▸ various 8.8- and 10-cm Flak [from the Germ. *Flieger-Abwehr-Kanone*, meaning "antiaircraft artillery"] shells; and
▸ the German WWII 8.8-cm Flak projectile (model KL8.8/5.27), which shows a substantial reduction in aerodynamic drag when a special nose geometry is adopted {HAACK ⇨ 1941}. |
| 1946 | *Ballistic Research Laboratory (BRL), Aberdeen Proving Ground, Maryland & National Defense Research Committee (NDRC), Office of Scientific Research and Development (OSRD), Washington, DC* | **James J. SLADE JR.,**[1906] a U.S. engineer and ballistician, **reviews the characteristics and effects of gun blasts, and their control:**
(i) Using sketches based on spark pictures taken of a 0.30-in. (7.62-mm)-caliber rifle firing ball ammunition using service charge, he gives an account of the muzzle blast phenomenon showing the blast at various stages of its evolution in outstanding detail.
(ii) Referring to a later stage in which the bullet is some 25 calibers from the muzzle and no longer interferes with the jet, he writes: "The principal expansion of the jet is confined to the 'bottle' *c*, which is bounded by the quasi-stationary normal shock *d* and oblique shock *e*. The flow within the bottle is practically adiabatic and starts at the muzzle with a Mach number close to one…" **The neck of this "shock bottle" is formed by the barrel end and its bottom by an axially centered Mach disc typically positioned from 10 to 16 calibers from the muzzle.** • Before this, most knowledge of the growth and decay of the muzzle blast was qualitative, obtained from spark photographs taken of small-caliber guns being fired {CRANZ ⇨ 1925; ⇨Fig. 4.5–L}. Beginning in the 1960s, more detailed investigations – often combined with muzzle flash studies and numerical hydrodynamic analyses – were resumed in England by F. SMITH[1907] and collaborators at the Royal Armament Research & Development Establishment (RARDE), in the United States by the aerospace engineer Edward M. SCHMIDT[1908] and collaborators at the Ballistic Research Laboratory (BRL), and by collaborating teams in Germany (EMI, Freiburg) and France (ISL, Saint-Louis).[1909] |

---

[1905] The diagram was reproduced in W. HAACK's paper presented at the 1959 Cranz Centenary Colloquium. See W. HAACK: *The calculation of the stability and damping of spin-stabilized projectiles*. In: (W.C. NELSON, ed.) *Selected topics on ballistics*. Cranz Centenary Colloquium [University of Freiburg, Germany, April 1958]. Pergamon Press, London *etc.* (1959), pp. 125-139.

[1906] J.J. SLADE JR.: *Muzzle blast: its characteristics, effects, and control*. Repts. NDRC -A391 (1946) and OSRD-6462 (1946).

[1907] F. SMITH: *Model experiments on muzzle brakes. Part I: Measurement of thrust. Part II: Measurement of blast*. Rept. No. 2/66. *Part III: Measurement of pressure distribution*. Rept. No. 3/68, RARDE, Fort Halstead, Kent, U.K. (1966).

[1908] E.M. SCHMIDT: *Muzzle devices, a state-of-the-art survey. Vol. I: Hardware study*. Memorandum Rept. No. 2276, BRL, Aberdeen Proving Ground, MD (Feb. 1973).

[1909] G. KLINGENBERG and J.M. HEIMERL: *Gun muzzle blast and flash*. Progr. Astronaut. Aeronaut., No. 139. AIAA, Washington, DC (1992).

(iii) Gun blasts can injure gun crews and damage neighboring structures. In flat-trajectory guns, dust is picked up from the ground by the action of high-speed gas. The dust raised may result in the obscuration of the target.

(iv) To a certain extent, muzzle blasts can be controlled by muzzle brakes {TRÉVILLE DE BEAULIEU ⇨1842} or simple cone ("blunderbuss") extensions to the muzzle in some antiaircraft guns. ▪ Muzzle brakes were first developed in Germany and used during World War II. The effective German two-baffle brake was copied extensively by other countries.

(v) In Appendix D of his report, he also discusses how the free-surface liquid analogy {JOUGUET ⇨1920; PREISWERK ⇨1938; H.A. EINSTEIN ET AL. ⇨1946/1947; GILMORE ET AL. ⇨1950} might be used to study ballistics, and in particular how an "open channel gun" could be used to get information about the deflection of the free jet.

| 1946 | CalTech, Pasadena, California | **Hsue-Shen TSIEN**,[1910] a Chinese-born U.S. aeronautical engineer, **coins the term *hypersonic*.**[1911] The new term implies that the flight velocity is very much greater than the ambient speed of sound. He does not give a precise definition of the velocity at which a supersonic flow becomes a hypersonic flow, because the onset of characteristic hypersonic flow effects is in fact gradual and varies with the geometry of the test body as well as with the nature of the surrounding atmosphere. ▪ The words *super* and *hyper* are of Latin and Greek origin, respectively, with both meaning "more than." Today, "hypersonics" designates flow or body velocities in excess of a Mach number of five. The choice of a Mach number of five as the lower bound in the definition of the hypersonic speed range can be derived from a variety of quite different considerations: |
|---|---|---|

▸ For a flight Mach number of five and above, kinetic heating is in excess of 800 °C, which demands the use of new materials.
▸ Above a Mach number of about four, new wind tunnel techniques are required due to the generation of detrimental condensation effects.
▸ Beginning at about a Mach number of five, the thermodynamic properties of gas are affected by kinetic heating, and molecular vibrational modes contribute to the internal energy – resulting in so-called "real gas effects."[1912]

To simulate flight enthalpies for hypersonic Mach numbers at a range of very large altitudes, special combustion-heated facilities – so-called **"high-enthalpy hypersonic wind tunnels"** – have been constructed.

In the same year, TSIEN[1913] also proposes the term *superaerodynamics*, to designate aerodynamics at very low ambient pressures when the dimensions of the moving body become small in relation to the length of the mean free path of the air molecules; *i.e.*, corresponding to very high altitudes where the density of the air is very small and the molecules are very far apart – a branch of fluid dynamics termed ***rarefied gas dynamics.*** ▪ Previously, G.A. CROCCO {⇨1932} had already proposed the term ***superaviation*** to designate this particular branch of aeronautics.

| 1946 | Institute of Chemical Physics, Moscow, Soviet Union | **Yakov B. ZEL'DOVICH**,[1914] an eminent Soviet shock wave and detonation physicist, **discusses the possibility that rarefaction shocks could happen near the critical point, where the differences between a vapor and a fluid are obliterated.** He theoretically predicts that rarefaction should propagate in a substance as a discontinuity under near-critical conditions, and |
|---|---|---|

---

[1910] H.S. TSIEN: *Similarity laws of hypersonic flows*. J. Math. Phys. **25**, 247-251 (1946).
[1911] The term *hypersonic* is somewhat ambiguous and has also come to be used to describe sound at frequencies ranging from 0.5 to 1,000 GHz; *i.e.*, beyond ultrasonics (20 kHz −500 MHz). *See* W. SCHAAFFS: *Molekularakustik*. Springer, Berlin *etc.* (1963).
[1912] A.R. COLLAR and J. TINKLER (eds.): *Hypersonic flow*. Butterworths, London (1960); *see Preface* by A.R. COLLAR, p. ix.
[1913] H.S. TSIEN: *Superaerodynamics, mechanics of rarefied gases*. J. Aeronaut. Sci. **13**, 653-664 (1946).
[1914] Y.B. ZEL'DOVICH: *On the possibility of rarefaction shock waves* [in Russ.]. Zh. Eksp. Teor. Fiz. (SSSR) **16**, 363-364 (1946).

| 1946/ | *Hydrodynamics* | compression should propagate as a continuous process. • Experimental evidence of this was first provided by Soviet researchers {KUTATELADZE ET AL. ⇨1978}. |
|---|---|---|

1946/
1947

*Hydrodynamics Laboratory, CalTech, Pasadena, California*

Hans Albert EINSTEIN,[1915] a Swiss-born U.S. hydraulic engineer and the first son of Albert EINSTEIN, **and Earl G. BAIRD,** who has been stimulated by previous hydraulic jump studies {PREISWERK ⇨1938}, **initiate systematic "water table" experiments to test the validity of the analogy between the flow of a liquid with a free surface and the flow of a compressible gas** {JOUGUET ⇨1920; PREISWERK ⇨1938; ⇨Figs. 4.12–I & 4.13–B}.

(i) In order to generate surface shock waves, they use a reservoir into which the water is drawn prior to its release by reducing the pressure using an aspirator – a simple but very effective method that had already been used in propagation studies on water waves {WEBER Bros. ⇨1825}.[1916]

(ii) These investigations, which also encompass Mach interactions of hydraulic jumps, are aimed at ascertaining the source of the discrepancy between experiment and theory, and will later be continued at CalTech by Harry E. CROSSLEY JR.[1917]

At the same time, similar water wave studies are also carried out in other parts of the United States:

▸ **W. James ORLIN**[1918] **and collaborators at NACA's Langley Memorial Aeronautical Laboratory apply the analogy between the flow of a compressible gas and the flow of a liquid with a free surface in order to better understand the nature of wind tunnel "choking."** This puzzling phenomenon was observed in NACA's 8-ft (2.44-m) transonic wind tunnel and which limits the speed achievable due to the presence of a test model. By setting up a small water channel which generates hydraulic jumps propagating at a velocity of about 3 ft/s (0.915 m/s), they obtain some interesting insights into the process of choking, including flow visualization, which agrees well with schlieren pictures taken in air.

▸ Stimulated by the 1946 Aleutian Earthquake {Unimak Island ⇨1946}, which produced devastating tsunamis at some lands facing or islands in the Pacific Ocean, **laboratory studies of the generation and reflection of water waves were carried out at the Hydraulic Laboratory of UC Berkeley** using a wave tank and a wave generator.[1919] The reflection of a solitary wave at oblique incidence to a wall revealed Mach reflection and the formation of strong eddies at the wall, aside from regular reflection.

Water table studies are very illustrative for student demonstrations: using a small water table {⇨Fig. 4.4–F} placed on a common viewgraph projector and slightly inclined so that the hydraulic jump propagates to a steadily decreasing depth, it is possible to slow down its speed to a full stop, thus enabling complex interaction phenomena – such as the Mach effect – to be observed with the naked eye.[1920]

---

[1915] H.A. EINSTEIN and E.G. BAIRD: *Analogy between surface shock waves on liquids and shocks in compressible gases.* Progr. Repts, CalTech Hydrodynamic Laboratory (Sept. 1946 and July 1947).

[1916] E.H. WEBER and W. WEBER: *Wellenlehre auf Experimente gegründet oder über die Wellen tropfbarer Flüssigkeiten mit Anwendung auf die Schall- und Lichtwellen.* Fleischer, Leipzig (1825), § 129.

[1917] H.E. CROSSLEY JR.: *Analogy between surface shock waves in a liquid and shocks in compressible gases.* Rept. N-54.1, CalTech Hydrodynamic Laboratory, Pasadena, CA (1949).

[1918] W.J. ORLIN, N.J. LINDNER, and J.G. BITTERLY: *Application of the analogy between water flow with a free surface and two-dimensional compressible gas flow.* Rept. NACA-TR 1185 (1947).

[1919] R.L. WIEGEL: *Research related to tsunamis performed at the Hydraulic Laboratory University of California, Berkeley.* In: (D.C. COX, ed.) *Proceedings of the Tsunami Meetings Associated with the 10th Pacific Science Congress* [Honolulu, HI, 1961]. Institut Géographique National, Paris (1963), pp. 174-197.

[1920] P. KREHL and M. VAN DER GEEST: *The discovery of the Mach reflection effect and its demonstration in an auditorium.* Shock Waves **1**, 3-15 (1991).

| 1947 | *Sikhote Alin Mountains [ca. 500 km northeast of Vladivostock], Soviet Union* | On February 12, **the Sikhote-Alin meteorite falls from the sky,** yielding one of the most spectacular meteorite falls in recorded history. Eyewitnesses observe a brilliant fireball which moves across the sky from north to south, leaving a thick smoke trail that remains visible for two hours. After the fireball disappears behind the hills, explosions like the firing of heavy-caliber guns are heard. ▪ The meteorite, a coarse iron octahedrite, appears to have broken up violently during its entry through the atmosphere. Numerous craters ranging from 26.5 to 0.5 m in diameter were later found, and hundreds of fragments of nickel-iron were collected.[1921] The speed of entry was estimated to be 14.5 km/s. |
|---|---|---|
| 1947 | *Texas City, Galveston Bay, Texas & Brest, northwest France* | On April 16, **the French cargo ship SS *Grandchamp*, moored in the harbor of Texas City and loaded with fertilizer-grade ammonium nitrate (FGAN), blows up in a prodigious explosion heard as far as 150 miles (241 km) away. An adjacent cargo ship, the SS *High Flyer*, laden with a 1,000 tons of FGAN, detonates 16 hours later in a colossal explosion** (468 people are killed, 3,500 are injured and about 100 are left missing). ▪ The ship, already loaded with 16 cases of small arms ammunition, as well as other assorted goods, was inordinately hot, and the bags of fertilizer started to burn. Then, at 9.12 A.M., an explosion ripped the ship apart, initiating a cascade of other explosions that destroyed the Monsanto plant near the dock, and numerous oil refineries. Impacted by the flames and debris, much of the town was wiped out. **The blast wave was strong enough to rip the wings off two planes flying overhead;** the 40,000-pound (18-ton) deck of the SS *Grandchamp* was catapulted half a mile away; and the entire harbor basin was briefly scooped out into a 20-ft (6-m)-high tidal wave that burst the wharf area.[1922]

In the same year, on July 28, a similar disaster occurs in Brest harbor, where the SS *Ocean Liberty* was carrying 3,300 tons of FGAN. ▪ Prior to these disasters, FGAN, a grained ammonium nitrate [$NH_4NO_3$] coated with 0.75% wax and conditioned with about 3.5% clay, was not considered to be an explosive. However, **subsequent investigations showed that FGAN was much more dangerous than previously thought, and more rigid shipment and storage regulations were promptly put into effect by the U.S. Government.** One major problem encountered when assessing the risks related to ammonium nitrate explosions lies in the fact that, for many scenarios, it is not clear which parameters are critical; for example, the specific conditions under which an explosion will result from heating in a confined space or shock caused by impact are often not very apparent.[1923] |
| 1947 | *Helgoland, North Sea, Germany* | On April 18, **6,700 tons of Axis munitions are blown up on the North Sea island of Helgoland by the British Navy.** ▪ This so-called "Helgoland blast" – then the largest non-nuclear man-made explosion in history – gave seismologists a unique opportunity to monitor a scheduled man-made earthquake. For example, the air shock was measured at Göttingen University after traveling a distance of about 324 km using an "undograph," a device for recording sound waves which was installed in the Gauss-House. The blast wave arrived about 14 minutes after the explosion – thus indicating a mean supersonic blast velocity of about 385 m/s.[1924] |

---

[1921] P.M. MILLMAN: *Earth, impact craters*. In: (S.P. MARAN, ed.) *The astronomy and astrophysics encyclopedia*. Van Nostrand & Reinhold, New York (1992), pp. 187-189.
[1922] B. MINUTAGLIO: *City on fire: the forgotten disaster that devastated a town and ignited a landmark legal battle*. Harper Collins, New York (2003).
[1923] R.J.A. KERSTEN and W.A. MAK: *Explosion hazards of ammonium nitrate. How to assess the risks?* 3rd National Research Institute of Fire and Disaster (NRIFD) Int. Symposium on Safety in the Manufacture, Storage, Use, Transport and Disposal of Hazardous Materials [Tokyo, Japan, March 2004]. NRIFD, Tokyo (2004).
[1924] W. KERTZ: *Ludger MINTROP, der die angewandte Geophysik zum Erfolg brachte*. Mitteil. Dt. Geophys. Gesell. **3**, 2-16 (1991).

| 1947 | U.S. Air Force Base, Edwards, California & Rogers Dry Lake, southern California | On October 14, **the 24-year old U.S. Air Force test pilot Charles ("Chuck") YEAGER becomes the first man to officially break through the "sound barrier"** {HILTON ⇨ 1935}. Due to limited propellant capacity, the small bullet-shaped aircraft that exceeded the barrier was carried by a Boeing B-29 ("Superfortress") to an altitude of about 25,000 feet (7,620 m) and then dropped.[1925] **At an altitude of 43,000 feet (13,106 m) over the Mojave Desert near Muroc Dry Lake, his research plane reached a velocity of $M = 1.06$ (700 mph or 313 m/s) in level flight.** ▪ His Bell XS-1 – a short for "Sonic Experimental 1" {⇨Fig. 4.20–F} and later known as the "X-1" – was a Sänger-type rocket plane {SÄNGER ⇨ 1933}. YEAGER christened it *Glamorous Glennis* in honor of his wife, following a tradition among American World War II fighter pilots.

YEAGER[1926] later described his first experience of breaking the sound barrier in his autobiography: "Suddenly the Mach needle began to fluctuate. It went up to .965 Mach, then tipped right off the scale. I thought I was seeing things. We were flying supersonic! And it was as smooth as a baby's bottom: grandma could be sitting up there sipping lemonade … I was thunderstruck. After all the anxiety, breaking the sound barrier turned out to be a perfectly paved speedway. There should've been a bump on the road, something to let you know you had just punched a nice clean hole through that sonic barrier. The unknown was a poke through Jello. Later-on, I realized that this mission had to end in a let-down, because the real barrier wasn't in the sky, but in our knowledge and experience of supersonic flight…"

In the following year, YEAGER collected the 1948 Collier Trophy {COLLIER ⇨ 1911} for achieving supersonic flight, along with John STACK of Langley Aeronautical Laboratory, who carried out research to determine the physical laws affecting supersonic flight, and Lawrence D. BELL, who supervised the manufacture of the Bell X-1. ▪ Today YEAGER's Bell X-1 is part of the collection at the Smithsonian National Air and Space Museum.[1927] The X-1 was only the first in a long series of supersonic research planes,[1928] which eventually accumulated in the development of the famous hypersonic plane X-15 {⇨1959 & 1961}.

The first Soviet aircraft to break the sound barrier in level flight was apparently the prototype fighter Lavochkin La-190, which reached a velocity of $M = 1.03$ in 1951 at an altitude of about 5,000 m. The first Soviet supersonic fighter put into service was the MiG-19 "Farmer" (max. speed $M = 1.35$); the prototype made its first flight in September 1953. |
|---|---|---|
| 1947 | NACA's Langley Memorial Laboratory [now NASA's Langley Research Center], Hampton, Virginia | On November 26, **the first American hypersonic-flow wind tunnel is operated successfully for the first time, at Mach 6.9** {⇨Fig. 4.9–I}. The primary objectives of the new tunnel are to obtain information about the effect of air at high speeds and temperatures on rockets and missiles, and to simulate space flight conditions.

(i) **The NACA blowdown facility was designed by John V. BECKER,**[1929] an aeronautical engineer and assistant chief in John STACK's Compressibility Research Division. The tunnel has a $11 \times 11$-in.$^2$ ($27.9 \times 27.9$-cm$^2$) test section. To reach high pressure ratios, air from a pressure tank is blown through the test section into an evacuated tank.[1930]

(ii) Detailed studies demonstrate that dry wind-tunnel air – *i.e.*, its components: nitrogen and oxygen – does not supersaturate. Consequently, the air supplied to a hypersonic wind tunnel, when taken from the atmosphere, must be preheated {5th Volta Conf. (PRANDTL) ⇨1935}. |

---

[1925] R.P. HALLION: *Supersonic flight: the story of the Bell X-1 and Douglas D-558*. Macmillan, New York (1972).
[1926] C.E. YEAGER: *YEAGER. An autobiography*. Hall, Boston, MA (1986), pp. 224-225.
[1927] *Bell X-1*. National Air & Space Museum, Smithsonian Institution, Washington, DC; http://www.nasm.si.edu/exhibitions/gal100/bellX1.html.
[1928] W.T. BONNEY: *High-speed research airplanes*. Scient. Am. **189**, 36-41 (Oct. 1953).
[1929] J.V. BECKER: *Results of recent hypersonic and unsteady flow research at the Langley Aeronautical Laboratory*. J. Appl. Phys. **21**, 619-628 (1950).
[1930] D.D. BAALS and W.R. CORLISS: *Wind tunnels of NASA*. NASA History Office, Washington, DC (1981), chap. 5: *NACA's first hypersonic tunnels*; http://www.hq.nasa.gov/office/pao/History/SP-440/cover.htm.

- More detailed investigations on this important subject of condensation were made in the early 1970s at Yale University by Peter WEGENER[1931, 1932] and Benjamin Ju-Cheng WU.

Aside from the work performed at NACA, **Antonio FERRI at the Polytechnic Institute of Brooklyn's Aerodynamic Laboratory had designed a hypersonic facility capable of achieving hypersonic flows of up to $M = 6$.** Using a large airflow preheating capacity ("pebble bed heater"), it allowed the high temperatures encountered in high-speed flows around re-entry vehicles from outer space to be simulated. This hypersonic facility was located at Freeport, LI, and was used as a model for other hypersonic facilities.

Hypersonic facilities played an integral role in the development of ballistic missiles and Apollo vehicles that could reenter the Earth's atmosphere without disintegrating.

| | | |
|---|---|---|
| 1947 | *U.S. Naval Research Laboratory (NRL), Washington, DC* | **The First Symposium on Shock and Vibration is held;** the general chairman is Elias KLEIN. The common objective of the meeting, which is devoted to Navy activities, is "to put into hands of the scientific and engineering personnel working in the field of shock and vibration all the extended knowledge which becomes available, so that each one may carry on his work most effectively." **Representatives from the David Taylor Model Basin propose the following definitions of the terms** *shock* **and** *vibration* **for discussion:** |

- ▸ "*Shock* **is a sudden and violent change in the state of motion** of the component parts or particles of a body or medium resulting from the sudden application of a relatively large external force, such as a blow or impact."
- ▸ "*Vibration* **is the periodic motion** of the component parts or particles of an elastic body or medium in alternately opposite directions from the position of equilibrium when the equilibrium has been disturbed."

The symposia, which traditionally include both classified (military) and unclassified sessions, subsequently became the leading forum for the structural dynamics and vibration community to present and discuss new developments and ongoing research. In Nov. 2007, the 78th Shock & Vibration Symposium took place at Philadelphia, PA.

| | | |
|---|---|---|
| 1947 | *Imperial College, London, England* | **Dennis GABOR,**[1933] a Hungarian-born British electrical engineer, **invents** *holography* [from the Greek *holos* "whole" and *gram* "message"], which he calls **"wavefront reconstruction."** This invention – a new lensless method which record in three spatial dimensions, sometimes called "lensless photography" – will earn him the 1971 Nobel Prize for Physics. • Originally proposed by GABOR as a new microscopic technique that could be used to improve imaging defects associated with the electron microscope, it has largely been applied to record macroscopic rather than microscopic objects instead. In his Nobel Lecture, delivered on December 11, 1971, GABOR[1934] also referred to progress in holographic interferometry.[1935] He showed an example of the dynamic holographic interferometry which Ralph F. WUERKER and coworkers obtained at TRW Physical Electronics Laboratory from an unusual shock wave interaction pattern when a train of head (shock) waves emerging from a supersonic bullet met another shock wave. |

With the advent of laser Q-switching {MCCLUNG & HELLWARTH ⇒1962; PERESSINI ⇒1963} and longitudinal mode control techniques {MCCLUNG & WEINER ⇒1965}, suitable light sources (*i.e.*, those of high intensity, nanosecond duration and sufficient coherence) for recording ultrahigh-speed events holographically became available {⇒Fig. 4.18–G}: events such as the head wave generated by a supersonic projectile {BROOKS ET AL.⇒1966; ⇒Fig. 4.6–I}.

---

[1931] P.P. WEGENER: *Nonequilibrium flow with condensation.* Acta Mech. **21**, 65-91 (1975).
[1932] P.P. WEGENER and B.J.C. WU: *Homogeneous and binary nucleation: new experimental results and comparison with theory.* Faraday Disc. Chem. Soc. **61**, 77-82 (1976).
[1933] D. GABOR: *A new microscopic principle.* Nature **161**, 777-778 (1948); *Microscopy of reconstructed wave-fronts.* Proc. Roy. Soc. Lond. **A197**, 454-487 (1949).
[1934] D. GABOR: *Holography, 1948–1971* [Nobel Lecture, presented on Dec. 11, 1971]; http://www.nobel.se/physics/laureates/1971/gabor-lecture.pdf.
[1935] L.O. HEFLINGER, R.F. WUERKER, and R.E. BROOKS: *Holographic interferometry.* J. Appl. Phys. **37**, 642-649 (1966).

| | | |
|---|---|---|
| 1947 | Polaroid Corporation, Boston, Massachusetts | **Edwin H. LAND,** a U.S. inventor and physicist, **first publicly demonstrates his so-called "Polaroid Land camera."** His "instant" camera takes only ten seconds to develop black-and-white film using developing agent carried in pods in the camera, and the full print process takes 60 seconds. ▪ The Land photographic process rapidly found numerous commercial, military and scientific applications. In particular, it proved extremely useful for snapshot photography of shock waves because it allowed a quick insight into the phenomenon of interest before installing bulky high-speed mechanical cameras. Another very important application was the recording of high-speed single-shot oscillograms. However, these speed advantages of Polaroid film have now been superseded by digital cameras and digitizers. |
| 1947 | Mont Lachat Observatoire, Dépt. Haute-Savoie, France | **Jacques MAURIN**[1936] **and Louis MÉDARD,** two French engineers at the Office Nationale d'Études et Recherches Aéronautique (ONERA) in Châtillon-sous-Bagneux, **measure the threshold overpressure of a shock wave generated by detonations used to trigger the freezing of natural supercooled fog droplets** {IVES ⇨ 1941}. They perform the measurements at the summit of Mont Lachat (alt. 2,077 m), and find a value of 0.3 psi (20.6 mbar) in natural fog at $-1.5\,°C$.<br><br>In the same year, MAURIN[1937] shows that a period of 0.2 s is necessary to completely freeze a 14-μm droplet at a temperature of $-0.5\,°C$. ▪ Their studies into triggered freezing were resumed by the 1964 Yellowstone Field Research Expedition {GOYER ⇨ 1964/1965}. |
| 1947 | Kaiser-Wilhelm-Institut (KWI) für Strömungsforschung, Göttingen, Germany | **Klaus OSWATITSCH,**[1938] an Austrian gas dynamicist and university lecturer, **reviews recent applications of the method of characteristics** {MONGE ⇨ 1770s} **in hydrodynamics.** In particular, simple methods are discussed for the two practical cases of an axially symmetric adiabatic supersonic flow and a nonstationary adiabatic streamline flow.<br><br>In the same year, **R. HILL**[1939] **and Donald C. PACK,** two British researchers at the Branch for Theoretical Research, Armament Research Dept. of the British Ministry of Supply, **calculate the pressure and velocity distributions in the gaseous products resulting from the detonation of an uncased two-dimensional explosive charge in a surrounding fluid medium (air).** The detonation products expand laterally behind the detonation wave as it travels down the charge. The boundary condition makes it necessary to find explicit theoretical formulae for the gas field near the charge. They solve the hydrodynamical equations numerically using the method of characteristics and calculate the position of the shock wave. |
| 1947 | Harvard College Observatory, Cambridge, Massachusetts | **Fred L. WHIPPLE,**[1940] a U.S. astronomer and one of the principal U.S. authorities on meteoroids, **proposes his so-called "Whipple shield" for protecting spacecraft from space debris.**<br>(i) It consists of simply placing a sacrificial bumper, usually aluminum, in front of the spacecraft, thus allowing it to absorb the initial impact {⇨Fig. 4.14−U}. The resulting debris cloud is diluted over a larger area and subsequently caught by a second shield, which is arranged in parallel to the primary shield.<br>(ii) In a short but remarkable note, he concisely writes, "Meteorites represent a potential hazard to a pressurized space vessel. Of fundamental interest is the value of the probability that the skin of the vessel will be punctured by a meteorite. In case this probability is appreciable the problem of protection from meteorites becomes important … A simple protection can be provided other than by the avoidance of known meteor streams. Considerations of the conservation |

---

[1936] J. MAURIN and L. MEDARD: *Congélation à distance d'un nuage surfondu par ondes de choc.* C. R. Acad. Sci. Paris **225**, 432-434 (1947).
[1937] J. MAURIN: *Durée de congélation d'une gouttelette d'eau surfondue.* C. R. Acad. Sci. Paris **225**, 814-816 (1947).
[1938] K. OSWATITSCH: *Über die Charakteristikenverfahren der Hydrodynamik.* ZAMM **25/27**, 195-208, 264-270 (1947).
[1939] R. HILL and D.C. PACK: *An investigation, by the method of characteristics, of the lateral expansion of the gases behind a detonating slab of explosive.* Proc. Roy. Soc. Lond. **A191**, 524-541 (1947); originally published in Jan. 1944 as a Ministry of Supply report.
[1940] F.L. WHIPPLE: *Meteorites and space travel.* Astron. J. **52**, 131 (1947). The citation given above was reproduced with the permission of the AAS.

of momentum and energy show that when a meteorite collides with a sheet of thickness comparable to the meteorite's diameter, the result is an explosion in which both the meteorite and the corresponding material of the sheet are vaporized and ionized at very high temperatures. Hence a **'meteor bumper'** consisting perhaps of a millimeter-thick sheet of metal surrounding the ¼-in. (6.35-mm) skin of the space vessel at a distance of an inch would dissipate the penetrating power of meteorites several times larger than one corresponding to an 8th-magnitude…"

In 1951, WHIPPLE[1941] was rather pessimistic about the chances of avoiding meteoroid penetration, and suggested the use of thick shielding on spacecraft to guard against structural damage. Quantitative data on the probability of meteoroid impact was first obtained by the first U.S. satellite Explorer 1, launched in January 1958: during the first month of its orbital life it recorded only seven hits by micrometeoroids – particles less than one millimeter in size impacting with velocities ranging from about 10 to 20 km/s.[1942] **Today WHIPPLE's principle of shielding is used in all kinds of spacecraft.** It has been largely modified, *e.g.*, by increasing the number of shielding stacks and/or by incorporating layers of Mylar® (DuPont), Kevlar® (DuPont) cloths and foam, in order to effectively absorb shock and debris.[1943]

| 1948 | DTMB, Carderock, Maryland | **Construction of a $7 \times 10$ ft² ($2.13 \times 3.05$ m²) sonic wind tunnel at the David Taylor Model Basin (DTMB).** Parts of the tunnel were obtained from a three-meter tunnel captured after the war at Ottobrunn, Germany.[1944] • In 1951, the tunnel was modified into a transonic wind tunnel of the same dimensions. |
|---|---|---|
| 1948 | Sternberg Astronomical Institute, Moscow, Soviet Union | The Soviet Academy of Sciences publishes the first **General Catalogue of Variable Stars (GCVS)**.[1945] • The 5th edition (five vols.) was published in the 1990s. Vols. I–III contain data for 40,215 individual variable objects discovered and named as variable stars by 1982 and located mainly in the Milky Way galaxy. Vol. V (extragalactic variable stars) consists of two catalogues: the *Catalogue of Variable Stars in External Galaxies* which contains 10,979 variable stars in 35 stellar systems, and the *Catalogue of Extragallactic Supernovae* which includes 984 confirmed or suspected supernovae.[1946] |

Astronomers have classified variable stars according to observable properties. There are two groups: ***intrinsic variable stars*** (here named classes I–III), which vary their light output and hence their brightness due to changes within the star itself; and ***extrinsic variable stars*** (classes IV & V), in which the light output varies due to either processes external to the star itself or the rotation of the star:

▸ **Class I – pulsating variables.** Their brightness variations are due to the periodic expansion and contraction of the surface layers of the stars; *i.e.*, the star actually increases and decreases in size periodically. For example, pulsating variables might swell and shrink due to internal forces {SEDOV ⇨ 1959}. Examples include Cepheids {GOODRICKE ⇨ 1784}, RR Lyrae, RV Tauri, and long-period variables such as Mira {FABRICIUS ⇨ 1596} and semiregular pulsating variables.

▸ **Class II – eruptive (or cataclysmic) variables.** These are unstable stars that experience in their lifetime one or more eruptions on their surfaces like flares or mass ejections. Examples encompass protostars, giants and supergiants, massive hot stars

---

[1941] F.L. WHIPPLE: *Meteoric phenomena and meteorites: the conquest of interplanetary space.* In: (C.S. White and O.O. BENSON JR., eds.) *Physics and medicine of the upper atmosphere.* University of New Mexico Press, Albuquerque, NM (1952), pp. 137-170.

[1942] L.S. SWENSON JR., J.M. GRIMWOOD, and C.C. ALEXANDER: *This new ocean: a history of project Mercury.* Rept. NASA SP-4201, NASA History Series (1989); http://history.nasa.gov/SP-4201/toc.htm.

[1943] *Shield development, basic concept.* NASA-JSC, Houston, TX; http://hitf.jsc.nasa.gov/hitfpub/shielddev/basicconcepts.html.

[1944] R.P. CARLISLE: *Where the fleet begins.* Naval Historical Center, Dept. of the Navy, Washington, DC (1998), pp. 483-484.

[1945] B.V. KUKARKIN and P.P. PARENAGO: *Obij katalog peremennych zvezd.* Izd. Akad. Nauk SSSR, Moskva (1948); English version of introduction and remarks from the *General catalogue of variable stars* (ed. by the International Astronomical Union). H.A. Kluyver, Leiden (1949).

[1946] *GCVS, description.* Sternberg Astronomical Institute, Moscow; http://www.sai.msu.su/groups/cluster/gcvs/gcvs/intro.htm.

{Wolf-Rayet Variables ⇨1976}, symbiotic variable (or binary) stars {Dual Hour-Glass Shaped Nebula ⇨1999}, and R Coronae Borealis variables (which decrease in brightness caused by the veiling of the star by expelled thick carbon clouds).

- **Class III – explosive (or catastrophic) variables.** They produce violent outbursts due to thermonuclear burst processes in their surface layers (novae) or deep in their interiors (supernovae). **The majority of explosive and nova-like variables are close binary systems.** It is often observed that the hot dwarf component of the system is surrounded by an accretion disk consisting of matter lost from the other, cooler, and more extended component. **Matter fed to the dwarf component from its distended companion appears to produce instabilities that result in a violent explosion,** rapidly raising the luminosity by as much as ten magnitudes. Some events – as implied by the term *catastrophic* (or *cataclysmic*) – result in the destruction of the star, whilst others can reoccur one or more times. Prominent examples include novae {ANDERSON ⇨1901; Nova CYGNI 1992 ⇨1992}, supernovae {Guest Star ⇨1006; Guest Star ⇨1054; BRAHE ⇨1572; KEPLER ⇨1604; BAADE & ZWICKY ⇨1931; I. SHELTON ⇨1987; DÍAZ ⇨1993}, and dwarf novae {HIND ⇨1855}.[1947]

- **Class IV – eclipsing binary variables.** Appearing as a single point of light to an observer, their variations in light intensity are caused by one star passing in front of the other relative to an observer; such stars include Algol {GOODRICKE ⇨1783} and Beta Lyrae {GOODRICKE ⇨1784}.

- **Class V – rotating variables.** The variation in brightness of a rotating variable derives from the presence of "starspots" on its surface, which can be either dimmer (such as the sunspots on our Sun) or brighter than the majority of the surface of the star. A few rotating neutron stars change in optical brightness (optically variable pulsars).

| | | |
|---|---|---|
| 1948 | U.S.A. & U.K. | In a joint study,[1948] **George D. BIRKHOFF** (Harvard University, Cambridge, MA), **Duncan P. MACDOUGALL** (Naval Ordnance Laboratory, Washington, DC), **Emerson M. PUGH** (Carnegie Institute of Technology, Pittsburgh, PA), **and Geoffrey I. TAYLOR** (Trinity College, Cambridge, U.K.) **present hydrodynamic theories of jet formation and target penetration by explosives with lined conical cavities.** |

(i) They assume that the pressures and energies involved are so great that the strength of the metals can be ignored and that, therefore, these **metals can be treated as compressible perfect liquids.**

(ii) The application of classical hydrodynamic theory, which ignores the role of shock waves within the flow, predicts a forward stream of material (a jet). The jet velocity becomes infinite as the angle between the colliding plates approaches zero. Their collision model shows good agreement with experiments.

(iii) Concerning the penetration phenomenon of a jet from a shaped charge impacting an armor plate, they make the following interesting note, "While a stream of water washes mud out of a mud bank, **the jet of metal does not wash or erode metal out of the target.** Careful weighings have shown that a metal jet is captured by a metal target, which loses no weight except a very small amount at the front surface. The hole is produced by plastic flow of the target material in a radial direction."

In the early 1950s, their theory was resumed by some Los Alamos shock physicists in order to determine the limiting conditions for jet formation in high-velocity collisions {WALSH, SHREFFLER & WILLIG ⇨1953}.

---

[1947] *Variable stars.* Australia Telescope National Facility, CSIRO; http://outreach.atnf.csiro.au/education/senior/astrophysics/variable_types.html#varextrinsic.

[1948] G. BIRKHOFF, D.P. MACDOUGALL, E.M. PUGH, and G.I. TAYLOR: *Explosives with lined cavities.* J. Appl. Phys. **19**, 563-582 (1948).

| 1948 | Trinity College, Cambridge, England | **Hermann Bondi**,[1949] an Austrian-British mathematician and astronomer, **and Thomas Gold**, an Austrian-born British astronomer at Royal Greenwich Observatory, **introduce a theory of "steady-state cosmology."** It is based on the idea that matter is continuously created in the Universe (causing it to expand), and always has been, and that there was no initial singularity to mark the beginning of its expansion.

In the same year, Sir Fred Hoyle,[1950] a British cosmologist at Cambridge and an early opponent of the Big Bang model {Alpher, Gamow & Herman ⇒1948}, supposes that matter (hydrogen) is created in the intergalactic vacuum at a small but steady rate, resulting in the expansion of the Universe but a constant mass density within it. • In a series of popular radio talks in Britain in the 1940s, entitled *Nature of the Universe*, **Hoyle coined the term *Big Bang*** to ridicule the rival concept of an explosive origin to the Universe {⇒c.14 Ga ago}, but this term is now widely used and the explosion theory is generally accepted. |
|---|---|---|
| 1948 | Harvard University, Cambridge, Massachusetts | **Percy W. Bridgman**,[1951] a U.S. professor of physics, **measures the equations of state for numerous elements** (*e.g.*, graphite, beryllium, aluminum, magnesium, titanium, uranium) **and substances** (*e.g.*, glass) **up to a static pressure *p* of 100 kbar**. His data can be simply described by the relation

$$(1 - \rho_0/\rho) = ap + bp^2,$$

where $\rho_0$ and $\rho$ are the densities of the initial and compressed states, respectively, and the constants $a$ and $b$ are characteristic parameters of the material. • Isothermal hydrostatic equations of state for isotropic solids that closely match Bridgman's simple relation have been proposed on the basis of

▸ an atomic model that can be applied up to almost 100 kbar, which was derived in 1944 by the German physicist Reinhold H. Fürth;[1952]

▸ quantum mechanical calculations {Feynman, Metropolis & Teller ⇒1949}; and

▸ finite elastic deformation {Murnaghan ⇒1937}.

**Since the change in entropy due to the change in pressure is only third-order for weak shock waves, it is possible to model metals behind the shock front as an ideal fluid, and to apply a hydrostatic equation of state** {Duvall ⇒1955}. |
| 1948 | Brown University, Providence, Rhode Island | **Robert Hugh Cole**,[1953] a U.S. professor of physics who acted as research supervisor at the Underwater Explosives Research Laboratory in Woods Hole, MA, during World War II, **publishes his book *Underwater Explosions*.** It is largely the result of research into underwater explosions carried out by many groups in the period 1941–1946 and his association with the work of the Underwater Explosives Research Laboratory.[1954] This wartime organization was established under contract to the Woods Hole Oceanographic Institution (WHOI) along with the Office of Scientific Research & Development (OSRD), and later to the U.S. Navy Bureau of Ordnance (NBO). • His book, reprinted several times, has become a classic on underwater explosion phenomena. Moreover, it is perhaps the most commonly consulted textbook on shock waves in liquids.

In 1950, the results of underwater explosion research carried out in the United States and the United Kingdom were published in a collection of papers {UNDEX Reports ⇒1950}. Pro- |

---

[1949] H. Bondi and T. Gold: *The steady-state theory of the expanding Universe*. Month. Not. Roy. Astron. Soc. **108**, 252-270 (1948).
[1950] F. Hoyle: *A new model for the expanding Universe*. Month. Not. Roy. Astron. Soc. **108**, 372-382 (1948).
[1951] P.W. Bridgman: *The compression of 39 substances to 100,000 kg/cm²*. Proc. Am. Acad. Arts Sci. **76**, 55-70 (1948).
[1952] R.H. Fürth: *On the equation of state for solids*. Proc. Roy. Soc. Lond. **A183**, 87-110 (1944).
[1953] R.H. Cole: *Underwater explosions*. Princeton University Press, Princeton, NJ (1948); Dover Publ., New York (1965).
[1954] W.G. Schneider, E.B. Wilson, and P.E. Cross: *Underwater explosives and explosions*. In: (M.P. White, ed.) *Effects of impact and explosion*. Tech. Rept. of Div. 2, NDRC, OSRD, Washington, DC (Sept. 1946), pp. 19-63.

gress in underwater explosion research up to 1956 has been reviewed by Hans G. SNAY,[1955] and progress up to 1977 by Maurice HOLT {⇨1977}. Compared to the very large number of studies that have been published on the effects of underwater explosions from scientific, technological and military points of view, rather few investigations into their effects on sea life have been carried out {LEWIS ⇨1996}.

| 1948 | Imperial Chemical Industries, Ltd., Stevenston, England |

J.H. COOK,[1956] a British explosives specialist, **first describes an explosively-driven *plane-wave generator*** {⇨Fig. 4.11−B}. Consisting of a lens-type combination of two explosives with a slow detonation velocity (a TNT/AN mixture) and a fast detonation velocity (RDX), it converts a point initiation into a planar detonation wave. • COOK's motivation for developing a plane-wave generator arose from the need to produce large-area explosive engravings {MUNROE ⇨1888}. A single ordinary (non-planar) explosive charge was satisfactory for small engravings, but this method produced increasingly vague patterns as the engravings became larger. With his plane-wave generator, COOK successfully produced large engravings up to a size of about $20 \times 20$ in.$^2$ ($50.8 \times 50.8$ cm$^2$).

COOK's ingenious concept was later adopted for early Hugoniot measurements {GORANSON ET AL. ⇨1955; DICK ⇨1969}. In addition, his method of tailoring explosives in order to shape detonation fronts is a crucial condition for the proper ignition of implosion-type nuclear weapons and of most *Explosively Formed Projectile (EFP)* techniques.

| 1948 | Ames Aeronautical Laboratory, Moffet Field, California |

George E. COOPER[1957] and George A. RATHERT, two U.S. aeronautical engineers, **first describe a method of making natural shadowgraphs to visualize shock waves traveling over straight-wing airplanes during transonic flight.** The method uses the refraction of the sunlight passing through the density discontinuity of the shock. • This technique was more recently resumed by David F. FISHER[1958] and collaborators at the NASA Dryden Flight Research Center in Edwards, CA, in order to study wing compression effects for a transport aircraft (Lockheed L-1011) during transonic flight up to a Mach number of $M = 0.85$ {⇨Fig. 4.14−L}.

| 1948 | Institute for Mathematics & Mechanics, University of New York, New York |

Richard COURANT[1959] and Kurt O. FRIEDRICHS, two German-born U.S. mathematicians, **publish their textbook *Supersonic Flow and Shock Waves.*** It describes the state of knowledge at the end of World War II, based on classical theory and wartime contributions mostly from England and the United States. • Their book has since become an important standard source of references and was (and probably still is) the most commonly consulted textbook on shock waves. However, as evidently from the title, it is limited to fluids only; *i.e.,* it does not treat shock waves in solids – a subject of research which has now grown into a huge field of its own.

Textbooks on gas dynamics addressing war and postwar research were written in Germany
- by Robert SAUER (1943): *Theoretische Einführung in die Gasdynamik* ("Theoretical Introduction into Gas Dynamics");
- by Klaus OSWATITSCH {⇨1952}: *Gasdynamik* ("Gas Dynamics");
- by Jürgen ZIEREP (1963): *Vorlesungen über theoretische Gasdynamik* ("Lectures on Theoretical Gas Dynamics");
- by Ernst BECKER (1966): *Gasdynamik* ("Gas Dynamics");

---

[1955] H.G. SNAY: *Hydrodynamics of underwater explosions.* In: (F.S. SHERMAN, ed.) *Symposium on Naval Hydrodynamics* [Washington, DC, Sept. 1956]. Nat. Research Council, Nat. Academy of Sciences, Washington DC (1957), pp. 325-352.
[1956] J.H. COOK: *Engraving on metal plates by means of explosives.* Research (Lond.) **1**, 474-477 (1948).
[1957] G.E. COOPER and G.A. RATHERT: *Visual observations of the shock wave in flight.* Rept. NACA-RM A8C25 (1948).
[1958] D.F. FISHER ET AL.: *Observations of shock waves on a transport aircraft.* In: (G.M. CARLOMAGNO and I. GRANT, eds.) *Proc. 8th Int. Symposium on Flow Visualization* [Sorrento, Italy, Sept. 1998]. CD Rom Proceedings, ISBN 0953399109, pp. 17.1-17.7.
[1959] R. COURANT and K.O. FRIEDRICHS: *Supersonic flow and shock waves.* Interscience, New York (1948), p. 334.

▸ by Robert SAUER (1966): *Nichtstationäre Probleme der Gasdynamik* ("Nonstationary Problems of Gas Dynamics"); and

▸ by Jürgen ZIEREP (1972): *Theoretische Gasdynamik* ("Theoretical Gas Dynamics").

In the Soviet Union, textbooks on gas dynamics were written

▸ by Yakov B. ZEL'DOVICH (1946): "Theory of Shock Waves and Introduction to Gas Dynamics" (in translation);

▸ by Kyrill P. STANYUKOVICH {⇨1955}: *Neustanovivshiesya dvizheniya sploshnoi sredy* ("Unsteady Motion of Continuous Media");

▸ by Genrich N. ABRAMOVIC (1958): *Prikladnaja gazovaja dynamika* ("Applied Gas Dynamics");

▸ by Aleksandr S. PREDVODITELEV (1962): *Gazodinamika i fizika gorenija* ("Gas Dynamics and Physics of Combustion");

▸ by Artases M. MKHITARYAN (1964): *Gidravlika i osnovy gazodinamiki* ("Hydraulics and Fundamentals of Gas Dynamics"); and

▸ by Yakov B. ZEL'DOVICH and Yuri P. RAIZER (1966 & 1967): *Fizika udarnykh voln i vysokotemperaturnykh gidrodinamicheskikh ëiiavleniæi* ("Physics of Shock Waves and High-Temperature Hydrodynamic Phenomena").

| | | |
|---|---|---|
| 1948 | *Engineering Laboratory, University of Cambridge, England* | **Rhisiart Morgan DAVIES,**[1960] a skilful experimenter and collaborator of Geoffrey I. TAYLOR, **improves the Hopkinson pressure bar** {B. HOPKINSON ⇨1914}. He devises a pressure bar – later known as the **"Davies bar"** {⇨Fig. 4.3–P} – in which the measurements are made electrically, thus giving a continuous record of the longitudinal displacement produced by the pressure pulse at the free end of the bar. By differentiating this curve, he obtains the pressure-time curve for this pulse. Based on his theoretical and experimental results, he concludes |

▸ that the pressure bar is incapable of accurately measuring pressures which are subject to rapid changes over timescales of the order of 1 µs; and

▸ that the pressure, deduced from the displacement at the measuring end, takes a finite time to rise to an approximately constant value – even when the force is applied instantaneously to the end of the pressure bar.

This rise time depends on the Poisson ratio as well as on the radius and length of the bar. For example, for a 1-in. (2.54-cm)-dia. 6-ft (189-cm)-long bar of annealed tool steel, the rise time amounts to about 5 µs.

| | | |
|---|---|---|
| 1948 | *Private laboratory, Hamburg, Germany* | **Frank B.A. FRÜNGEL,**[1961] a German physicist and inventor, **investigates the mechanical conversion efficiency of spark discharges in liquids** using small weights $m$ thrown up by the expanding spark plasma to a height $h$. He first compares the electrical energy $E_e = \frac{1}{2} C U_0^2$, initially stored in a capacitor of capacity $C$ charged up to a voltage $U_0$, with the mechanical potential energy $E_p = mgh$, and observes that efficiencies are only on the order of 1%, because it is impossible to establish an impedance match between the time-varying resistance of the liquid spark and the resistance of the critically damped condition in the discharge circuit (aperiodic limit resistance). However, since the conversion of energy is completed within a few microseconds, one can greatly accelerate the weight within this time period. |
| 1948 | *Dept. of Physics, George Washington University, Washington, DC* | **George GAMOW**, a Russian-born U.S. professor of physics, and his graduate student Ralph A. ALPHER prepare an article on the origins of the chemical elements, in which **they attempt to explain the distribution of chemical elements throughout the Universe via a primeval thermonuclear explosion: the Big Bang that gave birth to the Universe** {FRIEDMANN ⇨1922; LEMAÎTRE ⇨1927; BONDI ⇨1948}. According to this theory – the so-called **"Alpher-** |

---

[1960] R.M. DAVIES: *A critical study of the Hopkinson pressure bar*. Trans. Phil. Soc. **A240**, 375-457 (1948).
[1961] F.B.A. FRÜNGEL: *Zum mechanischen Wirkungsgrad von Flüssigkeitsfunken*. Optik **3**, 124-127 (1948).

Bethe-Gamow theory" (or "alpha, beta, gamma theory") – atomic nuclei were built up after the Big Bang by the successive capture of neutrons by the pairs and triplets of neutrons initially formed. • Also listing Hans A. BETHE from Cornell University as a coauthor of the article, GAMOW sent his "α-β-γ Paper" to the editor of the American journal *Physical Review*.[1962] In this famous scientific joke, often referred to as the "alphabetical article," GAMOW decided that it would be "unfair to the Greek alphabet to have the article signed by ALPHER and GAMOW only, and so the name of Dr. BETHE [*in absentia*] was inserted in preparing the manuscript for print." The paper was even published on April 1 (All Fools' Day)!

In the same year, ALPHER,[1963] Robert HERMAN (another graduate student of GAMOW) and GAMOW predict that **a residual blackbody radiation spectrum – the remnant from the primordial Big Bang event – should fill "empty" space.** They showed that the proportion of helium produced in the Big Bang depends on the temperature of the fireball in which the Universe was born. To match the observations that stars contain 25% helium, **they have to set the temperature of the Big Bang rather precisely, estimating that it has since been diluted to a few degrees Kelvin** – meaning that the black-body radiation should take the form of microwaves. • GAMOW, ALPHER and HERMAN proposed the hot Big Bang as a way to produce all of the elements observed in the Universe. However, elements heavier than helium are now thought to have originated in the interiors of stars which formed much later in the history of the Universe. However, discussions on early nucleosynthesis processes are still going on.

Seventeen years later, their prediction of residual microwave blackbody radiation was proved to be correct experimentally {PENZIAS & WILSON ⇨1965}, thus strongly supporting their *hot Big Bang model* which is now an essential pillar of modern cosmology {The Universe ⇨c.14 Ga ago; NASA-GSFC ⇨2003}.

The Big Bang model is somewhat misleading, for it suggests that the expansion of the cosmos is a result of the incredible initial temperature and pressure, rather like the explosion of a gigantic firecracker. Since the mass density (and therefore also gravitational forces) would also have been extremely high in the initial phase, which, according to EINSTEIN's General Theory of Relativity (1916), would lead to high positive pressures, this would contribute to an immediate collapse rather than to an expansion of the cosmos. However, the model has proven successful.[1964]

| | | |
|---|---|---|
| 1948 | *Mathematics Dept., Imperial College of Science and Technology, London, England* | Harry JONES,[1965] a British professor of mathematics, **and A.R. MILLER study the motion of the detonation products of a solid explosive (TNT) behind a planar detonation wave.** They calculate the changes in equilibrium compositions and energies as well as pressure and density during isentropic expansion after detonation.

(i) They state: "For a solid explosive the pressure, temperature and molecular volume in the detonation wave-front have values which lie far outside the range which has been studied in other connections, so that little guidance is available for the purpose of setting up the equation of state appropriate to the conditions in the detonation wave-front."

(ii) The detonation process for solid explosives depends on the loading density ($\rho_0$), so the calculation of this process for solid explosives presents greater difficulties than for explosive gases. They use a virial equation of state (EOS), given in the form of a pressure-density-temperature relation and in powers of the pressure, terminating at the term in $p^3$:

$$pv'/N' = RT + bp + cp^2 + dp^3,$$

---

[1962] R.A. ALPHER, H.A. BETHE, and G. GAMOW: *The origin of chemical elements.* Phys. Rev. **73** [II], 803-804 (1948).
[1963] R.A. ALPHER, R. HERMAN, and G. GAMOW: *The thermonuclear reactions in the expanding Universe.* Phys. Rev. **74** [II], 1198-1199 (1948).
[1964] E.P. TRYON: *Cosmic inflation.* In: (R.A. MAYERS, ed.) *Encyclopaedia of physical science and technology.* Academic Press, Orlando etc. (1992), vol. 3, pp. 709-743; J.P. HUCHRA: *Galactic structure and evolution.* Ibid. vol. 5, pp. 739-755.
[1965] H. JONES and A.R. MILLER. *The detonation of solid explosives: the equilibrium conditions in the detonation wave-front and the adiabatic expansion of the products of detonation.* Proc. Roy. Soc. Lond. **A194**, 480-507 (1948).

where $R$ is the gas constant, $v'$ is the volume and $N'$ is the number of moles at temperature $T$ and pressure $p$ of the gaseous products from the detonation of 1 mole of explosives. By comparing this relation with observed values for the detonation velocity, the virial coefficients in the equation of state can be determined. The virial coefficients $b$, $c$ and $d$ are functions of $T$ and of the mole fraction of the constituents of the gas. However, they assume that the virial coefficients are constants and determine values for them that provide agreement with the measured values of the detonation velocity for loading densities less than 1.5 g/cm$^3$.

(iii) For TNT at a loading density of 1.5 g/cm$^3$, they present pressure, temperature and volume data in a tabular form during adiabatic expansion of the detonation wavefront from 158.8 kbar to 2.8 bar.

In the 1960s, **the Jones-Miller equation of state (JM-EOS)** describing the adiabatic expansion of detonation products was further developed so that it could be applied to hydrodynamic calculations aimed at improving predictions of solid high explosive performance.

▸ At Lawrence Radiation Laboratory (LRL, Livermore, CA) Mark L. WILKINS,[1966] a detonation researcher, developed an equation of state based primarily on spherical metal expansion experiments; when used in hydrodynamic calculations, his equation accurately predicts results for experimental geometries, emphasizing the early stages of detonation product expansion.

▸ His colleague Edward Louis LEE[1967] (a U.S. physical chemist) and his collaborators at LRL extended this work using results from the *cylinder test*, in which a cylindrical charge expands a surrounding copper tube. **Their so-called "Jones-Wilkins-Lee equation of state (JWL-EOS)," which covers a wide range of thermodynamic states from very dense to highly expanded, is often used to model the behavior of explosive products.**

▸ Correlated with experiments for a number of explosives, the JWL-EOS can also be used for geometries involving a large expansion of the detonation products. The HOM EOS {MADER ⇒1962} and the JWL-EOS have been employed for the reactants and products when numerically modeling the reactions of homogeneous explosives (*e.g.*, TNT and RDX/TNT mixtures).[1968]

| | | |
|---|---|---|
| 1948 | *Randall Laboratories of Physics, University of Michigan, Ann Arbor, Michigan* | Charles W. MAUTZ,[1969] a U.S. physicist, **and his collaborators first demonstrate the suitability of a shock tube** {BLEAKNEY ⇒1946} **instead of a wind tunnel for studying supersonic flow patterns.** They point out that the shock tube, which acts as an intermittent wind tunnel for some 100 µs, should be particularly suited to studies of transonics, which was a regime of great concern to those involved in practical high-speed aeronautics at that time. They also present photos of bow wave effects on a wedge at $M = 2.42$. |
| 1948 | *Armament Research Establishment (ARE), Fort Halstead, Kent, England* | Philip L. OWEN[1970] and Charles K. THORNHILL, two British fluid dynamicists, **study the flow in an axially symmetric supersonic jet emerging from a nearly sonic orifice into a vacuum theoretically** {DE SAINT-VENANT & WANTZEL ⇒1839}.<br>(i) Using the method of characteristics, they entirely determine the flow numerically by an iterative process rather than graphically. In order to obviate the difficulties of a sonic orifice (at |

---

[1966] M.L. WILKINS: *The equation of state of PBX 9404 and LX04-01.* Rept. UCRL-7797, Lawrence Radiation Laboratory (LRL), UC Livermore, CA (1964).

[1967] E.L. LEE, H.C. HORNIG, and J.W. CURY: *Adiabatic expansion of high explosive detonation products.* Rept. UCRL-50422, Lawrence Radiation Laboratory (LRL), UC Livermore (1968).

[1968] L. DONAHUE and R.C. RIPLEY: *Simulation of cylinder expansion tests using an Eulerian multiple-material approach.* 22nd Int. Symposium on Ballistics [Vancouver, B.C., Nov. 2005].

[1969] C.W. MAUTZ, F.W. GEIGER, and H.T. EPSTEIN: *On the investigation of supersonic flow patterns by means of the shock tube.* Phys. Rev. 74 [II], 1872-1873 (1948).

[1970] P.L. OWEN and C.K. THORNHILL: *The flow in an axially symmetric supersonic jet from a nearly sonic orifice into a vacuum.* A.R.E. Rept. No. 30/48, Armament Research Establishment (ARE), U.K. (1948).

which the initial characteristics would be perpendicular to the flow and the potential equation would be parabolic), they assume a Mach angle of 85° in the plane of the circular orifice.

(ii) They state that within the area bounded by the orifice and the first wavefront from the orifice boundary – *i.e.*, within the first shock bottle {SLADE JR. ⇨1946} – the solution to the flow problem is a universal one and is independent of the external pressure.

(iii) Calculated axial distributions of the pressure along the axes of supersonic jets with sonic orifices show good agreement with Pitot and interferometric measurements.

In muzzle blast research, their method has frequently been applied in theoretical considerations of the shock bottle {ESCLANGON ⇨1925}.

| | | |
|---|---|---|
| 1948 | *Armament Research Dept., Ministry of Supply, London, England* | Donald Cecil PACK,[1971] a British applied mathematician, **and his collaborators publish the first paper on the experimental study of shock waves in solid matter.** Using a type of electric-pin method {G.I. TAYLOR ⇨1944} and a Baird-type microsecond chronometer to follow the propagation of shock waves through various lengths of steel and lead, **they first observe a two-wave structure indicating an elastic-plastic response ("elastic-plastic waves").** They measure the velocity of the plastic wave – a stress wave that results in the irreversible repositioning of atoms relative to their neighbors, and a type of shock wave – and find that its velocity in steel is less than (and in lead it is greater than) the velocity of planar elastic waves.<br><br>In his review of Soviet research into detonation and solid-state shock wave physics, Lev V. AL'TSHULER[1972] reported that he first encountered a two-wave configuration of elastic-plastic waves in the mid-1940s, while studying the effects of explosions on steel plates. |
| 1948 | *Physical Sciences Division, Stanford Research Institute (SRI), Menlo Park, California* | Thomas C. POULTER, a U.S. physicist and inventor, joins the Stanford Research Institute (SRI) and **establishes a laboratory dedicated to dynamic high-pressure research.** ▪ The laboratory – renamed the **Poulter Laboratory** in 1953 in acknowledgment of POULTER's numerous contributions to detonation and shock pulse phenomena – became widely known for its computational and experimental investigations on penetration, fracture and fragmentation, and the dynamic responses of shock-loaded materials and structures to impacts, explosions, fatigue, corrosion and fire.<br><br>Stanford Research Institute was founded in 1946 by the trustees of Stanford University. In 1977, it was renamed "SRI International." |
| 1948 | *Werkstoff-Hauptlaboratorium, Siemens & Halske AG, Berlin-Siemensstadt, Germany* | Werner SCHAAFFS[1973] and Ferdinand TRENDELENBURG, two German physicists and acousticians, **generate powerful electric discharges in thin layers of organic liquids.** Using flash soft radiography, they observe that the liquid surrounding the central discharge plasma is shock-compressed in the form of a **"compression ring."** ▪ Later studies performed at the Physics Department of the TU Berlin revealed that the ring structures of some shock-compressed organic liquids expand at only some tens of m/s – apparently indicating a shock-induced solidification of the liquid.[1974] This unusual phenomenon is accompanied by the explosion-like annular detachment of matter from the outer and inner periphery of the compression ring {⇨Fig. 4.14–T}, which is explained by the generation of *Schmelz-Stoßwellen* ("melting shock waves").<br><br>However, attempts made in the early 1970s to produce flash Debye-Scherrer diagrams from a $CCl_4$ compression ring – an unmistakable evidence for the existence of microcrystalline |

---

[1971] D.C. PACK, W.H. EVANS, and H.J. JAMES: *The propagation of shock waves in steel and lead*. Proc. Phys. Soc. (Lond.) **60**, 1-8 (1948).
[1972] L.V. AL'TSHULER: *Use of shock waves in high pressure physics*. Sov. Phys. Uspekhi **8**, 52-91 (1965), p. 85.
[1973] W. SCHAAFFS and F. TRENDELENBURG: *Röntgenographische Untersuchung der beim dielektrischen Funkenüberschlag auftretenden Schallwellen*. Z. Naturforsch. **3a**, 656-668 (1948).
[1974] W. SCHAAFFS and P. KREHL: *Röntgenblitzuntersuchungen über die Entstehung einer aus einer durch Stoßwellenprozesse zur Erstarrung gebrachten Flüssigkeit*. Acustica **23**, 99-107 (1970).

structure in the shock-solidified organic liquid – failed.[1975] One possible reason for the slow expansion velocity of the compression ring might be a strong increase in the dynamic viscosity in shock-compressed organic liquids, which means that the liquid does not change from an originally amorphous into a microcrystalline state. ▪ The relationship between the viscosity and changes in the phase state of shock-compressed substances has been discussed by various Soviet researchers {SAKHAROV ET AL. ⇨1963}.

| 1948 | Safety in Mines Research & Testing Branch, British Ministry of Fuel and Power, London, England | Wilfred Charles F. SHEPHERD,[1976] a British combustion researcher, **investigates ignition temperatures of gas mixtures** (for example methane/oxygen [$CH_4/O_2$] and ethylene/oxygen [$C_2H_4/O_2$]) **that can be detonated by an incident shock wave.** Using a "bursting diaphragm apparatus" – the term *shock tube* had not entered general use {BLEAKNEY ⇨1949} – with a buffer section to separate the driver gas from the combustible mixture, he observes the initiation of detonation using the schlieren technique and a high-speed drum camera. **His results show that ignition can be caused by shock waves of quite low intensity.** Furthermore, the photographic records show that the pressure effects are at their strongest fairly near to the diaphragm, and that ignition occurs here. The minimum igniting pressures for the most explosive mixtures of $CH_4/O_2$ are between 100 and 65 psi (6.9 and 4.5 bar), and for $C_2H_4/O_2$ they are below 50 psi (3.4 bar). |
|------|---|---|
| 1948 | University of Colorado at Boulder, Colorado | Harold Ward SIBERT,[1977] a U.S. mathematician and professor of aeronautical engineering, **publishes his textbook *High-Speed Aerodynamics*.** It is the first book dedicated to this particular branch of gas dynamics in which most of the studies performed had previously been classified due to its immediate military relevance. His book, provided with a note to instructors, also addresses supersonics and shock waves (*e.g.*, the Ackeret and Busemann approximate theories for supersonic flow, Prandtl-Meyer supersonic flow around a corner, and supersonic flow past a cone). ▪ Up to the end of World War II, the most advanced research on high-speed aerodynamics [Germ. *Hochgeschwindigkeits-Aerodynamik*] was carried out in Germany, where frontiers were pushed as far as hypersonics {ERDMANN ⇨1943/1944; ⇨Fig. 4.9−H}.<br><br>Four years later, **a textbook with the same title was also published in the United Kingdom** {HILTON ⇨1952}. The complex subject of high-speed aerodynamics was first treated thoroughly in the 12-volume compendium ***High Speed Aerodynamics and Jet Propulsion***, which was published in the United States in the period 1954–1959 {Aeronautics Publication Program ⇨1949}. |
| 1948 | University of New York, New York City, New York | James J. STOKER,[1978] a U.S. applied mathematician at the Institute of Mathematics and Mechanics, **uses mathematical analysis and the equations of motion to show that the wavefront of a strong hydraulic jump will grow steeper as it progresses.** However, unlike shocks in gases, the wavefront eventually leans forward, thus producing a "breaker" or "roller" {⇨Fig. 4.12−H}. |
| 1948 | Institute for Advanced Study (IAS), Princeton, New Jersey | Abraham H. TAUB,[1979] a mathematician and theoretical physicist, **works out the *relativistic Rankine-Hugoniot equations*.** ▪ **A number of astrophysical problems of current interest require the application of relativistic hydrodynamics to be able to treat them correctly.** This is particularly true for simulations of pulsar wind nebulae and the interactions of pulsar |

---

[1975] P. KREHL and W. SCHAAFFS: *Investigations concerning the applicability of flash X-ray interference and laser technology to shock-induced solidification of organic liquids.* In: (E. LAVIRON, ed.) *Proc. 10th Int. Congress on High-Speed Photography* [Nice, France, Sept. 1972]. Association Nationale de la Recherche Technique (ANRT), Paris (1973), pp. 295-300.

[1976] W.C.F. SHEPHERD: *The ignition of gas mixtures by impulsive pressures.* In: (B. LEWIS, ed.) *Proc. 3rd Symposium on Combustion, Flame & Explosion Phenomena* [Madison, WI, Sept. 1948]. Williams & Wilkins Co, Baltimore, MD (1949), pp. 301-316.

[1977] H.W. SIBERT: *High-speed aerodynamics.* Prentice-Hall, New York (1948).

[1978] J.J. STOKER: *The formation of breakers and bores.* Comm. Appl. Math. **1**, 1-87 (1948).

[1979] A.H. TAUB: *Relativistic Rankine-Hugoniot equations.* Phys. Rev. **74** [II], 328-334 (1948).

winds with old supernova remnants, the radial behavior of relativistic fireballs thought to give rise to gamma-ray bursts {Vela Satellites ⇨1960s; COLGATE ⇨1967}, and the dynamics of relativistic jets from active galactic nuclei (AGNs) and galactic "microquasars." Relativistic effects are also important at very high temperatures (*i.e.*, for large molecular velocities) and possibly when extrapolating equations of state for matter to very high pressures {FEYNMAN ET AL. ⇨1949}. Other applications are in nuclear physics, *e.g.*, in simulations of relativistic nuclear collisions.

After the war, TAUB also made other significant contributions to the relativistic theory of continua. Among his other achievements were the development of the Hamilton principle for a perfect fluid and other variational principles in general relativistic hydrodynamics, the circulation theorem, and the stability of fluid motions in general relativity.[1980]

| | | |
|---|---|---|
| 1948 | *Cavendish Laboratory, University of Cambridge, England* | **Geoffrey I. TAYLOR**,[1981] a British physicist and Yarrow Research Professor, **and A.C. WHIFFIN**,[1982] a British research engineer at the Road Research Laboratory of the Department of Scientific and Industrial Research, **report on theoretical and experimental investigations into the dynamic yield strengths of various materials at high strain rates associated with the impact and penetration of projectiles.** |

(i) In order to treat the normal collision of a circular cylinder of initial length $L$ and uniform cross-sectional area with a rigid target, they suggest a simple theoretical model where the rear portion of the projectile is taken to be completely rigid and to be traveling with instantaneous particle velocity. Furthermore, their model assumes a deformation process involving an unstressed rear portion of the cylinder of length $X$ and a discontinuity of length $L_1 - X$ in cross-section, which extends from the rigid target surface to the plastic wavefront; $L_1$ is the total length of the cylinder after impact.

(ii) Their method – known as the **"Taylor test"** {⇨Figs. 4.3–Q, R} – had already been developed in the late 1930s, but had not been published earlier due to World War II. Surprisingly, tests performed with mild steel show that the dynamic yield strength is considerably above the static value before plastic strain is suffered, an important result for shock-loaded armor plates. However, the momentary strain rate, which may vary along the impacted rod, is difficult to estimate from these measurements. • A method of equating the kinetic energy at impact with the plastic work provides a mean dynamic yield stress {HAWKYARD ET AL. ⇨1968}.

Since the mid-1980s, the Taylor test has usually been used to check constitutive models by comparing the shapes of recovered cylinders with computer predictions, rather than for its original purpose of providing the dynamic yield stresses of materials.[1983] The Taylor test is important because it helps to bridge the gap between the split-Hopkinson pressure bar and flyer plate experiments. André LICHTENBERGER[1984] and his associates in France discussed the applicability of the Taylor test at high strain rates of up to $5 \times 10^4 \, \text{s}^{-1}$ as a source of data for numerical models of dynamic materials behavior.

| | | |
|---|---|---|
| 1949 | *White Sands Proving Ground, New Mexico* | On February 24, a U.S. WAC-Corporal two-stage rocket carried atop a captured German V2 rocket {HVA Peenemünde ⇨1940/1941 & 1942; ⇨Fig. 4.20–C} modified as a booster sends a payload into space on a short suborbital flight. The V2 reaches an altitude of 102 km and a top |

---

[1980] A.H. TAUB: *Stability of general relativistic gaseous masses and variational principles*. Comm. Math. Phys. **15**, 235-254 (1969).

[1981] G.I. TAYLOR: *The use of flat-ended projectiles for determining dynamic yield stress. I. Theoretical considerations*. Proc. Roy. Soc. Lond. **A194**, 289-299 (1948).

[1982] A.C. WHIFFIN: *The use of flat-ended projectiles for determining dynamic yield stress. II. Tests on various metallic materials*. Proc. Roy. Soc. Lond. **A194**, 300-322 (1948).

[1983] S.M. WALLEY, P.D. CHURCH, R. TOWNSLEY, and J.E. FIELD: *Validation of a path-dependent constitutive model for FCC and BCC metals using "symmetric" Taylor impact*. In: *DYMAT 2000* [Krakow, Poland, Sept. 2000]. J. de Physique 4 (Proceedings) **10**, Pr9–69–74 (2000).

[1984] A. LICHTENBERGER, R. DORMEVAL, and P. CHARTAGNAC: *Test de TAYLOR*. Rept. No. RE/003/88, DYMAT, Establissement Technique Central de l'Armement, Arcueil, France (1988).

speed of 1,170 m/s, and the WAC an altitude of 400 km and a top speed of 2,300 m/s {⇨Fig. 4.20–D}.[1985] **The WAC-Corporal is the first man-made vehicle ever to achieve hypersonic speeds.** ▪ This unique concept involving the combination of a V2 missile with a WAC-Corporal missile – a vehicle known as the **"Bumper-WAC"** – is the world's first heavy two-stage rocket.

Between April 1946 and September 1952, 70 captured V2s were launched from the United States: 67 at White Sands Proving Ground (WSPG) in New Mexico, two at Cape Canaveral in Florida, and one from the deck of the aircraft carrier USS *Midway*.[1986]

| | | |
|---|---|---|
| 1949 | University of Toronto, Canada | In July, **foundation of the University of Toronto Institute of Aerophysics (UTIA).** The first director is Gordon N. PATTERSON.[1987] ▪ In 1964, the UTIA was renamed the UTIAS (University of Toronto Institute for Aerospace Studies).[1988] This Canadian institute has made significant contributions to shock wave physics, particularly to nonstationary gas dynamics. |
| 1949 | Institut d'Astrophysique, Paris, France | In August, **first Symposium on the Motion of Gaseous Masses of Cosmical Dimensions,** organized jointly by the International Astronomical Union (IAU) and the International Union of Theoretical and Applied Mechanics (IUTAM). Its main objective is "to bring about understanding between astrophysicists and aerodynamicists, and to formulate problems in such a way that they become amenable to mathematical treatment." The chairman is Johannes M. BURGERS. Astrophysicists meet the foremost representatives of the fields of aerodynamics, shock wave physics, detonics, plasma physics and theoretical physics (*e.g.*, H.O.G. ALFVÉN, W. HEISENBERG, F. HOYLE, Th. VON KÁRMÁN, J. VON NEUMANN, R.J. SEEGER, R.W. SPITZER, and C.F. VON WEIZSÄCKER) to discuss peculiarities in the motions and the evolution of stars and nebulae, and to consider theoretical models for a number of dynamic celestial problems of motion. **Further examples include gaseous masses with high relative velocities and with shock waves in a gravitational field, and turbulent gas motion in electromagnetic fields.**[1989] ▪ The proceedings of this conference were later published under the title *Problems of Cosmical Aerodynamics*.[1990] Beginning with the 2nd Symposium, the title of the conference was changed to **Symposium on Cosmical Gas Dynamics.** Subsequent conferences were held at Cambridge, MA (1953 & 1957); Varenna, Italy (1960); Nice, France (1965); and Yalta, Soviet Union (1969). |

At about the same time, *cosmic gas dynamics* (or *cosmic gasdynamics*) and *cosmic magnetohydrodynamics* evolved in the Soviet Union from astrophysics, plasma physics and shock wave physics.[1991] The presence of magnetic fields in a fast-moving, electrically conducting fluid yielded new supersonic phenomena due to the generation of the Alfvén wave, which is completely absent in pure electrodynamics and fluid dynamics. This led, particularly in cosmical gas dynamics, to spectacular hypotheses on supersonic motion {GOLD ⇨1953; HELFER ⇨1953; DUNGEY ⇨1958; PARKER ⇨1958 & 1961; AXFORD & KELLOGG ⇨1962} and unique discoveries {NEUGEBAUER & SNYDER ⇨1962; NESS ET AL. ⇨1964; Voyager 1 & 2 ⇨2003}.

---

[1985] M. WADE: *Encyclopedia astronautica*; http://www.astronautix.com/lvs/v2.htm.
[1986] E. STÜHLINGER and F.I. ORDWAY: *Wernher VON BRAUN: Aufbruch in den Weltraum*. Bechtle, Esslingen & München (1992), p. 163.
[1987] G.N. PATTERSON: *Pathway to excellence: UTIAS – the first twenty-five years*. University of Toronto Press, Toronto (1977).
[1988] I.I. GLASS: *Over forty years of continuous research at UTIAS on nonstationary flows and shock waves*. Shock Waves **1**, 75-86 (1991); see also UTIAS, http://www.utias.utoronto.ca/test/hist.html.
[1989] Papers directly relating to shock waves were presented by J.M. BURGERS (*Aerodynamical description of elementary expansion phenomena and shock waves; The properties of a shock wave in a gas with decreasing density*), and by E. SCHATZMAN (*The heating of the solar chromosphere by shock waves*).
[1990] J.M. BURGERS and H.C. VAN DE HULST (eds.): *Problems of cosmical aerodynamics: Proc. Symposium on the Motion of Gaseous Masses of Cosmical Dimensions* [Paris, France, Aug. 1949]. USAF Central Air Documents 254 Office, Dayton, OH (1951).
[1991] S.B. PIKELNER: *Cosmic gas dynamics*. In: (A.M. PROKHOROV, ed.) *Great Soviet encyclopedia*. Macmillan, New York & London, vol. 13 (1979), p. 173.

| | | |
|---|---|---|
| 1949 | *Soviet Nuclear Test Site Semipalatinsk-21, Kazakhstan, Soviet Union* | On August 29, **the first Soviet atomic bomb,** codenamed FIRST LIGHTNING (and dubbed JOE 1 in the Western World), **is exploded at a test site about 70 km south of Semipalatinsk-21.** It explodes at the top of a tower and has a yield of about 22 kilotons of TNT equivalent, similar to that of the first American atomic bomb {Trinity Test ⇨ 1945}.[1992, 1993]

(i) The fission bomb – an implosion-type device that uses two hemispheres of Pu-239 about 8 cm in dia. – a very similar design to that used for "The Gadget" {Trinity Test ⇨ 1945} and FAT MAN {Nagasaki ⇨ 1945} – was developed at the KB-11 Design Bureau by several theoreticians working in close cooperation, notably by Andrei D. SAKHAROV, Yakov B. ZEL'DOVICH and Evgeni I. ZABABAKHIN, as well as many top experimentalists.[1994] ▪ In response, Edward TELLER, a theoretical physicist at Los Alamos, proposed that the United States should develop H-bombs. At that time, the technical framework for their development was only embryonic; the crucial concepts evolved a year later due to work by S.M. ULAM and E. TELLER {MIKE Test ⇨ 1952}.

(ii) Vladimir S. KOMEL'KOV, who observed the explosion 15 km north from the tower, gave the following account,[1995] "On top of the tower an unbearably bright light blazed up. For a moment or so it dimmed and then with new force began to grow quickly. The white fireball engulfed the tower and the shop, and expanding rapidly, changing color, it rushed upwards. The blast wave at the base, sweeping in its path structures, stone houses, machines, rolled like a billow from the center, mixing up stones, logs of wood, pieces of metal, and dust into one chaotic mass ... Overtaking the firestorm, the shock wave, upon hitting the upper layers of the atmosphere, passed through several levels of inversion, and there, as in a cloud chamber, the condensation of water vapor began…"

In the following years, an increasing number of other countries began developing and testing their own nuclear weapons: the United Kingdom {Trimouille Island ⇨ 1952}, France {Reggan Test Site ⇨ 1960}, People's Republic of China (1964), India (1964), Israel (c.1967), Pakistan (1998), and North Korea (2006). According to U.S. statistics for the year 2002,[1996] the following nations – informally known in global politics as the "Nuclear Club" – have admitted to having the following approximate number of warheads under their control: United States (10,640), Russia (8,600), People's Republic of China (400), France (350), United Kingdom (200), Israel (estimated at 100–200),[1997] India (60–90), and Pakistan (24–48). |
| 1949 | *Princeton University, Princeton, New Jersey* | **Establishment of the Aeronautics Publication Program,** with the goal of publishing "a comprehensive and competent treatment of the fundamental aspects of the aerodynamics and propulsion problems of high-speed flight, together with a survey of those aspects of the underlying basic sciences cognate to such problems."[1998] The editorial board consists of Theodore VON KÁRMÁN, Hugh L. DRYDEN, and Hugh S. TAYLOR.

Within the period 1954–1964, a series of twelve volumes were published by Princeton University Press under the title **High Speed Aerodynamics and Jet Propulsion,** encompassing:
  ▸ vol. I: *Thermodynamics and Physics of Matter* (1955);
  ▸ vol. II: *Combustion Processes* (1956);
  ▸ vol. III: *Fundamentals of Gas Dynamics* (1958); |

---

[1992] *The Soviet Nuclear Weapons Program.* Nuclear Weapon Archive; http://nuclearweaponarchive.org/Russia/Sovwpnprog.html.
[1993] G.A. GONCHAROV and L.D. RYABEV: *The development of the first Soviet atomic bomb.* Phys. Uspekhi **45**, No. 2, 227-228 (2002).
[1994] L.V. AL'TSHULER: *Experiment in the Soviet atomic project.* In: (I.M. DREMIN and A.M. SEMIKHATOV, eds.) *Proc. 2nd Int. A.D. Sakharov Conference on Physics* [Moscow, Russia, May 1996]. World Scientific, Singapore (1997), pp. 649-655.
[1995] D. HOLLOWAY: *STALIN and the bomb.* Yale University Press, New Haven, CT & London (1994), p. 217.
[1996] *Archive of nuclear data.* Natural Resources Defense Council. New York (Nov. 2002); http://www.nrdc.org/nuclear/nudb/datainx.asp.
[1997] The Israeli government refuses to officially confirm or deny that it has a nuclear weapons program. However, in a December 2006 interview with a German TV station, Israeli Prime Minister Ehud OLMERT said that Iran aspires to possess nuclear weapons like the United States, France, Israel and Russia. Shortly after, however, he refused to confirm or deny Israel's nuclear weapon status.
[1998] From the *Preface* of vol. 1.

- vol. IV: *Laminar Flows and Transition to Turbulence* (1964);
- vol. V: *Turbulent Flows and Heat Transfer* (1959);
- vol. VI: *General Theory of High Speed Aerodynamics* (1954);
- vol. VII: *Aerodynamic Components of Aircraft at High Speeds* (1957);
- vol. VIII: *High Speed Problems of Aircraft and Experimental Methods* (1961);
- vol. IX: *Physical Measurements in Gas Dynamics and Combustion* (1954);
- vol. X: *Aerodynamics of Turbines and Compressors* (1964);
- vol. XI: *Design and Performance of Gas Turbine Power Plants* (1960); and
- vol. XII: *Jet Propulsion Engines* (1959).

Summarizing the state-of-the-art in theoretical high-speed aerodynamics and jet propulsion up to the late 1950s, this series became a useful resource for modern gas dynamicists because of its fundamental, classical treatment of compressible and rarefied flow.

1949 — *Oliver Machinery Company, Grand Rapids, Michigan*

Ralph B. BALDWIN,[1999] a U.S. astronomer and physicist, **publishes his book *The Face of the Moon*, in which he carefully discusses the impact theory with particular attention to meteoritic craters on the Earth** {GILBERT ⇨1893; GIFFORD ⇨1924}.

(i) He produces unassailable evidence that lunar craters were produced by explosions resulting from collisions with cosmic bodies.

(ii) Depicting the process of a large meteoritic impact and explosion {GIFFORD ⇨1924}, he writes, "When a meteorite strikes the Earth, it encounters tremendous resistance. This is the resultant of four factors. The first is the rigidity of the surface layers, particularly against shearing forces, the second and third are resistance to compression and the amount of heat produced, and the last is the inertia of the atoms and molecules in the earth as they fight to keep from being displaced. For a collision with a low-velocity body moving perhaps 5 miles per second [8,045 m/s] or slower, the resistance is largely controlled by rigidity; but once the velocity becomes greater than that of the waves created in the Earth's crust, then the resistance due to inertia becomes far and away the most powerful factor and all other resistances fade into insignificance. It is easy to visualize this condition. For an ordinary blow, such as a hammer striking an anvil, or even a bullet striking a piece of steel, the elastic waves travel in advance of the impinging body. Hence the molecules are set in motion and they have time to get out of the way. **On the other hand, a high-velocity meteorite strikes with a higher velocity than the shock waves it produces, the molecules are given no warning of its coming, and, consequently, they are trapped in front of the meteorite. The accelerations of the particles and the forces necessary to produce them become enormous** … At first the meteorite would plunge into the earth, moving faster than the shock waves and pushing ahead of it an ever increasing plug of compressed rock and probably a similar plug of compressed air. When the speed of the meteorite becomes less than that of the elastic waves, the vast amount of compression produced finds a shoulder against which to push, and the mass is soon stopped … At the instant of stopping, much of the transformed kinetic energy must be stored in the highly compressed matter ahead of the bolide as the elastic shock waves could not have transmitted any greater portion of it in the brief time and only a very little heat could have been conducted away. Conduction of heat is a rather slow process. With the stoppage of motion the meteorite is sitting on top of a tremendously compressed, tremendously hot plug of matter. Naturally, an explosion will be directed upward at first in the direction of least resistance, but as it develops and the explosive focus moves upwards, the blast will tend more and more horizontally."

**BALDWIN's concept has become the basis of the modern impact theory.** One riddle was the near-circular shape of lunar impact craters, which would obviously require relatively rare

---

[1999] R.B. BALDWIN: *The face of the Moon*. University of Chicago Press, Chicago, IL (1949), pp. 97-99.

vertical impacts. Laboratory experiments, however, showed that near-circular craters are also produced in high-velocity impacts where the angle of approach is far from the vertical.

| 1949 | Palmer Physical Laboratory, Princeton University, Princeton, New Jersey |

Walker BLEAKNEY,[2000] a U.S. gas dynamicist, **and Abraham H. TAUB,** a U.S. mathematician and theoretical shock wave physicist, **use the shock tube technique to study the reflection of shock waves.** To simplify triggering, they use a sliding pin passing through a seal in the wall, which pricks the center of the diaphragm, causing it to shatter. Using the Mach-Zehnder interferometer technique {L. MACH & ZEHNDER ⇨1891}, they measure the density field in the vicinity of a Mach reflection.

In the same year, **Walker BLEAKNEY**[2001] **coins the term** *shock tube* {⇨Fig. 4.10–B} in a paper published together with his collaborators David K. WEIMER and Charles H. FLETCHER. In order to visualize supersonic flow over bodies, they make the following very important proposition: "**The tube may be used as a wind tunnel with a Mach number variable over an enormous range,** it being a particularly advantageous arrangement for the investigation of phenomena of a transient nature such as the growth of boundary layers and the approach of a flow toward a steady state. It is uniquely suited to the study of flows in the transonic region…"
▪ The so-called **"cookie cutter,"** a particular shock tube technique, was also invented in BLEAKNEY's laboratory.[2002]

The shock tube proved to be a most worthwhile and economic diagnostic tool in supersonics and hypersonics.

| 1949 | Dept. of Mechanical Engineering, CalTech, Pasadena, California |

Donald S. CLARK,[2003, 2004] a U.S. physical metallurgist, **and David S. WOOD,** a U.S. materials scientist and assistant professor of mechanical engineering, **perform tension tests on solids in which the stress is applied rapidly up to values above the static elastic limit.** They observe that the transition from the elastic to the plastic regime and vice versa occurs at different stress levels – a phenomenon known as **"delayed yield."**

| 1949 | Los Alamos Scientific Laboratory (LASL), Los Alamos, New Mexico |

Richard P. FEYNMAN,[2005] Nicholas C. METROPOLIS and Edward TELLER, three theoretical physicists and contributors to the Manhattan Project {⇨1942}, **discuss the extrapolation of equations of state of matter to very high densities.** They introduce the simplification of describing the electron distribution in the matter by the Thomas-Fermi statistical model, and they improve the approximation by including relativistic effects in the calculation. They claim that their method can be applied to all atomic numbers by simply changing the linear dimensional scale used. ▪ The investigations, which had been initiated in 1943 during the Manhattan Project {⇨1942}, were related to the heavy elements U-235 {DEMPSTER ⇨1935; HEISENBERG ⇨1939; FRISCH & PEIERLS 1940; SEABORG ET AL. ⇨1941; Hiroshima Bombing ⇨1945} and Pu-239 {SEABORG ET AL. ⇨1941; Trinity Test ⇨1945; Nagasaki Bombing ⇨1945}.

At that time, both of the radioactive elements U-235 and Pu-239 were considered to be strong candidates for bomb material, but were not yet available in dense metallic forms and in sufficient quantities to experimentally determine their dynamic equations of state (Hugoniot curves). Theoretical studies on fast implosions indicated that shock pressures were in the Mbar range, which would highly compress the bomb material (Pu-239), thus decreasing the mean

---

[2000] W. BLEAKNEY and A.H. TAUB: *Interaction of shock waves*. Rev. Mod. Phys. **21**, 584-605 (1949).
[2001] W. BLEAKNEY, D.K. WEIMER, and C.H. FLETCHER: *The shock tube: a facility for investigations in fluid dynamics*. Rev. Scient. Instrum. **20**, 807-815 (1949).
[2002] R.J. EMRICH: *Walker BLEAKNEY and the development of the shock tube at Princeton*. Shock Waves **5**, 327-339 (1996).
[2003] D.S. CLARK and D.S. WOOD: *The time delay for the initiation of elastic deformation at rapidly applied constant stress*. Proc. Am. Soc. Test. Mat. (ASTM) **49**, 717-735 (1949).
[2004] T. VREELAND JR., D.S. WOOD, and D.S. CLARK: *A study of the mechanism of the delayed yield phenomenon*. Tech. Rept. NP-3433, CalTech, Pasadena, CA (Sept. 1951).
[2005] R.P. FEYNMAN, N. METROPOLIS, and E. TELLER: *Equation of state of elements based on the generalized Fermi-Thomas theory*. Phys. Rev. **75** [II], 1561-1573 (1949).

| | | |
|---|---|---|
| 1949 | Chair of Technical Mechanics and Thermodynamics, TH Stuttgart, Germany | free path of the neutrons and, therefore, the critical mass. **The so-called "Feynman-Metropolis-Teller (FMT) equation of state" for compressed atoms was subsequently used in all hydrodynamic calculations associated with nuclear weapons development.**[2006] |

1949     *Chair of Technical Mechanics and Thermodynamics, TH Stuttgart, Germany*     **Richard GRAMMEL,**[2007] a German professor of mechanical engineering, **and Konrad ZOLLER first use the Kucharski effect** {KUCHARSKI ⇨1941} **to calculate the dynamics of a whip lash,** particularly its motion at the turning point. Their analysis reveals that the tip can reach supersonic velocities, if the total mass of the lash is much larger than the mass of the tip – thus creating the typical sound of a cracking whip {LUMMER ⇨1905; CARRIÈRE ⇨1927; KREHL ET AL. ⇨1995}. ▪ It is interesting to note that GRAMMEL's studies were initiated by Prof. Heinrich ERGGELET at the ophthalmic hospital of the University of Göttingen, where strange eye injuries had been diagnosed in the late 1940s: small pieces of copper wire had penetrated deep into the eyes of patients, but their cornea had hardly suffered any injuries. These patients were coachmen who had used whip lashes made of stranded copper wire instead of textile or leather. This raised the question of whether these micro projectiles originated during whip cracking, and whether supersonic impact processes were involved.

1949     *Tohoku University, Japan*     **Akitune IMAMURA,**[2008] a Japanese tsunami researcher, **classifies *tsunami magnitude* in terms of maximum wave height measured at the coast.** However, his scale lacks a clear correlation between the magnitude and the intensity of the tsunami. ▪ The first primitive tsunami quantification scale was introduced in the 1920s by the German seismologist August SIEBERG,[2009] and (analogous to earthquake intensity scales) was based not on the measurement or estimation of transient physical parameters (*e.g.*, amplitude or wave height), but on residual macroscopic effects caused by tsunamis (*e.g.*, damage).

In the following years, the Japanese tsunami researcher Kumizi IIDA[2010, 2011] developed the concept of tsunami magnitude, $m$, which has been defined as $m = \log_2 H_{max}$, where $H_{max}$ is the maximum tsunami wave height (in meters) observed at the coast or measured with tide gauges. According to the so-called **"Imamura-Iida scale,"** tsunamis are classified into six grades in terms of *damage potential*:
- "–1" ($0.5 < H_{max} < 0.75$ m) – *no damage*
- "0" ($1$ m $< H_{max} < 1.5$ m) – *very little damage*
- "1" ($2$ m $< H_{max} < 3$ m) – *shore and ship damage*
- "2" ($4$ m $< H_{max} < 6$ m) – *some inland damage and loss of life*
- "3" ($8$ m $< H_{max} < 12$ m) – *severe destruction over 400 km of coast*
- "4" ($16$ m $< H_{max} < 24$ m) – *severe destruction over 500 km of coast*.

At the 2001 Tsunami Symposium held at Seattle, WA, Gerassimos A. PAPADOPOULOS,[2012] a tsunami researcher at the National Observatory of Athens, and Fumihiko IMAMURA, a professor of tsunami engineering at Tohoku University, proposed an improved, 12-grade tsunami intensity scale which is arranged according to
- effects on humans;
- effects on objects (including vessels of various size), and on nature; and
- damage to buildings.

---

[2006] L. HODDESON, P.W. HENRIKSEN, R.A. MEADE, and C. WESTFALL: *Critical assembly*. Cambridge University Press, Cambridge (1993), pp. 158-159.

[2007] R. GRAMMEL and K. ZOLLER: Zur Mechanik der Peitsche und des Peitschenknalls. Z. Phys. **127**, 11-15 (1949).

[2008] A. IMAMURA: *List of tsunamis in Japan* [in Japanese]. J. Seismol. Soc. Japan **2** [II], 23-28 (1949).

[2009] A. SIEBERG: *Geologische, physikalische und angewandte Erdbebenkunde*. G. Fischer, Jena (1923).

[2010] K. IIDA: *Earthquakes accompanied by tsunamis occurring under the sea off the islands of Japan*. J. Earth Sci. Nagoya University **4**, 1-43 (1956).

[2011] K. IIDA: *The generation of tsunamis and the focal mechanism of earthquakes*. In: (W.M. ADAMS, ed.) *Tsunamis in the Pacific Ocean*. East-West Center Press, Honolulu, HI (1970), pp. 3-18.

[2012] G.A. PAPADOPOULOS and F. IMAMURA: *A proposal for a new tsunami intensity scale*. In: *Proc. 20th Int. Tsunami Symposium* [Seattle, WA, Aug. 2001], pp. 569-577.

| | | |
|---|---|---|
| 1949 | *Imperial Chemical Industries Ltd., Welwyn, Hertfordshire, England* | Herbert KOLSKY,[2013] a British research scientist with a broad background in physics, chemistry and mechanical engineering, **introduces the *split Hopkinson pressure bar* in dynamic materials testing,** in order to measure the stress-strain behavior of shock-loaded disk-shaped specimens with round stress cycles over timescales on the order of 20 μs.[2014] ▪ His apparatus – later also known as the **"Kolsky bar"** {⇨Fig. 4.3–P} – is a modified Davies bar {DAVIES ⇨1948} which places the test specimen between the end faces of the main bar and an extension bar.<br><br>Almost 25 years later, KOLSKY[2015] presented a paper at the First Oxford Conference on Mechanical Properties of Materials at High Rates of Strain {⇨1974}, in which he reviewed previous research carried out during the last hundred years on the fractures produced by the propagation, reflection and diffraction of stress pulses of large amplitude. He also discussed recent brittle fracture investigations on the stress pulses generated in<br>▸ Hertzian impact experiments {HERTZ ⇨1882; ⇨Fig. 4.2–B};<br>▸ simple tensile tests; and<br>▸ flexure experiments on glass rods and bars.<br><br>KOLSKY's idea to use two instrumented bars with a wafer specimen sandwiched between them was so brilliant that his configuration has become the most popular one used to measure the responses of materials at high strain rates of up to about $10^3$/s.[2016] **A modified Kolsky bar was also used to investigate time-resolved friction; in particular, to measure the kinetic friction coefficient between two different materials (metal alloys) sliding against each other under impact.**[2017] |
| 1949 | *Dept. of Aeronautical Engineering and Applied Mechanics, Polytechnic Institute of Brooklyn, New York* | Morris MORDUCHOW,[2018] a Russian-born U.S. aeronautical engineer, **and Paul A. LIBBY,** a U.S. aeronautical engineer, **present "a general theory of the one-dimensional flow of a real continuous fluid in order to drive all of the mathematical implications inherent in the equations governing such flows."** Their computational results for a constant Prandtl number of ¾ show that the entropy within the thickness of the shock wave has a maximum at the center of the wave, which is positioned at the inflection point in the velocity distribution. **They consider that the presence of negative entropy gradients in separate sections does not violate the Second Law of Thermodynamics** – since, as they argue, the law applies to an entire system in thermal equilibrium. |
| 1949 | *United States* | Josiah Edward SPURR,[2019] a U.S. explorer and retired USGS geologist, **proposes in his three-volume work *Geology Applied to Selenology* a theory that the larger lunar craters were volcanic collapse structures (calderas).** ▪ Although results from the Apollo missions suggested that his geological interpretations of the Moon were wrong, volcanic activity does actually occur on the Moon {KOZYREV ⇨1958} and on other satellites such as Io {MORABITO ⇨1979}, and this has become a subject of modern geologic research. |

---

[2013] H. KOLSKY: *An investigation of the mechanical properties of materials at very high rates of loading*. Proc. Phys. Soc. (Lond.) **62B**, 676-700 (1949).

[2014] J.L. RAND and J.W. JACKSON: *The split Hopkinson pressure bar*. In: (M. ROY, ed.) *Behavior of dense media under high dynamic pressures*. In: *Symposium Hautes Pressions Dynamiques* [Paris, France, Sept. 1967]. Gordon & Breach, New York (1968), pp. 305-312.

[2015] H. KOLSKY: *Wave propagation effects and fracture*. In: (J. HARDING, ed.) *Mechanical properties at high rates of strain*. Conference Series No. 21, The Institute of Physics, London/Bristol (1974), pp. 199-214.

[2016] R.J. CLIFTON and J.R. KLEPACZKO: *On experimental investigation of the behavior of materials at high strain rates "Kolsky bar fifty years later."* In memoriam of the late Prof. H. KOLSKY. In: (L. LIBRESCU, ed.) *1999 ASME Mechanics & Materials Conference* [Blacksburg, VA, June 1999].

[2017] H.D. ESPINOSA, A. PATANELLA, and M. FISCHER: *A novel dynamic friction experiment using a modified Kolsky bar apparatus*. Exp. Mech. **40**, No. 2, 138-153 (2000).

[2018] M. MORDUCHOW and P.A. LIBBY: *On a complete solution of the one-dimensional flow equations of a viscous, heat conducting, compressible gas*. J. Aeronaut. Sci. **16**, 674-684, 704 (1949).

[2019] J.E. SPURR: *Geology applied to selenology*. The Science Press, Lancaster, PA (1949).

| | | |
|---|---|---|
| 1949 | Cavendish Laboratory, Cambridge, England | **Geoffrey I. TAYLOR,**[2020] a British applied physics professor, **investigates the density and particle velocity distributions behind planar and spherical detonation waves theoretically for gaseous explosives and TNT.** Describing the flow by a self-similarity variable, he obtains a solution for the waves trailing the detonation – **"Taylor expansion waves" or "Taylor waves."** ▪ These lateral rarefaction waves, also known as "release waves,"[2021] which follow a planar steady detonation and lead to a triangular region of wave interaction, are responsible for the immediate decay in pressure and particle velocity from the Chapman-Jouguet values. Thus, a test specimen in contact with a high explosive experiences a Taylor wave; *i.e.*, a *triangular-wave* rather than a *square-wave* loading profile. |
| 1949 | Dept. of Mathematics, University of Manchester, England | **Gilford Norman WARD,**[2022] a British mathematician, **calculates the drag force and lift for a body of revolution with wings of a small aspect ratio and moving supersonically.** He obtains a particularly simple expression for the lift in terms of the lift of the wing alone and that of the body alone which, as he suggests, may apply approximately to any wing system on a body of revolution. |
| 1949/ 1950 | University of Manchester, England | **M. James LIGHTHILL,**[2023] a British mathematics professor, **publishes a theoretical work on the diffraction of blast waves in air and the entire field in regular reflection.** It consists of computations for almost head-on collisions of shock waves on a wedge in the nonsteady state region, and the shape of the reflected shock. |
| 1950s | United States & Mexico | **George H. HERBIG,** a U.S. astronomer at Lick Observatory (UC Santa Cruz, CA), **and Guillermo HARO,** a Mexican astronomer at the Observatorio de Tacubaya (UNAM, Mexico City), **independently discover a number of compact nebulae with peculiar spectra near dark clouds** while investigating the neighborhoods of T-Tauri stars. ▪ These objects – called **"Herbig-Haro (HH) objects"** – are formed when highly energized particles (usually electrons and protons) are ejected from a young star and collide with nearby clouds of interstellar dust and gas. The jet particles stream out of the stars at speeds of more than 100 miles per second (160 km/s) and heat the surrounding clouds to an infrared glow that can be detected. Nearly 300 HH objects have been identified by astronomers around the world so far.<br><br>It took a further 25 years for the true nature of these objects and their role in the star formation process to be recognized: the U.S. astronomers Richard Dean SCHWARTZ[2024] at the Lick Observatory (UC Santa Cruz, CA) and John Charles RAYMOND[2025] at the Harvard College Observatory (Cambridge, MA) demonstrated that these objects were shock-excited nebulae. They are apparently excited by atomic collisions behind shock waves produced by supersonic collisions of ejecta moving away from young stars at speeds of from 50 to over 300 km/s. They appear as nonhomogeneous gas clouds at the leading edges of bipolar jets streaming away from extremely young stars, and have V- or U-shapes which are characteristic of bow shocks, similar to a supersonic bullet traveling through air {⇨Fig. 4.1–Y}.[2026]<br><br>In January 2006, at the 207th meeting of the American Astronomical Society in Washington, DC, **an image of a "cosmic tornado" was released, which was taken by U.S. astronomers with the Infrared Array Camera of the Spitzer Space Telescope managed by NASA-JPL.** This unique luminous tornado-shaped jet structure – known as "Herbig-Haro |

---

[2020] G.I. TAYLOR: *The dynamics of the combustion products behind plane and spherical detonation fronts in explosives.* Proc. Roy. Soc. Lond. **A200**, 235-247 (1949–1950).
[2021] This term was coined by the British explosives specialist Emerson M. PUGH and associates at Carnegie Institute of Technology, Pittsburgh, PA.
[2022] G.N. WARD: *Supersonic flow past slender pointed bodies.* Quart. J. Mech. Appl. Math. **2**, 75-97 (1949).
[2023] M.J. LIGHTHILL: *The diffraction of blast (I).* Proc. Roy. Soc. Lond. **A198**, 454-470 (1949); *The diffraction of blast (II).* Ibid. **A200**, 554-565 (1950).
[2024] R.D. SCHWARTZ: *T Tauri nebulae and Herbig-Haro nebulae – evidence for excitation by a strong stellar wind.* Astrophys. J. **195**, pt. 1, 631-642 (1975).
[2025] J.C RAYMOND: *Shock waves in the interstellar medium.* Astrophys. J. Suppl. **39**, 1-27 (1979).
[2026] *HH-47 star jet.* Astronomy picture of the day. NASA-GSFC (Oct. 12, 1995); http://antwrp.gsfc.nasa.gov/apod/ap951012.html.

|      |                        | 49/50" or "HH 49/50" – is about 0.3 light-years ($2.83 \times 10^{12}$ km) long and shows up in the infrared $\{\Rightarrow$ Fig. 4.1–Y$\}$.[2027] Shortly after, astronomers at the Harvard-Smithsonian Center for Astrophysics (CfA) gave an interesting preliminary interpretation, suggesting that
|      |                        | ▸ **the "tornado" is actually a shock front** created by a jet of material flowing downward through the field of view;
|      |                        | ▸ the shock may have developed instabilities as it plowed into surrounding material, creating eddies that give the "tornado" its distinctive spiral appearance; and
|      |                        | ▸ the triangular shape results from the wake created by the jet's motion, similar to the wake behind a speeding boat.[2028]

| 1950 | Washington, DC | On January 31, **U.S. President Harry S. TRUMAN directs the U.S. Atomic Energy Commission (AEC) to proceed with the development of the hydrogen bomb.** • At this time, the H-bomb was still far from being realized; it was still at the conceptual stage, the result of bold extrapolations by physicists from the energy processes in the Sun, and supported by some meager laboratory, experimental data on the reaction physics of mixtures of deuterium-deuterium and deuterium-tritium.[2029] |

| 1950 | Nevada Nuclear Test Site [about 70 miles northwest of Las Vegas], Nevada | On December 18, **U.S. President Harry S. TRUMAN approves the choice of a continental atomic proving ground in southern Nevada.**[2030]<br><br>The first atomic explosion in a series of atmospheric atomic tests at the newly christened Nevada Proving Ground was conducted on January 27, 1951 (Test ABLE, yield one kiloton) as part of Operation RANGER. Following the RANGER series, the U.S. Atomic Energy Commission (AEC) swiftly moved to turn the Nevada Nuclear Test Site into a permanent testing ground for nuclear weapons. |

| 1950 | U.S.A. & U.K. | **Both nations,** which cooperated in the period 1942–1945 in top secret research into problems posed by the conditions of undersea warfare, **publish their research results as UNDEX** (Underwater Explosions) **Reports,** a three-volume set of collected papers on underwater explosion phenomena.[2031] The editors are Gregory K. HARTMANN (U.S. Naval Ordnance Laboratory) and E.G. HILL (British Admiralty).<br><br>An update of the UNDEX Compendium on all kinds of underwater shock studies has recently been planned.[2032] The acronym "UNDEX" refers to simulations of underwater explosion effects on submerged and floating structures. A historical perspective on UNDEX analysis was given by Jeffrey CIPOLLA, a U.S. development engineer, at the 75th Shock & Vibration Symposium, held in 2004 at Virginia Beach, VA. |

| 1950 | Bureau of Ordnance and Hydrography, Washington, DC | Publication of the first two volumes of the ***Handbook of Supersonic Aerodynamics.***[2033] Containing contributions from the Applied Physics Laboratory of Johns Hopkins University, the U.S. Navy Dept. Bureau of Ordnance, and the U.S. Bureau of Naval Weapons, it is the first handbook to widely address the practical problems of the fast-growing branches of gas dynamics and supersonic aerodynamics. |

---

[2027] *Cosmic tornado HH 49/50.* Astronomy picture of the day. NASA-GSFC (Feb. 3, 2006); http://www.astronet.ru/db/xware/msg/1211349.

[2028] *Cosmic jet looks like giant tornado in space.* Press release No. 06-06 (Jan. 12, 2006) of the Harvard-Smithsonian Center for Astrophysics (CfA), Cambridge, MA; http://www.cfa.harvard.edu/press/pr0606.html

[2029] F.H. SHELTON: *Reflections of a nuclear weaponeer.* Shelton Enterprise Inc., Colorado Springs, CO (1988), pp. **4**:3-4.

[2030] *Nevada test site history – Project Nutmeg: the birth of the Nevada test site.* U.S. Dept. of Energy (DOE), National Nuclear Security Administration (NNSA), Las Vegas, NV (June 2004); http://www.nv.doe.gov/library/factsheets/DOENV_767.pdf.

[2031] *Underwater explosion research (UNDEX). A compendium of British and American reports.* The Library of Congress, Photoduplication Service, Washington, DC (1950). It consists of three volumes: vol. I: *The shock wave,* vol. II: *The gas globe,* and vol. III: *The damage process.*

[2032] R. SCAVUZZO: *UNDEX Compendium Panel.* In: *Proc. 69th Shock & Vibration Symposium* [Minneapolis/St. Paul, MN, Oct. 1998]. SAVIAC etc., Falls Church, VA (1998).

[2033] *Handbook of supersonic aerodynamics,* NavOrd Rept. No. 1488, published by the Bureau of Ordnance and Hydrography, Washington, DC. U.S. Govt. Print. Office, Washington, DC (1950–1959), vol. 1, p. 6.

| 1950 | Electrical Research Association Laboratories, Leatherhead, Surrey, England | Henry Wright BAXTER,[2034] a British researcher, **studies exploding wire phenomena of electric fuses** by simultaneously employing high-speed photography and oscilloscope recording of electrical discharge parameters (*e.g.*, voltage and current rise times, maximum current, pulse duration, *etc.*).

(i) **He observes that the wire explosion is accompanied by disintegration into a number of globules** {⇨Fig. 4.16–O}, which he correctly explains by the surface tension of molten wire material {VAN MARUM ⇨1799; SINGER & CROSSE ⇨1815}. When the globules have formed, the current has fallen to zero.

(ii) He also notices that the diameter of a large current-carrying wire becomes unstable, and that **"frozen unduloids"** are formed when melting ceases before the drops part from each other. He writes, "It is well known that, owing to surface tension, when the length of a cylinder is greater than its diameter it becomes unstable and tends to break up into drops. **During the passage from a cylinder to a succession of spheres, which is the stable state, the liquid column passes through a shape known as an "unduloid" – a solid of revolution, the outline of which is sinusoidal.** If melting ceases before the drops part from each other, a 'frozen' unduloid is obtained…"

Later, similar phenomena were also found by Heinrich ARNOLD and William M. CONN at the University of Würzburg.[2035] They placed a glass slide near an exploding wire on which the metallic vapor precipitated, and thus observed a characteristic pattern of **"striations"** rather than a homogeneous deposit {⇨Fig. 4.16–O}.[2036] The process of exploding-wire fragmentation was resumed by subsequent researchers {NASILOWSKI ⇨1964}. |
|---|---|---|
| 1950 | Harvard University, Cambridge, Massachusetts | Garrett BIRKHOFF,[2037] a U.S. applied mathematician, **discusses numerous hydrodynamic paradoxes of nonviscous flow in his book *Hydrodynamics*.** During World War II, he became interested in various hydrodynamic problems, such as shock waves around projectiles, shaped charges, bouncing bombs on water, and shock wave reflection phenomena. Addressing VON NEUMANN's result that regular reflection of weak shocks occurs for angles of incidence that are somewhat greater than those allowed by theory {VON NEUMANN ⇨1943; L.G. SMITH ⇨1945}, he writes: "The predicted limits for triple shocks differ grossly from those observed. This discrepancy, which may be called the ***triple shock paradox,*** was apparently discovered by John VON NEUMANN (1945). Repeated attempts have been made to resolve this paradox, which may be a 'singular point paradox,' due to an oversimplified guess as to the local behavior near the singular point. But no fully satisfactory explanation of it has yet been advanced."

BIRKHOFF's triple shock paradox – better known as the **"von Neumann paradox"** of oblique shock reflection – plays a central role in the transition between regular and irregular reflection of oblique shock waves on rigid walls and the diffraction patterns behind obstacles. The von Neumann paradox was studied by a number of Mach reflection researcher:
> In 1990, detailed experimental and numerical investigations were carried out by Phillip COLELLA,[2038] a Livermore physicist, and Le Roy F. HENDERSON, an Australian professor at the University of Sydney and an authority on Mach reflection. They arrived at the conclusion that "**when the von Neumann theory failed, the weak Mach reflection was transformed into a new type of irregular reflection which we called a 'von Neumann reflection'** [also called 'von Neumann-Mach reflection']. In this system, the incident and Mach shocks appear to form a single wave with a continuous |

---

[2034] H.W. BAXTER: *Electric fuses.* E. Arnold, London (1950), pp. 68-71, 77-79.
[2035] W.M. CONN: *Studien zum Mechanismus von elektrischen Drahtexplosionen (Metallniederschläge und Stoßwellen).* Z. Angew. Phys. **7**, 539-554 (1955).
[2036] For more about distances in "characteristic patterns" of exploding wires, *see* (W.G. CHACE and H.K. MOORE, eds.) *Exploding wires (II).* Plenum Press, New York (1962), pp. 77-86.
[2037] G. BIRKHOFF: *Hydrodynamics. A study in logic, fact and similitude.* Princeton University Press, Princeton, NJ (1950, 1960), pp. 25-26.
[2038] P. COLELLA and L.F. HENDERSON: *The von Neumann paradox for the diffraction of weak shock waves.* J. Fluid Mech. **213**, 71-94 (1990).

> turning point. The reflection is a smoothly disturbed and apparently self-similar pressure disturbance near its interaction region with the incident/Mach shocks, but it steepens into a shock as it retreats from them."
> - In 1994, the U.S. applied mathematicians Esteban G. TABAK[2039] at Princeton University and Rodolfo R. ROSALES at MIT studied the triple-point paradox in the context of an asymptotic model for the behavior of weak shock waves at almost glancing reflection on rigid walls and other cases of practical interest.
> - In 2001, Shigeru ITOH,[2040] a Japanese shock physicist at Kumamoto University, studied the propagation and reflection of underwater shock waves in converging systems for metalworking applications. ITOH observed that **von Neumann reflections are abundant in underwater explosions** – in particular that the underwater shock reflection pattern is not an ordinary Mach reflection, but rather a reflection with an extremely curved Mach stem.

| | | |
|---|---|---|
| 1950 | Metcalf Research Laboratories, Brown University, Rhode Island | George R. COWAN[2041] and Donald F. HORNIG, two U.S. chemists, **determine the thickness of a shock front in a gas by monitoring the magnitude of the optical reflectivity** {⇨Fig. 4.12−B}. They find that the thickness of the shock front is significantly greater than a few mean free paths as predicted by THOMAS {⇨1944}. ▪ The problem of determining the thickness of a shock front is of fundamental importance in shock wave physics and had already occupied a number of early pioneers {PRANDTL ⇨1909; Lord RAYLEIGH ⇨1910; BECKER ⇨1921}. |
| 1950 | Soviet Union | V.Y. DANILOVSKAYA,[2042] a Soviet scientist, **solves the famous (uncoupled) homogeneous thermoelastic problem of the half-space $x \geq 0$ subjected to a sudden step in surface temperature $T(0, t) = T_0 U(t)$**. This results in a propagating stress jump $$\sigma_x = E\alpha T_0/(1 - 2\nu),$$ where $\alpha$ is the coefficient of thermal expansion, $E$ is the Young modulus, and $\nu$ is the Poisson ratio. Recently, William J. PARNELL,[2043] an applied mathematician at the University of Manchester, extended the problem of fully coupled thermoelasticity to an inhomogeneous half-space, where the medium (a composite) has 1-D variations in its physical properties, and determined the magnitude of resulting discontinuities in field variables. |
| 1950 | Hydrodynamics Laboratory, CalTech, Pasadena, California | Forrest R. GILMORE,[2044] a U.S. physicist and applied mechanics instructor, **and his collaborators generate Mach disk-type hydraulic jump interactions and experimentally demonstrate the limitations of the analogy between hydraulic jumps in liquids and shock waves in gases.** Some sources of error are peculiar to hydraulic jumps and do not apply to compression shocks. They state, "One may conclude from the experimental observations discussed here that the deviations from the simple theory of hydraulic jump intersections limit the use and validity of the hydraulic analog as a means for studying compression-shock intersections in gases." ▪ In 1954, the correlation of water wave phenomena with certain phenomena in high-speed gas dynamics was reviewed by Donald R.F. HARLEMAN[2045] and Arthur T. IPPEN, two U.S. hydrodynamicists at MIT. |

---

[2039] E.G. TABAK and R.R. ROSALES: *Focusing of weak shock waves and the von Neumann paradox of oblique shock reflection*. Phys. Fluids **6**, 1874-1892 (1994).
[2040] S. ITOH: *Shock waves in liquids*. In: (G. BEN-DOR ET AL., eds.) Handbook of shock waves. Academic Press, San Diego etc. (2001), vol. 1, pp. 263-314.
[2041] G.R. COWAN and D.F. HORNIG: *The experimental determination of the thickness of a shock front in a gas*. J. Chem. Phys. **18**, 1008-1017 (1950).
[2042] V.Y. DANILOVSKAYA: *Thermal stresses in an elastic half-space due to a sudden heating of its boundary* [in Russ.]. Prikl. Mat. Mekh. **14**, 316-318 (1950).
[2043] W.J. PARNELL: *Coupled thermoelasticity in a composite half-space*. J. Engng. Math. **56**, 1-21 (2006).
[2044] F.R. GILMORE, M.S. PLESSET, and H.E. CROSSLEY JR.: *The analogy between hydraulic jumps in liquids and shock waves in gases*. J. Appl. Phys. **21**, 243-249 (1950).
[2045] D.R.F. HARLEMAN and A.T. IPPEN: *The range of application of the hydraulic analogy in transonic and supersonic aerodynamics*. In: *Mémoires sur la mécanique des fluides offerts à M. Dimitri P. RIABOUCHINSKY par ses amis, ses collègues et ses anciens élèves à l'occasion de son jubilé scientifique le 8 mai 1954*. Publ. Sci. Tech. Ministère de l'Air, Paris (1954), pp. 91-112.

| | | |
|---|---|---|
| 1950 | Los Alamos Scientific Laboratory (LASL), Los Alamos, New Mexico | **Frederic DE HOFFMAN**[2046] **and Edward TELLER,** two European-born U.S. theoretical physicists, **extend the theory of shock waves to electrically conducting fluids subjected to magnetic fields** and discover new phenomena. Performing a relativistic treatment of magnetohydrodynamic waves, they show that there are two limiting cases − namely hydrodynamic shock waves and electromagnetic waves. ▪ Their study was apparently an outgrowth of research on nuclear explosions during World War II.<br><br>In conventional shock waves, the material is transported through the discontinuity in the same direction as the wave propagation − resulting in so-called **"longitudinal shocks."** a magnetohydrodynamic shock wave propagating in a plasma in the presence of a magnetic field exhibits new types of discontinuities not observed in classical shock wave physics, such as **"transverse shocks,"** which occur when the material transport across the discontinuity is perpendicular to the direction of shock propagation.[2047] |
| 1950 | Polytechnic Institute, Leningrad, Soviet Union [now St. Petersburg, Russia] | **Lev Aleksandrovich JUTKIN**[2048] [or YUTKIN] **shows that underwater shock waves generated by strong electric spark discharges** {⇨Fig. 4.15−H} **are capable of destroying brittle materials** (such as plates of china and stones), but they have almost no affect on soft and elastic media. Furthermore, he notices that the primary shock is followed by a secondary shock when the cavity collapses, and that the shock strength increases with the capacity of the discharge capacitor and the discharge rate. ▪ JUTKIN later annotated autobiographically that he first observed the shaking action of a lightning strike in a log when he was still a boy, and that the shaking was most extreme in the parts of the log submerged in water.<br><br>JUTKIN's *electrohydraulic effect*[2049] was utilized in the 1950s by GOLDBERG {⇨1959} to destroy urinary calculi − a technique known as **"electrohydraulic shock lithotripsy."**[2050] However, it was also introduced into technology used to deform metal tubes and sheets (*hydrospark forming*), where the shock wave is generated by a powerful capacitor discharge, which is preferably initiated by an exploding wire in water.[2051] The high deformation rate of this technique results in a small amount of spring-back, which allows one to produce components to a high degree of accuracy. Furthermore, the applicability of this effect to the disintegration of rocks,[2052] as well as for the intensification of chemical-metallurgical and chemical engineering, has also been investigated.[2053] |
| 1950 | Institute for Advanced Study (IAS), Princeton, New Jersey & Los Alamos Scientific Laboratory (LASL), Los Alamos, New Mexico | **John VON NEUMANN,**[2054] a Hungarian-born U.S. mathematics professor at IAS, **and Robert D. RICHTMYER,** a U.S. physicist and leader of LASL's theoretical division, **propose to modify the equations of hydrodynamics through the inclusion of a purely mathematical, fictitious viscosity in order to numerically calculate hydrodynamic shocks using the digital computer ENIAC** (*see* Sect. 2.9.2):<br><br>(i) Summarizing the difficulties of the problem, they introductorily write, "[In the presence of shocks] the partial differential equations governing the motion require boundary conditions connecting the values of these quantities on the two sides of each such surface. The necessary boundary conditions are, of course, supplied by the Rankine-Hugoniot equations, but their application is complicated because the shock surfaces are in motion relative to the network of points in |

---

[2046] F. DE HOFFMAN and E. TELLER: *Magnetohydrodynamic shocks.* Phys. Rev. **80** [II], 692-703 (1950).
[2047] J.E. ANDERSON: *Magnetohydrodynamic shock waves.* MIT Press, Cambridge (1963).
[2048] L.A. JUTKIN: *Elektrogidravliceskij effekt.* Masgiz, Moskva (1955).
[2049] G. NESVETAILOV and E. SEREBRIAKOV: *Theory and practice of electrohydraulic effect* [in Russ.]. Minsk (1966).
[2050] V. GOLDBERG: *Zur Geschichte der elektrohydraulischen Lithotripsie (EHL).* In: *BMFT Symposium* [Meersburg, Germany, Juni 1976]. Wiss. Berichte, Dornier System GmbH, Friedrichshafen, Germany, pp. 130-131.
[2051] H.J. WAGNER and J.G. DUNLEAVY: *Hydrospark forming … evolution of the process.* Tool Engineer **44**, 83-86 (1960).
[2052] H.K. KUTTER: *The electrohydraulic effect: potential application in rock fragmentation.* Rept. of Investigations No. 7317. U.S. Bureau of Mines, RI (1969).
[2053] P.P. MALYUSHEVSKY: *Fundamentals of the discharge-pulsing technology* [in Russ.]. Naukova Dumka, Kiev (1983).
[2054] J. VON NEUMANN and R.D. RICHTMYER: *A method for the numerical calculation of hydrodynamic shocks.* J. Appl. Phys. **21**, 232-237 (1950).

space-time used for the numerical work, and the differential equations and boundary conditions are non-linear. Furthermore, **the motion of the surfaces is not known in advance but is governed by the differential equations and boundary conditions themselves. In consequence, the treatment of shocks requires lengthy computations (usually by trial and error) at each step, in time, of the calculation."**

(ii) **By giving the shocks a greater thickness than the spacing of the grid points, this procedure allows one to solve the difference equations directly, as if there are no shocks at all.**

(iii) In their artificial dissipation scheme, the "viscosity coefficient" is proportional to the square of the velocity gradient. This allows one to calculate the motions of shocks without postulating them explicitly, but by following the ordinary hydrodynamic equations step-by-step. They write, "When viscosity is taken into account, for example, the shocks are seen to be smeared out, so that the mathematical surfaces of discontinuity are replaced by thin layers in which pressure, density, temperature, *etc.* vary rapidly, but continuously. Our idea is to introduce (artificial) dissipative terms into the equations so as to give the shocks a thickness comparable to (but preferably somewhat larger than) the spacing of the points of the network. Then the differential equations (more accurately, the corresponding difference equations) may be used for the entire calculation, just as though there were no shocks at all. In the numerical results obtained, the shocks are immediately evident as near-discontinuities that move through the fluid with very nearly the correct speed and across which pressure, temperature, *etc.* have very nearly the correct jumps." ▪ Their proposed **"method of smearing out the shock"** over several cells through the introduction of an artificial dissipation – first successfully illustrated by them to one dimensional flows – evolved into a mathematical treatment that has been widely applied for shock waves.

Harold L. BRODE {⇨1959}, who applied the artificial viscosity technique of VON NEUMANN and RICHTMYER in order to calculate blast waves from spherical charges of TNT, summarized its advantages as follows:

▸ No limiting physical assumptions need be imposed on the differential equations to reduce them to ordinary differential equations or to make them more simply integrable – rather one deals with the partial differential equations sans similarity restrictions.

▸ **Shock waves are handled automatically,** and no special computational technique is required if the problem contains multiple shocks or when shocks impinge on the origin; *i.e.*, there are no floating boundary conditions that need to be satisfied.

▸ The complexity of computations is not seriously increased by allowing for the computation of detailed equations of state from either tabular data or from analytic fits.

▸ The method can be used over large ranges of time and space variables; *i.e.*, expansions around singular points are not necessary.

| 1950 | *Laboratoire de Recherches Techniques de Saint-Louis, Saint-Louis, France* | Rudi SCHALL,[2055] a German shock and detonation physicist, **presents a simple method for determining the shock adiabat (Hugoniot curve) of water up to 190 kbar.** Based on flash radiography, his method only requires a simple explosive arrangement for shock wave generation {⇨Fig. 4.15–B}. Using the equations of conservation of mass and momentum, he determines |
|---|---|---|

▸ the unknown compression ratio at the shock front by applying photodensitometry to the recorded flash radiograph; and

▸ the velocity of the shock front by using a closely positioned detonating Hg-fulminate detonation fuse of known detonation velocity as a "microsecond clock."

---

[2055] R. SCHALL: *Die Zustandsgleichung des Wassers bei hohen Drucken nach Röntgenblitzaufnahmen intensiver Stoßwellen*. Z. Angew. Phys. **2**, 252-254 (1950).

SCHALL originally had worked out his method in 1944 at the Physical Institute of Berlin University. The accuracy of his method was limited by the fact that the average density behind the shock front is reduced by rarefaction effects emanating from both sides, which rapidly decelerate the shock wave during its propagation into the cylindrical test sample {G.I. TAYLOR ⇨1949}. Nevertheless, his quantitative results allowed comparison with shock compression data obtained at Los Alamos Scientific Laboratory using free surface diagnostics {RICE ET AL. ⇨1958}, which were found to deviate from his data by up to (as little as) about 8%. His flash X-ray technique was also used to visualize shock waves in other low-density materials {SCHALL & THOMER ⇨1951}.

| 1950 | Private residence, Princeton, New Jersey | **Immanuel VELIKOVSKY,**[2056] a Russian-born American trained psychiatrist and author of a number of controversial books, **publishes his first book entitled *Worlds of Collision*** in which he proposes |
|---|---|---|

▸ that only a few thousand years ago our planetary system was unstable;
▸ that 3,500 years ago Venus was ejected from Jupiter as a comet, thereby affecting the Earth's orbit and rotation, and creating around 1450 B.C. the natural disasters described in Exodus;
▸ that more than one planet formerly moved on Earth-threatening courses; and
▸ that early civilizations experienced overwhelming catastrophes caused by planetary encounters.

He argues that many myths and traditions of ancient peoples and cultures are based on actual events: worldwide global catastrophes of a celestial origin, which had a profound effect on the lives, beliefs and writings of early mankind. He bases his claims on the basis of ancient cosmological myths rather than on astronomical evidence and scientific inference or argument.
▪ The Macmillan Company published the book in April 1950, but on the basis of persistent attacks from the scientific community was forced to transfer the rights to other publishers (Doubleday in the U.S., Gollancz in the U.K., Kohlhammer in Germany). After becoming a bestseller in the United States in 1952, his book which was banned from a number of academic institutions created an unprecedented scientific debacle, known as the "Velikovsky Affair." His book also appeared in condensed form in *Readers Digest*, and was several times reprinted in German translation.

| 1950 | University of Manchester, England | **Gerald B. WHITHAM,**[2057] a British applied mathematician, **presents a theory that satisfactorily describes the generation of shock waves around a supersonic aircraft and their attenuation through the atmosphere.** |
|---|---|---|

(i) He correctly predicts an N-shaped pressure signature for the far field, the so-called "N-wave" or "far-field N-wave" {DUMOND ET AL. ⇨1946}.
(ii) He develops a method of computing the sonic boom-producing shock fronts around bodies of revolution. ▪ WHITHAM[2058] applied his method to the particular case of a parabolic body of revolution. The resulting expression for the pressure rise across the shock in the far field has been called the "Whitham equation." It allows satisfactory estimates of the far-field pressures for complete airplanes.

His theory allowed him to establish an approximate lower bound for attainable sonic-boom overpressure, which depends on the airplane length, weight and volume, and on the flight conditions.[2059] Furthermore, his theory also gave aerodynamicists the opportunity to check his new

---

[2056] I. VELIKOVSKY: *Worlds of collision*. Macmillan Co., New York (1950), Gollancz, London (1950). *Welten im Zusammenstoß*. Kohlhammer, Stuttgart (1951); J. White, Wöllsdorf (2005).
[2057] G.B. WHITHAM: *The behavior of supersonic flow past a body of revolution, far from the axis*. Proc. Roy. Soc. Lond. **A201**, 89-109 (March 1950).
[2058] G.B. WHITHAM: *The flow pattern of a supersonic projectile*. Comm. Pure Appl. Math. **5**, 301-348 (1952).
[2059] H.W. CARLSON: *The lower bound of attainable sonic-boom overpressure and design methods of approaching this limit*. NASA TN D-1494 (Oct. 1962).

theoretical method experimentally in supersonic wind tunnels {⇨Fig. 4.6–K}. However, since the supersonic wind tunnels available at that time had only small test sections which could not be enlarged greatly because of the high expenditure required, only tiny models 0.25–1 in. (6.3–25.4 mm) in size were tested in the Ames and Langley Facilities. With the aid of this miniaturization, meaning that the tunnel walls were up to 150 body lengths away from the models, WHITHAM's theory of sonic booms could indeed be verified.[2060]

| 1951 | U.S. Office of Naval Research (ONR), Washington, DC |

In the period January 11–12, **the first Conference on the Chemistry and Physics of Detonation is held;** the general chairman is Calvin M. BOLSTER, chief of Naval Research. The purpose of this meeting is to bring together scientists from the government, contractors, and university laboratories to discuss the current status of and problems in this field of chemical physics.
▪ The two subsequent conferences, each renamed a **Symposium (ONR) on Detonation,** were held at Washington, DC (1955) and at Princeton, NJ (1960). Later conferences, each renamed a **Symposium (International) on Detonation,** were held at Silver Spring, MD (1965); Pasadena, CA (1970); Coronado, CA (1976); Annapolis, MD (1981); Albuquerque, NM (1985); Portland, OR (1989); Boston, MA (1993); Snowmass, CO (1998); San Diego, CA (2002); and Norfolk, VA (2006).

In 1997, the Los Alamos National Laboratory (LANL) published a general index of the first ten conferences, consisting of a title index, a topic phrase index, an author index, and an acronym & code index.[2061]

| 1951 | Daimler-Benz AG, Stuttgart, Germany |

On January 23, **Daimler-Benz engineer Béla BARÉNYI**[2062] **applies for a patent for a rigid passenger cell with crumple zones** [Germ. *Knautschzonen*] **used in the front and the rear of a car** – an effective way to dissipate the force of a crash before it reaches the passenger cell. The patent will be granted in the following year. ▪ His pioneering concept of *passive safety,* which involves absorbing the impact energy, completely changed the way cars were designed and built.

The world's first series of production vehicles with rigid passenger cages and integrated crumple zones were the Mercedes-Benz "fintail" types 220b, 220Sb and 220SEb (chassis model W111, from 1959), and the types 190c and 190Dc (chassis model W110, from 1962).[2063]

| 1951 | Edwards Air Force Base (AFB), California |

On June 20, **first flight of the Bell X-5, an experimental aircraft to determine the aerodynamic effects of *variable swept wing* in flight.**[2064] ▪ Using a system of electric motors the variable sweep wing could be adjusted in flight stepwise from 20, 40 to 60 degrees – thus reducing drag and increasing the aircraft's speed. As the X-5's wings were swept back, its center of gravity and center of pressure changed. To compensate, the entire wing assembly simultaneously moved forward on rails inside the fuselage. In later test flights the X-5 reached a maximum speed of about 1,100 km/h at an altitude of about 12 km.

The X-5 was largely based on the design of the German Messerschmitt P.1101 which was captured near the end of WWII and was brought to the United States for technical review and inspection. The P.1101 had variable swept wings which, however, could only be adjusted on the ground, prior to flight. ▪ **The idea of swept wings originated in Germany** {5th Volta Conference (BUSEMANN) ⇨1935}. **In 1942, Albert A. BETZ,**[2065] an eminent German professor of

---

[2060] H.W. CARLSON and O.A. MORRIS: *Wind-tunnel sonic-boom testing techniques.* J. Aircraft **4**, 245-249 (1967).
[2061] W.E. DEAL, J.B. RAMSAY, A.M. ROACH, and B.E. TAKALA: *Indexes of the proceedings for the ten International Symposia on Detonation 1951–1993.* Rept. LA-UR-97-1899, LANL, Los Alamos, NM (1997); http://www.intdetsymp.org/detsymp2002/fadeal.pdf.
[2062] B. BARÉNYI: *Kraftfahrzeug, insbesondere zur Beförderung von Personen* [Fahrzeug mit stabiler Fahrgastzelle und verformbarer Hinter- und Vorderfront (*Knautzone*)]. Germ. Patent No. 854,157 (Aug. 28, 1952).
[2063] H. NIEMANN: *Béla BARÉNYI – Sicherheitstechnik made by Mercedes-Benz.* Motorbuch-Verlag, Stuttgart (2002).
[2064] *Bell X-5.* Fact sheet. NASA Dryden Flight Researh Center; http://www.nasa.gov/centers/dryden/news/FactSheets/FS-081-DFRC.html.
[2065] A. BETZ: *Flügel mit veränderlicher Pfeilstellung (verstellbare Flügel).* Germ. Patent No. G799/42.

|      |                          |                                                                                                                                                                                                                                                                                                                                                                                                                                                                                                                                                                                                  |
|------|--------------------------|------|
|      |                          | aerodynamics at Göttingen University who studied the aerodynamic characteristics of swept wings, **obtained a secret patent on variable swept wings.** |
| 1951 | New York City, United States | On July 3, **the U.S. inventor Frank RIEBER**[2066] **obtains the first patent on an** *electrohydraulic shock-wave generator.* The shock wave is generated by a spark discharge between two closely-spaced electrodes immersed in an oil bath. ▪ RIEBER's invention is based on the *electrohydraulic effect*, which was discovered the year before in the Soviet Union {JUTKIN ⇨ 1950}. However, RIEBER had originally filed his patent application on May 24, 1947.<br><br>An electrohydraulic shock-wave generator was later used in medicine to disintegrate kidney stones {GOLDBERG ⇨ 1959} – known as **"electrohydraulic shock lithotripsy."** |
| 1951 | U.S. Naval Ordnance Laboratory (NOL), Silver Spring, Maryland | **A group of aerodynamicists,** supervised by Peter WEGENER, a former staff scientist at the Peenemünde Wind Tunnel Facility, **successfully operates the first U.S. hypersonic wind tunnel** ($12 \times 12$ cm$^2$), **in which disturbance-free flows are continuously produced over a large Mach-number range up to** $M = 8.25$, which is about equal to that achieved previously in Peenemünde {ERDMANN ⇨ 1943/1944}.[2067, 2068]<br><br>In the following years, the number of U.S. hypersonic wind tunnels increased enormously, particularly those used to test models of pilotless- and passenger-carrying reentry vehicles on a model scale. According to a partial list published in 1965, there were seven hypervelocity facilities operated by the government, and twenty-four operated by colleges and industry.[2069] |
| 1951 | Max-Planck-Institut (MPI) für Physik, Göttingen, Germany | **Ludwig F. BIERMANN,**[2070] a German astrophysicist, **postulates solar** *Korpuskularstrahlung* **("corpuscular radiation") based on an analysis of comet tail plasma deflection.** ▪ In the early 1600s, Johannes KEPLER, an eminent German astronomer, guessed that comet tails were driven by the pressure of sunlight, and his guess was found to be valid for the many comet tails that consist of dust. However, comets also have ion tails that may point in slightly different directions, and are sometimes observed to accelerate quite suddenly, causing them to become kinked or bent. The pressure of sunlight cannot explain such behavior, but in 1943 Cuno HOFFMEISTER, a German astronomer who observed an *aberration* of about 6° between the observed gas tail and the direction of the Sun, revived the idea of solar particle radiation. This was correctly interpreted in 1951 by BIERMANN in terms of an interaction between the ions in the tail and what later became known as the **"solar wind"** {PARKER ⇨ 1958}. |
| 1951 | Institute of Aerophysics, University of Toronto, Canada | **Gerald V. BULL**[2071] **investigates the starting process in an intermittent supersonic wind tunnel.** He divides the flow into five distinct phases. Based on the character of the flow along the wind tunnel axis, they are in cyclic order:<br>▸ the nonstationary starting phase;<br>▸ the quasi-stationary starting phase;<br>▸ the stationary phase;<br>▸ the quasi-stationary closing phase; and<br>▸ the nonstationary closing phase.<br><br>BULL's results stimulated further studies on the starting process, which revealed that during passage of the incident shock wave through the nozzle, an additional expansive wave, a secondary compression shock and several contact surfaces are formed, whose behavior per unit |

---

[2066] F. RIEBER: *Shock wave generator.* U.S. Patent No. 2,559,277 (July 3, 1951).
[2067] P.P. WEGENER and R.K. LOBB: *NOL Hypersonic Tunnel No. 4 Results II: diffuser investigation.* Rept. AD0895228, NOL, White Oak, MD (May 1952).
[2068] P.P. WEGENER: *The Peenemünde wind tunnels. A memoir.* Yale University Press, New Haven, CT etc. (1996), p. 174.
[2069] A. POPE and K.L. GOIN: *High-speed wind tunnel testing.* Wiley, New York etc. (1965), pp. 461-462.
[2070] L. BIERMANN: *Kometenschweife und solare Korpuskularstrahlung.* Z. für Astrophysik **29**, 274-286 (1951).
[2071] G.V. BULL: *Starting process in an intermittent supersonic wind tunnel.* UTIA Rept. No. 12, Institute of Aerophysics, University of Toronto (Feb. 1951); *Investigation into the operating cycle of a two-dimensional supersonic wind tunnel.* J. Aeronaut. Sci. **19**, 609-614 (1952).

of time has an important effect on the characteristics of the starting process. This problem is important not only in wind tunnels but also in shock tubes {AMANN ⇨ 1970}.

| 1951 | Harvard University, Cambridge, Massachusetts & Leiden Observatory, The Netherlands | Harold I. EWEN[2072] **and Edward M. PURCELL,** two U.S. astronomers at the Lyman Laboratory of Harvard University, **discover the 21-cm (1.42-GHz) spectral line due to neutral hydrogen in the Milky Way.** The same discovery is made almost simultaneously by the Dutch astronomers C. Alex MULLER[2073] and Jan H. OORT at Leiden Observatory, who publish their results in the same issue of the British journal *Nature*. ▪ This radiation permits observations of the previously inaccessible main component of interstellar gas, neutral hydrogen, thus stimulating the evolution of *cosmical aerodynamics* – renamed *cosmical gas dynamics* in 1951 {1st Symp. on the Motion of Gaseous Masses of Cosmical Dimensions ⇨ 1949}.

Observations of the 21-cm line are still a very important branch of radio astronomy to this day, resulting in ever more detailed understanding of the interstellar medium in our Galaxy and external galaxies. |

| 1951 | Cornell Aeronautical Laboratory (CAL), Buffalo, New York | Abraham HERTZBERG,[2074] a U.S. associate research aerodynamicist, **presents a new method of generating hypersonic flow in a shock tube by employing a diverging nozzle before the test section.** This permits the establishment of steady flow Mach numbers far exceeding those that could be produced in a constant area shock tube, which is only Mach 1.9 for perfect gases and about Mach 3 for air. ▪ HERTZBERG's *hypersonic shock tunnel* **was a real breakthrough,** because continuous hypersonic flow facilities, which had been available since 1945 at NACA-Langley's $11 \times 11$-in.$^2$ ($27.9 \times 27.9$-cm$^2$) test section and since 1946 at NACA-Ames' $10 \times 14$ in.$^2$ ($25.4 \times 35.6$-cm$^2$) test section, could not duplicate the thermal environment present at high Mach numbers typical of reentry.[2075]

The starting process that occurs in the nozzle of a simple expansion shock tunnel leads to a reduction in the duration of flow, which is already short. To overcome this limitation, tunnels operating in the "reflected shock" mode, in which the high-pressure and high-temperature gas behind a reflected shock was used as a reservoir, were constructed at CAL. |

| 1951 | The Leningrad Physico-Technical Institute, Leningrad, Soviet Union | Mark I. KORNFELD,[2076] a Soviet fluid dynamicist, **publishes his book entitled *Uprugost' i procnost' zidkostej* ("Elasticity and Strength of Fluids"), in which he also investigates what happens when a liquid jet is impacted by a metal striker in normal direction {⇨Fig. 4.3–V}.** At low striker velocity he observes a laminar deformation of the jet, while at high velocity (23 m/s), the liquid jet fractures in a brittle manner. ▪ In the late 1960s, KORNFELD's collision experiments were resumed and explained by a relaxation mechanism {KOZYREV & SHAL'NEV ⇨1970; ⇨Fig. 4.3–V}. |

| 1951 | Brown University, Providence, Rhode Island & Hofstra College, Long Island, New York | Erastus Henry LEE,[2077] a British-born U.S. applied mathematician, **and H. WOLF,** a German-born U.S. mathematician, **discuss the influence of plastic wave propagation effects in high-speed materials testing, a general problem which makes analyzing test results difficult.** They show that for tests at high strain rates – such as the Taylor test {G.I. TAYLOR & WHIFFIN ⇨1948} – the inertia of the specimen may become important, thus producing a nonuniform stress distribution along the specimen and introducing a spurious strain rate effect. They propose a comparatively simple calculation method that can be used to check the magnitude of |

---

[2072] H.I. EWEN and E.M. PURCELL: *Radiation from galactic hydrogen at 1,420 Mc/sec.* Nature **168**, 356 (1951).

[2073] C.A. MULLER and J.H. OORT: *The interstellar hydrogen line at 1,420 Mc/sec, and an estimate of galactic rotation.* Nature **168**, 357-358 (1951).

[2074] A. HERTZBERG: *A shock tube method of generating hypersonic flows.* J. Aeronaut. Sci. **18**, 803-804, 841 (Dec. 1951).

[2075] K. BURNS and A. BRUCKNER: *The history of aerospace research at Cornell Aeronautical Laboratory and Calspan.* 44th AIAA Aerospace Sciences Meeting and Exhibit [Reno, NV, Jan. 9-12, 2006]. Paper AIAA 2006-0335.

[2076] M.I. KORNFELD: *Uprugost' i procnost' zhidkosteæi.* Moskva (1951); Germ. translation: *Elastizität und Festigkeit der Flüssigkeiten.* Verlag Technik, Berlin (1952).

[2077] E.H. LEE and H. WOLF: *Plastic wave propagation effects in high-speed testing.* J. Appl. Mech. **17**, 379-386 (1951).

| | | |
|---|---|---|
| 1951 | Cornell University, Ithaca, New York | **Robert William PERRY,**[2078] a Ph.D. student under the supervision of Prof. Arthur KANTROWITZ, **reports on the generation of strong shock waves in a steadily convergent shock tube** {⇨Fig. 4.14–J}. He notices increasing instability at increasing strength. The cylindrical collapsing shocks produce luminous flashes at the center resulting from the high temperatures produced there.<br><br>In 1956, James A. FAY[2079] and Edward LEKAWA at Cornell University used the method of converging cylindrical shock waves to ignite hydrogen-oxygen and hydrogen-air mixtures in order to determine the minimum shock wave strength needed to produce ignition. |
| 1951 | Michelson Laboratories, U.S. Naval Ordnance Test Station, China Lake, California | **John S. RINEHART,**[2080] a U.S. research physicist, **develops a method for measuring the shape of the stress wave induced in a metal by an explosion.** He discovers that the change in hardness that results from the passage of the stress wave in mild steel decreases rapidly with increasing distance from the explosive. This trend, however, stops abruptly and remains constant for a considerable depth, but is still higher than the original hardness. On the other hand, for annealed copper, annealed brass and 24S-T4 aluminum alloy, the hardness curves do not show abrupt changes in slope. |
| 1951 | Soviet Union | **Andrei D. SAKHAROV,** an eminent Soviet physicist, proposes a potential way to convert the detonation energy of a high explosive or a nuclear bomb into magnetic energy and **discusses devices for producing very strong fields and currents by the explosive deformation of current-carrying conductors** {TERLETSKII ⇨1957}.[2081] Independent of these Soviet studies, various methods of magnetic flux compression were also investigated at Los Alamos {FOWLER ET AL. ⇨1960}. |
| 1951 | Laboratoire de Recherches Techniques de Saint-Louis, Saint-Louis, France | **Rudi SCHALL**[2082] **and Gustav THOMER,** two German physicists, **apply flash soft radiography to visualize shock waves in low-density materials,** such as in gaseous matter (air with a methyl iodide additive), in some liquids (water, acetone, alcohol, ether), and in low-density solids (Plexiglas, hard rubber, aluminum).<br>(i) In order to improve the X-ray contrast for shock waves in air, they<br>▸ operate the tube at a low voltage (max. 30 kV) and high pulse current (max. 50 kA);<br>▸ use a thin cellophane window at the tube output to minimize soft X-ray absorption;<br>▸ apply a flash X-ray tube with a copper anode to generate the soft $Cu$-$K\alpha$ radiation; and<br>▸ add a small quantity of gaseous methyl iodide [$CH_3I$] to the test gas.<br>(ii) In air they measure a compression ratio of $\rho/\rho_0 = 4.5$ in the shock wave. SCHALL {⇨1950} had previously used this diagnostic method to determine the Hugoniot of water.<br><br>Seven years later, HERRMANN {⇨1958} succeeded in taking the first flash radiographs of the shock front in gases, although without using any contrast-enhancing additives. |
| 1951 | U.S. Naval Ordnance Laboratory (NOL), Silver Spring, Maryland | **Raymond J. SEEGER,**[2083] a U.S. physicist, **and Harry POLACHEK,** a Polish-born U.S. mathematician, **perform theoretical studies on shock waves in water-like substances** in order to better predict the influence of thermodynamic factors on the interaction of shock waves in substances that behave like water. They define a *water-like substance* as a liquid for which the in- |

---

[2078] R.W. PERRY and A. KANTROWITZ: *The production and stability of converging shock waves.* J. Appl. Phys. **22**, 878-886 (1951).
[2079] J.A. FAY and E. LEKAWA: *Ignition of combustible gases by converging shock waves.* J. Appl. Phys. **27**, 261-266 (1956).
[2080] J.S. RINEHART: *Work-hardening of mild steel by explosive attack.* J. Appl. Phys. **22**, 1086-1087 (1951); *Some experimental indications of the stresses produced in a body by an exploding charge.* Ibid. **22**, 1178-1179 (1951).
[2081] A.D. SAKHAROV ET AL.: *Magnetic cumulation.* Sov. Phys. Dokl. **10**, 1045-1047 (1965).
[2082] R. SCHALL and G. THOMER: *Röntgenblitzaufnahmen von Stoßwellen in festen, flüssigen und gasförmigen Medien.* Z. Angew. Phys. **3**, 41-44 (1951).
[2083] R.J. SEEGER and H. POLACHEK: *On shock-wave like phenomena: waterlike substances.* J. Appl. Phys. **22**, 640-654 (1951).

trinsic energy can be separated into two terms: one term that is only dependent on density, the other only on entropy. They show that a number of intrinsic differences exist between the behavior of shock waves in water-like substances and their behavior in ideal gases, particularly in the theory of triple-shock intersections.

| | | |
|---|---|---|
| 1951 | NACA's Langley Aeronautical Laboratory [now NASA's Langley Research Center], Hampton, Virginia | **John STACK,** a U.S. aeronautical engineer and then assistant chief of research, **is awarded the 1951 Collier Trophy** {COLLIER ⇨1911} **"for the conception, development, and practical application of the transonic wind tunnel throat."** Based on the first experimental operation of a model slotted-throat wind tunnel performed towards the end of the 1940s by the physicist Ray H. WRIGHT and the aeronautical engineer Vernon G. WARD,[2084] **John STACK and associates incorporated a "slotted-throat" design into Langley's 8-ft (2.44-m) dia. high-speed tunnel,** a continuous transonic flow facility {⇨Fig. 4.9–F}.[2085] This was a landmark wind-tunnel technique that permitted the measurement of accurate data on airframe performance in the transonic range.<br><br>In the same year, **Richard T. WHITCOMB,**[2086] a U.S. aeronautical engineer at NACA's Langley Aeronautical Laboratory, **suggests an airplane configuration that minimizes drag at high speeds, which he terms the transonic** *area rule* {⇨Fig. 4.14–N} while collecting and analyzing data on lengthwise distributions of fuselage and wing volume. ▪ His discovery, made in connection with the development of the delta-winged jet fighter Convair F-102 ("Delta Dagger"), indicated that adoption of a "coke-bottle" (or wasp-waisted) shape can significantly increase the speed of a jet-propelled airplane. In the mid-1950s, the coke-bottle design was quickly applied to the fuselages of most U.S. jet interceptor configurations. In 1954, WHITCOMB received the prestigious Collier Trophy {COLLIER ⇨1911} "for discovery and experimental verification of the area rule, a contribution to base knowledge yielding significantly higher airplane speed and greater range with the same power." Compared with other significant advances in supersonic aerodynamics, this finding ranks in importance with swept airfoils {BUSEMANN ⇨1935} and supercritical wings {WHITCOMB ⇨1969}. ▪ **The area rule,** known as the "tadpole rule" in the Soviet Union, **was first applied by Andrei N. TUPOLEV's design bureau to the design of the bomber Tu-2** ("Bat," first flight in Oct. 1940), and was later refined in the jet bomber Tu-82 ("Butcher," first flight in March 1949).[2087] |
| 1951 | Palmer Physical Laboratory, Princeton University, Princeton, New Jersey | **Donald R. WHITE,**[2088] a Ph.D. student of Prof. Walker BLEAKNEY, **maps the density field in Mach reflection of strong air shocks on a wedge,** and studies the transition from Mach reflection to regular reflection. Stimulated by SMITH's previous studies on Mach reflection {L.G. SMITH ⇨1945}, which indicated a serious disagreement between measured data and the three-shock theory {VON NEUMANN ⇨1943}, **he discovers that "a new shock" develops behind the reflected shock** – a phenomenon characterized by two triple points and which later will be named **"double Mach reflection."** ▪ His report does not include an interferogram of double Mach reflection – as in the other examples of irregular reflection he studied – only a schematic {⇨Fig. 4.13–D}.<br><br>One year later, WHITE[2089] wrote in a conference paper, "The ability of the present shock tube to produce stronger shocks at higher densities than were previously available has led to |

---

[2084] R.H. WRIGHT and V.G. WARD: *NACA transonic wind-tunnel test sections.* Rept. NACA-RM L8J06 (1948).

[2085] *Eight-foot high speed tunnel.* In: H.A. BUTOWSKY: *Man in space. Excerpts from a national historic landmark theme study.* Natl. Park Service, U.S. Dept. of the Interior (1984); http://www.cr.nps.gov/history/online_books/butowsky4/space3.htm.

[2086] R.T. WHITCOMB: *A study of the zero-lift drag-rise characteristics of wing-body combinations near the speed of sound.* Rept. NACA-RM L52H08 (1952); *Recent results pertaining to the application of the area rule.* Rept. NACA-RM L53I15a (1953).

[2087] P. DUFFY and A. KANDALOV: *TUPOLEV: the man and his aircraft.* Airlife, Shrewsbury (1996), pp. 16-17.

[2088] D.R. WHITE: *An experimental survey of the Mach reflection of shock waves.* Tech. Rept. II-10, Dept. of Physics, Princeton University, NJ (1951).

[2089] D.R. WHITE: *An experimental survey of the Mach reflection of shock waves.* In: *Proc. 2nd Midwestern Conference on Fluid Mechanics* [Ohio State University, Columbus, OH, March 1952]. Ohio State University Studies. Engng. Series **XXI**, 253-262 (1952).

the discovery of hitherto unobserved features in the Mach reflection of strong shocks. At roughly the point where the gas velocity along the wedge becomes sonic with respect to the wedge the slipstream is observed to curl under itself in the immediate neighborhood of the wedge ... As the strength and angle increase, a reverse curvature of the reflected shock appears near the triple point. This becomes localized and **a *new shock* is observed between the wedge and the reflected shock, the resultant shock interaction giving rise to a second slipstream.**"

| | | |
|---|---|---|
| 1952 | *NATO Headquarters, Palais de Chaillot, Paris, France* | On January 24, **foundation of the Advisory Group for Aerospace Research & Development (AGARD),** a division of the North Atlantic Treaty Organization (NATO, established in 1949), with the goal of "bringing the treaty nations to a common level of knowledge of aerospace research." The first director is Theodore VON KÁRMÁN, who also inspired this organization. ▪ AGARD, the first scientific agency of NATO, provided scientific and technical advice and assistance to the NATO Military Committee in the field of aerospace research and development.<br><br>In 1966, AGARD became an agency under the Military Committee and in 1997 it was absorbed into the new NATO Research and Technology Organization (RTO).[2090] |
| 1952 | *UC Livermore, California* | In September, **establishment of the University of California Radiation Laboratory (UCRL) as part of the UC Berkeley Radiation Laboratory,** with a mission to ensure national security and to apply science and technology to the important issues of the time. The co-founders are the nuclear physicists Edward TELLER and Ernest O. LAWRENCE. The first director is Herbert F. YORK. ▪ In 1958, after LAWRENCE's death, the Livermore Laboratory was renamed the Lawrence Radiation Laboratory (LRL). In 1971, the name changed again to the Lawrence Livermore Laboratory (LLL), and eventually in 1981 to the **Lawrence Livermore National Laboratory (LLNL).** The Laboratory is presently managed by the University of California for the U.S. Department of Energy's National Nuclear Security Administration. |
| 1952 | *Myojin-sho Reef volcano [ca. 300 km south of Honshu], Pacific Ocean* | In the period September 16–26, **vigorous submarine volcanic eruptions occur at the reef, causing peculiar tsunami waves.**<br><br>(i) At the same time, these tsunami waves are recorded on Hachijo Island – about 130 km distant from the reef – by a self-registering wave gauge, a reconstruction of a statoscope for measuring microbarometric changes which is of the pressure recording type and uses a bellow.<br><br>(ii) One member of the investigation party who witnessed one of these eruptions while on a research vessel from Tokyo University, later reported: "When the eruption occurred, the sea surface, which had been even up to that time, heaved up suddenly into a water dome 300 m wide and less than 5 m high immediately after the eruption, 1,000 m wide after 10 seconds, and 1,500 m wide after 60 seconds."<br><br>(iii) Another crew member reported that "thin, black smoke rose on the water surface, sea water heaved up to more than 5–6 m high, and instantly an eruption took place at 13h 40m, when explosion noises were heard and slight vibration of the air was felt. The water which had heaved up collapsed into waves of wavelength 50–60 m, while the wave height decreased before our eyes and became only about 1 cm when the waves reached our ship, which was at a distance of about 1,600 m. Such waves came twice, and the third wave was not perceived. The period was 12–13 seconds." ▪ When volcanoes erupt in deep water, the weight of the overlying water prevents the explosive escape of volcanic gases. But when eruptions take place in shallower water, the water pressure can no longer confine the volcanic gases, thereby allowing them to escape. The presence of an unlimited supply of water also adds great quantities of |

---

[2090] *The foundation of AGARD 1950–1952.* NATO Research and Technology Organization (RTO); http://www.rta.nato.int/RTOHistory/AGARD.htm.

steam to the eruption. If the conditions are right, the result can be explosive, with rock debris and steam blasting out of the ocean.[2091]

During the same year, in Tokyo, Sanae UNOKI[2092] at the Central Meteorological Observatory and Masito NAKANO at the Meteorological Research Institute begin to analyze the measurements by applying the theory of Cauchy-Poisson waves {CAUCHY ⇨1815; POISSON ⇨1815}. They find that the characteristics of the tsunami waves – such as their period, variation in wave height and interference phenomena – are quite similar to those for the Cauchy-Poisson wave solution in the case when an initial impulse or an initial elevation is given to a finite area.

**These studies were the first to reveal that Cauchy-Poisson waves actually do exist in the ocean, and take a form very similar to that predicted by theory, thus allowing one to estimate the intensity and scale of a submarine eruption.**

| | | |
|---|---|---|
| 1952 | *Western shore of Trimouille Island, Monte Bello Islands, Australia* | On October 3, **Britain's first nuclear test is carried out in a lagoon.** The bomb, dubbed HURRICANE which has a yield of about 25 kilotons of TNT equivalent, is exploded 2.7 m below the water line, inside the hull of the frigate HMS *Plym* which is anchored in 12-m deep water about 400 m off-shore. The explosion mostly vaporizes the ship, except for scattered fragments of hot metal that set fire to the spinifex scrub covering Trimouille Island, and leaves a saucer-shaped crater on the seabed about 6 m deep and 300 m across.<br><br>The first British H-bomb, dubbed GRAPPLE X, was exploded over Christmas Island (Indian Ocean) on November 8, 1957. |
| 1952 | *Eluklab (or Elugelab) Island, Enewetak Atoll in the Marshall Islands, East Micronesia, Pacific Ocean* | On November 1, **the first American thermonuclear (or fusion) bomb – code-named MIKE – is exploded on Eluklab Island.** The detonation vaporizes the island, leaving a submerged crater about 1.9 km in dia. and 49 m deep. ▪ The detonation of MIKE was the first part of **Operation IVY,**[2093] a two-detonation atmospheric weapon test series conducted on Enewetak Atoll. The second detonation on November 16, code-named KING, was a 500-kilotons airburst from a stockpiled fission weapon (plutonium) dropped at 440 m from a Convair bomber B-36 ("Peacemaker") over a reef off Runit Island.<br><br>(i) The MIKE detonation produces a mushroom cloud 27 miles (43 km) high. Major F.E. MOORE and Lt. H.G. BECHANAN of the Joint Task Force (comprising 132 staff), who was aboard a vessel at sea, witnessed the shot as follows,[2094] "The heat wave was felt immediately at distances of 30–35 miles [48–56 km]. The tremendous fireball quickly expanded after a momentary hover time and appeared to be approximately a mile in diameter before the cloud-chamber effect and scud clouds partially obscured it from view … The shock wave and sound arrived at the various ships approximately 2½ minutes after the detonation, accompanied by a sharp report followed by an extended, broken, rumbling sound. The pressure pulse and the reduced pressure period as received by ear were exceptionally long."<br><br>(ii) Berlyn BRIXNER, a Los Alamos scientific photographer who had taken the first high-speed photographs of the explosion of the first fission bomb {Trinity Test ⇨1945}, also takes the first photographs of the first fusion bomb. He constructed a new camera capable of taking a sequence of 170, 12 × 14-mm² pictures at a rate of 3½ million frames per second for this specific purpose. ▪ Later, BRIXNER[2095] remembered: "I was drawn into the high-speed photography |

---

[2091] *Submarine eruptions – volcanoes on the rise*. USGS Hawaiian Volcano Observartory (HVO) (July 20, 2005); http://hvo.wr.usgs.gov/volcanowatch/2005/05_07_14.html.

[2092] S. UNOKI and M. NAKANO: *On the Cauchy-Poisson waves caused by the eruption of a submarine volcano*. Oceanogr. Mag. **4**, 119-141 (1952); Ibid. **5**, 1-13 (1953).

[2093] F.R. GLADECK ET AL.: *Operation IVY: 1952*. DNA Tech. Rept. 6036F. Defense Nuclear Agency (DNA), Washington, DC (Dec. 1982).

[2094] F.H. SHELTON: *Reflections of a nuclear weaponeer*. Shelton Enterprise Inc., Colorado Springs, CO (1988), pp. **5**:37-38.

[2095] B. BRIXNER: *High-speed photography of the first hydrogen-bomb explosion*. Proc. 20th Int. Congr. on High-Speed Photography and Photonics [Victoria, BC, Canada, Sept. 1992]. (J.M. DEWEY, ed.) *High-speed photography and photonics*. Proceedings of SPIE **1801**, 52-55 (1992).

of the first hydrogen-bomb experiment while attending a 1951 Los Alamos Coordinating Council meeting where plans for a future test at Enewetak atoll were being discussed. Edward TELLER suddenly pointed to me and said, in effect: 'I want the start of the explosion photographed every millionth of a second, or faster, if possible.' His request meant that I needed to have a camera running at one million frames per second, or faster, when the hydrogen bomb started to explode…"

(iii) **Edward TELLER**[2096] **later estimated the yield of his "super bomb" to be about 10 megatons of TNT equivalent,** of which 77% was due to fission.

The new thermonuclear bomb was a bulky experimental device rather than a compact weapon, because liquid deuterium was selected as the thermonuclear fuel, which required a cryogenics system. The bomb was a two-stage assembly and based upon an idea first conceived by the Polish-born U.S. mathematician Stanislaw M. ULAM. He proposed utilizing the tremendous flux of neutrons generated by the fission process to compress a light element fuel in order to achieve fusion. This concept was developed further in a collaboration between ULAM and TELLER in early 1950 – today known as the **"Teller-Ulam staged radiation implosion concept" or "Teller-Ulam design."** TELLER, generally considered to be the father of the hydrogen bomb,[2097] resumed ULAM's idea and solved the engineering problems.

The H-bomb concept was later conceived and realized independently in the former Soviet Union by the distinguished physicists Yakov B. ZEL'DOVICH and Andrei D. SAKHAROV {Soviet Superbomb Test ⇨ 1953; Test Site Semipalatinsk-21 ⇨ 1955}.

1952  *Washington, DC*  **First International Symposium on High-Speed Photography;** the general chairman is John H. WADDELL. It is organized by the American Society of Motion Pictures and Television Engineers (SMPTE, founded in 1916), although a few European members also participate.

The following symposia, each renamed an **International Congress on High-Speed Photography,** were held at Paris, France (1954); London, U.K. (1956); Cologne, Germany (1958); Washington, DC (1960), Scheveningen, The Netherlands (1962); Zurich, Switzerland (1965); Stockholm, Sweden (1968); Denver, CO (1970), Nice, France (1972); and London, U.K. (1974). The following symposia, each renamed an **International Congress on High-Speed Photography and Photonics,**[2098] were held at Toronto, Canada (1976); Tokyo, Japan (1978); Moscow, Soviet Union (1980); San Diego, CA (1982); Strasbourg, France (1984); Pretoria, South Africa (1986); Xi'an, Shaanxi, China (1988); Cambridge, U.K. (1990); Victoria, BC, Canada (1992); Taejon, South Korea (1994); Santa Fe, NM (1996); Moscow, Russia (1998); Sendai, Japan (2000); Beaune, France (2002); Alexandria, VA (2004); and Xi'an, China (2006). The symposia rapidly became an important international meeting place for shock physicists, who presented new diagnostic methods developed especially for the study of shock wave phenomena in many cases. **The conference proceedings, together with those of the Shock Tube Symposia {⇨1957}, represent one of the main sources of references on the diagnostics of shock wave, detonation and impact phenomena.**

In the United States, the Society of Photo Instrumentation Engineers (SPIE, founded in 1955) holds many annual conferences on photonic subjects and publishes the Proceedings of the International Congress on High-Speed Photography and Photonics. In the United Kingdom,

---

[2096] E. TELLER: *The work of many people.* Science **121**, 267-275 (Feb. 25, 1955).

[2097] In his book *In the shadow of the bomb: BETHE, OPPENHEIMER, and the moral responsibility of the scientist* [Princeton University Press, Princeton, NJ (2000), p. 166], Silvan S. SCHWEBER, a retired physics professor at Brandeis University, Waltham, MA, quoted Hans A. BETHE: "After the H-bomb was made, reporters started to call TELLER the father of the H-bomb. For the sake of history, I think it is more precise to say that ULAM is the father, because he provided the seed, and TELLER is the mother, because he remained with the child. As for me, I guess I am the midwife."

[2098] "*Photonics* is the technology of generating and harnessing light and other forms of radiant energy whose quantum unit is the photon. The science includes light emission, transmission, deflection, amplification and detection by optical components and instruments, lasers and other light sources, fiber optics, electro-optical instrumentation, related hardware and electronics…" From *Photonics dictionary.* Laurin Publ., Pittsfield, MA; http://www.photonics.com/dictHome.aspx.

| | | |
|---|---|---|
| | | the Association for High-Speed Photography and Photonics (AHSPP, founded in 1954) holds annual conferences on the applications of high-speed photography and the development of high-speed recording techniques.[2099] |
| 1952 | *Institut für Aerodynamik, ETH Zürich, Switzerland & ONERA, Paris, France* | Jakob ACKERET,[2100] a Swiss aerodynamicist, **and, independently, Maurice ROY,**[2101] a French physicist, **investigate the "double sonic boom,"** a curious acoustic phenomenon connected with diving stunts at transonic speeds close to Mach 1. Unlike an ordinary boom {E. MACH & SALCHER ⇨ 1886; DUMOND ET AL. ⇨ 1946}, it is noted as a "boom, boom!" when the aircraft moves towards the observer at variable speed. They explain this effect by the piling up of sound impulses – a process similar to the Doppler effect {DOPPLER ⇨ 1842}, but in this case produced by waves of finite amplitude such as those originated by engine noise or the emission of air shocks upon passing through the sound barrier. |
| 1952 | *NACA Ames Aeronautical Laboratory, Moffet Field, California* | H. Julian ALLEN and Alfred J. EGGERS, two U.S. aeronautical engineers studying reentry problems at $M < 20$, **find that increasing the drag of a high-speed vehicle causes much of the heat of reentry to be deflected away from the vehicle.** Therefore, the best designs for reducing heating by drag are not needle-shaped noses, but instead blunt noses: such a profile forms a thick shock wave ahead of the vehicle that both deflects the heat and slows the vehicle more rapidly, thereby protecting it. ▪ **The theory of ALLEN's so-called "blunt-body design" revolutionized the fundamental design of ballistic missile and spacecraft reentry shapes,** and heavily influenced the designs of the ablative heat shields that protected the Mercury, Gemini and Apollo astronauts as their space capsules reentered the Earth's atmosphere {⇨Fig. 4.6–N}.[2102]<br><br>In 1964, at an international symposium held on hypersonic flow at Cornell Aeronautical Laboratory, ALLEN[2103] reviewed the history of the aerodynamic heating of vehicles designed to reenter the Earth's atmosphere at hypersonic speeds. |
| 1952 | *Palmer Physical Laboratory, Princeton University, Princeton, New Jersey* | Walker BLEAKNEY,[2104] an applied physicist and the founder of Princeton's Shock Tube Laboratory, **uses the Princeton shock tube technique** {BLEAKNEY ⇨ 1946; BLEAKNEY, WEIMER & FLETCHER ⇨ 1949} **to investigate loading effects of planar blasts striking rigid two-dimensional (2-D) model structures.** He states, "In the general problem of predicting the damage to structures induced by a blast from some explosive source it is useful to divide the analysis into three separate although not independent parts. The first is concerned with the characteristics of the blast itself without reference to the structure, the second is the determination of the forces or loads on the target, and the third is the estimation of the response of the structure. Once approximate answers can be given to these separate questions, consideration may be given to their interactions on each other to give a better answer to the general problem…" ▪ The pressure of a nuclear explosion falls steadily behind the shock front, but the rate of decay is so slow that the first hundred feet of the wave profile can be considered to be flat-topped, as in the shock tube. This is the reason why shock tubes can be used to model blast loading on structures.[2105] |

---

[2099] Association for High Speed Photography and Photonics (AHSPP), U.K.; http://www.ahspp.com/ahspp.aspx.

[2100] J. ACKERET: *Akustische Phänomene bei hohen Fluggeschwindigkeiten*. Schweiz. Aero-Revue **11**, 429-430 (1952).

[2101] M. ROY: *A propos du gong sonique*. C. R. Acad. Sci. Paris **235**, 756-759 (1952).

[2102] H.J. ALLEN and A.J. EGGERS: *A study of the motion and aerodynamic heating of ballistic missiles entering the Earth's atmosphere at high supersonic speed*. Rept. NACA-TR 1381 (1958).

[2103] H.J. ALLEN: *The aerodynamic heating of atmosphere entry vehicles – a review*. In: (J.G. HALL, ed.) *Fundamental phenomena in hypersonic flow*. Cornell University Press, Ithaca, NY (1966), pp. 5-27.

[2104] W. BLEAKNEY: *A shock tube investigation of the blast loading of structures*. In: (C.M. DUKE and M. FEIGEN, eds.) *Proc. Symposium on Earthquake and Blast Effects on Structures* [Los Angeles, CA, 1952]. K.V. Steinbrugge, San Francisco (1957), pp. 46-73.

[2105] W. BLEAKNEY, D.R. WHITE, and W.C. GRIFFITH: *Measurements of diffraction of shock waves and resulting loading of structures*. ASME J. Appl. Mech. **17**, 439-445 (1950).

Combined with high-speed visualization, the shock tube became an extremely useful standard method for 2-D model blast loading studies, and is still applied in many laboratories around the world.

| | | |
|---|---|---|
| 1952 | Shock Wave Laboratory, Princeton University, Princeton, New Jersey | Wayland C. GRIFFITH,[2106] a U.S. fluid dynamicist, **first observes the existence of a boundary layer behind a moving shock wave using the schlieren technique.** ▪ With the expansion of shock tube technology, this phenomenon became a focal point for theoretical and experimental studies. The first theoretical studies on the development of laminar and turbulent boundary layer flow in a shock tube were performed by Harold MIRELS,[2107] a U.S. aeronautical engineer. |
| 1952 | National Physical Laboratory (NPL), London, England | William Frank HILTON,[2108] a British aerodynamicist who coined the term *sonic wall* {HILTON ⇨1935}, **publishes the textbook *High-Speed Aerodynamics*.** In the Foreword, Sir Leonard BAIRSTOW, then Great Britain's leading aerodynamicist and Emeritus Professor of aviation at the University of London, wrote, "The rate at which aviation has grown in the last few years is the reason for *High-Speed Aerodynamics*. Less than fifty years ago a speed of flight of 100 m.p.h. was noteworthy; aircraft now land at speeds of this order … During the whole of this fifty years speeds of flight have been increasing, but only in the last few years has a stage been reached which compels us, for flight purposes, to consider air as an essentially compressible gas. In the olden days, speeds greater than that of sound constituted the subject of external ballistics, and the bodies moving through the air were shells fired from gun. Now there is no such gap, for missiles have wings and aircraft do not confine themselves to speeds less than that of sound … The present volume is a fair collection of what is now known about high-speed aerodynamics. The designer will find information enabling him to use swept-back wings to delay the onset of the worst effects of shock waves. It may be some time yet before large numbers of man-carrying aircraft travel at supersonic speeds, but guided weapons are introducing us to the problem which will then be prominent." |
| 1952 | Aerodynamics Division, National Physics Laboratory (NPL), Teddington, Middlesex, England | Douglas W. HOLDER[2109] and R. John NORTH, two British aerodynamicists at the NPL Wind Tunnel Facility, **report on a new color schlieren system utilizing a constant-dispersion prism** {⇨Fig. 4.18–B}. They summarize the advantages of their new method as follows:<br>▸ "The color method is somewhat more sensitive than the black-and-white, owing to the ability of the eye to discriminate easier between colors than between shades of gray;<br>▸ solid models in the schlieren field appear black, while aerodynamic disturbances are colored, which simplifies the detection and analysis of boundary layer conditions; and<br>▸ color is more appealing to general and scientific audiences and permits easier reference to recorded details in the schlieren photograph."<br>In 1954, R. John NORTH[2110] suggested that the prism could be eliminated by placing three gelatin strips at the image source such that undisturbed light fell only on the central strip. In 1967, Paul H. CORDS JR.,[2111] a photographic science engineer and head of the photographic unit in the Aeroballistics Division at the U.S. Naval Ordnance Laboratory (NOL) in Silver Spring, MD, developed his "dissection" method – a system of colored filters at the source. It |

---

[2106] W. GRIFFITH: *Shock tube studies of transonic flow over wedge profiles.* J. Aeronaut. Sci. **19**, 249-264 (1952).
[2107] H. MIRELS: *Laminar boundary layer behind a shock advancing into stationary fluid.* Rept. NACA-TN 3401 (1955); *Boundary layer behind a shock or thin expansion wave moving into stationary flow.* Rept. NACA-TN 3712 (1956).
[2108] W.F. HILTON: *High-speed aerodynamics.* Longmans, Green & Co, London *etc.* (1952).
[2109] D.W. HOLDER and R.J. NORTH: *A schlieren apparatus giving an image in color.* Nature **169**, 466 (March 15, 1952); *Colour in the wind-tunnel.* The Aeroplane **82**, No. 2111 (Jan. 4, 1952); *Optical methods for examining the flow in high-speed wind tunnels.* AGARDograph No. 23, AGARD NATO, Paris (Nov. 1956).
[2110] R.J. NORTH: *A color schlieren system using multicolor filters of simple construction.* NPL Aero Note 266 (Aug. 1954).
[2111] P.H. CORDS JR.: *A high resolution, high sensitivity color schlieren method.* SPIE J. **6**, No. 3, 85-88 (Feb./Mar. 1968).

improved both resolution and sensitivity, but like all previous methods it was only capable of resolving index of refraction gradients in one dimension.

In 1971, **Gary S. SETTLES**,[2112] a mechanical engineer at the University of Tennessee, **improved the source-filter technique to make full two-dimensional color schlieren photographs.** Very recently, he reviewed the characteristics of the numerous color schlieren methods proposed so far in a book {SETTLES ⇨2001}.

| | | |
|---|---|---|
| 1952 | *Laboratoire de Recherches Techniques de Saint-Louis, Saint-Louis, Alsace, France* | **Bernhard KOCH**,[2113] a German research scientist, **uses a "radiofrequency interferometer" to record the velocity of a supersonic 37-mm caliber projectile during flight and its deceleration due to aerodynamic drag.** The precision of his instrument, which uses a 0.3-W Klystron oscillator ($\lambda = 14.6$ cm), is less than 0.3% – *i.e.*, it is superior to classical methods such as the Le-Boulengé chronograph {LE BOULENGÉ ⇨1860s; ⇨Fig. 4.19−B}, the electric spark photo-chronograph {⇨Fig. 4.19−J}, the Kerr-cell chronograph,[2114] and light barriers. ▪ The microwave reflection-type interferometer proved to be very useful for measuring high velocities of both mechanical objects (*e.g.*, projectiles, meteorites) and gaseous boundary layers (*e.g.*, nonionized and ionized shock waves, detonation waves, exploding wires, *etc.*).[2115] However, the VISAR interferometer {BARKER & HOLLENBACH ⇨1965} later superceded microwave displacement-time techniques. |
| 1952 | *Forschungsinstitut Weil am Rhein, Germany* | **Herbert OERTEL**,[2116] a German experimental physicist, **reports on a new gauge for measuring density-time profiles in shock waves with a rise time of about 1 μs.** Initially developed at the Faculty of General Sciences of the RWTH Aachen, the instrument is based on the corona discharge effect. Compared with standard quartz gauges, his new small-sized corona gauge allows better resolution of the weak cylindrically symmetric N-wave profiles emitted from small-caliber supersonic shots. |
| 1952 | *Michelson Laboratories, U.S. Naval Ordnance Test Station (NOTS), Inyokern, China Lake, California* | The U.S. physicists **John S. RINEHART**,[2117] head of the terminal ballistics branch, **and William Charles WHITE study the crater geometry generated by the oblique impact of high-speed pellets into plaster of Paris.**<br>(i) They produce the craters using 2-g steel pellets and 0.7-g aluminum pellets, both at 4.7 km/s; the angle of impact is varied from normal incidence to 75°.<br>(ii) They observe that for angles of incidence between 0° and 45°, the mouth of the crater is largely circular. At 60°, it is roughly arrowhead-shaped. At 75° obliquity, the pellet ricochets and produces an elliptical crater.<br>(iii) Studies of sectioned targets show that the pellet initially penetrates the target at the angle of incidence. There is then an undercutting before the pellet is deflected from the target.<br>(iv) In conclusion, in the case of an oblique impact, **the pellet impacts at the steep side of the crater and leaves at the other side.**<br>The idea of generating craters in model experiments in order to understand the shapes of the craters that could be produced by meteorites goes back to Robert HOOKE {⇨1665}, but had also been used by other researchers, notably the U.S. geologist Grove Karl GILBERT {⇨1893} and the German geophysicist Alfred WEGENER {⇨1920}. However, it is still difficult to simulate the high velocities of meteorites (which can enter the atmosphere with velocities up to 70 km/s) on a laboratory scale for substantial masses. |

---

[2112] G.S. SETTLES: *A two-dimensional color schlieren technique.* Image Technology **14**, 19-23 (June/July 1972).
[2113] B. KOCH: *Mesure radioélectrique de la vitesse des projectiles (effet DOPPLER-FIZEAU par réflexion de micro-ondes).* Onde Electrique **32**, 357-371 (1952).
[2114] H.M. SMITH: *Precision measurement of time.* J. Sci. Instrum. **32**, 199-204 (1955).
[2115] B. KOCH: *Radioelektrische Messverfahren in der Kurzzeitphysik.* In: (K. VOLLRATH and G. THOMER, eds.) *Kurzzeitphysik.* Springer, Wien etc. (1967), pp. 367-435.
[2116] H. OERTEL: *Knallwellenoszillographie mittels Koronasonde.* Z. Angew. Phys. **4**, 177-183 (1952).
[2117] J.S. RINEHART and W.C. WHITE: *Shapes of craters in plaster of Paris by ultra-speed pellets.* Am. J. Phys. **20**, 14-18 (1952).

| 1953 | Flight Research Division, NACA's Langley Field, Hampton, Virginia | On April 14, **first operation of the McDonnell XF-88B aircraft** {⇨Fig. 4.20–G}, an experimental turboprop version of the XF-88 ("Voodoo"), **built to conduct research into propellers for supersonic planes flying at speeds of up to just above Mach 1.** ▪ In order to delay the onset of shock waves, supersonic propellers have shorter and thinner blades than standard propellers; they are operated at a reduced blade angle in order to make the blades supercritical. In the late 1940s, NACA, together with the Air Force and Navy, began a high-speed propeller development program that culminated in the *supersonic propeller.* At a flight Mach number of 0.95, the propeller tip Mach number was 1.6. Although transonic propellers are attractive from an economic point of view, the rapid development of the jet engine doomed further research on this subject, and the NACA High-Speed Propeller Program eventually ended with the transition to NASA in 1958.[2118, 2119]

**In astrophysics, the term *supersonic propeller* refers to a model developed for the spindown process of an initially rapidly spinning magnetic neutron star** which is a member of a close binary system in which the companion is a massive main-sequence star losing mass by a stellar wind. This model describes how the interactions at the boundary of the magnetosphere affect the surrounding material originating from the stellar wind.[2120] In the so-called "propeller" stage of evolution of a rotating neutron star, the centrifugal force in the equitorial region is much larger than the gravitational force, and incoming matter, for example that originating from the wind of a binary companion, tends to be flung away from the neutron star by its rotating magnetic field. |
|---|---|---|
| 1953 | Soviet Nuclear Test Site, Semipalatinsk-21, Kazakhstan, Soviet Union | On August 12, **the first Soviet fusion bomb,** Test RDS-6s, the SLOIKA ("Layer Cake"), dubbed JOE-4 in the Western World, **is exploded.** The bomb, which detonates at the top of a tower, has a yield of 400 kilotons of TNT equivalent. ▪ The first Soviet fusion bomb was mainly the result of fundamental work on controlled nuclear fusion performed by Andrei D. SAKHAROV, an eminent Soviet physicist who today is considered to be the father of the Soviet H-bomb. The theoretical investigation was headed by Yakov B. ZEL'DOVICH. The bomb was developed and manufactured at the KB-11 Design Bureau, Arzamas-16 (currently Sarov, approximately 400 km from Moscow).

(i) Unlike the first American thermonuclear explosion {MIKE Test ⇨1952}, which had a yield 25 times larger, the first Soviet fusion bomb is a "dry" bomb which uses lithium-6 deuteride (Li-6D) fusion fuel, a whitish, slightly-blue powder pressed and shaped into a solid ceramic that does not require a cryogenics system.[2121]

(ii) Vladimir S. KOMEL'KOV,[2122] a detonics specialist and designer of precision detonator caps, witnesses the explosion from a distance of about 20 km as follows: "The intensity of the light was such that we had to put on dark glasses. The Earth trembled beneath us, and our faces were struck, like the lash of a whip, by the dull, strong sound of the rolling explosion. From the jolt of the shock wave it was difficult to stand on one's feet. A cloud of dust rose to a height of 8 km…"

The saga of the Soviet hydrogen weapons project has been chronicled more recently by the renowned physicist Yulii Borisovich KHARITON,[2123, 2124] then scientific leader of the Arzamas-16 Nuclear Weapons Center, and collaborators. |

---

[2118] J.B. HAMMACK, M.C. KURBJUN, and T.C. O'BRYAN: *On a propeller research vehicle at Mach numbers to 1.01.* NACA-RM L57E20 (July 1957); http://ntrs.nasa.gov/archive/nasa/casi.ntrs.nasa.gov/19930090296_1993090296.pdf.

[2119] J.B. HAMMACK and T.C. O'BRYAN: *Effect of advance ratio on flight performance of a modified supersonic propeller.* Rept. NACA-TN 4389 (Sept. 1958); http://naca.central.cranfield.ac.uk/reports/1958/naca-tn-4389.pdf.

[2120] R.E. DAVIES and J.E. PRINGLE: *Spindown of neutron stars in close binary systems. Part II.* Month. Not. Roy. Astron. Soc. **196**, 209-224 (1981).

[2121] *The Soviet nuclear weapons program.* Nuclear Weapon Archive; http://nuclearweaponarchive.org/Russia/Sovwpnprog.html.

[2122] D. HOLLOWAY: *STALIN and the bomb. The Soviet Union and atomic energy, 1939–1956.* Yale University Press, New Haven, CT *etc.* (1994), pp. 306-307.

[2123] Y.B. KHARITON and Y.N. SMIRNOV: *The Khariton version.* Bull. Atomic Scientists **49**, 20-31 (May 1993).

| | | |
|---|---|---|
| 1953 | *Ames Aeronautical Laboratory, Moffet Field, California* | **The Ames research staff publishes a report which presents equations, tables, and charts useful in the analysis of high-speed flows of compressible fluids.**[2125]<br><br>(i) The equations provide relations for continuous one-dimensional flow, normal and oblique shock waves, and Prandtl-Meyer expansions for both perfect and imperfect gases.<br><br>(ii) The tables present useful dimensionless ratios for continuous one-dimensional flow and for normal shock waves as functions of Mach number for air considered to be a perfect gas.<br><br>(iii) One series of charts presents the characteristics of the flow of air for oblique shock waves and for cones in a supersonic air stream; a second series shows the effects of caloric imperfections on continuous one-dimensional flow and on the flow through normal and oblique shock waves. |
| 1953 | *Medizinische Universitätsklinik und Poliklinik, Tübingen, Germany* | Fritz BECHER,[2126] a German neurologist, **investigates the hammering mechanism of woodpeckers** (of the family *Picidae*) **and how their anatomies cope with the enormous mechanical stresses on their skulls and brain percussion.** ▪ BECHER's motivation for studying this subject originated from a curiosity about how the central nervous systems of vertebrates are efficiently protected against concussion. Concussion can present problems for humans, but is not an issue for woodpecker species. BECHER appears to have been the first to tackle the riddle of efficient concussion absorption. Since then, numerous ornithologists, evolutionists and neurologists have dedicated their studies to this fascinating research subject.<br><br>The skull of the woodpecker {⇨Fig. 4.1–Z} has a remarkable suspension system that absorbs the force of its rapid hammer blows (performed at up to 20 Hz). Its forehead and some skull muscles are joined to its chisel-pointed beak, and the jaw joints are so robust that they help lessen the effect of the forceful strokes during pecking. The beak and brain itself are cushioned against impact. In most birds, the bones of the beak are joined to the bones of the cranium – the part of the skull that surrounds the brain. **But in the woodpecker, the cranium and beak are separated by a sponge-like tissue that absorbs the shock each time the bird strikes its beak against a tree. Coordination between strong neck muscles, which keeps the head perfectly straight, allows the bird to withstand the enormous shock resulting from hammering many thousands of times a day.** ▪ High-speed photography[2127] revealed that "in that millisecond before the strike the thickened nictitans close over the eye. This protects the eye from flying debris and chips, and also acts as a 'seat belt' to restrain the eyes from literally *popping out of the head*." |
| 1953 | *Société Française des Amortisseurs de Carbon SA, France* | Christian BOURCIER DE CARBON,[2128] a French shock researcher and inventor, **applies for a patent for his monotube high-pressure gas shock absorber.** In the same year, he founds the De Carbon Company. ▪ In the following year, the patent was granted to him. Soon after, a license was sold to the German Bilstein Company which perfected the design and revolutionized the shock absorber industry. Daimler-Benz AG, Stuttgart, Germany, put the first gas-charged shock absorbers on certain '58 models.<br><br>An ordinary shock absorber works by forcing fluid through various orifices and spring-loaded valves, thus damping oscillations (C.L. HOROCK 1901). However, as the rate of compression and rebound speeds up, cavitation (or "foaming") occurs in the hydraulic oil, which causes a loss of damping action. In a gas shock absorber, the reservoir of fluid is kept under |

---

[2124] Y.B. KHARITON, V.B. ADAMSKII, and Y.N. SMIRNOV: *On the making of the Soviet hydrogen (thermonuclear) bomb.* Phys. Uspekhi **39**, No. 2, 185-189 (1996).

[2125] *Equations, tables, and charts for compressible flow.* NACA Rept. 1135 (1953); http://ntrs.nasa.gov/archive/nasa/casi.ntrs.nasa.gov/19930091059_1993091059.pdf.

[2126] F. BECHER: *Untersuchungen an Spechten zur Frage der funktionellen Anpassung an die mechanische Belastung.* Z. Naturforsch. **8B**, 192-203 (1953).

[2127] I.R. SCHWAB: *Cure for a headache.* Brit. J. Ophthalmol. **86**, 843 (2002).

[2128] C. BOURCIER DE CARBON: *Amortisseur à liquide et gaz comprimé.* French Patent No. 1,082,032 (requested 1953, granted 1954).

| | | |
|---|---|---|
| | | high pressure by a sealed-in dose of nitrogen at pressures (in some single-tube versions) of up to 25 bar. This stops the bubbles from forming no matter how violently the piston is moving. Hence, the use of gas-charged shock absorbers, which suppress lag and react to the minutest movement, results in a smoother ride and less residual bounce.[2129] |
| 1953 | *Gates & Crellin Laboratories of Chemistry, Cal-Tech, Pasadena, California* | Tucker CARRINGTON,[2130] a chemist at the National Bureau of Standards, **and Norman DAVIDSON,** a biochemist at CalTech, **first apply the shock tube to study chemical kinetics in gases such as in nitrogen tetroxide** [$N_2O_4$], the storable liquid propellant of choice at that time. It is fairly simple kinetically and dissociates to nitrogen dioxide [$NO_2$], the simplest possible type of unimolecular chemical reaction. They use a shock wave to increase the translational and rotational temperature of a gas by a definite amount in a time of around a few collision times. Using a photoelectric method to follow the absorption bands, they observe the shock-induced rate of dissociation of nitrogen tetroxide optically by recording the change in the opacity of the gas with time.<br><br>The sudden heating of a gas by using a shock tube proved very useful to investigate how internal energy of rotation and vibration of the colliding molecules contributes to the energy available to produce thermal dissociation. In the following years, the rates of thermal dissociation of other polyatomic as well as diatomic molecules were studied. |
| 1953 | *R & D Dept., Norton Company, Worchester, Massachusetts* | Loring COES,[2131] a U.S. chemist, while subjecting samples of a mixture of sodium metasilicate [$Na_2SiO_3$] and diammonium phosphate [$(NH_4)_2HPO_4$] to a static pressure of 35 kbar at a temperature of 750 °C for a period of 15 hours, **synthesizes a new high-pressure silica polymorph: "coesite."** ▪ However, later attempts to produce coesite dynamically on a laboratory scale using shock waves failed, apparently due to the short compression period achievable in shock recovery experiments.<br><br>The structure of coesite is composed of $SiO_4$ tetrahedrons that are linked into four-membered rings which are then linked together into a chain-like structure {⇨Fig. 4.1–B}. It is more compact than the other members of the quartz group, except stishovite {STISHOV & POPOVA ⇨1961}. Because of the very high pressure necessary for its formation, it does not occur naturally in the Earth's crust. However, **coesite attained great importance in geology as a primary indicator of the shock-induced ultrahigh-pressure metamorphism of natural materials,** such as that provoked by meteoritic impact {CHAO ET AL. ⇨1960} – so-called "shock metamorphism" or "impact metamorphism." The Hugoniot of coesite, shocked into the stishovite regime, was determined up to 112 GPa (1.12 Mbar) by Russian researchers.[2132] |
| 1953 | *Los Alamos Scientific Laboratory (LASL), Los Alamos, New Mexico* | Russell E. DUFF,[2133] a U.S. physicist, **and Edwin HOUSTON determine the Chapman-Jouguet (CJ) pressure for a high explosive** by measuring the free-surface velocity as a function of thickness for metal plates in contact with the explosive.<br><br>(i) The theory behind their experimental method is based on an idea first advanced in 1945 by the Los Alamos scientist Roy W. GORANSON.[2134] He suggested that the length of the reaction zone of a detonating solid explosive – *i.e.*, which is the length of the steady-state reaction zone which follows the nonreactive shock and which is terminated at the CJ surface where the |

---

[2129] *Dampers.* Dept. of Mechanical Engineering, University of Bath, U.K.; http://www.bath.ac.uk/~en9apr/dampers.htm.
[2130] T. CARRINGTON and N. DAVIDSON: *Shock waves in chemical kinetics: the rate of dissociation of $N_2O_4$.* J. Phys. Chem. **57**, 418-427 (1953).
[2131] L. COES: *A new dense crystalline silica.* Science **118**, 131-132 (July 31, 1953).
[2132] M.A. PODURETS ET AL.: *Polymorphism of silica in shock waves, and equation of state of coesite and stishovite.* Izv. Acad. Sci. USSR Physics Solid Earth **17**, 9-15 (1981).
[2133] R.E. DUFF and E. HOUSTON: *Measurement of the Chapman-Jouguet pressure and reaction zone length in a detonating high explosive.* In: (L.D. HAMPTON ET AL., eds.) *Proc. 2nd Symposium (ONR) on Detonation* [White Oak, MD, 1953]. Rept. AD-052 145, Office of Naval Research (ONR), Arlington, VA (1955), pp. 225-239; J. Chem. Phys. **23**, 1268-1273 (1955).
[2134] R.W. GORANSON: *A method for determining equations of state and reaction zones in detonation of high explosives, and its application to pentolite, composition-B, baratol, and TNT.* Classified Los Alamos Rept. LA-487 (1945).

local flow velocity plus the speed of sound equals the detonation velocity in the ZND model – could be investigated by determining the initial free surface velocity imparted to thin metal plates as a function of plate thickness, and that the reaction zone length and CJ pressure could be estimated in this way.

(ii) They use aluminum plates ranging in thickness from 0.0085 to 0.2 in. (0.216–5.08 mm). Using a high-speed oscilloscope, they measure the free surface velocity by recording the arrival times of the metal surface at a series of contactors or pins.

(iii) **For Composition B (containing 63% RDX at a density of 1.67 g/cm$^3$), they estimate a CJ pressure of 272 kbar from a plot of measured surface velocity *vs.* plate thickness. They find a length of 0.13 mm for the reaction zone.**

(iv) Referring to the ***von Neumann spike*** {VON NEUMANN ⇨1942}, they annotate that "no pathological detonation in the von Neumann sense has yet been observed."

The subject was resumed two years later by their colleague William E. DEAL {⇨1957}.

---

**1953** — *Cambridge University, England*

At the Symposium on Cosmic Gas Dynamics, **Thomas GOLD,**[2135] an Austrian-born British astrophysicist, **is the first to interpret the sudden commencement of geomagnetic storms observed at the Earth approximately two days after a solar flare:**

(i) According to his theory, **this phenomenon is due to the increased mass flux behind an interplanetary shock impinging on the Earth's *magnetosphere*** – a term which he coins to designate the region around an astronomical object in which the motion of charged particles is dominated by the object's magnetic field.

(ii) He makes the seminal observation that, since the rise time for disturbances arriving at the Earth – signaled by the sudden start of the geomagnetic storm – is about five minutes, this would be compatible not only with the assumption of an extended interplanetary medium, but also with a steep-fronted wave.

(iii) Participating in a discussion panel on shock waves and rarefied gases, he says, "I should like to discuss, in connection with the subject of shock waves, some of the magnetic disturbances on the Earth that are caused by solar outbursts. The initial magnetic disturbance at a *sudden commencement* of a magnetic storm can be accounted for very roughly by an increase in pressure of the tenuous gas around the Earth. This increase in pressure may perhaps be described as the effect of a wave sent out by the Sun through the tenuous medium between Sun and Earth. In the complete absence of any such medium this description would then correspond to that of a stream of particles, while in the presence of a medium the correct description may lie anywhere between an acoustic wave, a supersonic shock wave or an unimpeded corpuscular stream. **It is known that some *commencements* occur on the Earth about 24 hours after the outburst has occurred on the Sun. Thus we know the velocity of the phenomenon which propagates itself, and this velocity is on the order of twenty or more times the velocity of sound in the medium.** Although we have a travel time of 24 hours or more, the initial increase in the magnetic force may reach a maximum in as little as two minutes. If this were to be attributed to a stream of particles which is unimpeded until it reaches the neighborhood of the Earth, then it would be necessary for this stream to have a quite unreasonably small velocity dispersion … **A much more reasonable interpretation of the phenomenon would be the arrival of a highly supersonic shock wave with the characteristic sharp wave front … The observation of magnetic storms may hence give us fairly direct proof of the existence of shock waves in the interplanetary medium.** The properties of waves mentioned by Arthur KANTROWITZ show that if the original outburst did not possess a very sharp beginning, a sharp front would be built up during the propagation through space. I know of no other theory that

---

[2135] T. GOLD: *Discussion on shock waves and rarefied gases*. In: (H.C. VAN DE HULST and J.M. BURGERS, eds.) *Proc. Symposium on Gas Dynamics of Cosmic Clouds* [Cambridge, U.K., July 1953]. Int. Astronautical Union (IAU) Symposium Series, North-Holland Publ. Co., Amsterdam (1955), pp. 103-104.

can reasonably provide the extreme suddenness of this phenomenon..." ▪ GOLD's far-reaching hypothesis, later resumed to explain the solar wind phenomenon {PARKER ⇒1958}, was confirmed nine years later experimentally {NEUGEBAUER & SNYDER ⇒1962}.

Historically, the evolution of a ***collision-free hydromagnetic shock theory*** was impeded by the difficulty with conceptualizing the kinetics of wave propagation in a rarefied cosmic gas characterized by an enormous mean free path. In an ionized gas, the electric and magnetic field fluctuations supported by the collective motions of the plasma cause the scattering and heating of the individual ions and electrons.[2136]

| | | |
|---|---|---|
| 1953 | U.S. Weather Bureau, Washington, DC | D. Lee HARRIS,[2137] a U.S. meteorologist, **examines the popular hypothesis that nuclear explosions are responsible for an increase in the number of tornadoes in considerable detail.** |

(i) Of the 532 tornadoes reported in 1953 in the United States, 294 were reported between March 17 and June 15 – the period when atomic weapons were being tested in Nevada. The coincidence of this increase in the number of reported tornadoes with the increased frequency of atomic explosions in 1953 led many people to believe that atomic explosions caused an increase in tornadoes.

(ii) HARRIS arrives at the conclusion that tornado reports have always been incomplete and that much of the recent upward trend in tornado frequency can be accounted for by improvements in the tornado reporting system. A comparison of the distribution of tornadoes to that of debris from an atomic explosion in time and space does not support the hypothesis that atomic explosions tend to increase the tornado frequency.

More recently, the interesting question of whether it is possible to tame destructive tornadoes, for example by "putting them out" using a controlled nuclear explosion,[2138] or by beaming microwave energy[2139] from a Thunderstorm Solar Power Satellite into the cold, rainy downdraft of a thunderstorm, where a tornado could originate, has been discussed.

| | | |
|---|---|---|
| 1953 | Yerkes Observatory, Williams Bay, Wisconsin | H. Lawrence HELFER,[2140] a predoctoral fellow of the National Science Foundation and coworker of Subrahmanyan CHANDRASEKHAR, an Indian-born U.S. astrophysics professor and astronomer at the University of Chicago, **provides an interpretation of the *de Hoffman-Teller shock-wave equations*** {DE HOFFMAN & TELLER ⇒1950} for an infinitely conducting medium analogous to the classical interpretation of the ordinary hydrodynamic shock wave equations of RANKINE {⇒1869} and HUGONIOT {⇒1887}. He shows that weak magnetic fields in interstellar clouds will be amplified by the compression of matter, so weak magnetic fields are always amplified by the passage of shock fronts: "The magnetic field acts as if glued to the matter and is compressed by the shock to exactly the same extent as the fluid." ▪ HELFER's prediction was fully confirmed with the advent of ISEE satellites in the 1970s, which allowed the magnetic profile of the Earth's bow shock to be probed for the first time {ISEE Program ⇒1977; ⇒Fig. 4.12–D}. An increase in the magnetic field was also observed during the Voyager mission and was taken as evidence that the space probe had passed through the *heliospheric termination shock* {Voyager 1 & 2 ⇒2003; Voyager 1 ⇒2004}. |

The Earth's bow shock is a so-called "fast shock" (*i.e.*, the plasma pressure and the field strength increase at the shock front), in contrast to a "slow shock" (characterized by an increase in plasma pressure and a decrease in magnetic field strength). Fast shocks are more common than slow shocks in Solar System plasmas.

---

[2136] M.A. LEE: *Shock waves, collisionless, and particle acceleration*. In: (S.P. MARAN, ed.) *The astronomy and astrophysics encyclopedia*. Cambridge University Press (1992), pp. 627-629.
[2137] D.L. HARRIS: *Effects of atomic explosions on the frequency of tornadoes in the United States*. Month. Weather Rev. **82**, No. 12, 360-369 (1954).
[2138] *Tornados*. D. BERGER, Bluffton University, Bluffton, OH; http://www.madsci.org/posts/archives/oct98/905866499.Es.r.html.
[2139] L. DAVID: *Taking the twist out of a twister*; http://www.space.com/scienceastronomy/planetearth/tornado_taming_000303.html.
[2140] H.L. HELFER: *Magneto-hydrodynamic shock waves*. Astrophys. J. **117**, 177-199 (Feb. 20, 1953).

| | | |
|---|---|---|
| 1953 | Shock Tube Laboratory, University of Michigan, Ann Arbor, Michigan | Otto LAPORTE,[2141] a German-born U.S. physicist, **and collaborators use the shock tube as a source of high-temperature thermal radiation.**<br><br>(i) They observe luminosity when the shock is reflected normally from a plate at the end of the shock tube. When xenon is used as the carrier gas, they estimate temperatures of up to 18,000 K.<br><br>(ii) Spectroscopic studies reveal that line broadening and shifting is possible for strong shocks.<br><br>(iii) They state, "Like the schlieren and shadowgraph methods, observations of luminous phenomena will provide information on the flow of a supersonic gas stream around a model. As the incoming flow is deflected or reflected by a model, the luminosity distribution will furnish a measure of the local temperature." |
| 1953 | G.H. Jones Chemical Laboratory, University of Chicago, Illinois | Stanley Lloyd MILLER,[2142] a graduate student of Prof. Harold C. UREY, **shows experimentally that organic compounds can form from nonbiological compounds, the famous Miller-Urey experiment** (or *Miller experiment*) {⇨Fig. 4.14–A}.<br><br>(i) To simulate the effects of lightning, he repeatedly generates a high-voltage spark discharge in an atmosphere of hydrogen, methane, ammonia and steam originating from purified water. Using paper chromatography, he shows that, after the experiment has been run for a week, the flask contains organic molecules as well as several left-handed amino acids (L-amino acids) that are essential constituents of organic life – a result of heat and shock and/or ultraviolet radiation.[2143, 2144]<br><br>(ii) However, he also finds equal amounts of right-handed amino acids (D-amino acids).<br><br>MILLER's paper was published in May 1953. Only two months before, an article was published in *The New York Times* which describes the work of William M. MACNEVIN, a chemist at Ohio State University, who passing 100,000-V sparks through methane and water vapor produced 'resinous solids' that were "too complex for analysis."[2145] ▪ MILLER was stimulated by a seminar given by Prof. UREY, winner of the 1934 Nobel Prize for Chemistry, who suggested that photochemical and charged particle reactions in the atmosphere caused by ultraviolet light from the Sun and by electrical discharges (lightning) could have led to the formation of the prebiotic compounds that eventually led to the development of living organisms. In the late 1960s, researchers at Cornell University in Ithaca, New York, synthesized organic compounds from mixtures of $CH_4$, $C_2H_6$, $NH_3$ and $H_2O$ – a reaction mixture of gases roughly simulating the primitive terrestrial atmosphere – in a shock tube {BAR-NUN ET AL. ⇨1970}. |
| 1953 | Nuclear Weapons Proving Ground, Nevada | Frank H. SHELTON,[2146] a U.S. nuclear weaponeer and physicist, **discovers the shock precursor phenomenon.** While analyzing recorded fireball data of the MIKE shot {Eniwetok Atoll ⇨1952} – a 10.4-megaton surface nuclear explosion – he noticed "a large jet that shot down the mile-long Ogle-Krause helium diagnostic box, ahead of the main blast wave.[2147] The jet down the helium box indicated that **a *precursor* can develop when the speed of sound of a medium significantly exceeds that of the surrounding air.** That is, the speed of sound in helium is larger than air, just as the thermally heated layer of air near the ground has a speed of |

---

[2141] O. LAPORTE, R.N. HOLLYER, A.C. HUNTING, and E.B. TURNER: *Luminosity generated by shock waves*. Nature **171**, 395-396 (1953).
[2142] S.L. MILLER: *Production of amino acids under possible primitive Earth conditions*. Science **117**, 528-529 (May 15, 1953).
[2143] S.L. MILLER and H.C. UREY: *Organic compound synthesis on the primitive Earth*. Science **130**, 245-251 (July 31, 1959).
[2144] S.L. MILLER: *The prebiotic synthesis of organic compounds as a step toward the origin of life*. In: (J.W. SCHOPF, ed.) *Major events in the history of life*. Jones & Bartlett, Boston, MA (1992), pp. 1-28.
[2145] *Looking back two billion years*. The New York Times (March 8, 1953), p. E9.
[2146] F.H. SHELTON: *Reflections of a nuclear weaponeer*. Shelton Enterprise Inc., Colorado Springs, CO (1988), p. **6**: 7.
[2147] The "Ogle-Krause box" was a 9-ft$^2$ (0.84-m$^2$) aluminum-sheathed plywood tunnel filled with helium ballonets which allowed gamma and neutron radiation from the blast to travel with minor absorption to test instruments on Bogon Island (close to Elugelab Island, the test site of the MIKE shot). Elugelab Island was completely destroyed by the enormous explosion.

sound that exceeds that of the ambient air above it, thus forming the conditions for a precursor in Nevada…" Based upon his unique observation, he writes the first "precursor" document.[2148]
• Two years later, E.J. BRYANT ET AL. {⇨1955} experimentally confirmed SHELTON's observation of a *precursor shock wave.*

In a *radiative-precursor shock wave,* radiation from the shock wave ionizes and heats the medium ahead of it. Many astrophysical systems, such as supernova remnants and jets, produce radiative-precursor shock waves. Laboratory shock waves driven by a laser-heated plasma are attractive for understanding radiation effects on the evolution of astrophysical shock waves.[2149]

| 1953 | *Los Alamos Scientific Laboratory (LASL), Los Alamos, New Mexico* | **Robert G. SHREFFLER**[2150] **and William E. DEAL,** two U.S. physicists, **demonstrate that considerably higher local energies could be obtained from an explosive system if the explosive was used to accelerate a thin metal plate over a distance of several centimeters.** Using brass plates of variable thickness $d$ driven by an $8 \times 8$-in.$^2$ ($20.3 \times 20.3$-cm$^2$) block of a high explosive (Composition B), they observe that the plate velocity increases with decreasing plate thickness, ranging from 1.636 mm/µs for $d = 1$ in. (25.4 mm) to 4.252 mm/µs for $d = 0.031$ in. (0.79 mm).
• Driver plates like this, moving at velocities of several km/s (mm/µs) – so-called **"flyer plate accelerators"** (or "flat plate accelerators") – were used to produce strong shock waves of up to 2 Mbar in stationary brass target plates {MCQUEEN & MARSH ⇨1960}.

In the same year, **John M. WALSH,**[2151] **Robert G. SHREFFLER and Frank J. WILLIG propose to divide collisions into jetless and jet-forming categories.**
(i) They report on a new theory which describes flow in the collision region for the jetless case. Contrary to a previously advanced theory {BIRKHOFF, MACDOUGALL & G.I. TAYLOR ⇨1948}, which ignored the role of shock waves in the collision, they describe flow in the collision region encountering a shock wave.
(ii) **They determine a *critical collision angle*** – as a function of material velocities and equation-of-state (EOS) properties of the materials – **above which a jet must result from the collision.** In order to also test their theoretical prediction, they carried out experiments on collisions of metal plates driven by high explosives, the impact of the plates being recorded with a high-speed smear camera. They observed that the critical angle that separates the two types of collision can be predicted sufficiently well by theory. • Explosion welding is based upon a jet-forming collision {CARL ⇨1944}. |
|---|---|---|
| 1954 | *U.S.A.* | In April, **the Supersonic Tunnel Association is founded,** with the objective "to provide at the operational level a means of interchange for ideas, techniques and solutions to problems." • In 1996, the name was extended to the **Supersonic Tunnel Association, International (STAI),** to better reflect the current worldwide makeup of the organization. |
| 1954 | *Pittsburgh, Pennsylvania* | On July 1, **foundation of The Combustion Institute** with the objective "to promote and disseminate research in combustion science carried out by its members." Its main activities are the **International Symposia on Combustion** {1st Symp. on Combustion ⇨1928}.[2152] *Combustion and Flame,* the monthly journal of The Combustion Institute, which covers the subjects of deflagration, detonation, and supersonic reacting flow, has been published since 1957. |

---

[2148] F.H. SHELTON: *The precursor, its formation, prediction, and effects.* Secret Rept. SC-2850, Sandia Corporation, Albuquerque, NM (July 1953).
[2149] T. DITMIRE ET AL.: *The production of strong blast waves through intense laser irradiation of atomic clusters.* Astrophys. J. (Suppl. Ser.) **127**, Part 1, 299-304 (2000).
[2150] R.G. SHREFFLER and W.E. DEAL: *Free surface properties of explosive-driven metal plates.* J. Appl. Phys. **24**, 44-48 (1953).
[2151] J.M. WALSH, R.G. SHREFFLER, and F.J. WILLIG: *Limiting conditions for jet formation in high velocity collisions.* J. Appl. Phys. **24**, 349-359 (1953).
[2152] *In celebration of the 50th anniversary of The Combustion Institute 1954–2004.* The Combustion Institute, Pittsburgh, PA; http://www.combustioninstitute.org/documents/HistoryFinal.pdf.

| | | |
|---|---|---|
| 1954 | RAND Corporation, Santa Monica, California | Harold L. BRODE,[2153] a U.S. theoretical physicist, **analytically treats the problem of the sudden expansion of a sphere of ideal gas initially at rest and high pressure** (a "spherical shock tube"), which produces a spherical shock wave that moves out into the surrounding medium.
(i) He employs an artificial viscosity in the numerical integration of the differential equations in Lagrangian form {VON NEUMANN & RICHTMYER ⇨1950}.
(ii) He considers three isothermal sphere problems: two of the spheres are at normal density, high initial temperature and initial pressures of 2,000 and 121 bar, respectively; the third sphere is at high initial density, normal temperature and an initial pressure of 121 bar. He finds that the shock overpressure from the hot spheres becomes coincident with the point source solution very rapidly.[2154]
(iii) Certain features of the isothermal sphere problem have much in common with phenomena that occur in the detonation problem.
Corresponding experiments on exploding thin glass spheres filled with a gas at high pressure were performed at UTIA, University of Toronto {GLASS & HALL 1957; ⇨Fig. 4.16–F}. |
| 1954 | Institute of Mathematical Studies, New York University, New York City, New York | Joseph B. KELLER,[2155] a U.S. applied mathematician, **derives the discontinuity conditions in an arbitrary continuous material for a general discontinuity surface;** *i.e.*, also including curved surfaces. He shows that only three types of discontinuities are possible: shocks, contact discontinuities, and phase change fronts. For the location of an acoustic shock front, he obtains a first-order partial differential equation which can be solved – as for optics – by means of rays. He determines the variation of shock strength along a ray and demonstrates his theory with an analysis of the shock tube. ▪ Two years before, KELLER[2156] had worked out a *geometrical theory of diffraction* (which was widely used later) to describe the propagation of waves, in particular
  ▸ to analyze radar reflection from objects;
  ▸ to calculate elastic wave scattering from flaws in solids; and
  ▸ to study acoustic wave propagation in the ocean.
In the following year, KELLER's geometrical method was applied to determine the reflection and refraction of weak spherical and cylindrical shock waves at a planar interface.[2157] |
| 1954 | Imperial Chemical Industries, Ltd., Welwyn, Hertfordshire, England | Herbert KOLSKY,[2158] a British research scientist at Butterwick Research Laboratories, **finds that a small electrical disturbance is generated, starting at the instant of detonation, while using small quantities of explosives to produce sharp stress pulses.** Using a 10-cm wire probe, a wide-band amplifier and a high-speed cathode-ray oscilloscope, he observes that the maximum potential is reached after a time of about 50 μs. For PETN, the potential acquired is negative, whereas for lead azide, silver acetylide and nitrogen tri-iodide it is positive. He explains this effect as being the result of the different mobilities of the highly ionized gases, thus creating an electrical dipole. |

---

[2153] H.L. BRODE: *Numerical solutions of spherical blast waves.* Rept. No. RM-1363-AEC, RAND Corporation, Santa Monica, CA (1954); *The blast from a sphere of high pressure gas.* Rept. P-582, RAND Corp., Santa Monica, CA (1955).
[2154] S.A. BERGER and M. HOLT: *Spherical explosions in sea water.* Tech. Rept. No. 19, Div. Appl. Math., Brown University, Providence, RI (1959).
[2155] J.B. KELLER: *Geometrical acoustics. I. The theory of weak shock waves.* J. Appl. Phys. **25**, 938-947 (1954).
[2156] J.B. KELLER. *The geometrical theory of diffraction.* In: (B.S. KARASIK, ed.) *The McGill Symposium on Microwave Optics* [Eaton Electronics Laboratory, McGill University, Montreal, Canada, June 1953]. Rept. AFCRC-TR-59-118, Air Force Cambridge Research Center, Bedford, MA (1959), vol. 1.
[2157] K.O. FRIEDRICHS and J.B. KELLER: *Geometrical acoustics. II. Diffraction, reflection, and refraction of a weak spherical or cylindrical shock at a plane interface.* J. Appl. Phys. **26**, 961-966 (1955).
[2158] H. KOLSKY: *Electromagnetic waves emitted on detonations of explosives.* Nature **173**, 77 (1954).

| | | |
|---|---|---|
| 1954 | *Graduate School of Aeronautical Engineering, Cornell University, Ithaca, New York* | **Shao-Chi LIN,**[2159] a Chinese-born U.S. physicist and hydrodynamicist, **expands previous work on point explosions** {G.I. TAYLOR ⇨1941} **and calculates the development of the shock wave in an ideal gas from the instantaneous energy release along a (straight) line.** He finds that the shock expands along the radius $r$ with a constant velocity, but that the pressure $p$ decays with distance $r$ according to $p \sim 1/r^2$. Applying the results of his analysis to the case of hypersonic flight, he can show that the shock envelope behind a meteor or a high-speed missile is approximately a paraboloid. ▪ Analyses of cylindrical explosions are also of interest for predicting the shock waves emitted from long sparks, lightning {UMAN ⇨1970}, and exploding wires. |
| 1954 | *Crimean Astrophysical Observatory, Nauchny, Ukraine, Soviet Union* | **Solomon Borisovich PIKELNER [or PIKEL'NER],**[2160] an eminent Soviet astronomer, **shows that the filaments in the Cygnus Loop are due to expanding blast waves from a supernova explosion propagating in the diffuse interstellar medium at about 100 km/s** {Cygnus Loop ⇨$c.$15,000 years ago}. In particular, in the doctoral dissertation he submits in the same year, he describes the mechanism of formation of the filament structure of the shells of supernova remnants and constructs a quantitative theory of the emission of filaments on the basis of shock wave fronts undergoing deexcitation in a heterogeneous medium.[2161] ▪ The expansion velocity of the ejecta is much larger than the speed of sound in the ambient gas, with the result that **the ejecta are preceded by a shock wave,** known among astronomers as a **"blast wave shock."** |
| 1954 | *U.S. Dept. of Commerce, Coast and Geodetic Survey, Washington, DC & Michelson Laboratory, U.S. Naval Ordnance Test Station (NOTS), China Lake, California* | **John S. RINEHART,**[2162] assistant director for research and development at the Coast & Geodetic Survey, **and John PEARSON,** a research scientist at the Michelson Laboratory, **publish their book** *The Behavior of Metals Under Impulsive Loads.* They state, "This book describes actions that occur under one type of loading, namely, the type which may develop when an explosive charge is detonated in intimate contact with a body or when one body impacts against another. Such load is termed *impulsive*. **The rational design of materials and systems that will not fail under impulsive loadings is, perhaps, the most challenging problem that faces the present day design engineer** … The basic character of the fracture and plastic flow processes that will be operative under such conditions is just beginning to be understood."<br><br>The dynamic behavior of materials encompasses a broad range of phenomena with technological applications in military and civilian sectors. **The field of dynamic behavior of materials comprises diverse phenomena such as processing (combustion synthesis, shock compaction, explosive welding and fabrication, shock and shear synthesis of novel materials),** and deformation, fracture, fragmentation, shear localization, and chemical reactions under extreme conditions. It has evolved considerably in the past twenty years and is now at a mature stage. This evolution has placed this field at a level of recognition comparable to fatigue, creep, and fracture. In extension of quasi-static deformation and fracture influences, the following effects play an increasingly important role in dynamic events:[2163]<br>▸ mass inertia which leads to the propagation of elastic, plastic, and shock waves;<br>▸ thermal inertia which leads to thermo-viscoplastic instabilities, most commonly known as "adiabatic shear bands;" and<br>▸ viscosity which can effect thermally activated processes. |

---

[2159] S.C. LIN: *Cylindrical shock waves produced by instantaneous energy release.* J. Appl. Phys. **25**, 54-57 (1954).

[2160] S.B. PIKELNER: *Spektrofotometričeskoe issledovanie mechanizma vozbužděnija voloknistnych tumannostej* ("Spectroscopic Investigation of Excitation Mechanism in Filamentary Nebulae"). Izvestija Krymskoj Astrofiziceskoj Observatorii (Izdatel'stvo Akademii Nauk SSSR, Moskva) **12**, 93-117 (1954).

[2161] Anonymous: *Solomon Borisovich PIKEL'NER (1921–1975).* Astron. Zh. **53**, 233-235 (Jan./Feb. 1976). Engl. translation by E.U. OLDHAM in Soviet Astronomy **20**, No. 1, 133-135 (1976); http://articles.adsabs.harvard.edu/cgi-bin/nph-iarticle_query?1976SvA....20..133.

[2162] J.S. RINEHART and J. PEARSON: *Behavior of metals under impulsive loads.* American Society for Metals (ASM), Clevland, OH (1954); *see Preface* and introduction of chap. 1.

[2163] Scope of the *Symposium Dynamic Behavior of Materials,* organized by M.A. MEYERS ET AL. and held at the *2007 TMS Annual Meeting & Exhibition* in Orlando, FL.

| | | |
|---|---|---|
| | | Today, so-called **"dynamic materials testing"** – covering a broad range of loading conditions with strain rates of up to some $10^9$ s$^{-1}$ – has become a huge field of research on its own, encompassing not only the investigation of metals and metal alloys, but increasingly also plastics, ceramics and composite structures. |
| 954 | *Paris, France* | At the 2nd International Congress on High-Speed Photography, **Hubert SCHARDIN**,[2164] an applied physics professor at Freiburg University and German director of the Laboratoire de Recherches Techniques de Saint-Louis in France, **reports on high-dynamic fracture studies in transparent solids.** Using his Cranz-Schardin multiple-spark camera {CRANZ & SCHARDIN ⇒1929; ⇒Fig. 4.19–M} and the Toepler schlieren method {A. TOEPLER ⇒1864} in reflected-light mode, he cinematographically visualizes the propagation of cracks in various shock-loaded samples, such as in explosive-loaded Plexiglas and glass specimens impacted by a bullet. He finds that cracks propagate faster than the shear (transverse) wave speed but slower than the longitudinal wave speed in the test sample. ▪ Intersonic crack speeds were also observed under extreme loading rate conditions {WINKLER, SHOCKEY & CURRAN ⇒1970}.<br><br>In another lecture presented at the same conference, Heinrich HAENSEL and Hubert SCHARDIN,[2165] using the same diagnostic method, report on the propagation of elastic wavefronts originating from longitudinal, transverse and Schmidt head waves {VON SCHMIDT ⇒1938}, and of the plastic deformation at the surface of an explosive-loaded steel plate. |
| 1955 | *Research Laboratory, General Electric Company (GEC), Schenectady, New York* | On February 16, **Francis P. BUNDY**,[2166] a U.S. physicist at GEC, **and collaborators announce the first reproducible transformation of graphite to diamond** using static pressures in excess of 100 kbar and temperatures of above 3,000 °C. ▪ The spectacular shock synthesis of diamond was first achieved six years later {DECARLI & JAMIESON ⇒1961}.<br><br>In 1963, BUNDY[2167] reported on the direct conversion of compressed graphite to diamond in a static pressure apparatus: he compressed the graphite to pressures in the range of > 130 kbar and then heated the graphite using a millisecond-duration capacitor discharge. His diamond was polycrystalline – very much like shock-synthesized diamond {DECARLI & JAMIESON ⇒1961} – but with a larger crystalline size of about 100 nm. |
| 1955 | *Nevada Proving Grounds, Nevada* | In the period from February 18 to May 15, **Operation TEAPOT is conducted** – a series of thirteen atmospheric nuclear tests ranging in yield from 1.2 to 43 kilotons of TNT equivalent. **E.J. BRYANT**,[2168] **Noel H. ETHRIDGE, and J.H. KEEFER,** three physicists and staff members at Sandia Corporation, measure the pressure-time profiles in the near-field from nuclear explosions and **observe air-blast disturbances arriving ahead of the main shock.**<br><br>This so-called **"shock wave precursor"** phenomenon {F.H. SHELTON ⇒1953} is caused by certain ground effects:[2169] in the case of nuclear explosions, the emitted thermal radiation preheats the air near the ground – particularly over dusty or heat-absorbing surfaces – which subsequently affects the passage of the blast wave {⇒Figs. 4.13–E & 4.16–Q}. Precursors are not necessarily associated with surface burst or low-level nuclear explosions; for example, they have also been observed in large-yield chemical explosions by the British-born Canadian |

---

[2164] H. SCHARDIN: *Application de la cinématographie par étincelles à l'examen des phénomènes de rupture.* Proc. 2nd Int. Congr. High-Speed Photography [Paris, France, Sept. 1954]. In: (P. NASLIN and J. VIVIE, eds.) *Photographie et cinématographie ultra-rapides.* Dunod, Paris (1956), pp. 301-314.

[2165] H. HAENSEL and H. SCHARDIN: *Ausbreitung elastischer und plastischer Oberflächendeformationen bei Metallen unter kurzzeitiger Belastung.* Proc. 2nd Int. Congr. High-Speed Photography [Paris, France, Sept. 1954]. In: (P. NASLIN and J. VIVIE, eds.) *Photographie et cinématographie ultra-rapides* Dunod, Paris (1956), pp. 315-326.

[2166] F.P. BUNDY, H.T. HALL, H.M. STRONG, and R.H. WENTORF JR.: *Man-made diamonds.* Nature **176**, 2-7 (July 9, 1955).

[2167] F.P. BUNDY: *Direct conversion of graphite to diamond in static pressure apparatus.* J. Chem. Phys. **38**, 631-643 (1963).

[2168] E.J. BRYANT, N.H. ETHRIDGE, and J.H. KEEFER: *Operation TEAPOT, Project 1.14b. Measurements of air-blast phenomena with self-recording gauges.* Rept. WT-1144, Sandia Base, Albuquerque, NM (1955).

[2169] F.H. SHELTON: *The precursor, its formation, prediction, and effects.* Tech. Rept. SC-2850, Sandia Corporation, Albuquerque, NM (1953).

physicist John M. DEWEY.²¹⁷⁰ ▪ Precursors of a quite different type can also be generated by the ballistic emission of supersonic fragments of a detonating explosive: their little head waves – *micro head waves* – are formed by individual fragments which outrun the main blast wave at some point. This phenomenon happens during the firing of most types of detonator caps {⇨Fig. 4.16–R}.²¹⁷¹ The interactions of closely neighboring micro head waves can even generate *micro Mach disks*.

| | | |
|---|---|---|
| 1955 | *Soviet Nuclear Test Site Semipalatinsk-21, Kazakhstan, Soviet Union* | On November 22, **the first Soviet Superbomb, a "true (staged)" H-Bomb, is detonated.**²¹⁷² Dubbed "RDS-37," this nuclear test explosion has a yield of about 1.6 megatons of TNT equivalent.<br><br>(i) This thermonuclear device, based upon the principle of "radiation implosion," is the Soviet Union's first test of a two-stage radiation implosion design, known as "Andrei D. SAKHAROV's Third Idea"²¹⁷³ in the Soviet Union, and as the "Teller-Ulam design" in the United States {MIKE Test ⇨1952}. The kernel of the principle consists in the radiation generated in a primary A-bomb explosion and confined by the radiation-opaque casing propagating throughout the interior casing volume and flowing around the secondary thermonuclear unit. The secondary unit experiences a strong compression under the irradiation, with a resulting nuclear and thermonuclear explosion.<br><br>(ii) In order to minimize the radioactive fallout, the Soviet "Super" is dropped from a bomber Tu-16 ("Badger") at high altitude – **the world's first air-dropped fusion bomb test.**<br><br>(iii) The bomb explodes underneath an inversion layer, which unexpectedly focuses the shock back toward the ground. The refracted shock wave produces unanticipated collateral damage, killing two people due to the collapse of a building.²¹⁷⁴<br><br>In 2005, on the 50th anniversary of the test, German A. GONCHAROV,²¹⁷⁵ a researcher at the Russian Federal Nuclear Center in Sarov, drew on documentary sources to gain insights into the historical origins and development of the design of the RDS-37. |
| 1955 | *Supersonic Naval Ordnance Research Track (SNORT), China Lake, California* | **Completion of a rail-guided rocket sled test facility for the acceleration of masses to supersonic velocities** in order to supplement aerodynamic wind tunnel and shock tube studies. The test facility is also equipped with methods that produce rapid deceleration. For example, masses of 91 kg and 9,080 kg can be accelerated up to velocities of 1,165 m/s and 333 m/s and decelerated at $30g$ and $5g$, respectively ($g$ = acceleration due to gravity).²¹⁷⁶ ▪ Similar rocket sled test beds were operated by the USAF at Edwards Air Force Base, CA, and Hurricane Mesa, UT, and by Sandia at Albuquerque, NM {⇨Fig. 4.20–M}.<br><br>**Sled vehicles accelerated by single- or multiple-stage rockets have proven to be very useful for a variety of hypersonic tests** {Holloman High-Speed Test Track ⇨2003}. |
| 1955 | *Springer-Verlag, Vienna, Austria* | **The international journal *Astronautica Acta* is founded** as an official organ of the International Astronautical Federation. The first editor is Friedrich HECHT. The first issue doesn't itemize any goals of this new journal. However, the German-born U.S. rocket engineer Wernher VON BRAUN uses a one-page statement to outline the feasibility of flight into outer space and logical extrapolations of the present art of rocketry, such as to realize voyages to the Moon and other planets. |

---

[2170] J.M. DEWEY: *Precursor shocks produced by a large-yield chemical explosion*. Nature **205**, 1306 (1965).
[2171] G. KAYAFAS (ed.): *Stopping time. Die Fotografie von Harold EDGERTON*. Stemmle, Photographie AG, Schaffhausen (1988), p. 143.
[2172] *The Soviet Nuclear Weapons Program*. Nuclear Weapon Archive; http://nuclearweaponarchive.org/Russia/Sovwpnprog.html.
[2173] The "First Idea" (A.D. SAKHAROV 1948) and the "Second Idea" (V.L. GINZBURG 1948) permitted to create the first Soviet H-bomb {Semipalatinsk ⇨1953}.
[2174] D. HOLLOWAY: *STALIN and the bomb*. Yale University Press, New Haven, CT & London (1994), pp. 315-316.
[2175] G.A. GONCHAROV: *The extraordinarily beautiful physical principle of thermonuclear charge design (on the occasion of the 50th anniversary of the test of RDS-37 – the first Soviet two-stage thermonuclear charge)*. Phys. Uspekhi **48**, No. 11, 1187-1196 (2005).
[2176] H. GARTMANN: *Raketen*. Kosmos, Stuttgart (1956).

| | | |
|---|---|---|
| | | In 1974, under the editorship of the Polish-born U.S. aeronautical engineer Antoni K. OPPENHEIM, the journal's name was changed (OPPENHEIM: "... for the sake of proper grammar") to *Acta Astronautica*, published by Pergamon Press, Oxford. |
| 1955 | *Santa Monica, California* | **The First Rand Symposium on High-Speed Impact is held.** The symposium arose out of a desire to "assess the vulnerability of the intercontinental ballistic missile to damage by high-speed fragments," and was initiated by the German-born U.S. General Bernard SCHRIEVER. The participants, stimulated by the the U.S. scientists Fred WHIPPLE and George GAMOW, recommend that the United States gather data on meteoroid impacts using sounding rockets, and develop three laboratory tools to achieve hypervelocity: light-gas guns {CROZIER & HUME ⇨1946; JONES ET AL. ⇨1966}; shaped charge jets {VON FÖRSTER ⇨1883; MUNROE ⇨1888; THOMANEK ⇨1938}; and electromagnetic guns {FAUCHON-VILLEPLÉE ⇨1916/1917; RASHLEIGH & MARSHALL ⇨1978}.<br><br>Subsequent symposia were held at Los Angeles, CA (1956); Washington, DC (1957); Chicago, IL (1958); Eglin AFB, FL (1960); Denver, CO (1961); Cleveland, OH (1963); and Orlando, FL (1964). After a decline in hypervelocity research in the 1970s due to various political reasons, this subject saw a renaissance in the early 1980s, eventually leading to the establishment of the **Hypervelocity Impact Symposium** {HVIS ⇨1986}. A historical perspective on hypervelocity research in the United States up to 1986 was given by Harry FAIR,[2177] a scientist at the Defense Advanced Research Projects Agency (DARPA) in Washington, DC. |
| 1955 | *Radiation Laboratory, UC Berkeley, California* | **A team led by Owen CHAMBERLAIN,**[2178] a U.S. physicist, **and Emile SEGRÉ,** an Italian-born U.S. experimental physicist, **observes that antiprotons** – *i.e.,* antiparticles each with the same mass as a proton but with an equal and opposite electrical charge – **are produced by high-energy collisions.** They use a high-energy particle accelerator ("Bevatron"), which allows them to accelerate protons up to a kinetic energy of 6.2 GeV. ▪ The experiment consisted mainly of a system of detectors and analyzers capable of accepting negatively charged particles whose masses were approximately equal to that of the proton. When the new negatively charged particles originating from a target bombarded with approximately $10^{10}$ protons per second came to rest inside a detector, a tremendous release of energy was observed – a process called "annihilation." **Their experiment provided conclusive proof that antiprotons exist.**<br><br>In the following year, CHAMBERLAIN and SEGRÉ also confirmed the existence of the antineutron. For their discovery of the antiproton, they jointly earned the 1959 Nobel Prize for Physics. ▪ Since then, matter and antimatter reactions {DIRAC ⇨1928}, which involve the total conversion of mass into radiation and hence are more potent than any nuclear explosions, have fascinated both scientists and science fiction writers. |
| 1955 | *Ottawa, Ontario, Canada* | At the 5th General Assembly of AGARD, **Hugh L. DRYDEN,**[2179] director of NACA, **and John E. DUBERG,** chief of the Structures Research Division at Langley Aeronautical Laboratory, **discuss the aeroelastic effects of aerodynamic heating at supersonic speed,** which provoke a complex interaction between heating, structural stiffness, and air forces. The elastic forces tend to resist distortion, while the aerodynamic forces tend to distort the structure. DRYDEN and DUBERG point to the aeroelastic consequences of aerodynamic heating in the design of high-speed aircraft, particularly the reduction in overall stiffness through the action of thermal stress, which can result in flutter for thin wing profiles and slender bodies. |

---

[2177] H. FAIR: *Hypervelocity – then and now.* Proc. First High Velocity Impact Symposium (HVIS) [San Antonio, TX, 1986]; *see* Int. J. Impact Engng. **5**, 1-11 (1987).

[2178] O. CHAMBERLAIN ET AL.: *Observation of antiprotons.* Phys. Rev. **100** [II], 947-950 (1955).

[2179] H.L. DRYDEN and J.E. DUBERG: *Aeroelastic effects of aerodynamic heating.* 5th General Assembly of AGARD (June 1955). NATO-AGARD AG20/P10; pp. 102-107.

| | | |
|---|---|---|
| 1955 | Poulter Laboratory, Stanford Research Institute (SRI), Menlo Park, California | George E. DUVALL,[2180] a U.S. shock wave physicist, **and Bruno J. ZWOLINSKI,** a U.S. physical chemist, **derive a hydrodynamic theory for the shock compression of solids:**<br>(i) They assume that under the impact of a very rapid compression, the solid material immediately transforms into a state in which it exhibits fluid-like behavior, and that the elastic properties of the material can be completely described by its hydrostatic equation of state.<br>• Principal experimental support came from the formation of jets by shaped charges {WALSH, SHREFFLER & WILLIG ⇨1953} and the associated penetration of matter {BIRKHOFF, MACDOUGALL, PUGH & G.I. TAYLOR ⇨1948}, as well as by equation-of-state (EOS) measurements obtained from either static compression {BRIDGMAN ⇨1948} or shock compression {WALSH & CHRISTIAN ⇨1955}.<br>(ii) DUVALL's hydrodynamic theory allows one to determine<br>▸ the amount by which the shock pressure exceeds the adiabatic pressure for a given compression;<br>▸ the temperature increase due to shock compression; and<br>▸ the entropy increase across the shock front. |
| 1955 | Washington University, Saint Louis, Missouri | Donald Herbert ELDREDGE,[2181] a research professor of otolaryngology (a medical branch dealing with the ear, nose, and throat) at the School of Medicine, **critically reviews the effects of blast phenomena on man** for the Committee on Hearing and BioAcoustics (CHABA) of the National Research Council (NRC). |
| 1955 | Cornell Aeronautical Laboratory (CAL), Buffalo, New York | Herbert S. GLICK,[2182] **William SQUIRE, and Abraham HERTZBERG,** three U.S. aerodynamicists, **first report on the construction of a** *chemical shock tube* **to solve problems in chemical kinetics.** Their device {⇨Fig. 4.10–D} has two membranes which burst – one after the other – in a controlled manner. Heating of the test gas up to the required temperature is achieved in two steps and is, therefore, more efficient. • Since then, studies into high reaction kinetics have became of great practical importance, *e.g.*, for the design of jet engines and when calculating the properties of hypersonic flow fields. The shock tube has the unique advantage that<br>▸ the gas sample can be heated up rapidly and homogeneously under accurately known enthalpy and pressure conditions; and<br>▸ the gas temperatures achievable are far in excess of those obtained in conventional reactors.<br>The *single-pulse shock tube*,[2183] a modern variant on GLICK's original design, is used to obtain information on the kinetics of decomposition of polyatomic molecules {CARRINGTON & DAVIDSON ⇨1953}. |
| 1955 | Los Alamos Scientific Laboratory (LASL), Los Alamos, New Mexico | Roy W. GORANSON,[2184] a high-pressure physicist and former student of Prof. Percy W. BRIDGMAN, **and coworkers provide the first review on techniques of generating large-amplitude planar shock waves in solids, and on corresponding diagnostic and analysis methods.** They present measured data on the dynamic compressibilities of metals in the range from 0.1 to 0.3 Mbar which they have combined with fluid mechanical theory to obtain high-pressure thermodynamic equation-of-state (EOS) data; the solids at extreme pressures are treated as fluids. They use a piezoelectric material (tourmaline) to measure shock pressures of up to 324 kbar in explosively loaded |

---

[2180] G.E. DUVALL and B.J. ZWOLINSKI: *Entropic equation of state and their application to shock wave phenomena in solids.* JASA **27**, 1054-1058 (1955).

[2181] D.H. ELDREDGE: *The effects of blast phenomena on man. A critical review.* Rept. No. 3. Committee on Hearing, Bioacoustics, and Biomechanics (CHABA), Armed Forces – National Research Council (NRC), St. Louis, MO (June 1955).

[2182] H.S. GLICK, W. SQUIRE, and A. HERTZBERG: *A new shock tube technique for the study of high temperature gas phase reactions.* In: *Proc. 5th Int. Symposium on Combustion* [Pittsburgh, PA, Aug./Sept. 1954]. Reinhold, New York (1955), pp. 393-402.

[2183] W. TSANG and A. LIFSHITZ: *Single-pulse shock tube.* In: (G. BEN-DOR ET AL., eds) *Handbook of shock waves.* Academic Press, New York (2000), vol. 3, pp. 107-210.

[2184] R.W. GORANSON, D. BANCROFT, B.L. BURTON, T. BLECHAR, E.E. HOUSTON, E.F. GITTINGS, and S.A. LANDEEN: *Dynamic determination of the compressibility of metals.* J. Appl. Phys. **26**, 1472-1479 (1955).

| | | iron. ▪ Their experimental studies, which they began at the end of World War II, were confined to velocity measurements of shocks and moving free surfaces using high-speed mechanical cameras and electrical contact pins {FRISCH ET AL. ⇨1944; ⇨Fig. 4.11−C}. |
|---|---|---|
| 1955 | Seismological Laboratory, Cal-Tech, Pasadena, California | Beno GUTENBERG,[2185] a German-born U.S. seismologist, **first identifies the effects of rupture propagation on seismograms** taken during the earthquake of magnitude 7.5 that happened in 1952 in Kern County, Central California. He determines the direction of rupture propagation from the azimuthal asymmetry of the surface-wave radiation. ▪ **His discovery that earthquake rupture propagates outward from the hypocenter implicitly recognized that earthquakes are caused by shear slip on faults.** |
| 1955 | Germany | H.V. HAJEK,[2186] a German military researcher, **discusses the various possibilities for initiating low-temperature thermonuclear reactions in shaped-charge configurations,** particularly in relation to the production of a simple nuclear weapon which he calls a "nuclear shaped charge" [Germ. *Atom-Hohlladung*].[2187]<br><br>In a paper published five years later, HAJEK[2188] addressed the subject in more detail:<br>▸ In order to produce temperatures of at least 400,000 °C to initiate a nuclear reaction, he proposes that a particular fuel (such as $^6$Li, $^9$Be or $^{233}$U) could be compressed by high-velocity jets from shaped charges which could be arranged in either a linear cascade or a twin configuration consisting of two shaped charges facing each other.<br>▸ He recommends the use of an extra neutron source to facilitate nuclear ignition.<br>▸ He notes that he performed his investigations for the French Ministry of Defense in the period 1946–1949, as summarized in a report entitled *Un explosif à désintégration atomique* ("An Explosive Arrangement for Nuclear Disintegration").<br>This report, however, was classified upon examination by the French Atomic Energy Commission of the French Ministry of War.<br><br>In September 1943, after becoming acquainted with the implosion bomb program at Los Alamos, John VON NEUMANN suggested that shaped charges would produce an appropriate spherical detonation wave. He pointed out that the method was not only likely to be faster than the gun-assembly (proposed in June 1942), but that it would produce higher pressures and reduce the amount of active material required, making the bomb more efficient.[2189] |
| 1955 | Combustion Colloquium, Liège, Belgium | Theodore von KÁRMÁN,[2190] a Hungarian-born U.S. engineering professor and pioneer of aeronautics and astronautics, **coins the term *aerothermochemistry*.** In a paper entitled *Fundamental Equations in Aerothermochemistry*, he states, "Aerothermochemistry deals with flow phenomena of compressible fluids in which chemical reactions take place. We want to restrict the definition to the case of chemical reactions. Changes in the chemical composition of a gas mixture may also take place without reactions (*e.g.*, due to diffusion); however, we want to include such phenomena in *aerothermodynamics* [CROCCO ⇨1931]. Of course, the line of de- |

---

[2185] B. GUTENBERG: *Magnitude determination for larger Kern County shocks, 1952*. In: *Effects of station azimuths and calculation methods*. Calif. Div. Mines and Geol. Bull. **171**, 171-175 (1955).

[2186] H.V. HAJEK: *Atom-Hohlladungen*. Explosivstoffe **3**, Nr. 5/6, 65-68 (1955).

[2187] In March 2005, the Berlin historian Rainer KARLSCH published the book *HITLER's Bombe* (Dt. Verlagsanstalt, München) in which he states that German physicists had conducted three nuclear weapons tests shortly before the end of World War II which claimed up to 700 lives. Although KARLSCH has no real proof to back up his spectacular theory, he presents a number of hitherto unknown documents on the history of nuclear science written in the 12-year period of the Third Reich, including a manuscript of one of Werner HEISENBERG's speeches. Already Prof. Erich SCHUMANN, until 1944 chief of research of the *Heereswaffenamt* (Army Ordnance Office), had the idea that two hollow charges aiming against each other may produce a sufficiently high temperature upon collision for initiating a nuclear fusion in an appropriate gaseous mixture.

[2188] H.V. HAJEK: *Die Möglichkeiten von Kernreaktionen mittels Hohlladungen*. Wehrtech. Monatshefte **57**, 8-21 (1960).

[2189] *Manhattan Project history – implosion takes center stage*. Manhattan Project Heritage Preservation Association (MPHPA), Inc., New York City, NY; http://www.childrenofthemanhattanproject.org/HISTORY/H-06c17.htm.

[2190] Th. VON KÁRMÁN: *Fundamental equations in aerothermochemistry*. *Selected combustion problems, Part 2*. Butterworths, London (1956); pp. 167-184.

marcation is somewhat arbitrary." ▪ The term *aerothermochemistry* was quickly adopted in the fluid dynamics community; it became a new field of research encompassing complex phenomena involving chemical changes that occur in gasdynamic systems.[2191]

In the following year, the first conference bearing this name in its title – the **Gas Dynamics Symposium on Aerothermochemistry** – was held at the Northwestern University in Evanston, Illinois; the chairmen were Donald K. FLEMING and J. Roscoe MILLER. The following symposia up to 1967 were each named the **Biennial Gas Dynamics Symposium** and held at Northwestern University, Evanston, IL. Each symposium focused on a particular topic: aerothermochemistry (**1**; 1955); transport properties in gases (**2**; 1957); dynamics of conducting gases (**3**; 1959); magnetohydrodynamics (**4**; 1961); physico-chemical diagnostics of plasmas (**5**; 1963); advances in plasma dynamics (**6**; 1965); and energy (**7**; 1967).

| | | |
|---|---|---|
| 1955 | *Chair of Applied Mathematics, University of Manchester, England* | **M. James LIGHTHILL**[2192] **and Gerald B. WHITHAM,** two British applied mathematicians and experts in fluid dynamics, **develop their famous model for** *traffic shocks* **– the so-called "Lighthill-Whitham model."** Since traffic jams display sharp discontinuities, there is a correspondence between traffic jams and shock waves. Such "traffic shocks" are a phenomenon encountered by many road users. |

(i) They present a model based on the analogy of vehicles in traffic flow with particles in a fluid, and show that
- the moving traffic shock waves are delimited by zones of slowing and zones of accelerating traffic;
- a moving stream of traffic may be treated as a fluid continuum; and
- on crowded freeways, extensive chain reaction collisions of individual motorcars may occur which, similar to the collision of gas molecules at the shock front, lead to traffic shocks.

(ii) Addressing the discontinuous nature of the traffic shock, they write, "Discontinuous waves are likely to occur on any stretch of road when the traffic is denser in front and less dense behind. For waves on which the flow is less dense travel forward faster than, and hence tend to catch up with, those on which the flow is denser. When this happens a bunch of continuous waves can coalesce into a discontinuous wave, or *shock wave*. When vehicles enter this, their mean speed is substantially reduced very quickly. The wave is not totally discontinuous of course, but its duration is not much longer than the braking time that each vehicle needs to make the required reduction of speed…"

Shortly after, **the U.S. mathematician Paul I. RICHARDS**[2193] **formulated, independently of LIGHTHILL and WHITHAM, a similar shock model of traffic flow obtained by means of a hydrodynamic approach.** Like LIGHTHILL and WHITHAM, he assumed that the velocity of the traffic is a function of density only. In this famous model – the so-called **"Lighthill-Whitham-Richards model"** – the one-way traffic flow through the system is represented by the motion of a one-dimensional wave representing a constant relationship between flow and density. When waves representing different traffic states intersect, shock waves are formed. These shock waves represent discontinuities in discharge flow across the wave boundary.

LIGHTHILL,[2194] later discussing the properties of a traffic shock in more detail, wrote, "Taking the case of traffic flow as an example, you can see that at low concentration the flow of cars is greater if you increase the concentration; but that at higher concentration the flow becomes less again. It follows that waves travel more slowly where the traffic becomes more congested. These waves are passing through you when you change your speed in response to

---
[2191] A.B. CAMBEL and B.H. JENNINGS: *Gas dynamics*. McGraw-Hill, New York *etc.* (1958), pp. 321-353.
[2192] M.J. LIGHTHILL and G.B. WHITHAM: *On kinematic waves. II: A theory of traffic flow on long crowded roads*. Proc. Roy. Soc. Lond. **A229**, 317-345 (1955).
[2193] P.I. RICHARDS: *Shock waves on the highway*. Operations Research **4**, 42-51 (1956).
[2194] M.J. LIGHTHILL: *Shock waves* [Ramsden Memorial Lecture]. Mem. Manchester Lit. Phil. Soc. **101**, 1-6 (1959).

the behavior of the drivers in front. They always pass backwards through the stream of cars, but, since the cars are going ahead at a good speed, they pass forwards along the road, until we get to those highly congested conditions in which a further increase in concentration decreases the flow, and then the waves pass backwards. It follows that a traffic 'hump' will have waves running together in the rear of it, and spreading out at the front. When you enter one of these regions of increased concentration, you know you have done so when you reach the shock wave caused by this running together, and have to brake rather suddenly."

Although this traditional traffic flow model {⇨Figs. 4.4–G, H} is now almost fifty years old, it is still accurate and has been the starting point for a number of more refined models, describing different aspects of traffic flow operation.[2195, 2196]

| 1955 | *Aerodynamische Versuchsanstalt (AVA) der Max-Planck-Gesellschaft (MPG), Göttingen, Germany* | Hubert LUDWIEG,[2197] a German fluid dynamicist, **proposes a new type of blow-down wind tunnel, which he names *Rohrwindkanal* ("tube wind tunnel")**. His idea earns him a prize from the German Ministry of Transport. ▪ The so-called **"Ludwieg tube"** {⇨Fig. 4.9–G} is a versatile facility for aerodynamic testing over the whole Mach range from subsonic to hypersonic speeds and also offers a very economical way to generate hypersonic flows of relatively long duration (about 0.3 seconds) in reasonably large test sections and at significant Reynolds numbers.

In 1968, the Ludwieg tube at AVA became operational with three different test sections for $3 < M < 11$. It was used for space-vehicle and missile research and development, for example ESA's Hermes space shuttle, ESA's Atmospheric Reentry Demonstrator (ARD), NASA/ESA's Crew Return Vehicle (CRV), and the NASA/ESA/DLR reentry vehicle X 38. In the upper Mach number range an externally, electrically heated supply tube was used.[2198]

Ludwieg tube tunnels were also operated in the United Kingdom (CABLE & COX 1963), and in the United States (DAVIS 1968; WARMBROD 1968; SHEERAN & WILSON 1969). Later, combusted Ludwieg tubes were also investigated for the production of supersonic flows in gasdynamic lasers.[2199] |
| 1955 | *U.S. Naval Ordnance Laboratory (NOL), White Oak, Silver Spring, Maryland* | Herbert Dean MALLORY,[2200] a U.S. physical chemist, **measures the velocity of shock waves in aluminum and the associated translational motions produced by metal-metal impact.** He uses electric pin contactors to measure the free surface velocity {MINSHALL ⇨1955}, from which he estimates the material velocity (or particle velocity) behind the shock front. Referring to a previous discussion applied to shock waves in water {COLE ⇨1948}, **he assumes that, to a first approximation, the particle velocity may be regarded as one-half the value of the free-surface velocity.** |
| 1955 | *Los Alamos Scientific Laboratory (LASL), Los Alamos, New Mexico* | Stanley MINSHALL,[2201] a U.S. shock wave experimentalist, **studies the elastic-plastic two-wave structure for two steel alloys and a sintered tungsten alloy.**

(i) His method, designed to yield the equation of state of any solid on the Hugoniot curve, consists of the measurement of the shock velocity and the particle velocity at the free surface of a test sample subjected to a contact explosion at the distal end. The particle velocity is measured by a group of judiciously spaced contactor pins {FRISCH ET AL. ⇨1944}. The shock velocity is measured by an image camera focused onto an argon-filled cavity formed by an in- |

---

[2195] Rachel SHINN, a Ph.D. student of Prof. G.B. WHITHAM, wrote a Ph.D. thesis entitled *Shocks and instabilities in traffic flows*. Applied Mathematics, CalTech, Pasadena, CA (1990).

[2196] S.P. HOOGENDOORN and P.H.L. BOVY: *State-of-the-art of vehicular traffic flow modeling*. J. Systems Control Engng. **215**, No. 4, 283-303 (2001).

[2197] H. LUDWIEG: *Der Rohrwindkanal*. Z. Flugwiss. **3**, 206-216 (1955); *Rohrwindkanal, eine Einrichtung zur Erzeugung eines kurzzeitigen, stationären Windstromes hoher Energie für aerodynamische Messzwecke*. Germ. Patent No. 944,334 (1956).

[2198] T. HOTTNER: *Der Rohrwindkanal der Aerodynamischen Versuchsanstalt (AVA) Göttingen*. Rept. 68A77, AVA, Göttingen (1968).

[2199] G.S. KNOKE: *Combusted Ludwieg tubes for gasdynamic laser research*. Ph.D. thesis, Washington University, Seattle, WA (Dec. 1975).

[2200] H.D. MALLORY: *The propagation of shock waves in aluminum*. J. Appl. Phys. **26**, 555-559 (1955).

[2201] S. MINSHALL: *Properties of elastic and plastic waves determined by pin contactors and crystals*. J. Appl. Phys. **26**, 463-469 (1955).

clined groove at the back side of the test sample and a Lucite block; this gap becomes luminous upon the arrival of the shock front at any point of the interface between metal and gas.

(ii) He finds that the elastic wave velocities are the same, within an experimental error of 2%, as the measured velocities of a longitudinal acoustic wave, and that the plastic wave velocities are about 15% less in steel and 10% less in tungsten than that of a longitudinal acoustic wave.

(iii) For steel, he calculates a dynamic elastic limit of 12 kbar, which he refers to "the stress on a dynamic stress-strain diagram where the relations between stress and strain cease to be linear."

(iv) He concludes that "no attempt has been made herein to correlate the dynamic elastic limit with values of yield strength or of ultimate strength, either static or dynamic. This is because there are fundamental differences in the experimental criteria by which these quantities are measured. It has been suggested that **Hugoniot elastic limit (HEL)** be substituted for *dynamic elastic limit* to emphasize the experimental differences between the phenomenon reported here and the engineering designations of yield stress and ultimate strength…"

The discovery of the "elastic limit" is a phenomenon closely related to the "yield point" of a stressed solid, which was first described in 1879 by Johann BAUSCHINGER,[2202] a German professor of technical mechanics at the TH Munich, who performed tensile tests on iron and mild steel.[2203]

| 1955 | *Laboratoire de Recherches Techniques de Saint-Louis, Saint-Louis, France* | Rudi SCHALL,[2204] a German research physicist, **proposes a direct method of determining the detonation pressure of a high explosive using flash radiography.**

(i) He measures the compression ratio $\rho/\rho_0$ by X-ray absorption using a calibrated step wedge and the detonation velocity $D$ from a series of flash radiographs taken at different moments in time. He then calculates the detonation pressure $p_D$ using the simple formula

$$p_D = \rho_0 D^2 (1 - \rho_0/\rho),$$

with $\rho_0$ as the initial density of the explosive.

(ii) Based upon his method, he finds that detonation pressures can range between 20 and 500 kbar, depending on the type of chemical explosive selected, its density and grain size. |
|---|---|---|
| 1955 | *Baumann Institute of Technology, Moscow, Soviet Union* | Kirill P. STANYUKOVICH,[2205] a Soviet professor of physics, **reviews wartime contributions to gas dynamics made by Soviet scientists in his book *Neustanovivshiesya dvizheniya sploshnoi sredy* ("Unsteady Motion of Continuous Media").**

(i) **According to his definition, "gas dynamics covers the flow of all compressible media, including liquids and even solids (the latter under conditions of high pressure)."**

(ii) Modestly he writes in his preface, "It is my sincere wish that English and American research workers in physics and mechanics will find this book useful, in spite of the availability of the excellent monograph *Supersonic Flow and Shock Waves* by R. COURANT and K.O. FRIEDRICHS [⇒1948]."

(iii) He restates not only classical shock wave theory, but also addresses many new applications related to explosions, fission or fusion reactors, and the motion of space vehicles.

In 1960, his textbook was translated into English by Maurice HOLT, a professor of applied mathematics at Brown University, and it soon became a widely used standard work on this subject. |

---

[2202] J. BAUSCHINGER: *Über die Quercontraction und -Dilatation bei der Längenausdehnung und -zusammendrückung prismatischer Körper.* Der Civilingenieur **25**, col. 81-124 (1879).

[2203] S.P. TIMOSHENKO: *History of strength of materials.* McGraw-Hill, New York (1953), pp. 280-281.

[2204] R. SCHALL: *Röntgenblitz-Untersuchungen bei Sicherheitssprengstoffen.* Nobel-Hefte **21**, 1-10 (1955).

[2205] K.P. STANYUKOVICH: *Neustanovivshiesya dvizheniya sploshnoi sredy.* Gostekhizdat, Moskva (1955); *Unsteady motion of continuous media.* Pergamon Press, London *etc.* (1960).

| 1955 | Los Alamos Scientific Laboratory (LASL), Los Alamos, New Mexico | **John M. WALSH**[2206] **and Russell H. CHRISTIAN**, two U.S. shock wave physicists, **pioneer a new method which allows the velocities of the shock front and free surface to be measured: the "flash gap technique"** {⇨Fig. 4.15–C}.
(i) It uses an argon flash gap approximately 0.1 mm in height placed over the metal surface to be monitored. The shock wave is generated by planar wave explosive lenses and explosive pads. Upon the arrival of the shock front, the flash gap emits an intense flash of light, which is recorded photomechanically using a streak camera.
(ii) Their method allows the first accurate determination of the dynamic equation-of-state data (EOS or Hugoniot data) for aluminum, copper, and zinc at pressures of between 100 and 500 kbar. Starting from a given Hugoniot curve, they also derive isotherms and adiabats neighboring the *principal Hugoniot curve* (*i.e.*, the Hugoniot curve which starts at 25 °C and 1 bar).
(iii) In their analysis, WALSH and CHRISTIAN treat metals as liquids. ▪ In 1961, George R. FOWLES,[2207] a shock physicist at SRI, Menlo Park, CA, was able to show that shear does exist behind the plastic wave by comparing shock measurements to BRIDGMAN's static measurements on aluminum. |
| 1956 | Rhode-Saint-Genèse [south of Brussels], Belgium | In October, **creation of the Von Karman Institute for Fluid Dynamics (VKI)**, an international institute for fluid dynamics, following Theodore VON KÁRMÁN's proposal.[2208] ▪ The **VKI** is a non-profit international educational and scientific organization, presently hosting three departments: (i) aeronautics and aerospace; (ii) environmental and applied fluid dynamics; and (iii) turbomachinery (*i.e.*, machines that transfer energy between a rotor and a fluid). |
| 1956 | University of Cambridge Press, Cambridge, England | **Foundation of the *Journal of Fluid Mechanics*** with the objective to exist "for the publication of theoretical and experimental investigations of all aspects of the mechanics of fluids." ▪ First editor was George K. BATCHELOR, a professor at Cambridge University and author of the monumental textbook *An Introduction to Fluid Dynamics* (1967). BATCHELOR also created the journal through which he established Fluid Mechanics as a discipline.
The journal is a rich mine not only for fluid dynamicists but also for shock physicists; the first issue already contained seven papers (of 39 in total) relating to shock waves. |
| 1956 | Rheinisch-Westfälische Technische Hochschule (RWTH) Aachen, Germany | **Foundation of the Stoßwellenlabor der RWTH Aachen** by Prof. Fritz SCHULTZ-GRUNOW, director of the Chair and Institute for Technical Mechanics. The Laboratory's main research topics encompass high temperature gas dynamics, the study of hypersonic/supersonic flows, and shock wave and detonation phenomena as well as the development and application of high-speed measurement techniques.[2209] ▪ Prof. SCHULTZ-GRUNOW was the first directeor of the Stoßwellenlabor (1956–1968); he was succeeded by Prof. Hans GRÖNIG (directorship 1968–1996) and Prof. Herbert OLIVIER (1996–present). |
| 1956 | Lawrence Radiation Laboratory (LRL), UC Livermore, California | **Berni J. ALDER**,[2210] a German-born U.S. physical chemist, **and Russell H. CHRISTIAN**, a U.S. shock wave physicist, **first observe a shock-induced increase in electrical conductivity in an insulator (alkali halide).** They interpret the increased conductivity as being due to a transition to a metallic state. ▪ Others, questioning this interpretation, explained the phenomenon on the basis of increased ionic conductivity (DORAN & MURRI 1964), or proposed that the conductivity was electronic in nature (KORMER ET AL. 1966). Shock-induced conduction is a commonly observed effect in good insulators, but is still poorly understood. |

---

[2206] J.M. WALSH and R.H. CHRISTIAN: *Equation of state of metals from shock wave measurements*. Phys. Rev. **97** [II], 1544-1556 (1955).
[2207] G.R. FOWLES: *Shock wave compression of hardened and annealed 2024 aluminum*. J. Appl. Phys. **32**, 1475-1487 (1961).
[2208] *Von Kármán Institut (VKI) history*. VKI for Fluid Dynamics, Rhode-Saint-Genèse, Belgium; http://www.vki.ac.be/public/history.htm.
[2209] *History of the Shock Wave Laboratory RWTH Aachen University*; http://www.swl.rwth-aachen.de/en/about-swl/history/.
[2210] B.J. ALDER and R.H. CHRISTIAN: *Metallic transition in ionic and molecular crystals*. Phys. Rev. **104** [II], 550-551 (1956).

| | | |
|---|---|---|
| 1956 | Los Alamos Scientific Laboratory (LASL), Los Alamos, New Mexico | **Dennison BANCROFT,**[2211] **Eric L. PETERSON, and Stanley MINSHALL,** three shock wave experimenters, **report on a shock-induced, high-pressure polymorphic transition in iron:** the body-centered cubic (bcc) phase or α-*phase* which exists at low pressure and room temperature changes into a new hexagonal close-packed (hcp) phase or ε-*phase* after undergoing a shock pressure of 130 kbar. They notice that recorded shock pressure profiles, reflected from a free surface, show three discrete shocks which correspond to the hypothesis that iron may transmit a stable compressive wave consisting of three successive discrete shocks. ▪ **The 130-kbar transformation in iron became the most widely studied shock-induced polymorphic phase transition,** and an understanding of its mechanism was also found to be important when analyzing the structures of iron meteorites {NININGER ⇨ 1956}. The progress in research from this newly discovered transition to a well-characterized phase transformation, known as the "**α → ε transition,**" was reviewed by DUVALL & GRAHAM {⇨ 1977}.<br><br>In the same year, Percy W. BRIDGMAN,[2212] a physics professor at Harvard College, attempts – albeit without success – to detect phase transitions in iron in static high-pressure experiments. Five years later, however, the gross difference between static and shock measurements was finally reconciled by resistance measurements performed by the U.S. chemical engineers Anthony S. BALCHAN[2213] and Harry G. DRICKAMER. |
| 1956 | Knob Lake Mine, Labrador, Canada | **Melvin A. COOK,** a U.S. physical chemist and explosives expert, **creates a new blasting agent using an unusual mixture of ammonium nitrate (AN), aluminum powder, and water** while consulting for Iron Ore Company of Canada.[2214] ▪ Two years later he founded the company IRECO Chemicals, which pioneered the production of the new blasting agent. Tests that followed resulted in the development of a new field of explosives: water-gel explosives, boosters, and pump trucks for their bulk delivery.<br><br>Water-gel explosives – widely known as **"slurry explosives"** – found worldwide application, particularly for large-scale operations. They can be used in wet conditions and are often cheaper and safer than nitroglycerin explosives. |
| 1956 | Dept. of Aerospace Engineering, University of Minnesota, Minneapolis | **August R. HANSON,**[2215] a U.S. aeronautical engineer, **and E.G. DOMICH are the first to study the break-up of drops of Newtonian fluids**[2216] **(such as water) at high subsonic and supersonic velocities using a shock tube.** ▪ The problem of the aerodynamic break-up of liquid drops has stimulated a large number of investigations, which were recently reviewed by Daniel D. JOSEPH,[2217] a professor of mechanical engineering at the University of Minnesota, and his collaborators. For example, the U.S. physical chemist Olive G. ENGEL[2218] made the important observation that water drops a millimeter in diameter would be reduced to mist by the flow behind an incident shock wave moving at Mach numbers ranging from 1.3 to 1.7. |

---

[2211] D. BANCROFT, E.L. PETERSON, and S. MINSHALL: *Polymorphism of iron at high pressures.* J. Appl. Phys. **27**, 291-298 (1956).
[2212] P.W. BRIDGMAN: *High pressure polymorphism of iron.* J. Appl. Phys. **27**, 659 (1956).
[2213] A.S. BALCHAN and H.G. DRICKAMER: *High pressure resistance cell, and calibration points above 100 kbars.* Rev. Scient. Instrum. **32**, 308-313 (1961).
[2214] *History of Dyno Nobel.* Dyno Nobel North America; http://www.dynonobel.com/dynonobelcom/en/northamerica/aboutus/ourhistory/.
[2215] A.R. HANSON and E.G. DOMICH: *The effect of viscosity on the breakup of droplets by air-blasts – a shock tube study.* Res. Rept. No. 130, Dept. of Aerospace Engineering, University of Minnesota, Minneapolis, MN (1956).
[2216] *Newtonian fluids* are those fluids that do not change their viscosity regardless of the agitation. They flow immediately upon the application of a force, and the rate of flow is directly proportional to the force being applied; *i.e.*, they behave according to NEWTON's Laws of Motion. In a Newtonian fluid the shear stress is proportional to the shear rate, and the viscosity remains constant as the shear rate is varied.
[2217] D.D. JOSEPH, J. BELANGER, and G.S. BEAVERS: *Breakup of a liquid drop suddenly exposed to a high-speed airstream.* University of Minnesota, Minneapolis, MN; http://www.aem.umn.edu/people/faculty/joseph/archive/docs/breakup99.pdf.
[2218] O.G. ENGEL: *Fragmentation of water drops in the zone behind an air shock.* J. Res. Natl. Bur. Stand. **60**, 245-280 (1958).

In a later study that used high-speed microcinematography to record drop diameters, HANSON[2219] and his collaborators found that the critical air velocity $U_C$ was related to the diameter $D$ of a single water droplet by the empirical equation

$$U_C^2 D = 6.21 \times 10^6,$$

where $U_C$ is the air velocity in ft/s, and $D$ is the droplet diameter in microns. For example, for a droplet diameter of 100 μm, the critical air velocity for the shattering of water droplets is 249 ft/s (75.9 m/s), which requires an air blast with an overpressure of 5.2 psi (335 mbar).

| 1956 | Lockheed Missile Systems Division, Palo Alto, California | At the Symposium on Magnetohydrodynamics, **Alan C. KOLB**,[2220, 2221] a U.S. theoretical spectroscopist at the U.S. Naval Research Laboratory (NRL) in Washington, DC, **reports on the construction of and experimental results from the first magnetically driven shock tube** {⇨Fig. 4.10−G} – the *electromagnetic shock tube*.

(i) The shock wave is generated by a capacitor discharge between two electrodes in a T-shaped *electric shock tube* which was first described by Richard G. FOWLER[2222] and collaborators (University of Oklahoma) in 1951. The generated plasma of ions is accelerated by the magnetic field from the return conductor, which is placed closely to the discharge tube.

(ii) **Using a strong external magnetic field, he succeeds in producing extraordinarily strong shocks in a deuterium layer with enormous Mach numbers ranging up to $M = 250$ and temperatures of up to one million degrees.**

In the same year, magnetically driven shock waves are independently studied at Lockheed Missile Systems Division, Palo Alto, CA, and at AVCO Manufacturing Corporation, Everett, MA. ▪ The theory behind the electromagnetically driven shock tube and the physical effects created by strong shock waves were reviewed by Robert A. GROSS,[2223] a U.S. plasma physicist at Columbia University, New York. Electromagnetic shock tubes were also investigated for possible use in controlled thermonuclear fusion, and for generating intense radiation in the visible and ultraviolet. |

| 1956 | Dept. of Applied Mathematics, University of Manchester, England | **M. James LIGHTHILL**,[2224] a British theoretical fluid dynamicist and applied mathematician, **presents a lengthy paper entitled *Viscosity Effects in Sound Waves of Finite Amplitude*.** Here the word "viscosity" refers to various thermodynamically irreversible processes, such as relaxation effects in sinusoidal and other waveforms, and nonlinear effects emerging from the theory of "planar waves of finite amplitude."

(i) The conflicting influences on wave propagation of heat and relaxation on the one hand and convection on the other reaches its height in shock waves, governing their formation, propagation and decay.

(ii) He gives an approximate solution for the velocity profile of a *dispersed shock wave* – one in which finite changes occur over distances large compared to the mean free path in the gas – which is valid for very weak waves. (In contrast a shock wave in air extends over only a few mean free paths.) |

---

[2219] A.R. HANSON, E.G. DOMICH, and H.S. ADAMS: *Shock tube investigation of the breakup of drops by air blasts*. Phys. Fluids **6**, 1070-1080 (1963).

[2220] A.C. KOLB: *Production of high energy plasmas by magnetically driven shock waves*. Phys. Rev. **107** [II], 345-350 (1957).

[2221] A.C. KOLB: *Magnetically driven shock waves, experiments at the U.S. Naval Research Laboratory*. In: (R.K.M. LANDSHOFF, ed.) *Symposium on Magnetohydrodynamics* [Lockheed Missile Systems Div., Palo Alto, CA, Dec. 1956]. Stanford University Press, Stanford, CA (1957), pp. 76-91.

[2222] R.G. FOWLER, J.S. GOLDSTEIN, and B.E. CLOTFELTER: *Luminous fronts in pulsed gas discharges*. Phys. Rev. **82** [II], 879-882 (1951).

[2223] R.A. GROSS: *Strong ionizing shock waves*. Rev. Mod. Phys. **37**, 724-743 (1965).

[2224] M.J. LIGHTHILL: *Viscosity effects in sound waves of finite amplitude*. In: (G.K. BATCHELOR and R.M. DAVIES, eds.) *Surveys in mechanics. A collection of surveys of the present position of research in some branches of mechanics, written in commemoration of the 70th birthday of Geoffrey Ingram TAYLOR*. Cambridge University Press (1956), pp. 250-351.

| | | |
|---|---|---|
| 1956 | American Meteorite Museum, Sedona, Arizona | **Harvey H. NININGER,**[2226] a renowned U.S. meteorite expert and museum director, **publishes his book *Arizona's Meteorite Crater. Past, Present, Future* in which he first addresses the possible significance of the new mineral coesite** {COES ⇨ 1953} **in a meteorite impact site.**

(i) Observing wide differences in the structure between Canyon Diablo meteorite specimens found on the rim of the Meteor Crater {⇨*c.*50,000 years ago} and on the plains surrounding the crater, he notes, "Recently, there has come to my attention the announcement of a new mineral, *coesite* [Science **118**, 131 (1953)]. This mineral, which has a specific gravity 16% higher than quartz, was produced by subjecting quartz to pressures of from 15,000 to 30,000 bars (atmospheres). The Coconino sandstone which is found in Meteor Crater at depths of from 350 to 1,000 feet [106–305 m] is composed of quartz grains and is very porous. In view of the excessively high impact pressures involved in the formation of this crater, a thorough search for this and possibly other new minerals, both within the crater and under the southern rim, might have significant results … **Along the same line, Harold UREY had suggested that the tremendous pressures presumed to accompany asteroidal impacts might accomplish transformation in terrestrial sediments beyond our capacity to estimate.**"

(ii) Becoming more specific, he proposes topics of future research, for example "… by a very thorough search, by drilling in the crater and under the south rim, for the newly created mineral *coesite* and any other possible alteration products." ▪ His suggestion was resumed and confirmed by CHAO, SHOEMAKER and MADSEN {⇨1960}.

In 1954, NININGER[2227] had reported on his findings of numerous glass-like fragments of fused limestone and sand in the environment of Meteor Crater, which typically contains small Fe/Ni-particles from the meteorite. Assuming that they had originated due to the heat generated by the impact of the meteorite, **NININGER classified these glassy to finely crystalline rock fragments as *impactites*** (or *impact glasses*) – a term coined by the U.S. geologist Henryk Bronislaw STENZEL in order to better differentiate such natural glasses from *tektites* – a material very similar in appearance to impactites but believed to be of extraterrestrial origin; *i.e.,* they were not produced on Earth by an impact or a volcanic eruption {C.R. DARWIN ⇨1844; O'KEEFE ⇨1967 & 1976; CHAPMAN ⇨1971}.[2228] |
| 1956 | Institute of Chemical Physics, Moscow, Soviet Union | **Yuri Nikolaevich RYABININ,**[2229] a Soviet chemical physicist, **reports on shock-induced chemical reactions in a number of chemical compounds up to 1 Mbar.** He placed various chemical substances in strong metallic cylinders which were encased in a high explosive. Upon recovering and opening the still-sealed cylinders, he found both gaseous and solid decomposition products intermixed with remnants of the original substances. For example, he observed the decomposition of rock salt into its elements according to the reaction: $2\,NaCl \rightarrow 2\,Na + Cl_2$. ▪ In his compression technique, the specimens were probably subjected to convergent shock waves. |

---

[2225] W.C. GRIFFITH and A. KENNY: *On fully-dispersed shock waves in carbon dioxide.* J. Fluid Mech. **3**, 286-288 (1957).

[2226] H.H. NININGER: *Arizona's meteorite crater. Past, present, future.* American Meteorite Museum, Sedona, AZ (1956), p. 50.

[2227] H.H. NININGER: *Impactite slag at Barringer Crater.* Am. J. Sci. **252**, 277-290 (1954).

[2228] The U.S. geologist Virgil E. BARNES wrote in his book *North American Tektites* [No. 3945, University of Texas Publ., Austin, TX (1940)], on page 558, "SPENCER's meteorite splash origin [SPENCER ⇨1932] … is valid for the formation of certain glasses. Glasses of this type will be distinguished in general from most of those now included under tektites. These meteorite splashes should be given a distinctive name such as 'impactites'. This name was suggested by Dr. H. B. STENZEL." ▪ In the period 1934–1954, Dr. Henryk Bronislaw STENZEL (1899–1980), a Polish-born U.S. paleontologist and stratigrapher, worked at the Bureau of Economic Geology at the University of Texas (UT) and in 1948 became professor of geology at UT Houston.

[2229] Y.N. RYABININ: *Sublimation of crystalline lattices under the action of a strong shock wave* [in Russ.]. Dokl. AN (SSSR) **109**, 289-291 (1956); *About various experiments on the dynamic compression of substances* [in Russ.]. Zh. Tekh. Fiz. (SSSR) **26**, 2661-2666 (1956).

BERTHELOT {⇨1881} had first demonstrated the shock-induced decomposition of a gaseous chemical compound.

| | | |
|---|---|---|
| 1956 | Stockholm, Sweden | **Nikolai N. SEMENOV** {⇨1934},[2230] a physical chemist and director of the Laboratory for Chemical Kinetics & Chemistry of Free Radicals at the University of Moscow, **and Sir Cyril HINSHELWOOD** {⇨1926},[2231] a chemistry professor at the University of Oxford, **jointly receive the 1956 Nobel Prize for Chemistry** "for their research into the mechanisms of chemical reactions, notably the mechanisms of chain and branched-chain chemical reactions." SEMENOV is the first Soviet physicist to win a Nobel Prize.<br><br>In the 1930s, SEMENOV and HINSHELWOOD had investigated the hydrogen/oxygen chain reaction that results in the formation of water. These studies initiated **Chemical Kinetics** – **a new branch of physical chemistry concerned with the study of the rates and mechanisms of chemical reactions.** Examples of research in this field and applications of it in modern chemical technology include:<br>▸ factors influencing reaction rates;<br>▸ predictions of rates of very rapid chemical reactions, such as detonation and supersonic combustion; and<br>▸ a fundamental understanding of the phenomena involved with starting and stopping chain reactions. |
| 1956 | Dept. of Mathematics, University of Manchester, England | Gerald B. WHITHAM,[2232] a British applied mathematician, **treats the problem of the propagation and ultimate decay of shock waves produced by explosions and by bodies in supersonic flight.**<br>(i) His **weak shock theory** which is of quite general application is an extension of his previous shock theory which was limited to directional symmetry only.[2233]<br>(ii) In order to check his new theory he applies it to numerous problems, such as<br>▸ to the outward propagation of spherical shocks;<br>▸ to the model of an unsymmetrical explosion;<br>▸ to steady supersonic flow past unsymmetrical bodies (*e.g.*, a flat plate delta wing at small incidence to the flow, and a thin wing having a finite curved leading edge); and<br>▸ to the calculation of wave drag on the wing. |
| 1957 | Kirtland Air Force Base, New Mexico | In the period February 26–27, **the First Shock Tube Symposium is held**; the general chairman is George Granger BROWN. ▪ The symposium was renamed three times:<br>▸ **International Shock Tube Symposium** (5.1965–10.1975);<br>▸ **Internatinal Symposium on Shock Tubes & Waves** (11.1977–17.1989); and<br>▸ **International Symposium on Shock Waves** (18.1991–present).<br>Further meetings were held at Palo Alto, CA (1958); Fort Monroe, VA (1959); Aberdeen, MD (1961); Silver Spring, MD (1965); Freiburg, Germany (1967); Toronto, Canada (1969); London, U.K. (1971); Palo Alto, CA (1973); Kyoto, Japan (1975); Seattle, WA (1977); Jerusalem, Israel (1979); Niagara Falls, NY (1981); Sydney, Australia (1983); Berkeley, CA (1985); Aachen, Germany (1987); Bethlehem, PA (1989); Sendai, Japan (1991); Marseille, France (1993); Pasadena, CA (1995); Great Keppel Island, Australia (1997); London, U.K. (1999); Fort Worth, TX (2001); Beijing, China (2003); Bangalore, India (2005), and Göttingen, Germany (2007). |

---

[2230] N.N. SEMENOV: *Einige Probleme der Kettenreaktionen und der Verbrennungstheorie* [Nobel Lecture, held on Dec. 11, 1956]. Angew. Chem. **69**, 767-777 (1957).

[2231] C.N. HINSHELWOOD: *Reaktionskinetik in den letzten Jahrzehnten*. Angew. Chem. **69**, 445-449 (1957).

[2232] G.B. WHITHAM: *On the propagation of weak shock waves*. J. Fluid Mech. **1**, 290-318 (1956).

[2233] G.B. WHITHAM: *The flow pattern of a supersonic projectile*. Comm. Pure Appl. Math. **5**, 301-348 (1952).

These symposia became the main international forum for reporting recent advances in shock wave research and applications in various fields, encompassing aeronautical and mechanical engineering, physics, chemistry, and, more recently, biology and medicine. Most of these symposia published proceedings which provide very useful sources of information on all kinds of shock wave problems.

| | | |
|---|---|---|
| 1957 | *Cosmodrome Baïkonour, Kazakhstan, Soviet Union* | On October 4, **Sputnik 1** [Russ. *Спутник*, meaning "traveling companion"], **the first artificial Earth satellite** (83 kg), **is launched.** Circling the Earth every 96 minutes at an altitude of between 231 km and 942 km, it provides data on air density, temperature, cosmic radiation and meteoroids. ▪ This important mission ended on January 4, 1958 when Sputnik 1 burned up in the atmosphere. The Sputnik missions, encompassing ten satellites in total (Sputnik 1 to Sputnik 10), were launched between October 4, 1957 and March 25, 1961. As well as inaugurating the Space Age, they also provoked a vigorous political and educational response in the Western World,[2234] known as the **"Sputnik shock."** |

The Soviets had announced their intention to launch an artificial satellite in August 1955, at the 6th International Astronautical Congress in Copenhagen. The chairman of the Soviet commission, Leonid I. SEDOV had said, "In my opinion, it will be possible to launch an artificial satellite of the Earth within the next two years, and there is the technological possibility of creating artificial satellites of various sizes and weights."

The first U.S. satellite was Explorer 1 (8.2 kg), which was launched on January 31, 1958. ▪ The idea of an artificial satellite was apparently first described by the U.S. science fiction writer Edward Everett HALE {⇨1870}.

▸ In order to launch a satellite into orbit around the Earth, a minimum velocity of 7.9 km/s (the ***first cosmic velocity*** or ***circular velocity***) is required [$v_{c1} = (g R_E)^{1/2}$, where $g$ is the acceleration due to gravity and $R_E$ is the Earth's radius].

▸ To escape from the Earth's gravitational field, an object must have a velocity of 11.2 km, termed the ***second cosmic velocity*** or ***escape velocity*** [$v_{c2} = (2gR_E)^{1/2}$].

▸ To explore outer space beyond the Solar System, a velocity of 16.6 km/s (the ***third cosmic velocity***) is required when the circular velocity of the Earth's orbit (29.8 km/s) is utilized; otherwise the escape velocity is 42.1 km/s.

These hypervelocities can currently only be achieved economically by rocket propulsion.

| | | |
|---|---|---|
| 1957 | *Europe* | **Reports are received that an *underwater explosive gun* – "firing a directional shock wave instead of a bullet" – has been seen somewhere in Europe.** It was reputedly being used for underwater fishing, and was said to be capable of stunning a large fish from a considerable distance. |

In the same year, A. COOMBS and Charles K. THORNHILL, two British scientists at the Armament Research and Development Establishment (ARDE), Fort Halstead, Kent, resume the idea and perform a theoretical study on an underwater explosive *shock gun*. It consists of a chamber in the form of a hollow circular cone, with a spherical sector of explosive charge fitted into the apex. They find that it is possible to design such a gun so that it is capable of projecting a high-intensity shock pressure beam over a considerable range, using only a small explosive charge. ▪ Their report was classified and was not published in the open literature until 1967.[2235]

---

[2234] F.J. KRIEGER: *Behind the Sputniks: a survey of Soviet space science.* Public Affairs Press, Washington, DC (1958).
[2235] A. COOMBS and C.K. THORNHILL: *An underwater explosive shock gun.* J. Fluid Mech. **29**, Part 2, 373-383 (1967).

A shock gun has also been proposed as a seismic wave source for use in hydrocarbon exploration; it was capable of producing peak amplitudes in water of 1.7 bar with a rise time of 6 μs and a cycle time of 10 s.[2236]

| 1957 | U.S. Ballistic Research Laboratory (BRL), Aberdeen Proving Ground, Maryland |

H.N. BROWN[2237] **first models the interaction and response of an elastic structure to an air blast** using the general equations for small elastic deflections of solids. ▪ His more general goal – to determine the law for large-deflection and elastic/plastic responses – became a subject of great practical concern, and still is today.

The inverse problem (*i.e.*, to be able to draw conclusions about the yield of an explosion by analyzing the structure damaged by it) is also important, and has been used by both the military and bureaus of criminal investigation; for example, to reconstruct the type of design, the applied explosive and yield used in bombs utilized in terrorism. This concept is even used in volcanology to some degree {NEWHALL & SELF ⇒ 1982}.

| 1957 | Dept. of Mathematics, University of Manchester, England |

Roy F. CHISNELL[2238] **treats the important practical problem of the motion of a shock wave in a channel.** His absorption model based on wave expansion will later be confirmed by shock tube measurements.[2239] ▪ Knowledge of the propagation of shock waves in channels is important in many engineering applications, *e.g.*, explosions of firedamp and/or coal dust in mine galleries, explosions provoked by the ignition of natural gas in pipelines, accidental detonations of munitions in underground shelters, accidental chemical explosions in long road and railway tunnels, reignitions of exhaust gases in exhaust pipes of reciprocating engines, *etc.*

Today a number of approximate analytical and numerical methods are available for studying the propagation of shock waves in ducts of arbitrary geometry and bifurcation.[2240]

| 1957 | Los Alamos Scientific Laboratory (LASL), Los Alamos, New Mexico |

William E. DEAL,[2241] a U.S. shock and detonation physicist, **reports on Chapman-Jouguet (CJ) pressure and detonation velocity measurements of various high explosives in order to characterize the CJ point at which the adiabatic expansion of the detonation products begins.** He uses an optical technique to measure the initial free-surface velocity as a function of thickness for aluminum alloy plates in contact with the detonation explosive – a method first suggested in 1945 by Roy W. GORANSON,[2242] a solid-state shock wave pioneer at Los Alamos. For RDX, TNT, Composition B, and 77/23 cyclotol the determined CJ pressures are 338, 189, 292, and 313 kbar, respectively. ▪ The ***Chapman-Jouguet pressure*** is defined as the maximum detonation pressure of the totally decomposed high explosive.

| 1957 | Poulter Laboratory, Stanford Research Institute (SRI), Menlo Park, California |

William E. DRUMMOND,[2243] a U.S. theoretical physicist, **presents a dynamic theory for the explosive production of multiple shocks in metals which undergo a phase transition.**

(i) He assumes that the metal can be treated hydrodynamically and is described by an isentropic equation of state with a kink located at the phase transition.

(ii) His remarkable results are as follows:

---

[2236] C. MACBETH and G. ARNOLD: *The shock gun: VSP experiments with a new seismic source.* The Leading Edge (Tulsa, OK) **17**, 183-188 (Feb. 1998).

[2237] H.N. BROWN: *Effect of scaling on the interaction between shock waves and elastic structures.* Appendix to BRL Rept. No. 1011, Aberdeen Proving Ground, MD (March 1957).

[2238] R.F. CHISNELL: *The motion of a shock wave in a channel, with applications to cylindrical and spherical shock waves.* J. Fluid Mech. **2**, 286-298 (1957).

[2239] H. REICHENBACH: *Funkenkinematographische Untersuchung der Stoßwellendämpfung durch Mehrfachreflexion an Blenden.* In: (N.R. NILSSON and L. HÖGBERG, eds.) Proc. 8th Int. Congress on High-Speed Photography [Stockholm, Sweden, June 1968]. Almquist & Wiksell, Stockholm, Sweden (1968), pp. 362-365.

[2240] W. HEILIG and O. IGRA: *Shock waves in channels.* In: (G. BEN-DOR ET AL., eds) Handbook of shock waves. Academic Press, New York (2000), vol. 2, pp. 319-396.

[2241] W.E. DEAL: *Measurement of Chapman-Jouguet pressure for explosives.* J. Chem. Phys. **27**, 796-800 (1957).

[2242] R.W. GORANSON: *A method for determining equations of state and reaction zones in detonation of high explosives, and its application to pentolite, composition-B, baratol, and TNT.* Classified Los Alamos Rept. LA-487 (1945).

[2243] W.E. DRUMMOND: *Multiple shock production.* J. Appl. Phys. **28**, 998-1001 (1957).

- If the pressure exceeds that at which a phase transition takes place, two compressive shocks are formed.
- If the maximum pressure is not too far above the transition pressure, these two shocks will separate.
- **A rarefaction shock will be formed which will overtake the second compressive shock after a period of time.**

(iii) He reports that his theoretical results were confirmed by experiments, although a rarefaction wave was not observed. He infers that the rarefaction shock exists by noting the sudden disappearance of the second compressive shock.

| 1957 | *Gruen Applied Science Laboratories, Hempstead, New York* | Antonio FERRI,[2244] an Italian-born U.S. professor of aeronautical engineering at Brooklyn Polytechnic Institute, **and Lewis FELDMAN and Walter DASKIN,** two U.S. aeronautical engineers, **describe a high-drag configuration of a lifting reentry vehicle – the so-called "Ferri sled."** Their spacecraft could produce a significant amount of lift when flying at an appropriate angle of attack. It would also have an open-top surface, and a bubble-canopy sealed cabin would be located in this open area. Reentry would be along a phugoid skip trajectory (where the nose rises and falls periodically, with a long period), with the lower surface of the vehicle acting as a heat sink. ▪ At the start of the following year, the Air Research and Development Command (ARDC) held a closed conference at Wright-Patterson Air Force Base near Dayton, OH, where eleven aircraft and missile firms outlined various classified proposals for a manned space vehicle to Air Force and NACA observers. |

The idea of the "Ferri sled" was resumed by the Republic Aviation Corporation (Farmingdale, Long Island, NY), which sketched a 1,800-kg vehicle of triangular planform with a 75-degree leading edge sweep. This "Republic Project 7969" was one of two approaches, where the vehicle would not be recovered; the pilot would eject from the spacecraft and parachute down to the ground.[2245] However, the concept was never realized.

In the same year, **Antonio FERRI**[2246] **and Paul A. LIBBY perform turbulent hypersonic flow studies.**

(i) They use a specially shaped nozzle designed to induce the required pressure distributions, and actual flight boundary and Reynolds number conditions around the test model, which influence aerodynamic heating. ▪ Later known as the **"shroud technique,"** it allows one to obtain large Reynolds numbers, thus overcoming the limitations of conventional hypersonic wind tunnels.

(ii) Also discussing air-breathing engines capable of operating at high Mach numbers ($M = 25$–$30$), they first propose the idea of injecting heat into a supersonic air stream by means of a mixing process.

| 1957 | *New York City, New York* | At the annual meeting of the American Physical Society (APS), **Frank C. GIBSON,**[2247] a physicist and explosives specialist at the Division of Explosives Technology of the Bureau of Mines (U.S. Dept. of the Interior, Pittsburgh, PA), **and his collaborators report on a new experimental method that can be used to determine detonation temperatures in high explosives.** |

(i) His method consists of embedding one end of a rod of Plexiglas into the explosive, the other end protruding from the charge. The light from the core of the explosive is transmitted along the rod and analyzed by a grating spectrograph, using four bands 100 Å wide and 600 Å

---

[2244] A. FERRI, L. FELDMAN, and W. DASKIN: *The use of lift for reentry from satellite trajectories*. Jet Propulsion **27**, 1184-1191 (Nov. 1957).
[2245] *Republic Project 7969*. Encyclopedia astronautica; http://www.astronautix.com/craft/rept7969.htm.
[2246] A. FERRI and P.A. LIBBY: *A new technique for investigating heat transfer and surface phenomena under hypersonic flight conditions*. J. Aeronaut. Sci. **24**, 464-465 (1957).
[2247] F.C. GIBSON, M.L. BOWSER, C.R. SUMMERS, F.H. SCOTT, and C.M. MASON: *Use of an electro-optical method to determine detonation temperatures in high explosives*. J. Appl. Phys. **29**, 628-632 (1958).

apart. The radiation temperatures are used to calculate the color temperatures within the detonating explosive.

(ii) For PETN, RDX, tetryl, and nitroglycerin, they determine detonation temperatures of 5,525 K, 5,135 K, 4,480 K, and 4,095 K, respectively.

Data on detonation temperatures were urgently required at this time in solid-explosive research in order to determine equations of state applicable to the Chapman-Jouguet (CJ) plane. GIBSON's successful technique was a milestone in this line of inquiry and it stimulated others to measure shock temperatures in transparent solids, such as in Plexiglas,[2248] and in single crystals of NaCl and KCl {KORMER ET AL. ⇨1965}.

| 1957 | Institute of Aerophysics, University of Toronto, Toronto, Canada |

**Irvine I. GLASS,**[2249] a professor of aeronautical engineering, **and J.G. HALL present some experimental results for the flow generated by the explosion of a sphere of high-pressure gas.** Glass spheres – one, two, and five inches (2.54, 5.08, and 12.7 cm) in dia. containing air, helium, or sulfur hexafluoride [$SF_6$] at overpressures of up to 21 bar – were used to produce a spherical analog of the shock tube. Their concept of a "spherical shock tube" shows that the experimental method can be applied to the study of explosions, implosions, and spherical wave interactions under controlled and known initial conditions.

In the years that followed, the experiments were continued at the Institute of Aerophysics by Donald W. BOYER[2250] {⇨Fig. 4.16–F} in order to compare the results from them with numerical solutions worked out in previous years by BRODE {⇨1954 & 1959} of the Rand Corporation. BOYER also extended his studies to spherical implosions. Using high-speed schlieren visualization, he noticed that the glass fragments form a jet in an implosion, similar to the phenomenon observed for a shaped charge.

| 1957 | Los Alamos Scientific Laboratory (LASL), Los Alamos, New Mexico |

**The U.S. Department of Defense (DOD) publishes the compendium *The Effects of Nuclear Weapons*, compiled and edited by Samuel GLASSTONE,**[2251] a U.S. physical chemist and an authority on the effects of nuclear weapons. It is the first official source of information on nuclear weapons phenomena and destruction effects resulting from blast and radiation, much of which had previously been classified.

In 1968, Harold L. BRODE,[2252] a U.S. theoretical physicist at RAND Corporation, Santa Monica, CA, reviewed nuclear explosion physics and provided (empirical) formulae for estimating the various transient phenomena. For example, in regard to mechanical effects resulting from surface bursts, he discussed

- the peak overpressure *vs.* distance;
- the shock radius and arrival time *vs.* peak overpressure;
- the shock temperature, peak dynamic pressure, shock velocity and maximum particle velocity at any point related to the peak overpressure at that point; and
- the properties of the air-blast-induced ground shock.

| 1957 | Dept. of Physic, Columbia University, New York City, New York |

**Gordon GOULD,** a Ph.D. student in physics under Charles TOWNES, the inventor of the *maser*, **conceives of a solution to the problem of how to produce a beam of coherent visible light,** a problem that has plagued researchers for several years. His solution is to excite atoms or molecules through either the use of bright light or atomic-level collisions in order to produce a

---

[2248] Y.B. ZEL'DOVICH, S.B. KORMER, M.V. SINITSYN, and A.I. KURYAPIN: *Temperature and specific heat of Plexiglas under shock wave compression*. Sov. Phys. Dokl. **3**, 938-939 (1958).

[2249] I.I. GLASS and J.G. HALL: *Shock sphere – an apparatus for generating spherical flows*. J. Appl. Phys. **28**, 424-425 (1957).

[2250] D.W. BOYER: *Spherical explosions and implosions*. Rept. No. 58, University of Toronto, Institute of Aerophysics, UTIA (1959). *See also* I.I. GLASS and J.P. SISLIAN: *Nonstationary flows and shock waves*. Clarendon Press, Oxford (1994), pp. 271-275.

[2251] S. GLASSTONE (ed.): *The effects of nuclear weapons*. U.S. Gvt. Print. Office, Washington, DC (1957); revised edition by U.S. Dept. of Defense, U.S.A.E.C. (April 1962).

[2252] H.L. BRODE: *Review of nuclear weapons effects*. Annu. Rev. Nucl. Sci. **18**, 153-202 (1968).

population inversion. At the same time, **he coins the name "laser" for this concept.** ▪ His work almost coincided with that of Arthur SCHAWLOW and Charles TOWNES' independent work on *optical masers* {SCHAWLOW & TOWNES ⇨1958}. **GOULD created the first written prototype for a laser** and, realizing its significance, took it to a neighborhood store to have his notebook notarized.[2253]

GOULD spent the following year refining and improving his model, but did not immediately file for a patent, believing that he had to build a prototype before filing. His early efforts to obtain patent protection for his invention were consistently rebuffed by the United States Patent and Trademark Office (USPTO). Interferences were declared between his applications, the first of which was filed on April 6, 1959, and the applications of other companies. Unfortunately, his technology was widely exploited by others. This resulted in a 20-year legal battle, which GOULD finally won in 1977, when the first of his laser patents was issued.[2254] ▪ He died on September 16, 2005, at the age of 85.

| | | |
|---|---|---|
| 1957 | *Physical Chemistry Division, Alcoa Research Laboratories, New Kensington, Pennsylvania* | George LONG,[2255] a U.S. research engineer, **reports on studies of aluminum/water steam explosions.** In his pioneering empirical experiments he poured various quantities of molten aluminum over coated and uncoated submerged surfaces; the suppression or occurrence of explosions was inferred empirically. For surface contact-initiated explosions he found – on an empirical basis – that certain surfaces (*e.g.,* rusted steel, gypsum, and lime) promote explosions. Other surfaces (*e.g.,* polished steel, aluminum) and those with organic coatings display relative inert to spontaneous explosions. ▪ Much of his work provided the first insights into these complex reactions, forming the basis for the methods used to prevent steam explosions in casting pits today. <br><br> More recently, Rusi TALEYARKHAN[2256] at the Oak Ridge National Laboratory (ORNL) studied explosive interactions between molten aluminum and water in order to determine the causes of explosion triggers and the extent of protection provided from various coatings. |
| 1957 | *Meppen Test Site, Ministry of Defense, Meppen, Lower Saxony, Germany* | Werner MÜLLER,[2257] a German governmental applied physicist who works closely with the ISL in France, **reports on his optical studies of exploding wires using a schlieren camera, which uses a Kerr cell** {KERR ⇨1875} **as an ultrafast shutter.** He observes that shortly after the wire begins to explode, a primary cylindrical shock wave is formed that separates from the expanding metal vapor. This first shock wave is followed by a secondary shock wave from the interior of the metal vapor cylinder, after the second energy release. |
| 1957 | *New York University, New York City, New York* | James J. STOKER,[2258] an applied mathematician at the Institute of Mathematical Sciences, **publishes his textbook *Water Waves*,** in which he mathematically analyzes water waves of all kinds – both continuous and discontinuous – such as water flow and flood waves of rivers and large reservoirs. ▪ His book became a classic on this subject. |

---

[2253] *Gordon GOULD (1920–2005).* Wikipedia; http://en.wikipedia.org/wiki/Gordon_Gould.
[2254] G. GOULD: *Optically pumped laser amplifiers.* U.S. Patent No. 4,053,845 (filed Aug. 16, 1974; issued Oct. 11, 1977); *Light amplifiers employing collisions to produce a population inversion.* U.S. Patent No. 4,704,583 (filed Aug. 11, 1977; issued Nov. 3, 1987).
[2255] G. LONG: *Explosions of molten aluminum and water – cause and prevention.* Metal Progress **71**, 107-112 (1957).
[2256] R. TALEYARKHAN: *Preventing melt-water explosions.* J. Miner. Metals Mater. Soc. **50**, No. 2, 35-38 (1998).
[2257] W. MÜLLER: *Der Ablauf einer elektrischen Drahtexplosion, mit Hilfe der Kerr-Zellen-Kamera untersucht.* Z. Phys. **149**, 397-411 (1957); *Studies of exploding wire phenomenon by use of Kerr cell schlieren photography.* In: (W.G. CHACE and H.K. MOORE, eds.): *Exploding wires (I).* Plenum Press, New York (1959), pp. 186-208.
[2258] J.J. STOKER: *Water waves.* Interscience, New York (1957).

| | | |
|---|---|---|
| 1957 | Institute of Nuclear Problems, Moscow State University | Yakov Petrovich TERLETSKII,[2259] a distinguished Soviet physicist, **gives a brief description of the principle of compressing magnetic fields by means of explosive-driven metallic conductors** {SAKHAROV ⇨1951} in order to build large energy-multiplying devices which can reach energy outputs that cannot be attained in conventional capacitor-bank systems. ▪ In 1972, TERLETSKII was awarded the Lenin Prize for his work in the field of "magnetic cumulation." |
| 1957 | Institute of Chemical Physics, Moscow, Soviet Union | Veniamin A. TSUKERMAN,[2260] a Soviet nuclear physicist, **and M.A. MANAKOVA report on a multishot flash X-ray system for recording successive phases of detonation processes and other fast phenomena, known as "cineradiography."** They use eight 2-MV pulsed X-ray tubes with needle anodes arranged in a three-by-three array such that each X-ray pulse projects an image of the object under study onto a separate fixed cassette containing an X-ray film. The arrangement of the flash X-ray tubes is very similar to the spark light sources in the Cranz-Schardin multiple-spark camera {CRANZ & SCHARDIN ⇨1929}. Using this recording technique, the number of frames is equal to the number of pulsed X-ray tubes, and the time interval between the frames can be as small as desired. ▪ Their method is very well-suited to a rotational target symmetry {CHANG ⇨1983}. However, in a *Cranz-Schardin arrangement* the space occupied by flash X-ray tubes only allows a small number of frames.<br><br>Multiple flash X-ray systems in a *linear arrangement* are particularly useful when the object to be studied moves along a trajectory, such as in ballistic ranges. In the 1960s, Gustav THOMER,[2261] a German applied physicist and flash X-ray pioneer {THOMANEK ⇨1938; SCHALL & THOMER ⇨1951} and Francis JAMET, a young French research physicist, developed at ISL (Saint Louis, Alsace, France) a multiple tube containing six small discharge gaps in a *circular arrangement* positioned on a 51°-segment in a single pump-driven vacuum chamber. They were pulsed successively at low voltages ranging between 20 and 40 kV; *i.e.*, producing soft flash X-rays that allow radiographs of high contrast to be obtained from low-density test samples. They used their device to study the fragmentation process of shaped charge jets cinematographically. |
| 1957 | Los Alamos Scientific Laboratory (LASL), Los Alamos, New Mexico | John M. WALSH, Melvin H. RICE, Robert G. MCQUEEN, and Frederick L. YARGER,[2262] four U.S. shock wave experimenters, **report on precise measurements of high-pressure Hugoniot data for 27 metals** obtained using a high explosive for sample compression. The shock wave velocity in the sample is measured using embedded gauges, and the particle velocity is determined from the velocity of the free back surface. ▪ Precise experimental data for equations of state at ultrahigh pressures are indispensable to nuclear weapons designers attempting to improve the fidelity of their computer simulations.<br><br>In the same year, RICE[2263] **and WALSH systematically study the optical transparency of fifteen shock-compressed liquids in the range from 50 to 150 kbar.** Only carbon tetrachloride [$CCl_4$] shows shock-induced freezing {SCHARDIN ⇨1940/1941}. However, for water shocked to 100 kbar, no sign of opacity due to freezing is observed. ▪ The results of RICE and WALSH were confirmed by Australian researchers.[2264] Soviet researchers {AL'TSHULER ET AL. ⇨1958}, however, first observed a marked change of slope in the measured Hugoniot curve at 115 kbar, of the kind expected from water about to freeze. They reported that they found a re- |

---

[2259] Y.P. TERLETSKII: *Production of very strong magnetic fields by rapid compression of conducting shells*. Sov. Phys. JETP **5**, 301-302 (1957).

[2260] V.A. TSUKERMAN and M.A. MANAKOVA: *Sources of short X-ray pulses for investigating fast processes*. Sov. Phys. Tech. Phys. **2**, 353-363 (1957).

[2261] G. THOMER and F. JAMET: *Ein Sechsfach-Röntgenblitzrohr für weiche Strahlung*. In: (N.R. NILSSON and L. HÖGBERG, eds.) *Proc. 8th Int. Congress on High-Speed Photography* [Stockholm, Sweden, June 1968]. Almqvist & Wiksell, Stockholm (1968), pp. 256-258.

[2262] J.M. WALSH, M.H. RICE, R.G. MCQUEEN, and F.L. YARGER: *Shock-wave compressions of twenty-seven metals. Equations of state of metals*. Phys. Rev. **108** [II], 196-216 (1957).

[2263] M.H. RICE and J.M. WALSH: *Dynamic compression of liquids from measurements on strong shock waves*. J. Chem. Phys. **26**, 815-823 (1957).

[2264] H.G. DAVID and A.H. EWALD: *Photographic observation on shock waves in liquids*. Austral. J. Appl. Sci. **11**, 317-320 (1960).

duction in the transparency of water above 115 kbar. But later experiments carried out by another Soviet group failed to confirm their results.

In the 1930s, Percy W. BRIDGMAN[2265] had discovered in his famous compression studies that water that is hydrostatically compressed at 25 kbar at room temperature transforms into a special ice phase which he called **"Ice VII"** (cubic phase).

| | | |
|---|---|---|
| 1958 | *Ripple Rock, Seymour Narrows, B.C., Canada* | On April 5, **the largest non-nuclear blast in history is detonated to demolish two underwater peaks, called "Ripple Rock."** ▪ In this unique project, 1,253 tons of explosives were used to displace 700,000 tons of rock that were a serious hazard to navigation.[2266] Initially, it was suggested that nuclear explosions could be used for this purpose. However, this idea was eventually dropped because of environmental problems {Plowshare Program ⇨1958; Test SEDAN ⇨1962; Test BUGGY ⇨1968; U.S. Atlantic-Pacific Interoceanic Canal Study Commission ⇨1970}.<br><br>The technique of using large chemical explosions for excavation work also gained increasing popularity in the former Soviet Union {Alma Ata ⇨1966}. |
| 1958 | *U.S. DOE Nevada Operations Office, Las Vegas, Nevada* | On June 6, **the Atomic Energy Commission (AEC) publicly announces the establishment of the Plowshare Program,** which has the objective to use nuclear explosives for civilian as opposed to military purposes. Two broad categories are being considered:<br><br>▸ **excavation applications** such as canals, harbors, highway and railroad cuts through mountains, open pit mining, construction of dams, and other quarry and construction-related projects {Test SEDAN ⇨1962; Test BUGGY ⇨1968; U.S. Atlantic-Pacific Interoceanic Canal Study Commission ⇨1970}; and<br><br>▸ **underground nuclear explosion applications,** including the stimulation of natural gas production, preparation of leachable ore bodies for *in situ* leaching, creation of underground zones of fractured oil shale for *in situ* retorting, and formation of underground natural gas and petroleum storage reservoirs {Test GASBUGGY ⇨1967}.<br><br>A series of tests to explore the feasibility of using nuclear explosions for excavation (to stimulate the production of natural gas from marginal fields) and for other peaceful uses began with the detonation associated with Project GNOME on December 10, 1961 in Carlsbad, NM, and concluded with those associated with Project RIO BLANCO on May 17, 1973, in Rifle, CO. Because of political reasons {Int. Nuclear Test Ban Treaty ⇨1963} as well as environmental and safety concerns (*e.g.,* fallout and concentration of radioactive material), the Plowshare Program[2267] was officially ended on June 30, 1975.<br><br>**The equivalent Soviet program on investigating peaceful nuclear explosions,** named "Nuclear Explosions for the National Economy" {Test CHAGAN ⇨1965}, **was many times larger than the U.S. Plowshare Program.**<br><br>The United States stopped performing nuclear tests in 1992. ▪ During the period between 1945 and 1992, the United States conducted more than 1,000 nuclear weapons tests, and the Soviet Union conducted more than 700 nuclear tests. |
| 1958 | *Nice, France* | In July, **first International Symposium on Rarefied Gas Dynamics;** the general chairman is Fernand Marcel DEVIENNE. It provides, for the first time, a focal point for the exchange of information between scientists with interests in kinetic theory, low-density gas dynamics, molecular interactions at surfaces, and other related fields. ▪ The first conference was followed by biennial symposia held at Berkeley, CA (1960); Paris, France (1962); Toronto, Canada (1964); Oxford, U.K. (1966); Cambridge, MA (1968); Pisa, Italy (1970); Palo Alto, CA (1972); Göt- |

---

[2265] P.W. BRIDGMAN: *The phase diagram of water to 45,000 kg/cm².* J. Chem. Phys. **5**, 964-966 (1937).
[2266] G. REYNOLDS: *Ripple Rock – the end comes with a bang.* DuPont Mag. (June/July 1958).
[2267] *Executive Summary – Plowshare Program.* Dept. of Energy (DOE), Nevada Operations Office (1994); https://www.osti.gov/opennet/reports/plowshar.pdf.

| | | |
|---|---|---|
| | | tingen, Germany (1974); Snowmass-at-Aspen, CO (1976); Cannes, France (1978); Charlotteville, VA (1980); Novosibirsk, Soviet Union (1982); Tsukuba, Japan (1984); Grado, Italy (1986); Pasadena, CA (1988); Aachen, Germany (1990); Vancouver, Canada (1992); Oxford, U.K. (1994); Beijing, China (1996); Marseille, France (1998); Sydney, Australia (2000); Whistler, B.C., Canada (2002); Bari, Italy (2004); St. Petersburg, Russia (2006); and Kyoto, Japan (2008). |
| 1958 | Lituya Bay, Glacier Bay National Park, southeastern coast of Alaska | On July 9, **a violent earthquake suddenly shakes the bay,** and 90 million tons of icy rock fall into the water from the northeastern wall of Gilbert Inlet, which forms part of a T-shaped inlet.[2268] **The splash of water it causes generates a *giant tsunami* that washes away trees to a maximum height of 520 meters at the entrance to the Gilbert Inlet** {⇨Fig. 4.1−Q}. A wave about 10−30 m high races at a speed of about 160 km/h (44 m/s) down the main bay. A day later, the only sign of the previous day's events is a shoreline that is stripped of soil and vegetation to a height of more than 30 m. ▪ George PARARAS-CARYANNIS,[2269] a tsunami researcher at Honolulu, suggested that a tsunami wave was formed by a rockslide impact similar to an asteroid impact. It formed a cavity in the ocean floor of the inlet and produced a giant wave that splashed up to tremendous heights. Charles L. MADER,[2270] a former Los Alamos detonation physicist and an authority on the numerical modeling of explosions, used the shallow-water code SWAN (Simulating Waves Nearshore)[2271] in order to numerically model the Lituya event, which he called a **"mega-tsunami."** SWAN can be used to solve long-wave, shallow-water, nonlinear equations of fluid flow.[2272]

During the last 150 years, five giant waves have occurred in Lituya Bay, leaving sharp trimlines marking the height below which the forest along the shores was totally or almost totally destroyed. The dates of occurrence of the five giant waves and the maximum altitudes of their trimlines are: 1853 or 1854 (395 ft or 120 m); *c.*1874 (80 ft or 24 m); *c.*1899 (about 200 ft or 70 m); 1936 (490 ft or 149 m); and 1958 (1,720 ft or 524 m).[2273] Fortunately, because Lituya Bay was an unpopulated area, therefore the loss of life was minimal. |
| 1958 | David Taylor Model Basin, Carderock, Maryland | On August 27, **Marshall TULIN,** a U.S. hydrodynamicist, **announces the development of a *supercavitating propeller*** which decreases the deleterious effects of cavitation by incorporating supercavitating blade sections.[2274] ▪ The design of a ship propeller that was able to run under supercavitating conditions has been compared in importance to the development of jet propulsion for aircraft.

***Supercavitation* is an extreme version of cavitation in which a single bubble is formed that envelops the moving object almost completely.** The subject was resumed in the 1990s in order to achieve supersonic velocities underwater, for example at velocities > 1,482 m/s (in pure water at 20 °C) and > 1,522 m/s (in sea water at 20 °C and a salinity of 3.5%). The goal was to develop two classes of supercavitating technologies: projectiles ("super penetrators") {⇨Fig. 4.14−E} and torpedoes such as the Russian "Shkval" and the German "Barracuda" {Naval Undersea Warfare Center ⇨1997}. |

---

[2268] *Lituya Bay close up.* Tsunami Research Group, University of Southern California; http://www.usc.edu/dept/tsunamis/alaska/1958/webpages/lituyacloseup.html.

[2269] G. PARARAS-CARAYANNIS: *Analysis of mechanism of tsunami generation in Lituya Bay.* Sci. Tsunami Hazards **17**, 193-206 (1999).

[2270] C.L. MADER: *Modeling the 1958 Lituya Bay mega-tsunami.* Sci. Tsunami Hazards **17**, 57-67 (1999).

[2271] SWAN was developed specifically for the nearshore and contains formulations for two physical processes: depth-induced wave breaking and triad wave-interaction. SWAN can also be used to solve long-wave, shallow-water, nonlinear equations of fluid flow; see C.L. MADER: *Numerical modeling of water waves.* CRC Press, Boca Raton, FL (2004).

[2272] C.L. MADER: *Numerical modeling of water waves.* CRC Press, Boca Raton, FL (2004).

[2273] D.J. MILLER: *Giant waves in Lituya Bay, Alaska.* Geological Survey, Professional Paper USGS 354-C. U.S. Gvt. Print. Office, Washington, DC (1960); https://www.uwsp.edu/geo/projects/geoweb/participants/dutch/LituyaBay/Lituya0.HTM.

[2274] R.P. CARLISLE: *Where the fleet begins.* Naval Historical Center, Dept. of the Navy, Washington, DC (1998), p. 487.

| 1958 | *National Aeronautics and Space Administration (NASA), Washington, DC* | On October 1, **NASA,** an independent U.S. governmental agency "for the research and development of vehicles and activities for the exploration of space within and outside of Earth's atmosphere," **is created.** The first president is T. Keith GLENNAN.

Building on the work done by NACA {⇨1915}, NASA contributed greatly to the advancement of hypersonic aerodynamics and our understanding of reentry phenomena through the use of wind tunnels and shock tunnels, real flight testing and fluid dynamics computer simulations.[2275, 2276] This resulted in a very successful series of milestone space programs:

- **Human space flight** – such as the *Mercury, Gemini,* and *Apollo* (lunar) missions {Apollo 11 ⇨1969; Apollo 14 ⇨1971; Apollo 16/17 ⇨1972}.
- **Hypersonic flight** – such as the **High-Altitude Mach 6 Aircraft X-15** {⇨1959 & 1961; ⇨Fig. 4.20–I}, the X-38 {LUDWIEG ⇨1955} and the *Columbia* **Space Shuttle** {⇨Fig. 4.20–K}.
- **Scientific space probes** – such as the *Explorer* Missions {Explorer 1, *see* Sputnik 1 ⇨1957; Explorer 18, *see* NESS ⇨1964; Sun-Earth Explorer Program ⇨1977}; the *Mariner* Missions {Mariner 2, *see* NEUGEBAUER ⇨1962; Mariner 4 ⇨1965; Mariner 9, *see* Mars Global Surveyor ⇨2002; Mariner 10 ⇨1973}; the Mars Global Surveyor orbiter {⇨2002}; the *Viking* Missions {Viking 2 Lander ⇨1975; Allan Hills Meteorite ⇨1984}; the *Pioneer* Missions {RUSSELL ⇨1979; ISSI Workshop ⇨1998; Cassini Spacecraft ⇨2004; Voyager 1 ⇨2004}; the *Voyager* Missions {MORABITO ⇨1979; Voyager 1 ⇨1980; Voyager 2 ⇨1989; Voyager 1/2 ⇨2003, 2004}; the *Hubble Space Telescope* {CYGNI 1992 ⇨1992; Supermassive Black Hole ⇨1994; LL Orionis ⇨1995; Southern Crab Nebula ⇨1999; NGC 6745 ⇨2004}; the *Chandra X-Ray Observatory* {SN 1987A ⇨2000; NGC 6240 ⇨2001}; the *Pioneer Venus Orbiter* {*see* Sect. 2.3.1}; the *Solar Radiation and Climate Experiment (SORCE)* **Satellite** {NASA-GSFC ⇨2003}; and the *High Energy Transient Explorer (HETE)* **Satellite** {GRB 030329 ⇨2003}.
- **Communications satellites** – such as Echo, Telstar, Syncom, Landsat, Seasat, and Intelsat.

**The most important NASA laboratories dedicated to supersonic and hypersonic research are:** Ames Research Center (ARC), Dryden Flight Research Center (DFRC), Goddard Space Flight Center (GSFC), Jet Propulsion Laboratory (JPL), Johnson Space Center (JSC), Langley Research Center (LaRC), Lewis Research Center (LeRC), Kennedy Space Center (KSC), and Marshall Space Flight Center (MSFC).

Shortly after the establishment of NASA, the NASA History Program was started.[2277] It serves three key functions:

- widely disseminating aerospace information, such as by the NASA History Series of publications (covering general histories, project histories, NASA Center histories, monographs in aerospace history, reference works, management histories, *etc.*), NASA Books and the NASA/NACA Online Archive[2278] (which provides an easy access to technical reports);
- helping NASA managers to understand and learn from past successes and failures; and
- interaction with the professional historian community, particularly with those researchers working in the history of science and technology fields. |

---

[2275] *NASA history in brief.* NASA History Office, Washington, DC; http://www.hq.nasa.gov/office/pao/History/brief.html.
[2276] R.E. BILSTEIN: *Orders of magnitude. A history of the NACA and NASA, 1915–1990.* The NASA History Series, NASA SP-4406, Office of Management, Scientific and Technical Information Division, Washington, DC (1989); http://www.hq.nasa.gov/office/pao/History/SP-4406/cover.html.
[2277] NASA History Division; http://history.nasa.gov/.
[2278] *NASA Technical Reports Server (NTRS);* http://ntrs.nasa.gov/search.jsp. ▪ NTRS searches three major STI (Scientific and Technical Information) collections: NACA (1915–1958), NASA (1958–present), and NASA Image eXchange (NIX) (1900–present).

| | | |
|---|---|---|
| 1958 | *Crimean Astrophysical Observatory, Ukraine, Soviet Union* | On November 3, **Nikolai A. KOZYREV,**[2279] a Soviet astronomer, **uses the Observatory's 50-in. (127-cm) reflector to photograph a gaseous eruption from the lunar surface, near the 119-km-dia. 2.7-km-deep crater Alphonsus: apparently some kind of volcanic activity.** He also records spectra from reddish glows that last for about an hour. ▪ Volcanic theories for the formation of the lunar craters have been around for a long time: internal volcanism, it has been argued, shaped the twisted surface, formed calderas, and filled the lunar plains with darkened ash or lava flows. Such theories were proposed as recently as the middle of the last century {SPURR ⇨1949}.<br><br>KOZYREV's claim that the Moon exhibits volcanic activity in the crater Alphonsus produced much controversy around the world:<br><br>(i) U.S. Nobel Prize winner Harold UREY was among the small group that believed that KOZYREV's theory of volcanic activity on the Moon was correct, and he urged NASA to conduct an investigation into the phenomenon.<br><br>(ii) As a direct result, NASA launched the enormous Operation "Moon Blink," which detected 28 lunar events in a relatively short period of time.[2280] This project later confirmed KOZYREV's assertions by detecting significant gas emissions on the Moon.<br><br>(iii) In 1963, astronomers at the Lowell Observatory in (Flagstaff, AZ) also saw reddish glows on the crests of ridges in the region of the crater Aristarchus (dia. 40 km, depth 3.7 km). These observations proved to be precisely identical and periodical, repeating themselves as the Moon moved closer to the Earth: when gravitational stresses are high, the crust shifts and gas escapes from the interior at regular intervals. The only transient lunar phenomena of short duration observed before this were seismic shocks, usually of very low magnitude, which happen quite frequently: they are known as **"moonquakes"** {Apollo 11 ⇨1969}.<br><br>There are some major physical differences between volcanism on the Earth and that on the Moon. Since lunar gravity is only one sixth that of the Earth's, lava flows spread out over large areas on the Moon, and explosive eruptions could throw debris further than on the Earth. Furthermore, **since the Moon essentially has no dissolved water, which plays a major role in driving violent steam eruptions on the Earth, explosive eruptions are much less likely on the Moon.**[2281] |
| 1958 | *American Institute of Physics (AIP), New York, New York* | **Foundation of the AIP journal *The Physics of Fluids*,** with the goal "to cover kinetic theory, statistical mechanics, structure and general physics of gases, liquids, and other fluids. The scope of these fields of physics includes magneto-fluid dynamics, ionized fluid and plasma physics, **shock and detonation phenomena, hypersonic physics,** rarefied gases and upper atmosphere phenomena, liquid state physics and superfluidity, as well as certain basic aspects of physics of fluids bordering geophysics, astrophysics, biophysics, and other fields of science." The editor-in-chief is François N. FRENKIEL. ▪ In 1989, the journal was divided into separate volumes: *Physics of Fluids A: Fluid Dynamics* {⇨1989}, and *Physics of Fluid Dynamics B: Plasma Physics*. The journal is devoted to the publication of original theoretical, computational, and experimental contributions to the dynamics of gases, liquids, and complex or multiphase fluids. |
| 1958 | *United Nations (UN), Geneva, Switzerland* | **Representatives of the United States, the United Kingdom and the Soviet Union convene to discuss techniques for monitoring compliance with a possible future Comprehensive Test Ban Treaty** {UN CTBT ⇨1996}. ▪ Following this conference, and in the ensuing ne- |

---

[2279] N.A. KOZYREV: *Spectroscopic proofs for existence of volcanic processes on the Moon.* In: (Z. KOPAL and Z.K. MIKHAILOV, eds.) *Symposium No. 14 of the International Astronomical Union (IAU)* [Pulkovo Observatory (near Leningrad), Dec. 1960]. Academic Press, London & New York (1962), pp. 263-272; *Relationships of tectonic processes of the Earth and Moon.* In: (J. GREEN, ed.) *Geological problems in lunar and planetary research.* AAS Science and Technology Series, vol. 25, distributed by the AAS Publ. Office, Tarzana, CA (1971), pp. 213-227.

[2280] J.J. GILHEANY: *Operation "Moon Blink."* The Strolling Astronomer **18**, 183-187 (1964).

[2281] R. WICKMAN: *Volcanism on the Moon.* VolcanoWorld; http://volcano.und.nodak.edu/vwdocs/planet_volcano/lunar/Overview.html.

| | | |
|---|---|---|
| 1958 | *Soviet Union* | **Lev V. AL'TSHULER,**[2282] a notable Soviet shock physicist, **and his associates report that they have recorded a shock-induced phase transition beginning at 115 kbar in water.** They also observe that the lead wave splits into two surfaces of discontinuity and that the optical transparency for shock pressures above the transition pressure is reduced. Using an impedance-match method with an aluminum driver, they measure the Hugoniot curve and notice a kink in the region of the phase transition.
Since the early 1940s, the interesting phenomenon of the shock-induced freezing of water, the most abundant of all liquids, has attracted the curiosity of physicists:
(i) Similar experiments to those reported by AL'TSHULER and his collaborators had already been performed previously in the United States {WALSH & RICE ⇨1957}. However, water shocked up to 100 kbar showed no evidence of a phase transition, probably because of an insufficient number of experimental readings.
(ii) Later experiments carried out in the Soviet Union by Yakov B. ZEL'DOVICH[2283] and collaborators to confirm AL'TSHULER's results failed as well. They showed that water remains transparent up to at least 300 kbar. From the recorded photochronograms recorded, they concluded that the nature of the reflection of light from the surface of a shock front corresponds to specular (not diffuse) reflection. This would seem to indicate a high degree of smoothness on the shock front.
(iii) After more than 40 years, the curious subject of the shock-induced freezing of water was resumed at Washington State University. These studies provided the first direct evidence for the freezing of water on a nanosecond timescale {DOLAN & GUPTA ⇨2004}. |
| 1958 | *Geneva, western Switzerland* | At the United Nations International Conference on the Peaceful Uses of Atomic Energy, **Lev A. ARTSIMOVICH** [or ARCIMOVIC],[2284] a Soviet physics professor and inventor of the Tokamak, reviews controlled fusion research in the Soviet Union. **He reports on the generation of $10^8$ neutrons per impulse by focusing converging spherical shock waves on a $UD_2T$ target.** • In 1963, using $UD_3$ and gaseous $D_2$ targets, this number was increased to $3 \times 10^{11}$. The converging shock wave was formed by the implosion of a spherical charge of explosive. In most of these experiments, the external diameter of the design was about 70 cm, and the mass of gaseous $D_2$ was about $3 \times 10^{-4}$ g.[2285] |
| 1958 | *U.S. Naval Research Laboratory (NRL), Washington, DC* | **Barry BERNSTEIN,**[2286] a U.S. mathematician and gas dynamicist, **and his collaborators discuss the production of a cracking sound by a bull whip.** They took motion pictures of the typical loop motion and also made shadowgraphs of the shock wave emitted from the tip using a still camera placed opposite a board covered with 3M Scotchlite™ in a darkened room and a spark light source triggered by a microphone placed before the camera. Similar to previous studies in France {CARRIÈRE ⇨1927}, they visualized the shock wave emitted from the whip tip {⇨Fig. 4.5−K}. Resuming a previous mathematical model for whip motion with fully elastic impact {GRAMMEL & ZOLLER ⇨1949}, they extended this model by assuming that the impact at the jump is inelastic and that there is an energy loss proportional to the square of the velocity. |

---

[2282] L.V. AL'TSHULER, A.A. BAKANOVA, and R.F. TRUNIN: *Phase transition of water compressed by strong shock waves.* Sov. Phys. Dokl. **3**, 761-763 (1958).
[2283] Y.B. ZEL'DOVICH, S.B. KORMER, M.V. SINITSYN, and K.B. YUSHKO: *A study of the optical properties of transparent materials under high pressure.* Sov. Phys. Dokl. **6**, 494-496 (1961).
[2284] L.A. ARTSIMOVICH: *Controlled fusion research in the U.S.S.R.* In: *Proc. 2nd United Nations Int. Conference on the Peaceful Uses of Atomic Energy.* Geneva (Sept. 1958), United Nations, Geneva (1958); see vol. 31: *Theoretical and experimental aspects of controlled nuclear fusion*, p. 2298.
[2285] A.S. KOZYREV, V.A. ALEKSANDROV, and N.A. POPOV: *Fusion first for U.S.S.R.* Nature **275**, 476 (Oct. 12, 1978).
[2286] B. BERNSTEIN, D.A. HALL, and H.M. TRENT: *On the dynamics of a bull whip.* JASA **30**, 1112-1115 (1958).

| | | |
|---|---|---|
| 1958 | NASA's George C. Marshall Space Flight Center (MSFC), Huntsville, Alabama | **The development of two- and three-stage Saturn launch vehicles intended for use in connection with the Apollo space program is started, led by the German-born U.S. rocket engineer Wernher VON BRAUN.** ▪ The three-stage, almost 110-m-long, nearly 2,800-ton Moon rocket Saturn V, first launched in November 1967, was used for manned Apollo lunar flights {U.S. President KENNEDY ⇨1961; Apollo 11 ⇨1969; Apollo 14 ⇨1971; Apollo 16/17 ⇨1972}. The first stage lifted the second and third stage, along with the spacecraft, to a speed of about 5,400 mph (2,413 m/s). The second stage accelerated the spacecraft to a speed of about 14,000 mph (6,257 m/s), and the third stage to about 24,300 mph (10,861 m/s) in order to put the spacecraft into a Moon orbit.[2287] |
| 1958 | Brooklyn Polytechnic Institute, New York City, New York | Robert F. CHAIKEN,[2288] a U.S. physicist, **investigates the initiation of detonation in nitromethane** [$CH_3NO_2$]. **He illustrates the initiation process using a time-distance diagram,** which can be directly obtained using ultrahigh-speed streak photography. ▪ His graphical method was resumed to illustrate the unsteady initiation process in homogeneous and heterogeneous explosives, such as<br>▸ the compression of a high explosive by a shock wave;<br>▸ the initiation of a thermal explosion after an ***induction period***;<br>▸ the evolution of a ***super detonation*** running forward into the precompressed explosive; and<br>▸ the decay of a super detonation into a steady detonation wave {CAMPBELL, DAVIS & TRAVIS ⇨1961; SHEFFIELD, ENGELKE & ALCON ⇨1989}. |
| 1958 | Dept. of Aeronautical Engineering, Princeton University, Princeton, New Jersey | Luigi Mario CROCCO,[2289] an Italian-born U.S. aeronautical engineer, **treats the problem of supersonic flow in a duct that is decelerated to subsonic velocity.**<br>(i) He shows that a series of planar shocks or ***lambda shocks*** are formed, which he names ***pseudo-shocks.*** They are generated as the result of an interaction of the incident shock wave with the wall boundary layer {FELDMANN & ACKERET ⇨1946; GRIFFITH ⇨1952}, which substantially retards the flow by viscous action.<br>(ii) CROCCO concludes: "Only for low supersonic velocities and thin boundary layers is a quasi-normal shock possible in a duct. Otherwise, a more complicated, non-one-dimensional pattern is produced, which, for sufficiently high Mach numbers, becomes multiple … As a result of the turbulence, the subsonic regions spread more and more into the supersonic region, until finally the latter disappears completely and the shock pattern terminates. The subsequent process is only one of adjustment of the subsonic velocity distribution, and does not produce any further appreciable increase in pressure. We shall call this process ***pseudo-shock.***" ▪ In practice, this effect is found in wind tunnel supersonic diffusers {⇨Fig. 4.14–P}, in the entrances to air-breathing engines, and near end-plates of shock tubes {MARK ⇨1958; ⇨Fig. 4.14–H}.<br>In 1970, Teiichi TAMAKI[2290] and collaborators at the Tokyo Institute of Technology (TIT) studied pseudo-shocks in a slightly divergent rectangular duct and gave a more specific explanation: "The flow separates from the wall with an adverse pressure gradient and forms a low speed layer near the wall. This layer interacts with the shock wave in the main flow and makes a series of shock waves (*upstream shock region*). The main flow is decelerated by these shock |

---

[2287] See *The new Encyclopædia Britannica, Micropædia*. Benton & Hemingway, Chicago *etc.*, vol. 8 (1974), p. 917.
[2288] R.F. CHAIKEN: *The kinetic theory of detonation of high explosives*. M.S. thesis, Brooklyn Polytechnic Institute, New York (June 1958).
[2289] L. CROCCO: *Shock waves and pseudo-shocks in ducts.* In: *High speed aerodynamics and jet propulsion*. Princeton University Press, Princeton, NJ; vol. III (1958): (H.W. EMMONS, ed.) *Fundamentals of gas dynamics*, pp. 110-130.
[2290] T. TAMAKI, Y. TOMITA, and R. YAMANE: *A study of pseudo-shock.* Bull. Jap. Soc. Mech. Eng. (JSME) **13**, 51-58 (1970).

| | | waves to sonic speed. After that, the mixing between the main flow and the low speed layer near the wall changes the complicated velocity profile to a smooth one (*downstream mixing region*)." |
|---|---|---|
| 1958 | *King's College, Newcastle upon Tyne, England* | James W. DUNGEY,[2291] a British lecturer in mathematics, **publishes a booklet entitled *Cosmic Electrodynamics*** in which he discusses a number of astrophysical problems from a theoretical point of view, focusing on solar phenomena in particular. In a section entitled *Shock Waves*, he reviews some of the few papers published on the influence of a magnetic field on a shock wave in a conducting fluid, and calculates the width of a shock in a stellar atmosphere. He also describes a process of "magnetic reconnection" which is expected to increase the flow of energy from the solar wind to the magnetosphere.

In the same year, his countryman A. Nicolai HERLOFSON[2292] extends the theory to a perfectly conducting compressible fluid, thereby analyzing three modes of propagation, some of them showing strong anisotropy. ▪ The theory of magnetohydrodynamic waves in a conducting incompressible liquid was studied in 1950 by the Swedish astrophysicist Hannes O.G. ALFVÉN,[2293] who applied it to the theory of sunspots. Ever since ALFVÉN had proposed an electromagnetic theory for the origin of the planets in 1954, it had generally been accepted that electromagnetic fields play an essential role in nearly all transient features of the Sun, and in stellar evolution.

The rapid progress in rarefied gas dynamics and magnetohydrodynamics since the 1950s further stimulated the evolution of ***cosmic gas dynamics*** {Symp. on the Motion of Gaseous Masses of Cosmical Dimensions ⇨1949}. |
| 1958 | *MIT, Cambridge, Massachusetts* | Harold E. EDGERTON,[2294] a U.S. professor of electrical engineering and renowned high-speed photographer, **describes a shadowgraph technique for photographing shock waves of large objects in daylight** {⇨Fig. 4.18–D}. Using a small-sized light source (such as a micro xenon flash lamp or a constrained air spark gap[2295]) and a retroreflective screen – such as 3M Scotchlite reflective foil, which has the remarkable ability to reflect the light from a source back to the source – he visualizes large-scale shock waves emerging from firecrackers or dynamite caps. ▪ EDGERTON's 1958 retroreflective shadowgraph method was further improved at the Gas Dynamics Laboratory, Penn State University, and has been used in modern full-scale cinematographic studies of explosions and gunshots that use video cameras.[2296] **Retroreflective screens also proved very useful for producing large color schlieren images** {SETTLES ⇨1983}. |
| 1958 | *Siemens & Halske AG, Berlin-Siemensstadt, Germany* | Karl-Heinz HERRMANN,[2297] a German industrial physicist, **measures the spatial density profile of shock waves propagating in a gas,** such as in pure argon, carbon dioxide and air {⇨Figs. 4.5–H & 4.18–D}. Using flash soft radiography he improves the radiographic contrast by applying the intense characteristic radiation (Cu-$K\alpha$) emitted from the conical copper anode of his vacuum-discharge-type flash X-ray tube. |

---

[2291] J.W. DUNGEY: *Cosmic electrodynamics*. Cambridge University Press (1958).

[2292] A.N. HERLOFSON: *Magneto-hydrodynamic waves in a compressible fluid conductor*. Nature **165**, 1020-1021 (1950).

[2293] H. ALFVÉN: *Cosmical electrodynamics*. Clarendon Press, Oxford (1950), chap. 4.

[2294] H.E. EDGERTON: *Shockwave photography of large subjects in daylight*. Rev. Scient. Instrum. **29**, No. 2, 171-172 (1958); *Electronic flash, strobe*. MIT Press, Cambridge, MA (1979), pp. 229-232, 343-344.

[2295] H.E. EDGERTON: *Small area flash lamps*. In: (W.G. HYZER and W.G. CHACE, eds.) *Proc. 9th Int. Congress on High-Speed Photography* [Denver, CO, Aug. 1970]. SMPTE, New York (1970), pp. 238-243.

[2296] G.S. SETTLES ET AL.: *Full-scale high-speed "Edgerton" retroreflective shadowgraphy of explosions and gunshots*. In: *CD Proceedings of PSFVIP-5 – 5th Pacific Symposium on Flow Visualization and Image Processing* [Great Barrier Reef, Australia, Sept. 2005], paper PSFVIP-5-251.

[2297] K.H. HERRMANN: *Röntgenblitzuntersuchungen an Funkenstoßwellen in Gasen*. Z. Angew. Phys. **10**, 349-356 (1958).

| | | |
|---|---|---|
| 1958 | *Detonation Physics Group, U.S. Naval Ordnance Test Station (NOTS), China Lake, California* | **Edward W. LaRocca**[2298] **and John Pearson,** a British-born U.S. research engineer, **report on the use of shock pressures of around a million psi** (68.9 kbar) **to compress powders to almost solid density.**<br><br>(i) They apply an *explosively activated double-piston press* – an indirect method consisting of two plungers facing each other, with the target powder in-between and cylindrical explosive discs positioned on the outsides of the plungers. This simple device for powder compaction shows excellent ability to produce high-density parts in a variety of flat configurations.<br><br>(ii) Because of the high strain rates and short times involved, they believe that sample heating is extremely localized, and that the integrated heating effect may be quite small. They note that if the plates are misaligned so their faces are not parallel, then the plates became explosively welded together. |
| 1958 | *NACA Flight Propulsion Research Laboratories, Cleveland, Ohio* | **Hans Mark,**[2299] a German-born U.S. physicist, **studies the head-on reflection of a planar shock in a shock tube.** He observes that the reflected shock interacts with the boundary layer which the flow, following the incident shock, has generated at the tube side walls. This leads to a bifurcation of the reflected shock in the vicinity of the boundary layer {⇨Fig. 4.14–H} – the so-called **"λ foot"** {Feldmann & Ackeret ⇨1946}. |
| 1958 | *Dept. of Physics, Astronomy and Astrophysics, University of Chicago, Illinois* | **Eugene N. Parker,**[2300] a U.S. astrophysicist, **first mathematically predicts that an ionized magnetized rarefied gas that fills the heliosphere continuously and supersonically emanates from the Sun in all directions.** He calls this the *solar wind.* Flowing away from the topmost layers of the Sun's corona, the solar wind is accelerated in the vicinity of the Earth's orbit to surprisingly high velocities of between 400 and 800 km/s. Four years later, his predictions were confirmed experimentally {Neugebauer & Snyder ⇨1962}. Already in 1951, "solar corpuscular radiation" had been postulated, based on an analysis of comet tail plasma deflection {Biermann ⇨1951}. In 1955, Leverett Davis Jr.,[2301] a theoretical physicist at CalTech, suggested the existence of a "heliosphere" – the region of space surrounding the Sun formed by the solar wind pushing back the interstellar medium.<br><br>In 1963, Parker[2302] devised his *solar wind theory.* It assumes that the solar wind – a steady stream of charged gas particles (mainly electrons and protons) that is continuously ejected from the solar corona in all directions and flows out into space – expands and accelerates, thereby reaching supersonic flow velocities (in relation to the speed of sound in the interplanetary medium) until it approaches an asymptotic velocity at some great distance from the Sun. In 1989, Parker received the U.S. National Medal of Science "for his fundamental studies of plasmas, magnetic fields, and energetic particles on all astrophysical scales, for his development of the concept of solar and stellar winds; and for his studies on the effects of magnetic fields on the solar atmosphere." |
| 1958 | *Cologne, Germany* | At the 4th International Congress on High-Speed Photography, **Ian Stewart Pearsall,**[2303] a British fluid dynamicist at the Mechanical Engineering Research Laboratory (Glasgow, Scotland), **reports on the use of high-speed photography in hydraulic research, particularly on the experimental difficulties involved in studying the short lives of cavitation bubbles.**<br><br>(i) The high collapse velocity involved requires the use of fast microcinematography with very short exposure times to minimize motion blur. Furthermore, cavitation bubbles are usually |

---

[2298] E.W. LaRocca and J. Pearson: *Explosive press for use in impulsive loading studies.* Rev. Scient. Instrum. **29**, 848-851 (1958).

[2299] H. Mark: *The interaction of a reflected shock wave with the boundary layer in a shock tube.* Rept. NACA-TM 1418 (1958).

[2300] E.N. Parker: *Dynamics of the interplanetary gas and magnetic fields.* Astrophys. J. **128**, 664-676 (1958).

[2301] L. Davis Jr.: *Interplanetary magnetic fields and cosmic rays.* Phys. Rev. **100** [II], 1440-1444 (1955).

[2302] E.N. Parker: *Interplanetary dynamical processes.* Wiley Interscience, New York (1963); *The solar wind.* Scient. Am. **210**, 66-76, 156 (April 1964).

[2303] I.S. Pearsall: *High-speed photography in hydraulic research.* Proc. 4th Int. Congress on High-Speed Photography [Cologne, Germany, Sept. 1958]. In: (O. Helwich and H. Schardin, eds.) *Kurzzeitphotographie.* Helwich, Darmstadt (1959), pp. 247-251.

very small, so considerable magnification is required as well as the greatest definition possible from the film, developer and camera.

(ii) In the subsequent discussion, Hubert SCHARDIN, a German professor of applied physics, asked PEARSALL whether he had also noticed the expansion of shock waves in his work on collapsing cavitation bubbles. PEARSALL answered, "I would ask Prof. SCHARDIN, whether we would be likely to see shock waves, as actually the bubbles collapse in a very short time and although there was a great amount of work done on their collapse, I have never heard of anybody seeing shock waves, and I would like to ask Prof. SCHARDIN, whether he thinks we could photograph any shock waves." SCHARDIN replied, "In other, similar cases, we have seen shock waves very often; for example, if you have bubbles, ordered air bubbles, and then the pressure is increased by, say, a projectile passing through the water (shock waves are not produced by a projectile with a velocity of less than the speed of sound), the pressure increase causes these air bubbles to collapse and give very, very intense shock waves. And the same must hold in the case of a water bubble during cavitation, but it is possible that the shock wave is not strong enough that we can see the shock wave. You certainly need a very good schlieren system to make these shock waves visible."

In fact, the first experimental evidence of a pressure pulse originating from strong bubble compression had already been observed by M. HARRISON in 1952 and W. GÜTH in 1954 {Lord RAYLEIGH ⇨ 1917}. But detailed investigations were not possible until the advent of enhanced diagnostics and the possibility of producing single bubbles of a known size at a fixed location by laser pulse focusing {OHL ET AL. ⇨ 1995}.

| | | |
|---|---|---|
| 1958 | *Los Alamos Scientific Laboratory (LASL), Los Alamos, New Mexico* | **Melvin H. RICE,**[2304] **Robert G. MCQUEEN, and John M. WALSH,** three U.S. shock wave physicists, **review current shock production methods and measurement techniques for determining pressure-compression states behind shock waves.** They summarize published data for solids and also extend experimental ($p,v,e$)-data used to describe thermodynamic states that neighbor experimental Hugoniot curves ($p$ = pressure, $v$ = specific volume, $e$ = specific internal energy). In order to transform experimental velocities to pressure-compression points, they use Hugoniot data from a well-studied reference material (24ST aluminum alloy), which permits an additional determination of shock wave particle velocity independent of the free-surface velocity approximation – the so-called **"impedance match method."** |
| 1958 | *Bell Telephone Laboratories, Murray Hill, New Jersey* | **Arthur L. SCHAWLOW**[2305] **and Charles H. TOWNES,** two U.S. physicists, **publish the idea that the maser principle could be extended to infrared and optical wavelengths.** They also calculate the fundamental limit for the linewidth of an optical maser: the so-called "Schawlow-Townes formula." ▪ The term *maser,* an acronym for "microwave amplification by stimulated emission of radiation," was coined in 1954 by TOWNES and associates, who also developed the first maser.[2306]<br><br>The first operational *laser,* an acronym for "light amplification by stimulated emission of radiation" {GOULD ⇨ 1957}, was constructed by Theodore H. MAIMAN {⇨ 1960}.[2307] |

---

[2304] M.H. RICE, R.G. MCQUEEN, and J.M. WALSH: *Compression of solids by strong shock waves.* In: (F. SEITZ and D. TURNBULL, eds.) *Solid state physics. Advances in research and applications.* Academic Press, New York & London (1958), pp. 1-63.

[2305] A.L. SCHAWLOW and C.H. TOWNES: *Infrared and optical masers.* Phys. Rev. **112** [II], 1940-1949 (1958).

[2306] W.T. SILFVAST: *Lasers.* In: (R.A. MEYERS, ed.): *Physical science and technology.* Academic Press, San Diego *etc.* (1987); vol. 8, chap. *Laser history,* p. 480.

[2307] M. BERTOLOTTI: *Masers and lasers: an historical approach.* A. Hilger, Bristol (1983, 1987).

| | | |
|---|---|---|
| 1958 | University of California, Los Alamos Scientific Laboratory (LASL), Los Alamos, New Mexico | Garry L. SCHOTT[2308] and Jimmy L. KINSEY, two U.S. combustion researchers using the shock-tube technique, **measure the formation of the hydroxyl radical [–OH] in the shock-wave-induced combustion of a $H_2/O_2$, mixture highly diluted in argon, by recording the absorption of ultraviolet OH line radiation oscillographically.**
(i) They demonstrate that the course of an exothermic reaction can be readily resolved by what is now the "standard shock tube" method:
▸ diluting the reaction mixture with a large excess of inert gas (nearly always argon);
▸ "overdriving" the test gas mixture; *i.e.*, supplying much more energy via the shock wave than is generated by the combustion reaction;
▸ and then probing the course of the reaction (for mixtures of varying composition and shock strengths covering wide temperature ranges), which is forced (by the dilution in inert gas) to expand spatially by an amount sufficient to permit the reaction profiles to be recorded.[2309]
(ii) **They measure ignition delay times of $H_2/O_2/Ar$ behind incident shock waves over a wide range of temperatures and $[H_2]:[O_2]$ concentration ratios,** and observe that the ignition is preceded by a short induction time ranging from a few microseconds to a few hundred microseconds during which chain branching occurs. ▪ Their landmark paper includes all of the basic experimental features of present-day shock tube studies of combustion; only the lasers and the computers that characterize modern shock tube studies of combustion are missing. An avalanche of investigations of a similar nature followed, and their technique became the standard shock tube method.
Reactions of the hydroxyl radical play a central role in combustion chemistry, atmospheric chemistry and other reactive environments. The hydroxyl radical is commonly found in space: it was first detected in interstellar matter in 1963, and in the inner magnetosphere of Saturn in 1993. |
| 1958 | University of Freiburg, Germany | At the Cranz Centenary Colloquium, **Hans G. SNAY,**[2310] a German-born U.S. shock physicist, **reports on model underwater explosions of a 0.2-g lead azide model charge in a vacuum tank.**
(i) He shows slides of the interesting behavior of the bubble at the minimum and comments: "At the moment of maximum [diameter], the bubble is an almost perfect sphere. But when the bubble contracts to the minimum, the sphere is distorted. The straight bottom seen on the second frame is due to the fact that the bubble surface is inverted here [*i.e.*, the bubble develops a dimple at the bottom]. The actual bubble surface is faintly visible in the interior of the bubble. **A short instant later, the two interfaces impinge on each other and the bubble becomes a torus** … The inversion of the bubble produces an impinging of the lower and upper bubble surfaces and causes a water-hammer effect which dissipates energy." The bubble energy consists of internal energy, kinetic energy, and potential energy.
(ii) He also discusses the difficulties involved with satisfying the scaling requirements of shock wave and bubble phenomena in an underwater explosion.
In 1966, the feasibility of a vortex ring combined with a kidney-shaped bubble generated by an underwater explosion was examined theoretically in the United States by the applied mathematician Maurice HOLT[2311] and the chemical engineer Raymond H. HEISKELL, who com |

---

[2308] G.L. SCHOTT and J.L. KINSEY: *Kinetic studies of hydroxyl radicals in shock waves. II. Induction times in the hydrogen-oxygen reaction.* J. Chem. Phys. **29**, 1177-1182 (1958).
[2309] From W.C. GARDINER's paper: *Shock tube studies of combustion chemistry, see* chap. 6.6.
[2310] H.G. SNAY: *Scaling problems in underwater ballistics.* In: (W.C. NELSON, ed.) *Selected topics on ballistics.* Cranz Centenary Colloquium [Freiburg, Germany, April 1958]. Pergamon Press, London *etc.* (1959), pp. 262-280.
[2311] M. HOLT and R.H. HEISKELL: *Vortex motion as related to migrated steam bubbles from underwater nuclear explosions.* U.S. Naval Radiol. Defense Lab. Curr. Rept. (1966).

| | | |
|---|---|---|
| | | pared the energy of a single pulsating bubble with that of a bubble combined with a vortex ring. Their theory revealed that the kinetic energy of the spherical vortex exceeded that of the pulsating bubble by up to 64%, which was in qualitative agreement with observation. |
| 1958 | *Institute of Mathematical Sciences, New York University, New York City, New York* | **Gerald B. WHITHAM,**[2312] a British professor of applied mathematics, **presents a simple rule for predicting the motion of a shock wave through regions of nonuniform flow** which change the speed and strength of the shock. He demonstrates the simplicity of his rule – so-called **"Whitham rule"** – by applying it to numerous practical examples:<br>▸ the propagation of a shock wave in a tube of nonuniform area;<br>▸ the interaction of an oblique shock wave with a shear layer in steady supersonic flow;<br>▸ the propagation of a shock wave through a stratified layer;<br>▸ the convergence of cylindrical and spherical shock waves; and<br>▸ the propagation of a bore arriving at the shore line of a sloping beach.<br>Two years later, his simple approximate formula for the bore height was confirmed by a numerical method which allowed the determination of the bore height and position, and the flow behind the bore.[2313] |
| 1958 | *Cornell Aeronautical Laboratory (CAL), Buffalo, New York* | **Merle R. WILSON,**[2314] a U.S. mechnical engineer, **and his collaborator Richard J. HIEMENZ report on a new high-speed multiple-spark light source suitable for use in photographic systems of shock tunnels, shock tubes and ballistic ranges.**<br>(i) Five spark assemblies are mounted in line with a lens system to provide a common point where all spark flashes are brought to focus; thus the new light source is applicable to any optical technique that requires a point light source, such as shadowgraphy {DVOŘÁK ⇨1880} and schlieren photography {A. TOEPLER ⇨1864}.<br>(ii) The duration of each spark flash is 100 ns; the delay between consecutive sparks can be as low as 100 ns.<br>(iii) They report that they have successfully synchronized their multiple-spark light source with a high-speed rotating drum camera.<br>Compared to the Cranz-Schardin multiple-spark camera {CRANZ ⇨1929}, their system is free of any detrimental parallax effects. However, a noticeable decrease in light intensity occurs between successive sparks due to losses in the lens system, which limits the applicability of their method to a relatively small number of sparks. |
| 1958 | *Institute of Chemical Physics, U.S.S.R. Academy of Sciences, Moscow, Soviet Union* | **Yakov B. ZEL'DOVICH**[2315] **and collaborators carry out the first temperature measurements in a shock-compressed solid.** For transparent bodies, one can measure the temperature directly from the brightness of the radiation emitted under shock compression. After a strong compression, where the temperature reaches a few thousand degrees as a result of shifting of electronic levels and electron excitation, a material which was originally transparent becomes nontransparent and strongly radiative. In polymethyl methacrylate or PMMA [$(C_5H_8O_2)_n$; tradenames are Plexiglas, Lucite, Perspex, *etc.*], shock-compressed from 1.18 g/cm$^3$ (initial density) to 3.15 g/cm$^3$, they photographically measure the light emission from the shock front in the blue (4,020 Å) and red (6,000 Å) regions of the spectrum, from which they estimate a brightness temperature in the red region of about 8,300 ± 500 K. |

---

[2312] G.B. WHITHAM: *On the propagation of shock waves through regions of non-uniform area or flow.* J. Fluid Mech. **4**, 337-360 (1958).
[2313] H.B. KELLER, D.A. LEVINE, and G.B. WHITHAM: *Motion of a bore over a sloping beach.* J. Fluid Mech. **7**, 302-316 (1960).
[2314] M.R. WILSON and R.J. HIEMENZ: *High-speed multiple-spark light source.* Rev. Scient. Instrum. **29**, 949-951 (1958).
[2315] Y.B. ZEL'DOVICH, S.B. KORMER, M.V. SINITSYN, and A.I. KURYAPIN: *Temperature and specific heat of Plexiglas under shock wave compression.* Sov. Phys. Dokl. **3**, 938-939 (1958).

| | | |
|---|---|---|
| 1958 | Deutsche Versuchsanstalt für Luftfahrt (DVL), Aachen, Germany | Jürgen ZIEREP,[2316] a German gas dynamicist at DVL's Institut für Theoretische Gasdynamik, **presents a new method for determining the geometry of a compression shock forming at a curved planar surface in a stationary flow.** This subject is important for local supersonic flow fields, such as aircraft structures in the transonic regime. For convex profiles, his theoretical results are in agreement with previous measurements obtained at ETH Zürich by Jakob ACKERET,[2317] Fritz FELDMANN, and Nicholas ROTT. |
| 1959 | Freiburg, Germany | In March, **foundation of the Ernst-Mach-Institut (EMI),** an institute of the Fraunhofer-Gesellschaft (FhG, "Fraunhofer Society") dedicated to applied research. Its headquarters are in Munich. The first director is Hubert SCHARDIN. ▪ In 1979, the title *Institut für Kurzzeitdynamik* ("Institute for High-Speed Dynamics") was incorporated into the name of the Ernst-Mach-Institut. a historical perspective of shock wave research carried out at the EMI up to 1990 was given by Dr. Heinz REICHENBACH,[2318] who was the director of the EMI in the period 1972–1990. |
| 1959 | U.S. Air Force Cambridge Research Center, Boston, Massachusetts | In April, **first Conference on the Exploding Wire Phenomenon,**[2319] the general chairmen are William G. CHACE and Howard K. MOORE. ▪ At that time, exploding wires were considered to be a very promising tool in applied physics, for example<br>▸ by spectroscopists to achieve extremely high temperatures;<br>▸ by high-speed photographers to provide flash light sources of high intensity and ultrashort duration;<br>▸ by shock physicists to produce mini shock fronts of arbitrary shape; and<br>▸ by thermonuclear laboratories to attempt to initiate fusion.<br>Two classification schemes for complex exploding wire phenomena, both involving four states of matter, were suggested:<br>▸ a first class for slow and fast wire explosions, and associated melting and ablation effects,[2320] and<br>▸ a second class for ultrafast explosions – *i.e.*, those where the rate of energy delivered to the wire is comparable to the time taken by a sonic wave to travel to the center of the wire and return.[2321]<br>After a further three conferences, all held at Boston, MA (1961, 1964 & 1967), interest from most scientists and engineers shifted increasingly towards pulse lasers, and exploding wires – with a few exceptions, such as in safety detonators {JOHNSTON ⇒1944} – are used only marginally in research and technology today. |
| 1959 | Dept. of Urology, First Municipal Clinic, Riga, Latvia | On July 5, **Viktor GOLDBERG,**[2322] a Latvian urologist and head physician of the clinic, **gives the first demonstration of the suitability of shock waves for disintegrating bladder calculus** *in vivo*.[2323] The phosphate stone of his first patient, a 58-year-old male, has a diameter of about 3 cm. All debris can be removed successfully, thus avoiding any subsequent surgical treatment. |

---

[2316] J. ZIEREP: *Der senkrechte Verdichtungsstoß am gekrümmten Profil*. ZAMP **IXb**, 764-776 (1958).

[2317] J. ACKERET, F. FELDMANN, and N. ROTT: *Untersuchungen an Verdichtungsstößen und Grenzschichten in schnell bewegten Gasen*. Mitteil. Inst. für Aerodyn. ETHZ, Nr. 10, Leemann, Zürich (1946).

[2318] H. REICHENBACH: *In the footsteps of Ernst MACH – a historical review of shock wave research at the Ernst-Mach-Institut*. Shock Waves **2**, 65-79 (1992).

[2319] W.G. CHACE and H.K. MOORE (eds.): *Exploding wires (I)*. Plenum Press, New York (1959).

[2320] W.G. CHACE and M.A. LEVINE: *Classification of wire explosions*. J. Appl. Phys. **31**, 1298 (1960).

[2321] F.H. WEBB ET AL.: *Submicrosecond wire exploding studies at electro-optical systems*. In: (W.G. CHACE and H.K. MOORE, eds.) *Exploding wires (I)*. Plenum Press, New York (1959), pp. 33-58.

[2322] V. GOLDBERG: *Zur Geschichte der Urologie: Eine neue Methode der Harnsteinzertrümmerung – elektrohydraulische Lithotripsie*. Der Urologe **19** [B], 23-27 (1979).

[2323] A. HUTTMANN: *Von der transurethralen elektrohydraulischen Lithotripsie zur extrakorporalen Stoßwellenzertrümmerung von Harnsteinen*. Der Urologe **28** [B], 220-225 (1988).

In the same year, on December 9, **GOLDBERG also successfully fragments ureteral calculus.** His method, based on the electrohydraulic effect {JUTKIN ⇨ 1950}, applies a high-voltage pulse discharge via a spark gap; its two electrodes are inserted via a cystoscope through the urethra and positioned at the surface of the bladder calculus. This technique is called "**electro-lithotripsy**" or, more precisely, "**transurethral shock lithotripsy**" {⇨Fig. 4.15–F}. ▪ His prototype apparatus, constructed at the Electrotechnical Works Riga by the engineer Leo W. ROSE according to instructions from Lev A. JUTKIN, was called an "**Electrohydraulic Lithotriptor (EHL)**" by urologists. First shown publicly in 1967, at the World Exposition in Montreal, it was commercialized as *Urat-1* by the Med-Export Company. The method was popularized by the urologists Olaf ALFTHAN in Finland and Hans J. REUTER in Germany.

Several drawbacks of JUTKIN's method, however, initiated the development of other methods of shock generation and shock coupling {EISENMENGER ⇨1959; HÄUSSLER & KIEFER ⇨1971; TAKAYAMA ET AL. ⇨1983}. A review of early lithotripsy up to the 1980s was given by Rainer M. ENGEL[2324] and Matthias A. REUTER.[2325]

| | | |
|---|---|---|
| 1959 | *Mt. Kilauea-Iki [ca. 30 km east of Mt. Kilauea volcano], southeastern Hawaii Island, Hawaii* | On November 14, **the volcano Mount Kilauea-Iki begins to erupt, producing spectacular lava fountains just over 500 m in height, creating a 120-m-deep lake of molten lava.**<br>So-called "**Hawaiian-type eruptions**" are the least violent explosive volcanic eruptions. They are characterized by a row of lava fountains from a crater or along a fissure. The lava bombs falling back to the ground are still fluid and flatten out on impact. This eruption type is common for Hawaiian volcanoes, but it is also the major type of eruptive behavior for almost all basaltic shield volcanoes around the world. |
| 1959 | *North American Aviation, Inc., Los Angeles, California* | **Beginning of a long series of test flights of the X-15, the first hypersonic rocket plane** – an outgrowth of a research program that NASA conducted with the U.S. Air Force, the U.S. Navy, and North American Aviation, Inc. ▪ The first hypersonic flight of the X-15 was performed two years later {X-15 ⇨1961; ⇨Fig. 4.20–I}.<br>The X-15 Program had a long 15-year lifetime.[2326]<br>▸ It used three piloted hypersonic rocket planes to fly as high as 67 miles (108 km) and as fast as almost Mach seven, thus becoming the world's fastest and highest-flying aircraft.<br>▸ Intended primarily as a hypersonic aerodynamic research tool, it also provided a wealth of information in many other areas too, including structures and materials, piloting problems, flight control system design and effectiveness, the interaction of aerodynamic and reaction control systems, guidance and navigation, and terminal area approach and landing behavior.<br>▸ The volumes of test data gleaned from the 199 X-15 missions helped shape the successful Mercury, Gemini, Apollo, and Space Shuttle human spaceflight programs. Two X-15s are displayed at the National Air and Space Museum (Washington, DC), and the National Museum of the USAF (near Dayton, OH).[2327, 2328] |

---

[2324] R.M. ENGEL: *De historia lithotomiae.* Curator of the William P. Didusch Museum of the American Urological Association (AUA), Baltimore, MD; http://www.urolog.nl/artsen/features/dehistoria.asp.
[2325] M.A. REUTER, R.M. ENGEL, and H.J. REUTER: *History of endoscopy.* Kohlhammer, Stuttgart (1999), p. 481.
[2326] J.V. BECKER: *The X-15 Program in retrospect* [3rd Eugen Sänger Memorial Lecture]. Presented at the First Annual Meeting, Deutsche Gesellschaft für Luft- und Raumfahrt. Bonn, Germany (Dec. 4–5, 1968); http://www.hq.nasa.gov/office/pao/History/x15lect/cover.html.
[2327] R.P. HALLION: *X-15: The perspective of history.* In: *Proc. X-15 First Flight 30th Anniversary Celebration* [NASA-ARC, Dryden Flight Research Facility, Edwards, CA, June 8, 1989]. NASA, Washington, DC (1991).
[2328] D.R. JENKINS: *Hypersonics before the shuttle: a concise history of the X-15 research.* Rept. NASA SP-2000-4518, Aerospace History No. 18 (2000).

| | | |
|---|---|---|
| 959 | *Applied Physics Laboratory, Johns Hopkins University, Baltimore, Maryland* | **Frederick BILLIG,** a U.S. mechanical engineer, **produces the first model engine based on supersonic combustion,** thereby coining the term *scramjet* for the new device – a condensation of *supersonic combustion ramjet* {LORIN ⇨1908}.[2329] To achieve quick combustion of the reaction products during the short time (< 100 μs) they pass through the combustion chamber (which only measures about a foot across), it is operated with triethyl-aluminum [$Al(C_2H_5)_3$], an exotic fuel which burns spontaneously on contact with air. ▪ The "air-breathing" hypersonic scramjet engine appears to be the most efficient engine cycle for hypersonic flight within the atmosphere; it can be operated with a variety of fuels and in a variety of configurations applicable to manned and unmanned vehicles {NASA-LaRC ⇨1998 & Hyper-X ⇨2004}.[2330] Although the potential high performance and general behavior of this engine were already well understood at that time, the first real scramjet flight was not achieved until 43 years later {HyShot ⇨2002}.

Scramjets could lead the way to a new generation of smaller and cheaper rockets. Conventional rockets must carry both fuel and oxidizer (such as liquid hydrogen and liquid oxygen) to burn, while scramjets only need to carry fuel – thus potentially halving a rocket's weight. However, the biggest problem with scramjet engines is that they only work at supersonic speeds, such speeds are needed to obtain the compression and oxygen densities required for good combustion. Therefore, conventional engines are needed to get the scramjet off the ground. |
| 1959 | *RAND Corporation Santa Monica, California* | **Harold L. BRODE,**[2331] a U.S. theoretical physicist, **reports on numerical calculations of the blast wave from the detonation of a spherical charge of TNT.**

(i) He determined pressures, densities, temperatures and velocities as functions of time and radius. His calculations show that – besides the blast wave or **main shock** – a **second shock** is originated as an imploding shock following inward rarefaction into the gases produced by the explosion.

(ii) Furthermore, a **series of subsequent minor shocks** are seen to appear in a similar manner, moving out in the negative phase behind the main shock until the energy in the gas is dissipated. Depending on their temperatures, these subsidiary shocks can overtake and join the main shock and so disappear. ▪ In 1954, **Hubert SCHARDIN,**[2332] a German physicist at the Laboratoire de Recherches Technique de Saint-Louis, France, **published schlieren pictures of the detonation of a small 1.4-g charge of lead azide in air which clearly show the formation of the first and second shock waves.**

A large number of experimental and numerical investigations on the shock propagation and reflection phenomena of detonating spherical blast waves in free air and in the vicinity of a solid boundary had been published since the late 1940s. This subject of great practical importance was reviewed recently by John M. DEWEY, a British-born Canadian physics professor at the University of Victoria (Victoria, BC, Canada).[2333] He compared BRODE's numerical results with blast measurements obtained from large-yield chemical explosions {CDRE Suffield ⇨1964}. |

---

[2329] A. NEWMAN: *Speed ahead of its time.* Johns Hopkins Magazine (Dec. 1988), pp. 27-54.
[2330] P.J. WALTRUP, G.Y. ANDERSON, and F.D. STULL: *Supersonic combustion ramjet (scramjet) engine development in the United States.* In: (D.K. HENNECKE, ed.) *Proc. 3rd Int. Symposium on Air Breathing Engines* [Munich, Germany, March 1976]. DGLR, Köln (1976), pp. 835-861.
[2331] H.L. BRODE: *Blast wave from a spherical charge.* Phys. Fluids **2**, 217-229 (1959).
[2332] H. SCHARDIN: *Measurement of spherical shock waves.* Comm. Pure Appl. Math. **VII**, 223-243 (1954).
[2333] J.M. DEWEY: *Expanding spherical shocks.* In: (G. BEN-DOR ET AL., eds.) *Handbook of shock waves.* Academic Press, San Diego etc. (2001), pp. 441-481.

| | | |
|---|---|---|
| 1959 | *Institute of Chemical Physics, U.S.S.R. Academy of Sciences, Moscow, Soviet Union* | **Yuri N. DENISOV**,[2334] **and Yakov K. TROSHIN,** two Soviet combustion researchers and students of Prof. Kirill I. SHCHELKIN, **first observe a multifront or "cellular" structure for detonation waves using foils covered with a thin layer of soot** {⇨Fig. 4.16–U}. ▪ Their experiments contradicted the model of a planar detonation wave, even in its simplest form {MIKHEL'SON ⇨1893; CHAPMAN ⇨1899; JOUGUET ⇨1904 & 1917; ⇨Fig. 4.16–S}. However, the first experimental evidence of irregularities at the detonation front had already been provided by the discovery of spinning detonation {CAMPBELL & WOODHEAD ⇨1926; ⇨Fig. 4.16–T}.

Their unusual method of recording using the simple soot method {ANTOLIK ⇨1874; E. MACH & WOSYKA ⇨1875} has stimulated many researchers of gaseous detonations to investigate the origins of periodic cell structures experimentally in more detail and increasingly also through numerical simulation since the 1990s. The cell pattern geometry is dependent on the chemical system, dilution, pressure level, and tube geometry.[2335] Recently, this diagnostic method even proved useful for also recording cell structures in dust-air detonations {⇨Fig. 4.16–U}.

In the following years, further experimental evidence of this curious phenomenon was also provided using other diagnostic methods, such as by interferometer,[2336] schlieren,[2337] and stroboscopic laser shadow {⇨Fig. 4.16–V} techniques. These studies demonstrated the existence of secondary waves propagating almost perpendicularly to the front, and a curved front structure with abrupt changes in slope at triple points where the secondary waves joined the main front. |
| 1959 | *Physics Dept., University of Göttingen, Germany* | **Wolfgang EISENMENGER,**[2338] a German applied physicist and acoustician, **describes an electromagnetic pulse generator for generating weak planar shock waves in liquids** {⇨Fig. 4.11–B}. The source consists of a flat solenoid and a metal membrane in close mechanical contact with the solenoid on one side and with a liquid contact on the other side. The solenoid is activated by the discharge of a pulse capacitor. His 5.3-cm-dia. source can generate shock pressures in water of up to 200 bar at a distance of 1 cm from the pump membrane. ▪ EISENMENGER's unique construction stimulated the Siemens AG (Erlangen, Germany) to combine such a shock wave generator with an acoustic lens for focusing purposes, and to use the system for **clinical extracorporeal shock wave lithotripsy** {CHAUSSY, EISENBERGER ET AL. ⇨1970}.[2339]

In the following year, John Brackett HERSEY,[2340] a U.S. physicist at the Woods Hole Oceanographic Institution (WHOI), and his collaborators, who were apparently not aware of EISENMENGER's shock generator, applied a very similar principle to the use of sonar in oceanography, and successfully realized a 40-cm-dia. source. |

---

[2334] Y. DENISOV and Y. TROSHIN: *Pulsating and spinning detonation of gaseous mixtures in tubes* [in Russ.]. Dokl. AN (SSSR) **125**, 110-113 (1959); *Structure of gaseous detonation in tubes.* Sov. Phys. Tech. Phys. **5**, No. 4, 419-431 (1960).

[2335] R.A. STREHLOW, R.E. MAURER, and R. RAJAN: *Transverse waves in detonation: I. Spacing in the hydrogen-oxygen system.* AIAA J. **7**, 323-328 (1969); R.A. STREHLOW and C.D. ENGEL: *Transverse waves in detonations. II. Structure and spacing in $H_2$–$O_2$, $C_2H_2$–$O_2$, and $CH_4$–$O_2$ systems.* Ibid. **7**, 492-496 (1969).

[2336] D.R. WHITE: *Turbulent structure in gaseous detonation.* Phys. Fluids **4**, 465-480 (1961).

[2337] R.I. SOLOUKHIN: *Multi-headed structure of gaseous detonation.* Combust. Flame **9**, 51-58 (1965).

[2338] W. EISENMENGER: *Elektromagnetische Erzeugung von ebenen Druckstößen in Flüssigkeiten.* In: (L. CREMER, ed.) *Proc. 3rd Int. Congress on Acoustics* [Stuttgart, Germany, 1959]. Elsevier, Amsterdam (1961), Part I, pp. 326-329; and Acustica **12**, 185-202 (1962).

[2339] H. REICHENBERGER and G. NASER: *Electromagnetic acoustic source for the extracorporeal generation of shock waves in lithotripsy.* Siemens F&E-Berichte (Berlin *etc.*) **15**, 187-194 (1986).

[2340] J.B. HERSEY, H.E. EDGERTON, S.O. RAYMOND, and G. HAYWARD: *Sonar uses in oceanography.* Proc. Fall Instrument-Automation Conference [New York, Sept. 1960]. In: *Conference Preprints of the Instrument Society of America (ISA).* ISA, Pittsburgh, PA (1960), pp. 21:60:1-9.

| | | |
|---|---|---|
| 959 | *Aerodynamic Laboratory, Polytechnic Institute of Brooklyn, New York* | Antonio FERRI,[2341] an Italian-born U.S. aeronautical engineer, **discusses conflicting characteristics observed in recent developments on hypersonic flow.**

(i) He states, "The phenomena to be investigated are much more complex and less amenable to simplified schemes of analysis and to experimental investigation than other fields of fluid dynamics, while at the same time much more precise detailed knowledge of the flow field is required in order to obtain the information necessary for practical applications. The complexity of hypersonic flow can be attributed to two main reasons: in very high speed flow fields of practical interest the fluid behaves as a non-ideal gas with variable properties; chemical-physical transformations can take place in limited regions of the flow; at the same time the flow to be investigated is highly nonlinear, rotational and, in important regions, of the transonic type."

(ii) Based on recent results obtained by large scientific groups and military organizations, he draws the following important conclusions:
- The results from theoretical simulations of flight conditions tend to indicate that, in the approximation within which the physical quantities can be determined, and for high flight Reynolds numbers, real gas effects are of relatively minor importance;
- Pressure distributions and forces on simple boundary conditions can already be obtained with simple approximations, which are often sufficient for engineering applications;
- The heat transfer on simple bodies can be obtained with sufficient approximation from existing theories for the laminar case, but cannot be obtained for turbulent cases; and
- Entropy gradients, which considerably influence the heat transfer, can be utilized to reduce aerodynamic heating at high speed. |
| 1959 | *University of Notre Dame, Indiana* | **Vincent Paul GODDARD**[2342] **and coworkers extend the low-speed** *smoke flow visualization method,* previously developed by the U.S. aeronautical engineer Frank N.M. BROWN, **to supersonic flow in a wind tunnel by introducing coherent smoke lines,** thus considerably enhancing preexisting high-speed optical methods and our understanding of complex supersonic flow phenomena.[2343] ▪ In fluid dynamics, the tracer method was first applied by Ludwig MACH,[2344] Ernst MACH's oldest son, in his wind tunnel in order to photograph the flow of air.

In the following years, GODDARD's technique was further developed in Canada at the Suffield Experimental Station in order to measure air velocities of blast waves generated by large-yield chemical explosions {CDRE Suffield ⇨ 1964}. |
| 1959 | *Aeronutronic Systems, Inc., Glendale, California* | **Harmon W. HUBBARD**[2345] **and Montgomery H. JOHNSON, two U.S. physicists, study the initiation of detonation.** By numerically integrating the one-dimensional hydrodynamic equations for several initial conditions, they establish a simple criterion for determining whether or not detonation will result from a given set of initial conditions. They find that that when an element of explosive has been subjected to a shock wave of appropriate strength, a time interval – referred to as a *time delay* or an *induction time* – ensues in which almost no chemical reaction takes place. The complete reaction then occurs in a very short time, and this reaction moves forward as a detonation wave. ▪ Their theory for the initiation of detonation was applied successfully to the process of initiation in nitromethane {CAMPBELL, DAVIS & TRAVIS ⇨ 1961}. |

---

[2341] A. FERRI: *A review of some recent developments in hypersonic flow.* Adv. Aeronaut. Sci. **1**, 723-770 (1959).
[2342] V.P. GODDARD, J.A. MCLAUGHLIN, and F.N.M. BROWN: *A visual supersonic flow by means of smoke lines.* J. Aerospace Sci. **26**, 761-762 (1959).
[2343] T.J. MUELLER: *Smoke visualization of subsonic and supersonic flows (The legacy of F.N.M. BROWN).* Rept. UNDAS TN-3412-1, University of Notre Dame, IN (1978).
[2344] L. MACH: *Sichtbarmachung von Luftstromlinien.* Z. Luftschiffahrt & Physik der Atmosphäre **15**, Heft 6, 129-139 (1896).
[2345] H.W. HUBBARD and M.H. JOHNSON: *Initiation of detonation.* J. Appl. Phys. **30**, 765-769 (1959).

| | | |
|---|---|---|
| 1959 | Evanston, Illinois | At the Third Biennial Gas Dynamics Symposium, **Theodore VON KÁRMÁN**,[2346] then Professor Emeritus at CalTech and chairman of AGARD, **gives an introductory lecture on the study of kinetic plasma flow phenomena.** He suggests that the general term *magnetofluidmechanics* should be used to encompass *magnetohydrodynamics* (in incompressible media) as well as two other terms suggested by him, *magnetogasdynamics* (in gaseous and compressible media) and *magnetoaerodynamics* (in air). ▪ The term *magnetohydrodynamic* (or *magneto-hydrodynamic*) had already been coined in the early 1940s by the Swedish plasma physicist Hannes O.G. ALFVÉN.[2347] |
| | | The three terms he proposed were indeed quickly adopted by the physics community. However, some encyclopedias consider *magnetohydrodynamics* to be the more general term, stating that it includes *magnetofluidmechanics* and *magnetogasdynamics* as well.[2348] |
| 1959 | Chair of Hydro-dynamics, Moscow State University Moscow, Soviet Union | **Leonid I. SEDOV**,[2349] a Soviet theoretical physicist and fluid dynamicist, **explains changes in the brightness of a variable star by changes in the temperature and radius of its photosphere, which also creates shock waves.** ▪ SEDOV's explanation refers to "pulsating variable stars." The different causes of light variation in variable stars provided the impetus to classify such stars into different categories.[2350] |
| | | In the same year, SEDOV[2351] **reviews previous attempts to describe the damping of shock waves in terms of the gas motion within the shock wave.** He presents a theory on one-dimensional unsteady gas motion within a shock wave. If the law of shock front motion is known – for example, found experimentally or on the basis of additional assumptions – his method allows one to determine the shock wave velocity as a function of the shock wave coordinates, the so-called **"asymptotic law of shock wave attenuation."** |
| 1959 | Cornell Aero-nautical Laboratory (CAL), Buffalo, New York | **Charles E. WITTLIFF**[2352] **and Merle R. WILSON,** two U.S. aeronautical engineers, **and Abraham HERTZBERG**, a U.S. aerodynamics professor and department head, **report that they successfully increased the testing time of their hypersonic wind tunnel by almost an order of magnitude by using a "tailored interface"** – a mode of operation where the reflected shock passes through the driver-driven interface without being reflected. ▪ Their **"shock tunnel"** was used to model hypersonic flight studies.[2353] |
| | | Shock tunnels operate at high Mach numbers (up to about 25) for time intervals of up to a few milliseconds by using air that has been heated and compressed in a shock tube. They are constructed to include
  ▸ a shock tube {VIEILLE ⇨ 1899; BLEAKNEY ⇨ 1946};
  ▸ a nozzle attached to the end of the driven section of the shock tube; and
  ▸ a diaphragm between the driven tube and the nozzle.
When the shock tube is fired and the shock generated reaches the end of the driven tube, the diaphragm at the nozzle entrance is ruptured. The shock is reflected from the end of the driven tube, and the heated and compressed air behind the reflected shock is available for the operation of the tunnel {HERTZBERG ⇨ 1951}.[2354] |

---

[2346] Th. VON KÁRMÁN: *Some comments on applications of magnetofluidmechanics.* In: (A.B. CAMBEL and J.B. FENN, eds.) *Dynamics of conducting gases.* In: *Proc. Third Biennial Gas Dynamics Symposium* [Evanston, IL, 1959]. Northwestern University Press, Evanston (1960), pp. ix-xi.
[2347] H. ALFVÉN: *On the cosmogony of the Solar System III.* Stockholms Observatoriums Annaler **14**, 9.1-9.29 (1942).
[2348] C. MORRIS (ed.): *Academic Press dictionary of science and technology.* Academic Press, San Diego etc. (1992), p. 1300.
[2349] L.I. SEDOV: *Similarity and dimensional methods in mechanics.* Infosearch Ltd., London (1959), pp. 305-353.
[2350] For the classification of variable stars *see* General Catalogue of Variable Stars ⇨ 1948.
[2351] L.I. SEDOV: *Similarity and dimensional methods in mechanics.* Academic Press, New York (1959), pp. 295-304.
[2352] C.E. WITTLIFF, M.R. WILSON, and A. HERTZBERG: *The tailored-interface hypersonic shock tunnel.* J. Aerospace Sci. **26**, 219-228 (1959).
[2353] A. HERTZBERG, C.E. WITTLIFF, and J.G. HALL: *Development of the shock tunnel and its application to hypersonic flight.* In: (F.R. RIDDELL, ed.) *ARS progress in astronautics and rocketry. Hypersonic flow research.* Academic Press, New York (1962), vol. 7, pp. 701-758.
[2354] A. POPE and K.L. GOIN: *High-speed wind tunnel testing.* Wiley, New York etc. (1965), pp. 453-455.

| 1960s | Dept. of Defense (DOD), Washington, DC | **A series of military Vela satellites, designed to spot clandestine nuclear detonations possibly carried out by the Soviet Union in outer space, is operated by the DOD in order to record the gamma-ray signatures of such events.** • The Advanced Vela series was to detect not only nuclear explosions in space but also in the atmosphere and carried improved detector packages, including optical nuclear flash instruments with sub-millisecond resolution – so-called **"bhangmeters."** They could determine the location of a nuclear explosion to within about 3,000 miles (4,800 km). • Atmospheric nuclear explosions produce a unique signature: a short and intense flash lasting around 1 ms, followed by a second much more prolonged and less intense emission of light taking a fraction of a second to several seconds to build up. This so-called "dual flash" phenomenon occurs because the surface of the early fireball is quickly overtaken by the expanding atmospheric shock wave. Acting as an optical shutter, it hides the small but extremely hot and bright early fireball behind an opaque ionized shock front which is comparatively quite dim. No natural phenomenon is known to produce this signature.

In the ten-year interval from July 1969 to April 1979, the four Vela satellites (5A, 5B, 6A & 6B; launched on July 20, 1965) recorded 73 gamma-ray bursts (GRBs), massive deep-space explosions. However, detailed studies revealed that the bursts, which typically lasted from a few seconds to a few thousand seconds, have nothing to do with military nuclear ambitions. Instead, in the most surprising discovery in the history of space-based astronomy, they found bursts of gamma rays coming from deep space.

The first GRB detection dates back to July 1967, however, because of security concerns, the Los Alamos scientists Ray W. KLEBESADEL,[2355] Ian B. STRONG and Roy A. OLSON did not publish their discovery until 1973. Their paper reported on 16 GRBs that had occurred between July 1969 and July 1972. They pointed out the lack of evidence for a connection between GRBs and supernovae, as proposed previously by COLGATE {⇨1967}, who had predicted that GRBs were caused by the emergence of a shock wave from a star during a supernova explosion. However, a quarter of a century later, observations of the low-energy afterglows of gamma-ray bursts provided evidence that at least some GRBs originated from a probably rare type of supernova.[2356]

GRBs – the most luminous explosive events in the Universe – release most of their energy as photons with energies of 30 keV to a few MeV. They are short nonthermal flashes with durations ranging between a few tens of and a few thousand seconds, and are caused by the ejection of ultra-relativistic matter from a powerful energy source and its subsequent collision with its surrounding environment. The kinetic energy of the explosive outflow from a long GRB is thought to be about ten times that of a supernova. A key question remains:[2357] why is much of this energy concentrated in a small mass ($10^{-5}$ solar masses) in a GRB, instead of the several solar masses associated with a supernova? |
|---|---|---|
| 1960 | Reggan Test Site, southwestern Algeria [then a protectorate of France] | On February 13, **France performs its first nuclear weapons test** (dubbed **"GERBOISE BLEUE"**) **in the Sahara desert**: a 60–70 kiloton atmospheric burst atop a 107-m-high tower. In April and December, two further tests (GERBOISE BLANCHE and GERBOISE ROUGE) are carried out, with the goal to develop a compact nuclear device appropriate for arming missiles. • The independence of Algeria in 1962 threatened further testing at the Algerian sites, and since then all French nuclear tests have been performed at the Mururoa (or Moruroa) and Fangatauta coral atolls in the Pacific. |

---

[2355] R.W. KLEBESADEL, I.B. STRONG, and R.A. OLSON: *Observations of gamma-ray bursts of cosmic origin.* Astrophys. J. **182**, L85-L88 (1973).

[2356] J. VAN PARADIJS, C. KOUVELIOTOU, and R.A.M.J. WIJERS: *Gamma-ray bursts afterglows.* Annu. Rev. Astron. Astrophys. **38**, 379-425 (2000); http://astronomy.sussex.ac.uk/~romer/Distant/annualreviews/2000/-annurev.astro.38.1.379.pdf.

[2357] A. MACFADYEN: *Long gamma-ray bursts.* Science **303**, No. 5654, 45-46 (2004).

| | | |
|---|---|---|
| 1960 | *Hughes Research Laboratories, Hughes Aircraft Company, Malibu, California* | On May 16, **Theodore H. MAIMAN,**[2358] a U.S. research physicist, **demonstrates the first working laser system – a *ruby laser*.** The laser pulse, which has a duration of about one millisecond, consists of a series of spikes, each lasting about one microsecond. However, the term *laser*, an acronym for "light amplification by stimulated emission of radiation," was not yet used in MAIMAN's classic paper. ▪ Laser radiation encompasses wavelengths ranging from 100 nm (UV-C) to 1 mm (IR-C). The principle behind the laser (as well as the name) had already been described previously {GOULD ⇒ 1957}.<br><br>Two years later, MAIMAN founded his own company – Korad Corporation at Santa Monica, CA – which was devoted to the research, development and manufacture of high-power pulsed lasers. Such lasers proved extremely useful as versatile tools for producing ultrashort shock wave pulses and for diagnosing shock waves. In 2000, SPIE published an article to honor the 40th anniversary of MAIMAN's epochal invention.[2359] |
| 1960 | *Valdivia [about 700 km south of Santiago], southern Chile* | On May 22, **the *Great Chilean Earthquake* (or Valdivia Earthquake) – the largest earthquake ever recorded – occurs off the coast of South America with a magnitude of 9.5;** its epicenter is located near Valdivia (number of casualties estimated between 2,200 and 6,000).[2360]<br><br>(i) A tsunami is generated along a fault that runs parallel to the coast about 20–25 km offshore for a distance of 1,000–1,200 km. ▪ An effective source length of 1,000 km and a rupture velocity of 3–4 km/s was inferred from the differential phase of the seismic surface waves recorded at distant stations.[2361]<br><br>(ii) The so-called **"Chilean Tsunami"** that propagates throughout the Pacific region results in severe damage not only to the Chilean coast, but also to remote coasts of Hawaii and Japan (*c*.5,700 casualties). ▪ Nobuo SHUTO, a Japanese professor of seismology at the Disaster Control Research Center (DCRC) of Tohoku University, Japan, numerically simulated the propagation and refraction of the Chilean Tsunami across the Pacific. His computer animation can be viewed via the Internet.[2362]<br><br>(iii) This disaster leads to a heightened interest in tsunamis and an increased recognition of our limited knowledge of these phenomena. In the same year, **at the meeting of the International Union of Geodesy and Geophysics (IUGG) in Helsinki the IUGG Tsunami Committee is established** by the International Association of Physical Oceanography (IAPO) and the International Association of Seismology and Physics of the Earth's Interior (IASPEI).<br><br>The Great Alaskan Earthquake {⇒1964} that happened only four years after the Great Chilean Earthquake also generated devastating tsunamis that affected many Pacific coasts, and so it again focused attention on the need for an efficient tsunami warning system in the Pacific {ITIC ⇒ 1965}. |
| 1960 | *Estes Park, Colorado* | In the period July 11–12, **a technical conference is held on the "Response of Metals to High Velocity Deformation,"** with the aim being to understand and digest the large body of data on effects obtained at high strain rates of shock-compressed metals. The general chairmen are the U.S. metallurgists Paul G. SHEWMON and Victor F. ZACKAY. The great interest in this subject is best illustrated by the fact that this conference is sponsored by the Physical Metallurgy Committee of the American Institute of Metals Division, the Mining and Metallurgical Society of |

---

[2358] T.H. MAIMAN: *Stimulated optical radiation in ruby*. Nature **187**, 493-494 (1960); *Optical and microwave-optical experiments in ruby*. Phys. Rev. Lett. **4**, 564-566 (1960).

[2359] G. FRIEDMAN: *Inventing the light fantastic: Ted MAIMAN and the world's first laser*. OE Reports (SPIE) No. 200 (Aug. 2000).

[2360] *Historic worldwide earthquakes* [ranging from the A.D. 1556 Shenshi Earthquake in China to the present time]. Compiled by the USGS Earthquake Magnitude Working Group; http://earthquake.usgs.gov/regional/world/historical.php.

[2361] F. PRESS, A. BEN-MENAHEM, and M.N. TOKSOZ: *Experimental determination of earthquake fault length and rupture velocity*. J. Geophys. Res. **66**, 3471-3485 (1961).

[2362] *Numerical models for 1960 Chilean Tsunami*. Data Control Research Center (DCRC), Tohoku University, Sendai; http://www.geophys.washington.edu/tsunami/general/historic/models_60.html.

America (MMSA), and the American Institute of Mining, Metallurgical, and Petroleum Engineers (AIME).

In the following decades, interest in the complex behavior of metals under conditions of intense and rapid dynamic loading increasingly extended to nonmetallic matter, stimulated by the need to develop efficient, lightweight spacecraft shields against micrometeoroids. Prominent examples include modern engineering polymers and fiber composites.

| 1960 | *Stockholm, Sweden* | On August 16, **the International Academy of Astronautics (IAA) is founded;** the first president is Theodore VON KÁRMÁN. The purposes of the IAA, as stated in the Academy's statutes, are |

▸ to foster the development of astronautics for peaceful purposes;

▸ to recognize individuals who have distinguished themselves in a branch of science or technology related to astronautics; and

▸ to provide a program through which the membership can contribute to international endeavors and cooperation in the advancement of aerospace science, in cooperation with national science or engineering academies.

Presently, approximately 65 countries are represented among the members. The journal *Acta Astronautica* {*Astronautica Acta* ⇒1955} is published monthly under the auspices of the IAA Publications Committee.

| 1960 | *Paris, France* | In October, **the 11th General Conference on Weights and Measures meets in Paris** – the birthplace of the metric system. A new international system of units is formulated, named ***Système International (SI),*** which adopts six base units: the meter (length); the kilogram (mass); the second (time); the ampere (electric current); the degree Kelvin (thermodynamic temperature); and the candela (light intensity). **Pressure, a derived SI unit, is defined in Newtons per square meter. In honor of PASCAL's pioneering work on pressure** {PASCAL ⇒1640s}, **1 N/m$^2$ is called a "Pascal;"** *i.e.,* **1 Pa** – a rather small unit. • In English-speaking countries, however, the unit "pound per square inch (psi)" is still in use in science and engineering, particularly in gas dynamics and aerodynamics (14.5038 psi = 10$^5$ Pa = 1 bar). |

| 1960 | *DTMB, Carderock, Maryland* | **The hypersonic wind tunnel at the David Taylor Model Basin (DTMB) becomes operational.** Used in the development and design of high-speed missiles, it also assists in the selection of materials for their construction. |

| 1960 | *Picatinny Arsenal, Dover, New Jersey* | **Publication of the first volume of the ten-volume** *Encyclopedia of Explosives and Related Items* (1960–1983) by the Large Caliber Weapons Systems Laboratory at Picatinny Arsenal. The editor is Basil T. FEDOROFF. |

| 1960 | *U.S. Geological Survey (USGS), Washington, DC* | Edward T.C. CHAO,[2363] Eugene M. SHOEMAKER and Beth M. MADSEN, three U.S. geologists, **report that they have identified** *coesite* {COES ⇒1953} – the high-pressure polymorph of quartz [SiO$_2$] known previously only as a synthetic compound – as being an abundant mineral in sheared Coconino sandstone at Meteor Crater {*c.*50,000 years ago; ⇒Fig. 4.1–B}. They conclude that this might have important connection with the recognition of meteorite impact craters in quartz-bearing geologic formations: **"The occurrence of coesite at Meteor Crater suggests that the presence of coesite may afford a criterion for the recognition of other impacts on the Earth and perhaps ultimately on the Moon and other planets."** • Their discovery confirmed previous assumptions on the origin of coesite made in the United States by the geologist Harvey H. NININGER and the chemistry professor Harold UREY {NININGER ⇒1956}. According to CHAO,[2364] however, the shock origin of coesite was proposed prior to |

---

[2363] E.C.T. CHAO, E.M. SHOEMAKER, and B.M. MADSEN: *First natural occurrence of coesite.* Science **132**, 220-222 (July 22, 1960).
[2364] E.C.T. CHAO: *Coesite.* In: (D.N. LAPEDES, ed.) *McGraw-Hill Encyclopedia of the geological sciences.* McGraw-Hill, New York *etc.* (1977), pp. 108-111.

the discovery of natural coesite by the U.S. chemistry professors Michael E. LIPSCHUTZ and Edward ANDERS {CARTER & KENNEDY ⇨1964}.

In the same year, **SHOEMAKER**[2365] **and CHAO, while performing geological studies in Germany in the Ries Basin** {⇨$c.$15 Ma ago; ⇨Fig. 4.1–C}, **discover traces of coesite** {COES ⇨1953} **in** *suevite* [Germ. *Schwabenstein*], a yellow-grayish fragmental rock with glassy inclusions – thus providing the first experimental evidence that the Ries Basin, a shallow, near-circular depression in northern Bavaria, is an impact crater rather than a volcanic crater, as previously supposed by some German geologists {WERNER ⇨1904; STUTZER ⇨1936}.
▪ Today the presence of coesite and stishovite {STISHOV & POPOVA ⇨1961; CHAO, FAHEY, LITTLER & MILTON ⇨1962} is established as being evidence of a meteoritic impact, and they are searched for when craters of unknown origin are examined – so-called **"shock metamorphism"** (or "impact metamorphism").

| | | |
|---|---|---|
| 1960 | *U.S. Naval Electronics Laboratory (NEL), Point Loma District, San Diego, California* | **Robert S. DIETZ,**[2366] a U.S. geologist, referring to shatter cones which had formed directly under the point of impact {BUCHER ⇨1933; BOON & ALBRITTON ⇨1938}, **suggests that ancient meteorite impact scars provide useful field criteria ("index fossils") for shock-wave fracturing in the geological past.** He coins the phrase *astrobleme* to describe impact structures created by extraterrestrial high-energy objects striking the Earth. He writes, "… bedrock shattering meteorite impacts are a process of some geologic consequence, so that one should find meteorite impact scars or 'astroblemes' in ancient formations if criteria can be developed for their recognition." |

In the following year, DIETZ[2367] explained this new term in more detail: "This newly coined word refers to ancient scars left in the Earth's crust by huge meteorites. The evidence for such impacts is largely the high-pressure mineral coesite and 'shatter cones' in the rocks … A few 'fossil' craters, scarcely discernible on the ground, have shown up in aerial photographs, appearing as faint circular features. Geological maps of surface and subterranean rock formations have revealed still other circular features, which geologists in the past have generally attributed to volcanic explosions. It now appears that many of these are the 'root' structures of ancient meteorite craters. **For those that prove to be obliterated craters made by a meteorite or the head of a comet I have proposed the term** *astrobleme*[2368] **from the Greek words for** *star* **and** *wound***.**" ▪ DIETZ used shatter cones to identify impact sites including the Ries and Steinheim Basins in Germany (1958) and the Vredefort Ring structure in South Africa (1961), a deformed and severely eroded impact crater. For example, the 10.5-km-dia. Ashanti Crater in Ghana, the approx. 300-km-dia. Vredefort Ring structure {⇨$c.$2 Ga}, and the oval 60×30-km² Sudbury Basin {⇨$c.$1.85 Ga} have been identified as most probable astroblemes resulting from meteoritic impacts.

| | | |
|---|---|---|
| 1960 | *Dust Explosion Research Section, U.S. Bureau of Mines, Pittsburgh, PA* | **Henry G. DORSETT JR.,**[2369] a U.S. physicist and section head, **and coworkers report on laboratory tests carried out on dried samples of fine dust in order to evaluate their explosivities.** |

(i) They compare the *explosion hazards* – classified as "moderate," "strong," and "severe" – of various dusty substances in terms of *ignition temperature of dust clouds* (in °C) and *minimum explosive concentration* (in oz./ft³).

---

[2365] E.M. SHOEMAKER and E.C.T. CHAO: *New evidence for the impact of the Ries Basin, Bavaria, Germany.* J. Geophys. Res. **66**, 3371-3378 (1961).
[2366] R.S. DIETZ: *Meteorite impact suggested by the orientation of shatter-cones at the Kentland, Indiana, disturbance.* Science **105**, 42-43 (Jan. 10, 1947); *Meteorite impact suggested by shatter cones in rock.* Ibid. **131**, 1781-1784 (June 17, 1960).
[2367] R.S. DIETZ: *Astroblemes.* Scient. Am. **205**, 50-58 (Aug. 1961).
[2368] Note that the prefix "astro" is used in a broad sense; it refers here not only to stars and star-like bodies (*i.e.*, to huge masses of glowing gases), but also to solid celestial bodies (planets, asteroids, comets, *etc.*), as used in the terms *astrobiology, astroblemes* and *astrogeology.*
[2369] H.G. DORSETT JR. ET AL.: *Laboratory equipment and test procedures for evaluating explosivity of dusts.* U.S. Bureau of Mines, Rept. of Investigation No. 5624 (1960).

(ii) Compared to strong hazards from coal dust (610 °C; 0.055 oz/ft$^3$ or 1.8 mg/cm$^3$), severe hazards exist for
- aluminum dust (650 °C; 0.055 oz/ft$^3$ or 1.8 mg/cm$^3$);
- magnesium dust (520 °C; 0.020 oz/ft$^3$ or 6.7 mg/cm$^3$);
- titanium dust (460 °C; 0.045 oz/ft$^3$ or 1.5 mg/cm$^3$); and
- uranium dust (20 °C; 0.060 oz/ft$^3$ or 2 mg/cm$^3$).

Besides uranium dust, zirconium and thorium dusts also have extremely low ignition temperatures.

| | | |
|---|---|---|
| 1960 | Soviet Union | E.A. FEOKTISTOVA[2370] **provides the first experimental evidence of** *steady Mach reflection in a detonating solid explosive* **with a cylindrical coaxial geometry** {⇨Fig. 4.13–K}.<br><br>In the same year, Lev V. AL'TSHULER[2371] and associates in the Soviet Union first measure bulk speeds of sound behind shock-compressed solids (Al, Fe, Cu and Pb) in the megabar range. They discover two sound wave propagation velocities: a lower velocity attributed to the plastic wave that characterizes the bulk compressibility, and a higher speed for the elastic longitudinal wave associated with one-dimensional compression. |
| 1960 | U.S. Naval Ordnance Laboratory (NOL), White Oak, Silver Spring, Maryland | William S. FILLER,[2372] a U.S. shock physicist, **performs measurements in a divergent, conical shock tube driven by a high explosive charge at the apex. This system provides a practical blast simulator for research purposes, and is termed the** *Filler tube.* His small-scale experiments indicate that a small charge of high explosive, when detonated while confined in an apex of angle $2\alpha$ at the end of a right circular tube, produces blast wave parameters characteristic of those produced by much larger weights of explosive detonated in free air, thereby achieving amplification by a factor of $1/\sin^2 \frac{1}{2}\alpha$. ▪ In the late 1950s, Otto LAPORTE[2373] and Alan Lee COLE at the University of Michigan, and R. Gordon CAMPBELL[2374] at the Rensselaer Polytechnic Institute had described conical or sector shock tubes driven by compressed gas that had been used to generate spherical or cylindrical shocks under laboratory conditions. Apparently, the idea of a conical shock tube originated from rumors spread in Europe of an underwater explosive gun {COOMBS & THORNHILL ⇨1957}.<br><br>Four years later, FILLER[2375] **also proposed a hydrodynamic conical shock tube for generating underwater shock waves** within a narrow conical of angle 7°. The shape and decay properties of the hydrodynamic shock wave as it progressed down the tube were very similar to those of a spherical shock wave generated in free water by a ¼-lb (113-g) sphere of TNT. |
| 1960 | Los Alamos Scientific Laboratory (LASL), Los Alamos, New Mexico | Clarence M. FOWLER,[2376] a LASL staff physicist, **and his collaborators perform the first highly dynamic** *magnetic flux compression* **experiments using high explosives.** Their set-up, a thin-walled metal cylinder imploded by a surrounding ring of charges of high explosive, compresses an initially strong magnetic field up to extremely strong magnetic flux densities in the 10–15 megagauss (1,000–1,500 Tesla) range. The magnetic pressure in a cylindrical liner of initial radius $R_0$, subjected to an initial magnetic field $B_0$, is given by<br><br>$$P_m(t) = B^2/2\mu_0 \text{ with } B = B_0 R_0^2/R_L(t)^2,$$<br><br>where $R_L(t)$ is the compressed liner radius at time $t$ and $\mu_0 = 4\pi \times 10^{-9}$ H/cm. Obviously, magnetic pressures of the order of megabars are developed using ultrahigh magnetic fields in the |

---

[2370] E.A. FEOKTISTOVA: *Experimental observation of Mach reflection of detonation waves in a solid explosive.* Sov. Phys. Dokl. **6**, 162-163 (1961).
[2371] L.V. AL'TSHULER, S.B. KORMER, M.I. BRAZHNIK, L.A. VLADIMIROV, M.P. SPERANSKAYA, and A.I. FUNTIKOV: *The isentropic compressibility of aluminum, copper, lead, and iron at high pressures.* Sov. Phys. JETP **11**, 766-775 (1960).
[2372] W.S. FILLER: *Measurements on the blast wave in a conical tube.* Phys. Fluids **3**, 444-448 (1960).
[2373] O. LAPORTE and A.L. COLE: *Construction of a sector shock tube.* University of Michigan, Engng. Res. Inst., Progr. Rept. No. 7 (1957).
[2374] R.G. CAMPBELL: *Initial wave phenomena in a weak spherical blast.* J. Appl. Phys. **29**, 55-60 (1958).
[2375] W.S. FILLER: *Propagation of shock waves in a hydrodynamic conical shock tube.* Phys. Fluids **7**, 664-667 (1964).
[2376] C.M. FOWLER, W.R.B. GARN, and R.S. CAIRD: *Production of very high magnetic fields by implosion.* J. Appl. Phys. **31**, 588-594 (1960).

range used by them. They find that the shaping and timing of the developing detonation front of the armature explosive which surrounds the liner is crucial for an efficient liner implosion.
▪ The method of flux compression, first proposed in the Soviet Union {SAKHAROV ⇨ 1951}, was later resumed in the United States[2377] and in the former Soviet Union[2378] in order to isentropically compress materials to several megabars.

Modern applications of magnetic flux compression (sometimes referred to as *magnetic cumulation*) also include **EMP (ElectroMagnetic Pulse) devices** – a generic term applied to any device that uses nuclear or conventional (chemical) explosives to generate a very intense but short electromagnetic field transient. In the literature, the term *E (Electromagnetic)-bomb* is used to describe both microwave-band and low-frequency weapons.[2379] The ***VIRCATOR (Virtual Cathode Oscillator)*** is a mechanically simple and robust one-shot device capable of producing a very powerful single pulse of radiation over a relatively broad band of microwave frequencies.[2380]

| | | |
|---|---|---|
| 1960 | UC Berkeley College of Engineering, Berkeley, California | Werner GOLDSMITH,[2381] a German-born U.S. professor of engineering mechanics and a pioneer in the field of collision mechanics, **publishes his monograph** *Impact: the Theory and Physical Behaviour of Colliding Solids*. Originated from a series of lectures given at UC Berkeley, it attempts to summarize the more important contributions to the subject of impact. |

(i) It is the first textbook to systematize the mechanics of collisions ranging from car crashes to refinery explosions, and it divides the subject into two separate phases:

▸ the dynamics of the process, encompassing **stereomechanics** (the classical rigid body theory employing a coefficient of restitution), vibrational aspects, contact phenomena, dynamic processes involving plastic strains; and

▸ information concerning material properties obtained by dynamic test methods, encompassing impact, tensile, compressive and torsion tests.

(ii) He writes, "During the last three decades, scientists and engineers have exhibited an increasing interest in the solution of problems concerned with the impact of solid bodies. Actual examples of this type of loading may be found in the fields of tool design, foundry and machine shop operations, protective ordnance, explosions, vehicle accidents, and in many other areas; however, the analysis of impact phenomena has been restricted to collisions involving only simple types of geometry. The paramount reason for this limitation is the severe mathematical complexity encountered in the theoretical development of the subject coupled with the relative ignorance of the behavior of materials under conditions of rapidly applied stress … The advent of high-speed calculators, which has alleviated the tedium of numerical computations, and the progress of modern methods of instrumentation, which has increased the scope and reliability of experimental data, are adding further impetus to the development of the field."

Prof. GOLDSMITH,[2382] an international authority on the mechanics of collision, also extended his studies to the biomechanics of head and neck trauma. He died on August 23, 2003, at the age of 79; his last paper, co-written with the American forensic pathologist John PLUN-

---

[2377] R.S. HAWKE, D.E. DUERRE, J.G. HUEBEL, H. KLAPPER, and D.J. STEINBERG: *Method of isentropically compressing materials to several megabars.* J. Appl. Phys. **43**, 2734-2741 (1972).

[2378] A.I. PAVLOVSKII, A.I. BYKOV, M.I. DOLOTENKO, N.I. EGOROV, and G.M. SPIROV: *Equation-of-state study using the method of isentropic compression by ultrahigh field pressure.* In: (V.M. TITOV and G.A. SHETSOV, eds.) *Megagauss fields and pulsed power systems.* Nova Science, New York (1990), pp. 155-161.

[2379] C. KOPP: *The E-bomb, a weapon of electrical mass destruction.* Dept. of Computer Science, Monash University, Clayton, Victoria, Australia; http://www.jya.com/ebomb.htm.

[2380] J. BENFORD and J. SWEGLE: *High-power microwaves.* Artech House, Boston, MA *etc.* (1992), pp. 307-337.

[2381] W. GOLDSMITH: *Impact: the theory and physical behaviour of colliding solids.* E. Arnold, London (1960); Dover Publ., Mineola, NY (2002).

[2382] W. GOLDSMITH and J. PLUNKETT: *A biomechanical analysis of the causes of traumatic brain injury in infants and children.* Am. For. Med. Path. **25**, No. 2, 89-100 (2004).

KETT, addressed the widely discussed ***shaken baby syndrome*** – a severe form of head injury caused by violently shaking an infant or child which may result in severe injuries to the infant, including permanent brain damage, and may even cause death.

| | | |
|---|---|---|
| 1960 | *Ballistic Research Laboratory (BRL), Aberdeen Proving Ground, Maryland* | George E. HAUVER,[2383] a U.S. experimenter, **exploits the piezoelectric nature of materials subjected to shock waves.** Using a thin disk of sulfur, he records the stress-time profile in a detonating charge of baratol (an explosive made of a mixture of TNT and barium nitrate, with a small quantity of wax used as a binder). At the same time, investigations are carried out independently in France by J. BERGER[2384] and collaborators at the Commissariat à l'Energie Atomique (CEA). They reveal that sulfur is useful only at stresses well above 70 kbar, because below this value the electrical resistivity is too large for practical use, and at about 70 kbar sulfur undergoes a dramatic decrease in resistivity.<br><br>Three years later, Ulf I. BERG,[2385] a scientist at the Research Institute of National Defense in Stockholm, reported on a sulfur transducer for measuring shock pressures within a narrow pressure region (100–180 kbar). He successfully recorded the arrival of relaxation and secondary shock waves in aluminum. |
| 1960 | *Bell Telephone Laboratories, Murray Hill, New Jersey* | Ali JAVAN,[2386] an Iranian-born U.S. physicist, **Donald R. HERRIOTT**, a U.S. engineer, **and William R. BENNETT**, a U.S. physicist, **jointly invent the first gas-discharge (helium-neon) laser** which produces – through stimulated emission – coherent, near-monochromatic light. • The term *laser* was not yet coined {GOULD ⇨ 1957}, and they termed their new light source an *optical maser*. The acronym *maser* stands for "microwave amplification by stimulated emission of radiation" {SCHAWLOW & TOWNES ⇨ 1958}. |
| 1960 | *Palomar Observatory, Pasadena, California* | Thomas A. MATTHEWS[2387] and Allan Rex SANDAGE, two U.S. astronomers, **find the first optical counterpart to a radio source that will later be identified as a** *quasar* {M. SCHMIDT ⇨ 1963}, a new class of astronomical objects. They identify a strange source of radio emission – later listed as "Quasar 3C-48" in the Third Cambridge ("3C") Catalogue – which emits more intense radio waves and ultraviolet radiation than an ordinary star. In visible light, this astronomical object looks like a faint star that exhibits a spectrum of broad spectral emission lines that cannot be identified. |
| 1960 | *Los Alamos Scientific Laboratory (LASL), Los Alamos, New Mexico* | Robert G. MCQUEEN[2388] and Stanley P. MARSH, two U.S. experimental shock wave physicists, **use an explosive-driven plate arrangement to generate very high shock pressures,** for example pressures as high as 2 Mbar in iron.<br>(i) Their technique – known as the **"flyer plate method"** (or "flat plate accelerator") {SHREFFLER & DEAL ⇨1953; ⇨Fig. 2.13; ⇨Fig. 4.11–C} – uses a flat projectile (a plastic cylinder tipped with a flat metal disk) and a plane-wave generator {J.H. COOK ⇨1948} which produces a planar detonation in a slab of explosive. Its diameter is several times larger than its thickness so that rarefactions arising from the sides do not seriously limit the area of planarity {G.I. TAYLOR ⇨1949}. |

---

[2383] G.E. HAUVER: *Pressure profiles in detonating solid explosive.* In: *Proc. 3rd Symposium on Detonation* [Princeton University, Princeton, NJ, Sept. 1960]. Office of Naval Research (ONR), White Oak, MD (1960), vol. 1, pp. 241-252.

[2384] J. BERGER, S. JOIGNEAU, and C. FAUQUIGNON: *Comportement du soufre sous l'action d une onde de choc.* In: *Les Ondes de Détonation. Colloques Internationaux du Centre National de la Recherche Scientifique (CNRS)* [Gif-sur-Yvette, France, Aug./Sept. 1961]. Edition N° 109 du CNRS, Paris (1962), pp. 353-361.

[2385] U.I. BERG: *Investigations on a very high pressure transducer.* Arkiv för Fysik **25**, No. 10, 111-122 (1963).

[2386] A. JAVAN, W.R. BENNETT JR., and D.R. HERRIOTT: *Population inversion and continuous optical maser oscillation in a gas discharge containing a He-Ne mixture.* Phys. Rev. Lett. **6**, 106-110 (1961).

[2387] T.A. MATTHEWS ET AL.: *First true radio star.* Sky & Telescope **21**, 148 (1961); T.A. MATTHEWS and A.R. SANDAGE: *Optical identification of 3c 48, 3c 196, and 3c 286 with stellar objects.* Astrophys. J. **138**, 30-56 (1963).

[2388] R.G. MCQUEEN and S.P. MARSH: *Equation of state for nineteen metallic elements from shock-wave measurements to two megabars.* J. Appl. Phys. **31**, 1253-1269 (1960).

    (ii) The flyer plate, which is initially in contact with the slab of explosive, is accelerated and, after passing a void, impacts the flat target specimen, with the face of the target being parallel to the face of the plate. Since the flyer plate accumulates momentum from the explosion products during its entire flight across the gap, it delivers to the target a higher pressure than the contact explosive alone {KORMER ET AL. ⇨1962}.

    This method of generating shock pressures by accelerating a flyer plate was modified by others; for example, the flyer plate has been driven by
- a conventional powder gun or a light-gas gun;[2389]
- an exploding foil {KELLER & PENNING ⇨1962; GUENTHER ET AL. ⇨1962; STEINBERG ET AL. ⇨1980};
- high explosives in special cascade hemispherical arrangements (KRUPNIKOV[2390] ET AL. 1963);
- intense 50-ns proton beams {BLUHM ET AL. ⇨1985};
- pulsed magnetic forces {FARBER ET AL. ⇨1969}; or
- an intense neutron pulse generated in an underground nuclear explosion {TRUNIN ET AL. ⇨1992; ⇨Fig. 4.11–F}.

| 1960 | Nobel Division of Imperial Chemical Industries Ltd., Stevenston, North Ayrshire, Scotland | **William T. MONTGOMERY[2391] and H. THOMAS publish a new explosive technique for compacting metal powder.** It does not use any plunger methods which require bulky anvils to avoid deformation and pressure reduction, but instead uses shock waves in a direct manner. The method uses a Cordex detonating fuse which is wrapped in a continuous spiral round the thin-walled metal container holding the powder. The container and charge are then placed in a suitable water bath and the explosive charge is detonated. ▪ This method was also successfully applied by Stanley W. POREMBKA[2392] and Charles C. SIMONS at the Batelle Institute (Columbus, OH) to explosively compact $UO_2$ fuel elements. |
| --- | --- | --- |
| 1960 | Los Alamos Scientific Laboratory (LASL), Los Alamos, New Mexico | **Robert D. RICHTMYER,[2393] a U.S. physicist, studies shock wave effects at an interface separating two fluids of different densities.** He theoretically predicts that such shock interactions may lead to instantaneous instabilities and a "mixing" of the two fluids, independent of the direction of the incident shock. ▪ Nine years later, RICHTMYER's theory was experimentally confirmed in the Soviet Union {MESHKOV ⇨1969}. |
| 1960 | Institute for Fluid Dynamics and Applied Mathematics, University of Maryland, College Park, Maryland | **Helmut D. WEYMANN,[2394] a German-born U.S. experimental physicist, reports on electric precursor signals in front of strong shock waves.** It was later recognized by others as being a detrimental effect associated with manned space vehicle reentry, due to the generation of "leading edge echoes." From electrostatic probe measurements, he concludes that electrons diffuse in the direction of the shock motion with velocities of up to several times the shock velocity.

At the same time, similar investigations are also carried out by Hans GRÖNIG[2395] at the Institut für Mechanik of the TH Aachen. Using a glow-discharge probe, he shows that in argon at medium Mach numbers ($M = 5-9$), the electric precursor signals are essentially due to an electron diffusion process. |

---

[2389] *Flat-plate accelerator.* NASA Johnson Space Center, Houston, TX; http://ares.jsc.nasa.gov/Education/websites/craters/fpa.htm.
[2390] K.K. KRUPNIKOV, A.A. BAKANOVA, M.I. BRAZNIK, and R.F. TRUNIN: *An investigation of the shock compressibility of titanium, molybdenum, tantalum, and iron.* Sov. Phys. Dokl. **8**, 205-208 (1963).
[2391] W.T. MONTGOMERY and H. THOMAS: *The compacting of metal powders by explosives.* Powder Metallurgy **6**, 125-128 (1960); J. TAYLOR, W.E. JOHNSTONE, and W.T. MONTGOMERY: *Compacting of metal powder.* Brit. Patent No. 833,673 (1960).
[2392] S.W. POREMBKA and C.C. SIMONS: *Explosively compacted $UO_2$.* The Dragon Project Fuel Element Symposium [Bournemouth, U.K., Jan. 1963].
[2393] R.D. RICHTMYER: *Taylor instability in shock acceleration of compressible fluids.* Comm. Pure Appl. Math. **13**, 297-319 (1960).
[2394] H.D. WEYMANN: *Electron diffusion ahead of shock waves in argon.* Phys. Fluids **3**, 545-548 (1960).
[2395] H. GRÖNIG: *Elektronenverteilung an starken Stoßwellen.* Ph.D. thesis, RWTH Aachen (1960).

| 1961 | Cosmodrome Baïkonour, Kazakhstan, Soviet Union | On April 12, **the first manned orbital flight is performed by the 27-year-old Soviet cosmonaut Major Yuri A. GAGARIN with the spacecraft Vostok 1** {⇨Fig. 4.20–J}. He orbits the Earth once in 1 hour 29 minutes at a maximum altitude of 301 km and a maximum speed of 28,968 km/h (8.047 km/s). **He is the first human to travel at hypersonic speeds.** His capsule achieves Mach 25 during reentry. ▪ The 4¾-ton Vostok 1 capsule, which consists of the pilot's spherical reentry cabin covered with ablative material and a cylindrical service module, was carried into orbit atop a modified R-7 missile. GAGARIN also proved that man could endure the rigors of lift-off, weightlessness and reentry, and yet still perform the manual operations essential to spacecraft flight. |
|---|---|---|
| 1961 | NASA, Cape Canaveral, Florida | On May 5, **Alan B. SHEPARD,** a 37-year old U.S. Navy lieutenant commander, **becomes the first American astronaut.** His spacecraft "Freedom 7" (Mercury-Redstone 3), launched by an enhanced version of the Redstone Ballistic Missile, enters space for a duration of 15 minutes and 28 seconds, but does not orbit the Earth.[2396] The main scientific objective of Project Mercury is to determine man's ability to cope in a space environment and in the other environments he encounters when traveling to and returning from space. ▪ His spacecraft was accelerated to a velocity of 5,134 mph (2,295 m/s). Upon reentry, SHEPARD experienced a force as high as eleven times the acceleration due to gravity. **The Mercury spacecraft, a "blunt body" design** {ALLEN ⇨1952}, **used a heat sink, but later versions used an ablative surface.** |
| 1961 | U.S. Congress, Washington, DC | On May 25, **U.S. President John F. KENNEDY,** speaking to U.S. Congress, **announces the NASA Project Apollo,** which is aimed at: "achieving the goal, before the decade is out, of landing a man on the Moon and returning him safely to Earth." ▪ This program stimulated hypersonic research tremendously in many respects, such as how to control, stabilize and test space vehicles during reentry. The ambitious project reached its climax with the first moonwalk {Apollo 11 ⇨1969}. |
| 1961 | University of Hawaii, Honolulu, Hawaii | In the period from August 21 to September 6, at the Tsunami Meeting (which is associated with the 10th Pacific Science Congress) **the term *tsunami* is adopted for general use** – instead of the term *seismic sea wave* or the (incorrect) term *tidal wave*.[2397] |
| 1961 | Novaya Zemlya, Arctic Sea, Soviet Union | On October 30, **detonation of the "Tsar Bomba"** [Russ. Царь-бомба] **at an altitude of 4,000 m, the largest nuclear explosion (about 50 megatons) ever set off.** ▪ The 8-m-long 27-ton bomb, codenamed IVAN, was originally conceived as a three-stage fission-fusion-fission design: the first and second stage producing a fission-fusion reaction and the third stage amplifying this reaction by fissioning an uranium fission tamper with fast neutrons from the fusion reaction. This concept, however, which would have resulted in a yield of about 100 megatons, was eventually given up by replacing the uranium tamper by lead, because most of the fallout from such a test would have fallen on populated Soviet territory.[2398]<br><br>The most powerful thermonuclear device ever detonated by the United States was a fusion bomb, lithium deuteride fueled and codenamed CASTLE BRAVO, a surface burst (2.13 m above surface) with a yield of 15 megatons of TNT equivalent (Bikini Atoll, March 1, 1954). "A perfectly circular shock wave tore the surface of the sea to frothy mist, shining like snow in the pitiless dazzling radiance of the still growing hemisphere of light. Limb darkening like that of a |

---

[2396] L.S. SWENSON JR., J.M. GRIMWOOD, and C.C. ALEXANDER: *This new ocean: a history of project Mercury.* Rept. NASA SP-4201. Washington, DC (1966); http://www.hq.nasa.gov/office/pao/History/SP-4201/cover.htm.

[2397] D.C. COX: *Status of tsunami knowledge.* In: (D.C. COX, ed.) *Proc. Tsunami Meeting associated with the 10th Pacific Science Congress* [University of Hawaii, Honolulu, HI, Aug./Sept. 1961]. Institut Géographique National, Paris (1963), pp. 1-6.

[2398] *The Tsar Bomba ("King of bombs"). The world's largest nuclear bomb.* Nuclear Weapon Archive; http://nuclearweaponarchive.org/Russia/TsarBomba.html.

| 1961 | U.S. Air Force Base, Edwards, California & Rogers Dry Lake, southern California | star outlined the fiery dome, with the shock wave at its base spreading as a sharply defined widening circle well beyond the explosion cloud."[2399]

On November 9, **Robert WHITE,** a U.S. test pilot, having dropped away from a Boeing B-52 ("Stratofortress") carrier aircraft, **flies the X-15 rocket plane at 4,093 mph (Mach 6), thus becoming the first pilot to reach hypersonic velocities with a winged craft** {Collier Trophy ⇨1911}.

In 1963, the X-15 reached an altitude of 67 miles (108 km), and performed a shuttle-style reentry from that altitude. During a test in 1967, it attained hypersonic speeds of up to $M = 6.72$. ▪ The test results for the X-15 {⇨1959; ⇨Fig. 4.20–I}, which proved the applicability of hypersonic theory and wind tunnel work to an actual flight vehicle, were important to the development of the Space Shuttle {⇨Fig. 4.20–K}. Even today, the X-15 remains the only aircraft capable of studying phenomena at hypersonic speeds, space-equivalent flight, and reentry flight. Hugh L. DRYDEN called it "the most successful research airplane in history." |
|---|---|---|
| 1961 | Seismology Division, U.S. Coast & Geodetic Survey (USC & GS), U.S.A. | **Establishment of the World-Wide Standardized Seismograph Network (WWSSN), the primary goal of which is to study the problems associated with the detection and identification of underground nuclear explosions.** ▪ It became the first truly global seismographic network. The network data obtained (from both short- and long-period seismometers) resulted in much more detailed and accurate catalogs of earthquake locations, and also promoted the development of the theory of plate tectonics {⇨Fig. 4.1–I}.[2400]

WWSSN operated from 1961 through 1996 and produced high-quality analog recordings of a very large number of interesting seismic events. The USGS Digital World Wide Standard Seismograph Network (DWWSSN) is the digital descendant of the WWSSN. |
| 1961 | United States | **Establishment of the Aeroballistic Range Association (ARA),** an international organization devoted to
  ▸ sharing research and development progress in the broad field of experimental ballistics;
  ▸ serving as a meeting point for institutions engaged in research and development using guns and related launchers; and
  ▸ fostering the infusion of young scholars into these exciting areas of research.

Membership from research facilities grew rapidly, and the ARA now contains 51 organizations. The 57th meeting of ARA was held in 2006 at Venice, Italy. |
| 1961 | Tektronix Corporation, Portland, Oregon | **The Tektronix model 519, the first ultrafast traveling-wave oscilloscope, is unveiled.** It has a minimum sweep rate of 2 ns/cm; *i.e.*, it provides a maximum writing speed of 5,000 km/s. ▪ Since the new GHz oscilloscope allowed single-sweep recording of ultrafast transient events, it
  ▸ **significantly promoted the development of high-speed strain and stress sensors for use in shock-loaded solids;**
  ▸ **stimulated investigations into shock effects in laser-irradiated targets** after the invention of Q-switching techniques {MCCLUNG & HELLWARTH ⇨1962; PERESSINI ⇨1963}; and
  ▸ enabled the recording of ultrafast photomultiplier signals in laser velocity interferometers {BARKER & HOLLENBACH ⇨1965}. |

---

[2399] *The day the Sun rose twice: the CASTLE BRAVO test March 1, 1954;* http://www.donaldedavis.com/CASTBRAV/CASLBRV.html.
[2400] J. OLIVER and L. MURPHY: *WWSSN: Seismology's global network of observing stations.* Science **174**, 254-261 (Oct. 15, 1971).

| | | |
|---|---|---|
| 1961 | Columbia University, New York City, New York & CalTech, Pasadena, California | Publication of the *Shock and Vibration Handbook*,[2401] a reference book for engineers edited by Cyril M. HARRIS and Charles E. CREDE aimed at "bringing together under one title classical vibration theory combined with modern applications of the theory to current engineering practice, including particularly the recently matured topics of mechanical shock and instrumentation for the measurement of shock and vibration." The editors provide a definition of the field of shock as treated in their handbook:<br><br>▸ **"Shock is a somewhat loosely defined aspect of vibration wherein the excitation is nonperiodic,** *e.g.* **in the form of a pulse, a step, or transient vibration. The word** *shock* **implies a degree of suddenness and severity.** These terms are relative rather than absolute measures of the characteristic; they are related to a popular notion of the characteristics of shock and are not necessary in a fundamental analysis of the applicable principles.<br><br>▸ From the analytical viewpoint, the important characteristic of shock is that the motion of the system upon which the shock acts includes both the frequency of the shock excitation and the natural frequency of the system. If the excitation is brief, the continuing motion of the system is free vibration at its own natural frequency…<br><br>▸ More frequently, shock and vibration are unwanted. Then the objective is to eliminate or reduce their severity or, alternatively, to design equipment to withstand their influences." |
| 1961 | Gibbs Chemical Laboratory, Harvard University, Cambridge, Massachusetts | John N. BRADLEY[2402] and George B. KISTIAKOWSKY, two physical chemists, **perform a mass spectroscopic analysis of shock-heated substances** in a reflected shock wave in order to study the decomposition rates of various gases up to 2,000 K, and the polymerization and oxidation of acetylene. Using a 13-mm-dia. shock tube and a Bendix time-of-flight mass spectrometer, **they sample the hot stationary gas** – *e.g.*, nitrous oxide [$N_2O$] – **behind a reflected shock,** which leaks through a tiny central hole (in the end plate of the shock tube) that leads directly into the ionization chamber of the spectrometer. Their set-up, which is capable of measuring spectra at time intervals of 50 µs, resolves individual masses of up to about 50. ▪ The method allows one to study a wide variety of very fast reactions in the gas phase and to derive possible chain-branching cycles. |
| 1961 | McGill University, Montreal, Canada & Flight range on the Island of Barbados, Caribbean Sea | Gerald V. BULL,[2403] a Canadian engineer and ballistician, **proposes a concept for a long-range gun** {RAUSENBERGER ⇨1916} **that could be used to study the ballistics of reentry vehicles at low cost.** ▪ This so-called "High Altitude Research Program (HARP)," an ambitious project (1961–1967) to study the upper atmosphere using instrumented projectiles shot from a "space cannon," was initially supported by the Ballistic Research Laboratory (BRL) of the U.S. Army and the Canadian Armaments and Research Development Establishment (CARDE). Later BULL also proposed to use this concept as a low-cost technology for launching considerable payloads into low-Earth orbit. In 1966, BULL and his team set a world record by shooting a specially designed 84-kg projectile 179 km into the sky using an old 16-in. (406-mm) U.S. Navy cannon.[2404] |

---

[2401] C.M. HARRIS and C.E. CREDE (eds.): *Shock and vibration handbook.* McGraw-Hill, New York *etc.* (1961); *see* vol. I: *Basic theory and measurements*; vol. II: *Data analysis, testing, and methods of control*; vol. III: *Engineering design and environmental conditions.*

[2402] J.N. BRADLEY and G.B. KISTIAKOWSKY: *Shock wave studies by mass spectrometry. (I) Thermal decomposition of nitrous oxide. (II) Polymerization and oxidation of acetylene.* J. Chem. Phys. **35**, 256-270 (1961).

[2403] G.V. BULL and C.H. MURPHY: *Paris Kanonen – the Paris guns (Wilhelmsgeschütze) and project HARP.* E.S. Mittler & Sohn, Bonn (1988).

[2404] *High Altitude Research Project (HARP).* Encyclopedia of Astrobiology, Astronomy & Spaceflight; http://www.daviddarling.info/encyclopedia/H/HARP.html.

Besides HARP, other promising concepts of ballistic launch to space have been proposed such as

- the *coil gun* {BIRKELAND ⇨ 1901};
- the *blast wave accelerator* {WILSON ET AL. ⇨ 1993}, a launcher analogous to the coil gun with the coils being replaced by rings of high explosive {BAKIROV & MITROFANOV ⇨ 1976; VOITENKO 1990; TARJANOV 1991; KRYUKOV 1995};[2405]
- the spiral *slingatron*,[2406] a device for launching objects from a curved track operating on the same principle as the ancient stone-throwing sling; and
- the *electromagnetic railgun* {HÄNSLER ⇨ 1944/1945; RASHLEIGH & MARSHALL ⇨ 1978}.

Recently, Edward M. SCHMIDT[2407] and Mark L. BUNDY, two research scientists at the U.S. Army Research Laboratory (ARL, Aberdeen Proving Ground, MD) critically reviewed these four possible launch technologies.

| | |
|---|---|
| 1961 *Los Alamos Scientific Laboratory (LASL), Los Alamos, New Mexico* | **Arthur Wayne CAMPBELL[2408] and collaborators study the process of initiating liquid explosives using strong planar shock waves with pressures of up to 100 kbar. The experiments demonstrate that thermal explosion occurs as a result of shock heating in the (homogeneous) explosive.** Typically, the detonation proceeds through the compressed explosive at a velocity greater than the steady state velocity in the uncompressed explosive, thereby overtaking the initial shock and overdriving detonation in the unshocked explosive – a process known as **"superdetonation"** (or "super detonation").[2409] ▪ According to Charles L. MADER,[2410] a former Los Alamos detonation physicist, "the term *super detonation* was used to describe the process of shocking an explosive up to a higher density and then having a second higher pressure shock initiate propagating detonation through the higher density explosive in the same direction as the initial shock – or as in the case of homogeneous shock initiation it occurred by thermal decomposition near the piston/explosive interface … We called such processes 'hydrodynamic tricks,' and they allowed us to double the CJ pressures of explosives in certain applications and to tailor how energy was delivered." |

In the same year, CAMPBELL[2411] and collaborators extend their study to solid explosives (cyclotol B, TNT, plastic-bonded HMX, and nitromethane-carborundum mixtures) using shock pressures of up to 200 kbar.

(i) They observe that the slightly reacting shock travels at increasing velocity for some distance (typically 1 cm), and then in a travel of perhaps 0.01 cm becomes full detonation, moving at full velocity.

(ii) In contrast to the behavior of liquid explosives, the velocity transition in solid explosives is not abrupt, and there is no strong overshoot in the detonation velocity after the transition.

(iii) They conclude that polycrystalline explosives, either cast or pressed, are inhomogeneous in nature and, therefore, produce a rough shock wave: in regions of convergent flow the local increase in compression results in "hot spots" which, because of the strong dependence of the reaction rate on the temperature, strongly influence initiation.

---

[2405] J. STARKENBERG: *Shock-physics simulation of blast wave accelerator launch dynamics.* Presented at the 40th AIAA Aerospace Sciences Meeting & Exhibit [Reno, NV, Jan. 2002], Paper AIAA-2002-682.

[2406] M.L. BUNDY, D.A. TIDMAN, and G.R. COOPER: *Sizing a Slingatron-based space launcher.* AIAA J. Propulsion & Power **18**, No. 2, 330-337 (March/April 2002).

[2407] E.M. SCHMIDT and M.L. BUNDY: *Ballistic launch to space.* In: *Proc. 22nd Int. Symposium on Ballistics* [Vancouver, BC, Canada, Nov. 2005].

[2408] A.W. CAMPBELL, W.C. DAVIS, and J.R. TRAVIS: *Shock initiation of detonation in liquid explosives.* Phys. Fluids **4**, 498-510 (1961).

[2409] R. ENGELKE and S.A. SHEFFIELD: *Explosives.* In: (G.L. TRIGG, ed.) *Encyclopedia of applied physics.* VCH, New York, vol. 6 (1997), pp. 327-357.

[2410] Dr. Charles L. MADER, Honolulu, HI; private communication (Jan. 2006).

[2411] A.W. CAMPBELL, W.C. DAVIS, J.B. RAMSAY, and J.R. TRAVIS: *Shock initiation of solid explosives.* Phys. Fluids **4**, 511-521 (1961).

| 1961 | Observatoire de Marseille, Université de Provence, France | **Georges Courtes**[2412] **and Paul Cruvellier,** two French astrophysicists, **discover an optical jet in the large spiral galaxy M106** (or NGC 4258).[2413] Deep Hα images show two faint structures coming from the nucleus perpendicular to the bar on a deprojected image of the galaxy [French *"... une longue bande d'émission qui apparaît symétriquement disposée par rapport au noyau central et qui a une forme vaguement spiralée peut-être due au mouvement de rotation de cette galaxie."*].

These Hα features of galaxy NGC 4258 – which have a similar size, width, and curvature to those of normal spiral arms, and so have been called "anomalous arms" – **have since been recognized to be an astrophysical jet.**[2414] • *Astrophysical jets* are highly collimated supersonic (relativistic) fluxes of plasma that some galaxies, quasars, microquasars, X-ray binaries, *etc.*, may produce due to the accretion of matter onto massive objects.

(i) This jet phenomenon, first observed at Lick Observatory {Curtis ⇒1918}, is the nearest extragalactic astrophysical jet to us, since it is "only" 24 million light-years away. Despite its low energy, it offers a prime laboratory for exploring the physics of astrophysical jets and their interactions with their environments.

(ii) The two-sided radiojet of the *Seyfert galaxy* M106 {Seyfert ⇒1943/1944} exits the galaxy's active nucleus traveling in two opposite directions and extends through the galactic disk for about 16,000 light-years before it deflects into the halo of the galaxy. The incline of M106 of about 72 degrees exposes the central region of the galaxy's barred core to Earth-bound telescopes, thus giving astronomers the opportunity to learn about the "jet" phenomenon in great detail.[2415]

(iii) Astrophysical jets – which are still one of the most poorly understood astrophysical phenomena – are collimated plasma flows of extraordinary stability which are ejected from some astronomical objects.

▸ Astrophysical jets have been observed in young stellar objects {Herbig & Haro ⇒1950s; ⇒Fig. 4.1–Y}, active galactic nuclei (AGNs) {Seyfert ⇒1943/1944}, and some binary stars.

▸ The scales of astrophysical jets vary widely: galactic jets are several orders of magnitude longer than stellar jets and can extend for some thousands of light-years from end to end.[2416]

▸ A number of stellar jets have a knot-like structure of bright emission along the jet axis that is reminiscent of the bright spots in the exhaust jet of a jet engine {⇒Fig. 4.20–F}. Time-dependent ejection of matter at the jet base and fluid instabilities are the most frequent explanations for knot production. |
|---|---|---|
| 1961 | Poulter Laboratory, Stanford Research Institute, Menlo Park, California | **Donald R. Curran,**[2417] a U.S. shock physicist, **discusses the possibility of second-order phase transitions in iron and Invar®** (an iron/nickel alloy with 64% Fe and 36% Ni). His analysis is directed toward the possibility that a multiple shock wave structure – a phenomenon previously observed in first-order transitions and characterized by a discontinuity in volume and entropy, leading to a cusp in the Hugoniot curve {Bancroft et al. ⇒1956; Drummond ⇒1957; Al'tshuler et al. ⇒1958} – would also be produced by a second-order transition |

---

[2412] G. Courtes and P. Cruvellier: *Détection de nouveaux nuages d'hydrogène ionisé dans les galaxies.* C. R. Acad. Sci. Paris **253**, 218–220 (1961).

[2413] The two-sided radiojet of the *Seyfert galaxy* M106 {Seyfert ⇒1943/1944} exits the galaxy's active nucleus in two opposite directions, and extends through the galactic disk for about 16,000 light-years before it deflects into the halo of the galaxy. The incline of M106 (about 72°) exposes the central region of the galaxy's barred core to earthbound telescopes, thus giving astronomers the opportunity to learn about the "jet" phenomenon in great detail. From Robert Gendler's Astroimaging Gallery; http://www.robgendlerastropics.com and http://www.robgendlerastropics.com/M106NM.html.

[2414] R.D. Blandford: *Astrophysical jets.* In: (P. Murdin, ed.) *Encyclopedia of astronomy and astrophysics.* 4 vols., Institute of Physics Publ., London *etc.* (2001); vol.1, pp. 135–138.

[2415] From Robert Gendler's Astroimaging Gallery; http://www.robgendlerastropics.com and http://www.robgendlerastropics.com/M106NM.html.

[2416] *Jets.* Cosmos: the Swinburne Astronomy Online (SAO) encyclopedia. Swinburne University, Hawthorn, Australia; http://www.cosmos.swin.edu.au/.

[2417] D.R. Curran: *On the possibility of detecting shock-induced second-order phase transitions in solids. The equation of state of Invar.* J. Appl. Phys. **32**, 1811–1814 (1961).

such as a ferromagnetic-to-paramagnetic transition in shock-compressed iron alloys. His experiments on Invar under explosive loading do not show a multiple wave structure; however, the stress-volume relation shows a gradual decrease in compressibility as the stress is increased.

| 1961 | Poulter Laboratory, Stanford Research Institute, Menlo Park, California & Dept. of Geophysical Sciences, University of Chicago, Illinois | Paul S. DeCarli,[2418] an experimental shock wave physicist at SRI, **and John C. Jamieson,** a professor of geophysics at the University of Chicago, **provide the first experimental evidence for a shock-induced transformation of graphite into diamond,** thus confirming previous speculation {Moissan & Sir Parsons; see Friedel ⇨ 1892}. By applying chemical separation on the residues followed by X-ray diffraction analysis, they find that all of the diamond was polycrystalline with a crystalline size of about 20 nm and particle sizes of below about 5 μm. • The idea of using shock waves to transform graphite into diamond was partly stimulated by the discovery that diamonds were found in minerals associated with meteorite craters {Friedel ⇨ 1892; Carter & Kennedy ⇨ 1964}.<br><br>In later studies, DeCarli[2419, 2420] showed that shock pressures as low as 100 kbar are sufficient to transform graphite into diamond, and that the shock synthesis can be interpreted in terms of a "hot spot" model. |
|---|---|---|
| 1961 | P.N. Lebedev Physical Institute, U.S.S.R. Academy of Sciences, Moscow, U.S.S.R. | **Vitaly Lazarevich Ginzburg**[2421] **and Sergei Ivanovich Syrovatskii,** two Soviet astrophysicists, **are apparently the first to suggest that cosmic rays are accelerated in supernova remnants.** They implicitly assume that the cosmic ray source is dominated by fresh nucleosynthetic material. • In the following year, Maurice M. Shapiro,[2422] a U.S. theoretical physicist who in 1949 founded a cosmic ray laboratory at the Naval Research Laboratory (NRL, Washington, DC) resumed the subject and assumed the same acceleration mechanism.<br><br>With subsequent developments in shock acceleration theory, it was realized that the most likely site for cosmic ray acceleration is the hot, low-density interstellar medium.[2423] The source of the particles that are accelerated is still the subject of much debate. Currently, a multi-institution, multi-disciplinary collaboration of U.S. researchers develops models to better understand the emission of intense bursts of gamma radiation during core collapse supernovae {Terascale Supernova Initiative ⇨ 2001}.<br><br>In November 2006, the Chandra X-ray Center released a map of acceleration of cosmic ray electrons in Cassiopeia A (or Cas A),[2424] a 325-year-old remnant in our Milky Way produced by the explosive death of a massive star. **The new map shows regions in Cassiopeia A, where the X-ray emission is generated by electrons spiraling along magnetic field lines and being accelerated as they pass across the remnant's shock front.**[2425] |
| 1961 | BRL, Aberdeen Proving Ground, Maryland | **George E. Hauver**[2426] **and Robert J. Eichelberger,** two U.S. high-pressure physicists at the U.S. Ballistic Research Laboratory (BRL), **first report on shockwave-induced polarization and charge release from unpoled dielectrics** (such as water). They suggest that these |

---

[2418] P.S. DeCarli and J.C. Jamieson: *Formation of diamond by explosive shock*. Science **133**, 1821-1822 (June 9, 1961).
[2419] P.S. DeCarli: *Method of making diamond*. U.S. Patent No. 3,238,019 (1966).
[2420] P.S. DeCarli: *Nucleation and growth of diamond in shock wave experiments: the evidence for the hot spot model*. In: (Y.M. Gupta, ed.) *Shock wave compression of condensed matter. Proc. Symposium in honor of G.E. Duvall* [Pullman, WA, Sept. 1, 1988]. Rept. 99164-2814, Shock Dynamics Laboratory (SDL), Washington State University, Pullman, WA (1988), pp. 16-19.
[2421] V.L. Ginzburg and S.I. Syrovatskii: *Present status of the question of the origin of the cosmic rays*. Sov. Phys. Uspekhi **3**, 504-541 (1961); *The origin of cosmic rays*. Pergamon Press, Oxford *etc.* (1964).
[2422] M.M. Shapiro: *Supernovae as cosmic-ray sources*. Science **135**, 175-193 (Jan. 19, 1962).
[2423] W.I. Axford: *The acceleration of cosmic rays by shock waves*. Ann. New York Acad. Sci. **375**, 297-313 (1981).
[2424] *Supernova remnants observed with the Chandra X-Ray Observatory*. Dept. of Astronomy, Pennsylvania State University (PSU), University Park, PA (July 2002); http://www.astro.psu.edu/users/green/Main/main5.html.
[2425] *Chandra discovers relativistic pinball machine*. CXC, Cambridge, MA (Nov. 15, 2006); http://chandra.harvard.edu/press/06_releases/press_111506.html.
[2426] G.E. Hauver and R.J. Eichelberger: *Solid state transducers for recording of intense pressure pulses*. In: *Les Ondes de Détonation. Colloques Internationaux du Centre National de la Recherche Scientifique* [Gif-sur-Yvette, France, Aug./Sept. 1961]. Edition N° 109 du CNRS, Paris (1962), pp. 363-381.

phenomena are "the result of dipole alignment or induction by the shock front." ▪ Based on an orientation polarization mechanism of water, a detailed analysis of their experimental work was carried out by Latvian and U.S. researchers.[2427] It revealed that the polarization characteristics of water, calculated from the polarization charge density, agree well with experimental data by EICHELBERGER and HAUVER. Finally, it was concluded that **the reorientation of water molecules is a dominating mechanism of shock-induced polarization.**

| 1961 | Institute of Physics, Hokkaido University, Sapporo, Japan | **Yoro ONO,**[2428] **Shiro SAKASHITA, and Noboru OHYAMA apply the gasdynamic equations to problems of cosmic dimensions.** In particular, they apply the shock relations to stellar interiors in order to evaluate cosmic shock waves, supernovae explosions and the origin of planetary nebulae. |
|---|---|---|
| 1961 | Dept. of Physics, Astronomy and Astrophysics, University of Chicago, Illinois | **Eugene N. PARKER,**[2429] a U.S. astrophysicist, **predicts that during periods of peak solar activity, the generation of large solar flares results in a sudden elevation of temperature.** ▪ A solar flare, which can reach heights of up to 100,000 km, can produce a *hydrodynamic blast wave* equivalent to a huge explosion. This sweeps outwards into space at some 1,000–2,000 km/s and reaches Earth within a day or two. ▪ Today solar flares are considered to be the biggest explosions that occur in the Solar System. |

In the same year,
- PARKER[2430] predicts that the solar wind changes from supersonic to subsonic flow at the outer limit of the Solar System {Voyager Missions ⇨2003}; and
- measurements performed by the two Cambridge astronomers Donald E. BLACKWELL[2431] and Michael F. INGHAM essentially confirm PARKER's model. In addition, the data revealed that the proton density in the hydrodynamic blast wave increases by a factor of about 30 compared to a "quiet-day solar wind," in which the speed of the solar wind heading toward Earth averages about 400 km/s.

| 1961 | USGS, Menlo Park & Hypervelocity Ballistic Range, Ames Research Center, Moffet Field, California | **Eugene M. SHOEMAKER,**[2432] **Donald E. GAULT, and Richard V. LUGN,** three U.S. impact researchers, **produce artificially shatter cones** {BUCHER ⇨1933} **in Kaibab dolomite** from Meteor Crater, Arizona {⇨c.50,000 years ago} by shooting a $^3/_{16}$-in. (4.8-mm)-dia. aluminum sphere from a light-gas gun at a velocity of 5.61 km/s. ▪ In the mid-1970s, similar impact experiments were carried out at Ernst-Mach-Institut (EMI) in Freiburg, Germany. Experimentally produced impact craters in limestone targets displayed millimeter-sized shatter cones within crater spallation zones. Variation of the impact velocity showed that at about 3 km/s shatter cone formation starts and is reproducible at any higher impact velocities. In most cases the cone apices were pointing in the direction of the impact center.[2433] |
|---|---|---|

The first man-made shatter cones were produced in 1959 during an underground nuclear explosion.[2434] In 1968, a series of 0.5-ton and 100-ton TNT explosion experiments were conducted in granitic rock near Cedar City, UT as part of a basic research program on cratering, shock wave propagation, and formation of shatter cones. The high explosion trials conducted demonstrated beyond any doubt, that shatter cones can be formed by shock wave processes

---

[2427] Y. SKRYL, A.A. BELAK, and M.M. KUKLJA: *Shock-induced polarization in distilled water*. Phys. Rev. B **76**, 064107 (2007).
[2428] Y. ONO, S. SAKASHITA, and N. OHYAMA: *On the mechanism of stellar explosion*. Progr. Theor. Phys. Suppl. **20**, 85-112 (1961).
[2429] E.N. PARKER: *Sudden expansion of the corona following a large solar flare and the attendant magnetic field and cosmic-ray effects*. Astrophys. J. **133**, 1014-1033 (1961).
[2430] E.N. PARKER: *The stellar-wind regions*. Astrophys. J. **134**, 20-27 (1961).
[2431] D.E. BLACKWELL and M.F. INGHAM: *Observations of the zodiacal light from a very high altitude station*. Month. Not. Roy. Astron. Soc. **122**, 113-127 (1961).
[2432] E.M. SHOEMAKER, D.E. GAULT, and R.V. LUGN: *Shatter cones formed by high-speed impact in dolomite*. Article 417: USGS Prof. Paper 424-D (1961), pp. 365-368.
[2433] E. SCHNEIDER and G.A. WAGNER: *Shatter cones produced experimentally by impacts in limestone targets*. Earth Planetary Sci. Lett. **32**, 40-44 (1976).
[2434] *The encyclopedia of astrobiology, astronomy and spaceflight*; http://www.daviddarling.info/encyclopedia/S/shatter_cone.html.

during cratering and that average formational pressures in these crystalline rocks are in the 2–6 GPa (20–60 kbar) range.[2435]

| 1961 | Dept. of Geochemistry, Moscow State University (MSU) & Institute of High Pressure Physics (HPPI), Troitsk, Moscow Region, Soviet Union | **Sergei M. STISHOV,**[2436] a graduate student at MSU, **and Svetlana V. POPOVA,** a mineralogist at HPPI, **statically synthesize** *stishovite* – a high-pressure polymorph of quartz, like ***coesite*** {COES ⇨1953}, with a rutile structure – **for the first time.** ▪ While coesite [$SiO_2$] is composed of very fine-grained polycrystalline aggregates (size 100–200 μm), stishovite [$SiO_2$] cannot be identified microscopically and requires X-ray diffraction analysis.<br><br>A few months later, **Ed CHAO, Eugene SHOEMAKER, and their coworkers discover this phase at Meteor Crater, Arizona** {⇨c.50,000 years ago}, **and name it** *stishovite* {CHAO, FAHEY, LITTLER & MILTON ⇨1962}. ▪ Stishovite attained greatest importance in geology as a primary indicator for natural shock metamorphism, such as provoked by meteorite or asteroid impact.<br><br>Later experiments revealed that stishovite could also be recovered from laboratory shock experiments.[2437] It is apparently formed during shock compression at lower pressures than for coesite, and coesite apparently crystallizes during pressure release. More recent shock wave experiments carried out on stishovite at up to 235 GPa (2.35 Mbar) by Shengnian LUO[2438] ET AL. at CalTech's Seismological Laboratory allowed the phase diagram of silica to be extended to the megabar regime, which might provide a more detailed interpretation of shock melting experiments on fused and crystalline quartz. |
|---|---|---|
| 1961 | Institute of Physical Chemistry, University of Göttingen, Germany | **Heinz Georg WAGNER,**[2439] a German physical chemist and combustion researcher, **reviews the present state of research into gaseous detonations.** In particular, he addresses<br>▸ the theory of stable detonation (CJ theory and ZND theory);<br>▸ a comparison between experimental and calculated values of the detonation velocity;<br>▸ the limits of detonability and governing factors (*e.g.*, tube diameter, wall effects, *etc.*);<br>▸ the system $C_2H_2$–$O_2$ (which shows a very rapid transition from deflagration to detonation);<br>▸ the emission spectra of detonations of different mixtures of hydrocarbons;<br>▸ the velocity of spherical detonation waves in comparison to planar detonation waves;<br>▸ the initiation of detonation and ignition by shock waves;<br>▸ spinning detonation (experimental results and theoretical approaches); and<br>▸ the reaction zone of a detonation (length and mechanism of the chemical reactions). |
| 1961 | ISL, Saint-Louis, France | **Franz WECKEN,**[2440] a German detonation physicist, **proposes several blast scaling laws for spherically symmetric explosions.** He also discusses the history of blast scaling, thereby attributing the Hopkinson law {B. HOPKINSON ⇨1915} to Carl CRANZ.[2441] |
| 1962 | Denver, Colorado | On April 24, **a series of earthquakes occur in the vicinity of a 12,045-ft (3,671-m)-deep well near Denver, into which liquid wastes are being pumped** (a direct result of military production and demilitarization programs).[2442] The seismic shocks are clearly associated with the |

---

[2435] D.J. RODDY and L.K. DAVIS: *Shatter cones formed in large scale experimental explosion craters.* Proc. Symposium on Planetary Cratering Mechanics [Flagstaff, AZ Sept. 1976]. In: (D.J. RODDY, R.O. PEPIN, and R.B. MERRILL, eds.) *Impact and explosion cratering: planetary and terrestrial implications.* Pergamon Press, New York (1977), pp. 715-750.

[2436] S.M. STISHOV and S.V. POPOVA: *A new modification of silica.* Geochemistry **10**, 923-926 (1961).

[2437] P.S. DECARLI and D.J. MILTON: *Stishovite: synthesis by shock wave.* Science **147**, 144-145 (Jan. 8, 1965).

[2438] S.N. LUO, J.L. MOSENFELDER, P.D. ASIMOW, and T.J. AHRENS: *Stishovite and its applications in geophysics: new results from shock-wave experiments and theoretical modeling.* Phys. Uspekhi **45**, No. 4, 435-439 (2002).

[2439] H.G. WAGNER: *Gaseous detonations and the structure of a detonation zone.* In: (A. FERRI, ed.) *Fundamental data from shock tube experiments.* NATO, AGARDograph No. 41. Pergamon Press, New York *etc.* (1961), pp. 320-385.

[2440] F. WECKEN: *Les lois de similitude dans les explosions à symétrie sphérique.* Mém. Artill. Franç. **35**, 438-459 (1961).

[2441] C. CRANZ: *Modellübertragungsregeln für die Wirkungen von Explosionen.* In: *Lehrbuch der Ballistik.* Springer, Berlin, vol. II (1926), pp. 181-182.

[2442] *Rocky Mountain Arsenal (RMA) History.* RMA (2001); http://www.rma.army.mil/site/hist-1.html.

pumping, allowing for a time lag. The largest shocks cause some damage in Denver, where perceptible earthquakes are usually infrequent. It is inferred that the pumping of fluids, acting as a lubricant and disturbing the equilibrium of the subsurface layers of rock, triggered the earthquake process.[2443]

**This amazing discovery led to speculations that earthquakes might be controllable** – for example that small artificial earthquakes created using high explosives could be used to gradually reduce stresses in the Earth's crust in order to prevent big earthquakes.

▸ Seismological studies performed in the 1960s in Nevada demonstrated that even underground nuclear explosions can only significantly influence seismicity in the near field of the explosion {BOUCHER ET AL. ⇨1969}.

▸ Experiments carried out in an oil field at Rangely, Colorado, demonstrated the feasibility of earthquake control.[2444] However, this hypothesis was not uniformly accepted by Earth scientists.

Since many large metropolitan areas around the world are close to faults (like Kobe, Mexico City, Los Angeles, San Francisco, Tangshan, Tokyo, *etc.*) any ability to reduce the likelihood of earthquakes, and therefore reduce the enormous number of casualties from them, would be potentially very attractive. On the other hand, however, artificially triggered earthquakes would also result in difficult legal problems, such as claims for material damage and personal injury.

**On December 8, 2006, drilling work for a planned geothermal power plant in Kleinhüningen,** a town near Basel, **triggered a small earthquake** that measured 3.4 on the Richter scale and caused minor damage to buildings. The canton Basel City prosecutor launched an investigation to find if the company behind the Swiss Deep Heat Mining Project[2445] should pay for repairs. This company plans to recover heat by pumping water deep into the Earth's crust and use it to generate electricity ▪ Basel is one of the areas in Switzerland most prone to seismic activity. In 1356, the city was almost entirely destroyed by a 6.5-magnitude earthquake.

| 1962 | Lawrence Radiation Laboratory (LRL), Livermore, California & Area 10, Nevada Nuclear Test Site, Nye County, Nevada | On July 6, **SEDAN – the first nuclear excavation experiment – is carried out by LRL in order to develop nuclear excavation technology,** a major objective of the AEC's Plowshare Program {⇨1958}. The test consists of detonating a 100-kiloton thermonuclear device at a depth of 625 ft (190 m) in a drill hole in desert alluvium at the Nevada Test Site. In just a few seconds, about 13 million cubic yards (10 million m³) of material are thrown upward. About half of this fell back, leaving a crater with a volume of about 6.5 million cubic yards (5 million m³) {⇨Fig. 4.16−J}. ▪ **SEDAN was the largest cratering test performed in the Plowshare Program.** Although SEDAN and similar tests performed thereafter clearly indicated the feasibility of using nuclear explosives for peaceful excavation applications, the Plowshare Program was eventually dropped in 1975 due to political and environmental problems. |
|---|---|---|
| 1962 | Johnston Island [a U.S. possession], North Pacific | On July 9, **a nuclear weapon with a yield of 1.4 megatons of TNT equivalent is detonated at 400-km altitude.** Ground zero is on Johnston Island, an atoll about 780 nautical miles (1,445 km) southwest of Hawaii. This nuclear test of Operation FISHBOWL – called **"STARFISH PRIME"** and announced publicly – is performed by order of the Defense Atomic Support Agency (DASA) and the Atomic Energy Commission (AEC), in order to observe an artificial radiation belt, and to test the confinement of charged particles in magnetic fields on a large scale. **It produces an auroral display over the Hawaiian Islands that covers most of the western sky, as well as EMP-induced failures in electrical systems, but no blast wave.**[2446] |

---

[2443] C.F. RICHTER: *Earthquakes.* In: *The new Encyclopædia Britannica, Macropædia.* Benton & Hemingway, Chicago *etc.*, vol. 6 (1974), p. 69.

[2444] J.H. HEALY, W.W. RUBEY, D.T. GRIGGS, and C.B. RALEIGH: *The Denver Earthquakes.* Science **161**, 1301-1310 (Sept. 27, 1968); C.B. RALEIGH, J.H. HEALY, and J.D. BREDEHOEFT: *An experiment in earthquake control at Rangely, Colorado.* Ibid. **191**, 1230-1237 (March 26, 1976).

[2445] M. HÄRING and R. HOPKIRK: *The Swiss Deep Heat Mining Project – the Basel exploration drilling.* GHC Bulletin **23**, No. 1, 31-33 (2002).

[2446] F.H. SHELTON: *Reflections of a nuclear weaponeer.* Shelton Enterprise Inc., Colorado Springs, CO (1988), pp. **11**:36-63.

• STARFISH PRIME soon revealed that relativistic electrons that had been released by the nuclear explosion and were partly (5–10%) trapped for a couple of years in the Earth's magnetic field had generated the **"STARFISH radiation belt,"** which damaged solar cells of various satellites and also endangered astronauts. Retrospectively, these nuclear explosion tests at large altitude can be considered to be the largest man-made pollution events ever to extend beyond terrestrial boundaries.

The STARFISH test was actually one of a series of tests, which were preceded by a similar nuclear test series codenamed **"Operation ARGUS"** (1958) secretly conducted in the South Atlantic (about 1,800 km southwest of Capetown, South Africa) using small nuclear weapons detonated at heights of 100–300 miles (160–482 km). These tests, which produced a transient belt or zone of plasma encompassing the Earth analogous to the Van Allen belts, were fairly large-scale demonstrations of the generation of a hot plasma in the Earth's magnetic field.

In the same year, on November 3, **the final atmospheric U.S. nuclear weapons test – codenamed TIGHTROPE – is conducted.** This low-yield test (< 20 kt), where a Nike Hercules missile launches the weapon to medium altitude, serves to document a *fireball blackout*, a temporary blocking of short wavelength radio and radar signals by the ionized gases of a fireball. • Between 1946 and 1962, the United States conducted over 190 atmospheric nuclear tests, 12 of them over Johnston Atoll. In the period 1955–1962, the United States carried out high-altitude nuclear explosions in order to study

> the physical phases of the interactions of the weapon outputs with the atmosphere (such as the formation of fireballs at low high-altitudes and the partition of energies and their distribution over very large spaces at the higher high-altitudes); and

> the effects of these explosions on the normal activities of populations and their protective measures.[2447]

| | | |
|---|---|---|
| 1962 | *Rheinisch-Westfälische Technische Hochschule (RWTH) Aachen, Germany* | In the period September 3–7, **the first Symposium Transsonicum is held** at RWTH Aachen, organized by the General Council of the International Union of Theoretical and Applied Mechanics (IUTAM). The general chairman is Klaus OSWATITSCH, an international authority on gas dynamics. • Subsequent symposia took place in the years 1975, 1988 and 2002 at Göttingen University, the "cradle of fluid mechanics." |
| 1962 | *Nevada Nuclear Test Site 7, Nye County, Nevada* | **An underground nuclear test is conducted to explore the feasibility of using *reactor-grade plutonium* as a nuclear explosive material.** • The explosion test, which had a yield of less than 20 kilotons of TNT equivalent, confirmed that reactor-grade plutonium could be used to make a nuclear explosive. This fact was declassified in July 1977. The release of additional information was deemed important in order to enhance public awareness of nuclear proliferation issues associated with reactor-grade plutonium, which can be separated out during the reprocessing of spent commercial reactor fuel.[2448] |

The Nuclear Control Institute (Washington, DC) arrived at the conclusion that "a Trinity-type device would be capable of bringing reactor-grade plutonium of any degree of burn-up to a state in which it could provide yields in the multi-ton range … The technical problems confronting a terrorist organization considering the use of reactor-grade plutonium are not different in kind from those involved in using weapons-grade plutonium, but only in degree."[2449]

---

[2447] H. HOERLIN: *United States high-altitude test experiences. A review emphasizing the impact on the environment.* LASL Monograph LA-6405, Los Alamos Scientific Laboratory, Los Alamos, NM (Oct. 1976); http://www.fas.org/sgp/othergov/doe/lanl/docs1/00322994.pdf.

[2448] *DOE facts: additional information concerning underground nuclear weapon test of reactor-grade plutonium.* Office of Public Affairs, U.S. Dept. of Energy; http://www.ccnr.org/plute_bomb.html.

[2449] J.C. MARK: *Reactor-grade plutonium's explosive properties.* Nuclear Control Institute (NCI), Washington, DC (Aug. 1990); http://www.nci.org/NEW/NT/rgpu-mark-90.pdf.

| | | |
|---|---|---|
| 1962 | *Arzamas-16, Soviet Nuclear Center, Sarov, Soviet Union* | **Lev V. AL'TSHULER,** the Soviet Union's leading solid-state shock physicist, **and his collaborators carry out unique shock studies in solids:** |

(i) **They investigate the shock compressibilities of various solids at ultrahigh pressures** (*e.g.*, of Ti, Mo, Ta, and Fe) using an advanced flyer plate method.[2450] In order to produce shock pressures of up to 10 Mbar (1 TPa), they use a thin steel disc which is accelerated by the expanding explosion products to as much as 14 km/s; *i.e.*, twice the velocity obtained with a two-stage gun and three times the velocity obtained with plane-wave explosive systems, as used in the United States.

(ii) **They first use Mach reflection to compress metals to ultrahigh pressures** (for example aluminum up to 4 Mbar and steel up to 7 Mbar).[2451] In particular, they study the oblique collision of shock waves and their irregular superposition using a symmetric twin flyer plate arrangement, and analyze the data obtained using shock polar curves. This unique experimental technique is also of great interest from a thermodynamic point of view, because in the region of two-stage compression the increase in entropy is smaller than during a single shock transition.

In the same year, **Samuil Borisovich KORMER**[2452] **and collaborators report on the generation of ultrahigh shock pressures of up to 9 Mbar in some porous metals** (Al, Cu, Pb, and Ni). The shock wave is produced by accelerating an iron flyer plate by the explosion products to a velocity of 8.64 mm/µs and impacting a cylindrical target. They present a new form of the equation of state (EOS) which takes the decreases in the specific heat and the Grüneisen coefficient with increasing temperature into account.

| | | |
|---|---|---|
| 1962 | *P.N. Lebedev Physics Institute, U.S.S.R. Academy of Sciences, Moscow, Soviet Union* | **Gurgen A. ASKAR'YAN**[2453] **and E.M. MOROZ,** two Soviet physicists, **suggest that the vaporization of material from the surface of a solid target due to pulsed irradiation from a high-powered laser results in the trasmission of a high-pressure pulse into the target** {READY ⇨ 1963}. This pressure pulse raises the boiling point of the underlying material, which becomes superheated as more heat is conducted into the interior. They conclude, "We note the possibility of the appearance of these effects in outer space as a pressure on the dust particles of a comet, on the surfaces of space vehicles, meteorites, *etc.* The direct pressure of solar radiation on space vehicles may displace their orbits quite considerably (up to 1 km per day). Therefore, these effects may be useful, *e.g.* for trajectory control by varying the evaporation pressure at the surface of such objects (by the use of special coatings, focusing of solar radiation to intensify evaporation, Venetian blinds to control the intensity of evaporation, *etc.*)…" ▪ Instead, this spectacular effect was widely discussed as a possible method of realizing fusion ("laser fusion") and to disable missiles – in the U.S. Strategic Defense Initiative, nicknamed "Star Wars" {SDI Concept ⇨1983}. Prof. ASKAR'YAN later also investigated the biological effects of laser radiation on human tissues. |

The laser enery is hydrodynamically coupled into the solid. The sudden evaporation of material generates a shock wave in the surrounding gas – among plasma physicists known as a **"hydrodynamic blast wave"** – which tends to be faster than the vapor jet emerging from the laser-irradiated surface. In addition, it creates a steep stress wave in the target which is not, however, the result of an ordinary shock steepening process along the Hugoniot curve, but is instead caused by the rapid recoil momentum of evaporated material and the pressure exerted on the surface by the hot plasma. Thus, this unique method also allows the generation of very steep stress pulses in the weak shock pressure regime (< 1 kbar).

---

[2450] L.V. AL'TSHULER, A.A. BAKANOVA, and R.F. TRUNIN: *Shock adiabats and zero isotherms of seven metals at high pressures.* Sov. Phys. JETP **15**, 65-74 (1962); K.K. KRUPNIKOV, A.A. BAKANOVA, M.I. BRAZNIK, and R.F. TRUNIN: *An investigation of the shock compressibility of titanium, molybdenum, tantalum, and iron.* Sov. Phys. Dokl. **8**, 205-208 (1963).

[2451] L.V. AL'TSHULER ET AL.: *Irregular conditions of oblique collision of shock waves in solid bodies.* Sov. Phys. JETP **14**, 986-994 (1962).

[2452] S.B. KORMER, A.I. FUNTIKOV, V.D. URLIN, and A.N. KOLESNIKOVA: *Dynamic compression of porous metals and equation of state with variable specific heat at high temperatures.* Sov. Phys. JETP **15**, 477-488 (1962).

[2453] G.A. ASKAR'YAN and E.M. MOROZ: *Pressure on evaporation of matter in a radiation beam.* Sov. Phys. JETP **16**, 1638-1639 (1963).

An understanding of radiation effects on the structure of the interstellar medium and on the evolution of the shock wave is of great importance to many problems in astrophysics.[2454] The effect of shock waves driven by a heated plasma also proved useful in micromechanics when testing the spallation of thin surface films and measuring their interfacial strength. Thin films are frequently used as multilayer components in microelectronics and microactuators.[2455]

| 1962 | *Defense Research Board, Canada & Ionosphere Research Laboratory, Pennsylvania State University, Pennsylvania* | **Independently, William Ian AXFORD,**[2456] an astrophysicist from New Zealand, **and Paul J. KELLOGG,**[2457] a U.S. theoretical physicist, **present their "bow shock picture of the Earth's vicinity"** {⇨Fig. 4.1–X} in the same issue of the *Journal of Geophysical Research*. Based on James W. DUNGEY's picture of the magnetosphere,[2458] they predict that the interaction of the solar wind with the Earth's magnetic field creates a MHD (magnetohydrodynamic) shock – the **"bow shock"** – similar to that ahead of a blunt object moving supersonically in air. They also propose that all planetary bodies that have either a magnetosphere or a highly conducting ionosphere also have bow shocks associated with the deflection of the solar wind around them too. Later investigations will show that the bow shock model as proposed by AXFORD and KELLOGG is indeed correct {ISEE Program ⇨1977; ⇨Fig. 4.12–D}. ▪ A similar model had already been suggested two years before: in a pioneering work, Soviet researchers considered the supersonic interaction of conducting fluid with the Earth's magnetic field.[2459]

**Nearly all of the planets in the Solar System possess magnetospheres and, therefore, exhibit *planetary bow shocks* or *planetary Mach cones*.**[2460] These provide insights into both the behavior of collisionless shocks and the nature of the planetary obstacle responsible for creating those bow shocks.[2461]

▸ Since Jupiter has a substantial magnetic field, which extends as much as 80 planetary radii or more from the planet, the magnetospheric obstacle is very large.[2462] Neptune {Voyager 2 ⇨1989} and Saturn {Cassini Spacecraft ⇨2004} also have vast magnetospheres and are therefore surrounded by bow shocks.

▸ For planets with no detectable intrinsic magnetic field, like Mars {Mariner 4 ⇨1965} and Venus {RUSSELL ET AL. ⇨1979}, the upper atmosphere/ionospheres may act as obstacles to the solar wind flow; these induce a magnetic field and produce a pseudo (or false) magnetopause, resulting in the formation of a detached bow shock. |
| 1962 | *P.N. Lebedev Physical Institute, U.S.S.R. Academy of Sciences, Moscow, Soviet Union* | **Nikolai G. BASOV**[2463] **and Anatolij N. ORAEVSKII, two Soviet physicists, propose that rapid cooling could produce population inversions in molecular systems.** Considering a system of three levels with different relaxation times, they point out that due to the different relaxation times for the various energy levels when thermodynamic equilibrium is established, rapid variation of the system temperature may create a negative temperature state for certain pairs of energy levels. **They conclude that "sufficiently fast heating of the system plays an important role in this method for increasing negative temperatures. In some cases, this could be** |

---

[2454] T. DITMIRE ET AL.: *The production of strong blast waves through intense laser irradiation of atomic clusters*. Astrophys. J (Suppl. Series) **127**, 299-304 (2000).
[2455] V. GUPTA ET AL.: *Measurement of interface strength by the modified laser spallation technique*. J. Appl. Phys. **74**, 2388-2410 (1993).
[2456] W.I. AXFORD: *The interaction between the solar wind and the Earth's magnetosphere*. J. Geophys. Res. **67**, 3791-3796 (1962).
[2457] P.J. KELLOGG: *Flow of plasma around the Earth*. J. Geophys. Res. **67**, 3805-3811 (1962).
[2458] J.W. DUNGEY: *The structure of the exosphere, or, adventures in velocity*. In: (C. DEWITT, J. HIEBLOT, and A. LEBEAU, eds.) *Proc. Les Houches Summer School* (1962). *Geophysics, the Earth's environment*. Gordon & Breach, New York (1963), pp. 505-550.
[2459] V.N. ZHIGULEV and E.A. ROMISHEVSKII: *Concerning the interaction of currents flowing in a conducting medium with the Earth's magnetic field*. Sov. Phys. Dokl. **4**, 859-862 (1959).
[2460] J.A. SLAVIN, R.E. HOLZER, J.R. SPREITER, and S.S. STAHARA: *Planetary Mach cones: theory and observation*. J. Geophys. Res. **89**, 2708-2714 (May 1, 1984).
[2461] C.T. RUSSELL: *Planetary bow shocks*. In: (B.T. TSURUTANI and R.G. STONE, eds.) *Collisionless shocks in the heliosphere: reviews of current research*. Geophysical Monograph **35**, American Geophysical Union (AGU), Washington, DC (1985), pp. 109-130.
[2462] W.S. KURTH: *Plasma waves associated with the bow shock of Jupiter*. Dept. of Physics & Astronomy, University of Iowa, Iowa City, IA; http://www-pw.physics.uiowa.edu/plasma-wave/tutorial/voyager1/jupiter/bowshock/text.html.
[2463] N.G. BASOV and A.N. ORAEVSKII: *Attainment of negative temperatures by heating and cooling of a system*. Sov. Phys. JETP **17**, 1171-1172 (1963).

accomplished by using fast chemical reactions or shock waves." ▪ In 1966, Vadim K. KONYUKHOV[2464] and Aleksandr M. PROKHOROV discussed the possibility of population inversion in $N_2/CO_2$ mixtures by rapid expansion through a supersonic nozzle.

Independently, a similar gasdynamic approach was also proposed in the United States {HURLE & HERTZBERG ⇨1965}, which eventually led to the construction of the first operational gasdynamic laser {KANTROWITZ ET AL. ⇨1966}.

| 1962 | U.S. Naval Radiological Defense Laboratory, San Francisco, California | Rodney R. BUNTZEN,[2465] a U.S. physicist, **demonstrates that, in some ways, a submerged exploding wire has a greater similarity to an underwater nuclear explosion than a conventional explosive.** He concludes:

(i) Advantageously, the steam bubble produced by an underwater exploding wire is transparent, thus facilitating the study of its interior using photographic techniques.

(ii) The explosion yield and detonation rate of an exploding wire are easily controlled by varying the discharge voltage and condenser bank capacitance.

(iii) The shape of the explosion can be easily altered according to the exploding wire geometry.

The study, however, showed that it is impossible to use submerged exploding wires to simulate the very short soft X-ray pulse of a nuclear explosion which is immediately absorbed, resulting in a maximum thermal emission at a wavelength of about 30 Å (3 nm), thus also contributing to the dynamics of the nuclear bubble. |

| 1962 | U.S. Geological Survey (USGS), Washington, DC | Edward C.T. CHAO,[2466] a Chinese-born U.S. geologist, **and collaborators report on their discovery of *stishovite*,** a high-pressure silica polymorph {STISHOV & POPOVA ⇨1961}, **at Meteor Crater In Arizona** {⇨c.50,000 years ago}. ▪ The discovery of stishovite in nature, together with the discovery of coesite {CHAO ET AL. ⇨1960}, led to a wide range of activities that have transformed the disciplines of high-pressure mineralogy and mineral physics into an essential component of Earth science research over the last forty years. |

| 1962 | Institute of Metals and Explosives Research, University of Utah, Salt Lake City, Utah | Melvin A. COOK,[2467] a U.S. physical chemist and explosives expert, **and his collaborators experimentally determine the equations of state for water and Lucite.**

(i) Their diagnostic method uses water as a "pressure gauge" based on the fact that water is transparent and therefore permits convenient and continuous optical observation of the propagating shock wave by a high-speed streak or framing camera – the **"aquarium technique"** previously developed by William C. HOLTON.[2468]

(ii) This technique allows the measurement of transient pressures, including the peak pressures in detonation waves of condensed explosives. The detonation pressures increase with the initial density of the high explosives tested, and they measure for granular TNT ($\rho = 0.86$ g/cm$^3$) and granular RDX (1.21 g/cm$^3$) detonation pressures of 50 kbar and 134 kbar, respectively. The peak pressures observed in detonation waves for explosives of various types are found to be the Chapman-Jouguet (CJ) or "detonation" pressures from thermohydrodynamic theory. |

| 1962 | Poulter Laboratory, SRI, Menlo Park, California | George E. DUVALL,[2469] a U.S. physicist, **discusses the general peculiarities of shock waves of planar geometry,** such as the decay of shock wave amplitude, the spreading of the wave, energy dissipations, and sources of instability in the shock transition in real materials. These |

---

[2464] V.K. KONYUKHOV and A.M. PROKHOROV: *Population inversion in adiabatic expansion of a gas mixture.* JETP Lett. **3**, 286-288 (1966).

[2465] R.R. BUNTZEN: *The use of exploding wires in the study of small-scale underwater explosions.* In: (W.G. CHACE and H.K. MOORE, eds.) *Exploding Wires (II)*. Plenum Press, New York (1962), pp. 195-205.

[2466] E.C.T. CHAO, J.J. FAHEY, J.J. LITTLER, and D.J. MILTON: *Stishovite, SiO₂, a very high pressure new mineral from Meteor Crater, Arizona.* J. Geophys. Res. **67**, 419-421 (1962).

[2467] M.A. COOK, R.T. KEYES, and W.O. URSENBACH: *Measurements of detonation pressure.* J. Appl. Phys. **33**, 3413-3421 (1962).

[2468] W.C. HOLTON: *The detonation pressures in explosives as measured by transmitted shocks in water.* NavOrd Rept. No. 3968, U.S. Naval Ordnance Laboratory (NOL), White Oak, MD (Dec. 1954).

[2469] G.E. DUVALL: *Concepts of shock wave propagation.* Bull. Seismol. Soc. Am. **52**, 869-893 (1962).

considerations are aimed at illuminating **the complex three-dimensional problem of shock wave research in seismology** – *i.e.*, to what degree is the far-out elastic wave generated by an explosion influenced by the details of the explosion itself and the features of the elastic wave? ▪ In principle, the propagation characteristics of the elastic wave are determined by the features of the initial phase of explosion and the complex propagation properties of real earth media.

| | | |
|---|---|---|
| 1962 | *Atomic Weapons Research Establishment (AWRE), Foulness, Essex, England* | Peter J.A. FULLER[2470] and John H. PRICE, two research physicists, **use manganin as a piezoresistive shock pressure gauge.** They observe that its response is linear with pressure for dynamic stresses of up to 300 kbar. Manganin is particularly well-suited to this task because its temperature coefficient is almost zero {BRIDGMAN ⇨1911}.<br><br>Two years later, FULLER[2471] and PRICE first reported on a ***manganin piezoresistive gauge.*** This type of pressure gauge was also developed independently, and at around the same time, in the United States {BERNSTEIN & KEOUGH ⇨1964}. |
| 1962 | *Sandia Laboratories, Albuquerque, New Mexico* | Orval E. JONES,[2472] a U.S. applied mechanical engineer, **and collaborators invent the so-called "Sandia quartz gauge,"** a pressure gauge of simple construction. It allows the measurement of shock wave profiles in solid specimens with nanosecond time resolution up to a pressure of about 25 kbar.<br><br>In the following years, this gauge technique was further improved, leading to the so-called **"Sandia shunted guard-ring gauge,"** which proved very useful for measuring not only shock-induced stresses in solid matter originating from impact and explosive loading, but also those arising from irradiation with ultrashort giant laser pulses.[2473] |
| 1962 | *Shock Dynamics Group, The Boeing Company, Seattle, Washington* | Donald V. KELLER[2474] and John R. PENNING, two U.S. research scientists, **report on the use of exploding foils to induce shock waves in solids.** The foils are applied either in direct contact with the solid, which results in maximum pressures of up to 10 kbar in the solid, or in order to accelerate a thin-plate projectile to high velocity before striking the target. By using very thin projectile plates, it is possible to generate extremely short pressure pulses. For example, a 5-mil (127-μm)-thick Mylar projectile produces an 80-kbar pulse about 100 ns in duration.<br><br>In the same year, similar experiments are also carried out by Arthur H. GUENTHER[2475] and collaborators at the Air Force Special Weapons Center of Kirtland AFB, NM, in order to determine the shock responses of various materials and structures to intense external loads. Reaching flyer velocities of up to 5 km/s, they study various types of damage exhibited by different material configurations, *e.g.*, by structural deformation, internal cracking, spallation, and delamination. |
| 1962 | *Los Alamos Scientific Laboratory (LASL), Los Alamos, New Mexico* | Charles L. MADER,[2476] a U.S. theoretical detonation physicist, **studies the shock initiation of detonation in nitromethane [$CH_3NO_2$], liquid TNT, and single-crystal PETN.**<br><br>(i) He uses a numerical method to solve the reactive fluid dynamics equations that later became known as "HOM," and obtains agreement with previously reported experimental results. |

---

[2470] P.J.A. FULLER and J.H. PRICE: *Electrical conductivity of manganin and iron at high pressures.* Nature **193**, 262-263 (1962).

[2471] P.J.A. FULLER and J.H. PRICE: *Dynamic pressure measurements to 300 kbars with a resistance transducer.* Brit. J. Appl. Phys. **15**, 751-758 (1964).

[2472] O.E. JONES, F.W. NEILSON, and W.B. BENEDICK: *Dynamic yield behavior of explosively loaded metal determined by a quartz transducer technique.* J. Appl. Phys. **33**, 3224-3232 (1962).

[2473] R.A. GRAHAM and J.R. ASAY: *Measurement of wave profiles in shock-loaded solids.* High Temperatures, High Pressures **10**, 355-390 (1978).

[2474] D.V. KELLER and J.R. PENNING JR.: *Exploding foils – the production of plane shock waves and the acceleration of thin plates.* In: (W.G. CHACE and H.K. MOORE, eds.) *Exploding wires (II).* Plenum Press, New York (1962), pp. 263-277.

[2475] A.H. GUENTHER, D.C. WUNSCH, and T.D. SOAPES: *Acceleration of thin plates by exploding foil techniques.* In: (W.G. CHACE and H.K. MOORE, eds.) *Exploding wires (II).* Plenum Press, New York (1962), pp. 279-298.

[2476] The first unclassified references to HOM are by C.L. MADER: *The hydrodynamic hot spot and shock initiation of homogeneous explosives.* Rept. LA-2703, LASL, Los Alamos, NM (1962); *Shock and hot spot initiation of homogeneous explosives.* Phys. Fluids **6**, 375-381 (1963). ▪ HOM was also described in MADER's book *Numerical modeling of explosives and propellants.* CRC Press, Boca Raton, FL (1998), pp. 309-311; http://www.chipsbooks.com/numodexp.htm.

(ii) He also performs studies which explain the mechanism of initiation of detonation at hot spots created by the interaction of a shock with an inhomogeneity in detail. Energy transfer is accomplished by shocks and rarefactions.

(iii) He shows how experimentally observed criticality in bubble size can be related to the divergence rate of the shock, and to other features of the fluid flow. He discusses how these results are related to the initiation of detonation in inhomogeneous explosives.

In January 2006, in a letter to the author, Dr. MADER[2477] explained the origin of HOM: "HOM and its many copies are still used by most numerical modelers to describe solid explosives, detonation products and any mixture of the two assuming pressure and temperature equilibrium … It was the first attempt (starting in 1960) to use the best numerical reactive hydrodynamics and as realistic equations of state as are available to obtain quantitative agreement for homogeneous shock initiation experiments. Since the realistic equation of state required electronic computers and considerable amounts of computer time, the use of so much computer time and money was considered by some to be a 'sin' and thus the hydrodynamic code with HOM was called SIN [[2478]]. In the 1960–1965 time frame, the first SIN and HOM codes were written in machine language for the IBM 709 and then for the IBM 7030. Then the codes were written in FORTRAN, the first computer language that could be used on different types of computers … Since much of the computer time and effort was expended in solving the equation of state using complicated interactive techniques, the equation of state was known as a 'Hell Of a Mess' or 'HOM.' Los Alamos scientists of the time did not take themselves very seriously. We could not have imagined that it would still be the explosive equation of state treatment of choice 45 years later … The other scientist who made a major contribution to the development of HOM was Charles FOREST [at Los Alamos] who later developed the heterogeneous shock initiation model which uses HOM called 'Forest Fire.' As a first rate mathematician, he was involved in the development of the interactive techniques and was the first to call the result a 'Hell Of a Mess.'"

| | | |
|---|---|---|
| 1962 | *Hughes Laboratories, Malibu, California* | **Frederick J. McCLUNG**[2479] **and Robert W. HELLWARTH,** two staff research physicists, **invent laser Q-switching through the application of a Kerr cell** {KERR ⇨ 1875}. ▪ The method of Q-switching allows laser pulses of only nanosecond duration and extreme intensity to be obtained – known as **"giant laser pulses."** These unique features enable shock waves to be visualized with practically no motion blur. These short, highly focused laser pulses of high intensity also proved to be very useful for generating shock waves in all states of matter within a well-defined area {ASKAR'YAN & MOROZ ⇨ 1963}. With the advent of *mode locking* {MCCLUNG & WEINER ⇨ 1965}, even shorter and more powerful laser pulses became available.<br><br>In the following year, Eugene R. PERESSINI[2480] at Aerospace Corporation, El Segundo, CA, also produced giant laser pulses, using a rotating prism as a gain switch instead of a Kerr cell. |
| 1962 | *NASA's Jet Propulsion Laboratory (JPL), CalTech, Pasadena, California* | **Marcia NEUGEBAUER**[2481] **and Conway W. SNYDER,** two U.S. plasma physicists, **confirm the existence of a steady, supersonic solar wind, based upon measurements carried out by the spacecraft Mariner 2 on its way to the Venus.** They analyze velocities ranging from 319 to 771 km/s, which are pretty close to PARKER's previous estimations of 400–800 km/s {PARKER ⇨ 1958}. |

---

[2477] Mader Consulting Co., Honolulu, HI; http://www.mccohi.com/.
[2478] C.L. MADER and W.R. GAGE: *FORTRAN SIN. A one-dimensional hydrodynamic code for problems which include chemical reactions, elastic-plastic flow, spalling, and phase transitions*. Rept. LA-3720, Los Alamos Scientific Laboratory (LASL), Los Alamos, NM (1967).
[2479] F.J. MCCLUNG and R.W. HELLWARTH: *Giant optical pulsations from ruby*. J. Appl. Phys. **33**, 828-829 (1962).
[2480] E.R. PERESSINI: *Ruby laser giant-pulse generation by gain-switching*. Appl. Phys. Lett. **3**, 203-205 (1963).
[2481] M. NEUGEBAUER and C.W. SNYDER: *Solar plasma experiment. Preliminary Mariner-2 observations*. Science **138**, 1095-97 (Dec. 7, 1962); *Interplanetary solar-wind measurements by Mariner-2*. Space Res. **4**, 89-113 (1964).

| | | |
|---|---|---|
| 1962 | College of Engineering, UC Berkeley, California | Antoni K. OPPENHEIM,[2482] a Miller Research Professor at the Dept. of Mechanical Engineering, and his collaborators apply optical diagnostics to detonation research in order "to unravel some of the mysteries still surrounding the formation of detonation in a gaseous medium." They report on the observation of the **onset of the *retonation wave*** in a stoichiometric hydrogen-oxygen mixture contained in a detonation tube with a rectangular cross-section of $1 \times 1.5$ in.$^2$ ($25.4 \times 38.1$ mm$^2$) by means of schlieren streak and instantaneous photographs. These pictures show a regime where a shock wave train originating from the "superdetonation" propagates in one direction and the retonation wave propagates in the opposite direction.<br><br>In the early 1950s, the U.S. physical chemist Bernhard GREIFER[2483] and coworkers at the U.S. Bureau of Mines investigated the initiation of detonation in a $CO/O_2$ explosion using schlieren photography and streak camera recording, and observed retonation. **Retonation has also been observed in solid explosives when detonation is initiated through shock waves.** The first explanation of the retonation wave was apparently proposed by Algot PERSSON,[2484] a Swedish explosives expert at Nitroglycerin AB in Stockholm, who measured the detonation velocity on the surface of the initiated charge and inside it. |
| 1962 | Astrophysics Dept., Sternberg State Astronomical Institute, Moscow, Soviet Union | Iosif Samuilovich SHKLOVSKII [or SHKLOVSKIJ],[2485] an eminent Soviet radio astronomy professor at Moscow State University, **recognizes the applicability of adiabatic blast wave theory {G.I. TAYLOR ⇨ 1941} to the problem of supernova remnant theory.** He proves that the ***self-similar solution***,[2486] previously developed independently by the British physicist Geoffrey I. TAYLOR (1950) and the Soviet physicist Leonid I. SEDOV (1959), is also applicable to a supernova explosion phenomenon. Therefore, this phase is known as the **"Taylor-Sedov phase."** ▪ Extensive reviews of blast wave physics and astrophysics were provided in the Soviet Union by L.I. SEDOV (1959, 1992), Y.B. ZEL'DOVICH & Y.P. RAIZER (1966) and G.S. BISNOVATYI-KOGAN & S.A. SILICH (1995), and in the United States by J.P. OSTRIKER & C.F. MCKEE (1988).[2487] |
| 1962 | ISL, Saint-Louis, France | Karl VOLLRATH[2488] and Rudi SCHALL, two German research physicists, **investigate the behavior of ferroelectrics under shock compression** and demonstrate that the initiated charge can be used to generate a spark discharge. They propose the application of this technique to optically measure the arrival time of a shock wave in a sandwich-type sample, or as a multiple pulsed light source of low weight and great simplicity in high-speed photography. ▪ Charge release measurements from shock-loaded ferroelectrics were apparently first reported by the U.S. physicist Frank W. NEILSON[2489] of Sandia Laboratories. |
| 1962 | Los Alamos Scientific Laboratory (LASL), Los Alamos, New Mexico | Jerry D. WACKERLE,[2490] a U.S. shock physicist and fluid dynamicist, **determines the stress-volume relation of x-, y-, z-cut quartz and fused quartz** (a noncrystalline form) **up to about 750 kbar.** He discovers a number of unusual effects, such as an extraordinarily high Hugoniot elastic limit (HEL). Recovery experiments reveal that at sufficiently high shock pressures quartz transforms from the crystalline to the fused state. |

---

[2482] A.K. OPPENHEIM, A.J. LADERMAN, and P.A. URTIEW: *The onset of retonation.* Combust. Flame **6**, No. 3, 193-197 (Sept. 1962).

[2483] B. GREIFER, F.C. GIBSON, and C.M. MASON: *Studies on gaseous detonation.* In: *Proc. 2nd Symposium (ONR) on Detonation* [White Oak, MD, Feb. 1955]. Office of Naval Research (ONR), Washington, DC (1955), pp. 281-294.

[2484] A. PERSSON: *The transmission of detonation from charges of TNT to LFB-dynamite, nitrolite or TNT.* Appl. Sci. Res. **6** (Sect. A), 365-371 (1956).

[2485] I.S. SHKLOVSKII: *Supernova outbursts and the interstellar medium.* Sov. Astron. **6**, 162-166 (1962).

[2486] "Self-similarity" means that the evolution of the blast wave is such that if some initial configuration expands uniformly, any subsequent configuration is an enlargement of the previous configuration.

[2487] J.K. TRUELOVE and C.F. MCKEE: *Evolution of non-radiative supernova remnants.* Astrophys. J. (Suppl. Ser.) **120**, 299-326 (1999); http://www.journals.uchicago.edu/ApJ/journal/issues/ApJS/v120n2/37202/37202.web.pdf.

[2488] K. VOLLRATH and R. SCHALL: *Piezoelektrische Funkengeneratoren.* In: (J.G.A. DE GRAAF and P. TEGELAAR, eds.) *Proc. 6th Int. Congress on High-Speed Photography* [Scheveningen, The Netherlands, Sept. 1962]. Willink & Zoon, Haarlem (1963), pp. 403-408.

[2489] F.W. NEILSON: *Effects of strong shocks in ferroelectric materials.* Bull. Am. Phys. Soc. **2** [II], No. 6, 302 (Sept. 1957).

[2490] J. WACKERLE: *Shock-wave compression of quartz.* J. Appl. Phys. **33**, 922-937 (1962).

| | | |
|---|---|---|
| 1963 | New York City, New York | On February 1, **foundation of the American Institute of Aeronautics and Astronautics (AIAA),** due to the increasing interest in space exploration. The goal of the AIAA is "to advance the arts, sciences, and technology of aeronautics and astronautics, and to promote the professionalism of those engaged in these pursuits." ▪ The AIAA was the result of a merger between the American Rocket Society (ARS, founded in 1930) and the Institute of Aerospace Studies (IAS, founded in 1932).<br><br>In the same year, **the *AIAA Journal* is established,** which is one of the major journals on aerodynamics, supersonics and gas dynamics. The first editors are William R. SEARS and Martin SUMMERFIELD. The new journal is the result of a merger of the *Journal of the Aerospace Sciences* (from 1958) with the *ARS Journal* (from 1959). |
| 1963 | U.S.S.R. Academy of Sciences, Moscow, Soviet Union | In May, **at a conference held at the Institute of Chemical Physics, Andrei D. SAKHAROV,**[2491] an eminent Soviet nuclear and shock physicist, **and collaborators report on a new method for investigating the viscosity of substances behind a shock-wave front** which is based on the experimental study of the development of small perturbations at the shock wave front. ▪ An analysis of experimental and theoretical data indicated that there is a relationship between the viscosity and phase transformations in shock-compressed water.[2492] |
| 1963 | Moscow, Soviet Union | On July 25, **the international Nuclear Test Ban Treaty is ratified,** with the intention of "banning nuclear weapons tests in the atmosphere, in outer space and under water." It comes into effect on October 10, 1963. ▪ In contrast to the Comprehensive Test Ban Treaty {CTBT ⇨1996}, this treaty – later known as the **"Limited Test Ban Treaty (LTBT)"** or **"Partial Test Ban Treaty (PTBT)"** – didn't attempt to stop the development of new nuclear weapons, but rather shifted research activity to extensive underground nuclear tests, which prompted the development of new testing methods with sophisticated new diagnostic techniques. The U.S. stopped performing nuclear tests completely in September 1992.[2493] |
| 1963 | Dept. of Physics, Harvard University, Cambridge, Massachusetts | Percy W. BRIDGMAN,[2494] the foremost pioneer and patriarch of static high-pressure research, **provides a general outlook on static high-pressure physics and also addresses the unique capabilities and perspectives of shock waves.** He writes, "The very highest pressures will doubtless continue to be reached by some sort of shock wave techniques. **It is conceivable that a way will be found of superposing shock wave pressures on static pressures,** although there are indications that the ordinary chemical type of detonation may be increasingly difficult to produce at higher pressures. **Perhaps some fortunate experimenters may ultimately be able to command the use of atomic explosives in studying this field.** It appears that there are still pressing questions in the ordinary range of detonation which must be solved before this powerful tool becomes capable of yielding results of precision comparable with those of static methods … Some way should be found of dealing more adequately with lack of isotropy, as in single crystals, which at present is almost smothered in a general amorphousness. At the same time, any unique results which this technique is capable of giving because of the very short time intervals involved should be further exploited … Under static conditions it is not possible to superheat a solid, but in the very short times involved in detonations this becomes possible … Here would appear to be a promising tool for studying the mechanism of transitions and of |

---

[2491] A.D. SAKHAROV, R.M. ZAĬDEL', V.N. MINEEV, and A.G. OLEĬNIK: *Experimental investigation of the stability of shock waves and the mechanical properties of substances at high pressures and temperatures.* Sov. Phys. Dokl. **9**, 1091-1094 (1965).

[2492] V.N. MINEEV and E.V. SAVINOV: *Relationship between the viscosity and possible phase transformations in shock-compressed water.* Sov. Phys. JETP **41**, 656-657 (1976).

[2493] *United States nuclear tests July 1945 through September 1992.* Rept. DOE/NV-209-REV 15, U.S. Dept. of Energy, Nevada Operations Office (Dec. 2000); http://www.nv.doe.gov/library/publications/historical/DOENV_209_REV15.pdf.

[2494] P.W. BRIDGMAN: *General outlook on the field of high-pressure research.* In: (W. PAUL and D.M. WARSCHAUER, eds.) *Solids under pressure.* McGraw-Hill, New York (1963), pp. 1-13.

melting." ▪ BRIDGMAN apparently never investigated high-pressure properties of matter under shock waves, his studies on explosives in the 1940s were rather directed towards their behavior under static pressure.[2495] However, the wealth of data he collected on static compressibility and phase transitions of a large number of elements and compounds were often the only – albeit crude and vague – guidance available to shock wave pioneers when studying the behavior of dynamically loaded materials.

| 1963 | *Institute of Physical & Chemical Research IPCR), Bunkyo-ku, Tokyo and Japanese Defense Academy (JDA), Yokosuka, Kanagawa, Japan* | **First successful shock syntheses of inorganic materials from the powdered forms of their elements:**<br>▸ Yoshikazu HORIGUCHI[2496] and Yokan NOMURA at IPCR report on the successful synthesis of titanium carbide [TiC], an extremely hard material with high thermal shock and abrasion resistance.<br>▸ Yasuyuki KIMURA[2497] at JDA reports that he shock-synthesized zinc ferrite [$ZnFe_2O_4$] by compressing a mixture of zinc oxide [ZnO] and iron oxide [$Fe_2O_3$] using a detonating explosive. Zinc ferrite is an interesting ceramic material which has low magnetization and high electrical resistivity.<br>In 1985, Sandia shock physicists analysed shock synthesized zinc ferrite using electron microscopy.[2498] Shock synthesized ferrite, having the spinel crystal structure, was found to occur as recrystallized regions within a heavily deformed matrix of well mixed reactants. |
|---|---|---|
| 1963 | *Aeroelastic and Structures Research Laboratory, MIT, Cambridge, Massachusetts* | Bo LEMBCKE,[2499] a Swedish research engineer, **reports on a new double-driver shock tube for simulating the blast loading of vehicles in supersonic flight – the so-called "shock-on-shock problem"** {⇨Fig. 4.14–R}. He uses a shock tube with a large-area driver separated from the shock tube by a diaphragm downstream of the area change {⇨Fig. 4.10–I}. This generates two self-timed shock waves at the end of the shock tube, which reach the test section one after the other.<br>At that time, important examples of the shock-on-shock problem encompassed:<br>▸ the interception of a reentry vehicle by the blast wave from a nuclear explosion; and<br>▸ the interaction between the shock waves from two supersonic airplanes, as seen by one of them.<br>In both of these cases, the transient interaction between a steady supersonic flow field and a moving shock wave is the issue in question.<br>Up to the present, the interaction of a shock wave with a supersonic body has been given a great deal of attention, both experimentally and numerically. Four types of shock-on shock (s-o-s) interactions have been proposed and reclassified:[2500]<br>▸ type I: regular concave s-o-s (or "intersecting tangents") interactions;<br>▸ type II: regular convex s-o-s (or "non-intersecting tangents") interactions;<br>▸ type III: irregular s-o-s (or "single tangent") interactions; and<br>▸ type IV: irregular s-o-s fan interactions. |

---

[2495] P.W. BRIDGMAN: *The effect of high mechanical stress on certain solid explosives.* J. Chem. Phys. **15**, 311-313 (1947).
[2496] Y. HOIGUCHI and Y. NOMURA: *An explosive synthesis of titanium carbide.* Bull. Chem. Soc. Japan **36**, 486 (1963).
[2497] Y. KIMURA: *Formation of zinc ferrite by explosive compression.* Jpn. J. Appl. Phys. **2**, 312 (May 5, 1963).
[2498] M.J. CARR and R.A. GRAHAM: *Analytical electron microscopy study of shock synthesized zinc ferrite.* In: *Proc. 4th APS Topical Conf. on Shock Waves in Condensed Matter* [Spokane, WA, July 1985]. Plenum Press, New York (1986), pp. 803-808.
[2499] B. LEMBCKE: *Double-shock shock tube for simulating blast loading in supersonic flow.* AIAA J. **1**, 1417-1418 (1963).
[2500] C. LAW, L.T. FELTHUN, and B.W. SKEWS: *Two-dimensional numerical study of planar shock-wave/moving-body interactions.* Shock Waves **13**, 381-394 (2003).

| | | |
|---|---|---|
| 1963 | *Los Alamos Scientific Laboratory (LASL), Los Alamos, New Mexico* | **Charles L. MADER,**[2501] a U.S. shock wave and detonation physicist, **improves the *Becker-Kistiakowsky-Wilson (BKW) equation of state*** {BECKER ⇨1922; KISTIAKOWSKY & WILSON ⇨1941} **of the detonation gas of an explosive** by including certain parameter values which he had derived from a comparison of calculated results with experimental pressures for various high explosives. His semi-empirical equation of state is based upon the application of a repulsive potential to the virial equation of state, and the assumption that the detonation products are in a chemical equilibrium. It has been the most extensively calibrated and used of the many equations of state that have been proposed to describe the detonation process.<br><br>In later years, MADER[2502] **devised a Fortran code for computing the detonation properties of explosives.** |
| 1963 | *Faculty of Engineering, University of Glasgow, Scotland, U.K.* | **Terence R.F. NONWEILER,**[2503] a professor of aerodynamics and fluid mechanics, **devises the concept of a *waverider* as a lifting system** – a vehicle with a hollow underside that allows hypersonic flight by riding a planar shock wave attached to the leading edges. ▪ In the late 1950s, NONWEILER – then working at Queen's University in Belfast – had designed a waverider atmospheric reentry vehicle that could act as a manned craft for the British space program. His waverider concept produces the fastest winged vehicles and is suited for hypersonic flight, and it has spawned a variety of waverider-based reentry vehicle concepts. |
| 1963 | *Honeywell Research Center, Hopkins, Minnesota* | **John Fetch READY,**[2504] a U.S. research scientist, **observes that a high-power laser pulse absorbed at an opaque metallic surface produces a luminous plume of vaporized material that has been blasted from the surface.** He concludes, "Material at some depth below the surface reaches its vaporization temperature before the material at the surface has absorbed its latent heat of vaporization. This leads to a pulse of high pressure and subsequent superheating of the underlying material until the temperature rises above the critical point. There is then no longer any distinction between the superheated solid and a highly condensed gas. **The emission of the vaporized material, which is delayed relative to the peak of the temperature pulse at the surface, then proceeds like a thermal explosion.**"<br><br>In the following years, this phenomenon was studied in great detail worldwide. It stimulated not only laser weapons technology, but also attained the greatest practical importance when working laser materials. Important medical applications using excimer lasers {SEARLES & HART ⇨1975} include their use in ophthalmology for reshaping or sculpturing the cornea surface {PULIAFITO ET AL. ⇨1987; ⇨Fig. 4.16−Z}, and in dermatology for the treatment of skin diseases and cosmetic purposes. |
| 1963 | *Colorado School of Mines, Golden, Colorado & U.S. Naval Ordnance Test Station (NOTS), China Lake, California* | **John S. RINEHART,**[2505] an applied physicist at the Mining Research Laboratory (Golden, CO), **and John PEARSON,** an explosives specialist at the Michelson Laboratory (China Lake, CA), **publish their book *Explosive Working of Metals.***<br><br>(i) They provide a definition of the term ***explosive working:*** "The term is used in an all-inclusive sense to include the change of shape of a metal part; the displacement, removal, or joining of metal; cutting and shearing; and changes in the engineering and metallurgical properties of a material all through the use of explosive energy." |

---

[2501] C.L. MADER: *Detonation properties of condensed explosives computed using the Becker-Kistiakowsky-Wilson equation of state.* Rept. LA-2900, LASL, Los Alamos, NM (1963). ▪ MADER's report is (still) classified. More information on his improved equation of state can be found, for example, in the book written by C.H. JOHANSSON and P.A. PERSSON: *Detonics of high explosives.* Academic Press, London & New York (1970), pp. 24-28.

[2502] C.L. MADER: *FORTRAN BKW (Becker-Kistiakowsky-Wilson): a code for computing the detonation properties of explosives.* Rept. LA-3704, LASL, Los Alamos, NM (1967).

[2503] T.R.F. NONWEILER: *Delta wings of shapes amenable to exact shock-wave theory.* J. Roy. Aeronaut. Soc. **67**, 39-40 (Jan. 1963).

[2504] J.F. READY: *Development of plume of material vaporized by giant pulse laser.* Appl. Phys. Lett. **3**, 11-13 (July 1, 1963); *Effects due to absorption of laser radiation.* J. Appl. Phys. **36**, 462-468 (1965).

[2505] J.S. RINEHART and J. PEARSON: *Explosive working of metals.* Pergamon Press, Oxford (1963); *see Preface*, p. vii.

(ii) They chronicle a host of impressive applications of shock waves in forming and joining metals and alloys. ▪ In the decades to follow, shock waves were used for a wide variety of industrial applications {MURR ⇒ 1988}.

In the 1950s, both authors (who had cooperated closely in detonics and materials dynamics research) had published the pioneering textbook *The Behavior of Metals Under Impulsive Loads*, which became a classic in this field {RINEHART & PEARSON ⇒ 1954}.

| 1963 | Mount Wilson Observatory & Palomar Observatory, CalTech, Pasadena, California |

Maarten SCHMIDT,[2506] a Dutch-American astronomer at CalTech, **discovers that quasars exhibit an extreme redshift.**

(i) He finds that the star-like object coincides with the position of a strange radio source designated No. 273 in the third catalog ("3C") compiled by radio astronomers at the University of Cambridge. The optical magnitude of this "quasar" (short for <u>qua</u>si-stell<u>ar</u> radio source) – the brightest of all known quasars and labeled 3C-273 – is 13 and has a redshift of 0.158. ▪ A subsequent study of quasar 3C-48 showed that it has even a redshift of 0.367.[2507]

(ii) A study of the strange spectrum of 3C-273 revealed that its light is shifted toward the red end of the spectrum by an amount that – according to the Hubble expansion law {HUBBLE ⇒ 1929} – indicates the object is receding at a velocity of 47,400 km/s (*i.e.*, almost one-sixth of the velocity of light). This implies that quasar 3C-273 must be extremely distant, approximately three billion light-years away.[2508]

(iii) **The fact that the quasars are visible at such distances implies that they emit enormous amounts of energy** and are certainly not stars but rather related to active galaxies.

(iv) He observes "forbidden" spectral lines of oxygen and neon in the quasar spectra. These forbidden lines are never experienced on Earth, and are observed only in the radiation from gaseous nebula (extremely thin clouds of ionized gas).

The first quasars, then believed to be stars, had already been found three years before during optical identifications of bright radio sources {MATTHEWS & SANDAGE ⇒ 1960}. There are at least a million stars of that magnitude across the sky. ▪ The term *quasi-stellar radio source* – later named QSR and shortened to *quasars* – came from the Goddard Institute for Space Studies, New York, and was coined by Hong-Yee CHIU,[2509] a NASA astrophysicist and space scientist. Since the discovery that the majority of these objects are not radio-emitters, the term *quasar* has fallen out of favor, with QSO (quasi-stellar object) now preferred.[2510]

In the years following, SCHMIDT and coworkers determined the velocities of recession of other quasars, among them **quasar 3C-9** with a redshift of ≈ 2 – **the farthest known object in the Universe in terms of both time and space, which appears to be receding at a velocity of about 240,000 km/s** (*i.e.*, four-fifths of the velocity of light). This implies that the light reaching Earth from quasar 3C-9 came from near the edge of the observable portion of the Universe, having left the quasar about ten billion years ago. ▪ By the early 1970s, several hundred radio sources had already been identified for star-like objects. In 1983, SCHMIDT[2511] gave a historical perspective on quasar research.

The nature of quasars is still the subject of debate. It has been proposed that quasars are in essence galaxies packed with supernovae, known as "active galactic nuclei (AGNs)." **The hydrodynamic collapse of the AGN results in a black hole, which powers its explosive en-

---

[2506] M. SCHMIDT: *A star-like object with large red-shift.* Nature **197**, 1040 (1963).
[2507] J.L. GREENSTEIN and M. SCHMIDT: *The quasi-stellar radio sources 3C-48 and 3C-273.* Astrophys. J. **140**, No. 1, 1-34 (July 1964).
[2508] M. SCHMIDT and F. BELLO: *The evolution of quasars.* Scient. Am. **224**, 54-69 (May 1971).
[2509] H.Y. CHIU: *Gravitational collapse.* Phys. Today **17**, 21-34 (May 1964).
[2510] P. ROBERTSON: *Beyond southern skies: radio astronomy and the Parkes telescope.* Cambridge University Press, Cambridge (1992), p. 234.
[2511] M. SCHMIDT: *Discovery of quasars.* In: (K. KELLERMANN and B. SHEETS, eds.) *Serendipitous discoveries in radio astronomy.* Proceedings of a Workshop [National Radio Astronomy Observatory (NRAO), Green Bank, West Virginia, May 1983]. NRAO, Green Bank, VA (1984).

| | | |
|---|---|---|
| 963 | U.S.S.R. Academy of Sciences, Moscow, Soviet Union | ergy output. The collapse is halted when nuclear densities are reached in the core. **The core bounces and sends a shock wave outward through the quasar.**

Kirill Ivanovich SHCHELKIN[2512] **and Yakov Kirillovich TROSHIN publish their book *Gazodinamika gorenija* ("Gas Dynamics of Combustion"),** in which they show a variety of photographs of surface traces on smoke-stained plates left by the detonation waves of explosive gaseous mixtures. They explain that the *periodic cell structure* {DENISOV & TROSHIN ⇨ 1959; ⇨Fig. 4.16–U} recorded is the result of periodic hydrodynamic instabilities which generate transverse waves, causing intermittent ignitions of the reactive gas.

**The mechanism of production for the periodic cell structure observed on smoke-coated surfaces was a mystery at first, but it soon became clear that the lines on the soot record are primarily produced by the triple points of Mach stems propagating across the detonation front** {⇨Fig. 4.16–V}. Referring to a triple point intersection in the detonation wave, the renowned U.S. combustion researcher Antoni K. OPPENHEIM[2513] wrote, "Besides the two head shock fronts – the weaker one referred to usually as the *incident wave*, while the stronger as the so-called 'Mach stem' – there is also a *reflected shock* generated by such an intersection in order to satisfy the dynamic compatibility requirement for the existence of a state of uniform pressure and particle path direction in the flow regime behind the two fronts. This regime is, moreover, divided into two parts by a slip-line that arises due to the fact that the gas particles were brought there by different routes and have, therefore, quite different flow velocity at its sides, the flow behind the Mach stem being locally subsonic, while that behind the reflection shock is supersonic. **This then gives rise to a considerable amount of shear that creates a concentrated vortex of high temperature gases. The net effect of these phenomena is quite spectacular: since such a vortex acts, in effect, as a *rotating stylus*, it can etch on a suitable material the trace of its path, and the detonation front is rendered thus the Biblical property of being able to write on the walls!** In the laboratory experiment such writing is obtained by having the wall of the detonation tube covered with a thin layer of carbon soot." ▪ An unmistakable experimental proof of his interesting hypothesis, however, is still pending.[2514] |
| 1963 | GEC Company, Palo Alto, California & Dept. of Electrical Engineering, UC Berkeley, California | Richard M. WHITE,[2515] a U.S. applied physicist and assistant professor at UC Berkeley, **reports on the production of high-frequency elastic waves by the impact of a pulsed beam of electrons and pulsed ruby laser light upon a solid target in a highly evacuated chamber.** The pressure gauge consists of a thin 0.5-in. (12.7-mm)-dia. piezoelectric barium titanate [BaTiO$_3$] crystal which is attached to the back side of the target.

(i) The 10-kV, 1 × 5-in.² (25.4 × 127-mm²) electron beam has a rectangular pulse duration of about 2 μs and produces a stress pulse of almost the same short time duration upon impact with the target.

(ii) Experiments with ruby laser irradiation show that a blackened target produces elastic waves of considerably higher amplitude. Similar to the results observed from the electron beam experiments, the stress pulse duration is comparable with that of the laser output pulse. |

---

[2512] K.I. SHCHELKIN and Y.K. TROSHIN: *Gazodinamika gorenija*. Izd. Akad. Nauk SSSR, Moskva (1963). Engl. translation: *Gas dynamics of combustion*. NASA TT-F-23 (1964); Mono Book Corporation, Baltimore, MD (1965), pp. 26-38.

[2513] A.K. OPPENHEIM: *Introduction to gas dynamics of explosions*. Springer, Vienna etc. (1970), pp. 27-29.

[2514] Mach reflection studies performed by the author in 1996 in thin layers of air enclosed between two parallel glass plates – an arrangement very similar to Ernst MACH's pioneering soot plate experiments {E. MACH & WOSYKA ⇨1875} but using a microscope and a picosecond pulse laser as a light source to minimize motion blur at high optical magnification – didn't show any micro vortices propagating along the triple point trajectory. Perhaps they are only formed when the glass plate is covered by a soot layer which acts as a rough surface, and therefore might produce turbulences upon transit of the triple point. ▪ Very recently, Japanese researchers gave a different explanation for the mechanism of shock-induced soot removal (*see* Sect. 2.8.3).

[2515] R.M. WHITE: *Elastic wave generation by electron bombardment or electromagnetic wave absorption*. J. Appl. Phys. **34**, 2123-2124 (1963); *Generation of elastic waves by transient surface heating*. Ibid. **34**, 3559-3567 (1963).

| 1963 | Institute of Chemical Physics, U.S.S.R. Academy of Sciences, Moscow, Soviet Union | **Yakov B. ZEL'DOVICH**[2516] **and Yuri P. RAIZER,** two leading Soviet shock and detonation physicists**, publish their two-volume textbook** *Fizika udarnych voln i vysokotemperaturnych gidrodinamiceskich javlenij* ("Physics of Shock Waves and High-Temperature Hydrodynamic Phenomena"). It widely treats high-rate phenomena associated with high concentrations of energy, very high temperatures and pressures, and extreme velocities such as those encountered in strong shock waves, explosions, hypersonic flight in the atmosphere, very strong electrical discharges, *etc.* ▪ Their book, which was translated into English a few years later (1966) and was recently reprinted, discloses many previously unknown research results obtained by former Soviet researchers to their colleagues in the Western World. It has become one of the main sources of references for shock physicists of astrophysical, nuclear, aerodynamic and seismological communities. |
|---|---|---|
| 1964 | Prince William Sound [ca. 120 km east of Anchorage], Alaska | On March 27, **Alaska is shaken by the second largest earthquake ever recorded and the largest recorded for the Northern Hemisphere:**<br>(i) This major earthquake – later called the **"Great Alaskan Earthquake"** – has a magnitude of 9.2; it lifts 25,000 square miles (64,700 km$^2$) of coast and causes extensive damage in Alaska.<br>(ii) Local tsunami waves triggered by this earthquake are extremely destructive in Prince William Sound and other areas of Alaska. However, the Pacific-wide tsunami generated by the earthquake also causes major damage in western Canada, Oregon, California, and even on the remote Hawaiian islands, where the tsunami transforms into a towering 20-ft (6-m) wall of water upon arrival.<br>(iii) There are 52 larger aftershocks, which are heavily concentrated on the northeast and the southwest of the uplifted coastal region which generated the tsunamis. The largest aftershock has a magnitude of 6.7. ▪ Aftershocks continued for more than a year, and thousands of small aftershocks were recorded in the months following the main earthquake.[2517]<br>The large number of casualties[2518] caused by tsunamis (122) compared to the relatively small number of casualties resulting from the earthquake (15) clearly demonstrated the need to improve existing tsunami warning systems and link them internationally {ITIC ⇨1965}. |
| 1964 | Canadian Defence Research Establishment (CDRE), Suffield [now Defence R & D Canada – Suffield], Alberta, Canada | In July, **blast wave researchers at the Experimental Proving Ground of CDRE monitor the surface detonation of a hemispherical charge of 500 U.S. tons (454,000 kg) of TNT – the largest detonation of conventional explosives ever.**<br>(i) The charge consists of 30,664 blocks of TNT with an average mass of 14.8 kg, arranged to form a hemisphere with the flat surface resting on the ground. The central booster charge consists of 14 blocks of 70/30 tetryl, each of mass 15 kg, to give a total charge mass of 454,000 kg.<br>(ii) The experiment is carried out in order to simulate the blast effects of a nuclear explosion. **John M. DEWEY,**[2519] a British-born Canadian physics professor at the University of Victoria and an authority on blast waves, **determines the physical properties of a spherically expanding shock wave from measured time-resolved particle trajectories within the wave – a method called "particle tracer photogrammetry."** He uses an array of smoke trails generated by a number of special smoke mortar shells simultaneously fired ahead of the incident blast wave {⇨Fig. 4.16–Q}. ▪ Prior to the Suffield Test, DEWEY[2520] used this diagnostic |

---

[2516] Y.B. ZEL'DOVICH and Y.P. RAIZER: *Fizika udarnych voln i vysokotemperaturnych gidrodinamiceskich javlenij.* Gos. Izd. Fiz. - Mat. Lit., Moskva (1963); Engl. translation: *Physics of shock waves and high-temperature hydrodynamic phenomena.* 2 vols., Academic Press, New York (1966–1977); 1 vol., Dover Publ., Mineola, NY (2002).

[2517] G. PARARAS-CARAYANNIS: *The March 27, 1964, Great Alaskan Earthquake;* http://www.drgeorgepc.com/Earthquake1964Alaska.html.

[2518] T.J. SOKOLOWSKI: *The Great Alaskan Earthquake & Tsunamis of 1964.* West Coast & Alaska Tsunami Warning Center (WCATWC), Palmer Alaska, AK; http://wcatwc.arh.noaa.gov/64quake.htm.

[2519] J.M. DEWEY: *The properties of a blast wave obtained from an analysis of the particle trajectories.* Proc. Roy. Soc. Lond. **A324**, 275-299 (1971).

[2520] J.M. DEWEY: *The air velocity in blast waves from TNT explosions.* Proc. Roy. Soc. Lond. **A279**, 366-385 (1964).

method to measure peak particle velocities in blast waves from surface-burst TNT charges of various weights, ranging from 30 to 200,000 pounds (13.6–90,718 kg). He found good agreement of his scaled measured particle velocity data with those calculated from the shock velocity, and demonstrated that Hopkinson scaling {B. HOPKINSON ⇨ 1915} applies over a very wide range of distances and explosive source energies.

In the following year, an identical charge was detonated by the U.S. Navy in the Pacific Ocean on Kahoolawe Island (the 8th largest Hawaiian Island). The study of large chemical explosions was resumed by CDRE in 1970. Of particular interest in such studies were blast overpressures ranging from about 7.5 to 0.5 bar and their response to structures. In this pressure range, the corresponding peak particle velocities amount to Mach 2 and 0.3, respectively.
▪ More recently, J.M. DEWEY[2521] reviewed the physical properties of expanding spherical shocks as well as their scaling, numerical modeling and experimental measurement techniques in a handbook article.

| 1964 | Lop Nur Test Ground, Province of Sinkiang, northwestern China | On October 16, **China detonates her first atomic bomb,** a pure-fission U-235 implosion device dubbed "596" (about 20 kilotons of TNT equivalent), thereby becoming the 5th Nuclear Power after the United States, the Soviet Union, Great Britain, and France.

Following the success of a boosted fission test (an uranium-lithium device) on December 18, 1966, China performed her first detonation of a two-stage thermonuclear weapon on June 17, 1967. The bomb exploded at an altitude of 9,472 ft (3,296 m) with a yield of 3.3 megatons. |
|---|---|---|
| 1964 | Poulter Laboratory, Stanford Research Institute (SRI), Menlo Park, California | David BERNSTEIN[2522] and Douglas D. KEOUGH, two U.S. shock wave physicists, **report on measurements of the change in resistance of manganin** {BRIDGMAN ⇨ 1911} **as a function of shock pressures** obtained from explosively generated planar shock waves in an insulating material (C-7 epoxy) with a known Hugoniot equation of state. The Hugoniot of manganin, an alloy consisting of 84% Cu, 12% Mn and 4% Ni, is measured up to 359 kbar. **They show that Ni does not have a great effect upon the piezoresistive response, and that the positive linear response is due to Mn.** ▪ Their studies were stimulated by project work that they finished in 1962 for the Defense Atomic Support Agency (DASA) in Washington, DC. At about that time, similar work on piezoresistive gauges was also initiated in the United Kingdom {FULLER & PRICE ⇨ 1962}. |
| 1964 | United States | Neville L. CARTER[2523] and George C. KENNEDY, two U.S. meteorite researchers at Texas A&M University (College Station, TX), **suggest that the formation of sand-grain-sized diamonds discovered in the Canyon Diablo meteorite** – the excavator of Meteor Crater {⇨ c.50,000 years ago} – **was the result of high gravitational pressures.**

Two years later, **Edward ANDERS,**[2524] a geophysics professor at the University of Chicago, **and Michael E. LIPSCHUTZ,** a chemistry professor at Purdue University, **explained the occurrence of diamond as being due to the shock-induced transformation of graphite.** This transformation likely occurred at the moment of terrestrial impact and disintegration of the projectile during crater formation.

Iron meteorites containing diamonds are very rare. For example, of the many meteorites recovered so far from the Allan Hills, Antarctica, only one meteorite, a reddish brown 10.51-kg object discovered in 1977, named ALHA 77283, contains diamonds produced by impact {CLARKE ET AL. ⇨ 1981}. In 1987, **a team of researchers at the University of Chicago reported the discovery of very fine-grained meteorite-embedded diamonds which contain an isotopic mix- |

---

[2521] J.M. DEWEY: *Spherical shock waves*. In: (G. BEN-DOR ET AL., eds.) *Handbook of shock waves*. Academic Press, New York (2001), vol. 2, pp. 441-481.
[2522] D. BERNSTEIN and D.D. KEOUGH: *Piezoresistivity of manganin*. J. Appl. Phys. **35**, 1471-1474 (1964).
[2523] N.L. CARTER and G.C. KENNEDY: *Origin of diamonds in the Canyon Diablo and Novo Urei meteorites*. J. Geophys. Res. **69**, 2403-2421 (1964); *Reply*. Ibid. **71**, 663-672 (1966).
[2524] E. ANDERS and M.E. LIPSCHUTZ: *Critique of paper by N.L. CARTER and G.C. KENNEDY, "Origin of diamonds in the Canyon Diablo and Novo Urei meteorites."* J. Geophys. Res. **71**, 643-661 (1966); *Reply*. Ibid. **71**, 673-674 (1966).

ture of xenon gas not found on Earth, thus indicating an extrasolar origin for the diamonds {LEWIS ET AL. ⇨1987}.[2525] The team proposed the idea that the lucent crystals formed in the atmosphere of a "red giant" or dying star before it collapsed and exploded billions of years ago, catapulting the diamond-studded material far out into space before eventually falling to Earth. If this scenario is correct, the team concluded, then interstellar dust may contain diamonds.

| | | |
|---|---|---|
| 1964 | Dept. of Physics, University of Göttingen, Germany | Wolfgang EISENMENGER,[2526] a German physicist, **measures the shock front thickness of weak shock waves in various liquids** (*e.g.*, water, acetone, carbon tetrachloride, toluene, methyl alcohol, and ethyl alcohol) **by analyzing the frequency spectrum of the propagating shock wave.** He applies an electromagnetic planar shock generator {EISENMENGER ⇨1959} which produces planar shock waves in the range 10–100 bar, and a shock wave microphone based on the principle of surface scanning of piezoelectric crystals, which gives rise times of less than 1 ns. Analyzing microphone signals in the range 280–950 MHz, he measures a shock front thickness of 5 μm in water at 10 bar, which decreases to about 1 μm at 60 bar. |
| 1964 | Chemistry Dept., Brown University, Providence, Rhode Island & Physical Institute, University of Bonn, Germany | Edward F. GREENE,[2527] a U.S. chemist at Brown University, **and J. Peter TOENNIES,** a German physicist at the University of Bonn, **publish their monograph *Chemical Reactions in Shock Waves,*** which will become a classic in this field. Emphasizing reactions in gases, it summarizes the experimental results reported in the literature up to the end of 1963 in tabular form.<br><br>In the following decades, research into shock-induced solid-state chemistry, particularly of inorganic powder materials, gained increasing interest in the Soviet Union[2528, 2529] and in the United States.[2530, 2531] ▪ While shock-induced chemical effects would not be expected in solids with perfect crystal lattices in thermodynamic equilibrium (*benign shock concept*) on the microsecond timescales of typical shock pulses, they are likely in shock-compressed defective lattices (*catastrophic shock concept*).[2532] |
| 1964 | Instytut Elektrotechniki, Warsaw, Poland | Jan NASILOWSKI,[2533] a Polish researcher, **studies melting effects of exploding wires and observes that only thin wires are transformed into unduloids** {BAXTER ⇨1950}. His photograph shows a 0.5-mm-dia. Cu wire during explosion, compared with a millimeter grid. ▪ In the 1970s, Walter LOCHTE-HOLTGREVEN,[2534] a German professor of experimental physics, resumed this subject. Photographs showed that the wire fractures at irregular distances along it, suggesting that the fractures occur at randomly distributed weak points along the wire.<br><br>**These experiments showed that thin exploding wires or fibers fracture into small solid pieces or segments before they can be vaporized by the electric current** – it is as if a negative pressure acts on the wires or fibers. However, many aspects of the process of wire frag- |

---

[2525] R.S. CLARKE JR., D.E. APPLEMAN, and D.R. ROSS: *An antarctic iron meteorite contains preterrestrial impact-produced diamond and lonsdaleite.* Nature **291**, 396-398 (June 4, 1981).

[2526] W. EISENMENGER: *Experimentelle Bestimmung der Stoßfrontdicke aus dem akustischen Frequenzspektrum elektromagnetisch erzeugter Stoßwellen in Flüssigkeiten bei einem Stoßdruckbereich von 10 ATM bis 100 ATM.* Acustica **14**, 187-204 (1964).

[2527] E.F. GREENE and J.P. TOENIES: *Chemical reactions in shock waves.* E. Arnold, London (1964). ▪ The book is a revised edition from the previous German edition *Chemische Reaktionen in Stoßwellen,* vol. 3 of *Fortschritte der Physikalischen Chemie.* Steinkopff-Verlag, Darmstadt (1959).

[2528] A.N. DREMIN and O.N. BREUSOV: *The chemistry of shock compression* [in Russ.]. Priroda **12**, 10-17 (1971); Engl. translation SAND-80-6003, Sandia National Laboratories, Albuquerque, NM (1980).

[2529] G.A. ADADUROV ET AL.: *Chemical conversions of condensed materials by shock waves.* Mendeleev Chem. J. **18**, 92-103 (1973); *Transformations of condensed substances under shock-wave compression in controlled thermodynamic conditions.* Russ. Chem. Rev. **50**, 948-957 (1981).

[2530] R.A. GRAHAM ET AL.: *Chemical reaction in shock compression of solids.* Rept. SAND-85-2411C, Sandia National Laboratories, Albuquerque, NM (1986).

[2531] R.A. GRAHAM: *Shock compression of solids as a physical-chemical-mechanical process.* In: (S.C. SCHMIDT and N.C. HOLMES, eds.) *Shock waves in condensed matter – 1987.* North-Holland, Amsterdam etc. (1988), pp. 11-18.

[2532] R.A. GRAHAM: *Shock-induced electrical and chemical activity in polymers and other materials.* Bull. Am. Phys. Soc. **25**, 495 (1980); *Shock-induced electrical activity in polymeric solids. A mechanically induced bond scission model.* J. Phys. Chem. **83**, 3048-3056 (1979).

[2533] J. NASILOWSKI: *Unduloids and striated disintegration of wires.* In: (W.G. CHACE and H.K. MOORE, eds.) *Exploding wires (III).* Plenum Press, New York (1964), pp. 295-313.

[2534] W. LOCHTE-HOLTGREVEN: *Nuclear fusion in very dense plasmas obtained from electrically "exploded" liquid threads.* Atomkern-Energie **28**, 150-154 (1976).

mentation still remain unclear. In 1983, Peter GRANEAU,[2535] a U.S. physicist at Northeastern University, Boston, argued that neither mechanical vibrations induced by the electromagnetic pinch force nor thermal expansion could have been responsible for the wire disintegration, because they were too weak. He speculated that the Ampere force law would lead to a longitudinal tension in the wire {Carl HERING 1907, see NORTHRUP ⇒1907}, which was later confirmed experimentally at Oxford University by his son Neal GRANEAU and collaborators.[2536] Researchers in the United Kingdom, using a simplified *magnetothermoelastic model* to study flexural vibrations induced by high pulsed currents in wires, have shown that the flexural vibrations induced are strong enough to lead to the breaking of the wire over a wide range of parameters.[2537]

| 1964 | NASA's Goddard Space Flight Center (GSFC), Greenbelt, Maryland | Norman F. NESS,[2538] a U.S. space scientist, **and his colleagues detect clear signs that a collisionless shock exists where the solar wind encounters the Earth's magnetic field.**<br><br>(i) They used magnetometer and solar wind detector data collected by the scientific satellite Explorer 18, also called "IMP 1" (short for Interplanetary Monitoring Platform).<br><br>(ii) **Their data analysis is the first to reveal the existence of a collision-free bow shock ahead of the magnetopause,** as predicted the year before {AXFORD ⇒ 1962; KELLOGG ⇒ 1962; ⇒Fig. 4.1–X}. ▪ In the 1950s, plasma physicists had theorized that – contrary to the expectations of many scientists – shock waves similar to aerodynamic shocks could form even in the near-vacuum of outer space, where particle collisions are extremely rare.<br><br>(iii) The data returned by IMP-1 appeared somewhat puzzling, because sometimes the shock appeared thin, while at other times it appeared thick. ▪ In 1971, Eugene W. GREENSTADT, a U.S. space scientist at TRW Defense & Space Systems Group (Redondo Beach, CA), and his colleagues assembled the first evidence that **the thickness of the Earth's shock varies with the direction of the magnetic field of the solar wind,** which constantly changes.[2539]<br><br>Data gathered from earlier Explorer missions, starting in 1958 and continuing into the 1970s, resulted in the discovery of the innermost of the two Van Allen radiation belts. **Caused by the solar wind, the radiation belt ends abruptly at a discrete outer edge – the front of the bow shock wave – usually at about 10 Earth radii.** The intensity of trapped particles decreases by a factor of about 1,000, and the Earth's magnetic field changes its character, within a layer less than 100 km thick {⇒Fig. 4.12–D}. |
|---|---|---|
| 1964 | U.S. Ballistics Research Laboratory (BRL), Aberdeen Proving Ground, Maryland | **Beauregard PERKINS JR.,** a U.S. geophysicist at BRL, **and Wendell F. JACKSON,** a U.S. chemist and ballistic engineer at the Explosives Dept. of E.I. du Pont de Nemours & Company, New Jersey, **publish their** *Handbook for Prediction of Air Blast Focusing* (U.S. Army, BRL Rept. 1,240). It shows how to calculate blast propagation and refraction effects under inhomogeneous atmospheric conditions. Unusual weather conditions (such as a low-level temperature inversion) may cause the blast to be focused on the ground some distance from the source. |
| 1964 | National Research Council (NRC) of Canada, Ottawa, Canada | S.A. RAMSDEN,[2540] a British applied physicist, **and P. SAVIC propose a detonation model for a laser-induced spark in air:**<br><br>(i) They write, "In the course of recent work on the spark produced in air by a focused ruby laser beam, the rather surprising result was obtained that, after breakdown, the spark developed asymmetrically, moving towards the lens with an initial velocity of about $10^7$ cm/s [100 km/s]. |

---

[2535] P. GRANEAU: *First indication of Ampère tension in solid electric conductors.* Phys. Lett. **A97**, 253-255 (1983).
[2536] N. GRANEAU, T. PHIPPS JR., and D. ROSCOE: *An experimental confirmation of longitudinal electrodynamic forces.* European Phys. J. D **15**, 87-97 (2005).
[2537] A. LUKYANOV and S. MOLOKOV: *Flexural vibrations induced in thin metal wires carrying high currents.* J. Phys. D (Appl. Phys.) **34**, 1543-1552 (2001).
[2538] N.F. NESS, C.S. SCEARCE, and J.B. SEEK: *Initial results of the IMP-1 magnetic field experiment.* J. Geophys. Res. **69**, 3531-3569 (1964).
[2539] R.Z. SAGDEEV and C.F. KENNEL: *Collisionless shock waves in interstellar matter.* Scient. Am. **106**, No. 4, 40-47 (April 1991); http://www.totse.com/en/fringe/fringe_science/shockwav.html.
[2540] S. RAMSDEN and P. SAVIC: *A radiative detonation model for the development of a laser-induced spark in air.* Nature **203**, 1217-1219 (1964).

In this article, this effect is discussed in terms of a new mechanism – that of a *radiation-supported shock wave.* It is assumed that after breakdown a shock wave propagates into the undisturbed gas, and that further absorption of energy from the laser beam then occurs behind the shock front traveling towards the lens, in the manner of a detonation wave. **After the end of the laser pulse, the heated gas then expands in the form of a blast wave."**

(ii) Replacing the reaction energy by the energy per unit mass absorbed behind the shock front from the laser beam, they treat the laser-supported shock wave in terms of the Chapman-Jouguet (CJ) hypothesis of detonation {JOUGUET ⇨1904–1906}. The later phase of shock wave emission appears to be best described by the blast wave theory for a point explosion {G.I. TAYLOR ⇨1941}.

(iii) Data measured when moving the focus towards the lens with a velocity of at least $10^5$ m/s, which they obtained using an image converter streak camera, show good agreement with the theoretically obtained 0.6th power law.

The radiation-induced breakdown within a sufficiently small volume of gas that is provoked by a high-energy pulsed laser can be considered to be almost a point explosion under laboratory conditions. This unique detonation phenomenon – later termed *laser-supported detonation (LSD)* {⇨Fig. 4.11–I} – was resumed in order to develop a "lightcraft" {MYRABO & MEAD ⇨1997; ⇨Fig. 4.11–J}, a laser-boosted rocket, as part of the research into the "Star Wars" anti-missile initiative {SDI Concept ⇨1983}.

Two years later, Yuri P. RAIZER,[2541] a Soviet plasma physicist, estimated the absorption and temperature of a laser-heated gas in more detail. He distinguished between three completely different and independent transfer mechanisms in the zone of light absorption:
▸ a hydrodynamic mechanism during which a shock wave is emitted in all directions;
▸ a "breakdown" mechanism during which the wave moves *against* the beam; and
▸ a radiative mechanism.

| | | |
|---|---|---|
| 1964 | *Private laboratory, Garden Grove, Orange County, California* | **Norman STINGLEY,**[2542] a U.S. chemist working for the Bettis Rubber Company at Whittier, CA, **uses his spare time to create a ball with unprecedented resilience by compressing a synthetic rubber material under 3,500 psi (241 bar).** His new material, which he names "Zectron®," contains the rubber polymer polybutadiene [$C_4H_6$], a synthetic rubber, and sulfur. It has about six times the bounce of ordinary rubber. ▪ Balls made from Zectron that were around the size and color of a plum were capable of bouncing back 92% of the height from which they were originally dropped – *i.e.*, the new material had a coefficient of restitution $$e = (h/h_0)^{1/2} = 0.92^{1/2} = 0.959$$ {THOMSON & TAIT ⇨1879}. The ball was immediately commercialized under the trade name *Super Ball*™ by Wham-O Manufacturing Company at San Gabriel, CA, which over the course of the decade sold some 20 million balls. It was America's most popular plaything in the summer and fall of 1965.<br><br>In the following years, collision experiments performed with a set of two superballs of different sizes showed unusual rebound effects {MELLEN ⇨1968}. |
| 1964 | *Kiew, Ukraine, Soviet Union* | **Alexander E. VOITENKO,**[2543] a Russian scientist, **proposes that the shaped charge originally developed to pierce thick steel armor could be adapted to the task of accelerating shock waves.** When the shaped charge detonates, most of its energy is focused onto a steel plate, which drives it forward and pushes the test gas ahead of it. The resulting device, which looks like a small wind tunnel, is called a **"Voitenko compressor."** ▪ The earliest studies into the capabilities of explosive shock tubes date back to mid-1950s. The first design simply detonated a |

---

[2541] Y.P. RAIZER: *Heating of a gas by a powerful light pulse.* Sov. Phys. JETP **21**, 1009-1017 (1965).
[2542] N. STINGLEY: *Highly resilient polybutadiene ball.* U.S. Patent No. 3,241,834 (March 22, 1966).
[2543] A.E. VOITENKO: *Generation of high-speed gas jets.* Sov. Phys. Dokl. **9**, No. 11, 860-862 (1965).

cylindrical high explosive inside a tube to produce a planar shock.[2544] A more advanced design is the multiple-charge blast wave accelerator proposed to propel a projectile through a gun tube {WILSON ET AL. ⇨1993}.

VOITENKO's idea was translated into a self-destroying shock tube ("suicidal wind tunnel") at the Hypersonic Free-Flight Facility (HFFF) at NASA's Ames Research Center. A 66-pound (30-kg) shaped charge accelerated the gas in a 3-cm-dia. glass-walled tube two meters in length. The velocity of the resulting shock wave was a phenomenal 220,000 ft/s (67 km/s). The **Voitenko compressor-driven shock tube** exposed to the detonation was, of course, completely destroyed, but not before useful data were extracted.[2545] ▪ Such ultrahigh velocities could be used to simulate phenomena associated with spacecraft entry into and impact with Jupiter's atmosphere.

| | | |
|---|---|---|
| 1964/ 1965 | National Center for Atmospheric Research (NCAR), Boulder, Colorado & Yellowstone National Park, Wyoming | In the winter of 1964/1965, **Guy G. GOYER**,[2546] a scientist at NCAR who conducted simulation tests on the freezing of super-cooled water droplets in the plume of Old Faithful Geyser, **discusses the effects of shock waves (such as those generated by a lightning discharge) on supercooled water droplets.**<br><br>(i) He writes, "Old Faithful Geyser was selected for these experiments since it offers the unique phenomenon of generating at ground-level, once every hour, a precipitating supercooled cloud comprising a wide range of water droplets. The lightning discharges were simulated by the detonation of Primacord, a detonation fuse manufactured by the Ensign-Bickford Company [now part of Dyno Nobel]. The explosive charge was suspended from a captive standard meteorological balloon anchored about 300 ft [91 m] above ground at a known distance from the cone of Old Faithful."<br><br>(ii) To realistically model a lightning discharge, he chose to use a Primacord containing 56 grains of PETN per foot (11.9 g/m), a value based on his previous calculations. Lengths of Primacords of between 6 and 140 ft (1.83 and 42.67 m) were detonated in the plume of the geyser.<br><br>(iii) One observation recorded by the 1964 Yellowstone Field Research Expedition[2547] is quoted here: **"Hail was always observed to fall immediately following the detonations. The hail was very noticeable by its bouncing off the clothing or the equipment carried by the observers."** ▪ Hail, rain drops, ice crystals, fog, clouds, *etc.*, are products of the condensation or sublimation of atmospheric water, which are known by the generic term **hydrometeors**.<br><br>In the same year, GOYER[2548] **reviews the various effects of lightning, particularly the mechanical effects provoked by the shock wave.** At that time, however, reliable data on the shock wave pressure of thunder as a function of distance were not yet available {UMAN ⇨1970; BASS ⇨1980; GRANEAU ⇨1989}. |
| 1965 | Balapan Test Area, Semipalatinsk Test Range, Kazakhstan, Soviet Union | On January 15, **the Soviet Union performs the first industrial nuclear explosion test – dubbed "CHAGAN"** – a 140-kiloton subsurface cratering experiment, similar to the U.S. test SEDAN {⇨1962}. The resultant crater has a dia. of 408 m and is 100 m deep. ▪ The Soviet program "Nuclear Explosions for the National Economy" was initiated to investigate peaceful nuclear explosions and ended in 1988. For example, the Soviet program encompassed:<br>▸ *Employment of Nuclear Explosive Technologies in the Interests of National Economy* (or *Program 6*), focusing on water reservoir development, canal and dam construction {Alma Ata ⇨1966}, and creation of underground cavities for toxic waste storage; and |

---

[2544] R.G. SCHREFFLER and R.H. CHRISTIAN: *Boundary disturbance in high-explosive shock tubes.* J. Appl. Phys. **25**, 324-331 (1954).
[2545] D.D. BAALS and W.R. CORLISS: *Wind tunnels of NASA.* Rept SP-440, NASA History Office, Washington, DC (1981); chap. 6: *Wind tunnels in the space age.*
[2546] G.G. GOYER: *Mechanical effects of a simulated lightning discharge on the water droplets of 'Old Faithful' Geyser.* Nature **206**, 1302-1304 (1965).
[2547] V.J. SCHÄFER: *Final report, fourth Yellowstone field research expedition.* Atmospheric Sciences Research Center, State University of New York (1964).
[2548] G.G. GOYER: *Effects of lightning on hydrometeors.* Nature **206**, 1203-1209 (1965).

|      |                                                                                                                              |                                                                                                                                                                                                                                                                                                                                                                                                                                                                                                                                                                                                                                                                                                                                                                                                                                                                                                                                                                                                                |
|------|------------------------------------------------------------------------------------------------------------------------------|---|

▸ *Peaceful Nuclear Explosions for the National Economy* (or *Program 7*), focusing on using nuclear explosions in geological exploration, breaking up ore bodies, stimulating the production of oil and gas, and forming underground cavities for storing the recovered oil and gas.

The Soviet program is the equivalent of the U.S. Plowshare Program {⇨1958}. However, it was many times larger than the U.S. Plowshare Program in terms of both the number of applications explored with field experiments and the extent to which they were introduced into industrial use.[2549]

| 1965 | *Planet Mars & NASA's Jet Propulsion Laboratoty (JPL), Pasadena, California* | In the period July 14 to 15, NASA's spacecraft Mariner 4 (also called "Mariner Mars") takes the first set (21 in total) of pictures of Mars in close-up, which show that the planet is heavily cratered. The probe also performs a number of programmed activities. Magnetometer data obtained by Mariner 4 in July 1965 for Mars are ambiguous and apparently show that the interactions of Mars with the solar wind are quite small.

Like Venus {RUSSELL ET AL. ⇨1979}, **Mars lacks measurable global magnetic fields, and the bow shocks observed for these planets are the result of interactions with each planet's ionosphere and atmosphere.** Various analyses of more detailed data obtained from further missions to Mars (such as the U.S. missions Mars 2, Mars 3, Mars 5, and the Soviet mission Phobos 2) show that the bow shock for Mars differs from that of Venus in that

▸ the Martian bow shock is located further from the planet (with respect to planetary size);

▸ the position of the Martian shock is far more variable and controlled by the interplanetary magnetic field; and

▸ the Mars shock location varies with solar activity.[2550] |

| 1965 | *Taal Volcano, southwestern Luzon Island [about 60 km south of the capital Manila], Philippine Islands* | In the period September 28–30, **Taal Volcano is violently shaken by phreatomagmatic eruptions.** A great column of steam, dust, ash, and cinders is blasted to a height of several thousand feet. Base surges from the volcano, which is situated in a highly populated region, travel 3 km across Taal Lake and blast the villages to the west of the vent at Volcano Island. This particular hazard results in much death and destruction on Volcano Island and in lakeside areas, as surges can propagate over the lake without any significant reduction in force (189 casualties). ▪ Taal Volcano, one of the world's lowest and deadliest volcanoes, is located at Taal Lake, inside a caldera approximately $25 \times 30$ km² in size. On the island is the 2-km-wide and 80-m-deep Main Crater Lake.[2551] Over 30 eruptions of Taal have been recorded since 1572, ranging from steam eruptions without lava ejection ("phreatic eruptions") to explosive eruptions resulting from water-magma interactions ("phreatomagmatic eruptions").

Base surges were first documented during this most famous 1965 Taal eruption by James G. MOORE,[2552] a USGS geologist, and his collaborators. **Base surges are flat, turbulent clouds of material resulting from water/magma interactions.** They form through the collapse of steam-saturated eruption columns and spread away from the volcano laterally at high velocities along the ground. The phenomenon of a base surge was first identified during the first U.S. underwater nuclear explosion {Operation CROSSROADS ⇨1946}. |

| 1965 | *Desolated parts of western Montana* | On October 12, **establishment of the Large Aperture Seismic Array (LASA). The primary objective of LASA is to monitor global underground nuclear explosions.** The online (real time) system consists of 525 seismometers organized in 22 subarrays and spread over an area of 30,000 km². ▪ In contrast to a single seismic station, which provides a point observation, an |

---

[2549] *Nuclear Explosions for the National Economy*. Wikipedia (May 9, 2007); http://en.wikipedia.org/wiki/Nuclear_Explosions_for_the_National_Economy.
[2550] D. VIGNES ET AL.: *Factors controlling the location of the bow shock at Mars*. Geophys. Res. Lett. **29**, No. 9, 42.1-42.4 (2002).
[2551] *Taal Volcano*. Philippine Institute of Volcanology, Quezon City, Philippines (2003); http://www.phivolcs.dost.gov.ph/Flyers/VOLCANO/Taal.doc.
[2552] J.G. MOORE ET AL.: *The 1965 eruption of Taal Volcano*. Science **151**, No. 3713, 955-960 (1966).

array yields an area observation and offers unique research opportunities in terms of seismologic quantification of a clearly 3-D Earth. **Array seismology also advanced research into the Earth's interior** {VIDALE ET AL. ⇨ 2000}.[2553]

Four years later, a second large array was established in Norway, the NORSAR (Norwegian Seismic Array), with 132 seismometers organized in 22 subarrays. While LASA was dismantled in 1978, NORSAR is still operating today. In terms of global nuclear explosion monitoring, these arrays provided significant improvements. However, they also demonstrated that the concept of a few large arrays would not provide adequate monitoring capabilities to check for compliance with the UN Comprehensive Test Ban Treaty {CTBT ⇨ 1996}.[2554]

| 1965 | *University of Honolulu, Hawaii* | In November, **establishment of the International Tsunami Information Center (ITIC),**[2555] with headquarters in Honolulu, by the Intergovernmental Oceanographic Commission (IOC) of UNESCO. Member states include the major Pacific rim nations in North America, South America and Asia. ▪ ITIC was initiated after the Great Alaskan Earthquake {⇨1964} in order to advise coastal communities of potentially destructive tsunamis. The primary basis for all tsunami warning systems is the close spatial and temporal association between tsunami generation and earthquake occurrence. Using a seismic network to locate epicenters of submarine quakes, this installation can predict the arrival of tsunamis at points around the Pacific Basin, often hours before the arrival of the waves at remote coasts.<br><br>**DART** (Deep-Ocean Assessment and Reporting of Tsunamis)[2556] **is an ongoing effort to maintain and improve the ability to detect tsunamis early and to report on tsunamis in the open ocean in real time.** Developed by the National Oceanic and Atmospheric Administration (NOAA), a U.S. governmental agency, and operated by its National Data Buoy Center (NDBC), DART's systems consist of an anchored seafloor bottom pressure recorder and a companion moored surface buoy for real-time communications via a Geodetic Earth-Orbiting satellite (GEOS) link to ground stations. |
|------|------|------|
| 1965 | *Le Creusot [Dépt. Bourgogne], France* | **First AIRAPT International High Pressure Conference;** the general chairman is Boris VODAR. The conference is organized by AIRAPT (Association Internationale pour L'Avancement de la Recherche et de la Technologie aux Hautes Pression), the International Association for the Advancement of High Pressure Science and Technology.[2557] ▪ Initially, the conferences were mainly devoted to static high pressures, but later conferences were also increasingly devoted to the field of high dynamic pressures, such as those achived by shock waves.<br><br>The following conferences were held at Schloss Elmau, Germany (1968); Aviemore, Scotland (1970); Kyoto, Japan (1974); Moscow, Soviet Union (1975); Boulder, CO (1977); Le Creusot, France (1979); Uppsala, Sweden (1981); Albany, NY (1983); Amsterdam, The Netherlands (1985); Kiev, Soviet Union (1987); Paderborn, Germany (1989); Bangalore, India (1991); Colorado Springs, CO (1993); Warsaw, Poland (1995); Kyoto, Japan (1997); Honolulu, HI (1999); Beijing, China (2001); Bordeaux, France (2003); Karlsruhe, Germany (2005); and Catania, Italy (2007). |
| 1965 | *Meersburg, Germany* | **The First International Conference on Rain Erosion and Associated Phenomena is held.** ▪ In the 1960s, the problem of rain erosion sustained by supersonic aircraft and missiles became of increasing concern in the aerospace industry. |

---

[2553] S. ROST and E.J. GARNERO: *Array seismology advances research into Earth's interior.* EOS **85**, No. 32, 301, 305-306 (2004).

[2554] E.S. HUSEBYE: *The role of small arrays and networks in seismology.* In: *Earthquake monitoring and seismic hazard mitigation in Balkan countries.* NATO Advanced Research Workshop [Borovetz, Rila Mountain, Bulgaria, Sept. 2005].

[2555] International Tsunami Information Centre, Honolulu, HI; http://www.prh.noaa.gov/itic/.

[2556] National Data Buoy Center (NDBC) of the National Oceanic and Atmospheric Administration (NOAA), Stennis Space Center, MS; http://www.ndbc.noaa.gov/Dart/dart.shtml.

[2557] For AIRAPT history, *see* http://www.unipress.waw.pl/airapt/.

The three subsequent conferences − with their coverage progressively broadening to include other types of erosion damage by solid particles and by cavitation − were held at Meersburg, Germany (1967); Elvetham Hall, U.K. (1970); and Meersburg, Germany (1974). The conference proceedings were edited by Andrew A. FYALL and Roy B. KING, and published by the Royal Aircraft Establishment (RAE), Farnborough, U.K. After that, **a series of international conferences on Erosion by Liquid and Solid Impact (ELSI) were held at Cavendish Laboratory** (Cambridge, U.K.) in 1979, 1983, 1987, 1994, and 1998.

**The first International Conference on Erosion, Abrasion and Wear (ICEA I) was incorporated into the 5th ELSI** and also held in 1998 at Cavendish Laboratory; ICEA II was held in 2003 at Churchill College, Cambridge, U.K. The scope of the ICEA meetings, which are in part sponsored by the Electric Power Research Institute (EPRI at Palo Alto, CA), encompasses all aspects of wear by hard particles, including erosion and abrasion, as well as erosion by cavitation in liquids, and erosion and damage via the impact of liquid jets and drops.

| | | |
|---|---|---|
| 1965 | *The Royal Society of London, Somerset House, London, England* | A discussion panel is held on the topic of the **Deformation of Solids by the Impact of Liquids, and its Relation to Rain Damage in Aircraft and Missiles, to Blade Erosion in Steam Turbines, and to Cavitation Erosion;** the organizer is Frank P. BOWDEN.[2558]<br><br>(i) BOWDEN, who also briefly addresses the **problem of rain erosion** in his introduction, notes that a raindrop striking a solid moving supersonically at about 500 m/s (about Mach 1.5) will exert a pressure of about 130 bar on the surface of the solid, thus producing an effect resembling a small explosion and causing damage by plastic flow or fracture.<br><br>(ii) On the other hand, **D.C. JENKINS**,[2559] a research engineer at Royal Aircraft Establishment in Farnborough who had previously studied droplet impact phenomena photographically,[2560] **points out that raindrops may be disintegrated by shock waves during supersonic flight, thus mitigating erosion damage on conical surfaces.**<br><br>The meeting, which was attended by about 150 scientists and engineers from nine different countries, summarized the present status of knowledge on this subject. Some presented milestone papers are discussed below {BOWDEN & MCONIE ⇨1965; BRUNTON & BOWDEN ⇨1965; LEACH & WALKER ⇨1965}. |
| 1965 | *U.S.S.R. Academy of Sciences, Novosibirsk, Soviet Union* | **Foundation of the Soviet journal** *Fizika Gorenija Vzryva*, published by the Siberian Division of the Academy of Sciences of the Soviet Union. First editor-in-chief is Mikhail A. LAVRENTYEV [or LAVRENT'EV]. • An English cover-to-cover translation ***Combustion, Explosion, and Shock Waves*** [vol. 1 (1966)] was also published (originally by Faraday Press, New York, but from 2004 onwards by Springer, New York). |
| 1965 | *Institute of Chemical Physics, Chernogolovka, Moscow region, Soviet Union* | **Gennady A. ADADUROV**[2561] **and collaborators discover the shock-induced polymerization of organic monomers.** The experiments reveal that various shock-induced reactions in organic substances are possible, such as<br>▸ chemical bond rupture (leading to destruction, a common feature);<br>▸ addition (the direct insertion of a small molecule into a double or triple carbon bond);<br>▸ substitution (the replacement of one atom or group by another); and<br>▸ isomerization (the rearrangement of atoms and bonds within a molecule without changing the empirical molecular formula). |

---

[2558] The 21 papers and discussions presented were published in the Proc. Roy. Soc. Lond. **A 260**, 73-315 (1966).
[2559] D.C. JENKINS: *Part C: Rain erosion of aircraft. Paper XI: Disintegration of raindrops by shock waves ahead of conical bodies.* Phil. Trans. Roy. Soc. Lond. **A260**, 153-160 (1966).
[2560] D.C. JENKINS and J.D. BOOKER: *A photographic study of the impact between water drops and a surface moving at high speed.* H.M.S.O., London (1960).
[2561] G.A. ADADUROV ET AL.: *Polymerization of condensed monomers in a shock wave* [in Russ.]. Dokl. AN (SSSR) **165**, 851-854 (1965).

| | | |
|---|---|---|
| 1965 | *Sandia Laboratories, Albuquerque, New Mexico* | **Lynn M. BARKER**, a U.S. applied solid-state physicist, **and Roy E. HOLLENBACH develop various ultrahigh-speed interference methods by analyzing the Doppler shift of reflected laser light.** At low free-surface velocities (< 100 m/s), they apply a modification of the Michelson interferometer in which fringe shift is made proportional to velocity instead of displacement.[2562] At higher velocity, they use a velocity interferometer that was later commercialized and named **VISAR** (Velocity Interferometer System for Any Reflector).[2563] ▪ The new instrumentation offers great advantages in terms of hypervelocity ballistics and impact applications over the older microwave displacement history techniques {KOCH ⇨1952}.<br><br>In the years following, the VISAR became a very powerful diagnostic tool in solid-state shock wave physics and has since been used preferentially either to measure the surface velocity of shock-loaded samples {⇨Fig. 4.12–G} or to monitor the velocity of an interface between a shock-loaded test sample and a transparent buffer.[2564] The VISAR technique has gained worldwide acceptance as the tool of choice for the measurement of shock and detonation phenomena.[2565] One limitation of the classical VISAR is that it can only measure one point on a surface. This disadvantage was overcome by the "line-imaging VISAR," a modification invented in the early 1990s by the Los Alamos scientist Willard F. HEMSING.[2566] This new instrument, however, is not yet commercially available. |
| 1965 | *Berkeley & Los Angeles, California* | **John W. BOND JR.**, a theoretical physicist at Aerospace Corporation,[2567] **Kenneth M. WATSON**, a physics professor at UC Berkeley, **and Jasper A. WELCH JR.**, a physicist at the U.S. Air Force in Los Angeles, **publish their textbook *Atomic Theory of Gas Dynamics*.** ▪ It is the first to describe, at the atomic level, both equilibrium and nonequilibrium phenomena in gases at high temperatures; this was a rapidly growing branch of gas dynamics that became important in aeronautics, rocketry and nuclear weaponry. |
| 1965 | *Surface Physics Division, Cavendish Laboratory, University of Cambridge, England* | **Frank Philip BOWDEN**,[2568] a Tasmanian-born British surface physicist, **and M.P. McONIE report on the detonation initiation of explosive liquids by impact or shock:**<br><br>(i) They argue that the initiation process is essentially a thermal process which requires some discontinuity, cavity or bubble to be be present in a liquid explosive. Cavities and cracks are present in solid explosives, and these play a similar role.<br><br>(ii) **An impact-generated shock wave that encounters a bubble or the curved surface of a cavity in a liquid** {⇨Fig. 4.8–J} **may produce a tiny Munroe jet** {MUNROE ⇨1888}, which they call a **"microjet."**[2569] In a liquid high explosive, such a microjet concentrates the energy into a localized "hot spot," which initiates an explosion {⇨Fig. 4.8–K}. ▪ BOWDEN performed studies on friction phenomena before he began to investigate explosives. Greatly interested in surface physics and hot-spot phenomena for many years, he had demonstrated that sliding produces friction over only a small fraction of the total area of the surface, but it can still produce exceedingly high temperatures and even induce melting in hot spots, thus decom- |

---

[2562] L.M. BARKER and R.E. HOLLENBACH: *Interferometer technique for measuring the dynamic mechanical properties of materials.* Rev. Scient. Instrum. **36**, 1617-1620 (1965).

[2563] L.M. BARKER: *Fine structure of compressive and release wave shapes in aluminum measured by the velocity interferometer technique.* Symposium International sur le Comportement des Milieux Denses sous Hautes Pressions Dynamiques [Paris, France, Sept. 1967]. In: (M. ROY, ed.) *Behavior of dense media under high dynamic pressures.* Gordon & Breach, New York and Dunod, Paris (1968), pp. 483-505.

[2564] L.M. BARKER and R.E. HOLLENBACH: *Shock-wave studies of PMMA, fused silica, and sapphire.* J. Appl. Phys. **41**, 4208-4226 (1970); *Laser interferometer for measuring high velocities.* Ibid. **43**, 4669-4675 (1972).

[2565] K.J. FLEMING ET AL.: *New innovations in shock diagnostics & analysis using high-speed multi-point velocimetry (VISAR).* In: (J.L. SHORT and J. MAIENSCHEIN, eds.) Proc. 12th Int. Detonation Symposium [San Diego, CA, Aug. 2002]. Office of Naval Research, U.S. Navy, Washington, DC (2005).

[2566] W.F. HEMSING: *Line-imaging laser interferometers for measuring velocities.* Los Alamos Science No. 21, 60-61 (1993).

[2567] J.W. BOND JR., K.M. WATSON, and J.A. WELCH JR.: *Atomic theory of gas dynamics.* Addison-Wesley, Reading, MA (1965).

[2568] F.P. BOWDEN and M.P. McONIE: *Cavities and micro jets in liquids, their role in explosion.* Nature **206**, 380-383 (1965).

[2569] F.P. BOWDEN: *Part A: The physics of impact and deformation: single impact. Paper III: The formation of microjets in liquids under the influence of impact or shock.* Phil. Trans. Roy. Soc. Lond. **A260**, 94-95 (1966).

posing the lubricant by heat at exactly the place where it is needed most. These studies led to a better understanding of the impact sensitivity of explosives and stimulated other researchers to investigate this subject of great practical importance further.

**In the same year, John Hubert BRUNTON**[2570] **and Frank Philip BOWDEN study the deformation of solids at high strain rates,** such as under the impact of short, high-velocity liquid jets with speeds of up to 1,200 m/s.

(I) They classify impact stresses which cause deformation into
- stresses produced by short-duration "explosive" compression of the surface arising from the water hammer effect; and
- stresses produced by the erosive scoring action of the high-speed radial flow.

(II) The impact of a water jet at 950 m/s against a plate of **hard polymer** (such as Plexiglas) causes an annular fracture separating a region of intense circumferential fracture from a central undamaged zone. The failure of the surface due to impact involves two processes:
- the production of fractures by radial tensile stresses; and
- the removal of material from the fractured regions due to the liquid flowing out at high speed over the surface. The impact is also accompanied by a light flash on the target surface – probably due to the adiabatic compression of an air layer trapped between the liquid and the solid. At velocities below 450 m/s the annular fracture is replaced by an annular depression.

(III) The impact of a water jet on **soft polymers** produces an outer ring of torn surface caused by the outward liquid flow, and a narrow central penetration with fracture.

Their liquid-jet technique for simulating single-drop impact proved very useful for rain erosion studies, and in the late 1980s was further developed into an automatic multiple impact jet apparatus.[2571]

| 1965 | *Poulter Laboratory, Stanford Research Institute (SRI), Menlo Park, California* | **George R. FOWLES,**[2572] a U.S. shock physicist and geophysicist, **and William M. ISBELL**, a U.S. physicist, **generate a Mach disk which moves steadily in time in a very simple coaxial explosive/target configuration** {⇨Fig. 4.13–L}. The idea of this arrangement, which easily produces in the central region of this configuration planar shocks at higher pressures (up to 4 Mbar) than those attainable by standard methods, was conceived by Prof. George E. DUVALL at Washington State University. Using an optical reflection technique, combined with a grid of alternate opaque and transparent lines on the face of a standard explosive argon light source and smear recording, they monitor the shock and free-surface arrivals. Their measurements of Hugoniot data for copper up to a shock pressure of 1.9 Mbar show good agreement with previously published data. ▪ Flash radiographic studies carried out in Germany proved that a coaxial explosive/target configuration is also most useful for generating high shock pressures in the Mbar region in water {⇨Fig. 4.13–H}.[2573] |
|---|---|---|

---

[2570] J.H. BRUNTON: *Part A: The physics of impact and deformation: single impact. Paper I: The physics of impact and deformation: single impact. High speed liquid impact.* Phil. Trans. Roy. Soc. Lond. **A260**, 79-85 (1966).

[2571] C.R. SEWARD, C.S.J. PICKLES, and J.E. FIELD: *Single- and multiple-impact jet apparatus and results.* Meet. Int. Soc. Opt. Engng. [San Diego, Ca, July 1990]. In: (P. KLOCEK, ed.) *Window and and dome technologies and materials II.* Proc. SPIE **1326**, 280-290 (1990).

[2572] G.R. FOWLES and W.M. ISBELL: *Method for Hugoniot equation-of-state measurements at extreme pressures.* J. Appl. Phys. **36**, 1377-1379 (1965).

[2573] It appears that Dr. Thomas C. POULTER, founder of SRI's Poulter Laboratory, had already considered using two such coaxial cylindrical arrangements, facing opposite each other, in order to generate high Mach numbers, possibly to initiate fusion in a target positioned along the cylinder axis. In October 1972, Dr. POULTER gave the author, then a Fulbright fellow at SRI, a list of his publications, which includes a reference entitled *Thermal fusion by opposing Mach 10 detonation fronts.* Tech. Rept. GU-960 (1958). Unfortunately, this report couldn't be located in SRI's archives (Dr. James D. COLTON, Poulter Laboratory; private communication, June 2004).

| | | |
|---|---|---|
| 1965 | *Verlag Chemie GmbH, Weinheim/Bergstr., Germany* | **Publication of the *Handbuch der Raumexplosionen* ("Handbook of Space Explosions").** The editor is Heinz Helmuth FREYTAG,[2574] a German safety engineer and retired senior government official. The book, which contains contributions from 18 authors, focuses on the following subjects:<br>▸ fundamentals of gaseous explosions and detonations [Wilhelm JOST and Heinz Georg WAGNER, University of Göttingen, Germany];<br>▸ properties and ignition of spherical explosions [Heinz FREIWALD, ISL, France];<br>▸ properties of combustible gases, vapors and dusts [Paul DITTMAR and Joachim ZEHR, Bundesanstalt für Materialprüfung (BAM), Berlin, Germany];<br>▸ influence of oxygen concentration on the explosivity of gaseous mixtures [Wolfgang WEGENER, BAM]; and<br>▸ the apparatus needed and administrative methods used to prevent explosion hazards in factories [Heinz Helmuth FREYTAG].<br>FREYTAG argues that explosion hazards are bound to localities; therefore, the specific term *Raumexplosion* ("space explosion") has been used in German safety regulations. For example, the term *Raumexplosionsgefahr* (literally meaning "space explosion hazard") refers to explosion hazards in closed localities caused by exothermal chemical reactions of explosive gases and dusts.<br>Two years later, FREYTAG[2575] reported on the hazards caused by ignition sources and suitable safety measures for these at the 15th ACHEMA. |
| 1965 | *Impulsphysik GmbH, Hamburg-Rissen, Germany* | Frank B.A. FRÜNGEL,[2576] a German physicist, inventor and industrialist widely known among pulsed-power engineers and physicists as "Mr. Pulse," **publishes the first two volumes of his four-volume compendium *High Speed Pulse Technology*.** In the first volume, he addresses the conversion of capacitively stored energy into acoustic impulses in detail, particularly<br>▸ electroacoustic converters and their applications;<br>▸ air impulse sound and air shock waves;<br>▸ shock waves generated by powerful underwater capacitor discharges, so-called "underwater sparks" (or "water arc explosions");<br>▸ applications of high-intensity shock waves (*e.g.*, in materials fabrication and manufacturing processes, metallurgy, seismology, biology, and pulsed power technology); and<br>▸ photography of shock waves resulting from underwater sparks.<br>His excellent compendium – an outgrowth of his German textbook *Impulstechnik* (Akad. Verlags-Gesell. Geest & Portig, Leipzig 1960) – is a rich mine of information for both pulsed-power engineers, acousticians, shock physicists, and high-speed photographers. |
| 1965 | *Midland Park, New Jersey* | Sighard F. HOERNER,[2577] a German-born U.S. fluid dynamicist, **publishes his book *Fluid Dynamic Drag: Practical Information on Aerodynamic Drag and Hydrodynamic Resistance*.** It addresses all kinds of drag (*e.g.*, those due to pressure, skin friction, lift, geometry, surface imperfections, *etc.*) at subsonic, transonic, supersonic, and hypersonic velocities. ▪ His book became the "Bible of Aerodynamic Drag." |

---

[2574] H.H. FREYTAG: *Handbuch der Raumexplosionen*. Verlag Chemie GmbH, Weinheim/Bergstr. (1965).

[2575] H.H. FREYTAG: *Gefahren durch Zündquellen und Schutzmaßnahmen* [presented at the 15th ACHEMA, Ausstellungs-Tagung für Chemisches Apparatewesen, Frankfurt/Main (1967)]. Herausgegeben von der Berufsgenossenschaft (BG) Chemie, Heidelberg (1967).

[2576] F.B.A. FRÜNGEL: *High speed pulse technology*. Academic Press, New York, vol. 1 (1965): *Capacitor discharges, magnetohydrodynamics, X-rays, ultrasonics*; vol. 2 (1965): *Optical pulses, lasers, measuring techniques*; vol. 3 (1976): *Capacitor discharge engineering*; vol. 4 (1980): *Sparks and laser pulses*.

[2577] S.F. HOERNER: *Fluid dynamic drag: practical information on aerodynamic drag and hydrodynamic resistance*. Hoerner Fluid Dynamics, Midland Park, NJ (1965).

| | | |
|---|---|---|
| 1965 | *Cornell Aeronautical Laboratory (CAL), Buffalo, New York* | **Ian R. HURLE**[2578] **and Abraham HERTZBERG suggest a new approach to realizing a "gas-dynamic laser"** {⇨Fig. 4.14–Q}. To create population inversion, they propose rapid cooling by nonequilibrium expansion of a three-level gas through a supersonic nozzle {HURLE[2579] ET AL. 1962; BASOV & ORAEVSKII ⇨1962}. ▪ First experiments, carried out using expansions of an initially hot xenon gas from a slit orifice placed at the termination of a shock tube, failed. However, the idea proved successful in the following year {KANTROWITZ ET AL. ⇨1966}. |
| 1965 | *Atomic Weapons Research Establishment (AWRE), Aldermaston, Berkshire, England* | **Alec E. HUSTON**,[2580] a British electronics engineer, **presents the first fully electronic high-speed multiframe image tube camera.** Frame frequencies and exposure times can be selected over a wide range. This new type of camera, which can be operated at low light levels and easily incorporated into most experimental set-ups, is a milestone in optical high-speed recording. ▪ HUSTON's camera, capable of recording in both streak and framing mode, was later commercialized by the British company John Hadland Photographic Instrumentation Ltd. under the tradename Imacon® 700 {⇨Fig. 4.19–P}.<br><br>Though completely electronic, HUSTON's camera was still an analog device and required special high-speed film (*e.g.*, Polaroid or Fuji) for instant recording. Modern electronic cameras, however, are pure digital devices {Hadland Photonics Ltd. ⇨1993}. Nowadays such cameras are used almost exclusively in shock wave and ballistic laboratories, and so the classical era of using mechanical high-speed cameras and film is definitely at an end. High-speed photography – in the past an art that was mastered by just a few experts who often invented and built their own instruments – has now turned into a quick, routinely used diagnostic tool. |
| 1965 | *Institute of Chemical Kinetics & Combustion, Novosibirsk, Soviet Union* | **Valery K. KEDRINSKII**,[2581] **M.I. VOROTNIKOVA, and Rem I. SOLOUKHIN report on a two-diaphragm hydrodynamic shock tube with a moving liquid piston for studying 1-D shock waves in liquids.** ▪ It was later applied to the study of the kinetics of chemical reactions in solutions; in particular for generating a temperature jump in order to determine the values of thermodynamic equilibrium shear for inverse chemical transformations in solutions.[2582] |
| 1965 | *Soviet Union* | **Samuil Borisovich KORMER**,[2583] a Soviet applied physicist, **and his associates measure the temperature of shock-compressed ionic single crystals up to 800 kbar.** Using photomultiplier tubes, together with two interference filters whose transmission maxima correspond to 4,780 Å and 6,250 Å, they compare the light emission from the shocked samples with that of a standard light source, and thus determine the melting curves for sodium chloride [NaCl] from 540 to 700 kbar, and for potassium chloride [KCl] from 330 to 480 kbar. |
| 1965 | *Safety in Mines Research Establishment (SMRE), Ministry of Power, Sheffield, England* | **S.J. LEACH**[2584] **and G.L. WALKER report on some aspects of rock cutting by high-speed (subsonic) water jets:**<br>(i) They use two types of experimental equipment: a water pump that produces continuous pressures of up to 600 bar and water jet velocities of up to 340 m/s, and a hydraulic intensifier that gives pulsed pressures of up to 5 kbar, resulting in higher water jet velocities of up to 1,000 m/s. |

---

[2578] I.R. HURLE and A. HERTZBERG: *Electronic population inversions by fluid-mechanical techniques.* Phys. Fluids **8**, 1601-1607 (1965).

[2579] I.R. HURLE, A. HERTZBERG, and J.D. BUCKMASTER: *The possible production of population inversion by gasdynamic methods.* Rept. RH-1670-A-1, Cornell Aeronautical Laboratory, Buffalo, NY (1962). ▪ This report is (still) classified and kept at Veridian Company in Buffalo, NY.

[2580] A.E. HUSTON: *A multi-frame image tube camera.* Proc. 7th Int. Congress on High-Speed Photography [Zurich, Switzerland, Sept. 1965]. In: (O. HELWICH and H. SCHARDIN, ed.) *Kurzzeitphotographie.* Helwich, Darmstadt etc. (1967), pp. 93-96.

[2581] V.K. KEDRINSKII, R.I. SOLOUKHIN, and M.I. VOROTNIKOVA: *Shock tube for studying one-dimensional waves in a liquid.* Fiz. Gor. Vzryva (Novosibirsk) **1**, 5-14 (1965).

[2582] V.K. KEDRINSKII, R.I. SOLOUKHIN ET AL.: *The study of high-rate reactions in solutions behind a powerful shock front.* Dokl. AN (SSSR) **187**, 130-133 (1969).

[2583] S.B. KORMER, M.V. SINITSYN, G.A. KIRILLOV, and V.D. URLIN: *Experimental determination of temperature in shock-compressed NaCl and KCl, and of their melting curves at pressures up to 700 kbar.* Sov. Phys. JETP **21**, 689-700 (1965).

[2584] S.J. LEACH and G.L. WALKER: *Part F: The application of high speed liquid jets to cutting. Some aspects of rock cutting by high speed water jets.* Phil. Trans. Roy. Soc. Lond. **A260**, 295-308 (1966).

(ii) The process by which several types of rock are penetrated by water jets is examined using optical and X-ray photography. They find that a 1-mm-dia. water jet drills an about 5-mm-dia. cylindrical hole.

Hydraulic mining is an old technology, and was first used during the Californian gold rush.[2585] For example, in 1852 a 40-ft (12.2-m)-long, 4-in. (10.2-cm)-dia. rawhide hose was used to direct a jet of water at the ore to remove the soft material. In the coal mining industry, the application of water jets not only substantially reduces the risk of firedamp explosions, but the water can also be used to transport the coal away from the working face in open troughs. In the 1960s, high-speed liquid jets were applied industrially, to wood and rock cutting, with some success. Today high-speed water jets are widely used in industry for cutting, drilling, splitting and notching.[2586]

| 1965 | Hughes Laboratories, Malibu, California | **Frederick J. MCCLUNG,**[2587] a U.S. physicist, **and Daniel WEINER invent *mode locking*,** a method of synchronizing the various modes involved in laser oscillations, which allows the generation of even shorter and more powerful laser pulses than obtained with Q-switching {MCCLUNG & HELLWARTH ⇨1962}. The laser pulses typically have a duration of only a few picoseconds, but beam powers of tens of terawatts. ▪ More recent developments in ultrafast laser sources, based on this principle of mode locking, can even provide ultrashort pulses in the femtosecond regime.[2588] Such laser pulses are most useful for generating ultrashort shock pulses in matter, which are suitable for many scientific, technical and medical applications {KÖNIG ET AL. ⇨1999}. |
|---|---|---|
| 1965 | Bell Telephone Laboratories, Crawford Hill, New Jersey | **Arno A. PENZIAS,**[2589] a German-born U.S. astrophysicist, **and Robert W. WILSON,** a U.S. radio astronomer, **measure a residual cosmic microwave background radiation of about 3.5 K** while working with a 7.35-cm (4.08-GHz) horn antenna. ▪ PENZIAS and WILSON (together with the Soviet nuclear physicist Peter L. KAPITZA for his fundamental work in low-temperature physics) earned the 1978 Nobel Prize for Physics for this important discovery.<br><br>(i) **Their discovery brilliantly confirms the Hot Big Bang Model, which predicts the presence in the Universe of a relic radiation field from the primordial fireball** {GAMOW ⇨1948} **at a temperature of 10 degrees Kelvin.**<br><br>(ii) Georges LEMAÎTRE {⇨1927}, the father of the Big Bang theory, then at the age of 71 is very pleased to hear that the background radiation has been discovered.<br><br>(iii) The renowned U.S. science journalist Walter S. SULLIVAN JR. immediately recognizes the importance of this discovery, and on May 21, 1965, he writes in the *New York Times*: "Scientists at the Bell Telephone Laboratories have observed what a group at Princeton University believes may be remnants of an explosion that gave birth to the Universe." ▪ In 1989, he received the first **Walter Sullivan Award for Excellence in Science Journalism,** an annually award for science feature writing, defined as "work prepared with a deadline of more than one week that makes geophysical information accessible and interesting to the general public."<br><br>The cosmic microwave background radiation plays a key role in our understanding of the evolution of the Universe, because it provides a picture of the Universe as it was only a hundred thousand years after the Big Bang {⇨c.14 Ga ago; NASA-GSFC ⇨2003; ⇨Fig. 4.1–W}. The temperature of the microwave background is remarkably uniform, thus confirming the *cosmological principle*, which asserts that the Universe is approximately homogeneous and isotropic when considered on a sufficiently large scale {COBE Satellite ⇨1992}. |

---

[2585] D.A. SUMMERS: *Hydraulic mining: jet-assisted cutting.* In: (H.L. HARTMAN, ed.) *SME mining engineering handbook.* Society for Mining, Metallurgy, and Exploration (SME), Inc., Littleton, CA (1992), vol. 2, pp. 1918-1929.

[2586] F. HAMMELMANN: *Die Anwendung der Hochdruckwassertechnik beim Bohren, Kerben und Spalten von Gestein* [with 243 refs.]. Shaker, Aachen (1997).

[2587] F.J. MCCLUNG and D. WEINER: *Longitudinal mode control in giant pulse lasers.* IEEE Trans. Quantum Electronics **1**, 94-99 (May 1965).

[2588] A. BALTRUSKA ET AL.: *Optical pulse compression to 5 fs at a 1-MHz repetition rate.* Optics Lett. **22**, 102-104 (Jan. 1997).

[2589] A.A. PENZIAS and R.W. WILSON: *A measurement of excess antenna temperature at 400 Mc/s.* Astrophys. J. **142**, 419-421 (1965).

| | | |
|---|---|---|
| 1965 | *Ernst-Mach-Institut (EMI), Freiburg, Germany* | **Heinz REICHENBACH,**[2590] a German research physicist, **reports on an "equal pressure shock tube" – a new blast wave simulator.** Invented jointly with Hubert SCHARDIN,[2591] an applied physics professor, it allows the duration of the pressure pulse to be adjusted independently of the peak pressure generated. Essentially a standard shock tube, but modified in the driver section to act as a baffle system and terminated in the driven section by an end-plate {⇨Fig. 4.10–E}, it is operated such that the reflected pressure equals the static pressure, thus generating a step pressure with a constant peak pressure (selectable between 4 and 12 bar) of unlimited duration. A Friedlander load function {FRIEDLÄNDER ⇨1946} at the end-plate can be achieved by inserting an outlet opening in the driven section. |
| 1965 | *Dept. of Physics, Australian National University (ANU), Canberra, Australia* | **Ray J. STALKER,**[2592] an Australian professor of aeronautical engineering, **reports on a new free-piston-driven shock tunnel** {⇨Fig. 4.10–K}. It uses a heavy piston in order to achieve isentropic compression of a helium driver gas, which results in hypersonic shock speeds. **His pioneering new type of shock tunnel substantially increases the velocities that can be achieved in wind tunnels.** STALKER first conceived his principle of shock tunnel operation in 1961, when he was employed at the National Research Council in Ottawa.[2593] • At that time, the **"Stalker tube"** – in modern terminology a *gun tunnel* rather than a *shock tunnel* – had already been recognized as being the most promising and economic tool for simulating high-enthalpy real gas effects in future hypersonic flow research of reentry vehicles and scramjet engines. The possibility of using a similar compression process had also been independently alluded to by Abraham HERTZBERG and associates. However, instead of using a piston accelerated in a tube for compression, they proposed a shock wave generated in a shock tube – the so-called **"shock tunnel"** {WITTLIFF, WILSON & HERTZBERG ⇨1959}.<br><br>Stalker tubes are now in use in many countries. The world's largest free-piston driven shock tunnel – the **High Enthalpy Shock Tunnel (HIEST)** – was constructed at the Kakuda Space Propulsion Laboratory of the National Aerospace Laboratory of Japan. A heavy piston, 0.6 m in dia. and ranging in weight from 300 to 560 kg, compresses helium in a 42-m-long compression tube. The 0.18-m-dia. 17-m-long shock tube is followed by a 1.2-m-dia. nozzle.[2594] HIEST is capable of producing a test flow with a maximum stagnation enthalpy of 25 MJ/kg and a maximum stagnation pressure of 150 MPa (1.5 kbar) in order to simulate the dissociation of molecules in the reentry flight path of the H2 Orbiting Plane (HOPE), a Japanese National Space Development Agency (NASDA) project of a reusable manned winged spacecraft. The facility allows model reentry vehicles as large as 50 cm in size to be tested, as well as scramjet models with a maximum length of 2.5 meters, at a dynamic pressure of up to 50 kPa (0.5 bar). The testing time is about 2 ms, and the air flow velocity can be varied between 4 and 7 km/s. |
| 1965 | *Los Alamos Scientific Laboratory (LASL), Los Alamos, New Mexico* | **Douglas VENABLE,**[2595] a staff physicist, **and Thomas J. BOYD use the PHERMEX** (Pulsed High Energy Radiographic Machine Emitting X-Rays) **facility at LASL to obtain precision radiographs of large explosive systems that contain high atomic number materials,** such as iron and, particularly, uranium. Using this new technique, they present clear flash radiographs at the 4th Symposium (International) on Detonation from the following phenomena: |

---

[2590] H. REICHENBACH: *The equal-pressure shock tube.* In: (Z.I. SLAWSKY, J.F. MOULTON, and W.S. FILLER, eds.) Proc. 5th Int. Shock Tube Symposium [White Oak, Silver Spring, MD, April 1965]. NOL, AD No. 484,600, pp. 543-556.

[2591] H. SCHARDIN and H. REICHENBACH: *Stosswellenrohr zur Erzeugung von Gleichdruckstoßwellen.* Germ. Patent No. 1,273,850 (1968).

[2592] R.J. STALKER: *Preliminary results with a free-piston shock tunnel.* AIAA J. **3**, 1170-1171 (1965); *The free piston shock tube.* Aeronaut. Quart. **17**, 351-370 (1966); *A study of the free-piston shock tunnel.* AIAA J. **5**, 2160-2165 (1967).

[2593] R.J. STALKER: *An investigation of free piston compression of shock tube driver gas.* Rept. MT-44, Division of Mechanical Engineering, National Research Council (NRC), Canada (1961).

[2594] K. ITOH ET AL.: *Design and construction of HIEST (High Enthalpy Shock Tunnel).* In: *Proc. Int. Conference on Fluid Engineering* [Tokyo, Japan, July 1997]. Japan Society of Mechanical Engineers (JSME), Tokyo (1997), vol. I, pp. 353-358.

[2595] D. VENABLE and T.J. BOYD JR.: *PHERMEX applications to studies of detonation waves and shock waves.* In: (S.J. JACOBS, ed.) *Proc. 4th Symposium (International) on Detonation.* Naval Ordnance Laboratory (NOL), White Oak, MD (1965), pp. 639-647.

- formation of Munroe jets {MUNROE ⇨ 1888};
- oblique collision of a flying plate with a stationary plate;
- collision of high-velocity metal jets;
- propagation of shock waves in explosively driven metal plates; and
- various spallation effects.

In 1969, VENABLE made a series of radiographic studies of two laterally colliding, diverging cylindrical detonation waves in the polymer-bonded explosive PBX 9404 (94/3/3 HMX/nitrocellulose/tris-β-chloroethyl phosphate). They showed that the Mach stems are curved and contain anomalous density regions. His results were reproduced numerically by use of the two-dimensionally Lagrangian hydrodynamic code 2DL with shock initiation burn model called "Forest Fire" {MADER ⇨ 1962}.[2596]

One main motivation for building PHERMEX was to study materials under a variety of shock conditions that could not be studied by other techniques. Developed at Los Alamos and built around 1963, PHERMEX – a pulsed 200-ns, 27-MeV linear electron accelerator – was capable of producing flashes of X-rays with an exposure of 168 Roentgens. Over 1,800 unclassified PHERMEX radiographs taken up to the year 1980 have been published in the *LASL PHERMEX Data* collection.[2597] ▪ A new milestone was reached in the mid-1990s by increasing the X-ray output of this machine to 380 Roentgens,[2598] a dose roughly equal to 20,000 chest X-rays and lethal to man.

In the 1990s, a new large two-flash 1,000-Roentgen X-ray machine, named **"DARHT"** (Dual Axis Radiographic Hydrodynamic Test), was developed at Los Alamos for use in so-called **"hydrodynamic tests,"** (or "hydros") because metals and other materials flow like liquids when driven by the high shock pressures and temperatures generated by the detonation of high explosives. Hydros are the most valuable experimental tool for diagnosing device performance of the primary stage in modern nuclear weapons. Hydros allow scientists to characterize the energy delivered from a layer of high explosives surrounding a mock pit, the response of this pit to hydrodynamic shocks, and the resulting distribution of pit materials when they are highly compressed. In this way, the experiments reveal the behavior of a nuclear weapon design from high-explosive detonation to the beginning of the nuclear chain reaction.[2599]

The DARHT facility, which permits three-dimensional pictures to be taken of a nuclear weapon's inner workings, performed its first hydrodynamic test in November 1999; the second axis was completed in 2002.[2600] **DARHT is supposed to eliminate the need for nuclear weapons testing by analyzing the effects of implosions during non-nuclear mock-up experiments** – simulations that will render actual weapons stockpile testing unnecessary {CTBT ⇨ 1996}.

| | | |
|---|---|---|
| 1965 | U.S. Naval Ordnance Laboratory (NOL), White Oak, Maryland | **W.A. WALKER**[2601] **and Hyman M. STERNBERG, a U.S. applied mathematician and physicist, discuss what is required to theoretically predict the underwater performances of solid high explosives,** which are important when calculating the peak pressure *vs.* distance and the pressure *vs.* time at fixed positions. |

---

[2596] C.L. MADER and D. VENABLE: *Mach stem formed by colliding cylindrical detonation waves.* Rept. LA-7869, LASL, Los Alamos, NM (Sept. 1979).
[2597] C.L. MADER, T.R. NEAL, and R.D. NICK: *LASL PHERMEX data.* 3 vols., University of California Press, Berkeley *etc.* (1980).
[2598] T. MARTINEZ: *PHERMEX reaches new milestone.* Newsbulletin (LANL) **16**, No. 13, 1 (1996).
[2599] *Ramrods shepherd hydrodynamic tests.* Sci. & Tech. Rev., LLNL (Sept. 2007); https://www.llnl.gov/str/Sep07/Bosson.html.
[2600] J. DANNESKIOLD: *DARHT second axis achieves final technical milestones.* LANL Daily NewsBulletin (Dec. 24, 2002); http://www.lanl.gov/orgs/pa/newsbulletin/2002/12/24/text03.shtml.
[2601] W.A. WALKER and H.M. STERNBERG: *The Chapman-Jouguet isentrope and the underwater shockwave performance of pentolite.* In: (S.J. JACOBS, ed.) *Proc. Fourth Symposium on Detonation* [White Oak, MD, Oct. 1964]. U.S. Naval Ordnance Laboratory, White Oak, MD (1965), pp. 27-38.

(i) They derive an explicit formula for the equation of state of water for use in shock hydrodynamic calculations – the **"Sternberg-Walker (SW) equation of state"** – which relates pressure $p$, specific volume $v$ and internal energy $E$ in the form

$$p = f(v, E) = f_1/v + f_2/v^3 + f_3/v^5 + f_4/v^7,$$

where the functions $f_1, f_2, f_3$, and $f_4$ are polynomial functions of $E$. They were found by fitting constant energy lines and matching measured shock Hugoniot data.

(ii) For pentolite (50/50 TNT/PETN), they initially calculated the detonation conditions using the Becker-Kistiakowsky-Wilson (BKW) equation of state {BECKER ⇨ 1922; KISTIAKOWSKY & WILSON ⇨ 1941}. However, a comparison of experimental data with results from hydrodynamic calculations revealed that it is necessary to use the Wilkins equation of state[2602] for the detonation product gases in the form

$$p = A\rho E + B\rho^4 + C\exp(-K/\rho),$$

which treats the expansion to low pressures differently to the BKW equation of state. Here $\rho$ is the density ($\rho = 1/v$), and $A$, $B$, $C$, and $K$ are constants. For pentolite with an initial density of 1.65 g/cm³ and a detonation energy of 1,280 cal/g, WALKER and STERNBERG obtained the following data: $A = 0.35$, $B = 0.002164$ mbar cm$^{12}$/g$^4$, $C = 2.0755$ mbar, and $K = 6$ g/cm³.

Based upon these data, STERNBERG[2603] and WALKER later calculated the partitions and distributions of kinetic energy and internal energies in the water and the gas sphere, and the energy dissipated by shock heating.

| | | |
|---|---|---|
| 1965 | *NASA's Flight Research Facility [now Dryden Flight Research Center], Edwards, California* | **Richard T. WHITCOMB**, a U.S. mechanical engineer and discoverer of the "area rule" {WHITCOMB ⇨ 1951}, having been stimulated by a number of previous studies on airfoil geometry and transonic drag, **begins to develop new tailored airfoils. His unique concept of "supercritical wings" delays the formation and reduces the strength of shock waves over the wing just below and above the speed of sound,** thus reducing drag during transonic flight.[2604]<br>• Real tests on supercritical wings began in 1969 on a Navy Chance-Vought F-8 ("Crusader"). Compared to conventional wings,<br>▸ the "NASA Supercritical Wing" is flatter on the top and rounder on the bottom with a downward curve at the trailing edge; and<br>▸ the test results showed an increase in transonic efficiency of as much as 15%.<br>NASA received a patent[2605] for this concept. In 1974, WHITCOMB[2606] reviewed the development of supercritical airfoils. |
| 1966 | *NASA's Goddard Space Flight Center (GSFC), Greenbelt, Maryland* | On April 16, **first Conference on Shock Metamorphism of Natural Materials,** which is intended to "emphasize the optical and petrographic effects of shock waves on rocks and minerals, and their rapidly increasing use as indicators of ancient terrestrial meteorite impacts." The general chairmen are Bevan M. FRENCH and Nicholas M. SHORT, two geologists at NASA's Goddard Space Flight Center.<br>(i) FRENCH[2607] provides definitions of the terms *shock metamorphism* and *impact metamorphism,* "created to describe all changes in rocks and minerals resulting from the passage of |

---

[2602] M.L. WILKINS, B. SQUIER, and B. HALPERIN: *Equation of state for detonation products of PB 9404 and LX 04-01.* In: *10th Symposium (Int.) on Combustion* [Cambridge, U.K., Aug. 1964]. The Combustion Institute, Pittsburgh, PA (1965), pp 769-778.

[2603] H.M. STERNBERG and W.A. WALKER: *Calculated flow and energy distribution following underwater detonation of a pentolite sphere.* Phys. Fluids **14**, 1869-1878 (1971).

[2604] J.V. BECKER: *The high speed frontier. Case histories of four NACA programs, 1920–1950.* Rept. NASA SP-445, NASA Scient. and Tech. Information Branch, Washington, DC (1980); http://www.hq.nasa.gov/office/pao/History/SP-445/cover.htm.

[2605] R.T. WHITCOMB: *Airfoil shape for flight at subsonic speed.* U.S. Patent No. 3,952,971 (April 27, 1976).

[2606] R.T. WHITCOMB: *Review of NASA supercritical airfoils.* In: (R.R. DEXTER, ed.) *Proc. 9th Congress Int. Council Aeronaut. Sciences.* Weizman Science Press, Jerusalem (1974); *see* Paper No. 74-10.

[2607] B.M. FRENCH: *Shock metamorphism as a geological process.* In: (B.M. FRENCH, ed.) *Proc. First Conference on Shock Metamorphism of Natural Materials* [NASA-GSFC, Greenbelt, MD, 1966]. Mono Book Corporation, Baltimore (1966), pp. 1-17.

transient, high-pressure shock waves. The great importance of such effects in geological studies arises from the fact that **the only known natural method of producing such effects is the hypervelocity impact of a large meteorite.** The terms *shock metamorphism* and *impact metamorphism* are thus closely related, and the unusual shock-metamorphic effects in rocks become important as criteria for the recognition of meteorite impact in structures so old that meteorite fragments no longer survive."

(ii) At the meeting, **three different scales for measuring the intensity of shock metamorphism are proposed.**[2608] These scales are largely based on the terminal response of quartz to shock.

Retrospectively, this meeting was a significant milestone in the understanding of shock-induced geologic effects which summarized the enormous progress made in this field since first impact hypotheses were advanced in the 1930s {SPENCER ⇒1932}. A more recent review of data from shock recovery experiments and natural impact sites with respect to their geological relevance was given by the two German mineralogists Dieter STÖFFLER and Falko LANGENHORST.[2609]

| | | |
|---|---|---|
| 1966 | Dept. of Engineering, University of Washington, Seattle, Washington | **Establishment of the Aerospace Research Laboratory (ARL);** the first director is Abraham HERTZBERG (who will stay in the position until his retirement in 1993). ▪ HERTZBERG initiated pioneering research in gas dynamics, ram accelerator technology, laser-controlled thermonuclear fusion, chemical lasers and other optical research. In 1975, the ARL was renamed the Aerospace and Engineering Research Building (AERB).[2610] |
| 1966 | Alma-Ata, south eastern Kazakhstan, Soviet Union | **Multiple chemical underground explosions are used on a large scale in order to generate an artificial dam to ward off mud flows.** The Soviet engineers V.V. GARNOV[2611] and A.G. FOMICHEV (O. Schmidt Institute of Physics of the Earth, USSR Academy of Sciences) apply high-speed stereoscopy in order<br>▸ to measure the initial displacement of soil and its initial velocity;<br>▸ to record the trajectories of the soil and dome outlines; and<br>▸ to register the overall pattern of the blast.<br>In the following years, this technique was refined in the Soviet Union and was also applied in hydraulic engineering projects {U.S. Plowshare Program ⇒1958}. After the Gulf War (1991), when most of Kuwait's oil wells were still burning, it was also considered as a way to quickly extinguish these fires through the **controlled movement of soil using multiple explosive charges** arranged concentrically around a burning well.[2612] However, since practical experience of this method was not available at the time, the fires were extinguished using established methods instead. |

---

[2608] E.C.T. CHAO: *Pressure and temperature histories of impact metamorphosed rocks – based on petrographic observations.* In: (B.M. FRENCH, ed.) *Proc. First Conference on Shock Metamorphism of Natural Materials* [NASA-GSFC, Greenbelt, MD 1966]. Mono Book Corporation, Baltimore (1968), pp. 135-158; W. VON ENGELHARDT and D. STÖFFLER: *Stages of shock metamorphism in the crystalline rocks of the Ries Basin, Germany*, Ibid., pp. 159-168; P.B. ROBERTSON, M.R. DENCE, and M.A. VOS: *Deformation in rock-forming minerals from Canadian craters*, Ibid., pp. 433-452; W. VON ENGELHARDT, F. HÖRZ, D. STÖFFLER, and W. BERTSCH: *Observations on quartz deformation in the breccias of West Clearwater Lake, Canada, and the Ries Basin, Germany*, Ibid., pp. 475-482.

[2609] D. STÖFFLER and F. LANGENHORST: *Shock metamorphism of quartz in nature and experiment: I. Basic observation and theory* [Barringer Award Address, presented on July 22, 1993 at Vail, CO]. Meteoritics **29**, 155-181 (1994).

[2610] J. LEE, D.S. EBERHARDT, R.E. BREIDENTHAL, and A.P. BRUCKNER: *A history of the University of Washington Department of Aeronautics and Astronautics 1917–2003.* University of Washington Aeronautics & Astronautics (UWAA) History (May 27, 2003).

[2611] V.V. GARNOW and A.G. FOMICHEV: *Application of photo recording methods in the observation of large-scale explosions.* In: (N.R. NILSSON and L. HÖGBERG, eds.) *Proc. 8th Int. Congress on High-Speed Photography* [Stockholm, Sweden, June 1968]. Almquist & Wiksell, Stockholm (1968), pp. 395-398; *Photorecording of large-scale underground explosions.* In: (W.G. HYZER and W.G. CHACE, eds.) *Proc. 9th Int. Congress on High-Speed Photography* [Denver, CO, Aug. 1970]. SMPTE, New York (1970), pp. 359-362.

[2612] *Zweifel an Peng und Puff.* Der Spiegel (Hamburg) **45**, Nr. 17, 180-186 (1991).

| | | |
|---|---|---|
| 1966 | Plenum Press, Data Division, New York City, New York | **Robert P. BENEDICT**[2613] at Westinghouse Electric Corporation (Monroeville, PA) **and William G. STELTZ** at Drexel Institute of Technology (Philadelphia, PA) **publish the *Handbook of Generalized Gas Dynamics*.** Addressing the large group of practicing engineers who must solve routine problems in gas dynamics daily, it focuses on fundamental one-dimensional gas dynamics and provides the user with numerical tables for various flow processes in terms of pressure ratios and Mach numbers. |
| 1966 | Physical Electronics Laboratory, TRW Systems, Redondo Beach, California | **Robert E. BROOKS,**[2614] **Lee O. HEFLINGER and Ralph F. WUERKER,** three physicists and staff members at TRW, **produce the first pulsed laser hologram of a shock wave. The example they chose for this first hologram was a head wave generated by a 0.22-in. (5.6-mm)-caliber bullet moving through air.** Using *double exposure holographic interferometry*, they demonstrate the advantages and three-dimensionality of holographic interferograms for visualizing high-speed aerodynamic phenomena. ▪ Since then, holographic studies on supersonic events have been carried out worldwide by a number of researchers, for example:<br>▸ in France at ISL by Paul SMIGIELSKI[2615] and Antoine HIRTH, who recorded a laser-generated shock wave and the head wave of a supersonic projectile flying at Mach 2.8 {⇨Fig. 4.6–I};<br>▸ in the former Soviet Union by Boris M. STEPANOV[2616] and his collaborator, who recorded the shock wave generated by an exploding wire in air and water;<br>▸ in the United Kingdom by J.W.C. GATES[2617] and collaborators, who recorded the shock wave generated by an electric spark;<br>▸ in the United States by Robert L. KURTZ[2618] and Hung-Yu LOH, who recorded the surfaces of hypervelocity projectiles moving at speeds in excess of 9,000 m/s; and<br>▸ in Germany by Werner LAUTERBORN {⇨1972} and his collaboratos when studying cavitation bubble collapse {OHL ET AL. ⇨1995}. |
| 1966 | Cornell Aeronautical Laboratory (CAL), Buffalo, New York | **John W. DAIBER,**[2619] **Abraham HERTZBERG, and Charles E. WITTLIFF,** three U.S. research aerodynamicists, **propose a new approach to heating a plasma to fusion regime temperatures based on the creation of an imploding shock wave driven by the rapid release of laser energy.**<br>(i) They describe an arrangement consisting of a number of pulsed high-energy lasers focused on a common point so that the entire solid angle is uniformly filled with radiation.[2620] Initially, some of the lasers are fired at an energy level sufficient to produce breakdown at the center, thus generating a spherical exploding blast wave.<br>(ii) After a certain time, when the blast wave has reached a desired radius, the remaining lasers are fired and absorbed by the periphery of the blast wave, thus generating a spherical imploding shock wave. For a deuterium-tritium gas mixture, they estimate the laser input energies required to reach temperatures $> 10^6$ K. |

---

[2613] R.P. BENEDICT and W.G. STELTZ: *Handbook of generalized gas dynamics.* Plenum Press, New York (1966).

[2614] R.E. BROOKS, L.O. HEFLINGER, and R.F. WUERKER: *Holographic photography of high-speed phenomena with conventional and Q-switched ruby lasers.* Appl. Phys. Lett. **7**, 92-94 (Aug. 1965); *Pulsed laser holograms.* IEEE J. Quant. Electron. **8**, 275-279 (1966).

[2615] P. SMIGIELKI and A. HIRTH: *New holographic studies of high-speed phenomena.* In: (W.G. HYZER and W.G. CHACE, eds.) *Proc. 9th Int. Congress on High-Speed Photography* [Denver, CO, Aug. 1970]. SMPTE, New York (1970), pp. 322-326.

[2616] B.M. STEPANOV and Y.I. FILENKO: *The use of holographic techniques for recording high-speed events.* In: (E. LAVIRON, ed.) *Proc. 10th Int. Congress on High-Speed Photography* [Nice, France, Sept. 1972]. Association Nationale de la Recherche Technique (ANRT), Paris (1973), pp. 240-243.

[2617] J.W.C. GATES, R.G.N. HALL, and I.N. ROSS: *High-speed recording of transilluminated events.* In: (W.G. HYZER and W.G. CHACE, eds.) *Proc. 9th Int. Congress on High-Speed Photography* [Denver, CO, Aug. 1970]. SMPTE, New York (1970), pp. 5-10.

[2618] R.L. KURTZ and H.Y. LOH: *A holographic technique for recording a hypervelocity projectile with front surface resolution.* Appl. Optics **9**, No. 5, 1040-1043 (1970).

[2619] J.W. DAIBER, A. HERTZBERG, and C.E. WITTLIFF: *Laser-generated implosions.* Phys. Fluid **9**, 617-619 (1966).

[2620] J.W. DAIBER, C.E. WITTLIFF, and A. HERTZBERG: *Shock-wave implosions by laser breakdown.* Bull. Am. Phys. Soc. **10** [II], 225 (1965).

(iii) They conclude, "The exploratory analysis discussed here indicates the very attractive possibility that lasers may be used to create by strong implosions a clean, high-density plasma from which a net fusion output may be obtained."

| 1966 | *Nobel's Explosives Co. Ltd. [a subsidiary of Imperial Chemical Industries Ltd.], U.K.* | **Stanley FORDHAM,**[2621] an explosives specialist, **publishes the book *High Explosives and Propellants*.** • This highly detailed and accurate work encompasses the following key subjects:<br>▸ high explosives (general principles, military high explosives, manufacture of commercial explosives, design of commercial explosives, assessment of explosives, permitted explosives);<br>▸ blasting accessories (initiating explosives, plain detonators, electric detonators, delay detonators, detonating & safety fuse);<br>▸ application of high explosives (commercial & military applications); and<br>▸ deflagrating and propellant explosives (black powder, manufacture of propellants, properties of propellants, design & application of propellants). |
|---|---|---|
| 1966 | *Lawrence Radiation Laboratory (LRL), UC Livermore, California* | **David W. GREGG**[2622] **and Scott J. THOMAS,** two research physicists, **measure the total momentum $M_{total}$ delivered by a focused giant laser pulse to a target using a simple ballistic pendulum** {J. CASSINI JR. ⇨1707; ROBINS ⇨1740}.<br>(i) The pendulum consists of a 1-cm-dia. sphere of the target material suspended by a 72-cm-long thread and is placed in a vacuum at $10^{-5}$ Torr. The laser beam is focused on the surface of the spherical target (Be, C, Al, Zn, Ag or W) and the pendulum's amplitude of swing viewed with a calibrated microscope through a window in the vacuum chamber.<br>(ii) **Their study showed that there is an optimum laser intensity for each material which gives a maximum amount of momentum transfer per joule of laser energy.**<br>(iii) Using the simple formula<br>$$P_{AV} = M_{total}/A\Delta t,$$<br>where $\Delta t$ is the width of the laser pulse and $A$ is the area over which the pressure is applied, they estimate that average shock pressures $P_{AV}$ ranging between 0.6 and 1 Mbar were generated, depending on the material.<br>The total momentum can also be determined by integrating the measured laser-induced pressure-time profile in the target {KREHL ET AL. ⇨1975}. |
| 1966 | *High Pressure Laboratory, CSIRO, Ryde, N.S.W., Australia* | **Sefton Davidson HAMANN,**[2623] an Australian applied chemist at the Division of Physical Chemistry of CSIRO (Australian Commonwealth Scientific & Industrial Research Organization), **reviews the effects of intense shock waves.** Addressing electrical and optical properties, phase transitions and chemical changes, he discusses ways of reaching higher shock pressures than currently available (about 10 Mbar), *e.g.*, by using nuclear explosions. |
| 1966 | *General Motors Defense Research Laboratories, Santa Barbara, California* | **Antrim Herbert JONES,**[2624] a U.S. analytical chemist, **in collaboration with William M. ISBELL,** a U.S. physicist, **and Colin J. MAIDEN,** a New-Zealand-born U.S. engineer, **first show that the light-gas gun is a very useful tool for equation-of-state (EOS) studies** {⇨Fig. 2.13}. They use a 20-mm-dia. light-gas gun to accelerate a driver plate against a target at velocities of up to 8 km/s and obtain Hugoniot data for Fansteel 77, tungsten and gold up to about 6 Mbar. An analysis of all possible errors shows that Hugoniot data can be obtained to an accuracy of about 1% with the present technique. |

---

[2621] S. FORDHAM: *High explosives and propellants*. Pergamon Press, Oxford *etc.* (1966, 1980); http://www.chimicando.it/e-book/%5Bebook%5DHigh%20Explosives%20And%20Propellants.pdf.

[2622] D.W. GREGG and S.J. THOMAS: *Momentum transfer produced by focused laser giant pulses*. J. Appl. Phys. **37**, 2787-2789 (1966).

[2623] S.D. HAMANN: *Effects of intense shock waves*. In: (R.S. BRADLEY, ed.) *Advances on high pressure research*. Academic Press, London & New York (1966), vol. 1, pp. 85-141.

[2624] A.H. JONES, W.M. ISBELL, and C.J. MAIDEN: *Measurement of the very-high-pressure properties of materials using a light-gas gun*. J. Appl. Phys. **37**, 3493-3499 (1966).

| 1966 | AVCO-Everett Research Laboratory, Everett, Massachusetts | **Arthur Robert KANTROWITZ,**[2625] a U.S. physics professor and vice-president of AVCO, **and a team of physicists and engineers successfully operate the first gasdynamic laser** {HURLE & HERTZBERG ⇨1965}. Using a hot high-pressure mixture of three gases ($CO_2/N_2/H_2O$) that flows through a supersonic nozzle and quickly expands in the subsequent laser cavity, they observe laser emission at a wavelength of 10.6 μm.<br><br>With the advent of the first gasdynamic laser, the pipe dream of producing laser beams that transfer tens or hundreds of kilowatts for durations of seconds or even minutes suddenly became reality. Compared to electric discharge lasers, which had problems associated with arcs discharging in large volumes, gasdynamic lasers can be scaled up to large sizes without any major physical complications. For a time, the gasdynamic laser was also seriously considered for use in the Strategic Defense Initiative or "Star Wars" program {SDI Concept ⇨1983} in order to allow high-power ground-based lasers to pass through the atmosphere without distortion, bounce off orbiting mirrors and shoot down intercontinental ballistic missiles. |
|---|---|---|
| 1966 | Centre for Prehistory and Palaeontology, Nairobi, Kenya | **Baroness Jane VAN LAWICK-GOODALL,**[2626] a British ethologist and anthropologist, **and her husband Baron Hugo VAN LAWICK,** a Dutch wildlife photographer, **report that they have observed an Egyptian vulture** [*Neophron percnopterus*] **attempting to break open ostrich eggs via percussion − by picking up a stone in its beak and then throwing it at the egg shell** {⇨Fig. 4.1−Z}. ▪ Other unusual examples of tool use in the animal world encompass:<br>▸ Chimpanzees have been observed to use rocks to break open sources of food, to insert sticks into honey, to use sticks and stems to feed on termites and ants, and to use leaves to pick up drinking water and to wipe their bodies.[2627]<br>▸ The woodpecker finch [*Cactospiza pallida*], one of the "DARWIN's finches," probes for insects in holes with spines or short twigs.<br>▸ The California sea otter [*Enhydra lutris nereis*] will place a slab of rock fifteen or twenty centimeters in diameter from the sea floor on its chest while floating on its back in the water. Then, holding a small mollusc shell in both forepaws, it will repeatedly strikes the shell on the stone with full swings until it is able to break it open {⇨Fig. 4.1−Z}. It is also known that the burrowing wasp [*Ammophila*] uses a small pebble as a hammer to pound the soil over its nest of eggs.[2628]<br>▸ John RAE,[2629] a Scottish Arctic explorer, reported that he was told by an eye-witness, an honest Eskimo from Repulse Bay, that polar bears use large blocks of ice as weapons: a polar bear [*Ursus maritimus*] was seen lying in wait at an elevated point above a beach for an unsuspecting walrus or sea calf to pass below. When one came within range, the polar bear grabbed a mass of ice in his paws and hit it with great force on the head of his victim by aiming his weapon very skillfully.<br><br>Prior to the discovery of tool use among animals, many anthropologists held that tools and weapons were cultural artifacts, and that man was the only "cultured" animal. **Culture is now defined as *learned behavior*, in contrast to the patterned *instinctive behavior* of animals.**[2630] |

---

[2625] J.D. ANDERSON JR.: *Gasdynamic lasers: an introduction*. Academic Press, New York *etc.* (1976).

[2626] J. VAN LAWICK-GOODALL and H. VAN LAWICK-GOODALL: *Use of tools by Egyptian vulture, Neophron percnopterus*. Nature **212**, 1468-1469 (Dec. 24, 1966); *Tool-using bird, the Egyptian vulture*. Nat. Geogr. Mag. **133**, No. 5, 631-651 (1968).

[2627] References given in VAN LAWICK-GOODALL's paper in the journal *Nature; see* preceding footnote.

[2628] K.P. OAKLEY: *Skill as a human possession*. In: (C. SINGER, E.J. HOLMYARD, and A.R. HALL, eds.) *A history of technology*. Clarendon Press, Oxford, vol. 1 (1954): *From early times to fall of ancient empires*, p. 2 and p. 5.

[2629] J. RAE: *Intelligence in animals*. Nature **27**, 366 (1883).

[2630] A.C. CUSTANCE: *Evolution or creation?* Zondervan Publ. House, Grand Rapids, MI (1976), chap. 5: *Man the culture maker*.

| | | |
|---|---|---|
| 1966 | ISL, Saint-Louis, Alsace, France | Herbert OERTEL,[2631] a German physicist, **publishes his book *Stoßrohre* ("Shock Tubes")**, certainly the most comprehensive work ever compiled on the theory and application of shock tubes. Primarily addressing physicists, chemists, gas dynamicists and aeronautical engineers, it provides the fundamentals of the design, construction and operation of shock tubes. ▪ His book became a classic on this subject. |
| 1966 | Institute for Fluid Dynamics and Applied Mathematics, University of Maryland, College Park, Maryland | Shih-I PAI,[2632] a Chinese-born aerodynamicist and research professor, **publishes the book *Radiation Gas Dynamics* which,** as well as furnishing readers with the basic elements of both radiative heat transfer and gas dynamics and their interactions, **discusses the influence of thermal radiation on the flow fields of high temperature gases.**<br><br>(i) In the introduction, he writes, "In ordinary gas dynamics, the effects of thermal radiation are always neglected. In the present Space Age, we are concerned with many technological developments in hypersonic flight, gas-cooled nuclear reactors, power plants for the needs of space exploration, fission and fusion reactions in which the temperature is very high and the density is rather low. As a result, the thermal radiation becomes an important mode of heat transfer. A complete analysis of a very high temperature flow field should be based upon a study of both the gasdynamic field and the thermal radiation field simultaneously. We use the term 'radiation gas dynamics' for this new branch of fluid mechanics." ▪ The term *radiation gas dynamics* was apparently coined in the early 1960s by Robert J. GOULARD,[2633] a gas dynamicist at Purdue University (West Lafayette, IN), who used it in a technical report.<br><br>(ii) **In radiation gas dynamics, the shock wave structure is more complicated and very similar to that of the shock structure in a chemically reacting medium** (such as a detonation wave) in which chemical equilibrium conditions are reached in a shock transition region: in addition to the viscosity and heat conductivity that occurs in the shock-structure problem in ordinary gas dynamics, where the mean free path of the gas plays an important role, the analysis of very high temperature flow requires the introduction of a new diffusive transport property – the so-called **"radiation mean free path."**<br><br>In most of the radiative transfer problems studied by astrophysicists, the interaction between the gasdynamic field and the radiation is negligibly small. However, in aerial nuclear explosions such interactions become significant, leading to synergistic effects {MORRIS ⇨1971}. |
| 1966 | Division of Engineering and Applied Physics, Harvard University, Cambridge, Massachusetts | Bradford STURTEVANT,[2634] a U.S. aeronautical engineer, **describes an "optical depth gauge" for the accurate local measurement of water waves** {⇨Fig. 4.12−I}. It is particularly appropriate for unsteady flow studies on bores, hydraulic jumps, *etc.* His optical gauge, which utilizes the absorption of infrared light by water, allows absolute depth measurements with a resolution of about 0.5 mm. ▪ The accurate determination of water depth behind a hydraulic jump – a task comparable to the determination of the density jump behind a shock wave – is essential for all discussions on the analogy between streaming and shooting liquid flow with a free surface (hydraulic jump) and two-dimensional compressible gas flow (shock wave). Here the accuracy of data obtained from the applied depth gauge is the crucial factor.<br><br>At that time, transformer-type wave gauges with variable coupling[2635] and wire-resistance-type wave gauges[2636] were among the most sensitive instruments available, but they suffered |

---

[2631] H. OERTEL SR.: *Stoßrohre. Mit einer Einführung in die Physik der Gase.* Springer, Wien *etc.* (1966).

[2632] S.I. PAI: *Radiation gas dynamics.* Springer, Vienna *etc.* (1966); *Fundamentals of radiation gas dynamics.* In: (W.H.T. LOH, ed.) *Modern developments in gas dynamics.* Plenum Press, New York & London (1969), pp. 255-310.

[2633] R.J. GOULARD: *Fundamental equations of radiation gas dynamics.* Rept. No. 62-4, School of Aeronautical & Engineering Sciences (A&ES), Purdue University, West Lafayette, IN (1962).

[2634] B. STURTEVANT: *Optical depth gauge for laboratory studies of water waves.* Rev. Scient. Instrum. **37**, 1460-1463 (1966).

[2635] M.J. TUCKER: *A ship-borne wave recorder.* In: (R.L. WIEGEL, ed.) *Proc. First Conference on Coastal Engineering Instruments* [UC Berkeley, 1955]. Council on Wave Research, The Engineering Foundation, Richmond, CA (1956), pp. 112-118.

from the difficulties common to all immersion devices: to the erratic dynamic behavior of the meniscus; in particular the existence of
- a viscous film of fluid on the gauge as the free surface recedes, and
- the occurrence of large disturbances around the gauge for large fluid velocities, *e.g.*, the upward-directed jet at the stagnation point and cavitation in the wake.

More recently, Sarma L. RANI[2637] and Margaret S. WOOLDRIDGE, two U.S. fluid dynamicists at the Dept. of Mechanical Engineering of Texas A&M University (College Station, TX), described the development of a non-intrusive optical method for the quantitative determination of water heights along a hydraulic jump in shooting water flows on a water table. Their technique which involves optically superimposing a series of alternating dark and clear fringes on the water flow is accurate to within ± 10%, thus meeting a fundamental need in the water table community.

| | | |
|---|---|---|
| 1967 | *Mullard Radio Observatory, Cambridge University, England* | In November, during a sky survey, **Susan Jocelyn BELL,** an Irish graduate student of Anthony HEWISH, A British radio astronomer and university lecturer, **discovers the first** *pulsar* – an important milestone in the history of astrophysics. Using a large-array multiple-dipole radio telescope tuned to 8.15 MHz (36.8 m), BELL and HEWISH observe an unusual radio signal coming from a small region in the constellation *Vulpecula* ("The Fox," a faint constellation of the Northern Hemisphere, near *Cygnus*) which consists of 37-ms pulses with a regular period of just 1.3 seconds. ▪ In March of the following year, the term *pulsar* was coined by Anthony MICHAELIS,[2638] a science correspondent for the British newspaper *The Daily Telegraph*, as a contraction of "pulsating star." |

In the following year, HEWISH and collaborators published their results, outlining the basic physical nature of the emitter and suggesting that only a collapsed star – such as a white dwarf {CLARK ⇨1862; CHANDRASEKHAR ⇨1931} or the hypothetical neutron star {BAADE & ZWICKY ⇨1934} – could offer a plausible explanation.[2639] ▪ One riddle to be solved was the origin of the mysterious pulsed electromagnetic radiation: in order to pulsate with a period of one second, the star would have to be very small and dense. Modern theories assume that **pulsars are highly magnetized rotating neutron stars** that emit dipole radiation in the form of radio waves along its magnetic axis. The signal appears to pulse on and off like a lighthouse beacon as the pulsar beam sweeps across the Earth. Although a basic physical model of pulsars was developed within a few years of their discovery, the source of the pulsar's radiation remains rather mysterious.

In 1974, **Sir Martin RYLE and Anthony HEWISH were jointly awarded the Nobel Prize for Physics, with HEWISH honored for the discovery of pulsars.** This was the first time the Prize was given for work in observational astronomy. ▪ The Nobel Prize announcement triggered public controversy. Sir Fred HOYLE, a renowned British astronomer, argued that BELL (since 1968 BELL-BURNELL), who subsequently discovered a further three pulsars, should have shared the Nobel Prize. In 1973, she received (jointly with HEWISH) the Michelson Medal from the Franklin Institute in Philadelphia.

As of 1997, thirty years after the discovery of the first pulsar, about 750 pulsars were known and had been catalogued. In 1999, there were just over 1,000 known radio pulsars, with pulse periods ranging from 1.557 milliseconds to over 8 seconds.[2640]

---

[2636] H.E. CROSSLEY JR.: *Analogy between surface shock waves in a liquid and shocks in compressible gases*. Rept. N-54.1, Hydrodynamics Laboratories, CalTech, Pasadena, CA (1949).
[2637] S.L. RANI and M.S. WOOLDRIDGE: *Quantitative flow visualization using the hydraulic analogy*. Experiments in Fluids **27**, 165-169 (2000).
[2638] A. MICHAELIS: *Pulsating star traced*. The Daily Telegraph, U.K. (March 6, 1968); http://www.garfield.library.upenn.edu/michaelis/title158.pdf.
[2639] A. HEWISH, S.J. BELL, J.D.H. PILKINGTON, P.F. SCOTT, and R.A. COLLINS: *Observation of a rapidly pulsating radio source*. Nature **217**, 709-713 (1968).
[2640] S.E. THORSETT: *Pulsars*. In: (P. MURDIN, ed.) *Encyclopedia of astronomy and astrophysics*. Institute of Physics Publ., London *etc.*, vol. III (2001), pp. 2177-2182.

| | | |
|---|---|---|
| 1967 | *Carson National Forest, Rio Arriba County, northwestern New Mexico* | On December 10, **the GASBUGGY Test − the first commercial test of a nuclear explosive − is carried out in order to spur the recovery of natural gas by *nuclear stimulation*,** a process by which nuclear bombs are used to crack open tight sandstone formations containing natural gas that cannot be accessed by wells. The 4-m-long fusion device has a yield of about 26 kilotons of TNT equivalent, and is detonated 1,292 m underground. ▪ The test explosion, initiated by efforts from the U.S. Atomic Energy Commission (AEC) to develop peaceful uses for atomic energy {U.S. Plowshare Program ⇒1958}, was financed by the El Paso Natural Gas (EPNG) Company and the AEC. The GASBUGGY well subsequently produced over five times as much gas as the well was expected to yield prior to the blast, but discussions about contamination problems (*e.g.*, by tritium and krypton-85) have continued ever since.<br><br>In the following year, the results from GASBUGGY encouraged EPNG to sign a contract with the AEC with the goal "to further develop the use of underground nuclear explosions to stimulate low permeability natural gas reservoirs." ▪ All such plans, however, were dropped upon the ratification of the Comprehensive Test Ban Treaty {CTBT ⇒1996}. |
| 1967 | *Free University of Brussels, Brussels, Belgium* | **First International Colloquium on the Gasdynamics of Explosions and Reactive Systems (ICOGERS);** the chairmen are Louis DEFFET and André L. JAUMOTTE. In his welcome address, Stark DRAPER,[2641] President of the International Academy of Astronautics (IAA), gives a definition of the subject matter of the Colloquium: **"*Gasdynamics of Explosions* is defined as being concerned primarily with the interrelation between rate phenomena associated with the deposition of energy at a high power density in a compressible medium and the concurrent nonsteady motion of this medium,"** ▪ The Colloquium originated due to the recognition by a group of visionary combustion specialists (Numa MANSON, Antoni K. OPPENHEIM, and Rem SOLOUKHIN) of the importance of the subject of the gasdynamics of explosions and reactive systems to the future of combustion technology and the control of global environmental emissions.<br><br>Subsequent meetings were held at Novosibirsk, Soviet Union (1969); Marseille, France (1971); La Jolla, CA (1973); Bourges, France (1975); Stockholm, Sweden (1977); Göttingen, Germany (1979); and Minsk, Soviet Union (1981). The conference, later renamed to the **International Colloquium on the Dynamics of Explosions and Reactive Systems (ICDERS)** and organized by The Institute for Dynamics of Explosions and Reactive Systems (IDERS),[2642] was held at Poitiers, France (1983); Berkeley, CA (1985); Warsaw, Poland (1987); Ann Arbor, MI (1989); Nagoya, Japan (1991); Coimbra, Portugal (1993); Boulder, CO (1995); Cracow, Poland (1997); Heidelberg, Germany (1999); Seattle, WA (2001); Hakone, Japan (2003); Montreal, Canada (2005); and Poitiers, France (2007). |
| 1967 | *University of Toronto, Canada* | **First International Symposium on Military Applications of Blast Simulation (MABS);** the general chairman is the fluid dynamicist Ashton Maynhard PATTERSON. ▪ This unclassified symposium (though primarily addressing military scientists, engineers and their consultants) is primarily focused on studies of the phenomenology and damage from shocks and blasts performed via computational methods and experimental simulation techniques, *e.g.*, explosive charges, shock tubes, blast simulators or other pressure-loading devices.<br><br>Beginning with the 14th Symposium held in 1994 at Albuquerque, NM, the conference was renamed to **Military Aspects of Blast and Shock (MABS).** Since then, the conference has been held at Banff, Alberta, Canada (1997); Oxford, U.K. (2000); Las Vegas, NV (2002); Oberjettenberg, Germany (2004); Calgary, Alberta, Canada (2006); and Oslo, Norway (2008). |

---

[2641] *Proceedings of the First International Colloquium on Gas Dynamics and Explosions* [Brussels, Belgium, Sept. 1967]. Astronaut. Acta **14**, No. 5, 385-584 (1968/1969); *see Preface.*

[2642] *History of ICDERS.* IDERS, University of Washington; http://www.engr.washington.edu/epp/iders/history.html.

| | | |
|---|---|---|
| 1967 | Sandia Corporation, Albuquerque, New Mexico | Lynn M. BARKER,[2643] a U.S. physicist, **obtains the full wave profile for the velocity of a particle in shock-loaded aluminum.** Using a new method of velocity interferometry – named **"VISAR"** {BARKER & HOLLENBACH ⇨1965} – he measures the maximum strain rate in plastic wavefronts, and suggests an approximate shock thickness relation. |
| 1967 | U.S. Naval Ordnance Laboratory (NOL), White Oak, Silver Springs, Maryland | Charles E. BELL[2644] and Jeremy A. LANDT, two NOL researchers, **generate and photograph strong shock waves in water by focusing Q-spoiled neodymium or ruby laser pulses of modest power (< 5 MW) into the water.** One part of the laser beam is reflected from a beam splitter and focused at the target, while the other part is optically delayed and used for single-shot shadowgraphy. Shock pressures of up to 233 kbar are deduced from the average shock front velocities, which are as high as 7,000 m/s; *i.e.*, about 4.8 times the sound velocity in water {COLLADON & STURM ⇨1828}.<br><br>Their method of generating spherical underwater micro-shock waves was later modified by transmitting the laser pulse through an optical fiber {NAKAHARA & NAGAYAMA ⇨1999}. |
| 1967 | Paris, France | At the IUTAM Symposium with the topic "High Dynamic Pressures," **David BERNSTEIN**,[2645] a shock physicist at Physics International Company (San Leandro, CA), **and his associates report on a new manganin-wire gauge for determining "deviatoric" stress:**<br><br>(i) They describe their goal as follows, "Present theories of stress-wave propagation in solids consider the stress tensor to be dependent on the strain tensor, the time rate of change of strain, the internal energy, and, in some cases, the past history of the material (*i.e.*, work hardening). In order to understand stress-wave propagation and the detailed relationship between the stress tensor and the various parameters that affect it, it would be desirable to measure within a material the deviatoric stresses during the passage of a stress wave. Such a gauge might be called a **'deviatoric stress gauge.'"**<br><br>(ii) They use two separate manganin gauges {FULLER & PRICE ⇨1962; BERNSTEIN & KEOUGH ⇨1964}: one gauge is used conventionally {BERNSTEIN & KEOUGH ⇨1964} to measure the stress ($\sigma_x$) in the direction in which the planar shock wave propagates through the target plate. The second gauge is embedded in a 10-mil (0.254-mm) layer of epoxy oriented at right angles in order to measure the lateral stress ($\sigma_y$).<br><br>In subsequent decades, some shock physics laboratories resumed their method of determining lateral stresses {GUPTA ET AL. ⇨1980}. More recent investigations have shown that the lateral stress has a positive dependence upon the impact stress.[2646] |
| 1967 | Calgary, southern Alberta, Canada | At the 10th International Conference on Cosmic Rays, **Stirling A. COLGATE**,[2647] a Los Alamos theoretical physicist, **proposes the first model of gamma-ray bursts (GRBs). His model invokes gamma-ray emission by particles accelerated during the breakout of a shock from a supernova progenitor's photosphere.** ▪ The finer details of his model later turned out to be incorrect, but long GRBs are indeed now believed to be caused by the death of massive stars. However, these stars do not explode in the way that ordinary stars do. They produce asymmet- |

---

[2643] L.M. BARKER: *Fine structure of compression and release wave shapes in aluminum measured by the velocity interferometer technique*. Symposium International sur le Comportement des Milieux Denses sous Hautes Pressions Dynamiques [Paris, France, Sept. 1967]. In: (M. ROY, ed.) *Behavior of dense media under high dynamic pressures*. Gordon & Breach, New York and Dunod, Paris (1968), pp. 483-504.

[2644] C.E. BELL and J.A. LANDT: *Laser-induced high-pressure shock waves in water*. Appl. Phys. Lett. **10**, No. 2, 46-48 (1967).

[2645] D. BERNSTEIN, C. GODFREY, A. KLEIN, and W. SHIMMIN: *Research on manganin pressure transducers*. Symposium International sur le Comportement des Milieux Denses sous Hautes Pressions Dynamiques [Paris, France, Sept. 1967]. In: (M. ROY, ed.) *Behavior of dense media under high dynamic pressures*. Gordon & Breach, New York and Dunod, Paris (1968), pp. 461-467.

[2646] Y. MEZIERE, J.C.F. MILLET, N.K. BOURNE, and A. WALLWORK: *Longitudinal and lateral stress measurements in NiTi under 1-D shock loading*. 14th APS Topical Conference on Shock Compression of Condensed Matter [Baltimore, MD (2005)]. In: (M.D. FURNISH, ed.) *Shock Compression of Condensed Matter 2005*. American Institute of Physics (AIP), Melville, NY (2005).

[2647] S.A. COLGATE: *Prompt gamma-rays and X-rays from supernovae*. Can. J. Phys. **46**, S476-S480 (1968); *Early gamma-rays from supernovae*. Astrophys. J. **187**, 333-336 (1974).

ric outflows traveling at almost the speed of light. Such ultrarelativistic outflows are required to produce the observed nonthermal spectrum and rapid variability. Speculative emission models have been developed to explain the afterglow of a GRB, such as the relativistic shock model,[2648] in which the prompt and afterglow emissions correspond to synchrotron radiation from shock-accelerated electrons – the so-called **"synchrotron shock model."** When COLGATE presented his paper in June 1967, GRBs were a still theoretical subject of research; they had not yet been discovered in nature. However, only one month later, in July 1967, the first gamma-ray burst was detected {Vela Satellites ⇨1960s}.

COLGATE **also created the** *AstroBlaster* – **a multiple-collision accelerator:** it consisted of four super-bouncy rubber balls held in alignment by a plastic rod. When the unit is dropped and the ball at the bottom (the largest one) hits the floor, the top ball shoots off at amazing speeds due to the successive elastic collisions of each ball with a smaller one. The top ball can reach heights of over five times the initial drop height. The Astroblaster was used by him to demonstrate the Laws of Conservation of Momentum and Energy during the creation of a supernova explosion: the shock wave accelerating outward through the star is analogous to the rapid departure of the top ball at high speed.[2649]

| 1967 | Los Alamos Scientific Laboratory (LASL), Los Alamos, New Mexico | William C. DAVIS,[2650] a U.S. shock physicist, **and Wildon FICKETT,** a U.S. physical chemist, **study detonation waves with improved and refined high-speed diagnostics.** Their results reveal that data derived from experiments on the basis of BECKER's classical theory of detonation {BECKER ⇨1921} correspond better to a weak detonation rather than to a Chapman-Jouguet detonation {CHAPMAN ⇨1899; JOUGUET ⇨1904–1906}. ▪ In 1987, the retired German detonation physicist Rudi SCHALL[2651] showed that this discrepancy essentially stems from the fact that the classical theory neglects the enthalpy of the unreacted explosive. |
|---|---|---|
| 1967 | NASA's Goddard Space Flight Center, Greenbelt, Maryland | John Aloysius O'KEEFE,[2652] a U.S. planetary scientist, **suggests that tektites** – small, pitted, pebble-like objects made of green-black high-quality glass – **are volcanic ejecta from the Moon** {O'KEEFE ⇨1976}. ▪ The U.S. aeronautical engineer Dean R. CHAPMAN {⇨1971} and others also supported the lunar theory.<br><br>Volcanic features observed within some layered tektites couldn't be explained by the terrestrial-impact theory. Furthermore, unlike all terrestrial impactite glasses, tektites are nearly free of internal water similar to lunar rocks. **Their origin remains an unsolved mystery after decades of debate** {C.R. DARWIN ⇨1844}. |
| 1967 | Explosives Research Center, Bureau of Mines, Pittsburgh, Pennsylvania | Richard W. WATSON,[2653] a U.S. engineer, **reports on a new and inexpensive pressure gauge with a submicrosecond rise time consisting of a standard 0.1-W carbon resistor.** In order to ensure good interfacial contact, the 5-mm-long, 1.7-mm-dia. resistor is coupled to a shock attenuator using a thin layer of petroleum jelly.<br>▸ For pressures ranging from 10 to 70 kbar, the gauge resistance drops from about 110 Ω to about 15 Ω. Since the resistance-pressure curve is hyperbolic and asymptotically approaches the abscissa in the high-pressure region, the gauge cannot be used to determine detonation pressures in powerful military explosives. |

---

[2648] P. MÉSZÁROS and M.J. REES: *Relativistic fireballs and their impact on external matter: models for cosmological gamma-ray bursts.* Astrophys. J. **405**, 278-284 (1993); P. MÉSZÁROS: *Theories of gamma-ray bursts.* Annu. Rev. Astron. Astrophys. **40**, 137-169 (2002).

[2649] S.A. COLGATE: *The physics of AstroBlaster*; http://fascinations.com/unique-toys-gifts-info/astro-blaster-science.htm.

[2650] W.C. DAVIS and W. FICKETT: *Detonation theory and experiment.* Symposium International sur le Comportement des Milieux Denses sous Hautes Pressions Dynamiques [Paris, France, Sept. 1967]. In: (M. ROY, ed.) *Behavior of dense media under high dynamic pressures.* Gordon & Breach, New York and Dunod, Paris (1968), pp. 1-11.

[2651] R. SCHALL: *On the hydrodynamic theory of detonation.* Propellants, Explosives, Pyrotechnics **14**, 133-139 (1989).

[2652] J.A. O'KEEFE: *Tektite sculpturing.* Geochemica et Cosmochimica **31**, 1931-1933 (1967); In: (V.E. BARNES and M.A. BARNES, eds.) *Tektites.* Dowden, Hutchinson & Ross, Stroudsburg, PA (1973), pp. 251-253.

[2653] R.W. WATSON: *Gauge for determining shock pressures.* Rev. Scient. Instrum. **38**, 978-980 (1967).

> However, he proposes that the new gauge could be used to rank explosives in terms of detonation pressure if a thin brass buffer is inserted between the gauge and the detonating explosive in order to reduce the magnitude of the pressure delivered to the gauge.

In the years following, carbon resistors were also applied in Germany at EMI (Freiburg), to measure shock waves in underwater explosions and at Batelle-Institut e.V. (Frankfurt/Main) to record the pressure in shock-loaded steel plates. Anton BUSSE,[2654] a researcher at the Universität der Bundeswehr (Munich), studied the applicability of carbon resistors for measuring blast waves.

| 1967 | *Atomic Weapons Research Establishment (AWRE), Aldermaston, Berkshire, England* | **Peter W. WRIGHT**[2655] **and collaborators perform experimental studies into the successive passage of multiple shocks through a material at intervals on the order of a microsecond, in order to obtain thermodynamic information on the behavior off the principal Hugoniot curve.** Hydrodynamic calculations that treat the metal as a fluid with the Grüneisen equation of state based on a known Hugoniot reference curve and using a 1-D hydrodynamic code are also performed in parallel with this work. For low-strength materials (*e.g.*, cadmium, tin, lead), the calculations accurately predict the rate at which the second shock catches up with the first one. However, for high-strength materials (*e.g.*, stainless steel, copper), such comparisons reveal considerable deviations which the authors explain by elastic-plastic material behavior. |
|---|---|---|
| 1968 | *Nuclear Test Site, Nevada* | On March 12, **the BUGGY Test – a ditch excavation experiment performed as part of the AEC's Plowshare Program – is carried out by the Lawrence Radiation Laboratory** {U.S. Plowshare Program 1958; ⇨Fig. 4.16–K}. Five nuclear devices, BUGGY-A to BUGGY-E, each with a yield of 1.08 kilotons of TNT equivalent, are detonated simultaneously about 150 ft (46 m) apart in separate 164-ft (50-m)-deep holes.[2656] This trench cratering experiment produces a crater 850 ft (259 m) long, 250 ft (76 m) wide, and 65 ft (20 m) deep. • One proposed application of the Plowshare Program was the widening of the Panama Canal {U.S. Atlantic-Pacific Interoceanic Canal Study Commission ⇨1970}, or even the construction of a new waterway through Nicaragua.

So-called "ditch blasting" is a quick and economic technique which was developed previously using chemical explosives and proved most useful in agriculture. There are two distinct methods of ditching with dynamite, differing in the means by which simultaneous detonation of a series of charges is produced.[2657]

> In the *propagation method*, a line of holes is put down and charged with straight dynamite, but only one hole is primed and fired either with an ordinary blasting cap and fuse or with an electric blasting cap method. **The shock from the explosion of the primed cartridge communicates itself through the soil to the other charges with sufficient strength to detonate them.**

> In the *electric method* each hole is primed with an electric blasting cap and connected in a circuit and fired with an electric blasting machine.

In the same year, Soviet researchers also develop a nuclear technique for generating rows of craters for canal construction. These excavation tests, named TEL'KEM-1 and TEL'KEM-2, were performed in autumn 1968. |

---

[2654] A. BUSSE: *Einsetzbarkeit von Kohle-Masse-Widerständen als Druckaufnehmer zur Messung von Stoßwellen in Gasen*. Universität der Bundeswehr, München (1993).

[2655] P.W. WRIGHT, G. EDEN, and B.D. LAMBOURN: *Behavior of various materials under multiple shocking*. Symposium International sur le Comportement des Milieux Denses sous Hautes Pressions Dynamiques [Paris, France, Sept. 1967]. In: (M. ROY, ed.) *Behavior of dense media under high dynamic pressures*. Gordon & Breach, New York and Dunod, Paris (1968), pp. 137-152.

[2656] BUGGY A, B, C, D, and E. U.S. Dept. of Energy; http://www.osti.gov/html/osti/opennet/document/nukeunan/sheets_b.html#ZZ10.

[2657] *Hercules dynamite on the farm: ditch blasting*. Hercules Powder Company, Wilmington, DE (1933); http://chla.library.cornell.edu/cgi/t/text/text-idx?c=chla;idno=3086190.

| | | |
|---|---|---|
| 1968 | *United Nations (UN) General Assembly, New York City, New York* | On July 1, **the Nuclear Non-Proliferation Treaty (NNPT) is opened for signature.** Its objectives are
▸ to prevent the spread of nuclear weapons and weapons technology;
▸ to promote cooperation in the peaceful uses of nuclear energy; and
▸ to further the goal of achieving nuclear disarmament and general and complete disarmament.
The Treaty was signed at Washington, DC, London, and Moscow on July 1, 1968 and entered into force on March 5, 1970. A total of 187 parties have joined the Treaty, including the five nuclear-weapons States: the United States (signed in 1968), the Soviet Union (1968), the United Kingdom (1968), France (1992), and the People's Republic of China (1992). India, Pakistan, and Israel have declined to sign the treaty. North Korea ratified the treaty in 1985, but withdrew from the treaty in 2003. |
| 1968 | *Zhukovski Test Center, U.S.S.R* | On December 31, **the Tupolev Tu-144** {⇨Fig. 4.20–H}, the Soviet SST (SuperSonic Transport) **flies for the first time.** The new plane, designed for $M = 2.4$ and 140 passengers by Aleksei Andreevich TUPOLEV, the son of the famous aircraft designer Andrei Nikolaevich TUPOLEV, was modeled in the large wind tunnel of the Central Aerohydrodynamic Institute [CAGI, Russ. abbrev. ЦАГИ] at Zhukovski Test Center, a large test airfield near Moscow. ▪ Flight velocities above Mach 1 were first reached with the Tu-144 on June 5, 1969, and speeds above Mach 2 on May 26, 1970. In 1973, Valentin BLIZNIUK became Chief Designer of the Tu-144 and Head of the the Soviet SST program. |
| 1968 | *Washington State University (WSU), Pullman, Washington* | **Foundation of the Shock Dynamics Laboratory (SDL)** by Prof. George E. DUVALL. Experimental shock wave research in condensed matter is initiated in the same year with support from the Department of Defense (DOD).[2658] ▪ Shock wave research at WSU already began with theoretical studies in the Dept. of Physics in the late 1950s.
In 1992, the laboratory was renamed the **Shock Dynamics Center (SDC)** to broaden its scope for research, and in 1997 it adopted its current name, the **Institute for Shock Physics (ISP).** Most American solid-state shock wave physicists studied at Washington State University. |
| 1968 | *NASA Goddard Institute for Space Studies (GISS), New York City, New York* | **First Pulsar Conference,** sponsored by NASA-GISS and the Yeshiva University (New York City, NY). ▪ Only a few months after the first publication of the newly discovered pulsating radio source {BELL & HEWISH ⇨1967}, further four pulsars were discovered. Many of their properties were already known, but at that time there was no comprehensive theory that could account for all the observed phenomena.[2659] |
| 1968 | *Boeing Corporation, U.S.A.* | **The Boeing Company announces the development of its High Speed Civil Transport (HSCT), the Boeing 2707-300 SST** {⇨Fig. 4.20–H}. ▪ This ambitious U.S. project, designed for $M = 2.65$ but intended to produce smaller sonic booms and lower engine noise than Concorde at its maximum speed ($M = 2.04$), was canceled in March 1971 when governmental funding was withdrawn.[2660]
In the years following, intensive research carried out by NASA and the U.S. industry showed that improved technology could make a commercially successful HSCT viable.[2661] Later, an even more ambitious project, a National Aero-Space Plane {NASP ⇨1985}, was started in the United States but was not completed either. |

---

[2658] *Institute for Shock Physics Applied Sciences Laboratory.* Washington State University (WSU); http://wsunews.wsu.edu/Shock%20Physics2.pdf.
[2659] S.P. MARAN and A.G.W. CAMERON: *Pulsars.* Phys. Today **21**, 41-49 (Aug. 1968).
[2660] W.T. GUNSTON: *The three supersonic transports.* Science J. **4**, No. 9, 32-39 (1968).
[2661] R. ROSEN: *The rebirth of the supersonic transport.* Technology Rev. **96**, No. 2, 22-29 (1993).

| | | |
|---|---|---|
| 1968 | KMS Industries, Inc., Ann Arbor, Michigan | **R.D. BUZZARD**[2662] **reports on a true 3-D schlieren system using pulsed laser holography that is suitable for measuring gas densities in wind tunnels and ballistic test ranges.** Since the recorded hologram stores all of the characteristics of the light wave from the test field, schlieren analysis can be performed on any or all of the different planes at various distances along the optical axis through the test section, thus distinguishing the effects of density gradients at different positions in the test section and providing data along the third dimension. |
| 1968 | U.S. Naval Ordnance Laboratory (NOL), White Oak, Maryland | **Nathaniel L. COLEBURN,**[2663] a U.S. research physical chemist, **and L.A. ROSLUND**[2664] **experimentally and theoretically study the collision of spherical shock waves in water.**<br>(i) They use 225-gram twin spheres of cast pentolite (50/50 TNT/PETN), which are fired simultaneously at their centers by electric detonators. The distance between the two spheres is varied to change the interaction pressures at the plane of collision.<br>(ii) They visualize the propagation and interaction of underwater shock waves and the associated spherical flow resulting from the simultaneous underwater detonations using high-speed framing camera shadowgraphs.<br>(iii) The critical angle for the onset of Mach reflection varies from 36° to 41° for incident pressures ranging from 6.5 kbar to 1.6 kbar, respectively.<br>(iv) The two intersecting underwater shock waves collide with the gas bubble from the other sphere. This results in two backward-facing rarefaction waves which produce a region of cavitation at the point of collision.<br>(v) Applying the theory of oblique shock reflection {VON NEUMANN ⇒ 1943} and an energy-dependent form of the equation of state, they calculate reflected shock states in water up to the critical point for Mach wave formation using measured peak pressure distance data.<br>(vi) They find that the critical angle for Mach wave formation increases as the incident pressure decreases. They attribute differences between data obtained by theory and that obtained by experiment to the fact that the theory does not consider the effects of rarefaction waves on the strength of the interacting shocks. |
| 1968 | Aeronautical Sciences Division, UC Berkeley, California | **Richard COLLINS,**[2665] a Canadian-born U.S. fluid dynamicist, **and Maurice HOLT,** a British-born U.S. applied mathematician, **systematically investigate the problem of an intense point-source explosion at the interface separating two different fluids.** In particular, they treat a nuclear explosion initiated on the surface of the ocean, and determine the position of the deformed ocean surface as a function of time during the initial phase of the explosion, and the profiles of shock and vacuum surfaces. |
| 1968 | University of Manchester Institute of Science and Technology (UMIST), England | **John Byrom HAWKYARD,**[2666] **David EATON,** a M.Sc. student, **and William JOHNSON,** a professor of mechanical engineering, **present a method of deducing mean yield strengths for some ductile metals from strain measurements on the "mushroomed" ends of flat-ended projectiles after impact on a flat, nominally rigid anvil – the "Taylor test"** {G.I. TAYLOR & WHIFFIN ⇒ 1948; ⇒ Fig. 4.3–Q}. The method applied, which assumes a homogeneous uniaxial compression of each cross-section and is based on an energy analysis that involves equating the kinetic energy at impact with the plastic work, provides a mean dynamic yield stress averaged over a wide strain range, thus improving correlation with the experimentally observed deformed shape. |

---

[2662] R.D. BUZZARD: *Description of three-dimensional schlieren system.* In: (N.R. NILSSON and L. HÖGBERG, eds.) *Proc. 8th Int. Congress on High-Speed Photography* [Stockholm, Sweden, June 1968]. Almquist & Wiksell, Stockholm (1968), pp. 335-340.

[2663] N.L. COLEBURN and L.A. ROSLUND: *Collision of spherical shock waves in water.* Tech. Rept. NOLTR 68-110, U.S. Naval Ordnance Laboratory (NOL), White Oak, MD (July 1968).

[2664] L.A. ROSLUND and N.L. COLEBURN: *Regular reflection of oblique shock waves in water.* Tech. Rept. NOLTR 68-209, U.S. Naval Ordnance Laboratory (NOL), White Oak, MD (Dec. 1968).

[2665] R. COLLINS and M. HOLT: *Intense explosion at the ocean surface.* Phys. Fluids **11**, 701-713 (1968).

[2666] J.B. HAWKYARD, D. EATON, and W. JOHNSON: *The mean dynamic yield strength of copper and low carbon steel at elevated temperatures from measurements of the "mushrooming" of flat-ended projectiles.* Int. J. Mech. Sci. **10**, 929-948 (1968).

| | | |
|---|---|---|
| 1968 | *Lovelace Foundation for Medical Education and Research, Albuquerque, New Mexico* | **Frederic G. HIRSCH,**[2667] a U.S. neurologist and otologist, **presents a review of what is known about otic blast trauma in view of recent advances in explosives technology and high-velocity ordnance.**<br>(i) Beginning in 1872 with the report from the American audiologist J.O. GREENE,[2668] he discusses<br>▸ previous studies on the vulnerability of the ear to blast overpressures under physical, anatomical and physiological considerations;<br>▸ quantitative considerations of eardrum rupture;<br>▸ hearing loss caused by blast trauma; and<br>▸ methods of preventing blast injury of the ear.<br>(ii) The vulnerability of an eardrum depends not only on the parameters of the blast wave at the site of the tympanic membrane (*e.g.*, peak overpressure, rise time, and duration of the pressure pulse), but it also varies with the age of the eardrum and depends on whether it has been subjected to previous trauma or disease.<br>(iii) He concludes, "Based upon previous quantitative studies, it can probably be safely said that the threshold pressure for damage to middle ear structures is about 5 psi [0.34 bar] and that overpressure near 15 psi [1.03 bar] will cause the rupture of 50% of the eardrums exposed to it." |
| 1968 | *NASA's Ames Research Center (ARC), Moffet Field, California* | **Enrique J. KLEIN**[2669] **first uses liquid crystals for boundary-layer flow visualization in aerodynamic measurements.** Based on his experiments, he concludes that liquid crystals are capable of showing the delineation of turbulent and laminar regions, discrete vortices, flow lines, and temperature distributions on the surface of wind tunnel test models. ▪ Color-change paints undergo irreversible changes in color and require that the model be cleaned and re-coated. On the other hand, liquid crystals – although fully reversible – are not resistant to abrasive forces due to shear stress when exposed to hypersonic flow, and so recoating is also necessary when using them. Encapsulated liquid crystals, however, overcome these restrictions, because the crystals are enclosed in plastic spheres 4–8 µm in diameter.<br>In the following years, this subject was resumed in Germany at the Deutsche Forschungs- und Versuchsanstalt für Luft- und Raumfahrt (DFVLR), Göttingen:<br>▸ **Arnulf KÜHN**[2670] **showed the feasibility of applying liquid crystals to quantitative heat transfer measurements in hypersonic flow;** and<br>▸ **Hennig SCHÖLER**[2671] **investigated the interference heating of fin/body junctions at Mach 6 using encapsulated liquid crystals.** He obtained reasonable agreement with thermocouple measurements.<br>An example of the use of ***liquid crystal surface thermography*** in British hypersonic flow studies is shown in Fig. 4.18–H. |
| 1968 | *Lowell Technological Institute, Lowell, Massachusetts* | **Walter Roy MELLEN,**[2672] a U.S. researcher, **reports in a short note on an unusual experiment involving elastic collision using "superballs"** {STINGLEY ⇒ 1964}. He dropped a small superball immediately after dropping a large one, one directly above the other. The small ball, upon rebounding from the large one, appeared to be shot into the air. **Under ideal elastic collisions the small ball should rise to about nine times the height from which it is dropped.** |

---

[2667] F.G. HIRSCH: *Effects of overpressure on the ear – a review.* Ann. New York Acad. Sci. **152**, 147-162 (1968).

[2668] J.O. GREENE: *Cases of injury to the ear from external violence.* Trans. Am. Otol. Soc. **5**, 88-98 (1872).

[2669] E.J. KLEIN: *Application of liquid crystals to boundary layer flow.* In: *AIAA 3rd Aerodynamic Testing Conference* [San Francisco, CA, April 1968]; see AIAA Paper No. 68-0376 (1968).

[2670] A. KÜHN: *Wärmeübergangsmessungen in einer Hyperschallströmung geringer Dichte mit Hilfe von flüssigen Kristallen.* Rept. IB 252-76 H06, DFVLR, Göttingen (1976).

[2671] H. SCHÖLER: *Application of encapsulated liquid crystals on heat transfer measurements in the fin/body interaction region at hypersonic speed*, AIAA Paper No. 78-777 (1978).

[2672] W.R. MELLON: *Superball rebound projectiles.* Am. J. Phys. **36**, 845 (1968).

Three years later, William G. HARTER,[2673] a U.S. physics instructor at the Physics Dept. of the University of Southern California at Los Angeles, and his students presented a qualitative interpretation of this unusual collision phenomenon obtained by developing an elastic continuous-force computer model which they also extended to multiple-stage collisions for an *n*-stage tower of balls.

| | | |
|---|---|---|
| 1968 | University of Kansas, Lawrence, Kansas | Raphael M. PAOLI[2674] and Russell B. MESLER, an assistant professor of chemical engineering, **explore the explosive mechanism that occurs when molten metal interacts with water.** This phenomenon, which has been known about for a long time {LEIBNIZ ⇨1704}, results in violent explosions and highly fragmented metal, and is extremely important in the nuclear reactor and metal industries. It is obviously not the result of chemical reactions, but rather of the rapid generation of steam. Using high-speed photography, they observe the appearance of ripples on the metal surface – comparable to the familiar occurrence of Helmholtz instabilities in fluid mechanics – which suggests a potential new way that water could become entrapped within the molten metal, thus confirming a recent new hypothesis based on the entrapment of quenched liquid.[2675] ▪ The principal hazards encountered in the foundry industry arise from the handling of molten metals. The aluminum industry in particular has made extensive efforts to gain an understanding of molten metal-water explosions, the factors that promote them, and how they might be prevented {LONG ⇨1957}. |

Severe accidents in the foundry industry caused by unexpected steam explosions have a long tradition, and generations of mechanical engineers and physicists attempted to uncover their causes. For example, in 1795 Charles HUTTON,[2676] a British professor of mathematics and ballistician at the Royal Military Academy in Woolwich, wrote, "… it has been inferred that aqueous steam is even vastly stronger than fired gun-powder. This is when water is thrown upon melted copper: for here the explosion is so strong as almost to exceed imagination; and **the most terrible accidents have happened, even from so slight a cause as one of the workmen spitting in the furnace where copper was melting; arising probably from a sudden decomposition of the water.** Explosions happen also from the application of water to other melted metals, though to a lower degree. When the fluid is applied in small quantities, and even to common fire itself, as every person's own experience must have informed him; and this seems to be occasioned by the sudden rarefaction of the water into steam. Examples of this kind often occur when workmen are fastening cramps of iron into stones; where, if the lead is poured, this will fly out in such a manner as sometimes to burn them severely. **Terrible accidents of this kind have sometimes happened in foundries, when large quantities of melted metal have poured into wet or damp moulds.** In these cases, the sudden expansion of the aqueous steam has thrown out the metal with great violence; and if any decomposition has taken place at the same time, so to convert the aqueous vapour into an aerial one, the explosion must be still greater."

| | | |
|---|---|---|
| 1968 | Los Alamos Scientific Laboratory (LASL), Los Alamos, New Mexico | At the Conference on Numerical Simulation of Plasma, **Carl R. SHONK**[2677] **and Richard L. MORSE discuss analytical approaches to the calculation of a shock wave in a collisionless plasma that propagates perpendicular to the magnetic field.**<br>(i) In order to numerically simulate a collisionless shock wave – which occurs when the mean free path of the particles in a plasma is longer than the shock width and the shock is car- |

---

[2673] W.G. HARTER ET AL.: *Velocity amplification in collision experiments involving superballs*. Am. J. Phys. **39**, 656-663 (1971).
[2674] R.M. PAOLI and R.B. MESLER: *Explosion of molten lead in water*. In: (N.R. NILSSON and L. HÖGBERG, eds.) *Proc. 8th Int. Congress on High-Speed Photography* [Stockholm, Sweden, June 1968]. Almquist & Wiksell, Stockholm (1968), pp. 463-466.
[2675] F.E. BRAUER, N. WAYNE, and R.B. MESLER: *Metal/water explosions*. Nucl. Sci. Engng. **31**, 551-554 (1968).
[2676] Ch. HUTTON: *A philosophical and mathematical dictionary*. Johnson & Robinson, London (1795), vol. 1, pp. 458-459.
[2677] C.R. SHONK and R.L. MORSE: *Two-dimensional simulation of shock waves in collisionless plasmas*. In: *Proc. of the APS Topical Conference on Numerical Simulation of Plasma* [Los Alamos, NM, Sept. 1968]. American Physical Society (APS), New York (1968), Paper C3.

ried by magnetohydrodynamic (MHD) effects only – they assume a dense cylindrical plasma sheet which is six times denser than the inside and outside. Using 64,000 particles in a 50 by 50 Eulerian mesh, each cell measuring $0.5\lambda$ on a side (with $\lambda$ as the ratio of the speed of light to the ion plasma frequency), they calculate the evolution of a cylindrical shock wave as driven radially inward by a magnetic field.

(ii) **Their computer simulation reveals various shock wave nonlinearities, such as a fin-like structure of the shock-compressed plasma** which is due to instabilities at the interface of the cylinder and the driving field. ▪ Similar shock instability phenomena have also been reported in photographs taken axially in theta-pinch shock devices which compress and heat the plasma.

| 1968 | Dept. of Physics, Princeton University, Princeton, New Jersey | **John A. WHEELER**,[2678] a U.S. theoretical physicist, **coins the term *black hole*** for a dense, compact object with such powerful gravity that nothing can escape from it, including light. ▪ **The idea of black holes had already been developed previously** {MICHELL ⇨1783; DE LAPLACE ⇨1796}. A black hole is a very massive (rotating or non-rotating) remnant of a star which, having completely burned its nuclear fuel, had collapsed under its own weight.

(i) Prior to this, discussions among theoretical physicists on the reality of an astronomical body consisting of a single, indefinitely dense point and the possibility of an infinite collapse had not yielded any clear answer. However, computer simulations of a collapsing star, adapted from simulations of fusion bombs, proved that the losses due to the ejection of matter, radiation *etc.* could not prevent the gravitational collapse.

(ii) In his paper, WHEELER states, "Light and particles incident from outside emerge and go down the black hole only to add to its mass and increase its gravitational attraction."

Black holes are regions of space into which matter "falls" and from which no material object (even light; which explains the name) can escape. There are two main types, classified by their mass:

▸ ***Stellar black holes*** have masses of between about 4 and 15 solar masses and result from the core-collapse of a massive star at the end of its life. ▪ In 1970, the first stellar black hole was discovered by Riccardo GIACCONI, an Italian-born U.S. physicist, the father of X-ray astronomy and the winner of the 2002 Nobel Prize for Physics.

▸ ***Supermassive black holes*** have masses of between about $10^6$ and $10^9$ solar masses and are found at the centers of most large galaxies. The gravitational pull would hurl gas and stars around it at almost the speed of light, and the violent clashing would heat the gas up to over a million degrees. ▪ **In 1994, the NASA/ESA Hubble Space Telescope detected the first supermassive black hole, a 2.4-billion-solar-mass object in the center of the giant elliptical galaxy M87** (or NGC 4486) which is located 50 million light-years away in the constellation *Virgo*.[2679] In 1997, a Hubble census of 27 nearby galaxies showed that supermassive black holes are common in large galaxies. |
| 1969 | Pueblito de Allende, Chihuahua, Mexico | On February 8, **a meteoroid enters the Earth's atmosphere, creating a huge fireball and exploding into a number of fragments which rain over an area of more than 60 square miles** (155.3 km$^2$). ▪ The weight of the ***Allende meteorite***, a carbonaceous chondrite, is estimated to be several tons. Two tons were collected after the meteorite fell, but specimens are still being found.

In 1997, **Robert L. FOLK**,[2680] a U.S. geology professor at the University of Texas, **and his coworker F. Leo LYNCH reported that they had seen and photographed tiny grape-like |

---

[2678] J.A. WHEELER: *Our Universe: the known and unknown* (reprint of a public lecture given on Dec. 29, 1967). Am. Sci. **56**, 1-20 (1968); see p. 9.

[2679] *Hubble confirms existence of massive black hole at heart of active galaxy*. Space Telescope Science Institute (STScI), Baltimore, MD (May 25, 1994); http://hubblesite.org/newscenter/archive/releases/1994/23/text/.

[2680] R.L. FOLK: *Nannobacteria: surely not figments, but what in heavens name are they?* naturalSCIENCE **1**, Article No. 3 (1997); http://naturalscience.com/ns/articles/01-03/ns_folk.html.

**clusters of nanobacterial cells** bridging a pore between two olivine crystals in the Allende meteorite, which closely resembled fossils of well-known terrestrial "nannobacteria." However, their interpretation − that these microfossils provided evidence of extraterrestrial lifeforms (or their remnants) − immediately provokes critical comments by other geologists.[2681]
▪ At around the same time, NASA researchers also studied the Martian meteorite ALH 84001 {⇨1984} and the Murchison meteorite {⇨1969} and found mineralized structures in these that may be the remains of extraterrestrial life.

The word "nanobacteria" was first published in 1988 by Richard Y. MORITA, a professor of microbiology and oceanography at Oregon State University who studied microbial lifeforms in oceanic environments. However, R.L. FOLK used the spelling "nanno-" instead of "nano-" to conform with geological usage (*e.g.*, "nannoplankton" and "nannofossil").

| 1969 | *Aérospatiale, Toulouse, France & British Aircraft Corporation (BAC), U.K.* | On March 2, **the British/French supersonic transport (SST) *Concorde* performs its maiden flight** − two months after that of the Soviet SST {Tupolev Tu-144 ⇨1968}.
(i) The Concorde {⇨Fig. 4.20−H}, a delta-winged supersonic jet airliner designed for $M = 2$ and 144 passengers, is powered by four jet engines (Rolls-Royce/Snecma Olympus 593s), the most powerful engines to be used for commercial flights.
(ii) The take-off velocity is 250 mph (402 km/h), compared with about 185 mph (298 km/h) for most subsonic aircraft.

On October 1 of the same year, **Concorde makes its first supersonic flight,** flying supersonically for nine minutes and attaining a maximum speed of Mach 1.5. ▪ In November 1970, **both Concorde prototypes reached Mach 2.** British Airways and Air France began operating Concordes in 1976. British Airways holds both the west-east and east-west transatlantic records: achieved on February 7, 1996, in 2 hrs, 52 mins and 59 secs from New York to London; and on October 8, 2003, in 3 hrs, 5 mins and 34 secs from London to Boston. However, due to a drop in demand, both airlines shut down their Concorde services in 2003.

Concorde was modeled in the large wind tunnel of the Royal Aircraft Establishment (RAE) at Bedford, and was produced jointly by the British Aircraft Corporation (BAC) and the French company Aérospatiale. ▪ A perspective on the early history of Concorde development was given by Francis G. CLARK[2682] and Arthur GIBSON, who were involved with BAC during the design and construction of the aircraft. |
| 1969 | *First U.S. Moon Expedition & NASA's Kennedy Space Center (KSC), Cape Canaveral, Florida* | In the period July 16−24: *Apollo 11* **mission to the Moon. During the mission, two American astronauts,** Neil A. ARMSTRONG (Lunar Module commander) and Edwin E. ALDRIN (Lunar Module pilot), **become the first men to land on the Moon.** Michael COLLINS, the third crewmember of Apollo 11, remains in the Command Module {⇨Fig. 4.20−J}, circling the Moon.
(i) On July 20, **ARMSTRONG,** an experienced hypersonic research pilot and veteran pilot of the hypersonic research plane X-15 {⇨1961}, **becomes the first human to set foot on the Moon's surface at** *Mare Tranquillitatis* ("Sea of Tranquility").
(ii) **A seismic instrument package powered by solar cells is left on the surface by the crew of Apollo 11 in order to measure** *lunar shock waves* **caused by "moonquakes," or impacts of meteoroids or of man-made objects on the Moon's surface.** ▪ Astronauts from subsequent lunar missions (Apollo 12/14/15/16) also placed seismometers at their landing sites. The Apollo instruments faithfully radioed data back to Earth until they were switched off in 1977. Between 1972 and 1977, 28 shallow moonquakes up to 5.5 on the Richter scale were de- |

---

[2681] R.P. HARVEY: *Nannobacteria: what is the evidence?* naturalSCIENCE **1**, Article No. 7 (1997); http://naturalscience.com/ns/articles/01-07/ns_rph.html.

[2682] F.G. CLARK and A. GIBSON: *Concorde: the story of the world's most advanced passenger aircraft.* Phoebus, London (1975).

tected (on Earth, quakes of magnitude 4.5 and above can cause damage to buildings and other rigid structures). According to NASA, there are at least four different kinds of moonquakes:
- deep moonquakes (~ 700 km below the surface, probably tidal in origin);
- meteoritic impact vibrations;
- thermal quakes (the frigid lunar crust expands when sunlight returns after the two-week lunar night); and
- shallow moonquakes (20 or 30 km below the surface).[2683]

(iii) During reentry into the Earth's atmosphere, the Command Module reaches a velocity of about Mach 30 and a shock-layer temperature of almost 12,000 K – *i.e.*, about twice the surface temperature of the Sun.[2684]

In the same year, ARMSTRONG, ALDRIN, and COLLINS receive the 1969 Collier Trophy {COLLIER ⇒ 1911} "for the epic flight of Apollo 11 and the first landing of man on the surface of the Moon."

**Rock samples collected from the Moon's surface during Apollo 11 and subsequently during Apollo 12** (November 14–24, 1969) **provide the first evidence of shock metamorphism in lunar matter:**
- Nicholas M. SHORT,[2685] a U.S. geologist at NASA's Goddard Space Flight Center (Greenbelt, MD) was one of the original principal investigators of the rocks brought back from the Moon. SHORT had previously specialized in the effects of shock metamorphism, such as those caused by underground nuclear explosions and the impacts of asteroids into rocks. He found that lunar microbreccias and loose regolith materials contain abundant evidence of shock metamorphism related to crater-forming meteorite impacts.
- A cooperative study carried out in the early 1990s at Yamaguchi University (Honshu, Japan) and Sandia National Laboratories (Albuquerque, NM) revealed that shocked lunar minerals of plagioclase (a series of tectosilicate minerals) $[(Na,Ca)(Si,Al)_4O_8]$ and silica $[SiO_2]$ show anomalous compositions and densities. The shock metamorphism showed evidence for two major impact processes on evolved and primordial lunar surfaces: (1) shocked silica phases with minor Al contents formed from plagioclase-rich primordial crusts of the Moon; and (2) shocked quartz formed by silica-rich target rocks, especially on evolved parts of the Moon.[2686]

| | | |
|---|---|---|
| 1969 | Murchison [ca. 100 km north of Melbourne], Victoria, Australia | On September 28, **a fireball explodes over the rural town of Murchison,** scattering about 100 kg of fragmented meteoritic material rich in carbon compounds over an area of five square miles (13 km$^2$); the largest stone weighs about 7 kg. ▪ The **"Murchison meteorite"** is classified as a carbonaceous chondrite, type II (CM2), and contains $SiO_2$ (29.07%), FeO (22.39%), MgO (19.94%), $H_2O^+$ (8.95%), FeS (7.24%), $Al_2O_3$ (2.15%), CaO (1.89%), C (1.85%), $H_2O-$ (1.14%), $Cr_2O_3$ (0.48%), $Na_2O$ (0.24%), *etc.*[2687] <br><br> The Murchison meteorite was suspected of having originated from a comet due to its high water content. An abundance of a mix of amino acids found within this meteorite – very similar to |

---

[2683] C. CHRISTENSEN: *Moonquake concerns: bases might need special construction.* Space.com (March 24, 2006); http://www.space.com/businesstechnology/060324_moonquakes.html.

[2684] J.D. ANDERSON JR.: *Aerodynamics: a tutorial discussion.* In: (E.A. THORNTON, ed.) *Thermal structures and materials for high speed flight.* Progr. Astronaut. Aeronaut. (New York) **140**, 3-57 (1992).

[2685] N.M. SHORT: *The nature of the Moon's surface: evidence from shock metamorphism in Apollo 11 and 12 samples.* Icarus (San Diego, CA) **13**, 383-413 (1971).

[2686] Y. MIURA and R.A. GRAHAM: *Shock metamorphism on the surface of the Moon.* Shock Waves **3**, No. 4, 293-298 (1994).

[2687] E. JAROSEWICH: *Chemical analyses of meteorites: a compilation of stony and iron meteorites analyses.* Meteoritics **25**, 323-337 (1990).

that produced in the Miller-Urey type experiment {S.L. MILLER ⇒1953; ⇒Fig. 4.14–A} – has led to intense study into its origins by researchers:[2688]

(i) Obviously, during its passage through the Earth's atmosphere the interior of this meteorite was never heated to temperatures above 120 °C – the degradation temperature of amino acids.

• **Hypervelocity impact studies** carried out by George COOPER[2689] and associates at NASA's Ames Research Center and NASA's Johnson Space Center **proved that complex organic compounds can survive heat and pressures such as those generated by impacts among asteroidal objects in the asteroid belt.** At laboratory shock pressures ranging from 100 to 200 kbar, they observed near-complete survival of sulfuric acids, while there was a significant drop in survival at approximately 300 kbar for all organic sulfur and phosphorus compounds.

(ii) **The discovery of amino acids in this meteorite suggests the idea that meteoroids, along with comets, may have deposited organic matter on the surface of the primordial Earth, and as a result, may have played an important role in providing the organic compounds needed to create the first terrestrial lifeforms.** The main argument against this so-called **"panspermia hypothesis"** is that these organic molecules could simply be terrestrial contaminants that found their way into the meteorite after it fell to Earth. However, of the more than 92 different amino acids identified within the Murchison meteorite, only 19 of these are found on Earth. This demonstrates that the remaining amino acids have no apparent terrestrial source.

(iii) The discovery of *enantiomeric excesses*[2690] in extraterrestrial molecules supports the hypothesis that exogenous delivery significantly contributed to the evolution of the organic chemicals necessary for life.[2691, 2692]

(iv) With few exceptions, notably benzene, the volatile products of the Murchison meteorite are substantially isotopically heavier than their terrestrial counterparts.[2693] In particular, the contents of individual amino acids found in the Murchison meteorite are enriched in the carbon isotope $^{13}C$ – signifying their extraterrestrial origin and suggesting the idea that a significant portion of prebiotic organic matter on the early Earth may have been introduced by carbonaceous asteroids and comets.[2694]

(v) In San Diego on July 29, 1997, Richard B. HOOVER[2695] of NASA's Marshall Space Flight Center announced the discovery of fossils in the Murchison meteorite that resemble microorganisms {⇒Fig. 4.1–D}.

*Astrobiology* is devoted to the scientific study of life in the Universe – its origin, evolution, distribution. This multidisciplinary field brings together the physical and biological sciences to address some of the most fundamental questions of the natural world:

---

[2688] K.A. KVENOLDEN ET AL.: *Evidence for extraterrestrial amino acids and hydrocarbons in the Murchison meteorite.* Nature **228**, 923-926 (1970).

[2689] G. COOPER, F. HÖRZ, A. OLEARY, and S. CHANG: *The survival of meteorite organic compounds with increasing impact pressure.* In: *Proc. First Astrobiology Science Conference* [NASA-ARC, Moffet Field, CA, April 2000]. NASA Ames Research Center (2000); published by NASA Astrobiology Institute in cooperation with the Carnegie Institution of Washington, *see* http://nai.arc.nasa.gov/team/index.cfm?page=pubs&teamID=14&year=7.

[2690] An *enantiomeric excess* exists when the proportion of one enantiomer of a particular organic molecule is greater than that of the other enantiomer of that molecule in sample. Enantiomers are compounds with identical chemical compositions that are mirror images of each other.

[2691] M.H. ENGEL and B. NAGY: *Distribution and enantiomeric composition of amino acids in the Murchison meteorite.* Nature **296**, 837-840 (1982).

[2692] J.R. CRONIN and S. PIZZARELLO: *Enantiomeric excesses in meteoritic amino acids.* Science **275**, 951-955 (Feb. 14, 1997).

[2693] G. YUEN, N. BLAIR, D.J. DES MARAIS, and S. CHANG: *Carbon isotope composition of low molecular weight hydrocarbons and monocarboxylic acids from Murchison meteorite.* Nature **307**, No. 5948, 252-254 (Jan. 1984).

[2694] M.H. ENGEL, S.A. MACKO, and J.A. SILFER: *Carbon isotope composition of individual amino acids in the Murchison meteorite.* Nature **348**, 47-49 (Nov. 1, 1990).

[2695] R.B. HOOVER: *Meteorites, microfossils, and exobiology.* In: (R.B. HOOVER, ed.) *Instruments, methods and missions for the investigations of extraterrestrial microorganisms.* Proc. SPIE **3111**, 115-136 (1997).

- How do living systems emerge?
- How do habitable worlds form and how do they evolve?
- Does life exist on worlds other than Earth?
- How could terrestrial life potentially survive and adapt beyond our home planet?

The term *astrobiology* has become the standard name applied to this new science, although an older term, *exobiology*, dates back to 1960. That term was invented by Joshua LEDERBERG, a U.S. biologist and winner of the 1958 Nobel Prize for Medicine/Physiology.

In 1998, NASA established the NASA Astrobiology Institute (NAI) as one element of its research program in astrobiology.[2696] In 2001, European scientists established the European Astrobiology Network Association (EANA).[2697]

| 1969 | U.S. Naval Research Laboratory (NRL), Washington, DC | **Foundation of the American journal *The Shock and Vibration Digest*** by NRL's Shock and Vibration Information Center, in order to present a digest of currently published material on shock and vibration for the shock and vibration engineer. The publisher of the journal is Sage Science Press, Thousand Oaks, CA. ▪ The *Shock and Vibration Bulletin* offers valuable information on the *Shock and Vibration Symposium* (the 78th Symposium was held in November 2007 at Philadelphia, PA). |
|---|---|---|
| 1969 | *Seismological Laboratory, Mackay School of Mines, University of Nevada, Reno, Nevada* | Gary Wynn BOUCHER,[2698] a U.S. assistant professor of geophysics, **Alan** RYALL, a U.S. research seismologist, **and Austin E. JONES investigate the important question of whether large man-made disturbances – such as those generated by underground nuclear explosions – have any effect on the seismicity in this region.** They analyzed seismic data collected at Tonopah and Unionville, two seismographic stations operated by the University of Nevada and located about 100 km and 370 km from the Nevada Test Site, respectively. In all cases of explosions with magnitude > 5.0, an increase in seismicity was observed for at least one day following the nuclear test. The results show that, for the most part, the explosion-induced seismic activity was confined to the test site, and was probably located within 20 km of the shotpoint, although Nevada is one of the most seismically active areas of the United States. |
| 1969 | *Research Division, Bell Laboratories, Murray Hill, New Jersey* | Willard S. BOYLE[2699] and George E. SMITH, two U.S. solid-state physicists, **invent a silicon chip composed of a grid of individual elements (pixels) that react to light in a similar way to the film grains in a traditional camera.** They term this chip a ***Charge Coupled Device (CCD)***. At a conference in New York, BOYLE unveils their new CCD invention and presents a paper entitled *The Future of Integrated Circuits*. ▪ The original motivation for this novel device was to find a silicon-based electronic analog for the magnetic bubble memory. However, it was soon recognized that the many advantages of CCDs – *e.g.*, their extremely high quantum efficiency, broad spectral bandwidth, very low readout noise, geometrical stability and fidelity, high resolution, large dynamic range, linear response and high photometric accuracy – were not only useful for TV cameras, but also for applications in high-speed diagnostics.[2700] Modern CCD chips often contain between one and five million pixels. For their important invention, BOYLE and SMITH received the 1973 Stuart Ballantine Medal of the Franklin Institute and the 1974 Morris Liebmann Award or the IEEE. |

---

[2696] NASA Astrobiology Institute (NAI); http://nai.nasa.gov/.
[2697] European Astrobiology Network Association (EANA); http://www.spaceflight.esa.int/users/index.cfm?act=default.page&level=1f&page=1856.
[2698] G. BOUCHER, A. RYALL, and A.E. JONES: *Earthquakes associated with underground nuclear explosions.* J. Geophys. Res. **74**, 3808-3820 (1969).
[2699] W.S. BOYLE and G.E. SMITH: *Charge coupled semiconductor devices.* Bell Syst. Tech. J. **49**, 587-593 (1970).
[2700] M.M. BLOUKE: *Charge-coupled devices.* In: (G.L. TRIGG, ed.) *Encyclopedia of applied physics.* VCH, New York *etc.*, vol. 3 (1992), pp. 241-272.

The combination of incorporated microchannel plates {FARNSWORTH ⇨ 1930} with CCDs led to the development of the first true digital cameras. Besides being incredibly popular consumer items, digital cameras have also become important research tools (this is true of both ordinary still cameras and ultrahigh-speed cameras). In particular, **multi-CCD cameras have become an extremely useful diagnostic tool for cinematographically recording shock waves and all kinds of high-velocity impact events** {Hadland Photonics Ltd. ⇨ 1993}.

| 1969 | *Sandia Laboratories, Albuquerque, New Mexico* | **Barry M. BUTCHER**[2701] **and C.H. KARNES,** two Sandia staff scientists, **present a theory on the propagation of a shock wave in a porous medium** in order to understand the mechanism of explosive powder compaction. Their theory is based on time-dependent pore closure during shock wave transmission through a ductile material with spherical voids. • Shock wave propagation in multiphase media has attained great attention in modern shock wave physics and encompasses not only porous media, but also granular media, bubbly liquids and liquid/gas suspensions. This complex field has recently been reviewed in several handbook articles.[2702] |

| 1969 | *U.S. Naval Weapons Laboratory (NWL), Dahlgreen, Virginia* | **D.W. CULBERTSON,**[2703] a U.S. engineer, **realizes a 2,400-ft (731.5-m)-long shock tube nuclear air blast simulator with a conical angle of 0.56°** {⇨Fig. 4.10–C}, as previously proposed by FILLER {⇨1960}. This design geometry results in an ideal amplification factor of about 160,000 which, assuming an "efficiency" of about 25%, yields an excellent 20-kiloton free air blast simulation from only a 100-pound (45.3-kg) TNT driver charge. • However, although the dimensions of the shock tube are enormous, the positive blast was still not long enough to simulate nuclear blast effects at larger yields. In order to reach this goal, other shock tube constructions were proposed by European researchers.[2704] In the 1990s, a large blast/thermal simulator was completed at White Sands Missile Range, NM to simulate the thermal pulse and air blast wave from a nuclear detonation and to investigate synergetic target effects. It allows to test nuclear survivability of full-scale vehicles.[2705] |

| 1969 | *Los Alamos Scientific Laboratory (LASL), Los Alamos, New Mexico* | **Richard D. DICK,**[2706] a U.S. experimental physicist, **reports on Hugoniot data measured up to several 100 kbar for a number of organic liquids,** such as benzene, carbon disulfide, carbon tetrachloride, and liquid nitrogen.<br>(i) In order to determine the Hugoniot curve, he uses a Cook-type plane-wave generator {J.H. COOK ⇨ 1948} and the "impedance match method" {RICE ET AL. ⇨ 1958}, which requires<br>▸ the measurement of the shock velocity;<br>▸ the initial density of the material being examined; and<br>▸ the shock velocity in a standard material of known Hugoniot (such as 2024 Dural, a standard aluminum alloy).<br>(ii) **He observes shock-induced phase transitions accompanied by a volume decrease of about 16%:** for benzene at 133 kbar, for carbon disulfide at 62 kbar, for carbon tetrachloride at 165 kbar, and for liquid nitrogen at 135 kbar. |

---

[2701] B.M. BUTCHER and C.H. KARNES: *Dynamic compaction of porous iron.* J. Appl. Phys. **40**, 2967-2976 (1969).

[2702] B.W. SKEWS ET AL.: *Shock wave propagation in multi-phase media.* In: (G. BEN-DOR ET AL., eds.) *Handbook of shock waves.* Academic Press, New York (2001), vol. 2, pp. 545-781.

[2703] D.W. CULBERTSON: *Description and performance of a conical shock tube nuclear air blast simulator.* Proc. 7th Int. Shock Tube Symposium [Toronto, Canada, June 1969]. In: (I.I. GLASS, ed.) *Shock tubes.* University of Toronto Press, Toronto (1970), pp. 396-409.

[2704] H. YU: *Recent developments in shock tube applications.* In: (K. TAKAYAMA, ed.) *Proc. 1989 National Symposium on Shock Wave Phenomena.* Shock Wave Research Center, Institute of Fluid Science, Tohoku University, Sendai (Sept. 1990), pp. 1-7.

[2705] *Large Blast / Thermal Simulator (LB/TS).* U.S. Army White Sands Missile Range, NM; see
http://www.wsmr.army.mil/capabilities/datts/testing/lab_fac/LargeBlastThermalSimulatorLBTS.html.

[2706] R.D. DICK: *Shock wave compression of benzene, carbon disulfide, carbon tetrachloride and liquid nitrogen.* J. Chem. Phys. **52**, 6021-6032 (1970).

| | | |
|---|---|---|
| 1969 | U.S. Air Force Weapons Laboratory (AFWL), Kirtland Air Force Base, New Mexico | Joseph FARBER,[2707] a U.S. plasma physicist, **and his collaborators first use a magnetically accelerated flyer plate to generate shock waves of nanosecond duration in solid targets** – a method which will become increasingly important for nuclear radiation testing and simulation of synergetic effects {MORRIS ⇨1971}. ▪ Today the method of accelerating conductors to high velocity is attributed to the Soviet researcher Vladimir N. BONDALETOV,[2708] who further extended this technique towards ultrahigh acceleration and hypersonic speeds.[2709] However, the principle of using eddy currents to accelerate conductors can be traced back as far as to the 19th century and the British electrician Elihu THOMSON, who first demonstrated that repulsive forces can be generated by pulsed magnetic fields in a classic experiment.[2710] |
| 1969 | Development Engineering Dept., Westinghouse Electric Corporation, Lester, Pennsylvania | Frank J. HEYMANN,[2711] a senior engineer at the Steam Division, **reviews the dynamics of high-speed impact between a compressible liquid drop and a solid surface.** <br><br>(i) The most-frequently used approximation to the pressure $P$ developed in liquid-solid impact is based on the one-dimensional water hammer $$P = \rho_0 C V_0,$$ where $\rho_0$ is the density of the undisturbed liquid, $V_0$ is the impact velocity, and $C$ is the velocity of propagation of the pressure wave or shock wave in the liquid, which emanates from the impact interface. One must distinguish between two cases: <br>▸ For small impact velocity the disturbance propagates with acoustic velocity $C_0$; i.e., $C \approx C_0$, hence $$P = \rho_0 C_0 V_0.$$ This relation is identical to ZHUKOVSKY's previous result derived for water hammer in water supply lines {KARELJSKICH ET AL. ⇨1898}. <br>▸ At high impact pressures, however, the compression propagates as a shock wave with a speed greater than the acoustic velocity ($C > C_0$). A more advanced one-dimensional analysis of water hammer for rigid solid targets incorporating a variable shock wave velocity was worked out by HEYMANN[2712] previously which resulted in the relation $$P = \rho_0 C V_0 = \rho_0 C_0 V_0 [1 + k(V_0/C_0)],$$ where $k$ is a constant. An analysis of experimental data shows that $k \approx 2$ up to about $V_0/C_0 = 1.2$. <br><br>(ii) Since the pressure or shock front generated by the impact is domed, not planar, simple water hammer pressure underestimates the actual peak impact pressure. In order to calculate the maximum pressure developed in the impact between a round liquid drop and a planar solid surface, **he presents a two-dimensional liquid-solid impact approximation assuming a linear relationship between shock velocity and particle velocity.** |
| 1969 | Denver Research Institute, Denver, Colorado | Chester R. HOGGATT[2713] and Rodney F. RECHT, two U.S. research engineers, **obtain dynamic uniaxial tensile stress-strain data at high strain rates of up to almost $10^5$ s$^{-1}$** by measuring the kinematics of thin-ring specimens expanding symmetrically by virtue of their own inertia. |

---

[2707] J. FARBER, G.W. SEMAN, and T. CIOLKOSZ: *Magnetic flyer combined response testing of composite specimens.* Rept AFWL-TR-69-18, Air Force Weapons Laboratory (AFWL), Kirtland Air Force Base, NM (Dec. 1969).

[2708] V.N. BONDALETOV: *Determination of electromagnetic forces, their working and electromechanical conversion efficiency in current-carrying contours* [in Russ.]. Elektrichestvo **87**, No. 1, 57-61 (1966).

[2709] V.N. BONDALETOV and E.N. IVANOV: *Contactless induction acceleration of conductors up to hypersonic speeds.* J. Appl. Mech. Tech. Phys. [Zh. Prikl. Mekh. Tekh. Fiz. (SSSR)] **16**, 765-766 (1975).

[2710] C. GERTHSEN: *Physik.* Springer, Berlin etc. (1958), p. 289.

[2711] F.J. HEYMANN: *High-speed impact between a liquid drop and a solid surface.* J. Appl. Phys. **40**, 5113-5122 (1969).

[2712] F.J. HEYMANN: *On the shock wave velocity and impact pressure in high-speed liquid-solid impact.* J. Basic Engng. **90**, 400-402 (Sept. 1968).

[2713] C.R. HOGGATT and R.F. RECHT: *Stress-strain data obtained at high rates using an expanding ring.* Exp. Mech. **9**, 1-8 (Oct. 1969).

(i) They use a specimen in the form of a thin ring which is shrunk-fit onto a hardened-steel core with a centrally located cavity containing a high explosive charge. Upon detonation of the explosive, a compressive shock wave moves radially outward through the steel.

(ii) Monitoring of the displacement-time history of the ring using a high-speed streak camera provides the information necessary to determine the stress-strain rate relationship.

(iii) For the materials tested (6061-T6 aluminum, 1020 cold-drawn steel, and 6Al-4V titanium) they observe parabolic displacement-time relationships, so exact differentiations can be performed to obtain the deceleration behavior of the ring and the resulting dynamic stress-strain relations.

In the 1970s, the method was resumed in Germany by Frank JÄHN,[2714] a mechanical engineer at Karlsruhe University, to determine yield stresses of metallic materials at ultrahigh strain rates and became known as the "ring expansion experiment" [Germ. *Ringaufweitungsversuch*].

| | | |
|---|---|---|
| 1969 | *Kobayashi Institute of Physical Research, Tokyo, Japan* | Heiji KAWAI,[2715] a Japanese physics professor and research director, **reports in a short note on his discovery of a strong piezoelectric effect in polyvinylidene fluoride [PVF$_2$] or PVDF**, a high molecular weight thermoplastic polymer [–(CH$_2$CF$_2$)$_n$–]. |

(i) He uses PVDF films stretched to several times their original length between 100 and 150 °C, and subsequently polarized by an electric field at about 300 kV/cm.

(ii) He observes piezoelectricity in polarized polycarbonate, polyethylene and polytetrafluorethylene, although the effect is smaller than in PVDF and decays with time.

KAWAI's discovery led not only to extensive research on PVDF itself, but also resulted in the discovery and study of many other new piezo/ferroelectric polymer materials. Since PVDF foils have a nanosecond response time, they are used in ultrasound transducers, underwater shock sensors, and high-speed accelerometers. **PVDF needle and membrane hydrophones were developed for lithotripter shock pressure field measurements**[2716, 2717] **and to investigate fragmentation effectiveness.**[2718]

Initially, commercially available PVDF materials did not exhibit reproducible properties due to the critical importance of the mechanical and electrical processing history. Early work carried out in France by François BAUER,[2719] an ISL staff physicist who explored the behavior of PVDF for dynamic high-pressure applications, led to a recognition of the need for such highly reproducible properties for PVDF. An international cooperative study between the Institut Franco-Allemand de Recherches de Saint-Louis (ISL) and Sandia National Laboratories (SNL) to develop the PVDF gauge has shown that, for example, 25-μm-thick PVDF film can be reliably used in a wide range of precise stress and stress-rate measurements.

| | | |
|---|---|---|
| 1969 | *Ernst-Mach-Institut (EMI), Freiburg & Chair of Early History, University of Tübingen, Germany* | Frank KERKHOF,[2720] a German physicist and an international authority on fracture mechanics, **and Hans Jürgen MÜLLER-BECK,** a German prehistorian, **discuss concepts of Hertzian fracture and its application to stone tools for the first time.** They investigate the fracture mechanics of point-loaded glass blocks cinematographically in order to better understand methods of prehistoric stone tool fabrication. Using the Cranz-Schardin multiple-spark camera {CRANZ & SCHARDIN ⇒1929}, they resolve the temporal evolution of the Hertzian cone geometry |

---

[2714] F. JÄHN: *Ein neues Verfahren zur Bestimmung der Fliessspannungen von metallischen Werkstoffen bei höchsten Dehngeschwindigkeiten*. Dissertation, Fakultät für Maschinenbau, Universität Karlsruhe (1979).

[2715] H. KAWAI: *The piezoelectricity of polyvinylidene fluoride*. Jpn. J. Appl. Phys. **8**, 975-976 (1969).

[2716] A.J. COLEMAN and I.E. SAUNDERS: *A survey of the acoustic output of commercial extracorporeal shockwave lithotripters*. Ultrasound Med. Biol. **15**, 213-227 (1989).

[2717] M. MÜLLER: *Comparison of Dornier lithotripters measurement of shock wave fields and fragmentation effectiveness*. Biomed. Technik **35**, 250-262 (1990).

[2718] W. FOLBERTH, G. KÖHLER, A. ROHWEDDER, and E. MATURA: *Pressure distribution and energy flow in the focal region of two different electromagnetic shock wave sources*. J. Stone Dis. **4**, 1-7 (1992).

[2719] F. BAUER and R.A. GRAHAM: *Very high pressure behavior of precisely-poled PVDF*. Ferroelectrics **171**, 95-102 (1995).

[2720] F. KERKHOF and H. MÜLLER-BECK: *Zur bruchmechanischen Deutung der Schlagmarken an Steingeräten*. Glastech. Ber. **42**, 439-448 (1969).

| | | |
|---|---|---|
| | | {HERTZ ⇨1882}. They visualize the growth, fracturing and shattering of the Hertzian cone {⇨Fig. 4.2−B} at increasing impact velocity. ▪ Soon after, their paper (now a classic among prehistorians) stimulated great interest in fracture mechanics.[2721] |
| 1969 | University of Michigan, Ann Arbor, Michigan | Yong Wook KIM[2722] and Otto LAPORTE, two shock physicists and spectroscopists, **report on the construction of a heated shock tube for studying shock-generated mercury plasmas of high electron densities** {⇨Fig. 4.10−J}. ▪ Their motivation for building a shock tube for metallic vapors was provided by previous spectroscopic studies using a shock wave as a spectroscopic source, which provides a well-defined discontinuity followed by a uniform flow at high temperature {LAPORTE ⇨1953}. |
| 1969 | Division of Applied Mathematics, Brown University, Providence, Rhode Island | Herbert KOLSKY,[2723] a British-born U.S. professor of applied physics, **discovers the formation of shock waves in the tensile deformation of rubber strings.**<br><br>(i) He determines the shapes of particle velocity *vs.* time profiles by observing the electrical outputs of light piano steel wires attached to the rubber as they cut the lines of force from magnetic fields of constant strength. He introduces a tensile pulse from the short section into the long stretched string section by suddenly volatilizing the steel wire through the application of a heavy electric current, and he analyzes wave profiles at several positions in the stretched section.<br><br>(ii) **He observes that the profiles of large pulses sharpen to form a shock wave after they have traveled about 2.74 m.**<br><br>(iii) His experiments first prove the theory that the tangent modulus of the material increases with increasing amplitude of deformation, thus steepening the head of the pulse during propagation.<br><br>Previously, KOLSKY[2724] had already suggested the idea that highly stretched vulcanized natural rubber becomes increasingly stiff with increasing tensile strain, stating "**There are some solids, however, such as rubbers and other high polymers, where large tensile strains result in an orientation of the long chain molecules** and this is accompanied by a large increase in the value of $d\sigma/d\varepsilon$. **It would therefore appear that when large deformations are propagated through these materials shock waves may develop.** Up to the present time no such experimental work appears to have been carried out to investigate whether this does in fact occur." His experiments on the deformation of rubber strings eventually confirmed his own hypothesis. |
| 1969 | McGill University, Montreal, Quebec, Canada | John H. LEE[2725] and Romuald KNYSTAUTAS, two U.S. professors of mechanical engineering, **report on experiments investigating the ignition of two chemically reactive gaseous mixtures ($C_2H_2/O_2$) using sparks introduced by lasers ("laser sparks"),** in this case a Q-switched ruby laser {⇨Fig. 4.11−G}. They find that, unlike electrical sparks, the minimum energy required to induce breakdown is of a sufficient magnitude to generate a strong spherical blast wave. They note that such information is of great importance when formulating safety criteria for the handling and storage of rocket propellants. |
| 1969 | Moscow, Soviet Union | Yevgeny Ye. MESHKOV,[2726] a Soviet fluid dynamicist, **first experimentally confirms RICHTMYER's theory of instability** {RICHTMYER ⇨1960}, which occurs when the interface between two fluids with different densities is accelerated by a shock wave striking the interface − the |

---

[2721] H. BERTOUILLE: *Théories physiques et mathématiques de la taille des outils préhistoriques.* Cahiers du quaternaire No. 15. CNRS, Paris (1989).

[2722] Y.W. KIM and O. LAPORTE: *Construction of a shock tube for metallic vapors.* Phys. Fluids **12**, I: 61-64 (1969).

[2723] H. KOLSKY: *Production of tensile shock waves in stretched natural rubber.* Nature **224**, 1301 (1969).

[2724] H. KOLSKY: *Stress waves in solids.* Clarendon Press, Oxford (1953). Dover Publ., New York (1963), p. 178.

[2725] J.H. LEE and R. KNYSTAUTAS: *Laser spark ignition of chemically reactive gases.* AIAA J. **7**, 312-317 (1969).

[2726] Y.Y. MESHKOV: *Instability of the interface of two gases accelerated by a shock wave* [in Russ.]. Izv. AN (SSSR) Mekhanika Zhidkosti i Gaza **4** (5), 151-157 (1969); Engl. translation in Sov. Fluid Dynamics (New York) **4**, 101-104 (1969); *Instability of a shock wave accelerated interface between two gases.* NASA Tech. Translation, NASA TTF-13074 [unpublished] (1970).

**"Richtmyer-Meshkov instability"** or "impulsive Rayleigh-Taylor instability" {G.I. TAYLOR ⇨ 1944}. His procedure is as follows:

(i) He uses a double-diaphragm shock tube consisting (in sequence) of a driver section, a thick membrane, a driven section, a very thin membrane, and a test chamber.

(ii) A shock wave is built up in the driven section after the rupture of the thick membrane, which travels towards the test chamber.

(iii) When the shock wave ruptures the thin membrane, a turbulent mixing process between the gas in the driven section and the gas in the test section is initiated, which can be visualized using schlieren or shadowgraph methods.

(iv) Shock refraction at the interface results in the formation of a vortex sheet that becomes turbulent.

Richtmyer-Meshkov instabilities (RMIs) combine compressible phenomena (such as shock interaction and refraction) with hydrodynamic instability across a wide range of Mach numbers. They are important in a large number of science and engineering applications. For example, **RMIs may play an important role**

- in astrophysics,[2727] *e.g.,* **in supernova explosions:** X-ray and gamma-ray observations of the famous supernova SN 1987A {I. SHELTON ⇨ 1987} revealed a considerable amount of radioactive cobalt among the explosive debris in the envelope, which suggests the occurrence of large-scale mixing in the ejecta during the explosion;[2728]
- in supersonic and hypersonic flows in linear and convergent shock tubes;
- in supersonic and hypersonic combustion performed by air-breathing vehicles;
- in inertial confinement fusion experiments in which laser energy is deposited on the outside of a fuel capsule which ablates and drives shocks in towards the deuterium/tritium fuel. RMIs may create "mixing" that introduces impurities into the fusion fuel which can dilute and cool the fusion fuel and quench the thermonuclear reaction,[2729] or may even break up the imploding shell, thus preventing the formation of a hot spot; and
- in small-scale laboratory experiments for validating computer simulations of nuclear weapons performance under various conditions.[2730] The current ban on underground nuclear testing severely limits the options for ensuring the safety, reliability, and performance of the nuclear stockpile {UN CTBT ⇨ 1996}.

In July 2002, Martin BROUILLETTE,[2731] an engineering professor at the Dept. of Mechanical Engineering of the Université de Sherbrooke (Sherbrooke, Québec, Canada), reviewed the basic physical processes underlying the onset and development of the RMIs in simple geometries and discussed principal theoretical, numerical and experimental methods of their validation.

In July 2004, at the 9th International Workshop on the Physics of Compressible Turbulent Mixing (held at Cambridge, U.K.), MESHKOV presented a review paper entitled *On New Possible Directions of Hydrodynamical Instabilities and Turbulent Mixing Investigations for the Solution of Some Practical Problems*.

---

[2727] N.A. INOGAMOV: *The role of Rayleigh-Taylor and Richtmyer-Meshkov instabilities in astrophysics*. Harwood, Basel (1999).
[2728] W.D. ARNETT, J.N. BAHCALL, R.P. KIRSCHNER, and S.E. WOOSLEY: *Supernova 1987A*. Annu. Rev. Astron. Astrophys. **27**, 629-700 (Sept. 1989).
[2729] N.E. LANIER ET AL.: *Multi-mode seeded Richtmyer-Meshkov mixing in a convergent, compressible, miscible plasma system*. Phys. Plasmas **10**, 1816-1821 (2003).
[2730] J.M. FISHBINE: *Code validation experiments – a key to predictive science*. Los Alamos Res. Quart. **1**, 6-14 (Fall 2002).
[2731] M. BROUILLETTE: *The Richtmyer-Meshkov instability*. Annu. Rev. Fluid Mech. **34**, 445-468 (2002).

| | | |
|---|---|---|
| 1969 | *Lehrstuhl für Zellenlehre & Organisch-Chemisches Institut der Universität Heidelberg, Germany* | Eberhard SCHNEPF,[2732] an eminent botanist, **and Wolf-Friedhelm WENNEIS and Hermann SCHILDKNECHT,** two biochemists, **discuss the explosion chemistry of the bombardier beetle** [*Brachinus crepitans*, family *Carabidae*], a ground beetle first described by the Swedish botanist and zoologist Carl VON LINNÉ in 1758. Its pygidial defense glands are capable of the synthesis and storage of 30% hydrogen peroxide, which then is mixed with hydroquinones (phenol-type aromatic organic compounds) and peroxidase (an oxidative enzyme) in a reaction chamber; this results in a violent reaction, forming a hot and almost explosive defensive discharge which is released through openings at the tip of the abdomen, causing a loud cracking sound similar to a bursting balloon. Using a piezoelectric microphone, a force transducer (piezoelectric crystal), and high-speed cinematography, the German researchers discovered that each discharge (lasting 2.6–24.1 ms) consists of 2–12 individual pulses, and that they are in reality **individual micro-explosions** repeating at 368–735 pulses per second within the reaction chamber.[2733] ▪ A very similar specy but less abundant than *Brachinus crepitans* is *Brachinus explodens*, described by the Austrian naturalist Caspar E. DUFTSCHMID[2734] in 1812.<br><br>The bombardier beetle has, in recent years, been a hot topic in the creation/evolution debate. It has been argued by creationists that the beetle serves as an excellent example of the kind of design that could not have formed through slow, random genetic mutations over time. |
| 1969 | *Ernst-Mach-Institut (EMI), Freiburg, Germany* | Fritz SCHULTZ-GRUNOW,[2735] a German professor of mechanical engineering, **studies the applicability of the soot method** {ANTOLIK ⇨1874; E. MACH & WOSYKA ⇨1875} **for recording shock waves in more detail.** He explains the regular cell structure of detonation patterns {SHCHELKIN & TROSHIN ⇨1963} as a self-sustaining process involving the formation of Mach stems by colliding blast waves which then initiate new explosion centers.<br><br>Three years later, **coworkers of SCHULTZ-GRUNOW experimentally confirmed his model concept using an arrangement of microdetonations all triggered simultaneously.** Using schlieren photography and high-speed cinematography, they visualized the superposition and Mach wave interactions of the individual detonation waves.[2736] |
| 1970 | *Ernst-Mach-Institut (EMI), Freiburg, Germany* | Hans O. AMANN,[2737] a German mechanical engineer, **investigates the starting process in a shock tube reflection nozzle.**<br><br>(i) In a conventional shock tube with constant cross-section, a continuous stationary flow (which is utilized for flow studies) occurs behind the shock front for only a few milliseconds. The Mach number of this flow cannot be increased arbitrary, but approaches a limiting value as the pressure ratio on the membrane increases. This limiting Mach number is 1.73 when air is used as the working gas, and the limiting value never exceeds 2, whatever the working gas. If higher Mach numbers need to be generated, then the gas behind the shock front can be expanded in a nozzle to get the desired supersonic or hypersonic Mach number. However, measurements on models behind the nozzle can only be performed after a steady flow has been established in the nozzle. |

---

[2732] E. SCHNEPF, W. WENNEIS, and H. SCHILDKNECHT: *Über Arthropoden-Abwehrstoffe XLI. Zur Explosionschemie der Bombardierkäfer (Coleoptera, Carabidae).* Z. Zellforsch. & mikroskopische Anatomie **96**, No. 4, 582-599 (1969).

[2733] J. DEAN, D.J. ANESHANSLEY, H.E. EDGERTON, and T. EISNER: *Defensive spray of the bombadier beetle: a biological pulse jet.* Science **248**, 1219-1221 (June 8, 1990).

[2734] C.E. DUFTSCHMID: *Fauna Austriae oder Beschreibung der osterreichischen Insecten fur angehende Freunde der Entomologie.* Verlag der k.u.k. privaten akademischen Kunst-, Musik- und Buchhandl., Leipzig, Part II (1812), p. 234.

[2735] F. SCHULTZ-GRUNOW: *Zellenstruktur von Detonationswellen.* Repts. Nos. 5/69 (1969) & 6/71 (1971), Ernst-Mach-Institut (EMI), Freiburg, Germany.

[2736] R. EWALD, E. SCHMOLINSKE, F. SCHULTZ-GRUNOW, and W.G. STRUCK: *Mach-stem-induced detonation and its relation to the structure of detonation waves.* Astronaut. Acta **17**, 467-473 (1972).

[2737] H.O. AMANN: *The process starting in a shock-tube reflection nozzle.* In: (W.G. HYZER and W.G. CHACE, eds.) *Proc. 9th Int. Congress on High-Speed Photography* [Denver, CO, Aug. 1970]. SMPTE, New York (1970), pp. 382-387.

(ii) Using a 24-spark camera {CRANZ & SCHARDIN ⇨ 1929}, he studies the process that occurs between the shock entering the nozzle and the point when a steady flow has been established. By uncovering the origins and the evolution processes of primary, secondary and reflected shocks, he confirms that the contact surfaces originate from multiple Mach reflections, and that their structure depends on the entrance geometry of the reflection nozzle used.

| | |
|---|---|
| 1970 *Dept. of Chemistry & Laboratory for Planetary Studies, Cornell University, Ithaca, New York* | Akiva BAR-NUN,[2738] a planetary scientist, **and collaborators report that energy injected via a shock (such as that generated in a shock tube) into a mixture of gases roughly simulating the primitive terrestrial atmosphere, can produce alpha-amino acids.**[2739]<br><br>(i) In their shock tube studies, they used different reaction mixtures, for example the mixture argon (80.1% by weight), methane (3.3%), ethane (11%) and ammonia (5.6%). The reaction mixture, together with water vapor, was introduced into the driven section and within less than half a minute the helium pressure in the driver section was increased until the diaphragm burst and the shock wave was initiated. They observed a high efficiency of shock-injected energy for production of alpha-amino acids.<br><br>(ii) Discussing potential high-energy sources on the primordial Earth that could trigger organic photochemistry, they conclude that "… energy sources of importance [[2740]] appear to be lightning discharges and volcanic thermal activity. In the first case, temperatures are generally too high for synthesis, and in the second the synthesis molecules are left for a significant time in the high-energy region. The time factor is probably also the principal reason for the low quantum yields for ultraviolet photoproduction. **We are thus led to the unexpected conclusion that cometary meteors, micrometeorites, and thunder were the principal energy sources for prebiological organic synthesis by a factor of perhaps $10^3$.**" ▪ In the early 1960s, Simon H. BAUER and Assa LIFSHITZ at Cornell University had (unsuccessfully) attempted to synthesize organic compounds from mixtures of $CH_4$, $NH_3$ and $H_2O$ in a shock tube. Their experimental results, however, were not published. This stimulated BAR-NUN and his colleagues to resume the matter.<br><br>In the years following, BAR-NUN[2741] **continued to investigate the synthesis of amino acids behind shock waves in methane, ethane, ammonia, and water using the single-pulse shock tube technique at the Hebrew University, Jerusalem** {⇨Fig. 4.14−B}. |
| 1970 *Dept. of Earth and Planetary Sciences, MIT, Cambridge, Massachusetts* | **Stephen Jay BLESS,**[2742] a Ph.D. student, **achieves high dynamic pressures in miniature samples by magnetic pinch compression** {NORTHRUP ⇨ 1907; TONKS ⇨ 1937}. Using small thin-walled metal tubes with an outer diameter of about 3 mm in which the sample under investigation is stored, he rapidly quenches the sample by applying a strong pulsed current of up to 1 MA from a capacitor discharge. He obtains evidence that this produces shock pressures of more than 130 kbar. ▪ The idea of using megagauss magnetic fields to produce high pressures in solids in order to study highly compressed matter such as that found inside stars and planets goes back to Francis BITTER,[2743] who was a physics professor at MIT's Magnet Laboratory (now the Francis Bitter Magnet Laboratory). He also conceived of the experiment and the original low-inductance circuit design.<br><br>In a recent joint study from LLNL, LANL and SNL, isentropic compression experiments were carried out by use of an imploding z-pinch using magnetic pressures of up to 1 Mbar gen- |

---

[2738] A. BAR-NUN, N. BAR-NUN, S.H. BAUER, and C. SAGAN: *Shock synthesis of amino acids in simulated primitive environments.* Science **168**, 470-473 (April 24, 1970).

[2739] An *alpha-amino acid* consists of an amino group, a carboxyl group, a hydrogen atom, and a distinctive R group bonded to a carbon atom.

[2740] S.L. MILLER and H.C. UREY: *Organic compound synthesis on the primitive Earth.* Science **130**, 245-251 (July 31, 1959).

[2741] A. BAR-NUN: *Shock synthesis of amino acids II.* Origins of Life **6**, 109-115 (1975).

[2742] S.J. BLESS: *The effect of magnetic pinch pressure on boron nitride, cadmium sulfide, graphite, silica glass, and some other materials.* Ph.D. thesis, MIT, Cambridge, MA (June 1970); *Production of high pressures by a capacitor discharge-powered linear magnetic pinch.* J. Appl. Phys. **43**, 1580-1585 (1972).

[2743] F. BITTER: *Ultrastrong magnetic fields.* Scient. Am. **213**, 64-73 (July 1965).

erated by SNL's powerful 2-MJ Z-machine.[2744] A theoretical study on pinch pressure generation in pulsed solid conductors performed at EMI, Freiburg, revealed that transient skin effects advantageously provoke an almost homogeneous compression of the interior of the conductor.[2745] The steep rise in pinch pressure $P_p(t)$ is not the result of shock front steepening, but is instead caused by the fact that the magnetic pressure is proportional to the square of the applied pulse current-time profile $I(t)$; i.e., $P_p \sim I^2$. Thus, steep current pulses produce much steeper pressure pulses.

| 1970 | Surface Physics Division, Cavendish Laboratory, University of Cambridge, England | John Hubert BRUNTON[2746] and Jean-Jacques CAMUS, two research physicists, **report on a new method of studying the impact of a liquid drop with a solid object** – an impact situation which is important in the problem of the rain erosion of aircraft – **in a two-dimensional (2-D) geometry.**<br>(i) In their sandwich geometry, a flat droplet that, enclosed between two glass plates, is impact-loaded by a metal "target" fired into the gap. The great advantage of this 2-D set-up is that it allows processes inside an impacted drop to be observed in detail without the refraction problems inherent with spherical drops.<br>(ii) Using high-speed microphotography, they observe that the damage mechanism consists of several stages:<br>▸ In the early stage of impact, where both solid and liquid are highly compressed in the contact region, there is no formation of a side spray of liquid ("jetting"). The contact region expands supersonically.<br>▸ The shock wave detaches from the contact area of impact and moves up through the drop, and jetting starts.<br>▸ When the shock is about to reflect at the upper surface of the drop, the jetting becomes more advanced.<br>▸ The jetting mechanism produces a central erosion pit in a soft metal or a ring crack in a brittle material.<br>Their experimental technique was later modified to use a "drop" shaped from a gel, which showed that **jetting can occur at a speed of about ten times the collision speed.**[2747] |
|---|---|---|
| 1970 | Depts. of Urology and Surgery, University of Munich, Germany | Christian CHAUSSY[2748] and Ferdinand EISENBERGER, two German urologists, **and their collaborators begin to investigate whether shock waves can be applied to fragment urinary tract stones.** Their concept – **"extracorporeal shock wave lithotripsy (ESWL)"** – uses shock waves generated at the first focus of an ellipsoidal cavity, which are then refocused at the second focus point, where the target stone is located. • The first commercially available apparatus for routine extracorporeal lithotripsy (*lithotripsers*) were developed by two German companies: Dornier Medizintechnik GmbH at Germering and Siemens AG at Erlangen. The application of extracorporeal shock waves for the noninvasive destruction of kidney stones in patients was not performed until 1980 by CHAUSSY[2749] and collaborators using Dornier's HM1-machine {FORSSMANN ⇒ 1976; ⇒ Fig. 4.15–F}.[2750] |

---

[2744] R. CAUBLE ET AL.: *Isentropic compression experiments to 1 Mbar using magnetic pressure.* J. Phys. Condens. Mat. **14**, 10821-10824 (2002).

[2745] P. KREHL: *Transient pressure effects in a pinch-compressed cylindrical metallic conductor.* J. Appl. Phys. **70**, 3488-3500 (1991); Ibid. **72**, 1206 (1992).

[2746] J.H. BRUNTON and J.J. CAMUS: *The flow of a liquid drop during impact.* In: (A.A. FYALL and R.B. KING, eds.) *Proc. 3rd Int. Conference on Rain Erosion & Associated Phenomena* [Elvetham Hall, U.K., Aug. 1970]. Royal Aircraft Establishment (RAE), Farnborough (1972), vol. 1, pp. 327-346.

[2747] M.B. LESSER and J. FIELD: *Studies in shock waves, liquid impact, jets and cavitation.* In: (K. TAKAYAMA, ed.) *Proc. 18th Int. Symposium on Shock Waves* [Sendai, Japan, July 1991]. Springer, Berlin etc. (1992), vol. 1, pp. 63-72.

[2748] C. CHAUSSY, F. EISENBERGER, ET AL.: *The use of shock waves for the destruction of renal calculi without direct contact.* Urological Res. **4**, 181 (1976).

[2749] C. CHAUSSY, W. BRENDEL, and E. SCHMIEDT: *Extracorporeally induced destruction of kidney stones by shock waves.* Lancet II, pp. 1265-1268 (1980).

[2750] A picture of the HM1-machine is shown in the catalogue *Deutsches Museum Bonn. Forschung und Technik in Deutschland nach 1945*. Raabe Verlag, Bonn (1999), p. 423.

Extracorporeal shock wave lithotripsy also proved useful for destroying gallstones and pancreatic duct stones[2751] as well as salivary gland stones.[2752]

> The historical development of this important medical therapy up to 1987 was reviewed by the German urologist Dieter JOCHAM.[2753]

> A more recent review was given by Achim M. LOSKE-MEHLING,[2754] a physicist at the Universidad Nacional Autónoma de México (UNAM) in Querétaro.

> A comprehensive list of papers on studies in extracorporeal shock wave therapy (ESWT) can be found on the Internet.[2755]

> Recently, Michel TANGUAY,[2756] a Ph.D. student at CalTech, studied the cloud of cavitation bubbles produced in the wake of the shock wave, which is a crucial element in the stone pulverization process {⇨Fig. 4.15−F}. In order to estimate the efficiency of this process as a function of the operational parameters (such as the intensity of the initial shock wave and the pulse frequency), he devised a numerical model for the two-phase flow inside an electrohydraulic lithotripter.

| | | |
|---|---|---|
| 1970 | Surface Physics, Cavendish Laboratory, University of Cambridge, England | **Graham Douglas COLEY**[2757] **and John E. FIELD,** two British physicists, **report on a study into the growth of explosions in various liquid explosives** (nitroglycerin, nitromethane, nitric acid, hydrogen peroxide, ethanol mixtures and diethylene glycol dinitrate). |

(i) They used a high-speed rotating-mirror framing camera with submicrosecond frame rates which allowed the burning and the transfer to detonation to be followed in detail, and they recorded the explosive reaction in transmitted light in 2-D (thin film) as well as in 3-D (araldite cell) configurations; the explosion was initiated by an electric spark discharge and an exploding nichrome bridgewire, respectively.

(ii) In the thin film configuration, precursor waves traveling in the confinement establish a circular cavitation field, giving rise to gas bubbles in the liquid explosive which can enhance the transition to a fast reaction

> due to the adiabatic collapse of the bubbles by compression shocks from the deflagration front, which gives rise to "hotspots" in the liquid;

> by presenting a larger surface area to the deflagration front as it enters the cavitation zone; and

> through the formation of ***micro-Munroe jets*** during bubble collapse {BRUNTON & CAMUS ⇨1970}, which may disperse the liquid as droplets in the heated cavity or produce high temperatures through impact.

| | | |
|---|---|---|
| 1970 | Institute of Physical Chemistry, University of Göttingen, Germany | **John Edward DOVE,**[2758] a visiting (U.S.) professor of chemistry from the University of Toronto, **and T.D. TRIBBECK numerically integrate the rate equations for $H_2/O_2$ reactions under the conditions of steady flow in a ZND detonation** {ZEL'DOVICH ⇨1940}. They use |

---

[2751] T. SAUERBRUCH ET AL.: *Fragmentation of gallstones by extracorporeal shock waves.* New Engl. J. Med. **314**, 818-822 (1986); Endoscopy **19**, 207-208 (1987).

[2752] H. IRO ET AL.: *Extracorporeal shock wave lithotripsy of salivary gland stones.* Lancet II, p. 115 (1989).

[2753] D. JOCHAM: *Historical development of ESWL.* In: (R.A. RIEHLE and R.C. NEWMAN, eds.) *Principles of extracorporeal shock wave lithotripsy.* Churchill Livingstone, New York (1987), pp. 1-11.

[2754] A.M. LOSKE-MEHLING: *Applications of shock waves in medicine.* In: (G. BEN-DOR ET AL., eds.) *Handbook of shock waves.* Academic Press, New York (2001), vol. 2, pp. 415-440.

[2755] *ESWT Literatur.* Schweizerische Gesellschaft für Stoßwellentherapie (SGST), Zürich, Schweiz; http://www.sgst.ch/Literatur/default.htm.

[2756] M. TANGUAY: *Computation of bubbly cavitating flow in shock wave lithotripsy.* Ph.D. thesis, Dept. of Mechanical Engineering, CalTech, Pasadena, CA (2004); http://etd.caltech.edu/etd/available/etd-05282004-130028/unrestricted/thesis.pdf.

[2757] G.D. COLEY and J.E. FIELD: *The initiation and growth of explosion in liquids.* In: (W.G. HYZER and W.G. CHACE, eds.) *Proc. 9th Int. Congress on High-Speed Photography* [Denver, CO, Aug. 1970]. SMPTE, New York (1970), pp. 466-473.

[2758] J.E. DOVE and T.D. TRIBBECK: *Computational study of the kinetics of the hydrogen-oxygen reaction behind steady state shock waves. Application to the composition limits and transverse stability of gaseous detonation.* Astronaut. Acta **15**, 387-397 (1970).

Arrhenius kinetic parameters {ARRHENIUS ⇨1889} measured in a shock tube for 28 elementary reactions involved in the reactive flow in the forward and reverse directions.

1970 — Bölkow Apparatebau GmbH, Schrobenhausen, Upper Bavaria, Germany

Manfred HELD,[2759] a German detonation physicist, **obtains a German patent for an arrangement used to protect an object against gun fire.** He refers to it as a drive-plate explosive sandwich that consists of two metallic plates with a high explosive in-between them – so-called **"Explosive Reactive Armor (ERA)."** ▪ The idea of applying a counterexplosion in an armor was first proposed by the Soviet gas dynamicist Bogdan V. VOITSEKHOVSKY in 1949.

Today ERAs play an important role as the main protection provided for battle tanks against both kinetic energy (KE) projectiles and shaped charge warheads. The most vulnerable areas of heavy U.S. tanks are partly protected by plates of reactive armor that are composed of military grade C4 and steel. ▪ Composition C4 is a plastic high explosive (91/9 RDX/polyisobutylene plasticiser) that is more powerful and has a higher detonation velocity of about 8,090 m/s than TNT {HAEUSSERMANN ⇨1891}.

1970 — Swedish Detonic Research Foundation, Vinterviken, Stockholm, Sweden

CARL H. JOHANSSON[2760] and Per A. PERSSON, two Swedish explosives specialists, **publish their book *Detonics of High Explosives*.** This outstanding book describes the behavior of high explosives, with emphasis on experimental data. In particular, it deals with

▸ the mechanism of detonation and initiation of detonation by means of strong shock waves or mechanical impact at low velocity;

▸ burning and the effects of heating;

▸ various light emission phenomena that occur during detonation; mechanical effects in surrounding media; and work performed by the reaction products of a contained charge (particularly the effects of a charge in rock); and

▸ methods of probing the physics of high dynamic pressures in condensed media in order to understand mechanical effects of detonation, and the results from such experiments.

1970 — Lawrence Radiation Laboratory (LRL), UC Livermore, California

Quintin JOHNSON,[2761] a U.S. physical chemist, **and his collaborators measure the shock-induced lattice compression of a polycrystalline lithium fluoride [LiF] sample using a TNT plane-wave generator for the first time.** Their shock experiment (which greatly enhanced our fundamental understanding of shock-compressed solids) allows one to observe the dynamic shift of the $\langle 200 \rangle$ diffraction line in a Debye-Scherrer flash X-ray diffraction diagram due to uniaxial shock compression at 130 kbar {⇨Fig. 4.12–E}. They obtain two important results:

(i) **The fact that diffraction takes place implies that crystalline order can exist behind the shock front,** and the required ordering takes place on a time scale which is short compared to 20 ns, the time duration of the applied X-ray pulse.

(ii) The location of the $\langle 200 \rangle$ reflection of shock-compressed LiF implies that, based on the unit cell, this **compression is isotropic;** *i.e.*, the shock compression is essentially hydrostatic. ▪ Their experiments stimulated similar studies abroad {JOHNSON & MITCHELL ⇨1972}.

It is interesting to note that **Veniamin A. TSUKERMAN[2762] and A.I. AVDEENKO in the former Soviet Union had already shown (in 1941) that a high-speed Laue diffraction pattern could be recorded using a single X-ray pulse of millisecond duration.** The idea of using flash X-ray diffraction to investigate the fine structure of shock-loaded crystal lattices was first realized in 1950 by the German physicist Rudi SCHALL.[2763] In his exploding foil experiments, however, the compression state was poorly defined, and this disadvantage could not be over-

---

[2759] M. HELD: *Schutzeinrichtung gegen Geschosse*. Germ. Patent No. 2,053,345 (issued Oct. 1970; publ. Feb. 1977).
[2760] C.H. JOHANSSON and P.A. PERSSON: *Detonics of high explosives*. Academic Press, London & New York (1970).
[2761] Q. JOHNSON, A. MITCHELL, R.N. KEELER, and L. EVANS: *X-ray diffraction during shock-wave compression*. Phys. Rev. Lett. **25**, 1099-1101 (1970).
[2762] V.A. TSUKERMAN and A.I. AVDEENKO: *The production of radiographs at very short exposure times* [in Russ.]. Zh. Tekh. Fiz. (SSSR) **12**, 185-196 (1942).
[2763] R. SCHALL: *Feinstrukturaufnahmen in ultrakurzen Zeiten mit dem Röntgenblitzrohr*. Z. Angew. Phys. **2**, 83-88 (1950).

| | | |
|---|---|---|
| | | come by subsequent researchers until high-explosive planar shock wave generators or high-velocity impact methods, such as the flyer plate method or the light-gas gun, were applied. |
| 1970 | *Dept. of Engineering Science, Tennessee Technological University, Cookeville, Tennessee* | **Publication of the book *High-Velocity Impact Phenomena*, edited by Ray KINSLOW,**[2764] a US. engineering professor. In this first compendium on **Terminal Ballistics,** fourteen contributors review experimental techniques and computational solutions to this complex problem. KINSLOW writes, "Terminal ballistics has been the subject of formal study for more than two centuries.[2765] The incentive for such a study was largely motivated by warfare, with its need for the development of faster projectiles and stronger armor. Current interest in high-speed impact is largely due to the need for information concerning the hazards posed by meteoroids to space vehicles. The speeds of this interplanetary debris exceed by an order of magnitude or more the impact velocities common to traditional ballistics, and the mechanics of the impact process are quite different … The reaction of materials to the extremely high pressures resulting from hypervelocity impact is of great fundamental importance to the physicist [HVIS ⇨1986]. The nature of the strong shock waves generated by such impact makes the subject one of interest to the seismologist. **These extremely high pressures and shock waves can be used to harden and form metals, bond dissimilar materials, and can be utilized by the engineer in many other ways in addition to the problem of spacecraft design.**" |
| 1970 | *Institute of the Problems of Mechanics, Moscow, Soviet Union* | **S.P. KOZYREV**[2766] **and K.K. SHAL'NEV resume previous Soviet liquid/solid interaction experiments** {KORNFELD ⇨1951}. They study the collision between a jet of a fluid, a mixture of rosin and mineral oil issued from a 8-mm-dia. nozzle under 5 bar of pressure, and a cylindrical 15-mm-dia. bullet impacting the jet in the normal direction {⇨Fig. 4.3–W}. Using high-speed photography, they observe that at low speed (9 m/s) the jet is plastically deformed, while at high speed (15 m/s) brittle fracture occurs – apparently due to a shearing process. They propose a relaxation hypothesis for the origin of the pressure peak and the duration of the pressure pulse in a collision between a fluid and a solid. |
| 1970 | *Convair Division, General Dynamics, San Diego, California* | **Richard J. MAGNUS**[2767] **and Hideo YOSHIHARA,** two U.S. aeronautical engineers and numerical fluid dynamicists, **use a nonlinear inviscid code to find the transonic solution for a practical lifting airfoil with embedded shock.** The results agree well with those from experiments, with local differences primarily accounted for by a Busemann-Guderley instability[2768,2769] rather than by viscous effects. |
| 1970 | *Physical Institute, University of Düsseldorf, Germany* | **Ferdinand MÜLLER**[2770] **and Erasmus SCHULTE,** two German physicists, **study the geometries of shock wavefronts generated in Plexiglas by electrically exploding copper foils.**<br><br>(i) Using a rotating mirror streak camera and an image tube camera to record the temporal and local development of the shock front, and a high-speed two-beam oscilloscope to record the electrical discharge parameters (current and voltage), they observe that groups of shock waves are initiated in the solid, depending on the coupling between the foil and the solid.<br><br>(ii) The two-dimensional shock front generated by the exploding foil is rather planar in the direction of the current, but it deviates considerably from planarity in the normal direction. |

---

[2764] R. KINSLOW (ed.): *High-velocity impact phenomena.* Academic Press, New York (1970).

[2765] *See,* for example, L.W. LONGDON: *Brief historical remarks on terminal ballistics.* In: (R. VINCENT, ed.) *Textbook of ballistics and gunnery.* H.M.S.O., London, vol. 1 (1987), pp. 644-647.

[2766] S.P. KOZYREV and K.K. SHAL'NEV: *Relaxation hypothesis of the fluid-solid collision mechanism.* Sov. Phys. Dokl. **15**, 513-515 (1970).

[2767] R.J. MAGNUS and H. YOSHIHARA: *Inviscid transonic flow over airfoils.* AIAA J. **8**, 2157-2162 (1970).

[2768] A. BUSEMANN: *The drag problem at high subsonic speeds.* J. Aeronaut. Sci. **16**, 337-344 (1949).

[2769] G. GUDERLEY: *Shocks in subsonic-supersonic flow patterns.* In: (R. VON MISES and TH. VON KÁRMÁN, eds.) *Advances in applied mechanics.* Academic Press, New York (1953), vol. 3, 145-184.

[2770] F. MÜLLER and E. SCHULTE: *Untersuchungen über das Verhalten durch explodierende Metallfolien erzeugter Stoßwellen in Festkörpern.* Bericht 4/70, Physikalisches Institut, Universität Düsseldorf, Germany (1970).

| 1970 | Westinghouse Research Laboratories, Pittsburgh, Pennsylvania | Martin A. UMAN,[2771] an electrical engineer and a fellow physicist at Westinghouse, **and collaborators perform laboratory experiments to simulate lightning.** By comparing their experimental results with previous theories, they provide the first quantitative information on thunder:

(i) In their studies they used a 6.4-MV impulse generator to produce 4-m-long spark discharges in air with an energy of 20 kJ, corresponding to a specific energy of 5 kJ/m.

(ii) Using calibrated piezoelectric microphones, they measured the pressure wave emerging from the spark as a function of distance up to 30 m.

▸ At a distance of 0.84 m, they measure an overpressure of $\Delta p/p_0 = 0.085$ at midgap height.

▸ **Within distances of 2 to 3 m from the spark, they mostly observe a single dominant shock wave.** At 3 m they measure $\Delta p/p_0 = 0.020$.

▸ **Further away, however (for example at 8 m), they observe several shock waves because of reflections from the ground.**

▸ **But at a distance of 16.5 m from the spark, the pressure waveform again resembles that of a single shock wave.** They measure an overpressure ratio of $\Delta p/p_0 = 0.005$ at this point; *i.e.*, the wavefront propagates at almost the speed of sound.[2772]

(iii) Their measured data are found to be consistent with a semiquantitative theory describing the generation of sound by lightning provided by Arthur A. FEW[2773] (a physicist and mathematician at Rice University, Houston, TX) the year before. However, close to the spark channel the measured overpressure and its duration fall significantly below the cylindrical theoretical curve, probably because the conditions of the spark do not satisfy the assumptions of the theory. • Since the shape of the waveform is related to the energy per unit length of the lightning flash, it is initially hoped that the energy in a real lightning channel can be estimated by analyzing the pressure-time profile (or "pressure signature") of a clap of thunder. However, the sound of thunder is the complex product of the energy released per unit length by the discharge and the tortuosity of the lightning channel. In addition, nonlinear wave propagation effects near the discharge and in the atmosphere (*e.g.*, attenuation, scattering, refraction and reflection) strongly influence the shapes of the waveforms recorded {BASS ⇒ 1980}.

(iv) UMAN's study was funded by federal agencies in order to gain insight into the in-flight explosion of a fuel tank on a Boeing 707 in 1963 due to a lightning strike. Furthermore, in 1969, Apollo 12 was struck by lightning twice after launch, which resulted in a momentary loss of electrical power and telemetry contact. However, power was then automatically switched to battery backup and the mission continued as planned.

The 1974 Encyclopædia Britannica,[2774] obviously referring to UMAN's results, notes that: "When air is crossed by a spark or a lightning flash, the air is heated rapidly and the cylindrical column expands at supersonic speed. **Within a metre or two the shock decays to a sound wave.** The sound heard as thunder comes from the entire channel length and is modified by the intervening medium. The result is a series of sounds that are variously described as peals, claps, rolls, and rumbles…" |

---

[2771] M.A. UMAN, A.H. COOKSON, and J.B. MORELAND: *Shock wave from a four-meter spark*. J. Appl. Phys. **41**, 3148-3155 (1970).
[2772] The ratio of the shock front velocity, $D$, and the sound velocity, $c_0$, is given by $D/c_0 = [(6 \Delta p/p_0 + 7)/ 7]^{1/2}$. For $\Delta p/p_0 = 0.005$ this gives $D/c_0 = 1.002$.
[2773] A.A. FEW JR.: *Power spectrum of thunder*. J. Geophys. Res. **74**, 6926-6934 (1969).
[2774] R.E. ORVILLE: *Lightning*. In: *The new Encyclopædia Britannica, Macropædia*. Benton & Hemingway, Chicago *etc.*, vol. 10 (1974), pp. 965-970.

In 2003, together with Prof. UMAN, Vladimir A. RAKOV[2775] (a Russian-born U.S. professor of electrical engineering at the University of Florida) **published the book *Lightning: Physics and Effects*** (Cambridge University Press), which is probably the most detailed and recent work of reference on this subject; it discusses the mechanism that produces the sound of thunder, gasdynamic modeling of lightning processes, ball lightning, extraterrestrial lightning, and the production of trace gases in the primordial atmosphere of Earth and the atmospheres of other planets by lightning.

| | | |
|---|---|---|
| 1970 | *Dept. of Solid Mechanics, Ernst-Mach-Institut (EMI), Freiburg, Germany* | **Siegfried WINKLER,**[2776] a German physicist, **and Donald A. SHOCKEY and Donald R. CURRAN,** two visiting U.S. physicists from Stanford Research Institute (SRI), Menlo Park, CA, **report on supersonic crack growth along weak crystallographic planes in anisotropic single crystals of potassium chloride [KCl].** They loaded the crack tip with an expanding plasma generated by a pulsed ruby laser. **Measured average crack velocities range from 20 to 60 km/s, which is about 4–15 times the velocity of the longitudinal elastic wave.** They attribute the supersonic speeds to their special experimental conditions, which allow energy to be supplied to the crack front at unusually high rates. ▪ These results contradicted the predictions of classical crack propagation theories: that crack velocities cannot exceed the speed of sound.<br><br>With the advent of supercomputers {ASCI WHITE ⇒ 1999} in the late 1990s, theoretical studies on supersonic crack propagation were resumed using atomistic computer modeling {ABRAHAM & GAO ⇒ 2000}. |
| 1971 | *Moon (Fra Mauro Site) & NASA's Kennedy Space Center (KSC), Cape Canaveral, Florida* | On January 31, **the Apollo 14 mission blasts off, heading to the Moon.** On February 5, the astronauts Alan B. SHEPARD JR. and Edgar D. MITCHELL land on Fra Mauro. They perform a number of scientific experiments before leaving on February 6.<br><br>(i) **They conduct the first active seismic experiment, which generates and monitors seismic waves in the Moon near the surface in order to study its internal structure to a depth of 460 m.**[2777]<br><br>(ii) The experimental equipment consists of three identical geophones, a thumper, a mortar package, a central electronics assembly, and interconnecting cabling.<br><br>▸ The geophones (miniature seismometers) are electromagnetic devices which are deployed on the lunar surface to translate surface movement into electrical signals.<br><br>▸ In order to create small seismic waves on the Moon, one of the crewmen uses the ***thumper,*** a hand-held tubular cartridge-firing device which contains 21 small explosive charges that provide seismic signals. These signals are generated by holding the thumper against the lunar surface at various locations along the row of geophones and firing explosive initiators located in the base of the thumper.<br><br>▸ The central electronics assembly consists of a three-channel amplifier with a log compressor to send the data to the Earth.<br><br>(iii) The mortar package, which was designed to launch four high-explosive grenades to distances of 5,000, 3,000, 1,000, and 500 feet (1,524, 914, 304, and 152 m), is planted 91 m from the lunar module, but the detonation signal (from Earth) will be postponed until the other experiments will be completed to avoid damaging them.<br><br>(iv) Thirteen initiators are fired successfully, five misfire, and three are deliberately not fired. The mission ended on February 9, 1971, with the safe return of the crew to Earth. |

---

[2775] V.A. RAKOV and M.A. UMAN: *Lightning: physics and effects (Encyclopedia of lightning)*. Cambridge University Press, Cambridge (2003).
[2776] S. WINKLER, D.A. SHOCKEY, and D.R. CURRAN: *Crack propagation at supersonic velocities, Part I.* Int. J. Fract. Mech. **6**, 151-158 (1970).
[2777] *Apollo 14 mission report.* Rept. NASA MSC-04112, NASA mission evaluation team, Manned Spacecraft Center, Houston, TX (May 1971), Appendix A.4.1: *Active seismic experiment*; http://www.hq.nasa.gov/office/pao/History/alsj/a14/a14mr.html.

(v) The thumper device provides data that indicate that at the Fra Mauro site two P-wave velocities were measured through the loose, fragmental material on the Moon's surface (so-called "regolith"): the near-surface has a seismic wave velocity of 104 m/s, and a sublayer starting at a depth of 8.5 m has a velocity of 299 m/s. Estimates of the thickness of this substratum range from 38 to 76 m, which is probably indicative of the depth of the Fra Mauro formation.[2778]

Active seismic experiments were also conducted in the following year {Apollo 16 & 17 ⇨1972}.

Prior to the Apollo 14 mission, **Robert L. KOVACH,**[2779] a geophysics professor at Stanford University and principal investigator of the Active Seismic Experiment, **and Thomas J. AHRENS,** a U.S. geophysicist and associate professor at CalTech, **performed model explosion experiments in order to design seismic experiments for the Apollo 14, 16 & 17 missions.** They studied the effect of vacuum on the transmission of seismic energy from charges detonated on or slightly above the surface of a thick Plexiglas plate. When using an untamped surface charge, the transmission of seismic energy under vacuum conditions – such as on the Moon – is different from its transmission in air. In vacuum, the explosive gas blast and the detonation products expand freely and continuously outward into the lunar vacuum, sweeping along the surface at a very high velocity of about 10 km/s. This may induce faulty seismic signals, *e.g.*, when the explosive gases

> ▸ interact directly with the seismometers; and/or
> ▸ act as a radially expanding source of seismic energy, inducing seismic signals that arrive at the seismometers well before the seismic signal originating at the point of detonation.

| | | |
|---|---|---|
| 1971 | U.S.A. | **The U.S. Atlantic-Pacific Interoceanic Canal Study Commission recommends the construction of a sea-level canal through Panama just north of the present canal.**[2780] ▪ Almost 30 routes across Central America have been proposed as possible paths for a large waterway since the 1960s, and the idea of **using nuclear explosives for earthmoving** {U.S. Plowshare Program ⇨1958; Test SEDAN ⇨1962; Test BUGGY ⇨1968} has also been put forward. While many such projects are feasible from an engineering economics standpoint, there are still major safety issues associated with them, including the generation of catastrophic seismic shocks, air blasts, and tsunamis in coastal operations. The biggest issue, however, would be the long-term radioactive contamination, which would preclude resettlement. |
| 1971 | NASA's Research Center (ARC), Moffet Field, California | **Dean R. CHAPMAN,**[2781] a U.S. aeronautical engineer and fluid dynamicist, **demonstrates that the aerodynamic ablation evidence on Australian tektites** (found in Southeast Asia, particularly on New Zealand and in Australia) **requires a lunar origin.** ▪ From the 1950s through the 1990s, CHAPMAN and others advanced the *lunar origin theory of tektites,* stating that tektites are molten parts from the Moon's surface which were thrown into space after the impact of large meteorites. CHAPMAN used complex orbital computer models and extensive wind tunnel tests to support the theory that the Australasian tektites originated from the Rosse ejecta ray of the large crater Tycho on the Moon's nearside. Until the Rosse ray is sampled, a lunar origin for these tektites cannot be ruled out. |

---

[2778] R.L. KOVACH: *Active seismic – Apollo 14 Lunar Module.* National Space Science Data Center (NSSDC) Master Catalogue Display Experiment. NASA-GSFC, Greenbelt, MD; http://nssdc.gsfc.nasa.gov/database/MasterCatalog?sc=1971-008C&ex=5.

[2779] R.L. KOVACH and T.J. AHRENS: *Explosive seismic sources for the Moon.* Geophys. **35**, No. 1, 33-44 (1970).

[2780] *Interoceanic canal studies, 1970.* U.S. Atlantic-Pacific Interoceanic Canal Study Commission, Depts. of State and Public Institutions, Washington, DC (Dec. 1970).

[2781] D.R. CHAPMAN: *Australasian tektite geographic pattern, crater and ray of origin, and theory of tektite events.* J. Geophys. Res. **76**, 6309–6338 (1971); In: (V.E. BARNES and M.A. BARNES, eds.) *Tektites.* Dowden, Hutchinson & Ross, Stroudsburg, PA (1973), pp. 328-357.

Today the terrestrial origin of tektites is accepted, based upon many geochemical and isotopic studies and the fact that most tektites are associated with known impact craters.[2782]

| 1971 | Dept. of Geophysical Sciences, University of Chicago, Illinois | **Tetsuya T. Fujita**,[2783] a Japanese-born U.S. meteorologist and discoverer of the downburst and microburst phenomena, **develops the international standard for measuring tornado severity.** Also popularly known as "Mr. Tornado," he classifies tornadoes into 13 numbers: $F = 0–12$, where $F = 6–12$ ranges from 319 mph (142 m/s) to the velocity of sound, approx. 760 mph (338 m/s). |
|---|---|---|

> **Tornadoes are powerful vortices** formed by severe thunderstorms and characterized by narrow, rapidly rotating funnel clouds. Since the actual wind velocity is the result of the inward spiraling of the air stream and the propagation velocity along its path, high velocities should be possible in extreme cases; rotational speeds are estimated to exceed 800 km/h (222 m/s) in extreme cases.[2784]

> However, **the existence of tornadoes that approach or even supercede the velocity of sound is obviously a myth** and couldn't be proved by measurements. ▪ In spite of this uncertainty, the so-called **"Fujita Tornado Scale"** was used routinely not only by the U.S. National Weather Service to estimate wind speeds in tornadoes, but also to classify buildings in terms of their ability to resist tornadoes (which depends upon their design and construction). Prior to 1970, U.S. nuclear power plants were designed (at considerable cost) to resist tornado wind speeds of > 600 mph (268 m/s).[2785]

> The starburst patterns of uprooted trees found in forests after the passages of tornadoes led FUJITA to his theory of "microbursts" – sudden, severe downdrafts from thunderstorms that can result in high-speed winds on or near the ground. He had seen similar patterns years before, when he had visited Nagasaki and Hiroshima just weeks after the atomic bombs were dropped in order to observe the effects of the shock waves on trees and structures in the devastated areas.

**Hurricanes and typhoons are spiraling storms;** they are the local names for large, traveling tropical cyclones. However, the wind velocities in strong hurricanes are smaller than those in strong tornados. The maximum hurricane wind speed ever measured – 165 mph (73.7 m/s) – was recorded for hurricanes "Camille" on August 17, 1969, and "Allen" on August 7, 1980, which were two of the strongest tropical cyclones in the Atlantic on record. However, although their wind speeds were far below the velocity of sound, they were nevertheless highly catastrophic. ▪ The *Safir-Simpson Hurricane Scale*, invented in the United States in the early 1970s by the engineer Herbert SAFFIR and the meteorologist Robert SIMPSON, **is a five-category scale used to quantify potential structural damage due to Atlantic hurricanes.** It ranges from category 1, with wind speeds of 74–95 mph (33–42.4 m/s) and surge heights of 4–5 ft (1.2–1.5 m), to **category 5 with wind speeds of > 155 mph (69 m/s) and surge heights of > 18 ft (5.5 m), which is the worst case.** The recent hurricane "Katrina" which devastated New Orleans, LA on August 29, 2005 was estimated to be close to a category-five hurricane.

| 1971 | University of Ulm, southern Germany | At the Spring Meeting of the Deutsche Physikalische Gesellschaft ("German Physical Society"), **Eberhard Häussler**,[2786] a German professor of applied physics and director of the Laboratorium für Kurzzeitphysik at the University of Homburg, **and his student W. Kiefer first report on a successful method for performing the contact-free disintegration of kidney stones using shock waves.** ▪ Working in close cooperation with the German company Dornier |
|---|---|---|

---

[2782] C. KOEBERL: *The geochemistry of tektites: an overview.* Tectonophys. **171**, 405-422 (1990).
[2783] T.T. FUJITA: *Tornadoes and downbursts in the context of generalized planetary scales.* J. Atmos. Sci. **38**, 1511-1534 (1981).
[2784] *The new Encyclopædia Britannica, Macropaedia.* Benton & Hemingway, Chicago etc. (1974), vol. 18, p. 514.
[2785] F. NATEGHI-A: *Assessment of wind speeds that damage buildings.* Natural Hazards **14**, 73-84 (1996).
[2786] E. HÄUSSLER and W. KIEFER: *Anregung von Stoßwellen in Flüssigkeiten durch Hochgeschwindigkeits-Wassertropfen.* Frühjahrstagung DPG, Fachausschuss Kurzzeitphysik, Ulm (1971). Verhandl. Dt. Physik. Gesell. **10** [VI], 786 (1971).

Medizintechnik GmbH (Germering, Upper Bavaria) on medical applications of shock waves, they generated the shock wave in a water basin by firing a water drop at high speed at the surface of the water above the stone.[2787]

| 1971 | Marseille, southern France | At the 3rd International Colloquium on Gas Dynamics of Explosions, **Hartmuth F. LEHR**,[2788] a German mechanical engineer and ballistician at ISL (Saint-Louis, France), **reports on experiments on shock-induced combustion in a stoichiometric hydrogen/air mixture, where the reaction was initiated by the bow wave generated by a supersonic blunt-nosed projectile.** |

(i) Using laser schlieren photography, he observes different blunt-body flow patterns at different projectile speeds: if the projectile is traveling at a velocity above the detonation velocity of the gas mixture, the reaction shows a coupled shock-deflagration system near the centerline of the projectile, but at projectile velocities below the Chapman-Jouguet (CJ) detonation velocity, the shock and the reaction in front of the projectile are separate.

(ii) There are three different regions which are clearly separated from each other:
▸ the undisturbed flow outside the bow shock wave;
▸ **a periodic tail structure of burned gas behind the projectile stern;** and
▸ a region of unburnt gas between the shock front of the bow wave and the tail structure.

(iii) LEHR shows that the pulsations are reaction instabilities which are mainly caused by the induction time of the gas mixture. ▪ These curious periodic features are apparently generated by a longitudinal detonation instability located at the front tip of the shock, thus leading to an approximately one-dimensional oscillatory detonation − known as an **"oscillatory galloping detonation."**[2789, 2790]

| 1971 | U.R.S. Research Company, San Mateo, California | **Philip J. MORRIS**,[2791] reviewing the effects of nuclear weapons in a secret report to the Defense Nuclear Agency (DNA), **shows that there is evidence that thermal loading of the target significantly enhances the damage produced by the blast wave which follows. These are called "synergetic effects."** ▪ When a nuclear weapon is detonated in the atmosphere, targets close by are first loaded with a thermal pulse propagating at the velocity of light. This is followed by the blast wave, which is many orders of magnitude slower. This curious phenomenon, where the effect of the thermal radiation and then the blast upon a target is significantly different to that achieved by the blast alone, was first observed in the 1950s during U.S. atmospheric nuclear weapons tests {F.H. SHELTON ⇒1953; BRYANT ET AL. ⇒1955}. Aircraft, light shelters and missiles would often be destroyed at overpressure levels that, in blast only tests, left the target minimally damaged. |

In the years following, these observations stimulated numerous thermal/blast synergism studies aimed at simulating these effects of combined loading in more detail in a laboratory using special thermal/blast simulators[2792] and by developing appropriate numerical codes.[2793]

---

[2787] E. HÄUSSLER and W. KIEFER: *Nierensteinzertrümmerung mit geführten Stoßwellen*. Annales Universitatis Saraviensis **11**, 150-159 (1974).
[2788] H.F. LEHR: *Experiments on shock-induced combustion*. Astronaut. Acta **17**, 589-597 (1972).
[2789] W. FICKETT and W.C. DAVIS: *Detonation*. University of California Press, Berkeley *etc.* (1979), pp. 231, 311-312.
[2790] P. CLAVIN: *Galloping detonations*. In: (M. CHAMPION and B. DESHAIES, eds.) *IUTAM Symposium on Combustion in Supersonic Flows*. Kluwer, Dordrecht (1997), pp. 335-346.
[2791] P.J. MORRIS: *A review of research on thermal radiation degradation of structural resistance to air blast*. Rept. URS 70-29-6 [DNA 2856F], U.R.S. Research Co., San Mateo, CA (1971).
[2792] B.S. KATZ and J.G. CONNOR: *Development of the NOL thermal blast simulator*. Rept. NOLTR 72-183, Naval Ordnance Laboratory (NOL), White Oak, Silver Spring, MD (July 1972).
[2793] D.M. WILSON: *A summary of methods for computing the degradation of structural elements to the thermal and thermal-blast effects of nuclear weapons*. Rept. NSWC/WOL/TR75-134. Naval Surface Weapons Center (NSWC), Silver Spring, MD (March 1976).

| 1971 | Boeing Scientific Research Laboratory, Seattle, Washington | **Earll M. MURMAN**[2794] **and Julian David COLE,** two U.S. aereodynamicists and experts in computational fluid dynamics, **use a simple and straightforward numerical procedure** to solve the "transonic small disturbance equation" in order **to determine the flow past thin airfoils, including cases with embedded shock waves.**
Two years later, Anthony JAMESON,[2795] a British-born U.S. aerodynamic engineer and numerical fluid dynamicist, devised a rotational difference scheme that extended the so-called **"Murman-Cole procedure"** so that it could be used to solve the transonic full potential equation in three dimensions, thus allowing alternative designs to be screened without the need to rely on a wind tunnel. The application of these two methods spread rapidly throughout the aerodynamic community, and airfoil sections were no longer designed experimentally by cutting and filing. |
|---|---|---|
| 1971 | Harry Diamond Laboratories, Washington, DC | **Robert B. OSWALD JR.,**[2796] a nuclear engineer, **and collaborators present a one-dimensional model for thermoelastic response, aimed at describing the free-surface motion of materials exposed to ultrashort pulsed energy deposition** – such as that resulting from exposure to high-power pulsed laser and electron beams. Using the uncoupled thermoelastic theory, in which the effects of heat conduction on the dynamics are neglected, their model assumes that the energy deposited in the sample by the radiation pulse is immediately coupled to the lattice phonons as thermal energy. For a single crystal of aluminum, they compare the measured time-dependent rear-surface displacement and velocity histories with the one-dimensional theory and find that they agree very well. |
| 1972 | Moon & NASA's Kennedy Space Center (KSC), Cape Canaveral, Florida | **The seismic studies performed during the Apollo 14 mission** {⇨1971} **are continued by two more missions to the Moon:**
▸ **Apollo 16** on April 16–27, 1972 (landing site at *Descartes Highlands*, spectacular mountain ranges) with the astronauts John W. YOUNG and Charles M. DUKE JR.; and
▸ **Apollo 17** on Dec. 7–19, 1972 (landing site at *Taurus-Littrow*, a deep narrow valley) with the astronauts Eugene A. CERNAN and Harrison H. SCHMITT.[2797]
The experiments involved seismic studies using artificial detonations from a series of small explosives.
(i) **In the *Active Seismic Experiments (ASEs)* performed during the Apollo 14 and 16 missions, explosive charges were detonated by an astronaut** using a device called a "thumper" along a 90-m-long geophone line. • The astronaut-activated thumper device was a staff with a large base containing 21 seismic charges, called "single bridgewire Apollo standard initiators." These were activated by a dial on the staff, and provided a small seismic shock for the geophones to monitor.[2798]
(ii) During Apollo 16, three mortar shells were also used to lob explosive charges to distances of up to 900 meters from the control unit. During Apollo 17, eight explosive charges were positioned at distances of up to 3.5 kilometers from the lunar module. These charges had masses of 57 g to 2.7 kg. Both the Apollo 16 mortar shells and the Apollo 17 explosives were detonated by radio control after the astronauts had left the lunar surface.
(iii) The seismic waves or ground motions caused by these explosions were measured by a network of geophones. **These experiments showed that the seismic velocity** (of the P-wave) |

---

[2794] E.M. MURMAN and J.D. COLE: *Calculation of plane steady transonic flow*. AIAA J. **9**, 114-121 (1971).

[2795] A. JAMESON: *Numerical calculation of the three-dimensional transonic flow over a yawed wing*. In: *Proc. AIAA Computational Fluid Dynamics Conference* [Palm Springs, CA, July 1973]. AIAA, New York (1973), pp. 18-26.

[2796] R.B. OSWALD, F.B. MCLEAN, D.R. SCHALLHORN, and L.D. BUXTON: *One-dimensional thermoelastic response of solids to pulsed energy deposition*. J. Appl. Phys. **42**, 3463-3473 (1971).

[2797] The missions Apollo 18 through 20, scheduled to fly to the Moon in the initial Apollo plan, were all cancelled due to budgetary constraints.

[2798] Lunar & Planetary Inst., Houston, TX: *Apollo 14 active seismic experiment*; http://www.lpi.usra.edu/expmoon/Apollo14/A14_Experiments_ASE.html.

is between 0.1 and 0.3 km/s in the upper few hundred meters of the crust at all three landing sites. These velocities are much lower than those observed for intact rock on Earth, but are consistent with a highly fractured or brecciated material produced by the prolonged meteoritic bombardment of the Moon.

(iv) At Taurus-Littrow — the Apollo 17 landing site, a mountainous region on the southeastern rim of the Serenitatis Basin — the surface basalt layer was thus determined to have a thickness of 1.4 km.

| | | |
|---|---|---|
| 1972 | Bölkow Apparatebau GmbH, Schrobenhausen, Upper Bavaria, Germany | Manfred HELD,[2799] a German detonation physicist, **shows that the light band produced by an explosive charge covered with transparent material** (such as a thick plate of glass or Plexiglas) **represents a compression of the air layer, where the products of the explosive form the "driving piston."**<br><br>His experimental results, obtained using an ultrahigh-speed image converter camera, confirm a previous theory advanced by Carl H. JOHANSSON[2800] and Per A. PERSSON, two prominent Swedish detonation researchers.<br>▸ They had predicted that the detonation gas expands from the end-surface of a cylindrical charge in a tube and acts like a piston moving with supersonic velocity, resulting in a shock-compressed gas layer ahead of the piston.<br>▸ They postulated that the velocity of the detonation front, and consequently the temperature and pressure at the front, must diminish in time, chiefly due to the growth of the luminous compression layer.<br>▸ Using a streak camera, they had observed that the light band, which is the heated compressed air in front of the detonation products, must become wider when the detonation way increases, which agrees with HELD's observation. |
| 1972 | Institut Franco-Allemand de Recherches de Saint-Louis (ISL), Saint-Louis, Alsace, France | Francis JAMET and Gustav THOMER,[2801] two staff research physicists working in flash X-ray diagnostics and detonics, **record a flash X-ray diffraction pattern for a potassium chloride [KCl] powder sample, shock-compressed to 12 kbar** due to the impact of a projectile at a velocity of about 800 m/s. An analysis of a Debye-Scherrer diagram obtained by the reflection method using pulsed Cu-K$\alpha$ radiation reveals that line shifting from reflections in the $\langle 200 \rangle$ and $\langle 220 \rangle$ directions corresponds to an isotropic compression of the cubic unit cell from the initial value of 0.628 nm to 0.611 nm.<br><br>The basic idea of flash X-ray diffraction is the following: a high-intensity flash X-ray source emits a strong characteristic radiation (a line) which is reflected from a family of lattice planes. The Bragg angle for these planes is $\theta_0$. Under the action of shock wave compression, this angle becomes $\theta_0 + \Delta\theta$. Thus, the recorded line is shifted on the plane of the detector (film). The measurement of this shift provides the variation $\Delta a$ between the planes, given by {RIGG & GUPTA ⇨1998}<br>$$\Delta a = a_0 [1 - \sin\theta_0 / \sin(\theta_0 + \Delta\theta)].$$<br>Here $a_0$ is the initial lattice spacing before the arrival of the shock wave. The measured value $\Delta\theta$ can be used to calculate the state of relative deformation $\Delta a/a_0$ of the crystal lattice. Depending on the lattice compression by the propagating shock wave, one obtains different density ratios:<br>▸ for 1-D (or *uniaxial*) compression: $\rho_0/\rho = 1 - \Delta a/a_0$, and<br>▸ for 3-D (or *isotropic*) compression: $\rho_0/\rho = 1 - 3\Delta a/a_0$. |

---

[2799] M. HELD: *Method of measuring the fine structure of detonation fronts in solid explosives*. Astronaut. Acta **17**, 599-607 (1972).

[2800] C.H. JOHANSSON and P.A. PERSSON: *Detonics of high explosives*. Academic Press, London & New York (1970), pp. 187-189.

[2801] F. JAMET and G. THOMER: *Enregistrement de diagrammes de diffraction par impulsions de rayons x de matériaux soumis à une compression par onde de choc*. In: (E. LAVIRON, ed.) *Proc. 10th Int. Congress on High-Speed Photography* [Nice, France, Sept. 1972]. Association Nationale de la Recherche Technique (ANRT), Paris (1973), pp. 292-294.

Flash X-ray diffraction is therefore a useful tool for deciding whether the compression behind the shock front is one- or three-dimensional.[2802]

| 1972 | Lawrence Radiation Laboratory (LRL), Livermore, California | Quintin JOHNSON[2803] and Arthur C. MITCHELL, two Livermore staff scientists, **provide the first microcrystalline evidence for the occurrence of a phase transition in a solid during the passage of a shock front using flash X-ray diffraction analysis** {⇨Fig. 4.12–E}. Pyrolytic boron nitride [BN], explosively shocked to 245 kbar, exhibits a diffraction pattern which indicates that the originally graphite-like crystal structure transforms during shock compression into a wurtzite-like structure.

In the years following, similar experiments were also performed in other laboratories, particularly aiming at uncovering shock compression effects in the microscopic domain in terms of crystal lattice deformation and structural changes (phase transformations) – e.g.,
▸ in the former Soviet Union by L.A. EGOROV[2804] ET AL.;
▸ in France at ISL by JAMET & THOMER {⇨1972};
▸ in Japan at TIT by Ken-ichi KONDO[2805] ET AL.;
▸ in Germany at the University of Düsseldorf by Ferdinand MÜLLER[2806] ET AL.;
▸ in the United Kingdom by WARK ET AL. {⇨1991}; and
▸ in the United States at Washington State University by RIGG & GUPTA {⇨1998}. |
|---|---|---|

| 1972 | Courant Institute, New York University, New York City, New York | David KORN,[2807] a U.S. applied mathematician, **and coworkers develop an approximate method for estimating the transonic performance of airfoils, and design a shock-free transonic airfoil** – the so-called **"Korn airfoil."** Using numerical methods, they predict the shape that produces a shock-free flow by applying the **Korn equation,** an empirical relation. ▪ Like all shock-free transonic designs, this is a single-point design and only valid for a Mach number of $M = 0.751$ and a zero angle of attack. For slightly different Mach numbers, however, a single shock pattern (such as for $M = 0.755$) will appear, or even a double shock pattern (for $M = 0.745$).

More recent studies suggest that shock-free airfoils can be designed for a band extending from a lift coefficient $C_L = 0.3$ at Mach 0.83 to $C_L = 10$ at Mach 0.72 with a thickness-to-chord ratio of around 10%.[2808] |

| 1972 | Third Physical Institute, University of Göttingen, Germany | Werner LAUTERBORN,[2809] a German research physicist, **and his collaborators visualize cavitation bubble collapse using high-speed holography.** This diagnostic method is particularly well suited for this purpose, because it provides a large depth of field and allows one to precisely determine the location of the collapse. They produce the bubbles by high-intensity acoustic waves using a piezoelectric transducer. The cavitation bubbles move with high speed, and in order to capture their motion on the film, they use a Q-switched ruby laser emitting pulses about 20 ns in duration. ▪ In 1976, LAUTERBORN received the Physikpreis of the Deutsche Physikalische Gesellschaft (DPG) for his works on cavitation. |

---

[2802] F. JAMET: *Flash radiography.* Conference of the NATO Advanced Study Institute (NATO-ASI) [Castelvecchio Pascoli, Italy, July 1983]. In: (J.E. THOMPSON and L.H. LUESSEN, eds.) *Fast electrical and optical measurements.* Nijhoff Publ., Dordrecht etc. (1986), pp. 845-861.

[2803] Q. JOHNSON and A. MITCHELL: *First X-ray diffraction evidence for a phase transition during shock-wave compression.* Phys. Rev. Lett. **29**, 1369-1371 (1972).

[2804] L.A. EGOROV, E.V. NITOCHKINA, and Y.K. OREKIN: *Registration of Debyegram of aluminum compressed by a shock wave.* Sov. Phys. JETP Lett. **16**, 4-6 (1972).

[2805] K. KONDO, A. SAWAOKA, and S. SAITO: *Microscopic observation of shock-compressed state of LiF by flash X-ray diffraction.* In: (K.D. TIMMERHAUS and M.S. BARBER, eds.) *Proc. 6th AIRAPT Int. High Pressure Conference* [Boulder, CO, July 1977]. Plenum Press, New York etc. (1979), vol. 2, pp. 905-910.

[2806] F. MÜLLER and E. SCHULTE: *Shock wave compression of NaCl single crystals observed by flash X-ray diffraction.* Z. Naturforsch. **33A**, 918-923 (1978).

[2807] F. BAUER, P. GARABEDIAN, and D. KORN: *A theory of supercritical wing sections, with computer programs and examples.* Springer, Berlin (1972).

[2808] M. HARBECK and A. JAMESON: *Exploring the limits of shock-free transonic air foil design.* Proc. 43rd Aerospace Sciences Meeting and Exhibit [Reno, NV, Jan. 2005]. AIAA, Reston, VA (2005), AIAA Paper 2005-1041; http://aero-comlab.stanford.edu/Papers/harbeck.aiaa.05-1041.pdf.

[2809] W. LAUTERBORN, K. HINSCH, and F. BADER: *Holography of bubbles in water as a method to study cavitation bubble dynamics.* Acustica **26**, 170-171 (1972).

Cavitation is a fast process which is connected with the production of shock waves that cause serious damage to material {Sir PARSON ⇒ 1884 & 1897; THORNYCROFT ⇒ 1895; Lord RAYLEIGH ⇒ 1917; S.S. COOK ⇒ 1928}. Since cavitation collapse is a fast microscopic process, optical magnification is required to record it, which practically speaking increases the speed of the cavitation phenomena in the detection plane (such as photographic film, the photocathode of an image tube, a microchannel-plate array). Thus, it can even require ultrahigh-speed cinematography, which, before the advent of ultrashort high-power pulsed lasers and electronic cameras, was a very challenging task for high-speed photographers.

| 1972 | Dept. of Geophysics & Astronomy, and Institute of Astronomy & Space Science, University of British Columbia, Vancouver, Canada | Michael W. OVENDEN,[2810] a Canadian astronomer, **proposes a theory that predicts a missing planet as well as the spacing of the planets and their major satellites.**

(i) He suggests the former existence of a planet of 90 Earth masses which supposedly filled the Titius-Bode gap in the asteroid belt and then suddenly disappeared 16 million years ago.

(ii) He believes that our present Solar System is currently evolving toward a minimum-interaction action configuration following the sudden disappearance of a planet that orbited in the same region of the Solar System now occupied by the asteroid belt.

(iii) He predicts that this planet must have been a giant planet – perhaps the size of Saturn, and much larger than all the minor planets combined {Solar System ⇒ c.3.2 Ma ago}.

In the late 1970s, Ernst J. ÖPIK,[2811] an astrophysicist at Armagh Observatory (Northern Ireland) refused OVENDEN's hypothesis. He showed that a removal by an explosion (nuclear, as the only possibility), however improbable, could have led to the formation of the asteroids, but that "life on Earth would have been completely destroyed by three successive blasts:
▸ one from the direct impact of the ejecta of the planet;
▸ another from the increased radiation suddenly emitted by the Sun when hit by the ejecta; and
▸ a third one (arriving, however, first) from the radiation emitted by the nuclear explosion."

He concluded that "the geological record of the continuity of life on Earth for the past $10^9$ years definitely excludes the possibility of such an explosion in the late Tertiary."

More recently, the U.S. astronomer Tom VAN FLANDERN[2812] cited many evidences in favor of OVENDEN's hypothesis, and predicted that future astrophysical discoveries, which he lists as tests of the theory, will tend to confirm it. |

| 1972 | Solar Physics Branch, NASA's Goddard Space Flight Center (GSFC), Greenbelt, Maryland | Charles L. WOLFF,[2813] a U.S. space scientist, **first suggests that solar flares – giant explosions on the Sun – may cause acoustic waves that travel through the Sun's interior, similar to the seismic waves on the Earth.**

(i) He assumes that a large solar flare can raise the temperature of the underlying photosphere by 10% which, due to thermal expansion, exerts a huge mechanical impulse upon the solar interior that stimulates free oscillation modes.

(ii) Because the sound speed increases with depth the waves are reflected in the deep layers of the Sun and appear back on the surface, forming expanding rings of the surface displacement.

With the advent of more advanced satellite diagnostics, first observations of the seismic waves caused by the flare of July 9, 1996 (classified as X2.6/1B) proved these predictions {KOSOVICHEV & ZHARKOVA ⇒ 1998}. |

---

[2810] M.W. OVENDEN: *Bode's law and the missing planet.* Nature **239**, 508-509 (Oct. 27, 1972).
[2811] E.J. ÖPIK: *The missing planet.* Earth, Moon, and Planets **18**, No. 3, 327-337 (May 1978).
[2812] See the book by Tom VAN FLANDERN {⇒1993}, chpt. 7: *Do planets explode?*
[2813] C.L. WOLFF: *Free oscillations of the Sun and their possible stimulation by solar flares.* Astrophys. J. **176**, 833 (Sept. 1972).

| 1973 | Vestmannaeyjar, Island of Heimaey, Vestmannaeyjar, southern Iceland | Early in the morning of January 23, **one of the most destructive volcanic eruptions in the history of Iceland begins near the town of Vestmannaeyjar,** Iceland's premier fishing port on Heimaey, the only inhabited island in the Vestmannaeyjar volcanic archipelago.[2814] During the eruption, an international commission suggests that an underwater lava flow threatening to close off the island's harbor might be dealt with using high explosives – it is hoped that the resulting mixing with water would thicken the lava and impede the flow. Shortly before the planned detonation, however, the awesome possibility that the mixing process might become self-sustaining is realized; in other words, the high-pressure steam produced could cause further mixing until all the lava had exchanged its heat with the water above it. |

In the same year,

- Stirling A. COLGATE,[2815] a U.S. physicist at the New Mexico Institute of Mining and Technology, Socorro, NM, and Thorbjörn SIGURGEIRSSON, an Icelandic geophysics professor at the University of Iceland, Reykjavik, who both belong to the Heimaey Eruption Commission, investigate the process of the dynamic mixing of water and lava. **They conclude that all cases of contact between a hot fluid (lava) and a cold vaporizable fluid (ocean water) – so-called "fuel-coolant interaction" – are potentially explosive due to a process of self-sustained mixing** which breaks up the liquid into small particles, provoking rapid heat transfer. Rayleigh-Taylor instabilities {G.I. TAYLOR ⇨1944} and a Helmholtz instability may favor the mixing of two initially separated fluids, thus giving rise to a violent explosive release of energy.
- D.J. BUCHANAN,[2816] a British researcher at UKAEA Research Group (Culham Laboratory, Abingdon, Berkshire), proposes a model for fuel-coolant interactions. The interaction is divided into five stages: (1) an initial perturbation which triggers the interaction and causes a vapour bubble to form at the fuel-coolant interface; (2) bubble expansion and collapse with jetting; (3) penetration of the fuel by the liquid jet; (4) heat transfer from the fuel to the jet; and (5) the formation of a new bubble.

Their results may explain previous disastrous industrial explosions that occurred when molten aluminum, iron, or salt suddenly came into contact with water.

| 1973 | Technical University of Denmark (TUD), Copenhagen, Denmark | In August, **Symposium on Finite-Amplitude Wave Effects in Fluids,** organized by the Physical Acoustics Group in the Dept. of Fluid Mechanics. The chairman is Leif BJOERNOE [BJØRNØ], a Danish acoustics professor. The meeting comprises three sessions: |

- nonlinear acoustics (general and air acoustics);
- nonlinear acoustics (underwater applications); and
- cavitation (general and basic aspects).

In the preface for the proceedings,[2817] BJOERNOE wrote: "Although the subject of finite-amplitude wave effects in fluids has a long history, the classical theory of finite-amplitude sound may be traced back to EULER's formulation of his equation of motion [EULER ⇨1755]: the last four decades and, especially, the last five years have seen a remarkable increase of interest in this subject among both theoreticians and experimenters." ▪ The term ***wave of finite amplitude*** was frequently used by early shock physicists {EARNSHAW ⇨1858; RIEMANN ⇨1859; MIKHEL'SON ⇨1893; Lord RAYLEIGH ⇨1910} to designate a longitudinal wave with a large amplitude compared to an acoustic wave with an almost infinitesimal amplitude. Whether

---

[2814] R.S. WILLIAMS JR. and J.G. MOORE: *Man against volcano: the eruption on Heimaey, Vestmannaeyjar, Iceland.* Published in 1976 by USGS as a booklet (USGS Information Center, Federal Center, Denver, CO) and in 1983 as a PDF file, *see* http://pubs.usgs.gov/gip/heimaey/heimaey.pdf.
[2815] S.A. COLGATE and T. SIGURGEIRSSON: *Dynamic mixing of water and lava.* Nature **244**, 552-555 (1973).
[2816] D.J. BUCHANAN: *A model for fuel-coolant interactions.* J. Phys. D: Appl. Phys. **7**, 1441-1457 (1974).
[2817] L. BJOERNOE (ed.): *1973 Symposium on finite-amplitude wave effects in fluids* [Copenhagen, Denmark, Aug. 1973]. IPC Science and Technology Press, Ltd., Guildford, Surrey (1974).

a wave of finite amplitude will develop into a shock wave depends on both the time, amplitude and geometry of the pressure wave.

| 1973 | Planets Mercury and Venus & NASA's Jet Propulsion Laboratory (JPL), Cal-Tech, Pasadena, California |
|---|---|

On November 3, **NASA launches the space probe Mariner 10 in order to perform three fly-bys of Mercury and Venus.** ▪ Mariner 10 returned 2,700 images, thereby photographing 45% of Mercury's terrain at a resolution of 1.5 km per pixel. The closest approach (705 km) was made on March 29, 1974. The most prominent feature observed by Mariner 10 was Caloris Basin, a giant impact crater about 1,300 km in diameter which was probably formed in Mercury's early history.

The region opposite the Caloris Basin – i.e., on the other side of Mercury – was photographed as well. **This spectacular "antipode" area, covered with a chaotic pattern of irregular hills and valleys, was apparently created when the seismic shock waves from the impact coincided at a focus.**[2818]

| 1973 | Nitro Nobel AB, Nora, Westmania, Sweden |
|---|---|

**Introduction of Nonel®, a *nonel*ectric system of initiating a detonation.** The Nonel fuse consists of a thick plastic tube with a bore dia. of about 1 mm, the inside surface of which is dusted with a small amount of powdered high explosive. If a shock wave is formed at one end of the tube, the explosive powder is raised to a dust and a stable detonation at a velocity of approximately 2,000 m/s (or 2 mm/μs) proceeds indefinitely along the fuse. The plastic itself is unaffected, and the only external effect is a flash of light seen through the tube walls. This therefore is an extremely safe method of propagating a detonation from one place to another.[2819]

| 1973 | Southwest Research Institute (SwRI), San Antonio, Texas & University of Tennessee, Knoxville, Tennessee |
|---|---|

**Wilfred E. BAKER,**[2820] **Peter S. WESTINE, and Franklin T. DODGE,** three U.S. engineers, **publish their textbook *Similarity Methods in Engineering Dynamics: Theory and Practice of Scale Modeling.***

(i) They state that they use
- the word **"similarity"** in the sense of Edgar BUCKINGHAM;[2821] i.e., to imply *physical* similarity between two systems; and
- the word **"model"** in the sense of Glenn MURPHY,[2822] "A *model* is a device which is so related to a physical system that observations on the model may be used to predict accurately the performance of a physical system in the desired respect."

(ii) Originating from work modeling the effects of weapons, their book addresses techniques for scaling high-rate phenomena – *e.g.*, blast waves, explosive forming, crater impact, and the response of structures subjected to transient loading – but it also covers a wide range of problems related to scaling at low and medium rates in engineering dynamics.[2823] ▪ Although numerous books on scaling and similarity have been published since, their book became the classic text on this subject. The second revised edition was published by Elsevier in 1991.

In the same year, BAKER,[2824] also associated with the Ballistic Research Laboratories (BRL) at Aberdeen Proving Ground, MD, **publishes his book *Explosions in Air*,** which is, in a sense, a counterpart to COLE's book on underwater explosions {COLE ⇨1948}. BAKER defines an explosion as **"a process by which a pressure wave of finite amplitude is generated in air by a rapid release of energy."** His examples of explosions in air encompass:

---

[2818] *Mercury – Mariner 10: jumbled terrain antipodal to the Caloris Basin, Mercury.* NASA GSFC (March 2003); http://nssdc.gsfc.nasa.gov/imgcat/html/object_page/m10_aom_11_20.html.

[2819] From Stanley FORDHAM's book *High explosives and propellants* {FORDHAM ⇨1966}, 2nd edn (1980), p. 125.

[2820] W.E. BAKER, P.S. WESTINE, and F.T. DODGE: *Similarity methods in engineering dynamics.* Hayden Book Co, Rochelle Park, NJ (1973), chap. 4: *Scaling of gas dynamics, with special attention to blast waves*, pp. 53-79.

[2821] E. BUCKINGHAM: *On physically similar systems; illustrations of the use of dimensional equations.* Phys. Rev. **4** [II], 345-376 (1914); *The principle of similitude.* Nature **96**, 396-397 (1915).

[2822] G. MURPHY: *Similitude in engineering.* Ronald Press, New York (1950).

[2823] F. BAUER, P. GARABEDIAN, and D. KORN: *A theory of supercritical wing sections, with computer programs and examples.* Springer, Berlin (1972).

[2824] W.E. BAKER: *Explosions in air.* University of Texas Press, Austin, TX & London (1973).

- chemical or nuclear materials capable of sustaining violent reactions;
- stored energy in a compressed gas or vapor, either hot or cold;
- the failure of a high-pressure gas storage vessel or a steam boiler;
- the rapid vaporization of a fine wire or thin metal film by a pulsed electric current;
- the release of electrical energy by discharge in a spark gap; and
- a muzzle blast in ballistics.

1973 — Institute for Materials Research, National Bureau of Standards (NBS) [since 1988 National Institute of Standards and Technology (NIST)], Washington, DC

**J. Dean BARNETT,**[2825] a visiting scientist from Brigham Young University, **and researchers at NBS invent the *ruby fluorescence method*.** Their so-called "ruby fluorescence pressure gauge" utilizes a pressure shift in the sharp R-line fluorescence spectrum of ruby. Applying this new technique to a diamond-anvil pressure cell, they report that "the precision of the pressure measurement in a hydrostatic environment up to 100 kbar is 0.5 kbar using ruby as the pressure sensor – *i.e.*, better than the accuracy of the present pressure scale above 40 kbar."
▪ Five years later, calibration of the ruby scale was extended into the very high-pressure domain by making specific volume measurements of four metals under static pressures simultaneously using X-ray diffraction, and referring these results to isothermal equations of state derived from shock-wave experiments.[2826]

The so-called "ruby standard," now commonly used throughout the world for pressure calibration in static experiments, **also proved useful in shock wave physics.** In 1986, Paul D. HORN[2827] and Yogendra M. GUPTA, two shock physicists at the Shock Dynamics Laboratory of Washington State University, used the ruby fluorescence method in impact experiments to measure the time-resolved, stress-induced wavelength shift of ruby fluorescence under shock loading up to 99 kbar. In the years following, instrumentation and data analysis were improved, and the fluorescence wavelength shift resulting from the ruby being shocked as well as that due to stress relief could be observed continuously over time.[2828] However, the ruby fluorescence wavelength shift is a function of temperature as well as stress, and this should be noted since it limits any comparison of measured shock data with hydrostatic results, particularly at shock pressures above 100 kbar.

1973 — Chemical Research Dept., University of Maryland, College Park, Maryland

**Julius William ENIG,**[2829] a U.S. mathematician, **works out a numerical solution for the unsteady phenomena of regular and Mach reflection that occur when two identical spherical explosive charges of pentolite** (50/50 TNT/PETN) **are detonated simultaneously under water** {COLEBURN & ROSLUND ⇨1968} – or, equivalently, when a single charge is detonated close to a rigid wall. Applying oblique shock theory {VON NEUMANN ⇨1943}, he finds that the use of the Sternberg-Walker equation of state (EOS) {WALKER & STERNBERG ⇨1965} for water and the HOM equation of state {MADER ⇨1962}, which combines a Mie-Grüneisen relationship for the condensed phase with a fitting EOS for the gas phase and is based upon a linear shock velocity *vs.* particle velocity relationship, correctly predicts the Mach stem pressure at the critical shock angle, while the Polachek-Seeger EOS {POLACHEK & SEEGER ⇨1951}, a γ-law type equation of state with γ = 7.15, does not.

---

[2825] J.D. BARNETT, S. BLOCK, and G.J. PIERMARINI: *An optical fluorescence system for quantitative pressure measurement in the diamond-anvil cell*. Rev. Scient. Instrum. **44**, 1-9 (1973).
[2826] H.K. MAO, P.M. BELL, J. SHANER, and D. STEINBERG: *Specific volume measurements of Cu, Mo, Pd, and Ag and calibration of the ruby R1 fluorescence pressure gauge from 0.06 to 1 Mbar*. J. Appl. Phys. **49**, 3276-3283 (1978).
[2827] P.D. HORN and Y.M. GUPTA: *Wavelength shift of the ruby luminescence R lines under shock compression*. Appl. Phys. Lett. **49**, 856-858 (1986).
[2828] P.D. HORN and Y.M. GUPTA: *Ruby fluorescence in shock wave experiments*. In: (E.R. MENZEL, ed.) *Fluorescence detection*. Proc. SPIE **743**, 42-48 (1987).
[2829] J.W. ENIG: *The unsteady regular and Mach reflection resulting from the interaction of spherical explosion shock waves in water*. Ph.D. thesis, Dept. of Mechanical Engineering, University of Maryland, College Park, MD (1973).

| | | |
|---|---|---|
| 1973 | Shock Dynamics Laboratory, Dept. of Physics, Washington State University, Pullman, Washington | **Dennis B. HAYES,**[2830] a Ph.D. student of Prof. George DUVALL, **experimentally investigates shock-induced polymorphic phase transitions in a single crystal of potassium chloride** [KCl]. He uses planar shock waves with stress amplitudes ranging from 17 to 31 kbar generated by planar impact, and a calibrated 50-$\Omega$ shorted X-cut quartz gauge located at the impact surface. He observes that the transformation rate depends upon the crystallographic orientation of the crystal:<br>(i) When the shock travels along a $\langle 111 \rangle$ crystallographic axis, phase transformation is observed to be completed in between 21 and 43 ns, for initial stresses from 25.8 to 23.0 kbar, respectively. Transformed states lie on the equilibrium surface.<br>(ii) However, when the shock travels along a $\langle 100 \rangle$ crystallographic axis, initial stress relaxation is not observed although partial transformation has occurred. Transformed states are metastable and decay toward the equilibrium surface after decay times of the order of 100 ns. |
| 1974 | Dept. of Engineering Science, University of Oxford, England | In the period April 2–4, **the first Oxford Conference on Mechanical Properties of Materials at High Rates of Strain is held.** The objectives are to discuss<br>▸ fundamental aspects of material behavior at high rates of strain;<br>▸ stress wave propagation effects and fracture;<br>▸ the responses of structures to dynamic loading; and<br>▸ applications to processes involving the dynamic straining of materials.<br>The chairman is John HARDING.<br>Subsequent conferences were held in the years 1979, 1984 and 1989 at Oxford University. Beginning in 1994, the Oxford Conferences merged with DYMAT, thus forming EURODYMAT {DYMAT 85 ⇒1985}. |
| 1974 | Orlando, Florida | In November, **first International Symposium on Ballistics.** The chairman is Richard S. DOWD of Martin Marietta Corporation, who also initiated the Symposium. ▪ In those days, with the Cold War at its peak, the science of ballistics was rarely publicized. Although great advances had happened during World War II, these were only known to a few people, and most literature on the topic had been written primarily by scholars in Germany pre-World War II. Therefore, most nations felt a need to exchange more basic scientific and technological information in the field. One of the main activities of the American Ordnance Association (AOA) [now the National Defense Industrial Association (NDIA)[2831]] was the organization of professional seminars, conferences and symposia on topics in armaments, ordnance and other defense-related technologies. The AOA organized several annual or semi-annual conferences related to weaponry and ballistic science and technology, but they were all purely national U.S. events. Mr. DOWD, chairing the Ballistic Committee of the AOA, suggested that an International Symposium should be arranged. This idea was checked with several NATO allies and supported by the U.S. Dept. of Defense (DOD), and it was decided that it should be organized at Orlando, FL, in 1974. Although essentially a NATO event when it started, other nations active in ballistics were later also admitted, such as the Soviet Union (1983) and China (1992).[2832]<br>Subsequent symposia were held at Daytona, FL (1976); Karlsruhe, Germany (1977); Monterey, CA (1978); Toulouse, France (1980); Orlando, FL (1981); The Hague, The Netherlands (1983); Orlando, FL (1984); Shrivenham, U.K. (1986); San Diego, CA (1987); Brussels, Belgium (1987); San Antonio, TX (1990); Stockholm, Sweden (1992); Quebec City, QC, Canada |

---

[2830] D.B. HAYES: *Experimental determination of phase transformation rates in shocked potassium chloride.* Rept. WSU SDL 73-01, Washington State University, Pullman, WA (May 1973).
[2831] *History of NDIA*; http://www.ndia.org/Content/NavigationMenu/Resources1/History_of_NDIA.htm.
[2832] Private communication by Dr. Bo JANZON, Director of Weapons & Protection; Swedish Defense Research Agency (FOI), Grindsjon, Sweden (Feb. 22, 2006).

| | | |
|---|---|---|
| 1974 | Pittsburgh, PA | **Foundation of the International Society of Explosives Engineers (ISEE).**[2833] The first president is Calvin J. KONYA (1975–1977). This professional society is dedicated to promoting the safe and controlled use of explosives in mining, quarrying, construction, manufacturing, demolition, aerospace, forestry, avalanche control, art, automotives, special effects, exploration, seismology, agriculture, law enforcement, and many other peaceful uses of explosives. ▪ The administrative offices and the international headquarters of the Society are located in Cleveland, OH. The bimonthly *Journal of Explosives Engineers,* founded in 1983 as the voice of ISEE, disseminates timely information to the entire explosives industry. |
| 1974 | *Messerschmitt Bölkow Blohm (MBB) GmbH, Schrobenhausen, Upper Bavaria, Germany* | **Manfred HELD,**[2834] a German research physicist, **presents three different methods of optical streak recording,** depending on the direction of the velocity vector of the event under observation *vs.* the streak slit position and the optical axis of the rotating mirror camera, thus demonstrating the high versatility of this method as a worthwhile diagnostic tool in detonics. ▪ Until the advent of electronic methods, the basic element of a mechanical high-speed streak camera was its rotating mirror {WHEATSTONE ⇒ 1834}.<br><br>In the following 25 years, HELD considerably extended the method of optical streak recording, thereby illustrating its great versatility with many new examples of application, such as in shock wave physics, detonics and ballistics.[2835] |
| 1974 | *ISL, Saint-Louis, Alsace, France & Dept. of Mechanics and Materials Science, Johns Hopkins University, Baltimore, Maryland* | **Francis JAMET**[2836] **and Gustav THOMER,** two research physicists at ISL, **first successfully record a flash X-ray diffraction pattern from a shaped charge jet under different experimental conditions.** For example, studying the jet of a shaped charge with an aluminum liner (charge weight 3.5 g, cone angle 90°, jet velocity about 5 km/s) and using a 50-ns flash X-ray generated in a vacuum-discharge-type flash X-ray tube operated at 80 kV, they observe a crystalline texture characterized by a grainy structure. From the shift in ring diameters in the Debye-Scherrer diffraction patterns recorded on film before and during detonation, they conclude that **the density of the jet is around 15% lower than that of the initial material.**<br><br>In the same year, **Robert E. GREEN,**[2837] a professor at Johns Hopkins University, **independently records flash X-ray diffraction patterns taken with an exposure time of about 70 ns from shaped charges** using a field-emission-type flash X-ray tube operated at 150 kV, and a two-stage image intensifier. Compared with the shaped charges applied at ISL, his shaped charges have also an aluminum liner, but a much larger charge weight (680 g), a smaller cone angle (60°), and a higher jet velocity (6.4 km/s). Upon analysis of the diffraction pattern, GREEN concludes that **the jet consisted of a particulate solid, with a grain size distribution ranging from about 1 mm down to about 0.01 mm.** |

---

[2833] For more about the history of the ISEE, *see* http://www.isee.org/about/history.htm.
[2834] M. HELD: *Streak technique as a diagnosis method in detonics.* In: (E.J. BRYANT, ed.) *Proc. First Int. Symposium on Ballistics* [Orlando, FL, Nov. 1974]. American Defense Preparedness Association, Washington, DC (1974); *see* Sect. IV: *Warhead mechanisms and effects*, pp. 177-210.
[2835] M. HELD: *High speed photography in detonics and ballistics.* In: (S.F. RAY, ed.) *High-speed photography and photonics.* Focal Press, Oxford *etc.* (1997), pp. 233-244.
[2836] F. JAMET and G. THOMER: *Diagramme de poudre d'un jet de charge creuse.* C. R. Acad. Sci. Paris **279** [B], 501-503 (1974).
[2837] R.E. GREEN JR.: *First X-ray diffraction photograph of a shaped charge jet.* Rev. Scient. Instrum. **46**, 1257-1261 (1975).

| | | |
|---|---|---|
| 1974 | *Poulter Laboratory, Stanford Research Institute (SRI), Menlo Park, California* | William MURRI,[2838] Donald R. CURRAN, Carl F. PETERSON, and Richard C. CREWDSON, four U.S. shock wave physicists, **review recent advances in shock wave physics,** thereby emphasizing the change in viewpoint that has been brought about by the appearance of in-material gauges of high temporal resolution.<br>(i) Introductorily, they specify the goal in more detail: "Whereas previous reviews have emphasized the concept of the Hugoniot equation of state (which describes the thermodynamic states attainable through a *steady* shock wave), **we shall emphasize the concept of *unsteady* waves and shall discuss the analytical procedures that have been developed to handle experimental data from unsteady waves in solid materials.**"<br>(ii) Addressing the problems of comparing static with dynamic results, they conclude: "One of the original motivations for shock wave experiments, using explosives, was to obtain pressure-volume data in high pressure regions that could not be reached with static pressure experiments … Where detailed comparisons between static and dynamic measurements can be made, it is usually found that agreement is good. There are many important differences between the two kinds of experiments, however, that must be considered in interpreting the experimental results:<br>▸ The most obvious difference is the length of time that a specimen can be subjected to high pressure.<br>▸ Another difference is in the stress distribution. In static work, the stress is more nearly hydrostatic. However, at pressures above about 30 kbar, the pressure-transmitting media solidify, causing a poorly defined shear stress to be present. Shock measurements, on the other hand, are generally conducted in a system of well-defined 1-D strain.<br>▸ Temperatures are more easily known and controlled in static experiments, but much higher temperatures can be reached in shock compression.<br>▸ Because of the different stress distribution, materials with a high Hugoniot elastic limit (HEL) may require a significant 'strength-of-materials' correction to properly compare static pressure results with dynamic stress results." |
| 1974 | *CalTech, Pasadena, California* | Gerald B. WHITHAM,[2839] a U.S. professor of applied mathematics, **publishes his book *Linear and Nonlinear Waves.*** Addressing in great detail the mathematical aspects of hyperbolic and dispersive waves, it provides many examples of specific wave types (*e.g.*, shock waves, sonic booms, blast waves, traffic flow, bores, waves in shallow water and in glaciers, and ship waves). ▪ The book became a classic on wave physics and was reprinted twice. |
| 1975 | *Roissy Airport, [ca. 23 km off Paris], France* | On May 28, **Concorde,** the British/French Supersonic Transport (SST) {Concorde ⇨1969}, **makes its first official demonstration flight.** Werner MEYER-LARSEN,[2840] a U.S. correspondent for the German news magazine *Der Spiegel*, describes his first supersonic flight as follows: "The 4,600 kilometers between Paris and Dakar, 6½ hours normal jet-time, were made by the 160-ton heavy bird at a stratospheric altitude of 19 km in 2 hours and 43 minutes. The time accelerator is represented by the formula 'Mach 2' (double the velocity of sound, about 2,400 km/h). For the civilians on board it remained a statistic value – readable only on a digital counter at the cabin wall [a so-called 'Machmeter,' ⇨Fig. 4.20–H]: the numbers on it started rolling up in ten minutes to a value of the velocity of sound and in twenty minutes to double it. However, the blood remained in the passengers' heads and the champagne in the glasses; the difference from the speed of a Boeing at Mach 0.85 remained unnoticed, and even **the dreaded sound barrier did not** |

---

[2838] W.J. MURRI, D.R. CURRAN, C.F. PETERSON, and R.C. CREWDSON: *Response of solids to shock waves.* Tech. Rept. No. 001-71, Poulter Laboratory of SRI., Menlo Park, CA (1971); In: (R.H. WENTORF JR., ed.) *Advances in High Pressure Research.* Academic Press, London & New York, vol. 4 (1974), pp. 1-163.
[2839] G.B. WHITHAM: *Linear and nonlinear waves.* Wiley, New York etc. (1974, 1978, 1999).
[2840] W. MEYER-LARSEN: *Zum Lunch nach Senegal.* Der Spiegel (Hamburg) **29**, Nr. 23, 16 (1975).

mar the horizontal flight." ▪ This account is a great illustration of the enormous progress made in supersonic aerodynamics in the space of 28 years, from the bumpy transition through the sound barrier of a little research aircraft (Bell X-1: length 9.4 m; wingspan 8.5 m; *see* YEAGER ⇨1947) to the smooth one of a huge airliner (Concorde: 62.2 m; 25.6 m).

Starting on January 1, 1976, Air France began to offer regular supersonic flights on Concordes taking the route Paris → Dakar → Rio de Janeiro. Due to a drop in demand, Air France shut down its regular passenger service in October 2003. Shortly after, Concorde memorabilia were sold at the British Airways Concorde Charity Auction in London and at the Air France Concorde Charity Auction in Paris for enormous prices: for example, the Machmeter of the pilot's control panel {⇨Fig. 4.20–H} was sold for more than $48,000 and the Radom (the nosecone) brought $552,600.[2841] ▪ A Machmeter (or Mach meter) is a sophisticated instrument based on the principle of the static Pitot probe {PITOT ⇨1732}. Combined with an analog or digital display, it indicates the velocity of a vehicle in terms of the Mach number {ACKERET ⇨1929}.[2842]

| 1975 | *ESA Headquarters, Paris, France* | On May 30, **the European Space Agency (ESA) is founded.** An organization encompassing 15 European countries (Austria, Belgium, Denmark, Finland, France, Germany, Ireland, Italy, The Netherlands, Norway, Portugal, Spain, Sweden, Switzerland, and the United Kingdom), the ESA is dedicated to space research and its related technologies.[2843] The first president is Roy GIBSON (1975–1980), from Great Britain. |
|---|---|---|

The ESA has developed the Ariane rocket (1979), which is used to launch most ESA satellites, the Spacelab scientific workshop, various scientific and communications satellites, and numerous space probes,[2844] such as

- ESA's ***COS-B gamma-ray satellites,*** which were designed to provide the first complete galactic high-energy gamma-ray survey (1975);
- NASA/ESA's ***ISEE*** *(International Sun-Earth Explorer)*, three heliocentric missions (1977, 1977 & 1978) performed to investigate solar-terrestrial relationships; in particular **to examine the structure of the solar wind near the Earth and the shock wave that forms the interface between the solar wind and the Earth's magnetosphere in detail** {ISEE Program ⇨1977; ⇨Fig. 4.1–V};
- ESA's ***Exosat*** (1983), designed to observe and detect high-energy sources, and to study the Universe at X-ray wavelengths;
- ESA's ***Giotto*** (1985), designed to help to solve mysteries about comet HALLEY by passing as close as possible to the comet's nucleus (achieved on March 13, 1986);
- ESA's ***Hipparchos*** (1989), designed to perform precise sky-mapping of the entire sidereal sphere, to measure fixed star luminosity and angular momentum, and to investigate the proper motions of the stars that make up the Milky Way;
- ESA/NASA's ***Ulysses*** (1990), designed to study the heliospheric magnetic field, to measure the solar wind over a wide range of solar activities, *etc.*;
- NASA/ESA's ***Hubble Space Telescope*** (1990), designed to orbit the Earth and to take pictures from all kinds of cosmic phenomena {Nova CYGNI 1992 ⇨1992; Supermassive Black Hole ⇨1994; LL Orionis ⇨1995};

---

[2841] *Concorde nosecone fetches more than $550,000 at auction.* VOA News (Dec. 01, 2003); http://www.theepochtimes.com/news/3-12-1/16312.html.

[2842] A Mach meter needs only Pitot static inputs to determine the Mach number of a vehicle, because the gases that make up our atmosphere behave in a manner consistent with the ideal gas equation. *See* the interesting note *Measuring Air Speeds* by Dave ESSER, a professor of aeronautical science at Embry-Riddle Aeronautical University, Daytona Beach, FL; http://www.womanpilot.com/past%20issue%20pages/2000%20issues/jan%20feb%202000/airspeed.htm.

[2843] J. KRIGE, A. RUSSO, and L. SEBESTA: *A history of the European Space Agency.* ESA SP-1235; vol. I: *The story of ESRO and ELDO, 1958–1987,* vol. II: *The story of ESA, 1973–1987.* ESA Publ. Div., Estec, Noordwijk, The Netherlands (2000).

[2844] *ESA science programme, list of operations.* ESA Science & Technology; http://sci.esa.int/science-e/www/area/index.cfm?fareaid=71.

|  |  |  |
|---|---|---|
|  |  | ► ESA/NASA's **SOHO** (1995), the space-based Solar Heliospheric Observatory designed to investigate the Sun from its deep core through its corona and the domain of the solar wind out to a distance of ten times beyond the Earth's orbit {LASCO ⇒2002};
|  |  | ► NASA/ESA/ASI's **Cassini-Huygens** (1997), designed to explore Saturn's ring systems and moons from orbit. In January 2005, the Huygens Probe successfully dived into Titan's thick atmosphere;
|  |  | ► ESA's **Mars Express/Beagle** (2003), designed to map the entire surface of Mars, its atmosphere, the mineral composition of the surface, *etc.*, and to determine the geology of the landing site and to search for life signatures (exobiology) with the **Beagle 2 Lander**;
|  |  | ► ESA's **SMART-1** (2003), the first of ESA's Small Missions for Advanced Research in Technology (SMART), designed to test solar-electric propulsion and gather more information about the Moon;
|  |  | ► ESA's **Rosetta** (2004), designed to undertake the long-term exploration of a comet at close quarters, and to release in 2014 a small lander on the icy nucleus of comet 67P/CHURYUMOV-GERASIMENKO; and
|  |  | ► ESA's **Planck Satellite** (2007), designed to make the most accurate maps yet of the microwave background radiation that fills space.
| 1975 | Poulter Laboratory, Stanford Research Institute (SRI), Menlo Park & Naval Postgraduate School (NPS), Monterey, California | In a joint study on the interaction of high-power pulsed laser irradiation with a solid target, **Peter KREHL** (SRI),[2845] **and Fred SCHWIRZKE and Alf W. COOPER** (NPS) **correlate recorded pressure-time profiles for the laser-induced stress wave in a metallic target with time-resolved plasma profiles in front of the target** using a Sandia-type quartz gauge {JONES ET AL. ⇒1962} and electrostatic double probes, respectively.<br><br>(i) The momentum $M_T(t)$ delivered by the focused laser beam to the target can be correlated to the impulse of the neutral and charged particles which, neglecting the small contribution by the electrons ($m_e \ll m_n$), is given by<br>$$M_T = A \int p(t)\,dt \approx m_i v_i(t)\, N_i(t) + m_n v_n(t)\, N_n(t),$$<br>where $A$ is the area of the focal spot, $p(t)$ is the stress in the target, $v_i$ and $v_n$, $N_i$ and $N_n$, and $m_i$ and $m_n$ are the velocities, numbers and masses of the ions and neutral atoms, respectively.<br><br>(ii) Their stress/plasma correlation analysis indicates that the target impulse is predominantly produced by fast-moving ionized particles which, however, represent only a small amount (about 10%) of the mass completely removed.<br><br>David W. GREGG and Scott J. THOMAS {⇒1966} had proposed to measure the total momentum delivered by a focused giant laser pulse to a target by using a simple ballistic pendulum. However, this overall method cannot provide true maximum shock pressures and data on the contribution of neutral and charged particles to the total momentum.
| 1975 | Thermo- and Gasdynamics Division, NASA's Ames Research Center (ARC), Moffet Field, California | Paul KUTLER,[2846, 2847] a U.S. aerospace engineer, **and collaborators determine the transient load encountered when a supersonic/hypersonic missile flies through a planar oblique blast wave traveling in the opposite direction.**<br><br>(i) In order to numerically solve this difficult **"shock-on-shock" problem** {LEMCKE ⇒1963; ⇒Fig. 4.14 R}, which contains complex multiple shock wave interactions, they apply a nonlinear inviscid code. Previously developed for hypersonic or purely supersonic flow, this code is based on a second-order, shock-capturing, finite-difference approach.

---

[2845] P. KREHL, F. SCHWIRZKE, and A.W. COOPER: *Correlation of stress-wave profiles and the dynamics of the plasma produced by laser irradiation of plane solid targets.* J. Appl. Phys. **46**, 4400-4406 (1975).

[2846] P. KUTLER, L. SAKELL, and G. AIELLO: *On the shock-on-shock interaction problem.* AIAA J. **13**, 361-367 (1975).

[2847] P. KUTLER and L. SAKELL: *Three-dimensional, shock-on-shock interaction problem.* AIAA J. **13**, 1360-1367 (1975).

(ii) They report that large shock tube and rocket-propelled test sled experiments performed at Holloman Air Force Base (Alamogordo, NM) have verified their numerical approach.

| | | |
|---|---|---|
| 1975 | Los Alamos Scientific Laboratory (LASL), Los Alamos, New Mexico | **Timothy R. NEAL,**[2848] a U.S. physicist and weapons technology and explosives expert, **uses a symmetric, two-column explosive arrangement to generate shock pressures in aluminum of up to 490 kbar using Mach reflection,** and examines the region of colliding shock waves using flash radiography {⇨Fig. 4.13–M}. Starting from a simple three-shock model of Mach waves with a slip line, he introduces several modifications in order to account for the regressive nature of the Mach disk; in gas dynamics this irregular reflection phenomenon is termed ***inverse Mach reflection.*** For large collision angles, his modified three-shock model is consistent with experimental data, but does not furnish a relationship between the collision angle and the stem growth angle. |
| 1975 | Optical Sciences Division, U.S. Naval Research Laboratory (NRL), Washington, DC | **Stuart K. SEARLES**[2849] **and George A. HART,** two NRL staff scientists at the Laser Physics Branch, **demonstrate the first excimer laser action obtained by bombarding a medium of xenon bromide [XeBr] with an electron beam gun.** ▪ Almost simultaneously, excimer laser action was also reported independently by three other U.S. research teams: at Northrop Research & Technology Center (Hawthorne, CA), AVCO-Everett Research Laboratory (Everett, MA), and Sandia Laboratories (Albuquerque, NM). |
| | | Excimer lasers are rare-gas halide or rare-gas metal vapor lasers emitting in the ultraviolet. They are primarily applied when machining a broad range of materials (*e.g.*, cutting, welding or surface modification operations), as well as in ophthalmology {PULIAFITO ET AL. ⇨1987; KRUEGER ET AL. ⇨1993; ⇨Fig. 4.16–Z}, dermatology, and biological research.[2850] When biological tissue is illuminated with an excimer laser, the relatively weak organic bonds are broken down. This creates a rise in pressure and a subsequent shock wave that removes material supersonically, with little heat transfer to the surrounding material, in a process called **"laser ablation."** |
| 1976 | Utopia Planitia, Planet Mars & NASA's Jet Propulsion Laboratory (JPL), CalTech, Pasadena, California | On September 3, **the lander from NASA's planetary probe Viking 2** (launched on Sept. 9, 1975) **touches down on Mars on Utopia Planitia,** a lava plain about 200 km west of Crater Mie.[2851] Among other tasks, seismic experiments are carried out to determine the level of seismic activity on Mars as well as its internal structure.[2852] The seismic technique applied is a passive one – *i.e.*, no artificial earthquakes are generated like during previous Moon missions {Apollo 14, 15, 16 & 17 ⇨1971 & 1972}. However, the simple device detects no natural seismic activity ("marsquakes") on Mars to the limits of its sensitivity. ▪ The interior of Mars shows very little movement and does not have quakes that have been measured. Natural marsquakes may be generated through the release of thermal stresses and by meteoritic impacts.[2853] |
| | | Seismology is one method used to search for the presence of water, a research task of the highest priority for NASA's Mars Exploration Program. On the other hand, the images from the Viking 2 orbiter seem to provide evidence of surface water. However, the biology experiment on the lander yielded no evidence of life at either landing site. |

---

[2848] T. NEAL: *Mach waves and reflected rarefactions in aluminum.* J. Appl. Phys. **46**, 2521-2527 (1975).
[2849] S.K. SEARLES and G.A. HART: *Stimulated emission at 281.8 nm from XeBr.* Appl. Phys. Lett. **27**, 243-245 (1975).
[2850] D. BASTING, K. PIPPERT, and U. STAMM: *History and future prospects of excimer laser technology.* Riken Rev. (Wako, Japan) **43**, 14-22 (Jan. 2002).
[2851] *Crater Mie,* a 95-km-dia. basin on Mars formed by asteroid or comet impact in Utopia Planitia, is named after Gustav MIE (1868–1957), a German physicist who in the 1920s, together with the German physicist Eduard GRÜNEISEN (1877–1949), worked out a kinetic theory for solid matter, thereby advancing the so-called "Mie-Grüneisen equation of state" which is used by modern solid-state shock wave physicists {GRÜNEISEN ⇨1926}.
[2852] D.L. ANDERSON ET AL.: *Seismology on Mars.* J. Geophys. Res. **82**, 4524-4546 (1977).
[2853] G.A. SOFFEN: *Scientific results of the Viking missions.* Science **194**, No. 4271, 1274-1276 (1976).

Ever since the earliest telescope observations made in the 1600s by the likes of GALILEI, HUYGENS, and CASSINI, people have wondered about the origins of some of the surface features on Mars. New high-resolution images from satellites have shown the Martian surface in unprecedented detail: unusual linear chains of pits may have been formed by active faults {FERRILL ET AL. ⇨ 2004}.

| 1976 | *Institut Kosmicheskikh Issledovanii, Moscow & Moskovskii Gosudarstvennyi Universitet, Soviet Union* | **Soviet astrophysicists develop the idea that in a binary star system (WR+OB) the fast WR wind collides with the companion's wind and generates a shock-heated plasma** with temperatures ranging from about $10^7$ to $10^8$ K, thus causing a bright, extensive X-ray emission.[2854, 2855] • Wolf-Rayet (WR) stars have fast stellar winds (1,000–3,000 km/s) with high mass loss rates. Young massive stars of the spectral type O and B also have supersonic winds and are the second components of such binary star systems.<br><br>In April 2005, NRAO astronomers, using the National Science Foundation's Very Long Baseline Array (VLBA) radio telescope, reported that they have tracked the motion of a violent region where the powerful winds of two giant stars – an about 20-solar-mass star and an about 50-solar-mass star, part of a system named WR 140 and circling each other in an elliptical orbit – slam into each other, producing bright radio emission.[2856] Astronomers have also identified the X-rays from about two-dozen of these binary systems in our Milky Way. Very recently, an international team led by Dr. Yael NAZE of the Université de Liege in Belgium has found such a system also outside of our galaxy: in the Small Magellanic Cloud, a dwarf galaxy located about 170,000 light-years from Earth.[2857] |
|---|---|---|
| 1976 | *Wiley-VCH (Verlag Chemie), Weinheim/Bergstr., Germany* | **Foundation of the international journal *Propellants and Explosives*,** in order "to provide a specialized vehicle of communication for scientists and technologists, whether in research, development, manufacture or applications, working in the area of propellants, explosives, primers and pyrotechnics together with those concerned with combustion and detonation processes." The first editor-in-chief is Hiltmar SCHUBERT, director of the Institut für Chemische Technologie (ICT) at Pfinztal-Berghausen, Germany, an institute of applied research belonging to the Fraunhofer-Gesellschaft (FhG), with headquarters in Munich.<br><br>Beginning with volume 7 (1982), the journal was renamed to ***Propellants, Explosives, Pyrotechnics.*** |
| 1976 | *Division of Engineering, Brown University, Providence, Rhode Island* | **Ahmed S. ABOU-SAYED,**[2858, 2859] a Ph.D. student of Prof. Rodney J. CLIFTON, **reports on a new technique developed for the dynamic pressure-shear loading of specimens. It involves using an oblique impact** in which the impact surface on the projectile nose is not perpendicular to the direction of motion. As a result, both a planar compressional shock and a planar shear wavefront can be introduced simultaneously into the specimen. They measure compression and shear waves using free-surface optical methods. • Their parallel-inclined impact technique was used to test piezoresistive pressure gauges for lateral stress measurements of shock-loaded specimens {GUPTA ET AL. ⇨ 1980}. |

---

[2854] O.F. PRILUTSKII and V.V. USOV: *X rays from Wolf-Rayet binaries.* Sov. Astron. **20**, 2-4 (1976).
[2855] A.M. CHEREPASHCHUK: *Detectability of Wolf-Rayet binaries from X rays.* Sov. Astron. Lett. **2**, 138-139 (1976).
[2856] S.M. DOUGHERTY ET AL.: *High-resolution radio observations of the colliding-wind binary WR 140.* Astrophys. J. **623**, 447-459 (April 10, 2005).
[2857] *First X-ray detection of a colliding-wind binary beyond Milky Way.* Space Daily (Feb. 23, 2007); http://www.spacedaily.com/reports/First_X_Ray_Detection_Of_A_Colliding_Wind_Binary_Beyond_Milky_Way_999.html.
[2858] A.S. ABOU-SAYED: *Analytical and experimental investigation of pressure-shear waves in solids.* Ph.D. thesis, Brown University, Providence, RI (1975/1976).
[2859] A.S. ABOU-SAYED, R.J. CLIFTON, and L. HERMAN: *The oblique-plate impact experiment.* Exper. Mech. **16**, 127-132 (1976); A.S. ABOU-SAYED and R.J. CLIFTON: *Pressure shear waves in fused silica.* J. Appl. Phys. **47**, 1762-1770 (1976).

One year later, Kyung-Suk KIM,²⁸⁶⁰ a collaborator of Prof. CLIFTON, reported on the development of a transverse displacement interferometer capable of monitoring normal and transverse motions in impact-loaded Y-cut quartz samples.

| 1976 | *Sandia Laboratories, Albuquerque, New Mexico* | **James R. ASAY,**²⁸⁶¹ a U.S. shock physicist, **and his collaborators first quantitatively study the physical mechanisms important in the degradation of surfaces during shock loading** – a phenomenon which can cause the development of either Rayleigh-Taylor instabilities at materials interfaces or the jetting of materials from surfaces {G.I. TAYLOR ⇨ 1944}. Using velocity interferometry and double-pulse holography, they measure a total specific mass of about 3 µg/cm² ejected from an aluminum sample when shocked to 250 kbar. • This complex phenomenon, which depends on the intensity of the incident shock wave, the nature and thickness of the sample and its roughness, was later investigated in more detail by a number of researchers and attributed to surface defect volume, microspallation and surface melting.²⁸⁶² |
|---|---|---|
| 1976 | *Institute of Hydrodynamics, Novosibirsk, U.S.S.R. Academy of Sciences, Soviet Union* | **I.T. BAKIROV**²⁸⁶³ **and Vladislav V. MITROFANOV,** two Soviet detonation researchers, **report that they have observed unusually high shock wave velocities in a steel tube lined with lead azide [Pb(N₃)₂] and filled with helium.** The shock wave initiated the lead azide and the combustion front accelerated to a velocity of 14 km/s; *i.e.*, more than three times the Chapman-Jouguet (CJ) velocity of lead azide. • In the previous two years, MITROFANOV and collaborators had already shown that a self-sustaining detonation process is possible in principle in a heterogeneous system consisting of alternate layers of a condensed explosive and a gas (for example, in a tube made of explosive and filled with gas), with a velocity which is several times the detonation velocity of a homogeneous charge of the same explosive.²⁸⁶⁴ |

Similar studies were performed in Canada at McGill University, but for safety reasons lead azide was replaced by a safer second explosive (PETN powder). They revealed that this curious phenomenon – known as the **"channel effect"** in the detonics community – is not caused by the precursor shock initiating detonation in the explosive; instead, **the shock wave precompresses the explosive to a higher initial density.** This can result in an acceleration of the detonation velocity, which varies almost linearly with density.²⁸⁶⁵

| 1976 | *Dornier System GmbH, NTF Medizintechnik, Friedrichshafen, southern Germany* | **Bernd FORSSMANN,**²⁸⁶⁶, ²⁸⁶⁷, ²⁸⁶⁸ a German physicist, **and collaborators report at a meeting in Meersburg that they have cinematographically visualized the fragmentation of an urinary stone by a shock wave for the first time** {⇨Fig. 4.15−F}. • This project, sponsored by the R&D Dept. of the German Government and carried out in collaboration with the Urological Clinic and the Institute for Surgical Research of the Ludwig-Maximilian University (both at Munich), resulted in a prototype lithotripter called "HM1" in February 1980, which prompted a |
|---|---|---|

---

²⁸⁶⁰ K.S. KIM, R.J. CLIFTON, and P. KUMAR: *A combined normal- and transverse-displacement interferometer with an application to impact of y-cut quartz.* J. Appl. Phys. **48**, 4132-4139 (1977).

²⁸⁶¹ J.R. ASAY, L.P. MIX, and F.C. PERRY: *Ejection of material from shocked surfaces.* Appl. Phys. Lett. **29**, 284-287 (1976).

²⁸⁶² P. CHAPRON and P. ELIA: *Surface phenomena of shock-loaded metallic samples.* In: (K. TAKAYAMA, ed.) *Proc. 18th Int. Symposium on Shock Waves* [Sendai, Japan, July 1991]. Springer, Berlin *etc.* (1992), vol. 1, pp. 435-440.

²⁸⁶³ I.T. BAKIROV and V.V. MITROFANOV: *High velocity two-layer detonation in an "explosive-gas" system.* Sov. Phys. Dokl. **21**, 704-706 (1976).

²⁸⁶⁴ V.V. MITROFANOV: *Detonation in two-layer system.* In: *Cinquième Colloque International sur la Dynamique des Gaz en Explosion et des Systèmes Réactifs* [Bourges, France, Sept 1975] Acta Astronaut. **3**, 995-1004 (Nov./Dec. 1976).

²⁸⁶⁵ V. TANGUAY, J. MAMEN, and A.J. HIGGINS: *Propagation of detonation by precursor shock wave in explosive lined channels.* In: (A.K. HAYASHI and D. DUNN-RANKIN, eds.) *19th International Colloquium on the Dynamics of Explosions and Reactive Systems (ICDERS)* [Hakone, Japan, July/Aug. 2003]. CD-ROM edition, ISBN 4-9901744-1-0, Paper No. 148.

²⁸⁶⁶ B. FORSSMANN, W. HEPP, G. HOFF, C. CHAUSSY, F. EISENBERGER, and K. WANNER: *Entwicklung eines Verfahrens zur berührungsfreien Zerkleinerung von Nierensteinen durch Stoßwellen.* Wissenschaftliche Berichte, Meersburg (June 1976).

²⁸⁶⁷ B. FORSSMANN: *Klinische Erprobung der berührungsfreien Nierensteinzertrümmerung durch Stoßwellen.* Fachinformationszentrum Energie, Physik, Mathematik, Eggenstein-Leopoldshafen (1983).

²⁸⁶⁸ B. FORSSMANN, W. HEPP, C. CHAUSSY, F. EISENBERGER, and K. WANNER: *Eine Methode zur berührungsfreien Zertrümmerung von Nierensteinen durch Stoßwellen.* Biomed. Tech. **22**, Heft 7, 166-168 (1977).

revolution in the removal of calculi {CHAUSSY, BRENDEL & SCHMIEDT 1980; see CHAUSSY & EISENBERGER ⇨1970}.

| 1976 | Dept. of Astronomy, University of Michigan, Ann Arbor, Michigan | **Jack G. HILLS**[2869] **and Carol A. DAY publish a theory that the remnant of a collision between two main-sequence stars could produce a** *blue straggler* − a star in a cluster that is hotter and bluer (*i.e.*, much younger) than normal globular cluster stars having the same luminosity. One of several ways to make such a star is through the direct physical stellar collision and merger of two less massive main-sequence stars. ▪ The cores of some globular clusters are dense enough that collisions between single stars are expected to happen at dynamically significant rates. Later, other astrophysicists proposed that binary star systems can enhance the number of stellar collisions, because they have a much larger collisional cross section than single stars.<br><br>Their collisional hypothesis for the blue stragglers was brilliantly confirmed in the 1990s, when large numbers of these stars were detected in the cores of dense star clusters with NASA/ESA's Hubble Space Telescope. An analysis of observational data of a blue straggler in the tumultuous center of the nearby globular cluster 47 TUCANAE revealed that blue stragglers are<br>▸ more massive and younger than normal globular cluster stars; and<br>▸ created by a slow coalescence of a gravitationally bound pair rather than by a violent collision of two unrelated stars.[2870]<br><br>The cluster 47 TUCANAE contains about 1 million stars and is located 15,000 light-years away in *Tucana* ("The Toucan," a constellation of the Southern Hemisphere).<br><br>In the following years, astrophysicists and computer specialists worked out mathematical models for hydrodynamic simulations of stellar collisions. In 2000, stellar collisions and their role in the evolution of stellar clusters were the subject of a conference {ASP Conference ⇨2000}. |
|---|---|---|
| 1976 | Tetra Tech, Inc., Pasadena, California | **Bernard LE MÉHAUTÉ,**[2871] a French-born U.S. hydrodynamicist and Senior Vice President of Tetra Tech, **publishes the book** *An Introduction to Hydrodynamics and Water Waves***, which also addresses various phenomena caused by underwater explosions:**<br>▸ Underwater explosions can cause significant blast wave effects both below and above the surface, but **the displacement of a great deal of water by bubble motion is also responsible for the propagation of large-amplitude surface waves,** which can cause enormous damage to shorelines and harbors under certain conditions. ▪ In 1996, together with Shen WANG, LE MÉHAUTÉ[2872] published a book on this particular subject.<br>▸ He discusses linear and nonlinear theories as well as experimental calibrations for both deep and shallow water explosions.<br>▸ Many of the theories and concepts presented are applicable to other types of water waves; in particular to tsunamis and waves generated by the fall of a meteorite.<br><br>In 1979, LE MÉHAUTÉ received the International Coastal Engineering Award from the American Society of Civil Engineering (ASCE), Washington, DC, for his contributions to the field of coastal engineering. |

---

[2869] J.G. HILLS and C.A. DAY: *Stellar collisions in global clusters.* Astrophys. Lett. **17**, L87-L93 (1976).

[2870] M.M. SHARA, R.A. SAFFER, and M. LIVIO: *The first direct measurement of the mass of a blue straggler in the core of a globular cluster: BSS 19 in 47 TUCANAE.* Astrophys. J. Lett. **489**, L59 (1997).

[2871] B. LE MÉHAUTÉ: *An introduction to hydrodynamics and water waves.* Springer, New York (1976).

[2872] B. LE MÉHAUTÉ and S. WANG: *Water waves, generated by underwater explosion.* Advanced Series on Ocean Engineering, vol. 10. World Scientific, Singapore etc. (1996).

| | | |
|---|---|---|
| 1976 | Sandia Laboratories, Albuquerque, New Mexico | **Darrell E. MUNSON,**[2873] a U.S. metallurgist and shock physicist, **and R.P. MAY measure the hypervelocity projectile velocity in a two-stage light gas gun as a continuous function of time** using the long focal length velocity interferometer VISAR {BARKER & HOLLENBACH ⇨1965}.<br><br>(i) The measurements show that the initial acceleration of the projectile is discontinuous due to the shock of the pressure applied to the projectile upon the rupture of the burst diaphragm.<br><br>(ii) Typically, all VISAR records show the same early shock acceleration and constant velocity steps. This behavior dominates the early response of the projectile and is a direct consequence of the physical arrangement of the gun.<br><br>(iii) The arrangement of the two-stage light gas gun is equivalent to a shock tube with a burst diaphragm separating a low-pressure and a high-pressure gas. They can explain the shock accelerations quantitatively using simple shock tube theory. |
| 1976 | NASA's Goddard Space Flight Center (GSFC), Greenbelt, Maryland | **John Aloysius O'KEEFE,**[2874] a planetary scientist at the Theoretical Division, **publishes the book *Tektites and Their Origin*,** an important reference work on this subject. • O'KEEFE who made field trips to Australia, Africa and various American localities proposed a theory that tektites – natural glass objects found in discrete strewn fields around the world {C.R. DARWIN ⇨1844} – are actually volcanic ejecta from the Moon {O'KEEFE ⇨1967; CHAPMAN ⇨1971}.<br><br>Since the late 1960s, scientists who have studied tektites have pointed to evidence that overwhelmingly suggests they were created as melt droplets from the impact of comets or small asteroids on Earth. However, O'KEEFE's most specific argument that such nearly pure glasses could not be made by any sudden process (*e.g.*, by a terrestrial impact) is still unanswered. O'KEEFE's work on tektites over the decades which includes many journal articles and two books stimulated research in several other fields, such as on the origin of the Moon, on the formation of the continental crust, and on the impact origin of Ontario's 65-km-long Sudbury Structure.[2875] |
| 1976 | University of Illinois at Urbana-Champaign (UIUC), Illinois & Dept. of Energetic Systems, Southwest Research Institute (SwRI), San Antonio, TX | **Roger A. STREHLOW**[2876] (College of Engineering, UIUC) **and Wilfred E. BAKER** (SwRI), two U.S. engineers, **give a general definition of an explosion in air:** "In general, an explosion is said to have occurred in the atmosphere if energy is released over a significantly small time and in a sufficiently small volume so as to generate a pressure wave of finite amplitude traveling away from the source. This energy may have originally been stored in the system in a variety of forms; these include nuclear, chemical, electrical or pressure energy, for example. **However, the release is not considered to be explosive unless it is rapid enough and concentrated enough to produce a pressure wave that one can hear.** Even though many explosions damage their surroundings, it is not necessary that external damage be produced by the explosion. All that is necessary is that the explosion is capable of being heard." |
| 1977 | Boulder, Colorado | In July, at the 6th AIRAPT Conference, **the International Association for the Advancement of High-Pressure Science and Technology** {AIRAPT ⇨1965} **establishes the *P.W. Bridgman Award*.** • Sir Percy Williams BRIDGMAN was a physics professor at Harvard University who invented the "Bridgman anvil" and investigated an enormous variety of materials at pressures of up to about 3 GPa (30 kbar) and mostly at room temperature. He was one of the founders of modern high-pressure research.<br><br>Harry DRICKAMER, a professor of chemical engineering at the University of Illinois, was the first recipient (1977). In the years following, further recipients of the Bridgman Award were |

---

[2873] D.E. MUNSON and R.P. MAY: *Interior ballistics of a two-stage light gas gun using velocity interferometry.* AIAA J. **14**, 235-242 (1976).
[2874] J.A. O'KEEFE: *Tektites and their origin.* Elsevier Scientific, Amsterdam & Oxford (1976).
[2875] P.D. LOWMAN and D.P. RUBINCAM: *John A. O'KEEFE (1916–2000).* EOS, Trans. Am. Geophys. Union **82**, No. 5, 55 (2001).
[2876] R.A. STREHLOW and W.E. BAKER: *The characterization and evaluation of accidental explosions.* Progr. Energy Combust. Sci. **2**, 27-60 (1976).

- **Boris VODAR,** C.N.R.S., Bellevue, France (1979);
- **Ernst U. FRANCK,** Institut für Physikalische Chemie, TH Karlsruhe, Germany (1981);
- **Francis BIRCH,** Dept. of Geology, Harvard University, Cambridge, MA (1983);
- **Nestor J. TRAPPENIERS,** Van der Waals Laboratory, University of Amsterdam, The Netherlands (1985);
- **Francis P. BUNDY,** Research Laboratory, GEC, Schenectady, NY (1987);
- **Ho-kwang D. MAO,** Geophysical Laboratory, Carnegie Institution of Washington (CIW), Washington, DC (1989);
- **Shigeru MINOMURA,** Dept. of Physics, Faculty of Science, Hokkaido University, Japan (1991);
- **Arthur L. RUOFF,** Dept. of Materials Science and Engineering, Cornell University, Ithaca, NY (1993);
- **Bogdan BARANOWSKI,** Institute of Physical Chemistry of the Polish Academy of Sciences, Warsaw, Poland (1995);
- **William A. BASSETT,** Dept. of Geological Sciences, Cornell University, Ithaca, NY (1997);
- **Vladimir E. FORTOV,** Moscow Institute of Physics and Technology, Russia (1999);
- **William J. NELLIS,** LLNL, Livermore, CA (2001);
- **Neil W. ASHCROFT,** Dept. of Physics, Cornell University, Ithaca, NY (2003); and
- **Sergei M. STISHOV,** Institute of High Pressure Physics, Moscow, Russia (2005); and
- **Takehuko YAGI,** University of Tokyo, Japan (2007).

1977  Washington, DC

On July 7, **the U.S. Government announces that it has tested a neutron bomb.** This new type of nuclear weapon – described as an ***enhanced radiation warhead (ERW)*** – is a very small hydrogen bomb that yields one tenth the usual blast, heat, and fallout. It is intended for use on enemy troops and is designed to confine damage from the blast and heat to a radius < 300 m. However, it produces massive amounts of neutron and gamma radiation, which kills people slowly and painfully within days.[2877] ▪ The strategic benefits of this new type of nuclear weapon were discussed in great depth in western Europe and considered to be very doubtful.

In the same year, George B. KISTIAKOWSKY,[2878] a Professor Emeritus of physical chemistry at Harvard University, who was involved in the development of the first atomic bomb in World War II {Manhattan Project ⇨1942}, comments: "If main reliance is placed on battlefield nuclear weapons, the concern about the dire consequences of even a short delay will certainly speed up the authorization of their use, hence lowering the threshold to nuclear warfare. **What will follow will be massive civilian casualties and other collateral damage in the environs of the battlefields and even these will be but a prelude to the nuclear devastation of Europe and then a general nuclear war.** These being the alternatives, it is impossible to escape the conclusion that NATO military efforts should emphasize various sophisticated 'precision guided' munitions and that *ERW*'s, as well as older battlefield nuclear weapons, be put into mothballs…" ▪ In the following year in April, the production of neutron warheads was canceled, but it was resumed on a small scale in 1981 under the orders of U.S. President Ronald REAGAN – such as the W70-3, a 1-kiloton-yield nuclear warhead for use on the surface-to-surface missile MGM-52 ("Lance") and the W79-0, an 8 in.-caliber nuclear artillery shell with enhanced radiation and three yields, ranging from 0.1 to 1.1 kiloton.

In 1999, Samuel COHEN,[2879] father of the neutron bomb, wrote, "… Having myself invented the neutron bomb in 1958 and closely following the subject ever since, I can say that the

---

[2877] *The Encyclopedia Americana.* Grolier, Danbury, vol. 20 (1997), p. 139.
[2878] G. KISTIAKOWSKY: *A chemist speaks out on the neutron bomb.* Chem. & Engng. News **56**, 3 (April 24, 1978).
[2879] S. COHEN: *The Cox Report – the surprising truth.* Insight Magazine – online (July 16, 1999); http://www.freerepublic.com/forum/a3790f7cc28ca.htm.

claim most likely is untrue that no nation has deployed this weapon. **The United States, the former Soviet Union, China, France (and very likely Israel) have developed such weapons for sound military and political reasons.** The United States, during the Reagan administration, built up a very substantial stockpile of neutron bombs with the stated intention of deploying them to NATO Europe; but in the face of European political (not military) opposition, Washington reversed its decision and stockpiled the warheads back home. After the Persian Gulf War, U.S. President George H.W. BUSH ordered the destruction of the neutron-bomb stockpile and Congress passed legislation forbidding the development of advanced versions of these warheads."

| | | |
|---|---|---|
| 1977 | *Tokyo, Japan* | In October, **first International Symposium on Flow Visualization (ISFV).** The general chairman is Tsuyoshi ASANUMA. ▪ This international meeting grew out of strong interest in the subject from the Japanese scientific and engineering communities after five years of domestic symposia on flow visualization, which also attracted a large number of researchers from other countries. The following meetings were held at Bochum, Germany (1980); Ann Arbor, MI (1983); Paris, France (1987); Prague, Czechoslovakia (1989); Yokohama, Japan (1992); Seattle, WA (1995); Sorrento, Italy (1998); Edinburgh, Scotland (2000); Kyoto, Japan (2002); Notre Dame, IN (2004); Göttingen, Germany (2006); and Nice, France (2008).<br><br>Flow visualization deals mostly with the development and application of optical methods for making fluid flows and waves visible for subsequent analysis, but other important techniques also encompass<br>▸ direct injection methods;<br>▸ tuft and wall tracing methods;<br>▸ chemical reaction and electrical control methods; and<br>▸ radiographic methods.<br>High-speed photography is used in most cases to record the visualized phenomena. Both the International Symposia on Flow Visualization and the International Congresses on High-Speed Photography {⇨1952} have become prime sources of valuable information for shock and detonation physicists. |
| 1977 | *NASA, United States & ESA, Europe* | On October 22, the Explorer-class spacecraft ISEE 1 and ISEE 2 are launched together at Cape Canaveral, FL. These craft carry a number of complementary instruments for performing measurements of plasmas, energetic particles, waves and magnetic fields. **Start of the International Sun-Earth Explorer (ISEE) Program,**[2880] an international cooperation between NASA and ESA in order to study the interaction of the solar wind with the Earth's magnetosphere.<br><br>(i) The work focused on<br>▸ investigating solar-terrestrial relationships at the outermost boundaries of the Earth's magnetosphere;<br>▸ **examining in detail the structure of the solar wind near the Earth and the shock wave that forms the interface between the solar wind and the Earth's magnetosphere** {⇨Fig. 4.12–D};<br>▸ investigating motions of and mechanisms operating in the plasma sheets; and<br>▸ continuing the investigation of cosmic rays and solar flare emissions in the interplanetary region near 1 AU.[2881]<br><br>(ii) An analysis of the ISEE data showed that the bow shock model {AXFORD & KELLOGG ⇨1962} is indeed correct and that the Earth's bow shock |

---

[2880] *International Sun-Earth Explorers (ISEE) – project information.* National Space Science Data Center (NSSDC), NASA-GSFC, Greenbelt, MD (Jan. 2003); http://nssdc.gsfc.nasa.gov/space/isee.html.

[2881] One astronomical unit (AU) is the distance between the Sun and Earth, equivalent to around 150 million kilometers.

| | | |
|---|---|---|
| | | ▸ approaches the Earth to about 92,000 km (14.4 Earth radii);
▸ varies its position depending on the dynamic pressure of the solar wind;
▸ has a thickness on the order of the ion gyroradius (*i.e.*, 100–1,000 km); and
▸ is followed by a sequence of complex plasma sheet structures.

In the following year, on August 12, 1978, the solar wind spacecraft ISEE 3 (renamed the International Cometary Explorer or "ICE" in 1982) was launched. After completing its original mission, the spacecraft was gravitationally maneuvered to intercept comet P/GIACOBINI–ZINNER.[2882] **On September 11, 1985, ICE flew through the tail of this comet, performing *in situ* cometary measurements and detecting the ions while crossing its "cometary bow shock."**[2883] Water and carbon monoxide ions were also identified, thus confirming the "dirty snowball" theory developed in 1949 by the U.S. astronomer Fred L. WHIPPLE. |
| 1977 | *Westwego, south-eastern Louisiana & Galveston, eastern Texas* | **On December 22** (Westwego) **and on December 27** (Galveston) **occur serious grain elevator explosions,** resulting in a large loss of life (54 casualties). ▪ Investigations revealed that the accidents were apparently caused by electric sparks, and that most of the damage occurred due to inadequate explosion-relief venting and secondary explosions from layers of dust throughout the plant.

**Further grain explosions in the American grain industry raised concern in the Federal Government about how to stop such disasters.** These accidents also revive international interest in the problems of dust explosions. ▪ In the following year, the U.S. Dept. of Agriculture (DOA) engaged the National Academy of Sciences (NAS) to conduct an international symposium on grain elevator explosions.[2884]

Following this symposium, the Occupational Safety and Health Administration (OSHA) of the U.S. Dept. of Labor (DOL) requested that the NAS forms a **Panel on Causes and Prevention of Grain Elevator Explosions.** Its membership comprised experts in systems analysis, explosion dynamics, investigations and prevention, instrumentation, grain handling and processing, agricultural insurance practices, employee relations, dust control methods, and aerodynamics. The Panel's work, completed in 1983, resulted in increased awareness in the grain handling industry of the hazards from grain dust, and it also provided specific methods for reducing these hazards.[2885] ▪ There are records of grain elevator explosions dating back over 120 years. Indeed, grain dust explosions {MOROZZO ⇨1785; ⇨Figs. 4.21–D, E} are still a major hazard, particularly in the modern grain industry with its huge elevator and silo facilities. Case histories of dust explosions, published throughout the world by governmental health and safety agencies, insurance companies and major chemical companies, represent the most useful sources of information for investigations into their complex causes.[2886] |
| 1977 | *U.S. Naval Weapons Center (NWC), China Lake, California* | **Dewey Philip ANKENEY,**[2887] a U.S. mechanical engineer, **investigates the wall response of an aircraft fuel cell during penetration by a high-speed projectile.** This failure is often accentuated, in brittle aluminum cell walls, by the bullet or high-speed projectile wound in the fuel cell walls, and a catastrophic brittle fracture failure can occur at low stress levels. This complex impact phenomenon – termed a ***hydraulic ram*** {CARRÉ ⇨1705} – is of particular importance |

---

[2882] Discovered in December 1900 by the French astronomer Michel GIACOBINI (1873–1938), rediscovered in October 1913 by the German astronomer Ernst ZINNER (1886–1970).

[2883] *ISEE-3/ICE – mission overview and instrumentation.* NASA-GSFC, Greenbelt, MD (June 26, 2003); http://heasarc.gsfc.nasa.gov/docs/heasarc/missions/isee3.html.

[2884] *Proc. on Grain Elevator Explosions, Int. Symposium* [Washington, DC, July 1978]. 2 vols., National Materials Advisory Board, U.S. Dept. of Agriculture, Washington, DC (1978).

[2885] Grain Elevator Explosion Investigation Team (V.L. GROSE, ed.): *Report on explosion of DeBruce grain elevator – Wichita, Kansas 8 June 1998.* OSHA Special Rept., U.S. Dept. of Labor, Kansas City, MO.

[2886] J. CROSS and D. FARRER: *Dust explosions.* Plenum Press, New York etc. (1982).

[2887] D.P. ANKENEY: *Physical vulnerability of aircraft due to fluid dynamic effects.* AGARD Advisory Rept. No. 106, NATO (1977).

to the survivability of military aircraft.[2888] ▪ ANKENEY's analysis built upon a hydraulic ram theory previously developed by his colleagues at NWC.[2889]

The potential hazard from meteoroids impacting into the liquid propellant tanks of a spacecraft is of particular concern because an impact of sufficient energy could not only puncture the tank, but also result in the catastrophic bursting or tearing of the tank wall.[2890]

| 1977 | Washington State University (WSU), Pullman, Washington & Sandia Laboratories (SL), Albuquerque, New Mexico | George E. DUVALL[2891] (WSU) **and Robert A.** GRAHAM (SL), two renowned shock physicists, **review first-order polymorphic, second-order melting, and freezing transitions induced by shock-wave loading.**<br>(i) They describe experimental techniques and review theories for the mechanics, thermodynamics, kinetics and shear strengths of shock-loaded materials.<br>(ii) In a detailed, 8-page table they summarize phase transition observations (such as stress and compression at transition) and applied experimental techniques for iron and iron alloys, elements, numerous compounds, and ferroelectric ceramics. |
|---|---|---|
| 1977 | Dept. of Mechanical Engineering, UC Berkeley, California | Maurice HOLT,[2892] a British-born U.S. professor of aeronautical sciences, **reviews underwater explosion research performed up to 1977.** In particular, he addresses<br>▸ the formation of spherical explosions and methods of calculating spherical explosion disturbances;<br>▸ bubble motion and the effect of gravity, especially concerning the migration of gas bubbles created by detonation of chemical explosives; and<br>▸ the influence of the ocean surface on underwater and near-surface explosions. |
| 1977 | Ernst-Mach-Institut (EMI), Freiburg, Germany | **Flash radiographic studies on the oblique collision of two shock waves in organic liquids such as carbon tetrachloride** [$CCl_4$], carried out by Peter KREHL,[2893] Ulrich HORNEMANN and Werner Heilig, **reveal two unusual phenomena:**<br>▸ **the liquid is thrown out by a two-jet process:** a primary fast jet is followed by a secondary slow jet {⇨Fig. 4.8–H}; and<br>▸ **the occurrence of** *regressive Mach reflection* – in gas dynamics known as **"inverse Mach reflection"**[2894] – *i.e.*, the Mach disk is well-developed at first, but steadily decreases in width over time until Mach reflection turns into regular reflection. Inverse Mach reflection is a fundamental type of shock reflection in truly unsteady flows. ▪ In most investigations of Mach reflection, experimental arrangements have been used in which the "Mach stem" (or the "Mach disk" in a symmetrical arrangement) grows with increasing time. The oldest examples of such ***progressive Mach reflections*** are Ernst MACH's recorded soot funnels {E. MACH & WOSYKA ⇨1875}. |

---

[2888] R. YURKOVICH: *Hydraulic ram: a fuel tank vulnerability study.* Rept. No. G964, McDonnell Douglas Co., St. Louis, MO (Sept. 1969).

[2889] E.A. LUNDSTROM and W.K. FUNG: *Fluid dynamic analysis of hydraulic ram IV.* [User's manual for pressure wave generation model]. Rept. JTCG/AS-74-T-018, Naval Weapons Center (NWC), China Lake, CA (Oct. 1976).

[2890] F.S. STEPKA, C.R. MORSE, and R.P. DENGLER: *Investigation of characteristics of pressure waves generated in water-filled tanks impacted by high-velocity projectiles.* NASA TN D-3143 (Dec. 1965).

[2891] G.E. DUVALL and R.A. GRAHAM: *Phase transitions under shock wave loading.* Rev. Mod. Phys. **49**, 523-579 (1977).

[2892] M. HOLT: *Underwater explosions.* Annu. Rev. Fluid Mech. **9**, 187-214 (1977).

[2893] P. KREHL, U. HORNEMANN, and W. HEILIG: *Flash radiography of unsteady regular and Mach reflection in a liquid.* Proc. 11th Symposium on Shock Tubes & Waves [Seattle, WA, July 1977]. In: (B. AHLBORN, A. HERTZBERG, and D. RUSSELL, eds.) *Shock tube and shock wave research.* University of Washington Press, Seattle & London (1978), pp. 303-312.

[2894] K. TAKAYAMA and G. BEN-DOR: *The inverse Mach reflection.* AAA J. **23**, 1853-1859 (Dec. 1985).

| | | |
|---|---|---|
| 1977 | *Los Alamos Scientific Laboratory (LASL), Los Alamos, New Mexico* | **Charles E. RAGAN III,**[2895] a U.S. nuclear physicist, **and collaborators report on the use of neutrons from an underground nuclear explosion to generate ultrahigh shock pressures of up to 2 TPa (20 Mbar).** They compressed a thin sheet of neutron-moderating material (Lucite) by rapidly fission-heating an adjacent slab of enriched uranium, which drives a shock wave into the target (a slab of molybdenum in their experiment). Particle velocities in this ultrahigh pressure region were measured using the Doppler shift technique. ▪ This new experimental technique, which uses underground nuclear explosions carried out in rocks, is the only way to legally elude the Moscow Nuclear Test Ban Treaty {⇨1963}. It permits an extension of previous dynamic methods {GORANSON ET AL. ⇨1955; WALSH ET AL. ⇨1955; BANCROFT ET AL. ⇨1956; WALSH ET AL. ⇨1957; MCQUEEN & MARSH ⇨1960} of determining data on the Hugoniot states of materials at extremely high shock pressures, which are of particular interest to those involved in the further development of nuclear weapons.<br><br>Similar experiments[2896] had already been conducted in the Soviet Union in the 1960s, but details of and results from extended studies performed using this method were not published until the early 1990s {TRUNIN ET AL. ⇨1992}. |
| 1977 | *Forschungsinstitut der Feuerfest-Industrie (FFI), Bonn & Ernst-Mach-Institut (EMI), Weil am Rhein, Germany* | **Hartmut SCHNEIDER,**[2897] a mineralogist at FFI, **and Ulrich HORNEMANN,** a shock physicist and ballistician at EMI (Weil Division), **report that single crystals of andalusite [$Al_2SiO_5$], when exposed to shock pressures in excess of 575 kbar, undergo chemical decomposition** into incoherently crystallized γ-$Al_2O_3$, well-crystallized α-$Al_2O_3$ and X-ray amorphous $SiO_2$. The X-ray amorphous $SiO_2$ is believed to be a pressure release product of stishovite {STISHOV & POPOVA ⇨1961}, the assumption being that stishovite [$SiO_2$] is produced along with γ-$Al_2O_3$ when andalusite disintegrates upon shock compression. |
| 1977 | *Lawrence Radiation Laboratory (LRL), UC Livermore, California* | **Mathias VAN THIEL,**[2898] an Indonesian-born U.S. physicist, **and his collaborators present a *Hugoniot Data Bank* containing data on hundreds of shock-compressed materials** obtained by various laboratories, previously scattered throughout a number of international publications. ▪ Since these data were obtained using different compression techniques, they require careful and critical review by the user.<br><br>In subsequent years, Los Alamos solid-state shock physicists published **a compendium of equation-of-state (EOS) data** {BENNETT ET AL. ⇨1978; MARSH ⇨1980} and Soviet colleagues **a collection of free-surface velocity profiles for commercial metals and alloys.**[2899] Later, the Soviet data were supplemented by experimental results obtained by other Soviet researchers for particle-velocity profiles from the shock-wave loading of condensed matter {KANEL' ET AL. ⇨2004}. A Russian *Shock Wave Database*, a collection of numerous shock-wave experimental points, can be found in the Internet.[2900] It includes measurements of thermodynamic properties of matter in shock and isentropic release waves. Collected are about 15,000 points from 300 references for 500 substances as ASCII; 80% of them are presented in this database. |

---

[2895] C.E. RAGAN III, M.G. SILBERT, and B.C. DIVEN: *Shock compression of molybdenum to 2.0 TPa by means of a nuclear explosion.* J. Appl. Phys. **48**, 2860-2870 (1977).

[2896] L.V. AL'TSHULER, B.N. MOISEEV, L.V. POPOV, G.V. SIMAKOV, and R.F. TRUNIN: *Relative compressibility studies of iron and lead at pressures of 31 to 34 Mbar.* Sov. Phys. JETP **27**, 420-422 (1968).

[2897] H. SCHNEIDER and U. HORNEMANN: The *disproportionation of andalusite ($Al_2SiO_5$) to $Al_2O_3$ and $SiO_2$ under shock compression.* Phys. Chem. Minerals **1**, 257-264 (1977).

[2898] M. VAN THIEL, A.S. KUSUBOV, and A.C. MITCHELL: *Compendium of shock wave data.* Rept. UCRL-50108, Lawrence Radiation Laboratory (LRL), Livermore, CA (1977).

[2899] G.I. KANEL' and S.V. RAZORENOV: *Shock-wave loading of metals: the motion of the surface of sample* [in Russ.]. Preprint of the Institute of Problems of Chemical Physics, Chernogolovka, U.S.S.R. (1989).

[2900] *Shock wave data base*; http://www.ficp.ac.ru/rusbank/index.php?text=2. See also P.R. LEVASHOV ET AL.: *Database on shock-wave experiments and equations of state available via Internet.* In: (M.D. FURNISH, ed.) *Shock compression in condensed matter – 2003.* AIP, Melville, NY. Conf. Proc. **706** [ISBN 0-7354-0181-0], 87-90 (2004).

| | | |
|---|---|---|
| 1977 | Harvard University & Smithsonian Astrophysical Observatory, Cambridge, Massachusetts | Steven WEINBERG,[2901] a U.S. professor of nuclear physics who shared the 1989 Nobel Prize in Physics, **publishes the book *The First Three Minutes: A Modern View of the Origin of the Universe*.**<br><br>(i) He divides the early Universe into six time frames, some lasting less than a second while others extend for more than a minute. In some phases energy dominates, while in others matter has the upper hand.<br><br>(ii) Based upon quantum and particle physics, he proposes a theory that "the fundamentals of the Universe were created in the first three minutes" of the Big Bang, suggesting that one-hundredth of a second after its birth, the Universe was characterized by a super-high temperature (an absolute temperature of 100 billion degrees Kelvin) and a super-high density. But after three minutes and 46 seconds, the temperature had fallen to 900 million degrees, and the fusion of nuclei of helium and hydrogen atoms had become stable.<br><br>(iii) The Universe continued to cool down and eventually the gas that formed the galaxies came into being. ▪ His book became a bestseller and has since been translated into 22 foreign languages. His "first three minutes" scenario has become the standard Big Bang model. |
| 1977/ 1978 | Institute of Cosmophysical Research & Aeronomics, Yakutsk, U.S.S.R. Academy of Sciences, Siberian Div., Soviet Union & Western World | Germogen F. KRYMSKY,[2902] a Soviet theoretical physicist **and independently theoretical physicists in Germany,[2903] the United States[2904] and England[2905] recognize that astrophysical shocks can accelerate particles by interacting with large-scale electric and magnetic fields.** One important condition is that the seed particles are sufficiently energetic to pass from the downstream to the upstream plasma against the oblique orientation of the magnetic field. ▪ Their theory of diffusive shock acceleration in shock waves associated with supernova remnants became the basic theory for the origin of cosmic rays {HESS ⇒ 1912}. It allows quantitative model calculations and appears capable of meeting many of the observational constraints on any cosmic ray acceleration theory.[2906] Two mechanisms are proposed for particle acceleration in shocks:<br><br>▸ *shock drift acceleration*, based on the energization of particles trapped in the vicinity of the shock front; and<br><br>▸ *diffusive acceleration*, based on the multiple scattering of particles moving long distances from the shock front.[2907] |
| 1978 | Karpacz, southwestern Poland | **First International Colloquium on Explosivity of Industrial Dusts,** organized by the Polish Ministry of Food Industry and Procurement. ▪ The following meetings, each renamed an **International Colloquium on Dust Explosions,** were initially held in Poland: at Baranów (1984); Jadwisin (1986); Szczyrk (1988); Porabka-Kozubnik (1990); and Pultusk (1993); and then abroad: at Shenyang, China (1994); Bergen, Norway (1996); Schaumburg, IL (1998); Tsukuba-Ibaraki, Japan (2000); Bourges, France (2002); Crakow, Poland (2004); and Halifax, Nova Scotia, Canada (2006). |

---

[2901] S. WEINBERG: *The first three minutes: a modern view of the origin of the Universe.* Basic Books, New York (1977, 1981, 1988, 1993).

[2902] G.F. KRYMSKY: *A regular mechanism for the acceleration of charged particles on the front of a shock wave* [in Russ.]. Dokl. AN (SSSR) **234**, 1306-1308 (1977); Engl. translation in Sov. Phys. Dokl. **23**, 327-328 (1977).

[2903] W.I. AXFORD, E. LEER, and G. SKADRON: *The acceleration of cosmic rays by shock waves.* In: *Proc. 15th Int. Cosmic-Ray Conference (ICRC)* [Plovdiv, Bulgaria, Aug. 1977]. Conf. Papers published by the Central Research Institute for Physics, Budapest (1977), vol. 11, pp. 132-137 (1977).

[2904] R.D. BLANDFORD and J.P. OSTRIKER: *Particle acceleration by astrophysical shocks.* Astrophys. J. **221**, L29-L32 (April 1, 1978).

[2905] A.R. BELL: *The acceleration of cosmic rays in shock fronts.* Month. Not. Roy. Astron. Soc. **182**, 147-156 (1978).

[2906] E.G. BEREZHKO: *Particle acceleration in supernova remnants.* In: *Proc. 27th Int. Cosmic Ray Conference (ICRC)* [Hamburg, Germany, Aug. 2001]. Conf. Papers published by the Copernicus Gesellschaft, Katlenburg-Lindau, Germany (2001), pp. 226-233; http://www.copernicus.org/C4/DATA/BEREZHKO.PDF.

[2907] M. GEDALIN: *Shock waves in space.* In: (G. BEN-DOR ET AL., eds.) *Handbook of shock waves.* Academic Press, New York (2001), vol. 1, pp. 455-483.

In the same year, an **International Symposium on Grain Elevator Explosions** is held at Washington, DC. • The meeting, organized by the National Academy of Sciences (NAS), was the result of several severe accidents in Louisiana and Texas {U.S. Grain Industry ⇨1977}.

| | | |
|---|---|---|
| 1978 | *Lawrence Berkeley Laboratory & Dept. of Geology and Geophysics, UC Berkeley, California* | Luis Walter ALVAREZ[2908] (a participant in the Manhattan Project {⇨1942} and winner of the 1968 Nobel Prize for Physics) accidentally discovers, along with his son Walter (a geologist), a band of sedimentary rock in Cretaceous and Tertiary limestone that contains unusually high levels of iridium, a rare element on Earth but one often found in asteroid matter, on a geology expedition to Gubbio, central Italy.<br><br>(i) **Luis W. and Walter ALVAREZ develop a theory that an asteroid with a diameter of $10 \pm 4$ km struck the Earth** {Chicxulub Crater ⇨*c.*65 Ma ago}, and that matter thrown into the stratosphere by the impact was distributed around the world. This airborne matter shielded the world from sunlight until it eventually fell back down to Earth after a few years, **resulting in dramatic biological effects.**<br><br>(ii) They write, "At this time, the marine reptiles, the flying reptiles, and both orders of dinosaurs died out, and extinctions occurred at various taxonomic levels among the marine invertebrates." • This "Alvarez impact hypothesis" is still the subject of much debate.<br><br>The attentions of many planetary scientists have recently turned to the problem of assessing the likelihood that our civilization may be threatened by a rogue comet or asteroid in the near future {Asteroid 2002 MN ⇨2002}. |
| 1978 | *Los Alamos Scientific Laboratory (LASL), Los Alamos, New Mexico* | **Bard I. BENNETT,**[2909] a staff physicist, **and collaborators present a compilation of equation-of-state (EOS) data and other material properties in a tabular form for practical application,** *e.g.*, to predict complicated hydrodynamic flow problems in the analysis of weapons effects, reactor safety, and laser fusion. Their so-called "Sesame Library" contains EOS data obtained by using various theoretical models to determine the function $p = p(\rho, T)$ and to predict real shock-induced melting conditions. • Today EOS research is necessary not only to develop a predictive computational capability, but also to assess the nuclear weapons stockpile {UN CTBT ⇨1996}. |
| 1978 | *Dept. of Engineering, University of Cambridge, England* | **William JOHNSON,**[2910] a British professor of Mechanical Engineering, **and Athanasios G. MAMALIS,** a Greek engineer and former technical manager at a steel and pipe plant in Thessaloniki, **publish their book** ***Crashworthiness of Vehicles.*** In this elementary monograph, they review vehicular impact under the headings of the four main classes of vehicle (motor cars, ships, aircraft, railway coaches), and lifts or elevators.<br><br>By definition, ***vehicle crashworthiness*** is the ability of a structure to plastically deform and yet maintain a sufficient survival space for its occupants during crashes.<br>▸ The internationally familiar term *crashworthiness* was apparently coined in association with aircraft safety. In the 1940s, the International Civil Aviation Organization (ICAO) Requirements and the British Civil Airworthiness Requirements already contained certain "crashworthiness" provisions, and RAF Wing Commander CARROLL suggested a *Certificate of Air and Crashworthiness.*[2911] |

---

[2908] L.W. ALVAREZ, W. ALVAREZ, F. ASARO, and H.V. MICHEL: *Extraterrestrial cause for the cretaceous-tertiary extinction.* Science **208**, 1095-1108 (June 6, 1980); W. ALVAREZ, L.W. ALVAREZ, F. ASARO, and H.V. MICHEL *The end of the Cretaceous: sharp boundary of gradual transition?* Ibid. **223**, 1183-1186 (March 16, 1984).

[2909] B.I. BENNETT (ed.), J.D. JOHNSON, G.I. KERLEY, and G.T. ROOD: *Recent developments in the Sesame equation-of-state library.* Rept. LA-713, LASL, Los Alamos, NM (1978); *Equations of state – theoretical formalism.* Los Alamos Science No. 26, 192 (2000).

[2910] W. JOHNSON and A.G. MAMALIS: *Crashworthiness of vehicles – an introduction to aspects of collision of motor cars, ships, aircraft, and railway coaches.* Mech. Engng. Publ. Ltd., London (1978).

[2911] *See* editor's note, J. Roy. Aeronaut. Soc. **52**, 572 (1948). • The *Oxford English dictionary* (Clarendon Press, Oxford) gives this reference as the oldest source of the term "crashworthiness," *see* vol. III (1989), p. 1121.

> In the 1950s, improvements in crashworthiness involved not only aircraft design and firefighting/emergency equipment at airports, but also automobile design in order to help passengers avoid injury or death in the event of a collision {BARÉNYI ⇨1951}. According to another source, the term *crashworthiness* was coined in the 1940s by the physician Dr. John Charles LANE (1918–1999), the father of aviation medicine in Australia and a road safety activist. LANE also demonstrated the efficiency of seatbelts before they became compulsory.[2912]

| | | |
|---|---|---|
| 1978 | Institute of Thermophysics, Novosibirsk, Soviet Union | **A group of Soviet scientists, headed by Samson Semenovich KUTATELADZE,**[2913] a Siberian thermophysicist, **gives the first experimental evidence for the existence of** *rarefaction shocks*. Using pressure gauges placed at increasing distances from the diaphragm of a shock tube, they demonstrate the evolution of a rarefaction shock wave in the gas Freon-13 [$CClF_3$] near the critical point. ▪ Rarefaction shock waves, which had already been predicted by ZEL'DOVICH {⇨1946}, can occur in gases with a sufficiently large specific heat near the critical point of liquid-vapor transition. |
| 1978 | Chair of Applied Mathematics, Cambridge University, England | **M. James LIGHTHILL**,[2914] a notable British professor of applied mathematics and a renowned fluid dynamicist, **publishes his textbook** *Waves in Fluids,* a thorough analysis of all kinds of waves in fluids (that is, in liquids and gases), such as sound waves, **shock waves and a treatment of the related subject of hydraulic jumps,** water waves, and internal gravity waves (*i.e.*, waves inside atmospheres and oceans resulting from density stratification). <br><br> In the following year, Donald Cecil PACK,[2915] a British mathematics professor at the University of Strathclyde who reviewed LIGHTHILL's book, wrote: "The book will become a standard work for reading and consultation by all interested in fluid flow. It will be invaluable to research students … Even among the most experienced readers few, if any, will fail to find in it new insights into some familiar formulae or results among the abundance of explanations, comments and comparisons that illuminate the chapters." The book, which went on to become a classic, had an enormous impact on all those working in fluid flows, and on waves in particular. |
| 1978 | Dept. of Engineering Physics, Research School of Physical Sciences, Australian National University (ANU), Canberra, Australia | **Scott C. RASHLEIGH**[2916] **and Richard A. MARSHALL report on the first operational inductively driven** *railgun* {⇨Figs. 4.11–D & 4.14–G}. The device allows a solid dielectric projectile (a 2.5-g, 12.7-mm cube made of Lexan®, a polycarbonate) to be accelerated up to a hypervelocity of 5.9 km/s. They use a high-current plasma arc supplied by the huge homopolar generator at ANU, which is capable of creating a pulsed current up to 1.6 MA, as the driving armature. ▪ This achievement was a significant milestone in the exciting history of electromagnetic gun inventions {FAUCHON-VILLEPLÉE ⇨1916/1917}, which reached its initial peak during World War II with the attempt to build an effective military weapon {HÄNSLER ⇨1944/1945}.[2917] <br><br> Research into railguns received a huge boost during the presidency of Ronald REAGAN {SDI Concept ⇨1983}. The biggest railgun facility was installed at the Air Force Armament Laboratory (Eglin Air Force Base, FL). While the military use of railguns in the near future appears very unlikely, they have proved to be a useful research tool, for example |

---

[2912] P. VULCAN: *A sad good-bye* [obituary on J. LANE]. Injury Issues Monitor (published by the Research Centre for Injury Studies, Flinders University, Adelaide, Australia) **16**, 4, 19 (March 1999); http://www.nisu.flinders.edu.au/monitor/monitor16.pdf.
[2913] S.S. KUTATELADZE, AL.A. BORISOV, A.A. BORISOV, and V.E. NAKORYAKOV: *Experimental detection of a rarefaction shock wave near a liquid-vapor critical point*. Sov. Phys. Dokl. **25**, 392-393 (1980).
[2914] M.J. LIGHTHILL: *Waves in fluids*. Cambridge University Press, Cambridge & New York (1978).
[2915] D.C. PACK: *Review of "Waves in Fluids" by James LIGHTHILL, Cambridge University (1978)*. J. Fluid Mech. **90**, 605-608 (1979).
[2916] S.C. RASHLEIGH and R.A. MARSHALL: *Electromagnetic acceleration of macroparticles to high velocities*. J. Appl. Phys. **49**, 2540-2542 (1978).
[2917] W. WITT and M. LÖFFLER: *Entwicklungsgeschichte der elektromagnetischen Beschleunigung: Die elektrische Kanone auf dem Weg zum Waffensystem*. Rheinmetall GmbH, Düsseldorf, und Technologie Zentrum Nord, Unterlüß, Germany. In: *Elektromagnetische Kanone*. Lehrgang der Carl-Cranz-Gesellschaft (CCG), Weil am Rhein, Germany (Dez. 1987).

- in hypersonic aerodynamics and reentry studies;
- in hypervelocity impact simulation studies;
- for obtaining equation-of-state (EOS) data;[2918]
- in high-pressure physics and plasma fusion research; and
- as a potential "space elevator" for moving people, payloads, power, and gases from the surface of the Earth to a geostationary Earth orbit at low cost.[2919, 2920]

In 2001, MARSHALL[2921] reviewed the art of modern railgunnery and its associated problems, encompassing armature transition, rail gouging, and pulsed power supplies.

**1979  New Orleans, Louisiana**

In January, at the 17th AIAA Meeting on Aerospace Sciences, **Dean R. CHAPMAN**,[2922] a NASA aeronautical engineer from NASA's Ames Research Center (Moffet Field, CA) **reviews historical progress in computational aerodynamics, thereby covering transonic, supersonic and hypersonic flows,** in his 1979 Dryden Lecture in Research. He differentiates four major stages of evolution: linearized inviscid, nonlinear inviscid, Reynolds-averaged Navier-Stokes, and full approximation to the Navier-Stokes equations.

Rapid progress in this field allowed flow simulations to be performed for successively more complex geometric configurations and physical phenomena. Aside from its use in experimental testing (such as in wind tunnels), computational aerodynamics has become a very important technology in many applications, and is used

- in the design of vehicles (spacecraft, missiles, cars, rotorcraft, submarines, ships, *etc.*);
- in performance analysis (estimation of drag, lift, and moment of the vehicle);
- in definitions of load for structural design (including structural deformation under load);
- in in aeroelastic analysis (including flutter and divergence); and
- in definitions of aerodynamic characteristics (stability, control, and handling).[2923]

**1979  NASA's Jet Propulsion Laboratory (JPL), Pasadena, California**

On March 9, **Linda MORABITO** [now KELLY], an engineer working in the Satellite Ephemeris Development and Orbit Determination section of JPL, **sees a large heart-shaped feature on Io,** one of Jupiter's satellites, in the color image taken on March 8 with a vidicon camera on board NASA's spacecraft Voyager 1. The image is presented the following day during the Science Imaging Team's press conference. **The image shows another example of an extraterrestrial volcanic eruption** – after that seen by KOZYREV {⇨1958}.[2924] ▪ One week prior to the encounter of Voyager 1 with Jupiter, Stanton J. PEALE,[2925] a U.S. physicist at UC Santa Barbara, published a paper suggesting that tidal heating might produce volcanic activity on Io.

A more detailed inspection showed that there were actually two eruptions occurring simultaneously from the volcanoes Pele and Loki; the former emitting giant ash clouds rising more than 150 miles (260 km) above Io's surface. Planetary geologists thought they were produced by explosive volcanic eruptions; however, it soon became apparent that this idea was flawed. On Earth, explosive eruptions are violent and short-lived, but on Io six of the eight plumes seen by Voyager 1 were still erupting when Voyager 2 flew past four months later. It was also calculated that explosive eruptions could not throw volcanic material that high.[2926] The huge

---

[2918] R.S. HAWKE: *A decade of railgun development for high pressure research*. In: (Y.M. GUPTA, ed.) *Shock wave compression of condensed matter* [Proc. Symposium held in honor of G.E. DUVALL (Sept. 1, 1988)]. Rept. 99164-2814, Shock Dynamics Laboratory, Washington State University, Pullman, WA (1988), pp. 53-57.

[2919] J. PEARSON: *Low-cost launch system and orbital fuel depot*. Astronaut. Acta **19**, No. 4, 315-320 (1989).

[2920] I. MCNAB: *Launch to space with an electromagnetic railgun*. IEEE Trans. Magn. **39**, No. 1, 295-304 (Jan. 2003).

[2921] R.A. MARSHALL: *Railgunnery: where we been? Where are we going?* IEEE Trans. Magn. **37**, 440-444 (Jan. 2001).

[2922] D.R. CHAPMAN: *Computational aerodynamics development and outlook*. AIAA J. **17**, 1293-1313 (1979).

[2923] W. MASON: *Applied computational aerodynamics*. Dept. of Aerospace and Ocean Engineering, Virginia Polytechnic Institute and State University; http://www.aoe.vt.edu/~mason/Mason_f/CAtxtTop.html.

[2924] *Volcanic eruption on Io*. Planetary Photojournal, NASA-JPL, Pasadena, CA; http://photojournal.jpl.nasa.gov/catalog/PIA00379.

[2925] S.J. PEALE, P. CASSEN, and R.T. REYNOLDS: *Melting of Io by tidal dissipation*. Science **203**, 892-894 (March 2, 1979).

[2926] C.J. HAMILTON: *Io's volcanic features*; http://www.solarviews.com/eng/iovolcano.htm.

plumes were interpreted as geyser-like eruptions of sulfur dioxide or sulfur gas which, because of the extremely cold conditions, immediately condensed into sulfurous snowflakes.[2927]

| 1979 | Washington State University, Pullman, Washington | In June, **first APS Topical Conference on Shock Waves in Condensed Matter,** a meeting organized by the American Physical Society (APS) with the aim "to provide a format for scientific interactions on questions of physico-chemical properties and processes of condensed matter under shock compression up to extreme pressures and temperatures." The general chairman is George E. DUVALL. ▪ The history of the APS Topical Group on Shock Compression of Condensed Matter was reviewed in 2002 by Jerry W. FORBES,[2928] a LLNL shock physicist. The APS Council officially accepted the formation of the Shock Compression of Condensed Matter Topical Group at its October 1984 meeting. This action firmly aligned the shock wave field with a major physical science organization.

The following conferences, which quickly developed into an important international forum for scientific exchange among solid-state shock physicists, were held at Menlo Park, CA (1981); Santa Fe, NM (1983); Spokane, WA (1985); Monterey, CA (1987); Albuquerque, NM (1989); Williamsburg, VA (1991); Colorado Springs, CO (1993); Seattle, WA (1995); Livermore, CA (1997); Snowbird, UT (1999); Atlanta, GA (2001); Portland, OR (2003); Los Angeles, CA (2005); and Kohala Coast, HI (2007). |
| 1979 | E.O. Hulburt Center for Space Research, NRL, Washington, DC & Astronomical Institute of Slovak Academy of Science, Tatranská Lomnica, Czechoslovakia | On August 30, **a brilliant, previously unreported sungrazing comet – labeled comet HOWARD-COOMEN-MICHELS (1979 XI) – actually collides with the Sun.** The comet merges into the corona that brightens substantially after the impact.

(i) A team of U.S. astronomers[2929] analyzed the data obtained from the Naval Research Laboratory's orbiting SOLWIND coronagraph and concluded: "… It appears from the data that the perihelion distance was less than 1 solar radius, so that the cometary nucleus encountered dense regions of the Sun's atmosphere, was completely vaporized, and did not reappear after the time of closest approach to the Sun. After this time, however, cometary debris, scattered into the ambient solar wind, caused a brightening of the corona over one solar hemisphere and to heliocentric distances of 5 to 10 solar radii [about 3.5 to $7 \times 10^6$ km]." ▪ Two years later, this team reported observations of two additional sungrazers that encountered the Sun on January 27, 1981 and July 20, 1981, respectively.[2930]

(ii) A team of Czechoslovakian astronomers[2931] analyzing coronagraph spectrographic records discovered faint emission features which could be ascribed to $Si_{II}$ and $Ni_{II}$. These emissions were obviously a transient phenomenon which was detected only 10 hours after the supposed fall of the comet in the Sun's photosphere. |
| 1979 | Physics Dept., University of Pennsylvania, Philadelphia, Pennsylvania | Howard BRODY,[2932] a physics professor and an enthusiastic tennis player, **reports on his investigations on the physics of a tennis racket** (as stimulated by the appearance of the then novel "oversized" racket). **He mainly discusses the center of percussion, the coefficient of restitution of the tennis ball, and the oscillations or vibrations of the racket.**

BRODY later became Science Advisor of the U.S. Professional Tennis Registry, and he continued his studies on hand-held sporting kits over the following 20 years: |

---

[2927] C. SAGAN: *Sulfur flows on Io.* Nature **280**, 750-753 (1979).
[2928] J.W. FORBES: *The history of the APS Topical Group on Shock Compression of Condensed Matter.* AIP Conference Proc. **620**, No. 1, 11-19 (July 2002).
[2929] D.J. MICHELS ET AL.: *Observations of a comet on collision course with the Sun.* Science **215**, 1097-1102 (Feb. 26, 1982).
[2930] N.R. SHEELEY JR. ET AL.: *Coronagraphic observations of two new sungrazing comets.* Nature **300**, 239-242 (Nov. 18, 1982).
[2931] D. CHOCHOL ET AL.: *Emission features in the solar corona after the perihelion passage of comet 1979 XI.* Astrophys. & Space Sci. **91**, 71-77 (1983).
[2932] H. BRODY: *Physics of the tennis racket.* Am. J. Phys. **47**, 482-487 (1979).

(i) He investigated the *sweet spot* of a tennis racket[2933] – a term used to describe the specific point or region of a tennis racket where the ball should be hit for optimum results. He noted three definitions of the sweet spot on the strings of a racket, each one based on a different physical phenomenon {⇨Fig. 4.3−X}:
- the point that gives a maximum coefficient of restitution {THOMSON & TAIT ⇨1879} – i.e., the point where the ball is returned fastest;
- the center of percussion {WALLIS ⇨1670/1671; ⇨Fig. 2.9} – i.e., the point at which the impact provides no reaction impulse at the racket handle; and
- the node – i.e., the point in the racket at which minimal vibrations result when the racket is struck by the ball.

(ii) He described methods of locating these points and defined a simple rigid model of a racket with strings interacting with a ball which was applicable to the newer, rigid, composite rackets.[2934]

(iii) He also extended his studies to the physics of a baseball bat, here locating three sweet spots for this as well {⇨Fig. 4.3−Y}. In particular, he located the impact point on the bat that leads to maximum "power" – i.e., to greatest batted ball speed.[2935]

| 1979 | Max-Planck-Institut (MPI) für Strömungsforschung, Göttingen, Germany | Georg DETTLEFF,[2936] a German fluid dynamicist, **and his collaborators experimentally verify the existence of a *liquefaction shock wave* –** i.e., a compression shock which completely converts a superheated vapor into a liquid across the shock front, as previously predicted theoretically.[2937] They produce the liquefaction shock as the reflected shock from the closed end of a shock tube, which is equipped with windows for optical measurement and photography as well as various transducers for measuring pressure and temperature. As test fluids they use various fluorocarbons such as perfluoro-1,2-dimethylcyclohexane [$C_8F_{16}$] and perfluoro-1-methyldecalin [$C_{11}F_{20}$]. Photographic observations revealing the presence of small two-phase toroidal rings near the shock front confirm the existence of a clear liquid phase. ▪ **The new phenomenon of *shockwave-induced liquefaction* differs distinctly from the well-known phenomenon of the *condensation shock*, first observed in the supersonic nozzle of wind tunnels {5th Volta Conf. (PRANDTL) ⇨1935}, and from *seismic shock-induced liquefaction*,** one of the most common and dangerous hazards arising from an earthquake {Kobe & Osaka Earthquake ⇨1995}.<br><br>Recently, Gerd A. MEIER,[2938] a German retired fluid dynamics professor, reviewed research on liquefaction shock phenomena, particularly the experimental details and results which show that the liquefaction shock wave does exist in a form consistent with classical shock-wave theory. |
|------|------|------|
| 1979 | Los Alamos Scientific Laboratory (LASL), Los Alamos, New Mexico | Wildon FICKETT[2939] and **William C. DAVIS,** two Los Alamos staff scientists, **publish their book *Detonation*, a tutorial and review on pertinent theories and experiments in this field.** It encompasses a survey of<br>- the simplest theory (Chapman-Jouguet model), along with applications of it using different equations of state for the detonation products (with or without explicit chemistry); |

---

[2933] H. BRODY: *Physics of the tennis racket II. The sweet spot.* Am. J. Phys. **49**, 816-819 (1981).
[2934] H. BRODY: *The physics of tennis. III. The ball-racket interaction.* Am. J. Phys. **65**, 981-987 (1997).
[2935] H. BRODY: *The sweet spot of a baseball bat.* Am. J. Phys. **54**, 640-643 (1986).
[2936] G. DETTLEFF, P.A. THOMPSON, G.E.A. MEIER, and H.D. SPECKMANN: *An experimental study of liquefaction shock waves.* J. Fluid Mech. **95**, 279-304 (1979).
[2937] P.A. THOMPSON and D.A. SULLIVAN: *On the possibility of complete condensation shock waves in retrograde fluids.* J. Fluid Mech. **70**, 639-649 (1975).
[2938] G.A. MEIER: *Liquefaction shock waves.* In: (M. VAN DONGEN, ed.) *Shock wave science and technology.* Springer, Berlin & Heidelberg. Vol. 1 (2007): *Multiphase flows I*, pp. 231-258.
[2939] W. FICKETT and W.C. DAVIS: *Detonation.* University of California Press, Berkeley etc. (1979).

- the simple theory (Zel'dovich-von Neumann-Döring model), illustrated using the examples of a gaseous and a solid explosive system;
- various types of one-dimensional steady solutions undergoing single or multiple reactions;
- unsteady solutions and the related problem of hydrodynamic stability; and
- knowledge of the structure of the detonation front.

| | | |
|---|---|---|
| 1979 | *U.S. Geological Survey (USGS), Menlo Park, California & Cal-Tech, Pasadena, California* | **Thomas C. HANKS,**[2940] a USGS geologist, **and Hiroo KANAMORI,** a Japanese-born seismologist at CalTech, **jointly develop the so-called "moment magnitude scale"** for a more precise study of great earthquakes – a successor to the Richter scale of magnitude {RICHTER & GUTENBERG ⇨1935}. ▪ The moment magnitude $M_W$, which is the most common one used today, is a physical quantity based on the concept of the seismic moment $M_0$, and is given by $$M_W = {}^2\!/_3 \log M_0 - 10.7.$$ $M_0$ is related to the physical size of the fault rupture and the displacement across the fault by $$M_0 = \mu A D,$$ where $\mu$ is the fault friction (which depends on the rigidity of the rock), $A$ is the area of rupture or the fault area, and $D$ is the fault displacement or slip distance, which can be estimated from seismograms. |
| 1979 | *Planet Venus, Solar System* | **Christopher T. RUSSELL,**[2941] a space physics professor at the Institute of Geophysics & Planetary Physics (UC at Los Angeles, CA), **and collaborators report that,** based on their analysis of Pioneer magnetometer data, **the Venus bow shock is not as strong as the terrestrial bow shock.** ▪ Venus has a bow shock in many respects similar to that of the Earth {GOLD ⇨1953; DUNGEY ⇨1958; AXFORD & KELLOGG ⇨1962; NESS ⇨1964; GREENSTADT 1971, see NESS ⇨1964; ISEE Program ⇨1977}. However, although Venus has no detectable intrinsic magnetic field, the solar wind is still deflected about the ionopause with the formation of a detached bow shock, because the diffusion time of the magnetized solar wind plasma into the ionosphere, under typical conditions at solar maximum, is very long.[2942]<br><br>Previously, the Venus bow shock was probed twice by NASA's Mariner 5 (flyby 1967) and once by Mariner 10 (1974) as well as by the Soviet probes Venera 4 (1967), Venera 6 (1969), and repeatedly by Venera 9 (1975) and Venera 10 (1975). |
| 1980s | *Western World* | **It is rumored that Soviet scientists are working on an explosive-pumped laser that can produce huge pulses that can be used to knock out boosters.** It is speculated that such weapons could be based on the use of conventional high-energy laser systems driven by explosive-pumped flux generators {TERLETSKII ⇨1957; FOWLER ⇨1960}. **One concept, advocated by Edward TELLER, would attempt to use a** *nuclear explosive-pumped laser* **as the kill mechanism of the interceptor.** However, this remained a paper concept as successive administrations focused on non-nuclear explosive principles. ▪ The High Energy Research and Technology Facility (HERTF) at Kirtland Air Force Base, NM, was established to study various kinds of high-power pulsed devices for military applications, including explosive-pumped lasers.[2943] |
| 1980 | *Mount St. Helens, Skamania, Washington* | On May 18, **the eruption of Mount St. Helens begins with an earthquake which triggers a massive debris avalanche** {⇨Fig. 4.1–G}. The ferocious blast, suffocating ash, searing heat |

---

[2940] T.C. HANKS and H. KANAMORI: *A moment magnitude scale.* J. Geophys. Res. **B84**, 2348-2350 (1979).
[2941] C.T. RUSSELL, R.C. ELPHIC, and J.A. SLAVIN: *Pioneer magnetometer observations of the Venus bow shock.* Nature **282**, 815-816 (Dec. 20, 1979).
[2942] T.L. ZHANG, C.T. RUSSELL ET AL.: *On the Venus bow shock compressibility.* Adv. Space Res. **33**, 1920-1923 (2004).
[2943] *High Energy Research and Technology Facility (HERTF).* Federation of American Scientists (FAS); http://www.fas.org/spp/military/program/asat/herft.htm.

and falling trees kill 57 people, several of whom tried in vain to outrun the blast in automobiles and trucks. **The event is accompanied by unusual lateral blast phenomena:**

(i) The sudden pressure release of the volcano – popularly known as **"uncorking"** – unleashes a tremendous, northward-directed lateral blast of rock, ash, and hot gases that devastates an area of about 230 square miles (595 km$^2$) in a fan-shaped sector north of the volcano. To the south, the devastated area is much smaller, extending only a small distance downslope of the summit. ▪ Calculations have shown that the blast's initial velocity of about 220 mph (98 m/s) quickly increased to about 670 mph (299 m/s). The average velocity did not surpass the speed of sound in the atmosphere, about 735 mph (330 m/s). This observation is consistent with a lack of reports of loud atmospheric shocks or "sonic booms" from nearby observers. In some areas near the blast front, however, the velocity may have approached, or even exceeded, the speed of sound for a few moments.[2944]

(ii) The lateral blast, which propels volcanic debris at an estimated 300 mph (134 m/s), is heard hundreds of miles away in British Columbia, Canada.

(iii) One eyewitness account,[2945] provided by W. and L. JOHNSON who were located 17 km NE on a ridge top with good view of Mount St. Helens, described the actions of the direct blast and ash cloud as follows: "Shortly after the vertical eruption began, a large **horizontal blast** occurred. Just before the top of the mountain became obscured, the south side of the summit crumbled into the hole formed by the avalanche. As the cloud grew, what appeared to be a shock wave similar to that associated with a nuclear explosion moved ahead of the cloud. **About 1½ minutes after the start of the avalanche, and perhaps 45 seconds after the start of the blast, a noise like a clap of thunder accompanied some sort of pressure change.** The initial noise was followed by a continuous rumbling, 'like a freight train.'"

(iv) The lateral blasts also produce curious erosional features on Mount St. Helens – so-called **"furrows"** – which will later be attributed to phenomena analogous to those observed on bodies during high-speed atmospheric flight {KIEFFER & STURTEVANT ⇨ 1984}.

(v) **John M. DEWEY**,[2946] an eminent Canadian shock physics professor at the University of Victoria, BC, who spent five days on the mountain within a week of the eruption, **couldn't find any evidence of a shock wave in air:** "I have estimated the speed of the flow in the so-called "blast" or "blow down region" to have been in the order of 100 km per hour [28 m/s]. No significant audible sound wave was produced near the eruption. a light airplane that was very close to the eruptions did not hear or feel a pressure pulse. Those on the ground close to the mountain, and who survived, heard the thunder from the lightning within the erupting material, but not the eruption itself. Nevertheless, a long wavelength sub-audible pressure pulse was produced and this was first heard, after passing through the upper atmosphere, at a distance of about 60 km, and reached its maximum intensity at about 150 km."

(vi) The lateral blast at Mount St. Helens has been interpreted as being the product of a single explosion by some stratigraphers and as two closely spaced explosions by others.[2947]

(vii) The eruption detaches a giant landslide (debris avalanche) into Spirit Lake, which produces an unusually large tsunami with a wave height of about 250 m when it slips off the volcano.[2948]

---

[2944] *Lateral "blast."* USGS; http://pubs.usgs.gov/publications/msh/lateral.html.

[2945] *Volcanoes* [contains four eyewitness accounts of the directed blast observed during the 1980 Mt. St. Helens explosive volcanic eruption]. USGS, Reston, VA (May 2005); http://mac.usgs.gov/isb/pubs/teachers-packets/volcanoes/lesson5/ms5-2c.pdf.

[2946] Private communication to the author (Dec. 7, 1998).

[2947] R.P. HOBLITT: *Was the 18 May 1980 lateral blast at Mount St Helens the product of two explosions?* Phil. Trans. Roy. Soc. Lond. **A358**, 1639-1661 (2000); http://www.journals.royalsoc.ac.uk/media/g3t6cgqvyk3xwxn7qu5m/contributions/d/l/c/v/dlcvfwec68b1880j.pdf.

[2948] *Geology of interactions of volcanoes, snow, and water: Mount St. Helens, Washington. Tsunami on Spirit Lake early during 18 May 1980 eruption.* USGS; http://vulcan.wr.usgs.gov/Projects/H2O+Volcanoes/Frozen/Geology/MSH/MSH.tsunami.html.

The violent eruption of Mount St. Helens – the first to be documented with modern observational methods – caused a revolution in the understanding of explosive volcanism, because it provided new data on pyroclastic flows and erosional and depositional effects of *lateral blasts* {Mt. Bandai-san ⇒1888; STURTEVANT ET AL. ⇒1991}. Since this eruption, blast waves induced by explosive-type volcanic eruptions have also become the subject of experimental field studies and numerical simulation. In particular, at Tohoku University, Sendai, Japan, such studies have been carried out on an interdisciplinary basis by geophysicists, volcanologists, fluid dynamicists, and shock wave physicists.[2949, 2950]

| | | |
|---|---|---|
| 1980 | *Saturn's moon Mimas & NASA's Goddard Space Flight Center (GSFC), Greenbelt, Maryland* | On November 11, the U.S. spacecraft Voyager 1 passes Mimas, one of the innermost moons of Saturn, at a distance of 425,000 km. **It takes pictures of Herschel Crater (about 100 km in dia. and 10 km deep), which has a central mountain that rises about 6 km above the crater floor.**[2951] This huge crater is about a quarter the diameter of Mimas itself, and it is believed that the impact that formed this complex crater with its central peak of uplifted rocks was almost large enough to destroy the satellite itself. The side of Mimas opposite Herschel Crater exhibits a trough system thought to have formed during the impact event that created Herschel. ▪ Mimas was discovered in 1789 by the German-born British astronomer Sir F. William HERSCHEL.<br><br>The two probes Voyager 1 and Voyager 2, which were launched in 1977 on a journey to the outer planets, are currently heading beyond the bounds of the Solar System at a velocity of about 3–4 AU per year (14.3–19 km/s) {Voyager 1 & 2 ⇒2003; Voyager 1 ⇒2004}. |
| 1980 | *Albuquerque, New Mexico* | **EXPLOMET '80 is held – the first International Conference on Metallurgical and Materials Applications of Shock-Wave and High-Strain Rate Phenomena in Materials.** The chairman and founder of EXPLOMET is Lawrence E. MURR {⇒1988}. ▪ Subsequent conferences were held at five-year intervals at Portland, OR (1985); La Jolla, CA (1990); and El Paso, TX (1995). EXPLOMET 2000 was held at Albuquerque, NM (2000); at this point the subtitle of the conference was changed to "International Conference on Fundamental Issues and Applications of Shock-Wave and High-Strain-Rate Phenomena." Further conferences were not held. |
| 1980 | *Dept. of Physics and Astronomy, University of Mississippi* | Henry E. BASS,[2952] a U.S. professor of physics, **examines current research into the relationship between the processes of thunder and lightning:**<br>(i) Since the geometry (tortuosity) of the discharge channel is not known exactly, and nonlinear effects strongly influence the propagation behavior, it is difficult to estimate the near-field pressure waveforms. BASS models the propagation of thunder from lightning as finite-wave propagation from a straight cylindrical discharge channel in the weak shock regime with overpressures $\Delta p$ ranging from 4.5 to 0.1 psi (310 to 7 mbar).<br>(ii) Assuming a typical lightning energy of 2.7 kJ/m, a current rise time of 1 μs, a decay time of 50 μs and an initial discharge radius of 0.5 mm, he finds that for distances greater than 10 cm (*i.e.*, where the weak shock limit is reached), the overpressure decays as $\Delta p \sim r^n$ with $n = -\frac{3}{4}$. For example, at a distance of $r = 0.15$ m from the lightning strike, the overpressure is |

---

[2949] H. TANIGUCHI and K. SUZUKI-KAMATA: *Direct measurement of overpressure of a volcanic blast on the June 1991 eruption at Unzen Volcano, Japan.* Geophys. Res. Lett. **20**, No. 2, 89-92 (1993); A. GOTO ET AL.: *Effect of explosion energy and depth to the pressure-wave form of a blast wave: field experimental study for the understanding of volcanic blast wave.* Ibid. **28**, 4287-4290 (2001).

[2950] T. EGUCHI, H. TANIGUCHI, T. SAITO, and K. TAKAYAMA: *Numerical simulations of the May 18, 1980 blast at Mount St. Helens.* In: *Japan Earth and Planetary Science Joint Meeting* [Tokyo, Japan, June 2000]. For abstract *see*
http://209.85.129.104/search?q=cache:O_jUAaGP3OkJ:www-jm.eps.s.u-tokyo.ac.jp/2000cd-rom/pdf/vb/vb-p017_e.pdf+EGUCHI+TANIGUCHI+SAITO+TAKAYAMA+helens&hl=en&ct=clnk&cd=6.

[2951] C.J. HAMILTON: *Mimas & Herschel Crater* (1999); http://www.solarviews.com/cap/sat/mimas2.htm.

[2952] H.E. BASS: *The propagation of thunder through the atmosphere.* JASA **67**, 1959-1966 (1980); *Atmospheric acoustics.* In: (G.L. TRIGG, ed.) *Encyclopedia of applied physics.* VCH, New York etc. (1991), vol. 2, pp. 145-179.

about 4.5 psi. **Propagation through the weak shock regime will (only) require tens of meters.**

(iii) Due to nonlinear effects, the positive phase duration of the pressure pulse increases, which shifts the acoustic spectrum to lower frequencies. This is the reason why the sound of thunder develops stronger rumbling as the distance from the lightning increases.

(iv) **His calculations clearly show that early speculations on the supersonic nature of thunder — particularly the overestimated high velocities of thunder previously claimed** {EARNSHAW ⇨ 1851; RAILLARD ⇨ 1860; MONTIGNY ⇨ 1860} **— are entirely unfounded.** ▪ For example, for a rather close distance to the strike of 0.5 m, BASS' theory would give an overpressure of about 0.03 $P_0$, with $P_0$ as the ambient atmospheric pressure. Using the Rankine-Hugoniot theory, and assuming a sound velocity at rest of 340 m/s, this would result in a shock front velocity of only 348 m/s.

| | | |
|---|---|---|
| 1980 | *Sandia National Laboratories (SNL), Albuquerque, New Mexico* | Robert A. GRAHAM[2953] and B.W. DODSON, two U.S. shock physicists, **present a bibliography on studies of shock-induced chemistry carried out in the Soviet Union up to 1980.** It encompasses 10 review articles and 110 research papers addressing<br>▸ methods of shock compression and sample recovery;<br>▸ shock activation of catalysts;<br>▸ shock-induced diffusion and shock-activated sintering;<br>▸ electrochemistry;<br>▸ polymerization, polymers and elastomers; and<br>▸ miscellaneous organic and inorganic materials.<br>The first experimental evidence for shock-induced chemistry was provided in the late 19th century in France {BERTHELOT ⇨ 1881}. However, although such studies have considerable technological potential, systematic investigations into the chemical effects of shock compression evolved only very slowly outside the Soviet Union. In 1993, Stepan S. BATSANOV,[2954] a Russian researcher at the Institute for Physical-Technical Measurements in Mendelejevo (Moscow region), reviewed the kinetics of chemical reactions in shock-compressed solids and discussed determinations of reaction times by chemical and physical methods. In his book ***Effects of Explosions on Materials***, BATSANOV[2955] discussed the modification and synthesis of materials under high-pressure shock compression. |
| 1980 | *Poulter Laboratory, SRI International, Menlo Park, California* | Yogendra M. GUPTA,[2956] an Indian-born U.S. high-pressure physicist, **and associates report on a new experimental technique of measuring lateral compressive stresses in impact-loaded solids using piezoresistive gauges** {BERNSTEIN ET AL. ⇨ 1967}. They applied an arrangement of four blocks of Plexiglas (polymethyl methacrylate, PMMA) provided with grooves in which 0.05-mm-thick ytterbium-foil gauges were embedded (two positioned in parallel and two perpendicular to the shock front). The projectile velocity was chosen so as to produce a stress amplitude of 5 kbar in the test samples. The goal of this study was<br>▸ to improve the determination of the shock state in a solid in order to provide a better understanding of the shock responses of solids; and<br>▸ to test the potentials and limitations of this new method.<br>Although their technique demonstrates the feasibility of correlating resistance changes with lateral stresses, they conclude that a more detailed understanding of piezoelectric gauge response is required to improve calibration. |

---

[2953] R.A. GRAHAM and B.W. DODSON: *Bibliography on shock-induced chemistry*. Rept. SAND80-1642, Sandia National Laboratories (SNL), Albuquerque, NM (Aug. 1980).
[2954] S.S. BATSANOV: *On the kinetics of chemical reactions in solids under shock compression (a review)*. Propellants, Explosives, Pyrotechnics **18**, 100-105 (1993).
[2955] S.S. BATSANOV: *Effects of explosions on materials*. Springer, New York (1994).
[2956] Y.M. GUPTA, D.D. KEOUGH, D. HENLEY, and D.F. WALTER: *Measurement of lateral compressive stresses under shock loading*. Appl. Phys. Lett. **37**, 395-397 (1980).

In the following years, GUPTA[2957] **and his team critically analyzed previous shock data obtained with manganin and ytterbium piezoresistive pressure gauges of various designs.** A detailed data analysis revealed the importance of gauge plasticity and gauge strain states in controlling the stresses in the gauge foils which, in turn, control the resistivity changes. Further studies have shown that the gauge data obtained must be evaluated with extensive computer simulations in order to understand the impact of the gauge on the stress field it is attempting to measure.[2958] Discussions about correct lateral stress measurements in shock-compressed materials are ongoing.

| 1980 | Los Alamos Scientific Laboratory (LASL), Los Alamos, New Mexico | Stanley P. MARSH,[2959] a Los Alamos staff member and experimental shock physicist, **publishes a compendium of equation-of-state (EOS) data from over 5,000 shock compression experiments.** His data collection includes experimentally obtained Hugoniot data for 450 solids, liquids, gases and explosives measured in the range from 500 bar to several Mbar.<br><br>In the same year, **Terry GIBBS**[2960] **and Alphonse POPOLATO publish the catalog *LASL Explosive Property Data*,** which contains data on physical and chemical properties, detonation velocities, shock initiation data, *etc.*, for 50 commonly used high-performance explosives that were studied at LASL. |
|---|---|---|
| 1980 | UC Berkeley (UCB) & NASA's Ames Research Center (ARC), Moffet Field, California | **Christopher F. MCKEE,**[2961] a physics and astronomy professor at UCB, **and David J. HOLLENBACH,** a theoretical astrophysicist at NASA-ARC, **discuss the structures of interstellar shocks driven by supernova remnants and by expanding H II regions**[2962] **around early-type stars.**<br>(i) They examine jump conditions, along with shock fronts, post-shock relaxation layers, collisional shocks, collisionless shocks, nonradiative shocks, radiative atomic shocks, and shock models of observed nebulae.<br>(ii) They also examine effects of shock waves on interstellar molecules, with reference to the chemistry behind shock fronts, infrared and vibrational-rotational cooling by molecules, and observations of shocked molecules.<br>(iii) They summarize some current problems in and applications of the study of interstellar shocks, including the initiation of star formation by radiative shock waves, interstellar masers, the stability of shocks, particle acceleration in shocks, and shocks in galactic nuclei. |
| 1980 | Lawrence Livermore National Laboratory (LLNL), UC Livermore, California | **Daniel J. STEINBERG,**[2963] a U.S. hydrodynamicist, **and his collaborators present a novel technique for generating shock pressures of greater than 1 TPa (10 Mbar).** In their system, the explosion of an electrically heated metal foil and the accompanying magnetic forces are used to accelerate a thin flyer plate to hypervelocities; this technique is known as an **"electric gun."** They suggest that this simple, low-cost device, driven by large pulsed laser systems or by a nuclear explosion, may be a promising alternative to equation-of-state (EOS) experiments in the 10–50 Mbar range. |

---

[2957] Y.M. GUPTA: *Analysis of manganin and ytterbium gauge data under shock loading.* J. Appl. Phys. **54**, 6094-6098 (1983).

[2958] R. FENG, Y.M. GUPTA, and M.K.W. WONG: *Dynamic analysis of the response of lateral piezoresistance gauges in shocked ceramics.* J. Appl. Phys. **82**, 2845-2854 (1997); *Determination of lateral stresses in shocked solids: simplified analysis of piezoresistance gauge data.* Ibid. **83**, 747-753 (1998); *See also* the response to these articles in Ibid. **86**, 3487-3489 (1999).

[2959] S.P. MARSH (ed.): *LASL shock Hugoniot data.* University of California Press, Berkeley *etc.*, vol. 5 (1980).

[2960] T. GIBBS and A. POPOLATO (eds.): *LASL explosive property data.* University of California Press, Berkeley (1980).

[2961] C.F. MCKEE and D.J. HOLLENBACH: *Interstellar shock waves.* Annu. Rev. Astron. Astrophys. **18**, 219-262 (1980).

[2962] The interstellar medium consists mainly of gas (about 99%) and dust (such as carbon and silicates particles), the gas consists mainly of hydrogen. Most of the gas in the galactic disk is present in an unionized or neutral state, and termed *H I region*. Under the right conditions, such as near a hot star or after a supernova explosion, the gas can become hot (~ $10^4$ K) and ionized, and these places are termed *H II region*. The gas in both states is more generally called the "Interstellar Medium (ISM)."

[2963] D. STEINBERG, H. CHAU, G. DITTBENNER, and R. WEINGART: *The electric gun: a new method for generating shock pressures in excess of 1 TPa.* In: (B. VODAR and P. MARTEAU, eds.) *High pressure science and technology.* Pergamon Press, Oxford, vol. 2 (1980), pp. 983-985.

| | | |
|---|---|---|
| 1981 | University of Tokyo, Japan | In April, **establishment of the Institute of Space and Astronautical Science (ISAS),**[2964] a joint research organization among Japanese universities, "with the aim of promoting further development of research in aeronautics and astronautics." It comprises the following branches related to fluid dynamics: aerodynamics and flight dynamics, gas dynamics, combustion, jet and rocket propulsion, heat transfer and internal fluid dynamics. ▪ ISAS is an outgrowth of the Institute of Space and Aeronautical Science, which was founded in 1964 at the University of Tokyo.<br><br>In 2003, ISAS was integrated into the Japan Aerospace Exploration Agency (JAXA) and became the Space Science Research Division of JAXA. |
| 1981 | CalTech, Pasadena, California | On May 16 (Alumni Seminar Day), **Susan Werner KIEFFER,**[2965] a U.S. geologist, **presents her supersonic flow theory regarding the last eruption of Mount St. Helens** {⇨1980}. Bradford STURTEVANT, a U.S. professor of aeronautics at CalTech's Graduate Aeronautical Laboratories (GAL), attending the seminar and stimulated by this hypothesis, initiates shock tube laboratory experiments on supersonic gas flow simulating pseudo-gas-erupting volcanic plumes. ▪ Important results from this work include the insight gained into the various shapes that volcanic plumes could assume depending on the composition of the volcanic gas, and its ratio of pressure to atmospheric pressure.[2966] |
| 1981 | Blacksburg, Virginia | In the period May 28–31, **the first International Conference on Physics in Collision is held,** with the aim of reviewing and updating key topics in elementary particle physics as well as encouraging informal discussions on new experimental results and their implications. Main topics are high-energy ee-, ep-, and pp-interactions. The chairman is W. Peter TROWER. ▪ The following symposia were held at Stockholm, Sweden (1982); Como, Italy (1983); Santa Cruz, CA (1984); Autin, France (1985); Chicago, IL (1986); Tsukuba, Japan (1987); Capri, Italy (1988); Jerusalem, Israel (1989); Durham, NC (1990); Colmar, France (1991); Boulder, CO (1992); Heidelberg, Germany (1993); Tallahassee, FL (1994); Krakow, Poland (1995); Mexico City, Mexico (1996); Bristol, U.K. (1997); Frascati, Italy (1998); Ann Arbor, MI (1999); Lisbon, Portugal (2000); Seoul, Korea (2001); Stanford, CA (2002); Zeuthen, Germany (2003); Boston, MA (2004); Prague, Czech Republic (2005); Búzios, Brazil (2006); Annecy, France (2007); and Perugia, Italy (2008). |
| 1981 | Snowbird, Utah | In October, **first International Meeting on Mass Extinctions and Global Catastrophes,** the remit of which includes the geological and biological consequences of large-scale events.<br><br>In the years following, symposia relating to this subject were held at Snowbird, UT {⇨1988}; Houston, TX (1994); and Vienna, Austria (2000). The main focus of these meetings was the question of whether (and if so how) short-term, high-energy events influence biological evolution on Earth. Mass extinctions of various scales mark some of the geological boundaries. |
| 1981 | Dept. of Physics and Astronomy, University of Victoria, B.C., Canada | **First International Mach Reflection Symposium.** The general chairman is John M. DEWEY, who also initiated the conference. There is a renewed interest in Mach reflection due to a new awareness of the importance of single and double Mach reflection on the height-of-burst (HOB) effects from nuclear air burst explosions. ▪ Later symposia were held at Sydney, Australia (1981); Melbourne, Australia (1982); Freiburg, Germany (1983); Tokyo & Sendai, Japan |

---

[2964] *History of ISAS.* Institute of Space and Aeronautical Science (ISAS); http://www.isas.ac.jp/e/about/index.shtml.
[2965] S.W. KIEFFER: *The blast at Mount St. Helens: what happened?* Engineering & Science (Alumni magazine publ. by CalTech) **XLV** (1), 6-12 (1981).
[2966] S.W. KIEFFER and B. STURTEVANT: *Laboratory studies of volcanic jets.* J. Geophys. Res. **89** (B10), 8253-8268 (1984).

(1984); Palo Alto, CA (1985); Beer Sheva, Israel (1986); Albuquerque, NM (1988); Freiburg, Germany (1990); Denver, CO (1992); Victoria, B.C. (1994); Pilanesberg, South Africa (1996); Beer Sheva, Israel (1998); Yonezawa & Sendai, Japan (2000); Aachen, Germany (2002); and Quelpart Island, South Korea (2004). Over the years, these meetings have become an international forum for discussions of ongoing work into Mach reflection phenomena, and an initiating point for collaborative research.

***Single Mach reflection,*** the most common wave configuration of ***irregular reflection*** (or ***irregular interaction***) of shock waves, is the archetypal form of Mach reflection (MR) {E. MACH & WOSYKA ⇨1875}. Depending on the shock interaction geometry, the size of the Mach disk or Mach bridge

- is invariant in time – *stationary MR* (*e.g.*, in cylindrical shock configurations; ⇨Figs. 4.13–H, L);
- increases in time – *progressive MR* (the common case; Figs. 4.5–D & 4.13–C); or
- decreases in time – *regressive MR* {KREHL ET AL. ⇨1977; ⇨Fig. 4.8–H}.

During the past 60 years of theoretical and experimental research on this subject, a number of new special cases of Mach reflection have been observed and described. This led to the definition of a number of new terms {⇨Fig. 2.14}, such as

- *compound MR* {L.G. SMITH ⇨1945};
- *inverse MR* {COURANT & FRIEDRICHS ⇨1948};
- *double MR* {WHITE ⇨1951};
- *irregular MR* (CARPENTER 1975);
- *transitional MR* (GLASS 1976); and
- *terminal double Mach reflection* {LEE & GLASS ⇨1984; ⇨Fig. 4.13–F}.

A review on the status of Mach reflection research up to 1992 was given by Gabi BEN-DOR[2967] and Kazuyoshi TAKAYAMA.

1981 | *Dept. of Mineral Sciences, National Museum of Natural History, Smithsonian Institution, Washington, DC* | **Roy S. CLARKE JR.,**[2968] **Daniel E. APPLEMAN, and Daphne R. ROSS,** three U.S. meteorite researchers, **referring to the Allan Hills meteorite ALH 77283 argue that diamond**[2969] **and lonsdaleite (a hexagonal form of diamond) were present in the meteoroid *before* its final ablative passage through the atmosphere and soft landing on the ground.** The shock event that produced these high pressure phases, therefore, must have taken place on its parent body or have been associated with the disruption of that body. They base their conclusion on metallographic and X-ray diffraction data.

Iron meteorites containing diamonds are very rare. For example, of the many meteorites recovered so far from the Allan Hills (Far Western Icefield, Antarctica), only one meteorite – a reddish brown 10.51-kg object discovered in 1977 and named ALH 77283 – contains impact-produced diamonds. The Canyon Diablo meteorite, the excavator of Meteor Crater in Arizona {⇨c.50,000 years ago}, is the only other iron meteorite known to contain the high-pressure minerals found in ALH 77283, and their occurrence in that meteorite has been explained as the result of shock-induced transformation of graphite, most probably at the moment of terrestrial impact and disintegration of the projectile during crater formation {FOOTE ⇨1891; FRIEDEL ⇨1892; CARTER & KENNEDY ⇨1964}.

---

[2967] G. BEN-DOR and K. TAKAYAMA: *The phenomena of shock wave reflection, a review of unsolved problems and future research needs.* Shock Waves **2**, 211-223 (1992).

[2968] R.S. CLARKE JR., D.E. APPLEMAN, and D.R. ROSS: *An Antarctic iron meteorite contains preterrestrial impact-produced diamond and lonsdaleite.* Nature **291**, 396-398 (June 4, 1981).

[2969] Diamonds crystallize in the isometric system, mostly in the octahedral form. The cubic form is less abundant.

| | | |
|---|---|---|
| 1981 | *Lawrence Livermore National Laboratory (LLNL), UC Livermore, California* | **Publication of the *LLNL Explosives Handbook*,** compiled by Brigitta M. DOBRATZ.[2970] It presents information and data for high explosives of interest to those involved in programs carried out at LLNL and other Dept. of Energy facilities, as well as to engineers and scientists working in other areas of research and development. The compilation is limited to the production of standard high explosives and their components; research explosives are not included. The data on high explosives encompass their physical, chemical, thermal, electrical, mechanical and performance properties (*e.g.*, detonation velocity, Hugoniot data, CJ detonation pressures). |
| 1981 | *School of Engineering, North Carolina State University, Raleigh, North Carolina* | Wayland C. GRIFFITH,[2971] a U.S. fluid dynamicist, **traces some of the principal lines of investigation on shock waves,** ranging from early motivations to the present state of understanding and application.<br><br>(i) "Motivation," he writes, "is not often consciously expressed in the scientific literature. Usually an external motivation in terms of identifiable needs for better understanding for the solution of practical problems can be identified; though much excellent work must be ascribed to that ubiquitous trait curiosity."<br><br>(ii) The topics covered in his article, chosen as representative of the basic elements of shock wave interactions and effects, encompass: shock structure, refraction, diffraction, shocks in liquid helium, condensation and liquefaction shocks, and approximate and computational methods developed for handling complex flow problems. |
| 1981 | *Institut für Didaktik der Physik, Universität Karlsruhe, Germany* | Friedrich HERRMANN,[2972] a professor of physics didactics, **and Peter SCHMÄLZLE,** a young physicist, **study the collision mechanics of "NEWTON's cradle"** {Royal Society of London ⇨1666; ⇨Fig. 4.4−B} − the well-known "multiple-ball percussion experiment − in more detail. They show that, contrary to common belief, the Laws of Conservation of Momentum and Energy alone are not sufficient to explain that the same number of balls move away to the other side as were initially displaced. Indeed, **the system must be *dispersion-free* −** *i.e.*, such that it transfers the total energy and momentum of the impacting balls to the same number of balls at the other side of the chain. In practice, this is accomplished by separating each ball from its neighbors by a small distance.<br><br>More recently, NESTERENKO {⇨2001} resumed this problem by treating chain percussion in NEWTON's cradle as a solitary wave {RUSSELL ⇨1834}. |
| 1981 | *Faculty of Engineering, Kyushu University, Fukuoka City, northern Kyushu, Japan* | Takefumi IKUI,[2973] a professor of mechanical engineering, **and his collaborators observe a new type of irregular shock wave reflection in low-gamma gases such as chlorofluorocarbon** [or Freon-12, $CF_2Cl_2$] with $\gamma = 1.141$. Using a planar shock wave interacting with a plane wedge and studying double-Mach reflection, they observe that, at increasing wedge angles and higher Mach numbers, the second triple point is pushed toward the rigid surface such that the reflected shock is reflected like a regular reflection, thereby forcing the second Mach stem to join the slipstream above the wedge surface. ▪ They name this phenomenon **"pseudostationary reflection."**<br><br>Pseudostationary reflection was subsequently studied in more detail at UTIAS by Prof. Irvine I. GLASS and collaborators.[2974, 2975] It only occurs in low-gamma gases at room temperature, such as in sulfur hexafluoride [$SF_6$] with $\gamma = 1.093$ and in isobutene [2-methyl pro- |

---

[2970] B.M. DOBRATZ: *LLNL explosives handbook: properties of chemical explosives and explosive simulants.* National Technical Information Service (NTIS), Springfield, VA (1981); With P.C. CRAWFORD, Ibid. Rept. UCRL-52997, LLNL, Livermore, CA (1985). ▪ *See also* the first edition of this compilation: *Properties of chemical explosives and explosive simulants.* Rept. UCRL-51319 (1974).
[2971] W.C. GRIFFITH: *Shock waves.* J. Fluid Mech. **106**, 81-101 (1981).
[2972] F. HERRMANN and P. SCHMÄLZLE: *Simple explanation of a well-known collision experiment.* Am. J. Phys. **49**, 761-764 (1981).
[2973] T. IKUI, K. MATSUO, T. AOKI, and N. KONDOH: *Mach reflection of a shock wave from an inclined wall.* Memoirs, Faculty of Engineering, Kyushu University **41**, No. 4, 361-370 (1981).
[2974] J.H. LEE and I.I. GLASS: *Pseudo-stationary oblique shock wave reflections in frozen and equilibrium air.* Progr. Aerospace Sci. **21**, 33-80 (1984).
[2975] T.C.J. HU: *Pseudo-stationary oblique shock-wave reflections in a polyatomic gas − sulfur hexafluoride.* UTIAS Tech. Note No. 253 (1985).

pane, CH(CH$_3$)$_3$] with $\gamma \approx 1.09$, and it significantly decreases at higher temperatures. GLASS proposed that pseudostationary reflection should be termed ***terminal double Mach reflection*** {⇨Fig. 4.13–F}.[2976]

| 1981 | *Aerodynamics Group, College of Aeronautics, Cranfield Institute of Technology, Bedford, eastern England* | **John Leslie STOLLERY**,[2977] a British professor of aerodynamics, **and collaborators use the hydraulic jump analogy as a simulation technique when investigating how to reduce the blast wave noise emanating from the rear of a shoulder-launched weapon.** Although the analogy to gas dynamic flow is not exact, and the water table can only simulate 2-D conditions, such hydraulic experiments provide a quick and inexpensive way to test a large range of silencer configurations under conditions of both confined and unconfined space. Such studies also promote an understanding of the various mechanisms that reduce the overpressure at the relevant "ear position."<br><br>(i) The sudden release of the gate of a water wave generator produces a hydraulic jump analogous to the blast wave from the rear of a shoulder-launched weapon. The water wave propagating down the channel through any silencer is then diffused from the mouth of the channel out over the bed of the water table.<br><br>(ii) The wave signature is measured at over 100 stations with a capacitance proximity probe and recorded. The maximum wave height $d_{max}$ is then identified from these records and converted into a maximum pressure $p_{max}$ using the water-to-gas analogy relationship given by<br>$$p_{max}/p_0 = (d_{max}/d_0)^2,$$<br>where $d_0$ is the height of the water at rest, and $p_0$ is the atmospheric pressure.<br><br>(iii) In addition, for comparison purposes, shock tube experiments have been carried out using an open-ended shock tube that is fitted with a variety of silencers and that exhausts into the atmosphere. |
|---|---|---|
| 1981 | *British Ministry of Defense & Dept. of Mechanical Engineering, University of Surrey, England* | **Leo H. TOWNEND**[2978] **and David G. EDWARDS**, two British researchers, **report on an unconventional blast suppression technique** based on inertial damping that is intended to reduce the initial expansion rate of a propellant gas cloud.<br><br>(i) **They modify the weapon such that it ejects a water sheet that surrounds the muzzle (or breech) region just before firing.** In the case of firing a 30-mm-caliber Aden cannon (a fully automatic, single-barrelled, five-chambered rotating cylinder gun used on many military aircraft), two liters of water around the muzzle produces a reduction in peak pressure of about 40%.<br><br>(ii) Using a water wall realized by a shallow trough consisting of a thin film of plastic supported on a wire frame in the near field at the exit of a shock tube, they successfully reduce the blast pressure level by about 14 dB (which corresponds to a factor of about five).[2979] |
| 1981 | *Dept. of Urology, Kyoto Prefectural University of Medicine, Kyoto, Japan* | **Hiroki WATANABE**,[2980] a Japanese professor of urology, **and collaborators report on the development of clinical *microexplosion lithotripsy*, used to destroy urinary stones.** A small dose of a high explosive (2–7 mg of lead azide) housed in a small steel chamber connected to the tip of a catheter and triggered electrically from outside is used *inside* the patient. • On the contrary, extracorporeal shock wave lithotripsy uses the shock wave source *outside* the patient {CHAUSSY ET AL. ⇨1970}. |

---

[2976] J.T. URBANOWICZ: *Pseudo-stationary oblique shock-wave reflections in low gamma gases isobutene and sulfur hexafluoride*. UTIAS Tech. Note No. 267 (1988).

[2977] J.L. STOLLERY, K.C. PHAN, and K.P. GARRY: *Simulation of blast fields by hydraulic analogy*. In: (C.E. TREANOR and J.G. HALL, eds.) *Proc. 13th Int. Symposium on Shock Tubes & Waves* [Niagara Falls, NY, July 1981]. State University of New York Press, Albany, NY (1982), pp. 781-789.

[2978] L.H. TOWNEND and D.G. EDWARDS: *The inertial damping of unsteady expansion*. In: (J.E. BACKOFEN JR., ed.) *Proc. 6th Int. Symposium on Ballistics* [Orlando, FL, Oct. 1981]. Batelle, Columbus Laboratories, Columbus, OH (1981), pp. 144-153.

[2979] Peak values of blast pressures in bar, $p_{max}$ [bar], are also given in decibels, $p_{max}$ [dB], according to the relation $p_{max}$ [dB] = 20 log ($p_{max}$ [bar]/$p_0$ [bar]) with $p_0 = 2 \times 10^{-10}$ bar; thus $p_{max} = p_0$ corresponds to 0 dB, $p_{max} = 1$ µbar to 74 dB, 1 mbar to 134 dB, 1 bar to 194 dB, *etc*. An increase of 6 dB corresponds to doubling the blast pressure.

[2980] H. WATANABE, M. UCHIDA, K. KITAMURA, Y. IMAIDE, and K. YONEDA: *Microexplosion lithotripsy for bladder and renal calculi*. In: (K. TAKAYAMA, ed.) *Proc. 18th Int. Symposium on Shock Waves* [Sendai, Japan, July 1991]. Springer, Berlin *etc.* (1992), vol. 2, pp. 1197-1200.

(i) Two blasting methods are applied which differ in terms of how the charge is brought in contact with the stone:
- in **"external charge" blasting** the explosive chamber is attached to the stone, and
- in **"confined" blasting** the explosive chamber is inserted into a hole made with an electric drill – a technique which has proved to be more effective than external blasting, because it needs less explosive and is capable of destroying even large and hard stones.

(ii) They report that 142 patients in total have been treated successfully using either one or the other of these methods.

In the following years, **this therapy was also extended to the treatment of kidney stones using microexplosive pellets outside the patient** {TAKAYAMA ⇨ 1983}. There are also other biomedical applications for detonation of micro-sized explosive charges such as localized destruction of pathological tissue.[2981] Reviews of the particular requirements in the design of micro-detonics devices as well as their as applications and prospects were given more recently {STEWART ⇨ 2002; JAGADEESH & TAKAYAMA ⇨ 2002}.

| | | |
|---|---|---|
| 1982 | *Honolulu, Hawaii* | **Establishment of the Tsunami Society** as an international society aimed at increasing and disseminating knowledge about tsunamis and their hazards. The Society is organized by William M. ADAMS, Augustine S. FURUMOTO, and George PARARAS-CARAYANNIS. ▪ The Tsunami Society publishes the journal ***The Science of Tsunami Hazards*** and conducts a ***Tsunami Symposium*** every three years at the East-West Center on the University of Hawaii campus. |
| 1982 | *NASA's Ames Research Center (ARC), Moffet Field, California* | Robert W. MACCORMACK,[2982] an aeronautical engineer and fluid dynamicist, **reports on a new numerical method for solving the equations of unsteady compressible viscous flow**, one of the main problems in aerodynamic design. He claims that his method<br>▸ is accurate to the second order in space and time, and unconditionally stable;<br>▸ preserves conservation form (homogeneous to the first degree);<br>▸ requires no block or scalar tridiagonal inversions;<br>▸ is simple and straightforward to program; and<br>▸ is more efficient than present methods.<br><br>His new technique – an outgrowth of the explicit predictor-corrector finite-difference method he developed previously[2983] – proved very useful for the numerical simulation of viscous, compressible fluid flows containing shock waves and strong shock wave/boundary layer interactions.[2984] The Flux Correction Transport (FCT)[2985] scheme was applied to the MacCormack method as a way to counter excessive "smearing" of systems with sharp fronts, such as shock waves, resulting in **"MacCormack FCT."**[2986]<br><br>In 1996, MACCORMACK received the Fluid Dynamics Award from the American Institute of Aeronautics and Astronautics (AIAA) "for pioneering work in computational fluid dynamics and the development of numerical algorithms which are used all over the world." |
| 1982 | *Los Alamos National Laboratory (LANL), Los Alamos, New Mexico* | Charles E. MORRIS,[2987] a U.S. solid-state physicist, **publishes a data collection which includes 300 shock wave profiles of various materials.** They are divided into six classes, similar to those used in the LASL Hugoniot Shock Data compendium {MARSH ⇨ 1980}: |

---

[2981] K. TAKAYAMA and T. SAITO: *Shock wave/geophysical and medical applications*. Annu. Rev. Fluid Mech. **36**, 347-379 (2004).

[2982] R.W. MACCORMACK: *A numerical method for solving the equations of compressible viscous flow*. AIAA J. **20**, 1275-1281 (1982).

[2983] R.W. MACCORMACK: *The effects of viscosity in hypervelocity impact cratering*. AIAA Paper 69-354 (1969).

[2984] C.M. HUNG ET AL.: *The MacCormack method – historical perspective*. In: (D.A. CAUGHEY and M.M. HAFEZ, eds.) *Frontiers of computational fluid dynamics 2002*. World Scientific, River Edge, NJ (2002).

[2985] J.P. BORIS and D.L. BOOK: *Flux-corrected transport. Parts I–III*. J. Comput. Phys. **11**, 38-69 (1973); Ibid. **18**, 248-283 (1975); Ibid. **20**, 397-431 (1976).

[2986] B.A. FINLAYSON: *Numerical methods for problems with moving fronts*. Ravenna Park Publ., Inc., Seattle, WA (1992).

[2987] C.E. MORRIS (ed.): *Los Alamos shock wave profile data*. University of California Press, Berkeley (1982).

- elements, alloys, minerals and compounds;
- rocks and mixtures of minerals;
- plastics;
- high explosives;
- high explosive simulants and propellants; and
- explosives/metal free-run systems.

The reproduced wave-time profiles were obtained using capacitor gauges, quartz gauges, manganin gauges, and electromagnetic probes.

**1982** — *Dept. of Earth Sciences, Dartmouth College, Hanover, New Hampshire*

**Christopher G. NEWHALL,**[2988] a U.S. geologist, **and Stephen SELF,** a British-born U.S. volcanologist, **propose a *volcanic explosivity index (VEI)*** in order to standardize the assignment of the magnitude of an explosive eruption. **The scale of their index ranges from 0 ("non-explosive eruption") to 8 ("very large explosive event").** It uses visually observable criteria such as

- the volume of ejected material;
- the height of the eruption column;
- the distances that blocks and fragments are hurled;
- the amount of aerosols injected into the upper atmosphere; and
- the duration of the eruption.

This index has also been applied for the general description and classification of prehistoric and historic volcanic eruptions.[2989, 2990]

- The largest explosive eruption with *VEI* = 8 happened in Colorado {⇨c.28 Ma ago}. The most recent event with *VEI* = 8 in prehistoric times was the ultra-Plinian mega-colossal eruption of Mt. Toba {⇨71,000 years ago}. • **A VEI-8 eruption from a *supervolcano*** – a term coined in 2000 by the producers of the BBC science program *Horizon* – **is a mega-colossal event** that produces at least 1,000 km$^3$ of magma and pyroclastic material.[2991]
- In the recent history of man, the largest explosive eruption happened in Indonesia {Tambora Eruption ⇨1815}. This had *VEI* = 7, which even surpassed the famous Krakatau eruption {⇨1883}, which had *VEI* = 6.

**1982** — *Lawrence Livermore National Laboratory (LLNL), UC Livermore, California*

**Craig M. TARVER,**[2992] a U.S. chemical physicist, **develops a *nonequilibrium Zel'dovich-von Neumann-Döring (NE-ZND) theory*** in order to explain the various nonequilibrium processes that precede and follow chemical energy release in self-sustaining detonation waves.

(i) In contrast to the classical one-dimensional (1-D) ZND model {ZEL'DOVICH ⇨1940}, which consists of a shock wave discontinuity followed by a zone of homogeneous chemical reactions, **his one-dimensional NE-ZND model consists of four zones:**

- a very thin leading shock front in which the unreacted explosive mixture is compressed and accelerated in the direction of shock propagation;
- a much thicker relaxation zone in which the rotational and vibrational modes of the unreacted explosive gases approach thermal equilibrium;

---

[2988] C. NEWHALL and S. SELF: *The Volcanic Explosivity Index (VEI): an estimate of explosive magnitude for historical volcanism.* J. Geophys. Res. **87**, No. C2, 1231-1238 (1982).
[2989] T. SIMKIN and L. SIEBERT: *Volcanoes of the world.* Geoscience Press, Tucson, AZ (1994).
[2990] For more information about VEI *see How big are volcanic eruptions?* Volcanoworld. Dept. of Geosciences, Oregon State University, Corvallis, OR; http://volcano.und.nodak.edu/vwdocs/eruption_scale.html.
[2991] *Supervolcano.* Wikipedia (April 10, 2006); http://en.wikipedia.org/wiki/Supervolcano#Known_eruptions.
[2992] C.M. TARVER: *Chemical energy release in one-dimensional detonation waves in gaseous explosives.* Combust. Flame **46**, 111-133 (1982); *Chemical energy release in the cellular structure of gaseous detonation waves.* Ibid. **46**, 135-156 (1982); *Chemical energy release in self-sustaining detonation waves in condensed explosives.* Ibid. **46**, 157-176 (1982).

> a relatively thin zone in which the chemical energy is released by rapid chain propagation and branching reactions into highly vibrationally excited reaction product gases; and

> another very thick relaxation zone in which the product gases expand.

(ii) TARVER demonstrates how to apply his NE-ZND model to the 3-D cellular structure of gaseous detonations, and to solid and liquid explosives.

| | | |
|---|---|---|
| 1982 | Theoretical Division, Los Alamos National Laboratory (LANL), Los Alamos, New Mexico | **Duane C. WALLACE,**[2993] a theoretical physicist, studies the thermodynamics of shock-compressed dense liquids (with densities > 1 g/cm$^3$), and **discovers that shocks in dense fluids are quite different in nature to those in gases.** He shows that, in addition to the condition for a compressive shock to propagate as a continuous steady wave {Lord RAYLEIGH ⇨1910}, two more limits exist for dense fluids, based on the fluid response at the leading edge of the shock:<br>> for shocks at the overdriven threshold and above, no solution is possible without heat transport; and<br>> for shocks near the viscous fluid limit (estimated to be 13 kbar for water and 690 kbar for mercury) and above, the fluid response at the leading edge of the shock is approximately that of a non-plastic solid. |
| 1983 | Washington, DC | On March 23, **U.S. President Ronald REAGAN announces his *Strategic Defense Initiative (SDI)*: a range of systems intended to protect against nuclear missiles.** ▪ The ambitious SDI Program — commonly called **"Star Wars"** after one of the popular science fantasy movies of the time — stimulated military-related research in many ways, particularly in shock wave physics, high-energy laser physics, detonics, and pulsed power technology. It also included several exotic projects which looked at the possibility of using antimatter to drive rockets and space-based weapons platforms. Since the contact of antimatter {DIRAC ⇨1928; CHAMBERLAIN & SEGRÉ ⇨1955} with ordinary matter results in complete conversion of the matter involved to energy, antimatter is the ultimate fuel.<br><br>Eventually, in May 1993, REAGAN's SDI concept was shelved due to its technical impracticability. Some feasibility studies, however, were continued. Military experts believed that aiming high-power laser systems at shells and rockets could dramatically alter the face of the battlefield. On November 2, 2002, **a high-energy laser was successfully used to destroy an artillery shell in supersonic flight for the first time.** The laser's tightly focused beam generated intense heat that caused the projectile to explode harmlessly in flight during the test. This operation, conducted by the U.S. Army at White Sands Missile Range, NM, was achieved with a test version of the Mobile Tactical High-Energy Laser developed by TRW Space & Electronics. |
| 1983 | Dept. of Mechanical Engineering, University of Liverpool, England | **First International Symposium on Structural Crashworthiness.** The chairman is Norman JONES. The purpose of this Symposium is "to bring research workers together from various branches of the structural crashworthiness field and provide an opportunity for the exchange of views and information in order to stimulate further progress." The structural crashworthiness of aircraft, cars, buses, trains, ships and offshore platforms are discussed as well as recent studies into the collapse and energy absorption of structural members. ▪ Subsequent symposia were held at Cambridge, MA (1988), and Liverpool, U.K. (1993). The 4th Symposium was renamed the *International Conference on Impact Loading of Lightweight Structures* and held at Florianopolis, Brazil (2005). |

---

[2993] D.C. WALLACE: *Theory of the shock process in dense fluids*. Phys. Rev. **A25** [III], 3290-3301 (1982).

| | | |
|---|---|---|
| 1983 | *University of Munich, Munich, Bavaria, Germany* | **Foundation of the *Deutsche Gesellschaft für Stoßwellenlithotripsie* ("German Society of Shock Wave Lithotripsy")** in order to discuss and clarify practical problems arising from this therapy. The first president is the urologist Egbert SCHMIEDT. ▪ This forum, composed of physicians, physicists and engineers, was the first in the world to be dedicated exclusively to this medical therapy. It soon stimulated the creation of similar organizations in other countries and symposia on this subject {DGST & IGESTO ⇨1995}. |
| 1983 | *Springer International, Heidelberg, Germany* | **Foundation of the international journal *Experiments in Fluids*** in order "to publish and disseminate research papers concerned with the development of new measuring techniques, and with their extension and improvement, for the measurement of flow properties necessary for the better understanding of fluid mechanics." The first editors are Wolfgang F. MERZKIRCH and James H. WHITELAW. ▪ Traditionally, many publications are also devoted to high-speed flow and shock waves – thus complementing the Springer journal *Shock Waves* {⇨1990}. |
| 1983 | *Pergamon Press, Oxford, England* | **Foundation of the *International Journal of Impact Engineering*** in order "to publish original research work concerned with the response of structures and bodies to dynamic loads arising from exposure to blast, collision or other impact events." The first editor-in-chief is William JOHNSON. ▪ The topics covered by this journal include the elastic and plastic response of structures and bodies to impact and blast loading, terminal ballistics, vehicle crashworthiness, containment, and other processes and phenomena in which effects due to impact predominate.<br><br>Impact physics and impact engineering – which both form an interesting link between classical "percussion mechanics" and modern shock wave physics – cover a broad range of impact loading phenomena that have gained increasing importance in mechanical engineering, aero- and astronautics, ballistics, geology, astrophysics, astrogeology, and planetology. |
| 1983 | *Castellvecchio, Tuscany, Italy* | At the meeting of the NATO Advanced Study Institute (NATO-ASI), **J. CHANG**,[2994] a researcher at Sandia National Laboratories (Albuquerque, NM), **reports on an unusual two- and three-frame flash radiography system for studying the implosions of hollow micro targets in particle beam inertial confinement fusion experiments cinematographically.**<br><br>(i) Each X-ray source has a spot size of 100 μm (realized by a 50-μm tungsten needle anode) and are driven by a 3-ns, 600-kV, 10-kA commercially available pulsed-power source.<br><br>(ii) The X-ray detectors use microchannel plates (MCPs) extensively as X-ray converters, gated shutters, and intensifiers to provide signal detection, background discrimination, and signal amplification {FARNSWORTH ⇨1930}. Curved MCP cameras were developed to permit easier shielding of the phosphor screen and the film package.<br><br>(iii) Cineradiographs of imploding hollow spherical targets (3 mm outer diameter, 25 μm wall thickness) and hollow cylindrical targets (6 mm in length, 16.5 μm wall thickness) made of high-Z materials (*e.g.*, Au or W) show that they implode with a high degree of symmetry and attain high average velocities ranging from 2,000 m/s for spherical targets to 10,000 m/s for cylindrical targets. |
| 1983 | *Luleå, Sweden & Cavendish Laboratory, Cambridge, England* | **Martin B. LESSER**,[2995] a mechanics professor at the University of Luleå, **and John E. FIELD**, a physics professor at the University of Cambridge, **discuss the problems of *liquid impact* in the intermediate speed range** (speeds above those obtainable by falling bodies but lower than supersonic speeds).<br><br>(i) They illustrate the general problems of liquid-solid impact using four configurations: |

---

[2994] J. CHANG: *3-ns flash X-radiography.* In: (J.E. THOMPSON and L.H. LUESSEN, eds.) *Fast electrical and optical measurements.* NATO-ASI Series, Series E: Applied Science, No. 109. Martinus Nijhoff Publ., Dordrecht etc. (1986), pp. 863-884.

[2995] M.B. LESSER and J.E. FIELD: *The impact of compressible liquids.* Annu. Rev. Fluid Mech. **15**, 97-122 (1983).

- a (spherical) liquid drop impacting a solid, the ideal model configuration;
- a normal impact of a liquid mass against a solid;
- a cylindrical jet impacting against a solid; and
- a two-dimensional impact {BRUNTON & CAMUS ⇨ 1970}.

(ii) The major features of such impacts are high edge pressures, side jetting with velocities far exceeding the impact velocity, and cavitation. These phenomena can be understood by employing a theoretical model of edge shock detachment and subsequent pressure relief leading to both jetting and cavitation.

(iii) They also discuss attempts to predict and measure the pressure distribution on the surface during liquid impact analytically, a difficult task because the pressures are intense, transient and highly localized.

| | | |
|---|---|---|
| 1983 | *Dept. of Engineering, University of Aberdeen, Scotland, U.K.* | **Stephen Robert REID,**[2996] a professor of mechanical engineering, **and his collaborators use high-speed cinematography to study a line of mild steel rings subjected to end impact.** In order to analytically model the deformation of the one-dimensional ring system, which proceeds from one ring to the next in a wave-like manner, they construct a shock wave theory analogous to that used for the propagation of longitudinal plastic waves along uniform metal bars {LEE & WOLF ⇨ 1951}, and obtain reasonable agreement with experimental observations. |
| 1983 | *Gas Dynamics Laboratory, Penn State University, Pennsylvania* | **Gary S. SETTLES,**[2997] a U.S. professor of mechanical engineering and director of the Gas Dynamics Laboratory, **reports on the use of retroreflective foils** {EDGERTON ⇨ 1958} **to produce large-field color schlieren images.** Using this technique, he designs the largest indoor schlieren system in the world, with a $2.9 \times 2.3$-m$^2$ field of view, based on the lens-and-grid principle {SCHARDIN ⇨ 1942}. ▪ **This technique has recently been applied in aviation security studies** to visualize the motion of blast waves generated by chemical explosions beneath full-size aircraft seats {⇨Fig. 4.5–N}, and inside luggage containers.[2998] |
| 1983 | *Institute of Fluid Science & School of Medicine, Tohoku University, Sendai, Japan* | **Kazuyoshi TAKAYAMA,**[2999] a fluid dynamics professor and one of Japan's foremost shock wave physicists, **and collaborators experiment with extracorporeal shock wave lithotripsy** {CHAUSSY, EISENBERGER ET AL. ⇨ 1970}. Although they initially use electric microdischarges for shock wave generation, they soon apply **laser-ignited microexplosives** (*e.g.*, 10 mg pellets of lead azide) instead – a technique known as **"microexplosive lithotripsy."** The new method provides better shock pressure reproducibility and does not require a high-voltage pulse generator. The microexplosives are used *outside* the patient, in contrast to a previous method developed in Japan, which used microexplosives *inside* the patient {WATANABE ET AL. ⇨ 1981}. ▪ Microexplosive lithotripsy, in combination with the use of ellipsoidal reflectors to concentrate the energy on the calculi, proved feasible and was first applied in 1985 to treat a patient sitting in a water tub. However, because of several drawbacks – in particular the need to replace the microexplosive after each shot – microexplosive lithotripsy didn't gain much popularity as a therapy. |

---

[2996] S.R. REID, W.W. BELL, and R.A. BARR: *Structural plastic shock model for one-dimensional ring systems.* Int. J. Impact Engng. **1**, 175-191 (1983).

[2997] G.S. SETTLES: *Large-field color schlieren visualization of transient fluid phenomena.* Bull. Am. Phys. Soc. **28**, 1404 (1983); *Full-scale schlieren flow visualization.* Proc. 3rd Symp. on Flow Visualization [Seattle, WA, Sept. 1995]. In: (J.P. CROWDER, ed.) *Flow visualization III.* Begell House, New York (1995), pp. 2-13.

[2998] G.S. SETTLES, B.T. KEANE, B.W. ANDERSON, and J.A. GATTO: *Shock waves in aviation security and safety.* Shock Waves **12**, 267-275 (2003).

[2999] K. TAKAYAMA, H. ESASHI, and N. SANADA: *Propagation and focusing of spherical shock waves produced by underwater microexplosions.* 14th Int. Symposium on Shock Tubes & Waves [Sydney, Australia, Aug. 1983]. In: (R.D. ARCHER and B.E. MILTON, eds.) *Proc. 14$^{th}$ Int. Sydney Symposium on Shock Tubes and Shock Waves.* University of New South Wales Press, Sydney (1984), pp. 553-562.

| | | |
|---|---|---|
| 1983 | Institute of Urology & Dept. of Lasers, University College Hospital, London, England | **Graham M. WATSON**[3000] **and John E.A. WICKHAM,** two renowned British urologists, **and collaborators fragment urinary calculi with laser-induced shock waves.** They use a flashlight-pulsed dye laser and transmit the energy through thin quartz fibers in direct contact with the stone, which results in a shock pulse with a duration of 15–20 ns. This method fragments stones safely, causing only 2% ureteral strictures and 7% perforations. It is not yet ready, however, for clinical application. ▪ Continued research into and improvements of their concept – **"laser-induced shock wave lithotripsy"** – resulted in a first series of successful treatments that were conducted at the Massachusetts General Hospital in Boston, MA. The laser light (pulse energy 30 mJ) was delivered through a fine quartz fiber only 0.25 mm in diameter; *i.e.*, considerably smaller than other devices previously used for fragmentation. Up to 4,000 pulses may be required to provoke the fragmentation of hard and large calculi. |
| 1983 | Dept. of Mechanical Engineering, Colorado State University, Fort Collins, Colorado | **Paul J. WILBUR,**[3001] a U.S. mechanical engineer, **and collaborators propose a new space launch concept – a ramjet –** as a possible alternative to proposed railguns and mass drivers for high-acceleration launch missions which are difficult to realize in practice because of the enormous current densities to be managed. Their method uses electrical heating in order to accelerate payloads up to 14–15 km/s. ▪ Their idea was later resumed by Abraham HERTZBERG and associates, although the necessary energy was provided by them chemically via combustion rather than electrically by making the projectile fly through a premixed fuel-oxidizer mixture {HERTZBERG ET AL. ⇨1986; ⇨Fig. 4.10–L}. |
| 1984 | Allan Hills, Far Western Icefield, Antarctica | On December 27, **a stony meteorite – later named the "Allan Hills meteorite" and labeled as ALH 84001 – is found by a team of meteorite hunters supported by ANSMET** (Antarctic Search for Meteorites), a program sponsored by the Polar Programs Office of the U.S. National Science Foundation (NSF).<br>    (i) The potato-sized 1.9-kg meteorite is one of only 12 meteorites identified so far that match the unique Martian chemistry measured by the Viking spacecraft that landed on Mars in 1976 {Viking 2 Lander ⇨1976}. ▪ Its possible Martian origins were not recognized until 1993. **ALH 84001 is believed to have originated from the impact of a huge comet or asteroid into Mars, which ejected a piece of Martian rock (the meteorite) with enough force to escape the planet** (> 5 km/s). This ejection mechanism has also been suggested to apply to other meteorites from Mars.[3002, 3003] After floating through space for about 16 million years (an estimated value based upon cosmic ray exposure data), ALH 84001 encountered the Earth's atmosphere about 13,000 years ago, until landing in Antarctica.<br>    (ii) ALH 84001 contains a variety of organic molecules and possibly also nanometer-scale structures that may have been produced by bacteria-like microorganisms. NASA researchers,[3004] studying these structures in meteorite ALH 84001, concluded: "Although inorganic formation is possible, formation of the globules by biogenic processes could explain many of the observed features, including the polycyclic aromatic hydrocarbons (PAHs). The PAHs, the carbonate globules, and their associated secondary mineral phases and textures could thus be fossil remains of a past Martian biota." |

---

[3000] G.M. WATSON, J.E.A. WICKHAM, ET AL: *Laser fragmentation of renal calculi*. Brit. J. Urology **55**, No. 6, 613-616 (Dec. 1983); *Initial experience with a pulsed dye laser for ureteric calculi*. Lancet I, pp. 1357-1358 (June 14, 1986).
[3001] P.J. WILBUR, C.E. MITCHELL, and B.D. SHAW: *The electrothermal RAMJET*. J. Spacecraft Rockets **20**, 603-610 (1983).
[3002] R.A. KERR: *Martian meteorites are arriving*. Science **237**, 721-723 (Aug. 14, 1987).
[3003] A.M. VICKERY and H.J. MELOSH: *The large crater origin of SNC meteorites*. Science **237**, 738-743 (Aug. 14, 1987).
[3004] D.S MCKAY ET AL.: *Search for past life on Mars: possible relic biogenic activity in Martian meteorite ALH 84001*. Science **273**, 924-930 (Aug. 16, 1996).

| | | |
|---|---|---|
| | | The Martian meteorite ALH 84001 reinvigorated discussions on the possibility that life may have existed at some point on Mars,³⁰⁰⁵ as well as the existence of extraterrestrial life in general {Allende Meteorite ⇨1969; Murchison Meteorite ⇨1969}. ALH 84001 is well on its way to becoming the single most important planetary specimen ever recovered, having been the focus of hundreds of scientific publications and discussed in thousands of others. |
| 1984 | *U.S. Geological Survey (USGS), Denver, Colorado* | Bruce F. BOHOR,³⁰⁰⁶ Peter J. MODRESKI, and Eugene E. FOORD, three U.S. geologists, **discover shock-metamorphosed quartz grains in a claystone at the Cretaceous-Tertiary (K/T) boundary near Brownie Butte in the Hell Creek area of east-central Montana.** This represents the first mineralogical evidence supporting the existence of an impact by an extraterrestrial body {Chicxulub Crater ⇨c.65 Ma ago}. **The shocked quartz grains exhibit planar deformation features – "shock lamellae."** • The principal evidence for this scenario before this discovery was the presence of anomalously high amounts of iridium and other siderophile elements measured in several boundary clay sites around the world {ALVAREZ ⇨1978}.<br><br>Three years later, BOHOR³⁰⁰⁷ and colleagues reported that they had found these characteristic planar features of shocked quartz at several sites worldwide, thus confirming that an impact event at the K/T boundary distributed ejecta products in a dust cloud that encircled the Earth, as postulated by the Alvarez impact hypothesis {ALVAREZ ⇨1978}. • The only other possible causes involved volcanism (but this had never been shown to produce sufficient shock pressures to give rise to more than one set of lamellae in quartz grains) and nuclear bomb explosions.³⁰⁰⁸ |
| 1984 | *Mount St. Helens, Skamania, Washington* | During a field trip to Mount St. Helens, **Susan W. KIEFFER,** a U.S. geologist, **and Bradford STURTEVANT,** a U.S. aeronautical engineer, **notice some unusual huge erosional "furrows" scoured into hill slopes by the lateral blast generated by the eruption four years ago** {Mt. St. Helens ⇨1980}. STURTEVANT interprets these furrows as being Taylor-Görtler vortices {GÖRTLER ⇨1940}. • His idea led to a detailed field program documenting, measuring and interpreting these furrows. These studies led to a new hypothesis: the vortices and furrows were produced by the geometry of a nose cone pointing obliquely into the oncoming supersonic flow of the blast {KIEFFER ⇨1981}. This mechanism confirmed the supersonic nature of the flow and allowed an estimate of flow velocities.³⁰⁰⁹ |
| 1985 | *Paris, France* | In September, **first International Conference on Mechanical and Physical Behavior of Materials under Dynamic Loading ("DYMAT 85").** The chairman is Jean PHILIBERT. • The following DYMAT conferences were held at Ajaccio, France (1988) and Strasbourg, France (1991).<br><br>**In 1994, the DYMAT Association** – originally a solely French organization – **was expanded to become a European organization, now known as "EURODYMAT."** The DYMAT Conferences were combined with the previous Oxford Conferences on Mechanical Properties of Materials at High Rates of Strain {⇨1974}. Subsequent EURODYMAT conferences |

---

³⁰⁰⁵ R.A. KERR: *Ancient life on Mars?* Science **273**, 738-743 (Aug. 16, 1996).

³⁰⁰⁶ B.F. BOHOR, E.F. FOORD, P.J. MODRESKI, and D.M. TRIPLEHORN: *Mineralogical evidenc for an impact event at the Cretaceous-Tertiary boundary layer.* Science **224**, 867-869 (May 25, 1984).

³⁰⁰⁷ B.F. BOHOR, P.J. MODRESKI, and E.E. FOORD: Shocked *quartz in the Cretaceous-Tertiary boundary clays – evidence for a global distribution.* Science **236**, 705-709 (May 8, 1987).

³⁰⁰⁸ In the late 1940s, U.S. scientists discovered that the quartz sand from the soil in the crater produced by a nuclear explosion looked different under a microscope than "normal" quartz: small parallel lines that intersected other parallel lines criss-crossed the face of the grains. These intersecting lines gave the impression of fractures in the quartz, but X-ray diffraction analysis of these grains showed that the crystalline lattice was only slightly deformed, and it was not fractured. These microscopic planar features were named "shock lamellae." See P. BIGELOW: *What is shocked quartz?* Hell Creek Life; http://www.scn.org/~bh162/shocked_quartz.html.

³⁰⁰⁹ S.W. KIEFFER and B. STURTEVANT: *Erosional furrows formed during the lateral blast at Mount St. Helens, May 18, 1980.* J. Geophys. Res. **93** (B12), 14793-14816 (1988).

were held at Toledo, Spain (1997); Crakow, Poland (2000); Porto, Portugal (2003); and Dijon, France (2006).

| 1985 | *University of Liverpool, northwestern England* | **Foundation of the Impact Research Centre (IRC)** at the Dept. of Mechanical Engineering, in order "to integrate several disciplines which contribute to our understanding of those problems involving the large dynamic loading response and failure of materials and structures which occur throughout the field of engineering." The first director is Norman JONES. |

| 1985 | *Institut für Neutronen- und Reaktorphysik (INR), Kernforschungszentrum Karlsruhe (KFK) [now Forschungszentrum Karlsruhe], Germany* | **Hans-Joachim BLUHM,**[3010] a German physicist, **and collaborators report on high-pressure shock experiments performed using the Karlsruhe Light Ion Facility (KALIF).** This light ion generator is capable of delivering intense 50-ns proton beams with power densities of up to 1 TW/cm$^2$ and depositing ion energies of up to about 40 kJ into a focal spot 6–8 mm in diameter. Methods involving either the direct interaction of the ion beam with test samples or indirect interaction achieved by impacting very thin flyer plates allow the realization of load durations in the nanosecond time range, thus considerably extending standard high-pressure shock experiments. Velocity measurements are made using laser Doppler velocimetry. |

In the years that followed, KALIF proved to be a versatile research tool in shock wave physics and materials dynamics. For example, in cooperation with Russian researchers from the High Energy Density Research Center at Izhorskaya, Moscow and the Institute for Chemical Physics Research (Chernogolovka, Moscow Oblast) {KANEL' ET AL. ⇨2004}, KALIF was used in

- equation-of-state (EOS) studies, including the interpretation of compression and relaxation processes;[3011, 3012]
- hypervelocity impact studies;[3013]
- Rayleigh-Taylor instability experiments;[3014] and
- investigations into materials dynamics at high loading rates of about 10$^{-7}$/s or less.[3015]

| 1985 | *Center of Electromagnetics Research, Northeastern University, Boston, Massachusetts & Physics Dept., King's College, London, England* | Peter GRANEAU[3016] (Boston) and Neal GRANEAU (London), two U.S. physicists, **carry out a current pulse experiment to look for longitudinal forces in water at various salinity levels.**

(i) They observe that in some cases the current discharges silently, while in others a luminous arc strikes between the two electrodes, a copper rod and a copper ring, accompanied by a hissing sound and shock waves in the water.

(ii) The boundary between the two kinds of discharge depends on the total charge in the discharge. This suggests that an arc is formed when the number of ions present in the solution is insufficient to discharge the capacitor. The energy not dissipated through the electrolytic current obviously goes into the arc discharge. With the same amount of energy, the discharge can either be silent or can cause an arc explosion, depending on the capacity and voltage of the capacitor.

(iii) In a typical experiment, the average force during the explosion is 21.6 N, throwing a 2.8-g weight floating on the surface about 20 cm up in the air; the current is about 94 A. The calculated thrust from pinch forces is 0.55 mN – too small to account for the force. Pinch thrust would also be present during the silent explosions, which proves that the observed force has something to do with the presence of an arc. ▪ Later, current pulses in the range of 10 to 25 kA |

---

[3010] H. BLUHM ET AL.: *Experiments on KFK's light-ion accelerator KALIF.* Proc. 5th IEEE Pulsed Power Conf. [Arlington, VA, 1985]. In: (M.F. ROSE and P.J. TURCHI, eds.) *Digest of technical papers.* IEEE, New York (1986), pp. 114-117.

[3011] G. KANEL' ET AL.: *Possible applications of the ion beams technique for investigations in the field of equation of state.* Nucl. Instrum. Meth. Phys. Res. **A415**, 509-516 (1998).

[3012] K. BAUMUNG ET AL.: *Shock-wave physics experiments with high-power proton beams.* Laser and Particle Beams **14**, No. 2, 181-209 (1996).

[3013] G.I. KANEL' ET AL.: *Applications of the ion beam technique for investigations of hypervelocity impacts.* Proc. Hypervelocity Impact Symp. HVIS'98 [Huntsville, AL, Nov. 1998]; see Int. J. Impact Engng. **23**, 421-430 (1999).

[3014] K. BAUMUNG ET AL.: *First proton-driven Rayleigh-Taylor experiments on KALIF.* Nucl. Instrum. Meth. Phys. Res. **A415**, 720-725 (1998).

[3015] K. BAUMUNG ET AL.: *Investigations of the dynamic strength variations in metals.* J. Phys. Paris **7** [IV], C3: 927-932 (1997).

[3016] P. GRANEAU and N. GRANEAU: *Electrodynamic explosions in liquids.* Appl. Phys. Lett. **46**, 468-470 (1985).

were used with another arrangement. The maximum force observed amounted to 430 kN, equivalent to a pressure of 27,000 atm.[3017]

(iv) Due to various reasons, they exclude thermal forces due to Joule heating as well as the production of superheated steam as being responsible for driving a thermal explosion, and explain the unusual phenomenon of *water arc explosions* by a change in the chemical bond energy. ▪ Similarly, **P. GRANEAU** suggested that thunder may be driven by *electrodynamic explosions*, **and not by thermic heating of the air** {GRANEAU ⇨ 1989}.

Later measurements carried out by P. GRANEAU and colleagues also indicated that the energy released substantially exceeds the energy supplied by the arc.[3018]

---

**1985** — *College of Engineering, UC Berkeley, California*

Antoni K. OPPENHEIM,[3019] a Polish-born U.S. professor of aeronautical sciences, **provides a unified treatment of the entire field of combustion dynamics,** thereby focusing on the fluid dynamics rather than on the chemical kinetics. He discusses four prominent cases relevant to a wide spectrum of combustion phenomena:

> ▸ **ignition** – the initiation of a self-sustained exothermic process, considered in the simplest case of a closed thermodynamic system and its stochastic distribution;
>
> ▸ **inflammation** – the initiation and propagation of self-sustained flames, presented in the most interesting case of turbulent flow;
>
> ▸ **explosion** – the dynamic effects caused by the deposition of exothermic energy in a compressible medium, illustrated by two classical examples: self-similar blast waves with energy deposition at the front, and the adiabatic non-self-similar wave; and
>
> ▸ **detonation** – the most comprehensive illustration of all the dynamic effects of combustion, exposed here by a phenomenological account of the development and structure of the wave.

---

**1985** — *Max-Planck-Institut für Aeronomie, Katlenburg-Lindau, southern Germany*

Klaus RINNERT,[3020] a German physicist, **discusses the necessary conditions for producing lightning discharges on other planets and the experimental possibilities for investigating extraterrestrial lightning.**

(i) From the knowledge on planetary atmospheres, he concludes that intensive lightning activity can be expected to exist in the Jupiter and Saturn cloud systems.

(ii) He reports that optical and radio frequency (RF) wave measurements from spacecraft have yielded evidence of possible lightning activity on Venus, Jupiter, and Saturn.

In the following years, RINNERT started investigating Jupiter's atmosphere in a long-lasting U.S. collaboration with Louis J. LANZEROTTI (Bell Particles Laboratory). This culminated in the descent of the Galileo probe into the Jovian atmosphere in 1995, when his lightning detector showed a much lower flash rate than expected.

Spacecraft in our Solar System have detected radio signals consistent with lightning on other planets, including Venus, Jupiter, Saturn, Uranus, and Neptune. Optical flashes from Jupiter were photographed recently by the Galileo orbiter.[3021] The generation of thunderclaps (shock waves), a result of strong lightning-strokes in Jupiter's dynamic gaseous atmosphere – which includes complex molecules (such as ammonia and methane) as well as simple molecules (such as helium, hydrogen, and sulfur) – is most likely, but has not yet been proved.

---

[3017] R. AZEVEDO, P. GRANEAU, C. MILLET, and N. GRANEAU: *Powerful water-plasma explosions.* Phys. Lett. **A117**, 101-105 (1986).

[3018] P. GRANEAU and N. GRANEAU: *Newtonian electrodynamics.* World Scientific, Singapore (1996).

[3019] A.K. OPPENHEIM: *Dynamic features of combustion.* Phil. Trans. Roy. Soc. Lond. **A 315**, 471-508 (1985).

[3020] K. RINNERT: *Lightning on other planets.* J. Geophys. Res. **90**, 6225-6237 (1985).

[3021] *Lightning on Jupiter.* Astronomy picture of the day (May 12, 1997); http://antwrp.gsfc.nasa.gov/apod/ap970512.html.

It is interesting to note that in antiquity the planet-god Jupiter was frequently shown with a thunderbolt in his hand. The electrical discharge coming from Jupiter is described in many ancient texts {PLINY the Elder ⇨A.D. 77}.

| | | |
|---|---|---|
| 1985 | Max-Planck-Institut (MPI) für Physik und Astrophysik, Institut für Astrophysik, Garching [near Munich], Germany | **Michael D. SMITH,**[3022] **Michael L. NORMAN, Karl-Heinz A. WINKLER, and Larry L. SMARR review the nonlinear features and extraordinary stability of astrophysical supersonic jets.**<br>(i) For the beam cap (or "working surface") of an axisymmetric jet ramming through its confining medium, they numerically derive brightness distributions using a high-resolution gas dynamics code.<br>(ii) Based upon these numerical studies, as well as on experimental results, they discuss the terminal shock systems formed at the working surfaces of laboratory jets {SALCHER & WHITEHEAD ⇨1889; L. MACH ⇨1897} and astrophysical jets {COURTÈS & CRUVELLIER ⇨1961}, which resembles the triple-shock configuration formed by Mach reflection in steady-state jets.<br>(iii) They show that significant insights into the physics of extragalactic radio sources can be obtained through numerical hydrodynamic simulations, and propose a model for the expected double-shock structure at the jet terminus where two shocks are formed {⇨Fig. 4.8–L}:<br>▸ the *terminal jet shock* or *Mach disk*, which effectively stops the incoming jet; and<br>▸ the *stand-off shock* or *bow shock*, which acts to accelerate and heat the ambient medium.<br>The two shocked fluids (jet and ambient) meet in pressure balance along a contact discontinuity.<br>Cygnus A – the closest ultraluminous radio galaxy located in *Cygnus* ("The Swan," a summer constellation of the Northern Hemisphere about 600 million light-years from Earth) – played a dominant role in the development of the theory of the jets that power the double structures in powerful radio galaxies.[3023] The jet from Cygnus A itself was discovered in Very Long Baseline Interferometry (VLBI) observations of the nucleus in the mid-1970s, and it was finally revealed in detail by the first high dynamic range images of the source taken with the Very Large Array (VLA) in the mid-1980s. Early aperture synthesis images of Cygnus A led to the discovery of radio "hotspots" at the extremities of the source. **The expanding radio source Cygnus A produces two bow shocks in the external medium.**[3024] |
| 1986 | Washington, DC | On February 4, in his State of the Union address, **U.S. President Ronald REAGAN announces efforts made by the National Aero-Space Plane (NASP) Program to develop the *X-30 NASP* {⇨Fig. 4.20–I}** – a civilian commercial transatmospheric hypersonic plane intended to be capable of cruising at a speed of 20,000 mph (32,000 km/h, about Mach 25) in Earth's orbit.<br>• After spending $2 billion on development work, the project was eventually stopped in 1993, and no such aircraft were ever built.<br>The successor of X-30 was the unmanned X-43, also named **"Hyper-X"** {⇨2004}, a small test vehicle which was developed by NASA's Langley Research Center {NASA-LaRC ⇨1998} in cooperation with Boeing, Microcraft, and the General Applied Science Laboratory (GASL) in order to demonstrate hydrogen-fueled scramjet engines {BILLIG ⇨1959}. |

---

[3022] M.D. SMITH, M.L. NORMAN, K.H.A. WINKLER, and L.L. SMARR: *Hotspots in radio galaxies: a comparison with hydrodynamic simulations.* Month. Not. Roy. Astron. Soc. **214**, 67-85 (1985). M.L. NORMAN and K.H.A. WINKLER: *Supersonic jets.* Los Alamos Science Magazine **12**, 38-70 (Spring/Summer 1985); http://www.fas.org/sgp/othergov/doe/lanl/pubs/00326955.pdf.

[3023] C.L. CARILLI and P.D. BARTHEL: *Cygnus A.* Astron. Astrophys. Rev. **7**, 1-54 (1996); http://www.astro.phys.ethz.ch/staff/schmid/private/PSagn/carilli96.pdf.

[3024] M.C. BEGELMAN and D.F. CIOFFI: *Overpressured cocoons in extragalactic radio sources.* Astrophys. J. **345**, L21-L24 (1989).

| 1986 | Chernobyl Nuclear Power Station at Pripyat [a town ca. 18 km northwest of the city of Chernobyl and 110 km north of Kiev], Ukraine, Soviet Union | On April 26, **the Power Unit IV reactor of the Chernobyl Nuclear Power Plant** (constructed in 1983 and known as "Chernobyl IV") **explodes, resulting in the largest ever nuclear power plant disaster** {⇨Fig. 4.21−F}. |

(i) Ironically, the disaster starts from a routine safety test which gets out of control: the operators wish to determine the length of time the reactor cooling water would continue to flow if steam to the turbine is cut off and all on-site electrical power is lost. Because of an unusual aspect of the reactor design, a sequence of unlikely operations results in an increase in reactivity, which destroys the reactor itself.[3025]

(ii) The explosion lifts the heavy cap off the top of the reactor, allowing air to enter and react with the hot graphite, resulting in a major fire and a nuclear meltdown.

(iii) 32 people die in the accident and from efforts to put out the fire. 38 more people die of acute radiation sickness in the following months. • A large number of clean-up workers received significant doses of radiation; several hundreds of thousands of people had to be relocated because of concerns about radioactive contamination.

(iv) The destruction of the reactor housing and the release of large quantities of radioactive material produces nuclear contamination across Europe. • Medical experts predict that the number of deaths by cancer will increase by at least 100,000 people due to the Cs-137 fallout. However, this is unlikely to be detectable because of the large number of cancers that arise from other causes.

Ten years later, in April 1996, an international seminar with the title *Chernobyl Lessons – Technical Aspects* was held in Russia at Desnogorsk (Smolensk Oblast). It was organized by Russian scientists in order to reconstruct the sequence of events leading to the Chernobyl accident.[3026] Based on their post-accident analysis, they concluded that "**the first explosion resulted in deformation of walls in the under-reactor compartment, and the second explosion centered under the central hall roof completely destroyed the reactor core. The shock wave and subsequent rarefaction wave that arose were the main reasons that the power unit building was destroyed.** Sharp decreases in pressure inside the steam drum compartments following ejection of the graphite stack and upper slab caused deformation of the steel thermal insulating lining in the steam drum compartments and the ejection of wall blocks of the steam drum compartments inside the central hall. Three wall blocks fell down in the reactor well. Some of the fuel that remained solid, together with graphite, was ejected in the central hall, onto the roof of Power Unit III, the roof of the unit holding subsidiary systems of reactor equipment and beyond the boundaries of the NPP [Nuclear Power Plant] building. Evaporated and finely dispersed fuel was carried out into the atmosphere to an altitude of up to 7,500 m."

In 2000, the Russian magazine *Ogonyok* [Russ. *Огонёк*] published an interview with Konstantin P. CHECHEROV, a physicist at the RRC Kurchatov Institute in Moscow who had been ordered by the Moscow Atomic Energy Institute to investigate the Chernobyl accident just one day after it had occurred.[3027] He also contributed to the report presented in 1996 at the international seminar mentioned above by developing a realistic model of the Chernobyl catastrophe. **CHECHEROV's main conclusion was that the accident was not a "partial breakdown" – the original official version of events – but rather an explosion that exhausted practically all of the nuclear fuel from the plant into the environment.**

---

[3025] *Chernobyl – 15 years later*. Canadian Nuclear Association (CNA), Ottawa, Ontario, Canada; http://www.cna.ca/english/Articles/CHERNO.pdf. • There are two conflicting official theories about the cause of the accident; *see* http://en.wikipedia.org/wiki/Chernobyl_accident.

[3026] Y.M. CHERKASHOV ET AL.: *Post-accident state of Chernobyl-4*. In: *Chernobyl Lessons – Technical Aspects* [Int. Seminar held at Desnogorsk, Smolensk NPP, Russia, in April 1996]. Moscow (1996); http://www.ignph.kiae.ru/ins/osrez/osrez.htm.

[3027] V.J. GOLOVANOV: *Interview with Konstantin CHECHEROV*. Ogonyok No. 15 (April 2000); http://www.ropnet.ru/ogonyok/win/200015/15-30-33.html.

| | | |
|---|---|---|
| 1986 | Southwest Research Institute (SwRI), San Antonio, Texas | **First Hypervelocity Impact Symposium (HVIS).** The general chairman is Charles E. ANDERSON JR. The objectives of this international meeting are<br>▸ **to promote a basic understanding of the physics of hypervelocity impact events;**<br>▸ to provide a comprehensive historical review of the research that had been conducted in the various hypervelocity impact disciplines; and<br>▸ to install a forum conducive to the presentation of research in the general field of hypervelocity impact.<br>HVIS was preceded by a number of workshops initiated by attempts to understand and assess the state of the art in hypervelocity impact phenomenology {1st Rand Symp. ⇨1955}. Subsequent symposia were held at San Antonio, TX (1989); Austin, TX (1992); Santa Fe, NM (1994); Freiburg, Germany (1996); Huntsville, AL (1998); Galveston, TX (2000); Noordwijk, The Netherlands (2003); Lake Tahoe, CA (2005); and Williamsburg, VA (2007). Since HVIS-1992 the proceedings are published in the British *International Journal of Impact Engineering* {⇨1983}.<br>***Hypervelocity impact*** is a field of research which began in the United States in the mid-1950s, focusing on the hazards to NASA spacecraft from meteorites and the feasibility of using solid projectiles to defend against incoming reentry vehicles. In the late 1970s, interest in this field was revitalized by an active program looking into systems that could provide non-nuclear kill defense against reentry vehicles. Interest in hypervelocity impact has grown considerably with<br>▸ the advent of space missions to comets {Deep Impact Mission ⇨1999; Stardust Mission ⇨1999};<br>▸ NASA's interest in the space station;<br>▸ the Strategic Defense Initiative {SDI Concept ⇨1983}; and<br>▸ the advent of theories of large meteoroid or cometary impacts against the Earth as well as other planets, moons, and asteroids.[3028] |
| 1986 | Seismological Laboratory, CalTech, Pasadena, California | **Thomas J. AHRENS,**[3029] a U.S. geophysicist, **and John D. O'KEEFE,** a U.S. planetary physicist, **discuss crater phenomena produced on Earth by the impact of meteorites into rock or into the ocean and meteorites that disintegrate in the air.** Using numerical techniques, they calculate the penetration, cratering and ejection process for various projectile and target material combinations. They believe that their numerical results, obtained via impact mechanics, support Luis and Walter ALVAREZ's model of mass extinction {⇨c.65 Ma ago; ALVAREZ ⇨1978}. |
| 1986 | Aerospace and Energetics Research Laboratory (AERL), Dept. of Aeroand Astronautics, University of Washington, Seattle, Washington | **Abraham HERTZBERG,**[3030] a U.S. physics professor and director of AERL, **Adam P. BRUCKNER,** a research associate professor of aeronautics and astronautics, **and David W. BOGDANOFF,** a research engineer, **report on a new propulsion technique by which relatively large masses** (up to hundreds of kilograms) **can, in principle, be accelerated efficiently to hypervelocities** (up to 12 km/s) **using chemical energy.** Their so-called **"Ram Accelerator (RAMAC)"** {⇨Fig. 4.10–L} consists of a tube filled with a combustible gas which thrusts a projectile with a special geometry forward during combustion. ▪ Initial aspects of this concept had already been envisioned in 1983 by Prof. HERTZBERG; the first accelerator of this particular design was installed at the University of Washington in 1985.[3031] |

---

[3028] C.E. ANDERSON JR., N.W. BLAYLOCK, and J.S. WILBECK: *Preface.* Proc. Hypervelocity Impact Symposium HVIS'86 [San Antonio, TX, Oct. 1986]; see: Int. J. Impact Engng. **5**, vii-viii (1987).

[3029] T.J. AHRENS and J.D. O'KEEFE: *Impact on the Earth, ocean and atmosphere.* Proc. 1st Hypervelocity Impact Symposium (HVIS'86) [San Antonio, TX, Oct. 1986]; see Int. J. Impact Engng. **5**, 13-32 (1987).

[3030] A. HERTZBERG, A.P. BRUCKNER, and D.W. BOGDANOFF: *The ram accelerator: a new chemical method for achieving ultrahigh velocities.* In: (G. DROUIN, ed.) *Proc. 37th Meeting of the Aeroballistic Range Association* [Quebec, Canada, Sept. 1986]. Defense Research Establishment Valcatier, Courcelette/Quebec, Canada (1986); published in AIAA J. **26**, No. 2, 195-203 (Feb. 1988).

[3031] A. HERTZBERG, A.P. BRUCKNER, D.W. BOGDANOFF, and C. KNOWLEN: *The ram accelerator and its applications.* In: (H. GRÖNIG, ed.) *Proc. 16th Int. Symposium on Shock Tubes and Waves* [Aachen, Germany, July 1987]. VCH-Verlag, Weinheim/Bergstr. (1988), pp. 117-128.

These efforts stimulated worldwide interest in hypervelocity launchers of this type. In order to provide a forum for exchanging information on the latest advances in the state of the art of this novel hypervelocity launcher technology, four International Workshops on Ram Accelerators (RAMAC I to RAMAC IV) have been held so far: at Saint-Louis, France (1993); Seattle, WA (1995); Sendai, Japan (1997); and Poitiers, France (2000). There are now six ram accelerator facilities of various sizes in the United States, France, and Japan.

**The ram accelerator, which is simpler to construct than rockets and also more cost-effective for the transport of materials insensitive to accelerations, could be the perfect method for delivering such materials into space.**[3032, 3033] Reviews of the development of ram accelerator technology up to 1997 and 2002 have been given by Günter SMEETS[3034] (ISL, Saint-Louis, France) and Adam P. BRUCKNER[3035] (University of Washington, Seattle, WA), respectively.

| | | |
|---|---|---|
| 1987 | Las Campanas Observatory [operated by the Carnegie Institute of Washington], La Serena, Chile | On February 23, **Ian SHELTON**, a young Canadian astronomer in the Dept. of Astronomy and Astrophysics at the University of Toronto, **reports on the discovery of a celestial object that could be seen with the naked eye** (*i.e.*, magnitude 5 or brighter), **ostensibly a supernova in the Large Magellanic Cloud,** a dwarf galaxy that is a satellite to our own Milky Way and about 163,000 light-years away.[3036, 3037] |

The catastrophic stellar explosion – later termed ***supernova SN 1987A*** – is a centenary event for astrophysicists:

- it is a new type of supernova in terms of its relatively compact progenitor, dim light curve, and some special spectral characteristics; and
- it emits neutrino bursts which are recorded independently by the Kamiokande II and IBM underground water Cherenkov detectors. These bursts mark the birth of a neutron star and confirm previous speculations {BAADE & ZWICKY ⇒ 1934}.

Subsequently, also gamma-rays, produced by radioactive decay, and hard X-rays are observed. The soft X-rays postulated to originate from the interaction of the explosion blast wave with circumstellar matter were not found until 1991, when they were discovered by ROSAT (short for Röntgensatellit), the German /British/American X-ray observatory. Since then the luminosity has steadily increased, indicating that the shock wave is propagating into the tail of the red supergiant wind, which the progenitor star had blown off at an earlier phase in its evolution.[3038] **Regions protruding outside the main shell of X-ray emission were later discovered and interpreted as being Mach cones** {E. MACH & SALCHER ⇒ 1886} **generated by rapidly moving hot star fragments** {ASCHENBACH ET AL. ⇒ 1995; ⇒ Fig. 4.1–Y}.

Today a supernova is understood as being the explosion of a massive star in which the core undergoes gravitational collapse, and the upper layers of its atmosphere are blown off. A supernova can be physically classified in terms of its layer characteristic (Type Ia: originates in a binary system; Type Ib: outer H layer stripped before collapse; Type Ic: outer H and He layers stripped before collapse; Type II: outer H layer remains at collapse). Supernovae are currently

---

[3032] M. COOKSEY: *Ram accelerator: space launch system of the future.* The Trend in Engineering **39**, No. 1, 1, 11 (1988).
[3033] H. WILLIAMS: *Shot into orbit.* The Seattle Times (May 23, 1988), pp. F1-F2.
[3034] G. SMEETS: *The ram accelerator: perspectives and experimental results already achieved.* In: (M. CHAMPION and B. DESHAIES, eds.) *IUTAM Symposium on Combustion in Supersonic Flows.* Kluwer, Dordrecht (1997), pp. 219-236.
[3035] A.P. BRUCKNER: *The ram accelerator: a technology overview* [invited paper]. 40th Aerospace Sciences Meeting and Exhibit [Reno, NV, Jan. 2002]. AIAA, Reston, VA (2002), *see* Paper AIAA 2002-1014.
[3036] D. HELFAND: *Bang: the supernova of 1987.* Phys. Today **40**, 25-32 (Aug. 1987).
[3037] A. BURROWS: *The birth of neutron stars and black holes.* Phys. Today **40**, 28-37 (Sept. 1987).
[3038] G. HASINGER, B. ASCHENBACH, and J. TRÜMPER: *The X-ray light curve of SN 1987A.* Astron. Astrophys. **312**, L9-L12 (1996).

| | | |
|---|---|---|
| 1987 | *Enrico Fermi Institute, Chicago, Illinois; NBS, Gaithersburg, Maryland & Scripps Institution of Oceanography, La Jolla, California* | **A team of U.S. researchers reports on the discovery of very fine-grained meteorite-embedded diamonds which contain an isotopic mixture of xenon gas not found on Earth, thus indicating an extra-solar origin for the diamonds.**[3040]<br><br>(i) The team proposes the idea that the lucent crystals formed in the atmosphere of a "red giant" (or "dying star") before it collapsed and exploded billions of years ago, sending the diamond-studded material far out into space and eventually falling to Earth.<br><br>(ii) They present evidence that part or all of the very fine-grained type of carbon is diamond which is not shock-produced but primary formed by stellar condensation as a metastable phase. If this scenario is correct, the team concludes, then interstellar dust may contain diamonds.<br><br>Three years later, **Gary R. Huss**,[3041] a researcher at the Enrico Fermi Institute (University of Chicago, IL), **showed that interstellar diamond and silicon carbide [SiC] were incorporated into all chondritic meteorites.** He concluded that his observations suggest that various classes of chondritic meteorites sampled the same Solar-System-wide reservoir of interstellar grains. Nanometer-sized diamonds (peaked around 2–5 nm) have been found not only in meteorites and interstellar dusts, but also in proto-planetary nebulae as well as in residues of man-made detonations {GREINER ET AL. ⇨1988} and diamond films. At LLNL, nanoparticles of diamonds for expeiments are obtained through synthesis from detonation.[3042] |
| 1987 | *Monterey, California* | At the 5th American Physical Society (APS) Conference on Shock Waves in Condensed Matter, **Melvin H. RICE, Robert G. MCQUEEN, and John M. WALSH,** three retired Los Alamos shock physicists, **receive the first APS Shock Compression Science Award** "for their outstanding and pioneering contribution to the field of shock wave physics" and their classical review article *Compression of Solids by Strong Shock Waves* {RICE, MCQUEEN & WALSH ⇨1958}, which summarizes early advances made in solid-state shock-wave techniques and theoretical analyses.<br><br>In the years following, recipients of this prestigious award were:<br><br>▸ **George E. DUVALL (1989)** "in recognition of his outstanding contributions to shock wave physics and his educational and organizational leadership in the shock physics community;"<br>▸ **Lev V. AL'TSHULER (1991)** "in recognition of his seminal and major contributions in the development of shock wave compression of condensed matter;"<br>▸ **Robert A. GRAHAM (1993)** "in recognition of his research achievements in shock compression of condensed matter and in particular the development of piezoelectric shock wave instrumentation;"<br>▸ **Thomas J. AHRENS (1995)** "in recognition of his outstanding contributions to the understanding of matter under shock compression and its application to problems in planetary physics;"<br>▸ **Arthur C. MITCHELL and William J. NELLIS (1997)** "in recognition of their pioneering experimental investigations of molecular and planetary fluids using shock compression;" |

discovered at the rate of about 20 or 30 per year, and several hundreds have already been cataloged.[3039]

---

[3039] *List of supernovae.* Harvard-Smithsonian Center for Astrophysics, Cambridge, MA; http://cfa-www.harvard.edu/cfa/ps/lists/Supernovae.html.
[3040] R.S. LEWIS, T. MING, J.F. WACKER, E. ANDERS, and E. STEEL: *Interstellar diamonds in meteorites.* Nature **326**, 160-162 (March 12, 1987).
[3041] G.R. HUSS: *Ubiquitous interstellar diamond and SiC in primitive chondrites: abundances reflect metamorphism.* Nature **347**, 159-162 (Sept. 13, 1990).
[3042] J.Y. RATY and G. GALLI: *Structural and electronic properties of isolated nanodiamonds.* Rept. UCRL-PROC-207103, LLNL, Livermore, CA (Oct. 8, 2004).

- **Lynn M. BARKER (1999)** "in recognition of his outstanding contributions to the temporal measurement and interpretation of nonlinear physical processes in shock-compressed matter;"
- **Yogendra M. GUPTA (2001)** "for many significant contributions to the mechanical, optical, and X-ray measurement of both continuum and microscopic aspects of shock waves in condensed matter;"
- **James R. ASAY (2003)** "in recognition of his contributions to understanding condensed matter and nonlinear physics through shock compression;" and
- **Vladimir E. FORTOV (2005)** "for pioneering research in high-energy density physics, strongly coupled plasmas, hot condensed matter, shock-compression science and their applications."

| 1987 | *Poulter Laboratory, SRI International, Menlo Park, California* | Donald R. CURRAN,[3043] Lynn SEAMAN, and Donald A. SHOCKEY, three U.S. shock physicists of the Shock Physics Group at Poulter Laboratory, **review the history of dynamic fracture work up to 1987, and describe the construction of mesomodels of microscopic failure, linking them to classical fracture mechanics.** • Shock-induced dynamic fracture of solids is of practical importance in many areas of materials science, chemical physics, engineering, and geophysics.<br><br>More recent developments in fracture mechanics were reviewed in the introduction of a recent monograph on "spall fracture" {ANTOUN ET AL. ⇨2003}. |
|---|---|---|
| 1987 | *Institute of Thermophysics, Novosibirsk, Siberia, Soviet Union* | Samson Semenovich KUTATELADZE,[3044] a Siberian thermophysicist, **and collaborators,** resuming previous studies on rarefaction waves {KUTATELADZE ET AL. ⇨1978}, **observe unique shock wave phenomena in a pure substance near to its thermodynamic critical point.** Since it is possible to obtain very low or even negative values of the fundamental gasdynamic derivative, this allows the demonstration of **rarefaction shock waves** – a wave phenomenon which early shock physicists had thought to be impossible {JOUGUET ⇨1904; ZEMPLÉN ⇨1905/1906}. |
| 1987 | *M.A. Lavrentyev Hydrodynamics Institute, Novosibirsk, Siberian Branch of the U.S.S.R. Academy of Sciences, Soviet Union* | L.A. MERZHIEVSKII[3045] and Vladimir M. TITOV present results from recent research related to the **collision of compact bodies with obstacles, with particular reference to the problem of the protection of spacecraft against meteoroid impact.**<br><br>(i) The characteristics of high-velocity impact in the case where the path of the impacting body is not normal to the obstacle surface are discussed using results obtained for ductile (metal), brittle, and porous obstacles.<br><br>(ii) Simple analytical expressions are presented which can be used to estimate the stability of structures against meteorite impact. |
| 1987 | *U.S. Naval Air Warfare Center (WAWC), China Lake, California* | Arnold T. NIELSEN,[3046] a U.S. chemist, **discovers the super explosive CL-20** – a high-density cyclic nitramine ($\rho$ = 2.04 g/cm$^3$). Its full chemical name is hexanitrohexaazaisowurtzitane [$C_6H_6N_{12}O_{12}$]. It is one of the world's most powerful non-nuclear explosives, with an extreme detonation pressure of about 450 kbar and a detonation velocity of 9,208 m/s.[3047] • Recently, this new highly energetic material has also been investigated for possible use as a gun propellant.[3048] |

---

[3043] D.R. CURRAN, L. SEAMAN, and D.A. SHOCKEY: *Dynamic failure of solids* [with 146 refs.]. Phys. Rept. **147**, 253-388 (1987).

[3044] S.S. KUTATELADZE, V.E. NAKORYAKOV, and A.A. BORISOV: *Rarefaction waves in liquid and gas-liquid media*. Annu. Rev. Fluid Mech. **19**, 577-600 (1987).

[3045] L.A. MERZHIEVSKII and V.M. TITOV: *High speed collision*. Combustion, Explosion, and Shock Waves (Fizika goreniya i vzryva) **23**, No. 5, 589-604 (Sept. 1987).

[3046] A.T. NIELSEN: *Polycyclic amine chemistry*. In: (G.A. OLAH and D.R. SQUIRE, eds.) *Chemistry of energetic materials*. Academic Press, San Diego, CA (1991), pp. 95-124.

[3047] R.L. SIMPSON, P.A. URTIEW, D.L. ORNELLAS, G.L. MOODY, K.J. SCRIBNER, and D.M. HOFFMANN: *CL-20 performance exceeds that of HMX and its sensitivity is moderate*. Propellants, Explosives, Pyrotechnics **22**, 249-255 (1997).

[3048] D. MUELLER: *New gun propellant with CL-20*. Propellants, Explosives, Pyrotechnics **24**, 176-181 (1999).

| | | |
|---|---|---|
| 1987 | *Institut für Werkstoffmechanik (IWM) der Fraunhofer-Gesellschaft (FhG), Freiburg, Germany* | **Rolf A. PRÜMMER,**[3049] an applied physicist at IWM and lecturer at Karlsruhe University, **publishes the textbook** *Explosiverdichtung pulvriger Substanzen* ("Explosive Compaction of Powdered Substances"). It reviews the various methods of producing shock waves appropriate for compacting powder samples, discusses the influence of pressure on the mechanical properties and structures of shock-compressed powders, and compares the peculiarities of dynamic compaction techniques with those for static techniques.<br><br>In 2004, PRÜMMER[3050] reviewed the prospects for ***hot explosive pressing*** of metal and ceramic powders and its combination with ***self-propagation high-temperature synthesis***, a new and promising method in powder metallurgy. |
| 1987 | *Laser Research Laboratory, Massachusetts; Eye and Ear Infirmary, Boston & Dept. of Ophthalmology, Harvard Medical School, Boston, Massachusetts* | **Carmen A. PULIAFITO,**[3051, 3052] a U.S. ophthalmic surgeon and co-inventor of optical coherence tomography, **and collaborators use laser-based high-speed photography to investigate "laser ablation of the cornea" caused by two different ultraviolet radiations from excimer lasers** {SEARLES & HART ⇨1975}. An analysis of the photographs obtained for the ablation plume taken 500 ns to 150 μs after laser irradiation at two wavelengths ($\lambda$ = 193 and 248 nm) shows that<br><br>▸ ejection of material from the cornea begins on a timescale of nanoseconds and continues for 5–15 μs following the excimer laser pulse;<br>▸ **material is ejected from the cornea at supersonic velocities but decelerates rapidly;**<br>▸ plume size and velocity increase with increasing fluence; and<br>▸ the ablation plume resembles a burst of smoke at 193 nm, and a spray of larger, discrete droplets at 248 nm.<br><br>**From these studies, they conclude that ultraviolet laser ablation of the cornea is an explosion-like process** {⇨Fig. 4.16–Z}.<br><br>The precise, submicron photoablation capabilities of the excimer laser were first demonstrated in 1981 by Rangaswamy SRINIVASAN, a researcher at IBM in Yorktown, NY. He discovered that an ultraviolet excimer laser could etch living tissue in a precise manner with no thermal damage to the surrounding area, and he named this phenomenon **"ablative photodecomposition."** His sensational experiments stimulated physicists, ophthalmologists and dermatologists to begin using excimer lasers in medical therapy. In 1983, the ophthalmologist Stephen L. TROKEL at Columbia University and R. SRINIVASAN at IBM first suggested the use of excimer laser surgery to reshape the cornea. Since then, multiple clinical trials have tested the use of excimer lasers for this task. |
| 1987 | *Institute of Urology, University of London, England* | **John WICKHAM,** a British urologist, **coins the term** *minimally invasive surgery* – a redefinition of technologically advanced surgery that minimizes risk and trauma for patients. Shock wave lithotripsy is one illustrious example of this new medical technology. • Retrospectively, WICKHAM[3053] wrote, "In 1982, Extracorporeal Shock Wave Lithotripsy was developed by EISENBERGER in Munich, and it became possible to disintegrate renal stones extracorporally – the focused shock wave being transmitted to the stone by way of a water coupling device to the soft tissues of the body … Then came Second Generation Lithotripsy. Multiple small focused shock waves are developed by the activation of 2,000 or more piezoelectric elements mounted on a focusing disc to produce an ultimate pressure of 100 kbars at a tight $2 \times 6$-mm² focus at the stone surface. **The shock waves,** having multiple points of entry through the body, **caused no perception of pain, the patient required no anesthetic, and the procedure became a 'walk** |

---

[3049] R.A. PRÜMMER: *Explosiverdichtung pulvriger Substanzen*. Springer, Berlin *etc.* (1987); R.A. PRÜMMER ET AL.: *Explosive compaction of powders and composites*. Science Publ., Enfield, NH (2006).
[3050] R.A. PRÜMMER: *Prospects of hot explosive pressing in powder metallurgy*. Mater. Sci. Forum **465/466**, 85-92 (2004).
[3051] C.A. PULIAFITO ET AL.: *High-speed photography of excimer laser ablation of the cornea*. Arch. Ophthalmol. **105**, No. 9, 1255-1259 (1987).
[3052] C.P. LIN, D. STERN, and C.A. PULIAFITO: *High speed photograph of Er:YAG laser ablation in fluid*. Invest. Ophthalmol. Vis. Sci. **32**, 2546-2550 (1990).
[3053] J.E.A. WICKHAM: *Minimally invasive therapy*. Health Trends **23**, No. 1, 6-9 (1991).

in walk out' phenomenon with negligible morbidity and an instant return to normal pursuits … We have thus moved rapidly into an entirely new era of interventional therapy which requires a name and some definition. The nearest that I could achieve for a name is the all embracing one of *Minimally Invasive Therapy*, with the process defined as the desire to minimize physical trauma in any field of interventional medical therapy."

To pursue this goal, a number of like-minded physicians founded the **Society for Minimally Invasive Therapy** in London in December 1989, in order "to bring practitioners of minimally invasive surgery and interventional radiology together with the instrument manufacturers concerned with interaction in these two areas." In 2000, the name of the society was formally changed to the **Society for Medical Innovation and Technology (SMIT).**

| | | |
|---|---|---|
| 1988 | Snowbird, Salt Lake Valley, Utah | In the period October 20–23, **first interdisciplinary Conference on Impacts, Volcanism, and Mass Mortality.**[3054] ▪ This meeting was held as the sequel to the **Conference on Large Body Impacts and Terrestrial Evolution: Geological, Climatological, and Biological Implications,** held at Snowbird, UT (1981). |
| 1988 | Cancer Center, New York City, New York; AZV, University of Amsterdam; UCLA School of Medicine & School of Medicine, University of Rochester, Rochester, New York | **Numerous researchers start experiments that attempt to use detrimental side effects from the interactions of shock wave with living tissue for the treatment of tumors,** *e.g.*, vessel wall damage, hemorrhages and venous thrombi. ▪ First results, however, were disappointing: the response to treatment[3055] was either limited to delayed tumor growth[3056, 3057] or it was insignificant.[3058, 3059] Investigations of this important medical problem are still performed today, and are partly directed at combining shock treatment with chemical therapy.<br><br>A review of the state of the art in the application of shock waves to medicine up to 1999 was given by Kazuyoshi TAKAYAMA,[3060] a renowned Japanese shock physics professor at Tohoku University in Sendai. Modern clinical applications of shock waves are becoming increasingly important in medicine. They are not only limited to the fields mentioned above, but they also encompass urology {Electrohydraulic Lithotripsy ⇨1959; Extracorporeal Lithotripsy ⇨1970; Laser Lithotripsy ⇨1983} and orthopedics {Shock-Induced Osteogenesis ⇨1995}. |
| 1988 | Cavendish Laboratory, Dept. of Physics, University of Cambridge, England | **John P. DEAR**[3061] **and John E. FIELD,** two British applied physicists, **observe that a shock wave that propagates in a liquid and strikes a cavity will result in the spalling of wall material.** It creates a jet inside the bubble {KORNFELD & SUVOROV ⇨1944} which travels in the direction of the shock, causing a ***shock-induced cavity collapse.*** ▪ Later studies revealed that this process is accompanied by luminescence – known as "sonoluminescence."<br><br>In his review of the different hypotheses for the origin of sonoluminescence, Lawrence A. CRUM,[3062] a physics professor and acoustician at the Applied Physics Laboratory of the University of Washington (Seattle, WA) also addressed the possible role of imploding shock waves. |

---

[3054] V.L. SHARPTON and P.D. WARD (eds.): *Global catastrophes in Earth history.* Special Paper No. 247, Geological Society of America (GSA), Boulder, CO (1990).
[3055] M. DELIUS ET AL.: *Tumor therapy with shock waves requires modified lithotripter shock waves.* Die Naturwissenschaften **76**, 573-574 (1989).
[3056] K. LEE ET AL.: *High energy shock waves enhance anti-tumor activity of cisalpine (DDP) murine bladder cancer.* J. Urology **139**, 326A (1988).
[3057] R.F. RANDAZZO ET AL.: *The in vitro and in vivo effects of extracorporeal shock waves on malignant cells.* Urol. Res. **16**, 419-426 (1988).
[3058] V.P. LAUDONE ET AL.: *Cytotoxicity of high energy shock waves: methodologic considerations.* J. Urology **141**, 965-968 (1989).
[3059] A.A. GELDORF ET AL.: *High energy shock waves do not affect either primary tumor growth or metastasis of prostate carcinoma.* Urol. Res. **17**, 9-12 (1989).
[3060] K. TAKAYAMA: *Application of shock wave research to medicine* [Plenary lecture]. In: (G.J. BALL, R. HILLIER, and G.T. ROBERTS, eds.) *Proc. 22nd Int. Symposium on Shock Waves* [Imperial College, London, U.K., July 1999]. CD Rom (ISBN 085432 706 1), University of Southampton (2000), pp. 23-32.
[3061] J.P. DEAR and J.E. FIELD: *A study of the collapse of arrays of cavities.* J. Fluid Mech. **190**, 409-425 (1988).
[3062] L.A. CRUM: *Sonoluminescence.* Phys. Today **47**, 22-29 (Sept. 1994).

| | | |
|---|---|---|
| 1988 | *High Explosives Application Facility, Lawrence Livermore National Laboratory (LLNL), Livermore, California* | Alan M. FRANK,[3063] a LLNL staff physicist, **reports on new laser-illuminated ultrafast microphotography equipment especially designed for use in detonics – the "LLNL Microdetonics Facility."** It combines high-speed electronic framing and streak cameras with appropriate specialized optics, spectrographs, and laser illumination to record events with temporal and spatial resolutions of a few nanoseconds and microns, respectively. **The new facility is particularly well suited to the study of rapidly evolving microstructures.** Examples encompass exploding bridge wires, foils, and their interactions with high explosives. Furthermore, it has been applied to investigate the triggering mechanism of aluminum metal combustion in water and the influence of fragmentation processes in sustaining this rapid reaction.[3064] |
| 1988 | *Los Alamos National Laboratory (LANL), Los Alamos, New Mexico & Institut für Treib- und Explosivstoffe (ICT) der Fraunhofer-Gesellschaft (FhG), Pfinztal/ Berghausen, Germany* | N. Roy GREINER,[3065] Dennis S. PHILLIPS, and James D. JOHNSON at LANL, **and** Fred VOLK at FhG-ICT **report on the preparation of diamonds by the application of explosive shocks to graphite loaded into the explosive.** The observed diamonds are 4–7 nm in diameter – "nanodiamonds" – and comprise 25 weight % of the soot. In terms of their size and infrared spectrum, they resemble diamonds similarly isolated from meteorites. ▪ Astrophysicists have long speculated that nanodiamonds were ancient products of stellar explosions that occurred before the Sun was born. In a joint study carried out by the Georgia Institute of Technology, the University of Washington, NASA's Goddard Space Flight Center and the Natural History Museum in London, it was discovered that nanodiamonds "are absent or very depleted in fragile, carbon-rich interplanetary dust particles" (*e.g.*, in cometary dust grains), which suggested the idea that some (perhaps most) nanodiamonds may have been produced in the inner Solar System.[3066]

In the same year, Soviet researchers at the Institute of Hydrodynamics, Siberian Division, Academy of Sciences of the U.S.S.R. (Novosibirsk) also reported on the manufacture of nanodiamonds by explosive detonation.[3067] ▪ Nanodiamond synthesis was discovered in the Soviet Union in 1963 at the Russian Federal Nuclear Center – All-Russian Scientific Research Institute of Technical Physics (RFNC-VNIITF) in Snezhinsk.[3068] |
| 1988 | *Max-Planck-Institut (MPI) für Strömungsforschung, Göttingen, Germany* | Gerd E.A. MEIER,[3069] a German physicist and head of the Transonic Aerodynamics Group, **reviews and classifies the wide field of unsteady transonic phenomena and their analytical treatment.** Using flow interferograms, he distinguishes between and illustrates four main categories:
▸ forced inviscid interactions due to gusts, shock waves, vortices and changes of quantities of gas;
▸ forced viscid interactions due to externally generated unsteady fluctuations, slipstreams and the decay of supersonic flow regimes; |

---

[3063] A.M. FRANK: *High-speed micro-photographic laboratory.* In: (W. DAHENG, ed.) *Proc. 18th Int. Congress on High-Speed Photography and Photonics* [Xian, China, Aug./Sept. 1988]. Proc. SPIE **1032**, 1026-1037 (1989).

[3064] W.C. TAO, A.M. FRANK, R.E. CLEMENTS, and J.E. SHEPHERD: *Aluminum metal combustion in water revealed by high-speed microphotography.* In: (L.L. SHAW ET AL., eds.) *Ultrahigh- and High-Speed Photography, Videography, Photonics, and Velocimetry '90* [held at SPIE's International Symposium on Optical and Optoelectronic Applied Science and Engineering in San Diego, CA, in 1990]. Proc. SPIE **1346**, 300-310 (1990).

[3065] N.R. GREINER, D.S. PHILLIPS, J.D. JOHNSON, and F. VOLK: *Diamonds in detonation soot.* Nature **433**, 440-442 (June 2, 1988).

[3066] Z.R. DAI ET AL.: *Possible in situ formation of meteoritic nanodiamonds in the early Solar System.* Nature **418**, 157-159 (July 11, 2002).

[3067] A.E. LYAMKIN ET AL.: *Production of diamonds from explosives.* Sov. Phys. Dokl. **33**, 705-706 (1988).

[3068] V.V. DANILENKO: *On the history of the discovery of nanodiamond synthesis.* Phys. Solid State **46**, 595-599 (April 2004).

[3069] G.E.A. MEIER: *Unsteady phenomena.* In: (J. ZIEREP and H. OERTEL, eds.) *Proc. Symposium Transsonicum III* [Göttingen, Germany, May 1988]. Springer, Berlin *etc.* (1989), pp. 441-458.

- self-excited inviscid flow oscillations due to flow interactions with flexible boundaries (transonic flutter), transonic cavity oscillations, the addition of heat by water condensation, vortex/vortex and shock turbulence interactions; and
- self-excited flow oscillations dominated by viscosity, which are due to vortex shedding, reattachment oscillations, "intake buzz," and "buffeting" of transonic airfoils.

Although the exact range of transonic Mach numbers will vary for different aircraft configurations and operating conditions, the transonic regime is generally said to lie between Mach 0.7 (or 0.8) and Mach 1.2 (or 1.3). **Modern airliners operate largely in the transonic range, typically flying at about $M = 0.85$, during which the flow over the wings is transonic or supersonic** {⇨Fig. 4.14–L}. This is particularly important, due to an effect known as "wave drag" {VON KÁRMÁN & MOORE ⇨1932} which is prevalent at these speeds.

| 1988 | Dept. of Materials Science and Engineering, Oregon Graduate Center, Beaverton, Oregon | Lawrence E. MURR,[3070] a U.S. professor of engineering, **edits the book *Shock Waves for Industrial Applications*,** a 13-chapter compendium written by 17 international contributors which encompasses the following broad topics:<br>▸ shock wave fundamentals: effects on materials behavior and properties;<br>▸ shock hardening and strengthening;<br>▸ explosion (or explosive) forming;<br>▸ explosion (or explosive) welding and cladding;<br>▸ shock compaction and consolidation of powered materials; and<br>▸ shock sensitization, new materials synthesis, and new and novel materials fabrication using shock waves.<br>The materials aspects include metals and alloys, ceramics, and polymers, particularly powdered materials in the case of polymers, and combinations of, or composites of, these materials in a variety of structural and microstructural forms. |
| --- | --- | --- |
| 1988 | I. Physikalisches Institut, Universität Stuttgart, Stuttgart, Germany | Joachim STAUDENRAUS,[3071] a Ph.D. student, **and Wolfgang EISENMENGER,** a physics professor, **report on a new *fiber optic probe hydrophone.*** It overcomes most of the problems involved with the use of piezoelectric measurement techniques in high-intensity sound fields.<br>They subsequently developed their hydrophone further into a compact transportable instrument, which gave a signal bandwidth of 20 MHz and a spatial resolution of 0.1 mm.[3072] Detectable pressures ranged from –15 MPa (–150 bar) to +100 MPa (1 kbar). Contrary to PVDF needle and membrane hydrophones {KAWAI ⇨1969}, which can only record the leading edge of the strong rarefaction phase following the compression shock in liquids, **the fiber optic probe hydrophone is capable of also registering a long negative pulse** – a feature which is very useful in cavitation research. |
| 1989 | Planet Neptune & NASA's Jet Propulsion Laboratory (JPL), Pasadena, California | On August 24, **Voyager 2 crosses Neptune's bow shock and then enters the magnetosphere.** ▪ In order to fully utilize all of the spacecraft observations, an improved nonlinear least squares Rankine-Hugoniot magnetohydrodynamic (MHD) shock-fitting technique was developed by Adam SZABO[3073, 3074] (MIT) and Ronald P. LEPPING (NASA-GSFC) in the early 1990s and applied to the Neptunian data set. It was found that the bow shock can be characterized as a low $\beta$ (the ratio of thermal to magnetic pressure), high Mach number, strong quasi- |

---

[3070] L.E. MURR (ed.): *Shock waves for industrial applications.* Noyes Publ., Park Ridge, NJ (1988).

[3071] J. STAUDENRAUS and W. EISENMENGER: *Optisches Sondenhydrophon.* Tagung der Deutsche Arbeitsgemeinschaft für Akustik (DAGA) [Braunschweig, März 1988]. Fortschritte der Akustik **14**, 467-470 (1988); W. EISENMENGER and J. STAUDENRAUS: *Sampling hydrophone.* U.S. Patent No. 5,010,248 (1991).

[3072] J. STAUDENRAUS and W. EISENMENGER: *Fibre-optic probe hydrophone for ultrasonic and shock-wave measurements in water.* Ultrasonics **31**, No. 4, 267-273 (1993).

[3073] A. SZABO: *The interaction of Neptune with the solar wind.* Ph.D. thesis, MIT, Cambridge, MA (1993); http://dspace.mit.edu/bitstream/1721.1/29865/1/29875844.pdf.

[3074] A. SZABO and R.P. LEPPING: *Neptune inbound bow shock.* J. Geophys. Res. **100**, No. A2, 1723-1730 (1995).

perpendicular shock moving outward, away from the planet towards the Sun, with a speed of 14 ± 12 km/s. The most likely reason for the bulk motion of the shock is the changing magnetic configuration of the planetary magnetosphere.

In the same year, on August 25, the **NASA spacecraft Voyager 2 makes its closest approach to Neptune, the outermost gas giant.**[3075]

(i) The probe discovers the Great Dark Spot, which may be similar to Jupiter's Great Red Spot. The Earth-sized spiral shape of both the dark boundary and the white cirrus suggests a storm system rotating counterclockwise. However, the size, shape, and location of the spot vary greatly over time; it even disappears and reappears occasionally. The atmosphere consists mostly of hydrogen (75%) and helium (24%). Icy particles of methane (1%) in its outer atmosphere give Neptune its deep blue color (methane absorbs red light).

(ii) Near Neptun's Great Dark Spot, Voyager 2 measures enormous wind speeds of up to 2,400 km/h (666 m/s) blowing opposite to its rotation. • **Neptune has the highest wind speeds measured on any planet in the Solar System – possibly even reaching supersonic velocities at some very cool spots.**[3076]

| | | |
|---|---|---|
| 1989 | Cambridge, Massachusetts | **First International Conference on Structures under Shock and Impact (SUSI),** which aims "to bring together workers in a wide range of engineering activities, who employ common analytical and experimental methods in their estimation of structural response." The general chairman is Philip Stanley BULSON. • Later conferences were held at Portsmouth, U.K. (1992); Madrid, Spain (1994); Udine, Italy (1996); Thessaloniki, Greece (1998); Cambridge, U.K. (2000); Montreal, Canada (2002); Crete, Greece (2004); The New Forest, U.K. (2006); and The Algarve, Portugal (2008). |
| 1989 | *American Institute of Physics (AIP), New York City, New York* | **Foundation of the American journal *Physics of Fluids A (Fluid Dynamics)*** in order "to cover special fields of applied physics such as magneto-fluid dynamics, shock and detonation phenomena, hypersonic physics, rarefied gas, hydrodynamics, and dynamics of compressible fluids." The first editor-in-chief is Andreas ACRIVOS. |
| 1989 | *Institute of Fluid Science, Tohoku University, Sendai, Japan* | **Foundation of the Shock Wave Research Center (SWRC),**[3077] with the goal "to establish a new interdisciplinary research field among existing sciences and technologies and medicine by reviewing these from common features of shock wave phenomena or short time physics." The first director is Kazuyoshi TAKAYAMA.<br><br>In the same year, the **International Workshop on Shock Wave Focusing** is held at Sendai; the chairman is K. TAKAYAMA. |
| 1989 | *Dept. of Pure and Applied Zoology, University of Leeds, West Yorkshire, England* | R. McNeill ALEXANDER,[3078] a British zoologist and an expert in biomechanics, **speculates that the long tapering tails of the giant sauropod dinosaurs of the families *Diplodocus* and *Apatosaurus* would have made formidable whips, which could have produced a loud "crack" when the tip of the whip exceeded the speed of sound** {LUMMER ⇨ 1905; CARRIÈRE ⇨ 1927; BERNSTEIN ET AL. ⇨ 1958; KREHL ET AL. ⇨ 1995}. Referring to the tails of *Diplodocus* and *Apatosaurus*, he notes that, "They could have been used to strike a predator, and I wonder whether they could also have been used to make a terrifying noise. When a circus ringmaster cracks his whip, he flicks it so as to make its tip move supersonically. Is it too wild a speculation to wonder whether *Diplodocus* could crack its tail?" |

---

[3075] *Voyager Neptune science summary*. NASA-JPL, Pasadena, CA (Dec. 20, 1989); http://www.solarviews.com/span/vgrnep.htm#nep.

[3076] The sound velocity in an ideal gas $c$ is given by $c = (\kappa RT)^{1/2}$, in which $R$ is the ideal gas constant per gram, $T$ is the absolute temperature, and $\kappa$ is the ratio of the specific heat of the gas at constant pressure to the specific heat at constant volume. For hydrogen ($\kappa = 1.41$) and helium ($\kappa = 1.66$), both at a temperature of $T = 55$ K, this results in sound velocities of 567 m/s and 436 m/s, respectively.

[3077] *Shock Wave Research Center*. Institute of Fluid Science, Tohoku University, Sendai, Japan (Sept. 1999); http://ceres.ifs.tohoku.ac.jp/~swrc/.

[3078] R.M. ALEXANDER: *Dynamics of dinosaurs and other extinct giants*. Columbia University Press, New York (1989).

*Diplodocus*, a relative of the more familiar *Brontosaurus*, was the longest land animal that ever lived; the longest known was 26.7 meters in length and must have approached a weight of about 80 tons. More recently, **computer models of the tail of the *Apatosaurus louisae* have shown that it could have reached supersonic velocities, thereby producing a noise analogous to the crack of a bullwhip.**[3079] Typically, the tail of this dinosaur, a specimen of which is kept at the Carnegie Museum in Pittsburgh, had a length of about 12.5 meters and an estimated mass of about 1.5 tons, while the mass of the whiplash part amounted to only 0.54% of the total tail mass.

| 1989 | Institute of Astronomy, University of Cambridge, England | **Houshang ARDAVAN,**[3080] a British mathematician and astrophysicist, **believes the pulsing of pulsars,** rapidly spinning neutron stars {BELL & HEWISH ⇨ 1967}, **could be "light booms;"** *i.e.*, **shock waves created by moving faster than light in vacuo,** analogous to the sonic booms created by a supersonic plane when it breaks the sound barrier. He points out that the rotating magnetic field surrounding the neutron star induces rotating patterns of electrical charges and currents that move faster than the speed of light. ▪ The idea of a light boom is not new {CHERENKOV ⇨ 1934}.<br><br>In 1995, H. ARDAVAN[3081] and John E. FFOWCS WILLIAMS, a professor of acoustical engineering at Cambridge University, discussed (as counterparts to acoustic volume sources) macroscopic electric charges and currents in empty space whose superluminal patterns move faster than light in vacuo. In 1998, H. ARDAVAN[3082] described a pulsed source of polarized electromagnetic radiation based upon the constructive interference of emitted waves and the formation of caustics with a polarization pattern that moves faster than light. He explained the mechanism pulsars use to generate regular, coherent radio wave pulses in a similar way. In the early 2000s, Arzhang ARDAVAN, supported by a team from Los Alamos and Oxford, built a "polarization synchrotron" based upon an idea put forward by his father, and claimed that the rotating directed superluminal polarization pattern could transform communications and radar systems.[3083] However, other researchers are skeptical that the polarization synchrotron, like a pulsar, emits radiation in a well-defined beam.[3084] |
|------|---|---|
| 1989 | Dept. of Aeronautical Engineering, Nagoya University, Honshu, Japan | **Toshi FUJIWARA**[3085] **and K. Viswanath REDDY first numerically simulate the structure of the three-dimensional reaction zone of a self-sustaining detonation propagating in a tube of circular cross-section.**<br><br>(i) Using the fundamental Euler equations and assuming a perfect gas (*i.e.*, ignoring real gas effects and transport phenomena), they use the numerical MacCormack FCT (flux-corrected transport) scheme, a finite difference algorithm capable of handling flow discontinuities {MACCORMACK ⇨ 1982}, and they calculate the pressure distributions for meridian planes and cross-sections of the tube.<br><br>(ii) Performing an analysis involving a million grid points on a supercomputer, **they obtain an insight into the unsteady mechanism of detonation propagation,** which is essentially a complex coupling of three different modes of shock waves: the frontal wave in the axial direction, the radial wave, and the azimuthal wave. |

---

[3079] N.P. MYRHVOLD and P.J. CURIE: *Supersonic sauropods? Tail dynamics in the diplodocids.* Paleobiology **23**, 393-409 (1997).

[3080] H. ARDAVAN: *The speed-of-light catastrophe.* Proc. Roy. Soc. Lond. **A424**, No. 1866, 113-141 (1989).

[3081] H. ARDAVAN and J.E. FFOWCS WILLIAMS: *Violation of the inverse square law by the emissions of supersonically or superluminally moving volume sources.* Astro-ph/9506023 (June 1995).

[3082] H. ARDAVAN: *Generation of focused, nonspherically decaying pulses of electromagnetic radiation.* Phys. Rev. **E58** [III], 6659-6684 (1998).

[3083] *Warp speed.* New Scientist No. 2288 (April 28, 2001); http://www.eurekalert.org/pub_releases/2001-04/NS-Ws-2404101.php.

[3084] E. CARTLIDGE: *Money spinner or loopy idea?* Science **301**, No. 5639, 1463-1465 (Sept. 12, 2003). ▪ For response *see* K.T. MCDONALD: *Synchrotron-Čerenkov radiation.* Science **303**, No. 5656, 310 (Jan. 16, 2004).

[3085] T. FUJIWARA and K.V. REDDY: *Propagation mechanism of detonation – three-dimensional phenomena.* Mem. Faculty Engng. (Nagoya University, Japan) **41**, No. 1, 93-111 (1989).

(iii) As had already been proven by the authors using a numerical analysis of two-dimensional nonsteady detonations, the mechanism that makes the detonation self-sustaining is also an interaction among shock waves. This interaction, however, is three-dimensional, and results in shock waves generating multiple Mach reflections.

| | | |
|---|---|---|
| 1989 | *Center for Electromagnetic Research, Northeastern University, Boston, Massachusetts* | Peter GRANEAU,[3086] a U.S. physics professor, **reviews previous theories of thunder and existing discrepancies with experiments performed in this field. Using numerous examples he attempts to prove that thunder is *not* caused by the thermal expansion of the lightning channel.** His theory is a contradiction to the widely accepted theory that thunder must begin with a (weak) shock wave in air due to the sudden thermal expansion of the plasma in the lightning channel {UMAN ⇨ 1970; BASS ⇨ 1980}. |

▸ Experiments with short atmospheric arcs that have similar strengths to lightning revealed average arc pressures in excess of 400 atm and peak pressures approaching 1,000 atm. These results demand much higher temperatures than those found by lightning spectroscopy (max. 36,000 K).

▸ When the strength of the short arc explosion was plotted against the action integral of the current pulse, it followed an electrodynamic law rather than a heating curve.

▸ Arc photography proved conclusively that the plasma did not expand thermally in all directions, but preferentially at right angles to the current, as if driven by organized electrodynamic action. **Electrodynamic forces that might drive the thunder shock wave include the Lorentz pinch force, the longitudinal Ampere force and the alpha-torque force of Ampere-Neumann electrodynamics.**

▸ The pinch force was found to be far too small and in the wrong direction to be the cause of thunder. Longitudinal and alpha-torque forces act in the correct direction but, so far, quantitative agreement has not been achieved. This may have to wait for a complete Ampère MHD.

More recently, GRANEAU[3087] **and collaborators suggested the idea that thunder is the result of an air explosion driven by the liberation of chemical bond energy from $N_2$ and $O_2$ molecules.** They also reported that laboratory experiments carried out with air arcs of similar strength to lightning confirmed their hypothesis. One interesting question is whether the water content in the air affects the intensity of thunder.

| | | |
|---|---|---|
| 1989 | *Institute of Fluid Science, Tohoku University, Sendai, Japan* | At the International Workshop on Shock Wave Focusing, Hans GRÖNIG,[3088] a German professor of gas dynamics at the Shockwave Laboratory (RWTH Aachen), discusses previous shock focusing research and its prospects for the future. |

(i) Shock wave focusing has its roots in early observations on acoustic phenomena: in antiquity, curious amplifications of sound waves were observed in particular architectural structures, such as in "whispering" galleries and later also in domes (*e.g.*, in St. Paul's Cathedral). Focusing effects were demonstrated in the laboratory in the 19th century with a mercury ripples dish, consisting of a heavy iron base with an elliptical dish holding a pool of mercury. An eye dropper was positioned over the focus of the ellipse and a drop was allowed to fall, setting up waves which converged to the other focus after reflection from the sides.

(ii) In the pioneering era of shock wave physics, the study of focusing spark waves generated by powerful electric discharges (which in many cases are actually weak shock waves) became a spectacular way to demonstrate the great potential of (then recently invented) schlieren and shadowgraph methods {ROSICKÝ ⇨ 1876; WOOD ⇨ 1900; M. TOEPLER ⇨ 1904}.

---

[3086] P. GRANEAU: *The cause of thunder.* J. Phys. D (Appl Phys.) **22**, 1083-1094 (1989).
[3087] P. GRANEAU, N. GRANEAU, and G. HATHAWAY: *Evidence of thunder being a chemical explosion of air.* J. Plasma Phys. **69**, 187-197 (2003).
[3088] H. GRÖNIG: *Past, present and future of shock focusing research.* In: (K. TAKAYAMA, ed.) *Proc. Int. Workshop on Shock Focusing* [Sendai, Japan, March 1989]. Shock Wave Research Center, Institute of Fluid Science, Tohoku University (1990), pp. 1-37.

(iii) During World War II, shock focusing was used in the implosion devices employed as the first fission bombs {Trinity Test ⇒1945; Hiroshima & Nagasaki Bombing ⇒1945}.

(iv) During and after World War II, shock focusing again attracted the attention of both theoretical and applied shock physicists {GUDERLEY ⇒1942; ⇒Fig. 4.14–K}, particularly in relation to advanced fusion weapons technology, fusion reactor research, and extracorporeal shockwave lithotripsy for treating stone diseases {CHAUSSY ET AL. ⇒1970}.

(v) More detailed investigations of shock focusing have revealed that shock focusing effects also play an important role in the initiation and propagation of explosions in combustible gaseous mixtures and in dust explosions {Int. Workshop on Shock Wave Focusing Phenomena in Combustible Media ⇒1998; ⇒Fig. 4.16–U}.

| 1989 | Impact Research Centre (IRC), University of Liverpool, England | Norman JONES,[3089] an eminent professor of mechanical engineering and director of the IRC {⇒1985}, **publishes the book *Structural Impact*.** In its eleven chapters he discusses
▸ the static plastic behavior of beams, plates and shells;
▸ the response when these structural members are loaded dynamically;
▸ the influence of transverse shear, rotary inertia and finite displacements on dynamic response;
▸ particular phenomena for impacted structural members, such as dynamic progressive and dynamic plastic buckling; and
▸ applicability and limitations of scaling laws to dynamic phenomena. |

| 1989 | Lunar & Planetary Laboratory, University of Arizona, Tucson, Arizona | Henry Jay MELOSH,[3090] a U.S. professor of planetary science, **publishes his book *Impact Cratering: A Geologic Process*.** Addressing physicists, geologists and planetary scientists, it contributes many and often novel ideas. In the preface of the book he summarizes its objectives: "This is an attempt to bring together the current knowledge on impact cratering in a single and comprehensive treatment. The field of impact cratering is so large at present, however, that I must restrict the scope to topics that seem to be of major importance to planetary geologists. Thus, I have excluded large amounts of material on microcraters, low velocity impacts, petrology, and geochemistry … **My primary focus is on the impact cratering process, and details of specific terrestrial and extraterrestrial craters are discussed only to the extent that they illuminate that process.** I have attempted to give the readers a feel for the essential physics of impact cratering and to establish a theoretical framework that he or she can apply to the particular cratering problem at hand." ▪ His book became a classic on the subject of impact cratering.

MELOSH also edited the books *Origin of Planets and Life* (1997), a reprint of articles published in *Annual Review of Earth & Planetary Sciences* and *Annual Review of Astronomy and Astrophysics* from 1988 to 1998. For his contributions to planetary sciences, MELOSH was elected to the National Academy of Sciences (NAS) in 2003, one of the most prestigious honors in American science. He is one of the twelve members of the science team for NASA's Deep Impact mission {⇒1999}. |

| 1989 | Los Alamos National Laboratory (LANL), Los Alamos, New Mexico | Stephen A. SHEFFIELD,[3091] Ray ENGELKE, and Robert R. ALCON, three detonation physicists, **report on studies of the initiation process of detonation in nitromethane** [$CH_3NO_2$], which is triggered by a shock wave generated by the impact of a projectile fired from a single-stage gas gun. They conducted ***homogeneous initiation experiments*** using nitromethane that |

---

[3089] N. JONES: *Structural impact*. Cambridge University Press, Cambridge *etc.* (1989); Russ. translation published by Mir, Moskva (1992).

[3090] H.J. MELOSH: *Impact cratering: a geologic process*. Oxford monographs on geology & geophysics No. 11, Oxford University Press, New York; Clarendon Press, Oxford (1989).

[3091] S.A. SHEFFIELD, R. ENGELKE, and R.R. ALCON: *In-situ study of the chemically driven flow fields in initiating homogeneous and heterogeneous nitromethane explosives*. In: *Proc. 9th Symposium (Int.) on Detonation* [Portland, OR, Aug./Sept. 1989]. OCNR 113291-7, Office of Naval Research (ONR), Arlington, VA (1989), pp. 39-49.

was chemically sensitized by an organic base and ***heterogeneous initiation experiments*** in nitromethane that was physically sensitized using silica particles.

(i) In contrast to classical shock wave trajectory recording using streak camera techniques, they perform *in situ* measurements using electromagnetic gauges to better resolve the evolution of the process of detonation.

(ii) Their homogeneous initiation experiments are only partly consistent with the classical homogeneous initiation model of detonation {CHAIKEN ⇨1958; CAMPBELL, DAVIS & TRAVIS ⇨1961}. The superdetonation {CAMPBELL ⇨1961} does not form immediately after an induction time; the reaction causes a wave to build up over a discernible length considerably behind the initial shock, and this wave evolves into a ***superdetonation*** which catches the initial shock.

(iii) Their heterogeneous initiation experiments indicate that the growth of the detonation wave occurs primarily in the shock front.

| | | |
|---|---|---|
| 1989 | *Kumamoto University and Mitsubishi Metal Corporation, Kitabukuro, Japan & Special Design Office of High Rate Hydrodynamics, Novosibirsk, Soviet Union* | **Kazuki TAKASHIMA,**[3092] a Japanese materials scientist at the Dept. of Mechanical Engineering & Materials Science of Kumamoto University, **and collaborators** at Mitsubishi Metal Corporation **report that they have successfully prepared a coil of high-temperature superconducting Y-Ba-Cu oxide using an explosive compaction technique** {⇨Fig. 4.15−G}.<br>(i) The shock compression of sample material, originally available only as powder, has two advantages: it increases<br>▸ the density of the ceramic oxide which also increases the critical current density; and<br>▸ the number of crystal defects. This possibly also promotes flux pinning − thus preventing "flux creep" which can create a pseudoresistance.<br>(ii) Subsequently, the coils were heat-treated in the temperature range 823−1193 K in a flowing oxygen gas atmosphere.<br>(iii) The zero resistance temperature for these coils is 90 K; *i.e.*, they can be operated at liquid nitrogen temperatures (−196 °C or 77 K).<br>Almost simultaneously, **similar successful experiments were also carried out by Andrei DERIBAS,**[3093] a Soviet researcher at Novosibirsk. He reported on the production of joints, tubes and coils from various types of metal/high-$T_c$ ceramics ($T_c$ = critical temperature). |
| 1989 | *Weizmann Institute of Science (WIS), Rehovot, Israel & Argonne National Laboratory (ANL), Argonne, Illinois* | **Zeev VAGER**[3094] **and Ron NAAMAN** (both at WIS), **and Elliot P. KANTER** (ANL), three particle physicists, **report on a new method that provides geometrical images of individual molecules.** This method takes advantage of the large Coulomb repulsion of the nuclei within molecules rapidly stripped of their electrons − so-called **"Coulomb explosion."** ▪ The break-up of small molecular ions impinging on an ultrathin target foil is usually well described by an explosion of the molecular ion due to the Coulomb forces between the atomic fragment ions that emerge are formed upon impact on the foil. Measuring the asymptotic velocity vectors of all the fragments using multi-particle 3D-imaging detectors, so-called "Coulomb explosion imaging," therefore allows to infer information about the initial geometrical structure of the molecular ion prior to its dissociation.[3095] |
| 1990 | *Canberra, southeastern Australia* | **The First Workshop on Shocktube Technology is held.** The goal is "to swap information and experiences about building and operating Free Piston Shock Tunnels (FPSTs), as well as |

---

[3092] K. TAKASHIMA ET AL.: *Preparation of oxide superconducting coils by explosive compaction.* In: (S.C. SCHMIDT, J.N. JOHNSON, and L.W. DAVISON, eds.) *Proc. 6th APS Topical Conference on Shock Compression of Condensed Matter* [Albuquerque, NM, Aug. 1989]. North-Holland, Amsterdam *etc.* (1990), pp. 591-594.

[3093] A. DERIBAS: *Use of explosive energy for the production of multi-layered composites from high $T_C$ superconductive ceramics and metals.* In: (S.C. SCHMIDT, J.N. JOHNSON, and L.W. DAVISON, eds.) *Proc. 6th APS Topical Conference on Shock Compression of Condensed Matter* [Albuquerque, NM, Aug. 1989]. North-Holland, Amsterdam *etc.* (1990), pp. 549-552.

[3094] Z. VAGER, R. NAAMAN, and E.P. KANTER: *Coulomb explosion imaging of small molecules.* Science **244**, 426-431 (April 28, 1989).

[3095] *Coulomb explosion imaging at the Test Storage Ring (TSR).* Heavy Ion Physics Group, MPI für Kernphysik, Heidelberg (Nov. 2004); http://www.mpi-hd.mpg.de/ato/molec/cei/.

performing research with them, and enhance collaboration and complementary research between Canberra and Brisbane." The chairman is Allan PAULL (University of Queensland).
▪ The first and second workshops were interstate rather than international meetings, between the FPST groups at the Dept. of Mechanical Engineering, University of Queensland at Brisbane, the Dept. of Physics and Theoretical Physics, The Faculties, Australian National University at Canberra, and the Dept. of Mechanical Engineering, University of New South Wales at the Australian Defence Force Academy.[3096]

The following meetings, renamed **International Workshops on Shocktube Technology (IWST),** were held at Brisbane, Australia (1991); Canberra, Australia (1992); Brisbane, Australia (1994); Göttingen, Germany (1996); Kakuda, Japan (1998); Ronkonkoma, U.S.A. (2000); Bangalore, India (2002); Woomera, Australia (2004); and Brisbane, Australia (2006).

| 1990 | *Springer International, Heidelberg, Germany* | **Foundation of the international journal *Shock Waves*.** The objectives are "to cover theoretical and experimental results on shock phenomena in gases, liquids, solids and two-phase media from both the fundamental research and the application points of view." The first editor-in-chief is Irvine I. GLASS (University of Toronto). |
|---|---|---|
| 1990 | *Lawrence Livermore National Laboratory (LLNL), UC Livermore, California* | John Edward OSHER,[3097] a U.S. plasma physicist, **and collaborators report on an improved ultrahigh-velocity electric gun design** {STEINBERG ET AL. ⇨1980} that allows velocities as high as 18 km/s to be attained. In order to accelerate objects ranging in size from tiny particles of meteoric dust (so-called "micrometeoroids") to 0.5-g projectiles, they constructed a 60-kV, 150-kJ, coaxial-feed electric gun that electrically explodes a 0.051-mm-thick aluminum bridge-foil load element in order to accelerate a 0.3-mm-thick Kapton® flyer. The pulse capacitors are arranged in a low-impedance, four-arm cross configuration. |
| 1990 | *Dept. of Engineering, University of Cambridge, England* | William J. STRONGE,[3098] a British impact researcher, **introduces an *energetic coefficient of restitution*** which, based on work done during contact by the normal component of contact force, is a measure of impact energy loss from internal sources. Contrary to the classic coefficient of restitution, a velocity ratio {Sir NEWTON ⇨1687; THOMSON & TAIT ⇨1879}, his new energy-based coefficient provides an opportunity to calculate the impulse ratio based on material properties, the impact configuration and the relative incident velocity at the impact point.<br>▪ His idea has been resumed by other impact researchers, for example<br>▷ for predicting the post impact velocities of the balls in NEWTON's cradle;[3099]<br>▷ for calculating planar impacts of slender bodies;[3100] and<br>▷ in multi-rigid-body systems with friction.[3101]<br>Very recently, **researchers at McGill University** (Montreal, Quebec, Canada) **established a new definition of the energetic coefficient of restitution** which is particularly useful for modeling collisions in multibody systems.[3102] Their new definition is based on a study of the absorption and restitution of the kinetic energy during contact involving multibody systems, which led to a generaliztion of the energetic coefficient of restitution. |

---

[3096] Private communication by Dr. Neil MUDFORD, Australian Defence Force Academy, Canberra, Australia (July 6, 2006).
[3097] J. OSHER, R. GATHERS, H. CHAU, R. LEE, G. POMYKAL, and R. WEINGART: *Hypervelocity acceleration and impact experiments with the LLNL electric guns.* J. Impact Engng. **10**, 439-452 (1990).
[3098] W.J. STRONGE: *Rigid body collisions with friction.* Proc. Roy. Soc. Lond. **A431**, 169-181 (1990).
[3099] V. CEANGA and Y. HURMUZLU: *A new look at an old problem: NEWTON's cradle.* J. Appl. Mech. **68**, No. 4, 575-583 (2001).
[3100] Y. HURMUZLU: *An energy based coefficient for planar impacts of slender bars with massive external surfaces.* ASME J. **65**, No. 4, 952-962 (1998).
[3101] W. YAO, B. CHEN, and C. LIU: *Energetic coefficient of restitution for planar impact in multi-rigid-body systems with friction.* Int. J. Impact Engng. **31**, No. 3, 255-265 (2005).
[3102] S.A. MODARRES NAJAFABADI, J. KÖVECSES, and J. ANGELES: *Energy analysis of contacts in multibody systems: a novel interpretation of the energetic coefficient of restitution.* 12th World Congress in Mechanism and Machine Science [Besançon, France, June 2007]. In: (J.P. MERLET and M. DAHAN, eds.) IFToMM 2007. Comité Français pour la Promotion de la Science des Mécanismes et des Machines, Besançon, France (2007).

| | | |
|---|---|---|
| 1991 | *Mt. Pinatubo, Luzon Island, Philippines* | On April 2, **a long series of heavy steam explosions begins from Mount Pinatubo,** a little-known and previously dormant volcano. Ten weeks later, they peak with the most violent shock eruptions recorded around the world over the last 50 years (resulting in about 1,000 casualties). The strongest eruption, on June 15, accelerated huge quantities of pulverized matter (particularly of sulfur) into the stratosphere in a column with a height of about 40 km and a diameter of about 200 km. The matter spread in the stratosphere and partially blocked sunlight – thus causing a temporarily cooling of the Earth.[3103] |
| 1991 | *Sendai, Japan* | In July, **a preliminary workshop on ram accelerators** {HERTZBERG ET AL. ⇨ 1986} **is held** in conjunction with the 18th International Symposium on Shock Waves. • In the following years, **International Workshops on Ram Accelerators and Related Phenomena (RAMAC)** were held at Saint-Louis, France (1993); Seattle, WA (1995); Sendai, Japan (1997); and Poitiers, France (1999). |
| 1991 | *Division of Engineering and Applied Sciences, CalTech, Pasadena, California* | **Bradford STURTEVANT,**[3104] a U.S. professor of aeronautical engineering, **and collaborators use a *vertical shock tube* to simulate physical processes of high-speed (supersonic) two-phase flows on a laboratory scale,** which (they assume) can occur in the vents and Plinian columns of volcanoes erupting explosively. The goal of such studies is to understand more about<br>▸ the source of pyroclastic flows;<br>▸ the mechanism by which magma fragments explosively; and<br>▸ the mechanisms by which pyroclastic flows are transported, in particular the processes of fluidization which allow them to propagate many kilometers.<br><br>Rapidly moving dense and hot pyroclastic flows are a major concern during explosive eruptions {Mt. Bandai-san ⇨ 1888; Mt. St. Helens ⇨ 1980; Mt. Pinatubo ⇨ 1991}. |
| 1991 | *Sendai, Japan* | At the 18th Internatonal Symposium on Shock Waves, **Justin S. WARK,**[3105] a British high-power laser physicist working at the Clarendon Laboratory at Oxford University, **and colleagues report on unique flash X-ray diffraction experiments involving laser-shocked single crystals of silicon ⟨111⟩.** They performed the study in collaboration with the U.S. Naval Research Laboratory (NRL) in Washington, DC, and the Blackett Laboratory at the Imperial College of Science, Technology & Medicine in London.[3106]<br><br>(i) **In contrast to previous flash X-ray diffraction experiments** {JOHNSON & MITCHELL ⇨ 1970 & 1972}, **they produced both the flash X-ray-pulse and the shock pressure pulse using high-power lasers.**<br>▸ They used the VULCAN laser system at the Central Laser Facility of the Rutherford Appleton Laboratory (RAL) in Oxfordshire, a high-intensity pulsed laser (100 J, 1 ns FWHM, wavelength 1.05 μm), for shock compression of the target. In order to generate a steep pressure pulse, they focused the laser beam onto a crystal coated with a thin aluminum layer at the front. Upon laser irradiation, the coating immediately evaporates, driving a shock wave into the crystal.<br>▸ A second laser – a high-intensity pulsed laser ($10^{14}$ W/cm$^2$, wavelength 0.53 μm) that is synchronous with but delayed with respect to the shock-driving first laser – was fo- |

---

[3103] *Restless Earth* (ed. by NGS). Book Division, National Geographic Society (NGS), Washington, DC (1997).

[3104] B. STURTEVANT, H. GLICKEN, L. HILL, and A.V. ANILKUMAR: *Explosive volcanism in Japan and the United States: gaining an understanding by shock tube experiments*. In: (K. TAKAYAMA, ed.) *Proc. 18th Int. Symposium on Shock Waves* [Sendai, Japan, July 1991]. Springer, Berlin *etc.* (1992), vol. 1, pp. 129-140.

[3105] J.S. WARK, N.C. WOOLSEY, W.J. BLYTH, and R.R. WHITLOCK: *Sub-nanosecond X-ray diffraction from laser-shocked crystals*. In: (K. TAKAYAMA, ed.) *Proc. 18th Int. Symposium on Shock Waves* [Sendai, Japan, July 1991]. Springer, Berlin *etc.* (1992), vol. 1, pp. 393-398.

[3106] J.S. WARK, D. RILEY, N.C. WOOLSEY, G. KEIHN, and R.R. WHITLOCK: *Direct measurements of compressive and tensile strain during shock breakout by use of subnanosecond X-ray diffraction*. J. Appl. Phys. **68**, 4531-4534 (1990).

cused on a separate 5-μm-thick titanium foil, the resultant laser-plasma serving as a small source of X-ray line radiation (titanium helium-alpha) which closely followed the laser pulse.

(ii) The X-rays were then Bragg-diffracted from the rear surface of the shocked crystal, and the line shift was recorded with an X-ray streak camera {⇒Fig. 4.12–F}.

(iii) They obtain the following remarkable results:

> measurement of the dynamic tension during shock breakout from the rear surface;
> observation of the onset of shock-induced plasticity, achieved by diffracting the X-rays from planes running perpendicular to the shock front; and
> first demonstration that **subnanosecond X-ray powder patterns of a laser-shocked polycrystalline material can be obtained with a single flash X-ray.**

| 1992 | *Cygnus ["The Swan," a summer constellation in the Northern Hemisphere]* | On February 19, **a tremendous explosion occurs in the constellation *Cygnus*. The event – labeled nova CYGNI 1992, an *explosive* (or *catastrophic*) *variable*[3107] – most probably occurred in an accretion disk binary system.** • Astronomers hypothesize that this system's white dwarf had so much gas (hydrogen) dumped onto its surface that conditions became ripe for nuclear fusion. The resulting thermonuclear detonation blasted much of the surrounding gas into a rapidly expanding shell of material around CYGNI 1992, which the NASA/ESA Hubble Space Telescope actually photographed in 1994. Nova CYGI 1992 was the brightest nova in recent history, and at its brightest it could even be seen with the naked eye. It was observed across the electromagnetic spectrum.[3108]<br><br>Novae occur suddenly and unexpectedly, showing a dramatic rise in luminosity over a period of only hours or days. The name originates from the Latin *nova stella* ("new star"), which is exactly what early astronomers thought they were seeing {BRAHE ⇒1572}. A nova is a binary, or double, star in which one member is a white dwarf and the other is a giant or supergiant. Matter from the large star falls onto the small star. **After a thick layer of the large star's atmosphere has collected on the white dwarf, the layer burns off in a sudden nuclear fusion reaction, producing a huge amount of energy.** Novae were distinguished from supernovae in the 1930s, and they are now subclassified, according to how their luminosities change over time, into:<br><br>> *classical novae*, which flare just once (the brightness of these novae can increase by 6–19 magnitudes before fading back to the original level);<br>> *recurrent novae*, which flare several times and are characterized by a runaway thermonuclear reaction which ejects the unburnt hydrogen into a rapidly-expanding shell around the white dwarf; and<br>> *dwarf novae*, which are intrinsically faint stars that undergo multiple flare events (the brightness of these novae can increase by 2–5 magnitudes).[3109] |
| --- | --- | --- |
| 1992 | *NASA's Goddard Space Flight Center (GSFC), Greenbelt, Maryland* | On April 23, in a press conference, **George F. SMOOT**, an astrophysics professor at UC Berkeley, and winner of the 2006 Nobel Prize in Physics (jointly with J.C. MATHER) "for their discovery of the blackbody form and anisotropy of the cosmic microwave background radiation," **announces that NASA's COBE** (Cosmic Background Explorer) **had observed "the oldest and largest structures ever seen in the early Universe** ... the primordial seeds of modern-day structures such as galaxies, clusters of galaxies, and so on ... huge ripples in the fabric of space-time left over from the creation period." And he uttered what became an instantly famous quote: "If you're religious, it's like seeing God."[3110] |

---

[3107] For the classification of variable stars *see* General Catalogue of Variable Stars ⇒1948.
[3108] *Nova Cygni 1992.* Astronomy picture of the day. NASA-GSFC (Dec. 27, 1995); http://antwrp.gsfc.nasa.gov/apod/ap951227.html.
[3109] *The Swinburne astronomy online encyclopedia.* Swinburne University of Technology, Hawthorn, Australia; http://www.cosmos.swin.edu.au/.
[3110] S.J. DICK: *Voyages to the beginning of time.* NASA (Oct. 16, 2006); http://www.nasa.gov/mission_pages/exploration/whyweexplore/Why_We_24_prt.htm.

(i) The first all-sky map of cosmic background radiation {GAMOW ⇨1948; PENZIAS & WILSON ⇨1965} are based on two years of data returned from COBE's differential mapping radiometer. These data show that **the cosmic microwave background spectrum matches that of a blackbody at a temperature of 2.726 K** to a precision of 0.03% of the peak intensity over a wavelength range of 0.1–5 mm. ▪ COBE was launched on November 18, 1989.[3111]

(ii) COBE's differential radiometer found anisotropies in the cosmic microwave background. The cosmic microwave background fluctuations which are extremely faint (only one part in 100,000 compared to the 2.73 K average temperature of the radiation field) are interpreted as being the imprints of the seeds that eventually grew under the influence of gravity into galaxies, clusters of galaxies, and clusters of clusters of galaxies.

| | | |
|---|---|---|
| 1992 | *Sudbury [a town located in the Sudbury Basin], Ontario, Canada* | In the period from August 31 to September 2, **first International Conference on Large Meteorite Impacts and Planetary Evolution,** organized by Burkhard O. DRESSLER. ▪ The Sudbury Basin is a huge oval crater produced by a meteorite impact {⇨c.1.85 Ga ago}.<br><br>The second conference was also held at Sudbury, Ontario, Canada (1997). The third conference, renamed the **International Conference on Large Meteorite Impacts,** was held at Nördlingen (2003), a German town located in the center of the Ries Basin, an impact crater about 20 km in diameter {Ries Basin ⇨c.15 Ma ago; ⇨Fig. 4.1–C}. |
| 1992 | *Tohoku University, Sendai, Honshu, Japan* | In the period November 2–4, **first International Symposium on Impact Engineering (ISIE).** The chairman is Ichiro MAEKAWA. The objective of the meeting is to promote investigations of common problems encountered in impact engineering by discussing fundamental topics, such as<br><br>▸ computational methods and materials behavior under high strain rates;<br>▸ structural behavior when subjected to an impact force;<br>▸ practical problems in the field of civil engineering; and<br>▸ new evaluation methods for impact strength.<br><br>Subsequent symposia were held in Beijing, China (1996); Singapore (1998); Kumamoto, Japan (2001); Cambridge, U.K. (2004); and Daejeon, South Korea (2007). |
| 1992 | *VNIIEF, Sarov [previously known as "Arzamas-16"], Russia* | **Ryurik F. TRUNIN**[3112] **and collaborators begin a series of underground nuclear explosion experiments aimed at generating ultrahigh shock pressures which are otherwise inaccessible in a laboratory** {⇨Fig. 4.11–F}. They determine the shock adiabat of iron up to 10 TPa (100 Mbar) in order to use iron as a standard material to implement a previous experimental method of shock compression in which unknown shock parameters of other materials were determined with respect to known Hugoniot data. ▪ Similar dynamic high-pressure studies using underground nuclear explosions were also performed at Los Alamos {RAGAN ET AL. ⇨1977}.<br><br>In the former Soviet Union, high-pressure studies were initiated in 1948 at the Federal Nuclear Center in Arzamas-16. Pioneering laboratory techniques increased the maximum shock pressures obtainable from 4 Mbar in 1958 to 25 Mbar in 1995. A review of previous Russian (Soviet) converging flyer shell devices used to generate ultrahigh shock pressures using the iron shock adiabat as a dynamical standard in megabar (or terapascal) compression studies on other substances was given by AL'TSHULER ET AL. {⇨1996}. |

---

[3111] *Cosmic Background Explorer (COBE).* Lawrence Berkeley National Laboratory (LBNL); http://aether.lbl.gov/www/projects/cobe/.

[3112] R.F. TRUNIN ET AL.: *Measurement of the compressibility of iron at 5.5 TPa.* Sov. Phys. JETP **75**, 777-780 (1992); *Determination of the shock compressibility of iron at pressures up to 10 TPa (100 Mbar).* Ibid. **76**, 1095-1098 (1993); *Shock compressibility of condensed materials in strong shock waves generated by underground nuclear explosions.* Phys. Uspekhi **37**, 1123-1145 (1994).

| | | |
|---|---|---|
| 1993 | World Trade Center (WTC), New York City, New York | On February 26, **a rented Ryder truck containing a homemade bomb and cyanide gas explodes in the underground garage of the WTC in an area adjacent to Tower 1 and under the Vista Hotel.** The blast from the bomb blows a hole five stories deep and half-a-football-field wide {⇨Fig. 4.21–G}. Six people are killed and over thousand are injured. ▪ The explosive device was made out of ordinary, commercially available materials including lawn fertilizer (urea nitrate) and diesel fuel, and cost less than $400 to make.[3113] |
| 1993 | Lugo, Galicia, northwestern Spain | On March 28, **Francisco Garcia DÍAZ,** a Spanish amateur astronomer and member of the Grupo de Busqueda Supernovas M1 ("Supernova Search Group M1"), **discovers the supernova SN 1993J** in M81, a spiral galaxy located in *Ursa Major* ("The Big Bear," a constellation in the Northern Hemisphere), **while observing with a 10-in. telescope.**[3114] ▪ SN 1993J was very unusual for the following reasons: <br>▸ This supernova appeared in a very close galaxy. Modern astronomers had only been able to observe one other supernova (SN 1987A, found in the Large Magellanic Cloud {SHELTON ⇨1987}) at such close range before. <br>▸ The time of explosion was known to within just a few hours. <br>▸ Supernova SN 1993J showed two maxima in its light curve. The first happened one day after the discovery, while the second occurred three weeks later. <br>▸ Supernova SN 1993J exhibited unique spectral behavior, changing from Type II when the explosion occurred to Type Ib some time later {CHANDRASEKHAR ⇨1931}. <br><br>**The images of SN 1993J represent the first long-term sequence of supernova images ever obtained.**[3115] They show the stellar explosion from its inception in 1993, when a powerful shock wave raced outward at more than 44 million mph (about 19,700 km/s). Five years later, the shock wave had slowed to less than half that speed as it fought increasing drag caused by particles in the interstellar medium.[3116] A sequence of images from very long baseline interferometry taken at wavelengths of 3.6 cm and 6 cm shows that the young radio supernova SN 1993J is expanding with near-circular symmetry. However, the circularly symmetric images show emission asymmetries: a scenario in which freely expanding supernova ejecta shock largely isotropic circumstellar material is strongly favored. **The sequence of images constitutes the first "movie" of a radio supernova.**[3117, 3118] <br><br>In April 2003, a large conference covering all theoretical and observational aspects of supernovae and gamma-ray bursts, as well as their impact on cosmology, took place in Valencia, Spain on the occasion of the 10th anniversary of SN 1993J. |
| 1993 | Okushiri [ca. 20 km west of Oshima Peninsula], southwestern Hokkaido, Japan | During the night of July 12, **a subduction seaquake with a magnitude of 7.6 occurs off the coast of Hokkaido at a depth of about 17 km.** 198 people are killed by the tsunami and earthquake, and hundreds more are injured. Because the quake occurs near the small island of Okushiri, huge tsunamis with heights of between 10 and 30 m attack the island within five minutes, taking most of their victims as they flee to higher ground and more secure locations after having survived the earthquake. ▪ Although the Japan Meteorological Agency (JMA), which monitors the seismic activity around Japan, issued a tsunami warning only five minutes after the earthquake, it comes after the first tsunami wave has already attacked the island. ▪ This severe disaster clearly demonstrated the limitations even of the excellent Japanese tsunami warning and evacuation system under such extreme conditions. |

---

[3113] E. BARNES ET AL.: *The $400 bomb.* Time (March 22, 1994).

[3114] *Chronology of the discovery of SN 1993J in M81.* Grupo de Busqueda Supernovas M1, Madrid, Spain; http://astrosurf.com/blazar/articulo/i1993j.htm.

[3115] N. BARTEL ET AL.: *The changing morphology and increasing deceleration of supernova 1993J in M81.* Science **287**, 112-116 (Jan. 7, 2000).

[3116] *Supernova shock wave captivates astronomers.* Press release (Jan. 6, 2000); http://www.space.com/scienceastronomy/astronomy/supernovae_000106.html.

[3117] J.M. MARCAIDE ET AL.: *Discovery of shell-like radio-structure in SN 1993J.* Nature **373**, 44-45 (1995).

[3118] *The remnant of supernova 1993J in M81.* Students for the Exploration and Development of Space (SEDS), MIT, Princeton, NJ (Sept. 2000); http://www.seds.org/messier/more/m081_snr.html.

***Subduction,*** meaning "sliding under," is a process which occurs when two sections of the Earth's crust collide, and one slab of lithosphere is forced back down into the mantle. Most volcanoes are products of lithosphere plate motions: the rocks thrust into the mantle become unstable, melt, migrate upward, and erupt from volcanoes. Subduction zones are also notorious for producing earth- and seaquakes {Sumatra-Andaman Islands Earthquake ⇨ 2004}.

| 1993 | *Wallops Island, Maryland & NASA's Langley Research Center (LaRC), Hampton, Virginia* | On December 13, above Wallops Islands, **Leonard M. WEINSTEIN,**[3119] a NASA Langley research scientist, **takes the first picture of the shock waves around a full-scale Northrop T-38 jet aircraft flying across the face of the Sun at** $M = 1.1$ **using the focusing schlieren system that he had developed.**[3120] His new schlieren camera, which he calls "SAF (Schlieren for Aircraft in Flight)" can photograph the shock waves of a full-sized aircraft in flight using the largest schlieren field of view (about $96 \times 71$ m$^2$) captured on film. ▪ The camera was then taken to the NASA Dryden Flight Research Center because of the high number of supersonic test flights performed there.[3121] The first aircraft flown with a modified SAF system was a McDonnell F-18 {⇨Fig. 4.6–J}. In 1995, the U.S. *Discover Magazine* gave WEINSTEIN the Award for Technical Innovation "for the development of an in-flight flow visualization technique. This process allows shock waves and turbulences from flying aircraft to be photographed and studied, thus enabling aircraft designers to better understand how their designs work at high, supersonic speeds." In 1999, he was chosen as the AIAA Engineer of the Year for his work on focusing schlieren and SAF. More recently, WEINSTEIN[3122] summarized his work on focusing schlieren and schlieren for rocket sleds and flight. |

First studies using sunlight shadowgraphy to reveal shock wave locations on high-speed aircraft wings were already carried out at Ames Aeronautical Laboratory {COOPER & RATHERT ⇨1948} – one year after the sound barrier was first officially broken by an airplane {YEAGER ⇨1947}. However, the complexity of transonic flow was both mysterious and dangerous, and tools to study it were scarce.

In 1999, **Gary S. SETTLES,**[3123] a U.S. professor of mechanical engineering at Penn State University, **gave a historical perspective of outdoor schlieren and shadowgraph imaging.** He also discussed the optical principles behind the sunlight shadowgraph method and schlieren observation by background distortion, and illustrated the versatility of outdoor visualization methods using numerous practical examples, encompassing thermal convection, combustion, explosion, and shock wave phenomena.

| 1993 | *John Wiley & Sons, Inc., New York City, New York* | **Foundation of the journal *Shock and Vibration*** in order "to provide a source for the publication of original, archival articles on shock, vibration, sound, structural dynamics, biodynamics, crashworthiness, earthquake engineering, gun dynamics, vehicle dynamics, and dynamics and vibration performance of civil engineering structures." The founding editor is Walter D. PILKEY (Dept. of Mechanical and Aerospace Engineering, University of Virginia, Charlottesville, VA). |

---

[3119] L.M. WEINSTEIN: *An optical technique for examining aircraft shock wave structures in flight.* Proc. 3rd High-Speed Research Program Sonic Boom Workshop [Hampton, VA, 1994]. In: (D.A. MCCURDY, ed.) *High-speed research: 1994 Sonic Boom Workshop – Atmospheric Propagation and Acceptability Studies.* Rept. NASA CP 3279 (1994), pp. 1-17.
[3120] B.D. NORDWALL: *NASA technique images supersonic shock waves.* Aviation Week & Space Technology **140**, 60 (Feb. 14, 1994).
[3121] L.M. WEINSTEIN, K. STACY, G.J. VIEIRA, E.A. HAERING JR., and A.H. BOWERS: *Imaging supersonic aircraft shock waves.* J. Flow Vis. Image Process. **4**, 189-199 (1997).
[3122] L.M. WEINSTEIN: *Schlieren.* In: (C. MERCER, ed.) *Optical metrology for fluids, combustion and solids.* Springer, Berlin (2003), pp. 1-36.
[3123] G.S. SETTLES: *Schlieren and shadowgraph imaging in the great outdoors.* In: (S. MOCHIZUKI, ed.) *Proc. 2nd Pacific Symposium on Flow Visualization & Imaging Processing (PSFVIP)* [Honolulu, HI, May 1999]. 1 CD-ROM (ISBN: 0-9652469-7-3), Honolulu, HI (1999), Paper PF302 (14 pp.).

| | | |
|---|---|---|
| 1993 | Christian Michelsen Research AS, Bergen, Norway | **Publication of the *Gas Explosion Handbook*.**[3124] Its editors are Dag BJERKETVEDT, Jan Roar BAKKE, and Kees VAN WINGERDEN. ▪ The purpose of this handbook is to provide a brief introduction to gas explosion safety, based on current knowledge of the subject and on experience of applying this knowledge to practical problems in industry. |
| 1993 | Hadland Photonics Ltd., Bovington [now DRS Hadlands Ltd, Tring], Hertfordshire, England | The British company **Hadland Photonics Ltd. presents the first commercially available multiple ultrahigh-speed ICCD (intensified CCD) camera – Imacon® 468 –** which is capable of digitally recording eight frames at a rate of up to 100 million frames per second.<br><br>(i) Using a proprietary pyramid beam splitter design, the Imacon 468 can, from a single optical input, provide high-resolution images at up to eight CCD sensors {BOYLE & G.E. SMITH ⇨ 1969} intensified by microchannel plates {FARNSWORTH ⇨ 1930}, which are also used for ultrafast shuttering purposes.<br><br>(ii) The ultrashort exposure time of only 10 ns is particularly suited to the visualization of ultrafast shock wave, detonation and impact phenomena.<br><br>In 1994, Joseph HONOUR,[3125] a research engineer at Hadland Photonics, discussed the advantages of multichannel CCD/MCP-based cameras compared to previous electronic image converter cameras {HUSTON ⇨ 1965}. The Imacon 468 awarded the 1999 Queens Award for Technology. |
| 1993 | Sandia National Laboratory (SNL) & University of New Mexico (UNM), Albuquerque, New Mexico | **James R. ASAY,**[3126] a shock physicist at SNL, **and Mohsen SHAHINPOOR,** a professor of mechanical engineering at UNM, **edit the textbook *High-Pressure Shock Compression of Solids*.**<br><br>In the years following, further volumes on this subject were published, forming an eight-volume compendium: the Springer series *High-Pressure Shock Compression of Condensed Matter*. This particular branch of high-pressure physics addresses shockwave-induced phenomena of the liquid and (mainly) solid states.[3127] |
| 1993 | Dept of Astrophysical Sciences, Princeton University, Princeton, New Jersey & Astronomy Dept., UC Berkeley, California | **Bruce T. DRAINE**[3128] **and Christopher F. MCKEE,** two U.S. professors of astrophysics, **outline the physical principles that underpin our theoretical understanding of the structures of interstellar shock waves.**<br><br>(i) Attention is given to single fluid shocks, with emphasis placed on fluid equations, jump conditions, and collisionless and radiative shocks.<br><br>(ii) Multifluid shocks in weakly ionized gas are examined with reference to fluid equations, atomic and molecular processes, the role of dust grains, length scales, critical ionization velocity, magnetic precursors to J (jump)-type shocks, and numerical modeling. |

---

[3124] D. BJERKETVEDT, J.R. BAKKE, and K. VAN WINGERDEN (eds.): *Gas explosion handbook*. Rept. CMR-93-A25034, publ. by Christian Michelsen Research, AS, Bergen, Norway (1993); http://www.gexcon.com/index.php?src=/handbook/GEXHBcontents.htm.

[3125] J. HONOUR: *Electronic camera systems take the measure of high-speed events*. Laser Focus World **30**, 121-127 (1994).

[3126] J.R. ASAY and M. SHAHINPOOR (eds.): *High-pressure shock compression of condensed matter*. Springer, Berlin (1993).

[3127] L. DAVISON, D.E. GRADY, and M. SHAHINPOOR (eds.): *High-pressure shock compression of condensed matter II – Dynamic fracture and fragmentation*. Springer, New York (1996);
L. DAVISON and M. SHAHINPOOR (eds.): *High-pressure shock compression of condensed matter III – High-pressure shock compression of solids*. Springer, New York (1998);
L. DAVISON, Y. HORIE, and M. SHAHINPOOR (eds.): *High-pressure shock compression of condensed matter IV – Response of highly porous solids to shock loading*. Springer, New York & London (1997);
L. DAVISON, Y. HORIE, and T. SEKINE (eds.): *High-pressure shock compression of condensed matter V – Shock chemistry with applications to meteorite impacts*. Springer, New York & London (2003);
Y. HORIE, L. DAVISON, and N.N. THADANI (eds.): *High-pressure shock compression of condensed matter VI – Old paradigms and new challenges*. Springer, New York & London (2003);
V.E. FORTOV ET AL. (eds.): *High-pressure shock compression of condensed matter VII – Shock waves and extreme states of matter*. Springer, New York & London (2004);
L.C. CHHABILDAS, L. DAVISON, and Y. HORIE (eds.): *High-pressure shock compression of condensed matter VIII – The science and technology of high-velocity impact*. Springer, Berlin (2005).

[3128] B.T. DRAINE and C.F. MCKEE: *Theory of interstellar shocks*. Annu. Rev. Astron. Astrophys. **31**, 373-432 (1993); ftp://ftp.astro.princeton.edu/draine/papers/pdf/ARAA_31_373.pdf.

| | | (iii) Multifluid shocks in ionized gases, instabilities in shock waves, dust in shocks, and shock chemistry are also discussed. |
|---|---|---|
| 1993 | *Meta Research, Inc., Sequim, Washington* | **Tom** VAN FLANDERN,[3129] a U.S. astronomer and president of Meta Research, Inc., **publishes the book *Dark Matter, Missing Planets and New Comets: Paradoxes Resolved, Origins Illuminated,*** in which he critically discusses many standard models of astronomy, such as the origin of the Solar System, planetary explosions, the Oort Cloud, the Dirty Snowball theory, and the Big Bang theory. |
| 1993 | *Sandia National Laboratory (SNL), Albuquerque, New Mexico* | **Robert A.** GRAHAM,[3130] a senior shock physicist, **publishes his book *Solids Under High-Pressure Shock Compression*** – the first comprehensive documentation of shock compression research in a book form. It describes shock-compressed matter derived from physical *and* chemical observations, and thus significantly differs from previous classical descriptions derived strictly from mechanical characteristics. |
| 1993 | *Cold Regions Research and Engineering Laboratory (CRREL), Fort Wainwright, Fairbanks, Alaska; Los Alamos National Laboratory (LANL), Los Alamos & Ktech Corporation, Albuquerque, New Mexico* | **Jerome B.** JOHNSON,[3131] a U.S. research physicist at the Snow and Ice Branch of the U.S. Army's CRREL, **and researchers from LANL and Ktech Corporation report on joint studies of shock wave loading, unloading and attenuation in snow.** The initial densities were ranging from 100 to 520 kg/m$^3$, temperatures from $-2$ to $-23$ °C, and stress levels from 20 to 40 MPa (200–400 bar). <br> (i) Using a 20-mm dia. gas gun, they fired flat-faced flyer plates into snow targets consisting of a copper cylinder sealed with aluminum end-plates to provide a vacuum-tight canister. <br> (ii) They measured stress-time records using carbon-film piezoresistive stress gauges. <br> (iii) Pressure-density data, derived from calculations using finite element and quasi-steady analysis, reveal a loading curve $P(\rho)$ which can be approximated in the low-pressure regime (2–20 MPa) by the simple law $$P(\rho) = AX^B,$$ where $P$ is the pressure in Pa, $X$ is the density change $(\rho - \rho_0)$ in kg/m$^3$, $\rho_0$ is the initial snow density, and $\rho$ is the snow density at pressure $P$. The parameters $A$ and $B$ are empirical functions of $\rho_0$ only. <br> (iv) The $P(\rho)$ plots have two modes: large deformations at low pressures, followed by a relatively abrupt increase in resistance, to further deformation (strain hardening) at higher pressures, which occurs at a critical density (porosity) for snow. <br> (v) **Shock loading causes significant permanent compaction of snow that severely attenuates shock waves traveling through it.** <br> Understanding the response of snow to shock loading is important when solving a number of problems in cold regions engineering, planetary sciences, seismic monitoring and shock isolation. According to JOHNSON, examples of such problems include:[3132] <br> ▸ the use of explosives to neutralize snow-covered land mines and other military uses of explosives in snow-covered terrain; <br> ▸ the use of explosives to release avalanches; <br> ▸ the determination of forces on structures caused by avalanche debris impacts and acting on snow removal equipment; <br> ▸ the determination of momentum and energy change for impacts between frost clouds and comets with celestial bodies; |

---

[3129] T. VAN FLANDERN: *Dark matter, missing planets and new comets: paradoxes resolved, origins illuminated.* North Atlantic Books, Berkeley, CA (1993).
[3130] R.A. GRAHAM: *Solids under high-pressure shock compression.* Springer, New York *etc.* (1993).
[3131] J.B. JOHNSON, D.J. SOLIE, J.A. BROWN, and E.S. GAFFNEY: *Shock response of snow.* J. Appl. Phys. **73**, 4852-4861 (1993).
[3132] *Shock waves in snow.* U.S. Army Corps of Engineers (USACE) Engineer Research and Development Center, Cold Regions Research and Engineering Laboratory, Hanover, NH; http://www.crrel.usace.army.mil/alaska-office/shokwave.html.

- the use of snow to isolate and protect objects from shocks; and
- the identification of the source signal and magnitude of large-explosive detonations on ice caps and in snow-covered terrain.

| | | |
|---|---|---|
| 1993 | D.A. McGee Eye Institute and Dept. of Ophthalmology, University of Oklahoma, Oklahoma City & University Center of Laser Research, Oklahoma State University, Stillwater, Oklahoma | Ronald R. KRUEGER,[3133] a U.S. ophthalmologist who specialized in refractive surgery, **and collaborators visualize the emission of shock waves during excimer laser ablation of the cornea using high-speed shadowgraphy at various time delays between 40 and 320 nanoseconds.**<br><br>(i) Their experimental set-up consists of a xenon-chloride excimer laser {SEARLES & HART ⇒1975} which emits 308-nm light at energy densities of 0.5 to 2 J/cm$^2$. This laser light ablates the specimen and excites a tunable dye laser emitting at 580 nm, which is then passed through an optical delay line before irradiating the specimen.<br><br>(ii) The semi-circular silhouettes of shock wavefronts in air, nitrogen, and helium are recorded by tangentially illuminating the ablated surface of the cornea with the dye laser pulse. The shadow of the shock wave produced during ablation is then cast onto a screen and photographed with a CCD video camera. The system is pulsed at 30 times per second to allow video recording of the shock wave at a fixed time delay.<br><br>(iii) They find that high-energy acoustic waves and gaseous particles are liberated during excimer laser corneal ablation, and these dissipate on a submicrosecond timescale.<br><br>(iv) **The velocity of the shock wave resulting from corneal ablation is on the order of km/s, depending on the atmospheric environment.** It can be increased two-fold when the ablation is performed in a helium atmosphere. Advantageously, this would reduce the exposure and transfer of heat to the surrounding tissue. |
| 1993 | North Carolina School of Science & Mathematics, Durham, North Carolina | Nicolas LEE,[3134] Spencer ALLEN, Elizabeth SMITH, and Loren M. WINTERS **investigate the problem of whether the tip of a snapped towel can exceed the speed of sound.** It is widely known that if a wet towel is flipped in just the right way, it will make a cracking sound. In order to obtain multiple-image photographs from videotape, they used a bank of four flash units triggered by a photo detector and controlled by a computer, as well as a video camera for recording. **Using this technique, they eventually succeeded in demonstrating that the towel tip can exceed the velocity of sound by about 20%.** |
| 1993 | Institut Saint-Louis (ISL), Saint-Louis, Alsace, France | At the first International Workshop on Ram Accelerators {HERTZBERG ET AL. ⇒1986; Workshop on Ram Accelerators ⇒1991}, **Dennis WILSON,**[3135] a professor of mechanical engineering (University of Texas, Austin, TX) **and collaborators present the concept for propelling a projectile through a gun tube – the so-called "blast wave accelerator" – for the first time.** They also present a preliminary quasi-steady analysis. Their concept involves the sequential detonation of high explosive charges behind a projectile. ▪ Later they showed that in theory launch velocities exceeding 6 km/s are possible.[3136]<br><br>The concept of a blast wave accelerator stimulated the interests of numerous researchers for quite different reasons:<br>- In previous years, NASA had already shown an interest in developing techniques allowing inexpensive access to space {HARP ⇒1961}.<br>- In the mid-1990s, the Space Systems Division at NASA's Langley Research Center identified a national need for a launcher capable of accelerating large packages to ve- |

---

[3133] R.R. KRUEGER, J.S. KRASINSKI, C. RADZEWICZ, K.G. STONECIPHER, and J.J. ROWSEY: *Photography of shock waves during excimer laser ablation of the cornea. Effect of helium gas on propagation velocity.* Cornea **12**, No. 4, 330-334 (1993).

[3134] N. LEE, S. ALLEN, E. SMITH, and L.M. WINTERS: *Does the tip of a snapped towel travel faster than sound?* The Physics Teacher **31**, 376-377 (1993); http://www.hiviz.com/PROJECTS/towel/towel.htm.

[3135] D. WILSON, Z. TAN, and P.L. VARGHESE: *Numerical simulation of the blast wave accelerator.* AIAA J. **34**, No. 7, 1341-1347 (1996).

[3136] D. WILSON and Z. TAN: *The blast wave accelerator – feasibility study.* AIP Conf. Proc. **552**, 589-598 (Feb. 2, 2001).

locities of 6 km/s in order to study hypervelocity aeroballistic and aerothermodynamic phenomena.[3137]

> The U.S. Army took an interest in the subject due to its ability to accelerate masses of several kilograms to velocities in excess of 3 km/s and its applicability to studies of large-scale hypervelocity impact and penetration dynamics.[3138]

1994    *Planet Jupiter*    Between July 16 and 22, **a comet – later labeled *SHOEMAKER-LEVY 9* (or *S-L 9*) – strikes Jupiter's atmosphere at an impact angle of about 45° with a velocity of over 60 km/s.** This Jupiter-family comet consists of about 20 fragments of stony and icy matter ranging from 1 to 4 km in diameter. The impact represents the most energetic event ever witnessed in the Solar System besides the daily output from the Sun; it produces huge plasma flares and sharp increases in radio wave emission.

(i) The impact of the fragments causes Jupiter to exhibit a number of highly visible big dark "scars." **Perhaps the largest impact is produced on July 18; it is estimated to pack the explosive power of 6 million megatons of TNT.**[3139]

(ii) On July 19, the ROSAT high-resolution imager detects intense X-ray emission immediately following the impact of fragment K from comet SHOEMAKER-LEVY 9. It is believed that the emissions result from the impacts of energetic electrons that were either accelerated along the magnetic field lines from the impact site to a conjugate region or precipitated by impact-induced pitch-angle scattering from Jupiter's trapped radiation belts.[3140]

Comet SHOEMAKER-LEVY 9 was discovered on March 25, 1993 by the U.S. geologist and planetary scientist Eugene M. SHOEMAKER, together with his wife Carolyn, an astronomer, and David H. LEVY,[3141] another U.S. astronomer and experienced comet discoverer, using a 450-mm Schmidt camera at the Mt. Palomar Observatory. The collision was an exceptional scientific event:

> Astronomers were able to watch in real time the response of a planetary atmosphere to a large meteorite.[3142]

> In order to predict and model the giant fireball plumes, Sandia scientists performed calculations for different size fragments and compared these numerical simulations with observed data. They used shock physics codes which were developed previously to calculate nuclear weapons effects up to yields of million megatons of TNT equivalent.[3143]

> A team of researchers from various American research institutes (UC Santa Cruz, Cornell University, and NASA-GSFC) performed a high-resolution, 3-D hydrodynamic simulation of the impact of porous ice comets into the atmosphere of Jupiter and calculated energy deposition profiles, depths of bolide penetration, and the disruptive impact process.[3144]

---

[3137] R. WITCOFSKI, W. SCALLION, and D. CARTER: *Advanced hypervelocity aerophysics facility: a ground-based flight test range.* AIAA Paper No. 91-0296 (Jan. 1991).

[3138] W.G. REINECKE and H.H. LEGNER: *Review of hypervelocity projectile aerophysics.* AIAA Paper No. 95-1853 (April 1995).

[3139] J.H. SHIRLEY: *Comet: impacts on Jupiter.* In: (J.H. SHIRLEY and R.W. FAIRBRIDGE, eds.) *Encyclopedia of planetary sciences.* Chapman & Hall, London *etc.* (1997), pp. 132-134.

[3140] *ROSAT X-ray images of fragment K impact.* NASA-JPL (July 19, 1994); http://www2.jpl.nasa.gov/sl9/image345.html.

[3141] E.M. SHOEMAKER, C.S. SHOEMAKER, and D.H. LEVY: *Discovering comet Shoemaker-Levy 9.* In: *Once in a thousand lifetimes: a guide to the collision of comet Shoemaker-Levy 9 with Jupiter.* The Planetary Society, Pasadena, CA (1994), pp. 2-3.

[3142] T. ENCRENAZ: *Shoemaker-Levy 9-Jupiter collision.* In: (P. MURDIN, ed.) *Encyclopedia of astronomy and astrophysics.* Institute of Physics Publ., London *etc.* (2001), pp. 2413-2420.

[3143] D. CRAWFORD, M. BOSLOUGH, A. ROBINSON, and T. TRUCANO: *Numerical simulations of comet Shoemaker-Levy 9 impact on Jupiter.* Comet Shoemaker-Levy 9 Impact Simulation Team, Sandia Laboratories, Albuquerque, NM; http://www.cs.sandia.gov/HPCCIT/jupiter.html.

[3144] D.G. KORYCANSKY ET AL.: *Shoemaker-Levy 9 impact modeling. I. High resolution three-dimensional bolides.* Astrophys. J. **646**, 642-652 (2006).

| | | |
|---|---|---|
| 1994 | *Space Telescope Science Institute (STScI), Baltimore, Maryland* | Using the NASA/ESA's Hubble Space Telescope, **U.S. and European astronomers find seemingly conclusive evidence for a *supermassive black hole* at the center of the giant galaxy M87 (or NGC 4486).**[3145] ▪ This elliptical galaxy, located about 50 million light-years away in the constellation *Virgo*, was discovered in 1781 by the celebrated French astronomer Charles MESSIER (M87 is short for Messier 87). By 1781, his catalogue of deep sky objects (such as nebulae and star clusters) had grown to 110 Messier Objects. |
| 1994 | *Scientific Research Institute of Technical Physics, Snezhinsk, Chelyabinsk Region, Russia* | **The first International Conference on Space Protection of the Earth ("SPE–1994") is held.** It is dedicated to the investigation and development of methods and facilities that prevent regional and global catastrophes caused by the approaches and collisions of asteroids, comets and meteorites with the Earth. Main organizer of the Conference is the Russian Federal Nuclear Center – Russian Scientific-Research Institute of Technical Physics (RFNC-VNIITF) ▪ Subsequent conferences were held in Russia at Snezhinsk, Chelyabinsk Region (Sept. 1996) and Evpatoriya, Ukraine (Sept. 2000).

In the same year, on December 9, a near-miss takes place when the asteroid 1994 XL1 (about 300 m in dia.) passes within 105,000 km of our planet. ▪ In the years that followed, the danger from and probability of a collision of a Near-Earth Object (NEO) with Earth was discussed with increasing frequency among high-velocity impact physicists, experts in the detection of NEOs and astrophysicists at newly founded conferences. Examples include the United Nations International Conferences: Near-Earth Objects (from 1995) and the AIAA Planetary Defense Conference: Protecting the Earth from Asteroids (from 2004). |
| 1994 | *Australia* | **Paul C.W. DAVIES,**[3146] former professor of physics at the University of Adelaide, cosmologist and author of numerous popular books, **publishes a book entitled *The Last Three Minutes: Conjectures About the Ultimate Fate of the Universe*** in which he discusses the latest theories on possible futures for the Universe. It states that the Universe could
   ▸ reach a steady state, neither expanding nor contracting, but staying the same forever;
   ▸ go on expanding indefinitely; or
   ▸ slow down and eventually collapse into a zero space, the so-called **"Big Crunch"** – the reverse of the Big Bang {ALPHER, HERMAN & GAMOW ⇨ 1948}. ▪ Edward R. HARRISON,[3147] a professor of physics and astronomy at the University of Massachusetts at Amherst, thought that the term ***Antibang*** was preferable to Big Crunch.

DAVIES' book was intended to be the companion to a previously published book on the Big Bang {WEINBERG ⇨ 1977}. |
| 1995 | *Kansai region, Honshu, south-central Japan* | On January 17, **a severe earthquake strikes the region of Kobe and Osaka,** shaking Japan's second-most heavily populated and industrialized area (about 5,500 people are killed and 35,000 injured). The shock occurs at a shallow depth on a fault running from Awaji Island through the city of Kobe, which has about 1.5 million inhabitants. ▪ The earthquake caused extensive ground failures, which destroyed more than 100,000 buildings and affected underground infrastructure, the port, highways, various facilities on soft or filled ground, and recovery efforts. **The ground failures occurred primarily due to *seismic shock-induced liquefaction* –** *i.e.*, the temporary transformation of loose, water-saturated sand into a semi-liquid state by the seismic shock. In the Kobe area, cretaceous granites are overlain by a relatively thick Plio-Pleistocene sedimentary unit which consists of alluvium interbedded with marine clays.[3148] |

---

[3145] F. MACCHETTO ET AL.: *The supermassive black hole of M87 and the kinematics of its associated gaseous disk.* Astrophys. J. **489**, 579-600 (1997).
[3146] P.C.W. DAVIES: The *last three minutes: conjectures about the ultimate fate of the Universe.* Weidenfeld & Nicolson, London (1994).
[3147] E.R. HARRISON: *Acceleration and dissolution of stars in the Antibang.* In: (G.O. ABELL and G.L. CHINCARINI, eds.) *Early evolution of the Universe and its present structure* [Symposium held at Kolymbari, Greece in Aug./Sept. 1982]. D. Reidel Publ. Co., Dordrecht & Boston (1983), pp. 453-455.
[3148] *The January 17, 1995 Kobe Earthquake.* Summary Rept. of EQE International, Inc., Oakland, CA (April 1995).

| | | |
|---|---|---|
| 1995 | *Ernst-Mach-Institut (EMI) der Fraunhofer-Gesellschaft (FhG), Freiburg & VKT Video Kommunikation GmbH, Pfullingen, Germany* | In March, **Peter KREHL**[3149] **and Stephan ENGEMANN** (EMI), **supported by Dieter SCHWENKEL** (VKT), **resume previous attempts to visualize the shock wave emitted during whip cracking** {CARRIÈRE ⇨ 1927; BERNSTEIN ET AL. ⇨ 1958}. To tackle this difficult task, they use large field-of-view shadowgraphy combined with a retroreflective screen and modern copper-vapor laser stroboscopy synchronized with a high-speed video camera. Correlating the motion of the tip of the whip lash (the "cracker") to the emitted shock wave, the movie clearly shows that at its final stage the tuft moves supersonically at more than twice the velocity of sound. However, the supersonic motion is only a *conditio sine qua non*: the decisive mechanism for generating a sharp cracking noise is the abrupt flapping of the tuft upon passing the turning point {⇨Fig. 4.5–K}. A whip lash with no tuft or an improperly-trimmed tuft cannot produce efficient cracking. ▪ The study was initiated by the *Zweites Deutsches Fernsehen* ("Second German Television"), which asked EMI to provide a movie on the mechanism of whip cracking that also showed the evolution of the shock wave.<br><br>In 2002, **Alan GORIELY**,[3150] a professor of mathematics at the University of Arizona in Tucson, **and Tyler MCMILLEN**, stimulated by these results, **presented a simple dynamic model for numerically calculating the propagation and acceleration of waves during the motion of whips.** Their model addresses<br>▸ the relationship between sonic boom and whip tip velocity;<br>▸ the effect of tapering;<br>▸ the boundary conditions; and<br>▸ the role of energy, linear momentum, and angular momentum.<br><br>In the following year, they used asymptotic analysis to show that a wave traveling along the whip increases its speed as the radius decreases. They also described the shape of the shock wave emitted by a material point on the whip that is traveling faster than the speed of sound.[3151] |
| 1995 | *Oklahoma City, Oklahoma* | On April 19, **the Alfred P. Murrah Federal Building is destroyed by a terrorist bomb attack** (resulting in 168 casualties).[3152] The bomb consisted of a large quantity (estimated at 4,800 pounds) of AN [ammonium nitrate, $NH_4NO_3$], a cheap, widely-used agricultural fertilizer, which was mixed with diesel fuel and carried in a truck. ▪ Since the occurrence of the disastrous accidental explosion in Texas City {SS *Grandchamp* ⇨1947}, it has been common knowledge that AN can explode and cause great destruction. Ammonium nitrate bombs are commonly used by terrorists across the world. Most countries, including the United States and Australia, do not regulate the sale of fertilizer-grade ammonium nitrate, but it is tightly restricted in the European Union. |
| 1995 | *Ecole Nationale Supérieure de Mécanique et d'Aérotechnique (ENSMA), Poitiers, France* | In October, **IUTAM Symposium on Combustion in Supersonic Flows.** It is organized by the International Union of Theoretical and Applied Mechanics (IUTAM) and hosted by the Laboratoire de Combustion et de Détonique at Chasseneuil du Poitou.[3153] The conference chairman is Michel CHAMPION. ▪ The presentations, related to combustion in supersonic streams and practical issues related to the development of new propulsion systems, focused on four main topics: |

---

[3149] P. KREHL, S. ENGEMANN, and D. SCHWENKEL: *The puzzle of whip cracking – uncovered by a correlation of whip-tip kinematics with shock wave emission.* Shock Waves **8**, 1-9 (1998).

[3150] A. GORIELY and T. MCMILLEN: *Shape of a cracking whip.* Phys. Rev. Lett. **88**, No. 24, Paper No. 244301 (2002).

[3151] T. MCMILLEN and A. GORIELY: *Whip waves.* Physica D **184**, 192-225 (2003); http://www.math.princeton.edu/~mcmillen/papers/WhipWaves2003.pdf.

[3152] *Alfred P. Murrah Federal Building bombing, April 19, 1995. Final report.* City of Oklahoma Document Management Team. Fire Protection Publication, Stillwater, OK (1996).

[3153] M. CHAMPION and B. DESHAIES (eds.): *IUTAM Symposium on Combustion in Supersonic Flows* [Poitier, France, Oct. 1995]. Kluwer Academic Publ., Dordrecht *etc.* (1997).

| | | |
|---|---|---|
| | | ▸ fundamental studies of premixed and unpremixed combustion; |
| | | ▸ aspects of supersonic combustion related to fluid dynamics; |
| | | ▸ applications of detonation to propulsion; and |
| | | ▸ practical systems, including scramjets, ram accelerators, and pulse detonation engines. |
| | | There were no further IUTAM Symposia held on this subject. |
| 1995 | *LL Orionis [in the constellation Orion]* | **NASA/ESA's Hubble Space Telescope reveals a bow shock around *LL Orionis* (or *LL Ori*), a very young star.** This *cosmic bow shock* or *cosmic Mach cone*, which is about half a light-year across, is created as the energetic stellar wind from *LL Orionis* collides with the Orion Nebula flow. As the fast stellar wind runs into slow moving gas a shock front is formed, analogous to an airplane traveling at supersonic speed. The slower gas is flowing away from *Theta Orionis* ("The Trapezium"), a young multiple star system in the heart of the Orion Nebula.[3154] |
| 1995 | *Berlin (DGST) and Mettmann North Rhine-Westphalia, (IGESTO), Germany* | **Foundation of two German societies for extracorporeal muscular skeletal shock wave therapy:**<br>▸ **DGST** (Deutsche Gesellschaft für Extrakorporale Stoßwellentherapie e.V.) and<br>▸ **IGESTO** (Internationale Gesellschaft für Extrakorporale Stoßwellentherapie e.V.)<br>with the objective to investigate, promote and develop this new therapy. ▪ Both societies later merged into **DIGEST** (Deutsche und Internationale Gesellschaft für extracorporale Stoßwellentherapie).[3155]<br><br>The growing interest in this therapy prompted international meetings, such as the annual International Congresses of the **ISMST** (International Society for Musculoskeletal Shockwave Therapy), previously held at London, U.K. (1998); Izmir, Turkey (1999); Naples, Italy (2000); Berlin, Germany (2001); Winterthur, Switzerland (2002); Orlando, FL (2003); Kaohsiung, Taiwan (2004); Vienna, Austria (2005); Rio de Janeiro, Brazil (2006); and Toronto, Canada (2007).<br><br>The suitability of shock waves for treating orthopedic diseases was discovered by chance by two independent groups of German physicians led by Jürgen GRAFF[3156] and Gerald HAUPT.[3157] Originally investigating possible detrimental side effects of shock waves on bones and tissue when using shock lithotripsy, they discovered positive effects such as shock-induced osteogenesis. However, the new shock therapy – which has been used on patients since 1992 – is not yet fully established and generally accepted, although numerous treatments of various orthopedic diseases have proved to be successful, *e.g.*,<br>▸ "heel spur" [*plantar fasciitis*];<br>▸ "tennis elbow" [*epicondylitis*]; and<br>▸ calcification of shoulder joints [*tendonitis calcarea*].<br>While the success of shock wave therapy in urology using shock lithotripsy is immediately evident due to the stone debris produced, the healing of orthopedic diseases is a more lengthy and complex process. In 2004, Ludger GERDESMEYER,[3158] a German orthopedic surgeon at the Dept. of Orthopedic Surgery and Sports Traumatology, TU Munich, edited the book *Extrakorporale Stoßwellentherapie* ("Extracorporeal Shock Wave Therapy"). Addressing to orthopedists it focuses on the use of "radial" (or spherical; *i.e.*, divergent in contrast to focused) shock waves. |

---

[3154] *LL Orionis: when cosmic winds collide*. Astronomy picture of the day. NASA-GSFC, Greenbelt, MD (Nov. 15, 2003); http://antwrp.gsfc.nasa.gov/apod/ap031115.html.
[3155] *Deutschsprachige internationale Gesellschaft für exraorporale Stoßwellentherapie (DIGEST) e.V.*, Berlin; http://digest-ev.de/.
[3156] J. GRAFF, K.D. RICHTER, and J. PASTOR: *Effect of high energy shock waves in bony tissue*. Urological Res. **16**, 252-258 (1988).
[3157] G. HAUPT ET AL.: *Enhancement of fracture healing with extracorporeal shock waves*. J. Urology **143**, 230A (1990).
[3158] L. GERDESMEYER (ed.): *Extrakorporale Stoßwellentherapie*. Books on Demand, Norderstedt (2004); *Extracorporeal shock wave therapy: technologies, basics, clinical results*. Data Trace Publ., Towson, MD (2006).

| | | |
|---|---|---|
| 1995 | *Aberdeen Test Center (ATC), Aberdeen Proving Ground, Maryland* | **The UNDEX (Underwater Explosions) Test Facility at ATC becomes fully operational.** This unique underwater shock facility – also known as "Superpond" – is a 150-ft (45-m)-deep body of water. ATC also operates a 55-ft (16.8-m)-deep test pond. It enables the U.S. Army to carry out large-scale underwater shock testing of submarine and surface ship structures, and to study the effects of underwater explosions on torpedoes, warheads, remotely operated underwater vehicles, *etc.*, as well as underwater gun firing.[3159] |
| 1995 | *Max-Planck-Insitut (MPI) für extraterrestrische Physik, Garching, Germany* | Bernd R. ASCHENBACH,[3160] Roland EGGER, and Joachim TRÜMPER, three German astrophysicists, **report on the discovery of six extended X-ray features well outside the blast-wave front from the *Vela supernova*,** based upon images taken with ROSAT (Roentgen Satellite), the German/-British/American space-based X-ray telescope {⇨Fig. 4.1–Y}. ▪ The Vela supernova occurred in our galaxy in the constellation *Vela*, resulting in the Gum Nebula {⇨*c*.9000 B.C.}, our closest supernova remnant.[3161] X-ray photography reveals a roughly spherical expanding shock wave.[3162]<br><br>(i) They propose that these X-ray features result from the passage of fragments formed by instabilities during the collapse and subsequent explosion of the progenitor star through the surrounding medium.<br><br>(ii) **In particular, they interpret the protruding objects as Mach cones extending back to the blast-wave front of the Vela, and estimate Mach numbers of between 2.4 and 4.0.**<br><br>(iii) They conclude that these low Mach numbers imply that the temperature of the ambient medium through which the objects are moving must be high, close to X-ray temperatures. **This would indicate that the Vela supernova exploded in a bubble of hot tenuous gas.** They calculate a velocity of 3,200 km/s for the blast wave emitted by the Vela supernova. |
| 1995 | *Institut für angewandte Physik, TH Darmstadt, Germany* | Claus-Dieter OHL,[3163] Achim PHILIPP, and Werner LAUTERBORN, three German physicists, **study cavitation phenomena produced in water by a focused pulsed laser beam at 20 million frames per second.**<br><br>(i) For the two most practically interesting cases of bubble dynamics – the generation of bubbles in a free liquid or near a solid boundary – they record the bubble collapse and shock wave emission in reflected light using a flash lamp for background illumination and the IMACON 700 {HUSTON ⇨1965; ⇨Fig. 4.19–P} high-speed image converter camera.<br><br>(ii) In an unbounded liquid, the initially spherical bubble changes to become ellipsoidal. After reaching its minimum size, the rebound process begins, accompanied by the emission of a spherical shock wave.<br><br>(iii) **The collapse of a cavitation bubble near a solid boundary is more complex, leading to the generation of a liquid jet and the emission of two shock waves, separated in time.** The first shock wave emerges when the jet hits the near side of the bubble wall; the second shock wave is emitted at minimum bubble size. |
| 1996 | *Cape Canaveral Air Station (CCAS), Florida* | On February 17, **launch of NASA's spacecraft NEAR** (Near-Earth Asteroid Rendezvous) **with the goal to collect data on the mass, structure, geology, and composition of asteroid 433 EROS.** ▪ This asteroid, discovered in 1898 at the Urania Observatory by the Berlin astronomer Carl Gustav WITT, is one of the larger Near-Earth Asteroids (NEAs): it has a size of about |

---

[3159] M. CAST: *Aquatic soldiers*. Soldiers. The U.S. Army Mag. (March 2001), pp. 46-48.
[3160] B. ASCHENBACH, R. EGGER, and J. TRÜMPER: *Discovery of explosion fragments outside the Vela supernova remnant shock-wave boundary*. Nature **373**, 587-590 (1995).
[3161] *Supernova remnants Vela and Puppis A*. X-ray Astronomy, Max-Planck-Institut für Extraterrestrische Physik (MPE), Garching, Germany; http://wave.xray.mpe.mpg.de/rosat/five_years/bild_38.
[3162] *Vela Supernova remnant in X-ray*. Astronomy picture of the day (June 12, 1996). NASA-GSFC, Greenbelt, MD;
[3163] C.D. OHL, A. PHILIPP, and W. LAUTERBORN: *Cavitation bubble collapse studied at 20 million frames per second*. Ann. Phys. **4** [VIII], 26-34 (1995).

$33 \times 13 \times 13$ km$^3$ and approaches to within 23 million kilometers of the Earth at times (during 1900/1901 and 1930/1931).

In February 2000 – after a four-year journey that included fly-bys of the Earth in January 1998, and of the two asteroids 253 MATHILDE in June 1997 and 433 EROS in December 1998 – NEAR began orbiting 433 EROS. In March 2000, NEAR was renamed NEAR Shoemaker in honor of the late U.S. geologist and planetary scientist Eugene M. SHOEMAKER {CHAO, SHOEMAKER & MADSEN ⇨1960; SHOEMAKER & CHAO ⇨1961}, who died in July 1997 in the Australian outback during an annual trip to search for asteroid craters.

**On February 12, 2001, NEAR Shoemaker touched down on 433 EROS after having taken close-up images of the asteroid's surface.** ▪ Asteroid 433 EROS exhibits a heavily cratered surface, with one side dominated by a huge, scallop-rimmed gouge, and the opposite side by a conspicuous crater with a sharply raised rim. Researchers from the Southwest Research Institute (SwRI), San Antonio, TX, carried out a conceptual seismic study of 433 EROS.[3164] Using the surface data geometry from the NEAR mission, they constructed a 3-D solid model of the asteroid and performed numerical computations of an explosive seismic source on the surface based on hypothetical material properties for 433 EROS.

| | | |
|---|---|---|
| 1996 | *United Nations (UN), New York City, New York* | On September 10, **the UN General Assembly votes 158-3 to approve a treaty prohibiting all nuclear tests.** Later known as the **"Comprehensive Test Ban Treaty (CTBT),"** it bans all nuclear test explosions (including those intended for peaceful purposes), and thereby prevents the development of new types of nuclear weapons.<br><br>(i) In paragraph 1 of Article I, each State Party to the Treaty undertakes, inter alia, "not to carry out any nuclear weapon test explosion or any other nuclear explosion." ▪ This means that each State Party is prohibited from carrying out a nuclear weapon test explosion or any other nuclear explosion anywhere in any environment, including underground. Thus, the Treaty completes the task that was begun in the Limited Test Ban Treaty {LTBT ⇨1963}, which prohibited nuclear weapon test explosions and any other nuclear explosions in three environments, but did not prohibit underground nuclear explosions.[3165]<br><br>(ii) A subject of particular concern is whether so-called "hydronuclear tests" would be sanctioned, because they offer a chance for established weaponeers to improve nuclear weapons and for newcomers from non-nuclear states to enter the field. In these shock experiments, nuclear charges with a small yield – less than four pounds of TNT equivalent – are used, but the chain reaction proceeds far more slowly upon detonation than in a full scale device. **For nuclear weapons designers,** *hydronuclear tests* **are a worthwhile supplement to** *hydrodynamic tests,* which can only provide data on materials in motion under severe compression; they do not entail the release of nuclear energy. ▪ Hydronuclear tests were also conducted in the United States during the 1958–1961 Test Moratorium in order to address stockpile safety concerns. In the former Soviet Union, similar tests to the U.S. hydronuclear tests – termed *nonexplosive chain reactions* at the time by Yulii B. KHARITON – provided the basis for the early phase of atomic bomb development and were used to study how the rate of neutron multiplication and hence the integral number of fissions in an explosive experiment is related to the maximum compression of the active zone of a shocked uranium or plutonium target.[3166] |

---

[3164] J.D. WALKER and W.F. HUEBNER: *Cracking a cosmic mystery.* Technology Today **25**, No. 1, 2-7 (Spring 2004).
[3165] Permanent electronic archive of the U.S. State Dept.; http://www.state.gov/www/global/arms/ctbtpage/treaty/artbyart.html.
[3166] L.V. AL'TSHULER, Y.B. ZEL'DOVICH, and Y.M. STYAZHKIN: *Investigation of isentropic compression and equations of state of fissionable materials.* Phys. Uspekhi **40**, No. 1, 101-102 (1997).

So far (as of Feb. 4, 2007), **the CBCT has been signed by 138 nations, including the United States, but it has not yet entered into force.**[3167] **The Treaty prompted the installation of an international monitoring network to search for evidence of clandestine nuclear explosions** {⇨Fig. 4.3–Z}.[3168]

| | | |
|---|---|---|
| 1996 | *Zhukovskii Air Base [near Moscow], Russia & NASA's Langley Research Center (LaRC), Hampton, Virginia* | **A team of American and former Soviet aircraft manufacturers establishes working relationships while using a supersonic Tupolev Tu-144LL jet as a flying laboratory** in order to conduct full-scale supersonic tests and to compare the results from them with data obtained from model tests in wind tunnels, computer-aided techniques and other flight tests.[3169] The aims of the experiments performed at high Mach numbers with the Tu-144LL – a modified Tu-144 {⇨1968} – primarily encompass<br>▸ the identification of areas of elevated temperature;<br>▸ measurements of temperatures within an engine compartment;<br>▸ boundary layer measurements of pressures and skin friction; and<br>▸ acoustic measurements of cabin noise.<br>Data collected from the flight and ground experiments during the NASA-funded Tu-144LL Flight Research Program are intended to aid the development of the technology base needed to realize a proposed second-generation American-built supersonic jetliner {High Speed Civil Transport ⇨1968}.<br>The aircraft's initial flight phase began in June 1996 and concluded in February 1998 after 19 research flights. Although the development of an advanced SST is currently on hold, commercial aviation experts in the United States estimate that a market for up to 500 such aircraft could develop by the third decade of the 21st century.[3170] |
| 1996 | *Centre National de la Recherche Scientifique (CNRS), Nice, France* | **Publication of the *Handbook of Computational Fluid Mechanics*.** The editor is Roger PEYRET[3171] at the Laboratoire de Mathématiques, Université de Nice. ▪ It reviews the state-of-the-art in computational fluid mechanics as it stands in the final years of the 20th century. Covering fluid motion in various situations, it also discusses finite-volume methods for compressible inviscid and viscous flows (Euler equations and Navier-Stokes equations) and the appropriate treatment of shock waves. |
| 1996 | *Pulsed Thermal Physics Research Center, Institute for High Temperatures, Russian Academy of Sciences, Moscow, Russia* | Lev V. AL'TSHULER,[3172] 83-year-old patriarch of Russian shock wave physics, **and his collaborators review the history of the development of explosive devices for producing ultrahigh shock pressures of up to 25 Mbar (2.5 TPa) at the Russian Federal Nuclear Center (Arzamas-16)** – which has previously been covered by a veil of secrecy.<br>(i) Independent of dynamic compressibility studies secretly conducted at the Los Alamos Scientific Laboratory (LASL) since the mid-1940s {GORANSON ⇨1944}, dynamic high-pressure studies were initiated at the Center in 1948, and even the earliest reports on that work contained data from the application of shock pressures of 4 Mbar (huge for the time), which were later published in the open literature.[3173] |

---

[3167] To enter into force, the CTBT must be signed and ratified by the 44 States listed in Annex 2 to the Treaty. These States formally participated in the 1996 session of the Conference on Disarmament and possessed nuclear power or research reactors at that time. On Feb. 4, 2007, only 34 of the Annex 2 countries ratified the Treaty.

[3168] K. WALTER: *Seismic monitoring techniques put to a test.* Sci. Technol. Rev. (Livermore, CA, April 1999); http://www.llnl.gov/str/pdfs/04_99.3.pdf.

[3169] *Tu-144LL experiments.* NASA-LRC, Hampton, VA; http://oea.larc.nasa.gov/PAIS/TU-144.html.

[3170] *Past projects – Tu-144LL Flying Laboratory.* NASA-DFRC, Edwards, CA (July 2006); http://www.nasa.gov/centers/dryden/history/pastprojects/TU-144/index.html.

[3171] R. PEYRET (ed.): *Handbook of computational fluid mechanics.* Academic Press, London etc. (1996, 2000).

[3172] L.V. AL'TSHULER, R.F. TRUNIN, K.K. KRUPNIKOV, and N.V. PANOV: *Explosive laboratory devices for shock wave compression.* Phys. Uspekhi **39**, No. 5, 539-544 (1996).

[3173] L.V. AL'TSHULER ET AL.: *Dynamic compressibility and equation of state of iron under high pressure.* Sov. Phys. JETP **7**, 606-614 (1958); *Dynamic compressibility of metals under pressures from 400,000 to 4,000,000 atmospheres.* Ibid. **7**, 614-619 (1958).

|  |  |  |
|---|---|---|
|  |  | (ii) Much larger shock pressures were generated using convergent overcompressed detonation waves in a hemispherical explosive charge. Uniformly ignited at its outer side, this drives a convergent detonation wave towards the center. The hemispherical charge contains an inner recess into which a thin-walled spherical metallic shell is inserted. This shell, acting as a flyer, is accelerated towards the center of the device, and its velocity increases with decreasing radius. The specimens to be studied can be placed at different depths (radii), allowing one to produce shock compressions at different pressures of up to 10–20 Mbar. |

(iii) Using cascade devices – multiple concentrically arranged hemispherical explosive charges, each with a metal shell backing – shock pressures of close to 13 Mbar were obtained at Arzamas-16 by the mid-1950s.

(iv) Extension to much higher pressures became possible when the iron shock adiabat was determined in nuclear underground explosion experiments for pressures of up to 100 Mbar {TRUNIN ET AL. ⇨1992}. This facilitated the interpretation of kinematic data obtained in multiple cascade devices.

| 1996 | *Aeronautical and Maritime Research Laboratory, Defence Science & Technology Organisation (DSTO), Melbourne, Victoria, Australia* | John A. LEWIS,[3174] a senior scientist at the Aeronautical and Maritime Research Laboratory, **publishes the technical report** *Effects of Underwater Explosions on Life in the Sea.* Although underwater explosions are used in a wide range of commercial and military operations – *e.g.,* to clear shipping lanes, for seismic investigations, military training exercises, shock testing of new naval vessels, *etc.* – relatively little is known by scientists, engineers and military people about their effects on nearby marine life. LEWIS' report
▸ discusses effects of underwater explosions on marine plants, invertebrates, fish, turtles, birds, and sea mammals;
▸ reviews relevant information on shock testing new vessels for defense purposes; and
▸ predicts safe ranges around explosives for humans and secondary effects which may impact marine communities.

The elastic waves that an earthquake generates in a lake or sea are also often fatal to fish life. Akitune IMAMURA,[3175] a Japanese seismologist, wrote the following in 1937: "With their air-bladders injured by the violent [seismic] shock, deep-sea fish rise to the surface, after which they do not seem able to descend. Some kinds of fish, upon being frightened by a shock, swim about in great shoals apparently looking for a place of safety. It is known that generally fish are caught only with the greatest difficulty a few days before a big earthquake. Perhaps they are sensible to the fore-shocks that do not affect human beings." |
| 1996 | *Dept. of Physics, University of Memphis, Memphis, Tennessee* | Gerrit L. VERSCHUUR,[3176] a U.S. radio astronomer and scientific writer, **publishes his book** *Impact: The Threat of Comets and Asteroids.* He discusses previous catastrophic collisions of comets and asteroids with the Earth and the possibility of such collisions in the future. ▪ About 10,000 tons of space debris fall to Earth every year, mostly in meteoric form. However, the major threats are asteroids and comets. Some 350 NEAs (Near-Earth Asteroids) have been located whose orbits cross the Earth's orbit. However, asteroid hunters have estimated that about 9,000 objects, 0.5 km in size or larger, are in near-Earth orbits.

Two years later, VERSCHUUR[3177] covered the range of issues surrounding the asteroid threat and defense issue including ways of mitigating public fears about threatening asteroids and the impact the asteroid threat will have on our cosmology. |

---

[3174] J.A. LEWIS: *Effects of underwater explosions on life in the sea.* Tech. Rept. DSTO GD-0080. Defence Science and Technology Organisation (DSTO), Melbourne, Victoria, Australia (1996).
[3175] A. IMAMURA: *Theoretical and applied seismology.* Maruzen Co., Tokyo (1937), p. 121.
[3176] G.L. VERSCHUUR: *Impact: the threat of comets and asteroids.* Oxford University Press, New York (1996).
[3177] G.L. VERSCHUUR: *Impact hazards: truth and consequences.* Sky & Telescope **96**, 27-34 (June 1, 1998).

| 1996 | Lawrence Livermore National Laboratory (LLNL), Livermore, California | **Samuel T. Weir,**[3178] **Arthur C. Mitchell, and William J. Nellis,** three Livermore experimental physicists, **provide the first evidence for "metallic hydrogen."**

(i) They use a reverberating shock wave rather than a single shock pulse, which less raises the temperature of the shock-compressed hydrogen.[3179]

(ii) They show that a thin layer of liquid hydrogen maintained between two hard sapphire plates and shocked up to 120 GPa (1.2 Mbar) reaches a minimum electrical resistivity of about 5,000 $\mu\Omega$cm, which corresponds to a semiconducting state. However, at a pressure of 140 GPa (1.4 Mbar), the resistivity drops to about 500 $\mu\Omega$cm – indicating that the hydrogen becomes metallic.[3180] The results for deuterium, a hydrogen isotope, follow the same pattern. • Five years later, **Nellis' team,** using essentially the same technique, **also found experimental evidence for a metallic phase in fluid molecular oxygen shock-compressed to 1.2 Mbar.**[3181]

The existence of metallic hydrogen, first predicted by the Princeton physicists Eugene P. Wigner and Hillard B. Huntington {⇨1935}, is of fundamental importance in both condensed matter physics and astrophysics.

▶ For example, it would help us to understand the dynamics of Jupiter, which is so massive that fluid hydrogen in its core is believed to be squeezed into the metallic state. This metallic hydrogen may be the source of Jupiter's very strong magnetic field, the largest of any planet in the Solar System.

▶ Hot metallic hydrogen is also believed to be present inside Saturn, and it may also occur in other large planets that have recently been discovered to exist outside the Solar System.

▶ An understanding of dense hydrogen might also become important in future technologies; for example, in the production of a room-temperature superconductor.[3182]

▶ In laser-driven inertial confinement fusion experiments, solid metallic hydrogen targets would produce substantially higher fusion yields than current forms of the hydrogen isotopes deuterium and tritium because it is ten times as dense.

▶ The stored energy released by reversion to the diatomic insulating fluid would also be very large, and so metastable metallic hydrogen would have widespread applications as fuels or super explosives. |
| 1997 | Earth's orbit | On June 25, during a docking test the unmanned Russian 7-ton resupply spacecraft Progress collides with the 100-ton Mir space station, leading to an almost fatal accident. **First collision of two spacecraft in space.** • The Progress collided with a solar array on the Spektr module. Then, the spacecraft hit Spektr itself, punched a hole in a solar panel, buckled a radiator, and breached the integrity of Spektr's hull. The collision had knocked Mir into a spin; and the power outage had shut down the gyrodynes so that the spin now went uncontrolled. The crew, consisting of two Russian cosmonauts (Vasily Tsibliev and Aleksandr Lazutkin) and one U.S. astronaut (Mike Foale), was not hurt by the accident and with the help of ground controllers managed to stabilize the Mir. The crew members continued their research program and later returned safely to Earth.[3183] |

---

[3178] S.T. Weir, A.C. Mitchell, and W.J. Nellis: *Metallization of fluid molecular hydrogen at 140 GPa*. Phys. Rev. Lett. **76**, 1860-1863 (1996).
[3179] W.J. Nellis: *Making metallic hydrogen.* Scient. Am. **282**, 60-66 (May 2000).
[3180] For comparison: copper at room temperature has a value of 1.68 $\mu\Omega$cm
[3181] M. Bastea, A.C. Mitchell, and W.J. Nellis: *High-pressure insulator-metal transition in molecular fluid oxygen*. Phys. Rev. Lett **86**, 3108-3111 (2001).
[3182] N.W. Ashcroft: *Dense hydrogen: the reluctant alkali*. Phys. World **8**, 43-48 (1995).
[3183] *Progess collision with Mir animation*. In: *Shuttle-Mir: the U.S. and Russia share history's highest stage*, NASA SP-2001-4225, NASA-Headquarters History Office; http://history.nasa.gov/SP-4225/multimedia/progress-collision.htm.

| | | |
|---|---|---|
| 1997 | *Great Keppel Island [at the rim of the Great Barrier Reef], Central Queensland, East Australia* | In July, at the occasion of the 21st International Symposium on Shock Waves, **Hans GRÖNIG** (RWTH Aachen, Germany), **Kazuyoshi TAKAYAMA** (Tohoku University, Sendai, Japan), **and Yasuyuki HORIE** (North Carolina State University), **three internationally renowned shock physicists, meet with Wolf BEIGLBOECK** (a mathematical physics professor at Heidelberg University and Senior Physics Editor at Springer Publishers, Heidelberg) **to discuss the possibility of editing a multi-volume *Encyclopedia of Shock Waves*.** ▪ After discussing over the years various concepts it was eventually decided rather to publish a collection of extensive, independent but topically interrelated surveys and reviews. The first two volumes which have recently been published are on Multiphase Flows[3184] and Solid-State Materials[3185] − to be followed by volumes on detonation dynamics, experimental techniques and instrumentation, numerical aspects as well as further volumes on applications to gaseous, fluid and solid-state media.<br><br>In the period 1993−2005, Springer published the eight-volume series *High Pressure Shock Compression of Condensed Matter* {ASAY ⇨1993}. |
| 1997 | *Black Rock Desert, northwestern Nevada* | On October 13, **Andy GREEN**,[3186] a British RAF Tornado pilot with 1,000 hours of flight experience in fast jets, **breaks the sound barrier** ($M = 1.007$) **in his jet car named *ThrustSSC*** (*Thrust*-powered *SuperSonic Car*) {⇨Figs. 4.6−J & 4.20−N}. He broke the barrier one day before the fiftieth anniversary of the successful attempt by YEAGER {⇨1947} to break the same barrier in an airplane.[3187] ▪ The ThrustSSC project started with a £40,000 of sponsorship from Castrol Ltd. and a two-year research program involving computational fluid dynamics and supersonic rocket model testing.<br><br>Two days later, **GREEN sets the supersonic world land speed record of 766.609 mph** ($342.63$ m/s), **corresponding to $M = 1.02$.** ▪ The sound barrier, which varies with temperature, corresponded to about 750 mph (335.21 m/s) during GREEN's record-breaking runs.<br><br>GREEN, when later being asked by a correspondent of the German newspaper *Die Zeit*,[3188] described his feat as follows: "When driving supersonically with the *ThrustSSC*, the problem is not the physical stress upon the body, but rather the shock wave which forms directly above the cockpit. It is a howling, shrieking noise, very unpleasant. Fortunately, I had a system of active noise suppression on board. This eliminates the noise electronically via earphones … **The shock wave at the sound barrier is of course not visible, however, one sees a haze above the car.** Twice I noticed how a fine fog formed in front of the cockpit between the power units. The same phenomenon can also be observed occasionally for supersonic planes [⇨Fig. 4.14−F]." |
| 1997 | *High Energy Laser Systems Test Facility (HELSTF), White Sands Missile Range, New Mexico* | On December 14, **Franklin MEAD**, an Air Force scientist **and Leik MYRABO**, an associate professor of engineering at Rensselaer Polytechnic Institute (RPI) in Troy, NY, **perform first flight tests of a scale model "Lightcraft"** − a vehicle driven by a train of high-energy laser pulses from the 10-kW $CO_2$ laser of the Pulsed Laser Vulnerability Test System {⇨Fig. 4.11−J}. The focused laser beam is absorbed by the air inside the engine, thereby creating a laser-supported detonation {RAMSDEN & SAVIC ⇨1964}. ▪ The concept of laser propulsion originated in the early 1970s and is credited to Arthur R. KANTROWITZ,[3189] while the idea of a light-powered craft was conceived by L. MYRABO in 1987. One grand vision is to integrate the Lightcraft concept into an Earth-to-orbit vehicle.[3190] |

---

[3184] M. VAN DONGEN (ed.): *Multiphase flows I*. Shock Waves Science and Technology Reference Library. Springer, Berlin & Heidelberg. Vol. 1 (2007).
[3185] Y. HORIE (ed.): *Solids I*. Shock Waves Science and Technology Reference Library. Springer, Berlin & Heidelberg. Vol. 2 (2007).
[3186] *ThrustSSC*, U.K.; http://www.thrustssc.com/thrustssc/contents_frames.html.
[3187] R. NOBLE and D. TREMAYNE: *Thrust: through the sound barrier* [with contributions from Andy GREEN and others]. Bantam, Sydney & London (1999).
[3188] U. BAHNSEN and U. WILLMANN: *Schockwelle in der Wüste*. Die Zeit (Hamburg), Nr. 7 (Feb. 5, 1998).
[3189] A. KANTROWITZ: *Propulsion to orbit by ground-based lasers*. Astronaut. Aeronaut. **10**, No. 5, 74-76 (May 1972).
[3190] L.N. MYRABO: *Transatmospheric laser propulsion: lightcraft technology demonstrator*. Final Tech. Rept, Strategic Defense Initiative Organization (SDIO) Laser Propulsion Program, Contract No. 2073803 (1989), pp. 117-142; *Highways of light*. Scient. Am. **280**, 68-69 (Feb. 1999).

In October 2000, a 4.8-in. (12.2-cm)-dia. laser-boosted rocket reached a record height of 233 ft (71 m) in a flight lasting 12.7 seconds.[3191]

| 1997 | *ESA's ASI Science Data Center, Frascati, central Italy & NASA's Goddard Space Flight Center (GSFC), Greenbelt, Maryland* | **Two important observations of cosmic Gamma-Ray Bursts (GRBs) are made using two satellites:** the Italian-Dutch **Beppo-SAX** ("Beppo" in honor of Giuseppe OCCHIALINI, SAX: Satellite per Astronomia X) and NASA's **CGRO** (Compton Gamma-Ray Observatory):[3192] |
|---|---|---|

(i) **The discovery of three GRB afterglows represents a major breakthrough in the field, since it provides the first direct evidence for the interaction between the explosive event and its environment,** and the first direct distance estimate for one of the bursts (GRB 970228).

▸ On February 28, a gamma-ray burst labeled **GRB 970228** is detected by Beppo-SAX for about 80 seconds. Its gamma-ray monitor establishes the position of the burst to within a few arc minutes in the constellation *Orion*, about halfway between the stars Alpha Tauri and Gamma Orionis – *i.e.*, the gamma-ray burst appears to originate from outside of our Galaxy. Within eight hours, operators in Rome turn the spacecraft around to look at the same region with an X-ray telescope. They find a source of X-rays that is fading fast and fix its location to within an arc minute.

▸ On May 8, Beppo-SAX operators locate a 15-second gamma-ray burst, labeled **GRB 970508.** Shortly after, Hawaiian astronomers take a spectrum of the burst with the Keck II telescope; it is strongest in the blue part of the spectrum and shows a few dark lines, apparently caused by iron and magnesium in an intervening cloud. Astronomers at CalTech find that the displacement of these absorption lines indicates a distance of more than seven billion light-years, the first distance measured for a GRB – *i.e.*, **the gamma-ray burst must have occurred at a cosmological distance. In that case, GRBs represent the most powerful explosions in the Universe.** Studies carried out with the Very Large Array (VLA) interferometer of the National Radio Astronomy Observatory (NRAO) located on the plains of San Agustin, NM showed that the fireball was about a tenth of a light-year in diameter a few days after the explosion, and that it was expanding at very nearly the speed of light.

▸ On December 14, **a GRB of extreme intensity, apparently originating from a *hypernova*, is emitted from the galaxy.** The burst, which is labeled **GRB 971214** and occurred about 12 billion light-years from the Earth, is detected by BeppoSAX and CGRO.[3193] Although it lasts only a few seconds, it is as luminous as the rest of the Universe: **it appears to have released several hundred times more energy than a supernova,** which had previously been considered to be the most energetic explosion phenomenon in the Universe (aside from the Big Bang).

(ii) The Space Telescope Science Institute (STScI) and NASA publish the first visible-light image of a GRB ever taken, which links the gamma-ray burst with a potential host galaxy.[3194]

In the following year, **Bohdan PACZYŃSKI,**[3195] a professor of astrophysics at Princeton University, **coined the term *hypernova*.** He used it to refer to the very bright events associated with the newly discovered optical counterparts to cosmic GRBs. ▪ The term *hypernova* is also used to refer to stellar explosions "with a power about 100 times that of the already astonishingly powerful 'typical' supernova."[3196]

---

[3191] D. FEIKEMA: *Analysis of the laser propelled lightcraft vehicle.* Rept. NASA TM-2000-210240, NASA-GRC, Cleveland, OH (June 2000).
[3192] G.J. FISHMAN and D.H. HARTMANN: *Gamma-ray bursts.* Scient. Am. **277**, 46-51 (July 1997); *Gamma-ray bursts: new observations illuminate the most powerful explosions in the Universe.* Scient. Am. Quart. **9**, No. 1, 68-73 (1998); http://www.mpe.mpg.de/~jcg/pola_030329/GRB_SciAm98.pdf.
[3193] S.R. KULKARNI ET AL.: *Identification of a host galaxy at redshift z = 3.42 for the γ-ray burst of December 1997.* Nature **393**, 35-39 (May 7, 1998).
[3194] *Gamma-ray bursts common to normal galaxies? Hubble data offer new clues and puzzles.* HubbleSite, Space Telescope Science Institute (STScI), Baltimore, MD (June 10, 1997); http://hubblesite.org/newscenter/newsdesk/archive/releases/1997/20/.
[3195] B. PACZYŃSKI: *Are gamma-ray bursts in star-forming regions?* Astrophys. J. **494**, Part II, L45-L48 (Feb. 10, 1998).
[3196] *A hypernova: the super-charged supernova and its link to gamma-ray bursts.* NASA's Imagine the Universe, NASA-GSFC, Greenbelt,MD; http://imagine.gsfc.nasa.gov/docs/science/know_l1/why_hyper.html.

It is now known that GRBs are caused by the ejection of ultra-relativistic matter from a powerful energy source and its subsequent collision with its environment. About three times a day, our sky flashes due to a powerful pulse of gamma-rays, which is invisible to the human eye but not to astronomical instruments. The sources of this intense radiation are likely to emit more energy than the Sun will in its entire lifetime (around ten billion years) within seconds or minutes. The GRB afterglow, which can linger for weeks or even months, gradually drops down the electromagnetic spectrum, initially emitting mostly gamma-rays, then peaking at X-rays, all the way down to radio waves.

| | | |
|---|---|---|
| 1997 | U.S. Naval Undersea Warfare Center (NUWC), Newport, Rhode Island | **First demonstration of a fully submerged launch of a supercavitating projectile with air injected in its nose.** With a muzzle velocity of 1,549 m/s, **it is apparently the first underwater weapon to break the sound barrier.** ▪ The idea of using *supercavitation* {⇨Fig. 4.14−E} to create supersonic underwater vehicles was developed in the United States in the late 1950s {TULIN ⇨1958} and later taken up by the Soviet ballistician Mikhail MERKULOV at the Hydrodynamic Institute in Kiev in order to create a super-fast torpedo.[3197] The VA-111 *Shkval* underwater rocket is a supercavitating torpedoe developed by the Russian Navy and capable of speeds in excess of 370 km/h (103 m/s).[3198]<br><br>The present state of supercavitation research, which is carried out by a number of countries, is difficult to estimate. New developments are kept classified because of their direct military relevance. Supersonic underwater vehicles and underwater supersonic ballistics using supercavitation could completely change the concepts of undersea warfare. |
| 1997 | Institute of Physics Publishing, Bristol, England | **Foundation of the British journal *Combustion Theory and Modelling*,** in order "to provide a clear focus for the publication of research relating to the development of fundamental understanding of all aspects of combustion from a theoretical and a mathematical modeling perspective." The editors are Bill DOLD and Mitch SMOOKE. ▪ A large number of the contributions published so far relate to detonation and deflagration phenomena. |
| 1997 | University of Westminster, London, England | **Publication of the multi-author textbook *High Speed Photography and Photonics*,** edited by Sydney F. RAY,[3199] a British lecturer in digital and photographic imaging. The book, originated and sponsored by the British Association for High Speed Photography and Photonics (AHSPP), is devoted to both historical and modern applications of this important diagnostic technique. In particular, it addresses scientists and engineers working in the fields of gas dynamics, aircraft engineering, shock wave physics, ballistics, the properties of materials, detonics, and combustion. ▪ The book was well received and awarded second prize in the 1998 Kraszna-Krausz Foundation Awards for the best books on photography published in the year 1997/1998.<br><br>A review of the gradual progress of high-speed photography was also given by Jeofry Stuart COURTNEY-PRATT, an Australian-born U.S. applied physicist and high-speed photography expert at Bell Telephone Laboratories (Murray Hill, NJ). His three review articles cover the early period up to 1957,[3200] the period from 1957 to 1972,[3201] and that from 1972 to 1982.[3202] |

---

[3197] D. GRAHAM-ROWE: *Faster than a speeding bullet.* New Scientist **167**, 26-29 (July 22, 2000).
[3198] *VA-111 "Shkval" underwater rocket.* Federation of American Scientists (FAS), Washington, DC (Sept. 3, 2000); http://www.fas.org/man/dod-101/sys/missile/row/shkval.htm.
[3199] S.F. RAY (ed.): *High speed photography and photonics*. Focal Press, Oxford *etc.* (1997).
[3200] J.S. COURTNEY-PRATT: *A review of the methods of high-speed photography.* Rept. Progr. Phys. **20**, 379-432 (1957).
[3201] J.S. COURTNEY-PRATT: *Advances in high-speed photography 1957–1972.* In: (E. LAVIRON, ed.) *Proc. 10th Int. Congress on High-Speed Photography* [Nice, France, Sept. 1972]. Association Nationale de la Recherche Technique (ANRT), Paris (1973), pp. 59-63.
[3202] J.S. COURTNEY-PRATT: *Advances in high-speed photography 1972–1982.* In: (J.E. THOMPSON and L.H. LUESSEN, eds.) *Fast electrical and optical measurements.* NATO Advanced Study Institute (ASI) [Castelvecchio Pascoli, Italy, July 1983]. Nijhoff Publ., Dordrecht *etc.* (1986), pp. 595-607.

| 1997 | Institute of Transportation Studies, UC Berkeley, Berkeley, California | Benjamin COIFMAN,[3203] a U.S. researcher studying congested traffic, **presents microscopic time-space diagrams for several shock waves** {LIGHTHILL & WHITHAM ⇨1955} **on Interstate 680** (a freeway in the San Francisco Bay Area) **and Highway 99** (also known as "The Main Street of California"). He videotaped several sections of highway and calibrated the mapping from an image plane of world coordinates at each site. His recorded data show that "slow" waves can transform into "stop" waves, and that a backward moving "stop" wave can smear over a "slow" wave due to particular maneuvers (such as a driver changing lane). By applying a graphical procedure to his recorded trajectory data, he finds that the disturbance travels at about 12 km/h. |
|---|---|---|
| 1997 | Los Alamos National Laboratory (LANL), Los Alamos, New Mexico | Richard L. GUSTAVSEN,[3204] Stephen A. SHEFFIELD, and Robert R. ALCON **estimate reaction zone parameters** – such as the *reaction zone time* and *reaction zone length*, and the *von Neumann spike pressure* – **in various kinds of HMX-based explosives.**<br>(i) They produce planar detonations by impacting the high explosive with projectiles launched in a gas gun, which generates a shock pressure $p$ in the specimen. They choose impactor, impact velocity and explosive thickness such that the run distance to detonation is always less than half the explosive thickness.<br>(ii) They measure the particle velocity $u(t)$ as function of time $t$ at an explosive/window interface using two VISAR interferometers {BARKER & HOLLENBACH ⇨1965} with different fringe constants in order to determine the velocity jump at the shock front.<br>(iii) Reaction zone parameters were estimated by analyzing recorded wave profiles $u(t)$ and by finding Hugoniot and isentrope intersections in the $(p, u)$-plane. For the high explosive PBX9501 (95 wt % HMX and 5% binders), the analysis leads to estimates for the von Neumann spike pressure of slightly over 50 GPa (500 kbar) and the reaction zone time of 15 ns, corresponding to a reaction zone length of 130 μm.<br>(iv) The reaction zone parameters could not be determined for porous HMX samples. |
| 1997 | Subatomic Physics Group, Physics Division, Los Alamos National Laboratory (LANL), Los Alamos, NM | **Los Alamos scientists have used for the first time protons from a linear accelerator to "photograph" the detonation wave from a small-scale explosion.** They detonated a small amount of high explosive inside a chamber specially designed to contain the explosion and let the 800-MeV proton beam at the Los Alamos Neutron Science Center enter and exit to produce a clear, high-resolution image of the explosion's shock wave. ***Proton radiography*** – which has several potential advantages over X-rays – could be used in future facilities for hydrodynamic testing of nuclear weapons mockups {VENABLE ⇨1965; UN General Assembly ⇨1996}.[3205] |
| 1997 | MOSI, Tampa; USFCAM, Tampa & ICLRT, Camp Blanding, Florida | Allan MCCOLLUM, an internationally acclaimed contemporary artist, **creates fulgurites that will form the centerpiece of an exhibit** organized by the Hillsborough County Museum of Science and Industry (MOSI) and the University of South Florida Contemporary Art Museum (USFCAM). His fulgurites consist of fused zircon sand {⇨Fig. 4.1–J} and are created artificially via a natural lightning bolt with the help of a small rocket. ▪ MCCOLLUM closely collaborated with researchers at the International Center for Lightning Research and Testing (ICLRT).[3206] Before the project, engineers at ICLRT regarded fulgurites as something of a curiosity, and once spent weeks excavating a 17-ft (5.18-m)-long fulgurite that is recognized by the Guinness Book of World Records as being the longest ever. |

---

[3203] B. COIFMAN: *Time-space diagrams for thirteen shock waves.* California Partners for Advanced Transit and Highways (PATH), Paper UCB-ITS-PWP-97-1 (1997); http://repositories.cdlib.org/its/path/papers/UCB-ITS-PWP-97-1.

[3204] R.L. GUSTAVSEN, S.A. SHEFFIELD, and R.R. ALCON: *Detonation wave profiles in HMX based explosives.* In: (M.D. FURNISH, ed.) *Shock compression of condensed matter – 1997.* Proc. 10th AIP Conference [Livermore, CA, 1997]. AIP Press, Woodbury, NY (1998), pp. 739-742.

[3205] *Los Alamos scientists photograph shock wave with protons.* News & Communications Office, LANL, Los Alamos, NM (Aug. 18, 1997); http://www.lanl.gov/news/index.php/fuseaction/home.story/story_id/1638.

[3206] A. HOOVER: *Learning about lightning.* Explore Mag. (University of Florida) **3**, No. 2 (Fall 1998).

Fulgurites have long fascinated researchers {WITHERING ⇒1790; BEUDANT ⇒1828; DARWIN ⇒1839; SPENCER ⇒1932}, but, until MCCOLLUM, no one had experimented with making the objects above ground with triggered lightning.

| 1997 | Dept. of Astronomy & Laboratory for Computational Science and Engineering (LCSE), University of Minnesota, Minneapolis, Minnesota |
|---|---|

**David H. PORTER,[3207] Steve E. ANDERSON, and Paul R. WOODWARD, three U.S. scientists, perform the first dynamical 3-D supercomputer simulation of a model red giant star.**

(i) They simulated the convection mechanism, which governs how energy produced in thermonuclear reactions in the core of a star is transported to the surface. Stellar convection of hot gases rising from the core and cold gases sinking from the surface produce complicated turbulence patterns which can cause gases to move at different speeds at the poles compared to the gases at the equator.

(ii) Their astrophysical gas dynamics simulation of the convection process in an entire rotating model star was extremely computationally intensive and carried out in a nine-day run at the National Center for Supercomputing Applications (NCSA) on a SGI/CRAY Origin-2000.[3208]

(iii) Their study also demonstrates that scaleable computers built from clusters of multiprocessor machines operated in parallel can work efficiently as single supercomputing systems.

**Stars pass through complex live cycles involving detonations and implosions until they develop into *red giant stars*.** Simplified, the cycles are as follows:[3209]
- When the star's core runs out of fuel (protons), it begins to cool and shrink. The outer layers of the star fall inwards under gravity, and they heat up as they fall.
- This causes the outer shell surrounding the central core to become hot enough to fuse protons into alpha particles. The star therefore gains a new source of energy.
- The core of the star is now hotter than it was before, and this heat causes the outer parts of the star to swell. The star becomes a giant.
- The radiation from the fusing shell has grown weak by the time it reaches the surface of the star. Weak radiation is red, so the star becomes a red giant.

Depending on its mass $M_S$, a star will become either a black dwarf, a neutron star, or a black hole. With $M_\odot$ = solar mass, one can distinguish three cases of life cycles:[3210]
- **Sun-like stars** ($M_S < 1.5 M_\odot$) →Red giant →Planetary nebula →White dwarf →Black dwarf;
- **huge stars** ($1.5 < M_S < 3 M_\odot$) →Red supergiant →Supernova →Neutron star;
- **giant stars** ($M_S > 3 M_\odot$) →Red supergiant →Supernova →Black hole.

*Red supergiants* are more massive than red giants. The red supergiant phase is extremely short-lived, lasting only a few 100 ka to 1 Ma at most before ending in a supernova. Betelgeuse,[3211] a variable star {General Catalogue of Variable Stars ⇒1948} located in the constellation *Orion* ("The Hunter"), about 427 light-years away, is the closest red supergiant star to the Solar System, and therefore the closest star that may end up as a supernova.

| 1998 | International Space Science Institute (ISSI), Bern, Switzerland |
|---|---|

In June, at the **ISSI Workshop on Corotating Interaction Regions,** space scientists involved in the Ulysses, Pioneer, Voyager, IMP 8, Wind, and SOHO missions exchange their data and interpretations with theorists in the fields of solar and heliospheric physics.[3212] • ***Corotating Interaction Regions (CIRs)*** are regions with transient and corotating flows in the solar wind,

---

[3207] D.H. PORTER, S.E. ANDERSON, and P.R. WOODWARD: *Simulating a pulsating red giant star.* Access, NCSA's general information magazine (1997); http://www.lcse.umn.edu/research/RedGiant/.

[3208] *NCSA and LCSE generate 3-D star simulation.* NCSA News, National Center for Supercomputing Applications (NCSA), Univerity of Illinois, Urbana-Champaign; http://access.ncsa.uiuc.edu/Releases/97Releases/970527.LCSE.html.

[3209] P.J. BROWN: *Global vision.* Penny, Coventry (1992); *See also History of the Universe: red giant*; http://www.historyoftheuniverse.com/starold.html.

[3210] *The death of stars.* Enchanted Learning, LLC, Mercer Island, WA; http://www.enchantedlearning.com/subjects/astronomy/stars/lifecycle/stardeath.shtml.

[3211] *Betelgeuse.* Astronomy picture of the day. NASA-GSFC, Greenbelt, MD (April 19, 1998); http://antwrp.gsfc.nasa.gov/apod/ap980419.html.

[3212] (A. BALOGH ET AL., eds.): *Corotating interaction regions: proceedings of an ISSI workshop, 6-13 June, 1998, Bern, Switzerland.* Kluwer, Dordrecht *etc.* (1999); *see also* http://www.springer.com/sgw/cda/frontpage/0,11855,5-10100-72-33624318-0,00.html.

where the magnetic field strength and pressure are high. CIRs form at the interface between the fast solar wind originating in the northern and southern coronal holes and the slow solar wind that originates near and within coronal streamers surrounding the heliomagnetic equator. CIRs and their successors play an important role in the structure and dynamics of the outer heliosphere. **CIRs consist of a forward and reverse shock pair in association with a region of enhanced solar wind density and temperature.** They are dominant structures in the solar wind at low heliographic latitudes and at heliocentric distances of between 2 and 8 AU (about 0.3–1.2 billion km).[3213]

| | | |
|---|---|---|
| 1998 | *Las Vegas, Nevada & Los Alamos, New Mexico* | On November 17, a Leonid meteor explodes over Las Vegas at an altitude of approximately 83 km. **A unique sequence of pictures is taken by Los Alamos astronomers using the ROTSE** (Robotic Optical Transient Search Experiment) **optical telescope.**[3214]<br><br>Leonid meteoroids, which are debris from comet TEMPEL-TUTTLE, hit the Earth's atmosphere at about 72 km/s – much faster than typical meteoroids. Leonids are believed to consist of sand-grain-sized particles, but they may also contain a significant number of larger-than-average meteoroids which fragment in the upper atmosphere, producing spectacular fireballs. If such a fireball penetrates into the stratosphere below an altitude of about 50 km and explodes, there is a chance that sonic booms will be heard on the ground below. Such events are rare, but sonic booms from Leonids have indeed been reported.[3215] |
| 1998 | *Rheinisch-Westfälische Technische Hochschule (RWTH) Aachen, Germany* | In the period December 15–16, **the International Workshop on Shock Wave Focusing Phenomena in Combustible Media is held at RWTH's *Stoßwellenlabor* ("Shock Wave Laboratory")**; the chairmen are Hans GRÖNIG and Boris GELFAND. ▪ The results presented provide new insights into the critical conditions for triggering explosions in industrial accidents. They illustrate the important role of focusing and reflection phenomena on the process of initiating gaseous explosions, such as those created in nonuniform pressure-temperature fields with local hot spots, contact surface, and sharp temperature gradients. |
| 1998 | *NASA's Langley Research Center (LaRC), Hampton, Virginia & NASA's Dryden Flight Research Center (DFRC), Edwards, California* | **In order to investigate the feasibility of ramjets/scramjets** (*i.e.*, air-breathing hypersonic propulsion technologies), **NASA establishes an experimental hypersonic ground and flight test program termed *Hyper-X*.** The purpose of the program (set to run for several years) is to explore an alternative to rocket power for vehicles used to access space. ▪ Air-breathing engines have an advantage over rockets: they require only fuel – no oxidizer – thus resulting in lighter, smaller and cheaper launch vehicles.[3216] ▪ In the previous year, Micro Craft Inc. (Tullahoma, TN) was selected to fabricate a series of small, unpiloted 3.6-meter-long aircraft, designated *X-43*. Tests of the Mach-10 engine, performed in the year 2000, demonstrated that it was capable of fulfilling the main performance goal required: that it would provide vehicle acceleration. The first successful flight test was performed in March 2004 {Hyper-X ⇨2004}. |
| 1998 | *U.S. Dept. of Energy (DOE) & IBM, Rochester, Minnesota* | **The Dept. of Energy turns to IBM in its quest to eliminate the need for live nuclear testing without compromising the nation's safety and security.** To replace the empirical conditions of live testing with predictive simulations, a computational system is required that is capable of calculating the reactions of billions of data points within small fractions of a second, using high-fidelity three-dimensional models. Using funds previously spent on the construction of nuclear weapons, the **DOE engages IBM to build the world's fastest supercomputer.** |

---

[3213] P.R. GAZIS: *A large-scale survey of corotating interaction regions and their successors in the outer heliosphere.* J. Geophys. Res. **105**, No. A1, 19-34 (2000).

[3214] *A Leonid meteor explodes.* Astronomy picture of the day. NASA-GSFC (Nov. 18, 2001); http://apod.nasa.gov/apod/ap011118.html.

[3215] *The 1998 Leonids: a bust or a blast?* NASA Space Science News (Nov. 27, 1998); http://spacescience.com/newhome/headlines/ast27nov98_1.htm.

[3216] C.R. MCCLINTON: *Air-breathing engines.* Scient. Am. **280**, 64-65 (Feb. 1999). ▪ Charles R. MCCLINTON was the Technology Manager for the Hyper-X Program at NASA's Langley Research Center (Hampton, VA) and responsible for X-43 vehicle definition, wind tunnel testing, and hypersonic technology development. He presented the 2006 Dryden Lecture in Research entitled *Hypersonic flight: recent successes and future opportunities.*

- When President Bill CLINTON signed the Comprehensive Test Ban Treaty {CTBT ⇨1996}, his decision was based on the Government's eventual ability to phase out live underground tests of nuclear weapons. In order to accomplish this and to meet the security needs of the 21st century, the DOE had begun to develop the Stockpile Stewardship Program[3217] and the Accelerated Strategic Computing Initiative (ASCI), thereby integrating the efforts of multiple vendors and three DOE laboratories (LLNL, LANL and SNL).[3218]

The ASCI project – formerly known as the Accelerated Strategic Computing Initiative, later renamed the Advanced Simulation and Computing Initiative – was established "to develop simulation capabilities needed to analyze and predict the performance, safety and reliability of nuclear weapons and certify their functionality."[3219] **The 12-teraflop ASCI WHITE, the world's fastest supercomputer, was completed in June 2000 and used at LLNL to study supersonic crack propagation in rapid brittle fracture** {ABRAHAM & GAO ⇨2000}.[3220]

| | | |
|---|---|---|
| 1998 | *Smithsonian Institution, Washington, DC* | Bevan M. FRENCH,[3221] a geologist at the Dept. of Mineral Sciences, **publishes his book *Traces of Catastrophe – a Handbook of Shock-Metamorphic Effects in Terrestrial Meteorite Impact Structures.*** Particularly addressing geochemists and shock physicists, it provides a detailed introduction and overview of impact processes, crater formation, and shock metamorphism. |
| 1998 | *Dept. and Laboratory of Biomechanics, University of Vienna, Austria* | Herbert HATZE,[3222] an Austrian professor and head of the Biomechanics Department, **notes that the center of percussion** {WALLIS ⇨1670/1671; ⇨Fig. 2.9} **is a concept of limited significance when an implement is held in the hand,** such as a tennis racket or a baseball bat {⇨Figs. 4.3–X, Y}.<br><br>Very recently, **Rod CROSS**[3223] (an Australian physics professor at the University of Sydney who performed a theoretical and experimental analysis of hand-held implements) **has shown that the impact point that feels best is usually the node of the fundamental vibration mode, not the center of percussion** – thus essentially confirming HATZE's hypothesis.<br><br>In 2005, **Robert F. ELLIOTT**,[3224] a British physicist, business man and croquet player, **reported that he had developed an effective croquet mallet.** It consists of a graphite chassis which allows almost all the required weight to be concentrated at the ends of the head and leads to a very high shock speed: the perfect characteristic seems to be such that the shock wave arrives back at the striking face quickly after it sets off towards the back of the mallet, so that the returning shock wave tends to push the ball off on its way. The sound of the strike is a click rather than the dull thud associated with a wooden mallet head. |

---

[3217] *The Stockpile Stewardship Program.* Fact Sheet released by the Bureau of Arms Control, U.S. Dept. of State, Washington, DC (Oct. 8, 1999); http://www.state.gov/www/global/arms/factsheets/wmd/nuclear/ctbt/fs_991008_stockpile.html.

[3218] About the IBM project *ASCI WHITE* see http://www-1.ibm.com/servers/eserver/pseries/hardware/largescale/supercomputers/asciwhite/.

[3219] *The mission of the Advanced Simulation and Computing Program (ASC).* LLNL, Livermore, CA; https://asc.llnl.gov/asc_history/asci_mission.html.

[3220] F.F. ABRAHAM ET AL.: *Simulating materials failure by using up to one billion atoms and the world's fastest computer: brittle fracture.* Proc. Natl. Acad. Sci. (U.S.A.) **99**, 5777-5782 (2002).

[3221] B.M. FRENCH: *Traces of catastrophe. A handbook of shock-metamorphic effects in terrestrial meteorite impact structures.* LPI Contribution No. 954, Lunar and Planetary Institute (LPI), Houston, TX (1998); http://www.lpi.usra.edu/publications/books/CB-954/CB-954.intro.html.

[3222] H. HATZE: *The center of percussion of tennis rackets: a concept of limited applicability.* Sports Engng. **1**, 17-25 (1998).

[3223] R. CROSS: *Center of percussion of hand-held implements.* Am. J. Phys. **72**, 622-630 (2004). • His paper contains 14 references relating to baseball-bat and tennis racket impact dynamics.

[3224] R.F. ELLIOTT: *In search of the perfect mallet*; http://www.insearchoftheperfectmallet.com/search.htm.

| | | |
|---|---|---|
| 1998 | Boston, Massachusetts | At the Spring Meeting of the American Geophysical Union (AGU), **Alexander G. Kosovichev,**[3225] a senior research scientist at the W.W. Hansen Experimental Physics Laboratory (HEPL) of Stanford University, CA, **and Valentina V. Zharkova,** a honorary research fellow at the Dept. of Physics and Astronomy of Glasgow University, Scotland, **report on their discovery that solar flares – giant explosions on the Sun – also affect the Sun's interior, generating seismic waves similar to earthquakes** {Wolff ⇨1972; ⇨Fig. 4.1–V}.

(i) They find a seismic signature in data on the Sun's surface collected by the Michelson Doppler imager on the SOHO (Solar and Heliospheric Observatory) spacecraft immediately following a moderate-sized flare on July 9, 1996.[3226] ▪ This solar quake was found to be 40,000 times more powerful than the Great San Francisco Earthquake of 1906; for example, it released enough energy to meet the power needs of the U.S.A. for 20 years.

(ii) In a press release from Stanford University issued on May 27, 1998, it reads as follows, **"The solar quake that the science team recorded looks much like ripples spreading from a rock dropped into a pool of water.** But over the course of an hour, the solar waves traveled for a distance equal to 10 Earth diameters before fading into the fiery background of the Sun's photosphere. Unlike water ripples that travel outward at a constant velocity, the solar waves accelerated from an initial speed of 22,000 miles per hour [9.8 km/s] to a maximum of 250,000 miles per hour [112 km/s] before disappearing … The solar seismic waves appear to be compression waves like the 'P' waves generated by an earthquake. They travel throughout the Sun's interior. In fact, the waves should recombine on the opposite side of the Sun from the location of the flare to create a faint duplicate of the original ripple pattern, Kosovichev predicts."

**Prior to their discovery, Kosovichev and Zharkova had predicted the existence of sunquakes. They had speculated that a shock wave created by energized electrons traveling faster than the speed of sound would hit the Sun's surface and cause a compression or dent.** This compression, they anticipated, would cause the Sun to vibrate like a plucked guitar string. ▪ A solar flare {Carrington ⇨1859} is a violent eruption of hydrogen gas from a localized region on the Sun and is named after its radiating arc of light, which creates a burst of cosmic radiation, a magnetic storm and shock waves that travel laterally through the photosphere and upward through the chromosphere and corona.

The science studying pressure wave oscillations throughout the Sun is called **"helioseismology."** It allows to measure, for the first time, the invisible internal structure and dynamics of a star. Helioseismology, a new branch of solar physics, got its start in the early 1960s when CalTech physicist Robert Leighton[3227] discovered the 5-minute oscillation on the surface of the Sun. |
| 1998 | Dept. of Aerospace Engineering Sciences, University of Colorado, Boulder, Colorado | **Culbert B. Laney,**[3228] an associate professor at the University of Colorado and senior scientist at ITT Industries, Advanced Engineering and Sciences Division (Colorado Springs, CO), **publishes his book *Computational Gasdynamics.*** ▪ This book concerns numerical methods that were developed between the mid-1950s and the late 1980s and are currently used in computational fluid dynamics. It contains an extensive survey of numerical methods for simulating unsteady, mainly one-dimensional high-speed flows of inviscid perfect gases, especially flows containing shock waves. |

---

[3225] A.G. Kosovichev and V.V. Zharkova: *X-ray flare sparks quake inside Sun.* Nature **393**, 317-318 (1998); *Observation of seismic effects of solar flares from the SOHO Michelson Doppler Imager.* In: (F.L. Deubner et al., eds.) *New eyes to see inside the Sun and stars – Pushing the limits of helio- and asteroseismology with new observations from the ground and from space.* IAU Symposia. Kluwer, Dordrecht, vol. 185 (1998), pp. 191-194.

[3226] SOHO, a joint project from the ESA and NASA, is part of the International Solar-Terrestrial Physics (ISTP) Program – a global effort to observe and understand our nearest star, and its effects on our environment.

[3227] R. Leighton, R. Noyes, and G. Simon: *Velocity fields in the solar atmosphere.* Astrophys. J. **135**, 474 (1961).

[3228] C.B. Laney: *Computational gasdynamics.* University of Cambridge Press, Cambridge, U.K. (1998).

| 1998 | *Departamento de Física Teórica, Universidad Complutense de Madrid, Spain* | **Antonio F. RAÑADA,**[3229] **Mario SOLER, and José Luis TRUEBA,** three Spanish theoretical physicists, **report on numerical simulation studies on ball lightning** {ARAGO ⇨1838; FARADAY ⇨1841; THOMSON ⇨1888} in which they have combined the Navier-Stokes equations {NAVIER ⇨1822} describing the motion of fluids with the Maxwell equations for magnetic fields. |
|---|---|---|

(i) They explain the stability of ball lightning as the coupling of an air ball to a magnetic knot (a magnetic field with linked magnetic lines) which is strong enough to confine a ball of glowing plasma. They describe the ball as "force-free."

(ii) They suggest that a ball starts its brief existence as a bolt of ordinary lightning that heats the gases in air to above 30,000 K. At that temperature, the plasma in the ball offers no electrical resistance, so the current continues to flow for a while. However, as the ball expands it would rapidly cool, losing its infinite conductivity as well as its electromagnetic knot, and thereby destroying itself. ▪ However, the theory put forward by the Spanish investigators cannot explain why lightning balls can float horizontally and are capable of passing through glass windows or even walls.

Ball lightning is still a mystery, and one which has attracted scientists from different fields. None of the current theories appear to account for all of the properties observed.

▸ In 1969, during a flight in an all-metal airliner, Roger C. JENNISON,[3230] a professor of physical electronics and radio astronomy at the Electronics Laboratory of the University of Kent (Canterbury, U.K.), observed a glowing ball of lightning 20 cm in diameter, "emerging from the pilot's cabin and passing down the aisle of the aircraft approximately 50 cm from me, maintaining the same height and course for the whole distance over which it could be observed."

▸ A 63-page compendium of ball lightning observations was published in the April 1990 issue of the British *Journal of Meteorology*. For example, it contains an unusual observation of **giant ball lightning,**[3231] reported by an officer at the coastguard station in Fishguard (Dyfed, Wales, U.K.) on June 1977: "The ball lightning phenomenon was very large and estimated to be about the size of a bus. It was described as a brilliant, yellow-green, transparent ball with a fuzzy outline which descended from the base of a towering cumulus … and appeared to 'float' down the hillside…"

▸ A collection of reported Russian and Ukrainian sightings of ball lightning was published in 2002 in the *Philosophical Transactions*.[3232]

▸ In 2000, **Graham K. HUBLER,**[3233] a nuclear physicist at the Naval Research Laboratory (NRL), Washington, DC, **had a rare color photograph of ball lightning taken in 1978 in Austria** {⇨Fig. 4.1–K}.

▸ On February 1, 2002, during one of the most severe thunderstorms seen in Melbourne (Victoria, Australia) for many years, a unique streak picture of giant ball lightning was captured on color film towards the end of an approx. 5–7 minute exposure.[3234]

**The International Symposia on Ball Lightning have since become an important meeting place for scientific exchange among ball lightning researchers.** They have so far been held in Tokyo, Japan (1988); Budapest, Hungary (1990); Los Angeles, CA (1992); Canterbury,

---

[3229] A.F. RAÑADA, M. SOLER, and J.L. TRUEBA: *A model of ball lightning as a magnetic knot with linked streamers*. J. Geophys. Res. **103**, No. D18, 23309-23313 (Sept. 27, 1998).

[3230] R.C. JENNISON: *Ball lightning*. Nature **224**, 895 (1969).

[3231] I. JONES: *Giant ball lightning or plasma vortex*. J. Meteorology **15**, 178 (1990).

[3232] J. ABRAHAMSON, A.V. BYCHKOV, and V.L. BYCHKOV: *Recently reported sightings of ball lightning: observations collected by correspondence and Russian and Ukrainian sightings*. Phil. Trans. Roy. Soc. Lond. **A360**, 11-35 (2002).

[3233] G.K. HUBLER: *Fluff balls of fire*. Nature **403**, 487-488 (2000).

[3234] *Ball lightning photograph, taken at Melbourne, Victoria, Australia, on Feb. 1, 2002*. ERN Mainka Photography, Kangaroo Ground, Victoria, Australia; http://www.ernmphotography.com/Pages/Ball_Lightning/Ball_Lightning_ErnM.html.

U.K. (1995); Tsugawa Town, Japan (1997); Antwerp, Belgium (1999); St. Louis, MO (2001); Chung-li, Taiwan (2004); Eindhoven, The Netherlands (2006); and Kaliningrad, Russia (2008).

| 1998 | *Institute for Shock Physics, Washington State University (WSU), Pullman, Washington* | **Paulo A. Rigg**[3235] **and Yogendra M. Gupta,** two U.S. shock physicists, **examine elastic-plastic deformation of low-atomic-number single crystals in laboratory impact experiments using flash X-ray diffraction.**

(i) In order to overcome several limitations in previously reported work {Johnson et al. ⇨1970; Johnson & Mitchell ⇨1972; Jamet & Thomer ⇨1972}, they improve the precision of the X-ray diffraction data measured by

▸ setting up a carefully aligned planar impact experiment to provide uniaxial strain loading of crystals along the $\langle 111 \rangle$ and $\langle 100 \rangle$ axes;
▸ backing the crystal samples with X-ray windows;
▸ confining the incident X-ray beam; and
▸ using a 50-ns burst of Cu-Kα radiation (0.154 nm or 1.5 Å) produced from a 300-kV flash X-ray tube and two-dimensional intensified CCD detectors in order to obtain high-quality diffraction signals.

(ii) **Shock compression of a LiF crystal below the known Hugoniot elastic limit (< 4 GPa) along $\langle 111 \rangle$ reveals that the unit cell undergoes uniaxial compression; in contrast, shock compression along $\langle 100 \rangle$ reveals isotropic compression of the unit cell** (*i.e.*, compression in all directions).

In the following years, further flash X-ray diffraction studies[3236] were carried out at Washington State University, for example in the close environment of a shock-induced phase transition in a KCl crystal shocked along $\langle 100 \rangle$ to 7 GPa (70 kbar). Interplanar spacing measurements revealed isotropic compression of the unit cell below the transition stress; above the transition stress, however, measured diffraction data show a rearrangement from the rock salt to the cesium chloride structure. **Subsequent development of a multiple diffraction technique permitted simultaneous determination of both the longitudinal and transverse lattice deformations, which allows the unit cell deformation to be monitored in real time.**[3237] |
| 1998 | *National Center for Supercomputing Applications (NCSA), University of Illinois at Urbana-Champaign, Illinois* | **F. Douglas Swesty,** a research scientist at NCSA, **and collaborators work out a computer animation simulating the collision of binary neutron stars.**[3238]

Systems of orbiting neutron stars are born when the cores of two old stars collapse in supernova explosions. Neutron stars have the mass of our Sun but are the size of a city, so dense that boundaries between atoms disappear. Einstein's Theory of General Relativity (1916) predicts that the orbit shrinks from ripples of space-time called "gravitational waves." After about 1 billion simulation years, the two neutron stars closely circle each other at 60,000 revolutions per minute. The stars finally merge in a few milliseconds, sending out a burst of gravitational waves.[3239]

On May 9, 2005 astronomers using the Swift X-ray Telescope observed an unusual gamma-ray burst (GRB 050509b) which lasted less than 1/30th of a second, but produced an X-ray afterglow. A team of astronomers and astrophysicists concluded that "there is now observational support for the hypothesis that short-hard bursts arise during the merger of a com- |

---

[3235] P.A. Rigg and Y.M. Gupta: *Real-time X-ray diffraction to examine elastic-plastic deformation in shock lithium fluoride crystals.* Appl. Phys. Lett. **73**, 1655-1657 (1998).

[3236] T. D'Almeida and Y.M. Gupta: *Real-time X-ray diffraction measurements of the phase transition in KCl shocked along ⟨100⟩.* Phys. Rev. Lett. **85**, 330-333 (2000).

[3237] P.A. Rigg and Y.M. Gupta: *X-ray diffraction measurements to determine longitudinal and transverse lattice deformation in shocked LiF.* In: (M.D. Furnish, L.C. Chhabildas, and R.S. Hixson, eds.) *Shock compression of condensed matter – 1999.* Proc. 11th AIP Conference [Snowbird, UT, 1999]. AIP, Melville, NY (2000)], pp. 1051-1056.

[3238] *Colliding neutron stars.* NCSA News, National Center for Supercomputing Applications (NCSA), University of Illinois (1998); http://access.ncsa.uiuc.edu/CoverStories/NeutronStar/neutron_1.html.

[3239] *Neutron star collision.* Scientific Visualization Studio, NASA-GSFC, Greenbelt, MD; http://svs.gsfc.nasa.gov/vis/a000000/a000500/a000560/.

pact binary (two neutron stars, or a neutron star and a black hole)." The key of their conclusion is the "localization" of the X-ray afterglow.[3240]

1999     *Cape Canaveral Air Station (CCAS), Florida*

On February 7, **NASA's spacecraft Stardust is launched aboard a Delta II rocket.** The primary goal of Stardust is to collect dust and carbon-based samples during its closest encounter with comet WILD 2, and to bring the samples back to Earth. • Comet WILD 2 is named after the Swiss astronomer Paul WILD who discovered the comet in January 1978. The comet, which orbits the Sun once every 6.39 years, never gets very close to the Sun. Therefore, it still has most of its ice.

After nearly five years of space travel, the rendezvous took place in January 2004 passing at about 6 km/s within 240 km of WILD 2. During the meeting, Stardust performed a variety of tasks, such as

- reporting counts of comet particles performed by the spacecraft with the Dust Flux Monitor;
- performing real-time analyses of the compositions of comet particles and volatiles taken by the Comet and Interstellar Dust Analyzer; and
- collecting cometary and interstellar dust (tiny particles in μm- and sub-μm size) using the 2,000-cm$^2$ Dust Collector Grid. The media used in the Grid are blocks of 1- and 3-cm-thick microporous silica aerogel, an ultra-lightweight material mounted in modular aluminum cells.

After capturing these samples and storing them for safe keeping on its long journey back to Earth, Stardust successfully arrived back on Earth on January 15, 2006.[3241] A preliminary analysis carried out by international scientists from the Stardust Examination Team revealed that dust from WILD 2 contains minerals rich in Ca, Al and Ti which only form at extremely high temperatures – temperatures that could not have existed where the comets formed.[3242] This suggests the idea that the dust originated from violent explosive events (*e.g.*, supernovae) which, occurring in the evolution of young Sun or in other stars, had transported ejected matter in the afterflow of shock waves to the Solar System. **Examinations of the first collected dust samples also showed that about 10% of comet WILD 2 is made of organic materials** (complex carbon molecules) – **thus strongly supporting the theory that the ingredients for life on Earth originated in space.**[3243]

1999     *Torino, north-western Italy*

In June, **the International Astronomical Union (IAU) adopts the Impact Hazard Scale.** • The so-called **"Torino Impact Hazard Scale,"** a risk assessment system, is a new tool to help scientists, the media, and the public assess the potential dangers from Near-Earth Objects or NEOs {Asteroid 2002 MN ⇨2002}. Created by Richard P. BINZEL,[3244] a U.S. professor of Earth, Atmospheric and Planetary Sciences at MIT, it is similar to the Richter magnitude scale {RICHTER ⇨1935} used for earthquakes. Based on the orbital trajectory for a given NEO (such as an asteroid or comet) a "Torino Impact Risk Value" or "Index" (0, 1, 2, 3…10) is assigned to the NEO, which takes into account the object's size, speed, and the probability of it striking the Earth. For example,

- *Index 0*: the likelihood of a collision is zero, or well below the chance that a random object of the same size will strike the Earth within the next few decades;
- *Index 2*: a somewhat close but not unusual encounter. Collision is very unlikely;

---

[3240] J.S. BLOOM ET AL.: *Closing in on a short-hard burst progenitor: constraints from early-time optical imaging and spectroscopy of a possible host galaxy of GRB 050509b*. Astrophys. J. **638**, 354-368 (2006).
[3241] *Stardust, NASA's comet sample return mission*. NASA; http://www.nasa.gov/mission_pages/stardust/main/index.html.
[3242] D. BROWNLEE: *Stardust analysis update*. NASA-JPL (May 12, 2006); http://stardust.jpl.nasa.gov/news/status/060512.html.
[3243] J. LEAKE: *Comet dust holds building blocks of life*. Timesonline (March 5, 2006); http://www.timesonline.co.uk/article/0,,2087-2070393,00.html.
[3244] R.P. BINZEL: *The Torino impact hazard scale*. Planet. Space Sci. **48**, 297-303 (2000); http://www.sizes.com/units/torino_scale.htm.

- *Index 4*: a close encounter, with a 1% or greater chance of a collision capable of causing regional devastation;
- *Index 6*: a close encounter by a large object posing a serious but still uncertain threat of a global catastrophe;
- *Index 8*: a collision capable of causing localized destruction. Such events occur somewhere on Earth between once every 50 years and once every 1,000 years;
- *Index 9*: a collision capable of causing regional devastation. Such events occur between once every 1,000 years and once every 100,000 years; and
- *Index 10*: a collision capable of causing a global climatic catastrophe. Such events occur once every 100,000 years, or less often than this.

The majority of known near-Earth asteroids (NEAs) have an Index of $0.^{3245}$ For example, the Torino Impact Risk Value for asteroid 2002 MN {⇨2002} was 1. **There is presently only one known asteroid – the 500-m-sized asteroid 2004 VD17 with a possible impact in 2102 – which reaches a level 2 on the Torino Scale.**

In 1995, P.V. KRYUKOV, [3246] a Russian scientist at NPP Ballistic Technologies (Korolev, Moscow Region), proposed ballistic throwing of kinetic penetrators towards an asteroid approaching the Earth – a method which he called "BALSAD (Ballistic System for Anti-Asteroid Defense)."

| | | |
|---|---|---|
| 1999 | *Laboratoire d'Automatique, Grenoble, France* | In the period from June 30 to July 1, at the **EUROMECH Colloquium 397, entitled "Impact in Mechanical Systems,"** five plenary lectures are given which are devoted to the study of rigid multi-body mechanical systems subjected to nonsmooth effects.[3247] These lectures focus |

- on mathematical analysis of Jean-Jacques MOREAU's sweeping process for higher order systems, a velocity-impulse formulation introduced in 1971 (M. KUNZE and M.D.P. MONTEIRO-MARQUES);
- on dynamic simulation of rigid bodies: modeling of frictional contact (M. ABADIE);
- on stability of periodic motions with impact (A.P. IVANOV);
- on contact problems for elasto-plastic impact (W.J. STRONGE); and
- on a new method of theoretically analyzing multiple impact problems without friction (Y. HURMUZLU and V. CEANGA).

The last two lectures listed above also discussed the response from impact in multi-body systems such as chain reactions in the case of NEWTON's cradle {The Royal Society of London ⇨1666; ⇨Fig. 4.4–B}.

| | | |
|---|---|---|
| 1999 | *MIT, Cambridge, Massachusetts* | In August, at the conference **Asymmetrical Planetary Nebulae II: From Origins to Microstructures,** pictures of the Southern Crab Nebula are presented that were taken with the NASA/ESA Hubble Space Telescope. They show a small, bright nebula embedded in the center of a larger nebula; this structure resembling an hourglass nestled within an hourglass, known as "Dual Hour-Glass Shaped Nebula."[3248] ▪ Astronomers speculate that this unusual structure is the result of two separate outbursts originating from the interaction of a red giant with a white dwarf – a so-called **"symbiotic system"** – that occurred several thousand years apart and created outbursts of material in the form of gaseous bubbles. The red giant, a bloated star that, exhausting its nuclear fuel, was shedding its outer layers in a powerful stellar wind. |

---

[3245] NASA published a table listing potential future Earth impact events that the JPL Sentry System (a highly automated collision monitoring system) has detected based on currently available observations. NASA-JPL, Pasadena, CA; *see* http://neo.jpl.nasa.gov/risk/.

[3246] P.V. KRYUKOV: *BALSAD – Ballistic system for anti-asteroid defense*. In: *Abstracts 2nd Int. Workshop RAM Accelerators (RAMAC-II)* [Seattle, WA, July 1995].

[3247] B. BROGLIATO (ed.): *Impacts in mechanical systems: analysis and modelling*. Lecture Notes in Physics, Springer, Berlin *etc.* (2000).

[3248] *Symbiotic star in the Southern Crab Nebula blows bubbles into space*. Office of Public Outreach, Space Telescope Science Institute, Baltimore, MD (1999); http://hubblesite.org/newscenter/newsdesk/archive/releases/1999/32/image/a.

| | | |
|---|---|---|
| | | **Symbiotic stars are variable stars that belong to the class of** *explosive* **(or** *catastrophic*) *variables*.[3249] The term *symbiotic star* – denoting a system with partially ionized gas and dust surrounding a binary containing a hot, compact star and a cool giant – was coined in 1950 by Paul W. MERRILL,[3250] a U.S. astronomer at Mt. Wilson Observatory, southern California. |
| 1999 | Northern Germany & KNMI, de Bilt, The Netherlands | On November 8, **a meteor explodes above northern Germany.** The event is detected and identified by the Seismological Division of the Royal Netherlands Meteorological Institute (KNMI), which operates an experimental infrasound array of 16 microbarometers.[3251] ▪ The explosive power of the meteor, which was estimated to be about 1.5 kilotons of TNT equivalent, was similar to that of a small nuclear explosion. Low-frequency sound (or infrasound) measurements have been selected by those involved with the Comprehensive Test Ban Treaty {CTBT ⇨1996} as a way to detect and identify possible nuclear explosions. |
| 1999 | NASA's Jet Propulsion Laboratory (JPL), Pasadena, California & Dept. of Astronomy, University of Maryland | **NASA begins to design and plan its discovery mission Deep Impact** – the first space mission to probe beneath the surface of a comet in order to reveal what lies inside.[3252]<br>(i) Deep Impact, a dual-element spacecraft, was built by Ball Aerospace & Technologies Corporation (Boulder, CO) and consists of the fly-by spacecraft and an instrumented 816-lb (370-kg) impactor probe, which is designed to crash into the comet.<br>(ii) **The spacecraft is intended to reach comet TEMPEL 1 in July 2005.** On impact, the crater produced is expected to range in size from that of a house to that of a football stadium, two to fourteen storeys deep. Ice and dust debris will be ejected from the crater, revealing fresh material beneath. Images and spectrometer readings will be taken from the Deep Impact fly-by spacecraft and sent to Earth during the approach, the impact and its aftermath.<br>(iii) The aim is to blast out material frozen inside the comet since the Solar System was formed 4.6 billion years ago. Further objectives of the Deep Impact Mission are to<br>▸ observe how the crater forms;<br>▸ measure the crater's depth and diameter;<br>▸ measure the composition of the interior of the crater and its ejecta; and<br>▸ determine the changes in natural outgassing produced by the impact.<br>On Monday July 4, 2005, at 05:52 GMT, at a distance of about 134 million km from Earth, Deep Impact successfully smashed its impactor probe into comet TEMPEL 1 at a relative velocity of about 23,000 mph (10.3 km/s). **The impact generated an immense flash of light,** which provided an excellent light source for the two cameras on the Deep Impact fly-by spacecraft. **The Deep Impact team found water ice on the mostly coal-black surface of comet TEMPLE 1. But these patches of surface ice only represent about 0.5% of the surface observed.**[3253] These preliminary results show that there is ice on the surface, but definitely not enough to account for the water in the cloud of gas and dust that surrounds the comet.[3254] |
| 1999 | Northwestern University & University of Illinois, Urbana-Champaign, Illinois | Based on extraordinarily bright X-ray emission data obtained from the German-British-American joint mission ROSAT (Roentgen Satellite), **U.S. astrophysicists obtain the first observational evidence for remnants from hypernovae explosions.** The two suspected hypernova remnants, coincident with very luminous X-ray sources and named MF 83 and NGC 5471B, are located in the nearby spiral galaxy M101 (or NGC 5457).[3255] |

---

[3249] For the classification of variable stars *see* General Catalogue of Variable Stars ⇨1948.
[3250] P.W. MERRILL: *Measurements in the combination spectra of RW HYDRAE, BF CYGNI, and CI CYGNI*. Astrophys. J. **111**, 484 (1955).
[3251] L.G. EVERS and H.W. HAAK: *Listening to sounds from an exploding meteor and oceanic waves*. Geophys. Res. Lett. **28**, 41-44 (2001).
[3252] *Deep Impact*. NASA-JPL, Pasadena, CA; http://deepimpact.jpl.nasa.gov/index.html and http://deepimpact.jpl.nasa.gov/mission/factsheet-color.pdf.
[3253] *Deep Impact: a smashing success*. NASA (Feb. 3, 2006); http://www.nasa.gov/mission_pages/deepimpact/main/.
[3254] M.F. A'HEARN ET AL.: *Deep Impact: excavating comet TEMPEL 1*. Science **310**, No. 5746, 258-264 (Oct. 14, 2005).
[3255] Q.D. WANG: *Detection of X-ray emitting hypernova remnants in M101*. Astrophys. J. **517**, L27-L30 (May 20, 1999).

Later investigations of the nature of these two hypernova remnant candidates at optical wavelengths with high angular and spectral resolutions revealed

- that MF 83 is possibly an X-ray-bright superbubble whose interior has been heated by recent supernovae (the thermal energy derived from the X-ray data is a few times $10^{52}$ ergs, which requires either a "hypernova" or 10 to 100 normal supernovae in the past $10^6$ years); and
- that the fast expanding shell in NGC 5471B is unmatched by any comparable shells in the Large Magellanic Cloud and that it is very likely that it was produced by a hypernova.[3256]

*Hypernovae* – explosions a hundred times more energetic than supernovae – release an enormous amount of energy, primarily in the form of gamma-rays, the most energetic form of electromagnetic waves (or photons). They may explain the mysterious phenomena known as "Gamma-Ray Bursts (GRBs)" {Beppo-SAX & CGRO ⇒ 1997}.

| | | |
|---|---|---|
| 1999 | *Faculty of Engineering, Kumamoto University, southern Japan* | **Establishment of the Shock Wave and Condensed Matter Research Center** [Jpn. *Shogeki Kyokugen Kankyo Kenkyu Senta*],[3257] with the objective "to generate multi-extreme conditions using various combinations of shock waves, static high pressures, mega-gravity fields, low temperatures, strong magnetic fields, strong electric fields, *etc*. The characterization of the properties of condensed matter under multi-extreme conditions and the creation of new materials under such specific conditions will be the focus of the Center." The first director is Akira CHIBA. ▪ The institute is an outgrowth of research activities made by the High Energy Rate Laboratory (established in 1971) and the Low Temperature Laboratory (established in 1984) at Kumamoto University. |
| 1999 | *Zentrum für Lasermikroskopie, Institut für Anatomie, Universität Jena, Thuringia, Germany* | **Karsten KÖNIG,**[3258] a German physicist, **and colleagues report on the use of high-intensity near-infrared femtosecond laser pulses to precisely dissect intercellular nanometer-sized regions – a so-called "nanoscalpel."** <br><br>(i) The destructive effects are due to the fact that the laser pulses cause molecular ionization and a rapidly expanding high-pressure microplasma. ▪ **Known as the "photodisruptive effect," this results in cavitation and destructive shock wave effects in the surrounding medium.** The effect, apparently first proposed in the former Soviet Union {ASKAR'YAN & MOROZ ⇒ 1963}, was first observed in the United States {BELL & LANDT ⇒ 1967} using nanosecond laser pulses. Advantageously, much shorter laser pulses of femtosecond duration allow the production of reduced shock waves and smaller cavitation bubbles, which enable their use in minimally invasive surgical procedures {WICKHAM ⇒ 1987}. <br><br>(ii) Using a titanium sapphire ultrafast tunable laser system (Coherent Mira model 900-F), they used this technique to knock out genomic regions within the nucleus of a living cell, focusing especially on the nanodissection of chromosomes. ▪ The new technique makes it possible to perform intracellular surgery, on chromosomes and other organelles for example – known as "nanosurgery." <br><br>In 2000, Dr. KÖNIG was awarded the International Robert Feulgen Prize of the International Society for Histochemistry (Zurich, Switzerland) for his innovative research in the field of gene diagnostics. |

---

[3256] C.H.R. CHEN ET AL.: *A critical examination of hypernova remnant candidates. Part I: MF 83.* Astrophys. J. **547**, 754-764 (2001); Ibid. *Part II. NGC 5471B.* Astron. J. **123**, 2462-2472 (May 2002).

[3257] *Shock Wave and Condensed Matter Research Center.* Kumamoto University, Japan; http://read.jst.go.jp/ddbs/plsql/KKN_EG_14?code=0384010000.

[3258] K. KÖNIG, I. RIEMANN, P. FISCHER, and K.J. HALBHUBER: *Intracellular nanosurgery with near infrared femtosecond laser pulses.* Cell. Mol. Biol. **45**, 195-201 (1999).

| 1999 | Wellman Laboratories of Photomedicine, Harvard Medical School, Boston, Massachusetts | Shun LEE,[3259, 3260] a research physicist, **and associates report on the use of a two-stage shock tube for the *in vivo* transdermal delivery of drugs,** which is more economical than generating the shock wave photomechanically using a Q-switched laser.[3261] A single shock wave is applied onto the skin which – prior to the treatment – is covered with a thin layer of an aqueous test drug solution (*e.g.*, rhodamine B dextran). The impacting shock wave permeabilizes the outermost layer of the skin [*stratum corneum*], thus allowing the drug solution to diffuse into the skin – also called **"needle-free drug delivery."** Fluorescence microscopy reveals that the drug solution can be delivered efficiently to a depth of 30–60 μm into viable skin.<br><br>Other devices for gene and drug delivery, which accelerate micro-particles in the transonic flow to a sufficient momentum in order to penetrate the outer layer of the skin, and target the tissue below, require a gas reservoir to be positioned coaxially inside the shock-tube driver.[3262] |
|---|---|---|
| 1999 | Institute of Mechanical Engineering, CalTech, Pasadena, California | Ares J. ROSAKIS,[3263] a U.S. professor of aeronautics and applied mechanics, **and coworkers report on the direct experimental observation of cracks moving faster than the shear wave speed.** They investigated shear-dominated crack growth along weak planes in a brittle polyester resin (Homalite H-100) that exhibits stress-induced birefringence. The resin sample was asymmetrically impacted by a projectile fired from a high-speed gas gun. The propagating shear crack emerging from the notch tip was visualized using dynamic photoelasticity. **The supersonically propagating crack tip produced a Mach cone with a Mach angle ranging from about 43° to 45°.** ▪ Supersonic cracks generated at an ultrahigh loading rate were apparently first observed by S. WINKLER, D.A. SHOCKEY and D.R. CURRAN {⇨1970}; however, their pictures, which were taken in transmitted light (shadowgraphy), didn't show a Mach cone.<br><br>In 2004, it has been demonstrated for the first time by CalTech seismologists that a very fast, spontaneously generated rupture known as "supershear" can take place on large strike-slip faults like the San Andreas Fault. They base their claims on a laboratory experiment designed to simulate a fault rupture.[3264] The results suggest that under certain conditions supershear rupture propagation can be facilitated during large earthquake events {San Francisco Earthquake ⇨1906; Kunlun Shan Earthquake ⇨2001}. **Theoretical, numerical, observational and experimental research indicates that shear ruptures can propagate faster than the shear wave speed $v_S$.** The rupture velocities $v_R$ of some fault segments in recent large strike-slip earthquakes have been interpreted as propagating at transonic ($v_S < v_R < v_L$) or even supersonic ($v_R > v_L$) velocities (in respect of the longitudinal velocity $v_L$). Recently, seismologists considered the observational constraints, from the field and the laboratory, of transonic rupture velocity estimates, as well as the potential ramifications of transonic rupture propagation for earthquake ground motions and earthquake fracture mechanics in moderate and large earthquakes.[3265] |

---

[3259] S. LEE, D.J. MCAULIFFE, T. FLOTTE, and A.G. DOUKAS: *Trancutaneous molecular delivery in vivo with a single mechanical pulse.* J. Invest. Dermatol. **112A**, 652 (1999), Abstract No. 776.

[3260] S. LEE, D.J. MCAULIFFE, T. KODAMA, and A.G. DOUKAS: *In vivo transdermal delivery using a shock tube.* Shock Waves **10**, 307-311 (2000).

[3261] S. LEE, N. KOLLIAS, D.J. MCAULIFFE, T.J. FLOTTE, and A.G. DOUKAS: *Topical drug delivery in humans with a single photomechanical wave.* Pharmacol. Res. **16**, 1717-1721 (1999).

[3262] M.C. MARRION, M.A.F. KENDALL, and Y. LIU: *The gas-dynamic effects of a hemisphere-cylinder obstacle in a shock-tube driver.* Experiments in Fluids **38**, No. 3, 319-327 (2005); N.K. TRUONG, Y. LIU, and M.A.F. KENDALL: *Gas and particle dynamics of a contoured shock tube for pre-clinical microparticle drug delivery.* Shock Waves **15**, No. 3-4, 149-164 (2006).

[3263] A.J. ROSAKIS, O. SAMUDRALA, and D. COKER: *Cracks faster than the shear wave speed.* Science **284**, 1337-1340 (May 21, 1999).

[3264] K. XIA, A.J. ROSAKIS, and H. KANAMORI: *Laboratory earthquakes: the sub-Rayleigh-to-supershear rupture transition.* Science **303**, No. 5665, 1859-1861 (March 19, 2004); with J.R. RICE: *Laboratory earthquakes along inhomogeneous faults: directionality and supershear.* Ibid. **308**, No. 5722, 681-684 (May 6, 2005).

[3265] R. ARCHULETA and M. BOUCHON: *Constraints on transonic rupture propagation.* 100th Anniversary Earthquake Conference Commemorating the 1906 San Francisco Earthquake [San Francisco, CA, April 2006]. Seismological Society of America (SSA) 2006 Annual Meeting, Special Sessions.

| 1999 | *Dept. of Physics & Astronomy, University of Iowa, Iowa City, Iowa & Max-Planck-Institut für extraterrestrische Physik (MPE), Garching, Germany* | Dmitry SAMSONOV,[3266] John A. GOREE, Zhi-Wei MA, and Amitava BHATTACHARJEE (all at the University of Iowa), **and Hubertus M. THOMAS and Gregor E. MORFILL** (MPE) **demonstrate the existence of Mach cones in a strongly coupled dusty plasma.** ▪ Dusty plasmas are defined as a mixture of partially or fully-ionized gases that contain micron-size particles of electrically charged solid material, either dielectric or conducting – all held in a suspension analogous to a colloidal crystal. |

(i) They place about 10,000 polymer spheres 8.9 μm in diameter in a weakly ionized krypton plasma, and then levitate these spheres above the lower electrode of a vertical diode biased to – 245 V. The spheres interact only with their nearest neighbors. This arrangement forms an almost stationary two-dimensional crystalline Coulomb lattice of repulsive charges with a particle spacing of 256 μm, and allows them to study the structure of and movements within the plasma at a microscopic level.

(ii) **By firing a supersonic particle into the dusty plasma and using a laser sheet visualization technique, they can observe the formation of a Mach cone,** a pattern produced by interacting shock waves {DOPPLER ⇨ 1847; E. MACH & SALCHER ⇨ 1886}.[3267]

(iii) They suggest that Mach cones might be found in planetary rings, produced by big boulders plowing through fields of charged dust. While the boulders keep pace with the Kepler orbital velocity, they argue, the Lorentz force of the planet's magnetic field could modify the orbital motion of the charged dust. A remote search for Mach cones in Saturn's rings with viewing instruments aboard the Cassini orbiter could yield information about the dusty plasma conditions in regions through which Cassini would not survive direct transit.

(iv) Their studies confirm a previous theory advanced by Ove HAVNES,[3268] a Norwegian physics professor at the University of Tromso, and collaborators. They first predicted that boulders moving through the dust of Saturn's rings should produce Mach cones and suggested that Mach cones should serve as valuable diagnostics since they can be directly observed through remote sensing and do not significantly affect the dusty plasma conditions.

Mach cones in dusty plasmas have since been observed in several experiments. They received further theoretical attention, and are now studied in many ongoing and planned experiments both on the ground and in space. **The V-shaped Mach cones, which occur if a localized body moves through a dusty plasma at a speed larger than the dust acoustic wave speed, are of a nature similar to those of cones occurring at supersonic speeds in gases, liquids, or some crystals.**

The rings of Saturn contain charged dust particles whose dynamics are controlled by electromagnetic and gravitational forces. In 2004, German researchers discussed the possibility of the Mach cone formation involving the modified dust-acoustic waves in a dusty magnetoplasma composed of electrons and positively charged dust grains.[3269] Dusty plasmas are common in astrophysical environments (*e.g.*, in the interstellar medium, cometary tails, planetary ring systems), and important in the laboratory (*e.g.*, fusion plasmas) and industry (*e.g.*, semiconductor manufacturing).[3270]

---

[3266] D. SAMSONOV, J. GOREE, Z.W. MA, A. BHATTACHARJEE, H.M. THOMAS, and G.E. MORFILL: *Mach cones in a Coulomb lattice and a dusty plasma.* Phys. Rev. Lett. **83**, No. 18, 3649-3652 (1999).
[3267] *Mach cones studied in plasmas.* CERN Courier **40**, No. 1 (2000); http://www.cerncourier.com/main/article/40/1/14.
[3268] O. HAVNES ET AL.: *Probing the properties of planetary ring dust by the observation of Mach cones.* J. Geophys. Res. **100**, 1731-1734 (1995).
[3269] A.A. MAMUM and P.K. SHUKLA: *Dust-acoustic Mach cones in magnetized electron-dust plamas of Saturn.* Geophys Res. Lett. **31**, L06808 (2004).
[3270] R.L. MERLINO and J.A. GOREE: *Dusty plasmas in the laboratory, industry, and space.* Phys. Today **57**, 32-38 (July 2004).

| | | |
|---|---|---|
| 2000 | *NASA's Goddard Space Flight Center (GSFC), Greenbelt, Maryland* | In February, **images taken by Chandra**[3271] – NASA's X-ray observatory named in honor of the late Indian-born U.S. Nobel laureate Subrahmanyan CHANDRASEKHAR {⇨1931} – **show for the first time the full impact of the blast wave from supernova SN 1987A** {SHELTON ⇨1987}.<br><br>(i) The observations are the first time that **X-rays from a shock wave have been imaged at such an early stage in a supernova explosion.** The X-rays are emitted by hot gas (at temperatures of tens of millions of degrees Celsius) following behind the shock wave emitted by the supernova explosion.<br><br>(ii) The X-ray observations appear to confirm the general framework of a model developed at the Dept. of Astrophysical & Planetary Sciences of the University of Colorado (Boulder, CO), which holds that a shock wave has been expanding ahead of the debris expelled by the explosion. As this shock wave collides with material outside the ring ejected thousands of years earlier in the death throes of the star, it heats it to millions of degrees. The shock wave is smashing into portions of the ring at a speed of 4,500 km/s.<br><br>(iii) In the next few years, the shock wave will light up still more material in the ring, and a ***reverse shock wave*** {ARDAVAN ⇨1973}, an inward-moving shock wave, will heat the material ejected in the explosion itself.[3272] |
| 2000 | *American Museum of Natural History, New York City, New York* | In the period from May 30 to June 2, **the ASP Conference on Stellar Collisions, Mergers and their Consequences is held** in celebration of the opening of the Rose Center for Earth and Space and the newly rebuilt Hayden Planetarium. The scientific conference is organized by the Astronomical Society of the Pacific (ASP), the chairman is Michael M. SHARA.<br><br>Close encounters and collisions between stars were long thought to be rare events. But while most stars live out their lives in relative isolation, stars in dense star clusters or in galactic nuclei do sometimes collide with each other. Such collisions may build up massive stars; also, encounters between stars could form exotic stellar systems. Results of hydrodynamic simulation studies are summarized in the conference proceedings.[3273] Illustrative examples are also shown in the Internet.[3274, 3275] |
| 2000 | *Engineering Dept., Cambridge University (CU), England* | In July, **publication of the book *Impact Mechanics* by William James STRONGE,**[3276] a renowned professor of applied mechanics at CU. • In his textbook which is concerned with the reaction forces that develop during a collision and the dynamic response of structures to these reaction forces, he develops several methodologies for analyzing collisions between structures, ranging from rigid body theory for structures that are stiff and compact, to vibration and wave analyses for flexible structures. His book encompasses<br>▸ an introduction to analysis of slow-speed impact;<br>▸ rigid body theory for collinear impact and planar (or 2-D) collisions, and 3-D impact of rough rigid bodies;<br>▸ rigid body impact with discrete modeling of compliance for contact region;<br>▸ continuum modeling for local deformation near contact area;<br>▸ axial impact on slender deformable bodies; |

---

[3271] *Chandra views a supernova remnant at its making.* NASA-GSFC, Greenbelt, MD (May 12, 2000); http://imagine.gsfc.nasa.gov/docs/features/news/12may2000.html.

[3272] E. MICHAEL ET AL.: *HST observations of high-velocity Lyα and Hα emission from supernova remnant 1987A – the structure and development of the reverse shock.* Astrophys. J. **593**, 809-830 (2003).

[3273] M.M. SHARA (ed.): *Stellar collisions, mergers and their consequences.* ASP Conf. Proceedings, vol. 263 (2002). Astronomical Society of the Pacific (ASP), San Francisco, CA.

[3274] J.E. BARNES: *Stellar collisions.* Institute for Astronomy, University of Hawaii (May 2003); http://www.ifa.hawaii.edu/~barnes/research/stellar_collisions/index.html.

[3275] *Stellar collisions and cosmic catastrophes.* Scientific visualizations. Space Telescope Science Institute (STScI), Baltimore, MD; http://terpsichore.stsci.edu/~summers/viz/scviz/scviz.html.

[3276] W.J. STRONGE: *Impact mechanics.* Cambridge University Press, Cambridge (2000).

- impact on assembly of "rigid" elements;
- collision against flexible structure;
- propagating transformations of state in self-organizing systems; and
- **a review on the role of impact in 17th and 18th century development of fundamentals of mechanics** (Appendix A), **and a glossary of terms** (Appendix B).

| | | |
|---|---|---|
| 2000 | *National Optical Astronomy Observatory (NOAO), Tucson, Arizona* | On November 2, in a HubbleSite press release **it is reported that an international team of NOAO astronomers,** while analyzing images taken in March 1996 with the Wide Field/Planetary Camera-2 of NASA/ESA's Hubble Space Telescope, **has discovered a striking example of what appears to be the aftermath of a rare physical collision between two galaxies.** |

(i) The interacting lenticular galaxy system, named NGC 6745, is located in *Lyra* ("The Lyre," a large constellation of the Northern Hemisphere), about 206 million light-years from Earth.[3277]

(ii) NOAO and NASA astronomers believe that these two galaxies did not merely interact gravitationally as they passed one another; they actually collided. ▪ Under the guidance of the instrument scientist Roger LYNDS, the NOAO team concluded that "in the rare case of NGC 6745, the gas and dust drifting between the stars in the two galaxies actually collided. Wherever the interstellar clouds of two galaxies collide, they do not freely penetrate each other, but rather, suffer something called **'inelastic collision.'** If the relative velocity in such collisions is sufficiently high, the pressure at the collision interface will produce densities of material extreme enough to trigger star formation through gravitational collapse, in much the same way that star formation is triggered by gravitationally induced shock waves."[3278]

Galaxies are mostly empty space, so when two galaxies collide, the chance of stars or planets colliding is very small. About the only objects that actually bump into each other are huge clouds of gas and dust. These collisions can cause new stars to form a million or so years after the collision, so the colliding galaxies would brighten up at that point. **Studies of colliding galaxies** – which reach back to before the 1950s and have been carried out by the Swedish astronomer Erik HOLMBERG, the Bulgarian-born U.S. theoretical physicist Fritz ZWICKY, and a few others – **have progressed rapidly in the last few years, driven by observations made with powerful new ground- and space-based instruments.**

Curtis J. STRUCK,[3279] a professor of physics and astronomy at the University of Iowa, recently reviewed the history of the study of galaxy collisions and discussed current attempts to integrate these studies into the key questions of modern astrophysics and cosmology. Examples include:

- What role do collisions play in the evolution of galaxies?
- What role does the environment play in galaxy collisions?
- How are star formation and nuclear activity orchestrated by large-scale dynamics, both before and during the merger?
- What is the relationship between ultraluminous X-ray sources and colliding galaxies?
- How can we improve galactic archaeology enough to allow us to determine the characteristics of precursor systems and the precise effects of their interactions?

The nearest large spiral galaxy to the Milky Way is the Andromeda Galaxy (NGC 224), which is about twice as big as the Milky Way and about 2.2 million light-years away from us in the constellation *Triangulum*, but the gap between us is closing at about 500,000 km/h (139 km/s). While most galaxies are rushing away as the Universe expands, **Andromeda is the only big spiral galaxy moving towards the Milky Way,** and it may collide with the Milky Way in

---

[3277] *A bird's eye view of a galaxy collision.* NASA HubbleSite (Nov. 2, 2000); http://hubblesite.org/newscenter/newsdesk/archive/releases/2000/34/.
[3278] *When galaxies collide.* Current science at NOAO (Nov. 2000); http://www.noao.edu/outreach/current/collide_hilite.html.
[3279] C.J STRUCK: *Galaxy collisions – dawn of a new era.* Dept of Physics and Astronomy, Iowa State University; E-Print Archive: Astro-ph/0511335 (Nov. 2005); http://arxiv.org/PS_cache/astro-ph/pdf/0511/0511335.pdf.

about three billion years. In 2000, John DUBINSKI,[3280] a professor in the Dept. of Astronomy & Astrophysics at the University of Toronto, numerically simulated this Milky Way-Andromeda interaction by following the motions of more than 100 million stars and dark matter particles as the gravitational forces of the two galaxies force them to collide. The simulation was a feat of parallel computing that took four days to complete on the San Diego Supercomputing Center's 1152-processor IBM SP3 "Blue Horizon" – one of a new class of supercomputers that can perform more than one trillion arithmetic operations per second.[3281]

| | | |
|---|---|---|
| 2000 | *Le Quartz, Brest, Brittany, north-western France* | On November 29/30, **Rogue Waves 2000 Workshop – the first international meeting on giant sea waves – is held** within the Brest SeaTech Week 2000. The chairmen are Michel OLAGNON and Gerassimos ATHANASSOULIS. ▪ One of the main aims of the meeting is to "to assess the state of the art as to conditions of occurrence of waves or groups of waves of unexpected severity responsible for ship wrecks and damage to offshore and gas production systems." Observations and measurements of rogue waves (sometimes also referred to as "freak" waves), as well as numerical modeling and laboratory-scale simulation in this area, have advanced various theories on the possible mechanisms and phenomena that cause these waves.<br><br>A general review of this unique subject – attracting fluid dynamicists and oceanologists as well as mariners and coastal engineers for scientific and practical reasons, respectively – was prepared by Kristian B. DYSTHE,[3282] a mathematician at the University of Bergen, Norway, and colleagues. Their summary refers to recent conference and journal articles, television programs, and research projects, including an animation of the evolution of a "breather" produced by the nonlinear superposition of two wave trains {⇨Fig. 4.1–T}. |
| 2000 | *IBM Almaden Research Center, San Jose & Stanford University (SU), Palo Alto, California* | Farid F. ABRAHAM,[3283] a research physicist and molecular dynamicist at IBM, **and Huajian GAO,** a professor of mechanical engineering at SU, **perform atomistic computer simulations of a crack propagating along a weak interface joining two harmonic crystals.**<br><br>(i) The aim of their investigation is to provide immediate insights into the nature of material failure on the atomic scale. In order to model the behavior of up to one billion atoms, ABRAHAM and his colleagues use the ASCI WHITE parallel supercomputer, which can complete more than $10^{13}$ computations a second {ASCI WHITE ⇨1999}.<br><br>(ii) They show that under special shear mode conditions, a crack can propagate at the longitudinal wave speed, thereby generating a Mach cone emanating from the crack tip.[3284] Their loading process seems to closely resemble the dynamic impact loading conditions in the experiments of ROSAKIS ET AL. {⇨1999}. **They suggest that their model study might provide an explanation for the *intersonic rupture* observed in shallow crust earthquakes.** ▪ A majority of earthquakes are caused by sudden rupture of the Earth's crust along a preexisting fault plane under the action of high ambient compressive and shear pre-stresses. For most shallow crustal earthquakes, average rupture speeds observed so far range from $0.7\beta$ to $0.9\beta$, where $\beta$ is the average shear (or S-) wave speed in the surrounding rock body. But – as ROSAKIS and col- |

---

[3280] J. DUBINSKI: *The merger of the Milky Way and Andromeda galaxies* (Jan. 2001). Canadian Institute for Theoretical Astrophysics (CITA), University of Toronto; http://www.cita.utoronto.ca/~dubinski/tflops/.

[3281] J. WONG: *Astrophysicist maps out our own galaxy's end.* Dept. of Public Affairs, University of Toronto (April 14, 2000); http://www.newsandevents.utoronto.ca/bin/000414b.asp.

[3282] K.B. DYSTHE, H.E. KROGSTAD, and K. TRULSEN: *Freak waves, extreme waves and ocean wave climate.* Universities of Bergen and Oslo, and Norges Teknisk-Naturvitenskapelige Universitet (NTNU), Trondheim, Norway; http://www.math.uio.no/~karstent/waves/index_en.html.

[3283] F.F. ABRAHAM and H. GAO: *How fast can cracks propagate?* Phys. Rev. Lett. **84**, 3113-3116 (2000).

[3284] *Simulations of fractures: supersonic crack propagation in brittle fractures.* IBM Almaden Research Center, San Jose, CA; http://www.almaden.ibm.com/st/computational_science/msmp/fractures/scp/?scp2.

laboratories later suggested – rupture is a highly transient process and rupture speeds could be *intersonic*; *i.e.*, between the shear wave speed and the longitudinal (or P-) wave speed.[3285]

Recent analyses of several large earthquakes gave indeed evidence of "supershear" rupture; *i.e.*, rupture which exceeds the shear velocity of the brittle part of the crust {San Francisco Earthquake ⇨1906; Kunlun Shan Earthquake ⇨2001}.

| | | |
|---|---|---|
| 2000 | *School of Chemistry, University of Illinois, Urbana-Champaign, Illinois* | **Dana D. DLOTT**,[3286] a U.S. professor of chemistry, **applies powerful laser-driven shock waves to produce large-amplitude compression in molecular materials on the picosecond timescale.**<br><br>(i) The miniature shock waves – which he calls **"nanoshocks"** – have a duration of a few nanoseconds and suddenly drive the irradiated material to extreme pressures and temperatures. Nanoshocks can be generated in almost any material of interest in chemistry, biology and medicine, and are promising tools for the study of fundamental processes at the molecular level.<br><br>(ii) Coupled with ultrafast molecular spectroscopy, he probes the molecular response to a nanoshock in detail. In order to characterize the nanoshock pulses, he uses simple molecular systems such as anthracene crystals [$C_{14}H_{10}$].<br><br>(iii) He applies well-characterized nanoshocks to study complex high-rate phenomena, *e.g.*,<br>▸ the properties of shock-induced chemical reactions;<br>▸ the shock-induced orientation of energetic solids; and<br>▸ the shock compression of organic polymers.<br><br>In the following years, DLOTT and his team began to extend shock wave spectroscopy to biological molecules, specifically to proteins.[3287] Using ultrafast coherent anti-Stokes Raman spectroscopy, they also studied variations in the velocity and pressure of repetitive 4 GPa (40 kbar) laser-driven shock fronts in a polymer thin film.[3288] |
| 2000 | *Dept. of Applied Physics, University of Twente, Netherlands & Dept. of Zoology, TU Munich, Germany* | **Detlef LOHSE**,[3289] **Michel VERSLUIS, and Anna VON DER HEYDT**, three Dutch applied physicists, **and Barbara SCHMITZ**, a German zoologist, **study the mechanism of how snapping shrimps snap.** Using high-speed videography, they observe that **the *Alpheus* species** {KRAUSS ⇨1843} – in Germany appropriately called *Pistolenkrebs* ("pistol shrimp") or *Knallkrebs* ("bang shrimp") – **generates weak shock waves in water by cavitation.** For example, at a distance of 4 cm from the snapping claw, a pressure of 80 bars was recorded {⇨Fig. 4.1–Z}. High-speed videos of the snapping shrimp's sound production mechanism can be found in the Internet.[3290]<br><br>In the following year, this team[3291] discovered that a short flash of sonoluminescent light is also generated during the rapid collapse of the cavitation bubble. They called this phenomenon "shrimpoluminescence." The light produced is of a lower intensity than the light produced by typical sonoluminescence, and is not visible to the naked eye. |

---

[3285] O. SAMUDRALA, Y. HUANG, and A.J. ROSAKIS: *Subsonic and intersonic shear rupture of weak planes with a velocity weakening cohesive zone*. J. Geophys. Res. (Solid Earth) **107**, Issue B8, Paper No. 2170 (Aug. 2002); http://www.solids.caltech.edu/~rosakis/pubs/2002-SubsonicIntrsonic.pdf.
[3286] D.D. DLOTT: *Nanoshocks in molecular materials*. Acc. Chem. Res. **33**, 37-45 (2000).
[3287] H. KIM, S.A. HAMBIR, and D.D. DLOTT: *Shock compression of organic polymers and proteins: ultrafast structural relaxation dynamics and energy landscapes*. J. Phys. Chem. **A104**, 4239-4252 (2000).
[3288] Y. YANG, S.A. HAMBIR, and D.D. DLOTT: *Ultrafast vibrational spectroscopy imaging of nanoshock planar propagation*. Shock Waves **12**, 129-136 (2000).
[3289] M. VERSLUIS, B. SCHMITZ, A. VON DER HEYDT, and D. LOHSE: *How snapping shrimps snap: through cavitating bubbles*. Science **289**, 2114-2117 (Sept. 22, 2000). See also K. BROWN: *For certain shrimp, life's a snap*. Science **289**, 2020-2021 (Sept. 22, 2000).
[3290] *Selection of high-speed video recordings of the snapping shrimp*. University of Twente, The Netherlands; http://stilton.tnw.utwente.nl/shrimp/video.html.
[3291] D. LOHSE, B. SCHMITZ, and M. VERSLUIS: *Snapping shrimp make flashing bubbles*. Nature **413**, 477-478 (Oct. 4, 2001).

| | | |
|---|---|---|
| 2000 | *Institute of Geophysics and Planetary Physics, UC Los Angeles & Lawrence Livermore National Laboratory, Livermore, California* | John E. VIDALE,[3292] and Paul S. EARLE (UCLA), **and** Dough A. DODGE (LLNL), three U.S. geophysicists, **find strong evidence for a comparatively slow rate of rotation of the Earth's inner core.**<br>(i) They use previously recorded seismic data collected with the Large Aperture Seismic Array {LASA ⇒1965} in Montana – one of the most sensitive instruments ever built, but now defunct – which came from two nuclear explosion tests conducted in 1971 and 1974 at Novaya Zemlya, in the former Soviet Union.<br>(ii) They find that the inner core rotates to the east at about 0.15° per year, which shifts its equator annually by about 3 km. |
| 2000 | *University of Texas at Austin, Texas* | J. Craig WHEELER, a U.S. astronomy professor, **publishes the popular book** *Cosmic Catastrophes: Supernovae, Gamma-Ray Bursts, and Adventures in Hyperspace* (Cambridge University Press). ▪ "The book follows the tortuous life of a star, from birth, evolution, and death, and leads on to ideas of complete collapse to a black hole, worm-hole time machines, the possible birth of baby bubble universes, and the prospect of a revolutionary view of space and time in a ten-dimensional string theory. Along the way we look at evidence that suggests that the Universe is accelerating, and recent developments in understanding gamma-ray bursts – perhaps the most catastrophic cosmic events of all. With the use of lucid analogies, simple language and crystal-clear cartoons, this book makes accessible some of the most exciting and mind-bending objects and ideas in the Universe." (Text taken from the book's inside cover.)<br>**Considering the enormous number of stars in the Universe,[3293] there is always a stellar explosion occurring at some point in the Universe.**<br>▸ Supernovae are the source of the heavier elements in the Universe: the material from which solid planets like the Earth and living organisms are made.<br>▸ Huge cosmic gamma-ray bursts, probably originating from supernovae and hypernovae, may reduce the shielding properties of the atmosphere and/or produce detrimental secondary effects which can affect terrestrial life. It is widely accepted that cosmic rays cause genetic mutation and are harmful to living organisms, but on the other hand, this may also stimulate biological evolution. |
| 2000 | *University of Chicago, Illinois & U.S. Naval Research Laboratory (NRL), Washington, DC* | Mao-Xi ZHANG, Philip E. EATON, and Richard GILARDI,[3294] three U.S. research chemists, **report on the synthesis of** *octanitrocubane* [$C_8N_8O_{16}$]. This new type of shock-insensitive high explosive, based on the building block hydrocarbon cubane, is more powerful than HMX, which is presently the standard military high explosive. ONC (octanitrocubane) is believed to be one of the most energetic non-nuclear substances known. ▪ In 1964, Philip E. EATON[3295] and Thomas W. COLE, two U.S. chemists at the University of Chicago, were the first to synthesize *cubane* [$C_8H_8$], one of the densest hydrocarbons ever synthesized. In octanitrocubane, the hydrogen atoms at the corner of cubane are all replaced by nitro groups [$-NO_2$]. |
| 2001 | *Physics Division, Oak Ridge National Laboratory (ORNL), Oak Ridge, Tennessee* | In February, **establishment of the Terascale Supernova Initiative (TSI) with the principal goal to simulate and, thereby, to understand the catastrophic end of the lives of massive stars, known as "core collapse supernova."** |

---

[3292] J.E. VIDALE, D.A. DODGE, and P.S. EARLE: *Slow differential rotation of the Earth's inner core indicated by temporal changes in scattering.* Nature **405**, 445-448 (May 25, 2000).

[3293] More recently, a team of Australian astronomers at the Australian National University (ANU), Canberra has calculated the number of stars in the visible Universe to be around $70 \times 10^{22}$; see http://www.cnn.com/2003/TECH/space/07/22/stars.survey/.

[3294] M.X. ZHANG, P.E. EATON, and R. GILARDI: *Hepta- and octanitrocubanes.* Angew. Chem. (Int. Ed.) **39**, 401-404 (2000).

[3295] P.E. EATON and T.W. COLE JR.: *The cubane system.* J. Am. Chem. Soc. **86**, 962-964, 3157-3160 (1964).

(i) Further goals are to understand all of the phenomena associated with these stellar explosions, such as their contribution to the synthesis of the chemical elements in the Periodic Table, their emission of radiation-like particles (known as "neutrinos"), of gravitational waves, and in some cases of intense bursts of gamma radiation.

(ii) TSI – a multi-million dollar, multi-year Department of Energy (DOE) initiative – is led by Anthony MEZZACAPPA, a U.S astrophysicist at ORNL who also conceived and proposed it. TSI involving several dozen researchers at a dozen international institutions is one of the world's largest computational astrophysics initiatives.[3296]

Core collapse supernovae of massive stars (of roughly more than ten times the mass of our Sun) are the most energetic explosions in the Universe with energy blasts > $10^{53}$ ergs and believed to be powered by the intense radiation (neutrinos) from the stellar core. They are the dominant source of elements in the Periodic Table between oxygen and iron, and there is growing evidence they are indeed responsible for producing half the elements heavier than iron. Moreover, there is now an indisputable connection between "peculiar" hyperenergetic core collapse supernovae (also known as "hypernovae") and one of two classes of gamma ray bursts in the Universe.

A 3-D visualization has revealed lopside blast waves of gas in the moments after a supernova explodes. When fusion stops with iron, the core collapse produces turbulent shock waves deep within the core leading to the synthesis of heavy elements. Within a fraction of a second, this collapse squeezes the inner part of the star's core into a volume just tens of kilometers wide that resists further crushing. The inner core rebounds like a piston, creating a shock wave that rifles back into the still-collapsing outer core.[3297]

| | | |
|---|---|---|
| 2001 | NASA's Goddard Space Flight Center (GSFC), Greenbelt, Maryland & Princeton University, Princeton, New Jersey | On June 30, **the Wilkinson Microwave Anisotropy Probe (WMAP) is launched.** Its mission is to provide a full sky map of the faint anisotropy or variations in the temperature of the cosmic microwave background radiation. The WMAP mission is scheduled to observe for four years at L2, a semi-stable region of gravity that is about four times further out than the Moon but follows the Earth around the Sun.[3298] ▪ WMAP is the result of a partnership between Princeton University and NASA-GSFC and was named in honor of the late David WILKINSON, a professor of physics at Princeton University and a renowned cosmologist who died in September 2002.<br><br>The cosmic background radiation {ALPHER, HERMAN & GAMOW ⇨1948; PENZIAS & WILSON ⇨1965} is the oldest light in the Universe; it has been traveling across the Universe for almost 14 billion years. The patterns in this light across the sky encode a wealth of details about the history, shape, content, and ultimate fate of the Universe. In February 2003, **the first picture of the background radiation, based on full sky data obtained by WMAP, was released by NASA, showing microwave light 379,000 years after the Big Bang** {⇨c.14 Ga ago; NASA-GSFC ⇨2003; ⇨Fig. 4.1–W}. |

---

[3296] *Shedding new light on exploding stars: terascale simulations of neutrino driven supernovae and their nucleosynthesis.* Scientific Discovery through Advanced Computing (SciDAC), DOD's Office of Science (Aug. 2006); http://www.scidac.gov/HENP/HENP_TSI/reports/TSI2003Annual1.html.
[3297] R. IRION: *The Terascale Supernova Initiative: modeling the first instants of a star's death.* SciDAC Rev. (IOP Publishing) No. 1, 26-37 (Spring 2006); http://www.scidacreview.org/0601/html/astro.html.
[3298] NASA: *Wilkinson Microwave Anisotropy Probe (WMAP) mission.* NASA-GSFC, Greenbelt, MD; http://map.gsfc.nasa.gov/index.html.

| | | |
|---|---|---|
| 2001 | *NASA's Chandra X-Ray Center (CXC); Smithsonian Astrophysical Observatory, Cambridge, MA & Max-Planck-Institut für Extraterrestrische Physik (MPE), Garching, Germany* | In July, **an international team of astrophysicists observes two supermassive black holes at the core of the same galaxy.** This extremely luminous galaxy – named **NGC 6240** – is located in *Ophiuchus* ("The Serpent Bearer," a constellation of the equatorial zone, near the center of the Milky Way), about 400 million light-years from Earth.<br><br>(i) **The peculiar butterfly shape of NGC 6240 was caused by the collision of two smaller galaxies.** The team, led by Stefanie KOMOSSA, an astrophysicist at MPE, could clearly distinguish the two X-ray emitting nuclei by using the Advanced CCD Imaging Spectrometer (ACIS) detector aboard NASA's Chandra X-Ray Observatory.[3299]<br><br>(ii) The black holes are orbiting each other and will collide and merge to create an even larger black hole – resulting in a catastrophic event that will unleash intense radiation and gravitational waves at the moment of impact. However, this will not happen for several hundred million years.[3300]<br><br>According to EINSTEIN's General Theory of Relativity (1916), gravitational waves are emitted by accelerating masses (such as electromagnetic waves are produced by accelerating charges). Consequently, colliding massive bodies produce gravitation waves, and colliding supermassive black holes are supposed to generate patterns of very strong gravitation waves. A first computer simulation of two black holes violently merging into one was carried out in the United States and Denmark in 2002 which supported this prediction. Further progress was achieved in the following years by NASA scientists using the Columbia supercomputer at NASA's Ames Research Center (Mountain View, CA).[3301, 3302] |
| 2001 | *World Trade Center (WTC), New York City, New York* | On September 11, **two Boeing 757 airliners heading for Los Angeles with full fuel loads (24,000 gallons or 90,847 liters) are hijacked by terrorists shortly after take-off. They order the pilots to crash the planes into the two 110-storey 416-m-high towers of the World Trade Center at about 500 mph (240 m/s).** Both towers collapse about an hour after impact {⇒Fig. 4.21–H}. It is the worst terrorist attack in recorded history, killing some 2,800 people.<br><br>Immediate investigations carried out at the Impact & Crashworthiness Laboratory of MIT revealed that the structural systems of both towers, essentially prefabricated steel lattices, were significantly weakened due to the impacts of the planes. The wings and fuselages of the planes sliced through the exterior columns of the façade (thin-walled box beams made of high strength steel) as if they were made of cardboard. The fuel in the wing tanks greatly increased the mass per unit length of the wings and added to their devastating power.[3303] In addition, the enormous quantities of burning fuel from the planes created extreme heat and ignited several floors in the building, causing the support columns to bend inward and buckle, and the steel flooring to separate from the columns. **The collapse of the upper portion of each building onto the floors below caused each floor to progressively collapse – a chain reaction of impacts which almost became a freefall.** • Since then, a number of more detailed explanations for the failure and collapse mechanisms have been developed by leading civil engineers from the United States, Australia and Germany.[3304, 3305, 3306, 3307] |

---

[3299] *NGC 6240: Never before seen: two supermassive black holes in same galaxy.* Chandra X-Ray Observatory. Harvard Smithsonian Center for Astrophysics, Cambridge, MA (revised Aug. 30, 2006); http://chandra.harvard.edu/photo/2002/0192/.

[3300] D. WHITEHOUSE: *Black holes on collision course.* BBC/News, world ed. (Nov. 19, 2002); http://news.bbc.co.uk/2/hi/science/nature/2493331.stm.

[3301] *NASA achieves breakthrough in black hole simulation.* NASA Exploring the Universe. (April 18, 2006); http://www.nasa.gov/vision/universe/starsgalaxies/gwave.html.

[3302] J.G. BAKER ET AL.: *Gravitational-wave extraction from an inspiraling configuration of merging black holes.* Phys. Rev. Lett. **96**, No. 11, Paper No. 111,102 (2006).

[3303] T. WIERZBICKI and X. TENG: *How the airplane wing cut through the exterior columns of the World Trade Center.* Int. J. Impact Engng. **28**, 601-625 (2003).

[3304] Z.P. BAŽANT: *Why did the World Trade Center collapse?* SIAM News **34**, No. 8, 1-3 (Oct. 2001). *Why did the World Trade Center collapse? – Simple analysis.* J. Engng. Mech. **128**, No. 1, 2-6 (2002); *Addenum.* Ibid. No. 3, 369-370 (2002).

[3305] S. ASHLEY: *When the Twin Towers fell.* Scient. Am. **285**, 9-10 (May 2001).

[3306] M. KISTLER: *Why the World Trade Center towers collapsed.* Illumin U.S.C. (University of Southern Cal.) No. 2, 4 pages (Fall 2002).

[3307] E. KAUSEL (ed.): *Towers lost and beyond.* WIT Press, Southampton (2003).

The U.S. National Institute of Standards and Technology (NIST) initiated investigations to uncover the complex factors contributing to the probable causes of the post-impact collapse of the WTC towers.[3308] The Federal Emergency Management Agency (FEMA), the Structural Engineering Institute of the American Society of Civil Engineers (SEI/ASCE), and several other relevant organizations formed a group of civil, structural, and fire protection engineers to study the performance of buildings at the WTC site, for example

- the immediate effects of the impacts of the aircraft on each tower;
- the spread of fire following the crashes;
- the fire-induced reduction in structural strength; and
- the mechanism that led to the collapse of each tower.

The most important findings of this FEMA report is an admission that no structures could have been reasonably expected to withstand the terrorist attack, and the building's robustness allowed for a large number of occupants to escape and survive.[3309]

| | | |
|---|---|---|
| 2001 | *Kunlun Fault, northern Tibetan Plateau, northwestern China* | On November 14, **a large earthquake with a magnitude of 8.1 occurs on the remote high plateau in the Kokoxili region near the Qinghai-Xinjiang border.** The main shock develops along a western segment of the Kunlun Fault system. ▪ The so-called "Kunlun Shan Earthquake" is the largest earthquake that has occurred in China in the past 50 years and an extraordinary event from the viewpoint of dynamic rupture mechanics: |

(i) Regional broadband recordings of this event provided an opportunity to accurately observe the speed at which a fault ruptures during an earthquake, which has important implications for seismic risk and for understanding earthquake physics.

(ii) **Michel BOUCHON**[3310] **and Martin VALLÉE**, two French seismologists at Grenoble, **determined that rupture propagated on the about 400-km long fault at an average speed of 3.7 to 3.9 km/s, which exceeded the shear velocity of the brittle part of the crust.** Rupture started at sub-Rayleigh wave velocity and became supershear (or intersonic) – *i.e.*, propagated faster than the shear (or S-) wave but slower than the longitudinal (or P-) wave – probably approaching 5 km/s after about 100 km of propagation.

Other events, such as the 1906 San Francisco Earthquake {⇨1906}, the 1979 Imperial Valley Earthquake, the 1992 Landers Earthquake, the 1999 Izmit Earthquake, and the 2002 Denali Earthquake, may also have featured supershear speeds.

| | | |
|---|---|---|
| 2001 | *University of California, San Diego, CA* | In the period November 18–20, at the 2001 APS Division of Fluid Dynamics 54th Annual Meeting, **Dutch fluid dynamicists**[3311] **at the University of Twente show a video of a steel ball which – dropping onto a deep layer of loose, very fine sand – creates an impressive granular jet.** The following phases can be distinguished: |

- upon impact, sand is blown away in all directions, forming a crown-like ejection similar to a drop impacting a liquid film with a relatively high impact velocity;[3312]
- the ball dives deep into the soft sand, creating a void (like a mine shaft);
- the void immediately collapses again: the "hydrostatic" pressure, focusing upon the axis of impact, pushes the sand straight up into the air.; and
- the resulting granular jet exceeds the release height of the ball.

---

[3308] *NIST and the World Trade Center.* National Institute of Standards and Technology (NIST), Gaithersburg, MD; http://wtc.nist.gov/.
[3309] FEMA: *World Trade Center building performance study: data collection, preliminary observations, and recommendations.* FEMA Rept. No. 403 (May 2002), Federal Emergency Management Agency (FEMA), Washington, DC; http://www.fema.gov/rebuild/mat/wtcstudy.shtm.
[3310] M. BOUCHON and M. VALLÉE: *Observation of long supershear rupture during the magnitude 8.1 Kunlunshan Earthquake.* Science **301**, No. 5634, 824-826 (2003).
[3311] R. MIKKELSEN, M. VERSLUIS, E. KOENE, G.W. BRUGGERT, D. VAN DER MEER, K. VAN DER WEELE, and D. LOHSE: *Granular eruptions: void collapse and jet formation.* Phys. Fluids **14**, No. 9, S14 (Sept. 2002); http://www.math.upatras.gr/~weele/media/GoFM2001_Granular_Jet.mpg.
[3312] I.V. ROISMAN and C. TROPEA: *Impact of a drop onto a wetted wall: description of crown formation and propagation.* J. Fluid Mech. **472**, 373-397 (2002).

This phenomenon of a granular jet is similar to the well-known *Worthington jet* which occurs when a drop impacts onto a pool of liquid {WORTHINGTON ⇨1894}. Their spectacular 3-minute video won the entry in the APS Gallery of Fluid Motion.

| | |
|---|---|
| 2001 *Academic Press, New York City, New York* | **Publication of the three-volume *Handbook of Shock Waves*** by Academic Press. The editors are Gabi BEN-DOR, Ozer IGRA, and Tov ELPERIN. Examples of addressed topics, reviewed in 17 chapters by 48 contributors, include |

- history of shock wave research (P. KREHL);
- general laws for propagation of shock waves through matter (L.F. HENDERSON);
- theory of shock waves in gases (G. EMANUEL), in liquids (S. ITOH), in solids (K. NAGAYAMA), and in space (M. GEDALIN);
- spherical shock waves: expansion (J.M. DEWEY) and attenuation (F. AIZIK et al.);
- rarefaction shocks (A. KLUWICK);
- shock waves in porous media (S. SKEWS, A. LEVY & D. LEVI-HEVRONI) and in inert reactive bubbly liquids (V. KEDRINSKII);
- shock waves in channels (W. HEILIG & O. IGRA);
- 1-D shock wave interaction (O. IGRA) and 2-D shock wave interaction (G. BEN-DOR, L.F. HENDERSON & J.M. DÉLERY);
- shock wave interactions with inert granular media (A. BRITAN & A. LEVY) and with liquid-gas suspensions (M.E.H. VAN DONGEN);
- shock wave focusing (F. HIGASHINO);
- geometrical shock dynamics (Z.Y. HAN & X.Z. YIN);
- stability of shock waves (N.M. KUZNETSOV) and shock-induced instability at interfaces (D. SHVARTS ET AL.);
- shock tubes (M. NISHIDA), blast tubes (R. ROBEY), and free piston-driven shock tubes/tunnels (R. MORGAN);
- supersonic and hypersonic wind tunnels (B. CHANETZ & A. CHPOUN);
- applications of shock waves in medicine (A.M. LOSKE);
- flow visualization (H. KLEINE) and spectroscopic diagnostics (D.F. DAVIDSON & R.K. HANSON);
- numerical methods (P. ROE);
- chemical and combustion kinetics (8 authors); and
- detonation waves in gaseous explosives (J.H.S. LEE).

| | |
|---|---|
| 2001 *Dept. of Mechanical and Aerospace Studies, UCSD, La Jolla, California* | **Vitali F. NESTERENKO,**[3313] a Russian-born U.S. physicist and professor of Materials Science, **publishes his book entitled *Dynamics of Heterogeneous Materials*.** |

(i) It deals with the behavior of essentially nonlinear heterogeneous materials (such as granular materials, powders, and laminates) under intense dynamic loading (such as from shock waves, solitons and localized shear), where microstructural effects play the main role. The topic is treated from the viewpoints of nonlinear phenomena and materials science rather than taking the traditional approach of the Hugoniot equations.

(ii) **Resuming the problem of chain percussion in NEWTON's cradle** {Sir NEWTON ⇨1687; HERRMANN & SCHMÄLZLE ⇨1981; ⇨Fig. 4.4−B}, **he argues that "the chain of contacting balls is not *dispersion-free*, on the contrary it is characterized by a strongly nonlinear dispersion resulting in a unique *solitary wave*** [or *soliton*; RUSSELL ⇨1834] … So this toy (if long enough, say, 30 balls) can be a simple and excellent example of complex highly-nonlinear wave dynamics, in addition to the demonstration of energy and momentum conservation for small numbers of balls. In the latter case there is too little time for solitons to be

---

[3313] V.F. NESTERENKO: *Dynamics of heterogeneous materials*. Springer, New York etc. (2001); for *solitons* and *NEWTON's cradle see* pp. 31-33.

formed simply due to a small propagation distance and not due to the absence of dispersion."
▪ It was already shown by NESTERENKO[3314] in 1983 that an impulse initiated at one end of a chain of elastic grains in mutual contact – *i.e.*, interacting via the nonlinear Hertz potential {HERTZ ⇨1882} – travels as a soliton-like object. Theoretical, experimental and numerical studies carried out in the 1990s by other researchers have validated his findings.

| 2001 | *Dept. of Physics, State University of New York, Buffalo, New York* | Surajit SEN,[3315] an Indian-born U.S. associate professor of physics, **and Felicia S. MANCIU and Marian M. MANCIU,** two Ph.D. students, **investigate the propagation of an impulse through a finite chain of elastic beads in which the grain diameters progressively shrink.** Using computer simulations, **they show that it should be possible to construct "tapered" chains that can effectively thermalize shock waves;** *i.e.*, **to turn the absorbed energy from the shock wave into heat.**[3316] |

(i) Granular materials, including sand and soil, have long been used to absorb impacts, but if the grains are all the same size, the shock waves are not always dispersed effectively. Instead, they simulate a shock wave traveling along a chain of several hundred spherical elastic beads of ever-decreasing size. The beads at one end of the chain are around ten centimeters in diameter, and become progressively smaller.

(ii) After the shock wave has passed through the large sphere at the beginning of the chain, it proceeds to the next – slightly smaller – sphere. But the wave cannot be transmitted symmetrically into this sphere. To ensure that its energy is conserved, the wave is forced to stretch out. Its leading edge accelerates away from its trailing edge, and this effect occurs every time the wave moves from one bead to the next. As the beads get smaller, the energy of the impulse is distributed and successive beads carry less and less kinetic energy.

(iii) They find that the smallest bead at the other end of the chain feels the initial large impact as a long series of very small shocks. The amplitudes of these mini-shocks are less than 10% of the original impulse.

(iv) They conclude that this very simple system of nonlinear wave propagation demonstrates
  ▸ that in theory a shock of any size can be absorbed with assemblies of appropriately tapered chains; and
  ▸ that this concept could one day be exploited to reuse the energy from unwanted man-made mechanical vibrations and even natural shocks from geological activity.

In the following years, **their predictions about the shock-absorbing capabilities of these "tapered chain shock absorbers" were experimentally confirmed** by independent researchers at the Colorado School of Mines in collaboration with a group at NASA's Glenn Research Center (GRC),[3317] as well as by researchers at the University of Santiago in Chile and the Institut Supérieur de Mécanique de Paris (SUPMECA).[3318]

| 2001 | *Gas Dynamics Laboratory, Penn State University, Pennsylvani* | Gary S. SETTLES,[3319] a U.S. professor of mechanical engineering and director of the laboratory, **publishes his book *Schlieren and Shadowgraph Techniques*** – still the two most commonly used visualization methods in gas dynamics and high-speed aerodynamics. ▪ His monograph not only provides modern fluid dynamicists and gas dynamicists with the necessary historical background knowledge, but it also addresses more recent developments and modifications in visualization and registration techniques. |

---

[3314] V.F. NESTERENKO: *Propagation of nonlinear compression pulses in granular media.* J. Appl. Mech. Tech. Phys. **24**, No. 5, 733-743 (1983).
[3315] S. SEN, F.S. MANCIU, and M.M. MANCIU: *Thermalizing an impulse.* Physica A **299**, 551-558 (Oct. 2001).
[3316] K. PENNICOTT (ed.): *Super shock absorber could protect buildings.* PhysicsWeb (Oct. 26, 2001); http://physicsweb.org/articles/news/5/10/15/1.
[3317] M. NAKAGAWA, J.H. AGUI, S. SEN, and D. WU: *Impulse dispersion of a tapered granular chain.* American Physical Society, Annual APS March Meeting, held at the Indiana Convention Center; Indianapolis, IN (March 2002).
[3318] F. MELO ET AL.: *Experimental evidence of shock mitigation in a Hertzian tapered chain.* Phys. Rev. E **73** [III], Paper No. 041305 (2006).
[3319] G.S. SETTLES: *Schlieren and shadowgraph techniques.* Springer, Heidelberg (2001).

|   |   |   |
|---|---|---|
|   |   | In 1999, SETTLES[3320] **gave a historical perspective of outdoor schlieren and shadowgraph imaging.** He discussed the optical principles of the sunlight shadowgraph method and schlieren observation by background distortion and illustrated the versatility of outdoor visualization methods at numerous examples (*e.g.*, thermal convection, combustion, explosion, and shock wave phenomena). |
| 2001 | *Agricultural Research Service (ARS), U.S. Dept. of Agriculture, Beltsville, Maryland* | Morse B. SOLOMON,[3321] a food researcher at USDA-ARS, **reports in a lecture entitled *Dynamite Recipe for Tendering and Sanitizing Meat* on a new method using hydrodynamic pressure shock waves** from the detonation of an underwater explosive charge in a containment vessel.[3322] In the new patented process, developed in cooperation with Hydrodyne, Inc. (Houston, TX), the shock waves cause muscle fiber disruptions which in turn result in instantaneous meat tenderization and a reduction in the inconsistency in tenderness of meat products. ▪ **The traditional method of tenderizing meat by disruption of its ultrastructure is based on the principle of percussion:** a "meat hammer" is used, which was originally a simple stone or wooden mallet but has since developed into a modern metallic tool. This tool is provided with a spiked surface on one side to tenderize the meat and a flat surface on the opposite side to pound meat into a uniform thickness to promote even cooking. |
| 2002 | *ESA, Europe & NASA, United States* | On January 4, **LASCO** (Large Angle and Spectrometric Coronagraph) – an instrument developed by German and U.S. scientists, and installed on SOHO (Solar and Heliospheric Observatory) – **observes an immense eruption of electrically charged gas, a so-called "coronal mass ejection (CME)."**[3323] These billion-ton eruptions of plasma are blasted from the Sun at high speeds, ranging from about 200 to 2,500 km/s. ▪ A CME can be regarded as a giant shock wave of electrified gases. This complex phenomenon, discovered in the 1970s and a subject of intensive research since then, is not necessarily caused by an impulsive "breaking loose" of a large-scale coronal structure. Various theoretical models have been proposed to explain the initiation and the enormous time-dependent CME flow in the presence of solar gravity.[3324]<br><br>In the following year, on November 4, 2003, the most powerful CME in modern times occurred, causing widespread spectacular auroras but no dramatic power outages.[3325] It was also the fastest CME on record, moving at 2,700 km/s. It would have hit the Earth within 15 hours had it been aimed at it. |
| 2002 | *Planet Mars & NASA's Jet Propulsion Laboratory (JPL), Pasadena, California* | In March, **NASA's Mars Global Surveyor records color pictures of the two Martian volcanoes Ceraunius Tholus and Uranius Tholus with its Mars Orbiter Camera.** The presence of impact craters on these volcanoes indicates that they are quite ancient and not active today.[3326] ▪ Ceraunius Tholus is a volcano in the Tharsis region that was first viewed in images obtained by Mariner 9 in 1972.<br><br>Mars supports some of the largest volcanoes in the Solar System. Most of them are classified as shields with effusive eruption styles, but **some Martian volcanoes exhibit explosive** |

---

[3320] G.S. SETTLES: *Schlieren and shadowgraph imaging in the great outdoors*. In: (S. MOCHIZUKI, ed.) *Proc. 2nd Pacific Symposium on Flow Visualization & Imaging Processing (PSFVIP)* [Honolulu, HI, May 1999]. 1 CD-ROM (ISBN: 0-9652469-7-3), Honolulu, HI (1999), Paper No. PF302 (p. 14).

[3321] M.B. SOLOMON: *Dynamite recipe for tenderizing and sanitizing meat.* USDA Beltsville Area Distinguished Lecture Series, U.S. Dept. of Agriculture (USDA); http://www.ba.ars.usda.gov/lectures2003/solomon.html.

[3322] J.R. CLAUS: *Hydrodyne process research.* Dept. of Animal Science and Food Science, College of Agricultural & Life Sciences, University of Wisconsin, Madison, WI (Feb. 3, 2003); http://www.ansci.wisc.edu/facstaff/Faculty/pages/claus/hyd/hydhome.htm.

[3323] *Sun erupts with an extraordinary mass ejection.* NASA-GSFC, Greenbelt, MD (Jan. 5, 2002); http://www.spaceflightnow.com/news/n0201/05soho/.

[3324] B.C. LOW: *Solar coronal mass ejection: theory.* In: (P. MURDIN, ed.) *Encyclopedia of astronomy and astrophysics.* Institute of Physics Publ., London *etc.*, vol. 3 (2001), pp. 2498-2502.

[3325] *Speed limit found for solar storms.* NASA-GSFC, Greenbelt, MD (June 14, 2004); http://www.space.com/scienceastronomy/solarstorm_speed_040614.html.

[3326] *Volcanoes Ceraunius Tholus and Uranius Tholus.* Malin Space Science Systems, San Diego, CA (April 18, 2002); http://www.msss.com/mars_images/moc/news2002/ceraunius/index.html.

eruption characteristics and probably consist of ash deposits. Examples include the about 135-km-dia. 5-km-high *Ceraunius Tholus,* the about 300-km-dia. 2-km-high *Tyrrhena Patera* and the about 60-km-dia. 1–2-km-tall *Hadriaca Patera.*[3327]

| | | |
|---|---|---|
| 2002 | Naval Sea Systems Command (NSSC), Washington, DC | On March 7, **David NAGLE**,[3328] chief journalist at NSSC Public Affairs, **reports on the development of *thermobaric bombs*** at the Naval Surface Warfare Center (NSWC), Indian Head Division, MD. ▪ Thermobaric weapons consist of a container of fuel and two explosive charges. When the first charge detonates, the fuel is very finely dispersed; a second charge then detonates the billowing cloud of fuel. Typically, the strong blast of the explosion is followed by a deep vacuum which is formed after the explosion cloud (a distinctive mushroom cloud commonly associated with a nuclear explosion) sucks out the air and solid particles. **Thermobaric bombs are closely related to *fuel-air explosives,*** because they are fuel-rich explosives that use oxygen from the target to create a longer burn time than conventional explosives, creating sustained temperatures and overpressures in confined structures.<br><br>Thermobaric bombs – known as "vacuum bombs" in Russia – were used in the early 2000s by the U.S. Army in the Afghanistan War to kill or injure people hiding in caves, bunkers, and similar shelters. According to Russian military sources, Russia tested the world's most powerful vacuum bomb in September 2007. The "super vacuum bomb" had the destructive power of a nuclear device, however, minus the long term radiation effects. |
| 2002 | *Institut für Kurzzeitdynamik, Ernst-Mach-Institut (EMI) & Institut für Werkstoffmechanik (IWM), Freiburg, Germany* | In April, **establishment of *crash*MAT, the Freiburg Center for Crash-Relevant Materials Characterization,** by Klaus THOMA (EMI) and Peter GUMBSCH (IWM).[3329] ▪ So far, *crash*MAT workshops have been held at Freiburg in 2002, 2004, and 2006.<br><br>*Crash*MAT is the result of a cooperation between EMI and IWM, two institutes of the Fraunhofer-Gesellschaft (FhG), a German society dedicated to contracted research in all fields of the engineering sciences, with headquarters at Munich.[3330] The objectives of the *crash*MAT workshops are<br>▸ to improve crash safety through the selective use of appropriate materials and numerical simulation, thereby covering a wide range of impact velocities (20–6,000 m/s); and<br>▸ to provide an open forum of discussion for industry (mainly for the automobile industry), as well as for institutions engaged in research and education. |
| 2002 | *Moon's orbit & Earth* | On June 14, **asteroid 2002 MN approaches the Earth at a distance of only 120,000 km at a velocity of about 10 km/s,** one of the closest approaches to Earth ever recorded, bringing it well inside the Moon's orbit (363,104–405,696 km). This near-Earth object (NEO), a giant rock about 100 meters in length, is first detected on June 17 by astronomers working at the Lincoln Laboratory in Socorro, NM, on the search project LINEAR (Lincoln Near-Earth Asteroid Research) – *i.e.,* three days *after* its closest approach.[3331] ▪ Asteroid 2002 MN is a lightweight among asteroids and is incapable of causing damage on a global scale. However, if it had hit the Earth, it may have caused local devastation similar to that produced by the Tunguska asteroid {⇨1908}. In recent decades, a close near-collision also took place in 1994, when asteroid 1994 XL1 passed within 105,000 km of our planet.<br><br>**The closest approach of an asteroid was discovered on March 31, 2004: the asteroid 2004 FU162 (about 10 m long) passed the Earth at a distance of only about 6,500 km.** It |

---

[3327] C.J. HAMILTON: *Martian volcanoes.* Views of the Solar System; http://www.solarviews.com/eng/marsvolc.htm.
[3328] D. NAGLE: *Navy scientists instrumental in developing new thermobaric weapons.* Navy NewsStand, Story No. NNS020307-03 (March 7, 2002); http://www.news.navy.mil/search/display.asp?story_id=999.
[3329] Fraunhofer Kompetenzzentrum crashMAT, Freiburg, Germany; http://www.emi.fraunhofer.de/EMI-Links/crashMAT/index.asp.
[3330] Fraunhofer-Gesellschaft (FhG) zur Förderung der angewandten Forschung e.V., Munich, Germany (Nov. 2005); http://www.fraunhofer.de/fhg/EN/index.jsp.
[3331] R. STENGER: *Surprise asteroid nearly hits home.* CNN Press release (June 21, 2002); http://archives.cnn.com/2002/TECH/space/06/20/asteroid.miss/.

would have exploded harmlessly in the upper atmosphere had it collided. ▪ The threshold asteroid diameter for an impact that would cause widespread global mortality and threaten civilization almost certainly lies between about 0.5 and 5 km, perhaps near 2 km. Impacts of objects this large occur several times or less per million years.

In 2003, after a year of analysis by scientists working on this issue, NASA released a technical report on potential future NEO search efforts.[3332] In 2005, Edward T. Lu[3333] and Stanley G. Love from NASA's Johnson Space Center presented a design concept for a spacecraft that would hover near the asteroid and, relying on gravitational force alone, would nudge the asteroid onto a path that avoids the Earth.

| | | |
|---|---|---|
| 2002 | *Woomera Test Facility, Woomera, South Australia* | On July 30, **the world's first successful supersonic combustion engine or scramjet test flight is performed** {BILLIG ⇨ 1959}, using gaseous hydrogen as a fuel. The scramjet engine was accelerated by a Terrier Orion Mk-70 two-stage rocket to a speed of Mach 7.6 to an altitude of 35 km, and the test took place within the last few seconds of the flight.[3334] ▪ The rocket flight was part of Australia's HyShot Flight Program, an effort involving in-flight tests of scramjet technology in order to validate experiments performed in ground test facilities such as the T4 shock tunnel at the Centre for Hypersonics of the University of Queensland. |
| 2002 | *Apopka and Sorrento, Florida* | On November 5, **Jay LIPELES**[3335] **and Glenn BROSCH,** two U.S. inventors, **obtain a patent on a *guided bullet.*** It relies on flexible circuit elements ("flaps") to serve as control surfaces to guide the bullet to its intended target. ▪ In the abstract of the patent, their invention is described as follows: "A projectile having a plurality of MicroElectroMechanical System (MEMS) devices disposed about the axis of flight for active control of the trajectory of the projectile. The MEMS devices each form an integral control surface/actuator. Control circuitry installed within the projectile housing includes both rotation and lateral acceleration sensors. Flap portions of the MEMS devices are extended into the air stream flowing over the projectile in response to the rate of rotation of the projectile, thereby forming a standing wave of flaps operable to impart a lateral force on the projectile…"<br><br>**There are various U.S. MEMS projects in transonic and supersonic aerodynamics.** They involve the development of sensors for the measurement of acceleration, velocity and orientation, with obvious military applicability to missile and projectile guidance:[3336]<br>▸ The University of Michigan investigated electrokinetic fluid microactuators for boundary layer drag reduction and on-demand leading edge vortex generation for control of aircraft.<br>▸ Stanford University also investigated MEMS-based transducers for active boundary layer control to reduce drag in turbulent flows.<br>▸ The Arizona State University worked on a project for a similar application, but using distributed arrays of micropistons to vary the roughness on the leading edges of swept wings to control the transition to turbulence. |
| 2002 | *Electric Power Research Institute (EPRI), Palo Alto, California* | **EPRI,** a U.S. non-profit energy research consortium, **undertakes an advanced computer modeling study to determine whether buildings at nuclear power plants can withstand the impact of an aircraft crash** similar to those seen in the September 11, 2001 terrorist at- |

---

[3332] NASA-JPL: *Study to determine the feasibility of extending the search for Near-Earth-Objects to smaller limiting diameters* (Aug. 22, 2003). Rept. of the Near-Earth Object Science Definition Team; http://neo.jpl.nasa.gov/neo/neoreport030825.pdf.
[3333] E.T. Lu and S.G. Love: *Gravitational tractor for towing asteroids*. Nature **438**, No. 7065, 177 (2005).
[3334] P. Woolnough: *HyShot program secures place in flight history*. Engineers Australia **74**, No. 9, 30-31 (2002).
[3335] J. Lipeles and G. Brosch: *Guided bullet*. U.S. Patent No. 6,474,593 (Nov. 5, 2002).
[3336] A. White: *A review of some current research in Microelectromechanical Systems (MEMS) with defence applications*. Rept. DSTO-GD-0316, Dept. of Defence, Defence Science & Technology Organisation, Australia; http://www.dsto.defence.gov.au/publications/2430/DSTO-GD-0316.pdf.

tacks {WTC ⇨2001}. ▪ Their building integrity study used advanced computer modeling and the most adverse assumptions:

> ▸ A Boeing 767 was selected as the aircraft because its maximum takeoff weight (450,000 pounds including 23,980 gallons of fuel) is greater than almost all other commercial jet airliners flown in the United States, and because over two-thirds of the commercial aircraft registered in this country are manufactured by Boeing.
>
> ▸ Both the engine and the aircraft fuselage were assumed to strike perpendicular to the surface of the wall.
>
> ▸ The study used a reasonable, controllable aircraft speed of 350 mph (563 km/h) for the strike analyzed.

**The analyses showed that nuclear plants should withstand aircraft crashes and would not be penetrated by an aircraft crash;** *i.e.*, that no radiation would leak from these structures. Because the structures used in nuclear plants are smaller than the WTC buildings, it is more difficult to aim the airplane such that it hits the structure at its most damaging point.[3337]

| 2002 | *Dept. of Aerospace Engineering, Indian Institute of Science (IISc), Bangalore, India & Shock Wave Research Center, Institute of Fluid Science, Tohoku University, Tohoku, Japan* | **Gopalan JAGADEESH,**[3338] an assistant professor at IISc, **and Kazuyoshi TAKAYAMA,** a fluid dynamics professor at Tohoku University and one of Japan's foremost shock wave physicists, **report on new applications of micro-shock waves** – *i.e.*, of spherical shock waves with typical radii of only a few millimeters, both in ambient air as well as in water with peak pressures in the range 1–100 MPa (10–1,000 bar).<br><br>(i) Introductorily they review the past and possible future applications of micro-shock waves in medical and biological sciences.<br><br>(ii) They briefly review the four important techniques commonly used for generating micro-shock waves which encompass<br><br>▸ pulsed laser beam focusing {BELL & LANDT ⇨1967; NAKAHARA & NAGAYAMA ⇨1999};<br><br>▸ generation of underwater shock waves by using micro-explosives (lead azide);<br><br>▸ generation of underwater shock waves by using electric discharges in water, so-called "electrohydraulic effect" {JUTKIN ⇨1950}; and<br><br>▸ application of an array of piezo-ceramic crystals, placed in a parabolic reflector and simultaneously subjected to a high-voltage pulse.<br><br>(iii) Various current applications of micro-shock waves are discussed, such as in food industry (destruction of *E. coli* bacteria), in agriculture (clearance of vessel blockages in timber), and in laser ablation-assisted micro-particle delivery systems (for gene therapy and needle-free drug delivery applications). |
| --- | --- | --- |
| 2002 | *Dept. of Theoretical & Applied Mechanics, University of Illinois, Urbana-Champaign, Illinois* | **D. Scott STEWART,**[3339] a professor of computational and applied fluid dynamics, **discusses theoretical, computational and experimental requirements of micro-detonics,** when conventional explosive systems with a charge on the order of a meter or a sizable fraction of a meter have to be scaled downwards by a factor of 100 or 1,000.<br><br>Miniaturization of exploding systems – so-called "micro-detonics" – is especially important for the design of new munitions for unmanned air vehicles and other military applications, but micro-sized explosive systems may also have applications in mining and demolition industries or be useful in very small scale materials processing (*e.g.*, surface treatment, and hardening of materials). The engineering and control of detonation fronts is a basic technology that requires the creation of new science and unique application of existing theory. An order-of magnitude |

---

[3337] *Security effectiveness: independent studies and drills.* Nuclear Energy Institute (NEI), Washington, DC; http://www.nei.org/index.asp?catnum=2&catid=279.
[3338] G. JAGADEESH and K. TAKAYAMA: *Novel applications of micro-shock waves in biological sciences.* J. Indian Inst. Sci. **82**, 1-10 (2002).
[3339] D.S. STEWART: *Towards miniaturization of explosive technology.* Shock Waves **11**, 467-473 (2002).

reduction of conventional explosive systems to micro systems with an overall size of about 1 cm presents extraordinary challenges to scientists and requires them to re-examine concepts established for large explosives when applied in the design of new miniaturized systems, such as

- the unsteady detonation shock transients associated with the initiation of a small explosive charge;
- the detonation acceleration (buildup) and steadiness of the detonation front during its propagation phase;
- the large energy loss in the detonation reaction zone due to lateral flow expansion;
- the requirement of providing explosive materials with a short detonation reaction zone (in relation to the overall dimension of the device), and the associated problems of safety; and
- the connection between unsteadiness and the high wave-front curvature in miniature explosive systems.[3340]

The development of miniaturized explosive systems that used condensed phase energetic materials can lead to a basic enabling technology with the potential for widespread applications. The output pressure and thermodynamic processing developed by micro-explosive systems are unique and can enable a new ubiquitous basic technology for materials processing and other applications at small scales.[3341]

| | | |
|---|---|---|
| 2003 | *NASA's Goddard Space Flight Center (GSFC), Greenbelt, Maryland* | On February 11, **NASA-GSFC releases a "baby" picture of the Universe** which captures the afterglow – the cosmic microwave background radiation of the Big Bang {The Universe ⇨ c.14 Ga ago, SMOOT ⇨ 1992; ⇨ Fig. 4.1 W}. |

(i) NASA scientists used NASA's Wilkinson Microwave Anisotropy Probe {WMAP ⇨ 2001} for a sweeping 12-month observation of the entire sky. By best-fitting a cosmological model to the cosmic microwave background, the WMAP Science Team[3342] reports in the *Astrophysical Journal* that **"the age of the best-fit Universe is $t_0$ = 13.7 ± 0.2 Gyr old"** (1 Gyr = 1 Giga-year = $10^9$ years).

(ii) WMAP's results also show that the cosmic background radiation was released about 389,000 years after the Big Bang (later than previously thought), and that the first stars formed about 200 million years after the Big Bang (earlier than anticipated). According to NASA's announcement in 2003,[3343] "the 'baby' picture recorded allows scientists to accomplish three main things:

- to reach back to earlier times to see what produced these patterns;
- to look forward from the time of the picture to predict how the Universe would develop; and
- to compare this to what is observed by other means (with galaxies, supernovae, *etc.*) to get the cosmic consistency."

ESA's Planck satellite, scheduled for 2007, will map the cosmic microwave background anisotropies as fully and accurately as possible and make very sensitive measurements of the temperature over the whole sky at nine frequencies between 30 and 900 GHz.[3344] **One of the key questions Planck will answer is: Will the Universe continue its expansion forever, or will it collapse into a "Big Crunch"?**

---

[3340] D. LAMBERT and D.S. STEWART: *Miniaturization of explosive systems technology*. AFRL Technology Horizons Magazine, Air Force Research Laboratory (AFRL), Wright-Patterson AFB, OH (August 2004); http://www.afrlhorizons.com/Briefs/Aug04/MN0313.html.

[3341] D.S. STEWART and R.J. ADRIAN: *Miniaturization of explosive systems and microdetonics*. 56th Ann. Div. Fluid Dynamics Meeting [East Rutherford, NJ Nov. 2003], Bull. Am. Phys. Soc. **48**, No. 10 (2003), Paper No. AD.001.

[3342] C.L. BENNETT ET AL.: *First year Wilkinson Microwave Anisotropy Probe (WMAP) observations: preliminary maps and basic results*. Astrophys. J. **148**, No. 1, 1-27 (Sept. 2003).

[3343] *The Wilkinson Microwave Anisotropy Probe (WMAP) charting the new cosmology. Big Bang and WMAP primer*. NASA-GSFC, Greenbelt, MD (Feb. 2003); http://www.gsfc.nasa.gov/gsfc/spacesci/pictures/2003/0206mapresults/MAPprimer.pdf.

[3344] *ESA Planck Spacecraft*. The University of Manchester, Jodrell Bank Observatory (Jan. 19, 2005); http://www.jb.man.ac.uk/research/cmb/planck.html.

| | | |
|---|---|---|
| 2003 | *Leo ["The Lion," a spring constellation of the Northern Hemisphere]* | On March 29, **a very bright burst of gamma-rays** – later designated GRB 030329 – **is observed by NASA's High Energy Transient Explorer** (HETE) **in the constellation *Leo*** at a distance of about 2,650 million light-years.[3345, 3346]<br><br>(i) Within 90 minutes, a new, very bright light source – the *optical afterglow* – is detected by telescopes in Australia and Japan.<br><br>(ii) Spectra taken of the optical afterglow of this gamma-ray burst show the gradual and clear emergence of a supernova spectrum of the most energetic class known – a hypernova caused by a huge explosion of a very heavy star (over 25 times heavier than the Sun). Measurements reveal that the explosion expands at a very high velocity (> 30,000 km/s), and that the total energy released was exceptionally high.<br><br>(iii) The observation of GRB 030329 evidently demonstrates that long-duration (> 2 s) gamma-ray bursts, the most luminous of all astronomical explosions, signal the collapse of massive stars in our Universe.<br><br>(iv) **The association of a hypernova with GRB 030329 strongly supports the collapsar model of gamma-ray bursts** {TeraScale Initiative ⇒ 2001} – *i.e.*, **shock breakout and relativistic shock deceleration in circumstellar material.**<br><br>The most popular model to account for the production of gamma-ray bursts involves the collapse of the core of a special kind of star to form a black hole. This core collapse occurs while the outer layers of the star explode in an especially energetic supernova dubbed a "hypernova" by astronomers. This model has been termed the ***collapsar model***.[3347] The term *collapsar* was coined in 1993 by Stanford E. WOOSLEY,[3348] an astronomy and astrophysics professor at UC Santa Cruz, CA who advocated such a model with great vigor. Since 1996 his work on the collapsar has been done jointly with his graduate student Andrew I. MACFADYEN.[3349] |
| 2003 | *The 846th Test Squadron, Holloman Air Force Base, Alamogordo, New Mexico* | On April 30, **a record-breaking hypersonic sled test is conducted at the Holloman High-Speed Test Track (HHSTT).** The sled train, accelerated by a Super Roadrunner rocket motor, delivers a 192-pound (87-kg) payload into a target at a velocity of 9,465 ft/s (2,885 m/s, about Mach 8.6).[3350] ▪ This test broke the previous world record, which was also established at HHSTT in 1982, when the sled train traveled at 8,974 ft/s (2,735 m/s) with a 25-pound (11.34-kg) payload.<br><br>**Sled vehicles, accelerated by single- or multiple-stage rockets, have proven very useful for a variety of hypersonic tests.** HHSTT at Holloman Air Force Base is presently the only ground facility in the world where full-scale test items can be studied at true supersonic or hypersonic conditions.[3351] |

---

[3345] *Cosmological gamma-ray bursts and hypernovae conclusively linked.* European Southern Observatory (ESO), Garching, Germany. Press release (June 18, 2003); http://www.eso.org/outreach/press-rel/pr-2003/pr-16-03.html.

[3346] J. HJORTH ET AL.: *A very energetic supernova associated with the γ-ray burst of 29 March 2003.* Nature **423**, 847-850 (June 19, 2003).

[3347] S.E. WOOSLEY, R.G. EASTMAN, and B.P. SCHMIDT: *Gamma-ray bursts and type Ic supernova SN 1998bw.* Astrophys. J. **516**, 788-796 (1999).

[3348] S.E. WOOSLEY: *Gamma-ray bursts from stellar mass accretion disks around black holes.* Astrophys. J. **405**, 273-277 (March 1993).

[3349] A.I. MACFADYEN and S.E. WOOSLEY: *Collapsars – gamma-ray bursts and explosions in "Failed supernovae."* Astrophys. J. **524**, 262-289 (1999).

[3350] J. BOWNE: *"Super Roadrunner" breaks land speed record.* RDECO Magazine, a publication of the U.S. Army Research Development Command (Jan. 2004); http://www.rdecom.army.mil/rdemagazine/200401/itl_amrdec_roadrunner.html.

[3351] L. WEINSTEIN and D. MINTO: *Focusing schlieren photography at the Holloman High Speed Test Track.* In: (D.L. PAISLEY, ed.) *22nd International Congress on High-Speed Photography and Photonics* [Santa Fe, NM, 1996]. Proc. SPIE **2869**, 865-873 (1996).

| | | |
|---|---|---|
| 2003 | *Khomas Highland [ca. 100 km from Windhoek], Namibia, southern Africa* | In the period from May to August, the HESS,[3352] the world largest gamma-ray telescope array, monitors the supernova remnant (SNR) RX J1713.7-3946. Based upon these data, **Astronomers discover a ring-like source of high-energy γ-rays of TeV energies in the sky in the Southern Hemisphere,** which is centered upon RX J1713.7-3946.[3353, 3354]<br><br>(i) RX J1713.7-3946 − one of the brightest galactic X-ray SNR known and located in *Scorpius* ("The Scorpion," a large constellation of the Southern Hemisphere) − was already discovered in 1995 during an all-sky survey by ROSAT (Roentgen Satellite), the German/British/American space-based X-ray telescope.[3355] The spatially resolved remnant has a shell morphology similar to that seen in X-rays, which indicates that very-high-energy particles are accelerated there.<br><br>(ii) RX J1713.7-3946 is believed to be about 1,000 years old, or to be the remnant of the A.D. 393 guest star {Guest Star ⇒ 393} as proposed by Chinese astronomers in 1997.[3356]<br><br>(iii) **The energy spectrum indicates efficient acceleration of charged particles** (mainly protons) **to energies beyond 100 TeV ($10^{14}$ eV), consistent with current theories for particle acceleration in young SNR shocks.** ▪ Electrons are known to be accelerated to cosmic-ray energies in supernova remnants, and the shock waves associated with such remnants.[3357] |
| 2003 | *Dept. of Physics, MIT, Cambridge, Massachusetts* | On May 23, **Evan J. REED**,[3358] a U.S. physicist and postdoctoral associate in MIT's Research Laboratory of Electronics, **and collaborators report on new physical phenomena which occur when light interacts with a shock wave propagating through a periodic optical (nano)structure − so-called "photonic crystal."** These new phenomena include<br><br>▸ the capture of light at the shock wave front and re-emission at a tunable pulse rate and carrier frequency across the bandwidth; and<br><br>▸ bandwidth narrowing, as opposed to the ubiquitous bandwidth broadening.<br><br>If the shock wave travels in the opposite direction to the light, the frequency of the light will get higher. If the wave travels in the same direction, the frequency of the light drops. By changing the way the crystal is constructed, researchers could control exactly which frequencies travel into the crystal and which come out − an effect which may have significant technological implications. |
| 2003 | *Guiana Space Centre [near Kourou], French Guiana, northern South America* | On September 27, **launch of ESA's spacecraft SMART-1** (Small Missions for Advanced Research in Technology). The spacecraft is designed to test solar electric propulsion and to perform scientific observations of the Moon.<br><br>On September 3, 2006, SMART-1 ended its mission in the course of its 2,890th orbit by being deliberately crashed onto the Moon's surface with a velocity of about 2,000 m/s.<br><br>(i) A preliminary analysis revealed that the spacecraft impacted the Moon in the ascending slope of a mountain about 1.5 km high, above the Lake of Excellence which is situated in the mid-southern region of the Moon. It produced an impact flash and subsequently a dust cloud of ejected material or debris. |

---

[3352] The name HESS (High Energy Stereoscopic System) honors Victor F. HESS {⇒ 1912}, the discoverer of cosmic rays {GINZBURG & SYROVATSKII ⇒ 1961; ISEE Program ⇒ 1977}. HESS which allows to explore cosmic gamma rays in the 100 GeV energy range consists of four 13-m-dia. telescopes spaced at the corners of a square of side length 120 m. HESS was awarded the 2006 Descartes Prize for Research of the European Commission.

[3353] D. BERGE ET AL.: *Observations of SNR RX J1713.7-3946 with H.E.S.S.* In: (F.A. AHARONIAN, ed.) *High energy gamma-ray astronomy: 2nd Int. Symposium on High Energy Gamma-Ray Astronomy* [Heidelberg, Germany, July 2004]. American Institute of Physics (AIP), Melville, NY (2005), pp. 263-268.

[3354] F.A. AHARONIAN ET AL.: *High-energy particle acceleration in the shell of a supernova remnant.* Nature **432**, 75-77 (Nov. 4, 2004).

[3355] E. PFEFFERMANN and B. ASCHENBACH: *ROSAT observation of a new supernova remnant in the constellation Scorpius.* In: (H.U. ZIMMERMANN ET AL., eds.) *Rontgenstrahlung from the Universe.* Proc. Int. Conf. on X-Ray Astronomy and Astrophysics [Würzburg, Sept. 1995]. Max-Planck-Institut für Extraterrestrische Physik (MPE), Garching (1996), Rept. 263, pp. 267-268.

[3356] Z.R. WANG, Q.Y. QU, and Y. CHEN: *Is RX J1713.7-3946 the remnant of the A.D. 393 guest star?* Astron. Astrophys. **318**, L59-L61 (Feb. 1997).

[3357] K. KOYAMA ET AL.: *Evidence for shock acceleration of high-energy electrons in the supernova remnant SN 1006.* Nature **378**, 255-258 (1995).

[3358] E.J. REED ET AL.: *Color of shock waves in photonic crystals.* Phys. Rev. Lett. **90**, Paper No. 203904 (2003).

(ii) The impact flash was observed in the infrared (at 2.2 μm) by the Canada-France-Hawaii Telescope (CFHT).[3359] In order to determine what part of the flash came from the lunar rock heated at impact or from the volatile substances released by the probe, it is important to also consider measurements in several optical and other infrared wavelengths which requires an international cooperation of both professional and amateur astronomers.[3360]

(iii) A new map of the lunar north pole was obtained with images taken by the Advanced Moon Imaging Experiment (AMIE) on board of SMART-1. Peary, an about 73-km-wide impact crater, is closest to the north pole – an area most likely to be permanently sunlit and therefore the best spot to settle on the Moon.

| | | |
|---|---|---|
| 2003 | *Outer limit of the Solar System & NASA's Jet Propulsion Laboratory (JPL), Pasadena, California* | In November, **NASA's space probe Voyager 1 approaches a hypothetical, quasi-perpendicular boundary – the so-called "heliospheric termination shock"** (or "solar wind termination shock"). ▪ **Voyager 2 is also escaping the Solar System** at a speed of about 3.3 AU per year (15.7 km/s). In 2006, Voyager 2 found that the shock in the south is a source of low energy ions as was discovered by Voyager 1 in the north.[3361] |

(i) At this region, often considered to mark the "edge" of the Solar System,
- the supersonic particles of the solar wind, which continuously stream out from the Sun with an average supersonic speed of 300–700 km/s {PARKER ⇨ 1961; NEUGEBAUER & SNYDER ⇨ 1962}, first start to plough into the thin interstellar gas; and
- the solar wind, which is therefore slowed down to subsonic velocities, becomes denser and hotter, causing an increase in the strength of the magnetic field carried by the solar wind.

(ii) Quantitative information on the distance of this termination shock from the Sun, which also depends on the Sun's activity, would allow scientists to estimate the size of the ***heliosphere***, the immense magnetic bubble containing the Solar System, the solar wind, and the entire solar magnetic field.

Astrophysicists expected that the termination shock may be responsible for the acceleration of interstellar particles which are ionized in the heliosphere and become charged with energies of the order of 20–300 MeV, a hypothesis known as the "anomalous cosmic-ray component" (*i.e.*, cosmic rays with unexpectedly low energies).[3362] ▪ Contrary to earlier predictions, however, neither Voyager 1 nor 2 have found the source of higher energy anomalous cosmic rays.

In the summer of 2002, when Voyager 1 – the "ultimate" man-made hypersonic craft – was about 85 AU (ca. 12.75 billion km) from the Sun, it measured a sharp increase in the number of energetic particles over a period of six months – possibly measuring precursor energetic ions and electrons from the heliospheric termination shock. In the following year, this phenomenon was interpreted by a team of astronomers[3363] at the Applied Physics Laboratory of Johns Hopkins University (Laurel, MD) as the termination shock. However, another team[3364] of astronomers at College Park, University of Maryland, believed that Voyager 1 had not yet reached the termination shock, pointing out that the probe did not register large increases in the local magnetic field. At that time, estimations of the distance to the termination shock from the Sun ranged from 85 to 120 AU (ca. 12.75 to 18 billion km).

In December 2004, Voyager 1 observed the magnetic field strength increasing by a factor of 2.5, as expected when the solar wind slows down {Voyager 1 ⇨ 2004}.

---

[3359] C. VEILLET: *Observation of the impact of SMART-1*. Canada-France-Hawaii Telescope (CFHT) Corporation, Kamuela, HI (2006); http://www.cfht.hawaii.edu/News/Smart1/#Impact.

[3360] *SMART-1 impact flash and debris: crash scene investigation*. ESA News (Sept. 7, 2006); http://www.esa.int/esaCP/SEMWX03VRRE_index_0.html.

[3361] *New discoveries at the edge of the Solar System*. NASA News Release (May 23, 2006); http://www.nasa.gov/vision/universe/solarsystem/voyager_2006agu.html.

[3362] M.E. PESSES, D. EICHLER, and J.R. JOKIPII: *Cosmic ray drift, shock wave acceleration, and the anomalous component of cosmic rays*. Astrophys. J. Lett. **246**, L85-L88 (June 1, 1981).

[3363] S.M. KRIMIGIS ET AL.: *Voyager 1 exited the solar wind at a distance of* ≈ 85 AU *from the Sun*. Nature **426**, 45-48 (2003).

[3364] F.B MCDONALD ET AL.: *Enhancement of energetic particles near the heliospheric termination shock*. Nature **426**, 48-50 (2003).

| | | |
|---|---|---|
| 2003 | NASA's Goddard Space Flight Center (GSFC), Greenbelt, Maryland | On November 4, **an X28 class solar flare which is the largest flare on record – so-called "monster blast" – erupts at the right edge of the Sun.**[3365] ▪ The giant flare, detected by NASA's Solar Radiation and Climate Experiment (SORCE) satellite, was an explosion of particles and radiation that zapped Earth. It disrupted satellites and other transmissions and triggered an intense geomagnetic storm, enabling the northern lights to be seen as far south as Arkansas, Texas and Oklahoma.<br><br>In the following year, on December 27, **the magnetar SGR 1806–20** – a rapidly spinning neutron star with an extremely powerful magnetic field which is about 50,000 light-years away in the constellation *Sagittarius* ("The Archer") – **produced a monster blast of radiation** that made it, for a tenth of a second, brighter than the full Moon, the brightest object ever seen outside the Solar System. |
| 2003 | Dept. of Mechanical Engineering, University of Cape Town, South Africa | **Establishment of the Blast Impact and Survivability Research Unit (BISRU)** with the objective "to reduce risk of injury and save lives through fundamental principles of science and engineering, using experimental, analytical and computational tools and techniques to understand the mechanics and dynamics of blast and impact loads."[3366] ▪ Research at the University of Cape Town in the field of blast, impact and survivability already began in the mid-1970s. BISRU has also developed collaborative links with industry and academia on national and international levels. |
| 2003 | United States & Russia | **Tarabay ANTOUN** (LLNL, Livermore, CA), **Lynn SEAMAN and Donald R. CURRAN** (Poulter Laboratory of SRI International, Menlo Park, CA), **Gennady I. KANEL'** (High-Energy Density Research Center, United Institute of High Temperatures, Moscow, Russia), **and Sergey V. RAZORENOV** (Institute of Problems of Chemical Physics, Chernogolovka, Russia) **jointly publish a book entitled *Spall Fracture*.**[3367] An outgrowth of American-Russian cooperation over the years, it reviews recent investigations into ***spall* – a term** referring to flakes of a material that are broken off a larger solid body such as produced by projectile impact; *i.e.*, **related to a high-speed dynamic fracture phenomenon** {B. HOPKINSON ⇒1912}. The book summarizes experimental techniques, mesomechanical constitutive modeling of the failure process, and provides a library of data and constitutive model parameters for numerous important engineering materials. |
| 2003 | Dept. of Mechanical Engineering, Université de Sherbrooke, Québec, Canada | **Martin BROUILLETTE,**[3368] a professor of mechanical engineering, **discusses what is required for the miniaturization of new devices which, ranging in the micrometer to millimeter range, make use of shock waves and shock-assisted combustion.**<br><br>(i) He presents a model for the effects of scale, via molecular diffusion phenomena, on the generation and propagation of shock waves. A simple parametrization of shear stress and heat flux at the wall leads to the determination of new jump conditions, which show that, for a given wave Mach number at small scales, the resulting particles velocities are lower but the pressures are higher. The model predict<br>▸ that, for a given wave Mach number at small scales, the resulting particle velocities are lower but the pressures are higher; and<br>▸ that the flow at small scale is isothermal and that the minimum wave velocity can be subsonic.<br><br>(ii) In his experimental study he uses a miniature shock tube (inner dia. 5.3 mm, lengths of driver section 8 cm and of driven section 45 cm). With various combinations of driver pressures (3–30 bar) and initial driven pressures (0.001–1 bar), diaphragm pressure ratios ranging |

---

[3365] *Monster blast.* StarDate online, Austin, TX (Nov. 24, 2005); http://stardate.org/resources/gallery/gallery_detail.php?id=567.
[3366] Dept. of Mechanical Engineering, University of Cape Town, South Africa; http://www.meceng.uct.ac.za/research/research.php#bisru.
[3367] T. ANTOUN: *Spall fracture.* Springer, New York *etc.* (2003).
[3368] M. BROUILLETTE: *Shock waves at microscales.* Shock Waves **13**, No. 1, 3-12 (2003).

from 10 to 10,000 could be produced, thus spanning a wide range of Reynolds numbers (100–100,000). Tests using low pressures to simulate the effect of small scale have shown qualitative agreement with the proposed model.

(iii) From his applied model and experiments he concludes that – contrary to conventional large-scale shock tubes – shock tubes at microscales are very inefficient to generate high flow velocities and temperatures, but useful to produce a large pressure increase over a short period of time.

**Miniature shock-tubes, ranging from some centimeter to submillimeter diameters, have been used for quite different purposes:**

- for gene and drug delivery in hand-held clinical systems {LEE ET AL. ⇨ 1999};
- for visualizing the collapse of a confined accelerated air cavity in water;[3369]
- for studying the temporal evolution of the nonlinear growth of Richtmyer-Meshkov instabilities;[3370] and
- for investigating the nonlinear growth of an unstable density interface from single-mode initial perturbations at high Mach number[3371] and hydrodynamic instabilities in doubly shocked systems[3372] using the high-power Nova laser at LLNL.

The creation and the propagation of a shock wave in a micro shock tube has been investigated numerically for different initial pressure ratios and diameters. Due to the formation of a boundary layer the attenuation of the shock wave increases with decreasing tube diameter. The shock wave can even vanish and transform into a compression wave along the tube.[3373]

2003 — Max-Planck-Institut für Metallforschung (MPI-MF), Stuttgart, Germany & IBM Research Division, Almaden Research Center, San Jose, California

In a cooperative study, **Markus J. BUEHLER** (MPI),[3374] **Farid F. ABRAHAM** (IBM), **and Huajian GAO** (MPI) **derive the conditions under which hyperelasticity governs dynamic fracture by combining theoretical considerations and large-scale molecular dynamics simulations** {ABRAHAM & GAO ⇨ 2000}.

(i) It is only when the deformation is infinitesimally small that elastic moduli can be considered constant, and hence the elasticity linear. Yet, many existing theories model fracture using linear elasticity, despite the fact that materials will experience extreme deformations at crack tips. In real solids, however, the relation between stress and strain is strongly nonlinear due to large deformation near a moving crack tip, a phenomenon referred to as *hyperelasticity* (or *nonlinear elasticity*).

(ii) Using large-scale atomistic simulations they discover that cracks can propagate supersonically when the elasticity at large strains becomes dominant within a zone of high energy transport near the crack tip. **The supersonically propagating crack produces a Mach cone associated with the shear wave speed** {ROSAKIS ET AL. ⇨ 1999}.[3375] They state that this phenomenon may be important in understanding the dynamics of large earthquakes {San Francisco Earthquake ⇨ 1906; Kunlun Shan Earthquake ⇨ 2001}, or the nucleation and propagation of cracks in aircrafts and space shuttles.

(iii) The results are in clear contrast to classical theories in which the speed of elastic waves was considered the limiting speed of fracture, analogous to the speed of light in the theory of relativity.

---

[3369] F.K. LU and X. ZHANG: *Visualization of a confined accelerated bubble*. Shock Waves **9**, No. 5, 333-339 (1999).

[3370] L.M. LOGORY ET AL.: *An experimental study of the Richtmyer-Meshkov instability, including amplitude and wavelength variations*. In: (G. JOURDON and L. HOUAS, eds.) *Proc. 6th Int. Workshop on the Physics of Compressible Turbulent Mixing* [Marseille, France June 1997], Imprimerie Caractère, Marseille (1997); Preprint UCRL-JC-127665.

[3371] D.R. FARLEY and L.M. LOGORY: *Single-mode, nonlinear mix experiments at high Mach number using Nova*. Astrophys. J. (Suppl. Ser.) **127**, Part 1, 311-316 (2000).

[3372] D.J. WARD ET AL.: *Doubly shocked Richtmyer-Meshkov instability experiments at Nova*. Program 44th Annual Meeting of the Division of Plasma Physics [Orlando, FL Nov. 2002]. American Physical Society (APS), Melville, NY (2002); *see* Abstract No. KP1.139.

[3373] D.E. ZEITOUN and Y. BURTSCHELL: *Navier-Stokes computations in micro shock tubes*. Shock Waves **15**, 241-246 (2006).

[3374] M.J. BUEHLER, F.F. ABRAHAM, and H. GAO: *Hyperelasticity governs dynamic fracture at a critical length scale*. Nature **426**, 141-146 (2003).

[3375] *See* Fig. 4 of their paper; http://www.nature.com/nature/journal/v426/n6963/fig_tab/nature02096_F4.html.

| | | |
|---|---|---|
| 2003 | Nottingham Trent University (NTU), Nottinghamshire, England | Trevor PALMER,[3376] a professor of life sciences at NTU, **publishes his book *Perilous Planet Earth: Catastrophes and Catastrophism Through the Ages*,** in which he discusses not only how natural catastrophes (such as meteorite impacts) in the past may have influenced the course of life on Earth, but also present-day extraterrestrial threats to Earth. According to the catastrophists, new species do not evolve gradually; they proliferate following sudden mass extinctions. |
| 2003 | Dept. of Physics, University of Nevada, Reno, Nevada | Friedwardt WINTERBERG,[3377] a 74-year-old German-born U.S. professor of theoretical physics and a previous student of Werner HEISENBERG, **publishes an idea for producing energy via thermonuclear fusion.** He proposes a system of high thermonuclear gain consisting of two electric transmission lines (charged by Marx generators), which are discharged almost simultaneously. The first discharge is used for the compression and confinement of a small, conically shaped target of deuterium-tritium, while the second discharge is used for ignition.[3378]<br><br>In the early 1960s, the British cosmologist **Edward R. HARRISON**[3379] **at the Rutherford High Energy Laboratory** (Chilton, England), **and WINTERBERG**[3380] **had proposed "impact fusion" to achieve controlled thermonuclear power:**<br>▸ the acceleration of small projectiles of appropriate mass, material and design to a velocity of about 1,000 km/s;<br>▸ their collision with a deuterium/tritium target; and<br>▸ the abruptly conversion of their kinetic energy into thermal energy (temperature about $10^8$ K) that is inertially confined in the shocked region. |
| 2004 | Kumamoto University, Kumamoto, Japan | In the period March 15–17, **the first International Symposium on Explosion, Shock Wave and Hypervelocity Phenomena** (ESHP Symposium) **is held,** with the objective "to discuss the future trends in the related research field as a purpose to find our new goals as well as to establish a worldwide network." The chairman is Shigeru ITOH. ▪ The 2nd Symposium (2007) was also held at Kumamoto University. |
| 2004 | NASA's Dryden Flight Research Center (DFRC), Edwards, California | On March 27, **the first test flight of the Hyper-X** {U.S. President REAGAN ⇨1986; NASA-LaRC ⇨1998} **is carried out successfully.**<br><br>(i) The hypersonic vehicle carrying the air-breathing X-43A scramjet is released from a B-52 ("Stratofortress") about 40,000 ft (12.2 km) over the Pacific Ocean. Using a rocket motor as a booster, the Hyper-X is accelerated to about Mach 7. Then the scramjet engine, which compresses oxygen from the atmosphere using the shape of the vehicle's airframe, is activated at an altitude of about 95,000 ft (28.9 km).<br><br>(ii) Separated from the booster, the Hyper-X executes a number of aerodynamic maneuvers during its eight-minute flight, before finally crashing into the ocean approximately 450 miles (724 km) from the launch point.[3381]<br><br>On November 16, 2004, during the last and fastest of three unpiloted flight tests in NASA's Hyper-X Program, **the X-43A research vehicle reached a velocity of nearly 7,000 mph (3,129 m/s) or Mach 9.8 at an altitude of about 110,000 ft (33.5 km).** The record flight took place in restricted airspace over the Pacific Ocean northwest of Los Angeles. |

---

[3376] T. PALMER: *Perilous planet Earth: catastrophes and catastrophism through the ages.* Cambridge University Press, Cambridge (2003).

[3377] F. WINTERBERG: *Ignition of a thermonuclear detonation wave in the focus of two magnetically insulated transmission lines.* Z. Naturforschung **58a**, 197-200 (2003).

[3378] WINTERBERG had already proposed a similar concept previously which consisted of just one Marx generator for compression and ignition; *see* his paper *The possibility of producing a dense thermonuclear plasma by an intense field emission discharge.* Phys. Rev. **174** [II], 212-220 (1968).

[3379] E.R. HARRISON: *Alternative approach to the problem of producing controlled thermonuclear power.* Phys. Rev. Lett. **11**, 535-537 (1963).

[3380] F. WINTERBERG: *On the attainability of fusion. Temperatures under high densities by impact shock waves of small solid particles accelerated to hypervelocities.* Z. Naturforsch. **19A**, 231-239 (1964).

[3381] *Orbital's Hyper-X rocket successfully launches NASA's X-43A hypersonic scramjet.* Press release from SpaceRef Interactive Inc., Reston, VA (March 27, 2004); http://www.spaceref.com/news/viewpr.html?pid=13926.

NASA has been officially recognized for setting the speed record for a jet-powered aircraft by Guinness World Records.[3382]

2004 — *Planet Saturn* — On June 27, **the Cassini Spacecraft crosses the bow shock of Saturn** at a distance of 2.95 million km (49.2 Saturn radii) from Saturn.[3383] ▪ Cassini-Huygens, a joint U.S./European mission to Saturn and its moon Titan, was launched on October 15, 1997. After flying past the planets Venus, Earth, and Jupiter, it entered Saturn orbit on July 1, 2004.

Before Cassini, the brief encounters of Pioneer 11 in 1979 and the two Voyager spacecraft in 1980/1981 had provided most of the information available about the structure and dynamics of Saturn's magnetosphere, such as the **first evidence of Saturn's bow shock** at a distance of between 19 and 35 Saturn radii and a bow shock thickness of about 2,000 km. The first results for Saturn's bow shock were published recently by a team of German, British and U.S. astrophysicists.[3384]

2004 — *Beijing, northern China* — In July, at the 24th meeting of the International Symposium on Shock Waves (ISSW24), **the concept for the formation of an International Shock Wave Institute (ISWI) is generally accepted.** ▪ For several years, the idea of an association to promote and integrate shock wave research at an international level has been investigated. Original proposals were for several geographically defined groups (*e.g.*, an Asian-Pacific Association), but the idea of a full international group proved to be attractive.

On March 3, 2005, at the 2nd Int. Symposium on Interdisciplinary Shock Wave Research (ISISWR-2) at Sendai, Japan, a meeting of all delegates was held to discuss the association. As the meeting was fully representative of the wide discipline areas of shock wave research, **a resolution was passed that the ISWI be formed from that date.**[3385] The first president is Kazuyoshi TAKAYAMA.

2004 — *Outer limit of the Solar System; NASA's Jet Propulsion Laboratory (JPL), Pasadena, California & NASA's Goddard Space Flight Center (GSFC), Greenbelt, Maryland* — On December 15, at a distance of about 94 AU (ca. 14.1 billion km) from the Sun and no longer capable to observe the solar wind, the space probe Voyager 1 observes an interplanetary magnetic field of 0.05 nT or less (for comparison, the field in the solar wind near Earth can be 5–10 nT). On December 17, its intensity rises to 0.12–0.15 nT, and stays at this higher level. Apparently, at some point between those observations, **Voyager 1 crossed a transition known as the "heliospheric termination shock"** {Voyager 1 & 2 ⇨2003} **at which the solar wind velocity dropped steeply, as also signaled by an increase in the magnetic fluctuation level ("turbulence").**[3386] ▪ In 1994, the New Zealand astrophysicist William Ian AXFORD[3387] had speculated that "with Pioneer 10 out at about 60 AU, I have taken to suggesting that it [the termination shock] might be at about 90–100 AU."

The magnetic field remained at these high levels for the next half-year, the strongest evidence that Voyager 1 had passed through the termination shock into the slower, denser wind. However, scientific debate on the phenomena observed and data recorded, and what they may infer about the approach to the termination shock, continued into 2005.[3388, 3389] When Voyager

---

[3382] *Faster than a speeding bullet: Guinness recognizes NASA scramjet.* NASA press release No. 05-156 (June 20, 2005); http://www.nasa.gov/home/hqnews/2005/jun/HQ_05_156_X43A_Guinness.html.

[3383] *Cassini encounters Saturn's bow shock.* The Radio and Plasma Wave Group, Dept. of Physics & Astronomy, College of Liberal Arts & Sciences, University of Iowa; http://www-pw.physics.uiowa.edu/space-audio/cassini/bow-shock/.

[3384] S. HENDRICKS ET AL.: *Variability in Saturn's bow shock and magnetopause from Pioneer and Voyager: probabilistic predictions and initial observations by Cassini.* Geophys. Res. Lett. **32**, No. 20, L20S08 (Aug. 2005); for abstract see http://www.agu.org/pubs/crossref/2005/2005GL022569.shtml.

[3385] Memorandum of the International Shock Wave Institute (ISWI); http://www.iswi-online.org/institute.html.

[3386] L.F. BURLAGA ET AL.: *Crossing the termination shock into the heliosheath: magnetic fields.* Science **309**, No. 5743, 2027-2029 (Sept. 23, 2005).

[3387] W.I. AXFORD: *The good old days.* J. Geophys. Res. **99**, No. A10, 19.199-19.212 (Oct. 1, 1994).

[3388] *Voyagers surpass 10,000 days of operation.* NASA-JPL (Jan. 20, 2005); http://voyager.jpl.nasa.gov/.

[3389] *Voyager enters Solar System's final frontier.* NASA-GSFC (May 24, 2005); http://www.nasa.gov/vision/universe/solarsystem/voyager_agu.html.

2, which is trailing Voyager 1 by about 20 AU (ca. 3 billion km), reaches the possible location of the termination shock it should provide more data and hopefully definitive answers.

At around 2020, Voyager 1 is expected to reach the heliopause at 135 AU (ca. 20 billion km), a region where the Sun's influence fades away entirely and interplanar space begins.

2004    *Sumatra-Andaman Islands, Indian Ocean*

On Sunday, December 26, at 7:58 A.M. local time, **thrust-faulting produces a violent undersea earthquake – *i.e.*, a seaquake – in the Indian Ocean with a magnitude of 9.0**. The epicenter is located about 100 km off the west coast of Sumatra at a shallow depth below the ocean floor, about 30 km below sea level. It is the biggest earthquake worldwide for forty years {Great Alaskan Earthquake ⇨1964} and the fourth largest in a century.[3390]

(i) The quake triggers a series of lethal tsunamis up to about ten meters high – popularly known as "killer waves" – that crash onto the shores along some coastal regions of West Sumatra near the fault region. Walls of water up to about four meters high are reported at remote coastlines such as those in Sri Lanka and Thailand.

(ii) Witnesses report that the sea level first receded, producing what appeared to be an extremely low tide, only to return with a destructive, towering "wall of water." ▪ This phenomenon of receding water is typical of tsunamis approaching shallow water, and had been observed several times previously {MARCELLINUS ⇨A.D. 365; Lisbon Earthquake ⇨1755; Arica Earthquake ⇨1868}.

(iii) The tsunami arrives without any warning – the main reason for the large number of casualties. Having survived the earthquake, people are killed by the subsequent tsunamis; they are either drowned by the incoming wave, smashed against buildings and debris, buried under crashing houses, or dragged along recurrent waters and drowned in the ocean. ▪ Tsunamis are rare events in the Indian Ocean, and, in contrast to the tsunami warning system that covers the Pacific Ocean {Aleutian Earthquake ⇨1946; ITIC ⇨1965}, there is no international tsunami warning system for Indian Ocean rim nations.

(iv) The tsunami, which is more devastating than any other in recorded history, kills a large number of people not only near the epicenter, such as in western Sumatra, but also in western Thailand, on the Nicobar and Andaman Islands, in Sri Lanka, India, Bangladesh, on the Maldives, and even as far away as the coast of Africa (in Somalia, Tanzania, Kenya). The number of casualties from the earthquake and tsunami is estimated to be over 283,000 in eleven countries.[3391] The tsunami also displaces over one million people, and devastates innumerable buildings, harbors and coastal infrastructures {⇨Fig. 4.1–S}. The seawater also floods fields in coastal areas, completely destroying crops.

Shortly after the disaster – known as the **"Boxing Day Tsunami"** in some Commonwealth countries because it occurred on Boxing Day (December 26), a public holiday – the USGS released the following preliminary tectonic analysis: "**The devastating *megathrust earthquake* occurred at the interface of the India and Burma plates off the west coast of northern Sumatra at a depth of about 10 km** [later estimated at 30 km] **and was caused by the release of stresses that develop as the India plate subducts beneath the overriding Burma plate** [⇨Fig. 4.1–N]. Preliminary locations of larger aftershocks following the megathrust earthquake show that approximately 1,200 km of the plate boundary slipped as a result of the earthquake. By comparison with other large megathrust earthquakes, the width of the causative fault-rupture was likely over one-hundred km. From the size of the earthquake, it is likely that

---

[3390] These are: the Chilean Earthquake {⇨1960} with moment magnitude $M_W$ = 9.5, the Prince William Sound, Alaskan Earthquake {⇨1964} with $M_W$ = 9.2, and the Alaskan Earthquake (1957) with $M_W$ = 9.1. The Kamchatka Earthquake (1952) had the same moment magnitude ($M_W$ = 9) as the recent Sumatra-Andaman Islands Earthquake {⇨2004}. However, the tsunami created by the Kamchatka Earthquake (which had waves up to 13 meters high) happened in a much more sparsely populated area.

[3391] A more recent analysis compiled by the United Nations (UN) lists a total of 229,866 people lost, including 186,983 dead and 42,883 missing. UN Office of the Special Envoy for Tsunami Recovery; http://www.tsunamispecialenvoy.org/country/humantoll.asp.

the average displacement on the fault plane was about fifteen meters. The sea floor overlying the thrust fault would have been uplifted by several meters as a result of the earthquake. The above estimates of fault-dimensions and displacement will be refined in the near future as the result of detailed analyses of the earthquake waves."

National and international teams of tsunami researchers began to collect important data a few days after the tsunami event, because evidence of tsunami flow depth, flow direction, run-up elevation, and maximum inundation is easily destroyed or altered by weather and clean-up efforts. For example, the Sri Lanka International Tsunami Survey Team (ITST) studied tsunami sand deposits at numerous locations along the coasts of Sri Lanka, which are important for tsunami risk assessment. The Sri Lanka ITST measured water levels near the shoreline that varied between under three meters to more than ten meters.[3392]

In the following year, Xiaoming WANG[3393] and Philip L.F. LIU, two tsunami researchers at Cornell University, began to simulate the tsunami computationally using COMCOT (Cornell Multigrid Coupled Tsunami), a numerical model. Although the length of the entire rupture zone was up to 1,300 km long for this earthquake, the duration of slip and the rupture speed were still relatively short in comparison with the timescale and the propagation speed of the tsunami. Therefore, the impulsive fault plane model simulates the tsunami in the deep ocean reasonably well. Their video simulation shows in detail how the massive wave system spread outward from the epicenter of an undersea earthquake northwest of Sumatra.[3394]

| 2004 | *Institute for Shock Physics & Dept. of Physics, Washington State University, Pullman, Washington* | **Daniel H. DOLAN,**[3395] a young U.S. physicist and postdoctoral student, **and Yogendra M. GUPTA,** an Indian-born U.S. professor of physics, **report the first direct evidence of shock-induced water freezing on a nanosecond timescale.** This research area has attracted the curiosity of shock physicists and produced controversial discussions since the early 1940s {SCHARDIN ⇒1941; DÖRING 1944, *see* SCHARDIN ⇒1941; WALSH & RICE ⇒1957; AL'TSHULER ET AL. ⇒1958; ZEL'DOVICH 1961, *see* AL'TSHULER ET AL. ⇒1958; HAMANN ⇒1966}.

(i) Using multiple shock compression to compress liquid water to 2–5 GPa (20–50 kbar), which is well inside the stability field of ice VII {BRIDGMAN 1937, *see* SCHARDIN ⇒1941}, and optical transmission and imaging measurements, **they watch water freezing over timescales of 20–200 ns.**

(ii) **They observe that when water is compressed in silica [$SiO_2$] windows it shows a large drop of optical transmission, while minimal or no extinction is observed at the same pressure if sapphire [$Al_2O_3$] windows are used.** Discussing this difference, they conclude that, "A possible explanation for the difference between silica and sapphire windows is the surface charge of each substrate. Sapphire develops a positive surface charge in the presence of solutions with $pH < 9$, whereas silica surfaces become negatively charged for solutions with $pH > 2$. The sample $pH$ in this work was in the range 6–7, which would lead to different signs and magnitudes of surface charge, and thus very different electric fields at the water/window interface." |

---

[3392] *The December 26, 2004 Indian Ocean tsunami: initial findings on tsunami sand deposits, damage, and inundation in Sri Lanka.* Western Coastal & Marine Geology, USGS; http://walrus.wr.usgs.gov/tsunami/srilanka05/.

[3393] X. WANG and P.L.F. LIU: *An analysis of 2004 Sumatra Earthquake fault plane mechanisms and Indian Ocean tsunami.* J. Hydraulic Res. **44** (2), 147-154 (2006).

[3394] *Tsunami video.* Cornell News; http://www.news.cornell.edu/releases/Jan05/tsunamiVid320.html.

[3395] D.H. DOLAN: *Time dependent freezing of water under multiple shock wave compression.* Ph.D. thesis, Dept. of Physics, Washington State University, Pullman, WA (2003); D.H. DOLAN and Y.M. GUPTA: *Nanosecond freezing of water under multiple shock wave compression: optical transmission and imaging measurements.* J. Chem. Phys. **121**, No. 18, 9050-9057 (2004); D.H. DOLAN, J.N. JOHNSON, and Y.M. GUPTA: *Nanosecond freezing of water under multiple shock wave compression: continuum modeling and wave profile measurements.* Ibid. **123**, No. 6, Paper No. 064702 (2005).

| | | |
|---|---|---|
| 2004 | High-Pressure Group, Max-Planck-Institute (MPI) for Chemistry, Mainz, Germany | Mikail EREMETS,[3396] and collaborators report that they have synthesized a polymeric cubic form of nitrogen where the atoms are linked by single covalent bonds, similar to the carbon atoms in diamond. They therefore call this phase "nitrogen diamond." This cubic phase has not been observed before in any element. • Nitrogen usually consists of unreactive molecules in which the atoms are linked together in triple bonds. These triple bonds are very stable, which is why the molecules are also very stable. For the past twenty years scientists have been searching for a form of nitrogen where the atoms form a simple polymer net and which therefore has a high energy density.<br><br>(i) By using a diamond high-pressure cell heated by a laser, the scientists succeeded in synthesizing this form of nitrogen.<br><br>(ii) They plan to investigate whether polymeric nitrogen – a three-dimensional crystal – can be used as an environmentally friendly fuel or as an explosive. **This theoretically predicted cubic gauche structure possesses unique properties; for example, its energy capacity is more than five times that of the most powerful explosives.** |
| 2004 | Southwest Research Institute (SwRI), San Antonio, Texas | David FERRILL,[3397] director of the SwRI Earth, Material and Planetary Sciences Department, **and his colleagues at the University of Texas** (San Antonio, TX) **report on the origin of unusual linear chains of pits that are common on Mars.** Using a combination of laboratory simulations, mathematical modeling to take into account the decreased gravity field on Mars, and comparison with modern examples in Iceland, the authors show that **these features are formed by active faulting.** As the faults move, unconsolidated loose material above the fault collapses into the fault zone, forming distinctive conical depressions aligned along the trace of the fault. The pristine morphology of the pits is taken to indicate that the pits formed recently and thus, like Earth, Mars may have active faults. When the planet was more tectonically active, faulting could have produced "marsquakes" similar to the earthquakes on Earth.<br><br>In 2002, researchers at Imperial College London begun to design and build tiny earthquake-measuring devices to send to Mars on the 2007 NetLander mission led by the French space agency CNES (Centre National d'Études Spatiales) and ESA. Although the NetLander project was officially stopped since May 2003 by CNES, it still arouses a great interest in the installation of a geophysical and meteorological network on Mars. Both agencies have planned to send other orbiters and landers for missions like ExoMars. |
| 2004 | Institute for Chemical Physics Research, Chernogolovka, Russia | Gennadij I. KANEL',[3398] Sergej V. RAZORENOV, and Vladimir E. FORTOV, three Russian shock physicists, **publish their book *Shock-Wave Phenomena and the Properties of Condensed Matter.*** • The authors report on international investigations of shock-wave phenomena in condensed matter and methods developed to predict the effects of explosions, high-velocity collisions, and other kinds of intense dynamic loading of materials and structures conducted over the past 30 years.<br><br>**Shock-wave phenomena in condensed matter are accompanied by very high strain rates, phase transitions, fast fracture, and other relaxation processes** which are investigated through the analysis of wave profiles. A great number of experiments measuring pressure or particle velocity histories in planar shock waves have been performed in many laboratories for metals and alloys, ceramics, rocks, polymers, high explosives and other materials, and a huge amount of experimental data has been accumulated. More advanced models of material response to intense pulse loading have to be based on experimental data or verified through comparison with experimental data.[3399] |

---

[3396] M.I. EREMETS ET AL.: *Single-bonded cubic form of nitrogen.* Nature Materials **3**, 558-563 (2004).
[3397] D.A. FERRILL ET AL.: *Dilational fault slip and pit chain formation on Mars.* GSA Today **14**, No. 10, 4-12 (2004).
[3398] G.I. KANEL', S.V. RAZORENOV, and V.E. FORTOV: *Shock-wave phenomena and the properties of condensed matter.* Springer, New York (2004).
[3399] G.I. KANEL', S.V. RAZORENOV, A.V. UTKIN, and K. BAUMUNG: *Shock wave profile data.* IVTAN (Institute for High Temperatures), Russ. Academy of Sciences, Moscow (1996).

| 2004 | Institute of Earth Sciences (IES) and Racah Institute of Physics (RIP), Hebrew University of Jerusalem, Jerusalem, Israel | Amir SAGY (IES),[3400] Jay FINEBERG (RIP), and Ze'ev RECHES (IES), three Israeli professors, **characterize the structure of shatter cones by field and microanalyses and explain their formation by dynamic fracture mechanics.**

(i) Their analyses reveal that shatter cones always occur as multilevel, 3-D networks, ranging in size from centimeter to hundred of meters, with hierarchal branched fractures.

(ii) Their model assumes that shatter cone formation is driven by the tensile stresses that develop in the tail of the shock wave. It explains all of the structural features of shatter cones (such as curved surfaces, cone directivity, unique striations, hierarchic, and multilevel structure) and their exclusive occurrence at impact sites.

Shatter cones are rock discontinuities. They were first described in the Steinheim impact structure {Ries Basin & Steinheim Basin ⇨c.15 Ma ago; ⇨Fig. 4.1−C} and (incorrectly) attributed to a "cryptovolcanic" explosion.[3401] Robert S. DIETZ first (correctly) argued that shatter cones develop under the "mechanical shock of explosive violence" associated with extraterrestrial impacts {DIETZ 1947, see DIETZ ⇨1946}. |
|---|---|---|
| 2004 | Ritter Astrophysical Research Center (RARC), University of Toledo, Toledo, Ohio & Steward Observatory, Center for Astronomical Adaptive Optics (CAAO), University of Arizona, Tucson, Arizona | Uma P. VIJH (RARC),[3402] Adolf N. WITT (RARC), and Karl D. GORDON (CAAO), three U.S. astrophysicists using telescopes in Chile and Arizona, **discover blue luminescence in the spectrum of the Red Rectangle nebula** which they identify as fluorescence by small three- to four-ringed polycyclic aromatic hydrocarbon (PAH) molecules. ▪ PAH molecules and more complex aromatic networks make up the bulk of carbon in carbonaceous meteorites and interplanetary dust particles. **PAHs are formed in the outflows of carbon-rich stars or possibly by shock fragmentation of interstellar grains and considered of being important for early life.**

(i) Red Rectangle,[3403] catalogued as HD 44179, is an infrared source. This most unusual X-shaped nebula in our galaxy lies about 2,300 light-years from Earth in the direction of *Monoceros* ("Unicorn," a constellation of the Equatorial Zone). The central star in Red Rectangle is dying, and substantial mass is being blown off the star in a bipolar outflow, creating a large amount of dust particles which are visible as they reflect light from the central star.

(ii) Using spectroscopy **they detected traces of anthracene [$C_{14}H_{10}$] and pyrene [$C_{16}H_{10}$].** These 3- and 4-fused benzene rings, respectively, are the most complex organic molecules, to that date, found in space.

Already in December 1998, scientists from the NASA Ames Research Center (Moffett Field, CA) and the Astrophysical Institute & University Observatory (Jena, Germany) reported finding the spectroscopic signatures of PAHs in interstellar space.[3404] ▪ The result of this collaboration was the conclusion that previously unidentified absorption features in the spectra of the interstellar medium are due to PAHs in space which originated in the atmospheres of red giants. **When a red giant has shed its outer envelopes explosively as a nova or supernova, it returns to the interstellar medium elements heavier than hydrogen that it has synthesized in its interior.** |

---

[3400] A. SAGY, J. FINEBERG, and Z. RECHES: *Shatter cones: branched, rapid fractures formed by shock impact.* J. Geophys. Res. **109**, B10209 (2004).

[3401] W. BRANCO and E. FRAAS: *Das kryptovulcanische Becken von Steinheim.* Abhandl. Königl. Preuss. Akad. Wiss. Berlin I, 64 Seiten (1905).

[3402] A.N. WITT, U.P. VIJH, and K.D. GORDON: *Discovery of blue luminescence in the Red Rectangle: possible fluorescence from neutral polycyclic aromatic hydrocarbon molecules?* Astrophys. J. **606**, Part 2, L65-L68 (2004); *Small polycyclic aromatic hydrocarbons in the Red Rectangle.* Ibid. **619**, Part 1, 368-378 (2005).

[3403] *Rungs of the Red Rectangle.* Astronomy picture of the day, NASA-GSFC (May 13, 2004); http://antwrp.gsfc.nasa.gov/apod/ap040513.html.

[3404] T. HENNING and M. SCHNAITER: *Carbon − from space to laboratory.* In: (P. EHRENFREUND ET AL., eds.) *Laboratory analysis and space research.* Astrophysics and Space Science Library (Kluwer Acad. Publ., Dordrecht & London), vol. **236**, p. 249 (1999); T. HENNING: *Laboratory astrophysics of cosmic dust analogues.* In: (T. HENNING, ed.) *Astromineralogy.* Lecture Notes in Physics. Springer, Berlin & Heidelberg; vol. 609 (2003), pp. 266-281.

# 4

## PICTURE GALLERY

# 4 PICTURE GALLERY

## 4.1 SHOCK AND PERCUSSION IN NATURE – Lunar Surface, a Result of Meteorite Impacts

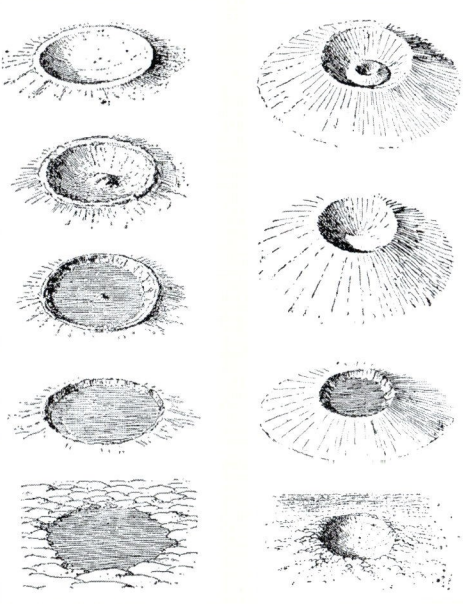

**4.1−A** *Top:* Galileo GALILEI, inspired in 1609 by the account of a Dutch invention of a telescope, built his own instrument. Using the telescope to observe the surface of the Moon, Galileo discovered new spots in addition to those already visible to the naked eye. From their change of light at different phases he concluded that the surface of the Moon must be rough with deep depressions and high mountains, contrary to the opinions of the ancients who assumed a smooth and polished surface. Those parts which remained or became brilliant he inferred were land, while those which remained obscure – the permanent spots – were water. These sketches are from his memoir *Sidereus Nuncius* (1610). [G. GALILEI. Opere. R. Ricciardi, Milano (1953)] *Center:* Pen-and-ink drawing of lunar crater Clavius (diameter c.225 km, depth c.3.5 km) by U.S. geologist Grove Karl GILBERT showing a grouping of craters. He used this crater as an example to illustrate that younger and typically smaller craters occurred on and within larger craters. [Bull. Phil. Soc. Wash. **12**, 241 (1893)] *Bottom, left:* This is a more recent photo of Clavius. [Photo taken on Feb. 2, 2003. © 2007 Markus WEBER, Verein Sternwarte Trier e.V.] *Bottom, right:* In 1893, GILBERT gave cogent argument for the impact origin of craters on the Moon. He concluded that "all features of the typical lunar crater and of its varieties may be explained as the result of impact." His hypothesis has been the basis of all subsequent thinking along these lines. [Reprinted from Bull. Phil. Soc. Wash. **12**, 241 (1893)]

*Left:* Classifying lunar crater structures according to size, GILBERT noticed that the smallest craters have a simple concave geometry (*uppermost sketch*), while with increasing size a central hill in the inner plain becomes typical, which, again, does not exist in craters of the largest diameters (*lowermost*). *Right:* He also studied terrestrial crater forms like craters of the Vesuvius type with central hills (*uppermost*) and without such hills, craters of the Hawaiian type, and craters of the mare type (*lowermost*).

## 4.1 SHOCK AND PERCUSSION IN NATURE – Meteor Crater and Shock Metamorphism

**4.1–B** *Top:* Aerial view of Meteor Crater (dia. 1.2 km, depth 200 m) – now officially called "Barringer Crater" – that is about 50,000 years old. In 1905, the U.S. mining engineer Daniel M. BARRINGER first suggested its extraterrestrial origin. U.S. geologist Harvey H. NININGER, following him along these lines, proposed in 1956 the natural occurrence of coesite in Coconino sandstone at the crater bottom. [Photo by David J. RODDY and Karl ZELLER, courtesy USGS; http://www.lpi.usra.edu/publications/slidesets/craters/slide_10.html] *Bottom:* Using X-ray diffraction (*left*), U.S. geologist Edward C.T. CHAO and collaborators demonstrated that the polymorphic transformation from quartz to coesite, a high-pressure modification of quartz, may have occurred under shocks generated from meteorite impact, thus providing a reliable criterion for the recognition of other impact sites on Earth and other planets. [In the public domain. Reprinted from Science **132**, 220 (July 22, 1960)] Schematics of crystal models of natural quartz (*center*) and coesite (*right*). The difference in crystal structure cannot be seen under a microscope but requires X-ray diffraction analysis. [Courtesy Prof. Caroline RÖHR, Dept. of Chemistry and Pharmacy, Univ. of Freiburg]

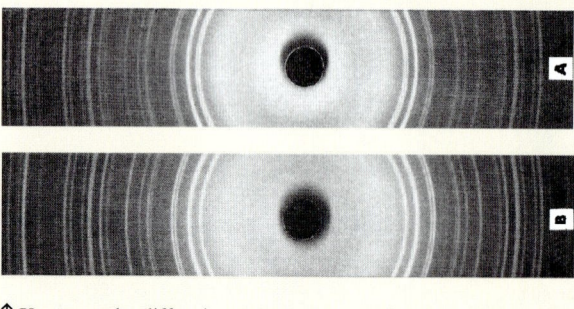 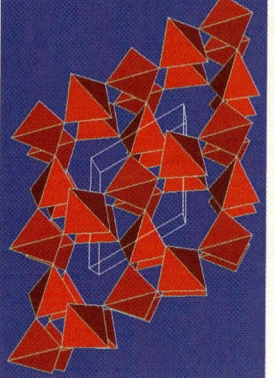 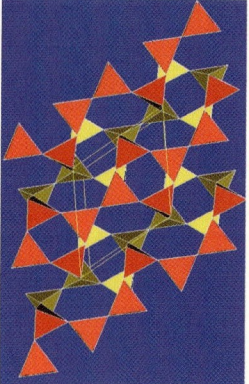

↑ X-ray powder diffraction patterns: Pattern of natural coesite with minor amounts of quartz (*A*) from sheared Coconino sandstone. It is identical to the X-ray powder diffraction pattern (*B*) of coesite synthesized by Francis R. BOYD at the Geophysical Laboratory, Carnegie Institute, Washington, DC.

# 4 PICTURE GALLERY

## 4.1 SHOCK AND PERCUSSION IN NATURE – Asteroid Impact: Ries Basin and Steinheim Basin

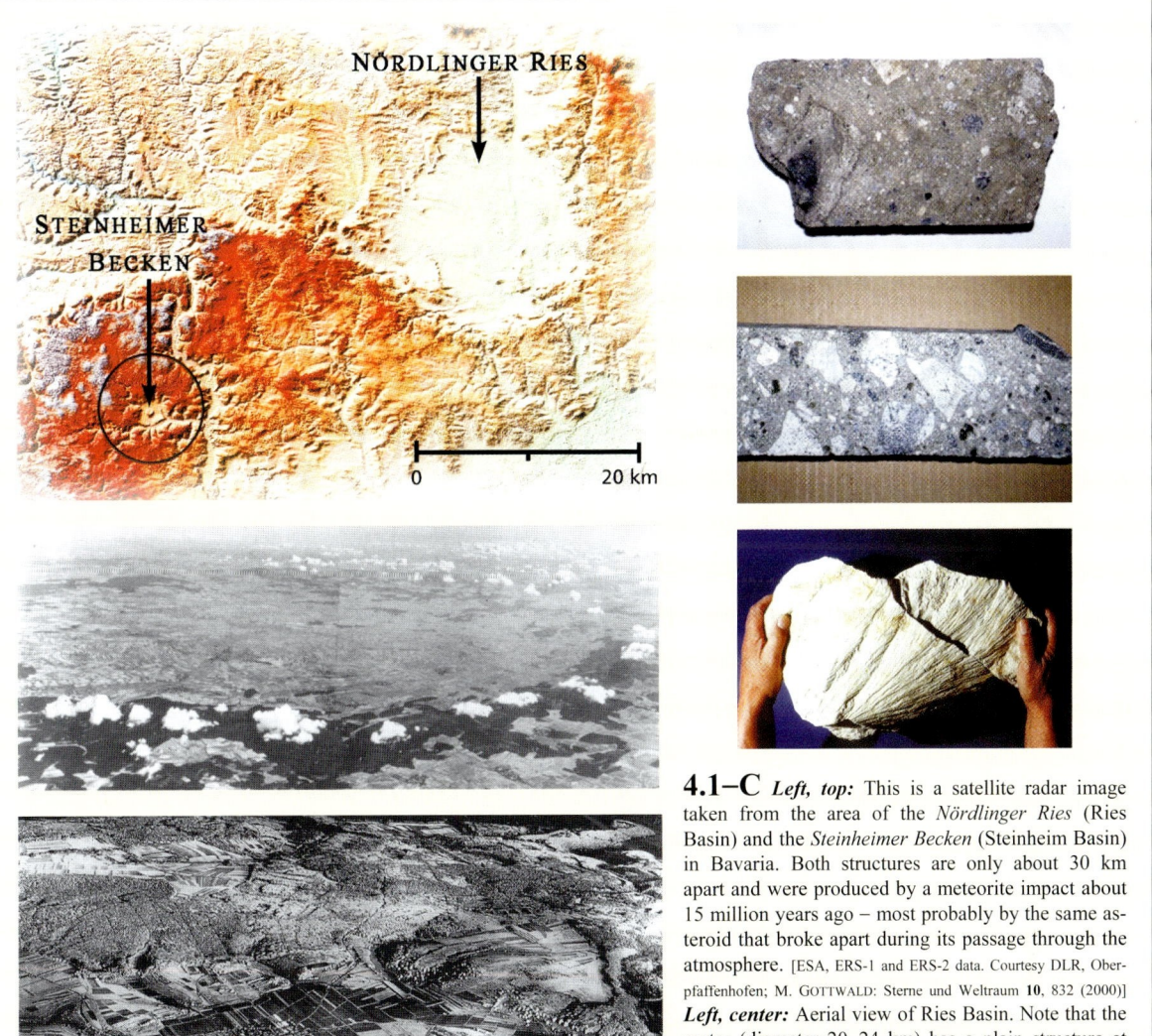

**4.1–C** *Left, top:* This is a satellite radar image taken from the area of the *Nördlinger Ries* (Ries Basin) and the *Steinheimer Becken* (Steinheim Basin) in Bavaria. Both structures are only about 30 km apart and were produced by a meteorite impact about 15 million years ago – most probably by the same asteroid that broke apart during its passage through the atmosphere. [ESA, ERS-1 and ERS-2 data. Courtesy DLR, Oberpfaffenhofen; M. GOTTWALD: Sterne und Weltraum **10**, 832 (2000)] *Left, center:* Aerial view of Ries Basin. Note that the crater (diameter 20–24 km) has a plain structure at its bottom. [Courtesy Rieskrater-Museum, Nördlingen, Germany] *Left, bottom:* An aerial view of the Steinheim Basin (diameter 3.5 km, depth 120 m) clearly reveals a central hill. *Right:* The findings of *Schwabenstein* or suevite (*top* and *center*) and of coesite in the Ries Crater as well as of so-called "shatter cones" (*bottom*) in the Steinheim Basin gave clear evidence that these geological structures were indeed generated by a meteorite impact. [Courtesy Prof. Eberhard STABENOW, Heidenheim]

## 4.1 SHOCK AND PERCUSSION IN NATURE – Two Famous Meteorites: Ensisheim and Murchison

**4.1–D** *Top, left:* On November 7, 1492, after a very loud explosion, a 127-kg meteorite fell in a wheat field near the German village of Ensisheim, in the province of Alsace (now France), where it produced an approx. 1-m-deep crater. An old woodcut depicting the scene shows the fall watched by two people emerging from a forest. In fact, the event was only witnessed by a young boy who led the townsfolk to the impact site. [From leaflet by Sébastien BRANT (1492); http://www.educnet.education.fr/planeto/pedago/systsol/astero.htm] *Top, right:* The Ensisheim meteorite is perhaps the most famous of all historic falls because it is the oldest witnessed fall in the Western Hemisphere from which specimens are preserved in many museums around the world. An approx. 55-kg residue, almost without fusion crust, is preserved and displayed at the Regency Museum in the Town Hall of Ensisheim. [Photo by Bernd GRÜNEWALD, EMI, Freiburg; with kind permission of Regency Museum, Ensisheim] *Bottom, left:* The Murchison meteorite fall occurred on September 28, 1969 over Murchison, a town in the state of Victoria, Australia. Over 100 kg of this meteorite have been found. Classified as a carbonaceous chondrite, type II (CM2), this meteorite is of possible cometary origin, based on its high water content of 12%. An abundance of amino acids found within this meteorite has led to intense study by researchers as to its origins. More than 92 different amino acids have been identified within the Murchison meteorite to date; nineteen of these are found on Earth. The remaining amino acids have no apparent terrestrial source. This picture shows a 5.28-g fragment which has a size of about $16 \times 15$ mm$^2$. [Photo courtesy Pierre-Marie PELÉ, France; http://www.meteor-center.com/collection/chondrites.asp] *Bottom, right:* In 1997, Richard B. HOOVER, a scientist at NASA's Marshall Space Flight Center, announced that he had seen and photographed in the Murchison meteorite microfossils that resembled microorganisms. The fossils were seen in freshly broken pieces of the meteorite, so the chance that they were earthly contaminants was low. The chemical evidence around the microfossils was most readily explained as the result of biological activity. The most interesting and unusual form, seen in many examples, curls to a tapered end. [Courtesy R.B. HOOVER, NASA/MSFC/NSSTC; Proc. SPIE **3111**, 115 (1997)]

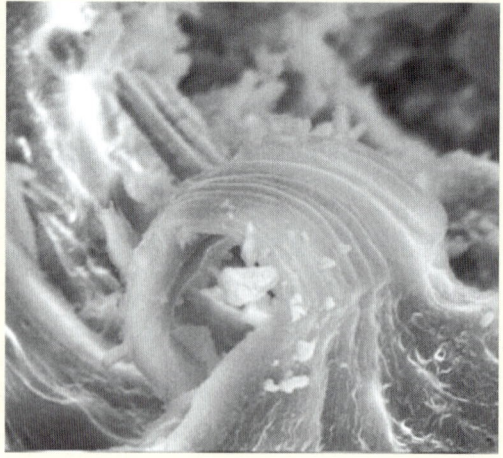

# 4 PICTURE GALLERY

## 4.1  SHOCK AND PERCUSSION IN NATURE – Great Earthquakes: Lisbon (1755)

**4.1–E** These two contemporary pen-and-ink drawings illustrate the terrible Lisbon Earthquake that occurred on the morning of November 1 (All Saints' Day), 1755 and largely destroyed the Royal Portuguese residential town (*bottom*). Thousands of Portuguese who survived the earthquake were killed by a tsunami that followed a few minutes later. Before the great wave hit, the harbor waters retreated, revealing lost cargo and forgotten shipwrecks. The disaster, which in Lisbon killed about 10,000 people and injured 40,000 to 50,000, challenged the well-established Baroque philosophy that hitherto regarded the Universe as a perfect creation. It was this catastrophe that is said to have caused VOLTAIRE, one of the greatest French authors and philosophers, to abandon religion. However, the disaster also stimulated the scientific community to investigate its possible causes. [*Kurz-verfaste Beschreibung der vortrefflichen, mächtigen und reichen Haupt- und Residenzstadt Lissabon im Königreiche Portugall (1756).* Courtesy Bayerische Staatsbibliothek, Munich]

## 4.1 SHOCK AND PERCUSSION IN NATURE – Great Earthquakes *(cont'd)*: Lisbon (1755)

Interrogatorios para a organização
do «Diccionario Geographico» do P.° Luis Cardoso

(Mandados pelo Governo aos parochos depois do terremoto de 1755)

O QUE SE PROCURA SABER D'ESSA TERRA É O SEGUINTE

*Venha tudo escrito em lettra legivel, e sem breves*

1. Em que provincia fica, a que bispado, comarca, termo e freguesia pertence?

2. Se é d'el-rei, ou de donatario, e quem o é ao presente?

3. Quantos vizinhos tem [*e o numero das pessoas*]?

4. Se está situada em campina, valle, ou monte, e que povoações se descobrem d'ella, e quanto dista?

5. Se tem termo seu, que lugares, ou aldeias comprehende, como se chamam, e quantos vizinhos tem?

6. Se a parochia está fóra do lugar, ou dentro d'elle, e quantos lugares, ou aldeias tem a freguesia, todos pelos seus nomes?

7. Qual é o seu orago, quantos altares tem, e de que santos, quantas naves tem; se tem irmandades, quantas, e de que santos?

8. Se o parocho é cura, vigario, ou reitor, ou prior, ou abbade, e de que apresentação é, e que renda tem?

9. Se tem beneficiados, quantos, e que renda tem, e quem os apresenta?

10. Se tem conventos, e de que religiosos, ou religiosas, e quem são os seus padroeiros?

11. Se tem hospital, quem o administra, e que renda tem?

12. Se tem casa de misericordia, e qual foi a sua origem, e que renda tem; e o que houver notavel em qualquer d'estas cousas?

13. Se tem algumas ermidas, e de que santos, e se estão dentro, ou fóra do lugar, e a quem pertencem?

14. Se acode a ellas romagem, sempre, ou em alguns dias do anno, e quaes são estes?

15. Quaes são os fructos da terra, que os moradores recolhem em maior abundancia?

16. Se tem juiz ordinario, etc., camara, ou se está sujeita ao governo das justiças de outra terra, e qual é esta?

17. Se é couto, cabeça de concelho, honra, ou behetria?

18. Se ha memoria de que florecessem, ou d'ella sahissem, alguns homens insignes por virtudes, lettras, ou armas?

19. Se tem feira, e em que dias, e quantos dura, se é franca ou cativa?

20. Se tem correio, e em que dias da semana chega, e parte; e, se o não tem, de que correio se serve, e quanto dista a terra aonde elle chega?

21. Quanto dista da cidade capital do bispado, e quanto de Lisboa, capital do reino?

22. Se tem alguns privilegios, antiguidades, ou outras cousas dignas de memoria?

23. Se ha na terra, ou perto d'ella alguma fonte, ou lagoa celebre, e se as suas aguas tem alguma especial qualidade?

24. Se for porto de mar, descreva-se o sitio que tem por arte ou por natureza, as embarcações que o frequentam e que póde admittir?

25. Se a terra for murada, diga-se a qualidade de seus muros; se for praça de armas, descreva-se a sua fortificação. Se ha nella, ou no seu districto algum castello, ou torre antiga, e em que estado se acha ao presente?

26. Se padeceu alguma ruina no terremoto de 1755, e em que, e se está reparada?

27. E tudo o mais, que houver digno de memoria, de que não faça menção o presente interrogatorio.

**4.1–E** *(cont'd)* **Left:** POMBAL's questionnaire, entitled *Interrogatórios enviados aos párocos depois do Terramoto de 1755* ("Interrogations Sent to the Parishes After the Earthquake of 1755"), was divided into three parts: questions regarding information about the land (27 questions, *left*), about the mountains (13 questions), and about the rivers (20 questions). For example, his questionnaire asked whether dogs or other animals behaved strangely prior to the earthquake. Was there a noticeable difference in the rise or fall of the water level in wells? How many buildings were destroyed and what kind of destruction occurred? Their answers – today kept in the Arquivos Nacionais, Torre do Tombo, Alameda da Universidade, Lisboa – enabled modern seismologists to precisely reconstruct the course of the earthquake and its severity at different localities. The response of the Portuguese authorities to the earthquake represents one of the first historic examples of integrated post-disaster recovery planning. Under the direction of king JOSEPH's first minister, Sebastião José DE CARVALHO E MELO (later Count of Oeiras and Marquês DE POMBAL), data were collected in written form and in the form of city plans and maps. [Archeólogo Português (Lisboa) 1, 268 (1895)] **Right:** Oil painting showing the Marquis DE POMBAL (1699–1782). He commissioned the military engineers General Manuel DA MAIA, Captain Eugénio DOS SANTOS, and Lieutenant Colonel Carlos MARDEL not only to remap the city but also to make a complete inventory of damage. This provided a worthwhile database for earth scientists and hazard analysts prior to the advent of the seismograph in the late 19th century. [Portrait of Sebastião José DE CARVALHO E MELO – oil on canvas. Unknown artist, middle of XVIII century. Courtesy Lisbon City Museum]

# 4 PICTURE GALLERY

## 4.1 SHOCK AND PERCUSSION IN NATURE – Great Earthquakes *(cont'd)*: San Francisco (1906)

**4.1–E** *(cont'd):* On April 18, 1906, at 5:13 A.M., San Francisco was wrecked by a series of heavy earthquake shocks of magnitude 8.3 on the Richter scale. Many buildings immediately caught fire, and trapped victims could not be rescued. A major aftershock struck at 8:14 A.M. causing the collapse of many damaged buildings (the death toll was more than 3,000 from all causes). The earthquake destroyed 3,000 acres (ca. 12 km²) in the heart of the city. ***Top, left:*** The steel skeleton of the City Hall of San Francisco was stripped of masonry by the seismic shocks. ***Top, right:*** The San Andreas fault ruptured 430 km from San Juan Bautista to the Cape Mendocino triple junction (with San Francisco in between). [Courtesy Zpub, San Francisco; http://www.zpub.com/sf/history/1906earth.html] ***Bottom:*** In the largest maritime rescue in U.S. history more than 30,000 people were taken from the shoreline between Fort Mason and the foot of Lombard Street. The painting by William Alexander COULTER (1849–1936) depicts the flotilla of rescue vessels ferrying survivors from the burning city to Sausalito. [Courtesy Virtual Museum of the City of San Francisco; http://www.sfmuseum.org]

## 4.1 SHOCK AND PERCUSSION IN NATURE – Explosive Volcanic Eruption: Krakatau (1883)

← Map of present Krakatau Island and vicinity. The volcano is still active and formed Anak Krakatau, a small island in the middle of the ocean-filled caldera. [By Lyn TOPINKA, USGS/Cascades Volcano Observatory, Vancouver, WA. Modified from T. SIMKIN and R.S. FISKE: *Krakatau 1883: the volcanic eruption and its effects*. Smithsonian Institution Press, Wash., DC (1983)]

**4.1–F** The famous Krakatau (or Krakatoa) eruption was one of the largest in history (with an Volcanic Explosivity Index of $VEI = 6$). ***Top, left:*** View of Krakatau during modest, early stage of eruption taken on May 27, 1883. [G.J. SYMONS (ed.): *The eruption of Krakatoa and subsequent phenomena*. Trübner, London (1888)] ***Top, right:*** The Krakatau volcano lies in the Sunda Strait between Java and Sumatra (*top*). Since 1927, small but frequent eruptions have constructed a new island, Anak Krakatau (*bottom*), shown here with a column of ash rising above it. Remnants of the northern part of Krakatau are visible in the background. [Reprinted with permission from J. Geophys. Res. **66**, 3497 (1961)] ***Bottom:*** Drawing of northern remnants of Krakatau island made 2 months after the final, most violent outbursts, which occurred on August 26 and 27 of the same year. They destroyed a large northern portion of the basaltic cone of the former Rakata Peak (2,623 ft) and formed a nearly vertical cliff, thus giving rise to a magnificent section that afforded a perfect view of the interior structure of the volcano. [R.D.M. VERBEEK: *Krakatau, Album van Straat Soenda*. Landsdrukkerij, Batavia (1885/1886)]

## 4.1  SHOCK AND PERCUSSION IN NATURE – Explosive Volcanic Eruptions: Krakatau (1883) *(cont'd)*

Examples of accounts of observed sound phenomena given in English miles at the following locations:

1 – **Merak** (Java; 38 miles from Krakatau) "heavy detonations and violent shocks, but no earthquake."
2 – **Batavia** (Java; 94 miles) "a series of detonations, towards night they grew louder, till in the early morning (August 27, 1883) the reports and concussions were simply deafening."
3 – **Brig** *Airlie* at sea (390 miles; Lat. 0° 30' S, Lon. 105° 54' E) "sounds like those of heavy artillery… The last report made the ship tremble all over."
4 – **Hambantota** (Ceylon; 1,866 miles) "a steady sequence of reports, and then a rapid succession of them, ending, very often, in a loud burst of two or three, or half a dozen, almost together, which was generally followed by a lull. The intensity of the sounds greatly varied."
5 – **Alice Springs** (Central Australia; 2,233 miles) "two distinct reports, similar to the discharge of a rifle."
6 – **Rodriguez** (island near Mauritius, Indian Ocean; 2968 miles) the most distant location at which sounds originating from the eruption were perceived, "reports like the distant roars of heavy guns."

**4.1–F** *(cont'd)* The illustrated map, with Krakatau at its center, shows the places at which the sounds of the explosive eruptions were heard on August 26–27, 1883. The large number of accounts of observed sound phenomena were carefully collected and analyzed by the Krakatoa Committee established in 1884 by The Royal Society of London. The *red line* indicates the approximate area over which the sounds were heard; the most distant locations were more than 4,000 km away. It is remarkable that at many places in the more immediate neighborhood of the volcano the sounds ceased to be heard. Very probably this peculiar phenomenon was caused by the large amount of solid matter that, emitted into the atmosphere by the volcano, formed in the lower strata of the air a screen of sufficient density to prevent the sound waves from penetrating to those places over which it was more immediately suspended. The eruptions caused atmospheric fluctuations that were also recorded in Europe. Furthermore, the tremendous eruptions caused tsunamis that reached heights of 40 m, killing tens of thousands of people on the low shores of Java and Sumatra. [After G.J. SYMONS (ed.): *The eruption of Krakatoa and subsequent phenomena*. Trübner, London (1888)]

## 4.1 SHOCK AND PERCUSSION IN NATURE – Explosive Volcanic Eruption: Mount St. Helens (1980)

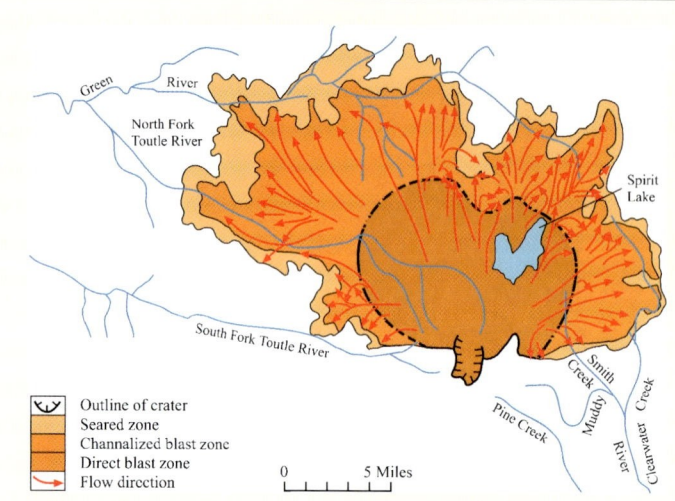

↑ The inner zone – the *direct blast zone* – averaged about 8 miles in radius in which virtually everything, natural or manmade, was obliterated or carried away. For this reason, this innermost zone has also been called the "tree-removal zone." The flow of the material carried by the blast was not deflected by topographic features in this zone. The *channelized blast zone*, an intermediate zone, extended out to distances as far as 19 miles from the volcano, an area in which the flow flattened everything in its path and was channeled to some extent by topography. In this zone, the force and direction of the blast are strikingly demonstrated by the parallel alignment of toppled large trees, broken off at the base of the trunk like blades of grass mown by a scythe. This zone was also known as the "tree-down zone." The *seared zone*, also called the "standing dead zone," is the outermost fringe of the impacted area, a zone in which trees remained standing but were singed brown by the hot gases of the blast.

**4.1–G** On May 18, 1980, a magnitude 5.1 earthquake shook Mount St. Helens in the State of Washington, and the volcano erupted explosively. *Right:* View of eruption taken on this day. The bulge and surrounding area slid away in a gigantic rockslide and debris avalanche, releasing pressure, and triggering a major pumice and ash eruption of the volcano. The eruptive cloud rose to an altitude of more than 12 miles in 10 min. The swirling ash particles in the eruptive cloud generated lightning, which in turn started forest fires. Other fires were ignited by the initial blasts and later pyroclastic flows. Nearly 550 million tons of ash fell over an area of approx. 60,000 km$^2$. Note the erosional "furrows" below the mouth of the volcano. [Courtesy USGS; photo by Austin POST.] *Left, top:* The area affected by the blast from the explosive eruption can be subdivided into three roughly concentric zones. [USGS, http://pubs.usgs.gov/publications/msh/lateral.html] *Left, bottom:* Oblique aerial view of a devastated area in Skamania and Cowlitz Counties, showing uprooted trees blown down by the blast wave. Note how the blast followed the contours of the mountainside. [Courtesy USGS, Branch of Exhibits]

# 4 PICTURE GALLERY

## 4.1 SHOCK AND PERCUSSION IN NATURE – Plate Tectonics

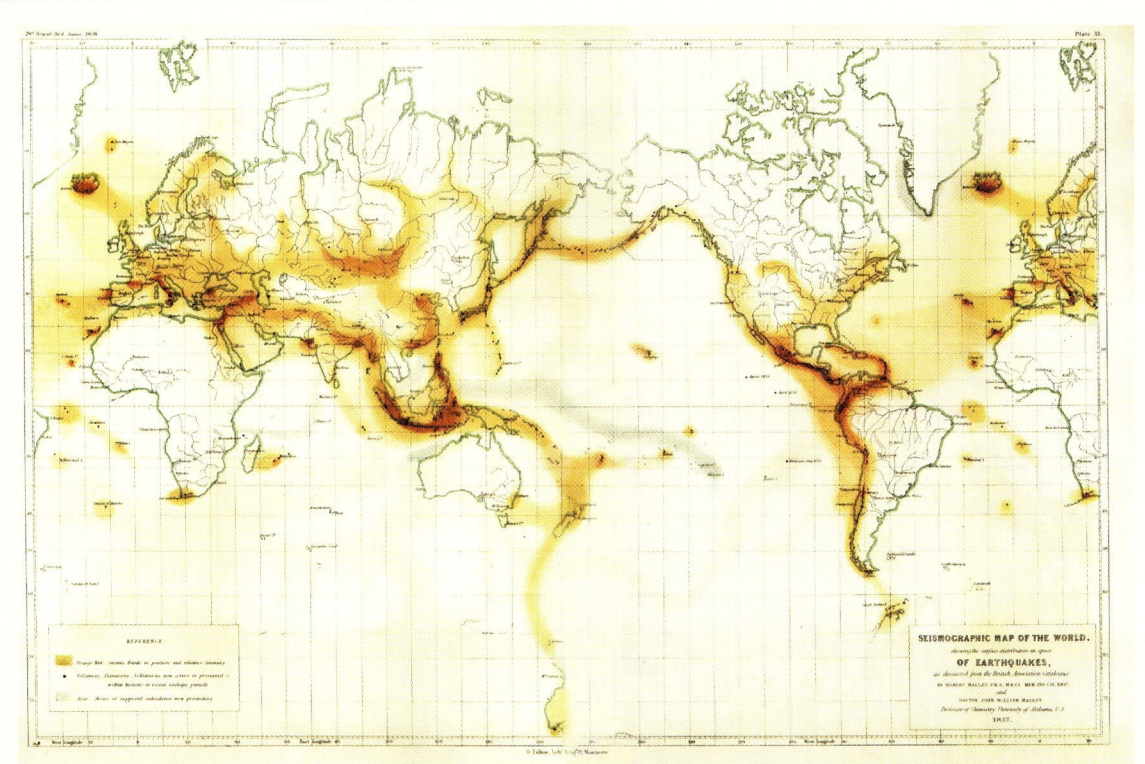

**4.1−H** The Irish engineer Robert MALLET compiled a "Seismographic Map" (1857) of the world for the British Association for the Advancement of Science using his great "Earthquake Catalogue" (1852–1854), which contains 6,831 earthquakes reported between 1606 B.C. and A.D. 1858. In addition, he used data from previous attempts such as the "Physikalischer Atlas" (1845) of the German geographer Heinrich BERGHAUS and the "Physical Atlas" (1848) of the Scottish cartographer and geographer royal Alexander Keith JOHNSTON. By mapping earthquakes in terms of disturbed areas rather than of epicenters (the modern and more accurate method), MALLET divided earthquakes into three classes – great, mean, and minor. His map (original size $75 \times 48$ in.$^2$) is not only a very attractive piece of early color-printed cartography, but also illustrates the importance of seismic bands, their color varying according to the intensity of the shocks experienced, the volcanoes active in historic times, and the areas of supposed oceanic subsidence. These seismic bands generally follow the lines of elevation that divide the great oceanic or terr-oceanic basin of the earth's surface; insofar as these are frequently the lines of mountain-chains, and these latter of volcanic vents, so the seismic bands are found to follow them as well. The regions of least or no disturbance are the central areas of great oceanic or terr-oceanic basins and the greater islands existing in shallow seas. The great seismic band, running from the Aleutian and Kurile Islands through Japan and the Philippines to the East Indies, is clearly marked. By delimiting the lithosphere plates, MALLET unconsciously anticipated the development of plate tectonic theory in the 1960s; *see also* C. DAVISON: *The Founders of Seismology.* New York, pp. 73–75 (1978). [Rept. Meet. Brit. Assoc. **28**, 1 (1858); plate 11]

## 4.1 SHOCK AND PERCUSSION IN NATURE – Plate Tectonics *(cont'd)*

**4.1–I** *Left, top & center:* The location of plate boundaries coincides in most cases with the loci of volcanoes and the epicenters of earth- and seaquakes *(top)*, *i.e.*, with the loci of naturally generated discontinuities such as seismic shocks and lateral blasts. Global distribution of volcanoes (▲) and earth- and seaquakes (· . · . ·). Major tectonic plates of the world are separated by faults (⎯⎯), major fractures of the Earth's crust *(bottom)*. Geologists define 7 large plates and 20 smaller plates. Most of the plates consist of both oceanic and continental lithosphere. [Courtesy Dept. of Space Studies, University of North Dakota; http://volcano.und.nodak.edu/vwdocs/vwlessons/plate_tectonics/part12.html]
*Right, top:* The San Andreas fault zone, which is about 1,300 km long and in places tens of kilometers wide, slices through two thirds of the length of California. Aerial view of the San Andreas fault, slicing through the Carrizo Plain in the Temblor Range east of the city of San Luis Obispo (about 320 km south of San Francisco). [Photo by Robert E. WALLACE, USGS; http://pubs.usgs.gov/publications/text/San_Andreas.html]
*Left, bottom:* Example of a left-lateral strike-slip fault, generated during the 1976 Guatemala Earthquake in the cultivated field west of El Progreso (Dept. Jutiapa). The thick, saturated, unconsolidated deposits have yielded by plastic deformation rather than rupture along the fault. [Courtesy NGDC, NOAA; http://www.ngdc.noaa.gov/seg/hazard/slideset/10/10_slides.shtml] *Right, bottom:* Fault trace northwest of Olema, generated during the 1906 San Francisco Earthquake. The greatest displacement was 21 ft (6.4 m) about 30 miles (48 km) northwest of San Francisco. [Courtesy J.B. Macelwane Archives, Saint Louis University, St. Louis, MO; http://www.eas.slu.edu/Earthquake_Center/1906EQ/olema/m40A.html]

## 4 PICTURE GALLERY

## 4.1  SHOCK AND PERCUSSION IN NATURE – Lightning and Thunder

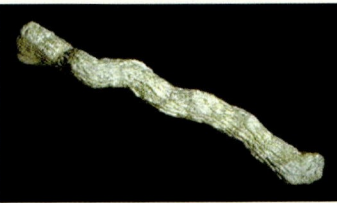

**4.1–J** In ancient times, the hammer and mallet – the basic tools to produce percussion and shock phenomena – were considered to be of divine origin and to possess marvelous, supernatural abilities. **Left, top:** View of a small, 6.4-cm bronze statuette of THOR (ca. A.D. 1000), the god of thunder, which he produced with his short-handed hammer *Mjolnir* (meaning "The Destroyer"). The hammer, here shown on THOR's knees, had the marvelous quality of returning to the thrower like a boomerang. THOR was a deity common to all early Germanic peoples. [Thor (Natmus #10880), photo by Ivar BRYNJOLFSSON. Courtesy National Museum of Iceland, Reykjavik] **Right, top:** Corinthian bronze statuette (ca. 470 B.C.) found at Dodōna in the ancient Epirus (now Dodoni, Albania). It shows the god ZEUS hurling a thunderbolt. He was regarded as the sender of lightning, thunder, rain, and winds. His most prominent symbol was the thunderbolt, resembling a mallet. In Teutonic mythology, thunderbolts were ascribed to THOR, in Germany called DONAR, while in the Hindu cosmology, SHIVA, one of the great trinity of Hindu deities, was represented with a bow, a thunderbolt, and an axe. [Photo by Bildarchiv Preußischer Kulturbesitz. Berlin] **Right, center:** Some lightning phenomena are known as "lightning tubes" ore "fulgurites," which are branched, irregular tubular bodies with a glassy structure. They are produced by lightning in loose, unconsolidated sand (mostly quartz). The interior is normally very smooth or lined with fine bubbles. [From P. MENZEL, National Geographic World No. 250 (June 1996)] **Right, bottom:** Obviously, the expanding shock wave, generated during a lightning strike, pushed and compressed the sand grains into a thin shell of compacted sand. The silicate fulgurites shown below were found in Egypt. Although of very narrow cross-section and short length, some specimens have been found to exceed several centimeters in diameter and 20 m in length. [Courtesy Mineralogical Research Company, San Jose, CA; http://www.minresco.com/]

**Left, bottom:** Damage by lightning strikes includes explosion and melting effects; the best known examples are split tree trunks, damaged brickwork buildings, and destroyed electrical equipment by induced short-circuits. This rare snapshot of a lightning strike was taken in 1996 in Alabama by Johnny AUTERY from his pickup truck. A 65-foot-tall sycamore tree is lit from top to bottom by a direct lightning strike. Lightning often damages or destroys trees. [A. FALLOW: *Powers of nature: lightning*. National Geographic World No. 250 (June 1996)]

## 4.1 SHOCK AND PERCUSSION IN NATURE – The Riddle of Ball Lightning

**4.1–K** *Left:* A 19th-century French wood engraving showing the propagation of ball lightning in a barn. Since antiquity this curious but rare phenomenon, which can end in a violent explosion, has often been observed and described, but it remains an enigma to modern science. Note that the size of ball lightning, ranging in diameter between a golf ball and a beach ball as reconfirmed also by numerous more recent accounts, has been reproduced correctly. [Courtesy Bildarchiv Preußischer Kulturbesitz (bpk), Berlin, image no. 20.031.187] *Center:* Although about 1% of the population reports having seen ball lightning, clear photographs of this phenomena are extremely rare. This picture was taken by D. KUHN in Germany at Ludwigshafen/Rhein during a heavy thunderstorm. According to his account, the phenomenon had a diameter of about 50 cm and passed by silently. [Die Naturwissenschaften **38**, 518 (1951)] *Right:* This spectacular example of a fireball, which may be the rare phenomenon of ball lightning, was taken by Werner BURGER on a summer night in 1978 at St. Gallenkirch in Vorarlberg, western Austria. It had a whitish center with a blue surround and a luminous tail. [Courtesy Fortean Picture Library, Ruthin, U.K.]

## 4.1 SHOCK AND PERCUSSION IN NATURE – Tidal Bores

**4.1–L** Tidal bores are hydraulic jumps propagating upstream and frequently can be observed in shallow estuaries. *Left:* In China tidal-bore watching at the Qiantang River has a history of over 2,000 years. This picturesque scene of a bore propagating up the Qiantang River was recorded by the Chinese artist Li SUNG (1166–1243). [Collection of the National Palace Museum. Taiwan, Republic of China] *Right:* The tidal bore of the Qiantang River, the greatest such bore in the world reaching heights of up to almost 9 m, is caused by the horn-shaped topography of the Hangzhou Bay, which concentrates the bore's energy. [Photographer unknown. Linden Software, North Ferriby, U.K.; http://www.linden-software.com/china.html]

# PICTURE GALLERY

## 4.1 SHOCK AND PERCUSSION IN NATURE – Hydraulic Jumps

← The flowing water, when hitting the floor of the sink, spreads out at a speed that is higher than the local wave speed, forming a disk of shallow, rapidly diverging flow.
• A ring-shaped discontinuity can also be observed in space: it arises when the supersonic solar wind, interacting with the interstellar medium, has slowed down to subsonic velocity. Compression and heating effects produce a ring-shaped "termination shock" surrounding the Sun at a distance of about 75–90 AU ($c$.11–13 billion km).

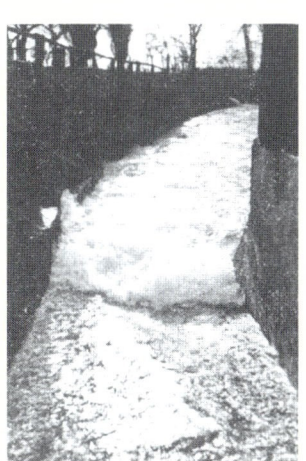

**4.1–M** Hydraulic jumps are a curious flow phenomenon. They are characterized by sudden changes in depth and velocity, separated by an intermediate region of turbulence, and can occur both as stationary and traveling discontinuities. ***Top, left:*** This illustrious example of a stationary hydraulic jump can easily be observed in a kitchen sink when the tap is left running, which produces a ring-shaped discontinuity. A bore, which in nature is a traveling discontinuous phenomenon and is called a "hydraulic jump," can be demonstrated by turning off the faucet. Then the front moves inward. [Courtesy Prof. Philippe BELLEUDY, Laboratoire d'études des Transferts en Hydrologie et Environnement (LTHE), Grenoble, France; http://www.eng.vt.edu/fluids/msc/gallery/waves/sink.htm] ***Top, left:*** View of a shooting flow in a narrow channel, a so-called "hydraulic jump" that can be generated, for example, by suddenly opening a water reservoir at the other end of the channel. [E. PREISWERK: *Anwendungen gasdynamischer Methoden auf Wasserströmungen mit freier Oberfläche*. Ph.D. thesis, ETH Zurich (1938), p. 65] ***Bottom:*** Mach reflection can also be seen in nature. The British scientist Vaughan CORNISH first photographed the interaction of two hydraulic jumps approaching each other in very shallow water: regular reflection (*left*) and Mach reflection (*right*). Unfortunately, CORNISH didn't report on the scale of his observed interaction phenomena. The pictures might suggest aerial views, *i.e.*, that the interactions took place on a rather large scale. However, it is more probable that the photos are close-ups. On a small scale, say in a field of about 1×1 m$^2$, Mach reflection can be observed on a shallow beach quite frequently, particularly when the angle between the incident wave front and the beach is large. [V. CORNISH: *Waves of the sea and other water waves*. Fisher Unwin, London (1910), frontispiece and p. 173]

## 4.1 SHOCK AND PERCUSSION IN NATURE – Tsunami Caused by Submarine Volcanic Eruption

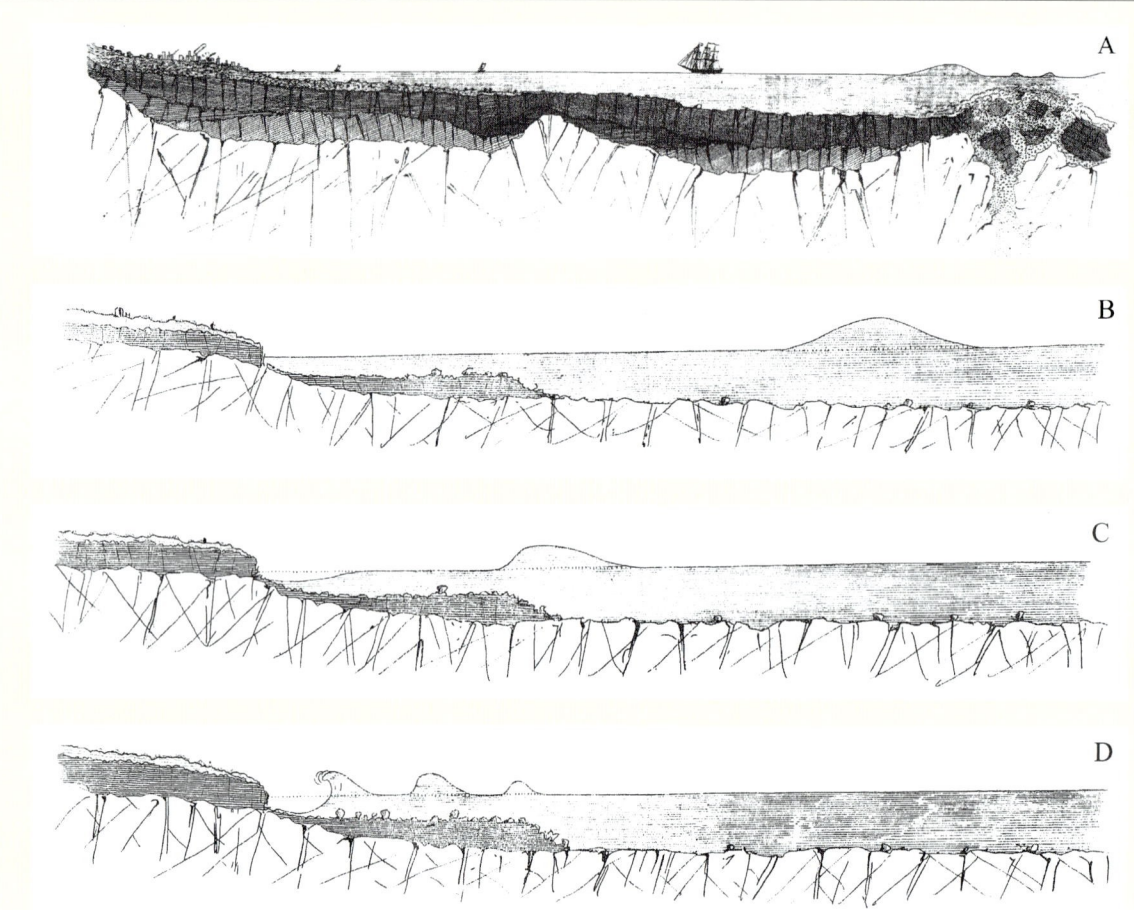

**4.1–N** The Irish engineer Robert MALLET considered an earthquake to be "the transit of a wave of elastic compression in any direction, from vertically (as shown here) to horizontally, through the surface and crust of the Earth…" The causes of tsunamis – or great sea waves as they were still called in the 19th century – are very complex and were a puzzle to early seismologists. Particularly the Lisbon Earthquake (1755) had clearly demonstrated a correlation between earthquakes and tsunamis. Although the origin of seismic waves by earth- and seaquakes was already recognized by the German natural philosopher Immanuel KANT (1756) and the British geologist Reverend John MICHELL (1760), the origin of these vertical displacements of the sea floor was disputed, and most contemporary naturalists favored the hypothesis of submarine explosions. In his memoir *On the Dynamics of Earthquakes* (1846), MALLET illustrated earthquake motion in the Earth's crust and the seismic origin of sea waves. His schematics illustrate the general relation of earthquake phenomena in their successive occurrence. *(A)* Following a submarine eruption, the fast Earth wave, prior to the great sea wave, after passing the ship at sea, has arrived at land, marked by a tower falling upon the shore (*left*). The great sea wave and its minor successors are still on their way toward land. This sequence of the following three schematics shows the effects upon the great sea wave of its arrival from deep water upon a shore, which suddenly shelves by steep escarpments. *(B)* The great sea wave advances as a solitary mass with equal front and rear slopes over the deep sea. *(C)* After reaching the sounding edge, the front face of the wave becomes steep while the rearward slope flattens and the water at the beach is in the process of receding. *(D)* Shortly thereafter, the solitary wave is broken into several smaller ones, at heights bearing relation to the shallow water beneath, and the leading wave is about to form a great "breaker" upon the shore, no longer having any depth to remain unbroken. [Trans. Roy. Irish Acad. **21**, 51 (1846)]

# 4 PICTURE GALLERY

## 4.1 SHOCK AND PERCUSSION IN NATURE – Tsunami Caused by Subduction

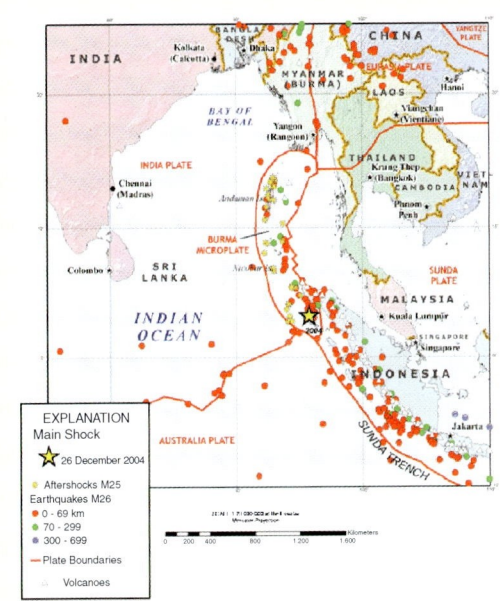

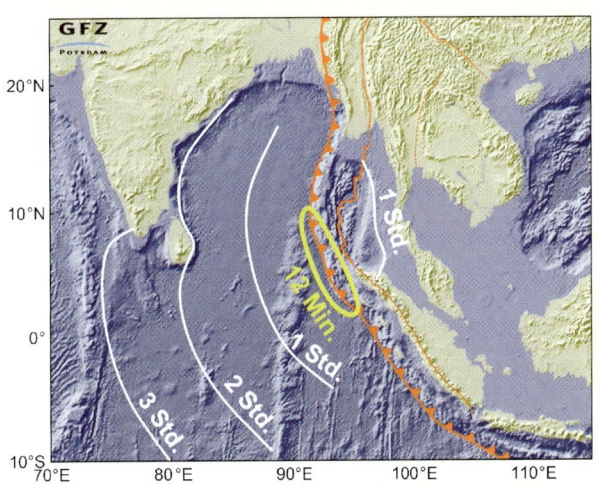

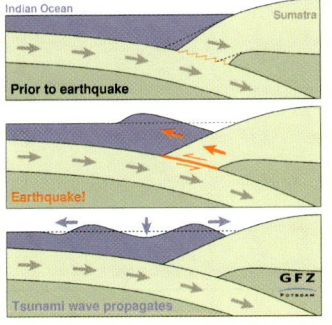

The velocity of an ocean wave $v_O$ whose length $L$ is sufficiently large compared to the still water depth $H$ (i.e., for $L > 25\,H$) can be approximated by $v_O = (gH)^{1/2}$, where $g$ is the acceleration of gravity. Thus, the velocity of such "shallow-water waves" is independent of wavelength $L$. As the wave approaches the coast, the wavelength decreases and the wave height increases. The rupture, triggered by frictional instability and initially affecting only a relatively small area, propagated outside the hypocenter northwestward about 1,200 km along the fault with the shear-wave velocity $v_S$ (presumably ranging between 2 and 4 km/s). The required time for the rupture is then of the order of only 5 to 10 minutes. Even in deep water ($L \gg H$) $v_O$ is much smaller (about 5%) than the rupture velocity, and the fault displacement was quasi-instantaneous.
• Theoretical, numerical, observational and experimental research indicates that shear ruptures can propagate faster than the shear wave speed {ROSAKIS ET AL. ⇨1999}. The rupture velocities $v_R$ of some fault segments in recent large strike-slip earthquakes have been interpreted as propagating at transonic ($v_S < v_R < v_L$) or even supersonic ($v_R > v_L$) velocities (in respect of the longitudinal velocity $v_L$).

**4.1−N** *(cont'd)* **Left, top:** On December 26, 2004, the world's largest earthquake in 40 years hit southern Asia (about 230,000 casualties). Known as the "2004 Sumatra-Andaman Earthquake" it occurred at the interface between the India and Burma plates and was caused by the release of stresses that develop as the India plate subducts beneath the overriding Burma plate. [Courtesy USGS] **Right:** The magnitude 9 quake, which occurred under the sea floor at a depth of more than 10 km and had its epicenter about 200 km off the west coast of northern Sumatra, produced a destructive tsunami that traveled at speeds of up to 800 km/s and arrived at all coasts along the Indian Ocean. Countries in the vicinity of the quake such as Indonesia, Thailand, Malaysia, Sri Lanka, and the Maldives were particularly affected by the tsunami. Note that the tsunami arrived at the west coast of Sumatra only 13 min after the quake but needed about 1 h to reach the west coast of Thailand and 2 h to reach the east coast of Sri Lanka. Many lives could have been saved by a tsunami warning system, which, however, did not exist. Just three months apart, in March 2005, another earthquake happened along the Sunda megathrust and produced a smaller but still considerable tsunami. **Left, bottom:** Schematic illustrating the tsunami wave development mechanism by the sudden release of strain that shifted the Burma plate upward by ca. 10 m, thus causing the overlying water to move up and down and generating the tsunami. [Courtesy Geoforschungszentrum Potsdam; http://www.gfz-potsdam.de/news/recent/archive/20041226/Downloads/index.html]

## 4.1 SHOCK AND PERCUSSION IN NATURE – Examples of Early Tsunami Research

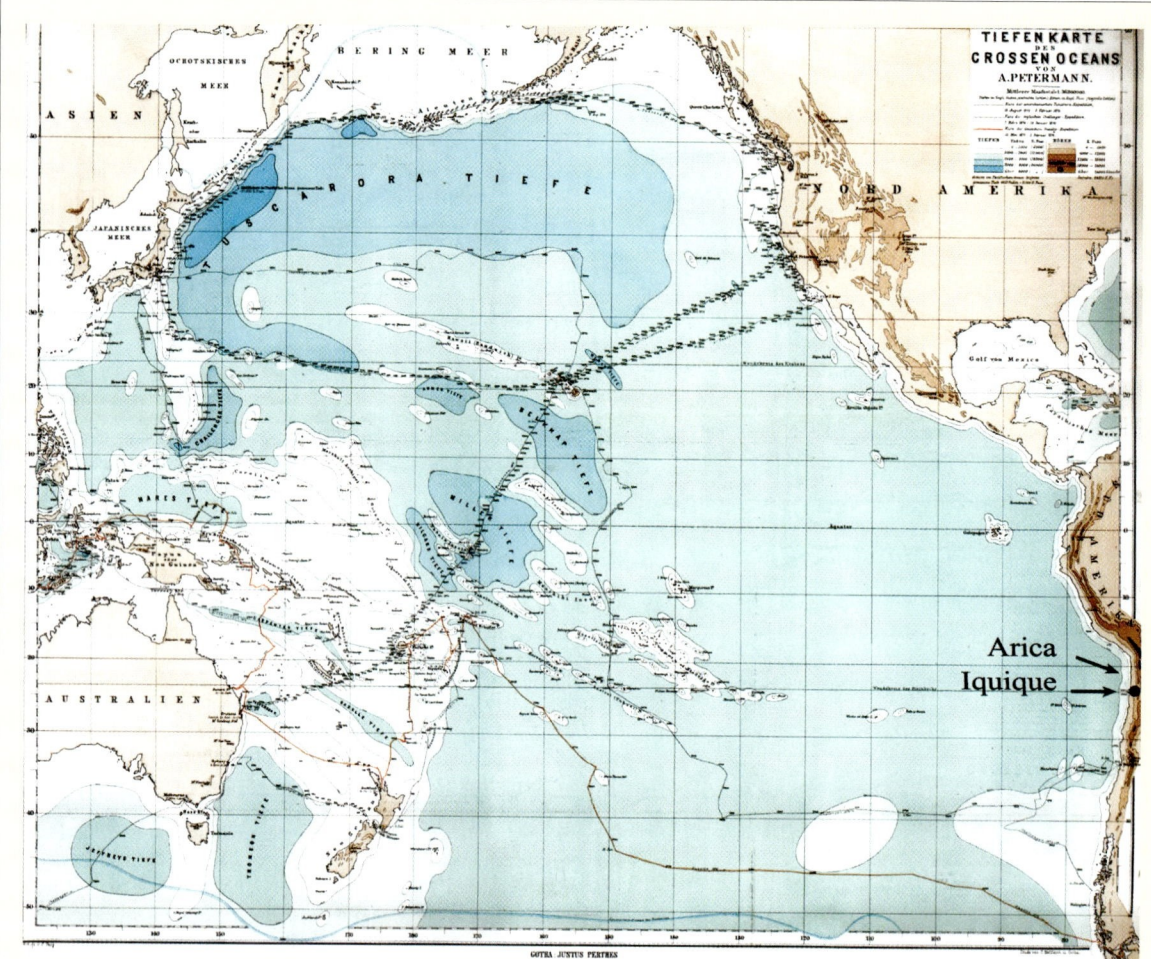

**4.1–O** The German geographer August PETERMANN published in 1877 the first chart of bathymetric data of the Pacific Ocean, based on 738 deep-sea soundings performed during the expeditions of the HMS *Challenger* (1872), the USS *Tuscarora* (1873–1876), and the German SMS *Gazelle* (1874–1876). It first revealed the existence of submarine ridges of considerable depth, later identified as sources of submarine landslides causing dangerous tsunami effects at remote coasts. However, as shown in the map, the ridge west of the South American coast, a seismically active region, was not yet fully known. Inspired by the Arica Earthquake (1868), which created destructive tsunami effects at remote coasts of the Pacific Ocean, German geologist Ferdinand VON HOCHSTETTER proposed to calculate the average velocity of the seismic sea wave $v_{AV}$ by the formula $v_{AV} = (h_{AV} g)^{1/2}$ and to compare the result with an average velocity $w_{AV} = s/(t_2 - t_1)$. Here $h_{AV}$ is an average depth calculated from bathymetric data, $g$ the gravitational acceleration, $t_1$ the time at which the earthquake started, and $t_2$ the arrival time of the seismic sea wave at remote distance $s$. He found reasonable agreement for waves traveling freely between Arica and remote coasts ($v_{AV} \approx w_{AV}$). His hypothesis was also confirmed by his colleague F. Eugen GEINITZ on the occasion of the Iquique Earthquake (1877), which caused tsunami effects as destructive as the Arica Earthquake. [Mitth. aus Justus Perthes' Geograph. Anstalt **23**, 125 (1877)]

## 4.1 SHOCK AND PERCUSSION IN NATURE – Examples of Early Tsunami Research *(cont'd)*

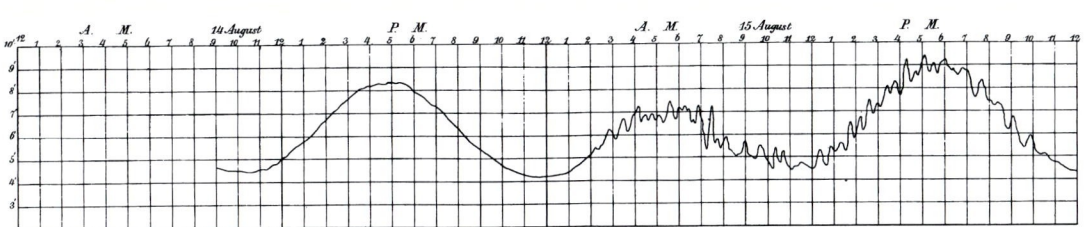

**4.1–P** *Top:* At remote coasts a water-depth gauge records a tsunami as a number of spikes superimposed on the daily fluctuations caused by the tides. This record of a self-registering gauge was taken in August 1868 at Sydney harbor after the violent Arica Earthquake (1868) with its epicenter near the coastal town of Arica, Chile. A similar record was obtained at South Georgia (an island in the South Atlantic Ocean) after the explosive eruption of Krakatau volcano. It first showed that, despite different causes, similar tsunami effects are generated at remote coasts of the Pacific Ocean. [Sitzungsber. Akad. Wiss. Wien **60** (II), 818 (1869)] ***Bottom:*** This map shows schematically epicenters of earthquakes that occurred in Japan during the period 684–1960 and were accompanied by tsunamis. [Proc. Tsunami Meetings at the 10th Pacific Science Congr., University of Hawaii, Honolulu (1961). Institut Géographique National, Paris (1963), p. 260]

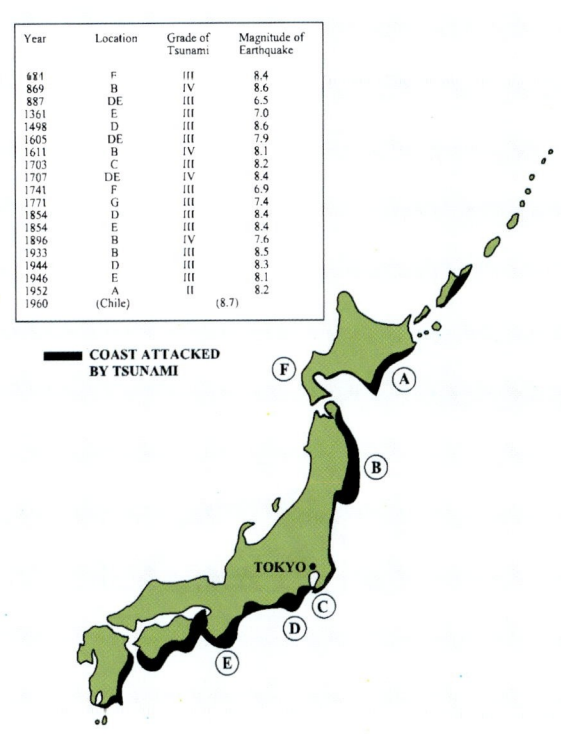

← Japan has frequently been pounded by tsunamis that caused tremendous loss of life and damage on the coast. Since tsunami waves follow an earthquake within a few minutes, it was recognized early that a causal connection must exist between earthquakes and tsunamis. However, the complex mechanism of tsunami generation has been a subject of debate among scholars that has continued to this day. Historical tsunami research may prove helpful in analyzing the frequency of occurrence of tsunamis and their relationship to large earthquakes.

The islands of Japan lie along the circum-Pacific seismic zone. This figure shows the locations and magnitudes of large earthquakes that had their epicenters on the sea bottom very close to the coasts along which the great tsunamis were generated. In the table the magnitudes of earthquakes are given in terms of the Richter-Gutenberg scale and the grades of tsunamis in terms of total energy:

Grade II: $4 \times 10^{22}$ ergs;
Grade III: $16 \times 10^{22}$ ergs, and
Grade IV: $64 \times 10^{22}$ ergs.

The most powerful tsunamis over the last 100 years were the Sanriku tsunamis (location B), which occurred in 1896 (Grade IV) and 1933 (Grade III). On the Sanriku Coast many tsunami monuments have been erected bearing the warning "Expect a tsunami if you feel an earthquake!" However, dangerous tsunamis generated off the coast at very remote distances – so-called "teletsunamis," such as those resulting from the Chilean Earthquake (1960), topographically diametrically opposite Japan – also proved most destructive to Japan's coasts.

## 4.1 SHOCK AND PERCUSSION IN NATURE – The Tsunami in Indian Mythology

**4.1–Q** Rockslides in the Lituya Bay (southern Alaska) can transform into dangerous tsunamis. ***Left, top:*** In July 1958, a rockslide, triggered by an earthquake, occurred along the eastern wall of the Gilbert Inlet. The mass of rock striking the surface of the bay created a giant splash, which sent water surging to a height of 1,720 ft (524 m) across the point opposite the inlet. The giant local tsunami sweeping across the bay inundated approx. 5 square miles (15.5 km$^2$) of land along the shores of Lituya Bay, sending water as far as 3,600 ft (1,027 m) inland. This map shows a close-up of Lituya Bay. [Tsunami Research Group, University of Southern California; http://www.usc.edu/dept/tsunamis/alaska/1958/webpages/lituyacloseup.html] ***Center:*** This aerial photo was taken in August 1958, one month after the earthquake. [Courtesy USGS Photographic Library] ***Bottom, left:*** A unique example of illustrating wave discontinuities is kept at the Museum of the National American Indian, Heye Foundation, New York. [Courtesy National Museum of the American Indian, Smithsonian Institution. NMAI photo no. T009205] ***Bottom, right:*** Reproduction of a carving on an about 500-year-old wooden box excavated from an archaeological site on Kodiak Island, Alaska – the oldest known illustration of a volcanic eruption and volcanic tsunami in the Western Hemisphere. [Courtesy Prof. James E. BEGÉT, Alaska Volcano Observatory; *Encyclopedia of volcanoes.* Academic Press, San Diego (2000), p. 1007]

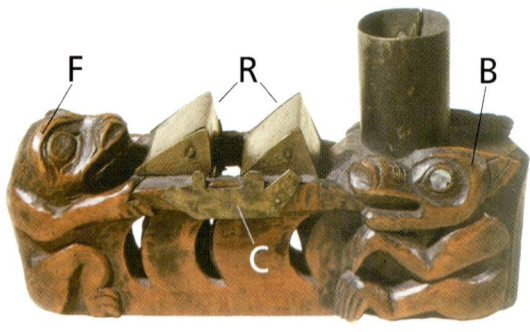

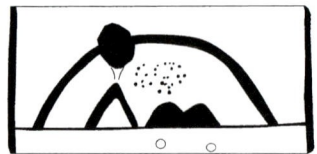

← According to an old Indian legend and illustrated on this beautiful Tlingit Indian pipe, a froglike monster *F* – today identified as a large geological fault running across the mouth of the bay – sits opposite a bearlike demon *B* who dwells in the underwater caverns close to the bay entrance. Note the two enormous ridges *R* created by the froglike monster swamping a two-man canoe, shown here in the form of a brass plate *C*.

## 4.1 SHOCK AND PERCUSSION IN NATURE – Destructive Tsunami Effects

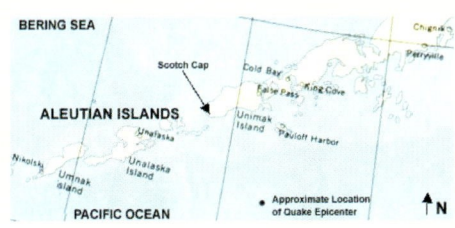

**4.1–R** *Top:* On April 1, 1946, a magnitude 8.0 earthquake with the source about 145 km south of Unimak Island and a focal depth of 25 km occurred in the Aleutian Islands of Alaska. It triggered a Pacific-wide tsunami which had a surface-wave magnitude of 7.8. The tsunami, then bounding 3,800 km to the Hawaiian Island, stroke its coasts 4.9 hours later, in some areas producing waves over 6 meters high. [USC Tsunami Research Center, Los Angeles, CA]. **Bottom:** Destructive effects on Unimak Island by the tsunami [National Geophysical Data Center (NGDC)/USGS; http://www.ngdc.noaa.gov/nndc/struts/results?eq_1=25&t=101634&s=0&d=4&d=44]

←← Scotch Cap Lighthouse on Unimak Island, Alaska as it looked before the earthquake and tsunami. The structure was built in 1940. It was located 9.8 m above sea level and five stories high.

← Remains of Scotch Cap Lighthouse after the tsunami. All five occupants were killed. Only the foundation and part of the concrete sea wall remained. The tsunami also wiped out the 31-m high radio antenna.

**4.1–S** On 26 December 2004, at 8 a.m. local time, a magnitude 9.1 earthquake happened off the coast of Sumatra which triggered a huge tsunami. Because of the fault geometry, the waves propagating to the East (towards Thailand and Myanmar) began with a receding wave which explains why the sea started to retreat minutes before flooding the coast. On the opposite, to the West (towards India and Sri Lanka) a large wave suddenly hit the coast without warning. **Left:** View of the tsunami in the moment of hitting the famous Hat Railay Beach near Krabi in southern Thailand, a tourist resort located about 860 km distant from the epicenter of the quake. The woman in the foreground, running toward the wave, is a Swedish mother attempting to save her three children. They somehow managed to survive. Also the other tourists, running for their lives and rushing to safety, reportedly survived. The water had receded before the first of six tsunami waves struck the beach. About 200 people died in this area. [© Getty Images Deutschland GmbH; Photo by Agence France Presse (AFP), Paris]

## 4.1 SHOCK AND PERCUSSION IN NATURE – Destructive Tsunami Effects *(cont'd)*

Aftershocks, defined as earthquakes that follow the largest earthquake of an earthquake sequence, are smaller than the main shock and within 1–2 fault lengths distance from the main shock fault. They can continue over a period of weeks, months, or years. In general, the larger the main shock, the larger and more numerous the aftershocks, and the longer they will continue.

**4.1–S** *(cont'd)* The 26 December 2004 Earthquake – actually a seaquake – happened on the interface of the India and Burma plates and extended along the subduction zone from western Sumatra to the Andaman Islands {⇨Fig. 4.1–N} – thus named "Sumatra-Andaman Islands Earthquake". Resulting from thrust-faulting in a shallow depth of about 30 km, it produced a destructive tsunami causing more casualties (estimated 230,000) than any other in recorded history. But also remote coasts of other nations bordering the Indian Ocean were affected such as in India, Sri Lanka, Thailand, Malaysia, Kenya, Somalia, Tanzania and on Mauritius, Madagascar, the Nicobar Islands and Andaman Islands, and the Maldives. *Left:* Map showing the epicenters of the 26 December 2004 Earthquake (moment magnitude 9.1) which produced a large tsunami and of the 28 March 2005 Earthquake (moment magnitude 8.7) which, probably a result of stress placed on the fault by the first quake, only produced a small tsunami. Both earthquakes happened along the same Sunda trench fault line and at about the same focal depth of about 30 km. [Courtesy USGS] *Right:* The epicenter of the 26 December 2004 was off the west coast of western Sumatra, only about 250 km apart from Banda Aceh, capital of the Indonesian Province of Aceh. Banda Aceh, located at the northernmost part of Sumatra, was the closest major city to the earthquake's epicenter. The two satellite photos taken with QuickBird show a shore detail of Banda Aceh on June 23, 2004 *(top)* and on December 28, 2004 *(bottom)*, only two days after being struck by the tsunami. [© Getty Images Deutschland GmbH; Satellite Photography, reproduced with permission from DigitalGlobe] This comparison illustrates the enormous destructive power which the tsunami caused in coastal regions of western Sumatra. A considerable portion of the town of Banda Aceh was totally destroyed, but much of downtown, the airport and roads from outside survived so that recovery supplies could get in more systematically than in other parts of the province. According to the U.S. Agency for International Development (fiscal year 2005) over 104,000 people were killed, over 10,000 were missing and over 650,000 were displaced in Indonesia alone.

# 4 PICTURE GALLERY

## 4.1 SHOCK AND PERCUSSION IN NATURE – Rogue or Freak Waves

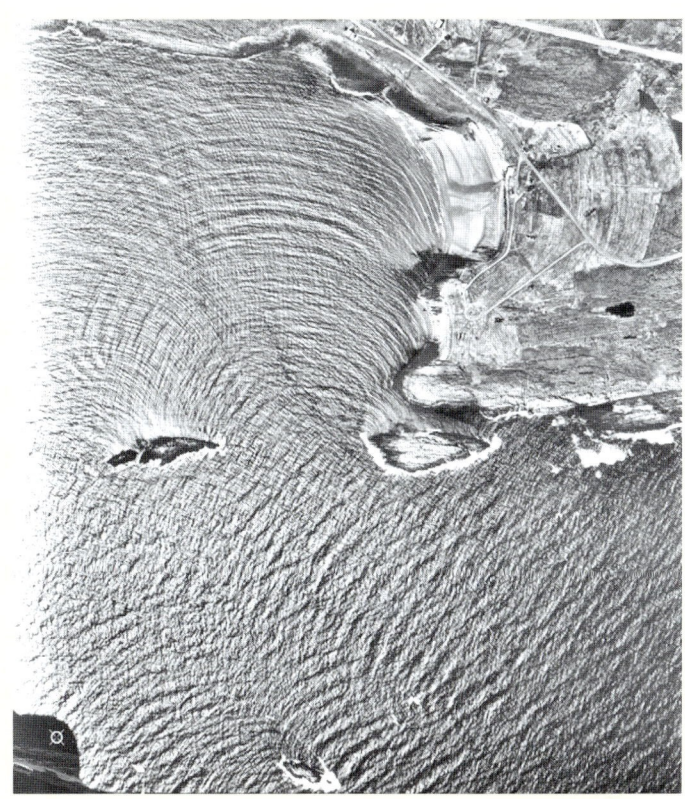

**4.1–T** Rogue waves, also known as "freak waves," are giant sea waves. Presently, there is no consensus about whether they are "normal" extremes of a superposition of waves or the result of totally different generation mechanisms, such as nonlinear interaction and phase locking of wave trains. *Top:* Aerial view of an area near Kiberg on the coast of Finmark, a county in the extreme northeast of Norway, taken on June 12, 1976. Aerial photography allows one to localize dangerous areas where in shallow waters refraction and diffraction of surface waves can cause focusing of wave energy, resulting in waves of unusual amplitudes. In addition, radar images taken by satellites allow one to measure the wave length, crest height, and direction of propagation of ocean surface waves. [Courtesy Vardø Kommune, Norway. Photo by Fjellanger Widerøe Foto AS, Trondheim & Oslo, Norway] *Bottom, left:* The unusual record of an extreme wave was measured on January 1, 1995 with a depth gauge under an oil platform in the North Sea, proving that such waves do indeed exist. [http://www.math.uio.no/~karstent/waves/index_en.html] For centuries sailors have blamed mysterious surges of water for unexplainable sinkings or damage, but the claims have always attracted plenty of skepticism. *Bottom, right:* View of a giant wave shortly before striking a cargo ship with huge force. Extreme waves are studied jointly by Norwegian researchers at the Universities of Oslo and Bergen and at Norges Teknisk-Naturvitenskapelige Universitet (NTNU). [Courtesy BBC, London http://news.bbc.co.uk/1/hi/sci /tech/2450407.stm]

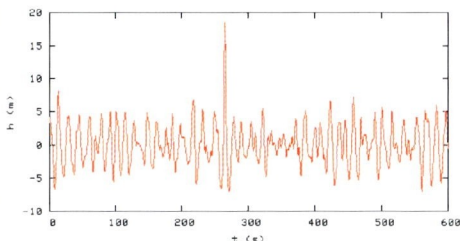

↑ The maximum amplitude of 18.5 m is more than three times the significant amplitude for the wave train. This wave is known in the international scientific community as the "New Year Wave." [Courtesy Prof. Karsten TRULSEN, University of Oslo. Data provided by Drs. Ove T. GUDMESTAD and Sverre HAVER at Statoil, Stavanger, Norway]

870   4 PICTURE GALLERY

## 4.1  SHOCK AND PERCUSSION IN NATURE – Sunspots, Solar Flares, and Prominences

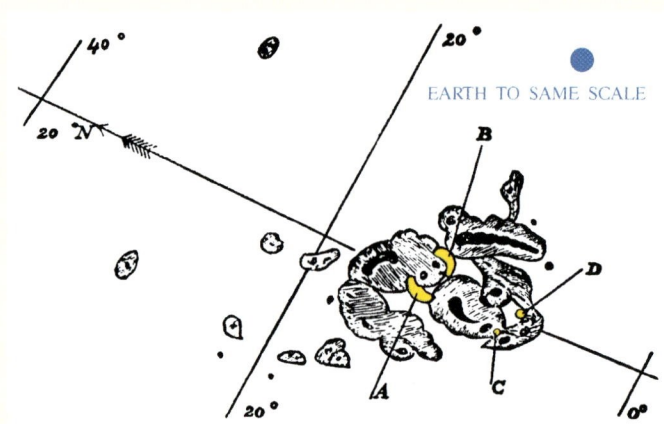

← CARRINGTON observed near a large sunspot (an area of cooler gas on the Sun's surface) two patches of intensely bright light, *A* and *B*, which moved within 5 min to positions *C* and *D*, respectively. Corresponding to a distance of 35,000 miles (about 56,000 km), this indicated an average lateral velocity of about 116 miles/s (187 km/s). The diminished size of spots *C* and *D* clearly shows one of the characteristic features of a solar flare, *i.e.*, that its intensity strongly decreases within a short time. His famous discovery first revealed the enormous dimensions and velocities of solar-surface phenomena. The techniques of CARRINGTON and his contemporaries gave birth to the new science of solar physics, which played a key role in the evolution of modern astrophysics.

**4.1–U**  *Top:* In 1860, the English astronomer Richard C. CARRINGTON customarily observed sunspots using a telescope that projected the image of the Sun's disk onto a plate of glass coated with distemper of a pale straw color. He witnessed unusual violent eruptions, so-called "flares." At the same time, similar observations were made independently by his countryman Richard HODGSON. [Month. Not. Astron. Soc. **20**, 14 (1860)]  **Bottom, left:** Prominences are beautiful solar phenomena with enormous horizontal and vertical dimensions attaining in some cases some 100,000 km. They can be active or quiescent and encompass different types of structures. These prominences were observed by the French artist and amateur astronomer E. Leopold TROUVELOT on April 15, 1872 (*top*) and April 29, 1872 (*bottom*). [*Meyers Konversations-Lexikon*. Bibliographisches Institut, Leipzig (1897), vol. 16, plate II]  **Bottom, right:** This ultraviolet photograph was acquired from NASA's Skylab space station on December 19, 1973. It shows one of the most spectacular solar flares ever recorded that, propelled by magnetic forces emitted by the Sun, spans more than 588,000 km of its surface. [Courtesy NASA-GSFC]

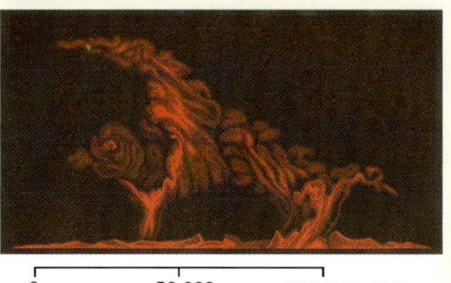

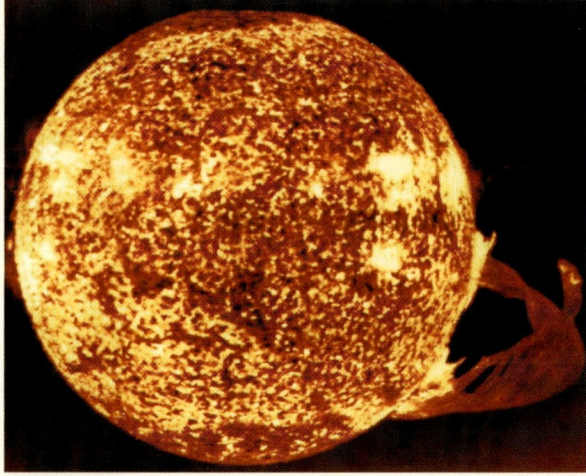

← In 1876, TROUVELOT's pictures were issued as lithographic color prints by Harvard College Observatory and became widely known. Not only are they aesthetic, but they also give a vivid impression of the enormous dimensions of surface phenomena which happen on the Sun, our closest star.

# 4 PICTURE GALLERY

## 4.1 SHOCK AND PERCUSSION IN NATURE – Solar Flares and Solar Quakes

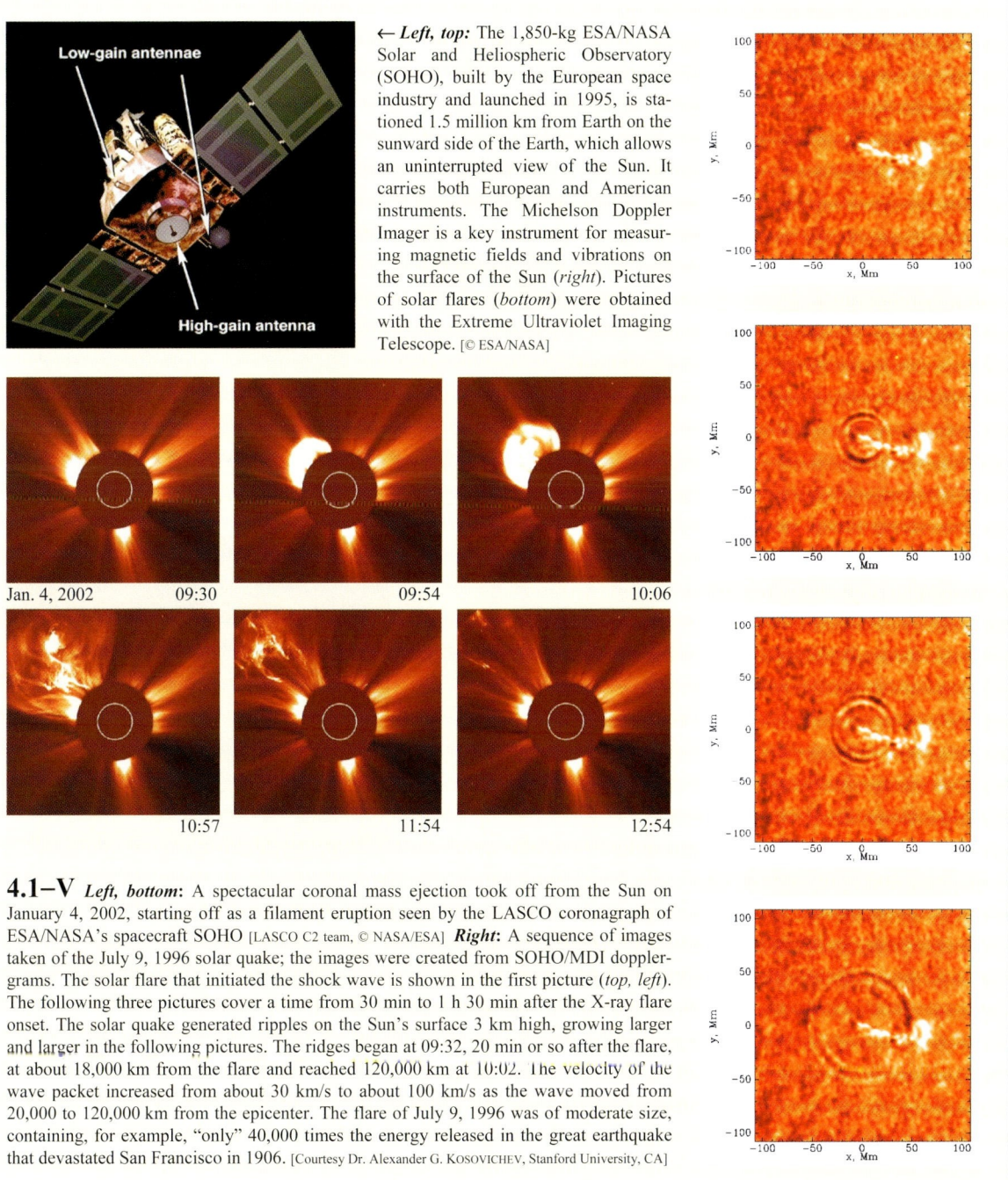

← *Left, top:* The 1,850-kg ESA/NASA Solar and Heliospheric Observatory (SOHO), built by the European space industry and launched in 1995, is stationed 1.5 million km from Earth on the sunward side of the Earth, which allows an uninterrupted view of the Sun. It carries both European and American instruments. The Michelson Doppler Imager is a key instrument for measuring magnetic fields and vibrations on the surface of the Sun (*right*). Pictures of solar flares (*bottom*) were obtained with the Extreme Ultraviolet Imaging Telescope. [© ESA/NASA]

**4.1–V** *Left, bottom*: A spectacular coronal mass ejection took off from the Sun on January 4, 2002, starting off as a filament eruption seen by the LASCO coronagraph of ESA/NASA's spacecraft SOHO [LASCO C2 team, © NASA/ESA] *Right*: A sequence of images taken of the July 9, 1996 solar quake; the images were created from SOHO/MDI dopplergrams. The solar flare that initiated the shock wave is shown in the first picture (*top, left*). The following three pictures cover a time from 30 min to 1 h 30 min after the X-ray flare onset. The solar quake generated ripples on the Sun's surface 3 km high, growing larger and larger in the following pictures. The ridges began at 09:32, 20 min or so after the flare, at about 18,000 km from the flare and reached 120,000 km at 10:02. The velocity of the wave packet increased from about 30 km/s to about 100 km/s as the wave moved from 20,000 to 120,000 km from the epicenter. The flare of July 9, 1996 was of moderate size, containing, for example, "only" 40,000 times the energy released in the great earthquake that devastated San Francisco in 1906. [Courtesy Dr. Alexander G. KOSOVICHEV, Stanford University, CA]

## 4.1 SHOCK AND PERCUSSION IN NATURE – Big Bang Portrait

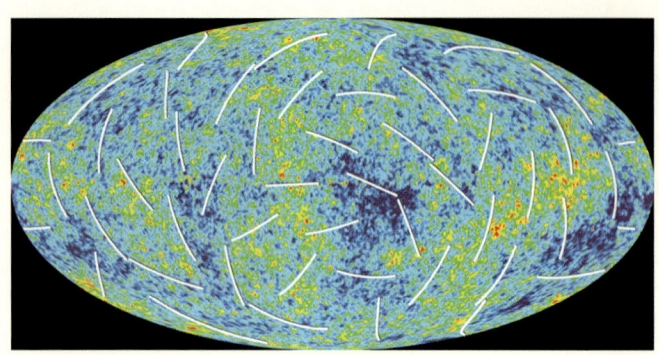

**4.1–W** Cosmic portrait showing afterglow – so-called "cosmic microwave background" – of the Big Bang that occurred over 13 billion years ago. It was captured using NASA's Wilkinson Microwave Anisotropy Probe (WMAP) during a sweeping 12-month observation of the entire sky. This is a full-sky map of the oldest light in the Universe released by NASA-GSFC on Feb. 11, 2003. The microwave light captured in this picture is from 379,000 years after the Hot Big Bang, so to speak a "baby" picture of the Universe in galactic coordinates. The oval shape is a projection to display the whole sky, similar to the way the globe of the Earth can be represented as an oval.
[Courtesy NASA/WMAP Science Team, NASA-GSFC; see http://map.gsfc.nasa.gov/m_mm.html]

↑ Colors indicate "warmer" (*red*) and "cooler" (*blue*) spots. The *white bars* show the "polarization" direction of the oldest light

## 4.1 SHOCK AND PERCUSSION IN NATURE – Earth's Bow Shock

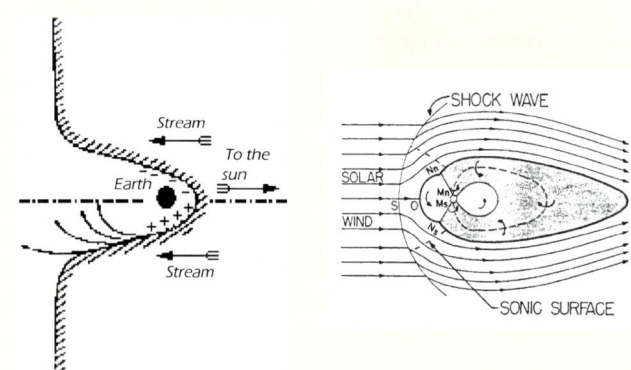 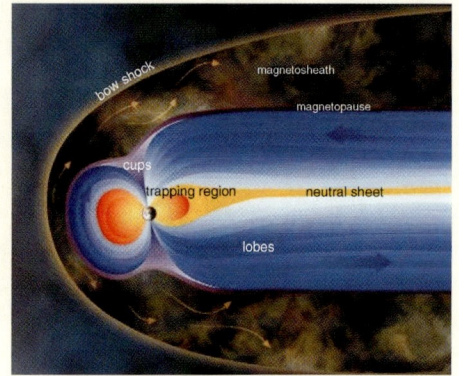

**4.1–X** *Left:* In 1931, the British astrophysicists Sydney CHAPMAN and Vincent FERRARO proposed that the Sun sent out huge clouds of electrically neutral plasma and that magnetic storms arose when those clouds enveloped the Earth. Many magnetic storms were observed to begin with a small steplike jump in the magnetic field, taking just a minute or so. CHAPMAN and FERRARO also proposed that such jumps marked the cloud's arrival. They realized that the strong field of the Earth would hold off the cloud, carving a cavity in the cloud in which the Earth and its magnetic field would be confined. Their theory proved to be prophetic, except for one important detail: the flow of plasma from the Sun was not confined to isolated clouds, but went on all the time, in the form of the solar wind. [Terr. Magnet. Atmos. Electric. **36**, 171 (1931). See also D.P. STERN and M. PEREDO: *The magnetopause – history*; http://www-spof.gsfc.nasa.gov/Education/whmpause.html] ***Center:*** In 1962, the astrophysicists William I. AXFORD and Paul J. KELLOG predicted independently of each other that a shock wave must exist in the interplanetary medium a short distance upstream of the Earth's magnetosphere. This discontinuity is created when the supersonic solar wind ($M = 3$–$20$) encounters the magnetic field of the Earth. [J. Geophys. Res. **67**, 3791, 3805 (1962) © American Geophysical Union (AGU). Reproduced by permission of AGU] ***Right:*** Based on an increasing amount of data obtained by various satellite missions the schematic of the magnetosphere underwent numerous refinements. The solar wind, a stream of charged particles continuously being ejected by the Sun, flowing past the Earth, and interacting with the Earth's magnetic field, generates a bow shock similar to an object exposed to a supersonic flow. [Courtesy NASA-GSFC; http://www.eng.vt.edu/fluids/msc/gallery/shoks/earth.htm]

# 4 PICTURE GALLERY

## 4.1 SHOCK AND PERCUSSION IN NATURE – Cosmic Jets, Shock Waves, and Mach Cones

**4.1–Y** *Left, top:* Astrophysical jets are highly collimated beams of matter and energy that are ejected from some astronomical objects. The Herbig-Haro object HH 47, a small bright nebula in a star-forming region, expels an energetic beam of charged particles into interstellar space. The jet, here moving from left to right, moves at a velocity of nearly 300 km/s. [Courtesy NASA-GSFC; http://antwrp.gsfc.nasa.gov/apod/ap951012.html] *Left, center:* Cosmic tornadoes – here shown using the example of HH 49/50 – are mysterious astrophysical jets with a spiral structure. They are energetic outflows associated with the formation of young stars. [J. BALLY (Univ. of Colorado) ET AL., JPL-CalTech, NASA; http://www.astronet.ru/db/xware/msg/1211349] *Left, bottom:* Astrophysical shock waves can arise in a number of different situations. For example, an explosive event may generate supersonic flows that, impacting the surrounding gas, generates a cosmic bow shock. BZ Cam is a binary star system that contains an accreting white dwarf and a 0.3 to 0.4 solar mass main-sequence donor. It lies about 2,500 light-years away toward *Camelopardalis* ("The Giraffe," a constellation of the Northern Hemisphere) and expels an unusually large wind of particles that creates a large bow shock as the system moves through the surrounding interstellar gas. [R. CASALEGNO, C. CONSELICE ET AL.; WIYN, NOAO, MURST, NSF; http://antwrp.gsfc.nasa.gov/apod/ap001128.html] *Right:* Based on X-ray images taken by the Roentgen satellite ROSAT in the spectral range 0.1 to 2.4 keV, astrophysicists at the MPE (Garching, Bavaria) discovered explosion fragments (labeled *A–F*) well outside the *Vela* supernova remnant shock-wave boundary. The area containing feature *D* shows evidence for the superposition of two bow-shaped objects. They suggested that X-ray emission associated with the protruding objects was produced by shock-heating of the ambient medium by supersonic motion of the objects. For the protruding objects *A–E*, which they interpreted as Mach cones, they calculated Mach numbers ranging from 4 (for object *A*) to 2.4 (for object *E*). [Nature **373**, 587 (1995); Reprinted with permission of Dr. Bernd ASCHENBACH, MPI für extraterrestrische Physik (MPE), Garching]

## 4.1 SHOCK AND PERCUSSION IN NATURE – Animal World

**4.1–Z** Only a few animal species use the principle of percussion as a tool. ***Top, left:*** The Egyptian vulture [*Neophron percnopterus*] uses a stone, ranging in weight from 100 to 300 g, as a tool to fracture the shell of an ostrich-egg. He picks the stone in his beak and hammers or throws it against the shell. [Photo by H. VAN LAWICK and J. VAN LAWICK-GOODALL. Reprinted with permission from Nat. Geographic Mag. **133**, No. 5, 631 (1968). ©2007, National Geograhic Image Collection] ***Top, right:*** Tool-using is also practiced by the California sea otter [*Enhydra lutris nereis*]. It applies small stones to detach and open shellfish, particularly abalones which are often tightly attached to the rocks. It usually eats while floating on its back as shown here. If it cannot get at the fleshy animal inside the shell, it will hold the shell against its chest with one paw and pound it with a stone. It often tucks a good stone under an armpit as it swims or dives. [Photo by W.F. BRYAN, reprinted from D.R. GRIFFIN, Am. Scient. **72**, 456 (1984)]

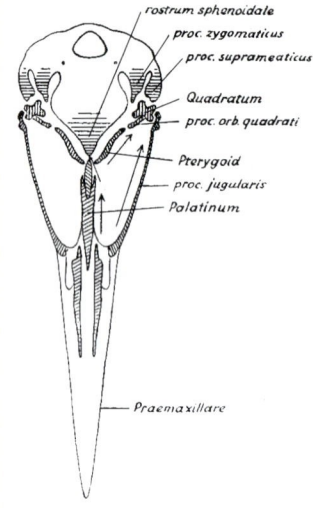
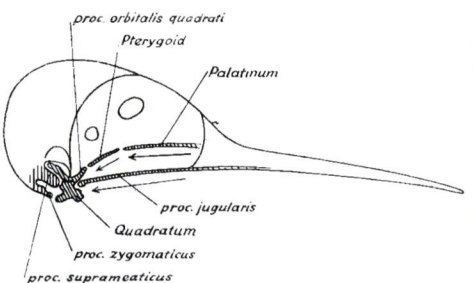

***Bottom, left & center:*** Woodpeckers hammer their lives away for feeding, nest construction and drumming, up to 20 times a second. Their thick bony skulls with relatively spongy bones partially cushion the incessant blows, thus protecting the brain. In the early 1950s, the German neurologist Fritz BECHER studied the skull and beak anatomy, and how the impact force is transmitted and absorbed. In his schematic of the skull of a woodpecker the arrows indicate the direction of the impact force which splits into several branches, thus favorably bypassing the shock-sensitive brain. The hatched zones indicate those parts of the skull which are most stressed during hammering action. Inside the skull there is almost no cerebrospinal fluid which further prevents shock transmission to the brain. [Reprinted from Z. Naturforschung **8B**, 192-203 (1953)] ***Bottom, right:*** Woodpeckers [family *Picidae*] are found worldwide including over 200 species. View of the German "Buntspecht" [*Dendrocopos mayor*] which lives in central Europe. [Reprinted from J.F. NAUMANN: *Naturgeschichte der Vögel Mitteleuropas*. Köhler, Gera (1901), Bd. IV, Tafel 31]

# 4 PICTURE GALLERY

## 4.1 SHOCK & PERCUSSION IN NATURE – Animal World *(cont'd)*

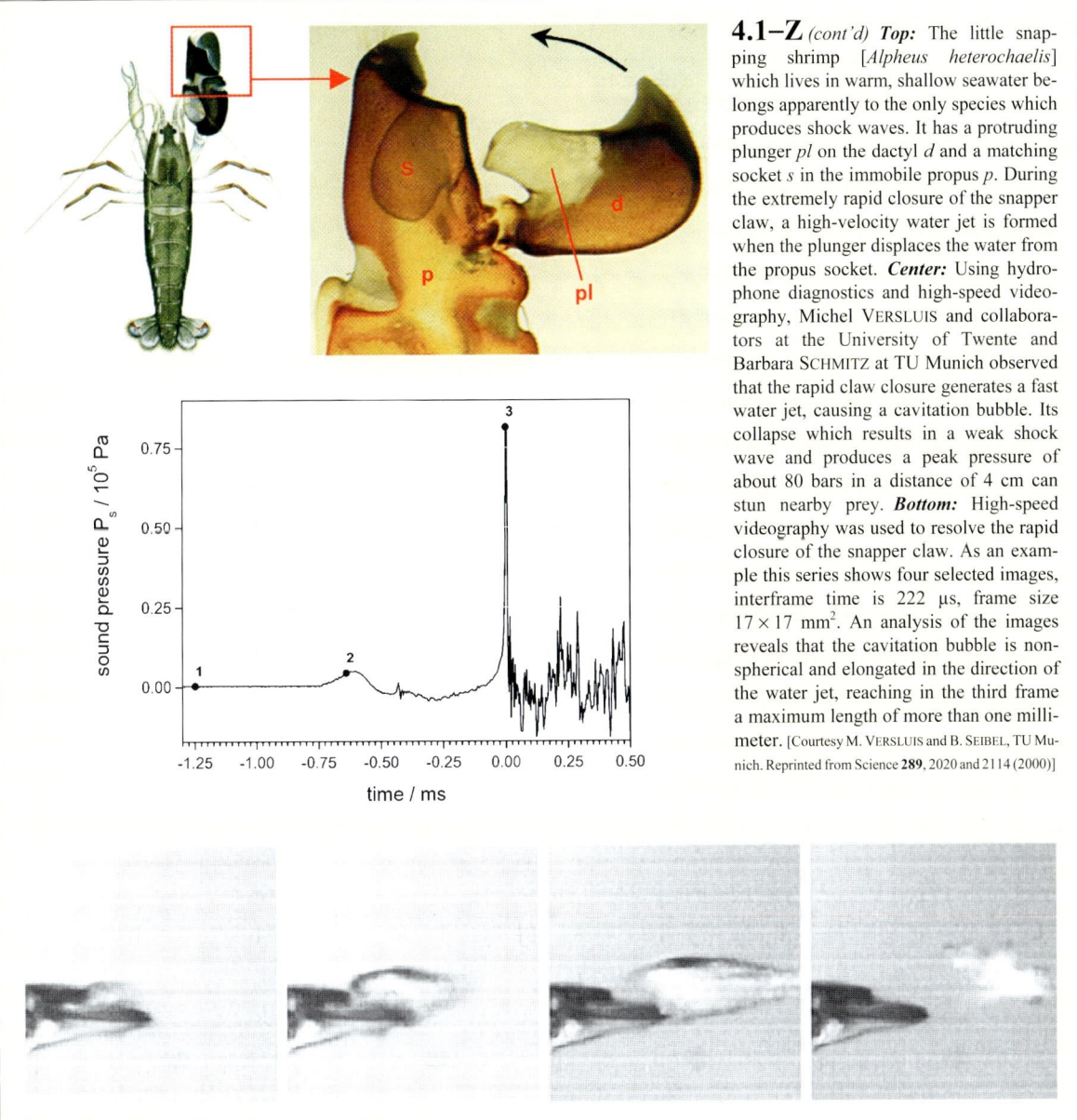

**4.1–Z** *(cont'd)* ***Top:*** The little snapping shrimp [*Alpheus heterochaelis*] which lives in warm, shallow seawater belongs apparently to the only species which produces shock waves. It has a protruding plunger *pl* on the dactyl *d* and a matching socket *s* in the immobile propus *p*. During the extremely rapid closure of the snapper claw, a high-velocity water jet is formed when the plunger displaces the water from the propus socket. ***Center:*** Using hydrophone diagnostics and high-speed videography, Michel VERSLUIS and collaborators at the University of Twente and Barbara SCHMITZ at TU Munich observed that the rapid claw closure generates a fast water jet, causing a cavitation bubble. Its collapse which results in a weak shock wave and produces a peak pressure of about 80 bars in a distance of 4 cm can stun nearby prey. ***Bottom:*** High-speed videography was used to resolve the rapid closure of the snapper claw. As an example this series shows four selected images, interframe time is 222 µs, frame size $17 \times 17$ mm$^2$. An analysis of the images reveals that the cavitation bubble is non-spherical and elongated in the direction of the water jet, reaching in the third frame a maximum length of more than one millimeter. [Courtesy M. VERSLUIS and B. SEIBEL, TU Munich. Reprinted from Science **289**, 2020 and 2114 (2000)]

## 4.2 PERCUSSION IN THE EVOLUTION OF TECHNOLOGY – Basic Tool of Civilization

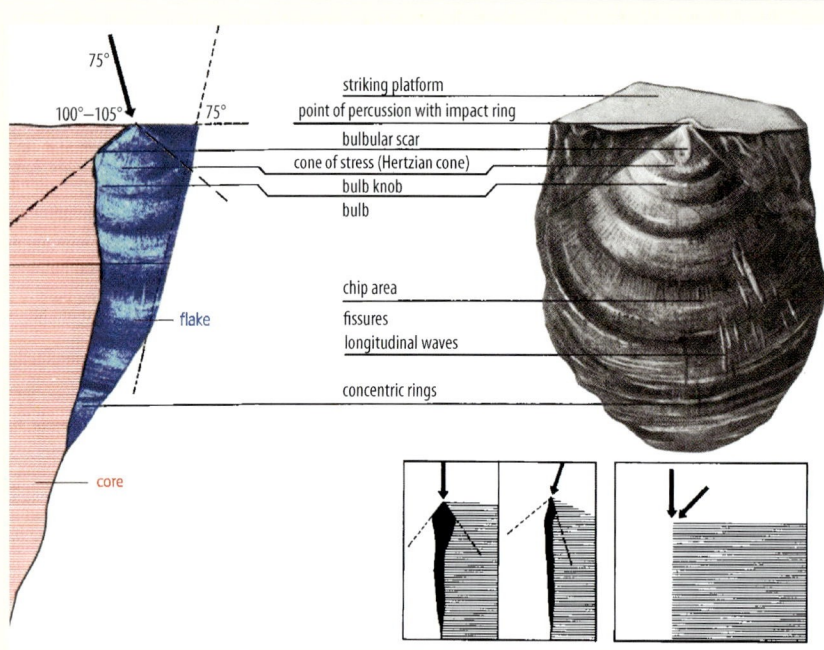

← The flake (*left*) was processed into a handaxe, a knife, a saw, an arrow point, a scraper, a bore, *etc.* The core (*right*) was discarded, processed into further flakes, or used as a heavy-duty chopping tool. [After L. PFEIFFER: *Die steinzeitliche Technik*. Fischer, Jena (1912); and P. HONORÉ: *Es begann mit der Technik*. Deutsche Verlagsanstalt, Stuttgart (1969)]

**4.2–A** *Left:* The basic method of stone fracture and disintegration, used by primitive humans throughout a period of at least 2 million years, is based on the generation of a *percussion bulb* extending from the point of percussion into the stone. This allowed the user to manufacture crude, but very effective, stone tools and weapons of great diversity. **Right:** Drawing of an artificial cone of flint produced by a single blow of a hammer. [J. EVANS: *The ancient stone implements*. Longman & Green, London (1872), p. 247] The term "bulb of percussion" was apparently coined in the middle of the 19th century by the Scottish palaeontologist Hugh FALCONER (1808–1865).

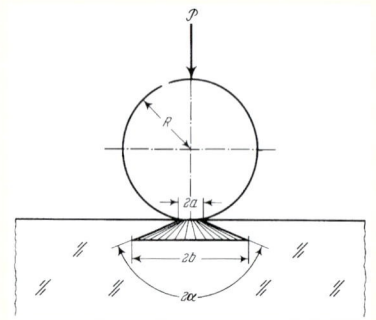

← Schematic of the formation of *Hertzian cone* with apex angle $2\alpha$ when a glass cube is loaded by a steel sphere under pressure $P$, such as by a static load or dynamically by impact. The circular contact area between sphere and glass cube has the diameter $2a$.

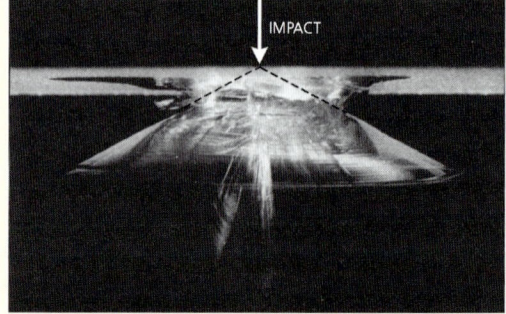

**4.2–B** Frank KERKHOF, a German applied physics professor and expert on fracture mechanics at EMI, Freiburg, and Hans Jürgen MÜLLER-BECK, a prehistorian at the University of Tübingen, first recorded cinematographically the growth of the "bulb of percussion" – a fracture residue of the "Hertzian cone," which they generated by impacting a glass cube at the top with a 30-mm-dia., 3.25-kg semispherical steel anvil dropped from a height of 23 cm. *Left:* Schematic illustrating the generation of the Hertzian cone. **Right:** This photo was taken in reflected light and shows the fully grown Hertzian cone. Its surface represents the fracture area that, from the point of impact, extends into the interior at an angle of $2\alpha \approx 135°$. [Glastech. Berichte **42**, 439 (1969)]

## 4.2 PERCUSSION IN THE EVOLUTION OF TECHNOLOGY – Basic Tool of Civilization *(cont'd)*

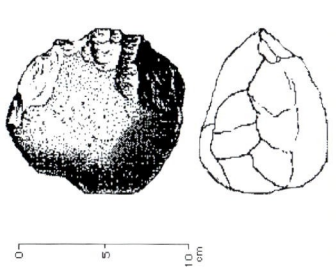

↑ Schematic of side chopper found at Olduvai Gorge. Flakes, provided with razor-sharp edges, were most effective at slicing through animal hide, cutting up and defleshing carcasses. Cores were most suited to removing dried flesh and smashing bones to get access to brain and bone marrow.

**4.2–C** *Left:* The first stone tools appearing in the fossil record around 2.6 million years ago were discovered in East Africa in 1959 by the British archeologists Mary and Louis LEAKEY at Olduvai Gorge, northern Tanzania. They have been taken as a key indicator of an early human species that anthropologists have called "australopithecines." [Photo by Guston SONDIN-KLAUSNER taken in 2006; http://commons.wikimedia.org/wiki/Image:Olduvai_Gorge.jpg] *Right:* Typically, these hominoids applied lava cobbles (cores) with flakes chipped away simply by hitting with another stone, the so-called "direct percussion technique" (Oldowan stone tools, 2.6 to 1.5 million years ago). [Dept. of Anthropology, UC Santa Barbara; http://www.mc.maricopa.edu/dept/d10/asb/anthro2003/archy/lithictech/lithictech5.html]

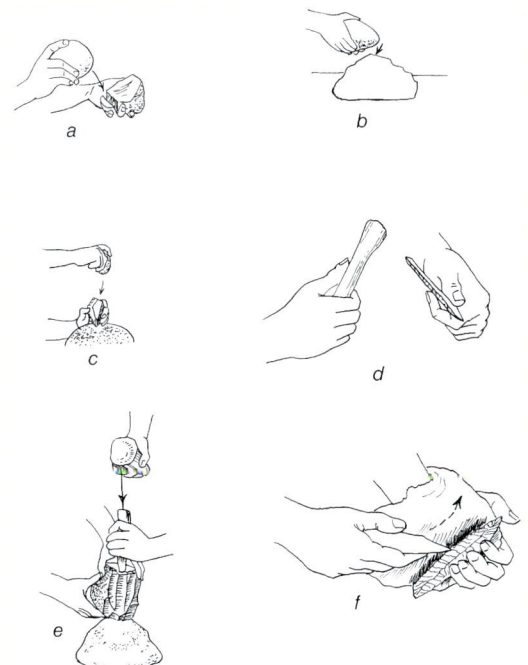

**4.2–D** Schematics of basic examples of major dynamic and static high-pressure techniques as practiced in the Old Stone Age (about 2,500,000–10,000 years ago) for the manufacture of stone tools and weapons. The various dynamic techniques, using percussion in a direct or indirect manner, are the oldest ones and probably all originated in East Africa.

(1) In the "direct percussion method," the core was held in the hand and struck with a hammer of stone, wood, or bone. *(a)* hard-hammer, free-hand percussion; *(b)* anvil technique; *(c)* bipolar technique; and *(d)* soft-hammer percussion.

(2) In the "indirect percussion method," *(e)* the core was placed on a large stone on the ground and the flake detached with a hammer and punch. Or the core was placed on a pointed anvil stone and struck at the top with a hammer (not shown here). This indirect percussion method was refined in the Late Pleistocene (100,000–35,000 years ago) and produced very delicate and fine blades.

(3) Blades could also be detached by generating the Hertzian cone statically, for example *(f)* by exerting a static pressure on the core with a pointed tool such as a sharpened piece of antler or bone, a technique known as "pressure flaking." [I. TATTERSALL ET AL. (eds.): *Encyclopedia of human evolution and prehistory.* Garland, New York (1988), p. 545]

## 4.2 PERCUSSION IN THE EVOLUTION OF TECHNOLOGY – Basic Tool of Civilization *(cont'd)*

**4.2–E** *Left, top:* The handaxe, based on the "direct percussion method" and developed from the crude pebble chopper, was one of the principal tools in ancient Egypt for working stone surfaces. Many were made of dolerite, a very hard volcanic stone, and have been found at Giza, closely to the great pyramids. Another important tool based on the principle of direct percussion was the adze with a short handle, made with a metal blade of bronze or copper and used in the quarries of ancient Egypt. *Left, bottom:* An example of a tool based on the "indirect percussion method" is the chisel, which was applied together with a wooden mallet. It was the basic tool for most crude and fine masonry of ancient cultures. Typically, mallets have short handles, and the amplitude of swing is small, allowing a succession of rapid blows without undue operator fatigue. To provide energy and momentum, the mallet head is heavy. Being of wood, it does not rebound in the same manner as a metal head but stays on the chisel, which transmits the blow to the cutting edge that focuses it into a small area of stone to be spalled off. Ancient Egyptian chisels consisted of copper, arsenical copper, tin bronze, and leaded tin bronze. In the 26th Dynasty (664–525 B.C.), iron became as common as bronze. The tools shown here were used by the pyramid builders and found in the quarries north of the great pyramids at Giza. [Courtesy Dr. Rosemarie KLEMM, Univ. of Munich; *see also* her book *Steine und Steinbrüche im alten Ägypten.* Springer, Berlin (1993)] **Right, top:** Reconstruction of a scene of a man dressing a limestone block with chisel and mallet. The painting is from the Theban tomb of the Vizier REKHMIRA serving under Pharaoh THUTMOSE III who ruled from 1504–1450 B.C. [Courtesy W. MARTINI, EMI, Freiburg] **Right, bottom:** The Great Pyramid at Giza, on which construction began at about 2,590 years B.C., was built in a relative short period of less than 25 years. It consists of about 2.3 million blocks of white limestone. These blocks, which have an average weight of about 2.5 tons, were essentially produced using only simple tools based on the principle of percussion. [http://www-lib.Haifa.ac.il/www/art/construction_lower-closer.gif]

## 4.2 PERCUSSION IN THE EVOLUTION OF TECHNOLOGY – Early War Machines

 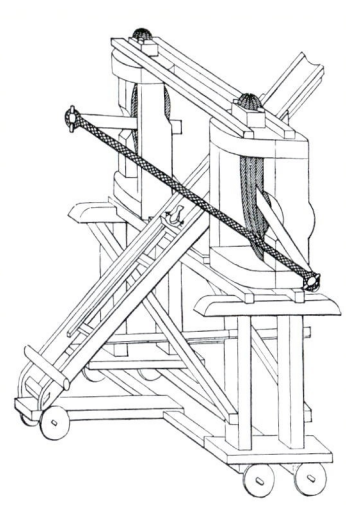

**4.2–F** *Left:* Illustration on an Attic drinking bowl (about 540 B.C.) showing a war galley (*right*) that is going to attack a trading vessel with reefed sail (*left*). Note the ram at the galley's bow, which was commonly fortified by metal. [Courtesy British Museum, London]
*Right:* The catapult or *ballista* was used on a large scale in the Punic Wars (246–146 B.C.). Being a Greek invention, it was capable of bestowing the projectile with a considerable kinetic energy that upon impact was transformed via percussion into destructive power.
[*Meyers Konversationslexikon.* Bibliogrphisches Institut, Leipzig (1894), vol. 2, p. 291]

**4.2–G** *Left:* The "battering ram" (Lat. *aries*), a wooden pole or plank usually tipped with a bronze or iron ram, is based on the principle of percussion and was used in Antiquity until the Middle Ages to break down walls and gates. It was up to 30 m long and was operated by a crew of up to 100 men. To better protect from battlement attacks, a transportable shelter was applied, the ram being suspended from its roof on chains. [F. GROSE: *Military antiquities respecting a history of the English army.* S. Hooper, London (1786)] *Right:* This bronze battering ram is the only surviving besieging instrument of its kind from Antiquity (5th century B.C.). On both sides are symbolic depictions of ram heads, whence its name. [Courtesy Archaeological Museum, Olympia, Greece]

## 4.2 PERCUSSION IN THE EVOLUTION OF TECHNOLOGY – Devices Based on Rapid Expansion

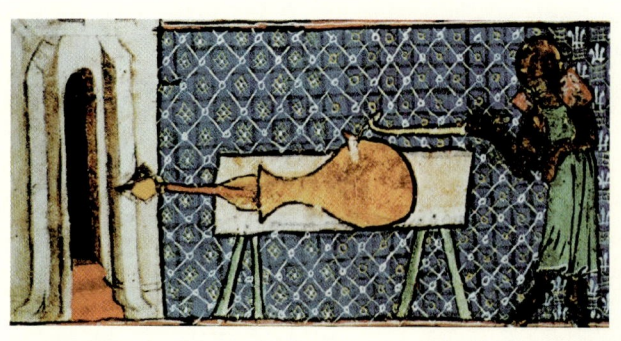 

**4.2−H** *Left:* This is the oldest known illustration (A.D. 1320) of a cannon shot, found in a miniature of the English cleric Walter DE MILIMETE. It shows a subcaliber incendiary dart projectile at the moment of issue from the muzzle. [Courtesy Library of Christ Church College, Oxford] *Right:* In 1861, a bronze cannon was found in the south Swedish town of Loshult. Dating back to the early 14th century, it is the oldest known firearm. Note that its vaselike combustion chamber is similar to the geometry shown in the left picture. [Courtesy Statens Historiska Museum, Stockholm]

 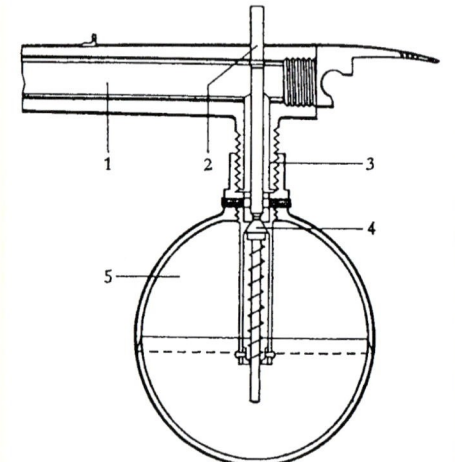

← 
*1* – Barrel 
*2* – Tappet and trigger 
*3* – Joint 
*4* – Spring-loaded cone valve 
*5* – Air pressure reservoir

**4.2−I** *Left:* The "blowgun" was apparently invented by Malaysians and already in pre-Columbian times imported to America and Europe. Applying the human breath to propel a projectile, it is a high-tech pneumatic tubular weapon still in use by some primitives. Veterinarians in zoos and elsewhere also use the blowgun to narcotize wild and large animals before treatment. Blowguns vary in length from 0.45 to 7 m and allow surprisingly precise aiming. The picture shows an Indian from the Amapá Territory, northern Brazil, while loading his approx. 2-m-long blowgun with a thin, pointed arrow. [Orion **12**, 423 (1957)] *Right:* The "wind-gun" or "air-gun," a derivative of the blowgun, was invented in Germany in the 15th century. This schematic shows the construction of a wind-gun dated A.D. 1606. Provided with an attached spherical reservoir of compressed air, it was capable of firing only a single shot, but more advanced gun constructions allowed one to fire a number of shots without recharging. Since the muzzle blast is modest compared to common firearms, it was favorably used by poachers and political assassins and, therefore, outlawed in numerous countries. Modern commercially available wind-guns reach muzzle velocities of up to 280 m/s. [H. MÜLLER: *Gewehre, Pistolen, Revolver*. Kohlhammer, Stuttgart (1979)]

## 4.2 PERCUSSION IN THE EVOLUTION OF TECHNOLOGY – Devices Based on Rapid Expansion
*(cont'd)*

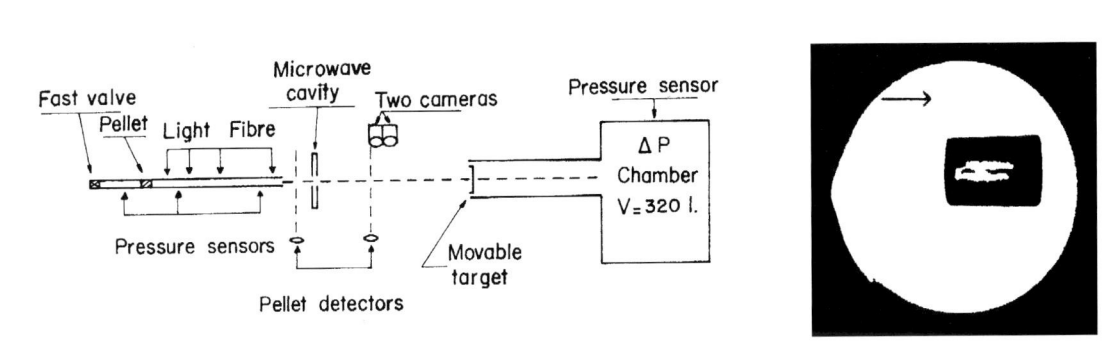

**4.2–J** The principle of the "pneumatic gun" is very versatile. **Left:** It was resumed in the 1980s in France at the Centre d'Etudes Nucléaires de Grenoble in order to accelerate deuterium pellets for possibly refueling fusion plasmas in a tokamak reactor. **Right:** The shadowgraph shows a 6-mm-dia. solid deuterium pellet at a velocity of 1,390 m/s that, before firing, was produced in a cryostat from deuterium gas frozen to 14 K. [Proc. 14th Symp. on Fusion Technology, Avignon, France (1986). Euratom Publ. 10936 EN, Pergamon, Oxford (1986)]

## 4.2 PERCUSSION IN THE EVOLUTION OF TECHNOLOGY – Devices Based on Rapid Compression

**4.2–K** In the 16th century, the adiabatic nature of the rapid compression of air was well known and apparently has been used for a long time in southeast Asia – from Burma to Flores and Sumatra to Mindanao – in daily life in so-called "pneumatic lighters," also known as "compression lighters" or "fire pumps." In the early 19th century, the device was reinvented almost simultaneously in England and France. The application of the device was very simple: a small piece of dry tinder was fixed on a little hook attached to the bottom of the piston, which was rapidly pushed into the cylinder. After compression the piston was immediately pulled out to provide the ignited tinder with fresh air. **Top:** An early pneumatic lighter from the Isle of Luzon, the Philippines. The cylinder and the piston were made from natural material such as wood and buff horn. **Center:** Advanced European model as used in the 19th century, consisting of a glass cylinder and a metal piston. **Bottom:** Pneumatic lighter from China. [Courtesy Technisches Museum, Vienna]

## 4.2 PERCUSSION IN THE EVOLUTION OF TECHNOLOGY – Periodically Operating Devices

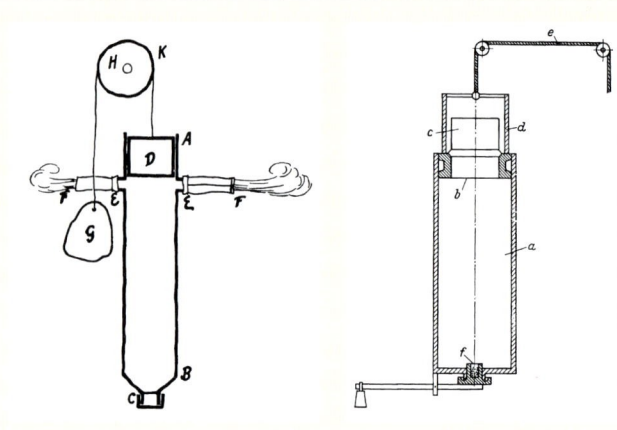

**4.2–L** *Left:* Christiaan HUYGENS' sketch of his "ballistic pump," dated September 22, 1673. Gunpowder, placed in a cylinder and ignited at the bottom, moves a piston *P* upward. The explosive products leave the cylinder at valves arranged at the top. This creates a vacuum that moves the piston down, thus creating work by atmospheric pressure such as for lifting a weight. [*Œuvres Complètes de C. HUYGENS*. Martinus Nijhoff, La Haye (1897), vol. 7: *Correspondance*, p. 357; Deutsches Museum München] *Right:* Denis PAPIN, a French-born English physicist, incorporated the outlet valves in the piston head and facilitated refueling by using a lever arrangement at the piston bottom. He proposed steam instead of gunpowder, thus anticipating Thomas NEWCOMEN's atmospheric steam engine. [E. GERLAND: *LEIBNIZens und HUYGENS' Briefwechsel mit PAPIN, nebst der Biographie Papin's*. Verlag der Königl. Akad. Wiss., Berlin (1881), p. 44]

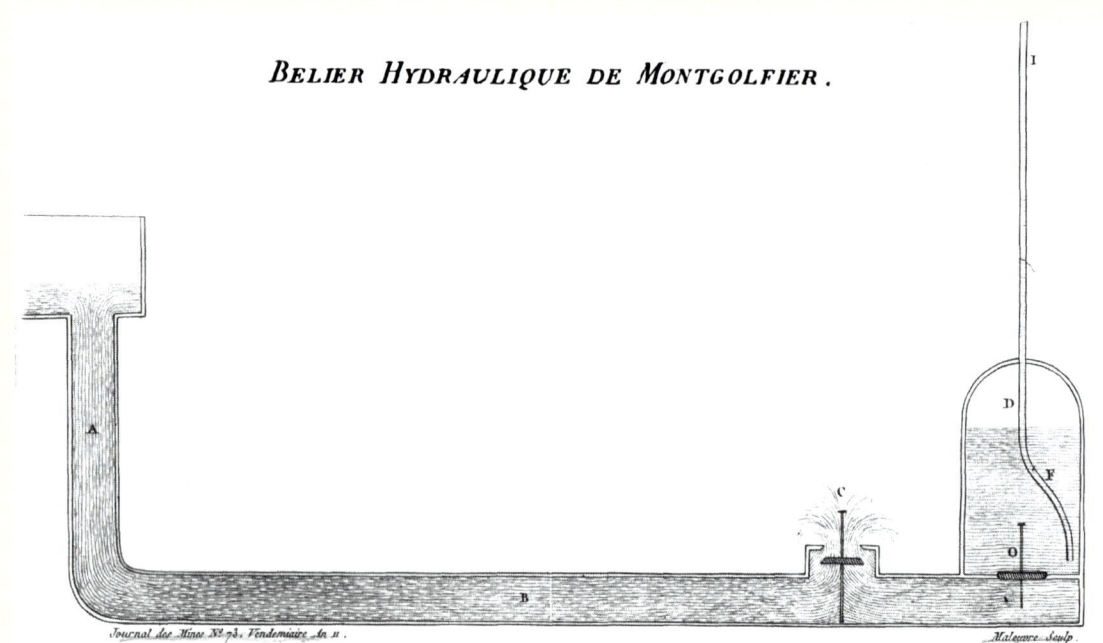

**4.2–M** In the late 1790s, the French manufacturer Joseph Michel MONTGOLFIER, today widely known for his hot-air balloon rides, devised the first "hydraulic ram," a simple water pump. It uses a copious flow of water to drive running water to a higher level that, emerging from an elevated water source *A* and propagating in a pipe *B*, leaves at an escape valve *C*. When *C* is suddenly closed, the resulting pressure increase forces the water through a delivery valve *O* into a chamber *D*, which compresses the air that partially fills that chamber. After valve *C* shuts, the air pressure pushes water up the vertical outlet pipe *F* to a height *I*. When the escape valve *C* drops open, the cycle begins to repeat. Because of its simplicity, the hydraulic ram is still in use today. [*J. des Mines* **XIII**, 42 (1802/1803)]

## 4.2 PERCUSSION IN THE EVOLUTION OF TECHNOLOGY – Pile Driver

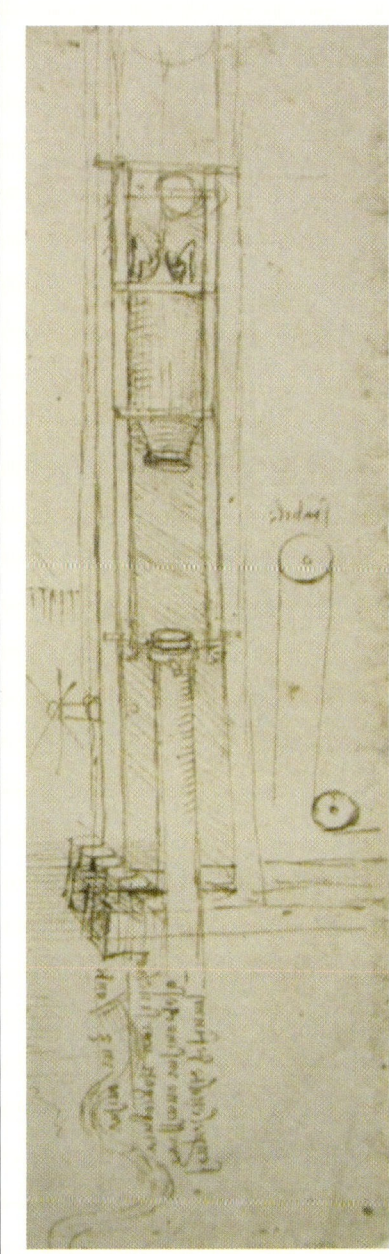

**4.2–N** Piles are large stakes or beams, sharpened at the end to be driven into the ground, for a foundation to build upon in marshy places. The effect of the blow is proportional to the height fallen by the hammer or ram, to the head of the pile, and the weight of the hammer or ram. **Left:** Pencil drawing of a so-called "pile engine" made by Leonardo DA VINCI around 1490. Such machines were then used for changing the course of the Arno River. [Il *codice atlantico di Leonardo di Vinci*. Giunti-Barbera, Florence (1975), vol. IX, p. 785] **Right:** In the 1750s, the most advanced machine for driving piles was based on an invention made by the watchmaker James VAULOUE: it consisted of a high frame with appliances for raising and dropping the pile hammer, a heavy weight that was lifted by horsepower. The whole engine was installed on a pontoon and used, for example, in driving piles in the foundations of Westminster Bridge in London.
[J.T. DESAGULIERS: *A course of experimental philosophy*. W. Innys et al., London (1745), vol. 2, plate 26]

## 4.2 PERCUSSION IN THE EVOLUTION OF TECHNOLOGY – Forge Steam Ram and Explosion Ram

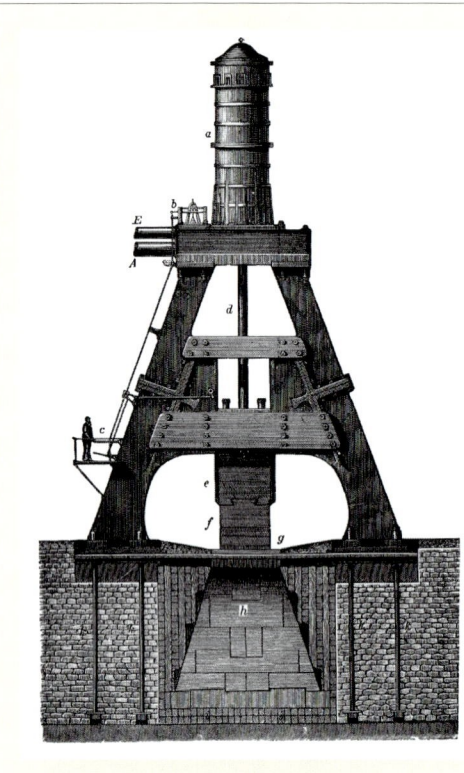 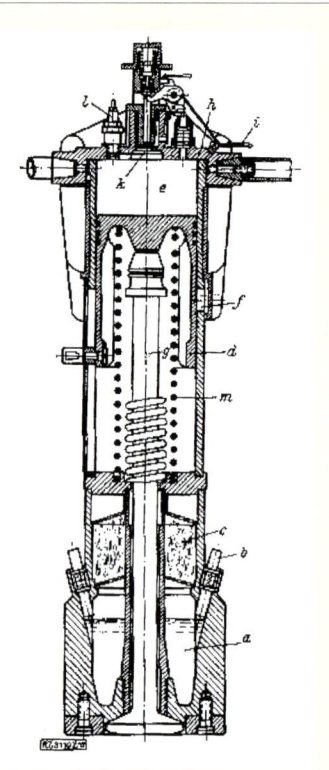

**4.2–O** *Left:* With the invention of the steam engine in the latter part of the 18th century, forces of men and horses were increasingly replaced by steam engines, and bigger pile engines could be constructed. This huge, single-action steam hammer was used for forging operations in the late 1880s at the Schneider & Co. ironworks in Le Creusot, France. The hammer weighed 80 tons and was dropped from a height of 5 m. [*Brockhaus' Konversations-Lexikon*, Brockhaus, Leipzig (1908), vol. 4, p. 666] *Right:* Example of a handheld explosion ram, first built in 1926 by the Deutsche Elektromaschinen- und Motoren-Bau AG (DELMAG Co.). The main parts of an explosion ram are: $a$ – fuel chamber, $b$ – air intake, $c$ – carburetor, $d$ – piston, $e$ – cylinder, $f$ – exhaust slot, $g$ – free piston shaft, $h$ – cylinder head, $i$ – hand lever, $k$ – main valve, $l$ – spark plug, and $m$ – piston spring. The machine was used mainly for paving but also for pile driving. [Z. VDI **73**, 1273 (1929)]

## 4.2 PERCUSSION IN THE EVOLUTION OF TECHNOLOGY – Explosion Tamper and Wrecking Ball

**4.2–P** *Left:* This 2.5-ton DELMAG explosion thumper, popularly known as the "frog tamper" [Germ. *Froschstampfer*] was built in 1939 for soil condensation purposes. A large explosive percussion force is generated by igniting a diesel fuel/air mixture in a cylinder that drives a piston connected with the ram. The machine was widely used in the construction of the German Autobahn. [Courtesy S.A. MEWES, DELMAG, Esslingen, Germany]

*Right:* View of a 2-ton wrecking ball for building demolition purposes. [Courtesy LST, Herrsching, Germany]

## 4.2 PERCUSSION IN THE EVOLUTION OF TECHNOLOGY – Percussion Boring

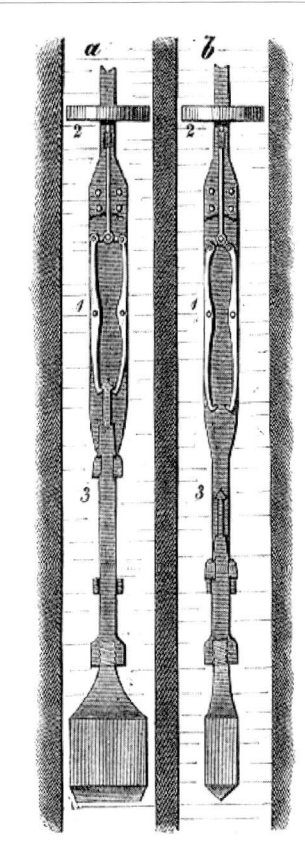

← In early mining operations, manual percussion drilling performed by two men was the common method until the advent of pneumatic drilling machines: one man turns the drill while the second swings the hammer. The blows are applied successively, and with each blow the tool is rotated slightly so that a new portion of the face is attached at each blow. The origin of this method is unknown, but since it requires a metal chisel, it could have been invented at the earliest in the Bronze Age (in Europe ca. 2000–700 B.C.).

**4.2–Q** *Left:* In 1834, the German mining engineer Karl G. KIND invented "percussion boring," the first effective technique for drilling deep holes. It uses a gliding jar *1* that is connected to the end of the bore rod and centered in the bore hole by a disk *2*. A heavy tool *3* that is connected to the jar *1* is repeatedly dropped on the same spot from a suitable height, thereby pulverizing the rock by percussion and gradually penetrating the spot. Illustration *a* shows the apparatus in the starting position when the percussion tool *3* is lifted. Illustration *b* shows the triggered jar after provoking the dropping of tool *3*. Then the bore hole is flushed by water, which not only washes away the pulverized rock but also triggers *3* via *1* by suddenly shifting the bore rod. [E. BOBRIK (ed.): *Das neue Buch der Erfindungen, Gewerbe und Industrien. III. Die Gewinnung der Rohstoffe.* Spamer, Leipzig (1864), vol. 3]
*Right:* This 19th-century pen-and-ink drawing illustrates the "two-men percussion drilling method" which occasionally is still in use because of its simplicity.

## 4.2 PERCUSSION IN THE EVOLUTION OF TECHNOLOGY – Ricocheting

**4.2–R** When a moving body obliquely strikes the surface of a solid or a liquid at a slight angle, it rebounds in a succession of skips. This unique phenomenon, called "ricochet" and invented in the 17th century by the French military engineer Marshal Sébastien LE PRESTRE DE VAUBAN, is a complex process of gliding and percussion that generates finite-amplitude waves in the struck body. In the 1920s, it attracted renewed interest with the advent of hydroplanes. Ricochet fire was delivered by guns, howitzers, and sometimes mortars, using small charges of powder. This particular artillery technique also allows projectiles to be skipped over the surface of the water to better ensure a hull hit. However, ricochet firing requires a perfectly smooth sea. The picture shows a "ricochet fire" of a 16-in. (40.64-cm) projectile in water. [J. Appl. Phys. **15**, 264 (1944)]

## 4.2 PERCUSSION IN THE EVOLUTION OF TECHNOLOGY – Big Guns of the 19th Century

**4.2–S** In the 1860s, the construction and use of big guns stimulated not only classic ballistics and the invention of new propellants but also aroused the fantasy of science fiction writers. ***Left, top:*** During the American Civil War (1861–1865), Lieut. Thomas J. RODMAN of the Union Artillery initiated the production of huge smooth-bore guns using his method of hollow casting. Cooling occurred from the interior by introducing a stream of water into the core. His guns were cast by Knapp, Rudd & Co. of Pittsburgh, PA. View of a Rodman 20-in. (50.8-cm)-caliber monster gun (1864), then the world's biggest and heaviest gun weighting 80 tons. However, it did not attain military relevance, and the few pieces that were made were fired only a few times. [Collection of the Pennsylvania Dept., The Carnegie Library of Pittsburgh, PA] ***Left, bottom:*** This sketch illustrates the testing of a Rodman 15-in. (38.1-cm) caliber columbiad at Fort Monroe, VA. Slightly smaller than the monster gun shown above, this gun type was installed in large numbers for seacoast defense. Today Fort McHenry near Baltimore has the largest surviving collection of 15-in. Rodman guns. [Harper's Weekly: J. Civilizat. **5**, No. 222 (March 30, 1861)] ***Right:*** RODMAN's huge guns inspired the French science fiction writer Jules VERNE. In his novel *De la terre à la lune* (1865), he discussed the ballistic requirements of a cannon to transport a crew to the Moon. To achieve the necessary velocity to escape from Earth, the so-called "second cosmic velocity" of 11,200 m/s, his 10-ton 108-in. (2.74-m)-dia. aluminum spaceship was fired like a projectile (*top*) from a 270-m-long cast-iron cannon (*bottom*) propelled by 400,000 pounds of gun-cotton. The cannon was positioned vertically and buried in the ground. The crew of VERNE's spacecraft consisted of two Americans, one Frenchman, and two dogs. [J. VERNE: *Von der Erde zum Mond. Reise um den Mond.* Bärmeier & Nikel, Frankfurt/Main (1966)]

## 4.2 PERCUSSION IN THE EVOLUTION OF TECHNOLOGY – Superguns of the 20th Century

**4.2−T** The motivation for constructing huge-caliber guns by the German military was the experience from World War I that modern concrete-reinforced fortifications (such as those of Verdun in France) were extremely strong and could not be efficiently destroyed by standard artillery munitions. *Top, left:* The German Krupp works built a 42-cm howitzer that was nicknamed *Dicke Bertha* ("Big Bertha") after Mrs. Bertha KRUPP VON BOHLEN UND HALBACH. Capable of shooting a 1-ton shell over a short range (14–15 km) in a high arched trajectory, it was used during World War I against Belgian fortifications. [http://www.waffenhq.de/panzer/dickeberta.html] *Top, right:* Another big gun, also used in WW I, was the Krupp "Paris Gun" – officially called "Kaiser-Wilhelm-Geschütz" – a 140-ton, 21-cm caliber gun. It was capable of reaching extraordinarily long distances, up to 122 km. To hold the 33.5-m-long barrel straight, it needed a special support. Four such guns were used in the bombardment of Paris. [http://www.westfront.de/00291.htm] *Bottom, left:* During World War II, the German 80-cm railroad gun *Dora* was built and had a weight of 1,350 tons and a barrel length of 32 m. *Bottom, right: Dora* was capable of firing 7-ton armor piercing shells (*right*) with an initial velocity of 720 m/s ($M = 2.15$) over a distance of up to 38 km. Transportation of this supergun required 5 trains with a total of 99 cars. Being the world's largest artillery gun ever built, it was successfully used in June 1942 for destroying the strong fortifications of Sevastopol. However, further plans for building even larger and heavier guns – *e.g.,* the *Schwerer Gustav* ("Heavy Gustav"), a 80-cm gun with a barrel length of 84 m and an anticipated firing distance of at least 150 km – could only partly be realized. [http://www.waffenhq.de/panzer/dora.htm] Although remarkable high-tech products of the German war industry, all these superguns eventually proved to be faulty developments, from both a military and an economic point of view.

## 4.3 PERCUSSION STUDIES – 17th Century: The Pioneering Era of Percussion Research

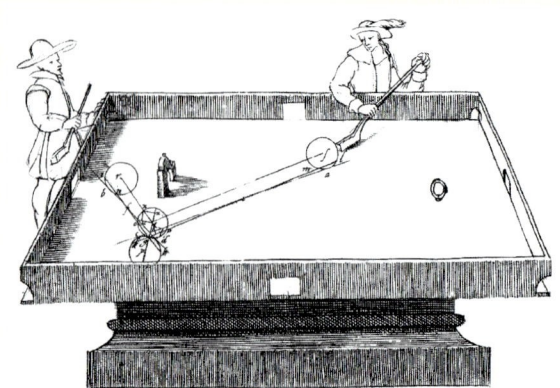

**4.3–A** Billiards was probably invented in Italy in the 15th century, but reliable references are rare. Well suited to study elastic collision of both centrally and eccentrically hit balls, it has since inspired physicists and mathematicians to uncover the laws of motion and collision of bodies. *Left:* Copper engraving by Antoine TROUVAIN (1694) showing king LOUIS XIV playing billiards in his palace at Versailles. [*Grand Larousse encyclopédique*. Larousse, Paris (1960), p. 137] *Right:* Marcus MARCI VON KRONLAND, a Bohemian natural philosopher, found that when a ball strikes a plane wall obliquely, the angle of incidence is equal to the angle of reflection. He applied his observation to the game of billiards and attempted to tackle the difficult "carom" problem of French billiards: three balls of equal size and elasticity – $p$, $r$, and $s$ – are arranged such that their centers of gravity are not positioned along a straight line. How should the player hit $s$ such that after collision with $p$, $p$ is set in motion and hits $r$? The problem, however, was too complex to be solved with available mathematical means. MARCI tried to apply his law of optical reflection, which he derived under the assumption of a plane wall, however, overlooking that the ball's motion is a superposition of translation and rotation. [I.M. MARCI: *De proportione motus*. Typis I. Bilinae, Pragae (1639)]

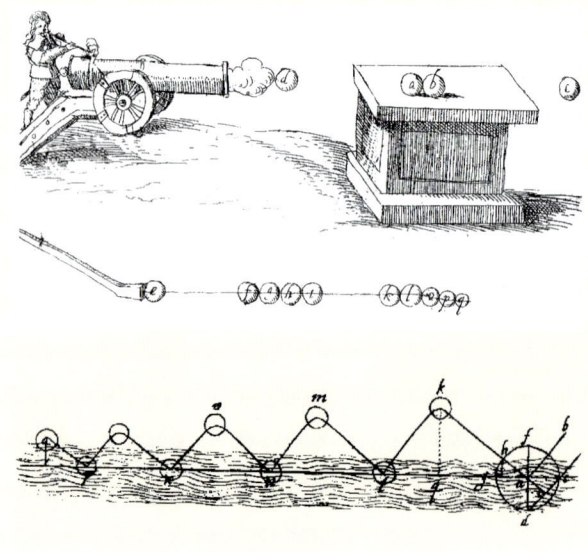

**4.3–B** Marcus MARCI VON KRONLAND investigated straight collision of elastic bodies and also experimented with stone cannonballs and ivory billiard balls. *Top:* In his treatise *De proportione motus* (Prop. XXXVII, Por. I) MARCI stated that a ball $d$, striking a row of two balls $a$ and $b$ at rest and each having the same mass as $d$, fully transmits its motion to the last ball $b$ in the row, which, illustrated here as ball $c$, flies away with the same velocity as that with which ball $d$ had initially struck the row of balls $a$ and $b$. The principle of NEWTON's cradle is essentially based upon this experiment. *Bottom:* MARCI's illustration of a body (*e.g.*, a pebble) impinging under oblique incidence on a water surface. Being reflected along the surface, it jumps from right to left. This phenomenon, explained by MARCI using the simple law of reflection, is actually a complex combination of gliding and percussion. It attracted renewed interest in the 1930s by aeronautical engineers for optimizing the hull geometry of hydroplanes. [I.M. MARCI: *De proportione motus*. Typis I. Bilinae, Pragae (1639)]

## 4.3 PERCUSSION STUDIES – 17th Century: The Pioneering Era of Percussion Research *(cont'd)*

APROINO: I shall mention the first experiment that our friend essayed in order to get to the heart of this admirable problem of impact: What is sought is the means of finding and measuring its great force, and if possible simultaneously of resolving the essence of impact into its principles and prime causes... It was the Academician's first idea to try to find out what part in the effect and operation of impact belonged, for example, to the weight of a hammer, and what part belonged to the greater or lesser speed with which it was moved. He wanted if possible to find out one measure that would measure both of these, and would assign the energy of each.; and arrive at this knowledge, he imagined what seems to me to be an ingenious experiment.

He took a very sturdy rod, and about three braccia long, pivoted like the beam of a balance, and he suspended at the ends of these balance-arms two equal weights, very heavy. One of these consisted of copper containers; that is, of two buckets, one of which hung at the said extremity of the beam and was filled with water. From the handles of this bucket hung two cords, about two braccia each in length, to which was attached by its handles another like bucket, but empty; this hung plumb beneath the bucket already described as filled with water. The bottom of the upper bucket had been pierced by a hole the size of an egg or a little smaller, which hole could be opened and closed.

Our first conjecture was that when the balance rested in equilibrium, the whole apparatus having been prepared as described, and then the hole in the upper bucket was unstoppered and the water allowed to flow;, this would go swiftly down to strike in the lower bucket; and we conceived that the adjoining of this impact must add to the static moment on that side, sol that in order to restore equilibrium it would be necessary to add more weight to that of the counterpoise on the other arm. This addition would evidently restore and offset the new force of impact of the water, so that we could say that its momentum was equivalent to the weight of the ten or twelve pounds that it would have been necessary to add to the counterpoise.

SAGREDO: This scheme seems to me really ingenious, and I am eager waiting to hear how the experiment succeeded.

APROINO: The outcome was no less wonderful than it was unexpected by us. For the hole being suddenly opened, and the water commencing to run out, the balance did indeed tilt toward the side with the counterweight; but the water had hardly begun to strike against the bottom of the lower bucket when the counterweight ceased to descend, and commenced to rise with very

tranquil motion, restoring itself to equilibrium while water was still flowing; and upon reaching equilibrium it balanced and came to rest without passing a hairbreadth beyond.

SALVIATI: This clever contrivance much pleases me, and it appears to me that without straying from that path, in which some ambiguity is introduced by the difficulty of measuring the amount of this falling water, we might by a not unlike experiment smooth the road to the complete understanding which we desire.

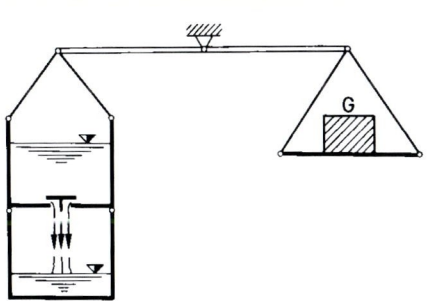

**4.3–C** *Top:* In his *Discorsi e dimostrazioni matematiche* (Leiden 1632), GALILEI pondered on an experiment on how to measure the force of percussion, the first scientific approach to a phenomenon used in stone tools and weapons throughout the evolution of man. He developed his thoughts in a platonic dialog between SALVIATI (a Copernican, representing GALILEI), SAGREDO (an educated layman), and APROINO (an intelligent Aristotelian, his oldest pupil). The experimenters expected some constant effect as long as the flow of water continued, enabling them to reestablish equilibrium by adding weight to the counterpoise. [G. GALILEI: *Two new sciences*. Univ. of Wisconsin Press, Madison, WI (1974), pp. 283-284] The copper engraving made by Stefano DELLA BELLA for the first edition of GALILEI's *Discorsi* shows (from *left* to *right*) the natural philosophers ARISTOTLE, PTOLEMAEUS, and COPERNICUS in a discussion. **Bottom:** Schematic of GALILEI's intellectual experiment. [I. SZABÓ: *Geschichte der mechanischen Prinzipien*. Birkhäuser, Basel (1987), p. 428]

## 4.3 PERCUSSION STUDIES – 17th century: The Pioneering Era of Percussion Research *(cont'd)*

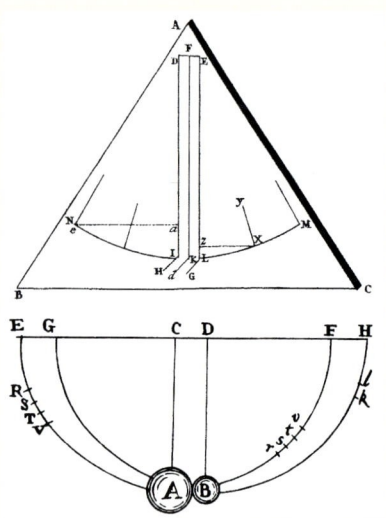

**4.3–D** Evolution of the percussion machine. *Top:* Instead of studying elastic percussion on a plane table like Marcus MARCI (1639) did in his experiments {⇨Fig. 4.3–B}, Edmé MARIOTTE (1676) used a two-sphere percussion pendulum that he applied successfully to experimentally confirm the laws of percussion previously derived by Sir Christopher WREN (1668), John WALLIS (1668), and Christiaan HUYGENS (1669). [*Œuvres de Mariotte,* The Hague (1740), Table I, Fig. 3] *Bottom, left:* In 1687, Sir Isaac NEWTON recognized that the ratio $e$ of the relative velocity of two bodies just after a collision to the relative velocity just before the collision is an important quantity to characterize the elasticity in a 1-D collision. To measure the "percussion number," $e$, of different materials, defined by him as the ratio $(C - c)/(v - V)$, NEWTON used two spheres of two such materials and suspended them on long threads. He determined the velocities of two balls ($V$, $C$) before collision and their velocities ($v$, $c$) after collision by releasing them from a given height and measuring their heights to which each ball climbed after collision. For example, the big ball $A$ was released at height $H_A$ and after collision climbed to height $h_A$. The heights are proportional to the square of the velocities $V$ and $C$, and $v$ and $c$, before and after collision, respectively. He found that his "percussion number" ranged between 0 for fully inelastic collision and 1 for perfectly elastic collision. [I. NEWTON: *Principia.* S. Pepys, Londini (1687); Corollary VI, p. 22]

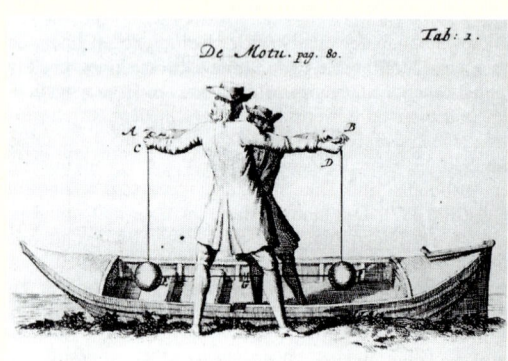

**4.3–E** Christiaan HUYGENS first pondered the relativity of percussion phenomena in his famous thought experiment. He stated in his first tenet that two equal elastic bodies, impacting with equal velocities, rebound from each other with exactly the same (but opposite) velocities. When this happens in a boat moving with velocity $v$, he showed that for a spectator in the boat the balls approach with velocities $+v$ and $-v$ and rebound with $-v$ and $+v$. But for a spectator on the shore the velocities of the balls before impact are $2v$ and 0 and after impact 0 and $2v$, respectively. Thus, an elastic body, impinging on another of equal mass at rest, communicates to the latter its entire velocity and remains after the impact itself at rest. [C. HUYGENS: *De motu corporum ex percussione.* Lugduni Batavorum, Leiden (1703)]

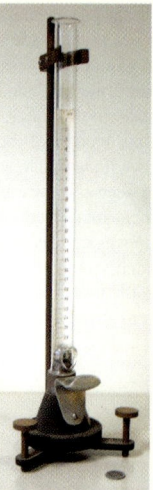

**4.3–F** NEWTON's percussion number $e$ can be found easily by dropping a small ball on the horizontal surface of a large mass and by comparing the height $h$ to which the ball bounces with the height $H$ from which it is dropped, hence $e = (h/H)^{1/2}$. In 1879, William THOMSON and Peter G. TAIT called this ratio $e$ the "coefficient of restitution." View of an apparatus for measuring $e$ as used at the University of California at San Diego. [Courtesy Physics Dept., UC San Diego, CA]

## 4 PICTURE GALLERY

### 4.3 PERCUSSION STUDIES – 18th Century: Percussion Machine for Demonstrating Central Percussion

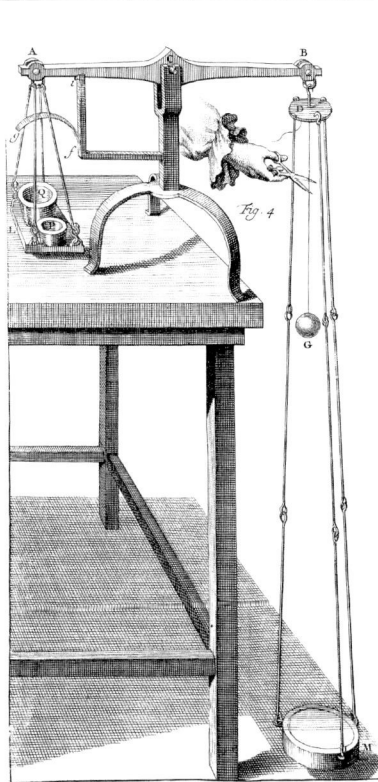

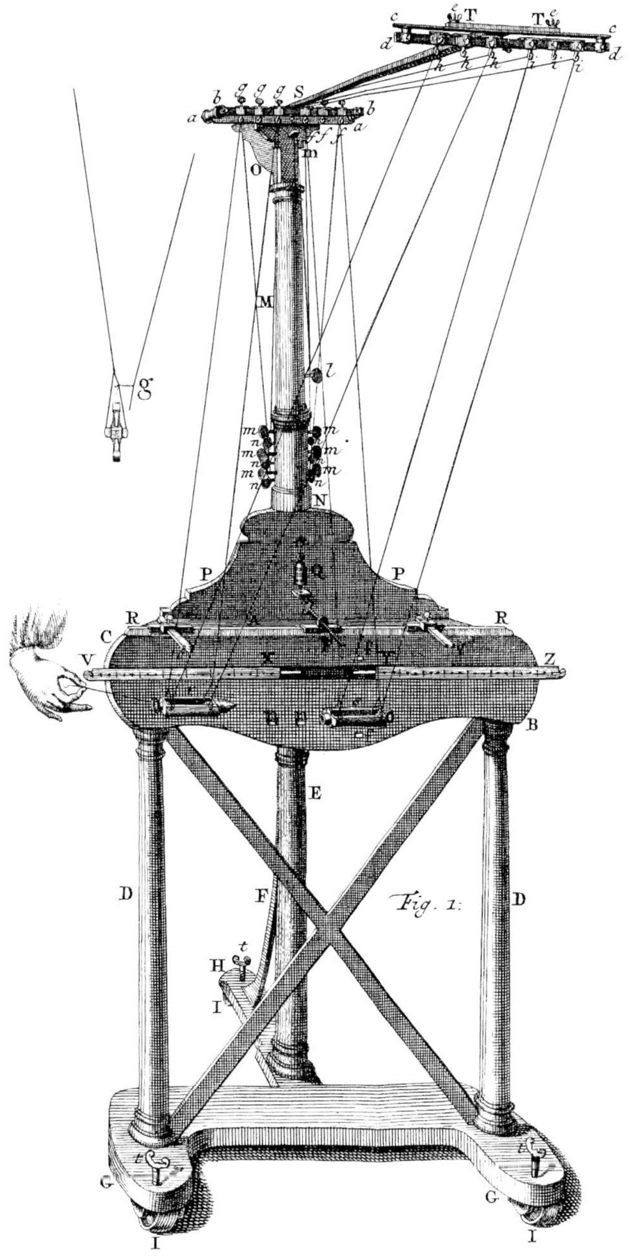

**4.3–G** The Dutch natural philosopher Willem Jacob's GRAVESANDE constructed throughout his teaching period at the University of Leiden many unique physical apparatus for demonstration purposes. *Left:* For example, he modified GALILEI's thought experiment {⇨Fig. 4.3–C} by using a ball pendulum instead of a water jet. The experiment was started by cutting the thread with scissors. *Right:* In the early 1720s, GRAVESANDE constructed a percussion machine for quantitatively studying the laws of elastic and inelastic percussion. Note the bifilar suspension of his test bodies, which could be provided at their heads with different materials for studying their elastic properties. It was also possible to mount leaf springs on the heads to simulate a linear chain of mass/spring oscillators. [W.J.'s GRAVESANDE: *Physices elementa mathematica, experimentis confirmata.*. Langerak & Verbeeck, Leiden (1742), plates 27 & 35]

## 4.3 PERCUSSION STUDIES – 18th Century: Percussion Machine for Demonstrating Oblique Percussion

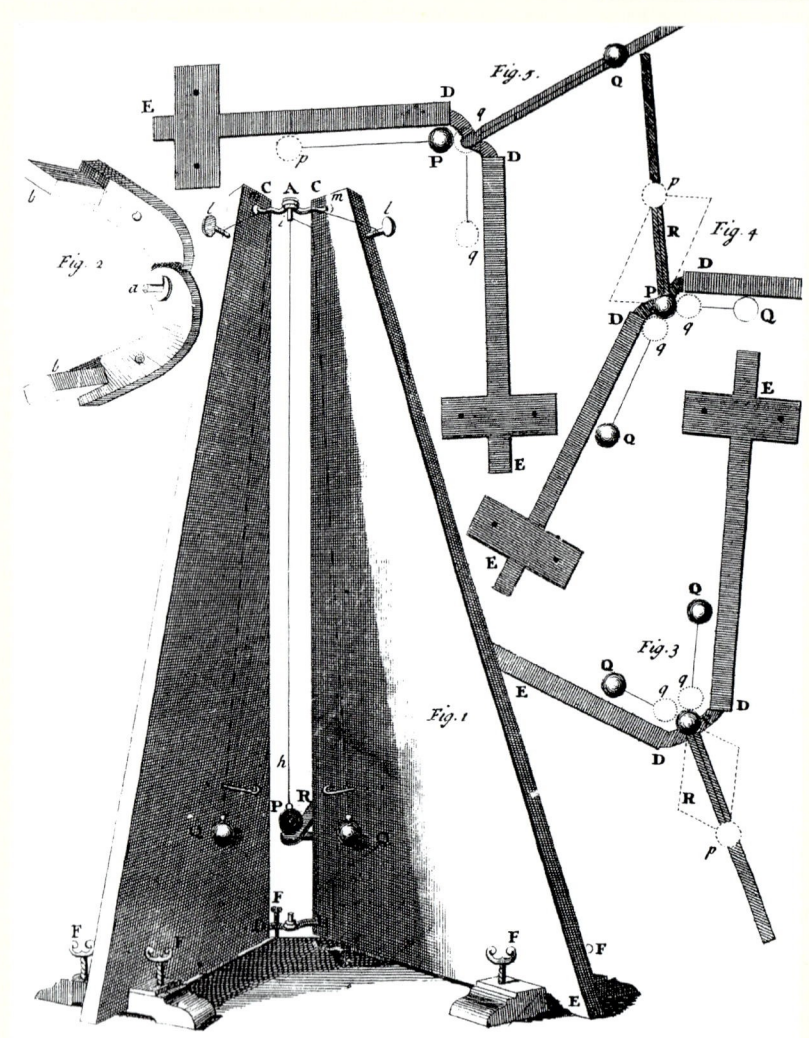

*←←← Fig. 2:* "In the Center of the upper Hinge there is a small Cylinder *a*, in whose Case there is a Hole, which meets another in the Side, thro' which the Thread *ih* is to run; at one End of this Thread a Ball, as *P* hangs, and the other End is join'd to the Key *l*. At *m, m*, there are two Pins fix'd to the two Planes, from which Pins the Balls *Q, Q*, hang, at such a Distance from the Planes, that they may almost touch them; so that if you suppose a Line to pass thro' the Centers of the Balls *P* and *Q*, it shall be parallel to the Plane on that Side: besides, it is required, that, when those Balls hang at the same Height, they touch one another."

*←← Fig. 5:* "Let the Ball *Q* and *P* hang; having set the Planes at Right Angles let the Body *Q* with any Direction, and from any Height, come down upon *P*, and strike against it: after the Stroke the Bodies will follow the Directions of the Planes, and rise to Heights, which may be determined by what has been said hitherto."

*← Figs. 3 and 4:* "The Body *P* may be driven by either of the Bodies *Q*, with any Direction and Velocity. If the Bodies *Q* and *Q* are let fall at the same time, the Body *P* has two Motions impress'd upon it at the same time, and therefore runs in the Diagonal *Pp* of the Parallelogram, to express those two Motions, and runs up to an Height proportionable to the Length of that Diagonal."

**4.3−H** In order to quantitatively study oblique percussion phenomena as well, GRAVESANDE constructed a unique machine that he described as follows: "It consists of two Boards, or wooden Planes, *CDE, CDE* of the Figure of a right-angled Triangle, whose Side *CD* is in Length about 3 Foot and a half, and the Side *DE* about 1 Foot and a half; these Boards are fixed so as to move in a vertical Situation about the Hinges *A* and *B*. The Experiments upon this Machine are made with Ivory Balls of an Inch and a half Diameter [38.1 mm] … By Help of the Screws *F, F, F, F, F*, his Machine is set perpendicular; so as to have the Thread *hi* hang in the Axis of the Machine … Each Ball *Q*, when it swings, moves along the Plane to which it is applied; and the Height, from which it is made to fall, is shewn by an Index fixed to the Plane." [W.J.'s GRAVESANDE: *Mathematical elements of natural philosophy confirmed by experiments, or, an introduction to Sir Isaac NEWTON's philosophy* (Engl. translation by J.T. DESAGULIERS). Senex & Taylor, London (1726), Book I, plate 12]

## 4.3 PERCUSSION STUDIES – 18th Century: First Ballistic Pendulum

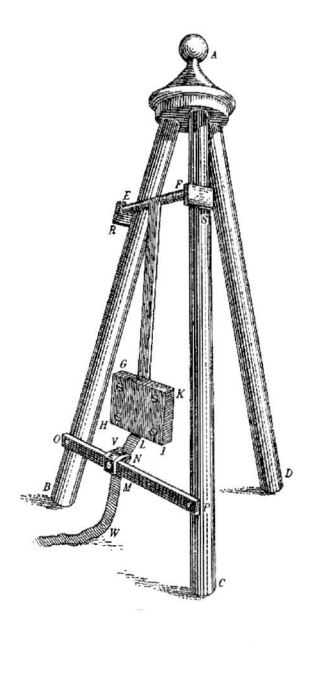

**4.3–I** In the period 1746–1791, the Englishmen Benjamin ROBINS and Charles HUTTON performed ballistic drag studies that covered velocities up to 2,030 ft/s (626 m/s; $M = 1.85$). *Top:* Schematic of ROBINS' so-called "ballistic pendulum," developed by him while at the Royal Military Academy in Woolwich. It essentially consisted of an iron plate, $G$–$K$–$H$–$I$, faced with wood to catch and retain the bullet. The deflection was determined by a thin metal tape $W$, connected at the outer end of the pendulum, $H$–$I$, and passed through two steel edges, $V$–$N$, which were pressed lightly together. He derived a formula to estimate the muzzle velocity from the pendulum's swing angle and the mass of the projectile. The device was used by him to measure the muzzle velocity of small shot, like the ball of a musket. By moving the instrument at progressively greater distances from the muzzle, he also studied how aerodynamic drag reduces the velocity of projectiles. [B. ROBINS: *New principles of gunnery*. Nourse, London (1742)] *Bottom:* In the 1780s, ROBINS' pendulum was further developed by HUTTON for use of large shot, the so-called "gun pendulum." At Woolwich he combined two similar methods of measuring the velocity of a cannonball: the schematic on the *left* shows a side view of his ballistic pendulum into which the balls were fired. Compared to ROBINS' pendulum, it was reinforced by thick bars of iron. The schematic on the *right* shows a second pendulum for observing the arch of recoil of the gun that was attached to the pendulum. From the deflection angle, measured by this unusual pendulum construction, he could also determine the initial velocity of the cannonball, thus allowing a comparison of both methods. [C. HUTTON: *Tracts on mathematical and philosophical subjects*. T. Davison, London (1812), vol. II, plate VI] Later a pendulum with a 5-ton cannon of 13-in. (33-cm) caliber was installed at Woolwich Arsenal that was described by Fritz HEISE in his book *Sprengstoffe und Zündung der Sprengschüsse, mit besonderer Berücksichtigung der Schlagwetter- und Kohlenstaubgefahr* (Springer, Berlin 1904). [C. HUTTON: *Tracts on mathematical and philosophical subjects*. T. Davison, London (1812), vol. II, plate VI]

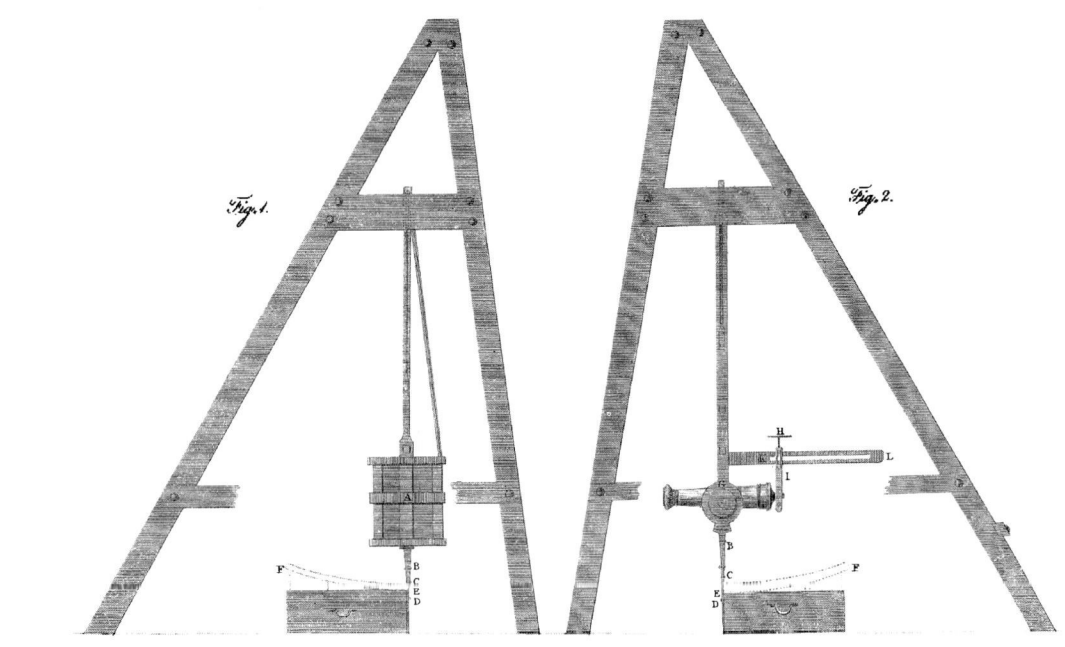

## 4.3 PERCUSSION STUDIES – 19th Century: Measurement of Shock Duration

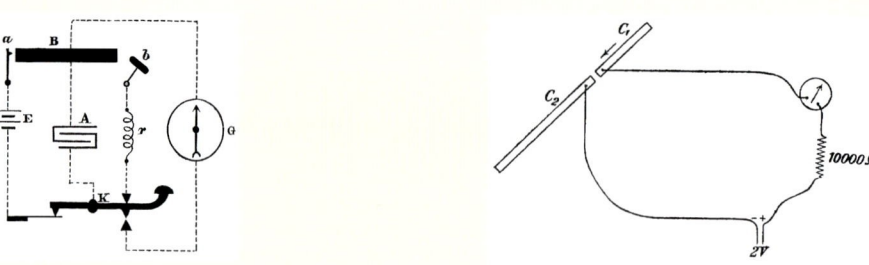

**4.3–J** In 1844, Claude S.M. POUILLET in France worked out an electric method for measuring the contact time of impacting bodies that was later improved by Robert SABINE in England and Carl W. RAMSAUER in Germany. *Top, left:* SABINE's electric apparatus, also known as the "ballistic galvanometer," was capable of measuring contact times of impacting bodies in the microsecond regime. A charged capacitor $A$ was partly discharged when the impactor $b$ hit an anvil $B$. The contact time was analytically determined from galvanometer readings $G$ before and after impact and from the circuit parameters. [Phil. Mag. **1** (V), 337 (1876)] *Top, right:* RAMSAUER, improving SABINE's method, succeeded in resolving contact times as small as 1 µs. In contrast to SABINE's circuit, however, it used a battery instead of a charged capacitor. During the contact time of the two impacting rods $C_1$ and $C_2$, a certain quantity of electric charge provokes a certain deflection of a galvanometer that is proportional to the contact time and can be calibrated, for example, using the Helmholtz pendulum interrupter. [Ann. Phys. **30** (IV), 417 (1909)] *Bottom, left:* Peter G. TAIT, a Scottish physics professor, constructed an apparatus for measuring the duration of percussion phenomena that he humorously called "my guillotine." It used a free-falling bloc $A$ that, sliding between vertical guide rails, collided with a test sample $B$ at rest. The temporal movement and rebound of $A$ was recorded on a soot-covered revolving disk $D$ via a needle $E$ that was connected with $A$. The contact time was evaluated by simultaneously recording the oscillations of a tuning fork $F$ bearing a needle $G$ writing on the soot disk $D$. *Bottom, right:* Percussion diagram of recorded soot traces taken with TAIT's apparatus and showing the time-resolved behavior of percussion and rebound of two elastic bodies, $A$ and $B$. [Rev. Gén. Sci. Pures Appl. **3**, 777 (1892)]

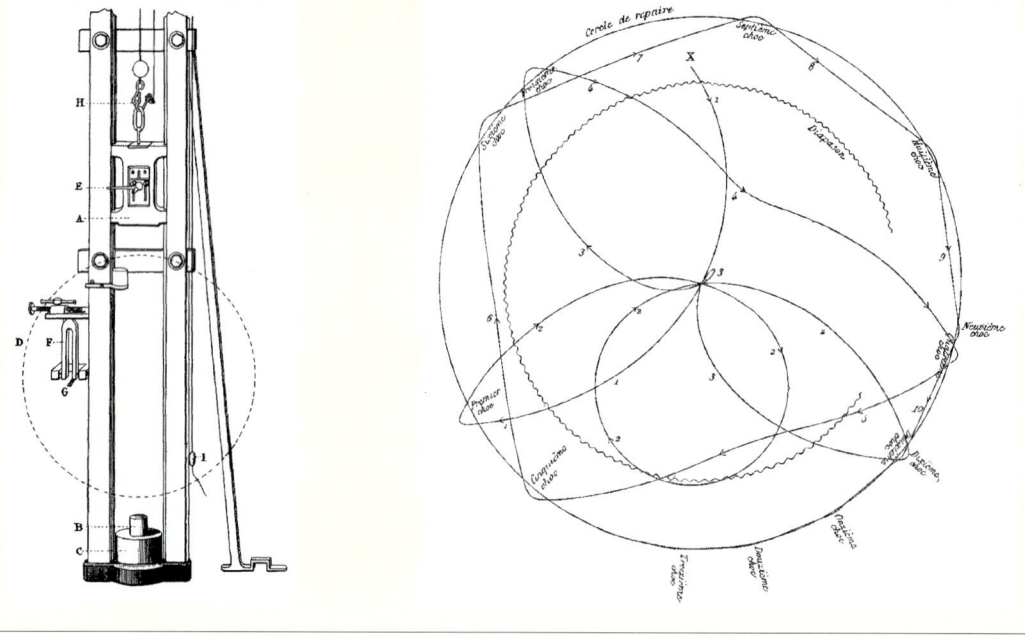

## 4.3 PERCUSSION STUDIES – 19th Century: DE CORIOLIS' Mathematical Studies on Billiards

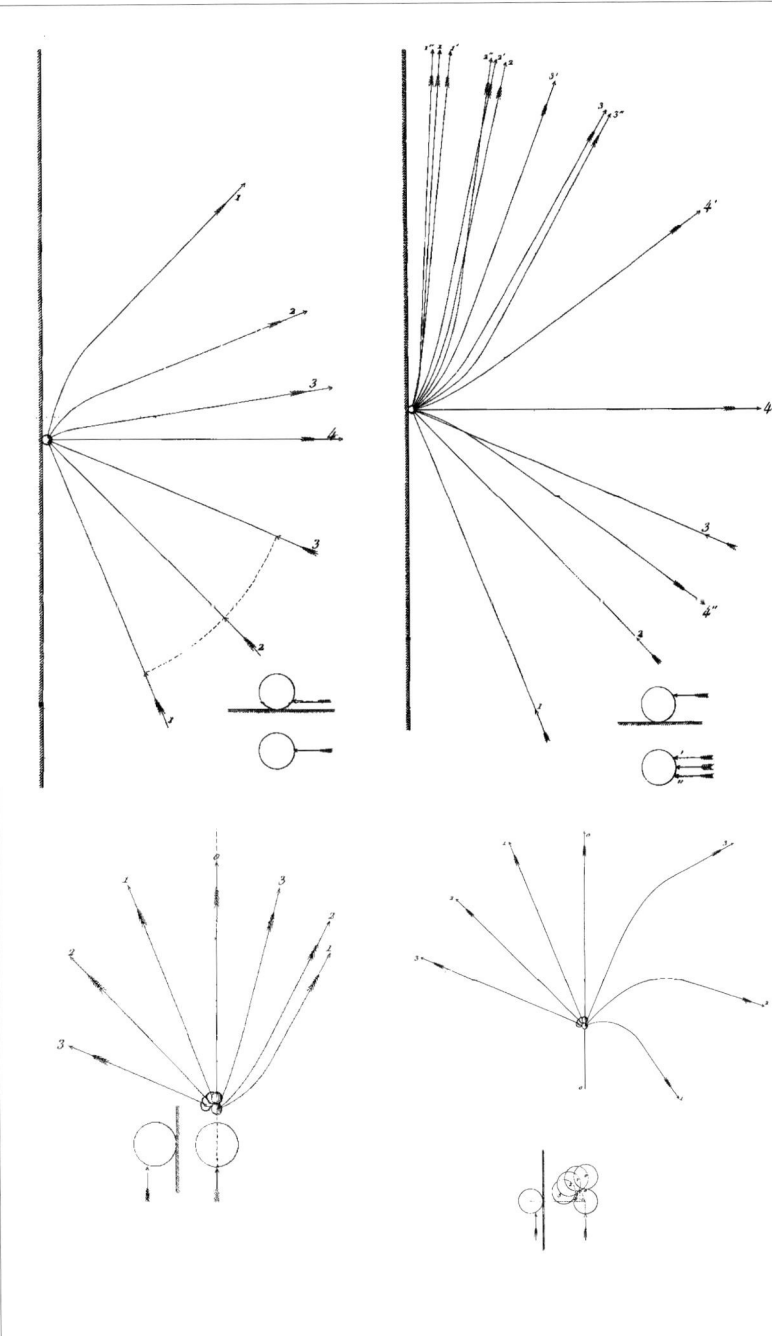

**4.3−K** In the early 1830s, Gaspard G. DE CORIOLIS, a French mathematics and mechanics professor at the Ecole Centrale des Arts et Manufactures in Paris, first analyzed the trajectories of billiard balls with eccentric twists considering friction between ball and cloth. *Top:* Deflection of cue ball at cushion. When a ball is hit below (*left*) or above (*right*) its center of gravity, the induced spin and its direction influence the ball trajectory after deflection, which results in a convex (*left*) or concave (*right*) curve, respectively. This phenomenon also depends on the angle of incidence. Obviously, the simple optical Law of Reflection that the angle of incidence equals the angle of reflection as assumed by Marcus MARCI in 1639 is no longer valid. When playing to the "side" [French *effet*] as illustrated in the right picture at bottom, which provides the ball with an additional spin around its vertical axis, the ball's trajectory after deflection can be influenced to some extent. *Bottom:* Collision of cue ball with an object ball. He also studied the difficult problem of when a ball, being hit above (*left*) or below (*right*) its center of gravity, strikes a second, resting ball centrally or eccentrically − a difficult task that MARCI in 1639 had tried to solve in vain. When the second ball is hit centrally (case *0−0*), it moves along a straight line in the direction of the first, striking ball. However, when the first, striking ball hits the cue ball above (*left*) or below (*right*) its center of gravity, it collides eccentrically with the second, resting ball, and the striking ball moves after collision along concave (*left*) or convex (*right*) trajectories, respectively. Note that, after collision, in both cases the trajectories of the first, striking, ball are curved (right family of curves *1, 2, 3*) while the trajectories of the second, struck, ball are always straight lines (left family of curves *1, 2, 3*), which are inclined in the direction of the striking ball. [G. CORIOLIS: *Théorie mathématique des effets du jeu de billard.* Carilian-Goeury, Paris (1835)]

## 4.3 PERCUSSION STUDIES – Oldest Known Seismoscope and 19th-Century Seismography

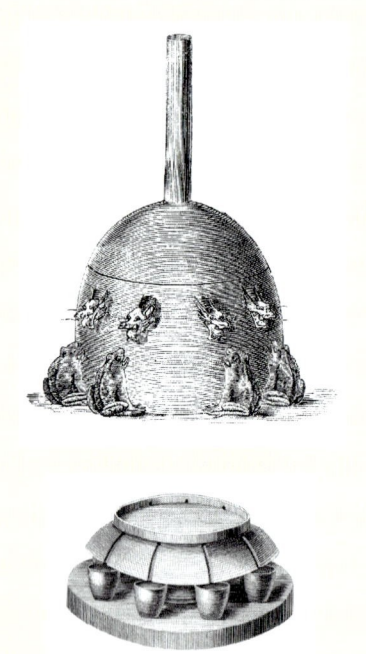

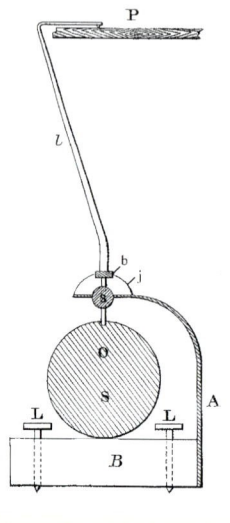

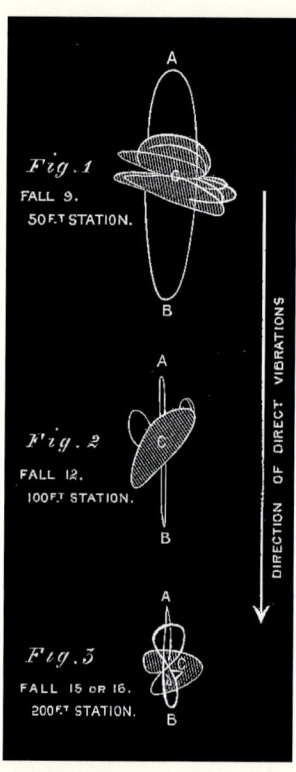

**4.3–L** Early seismoscopes were primitive but ingenious mechanical instruments supposed to record the direction and strength of seismic shocks. *Top:* The oldest known seismoscope was invented in China in A.D. 136. On the outside there are eight dragon heads, each of which contained a ball. Underneath these heads were eight frogs positioned in such a way that they were ready to receive the ball if it were shaken out of place by an earthquake, thus recording the direction of the seismic shock. [Nature **26**, 627 (1882)] *Bottom:* To record the strength and direction of seismic shocks, Gaetano CACCIATORE improved the mercury seismometer that Jean DE HAUTE FEUILLE had invented in 1703. CACCIATORE's mercury-horizontal seismometer consisted of a shallow, circular 10-in. (25.4-cm)-dia. bowl. In level height, its side wall was bored through eight equidistant holes. The bowl was surrounded by an annular bulge containing eight channels, corresponding to the eight holes and leading to eight small cups. The instrument had to be installed on a level pedestal in an area free of human-induced vibrations. After careful alignment of the eight holes with the four cardinal points, the bowl was filled up with mercury. Used by CACCIATORE in Palermo during the severe earthquake in 1823, he was able to prove that the seismic shocks came from the direction of the volcano Mount Etna. [Ann. Phys. Chem. **24** (II), 62 (1832)]

**4.3–M** The Englishmen John MILNE and Thomas GRAY, working as visiting professors in the 1880s at the Imperial College in Tokyo, investigated the dynamics of earthquakes. They constructed new seismographs to study the motions of the ground originating from natural and artificial earthquakes. *Left:* To better observe earthquakes in Japan, GRAY constructed a "rolling-sphere seismograph" that writes the Earth's vibrations directly upon a smoked glass plate by means of a pointer. Later installed at a number of different stations, it was the first seismograph to allow a clear distinction between normal and transverse vibrations, together with the maximum amplitudes of each of these two distinct movements. [Phil. Mag. **12** (V), 199 (1881)] *Right:* Examples of three records taken with GRAY's seismograph; each shows the motion of the ground magnified 85 times. These figures are arranged such that their greatest lengths are parallel, and the direction of greatest motion is that of a line joining the instrument and the point where the ball fell. At the commencement of a seismic shock, the needle at the pointer rested at or very near the center of curve C. The deflections occurred first in directions $A \to C \to B \to C$, but suddenly, when the shear waves arrived, the deflections occurred at nearly right angles to the first motion. [Phil. Trans. Roy. Soc. Lond. **173**, 863 (1883)]

## 4.3 PERCUSSION STUDIES – 19th-Century Seismography *(cont'd)*

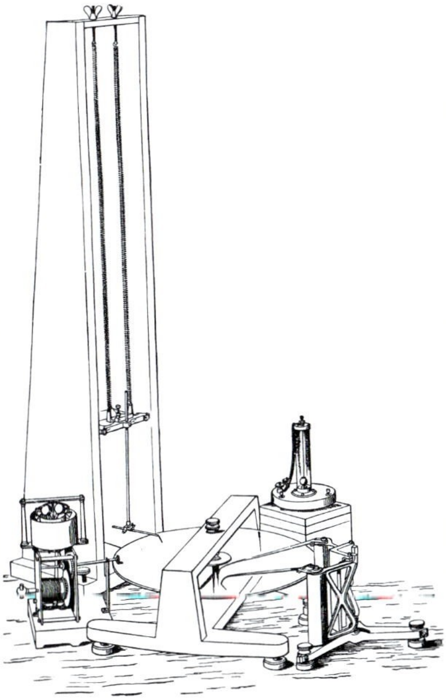

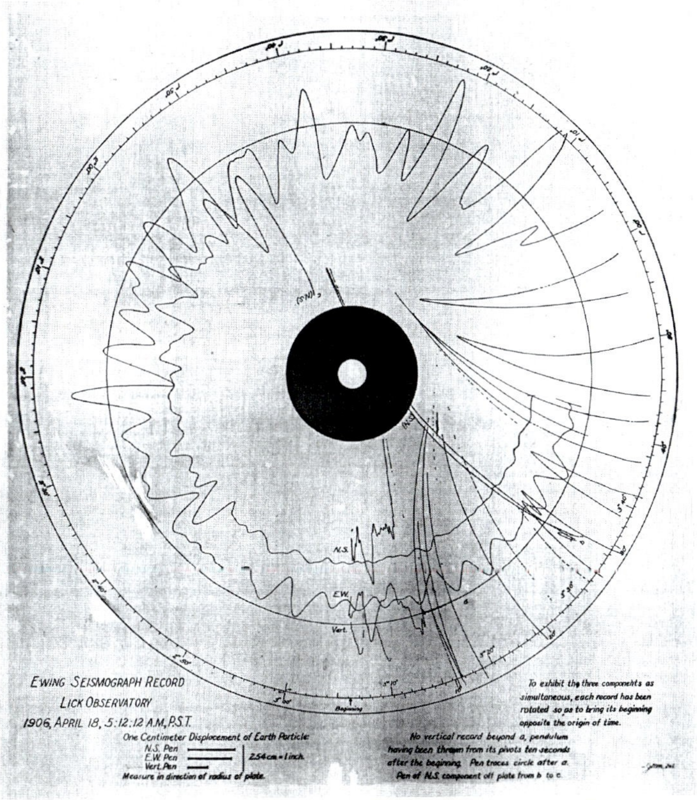

**4.3–M** *(cont'd)* **Top, left:** The first seismograph in the United States was constructed by the Scottish engineering and physics professor Sir James A. EWING and installed in 1897 at the Lick Observatory on Mount Hamilton, about 50 miles southeast of San Francisco, CA. This three-component instrument for recording motions in all directions, provided with clock and driving plate, faithfully recorded earthquake disturbances throughout the period 1887–1926. The clock was set in motion by means of a Palmieri seismograph, a small common pendulum that appears on the right in the figure behind the plate. This occurred during the preliminary tremors usually found in advance of the main movements of an earthquake. **Top, right:** Reproduction of the EWING seismograph record from Lick Observatory taken in the morning of April 18, 1906 at 5:12 A.M. P.S.T., when San Francisco and a wide area around the city were hit by a heavy earthquake. Ten seconds after the beginning of a series of violent shocks, the pendulum was thrown from its pivots. **Bottom:** Picture of Lick Observatory taken in November 1881. [Courtesy Mary Lea Shane Archives of the Lick Observatory, University of California, Santa Cruz]

## 4.3 PERCUSSION STUDIES – 19th Century: Percussion Figures

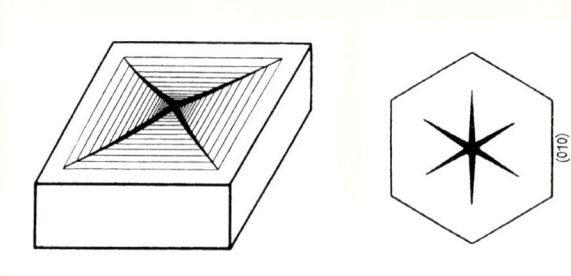

**4.3–N** When a crystal is hit by a pointed instrument normal to its surface, a starlike fracture pattern is produced. This so-called "percussion figure," first discovered by the German physicist Friedrich E. REUSCH (1867), is characterized by radiating lines from the point of impact that are parallel to the plane of symmetry of the crystal. For example, when a rock salt crystal is struck by a sharp blow, a four-rayed star (*left*) is formed, while a cleavage flake of mica typically shows a six-rayed star (*right*). [P. NIGGLI: *Lehrbuch der Mineralogie*. Gebr. Borntraeger, Berlin (1920), vol. I, p. 268]

## 4.3 PERCUSSION STUDIES – 20th Century: Measurement of Deformation and Force

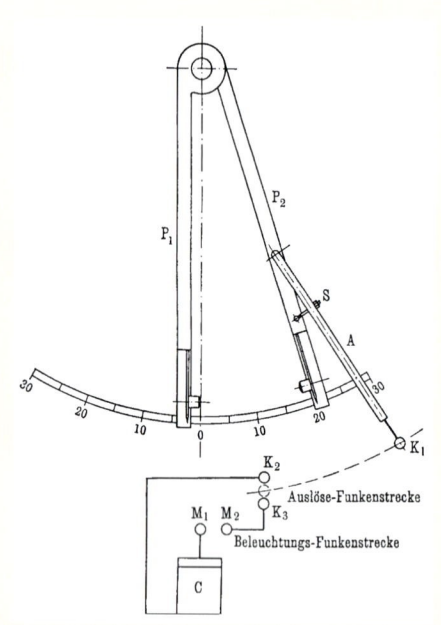

**4.3–O** In the early 1920s, Franz BERGER, an Austrian engineer at Vienna University, first recorded force-time profiles of colliding bodies using two different percussion apparatus: one for a stepwise recording as shown here, and a second, more advanced, one for a complete recording of the force-displacement curve. **Left:** Schematic of his setup. Using a twin pendulum $P_1$–$P_2$, each pendulum carries at its lower end a test cylinder, both of which collide with each other head-on. To measure the force of percussion, BERGER inserted a leaf spring between the test body and the pendulum end. The momentary displacement of the leaf spring was illuminated by a triggered spark light source $M_1$–$M_2$ and photographed using a still camera. The time delay of the flash could be adjusted at the screw $S$ of the contact arm $A$ that, triggering the spark gap $K_2$–$K_3$ via ball $K_1$, allowed one to observe different time instants within the percussion cycle. From these displacement-time data BERGER calculated the force-time profile and determined how the maximum percussion force depends on the impact velocity. **Right:** Photo of BERGER's percussion apparatus. [F. BERGER: *Das Gesetz des Kraftverlaufs beim Stoß*. Vieweg, Braunschweig (1924)]

## 4.3   PERCUSSION STUDIES – 20th Century: Measurement of Deformation and Force *(cont'd)*

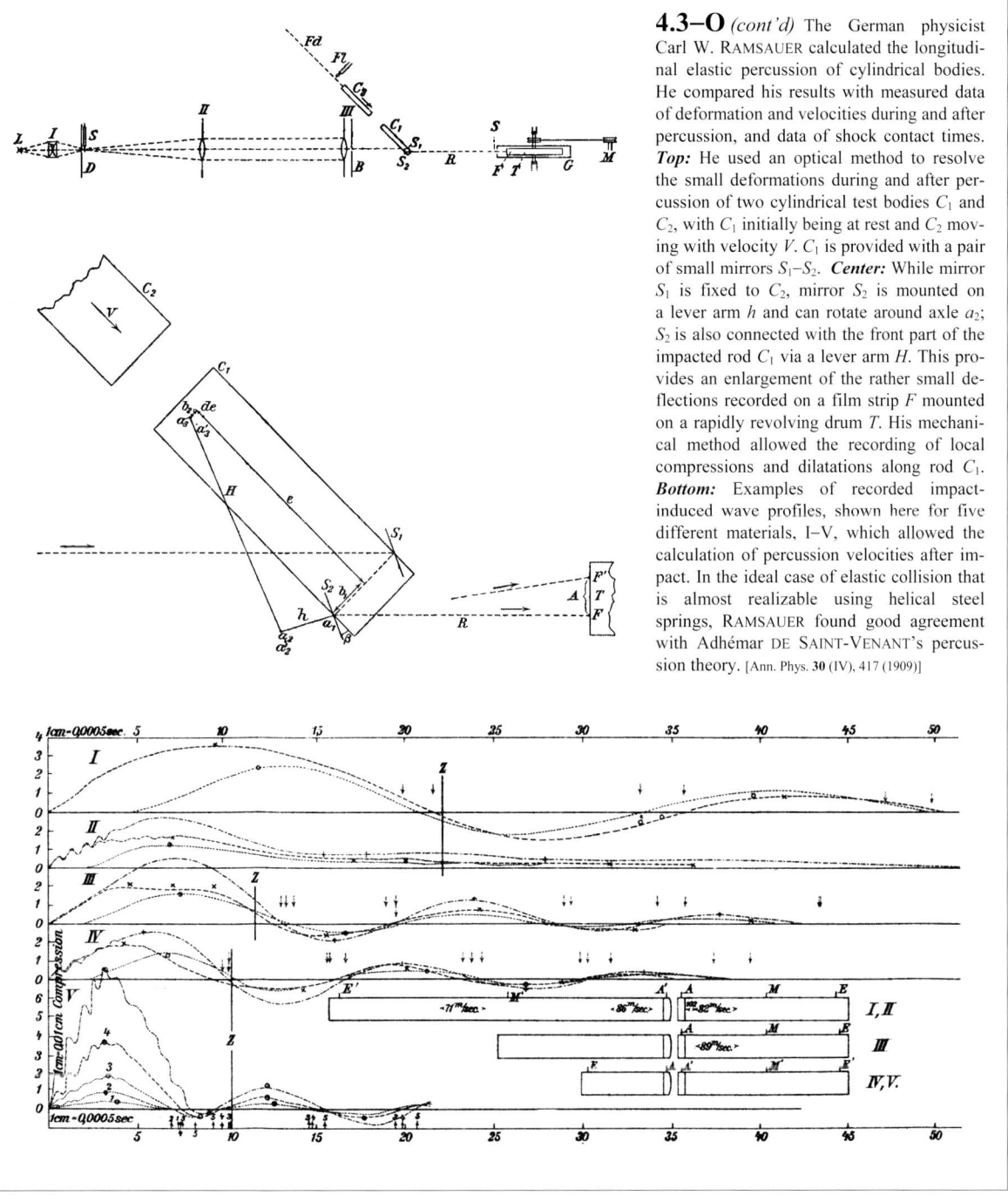

**4.3–O** *(cont'd)* The German physicist Carl W. RAMSAUER calculated the longitudinal elastic percussion of cylindrical bodies. He compared his results with measured data of deformation and velocities during and after percussion, and data of shock contact times. *Top:* He used an optical method to resolve the small deformations during and after percussion of two cylindrical test bodies $C_1$ and $C_2$, with $C_1$ initially being at rest and $C_2$ moving with velocity $V$. $C_1$ is provided with a pair of small mirrors $S_1$–$S_2$. *Center:* While mirror $S_1$ is fixed to $C_2$, mirror $S_2$ is mounted on a lever arm $h$ and can rotate around axle $a_2$; $S_2$ is also connected with the front part of the impacted rod $C_1$ via a lever arm $H$. This provides an enlargement of the rather small deflections recorded on a film strip $F$ mounted on a rapidly revolving drum $T$. His mechanical method allowed the recording of local compressions and dilatations along rod $C_1$. *Bottom:* Examples of recorded impact-induced wave profiles, shown here for five different materials, I–V, which allowed the calculation of percussion velocities after impact. In the ideal case of elastic collision that is almost realizable using helical steel springs, RAMSAUER found good agreement with Adhémar DE SAINT-VENANT's percussion theory. [Ann. Phys. **30** (IV), 417 (1909)]

## 4.3 PERCUSSION STUDIES – 20th Century: Pressure-Bar Devices

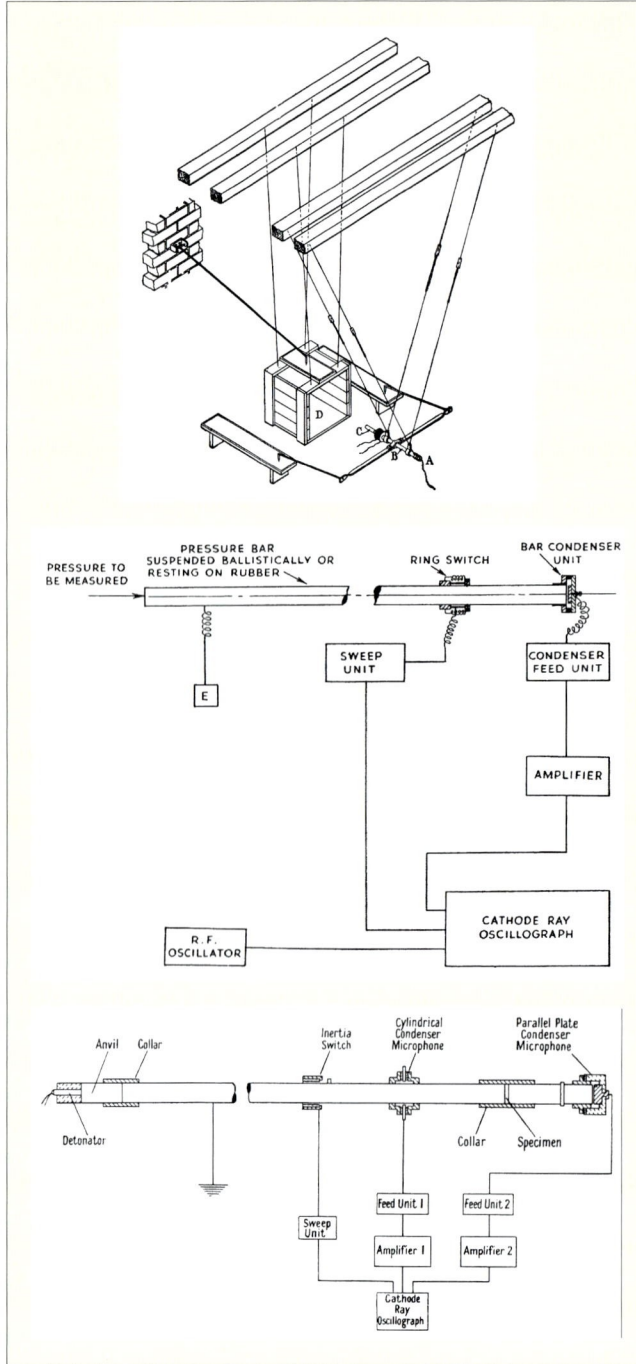

**4.3–P** Throughout the 20th century, various pressure-bar methods were devised for the generation of well-defined pressure pulses in order to derive dynamic stress/strain materials data. *Top:* Schematic of "Hopkinson pressure bar," a variant of the ballistic pendulum invented in the 1910s by the British engineer Bertram HOPKINSON. It consists of a long steel bar $B$, about 1 in. in diameter, suspended in a horizontal position by four threads so that it can swing in a vertical plane. At one end of the bar a short cylindrical pellet $C$, known as the "time piece," is wrung on. The transient pressure is applied at the other end. When a bullet impinges on the firing end of the long bar or a pressure pulse is initiated by a detonation $A$, the compression pulse is transmitted through the joint between the long and the short bar. But after reflection at the free end of the short bar this pressure pulse is converted into a pulse of tension. This causes the short rod $C$ to fly off with the momentum trapped in it, which is measured with a conventional ballistic pendulum $D$. [Proc. Roy. Soc. Lond. **A89**, 411 (1914)] *Center:* In 1948, the British engineer Rhisiart M. DAVIES devised an electrical modification of the Hopkinson pressure bar for measuring the relation between pressure and time in experiments on high pressures of short duration. His experimental method, the so-called "Davies bar," applies a long bar to propagate the pressure pulse generated at the entrance face to the end face, which is equipped with a displacement sensor, a condenser microphone. Under the assumption that the pressure waves are elastic waves, the unknown stress-time curve can be determined by differentiating the measured displacement-time profiles. [Phil. Trans. Roy. Soc. Lond. **A240**, 375 (1948)] *Bottom:* In the Davies method, the pulse is propagated down the bar, only without an appreciable change in form when the pulse is long in comparison with the diameter of the bar and when there are no sudden changes in pressure. The method proposed by the British researcher Herbert KOLSKY avoids this limitation. The so-called "Kolsky pressure bar" – also known as the "split Hopkinson pressure bar" – is similar to the Davies pressure bar, except that the bar is in two parts, and a second condenser microphone and amplifier are introduced. The shock pressure pulse is produced by a detonator held against a replaceable steel anvil at the firing end of the bar. KOLSKY's method allows one to measure the stress-strain behavior of disk-shaped specimens. If a sufficiently thin specimen is inserted, the pressure pulse is effectively the same throughout the specimen during the passage of the compression pulse. This pressure is communicated along the extension bar to the condenser microphone, which delivers a displacement-time signal. The stress-time relation for the specimen is obtained by differentiation of this curve. [Proc. Phys. Soc. (Lond.) **62B**, 676 (1949)]

## 4.3 PERCUSSION STUDIES – 20th Century: Taylor Test

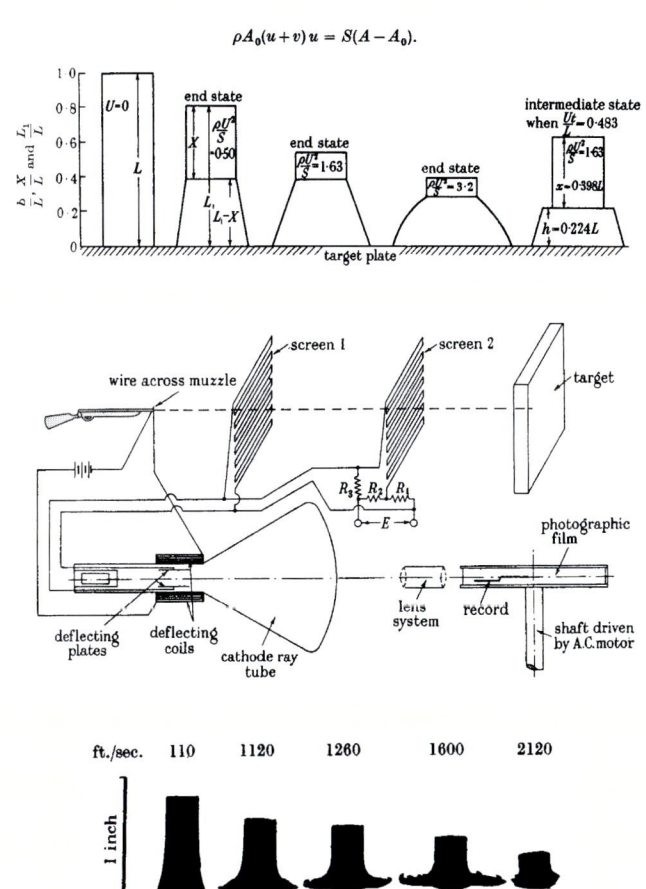

**4.3–Q** It has long been known that metals may be subjected momentarily to stresses far exceeding their static yield stress without suffering plastic strain. The theoretical concept for determining the dynamic yield stress is the so-called "Taylor test" devised by Geoffrey I. TAYLOR, a British applied mathematician at the Cavendish Laboratory of the University of Cambridge. It was first tested in practice by his coworker A.C. WHIFFIN, who found good agreement with theory. *Top:* The method allows one to determine the dynamic yield strength $S$ of a material by using a rod made of this material and impacting it end-on at a known velocity $U$. TAYLOR derived a simple formula, given by

$$S = C \times U^2 (L-X)/(L-L_1),$$

with $C$ being a constant, $L$ the initial overall length of the test cylinder, $X$ the unstrained length of the specimen, and $L_1$ the overall length of the test cylinder after impact. *Center:* Since the dynamic strength is proportional to the square of the striking velocity, a careful measurement of the velocity is absolutely necessary. WHIFFIN used two wire screens to subsequently deflect the bright spot of a cathode ray tube which he recorded on a film strip mounted on a rapidly revolving drum. *Bottom:* This picture shows examples of mushroomed steel cylinders after the test that were impacted with velocities ranging from 110 to 2,120 ft/s (33.5–646 m/s). The method of determining dynamic yield stress proved very useful for strain rates up to about $10^4$ in./in./s. [Proc. Roy. Soc. Lond. **A194**, 289 (1948)]

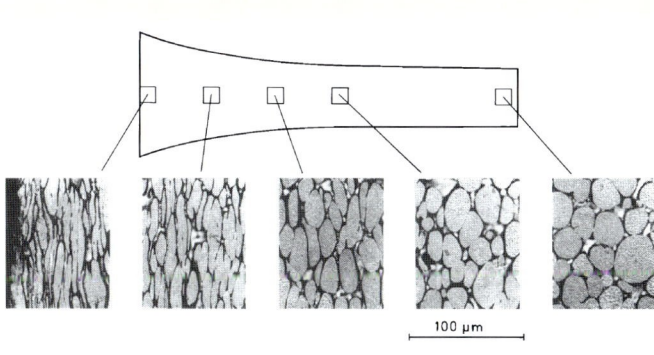

**4.3–R** Reinhard THAM and Alois J. STILP (EMI, Freiburg) studied the penetration depth of the plastic wave front in metal alloys by analyzing their microstructure after impact. Using 5.8-mm-dia. rods with an initial length of 58 mm made of DF-17 (a W-Ni-Fe compound made by Plansee GmbH, Germany) and impacting them on a rigid wall at 203 m/s, they measured the grain deformation and therefrom determined the dynamic compression strength. Note that deformations of the tungsten grains are particularly pronounced in the impact region on the left-hand side of the rod. [J.R. ASAY ET AL. (eds) *Shock waves in condensed matter – 1983*. North-Holland, Amsterdam (1984), p. 167]

## 4.3 PERCUSSION STUDIES – 20th Century: Examples of Ball Percussion Studies

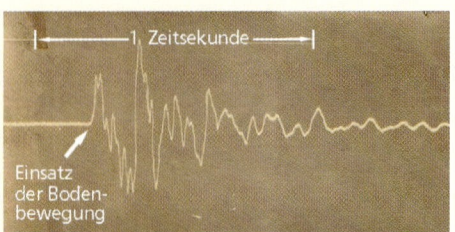

**4.3–S** Example of inelastic percussion at low velocity (< 17 m/s): *Left:* View of the "Mintrop ball," a 4,000-kg, approx. 1-m-dia. iron drop weight used in 1908 by Ludger MINTROP at Göttingen University to generate artificial earthquakes. The dropping tower, a 14-m-high steel construction, and the ball were donated by the Krupp Company. MINTROP recorded the seismic shocks at different distances from the impact site using a transportable, high-sensitivity seismograph with a magnification of 50,000. From that he derived the velocity of the earth wave. [Courtesy Institute of Geophysics, University of Göttingen] *Right:* Partial view of a seismogram taken in 1908 by MINTROP at a distance of 120 m from the ball impact using the Wiechert seismograph. At this close distance, the separation of ground motion into P- and S-waves is not yet fully accomplished. [From L. MINTROP's Ph.D. thesis, University of Göttingen. Girardet, Essen (1911)]

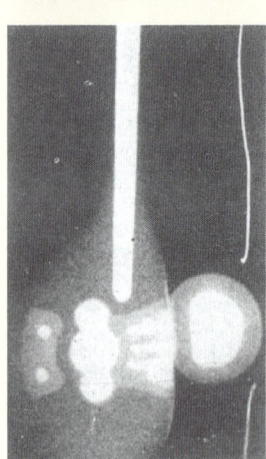

**4.3–T** Examples of freezing the moment of violent interaction of a golf ball with a club. *Top:* In 1935, Harold E. EDGERTON at MIT (Cambridge, MA), using a xenon flash lamp of short duration (< 1 μs) and a still camera, took unique reflected-light photographs of this high-speed percussion event. Note the strong elastic deformation of the ball. Because of the effect of inertia, the lower, undeformed part of the ball is still fully resting on the tee, while a tremendous percussion force is already acting on its central part. [H. EDGERTON: *Stopping time*. Stemmle, Schaffhausen (1987), p. 54] *Bottom:* With the advent of flash radiography in 1938 it became possible to also study the interior of shock-compressed opaque objects. The picture was taken by Charles M. SLACK at Westinghouse Electric Corporation. Note the compressed core of the golf ball. [J. Soc. Mot. Pict. Eng. **52**, No. 3, p. 63 (1949)]

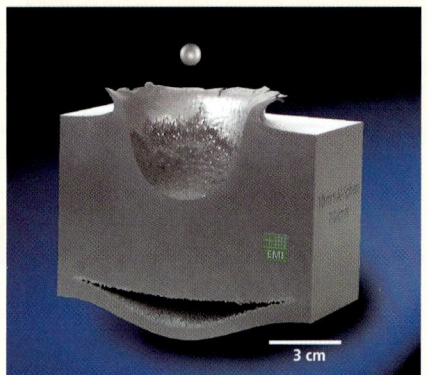

**4.3–U** Example of a hypervelocity impact: a 10-mm-dia. aluminum sphere impinged a plane aluminum plate at hypervelocity ($M \approx 14$). This picture well illustrates that a hypervelocity impact produces a crater that is much larger in diameter than the size of the impacting body, in this example by a factor of about 7. This is an important phenomenon for correctly interpreting meteorite impact craters. Note the well-marked spallation zone. [Dept. of Terminal Ballistics & Impact Physics, EMI, Freiburg]

## 4.3 PERCUSSION STUDIES – 20th Century: Liquids under Impact

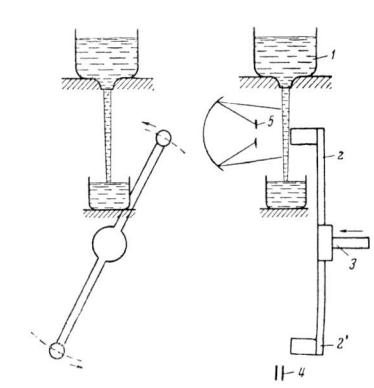

**4.3–V** At high impact velocities liquids can behave like solid bodies. Depending on the viscosity of the liquid and the rate of deformation, the same liquid can show plastic flow as well as brittle fracture, which was first demonstrated by Mark I. KORNFELD. *Left:* He generated a vertical jet of liquid by providing a hole in the bottom of a vessel *1* and used a rotating arm with two metallic segments *2* and *2'* facing each other. The first one served as an impactor and the second as a trigger of a spark light source *5*. By suddenly pushing forward the rotating arm on the axis *3*, the segment *2* strikes the liquid jet, thereby simultaneously triggering the spark light source. A still camera with open shutter was used to photograph the impact phenomenon in reflected light. He used a liquid that had a viscosity of about 50 kg s/m², corresponding to the consistency of honey or syrup. *Bottom:* "Brittle fracture" of liquid can be observed also at lower velocities when a liquid with a higher viscosity is used. [M.I. KORNFELD: *Elastizität und Festigkeit von Flüssigkeiten*. Verlag Technik, Berlin (1952), p. 74]

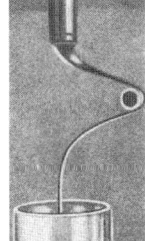

← At a low impact velocity (< 23 m/s), the liquid jet only reveals a plastic (laminar) deformation.

→ A brittle-type destruction of the liquid jet can be observed at an impact velocity of > 23 m/s.

 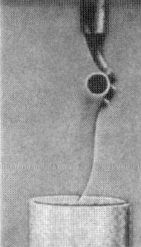 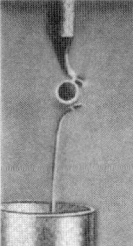 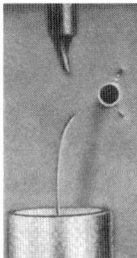

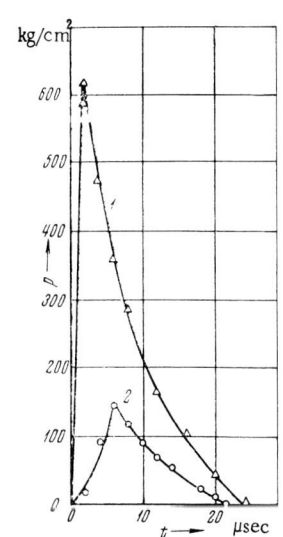

**4.3–W** *Left:* In the former Soviet Union, S.P. KOZYREV and K.K. SHAL'NEV studied the impact of liquid jets on a solid. They observed that the short duration of the initiated pressure pulse depends on the shear viscosity and impact velocity [*curve 1*, 64 m/s; *curve 2*, 32 m/s]. *Right:* To explain this phenomenon, they proposed a relaxation model and tested it by ballistic impact of a cylindrical projectile with a 5-bar liquid jet, a solution of rosin in mineral oil, issued from an 8-mm-dia. nozzle. Like M.I. KORNFELD {⇨Fig. 4.3–V}, they observed plastic deformation of the liquid jet at low velocity (9 m/s, *left*) and brittle fracture at a higher velocity (15 m/s, *right*). [Sov. Phys. Dokl. **15**, 513 (1970)]

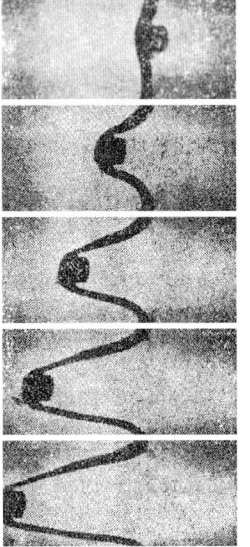 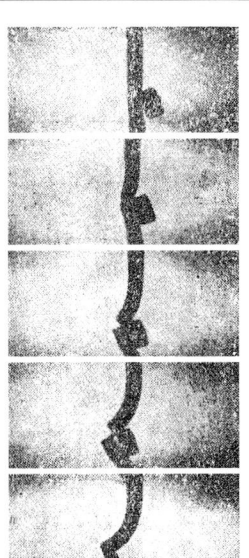

## 4.3 PERCUSSION STUDIES – 20th Century: "Sweet Spots" of Sports Equipment

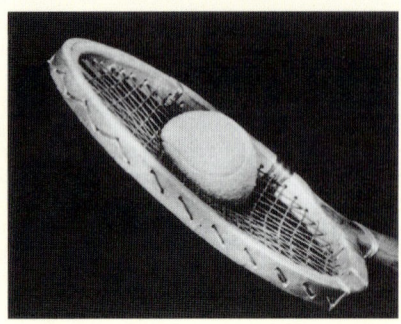
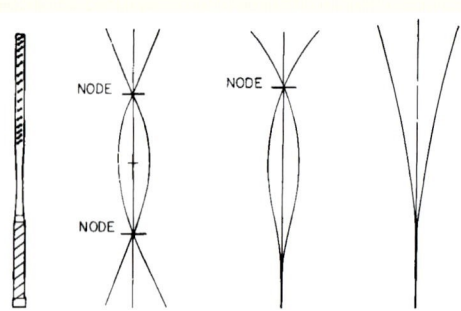
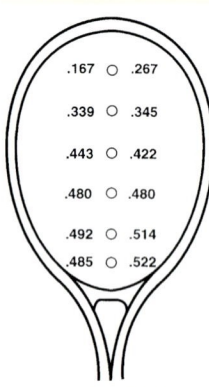

**4.3–X** Recent studies performed in Australia, Austria, and the United States on the impulsive force generated in handheld implements (such as tennis rackets and baseball bats) have shown that the impact point that feels best – commonly called the "sweet spot" – is usually the node of the fundamental vibration mode, not the center of percussion (COP), a well-defined spot for a rigid body rotating around a fixed axis {⇒Fig. 2.9}. The Austrian biomechanics professor Herbert HATZE {⇒1998} concluded that the concept of the COP is of limited significance when an implement is handheld because the force exerted by the hands will affect the location of the axis of rotation, which in turn affects the location of the COP. ***Top, left:*** Photo of a tennis ball in compression taken by MIT professor Harold E. EDGERTON using a xenon flash light source [Proc. 4th Int. Congress on High-Speed Photography. Helwich, Cologne (1958), p. 14] ***Top, right:*** When a ball hits a racket, it produces a transverse wave that travels along the racket and is then reflected from both the tip and the butt end. The U.S. physics professor Howard BRODY studied the vibrations of a tennis racket for the free-free (*left*) and the clamped-free (*center and right*) modes. For an impact at the vibration node, the vibrational component becomes zero. [Am. J. Phys. **49**, 816 (1981)]. ***Bottom:*** Positions of experimentally determined ball speed ratios (*left*) and calculated ball speed ratios (*right*) for a free racket according to BRODY. This ratio falls off as the impact point moves from the center of mass toward the tip. [Am. J. Phys. **65**, 981 (1997)]

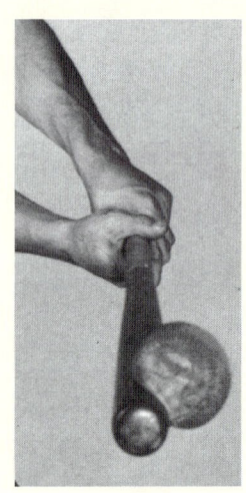
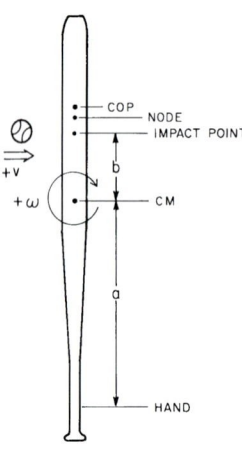

**4.3–Y** ***Left:*** High-speed flash photograph of a bat striking a baseball, taken by H.E. EDGERTON at MIT. The large deformation of the ball illustrates well the enormous magnitude of the impulsive force at this instant. [From his book *Stopping time*. Stemmle, Schaffhausen (1988), p. 55] ***Right:*** According to studies by Howard BRODY, there exist three "sweet spots" on a baseball bat:

(1) The center of percussion (COP), which is located ca. 17 cm from the fat or distal end of the bat. When the hands grip the bat at a distance *a* from the center of mass (CM), there is no net translational motion at that point of the bat.

(2) The node of oscillation, which is located ca. 19 cm from the end of the bat, approx. one quarter of the length of the beam. When the ball is hit at the location of the node, the hands will not "sting."

(3) The impact point, which is located at a distance *b* from the CM. When the ball is hit at an impact point that is not the COP, this leads to maximum "power," *i.e.*, to the fastest batted ball speed. The distance *b* varies between 18.2 and 8.8 cm for a bat/ball mass ratio varying between 5 and 7, respectively. [Am. J. Phys. **54**, 640 (1986)]

## 4.3 PERCUSSION STUDIES – 20th Century: Seismology of Nuclear Explosions

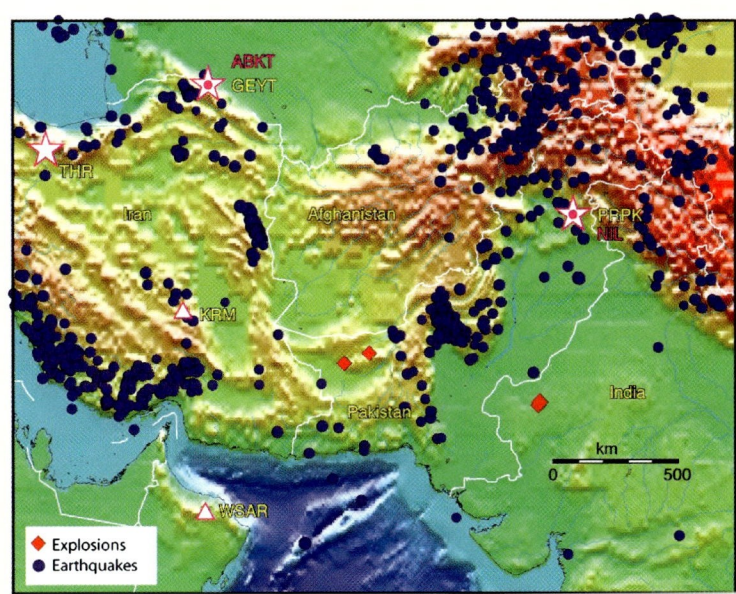

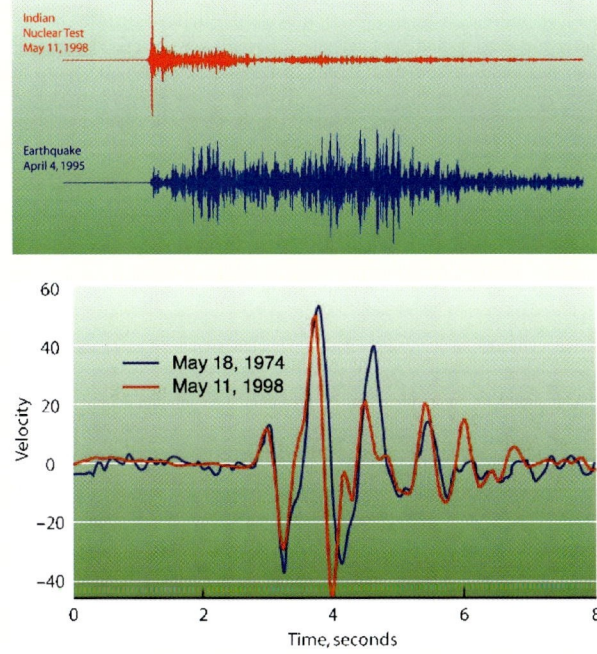

**4.3–Z** The Nuclear Test Ban Treaty that was signed in 1963 banned nuclear weapons tests in the atmosphere, in outer space, and under water, but permitted underground testing. The Comprehensive Test Ban Treaty (CTBT), which has hitherto been signed by 174 nations, bans *all* nuclear test explosions, including underground tests and those intended for peaceful purposes. The CTBT provides for an International Monitoring System (IMS) of automated seismic stations to record any evidence of clandestine nuclear explosions that transmit data via satellite to the International Data Center in Vienna, Austria. Teams in various countries are working to improve ways to seismically differentiate clandestine underground nuclear explosions from other sources of seismicity (*e.g.*, earthquakes and mining explosions) and to characterize them in terms of their locations, depths, and yields.

Under the auspices of the U.S. Dept. of Energy (DoE), William R. WALTER and collaborators at LLNL (Livermore, CA) worked out methods of analyzing seismograms to differentiate between nuclear explosions and other types of seismic sources. They discovered that differences in seismic P- and S-wave energy provide one method of discriminating explosions from earthquakes: seismic P-waves are compressional waves; therefore explosions should have higher P/S ratios than earthquakes. *Top:* This topographic map shows seismic locations both of nuclear tests (*red diamonds*) performed in India (in 1974 and on May 11 and May 13, 1998) and in Pakistan (on May 28 and 30, 1998), and earthquakes recorded in the region between 1995 and 1997 (*blue circles*), taken from the PDE (Preliminary Determination of Epicenters) catalog. Also shown are planned locations of IMS primary seismic stations (*stars*) and auxiliary seismic stations (*triangles*). *Center:* The seismogram of the Indian nuclear test, performed on May 11, 1998, clearly differs from that of a representative earthquake. *Bottom:* Seismic waveforms from different nuclear explosions show a remarkable similarity. [From LLNL's Science & Technology Review (April 1999). Courtesy University of California, LLNL; http://www.llnl.gov/str/Walter.html]

## 4.4 PERCUSSION AND SHOCK WAVE MODELS – Corpuscular Models and NEWTON's "Cradle"

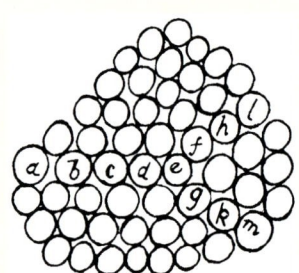

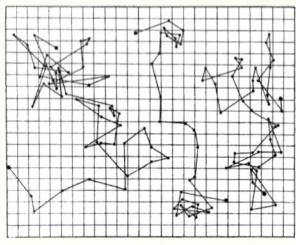

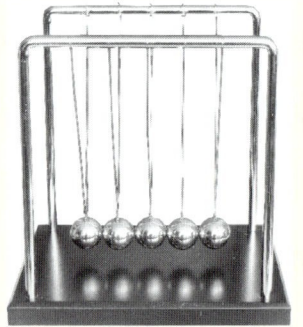

**4.4–A** Historic 2-D collision models. ***Top:*** Simple model of an elastic fluid used by Sir Isaac NEWTON to illustrate the transport of impulse from one individual particle to another. He writes, "If the particles $a, b, c, d, e$ lie in a right line, the pressure may be indeed directly propagated from $a$ to $e$, but then the particle $e$ will urge the obliquely posited particles $f$ and $g$ obliquely … the pressure begins to deflect towards one hand and the other, and will be propagated obliquely in infinitum…"
[I. NEWTON: *Principia*. S. Pepys, London (1687), Lib. II, Prop. XLI]
***Center:*** In 1992, the U.S. researcher Brian J. FORD reproduced some of Robert BROWN's famous experiments on "Brownian movement" (1827) using BROWN's own microscope. FORD's video recording on the movement of tiny particles (here fat droplets of about 1 µm in diameter in cow's milk) can be viewed on the Internet. Note that the blurring in FORD's video is caused by the primitive, single-lens construction of BROWN's microscope.
[http://www.sciences.demon.co.uk/wbbrowna.htm]
***Bottom:*** The French physicist Jean-Baptiste PERRIN, studying the successive movements of small particles, quantitatively confirmed the kinetic theory of heat. Shown here are three examples of observed temporal positions of an individual grain taken through out 50 time intervals, each lasting 30 s. The particles quickly change direction, and in the microscope only a mean shift is observable.
[Ann. Chim. Phys. **18** (VIII), 5 (1909)]

**4.4–B** The Newtonian demonstrator, popularly known as "NEWTON's cradle" or "balance ball," has become a widely used apparatus for demonstrating elastic percussion, the Third Law of Motion, and the Laws of Conservation of Momentum and Energy. Suspension of the balls on twin strings ensures that oscillations of all balls remain in a fixed plane, even after passing several cycles. To verify that during each collision cycle the impulse is fully propagated, the balls are suspended such that they barely touch in their positions at rest. The instrument shown here can be bought at the Deutsches Museum Shop GmbH, Munich.

## 4.4 PERCUSSION AND SHOCK WAVE MODELS – Shock Wave Demonstration Apparatus

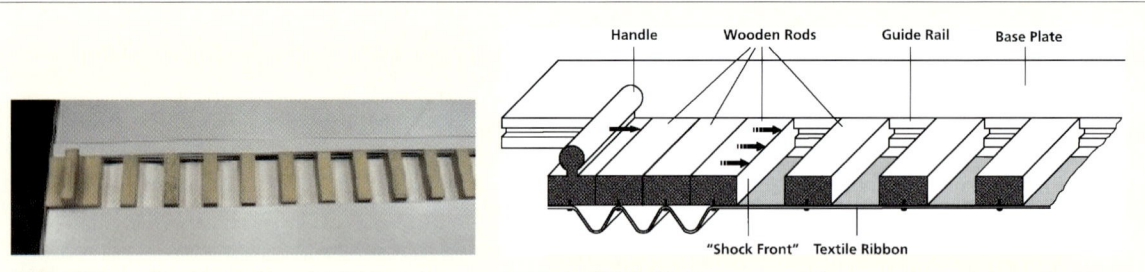

**4.4–C** This simple shock wave model is used to demonstrate to visitors of the Rieskrater-Museum the "shock wave" phenomenon. It consists of a couple of wooden rods mounted on a textile ribbon. Initially separated by each other, the rods, beginning from the left, are successively pushed together. In this way, the leading edge of collision proceeds successively from one rod to the other, thus illustrating the successively acceleration of particles when being hit by the shock front. [Courtesy Gisela PÖSGES, Rieskrater-Museum, Nördlingen, Bavaria]

## 4.4 PERCUSSION AND SHOCK WAVE MODELS – One-Dimensional Shock Wave Models

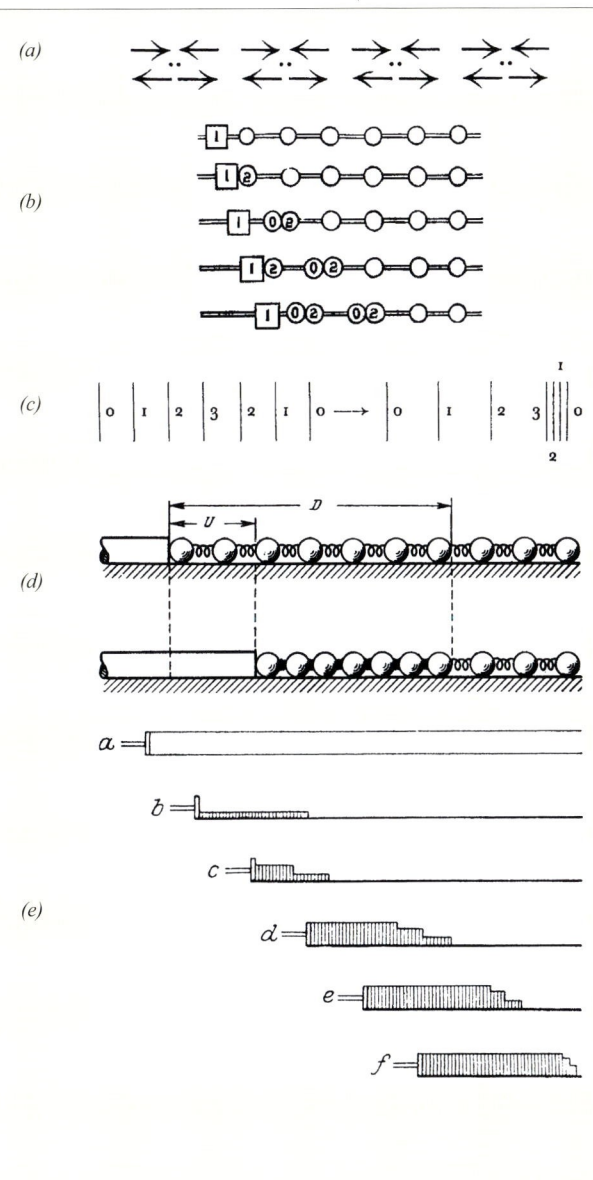

**4.4–D** Shock wave models arose with the discovery of the phenomenon itself, but they are also used in modern textbooks. *(a)* Ernst MACH's sketch of his percussion model, illustrating that a violent disturbance must propagate with a higher velocity than the sound velocity *c*. MACH assumed a linear row of gas molecules arranged two by two and colliding with each other (*upper row of arrows*) with velocity $u \gg c$. During collision the molecules gain a velocity increase by $u + c$. After collision they rebound in the opposite direction (*lower row of arrows*). [Sitzungsber. Akad. Wiss. Wien **75** (II), 101 (1877)] *(b)* The British physicist Charles V. BURTON used a linear shock wave model that assumes a linear row of balls separated by an equal number of weak springs (not shown here). A piston, coming from the left side, hits the first ball and transmits its impulse to the next one. [Phil. Mag. **35** (V), 317 (1893)] *(c)* The shock wave model of explaining shock wave formation proposed by the British fluid dynamicist Sir M. James LIGHTHILL is very simple and plausible. LIGHTHILL says, "Note that the running of waves in the region of rising pressure, which produces a local ordering of the pressure rise into a discontinuous jump, is balanced by the gradual separation of waves from one another in any region of falling pressure, where the waves carrying the lower pressures lag behind those in front and spread the wave out as illustrated. The numbers on the waves might represent percentage excess pressure, and the right-hand figure shows the waves in the region of the rising pressure just before they have finally run together." [Mem. Manch. Lit. Phil. Soc. **101**, 1 (1959)] *(d)* The 1-D model of shock wave propagation of a shock wave in an elastic medium proposed by the Russian physicist Lev V. AL'TSHULER is similar to BURTON's model: The propagation of a shock wave in a substance can be modeled by a series of spheres that are elastically coupled to one another. The rate of displacement of the spheres, the "mass velocity" of substance *U*, is equal in this model to the velocity of a piston that causes the spheres to move. This velocity is always lower than the velocity *D* of the perturbation boundary separating the resting and separated spheres from those moving and gathered into a more compact mass. [Sov. Phys. Uspekhi **8**, 52 (1965)] *(e)* The famous "shock wave piston model," proposed by the German physicist Richard A. BECKER, explains the formation of shock waves by coalescence of discontinuous pressure pulses and by means of the formula for the sound velocity $c = (k\, \partial p/\partial \rho)^{1/2}$. The motion of the piston is thought of as being divided into a large number of small, successive motions, each separate motion producing a pressure pulse that propagates through the medium ahead at sonic velocity. Since each pressure pulse heats the gas adiabatically, which also increases the sound velocity, later pulses will tend to overtake the preceding ones. [Z. Phys. **8**, 321 (1922)]

## 4.4 PERCUSSION AND SHOCK WAVE MODELS – Apparatus for Demonstrating Hydraulic Jumps

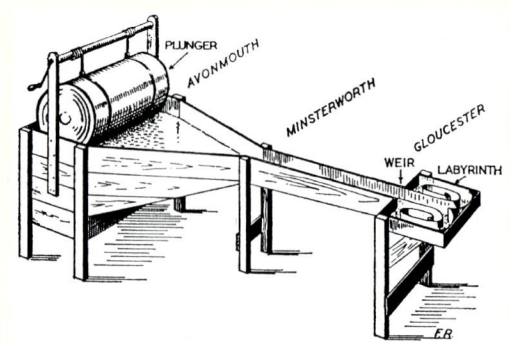

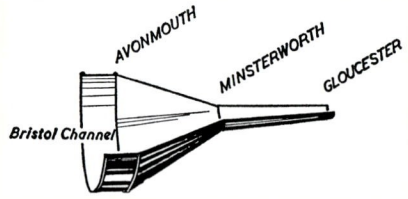

**4.4–E** *Left:* The Severn is Britain's longest river (about 290 km). The shape of its estuary is such that the water is funneled into an increasingly narrow channel as the tide rises. Between Minsterworth and Gloucester the river is less than 100 yards across, thus forming the famous "Severn Bore," one of the most studied surge waves in the world. *Right:* The generation of a bore can be demonstrated using this simple apparatus. It uses a funnel-shaped channel to simulate the estuary and a plunger to drive a model bore into the straight part of the funnel. At its end the wave is absorbed in a labyrinth. [F. ROWBOTHAM: *The Severn bore*. Macdonald, London (1964), p. 23]

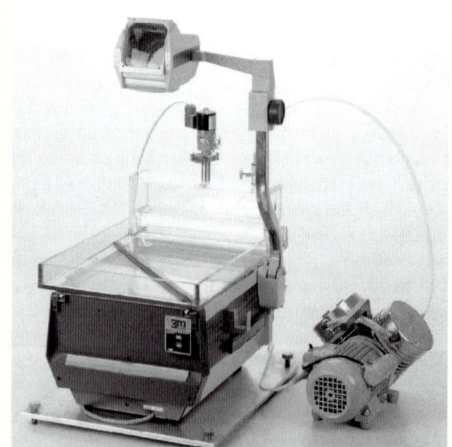

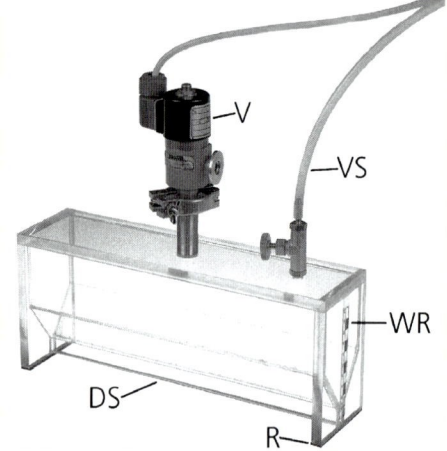

**4.4–F** *Left:* The propagation and reflection of hydraulic jumps was demonstrated by the author in an auditorium using a small water table, a wave generator, and a common viewgraph projector. A plane wave leaving the slit of a wave generator can be used to study the interaction of a hydraulic jump with a solid boundary of any desired geometry. In the example shown here, a brass bar, positioned at an oblique angle to the incident wave front, was used to illustrate regular and Mach reflection phenomena. *Right:* View of the used plane wave generator: *WR* – water reservoir, *DS* – discharge slot; *R* – rubber strip; *V* – air inlet valve; and *VS* – connection to vacuum pump. The strength of the hydraulic jump depends on the water level to which the water has been raised by generating a negative pressure via a vacuum pump prior to discharge. [Shock Waves **1**, 3 (1991)]

## 4.4 PERCUSSION AND SHOCK WAVE MODELS – Traffic Shocks

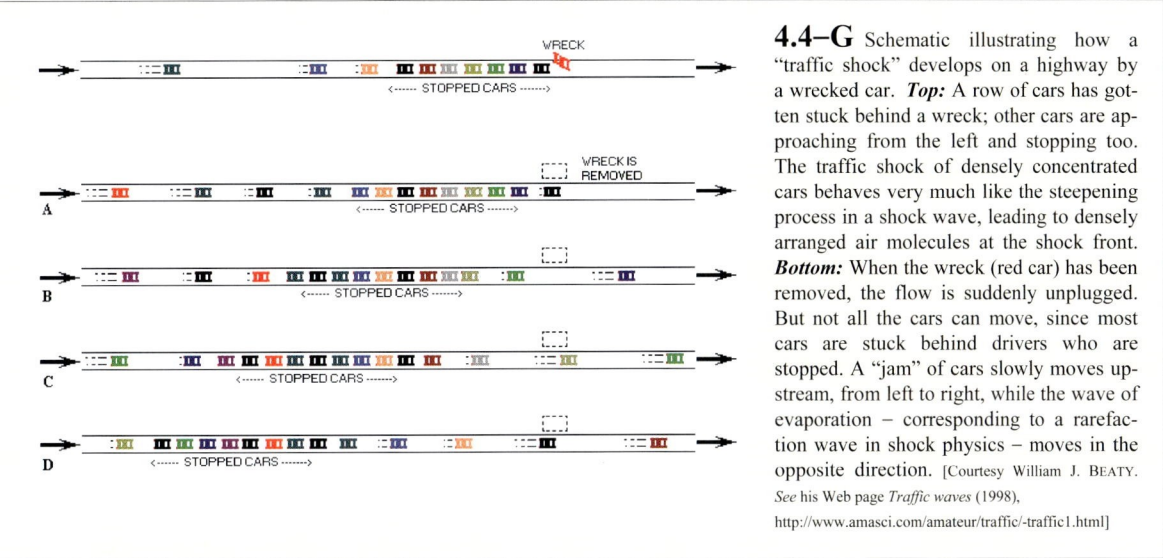

**4.4–G** Schematic illustrating how a "traffic shock" develops on a highway by a wrecked car. *Top:* A row of cars has gotten stuck behind a wreck; other cars are approaching from the left and stopping too. The traffic shock of densely concentrated cars behaves very much like the steepening process in a shock wave, leading to densely arranged air molecules at the shock front. *Bottom:* When the wreck (red car) has been removed, the flow is suddenly unplugged. But not all the cars can move, since most cars are stuck behind drivers who are stopped. A "jam" of cars slowly moves upstream, from left to right, while the wave of evaporation – corresponding to a rarefaction wave in shock physics – moves in the opposite direction. [Courtesy William J. BEATY. *See* his Web page *Traffic waves* (1998), http://www.amasci.com/amateur/traffic/-traffic1.html]

## 4.4 PERCUSSION AND SHOCK WAVE MODELS – Amusing Cartoons

**4.4–H** *Right:* This is a picturesque example of a "traffic shock," illustrating the traffic situation when a preceding car suddenly slows down, such as by a breakdown, and drivers of subsequent cars cannot react to the new situation quickly enough. This chain-reaction-type collision, a frequently occurring phenomenon on highways, often results in dreaded mass collisions. The same mechanism that occurs in a traffic flow on long crowded roads causes the steepening of a shock wave in fluid dynamics, here each car resembling a decelerated wave. The shock front is identified with the point of collision that moves with increasing time counter to the flow of traffic. [P.A. THOMPSON: *Compressible-fluid dynamics*. McGraw-Hill, New York (1972), p. 306] *Left:* A similar situation might also happen to a group of skiers when a skier in front is suddenly slowed down, for example by head-on collision with an obstacle. [G.E. DUVALL and G.R. FOWLES: *Shock waves*. In: (R.S. BRADLEY, ed.) *High pressure physics and chemistry*. Academic, New York (1963), p. 213]

## 4.5 SHOCK WAVE VISUALIZATION – Toepler's Stroboscopy of Propagating Shock Waves

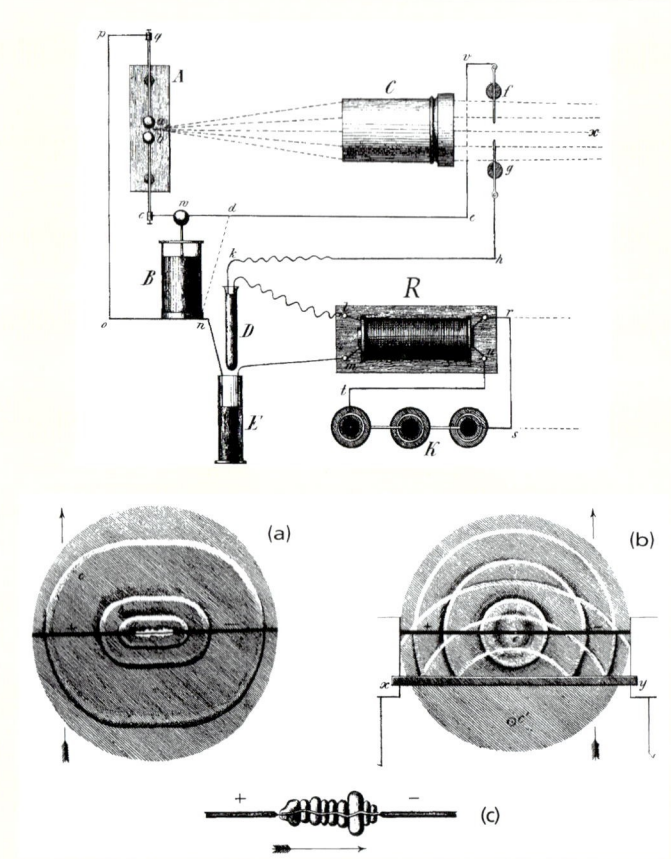

**4.5–A** *Top:* This experimental setup was used by the German physicist August Toepler in the mid-1860s to visualize the propagation and reflection of shock waves subjectively, using his schlieren method {⇨Fig. 4.18–A} and an electric circuit to delay the illumination spark *a–b* relative to the shock wave that he generated by the spark *f–g*. The principle involved uses two capacitor discharges that are coupled to each other via the spark gap *f–g*, the so-called "Knochenhauer circuit" {⇨Fig. 4.19–C}. The delay time between the two sparks can be varied simply by changing the immersion depth of the capacitor *D/E*. *Bottom:* Toepler's pen-and-ink drawings of what he observed as flashing pictures through his telescope. Illustrated are shock wave fronts at increasing time instants originating from a long spark *(a)*; and reflected at a solid boundary *(b)*. He also observed the spark channel and found a pinch-type plasma structure *(c)*. His drawings have often been taken erroneously as photographs, but the first photograph of a propagating shock wave was not taken until 1885 by Ernst Mach and his son Ludwig {⇨Fig. 4.5–G}. Since Toepler was actually the first to "see" a shock wave rather than to record it on film, his widow provided his tombstone with the following epitaph: *August Toepler – Er sah als Erster den Schall* ("August Toepler – He was the first to see the sound"). She erroneously chose the term "sound" instead of "shock wave." [A. Toepler: *Beobachtungen nach einer neuen optischen Methode.* Max Cohen & Sohn, Bonn (1864); plate III, figs. 1 and 10, and plate IV, figs. 5 and 11]

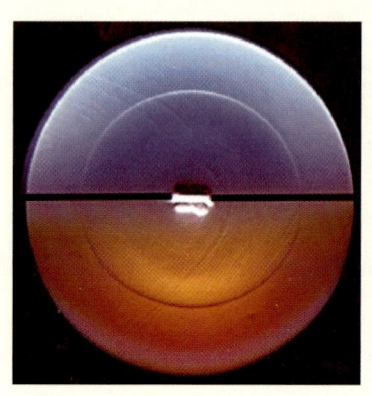

**4.5–B** Toepler's first schlieren experiments reproduced by P. Krehl (EMI, Freiburg) on the example of an air shock wave emerging from a short spark. They illustrate the propagation of the shock wave *(left)* and its reflection at a solid boundary *(right)*. Since Toepler did not use a schlieren head corrected for chromatic aberration, he must have seen the shock wave on a colored background divided into two halves of different colors as shown here. Note the amazing relieflike structure of the shock wave front, which is typical for the schlieren method of reproducing steep density gradients. [Shock Waves **5**, 1 (1995)]

## 4.5 SHOCK WAVE VISUALIZATION – ANTOLIK's Soot Method

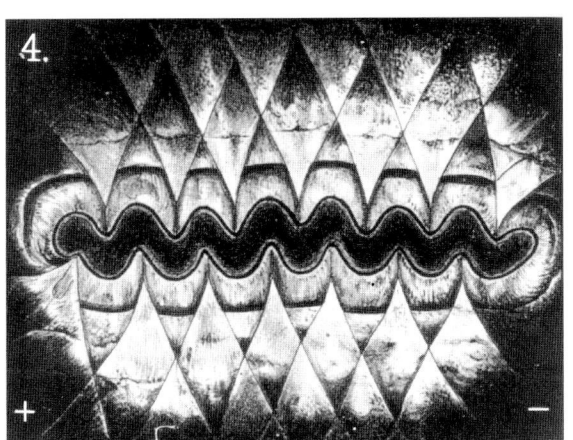

**4.5–C** In the mid-1870s, the Hungarian physicist and schoolmaster Károly ANTOLIK (Kaschau, Austro-Hungarian Empire) invented a curious technique of visualizing the mechanical effects originating from electric spark discharges. He generated a zigzag-shaped gliding spark between two closely spaced glass plates. When coating one plate facing the spark channel with a soot layer, he observed on this plate that conical-shaped branches were recorded in the soot layer that emerged from each concave part of the spark path. He attributed this phenomenon to electrical rather than to mechanical reasons. In reality, however, ANTOLIK recorded with his so-called "soot method" the very first irregular interaction patterns of shock waves. This was first recognized by Ernst MACH immediately after the publication of ANTOLIK's paper. [Ann. Phys. **154** (II), 14 (1875)]

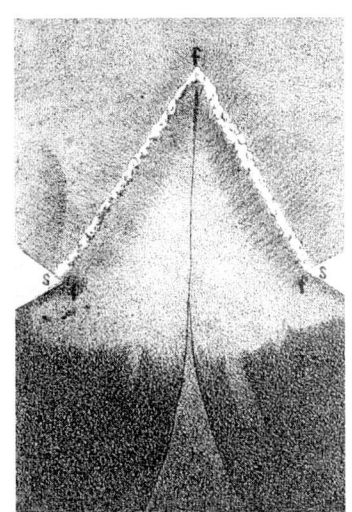

**4.5–D** In 1875, Ernst MACH at the German Charles University in Prague began to tackle one of the most challenging problems in shock wave physics: the oblique interaction of shock waves. *Top:* Ernst MACH and his student Jaromir WOSYKA, repeating ANTOLIK's experiments, reduced his complicated gliding spark geometry to a single zigzag element ("V-spark"), thus confirming that the production of conical branches is of a purely mechanical origin and that the area between them is swept clean of soot upon discharge, which they called "V-shaped propagation" [Germ. *V-förmige Ausbreitung*], later also called "soot funnel." They correctly interpreted its origin by a superposition of mechanical waves. MACH later recognized that the puzzling phenomenon was caused by aerial waves of finite amplitude – then also called "Riemann waves" [*Riemann'sche Wellen*] – the term "shock wave" [*Stoßwelle*] was not used by him until 1885. *Bottom:* MACH and WOSYKA, using a geometry of two opposing V-sparks (*left*), proved the existence of a straight part of the resulting interfering wave pattern, later called "Mach disk" by John VON NEUMANN. The soot transport emerging simultaneously from both sides is stopped along a straight central line where the afterflows of both shocks eliminate each other. This ingenious setup converted the transient shock wave interaction pattern *during* the discharge to an irreversible soot pattern *after* discharge. A superposition of three Mach disks (*center*), one coming from the top and two from the sides, results in an interesting Y-configuration, *cf.* enlargement (*right*).
[Sitzungsber. Akad. Wiss. Wien **72** (II), 44 (1875)]

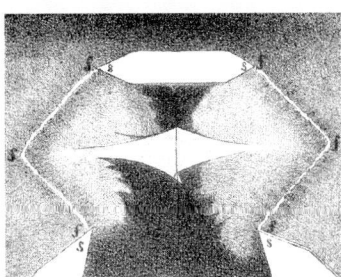 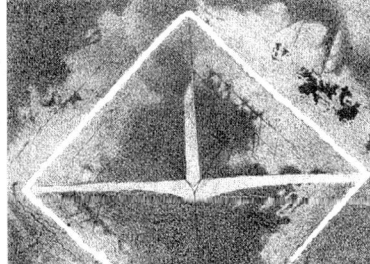 

## 4.5 SHOCK WAVE VISUALIZATION – ANTOLIK's Soot Method *(cont'd)*

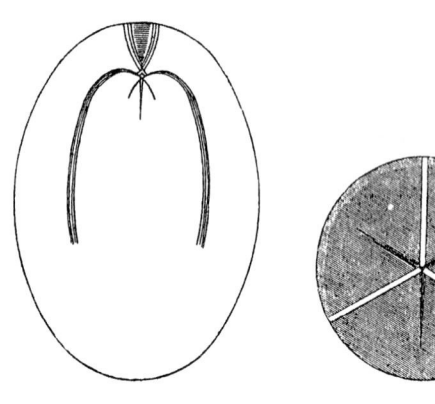

**4.5–E** Wenzel ROSICKÝ, one of Ernst MACH's assistants at the German Charles University in Prague, visualized focusing effects of weak shock waves in an elliptic reflector backed by a soot-coated glass plate. *Left:* He generated the shock wave by an electric spark discharge in one focus and observed a star-shaped interference pattern in the other focus. *Right:* A close-up inspection revealed that this other focus had a tri-star formation with a center completely free of soot. [Sitzungsber. Akad. Wiss. Wien **73** (II), 629 (1876)]

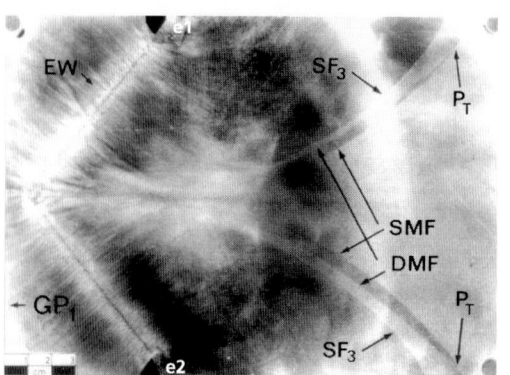

**4.5–F** Recording of shock interactions at higher pressure levels is possible using sand-blasted rather than plain glass plates, a discovery made by P. KREHL (EMI, Freiburg). A strong V-shaped electric gliding spark $e_1$–$e_2$ produces two concentric Mach funnels *SMF* and *DMF*, resulting from single and double Mach reflections, respectively. The third funnel $SF_3$ was apparently created by the interaction of primary shock waves with secondary ones from the oscillatory capacitor discharge. [*Proc. 18th Int. Symposium on Shock Waves.* Springer, Berlin (1992), p. 221]

## 4.5 SHOCK WAVE VISUALIZATION – Examples of Shock Wave Photography in Gases

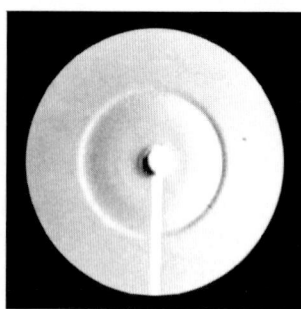

**4.5–G** The first schlieren photograph of a shock wave was taken in 1885 by Ernst MACH and his son Ludwig using high-sensitivity silver bromide gelantin dry plates. They generated the shock wave by discharging a Leiden jar between two closely spaced glass plates. Note the relief-type structure of the shock front, which is typical for the schlieren method of reproducing steep density gradients. [Sitzungsber. Akad. Wiss. Wien **98** (II), 1333 (1889)]

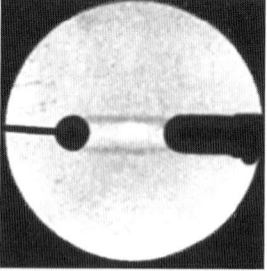 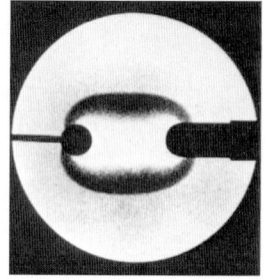

**4.5–H** Karl-Heinz HERRMANN at the Freie Universität Berlin took flash radiographs of shock waves emerging from a spark discharge in gases using soft Cu-K radiation ($\lambda \approx 1.5$ Å), which is essential for getting a high radiographic contrast. The two flash radiographs of the shock wave in air were taken 3 µs (*left*) and 9.5 µs (*right*) after the onset of discharge. Densitometry revealed that the density jump in the emitted wave profile is already established after the wave has traveled a few centimeters from the spark, indicating that the steepening process of the shock front is almost completed. [Z. angew. Phys. **10**, 349 (1958)]

4 PICTURE GALLERY 913

## 4.5 SHOCK WAVE VISUALIZATION – Shock Tube Studies

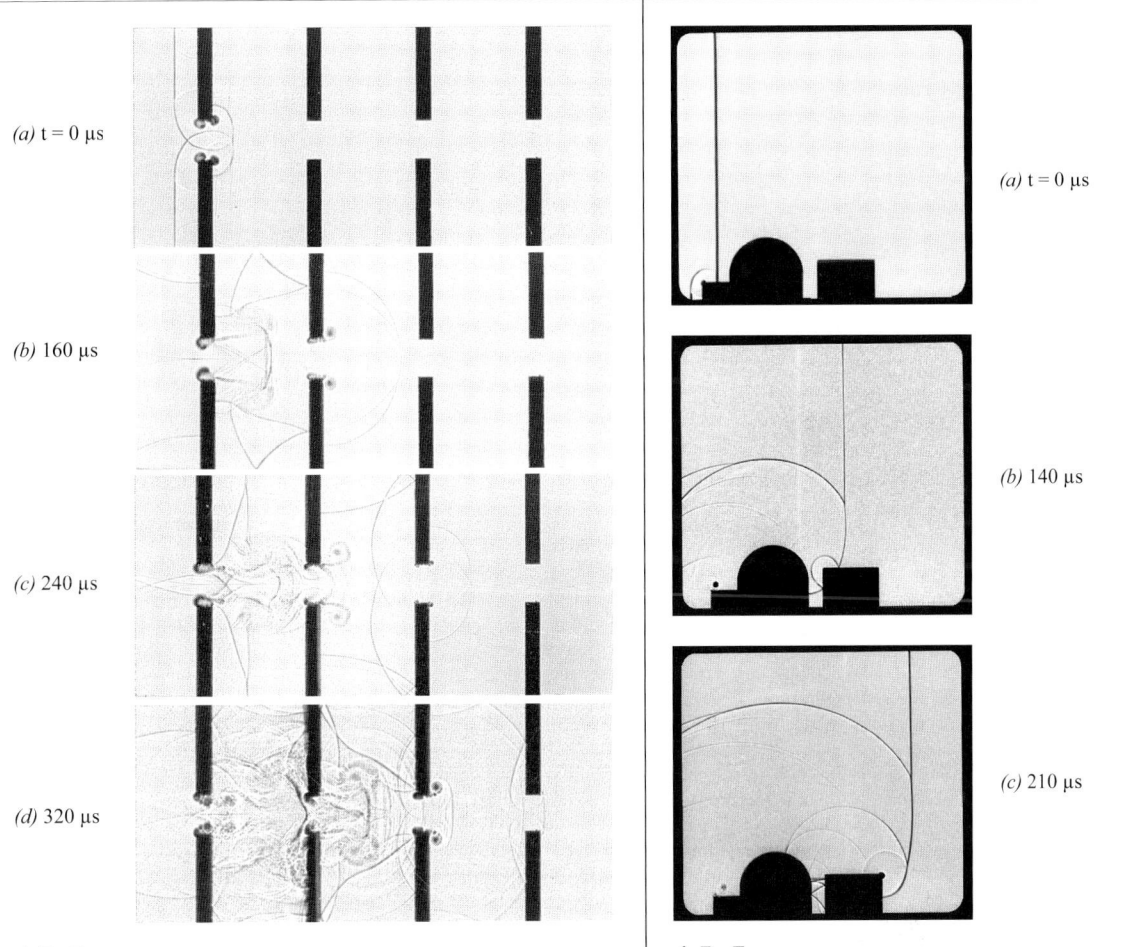

(a) t = 0 μs

(b) 160 μs

(c) 240 μs

(d) 320 μs

(a) t = 0 μs

(b) 140 μs

(c) 210 μs

**4.5–I** Using the EMI Shock Tube Facility, Heinz REICHENBACH and collaborators investigated the shock attenuation in a symmetric baffle system (baffle height 110 mm). Applying the 24-spark Cranz-Schardin camera and shadowgraphy, they visualized the propagation of the shock front and the interaction with the baffle elements, cf. frames *(a) – (d)*. An analysis revealed that a shock wave, here moving with $M = 1.5$ from left to right, is already absorbed by 90% after the passage of only four baffles – a surprising phenomenon due mainly to expansion. Note the vortex formation at the aperture edges that begins immediately after passage of the incident shock wave. [*Proc. 8th Int. Congr. on High-Speed Photography*, Almquist & Wiksell, Stockholm (1968), p. 362.]

**4.5–J** Cinematography of a propagating plane shock wave striking a model nuclear power plant. Initially, the shock wave is reflected regularly *(a)*, but after passing the left building block *(b)* it generates Mach reflection. Subsequently, the resulting Mach stem interacts with the right building model, creating for its part a second Mach reflection. Note the trapping of shock waves between the two model buildings *(b, c)*. This series of shadowgraphs was also recorded with the Cranz-Schardin multiple-spark camera. [Photos by W. SCHÄTZLE and W. GEHRI; EMI-Archives, Freiburg]

## 4.5 SHOCK WAVE VISUALIZATION – Whip Cracking

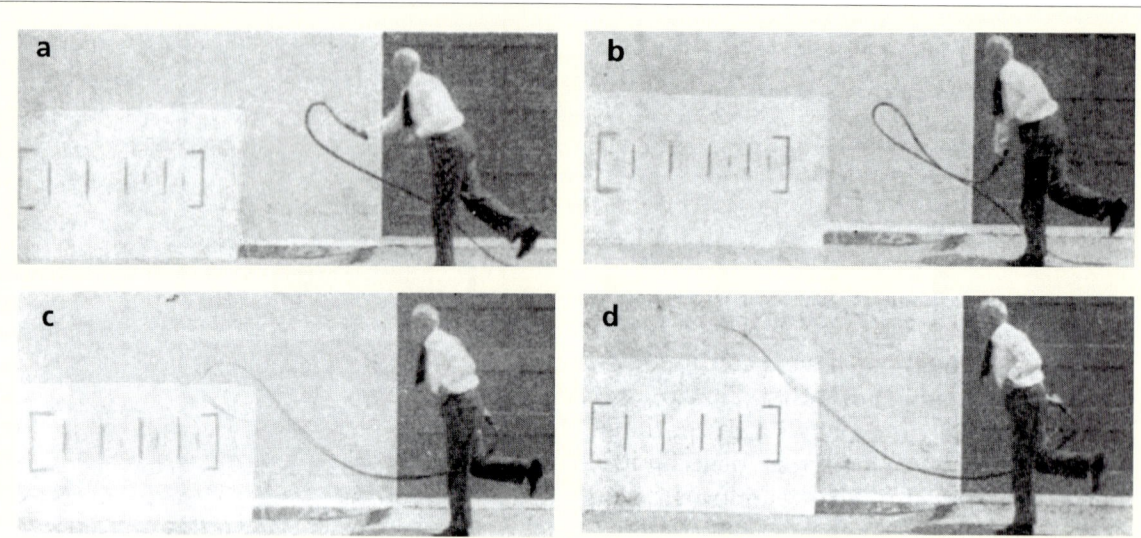

**4.5–K** *Top:* Barry BERNSTEIN and collaborators at the U.S. Naval Research Laboratory (Washington, DC) visualized the motion of the whip lash before, during, and after whip cracking. Frames *a–d* were taken with a movie camera at 4,000 frames/s, the heavy vertical lines are one foot apart. Frames *c* and *d* were taken before and shortly after the whip cracked. Note that the loop starts near the handle and propagates down until it reaches the whip tip where it generates the sharp cracking sound. [JASA **30**, 1112 (1958)] *Bottom, left:* In 1927, Zéphirin CARRIÈRE, a French physicist and clergyman at the Institut Catholique (Toulouse, France), was the first to succeed in visualizing the shock wave emerging from the whip tip of a simulated *laboratory whip*. He performed this difficult task by using schlieren photography and a spark point light source. Almost 30 years later, he resumed his studies and was able to improve the contrast of his photos. [Cahier de Physique No. 63, p. 5 (1955)] *Bottom, center:* BERNSTEIN ET AL. (*see above*) were the first to successfully photograph the shock wave emerging from the tip of a *real whip*, in their studies a 12-ft (3.66-m)-long bull whip. Using a still camera and flash shadowgraphy, they recorded snapshots of the propagating shock wave at different time instants. [JASA **30**, 1112 (1958)] *Bottom, right:* More recent experiments on whip cracking, carried out at EMI, Freiburg by Peter KREHL and collaborators in March 1995 using high-speed videography and laser stroboscopy, revealed that the supersonic motion of the tuft is only a *conditio sine qua non*, but that the essential mechanism of shock generation occurs in the final stage of acceleration and is due to the abrupt flapping of the tuft at the turning point – *i.e.*, when, as illustrated in the schematic, the tuft flaps from position "8" into position "9". [Shock Waves **8**, 1 (1998)]

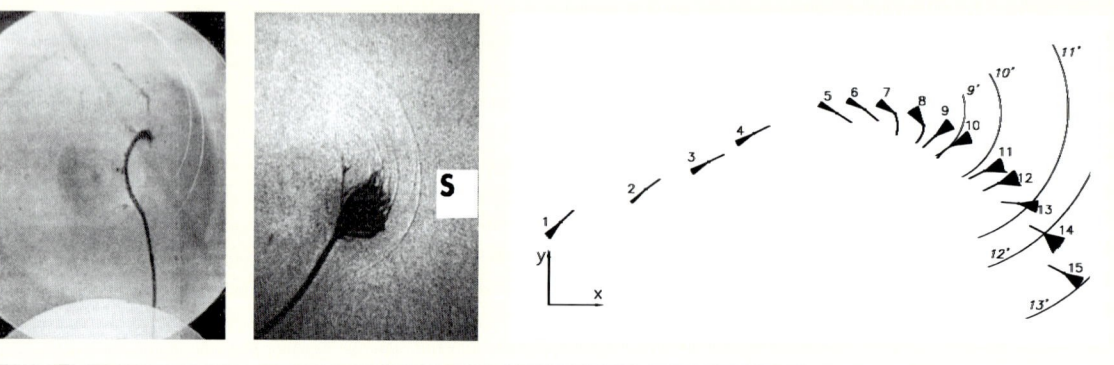

4 PICTURE GALLERY

## 4.5 SHOCK WAVE VISUALIZATION – Muzzle Blast and Head Wave

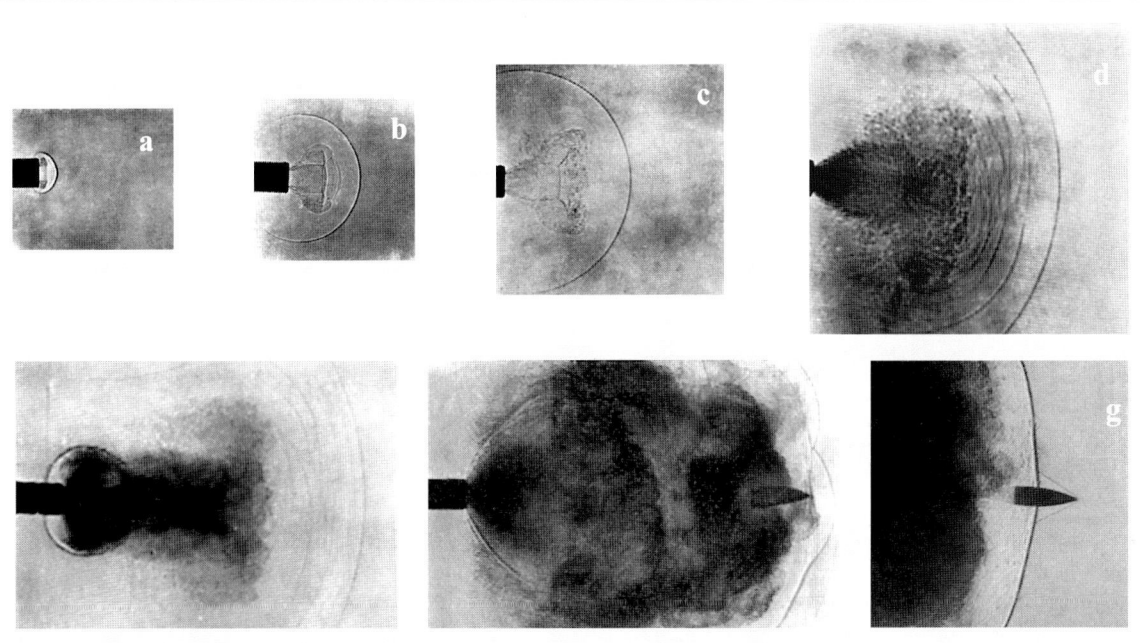

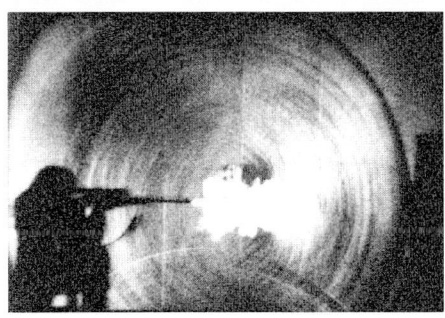

**4.5–L** High-speed cinematography revealed hitherto unknown flow phenomena in the environment of the muzzle and established *Intermediate Ballistics*, a new branch of ballistics. ***Top:*** Muzzle phenomena observed at the German rifle Mauser M98 while firing an S-bullet ($v_0$ = 880 m/s): *(a)* the air, compressed in the barrel in front of the accelerating projectile, creates a weak shock wave that has already partly left the muzzle; *(b), (c)* the shock wave expands spherically, and an air cushion has piled up between muzzle and shock front; *(d)* some powder smoke, leaking between barrel and projectile, already begins to leave the muzzle; *(e)* in this moment, the projectile base has just passed the muzzle, and the main part of the powder smoke can leave the barrel, which, impinging on the surrounding air, produces a second, stronger shock wave – the so-called muzzle blast; *(f)* the powder smoke, initially emerging from the muzzle with velocity $v_0$, decreases in velocity because of aerodynamic drag effects; and *(g)* after a certain distance from the muzzle, the head wave overtakes the muzzle blast, which quickly loses in strength because of its 3-D expansion. Note that the head wave, sitting rooflike above the muzzle blast, does not extend beyond the front of the muzzle blast. [C. CRANZ: *Lehrbuch der Ballistik*. Springer, Berlin; vol. II (1926): *Innere Ballistik*, pp. 443-449] ***Bottom:*** These schlieren pictures show the formation of the muzzle blast and the head wave upon firing a Remington 0.30-06 high-powered rifle. Taken on a scale an order of magnitude larger than previously done by others with the Full-Scale Schlieren System at Penn Sate University {⇨Fig. 4.5–N}, they allow one to observe and study the entire process, including shock wave reflections, bullet impact upon a target, and hearing protection for the shooter. Note that near the muzzle the head wave propagates almost twice as fast as the muzzle blast. The images appear somewhat grainy and indistinct because they are individual frames from a high-speed movie shot at 30,000 frames/s by way of a drum camera. [Courtesy Prof. Gary S. SETTLES, Gas Dynamics Laboratory, Penn State University, PA]

## 4.5 SHOCK WAVE VISUALIZATION – Color Schlieren Photography

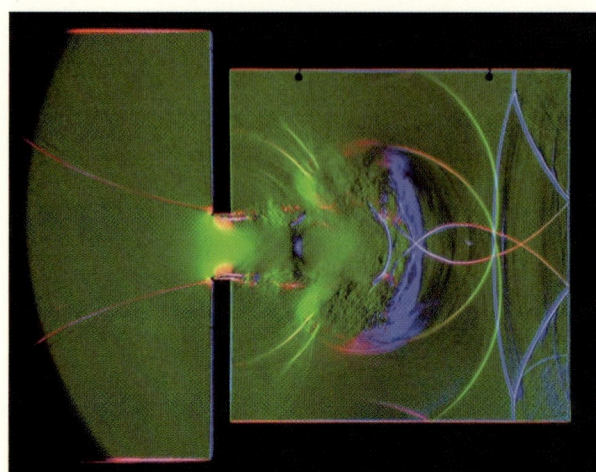

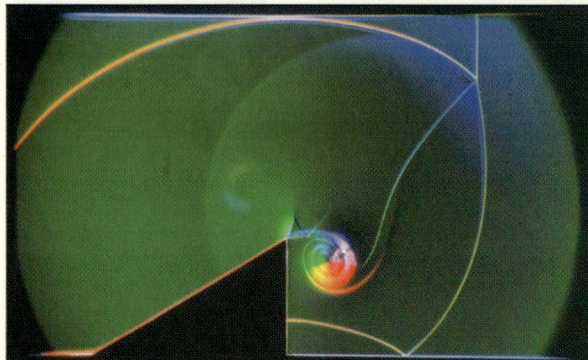

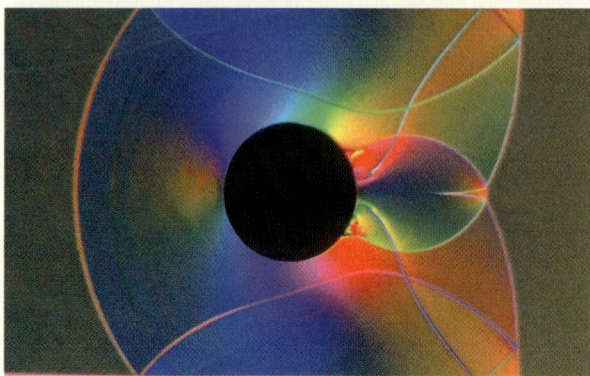

**4.5–M** Color schlieren photography is an extension of the original black-and-white schlieren photography and has the advantage that a specific hue can be better distinguished by the eye than a corresponding luminance within a gray-scale on a black-and-white schlieren photograph. The shock-tube pictures presented here were made with a multi-color filter in place of the knife edge. In all three cases shown here the incident shock wave propagates from left to right. *Top:* Schlieren photo of an aerial shock wave of strength $p_1/p_0 = 2.15$, propagating in air with a Mach number of $M = 1.41$ and penetrating through an aperture into a dead-end chamber. The picture was taken at a time instant shortly after reflection of the incident shock wave at the end of the chamber. Note that a complex reflection pattern is created, including both regular reflection and irregular (or Mach) reflection. A symmetric four-sector bicolor (blue/green) filter was used that reproduced the pressure gradients in different colors. The reflected shock wave, propagating from right to left, is reproduced in a bluish color, while several subsequent weaker shocks, still propagating from left to right, are reproduced in a greenish color. The following two pictures were taken with a three-sector tricolor (blue/green/red) filter. *Center:* Interaction of an aerial shock wave ($p_1/p_0 = 2.27$; $M = 1.45$) with a 30° wedge that generates Mach reflection. The picture was taken at a moment after the Mach stem had already passed the wedge top. The flow behind the Mach stem, after having entered the undisturbed corner below the edge, has created a counter-clockwise rotating vortex near the edge of the wedge. This motion has also been communicated to the slip line. Shock wave diffraction over a knife edge was first studied in 1958 by Hubert SCHARDIN at ISL (Saint-Louis, France) using black-and-white schlieren photography. [Courtesy Dr. P. NEUWALD, EMI, Freiburg] *Bottom:* This picture shows a complex wave interaction pattern generated by a plane shock wave ($p_1/p_0 = 2.39$; $M = 1.48$), propagating into nitrogen at atmospheric pressure and being diffracted at a circular cylinder. The single wave front on the left-hand side is the reflected wave. The two diffracted waves, propagating from left to right, have interacted and created a Mach disk that propagates along the axis of symmetry and varies in width. The diffracted waves, after being reflected at the tube walls, have created two Mach stems. [Courtesy Dr. Harald KLEINE, Stoßwellenlabor of RWTH, Aachen]

## 4.5 SHOCK WAVE VISUALIZATION – Color Schlieren Photography *(cont'd)*

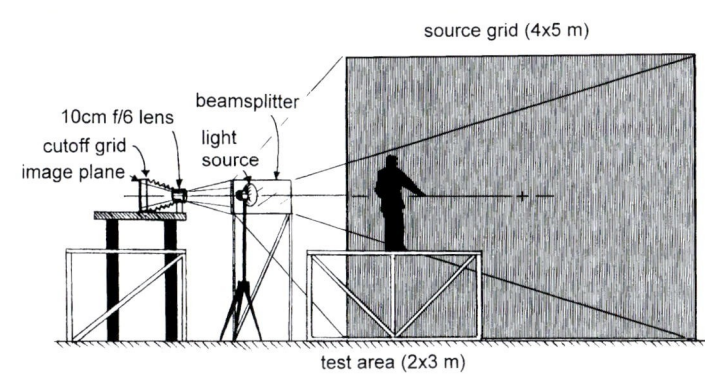

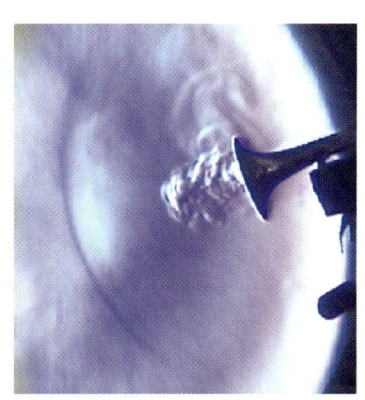

**4.5–N** Gary S. SETTLES, a professor at the Gas Dynamics Laboratory of Penn State University (University Park, PA), developed a sensitive large-aperture schlieren optical instrument that proved very useful in studying at full scale both heat- and shock-wave-generated schlieren flows. *Left, top:* Schematic of his Full-Scale Schlieren System, the largest indoor schlieren system in the world; the field-of-view is $2.9 \times 2.3$ m². Based on the lens-and-grid principle first developed in Germany by Hubert SCHARDIN in the late 1930s, it essentially consists of (1) a high-intensity light source of small dimension, (2) a large front-lit retroreflective source grid that covers one wall of the laboratory, (3) a cutoff grid that is a precise negative image of the source grid, and (4) for recording purposes either a high-speed movie camera system or a still camera for operation in single-frame mode. Contrary to classic schlieren methods using transparent light, the lens-and-grid method views the object in reflected light, which is more informative and realistic than imaging of contours only. *Left, center & bottom:* To study onboard explosions in an aircraft, full-scale schlieren flow visualizations of an explosion in a mockup aircraft cabin were performed. The blast waves, with Mach numbers ranging between 1.0 and 1.2 at a distance of about 1 m from the blast center, were generated by the detonation of an oxygen-acetylene gas mixture in a small toy balloon which, depending on its size, were equivalent to the yield of about 1 to 10 grams of TNT. The schlieren images show the blast propagation beneath full-size aircraft seats occupied by mannequins, both in side view (*center*) and front view (*bottom*). [Shock Waves **12**, 267 (2003)] *Right:* B.H. PANDYA, G.S. SETTLES, and J.D. MILLER at Penn State University visualized gasdynamic phenomena at the exit of a trumpet. Especially for loud, high-pitched trumpet notes they observed the emission of weak shock waves, which are the result of cumulative nonlinear acoustic propagation inside the trumpet bore. Though of weak intensity, ranging from 118 to 124 peak dB (A) and thus only marginally propagating above the sound speed, they can clearly be visualized with SETTLES' Full-Scale Schlieren System. The schlieren pictures shown here were made in single-frame mode using a xenon flash tube and 120-size ISO 800 color negative film. [JASA **114**, 3363 (2003)]

## 4.6 HEAD WAVE STUDIES – HUYGENS' Principle of Wave-Front Construction

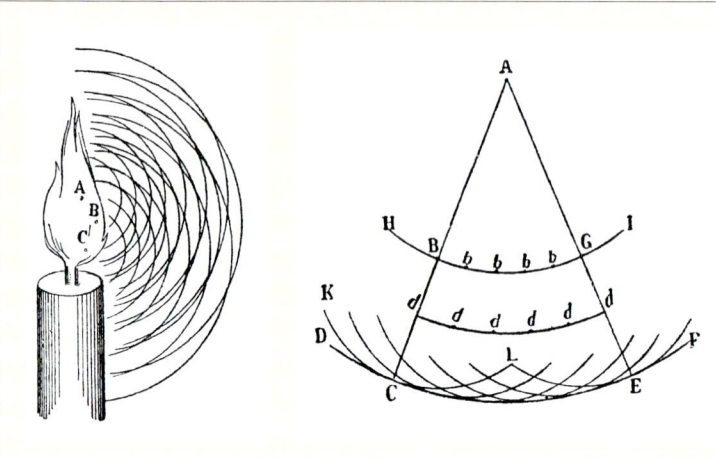

**4.6–A** The Dutch natural philosopher Christiaan HUYGENS suggested (i) that each point on the surface of a wave front acts as a point source for outgoing spherical waves ("wavelets") which advance with a speed and frequency equal to that of the primary wave at each point in space, and (ii) that the sum of the wavelets produces a new wavefront – so-called "Huygens principle." To some extent the method is also useful for constructing the front of a shock wave. HUYGENS, not yet knowing the features of a shock wave, assumed that the wavelets had very small amplitude and that the wavefronts were a linear superposition of wavelets. [C. HUYGENS: *Traité de la lumière*. Vander Aa, Leiden (1690)]

## 4.6 HEAD WAVE STUDIES – DOPPLER, Father of the Head Wave Phenomenon

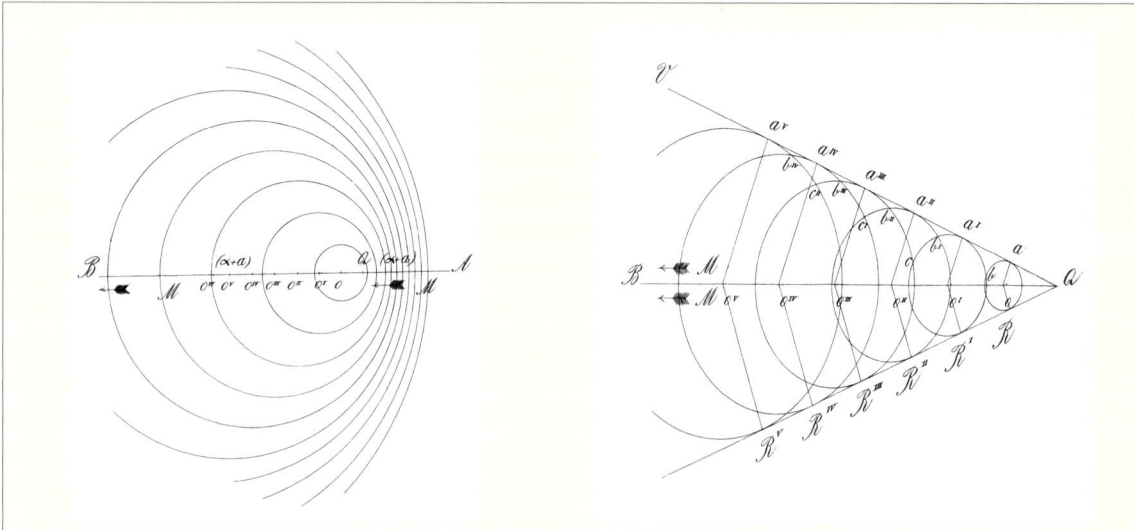

**4.6–B** Christian DOPPLER's schematics of a wave phenomenon arising from a disturbance moving along a straight line with a constant velocity $u$ with respect to the sound velocity $a$ of the surrounding medium. ***Left:*** Formation of waves in front of an object moving with subsonic velocity. ***Right:*** For supersonic motion the disturbances emitted from the moving object form the typical "Mach cone." If the body's speed is not constant along a straight course, the region of disturbance would be bounded by curved lines and its shape would also change with time. DOPPLER also derived the famous formula for the half-cone angle, given by $\sin\alpha = a/u$. In the period 1886–1887, Ernst MACH and Peter SALCHER first proved his theoretical model by visualizing the "head wave" in their supersonic ballistic experiments. [Abhandl. Böhm. Gesell. Wiss. Prag **5** (V), 293 (1848)]

4.6   **HEAD WAVE STUDIES** – DOPPLER, Father of the Head Wave Phenomenon *(cont'd)*

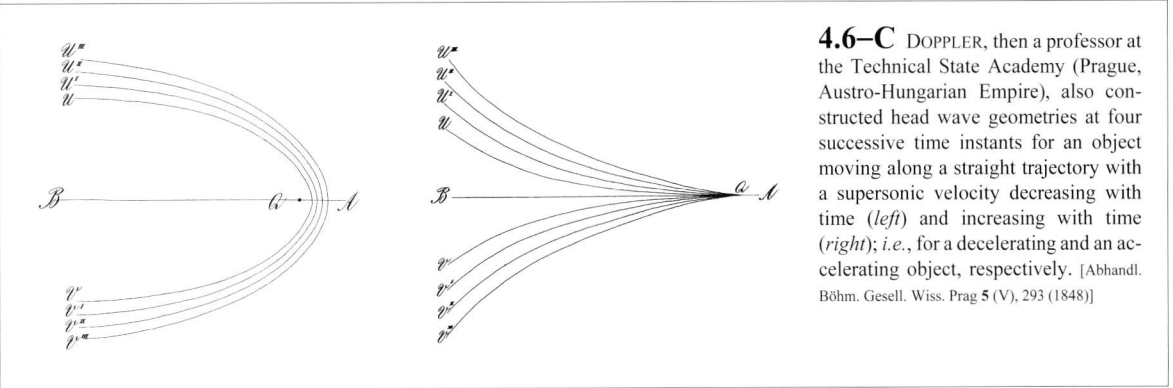

**4.6–C** DOPPLER, then a professor at the Technical State Academy (Prague, Austro-Hungarian Empire), also constructed head wave geometries at four successive time instants for an object moving along a straight trajectory with a supersonic velocity decreasing with time *(left)* and increasing with time *(right); i.e.*, for a decelerating and an accelerating object, respectively. [Abhandl. Böhm. Gesell. Wiss. Prag **5** (V), 293 (1848)]

4.6   **HEAD WAVE STUDIES** – Surface Wave Pattern Produced by a Moving Body in Water

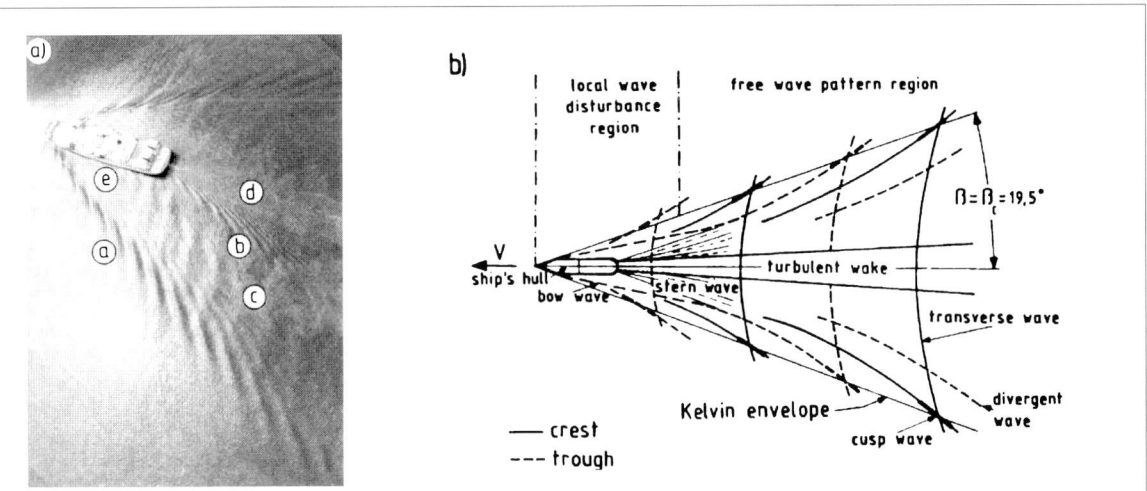

**4.6–D** *Left:* Aerial photograph showing the components of a ship wake pattern: *a* – bow wave, *b* – stern wave, *c* – transverse wave, *d* – turbulent wake, and *e* – turbulent region, adjacent to ship's hull. *Right:* When a ship travels through water, it generates a conical wave pattern. A closer inspection of the wave drawn in water, the so-called "bow wave," was carried out in 1871 by William THOMSON, the later Lord KELVIN [Phil. Mag. **42** (IV), 368 (1871)]. His study revealed that the wave front is not made up of only a single wave front, as in the case of the Mach head wave generated by a supersonic projectile, but is rather composed of a number of short waves (or cusp waves). They typically have wavelengths between 10 and 40 m and amplitudes between 0.2 and 1 m. They are not arranged parallel to the two lines emanating from the bows but are inclined at a certain angle to them. These component waves travel forward through the lines from the bows, at the same time as they travel outward with them, as the lines get gradually farther and farther apart. The envelope of these short waves is called the "Kelvin envelope" or "Kelvin wake." The angle between the waves in the "Kelvin arms" of a wake is independent of the ship's velocity and, contrary to a head wave generated by a supersonically moving object, is always 39°, *i.e.*, the Mach number for motion on water is $1/\sin 19.5° = 3$. [*Ship wakes*. ESA, Earth Observation Applications, Earthnet Online; http://earth.esa.int/applications/ERS-SARtropical/oceanic/shipwakes/intro/index.html. Courtesy R. DOERFFER, GKSS, Geesthacht, Germany]

## 4.6 HEAD WAVE STUDIES – MACH and SALCHER: Prelude to a Pioneering Ballistic Experiment

*Dear Professor [Salcher],*

*I am very grateful to you for your kind intention to do the experiment, and if you don't mind I communicate right away some experiences and remarks.*

*I believe that an infantry rifle will be sufficient to do the experiment. Most important is that the projectile velocity surmounts the sound velocity. It will be sufficient at first to observe the phenomenon optically and later to photograph it. For its representation an additional illumination is certainly not required. This setup should work with this type of experiment as well as when photographing the bullet. The closing spark of a battery of jars, corresponding to a capacity of a free 12-m-diameter ball and in my case charged up to a voltage of about 1-cm gap length, has two gaps, I and II. Passing in front of the lens O, the bullet passes II, which triggers the illumination spark I.*

*The rifle will be positioned several meters away from II in order to prevent smoke from entering the field of view.*

*As electrodes in II I have applied end-melted glass tubules into which very thin wires were inserted. The straight parts stand vertically with respect to the trajectory of the bullet, and their distance is slightly smaller than the diameter of the projectile. Annoying are the small glass fragments that could easily scratch the lens. Besides, using another method I could not provoke precise triggering and have preferred to position II farther away from the lens.*

*The setup would differ from that for photographing the bullet only by insertion of the aperture B, which at its border retains the picture imaged by the lens O from the telescope focused on II. For this experiment Toepler's time delay method would be too unreliable. If necessary I could supply you with a better triggering that is easy to produce.*

*I expect that the projectile will carry an envelope of compressed air of an approximate geometry as shown opposite. The apex of the truncated cone will doubtless depend on the ratio of the sound velocity to the projectile velocity.*

*I would be very happy if you succeeded with the experiment. I am entirely satisfied with having stimulated it.*

*Very sincerely, yours*                     E. MACH

**4.6-E** *Top, left & right:* Ernst MACH assumed that a head wave should surround a projectile flying at supersonic speed. In his letter to Peter SALCHER dated February 16, 1886, he specified his idea for an appropriate setup to visualize the head wave. This historically important letter, until 1990 assumed lost, was rediscovered by the author in private archives of the Salcher family. [Courtesy Dr. Günther SALCHER, Hermagor, Austria] *Bottom:* In a letter (*left*) to MACH dated May 21, 1886, SALCHER fully confirmed MACH's speculation by including a sketch of the head wave. It also shows a circular wave emitted from a spark gap that, triggered by the projectile, activated the spark light source. In a further letter (*right*) to MACH dated May 23, 1886, SALCHER also visualized so-called "Mach lines" that arise from the rough surface of a supersonic projectile. He wrote, "A stripe (2) – if I'm not mistaken – goes out exactly from the place where the nose of the bullet begins." Both of SALCHER's letters mentioned above belong to a collection of 140 letters that he had sent to MACH at a time when they were both involved in the visualization of head waves and free-air jet phenomena (1886–1889). They are now kept at the Archives of the Deutsches Museum, Munich. MACH's letters to SALCHER are kept at Dr. Günther SALCHER's private archive at Hermagor.

# 4 PICTURE GALLERY

## 4.6 HEAD WAVE STUDIES – First Experimental Evidence

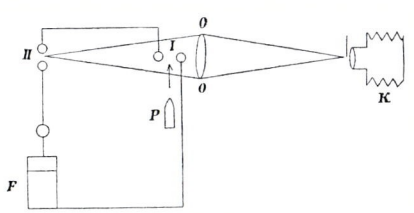 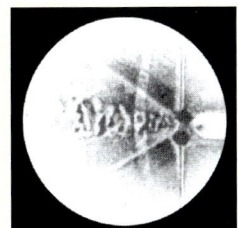 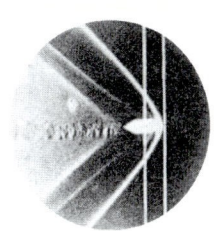

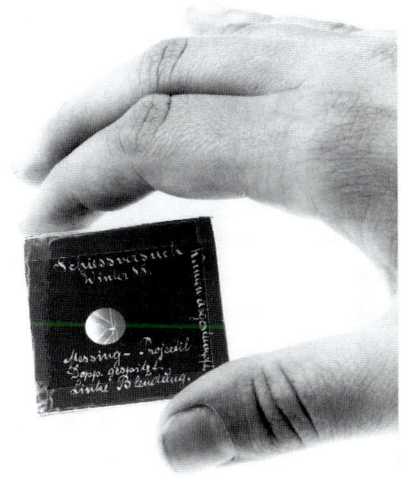

**4.6−F** In May 1886, E. MACH and P. SALCHER first reported in a short note on their observed head wave phenomenon. More detailed data were published by them in 1887. *Left, top:* Schematic of their experimental setup to capture a supersonic projectile *P* in flight using TOEPLER's schlieren method. Note the great similarity with MACH's previous sketch {⇨Fig. 4.6−E}. They took the picture in a darkened room about 2 to 4 m away from the muzzle. After opening the camera shutter and firing the rifle, the light source was triggered when the projectile *P* entered the electrode spacing of the spark gap *I*, which, causing a breakdown, triggered the illumination spark gap *II*. [Sitzungsber. Akad. Wiss. Wien **95** (IIa), 764 (1887)] *Left, bottom:* Since the intensity of their spark point light source, as well as the sensitivity of the first commercially available silver bromide gelatin dry plates, was limited, they reduced the schlieren picture to a small size of only some millimeters in diameter, thus obtaining sufficient exposure density on film. *Center & right:* These two schlieren pictures of the head wave, generated by a supersonic projectile flying from left to right, are among the very first ones they took. The vertical line (*left*) is a wire with a spark gap inserted for triggering the spark light source. They observed that triggering is more reliable when using two parallel lines acting as a spark gap (*right*). SALCHER used a Guedes infantry rifle ($v_0 = 530$ m/s), which resulted in a Mach cone angle of $2\alpha \approx 80°$. [EMI-Archives, Freiburg]

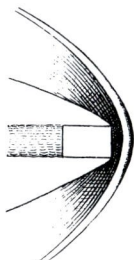 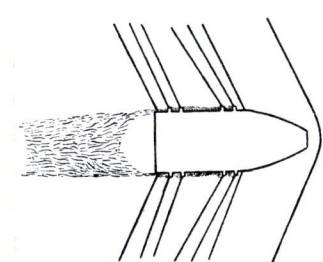 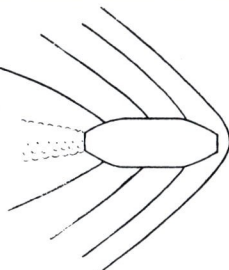

**4.6−G** MACH and SALCHER visualized head wave phenomena not only of small shots, but also of 9-cm-caliber projectiles – a difficult experimental task that had to be done in the harsh environment of a shooting range. Observing similar head wave phenomena as in the case of small shots, they correctly concluded that large shots could be simulated also in small-scale model experiments. *Left:* Blunt-headed projectiles typically reveal a double-layer structure of the head wave. *Center and right:* In the case of a projectile with guiding rings, vortices are already created at the first ring. Note the series of secondary head waves emerging from the periphery of the projectile, so-called "Mach lines." For projectiles provided with guiding rings they emerge under different Mach angles. In all cases, a train of vortices is followed behind the projectile stern. These vortices produce additional aerodynamic drag. [Sitzungsber. Akad. Wiss. Wien **98** (IIa), 41 (1889)]

## 4.6 HEAD WAVE STUDIES – Other Optical Methods

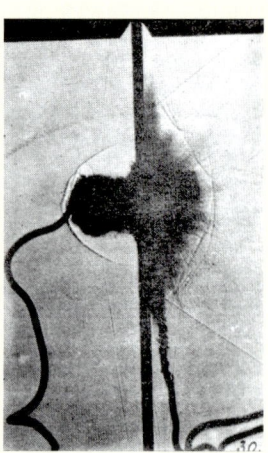

**4.6–H** In the early 1890s, Charles V. BOYS in England repeated MACH and SALCHER's ballistic photography of supersonic projectiles in free flight, but he used the silhouette rather than the schlieren method. *Left:* This photo shows BOYS' laboratory at the Royal College of Science in London. The rifle, operated by an assistant at the very left, was fired into a "recording box" that, shown here in the center with its open cover, contained the film plate and the spark light source. Its pulse capacitor was charged up via a Whimphurst influence machine. After leaving the recording box, the bullet was caught behind in a second box. *Right:* BOYS also studied the penetration and perforation phenomena of a glass plate struck by a projectile flying here from right to left. The expansion of the cloud of debris is clearly resolved. The two wires were used for triggering purposes of the spark light source. [Rev. Gén. Sci. Pures Appl. **3**, 661 (1892)]

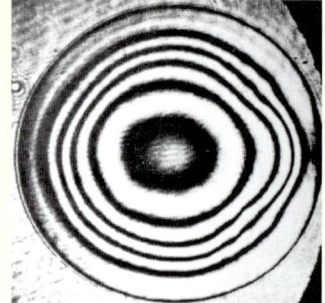
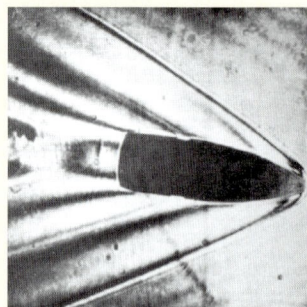

**4.6–I** *Left:* This is Hubert SCHARDIN's famous color schlieren photo of a supersonic projectile penetrating the flame of a candle. Outside of the Mach cone, in the so-called "zone of silence," the candle schlieren are not yet disturbed by the head wave. Note the train of vortices behind the projectile tail. (EMI-Archives, Freiburg) *Right:* Flash holography allows one to record a shock wave in three dimensions and to subsequently analyze details according to the required sensitivity such as corresponding to the schlieren method, shadowgraphy, or differential interferometry. Example of a flash hologram (*left*) of a supersonic projectile using a Q-switch ruby laser and applying holographic interferometry, and a reconstruction of the hologram (*right*). [Courtesy Drs. A. HIRTH and P. SMIGIELSKI, Rept. T16/69, ISL, Saint-Louis, France (1969)]

## 4.6 HEAD WAVE STUDIES – Other Optical Methods (cont'd)

↑ Although Andy GREEN, an RAF fighter pilot in his real job, still drove the 10-ton jet-mobile at subsonic speed ($M = 0.95$), close to the ground it produced a bow shock that extended out about 150 ft on either side of the vehicle.

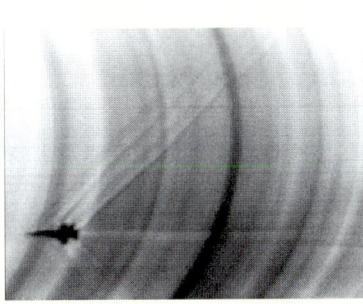

**4.6–J** The visualization of head waves produced by a full-sized supersonic vehicle is a particular challenge for high-speed photographers because it requires monitoring of a large field of view that is not directly accessible using conventional optical laboratory methods. *Top, left:* Aerial photo of shock bow wave generated by British supersonic jet car "ThrustSSC" {⇨Fig. 4.20–N} in the Black Rock Desert of Nevada. The photo was taken on October 8, 1997 from a microlight spotter aircraft, shortly before reaching Mach 1. Only 3 days later, GREEN broke the sound barrier. [Courtesy British Microlight pilot Richard MEREDITH-HARDY] *Top, right:* An array of head waves, emerging from different parts along the supersonic jet car, is visible against the background. [Courtesy U.S. photographer Chris ROSSI; http://ourworld.compuserve.com/homepages/Andy_Graves/SSC_pics.html] **Bottom:** The largest field-of-view schlieren photo ($96 \times 71$ m$^2$) ever captured on film, taken of a full-sized aircraft Northrup T-38 at supersonic flight against the Sun. The plane (total length 14.12 m) was at 11.6 km altitude and at a slant distance of 19.5 km. [Courtesy Dr. Leonard M. WEINSTEIN, NASA Langley Research Center, Hampton, VA]

## 4.6 HEAD WAVE STUDIES – Model Sonic Boom Studies

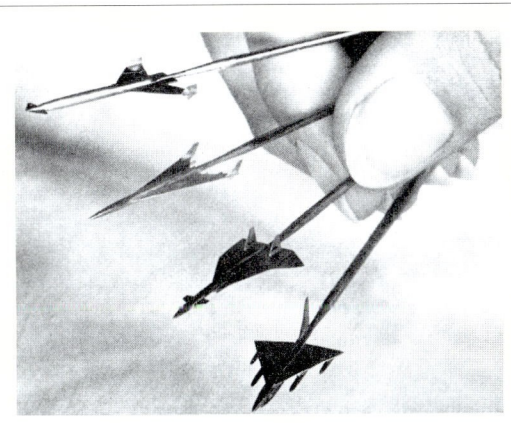

**4.6–K** To study sonic boom effects in the Langley $4 \times 4$-ft$^2$ ($1.22 \times 1.22$-m$^2$) supersonic wind tunnel, Harry W. CARLSON and Odell A. MORRIS at NASA's Langley Research Center (Hampton, VA) used precision miniature models of supersonic transport and bomber configurations, only about 1 in. in overall length. They visualized the resulting head wave and flow field up to $M = 2$ using a schlieren technique and measured the model-induced pressure-time profile, the so-called "N-wave," using a differential pressure gauge. Model vibration as well as nonuniform and nonsteady tunnel flow were the prime difficulties encountered in this method of sonic boom testing. [J. Aircraft **4**, 245 (1967)]

## 4.6 HEAD WAVE STUDIES – Pressure Measurements Around a Flying Projectile

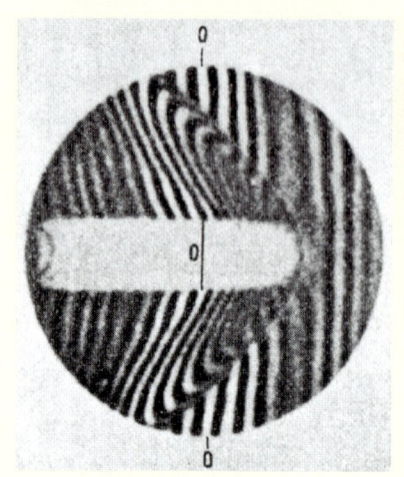

**4.6–L** Flash interferogram of a head wave generated by a supersonic steel projectile ($v_0 = 620$ m/s) taken by Ludwig MACH on August 15, 1893 with the Mach-Zehnder interferometer. To provide sufficient light intensity – the crucial condition particularly for this optical method – MACH used the continuous spectrum of an electric point spark and positioned a concave mirror at the back of the spark. He used high-sensitivity dry plates of the Schleussner Co., which had to be developed for a period of 1.5 h. Note that his original photo had a diameter of only 4 mm. From the shift of the system of interference lines he also tried to estimate the density and pressure behind the projectile head, a delicate problem. Using the Fresnel principle and converting densities into pressures, he estimated for a distance of 19 mm behind the projectile tip a maximum compression of 0.2 bar and at the tail a rarefaction of 0.07 bar. He correctly concluded that the projectile, pushing the air to the sides, had only slightly compressed the surrounding air. His studies provided the first quantitative evidence that Louis H.F. MELSENS' hypothesis (1872), that a supersonic projectile carries with it large masses of compressed air, was not correct. [Sitzungsber. Akad. Wiss. Wien **105** (IIa), 605 (1896)]

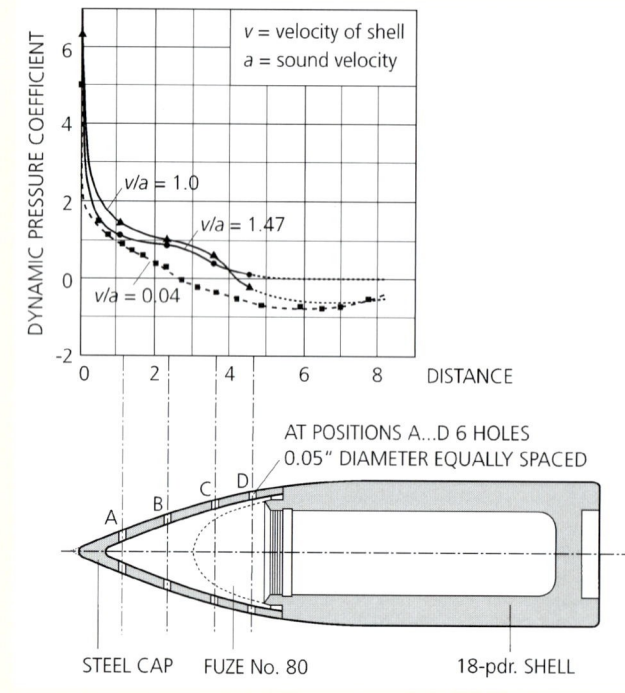

**4.6–M** In England, Leonard BAIRSTOW, Ralph H. FOWLER, and Douglas R. HARTREE used an ingenious method of measuring the pressure at different locations along the head of an 18-pdr. shell moving at high velocity. Each location of measurement was connected with a time fuse being affected by the actual pressure during flight. It is well known that the rate of burning of a time fuse depends on pressure. The measurements were carried out in two steps: (i) At first this dependency between the external pressure and the time was determined by laboratory experiments. If the circumstances of the shell's motion are known, this leads at once to a relation between the dynamic pressure, the velocity of the shell, and the prevailing atmospheric conditions. Depending on the projectile velocity, the time fuse burns slower or faster, eventually igniting the charge. (ii) Then the observers measured the time and location of the explosion. The results obtained for the derived dynamic pressure coefficient, $p/\rho v^2$, are shown here schematically for different velocity ratios $v/a$, with $v$ as the projectile velocity, $a$ as the sound velocity, $\rho$ as the density, and $p$ as the overpressure (or underpressure) with regard to the quiescent air. The curve for $v/a = 0.04$ (*broken line*) was obtained in wind-tunnel experiments. Note that in 1920 the term *Mach number*, here the ratio $v/a$, had not yet been coined. [Drawings by EMI, Freiburg; after Proc. Roy. Soc. Lond. **A97**, 202 (1920), Figs. 1 and 3]

## 4.6 HEAD WAVE STUDIES – Blunt Body Concept

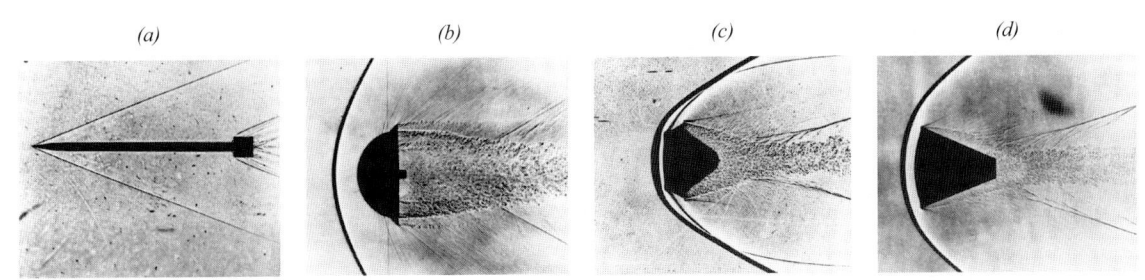

**4.6–N** Example of aerodynamic research contributing to the U.S. Project Mercury, the first series of manned space flights conducted in the period 1961–1963. In 1952, H. Julian ALLEN at NACA's Ames Aeronautical Laboratory pioneered the Blunt Body Theory, which made possible the heat shield designs embodied in the Mercury, Gemini, and Apollo space capsules, enabling astronauts to survive the fiery reentry into the Earth's atmosphere. A blunt body produces a shock wave in front of the vehicle that actually shields the vehicle from excessive heating. As a result, blunt body vehicles can stay cooler than pointy, low-drag vehicles. *(a)* Initial concept. *(b)* Blunt body concept 1953. *(c)* Missile nose cones 1953–1957. *(d)* Manned capsule concept 1957. Later versions of spacecraft used the blunt body design and an ablative surface [Courtesy NASA; http://www.centennialofflight.gov/essay/Evolution_of_Technology/reentry/Tech19G3.htm]

## 4.6 HEAD WAVE STUDIES – Phenomena at Hypersonic Velocities

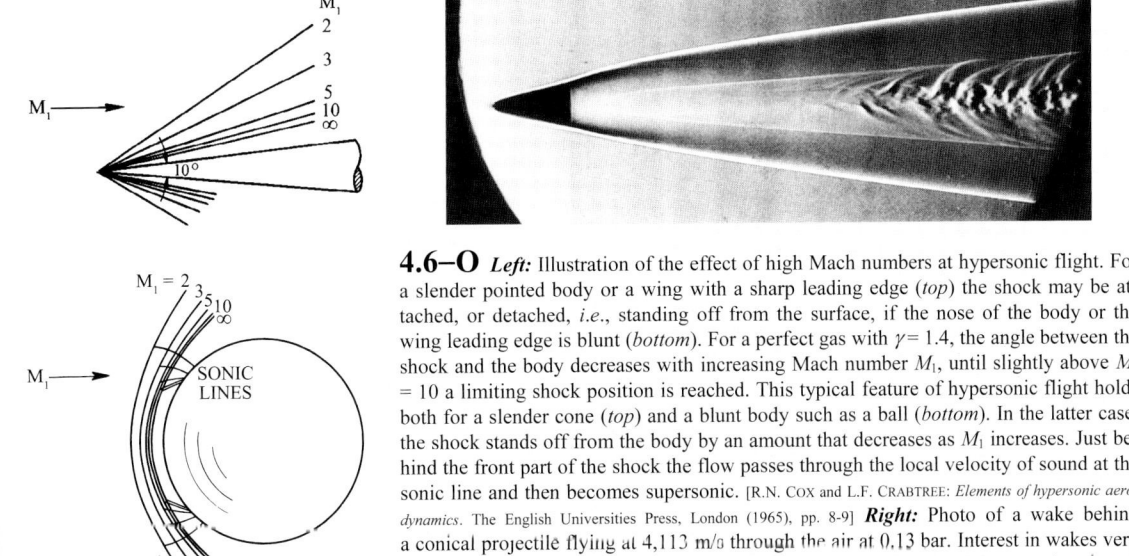

**4.6–O** *Left:* Illustration of the effect of high Mach numbers at hypersonic flight. For a slender pointed body or a wing with a sharp leading edge (*top*) the shock may be attached, or detached, *i.e.*, standing off from the surface, if the nose of the body or the wing leading edge is blunt (*bottom*). For a perfect gas with $\gamma = 1.4$, the angle between the shock and the body decreases with increasing Mach number $M_1$, until slightly above $M_1 = 10$ a limiting shock position is reached. This typical feature of hypersonic flight holds both for a slender cone (*top*) and a blunt body such as a ball (*bottom*). In the latter case, the shock stands off from the body by an amount that decreases as $M_1$ increases. Just behind the front part of the shock the flow passes through the local velocity of sound at the sonic line and then becomes supersonic. [R.N. COX and L.F. CRABTREE: *Elements of hypersonic aerodynamics*. The English Universities Press, London (1965), pp. 8-9] *Right:* Photo of a wake behind a conical projectile flying at 4,113 m/s through the air at 0.13 bar. Interest in wakes very far downstream has been stimulated by observations of entry vehicles entering the atmosphere, which may leave ionized trails thousands of feet long. In the 1950s, various criteria were worked out to calculate the transition point from laminar to turbulent flow. [B.C. JAEGGY: Rept. CO 229/85, ISL, Saint-Louis, France (1985)]

## 4.7 NOZZLE STUDIES AND APPLICATIONS – Early Safety Valve Constructions

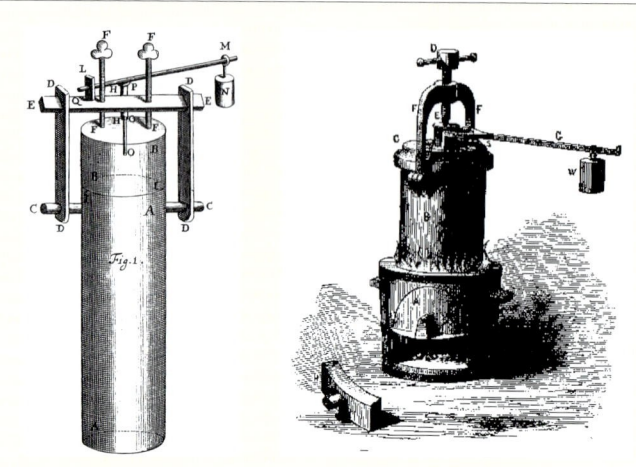

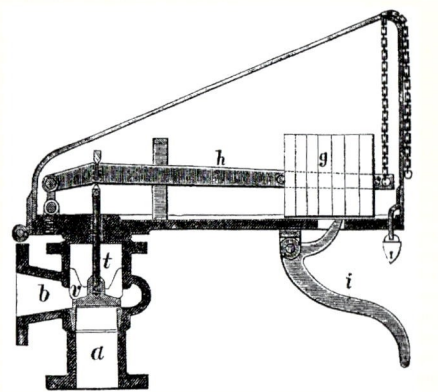

**4.7–A** *Left:* In 1679, the French-born British physicist Denis PAPIN designed the first safety valve to prevent an explosion of his "steam digester," the first pressure cooker. It consisted of a small plate $P$ covering the exit hole of a pipe $H$-$H$ in the cooker lid and loaded by a weight $N$. When the pressure exceeded a certain value, $P$ was lifted so that the steam could escape. [D. PAPIN: *A new digester or engine for softening bones*. Bonwicke, London (1681)] *Right:* View of PAPIN's digester. [http://www.ledenispapin.com/digesteur.htm]

**4.7–B** PAPIN's valve was also used in steam boilers and modified in many ways. To avoid unauthorized manipulations of the valve – one of the numerous reasons for disastrous steam boiler explosions – it was enclosed by a case and secured by a locker. The so-called "safety valve" became one of the most delicate elements of steam engines and stimulated nozzle outflow research. [*Brockhaus Enzyklopädie*. Brockhaus, Leipzig (1898), vol. IV, p. 728]

## 4.7 NOZZLE STUDIES AND APPLICATIONS – Predecessors of Steam Turbines

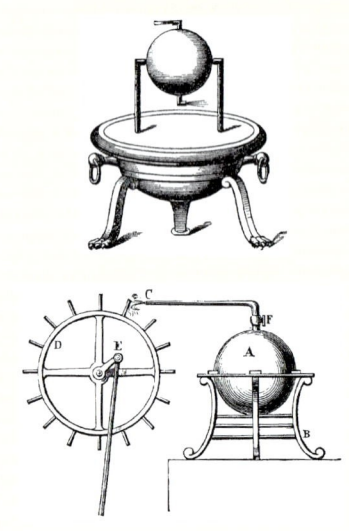

**4.7–C** Early steam engines were used for demonstration purposes or as children's toys, but never gained any technical importance. *Top:* The "steam ball" [*aeolipile*], the first steam turbine, was described by HERON of Alexandria in his *Pneumatica* (A.D. 100). Based on Sir Isaac NEWTON's Third Law of Motion, his mechanical principle of reaction, it consists of a ball filled with steam and is provided with a number of fine nozzles arranged in a tangential direction around the ball. Steam emerging from these nozzles sets the steam ball in rotation. [*Meyers Konversations-Lexikon*. Bibliogr. Inst., Leipzig (1894), vol. 4, p. 525] In earlier centuries, HERON's small steam turbine was often demonstrated in physical cabinets. The reaction principle was later resumed and successfully applied on an industrial scale in Andreas SEGNER's water reaction turbine (1750) and in Charles PARSON's steam reaction turbine (1884). *Bottom:* Schematic of Giovanni BRANCA's "steam wheel" (1629), which is based on the impulse (or action) principle. The steam emerging from a narrow nozzle is directed toward the blades of a small paddle wheel. The principle of action was later resumed in DE LAVAL's impulse turbine, which used a special nozzle of convergent-divergent geometry, the so-called "Laval nozzle." The very high steam speed at the exits of Laval nozzles, impinging on blades projecting from a wheel, resulted in a high turbine efficiency (1888). [*Meyers Konversations-Lexikon*. Bibliogr. Inst., Leipzig (1875), vol. 4, plate *Dampfmaschine I*]

## 4.7 NOZZLE STUDIES AND APPLICATIONS – Forerunners of the Laval Nozzle

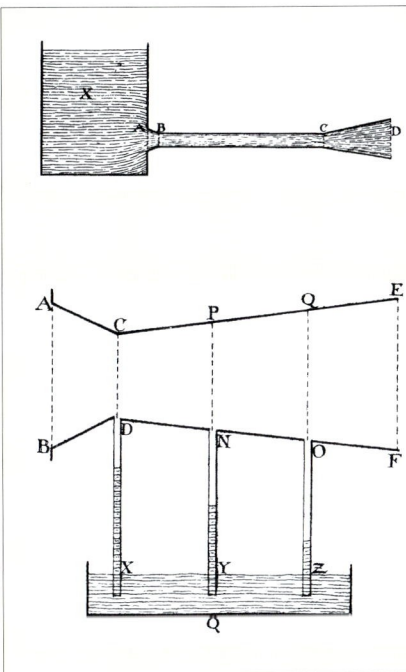
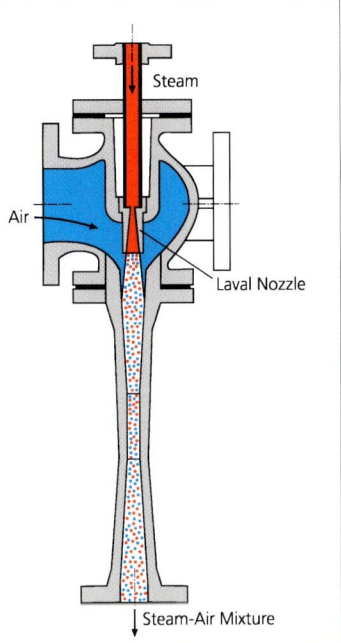

**4.7–D** *Left, top:* In the 1790s, the Italian physicist Giovanni B. VENTURI, experimenting with the outflow of water from vessels, observed that the rate of outflow could be increased by attaching to the end of a drain pipe a conical diverging tube, thus inventing the diffuser principle. *Left, bottom:* While studying the flow in a convergent-divergent tube, VENTURI noticed that in the constriction the flow speed increased and the pressure decreased, an important phenomenon used in the so-called "Venturi tube" to measure fluid flows, and as a pump. [Ann. Phys. **2**, 418 (1799)] *Right:* In the mid-1870s, the German KÖRTING Brothers at Hannover used a divergent nozzle in their steam injection pump, even before Gustaf DE LAVAL. However, their patent application was rejected. This is a more recent schematic of a steam jet vacuum pump. A supersonic jet velocity improves the pump efficiency. [After R. VON MILLER (ed.) *Lexikon der Energietechnik und Kraftmaschinen.* Deutsche Verlagsanstalt, Stuttgart (1965), vol. 6, p. 188]

## 4.7 NOZZLE STUDIES AND APPLICATIONS – Use of the Venturi Nozzle in the First Wind Tunnel

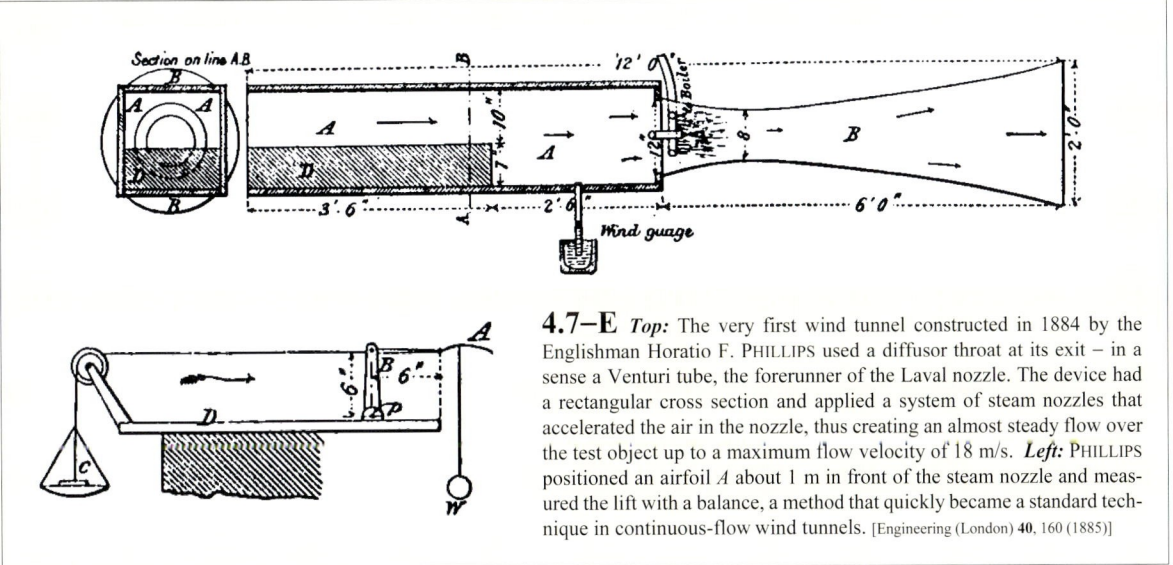

**4.7–E** *Top:* The very first wind tunnel constructed in 1884 by the Englishman Horatio F. PHILLIPS used a diffusor throat at its exit – in a sense a Venturi tube, the forerunner of the Laval nozzle. The device had a rectangular cross section and applied a system of steam nozzles that accelerated the air in the nozzle, thus creating an almost steady flow over the test object up to a maximum flow velocity of 18 m/s. *Left:* PHILLIPS positioned an airfoil *A* about 1 m in front of the steam nozzle and measured the lift with a balance, a method that quickly became a standard technique in continuous-flow wind tunnels. [Engineering (London) **40**, 160 (1885)]

## 4.7 NOZZLE STUDIES AND APPLICATIONS – First Use of Laval Nozzle in a Steam Turbine

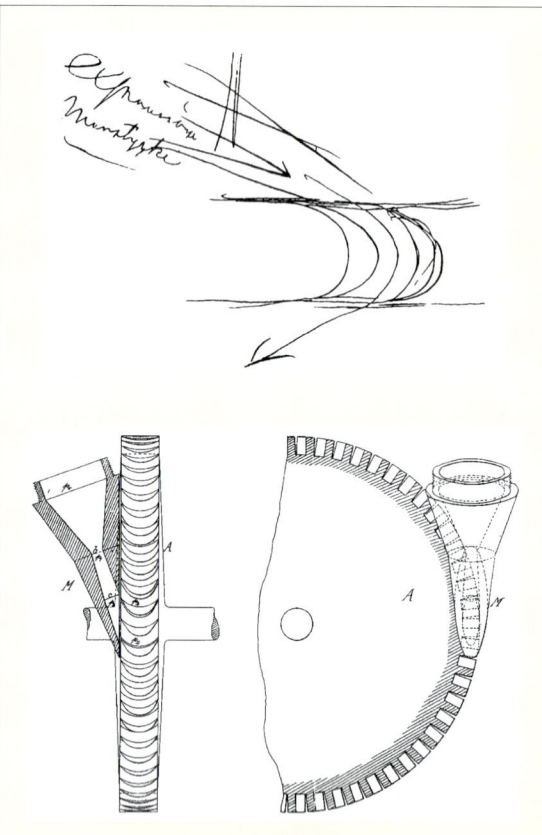

**4.7–F** Gustaf DE LAVAL, a Swedish mechanical engineer, dedicated much time to optimizing the geometry of steam nozzles for use in steam turbines. He correctly recognized by experimentation that a divergent geometry significantly increased the velocity of the steam jet, thus improving the efficiency of steam turbines. *Left, top:* His sketch of the first axial impulse blade with a rather modern aerodynamic blade form and a supercritical nozzle, dated 1888. His Swedish handwriting reads *expansion munstycke*, meaning "expansion nozzle." [I. JUNG: *De Laval Memorial Lecture 1973*, Roy. Swed. Acad. Eng. Sci. (1973)] *Left, center:* Schematic of the Laval nozzle as used in his steam turbine. Note the divergent nozzle exit geometry [C.G.P. DE LAVAL: *Steam inlet channel for rotating engines*. Swed. Patent No. 1,902 (1888)] *Right:* Photograph of Laval's steam turbine from 1888, the first to employ Laval nozzles. *Left, bottom:* Enlargement of a cut through the injection system, showing the turbine wheel and two Laval steam nozzles in more detail. [Courtesy Deutsches Museum, Munich]

## 4.7 NOZZLE STUDIES AND APPLICATIONS – Steam Flow in a Divergent Nozzle

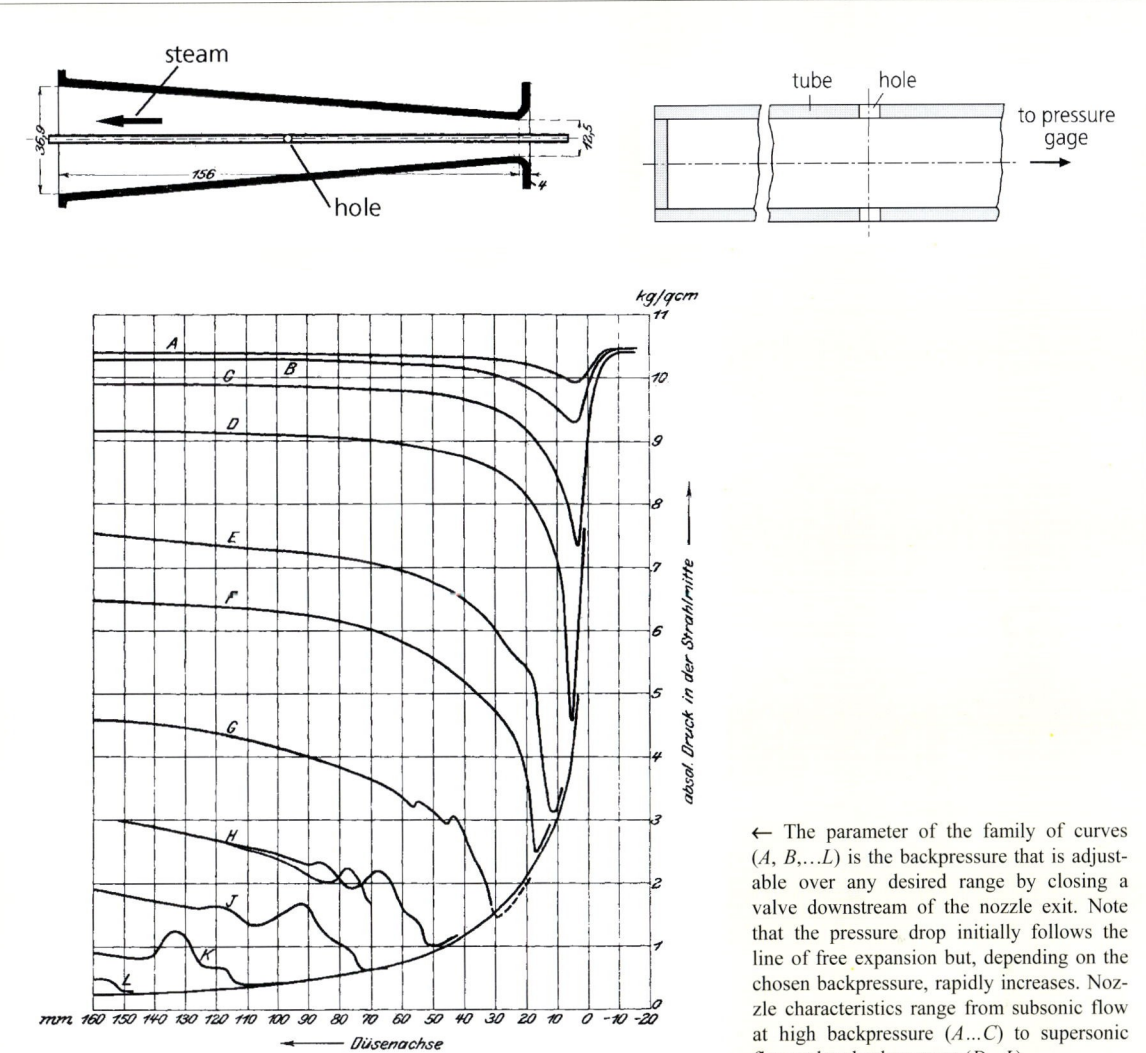

← The parameter of the family of curves (*A, B,...L*) is the backpressure that is adjustable over any desired range by closing a valve downstream of the nozzle exit. Note that the pressure drop initially follows the line of free expansion but, depending on the chosen backpressure, rapidly increases. Nozzle characteristics range from subsonic flow at high backpressure (*A...C*) to supersonic flow at low backpressure (*D...L*).

**4.7–G** Aurel B. STODOLA, a Hungarian-Swiss mechanical engineer at ETHZ, made the first axial pressure measurements of supersonic flow inside a Laval nozzle. **Left, top:** Using a 156-mm-long divergent nozzle with a steam entrance diameter of 12.5 mm and an exit diameter of 36.9 mm, STODOLA measured the axial pressure by moving a tubular gauge in the axial direction. The thin tube, closed at the left end and connected to a mechanical pressure gauge at its right end, allowed one to measure the pressure variation along the nozzle axis. **Right:** The tube was provided with a small hole drilled perpendicular to the tube axis and connected with a manometer. [By the author] **Left, bottom:** Diagram showing curves of measured absolute steam pressures along nozzle axis. STODOLA correctly interpreted the strong increases in pressure as a "condensation shock" [Germ. *Verdichtungsstoß*], such as was derived theoretically in 1859 by the German mathematician G.F. Bernhard RIEMANN. At low backpressures (curves *H* to *L*), the local pressure shows small oscillations that, however, are quickly used up by friction when approaching the nozzle exit. [A.B. STODOLA: *Die Dampfturbinen*. Springer, Berlin (1903), p. 19]

## 4.7 NOZZLE STUDIES AND APPLICATIONS – Laval Nozzles as Power Generators in Aeronautics

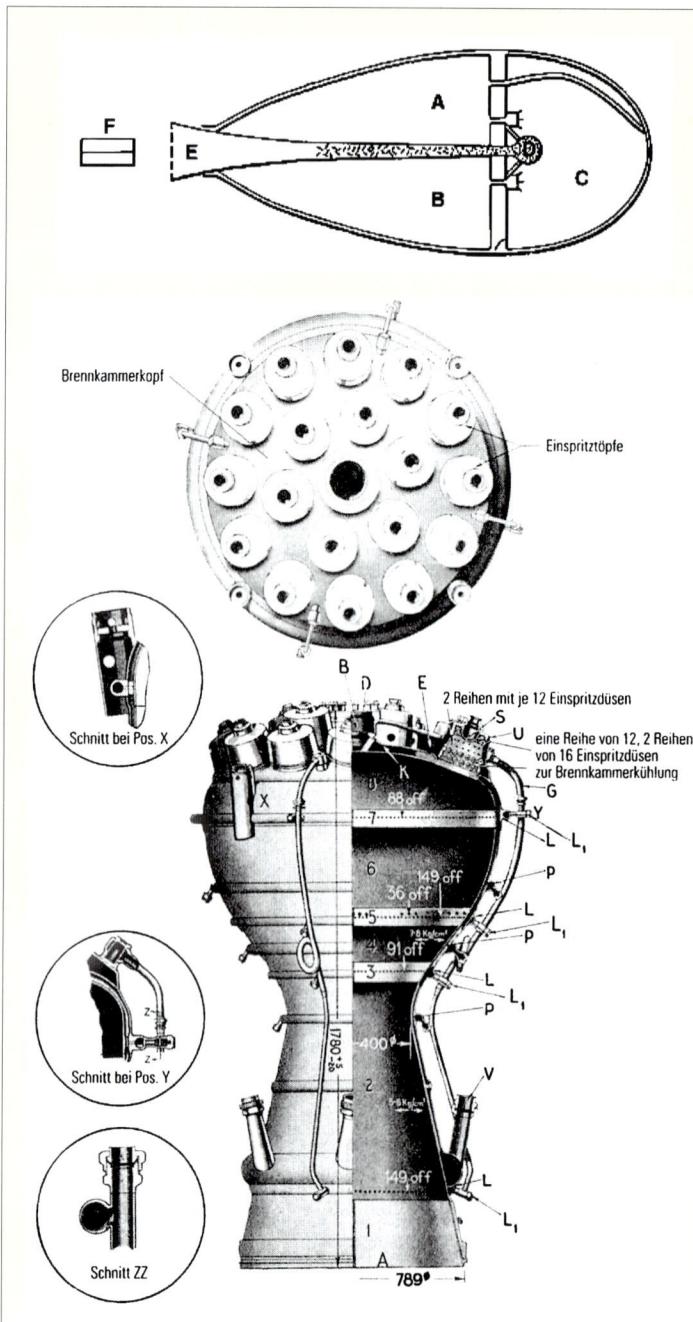

**4.7–H** The concept of a convergent-divergent nozzle geometry, the so-called "Laval nozzle," had also been proposed for propulsion purposes, particularly to realize in this way a high exit velocity of the hot exhaust gases. *Top:* The Russian rocket pioneer Konstantin E. TSIOLKOVSKY proposed manned rocketry in the 1890s. This is a conceptual sketch of a liquid fuel rocket from his 1903 milestone paper entitled *Exploration of Cosmic Space by Means of Reaction Devices* (Nauchnoe Obozrenie). The letters, here added later, designate: $A$ – tank with freely evaporating liquid oxygen kept at very low temperature; $B$ – tank with liquid hydrogen (or hydrocarbon); $C$ – crew and breathing equipment; $D$ – burning chamber; $E$ – exhaust nozzle; $F$ – rudders, positioned in the stream of the exhaust gases, serving for flight control (a method which was later used in the German V2). The rocket thrust chamber and a partition wall separate the fuel and oxidizer.
[Uranos Group, *see* http://www.uranos.eu/biogr/ciolke.html]
*Bottom:* View of the 1.83-m-long engine of the German WWII rocket A-4, in 1943 renamed V2 (meaning "Vengeance Weapon 2"). The rocket engine consisted mainly of an 18-nozzle injection system, a spherical combustion chamber, and a common exhaust nozzle, a large 25° conical Laval nozzle. A small, only 30-cm-diameter two-stage 500-hp turbo-centrifugal pump (not shown here) – in WW II a riddle to allied secret services – injected the propellant (liquid oxygen and a fuel mixture consisting of 75% ethyl alcohol and 25% water) into the combustion chamber. The hot combustion gases, leaving the combustion chamber at a temperature of about 2,400 °C and expanding to ambient pressure, were accelerated up to 2,000 m/s, thus producing by the exhaust blast a thrust of 55,000 pounds for about 1 min. The A-4, weighing about 13.5 tons at takeoff, reached a maximum speed of over $M = 4$ {⇨Fig. 4.20–C}. Its maximum range was about 320 km. [M.J. NEUFELD: *The rocket and the Reich*. Free Press, New York (1995), p. 98]

## 4.8 SUPERSONIC JET PHENOMENA – Salcher's and Mach's First Free Air Jet Studies

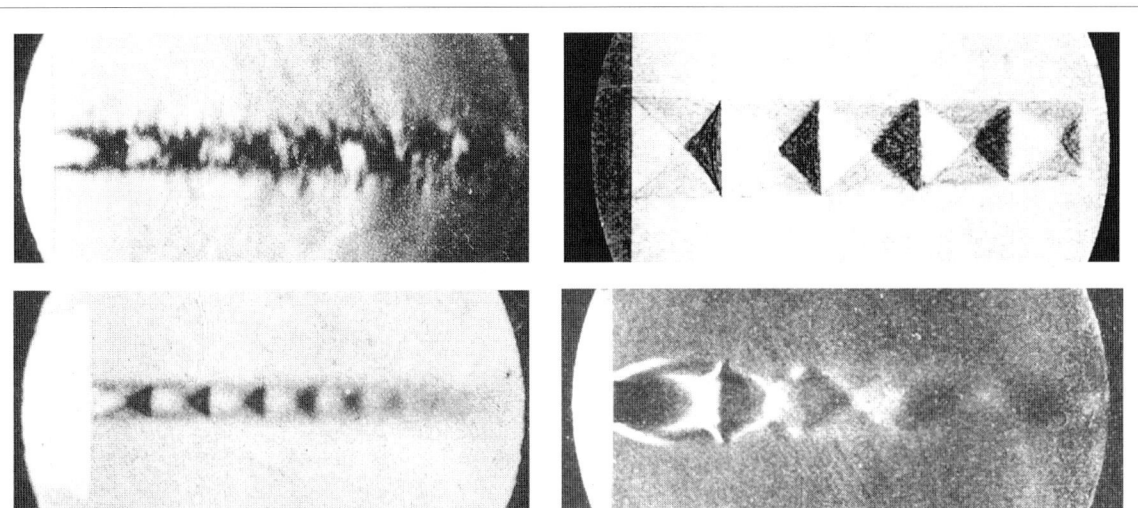

**4.8–A** Ernst MACH at the Charles University in Prague inspired a number of studies on the rapid outflow of gases from small orifices that brought to light many new supersonic flow phenomena. ***Left:*** Together with Peter SALCHER, an Austrian physics professor at the Royal Naval Academy in Fiume, he made the first schlieren photos of a supersonic air jet emerging from a 4.5-mm-dia. nozzle operated at a pressure of 37 bar. They used two different methods of illumination: flash illumination (*top*) provided by an electric spark is useful to visualize momentary nonstationary phenomena, such as vortices. On the other hand, they are suppressed when a long-duration illumination is used, such as a Geissler-discharge tube (*bottom*), which shows more clearly the stationary wave pattern. [Sitzungsber. Akad. Wiss. Wien **98** (IIa), 1303 (1889)] ***Right:*** Ludwig MACH, resuming these experiments in the mid-1890s, studied more closely the wave pattern inside supersonic free air jets resulting from different reservoir pressures and nozzle shapes. At low pressure he observed a crossed wave pattern in the case of regular shock reflection (*top*). At higher pressures, the crossed wave pattern transformed into a new wave front at the point of intersection (*bottom*) – the so-called "Mach disk," a result of a symmetric irregular shock interaction. [Sitzungsber. Akad. Wiss. Wien **106** (IIa), 1025 (1897)]

**4.8–B** Flash interferogram of an air jet, emerging under 20 bar from a cylindrical nozzle with a 7.5-mm dia. and propagating from left to right. The mouth of the nozzle is on the far left side and not visible in the picture. This interferogram was taken in July 1893 by Ludwig MACH using the famous "Mach-Zehnder interferometer," a construction invented in 1891 by him and, independently, by Ludwig ZEHNDER at the University of Würzburg, a Ph.D. student of Prof. Wilhelm Conrad RÖNTGEN. The picture clearly shows the formation of a steady "Mach disk" in front of the nozzle exit.
[Sitzungsber. Akad. Wiss. Wien **106** (IIa), 1025 (1897)]

## 4.8 SUPERSONIC JET PHENOMENA – EMDEN's First Steam Jet Studies

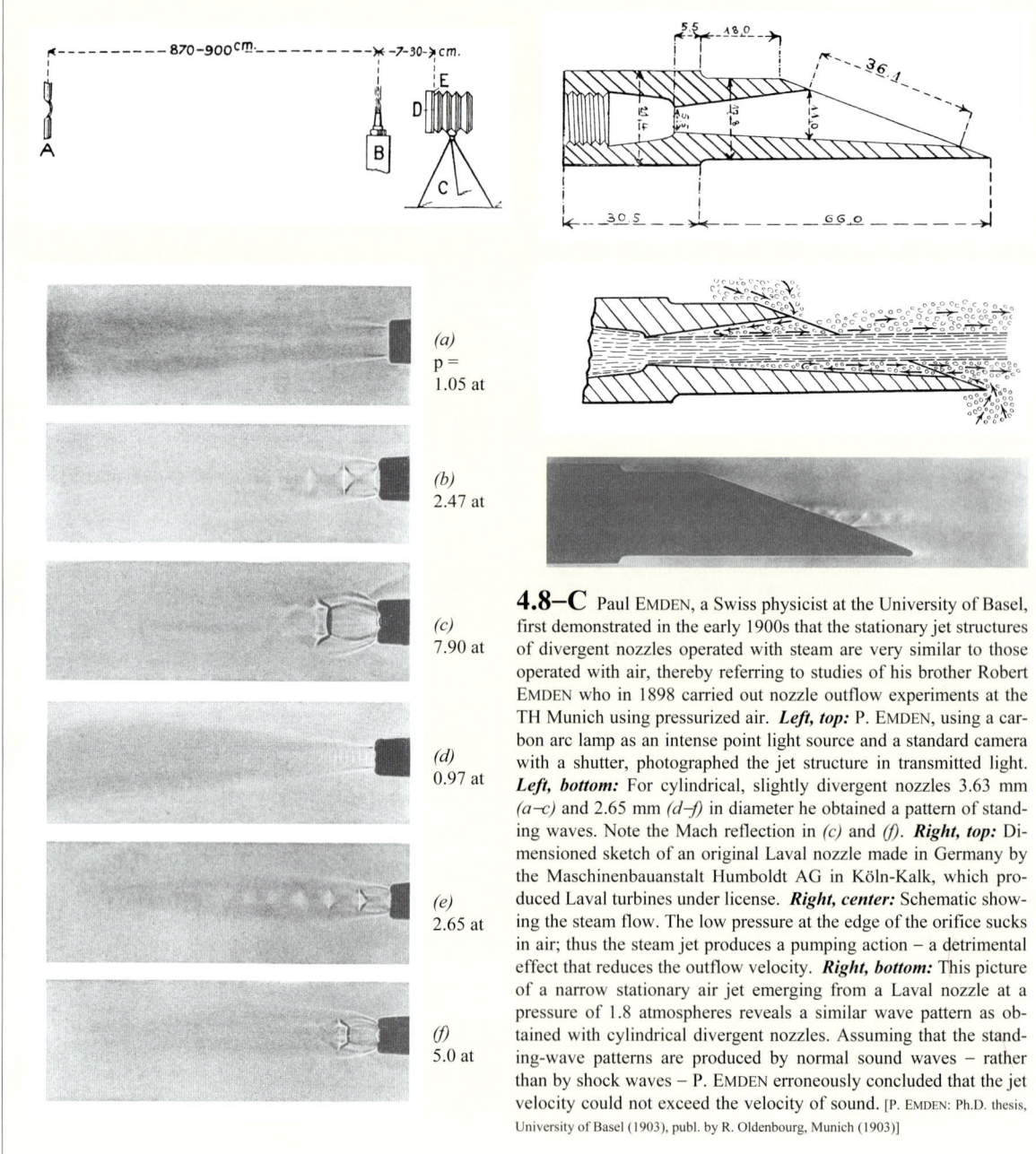

*(a)* p = 1.05 at

*(b)* 2.47 at

*(c)* 7.90 at

*(d)* 0.97 at

*(e)* 2.65 at

*(f)* 5.0 at

**4.8−C** Paul EMDEN, a Swiss physicist at the University of Basel, first demonstrated in the early 1900s that the stationary jet structures of divergent nozzles operated with steam are very similar to those operated with air, thereby referring to studies of his brother Robert EMDEN who in 1898 carried out nozzle outflow experiments at the TH Munich using pressurized air. *Left, top:* P. EMDEN, using a carbon arc lamp as an intense point light source and a standard camera with a shutter, photographed the jet structure in transmitted light. *Left, bottom:* For cylindrical, slightly divergent nozzles 3.63 mm *(a−c)* and 2.65 mm *(d−f)* in diameter he obtained a pattern of standing waves. Note the Mach reflection in *(c)* and *(f)*. *Right, top:* Dimensioned sketch of an original Laval nozzle made in Germany by the Maschinenbauanstalt Humboldt AG in Köln-Kalk, which produced Laval turbines under license. *Right, center:* Schematic showing the steam flow. The low pressure at the edge of the orifice sucks in air; thus the steam jet produces a pumping action − a detrimental effect that reduces the outflow velocity. *Right, bottom:* This picture of a narrow stationary air jet emerging from a Laval nozzle at a pressure of 1.8 atmospheres reveals a similar wave pattern as obtained with cylindrical divergent nozzles. Assuming that the standing-wave patterns are produced by normal sound waves − rather than by shock waves − P. EMDEN erroneously concluded that the jet velocity could not exceed the velocity of sound. [P. EMDEN: Ph.D. thesis, University of Basel (1903), publ. by R. Oldenbourg, Munich (1903)]

## 4.8 SUPERSONIC JET PHENOMENA – PRANDTL's and MEYER's Nozzle and Jet Studies

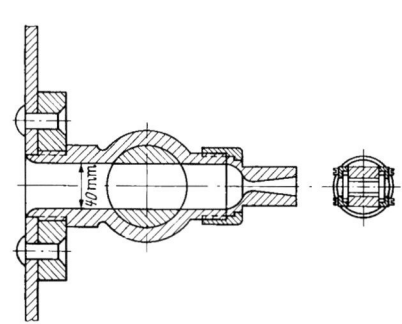

↑ The test nozzle was attached to a 40-mm-dia. ball valve.

**4.8–D** In the early 1900s, Ludwig PRANDTL and Theodor MAYER at Göttingen University studied steady air flow fields inside and outside of plane nozzles enclosed between two parallel glass plates using the schlieren method which reproduces regions of compression and expansion as dark and bright areas, respectively. *Left, top:* Experimental setup. *Left, bottom:* Schematics of three main jet types: jet emerging *(a)* from a parallel nozzle with supersonic velocity; *(b)* from a convergent nozzle with sound velocity; and *(c)* from a divergent nozzle showing a more simplified jet structure. *Arrows* in the schematics indicate the direction of density gradients. *Right:* Schlieren pictures of jet structures: *(a)* jet emerging supersonically from a divergent nozzle, exhibiting in the case of overpressure ($P_{JET} > P_0$) a series of crossing fans of expansion and compression waves; *(b)* supersonic jet in the case of equal pressure ($P_{JET} = P_0$), revealing a more uniform structure; and *(c)* jet emerging with sound velocity from a parallel nozzle, showing disturbances. In all cases the flow propagates from left to right. [Physik. Z. **8**, 23 (1907)]

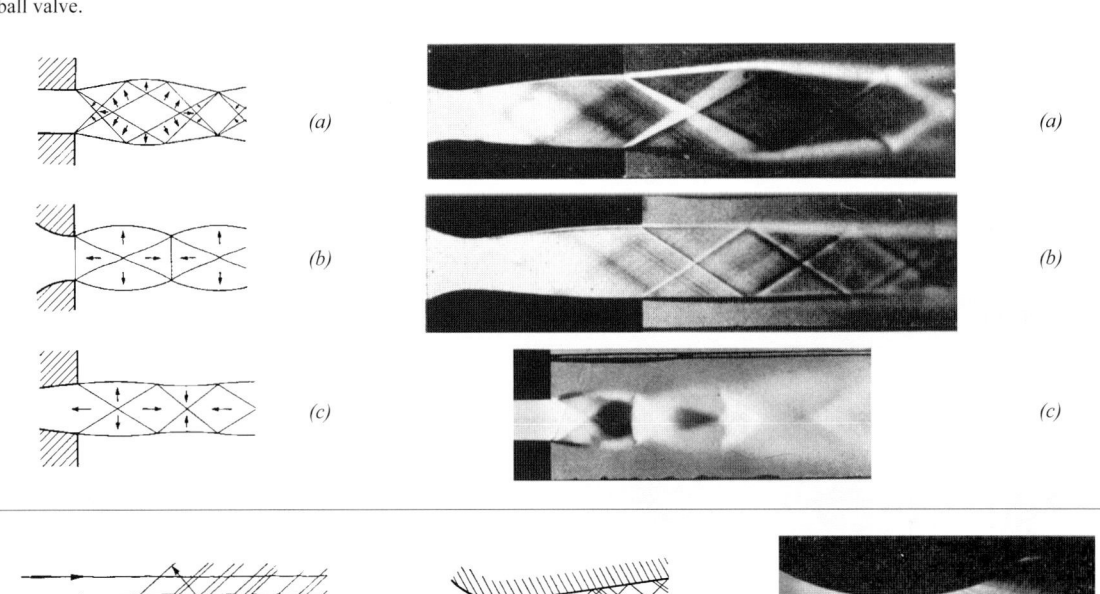

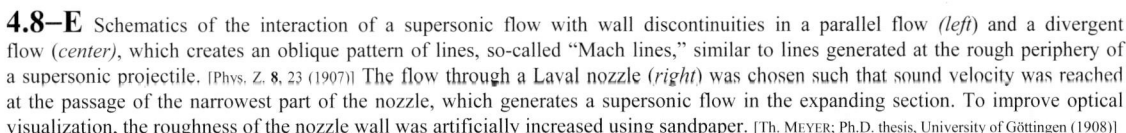

**4.8–E** Schematics of the interaction of a supersonic flow with wall discontinuities in a parallel flow *(left)* and a divergent flow *(center)*, which creates an oblique pattern of lines, so-called "Mach lines," similar to lines generated at the rough periphery of a supersonic projectile. [Phys. Z. **8**, 23 (1907)] The flow through a Laval nozzle *(right)* was chosen such that sound velocity was reached at the passage of the narrowest part of the nozzle, which generates a supersonic flow in the expanding section. To improve optical visualization, the roughness of the nozzle wall was artificially increased using sandpaper. [Th. MEYER; Ph.D. thesis, University of Göttingen (1908)]

## 4.8 SUPERSONIC JET PHENOMENA – THOMER's First Radiographs of Detonating Shaped Charges

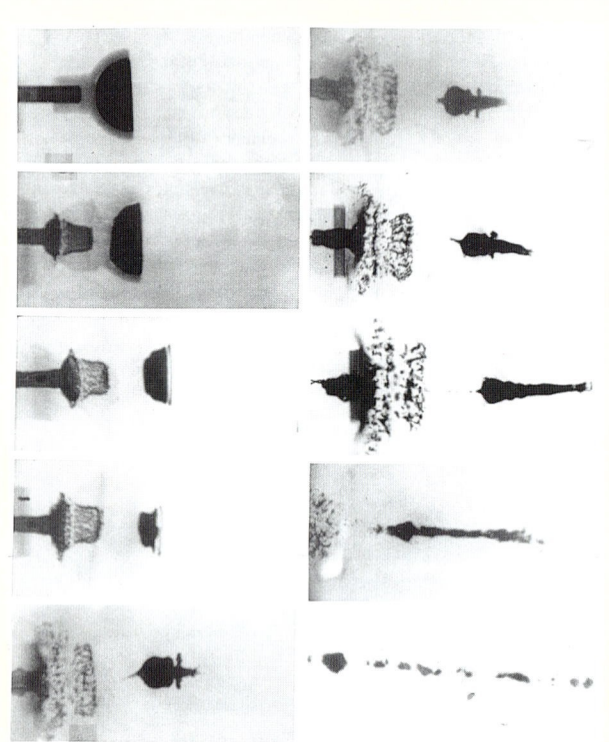

**4.8–F** This series of historic flash radiographs of hemispherical shaped charges was made in 1941 by Gustav THOMER at Prof. Hubert SCHARDIN's Ballistic Research Laboratory of the Technische Akademie der Luftwaffe (Air Force Technical Academy) in Berlin-Gatow and published in November 1941 in a secret institute report. In order to illustrate the evolution of the jet phenomenon, the pictures were taken from different charges at different time instants and assembled to a chronology of a detonating hollow charge, a technique called "pseudocine radiography." The time proceeds from *top* to *bottom*, and from *left* to *right*. The jet formation and propagation as well as its subsequent rupture into a number of droplets is clearly resolved.
[Courtesy Dr. Pascale LEHMANN, ISL, France; Proc. Flash Radiography Symp.. 36th Natl. Fall Conf. of American Society for Nondestructive Testing (ASNT), Houston, TX (1976), pp. 1-14]

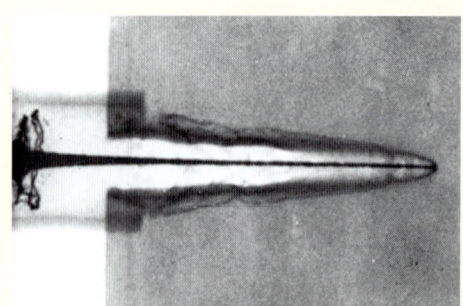 

**4.8–G** *Left:* Flash radiograph of a supersonic jet resulting from a shaped charge and penetrating a duralumin target. Note that the shock wave initiated in the target is not visible. In detonating shaped charges, most of the jet formed moves at hypersonic speed, the tip traveling in the 10–14 km/s region and the jet tail at a lower velocity. [F. JAMET and G. THOMER: *Flash radiography*. Elsevier, Amsterdam (1976), p. 137]
*Right:* View of a "slug," the back part of the collapsing liner that travels along the same direction as the jet but at much lower velocity than the jet tail. In this well-preserved form, it is only occasionally found after detonation of shaped charges [Courtesy Dr. Norbert HEIDER, EMI, Freiburg]

## 4.8 SUPERSONIC JET PHENOMENA – Formation and Structure of Liquid Jets

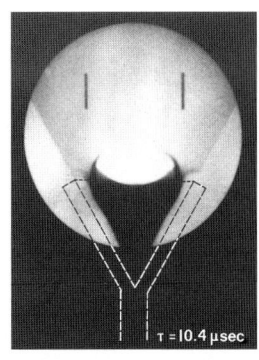 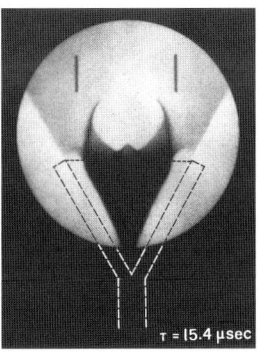 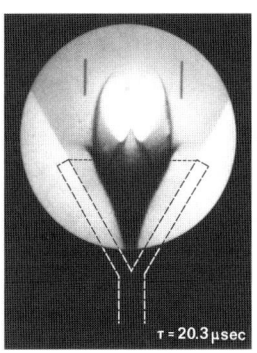 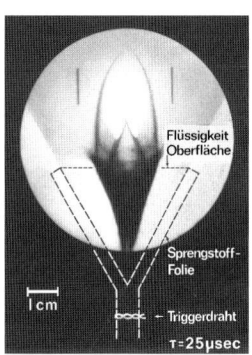

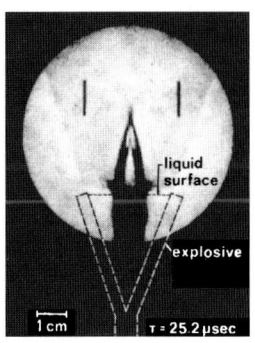

**4.8–H** Peter KREHL, Werner HEILIG, and Ulrich HORNEMANN at EMI, Freiburg, studied unsteady Mach reflection in organic liquids. *Top:* This series of flash radiographs from the formation of supersonic liquid jets was obtained when a plane hollow 60° cone of a solid explosive (nitropenta), filled with a liquid (carbon tetrachloride, $CCl_4$), was ignited at its cone tip. The region of interaction of the two head waves, indicated by the two detonation waves in the liquid, results in "inverse Mach reflection," a regressive type of irregular reflection. The liquid, squeezed and accelerated along the cone axis, is thrown out as a two-jet phenomenon. The two outer jets are formed first, followed by the generation of the two inner jets. Both branches of outer and inner jets move at a velocity of about 1,000 m/s. *Bottom:* At decreasing cone angle, here shown for a 40° cone geometry, the intersection of the two colliding head waves just before reaching the surface is regular, but the typical ejection phenomenon is again obvious. However, the jet formation becomes now increasingly focused in the axial direction, and the jet velocity increases. [Proc. 11th Int. Symp. on Shock Tubes and Waves, Seattle (July 1977), p. 303]

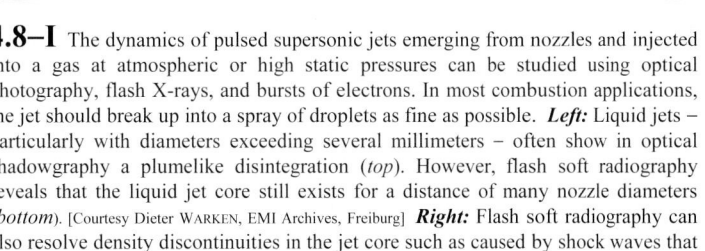

↑ Optical shadowgraph of a petroleum jet (*top*) and X-ray shadowgraph of a nitromethane-methanol jet (*bottom*), emerging from a 2-mm- and a 4-mm-dia. cylindrical nozzle, respectively, at a velocity of about 410 m/s into the atmosphere.

**4.8–I** The dynamics of pulsed supersonic jets emerging from nozzles and injected into a gas at atmospheric or high static pressures can be studied using optical photography, flash X-rays, and bursts of electrons. In most combustion applications, the jet should break up into a spray of droplets as fine as possible. *Left:* Liquid jets – particularly with diameters exceeding several millimeters – often show in optical shadowgraphy a plumelike disintegration (*top*). However, flash soft radiography reveals that the liquid jet core still exists for a distance of many nozzle diameters (*bottom*). [Courtesy Dieter WARKEN, EMI Archives, Freiburg] *Right:* Flash soft radiography can also resolve density discontinuities in the jet core such as caused by shock waves that propagate in the core as in a wave guide. [Picture by Peter KREHL (EMI Archives, Freiburg). Proc. 12th Int. Symp. on Ballistics, San Antonio, TX (1990); Proc. 19th Int. Congr. on High-Speed Photography and Photonics, Cambridge, U.K. (1990)]

## 4.8 SUPERSONIC JET PHENOMENA – Generation of Microjets

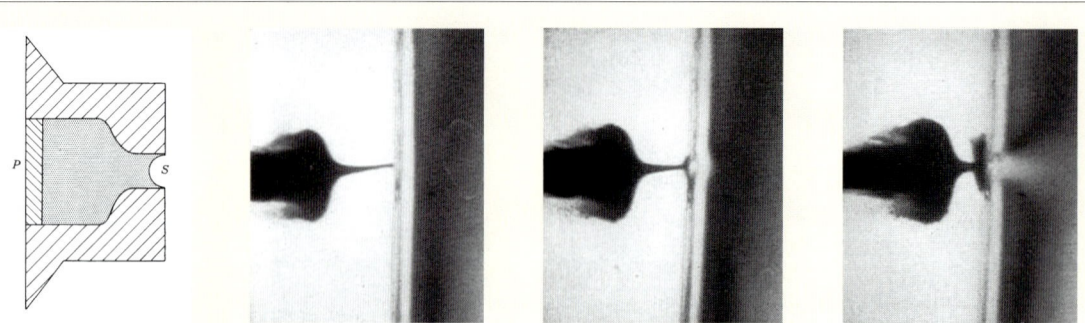

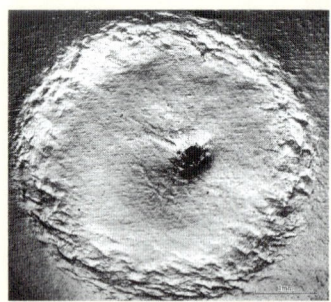

**4.8–J** In the 1960s, Frank P. BOWDEN and John H. BRUNTON at Cavendish Laboratory (Cambridge, U.K.) studied the formation of microjets in liquids under the influence of impact or shock. *Top, left:* Schematic of generating small high-velocity liquid jets. A small amount of water is held in a strong steel container with a small hole in it. A sudden pressure pulse is generated, *e.g.*, by firing a bullet on the rubber plug *P*. When the shock wave reaches the curved free surface *S*, the water emerges as a tiny Munroe jet at much higher velocity. *Top, right:* A three-frame series showing the microjet formation process and its interaction with a steel surface; interframe time is 0.8 µs. This technique of liquid jet generation was later refined and used in rain erosion studies. *Bottom:* Example of a 4-mm-dia. crater formed by liquid impact on stainless steel. Near its center a second small but deep crater has been formed by a microjet. [Phil. Trans. Roy. Soc. Lond. **A260** (1966), plates 10 and 11]

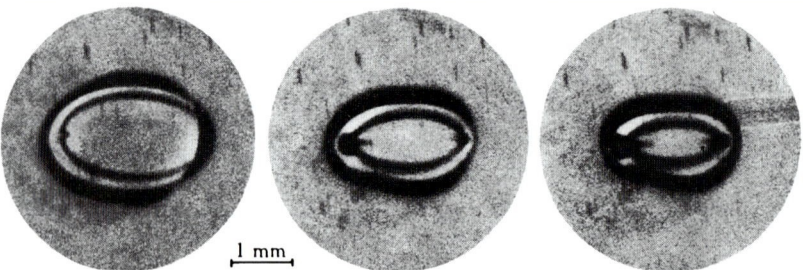

**4.8–K** There is strong evidence that microjets play a significant role in the initiation process of liquid high explosives. The initiation by impact or shock is essentially a thermal process in which a small region of the explosive is heated to a temperature at which reaction can occur. Shock heating requires very high pressures (ca. 60 to 100 kbar). However, if a discontinuity – such as a bubble or cavity – is present in the liquid explosive, the energy of the shock can be concentrated in a small region, and a localized "hot spot" can be formed that initiates detonation. Frank P. BOWDEN and M.P. MCONIE at Cavendish Laboratory demonstrated that an annulus of nitroglycerin, which had trapped a single cavity containing air and was struck by a falling hammer, generated tiny jets in the regions of maximum curvature. These so-called "microjets" project at about 30 to 50 m/s into the central cavity as the cavity is compressed. *Left:* Beginning of jet formation, here starting at the left inner wall of the bubble. *Center:* 10 µs later, a second jet is formed at the opposite end of the cavity. *Right:* After another 5 µs, two jets have formed that later collide head on. The initiation of microjets will aid adiabatic heating of the entrapped gas and initiation of detonation. [Nature **206**, 380 (1965); Phil. Trans. Roy. Soc. Lond. **A260**, 94 (1966)]

# 4.8 SUPERSONIC JET PHENOMENA – Astrophysical Jets

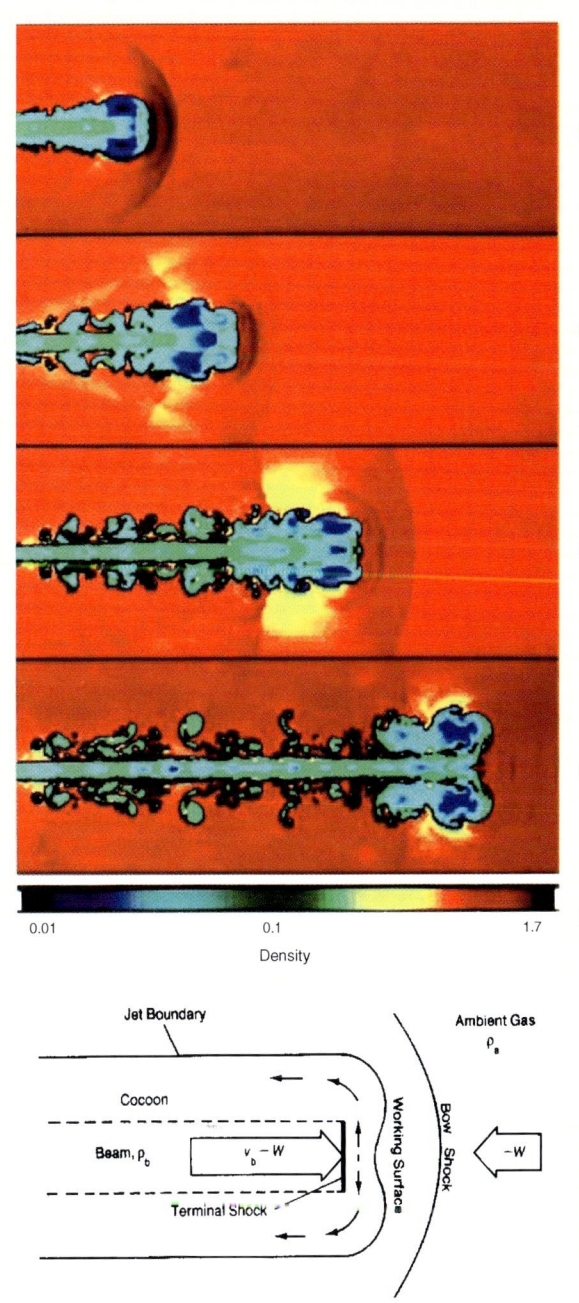

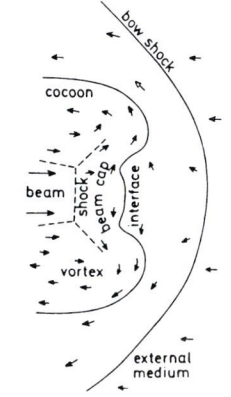

← Detail of an astrophysical *terminal jet shock* structure originating from a powerful radio source. The beam gas does not collect at the region of impact, the so-called "beam cap," but is shock-decelerated and deflected sideways to form a low-density backward-flowing cocoon around the forward-moving beam. At the terminal shock the jet gas is heated and flows into the cocoon. The interface is the contact discontinuity between the shocked jet material and the shocked intracluster material.

**4.8–L** Astrophysical jets are mysterious phenomena that can extend in a remarkably collimated manner from the center of a radio galaxy to distances of thousands or millions of light-years. In order to study nonlinear features of supersonic jet flow that may explain the unusual stability of extragalactic jets, astrophysicists at the MPI für Physik und Astrophysik (Garching, Germany) performed numerical studies using a high-resolution gasdynamic code. **Left, top:** Numerical simulation of the evolution of a supersonic jet ($M = 3$) expanding from a narrow orifice into a quiescent gas; the calculated gas density varies according to the color code *(e)*. Stages of jet formation: *(a)* & *(b):* Establishment and turnover of jet cocoon, which is composed of gas that has passed through the terminus of the jet. *(c)* & *(d):* Subsequent mixing of cocoon and ambient gas, which leaves intact only a lobe of cocoon gas at the jet head. The bow shock, driven by the supersonically advancing jet head, is revealed by the jump in density it produces. [M.D. SMITH, M.L. NORMAN, K.H.A. WINKLER, L. SMARR, MNRAS **214**, 67 (1985); Los Alamos Sci. Mag. **12**, 38 (Spring/Summer 1985)] **Left, bottom:** Flow schematic of the jet head. The working surface or "beam cap" maintains a high pressure of order $M^2$ higher than the static pressure of the external medium. **Right:** The above jet morphology was used to explain the possible physical conditions in extragalactic radio sources. This schematic shows a detail of the expected double-shock structure at the jet terminus as the result of an expanding radio source on the external medium. The Mach disk or "jet shock" effectively stops the incoming jet, and the bow shock accelerates and heats the ambient medium. The beam shock and cap correspond to observed radio hotspots, and the cocoon corresponds to the radiolobe of a radio galaxy.

**4.9  WIND TUNNELS** – Pioneering Supersonic Devices in France and England

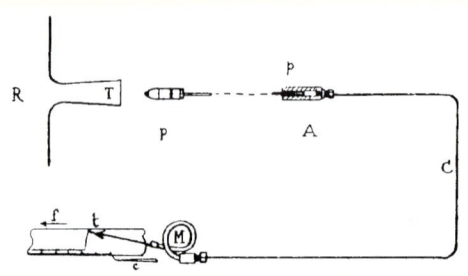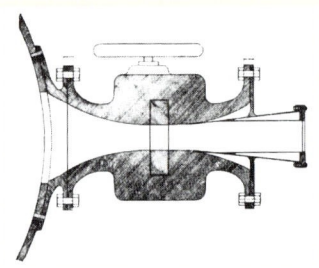

**4.9–A** *Left:* Eugène HUGUENARD and Jean André SAINTE-LAGUË, two French researchers at Paris, apparently built the first supersonic test facility for aeroballistic studies. They generated a supersonic flow emerging from a Laval nozzle $T$ that was operated by discharging a high-pressure air reservoir $R$ via $T$ and reached flow velocities up to 450 m/s. To record the drag force of a projectile $P$ they used a piston gauge $A$ coupled to an oil-pressurized manometer $M$. Note that the projectile was positioned in front of the Laval nozzle and not within it. *Right:* This is a close-up view of their Laval nozzle illustrating its geometry. [La Technique Aéronautique **15**, 346 (1924)]

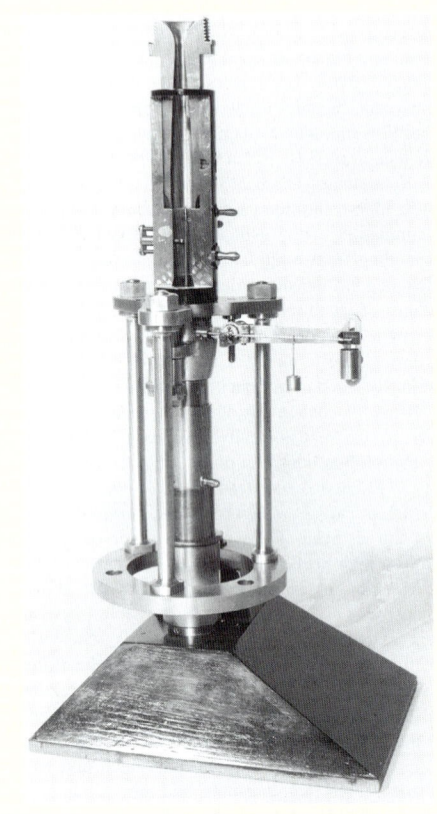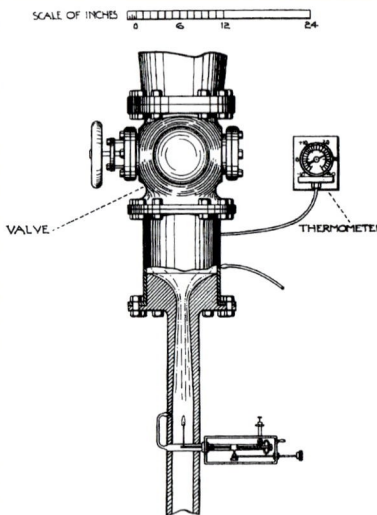

← The NPL supersonic wind tunnel employed a converging/diverging nozzle ranging from 0.515 to 0.8 in. (13 to 20.6 mm) in diameter; the test body was fully emerged in the Laval nozzle. To minimize any wall effects on the air flow around the model, only tiny model projectiles with a diameter not exceeding 0.09 in. (2.3 mm) could be used. The drag force was measured with a small balance. To provide a uniform flow, a honeycomb and guide-blade system were later inserted between stop valve and nozzle. [Proc. R. Soc. Lond. **A131**, 122 (1931)]

**4.9–B** *Left:* View of world's first supersonic wind tunnel ($M = 2$), a continuously working miniature facility devised by the British engineer Thomas E. STANTON and collaborators at NPL (Teddington, U.K.). The apparatus, first operational in 1921, was applied on ballistic drag studies. [Crown Copyright 1980. Reproduced by permission of controller of H.M.S.O. Courtesy National Physical Laboratory (NPL), Teddington, U.K.] *Right:* Schematic of STANTON's improved wind tunnel. A considerable increase in the scale of experiments became possible by equipping the laboratory with a large air-compressing plant (2,000 m$^3$/min) that allowed one to increase the speed up to $M = 3.25$ at a 3.07-in. (79.5-mm)-dia. test section.

4 PICTURE GALLERY

## 4.9 WIND TUNNELS – Prewar Supersonic Facility at TH Aachen

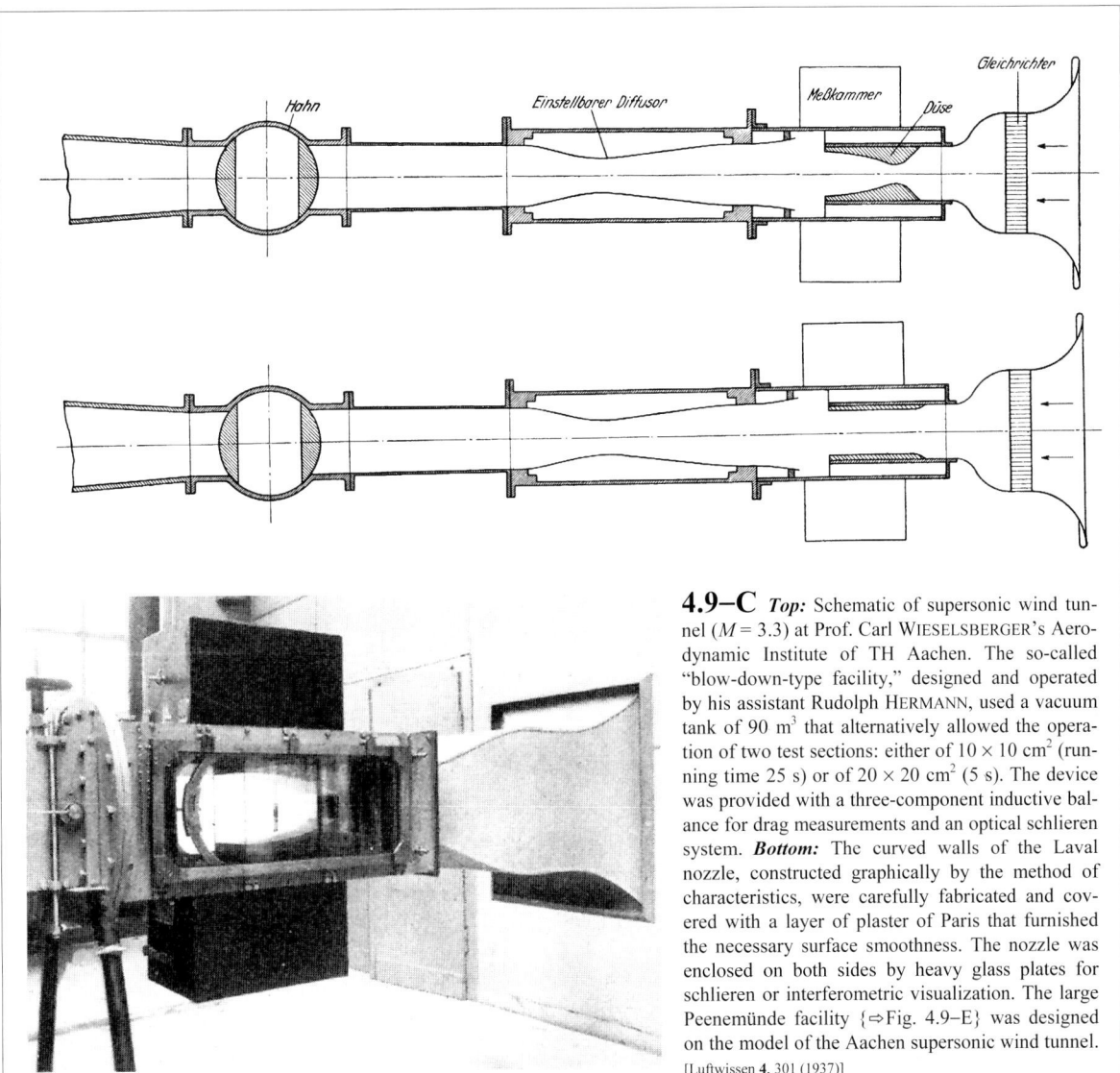

**4.9–C** *Top:* Schematic of supersonic wind tunnel ($M = 3.3$) at Prof. Carl WIESELSBERGER's Aerodynamic Institute of TH Aachen. The so-called "blow-down-type facility," designed and operated by his assistant Rudolph HERMANN, used a vacuum tank of 90 m³ that alternatively allowed the operation of two test sections: either of $10 \times 10$ cm² (running time 25 s) or of $20 \times 20$ cm² (5 s). The device was provided with a three-component inductive balance for drag measurements and an optical schlieren system. *Bottom:* The curved walls of the Laval nozzle, constructed graphically by the method of characteristics, were carefully fabricated and covered with a layer of plaster of Paris that furnished the necessary surface smoothness. The nozzle was enclosed on both sides by heavy glass plates for schlieren or interferometric visualization. The large Peenemünde facility {⇨Fig. 4.9–E} was designed on the model of the Aachen supersonic wind tunnel.
[Luftwissen **4**, 301 (1937)]

## 4.9 WIND TUNNELS – Continuous-Flow Closed-Circuit Supersonic Facility at ETH Zurich

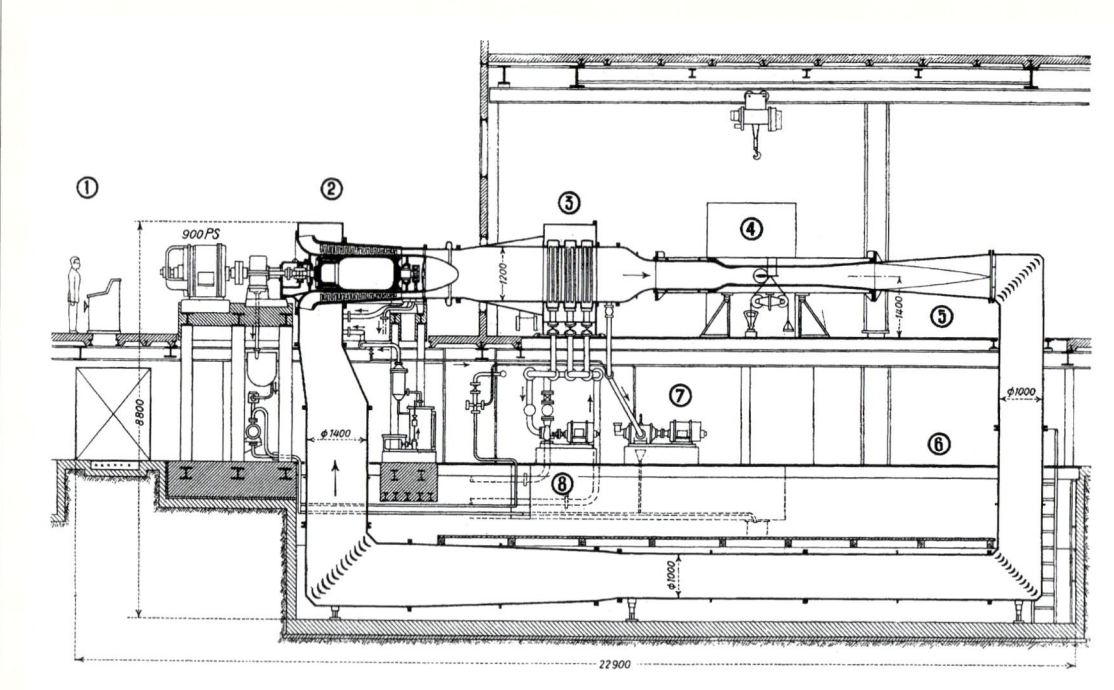

← View of open test section showing Laval nozzle geometry. Note that the flow direction is from left to right. The upper porthole is the field of view for schlieren, shadow, or interferometer visualization; the lower porthole is used to accept the reference beam of a Mach-Zehnder interferometer. [Courtesy Prof. Leonhard KLEISER, Institut für Fluiddynamik, ETH Zürich]

**4.9–D** *Top:* Schematic of Prof. Jakob ACKERET's famous continuously operated closed-circuit supersonic wind tunnel (1933), the first of its kind in the world, which allowed Mach numbers up to two and stationary flow visualization. The device allowed one to vary the Mach number and, independently, the Reynolds number. *1* – power house; *2* – axial fan with 13 steps ranging from 0 to 3,900 rpm; *3* – air cooler with 750,000 kcal/h; *4* – test section $40 \times 40$ cm$^2$; *5* – ground floor; *6* – basement; *7* – vacuum pump; and *8* – cooling water pump. [Interavia **1**, 1 (1946)] The facility, further developed, is still in operation at ETH Zurich. *Bottom:* View of Laval nozzle.

# 4 PICTURE GALLERY

## 4.9 WIND TUNNELS – Supersonic Intermittent Indraft Facility at Heeresversuchsanstalt Peenemünde

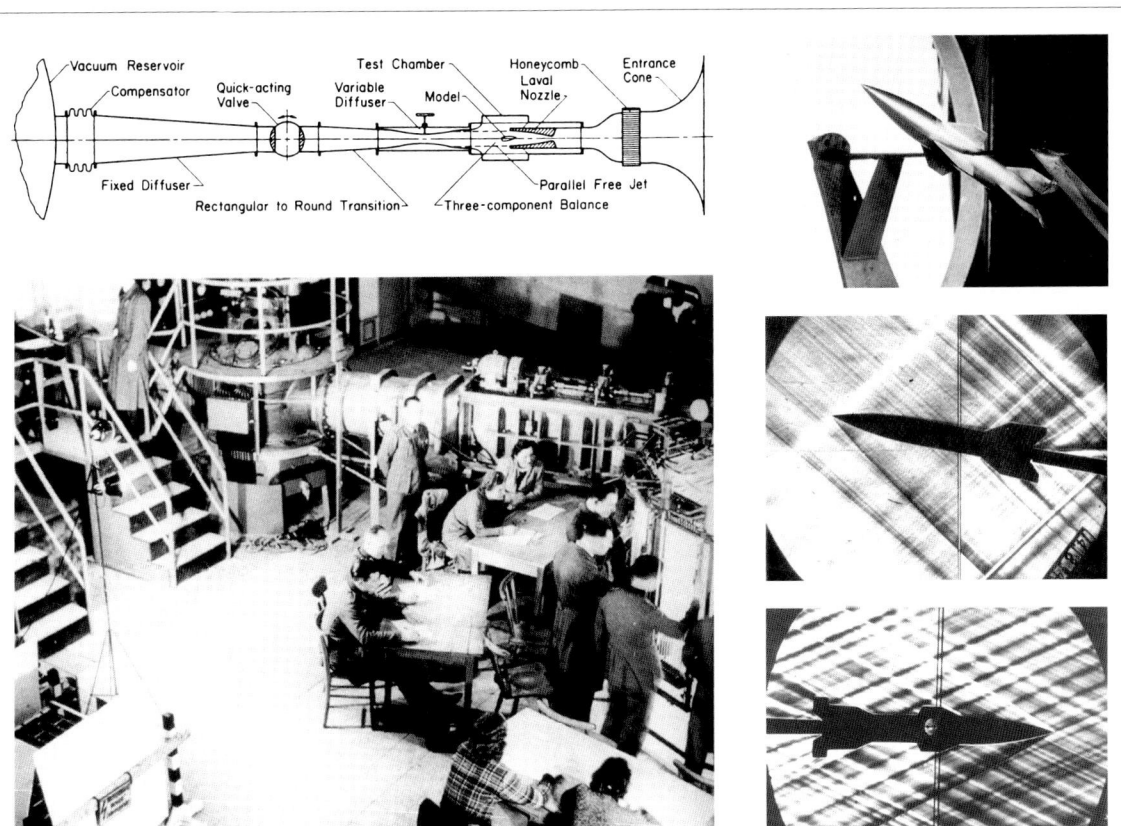

**4.9–E** *Left, top:* Schematic of supersonic intermittent indraft wind tunnel at the Heeresversuchsanstalt (HVA) Peenemünde, designed in 1937 by Dr. Rudolph HERMANN for model testing of rockets, gun-fired projectiles, and missiles up to velocities of Mach 5. It had a $40 \times 40$-cm$^2$ test section. A system of large rotary vacuum pumps was provided to evacuate a huge steel sphere with a diameter of more than 12 m. The blow-down operation was initiated by opening a special valve leading to the wind tunnel, which caused air to rush through the test section into the vacuum sphere throughout a running time of 20 to 25 s, depending on the chosen Mach number. A set of differently shaped Laval nozzles was used to obtain a supersonic flow, free of shocks, ranging from $M = 1.22$ to 5.18. Since Peenemünde is located on the Baltic Sea and has a humid climate, the air entering the wind tunnel from outside had to pass an extensive drying system at low speed (installed in 1940). This provided a shock-free flow up to $M = 5.2$. [Proc. 32nd Int. Astronautical Congr., Rome (1981). Pergamon, Oxford (1982), p. 436] *Left, bottom:* View of the test section of one of Peenemünde's wind tunnel testing facilities – then the most advanced in the world. Aerodynamic pressure loading at many locations along the surface of rocket models was measured with the help of mercury manometers attached to tiny tubes leading to small holes in the surface of the test model. To observe and record the gauge data during the short operation time of the wind tunnel, each gauge was assigned to a technician. The schlieren-optical system had two nearly parabolic mirrors, 50 cm in diameter each, with a focal length of 10 m. Schlieren pictures were taken with various types of high-speed cameras and light sources. Aerodynamic forces like drag, lift, and pitching were measured by a three-component balance. [P.P. WEGENER: *The Peenemünde wind tunnels*. Yale University Press, New Haven etc. (1996)] *Right:* One of the main tasks of HERMANN's group was the model study of the aerodynamic behavior of the supersonic rocket A-4 or "V2" (*center*), the first ballistic missile ($M_{max} > 4$), and the supersonic winged "Wasserfall" (*top & bottom*), an anti-aircraft missile ($M_{max} \approx 2$). The models were mounted on a rod and could be exposed under different yaw angles to the supersonic flow. [Courtesy Historisch-Technisches Informationszentrum Peenemünde]

## 4.9 WIND TUNNELS – Slotted Throat of Supersonic Facility at NACA, Hampton, VA

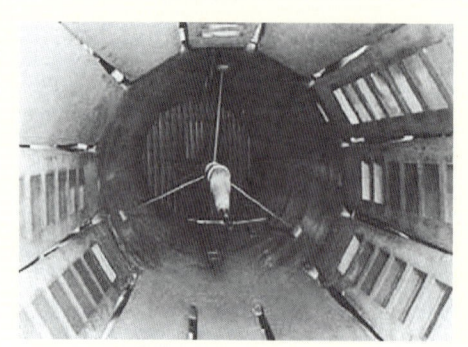

**4.9–F** To eliminate any net effects of the walls on wind tunnel test results, a research program was devised at NACA in the mid-1940s with the idea of absorbing shock waves by means of longitudinal openings, or slots, in the test section – a combination of opposite effects of open and closed walls. This, however, required an increasing power of the air compression plant. The slotted throat avoided the so-called "choking effect" on the achievable speed due to the presence of a test model and permitted a full spectrum of transonic flow studies, thus becoming a milestone in the evolution of high-speed wind tunnels. The first successful slotted tunnel was a 12-in. (30.48-cm) model tunnel with a maximum useful Mach number of 1.26, which was applied only for tunnel development (1947). The view of the NACA 8-ft (2.44-m) slotted throat shows the diffuser-entrance flaps. [NASA Rep. SP-445 (1980)]

## 4.9 WIND TUNNELS – Ludwieg Tube Facility at AVA, Göttingen

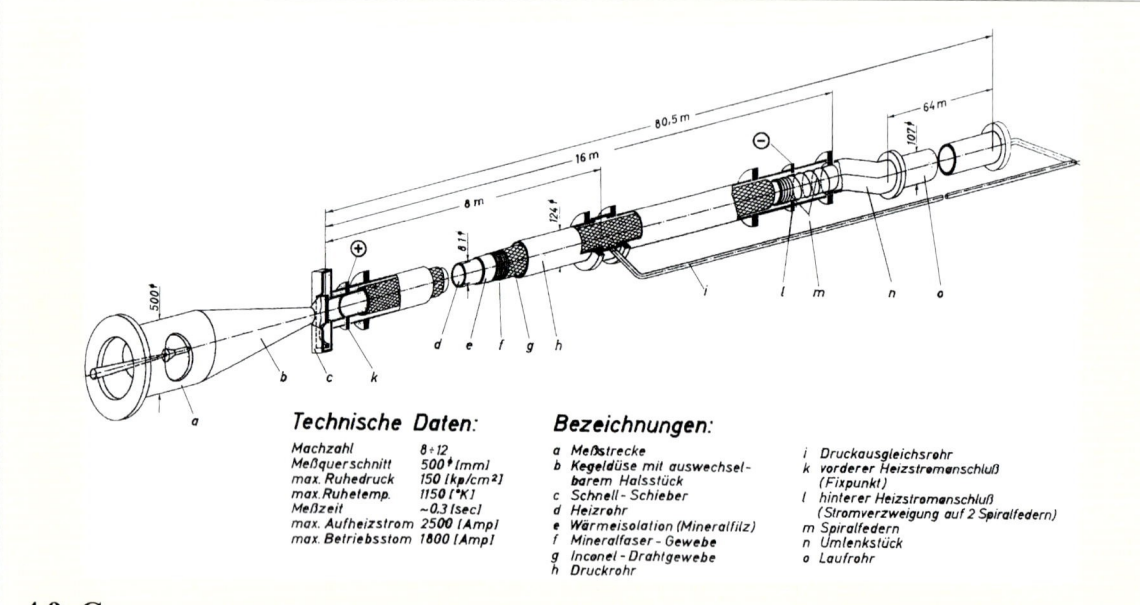

**Technische Daten:**

| | |
|---|---|
| Machzahl | 8÷12 |
| Meßquerschnitt | 500² [mm²] |
| max. Ruhedruck | 150 [kp/cm²] |
| max. Ruhetemp. | 1150 [°K] |
| Meßzeit | ~0.3 [sec] |
| max. Aufheizstrom | 2500 [Amp] |
| max. Betriebsstom | 1800 [Amp] |

**Bezeichnungen:**

- a Meßstrecke
- b Kegeldüse mit auswechselbarem Halsstück
- c Schnell-Schieber
- d Heizrohr
- e Wärmeisolation (Mineralfilz)
- f Mineralfaser-Gewebe
- g Inconel-Drahtgewebe
- h Druckrohr
- i Druckausgleichsrohr
- k vorderer Heizstromanschluß (Fixpunkt)
- l hinterer Heizstromanschluß (Stromverzweigung auf 2 Spiralfedern)
- m Spiralfedern
- n Umlenkstück
- o Laufrohr

**4.9–G** Schematic of a particular type of hypersonic wind tunnel – the so-called "Ludwieg tube" – at the Aerodynamische Versuchsanstalt (AVA) in Göttingen. The principle, invented in 1955 at AVA by the German fluid dynamicist Hubert LUDWIEG, allows a wide range of operation, ranging from Mach 3 to 12. The facility, which had a test cross section $a$ of $50 \times 50$ cm², had a pressure storage, designed as a pressure tube $h$, which was connected to test chamber $a$ via a fast-closing valve $c$. The inserted conical nozzle $b$ provided a parallel flow in $a$. [AVA Rep. 68A77, Göttingen (1968)]

## 4.9  WIND TUNNELS – First Hypervelocity Facilities

↑ ERDMANN used a plane nozzle with an extremely narrow slit at the throat, followed by two symmetrical, sharply opening nozzle walls enclosed in parallel plate-glass sidewalls.

**4.9–H** *Left:* First schlieren photo of a hypersonic flow close to $M = 9$, taken in 1943/1944 by Siegfried F. ERDMANN at Heeresversuchsanstalt (HVA) Peenemünde shortly before evacuation of the facility to Kochel in the Bavarian Alps. *Right:* Model of the hypersonic wind tunnel ($M = 10$) as planned in 1944 at the German Rocket Center. It was considered to operate this huge wind tunnel facility by the nearby 120-MW Walchensee Hydroelectric Power Plant. [P.P. WEGENER: *The Peenemünde wind tunnels. A memoir.* Yale University Press, New Haven, CT (1996)]

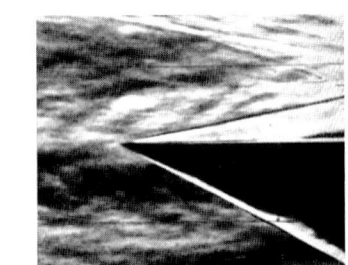

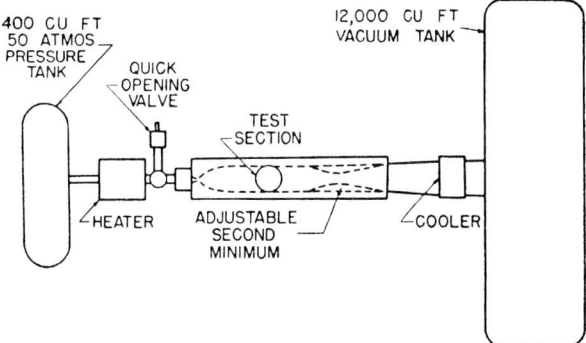

**4.9–I** After World War II, hypersonic wind tunnel studies were resumed first in the United States. Dr. John V. BECKER at Langley Aeronautical Laboratory built an 11-in. (279-mm) dia. hypersonic tunnel that first became operational in November 1947 – the first operation of a hypersonic tunnel in the United States. *Left, top:* Schlieren photograph of the boundary layer and shock on a flat surface aligned with the flow at $M = 6.9$. *Right:* Schematic, shown here with a conventional single step nozzle that produced uniform flow up to $M = 6.9$. [J. Appl. Phys. **21**, 619 (1950)] *Left, bottom:* View of NACA's 11-in. dia. hypersonic wind tunnel. An electric heater was incorporated in front of the test section to prevent liquefaction. [NASA SP-4305 (1987)]

## 4.10 SHOCK TUBES – VIEILLE's Pioneering Setup

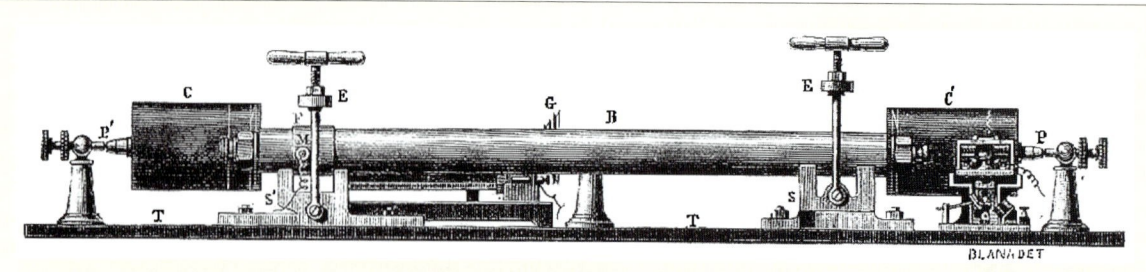

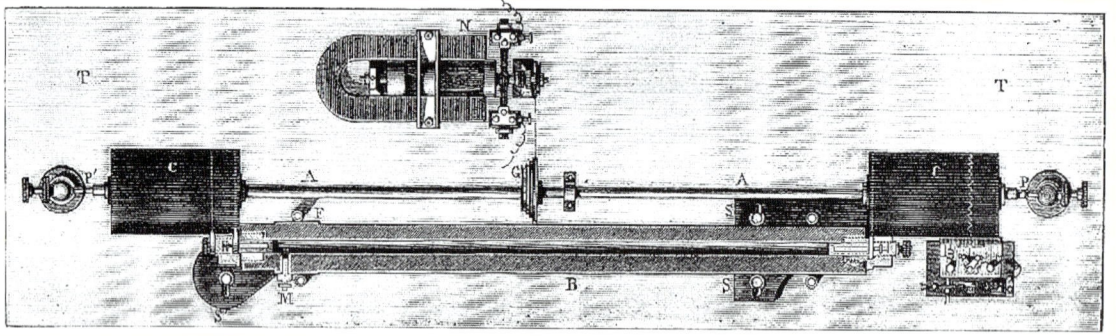

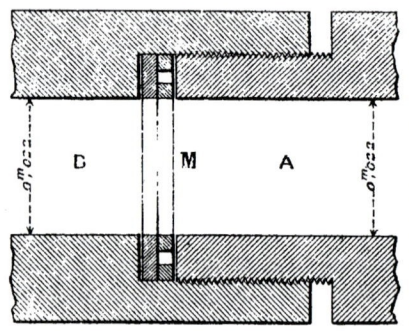

**4.10–A** *Top & center:* Side view and top plan view of the first "shock tube" invented in the late 1890s by the French physicist Paul VIEILLE at the Laboratoire Central of the Service des Poudes et Salpêtres in Paris. It allowed gas dynamicists to study in detail the propagation and reflection of shock waves. Its length ranged from 1 to 32 m. To measure pressure-time profiles at two different locations, two small pressure pistons, spring-loaded and positioned perpendicular along the tube axis, were applied. Their temporal displacements were recorded on two synchronously rotating, smoke-covered drums [Mém. Poudres Salpêtres **3**, 177 (1890)]. **Bottom:** Drawing of his membrane holder. The 22-mm-diameter membrane $M$ ruptured after reaching a critical burst pressure, which generated the shock wave. VIEILLE tested various membrane materials and measured corresponding burst pressures $p_R$, *e.g.*, for a 0.6- to 1.5-mm-thick glass plate: $p_R = 5-35$ bar; for a 0.27-mm-thick collodion foil: $p_R = 25-30$ bar; and for a sheet of paper: $p_R = 0.95-1$ bar [C. R. Acad. Sci. Paris **129**, 1228 (1899); Mém. Poudres Salpêtres **10**, 177 (1900)].

## 4.10 SHOCK TUBES – BLEAKNEY's Triggerable Shock Tube

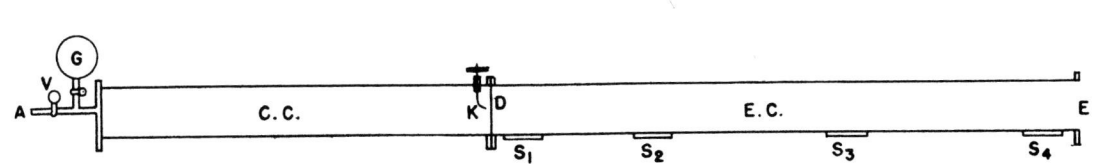

**4.10–B** During World War II, Walker BLEAKNEY, a U.S. professor of applied physics, and his team at the Shock Wave Laboratory of Princeton University (Princeton, NJ) took up and refined VIEILLE's shock tube technique – thereby also coining the term "shock tube." They introduced a sliding pin $K$ for rupturing the diaphragm $D$, which allowed for an easier triggering of attached diagnostic instrumentation. Their pioneering setup basically consisted of a compression chamber $CC$ and an expansion chamber $EC$. Air was pumped into $CC$ through the inlet $A$, pressure read by gauge $G$, valve $V$ closed, and then knife $K$ pushed into diaphragm $D$. The shock wave, propagating down the tube $EC$, was monitored by pressure gauges $S_1...S_4$ positioned sideways along the tube wall before escaping through the open end $E$ of the shock tube. [Rept. OSRD No. 1519 (1943), Princeton University Station]

## 4.10 SHOCK TUBES – Special Types

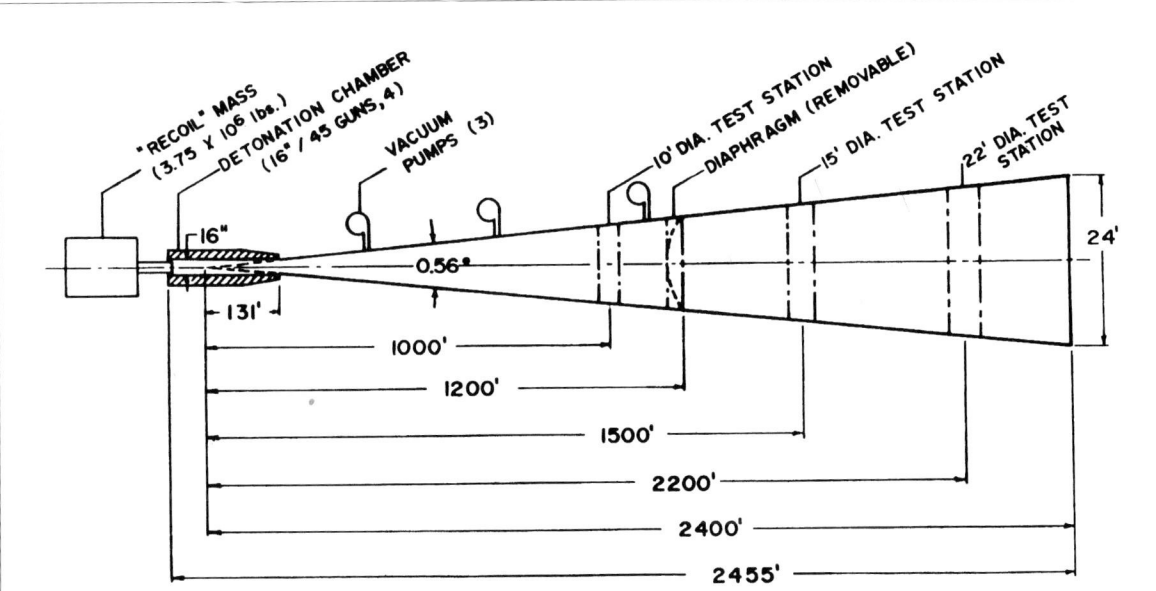

**4.10–C** Based upon the "Filler tube" (1960), a divergent conical blast simulator driven by a high explosive charge at the apex, the U.S. engineer D.W. CULBERTSON and collaborators at the Naval Weapons Laboratory (Dahlgreen, VA) constructed a large-scale, explosive-driven, conical shock tube to simulate free-air blast waves from a nuclear explosion in the kiloton-yield range. Their simulator applying TNT in a detonation chamber was set up at Dahlgreen. [Proc. 7th Int. Shock Tube Symp., University of Toronto Press, Toronto (1970), p. 396]

## 4.10 SHOCK TUBES – Special Types *(cont'd)*

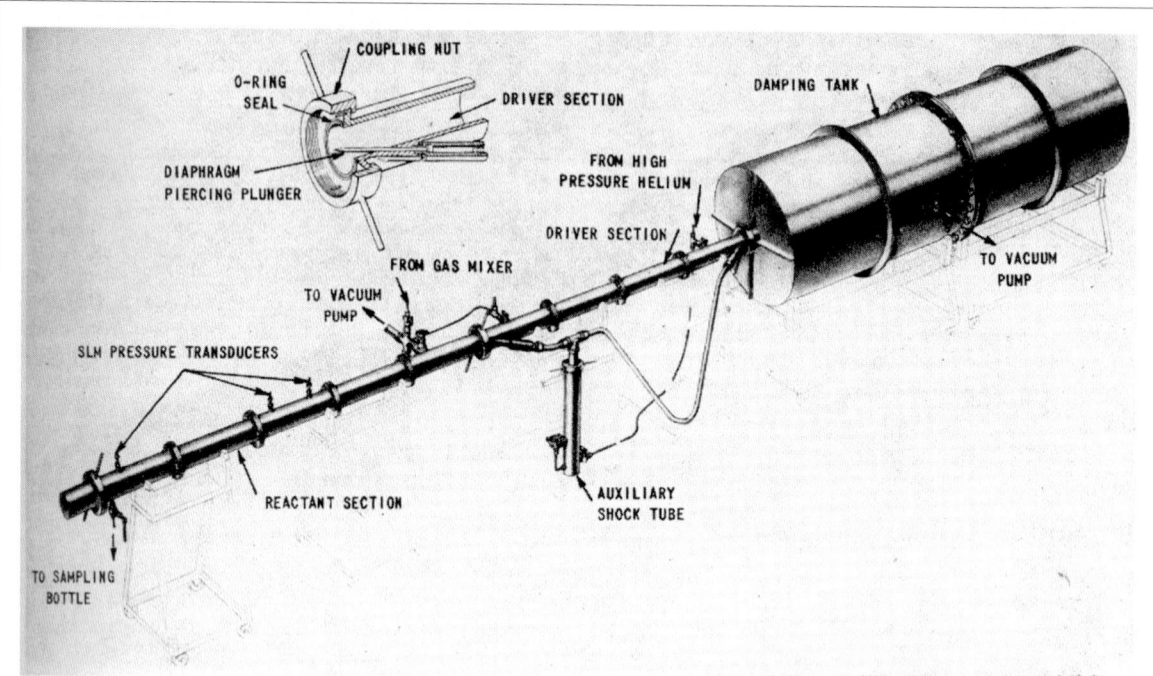

**4.10–D** Herbert S. GLICK and collaborators at the Aeronautical Laboratory of Cornell University (Ithaca, NY) designed the first single-pulse chemical shock tube. The reactant gas was first compressed by the incident shock. After being reflected at the closed end of the tube, the test gas was further compressed and heated, thus allowing gas dynamicists to process a reactant gas sample with a single closely controlled, high temperature pulse in the order of several milliseconds. [J. Chem. Phys. **27**, 850 (1957)]

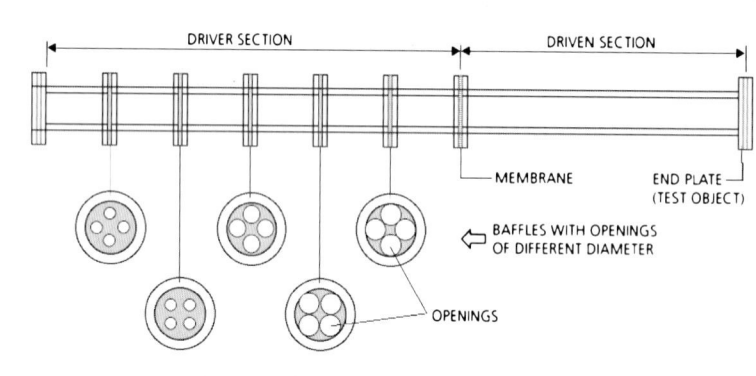

**4.10–E** In the 1960s, Hubert SCHARDIN and Heinz REICHENBACH at EMI, Freiburg designed an "equal-pressure shock tube" for testing objects subjected to a pressure jump of a long, even infinitely long, duration [Germ. Patent No. 1,273,850 (1968)]. It differed from a common shock tube in that the driver section contained a series of baffles for attenuation of transient shock and rarefaction waves. The test object was placed in front of the end plate, or served to hermetically terminate the end of the shock tube. [EMI-Archives, Freiburg]

## 4.10 SHOCK TUBES – Special Types *(cont'd)*

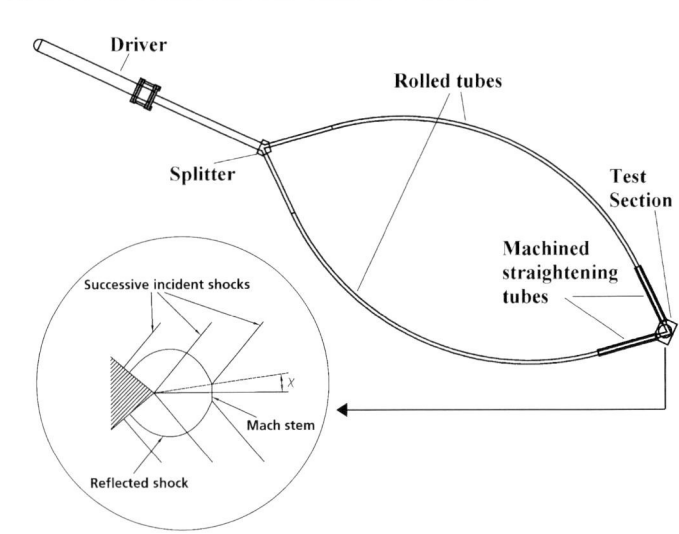

**4.10–F** Beric W. SKEWS, a gas dynamics professor at the University of Witwatersrand (Johannesburg, South Africa) and collaborators constructed a unique bifurcated shock tube to study the interaction of two-plane shock waves. An initially plane shock wave is split symmetrically into two equal shock waves and then recombined at the trailing edge of a wedge. The plane of symmetry acts as an ideal rigid wall, thus eliminating any thermal and viscous boundary layer effects. [Courtesy Prof. B.W. SKEWS; Proc. 22nd Int. Congr. on High-Speed Photography and Photonics, Santa Fe, NM (1996). SPIE vol. 2869, SPIE, Bellingham, WA (1997), p. 623]

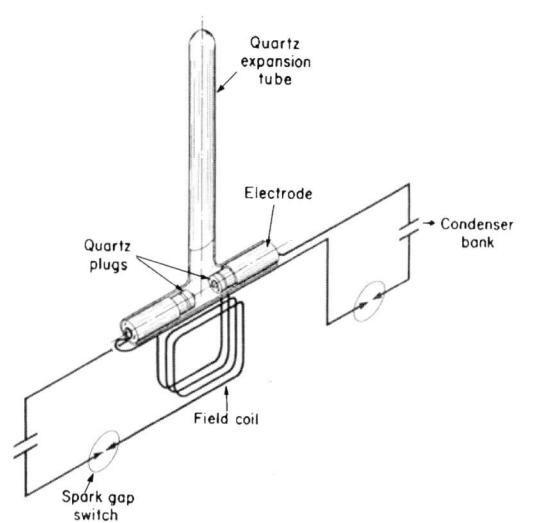

**4.10–G** A.C. KOLB at the U.S. Naval Research Laboratory (Washington, DC) built a T-tube arrangement for electromagnetic generation of very strong shock waves up to Mach numbers > 100 and temperatures on the order of 100,000 K. [3rd Lockheed Symp. on Magnetohydrodynamics, Palo Alto, CA (1956). Stanford University Press (1958), p. 76]

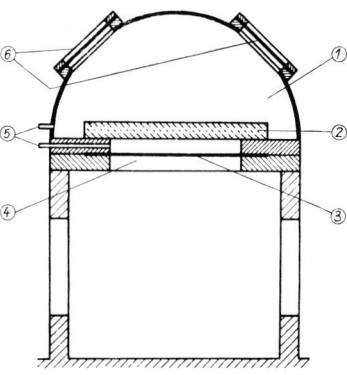

**4.10–H** In 1960, Hubert SCHARDIN and Manfred FROBÖSE at ISL (Saint-Louis, France) designed a blast simulator for full-scale testing of structural elements (*e.g.*, doors, windows, wall panels), consisting of a hemispherical pressure vessel *1* that contained the test structure *2* at its bottom and a lower chamber *4*, both being separated by a membrane *3*. The test structure was shock-loaded by suddenly destroying the membrane *3*. The shape of the pressure-time profile could be additionally controlled by destroying sequentially further membranes *6*. [ISL-Bericht 6/62 (1962)]

## 4.10 SHOCK TUBES – Special Types (cont'd)

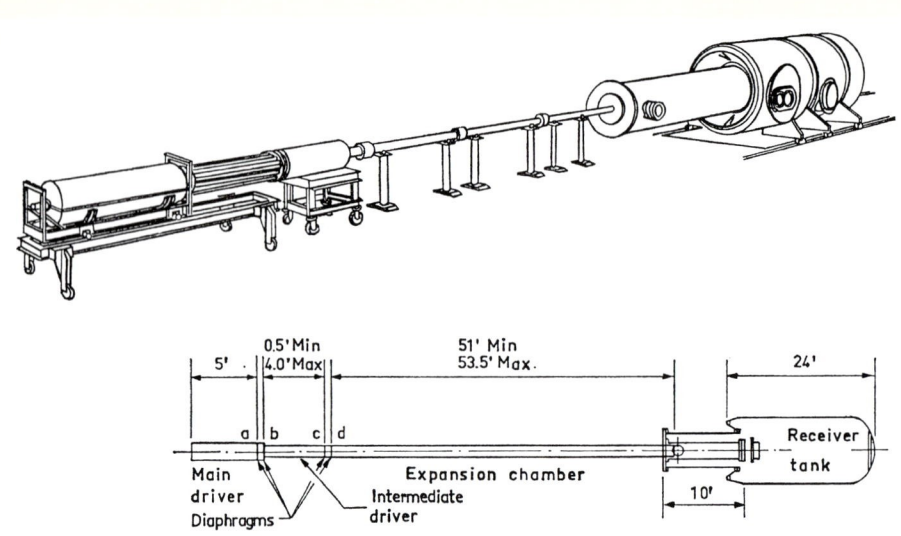

**4.10–I** Bo LEMBCKE, a Swiss mechanical engineer at MIT (Cambridge, MA) observed that when a shock tube configuration with a large area driver, separated from the shock tube by a diaphragm downstream of the area change, is used, two shock waves appear at the end of the shock tube. The double-driver self-timing shock tube at Cornell Aeronautical Laboratory, closely following this principle, applied a 5-in. (12.7-cm)-diameter main driver that could heat the driver gas (*e.g.*, helium or hydrogen at 2 kbar) up to a temperature of 700 K. The two shock waves reach the test section one after the other before entering the receiver tank. So-called "double-shock shock tubes" are useful for simulating blast loading effects on supersonically moving bodies. [AIAA J. **1**, 1417 (1963); FFA-Rept. 109, Aeronautical Research Institute of Sweden, Stockholm (1967)]

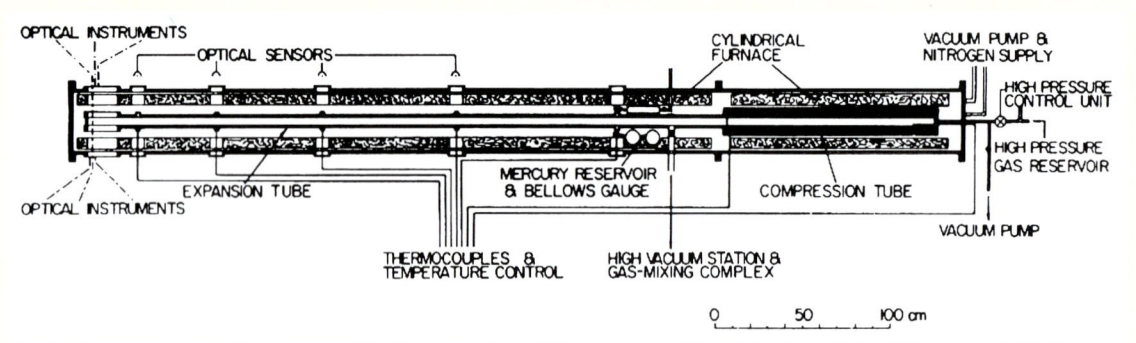

**4.10–J** In the 1960s, Yong W. KIM and Otto LAPORTE, two shock physicists at the University of Michigan (Ann Arbor, MI), constructed a heated shock tube as a spectroscopic source for studying metallic vapors at high temperatures. Their "mercury shock tube" consisted of three major parts: the main shock tube, a cylindrical furnace surrounding it, and a large outer tube within which both were contained. It provided a thermodynamically well-defined discontinuity followed by a uniform flow at high temperature. The shock-tube temperature could be raised up to 400 °C. [Phys. Fluids **12**, I:61 (1969)]

## 4.10 SHOCK TUBES – Special Types *(cont'd)*

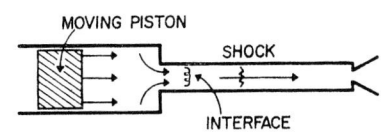

**4.10–K** Ray J. STALKER, an aeronautical engineering professor at the Australian National University (Canberra) proposed his "free-piston shock tunnel" – later known as the "Stalker tube." It uses a heavy piston to achieve isentropic compression of a helium driver gas. The shock-tube diaphragm ruptures before the piston compression stroke is complete. The resulting very high shock speeds allow model testing at hypersonic speeds. With his prototype gun STALKER achieved a shock wave velocity of 5,334 m/s. [AIAA J. **3**, 1170 (1965)]

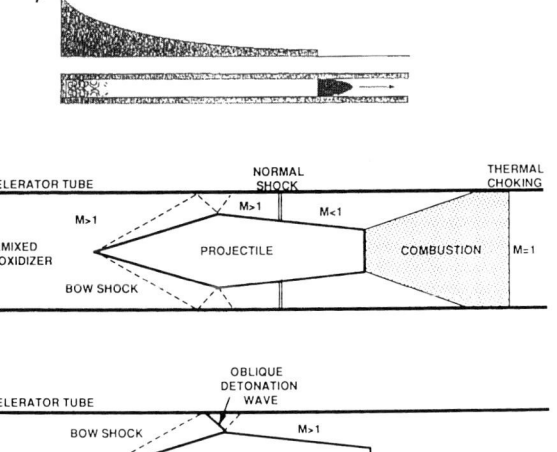

**4.10–L** *Top:* Principle of the ram accelerator (*right*) when compared to classical interior ballistics (*left*). In a ram accelerator the projectile, behaving like a surfboard riding on a wave propagating in a tube, is continuously accelerated by the combustion generated overpressure at its rear end. [G. SMEETS: IUTAM Symp. on Combustion in Supersonic Flows. Kluwer, Dordrecht (1997), p. 228] ***Center & bottom:*** The first ram accelerator, devised and built by Abraham HERTZBERG and collaborators at the University of Washington (Seattle, WA) used the chemical energy of a driving gas for accelerating projectiles to hypersonic velocities (up to 12 km/s). It consisted of a steel tube filled with a gaseous mixture of fuel, oxidizer, and diluent. The most popular gases were methane, oxygen, and nitrogen. To reduce the length of the tube, the mixture was pressurized. A projectile resting on a sabot was fired from a conventional powder gun into the ram accelerator. The projectile compressed the mixture to the point of ignition. Thrust was generated by the mixture expanding behind the projectile. The two schematics illustrate two different drive modes: subsonic combustion wave (*top*) and overdriven detonation wave (*bottom*), which enabled efficient acceleration to higher velocities. [Proc. 37th Meeting of Aeroballistic Range Association (ARA), Québec (1986). Defense Research Establishment Valcatier, Québec (1986)]

## 4.11 SHOCK WAVE GENERATION – Snapping Belts and Whip Cracking

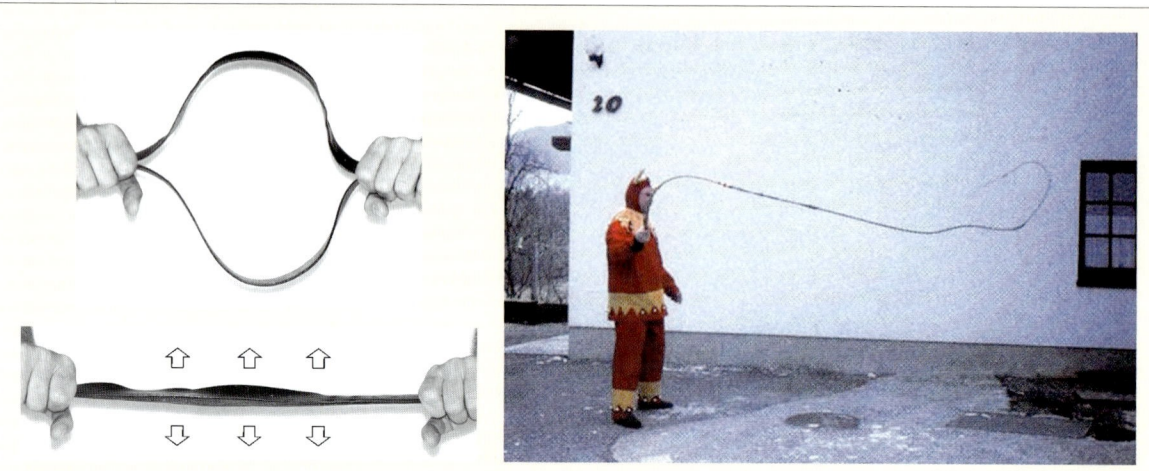

**4.11–A** *Left:* The "snapping belts" are a very primitive way of generating aerial waves of final amplitude: two belts' ends are firmly held together at both sides at the starting position (*top*) and then rapidly moved apart (*bottom*). This results in a peak pressure of about 130 dB ref. $2 \times 10^{-4}$ µbar. However, although the ear has the impression of a sharp report, it is in the strict sense not a shock wave because the pressure front does not have the typical steplike rise. [EMI-Archives, Freiburg] *Right:* View of a Black Forest whip cracker in his traditional costume at Shrovetide. The cracking sound of a whip is indeed a shock wave, albeit a weak one. [Shock Waves **8**, 1 (1998)]

## 4.11 SHOCK WAVE GENERATION – Plane-Wave Generators

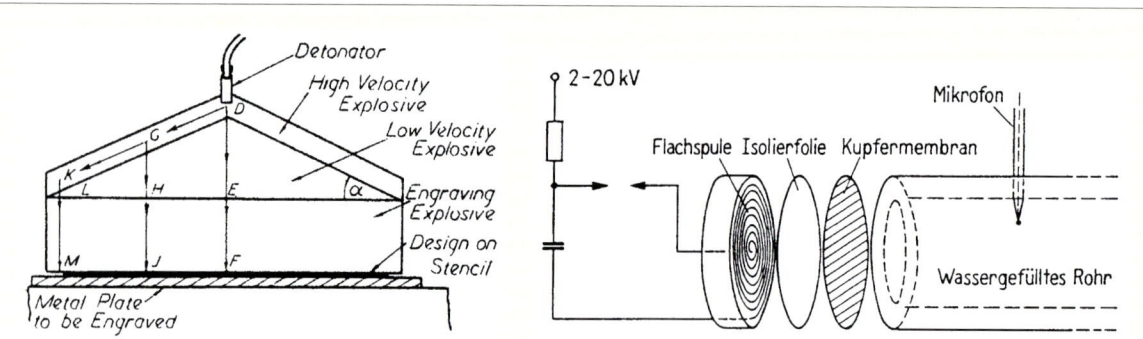

**4.11–B** Examples of plane-wave generators: *Left:* J.H. COOK at Imperial Chemical Industries (Stevenston, U.K.) used a novel explosive arrangement for engraving purposes. It applied a conical geometry assembled from explosives of different detonation velocities to tailor a plane detonation wave. His method of generating plane shock fronts was widely adopted in shock wave research. [Research **1**, 476 (1948)] *Right:* Wolfgang EISENMENGER, a German physics professor at Göttingen University, used an electromagnetic shock wave generator that could produce plane shock fronts in liquids with a diameter of 5 cm and shock pressures up to 700 bar. Today this principle is used in shock lithotripters. [Acustica **12**, 185 (1962)]

# 4 PICTURE GALLERY

**4.11 SHOCK WAVE GENERATION** – Plane-Wave Generators *(cont'd)*

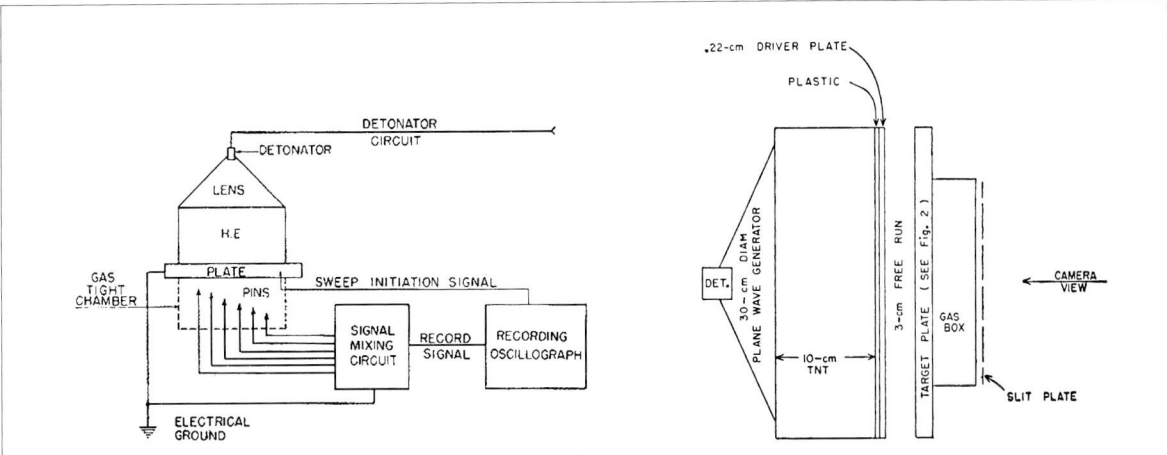

**4.11–C** *Left:* Roy W. GORANSON and collaborators at LASL (Los Alamos, NM) introduced the "electrical pin method" to measure free-surface velocities of shocked solid samples. Together with measured shock front velocities, it allowed for the determination of Hugoniot data. [J. Appl. Phys. **26**, 1475 (1955)] *Right:* Schematic of the "flyer plate method" as used by Robert G. McQUEEN and Stanley P. MARSH at LASL to determine Hugoniot curves of solids. The method applied an explosive plane-wave generator {⇨Fig. 4.11–B} to accelerate a thin plate that, upon impact, generated an intense plane shock wave in the test specimen. [J. Appl. Phys. **31**, 1253 (1960)]

**4.11 SHOCK WAVE GENERATION** – Gun-Type High-Velocity Accelerators

Behind piston *P* traveling in a bore *H* of a cylinder *D* is a conventional powder chamber *C*, loaded with nitrocellulose propellant.

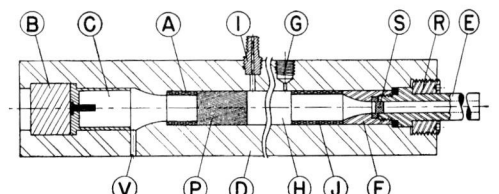

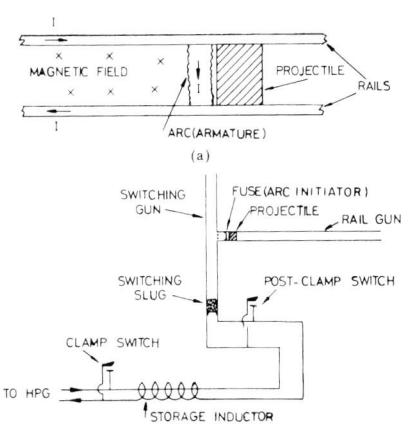

**4.11–D** Schematics of high-velocity guns. *Left:* The "rail gun" was invented in France by André FAUCHON-VILLÉPLÉE (1916). In Germany, Joachim HÄNSLER (1944) built the first practical device, his objective was to use rail guns in the battlefield. The method was later taken up by Scott C. RASHLEIGH and Richard A. MARSHALL. Their setup shown here could accelerate 2.5 g up to 5,900 m/s. *Right:* William D. CROZIER and William HUME (Socorro, NM) first proposed a "light gas gun," using a column of hydrogen or helium instead of conventional powder gas. [J. Appl. Phys. **28**, 892 (1957), Ibid. **49**, 2540 (1978)]

## 4.11 SHOCK WAVE GENERATION – Other Methods

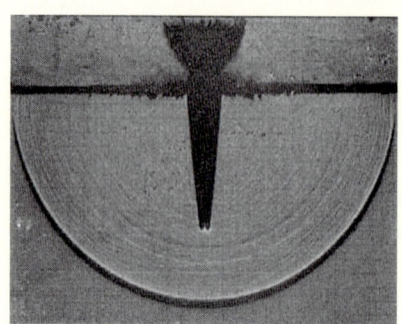

**4.11–E** J. Howard MCMILLEN and E. Newton HARVEY at Princeton University (Princeton, NJ) studied "water impact waves" generated by fast moving objects striking a water surface. Using flash shadowgraphy they visualized the shock wave produced by a vertical shot of a $^4/_{32}$-in. (3.2-mm)-dia. steel sphere and noticed that the wave front geometry was not hemispherical but rather ellipsoidal: since the water pressure was higher in shot direction, the wave propagated there supersonically but in perpendicular direction only sonically. [J. Appl. Phys. **17**, 541(1946)]

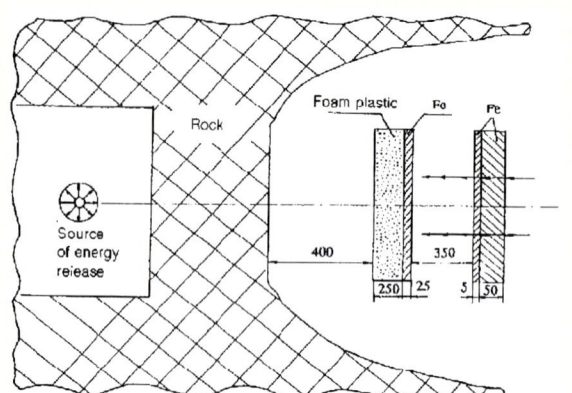

**4.11–F** Ryurik F. TRUNIN and collaborators at VNIIEF (Sarov, Russia) determined the shock compressibility of iron up to 100 Mbar using the speed of flight of an iron striker that was set off with the help of the energy of an underground nuclear explosion. The mean velocity of the shock wave in the target was measured using a system of electric-contact sensors. The mass velocity was close to half the velocity of the striker iron plate, which was provided at its front side with a foam to absorb the neutron flux of the explosion and to drive the iron striker by its explosion products. [Sov. Phys. JETP **76**, 1095 (1993)]

## 4.11 SHOCK WAVE GENERATION – Laser-Induced Spark

**4.11–G** Russell G. MEYERAND JR. and collaborators at United Aircraft Research Laboratories (East Hartford, CT) studied the interaction of extremely high-intensity laser radiation with air. *Left:* A laser pulse from a Q-switched ruby laser, focused by a simple lens in air at atmospheric pressure – a so-called "laser spark" – produced a bright flash, resulting in the emission of a blast wave, causing a sharp report. *Right:* The breakdown that developed in an elongated, egg-shaped plasma region showed a curious structure: it was rounded on the left side, the direction from which the laser beam was incident, and lobed on the right or downstream side.
[Proc. 6th Int. Conf. on Ionization Phenomena in Gases. S.E.R.M., Paris (1963), vol. II, p. 479]

## 4.11 SHOCK WAVE GENERATION – Laser-Supported Detonation (LSD)

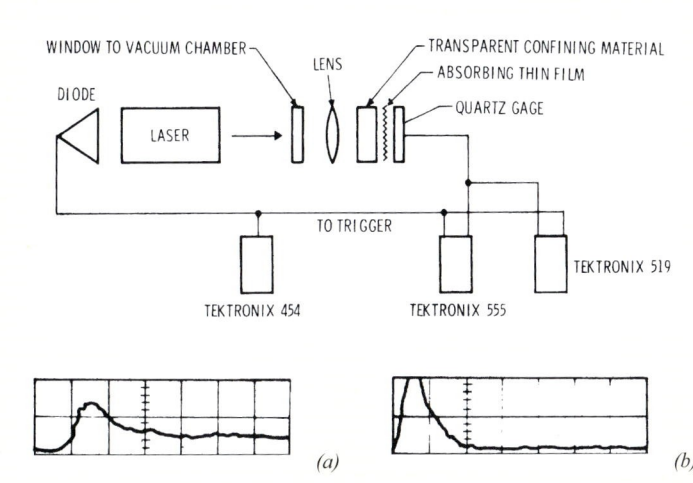

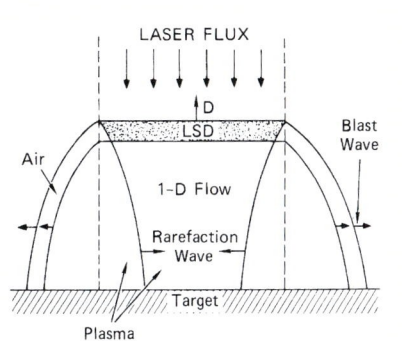

**4.11–H** N.C. ANDERHOLM at Sandia Laboratories (Albuquerque, NM) reported on a method of generating ultrashort stress pulses of 34 kbar using a pulsed ruby laser (7 J in 12 ns). *Top:* Schematic of experimental setup. The laser beam is focused into a 6-mm-dia. region. A transparent material is used to impede the expansion of the vaporized absorber. *Bottom:* The high-resolution stress pulse *(a)* was recorded with a Sandia-type quartz gauge (horiz. scale 20 ns/div); its rise time was nominally the laser pulse width *(b)*. [Appl. Phys. Lett. **16**, 113 (1970)]

**4.11–I** When matter is irradiated by an ultrashort high-intensity laser pulse, it provokes a "laser-supported detonation" (LSD) wave. Samuel HOLMES at SRI International (Menlo Park, CA) proposed a detonation model for the plasma zone. Using ytterbium pressure gauges, he measured the surface pressures produced by a pulsed 200-J $CO_2$-laser beam on an Al target to be around 140 bar. [*Shock waves in condensed matter – 1983.* North-Holland, Amsterdam (1984), p. 339]

## 4.11 SHOCK WAVE GENERATION – Laser-Propelled "Lightcraft"

**4.11–J** In 1987, Leik MYRABO, a U.S. professor of engineering at Rensselaer Polytechnic Institute (Troy, NY) had the idea of a "lightcraft" – a vehicle driven by a train of high-power laser pulses. Experiments using a 10-kW $CO_2$ laser were carried out at the USAF Research Laboratory (Edwards AFB, CA). *Left:* The optical surface of the lightcraft was used to focus a horizontal laser beam into the rear of the vehicle, where it was absorbed by the air inside the engine, thus creating a laser-supported detonation. This picture shows the plasma induced in the 14-cm-dia., 50-g Al lightcraft model from a single laser pulse. *Right:* Superposition of four pictures illustrating the starting phase. A high-pressure, high-temperature plasma is used to create the thrust that propels the lightcraft into the sky. In October 2000, a spin-stabilized lightcraft reached a record height of 71 m. [AIAA 98-1001 (1998)]

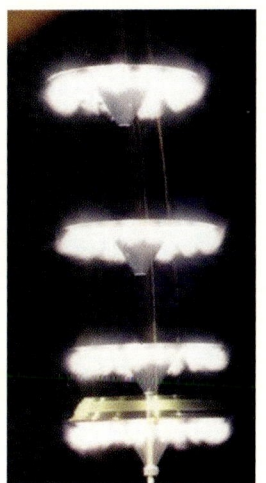

## 4.12 SHOCK FRONT ANALYSIS – In Gaseous Matter

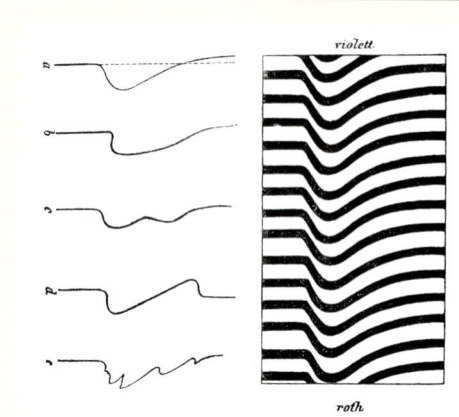

**4.12–A** Ernst MACH and J. VON WELTRUBSKY (Charles University, Prague), using a Jamin interferometer and a spark light source, observed the shifting of diffraction lines behind a blast wave that they generated by repeatedly operating an electric spark discharge. They visualized interferograms of the blast wave at different time instants (*right*) and then determined the corresponding density profiles (*left*). Note that the interferograms are drawings of what they observed and not photographic records. High-sensitive film was not yet available. [Sitzungsber. Akad. Wiss. Wien **78** (II), 551 (1878)]

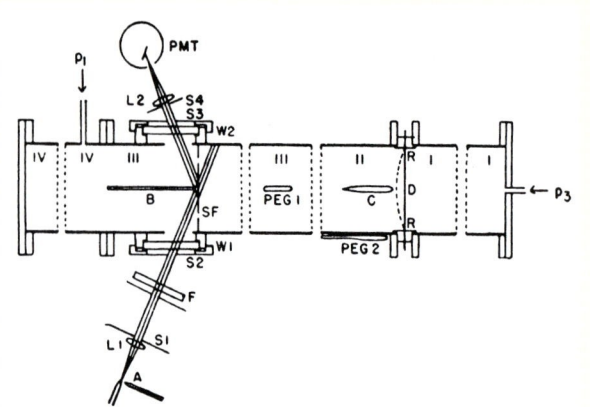

**4.12–B** George R. COWAN and Donald F. HORNIG at Brown University (Providence, RI) tackled the difficult task of measuring the shock front thickness in gases. Using an optical reflection method, they obliquely directed an intense collimated beam of nearly monochromatic light onto the shock front *SF* and recorded the weakly reflected light (reflectivity $4 \times 10^{-6}$ in $N_2$ at $M = 1.3$) by a photomultiplier tube *PMT*. The main problem was the detection of the weakly reflected light against a background of scattered light. They found the shock front thickness to be significantly greater than hitherto predicted theoretically based on the Navier-Stokes equations. [J. Chem. Phys. **18**, 1008 (1950)]

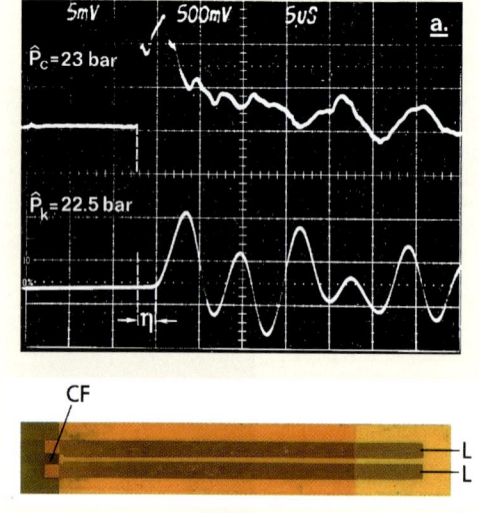

**4.12–C** In gas dynamics, piezoelectric gauges are widely used to record shock-pressure profiles. However, their rise time is not short enough to truly record the steep rise at the shock front. On the contrary, piezoresistive gauges have rise times of < 100 ns and since the early 1960s have been advantageously applied to high pressures in solid-state shock wave physics. To extend the use of piezoelectric gauges to very low pressure levels, P. KREHL (EMI, Freiburg) used a pulsed, double-compensated Wheatstone bridge in order to compensate for the enormous offset originated from the slightest changes in temperature. To demonstrate the feasibility of piezoresistive carbon gauges for laboratory studies, head-on and side-on shock-wave-collision experiments were carried out in a shock tube. *Top:* This picture compares head-on pressure-time profiles recorded with a carbon gauge (*upper trace*) and a piezoelectric gauge (Kistler model 603 E, *lower trace*). Obviously, a piezoelectric gauge has a much larger rise time and requires more time to reach stress equilibrium. *Bottom:* View of a thin 50-Ohm piezoresistive carbon shock pressure gauge, Dynasen model 15-300-E. The sensitive element, a $1.5 \times 1.5$-$mm^2$-sized carbon film *CF* manufactured by Dynasen Inc. (Goleta, CA) is connected to two electrical leads *L* that are encapsulated between two layers of thin Kapton foils. [Rev. Sci. Instrum. **49**, 1477 (1978)]

## 4.12 SHOCK FRONT ANALYSIS – In Space

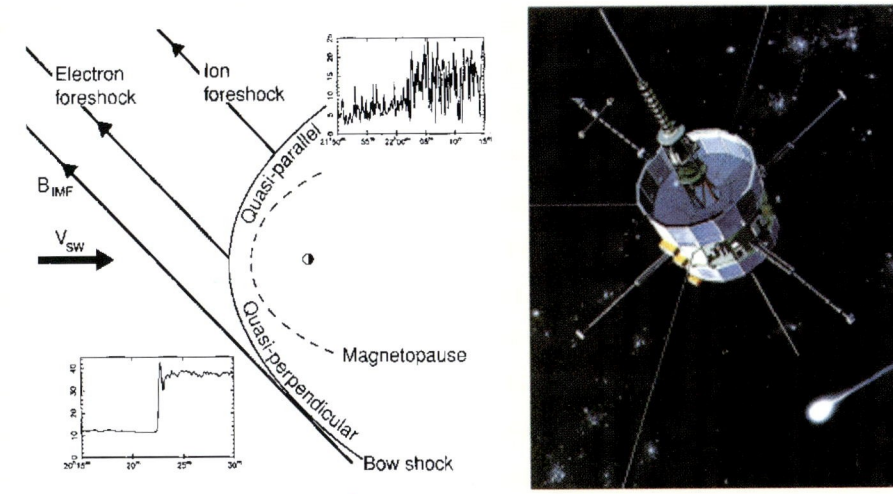

← An artist rendering showing the 469-kg ISEE space probe in orbit. This Explorer-class heliocentric satellite was used in the ISEE program to investigate the structure of the solar wind near the Earth and the shock wave that forms the interface between the solar wind and the Earth's magnetosphere. [Courtesy NSSDC and NASA]

**4.12–D** *Left:* Schematic of Earth's bow shock in equatorial plane as given by David BURGESS. Since in space plasmas the mean free path between collisions is very large, plasma processes, related to the changes in electric and magnetic fields, govern the width and internal structure of the shock layer. The Earth's bow shock has a width of between roughly 100 km and 1,000 km, depending on the shock and plasma parameters. The direction of the upstream solar wind flow is indicated schematically by $V_{SW}$. Examples of different types of shock crossing – *e.g.*, the "quasi-perpendicular bow shock" and the "quasi-parallel bow shock" – are shown as time series of the magnetic field magnitude in nanoTesla (nT). [*Encyclopedia of astronomy and astrophysics*. Institute of Physics, Bristol (2001), vol. 2, p. 1565]
*Right:* View of ISEE spacecraft that took the magnetic profiles of the shock. [http://www.friends-partners.org/mwade/craft/isee.htm]

## 4.12 SHOCK FRONT ANALYSIS – In Solid Matter

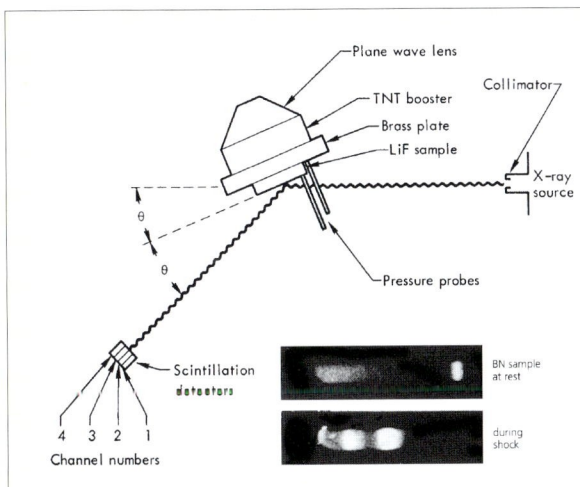

**4.12–E** To study the microcrystalline state of shock-compressed lithium fluoride (LiF) behind the shock front, Quintin JOHNSON and collaborators at LLNL (Livermore, CA) applied the flash X-ray diffraction technique. Their experimental setup is shown here schematically. The test sample was compressed using a COOK-type, planar shock wave generator {⇒ Fig. 4.11–B}. Either film or a detector array was applied to record the shock-induced shift of diffraction lines, which allowed for the estimation of the compression of the unit cell. [Phys. Rev. Lett. **25**, 1099 (1970)] In another study, using shock waves to compress boron nitride (BN) up to 245 kbar, they obtained the first evidence of a shock-induced phase transition that showed up in the origin of new lines in the flash X-ray diffraction pattern. [Phys. Rev. Lett. **29**, 1369 (1972)]

## 4.12 SHOCK FRONT ANALYSIS – In Solid Matter *(cont'd)*

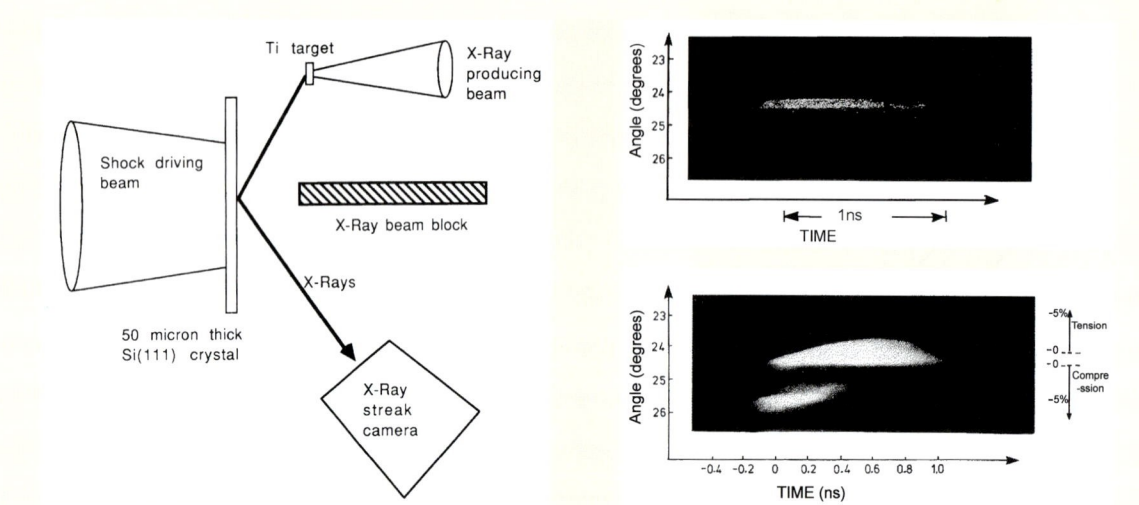

**4.12–F** Justin S. WARK at Oxford University, together with researchers at Imperial College (London, U.K.) and the Naval Research Laboratory (Washington, DC) measured the compressive and tensile strain during breakout of a 100-kbar shock wave which was launched into a 50-μm-thick single crystal of silicon <111>. *Left:* As shown in the schematic of their experimental setup, they used a 100-J 1.05-μm laser pulse which they split into two arms. One arm was focused to a small focal spot generating the shock wave; the second laser beam, synchronous but delayed with respect to the shock driving beam, was focused to a tight spot on a separate target, thus acting as an intense line X-ray source emitting at 23.6 Å. *Right:* Changes in interatomic spaces were deduced from a resultant shift in Bragg angle. The upper trace is the diffraction from the crystal at rest, the lower trace from the shocked crystal.
[J. Appl. Phys. **68**, 4531 (1990); Proc. 18th Int. Symposium on Shock Waves, Springer, Berlin (1992), p. 393]

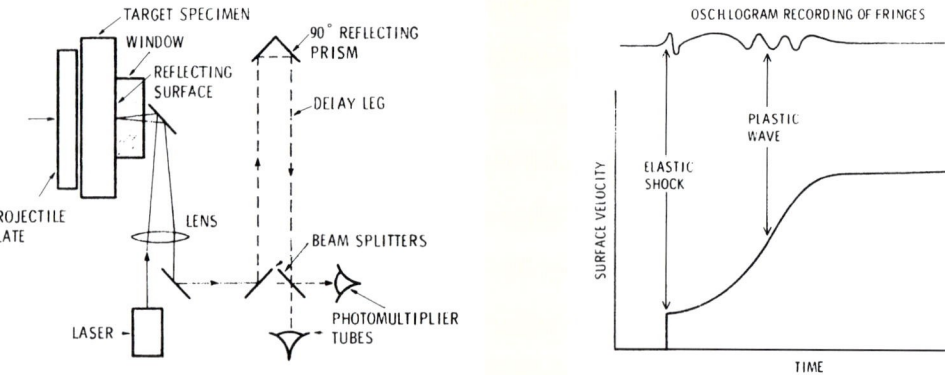

**4.12–G** James N. JOHNSON and Lynn M. BARKER at Sandia Laboratories (Albuquerque, NM) first used laser interferometry to measure the free rear surface velocity *vs.* time of an impact-loaded metal plate. *Left:* Schematic of measurement system. *Right:* The evaluated velocity-time profile, here shown in aluminum, allows one to determine the thickness of the plastic wave and to compare it with the dislocation model of plastic flow. [J. Appl. Phys. **40**, 4321 (1969)]

## 4.12 SHOCK FRONT ANALYSIS – Hydraulic Jumps in Water

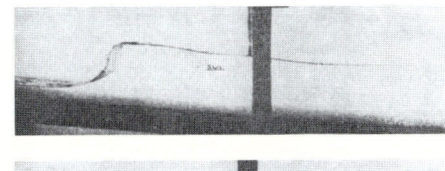

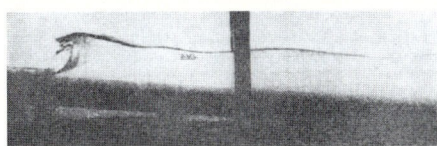

↑ Photographs of the breaking and formation of what looks like a jet (*bottom*) at the summit of the wave.

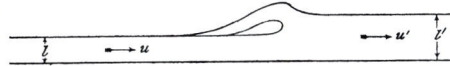

**4.12–H** Since the propagation velocity $v$ of a hydraulic jump in shallow water of depth $h$ is approximately given by $v = (gh)^{½}$ with $g$ as the gravitational acceleration, the wave crest in shallow water travels faster than the wave trough, and the front, leaning forward, generates a "breaker" or "roller". Obviously, the front of a water wave can never be as steep as in the case of a shock wave. ***Left, top & center:*** Propagation of a long wave in shallow water. [J.J. STOKER: *Water waves*. Interscience Publ., New York (1957), p. 373] ***Left, bottom:*** Sketch of the wave front of a tidal wave made by Lord RAYLEIGH. [Reprinted with permission from Proc. Roy. Soc. Lond. **A81**, 448 (1908). © 2007, The Royal Society] ***Right:*** Two tsunami logos (*left* and *center*) adopted by the 19th Session of ITSU (2003); and international tsunami hazard sign (*right*).

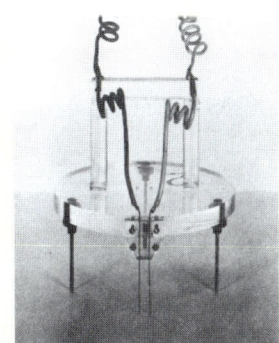

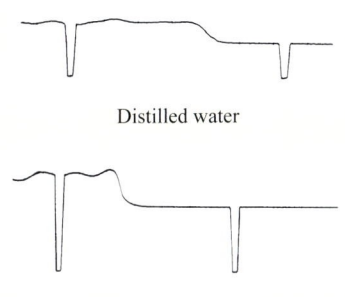

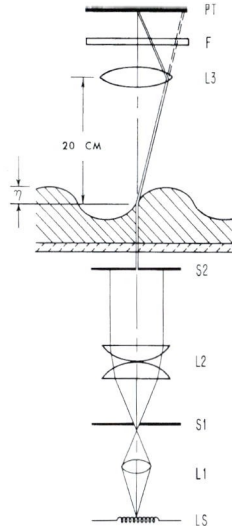

**4.12–I** In the 1940s, Harry E. CROSSLEY JR. at CalTech (Pasadena, CA) performed so-called "water table" studies of hydraulic jumps. ***Left:*** He used an electrolytic gauge for measuring the height of a propagating hydraulic jump. The change in the electrical resistance between two metal wires immersed in the working fluid served as a measure for the instantaneous water height. The short interruptions in the recorded wave profiles were due to an automatic current interrupter that broke the gauge circuit at intervals of approx. 0.6 s. ***Center:*** The wave profile of a hydraulic jump propagating in pure water was not as discontinuous as desired. CROSSLEY observed that by adding a detergent (such as isoquinolium bromide) to the distilled water the slope of the wave front substantially increased. [Hydrodynamics Laboraory Rept. N-54.1, CalTech (1949)] ***Right:*** Immersion gauges suffer from difficulties attributable to the erratic dynamic behavior of the meniscus and large disturbances around the gauge when fluid velocities become large. Bradford STURTEVANT at Harvard University (Cambridge, MA) developed an optical depth gauge with fast response time for laboratory studies of water waves using the absorption of infrared light by water. [Rev. Sci. Instrum. **37**, 1460 (1966)]

↑ Schematic of STURTEVANT's optical-depth gauge: A beam of light from a light source *LS*, collimated by slits *S1* and *S2*, is projected from below through the glass bottom of a water table. After passing through the water, the attenuated beam is focused by a lens *L3* through an infrared filter *F* onto an infrared-sensitive phototube *PT*.

## 4.13 MACH EFFECT – Interactions of Hydrodynamic Jumps

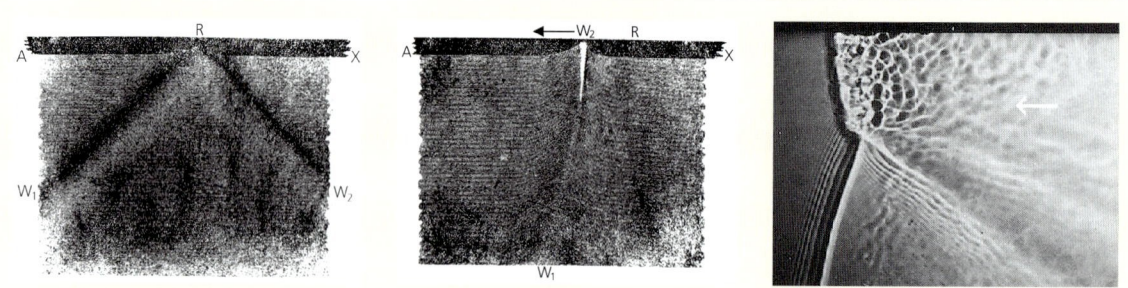

**4.13–A** Early studies of hydraulic jumps in shallow water. *Left:* In the period 1833–1840, the English engineer John Scotch RUSSELL studied the oblique reflection of solitary waves on a vertical plane surface immovable at R. For an incident water wave ($W_1$–R) with an angle less than 30° he observed regular reflection, the reflected wave ($W_2$–R) being equal in angle and quantity. Note that his pictures are copper-plate impressions and not photographic snapshots. They were prepared after drawings that he made immediately after observation. *Center:* The magnitude of the reflected wave diminishes as the angle of incidence increases. Then the velocity of the incident wave ($W_1$–R) increases near the wall and moves forward rapidly, thus forming a wave front with a high crest ($W_1$–$W_2$) at right angle to the resisting surface, also accompanied by the disappearance of the reflected wave. His observation of irregular reflection which he called a "lateral accumulation" is actually the "Mach effect," which Ernst MACH and Jaromir WOSYKA recovered in 1875 by studying the interaction of aerial weak shock waves originating from an electric gliding spark. [J.S. RUSSELL: *Report on waves (1833–1840)*. Rep. Meet. Brit. Assoc. **14**, 311 (1844)] *Right:* Reflection of a hydraulic jump at a solid boundary under an angle of about 75°. Very similar to RUSSELL's result (*center*), the Mach stem emanating almost perpendicularly from the wall is well pronounced, but the reflected wave diminishes in amplitude and at a large angle of incidence becomes barely visible. [Photo taken in 1990 by the author]

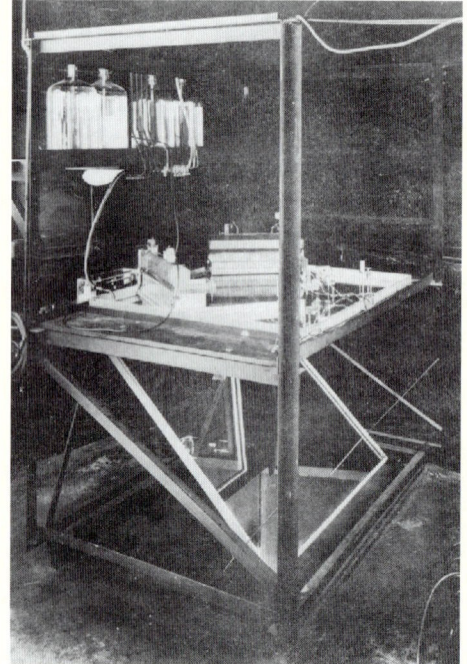

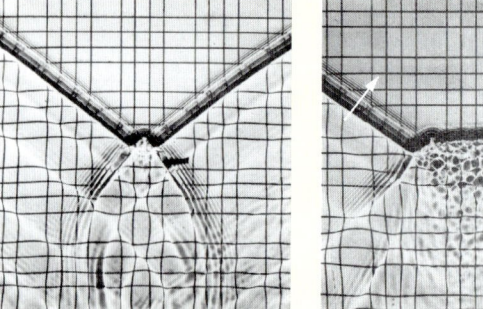

**4.13–B** *Left:* View of CalTech's ripple tank as used in the 1940s to model the interaction of hydraulic jumps. It consisted of a shallow glass-bottom tank; the hydraulic jumps or "surface shock waves" were produced by two 24-in. (61-cm)-long slit generators. Snapshot silhouette photography was used to visualize the wave interference phenomena. *Center & right:* Two shadowgraphs showing the interaction of two upward moving hydraulic jumps in shallow water under an angle of 56°, taken at two time instants $\tau$ after beginning of interaction. The strength of a hydraulic jump, defined as $\xi = (h_1/h_2)^2$, is analogous to the pressure ratio across a compression shock in a perfect gas with $\gamma = 2$. At $\tau = 0.45$ s (*center*), the region of interaction shows the beginning of Mach reflection, while at $\tau = 1.25$ s (*right*) the Mach disk has been clearly established. [H.E. CROSSLEY JR.: *Analogy between surface shock waves in a liquid and shocks in compressible gases.* CalTech Rept. N-54.1 (1949)].

### 4.13 MACH EFFECT – Shock Interactions in Gases

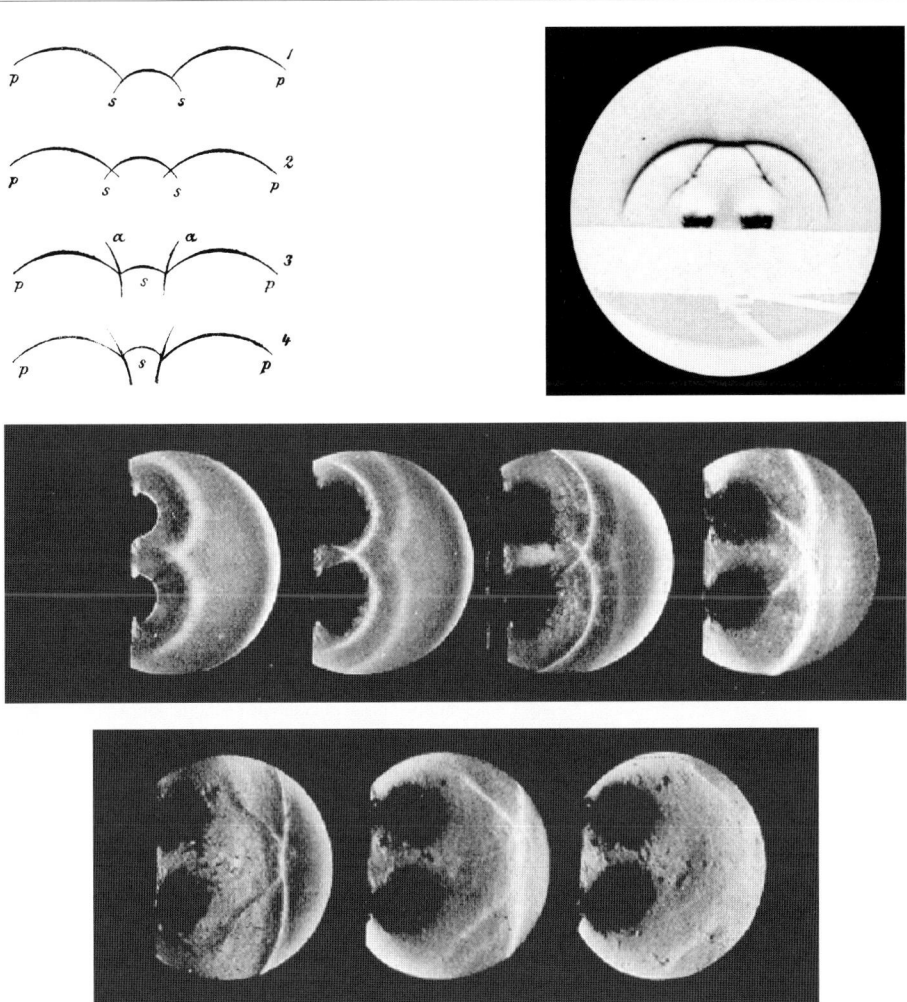

**4.13–C** Steps of disclosing the riddle of "Mach reflection." *Top, left:* Ernst MACH (German Charles University, Prague) studied the intersection of two spherical spark (weak shock) waves *p* in air. Since high-sensitivity photo plates were not yet available to him, he visualized the schlieren pictures subjectively. His schematic clearly shows that he correctly interpreted the newly formed secondary wave front *s–s* as the result of increased density at the point of interaction of the two primary waves *p p*. Note that the distance *s–s* increases in size with increasing time ($t_1 > t_2 > t_3 > t_4$). [Sitzungsber. Akad. Wiss. Wien **78** (II), 467 (1878)] *Top, right:* First published photo of "Mach reflection," taken by Ernst and Ludwig MACH. The two closely spaced point sparks emit shock waves that, obliquely interacting under a sufficiently large angle of incidence, create the so-called "Mach bridge" or "Mach disk." They also obtained quite similar results by simultaneously igniting two small quantities of a high explosive, thus proving that the effect is indeed of mechanical rather than of electric nature. [Sitzungsber. Akad. Wiss. Wien **98** (IIa), 1333 (1889)] *Center & bottom:* Shortly after their invention of the "multiple-spark camera," Carl CRANZ and Hubert SCHARDIN (Institut für Technische Physik, Berlin) made the first series of oblique shock interactions. This series of shadowgraphs showing the evolution of the Mach disk fully confirmed MACH's interpretation of the origin of this nonlinear superposition phenomenon. The aerial shock waves were generated by two small piles of detonating silver fulminate. The spherical waves, interacting with each other, formed the typical "Mach disk," here beginning in the fourth frame (*left* to *right*). [Z. Phys. **56**, 147 (1929)]

## 4.13 MACH EFFECT – Shock Interactions in Gases *(cont'd)*

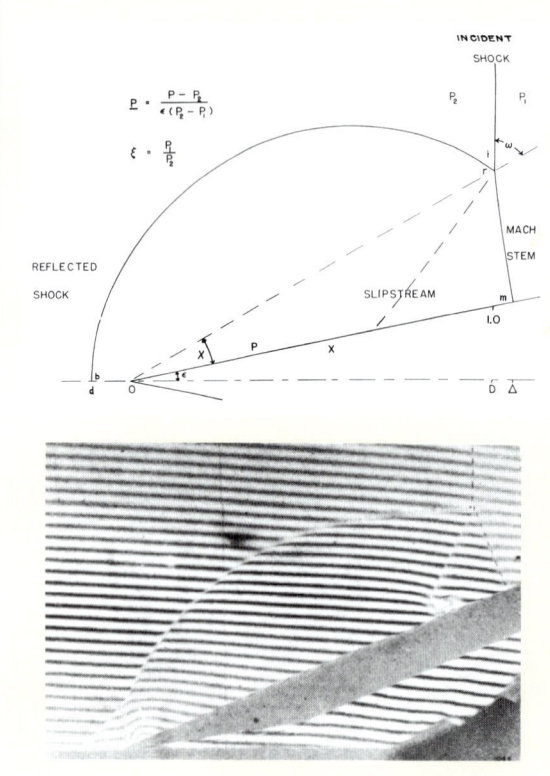
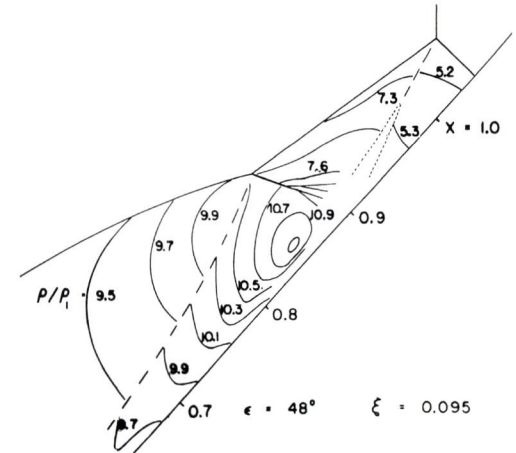

**4.13–D** *Left:* Schematic (*top*) and corresponding photograph (*bottom*) of "single Mach reflection," taken by Donald R. WHITE at Palmer Physical Laboratory of Princeton University (Princeton, NJ) using a shock tube and Mach-Zehnder interferometry. *Right:* WHITE's schematic of a strong shock wave interaction at a wedge, giving rise to a second slipstream and a second triple point. This unique shock interaction phenomenon was later called the "double Mach reflection." [D.R. WHITE: *An experimental survey of the Mach reflection of shock waves*. Tech. Rep. II-10, Dept. of Physics, Princeton Univ. (1951)]

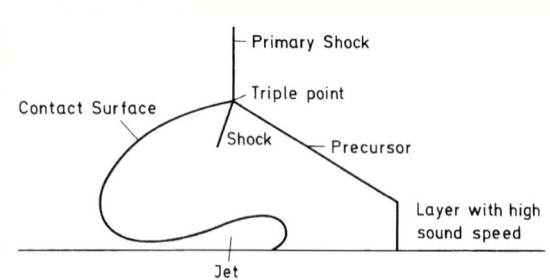
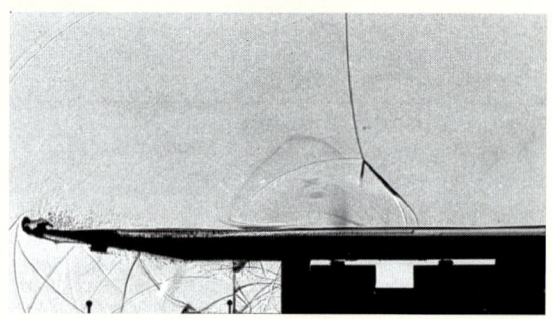

**4.13–E** Heinz REICHENBACH and collaborators (EMI, Freiburg) performed laboratory-scale air-blast precursor experiments in a shock tube. *Left:* The primary shock ($M = 1.66$), here moving from left to right over an electrically heated surface, propagates faster in the hot gas layer than in the rest of the atmosphere, thus creating a "precursor" wave that encounters the primary shock in the cool gas at an angle different from 180°. *Right:* At the same time, a complex three-shock configuration is created, similar to that created by classical Mach reflection. The contact surface is marked by a shear flow, which causes the cool gas to move towards the ground in a forward direction. [Rept. E22, EMI, Freiburg (1985)]

## 4.13 MACH EFFECT – Shock Interactions in Gases (cont'd)

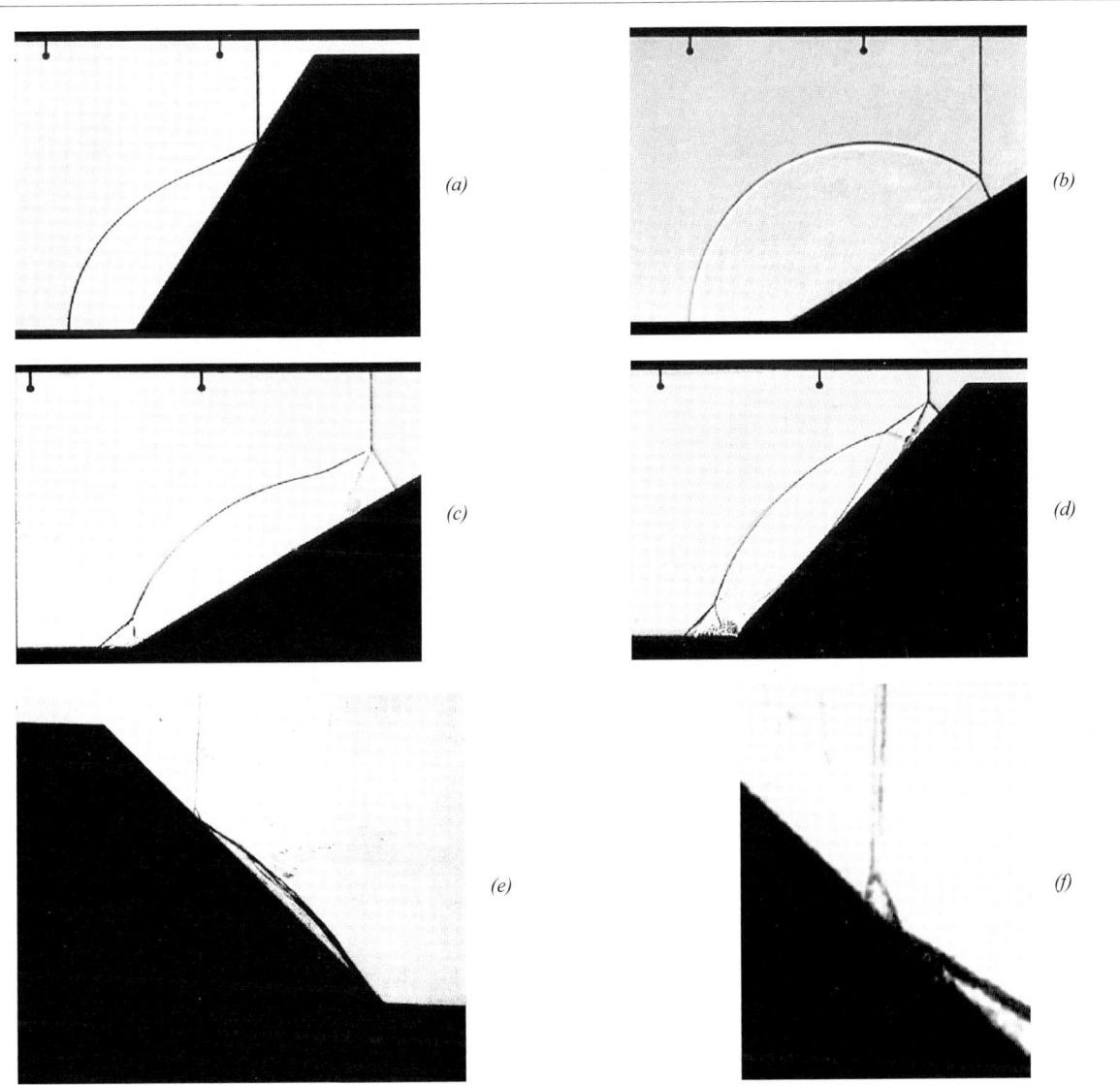

**4.13–F** Types of oblique shock wave reflections in a pseudo-stationary gasdynamic flow {⇒Fig. 2.14}, generated in a shock-tube arrangement and visualized using the shadowgraph method. *(a)* Regular reflection. *(b)* Single Mach reflection. *(c)* Complex Mach reflection. *(d)* Double Mach reflection. [Courtesy Dr. Werner HEILIG, FMI, Freiburg] *(e)* Terminal double-Mach reflection. *(f)* Enlargement of *(e)* showing the region of the two triple points in more detail. This unusual shock wave reflection configuration as shown here is only possible in low-gamma gases, e.g., in isobutane and sulfur hexafluoride ($\gamma \approx 1.094$ at room temperature). [J.T. URBANOWICZ, UTIAS Tech. Note No. 267, University of Toronto (1985)] Note that in *(a)–(d)* the incident shock wave propagates from left to right; in *(e)* and *(f)*, however, from right to left.

## 4.13 MACH EFFECT – Shock Interactions in Liquids

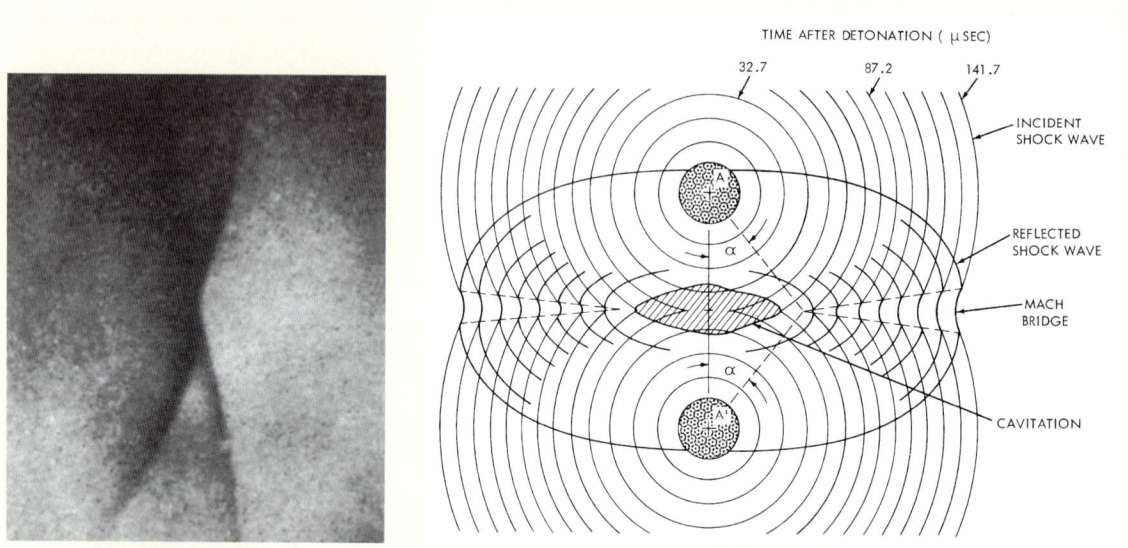

**4.13–G** *Left:* In 1943, Ralph W. SPITZER and Robert S. PRICE at the Underwater Explosives Research Laboratory (Woods Hole, MA) first confirmed John VON NEUMANN's assumption that the Mach reflection also exists in water. Using two spherical 50-g tetryl charges, 8.4 in. (21.3 cm) apart and fired simultaneously under water, they photographed the Mach intersection at a distance of 12 in. (30.5 cm) from the charges, where the fronts of the interacting shock waves formed an angle of 142°. However, at the low pressure levels they did not observe any slipstreams. [R.H. COLE: *Underwater explosions*. Dover, New York (1948), plate III] *Right:* Time profiles of collision of two equal spherically expanding underwater shock waves emerging from two equal charges $A$ and $A'$, and the formation of the so-called "Mach disk" (or "Mach bridge"). Note that a lens-shaped cavitation zone is created between both detonating charges.
[N.L. COLEBURN and L.A. ROSLUND: *Collision of spherical shock waves in water*. Rept. NOLTR 68-110, U.S. Naval Ordnance Laboratory, White Oak, MD (1968)]

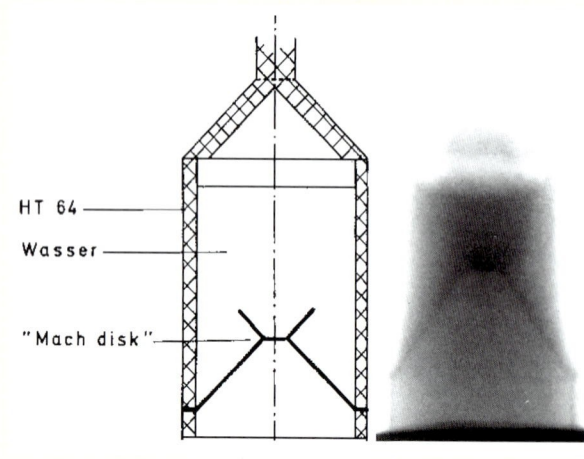

**4.13–H** Flash radiograph of Mach reflection in water taken by Klaus HOLLENBERG and Ferdinand MÜLLER (Physics Dept., University of Düsseldorf). The 40-mm-dia. water column is surrounded by a hollow cylinder of the high explosive HT64 (a mixture of hexogen and TNT), which has an outer diameter of 50 mm. After ignition at the top, the coaxial detonation wave generates a "Mach disk," which steadily moves in axial direction downwards with constant width. This simple configuration allows the generation of axial pressures in water up to 2 Mbar. Since the X-ray absorption of water is rather low, it is very difficult to obtain high-contrast flash radiographs, and a special "flash soft X-ray tube," operated at high current but low voltage ($< 30$ kV), is required to resolve the density jump at the shock front.
[Bericht 1/7, Arbeitsgruppe für Physik der Hochdruckplasmen und Impulsentladungen (AGD), Fraunhofer-Gesellschaft e.V. Physikalisches Institut I, Universität Düsseldorf (1971)]

## 4.13 MACH EFFECT – Shock Interactions in Liquids *(cont'd)*

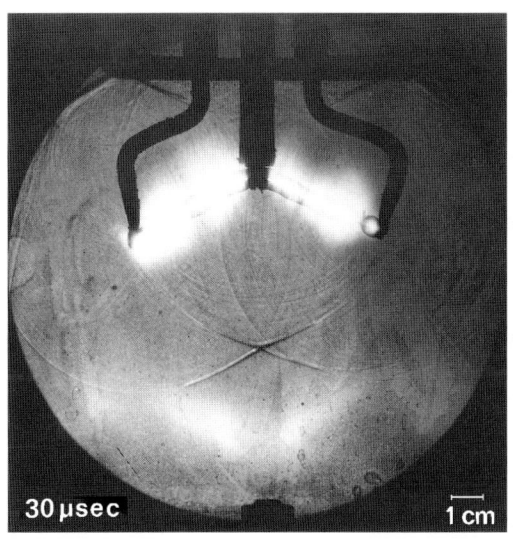
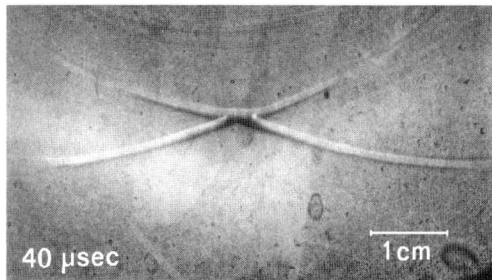
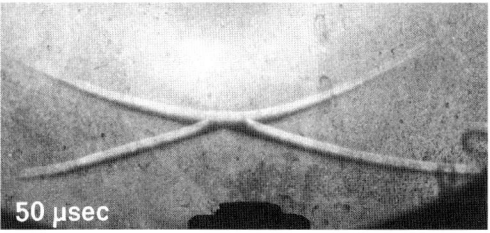

**4.13–I** *Left:* Optical snapshot shadowgraphy of two interacting cylindrical shock waves in water, generated by two 4-cm-long exploding Cu wires arranged in a "V-geometry" at an angle of 130°. At a distance of about 4 cm from the wire, the geometry of the shock waves begins to transform into a spherical one. This increases the angle of incidence – thus favoring Mach reflection, which in this example begins at about 30 μs. *Right:* At 40 μs *(top)* and 50 μs *(bottom)*, the Mach disk has more grown in width. [Photos by the author]

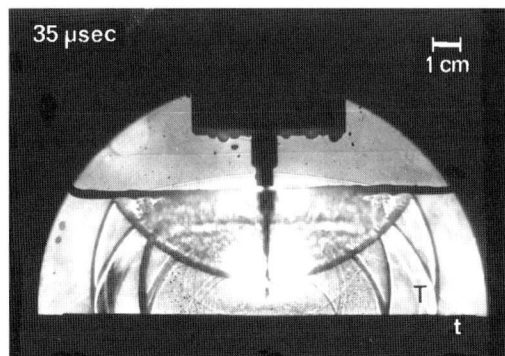

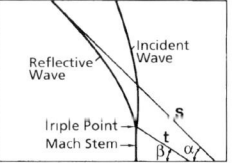

**4.13–J** *Top:* In the 1930s, the German physicist Oswald VON SCHMIDT at Humboldt University in Berlin demonstrated that head waves arise when a disturbance is generated at or near the interface of two media with different sound (or shock) wave propagation velocities {⇨Fig. 4.14–O}. This so-called "Schmidt head wave" (SHW) commonly encompasses a family of such head waves that emerge under different angles, $\alpha_L$, $\alpha_T$, and $\alpha_S$, according to the relation $\alpha_{L,T,S}$ = arc sin $u_1/(u_2)_{L,T,S}$. Here $u_1$ denotes the sound velocity in the upper (here liquid) half space; the velocities $(u_2)_{L,T,S}$ are, respectively, the longitudinal, transverse, and surface wave velocities in the lower (here solid) half space. *Bottom:* In the case of Mach reflection, a second family of SHWs is generated: at the transition from regular reflection into Mach reflection, a disturbance is produced that propagates along the interface of the two media, thereby according to HUYGENS' principle sending secondary waves into the upper half space and additionally generating a second family of SHWs under angles $\beta_L$, $\beta_T$, and $\beta_S$, which all run into the triple point. However, in the case of water as the upper medium ($u_W$ = 1,483 m/s at 20 °C) and lead (longitudinal velocity $u_L$ = 1,960 m/s, $u_T$ = 690 m/s) as the lower medium, only the longitudinal Schmidt head waves of both families can arise. [P. KREHL ET AL.: *Erzeugung von Mach-Reflexion bei schwachen Stoßwellen in Wasser.* Frühjahrstagung "Kurzzeitphysik," Verhandl. Dt. Phys. Gesellsch., Hannover (1976)]

## 4.13 MACH EFFECT – Shock Interactions in Solids

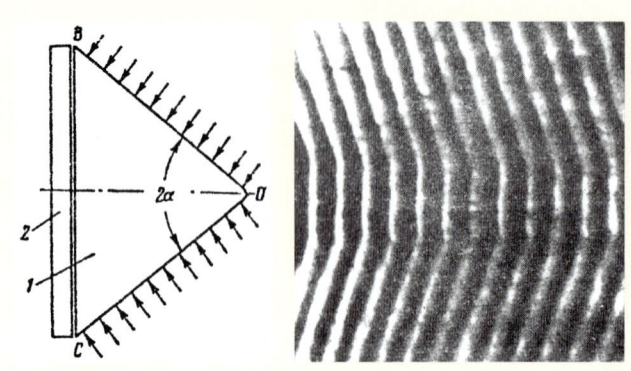

**4.13–K** In the former Soviet Union, E.A. FEOKTISTOVA first visualized Mach reflection in a detonating solid explosive. *Left:* The schematic shows the interaction of detonation waves emerging from the lateral prism faces *OB* and *OC* of a solid explosive *1* and being recorded through a Plexiglas window *2* in the plane of the prism base *BC* by a high-speed photochronograph. A series of slits, located at the focus of the chronograph optical system, were arranged perpendicular to the direction of the development of the interaction process. *Right:* The photochronogram, taken along the axis of symmetry, clearly shows the Mach reflection regime. [Sov. Phys. Dokl. **6**, 162 (1961)]

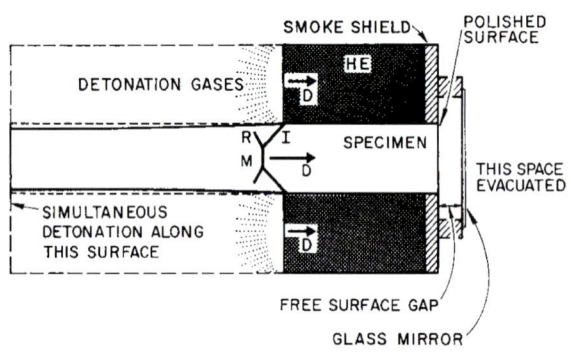

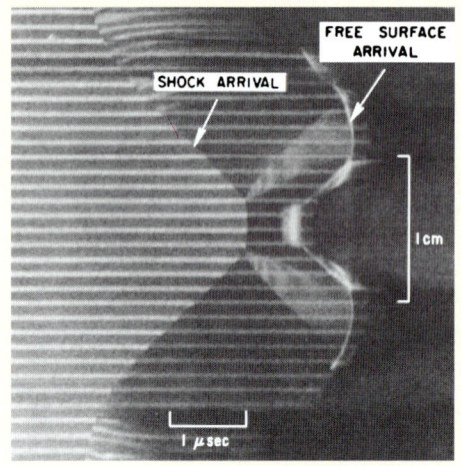

**4.13–L** In the mid-1960s, George R. FOWLES and William M. ISBELL at Pouter Laboratory of Stanford Research Institute (Menlo Park, CA) carried out optical studies on the interaction of shock waves in non-transparent solids. *Top:* At the example of a cylindrical copper sample, compressed up to 1.9 Mbar by a surrounding hollow cylinder of a high explosive, they visualized the shock interaction phenomena by a partially reflecting glass plate supported at a known distance from the surface. *Bottom:* Using an ultrahigh-speed streak camera, they recorded images of point light sources reflected in the glass and in the polished metal surface. A record of the shock and free-surface arrival shows a distinct region of Mach reflection that is accessible to Hugoniot equation-of-state measurements. The size of the Mach disk depends on the type of explosive employed and the specimen material. [J. Appl. Phys. **36**, 1377 (1965)]

## 4.13 MACH EFFECT – Shock Interactions in Solids *(cont'd)*

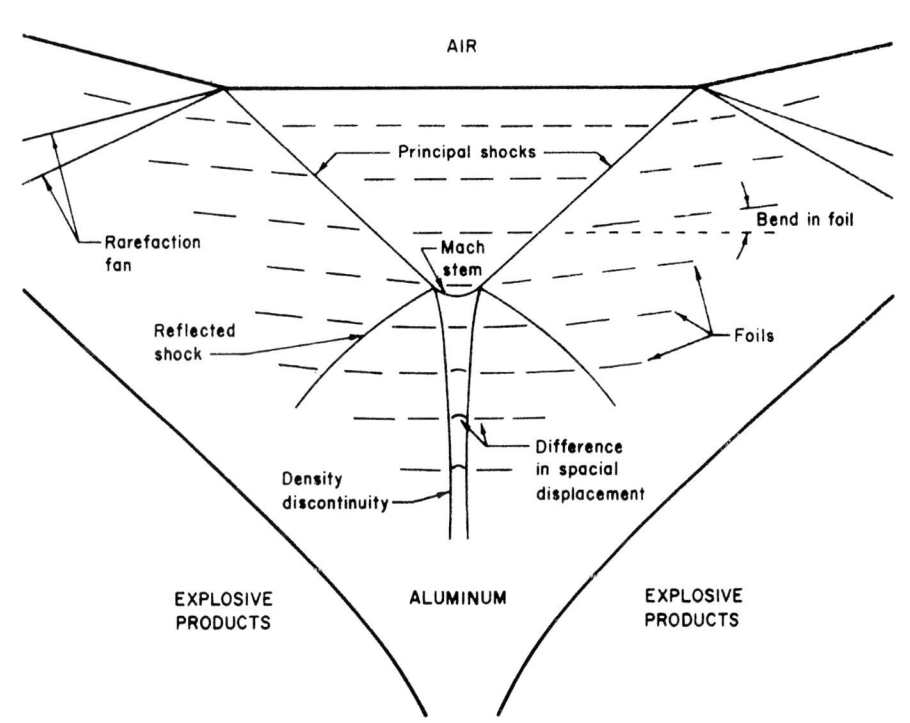

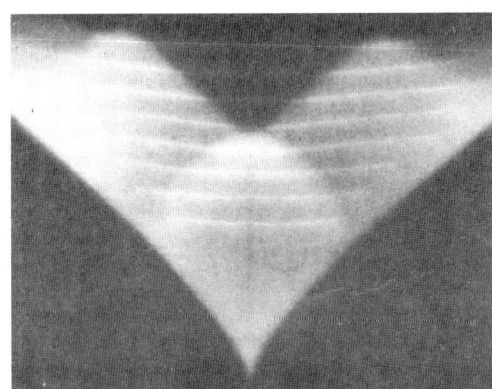

**4.13–M** Timothy R. NEAL at LASL (Los Alamos, NM) studied Mach reflection in shock-compressed aluminum using flash radiography. *Top:* To monitor the material flow and to provide a high radiographic contrast, he used a parallel, equidistant arrangement of 0.0127-mm-thick tantalum foils embedded in the aluminum test sample. By measuring the bents in the foils he determined the density behind the principal shocks from which he calculated a shock pressure of 277 kbar. From his Mach reflection model, which closely follows John VON NEUMANN's first proposed three-shock model (1943), he estimated a shock pressure behind the Mach disk of 887 kbar. *Bottom:* Example of a flash radiograph of two interacting shock waves. Driven by the explosive products of two columns of solid explosive (Composition B3), they obliquely interfere and produce Mach reflection. Note that the two reflected shocks and the slip lines, separating regions of the same pressure but different densities, are clearly visible. [J. Appl. Phys. **46**, 2521 (1975)]

## 4.14 SHOCK WAVE EFFECTS – Shock-Induced Creation of Prebiotic Substances

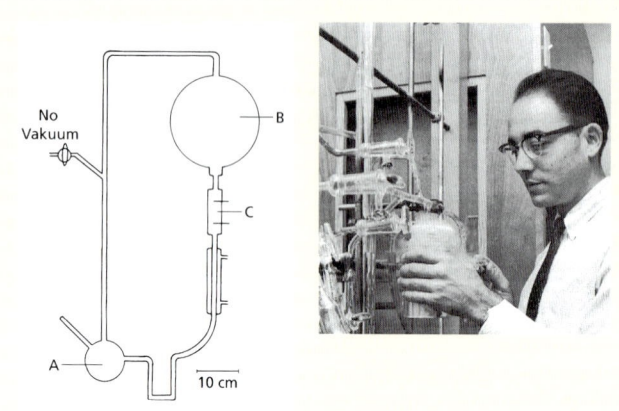

**4.14–A** In 1953, Stanley L. MILLER, a chemistry student of Prof. Harold UREY at the University of Chicago, devised a spectacular experiment producing amino acids by repetitively generating a spark discharge in a gaseous "atmosphere" of hydrogen, methane, ammonia, and purified water, thus simulating the influence of heat and shock on a primitive Earth atmosphere such as generated by lightning and thunder. **Left:** Schematic of his experimental setup. A steady stream of steam, generated in boiling flask *A* and mixed with the gases in flask *B*, passed the spark gap *C* fed by an induction coil, and returned to *A*. After about 1 week of operation, amino acids – essential building blocks of organic life – accumulated in the small flask *A*. [Science **117**, 528 (1953)] **Right:** This photo shows MILLER together with his famous so-called "Miller-Urey spark-discharge apparatus." [http://www.astro.virginia.edu/~eww6n/bios/Miller.html]

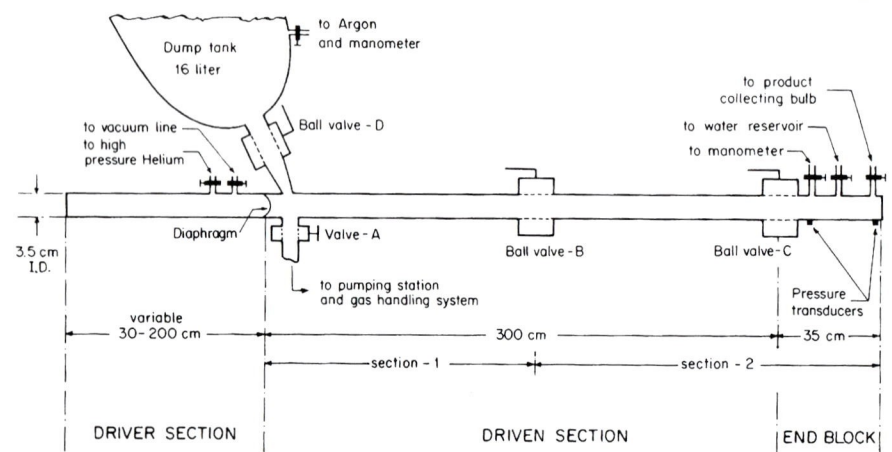

← Two different shock tubes of similar design were used: a 50-mm-inner-dia. Pyrex tube and a 35-mm-inner-dia. stain-less steel tube as shown here. Both driven sections, which could be heated to 80 °C, were pumped to $< 10^{-4}$ Torr, and water vapor and the reaction mixture were introduced into the entire driven section. The dwell times at the high temperature ranged from 200 to 500 µs.

**4.14–B** Akiva BAR-NUN at Hebrew University in Jerusalem synthesized amino acids behind high-temperature shock waves in methane, ethane, ammonia, and water. Aldehydes and HCN were formed separately during the short-duration high-temperature period, which recombined with ammonia to form α-amino nitriles. **Left:** Schematic of his single-pulse shock-tube technique. **Right:** His results obtained on the 50-mm-dia. shock tube. The reaction products were analyzed using paper, column, and gas chromatography. He concluded that thunder shock waves might be a suitable source of energy for the production of amino acids. [Origins Life **6**, 109 (1975)]

TABLE II
Summary of products analysis, µmoles amino acid[a]

| Run | Glycine | Alanine | Valine | Leucine | Isoleucine[b] |
|---|---|---|---|---|---|
| 2 | 0.98 | 0.08 | 0.06 | 0.22 | – |
| 3 | 50.50 | 27.50 | 10.20 | 9.50 | <0.01 |
| 5 | 121.30 | 60.10 | 4.40 | 12.60 | 0.30 |

[a] Calculated from the results of column chromatography and gas chromatography.
[b] Isoleucine was identified only by its elution time in column chromatography. In the gas chromatography and paper chromatography it was not separated from the leucine.

## 4.4   SHOCK WAVE EFFECTS – Cavitation

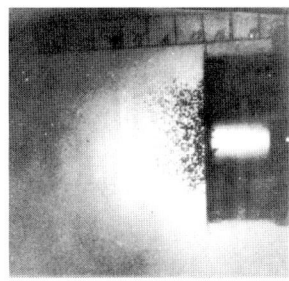
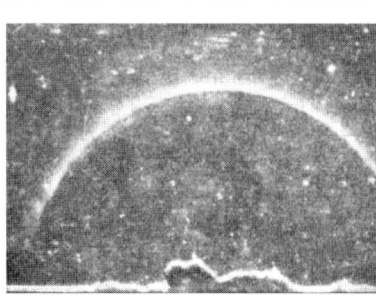
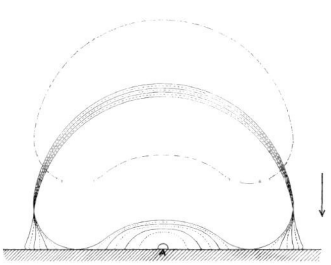

**4.14−C** *Left:* Stanley S. COOK (Brit. Committee of Erosion Research) observed that erosion at steel blade tips (*top*) of a turbine wheel only occurred at the moving blades, which he attributed to water hammer: drops of water, swept off the fixed blades into the path of the moving blades, impacted the blade (*bottom*). [Proc. Roy. Soc. Lond. **A119**, 481 (1928)] *Center:* In 1946, John E. ELDRIDGE and collaborators at the U.S. Navy Bureau of Ordnance (Washington, DC) observed that an underwater shock wave, reflected at a model hull (a plate immersed in water and backed by air), caused tension in the water, thus producing a zone of cavitation bubbles near the surface of the plate. The large amount of shock wave energy, trapped in this way in the cavitated water, could later expand in considerable plastic work of deformation. [R.H. COLE: *Underwater explosions*. Dover Publ., New York (1965), plate XII] *Right:* Wernfried GÜTH (Dept. of Physics, University of Göttingen), using the schlieren method and snapshot reflected-light photography, recorded the shock wave emerging from the collapse of a single cavitation bubble. Cavitation was generated at the surface of a metal piece – here located at the bottom of the picture – which, excited by a violent blow, was set into strong oscillations. [Acustica **6**, 526 (1956)]

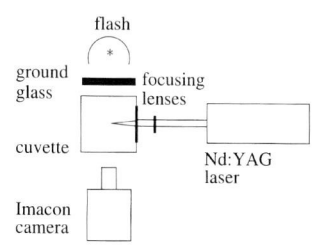
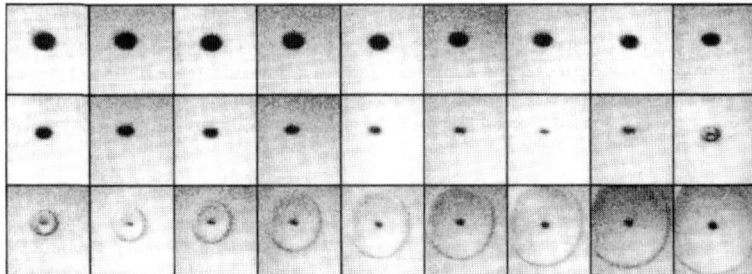
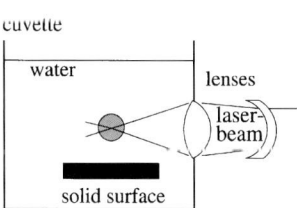

**4.14−D** *Left, top:* Claus-Dieter OHL and collaborators (Institute of Applied Physics, TH Darmstadt) studied the collapse of laser-generated cavitation bubbles using high-speed cinematography. *Right:* Example of bubble collapse in a free liquid photographed at 20.8 million frames/s. The maximum bubble diameter before collapse was 2.2 mm. Approaching the very moment of collapse (frame 16), the bubble reveals an ellipsoidal shape. Therefore, the shape of the subsequently rebounding bubble is not spherical, either. The emitted shock wave is first clearly visible in frame 19. *Bottom:* When the bubble collapses and rebounds in close proximity to a solid boundary, a jet inside the bubble is created. An additional shock wave, somewhat delayed after the first shock wave due to bubble collapse, is generated, which the researchers sucessfully resolved by cinematography as well. [Ann. Phys. **4** (8), 26 (1995)]

## 4.14 SHOCK WAVE EFFECTS – Supercavitation

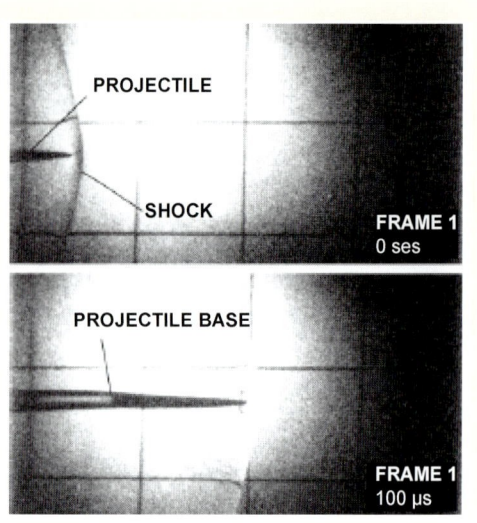

**4.14–E** *Left:* James D. HRUBES at NUWC used slender, 131-mm-long projectiles with a blunt nose to generate a single bubble entirely surrounding the projectile when shot into water. This effect, called "supercavitation," considerably reduces hydrodynamic drag and even allows supersonic projectile velocities under water. *Right:* Note the formation of the Mach head wave ($M = 1.03$). Supersonic velocities could be sustained up to a maximum distance of 2.5 m. [Courtesy Dr. J.D. HRUBES, U.S. Naval Undersea Warfare Center (NUWC), Newport, RI; Exp. Fluids **30**, 57 (2001)]

## 4.14 SHOCK WAVE EFFECTS – Condensation

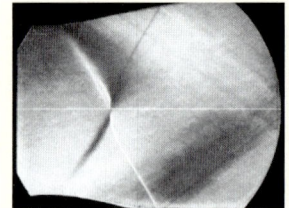
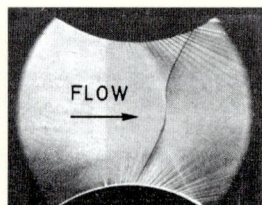

**4.14–F** *Top, left:* Formation of a "condensation shock" in a Laval nozzle, also called "X-shock" or "ghost." This phenomenon, first observed in the 1930s by Carl WIESELSBERGER (TH Aachen) in wind tunnels operated at high Mach numbers, occurs when atmospheric (*i.e.*, moist) air is used. [*50 Jahre Aerodynamisches Institut RWTH Aachen* Abhandl. Aerodyn. Inst. RWTH Aachen, Heft 17 (1963)] *Top, right:* Another schlieren picture of a condensation shock in a supersonic nozzle that L. PRANDTL (University of Göttingen) showed in 1935 at the 5th Volta Congress on High Velocity in Aviation. [Acta Mechanica **21**, 77 (1975)] *Bottom:* Snapshot of shock condensation effect around the jet fighter Boeing F-18 ("Hornet") at transonic flight in humid air. It apparently arises during acceleration when the air flow at some parts of the fuselage reaches supersonic speeds. When the resulting shock wave detaches, it builds up a sudden rarefaction that, lowering the temperature, causes condensation of the ambient water vapor. This unique picture was published in many journals. [Courtesy J. GAY, U.S. Navy] A similar picture was recently taken from an F-4 ("Phantom II") during an air show at Point Magu Naval Air Station, CA. [*The Military Aircraft Archive*, http://www.milair.simplenet.com/]

## 4.14 SHOCK WAVE EFFECTS – Aerodynamic Shock Heating

 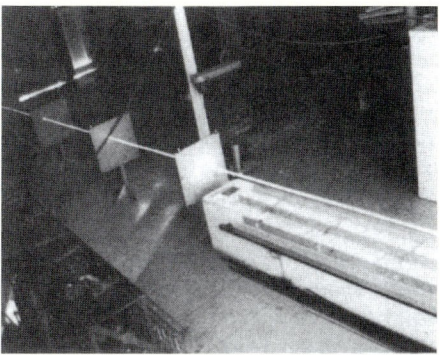 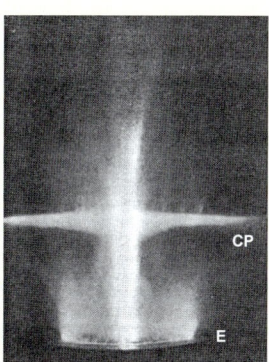

**4.14–G** *Left:* A meteoroid entering the Earth's atmosphere at hypervelocity produces a brilliantly luminous cap of ionized air at its nose, leaving a luminous streak or train of light along its path. Most stony meteorites break up explosively in flight. This photograph of an exploding meteorite was taken by Charles P. BUTLER on November 23, 1895. [Courtesy Science Museum Pictorial, U.K. Image-No. 1901-0146] *Center:* This is not a laser beam, but rather a time-exposure photograph of a small cubic polycarbonate (Lexan) projectile fired from an inductively driven rail-gun accelerator and flying at hypervelocity ($M \approx 17$) down the laboratory in the open air. Note the meteorite-like streak that, using a still camera, was taken by Scott C. RASHLEIGH and Richard A. MARSHALL at the University of Canberra. [J. Appl. Phys. **49**, 2540 (1978)] *Right:* This unique example of shock heating, caused by shock wave interaction, was taken by Henri MURAOUR in France. It shows the detonation of a bloc of explosive $E$ fired at the bottom. About 40 mm above the surface of the explosive a sheet of cigarette paper $CP$ was positioned at which the shock wave was reflected, thus creating a strong luminous zone. Initially withstanding the shock by its inertia, the paper was pulverized very shortly afterwards. [L'Astronomie **50**, 153 (1936)]

## 4.14 SHOCK WAVE EFFECTS – λ-Shock Configuration

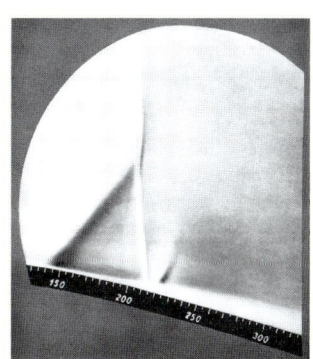 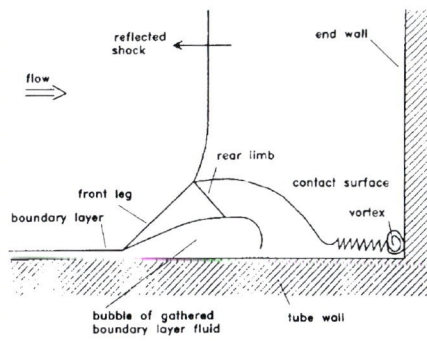 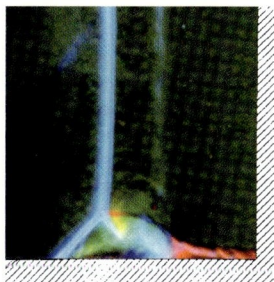

↑ λ-shock in $CO_2$, incident shock Mach number $M_{SI} = 2.45$.

**4.14–H** *Left:* Jakob ACKERET and Fritz FELDMANN (ETH Zurich) observed that the interaction of a shock wave with a boundary layer results in a new wave configuration, which they called "λ-shock" because of its striking similarity with the Greek letter λ. This schlieren picture shows a λ-shock formed above the surface of an airfoil resulting from an oblique interaction of an incident shock ($M = 1.225$) with a laminar boundary layer. Note that the flow direction is here from left to right. [Interavia **1**, 1 (1946)] In 1958, Hans MARK (NACA's Flight Propulsion Research Laboratory, Cleveland, OH) observed that the interaction of a reflected shock wave with a boundary layer produces a bifurcation. [NACA-TM1418 (1958)] *Center & right:* Harald KLEINE (RWTH Aachen) used color schlieren photography to resolve this characteristic wave pattern – a corkscrew-like vortex structure emerging from the corner. [Proc. 18th Int. Symp. on Shock Waves, Sendai (1991). Springer, Berlin (1992), p. 261]

## 4.14 SHOCK WAVE EFFECTS – Shock Focusing

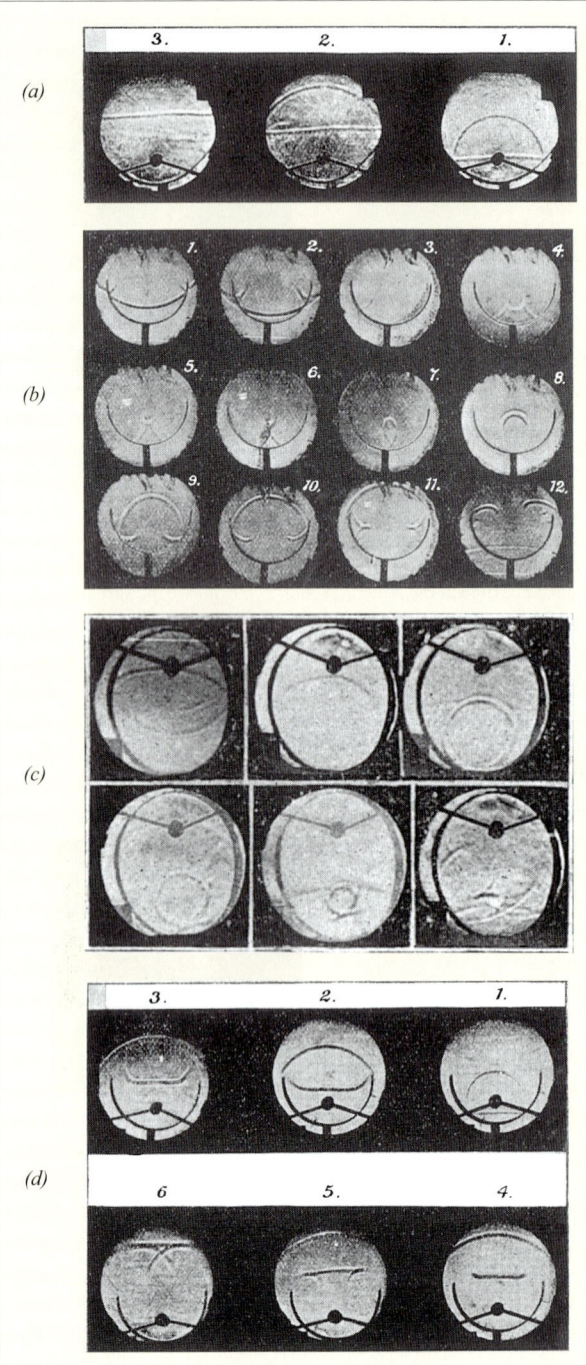

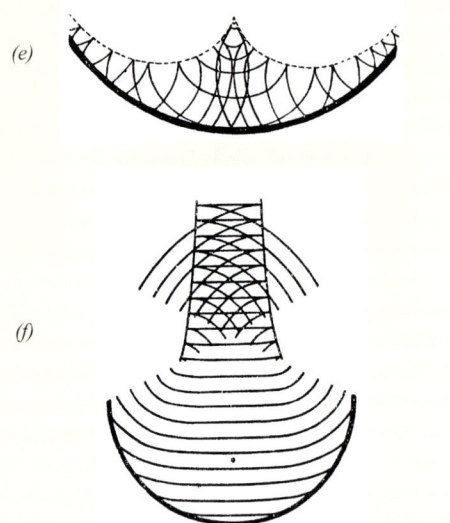

**4.14–I** In the late 1890s, the U.S. physics professor Robert W. WOOD (University of Wisconsin) resumed the pioneering schlieren studies of August TOEPLER, who in the mid-1860s was the first to visualize stroboscopically the propagation and reflection of weak shock waves emerging from electric sparks {⇨Figs. 4.5–A, B}. WOOD's original motivation was to illustrate in his lectures wave phenomena, in particular certain optical wave phenomena, by the "wave-front method" rather than by the "ray method" – today a standard technique in shock wave physics of interpreting wave phenomena. *Left:* He photographed a number of unusual wave focusing phenomena. *(a)* Transformation of a spherical into a plane wave by a parabolic mirror. *(b)* Generation of a cusp that always lies on the caustic surface when a plane wave enters a hemispherical mirror. *(c)* Demonstration of conjugated foci of an elliptical mirror that transforms by reflection a spherical wave diverging from one focus into a converging sphere, shrinking to a point at the other focus. *(d)* Transformation of a wave, starting at the principal focus of a hemispherical mirror and being reflected, into a nearly plane wave front in the vicinity of the axis, which curls up at the edges. *Right:* From some of the photographed wave focusing phenomena he drew schematics *(e)* and *(f)* to better illustrate their temporal evolution. [Proc. Roy. Soc. Lond. 66, 283 (1900); Nature 62, 342 (1900)]

## 4.14  SHOCK WAVE EFFECTS – Shock Focusing (cont'd)

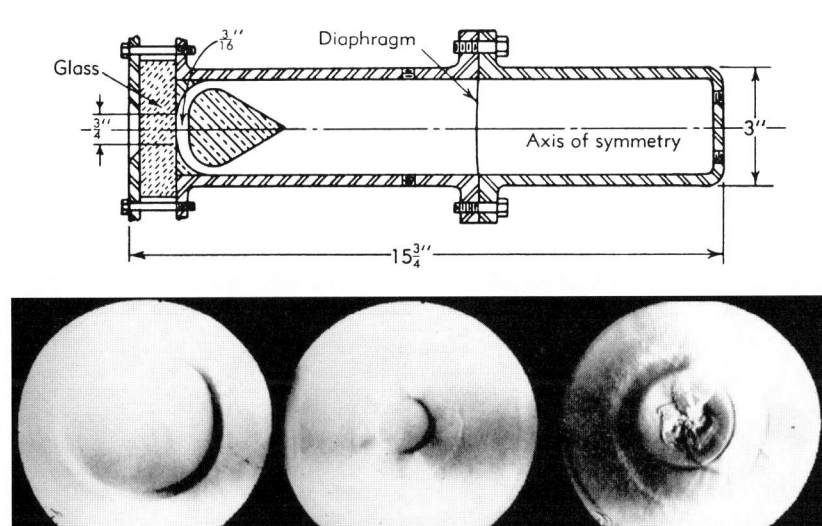

**4.14–J** *Top:* Robert W. PERRY and Anton KANTROWITZ (Cornell University, Ithaca, NY) used a special shock-tube geometry for producing converging cylindrical shock waves. They studied the implosion process through a glass plate "end-on." *Bottom:* High-speed cinematography revealed in more detail the conversion process of plane air shock waves with $M = 1.8$ into convergent cylindrical shock waves. The sequence of shadowgraphs was taken at increasing time instants (A → C → E). They noticed that the convergence stability became poorer with increasing shock strength. [J. Appl. Phys. **22**, 878 (1951)]

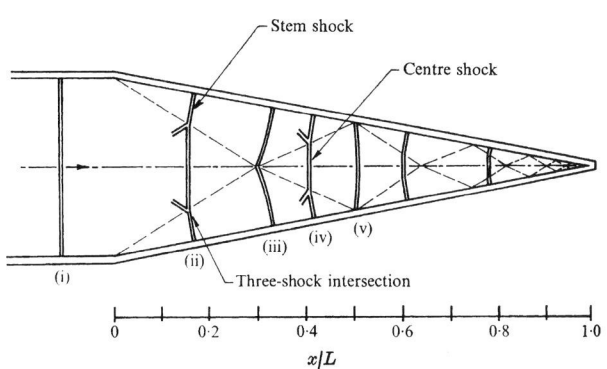

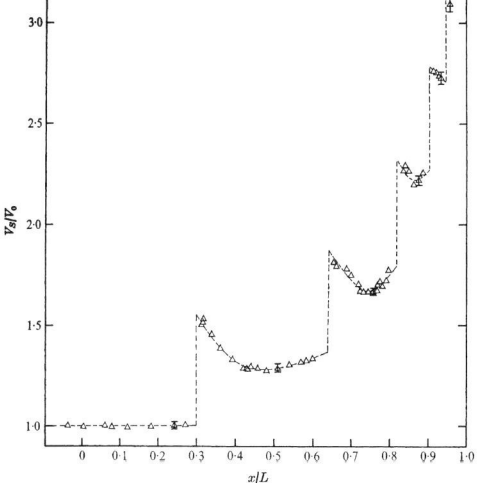

**4.14–K** *Left:* Robert E. SETCHELL, Erik STORM, and Bradford STURTEVANT (CalTech's Aeronautical Laboratory, Pasadena, CA) generated in a shock tube in argon at 1.5 Torr plane shock waves with high initial Mach numbers ($M = 6$–$10.2$) and investigated shock strengthening in a 10° half-angle conical convergent channel. *Right:* The profile of the shock wave velocity along the cone center line, measured with a new piezoelectric shock velocity probe, revealed that the shock velocity does not display the gradual monotonic increase as predicted by the 1-D Chester-Chisnell-Whitham theory, but rather shows a number of short intervals with sudden jumps. [J. Fluid Mech. **56**, 505 (1972)]

↑ The repeated cycles of Mach reflection on the cone wall, followed by Mach reflection of the stem shock on the cone axis, produce velocity jumps of the shock front along the cone center line, each jump being followed by a rapid decline and a gradual acceleration.

## 4.14 SHOCK WAVE EFFECTS – Transonic Shock Phenomena

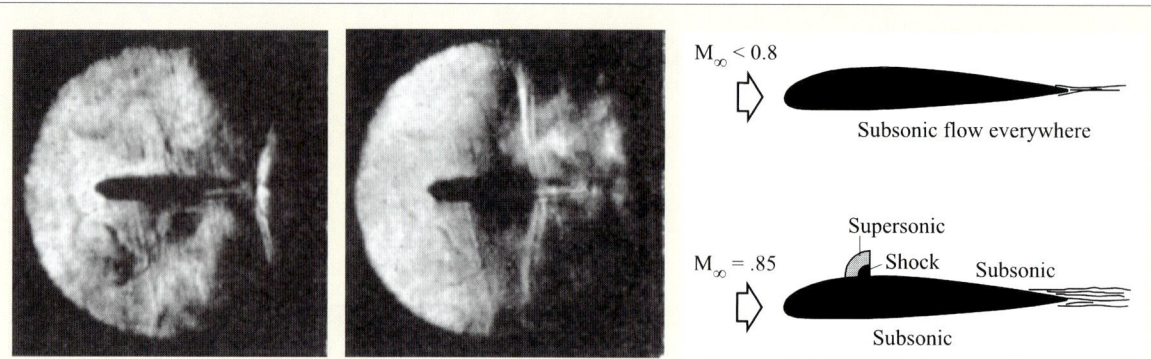

**4.14–L** When an aircraft approaches the speed of sound, the airflow over the wing reaches supersonic speed before the airplane itself does, and a shock wave forms on the wing. ***Top, left & center:*** In 1935, the U.S. aerodynamicist Eastman N. JACOBS showed pictures taken in NACA's wind tunnel from a transonic flow over a model wing, a NACA airfoil No. 0012-63. For $M < 0.79$ no shock has formed (*left*), but at reaching a so-called "critical Mach number," here $M_c = 0.79$, a shock is created (*center*) at the upper surface of the wing. [Proc. V. Convegno Volta, Rome (1935). Reale Accademia d'Italia, Rome (1936), pp. 383] ***Top, right:*** Schematic of shock wave formation in the transonic regime, when the Mach stream number $M_\infty$ is increased from $< 0.8$ (subsonic flow above the wing) to 0.85 (supersonic flow). This leads to a sharp rise in drag and a loss of lift – so-called "shock stall." [Courtesy NASA] ***Bottom:*** Two examples of wing compression shock shadowgraphs taken during banked turns of a Lockheed L-1011 at $M = 0.85$ and an altitude of 10,700 m. Note that the left photo was taken looking toward the Sun. The right photo shows the wing-tip shock, which appears to be approx. 2 m high. [Courtesy Carla THOMAS, NASA Dryden Flight Research Center; Proc. 8th Int. Symp. on Flow Visualization, Sorrento (1998). IOS, Amsterdam (1999), p. 17.1]

## 4.14 SHOCK WAVE EFFECTS – Aerodynamic Drag

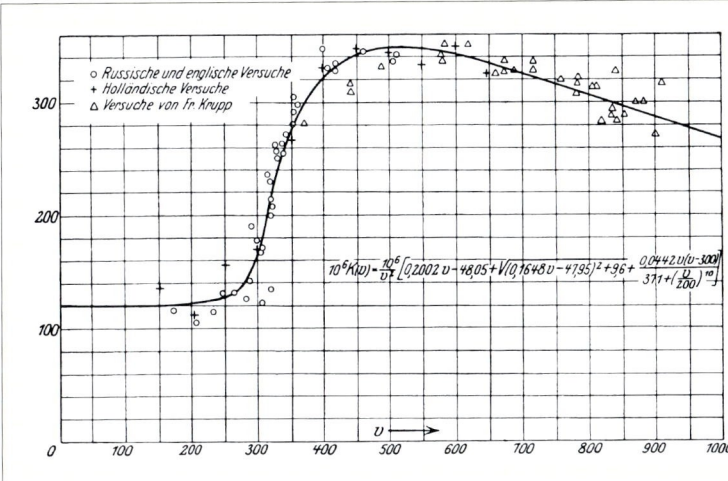

**4.14–M** This diagram is a result of numerous ballistic tests performed in Russia (MAIYEVSKII 1869), England (BASHFORTH 1866–1870), The Netherlands (HOJEL 1884), and Germany (Krupp Co. 1912). It shows that aerodynamic drag, also called "wave drag," decreases after passing the sound barrier {⇒Fig. 4.19-J}. From these data the Italian ballistician Francesco SIACCI derived in 1896 a standardized law of drag at high velocities. The projectile velocities $v$ are in (m/s). [C. CRANZ: *Lehrbuch der Ballistik*. Springer, Berlin (1925), vol. 1, p. 65; Rivista di Artiglieria e Genio (Rome), vol. 1 (1896)]

## 4.14 SHOCK WAVE EFFECTS – WHITCOMB's Area Rule

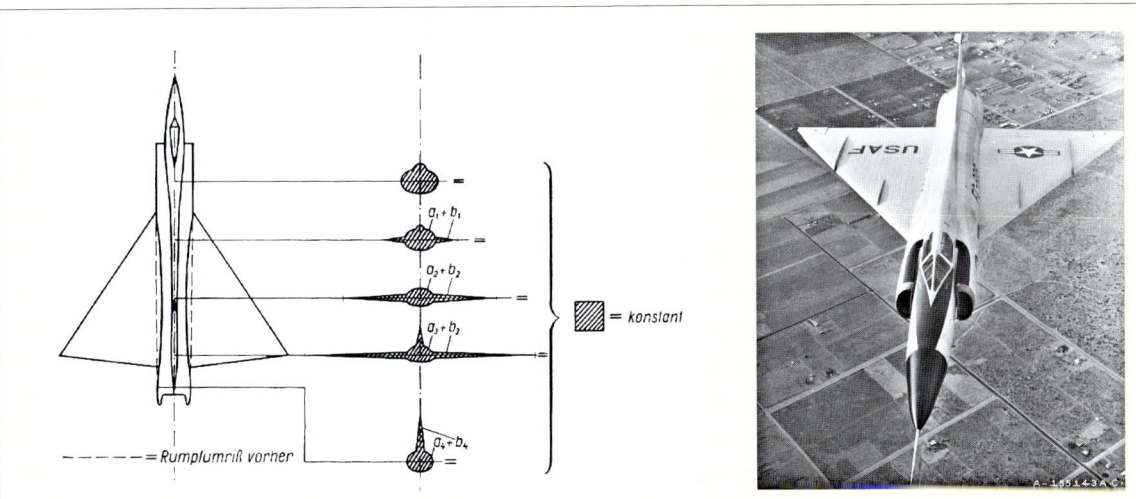

**4.14–N** *Left:* In 1951, Richard T. WHITCOMB at NACA discovered the "area rule," which postulates that, in order to avoid detaching of shock waves at transonic flight, all stepwise changes in cross sections of wings and the fuselage should be avoided and their total areas be kept constant along the plane's axis. In the case of delta wings, this results in a "coke-bottle" (or "Marilyn Monroe") shaped design of the fuselage (*broken lines*). Favorably, this reduces drag – thus allowing higher flight velocities and saving fuel. [Orion 11, 974 (1956)] *Right:* The Convair F-102 ("Delta Dagger") was the first aircraft constructed after the area rule such that the total cross-sectional area of wings, fuselage, and tail should be that of an ideal streamlined body, *i.e.*, with a wasplike waist and a bulging tail. It was the world's first supersonic single-seat all-weather jet interceptor and the USAF's first operational (knife edge) delta-wing aircraft. The F-102 made its initial flight on October 24, 1953; the maximum flight velocity was 810 mph (1,303 km/h). [Courtesy USAF Museum Photo Archives]

## 4.14 SHOCK WAVE EFFECTS – Pseudo Supersonic Wave Effects

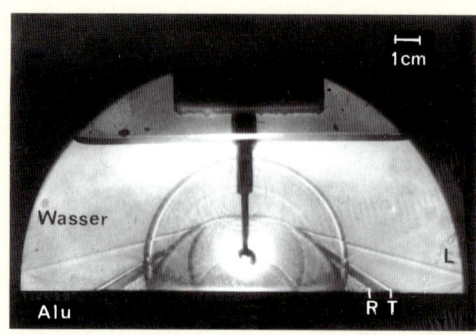

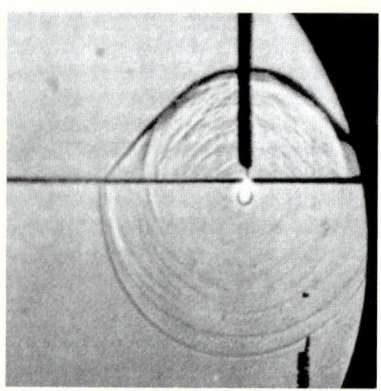

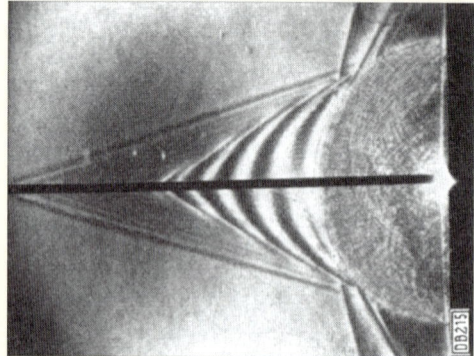

**4.14–O** Oswald VON SCHMIDT, a physicist at Humboldt University in Berlin, showed that any disturbance produced at the boundary of two media with different sound velocities generates a head wave similar to that of a supersonic projectile – actually a pseudo supersonic wave phenomenon that occurs for both acoustic and shock waves. ***Right:*** An electric spark generated at the interface of two liquids (*above:* an aqueous NaCl solution, sound speed 1,600 m/s; *below:* xylol with 1,175 m/s) produces two hemispherical waves. Waves, traveling some horizontal distance along the interface with 1,600 m/s, return to the upper half space, thereby according to HUYGENS' principle producing a straight "Schmidt head wave" (SHW). [Physik. Z. **39**, 868 (1938)] ***Left, top:*** A thin copper wire, exploding in water (sound speed $u_1$) above an aluminum surface (sound speed $u_2$), generates the three main types of SHWs, which emerge under different angles $\alpha_L$, $\alpha_T$, and $\alpha_S$ according to the relation

$$\alpha_{L,T,S} = \arcsin u_1/(u_2)_{L,T,S}.$$

Here $(u_2)_{L,T,S}$ are the longitudinal, transverse, and surface wave speeds in the lower half space, respectively [Photo by author]. ***Left, bottom:*** At Göttingen University, Erwin MEYER showed that flexural waves also generate SHWs. [Z. Tech. Phys. **19**, 554 (1938)]

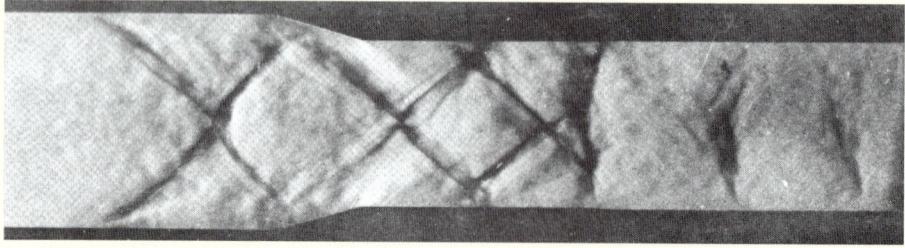

**4.14–P** E.P. NEUMANN and F. LUSTWERK at MIT (Cambridge, MA) observed the development of a pseudoshock in the presence of thick boundary layers. The pseudoshock taken in the throat of a diffuser actually begins for $M = 2.55$ at the entrance of the convergent portion (*left*). [J. Appl. Mech. **17**, 195 (1949)]

## 4.14 SHOCK WAVE EFFECTS – Gasdynamic Laser

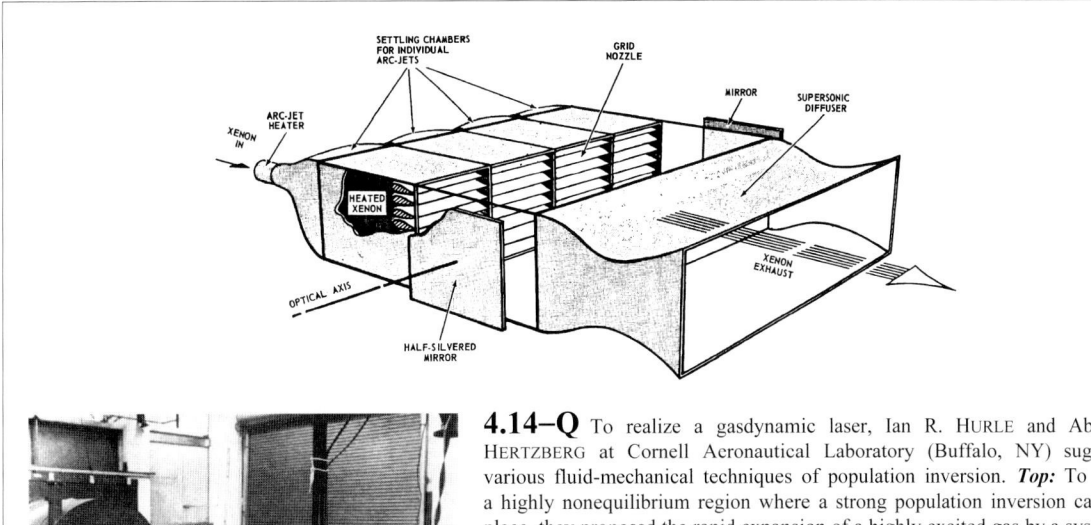

**4.14–Q** To realize a gasdynamic laser, Ian R. HURLE and Abraham HERTZBERG at Cornell Aeronautical Laboratory (Buffalo, NY) suggested various fluid-mechanical techniques of population inversion. *Top:* To create a highly nonequilibrium region where a strong population inversion can take place, they proposed the rapid expansion of a highly excited gas by a system of grid nozzles. However, the immediate experimental proof failed. [Phys. Fluids **8**, 1601 (1965)] In 1966, the first gasdynamic laser, using a hot gas mixture of $CO_2/N_2/H_2O$ expanding through a supersonic nozzle, was successfully operated at AVCO Everett Research Laboratory (Everett, MA). ***Bottom:*** This large-scale gasdynamic laser, developed there and operated with CO as the fuel, was among the very first high-power lasers. It produced 60 kW of multimode power. [IEEE Spectrum **7**, No. 11, 51 (1970)]

## 4.14 SHOCK WAVE EFFECTS – Shock-On-Shock Problem

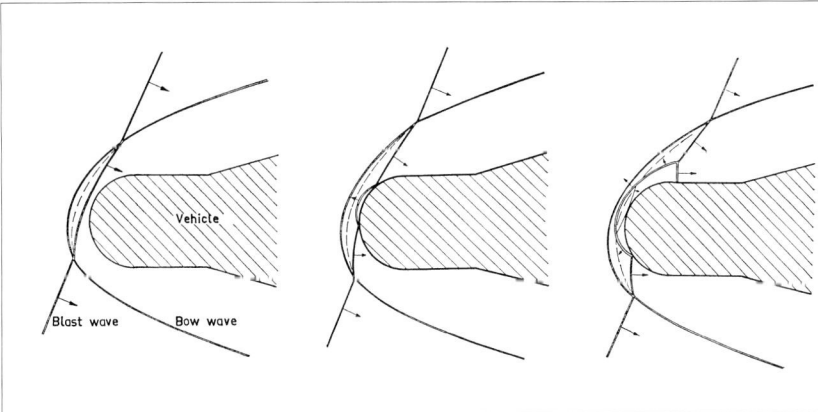

**4.14–R** Schematic of a "shock-on-shock" (SOS) interaction, illustrated here by Bo LEMBCKE, a Swedish mechanical engineer, at the example of a supersonic body hit obliquely by a plane blast wave (time proceeds from *left* to *right*). A complex wave pattern of reflecting and interacting shocks is produced, including Mach reflection. [FFA-Rep. 109, Aeronaut. Res. Inst. of Sweden, Stockholm (1967)] 2-D and 3-D SOS interactions stimulated many numerical studies.

## 4.14 SHOCK WAVE EFFECTS – Shock Wave Interactions in Metals

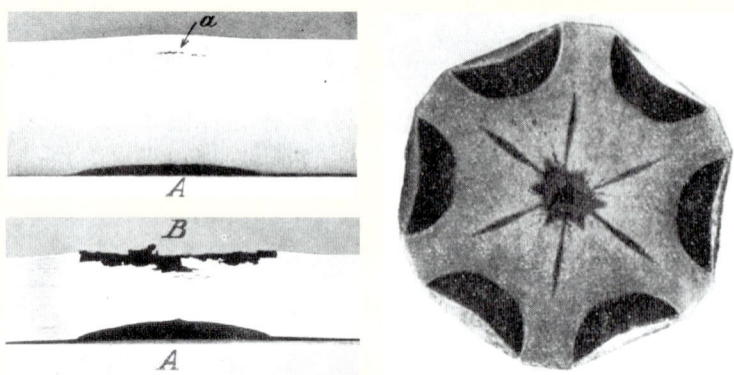

→ In 1962, Lev V. AL'TSHULER in the Soviet Union studied the interaction of colliding shock waves in metals subjected to the explosive action of several charges. The dark etched zones of the polished sample, stretching along the radii, are the results of oblique collision of two neighboring shock waves, which produce phase transformation induced by pressures > 130 kbar. The central spot is the result of a superposition of six waves. [Sov. Phys. Uspekhi **8**, No. 1, 72 (1965)]

**4.14–S** *Left:* In the early 1900s, Bertram HOPKINSON at Cambridge University observed that mild steel plates, being exposed to detonating guncotton at their bottom side *A*, revealed either a crack *a* in the case of a 1.25-in. (31.8-mm)-thick plate (*top*) or a scab torn off from *B* in the case of an only 0.75-in. (19-mm)-thick plate (*bottom*). He correctly interpreted that this "spalling effect" is caused by the reflected stress wave, leading to tension and fracture in the shocked material. [*Scientific Papers of B. HOPKINSON*. Cambridge University Press, Cambridge (1921)] *Right:* Macrostructure of a steel cylinder in the cross section when six charges are blasted simultaneously

## 4.14 SHOCK WAVE EFFECTS – Shock-Induced Solidification

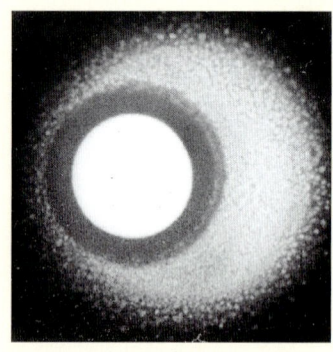
$\tau = 77$ μs; $d = 0.15$ mm

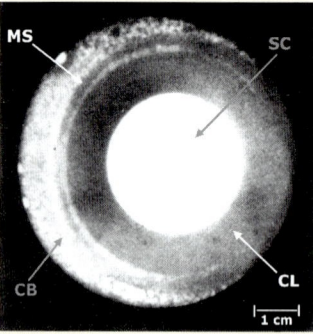
$\tau = 243$ μs; $d = 2.0$ mm

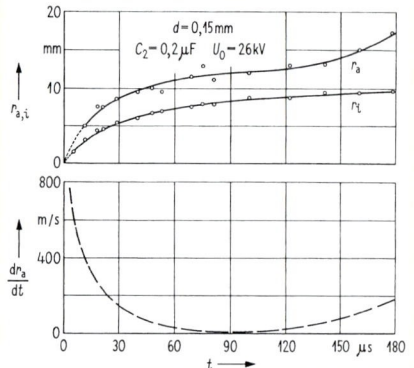

**4.14–T** *Left & center:* Flash radiographs taken at different time instants τ after the onset of discharge of a melting shock wave *MS*, propagating by stepwisely melting from the periphery of a shock-solidified organic liquid (bromobenzene, $C_6H_5Br$). This phenomenon, predicted by Werner SCHAAFFS at TU Berlin in 1948 and first photographed by Peter KREHL in 1968, was generated in a thin liquid layer of thickness *d* by an expanding plasma *SC* from a capacitor discharge between two plane electrodes. The surrounding compressed liquid *CL* contains numerous cavitation bubbles *CB* initiated by electrode vibrations during discharge. *Right:* Diagrams showing the radial expansion of the plasma channel, $r_i(t)$, and of the compression ring, $r_a(t)$. For a 0.15-mm-thick liquid layer, the expansion velocity of the compression ring, $dr_a/dt$, reaches at about 90 μs a minimum of only a few meters per second, apparently due to the shock-induced solidification of the liquid that solidifies at room temperature under static pressures already at about 3 kbar. [*Acustica* **23**, 99 (1970)]

## 4.14 SHOCK WAVE EFFECTS – Other Phenomena

**4.14–U** Superposition of three flash radiographs showing a sequence of a 30-06 bullet in air penetrating a so-called "Whipple shield" – a configuration invented in 1946 by the astronomer Fred L. WHIPPLE. It is based on the principle that small meteoroids and orbital debris explode when they strike a solid surface; therefore if a spacecraft is protected by an outer skin about a tenth of the thickness of its main skin, an impinging body will be destroyed before it can cause any real damage. The simplest kind of Whipple shield consists of a bumper, such as a thin sheet of aluminum, and a standoff, or open space, between the bumper and the wall of the spacecraft to allow any remaining fragments that make it past the bumper to spread out. Here the shield consists of two $1/8$-in. (3.2-mm) lead plates, arranged in parallel at a small distance. [Picture taken by John P. BARBOUR and associates at Field Emission Corporation in McMinnville, OR. See Proc. 7th Int. Congr. on High-Speed Photography, Zurich (1965). Helwich, Darmstadt (1967), p. 292]

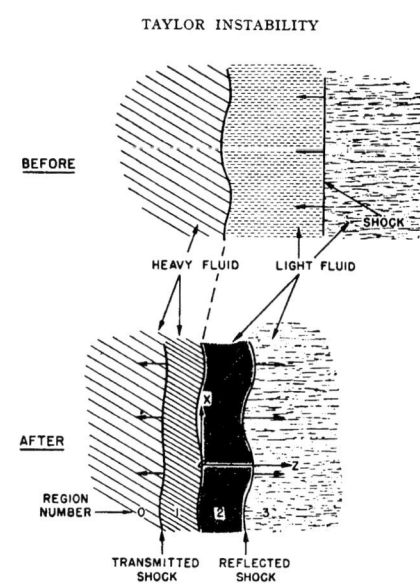

**4.14–V** The British physicist Geoffrey I. TAYLOR at Cambridge University developed a theory of the growth of irregularities on the interface between two fluids of different densities when they are in accelerated motion, such as illustrated by him in the case of a shock wave striking a boundary. This leads to damped oscillations in the shock wave, which influence the motion of the interface – a phenomenon called "Taylor instability." [Comm. Pure Appl. Math. **13**, 299 (1960)]

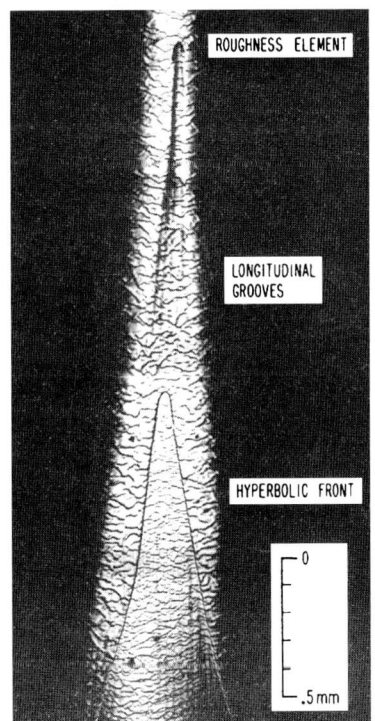

← CANNING and collaborators observed that "longitudinal grooves" and a "hyperbolic front" were superimposed on a pattern of "long- and short-wavelength ripples," overlying the entire surface. The origin of these grooves, etched in the surface within the wedge area, were later interpreted as a by-product of longitudinal vortices caused by Taylor-Görtler vortices in the boundary layer.

**4.14–W** In the 1960s, Thomas N. CANNING and collaborators at NASA's Ames Research Center (Moffet Field, CA) studied ablation patterns on cones of 1-cm-dia. plastic (Lexan or Delrin) models at hypersonic velocities up to 7 km/s and noticed curious patterns on the cone surface [AIAA J. **6**, 174 (1968)]

## 4.15 SHOCK WAVE APPLICATIONS – Miscellaneous

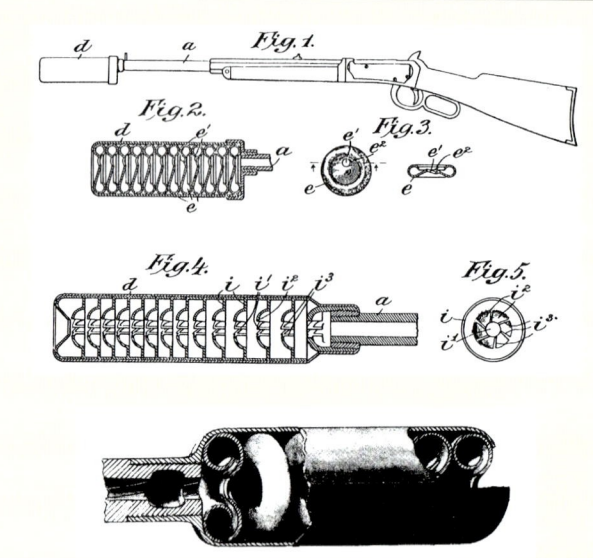

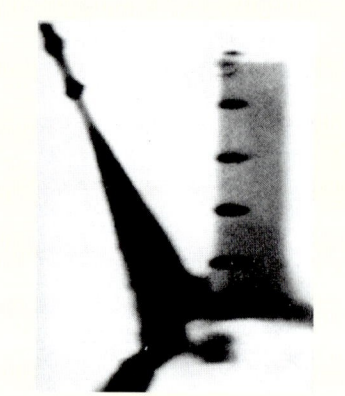

**4.15–A** The silencer for firearms was invented in New York around 1905 by Hiram Percy MAXIM (son of Sir Hiram Stevens MAXIM – who in 1889 developed the first fully automatic machine gun). *Top:* He already used the modern concept of a multiple-baffle system. This drawing was reproduced from one of his early patents granted to him in 1910 by the German Patent Office. [K.R. PAWLAS: *Waffengeschichte*. Chronica-Reihe, Folge **W 123**, Publ. Archiv, Nürnberg (1983); DRP Nr. 220,470 (1910)] *Bottom:* View of his perforated deflector system, which reversed the gases stepwise, from diaphragm to diaphragm.

**4.15–B** Rudi SCHALL and Gustav THOMER, two German physicists at ISL (Saint-Louis, France), proposed a simple method to derive Hugoniot data from flash radiographs of shock-compressed materials. Flash radiograph of a shock wave generated from a solid explosive at the bottom of a water column (*right*) moving upwards. Lead marks positioned along the water column are used for distance calibration. The shock velocity in the water was determined by comparison with a detonating fulminate fuse (*left*) of known detonation velocity and the shock compression ratio $\rho/\rho_0$ at the shock front from X-ray densitometry data. [Z. angew. Phys. **2**, 252 (1950)]

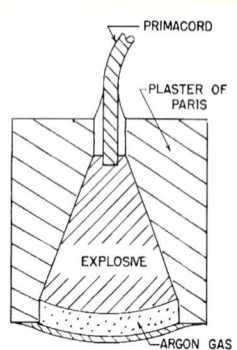
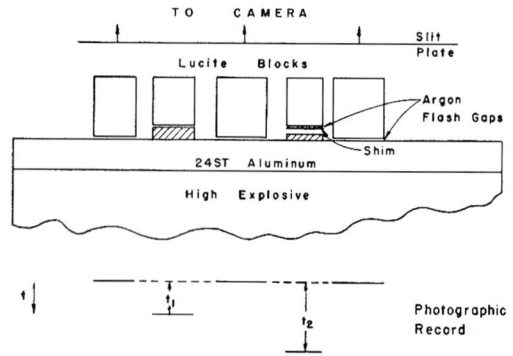

**4.15–C** A brilliant flash of light is generated when a strong shock wave strikes a layer of a noble gas. The duration of intensity can be made very short by decreasing the thickness of the gas layer. *Left:* Example of an "explosive flash charge" designed as a flash light for underwater photography using a thin layer of argon. [R.H. COLE: *Underwater explosions*. Dover, New York (1965), p. 214]  *Right:* This phenomenon was also applied in the so-called "flash gap technique" developed in the late 1950s by John M. WALSH and Russell H. CHRISTIAN at LASL (Los Alamos, NM) in order to determine the shock-wave and free-surface velocities from measured flash time arrivals, known specimen thickness, and free-run distance. [F. SEITZ and D. TURNBULL (eds.): *Solid state physics*. Academic Press, New York (1958), vol. 6, p. 21]

### 4.15 SHOCK WAVE APPLICATIONS – Determination of Seismic Wave Velocities

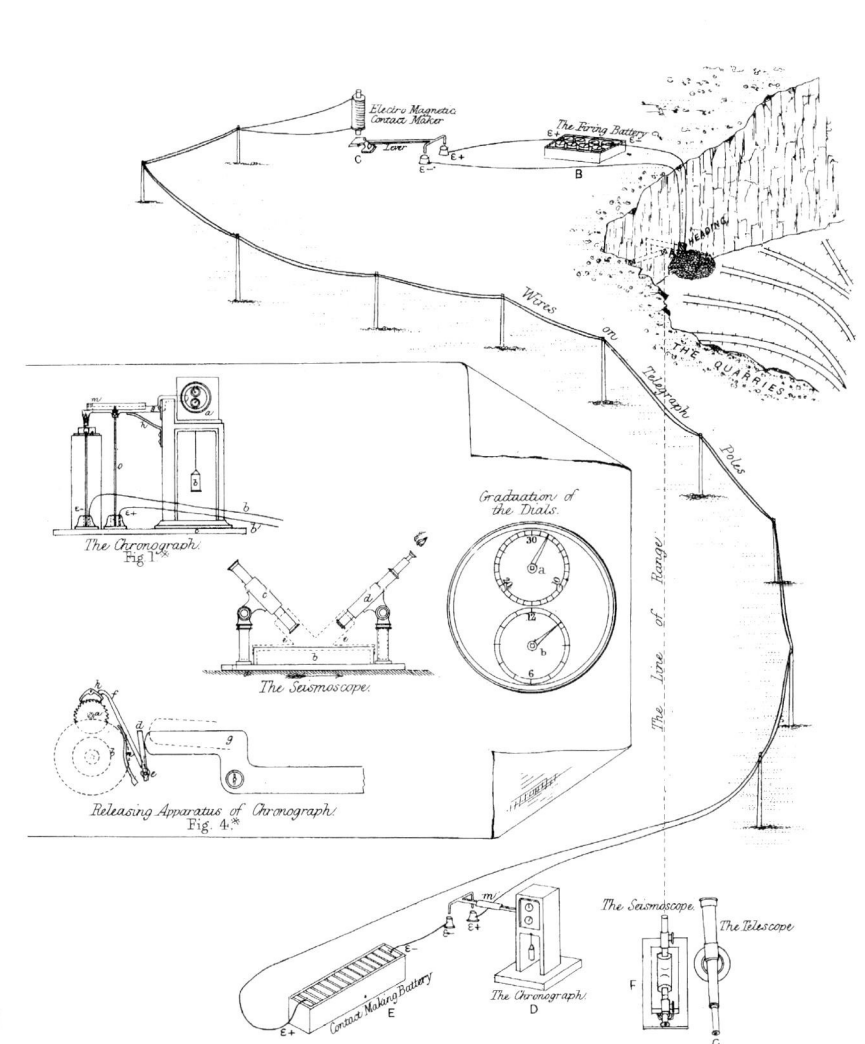

← MALLET made his seismic experiments at Holyhead in North Wales in order to ascertain the transit velocity of waves, analogous to earthquake waves, through the local rock formations. Supported in his study by Charles WHEATSTONE and Thomas R. ROBINSON, he generated artificial earthquakes by applying black powder, a low explosive not very appropriate for this purpose; however, dynamite had not yet been invented. He positioned a mercury horizontal seismometer at a large distance from the explosion and observed the arrival of the seismic wave that he assumed would show up based on the sudden appearance of ripples at the mercury surface. Note that he used the Hipp-Wheatstone chronometer {⇨Fig. 4.19–A} to measure the delay between triggering the charge and arrival of the longitudinal (P-) wave.

**4.15–D** In the period 1856–1857, the Irish engineer Robert MALLET performed seismic experiments to study the propagation and velocity of elastic waves in geologic layers, thus hoping to get information on the geologic nature of subterraneous layers. He correctly assumed that the longitudinal (elastic) waves precede the slower transversal waves; however, he did not assume any surface waves. Although he used huge single charges of black powder up to 5,000 kg, he was unable to detect the precursor in a distance beyond 2 km from the origin of explosion. MALLET's pioneering experiments, albeit unsuccessful, were a milestone toward geophysical prospecting. In 1876, the U.S. military engineer Henry L. ABBOT, on the occasion of blowing off the gneiss river bed of the East River at Hallet's Point, NY by using huge charges of high explosives (up to 22,000 kg of dynamite), was able to clearly resolve the arrival of the longitudinal wave at a distance of up to 220 km. [Phil. Trans. Roy. Soc. Lond. **150**, 655 (1860), plate XXIII]

## 4.15 SHOCK WAVE APPLICATIONS – Explosion Seismology: the "Mintrop Wave"

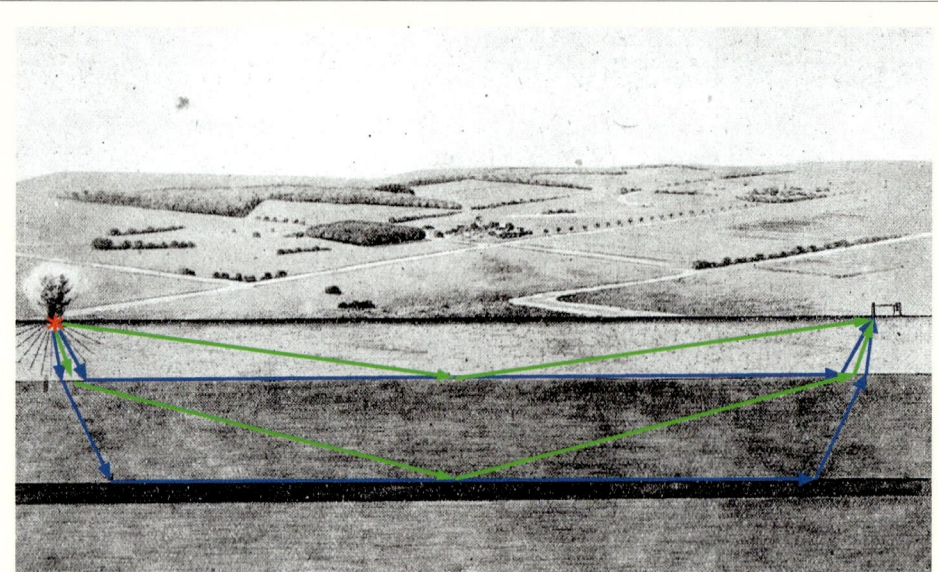

← The two principal seismic survey techniques in simplified form: in the *reflection method* the seismic waves take the green paths; in MINTROP's *refraction method* they take the blue paths. At greater distances the seismic pulse travels faster by the refraction path because its velocity is greater along the boundary than it is through the upper layer, a pseudo-supersonic effect.

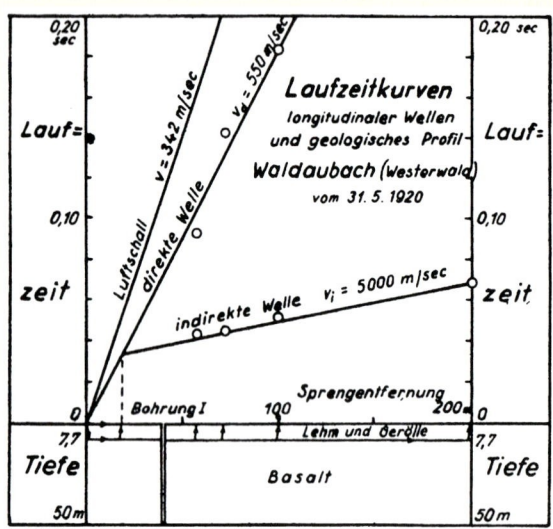

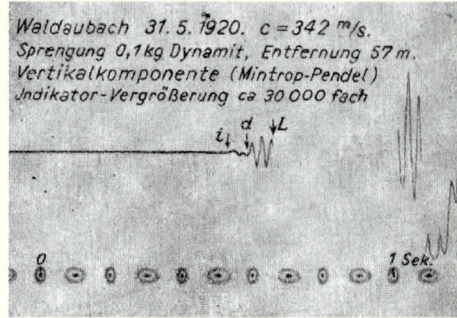

← The arrival times of the direct wave and the indirect wave plotted in a time-distance diagram are located at two straight lines of different slopes that intersect at the so-called "bent distance" from which the unknown depth of the lower stratum can easily be inferred. In the illustrated example, the upper layer consisted of a 7.7-m-thick clay stratum, followed by a thick basalt stratum.

**4.15–E** *Top:* Schematic of the generation, propagation, and recording of elastic seismic waves using MINTROP's refraction or "boundary wave" method. Generally, the density of rocks near the surface of the Earth increases with depth. An explosion generated at the surface induces elastic seismic waves in the soil. When arriving at the boundary of a deeper layer of higher acoustic impedance, the waves are transmitted along the boundary with the sound velocity of the lower layer, *i.e.*, supersonically with respect to the sound velocity of the upper layer. *Bottom, left:* Diagram of seismic wave propagation. *Bottom, right:* MINTROP's famous seismogram of an "artificial" earthquake taken on May 31, 1920 in Waldaubach, Germany by detonating 0.1 kg of dynamite. [Die Naturwissenschaften **34**, 257 (1947)]

## 4.15 SHOCK WAVE APPLICATIONS – Medical Therapy

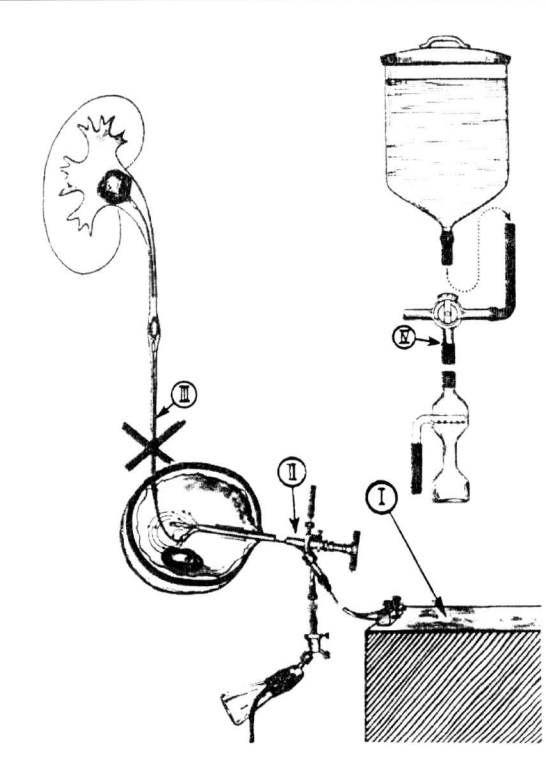
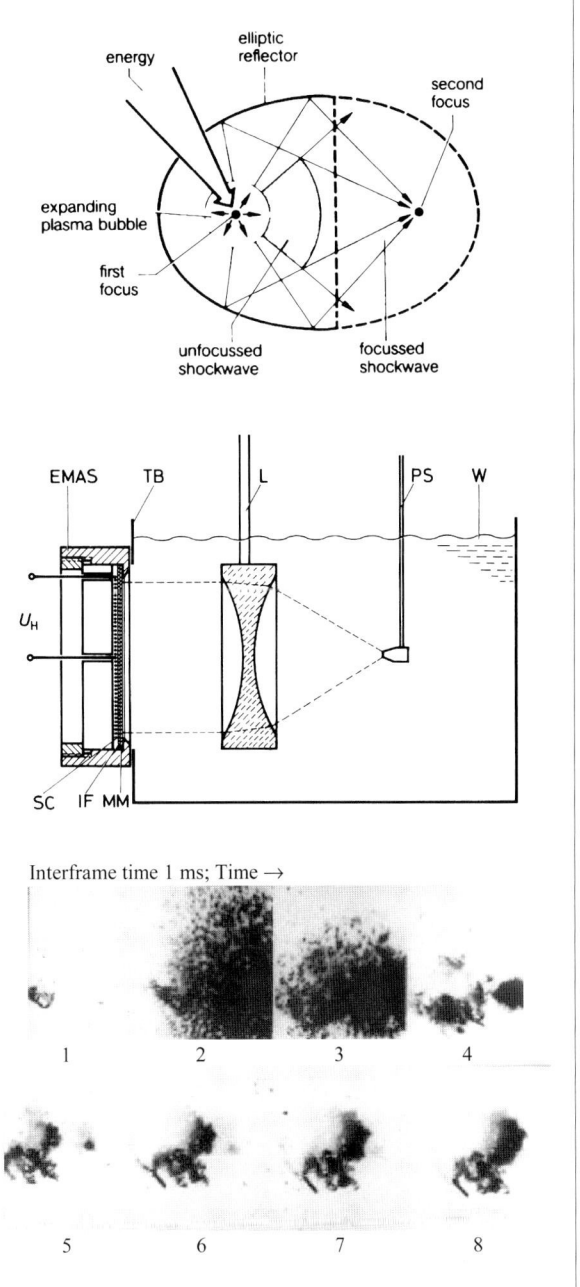

**4.15–F** *Left:* Victor GOLDBERG's "electro-lithotripsy" – the archetype method of modern shock lithotripsy. It was the first successful method of so-called "intracorporeal shock lithotripsy." His first patient was a 58-year-old man. [Der Urologe **B19**, 23 (1979)] *Right, top:* So-called "extracorporeal shock lithotripsy," a more advanced method. Developed in Germany by Dornier Medizintechnik, it employs an exploding plasma emitter as a point source together with an elliptical reflector. [Proc. IEEE **76**, 1236 (1988)] *Right, center:* Another method, commercialized by Siemens AG, applies an electromagnetic plane shock wave generator with a metallic membrane *MM*, combined with an acoustic lens *L*. [Siemens Forsch.- Entwickl.-Ber. **15**, 187 (1986)] *Right, bottom:* The first motion pictures of fragmentation of a kidney stone, taken at 1,000 frames per second by Bernd FORSSMANN and collaborators at Dornier System, NTF Medizintechnik (Friedrichshafen, Germany). The shock wave was generated by a spark discharge and focused by an elliptical reflector. The movie revealed that the stone disintegration was invisible until about 1 to 2 ms after interaction with the shock wave. [BMFT Symp. "Biophysikalische Verfahren zur Diagnose und Therapie von Steinleiden der Harnwege," Meersburg, Germany (Juni 1976). Wiss. Berichte, Dornier System GmbH, Friedrichshafen]

## 4.15 SHOCK WAVE APPLICATIONS – Materials Research and Metal Working Industry

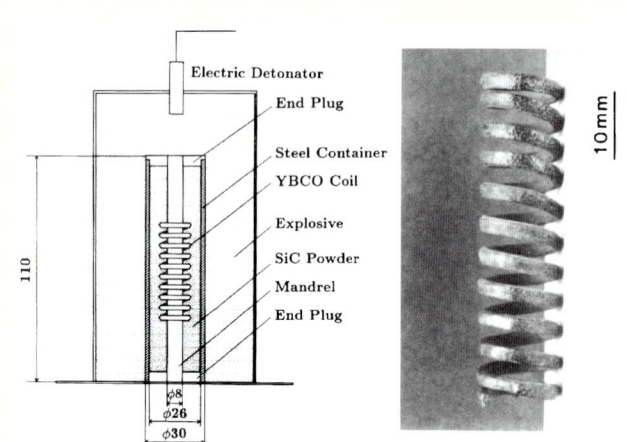

**4.15–G** Explosive setup used by Kazuki TAKASHIMA and collaborators in Japan for producing a coil of high-density superconductive material from Y-Ba-Cu-oxide powder. [*Proc. 6th APS Topical Conf. on Shock Compression of Condensed Matter.* Albuquerque, NM (1989). North-Holland, Amsterdam (1990), p. 591]

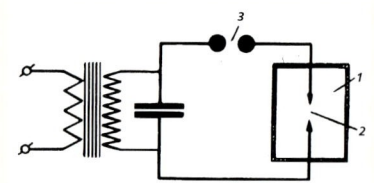

**4.15–H** Lev A. YUTKIN's circuitry to generate intense shock waves by a spark discharge (*2*) immersed in a non-conducting liquid (*1*). It avoids high-voltage DC potentials by using a pulse transformer. YUTKIN's so-called "electrohydraulic effect" was later applied to industrial metal forming ("hydrospark forming") and intensification of chemical-metallurgical processes, but also appears promising for well-drilling and grinding of materials. The urologist V. GOLDBERG first used this method for disintegrating bladder stones {⇨Fig. 4.15–F}. [L.A. YUTKIN: *Elektrogidrav-liceskij effekt.* Masgiz, Moscow (1955)]

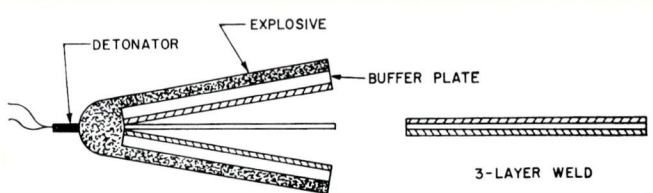

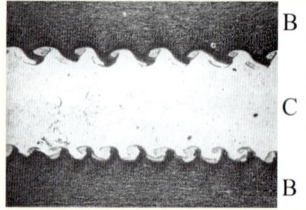

← This photomicrograph shows multiple welds between brass *B* and copper *C*. The surface waves that arose across while the metals were melted have frozen into a pattern of tiny hooks that strengthen the two bonds considerably.

**4.15–I** The most common explosive metalworking technique is "explosive welding," also known as "explosive bonding." When two dissimilar metal plates are driven together by detonation of a high explosive, they impact with each other at high velocity and may bond along the interface. ***Top:*** Experimental arrangement for a three-layer weld. As the detonation wave propagates along the explosive layers, the upper and lower plate violently collapse against the central plate. ***Bottom:*** The sequential collapse of the plate elements causes the metal close to the interacting surface to participate in the surface jetting effect. [J.S. RINEHART and J. PEARSON: *Explosive working of metals.* Pergamon, Oxford (1963), p. 312]

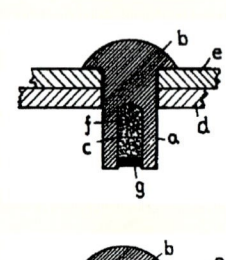

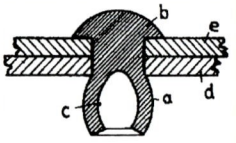

**4.15–J** The explosive rivet was invented by the German BUTTER Brothers. ***Top:*** Its shank *a* is filled with an explosive compound *c*, and the head of rivet *b* is heated to initiate the explosive. ***Bottom:*** The explosion forces the bottom of the rivet that extends past the edge of the material to expand to be larger than the diameter of the drilled hole. Blind rivets are needed when space limitations make conventional rivets impractical. They were first largely used during World War II in the U.S. aircraft industry. [Germ. Patent Nr. 655,669 (1937)]

## 4.15 SHOCK WAVE APPLICATIONS – Oil Production Industry

**4.15–K** *Left:* Photo of a burning oil well taken in 1913 at KT&O, Midway Fields (near Taft, CA). It shows smoke and flames ringing straight up in a column over 300 m high. The American Karl KINLEY had the idea of putting out the fire using for the first time nitroglycerin, appropriately calling the blast's action on the flame torch "snuffing it out." *Right:* Photo showing KINLEY standing far right, his right hand resting on the box of explosives which he used to put out the fire. [Courtesy Karl KINLEY Jr., Kinley Corp., Houston, TX]

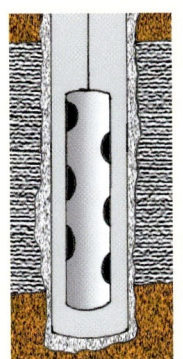
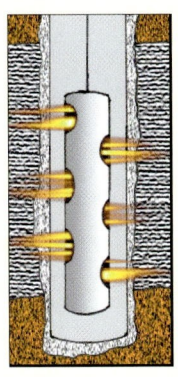
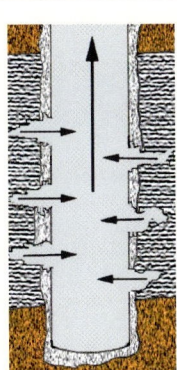
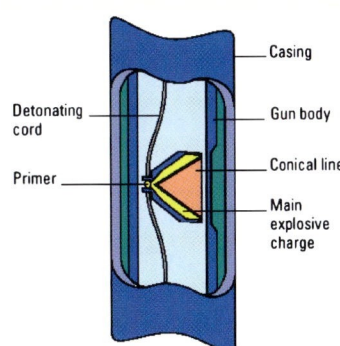

**4.15–L** Shaped charges accomplish penetration by creating jets of high-pressure, high-velocity gas. When the petroleum industry discovered the outstanding utility of shaped charges, they were used in the perforation of steel casings of well bores at depths thousands of feet below the surface. *Left:* The shaped charges, similar to those used in armor-piercing shells, are arranged in a tool, called a "gun," which is lowered into the well on a wire line and positioned opposite the producing zone (*left*). Then the charges are fired by electronic means from the surface (*center*), which produces holes extending some distance into the geologic formation. After the perforations are made, the tool is retrieved (*right*). Oil or gas now flows easier into the well bore. The use of these so-called "shaped-charge perforating guns" became the standard method of completing oil- and gas-producing wells. [Maverick Energy, Robinson, IL, http://www.maverickenergy.com/fundamentals.html] *Right:* A close-up of the shaped-charge perforating gun showing a shaped charge with primer belonging to it. A number of shaped charges are connected in a series via a detonation cord which provokes that the shaped charges are fired in a sequence. [Schlumberger Oil Field Glossary; http://www.glossary.oilfield.slb.com/DisplayImage.cfm?ID=474]

## 4.16 EXPLOSION, IMPLOSION, AND DETONATION – Early Use of Black Powder

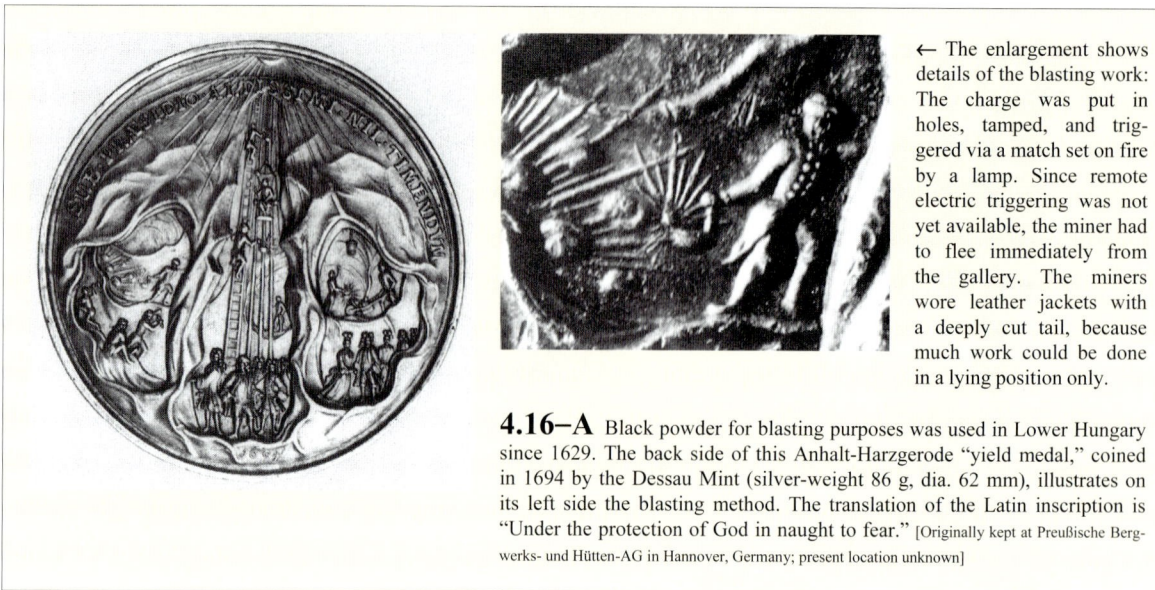

← The enlargement shows details of the blasting work: The charge was put in holes, tamped, and triggered via a match set on fire by a lamp. Since remote electric triggering was not yet available, the miner had to flee immediately from the gallery. The miners wore leather jackets with a deeply cut tail, because much work could be done in a lying position only.

**4.16–A** Black powder for blasting purposes was used in Lower Hungary since 1629. The back side of this Anhalt-Harzgerode "yield medal," coined in 1694 by the Dessau Mint (silver-weight 86 g, dia. 62 mm), illustrates on its left side the blasting method. The translation of the Latin inscription is "Under the protection of God in naught to fear." [Originally kept at Preußische Bergwerks- und Hütten-AG in Hannover, Germany; present location unknown]

## 4.16 EXPLOSION, IMPLOSION, AND DETONATION – Hollow Charge Effect

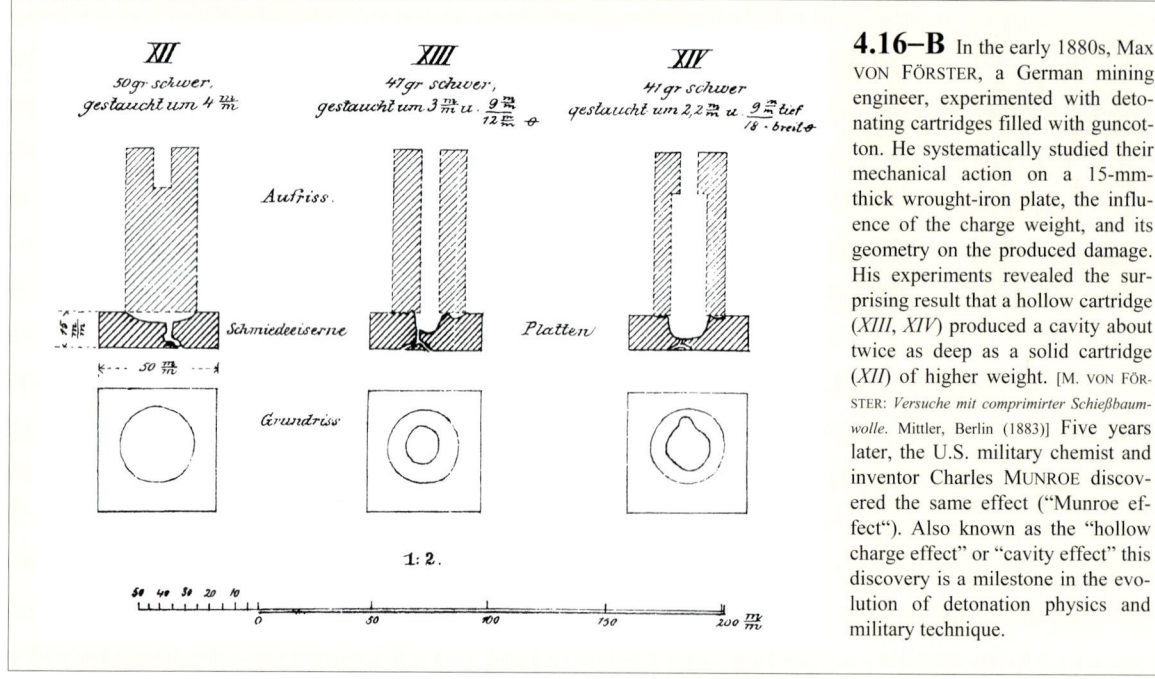

**4.16–B** In the early 1880s, Max VON FÖRSTER, a German mining engineer, experimented with detonating cartridges filled with guncotton. He systematically studied their mechanical action on a 15-mm-thick wrought-iron plate, the influence of the charge weight, and its geometry on the produced damage. His experiments revealed the surprising result that a hollow cartridge (*XIII, XIV*) produced a cavity about twice as deep as a solid cartridge (*XII*) of higher weight. [M. VON FÖRSTER: *Versuche mit comprimirter Schießbaumwolle*. Mittler, Berlin (1883)] Five years later, the U.S. military chemist and inventor Charles MUNROE discovered the same effect ("Munroe effect"). Also known as the "hollow charge effect" or "cavity effect" this discovery is a milestone in the evolution of detonation physics and military technique.

## 4.16  EXPLOSION, IMPLOSION, AND DETONATION – Underwater Explosions

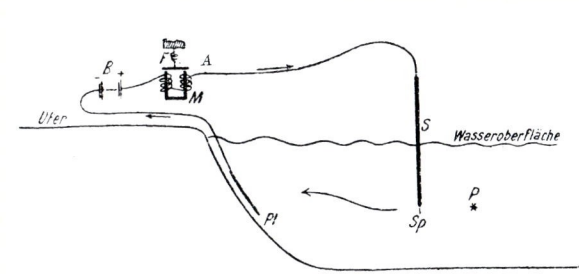

**4.16–C** Carl W. RAMSAUER, a German physicist at Danzig University, used an electrolytic method to measure the expansion dynamics of the gas sphere ("bubble") of an underwater explosion. Illustrated here on the example of a single-probe circuit, it applies a battery $B$, an electromagnet $M$, an isolated steel rod $S$ with a bare tip $Sp$, and a plate electrode $Pl$. The sea water, acting as an electrolyte, forms a conducting circuit and activates $M$. When an explosive is initiated at $P$, the bubble expands and at the arrival of $Sp$ interrupts the current, thereby deactivating $M$. Using a chronograph and many such probes arranged at different distances from $P$, it is possible to resolve in time both the growth of the bubble and its upward motion. [Ann. Phys. **72** (IV), 265 (1923)]

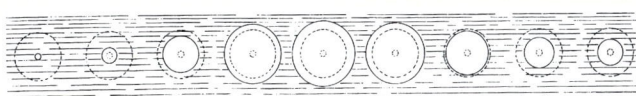

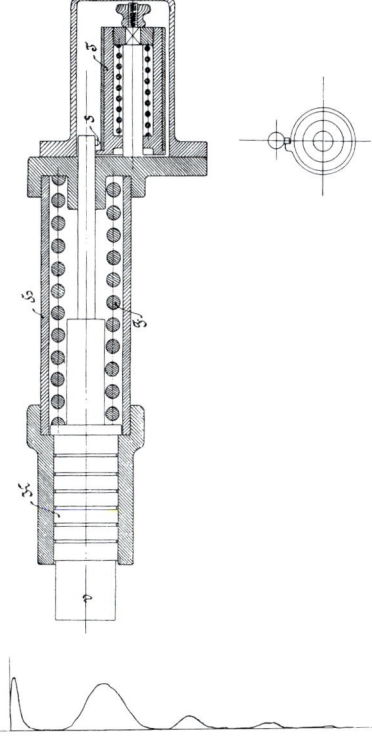

**4.16–D** In the 1890s, G.F. Rudolph BLOCHMANN, a military engineer at the German Imperial Navy, first studied the gas globe oscillation of an underwater explosion (*top*). To measure the pressure-time profiles in the water he used a "dynamometer" (*center*), a pressure gauge with a spring-loaded piston, and recorded the piston displacement by a rotating-drum chronograph. He produced small-scale underwater explosions by detonating small charges of guncotton (15–60 g) and first observed a train of pressure pulses (*bottom*) which he correctly correlated to the oscillating gas sphere, the so-called "bubble pulsations." Working out a first theory of underwater explosions, he compared the gas globe oscillation with the vibration of a simple spring-mass system. [Marine-Rundschau **9** (1. Teil), 197 (1898)]

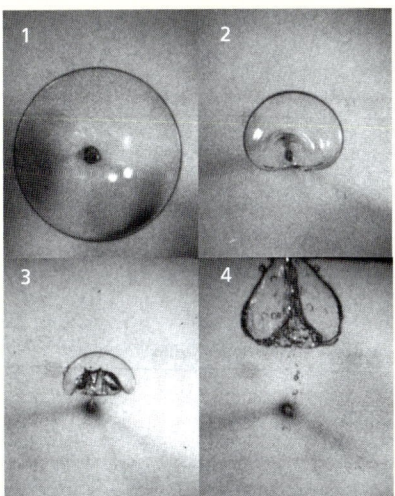

**4.16–E** Georges L. CHAHINE at Dynaflow and researchers at the Naval Surface Weapons Center used spark-generated bubbles for modeling underwater explosion bubble dynamics. This allows clear observation of reentrant jet formation inside the bubble, because the explosive detonation products generally occlude the view of the interior of the bubble. This series, taken with a Hycam camera at 10,000 frames/s, illustrates bubble dynamics in a gravity field, the reentrant jet dynamics, and the toroidal bubble rebound. [Courtesy Dr. Georges L. CHAHINE, Dynaflow, Fulton, MD. Proc. 66th Shock & Vibration Symp., Biloxi, MS (1995). SAVIAC, Arlington, VA (1995), vol. 2, p. 265]

## 4.16 EXPLOSION, IMPLOSION, AND DETONATION – Implosion and Explosion in a Gas

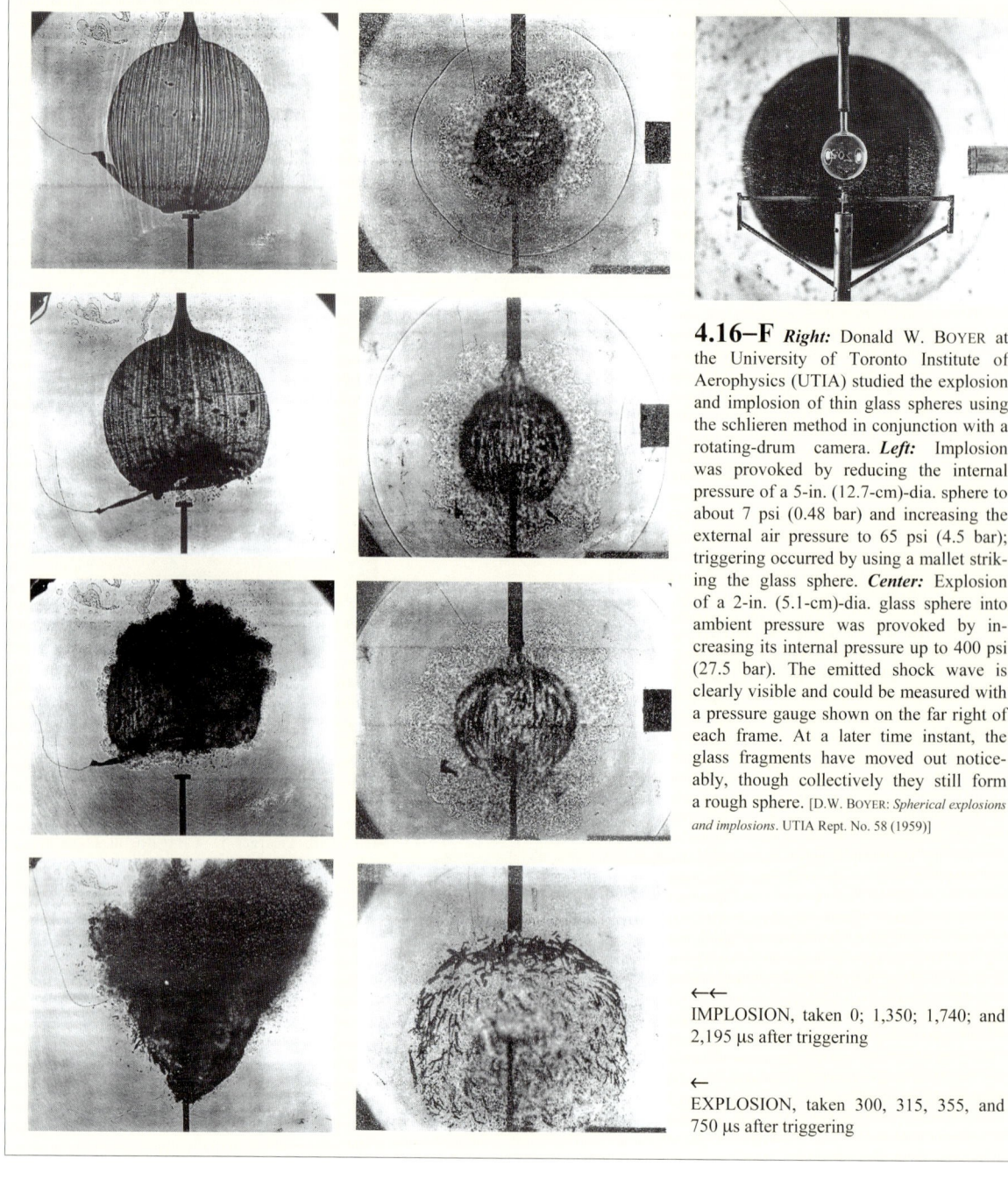

**4.16–F** *Right:* Donald W. BOYER at the University of Toronto Institute of Aerophysics (UTIA) studied the explosion and implosion of thin glass spheres using the schlieren method in conjunction with a rotating-drum camera. *Left:* Implosion was provoked by reducing the internal pressure of a 5-in. (12.7-cm)-dia. sphere to about 7 psi (0.48 bar) and increasing the external air pressure to 65 psi (4.5 bar); triggering occurred by using a mallet striking the glass sphere. *Center:* Explosion of a 2-in. (5.1-cm)-dia. glass sphere into ambient pressure was provoked by increasing its internal pressure up to 400 psi (27.5 bar). The emitted shock wave is clearly visible and could be measured with a pressure gauge shown on the far right of each frame. At a later time instant, the glass fragments have moved out noticeably, though collectively they still form a rough sphere. [D.W. BOYER: *Spherical explosions and implosions*. UTIA Rept. No. 58 (1959)]

←←
IMPLOSION, taken 0; 1,350; 1,740; and 2,195 μs after triggering

←
EXPLOSION, taken 300, 315, 355, and 750 μs after triggering

## 4.16 EXPLOSION, IMPLOSION, AND DETONATION – Nuclear Implosion Device: The "Gadget"

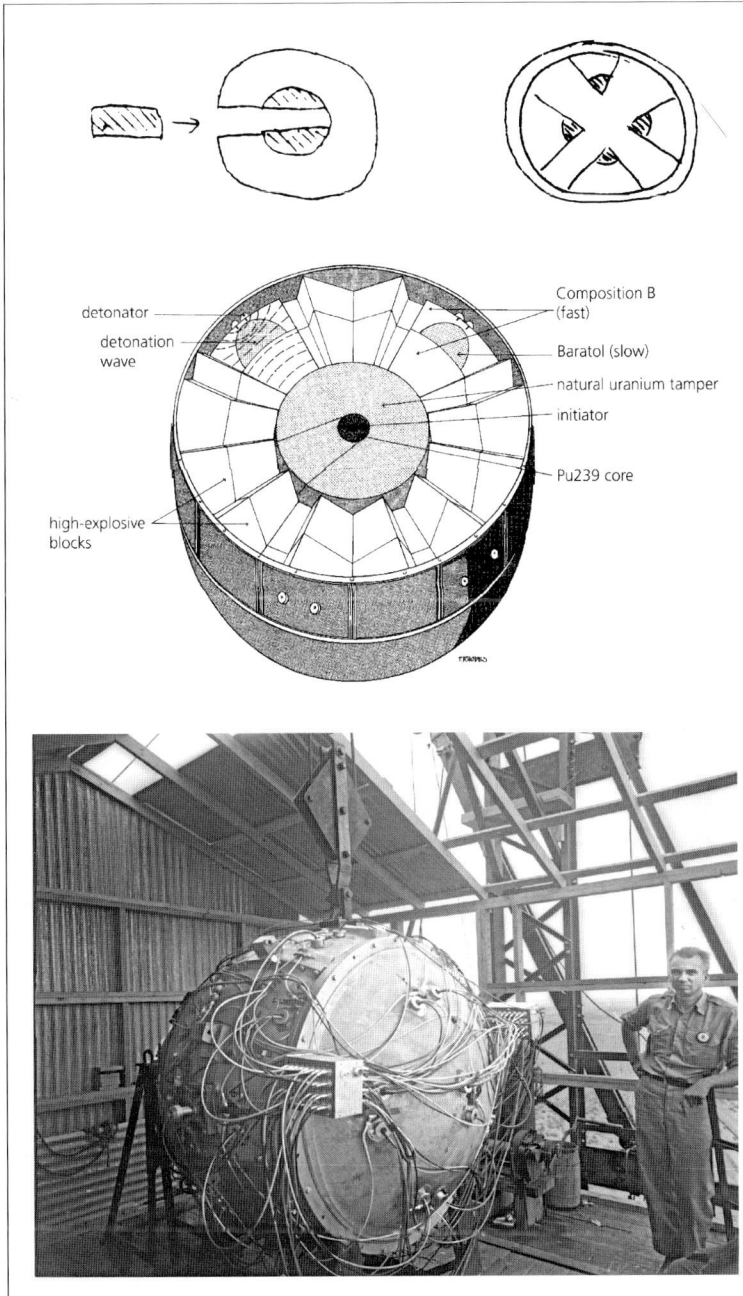

**4.16–G** *Top:* In April 1943, the physicist Robert SERBER delivered five top-secret lectures on how to build an atomic bomb in order to introduce scientists arriving at Los Alamos to the current state of nuclear weapons research. He showed sketches of the gun assembly (*left*) and the implosion-assembly device (*right*), the "Gadget," which became the first atomic bomb. [R. SERBER: *The Los Alamos Primer.* Univ. of Cal. Press, Berkeley 1992)] **Center:** Schematic of implosion bomb showing fast and slow explosive lens components and the two subcritical hemispheres of plutonium Pu-239 (diameter approx. 8.1 cm) that were driven inward and compressed to supercriticality by the implosion. In the center of the plutonium sphere, which had approximately the diameter of an orange, an initiator was placed in order to release a burst of neutrons when struck by the pressure wave. The bomb initiator was based on the ($\alpha$, n)-reaction of Po-210 and Be-9. The arrival of the shock wave caused mechanical mixing that allowed $\alpha$-particles to impinge on the beryllium and to yield neutrons. One of the greatest problems of any implosion device using high explosives is to ensure a symmetric collapse and to avoid any jet formation and instabilities that, provoking a predetonation, would result in a low energy release (yield). The spherical implosion charge consisted of 96 segments cast from slurries of explosives. Set such that detonation started simultaneously at certain points on the surface of the charge, the individual detonation waves were converted by high explosive lenses into a single convergent wave front compressing the natural uranium tamper with the contained plutonium. The tamper served to reflect neutrons back into the core, thus improving the chain reaction rate. [R. RHODES: *The making of the atomic bomb.* Simon & Schuster, New York (1986), p. 575; A more detailed schematic of "The Gadget" is shown in F.H. SHELTON's book *Reflections of a nuclear weaponeer.* Shelton Enterprise, Colorado Springs, CO (1988), p. 3-36] **Bottom:** View of the completely assembled "gadget" atop the 100-ft tower, with physicist Norris BRADBURY. The photo was taken the day before the Trinity Test, which occurred on July 16, 1945. [From SERBER's book cited above.]

## 4.16 EXPLOSION, IMPLOSION, AND DETONATION – Examples of Nuclear Explosions

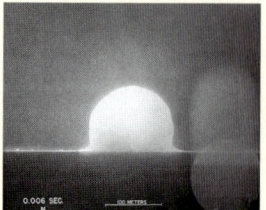
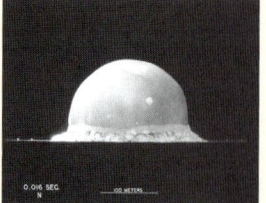

**4.16–H** *Left:* On July 16, 1945 the first atomic explosion, the so-called "Trinity Test," took place at Trinity Site in the Alamogordo Desert of New Mexico. The plutonium bomb, an implosion-type device, detonated on top of a 100-ft (30-m)-high tower and had a yield of about 19 kilotons TNT equivalent. [Courtesy LLNL, Livermore, CA]
*Right:* These are four pictures from a movie of the explosion taken by LASL staff scientist Berlyn BRIXNER, a professional high-speed photographer and inventor. His Fastax camera was positioned in the North Shelter, at a distance of 10,000 yards (9,144 m) from the explosion. The pictures were taken 6 ms (*top*), 16 ms (*center*), and 10 s (*bottom*) after ignition. A movie of the first nuclear explosion in real time can be viewed on the Internet. [LLNL Archives, Livermore, CA; http://www.cddc.vt.edu/host/atomic/trinity/tr_test.html]

**4.16–I** The first nuclear underwater explosion – officially called Test "BAKER" of operation "CROSSROADS" – and the fifth nuclear explosion in history took place at Bikini Atoll on July 25, 1946. The weapon was suspended beneath the landing craft LSM-60 (a World War II amphibious assault ship) anchored in the midst of the target fleet and ignited within the atoll 90 ft (27.4 m) beneath the surface of the lagoon. A target armada consisting of nine ships was exposed to the explosion in order to study nuclear vulnerability effects under combat conditions and to determine the hull damage caused by the hydraulic effect of the underwater nuclear detonation. It produced a huge moisture cloud typical for underwater nuclear explosions. Eight ships sunk, the USS *Prinz Eugen* (a captured heavy cruiser of the German Navy), survived the test but was too radioactive to have leaks repaired. [F.H. SHELTON: *Reflections of a nuclear weaponeer*. Shelton Enterprise, Colorado Springs, CO (1988), p. 2–48]

## 4.16 EXPLOSION, IMPLOSION, AND DETONATION – Examples of Nuclear Explosions *(cont'd)*

**4.16–J** Nuclear excavation experiments were conducted by the U.S. Atomic Energy Commission (AEC) in order to develop possible peaceful uses for nuclear explosives. In the so-called "AEC's Plowshare Program," various earth-moving projects were under study. For example, on July 6, 1962, the experiment SEDAN was carried out. The 104-kT nuclear device was detonated at the Nevada Test Site, buried under 635 ft (193 m) of desert alluvium. *Top:* Three pictures from a series of the moving earth bubble taken at 1.9 s (*left*), 2.8 s (*center*), and 5 s (*right*) after ignition of the nuclear device. *Bottom:* The detonation produced a crater 1,280 ft (390 m) in diameter and 320 ft (97.5 m) deep. [Courtesy LLNL Archives, Livermore, CA]

Open ditches are necessary for farm drainage, mosquito and flood control, pipelines, highway construction, land reclamation, stream corrections, and various other purposes. They can be dug quickly in many types of soil by using dynamite – so-called "blast ditching." In the 1970s, an idea was proposed of using this technique to construct a new, wider, and deeper Atlantic-Pacific canal north of the present Panama Canal, but using nuclear instead of chemical explosives.

**4.16–K** *Left:* This photo shows a well-executed ditch blast in action, resulting from a chemical explosive. In wet soil, ditches may be blasted by the propagation method, *i.e.*, only one hole is primed, the concussion from the explosion of this one charge being sufficient to propagate the detonation through the soil and set off the whole line of charges. The other method of initiating applies either electric caps or "Primacord" to each hole. [*Blasters' handbook*. E.I. du Pont de Nemours, Wilmington, DE (1967), p. 368] *Right:* The "BUGGY" Excavation Test performed in 1968 used five nuclear devices in a row, each about 50 m apart, that were detonated simultaneously. It was part of the U.S. Plowshare Program, an initiative to develop peaceful uses of nuclear weapons. [Courtesy LLNL Archives, Livermore, CA]

## 4.16 EXPLOSION, IMPLOSION, AND DETONATION – Electric Guns

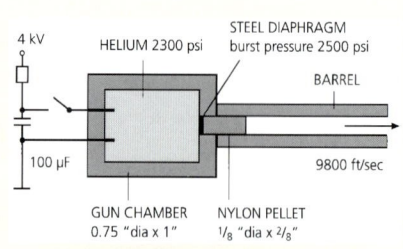

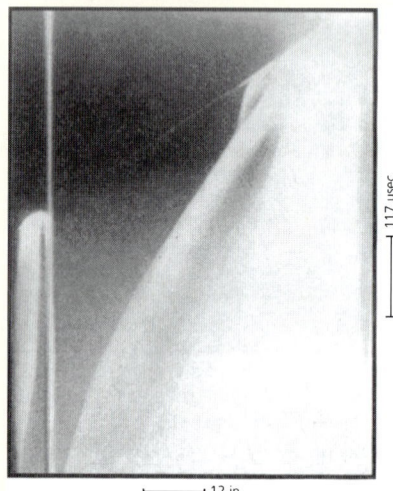

**4.16–L** This model "electric gun" was used in the 19th century to demonstrate the action of suddenly heated air by discharging a battery of Leiden jars via two electrodes into a wooden discharge chamber. The expanding hot gases blow out the mortar – here an ivory ball. [G.L.E. TURNER: *19th-century scientific instruments*. Sotheby, London (1983), p. 193]

**4.16–M** The principle of the electric gun, in modern terms an "electrothermal plasma gun," saw a renaissance after World War II. ***Top:*** Daniel E. BLOXSOM JR., a researcher at Rhodes & Bloxsom Co. (Canoga Park, CA), used pressurized helium gas that was suddenly heated in a small gun chamber by a capacitor discharge. Using energies up to 8 kJ, the helium could be heated up to approx. 21,000 K. The high dynamic pressure in the chamber ruptured a thin steel diaphragm, thereby accelerating a small pellet freely moving in a barrel up to hypersonic velocities. ***Bottom:*** Example of an optical streak record – a diagram showing time *vs.* distance – taken from a $1/3$-in. (8.47-mm)-dia. nylon pellet fired into laboratory air at 9,800 ft/s (2,987 m/s). The flash on the left shows the nylon pellet disintegrating upon striking a thin aluminum foil. Ablation of the nylon pellet may also be seen in the first 3 in. of flight. [J. Appl. Phys. **29**, 1049 (1958)]

## 4.16 EXPLOSION, IMPLOSION, AND DETONATION – Exploding Wires

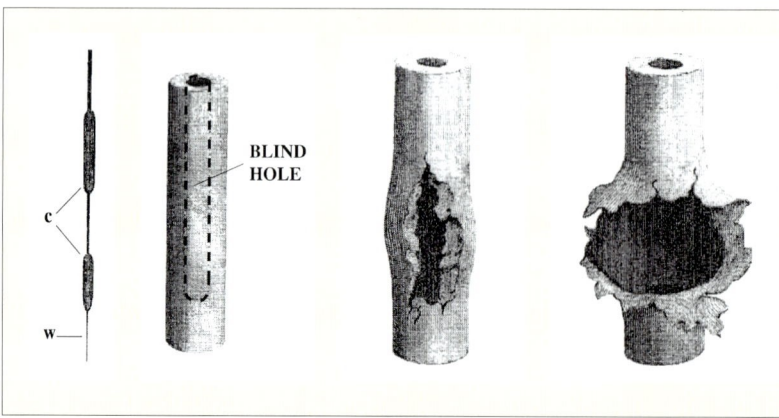

**4.16–N** *Left:* In 1815, the English electricians George J. SINGER and Andrew CROSSE studied the destructive effects of exploding wires by inserting a leaden wire *w* to the lower end of a needle, which was provided with insulating space holders *c* positioned on the axis of a water-filled metallic cylinder, the wire resting with its point on the bottom thereof. *Center & right:* The discharge of Leiden jars through the wire *w* provoked ruptures even of thick-walled tubes such as of a musket barrel. [Phil. Mag. **46** (I), 161 (1815)]

## 4.16 EXPLOSION, IMPLOSION, AND DETONATION – Exploding Wires (cont'd)

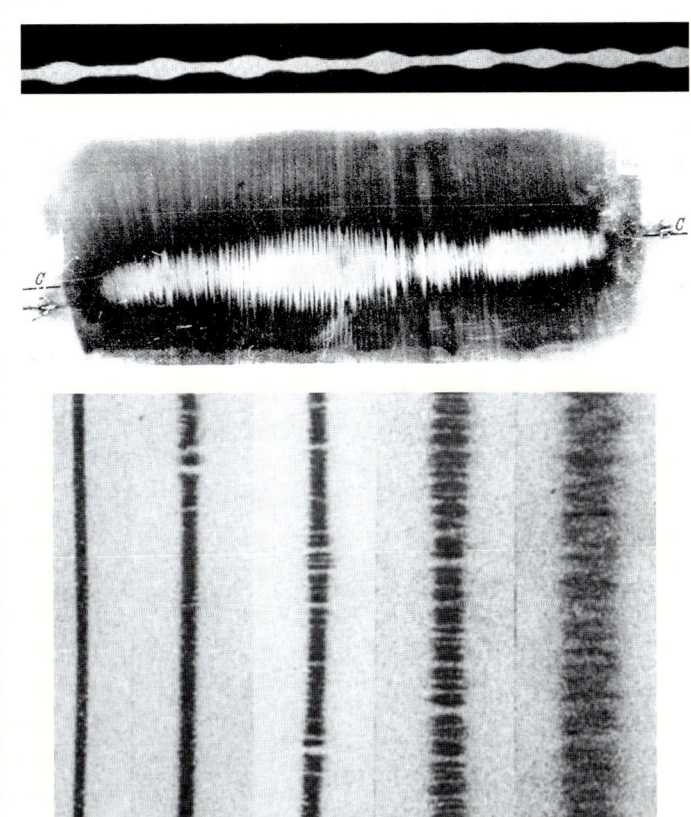

**4.16−O** *Top:* Silhouette of a long copper wire pulsed by a high current. Shortly before the formation of molten wire droplets begins, this curious wire geometry arises, which Henry W. BAXTER, a British researcher, first called "unduloids." He correctly interpreted this curious phenomenon as the result of surface tension. [H.W. BAXTER: *Electric fuses*. Arnold, London (1950), p. 69] *Center:* Heinrich ARNOLD and William M. CONN, two German researchers at Würzburg University, placed a glass slide at a short distance from and parallel to an exploding wire and found in the deposits a characteristic "striation" structure. This picture was obtained from a 0.16-mm-dia. silver wire; the distance from wire to glass plate was 3 mm. [W.G. CHACE and H.K. MOORE (eds.) *Exploding wires*. Plenum, New York (1962), vol. II, p. 77] *Bottom:* The German physicists Francis JAMET and Gustav THOMER, two research physicists at ISL (Saint-Louis, France), using flash radiography resolved this striation process in time and found that the disintegration of the wire occurs shortly after the current pause. At this characteristic time instant – the so-called "dwell" – the wire material, completely evaporating and becoming non-conducting, expands at high speed in a radial direction. The pictures illustrate (from left to right) the evolution of the explosion process of a 0.1-mm-dia. copper wire, taken at a rate of $10^5$ frames/s. [F. JAMET and G. THOMER: *Flash radiography*. Elsevier, Amsterdam (1976), p. 153]

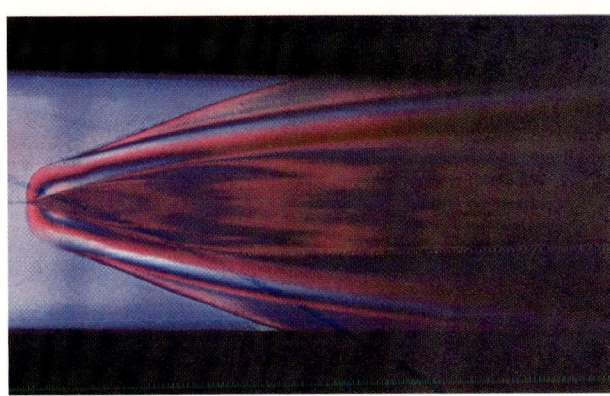

**4.16−P** Frederick D. BENNETT and D.D. SHEAR at BRL (Aberdeen, MD) investigated the mechanism of shock wave production from exploding wires. They found that shock wave production does not proceed from the expansion of the heated metal vapor alone, but rather depends on a two-stage process in which the early expansion of a plasma created by the peripheral arc plays an important part. This single-fringe, streak interferogram of an exploding 4-mil (0.1-mm) copper wire in argon at $1/8$ bar was taken in white light. Note that the complex jagged fringes near the axis attest to a high degree of cylindrical symmetry in the expanding metal. The outermost boundary is the head shock trajectory, while the rounded tip represents the peripheral arc. [W.G. CHACE and H.K. MOORE (eds.) *Exploding wires*. Plenum, New York (1962), vol. II, p. 181]

## 4.16 EXPLOSION, IMPLOSION, AND DETONATION – Large Yield Surface Detonations

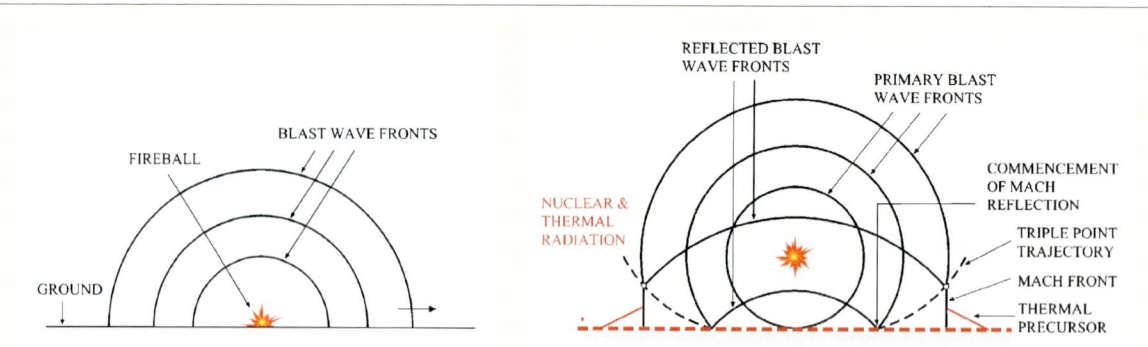

**4.16–Q** *Top:* Schematic of blast wave propagation from a chemical explosive detonated at the surface (*left*) and at a certain altitude above ground (*right*). In the case of a nuclear explosion, a thin layer of high-temperature gas above the ground surface is produced by adsorption of the thermal radiation (*broken red line*), creating ahead of the Mach front a "thermal precursor" – a shock wave (*solid red line*). The rapid expansion of the fireball also initiates seismic shock waves in the ground that generally precede the blast wave. [By author] *Bottom:* Refractive photography of the blast wave generated at the surface by detonation of a hemispherical 500-ton TNT charge, taken in July 1964 at the Canadian Defence Research Establishment (CDRE), Suffield, Alberta. The velocity of the blast wave was measured using cinematography, and the particle velocity using the smoke trail displacement technique. The white smoke trails were formed using cylindrical containers filled with titanium tetrachloride, which were projected from simple mortars. To improve the contrast of the photographic records, a black-smoke background was formed by burning crude oil. These data then allowed the determination of all other blast parameters as functions of time and radial distance. [Courtesy Defence R&D Canada, Suffield, Canada] The analysis of data was carried out and described in more detail by J.M. DEWEY in Proc. Roy. Soc. Lond. **A324**, 275 (1964).

## 4.16 EXPLOSION, IMPLOSION, AND DETONATION – Precursor Detonation Phenomenon

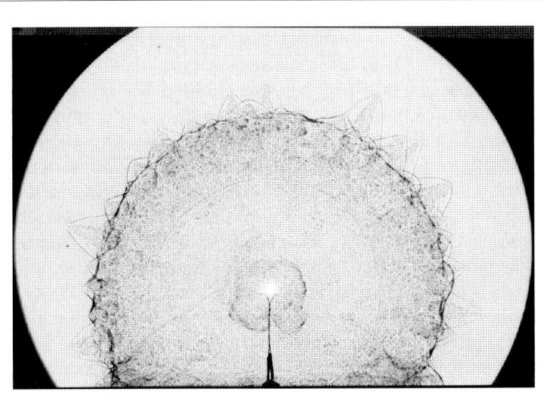
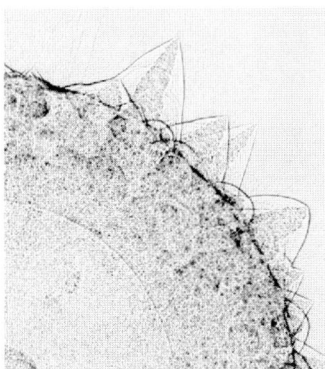

← Magnification of mini precursor shock waves reveals that a conical flow region of turbulent detonation reaction products, following each supersonic particle, is clearly visible. This "head wave" phenomenon is well known from supersonic ballistics and was first observed by Peter SALCHER and Ernst MACH in 1886.

**4.16–R** Precursor shocks are generated in the early stage of a detonating explosive, shown here for a 1-g nitropenta detonator. Small particles of still detonating explosive or paint particles have separated from the main body and, flying with higher velocity than the blast wave, produce small head waves. [Courtesy Dr. Peter NEUWALD, EMI, Freiburg]

## 4.16 EXPLOSION, IMPLOSION, AND DETONATION – One-Dimensional Detonation Front Models

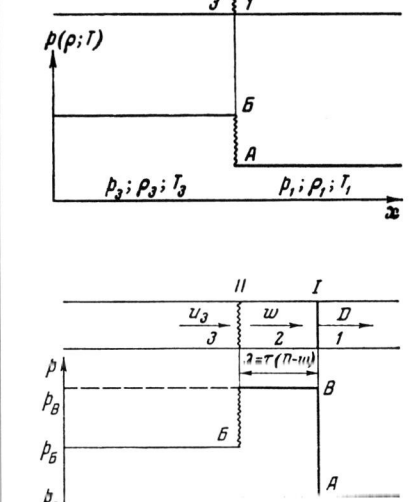
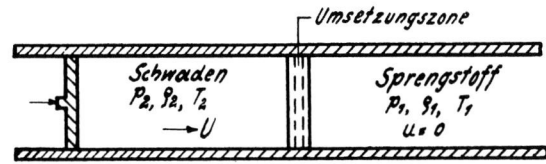

**4.16–S** *Left, top:* MIKHEL'SON (1893), CHAPMAN (1899), and JOUGUET (1904, 1917) assumed that in a detonation the distribution of pressure, density, and temperature occurs instantaneously at the "detonation front." However, combustion is not an instantaneous process, and it takes a certain time until the compressed mixture reacts. *Left, bottom:* In the early 1940s, Yakov ZEL'DOVICH in the U.S.S.R., John VON NEUMANN in the United States, and Werner DÖRING in Germany independently introduced a chemical reaction zone, here denoted by I–II, to take into account the kinetics of the chemical reaction in a detonation. Both models proved useful for a large number of applications. However, experimental facts of irregularities in the detonation front – such as spinning detonation {⇨Fig. 4.16–T} and periodic cell structure {⇨Fig. 4.16–U} – contradict both models: instead of a propagating plane detonation front, gas ignition rather occurs at separate centers, and the plane front disintegrates. [K.I. SHCHELKIN and Y.K. TROSHIN: *Gasdynamics of combustion*. Mono, Baltimore (1965), p. 12] *Right:* Schematic of DÖRING's detonation model as illustrated in his secret report "Beiträge zur Theorie der Detonation." [Forschungsbericht Nr. 1939, Deutsche Luftfahrtforschung, Berlin-Adlershof (1944)]

## 4.16  EXPLOSION, IMPLOSION, AND DETONATION – Spinning Detonation

**4.16–T** In 1926, the British chemists Colin CAMPBELL and Donald W. WOODHEAD at the University of Manchester studied the ignition of explosive gaseous mixtures and measured the velocity of the detonation wave. Their apparatus consisted of a series of metal and glass tubes, forming a continuous horizontal gallery through which the flame passed in front of a drum camera that had a peripheral speed of about 45 m/s. The flame in the test mixture was initiated by the explosion of a hydrogen-oxygen mixture contained in a similar tube in a coaxial position; the complete separation of the two mixtures until shortly before firing was effected by a metal shutter. *Top:* This is a "moving-film" or "smear" record, a diagram time *vs.* distance, of the explosion-wave passing through a moist mixture of $2\,CO + O_2$. The *vertical black lines* are for reference purpose and have intervals of 10 cm. The researchers noticed that the burning gases behind the wave front showed marked horizontal bands. In the following year, a more detailed study of this apparently undulatory form of the detonation wave suggested the idea that the detonation wave propagated along a helical path, a unique phenomenon called "spinning detonation." [J. Chem. Soc. **130**, 1573 (1927)]

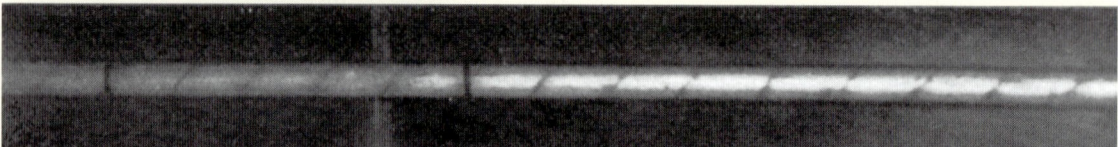

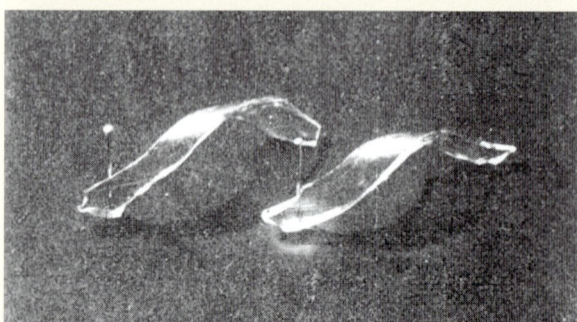

*Center:* William A. BONE and Reginald P. FRASER, two combustion researchers at Imperial College (London, U.K.), resumed Colin CAMPBELL's studies on spinning detonation. They provoked detonation in a lead tube, followed by a glass tube. They photographed the helical track formed by the "head" of detonation in the glass tube as a thin gray film of lead. [Phil. Trans. Roy. Soc. Lond. **A228** (1929), plate 11] *Bottom:* Spiral fragments of a 1.3-cm-inner-dia. glass tube that had been shattered by the detonation of a $C_2N_2 + O_2$ medium in it, as though a spiraling compression wave had passed through the glass and sheared it. [Phil. Trans. Roy. Soc. Lond. **A230** (1932), plate 20]

## 4.16 EXPLOSION, IMPLOSION, AND DETONATION – Periodic Cell Structure

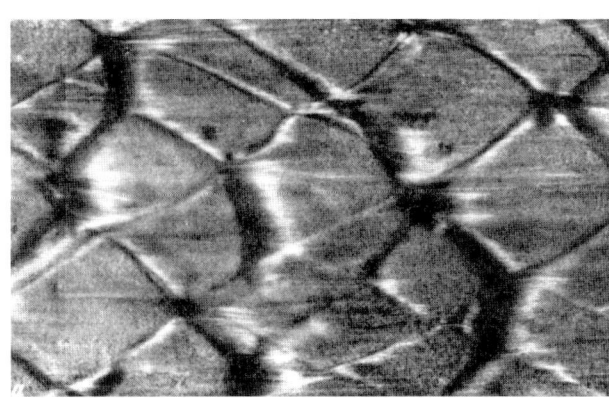
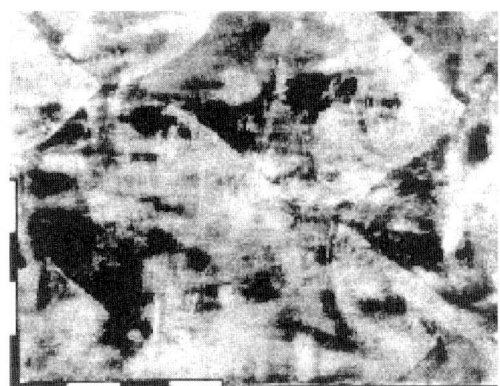

**4.16–U** *Left:* In 1959, the Soviet combustion researchers Yuri N. DENISOV and Yakov K. TROSHIN first reported on the discovery of a cellular structure of detonation waves using smoked foils exposed to the detonation, a recording technique based upon ANTOLIK's soot method {⇒Fig. 4.5–C}. Depending on the ambient pressure and mixing ratio, they observed curious soot patterns of an almost periodic cell structure. [Dokl. Akad. Nauk **125**, 110 (1959)] *Right:* More recently, Fuqing ZHANG and collaborators at RWTH Aachen, using the smoked-foil technique, recorded from a dust detonation traces of triple points on the tube wall, thus first showing that cell structures also exist in dust explosions. [Shock Waves **11**, 53 (2001). Courtesy Prof. Hans GRÖNIG, Stoßwellenlabor der RWTH Aachen]

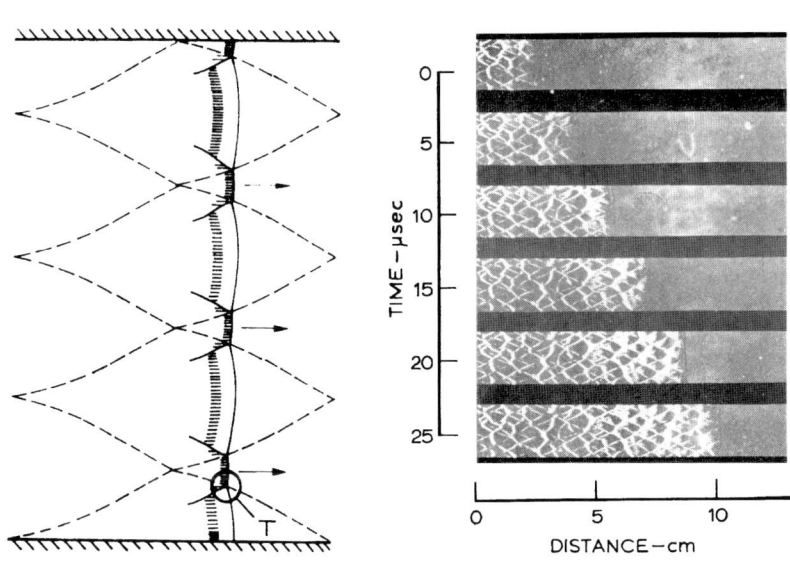

**4.16–V** Schlieren photography has shown that the detonation front actually consists of a system of intersecting shock fronts that produce Mach reflection. *Left:* Schematic diagram of a detonation wave. The *solid lines* represent curved shock fronts, the *shaded regimes* are the combustion zones, and the *broken lines* depict the traces of the wave intersection points. Vortices in the triple points act as rotating styli and generate the diamond-shaped cellular pattern. The *arrows* indicate the positions where the wave system may receive the propulsive drive. *Right:* Sequence of stroboscopic laser-shadow photographs of a gaseous detonation in a hydrogen-oxygen mixture. [A.K. OPPENHEIM: *Introduction to gasdynamics.* Springer, Vienna (1970), pp. 27, 31]

## 4.16 EXPLOSION, IMPLOSION, AND DETONATION – Other Explosion Phenomena *(cont'd)*

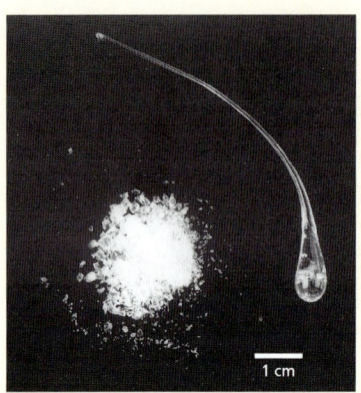

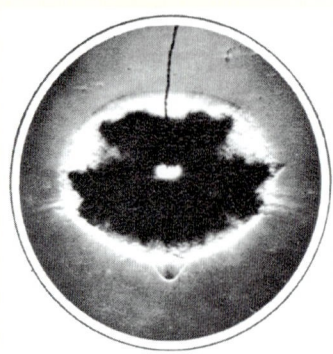 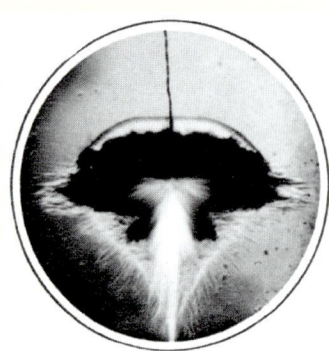

**4.16–W** So-called "Prince RUPERT's drops" are formed by dropping a small gob of hot, molten glass into cold water. They were used from the 17th century to demonstrate the high internal stress frozen inside of the tadpole-shaped glass. Upon breaking the thin tail, the glass releases the internal stress explosively so that the entire piece shatters into a fine powder. [Courtesy Dr. Heinz REICHENBACH, EMI, Freiburg]

**4.16–X** In England, William H. PAYMAN and collaborators at the British Safety in Mines Research Establishment (SMRE) studied near-field explosion phenomena of plain No. 6 detonator caps using a schlieren camera and making snapshot spark photographs. The solid particles that appear to play an important part in the initiation of explosion by a detonator may be composed either of metal torn from the case or of unconsumed detonating composition, or of both. ***Left:*** For a copper-cased fulminate-chlorate detonator bursting occurs quicker in the lateral than in the axial direction, leading to an ellipsoidal wave geometry, and no focusing effect happens. ***Right:*** For a compound aluminum-cased lead azide detonator the researchers observed jetting of gases and particles sent out in advance of the main shock. This type of detonator will ignite firedamp-air mixtures, unlike the copper-cased fulminate-chlorate detonator. [Proc. Roy. Soc. Lond. **A148**, 604 (1935)]

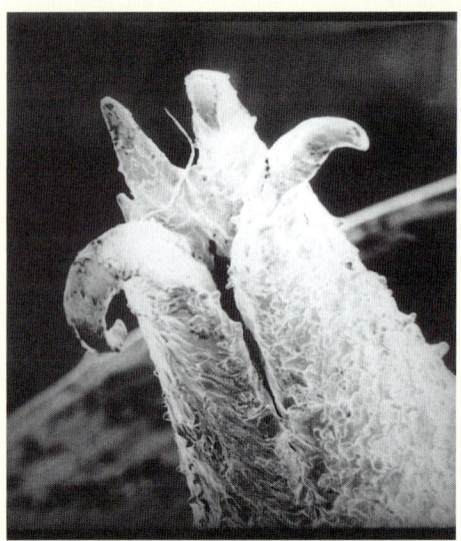

**4.16–Y** In flash X-ray tubes high-velocity electrons impinging as a beam on a metallic surface are scattered, slowed down, and stopped within a short depth below the surface. Most of the kinetic energy of the beam is transformed into heat and only about 1‰ goes into X-rays. Depending on the electron energy and the absorption and melting characteristics of the applied anode material, the impact of fast electrons can be accompanied by subsurface mini explosions. The SEM picture, taken by Reinhard THAM (EMI, Freiburg) in 1984, shows the tip of a conical Mo-anode from a vacuum-discharge flash X-ray tube, after being pulsed several hundred times via a 100-kV low-impedance transmission line. [P. KREHL, Proc. 13th Int. Congr. High-Speed Photography & Photonics, Tokyo (1978). Japan Society of Precision Engineering, Tokyo (1979), p. 409] For a photon energy of 100 keV the penetration depth in molybdenum is in the order of 1 mm and quickly diminishes with decreasing photon energy. The high-velocity bombardment of electrons generates a curious, "banana-split"-type erosion pattern. [EMI Archives, Freiburg, Germany]

## 4.16 EXPLOSION, IMPLOSION, AND DETONATION – Explosive Ablation of Biological Tissue

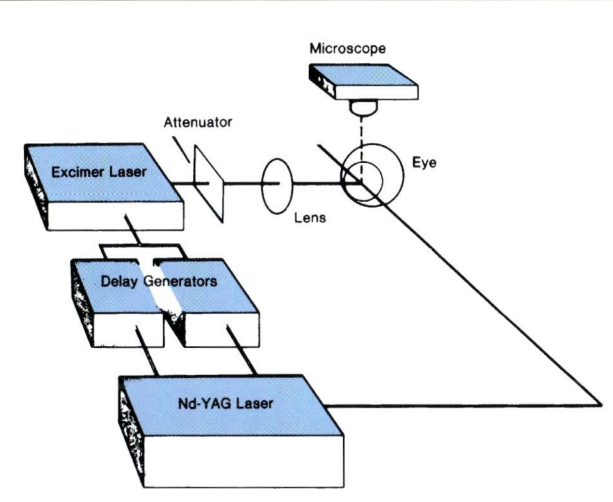

**4.16–Z** The irradiation of biological tissue by an ultrashort laser pulse is an explosionlike process similar in appearance to other laser-induced explosive phenomena. This so-called "ablative photo decomposition effect" is used in medical therapy, for example in dermatology and ophthalmology. In the mid-1980s, the U.S. ophthalmologist Carmen A. PULIAFITO and collaborators at Boston's Laser Research Laboratory and Harvard's Medical School first studied laser ablation of the cornea using high-speed photography. *Top:* They used a 0.9-J/cm$^2$ 15-ns FWHM excimer laser at 193 nm, which was focused on the cornea to a spot size of 0.44 mm in diameter. A second laser, triggered after a variable delay time, was used as a flash light source to illuminate the near environment of the cornea and to visualize the ablation process perpendicularly to the plume propagation. Pictures of the ablation process were recorded using a 35-mm still camera (Nikon) attached to an operating microscope. [Arch. Ophthalmol. **105**, 1255 (1987)]

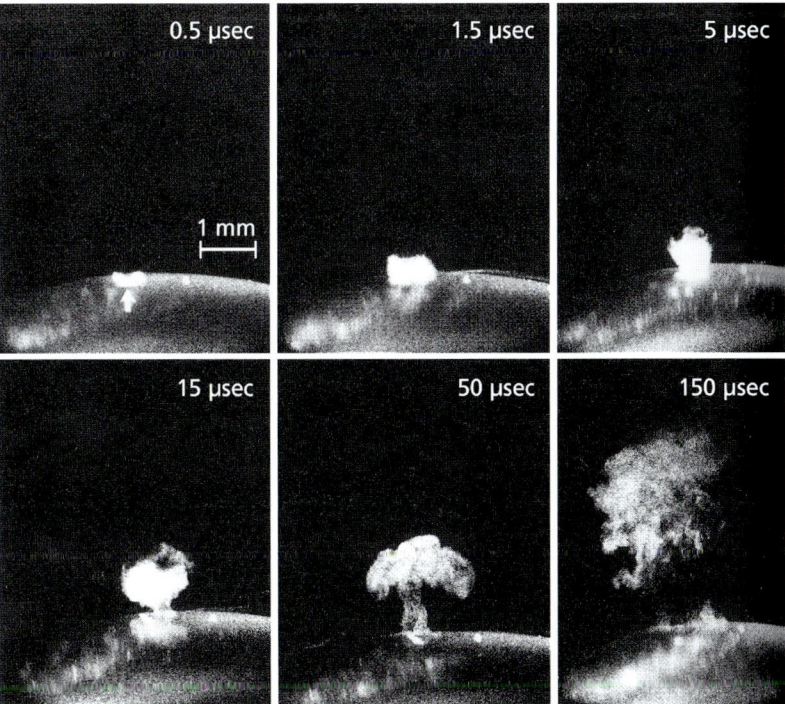

*Bottom:* This is a series of single-shot shadowgraphs (or pseudo cinematography) of the ablation process taken at six different time instants after firing the excimer laser. The laser beam comes from above the photograph and strikes the cornea near the point indicated by the *white arrow* (*top, left*). At 0.5 μs after firing of the excimer laser pulse the plume of ablated material extends 0.2 mm from the cornea, indicating an initial average supersonic velocity of 400 m/s. The ablation process is essentially completed between 5 and 15 μs after the excimer pulse. Note the formation of a mushroom cloud with a long stem at 50 μs. By 150 μs the effects of diffusion and turbulence have become dominant. Today ultraviolet pulsed radiation from excimer lasers is routinely applied to the surface of the cornea in order to correct low to high levels of nearsightedness, farsightedness, and astigmatism.

## 4.17  EXPLOSION AND DETONATION DIAGNOSTICS – Maximum Pressure of Fired Gunpowder

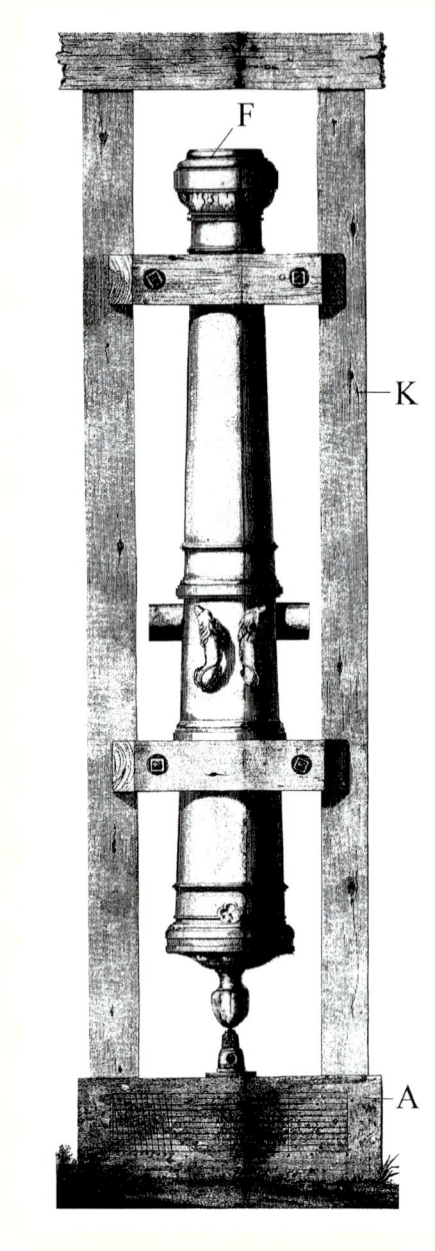

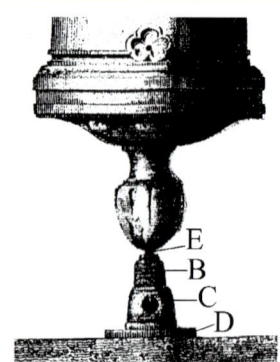

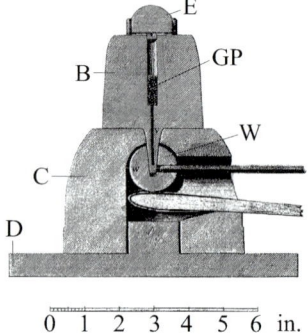

← Cross section of his setup for testing the force of fired gunpowder:
A – solid block of very hard stone
B – barrel of hammered iron
C – support of cast brass
D – circular plate of hammered iron
E – hemisphere of hardened steel
F – 24-pdr cannon, used as a heavy weight
GP – gunpowder test sample
K – wooden frame
W – heated iron ball for igniting GP

**4.17–A**  In the 1790s, Sir Benjamin THOMPSON, Count VON RUMFORD, then Bavarian Minister of War, developed an experimental method to systematically determine the force of fired gunpowder. *Right:* His "test bomb" *B*, a short iron gun barrel with thick walls, was closed at the top by a movable hemisphere *E*. *Left*: To counterbalance the pressure of the generated "elastic fluid" (*i.e.*, of the hot explosion products), he used a heavy 24-pdr cannon *F* that rested on *E* and was movable in a vertical position within a wooden frame *K*. The bottom of *B* rested on a solid block *A* via a steel disk *D*. In order to avoid erosion of the vent and a loss of elastic fluid, he ignited the gunpowder *GP* inside *B* by bringing a red-hot ball *W* in contact with the bottom of *B*. When the force of the generated elastic vapor was just sufficient to raise the weight, the explosion was attended by a very sharp report. In the special case that the chamber was completely filled, which required 1.8 g of powder, he estimated a force of fired gunpowder of about 29,000 at. However, he observed that the force of fired gunpowder also depended on its moisture content. From an experiment resulting in a destroyed chamber he (erroneously) estimated the enormous pressure of 101,021 atm – a value very different from 5,500 atm as later determined by Sir Frederick A. ABEL and Sir Andrew NOBLE in careful measurements. [Phil. Trans. Roy. Soc. Lond. **87**, 222 (1797)]

## 4.17 EXPLOSION AND DETONATION DIAGNOSTICS – Maximum Pressure and Temperature

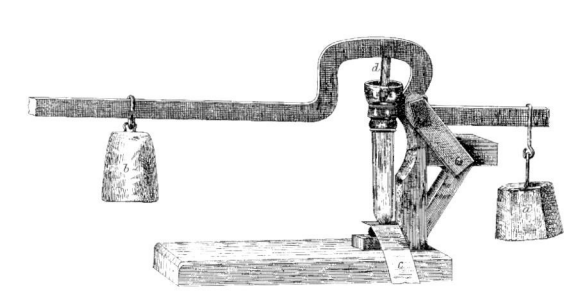

**4.17–B** View of Robert W. BUNSEN's exploding vessel, which he used in his laboratory at the University of Heidelberg in an attempt to measure the maximum pressure caused by the detonation of explosive gaseous mixtures. His handy instrument consisted of a closed air-tight tube with a weight-loaded valve and allowed one to determine the temperature of detonation. The ignition of the test gas occurred by an electric spark via electrodes $c$ and $d$ through the entire length of the tube. He observed which weight $b$ could still be raised by the detonating gases. [Phil. Mag. **34** (IV), 489 (1867)]

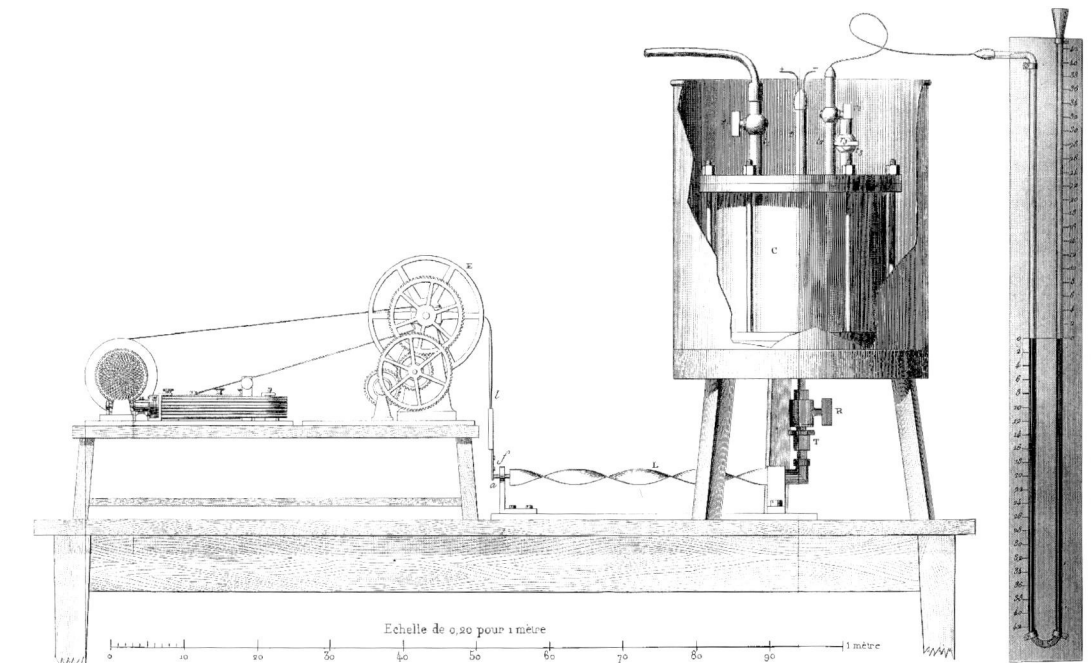

**4.17–C** François E. MALLARD and Henry L. LE CHÂTELIER at the Ecole des Mines in Paris determined the temperature of detonation of firedamp and other combustible gaseous mixtures indirectly by measuring the explosion pressure in a closed reaction vessel. They used a calibrated Bourdon gauge connected via a stiff axle $L$ to a Deprez chronograph, which was provided with a needle $l$ at its end, and registered the pressure on a revolving cylinder covered with black paper. The gaseous mixture was fed to the reaction vessel $C$ via a pipe $t_1$. Another pipe $t$ was a feed-through for two electric wires, which ignited the test gaseous mixture via an electric spark. While studying the specific heat and dissociation of gases at elevated temperatures, they made the curious observation that ignition of firedamp only occurs when the gaseous mixture is kept a certain time above its ignition temperature, which is approx. 650 °C for firedamp. This delay of ignition allows the application of special explosives of even high combustion temperatures for shooting purposes in the coal-mining industry, such as ammonium nitrate. Although its combustion temperature is about 1,100°C – *i.e.*, well above the ignition temperature of firedamp – it cannot ignite it, because the duration of this high combustion temperature is too short. [Ann. des Mines **4** (IV), 274 (1883)]

## 4.17 EXPLOSION AND DETONATION DIAGNOSTICS – Chamber Pressure of a Detonating Explosive

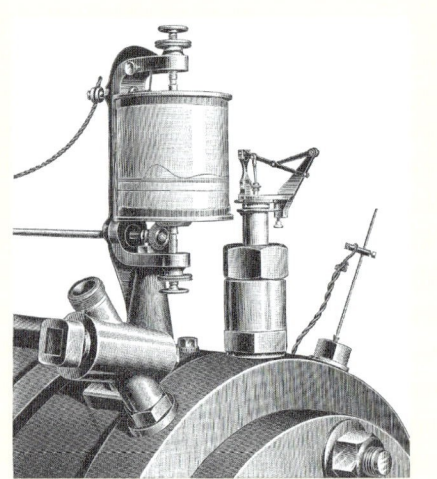

**4.17−D** *Left:* At the Sprengstoff-AG Carbonit (Schlebusch/Köln), Christian E. BICHEL, a German combustion researcher, determined the pressure in a closed chamber generated by the detonation of various explosives. *Right:* His mechanical "brisance gauge," a steam-engine indicator coupled with a rotating drum, provided crude pressure-time profiles from which he tried to determine the maximum pressure and the performed work. From the rise of pressure he estimated the brisance of an explosive. [C.E. BICHEL: *Experimentelle Untersuchung von Sprengstoffen*. Mittler, Berlin (1898), plates V and VI]

## 4.17 EXPLOSION AND DETONATION DIAGNOSTICS – Brisance Test of Gunpowder

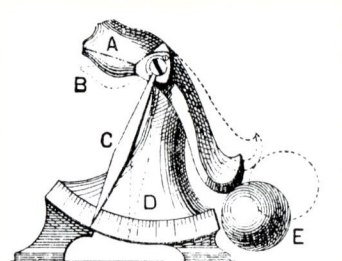

**4.17−E** *Left & center:* A 19th-century éprouvette, probably of British origin, in the form of a small pistol (overall length about 15 cm). Upon firing, the force of gunpowder turns a friction-restrained wheel numbered 1,2…,8 for calibration purposes [Courtesy R. WATSON, Manitou, Manitoba, Canada] *Right:* The éprouvette was used for quickly testing the brisance quality of gunpowder for use in cannons. It consists of a lever that carries on one end a lid $B$ covering a small explosion chamber $A$ and at the opposite end a counterweight $E$. The explosion of a test quantity of gunpowder in $A$ moves lever $B$ and the trailing pointer $C$ into the *dotted position*. The amount of shift read at scale $D$ was a measure of the brisance of the tested gunpowder. [*Larousse du XX$^e$ Siècle*. Larousse, Paris (1930), vol. 3, p. 231] However, the efficiency of gunpowder applied in firearms depends on a variety of firearm parameters and cannot simply be determined by these primitive testing methods. For example, Carl CRANZ appropriately wrote, "The term 'brisance' is not an established one. For one firearm a powder may be too explosive, while for another it may be too sluggish…" [C. CRANZ: *Lehrbuch der Ballistik*. Springer, Berlin (1927), vol. 2, p.116]

## 4.17 EXPLOSION AND DETONATION DIAGNOSTICS – Brisance Test of a High Explosive

← View of ten cut-open Trauzl probes of different high explosives but of the same weight. *Left* to *right*: Nine explosives of German production (Dynamit, Roburit, Dahmenit, Dahmenit A, Westfalit, Köln-Rottweiler Sicherheitssprengpulver, Progressit, Roburit I, Kohlenkarbonit), followed by conventional gunpowder (*far right*). All high explosives produce a pear-shaped bulge, while only gunpowder produces a cylindrical cavity.

**4.17−F** In the 1880s, Isidor TRAUZL, an Austrian blasting engineer, devised a simple brisance test, the so-called "Trauzl lead block test." First introduced in Germany in 1903 as a standard test method, it allowed one to compare the brisance of new explosives in a simple manner. For the standard test, the blocks are cast from pure lead, 20 cm in height and 20 cm in diameter, with a central hole made by the mold, 12.5 cm deep and 2.5 cm in diameter. The Trauzl lead block test was later adopted worldwide and is still in use today. [F. HEISE: *Sprengstoffe und Zündung der Sprengschüsse*. Springer, Berlin (1904), p. 36]

## 4.17 EXPLOSION AND DETONATION DIAGNOSTICS – Test of Explosives Used in Mining

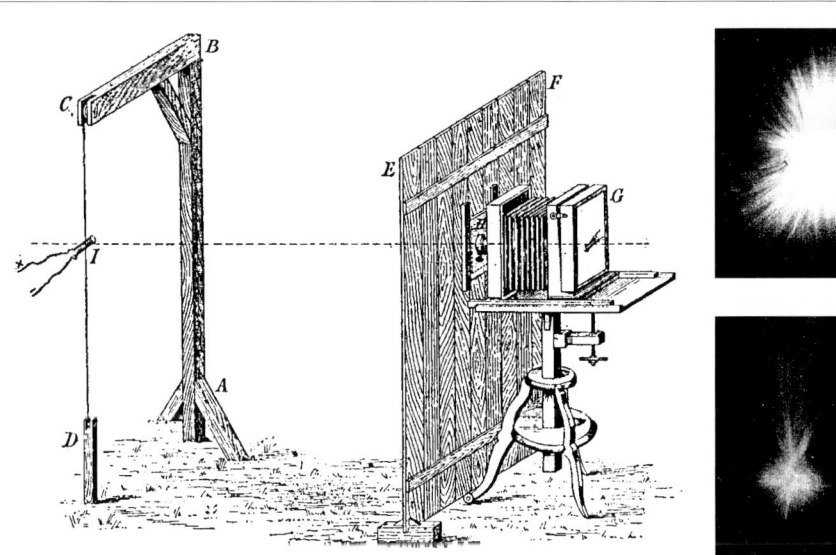

← SIERSCH at Dynamit AG in Vienna tested a large number of explosives with his photographic method. For example, he observed that a 100-g charge of Guhr Dynamite (*top*), housed in a freely suspended cartridge, produces a brilliant light flash that might ignite firedamp or coal dust in a mine, while the explosion of the same amount of nitroglycerine (*bottom*) is only accompanied by a faint flash of light.

**4.17−G** The Austrian explosives specialist Alfred SIERSCH used photography as a tool to classify "safe" explosives for use in mining. *Left:* He photographed the test explosive *I* in its own light with a still camera *G* through a shield *E−F*. Since he worked with an open shutter *H*, he had to conduct his studies at night. *Right:* According to his classification, safe explosives were those that emitted only a little light and, therefore, only a little heat to provoke firedamp and/or coal dust explosions. [Trans. Fed. Inst. Min. Eng. **11**, 2 (1896)]

## 4.17 EXPLOSION AND DETONATION DIAGNOSTICS – Dust Explosion Tester

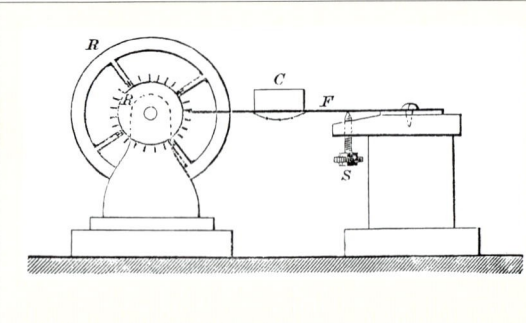

**4.17–H** Rudolph WEBER at the Polytechnic Institute in Berlin studied the causes of flour-dust explosions. To generate a steady stream of dust for experimentation with an open flame or an electric spark, he invented a simple dust generator. It consisted of a wheel $R$ provided with 24 spikes that successively struck a leaf spring $F$ carrying a cylinder $C$. This cylinder contained the flour dust sample to be tested and was provided at its bottom with a fine grid. Upon rotation of $R$, the spikes shook the cylinder, thus producing a constant flow of flour dust. A screw $S$ allowed one to adjust the dust intensity. WEBER tested the trigger ability of a flour-dust explosion by positioning the heat source close to the generator and found that explosions occurred at concentrations of 20–30 mg dust in 1 liter of air. [Z. Tech. Hochschulen **3**, 51 (1878)]

## 4.17 EXPLOSION AND DETONATION DIAGNOSTICS – Detonation Velocity of a High Explosive

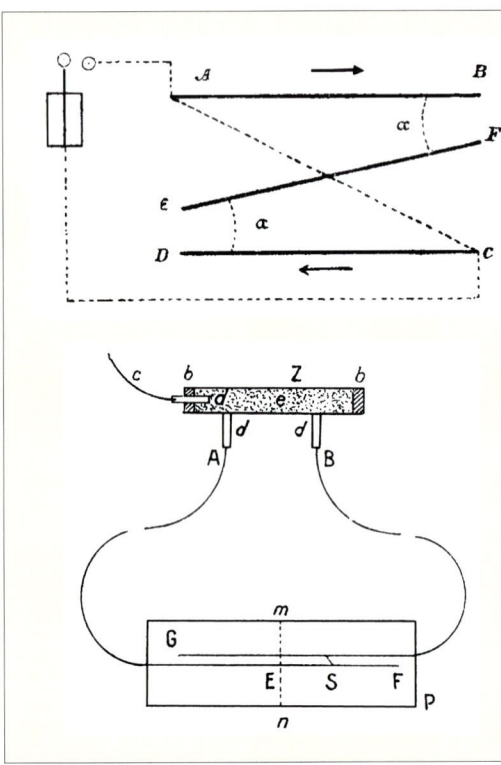

**4.17–I** Illustration of two examples of "antiparallel" testing methods as used in early detonation diagnostics: *Top:* Ernst MACH and Josef WENTZEL at the German Charles University in Prague used an ingenious setup to determine the detonation velocity of silver fulminate: two parallel line charges, $AB$ and $CD$, triggered simultaneously by discharging a Leiden jar, produce an interference line $EF$, which is recorded on a fixed soot-covered plate positioned in the center. They determined the detonation velocity $D$ from the inclination of the line $EF$ by $\sin\alpha = c/D$, with $c$ as the sound velocity in air. [Sitzungsber. Akad. Wiss. Wien **92** (II), 625 (1885)] *Bottom:* The Frenchman Henri J. DAUTRICHE also invented an antiparallel method of determining the detonation velocity that, like MACH and WENTZEL's setup does not require the use of a chronograph. However, his method depends upon a comparison of the velocities of the unknown explosive $Z$ with a standard explosive of known velocity of detonation, here given in the form of a cordeau of length $A–B$. The point $S$ where the two explosive waves in the cordeau meet is marked on the surface of a lead plate upon which the cordeau is resting. The shift $E–S$ of this mark from the midpoint of cordeau allows one to determine the detonation velocity by a simple formula. [Mém. Poudres Salpêtres **14**, 216 (1906–1907)]

## 4.17 EXPLOSION AND DETONATION DIAGNOSTICS – Detonation Velocity of a Gaseous Explosion

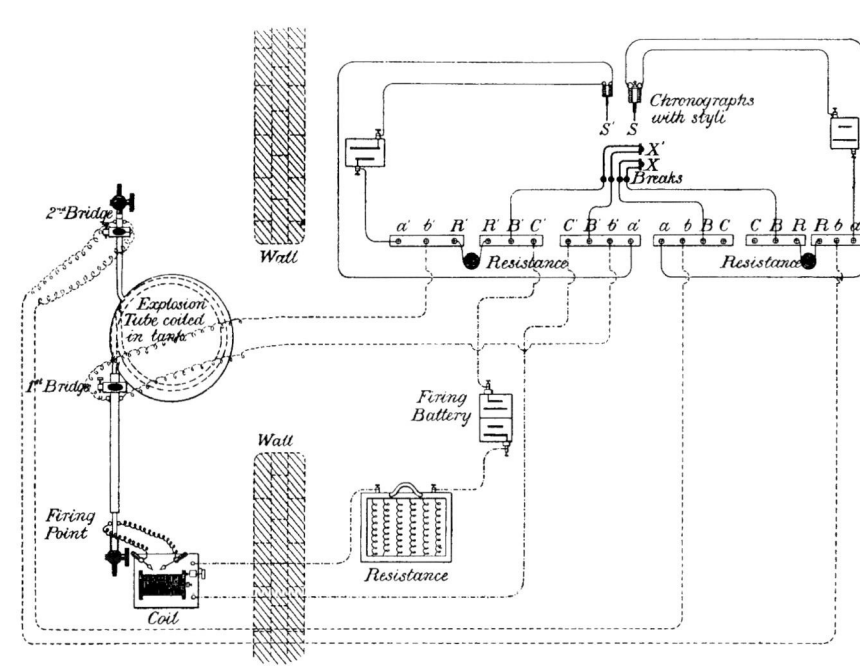

← Two styli, $S$ and $S'$, were put in a circuit through resistances $RR$ and $R'R'$ and the breaks $X$ and $X'$, respectively. The explosion wave traveling down the tube first breaks the bridge wire at the entrance of the coil and then breaks a second wire upon reaching the coil exit. The time difference between these activations of the two styli, a measure of the flame needed to propagate through the coiled tube, was recorded with an Elliot chronograph – a heavy pendulum carrying a smoked-glass plate. On letting the pendulum fall, the two styli were released, thus registering their marks on the moving glass plate.

**4.17–J** Schematic of a setup used by the British physicist Harold B. DIXON to measure the detonation velocity of gaseous mixtures. He used a long leaded tube, mostly 100 m long and 9 mm in diameter. Wound up to a coil and filled with the electrolytic gas, it was fired at the fixing point by interrupting the current flowing through an ignition coil via the break contact $X'$. For an oxyhydrogen detonation he found a detonation velocity of 2,821 m/s. [Phil. Trans. Roy. Soc. Lond. **A184**, 97 (1893)]

## 4.17 EXPLOSION AND DETONATION DIAGNOSTICS – Interior Ballistic Studies

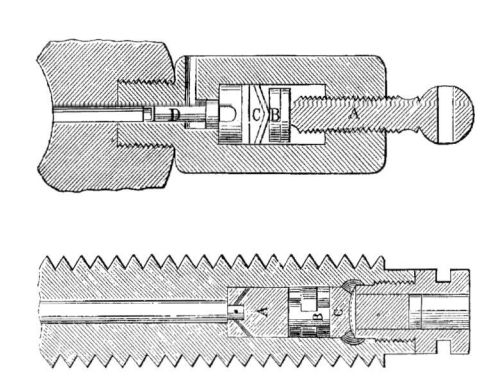

**4.17–K** *Top:* In 1857, Captain Thomas J. RODMAN of the U.S. Army (Allegheny Arsenal, PA) used an indentation gauge to measure the maximum gas pressure in the bore of a gun, the so-called "Rodman gauge." He observed that the pressure diminishes with increasing grain size of the gunpowder, which is particularly advantageous for the operation of large-caliber guns. *Bottom:* For the same purpose, Sir Andrew NOBLE at W.G. Armstrong & Co. (Elswick, U.K) invented in 1862 a crusher gauge, the so-called "Noble gauge." Based on the analysis of the deformation of a metal cylinder, it provides more precise pressure data than the Rodman gauge. Both gauges were calibrated using a static pressure loading. [*Meyers Konversations-Lexikon*, Bibliographisches Institut, Leipzig (1897), vol. 15, p. 434] NOBLE, measuring the "elastic force of fired gunpowder" by using his crusher gauge, noticed that RODMAN had measured erroneously a too high pressure of 12,400 atmospheres, a value almost twice as much as he obtained.

## 4.17 EXPLOSION AND DETONATION DIAGNOSTICS – Interior Ballistic Studies *(cont'd)*

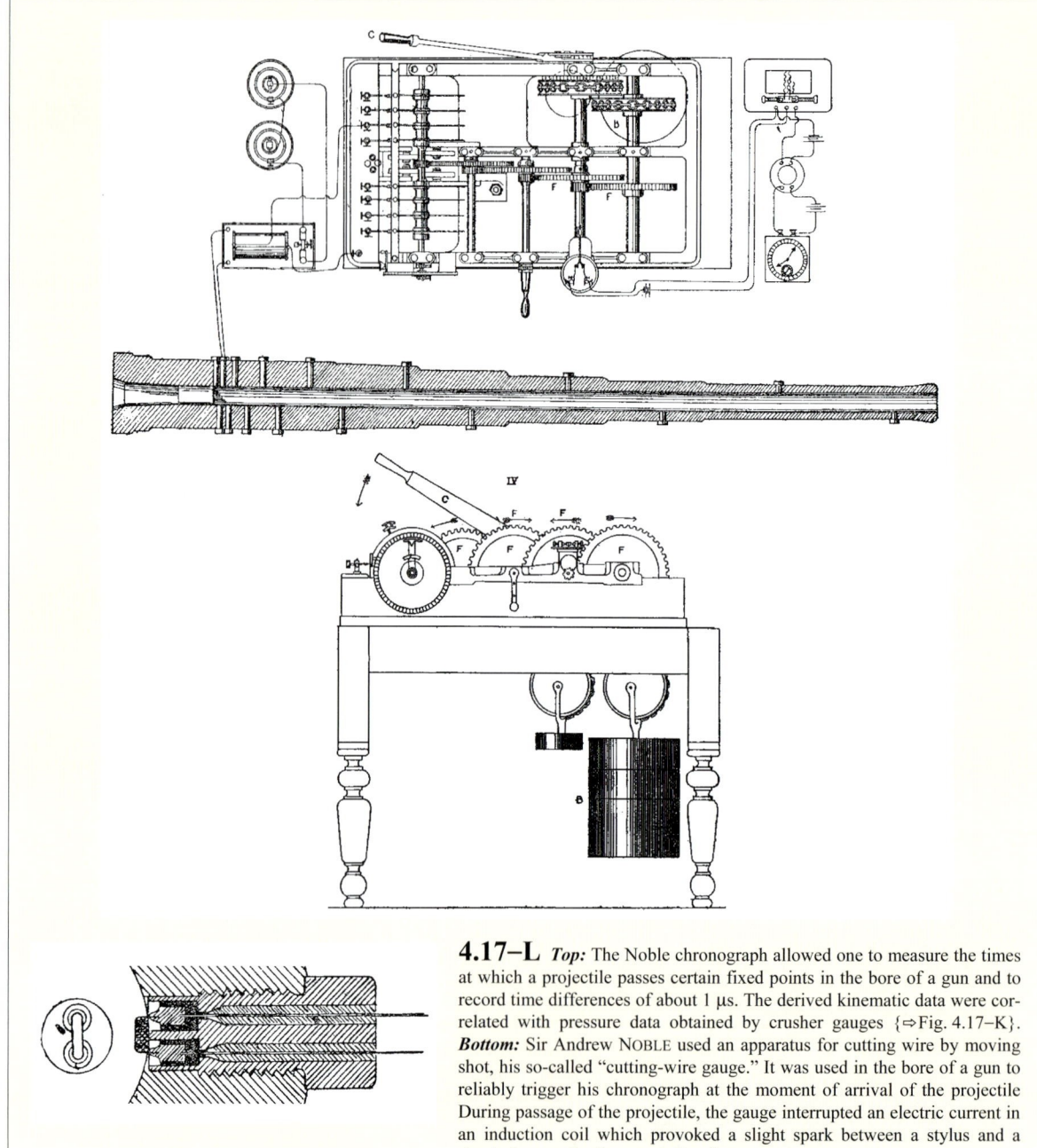

**4.17–L** *Top:* The Noble chronograph allowed one to measure the times at which a projectile passes certain fixed points in the bore of a gun and to record time differences of about 1 µs. The derived kinematic data were correlated with pressure data obtained by crusher gauges {⇨Fig. 4.17–K}.
*Bottom:* Sir Andrew NOBLE used an apparatus for cutting wire by moving shot, his so-called "cutting-wire gauge." It was used in the bore of a gun to reliably trigger his chronograph at the moment of arrival of the projectile During passage of the projectile, the gauge interrupted an electric current in an induction coil which provoked a slight spark between a stylus and a sheet of paper fixed on a rotating drum. [Brit. Assoc. Rept. **64**, 523 (1894)]

## 4.17 EXPLOSION AND DETONATION DIAGNOSTICS – Blast Wave Recording

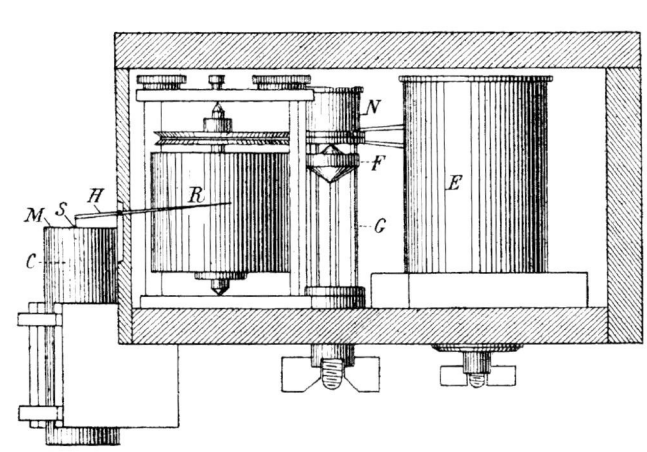

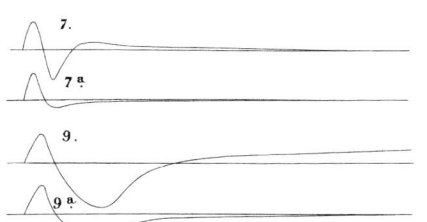

↑ $C$ – brass cylinder, $E$ – small electric motor, $F$ – dropping weight, $G$ – gliding bars, $H$ – lever, $M$ – caoutchouc membrane, $N$ – electromagnet, $R$ – recording drum, $S$ – stylus

**4.17–M** W. WOLFF, a military engineer at the German Army Ballistic Test Site in Berlin-Cummersdorf, used a mechanical membrane-type pressure gauge and a rotating-drum chronograph to record free-field pressure-time profiles of large-scale chemical explosions in air. He compared blast effects of charges up to weights of 1,500 kg using two different explosives, black powder and trinitrophenol. Note that the typical negative pressure phase of a blast wave is correctly reproduced. [Ann. Phys. Chem. **69** (III), 329 (1899)]

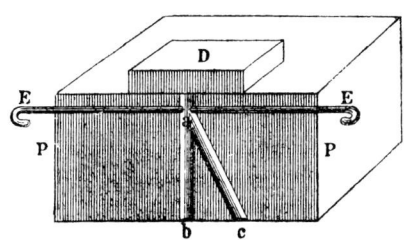

**4.17–N** Schematic of Ernst MACH's ingenious arrangement for measuring the propagation velocity of a blast wave, which he generated by an electric spark $a$ between two electrodes $E$–$E$. The blast wave, split into two shock waves and propagating in the two bores $a$–$b$ and $a$–$c$ of different lengths arranged in a wooden block $P$–$P$, arrived at two different time instants at openings $b$ and $c$. Here their arrivals were recorded as angular marks on a soot-coated, fast-revolving disk. Using blocks $P$-$P$ of different thickness that provided different channel lengths $a$–$b$ and $a$–$c$, MACH also applied this simple method to determine point-by-point how the shock-front velocity decays with distance from the explosion source. [Sitzungsber. Akad. Wiss. Wien **77** (II), 819 (1878)]

**4.17–O** Wolfgang PARR and collaborators at EMI developed a ballistic "roller pendulum" to measure the combined blast and fragment loading on a vertical surface near the ground. The instrument had a target area of 0.2 m² and allowed one to measure with sufficient accuracy impulsive loads ranging from 0.2 to 200 kg m/s. The picture shows a setup of four roller pendulums positioned in the near-field around a detonating model shell consisting of 1 kg Composition B and a 2-kg steel jacket. In order to avoid gliding, the center of percussion in this pendulum type must be located close to the point of contact. [Bericht E20 (1983), EMI, Freiburg]

## 4.17 EXPLOSION AND DETONATION DIAGNOSTICS – Streak Photography in Detonics

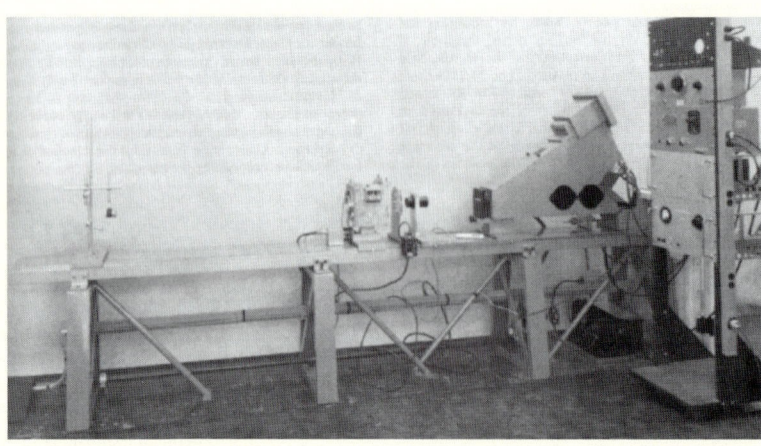

**4.17–P** Berlyn BRIXNER, a high-speed photographer at LASL (Los Alamos, NM), developed a high-speed sweeping image camera for routine explosive testing. The rotating mirror, revolving at 2,000 rps, sweeps the image of a 3-in. (76-mm) long slit at a velocity of 13 mm/µs. The photo shows a general view of his camera assembly used for studying miniature explosive events such as sparks, exploding wires, detonators, and other miniature explosive charges contained in the box shown at the center; a welded-steel construction provided with Lucite windows as viewing ports on two sides was also used to permit use of backlighting when desired. [J. SMPTE **70**, 180 (1961)]

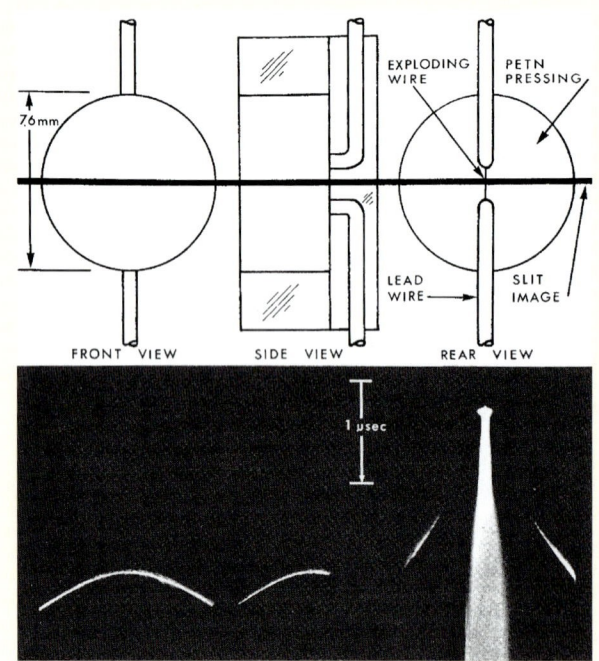

**4.17–Q** In the 1960s, James H. BLACKBURN and Robert J. REITHEL, two detonation researchers at LASL, used BRIXNER's high-speed sweeping image camera to study the initiation of detonation in PETN by an exploding wire. They took photographs through a modified, transparent head of a detonator. ***Top:*** Schematic of their test setup showing three views of a modified exploding wire detonator with two 0.8-mm-dia. lead wires embedded in a Lucite insulator. ***Bottom:*** Streak photographs taken in front, side, and rear view. The first light recorded has the image of an arrowhead, the tip designated as the wire light and the blade as the flare light. The two traces proceeding outward to the sides denote the initiated detonation wave. Note that detonation does not occur immediately after wire explosion begins, but rather that a buildup phase with a duration of about 1 µs is required.
[W.G. CHACE and H.K. MOORE (eds.) *Exploding wires*. Plenum, New York (1964), vol. 3, p. 153]

## 4.17 EXPLOSION AND DETONATION DIAGNOSTICS – Reflected-Light Photography in Detonics

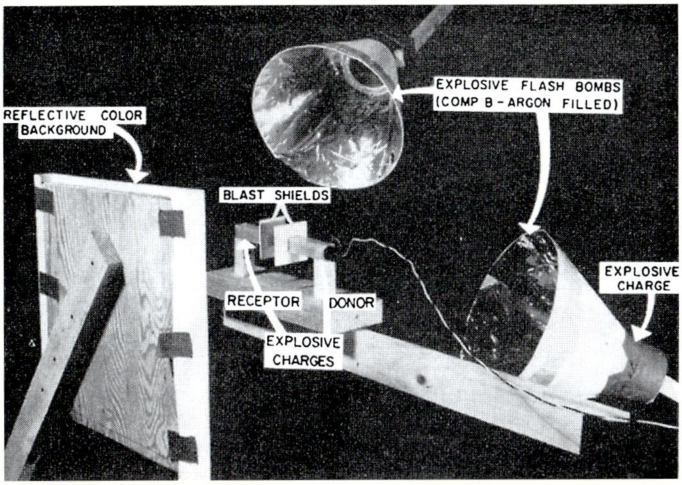

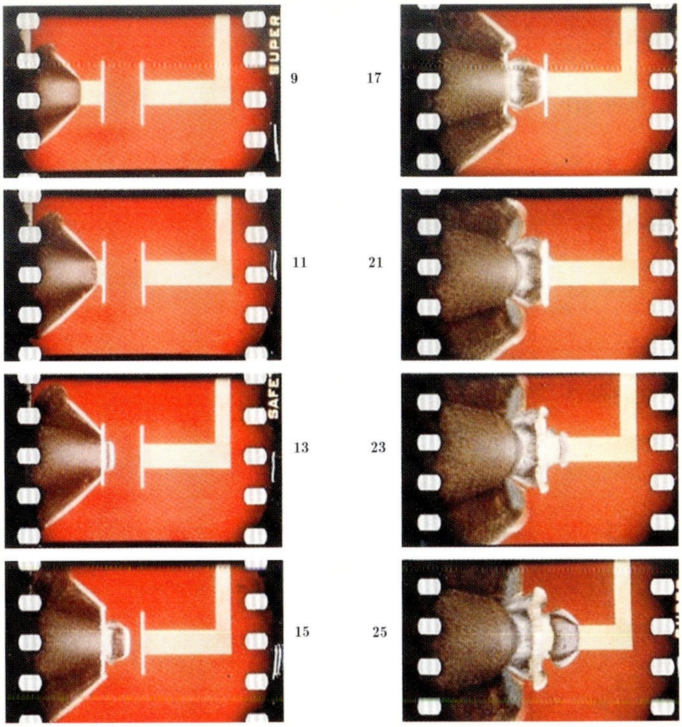

**4.17–R** Morton SULTANOFF and Robert L. JAMESON at BRL (Aberdeen, MD) used submicrosecond color cinematography to investigate the influence of the air shock sent out from the end face of a detonating explosive on a second high explosive positioned close by. This strong blast interaction did not cause immediate reaction in the receptor explosive, but rather produced a discontinuity to the detonation of the receptor charge. *Top, left:* View of experimental setup. Lighting was provided by two argon flash bombs positioned at 45° from the event (flash duration ca. 30 µs). Both end faces were provided with "blast shields" to prevent the detonation gases from spilling over and obscuring the receptor stick until the appearance of the detonation in the receptor stick. *Top, right:* View of the Beckman & Whitley camera model 189, a 25-frame reimaging high-speed camera. Operated at a rate of up to 1.2 million frames/s, this resulted in a full recording time of 20 µs and an exposure time of 0.1 µs. [W.G. HYZER: *Engineering and scientific high-speed photography*. Macmillan, New York (1962), p. 127] **Bottom:** The receptor charge (*right*) was separated from the donor charge (*left*) by a 5-mm air gap. Using 35-mm film Super Anscochrome, they recorded the detonation events in reflected light. The obvious preliminary shock in the receptor explosive stick, which explains the heretofore incompatible physical discontinuity to detonation, can be seen in the selected frames. [J. SMPTE **69**, 113 (1960)]

## 4.18 OPTICAL METHODS FOR FLOW VISUALIZATION – Schlieren Photography

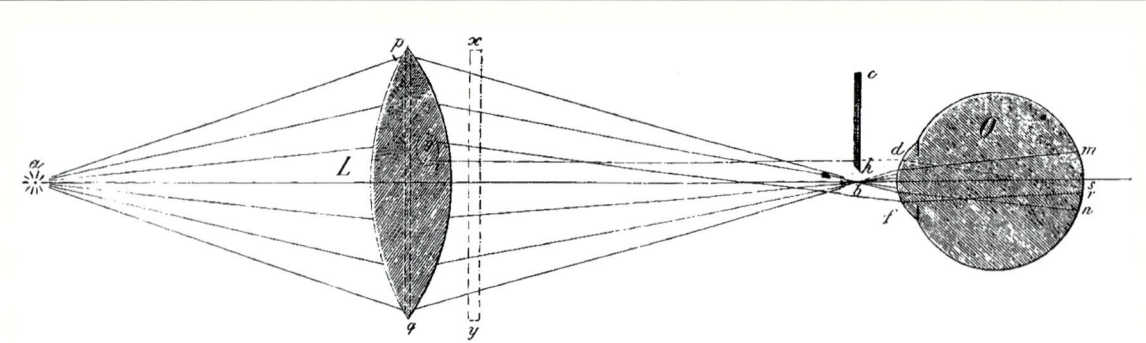

**4.18–A** Schematic of August TOEPLER's schlieren method. It essentially consists of an electric spark *a* acting as a high-intensity short-duration flash point light source, a lens $L$, and a knife edge $c$–$h$. TOEPLER positioned the object $x$–$y$ to be studied behind $L$ and used the edge $c$–$h$ close to the eye $O$ such that the central beam was cut off. The object, for example the density jump at the front of an aerial shock wave, is then visible on a dark background. [A. TOEPLER: *Beobachtungen nach einer neuen optischen Methode*. Cohen, Bonn (1864), plate I, fig. 1]

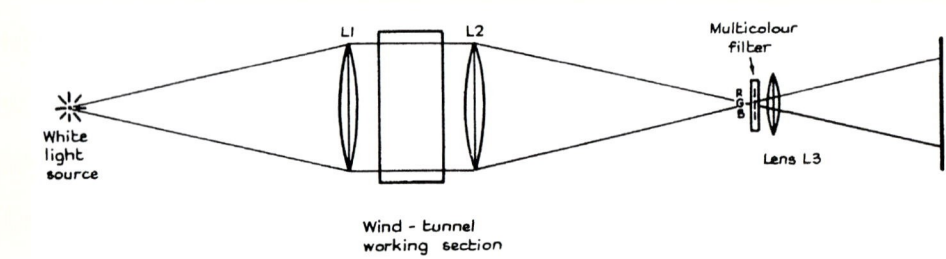

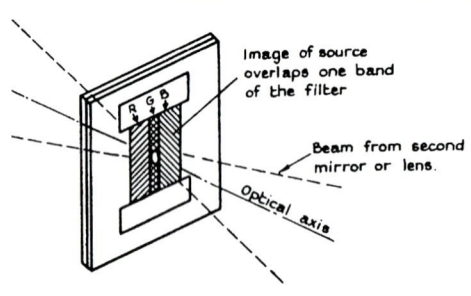

**4.18–B** Color schlieren methods are derivatives of TOEPLER's classic schlieren method and have been known since the end of the 19th century. Not only are hue, saturation, and color intensity useful as aids in a quantitative analysis, but color schlieren pictures also result in more aesthetic pictures than black-and-white images. There exists a wealth of methods for color coding of the refraction magnitude and direction, *e.g.*, by using prisms, diffraction gratings, interference filters, and chromatic aberration. ***Top:*** The color schlieren method shown here schematically was devised by the British researchers R. John NORTH and Douglas W. HOLDER at the former Aerodynamics Division of the NPL (Teddington, U.K.). He applied a white light source (such as an electric spark or a xenon flash tube) in the source plane and a multicolor filter in the cutoff plane (instead of the classic knife edge). Today 1- and 2-D color schlieren techniques are widely applied in high-speed flow visualization studies, using colored filter masks of different designs and gradient directions especially tailored to a certain problem. [NPL Aero Rep. No. 266 (1954). Courtesy British Marine Technology, Teddington, U.K.]

***Bottom:*** A tricolor filter can consist of a simple arrangement of narrow strips of colored gelatin film laid side by side and clamped between two thin glass plates. Schematic showing position and orientation of color filter: $R$ – red, $G$ – green, and $B$ – blue.

## 4.18 OPTICAL METHODS FOR FLOW VISUALIZATION – Shadowgraphy

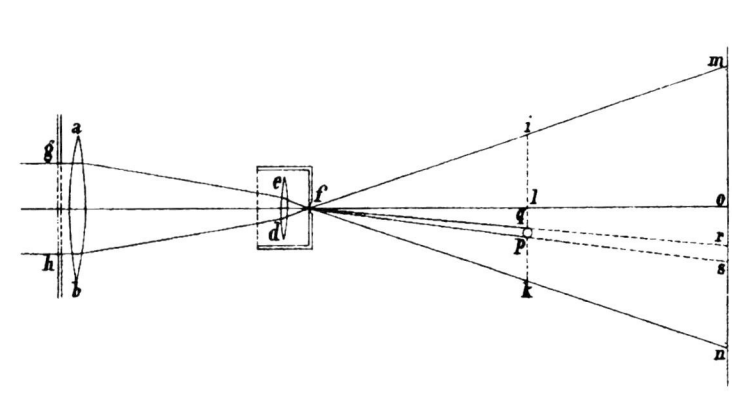

**4.18–C** The shadow method was invented in the late 1870s by Vincent DVOŘÁK at Agram University. It only requires a point light source and a white screen for projection purposes. The light, passing the test object $p$–$q$ (*e.g.*, a small inhomogeneity in glass or a "schliere" in air), is deviated and produces its image $r$–$s$ on the screen. The aperture $i$–$k$ determines the field of view $m$–$n$ on the screen. DVOŘÁK realized a point light source by using a heliostat (*i.e.*, using sunlight) and focusing the light onto a pinhole $f$. [Ann. Phys. Chem. **9** (III), 502 (1880)]

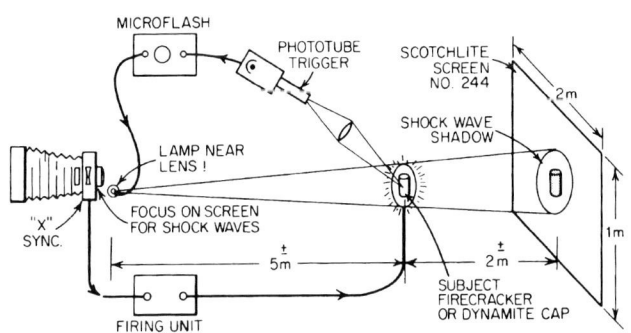

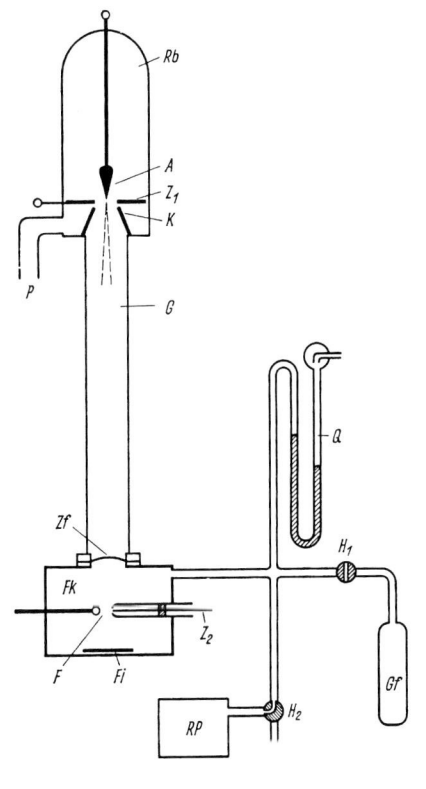

**4.18–D** The shadow method is most useful for visualizing shock waves and can be modified in a manifold manner. *Left:* Harold E. EDGERTON at MIT (Cambridge, MA) used an arrangement that employs a retroreflective screen (*e.g.*, Scotchlite, a 3M product) to photograph shock waves in large-scale experiments. The point light source should be placed as close to the lens as practical, but with a shield to keep the direct light from the camera lens. [H.E. EDGERTON: *Electronic flash, strobe*. MIT Press, Cambridge, MA (1970), p. 344] *Right:* Radiography is also a shadow method but uses X-rays instead of visible light. Karl-Heinz HERRMANN at the Freie Universität Berlin was the first to make a radiograph of a propagating shock wave in air and other gases without using any high-atomic-number additives in the test gas {⇨Fig. 4.5–H}. In order to get a sufficient contrast, he applied a flash soft X-ray tube $Rb$ pulsed from a low-voltage high-current capacitor discharge, and included a vacuum tube $G$ between $Rb$ and object $F$ to minimize the absorption of soft X-rays. The shock wave emerging from a spark gap $F$ was recorded on film $Fi$. Photo densitometry of the radiograph first allowed for the determination of the density jump and the steepening process at the shock front. [Z. angew. Phys. **10**, 349 (1958)]

## 4.18 OPTICAL METHODS FOR FLOW VISUALIZATION – Interferometry

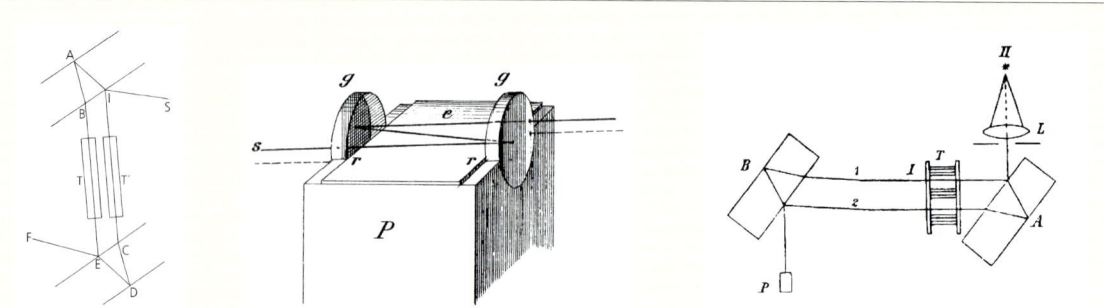

**4.18–E** The archetype of most interferometer constructions currently used in fluid and gas dynamics is the "Jamin interferometer." *Left:* Invented in 1856 in France by Jules C. JAMIN, it consists of two equally-thick plane parallel glass plates of the same refractive index, opaquely silvered on their back surfaces, thus combining mirror and beam splitter into one element. A beam of light $S \rightarrow I$, incident on the upper shown plate at about 45°, gives rise to two beams: one reflected from the front surface and the rear surface of the second plate ($I \rightarrow C \rightarrow D \rightarrow E$), the other reflected from the rear surface of the first plate and the front surface of the second plate ($I \rightarrow A \rightarrow B \rightarrow E$). Both beams are recombined to give an interference pattern in the axis $E \rightarrow F$ of a telescope. JAMIN positioned two tubular gas chambers, $T$ and $T'$, in the two beams in order to measure the difference in the refractive index. The instrument, however, only allows a small distance between the two beams, normally not exceeding a few centimeters. [Ann. Phys. **98** (II) 345 (1856)] *Center:* August TOEPLER and Ludwig E. BOLTZMANN used the Jamin interferometer to measure the amplitude of sound at the threshold of hearing. They closely attached two circular parallel glass plates $g$–$g$ at the side walls of a covered wooden pipe $P$ such that they touched the two side walls $r$–$r$ of an iron plate $e$ and parallel faced each other, thus forming a Jamin interferometer. One light beam (*solid line*) ran outside of the pipe, the other (*broken line*) inside, being exposed to the periodic changes in refractive index. A stroboscopic illumination allowed one to observe the fringes as a stationary picture. [Ann. Phys. **141** (II), 321 (1870)] *Right:* Ernst MACH and J. VON WELTRUBSKY first visualized the local density profile through the front of an aerial shock wave. $A$ and $B$ denote the two Jamin plates. The shock wave was generated by an electric spark $I$ in a cuvette $T$ and interacted with the light beam $1$; beam $2$ was shielded from the shock wave. A second spark $II$, serving as a flash light source, was triggered in delay to spark $I$. Observation of the fringes occurred through a prism $P$ such that the eye rested on $T$ as well as on a slit positioned behind the condenser $L$. [Sitzungsber. Akad. Wiss. Wien **78**, 551 (1878)]

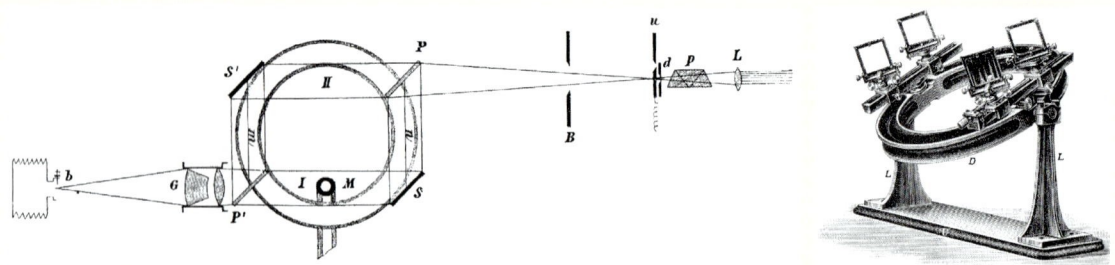

**4.18–F** *Left:* In 1891, Ludwig MACH at the German Charles University in Prague and, independently, Ludwig ZEHNDER at the University of Würzburg invented the so-called "Mach-Zehnder interferometer." It differs from JAMIN's interferometer in that the two beams $I$ and $II$ are widely separated so that one beam can be exposed to a violent environment, such as the trajectory of a projectile, and the other beam can be shielded from any aerial disturbances. [Sitzungsber. Akad. Wiss. Wien **106** (Abt. IIa), 1057 (1897)] *Right:* L. MACH applied his interferometer construction in nonstationary gas dynamics. The two plane mirrors and the two beam splitters, both $10 \times 10$ cm² in size, are altogether mounted on a heavy annular support of about 40 cm in diameter, which provides a central area between these four optical elements free of any mechanical support. This allowed L. MACH to position the interferometer along the trajectory of a flying projectile {⇨Fig. 4.6–L}. [Z. für Instrumentenkunde **12**, 89 (1892)]

## 4.18  OPTICAL METHODS FOR FLOW VISUALIZATION – Holography

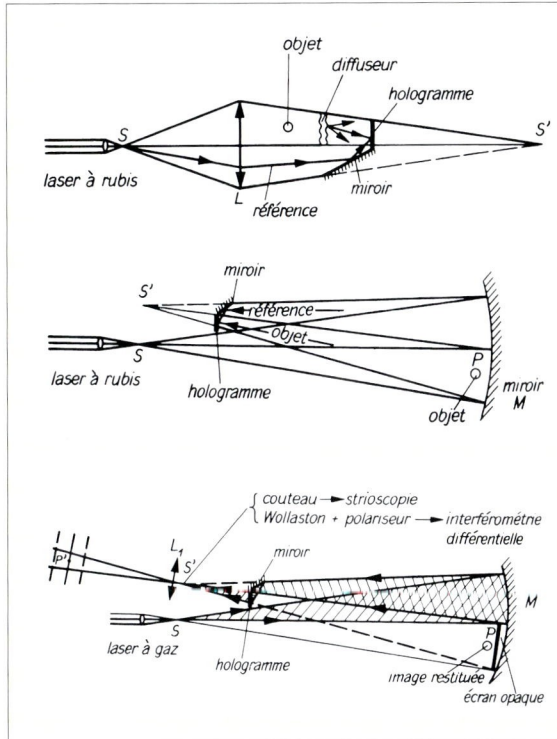

**4.18–G** In 1948, the Hungarian-born British engineer Dennis GABOR invented holography, which was extended in the 1960s to the recording of fast transient events. So-called "high-speed holography" allows a 3-D reconstruction of a fast moving object, *e.g.*, a propagating shock wave or a fast-flying projectile and its aerodynamic flow environment. When a flash hologram of a high-speed event has been taken at a desired time instant, it can be submitted successively to various standard optical methods such as to shadowgraphy, schlieren, and differential interferometry. This allows one to solve a variety of problems with a single shot only. The schematics given here illustrate the high-speed holographic technique in more detail. ***Top:*** Schematic arrangements for taking holograms in diffuse light as used by Paul SMIGIELSKI and Henri ROYER at ISL (Saint-Louis, France). The light source is a pulsed laser of high coherence and high energy per pulse, such as a ruby laser. ***Center:*** Arrangement for taking holograms in reflected light. ***Bottom:*** For reconstruction purposes, the same arrangement as sketched above is used but with a continuously operated laser instead of a pulsed laser. Furthermore, the object beam is screened. An example of a holographic image of a head wave is shown in Fig. 4.6–I. Using a repetitive Q-switched ruby laser and a rotating photographic plate – or different lasers pulsed at different times and a steady plate – they were able to realize cine-holography, which significantly extends holography for the study of high-speed events.
[Courtesy ISL; Proc. 8th Int. Congr. on High-Speed Photography, Stockholm (1968). Almquist & Wiksell, Stockholm (1968), p. 324].

## 4.18  OPTICAL METHODS FOR FLOW VISUALIZATION – Surface Thermography

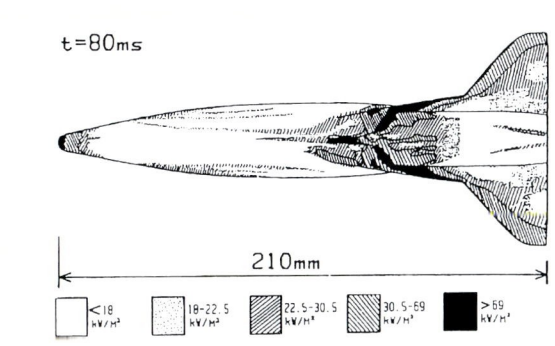 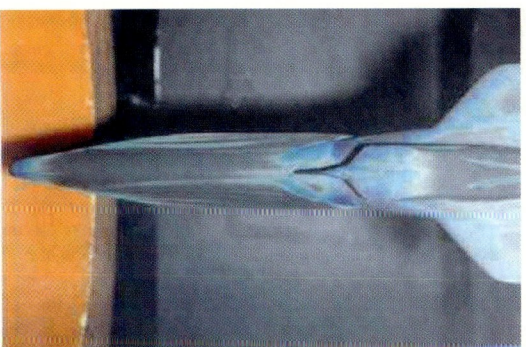

**4.18–H** Robin A. EAST at the Dept. of Aeronautics and Astronautics (University of Southampton, U.K.) applied "liquid crystal surface thermography" to the study of hypersonic flow with heat transfer rates ranging from 1 to 500 kW/m². Using a coating of micro-encapsulated liquid crystals he obtained quantitative heat transfer data for an aerospace plane model at $M = 6.85$ and $Re_\infty = 3 \times 10^7$/m in running times as short as 20 ms. [Courtesy Prof. R.A. East; Proc. 18th Int. Symp. on Shock Waves, Sendai, Japan (1991). Springer, Berlin (1992). p. 643]

## 4.19 HIGH-SPEED DIAGNOSTICS – Chronoscopes and Chronographs

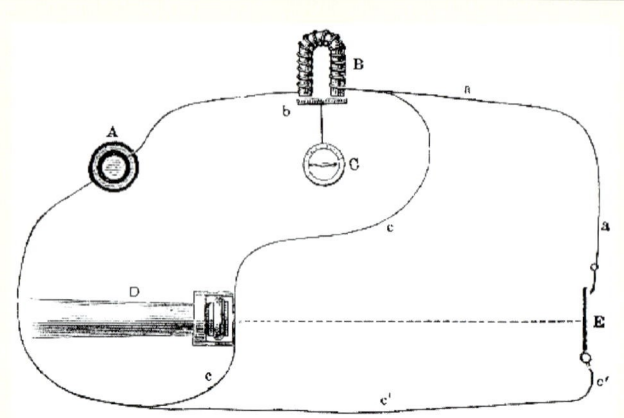

← Matthias HIPP, a German-born Swiss watchmaker, improved the accuracy of the Wheatstone chronoscope by using a clock with two faces, each moving over a scale divided into intervals of 10 ms and 1 ms.

**4.19–A** *Top:* In 1840, the English physicist Charles WHEATSTONE devised the so-called "Wheatstone chronoscope," an electrically activated stopwatch. The principle is illustrated here on the example of measuring the muzzle velocity of a gun *D*. The projectile first deactivates a magnetic clutch *B*, which starts a watch *C*. After traveling a known distance in free flight, it closes a contact *E*, which activates a magnet *B*, stopping the watch. [*Meyers Konversations-Lexikon*. Bibliogr. Inst., Leipzig (1894), vol. 4, p. 154] *Bottom:* View of a "Hipp-Wheatstone clock" made around 1900 by the renowned German instrument maker Ernst ZIMMERMANN at Leipzig. [Courtesy Institut für Psychologie, Karl-Franzens-Universität, Graz, Austria]

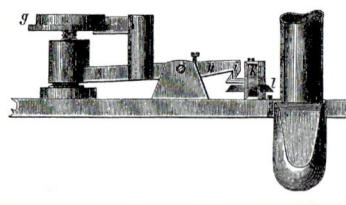

**4.19–B** In the early 1860s, the Belgian artillerist Paul LE BOULENGÉ devised the electrically triggerable so-called "Le-Boulengé chronograph," which became the standard instrument in ballistics for measuring muzzle velocities. Using two wire screens along the trajectory, the bullet, interrupting the first screen, deactivated a magnet, thus starting the free-fall dropping of a first rod *c*. Upon passage of the second screen another suspended rod *f* fell, thereby releasing a spring-loaded knife (*bottom*) marking rod *c*. Thus, the height of fall is known, and, likewise, the time of fall. [*Brockhaus' Konversations-Lexikon*. Brockhaus, Leipzig (1908), vol. 4, p. 241]

## 4.19 HIGH-SPEED DIAGNOSTICS – Cathode-Ray Oscilloscopes

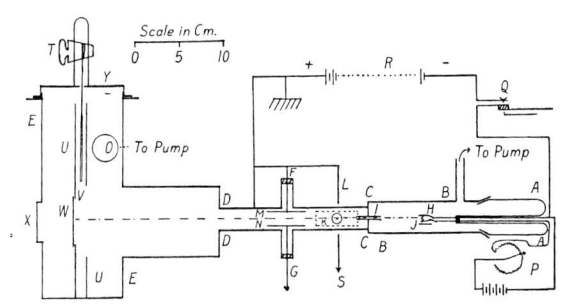

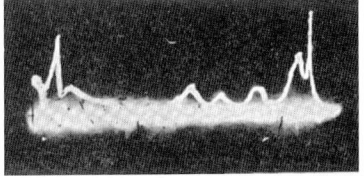

↑ The Wood oscillograph, devised by the British physicist Joseph J. THOMSON, used a cathode-ray tube (or Braun tube) with a hot cathode and two pairs of deflection plates.

**4.19–C** In 1921, the British physicist David A. KEYS at McGill University (Montreal, Canada) applied the so-called "Wood oscillograph" to record pressure-time profiles of underwater explosions. *Left:* Schematic (*top*) of his experimental setup and a pressure-time profile (*bottom*) of an explosion of guncotton in water, recorded with a tourmaline gauge. *Right:* View of the Wood oscillograph constructed and tested by the British physicist Albert B. WOOD at the Admiralty Research Laboratory (Teddington, U.K.). The film plate, placed inside the vacuum, was directly exposed by the 3-kV cathode ray. [J. Franklin Inst. **196**, 576 (1923)]

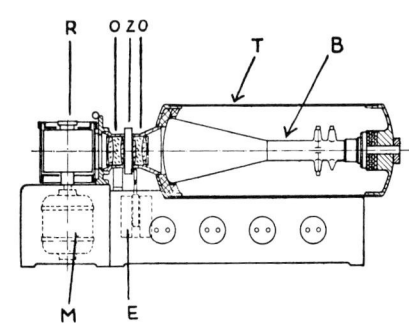

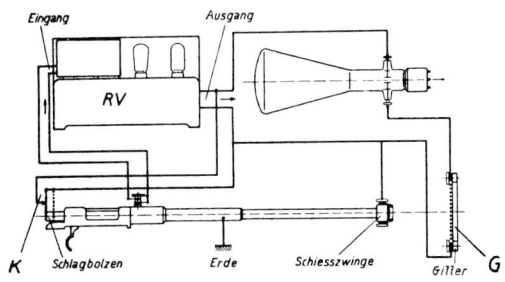

**4.19–D** In 1932, the physicists Hans JOACHIM and Hans ILLGEN at Zeiss-Ikon AG (Dresden, Germany) measured the dynamic gas pressure in firearms using a quartz gauge. *Right:* The signal of the pressure gauge was amplified (*RV*) and via a grid *G* fed to the vertical pair of deflection plates of a Braun tube *T*. Triggering occurred via contact *K*, which opened at the moment the gun was fired. The voltage at the deflection plates was interrupted when the bullet destroyed *G*. *Left:* Schematic (*top*) and view (*bottom*) of their electronic-mechanical oscillograph: time display was achieved mechanically by sweeping the image of *T* via lenses *O* and a mechanical shutter *Z* onto a film fixed on a rotating drum *R* driven by a synchronous motor *M* at 3,000 rpm. They found that gas pressures in gun barrels typically range between 3 and 4 kbar, which, however, only last about 1 ms FWHM. [Z. ges. Schieß- & Sprengstoffwesen **27**, 121 (1932)]

## 4.19 HIGH-SPEED DIAGNOSTICS – Time-Delay Generators

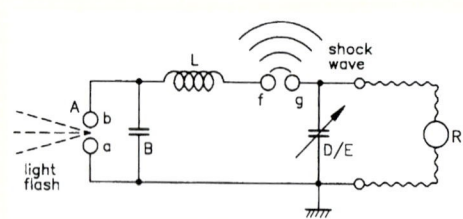
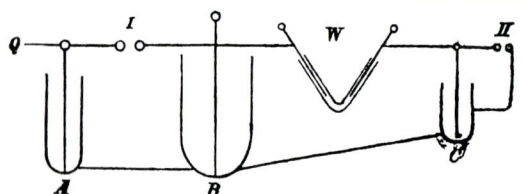

**4.19–E** In 1858, Karl-Wilhelm KNOCHENHAUER, a high-school teacher (Meiningen, Thuringia, Germany) devised a high-voltage delay circuit based on the coupling of two capacitor discharges. *Left:* The so-called "Knochenhauer circuit" was used in 1864 by August TOEPLER {⇨Fig. 4.5–A} to generate a shock wave by a first spark $f$–$g$ and a light flash by a second, delayed spark $a$–$b$ for visualizing the propagating shock wave at different time instants, which were varied by trimming the capacitor $D/E$. *Right:* In 1878, this circuit was improved by Ernst MACH, the so-called "Mach circuit." The spark generating the shock wave is designated by $I$ and the illumination spark, fired delayed, by $II$. [Shock Waves **5**, 1 (1995)]

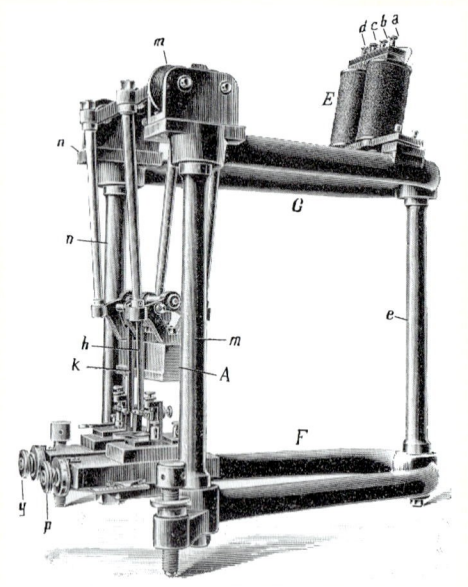
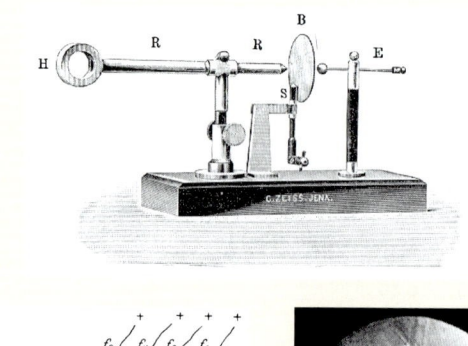
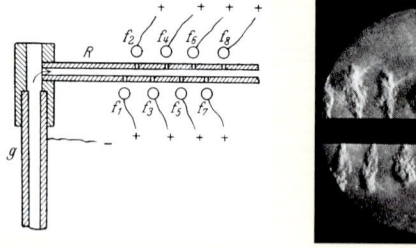

**4.19–F** The "Helmholtz pendulum," a precise contact breaker, was originally constructed for physiological studies by the German professor Max T. EDELMANN. But it was also used in ballistics and detonics for triggering purposes to derive a delayed electric pulse from the high-speed event to be studied. Provided with two pairs of contacts for opening and closing actions and adjustable against each other via micrometric screws, it allowed the generation of current pulses of variable duration. One full turn of the micrometric screw corresponded to a time difference of 156 μs. [Z. f. Instumentenkunde **21**, 124 (1901)]

**4.19–G** *Top:* Ludwig MACH (German Charles University, Prague) applied the head wave to trigger a spark outside of his interferometer such that the head wave upon entering the interferometer was illuminated by a spark flash after a certain delay time. Using a ring $H$ connected to a tube $R$, the head wave, generated by the projectile flying through $H$, coupled a shock wave to $R$, which, arriving at the end and passing a hole in an aperture $B$, triggered a spark gap $S$–$E$. [Sitzungsber. Akad. Wiss. Wien **105** (IIa), 605 (1896)] *Bottom:* Carl CRANZ and Hubert SCHARDIN (Institut für Technische Physik, Berlin) taking up this method in their multiple-spark camera, generated a series of eight flashes up to a rate of 50,000/s. They used a gun barrel $G$ connected to a pipe $R$ along which eight trigger spark gaps, $f_1 \ldots f_8$, were arranged (*left*). Illustration of the successive escape of hot gases from the holes after passage of the shock wave (*right*). [Z. Phys. **56**, 147 (1929)]

## 4.19 HIGH-SPEED DIAGNOSTICS – Triggered Snapshot Photography

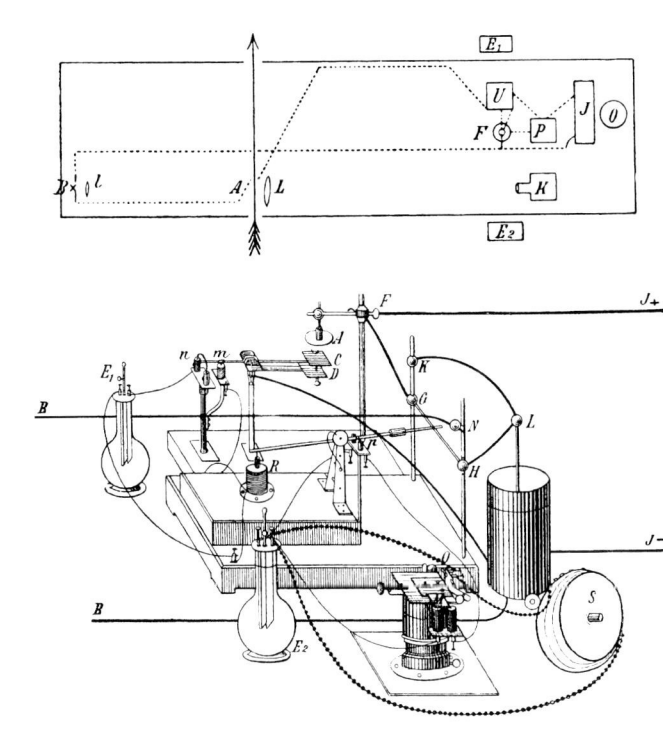

**4.19–H** The "snapshot" or "single-shot" photography was first used in 1851 by the Englishman W.H. Fox TALBOT. It required the least expenditure of equipment but only provided a single picture at a preselected time instant. With the advent of high-sensitive gelatin dry plates, snapshot photography was increasingly used to freeze high-speed events. ***Top:*** Schematic of an experimental setup that Ernst MACH and Peter SALCHER used in the late 1880s at the Meppen Ballistic Test Site in northern Germany for outdoor photography of large-caliber supersonic projectiles in free flight, about 12 m from the muzzle. Inside a wooden cabin they installed an optical setup ($B$ – spark light source, $l$ and $L$ – schlieren optics, $K$ – camera, an electric circuitry ($F$ – Leiden jar, $J$ – influence machine, $P$ – voltage controller, $U$ – commutator), and an external power supply ($E_1$ and $E_2$ – Bunsen batteries). The cabin was heated to minimize any loss of charge caused by corona discharges. The projectile, entering and leaving the cabin through small portholes to minimize detrimental daylight exposure, passed a fluid-dynamic trigger device $A$ {⇨Fig. 4.19–G}. ***Bottom:*** To ensure precise triggering and safe operation it required precise circuitry: When the full potential of the Leiden jar was reached, the circuit automatically activated the camera shutter and a bell $S$. Upon the bell signal, the gun had to be fired and the camera shutter closed immediately after, both by hand. Precise triggering of the illumination spark $B$ was achieved automatically upon projectile passage of the trigger spark gap $A$; both spark gaps were connected in series and with the circuitry via lines $B$. The HV circuitry was charged by the influence machine $J$ via lines $J+$ and $J–$.
[Sitzungsber. Akad. Wiss. Wien **98** (IIa), 41 (1889)]

## 4.19 HIGH-SPEED DIAGNOSTICS – Rotating Mirror

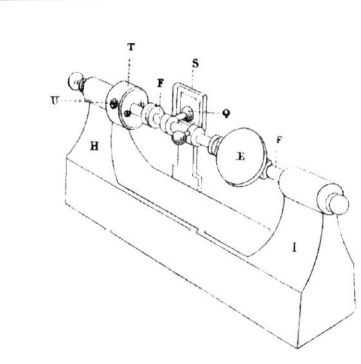

**4.19–I** In 1834, the English physicist Sir Charles WHEATSTONE first proposed the "rotating mirror" as a diagnostic tool to measure time spans to the millionth of a second. The device that he used to determine the velocity of electricity and light consisted of a brass frame $H$-$I$, a plane circular mirror $E$ of polished steel fixed to the horizontal axis $F$–$F$ such that the axis of rotation was in the plane of the mirror. Motion up to 800/s was communicated to the axis by means of a thread passing to a hand wheel. The arm $Q$, connected with a discharger $S$ and mounted on $F$–$F$, provided a stroboscopic switch for triggering purposes. [Proc. Roy. Soc. Lond. **3**, 583 (1834)]. WHEATSTONE used the rotating mirror in the first measurement of the velocity of an electrical current passing through a wire. He later suggested that the same device could be used to also measure the speed of light. WHEATSTONE, together with the Irish natural philosopher John TYNDALL, also used his rotating mirror to show that Michael FARADAY's hypothesis on the "chemical harmonica" (or "burning harmonica"), a special case of the singing flame, was indeed correct.

## 4.19 HIGH-SPEED DIAGNOSTICS – Spark Chronography of a Flying Projectile

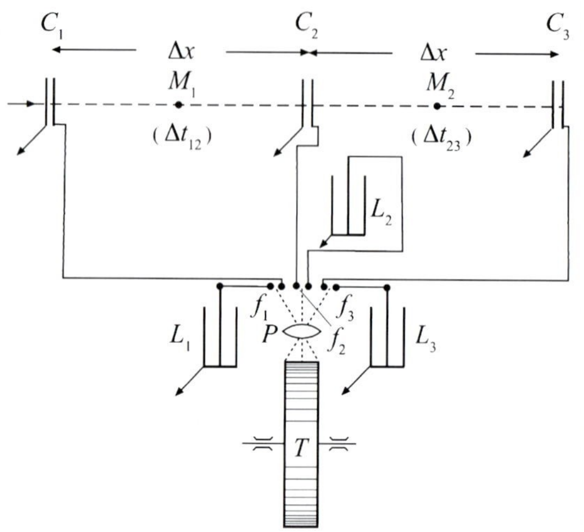

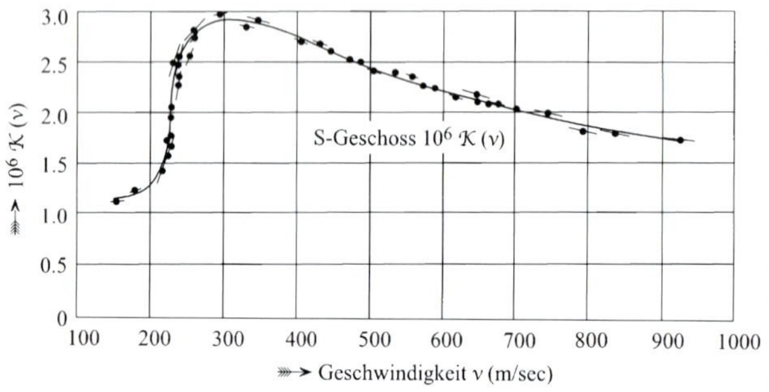

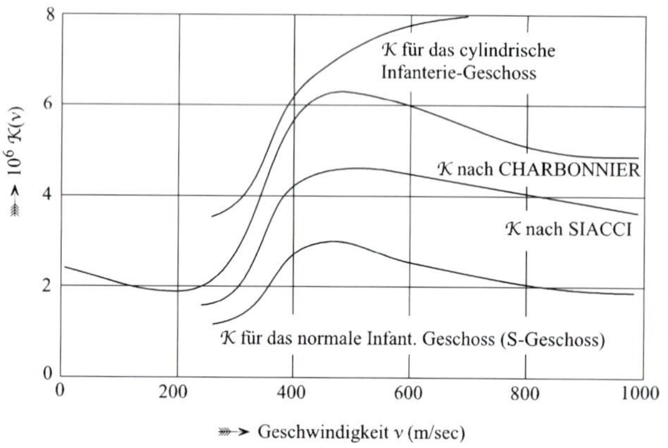

**4.19–J** In the late 1900s, Karl BECKER and Carl CRANZ, two German ballisticians at the Militärtechnische Akademie Berlin, began to develop a method to measure aerodynamic drag of infantry bullets as a function of the projectile velocity $v$ with an accuracy of ca. 1%. **Left, top:** Their electric-spark photochronograph, the so-called "Cranz-Becker chronograph," consisted of a rapidly revolving drum $T$ covered with a photographic film onto which, via a focusing lens $P$, the light of three spark gaps, $f_1$, $f_2$, and $f_3$, was projected. These spark gaps were triggered subsequently when the projectile had passed the contact plates $C_1$, $C_2$, and $C_3$, thus discharging the Leiden jars $L_1$, $L_2$, and $L_3$ via $f_1$, $f_2$, and $f_3$, respectively. Each contact plate $C$ consisted of two pieces of aluminum foil, closely facing each other, which became conductive upon penetration of the bullet. The distance $\Delta x$ between $C_1$ and $C_2$ and $C_2$ and $C_3$ was 20 m. Prior to firing the rifle, the Leiden jars were charged up to a high voltage by an influence machine and then disconnected. **Left, center:** From the loss of velocity BECKER and CRANZ determined the drag coefficient $K(v)$. **Left, bottom & right:** They obtained the remarkable result that, compared to a cylindrical bullet with a flat head, a pointed bullet with an ogival head – the so-called "S-bullet" [Germ. *Spitzgeschoss*] – has a much lower aerodynamic drag. In addition, it was found that S-bullets shoot more accurately than those with round or truncated noses and retain a higher velocity when reaching the remote target. As early as before World War I, most European armies began to replace hitherto used round ammunition by the more efficient pointed one. [Artilleristische Monatshefte **69**, 189 (1912); Ibid. **71**, 333 (1912)]

## 4.19 HIGH-SPEED DIAGNOSTICS – Rotating Mirror Streak Cameras

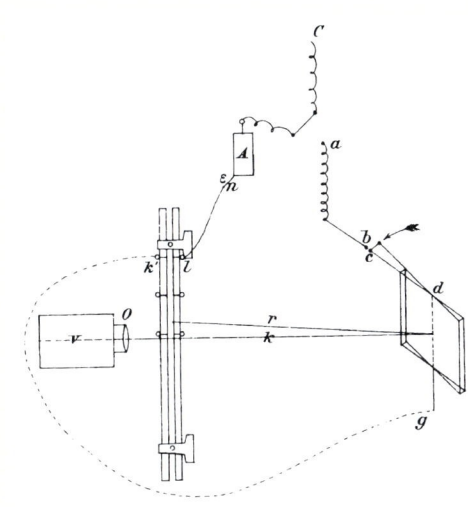

← The propagation of the front of the detonation wave was visualized through a 3-mm-wide slit glued along the tube axis and imaged via the rotating mirror $d$ into a conventional still camera $F$. The mirror was provided with a pointer $c$ in order to trigger the detonation in the glass tube by discharging a Leiden jar $A$ along the electric path $n$–$l$–$k$–$g$–$d$–$c$–$b$–$a$.

→

For oxyhydrogen they determined from the inclinations of the streak records a detonation velocity ranging from approx. 2,200 m/s at the beginning to approx. 850 m/s at the end of the eudiometer tube.

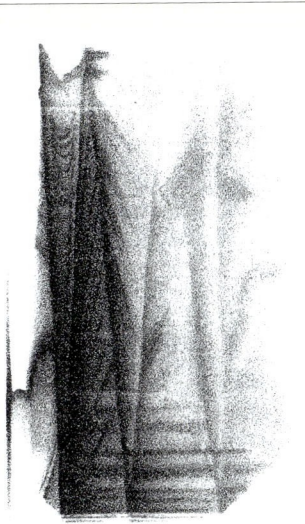

**4.19–K** *Left:* In 1888, Arthur VON OETTINGEN and Arnold VON GERNET, two German physicists at the University of Dorpat (West Russia), studied the propagation and reflection of detonation waves of gaseous explosive mixtures in 400-mm-long glass tubes using a rotating mirror. *Right:* Example of a streak record of an oxyhydrogen explosion. The streak record, beginning at the left side with the triggering spark discharge and covering a time span of about 3.7 ms, shows the propagation and three reflection cycles of the detonation front. [Ann. Phys. Chem. **33** (III), 586 (1888)]

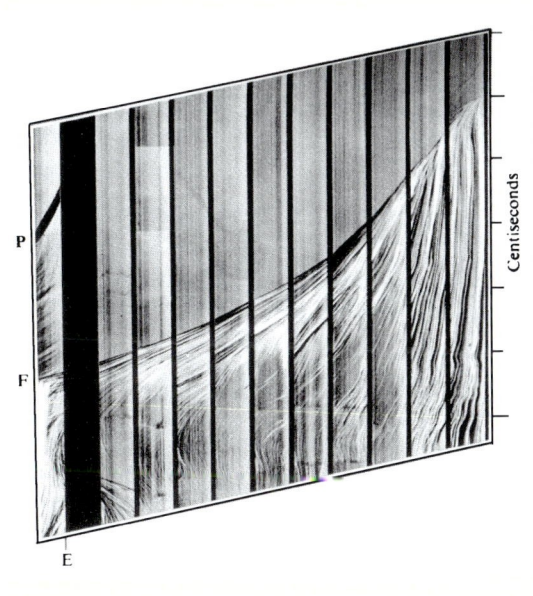

**4.19–L** *Right:* William H. PAYMAN and collaborators at the British Safety in Mines Research Establishment (SMRE) constructed a "wave speed camera," a streak camera, to record detonation phenomena in gaseous mixtures [Proc. Roy. Soc. Lond. **A132**, 200 (1931)] *Left:* Example of a typical composite streak record $s(t)$ showing the propagation of the detonation front in an air/methane mixture as a function of time $t$. From the slope of the streak record PAYMAN could analyze the starting process of detonation. [Proc. Roy. Soc. Lond. **A158**, 348 (1937)]

## 4.19 HIGH-SPEED DIAGNOSTICS – High-Speed Cinematography

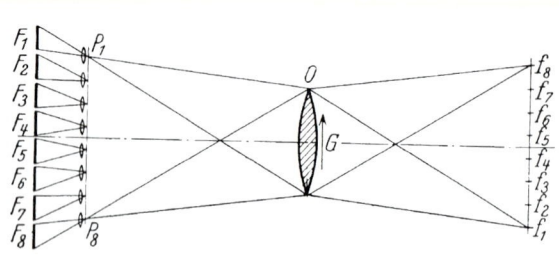

**4.19–M** *Top:* Schematic of the "Cranz-Schardin multiple-spark camera" invented by Carl CRANZ and Hubert SCHARDIN (Institut für Technische Physik, Berlin) in the late 1920s. It consists of a bundle of single-shot spark light source, $f_1 \ldots f_8$, arranged around a common axis and triggerable at any desired time instant. The camera gives from an object, positioned at $G$ and viewed through a common lens $O$, a limited number of slightly oblique views $F_1 \ldots F_8$, which are imaged via individual lenses as pictures $P_1 \ldots P_8$ on a common film plane. Since the film is stationary, the picture frequency can be very high, and the quality is as good as that from a single flash system. [Z. Phys. **56**, 147 (1929)] *Bottom:* This photo shows Hubert SCHARDIN while recording ballistic experiments cinematographically using his multiple-spark camera. The camera prototype, built in the period 1927–1928, had only nine spark/frame units, but in 1936 this was extended to 24 frames. Note the spark head $SH$ with the sparks $f_1 \ldots f_9$ and the camera head $CH$. Each spark was fed by a pair of cylindrical, low-inductance capacitors (shown in the background). This camera type was later used at EMI, ISL, MIT, and other laboratories. [Proc. 4th Int. Congr. on High-Speed Photography, Cologne (1958). Helwich, Darmstadt (1959), p. 139. Courtesy EMI-Archives, Freiburg]

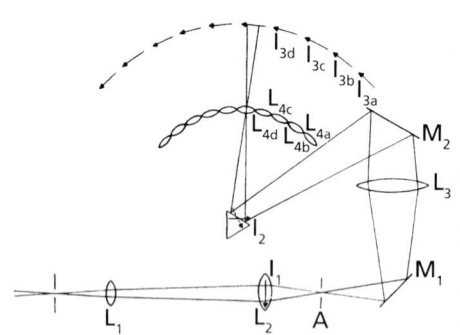
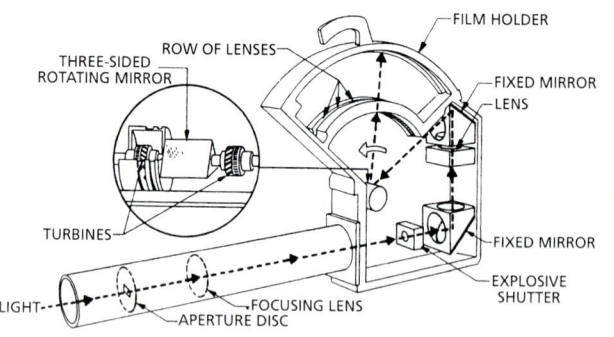

**4.19–N** Berlyn BRIXNER at LASL (Los Alamos, NM) developed an ultrahigh-speed framing camera capable of taking 96 consecutive motion pictures of explosive phenomena and shock wave actions (particularly of nuclear detonations) at a rate of 15 million frames/s. *Left:* The optical system in this camera is based on the use of a rotating mirror and refocused revolving beams as pioneered by Cearcy D. MILLER in the mid-1930s. An image $I_1$ of the event being studied is formed at field lens $L_2$ by the objective $L_1$. This image is relayed to the rotating-mirror surface at $I_2$ by means of lens $L_3$. Folding mirrors $M_1$ and $M_2$ are used to make a compact camera. As the mirror rotates, the reflected light beam passes through the final framing lenses $L_{4a}$, $L_{4b}$, *etc.* to form images $I_{3a}$, $I_{3b}$, *etc.* on the film plane. [U.S. Patent No. 2,400,887 (1946)] *Right:* Schematic of the Los Alamos camera, showing path of light rays and direction in which the light beam is swept across secondary lenses. [Proc. 2nd Int. Congr. on High-Speed Photography, Paris (1954). Dunod, Paris (1956), p. 108]

## 4.19 HIGH-SPEED DIAGNOSTICS – High-Speed Cinematography *(cont'd)*

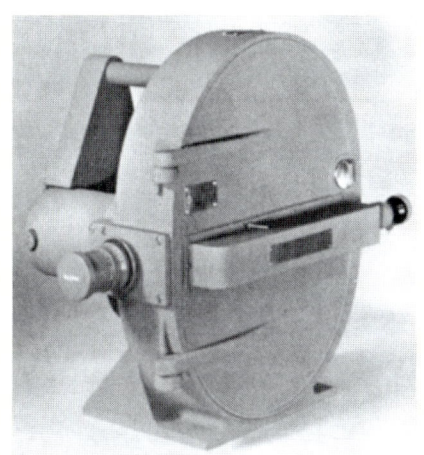

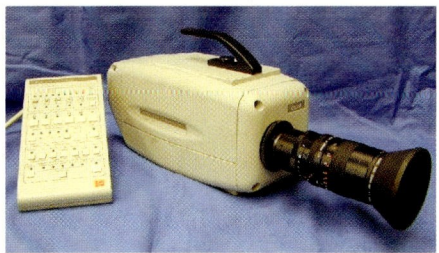

**4.19–O** *Top:* The Fastax, a high-speed rotating-prism camera (max. 16,000 frames/s), was a milestone in photo instrumentation and was also successfully used at Trinity Test (1945) to record minute details of the nuclear fireball growth. Today mechanical cameras at this frame rate are superseded by high-speed video cameras. [R.F. SAXE: *High-speed photography*. Focal Press, London (1966), p. 25] ***Bottom:*** In 1991, Photron Ltd. at Tokyo, Japan designed the world's fastest high-speed video system (max. 40,500 frames/s), which was commercialized by Kodak as model 4540. One important advantage of this digital camera is that it permanently records, and a trigger pulse stops recording. This enables the experimenter to also capture pretrigger events, such as in failure studies, which otherwise would not be accessible to observation. Contrast and illumination can be improved by using a copper vapor laser as a stroboscopic light source: Because of the ultrashort duration of laser flashes, image blurring is practically eliminated, regardless of the speed of the event. [Courtesy Photron Ltd., Tokyo]

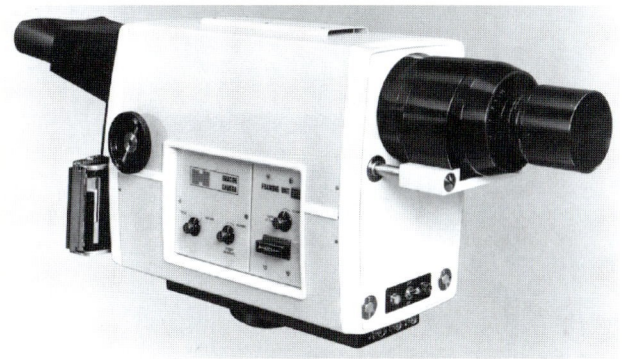

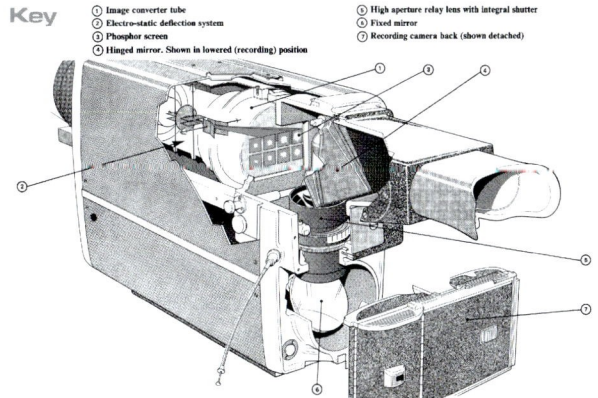

**4.19–P** A framing operation with image converters is usually achieved by (1) interrupting the electron beam in the tube to effect repetitive shuttering and (2) shifting the interrupted beam to different areas of the screen – thereby producing a pattern of discrete images on screen. ***Top:*** View of the Imacon 700, the first commercial electronic high-speed framing camera, which allowed one to take up to 16 images at a rate ranging from 25,000 to 20,000,000 frames/s. Based on original ideas developed by the Englishman Alec E. HUSTON at AWRE (Aldermaston, U.K.); it was commercialized by John Hadland and first became available in May 1967 ***Bottom:*** Schematic of the camera as seen from the back. The latent picture on the phosphorus screen was recorded on Polaroid film, *i.e.*, recording of a complete scene was available on film only 10 s after the experiment. Lens coupling of the film was later replaced by fiber optic coupling. The camera concept opened the door to a new generation of electronic digital multi-frame cameras, which today considerably facilitates high-speed recording of shock waves, detonation waves, exploding wires, fracture phenomena, and other ultra-high-speed events. [Courtesy B. SPEYER, DRS Hadland, Tring, U.K.]

## 4.20  HIGH-SPEED VEHICLES – Mythologies

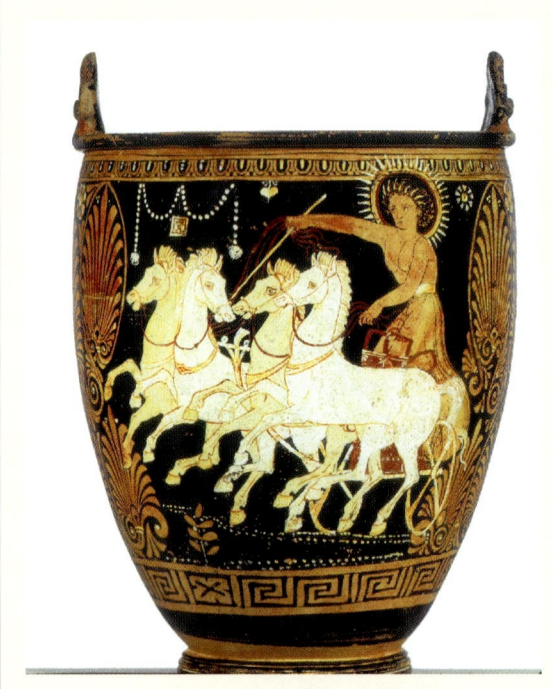

**4.20–A** Covering large distances in a short time has been humanity's dream since time immemorial. High-speed vehicles, in antiquity light two-wheeled chariots drawn by up to four horses, were a privilege and only reserved to deities and high-ranked persons such as emperors, kings, generals, *etc*. In Greek mythology, the Sun-god HELIOS was conceived as driving a glowing chariot daily across the sky from the morning's eastern portals to the evening's western gates and as sailing nightly in a golden boat on the ocean along the Earth's northern border back to his palace. This picture shows a *situla* made of pottery (about 330 B.C.) – a ceremonial bucket-shaped vessel probably used to hold holy water; the total height is 33.5 cm. It was found in a tomb in Apulia, southeastern Italy, then a Greek colony. HELIOS is shown here driving a quadriga and wearing a radiative crown. His "supersonic" vehicle is a delicate chariot drawn by four fiery stallions. [From the Collection W. KROPATSCHECK, Helgoland. Courtesy Museum für Kunst und Gewerbe, Hamburg]

**4.20–B** *Left:* According to legend, a Chinese minor official named WAN-HOO attempted in the early 16th century a flight to the Moon using a large wicker chair to which were fastened 47 large rockets. However, his rocket experiment failed, and he disappeared in a huge cloud of smoke. [http://encyclopedia.thefreedictionary.com/_/viewer.aspx?path=5/5c/&name=Wan_Hu_large.png]
*Right:* In the 12th century, the Chinese invented rockets and began to use military rockets in the battlefield. This is a drawing of a Chinese soldier launching a fire arrow. [NASA-MSFC; http://history.msfc.nasa.gov/rocketry/03.html]

## 4.20 HIGH-SPEED VEHICLES – First Supersonic Rocket Flight

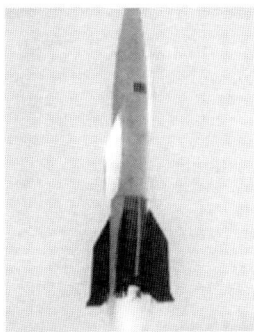

**4.20–C** *Left:* Launch of the German rocket A-4 (in 1944 renamed V2) at the Peenemünde Test Site on October 3, 1942, reaching a record distance of 191 km, a velocity of M > 4, and a height of 84.5 km. It was the first time that a large structure (length 14.3 m, takeoff weight 12.8 tons) broke the sound barrier. *Center:* A rare picture of a successful takeoff of an A-4b, the winged long-distance version of the A-4, taken on January 24, 1945. [Courtesy Deutsches Museum, Munich] *Right:* Schlieren photo of a model of the A-4b taken at a velocity of about $M = 2.1$ in the Peenemünde wind tunnel. It was anticipated that this so-called "glider" could increase the distance from firing location to the target. Note that up to about $M = 2.6$ the wing tips would still remain within the Mach cone, thus considerably reducing aerodynamic drag.
[P.P. WEGENER: *The Peenemünde Wind Tunnels*. Yale University Press, New Haven, CT (1996)]

## 4.20 HIGH-SPEED VEHICLES – First Hypersonic Rocket Flight

**4.20–D** *Left:* The U.S. WAC-Corporal, a 16-ft (4.88-m)-long rocket with a thrust of 1,500 pounds, was developed by Theodore VON KÁRMÁN's research team at CalTech near the end of World War II. The first WAC-Corporal was launched in October 1945 at White Sands Proving Ground, NM. It reached an altitude of about 45 miles (75 km). The WAC-Corporal, an unguided rocket, was applied for high-altitude research and for providing information and engineering experience to be used in future surface-to-surface missile programs. *Right:* On February 24, 1949, a WAC-Corporal, boosted by a captured German V2, reached a record altitude of about 400 km and a hypervelocity of about 2,300 m/s
[National Air and Space Museum, Smithsonian Institution, Washington, DC (NASM, image nos. A-5048-B and SI 76-15531)]

## 4.20 HIGH-SPEED VEHICLES – First Transonic Rocket Plane

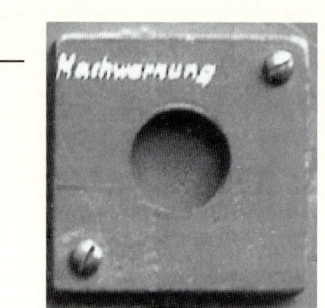

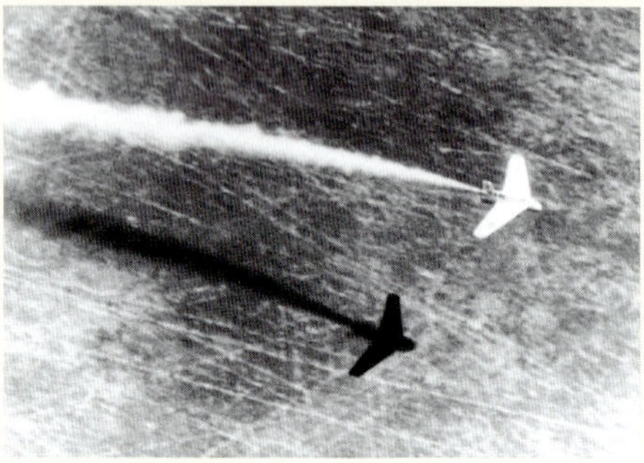

**4.20–E** The Messerschmitt Me 163 or "Komet" was the first rocket plane and the most advanced transonic aircraft of its time. It was designed by the famous German aeronautical engineer Alexander LIPPISCH. Piloted by Heini DITTMAR, it reached on October 2, 1941 a record speed of 1,003 km/h ($M = 0.84$), thus remaining the fastest jet fighter until the end of World War II. However, it arrived too late to be of any military relevance. **Left, top:** A warning lamp was installed at the upper far right side of the cockpit panel. This lamp, labeled *Machwarnung* ("Mach warning"), indicated to the pilot the approach to the sound barrier. Apparently, it was the very first "Mach indicator" ever installed in an airplane. **Left, center:** This picture was taken of the Komet during flight. Note the swept-back wings, an idea of the German aerodynamicist Adolf BUSEMANN. First proposed by him at the 5th Volta Conference held in Guidonia/Rome in 1935, this wing concept revolutionized modern high-speed aeronautics. **Left, bottom:** View of the small aircraft, which had only a length of 5.7 m and a wing span of 9.3 m. Powered by a 3,300-pound thrust liquid-fuel rocket motor, it was also popularly called *"Kraftei"* – a German word meaning "power egg" – because of its compact fuselage and enormous thrust. [Courtesy Deutsches Museum, Munich] A Messerschmitt Me 163 B-1a (built in 1943) is on display at the Deutsches Museum, Munich.

## 4.20 HIGH-SPEED VEHICLES – First Supersonic Rocket Plane

**4.20–F** View of the Bell XS-1 (later designated X-1) with which the U.S.A.F. research pilot Capt. Charles E. YEAGER first broke the sound barrier ($M = 1.07$) in level flight on October 14, 1947. The little rocket-powered airplane was launched in the air from the bomb-bay of a Boeing B-29 ("Superfortress"), the same kind of airplane that dropped the atomic bombs on Japan in August 1945. The plane, which did not yet have swept-back wings for easing the shock wave problem, measured 31 ft (9.45 m) long with a wingspan of just 28 ft (8.53 m). The maximum speed attained by the X-1 was Mach 1.45 at 40,130 ft (12,232 m) during a flight by YEAGER on March 26, 1948. *Left:* Note the stationary, axisymmetric diamondlike shock wave pattern in the jet exhaust of the rocket engine. *Right:* The small aircraft is now on display in the Milestones of Flight Hall at the Smithsonian National Air and Space Museum. [Smithsonian NASM, Washington, DC, image nos. SI 97-17485 and SI 2004-50572]

## 4.20 HIGH-SPEED VEHICLES – Supersonic Propeller

**4.20–G** The XF-88B (McDonnell 1953) was a turboprop experimental version of the XF-88 ("Voodoo"). Built to conduct propeller research for supersonic planes, the XF-88 was first operated in 1953. At that time, the theory on thin "supersonic" propellers predicted that at low supersonic speeds a specially designed propeller could be more efficient under some circumstances than a jet engine. Until 1956 various propellers were tested at flight speeds up to slightly above Mach 1. At Mach 0.95 a peak efficiency of 80% had been measured. After being kept at Langley Air Force Base in Virginia for several years, the XF-88B was eventually scrapped. [© Boeing, image no. 5212]

## 4.20 HIGH-SPEED VEHICLES – Supersonic Transport (SST)

**4.20–H** In the early 1960s, aeronautical engineers of Great Britain, France, the Soviet Union and the United States developed plans to build a commercial supersonic airliner – so-called "SST" (SuperSonic Transport). All three SST liners were designed to be powered by four turbojet engines and have a fixed delta wing; the concept of using a variable geometry (so-called "swing wing") was abandoned because of weight and complexity. However, the Concorde was designed for lowering its nose during takeoff and landing to improve visibility. Like the Concorde, the Boeing SST had a variable nose geometry to improve flight deck forward views on approach. The primary structural material of SST planes were aluminum alloys, but the leading edges of the nose and wings, where supersonic flight generates high temperatures up to about 150 °C, were manufactured of stainless steel and titanium. *Top:* The Soviet Tupolev Tu-144 turboprop airliner was designed for $M = 2.4$ and 140 passengers. This picture was taken on 3 June 1973 at the Paris Air Show, shortly before the aircraft crashed in Goussainville; the six people on board died. In the 1990s, a modified version of the Tu-144 – named Tu-144LL – was operated as a "flying laboratory" in a joint U.S./Russian research program on supersonic flight. [Photo GettyImages] *Center:* The Concorde, a British-French development of BAC and Aerospatiale designed for $M = 2$ and 144 passengers, was the only operable SST civil airliner and in service until October 2003. [Photo taken by Paul JONGENEELEN, Hoboken, Belgium] *Bottom:* Artist's concept of the American SST prototype, the Boeing 2707-300. It was designed for $M = 2.7$ and 234 passengers. However, governmental funding was withdrawn in 1971 before even the prototype was finished. The original mockup of this unusual aircraft is on display at Hiller Aviation Museum in San Carlos, CA. [© Boeing, image no. 7049]

## *M 2.02*

Both the Tu-144 Machmeter (*left*) and the Concorde Machmeter (*center*) were analog instruments installed on the pilot's control panel. The Concorde also had a $23.5 \times 7.3$-cm$^2$ wall-mounted digital Machmeter (*right*) installed in the passenger cabin. [*Left & right:* Courtesy Auto & Technik Museum Sinsheim, Germany; *center:* Photo by British Airways]

## 4.20 HIGH-SPEED VEHICLES – Hypersonic Aircraft

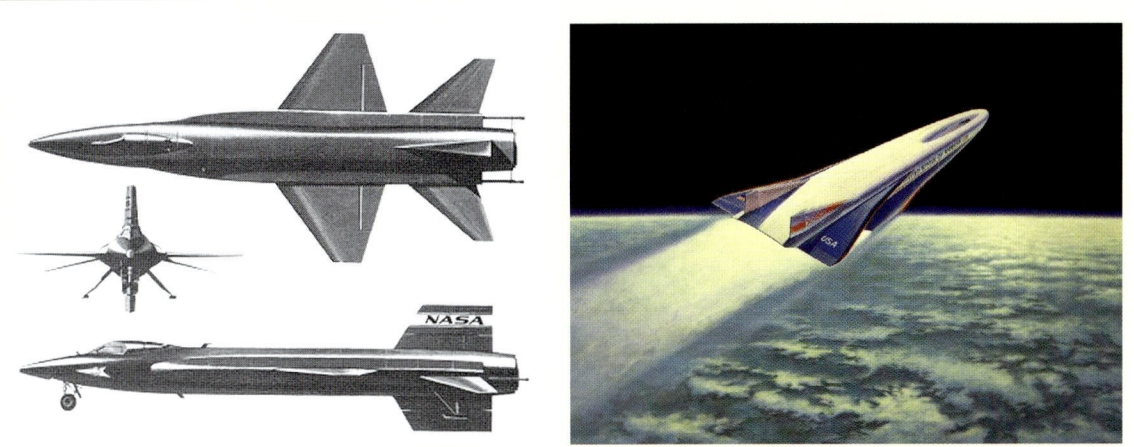

**4.20–I** *Left:* Three views of the North American X-15's original configuration with which it achieved on November 9, 1961 a maximum speed of $M = 6.06$ and a maximum altitude of 354,200 ft (108 km). [W.H. STILLWELL: *X-15. Research results*. NASA Rep. SP-60. NASA Scient. & Tech. Information Office, Washington, DC (1964)] *Right:* Artist's concept of the American National Aerospace Plane, X-30 NASP, supposed to take off from a conventional airfield and fly to orbit at a maximum speed of $M = 25$. This unusual aircraft was already planned in the 1980s but never realized. [NASP program, NASA-LRC, image no. EL-2001-00432]

## 4.20 HIGH-SPEED VEHICLES – Manned Spacecraft: Reentry Capsules

← Layers of special "ablative" material (*e.g.*, phenolic epoxy resin) on the shield, varying in thickness from 1.8 to 6.9 cm, were purposely allowed to burn away during reentry to help dissipate the extremely high temperatures caused by atmospheric friction. The blunt-end design for the Command Module was chosen to build upon experience with the similarly shaped Mercury and Gemini spacecraft.

**4.20 J** *Left.* View of the 2.5-m-dia. reentry capsule of the Soviet Vostok-1 mission in which the first man in space, Soviet research pilot Yuri A. GAGARIN, entered space and safely returned to Earth, thereby entering the atmosphere with the so-called "first cosmic velocity" of about 8 km/s, that is, $M > 20$. [Photo by Anatoly ZAK, Moscow; http://www.russianspaceweb.com/spacecraft_manned-first.html] *Right:* View of the American Apollo 11 Command Module Columbia, which, carrying on July 16-24, 1969 three astronauts on their historic voyage to the Moon and back, reentered the atmosphere with its protective heat shield facing forward. Since Columbia returned from the Moon to the Earth, the reentry velocity was higher than in the case of Vostok's reentry and amounted to $M > 30$. [Photo by Eric F. LONG, National Air and Space Museum, Smithsonian Institution (SI 98-16042)]

## 4.20 HIGH-SPEED VEHICLES – Manned Spacecraft: Shuttles

**4.20–K** The space shuttle Columbia was designed for scientific missions in the fields of space research and exploration as well as for technical and scientific applications. The speed of a space shuttle in low Earth orbit is about 7.8 km/s. An orbiter can fly at various altitudes, from under 300 to about 560 km above sea level, depending on the mission requirements. In 1981, Columbia became the first space shuttle to fly into Earth's orbit. *Left:* Launch view of the Columbia on April 12, 1981 at Kennedy Space Center, FL for the first shuttle mission (STS-1). *Right:* After 36 orbits during two days in space, Columbia landed at Edwards Air Force Base, CA in a manner similar to that of an aircraft. [NASA-JSC, Digital Image Collection; http://images.jsc.nasa.gov/index.html]

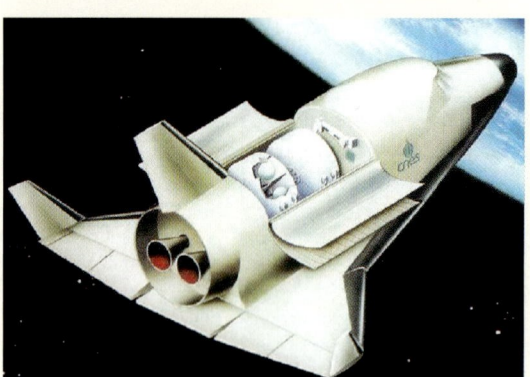

**4.20–L** *Left:* Artist's conception of the Hermes Space Plane (length 19.0 m, weight 23 tons). Designed in the 1980s by the European Space Agency (ESA), it was cancelled in the early 1990s for economic reasons. The purpose of Hermes was for (1) transferring crews and equipment to space stations, (2) servicing unpiloted platforms, (3) repairing satellites in orbit, and (4) putting scientific experiments in orbit during unpiloted flights. [ESA via Marcus LINDROOS at ESA; http://www.astronautix.com/craft/hermes.htm] *Right:* Aerodynamic model testing was partly performed at the Hypersonic Shock Tube Facility of the RWTH Aachen. The Hermes model was exposed to various hypersonic flow conditions, here shown at a Mach number of 7.5 and a yaw angle of 30°. Hypersonic flow phenomena were visualized using a color schlieren method. Note that the flow striking the fixed model propagates from right to left. [Courtesy Prof. Hans GRÖNIG, Stoßwellenlabor der RWTH Aachen]

## 4.20 HIGH-SPEED VEHICLES – Rocket Sleds

**4.20–M** Modern supersonic rocket sleds, filling the gap between wind-tunnel and airborne (free-flight) testing, are used to simulate flight conditions and study terminal ballistic phenomena at high speeds. *Top:* View of an early wheeled rocket sled during tests performed in 1948 on the Transonic (B-4) Track at the Naval Ordnance Research Track (SNORT), CA. [*China Lake Historical Overview*, http://www.nawcwpns.navy.mil/clmf/oldsled.html] The 50,988-ft (15,541-m)-long Holloman High-Speed Test Track (HHSTT), NM, the largest rocket sled facility in the world, can accelerate test objects from high subsonic speeds to Mach 8. *Bottom, left:* An array of head waves, emerging from different parts along the rocket sled, are visible against the background. [By courtesy of Dr. Leonard M. WEINSTEIN, NASA-LRC, Hampton, VA] *Bottom, right:* Reflected-light photography of a rocket sled traveling at 4,800 ft/s (M ≈ 4.3). [Photo by A. SEHMER, Sandia National Laboratories, Albuquerque, NM]

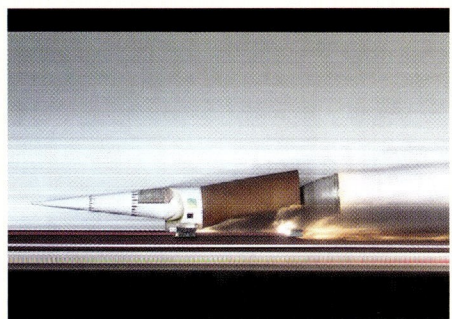

## 4.20 HIGH-SPEED VEHICLES – Supersonic Car

**4.20–N** *Left:* View of the British "ThrustSSC" supersonic car which measures about 16.5 m in length and is powered by a pair of Rolls-Royce jet engines as used in the Phantom fighter, each producing 20,000 pounds of thrust. To avoid lateral deviations from the race track, their individual thrusts are controlled electronically. The wheels consist of forged aluminum discs. *Right:* From left to right: The British engineer Richard NOBLE, holder of the 1983 Land Speed Record (1018 km/h), was project director of the ThrustSSC. The car was designed by Glynne BOWSHER. Ron AYERS was chief aerodynamicist of the ThrustSSC project, he only used computational fluid dynamics; no wind tunnel experiments were conducted. The British test pilot Andy GREEN, a member of Richard NOBLE's ThrustSSC team, broke the sound barrier on 17 October 1997 at Black Rock Desert in Nevada {⇨Fig. 4.6–J}: flying mile 763.035 mph (1,227.985 km/h). In setting the record, the sound barrier was broken in both north and south runs. [© Jeremy DAVEY; ThrustSSC team]

## 4.21 MAN-MADE DISASTERS – Steam-Boiler Explosions

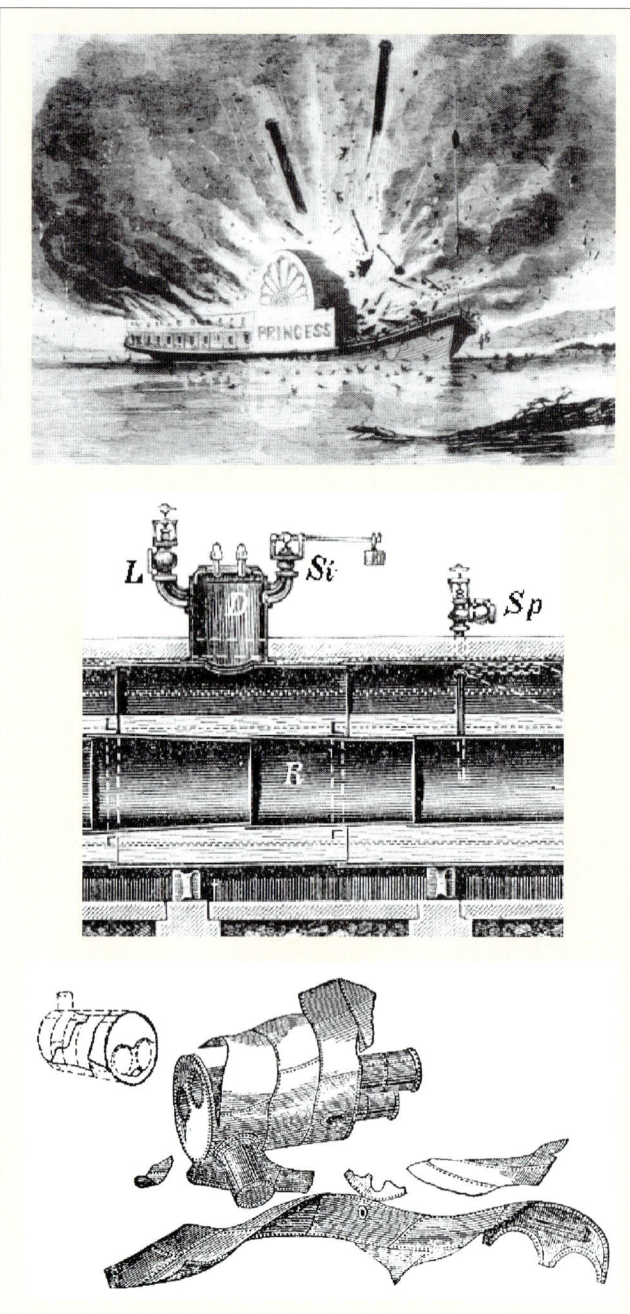

**4.21–A** In the main era of steam engines, beginning in the early 1700s and reaching its peak at the end of the 19th century, boiler explosions were an omnipresent hazard to man and equipment. They were considered the results of (1) complex thermodynamic phenomena in the steam/water mixture; (2) unfavorable boiler construction, *e.g.*, caused by poor production technology and the use of inappropriate materials; and (3) malfunctions of the safety valve and undue manipulations of it. The numerous accidents not only gave rise to the foundations of the first safety inspection authorities, but also initiated theoretical fluid dynamic studies on the discharge mechanism of a pressurized fluid through a small opening. They revealed that dangerous overpressures in steam boilers could not be reduced immediately, but only with a certain delay. ***Top:*** Contemporary drawing of a steam-boiler explosion on the side-wheeler SS *Princess* with 200 casualties, which occurred in 1859 on the Mississippi River near Baton Rouge, LA. This disaster was typical of many similar tragedies accompanying the use of steam engines in industry and traffic. [F. LESLIE's *Illustrated Newspaper* (March 19, 1859)] ***Center:*** Partial view of a British steam boiler of the Cornwall type, a construction provided with a single fire tube only. The safety valve *Si* was installed on top of the dome *D*; fresh water was supplied via valve *Sp* and the steam taken out via valve *L*. [*Brockhaus' Konversations-Lexikon*, Brockhaus, Leipzig (1908), vol. 4, p. 668] ***Bottom:*** Since high-speed diagnostics was then still in its infancy, it was difficult to uncover the puzzle of boiler explosions. Physical, chemical, and material reasons were likewise discussed. To get insight into possible causes, data from boiler explosions were meticulously collected and published in England in so-called "Boiler Records." From a visual inspection of boiler debris it was hoped that future boiler design could be improved. The picture below shows debris of the exploded boiler of the Bingley-type (1869). [E.B. MARTEN: *Records of steam-boiler explosions*. Stourbridge, London (1869)]

## 4.21 MAN-MADE DISASTERS – Firedamp Explosions

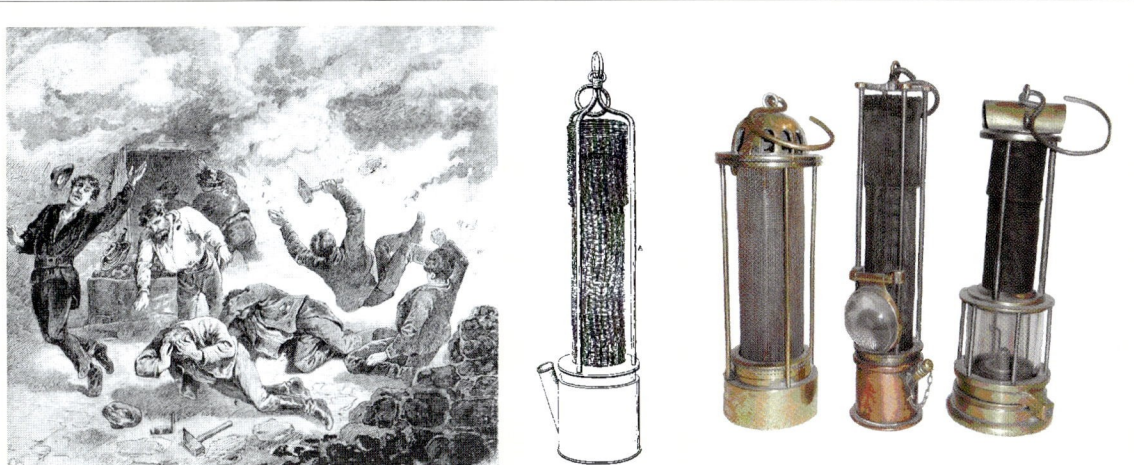

**4.21−B** *Left:* Drawing of a firedamp explosion in the soft-coal mining industry of Bilin [now Bílina, Czechia], killing 3 and seriously injuring 5 miners. Besides fatal blast effects, the sudden deficiency of oxygen often caused death from suffocation. [Das interessante Blatt **8**, No. 9, 1 (1889); courtesy Österreichische Nationalbibliothek, Vienna] *Center:* The English chemist Humphry DAVY studied the ignition of firedamp and the propagation of flames in tubes, which led to his invention of a safety mining lamp, the so-called "Davy lamp." A gauze mesh surrounding the flame of the lamp allows light to pass, but not the flame front of ignited firedamp to escape out in the mine atmosphere. [Phil. Trans. R. Soc. Lond. **107**, 84 (1817)] *Right:* Early flame safety lamps designed by the railroad engineer George STEPHENSON (*left*), by H. DAVY (*center*), and by the physician William R. CLANNY (*right*). [Courtesy of David S. BARRIE; http://www.thewandofscience.co.uk/]

## 4.21 MANMADE DISASTERS – Gun Barrel Bursts

← View of German heavy 15-cm field howitzers, model 14 (*left*), and model 02 (*right*) which became operational in 1914 and 1902, respectively.

**4.21−C** Barrel bursts of fire arms are rare phenomena in peacetime but frequently occur in times of war, where they can kill more soldiers and destroy more guns than can enemy fire. In the great battle of Verdun (1916), barrel bursts [Germ. *Rohrkrepierer*] were observed particularly on the German heavy 15-cm field howitzer. [Reprinted with permission from H. SCHIRMER: *Das Gerät der schweren Artillerie vor, in und nach dem Weltkrieg. Bilderband.* Bernhard & Graefe, Berlin (1937)]

## 4.21 MANMADE DISASTERS – Grain Dust Explosions

**4.21–D** Fine dust, in particular grain dust, caused many accidents in the past. It becomes extremely explosive when the dust gets in with the right mixture of oxygen and then comes in contact with a spark or another source of heat. This pen-and-ink drawing shows the mills at Minneapolis after the great flour dust explosion on May 2, 1878. Similar accidents happened there quite frequently in the 1870s, when the center of wheat production moved to Minnesota and Dakota Territory. [Courtesy Minnesota Historical Society, St. Paul, MN]

**4.21–E** Despite various countermeasures, dust explosions are still an omnipresent hazard to modern grain industry. They often have complex causes and are still a matter of present-day investigations. *Left:* In June 1998, a serious grain dust explosion happened at the half-mile-long grain elevator complex of DeBruce Inc., near Wichita, KS. The accident caused a series of explosions throughout the world's largest grain elevator, killing 7 workers and injuring 10 others. The facility mainly consisted of 246 circular grain silos and a central headhouse containing 4 elevator legs. Elevated grain was carried horizontally by belt from the headhouse to a selected silo. The accident was probably caused by a locked roller of a belt, which raised its temperature well beyond the ignition temperature of layered grain dust, which is only around 220 °C. *Right:* Aerial view of a part of the elevator complex taken within 1 h after the explosion. [Courtesy U.S. Department of Labor, Occupational Safety and Health Administration (OSHA); http://www.osha.gov/as/opa/foia/hot_6.html]

## 4.21 MAN-MADE DISASTERS – Nuclear Reactor Explosion

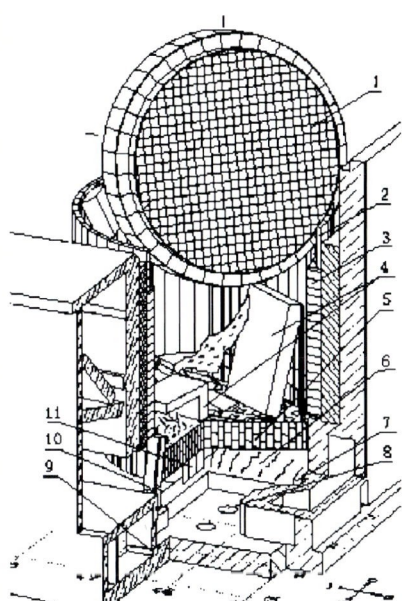

**4.21–F** On April 26, 1986, the nuclear power plant at Chernobyl (northern Ukraine) in the former Soviet Union was largely destroyed by a huge steam explosion, thereby setting free large quantities of radioactive matter. Ironically, the accident – the most severe ever to have occurred in the nuclear industry – was triggered by a safety test that got out of control. Actions taken during this exercise resulted in a rapid increase in the power level of the reactor, causing a destructive shock wave in the cooling water. Access to the reactor well allowed researchers to switch themselves from hypothetical models describing postaccident reactor state to studying a real state of reactor core, metal structures, and systems of the reactor. There was no core found in the reactor well. Core portions, such as graphite blocks, fragments of fuel channels, and assemblies, were found on the roof of nearby buildings, and individual graphite blocks and fragments of fuel assemblies and fuel rods were found close by, up to 150 m away. **Left:** Aerial view of the destroyed Block IV of the nuclear power plant. [From Wikipedia: *Chernobyl disaster*, source unknown] **Right, top:** Schematic of the reactor section of Unit IV after the accident, showing the empty core region. [Reprinted from Y.M. Cherkashov et al.: *Post-accident state of Chernobyl-4*. In: "Chernobyl Lessons – Technical Aspects." Int. Seminar held at Desnogorsk, Russia (April 1996); http://www.ignph.kiae.ru/ins/osrez/osrez.htm] **Right, center & bottom:** Partial views of destructions in the turbine hall *(center)* and adjacent reactor building *(bottom)*. [From Ogonyok (Moscow), No. 15 (April 2000); http://www.ropnet.ru/ogonyok/win/200015/15-30-33.html]

## 4.21 MAN-MADE DISASTERS – World Trade Center, NY: Terrorist Bomb Attack

**4.21–G** *Left, top:* On February 26, 1993 a massive explosion occurred in a subterranean garage below the World Trade Center (WTC) Plaza. The bomb, placed by terrorists in a truck parked on level 2, had an estimated yield equivalent to approx. 1,200 lb (544 kg) TNT, blowing a hole half the size of a football field in the basement. The explosions and subsequent fires caused extensive structural damage on several basement levels. Six people were killed and more than 1,000 injured, mostly from smoke inhalation, but the building did not collapse as the terrorists had hoped. The explosion caused nearly $600 million in property damage. Analysts later determined that, had the terrorists not made a minor error in the placement of the bomb, both towers could have fallen. [Courtesy U.S. Bureau of Alcohol, Tobacco, and Firearms; http://www.nycop.com/Stories/Dec_00/World_Trade_Center_Bombing/body_world_trade_center_bombing.html]

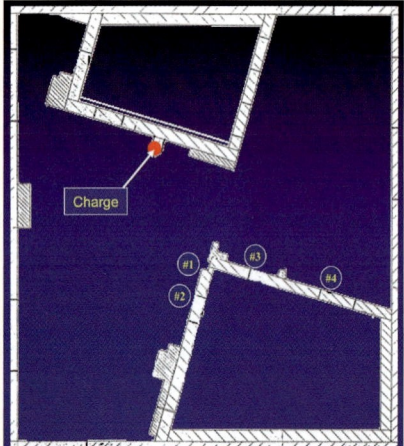

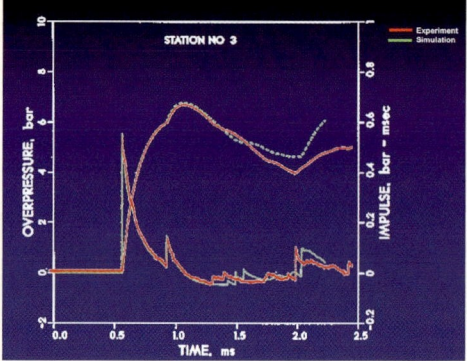

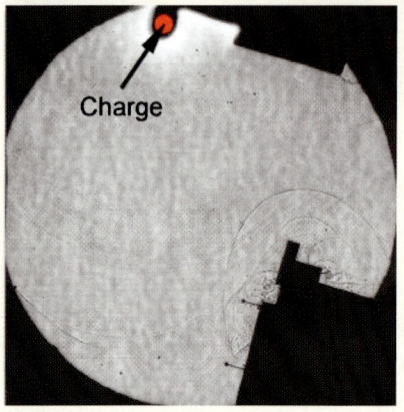

*Left, center:* Schematic of the floor plan of level 2 of the parking garage. The position where the truck with the bomb was parked is indicated by a *red dot*. *Left, bottom:* Heinz REICHENBACH and collaborators at EMI, Freiburg performed 2-D model blast propagation and reflection studies using small explosive charges and optical shadowgraphy. Pressure-time profiles were recorded at various distances from the model explosion center using Kistler pressure gauges. The blast wave, emerging from the charge (*top*), is reflected at the opposite walls and focused in corners (*bottom*), leading at those locations to significant increases in peak pressures. In such areas, the observed damage to parked cars was even greater than in the close environment of the explosion center. [Rept. E2/94 (1994), EMI, Freiburg] *Right:* Joseph D. BAUM at Science Applications Int. performed a numerical simulation of the blast wave propagation. A comparison between EMI's measured pressure-time profiles and calculated ones at different locations showed a fairly good agreement. Based on these promising results, BAUM and colleagues at George Mason University, Fairfax, VA, initiated a 3-D numerical modeling of the blast wave propagating in the parking deck. They used the FEFLO96 code, which is based on a 3-D finite-element, shock-capturing methodology. [Courtesy Dr. J.D. BAUM, Science Applications International Corporation, McLean, VA; Proc. 33rd Aerospace Sciences Meeting & Exhibit, Reno, NV (Jan. 1995)]

## 4.21  MAN-MADE DISASTERS – World Trade Center, NY: Terrorist Aircraft Attack

**4.21–H** On September 11, 2001, two Boeing 757s hijacked by terrorists crashed into the two 417-m-high towers of the World Trade Center. It was the worst disaster in recorded history, killing some 2,800 people. The North Tower and the South Tower collapsed 105 and 62 min after impact, respectively. The Twin Towers were designed to withstand as a whole the forces caused by a horizontal impact of a large aircraft and strong winds. Unfortunately, however, the towers were vulnerable to fire damage due to their primarily steel construction. In the floors impacted by the aircraft the tens of thousands of gallons of aviation fuel that spilled from the doomed airliners caused such intense heat that the supporting steel structures were weakened. This eventually resulted in a collapse of these floors. When the mass of the upper part of the building began to fall, it gained momentum that exceeded the force that the structure below could resist. The resulting collapse then continued down the building as an unstoppable chain reaction borne along by the force of gravity.

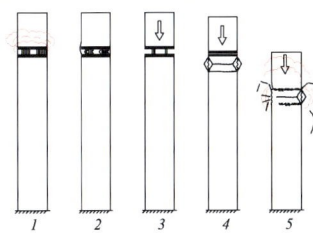

← Photographs showing the collapse of the South Tower at about 5, 5.9, and 7.5 seconds after initiation.

***Top:*** This unique picture was taken shortly before the second plane crashed into the WTC's North Tower. [Reprinted with permission of Robert CLARK, New York City; who took the photo] ***Bottom, left & center:*** View of the South Tower collapsing in a plume of ash and debris. [Images taken from *9-11 Research: South Tower Collapse*: http://911research.wtc7.net/wtc/evidence/photos/wtc2exp9.html; image source unknown] ***Bottom, right:*** Prof. Zdeněk P. BAŽANT and collaborators at the Dept. of Civil Engineering, Northwestern University in Evanston, IL, suggested that the gravitational collapse was the dynamic consequence of the prolonged heating of the steel columns to very high temperatures, thus causing creep buckling and a loss of strength. This is a schematic showing five stages of their failure scenario. The vertical impact of the mass of the upper part onto the lower part produced an enormous vertical dynamic load on the underlying structure. This caused buckling and plastic deformation of the steel structure elements. The collapse of the tower was an almost free fall because the observed duration of 9 s was pretty close to the free fall time of 8.93 s from the tower top (416 m) to the top of the final heap (ca. 25 m above ground). [Reprinted from J. Eng. Mech. ASCE **128**, 2, 369 (2002). With permission of Prof. Z.P. BAŽANT, Northwestern University, Evanston, IL]

# 5

# BIOGRAPHIES INDEX

# ABEL, Sir Frederick Augustus (1827–1902); from 1893 first Baronet

- English chemist and military explosives specialist

Sir Frederick A. ABEL was born in Woolwich, southeast London. He was the eldest son of Johann Leopold ABEL, a music master in Kennington of German descent. After attending high school at the Johanneum in Hamburg, he studied chemistry at the London Polytechnic Institute and the Royal College of Chemistry. He first worked on aniline derivatives and then began offering instruction in practical chemistry to artillery officers at the Royal Military Academy in Woolwich, south-east London (1849). Succeeding Michael FARADAY as professor of chemistry at this prestigious institution (1851), he was appointed Ordnance Chemist (1854), then Great Britain's first Chemist to the War Department (1854–1888), and later also Chemical Referee to the British Government.

In conjunction with Sir Charles WHEATSTONE, ABEL investigated the applicability of electricity to military purposes (1856–1861) and worked as well on the detonation of explosives by electrical means. In 1866, he developed a process to prevent guncotton from exploding spontaneously by reducing it to a fine pulp that could be worked and stored with little danger. He measured the detonation velocity of nitroglycerin, then a subject of great controversy, and obtained a velocity of 1,525 m/s in tubes of 3 mm inner diameter (1867). Together with the British chemist Sir Andrew NOBLE he investigated the nature of chemical changes that result from firing explosives and measured the temperature of fired gunpowder (1875–1880). Together with (later Sir) James DEWAR, a Scottish professor of chemistry at Cambridge University, he invented and developed *cordite* (1889), a mixture made from purified ingredients of nitroglycerin, nitrocellulose, and petroleum jelly. This new safe and smokeless explosive, later adopted as the standard explosive of the British Army, was of vital importance in the First World War. One of ABEL's inventions was the so-called "Abel Heat Test" (1875) for checking the stability of a heated sample of cordite and other nitroglycerin and nitrocellulose explosives. It consists of heating a sample of the explosive in a test tube under rigid temperature control and estimating the degree of stability from the time taken to develop a brown color on a special paper suspended in the tube over the cordite. In the early 1880s, he studied the causes of firedamp explosions at the Seaham Colliery, Durham and their connections with the presence of finely divided coal dust in the air. This work also shed light on previously inexplicable but disastrous explosions in flour mills. ABEL also sat on the Royal Commission on Accidents in Mines (1883).

ABEL was knighted in 1883 and became a baronet in 1893. He received the Royal Medal (1887) for his research on explosives and the Bessemer Gold Medal (1897). He was President of the Chemical Society (1875–1877) and second holder of the Presidential Chair of the Institute of Chemistry (1880–1883) and the Committee on Explosives (1888–1891). During his career he also presided over other scientific organizations such as the Institution of Electrical Engineers (IEE), the Society of Chemical Industry (SCI), and the Iron and Steel Institute (ISI) and participated in the foundation of the Imperial Institute in London. In 1888, he was made Doctor of Science by Cambridge University.

**ORIGINAL WORKS.** With C.L. BLOXAM: *Handbook of chemistry: theoretical, practical, and technical*. Churchill, London (1858) — With C. WHEATSTONE: *Result of investigation conducted at Woolwich and Chatham on the application of electricity from different sources to the explosion of gunpowder*. Roy. Eng. Papers **10**, 89-115 (1861) — *On some phenomena exhibited by gun-cotton and gunpowder under special conditions of exposure to heat*. Proc. Roy. Soc. Lond. **13**, 204-217 (1864) — *Improvements in the preparation and treatment of gun-cotton*. Brit. Patent No. 1,102 (1865) — *On the chemical history and application of gun-cotton*. Proc. Roy. Inst. **4**, 245-263 (1866) — *On recent progress in the history of proposed substitutes for gunpowder*. Proc. Roy. Inst. **4**, 616-626 (1866) — *On the manufacture and composition of gun-cotton*. Phil. Trans. Roy. Soc. Lond. **156**, 269-308 (1866) — *Researches on gun-cotton. Second memoir. On the stability of gun-cotton* [Bakerian Lecture]. Ibid. **157**, 181-254 (1867) — *On the chemical composition of the great cannon of* MUHAMMED II, *recently presented by the Sultan Abdul* AZIZ KHAN *to the British Government*. Rept. Meet. Brit. Assoc. **38**, 34-35 (1868) — *Nouvelles études sur les propriétés des corps explosibles*. C. R. Acad. Sci. Paris **119**, 105-121 (1869) — *Contributions to the history of explosive agents (First memoir)*. Phil. Trans. Roy. Soc. Lond. **159**, 489-516 (1869) — *Explosive agents applied to industrial purposes*. Proc. Inst. Civ. Eng. **33**, 327-371 (1872) — *On the more important substitutes for gunpowder*. Proc. Roy. Inst. **6**, 517-533 (1872) — *The rapidity of detonation*. Nature **8**, 534 (1873) — *Contributions to the history of explosive agents (Second memoir)*. Phil. Trans. Roy. Soc. Lond. **164**, 337-395 (1874) — *Studien über Eigenschaften explosiver Stoffe*. Dingler's Polytech. J. **213** [V], 428-430 (1874) — *Accidental explosions*. Nature **11**, 436-439, 477-478, 498-499 (1875); Roy. Inst. G.B., London (1875) — With A. NOBLE: *Researches on explosives. Fired gunpowder. Part I*. Phil. Trans. Roy. Soc. Lond. **165**, 49-155 (1875); *Part II*. Ibid. **171**, 203-279 (1880) — *Recent contributions to the history of detonating agents*. Chem. News **39**, 165-166, 177-179, 187-188, 198-200, 209-212 (1879) — *Explosive agents applied to industrial purposes*. Proc. Inst. Civ. Eng. **61**, 58-104 (1880) — *On colliery*

*explosions* [Report on experiments with dust from Seaham Colliery]. Chem. News **44**, 16-18, 27-31, 39-42 (1881) — *On electricity applied to explosive purposes*. Nature **28**, 66-68 (1883) — *Explosions in coal mines*. Ibid. **33**, 108-112, 138-142 (1886) — *Coal dust and explosions*. Ibid. **33**, 417 (1886) — *Accidents explosions produced by non-explosive liquids*. Proc. Roy. Inst. **11**, 218-242 (1887) — *Accidents in mines*. Proc. Inst. Civ. Eng. **90**, 160-200 (1887); Ibid. **91**, 36-83 (1888) — *Mining accidents and their prevention*. Scientific Publ., New York (1889) — *Smokeless explosives*. Proc. Roy. Inst. **13**, 7-23 (1893).
**SECONDARY LITERATURE**. *Leading men of London*. Brit. Biogr. Company, London (1895) — A.T.C. PRATT: *People of the period*. Beemann, London (1897) — V.G. PLARR: *Men and women of the time*. Routledge, London (15th edn., 1899) — W.R.D. (anonymous): *Sir Frederick ABEL*. Nature **66**, 492-493 (1902) — J. SPILLER: *Frederick August ABEL*. J. Chem. Soc. (Lond.) **87**, 565-570 (1905) — H.B. MASON (ed.) *Encyclopedia of ships and shipping*. Shipping Encyclopaedia, London (1908) — L. STEPHEN (ed.) *The dictionary of national biography. Suppl. 2 (1901–1911)*. Milford, London (1912) — F.D. MILES: *A history of research in the Nobel division of I.C.I.* Imperial Chemical Industries, Glasgow (1955) — W.G. NORRIS: *Chemical service in defense of the realm. One hundred years of chemical inspection. The story of the chemical inspectorate*. Ministry of Supply, London (1957) — K.R. WEBB: *Former Presidents of the Institute: Sir Frederick ABEL, 1827–1902. Second President 1880–1883*. J. Roy. Inst. Chem. **82**, 147-150 (1958) — R. STEELE and K.D. WATSON: *ABEL, Sir Frederick Augustus*. In: (H.C.G. MATTHEW, ed.) *Oxford dictionary of national biography*. Oxford University Press, Oxford (2004), vol. 1, pp. 62-63.
**PICTURE**. Courtesy Library and Information Centre, Royal Society of Chemistry, Cambridge, U.K.

# ACKERET, Jakob (1898–1981)

- Swiss aerodynamicist; pioneer of aerodynamics

The son of Jakob ACKERET, a master locksmith, Jakob ACKERET was born in Zurich. After schooling at the Industrieschule Zürich, he studied mechanical engineering at the Eidgenössische Technische Hochschule (ETH) in Zurich (1916–1920) and became assistant to Prof. Aurel STODOLA, one of the most eminent representatives of mechanical engineering of his time. On STODOLA's recommendation he went to the University of Göttingen to study under Prof. Ludwig PRANDTL at the Aerodynamische Versuchsanstalt (AVA) to learn about the aerodynamics of aircraft. During the period 1921–1927 he worked out the essential theoretical fundamentals of supersonic flight and developed his famous linearized wing theory of thin sharp-edged supersonic airfoils – the so-called "Ackeret theory" (1925) – which the British physicist Geoffrey I. TAYLOR restated for the benefit of English-speaking readers (1932).

In the period 1925–1926, he directed the extension of PRANDTL's institute, which was financed by the German government and became the famous Kaiser-Wilhelm-Institut für Strömungsforschung. The new installations were used to study problems in gas dynamics and cavitation and to test new aerodynamic theories of flight at high speeds in practice. Subsequently, he became head of the hydraulic laboratory at the machine factory Escher-Wyss AG in Zurich and established testing procedures for hydraulic machines as well as for gas and steam turbines (1927–1931). In 1928, he qualified as a university lecturer at the ETHZ with a paper entitled *Über Luft-Kräfte bei sehr großen Geschwindigkeiten insbesondere bei ebenen Strömungen* ("On Air Forces at Very High Velocities, Particularly for Two-Dimensional Flows"), in which he coined the term *Mach number*. In 1931, he became *Privatdozent* (university lecturer) and 3 years later was appointed full professor of aerodynamics at the ETHZ and director of the newly founded Institut für Aerodynamik (IfA). He investigated the aerodynamic lift of rotating cylinders, known as the "Magnus effect," which led to the development of the Flettner-Rotor (1925) for possible driving and modernizing of sailing ships.

ACKERET's transfer of scientific know-how of modern aerodynamics to the construction and economic operation of steam and gas turbines was particularly successful and acknowledged worldwide. To increase the lift or to reduce the length of diffusers, he studied the removal of boundary layers by suction (1925–1927). He constructed the first closed-loop supersonic wind tunnel (1933–1934) which, built in cooperation with Brown Bovery & Co. (BBC) in Baden and Escher-Wyss AG in Zurich, was installed at ETH's new Maschinenlaboratorium. A similar wind tunnel was built under his direction at the Italian research center at Guidonia. He also treated the problem of cavitation (1938), which is essential for the trouble-free operation of steam turbines.

ACKERET was editor (1934–1961) of the bulletin Mitteilungen aus dem Institut für Aerodynamik ETHZ, where his most important results were also published.

**ORIGINAL WORKS**. *Neue Untersuchungen der Aerodynamischen Versuchsanstalt*. VDI-Z. **68**, 1087, 1155-1156 (1924) — *Das Rotorschiff*. Vandenhoeck & Ruprecht, Göttingen (1925) — *Luftkräfte auf Flügel, die mit* größerer als Schallgeschwindigkeit bewegt werden. Z. Flugtech. u. Motor-Luftschiffahrt **16**, 72-74 (1925) — *Über Grenzschichtabsaugung*. VDI-Z. **70**, 1153-1158 (1926) — *Gasdynamik*. In: (H. GEIGER and K. SCHEEL, eds.) *Handbuch der Physik*. Springer, Berlin, vol. VIII (1927), pp. 289-342 — *Ausbreitungsgeschwindigkeit von Schallwellen mit endlicher Amplitude (RIEMANN'scher Verdichtungsstoß)*. Z. math.-naturwiss. Unterricht **59**, 307-314 (1928) — *Über Luftkräfte bei sehr großen Geschwindigkeiten, insbesondere bei ebenen Strömungen*. Helv. Phys. Acta **1**, 301-322 (1928) — *Der Luftwiderstand bei sehr großen Geschwindigkeiten*. Schweiz. Bauz. **94**, 179-183 (1929) — *Experimentelle und theoretische Untersuchungen über Hohlraumbildung (Kavitation) im Wasser*. Dissertation, ETH Zürich (1930); Tech. Mech. Thermodyn. (VDI, Berlin) **1**, 1-22, 63-72 (1930) — *Kavitation*. In: (L. SCHILLER, ed.) *Handbuch der Experimentalphysik*. Akad. Verlagsgesell., Leipzig (1931), vol. IV, 1 (1931), pp. 461-486 — With P. DE HALLER: *Untersuchungen über Korrosion durch Wasserstoß*. Schweiz. Bauz. **98**, 309-310 (1931) — *Einfluß hoher Umfangsgeschwindigkeiten auf den Wirkungsgrad von Luftschrauben*. Ibid. **101**, 11-12 (1933) — *Über die Bildung von Wirbeln in reibungslosen Flüssigkeiten*. ZAMM **15**, 3-4 (1935) — *Der Überschallkanal des Instituts für Aerodynamik an der ETH*. Schweiz. Aero-Rev. **10**, 112-114 (1935) — *Short autobiography*. In: *V Convegno Volta su "Le alte velocità in aviazione"* [Rome, Sept./Oct. 1935]. Reale Accademia d'Italia, Roma (1936), p. 661 — *Aerodynamischer Auftrieb bei Überschallgeschwindigkeit*. Ibid., pp. 328-360; Luftfahrtforsch. **12**, 210-220 (1935) — *High-speed wind tunnels*. NACA-TM 808 (1936) — With P. DE HALLER: *Über die Zerstörung von Werkstoffen durch Tropfenschlag und Kavitation*. Schweiz. Bauz. **108**, 105-106 (1936) — With P. DE HALLER: *Über Werkstoffzerstörung durch Stoßwellen in Flüssigkeiten*. Arch. Angew. Wiss. Tech. **4**, 293-294 (1938) — With M. RAS: *Über die Verhinderung der Grenzschicht-Turbulenz durch Absaugung*. Helv. Phys. Acta **14**, 323-325 (1941) — *Zur Theorie der Raketen*. Ibid. **19**, 103-112 (1946) — With F. FELDMANN and N. ROTT: *Untersuchungen an Verdichtungsstößen und Grenzschichten in schnell bewegten Gasen*. Mitt. Inst. Aerodyn. ETH, Nr. 10, Leemann, Zürich (1946); *Investigation of compression shocks and boundary layers in gases moving at high speeds*. NACA-TM 1113 (1947) — *Elementare Betrachtungen über die Stabilität der Langgeschosse*. Helv. Phys. Acta **22**, 127-134 (1949) — *Untersuchungen an gepfeilten und ungepfeilten Flügeln bei hohen Unterschallgeschwindigkeiten*. ZAMP **1**, 32-42 (1950) — *Akustische Phänomene bei hohen Fluggeschwindigkeiten*. Schweiz. Aero-Revue **11**, 429-430 (1952) — *Über exakte Lösungen der Stokes-Navier-Gleichungen inkompressibler Flüssigkeiten bei veränderten Grenzbedingungen*. ZAMP **3**, 259-271 (1952).

**SECONDARY LITERATURE.** G.I. TAYLOR: *Applications to aeronautics of ACKERET's theory of airfoils moving at speeds greater than that of sound*. Brit. Aeronaut. Research Committee (BARC), R&M No. 1467, WA-4218-5a (1932) — *Festschrift Jakob ACKERET zum 60. Geburtstag am 17. März 1958*. ZAMP **IX b**, Birkhäuser, Basel (1958); F. TANK: *Jakob ACKERET*, Ibid., pp. 9-16; L. MEYER and H. SPRENGER: *Bibliographie ACKERET*, pp. 17-25; A. BETZ: *ACKERET in Göttingen*, pp. 34-36 — F. SCHULTZ-GRUNOW: *J. ACKERET: Persönliche Erinnerungen*. Schweiz. Ing. Arch. **21**, 587-590 (1983) — N. ROTT: *Jakob ACKERET and the history of the Mach number*. Annu. Rev. Fluid Mech. **17**, 1-9 (1985) — Z. PLASKOWSKI: *Hochgeschwindigkeitskanal* [description of the world's first closed-loop supersonic wind tunnel]. Pro Technorama **4**, 5-12 (1988) — G. BRIDEL: *Jakob ACKERET (1898-1981)*. Schweiz. Pioniere Wirtschaft Tech. **67**, 73-90 (1998).

**PICTURE.** Courtesy of Prof. Leonhard KLEISER, Institut für Fluiddynamik, ETH Zurich, Switzerland.

**NOTE.** *Jakob ACKERET (1898-1981), Pionier der Aerodynamik.* Präsentiert zum 20. Todestag am 27. März 2001 von der Bibliothek der ETH Zürich; http://www.ethbib.ethz.ch/exhibit/ackeret/ackeret_frame.html. *See also* 13 pictures of Jakob ACKERET, "bildarchivonline" der ETH Bibliothek, Zürich; http://ba.e-pics.ethz.ch/ETH_Bibliothek/Standard/.

# AIRY, Sir George Biddell (1801–1892)

- British mathematician and Astronomer Royal

George B. AIRY was born at Alnwick in Northumberland to William AIRY, a farmer and erstwhile collector of excise, and received his early education at private academies in Hereford and Colchester. With the support of his uncle he entered Trinity College in Cambridge (1819), took a scholarship (1822), and graduated as senior wrangler and first Smith's prizeman (1823). On his election to a fellowship at Trinity College, he became assistant mathematical tutor (1824) and Lucasian professor of mathematics at Cambridge (1826); illustrious philosophers such as Isaac BARROW and Sir Isaac NEWTON had preceded him as occupants of that traditional chair.

AIRY's scientific contributions in the fields of mathematics, physics, and astronomy, commencing in 1824, have been numerous and of high merit. His *Mathematical Tracts on Physical Astronomy* (1826) became a standard textbook used in the university. When the Plumian professorship of astronomy became vacant, he was appointed to the chair (1828), which was also connected to the directorship of the Cambridge Observatory. He gave popular lectures on statics, dynamics, hydrostatics, geometrical optics, and on the theory of undulations, a chief subject of interest throughout his professional career. When he became the seventh Astronomer Royal (1835–1881), he reorganized the Royal Greenwich Observatory and installed modern equipment; some pieces were of his own invention. Based upon the results of his astronomical and geodetic work, the Greenwich meridian was accepted as the international zero longitude and prime meridian of the world (1884).

AIRY detected a "long inequality" in the orbits of Venus and the Earth and improved the theory of their orbital motions. He determined the orbits of comets from observations (1839), introduced observation of sunspot phenomena (1873), and invented instruments for lunar observations. He measured gravity by swinging the same pendulum at the top and bottom of a deep mine and thus computed the density of the Earth (1854). His labors in connection with the Royal Observatory of

Greenwich, both in its development and in his personal scientific work, gave him a position of rare eminence.

In around 1841, AIRY turned his attention to the theory of tides. He wrote several papers on this subject, discussing separately the tides in the Thames, at Ipswich, Southampton, the coast of Ireland, and, later on, the tides at Malta. His chief work on this subject is his essay on *Tides and Waves* (1845) in which, for example, he discusses the broken water seen on the edge of a shoal, why the rise of tide takes less time than the fall, the solitary wave, the breaking of waves, the effect of the wind, and the effect of friction. Based on his theory of river tides, AIRY gave a general explanation that the retardation of the rotation of the Earth is caused by tidal friction. He also theoretically studied the motion and form of "waves of finite amplitude" in a broad, uniform canal of rectangular section and found, by methods of successive approximation, that in a progressive wave different parts will travel at different velocities. This pioneering analytical treatment of river tides might have motivated him to treat also sounds of finite amplitude in air, which, however, occupied him only a short period (1848–1849), while he maintained an interest in tidal waves through his later years. The same subject also stimulated James CHALLIS and George STOKES to turn to this problem on which they carried on a prolonged dispute.

AIRY made a number of worthwhile contributions to applied optics. For example, in the 1830s he carried out various experiments on the diffraction of Newton rings and on the intensity of light in the neighborhood of a caustic, and he attempted to theoretically explain the polarization of light. He used a water-filled telescope to test the effect of the Earth's motion on the aberration of light (1871). He studied astigmatism and was the first to correct astigmatism in the human eye (his own) by use of a cylindrical eyeglass lens (1825), a method that is still used. He was also actively involved in the improvement of lighthouses.

AIRY's continuous interest in the application of mathematics led to the solution of many physical and engineering problems. For example, his paper on strains in the interior of beams (1863) stimulated James C. MAXWELL to develop a general theory of stress diagrams for 3-D stress systems, while his studies on the causes of destructive steam-boiler explosions (1863) served immediate practical needs of greatest relevance. He was called in as a consultant on a project involving the removal of the magnetic-compass disturbance in iron-built ships and set up a magnetic department (1838), discovering that the deviation of the compass is accounted for almost entirely by the permanent magnetism of the hull (1840, 1856). He was also consulted on the launch of the SS *Great Eastern* (1858), the design of Big Ben (1850s), and the laying of the Atlantic telegraph cable (1866).

AIRY's scientific output was enormous; he wrote over 500 published papers and numerous books on mathematical physics and essays on history. From the Royal Society he received the Copley Medal (1831) for his successful optical theories and the Royal Gold Medal (1845) for his tidal investigations. He was awarded the Lelande Prize (1834) by the French Institute in honor of his discoveries in astronomy and the Telford Medal (1867) from the Institution of Civil Engineers. Twice the Gold Medal (1833, 1846) of the Royal Astronomical Society was awarded to him for his discovery of an inequality of long period in the movement of Venus and for his reduction of planetary observations. He was an honorary member of many scientific societies both at home and abroad. AIRY was knighted in 1872. He held honorary degrees from three great universities of Great Britain and was a foreign correspondent of various scientific societies.

The *Airy spiral* (an optical phenomenon visible in quartz) and the *Airy disc* (the central spot of light in the diffraction pattern of a point light source) are named after him. Astronomers named a crater on the near side of the Moon and a crater on Mars after him.

**ORIGINAL WORKS.** *On the laws of the rise and fall of the tide in the river Thames.* Phil. Trans. Roy. Soc. Lond. **132**, 1-8 (1842) — *On the laws of individual tides at Southampton and Ipswich.* Ibid. **133**, 45-54 (1843) — *An account of the results of the tide observations on the coast of Ireland.* Rept. Meet. Brit. Assoc. **14**, 4-6 (1844) — *On the laws of the tides on the coasts of Ireland, as inferred from an extensive series of observations made in connection with the Ordnance Survey of Ireland.* Phil. Trans. Roy. Soc. Lond. **135**, 1-124 (1845) — *Tides and waves* [1842]. In: (E. SMEDLEY and H.J. ROSE, eds.) *Encyclopaedia Metropolitana*. Fellowes, London, vol. 5 (1845), Art. 208, pp. 241-396. The section on river tides was translated into French by P. GUIEYSSE and printed in J. Math. Pures Appl. **1** [III], 399-450 (1875) — *The Astronomer Royal's remarks on Prof. CHALLIS's "Theoretical determination of the velocity of sound."* Phil. Mag. **32** [III], 339-343 (1848) — *On a difficulty in the problem of sound.* Ibid. **34** [III], 401-405 (1849) — *On the strains in the interior of beams.* Ibid. **23** [IV], 25-28 (1863) — *Report on steam-boiler explosions.* Rept. Meet. Brit. Assoc. **33**, 686-688 (1863) — *On the numerical expression of the destructive energy in the explosions of steam-boilers, and on its comparison with the destructive energy of gunpowder.* Phil. Mag. **26** [IV], 329-336 (1863) — *On the supposed possible effect of friction in the tides, in influencing the apparent acceleration of the Moon's mean motion in longitude.* Month. Not. Roy. Astron. Soc. **26**, 221-235 (1866) — *An elementary treatise on partial differential equations.* Macmillan, London (1866) — *On sound and atmospheric vibrations, with the mathematical elements of music.* Macmillan, London etc. (1868) — *On a controverted point in LAPLACE's theory of the tides.* Phil. Mag. **50** [IV], 277-279 (1875) — *On the tides at Malta.* Proc. Roy. Soc. Lond. **26**, 485-487 (1878) — (W. AIRY, ed.) *The autobiography of Sir George Biddell AIRY.* Cambridge University Press, Cambridge (1896).

**SECONDARY LITERATURE.** M. FAYE: *Notice sur Sir George Biddell AIRY, Associé étranger de l'Académie.* C. R. Acad. Sci. Paris **114**, 91-93 (1892); Ibid. **115**, 1117-1118 (1892) — E. BUDDE: *George Biddell AIRY.* Ann. Phys. **45** [III], 601-604 (1892) — Anonymous: *Sir George Biddell AIRY.* Nature **45**,

232-233 (1892) — *Miscellaneous intelligence.* Am. J. Sci. **43** [III], 248 (1892) — E.J.R. [E.J. ROUTH]: *Obituary note of Sir G. AIRY.* Proc. Roy. Soc. Lond. **51**, i-xxi (1892) — H.H. TURNER: *Obituary notice of Sir George B. AIRY.* Month. Not. Roy. Astron. Soc. **52**, 212-229 (1892) — A. HILL: *Sir George Biddell AIRY.* Science No. 19, 64-65 (Jan. 29, 1892); Proc. Roy. Soc. Edinb. **19**, i-viii (1893) — O.J. EGGEN: *G.B. AIRY.* In: (C.C. GILLESPIE, ed.) *Dictionary of scientific biography.* Scribner, New York, vol. 1 (1970), pp. 84-87 — A. CHAPMAN: *AIRY, Sir George Biddell.* In: (H.C.G. MATTHEW, ed.) *Oxford dictionary of national biography.* Oxford University Press (2004), vol. 1, pp. 521-524.
**PICTURE.** Photo courtesy Deutsches Museum, Munich, Germany.

## AL'TSHULER [Russ. *АЛЬТШУЛЕР*], Lev Vladimirovich (1913–2003)

- Russian physicist, dean of Russian shock wave physics

Lev V. AL'TSHULER was born in Moscow in a highly educated family; his father, a lawyer and active revolutionary, was later appointed to a post in the Soviet Ministry of Finance. After graduating from ordinary school in 1930, he worked for 2 years in the countryside. In 1932, he entered the Roentgen Laboratory of the Evening Institute of Mechanical Engineering in Moscow, which was headed at that time by Prof. E.F. BAKHMETEV, a recognized expert in X-ray analysis. BAKHMETEV cultivated in him a love for science, and studying under the famous scientist greatly influenced AL'TSHULER throughout his career. In 1933, AL'TSHULER entered the Faculty of Physics at Moscow State University. Being an external student in several subjects, he graduated from the University in 1936 ahead of schedule with a specialization in metal physics. He continued working at the Roentgen Laboratory till 1940; in 1939 the Laboratory was incorporated into the Institute of Engineering Science of the U.S.S.R. Academy of Sciences. In 1940, he was drafted and took part in World War II as a sergeant-mechanic and military technician of the Soviet Army, but in 1942 he was called back from the front to work in the U.S.S.R. Academy of Sciences. From 1942 to 1946, AL'TSHULER was Senior Research Associate at the Institute of Engineering Science. Together with Veniamin A. TSUKERMAN he developed pulsed radiography for studying shaped-charge effects on tank armor.

In the following years, the next important stage of his career began when he and TSUKERMAN were invited by Yulii KHARITON, then scientific director of the Soviet Nuclear Center Arzamas-16, to participate in the Soviet Atomic Project and to diagnose experimentally what happens to metals placed inside an explosive system. He worked at Arzamas-16 – now the All-Russia Research Institute of Experimental Physics (VNIIEF) located in the town of Sarov – from 1946 to 1969 and carried out experimental work in close cooperation with Yakov ZEL'DOVICH, Andrei SAKHAROV, and other prominent scientists. For example, he developed a method to measure the pressure of plane detonation waves; the basic results were obtained, together with Konstantin K. KRUPNIKOV, in 1948. In addition, equations of the state of compressed and heated explosion products were determined, dynamic methods for compressibility study developed, and the compressibility of fissile materials (uranium and plutonium) in the megabar range investigated. Together with Yakov B. ZEL'DOVICH and Yu.M. STYAZHKIN he investigated at multiple megabars the compressibility of fissile materials and polymorphic transitions during shock compression. Non-monotonous change of compressibility, corresponding to the reconstruction of energetic electronic spectra, was revealed for the first time by him in conjunction with A.A. BAKANOVA for rare-earth and alkaline metals. Together with M.N. PAVLOVSII, he studied phase transitions of the IV-group elements and ionic compositions, as wells as of minerals and rocks. Shock wave experiments, performed in conjunction with Aleksei M. PODURETS and Ryurik F. TRUNIN, showed that nearly all minerals and rocks form dense modifications under pressures above critical pressures. This significantly advanced the knowledge of the Earth's structure.

Returning to Moscow in 1969, AL'TSHULER headed a laboratory at the All-Union Institute of Optical-Physical Measurements. From 1989 he worked as Principal Research Associate at the Institute of High Temperatures of the Russian Academy of Sciences. Here he initiated efforts associated with the development of wide-range equations of state based on the joint interpretation of theoretical data and shock wave experiments. As founder of the Russian school of dynamic researches on the properties of shock-compressed materials, AL'TSHULER made a great contribution by teaching experts at this school. He is the author of over 60 scientific publications and coeditor of the book *Shock Waves and Extreme States of Matter* (New York, 2004).

He received the Stalin Prize in 1946 and the Order of Lenin in 1949. For a series of research efforts at VNIIEF, AL'TSHULER was awarded two more Stalin Prizes (1949, 1953), the Lenin Prize (1962), and three Orders of Lenin. He received the Shock Compression Science Award (1991) of the American Physical Society "in recognition of seminal and major contributions in the development of the field of shock wave compression of condensed matter." AL'TSHULER died on Dec. 23, 2003.

ORIGINAL WORKS. With A.A. BAKANOVA and R.F. TRUNIN: *Phase transformations of water compressed by strong shock waves*. Sov. Phys. Dokl. **3**, 761-763 (1958) — With K.K. KRUPNIKOV, B.N. LEDENEV, V.I. ZHUCHIKHIN, and M.I. BRAZHNIK: *Dynamic compressibility and equation of state of iron under high pressure*. Sov. Phys. JETP **7**, 606-614 (1958) — With K.K. KRUPNIKOV and M.I. BRAZHNIK: *Dynamic compressibility of metals under pressures from 400,000 to 4,000,000 atmospheres*. Ibid. **7**, 614-619 (1958) — With S.B. KORMER, A.A. BAKANOVA, and R.F. TRUNIN: *Equation of state for aluminum, copper, and lead in the high pressure region*. Ibid. **11**, 573-579 (1960) — With S.B. KORMER, M.I. BRAZHNIK, L.A. VLADIMIROV, M.P. SPERANSKAYA, and A.I. FUNTIKOV: *The isentropic compressibility of aluminum, copper, lead, and iron at high pressures*. Ibid. **11**, 766-775 (1960) — With L.V. KULESHOVA and M.N. PAVLOVSKII: *The dynamic compressibility, equation of state, and electrical conductivity of sodium chloride at high pressures*. Ibid. **12**, 10-15 (1961) — With A.P. PETRUNIN: *An X-ray investigation of the compressibility of light materials under the action of shock-wave impacts*. Sov. Phys. Tech. Phys. **6**, 516-522 (1961) — With D.M. TARASOV and M.P. SPERANSKAJA: *The deformation of steel under the action of shock waves from explosion* [in Russ.]. Fis. Met. **13**, 738-743 (1962) — With S.B. KORMER, A.A. BAKANOVA, A.P. PETRUNIN, A.I. FUNTIKOV, and A.A. GUBKIN: *Irregular conditions of oblique collision of shock waves in solid bodies*. Sov. Phys. JETP **14**, 986-994 (1962) — With A.A. BAKANOVA and R.F. TRUNIN: *Shock adiabats and zero isotherms of seven metals at high pressures*. Ibid. **15**, 65-74 (1962) — With M.N. PAVLOVSKII, L.V. KULESHOVA, and G.V. SIMAKOV: *Investigation of alkali-metal halides at high pressures and temperatures produced by shock compression*. Sov. Phys. Solid State **5**, 203-211 (1963) — *Use of shock waves in high pressure physics*. Sov. Phys. Uspekhi **8**, 52-91 (1965) — With A.A. BAKANOVA and I.P. DUDOLADOV: *Peculiarities of shock compression of lanthanides*. Sov. Phys. JETP Lett. **3**, 315-317 (1966) — With S.A. NOVIKOV and I.I. DIVNOV: *The relationship between the critical breaking stresses and time of failure as a result of explosive stressing of metals*. Sov. Phys. Dokl. **11**, No. 1, 79-82 (1966) — With M.N. PAVLOVSKII and V.P. DRAKIN: *Peculiarities of phase transitions in compression and rarefaction shock waves*. Sov. Phys. JETP **25**, 260-265 (1967) — With A.A. BAKANOVA and I.P. DUDOLADOV: *Effect of electron structure on the compressibility of metals at high pressure*. Ibid. **26**, 1115-1120 (1968) — With B.N. MOISEEV, L.V. POPOV, G.V. SIMAKOV, and R.F. TRUNIN: *Relative compressibility studies of iron and lead at pressures of 31 to 34 Mbar*. Ibid. **27**, 420-422 (1968) — With M.I. BRAZHNIK, V.N. GERMAN, and L.I. MIRKIN: *Explosive deformation of single crystals*. Sov. Phys. Solid State **9**, 2417-2421 (1968) — With I.I. SHARIPDZHANOV: *Additive equations of state of silicates at high pressures*. Earth Phys. **3**, 11-28 (1971) — With V.A. SIMONENKO: *History and prospects of shock wave physics*. High Pressure Research **5**, 813-815 (1990) — *Next to SAKHAROV*. In: (B.L. AL'TSHULER ET AL., eds.) *Andrei SAKHAROV. Facets of a life*. Editions Frontieres, P.N. Lebedev Physics Institute, Moscow (1991), pp. 44-52 — *Shock waves and extreme states of matter*. In: (S.C. SCHMIDT ET AL., eds.) *Proc. 7th Conference on Shock Waves in Condensed Matter* [Williamsburg, VA, June 1991]. North-Holland, Amsterdam (1992), pp. 3-14 — *Experiment in the Soviet atomic project*. In: (I.M. DREMIN and A.M. SEMIKHATOV, eds.) *Proc. 2nd Int. A.D. Sakharov Conf. on Physics* [Moscow, Russia, May 1996]. World Scientific, Singapore (1997), pp. 649-655 — With R.F. TRUNIN, K.K. KRUPNIKOV, and N.V. PANOV: *Explosive laboratory devices for shock wave compression studies*. Phys. Uspekhi **39**, No. 5, 539-544 (1996) — With Y.B. ZEL'DOVICH and Y.M. STYAZHKIN: *Investigation of isentropic compression and equations of state of fissionable materials*. Ibid. **40**, No. 1, 101-102 (1997) — With R.F. TRUNIN, V.D. URLIN, V.E. FORTOV, and A.I. FUNTIKOV: *Development of dynamic high-pressure techniques in Russia*. Ibid. **42**, No. 3, 261-280 (1999).

SECONDARY LITERATURE. A.A. BRISH, V.L. GINZBURG, R.I. IL'KAEV, N.N. KALITKIN, G.I. KANEL', K.K. KRUPNIKOV, B.V. LEVIN, A.Y. RUMYANTSEV, L.D. RYABEV, V.M. TITOV, R.F. TRUNIN, and V.E. FORTOV: *In memory of Lev Vladimirovich AL'TSHULER*. Phys. Uspekhi **47**, No. 3, 319-320 (2004) — B.L. AL'TSHULER, V.E. FORTOV, and W.J. NELLIS: *AL'TSHULER, Lev Vladimirovich 1913–2003; see* http://www.unipress.waw.pl/airapt/altshuler.pdf.

PICTURE. Courtesy Prof. Anatoly MIKHAYLOV, director of the Institute of Experimental Gasdynamics and Physics of Explosion at VNIIEF, Sarov, Russia, who also contributed to the biography.

NOTE. His name has also been transliterated as AL'TŠULER.

## ANTOLIK, Károly (1843–1905)

- Hungarian physicist and schoolmaster, inventor of the sooth method

Károly ANTOLIK was born in Kolbach (now Studenec) in East Slovakia, then a part of the Austro-Hungarian Empire, and studied physics and mathematics at the University of Budapest. There he became deeply influenced by Ányos JEDLIK, a prominent Hungarian physicist. After graduation, he taught at several Hungarian secondary schools. As a teacher at the Realschule Kaschau, he performed experiments with gliding sparks in order to analyze the mechanism of electric breakdown and tested various methods to record exactly the path of a spark discharge (1873–1874). Initially he experimented with various dust methods to mark spark traces such as LICHTENBERG's and CHLADNI's dust patterns. ANTOLIK was awarded a scholarship in Germany (1874–1875), where he spent the first semester in Berlin in the laboratory of Hermann VON HELMHOLTZ and the second semester under Robert W. BUNSEN and Georg H. QUINCKE at the University of Heidelberg. Here he resumed his studies of dust patterns.

Today it is not known precisely where he invented his soot method. In his paper *Das Gleiten elektrischer Funken* ("The Gliding of Electric Sparks") published in 1875, he noted that he had discovered by chance the suitability of soot as a recording medium: bringing a small soot-coated glass balloon close to the spark of an influence machine, he noticed a well-marked trace of a spark in the soot. When he had sufficiently developed his new soot recording technique, he observed a significant phenomenon: soot-covered glass plates, brought close to crooked gliding sparks, showed complicated "V-shaped patterns" that, however, disappeared when the air between the plates was evacuated (1875). This suggested that the phenomenon must be of an acoustical nature, but ANTOLIK was too caught up in the interpretation of electrostatic phenomena and his hypothesis of the existence of a point of encounter between two types of electricity. Prof. Roland VON EÖTVÖS, the most prominent Hungarian physicist of the time, had in one of his private communications compared ANTOLIK's experiments to children's games. However, his experiments stimulated Ernst MACH, who after reading ANTOLIK's paper immediately turned to the study of his spark patterns. Yet in 1875 MACH scientifically proved that the V-shaped contours are of mechanical (acoustic) and not electrical origin and the result of a peculiar (irregular) interaction of strong acoustic waves (the term *shock wave* had not yet been established). Curiously enough, ANTOLIK, being familiar with MACH's work and also corresponding with him, remained unimpressed by MACH's skepticism on using soot patterns in the field of electrical research.

ANTOLIK eventually became director of the Staatliche Oberrealschule in Pressburg (now Bratislava, Czechia). Nicknamed the "Hungarian pioneer of spark patterns" by his contemporaries because of his numerous studies of spark discharges, he is considered today an early pioneer of 19th-century Hungarian physics.

ORIGINAL WORKS. *Das Gleiten elektrischer Funken.* Ann. Phys. **151** [II], 127-130 (1874); Ibid. **154** [II], 14-37 (1875) — *Über elektrische Rauchfiguren.* Ann. Phys. **1** [III], 310-314 (1877) — *Über das Gleiten elektrischer Funken und über die Ausgleichungsstelle in der Schlagweite.* Ibid. **3** [III], 483-489 (1878) — *Über neue elektrische Figuren und über das Gleiten elektrischer Funken.* Ibid. **15** [III], 475-491 (1882) — *Les figures électriques.* Lumière électrique **11** [I], 310-315 (1884) — *Über Klangfiguren gespannter Membranen und Glasplatten.* Verhandl. des Vereins für Natur- und Heilkunde zu Pressburg **15** [II], 71-139 (1903).
SECONDARY LITERATURE. L. WAGNER: *Károly ANTOLIK 1843–1905* [nekrológ]. A Pozsonyi Orvos, Természettudományi E gycsület Közle-ményei **XVII** (1905) — M.T. MOROVICS: *Egy múlt századi kisérletező fizikus, ANTOLIK Károly életmüvéröl.* Fizikai Szemle (Budapest) **34**, 222-227 (1984) — J. TIBENSKY: *Karol ANTOLIK (1843–1905).* Prie-kopníci vedy a techniky na Slovensku (Bratislava) **2**, 431-437 (1988) — M.T. MOROVICS: *The spark patterns of Karol ANTOLIK, a source of MACH's work in the field of aerodynamics.* In: (V. PROSSER and J. FOLTA, eds.) *Ernst MACH and the development of physics* [Int. Conf., University of Prague, Sept. 1988]. Universitas Carolina Pragensis (1991), pp. 205-216.
PICTURE. Courtesy Central Library of the Slovak Academy of Sciences (SAS), Bratislava, Slovak Republic.

## BECKER, Richard Adolf (1887–1955)

• German theoretical physicist

Richard Adolf BECKER was born in Hamburg and attended high school at the famous Johanneum in Hamburg. He studied first zoology and at Freiburg wrote his Ph.D. thesis on the fly [*Diptera*] larva. Inspired by a lecture by Prof. Arthur SOMMERFELD, he decided to study physics. After graduation he worked as an assistant at the Institute of Physics of the TH Hannover and at the Kaiser-Wilhelm-Institut für Physikalische Chemie in Berlin. After carrying out research on explosives at the Sprengstoff AG Carbonit in Schlebusch, Cologne region (1913–1916), he worked for three years in Berlin as an assistant referee at the Waffen- und Munitionsbeschaffungsamt (Weapons and Munitions Supply Bureau). As a coworker of the chemist Friedrich BERGIUS at Heidelberg he got involved in coal liquefaction and the use of soft coal. He matriculated in Berlin at the Friedrich-Wilhelms-Universität with a thesis entitled *Stoßwelle und Detonation* ("Shock Wave and Detonation" 1922), a classical memoir in elementary shock wave physics famous for its lucidity in which he discussed in detail the effects of heat conduction and viscosity. After 4 years in industry he was appointed full professor of theoretical physics at the TH Berlin-Charlottenburg. In 1936, he took over the chair of theoretical physics at the University of Göttingen, where he remained until his death.

During World War II, BECKER managed the Erfahrungsgemeinschaft Hohlladungen (Shaped Charges Experience

Community) in Göttingen, which was founded in order to coordinate and exchange experimental work on the shaped charge effect at various German research institutes. His most renowned disciple and assistant was Werner DÖRING, cofounder of the *Zel'dovich-von Neumann-Döring detonation theory* (or *ZND theory*). Throughout his professional career BECKER worked in various fields of basic research such as detonation, shock waves, plasticity, supra conductivity, nucleus formation, and ferromagnetism. His thorough contributions, also to difficult subjects, were characterized by great clarity of expression.

**ORIGINAL WORKS.** *Zur Theorie der Detonation.* Z. Elektrochem. **23**, 40-49, 93-95 (1917) — *Über die Fortleitung einer Detonation in flüssigen Sprengstoffen.* Ibid. **23**, 304-308 (1917) — *Eine Zustandsgleichung für Stickstoff bei großen Dichten.* Z. Phys. **4**, 393-409 (1921) — *Stoßwelle und Detonation.* Ibid. **8**, 321-362 (1922); Engl. translation: *Impact, waves and detonation.* NACA-TM 505 (1929) and NACA-TM 506 (1929) — *Physikalisches über feste und gasförmige Sprengstoffe.* Z. Tech. Phys. **3**, 152-159, 249-256 (1922) — *Zur Thermodynamik der Ionisierung einatomiger Gase.* Physik. Z. **24**, 485-486 (1923) — *Explosionsvorgänge.* In: (G. GEHLHOFF, ed.) *Lehrbuch der Technischen Physik.* Barth, Leipzig, vol. 1 (1924): *Masse und Messen, Mechanik, Akustik und Thermodynamik,* pp. 376-386 — *Über Plastizität, Verfestigung und Rekristallisation.* Z. Tech. Phys. **7**, 547-555 (1926) — With G. SCHWEIKERT: *Theorie der Zustandsgleichungen.* Z. Phys. **92**, 680-82, 689 (1934) — *Über Detonation.* Z. Elektrochem. **42**, 457-461 (1936) — With W. DÖRING: *Kinetische Behandlung der Keimbildung in übersättigten Dämpfen.* Ann. Phys. **24** [V], 719-752 (1935) — *Die Zusammenhänge zwischen den Eigenschaften der Knallwelle und der Detonationswelle.* In: *Probleme der Detonation.* Schriften Dt. Akad. Luftfahrtforsch. Nr. 1023/41G, 15-28 (1941) — *Theorie der Wärme.* Springer, Berlin (1955).

**SECONDARY LITERATURE.** (Ed.) *Richard Adolf BECKER.* Jb. Dt. Akad. Luftfahrtforsch. (1939/1940), p. 211 — L.H. THOMAS: *Note on BECKER's theory of the shock front.* J. Chem. Phys. **12**, 449-457 (1944) — W. DÖRING: *Richard BECKER 60 Jahre.* Physik. Blätter **3**, 393 (1947) — (Ed.) *R. BECKER zum 60. Geburtstag.* Z. Phys. **124**, 441 (1948) — G. LEIBFRIED: *Richard BECKER †.* Physik. Blätter **11**, 319-320 (1955) — M. KERSTEN: *Richard BECKER 1887–1955.* Ibid. **34**, 379-382 (1978).

**PICTURE.** Courtesy Staats- und Universitätsbibliothek (SUB) Göttingen, Germany; Sammlung Voit: R. BECKER, Nr. 4.

**NOTE.** On Dec. 3, 1987, a Ballistic Seminar was given at the Ernst-Mach-Institut (EMI), Abteilung Weil am Rhein, Germany, on the occasion of R. BECKER's 100th birthday.

# BERNOULLI, Daniel (1700–1782)

- Swiss mathematician, physicist, physician, and philosopher; cofounder of theoretical fluid dynamics

Daniel BERNOULLI was born in Groningen, Holland, in a renowned family of scholars that had emigrated in 1622 from Antwerp, Holland to Basel, Switzerland. He was the

second son of Johann BERNOULLI (1667–1748), who occupied the chair of mathematics at the University of Basel, and a nephew of Jakob BERNOULLI (1665–1705), who also became a famous mathematician. His brother Nikolaus BERNOULLI (1695–1726) became a mathematician and physicist.

At first Daniel BERNOULLI was slated for a career in business, but he eventually turned to medicine instead. After completing his M.D. at the University of Basel on the action of the lungs (1721), he briefly worked with the physician Pierre MICHELOTTI in Venice but soon reverted to the family tradition of mathematics and intensified his studies in mathematics, receiving instruction from his father and his beloved brother Nikolaus.

His first paper on mathematics, *Exercitationes mathematicae* (Venice 1724), won him much praise and an appointment as professor of mathematics at the Academy of St. Petersburg (1725–1733), joining there the young Swiss mathematician Leonard EULER, 7 years his junior, who had studied mathematics under Johann BERNOULLI and, largely as the result of Daniel's influence, was also invited to St. Petersburg by Catherine I, Empress of Russia. During his stay in St. Petersburg he collaborated with EULER and in 1729 began writing his famous treatise *Hydrodynamica* (1738), which deals with the behavior of fluids; it was in this work that BERNOULLI introduced the term *hydrodynamics*. This most important work on pure and applied fluid motion encompasses (1) a history of hydraulics, followed by a brief presentation of hydrostatics; (2) formulas for the outflow of a fluid from the opening of a container; (3) oscillations and energy loss of water in a tube immersed in a water tank; (4) a theory of the performance of hydraulic machinery and windmills; (5) discussions on the properties and motion of "elastic fluids" (*i.e.*, gases), such as the flow velocity of air streaming from a small opening; (6) the first formulation of the kinetic theory of gases; and (7) applications of the principle of conservation of energy to problems of fluid flow. He discovered that the total mechanical energy of a flowing fluid, which comprises the energy associated with fluid pressure, the gravitational potential energy of elevation, and the kinetic energy of fluid motion, remains constant – the so-called "Bernoulli equation" (or "Bernoulli theorem"). A manuscript draft of this famous *Hydrodynamica*, which already contains most of the later published book and was left behind by him in St. Petersburg, is still preserved in the archives at the St. Petersburg Academy.

BERNOULLI left Russia and returned to Basel as professor, first of anatomy and botany (1733) and later of natural philosophy; *i.e.*, philosophy and physics (1750). He also treated special cases of percussion of bodies of non-symmetric geometry, demonstrated the propulsion of a small vessel by means of an ejected jet of water from the stern, invented a clepsydra (water clock) designed for the more accurate measurement of time on sea, investigated problems of friction and vibrating strings, and studied the nature and cause of ocean currents and tides of the sea.

BERNOULLI defined the "simple modes" and the frequencies of oscillation of a system with more than one degree of freedom, the points of which pass their positions of equilibrium at the same time, and demonstrated his concept on an arrangement consisting of a hanging rope loaded with several bodies. He determined their amplitude rates and frequencies and found that the number of simple oscillations equals the number of bodies; *i.e.*, the degrees of freedom. He was also the first to theoretically investigate the influence of elastic vibrations and translation on the percussion process. For a homogeneous straight elastic rod struck in its center of gravity by a percussive force, he calculated the loss of *vis viva* due to elastic vibrations (1770).

In addition to his continuous interest in applied mechanics and mathematics, BERNOULLI has been called the founder of mathematical physics. Later he studied differential equations, as well as probability and its applications to statistics. He also combined mathematics with medicine and, studying the rate of mortality resulting from smallpox in various age groups, introduced medical statistics.

For his numerous ingenious contributions to the natural sciences BERNOULLI won or shared the annual prize of the Paris Académie des Sciences ten times (as did his friend EULER) for the solution of designated problems.

**ORIGINAL WORKS.** *Theremata de osccillattonibus corporum filo flexili connexorum et catenae verticaliter suspensae.* Comm. Acad. Sci. Imp. Petropol. **VI**, 108-122 (1732/1733) — *De variatione motuum a percussione excentrica.* Ibid. **IX**, 189-206 (1737) — *Hydrodynamica, sive de viribus et motibus fluidorum commentarii.* Decker, Strassburg (1738); *Hydrodynamics, or commentaries on forces and motions of fluids* (translated from Latin by T. CARMODY and H. KOBUS). With a *Preface to the English translation* by H. ROUSE. Dover Publ., New York (1968) — *Examen physico-mechanicum de motu mixto qui laminis easticis a percussione simul imprimitur.* Nov. Comm. Acad. Sci. Imp. Petropol. **XV**, 361-380 (1770) — (D. SPEISER, ed.) *Die Werke von Daniel BERNOULLI.* Hitherto published: vol. 1 (1996): *Medizin und Physiologie*; vol. 2 (1982): *Analysis, Wahrscheinlichkeitsrechnung*; vol. 3 (1987): *Mechanik*; vol. 4 (1993): *Reihentheorie*; vol. 5 (2002): *Hydrodynamik II*; vol. 7 (1994): *Magnetismus und Technologie I*; vol. 8 (2004): *Technologie II*. Birkhäuser, Basel *etc.*

**SECONDARY LITERATURE.** D. BERNOULLI JR.: *Vita Danielis BERNOULLI.* Helvetica **9**, 1-32 (1787) — Various authors: *Die Basler Mathematiker D. BERNOULLI und L. EULER. 100 Jahre nach ihrem Tode gefeiert von der Naturforschenden Gesellschaft* [Festschrift]. Georg, Basel (1884) — C.A. TRUESDELL III: *Daniel BERNOULLI's work on the propagation of sound before 1762.* In: (C.A. TRUESDELL III, ed.) *Leonardi EULERI Opera Omnia* **XIII** [II]. Teubner, Leipzig (1926). *See also* editor's introduction, pp. xxxiv-xxxv; *The hydrodynamics of Daniel BERNOULLI and John BERNOULLI (1727–1740).* In: *Leonardi Euleri Opera Omnia* **XII** [II]. Teubner, Leipzig *etc.* (1954). See editor's introduction, pp. xxiii-xxxviii — O. SPIESS: *Daniel BERNOULLI aus Basel, 1700–1782.* In: (E. FUETER, ed.) *Große Schweizer Forscher.* L'art ancien, Zurich (1939) — H. ROUSE and S. INCE: *History of hydraulics.* Dover Publ., New York (1957), chpt. VIII — K. FLIERL: *Anmerkungen zu Daniel BERNOULLI's Hydrodynamik.* Veröffentlichungen des Forschungsinstituts des Deutschen Museums für die Geschichte der Naturwissenschaften und der Technik. Reihe C, Nr. 1b, Munich (1965) — H. STRAUB: *D. BERNOULLI.* In: (C.C. GILLESPIE, ed.) *Dictionary of scientific biography.* Scribner, New York, vol. 2 (1970), pp. 36-46 — I. SZABO: *Geschichte der mechanischen Prinzipien und ihre wichtigsten Anwendungen.* Birkhäuser, Basel (1977), pp. 465-470 — A.T. GRIGOR'YAN and B.D. KOVALEV: *Daniil BERNULLI, 1700–1782* [in Russ.]. Scientific-biographic literature. Nauka, Moscow (1981) — W. JOHNSON: *Encounters between ROBINS, and EULER and the BERNOULLIS; Artillery and related subjects.* Int. J. Mech. Sci. **34**, 651-679 (1992).

**PICTURE.** Photo courtesy Deutsches Museum, Munich, Germany.

# BERTHELOT, Pierre Eugène Marcelin [Marcellin] (1827–1907)

- French organic and physical chemist; founder of thermochemistry

P.E. Marcelin BERTHELOT was born in Paris to Dr. Jacques Martin BERTHELOT, a physician. After attending courses in the Paris Faculty of Medicine and the Faculty of Science (1847–1849), he became a staff assistant to Antoine-Jérôme BALARD and demonstrator at the Collège de France (1851), where he took his Ph.D. in chemistry (1854). After carrying out further studies at the Ecole de Pharmacie, he graduated as a pharmacist (1858) and was appointed to the newly created chair of organic chemistry at the Ecole Supérieure de Pharmacie in Paris (1859). He became professor at the Collège de France (1864) and began numerous researches on the acetylides of silver and copper (1862–1866), which inspired him to study other explosives as well. Following this period, BERTHELOT investigated the explosive force of gunpowder using his new calorimeter (1870). In the defense of Paris during the Franco-Prussian War (1870–1871), he investigated the possibility of extracting saltpeter for producing gunpowder within the city.

BERTHELOT studied the combustion of explosive mixtures of gases (1871), made measurements of the heat resulting from the formation of nitroglycerin (1874–1876), and extended his studies to determine their combustion temperature and velocity (1877). He also examined explosive mixtures of dust with air (1878) and fulminating mercury (1880). Investigations performed in collaboration with Paul VIEILLE on the velocity of the flame speed in explosive mixtures of gases led to the important discovery that the rate of explosion rapidly increases from the point of origin until it reaches a maximum, which remains constant, however long the column of gases might be, thus forming a new physico-chemical constant important for theoretical and practical applications (1881). He called this rapidly propagating flame front *l'onde explosif* ("explosive wave"). For example, for oxyhydrogen BERTHELOT and VIEILLE measured a velocity of 2,841 m/s. BERTHELOT concluded that the explosion wave propagates upon impact of the products of combustion of one layer on the unburnt gases in the next layer and so on to the end of the tube at the rate of movement of the products of combustion themselves, thus identifying the maximum velocity of the flame with the mean translational velocity of the molecules themselves.

Besides his interest in the nature of explosion, BERTHELOT made major contributions to the synthesis of organic compounds and expanded our present knowledge of the history of alchemy. In 1868, he analyzed samples of the Orgueil Meteorite, a shower of stony meteorites of the rare carbonaceous chondrite type that fell on May 14, 1864 in southern France near the town of Peillerot. He reported finding in them hydrocarbons of the alkane family comparable to the oils of petroleum – thus confirming previous findings of Jöns J. BERZELIUS (1834) and Friedrich WÖHLER (1858) that some meteorites may contain complex organic matter, which gave rise to the question of possible extraterrestrial life.

BERTHELOT was Secretary of the Academy of Sciences and also a prominent politician, acting temporarily as Senator of France, Minister of Public Instruction, and Minister of Foreign Affairs. He is buried in the Panthéon in Paris.

**ORIGINAL WORKS.** *Sur quelques phénomènes de dilatation forcée des liquides.* Ann. Chim. Phys. **30** [III], 232-237 (1850) — *Sur la matière charbonneuse des météorites.* C. R. Acad. Sci. Paris **67**, 849 (1868) — *Sur la force de la poudre et des matières explosives.* Ibid. **71**, 619-625, 667-677, 709-728 (1870); Ann. Chim. Phys. **23** [IV], 223-273 (1871) — *Sur la force des mélanges gâteaux détonants.* C. R. Acad. Sci. Paris **72**, 165-168 (1871) — *Sur la force de la poudre et des matières explosives.* Gauthier-Villars, Paris (1871) — With P. VIEILLE: *Sur les températures de combustion.* Ann. Chim. & Phys. **12** [V], 302-310 (1877) — *Observation sur le mémoire de M.M. NOBLE et ABEL relative aux matières explosives.* C. R. Acad. Sci. Paris **89**, 192-196 (1879) — *Sur la chaleur de formation des oxydes de l'azote.* Ibid. **90**, 779-785 (1880) — With P. VIEILLE: *Etude des propriétés explosives du fulminate de mercure.* Ibid. **90**, 946-952 (1880) — *Appareil pour mesurer la chaleur de combustion des gaz par détonation.* Ibid. **91**, 188-191 (1880) — *Méthode pour mesurer la chaleur de combustion des gaz par détonation.* Ann. Chim. Phys. **23** [V], 160-187 (1881) — *Sur la vitesse de propagation des phénomènes explosifs dans les gaz.* C. R. Acad. Sci. Paris **93**, 18-22 (1881) — *Détonation de l'acétylène, du cyanogène et des combinaisons endothermiques en général.* Ibid. **93**, 613-619 (1881) — With P. VIEILLE: *Sur la vitesse de propagation des phénomènes explosifs dans les gaz.* Ibid. **94**, 101-108, 822-823 (1882); *Sur l'onde explosive.* Ibid. **94**, 149-152 (1882) — With P. VIEILLE: *Nouvelles recherches sur la propagation des phénomènes explosifs dans les gaz.* Ibid. **95**, 151-157 (1882); *Sur la période d'état variable qui précède le régime de détonation et sur les conditions d'établissement de l'onde explosive.* Ibid. **95**, 199-205 (1882) — *Sur la vitesse de l'onde explosive.* Ibid. **96**, 672-673 (1883) — *Sur quelques relations entre les températures de combustion, les chaleurs spécifiques, la dissociation et la pression des mélanges tonnants.* Ibid. **96**, 1186-1191 (1883) — *Sur la force des matières explosives.* Ibid. **97**, 767-768 (1883) — *Sur la force des matières explosives d'après la thermochimie.* Gauthier-Villars, Paris (1883); *Explosives and their power.* Murray, London (1892) — With P. VIEILLE: *L'onde explosive.* Ann. Chim. Phys. **28** [V], 289-332 (1883); *Recherches sur les mélanges gazeux détonants.* C. R. Acad. Sci. Paris **98**, 545-550 (1884); *Mélanges gazeux détonants. Calcul des températures et des chaleurs spécifiques.* Ibid. **98**, 601-606 (1884); *Vitesse relative de combustion des mélanges gazeux détonants.* Ibid. **98**, 646-651 (1884); *Sur la chaleur spécifique des éléments gazeux, à très hautes températures.* Ibid. **98**, 770-775 (1884); *Sur les chaleurs spécifiques de l'eau et de l'acide carbonique à de très hautes températures.* Ibid. **98**, 852-858 (1884); *Nouvelle méthode pour la mesure de la chaleur de combustion du charbon et des composés organiques.* Ibid. **99**, 1097-1103 (1884); Ann. Chim. Phys. **6** [VI], 546-556 (1885) — *Sur la vitesse de propagation de la détonation dans les matières explosives solides et liquides.* C. R. Acad. Sci. Paris **110**, 314-320 (1885) — *Sur l'onde explosive, sur les données caractéristiques de la détonation et sa vitesse de propagation dans les corps solides et liquides, et spécialement dans le nitrate de méthyle.* Ibid. **112**, 16-27 (1891) — With P. VIEILLE: *Sur les dissolutions d'acétylène et sur leurs propriétés explosives.* Ibid. **124**, 988-996 (1897) — With H.L. LE CHATELIER: *Sur la vitesse de détonation de l'acétylène.* Ibid. **129**, 427-434 (1899); Ann. Chim. Phys. **20** [VII], 15-26 (1900).

**SECONDARY LITERATURE.** G. BREDIG: *Marcellin BERTHELOT.* Z. Angew. Chem. **20**, 689-694 (1907) — W.R. (anonymous): *Pierre Eugène Marcellin BERTHELOT (1827–1907).* Proc. Roy. Soc. Lond. **A80**, iii-x (1908) — C. GRAEBE: *Marcellin BERTHELOT.* Ber. Dt. Chem. Gesell. **41**, 4805-4872 (1908) — E. JUNGFLEISCH: *Notice sur la vie et les travaux de Marcellin BERTHELOT.* Bull. Soc. Chim. France **13**, i-cclx (1913) — H.B. DIXON: *Berthelot memorial lecture.* In: *Memorial lectures delivered before the Chemical Society 1901-1913.* Gurney & Jackson, London (1914), vol. II, pp. 167-185 — A. BOUTARIC: *Marcellin BERTHELOT.* Payot, Paris (1927) — H. LE CHATELIER: *Souvenir du centenaire de Marcellin BERTHELOT.* N.P., Paris (1927) — L. VELLUZ: *Vie de BERTHELOT.* Plon, Paris (1964) — G. MENDELEVICH: *What Marcellin BERTHELOT could not foresee* [in Russ.]. Technika Molodeži **32**, No. 4, 16-17 (1964) — R. VIRTANEN: *Marcellin BERTHELOT. A study of a scientist's public role.* University of Nebraska Studies No. 31, Lincoln, Nebraska (1965) — M.P. CROSLAND: *P.E.M. BERTHELOT.* In: (C.C. GILLESPIE, ed.) *Dictionary of scientific biography.* Scribner, New York, vol. 2

(1970), pp. 63-72 — A. HOREAU: *Marcellin BERTHELOT, savant, philosophe et homme politique.* L'Actualité Chimique **9**, 33-38 (Feb. 1981) — E. FÄRBER: *BERTHELOT.* In: (G. BUGGE, ed.) *Das Buch der großen Chemiker.* Verlag Chemie, Weinheim/Bergstr. *etc.* (1984) — N. MANSON: *Notes et remarques sur la contribution de M. BERTHELOT à l'étude des matières explosives.* Présenté au colloque: (J. DHOMBRES and B. JAVAULT, eds.) *Actes du Colloque Marcellin BERTHELOT: Une vie, une époque, un mythe.* Saint-Maur, France (1988). Soc. Franc. Hist. Sci. & Tech., Paris (1992) — L. MEDARD and H. TACHOIRE: *Histoire de la thermochimie.* Publications de l'Université de Provence, Aix-en-Provence (1994); *Sur la légende de la bombe calorimétrique de BERTHELOT,* pp. 209-211.
**PICTURE.** In the public domain, taken from Wikipedia, the free encyclopedia; http://de.wikipedia.org/wiki\bild:Marcellin_Berthelot.jpg.

## BETHE, Hans Albrecht (1906–2005)

- German-born U.S. theoretical physicist; father of nuclear astrophysics

The son of a university professor, Hans A. BETHE was born in Strassburg, Germany (now Strasbourg, Dépt. Alsace-Lorraine, France). He studied physics at the University of Frankfurt am Main. After taking his doctorate at the University of Munich under the supervision of Prof. Arthur SOMMERFELD (1928), BETHE began to teach theoretical physics at various German universities until 1933, when he fled Germany's growing Nazi regime. He went to England for the next 2 years, working in nuclear physics. He held a temporary position as lecturer at the University of Manchester for the period 1933–1934 and a fellowship at the University of Bristol in the fall of 1934. Together with the German-British physicist Rudolf E. PEIERLS he developed a theoretical model of the deuteron shortly after its discovery (1932). In February 1935 he was appointed assistant professor at Cornell University in Ithaca, NY (1935–1975). BETHE became renowned for his theory of how the Sun and stars use nuclear reactions to supply the energy they radiate. Partly in collaboration with U.S. physicist Charles L. CRITCHFIELD, he proposed the proton-proton chain (1938) and the carbon-nitrogen cycle (1939), the latter dominating in hotter stars. The end result of the reactions is the fusion of hydrogen nuclei to form helium nuclei. This process was independently suggested at the same time by Carl F. VON WEIZSÄCKER in Germany – so-called "Bethe-Weizsäcker cycle."

At the beginning of World War II, BETHE, on his own, formed a theory of the penetration of armor by projectiles (publ. 1945). Together with Edward TELLER he showed that behind strong shock waves a fluid (such as air) is in non-equilibrium because it takes a certain amount of time for the energy of the molecules corresponding to the vibrational degrees of freedom to arrive at equilibrium with the molecular energies corresponding to the translational and rotational degrees of freedom (1941). He began to work out a theory of shock wave propagation in dissociated gases, which became an important effect in the case of the detonation of nuclear weapons. During the war he worked for the Manhattan Project, leading the Theoretical Division of the Los Alamos Scientific Laboratory (1942–1945). In 1947, he edited the Los Alamos report *Blast Wave* (Rept. LA-2000), one of his most cited works among shock physicists, which describes how a nuclear weapon blast wave develops over time and distance. For an arbitrary equation of state he discussed theoretically the shock wave velocity in comparison to the sound velocity behind the shock front (1942). Together with John G. KIRKWOOD, he worked out an analytical approach to shock wave propagation in water for a number of explosives, the so-called "Kirkwood-Bethe propagation theory." Later he also made contributions to the design of the heat shield for ballistic missiles when they reenter the atmosphere.

After World War II, BETHE worked extensively on the collision of charged particles with atoms and pursued his earlier research on stellar nuclear energy for which he received the 1967 Nobel Prize for Physics "for his contribution to the theory of nuclear reactions, especially his discoveries concerning the energy production in stars." He also derived an equation of state of matter at supernuclear density ($\rho > 10^{16}$ g/cm$^3$) in order to determine the moment of inertia and mass of neutron stars (1974). In 1978, he began in a team to investigate the behavior of matter in a collapsing giant star that, due to gravitation, is supposed to give rise to a supernova explosion. Throughout his life BETHE remained vigorously opposed to U.S. H-bomb research, although after World War II he returned to Los Alamos as a consultant. He was a member of the President's Science Advisory Committee (1956–1959). At the age of 88, he called in an open letter on all scientists of all nations "to cease and desist from work creating, developing, improving, and manufacturing further nuclear weapons." In 1997, BETHE, at the age of 91, became even more specific and wrote to U.S. President CLINTON that "the time has come to stop sponsorship also of *computational experiments*, or even create thought designed to produce new categories of nuclear weapons."

From 1944 he was a member of the U.S. Academy of Sciences and received its Draper Medal (1948). He was president of the American Physical Society (1954) and was awarded the U.S. Medal of Merit (1946), the Planck Medal (1955) of the German Physical Society, and the Eddington Medal (1963) of the Royal Astronomical Society. He also won the Enrico Fermi Prize (1961) of the U.S. Atomic Energy Commission. His main work was published in more than 250 papers in scientific journals. He also wrote many superb review articles or books on the theory of metals, quantum mechanics, atomic systems, nuclear physics, and field theory. BETHE died on March 6, 2005 at the age of 98 at his home in Ithaca, NY, near Cornell University.

**ORIGINAL WORKS.** With A. SOMMERFELD: *Elektronentheorie der Metalle.* In: (H. GEIGER and K. SCHEEL, eds.) *Handbuch der Physik.* Springer, Berlin, vol. 24 (1933), Part II, pp. 353-622 — *Energy production in stars.* Phys. Rev. **55** [II], 103, 434-56 (1939) — With E. TELLER: *Deviations from thermal equilibrium in shock waves.* Rept. X-117, BRL, Aberdeen Proving Ground, MD (1941) — With J.G. KIRKWOOD and E. MONTROLL: *The pressure wave produced by an underwater explosion (I, II).* Rept. OSRD-114, NDRC Div. B (1941) — With J.G. KIRKWOOD: *The pressure wave produced by an underwater explosion.* Progr. Rept. OSRD-588, NDRC Div. B (1942) — *Energy production in stars.* Am. Scientist **30**, 243-264 (1942) — *The theory of shock waves for an arbitrary equation of state.* Rept. OSRD-545, NDRC, Div. B (1942) — *Specific heat of air up to 25,000 K.* Rept. OSRD-369, NDRC, Div. B (1942) — With K. FUCHS, J. VON NEUMANN, R. PEIERLS, and W.G. PENNEY: *Riemann method; shock waves and discontinuities (1-D), 2-D hydrodynamics. Lectures in shock hydrodynamics and blast waves.* Rept. AECD-2860 (1944) — *An attempt at a theory of armor penetration.* Ordnance Laboratory, Frankford Arsenal (1945) — With K. FUCHS, J.O. HIRSCHFELDER, J.L. MAGEE, R. PEIERLS, and J. VON NEUMANN: *Blast wave.* Rept. LASL-2000, LASL, Los Alamos, NM (Aug. 1947, distributed March 1958) — *The hydrogen bomb, II.* Scient. Am. **182**, 18-23 (April 1950) — *Brighter than a thousand suns.* Bull. Atom. Sci. **14**, 426-428 (Dec. 1958) — With M.C. ADAMS: *A theory for the ablation of glassy materials.* J. Aerospace Sci. **26**, 321-328, 350 (1959) — *Theory of the fireball.* Rept. LA-3064, LASL, Los Alamos, NM (Feb. 1964) — *The fireball in air.* J. Quant. Spectrosc. Rad. Transfer **5**, 9-12 (1965) — *Energy production in stars* [Nobel Lecture]. Science **161**, 541-547 (Aug. 9, 1968) — *Robert J. OPPENHEIMER.* Biogr. Mem. Fell. Roy. Soc. (Lond.) **14**, 391-416 (1968) — *Introduction.* In: (S. FERNBACH and A. TAUB, eds.) *Computers and their role in the physical sciences.* Gordon & Breach, New York (1970), pp. 1-9 — *Equation of state at densities greater than nuclear density.* In: (C.J. HANSEN and L.H. VOLSKY, eds.) *Physics of dense matter.* Reidel, Dordrecht (1974), pp. 27-46 — With M.I. SOBEL, P.J. SIEMENS, and J.P. BONDORF: *Shock waves in colliding nuclei.* Nucl. Phys. **A251**, 502-529 (1975) — *Equation of state in the gravitational collapse of stars.* Ibid. **A324**, 487-533 (1979) — *Comments on the history of the H-bomb.* Los Alamos Science **3**, 42-53 (Fall 1982) — With G.E. BROWN, J. COOPERSTEIN, and J.R. WILSON: *A simplified equation of state near nuclear density.* Nucl. Phys. **A403**, 625-648 (1983) — With J. COOPERSTEIN and G.E. BROWN: *Shock propagation in supernovae: concept of net ram pressure.* Ibid. **A429**, 527-555 (1984) — *How a supernova explodes.* Scient. Am. **252**, 60-68 (May 1985) — *Weapons in space.* 2 vols., American Academy of Arts and Sciences, Cambridge, MA (1985) — *Supernova theory.* In: (N. METROPOLIS, ed.) *New directions in physics.* The Los Alamos 40th Anniversary Volume (1987), pp. 235-256 — *SAKHAROV's H-bomb.* Bull. Atom. Sci. **46**, 8-9 (Oct. 1990) — *Supernovae. By what mechanism do massive stars explode?* Phys. Today **43**, 24-27 (Sept. 1990) — *The road from Los Alamos.* Masters of Modern Physics series, American Institute of Physics (AIP), New York (1991) — *Preservation of the supernova shock.* Astrophys. J. **419**, 197 (1993) — *Bombs after Hiroshima* [Book review of R. RHODES: *Dark Sun: the making of the Hiroshima bomb.* Simon & Schuster, New York (1995)]. Science **269**, 1455-1457 (Sept. 8, 1995) — *The supernova shock.* Astrophys. J. **449**, 714-726 (1995) — *Breakout of the supernova shock.* Ibid. **469**, 737-739 (1996) — *J. Robert OPPENHEIMER.* Biogr. Mem. Natl. Acad. Sci. (U.S.A.) **71**, 175-220 (1997) — *Selected works of Hans A. BETHE* [with historical notes written by BETHE and a list of his 290 publications]. World Scientific, Singapore (1997) — *The German uranium project.* Phys. Today **53**, 34-36 (July 2000).

**SECONDARY LITERATURE.** (Ed.) *Henry Draper Medal award to H. BETHE.* Nature **161**, 861 (1948) — (Ed.) *H. BETHE. New foreign member of the Royal Society.* Nature **179**, 1221 (1957) — H.C. ALLISON: *Detection of underground explosions* [A censored version of the Hearing before a subcommittee of the Committee on Foreign Relations], U.S. Senate, 86th Congress, first session]. Bull. Atom. Sci. **15**, 257-259 (1959) — R.O. REDMAN: *On the award of the Eddington Medal to Professor H.A. BETHE* [Presidential address]. Quart. J. Roy. Astron. Soc. **2**, 107-108 (1961) — R.F. BACHER and V.F. WEISSKOPF: *The career of Hans BETHE.* In: (R.E. MARSHAK, ed.) *Perspectives in modern physics: essays in honor of Hans A. BETHE on the occasion of his 60th birthday, July 1966.* Wiley, Interscience, New York (1966), pp. 1-8 — P.P. EWALD: *BETHE, Hans Albert.* In: *Encyclopædia Britannica, Macropædia.* Benton & Hemingway, Chicago, vol. 2 (1974), pp. 871-872 — J. BERNSTEIN: *Hans BETHE, prophet of energy.* Basic Books, Inc., New York (1980) — S.P. PARKER (ed.) *McGraw-Hill modern scientists and engineers.* McGraw-Hill, New York (1980), vol. 1, pp. 90-91 — R.L. WEBER: *Pioneers of science.* Institute of Physics, Nobel Prize winners in physics. Institute of Physics, Bristol (1980), pp. 210-211 — On BETHE's contribution to the Manhattan Project *see* L. HODDESON ET AL.: *Critical assembly.* Cambridge University Press (1993) — I. GOODWIN: *Bethe Fest: a tribute to a titan of modern physics.* Phys. Today **48**, 39-40 (June 1995) — J.M. BONNET-BIDAUD: *Hans BETHE: Des étoiles à la bombe.* Ciel et Espace **314**, 32-38 (1996) — S.S. SCHWEBER: *Hans BETHE: Superlative scientist, incomparable mentor, humanitarian, friend.* Phys. Today **58**, 12-13 (May 2005) — W.J. BROAD: *Hans BETHE, father of nuclear astrophysics, dies at 98.* The New York Times (March 7, 2005) — F. DYSON: *Hans A. BETHE (1906–2005).* Science **308**, 219 (April 8, 2005) — K. GOTTFRIED and E.E. SALPETER: *Obituary: Hans A. BETHE (1906–2005).* Nature **434**, 970-971 (April 21, 2005).

**PICTURE.** © The Nobel Foundation, Stockholm, Sweden.

# BLEAKNEY, Walker (1901–1992)

- U.S. physicist and gas dynamicist; pioneer of shock tube research

Walker BLEAKNEY was born in a country farmhouse near Echo, OR, and studied physics at Whitman College in Walla Walla (Washington State) and at Harvard. After completing his Ph.D. on the ionization of gases at the University of

Minnesota, he won a National Research Council Fellowship to continue research at Princeton University, where he remained for his entire career. He began to work in atomic and molecular ionization processes and built the most modern mass spectrometer of his time. During World War II, he became leader of a group to study and advise the U.S. Government on terminal ballistics. He began to improve the shock tube in order to generate well-defined shock profiles for a quantitative study of shock reflection, thereby coining the term "shock tube" (1946). He became director of the Princeton University Station, a division of the National Defense Research Committee (NDRC) of the Office of Scientific Research and Development (OSRD), and founded the Shock Wave Laboratory at Princeton University. One of his graduate students, Donald R. WHITE, discovered in 1951 "compound Mach reflection," later renamed "double Mach reflection." In the following year, his team began to apply the shock tube to study blast effects on model structures, then a field of growing interest because of the increasing menace of possible nuclear conflicts.

One of his team's major contributions was the study of real gas effects (*e.g.*, vibrational modes of diatomic molecules, dissociation, and ionization) that occur as the Mach number of the flow increases, and the extension and application of optical visualization techniques (*e.g.*, shadowgraphy, schlieren, and Mach-Zehnder interferometry) on shock reflection and shock wave superposition phenomena.

**ORIGINAL WORKS.** *A new method of positive ray analysis and its application to the measurement of the probability and critical potentials for the formation of multiply charged ions in Hg vapor by electron impact.* Ph.D. thesis, University of Minnesota (1930); reprint from Phys. Rev. **34** [II], No. 1, 157-160 (1929) and Ibid. **35** [II], No. 2, 139-148 (1930) — With K.O. FRIEDRICHS: *Interaction of shock and rarefaction waves in one-dimensional motion.* NDRC, Applied Math. Panel Rept. 38.1R, Appl. Math. Group, NYU No. 1 (1943) — With A.H. TAUB: *Remarks on fortification design.* Interim Memorandum M-10, CFD Natl. Res. Council (Nov. 1944) — *Study of the physical vulnerability of aerial bombardment.* Rept. OSRD-6444, NDRC A-385 (Jan. 1946) — *Shock waves in a tube.* Phys. Rev. **69** [II], 678 (1946) — With F.B. HARRISON: *Remeasurement of reflection angles in regular and Mach reflection of shock waves.* Physics Dept., Princeton University (March 1947) — With R.G. STONER: *The attenuation of a spherical wave.* Phys. Rev. **72** [II], 170 (1947); *The attenuation of spherical shock waves in air.* J. Appl. Phys. **19**, 670-678 (1948) — With A.H. TAUB: *Interaction of shock waves.* Rev. Mod. Phys. **21**, 584-605 (1949) — With D.K. WEIMER and C.H. FLETCHER: *The shock tube: a facility for investigation in fluid dynamics.* Rev. Scient. Instrum. **20**, 807-815 (1949) — With D.K. WEIMER and C.H. FLETCHER: *A shock tube for the study of transient gas flow; Experimental measurement of the density field in the Mach reflection of shock waves.* Phys. Rev. **75** [II], 1294-1295 (1949) — With D.K. WEIMER and C.H. FLETCHER: *Transonic flow in a shock tube.* J. Appl. Phys. **20**, 418 (1949) — With D.R. WHITE and D.K. WEIMER: *Diffraction of shock waves around obstacles and the transient loading of structures.* Tech. Repts. II-3 and II-6. Dept. of Physics, Princeton University (1950) — With D.R. WHITE and W.C. GRIFFITH: *Measurements of diffraction of shock waves and resulting loading of structures.* ASME J. Appl. Mech. **17**, 439-445 (1950) — *The effect of Reynolds number on the diffraction of a shock wave.* Dept. of Physics, Tech. Rept. II-8, Princeton University (1951) — With C.H. FLETCHER and A.H. TAUB: *Reflection of shock waves at nearly glancing incidence.* Rev. Mod. Phys. **23**, 271-286 (1951) — *A shock tube investigation of the blast loading of structures.* In: (C.M. DUKE and M. FEIGEN, eds.) Proc. Symp. on Earthquake & Blast Effects on Structures. University of California, Los Angeles (1952). Earthquake Engineering Research Institute & University of California (1952), pp. 46-73 — *Shock loading of rectangular structures.* Tech. Rept. II-11. Dept. of Physics, Princeton University, Princeton, NJ (1952) — *Review of significant observations on the Mach reflection of shock waves.* Proc. Symp. on Applied Mathematics (A.M.S.) **5**, 41-47 (1954) — With W.C. GRIFFITH: *Shock waves in gases.* Am. J. Phys. **22**, 597-612 (1954) — With A.B. ARONS: *Pressure measuring manometers and gauges.* In: *High speed aerodynamics and jet propulsion.* Princeton University Press, Princeton, NJ; vol. IX (1954): (R. LADENBURG, B. LEWIS, R.N. PEASE, and H.S. TAYLOR, eds.) *Physical measurements in gas dynamics and combustion*, pp. 124-135 — With A.B. ARONS: *Velocity of shock waves by the light screen technique.* Ibid., pp. 159-163 — With R.J. EMRICH: *The shock tube.* In: *High speed aerodynamics and jet propulsion.* Princeton University Press, Princeton, NJ; vol. VIII (1961): (E. GODDARD, ed.) *High speed problems of aircraft and experimental methods*, pp. 596-647.
**SECONDARY LITERATURE.** J.H. VAN VLECK: *Reminiscences from the youth of Walker BLEAKNEY and Alfred KASTLER: two unusual autobiographical chapters* [contributions by A. KASTLER and W. BLEAKNEY]. Am. J. Phys. **40**, 950-958 (1972) — R.J. EMRICH: *Walker BLEAKNEY and the development of the shock tube at Princeton.* Shock Waves **5**, 327-339 (1996) — A. HERTZBERG: *Shock tube research, past, present and future: a critical survey.* Proc. 7th Int. Shock Tube Symp., Toronto (June 1969). In: (I.I. GLASS, ed.) *Shock tubes*, University of Toronto Press (1970), pp. 3-5.
**PICTURE.** Reprinted from Shock Waves **5**, 328 (1996).

# BOYS, Sir Charles Vernon (1855–1944)

- British experimental physicist and inventor

The son of Rev. Charles BOYS, Rector of Wing, Sir Charles Vernon BOYS was born at the Rectory of Wing, Uppingham, in Rutland. He graduated in mining and metallurgy at the Royal School of Mines, London. Later he became assistant professor at the Royal College of Science in South Kensington in London (1889–1897), now a constituent of the Imperial College, London. In 1886, BOYS took on the demonstratorship of the

Physical Society and also took over temporarily as acting professor. In 1897, he gave up his assistant professorship, also resigning his honorary positions as demonstrator and librarian for the Physical Society, and took up an applied science post at the Metropolitan Gas Referee in Westminster (1897–1939), which enabled him to continue his inventive activity without the onerous duties of teaching. During World War I, BOYS gave military advice on ballistics. In the period 1916–1917, he served as president of the Physical Society of London.

BOYS invented various scientific instruments such as (1) an *integraph* (1881), an integral machine for drawing the integral curve corresponding to any given curve (1881); (2) a *radio micrometer* (1883) for measuring minute streams of radiation such as the heat radiation from the Moon and other planets; (3) a *high-speed rotating mirror streak camera* (1892) to measure the pulse duration of submicrosecond spark light sources; (4) an improved *gas calorimeter* (1906) for testing the caloric value of gaseous fuels; and (5) a *high-speed rotating lens camera* – so-called "Boys camera" (1900).

His work on quartz fibers, utilizing their torsion for the measurement of minute forces, led to his conducting Cavendish-type experiments to improve the precision of the value obtained for the gravitational constant (1895), which was unsurpassed for decades.

Another important contribution to high-speed diagnostics and ballistics is his snapshot photography of supersonic bullets (1890–1893). Repeating Ernst MACH and Peter SALCHER's ballistic experiments (1886–1887), he simplified their experimental setup by using DVOŘÁK's method of shadowgraphy instead of TOEPLER's schlieren method and visualized the interaction and reflection of head waves. BOYS produced the first system to measure the spin rate of supersonic shot using specially prepared bullets and back light illumination along the trajectory. His streak camera, using a mirror of hardened steel rotating at 500 revolutions per second and a scan radius of 6 m, attained a writing speed of 37.5 mm/µs; time calibration occurred with a tuning fork. He measured the flash duration of various spark configurations, which resulted in a duration of between 0.1 and 1 µs.

As early as 1889 he suggested that the flickering of many lightning flashes indicated that they comprised several discharges down the same channel, which he eventually proved with his two-lens revolving camera. It had a fixed plate and two lenses mounted 180° apart on a rapidly rotating disk and allowed one to take two pictures with a minimum interframe time of 25 µs. From a comparison of the two pictures and a knowledge of the velocities of the lenses, he deduced the direction and speed of the developing discharge. Although he was in the habit of carrying this camera around for 26 years without obtaining a photograph, he eventually succeeded in photographing this progressive lightning during a visit of Mr. LOOMIS, a wealthy banker and amateur scientist who had a private laboratory at Tuxedo Park, NY (1928). His conclusion that the flash starts at the ground stimulated Basil F. SCHONLAND in South Africa to resume his technique (1934). BOYS also described a beautiful stereoscopic method of studying pictures in which they appear in two space dimensions and one time dimension.

BOYS was honored with the Royal Medal (1896), the Rumford Medal (1924) of the Royal Society, the Duddell Medal (1925) of the Physical Society, and the Cresson Medal (1939) of the Franklin Institute, Philadelphia. From 1888 he was a member of the Royal Society. He acted as president of the Physical Society of London (1916–1918) and was knighted in 1935. He bequeathed £1,000 to the Physical Society in 1944 "to be used to found a Boys Prize or a Boys Lecture or to be used in other manner at the discretion of the Council of the Society to further interest in experimental physics." In 1992, the Council of the Institute decided that the Charles Vernon Boys Prize should be changed to a medal and prize. This prestigious award recognizes outstanding contributions to experimental physics by a scientist under the age of 35.

**ORIGINAL WORKS.** *Notes on photographs of rapidly moving objects, and on the oscillating electric spark.* Phil. Mag. **30** [V], 248-260 (1890); Proc. Phys. Soc. (Lond.) **11**, 1-15 (1892) — *Les projectiles pris au vol. Méthode pour l'étude des mouvements dans les gaz.* Rev. Gén. Sci. Pures Appl. **3**, 661-670 (1892) — *Photography of flying bullets, etc.* [with discussion]. Photogr. J. **16**, 199-209 (1892) — *On electric spark photographs; or, Photography of flying bullets, etc., by the light of the electric spark.* Nature **47**, 415-421, 440-446 (1893) — *The visibility of a sound shadow.* Ibid. **56**, 173-174 (1897) — *Soap-bubbles. their colors and the forces which mould them.* S.P.C.K., London (1920) — *Progressive lightning (I).* Nature **118**, 749-750 (1926) — *Solid dipleidoscope.* Proc. Roy. Soc. Lond. **A119**, 489-505 (1928); Ibid. **A121**, 1-8 (1928) — *Progressive lightning (II).* Ibid. **A122**, 310-311 (1928) — *A new transit instrument.* Ibid. **A122**, 977 (1928) — *On rotating mirrors at high speeds.* Proc. Roy. Soc. Edinb. **57**, 377-378 (1936/1937) — *Weeds, weeds, weeds.* Wightman, London (1937).

**SECONDARY LITERATURE.** A. MALLOCK: *On the resistance of air.* Proc. Roy. Soc. Lond. **A79**, 262-276 (1907) — B.F.J. SCHONLAND and H. COLLENS: *Progressive lightning.* Ibid. **A143**, 654-674 (1934) — A.G. LOWNDES:

*Sir Charles* BOYS. Nature **133**, 677 (1934) — J.S.G.T. (anonymous): *Dr.* BOYS *on gas calorimetry.* Ibid. **133**, 710-711 (1934) — R.A.S. PAGET: *Sir Charles* BOYS. Proc. Phys. Soc. (Lond.) **56**, 397-403 (1944) — R.J. STRUTT: *C.V.* BOYS. Obit. Not. Fell. Roy. Soc. (Lond.) **4**, 771-788 (1944) — C.T.R. WILSON: *Sir Charles V.* BOYS, *F.R.S.* Nature **155**, 40-41 (1945) — R.V. JONES: *C.V.* BOYS. In: (C.C. GILLESPIE, ed.) *Dictionary of scientific biography.* Scribner, New York, vol. 15 (1978), pp. 59-61 — J.T. STOCK: *Sir C.V.* BOYS, *guardian of the flame.* Bull. Sci. Instrum. Soc. **23**, 2-6 (1989) — R. HILLIER: *High speed aerodynamics, be prepared for shocks.* Imperial College Inaugural Lecture (May 20, 1997) — G.J.N. GOODAY: BOYS, *Sir Charles Vernon.* In: (H.C.G. MATTHEW, ed.) *Oxford dictionary of national biography.* Oxford University Press, Oxford (2004), vol. 7, pp. 116-118.
**PICTURE.** Courtesy Science Museum Library, London, U.K.
**NOTE.** Institute of Physics: *The Charles Vernon Boys Medal and Prize – previous winners*; http://about.iop.org/IOP/Awards/vernon.html.

# BRIDGMAN, Sir Percy Williams (1882–1961)

- U.S. physicist and philosopher of science; dean of static high-pressure physics

Percy W. BRIDGMAN was born in Cambridge, MA, as son of Raymond Landon BRIDGMAN, a newspaperman and author. He received his early education in public schools in the nearby city of Newton until 1900, when he entered Harvard University. He studied physics at Harvard College and remained there his entire career: MA (1905), Ph.D. (1908), physics instructor (1910), assistant professor (1919), Hollis Professor of Mathematics and Natural Philosophy (1926), Higgins University Professor (1950), and professor emeritus (1954).

BRIDGMAN focused his researches on the effects of static high pressures on materials and their thermodynamic behavior. In 1905, he commenced his high-pressure studies which he continued throughout his whole career. His essential invention was the self-sealing feature of his first high-pressure packing, the so-called "Bridgman unsupported area seal" (1909), an eminent discovery for pioneering the field of high-pressure physics that, in his own words, "had a strong element of accident." It was the starting point of many successive apparatus to produce high hydrostatic pressure without leak. Using initially a screw compressor (1908–1909), which allowed maximum pressures up to 6 kbar, he succeeded in pushing pressures up to 20 kbar by employing a hydraulic press (1910).

Further technical improvements, which allowed for gradual pressure increases up to 30 kbar, facilitated the first routine measurements of the mechanical, electrical, and thermal properties of matter. Still higher pressures up to about 400 kbar using his so-called "Bridgman opposed-anvil apparatus" were finally obtained in quasi-fluid systems (1930). He found that all liquids, with the exception of water, behaved qualitatively alike, the viscosity increasing with pressure at a rapidly increasing pressure. Subjecting water to high pressure, he found new crystalline forms; one, produced above 40,000 bars, was the so-called "hot ice" with a melting point of about 200 °C. In 1932, he took an active part in initiating a program at Harvard for high-pressure studies devoted to geophysical behavior. One of BRIDGMAN's most famous undergraduate students was J. Robert OPPENHEIMER, a U.S. theoretical physicist who profited much from his contact with BRIDGMAN and later became scientific director of the Manhattan Project.

During World War II, BRIDGMAN participated in the Manhattan Project and measured the static compressibility of uranium and plutonium. He also studied the plastic flow of steel under high pressure as related to the strengthening of amour plate. It was at this time that he developed a method for increasing the yield point of artillery gun barrels by preliminary stretching with internal hydrostatic pressure. Although he never studied the behavior of matter under shock compression, his numerous papers on high-pressure research and his famous textbook *The Physics of High Pressures* (1931) have become standard sources of reference for shock physicists as well. His carefully measured data on static compression and on the state of matter (such as polymorphism and phase transitions) were often the only available source of information when physicists and military engineers began to investigate the behavior of solids and liquids under shock loading. He first suggested the use of the piezoresistive effect to measure static pressures and applied manganin wires (1950), a method which was resumed later in solid-state shock physics experiments. BRIDGMAN won the 1946 Nobel Prize for Physics "for the invention of an apparatus to produce extremely high pressures, and the discoveries he made therewith in the field of high-pressure physics." In the early 1950s, he performed experiments on the compressibility of glass up to static pressures of 200 kbar – approximately the upper limit accessible with his anvil compression technique.

His writings on high-pressure physics, spanning the years 1909–1963, combine experimental expertise with philosophic insight. He is renowned not only for his numerous contributions to static high-pressure research, but also for his brilliant mechanical constructions of high-pressure apparatus and accessories. In his book *Dimensional Analysis* (1931) he treated mathematically the Buckingham Pi Theorem (1914–1915), which has become the basis of most dimensional analyses in engineering dynamics. BRIDGMAN's interest in natural philosophy, pointing in the direction of the positivism of Ernst MACH and the Vienna Circle, was recorded in various books.

BRIDGMAN was a member of the National Academy of Sciences and various foreign academies. His high appreciation to the contribution of science is best reflected by the large number of awards bestowed upon him, such as the Rumford Medal (1917) by the American Academy, the Cresson Medal (1932) of the Franklin Institute, the Roozeboom Medal (1933) of The Netherlands Royal Academy, the Comstock Prize of the National Academy, the Research Corporation Award (1937), and the Bingham Medal (1951) of the Society of Rheology.

Since 1977 the *Bridgman Award* of the International Association for the Advancement of High Pressure Science and Technology (AIRAPT) is given biennially to researchers who make significant contributions to the progress of high-pressure physics.

A crater on the far side of the Moon is named for him.

**ORIGINAL WORKS.** *The measurement of high hydrostatic pressure. (I) A simple primary gauge.* Proc. Am. Acad. Arts Sci. **44**, 201-217 (1909) — *(II) A secondary mercury resistance gauge.* Ibid. **44**, 221-251 (1909) — *The measurement of hydrostatic pressures up to 20,000 kilograms per square centimeter.* Ibid. **47**, 321-343 (1911) — *Water, in the liquid and five solid forms.* Ibid. **47**, 441-558 (1911) — *Change of phase under pressure.* Phys. Rev. **3** [II], 153-203 (1914) — *The effect of tensions on the transverse and longitudinal resistance of metals.* Proc. Am. Acad. Arts Sci. **60**, 423-449 (1925) — *The physics of high pressure.* Bell, London (1931) — *Dimensional analysis.* Yale University Press, New Haven, CT (1931) — *Theoretically interesting aspects of high pressure phenomena.* Rev. Mod. Phys. **7**, 1-33 (1935) — *Effects of high shearing stress combined with high hydrostatic pressure.* Phys. Rev. **48** [II], 825-847 (1935) — *The pressure-volume-temperature relations of the liquid, and the phase-diagram of heavy water.* J. Chem. Phys. **3**, 597-605 (1935) — *The phase diagram of water to 45,000 kg/cm$^2$.* Ibid. **5**, 964-966 (1937) — *Flow phenomena in heavily stressed metals.* J. Appl. Phys. **8**, 328-336 (1937) — *The measurement of hydrostatic pressure to 30,000 kg/cm$^2$.* Proc. Am. Acad. Arts Sci. **74**, 1-10 (1940) — *The compression of 46 substances to 50,000 kg/cm$^2$.* Ibid. **74**, 21-51 (1940) — *Freezing and compressions to 50,000 kg/cm$^2$.* J. Chem. Phys. **9**, 794-797 (1941) — *Compression and polymorphic transitions of 17 elements to 100,000 kg/cm$^2$.* Phys. Rev. **60** [II], 351-354 (1941) — *Freezing parameters and compression of 21 substances to 50,000 kg/cm$^2$.* Proc. Am. Acad. Arts Sci. **74**, 399-424 (1942) — *Pressure-volume relations for seventeen elements to 100,000 kg/cm$^2$.* Ibid. **74**, 425-440 (1942) — *The nature of thermodynamics.* Harvard University Press, Cambridge, MA (1943) — *General survey of certain results in the field of high-pressure physics* [Nobel Lecture, given on Dec. 11, 1946]. In: *Nobel lectures, physics 1942–1962.* Elsevier, Amsterdam (1964), pp. 53-70 — *The effect of high mechanical stress on certain solid explosives.* J. Chem. Phys. **15**, 311-313 (1947) — *The compression of 21 halogen compounds and 11 other simple substances to 100,000 kg/cm$^2$.* Proc. Am. Acad. Arts Sci. **76**, 1-7 (1948) — *The compression of 39 substances to 100,000 kg/cm$^2$.* Ibid. **76**, 55-70 (1948) — *Linear compressions to 30,000 kg/cm$^2$, including relatively incompressible substances.* Ibid. **77**, 187-234 (1949) — *Physics above 20,000 kg/cm$^2$* [Bakerian Lecture]. Proc. Roy. Soc. Lond. **A203**, 1-17 (1950) — *Studies in large plastic flow and fracture, with special emphasis on the effects of hydrostatic pressure.* McGraw-Hill, New York (1952) — *High-pressure instrumentation.* Mech. Engng. **75**, 111-113 (1953) — With I. SIMON: *Effects of very high pressures on glass.* J. Appl. Phys. **24**, 405-413 (1953) — *The effect of pressure on the tensile properties of several metals and other materials.* Ibid. **24**, 560-570 (1953) — *Synthetic diamonds.* Scient. Am. **193**, 42-46 (Nov. 1955) — *High pressure polymorphism of iron.* J. Appl. Phys. **27**, 659 (1956) — *Significance of the Mach principle.* Am. J. Phys. **29**, 32-36 (1961) — *General outlook on the field of high-pressure research.* In: (W. PAUL and D.M. WARSCHAUER, eds.) *Solids under pressure.* McGraw-Hill, New York (1963), pp. 1-13 — *Collected experimental papers by P.W. BRIDGMAN.* 7 vols., Harvard University Press, Cambridge, MA (1964).

**SECONDARY LITERATURE.** (Ed.) *Percy W. BRIDGMAN.* Phys. Today **14**, 78 (Oct. 1961) — F. BIRCH, R. HICKMAN, G. HOLTON, and E.C. KEMBLE: *P.W. BRIDGMAN.* Harvard University Gazette (March 31, 1962) — D.M. NEWITT: *Percy Williams BRIDGMAN.* Biogr. Mem. Fell. Roy. Soc. (Lond.) **8**, 27-40 (1962) — E.C. KEMBLE, F. BIRCH, and G. HOLTON: *P.W. BRIDGMAN.* In: (C.C. GILLESPIE, ed.) *Dictionary of scientific biography.* Scribner, New York, vol. 2 (1970), pp. 457-461 — E.C. KEMBLE and F. BIRCH: *Percy Williams BRIDGMAN.* Biogr. Mem. Natl. Acad. Sci. (U.S.A.) **41**, 23-67 (1970) — R.S. BRADLEY: *BRIDGMAN, Percy Williams.* In: *Encyclopædia Britannica, Macropædia.* Benton & Hemingway, Chicago (1974), vol. 3, pp. 191-192 — R.L. WEBER: *Pioneers of science.* Institute of Physics, Nobel Prize winners in physics. Institute of Physics, Bristol (1980), pp.127-128.

**PICTURE.** © The Nobel Foundation, Stockholm, Sweden.

**NOTE.** AIRAPT, Harvard University, Cambridge, MA: *List of Bridgman Award recipients (1977–2005);* http://www.unipress.waw.pl/airapt/.

# BUNSEN, Robert Wilhelm Eberhard (1811–1899)

- German chemist and combustion researcher

Robert Wilhelm Eberhard BUNSEN was born in Göttingen to Christian BUNSEN, a professor of philology at the University of Göttingen. He studied chemistry, physics, mineralogy, and mathematics. In 1830, at the age of 19, BUNSEN earned a Ph.D. at the University of Göttingen under chemistry professor Friedrich STROMEYER and set off to study in Paris, Berlin, and Vienna. After touring Europe and visiting various factories, laboratories, and places of geological interest

(1830–1833), he became lecturer at the University of Göttingen (1833) and the Polytechnic School at Kassel (1836–1838). In Kassel he succeeded Friedrich WÖHLER, a pioneer in organic chemistry. Thereafter, BUNSEN accepted a position as an *Extraordinarius* (associate professor) of chemistry (1838–1842) at the University of Marburg, and in 1852 he succeeded Leopold GMELIN, a physiological chemist, at the University of Heidelberg as *Ordinarius* (full professor), where he stayed until his retirement in 1889. Most of his research he carried out at his new laboratory at Heidelberg University, which was constructed for him by the government of Baden (1855).

BUNSEN invented various instruments, such as the carbon-zinc battery or *Bunsenbatterie*, a filter pump for washing precipitates, and a thermopile. He devised a sensitive ice calorimeter and a vapor calorimeter, used magnesium to provide a brilliant light source, and improved a gas burner for use in spectroscopy – the famous *Bunsenbrenner* (Bunsen burner). Together with Gustav KIRCHHOFF he worked out the spectral analysis that later allowed BUNSEN to discover cesium and rubidium. While investigating compounds of cacodyl, an arsenic-containing organic compound, he lost the use of his right eye in an explosion of cacodyl cyanide. Most of his life he dedicated to the study of gases, particularly to the combustion process of gaseous mixtures and to the temperature of flames, which he determined by measuring the maximum pressure produced at the moment of explosion using a reaction vessel that, closed by a valve, was coupled to a weight-loaded lever arrangement – in construction similar to a safety valve. He investigated the ignition process of an oxyhydrogen atmosphere by using a rotating stroboscopic disk that had a known rate of rotation and that was comprised of radiating segments. Looking through his rotating disk at a white surface illuminated by the light from exploding gases, he could measure the duration of illumination. Together with the Russian chemist Léon SCHISCHKOFF he was the first to investigate exactly the chemical reactions of exploding gunpowder (1857). He compiled his research on the phenomena of gases into his only book, *Gasometrische Methoden* (1857). He observed that an explosion does not occur simultaneously in the whole test chamber but rather has a finite propagation velocity. Using a narrow tube he attempted to determine the rate at which an explosion is propagated and (erroneously) came to the conclusion that for a mixture of hydrogen and oxygen this was 34 m/s and for a mixture of hydrogen and carbon monoxide about 1 m/s (1867). However, subsequent studies carried out in France {BERTHELOT ⇨1881; BERTHELOT and VIEILLE ⇨1882; MALLARD and LE CHÂTELIER ⇨1883} and in England {HARCOURT ⇨1880; DIXON ⇨1877 & 1881} showed that this low rate of explosion only forms during the initial period of the combustion before the explosion wave attains its maximum velocity, which, in the case of a detonating mixture of hydrogen and oxygen, amounts to nearly 3,000 m/s.

BUNSEN was elected Foreign Fellow of the Royal Society (1858) and honored by several European societies. The Copley Medal (1860) was awarded to him, and, together with his German colleague KIRCHHOFF, he received the first Davy Medal (1877).

Astronomers named a crater on the near side of the Moon after him. The *Deutsche Bunsen-Gesellschaft für Physikalische Chemie e.V.* (German Bunsen Society for Physical Chemistry) at Frankfurt/Main, established in 1894, promotes the scientific and technical aspects of physical chemistry, with particular emphasis on the interaction between science and technology. The *Robert Wilhelm Bunsen Medal* has been established by the European Geosciences Union (EGU), Division of Geochemistry, Mineralogy, Petrology and Volcanology, in recognition of BUNSEN's scientific achievements. The first winners were the Canadian experimental chemist Terry M. SEWARD (2005) and the French physicist and geochemist Pascal RICHET (2006).

**ORIGINAL WORKS.** *Untersuchungen über die Kakodylreihe.* Ann. Chem. Pharm. **37**, 1-57 (1841); Ibid. **42**, 14-46 (1842); **46**, 1-48 (1843) — *On the radical of the Kakodyl series.* Phil. Mag. **20** [III], 382-393 (1842) — *Untersuchungen über die chemische Verwandtschaft.* Ann. Chem. Pharm. **85**, 137-155 (1853) — *Gasometrische Methoden.* Vieweg, Braunschweig (1857, 1877); Engl. translation by H.E. ROSCOE: *Gasometry. Comprising the leading physical and chemical properties of gases.* Walton & Maberly, London (1857) — With L. SCHICHKOFF: *Chemische Theorie des Schießpulvers.* Ann. Phys. **102** [II], 321-353 (1857); *On the chemical theory of gunpowder.* Phil. Mag. **15** [IV], 489-512 (1858) — *Über die Temperatur der Flammen des Kohlenoxyds und Wasserstoffs.* Ann. Phys. **131** [II], 161-179 (1867) — *On the temperature of a flame of carbonic oxide and hydrogen.* Phil. Mag. **34** [IV], 489-502 (1867) — W. OSTWALD and M. BODENSTEIN (eds.) *Gesammelte Abhandlungen von Robert BUNSEN.* 3 vols., Engelmann, Leipzig (1904).

**SECONDARY LITERATURE.** R. RATHKE: *Robert Wilhelm BUNSEN.* Z. Anorg. Chemie **23**, 393-438 (1900) — H.E. ROSCOE: *Bunsen Memorial Lecture.* J. Chem. Soc. (Lond.) **77**, 513-554 (1900) — W. OSTWALD: *Gedenkrede auf Robert BUNSEN.* Z. Elektrochemie **7**, 608-618 (1900) — H.B. DIXON: *VIII. On the movements of the flame in the explosion of gases.* Phil. Trans. Roy. Soc. Lond. **A200**, 315-352 (1903) — H.E.R. (H.E. ROSCOE): *Robert Wilhelm BUNSEN.* Proc. Roy. Soc. Lond. **75**, 46-49 (1905)

— W. OSTWALD: *Robert W. BUNSEN.* L. Joachim, Leipzig (1923) — K. FREUDENBERG: *150. Geburtstag R.W. BUNSEN.* Physik. Blätter **17**, 111-113 (1961) — S.G. SCHACHER: *R.W. BUNSEN.* In: (C.C. GILLESPIE, ed.) *Dictionary of scientific biography.* Scribner, New York, vol. 2 (1970), pp. 586-590 — K. HENTSCHEL: *Gustav Robert KIRCHHOFF (1824–1887) und Robert Wilhelm BUNSEN (1811–1899).* In: (K. VON MEŸENN, ed.) *Die großen Physiker.* C.H. Beck'sche Verlagsbuchhandl., München (1997), vol. 1, pp. 416-430. **PICTURE.** Courtesy Universitätsarchiv Heidelberg, Germany. **NOTE.** European Geosciences Union (EGU), Katlenburg-Lindau, Germany: *Robert Wilhelm Bunsen Medal and list of Robert Wilhelm Bunsen medallists*; http://www.copernicus.org/EGU/awards/robert_w_bunsen_overview.html.

## BUSEMANN, Adolf (1901–1986)

- German aeronautical engineer; father of the swept-wing concept

Adolf BUSEMANN was born in Lübeck and studied mechanical engineering at the TU Braunschweig (1920–1924), where he also earned his Ph.D. under the supervision of Prof. Otto FÖPPL (1924). In Göttingen he began working in aeronautical research at the Kaiser-Wilhelm-Institut (KWI) [now Max-Planck-Institut (MPI)] für Strömungsforschung under the guidance of Prof. Ludwig PRANDTL (1925). This period of his life significantly determined his future scientific career. Together with PRANDTL he worked out a method of characteristics for 2-D supersonic flow, the so-called "Prandtl-Busemann method," and developed the shock polar diagram (1928). When Jakob ACKERET left Göttingen, he succeeded him as department head (1927–1931). He habilitated on a subject of applied mechanics and became *Privatdozent* (university lecturer) at the University of Göttingen (1930–1931) and at the University of Dresden (1931–1935), teaching fluid mechanics and thermodynamics.

At the 5th Volta Conference in Rome (1935), which was dedicated to the topic *Le alte velocità in aviazione* ("High Velocities in Aviation"), he suggested "sweepback wings" (or "swept wings") [Germ. *gepfeilte Tragflügel*] for high-speed aircraft and showed how their properties might solve many aerodynamic problems at speeds just below and above the speed of sound. Sweepback wings, resembling the tip of an arrow, hinder bow waves from striking the wings, thus reducing aerodynamic drag and preventing shock stall. One year later, the German Air Force made his concept of swept wings a state secret. In the 1930s, he also suggested a supersonic airframe design, later much discussed and known as the "Busemann biplane." It uses two identical wing surfaces facing each other, which results in zero wave drag – thus prohibiting the formation of a sonic boom – but, unfortunately, also in zero lift. His concept was investigated vigorously in Japan for the possible application to Boomless Supersonic Transport (Paper AIAA-2006-654).

In 1936, he became professor and director of the newly founded Institut für Gasdynamik of the Luftfahrtforschungsanstalt Hermann Göring at Braunschweig, a huge camouflaged aerodynamic research facility that remained unnoticed by the Allies throughout the war. He began to analyze the nature of supersonic lift, which laid the foundation for the aerodynamic design of German jet planes of World War II. BUSEMANN's idea could not be used until engines were developed to provide the high speed. In Germany, the Messerschmitt Me-262 (the first operational jet airplane) and the Me-163 (the fastest flying WWII aircraft and the only operational rocket-powered fighter) had sweepback wings. The German A4b, a long-distance version of the rocket V2 designed at the Heeresversuchsanstalt (HVA) Peenemünde under the leadership of Walter DORNBERGER and Werner VON BRAUN, had also sweepback wings.

After the war BUSEMANN worked in England at Farnborough, but after one year he decided to work for the NASA at Langley Research Center in Hampton, VA (1947–1964). In 1963, he became professor of fluid mechanics at the University of Colorado. After his retirement he lived in Boulder, CO.

Still during PRANDTL's lifetime, he became foreign scientific member of the Max-Planck-Institut für Strömungsforschung (1950). He received the Ludwig-Prandtl-Ring (1966), the highest award of the Wissenschaftliche Gesellschaft für Luft- und Raumfahrt, and shortly thereafter became *doctor honoris causa* of the University of Aachen. He was a Fellow of the American Astronautical Society (from 1968) and elected member of the National Academy of Engineering (1970).

**ORIGINAL WORKS.** With O. FÖPPL: *Physik. Grundlagen der Elastomechanik.* In: (H. GEIGER, ed.) *Handbuch der Physik.* Springer, Berlin; vol. 6 (1928): *Mechanik der elastischen Körper*, pp. 1-46 — *Profilmessungen bei Geschwindigkeiten nahe der Schallgeschwindigkeit (im Hinblick auf Luftschrauben).* Jb. Wiss. Gesell. Luftfahrt 1928. Vieweg, Braunschweig (1928), pp. 95-99 — *Überschallgeschwindigkeit in zylindrischen Rohren.* ZAMM **8**,

419-420 (1928) — *Zeichnerische Ermittlung von ebenen Strömungen mit Überschallgeschwindigkeit.* Ibid. **8**, 423-424 (1928) — *Drücke auf kegelförmige Spitzen bei Bewegung mit Überschallgeschwindigkeit.* Ibid. **9**, 496-498 (1929) — With L. PRANDTL: *Näherungsverfahren zur zeichnerischen Ermittlung von ebenen Strömungen mit Überschallgeschwindigkeit* [Festschrift zum 70. Geburtstag von Prof. A. STODOLA]. Füssli, Zurich (1929), pp. 499-509 — *Widerstand bei Geschwindigkeiten nahe der Schallgeschwindigkeit.* In: (C.W. OSEEN, ed.) *Proc. 3rd Int. Congress on Applied Mechanics.* Sveriges Litografiska Tryckerier, Stockholm (Aug. 1930), pp. 282-286 — *Verdichtungsstöße in ebenen Gasströmungen.* In: (A. GILLES, L. HOPF, and T. VON KÁRMÁN, eds.) *Vorträge aus dem Gebiet der Aerodynamik und verwandter Gebiete* (Aachen 1929). Springer, Berlin (1930), pp. 162-169 — *Gasdynamik.* In: (L. SCHILLER, ed.) *Handbuch der Experimentalphysik.* Akad. Verlagsgesell., Leipzig (1931), vol. IV, Teil 1 (1931), pp. 343-460 — With O. WALCHNER: *Profileigenschaften bei Überschallgeschwindigkeiten.* Forsch. Ing. Wes. **4A**, 87-92 (1933) — *Flüssigkeits- und Gasbewegung.* In: (E. KORSCHELT, ed.) *Handwörterbuch der Naturwissenschaften.* Fischer, Jena, vol. 4 (1933), pp. 244-279 — *Aerodynamischer Auftrieb bei Überschallgeschwindigkeit.* In: *V Convegno Volta su "Le alte velocità in aviazione"* [Rome, Sept./Oct. 1935]. Reale Accademia d'Italia, Roma (1936), pp. 328-360, and short autobiography. Ibid., p. 667; Luftfahrtforsch. **12**, 210-220 (1935) — *Gasströmung mit laminarer Grenzschicht entlang einer Platte.* ZAMM **15**, 23-25 (1935) — *Bericht über den Paul SCHMIDTschen Strahlrohrantrieb.* Bericht FB530, DVL, Berlin-Adlershof (1936) — *Hodographenmethode der Gasdynamik.* ZAMM **17**, 73-79 (1937) — *Aufgaben der Hochgeschwindigkeitstechnik.* Luftfahrtforsch. **30**, 17-36 (1940) — *Lavaldüsen für gleichmäßige Überschallströmungen.* VDI-Z. **84**, 857-862 (1940) — *Zeichnerische Behandlung nichtstationärer Gasströmungen.* In: *Probleme der Detonation.* Schriften Dt. Akad. Luftfahrtforsch. Nr. 1023/41G, 59-89 (1941) — *Die achsensymmetrische keglige Überschallströmung.* Luftfahrtforsch. **19**, 137-144 (1942) — *Infinitesimale keglige Überschallströmung.* Jb. Dt. Akad. Luftfahrtforsch. (1942/1943), pp. 455-470; Engl. translation NACA-TM 1100 (1947) — With G. GUDERLEY: *The problem of drag at high subsonic speeds.* MOS Reports and Translations No. 184 (1947) — *A review of analytical methods for the treatment of flows with detached shock waves.* NACA-TN 1858 (1949) — *The drag problem at high subsonic speeds.* J. Aeronaut. Sci. **16**, 337-344 (1949) — *The relation between minimizing total noise and noise at supersonic speeds.* In: (A. FERRI, N.J. HOFF, and P.A. LIBBY, eds.) *Proc. of Conf. on High-Speed Aeronautics* [New York, Jan. 1955]. Polytechnic Institute of Brooklyn, New York (1955), pp. 133-144 — *Compressible flow in the thirties.* Annu. Rev. Fluid Mech. **3**, 1-12 (1971) — With R.D. CULP and N.X. VINH: *Optimum altitude for coasting flight of a hypervelocity vehicle.* J. Astronaut. Sci. **21**, No. 1, 32-48 (July/Aug. 1973); *Optimum three-dimensional atmospheric entry.* Acta Astronautica **2**, No. 7-8, 592-611 (July/Aug. 1975); *Hypersonic and planetary entry flight mechanics.* University of Michigan Press, Ann Arbor, MI (1980).
**SECONDARY LITERATURE.** (Ed.) *Adolf BUSEMANN.* Jb. Dt. Akad. Luftfahrtforsch. (1939/1940), pp. 134-135 — K. OSWATITSCH: *Adolf BUSEMANN, 65 Jahre.* Z. Flugwiss. **14**, 206 (1966) — Th. VON KÁRMÁN and L. EDSON: *The wind and beyond. Theodore VON KÁRMÁN, pioneer and pathfinder in space.* Little, Brown, Boston (1967) — K.G. GUDERLEY: *Adolf BUSEMANN, 70 Jahre.* Z. Flugwiss. **19**, 191 (1971) — N.X. VINH: *An appreciation of Prof. A. BUSEMANN.* J. Optimization Theory & Applications **17**, Nos. 5-6, 353-359 (1975) — R.D. CULP: *The swept wing.* Aviation Quart. **5**, 276-287 (1979) — W. TILLMANN: *Adolf BUSEMANN: 20.4.1901-3.11.1986.* Max-Planck-Gesellschaft (MPG). Berichte Mitteilungen **4**, 74-75 (1987) — R.T. JONES: *Adolf BUSEMANN 1901-1986.* Mem. Tributes: Natl. Acad. Engng. **3**, 62-67 (1989) — J.D. ANDERSON JR.: *Modern compressible flow, with historical perspective.* McGraw-Hill, New York (1990), pp. 283-286, 355-356.
**PICTURE.** Reprinted, with permission, from the *Annual Review of Fluid Mechanics*, Volume 3. © 1971 by Annual Reviews, Palo Alto, CA; http://www.annualreviews.org.

# CARRÉ, Louis (1663–1711)

- French mathematician, physicist, and science writer

Louis CARRÉ was born at Clofontaine (Dépt. Seine-et-Marne) to a laborer and was set to become a priest. However, after finishing his studies in theology, he refused to take orders and became secretary of the famous philosopher and theologian Nicolas MALEBRANCHE (1638–1715), where he remained and continued his studies. In 1697, he was admitted to the Paris Academy of Sciences. His fields of interest included mathematics, geometry, fluid dynamics, and acoustics, particularly musical acoustics. In 1700, he published the first complete work on the integral calculus under the title *A Method of Measuring Surfaces and Solids, and Finding their Centers of Gravity, Percussion, and Oscillation.*

In 1702, CARRÉ reported on curious ballistic experiments in which he observed the destructive action of bullets being fired in a wooden tank filled with water, a phenomenon later termed *hydraulic ram* or *hydrodynamic ram*. It was certainly known to have happened in armed conflicts but had hitherto never been perceived or even analyzed scientifically. Hydrodynamic ram has become a present-day problem related to protecting aircraft fuel tanks against ballistic impact. CARRÉ also speculated on the trajectory and resistance that a bullet undergoes when striking a surface of water at an oblique angle, mentioning in connection with this the phenomenon of water ricocheting (1705). In 1707, he reported to the Academy on the first ballistic pendulum experiments made by Jacques CASSINI {⇨1707}. His memoirs were printed in the volumes of the Memoirs of the French Academy of Sciences, from 1701 to 1710.

**ORIGINAL WORKS.** *Méthode pour la mesure des surfaces, la dimension des solides, leurs centres de pesanteur, de percussion, d'oscillation par l'application du calcul intégral.* J. Boudot, Paris (1700) — *Expériences physiques sur la réfraction des balles de mousquet dans l'eau et sur la résistance de ce fluide.* Mém. Acad. Sci. Paris (1705), pp. 277-287; J. Corréard, Paris (1846) — *Sur la théorie générale du son, sur les différents accords de la musique et sur le monochorde.* Supplément au J. des Savants **35**, 547-578 (1707) — *Démonstrations simples et faciles de quelques propriétés qui regardent les pendules, avec quelques nouvelles propriétez de la Parabole.* Mém. Acad. Sci. Paris (1707), pp. 61-72 — *Problèmes de hydrostatique.* Ibid. (1707), pp. 363-366 — *Expériences sur le ressort de l'air.* Mém. Acad. Sci. Paris (1710), pp. 1-9 — Campbell translation of Louis CARRÉ. Reference code GB 0103 MS GRAVES 34. University College London.
**SECONDARY LITERATURE.** B.L.B. FONTENELLE: *Eloge de M. CARRÉ.* Hist. Acad. Sci. Paris (1711), pp. 132-139 — Anonymous: *CARRÉ (Lewis).* In: (A. CHALMERS, ed.) *The general biographical dictionary: containing an historical and critical account of the lives and writings of the most eminent persons in every nation.* Nichols, London, vol. VIII (1813), pp. 283-285.

# CAVENDISH, Henry (1731–1810)

- British experimental chemist and physicist

Henry CAVENDISH was born in Nice, southern France (Dépt. Alpes-Maritimes) to Lord Charles CAVENDISH, third son of William, second duke of Devonshire. He studied mathematics at St. Peter's College (now Peterhouse), Cambridge (1749–1753), but left the university without a degree. After touring Europe with his brother Frederick, he lived a frugal life at 13 Great Marlborough Street in Soho, London (1753–1784), and decided to experiment in physics and chemistry in his private laboratory. In 1771, at the age of 40, he became a millionaire through the inheritance of a fortune that made him, according to Jean-Baptiste BIOT, a contemporary French scientist, "the richest of all learned men, and very likely also the most learned of all the rich." In the late 1780s, he moved his laboratory to a large villa in Clapham Common, a fashionable suburb of London, where he set up a well-equipped laboratory and library, and became an eccentric scientist.

CAVENDISH – throughout his research inspired by Sir Isaac NEWTON's *Principia* (1687), his model of exact science – searched for the forces of particles. However, contrary to NEWTON, who assumed the existence of extended corpuscles, CAVENDISH preferred point-particles as proposed by some contemporary natural philosophers (*e.g.*, John MICHELL and Roger J. BOSCOVICH). CAVENDISH divided mechanical effects into visible motions and invisible vibrations, the latter being linked with active and latent heat. He planned to work out a theory of motion using the principle of conservation of mechanical momentum, the product of mass and velocity. He included the hidden momenta contributed by elasticity and gravitation and identified heat with the mechanical momentum of the vibrations of invisible particles. Mechanical momenta, he argued, must be conserved in heat exchanges and were applied to sound, water waves, and the heating effects of light. However, he could not yet explain basic phenomena such as the relation between mechanical heat and measurable temperature changes when bodies were in different physical states. In the 1760s, he started experiments on heat, and in the 1780s, responding to the caloric theory of the prominent French scientists Antoine-Laurent LAVOISIER and Pierre-Simon DE LAPLACE, drafted a long paper applying his early theory of mechanical momentum to thermal phenomena, attempting to extend his thermal mechanics to percussion, expansion, the electrical heating of wires, and the conversion of mechanical into thermal effects.

His ingenious research in chemistry and physics is characterized by a quantitative approach. He performed numerous scientific investigations but published only 20 articles and no books. He tackled many puzzling phenomena of his time, *e.g.*, the composition of air, the nature and properties of hydrogen, the specific heat of certain substances, the composition of water, and various properties of electricity. In order to establish that hydrogen gas was a substance entirely different from ordinary air, he calculated their densities as well as the densities of several other gases. He found that common air, as well as air brought by a balloon from the upper atmosphere, is made up of nitrogen in a 4:1 ratio by volume. He also showed that water is composed of oxygen and hydrogen.

Apparently, his only contribution relating to explosions was his investigation of the nature of hydrogen. He isolated carbon dioxide, then called "fixed air," and hydrogen (1766), which he called "inflammable air," and measured their specific weights. To uncover also the properties of hydrogen, he studied various mixtures of hydrogen and air. He subjectively measured the combustibility of air by the loudness of the explosion when detonated with inflammable air – a dynamic pressure gauge of sufficiently short response was not yet available to him. To find the volume ratio leading to the strongest explosions, he constructed a "measurer of explosions of inflammable air," which was the first instrument to quantitatively measure the efficiency of a gas explosion. Stimulated by experiments on an exploding mixture of air and hydrogen carried out in previous years by the Birmingham scientist John WARLTIRE, he made the important discovery that the combustion product of hydrogen and oxygen, deposited as dew on the walls of the reaction vessel, is pure water, thus showing that water is not a basic chemical element (then the common opinion) but rather a chemical compound formed from hydrogen and oxygen.

CAVENDISH also performed careful eudiometric measurements of the constitution of atmospheric air and discovered that after combustion of substances the residual air contains not only nitrogen but also another indifferent component (1783–1788). Later experiments by Sir William RAMSAY and Lord RAYLEIGH showed that this inert component is a new element, which they called "argon" (1895). At the age of 67, CAVENDISH performed one of the most difficult measurements in the history of physics, which allowed him to calculate the mass of the Earth (1798): the determination of the gravitational constant between small bodies – since then known as the "Cavendish experiment."

CAVENDISH was also interested in applied science. For example, in the early 1770s he served as member of a committee charged with the practical task of devising the best method of protecting the Purfleet powder magazine from lightning. In the 18th century, Purfleet, a town in Essex, was a major storage location for gunpowder in outer London that had burned down after a lightning strike. CAVENDISH took a leading role in the long-running debate about the appropriate shape of lightning rods. Eventually, the committee recommended the erection of pointed rather then blunted rods, with a minute description of their construction.

Astronomers named a crater on the near side of the Moon after him. The Cavendish Laboratory at Cambridge University was not named after him, but rather after the seventh duke of Devonshire, Spencer Compton CAVENDISH, who was chancellor of the University and who in 1879 donated the necessary funds to establish the laboratory.

**ORIGINAL WORKS.** *Three papers, containing experiments on factitious air* [1766]. Phil. Trans. Roy. Soc. Lond. **56**, 141-184 (1767) — With W. WATSON, B. FRANKLIN, and J. ROBERTSON: *Report on lightning conductors for the powder magazines at Purfleet. To the president and council of the Royal Society* (Aug. 21, 1772). In: (B. FRANKLIN and W.T. FRANKLIN, eds.) *Memoirs of the life and writings of Benjamin FRANKLIN.* H. Colburn, London (1818/1819), pp. 197-202 — *An account of a new eudiometer.* Phil. Trans. Roy. Soc. Lond. **73**, 106-135 (1783) — *Experiments on air.* Ibid. **74**, 119-153, 170-177 (1784); Ibid. **75**, 372-384 (1785) — *A measurer of explosions of inflammable air* [laboratory note]. In: (E. THORPE, ed.) *The scientific papers of the honourable Henry CAVENDISH, F.R.S.* Cambridge University Press, Cambridge, U.K., vol. II (1921): *Chemical & dynamical*, pp. 318-320.
**SECONDARY LITERATURE.** H. BROUGHAM: *CAVENDISH.* In: (H. BROUGHAM, ed.) *Lives of men of letters and science: who flourished in the time of GEORGE III.* C. Knight, London (1845), vol. I, pp. 250-259 — A.J. BERRY: *Henry CAVENDISH: his life and scientific work.* Hutchinson, London (1960) — R. MCCORMMACH: *Henry CAVENDISH: a study of rational empiricism in 18th-century natural philosophy.* Isis **60**, 243-306 (1969); *H. CAVENDISH.* In: (C.C. GILLESPIE, ed.) *Dictionary of scientific biography.* Scribner, New York, vol. 3 (1971), pp. 155-159 — S. SCHAFFER: *CAVENDISH, Henry.* In: *The new Encyclopædia Britannica, Macropædia.* Benton & Hemingway, Chicago, vol. 3 (1974), pp. 1018-1020 — C. JUNGNICKEL and R. MCCORMMACH: *CAVENDISH, the experimental life.* Bucknell, Cranbury, NJ (1999) — R. MCCORMMACH: *Speculative truth: Henry CAVENDISH, natural philosophy, and the rise of modern theoretical science.* Oxford University Press, Oxford & New York (2004) — S. SCHAFFER: *CAVENDISH, Henry.* In: (H.C.G. MATTHEW, ed.) *Oxford dictionary of national biography.* Oxford University Press (2004), vol. 10, pp. 621-627 — F. SEITZ: *Henry CAVENDISH: the catalyst for the chemical revolution.* Proc. Am. Phil. Soc. **148**, No. 2, 151-179 (June 2004); http://www.aps-pub.com/proceedings/1482/480201.pdf.

**PICTURE.** Drawing probably by Charles ROSENBERG, after William ALEXANDER, early 19th century. © National Portrait Gallery, London, U.K.
**NOTE.** The vast bulk of CAVENDISH's manuscript papers and correspondence has not been published; it is deposited in Chatsworth House (Derbyshire), in the possession of the duke of Devonshire.

## CHALLIS, James (1803–1882)

- British clergyman, physicist, and astronomer

James CHALLIS was born at Braintree in Essex to John CHALLIS, a stone mason. Following his studies at Trinity College in Cambridge (1821–1825), he was elected Fellow of Trinity (1826) and resided there until ordained (1830). When George AIRY was elected Astronomer Royal (1835), he was succeeded by CHALLIS as Plumian Professor of astronomy and experimental philosophy (1836), and CHALLIS became director of the Cambridge Observatory (1836). In the first years of his professorship CHALLIS lectured on pneumatics, hydrodynamics, and geometrical and physical optics with special reference to the mathematical theories of light and sound. He published numerous papers on hydrodynamics, heat, light, the theory of colors, and astronomy. By invitation of the British Association for the Advancement of Science he wrote two reports of the current state of research on hydrodynamics. His *Report on the Present State of the Analytical Theory of Hydrostatics and Hydrodynamics* (1833) was the best known of his mathematical papers among contemporary scientists, which he followed with a substantial *Supplementary Report* (1836).

Attempting to treat the sound of finite longitudinal disturbance mathematically (1845–1851), he attacked his countrymen Samuel EARNSHAW, Sir George Biddell AIRY, and Sir George G. STOKES, later also crossing swords with them on other subjects. However, his disputes on the problem of how to correctly treat analytically *waves of finite amplitude* (*i.e.*, of shock waves) kept discussions among early shock wave pioneers in full swing.

In the astronomy community CHALLIS soon became famous for his failure to discover the planet Neptune of which John C. ADAMS, another Cambridge astronomer, had already calculated the position and which was eventually discovered by Johann G. GALLE at the Berlin Observatory (1846) through Urban J. LEVERRIER's prediction. CHALLIS published some 60 papers reporting observations of comets and asteroids. Stress, due to arrears of reduction derided by AIRY, compelled him to give up his occupations at Cambridge Observatory (1861), but he retained the Plumian chair until his death.

CHALLIS contributed almost 250 mathematical, physical, and astronomical papers to scientific journals, partly as a coauthor, and published in 12 volumes the *Astronomical Observations Made at the Observatory of Cambridge* (1832–1864), the chief result of his work. In his book *Notes on the Principles of Pure and Applied Calculations, and Applications of Mathematical Principles to Physics* (1869), a large volume of 700 pages, he also addressed applications of his hydrodynamical research. He was a Fellow of the Royal Astronomical Society (from 1836), a Fellow of the Royal Society of London (from 1848), and president of the Cambridge Philosophical Society (1845–1847).

Astronomers named a crater on the near side of the Moon after him.

**ORIGINAL WORKS.** *On the integration of the general equations of the motion of incompressible fluids.* Phil. Mag. **6** [II], 123-133 (1829) — *On the determination of the forms of the arbitrary functions which occur in the integrals of partial differential equations.* Ibid. **6** [II], 296-301 (1829) — *On the theory of the small vibratory motions of elastic fluids.* Trans. Cambr. Phil. Soc. **III**, 269-320 (1830) — *On the general equations of the motion of fluids, both incompressible and compressible, and on the pressure of fluids in motion.* Ibid. **III**, 383-416 (1830) — *On the theoretical determination of the motion of fluids.* Phil. Mag. **9** [II], 7-11 (1831) — *On the theory of the compressibility of the matter composing the nucleus of the Earth, as confirmed by what is known of the ellipticities of the planets.* Ibid. **10** [II], 200-204 (1831) — *Report on the present state of the analytical theory of hydrostatics and hydrodynamics.* Rept. Meet. Brit. Assoc. **3**, 131-151 (1833) — *Researches in the theory of the motion of fluids.* Trans. Cambr. Phil. Soc. **V**, 173-204 (1835) — *Supplementary report on the mathematical theory of fluids.* Rept. Meet. Brit. Assoc. **6**, 225-252 (1836) — *A general investigation of the differential equations applicable to the motion of fluids.* Trans. Cambr. Phil. Soc. **VII**, 371-396 (1842) — *On the necessity of three fundamental equations for the general analytical determination of the motion of fluids.* Phil. Mag. **26** [III], 425-431 (1845) — *Theoretical determination of the velocity of sound.* Ibid. **32** [III], 276-284 (1848) — *On the velocity of sound, in reply to the remarks of the Astronomer Royal.* Ibid. **32** [III], 494-499 (1848) — *Additional analytical considerations respecting the velocity of sound.* Ibid. **33** [III], 98-101 (1848) — *On the nature of aerial vibrations.* Ibid. **33** [III], 462-466 (1848) — *On the mathematical theory of aerial vibrations.* Ibid. **34** [III], 88-98 (1849) — *On the theoretical value of the velocity of sound, in reply to Mr. STOKES.* Ibid. **34** [III], 284-286 (1849) — *Determination of the velocity of sound on the principles of hydrodynamics.* Ibid. **34** [III], 353-366 (1849) — *On spherical waves in an elastic fluid, in reply to Mr. STOKES.* Ibid. **34** [III], 449-450 (1849) — *On some points relating to the theory of fluid motion.* Ibid. **34** [III], 512-520 (1849) — *On the modification of sounds by distance of propagation.* Ibid. **35** [III], 241-244 (1849) — *On the principles of hydrodynamics.* Ibid. **1** [IV], 26-38, 231-241, 477-478 (1851) — *On the theory of the velocity of sound.* Ibid. **1** [IV], 405-408 (1851) — *On the principles of hydrodynamics.* Ibid. **4** [IV], 438-450 (1852); Ibid. **5** [IV], 86-102 (1853) — *On some theorems in hydrodynamics.* Ibid. **6** [IV], 338-344 (1853) — *Theoretical considerations respecting the relation of pressure to density.* Ibid. **17** [IV], 401-404 (1859) — *On the general differential equations of hydrodynamics.* Ibid. **23** [IV], 436-445 (1862) — *On the principle of discontinuity in solutions of problems in the calculus of variations.* Ibid. **24** [IV], 196-201 (1862) — *Notes on the principles of pure and applied calculations, and applications of mathematical principles to physics.* Publ. privately (1869) — *A theory of the effects produced by fog and vapor in the atmosphere on the intensity of sound.* Ibid. **47** [IV], 277-281 (1874) — *On the mathematical principles of LAPLACE's theory of tides.* Ibid. **50** [IV], 544-548 (1875).
**SECONDARY LITERATURE.** J.W.L.G. [J.W.L. GLAISHER]: *James CHALLIS*. Month. Not. Roy. Astron. Soc. **43**, 160-179 (1883) — O.J. EGGEN: *J. CHALLIS*. In: (C.C. GILLESPIE, ed.) *Dictionary of scientific biography.* Scribner, New York, vol. 3 (1971), pp. 186-187 — M. YAMALIDOU: *Molecular ideas in hydrodynamics.* Ann. Sci. **55**, 369-400 (1998) — A.M. CLERKE and D.B. WILSON: *CHALLIS, James.* In: (H.C.G. MATTHEW, ed.) *Oxford dictionary of national biography.* Oxford University Press, Oxford (2004), vol. 10, pp. 859-862.
**PICTURE.** © The Royal Society of London, London, U.K.

## CHAPMAN, David Leonard (1869–1958)

• British physical chemist, cofounder of the first theory on detonation

Born in Wells, Norfolk, David L. CHAPMAN's father, David CHAPMAN, was a builder in Manchester. After his education at Manchester Grammar School he won an Open Exhibition at Christ Church, Oxford, and was placed in the First Class of the Final Honor School of Natural Science (1893), being awarded a Second Class for Physics. When he took up a post as science master at Gigglewick (1894), he was soon picked by Harold B. DIXON, a professor of chemistry, to join his staff at Owens College, University of Manchester (1897–1907). In his Bakerian Lecture (1893), DIXON had reported about his observation on the velocities of explosion in gases, to which Prof. Arthur SCHUSTER made the comment that on either side of the explosive wave, later to be called the "detonation wave," the exploded and unexploded gases might have uniform densities and velocities. Based upon SCHUSTER's hypothesis and DIXON's experimental data of detonation velocities, CHAPMAN theoretically investigated in his paper *The Rate of Explosions in Gases* (1899) the phenomenon of detonation on a rational basis. First applying Bernhard RIEMANN's formula to both sides of the detonation front, which separates two chemically different states, he made the important assumption that the detonation moves steadily and – consistent with the condition of maximum entropy – possesses a minimum velocity, the so-called "Chapman equation." With data thus obtained, he was able to calculate the explosion velocities for some 40 other mixtures. Independently, a very similar assumption was later made in France by Emile JOUGUET, the so-called "Chapman-Jouguet (CJ) hypothesis" and the conditions behind an advancing explosive wave are now referred to as the Chapman-Jouguet (CJ) state (1899–1905). Detonations with fronts advancing at sonic speeds – so-called "Chapman-Jouguet detonations" – are the most common.

After his famous studies on explosives at Owens College, CHAPMAN accepted a Fellowship at Jesus College, Oxford, and began to investigate the kinetics of chemical reactions in more detail, partly with the assistance of his wife and C.H. BURGESS, two chemists. These fundamental studies (1899–1937), including also photochemically induced explosive reactions in gases, opened a new field of research and stimulated Max BODENSTEIN and Walther H. NERNST in Germany on their concept of chain reactions. Together with his students CHAPMAN published over 40 papers, reporting on studies of the decomposition of water vapor by electric spark (1902), retarding reaction effects (1909), and photochemical reactions of gases subjected to ultraviolet radiation and their dependency on light intensity, here particularly focusing on the puzzling chlorine-hydrogen reaction (1909–1933). As a forerunner of modern reaction kinetics he was already engaged in discussions about problems of the decomposition of ozone (1910–1911) and the role of metals in the catalysis of gases (1929–1936). CHAPMAN remained at Jesus College as College tutor in charge of the laboratories until his retirement (1944).

He was a Fellow of the Royal Society (from 1913) and served on its Council (1934–1936).

**ORIGINAL WORKS.** *On the rate of explosion in gases.* Phil. Mag. **47** [V], 90-104 (1899) — With F.A. LIBDURY: *Decomposition of water vapor by the electric spark.* Proc. Chem. Soc. Lond. **18**, 183 (1902) — With C.H. BURGESS: *Note on the cause of the period of chemical induction in the union of hydrogen and chlorine.* Proc. Roy. Soc. Lond. **74**, 400 (1905) — With C.H. BURGESS: *The interaction of chlorine and hydrogen.* J. Chem. Soc. (Lond.) **89**, 1399-1434 (1906) — With P.S. MACMAHON: *The interaction of hydrogen and chlorine. The nature of photochemical inhibition.* Ibid. **97**, 845-851 (1910) — With L.K. UNDERHILL: *The interaction of chloride and hydrogen. The influence of mass.* Ibid. **103**, 496-508 (1913) — *General and physical chemistry.* Annu. Repts. Progr. Chem. **11**, 1-33 (1914) — With M.P. APPLEBEY: *On the equation of state.* Phil. Mag. **40** [VI], 197-200 (1920) — With J.E. RAMSBOTTOM and C.G. TROTMAN: *The union of hydrogen and oxygen in presence of silver and gold.* Proc. Roy. Soc. Lond. **A107**, 92-100 (1925) — With F. BRIERS and E. WALTERS: *The influence of the intensity of illumination on the velocity of photochemical changes.* J. Chem. Soc. (Lond.) **128**, 562-569 (1926) — *Some conclusions from recent work on photochemistry.* Trans. Faraday Soc. **21**, 547-553 (1926) — With P.P. GRIGG: *Note on the rate of photochemical combination of chlorine and hydrogen in glass capillary tubes.* J. Chem. Soc. (Lond.) **131**, 3233-3235 (1928); *The mean life of the catalyst postulated in the photochemical union of chlorine and hydrogen.* Ibid. **132**, 2426-2432 (1929); *Photochemical interactions of hydrogen with chlorine and bromine.* Nature **127**, 854 (1931) — With J.S. WATKINS: *The photochemical union of chlorine and hydrogen in the presence of oxygen, and the relative rates of formation of water and hydrogen chloride in illuminated mixtures of the three gases rich in oxygen.* J. Chem. Soc. (Lond.) **136**, 743-745 (1933) — With P.W. REYNOLDS: *The catalytic combination of hydrogen and oxygen at the surface of platinum.* Proc. Roy. Soc. Lond. **A156**, 284-306 (1936).
**SECONDARY LITERATURE.** E.J. BOWEN: *David Leonard CHAPMAN.* Biogr. Mem. Fell. Roy. Soc. (Lond.) **3**, 35-44 (1958) — D.L. HAMMICK: *David Leonard CHAPMAN 1869–1958.* Proc. Chem. Soc. (1959), pp. 101-103 — E.J. BOWEN: *David Leonard CHAPMAN.* In: (C.C. GILLESPIE, ed.) *Dictionary of scientific biography.* Scribner, New York, vol. 3 (1971), p. 197 — R. CHÉRET: *Chapman-Jouguet hypothesis 1899–1999: one century between myth and reality.* Shock Waves **9**, 295-299 (1999) — E.J. BOWEN and K.J. LEIDLER: *CHAPMAN, David Leonard.* In: (H.C.G. MATTHEW, ed.) *Oxford dictionary of national biography.* Oxford University Press, Oxford (2004), vol. 11, pp. 37-38.
**PICTURE.** Taken in 1920. Courtesy Manchester Archives and Local Studies, Manchester, U.K.

# CORIOLIS, Gaspard Gustave DE (1792–1843)

- French mathematical physicist and engineer

G. Gustave DE CORIOLIS was born in Paris to Jean-Baptiste-Elzéar DE CORIOLIS, an officer of king Louis XVI. He was brought up in Nancy to where his family had taken refuge and where his father became an industrialist. He entered the Ecole Polytechnique (1810), and after graduation at the Ecole des Ponts et Chaussées (1813) he served in its Corps of Engineers, spending several years in the Dépt. Meurthe-et-Moselle and in the Vosges Mountains. Recommended by the

 eminent mathematician Augustin-Louis CAUCHY, he became in 1816 tutor in mathematical analysis at the Ecole Polytechnique. In 1829, he accepted the position of chair of mechanics at the newly founded Ecole Centrale des Arts et Manufactures in Paris, and in 1832 assisted Louis NAVIER in applied mechanics at the Ecole des Ponts et Chaussées where he succeeded him in 1836. CORIOLIS ended his teaching at the Ecole Polytechnique in 1838 and succeeded Pierre L. DULONG as director of studies in the same school.

CORIOLIS' whole life, although overshadowed by a delicate constitution, was devoted to research and methods of teaching mechanics, which, together with the mathematician Jean PONCELET, he successfully reformed. Inspired by the writings of Nicolas CARNOT, he worked out his theory of machines, which, addressing the "economy" of mechanical power and introducing the important term *travail* ("work"), was published in his first book *Du calcul de l'effet des machines* (1829); its second edition was published posthumously under the new title *Traité de Mécanique* (1844). In this important work CORIOLIS proposed a unit of measurement of work, the *dynamode* (corresponding to 1,000 kg-meters). While this term was not adopted by his contemporaries, his term *force vive* ("kinetic energy") for one half the product $mv^2$ was generally accepted. Today CORIOLIS is best remembered for the composed centrifugal force [French *force centrifuge composée*], the so-called "Coriolis force," which first appeared in the paper *Sur les équations du mouvement relatif des systèmes de corps* (1835). In this paper he showed that the Laws of Motion could be used in a rotating frame of reference if an extra force is added to the equations of motion *(Coriolis theorem)*. The so-called "Coriolis effect" denotes the effect of the Coriolis force to deviate a moving body perpendicular to its velocity vector. Enunciated by him regarding relative motions, it has found numerous applications, particularly in the case of motion on, above, or below the surface of the Earth, for example (1) the deviation toward the east of falling bodies, an observation that had already been cited by Robert HOOKE and Sir Isaac NEWTON as a possible experimental proof of the Earth's rotation; (2) the tendency of an ocean current, a wind system, or an artillery round to drift sideways from its course; and (3) the apparent rotation of the plane of vibration of a Foucault pendulum. In 1873, the French mechanics professor Henry A. RÉSAL used the Coriolis theorem to calculate the vibrational motions of molecules.

In the early 1830s, CORIOLIS began to tackle the difficult task of analytically investigating in the game of billiards the various modes of collision and the influence of friction. In the same year in which he published his Coriolis effect, CORIOLIS also presented his famous mathematical theory of billiards, certainly the most prominent and spectacular example of elastic collision. After watching for a while the famous French player Captain François MINGAUD who first introduced new, highly surprising *effets* ("side shots") into the game, he worked out an analytical solution to these complicated collision processes, thereby considering not only the translational and rotational kinetic energy of the hit ball but also friction effects between ball and cloth, and modifications of the ball trajectory in the special case of side shots. CORIOLIS himself considered his contribution to billiards as his greatest work, but his fame from mathematically analyzing billiards was soon surpassed by his *Coriolis effect*.

In his book on classical mechanics, *Traité de la mécanique des corps solides et du calcul de l'effet des machines* (1844), which was published posthumously, he introduced a moving frame of reference that allows one to study the relative motions of three bodies, known as the "three-body problem" – a particularly challenging task that occupied generations of mathematicians.

CORIOLIS was a member of the French Academy at Paris (from 1836). The *Coriolis Data Center* is a French contribution to operational oceanography. In honor of his lasting contributions to theoretical and applied mechanics, a French oceanographic research vessel was named for him. Its scientific team discovered the *Coriolis Troughs* which, located in the south of the New Hebrides, are one of the world's most youthful, magmatically-active back arc basins. The *Coriolis parameter* is a measure of planetary rotation as a function of latitude.

A crater on the far side of the Moon is also named after him.

**ORIGINAL WORKS.** *Sur l'influence du mouvement d'inertie du balancier d'une machine à vapeur et de sa vitesse moyenne, sur la régularité du mouvement de rotation que le va-et-vient du piston communique au volant.* J. Ecole Polytech. **13**, 228-268 (1832) — *Sur le principe des forces vives dans les mouvements relatifs des machines.* Ibid. **13**, 268-302 (1832) — *Sur la manière d'établir les différents principes de mécanique pour des systèmes de corps, en les considérant comme des assemblages de molécules.* Ibid. **15**, 93-125 (1835) — *Sur la théorie des pertes de travail dues aux frottements dans les engrenages coniques.* Ibid. **15**, 126-133 (1835) — *Sur les équations du mouvement relatif des systèmes de corps.* Ibid. **15**, 142-154 (1835) — *Théorie mathématique des effets du jeu de billard.* Carilian-Gœury, Paris (1835); Engl. translation by D. NADLER: *Mathematical theory of spin, friction,* *and collision in the game of billiards*. D. Nadler, San Francisco, CA (2006) — *Sur un moyen de tracer des courbes données par des équations différentielles*. J. Math. Pures Appl. **1** [I], 5-9 (1836) — *Mémoire sur le degré d'approximation qu'on obtient pour les valeurs numériques d'une variable qui satisfait à une équation différentielle, en employant pour calculer ces valeurs, diverses équations aux différences plus ou moins approchées*. Ibid. **2** [I], 230-245 (1837) — *Note sur une manière simple de calculer la pression produite contre les parois d'un canal dans lequel se meut un fluide incompressible*. Ibid. **2** [I], 130-132 (1837) — *Note sur un théorème de mécanique*. C. R. Acad. Sci. Paris **12**, 267-268 (1841) — *Traité de la mécanique des corps solides et du calcul de l'effet des machines*. Carilian-Gœury, Paris (2nd edn. 1844); *Lehrbuch der Mechanik fester Körper und der Berechnung des Effektes der Maschinen*. Meyer, Braunschweig (1846).

**SECONDARY LITERATURE.** *Mémoires et communications*. C. R. Acad. Sc. Paris **17**, 555, 559, 1327 (1843) — J. BINET: *Discours prononcé aux funérailles de CORIOLIS le 20 septembre 1843*. F. Didot Frères, Paris (1843) — N.A. RENARD: *Notice historique sur la vie et les travaux de G. CORIOLIS*. Raybois, Nancy (1862), Extrait des Mémoires de l'Académie de Stanislas (1861) — H.A. RESAL: *De l'équation des forces appliquée à un système de corps solides, en ayant égard aux ébranlements des molécules* (*théorème de CORIOLIS*). In: *Traité de mécanique générale*. Gauthier-Villars, Paris (1873), pp. 446-449 — A. DE LAPPARENT: *CORIOLIS (1792–1843)*. In: *Livre du Centenaire de l'Ecole Polytechnique 1794–1894*. Gauthier-Villars, Paris (1895). Reprinted as Preface in: G.G. CORIOLIS: *Théorie mathématique des effets du jeu de billard, suivi par des deux célèbres mémoires publiés en 1832 et 1835…*. J. Gabay, Sceaux (1990) — R. DUGAS: *Sur l'origine du théorème de CORIOLIS*. La Rev. Sci. **79**, 267-270 (1941) — L.S. FREIMAN: *Gaspard Gustave CORIOLIS*. Moscow (1961) — P. COSTABEL: *Gaspard Gustave DE CORIOLIS*. In: (C.C. GILLESPIE, ed.) *Dictionary of scientific biography*. Scribner, New York, vol. 3 (1971), pp. 416-419 — H.M. STOMMEL and D.W. MOORE: *An introduction to the Coriolis force*. Columbia University Press, New York (1989).
**PICTURE.** © Collections Ecole Polytechnique, Paris, France.

# COURANT, Richard (1888–1972)

- German-born U.S. mathematician and fluid dynamicist

Richard COURANT was born at Lublinitz in Upper Silesia (now Lubliniec, southern Poland) to Siegmund COURANT, an unsuccessful Jewish businessman. He attended the gymnasium in Breslau and studied mathematics at the Universities of Breslau (1906), Zürich (1907), and Göttingen (1908–1910). He began to teach mathematics as *Privatdozent* (university lecturer) at Göttingen (1910–1920), where he succeeded Prof. Christian Felix KLEIN, an eminent German mathematician. After serving in the German Army (1914–1918) and becoming a highly decorated officer, he resumed his teaching activity in Göttingen and became professor of mathematics and director of the Mathematical Institute at the University of Göttingen (1920–1933). In a joint study with Kurt O. FRIEDRICHS and Hans LÉWY, two German mathematicians, he discovered the conditional stability of the difference-equation integration method for partial differential equations, the so-called "Courant-Friedrichs-Léwy condition" (1928), which, rediscovered in the 1940s by John VON NEUMANN, who called it the "Courant criterion," became important in the pioneering era of digital computers for the numerical simulation of hydrodynamic shocks. The "Courant number," the ratio of a time step to a cell residence time, representing a stability factor, is named for him. Due to increasing political pressure, COURANT emigrated with his family first to the United Kingdom (1933) and then to the United States, where he became professor and head (1936–1958) of the department of mathematics at New York University (NYU), New York City.

In 1943, COURANT first developed Finite Element Analysis (FEA) utilizing the Ritz method of numerical analysis and minimization of variational calculus to obtain approximate solutions to vibration systems. During World War II, he got involved in shock waves and associated interaction phenomena. COURANT also directed theoretical studies of pulsejet engines (1944), which became a subject of immediate interest due to the increasing employment of the German Schmidt tube in the first cruise missile, the V1. Based on the Göttingen approach, he established at NYU the Institute for Mathematics and Mechanics and became its first director (1947–1958). Together with Kurt O. FRIEDRICHS, his former student from Göttingen, and James J. STOKER, an American originally trained in engineering, he transformed his small institute into one of the world's largest institutes of applied mathematics. The mathematical center was renamed the Institute of Mathematical Sciences (1953), a name that was also adopted by other leading mathematical institutions. In 1958, the institute was renamed the Courant Institute of Mathematical Sciences (CIMS) and became a center for research and advanced training in mathematics and computer science. It is attached to New York University and is located in Greenwich Village in lower Manhattan.

After his retirement (1958) COURANT served as consultant to both governmental agencies and private industry and was also active on various scientific commissions. Throughout his life he was devoted to the applicability of pure mathematics, particularly to quantum mechanics and gas dynamics. He is also credited with paving the way for the use of electronic computers in applied science. COURANT worked on mathematical analysis and physics, the theory of functions, and the calculus of variations. He wrote several best-sellers on mathematics such as the book he cowrote with Herbert ROBBINS, *What is Mathematics?* (1941). In the gas dynamics community he is best known for his book *Supersonic Flow and Shock Waves* (1948), which he cowrote with FRIEDRICHS.

**ORIGINAL WORKS.** With D. HILBERT: *Methoden der mathematischen Physik.* 2 vols., Springer, Berlin (1924); *Methods of mathematical physics.* Interscience, New York (1937) — *Bernhard RIEMANN und die Mathematik der letzten hundert Jahre.* Die Naturwissenschaften **14**, 813-818 (1926) — With K. FRIEDRICHS and H. LÉWY: *Über die partiellen Differentialgleichungen der mathematischen Physik.* Math. Ann. **100**, 32-74 (1928) — *Differential and integral calculus.* 2 vols., Blackie, London (1934, 1936) — With H. ROBBINS: *What is mathematics?* Oxford University Press, London (1941) — With K. FRIEDRICHS: *Interaction of shock and rarefaction waves in one-dimensional motion.* Div. 8 and Appl. Math. Panel, NDRC Rept. OSRD-1567 (1943) — *Technical conference on supersonic flow and shock waves.* Nav. Rept. 203-45, Bureau of Ordnance (1945) — With K.O. FRIEDRICHS: *Supersonic flow and shock waves.* Interscience, New York (1948) — With P. LAX: *On nonlinear partial differential equations with two independent variables.* Comm. Pure Appl. Math. **2**, 255-273 (1949) — With E. ISAACSON and M. REES: *On the solution of nonlinear hyperbolic differential equations by finite differences.* Ibid. **5**, 243-255 (1952) — With D. HILBERT: *Methods of mathematical physics.* Interscience, New York (1953) — *Die Bedeutung der modernen Rechenmaschinen für mathematische Probleme der Hydrodynamik und Reaktortechnik.* Arbeitsgemeinschaft für Forschung des Landes Nordrhein-Westfalen, Heft 59. Westdt. Verlag, Köln/Opladen (1958), pp. 7-17 — *Mathematics in the modern world.* Scient. Am. **211**, 41-49 (March 1964).
**SECONDARY LITERATURE.** K.O. FRIEDRICHS (ed.) *Studies and essays* [presented to R. COURANT on his 60th birthday on Jan. 8, 1948]. Interscience, New York (1948) — C. REID: *Richard COURANT, 1858–1972.* Springer, Berlin (1979); *COURANT in Göttingen and New York.* Springer, New York (1986) — P.D. LAX: *Richard COURANT. Jan. 8, 1888 – Jan. 27, 1972.* Biogr. Mem. Natl. Acad. Sci. (U.S.A.) **82**, 78-97 (2003) — *Richard COURANT (1888–1972).* Math. Fakultät der Georg-August-Universität Göttingen; http://www.math.uni-goettingen.de/Personen/bedeutende_Mathematiker/courant/courant.html.
**PICTURE.** Courtesy New York University Archives, New York City, NY.

# CRANZ, Carl Julius (1858–1945)

- German physicist, dean of German ballistics

Carl J. CRANZ was born at Hohebach, Oberamt Künzelsau, in Württemberg, to a Protestant minister. Initially his family intended for him to also embark upon a theological career, and

in 1877 he began to study theology and philosophy at Tübingen University. But in 1879 he left Tübingen and began to study physics at the Friedrich-Wilhelms-Universität (since 1945 Humboldt-Universität) in Berlin under the famous physicists Gustav KIRCHHOFF and Hermann VON HELMHOLTZ and the electrophysiologist Emil DUBOIS-REYMOND. Returning to Tübingen in 1880, he continued his study of mathematics under the physicist Friedrich Eduard REUSCH and the mathematician Paul DUBOIS-REYMOND. Under the latter he took his doctorate with a thesis on the ideal shape of a projectile to minimize deviations (1883), a subject that he returned to in later years. In the period 1882 to 1883, he passed his examination for professorship at the TH Stuttgart, where he became an instructor of mathematics and physics. Besides teaching in Stuttgart at the Friedrich-Eugen-Lyceum (1882–1903), a secondary school for girls, he started various research activities, comprising mathematical, ballistic, physical, and geophysical problems.

Owing to his expertise in ballistics he was offered a chair at the Technische Hochschule (TH) in Berlin-Charlottenburg and became first director of the Militärtechnische Akademie (1903–1918), which was founded in 1903 close to the campus of the TH Berlin. After World War I, the military academy was shut down, but it reopened in 1919 under the demilitarized designation "Institut für Technische Physik" with CRANZ as its director (1919–1935). It was shut down permanently at the end of World War II.

His contributions are manifold: (1) measurement of projectile drag at supersonic speed; (2) barrel vibration and flow phenomena at the muzzle of a fired gun; (3) generation and propagation of shock waves in gaseous, liquid, and solid matter; (4) interaction of shock waves with objects; (5) graphical methods for use in interior and exterior ballistics; (6) similarity laws in ballistics; and (7) development of high-speed diagnostic instrumentation. His most prominent disciple was Hubert SCHARDIN, who was his close assistant for 10 years (1926–1936) and with whom he went to Nanking to establish the first ballistic research institute in China (1934–1936). Together with SCHARDIN he developed in the 1930s the so-called "Cranz-Schardin multiple spark camera," which allowed one to optically resolve high-speed events at frame rates up to some $10^6$/s. Their principle was also used by others to study self-luminous phenomena using an array of flash X-ray tubes {TSUKERMAN & MANAKOWA ⇒1957}.

For 62 years CRANZ contributed largely to theoretical ballistics and methods of high-speed diagnostics related to solving problems in ballistic research. During his life he saw great changes in ballistics, ranging from classical ballistics of firearms and the development of huge guns (e.g., *Paris-geschütz, Dicke Bertha, Dora*) up to the development and immediate applications of new types of weapons that dramatically changed military strategy down to our own times, such as cruise missiles (V1) and ballistic rocketry (A-4 or V2, and A-4b). Shortly after World War II, CRANZ died in Esslingen, Germany at the age of 87.

**ORIGINAL WORKS.** *Theoretische Untersuchungen über die regelmäßige Abweichung der Geschosse und die vortheilhafteste Gestalt der Züge.* Ph.D. thesis, Universität Tübingen (1883); Archiv für die Artill.-Ingenieuroffiziere **90**, 477-543 (1883) — *Zur Bewegung der Geschosse.* Civilingenieur **31**, 103-126 (1885) — *Theoretische Studien zur Ballistik der gezogenen Gewehre.* Helwing, Hannover (1887) — *Gemeinverständliches über die sogenannte vierte Dimension.* Verlags-Anstalt & Druckerei-AG, Hamburg (1890) — *Compendium der theoretischen äußeren Ballistik.* Teubner, Leipzig (1896) — *Theoretische und experimentelle Untersuchungen über die Kreisbewegungen der rotierenden Langgeschosse während ihres Fluges.* Z. Math. Phys. **43**, 133-162, 169-215 (1898) — With K.R. KOCH: *Untersuchung über die Vibration des Gewehrlaufs. (A.) Gewehre vom Typ des Mausergewehres: Schwingungen in verticaler Ebene bei horizontal gehaltenem Gewehr.* Abhandl. Königl. Bayr. Akad. Wiss. **19** [Cl. II], 747-775 (1899); (*B.*) *Versuche mit kleinkalibrigen Gewehren: I. Schwingungen in verticaler Ebene.* Ibid. **20** [Cl. II], 591-611 (1900); *II. Schwingungen in horizontaler Ebene.* Ibid. **21** [Cl. II], 525-574 (1901) — With K.R. KOCH: *Über die explosionsartige Wirkung moderner Infanteriegeschosse.* Ann. Phys. **3** [IV], 247-273 (1900) — *Anwendung der elektrischen Momentphotographie auf die Untersuchung von Schusswaffen.* Carl Cranz, Halle a.S. (1901) — *Ballistik.* In: (F. KLEIN and C. MÜLLER, eds.) *Enzyklopädie der mathematischen Wissenschaften.* Teubner, Leipzig, vol. IV (1904): *Mechanik*, Part 18, pp. 185-279 — *Über die Messung der Verbrennungsdauer des Pulvers beim Schuß.* Z. ges. Schieß- u. Sprengstoffwesen **2**, 321-324 (1907) — *Zur Messung des Abgangsfehlerwinkels bei Gewehren. Zur Theorie und Praxis des Gewehr-Rücklaufmessers.* Ibid. **2**, 345-348 (1907) — *Über die relative und absolute Genauigkeit verschiedener Flugzeitmesser.* Ibid. **3**, 7-10 (1908) — With R. ROTHE: *Temperatur- und Wärmemessungen am Infantriegewehr M.98 S.* Ibid. **3**, 301-305, 327-332, 474 (1908) — *Über einen ballistischen Kinematographen mit 5000 bis 10000 Bildern je Sekunde.* Ibid. **4**, 321-323 (1909) — *Versuche über den vertikalen Schuß.* Schuß u. Waffe **2**, 413-420 (1909) — *Ballistische Bemerkungen [Über die Berechnung steiler Flugbahnen].* Artillerist. Monatshefte **30**, 401-415 (1909) — *Ballistische Bemerkungen [Näherungsformel für die Scheitelhöhe einer Flugbahn].* Ibid. **34**, 241-247 (1909) — With A. BENSBERG: *Über eine photographische Methode zur Messung von Geschwindigkeiten und Geschwindigkeitsverlusten bei Infanteriegeschossen.* Ibid **41**, 333-346 (1910) — With K. BECKER: *Über ein Vergleichspendel zur Feststellung größerer Zeitintervalle.* Z. ges. Schieß- u. Sprengstoffwesen **6**, 1-4, 22-28 (1911) — *Über die empirischen Luftwiderstandsgesetze und über den gegenwärtigen Stand der theoretischen äußeren Ballistik.* Artillerist. Monatshefte **56**, 85-115 (1911) — With K. BECKER: *Messungen über den Luftwiderstand bei großen Geschwindigkeiten.* Ibid. **69**, 189-196; Ibid. **71**, 333-368 (1912) — With P.A. GÜNTHER: *Einige Versuche über die sogen. Explosionswirkung (Dumdumwirkung) moderner Infanteriegeschosse.* Z. ges. Schieß- u. Sprengstoffwesen **7**, 317-319 (1912) — *Die Arbeitsleistung der Sprengstoff- und Geschossmittel.* Z. Elektrochemie **19**, 607, 731-738 (1913) — With P.A. GÜNTHER and F. KÜLP: *Photographische Aufnahme von sehr rasch verlaufenden Vorgängen, insbesondere von Schussvorgängen, mittels Vorderbeleuchtung durch das Licht elektrischer Funken* [mit 4 Bildtafeln]. Z. ges. Schieß- u. Sprengstoffwesen **9**, 1-4 (1914) — *Beiträge zur Waffenuntersuchung.* Artillerist. Monatshefte **88**, 252-262 (1914) — With B. GLATZEL: *Die Ausströmung von Gasen bei hohen Anfangsdrucken.* Ann. Phys. **43** [IV], 1186-1204 (1914) — With R. ROTHE: *Zur Lösung des Hauptproblems der äußeren Ballistik für ein beliebiges Luftwiderstandsgesetz.* Artillerist. Monatshefte **125/126**, 197-239 (1917) — *Eine graphische Lösung des innerballistischen Hauptproblems auf thermodynamischer Grundlage.* Ibid. **142/143**, 89-126 (1918) — With P. VON ZECH: *Aufgabensammlung zur theoretischen Mechanik.* Metzler, Stuttgart (1920) — *Über die WIENERsche Methode der Flugbahnberechnung.* Artillerist. Monatshefte **183/184**, 115-122 (1922) — With E. BAMES: *Über Hochfrequenz-Schlierenkinematographie und ihre Verwendung zur Untersuchung von Explosionserscheinungen und anderen sehr rasch verlaufenden Vorgängen.* Z. Angew. Chem. **36**, Nr. 11, 76-80 (1923) — With O. VON EBERHARD: *Ballistik.* In: (A. BERLINER and K. SCHEEL, eds.) *Physikalisches Handwörterbuch.* Springer, Berlin (1924), pp. 74-75 — With W. SCHMUNDT: *Berechnung einer Geschoß-Steilbahn unter Berücksichtigung des Kreiseleffekts und des Magnuseffekts.* ZAMM **4**, 449-464 (1924) — *Lehrbuch der Ballistik.* Springer, Berlin; vol. I (1925): *Äußere Ballistik;* vol. II (1926): *Innere Ballistik;* vol. III (1927): *Experimentelle Ballistik;* vol. IV (1936): *Supplement* — With K. SCHEEL: *Ballistische Paradoxa.* Z. Tech. Phys. **8**, 359-362 (1927) — With H. SCHARDIN: *Kinematographie auf ruhendem Film mit extrem hoher Bildfrequenz.* Z. Phys. **56**, 147-183 (1929) — With O. VON EBERHARD: *Die neuzeitliche Entwicklung der Schusswaffen.* VDI-Verlag, Berlin (1931) — With H. SCHARDIN: *Eine neue Methode zur Messung des Geschosswiderstandes im Rohr.* Z. Tech. Phys. **13**, 124-132 (1932) — With R.E. KUTTERER and H. SCHARDIN: *Der Kerreffekt-Chronograph, ein neuer Geschossgeschwindigkeitsmesser.* Wehr u. Waffen **10**, Nr. 9, 385-394 (1932) — With R.E. KUTTERER: *Einiges aus der Kriegstechnik des Altertums und des Mittelalters im Lichte der Gegenwart.* In: (E. VON FRAUENHOLZ) *Wehrpolitik und Wehrwissen.* Quelle & Meyer, Leipzig (1935) — *Entwicklung der Funkenkinematographie.* Schriften Dt. Akad. Luftfahrtforsch. Nr. 40, 1-18 (1941).

**SECONDARY LITERATURE.** V. VON NIESIOLOWSKI-GAWIN: *Fünfzig Jahre ballistische Forschung.* Die Naturwissenschaften **16**, 269-280 (1928) — K. BECKER and C.W. RAMSAUER: *Carl CRANZ 70 Jahre.* Z. Tech. Phys. **9**, 1-7 (1928) — K. BECKER: *Carl CRANZ 80 Jahre.* Vortrag gehalten auf der Festsitzung der Wehrtechnischen Fakultät der TH Berlin am 10. Jan. 1938 — (Ed.) *Carl CRANZ.* Jb. Dt. Akad. Luftfahrtforsch. (1939/1940). For his bibliography see pp. 98-101, for his biography pp. 112-113 — H. SCHARDIN: (i) *Beiträge zur Ballistik und technischen Physik: verfasst von Schülern des Herrn Geheimrat Professor Dr. phil. Dr.-Ing. E.h. Carl CRANZ anlässlich seines 80. Geburtstages am 2. Januar 1938.* J.A. Barth, Leipzig (1938), pp. 1-9. (ii) *Carl CRANZ.* Physik. Blätter **14**, 463-466 (1958). (iii) *C. CRANZ als Mitbegründer der Kurzzeitphotographie.* In: (O. HELWICH and H. SCHARDIN, eds.) *Proc. 4th Int. Congress on High-Speed Photography* [Cologne, Germany, Sept. 1958]. Helwich, Darmstadt (1959), pp. 1-8. (iv) *La vie et l'œuvre de C. CRANZ.* In: (W.C. NELSON, ed.) *Selected topics on ballistics* [Cranz Centenary Colloquium, Freiburg, Germany, April 1958]. Pergamon, Press London (1959), pp. 1-7 — H. EBERT and H.J. RUPIEPER: *Technische Wissenschaft und nationalsozialistische Rüstungspolitik: Die Wehrtechnische Fakultät der TH Berlin 1933–1945.* In: (R. RÜRUP, ed.) *Wissenschaft und Gesellschaft: Beiträge zur Geschichte der Technischen Universität Berlin 1879–1979.* Springer, Berlin (1979), pp. 469-481 — P. BERZ: *08/15. Ein Standard des 20. Jahrhunderts.* Ph.D. thesis, Phil. Fak. III, Humboldt-Universität, Berlin (1998) — P.W.W. FULLER: *Carl CRANZ, his contemporaries, and high-speed photography.* In: (D.L. PAISLEY ET AL., eds.) *Proc. 26th Int. Congress on High-Speed Photography and Photonics* [Alexandria, VA, Sept. 2004]. Proc. SPIE **5580**, 250-260 (2005).

**PICTURE.** Reprinted from Z. Tech. Phys. **9**, 1 (1928).

## Crocco, Luigi Mario (1909–1986)

- Italian-born U.S. aeronautical engineer

Luigi M. Crocco was born in Palermo, Sicily to General Gaetano Arturo Crocco (1877–1968), a renowned Italian pioneer of aeronautics who later became professor at the School of Aeronautical Engineering of the University of Rome. At this school Luigi Crocco studied mechanical engineering. Simultaneously with his studies, together with his father he conducted research in the field of rocketry (1927–1931), particularly to determine the combustion laws of double-base powder propellants (1927–1929) and liquid propellants (1929–1931). In addition, he was very attracted to the aerodynamic theory of high speed, a field that he pursued throughout his entire career. As an assistant professor of aviation engines at the School of Aeronautical Engineering of the University of Rome (1937), he became increasingly attracted to high-speed aerodynamics and in 1939 was appointed Chair of Aviation Engines (1939–1949).

His first publication (1931) was on high-speed boundary layers, and his approach led to what later became known as the "Crocco energy integral." Following graduation as a mechanical engineer (1931) and military service (1931–1933), he continued his rocket research and became involved in non-atmospheric propulsion. He also reviewed the characteristics of different types of supersonic wind tunnels of that time, later renowned as *Crocco's bible* (1935). His other major contributions were the *Crocco vorticity law* (1937) and studies of compressibility effects of laminar boundary layers. In 1939, he proposed a boundary-layer equation (*Crocco equation*) in a form that could be converted into an integral equation for constant pressure flow, and six years later he showed that the integral equation could be conveniently solved by iteration. Later his research focused on liquid monopropellants, mainly on mononitromethane, which does not require the use of air as an oxidizer and, therefore, appeared to be a promising propellant for underwater and stratospheric propulsion (1933–1937).

During World War II his scientific activities shifted toward aviation motors and jet engines. In a joint research project with the French War Ministry he resumed his studies on the use of nitromethane in rocket motors (1947). In 1949, Harry F. Guggenheim created two jet propulsion centers, one at CalTech in Pasadena and one at Princeton University. Crocco accepted an invitation to join the staff at the Princeton center as a visiting professor. He became a member of the newly founded Department of Aeronautical Engineering at Princeton University (1949–1970) and Goddard Professor to supervise the Guggenheim Jet Propulsion Center. Besides his teaching duties he was deeply involved in the investigation of combustion instability in liquid-propellant rocket motors and their elimination. In particular, he provided the first theoretical explanations of high-frequency instability, then a serious problem and of immediate practical interest in many NASA rocket-development programs. Returning to Europe in 1970, he taught at the University of Rome, at the Ecole Polytechnique in Paris, and worked as a consultant to ONERA. He retired in 1978.

Crocco was a member of various national academies. He received the Pendray Award (1965), the Wild Award (1969), and the Columbus International Prize and Gold Medal (1973). His high reputation and great intellect are best reflected by the fact that he was fluent in six languages and sought out as a consultant by major aeronautical and aerospace companies and governmental agencies throughout the Western World.

The *Luigi Crocco Teaching Prize*, a cash prize, was first awarded by the Dept. of Mechanical and Aerospace Engineering (MAE) of Princeton University to a student in the fall of 1988. The Cluster for Research on Complex Computations (CROCCO) at MAE was named for Luigi Crocco, because he was among the first in his generation to seriously devote efforts on numerical techniques to solving practical and complex fluid flows. The *Luigi Crocco Colloquia* bring outstanding engineers and scientists from other research centers to Princeton throughout the year and are considered an important feature of the department's graduate program.

**ORIGINAL WORKS.** *Flying in the stratosphere. A theoretical examination of the possibilities of achieving great speeds at very high altitudes.* Aircraft Engng. **4**, 171-175, 204-209 (1932) — *Gallerie aerodinamiche per alte velocità.* L'Aerotecnica **15**, 237-275, 735-778 (1935); *Tunnels aérodynamiques pour grandes vitesses.* Extrait Mém. Artill. Franç., Imprim. Nat., Paris (1938), pp. 358-442 — Short autobiography in: *V Convegno Volta su 'Le alte velocità in aviazione'* [Rome, Sept./Oct. 1935]. Reale Accademia d'Italia, Rome (1936), p. 671 — *Singolarità della corrente gassosa iperacustica nell'intorno di una prora a diedro.* L'Aerotecnica **17**, 519-534 (1937) — *Eine neue Stromfunktion für die Erforschung der Bewegung der Gase mit Rotation.* ZAMM **17**, 1-7 (1937) — *Una caracteristica transformazione delle equazioni dello strato limite nei gas.* Atti di Guidonia **7**, 105-120 (1939); Engl. translation: *A characteristic transformation of the boundary layer equations in gases.* Rept. ARC 4582 (1940) — *Lo strato limite laminare nei gas.* Monografie Scientifiche di Aeronautica No. 3 (1946); Engl. translation by I. Hodes and J. Castelfranco: *The laminar boundary layer in gases.* Rept. AL-684, North American Aviation, Los Angeles (1948) — *Il turboreattore a due* *flussi*. Monografie Sci. Aeronaut. No. 2 (March 1947) — *Diagrammi termodinamici dei gas di combustione*. In: *Monografia Scientifiche di Aeronautica*. Associazione culturale aeronautica, Roma No. 2 (Marzo 1947) — *Una nuova funzione potenziale per lo studio del moto bidimensionale non isentropico die gas*. L'Aerotecnica **29**, 347-355 (Dec. 1949) — With L. LEES: *A mixing theory for the interaction between dissipative flows and nearly isentropic streams*. J. Aeronaut. Sci. **19**, 649-676 (1952) — With R.F. PROBSTEIN: *The peak pressure rise across an oblique shock emerging from a turbulent boundary layer over a plane*. Rept. No. 254, Aeronaut. Engng. Dept. of Princeton University, Princeton, NJ (1954) — With C.B. COHEN: *Compressible laminar boundary layer with heat transfer and pressure gradient*. In: *50 Jahre Grenzschichtforschung*. Vieweg, Braunschweig (1954), pp. 280-293 — *Considerations on the shock-boundary layer interaction*. In: *Proc. Conf. on High-Speed Aeronautics* [Brooklyn Polytechnic Institute, NY, Jan. 1955], pp. 75-112 — With S. CHENG: *Theory of combustion instability in liquid propellant rocket motors*. AGARDograph No. 8, NATO, Butterworths, London (1956) — *One-dimensional treatment of steady gas dynamics*. In: *High speed aerodynamics and jet propulsion*. Princeton University Press, Princeton, NJ; vol. III (1958); (H.W. EMMONS, ed.) *Fundamentals of gas dynamics*, pp. 64-349 — With S.H. LAM: *Note on the shock-induced unsteady laminar boundary layer on a semi-infinite flat plate*. J. Aerospace Sci. **26**, 54-56 (1959) — *A suggestion for the numerical solution of the steady Navier-Stokes equations*. AIAA J. **3**, 1824-1832 (1965) — With W.A. SIRIGNANO: *Behavior of supercritical nozzles under three-dimensional oscillatory conditions*. AGARDograph No. 117, NATO (1967) — *Coordinate perturbation and multiple scale in gasdynamics*. Phil. Trans. Roy. Soc. Lond. **A272**, 275-301 (1972).

**SECONDARY LITERATURE.** C. CASCI and C. BRUNO (eds.): *Recent advances in aerospace sciences. In honor of Luigi CROCCO on his 75th birthday* [with biography and bibliography]. Plenum Press, New York (1985) — S.M. BOGDONOFF: *Luigi CROCCO 1909-1986*. Mem. Tributes: Natl. Acad. Engng. **3**, 100-104 (1989) — I. GLASSMAN: *Luigi CROCCO*. In: (P.H. MARKS, ed.) *Luminaries; Princeton faculty remembered*. Princeton University Press, Princeton, NJ (1996), pp. 67-73.

**PICTURE.** Courtesy Prof. Irvin GLASSMAN, Dept. of Mechanical & Aerospace Engineering, Princeton University, NJ.

# DAVY, Sir Humphry (1778–1829); from 1818 Baronet

- English chemist and inventor

Sir Humphry DAVY, born at Penzance in Cornwall, was the eldest son of Robert DAVY, a woodcarver. His first experience with chemistry was when he made fireworks with his sister. During his apprenticeship to an apothecary-surgeon (1795–1798) he discovered his interest in chemistry and drew up a formidable program of self-education. His scientific career began when he was appointed superintendent in Thomas BEDDOES' Pneumatic Institution at Clifton. Here he became interested in gases and discovered in self-tests the anesthetic properties of nitrous oxide. His book on this specific gas, entitled *Researches, Chemical and Philosophical...* (1799), made his reputation. He was appointed to a lectureship at the Royal Institution in London (1801) and became professor of chemistry (1802). His lectures were famous for their brilliant presentation. He employed Michael FARADAY as amanuensis, and on DAVY's recommendation FARADAY was given a job at the Royal Institution of Great Britain (1812).

His contributions to both organic and inorganic chemistry are enormous; prominent examples include (1) experimental evidence that oxygen and hydrogen are the only product of the electrolysis of pure water (1800); (2) improvement of the Volta cell by making piles with charcoal replacing one metal, and with two fluids and one metal (1801–1802); (3) observations on the processes of tanning and lectures on the chemistry of this subject (1801); (4) discovery of boron, hydrogen telluride, and hydrogen phosphide (or phosphine) [$PH_3$], and isolation of the metals sodium and potassium from their compounds by means of electricity (1807); (5) analysis of the alkaline earth metals and isolation of magnesium, calcium, strontium, and barium (1808); (6) experimental evidence that chlorine is an element and hydrochloric acid free of oxygen (1807); (7) investigations on nitrogen trichloride [$NCl_3$] and its detonating properties (1812–1813); (8) experiments on the combustion of the diamond (1814–1818); (9) researches on the preservation of metals by electrochemical means (1824–1825); and (10) studies on the nature of the electrical action of living animals (1829) such as the organs of the electric ray [*Torpedinidae*] and the electric eel [*Gymnotidae*].

Throughout his life DAVY was also interested in geology and also studied volcanism using Vesuvius as an example (1827). He suggested that volcanoes might have a core of molten alkali metal, acted on by water to cause eruption. But an analysis of lava did not confirm his hypothesis, and he eventually dropped it.

Spurred by a number of disasters in coal mines, he turned his attention to the problem of explosive firedamp (1813). Analyzing gas samples from English coal mines, he concluded that methane is the main constituent and confirmed previous observations that this gas could be ignited only at a high temperature. He constructed a lamp in which the air intake and the chimney were exposed to narrow tubes and found that they did not explode firedamp. Later he found that wire gauze surrounding the flame was equally efficient – his famous *Davy lamp* (1815). He did not ask for a patent and in 1816 wrote, "I never thought of such a thing, my sole object was to serve the cause of humanity, and if I succeeded I am amply rewarded in the gratifying of having done so." He was the first to observe the rate at which an explosion of gases was propagated in a tube, and he also made the first rough experiment on the temperature reached in a gaseous explosion (1816).

DAVY's safety lamp proved extremely useful. From miners he received numerous letters of thanks and from the Association of Coal Owners a service of plates valued at 2,500 pounds, which was eventually sold to found the *Davy Medal* – an award conferred annually by the Royal Society of London, "for an outstanding important recent discovery in any branch of chemistry." When first awarded in 1877, the medal was jointly presented to Robert BUNSEN and Gustav KIRCHHOFF for their researches and discoveries in spectrum analysis.

DAVY was knighted (1812) and made a baronet (1818). He was admitted to be a Fellow of the Royal Society of London (1803), elected secretary (1807), and later became its president (1820–1827), then succeeding the physician and scientist William H. WOLLASTON. After spending the winter of 1828/1829 in Italy inquiring into volcanic action, he had a stroke that resulted in paralysis on his right side and during his return to England died in Geneva, Switzerland. DAVY received the Copley Medal (1805) for his researches on voltaic cells, tanning, and mineral analysis, the Napoleon Prize for his researches on electrolysis (1807), the Rumford Medals (gold and silver) for his researches on flame (1816), and the Royal Medal for his ideas on electrochemistry (1827). DAVY was a cofounder of the Geological Society of London (founded in 1807), the oldest association of its kind in the world.

Astronomers named a crater on the near side of the Moon after him.

**ORIGINAL WORKS.** *Researches, chemical and philosophical, chiefly concerning nitrous oxide and its respiration.* 2 vols, Biggs & Cottle, Bristol (1800) — *Observations on the appearances produced by the collision of steel with hard bodies.* J. Roy. Inst. **50**, 264-268 (1802) — *On a new detonating compound.* Phil. Trans. Roy. Soc. Lond. **103**, 1-7 (1813); *Some further observations on a new detonating substance.* Ibid. **103**, 242-251 (1813) — *Some experiments on the combustion of the diamond and other carbonaceous substances.* Ibid. **104**, 557-570 (1814) — *(I) On the fire-damp of coal mines, and methods of lighting the mines so as to prevent its explosion.* Ibid. **106**, 1-22 (1816) — *(II) An account of an invention for giving light in explosive mixtures of Firedamp in coal mines, by consuming the Fire-damp.* Ibid. **106**, 23-24 (1816) — *A few additional practical observations on the wire-gauze safety lamps for miners.* Phil. Mag. **48** [I], 51-59 (1816) — *Suggestions arising from inspections of wire-gauze lamps, in their working state.* Ibid. **48** [I], 197-200 (1816) — *Some researches on flames.* Phil. Trans. Roy. Soc. Lond. **107**, 45-76 (1817) — *Some new experiments and observations on the combustion of gaseous mixtures, with an account of a method of preserving a continued light in mixtures of inflammable gases and air without flame.* Ibid. **107**, 77-85 (1817) — *On the safety lamp for coal miners, with some researches on flame.* Hunter, London (1818) — *Combustion of the diamond.* Quart. J. Sci. **IV**, 155 (1818) — *On the phenomena of volcanoes.* Phil. Trans. Roy. Soc. Lond. **118**, 241-250 (1828) — *An account of some experiments on the torpedo.* Ibid. **119**, 15-18 (1829) — (J. DAVY, ed.) *The collected works of Sir Humphry DAVY, Bart.* Smith, Elder, London (1839/1840) — J.Z. FULLMER: *Sir Humphry DAVY's published works.* Harvard University Press, Cambridge, MA (1969).

**SECONDARY LITERATURE.** J.A. PARIS: *The life of Sir Humphry DAVY.* Colburn & Bentley, London (1831) — J. DAVY: *Memoirs of the life of Sir Humphry DAVY.* 2 vols., Longman, London (1836) — H.B. DIXON: *VIII. On the movements of the flame in the explosion of gases.* Phil. Trans. Roy. Soc. Lond. **A200**, 315-352 (1903) — W. PRANDTL: *Humphry DAVY, Jöns Jacob BERZELIUS: zwei führende Chemiker aus der ersten Hälfte des 19. Jahrhunderts.* Wiss. Verlags-Gesell., Stuttgart (1948) — G. WILSON: *The life of the Honourable Henry CAVENDISH.* Cavendish Society, London (1951) — R. SIEGFRIED: *Sir Humphry DAVY on the nature of the diamond.* Isis **57**, 325-335 (1966) — H. HARTLEY: *Humphry DAVY.* Nelson, London (1966) — D.M. KNIGHT: *H. DAVY.* In: (C.C. GILLESPIE, ed.) *Dictionary of scientific biography.* Scribner, New York, vol. 3 (1971), pp. 598-604 — F.W. GIBBS: *DAVY, Sir Humphry.* In: *The new Encyclopædia Britannica, Macropædia.* Benton & Hemingway, Chicago (1974), vol. 5, pp. 523-524 — R. SIEGFRIED and R.H. DOTT JR.: *Humphry DAVY on geology: the 1805 Lectures for the general audience.* University of Wisconsin Press, Madison (1980) — B. BOWERS and L. SYMONS (eds.) *Curiosity perfectly satisfied: FARADAY's travels in Europe, 1813–1815.* Peregrinus, London (1991) — D.M. KNIGHT: *Humphry DAVY: science & power.* Blackwell, Oxford (1992) — C. JUNGNICKEL and R. MCCORMMACH: *CAVENDISH, the experimental life.* Bucknell, Cranbury, NJ (1999) — D. KNIGHT: *DAVY, Sir Humphry, baronet.* In: (H.C.G. MATTHEW, ed.) *Oxford dictionary of national biography.* Oxford University Press (2004), vol. 15, pp. 506-512.

**PICTURE.** Engraving after a painting by James LONSDALE (after 1800). In the public domain, taken from Wikipedia, the free encyclopedia; http://en.wikipedia.org/wiki/Image:Sir_Humphry_Davy.jpg.

**NOTE.** *The Royal Society: The Davy Medal (1877). The Davy archive winners 1899–2005;* http://www.royalsoc.ac.uk/page.asp?tip=1&id=1754.

# DIXON, Harold Baily (1852–1930)

• British chemist, founder of the Manchester School of Combustion Research

Harold B. DIXON was born in London to William Hepworth DIXON, a traveler, historical writer, and editor of the *Athenaeum,* a literary magazine published in London (1828–1921). Initially he intended to follow a literary career and entered Christ Church College at Oxford (1870); however, he transferred from classical studies to natural science and

after graduation was elected to a fellowship at Trinity College, Cambridge. It was at the instigation of A. Vernon HARCOURT, a reader in chemistry at Christ Church, that DIXON commenced studying chemistry (1873) and later, in particular, gaseous explosions (1876). He was a lecturer at Balliol College and Trinity College, Cambridge, and in 1886 became professor of chemistry at Owens College of Victoria University, Manchester, where he stayed until his retirement. There he founded his Manchester School of Combustion Research. His research assistants were Harry Wood SMITH and George S. TURPIN (1888), then William Arthur BONE and Bevan LEAN (1891), Edward Halford STRANGE and Edward GRAHAM (1894), Edward John RUSSELL (1896), and later R.H. JONES and L. BOWER, all of whom collaborated with him during the period 1888–1903 in experimental work on the ignition and detonation properties of gaseous mixtures. In addition, he had a number of students working on this subject along the same lines as he – the most prominent being David Leonard CHAPMAN {⇨1899}, cofounder of the theory of detonation, and Colin CAMPBELL, codiscoverer of "spinning detonation" {CAMPBELL and WOODHEAD ⇨1926/1927}.

DIXON's combustion research followed three principal lines. (1) In his early Oxford researches on gaseous explosions he demonstrated that purified and dried explosive gases could not be ignited by an electric spark but detonated when a small amount of water was added (1877–1880), thus proving – in opposition to the German chemist Robert W. BUNSEN {⇨1853} – the validity of BERTHOLLET's Law of Mass Action for chemical explosion. (2) In 1880 he began studying the rate of gaseous explosions, instigated by Prof. Vernon HARCOURT on the occasion of a disastrous explosion of a gas main line that happened in London in the same year. He photographed the flame movements in an explosion using a streak method very similar to that applied by the French chemists François Ernest MALLARD and Henri Louis LE CHÂTELIER (1880–1883) and the German physicists Arthur VON OETTINGEN and Arnold VON GERNET (1888), but refining the technique and improving the resolution (1890–1900). (3) DIXON discovered the backward traveling "retonation wave" and also observed "reflection waves" arising when a detonation wave is either reflected at the closed end of a tube or on passing a restriction in it. (4) He first determined the ignition temperatures of explosive gaseous media, particularly when they contain small amounts of impurities (1903–1930). Thus he found that the presence of small quantities of oxide of nitrogen lowers, whereas that of iodine vapors materially raises, the ignition temperatures.

DIXON was elected president of the Manchester Literary and Philosophical Society (1907–1909) and the Chemical Society of London (1909–1911). In his last paper published in 1929 in the journal *Nature* (London), one year before his death, DIXON proudly wrote: "If the highest reward a teacher can reach is to start a school which will carry on his lines of research, improving his technique, extending his data and enlarging his horizon, I may well claim that my lines have fallen on pleasant and fruitful places."

**ORIGINAL WORKS.** *On the influence of water on the union of carbonic oxide with oxygen at high temperature.* Rept. Meet. Brit. Assoc. **50**, 503-504 (1880) — *The velocity of explosion of a mixture of carbonic oxide and oxygen with varying quantities of aqueous vapor.* Ibid. **52**, 487 (1882) — *The influence of aqueous vapor on the explosion of carbonic oxide and oxygen.* Chemical News **46**, 151 (1882) — *On the incomplete combustion of gases.* Rept. Meet. Brit. Assoc. **54**, 671-672 (1884) — *Conditions of chemical change in gases: hydrogen, carbonic oxide, and oxygen.* Phil. Trans. Roy. Soc. Lond. **175**, 617-684 (1885) — *The rate of explosion of hydrogen and oxygen.* Rept. Meet. Brit. Assoc. **55**, 905 (1885) — *The combustion of cyanogens.* J. Chem. Soc. (Lond.) **49**, 94-112 (1886) — *On the union of hydrogen and nitrogen.* Mem. Proc. Manch. Lit. Phil. Soc. **31**, 91-92 (1888) — With H.W. SMITH: *The incompleteness of combustion in explosions.* Rept. Meet. Brit. Assoc. **58**, 632 (1888) — (Ed.) *The explosion of a mixture of hydrogen, chlorine and oxygen.* Nature **40**, 587-588 (1889) — *The mode of observing the phenomena of earthquakes.* Ibid. **42**, 491 (1890) — *The rate of explosion in gases.* Trans. Am. Inst. Min. Eng. **3**, 312-321 (1892); Proc. Roy. Inst. **13**, 443-450 (1893); Phil. Trans. Roy. Soc. Lond. **A184**, 97-188 (1894) — (Ed.) *Explosions in mines, with special reference to the dust theory.* Nature **48**, 530 (1893) — With J.C. CAIN: *On the instantaneous pressures produced in the explosion-wave.* Mem. Proc. Manch. Lit. Phil. Soc. **38**, 174-180 (1894) — *The explosion of gases in glass vessels.* Nature **51**, 151-152 (1894) — (Ed.) *Gaseous explosions.* Ibid. **56**, 463 (1897) — *VIII. On the movements of the flame in the explosion of gases.* Phil. Trans. Roy. Soc. Lond. **A200**, 315-352 (1903) — *Über Explosionswellen.* Ber. Dt. Chem. Gesell. **38**, 2419-2446 (1905) — With L. BRADSHAW: *On the explosion of pure electrolytic gas.* Proc. Roy. Soc. Lond. **A79**, 234-235 (1907) — With L. BRADSHAW: *Über die Explosion von reinem Knallgas.* Z. Phys. Chem. **61**, 373-375 (1908) — With H.M. LOVE: *Experiments on ABEL's theory that incombustible dusts act catalytically in igniting weak mixtures of methane and air.* Mem. Proc. Manch. Lit. Phil. Soc. **57**, No. 15 (1913/1914) — With C. CAMPBELL and W. SLATER: *Photographic analysis of explosions in the magnetic field.* Proc. Roy. Soc. Lond. **A90**, 506-511 (1914) — *Berthelot Memorial Lecture.* In: (Brit. Chem. Soc., ed.) *Memorial lectures delivered before the Chemical Society 1901-1913.* Gurney & Jackson, London, vol. II (1914), pp. 167-185 — With C. CAMPBELL and A. PARKER: *On the velocity of sound in gases at high temperatures, and the ratio of the specific heats.* Proc. Roy. Soc. Lond. **A100**, 1-26 (1921) — With N.S. WALLS: *On the propagation of the explosion wave. I. Hydrogen and carbon monoxides.* J. Chem. Soc. (Lond.) **123**, 1025-1037 (1923) — *The movement of flame in carbonic oxide-oxygen explosions.* Nature **124**, 580-584 (1929).

**SECONDARY LITERATURE.** J.F. THORPE: *Presidential address* [delivered March 27, 1930]. J. Chem. Soc. **133**, Part I, 872-887 (1930) — W.A. BONE: *Prof. H.B. DIXON.* Chem. Ind. Rev. **49**, 805-806 (1930) — A. LAPWORTH: *Prof. H.B. DIXON.* Ibid. **49**, 826 (1930) — G.J. FOWLER: *Prof. H.B. DIXON* [Obituary notice]. Nature **126**, 958 (1930) — H.B. BAKER and W.A. BONE: *Harold Baily DIXON 1852–1930.* J. Chem. Soc. (Lond.) **134**, Part II, 3349-3368 (1931); Proc. Roy. Soc. Lond. **A134**, i-xvii (1932) — A.B. COSTA: *H.B. DIXON.* In: (C.C. GILLESPIE, ed.) *Dictionary of scientific biography.* Scribner, New York, vol. 4 (1971), p. 130.

**PICTURE.** Courtesy Library and Information Centre, Royal Society of Chemistry, Cambridge, U.K.

## DOPPLER, Christian Andreas (1803–1853)

- Austrian physicist

Christian A. DOPPLER, son of noted stonemason Johann E. DOPPLER, was born in Salzburg, Austro-Hungarian Empire. His poor health, however, did not allow him to follow his father's profession. Because of his great talent in mathematics he attended the Polytechnic Institute in Vienna (1822–1825), where he was employed later as a mathematical assistant (1829–1833). In Prague he became professor of mathematics and accounting at the State Secondary School (1835), followed by a professorship of elementary mathematics and geometry at the Technical State Academy (1841). For a short time, he stayed at the Mining Academy of Schemnitz (now Banská Štiavnica, Slovakia) as *Bergrat* and professor of mathematics, physics, and mechanics (1847–1850). Forced by political turbulence, he returned to Vienna and became the first director of the new Physical Institute (1850–1853) and full professor of experimental physics at the Royal Imperial University of Vienna. Suffering from a lung disease since his period at Prague, he traveled in November 1852 to Venice in the hope that the warmer climate would bring about some improvement. He died there in March 1853 at the age of 49.

His scientific fame is based on his *Doppler effect* (1842), an apparent shift in the frequency of waves received by an observer, depending on the relative motion between the observer and the source of waves. DOPPLER, trying to explain the colored light of double stars, correctly recognized the importance of his *Doppler principle*, in acoustics, optics, and astronomy. The Dutch meteorologist Christophorus BUYS-BALLOT first proved the correctness of the Doppler effect in a material medium by acoustic experiments. From the red shift of spectroscopic lines in the light from the stars, the British astronomer William HUGGINS first determined the velocities of these stars relative to our Sun (1868). More than 60 years later, the U.S. astronomer Edwin P. HUBBLE, based on the Doppler effect, concluded that galaxies are receding from us with relative velocities that increase in proportion to the distance (1929).

To illustrate the consequences of his principle in acoustics, DOPPLER assumed explosion-like disturbances moving on a straight or curved line with a velocity slower, equal to, or faster than the sound velocity of the surrounding medium (1846). For disturbances propagating along a straight line with a constant supersonic velocity $u$ in a medium of sound velocity $a$, DOPPLER, using HUYGENS' principle of wave propagation, constructed the envelopes of all elementary disturbances, which resulted in a straight cone geometry with a half cone angle $\alpha = \arcsin a/u$. His supersonic wave model, later taken up by Ernst MACH to explain his head wave phenomenon of supersonic shots (1887), was named *Mach cone* by 20th-century scientists, although credit should have been given to DOPPLER rather than to MACH. In 1846, DOPPLER, using HUYGENS' principle of constructing wave fronts by sources of secondary spherical wavelets, treated also two cases of disturbances moving with accelerated and decelerated supersonic velocities along a straight line, which resulted in a concave and convex cone geometry, respectively. DOPPLER's third example describes a supersonic disturbance moving on a circle that, seen from the center of rotation, results in a concave shaped head wave geometry. This case became of particular interest for aircraft designers at the end of World War I when propeller tips began to exceed sound velocity.

Other examples of applications of the Doppler effect are in high-speed diagnostics (*e.g.*, laser anemometry, laser velocimetry), meteorology, radar, navigation, medical ultrasonic diagnostics, and Mößbauer-effect studies.

The *Christian-Doppler-Fonds* (Christian-Doppler Foundation) and the *Doppler-Institut für medizinische Wissenschaft und Technologie* (Christian-Doppler-Institute for Medical Science and Technology), both located in Salzburg, Austria, support scientific research and publications related to the Doppler effect.

A crater on the far side of the Moon and a minor planet (asteroid 3905 DOPPLER) are named for him.

**ORIGINAL WORKS.** Über das farbige Lichtes der Doppelsterne und einiger anderer Gestirne des Himmels. Abhandl. Königl. Böhm. Gesell. Wiss. (Prag) **2** [V], 465-482 (1841/1842) — Über eine bei jeder Rotation des Fortpflanzungsmittels eintretende eigenthümliche Ablenkung der Licht- und Schallstrahlen. Ibid. **3** [V], 417-430 (1843/1844) — Über ein Mittel, periodische Bewegungen von ungemeiner Schnelligkeit noch wahrnehmbar zu machen und zu bestimmen. Ibid. **3** [V], 779-782 (1843/1844) — Drei Abhandlungen aus dem Gebiete der Wellenlehre, nebst Anwendungen aus Acustik, Optik und Astronomie. Ibid. **4** [V], 497-523 (1845/1846) — Beiträge zur Fixsternkunde. Ibid. **4** [V], 621-646 (1845/1846) — Bemerkungen zu meiner Theorie des farbigen Lichts der Doppelsterne, &c. mit vorzüglicher Rücksicht auf die von Hrn. Dr. BALLOT zu Utrecht dagegen erhobenen Bedenken. Ann. Phys. **68** [II], 1-34 (1846) — Über den Einfluß der Bewegung des Fortpflanzungsmittels auf die Erscheinungen der Äther-, Luft- und Wasserwellen. Abhandl. Königl. Böhm. Gesell. Wiss. (Prag) **5** [V], 293-306 (1848) — Über ein Mittel, die Brechung der Schallstrahlen experimentell nachzuweisen und numerisch zu bestimmen. Sitzungsber. Akad. Wiss. Wien **2** (Abth. II), 322-329 (1849) — Einige weitere Mittheilungen und Bemerkungen meiner Theorie des farbigen Lichtes der Doppelsterne betreffend. Ann. Phys. **81** [II], 270-275 (1850) — Über den Einfluss der Bewegung auf die Intensität der Töne. Ibid. **84** [II], 262-266 (1851) — Weitere Mittheilungen meiner Theorie des farbigen Lichtes. Ibid. **85** [II], 371-378 (1852).

**SECONDARY LITERATURE.** B. BOLZANO: Christian DOPPLER's neueste Leistungen auf dem Gebiete der physikalischen Apparatelehre, Akustik, Optik und optischen Astronomie. Ann. Phys. **72** [II], 530-555 (1847) — A. SCHRÖTTER: Bericht des Sekretärs [Nachruf auf DOPPLER]. Almanach Kaiserl. Akad. Wiss. **4**, 112-120 (1854) — J. SCHEINER: Johann Christian DOPPLER und das nach ihm benannte Prinzip. Himmel und Erde **8**, 260-271 (1896) — A.E. WOODRUFF: J.C. DOPPLER. In: (C.C. GILLESPIE, ed.) Dictionary of scientific biography. Scribner, New York, vol. 4 (1971), pp. 167-168 — G. JÄGER: Christian DOPPLER. In: (A. BETTELHEIM, ed.) Neue österreichische Biographie: 1815-1948. Wiener Drucke, Wien, Bd. 3 (1926), pp. 72-81 — A. EDEN: Christian DOPPLER – thinker and benefactor. Christian-Doppler-Institut for Medical Science and Technology, Salzburg (1988) — H. GRÖSSING and K. KADLETZ: Christian DOPPLER (1803-1853), Band I, 1. Teil: Wissenschaft, Leben, Umwelt, Gesellschaft. 2. Teil: Quellenanhang; P. SCHUSTER: Christian DOPPLER (1803-1853), Band II, 3. Teil: Das Werk. Böhlau, Wien etc. (1992) — I. STOLL ET AL.: The phenomenon of Doppler. The Czech Technical University, Faculty of Nuclear Science & Physical Engineering, Prague (1992) — A. EDEN: The search for Christian DOPPLER. Springer, Vienna (1992) — J. DORSCHNER: 150 Jahre Doppler-Effekt. Die Sterne **69**, 311-312 (1993) — P.M. SCHUSTER: Weltbewegend und doch unbekannt. Zum 200. Geburtstag von Christian Andreas DOPPLER. Physik J. **2**, Nr. 10, 47-51 (2003); Weltbewegend – unbekannt. Leben und Werk des Physikers Christian DOPPLER und die Welt danach. Living Edn., Pöllauberg (2003); Moving the stars – Christian DOPPLER: his life, his works and principle, and the world after. Living Edn., Pöllauberg (2005).

**PICTURE.** This daguerreotype, taken in 1844, is the only one of DOPPLER that has survived. Courtesy Dr. Peter M. SCHUSTER (Pöllauberg, Austria), who discovered it in the personal property of Mrs. Dorothea MERSTALLINGER, DOPPLER's great-granddaughter. The picture is now kept at the Archives of the Christian Doppler Foundation in Salzburg, Austria.

**NOTE.** In some biographies DOPPLER's first names were given incorrectly as Johann Christian rather than Christian Andreas; see SCHUSTER's book Moving the stars – Christian DOPPLER (cf. above), pp. 18-19.

## DÖRING, Werner Siegfried (1911–)

- German theoretical physicist, cofounder of modern detonation theory

Werner S. DÖRING was born in Berlin-Tegel to chief engineer Gebhard DÖRING and studied technical physics at the TH Stuttgart (1930–1932) and the TH Berlin (1933–1935). Under the supervision of Prof. Richard A. BECKER, a theoretical physicist at the TH Berlin, he received his Ph.D. with an experimental study on the temperature dependency of magnetostriction in nickel (1936). He wrote his Habilitationsschrift (habilitation thesis) on reversible processes in magnetic materials with small inner stress which qualified him for being admitted as a university professor. In 1939, he became Privatdozent (university lecturer) at the University of Göttingen and in 1942 associate professor at the Reichsuniversität Posen. After World War II, he served as lecturer (Diätendozentur) of theoretical physics at the TH Braunschweig (1946–1949). In this period he also did research for a short time in France at the ISL, Saint-Louis, and in Switzerland at the Zurich IBM Research Laboratories. Thereafter, he became full professor and taught theoretical physics at the University of Gießen (1949–1963). In 1963, he was called to the chair of the Theoretical Physics Department at the University of Hamburg, where he stayed until his retirement (1976).

His work contributed mainly to three fields: He dedicated much of his research efforts to the study of magnetic properties of matter, particularly of ferromagnetism and micro magnetism. In 1941, at a secret workshop held in Berlin, DÖRING first reported on a theory of how to determine the pressure distribution behind a detonating explosive, thereby developing the concept that a detonation wave can be described as a shock wave immediately followed by a flame, whereby the reaction rate is finite, a hypothesis suggested independently by Yakov B. ZEL'DOVICH (1940) in the Soviet Union and John VON NEUMANN (1942) in the United States. This so-called "Zel'dovich-von Neumann-Döring (ZND) model" – actually named correctly the ZDN model – stimulated numerous researchers to study the influence of finite reaction rates on the structure of a 1-D detonation wave in a compressible medium. His interests in ferromagnetism and detonation were already stimulated in the late 1930s by his previous cooperation with Prof. R.A. BECKER. DÖRING also developed a theory of germ formation in supersaturated phases.

Besides teaching physics to students, DÖRING dedicated much of his time to training physics teachers for teaching at the high-school level. He wrote several textbooks on special fields of physics such as the electromagnetic field (1955), thermodynamics (1956), statistical mechanics (1957), theoretical physics (1960), ferromagnetism (1966), atomic physics, and quantum mechanics (1973) and is the author of several related handbook articles.

**ORIGINAL WORKS.** With R.A. BECKER: *Kinetische Behandlung der Keimbildung in übersättigten Dämpfen.* Ann. Phys. **24** [V], 719-752 (1935); *Ferromagnetismus.* Springer, Berlin (1939) — *Grundlegende Betrachtungen über den Detonationsvorgang.* Berichte des Ballistischen Instituts der Technischen Akademie der Luftwaffe (TAL, Berlin) **1a** (1941) — *Zur Theorie der Detonation.* Ibid. **5a** (1941) — *Der Druckverlauf in den Schwaden und im umgebenden Medium bei der Detonation.* In: *Probleme der Detonation.* Schriften Dt. Akad. Luftfahrtforsch. Nr. 1023/41G, 31-45 (1941) — *Über den Detonationsvorgang in Gasen.* Ann. Phys. **43** [V], 421-436 (1943) — With G. BURKHARDT: *Beiträge zur Theorie der Detonation: (I) Die Stoßwelle in verschiedenen Medien,* pp. 1-62; *(II) Die Stoßwelle in Luft unter Berücksichtigung der Dissoziations- und Ionisationsvorgänge,* pp. 63-80; *(III) Die Stoßwelle in Wasser,* pp. 81-101; *(IV) Das Verhalten einer Stoßwelle an Materiegrenzen,* pp. 103-154; *(V) Die Detonationswelle,* pp. 155-196; *(VI) Die Druckverteilung in den Schwaden,* pp. 197-220; *(VII) Die Anfangsintensität der Stoßwellen in der Umgebung eines detonierenden Körpers,* pp. 221-236; *(VIII) Die Druckverteilung in der Umgebung eines detonierenden Sprengstoffes,* pp. 237-277; Anhang: *Die Charakteristikenmethode zur numerischen Ermittlung der Druckverteilung in nichtstationären, eindimensionalen Strömungen,* pp. 278-311. Zentrale für wiss. Berichtswesen (ZWB) der Luftfahrtforschung des Generalluftzeugmeisters, Berlin-Adlershof, Forschungsbericht Nr. 1939 (1944); (Engl. translation by Brown University) *Contributions to the theory of detonation.* Tech. Rept. No. F-TS-1227-IA (GDAM A9-T-45), Wright-Patterson Air Force Base, Dayton, OH (1949) — *Die Geschwindigkeit und Struktur von intensiven Stoßwellen in Gasen.* Ann. Phys. **5** [VI], 134-150 (1949) — With G. SCHÖN: *Über die Detonationsgeschwindigkeit des Methans und Dizyans im Gemisch mit Sauerstoff und Stickstoff.* Z. Elektrochem. Angew. Phys. Chem. **54**, 231-239 (1950) — With H. SCHARDIN: *Detonationen.* In: *FIAT Review of German Science 1939–1946.* Erich Verlag, Wiesbaden (1948); *see* vol. 1.5: (A. BETZ, ed.) *Hydro- and aerodynamics,* pp. 97-125 — *Einführung in die Quantenmechanik.* Vandenhoeck & Ruprecht, Göttingen (1962) — *Atomphysik und Quantenmechanik.* De Gruyter, Berlin (1973) — *Modernisierung des Physikunterrichts.* Physik Didaktik **3**, 239-244 (1974).
**SECONDARY LITERATURE.** G. SIMON and V. ZEHLER: *Werner DÖRING 70 Jahre.* Physik. Blätter **37**, 300 (1981) — H. SCHMIDT: *Werner DÖRING 75 Jahre alt.* Universität Hamburg-Harburg: Berichte und Meinungen aus der Universität Hamburg **17**, Nr. 5, 80-81 (1986).
**PICTURE.** Courtesy Archives of the University of Hamburg, Germany.
**NOTE.** DÖRING's classic report "*Beiträge zur Theorie der Detonation*" [Forschungsbericht Nr. 1939 (1944)], in which he specified his detonation theory, was classified during World War II and distributed only in very limited numbers. For example, one copy is kept at the Archives of the Deutsches Museum, Munich. DÖRING's personal copy, transferred by him to the author in Nov. 2002, is now kept at EMI library, Freiburg.

# DRYDEN, Hugh Latimer (1898–1965)

- U.S. physicist and aerodynamicist, and administrator

Hugh L. DRYDEN was born in Pocomoke City, MD, to Samuel Isaac DRYDEN, a schoolteacher who ran a general store. He obtained his early education in the Baltimore public schools. He entered Johns Hopkins University, Baltimore, MD at the early age of 15. While still working on his Ph.D. under the supervision of Joseph S. AMES, a chairman of NACA, he began his aeronautical research at the newly installed wind tunnel facility of the National Bureau of Standards (NBS) (1918–1944). After receiving his Ph.D. in physics, with a dissertation entitled *Air Forces on Circular Cylinders* (1919), he became technical director of the aerodynamics division at NBS (1920), where he began his research on boundary layer control and the origin of turbulence. He also developed methods of accurately measuring turbulence in wind tunnels, contributed to the design of wind tunnels with very low turbulence, and studied the influence of turbulence on aerodynamic forces imposed upon models tested in wind tunnels at high speeds. During World War II, DRYDEN and his associates Galen B. SCHUBAUER and Harold K. SKRAMSTAD successfully verified experimentally previous theoretical predictions on the onset of instability made by Hermann SCHLICHTING and Walter TOLLMIEN, who had observed in the early 1930s that the transition to turbulence is initiated and continuously generated at a multiplicity of spots on the surface of an object exposed to a high-speed flow at the upstream edge. During World War II, he also headed the development of the Bat air-o-surface missile, which earned him the Presidential Certificate of Merit.

In 1947, he became director of aeronautical research of the National Advisory Committee on Aeronautics (NACA). Two years later he was appointed director of NACA, the highest career position of this agency, which he held until 1958 when he became deputy administrator of the National Aeronautics and Space Administration (NASA). In the period 1954 to 1957, he was chairman of the Air Force-Navy-NACA Research Airplane Committee, which was installed in order to supervise the design and production of the X-15. This hypersonic plane, which shortly before DRYDEN's death was the first to reach Mach 6 and an altitude of nearly 70 miles (112 km), provided most worthwhile full-scale flight data at hypersonic speeds and high altitudes for future spacecraft. DRYDEN also played a key role in other aircraft testing, such as the X-1, D-558, X-3, X-4, X-5, and the XB-70.

In 1958, he was named deputy administrator of NASA and participated in the organization of the projects of manned space flights of Mercury, Gemini, and Apollo and in the decision to mount a lunar exploration mission. In 1959, he was appointed one of two men to assist Ambassador Henry Cabot LODGE JR. at the first meeting of the United Nations Committee on the Peaceful Uses of Outer Space (1959). Together with Theodore VON KÁRMÁN he edited the series *Advances in Applied Mechanics* (vols. IV–IX, 1956–1966), published by Academic Press, which contains review articles of internationally renowned experts in the field of fluid mechanics and shock waves. DRYDEN's numerous contributions to aeronautics as well as his great ability in organizing international cooperation in aeronautical and aerospace research were acknowledged by many national and international organizations, such as by awarding him 16 honorary degrees. DRYDEN was also recipient of the Daniel Guggenheim Medal (1950), and the Robert H. Goddard Memorial Trophy (1964) which he received together with James E. WEBB for "representing the Gemini Program teams which significantly advanced human experience in space flight," and the National Medal of Science (1966). The greatest honor came posthumously in 1976 when NASA renamed its Flight Research Center in Edwards, CA the *NASA Hugh L. Dryden Flight Research Center (NASA-DFRC)*. The *Dryden Flow Visualization Facility* at NASA-DFRC, a water tunnel facility that became operational in 1983, is used primarily as a low-cost diagnostic tool for visualizing and analyzing vortical flows on aircraft and other shapes at high incident angles. The *AIAA Dryden Lectureship in Research* was named in honor of him in 1967, succeeding the Research Award established in 1960. The lecture emphasizes the great importance of basic research to advancements in aeronautics and astronautics and is a salute to research scientists and engineers.

A crater on the far side of the Moon is also named for him.

**ORIGINAL WORKS.** With L.J. BRIGGS and G.F. HULL: *Aerodynamic characteristics of airfoils at high speeds*. NACA-TR 207 (1925); *Pressure distribution over airfoils at high speeds*. NACA-TR 255 (1927); *Aerodynamic characteristics of twenty-four airfoils at high speeds*. NACA-TR 319 (1929); *Aerodynamic characteristics of circular-arc airfoils at high speeds*. NACA-TR 365 (1930); *The effect of compressibility on the characteristics of airfoils*. In: (C.W. OSEEN, ed.) *Proc. 3rd Int. Congress on Applied Mechanics*. Sveriges Litografiska Tryckerier, Stockholm (Aug. 1930), pp. 417-422 (1930) — With F.D. MURNAGHAN and N.H. BATEMAN: *Report of the Committee on hydrodynamics*. NRC Bulletin 84 (1932) — *Computation of the two-dimensional flow in a laminar boundary layer*. NACA-TR 497 (1934) — *Air flow in the boundary layer near a plate*. NACA-TR 562 (1936) — *Turbulence and the boundary layer* [2nd Wright Bros. Lecture]. J. Aeronaut. Sci. **6**, 85-105 (1939) — *Some recent contributions to the study of transition and turbulent boundary layers*. In: *Proc. 6th Int. Congr. on Applied Mechanics* [Paris, Sept. 1946]; NACA-TN 1168 (1947) — With G.B. SCHUBAUER: *The use of damping screens for the reduction of wind tunnel turbulence*. J. Aeronaut. Sci. **14**, 221-228 (1947) — *Recent advances in the mechanics of boundary layer flow*. In: (R. VON MISES and Th. VON KÁRMÁN, eds.) *Advances in applied mechanics*. Academic Press, New York (1948), vol. I, pp. 1-40 — *The dawn of the supersonic age*. U.S. Air Services (June 1948), pp. 11-14, 24 — *Faster than sound*. Phys. Today **1**, 6-10 (Oct. 1948) — *The aeronautical research scene – goals, methods, and accomplishments* [37th Wilbur Wright Memorial Lecture]. J. Roy. Aeronaut. Soc. **53**, 623-666 (1949) — With I.H. ABBOTT: *The design of low-turbulence wind tunnels*. NACA-TR 940 (1950) — *Review of published data on the effect of roughness on transition from laminar to turbulent flow*. J. Aeronaut. Sci. **20**, 477-482 (1953) — *Supersonic travel within the last two hundred years*. Sci. Month. **78**, 289-295 (May 1954) — With F.D. MURNAGHAN and N.H. BATEMAN: *Hydrodynamics*. Dover Publ., New York (1956) — *Scientists intensify study of thermal barrier*. The Legion Air Rev. **7**, No. 8, 1-3 (16 Aug. 1956) — *Contribution to the subject "Transition."* In: *Symposium Grenzschichtforschung* [Freiburg, Germany, Aug. 1957]. Springer, Berlin (1958), pp. 140-141 — *Gegenwartsprobleme der Luftfahrtforschung* [Zweite Ludwig-Prandtl-Gedächtnis-Vorlesung, 7. Mai 1958]. Z. Flugwiss. **6**, 217-233 (1958) — *Transition from laminar to turbulent flow*. In: *High speed aerodynamics and jet propulsion*. Princeton University Press, Princeton, NJ; vol. V (1959): (C.C. LIN, ed.) *Turbulent flows and heat transfer*, pp. 3-74 — *Amerikas internationale Zusammenarbeit in der Raumforschung*. Weltraumfahrt **12**, 78-80 (1961) — *Towards the new horizon of tomorrow* [1st Annual ARS von Kármán Lecture]. Astronautics **8**, 14-19 (Jan. 1963) — (R.K. SMITH, ed.) *The Hugh L. DRYDEN papers, 1898–1965* [A preliminary catalogue of the basic collection]. Milton S. Eisenhower Library, Johns Hopkins University, Baltimore (1974).

**SECONDARY LITERATURE.** *Dr. Hugh DRYDEN of NASA is dead*. New York Times (Dec. 3, 1965) — H. SCHLICHTING: *Hugh L. DRYDEN †*. Jb. Wiss. Gesell. Luft- u. Raumfahrt (1965), p. 552 — W.J. COUGHLIN: *The man from Pocomoke City*. Missiles Rockets **17**, 46 (1965) — (Ed.) *Hugh Latimer DRYDEN 1898–1965*. Astronaut. Acta **12**, 192 (1966) — R.L. BISPLINGHOFF: *Hugh Latimer DRYDEN*. Appl. Mech. Rev. **19**, 1-5 (1966); Phys. Today **19**, 153 (Jan. 1966); Astronaut. Acta **12**, 192 (1966); Z. Flugwiss. **14**, 542-543 (1966) — S. THOMAS: *Hugh DRYDEN*. In: *Men of space: profiles of the leaders in space research, development, and exploration*. (Chilton, Philadelphia) **2**, 65 (1961) — J.C. HUNSAKER and R.C. SEAMANS JR.: *Hugh Latimer DRYDEN* [with comprehensive bibliography]. Biogr. Mem. Natl. Acad. Sci. (U.S.A.) **XL**, 35-68 (1969) — R.K. SMITH: *Biographical sketch*. In: (R.K. SMITH, ed.) *The Hugh L. DRYDEN papers*. Milton S. Eisenhower Library, Johns Hopkins University, Baltimore (1974), pp. 19-33 — (Ed.) *Buchbesprechungen* (R.K. SMITH: *The DRYDEN papers*). Z. Flugwiss. **23**, 336 (1975) — E.M. EMME: *Astronautical biography: Hugh Latimer DRYDEN*. J. Astronaut. Sci. **25**, 151-171 (1977) — *Hugh Latimer DRYDEN 1898–1965*. Mem. Tributes: Natl. Acad. Engng **1**, 32-42 (1979) — M.H. GORN: *Hugh L. DRYDEN's career in aviation and space*. Monographs in Aerospace History No. 5, NASA History Office Code ZH, NASA Headquarters, Washington, DC (1996); http://www.dfrc.nasa.gov/History/Publications/PDF/Dryden.pdf.

**PICTURE.** Taken on October 28, 1958. Courtesy NASA Dryden Flight Research Center, Edwards, CA (image No. E4248).

# DUHEM, Pierre Maurice Marie (1861–1916)

- French physicist and philosopher of science

Pierre M.M. DUHEM was born in Paris to Pierre-Joseph DUHEM, a commercial traveler. After schooling at the Collège Stanislas in Paris (1872), where he surprised his teachers by his great talent for mathematics, he studied science at the Ecole Normale Supérieure. In his Ph.D. thesis on thermodynamic potentials in chemistry and physics, following Josiah Willard GIBBS and Hermann VON HELMHOLTZ, he defined the criterion for the spontaneity of chemical reactions in terms of free energy rather than following the conception of maximum work, which had been stated 20 years before by P.E. Marcellin BERTHELOT. This disagreement with BERTHELOT, who was a leading authority of great influence, led to a long-lasting enmity and forced DUHEM to write his thesis on another subject. It also pushed his life into an increasing academic isolation. He rewrote his thesis on magnetism (1886) and taught at the universities of Lille (1887–1893), Rennes (1893–1894), and Bordeaux (1894–1916).

His fields of interest included thermodynamics, electromagnetism, hydrodynamics, and the history and philosophy of science. Today his contributions to shock wave physics are almost forgotten, although laid down in detail in his two books *Hydrodynamique, élasticité, acoustique* (1891) and *Recherches sur l'hydrodynamique* (1903–1904). However, they had an important influence on contemporary physicists and mathematicians, because they called attention to Pierre-Henri HUGONIOT's work on waves of finite longitudinal disturbances. The first book stimulated Jacques HADAMARD in his own work on wave propagation, a colleague at Bordeaux University for a period of 3 years who became his lifelong friend. One of his best-known disciples was Emile JOUGUET. DUHEM was also the first to show that true shock waves could only propagate in perfect fluids, which he called "true Hugoniot waves," while in real fluids only "quasi-shock waves" are possible, as he argued (1904).

His profound scientific output is enormous and comprises 45 books and nearly 400 papers. DUHEM, interested in many fields of physics of his time, was also a positivist like Ernst MACH, relying heavily on historical examples in presenting his philosophy of science. Controversies over scientific and political matters with P.E.M. BERTHELOT and others, however, resulted in a partial suppression and ignorance of his work. Today his numerous contributions, all of which were published in French, are being increasingly rediscovered and acknowledged. However, a detailed discussion of his contributions to shock wave physics from the modern point of view is still pending.

**ORIGINAL WORKS.** *Le potentiel thermodynamique et ses applications à la mécanique chimique et à la théorie des phénomènes électriques.* 2 vols., A. Hermann, Paris (1886) — *Hydrodynamique, élasticité, acoustique.* 2 vols., A. Hermann, Paris (1891) — *Théorie thermodynamique de la viscosité, du frottement et des faux équilibres chimiques.* Mém. Soc. Sci. Phys. Nat. (Bordeaux) **2** [V], 1-208 (1896) — *Traité de mécanique chimique fondée sur la thermodynamique.* 4 vols., Hermann, Paris (1897–1899) — *Sur le théorème d'HUGONIOT et quelques théorèmes analogues.* C. R. Acad. Sci. Paris **131**, 1171-1173 (1900) — *Sur la condition supplémentaire en hydrodynamique.* Ibid. **132**, 117-120 (1901) — *De la propagation des ondes dans les fluides visqueux.* Ibid. **132**, 393-396 (1901) — *Sur les ondes du second ordre par rapport aux vitesses, que peut présenter un fluide visqueux.* Ibid. **132**, 607-610 (1901) — *De la propagation des discontinuités dans un fluide visqueux.* Ibid. **132**, 658-662 (1901) — *De la propagation des discontinuités dans un fluide visqueux. Extension de la loi d'HUGONIOT.* Ibid. **132**, 944-946 (1901) — *Sur les théorèmes d'HUGONIOT, les lemmes de M. HADAMARD et la propagation des ondes dans les fluides visqueux.* Ibid. **132**, 1163-1167 (1901) — *Sur les ondes longitudinales et transversales dans les fluides parfaits.* Ibid. **132**, 1303-1306 (1901) — *Des ondes qui peuvent persister en un fluide visqueux.* Ibid. **133**, 579-580 (1901) — *Sur les ondes-cloisons.* Ibid. **137**, 237-240 (1903) — *L'évolution de la mécanique.* Joanin, Paris (1903). Germ. translation: *Die Wandlungen der Mechanik und der mechanischen Naturerklärung.* Barth, Leipzig (1912); VDM Verlag Dr. Müller, Saarbrücken (2006). Engl. translation: *The evolution of mechanics.* Sijthoff & Noordhoff, Alphen aan den Rijn (1980) — *Recherches sur l'hydrodynamique.* Ann. Fac. Sci. Toulouse **1-3** [II] (1901–1903); 2 vols., Gauthier-Villars, Paris (1903/1904); Service de Documentation & d'Information Technique de l'Aéronautique, Paris (1961) — *Sur l'impossibilité des ondes de choc négatives dans les gaz.* C. R. Acad. Sci. Paris **141**, 811 (1905) — *Sur les quasi-ondes de choc et la distribution des températures en ces quasi-ondes.* Ibid. **142**, 324-327 (1906) — *Quelques lemmes relatifs aux quasi-ondes de choc.* Ibid. **142**, 377-380 (1906) — *Sur une inégalité importante dans l'étude des quasi-ondes de choc.* Ibid. **142**, 491-493 (1906) — *Sur les quasi-ondes de choc au sein des fluides mauvais conducteur de la chaleur.* Ibid. **142**, 612-616 (1906) — *Sur les quasi-ondes de choc au sein des fluides bon conducteur de la chaleur.* Ibid. **142**, 750-752 (1906) — *Recherches sur l'élasticité.* Gauthier-Villars, Paris (1906) — *La théorie physique, son objet et sa structure.* Chevalier & Rivière, Paris (1906, 1914); Vrin, Paris (1981, 1989). Germ. translation: *Ziel und Struktur der physikalischen Theorien* [mit einem Vorwort von Ernst MACH]. Barth, Leipzig (1908); Meiner, Hamburg (1978, 1998). Engl. translation: *The aim and structure of physical theory* [with a foreword by Prince L. DE BROGLIE]. Princeton University Press (1954); Atheneum, New York (1962) — *Sur la propagation des quasi-ondes de choc.* C. R. Acad. Sci. Paris **144**, 179-181 (1907) — *Sur la propagation des ondes de choc au sein des fluides.* Z. Phys. Chem. **69**, 169-186 (1909) — *Traité d'énergétique ou de thermodynamique générale.* Gauthier-Villars, Paris (1911) — *Le système du monde: histoire des doctrines cosmologiques de PLATON a COPERNIC.* 10 vols., Hermann, Paris (1913-1959); Engl.

translation: *Medieval cosmology: theories of infinity, place, time, void, and the plurality of worlds*. University of Chicago Press, Chicago (1985, 1987, 1990).

**SECONDARY LITERATURE.** E. JOUGUET: *L'œuvre scientifique de Pierre DUHEM*. Rev. Gén. Sci. Pures Appl. **28**, 40-49 (1917) — E. PICARD: *La vie et l'œuvre de Pierre DUHEM*. Gauthier-Villars, Paris (1922) — O. MANVILLE: *L'œuvre scientifique de Pierre DUHEM*. Mém. Soc. Sci. Phys. Nat. (Bordeaux) **1** [VII], 437-464 (1927) — J. HADAMARD: *L'œuvre de DUHEM sous son aspect mathématique*. Ibid. **1**, 637-665 (1927) — C. TRUESDELL and R. TOUPIN: *The classical field theories*. In: (S. FLÜGGE, ed.) *Handbuch der Physik*. Springer, Berlin (1960), vol. III/1 — D.G. MILLER: *Pierre DUHEM, ignored intellect*. Phys. Today **19**, 47-53 (Dec. 1966) — D.G. MILLER: *P.M.M. DUHEM*. In: (C.C. GILLESPIE, ed.) *Dictionary of scientific biography*. Scribner, New York, vol. 4 (1971), pp. 225-233 — P. BROUZENG: *DUHEM (1861–1916): un savant, une époque*. Belin, Paris (1987) — S.L. JAKI: *Uneasy genius: the life and work of Pierre DUHEM*. Nijhoff, Dordrecht (1987) — G. GENDRON: *English translation of three milestone papers* [by DUHEM (1909), JOUGUET (1901), and ZEMPLÉN (1905)] *on the existence of shock waves*. Rept. VPI-E-89-12, Virginia Polytechnic Institute, Blacksburg, VA (1989) — A. BRENNER: *DUHEM: science, réalité et apparence; la relation entre philosophie et histoire dans l'œuvre de Pierre DUHEM*. Vrin, Paris (1990) — R.N. MARTIN: *DUHEM and the origins of statics: ramifications of the crisis of 1903–04*. Synthese **83**, 337-355 (June 1990); *Pierre DUHEM: philosophy and history in the work of a believing physicist*. Open Court, La Salle, IL (1991) — S.L. JAKI: *Scientist and catholic: an essay on Pierre DUHEM*. Christendom Press, Front Royal, VA (1991) — *Pierre DUHEM* [various papers in Engl and French], Rev. Int. Phil. (Presses Univ. de France, Evry) **46**, No. 3, 290-409 (1992).

**PICTURE.** DUHEM at age 50–54. Courtesy Académie des Sciences, Archives et Patrimoine Historique, Paris, France.

# DUVALL, George Evered (1920–2003)

- U.S. physicist, dean of U.S. shock wave science

George E. DUVALL, son of George W. DUVALL, was born in Leesville, LA. After his junior year at Oregon State College (now Oregon State University), he left to work as an associate physicist on problems of underwater sound at the University of California Division of War Research in San Diego (1941–1945). He wrote or contributed to over 20 research reports on acoustics covering wave propagation, scattering, echo response from submerged objects, and piezoelectric transducers. After his return to Oregon State College, he completed his B.S. (1946) and then went to MIT, where he completed his Ph.D. in physics (1948). He worked on nuclear reactor problems at General Electric in Richland, WA (1948–1953), from 1950 heading the theoretical group there. In late 1953, he was hired by Dr. Thomas POULTER to work on shock wave problems at Stanford Research Institute (SRI) in Menlo Park, CA as a senior physicist (1954–1957). He became scientific director of SRI's Explosives and Extreme Pressure Laboratory (1957–1964), later named the Poulter Laboratory. Initially he concentrated his efforts on the thermodynamic and hydrodynamic aspects of shock wave propagation as needed for ballistic and metallurgical applications.

His work was primarily responsible for clarifying many fundamental theoretical issues related to shock wave propagation. DUVALL and his collaborators contributed to some of the earliest theoretical and experimental advances in the field, including such topics as hydrodynamic attenuation, material strength effects, optical techniques for free surface measurements, instabilities and multiple wave interactions, rate effects, studies on porous materials, impact welding, and detonation studies. Since 1964 he worked as a professor of physics at Washington State University (WSU), directing the university's Shock Dynamics Laboratory (SDL), now the Institute for Shock Physics (ISP). During this time he and his students made noteworthy contributions to the following problems: (1) shock-induced phase transitions and the incorporation of kinetic effects; (2) systematic calculations of the effects of transformation rates on the details of shock evolution; (3) nonlinear wave propagation in lattices; (4) understanding of atomic mechanisms controlling inelastic deformation in shocked crystals; (5) equations of state of solids and liquids; (6) electrical and thermoelectric measurements under shock loading; and (7) time-resolved spectroscopic studies to understand shock-induced chemical reactions.

He is author or coauthor of almost 170 articles or technical reports and supervised Ph.D. theses of over 25 students. He became the mentor for a whole generation of research scientists who themselves made significant contributions to shock wave physics and had distinguished careers at private and governmental laboratories. DUVALL was one of the founding members of the Association Internationale pour L'Avancement de la Recherche et de la Technologie aux Hautes Pression (AIRAPT). At international and national meetings he gave numerous lectures on the progress of shock wave physics in condensed matter and prepared worthwhile review articles, summarizing recent milestone achievements of this rapidly growing discipline. Chairing the National Materials Advisory Board (NMAB), he was mainly responsible for the report on Shock Compression Chemistry in Materials Synthesis and Processing.

In 1989, he received the second Shock Compression Science Award of the American Physical Society (APS) "in recognition of his outstanding contributions to shock wave physics and his educational and organizational leadership in the shock physics community." In honor of DUVALL's great contributions to academic education, Washington State University offers the *George E. Duvall Scholarship*. This new scholarship recognizes outstanding achievements in graduate research in the area of shock compression science.

DUVALL died in Vancouver at the age of 82 after a long illness.

**ORIGINAL WORKS.** With B.J. ZWOLINSKII: *Entropic equation of state and their application to shock wave phenomena in solids.* JASA **27**, 1054-1058 (1955) — *Rock breakage by explosion.* U.S. Bureau of Mines Rept. Inv. 5396 (Sept. 1957) — *Some properties and applications of shock waves.* In: (P.G. SHEWMON and V.F. ZACKAY, eds.) *Response of metals to high velocity deformation.* Interscience, New York (1961), pp. 165-203 — *Stability of shock waves.* In: *Sur les ondes de détonation.* Colloque International C.N.R.S., Paris (1961) — *Shock waves in the study of solids.* Appl. Mech. Rev. **15**, 849-854 (1962) — *Concepts of shock wave propagation.* Bull. Seismol. Soc. Am. **52**, 869-893 (1962) — *Shock waves in solids.* Int. Sci. Tech. **16**, 45-52 (1963) — With G.R. FOWLES: *Shock waves.* In: (R.S. BRADLEY, ed.) *High pressure physics and chemistry.* Academic Press, New York (1963), vol. 2, pp. 209-291 — *Shock waves and equations of state. Applications.* In: (P.C. CHOU and A.K. HOPKINS, eds.) *Dynamic response of materials to intense impulsive loading.* Air Force Materials Laboratory, OH (1964), Ch. 4, pp. 89-122, 481-516 — *Propagation of plane shock waves in a stress relaxing medium.* In: (H. KOLSKY and W. PRAGER, eds.) *Proc. IUTAM Symp.* [Providence, RI, April 1963]. *Stress waves in anelastic solids.* Springer, Berlin (1964), pp. 20-32 — With Y. HORIE: *Shock-induced phase transitions.* In: (S.J. JACOBS, ed.) *Proc. 4th Symposium on Detonation.* ACR-126, Office of Naval Research, Dept. of the Navy, Washington, DC (Oct. 1965), pp. 248-257 — With R.M. THOMSON: *Contact potential of a shocked metal.* In: (C.T. TOMIZUKA and R.M. EMRICK, eds.) *Physics of solids at high pressures.* Academic Press, New York (1965), pp. 196-212 — With T.J. AHRENS: *Stress relaxation behind elastic shock waves in rocks.* J. Geophys. Res. **71**, 4349-4360 (1966) — With Y. HORIE: *Shock-induced phase transition in iron.* Proc. Symp. on High Dynamic Pressure. IUTAM [Paris, Sept. 1967]. In: (M. ROY, ed.) *Behavior of dense media under high dynamic pressure.* Gordon & Breach, New York (1968), pp. 355-359 — *Shock waves in solids.* In: (B.M. FRENCH and N.M. SHORT, eds.) *Shock metamorphism of natural materials.* Mono Book, Baltimore (1968), pp. 19-29 — With Y. HORIE: *Shock waves and kinetics of solid-solid transition.* In: *Proc. Army Symp. on Solid Mechanics* [Johns Hopkins University, Baltimore, MD, Sept. 1968]. Applied Materials & Mechanics Research Center, Watertown, MA (1968) — With D.E. GRADY and E.B. ROYCE: *Shock-induced anisotropy in ferromagnetic materials. II. Polycrystalline behavior and experimental results for YIG.* J. Appl. Phys. **43**, 1948-1955 (1972) — *Problems in shock wave research* [invited paper]. In: (R.W. ROHDE, B.M. BUTCHER, J.R. HOLLAND, and C.H. KARNES, eds.) *Conf. on Metallurgical Effects at High Strain Rates* [Sandia Labs, Albuquerque, NM, Nov. 1973]. Plenum Press, New York (1973), pp. 1-13 — *Shock-induced phase transitions in solids.* Applied Mechanics Conf. [Salt Lake City, UT, June 1976]. In: (E. VARLEY, ed.) *Propagation of shock waves in solids.* Applied Mechanics Division (AMD Series), ASME (New York) **17**, 97-114 (1976) — With R.A. GRAHAM: *Phase transitions under shock wave loading.* Rev. Mod. Phys. **49**, 523-579 (1977) — With J.B. FORBES: *Theoretical basis for understanding the mixed phase region in shock-compressed iron.* Proc. 6th AIRAPT Conference [Boulder, CO, July 1977]. In: (K.D. TIMMERHAUS and M.S. BARBER, eds.) *High-pressure science and technology* Plenum Press, New York (1979), vol. 1, pp. 210-216 — With D.A. CREMERS and P.L. MARTSON: *Sources of light emission during shock experiments.* High Temperatures, High Pressures (U.K.) **12**, 109-112 (1980) — With G. ROSENBERG: *Precursor amplitudes in LiF from shocks propagating in ⟨111⟩ directions.* J. Appl. Phys. **51**, 319-330 (1980) — With J.E. VORTHMAN: *Effects of temperature on attenuation of the shock wave precursor in ⟨100⟩ LiF.* Ibid. **52**, 764-771 (1981) — With R.S. HIXON and P.M. BELLAMY: *Effect of shock waves on the absorption spectrum of ruby.* In: (W.J. NELLIS, L. SEAMAN, and R.A. GRAHAM, eds.) *Proc. AIP Conf. on Shock Waves in Condensed Matter* [Menlo Park, CA, June 1981]. AIP, New York (1982), pp. 282-286 — With K. OGILVIE: *Time resolved spectroscopy of shock compressed liquids.* Ibid. pp. 292-295 — With C.R. WILSON and K. OGILVIE: *The resistivity of liquid carbon disulfide during shock compression.* Ibid., pp. 296-298 — With J.E. VORTHMAN: *Dislocations in shocked and recovered LiF.* J. Appl. Phys. **53**, 3607-3615 (1982) — With K.M. OGILVIE: *Shock-induced changes in the electronic spectra of liquid $CS_2$.* J. Chem. Phys. **78**, 1077-1087 (1983) — With S.A. SHEFFIELD: *Response of liquid carbon disulfide to shock compression: equation of state at normal and high densities.* Ibid. **79**, 1981-1990 (1983) — *Shock wave research: yesterday, today and tomorrow.* In: (Y.M. GUPTA, ed.) *Proc. 4th APS Topical Conf. on Shock Waves in Condensed Matter* [Spokane, WA, July 1985]. Plenum Press, New York (1986), pp. 1-12 — With R.H. GRANHOLM, P.M. BELLAMY, and J.E. HEGLAND: *Effects of temperature on the UV-visible spectrum of dynamically compressed $CS_2$.* Ibid., pp. 213-219.

**SECONDARY LITERATURE.** Y.M. GUPTA (ed.) *Shock wave compression of condensed matter.* Proc. Symposium in Honor of George. E. DUVALL. Shock Dynamics Laboratory, Washington State University (WSU), Pullman, WA (Sept. 1988) — (Ed.) *George E. DUVALL, WSU shock physics pioneer dies at 82.* WSU News (Jan. 9, 2003); http://wsunews.wsu.edu/detail.asp?StoryID=3532.

**PICTURE.** From *Shock compression of condensed matter – 1989.* North-Holland, Amsterdam (1990), p. viii.

# EARNSHAW, Samuel (1805–1888)

• British clergyman and mathematician

Samuel EARNSHAW was born in Sheffield to Joseph EARNSHAW, a file-cutter who had been in Charity School as a boy. He began early to hold various teaching positions at Carver Street Schools in Sheffield (1819–1824). The Revd William H. BULL, curate of the Parish Church who had perceived young EARNSHAW's singular aptitude for mathematics, encouraged him to continue higher education. He entered St. John's College at Cambridge (1827), graduated as Senior Wrangler (1831), and was also first Smith's prizeman. He remained at Cambridge University as tutor and coach (1831–

1847), which brought him great renown, and became ordained deacon and priest. The Church Burgesses, recognizing his great learning, appointed him chaplain of Queen Mary's Foundation in the church and parish of Sheffield (1847–1888), which came with a considerable annual stipend. During his ministry he also took an active part in all religious, educational, and philanthropic movements and also published several books on mathematics and philosophy, including treatises on statics and dynamics, differential equations, and similar subjects, some of which went through several editions. He delivered many learned papers before the Sheffield Literary and Philosophical Society such as *On the Theory of Heat*, *What Geometry Says to Evolution*, and *The Arithmetic of Infinities*. He was the author of several books on mechanics, mathematics, and philosophy, some of which went through several editions.

In the 1840s, he got involved in fluid dynamics and related problems of solving partial differential equations. At first he investigated fluid motions such as solitary wave propagation and, in the 1850s, turning to acoustic waves, worked out a mathematical theory of sound. His interest in sounds of finite amplitude (*i.e.*, in shock waves) arose during his period at Sheffield. In the spring of 1851, he observed that a thunderstorm was terminated by a flash of lightning of great vividness, which was instantly followed by an awful crash. However, no damage was done at the locus of observation, but rather at a distance of more than a mile away. EARNSHAW, speculating on this phenomenon, concluded that violent sounds, for example emitted from a thunder clap, would propagate faster than gentle sound. In 1858, he first reported on this phenomenon at the Meeting of the British Association at Leeds and his attempt to treat this problem of sounds of finite amplitude on a mathematical basis. His interesting contributions certainly stimulated his contemporaries to turn to this unique branch of nonlinear acoustics.

**ORIGINAL WORKS.** *Dynamics, or an elementary treatise on motion.* Deighton, Cambridge (1832) — *On fluid motion, so far as it is expressed by the equation of continuity* [delivered March 21, 1836]. Trans. Cambr. Phil. Soc. **6**, 203-233 (1838) — *The mathematical theory of the two great solitary waves of the first order* [delivered Dec. 8, 1845]. Ibid. **8**, 326-341 (1849) — *On the transformation of linear partial differential equations with constant coefficients to fundamental forms.* Phil. Mag. **35** [III], 24-28 (1849) — *On the mathematical theory of sound.* Rept. Meet. Brit. Assoc. **28**, 34-35 (1858); Proc. Roy. Soc. Lond. **9**, 590-591 (1858) — *On a new theoretical determination of the velocity of sound.* Phil. Mag. **19** [IV], 449-455 (1860); Ibid. **20** [IV], 37-41 (1860); Rept. Meet. Brit. Assoc. **30**, 58 (1860) — *On the triplicity of sound.* Phil. Mag. **20** [IV], 186-192 (1860); Rept. Meet. Brit. Assoc. **30**, 58-59 (1860) — *On the mathematical theory of sound.* Phil. Trans. Roy. Soc. Lond. **150**, 133-148 (1860) — *Reply to some remarks of Dr. LE CONTE in his paper on the problem of the velocity of sound.* Phil. Mag. **27** [IV], 98-104 (1864) — *Partial differential equations: an essay towards an entirely new method of integrating them.* Macmillan, London (1871) — *The finite integrals of certain partial differential equations which present themselves in physical investigations.* Phil. Mag. **4** [V], 213-215 (1877) — *The doctrine of germs, or The integration of certain partial differential equations which occur in mathematical physics.* Deighton, Bell, Cambridge (1881).

**SECONDARY LITERATURE.** (Ed.) [Obituary note]. Sheffield Daily Telegraph (Dec. 7, 1888) — F. BOASE (ed.): *Modern English biography.* 6 vols., Cass, London (1892–1921), vol. 1, p. 951 — J.F. KIRK: *A supplement to ALLIBONE's critical dictionary of English literature.* 2 vols., Lippincott, Philadelphia (1897) — W. ODOM: *Hallamshire worthies. Characteristics and work of notable Sheffield men and women.* J.W. Northend, Sheffield (1926) — *Alumni Cantabrigienses.* Compiled by J.A. VENN, Cambridge University Press, Cambridge, U.K. (1944), p. 374.

**PICTURE.** Courtesy Sheffield City Libraries, Sheffield, U.K.

# EULER, Leonhard (1707–1783)

- Swiss theoretical and applied mathematician, physicist, and astronomer; cofounder of modern theoretical fluid dynamics

Leonhard EULER was born in Basel, Switzerland. He was the eldest son of Paul EULER, a Protestant minister, and brought up in Riehen, a small town near Basel. His father wanted to follow him into the church and sent him to the University of Basel to prepare for the ministry. He entered the university in 1720, first to obtain a general education before going on to more advanced studies. After receiving his master's degree in philosophy (1723), he turned to theology, thus fulfilling his father's wish. Johann BERNOULLI, a dear friend of his father, soon discovered EULER's great potential for mathematics, and during his years of study he attended lectures on mathematics given by J. BERNOULLI. In 1727, he accepted an invitation by the newly organized St. Petersburg Academy of Sciences. He became professor of physics (1730) and professor of mathematics (1733). Besides his diverse activities in scientific and technical fields, he reorganized the university and the curriculum.

Owing to political problems in the Russian capital, he left Petersburg (1741) and accepted the invitation of the Prussian king FREDERICK II (the Great) to join the Berlin Society of Sciences (1741–1766). He was appointed director of the mathematical class of the Royal Prussian Academy (1744) and member of the board. The Prussian king charged him also with practical problems, such as supervision of the pumps and pipes of the hydraulic system at castle Sanssouci in Potsdam, the royal summer residence, or the translation of the book *New Principles of Gunnery* (1842) by the British ballistician Benjamin ROBINS. Here EULER added important enhancements and expanded ROBINS' work to a length many times that of the original text (1745). EULER's translation occupies an important place in the history of ballistics. It was retranslated into English by Hugh BROWN (1784) and Charles HUTTON (1805), both providing worthwhile additional notes. EULER's contribution to the evolution of theoretical fluid mechanics are significant. The momentum equation in the form we frequently use in modern compressible flow was derived by him (1748), as was the continuity equation in the general form (1757). Due to serious conflicts with king FREDERICK II over financial matters and his management of the Academy, EULER left Berlin (1766) and returned to St. Petersburg, where he continued working in leading positions until his death.

EULER's contributions to modern mathematics, encompassing also mathematical nomenclature, are enormous. For example, he defined a function as an analytic expression and introduced $f$ for the mathematical term *function* – a word used by the Swiss mathematician Johann BERNOULLI in 1698 in an article on the solution to a problem involving curves – and included brackets for $f(x)$. In the 1750s, studying the motion of a vibrating string like some of his contemporaries, he found it useful to expand certain functions in terms of simpler ones, thus considering what we call today the "Fourier series solution of the wave equation." His work sparked a tremendous debate among the leading mathematicians of the day over the definition of the function concept, one of the greatest contributions of 19th-century mathematics. In the early 1750s, EULER worked out the equations for the motion of an inviscid compressible fluid: introducing the differential element of fluidic matter, he derived the differential forms of the mass (or continuity) equation and the momentum equation, the so-called "Euler equations." They were later extended by others by including the effect of viscosity (the Navier-Stokes equations).

In 1759, in a letter to Joseph L. DE LAGRANGE, EULER speculated on the possibility that the propagation of sound might depend on the "size of disturbances;" *i.e.*, on the intensity of sound – thus anticipating the shock wave problem.

During his lifetime EULER published about 560 books and articles, and many manuscripts were found unpublished among his personal belongings after his death. In September 1783 EULER died in St. Petersburg, where he is buried in the town's necropolis.

Astronomers named a crater on the near side of the Moon and a minor planet (asteroid 2002 EULER) after him.

**ORIGINAL WORKS.** *De communicatione motus in collisione corporum* ("On the communication of motion in collisions") [1730]. Comm. Acad. Scient. Imp. Petropol. **V**, 159-168 (1738) — *De communicatione motus in collisione corporum sese non directe percutientium* ("On the imparting of motion from a collision of bodies not striking each other directly") [1737]. Ibid. **IX**, 50-76 (1744) — *Neue Grundsätze der Artillerie enthaltend die Bestimmung der Gewalt des Pulvers nebst einer Untersuchung über den Unterschied des Widerstands der Luft in schnellen und langsamen Bewegungen.* [Germ. translation of Benjamin ROBINS' book *New principles of gunnery*]. Haude, Berlin (1745); Reprinted in: (F.R. SCHERRER, ed.) *Leonardi Euleri opera omnia* **14** [II]. Teubner, Leipzig (1922); Retranslated into English by Hugh BROWN: *The true principles of gunnery investigated and explained; comprehending translations of Prof. EULER's observations upon the new principles of gunnery, published by the late Mr. Benjamin ROBINS, and that celebrated author's Discourse upon the Track described by a Body in a resisting medium, inserted in the memoirs of the Royal Academy of Berlin, for the year 1753...* Nourse, London (1777) — *De la force de percussion et de sa véritable mesure.* Mém. Acad. Berlin **1**, 21-53 (1745, publ. 1746) — *Recherches sur la véritable courbe que décrivent les corps jetés dans l'air ou dans un autre fluide quelconque.* Hist. Acad. Roy. Sci. Belles Lettres (Berlin) **9**, 321-352 (1753) — *Principes généraux de l'état d'équilibre des fluides.* Mém. Acad. Sci. Berlin **11**, 217-273 (1757) — *Principes généraux du mouvement des fluides.* Ibid. **11**, 274-315 (1757) — *Continuation des recherches sur la théorie du mouvement des fluides.* Ibid. **11**, 316-361 (1757) — *De collisione corporum gyrantium* ("On collision of gyrating bodies") [1772]. Comm. Acad. Scient. Imp. Petropol. **XVII**, 272-314 (1773) — *De collisione corporum pendulorum, tam oblique, quam motu gyratorio perturbata* ("On collisions of swinging bodies, even if oblique, when the motion is perturbed by gyrations") [1772]. Ibid. **XVII**, 315-332 (1773) — *De natura et propagatione.* In: (F. RUDIO, ed.) *Leonardi EULERI opera omnia* **1** [III]. Teubner, Leipzig (1926), pp. 183-187. *De productione soni*, Ibid. pp. 188-196; *De la propagation du son*, Ibid. pp. 428-451. *Supplément aux recherches sur la propagation du son*, Ibid. pp. 452-483. *Continuation des recherches sur la propagation du son*, Ibid. pp. 484-507.

**SECONDARY LITERATURE.** O. SPIESS: *Leonard EULER. Ein Beitrag zur Geistesgeschichte des XVIII. Jahrhunderts.* Frauenfeld, Leipzig (1929) — A. SPEISER: *Die Basler Mathematiker.* Helbing & Lichtenhahn, Basel (1939) — F.I. FRANKL: *On the priority of EULER in the discovery of the similarity law for the resistance of air to the motion of bodies at high speeds* [in Russ.]. Dokl. AN (SSSR) **70**, 39-42 (1950) — J.E. HOFMANN: *Leonard EULER.* Physik. Blätter **14**, 117-122 (1958) — C. BLANC (ed.) *Leonhardi EULERI commentationes mechanicae (Ad theoriam corporum pertinentes).* Füssli, Basel (1968) — I. GRATTAN-GUINNESS: *The development of the foundations of mathematical analysis from EULER to RIEMANN.* MIT Press, Cambridge, MA (1970) — A.P. YOUSCHKEVITCH: *L. EULER.* In: (C.C. GILLESPIE, ed.)

*Dictionary of scientific biography.* Scribner, New York, vol. 4 (1971), pp. 467-484 — I. SZABÓ: *Geschichte der mechanischen Prinzipien und ihre wichtigsten Anwendungen.* Birkhäuser, Basel (1977), pp. 452-457 — W. JOHNSON: *Encounters between ROBINS, and EULER and the BERNOULLIS; Artillery and related subjects.* Int. J. Mech. Sci. **34**, 651-679 (1992) — H.H. VON BORZESZKOWSKI and R. WAHSNER: *Leonard EULER (1707–1783) und Joseph Louis LAGRANGE (1736–1813).* In: (K. VON MEŸENN, ed.) *Die großen Physiker.* C.H. Beck'sche Verlagsbuchhandl., München (1997), vol. 1, pp. 229-242. **PICTURE.** Engraving by Joseph-Friedrich-August DARBES. Courtesy Dibner Library of the History of Science and Technology, Smithsonian Institution Libraries. National Museum of American History, Washington, DC. **NOTE.** EULER's original works are available online; *see The Euler Archive*; http://www.math.dartmouth.edu/~euler/.

## FERRI, Antonio (1912–1975)

- Italian-born U.S. aeronautical engineer and inventor; pioneer of supersonic testing

Born in Norcia, Italy, Antonio FERRI studied electrical and aerospace engineering. After earning doctorates in industrial electrical engineering from the University of Rome (1934) and in aeronautical engineering from the Graduate School of Aeronautics at Rome (1936), he initially worked as an aerodynamic research scientist at the Italian Air Ministry's aeronautical research center in Guidonia (1935–1937), where he soon became head of the *Galleria Ultrasonora* (Supersonic Wind Tunnel) (1937–1940). In Guidonia he also studied various supersonic biconvex airfoil geometries, including a 10% thick biconvex airfoil at a Mach number of 2.13 – a concept that was later taken up in the U.S.A. in the design of the Bell X-2, a Mach 3 research rocket plane.

During World War II his supersonic wind tunnel was taken to Germany, where he temporarily continued his aerodynamic research (1940–1943). Returning to Italy, he joined the underground and became leader of the partisan brigade Spartaco against the Nazis (1943–1944). He was brought to the U.S.A. by the War Department, and he joined the National Advisory Committee for Aeronautics (NACA) and was assigned to the Langley Aeronautical Laboratory in Virginia. Initially working as a senior scientist (1944–1949) in wind tunnel testing of war planes, he became head of the gas dynamics branch (1949–1951).

Recruited by the Aerospace Department of the Polytechnic Institute of Brooklyn, FERRI was appointed professor of aerodynamics (1951–1964). During this period he installed at Freeport, Long Island the first American hypersonic wind tunnel, which, being from the blow-down type, allowed simulation of the high temperatures encountered in high-speed flow up to $M = 6$. Unlike with hypersonic shock tubes, the long operation time was particularly attractive to simulate the time-history of reentry phenomena of spacecraft and intercontinental missiles, which also helped to effectively design the nose cone to withstand the combined action of heat and aerodynamic loading during reentry. In 1964, FERRI joined the University of New York (NYU) as Astor professor of aerospace sciences (1964–1975). Later employed at NYU's Aerospace Energetics Laboratory in Westbury, NY on Long Island, he directed studies on sonic boom effects and air pollution problems originating from supersonic aircraft engines (1967–1975).

FERRI has been considered one of the most innovative experimentalists in aeronautics in the United States. He was always interested in propulsion and did pioneering work on supersonic inlets and supersonic compressors and also held numerous U.S. patents in these fields. He helped design a 2,300-mph (1,028 m/s) ramjet-powered fighter, the Republic XF-103, a high-speed, high-altitude aircraft specifically designed to intercept incoming enemy bombers. However, in 1957 the project was cancelled by the USAF in favor of the competing Convair F-102. His most recognized contributions were the "Ferri sled" (1957), a proposed spaceship design to better meet problems of reentry from outer space, and his "shroud technique" (1957), which allows one to channel the flow around large models in order to reproduce actual flight boundary layer and Reynolds number conditions. He was one of the leading pioneers of low-sonic-boom technology relevant to the design of supersonic-cruise aircraft.

Together with Nicholas J. HOFF and Paul A. LIBBY FERRI edited the Proceedings of the Conference on High-Speed Aeronautics, held in January 1955 at the Polytechnic Institute of Brooklyn. Together with Dietrich KÜCHEMANN and Laurence H.G. STERNE he also edited the 2-volume book *Progress in Aeronautical Sciences* (Pergamon Press, Oxford 1961/1962).

FERRI was cofounder and president of the General Applied Science Laboratories, Inc. (GASL) in Westbury, NY and a consultant to the aircraft industry. He received various scientific awards from Italian, American, and British professional societies.

**ORIGINAL WORKS.** *Experimental results with airfoils tested in the high-speed tunnel at Guidonia.* NACA-TM 946 (1940), an Engl. translation of Atti di Guidonia Tech. Rept. No. 17 (1939) — *Esperienze su di un biplano iperacustico tipo Busemann.* Atti di Guidonia Tech. Rept. No. 37-38 (1940) — *Application of the method of characteristics to supersonic rotational flow.* NACA-TN 135 (1946) — *Elements of aerodynamics of supersonic flows.* Macmillan, New York (1949) — *Method for evaluating from shadow or schlieren photographs the pressure drag in two-dimensional or axially symmetric flow phenomena with detached shock.* NACA-TN 1808 (1949) — *The method of characteristics for the determination of supersonic flow over bodies of revolution at small angles of attack.* NACA-TN 1809 (1949) — *Supersonic flow around circular cones at angles of attack.* NACA-TN 2236 (1950) — With L.N. NUCCI: *Preliminary investigation of a new type of supersonic inlet.* NACA-TN 2286 (1951) — *The linearized characteristics method and its application to practical nonlinear supersonic problems.* NACA-TN 2515 (1951) — With P.A. LIBBY and V. ZAKKAY: *A theoretical analysis of a new type of heater suitable for intermittent hypersonic wind tunnel operation.* PIBAL Rept. No. 201, AD-130911 (1952) — With N. NESS and T. KAPLITA: *Supersonic flow over conical bodies without axial symmetry.* J. Aeronaut. Sci. **20**, 563-571 (1953) — With S.M. BOGDONOFF: *Design and operation of intermittent supersonic wind tunnels.* AGARDograph No. 1 (1954) — *The method of characteristics.* In: (W.R. SEARS, ed.) *General theory of high-speed aerodynamics.* Princeton University Press, Princeton, NJ (1954) — With P.A. LIBBY: *Note on the interaction between the boundary layer and the inviscid flow.* J. Aeronaut. Sci. **21**, 130 (1954) — With P. LIBBY, M. BLOOM, and V. ZAKKAY: *Development of the Polytechnic Institute of Brooklyn Hypersonic Facility.* Rept. WADC-TN 55-695, Wright Air Development Center (1955) — *Note on the flow on the rear part of blunt bodies in hypersonic flow.* Rept. WADC-TN 56-294 (1956) — *The method of characteristics.* In: *High speed aerodynamics and jet propulsion.* Princeton University Press, Princeton, NJ; vol. VI (1956): (W.R. SEARS, ed.) *General theory of high speed aerodynamics,* pp. 583-669 — *Supersonic flows with shock waves.* Ibid., pp. 670-747 — *Cooling by jets directed upstream in hypersonic flow.* Rept. WADC-TN 56-382 (1956) — With J.H. CLARKE: *On the use of interfering flow fields for the reduction of drag at supersonic speeds.* J. Aeronaut. Sci. **24**, 1-18 (1957) — With P.A. LIBBY: *A new technique for investigating heat transfer and surface phenomena under hypersonic flight conditions.* Ibid. **24**, 464-465 (1957) — *Achievements of high heat fluxes in a wind tunnel.* Ibid. **24**, 772-773 (1957) — With L. FELDMAN and W. DASKIN: *The use of lift for reentry from satellite trajectories.* Jet Propulsion **27**, 1184-1191 (1957) — *A review of some recent developments in hypersonic flow.* In: *Proc. 1st Int. Congress in the Aeronautical Sciences* [Madrid, Spain, Sept. 1958]. Adv. Aeronaut. Sci. **1**, 723-770 (1959) — *Some heat transfer problems in hypersonic flow.* WADC-TN 59-308 (1959) — *Future research in high Mach number air breathing engines.* In: (A.L. JAUMOTTE ET AL., eds.) *4th AGARD Colloquium: Combustion and Propulsion* [Milan, Italy, April 1960]. Pergamon Press, Oxford (1961), pp. 3-15 — With L.G. NAPOLITANO: *Fluid dynamics of nonsteady flows.* In: (A. FERRI, ed.) *Fundamental data obtained from shock-tube experiments.* AGARDograph No. 41. Pergamon Press, New York (1961), pp. 1-46 — With V. ZAKKAY and L. TING: *Blunt body heat transfer at hypersonic speed and low Reynolds numbers.* J. Aerospace Sci. **28**, 962-972 (1961); Ibid. **29**, 882-883 (1962) — With P.A. LIBBY and V. ZAKKAY: *Theoretical and experimental investigation of supersonic combustion.* In: (M. ROY, ed.) *Proc. 3rd Congr. of International Council of the Aeronautical Sciences (ICAS)* [Stockholm, Sweden, 1962]. Spartan Books, Washington, DC (1964), pp. 1089-1155 — *Hypersonic flight testing.* Int. Sci. Technol. (New York) **28**, 64-76 (April 1964) — *Review of problems in application of supersonic combustion* [7th Lanchester Memorial Lecture]. J. Roy. Aeronaut. Soc. **68**, 575-597 (1964) — *Supersonic combustion progress.* Astronaut. Aeronaut. **2**, 32-37 (1964) — *Review of scramjet technology.* J. Aircraft **5**, 3-10 (1968) — With A. ISMAIL: *Effects of lengthwise lift distribution on sonic boom.* AIAA J. **7**, 1538-1541 (1969) — With H.C. WANG and H. SORENSON: *Experimental verification of low sonic boom configuration.* NASA Rept. CR-2070 (June 1972) — *Mixing-controlled supersonic combustion.* Annu. Rev. Fluid Mech. **5**, 301-338 (1973) — *Improved nozzle testing techniques in transonic flows.* AGARDograph No. 288. NATO, Neuilly-sur-Seine (1975) — *Selected papers on advanced design of air vehicles.* AGARDograph No. 226. NATO, Neuilly-sur-Seine (1977) — *Selected collection of Antonio FERRI's scientific work.* Associazione Italiana di Aeronautica e Astronautica (AIDAA), Rome (1981).

**U.S. PATENTS.** *Air inlet for supersonic airplane or missile.* No. 2,772,620 (1956) — *Multi-scoop supersonic inlet.* No. 2,788,183 (1957) — With S. SLUTSKY: *Water spray cooling method and apparatus for supersonic nozzle.* No. 2,892,308 (1959) — *Apparatus for cooling of supersonic aircraft.* No. 3,005,607 (1961) — *Scoop-type supersonic inlet with precompression surface.* No. 2,990,142 (1961) — *Engine for supersonic flight.* No. 2,989,843 (1961) — *Ram-jet engine.* No. 3,158,990 (1964) — *Vertical takeoff aerial lifting device.* No. 3,170,285 (1965) — *Wind tunnel nozzle structure.* No. 3,353,405 (1965) — *High velocity fluid accelerator.* No. 3,302,866 (1967) — *Supersonic engine.* No. 3,363,421 (1968) — *Wind tunnel fabrication method.* No. 3,523,350 (1970) — With P.A. LIBBY and M.H. BLOOM: *Sonic boom pressure wave simulation method and apparatus.* No. 3,555,878 (1971).

**SECONDARY LITERATURE.** R.H. CUSHMAN: *Hypersonic tunnels yield practical data.* Aviation Week (Oct. 21, 1957) — Anonymous: *Dr. Antonio FERRI, aerospace expert* [Obituary note]. New York Times (Dec. 30, 1975) — M. INGRASSIA: *Antonio FERRI, pioneer of supersonic testing.* Long Island newspaper *Newsday* (Dec. 29, 1975) — Anonymous: *A. FERRI, founded aero labs.* Polytecnic Cable, N.Y.U. (Jan. 1976) — A. BUSEMANN: *Antonio Ferri 1912–1975.* Mem. Tributes: Natl. Acad. Engng. **1**, 56-60 (1979) — J.R. HANSEN: *Engineer in charge. A history of the Langley Aeronautical Laboratory, 1917–1958.* NASA Rpt. SP-4305, Washington, DC (1987) — P.A. LIBBY: *Observations concerning supersonic combustion* [with a review of the early history of supersonic combustion]. In: (M. CHAMPION and B. DESHAIES, eds.) *IUTAM Symp. Combustion in Supersonic Flows.* Kluwer, Dordrecht (1997), pp. 1-11 — P.J. BOBBITT and D.J. MAGLIERI: *Dr. Antonio FERRI's contribution to Supersonic Transport (SST) sonic-boom.* J. Spacecraft Rockets **40**, 459-466 (2003).

**PICTURE.** From Aeronaut. Engng. Rev. **10**, 21 (Nov. 1951).

# FRIEDMANN [or Fridman; Russ. *ФРИДМАН*], Aleksandr Aleksandrovich (1888–1925)

• Russian mathematician, fluid dynamicist, and cosmologist; cofounder of Big Bang theory

Aleksandr A. FRIEDMANN was born in St. Petersburg (renamed Petrograd in 1920, renamed Leningrad in 1924). His father, Aleksandr FRIEDMANN, was a composer and ballet dancer and his mother a pianist. He graduated from the

gymnasium with the gold medal and entered the University of St. Petersburg (1906), where he became a student of Prof. Vladimir Andreevich STEKLOV, who founded the school of mathematical physics that later achieved considerable distinction. Together with the Russian mathematician Jakow D. TAMARKIN, FRIEDMANN wrote an unpublished paper on second-degree indeterminate equations that earned him a gold medal from the Department of Physics and Mathematics. After graduating from the university (1910) he was retained in the department to prepare for the teaching profession.

In 1913, he passed the examinations for the degree of master of pure and applied mathematics and was appointed to a position in the Aerological Observatory in Pavlovsk, a suburb of St. Petersburg, which involved him in meteorology. In the same year, he published a paper on meteorology, entitled (in translation) *On the Relationship of Temperature to Altitude*, in which he discusses the possible existence of an upper temperature inversion point in the stratosphere. During World War I, he served in an aviation detachment and later got involved in supervising the manufacture of measuring instruments in aviation. Thereafter, he became professor in the department of theoretical mechanics at Perm University (1918–1920) and worked as head, and later as director, of the Department of Theoretical Meteorology at the Geophysical Central Observatory of the Academy of Sciences in Petrograd (1920–1925). He studied a number of fundamental aerodynamic and hydrodynamic physical processes such as the motion of a compressible fluid under the influence of given forces with respect to some problems of treating cyclones and anticyclones, thereby creating "dynamical meteorology" – a new branch of atmospheric science.

In the early 1920s, he took up an interest in EINSTEIN's General Theory of Relativity (1916). FRIEDMANN's most prominent disciple was probably the Russian-born U.S. physicist George GAMOW, spiritual father of the Hot Big Bang Theory who studied, though only briefly, relativity under FRIEDMANN in 1925, the year of FRIEDMANN's death. FRIEDMANN published two classical papers in the prestigious German journal *Zeitschrift für Physik* on a dynamic model describing the evolution of the Universe in mathematical terms (1922, 1924). In his first paper, entitled (in translation) *On the Curvature of Space*, he found that even without the cosmological term there are still solutions of the field equations where matter has a finite density everywhere in space, provided this density is not time-independent. Supposing a closed Universe – so-called "Friedmann closed Universe" – his first solution showed that the radius of curvature of the Universe can be either increasing or a periodic function of time; *i.e.*, a Big Bang followed by expansion, then contraction and an eventual Big Crunch. Prior to FRIEDMANN, the German theoretical physicist Albert EINSTEIN had proposed a static, finite spherical Universe (1917). EINSTEIN, at first claiming that FRIEDMANN's solution does not satisfy the field equations, conceded after further study that FRIEDMANN was indeed right.

In July 1925, FRIEDMANN made a record-breaking ascent in a balloon to 7,400 m to make meteorological and medical observations. A few months later, shortly after falling ill on typhoid fever, he died at Leningrad at the early age of 37. He was awarded the George Cross (1915) for bravery in his flights during World War I and posthumously the Lenin Prize (1931) for his scientific work. Prof. Eduard A. TROPP and colleagues at Ioffe Technico-Physical Institute, summing up FRIEDMANN's contributions, appropriately wrote, "Just as COPERNICUS made the Earth orbit round the Sun, so FRIEDMANN made the Universe expand." Today, FRIEDMANN, together with the Belgian mathematician and astronomer Georges LEMAÎTRE, is considered the main founder of the Big Bang model.

The *Alexander Friedmann Seminars*, which have been held in St. Petersburg (1988, 1993, 1995, 1998), João Pessoa, Brazil (2002) and on Corsica, France (2004), became the international forum for astrophysicists to discuss actual topics of cosmology such as problems of gravitational theory, inflationary Universe, primordial radiation, dark matter, quantum effects in curved space-time, and observational cosmology. The first Friedmann Seminar held in 1988 (*see below*) was dedicated to the centenary anniversary of Alexander FRIEDMANN's birth. The *Alexander Friedmann Laboratory for Theoretical Physics* in St. Petersburg bears his name.

A crater on the far side of the Moon is named for him. The *Cosmonautics Day* is a Russian holiday celebrated every April 12 to commemorate GAGARIN's first manned space flight.

**ORIGINAL WORKS.** *Zur Theorie der Vertikaltemperaturverteilung.* Meteorol. Z. **31**, 154-156 (1914) — *Sur la recherche des surfaces isodynamiques.* C. R. Acad. Sci. Paris **154**, 864-865 (1912) — *Sur les tourbillons dans un liquide à température variable.* Ibid. **163**, 219-222 (1916) — *The hydrodynamics of a compressible fluid* [in Russian. The first part is on the kinematics of vortices, the second part on the dynamics of a compressible fluid]. Ph.D. thesis, Leningrad University (1922); Opyt Gidtmekhaniki Szhimaemoy Zhidkosti ("Essay on the hydrodynamics of a compressible fluid"). ONTI-GTTI, 370 pp. (1934) — *Über die Krümmung des Raumes.* Z. Phys. **10**, 377-386 (1922) — *Die Welt als Raum und Zeit* (1923, 1965) [Germ. translation by G. SINGER]. In: *Ostwald's Klassiker der exakten Wissenschaften.* Nr. 287, Deutsch, Thun (2000, 2002) — *Über die Möglichkeit einer Welt mit konstanter negativer Krümmung des Raumes.* Z. Phys. **21**, 326-332 (1924) — With J. TAMARKIN: *Über eine Methode der Bestimmung der vertikalen Windgeschwindigkeit.* Meteorol. Z. **41**, 90-91 (1924) — *On the extent of discontinuity in a compressible liquid* [in Russ.]. Zh. Russkago fiziko-khimicheskago obshchestva **56**, No. 1, 40-58 (1924) — *Über Wirbelbewegung in einer kompressiblen Flüssigkeit.* ZAMM **4**, 102-107 (1924) — With L. KELLER: *Differentialgleichungen für die turbulente Bewegung einer kompressiblen Flüssigkeit.* In: (C.B. BIEZENO and J.M. BURGERS, eds.) *Proc. 1st Int. Congr. Applied Mechanics* [Delft, Netherlands, 1924]. Waltman, Delft (1925), pp. 395-405 — *Hydrodynamische Arbeiten russischer Gelehrter* [Zusatz des Herausgebers]. Ibid., pp. 409-410 — *Théorie du mouvement d'un fluide compressible et ses applications aux mouvement de l'atmosphère.* Recueil de Géophysique. Observatoire physique centrale de Russie (Leningrad) **5**, 16-56 (1927).

SECONDARY LITERATURE. A.F. VANGENGEIM: *A.A. FRIEDMANN* [in Russ.]. Klimat i pogoda, No. 2/3, 5-7 (1925) — H. VON FICKER: *A.A. FRIEDMANN †.* ZAMM **5**, 526-527 (1925); Meteorol. Z. **42**, 440 (1925) — N.M. GYUNTER: *The scientific works of A.A. FRIEDMANN* [in Russ.]. Zh. Leningradskogo fiziko-matematicheskogo obshchestva **1**, No. 1, 5-11 (1926) — A.F. GAVRILOV: *In memory of A.A. FRIEDMANN* [in Russ.]. Uspekhi Fiz. Nauk **6**, 73-75 (1926) — I.V. MESHCHERSKY: *A.A. FRIEDMANN's works on hydrodynamics* [in Russ.]. Geofizichesky sbornik (Leningrad) **V**, part I, 57-60 (1927) — M.A. LORIS-MEKHOV: *FRIEDMANN's works on the theory of relativity* [in Russ.]. Ibid. **V**, part I, 61-63 (1927) — L.G. LOYTSYANSKY and A.I. LURIE: *A.A. FRIEDMANN* [in Russ.]. Trudy Leningradskogo politekhnicheskogo instituta M.I. Kalinina No. 1, 83-86 (1949) — A.T. GRIGORIAN: *Aleksandr Aleksandrovich FRIEDMANN.* In: (C.C. GILLESPIE, ed.) *Dictionary of scientific biography.* Scribner, New York, vol. 5 (1971), pp. 187-189 — M.A. MARKOV ET AL.: *A. A. FRIEDMANN, centenary volume.* In: (M.A. MARKOV, ed.) *Proc. Friedmann Centenary Conf.* [Leningrad, U.S.S.R., June 1988]. World Scientific, Singapore (1990) — E.A. TROPP, V.Y. FRENKEL, and A.D. CHERNIN: *Alexander A. FRIEDMANN: the man who made the Universe expand.* Translated by A. DRON and M. BUROV, Cambridge University Press (1993) — A detailed biography on FRIEDMANN, written by J.J. O'CONNOR and E.F. ROBERTSON, has been published by the MacTutor History of Mathematics Archive, University of St. Andrews, Scotland; see http://www-history.mcs.st-andrews.ac.uk/history/Mathematicians/Friedmann.html.

PICTURE. From the MacTutor History of Mathematics Archive (*see above*).

# FRIEDRICHS, Kurt Otto (1901–1982)

- German-born U.S. applied mathematician and theoretical fluid dynamicist

Kurt O. FRIEDRICHS was born in Kiel and attended elementary and high school in Düsseldorf, where he also began his univer-

sity studies. After spending several years at different German universities, he completed his studies in Göttingen. There he wrote his Ph.D. thesis (1927) on the theory of elastic plates under Richard COURANT's supervision, then director of the Institute of Mathematics. FRIEDRICHS became his assistant and helped with the Courant-Hilbert book (1927). He went to the University of Aachen (1929) to join Theodore VON KÁRMÁN who had become the first professor of aeronautical engineering. In 1931, at 30 years of age, he was called to the TH Braunschweig as a full professor. In 1937, he emigrated to America and at New York University (NYU) joined Prof. COURANT, who already had emigrated by that time, in teaching and research.

Until the end of World War II his main contributions were in fluid dynamics and in elasticity with the U.S. applied mathematician James J. STOKER. Together with his colleague Joseph B. KELLER, an applied mathematician, he investigated in the 1950s the propagation, reflection, and refraction properties of weak shock waves. Conducting here groundbreaking work in pure and applied mathematics, he was author of several books. His main work was on partial differential equations in mathematical physics, using finite differences to prove the existence of solutions. In the shock physics community his name is well known as being the coauthor of COURANT's classical textbook *Supersonic Flow and Shock Waves.* He made many contributions to fluid dynamics; several of them, resulting from wartime work of the 1940s, were not published. These include work on fluid flow through nozzles, over surfaces of revolution, and in detonations and deflagrations.

Later FRIEDRICHS directed the Courant Institute (1966–1967) at NYU. He was a member of the National Academy of Sciences (from 1959) and recipient of the National Medal of Science (1977) "for bringing the powers of modern mathematics to bear on problems in physics, fluid dynamics and elasticity."

ORIGINAL WORKS. With H. LÉWY: *Das Anfangsproblem einer beliebigen nichtlinearen hyperbolischen Differentialgleichung beliebiger Ordnung in zwei Variablen. Existenz, Eindeutigkeit und Abhängigkeitsbereich der Lösung.* Math. Ann. **99**, 200-221 (1928) — With R. COURANT and H. LÉWY: *Über die partiellen Differentialgleichungen der mathematischen Physik.* Ibid. **100**, 32-74 (1928) — With J.J. STOKER: *The non-linear boundary* value problem of the buckled plate. Am. J. Math. **63**, 839-888 (1941) — *Fluid dynamics*. Bergmann, Providence, RI (1942) — With J.J. STOKER: *Buckling of the circular plate beyond the critical thrust*. J. Appl. Mech. **9**, 7-14 (1942) — *Remarks on the Mach effect*. Repts. Nos. 38.4M and 38.5M, NDRC, Appl. Math. Panel Memos; Repts. Nos. 5 and 6 Appl. Math. Group, New York University, New York (1943) — *Theoretical studies on the flow through nozzles and related problems*. Rept. 82.1R, NDRC, Appl. Math. Panel; Rept. No. 3, Appl. Math. Group, N.Y. University (1944) — *Lectures on nonlinear elasticity at New York University*, New York (1945) — *On the mathematical theory of gas flow, flames, and detonation waves*. Five lectures presented at the University of Michigan (1946) — *On the mathematical theory of deflagrations and detonations*. Rept. No. 79-46, Naval Bureau of Ordnance (1946) — *On the non-occurrence of a limiting line in transonic flow*. Rept. No. 165, Inst. for Math. & Mech., N.Y. University (1947) — *On the boundary-value problems of the theory of elasticity and KORN's inequality*. Ann. Math. **48**, No. 2, 267-297 (1947) — With R. COURANT: *Supersonic flow and shock waves*. Interscience, New York (1948) — With R. COURANT: *Interaction of shock and rarefaction waves in one-dimensional motion*. Rept. OSRD-AMP 38.1R (1948) — *On the derivation of the shallow water theory*. Appendix to the paper by J.J. STOKER: *The formation of breakers and bores*. Comm. Pure Appl. Math. **1**, 81-85 (1948) — With D.H. HYERS: *The existence of solitary waves. Water waves on a shallow sloping beach*. Ibid. **1**, 109-134 (1948) — *Formation and decay of shock waves*. Ibid. **1**, 211-245 (1948) — With D.A. FLANDERS: *On the non-occurrence of a limiting line in transonic flow*. Ibid. **1**, 287-301 (1948) — *Nonlinear hyperbolic differential equations for functions of two independent variables*. Am. J. Math. **70**, 555-589 (1948) — *Symmetric hyperbolic linear differential equations*. Comm. Pure Appl. Math. **7**, 345-392 (1954) — With D.H. HYERS: *The existence of solitary waves*. Ibid. **7**, 517-550 (1954) — With J.B. KELLER: *Geometrical acoustics. Part II. Diffraction, reflection, and refraction of a weak spherical or cylindrical shock at a plane interface*. J. Appl. Phys. **26**, 961-966 (1955) [Part I, entitled *The theory of weak shock waves*, was written by J.B. KELLER, J. Appl. Phys. **25**, 938-947 (1954)] — *Mathematical aspects of flow problems of hyperbolic type*. In: *High speed aerodynamics and jet propulsion*. Princeton University Press, Princeton, NJ; vol. VI (1954): (W.R. SEARS, ed.) *General theory of high speed aerodynamics*, pp. 31-60 — *Asymptotic phenomena in mathematical physics*. Bull. Am. Math. Soc. **61**, 485-504 (1955) — *Nonlinear wave motion in magnetohydrodynamics*. Rept. No. 1845, Los Alamos Development Center (1955) — *Nichtlineare Differentialgleichungen. Stoß- und Expansionswellen*. Vorträge gehalten in Göttingen (Juli 1955) — *Nonlinear waves in magnetohydrodynamics*. Rept. MH-8, Institute of Mathematical Sciences, New York University (1958) — *Symmetric positive linear differential equations*. Comm. Pure Appl. Math. **11**, 333-418 (1958) — *Special topics in fluid dynamics* [with notes by S. CIOLKOWSKI]. Gordon & Breach, New York (1967) — With R. VON MISES: *Fluid dynamics*. Springer, New York (1971) — C.S. MORAWETZ, (ed.): *Kurt Otto FRIEDRICHS, Selecta*. 2 vols., Birkhäuser, Boston (1986).

SECONDARY LITERATURE. C.S. MORAWETZ: *Kurt Otto FRIEDRICHS*. In: (G.C. ROTA, ed.) *Contemporary mathematicians*. Birkhäuser, Boston (1986). Vol. 1 contains a biography of K.O. FRIEDRICHS written by C. REID — C.S. MORAWETZ: *Kurt Otto FRIEDRICHS*. Biogr. Mem. Natl. Acad. Sci. (U.S.A.) **67**, 131-146 (1995) — The FRIEDRICHS family (with a list of K.O. FRIEDRICH's publications, awards and honors); http://www.friedrichs.us/x-history.html.

PICTURE. Courtesy Universitätsarchiv der Technischen Universität Braunschweig, Germany; Signatur UniA BS JIF:15.

# GAGARIN [Russ. *ГАГАРИН*], Yuri Alekseyevich (1934–1968)

• Soviet cosmonaut; "First Man in Space" and first human to fly hypersonically

Yuri A. GAGARIN was born near Gzhatsk (now Gagarin) in the Smolensk region, west of Moscow. His father was a carpenter on a collective farm. After graduation as a molder from a trade school near Moscow (1951) he continued his studies at the industrial college at Saratov, where he took a course in flying. He entered the Soviet Air Force cadet school at Orenburg, where he graduated with top honors (1957). Selected with the first group of cosmonauts, he was eventually chosen for the historic orbital flight of 1961. His spacecraft Vostok 1 was launched at 9:07 A.M. Moscow time on April 12, 1961, orbited the Earth once in 1 h 29 min at a maximum speed of 28,968 km/h (about 8 km/s) and at a maximum altitude of 301 km, and landed safely at 10:55 A.M. in Siberia in the Sarov region. This famous flight during which GAGARIN experienced 8 g's during reentry was the first demonstration that problems of weightlessness and safe reentry into the Earth's atmosphere could be solved successfully, thus opening the door for all succeeding space flights. Hitting Mach 25 during reentry, he was the first human to fly at hypersonic velocity.

GAGARIN was awarded the Order of Lenin and given the titles of Hero of the Soviet Union and Pilot Cosmonaut of the Soviet Union. He was killed with another pilot in 1968 in the crash of a MiG-15 on a routine training flight near Moscow. The Cosmonaut Training Center, established in 1960 near Moscow, was renamed in 1968 the *Yuri Gagarin Cosmonaut Training Center* in memory of his famous space flight.

A crater on the far side of the Moon is also named for him.

ORIGINAL WORKS. *Road to the stars: notes by cosmonaut No. 1*, Foreign Languages, Moscow (1962) — *Ich war der erste Mensch im Weltall: Psychologie und Kosmos*. Goldmann, München (1970).
SECONDARY LITERATURE. *The first man in space: the record of GAGARIN's historic first venture into cosmic space: a collection of translations from Soviet press reports*. Crosscurrents Press, New York (1961) — W. BURCHETT and A. PURDY: *Cosmonaut Yuri GAGARIN, first man in space*. Gibbs & Phillips, London (1961) — (Ed.) *Yuri Alexeievitch GAGARIN*

*1934–1968.* Astronaut. Acta **14**, 690 (1968/1969) — *Encyclopædia Britannica, Micropædia.* Benton & Hemingway, Chicago (1974), vol. V, p. 377 — N. TSYMBAL (ed.) *First man in space: the life and achievement of Yuri GAGARIN. A collection.* Progress, Moscow (1984) — G.P. KENNEDY: *The first men in space.* Chelsea, New York (1991) — J. DORAN and P. BIZONY: *Starman: the truth behind the legend of Yuri GAGARIN.* Bloomsbury, London (1998) — R. HALL and D.J. SHAYLER: *The rocket men: Vostok & Voskhod, the first Soviet manned space flights.* Springer, London (2001).
**PICTURE.** Created in 1961. In the public domain, taken from Wikipedia, the free encyclopedia; http://scn.wikipedia.org/wiki/Yuri_Gagarin.

## GAMOW, George Anthony [born Georgiy Antonovich] (1904–1968)

- Russian-born U.S. physicist and cosmologist; father of the Hot Big Bang theory

George A. GAMOW was born in Odessa, Russia, to Anton GAMOW, a teacher of Russian language and literature. He was educated at the University of Leningrad (now St. Petersburg), where he briefly studied relativistic cosmology under Alexander FRIEDMANN before turning to quantum theory. After receiving his Ph.D. (1928) from the University of Leningrad, he worked for brief periods at the universities of Göttingen, Cambridge, Copenhagen, Paris, and London. In 1933, he left Europe and emigrated to the U.S.A., where he held professorships at George Washington University (1934–1956) and the University of Colorado (1956–1968). During World War II, he served as a consultant to the Division of High Explosives in the Bureau of Ordnance of the U.S. Navy and studied the propagation of shock and detonation waves in various conventional high explosives. Later he cooperated with Edward TELLER and Stanislaw ULAM on the hydrogen bomb at Los Alamos. Together with TELLER he formulated the so-called "Gamow-Teller rules" for classifying subatomic particle behavior in radioactive decay, and attempted to apply the new understanding of atomic phenomena to astrophysics.

Being interested in astronomy since his childhood, he later turned to astrophysical problems and worked on the theory of the internal structure of red giant stars (1939). He was a major proponent of the Big Bang cosmological theory of the origin of the Universe, which was previously developed by the Russian mathematical physicist and fluid dynamicist Alexander FRIEDMANN (1922–1924) and, independently, by the Belgian cosmologist and priest Georges LEMAÎTRE (1927). GAMOW and collaborators suggested the so-called "Hot Big Bang model" (1946): they argued that at the beginning of the expansion of the Universe, matter was not only very dense but also very hot: as a result, thermonuclear reactions would have taken place that promoted the formation of all elements heavier than hydrogen, especially of helium. In 1948, they proposed that low-temperature radiation has survived as a vestige of the Big Bang and that by the present time it would have been diluted to a temperature of only about 10 K. This was confirmed in 1965 by Arno A. PENZIAS and Robert W. WILSON, two U.S. physicists at Bell Telephone Laboratories, who discovered cosmic microwaves at a wavelength of 7 cm. This radiation, which exceeded the radiation coming from known sources by a factor of about 100, corresponds to an approx. 3-K black-body radiation – thus strongly supporting the Lemaître Big Bang model of an expanding Universe. GAMOW also worked out a theory for the origin of the elements in the Big Bang, his Alpha-Beta-Gamma Theory (1948), and speculated on the question of whether physical constants change over time.

In his later life, GAMOW became interested in biology and genetics and correctly assumed that the DNA structure forms a code that directs protein synthesis. He wrote almost 140 scientific articles and 28 books, mostly for nonscientists, which earned him the UNESCO Kalinga Prize (1956). In 1994, the 90th anniversary of his birth, the International Conference *Astrophysics and Cosmology after Gamow* was held in Odessa and St. Petersburg to discuss actual problems in astrophysics and cosmology, followed 5 years later by the *Gamow Memorial International Conference* (GMIC-99). The *George Gamow Lecture* is a prominent lecture series given at the University of Colorado at Boulder that brings renowned scientists to campus to address a general audience of nonscientists. At George Washington University (Washington, DC), the *George Gamow Undergraduate Research Fellowship* is designed to give promising undergraduates the opportunity to engage in a well-defined research project under the guidance of a faculty member in the chosen field of study.

A crater on the far side of the Moon is named for him.

**ORIGINAL WORKS.** With D.D. IVANENKO: *Zur Wellentheorie der Materie.* Z. Phys. **39**, 865-868 (1926) — *Tentative theory of novae.* Phys. Rev. **54** [II], 480 (1938) — With E. TELLER: *On the origin of great nebulae.* Ibid. **55** [II], 654-657 (1939) — *Physical possibilities of stellar evolution.* Ibid. **55**, 718-725 (1939) — With E. TELLER: *Energy production in red giants* [letter to the ed.]. Ibid. **55**, [II], 791 (1939); *The expanding Universe and the origin of the great nebulae*. Nature **143**, 116-117, 375 (1939) — With M. SCHOENBERG: *Neutrino theory of stellar collapse*. Phys. Rev. **59** [II], 539-547 (1941) — *The evolution of contracting stars*. Ibid. **65** [II], 20-32 (1944) — *Rotating Universe?* Nature **158**, 549 (1946) — *Expanding Universe and the origin of elements*. Phys. Rev. **70** [II], 572-573 (1946) — With R. FINKELSTEIN: *Theory of the detonation process*. NavOrd Rept. No. 90-46 (103), Navy Dept., Bureau of Ordnance (1947) — *The evolution of the Universe*. Nature **162**, 680-682 (1948) — *The origin of elements and the separation of galaxies*. Phys. Rev. **73** [II], 505-506 (1948) — With H. BETHE and R.A. ALPHER: *The origin of chemical elements*. Ibid. **73**, 803-804 (1948); *The thermonuclear reactions in the expanding Universe*. Ibid. **74**, 1198-1199 (1948) — *Supernovae*. Scient. Am. **181**, 18-21 (Dec. 1949) — *Hydrogen exhaustion and explosions of stars*. Nature **168**, 72-73 (July 14, 1950) — *The creation of the Universe*. Viking, New York (1952) — *The role of turbulence in the evolution of the Universe*. Phys. Rev. **86** [II], 251 (1952) — *The creation of the Universe*. Viking Press, New York (1952) — *Evolutionary Universe*. Scient. Am. **195**, 136-140 (Sept. 1956) — *Physics: foundations and frontiers*. Prentice Hall, Englewood Cliffs, NJ (1960) — *The great physicists from GALILEO to EINSTEIN*. Harper & Bros., New York (1961). Dover Publ., New York (1988) — *Thirty years that shook physics: the story of quantum theory*. Doubleday, Garden City, NY (1966). Dover Publ., New York (1985) — With M. YCAS: *History of the Universe*. Science **158**, 766-769 (Nov. 10, 1967) — *My world line, an informal autobiography* [with a bibliography of GAMOW's scientific and popular writings]. Viking, New York (1970) — (F. REINES, ed.) *Cosmology, fusion and other matters*. George Gamow memorial volume. Hilger, London (1972).

**SECONDARY LITERATURE.** P.G. ROLL and D.T. WILKINSON: *Measurement of cosmic background radiation at 3.2 cm wavelength*. Ann. Phys. **44**, 289-321 (1967) — *The swashbuckling physicist. A talk with George GAMOW*. Phys. Today **21**, 101-103 (Feb. 1968) — R.A. ALPHER and R. HERMAN: *G. GAMOW dies; nuclear and astrophysicist was popular writer*. Ibid. **21**, 102-103 (Oct. 1968) — Anonymous: *G. GAMOW*. Nature **220**, 723 (1968) — (Ed.) *George GAMOW 1904-1968*. Astronaut. Acta **14**, 690 (1968/1969) — R.H. STUEWER: *G. GAMOW*. In: (C.C. GILLESPIE, ed.) *Dictionary of scientific biography*. Scribner, New York, vol. 5 (1972), pp. 271-273 — A.M. BYKOV (ed.): *George GAMOW and astrophysics*. In: *Jubilee Gamow Seminar* [St. Petersburg, Russia, 1994]. Kluwer, Dordrecht (1995) — *Celebration of GAMOW's birth: 90 years later*. In: *Selected papers of an international conference dedicated to the memory of George GAMOW and to the 20th anniversary of the city of Odessa* [Odessa, Russia, Sept. 1994]. Gordon & Breach, Amsterdam (1996) — H. BETHE: *Influence of GAMOW on early astrophysics and on early accelerators in nuclear physics*. In: (E. HARPER, W.C. PARKE, and G.D. ANDERSON, eds.) *The George Gamow Symp.* [Washington, DC, April 12, 1997]. Astronomical Society of the Pacific (ASP), San Francisco (1997). Conf. series, vol. 129, p. 44. — E.P. HARPER: *Getting a bang out of GAMOW*. GW Magazine (George Washington University), p. 14 (Spring 2000).

**PICTURE.** Photo Deutsches Museum, Munich, Germany.

# GLASS, Irvine Israel (1918–1994)

- Polish-born Canadian aeronautical engineer and aerospace scientist

Irvine I. GLASS was born at Slupia Nowa (near Kielce, Poland) and emigrated with his parents to Canada at the age of 12. At

the University of Toronto he studied aeronautical engineering, where he received his B.Sc. in engineering physics (1947) and his M.Sc. in aeronautical engineering (1948). As a Ph.D. candidate he entered the newly founded University of Toronto Institute of Aerophysics (UTIA), later renamed the University of Toronto Institute for Aerospace Studies (UTIAS). After obtaining his Ph.D. (1950) in aerophysics, he established and headed the Gas Dynamics and Shock-Wave Phenomena Group until his retirement (1983), and thereafter worked as professor emeritus.

Together with his numerous students and research associates, many also from abroad, he contributed within a period of more than 40 years to a wide field in shock wave physics: (1) Theoretical and experimental study of shock tubes; (2) fundamentals of shock wave and rarefaction wave dynamics; (3) nonstationary oblique shock wave interactions in perfect, imperfect, and dusty gases with various boundaries and establishment of gasdynamic criteria for transition between various reflection configurations; (4) high-temperature non-equilibrium supersonic plasma flow phenomena behind strong shock waves; (5) shock-induced chemical kinetics; (6) development of hypersonic launchers; (7) study and simulation of sonic boom phenomena and their effects on humans and animals; (8) synthesis of new materials using shock implosion techniques; and (9) shock effects in dusty gases. He wrote the popular book *Shock Waves and Man* (1974), which interprets the rapidly growing field of shock wave research to a wide public. His textbook *Nonstationary Flows and Shock Waves* (1994), written together with Jean P. SISLIAN, covers the use of shock tubes to investigate physical and chemical-reactive effects in supersonic and hypersonic flows.

Prof. GLASS is author or coauthor of more than 200 papers and became a leading authority on shock waves. He served at the University of Toronto for more than 45 years and was named a Distinguished University Professor (1981), the highest honor that the university can bestow on one of its faculty members. He was also a visiting professor at Imperial College, London (1957–1958), Kyoto University (1975), and Haifa's Institute of Technology and awarded an honorary professorship from the Chinese Aeronautical Institute at Nanjing

(1985). He was a cofounder of the journal *Shock Waves* (Springer, Berlin Heidelberg New York) and its first editor-in-chief (1990–1994). In the first issue of this journal (1991), he reviewed the research activities and problems tackled by him and his team at UTIAS over the preceding four decades.

**ORIGINAL WORKS.** *The design of a wave interaction tube.* UTIA Rept. No. 6 (1950) — *On the speed of sound in gases.* J. Aeronaut. Sci. **19**, 286-287 (1952) — With W.A. MARTIN and G.N. PATTERSON: *A theoretical and experimental study of the shock tube.* UTIA Rept. No. 2 (1953) — *On the interaction of two similarly facing plane shock waves.* J. Appl. Phys. **25**, 1549-1550 (1955) — *Experimental and theoretical aspects of shock wave attenuation.* Ibid. **26**, 113-120 (1955) — With G.N. PATTERSON: *A theoretical and experimental study of shock tube flows.* J. Aeronaut. Sci. **22**, 73-100 (1955) — With J.G. HALL: *Determination of the speed of sound in sulfur hexafluoride in a shock tube.* J. Chem. Phys. **27**, 1223 (1957) — With J.G. HALL: *Shock sphere, an apparatus for generating spherical flows.* J. Appl. Phys. **28**, 424-425 (1957) — With L.E. HEUCKROTH: *An experimental investigation of the head-on collision of spherical shock waves.* Phys. Fluids **2**, 542-546 (1959) — With G. HALL: *Shock tubes.* In: *Handbook of supersonic aerodynamics.* NavOrd Rept. No. 1488, U.S. Govt. Printing Office, Washington, DC (1959), vol. 6, sect. 18 — *Aerodynamics of blasts.* Can. Aeronaut. Space J. **7**, No. 3, 109-135 (1961) — *On the one-dimensional overtaking of a shock wave by a rarefaction wave.* ARS J. **31**, 1453-1454 (1961) — With L.E. HEUCKROTH: *The hydrodynamic shock tube.* Phys. Fluids **6**, 543-547 (1963) — With A. TAKANO: *Non-equilibrium expansion flows of dissociated oxygen and ionized argon around a corner.* Progr. Aeronaut. Sci. **6**, 163-249 (1965) — *Research frontiers at hypervelocities.* Can. Aeronaut. Space J. **13**, 348-367, 401-425 (1967) — With R.F. FLAGG: *Explosive-driven spherical implosion waves.* Phys. Fluids **11**, 2282-2286 (1968) — With C.K. LAW: *Diffraction of a strong shock wave by a sharp corner.* CASI Transactions **4**, 2-12 (1971) — With H.L. BRODE and A.K. OPPENHEIM: *Gasdynamics of explosions today.* In: (J.L. STOLLERY ET AL., eds.) *Shock tube research.* Chapman & Hall, London (1971), pp. 2:1-56 — *Appraisal of UTIAS implosion-driven hypervelocity launchers and shock tubes.* Progr. Aerospace Sci. **13**, 223-291 (1972) — With H.S. RIBNER and J.J. GOTTLIEB: *Canadian sonic-boom simulation facilities.* Can. Aeronaut. Space J. **18**, 235-246 (1972) — With H.L. BRODE and S.K.L. CHAN: *Strong planar shock waves generated by explosively-driven spherical implosions.* AIAA J. **12**, 367-374 (1974) — *Shock waves and man.* University of Toronto Press, Toronto (1974) — With S.P. SHARMA: *Production of diamonds from graphite using explosive-driven implosions.* AIAA J **14**, 402-404 (1976) — *Terrestrial and cosmic shock waves.* Am. Sci. **65**, 473-481 (1977) — With W.S. LIU and F.C. TANG: *Effects of hydrogen impurities on shock structure and stability in ionizing monatomic gases. Part 1: Argon.* J. Fluid Mech. **84**, 55-77 (1978); *Part 2: Krypton.* Can. J. Phys. **55**, 1269-1279 (1977) — With G. BEN-DOR: *Nonstationary oblique shock-wave reflections: actual isopycnics and numerical experiments.* AIAA J. **16**, 1146-1153 (1978) — With W.S. LIU and F.C. TANG: *Radiation-induced shock tube flow nonuniformities in ionizing argon.* Phys. Fluids **23**, 224-225 (1980) — With M. SHIROUZU: *An assessment of recent results on pseudo-stationary oblique-shock-wave reflections.* UTIAS Rept. No. 264 (1982) — With J.H. LEE: *Pseudo-stationary oblique-shock-wave reflections in frozen and equilibrium air.* Progr. Aerospace Sci. **21**, 33-80 (1984) — *Some aspects of shock wave research* [AIAA Dryden Lecture in Research]. In: *AIAA 24th Aerospace Sciences Meeting* [Reno, NV, Jan. 1986]; see AIAA J. **25**, 214-229 (1987) — *Over forty years of continuous research at UTIAS on nonstationary flows and shock waves.* Shock Waves **1**, 75-86 (1991) — With J.P. SISLIAN: *Nonstationary flows and shock waves.* Clarendon Press, Oxford (1994).
**SECONDARY LITERATURE.** *Who's Who in America.* Marquis, New Providence, NJ (1947–1949) — Annu. Progr. Rept. [with short biographies of academic staff]. UTIAS (1991/1992), p. 14 — Anonymous: *Irvine GLASS, leading expert on shock waves.* Toronto Star, P A13 (Oct. 8, 1994) — O. IGRA: *In memoriam Prof. Irvine Israel GLASS (1918–1994).* Shock Waves **4**, 169-170 (1995) — A. HERTZBERG: *Irvine I. GLASS Memorial Lecture.* University of Toronto (1996).
**PICTURE.** From: Shock Waves **4**, 169 (1995).

# GLAUERT, Hermann (1892–1934)

- British aeronautical engineer

Hermann GLAUERT was born in Sheffield to Louis GLAUERT, a naturalized British citizen of German birth who settled in England as a young man. Hermann GLAUERT was educated at the local King Edward VII School and at Trinity College in Cambridge, where he gained the Tyson Medal for astronomy (1913), the Isaac Newton Studentship in astronomy and physical optics (1914), and the Rayleigh Prize for mathematics (1915). At the outbreak of World War I he studied astronomy at Cambridge but later turned to aeronautics, accepting an appointment on the staff of the Royal Aircraft Establishment (RAE) at Farnborough (1916–1934). GLAUERT entered the aerodynamics research division and later became Principal Scientific Officer and head of the aerodynamic department. His initial studies related to the analysis of experimental work resulting from the engineering design of aeroplanes and were laid down in numerous reports and memoranda published by the Aeronautical Research Committee (1917–1920). This stimulated him to further develop existing aerodynamic theories by Frederick LANCESTER, Wilhelm KUTTA, and Ludwig PRANDTL and to apply them to all problems of flight, particularly with a view to its use by engineers. His book *The Elements of Aerofoil and Airscrew Theory* (1926), which quickly disseminated PRANDTL's airfoil and wing theory around the English-speaking world, became the standard reference source on incompressible flow and was later also translated into German

(1929). He was one of the first in England who contributed to the theory of flight at high speed and the effect of compressibility on the lift force of an airfoil (1928), thereby, independently of PRANDTL in Göttingen, working out a rule for subsonic airflow that describes the compressibility effects of air at high speeds, the so-called "Prandtl-Glauert rule."

GLAUERT was killed by a falling tree in an explosion accident at Farnborough while walking in Fleet Common Park with his children and watching the demolition of a tree from a position that he was told was safe.

**ORIGINAL WORKS.** With R. WOOD: *Preliminary investigation of multiplane interference applied to propeller theory*. Aeronautical Research Commission, Reports & Memoranda (ARC-R&M) No. 620 (1918) — *Aerofoil theory*. ARC-R&M No. 723 (1921) — *An aerodynamic theory of the airscrew*. ARC-R&M No. 786 (1922) — *Note on the vortex theory of airscrews*. ARC-R&M No. 869 (1922) — *The elements of aerofoil and airscrew theory*. Cambridge University Press (1926); Germ. translation by H. HOLL: *Die Grundlagen der Tragflügel- und Luftschraubentheorie*. Springer, Berlin (1929) — *A non-dimensional form of the stability equations of an airplane*. ARC-R&M No. 1093 (1927) — *Theoretical relationship for an airfoil with hinged flap*. ARC-R&M No. 1095 (1927) — *The effect of compressibility on the lift of an aerofoil*. Proc. Roy. Soc. Lond. **A118**, 113-119 (1928) — *The characteristics of a Kármán vortex street in a channel of finite breadth*. Ibid. **A120**, 34-46 (1928) — *The force and moment of an oscillating airfoil*. ACR-R&M No. 1242 (1929) — *Wind tunnel interference on wings, bodies and air screws*. ACR-R&M No. 1566, H.M.S.O., London (1933).
**SECONDARY LITERATURE.** W.S. FARREN and H.T. TIZARD: *Hermann GLAUERT 1892–1934*. Obit. Not. Fell. Roy. Soc. (Lond.) **1**, 607-610 (1935) — (Ed.) *Hermann GLAUERT*. Z. Flugwiss. **4**, 396 (1956) — J.D. ANDERSON JR.: *Modern compressible flow, with historical perspective*. McGraw-Hill, New York (1990), pp. 289-290.
**PICTURE.** © The Godfrey Argent Studio, London, U.K.

# GOLDBERG, Viktor (1903–1978)

- Latvian urologist

Viktor GOLDBERG was born in Riga, then capital of the Russian province of Latvia. After attending the gymnasium in Riga he began to study medicine at the University of Würzburg, Germany, and took his M.D. at the University of Vienna (1927). His first appointment was a position as an assistant physician at the Department of Surgery of the Bikur-Chalim Hospital in Riga (1927–1929). Thereafter he went to Berlin where he worked at the department of urology of the St. Hedwig Hospital under the guidance of the German surgeon Alexander VON LICHTENBERG, who introduced intravenous urography (pyelography) in order to diagnose nephritic disorder.

GOLDBERG studied the connection between the specific density of urine and the physiological processes in the kidney, and, together with Isidor TRAUBE and Konstantin SKUMBURDIS from the Technische Hochschule Berlin-Charlottenburg, he investigated the colloid-chemical generation of kidney stones. After suffering a period of racial persecution because of his Jewish ancestry, GOLDBERG resumed his research only at the end of World War II. In Riga he became head physician at the First Municipal Clinical Hospital, directing also the department of urology of the Medical Institute Riga. He also invented a new radiographic method for diagnosing the lower urethra system by incorporating a contrast liquid (genitography, 1946) and constructed a new apparatus for removing calculi debris after shock wave treatment, which he called the "hydrodynamic evacuator" (litholapaxie, 1977). His greatest contribution to medicine, however, was the first demonstration of transurethral shock lithotripsy and also its further clinical development for routine use. During his retirement he moved to Stuttgart (1973–1978), where he worked at Prof. Hans J. REUTER's hospital until his death.

**ORIGINAL WORKS.** With I. TRAUBE and K. SKUMBURDIS: *Beiträge zum Problem der Nierensteinbildung*. Münchener Medizinische Wochenschrift **79**, 1083-1085 (1932) — *Studium über Konkrementbildung und Verhinderung der Steinentstehung in den Harnwegen*. Z. Urolog. Chir. **35**, 347-358 (1932) — *Zur Geschichte der elektrohydraulischen Lithotripsie (EHL)*. In: (E. SCHMIEDT, ed.) Symp. Biophysikalische Verfahren zur Diagnose und Therapie von Steinleiden der Harnwege. BMFT Symp., Meersburg, Germany (1976). Wissenschaftliche Berichte, Dornier System GmbH, Friedrichshafen, Germany, pp. 130-131.
**SECONDARY LITERATURE.** (S. TAYLOR, ed.) *Who's who in central and east Europe*. Central European Times, Zurich (2nd edn., 1937) — M. KLAUSS-GOLDBERG: *Zur Geschichte der elektrohydraulischen Lithotripsie: V. GOLDBERG: Eine neue Methode der Harnsteinzertrümmerung, elektrohydraulische Lithotripsie*. Der Urologe **19B**, 23-27 (1979) — A. HUTTMANN: *Von der transurethanen elektrohydraulischen Lithotripsie zur extrakorporalen Stoßwellenzertrümmerung von Harnsteinen*. Ibid **28B**, 220-225 (1988) — *Viktor GOLDBERGs literaturas saraksts*. This list of GOLDBERG's publications was compiled by the Medicīnas Zinātniskā Bibliotekā, Rīga, Latvia (1999) — M.A. REUTER (with R.M. ENGEL and H.J. REUTER): *History of endoscopy*. Kohlhammer, Stuttgart (1999), p. 481 — R.M. ENGEL: *De historia lithotomiae* [includes electrohydraulic and laser lithotripsy]. Curator of the William P. Didusch Museum of the American Urology Association, Baltimore, MD; http://www.urolog.nl/urolog/php/content.php?doc=lithotomia&profmenu=yes.
**PICTURE.** Courtesy Mrs. G. TIFONOWA, Medicīnas Zinātniskā Bibliotēka, Riga, Republic of Latvia.

# HADAMARD, Jacques Salomon (1865–1963)

- French mathematician and theoretical fluid dynamicist

Jacques S. HADAMARD was born in Versailles. His father Amadée HADAMARD was a professor of Latin and his mother taught piano. After studying at the Ecole Normale Supérieure (1884–1888), he taught at the Lycée Buffon in Paris (1890–1893), where he also received his Ph.D. in sciences (1892). He began as a lecturer at the University of Bordeaux (1893–1897) and the Sorbonne (1897–1909). Returning to Paris, he became entrance examiner (1910–1911) and then professor (1912–1937) at the Ecole Polytechnique and the Ecole Centrale des Arts et Manufactures (1920–1937). He also taught at the Collège de France (1907–1937), where he succeeded Prof. Camille JORDAN as chair of analysis (1912). Here he established a seminar (1913) that soon became a favorite meeting place for leading mathematicians.

In mathematics, HADAMARD contributed to (1) the analytic continuation of a Taylor series and the distributions of the singularities of the series in terms of the nature of its coefficients; (2) the theory of functions of a complex variable; (3) the famous problem concerning the distribution of the prime numbers; (4) differential geometry; (5) the solution to his maximum determinant problem, the so-called "Hadamard matrix" of order $n$ which can be used to make error-correcting codes; and (6) the problem of solving equations with partial derivatives. He also addressed various problems of hydrodynamics, mechanics, and probability theory of his time. In his book *Leçon sur la propagation des ondes et les équations de l'hydrodynamique* (1903), he treats waves of finite amplitude in a general manner using his "method of operator differences." This book, a result of his teaching activities in the period 1898–1900 and fallen almost into oblivion, is now increasingly being cited by present shock physicists as an important early source of the theoretical treatment of shocks. The starting point of getting involved in shock waves was Pierre-Henri HUGONIOT's work on the rectilinear movement of a gas, or rather Pierre DUHEM's exposition of it in his lectures and in conversations that took place in the years 1893 to 1897. He termed shock discontinuities *ondes d'accélération* ("acceleration waves") to underline the steepening process, but he also used the designation *onde de choc* ("shock wave") to illustrate the wave nature of the propagating discontinuity. HADAMARD theoretically investigated vorticity effects in shock waves and discovered that curved shock waves produce vorticity, while a vortex-free flow, passing through a plane shock wave, remains vortex-free – the so-called "Hadamard theorem."

HADAMARD also first investigated analytically the formation of shock waves in elastic solids. In 1923, in a series of lectures given at Yale University, he formulated in mathematical terms three different meanings of the "Huygens principle" he found in the literature of his time. The so-called "Hadamard problem" consists in classifying all second-order hyperbolic operators that obey the Huygens principle, up to trivial relations (1932). A *Hadamard material* is one in which longitudinal waves may propagate in every direction when the material is homogeneously deformed.

HADAMARD wrote about 300 scientific papers and several books for a wider audience. He was an associate member of several foreign academies and held honorary doctorates from many foreign universities.

**ORIGINAL WORKS.** *Leçon sur la propagation des ondes et les équations de l'hydrodynamique.* A. Hermann, Paris (1903); Chelsea, New York (1949) — *Remarque au sujet de la note de M. Gyözö ZEMPLEN.* C. R. Acad. Sci. Paris **141**, 713 (1905) — *Recherches sur les solutions fondamentales et l'intégration des équations linéaires aux dérivées partielles (2e Mém.).* Ann. Ecole Norm. Sup. **22**, 101-141 (1905) — *Sur les ondes liquides.* C. R. Acad. Sci. Paris **150**, 609-611, 772-774 (1910) — *Cours d'analyse professé à l'Ecole Polytechnique.* A. Hermann, Paris (1927) — *Lectures on CAUCHY's problem in linear partial differential equations* [Silliman Lecture]. Yale University Press, New Haven, CT (1928) — *Le problème de CAUCHY et les équations aux dérivées partielles linéaires hyperboliques.* A. Hermann, Paris (1932) — *An essay on the psychology of invention in the mathematical field.* Princeton University Press, Princeton, NJ (1949) — *La théorie des équations aux dérivées partielles.* Gauthier-Villars, Paris (1964) — *Œuvres de Jacques HADAMARD.* 4 vols., Editions du CNRS, Paris (1968).

**SECONDARY LITERATURE.** C. TRUESDELL and R. TOUPIN: *The classical field theories.* In: (S. FLUEGGE, ed.) *Handbuch der Physik.* Springer, Berlin (1960), vol. III/1 — A.S. PREDVODITELEV: *HADAMARD's method of operator differences, and applications.* In: (A.S. PREDVODITELEV, ed.) *Gas dynamics and physics of combustion.* Israel Program for Scientific Translations, Jerusalem (1962), pp. 3-39 — H. HEILBROWN and L. HOWARTH: *Jacques HADAMARD.* Nature **200**, 937-938 (1963) — M.L. CARTWRIGHT: *J. HADAMARD.* J. Lond. Math. Soc. **40** [I], 722-748 (1965); Biogr. Mem. Fell. Roy. Soc. (Lond.) **11**, 75-99 (1965) — P. LEVY, S. MANDELBROJT, B. MALGRANGE, and P. MALLIAVIN: *La vie et l'œuvre de Jacques HADAMARD.* L'Enseignement Mathématique, No. 16. Université de Genève (1967) — P. CURRIE and M. HAYES: *Longitudinal and transverse waves in finite elastic strain. Hadamard and Green materials.* IMA Appl. Math. **5**, No. 2, 140-161 (1969) — P. CURRIE: *A note on shock waves in elastic Hadamard and Green materials.* ZAMP **22**, 355-359 (March 1971) — S. MANDELBROJT: *J. HADAMARD.* In: (C.C. GILLESPIE, ed.) *Dictionary of scientific biography.* Scribner, New York, vol. 6 (1972), pp. 3-5 — A.J. WILLSON: *Surface waves in restricted*

*Hadamard materials*. Pure Appl. Geophys. **110**, No. 1, 1967-1976 (Dec. 1973) — V. MAZ'YA and T. SHAPOSHNIKOVA: *Jacques HADAMARD, a universal mathematician*. Am. Math. Soc. & Lond. Math. Soc., Providence, RI (1990).
**PICTURE.** In the public domain, taken from Wikipedia, the free encyclopedia; http://fr.wikipedia.org/wiki/Image:Hadamard.jpg.

# HAEUSSERMANN, Carl (1853–1918)

- German chemist

The son of a Protestant minister, Carl HAEUSSERMANN was born in Stuttgart. After attending the gymnasium, he finished an apprenticeship in a local pharmacy, whereupon he decided to study chemistry at the Polytechnikum Stuttgart (1869) and the TH Munich (1871–1873), where he became private assistant of Prof. Emil ERLENMEYER, an experimental chemist. Working in the first instance as a chemist in the chemical industry and being involved in the fabrication of nitrobenzene and aniline dyes and in oil refinery, he took his Ph.D. at the University of Heidelberg (1876) and shortly afterwards habilitated in Stuttgart (1877) on the fabrication of aniline dyes. He also worked temporarily in the French dye industry at Paris (1878–1880) and several years as a consultant to the Chemische Fabrik Griesheim-Elektron, which offered him a leading position. He became assistant director and later member of the board, supervising the fabrication of aniline dyes and derivatives (1885–1891). In addition, with the permission of the Prussian War Ministry he acted as consulting chemist to the Royal Powder Mills in Hanau. This activity brought him into close contact with actual problems of nitroexplosives, particularly with their manufacturing techniques and explosive properties.

During his stay at Chemische Fabrik Griesheim-Electron he became aware of trinitrotoluol or TNT (hexanitrodiphenylamine) and began to work out a method of manufacturing that became the standard method for many years. In 1891, he first discovered the explosive properties of trinitrotoluol.

TNT is very stable, rather insensitive to shock, almost insoluble in water, neutral, and does not attack metals. The commercial production of TNT began in Griesheim around 1893, but production on a large scale did not start until the early 20th century. He first suggested the military use of TNT and recommended it as a filling for shells, a practice that was followed as early as 1901. Because of these outstanding features it became by far the most important explosive for blasting charges of all kinds of weapons in both world wars.

HAEUSSERMANN held the Chair of Chemical Technology at the Technische Hochschule Stuttgart (1891–1906). He died of a heart attack on a business trip to the moorlands of Günzburg/Ulm, which he intended to inspect as a possible new fuel resource. His pioneering contributions to the chemistry of nitro explosives, nitrocelluloses, and combustibles made him internationally renowned.

**ORIGINAL WORKS.** *Über die explosiven Eigenschaften des Trinitrotoluols*. Z. Angew. Chem. **4**, 508-511 (1891) — *Verfahren zur Herstellung von Trinitrotoluol*. Ibid. **4**, 661-662 (1891) — *Les matières explosives en 1892*. Moniteur Sci. **7**, 861-863 (1893) — *Sprengstoffe und Zündwaren*. Metzler, Stuttgart (1894) — *Brenn- und Explosivstoffe* [Jahresfachreferat]. In: (R. MEYER, ed.) Jb. Chemie. Vieweg, Braunschweig (1897–1914), vol. VII-XXIV — *Wesen und Wirkungsweise der modernen Explosivstoffe*. Verein vaterländ. Naturkunde (Württ.) **59**, 328-335 (1903) — *Zur Kenntnis der Nitrocellulose*. Ber. Dt. Chem. Gesell. **36**, 3956 (1903), Ibid. **37**, 1624-1625 (1904) — *Alfred NOBEL und die Erfindung der Nitroglyzerinpulver*. Union Dt. Verlagsgesell., Stuttgart (1904) — *Über die Denitrierung der Pyroxyline*. Chemiker-Z. **29**, 420-422 (1905) — *Zur Kenntnis der Pyroyline*. Z. ges. Schieß- u. Sprengstoffwesen **1**, 305 (1906) — *Zur Kenntnis der Xyloidine*. Ibid. **1**, 39 (1906); Ibid. **2**, 426 (1907); Ibid. **3**, 305-306 (1908) — *Über die Einwirkung der Salpetersäure auf Cellulose*. Ibid. **3**, 123-124 (1908) — *Gedächtnisrede auf Christian Friedrich SCHÖNBEIN*. Z. ges. Schieß- u. Sprengstoffwesen **4**, 433-434 (1909) — *Über das Verhalten der Zellulose gegen reine Salpetersäure*. Z. Angew. Chem. **23**, 1761-1763 (1910); Ibid. **26**, 456 (1913) — *Über die explosiven Eigenschaften des Trinitrotoluols*. Z. ges. Schieß- u. Sprengstoffwesen **8**, 378-379 (1913) — *Die Nitrozellulosen, ihre Bildungsweisen, Eigenschaften und Zusammensetzung*. Vieweg, Braunschweig (1914).
**SECONDARY LITERATURE.** (Ed.) *Carl HAEUSSERMANN †*. Z. ges. Schieß- u. Sprengstoffwesen **13**, 325-327 (1918) — C. HELL: *Carl HAEUSSERMANN †*. Z. Angew. Chem. **31**, 413-414 (1918) — A. SCHWEITZER: *Carl HAEUSSERMANN †*. Chemiker-Z. **42**, 397-398 (1918) — B. LEPSIUS: *Carl HAEUSSERMANN*. Ber. Dt. Chem. Gesell. **51**, 1683-1685 (1918).
**PICTURE.** Photo Deutsches Museum, Munich, Germany.

# HERMANN, Rudolph (1904–1991)

- German-born U.S. physicist and aerospace engineer

Rudolph HERMANN was born in Leipzig, where he studied physics, mathematics, and astronomy at the university and

also received his Ph.D. (1929). After serving there as an assistant in the Department of Applied Mechanics and Thermodynamics (1929–1933), he accepted a research fellowship offered by the TH Aachen (1933–1934), where he became assistant at Prof. Carl WIESELSBERGER's Institute of Aerodynamics. Heading there the Supersonic Wind Tunnel Division (1934–1937), he began much pioneering work at the $10 \times 10$ cm$^2$ supersonic wind tunnel ($M = 3.3$), performing for the Heereswaffenamt (Army Ordnance Office) first aerodynamic tests of the A-3 rocket, a fore-runner of the A-4 (later known as the "V2"). In addition, he habilitated (1935) and lectured in supersonic aerodynamics (1936–1937). Selected by the German rocket pioneer Werner VON BRAUN to install the new supersonic aerodynamic institute at the Heeresversuchsanstalt (HVA) Peenemünde-Ost, the Rocket Research & Test Facility of the German Army, he became its first and only director (1937–1945). Compared to the Aachen facility, his team began to install a more advanced, $40 \times 40$ cm$^2$ wind tunnel that, depending on the selected nozzle geometry, later allowed one to choose different Mach numbers ranging from 1 to 4.4. His major research activities were devoted to the model studies of the A-4 and A-5 rockets in the supersonic wind tunnel, particularly to (1) the investigation of the dependency of the trajectory on the inclination angle of the launching track and the acceleration due to thrust; (2) three-component measurements of aerodynamic forces; (3) the determination of the aerodynamic characteristics of missiles and location of the center of pressure; and (4) calculations of the skin and warhead temperature. After evacuation of the test facility to the small town of Kochel in the Bavarian Alps (1943), HERMANN also directed, until the end of war, the Wasserbauversuchsanstalt or WVA (Hydraulic Engineering Testing Facility) – a code name for the secret aerodynamic institute. He supervised plans to install a more advanced, huge hypersonic wind tunnel ($1 \times 1$ m$^2$, $M = 10$), which, however, never opened because of Germany's defeat.

After the war HERMANN served as consultant for hypersonic wind tunnels and ramjet engines to the Wright-Patterson Air Force Base in Dayton, OH (1945–1950). He was professor of aeronautical engineering at the University of Minnesota, heading there also the hypersonic facilities at Rosemount Aeronautical Laboratories (1950–1962). He eventually transferred to the University of Alabama in Huntsville and was appointed professor of physics and aerospace engineering at the University of Alabama, directing also its research institute. HERMANN contributed to various fields of fluid dynamics, such as pipe friction, free convection heat transfer, viscosity of non-Newtonian fluids, fin-stabilized projectiles, guided missiles, supersonic and hypersonic wind tunnels and diffusers, film cooling, hypersonic physics, and reentry of satellite vehicles.

After his retirement (1969) he gave guest lectures at various international research institutions and dedicated himself to the history of science and technology. He published one book and over 100 scientific papers.

**ORIGINAL WORKS.** *Experimentelle Untersuchungen zum Widerstandsgesetz des Kreisrohres bei hohen Reynolds'schen Zahlen und großen Anlauflängen.* Dissertation, Universität Leipzig (1929); Akad. Verlagsgesell. Leipzig (1930) — With T. BURBACH: *Strömungswiderstand und Wärmeübergang in Rohren.* Akad. Verlagsgesell., Leipzig (1930) — *Turbulenzentstehung bei Wärmeübergang durch freie Konvektion an senkrechter Platte und waagerechtem Zylinder.* ZAMM **13**, 433-434 (1933) — *Wärmeübergang bei freier Konvektion am waagerechten Zylinder in zwei-atomigen Gasen.* Habilitationsschrift, T.H. Aachen (1935); VDI-Forschungsheft Nr. 379 (July/Aug. 1936) — *Der Kondensationsstoß in Überschall-Windkanaldüsen.* Luftfahrtforsch. **19**, 201-209 (1942) — *Entwicklung flügelstabilisierter Geschosse zum Zwecke der Leistungssteigerung.* Schriften Dt. Akad. Luftfahrtforsch. Nr. 1059/43 [gKdos, geheime Kommandosache], 9-31 (1943/1944) — *The supersonic wind tunnel of the Heereswaffenamt and its application in external ballistics.* Interner Forschungsbericht, Kochel, Bavaria (June 16, 1945). Microfilm Mi 56-4553 [see Natl. Union Cat., Library of Congress, Washington, DC (1972), vol. 242, p. 259] — *Theoretical calculations of the diffuser efficiency of supersonic wind tunnels with free jet test section.* Heat Transfer and Fluid Mechanics Institute, Berkeley, CA (1949). ASME, New York (May 1949), pp. 255-270 — *Diffuser efficiency and flow process of supersonic wind tunnels with free jet test section.* Air Force Tech. Rept. No. 6334, Wright Field, OH (Dec. 1950) — *Diffuser efficiency of free-jet supersonic wind tunnels at variable test chamber pressure.* J. Aeronaut. Sci. **19**, 375-384 (1952) — *Supersonic diffuser problems for inlet ducts and wind tunnels in one-dimensional analysis.* Proc. 2nd Midwestern Conf. Fluid Mechanics [Ohio State Univerity, Columbus, OH, March 1952]. In: (A. TIFFORD, ed.) *Ohio State University studies. Engineering series.* College of Engineering, Ohio State University, Columbus, OH (1952), vol. 21, No. 3, pp. 231-242 — *Supersonic inlet diffusers and introduction to internal aerodynamics.* Minneapolis-Honeywell, MN (1958) — *Supersonic inlet diffusers and introduction to internal aerodynamics.* Minneapolis-Honeywell Regulator, Minneapolis, MN (1958) — *The supersonic wind tunnel installations at Peenemünde and Kochel, and their contributions to the aerodynamics of rocket-powered vehicles.* Selected papers from the 32nd Int. Astronautical Congr. [Rome, Sept. 1981]. In: (L.G. NAPOLITANO, ed.) *Space: mankind's fourth environment.* Pergamon Press, Oxford (1982), pp. 435-446.

**SECONDARY LITERATURE.** W. KRAUS: *Der Überschall-Windkanal von Peenemünde.* Interavia **6**, 558-561 (1951) — H. KURZWEG: *The aerodynamic development of the V2.* AGARD First Guided Missiles Seminar [Munich, April 1956]. In: (T. BENECKE and A.W. QUICK, eds.) *History of German guided missile development.* Appelhaus, Brunswick (1957), pp. 50-69 —

H.O. RUPPE: *Rudolf HERMANN 65 Jahre. Z. Flugwiss.* **17**, 459 (1969) — (Ed.) *Zum 70. Geburtstag von R. HERMANN.* Ibid. **22**, 436 (1974) — P.P. WEGENER: *The Peenemünde wind tunnels.* Yale University Press, New Haven, CT (1996).
**PICTURE.** Photo Deutsches Museum, Munich, Germany.
**NOTE.** Shortly after World War II, the original technical reports of the Aerodynamic Institute, covering both the periods at Peenemünde and at the WVA in Kochel, were transferred to the U.S. Army and compiled by its Ordnance Dept. as "German Documents (GD);" see WEGENER's book cited above, pp. 90-96.

# HERTZ, Heinrich Rudolf (1857–1894)

- German physicist; pioneer of theoretical collision mechanics

Heinrich R. HERTZ was born in Hamburg into a prosperous and cultured family. His father, Gustav HERTZ, was a barrister and later a senator. Educated in modern and ancient languages, he passed his Abitur at the Johanneum Gymnasium (1875) in Hamburg. He prepared for an engineering career in Frankfurt, Dresden, and Munich, but finally decided to attend a university instead of a polytechnic school. After spending a year in Munich, he continued his studies in Berlin (1878), in addition to establishing close contact with Hermann VON HELMHOLTZ and Gustav KIRCHHOFF and winning a prize awarded by the Berlin Philosophical Faculty on an experimental problem of electrical inertia (1879). He wrote his Ph.D. thesis on electromagnetic induction (1880) and worked as an assistant to VON HELMHOLTZ at the Berlin Physical Institute.

At this time he became interested in the theory of compression of elastic bodies, and in January 1881 he presented his famous paper on the classical problem of collision (publ. 1882). HERTZ not only offered the general solution of the problem, but also applied it to particular cases and even prepared a numerical table to facilitate practical applications. Extending his theory to the impact of two spheres, he derived formulas for calculating the stress and duration of impact. Subsequently he studied the hardness of materials, which is the crucial point in treating the complicated phenomenon of impact.

Followed by a short period as a lecturer at the University of Kiel (1883), he became professor of physics at the Technische Hochschule Karlsruhe (1885–1889), where he performed his famous experiments of broadcasting and receiving radio waves and measuring their length and velocity. Accepting an offer from the Prussian Ministry of Culture, he moved in 1888 to Bonn University and became professor of physics and director of the Physics Institute as successor to Rudolf CLAUSIUS. But during the almost 5 years that he spent in Bonn, HERTZ abandoned almost all experimental work and devoted 3 years to difficult theoretical work on mechanics, which culminated in the posthumous publication in 1894 of his book *Die Prinzipien der Mechanik in neuem Zusammenhange dargestellt* (The Principles of Mechanics Presented in a New Form, London 1899). On January 1, 1894 he died of blood poisoning at the early age of 36.

His name was given to the unit of frequency (hertz, abbreviated Hz), which replaced the use of cycles per second for the unit of frequency in the late 1960s, and electromagnetic waves in the radio and radar spectrum (*Hertzian waves*) are named in honor of him. In mechanics, the *Hertzian crack* is a localized cone-shaped crack (*Hertzian cone*) that appears at the point of contact or low-velocity impact, and in the *Hertzian fracture* test the fracture of a brittle solid is studied under a spherical indenter.

A crater on the far side of the Moon is named for him.

**ORIGINAL WORKS.** *Über die Berührung fester elastischer Körper.* J. Reine u. Angew. Math. **92**, 156-171 (1882) — *Über die Berührung fester elastischer Körper und über die Härte.* Verhandl. des Vereins zur Beförderung des Gewerbefleißes (Berlin) **61**, 449-463 (Nov. 1882) — *Gesammelte Werke.* A. Barth, Leipzig; Bd. III (1894): *Die Prinzipien der Mechanik in neuem Zusammenhang dargestellt.* See chap. *Von den Unstetigkeiten der Bewegung*, pp. 286-306 — *The principles of mechanics: presented in a new form.* Dover Publ., New York (1956).
**SECONDARY LITERATURE.** M. PLANCK: *Gedächtnisrede auf Heinrich HERTZ.* Verhandl. Phys. Gesell. Berlin **13**, 9-29 (1894) — S.P. TIMOSHENKO: *History of strength of materials.* McGraw-Hill, New York (1953) — F. BOPP and W. GERLACH: *Heinrich HERTZ zum 100. Geburtstag am 22.2.1957.* Die Naturwissenschaften **44**, 49-52 (1957) — C. SUSSKIND: *Heinrich HERTZ: a short life.* San Francisco Press (1995) — R. MCCORMMACH: *H. HERTZ.* In: (C.C. GILLESPIE, ed.) *Dictionary of scientific biography.* Scribner, New York, vol. 6 (1972), pp. 340-350 — I. SZABO: *Geschichte der mechanischen Prinzipien und ihre wichtigsten Anwendungen.* Birkhäuser, Basel (1977), pp. 472-477 — J.F. MULLIGAN: *Heinrich HERTZ and the development of physics.* Phys. Today **42**, 50-57 (March 1989) — M.G. DONCEL: *Heinrich Rudolf HERTZ (1857–1894).* In: (K. VON MEŸENN, ed.) *Die großen Physiker.* C.H. Beck'sche Verlagsbuchhandlung, München (1997), vol. 2, pp. 121-142.
**PICTURE.** Archiv der Rheinischen Friedrich-Wilhelms-Universität Bonn, Germany.

# HERTZBERG, Abraham ("Abe") (1922–2003)

- U.S. aerospace engineer, gas dynamicist and inventor; the "Idea Man"

Abraham HERTZBERG was born in the Bronx, NY. Shortly thereafter his family moved to Richmond, VA. He graduated from Virginia Polytechnic Institute with a Bachelor's degree in 1943 and started working at Curtiss-Wright Corporation that same year as an aerodynamicist. Near the end of World War II, he became a flight test engineer for the Army Air Force. Following his discharge from the military in 1946, he returned to school as a student of Prof. Arthur KANTROWITZ at Cornell University, Ithaca, NY, and received the M.S degree in 1949, writing a dissertation on hypersonic theory. Between 1950 and 1954 he took additional graduate-level courses in physics, chemistry, and gas dynamics at the University of Buffalo, NY, but because of his intense commitment to his work (*q.v.*) never pursued a Ph.D.

In 1949, HERTZBERG joined the Cornell Aeronautical Laboratory (CAL), Buffalo, NY. He was promoted to Assistant Head (1957) and Head of the Aerodynamics Research Department (1959–1965). During his sixteen years at CAL, HERTZBERG and his group made basic contributions in the fields of supersonic and hypersonic aerodynamics and hypersonic test facilities. In the 1950s, HERTZBERG proposed and developed a shock tunnel facility, which became accepted throughout the world as a standard hypersonic research tool. He was also responsible for the wave superheater, used for studies of ablative phenomena. Aerodynamic contributions included basic studies of viscous effects at hypersonic speeds and nonequilibrium flow phenomena. In the early 1960s, HERTZBERG carried out pioneering work in the development of gasdynamic lasers and laser-induced fusion systems. In addition, he studied methods of developing high efficiency thermal power plants as well as the use of shock wave and expansion waves in novel propulsion and chemical production systems. Walter KISTLER, a Swiss-born U.S. physicist and inventor, and HERTZBERG developed miniature high-frequency acceleration-compensated quartz pressure sensors with microsecond response time. This research spearheaded the development of shock tube technology crucial to studying the sort of aerodynamic shock waves that spacecraft can encounter during reentry.

In 1966, HERTZBERG was appointed as professor of aeronautics and astronautics, and director of the Aerospace Research Laboratory (ARL) at the University of Washington (ARL was later renamed as the Aerospace and Energetics Research Program, or AERP). In addition to continuing his pioneering work on high power lasers, he engaged in research on laser applications, the study of novel propulsion systems, advanced fusion concepts, and expanded his work on the use of shock waves in chemical production systems. In addition, he developed new concepts in areas such as hypervelocity launchers (the ram accelerator) and automobile propulsion (cryogenic engines). He retired in 1993 but remained active in research until shortly before his death in March 2003. HERTZBERG was the author of more than 100 papers in the fields of gas dynamics, physics of high temperature gases, lasers, space launchers, cryogenic automobile propulsion, and other novel concepts. He also held numerous U.S. patents in these fields.

In addition to his research efforts at the University of Washington, HERTZBERG also immersed himself into the academic life of the department, teaching a variety of courses, serving on numerous committees, and engaging in public outreach. In 1970, he began to offer a yearly special topics graduate course in which he and his students explored many areas of emerging interest, such as the energy crisis, the green revolution, developments in automobile technology, space power systems, efficient energy conversion technologies, beamed power via lasers, *etc*. This was the course for which he was best known, and from which several major funded research programs sprang. In 1979, HERTZBERG initiated the undergraduate senior capstone design course in space systems engineering. Generations of students, both graduate and undergraduate, benefited from his unique teaching style, which combined creativity and humor with rigor and high expectations.

HERTZBERG was also a prolific and much sought-after consultant to industry and government. Among the companies and institutions for which he consulted over the years were Boeing Scientific Research Laboratory; G.E. Missiles & Space Division; Aerospace Corporation; S.T.I. Optronics, Inc. (formerly Mathematical Sciences Northwest, Inc. which he helped found in 1969); Rockwell International; Lockheed Missiles & Space Co.; Brookhaven National

Laboratory; Los Alamos National Laboratory (LANL); Pratt & Whitney; Olin/Rocket Research; and Kistler Aerospace Corporation.

HERTZBERG served on numerous professional and governmental committees and also on the NASA Research and Technology Advisory Council and the USAF Scientific Advisory Board. He chaired the Fluids Subcommittee of the National Research Council Survey of Plasma Physics and Fluids (1983–1984) and was a member of the LANL Advisory Committees (1984–1993). He was a member of the American Physical Society Directed Energy Weapons Study Group (1984–1986), and, beginning in 1978, served on the NASA Space Systems and Technology Advisory Committee. He was also a Fellow of the American Institute of Aeronautics and Astronautics (1976), an elected Fellow of the International Academy of Astronautics (1987), Fellow of the American Association for the Advancement of Science (1995), an elected member of the National Academy of Engineering (1976), and a member of the American Physical Society and Sigma Xi. He was the AIAA Dryden Medallist and received the AIAA Plasmadynamics and Lasers Award (1992). He was Lecturer (1977) and Visiting Lecturer (1983, 1988) at the Chinese Academy of Sciences, Beijing, and Paul Vieille Lecturer at the International Symposium on Shock Tubes and Waves (1969, 1989). In 1996, he presented the First Memorial Lecture to honor Prof. Irvine I. GLASS in Toronto, Canada.

**ORIGINAL WORKS.** With A. KANTROWITZ: *Studies with an aerodynamically instrumented shock tube*. J. Appl. Phys. **21**, 874-878 (1950) — *A shock tube method of generating hypersonic flows*. J. Aeronaut. Sci. **18**, 803-805 (1951) — *Shock tubes for hypersonic flow*. Rept. AF-702-A-1, Cornell Aeronautical Laboratory, Buffalo, NY (1951) — With W.E. SMITH: *A method for generating strong shock waves*. J. Appl. Phys. **25**, 130-131 (1954) — With W. SQUIRE and W.E. SMITH: *Real gas effects in a hypersonic shock tunnel*. CAL Rept. AD-789-A-1, AEDC-TN-55-14, AD 56189 (March 1955) — With H.S. GLICK and W. SQUIRE: *A new shock tube technique for the study of high-temperature gas-phase reactions*. In: *5th Symp. (International) on Combustion* [Pittsburgh, PA, Sept./Aug. 1954]. Reinhold, New York (1955), pp. 393-402 — With J.G. LOGAN: *The application of the shock tube to the study of aerothermal problems of high-speed flight*. Proc. Conf. Chem. Aeronomy [Cambridge, MA, June 1956]. In: (M. ZELIKOFF, ed.) *Threshold of space*. Pergamon Press, New York etc. (1957), pp. 276-287 — *The application of the shock tube to the study of the problems of hypersonic flight*. Jet Propulsion **26**, 549-554 (1956) — *The application of the shock tube to the study of high-temperature phenomena in gases*. Appl. Mech. Rev. **9**, 505-509 (1956) — With A. RUSSO: *A method for improving the performance of shock tubes*. Jet Propulsion **27**, 1191-1193 (1957) — With R.C. WEATHERSTON: *Investigation of rocket flow problems by means of short-duration flow devices*. ARS J. **31**, 1149-1151 (1961) — With I.R. HURLE and J.D. BUCKMASTER: *The possible production of population inversions by gasdynamic methods*. CAL Rept. RH-1670-A-1 (Dec. 1962) — With W.J. RAE: *On the possibility of simulating meteoroid impact by the use of lasers*. CAL Rept. AI-1821-A-1 (April 1964) — With I.R. HURLE: *Electronic population inversions by fluid-mechanical techniques*. Phys. Fluids **8**, 1601-1607 (1965) — With J.W. DAIBER and C.E. WITTLIFF: *Laser-generated implosions*. Ibid. **9**, 617-619 (1966) — With J.M. DAWSON ET AL.: *Controlled fusion using long-wavelength laser heating with magnetic confinement*. Proc. Esfahan Symposium on Fundamental and Applied Laser Physics [Esfahan, Iran, Aug./Sept. 1971]. In (M.S. FELD, A. JAVAN, and N.A. KURNIT, eds.) *Fundamental and applied laser physics*. Wiley, New York (1973), pp. 119-140 — With A.T. MATTICK: *Liquid droplet radiators for heat rejection in space*. J. Energy **5**, 387-393 (1981) — With A.P. BRUCKNER: *Ram accelerator direct launch system for space cargo*. 38th Congr. Int. Astronaut. Fed. [Brighton, U.K., Oct. 1987], Paper No. IAF-87-211 — With A.P. BRUCKNER and D.W. BOGDANOFF: *The ram accelerator: a new chemical method of achieving ultrahigh velocities*. Proc. 37th Meet. Aeroballistic Range Association (ARA), [Quebec, Canada, Sept. 1986]; *Ram accelerator: a new chemical method for accelerating projectiles to ultrahigh velocities*. AIAA J. **26**, 195-203 (1988) — With A.P. BRUCKNER and C. KNOWLEN: *Experimental investigation of ram accelerator propulsion modes*. Shock Waves **1**, 17-25 (1991) — With A.T. MATTICK and D.A. RUSSELL: *Shock controlled reactors*. Proc. 18th Int. Symp. on Shock Waves [Sendai, Japan, July 1991]. In (K. TAKAYAMA, ed.) *Shock Waves*. Springer, Berlin etc. (1992), pp. 1289-1294.

**U.S. PATENTS.** With H.S. GLICK and W. SQUIRE: *Process for rapidly heating and cooling gases and apparatus therefore*. No. 601,642 (1960) — With J.W. DAIBER and C.E. WITTLIFF: *Laser-driven shock tube*. No. 3,410,142 (1968) — With F.J. STODDARD and J.G. HALL: *Isentropic compression tube*. No. 3,415,442 (1968) — With I.R. HURLE: *Method and apparatus for creating electronic population inversions leading to laser action by convective fluid mechanical techniques*. No. 3,487,333 (1969) — With D.A. RUSSELL: *Weapons system* [an explosive shell precisely guided to a target to ensure a sure hit and kill capability]. No. 4,170,330 (1979) — With A.P. BRUCKNER and D.W. BOGDANOFF: *Apparatus and method for the acceleration of projectiles to hypervelocities*. No. 4,938,112 (1990) — With A.P. BRUCKNER, D.W. BOGDANOFF, and C. KNOWLEN: *Method and apparatus for initiating stable operation of a ram accelerator*. No. 4,982,647 (1991) — With A.P. BRUCKNER, C. KNOWLEN, and K.A. MCFALL: *Method and apparatus for zero velocity start ram acceleration*. No. 5,097,743 (1992) — With A.T. MATTICK and D.A. RUSSELL: *Apparatus for initiating pyrolysis using a shock wave*. No. 5,219,530 (1993); *Method for initiating pyrolysis using a shock wave*. No. 5,300,216 (1994).

**SECONDARY LITERATURE.** J. LEE, D.S. EBERHARDT, R.E. BREIDENTHAL, and A.P. BRUCKNER: *A history of the University of Washington Department of Aeronautics and Astronautics 1917–2003*. University of Washington Aeronautics & Astronautics (UWAA) History (May 27, 2003); http://www.aa.washington.edu/about/history/AA_History.pdf#search=%22abraham%20hertzberg%20died%20cornell%22 — K. BURNS: *The history of aerospace research at Cornell Aeronautical Laboratory (Part I: The FURNAS years – 1946 to 1954)*. Proc. Space 2004 Conference and Exhibition [San Diego, CA, Sept. 2004]. Paper AIAA 2004-5884 (2004); http://pdf.aiaa.org/preview/CDReadyMSPACE2004_1014/PV2004_5884.pdf — (Ed.) *Abe HERTZBERG: the reluctant "Idea Man."* Columns Magazine, The University of Washington Alumni Magazine (Dec. 2004); http://www.washington.edu/alumni/columns/dec97/car5.html — K. BURNS and A. BRUCKNER: *The history of aerospace research at Cornell Aeronautical Laboratory and Calspan (Part III: The life of Abe HERTZBERG and his contributions to hypersonic research)*. Proc. 44th AIAA Aerospace Sciences Meeting and Exhibit [Reno, NV, Jan. 2006]. Paper AIAA 2006-0335 (2006); http://pdf.aiaa.org/preview/CDReadyMASM06_778/PV2006_335.pdf.

**PICTURE.** Taken in 1994. Courtesy Dept. of Aeronautics and Astronautics, University of Washington, Seattle, WA.

**NOTE.** The biography was kindly provided by Prof. Adam P. BRUCKNER of the University of Washington. A more detailed biography is available in Paper AIAA 2006-0335, see above.

## HOPKINSON, Bertram (1874–1918)

• British mechanical and aeronautical engineer; pioneer of dynamic materials testing

Bertram HOPKINSON was born in Birmingham, U.K. He was the eldest son of John HOPKINSON (1849–1898), an engineer, manager, and inventor renowned for the construction of dynamo machines. Stimulated at a young age by his father's professional activities and ways of thinking, he studied mathematics and engineering at Trinity College (1891–1896). He was trained as a patent lawyer until his father's death (1898), at which point he decided to carry on his father's work in engineering and technological education. However, in later years he continued his practice in the law courts together with his research.

When he was appointed chair of the Mechanism and Applied Mechanics Department at Cambridge University (1903–1918), he also took over the supervision of the Cambridge Engineering School, which was part of the University, and became editor of the serial *Cambridge Engineering Tracts*. His main interest was in developing research in the department, with the aim of making it comparable with that of experimental physics at the Cavendish Laboratory. He built up a team of researchers looking at the science of flames and explosions as well as the impact of bullets on steel plates. Among his research students was Harry RICARDO, an engineer who made a name for himself with his pioneering work on internal combustion engines. It was HOPKINSON who encouraged RICARDO to work on engines, turning him from the more traditional pursuit of civil engineering at that time. Starting out with investigations in gas engines and petrol motors, HOPKINSON studied gas explosion phenomena and invented a recording calorimeter for explosions (1906) and an electrical thermometer for measuring gas engine temperatures (1907). He designed instruments to measure the rate of loss of heat through the walls of the reaction vessel (1906) and the emitted radiation (1910). Based on his expertise in explosion he became, together with Sir Dugald CLERK, secretary of the British Association Committee on Gaseous Explosions.

Accepting a fellowship offered to him by King's College, Cambridge (1914), HOPKINSON studied the mechanical strength of metals and designed a high-speed fatigue tester for investigating metal alloys under alternating stresses. Later he extended his studies to high-rate loading, such as that produced by placing high explosives in close contact with the test sample or impacting it with supersonic bullets. To determine the nature of the pressure-time profile when an explosive is detonated or when a projectile impinges on a hard surface, he invented an apparatus that has become known as the "Hopkinson pressure bar" (1914) – a derivative of the ballistic pendulum – which became a standard piece of equipment in the dynamic testing of materials. It uses a cylindrical bar where the length of the pulse is great compared with the radius of the bar. This method was further developed into the so-called "split Hopkinson bar" (1949) by Herbert KOLSKY, a U.S. professor of applied physics, who used two bars with the sample under investigation situated in between. On the outbreak of World War I, HOPKINSON, who obtained a commission in the Royal Engineers, applied his knowledge of explosions to problems of both attack and defense and worked on the best form of bomb to drop from aircraft. He also suggested an additional outer shell to the hull – a so-called "blister" – for the protection of warships from the effects of mines and torpedoes. To test his concept, he began to model dynamic phenomena in small-scale experiments, leading to his so-called "Law of Comparison." As an important result of such studies he formulated his famous "cube-root scaling law" (1915) – also known as the "Hopkinson scaling law." He established an experimental station for the Royal Flying Corps, where testing of aircraft was under his control. For example, he performed for the Air Force model tests on a one-sixth scale to optimize proportions of bomb-case weight to weight of explosive as well as the best material for bomb cases (1915).

HOPKINSON piloted a small plane himself to quicker communicate between Cambridge and his experimental station on the east coast, and he died in a flying accident in bad weather near London (1918). He was a Fellow of the Royal Society (from 1910) and a professional Fellow of King's College (from 1914).

**ORIGINAL WORKS.** *Discontinuous fluid motions involving sources and vortices.* Proc. Lond. Math. Soc. **29**, 142-164 (1898) — *The effects of momentary stresses in metals.* Proc. Roy. Soc. Lond. **74**, 498-506 (1905) — *Explosions of coal-gas and air.* Ibid. **A77**, 387-413 (1906) — *A recording calorimeter for explosions.* Ibid. **A79**, 138-154 (1907) — *On radiation in a gaseous explosion.* Ibid. **A84**, 155-172 (1910) — *A high-speed fatigue tester, and the endurance of metals under alternating stresses at high frequency.* Ibid. **A86**, 131-149 (1912) — With G.T. WILSONS: *The elastic hysteresis of steel.* Ibid. **A87**, 502-511 (1912) — *The pressure of a blow* [evening discourse on Jan. 26, 1912, at the Royal Institution with Lord RAYLEIGH as chair]; *see Collected works*, pp. 423-437 — *On holes and cracks in plates*. Trans. Inst. Naval Architects **LV**, 232-234 (1913) — *The effects of the detonation of gun-cotton.* Proc. North-East Coast Inst. Engineers & Shipbuilders **30** (1913/1914); *see Collected works*, pp. 461-474 — *A method of measuring the pressure produced in the detonation of high explosives, or by the impact of bullets.* Proc. Roy. Soc. Lond. **A89**, 411-413 (1914); Phil. Trans. Roy. Soc. Lond. **A213**, 437-456 (1914) — [On HOPKINSON's scaling law]. Brit. Ordnance Board Minutes 13,565 (1915) — *The scientific papers of Bertram HOPKINSON. Collected and arranged by Sir J.A. EWING and J. LAMOR.* Cambridge University Press, Cambridge, U.K. (1921).

**SECONDARY LITERATURE.** A.V. HILL: *Colonel Bertram HOPKINSON, an appreciation.* Alpine J. **32**, 353-356 (1918/1919) — J.A. EWING: *Bertram HOPKINSON, 1874–1918.* Proc. Roy. Soc. Lond. **A95**, xxvi-xxxvi (1919) — J.W. LANDON and H. QUINNEY: *Experiments with the Hopkinson pressure bar.* Ibid. **A103**, 622-643 (1923) — H. KOLSKY: *An investigation of the mechanical properties of materials at very high rates of loading.* Proc. Phys. Soc. (Lond.) **62B**, 676-700 (1949) — H. KOLSKY: *Stress waves in solids.* Clarendon Press, Oxford (1953) — T.M. CHARLTON: *Professor Bertram HOPKINSON, C.M.G., M.A., B.Sc., F.R.S. (1874–1918).* Notes Rec. Roy. Soc. **29**, 101-109 (1974/1975) — J. HEYMAN: *HOPKINSON, Bertram.* In: (H.C.G. MATTHEW, ed.) *Oxford dictionary of national biography.* Oxford University Press, Oxford (2004), vol. 28, p. 72.

**PICTURE.** From *The scientific papers of Bertram HOPKINSON.* Cambridge University Press (1921). Reprinted with permission of Cambridge University Press, Cambridge, U.K.

## HUBBLE, Edwin Powell (1889–1953)

- U.S. astronomer; founder of modern extragalactic astronomy

Edwin P. HUBBLE was born in Marshfield, MO to John Powell HUBBLE, a lawyer. His early interest in astronomy was sparked at the University of Chicago by the astronomer George E. HALE. After having earned an undergraduate degree in both mathematics and astronomy (1910), he turned away from astronomy and studied law at the Oxford University (1912). However, soon after he dissolved his practice and returned to the University of Chicago, where he again focused on astronomy at Yerkes Observatory in Wisconsin. In 1917, he earned a Ph.D. in astronomy with a focus on faint nebulae using photography as a diagnostic tool. After serving in World War I, he settled down to work at Mount Wilson Observatory (now part of the Hale Observatories) and began studying planetary nebulae, nebulous stars, novae, and stars variable in light with a 60-in. telescope, objects that are all within our own galaxy. Using the Hooker 100-in. telescope at Mount Wilson Observatory, then the largest optical instrument in the world, he observed that not all nebulae in the sky are part of the Milky Way, but rather that some are located outside – at least out to the extreme range of the largest telescopes, a billion or more light-years away. In 1924, HUBBLE announced the discovery of the presence of Cepheid variables in *extragalactic nebulae*, a term that he coined, later called by astronomers "galaxies." He established that the majority of observed nebulae are very far away from our Milky Way galaxy and that most likely they, too, are galaxies. After their classification into four principal types according to luminosity, degree of concentration, degree of diffuseness, and form (1925) – a classification still widely used today – he recognized that rotational symmetry about a dominating nonstellar nucleus was an almost universal characteristic of extra-galactic nebulae. Using spectroscopy to determine the motion in the line of sight, he observed, with the cooperation of Milton L. HUMASON, a systematic displacement toward longer lines in the spectra of distant objects. The so-called "Hubble relation" correlates spectral red shift with distance. The two astronomers made the surprising observation that these galaxies are apparently receding from ours and that the further away they are, the faster they are receding (1929). According to this so-called "Hubble law," the velocity of recession $v$ is directly proportional to distance $D$; i.e., $v = H_0 D$, where $H_0$ is the "Hubble constant" or "Hubble parameter" at the present epoch in the history of the Universe. In his book *The Realm of Nebulae* (1936) HUBBLE described in a semipopular form the results of his special researches. With the availability of more powerful telescopes in the late 1940s, it became obvious that HUBBLE's cosmic distance scale had to be stretched by at least a factor of 5.

During World War II, HUBBLE, who had already served as a line officer in World War I, was chosen by the Army Ordnance because it was believed that "ballistics has a curious affinity with astronomy" as he himself once described his assignment. He became chief of the Exterior Ballistics Branch of the Ordnance Research Laboratory at Aberdeen Proving Ground in Maryland and served as director of the Supersonic Wind Tunnel Laboratory. He "found out … ballistics was both underdeveloped and highly classified … The place is not the home of genius, but it knows the answers to many problems and how to get the answers for others" (Biogr. Mem. Natl. Acad. Sci. **41**, 182). He remained at BRL until

1946, after which time he returned to the Mt. Wilson Observatory and resumed his astronomical research. He greatly assisted in the design of the 200-in. Hale Telescope and served as chair of the Mt. Wilson Observatory Advisory Committee planning the building of the Palomar Observatory.

HUBBLE did not contribute to shock wave physics directly, but his observed phenomenon of expansion spurred others to propose expanding world models (*e.g.*, Alexander FRIEDMANN and Georges LEMAÎTRE), on calculating the age of the Universe (Edward A. MILNE), and the mechanism of its origin (George GAMOW). The Big Bang theory of other scientists, which he supported by his research results, stimulated the theoretical treatment of huge astrophysical explosions and shock waves resulting in the new branch of cosmic gas dynamics, and the numerical modeling of cosmogony.

HUBBLE was a member of many astronomical societies and the recipient of numerous honorary degrees. For his contributions to cosmogony he was awarded numerous gold medals, and for his defense work in World War II he received the Medal of Merit. The NASA/ESA *Edwin P. Hubble Space Telescope,* a giant 11.5-ton telescope placed in low-Earth orbit in order to make observations above the turbulent atmosphere, is named in honor of him. The *Hubble Award* established by the Advanced Imaging Conference (AIC) recognizes outstanding contributions to the art and science of astronomical imaging (since 2006).

Astronomers named a crater on the near side of the Moon and a minor planet (asteroid 2069 HUBBLE) after him.

**ORIGINAL WORKS.** *Photographic investigation of faint nebulae.* Ph.D. thesis, University of Chicago (1917); Publs. Yerkes Observatory **4**, 69-85 (1920) — *A general study of diffuse galactic nebulae.* Astrophys. J. **56**, 162-199 (1922) — *Density distribution in the photographic images of elliptic nebulae.* Popular Astron. **31**, 644 (1923) — *Cepheids in spiral nebulae.* Observatory (Lond.) **48**, 139-142 (1925) — *Extra-galactic nebulae.* Astrophys. J. **64**, 321-369 (1926) — *A relation between distance and radial velocity among extra-galactic nebulae.* Proc. Natl. Acad. Sci. **15**, 168-173 (1929) — With M.L. HUMASON: *The velocity-distance relation among extra-galactic nebulae.* Astrophys. J. **74**, 43-80 (1931) — *Red shifts in the spectra of nebulae.* Clarendon Press, Oxford (1934) — With R.C. TOLMAN: *Two methods of investigating the nature of the nebular red-shift.* Astrophys. J. **82**, 302-337 (1935) — *Effects of red shifts on the distribution of nebulae.* Ibid. **84**, 517-554 (1936) — *The realm of the nebulae.* Yale University Press, New Haven, CT (1936) — *The observational approach to cosmology.* Clarendon Press, Oxford (1937) — *The problem of the expanding Universe.* Am. Scient. **30**, 99-115 (1942) — *The law of red-shifts.* Month. Not. Roy. Astron. Soc. **113**, 658-666 (1953) — With A. SANDAGE: *The brightest variable stars in extragalactic nebulae. I. M31 and M33.* Astrophys. J. **118**, 353-361 (1953) — *The law of red-shifts* [George Darwin Lecture]. Month. Not. Roy. Astron. Soc. **113**, 658-666 (1954).

**SECONDARY LITERATURE.** (Ed.) *Personalities in science: Edwin Powell HUBBLE.* Scient. Am. **161**, 205 (July 1939) — I.S. BOWEN: *Edwin P. HUBBLE: 1889-1953.* Science **119**, 204 (Feb. 12, 1954) — M.L. HUMASON: *Edwin HUBBLE.* Month. Not. Roy. Astron. Soc. **114**, 291-295 (1954) — W.S. ADAMS: *Obituary: Dr. Edwin P. HUBBLE.* Observatory (Lond.) **74**, 32-35 (1954) —

H.P. ROBERTSON: *Edwin Powell HUBBLE: 1889–1953.* Publs. Astron. Soc. Pacific **66**, 120-125 (1954) — N.U. MAYALL: *Edwin Powell HUBBLE.* Biogr. Mem. Natl. Acad. Sci. (U.S.A) **41**, 175-214 (1970) — G.J. WHITROW: *E.P. HUBBLE.* In: (C.C. GILLESPIE, ed.) *Dictionary of scientific biography.* Scribner, New York, vol. 6 (1972), pp. 528-533 — R.W. SMITH: *Edwin P. HUBBLE and the transformation of cosmology.* Phys. Today **43**, 52-58 (April 1990) — A. SANDAGE: *The deep Universe.* Springer, Berlin (1995) — *HUBBLE, Edwin Powell (1889–1953).* In: (P. MURDIN, ed.) *Encyclopedia of astronomy and astrophysics.* Institute of Physics, London (2001), p. 1178.

**PICTURE.** Courtesy The Observatories of the Carnegie Institution of Washington, Pasadena, CA.

## HUGONIOT, Pierre-Henri (1851–1887)

- French physicist and applied mathematician; cofounder of modern shock wave theory

Pierre-Henri HUGONIOT was born in Allenjoie (Dépt. Doubs). The second son of Pierre HUGONIOT, a metallurgist, he showed an early talent for mathematics. At the age of 17 he was nominated *Préparateur de Physique* in the Strasbourg Faculty of Science. After graduating first in his class at the Ecole Normale Supérieure in Paris, he studied at the Ecole Polytechnique (1870–1872). Thereafter, he was accepted into the marine artillery service and became professor of mechanics and ballistics at the Ecole d'Artillerie de la Marine at Lorient, Brittany (1879–1882) and assistant director of its Central Laboratory (1882–1884). In 1884, he was appointed captain. His first research, done in collaboration with Hippolyte SÉBERT and relating to the effect of powder gases on the bore of a weapon (1882), was still based on treating the discontinuous flow as an adiabatic process using POISSON's law. Based on his assistance to Prof. Felix HÉLIE on the latter's book *Traité de balistique expérimentale* (Paris 1864), a report on ballistic experiments carried out by the French artillery at Gâvres, Brittany in the period 1830–1864 which earned them an award from the Paris Academy of Sciences, HUGONIOT was appointed *Répétiteur de mécanique* at the

Ecole Polytechnique (1884–1887). There he established the fundamentals of the theory of compressible, discontinuous flows (*i.e.*, of shock waves). In subsequent years, he treated the problem of discontinuous flow on a more general basis and, applying the law of conservation of energy to the pressure jump, he obtained for the first time a dynamic pressure-density relation for shock compression, a "dynamic adiabat," later called the "Hugoniot curve" or, "Hugoniot" for short. Furthermore, he demonstrated that for a perfect gas of constant ratio of specific heats $\gamma$, the maximum possible compression by a shock wave is given by the quotient $(\gamma+1)/(\gamma-1)$.

Little is known about HUGONIOT's personal life. According to the *Grand Larousse Encyclopédique* (1963 edition), he also worked on the mechanics of steam turbines, and he apparently also served as a consultant on this subject. The French mathematician Roger LIOUVILLE edited posthumously HUGONIOT's famous paper *Mémoires sur la propagation du mouvement dans un fluide indéfini*. In his obituary LIOUVILLE reported that HUGONIOT died on a business trip to the Compagnie des Tramways de Nantes. HUGONIOT, possibly consumed by his numerous tasks, passed away at the early age of 36.

**ORIGINAL WORKS.** With F. HELIE: *Traité de balistique expérimentale*. Gauthiers-Villars, Paris (1865, 1884) — With H. SEBERT: *Etude des effets de la poudre dans un canon de 10 cm.* J. Dumaine, Paris (1882); *Sur les vibrations longitudinales des barres élastiques dont les extrémités sont soumises à des efforts quelconques.* C. R. Acad. Sci. Paris **95**, 213-215, 278-281, 338-340 (1882); *Sur le choc longitudinal d'une tige élastique fixée par l'une de ses extrémités.* Ibid. **95**, 381-384 (1882); *Sur les vibrations longitudinales des verges élastiques et le mouvement d'une tige portant à son extrémité une masse additionnelle.* Ibid. **95**, 775-777 (1882); *Sur la propagation d'un ébranlement uniforme dans un gaz renfermé dans un tuyau cylindrique.* Ibid. **98**, 507-509 (1884) — With F. HELIE: *Traité de balistique expérimentale.* Gauthiers-Villars, Paris (1884) — *Sur la propagation du mouvement dans les corps et spécialement dans les gaz parfaits.* C. R. Acad. Sci. Paris **101**, 794-796 (1885) — *Sur un théorème général relatif à la propagation du mouvement.* Ibid. **102**, 858-860 (1886) — *Sur la propagation du mouvement dans un fluide indéfini.* Ibid. **102**, 1118-1120, 1229-1232 (1886) — *Sur l'écoulement des gaz dans le cas du régime permanent.* Ibid. **102**, 1545-1547 (1886) — *Sur la pression qui existe dans la section contractée d'une veine gazeuse.* Ibid. **103**, 241-243 (1886) — *Sur l'écoulement d'un gaz qui pénètre dans un récipient de capacité limitée.* Ibid. **103**, 922-925 (1886) — *Sur le mouvement varié d'un gaz comprimé dans un réservoir qui se vide librement dans l'atmosphère.* Ibid. **103**, 1002-1004 (1886) — *Sur un théorème relatif au mouvement permanent et à l'écoulement des fluides.* Ibid. **103**, 1178-1181 (1886) — *Sur l'écoulement des fluides élastiques.* Ibid. **103**, 1253-1255 (1886) — *Sur la vitesse limitée d'écoulement des gaz.* Séances Soc. Franç. Phys. (Paris), pp. 120-124 (1886) — *Remarques relatives aux observations de M. HIRN sur l'écoulement des gaz.* C. R. Acad. Sci. Paris **104**, 46-49 (1887) — *Plusieurs conséquences singulières et anomales auxquelles conduisent les formules théoriques relatives à l'écoulement des fluides élastiques.* Séances Soc. Franç. Phys. (Paris), pp. 7-10 (1887) — *Mémoire sur la propagation du mouvement dans les corps et plus spécialement dans les gaz parfaits. Part I.* J. Ecole Polytech. (Paris) **57** [I], 3-97 (1887); *Part II.* Ibid. **58** [I], 1-125 (1889) — *Mémoire sur la propagation du mouvement dans un fluide indéfini. Part I.* J. Math. Pures Appl. **3** [IV], 477-492 (1887); *Part II.* Ibid. **4** [IV], 153-167 (1888).

**SECONDARY LITERATURE.** P.M.M. DUHEM: *Sur le théorème d'HUGONIOT et quelques théorèmes analogues.* C. R. Acad. Sci. Paris **131**, 1171-1173 (1900); *De la propagation des discontinuités dans un fluide visqueux. Extension de la loi d'HUGONIOT.* Ibid. **132**, 944-946 (1901) — Z. ADAMAR: *HUYGENS' principle and HUGONIOT's theory* [in Russ.]. Trudy pervogo Vsesoyuznogo sezda matematikov, Kharkov (1930), Moscow & Leningrad (1936) — R. LIOUVILLE: *Notice sur la vie et les travaux d'HUGONIOT.* J. Ecole Polytech. (Paris) **28** [II], 1-14 (1931) — R. DUGAS: *Histoire de la mécanique.* Griffon, Neuchâtel (1950); see chpt. IX: *HUGONIOT et la propagation des mouvements dans les milieux continus*, pp. 407-418 — *Grande Larousse Encyclopédique.* Librairie Larousse, Paris, vol. 5 (1962), pp. 983-984 — N.M. MERKOULOVA: *P.-H. HUGONIOT.* In: (C.C. GILLESPIE, ed.) *Dictionary of scientific biography.* Scribner, New York, vol. 6 (1972), pp. 545-546 — R. CHÉRET: *The life and work of Pierre-Henri HUGONIOT.* Proc. 6th Conf. Shock Compression of Condensed Matter [Albuquerque, NM, Aug. 1989]. In: (S.C. SCHMIDT, J.N. JOHNSON, and L.W. DAVIDSON, eds.) *Shock compression of condensed matter – 1989.* North-Holland, Amsterdam (1990), pp. 11-19; Shock Waves **2**, 1-4 (1992) — Y. BIELINSKI: *Le génie méconnu d'Allenjoie.* Publ. in journal Le Pays (1993), Montbéliard, France; reprinted in report *Allenjoie – Bulletin Municipal* No. 17 (April 1993).

**PICTURE.** © Collections Ecole Polytechnique, Paris. The picture is part of a group photo showing HUGONIOT among his fellow students; it was taken on the occasion of their promotion in 1870. There also exists another picture of HUGONIOT (possibly the only other that has survived), showing him in a more advanced age and reproduced in the *Grand Larousse Encyclopédique*. Librairie Larousse, Paris (1962), vol. 5, p. 983. Unfortunately, its source is not given there.

**NOTE.** The library of the Ecole Polytechnique keeps 43 letters received by HUGONIOT in the period 1883–1886.

# HUTTON, Charles (1737–1823)

- British mathematician, military engineer, and scientific writer

Charles HUTTON was born in Newcastle upon Tyne. He was the youngest son of Henry HUTTON, an overseer in a local colliery. He worked for a short time as a hewer in a pit at Longbenton in Northumberland, but since he acquired a taste for books, it was decided that teaching was his proper occupation. Largely self-educated, he began to teach mathematics and natural philosophy at his own Writing and Mathematical School, which he established in Newcastle (1760). Contributing to engineering problems, land survey, and mathematical education, he wrote several tracts and textbooks such as *The Schoolmaster's Guide* (1764), an elementary textbook on arithmetic, his first publication; *A Treatise on Mensuration* (1767–1770), illustrated by the famous Thomas BEWICK, who established wood engraving as a major printmaking technique; *Plan of Newcastle and Gateshead* (1770), a local land survey of the city and its suburbs; and *The Principle of Bridges* (1772), a tract on the equilibrium of bridges.

Obviously based on his wide scope of interests in educational and scientific affairs, he was appointed professor of mathematics at the Royal Military Academy in Woolwich, southeast London (1773–1807). His researches centered on the convergence of series of experiments in ballistics, the building of bridges, and measurement of the mean Earth's density. In this fruitful period he wrote other renowned books on mathematics such as *A Course of Mathematics for the Cadets of the Royal Military Academy* (1798–1801), a two-volume textbook for his students at Woolwich; and the historical introduction to *Mathematical Tables* (1785), which contains the common, hyperbolic, and logistic logarithms. His *Mathematical and Philosophical Dictionary* (1795–1796), probably his bestknown work, contains a glossary of terms used in mathematics, astronomy, and natural philosophy as well as an interesting historical account of the rise, progress, and state of contemporary science. It has recently been reprinted in Germany (1973). He also edited a great many almanacs, including the *Ladies' Diary* (1773–1818), dealing with the popular mathematical and poetic sections of such books.

HUTTON contributed many papers to the *Philosophical Transactions*. For his investigations on *The Force of Fired Gunpowder and the Velocities of Cannon Ball* (publ. 1778) – along with the work of his countryman Benjamin ROBINS a milestone of modern internal ballistics as well as of supersonic aeroballistics – he received the Copley Medal (1778) of the Royal Society. Following in ROBINS' footsteps, he was the first to extend the ballistic pendulum technique to large-caliber (cannon) shots, as well as the first to verify the existence of supersonic muzzle velocities (1783). In addition, he first measured the aerodynamic drag ranging from subsonic to supersonic velocities and noticed that the power of the resistance-velocity law is not a constant. In the early 1800, HUTTON speculated on the origin of meteors ("stones that have fallen from the atmosphere…").

HUTTON also became renowned for his computation of the mean density of the globe (1778). After the British astronomer Nevil MASKELYNE had completed his series of observations at Mount Schiehallion, North Perthshire, to measure the attraction of mass by the deflection of a plumb line, HUTTON was chosen to deduce the corresponding estimate of the mean density of the globe. It was found that the mean Earth's density is 4.481 times that of water (modern value 5.517 g/cm$^3$). The French physicist and mathematician Pierre S. DE LAPLACE acknowledged the value of HUTTON's work in computing the density of the Earth in an article published in the French journal *Connaissance des Temps* (1823).

HUTTON retranslated into English EULER's German translation *Neue Grundsätze der Artillerie* (1745) of ROBINS' book *New Principles of Gunnery* (1742), which EULER had extended by numerous valuable comments. Corrected and enlarged, HUTTON's new edition (1805) of ROBINS' book became a basic source for most subsequent work on the theory of artillery and projectiles. His three-volume *Tracts on Mathematical and Philosophical Subjects* (1812) summarizes his most important contributions: it contains a treatise on various bridges and mathematical tables (vol. 1); calculations of the Earth's density (vol. 2); and a description of a new gunpowder eprouvette, new experiments to determine aerodynamic drag using a whirling machine, and a treatise on the theory and practice of gunnery (vol. 3).

HUTTON was made Fellow of the Royal Society (1774) and later served as its foreign secretary (1779–1783). He resigned from office upon the request of Sir Joseph BANKS, then president of the Society, who reproached him for not carrying out his duties efficiently. The University of Edinburgh awarded him the degree of doctor of law (LL.D. 1783).

**ORIGINAL WORKS.** *The force of fired gunpowder and the velocities of cannon balls.* Phil. Trans. Roy. Soc. Lond. **68**, 50-85 (1778) — *An account of the calculation made from the survey and measures taken at Mount Schiehallion, in Perthshire, in order to ascertain the mean density of the Earth.* Ibid. **68**, 689-778 (1778) — *Abstract of experiments made to determine the true resistance of the air to the surfaces of bodies, of various figures, and moved through it with different degrees of velocity.* Trans. Roy. Soc. Edinb. **II**, 29-36 (1790) — *A mathematical and philosophical dictionary.* 2 vols., Johnson & Robinson, London (1795–1796); reprinted by Olms, Hildesheim (1973) — *On the origin of stones that have fallen from the atmosphere.* Phil. Mag. **22** [I], 71-79 (1805) — *New experiments in gunnery, for determining the force of fired gunpowder, the initial velocity of cannon ball, the ranges of projectiles at different elevations, the resistance of the air to projectiles, the effect of different lengths of guns, and of different quantities of powder, &c, &c.* In: C. HUTTON: *Tracts on mathematical and philosophical subjects.* 3 vols., T. Davison, London (1812), vol. 2, Tract XXXI, 306-384 — *On a new gunpowder eprouvette.* Ibid., vol. 3, Tract XXXV, 153-163 — *Resistance of the air determined by the whirling machine.* Ibid., Tract XXXVI, 163-208 — *Theory and practice of gunnery, as dependent on the resistance of the air.* Ibid., Tract XXXVII, 209-315 — *To determine the effects of pile-engines.* Ibid., Tract XXXVIII, Problem 2, 317-321; *To assign the velocity which water, or other fluids, spouts out from the bottom of a vessel.* Ibid., Problem 24, 352-356.

**SECONDARY LITERATURE.** A.D. (anonymous): *Dr. Charles HUTTON, F.R.S.* Brit. Public Characters, London (1799–1809), vol. 10 — J. BRUCE: *A memoir of Charles HUTTON.* Hudgson, Newcastle (1823) — R.E. ANDERSON: *Charles HUTTON.* Dict. Natl. Biogr. **XXVIII**, 351-353 (1891) —

O. GREGORY: *Brief memoir of the life and writings of Charles* HUTTON, *LL.D., F.R.S.* The Imp. Mag. **5**, 202-227 (1823) — M.E. BARON: *Ch.* HUTTON. In: (C.C. GILLESPIE, ed.) *Dictionary of scientific biography.* Scribner, New York, vol. 6 (1972), pp. 576-577 — A.G. HOWSON: *A history of mathematics education in England.* Cambridge University Press, Cambridge, U.K. (1982), pp. 59-74 — D.P. MILLER: *The revival of the physical sciences in Britain, 1815-1840.* Osiris (Chicago) **2** [II], 107-134 (1986) — W. JOHNSON: *Charles* HUTTON *1737-1823: The prototypical Woolwich professor of mathematics.* Int. J. Mech. Work. Technol. **18**, 195-230 (1989) — W. JOHNSON: *Collected works on Benjamin* ROBINS *& Charles* HUTTON. Phoenix, New Delhi (2001) — N. GUICCIARDINI: *Hutton, Charles.* In: (H.C.G. MATTHEW, ed.) *Oxford dictionary of national biography.* Oxford University Press, Oxford (2004), vol. 29, pp. 51-53.

**PICTURE.** From HUTTON's *Tracts on mathematical and philosophical subjects.* T. Davison, London (1812), vol. 1, frontispiece.

# HUYGENS [or Huyghens, Lat. *HUGENIUS*], Christiaan (1629–1695)

- Dutch physicist, mathematician, geometrician, optician, astronomer, and inventor; main contributor to the classical theory of percussion

Born in The Hague, The Netherlands, Christiaan HUYGENS belonged to a prominent family; his father Constantijn HUYGENS was a secretary to Prince Frederic Henry. He was supposed to continue the diplomatic tradition of his family and studied law at the University of Leiden (1645–1647) and at the College of Orange (1647–1649). However, influenced by the French philosophers Marin MERSENNE and René DESCARTES, who exchanged letters with his father, he began to study mathematics (1645) and eventually turned to the private study of nature (1650–1666). In 1666, he became one of the founding members of the Académie Royale des Sciences, and from that year lived mainly in Paris (1666–1681), where he met the German philosopher and mathematician Gottfried W. LEIBNIZ. Due to illness and political reasons he returned to The Hague, where he died in 1695.

HUYGENS made major contributions to mathematics, optics, and mechanics (particularly to statics and hydrostatics, and impact). He developed new optical techniques together with his brother Constantijn, formulated a wave theory of light, and applied geometrical optics to a number of optical systems. He improved the microscope and telescope and in 1655 discovered that the planet Saturn also has a large moon, now known as "Titan." Four years later he also discovered the true shape of the rings of Saturn. Studying fall and projectile motion in resisting media, he became convinced by experiments that the resistance in such media as air and water is proportional to the square of their flow velocity. He also cooperated with Denis PAPIN in building a *moteur à explosion* (1673).

Obviously dissatisfied with DESCARTES' treatment of percussion published in his *Principia Philosophiae* ("Principles of Philosophy," 1644), HUYGENS dedicated much of his time to studying the percussion of elastic bodies. Applying a geometrical treatment, HUYGENS worked out new rules of percussion that, confirmed in the course of discussions in Paris by repeated experiments and recorded in 1656 in his treatise *Tractatus de motu corporum ex percussione* ("On the Motion of Bodies by Percussion"), were published posthumously in 1703. His results were partly based on the principle that in any system of bodies the center of gravity could never rise of its own accord above its initial position. Asked in 1667 by Henry OLDENBURG, editor of *Philosophical Transactions*, to contribute to the problems of percussion, he submitted in 1669 a paper on this subject in which he briefly summarized his previous results. Most importantly, he found that during elastic percussion the sum of the products of the quantity of matter and the squares of the velocity is conserved, which in modern terms is the Law of Conservation of Kinetic Energy.

In his treatise *Horologium oscillatorium sive de motu pendulorum* ("The Pendulum Clock, or On the Motion of Pendulums," 1673), he thoroughly investigated the theory of the center of oscillation of compound pendulums – a difficult task that, already posed by the French mathematician and theologian Marin MERSENNE to the 17-year-old HUYGENS, had been previously grappled with in vain by renowned scholars of his time, such as the French mathematicians René DESCARTES and Honoratus FABRI. HUYGENS succeeded in formulating a general computation rule for determining the center of oscillation, applicable to all sorts of compound pendulums. The "center of oscillation" is identical to the "center of percussion," as was first noticed by the British mathematician John WALLIS.

In his book *Traité de la lumière* ("Treatise on Light," 1690), HUYGENS considered light as an irregular series of mechanical disturbances that propagate with very great, but finite, velocity through the aether, a medium that supposedly consisted of uniform, minute elastic particles filling all of space. He considered light propagation therefore as a serial longitudinal displacement similar to a collision that, according to his concept, proceeds through a row of billiard balls: colliding particles produce around each particle new wave fronts – secondary spherical "wavelets" – that have the same speed as the overall wave. Thus, the observed wave front is the envelope of all fronts of the individual particles, the so-called "Huygens principle." This mechanistic wave model proved to be very successful in describing acoustic wave phenomena as was first shown geometrically by Christian A. DOPPLER in the case of a subsonically, sonically, and supersonically moving object (1846) and first proved experimentally with the discovery of the "head wave" phenomenon by Ernst MACH and Peter SALCHER (1886–1887). The Huygens principle was later extended by Augustin FRESNEL and Gustav KIRCHHOFF to explain interference and diffraction phenomena.

In the final years of his life, HUYGENS also discussed extraterrestrial life in a letter to his brother Constantijn, speculating that rational creatures live on each planet. This letter was published in Latin 3 years after his death as the *Cosmotheoros* (1698), further entitled (in translation) *The Celestial Worlds Discover'd: or, Conjectures Concerning the Inhabitants, Plants and Productions of the Worlds in the Planets*.

HUYGENS' output in mathematics (particularly in geometry, algebra, and calculus) and mechanics, optics, astronomy, and chronometry was enormous. Throughout his life he maintained an extensive correspondence with renowned contemporaries such as DESCARTES, LEIBNIZ, MERSENNE, OLDENBOURG, and WALLIS.

The *Christiaan Huygens Wetenschapsprijs* ("Christiaan Huygens Science Prize") of the Koninklijke Nederlandse Akademie van Wetenschappen (KNAW), the Dutch Royal Society, was established in 1998 and is awarded annually to researchers who have made a highly original contribution to a certain discipline.

Almost 350 years after HUYGENS' discovery of Titan, the scientific probe *HUYGENS*, since September 2004 orbiting Saturn on board the NASA/ESA/ASI CASSINI spacecraft, was released on December 25, 2004: after a 22-d journey to Titan it parachuted through its atmosphere and safely landed on its surface. Astronomers named a crater on Mars (*Huygens Crater*), a mountain on the near side of the Moon (*Mons Huygens*), and a minor planet (asteroid 2801 HUYGENS) after him.

**ORIGINAL WORKS.** *De motu corporum ex percussione* [manuscript of 1656]. In: (B. DE VOLDER and B. FULLENIUS, eds.) *C. HUYGENII ... opuscula postuma*. Boutesteyn, Lugduni Batavorum (1703); also publ. in *Œuvres complètes*; vol. XVI (1929): *Percussion — De vi centrifuga*. Manuscript originating from 1659 — *Question de l'existence et de la perceptibilité du mouvement absolu. Force centrifuge. Travaux divers de statique et de dynamique de 1659 à 1666*. Germ. translation: *Über die Bewegung der Körper durch den Stoß. Über die Centrifugalkraft*. In: *Ostwald's Klassiker der exakten Wissenschaften*. Nr. 138, Engelmann, Leipzig (1903) — *Règles du mouvement dans la rencontre des corps* [Extrait d'une lettre de M. HVGENS a l'Auteur du Iournal]. J. Sçavans (Paris) **5**, 22-24 (March 18, 1669). Also publ. in *Œuvres complètes*, vol. VI (1895), 383-385 — *The laws of motion on the collision of bodies*. Phil. Trans. Roy. Soc. Lond. **4**, No. 46, 925-928 (April 12, 1669) — *Horologium oscillatorium, sive de motu pendulorum ad horologia aptato demonstrationes geometricae*. Muguet, Parisii (1673). Germ. translation: (A. HECKSCHER and A. VON OETTINGEN, eds) *Die Penduluhr*. In: *Ostwald's Klassiker der exakten Wissenschaften*. Nr. 192, Engelmann, Leipzig (1913) — *Traité de la lumière*. Vander Aa, Leiden (1690). Germ. translation: *Abhandlung über das Licht*. In: (E. LOMMEL, ed.) *Ostwald's Klassiker der exakten Wissenschaften*. Nr. 20, Engelmann, Leipzig (1890) — *Nouvelle force mouvante par le moyen de la poudre à canon et de l'air* [1673]. *Divers ouvrages de mathématique et de physique, par Messieurs de l'Académie Royale des Sciences*. Imprimerie Royale, Paris (1693) — *Christiani Hugenii Cosmotheoros, sive de terris coelestibus, earumque ornatu coniecturae*. Hagae-Comitum (1698); Germ. translation: *Christian HUGENS Cosmotheoros oder Welt-betrachtende Muthmassungen von denen himmlischen Erd-Kugeln und deren Schmuck* [geschrieben an seinen Bruder Constantijn HUGENS]. F. Lanckischens Erben, Leipzig (1703) — (Ed. by W.J.'s GRAVESANDE) *Opera varia*. 4 vols., J. vander Aa, Lugduni Batavorum (1724–1728) — *Œuvres complètes de Christiaan HUYGENS. Publiées par la Société Hollandaises des Sciences*. 22 vols., M. Nijhoff, La Haye (1888–1950).

A list of HUYGENS' work published during his lifetime or as a consequence of his last will was provided by the Dutch Huygens Web, University of Utrecht; http://www.phys.uu.nl/~huygens/hug_biblio1_en.htm#top.

**SECONDARY LITERATURE.** P.G. TAIT: *Note on a singular passage in the 'Principia.'* Proc. Roy. Soc. Edinb. **13**, 72-78 (1886) — M. ZWERGER: *Der Schwingungsmittelpunkt zusammengesetzter Pendel: historisch-kritische Untersuchung nach den Quellen bearbeitet*. Lidauscher, München (1889) — H.L. BRUGMANS: *Le séjour de Christian HUYGENS à Paris et ses relations avec les milieux scientifiques français*. E. Droz, Paris (1935) — Z. ADAMAR: *HUYGENS' principle and HUGONIOT's theory* [in Russ.]. Trudy pervogo Vsesoyuznogo sezda matematikov, Kharkov (1930), Moscow & Leningrad (1936) — G. LORIA: *La vita di Cristiano HUYGENS quale si desume del suo carteggio*. Comm. Pont. Acc. Scient. (Città del Vaticano) **6**, 1079-1138 (1942) — A.E. BELL: *Christian HUYGENS and the development of science in the 17th century*. E. Arnold, London (1947) — J.A. VOLLGRAFF: *Biographie de Christiaan HUYGENS*. In: *Œuvres complètes*. Société Hollandaises des Sciences. M. Nijhoff, La Haye, vol. 22 (1950), pp. 383-771 — B.B. BAKER and E.T. COPSON: *The mathematical theory of HUYGENS' principle*. Clarendon Press, Oxford (1950) — E.J. DIJKSTERHUIS: *Christiaan HUYGENS*. Bohn, Haarlem (1951) — A.R. HALL: *Mechanics and the Royal Society, 1668–1670*. Brit. J. Hist. Sci. **3**, 24-38 (1966/1967) — A.R. HALL and M.B. HALL (eds.) *The correspondence of Henry OLDENBURG*. University of Wisconsin Press, Madison, WI, vol. 5 (1968): *1668–1669*, pp. 127-128, 283-284, 373-375, 452-453, 465-467, 502-504, and 582-584 — R.S. WESTFALL: *Force in NEWTON's physics*. MacDonald, London (1971), see *Christiaan HUYGENS' kinematics*, pp. 146-193 — H.J.M. BOS: *Ch. HUYGENS*. In: (C.C. GILLESPIE, ed.) *Dictionary of scientific biography*. Scribner, New York, vol. 6 (1972), pp. 597-613 — M. FIERZ: *Vorlesungen zur Entwicklungsgeschichte der Mechanik*. Lecture notes in physics No. 15. Springer, Berlin (1972), pp. 73-79 — J. HERIVEL: *HUYGENS, Christiaan*. In: *Encyclopedia Britannica. Macropaedia*. Benton & Hemingway, Chicago, vol. 9 (1974), pp. 74-75 — I. SZABÓ: *Geschichte der mechanischen Prinzipien und ihre wichtigsten Anwendungen*. Birkhäuser, Basel (1977), pp. 446-452 — A. D'ELIA: *Christiaan HUYGENS: Una biografia intellettuale*. Franco Angeli, Milano (1985) — Y.Y. BEREST and A.P. VESELOV: *Huygens' principle and integrability.* Russ. Math. Surv. **49**, No. 6, 5-77 (1994) — V. SCHÜLLER: *Christiaan Huygens (1629–1695).* In: (K. VON MEŸENN, ed.) *Die großen Physiker.* C.H. Beck'sche Verlagsbuchhandlung, München (1997), vol. 1, pp. 185-193 — *Christiaan Huygens: a biographical sketch.* Dutch Huygens Web, University of Utrecht; http://www.phys.uu.nl/~huygens/bio1_en.htm.

PICTURE. By the Dutch print artist Frederik OTTENS, based on HUYGENS' portrait made by Gerard EDELINCK (1687) and prepared for publication of HUYGENS' *Opera varia* (1724). Courtesy Dibner Library of the History of Science and Technology, Smithsonian Institution Libraries. National Museum of American History, Washington, DC.

NOTE. In summer 1995, an International Congress on Christiaan HUYGENS was held in Leyden and Voorburg, The Netherlands, under the auspices of the Royal Netherlands Academy of Arts and Sciences and the Académie des Sciences of the Institut de France, Paris. The purpose of this major international Congress was to commemorate HUYGENS' life and work on the 300th anniversary of his death. *See* http://web.clas.ufl.edu/users/rhatch/pages/03-Sci-Rev/SCI-REV-Home/resource-ref-read/major-individuals/huygens/.

## JOHNSON, William (1922–)

• British mechanical engineer; founder of the *International Journal of Mechanical Sciences* and the *International Journal of Impact Engineering*

William JOHNSON was born in Manchester, U.K. He was the elder son of James JOHNSON, a foreman in a wire-drawing plant. His early working-class education was at a high school and later in a mechanical engineering course at the Manchester College of Technology (later named UMIST), from which he graduated, B.Sc. (Hons), in 1943. After some minor work in engineering he was called up for service in the army in World War II, and in due course he became an officer in the Corps of Electrical and Mechanical Engineers serving until September 1947 in Italy and later in Austria. After demobilization he spent 2 years in the Civil Service but moved on to a lectureship in mechanical engineering and solid mechanics, first in Northampton Polytechnic, London, and later at the University of Sheffield, specializing in metal-forming plasticity theory in 1952. JOHNSON became a senior lecturer at Manchester University (1956–1960) and there continued his career work as a professor of mechanical engineering on slip-line theory with associated experiments. He also developed historical interests in science in these years, after attending courses at the University College, London, in the history and philosophy of science.

In 1959, he was invited to become founder and editor of the *International Journal of Mechanical Sciences* after suggesting the idea to the late Robert MAXWELL of Pergamon Press, Oxford; this monthly journal still flourishes today. For many years JOHNSON was also editor of the quarterly *Bulletin of Mechanical Engineering Education*. In 1984, at his suggestion, the *International Journal of Impact Engineering* was created, which is specifically devoted to problems in impact engineering. These journals are prominent international ones today, run by his former younger colleagues and researchers.

In 1975, JOHNSON was invited to a chair in mechanics in the Engineering Department at Cambridge University, retiring from there in 1982 but moving to fill a similar appointment in the United States at Purdue University in Indiana (1983–1989). Thereafter he visited and worked at many universities across the world, mostly with former colleagues. He was elected to a fellowship of the British Royal Society (1982) and later to the academies of India and Greece.

JOHNSON published (with coauthors) about 500 research papers and is the author of ten books, eight technical, one being an autobiography and the other a volume on the historical works of the British mathematicians and military engineers Benjamin ROBINS and Charles HUTTON, both of the latter appearing in 2003. The first of the eight editions above include *Plasticity for Mechanical Engineers* (1962, 1966) with Peter B. MELLOR; *The Mechanics of Metal Extrusion* (1962) with Hideaki KUDO; *Plane-Strain Slip-Line Fields* (1970, 1982) with Robert SOWERBY, James B. HADDOW, and Ronald D. VENTER; *Impact Strength of Materials* (1972); *Engineering Plasticity* (1973, 1986) with Peter B. MELLOR; and *Plasticity and Metal Forming* (1978) and *Crash-Worthiness of Vehicles* (1978), both with Athanasios G. MAMALIS. Many of the latter volumes have appeared in translation.

In the late 1950s, JOHNSON helped start the Department of the History of Science and Technology at Manchester University and a Medical Engineering Unit in 1973. He became attached to researching historically a number of British and continental scientists especially in his later years, particularly Jacques CASSINI, Martin FOLKES, James GLENIE, Charles HUTTON, Alfred MORDECAI, Benjamin ROBINS, Isaac TODHUNTER, and VOLTAIRE, about some of whom he wrote several papers. He also addressed the early history of the ballistic pendulum and some monster guns and reviewed the contributions of American ballisticians, such as John A.B. DAHLGREN and Thomas J. RODMAN, to ballistic research.

Between 1965 and 1990 Prof. JOHNSON received several honorary degrees, prizes, and medals, most recently the History and Heritage Engineer-Historian Award of the American Society of Mechanical Engineers (ASME) for the year 2000 "for his many publications on a wide variety of technological-history subjects, including projectiles, the life and works of Benjamin ROBINS, manufacturing technology, and steam hammer forging."

**ORIGINAL WORKS.** *Research into metal forming and shaping operations.* J. Inst. Metals **84**, 165-179 (1956) — *Indentation and forging and the action of NASMYTH's anvil.* Engineer (Lond.) **205**, 348-350 (1958) — With R.I. TANNER: *Temperature distribution in fast metal working operations.* Int. J. Mech. Sci. **1**, 28-44 (1960) — With H. LIPPMANN: *Temperature development based on technological analysis: fast rolling as an example.* Appl. Sci. Res. **AIX**, 345-356 (1960) — With J.B. HADDOW: *Experiments in the piercing of soft metals.* 2nd Int. Machine Tool Design and Research Conf. [Manchester, U.K., Sept. 1961]. Int. J. Machine Tool Des. Res. **2**, No. 1, 1-18 (Jan. - March 1962) — With R.A.C. SLATER and E. LAITHEWAITE: *Impact extrusion using a linear induction motor.* Proc. Inst. Mech. Eng. **179**, 15-35 (1964) — With F.W. TRAVIS: *Explosive fracturing.* In: (S.A. TOBIAS, ed.) *Advances in machine tool design and research 1965: Proc. 6th Int. Machine Tool Design and Research Conf.* [Manchester, U.K., Sept. 1965]. Pergamon Press, Oxford (1966), pp. 741-764 — With D.G. DALYMPLE: *A study of thin tube forming using non-uniform explosive charges.* Int. J. Mech. Sci. **8**, 353-360 (1966) — With F.W. TRAVIS: *Explosive hydrodynamic extrusion.* Proc. Inst. Mech. Eng. **182** (Part 3C), 231-238 (1967/1968) — With J.B. HAWKYARD: *An analysis of the changes in geometry of a short hollow cylinder during axial compression.* Int. J. Mech. Sci. **9**, 163-182 (1967) — With S.R. REID: *Amplitude of interface waves in exploding welding.* Nature (Phys. Sci.) **231**, No. 26, 205-206 (June 28, 1971) — With P.D. SODEN and E.R. TRUEMAN: *A study in jet propulsion: an analysis of the motion of squid, Loligo vulgaris.* J. Exp. Biol. **56**, 155-165 (1972) — With S.T.S. AL HASASINI and J.L. DUNCAN: *On the parameters of the magnetic forming process.* J. Mech. Engng. Sci. **16**, 1-9 (1974) — With S.R REID: *Ricochet of spheres off water.* Ibid. **17**, 71-81 (1975) — With A.S. SOLIMAN and S.R. REID: *The effect of spherical projectile speed in ricochet off water and sand.* Ibid. **18**, 279-284 (1976) — *'Simple' linear impact.* Int. J. Mech. Engng. Educ. (IJMEE) **4**, 167-181 (1976) — With A.G. MAMALIS: *Aspects of mechanics in some sports and games: the elementary mechanics of golf, football, cricket.* VDI-Verlag, Düsseldorf (1977); *The perforation of circular plates with four-sided pyramidally-headed square-section punches.* Int. J. Mech. Sci. **20**, 849-866 (1978) — With T. YELLA-REDDY and S.R. REID: *Model road-tank vessels subject to internal explosion.* J. Strain Analysis **15**, 225-233 (1980) — *Benjamin ROBINS' New Principles of Gunnery.* Int. J. Impact Engng. **4**, 205-219 (1986) — *Henri TRESCA as the originator of adiabatic heat lines.* Int. J. Mech. Sci. **29**, 301-310 (1987) — *Admiral John A.B. DAHLGREN (1809–1870), his life times and technical work in the U.S. Naval Ordnance.* Int. J. Impact Engng. **8**, 355-387 (1990) — *The origin of the ballistic pendulum: the claims of Jacques CASSINI and Benjamin ROBINS.* Int. J. Mech. Sci. **32**, 345-374 (1990) — *Elements of crashworthiness: scope and actuality.* Proc. Inst. Mech. Eng., Part D: J. Automobile Engng. **204**, No. D4, 255-273 (1990) — *T.J. RODMAN: mid-19th century gun barrel design and research for the U.S. Army.* Int. J. Impact Engng. **9**, 127-159 (1990) — *Some monster guns and unconventional variations.* Int. J. Impact Engng. **11**, 401-439 (1991) — With S. CHANDRASEKAR: *Volcanic bombs – their form and physical structure: comparison with cast iron cannon balls.* Ibid. **12**, 459-467 (1992); *RUPERT's glass drops: residual stress measurements and calculations, and hypotheses for explaining disintegrating fracture.* J. Mater. Process. Technol. **31**, 413-440 (1992) — *The ricochet of spinning and non-spinning spherical projectiles, mainly from water.* Int. J. Impact Engng. **21**, 15-34 (1998) — *Collected works on Benjamin ROBINS & Charles HUTTON.* Phoenix, New Delhi (2001).

**SECONDARY LITERATURE.** S.R. REID (ed.) *Metal forming and impact mechanics: William JOHNSON commemorative volume.* Pergamon Press, Oxford (1985) — W. JOHNSON: *Record and services, satisfactory* [an autobiography]. Memoir Club, Whitworth Hall, Spennymoor, County Durham (2003); ISBN 1-84104-059-2. • A full biography and record of JOHNSON's publications and honors will be found in the library of the Royal Society of London and other autobiographical material in his memoir.

**PICTURE.** Courtesy Prof. emeritus W. JOHNSON, Cambridge University, Cambridge, U.K.

**NOTE.** This short biography was composed by Prof. JOHNSON and sent to the author in September 2005.

## JOUGUET, Jacques Charles Emile (1871–1943)

• French engineer and mathematician; cofounder of the first theory on detonation

J.C. Emile JOUGUET was born in Bessèges (Dépt. Languedoc). His father was a mining engineer and director in the steel and iron industry. After schooling in Nîmes (1891) he studied at the Ecole Polytechnique and graduated as an engineer (Ph.D. 1889). He started his professional career as a railroad supervisor at Bordeaux (1895–1898), where he came under the influence of Prof. Pierre DUHEM and increasingly became involved in teaching. He was professor of general theoretical and applied mechanics (1898–1907) at the Ecole des Mines in Saint-Étienne (Dépt. Loire), *répétiteur de mécanique* (1809) at the Ecole Polytechnique, and professor of analysis, descriptive geometry, and topography (1910–1914) at the Ecole des Mines in Paris. In World War I, he served in the French Artillery as a lieutenant-colonel. In the period 1920–1939, he taught in Paris and was professor of machines at the Ecole des Mines, professor of thermodynamics at the Ecole du Génie rural, and professor of mechanics at the Ecole Polytechnique, where he extended his teaching to mathematical analysis, thermodynamics, mechanical engineering, and topography. He also taught courses on theoretical and applied mechanics at the Ecole Nationale du Génie Rural in Paris.

Together with his friends and colleagues Jacques HADAMARD and Pierre DUHEM he belonged to the small group of leading French shock physicists who tackled many problems posed by this new discipline. Examples include various problems of similarity, the analogy between shocks in gases and hydraulic jumps, and propagation effects of shock waves in solids. Independently of the English physical chemist David Leonard CHAPMAN, he analytically formulated a theory of detonation, the so-called "Chapman-Jouguet (CJ) hypothesis" (1899–1905). Detonations with fronts advancing at sonic speeds – so-called "Chapman-Jouguet detonations" – are the most common. The region immediately behind a detonation wave is still referred to as the "Chapman-Jouguet zone."

Together with his colleague Louis CRUSSARD he investigated the stability of detonation waves and applied the Law of Similarity to its propagation (1907–1908). In his book *Mécanique des explosifs, étude de dynamique chimique* ("Mechanics of Explosives, a Study of Chemical Dynamics"), published in 1917, JOUGUET treated the fluid dynamical aspects of fast chemical reactions, such as detonation and deflagration, theoretically and in a very general manner. He also kept a steady interest in applying thermodynamics to mechanical engineering and contributed to the development of steam turbines and thermal engines.

JOUGUET became a member of the Académie des Sciences (1930) and commander of the Légion d'honneur (1936), and retired as Chief Engineer of Mines. He was a major contributor to the theory of detonation and the theory of shock waves.

**ORIGINAL WORKS.** *Sur la propagation des discontinuités dans les fluides.* C. R. Acad. Sci. Paris **132**, 673-676 (1901) — *Mécanique des fluides.* J. Thomas, Saint-Etienne (1904) — *Remarques sur la propagation des percussions dans les gaz.* C. R. Acad. Sci. Paris **138**, 1685-1688 (1904) — *Sur l'onde explosive.* Ibid. **139**, 121-124 (1904) — *Remarques sur la loi adiabatique d'HUGONIOT.* Ibid. **139**, 786-789 (1904) — *Sur l'onde explosive.* Ibid. **140**, 711-712 (1905) — *Sur la propagation des réactions chimiques dans les gaz.* J. Math. Pures Appl. **1** [VI], 347-425 (1905); Ibid. **2** [VI], 5-86 (1906) — *Sur la similitude dans le mouvement des fluides.* C. R. Acad. Sci. Paris **141**, 346-348 (1905) — *Sur l'accélération des ondes de choc planes.* Ibid. **142**, 831-833 (1906) — *Sur l'accélération des ondes de choc sphériques.* Ibid. **142**, 1034-1036 (1906) — *Remarque sur les ondes de choc. Application à l'onde explosive.* Ibid. **144**, 415-417 (1907) — With L. CRUSSARD: *Sur les ondes de choc et combustion. Stabilité de l'onde explosive.* Ibid. **144**, 560-563 (1907) — *Sur les ondes de choc et de combustion sphériques.* Ibid. **144**, 632-633 (1907) — *Sur la résistance de l'air.* Ibid. **145**, 500-502 (1907) — *Application des lois de la similitude à la propagation des déflagrations.* Ibid. **146**, 915-917 (1908) — With L. CRUSSARD: *Application des lois de la similitude à la propagation des détonations.* Ibid. **146**, 954-956 (1908) — *Lectures de mécanique.* Gauthier-Villars, Paris (1908/1909) — *Sur la vitesse des ondes de choc et combustion.* C. R. Acad. Sci. Paris **149**, 1361-1364 (1910) — *Impossibilité de certaines ondes de choc et combustion.* Ibid. **150**, 91-93 (1910) — *Loi adiabatique dynamique dans le mouvement des fils.* Ibid. **153**, 761-764 (1911) — *Sur l'accélération des ondes de choc dans les fils.* Ibid. **153**, 933-936 (1911) — *Sur la vitesse et l'accélération des ondes de choc de seconde et troisième espèce dans les fils.* Ibid. **153**, 1062-1064 (1911) — *Sur la propagation des déflagrations dans les mélanges gazeux.* Ibid. **156**, 872-875 (1913) — *Sur la propagation des déflagrations et sur les limites d'inflammabilité.* Ibid. **156**, 1058-1061 (1913) — *Sur quelques propriétés d'onde de choc et combustions.* Ibid. **157**, 545-547 (1913) — *Mécanique des fluides.* Théolier, St.-Etienne (1914) — *Mécanique des explosifs, étude de dynamique chimique.* 2 vols., Octave Doin et Fils, Paris (1917) — With A. RATEAU and M. DE SPARRE: *Étude théorique et expérimentale sur les coups de bélier dans les conduites forcées.* In: Compte rendu du 2e Congrès de la Houille Blanche (Lyon, 1914). Dunod & Pinat, Paris (1917) — *Notice sur les travaux scientifiques de M. E. JOUGUET.* Gauthier-Villars, Paris (1918) — With L. CRUSSARD: *Sur la célérité des déflagrations.* C. R. Acad. Sci. Paris **168**, 820-822 (1919) — *Sur une problème d'hydraulique généralisée. Ecoulement d'un mélange gazeux en combustion.* Ibid. **169**, 326-328 (1919) — *Remarques sur les lois de la résistance des fluides.* Ibid. **171**, 96-99 (1920) — *Sur les ondes de choc dans les corps solides.* Ibid. **171**, 461-464 (1920) — *Sur la célérité des ondes de choc dans les solides élastiques.* Ibid. **171**, 512-515 (1920) — *Sur la variation d'entropie dans les ondes de choc des solides élastiques.* Ibid. **171**, 789-791 (1920) — *Application du principe de CARNOT-CLAUSIUS aux ondes de choc des solides élastiques.* Ibid. **171**, 904-907 (1920) — *Quelques problèmes d'hydrodynamique générale.* J. Math. Pures Appl. **3** [VIII], 1-63 (1920) — *Note de mécanique chimique. Sur les lois de la dynamique chimique.* J. Ecole Polytech. (Paris) **21** [II], 181-194 (1921) — *Sur la célérité des déflagrations.* C. R. Acad. Sci. Paris **179**, 454-457 (1924) — *Comparaison de la théorie de l'onde explosive avec quelques expériences récentes.* Ibid. **181**, 546-548 (1925) — *Ondes de choc et combustion avec combustion résiduelle irréversible.* Ibid. **181**, 658-660 (1925) — *La théorie thermodynamique de la propagation des explosions.* In: (E. MEISSNER, ed.) Verhandl. des 2. Int. Kongresses für Technische Mechanik [Zurich, Switzerland, Sept. 1926]. Füssli, Zürich (1927), pp. 12-22 — *Cavitation et similitude.* Bull. Tech. Suisse Romande (18 juin 1927) — *Remarques sur un théorème d'HUGONIOT relatif à l'écoulement des fluides.* C. R. Acad. Sci. Paris **194**, 141-146 (1932) — *Résumé des théories sur la propagation des explosions.* La Science Aérienne **3**, 138-155 (1934) — *Sur les ondes de choc produites dans un gaz par un explosif solide.* C. R. Acad. Sci. Paris **202**, 1225-1229 (1936) — *Commentaire sur la théorie des ondes de choc produites dans une atmosphère gazeuse par un explosif solide.* Ibid. **202**, 1320-1322 (1936) — Preface in the book by P.F. LAFFITTE: *La propagation des flammes dans les mélanges gazeux.* Hermann, Paris (1939).

**SECONDARY LITERATURE.** G. BERTRAND: *Notice nécrologique sur M. JOUGUET.* C. R. Acad. Sci. Paris **216**, 513-515 (1943) — M. LIENARD: *Emile JOUGUET 5 janvier 1871 – 2 avril 1943.* Ann. Mines & Carburants **2** [XIV], 129-132 (1943) — M. ROY: *L'œuvre d'Emile JOUGUET.* Ibid. **2** [XIV], 133-149 (1943) — M. LATOURTE: *JOUGUET, Professeur.* Ibid. **2** [XIV], 150 (1943) — *Liste chronologique des publications.* Ibid. **2** [XIV], 151-154 — M. ROY: *Un grand mécanicien français. Emile JOUGUET.* Bull. Soc. Franç. Mécan. **28**, 165-167 (1944) — G. GENDRON: *English translation of three milestone papers* [publ. by DUHEM (1909), JOUGUET (1901), and ZEMPLÉN (1905)] *on the existence of shock waves.* Rept. VPI-E-89-12, Virginia Polytechnic Institute, Blacksburg, VA (1989) — R. CHÉRET: *Chapman-Jouguet hypothesis 1899–1999: one century between myth and reality.* Shock Waves **9**, 295-299 (1999).

**PICTURE.** Bibliothèque de l'Ecole des mines de Paris, France.

# KANT, Immanuel (1724–1804)

- German philosopher, physicist and cosmologist

Immanuel KANT was born in Königsberg, East Prussia (now Kaliningrad, Russia). He was the fourth of nine children of Johann Georg CANT, a harness maker of modest means. He attended the famous pietistic Collegium Fredericianum in Königsberg and thereafter began to study mainly physics and mathematics at the University of Königsberg (1740). Forced by financial circumstances, he interrupted his study and worked as a private tutor for a period of nine years. He resumed his study at the University of Königsberg, where he received his doctorate in philosophy and habilitated (1755). Working there first as an underpaid instructor (1755–1770), he became professor of logic and metaphysics (1770), a position he held until his retirement (1797).

KANT, more known for his many philosophical works, particularly his *Kritik der reinen Vernunft* ("Critique of Pure Reason," 1781), also speculated on the origin of earthquakes. From a historical point of view his contributions can be regarded as milestones in the evolution of seismology. One year before the 1755 Lisbon Earthquake, KANT finished his long memoir *Allgemeine Naturgeschichte und Theorie des Himmels* ("Universal Natural History and the Theory of the Heavens," published in 1755) in which he treats cosmology according to Newtonian principles. He speculated that the spiral nebulae are distant stellar systems comparable in size to our own stellar system and interpreted the Milky Way as a great disklike structure of a vast swarm of stars, all orbiting some common center or centers in a manner very similar to the planets orbiting around the Sun. KANT also suggested that the nebulae were other disk systems outside our Milky Way, which he called "island universes" – now known as "galaxies." His representation of our Milky Way was basically correct, and in 1993 a first direct proof was given by NASA's Cosmic Background Explorer (COBE) using high-resolution infrared digital imaging.

His theory of nebulae was taken up by Pierre-Simon DE LAPLACE in his Nebular Hypothesis of Earth's Creation (1796), which postulates that the Solar System was formed from a spinning cloud of gas. In the late 1920s, in a series of observations carried out by the U.S. astronomer Edwin HUBBLE, it was conclusively proved that indeed separate galaxies exist beyond our own – thus increasing the size of the Universe by a factor of more than 100. KANT tried to explain the origin of the Universe from a primordial chaotic state of matter, contrary to Sir Isaac NEWTON who had explained the Solar System in terms of a stationary state.

In 1756, KANT also wrote three papers on the Great Lisbon Earthquake that occurred on November 1 (All Saints' Day), 1755. He did not speculate on moral or theological aspects like many of his contemporaries such as François VOLTAIRE and Jean-Jacques ROUSSEAU but rather thoroughly collected all reported data on earthquakes available to him (he never left East Prussia throughout his life), compared existing earthquake theories, and analyzed them soberly also in regard to their possible quantitative effects. Despite these efforts, he eventually ended up, like some of his contemporaries, with an attempt to explain them by subterraneous fires and explosions quoting Nicolas LÉMERY's famous "model volcano" experiment. Since he was not an experimentalist, he probably never considered the simple practical aspect of how nature could have provided such huge quantities of pure sulfur and iron, closely positioned and initially separated, to create an oxyhydrogen explosion with dimensions of destruction on the scale of the Lisbon event. On the other hand, he first attributed the origin of water movements of those inland lakes that are not connected with the sea to either varying atmospheric pressure (a phenomenon later studied in more detail by the Swiss scientist François A. FOREL and called "seiches") or to ground motions as observed during the Lisbon Earthquake in several European lakes far from the epicenter. He also correctly attributed the origin of large surge waves observed during the Lisbon Earthquake that killed many people, later known as "tsunamis," to earthquake-induced displacements of the sea floor.

Although philosophical and theological aspects predominated KANT's later works, he maintained a close affiliation with the natural sciences. His treatment *Metaphysische Anfangsgründe der Naturwissenschaft* ("Metaphysical Foundations of Natural Science," 1786), reprinted three times until 1800, found a wide audience among natural scientists of all fields. Today he is widely considered the foremost philosopher since classical antiquity, effecting a revolution in philosophy and influencing the development of science, theology, and philosophy in a diverse manner.

Astronomers named a crater on the near side of the Moon after him.

**ORIGINAL WORKS.** *Allgemeine Naturgeschichte und Theorie des Himmels.* Petersen, Königsberg (1755) — *Von den Ursachen der Erschütterungen bei Gelegenheit des Unglücks, welches die westlichen Länder von Europa gegen Ende des vorigen Jahres getroffen hat.* Königsbergische wöchentliche Frag- und Anzeigungs-Nachrichten Nr. 4 & Nr. 5 (Jan. 1756) — *Fortgesetzte Betrachtung der seit einiger Zeit wahrgenommenen Erderschütterungen.* Ibid. Nr. 15 & Nr. 16 (April 1756) — *Geschichte und Naturbeschreibung der merkwürdigen Vorfälle des Erdbebens, welches an dem Ende des 1755sten Jahres einen großen Teil der Erde erschüttert hat.* J.H. Hartung, Königsberg (März 1756) — *Metaphysische Anfangsgründe der Naturwissenschaft.* Hartknoch, Riga (1786).

**SECONDARY LITERATURE.** G. RABEL: *KANT* [biography]. Clarendon Press, Oxford (1963) — W. LEY (ed.) *KANT's cosmonogy.* Greenwood, New York (1968) — J.W. ELLINGTON: *I. KANT.* In: (C.C. GILLESPIE, ed.) *Dictionary of scientific biography.* Scribner, New York, vol. 7 (1973), pp. 224-235 — P. BORMANN: *Der Beitrag Immanuel KANT's zur Entwicklung wissenschaftlicher Vorstellungen über die Natur der Erdbeben.* In: (P. SCHMIDT, ed.) *Geschichte der Seismologie, Seismik und Erdbebenforschung* [Meeting in Eisenach, Germany, Dec. 1979]. Veröffentlichung des Zentralinstituts für Physik der Erde (Potsdam) Nr. 64, 17-24 (1981) — O. REINHARDT and D.R. OLDROYD: *KANT's theory of earthquakes and volcanic action.* Ann. Sci. **40**, 247-272 (1983) — E. OESER: *Historical earthquake theories from ARISTOTLE to KANT.* Abhandl. Geol. Bundesanstalt Wien **48**, 11-31 (1992) — W. BREIDERT: *Die Erschütterung der vollkommenen Welt.* Wiss. Buchgesell., Darmstadt (1994); *Das Erdbeben von Lissabon und die Erschütterung seiner Zeitgenossen.* Lichtenberg-Jb., SDV, Saarbrücken (1994), pp. 56-67.

**PICTURE.** Steel engraving by Johann L. RAAB (around 1860), after a painting by Gottlieb DÖBLER (1791). In the public domain, taken from Wikipedia, the free encyclopedia;
http://en.wikipedia.org/wiki/Image:Immanuel_Kant_%28portrait%29.jpg.

## KÁRMÁN, Theodore VON [born VON SKOLLOSKISLAKI KÁRMÁN Tódor] (1881–1963)

- Hungarian-born U.S. physicist, aerodynamicist, and applied mathematician; father of supersonic flight

VON SKOLLOSKISLAKI KÁRMÁN Tódor – who came to be known as "Theodore VON KÁRMÁN" – was born in Jozsefvaros, a suburb of Budapest, as the third son of VON SKOLLOSKISLAKI KÁRMÁN Mór (Maurice), a university professor of education. He already showed in his childhood a surprising talent for mathematics. Before entering university he won the Eötvös Prize for Hungarian secondary students in science and mathematics. After studying engineering at the Múegyetem (Royal Josephs Polytechnic) in Budapest (1903–1906), he received a 2-year fellowship from the world famous University of Göttingen (1906). Under the influence

of the mathematicians Christian Felix KLEIN and David HILBERT he began to collaborate with Ludwig PRANDTL and Max BORN on various problems of aerodynamics and thermodynamics.

After his habilitation (1910) at the University of Göttingen with a thesis on solid-state mechanics, entitled *Untersuchungen über die Bedingungen des Bruches und der plastischen Deformation, insbesondere bei quasi-isotropen Körpern* ("Investigations on the Conditions of Fracture and Plastic Deformation, Particularly of Quasi-Isotropic Bodies"), he taught as a *Privatdozent* (university lecturer) in Göttingen (1910–1912) and at the Hungarian Mining College (1912–1913) in Schemnitz (now Banská Štiavnica, Slovakia). He began working on the mechanics of solid continua, in particular on the strength and elasticity of materials. For example, he studied the brittle-ductile transition in rock deformation and demonstrated in his famous experiments that sandstone (usually a brittle material) becomes ductile when subjected to uniform compression and deformation (1911).

Returning to Germany, he was appointed chair of the Mechanics and Aeronautics Department at the RWTH Aachen and became director of the Institute of Aerodynamics (1913–1934), which was interrupted only by his military service throughout World War I.

In 1926, he first made plans for setting up a laboratory for the study of supersonic motion at Aachen, which was opened in 1929. During the period 1926–1927, he spent some time at CalTech in Harry F. GUGGENHEIM's new aeronautical laboratory. In Kobe, Japan, he built a new wind tunnel, similar to the one in Aachen, for the aircraft division of the Kawanishi Works (1927). Upon returning to Aachen, he accepted an invitation from the president of CalTech to advise on the design of a wind tunnel (1926). In the following two years, he spent half his time at CalTech but in addition remained director of the Institute of Mechanics at the TH Aachen. In 1930, he accepted an offer from the U.S. physicist Robert A. MILLIKAN to become director of the Guggenheim Aeronautics Laboratory at CalTech (1930–1949). At CalTech he began to study supersonic air flows around projectiles of various shapes. Together with Norton B. MOORE, one of his doctoral candidates, he published a remarkable paper on these results (1932) that significantly promoted research on supersonic aerodynamics in the United States. The method they developed was applied not only to projectiles but later also to the design of airplanes and all kinds of airplane components.

In 1936, VON KÁRMÁN became a U.S. citizen. His contributions, covering both sub- and supersonic flight and treating the theory of elasticity, boundary layer theory, heat transfer, turbulence, compressible fluids, and airfoil and propeller profiles, laid the theoretical foundations for modern aerodynamics. His ideas also influenced the design of the first aircraft to break the sound barrier, the Bell X-1. Probably his best-known achievement is the discovery of a special type of periodic vortex pattern of unsteady flow separation over bluff bodies, the so-called "von Kármán vortex street," which can cause detrimental periodic vibration effects on aircraft structures.

VON KÁRMÁN was a cofounder of the Aerojet Engineering Corporation (1942), the RAND Corporation (1948), and CalTech's Jet Propulsion Laboratory, now NASA-JPL. He also gave direction to the early stages of the American rocket and space program. Under his leadership NATO's Advisory Group on Aeronautical Research and Development (AGARD) established the Training Center for Experimental Aerodynamics (1956), located in Belgium at Rhode-Saint-Genèse. After his death, it was renamed the *Von Kármán Institute (VKI) for Fluid Dynamics*; the motto of the Institute, which houses three departments (aeronautics & aerospace, environmental & applied fluid dynamics, and turbomachinery & propulsion), is "Training in research through research." In 1959, serving as chief scientific advisor to the U.S. Air Force, the Gas Dynamics Facility at Arnold Engineering Development Center (AEDC), located in southern Middle Tennessee, was renamed *Von Kármán Gas Dynamics Facility (VKF)*.

VON KÁRMÁN edited the journal *Advances in Applied Mechanics* (vols. I–VIII, 1956–1964, publ. by Academic Press), which contains review articles of internationally renowned experts in the field of fluid mechanics and shock waves. He was awarded the Wright Brothers Trophy (1954), and President John F. KENNEDY awarded him the first Medal of Science (1963) to honor his contributions to science, technology, and education. In addition, he received many other honors and 29 honorary doctorates from various national and foreign universities and colleges.

*Theodore von Kármán Medal*, established and endowed in 1960 by the Engineering Mechanics Division of the American Society of Civil Engineers (ASCE), is awarded to an individual in recognition of distinguished achievements in engineering mechanics. The *von Kármán Award*, instituted in 1987, is the premier award of the International Academy of Astronautics (IAA) given annually to recognize outstanding lifetime achievements in any branch of science. The *von Kármán Lecture Series* is held at VKI and given by active international experts from universities, research establishments, and industry. In the 1990s, the *Kármán Tódor Wind Tunnel Laboratory (KTWTL)* at the Department of Fluid Mechanics of the Budapest University of Technology and Economics was established in his honor.

A crater on the far side of the Moon and a crater on Mars are named for him. On August 31, 1992, the U.S. Post Office issued the 29-cent von Kármán postage stamp featuring space exploration.

ORIGINAL WORKS. *Über stationäre Wellen in Gasstrahlen.* Phys. Z. **8**, 209-211 (1907) — *Recent investigations regarding the flow phenomena of vapors and gases.* J. Soc. Hungarian Eng. Architect. Nos. III and IV (1908) — *Festigkeitsversuche unter allseitigem Druck.* VDI-Z. **55**, 1749-1757 (1911) — With H. RUBACH: *Über den Mechanismus des Flüssigkeits- und Luftwiderstandes.* Physik. Z. **13**, 49-59 (1912) — *Elastizität.* In: (E. KORSCHELT, ed.) *Handwörterbuch der Naturwissenschaften.* Verlag G. Fischer, Jena (1913), vol. 3, pp. 165-193; *Festigkeit.* Ibid., vol. 3, pp. 1014-1030; *Gleichgewicht.* Ibid., vol. 4, pp. 245-261; *Härte und Härteprüfung.* Ibid., vol. 5, pp. 198-202 — *Das Gedächtnis der Materie.* Die Naturwissenschaften **4**, 489-494 (1916) — *Über laminare und turbulente Reibung.* ZAMM **1**, 233-252 (1921) — *Gastheoretische Deutung der Reynolds'schen Kennzahl.* Ibid. **3**, 395-396 (1923) — *Zur Theorie der Luftschrauben.* VDI-Z. **68**, 1237-1242, 1315-1318 (1924) — With J.M. BURGERS: *General aerodynamic theory.* 2 vols., Springer, Berlin (1924) — *Berechnung der Druckverteilung an Luftschiffkörpern.* Abhandl. Aerodyn. Inst. TH Aachen **6**, 3-17 (1927) — With F.L. WATTENDORF: *The impact on seaplane floats during landing.* NACA-TN 321 (1929) — *Eine praktische Anwendung der Analogie zwischen Überschallströmung in Gasen und überkritischer Strömung in offenen Gerinnen.* ZAMM **10**, 334-345 (1930) — With N.B. MOORE: *The resistance of slender bodies moving with supersonic velocities, with special reference to projectiles.* Trans. ASME **54**, 303-310 (1932) — *Turbulence and skin friction.* J. Aeronaut. Sci. **1**, 1-20 (1934) — *General aerodynamic theory: perfect fluids.* In: (W.F. DURAND, ed.) *Aerodynamic theory.* Springer, Berlin (1935), vol. 2 — *The problem of resistance in compressible fluids.* In: *V Convegno Volta su "Le alte velocità in aviazione"* [Rome, Sept./Oct. 1935]. Reale Accademia d'Italia, Roma (1936), pp. 222-276 — [Short autobiography.] Ibid., p. 676 — *Eine praktische Anwendung der Analogie zwischen Überschallströmung in Gasen und überkritischer Strömung in offenen Gerinnen.* ZAMM **18**, 49-56 (1938) — With H.S. TSIEN: *Boundary layer in compressible fluids.* J. Aeronaut. Sci. **5**, 227-232 (1938) — With W.R. SEARS: *Airfoil theory for nonuniform motion.* Ibid. **5**, 379-390 (1938) — *The engineer grapples with nonlinear problems.* Bull. Am. Math. Soc. **46**, 615-683 (1940) — *Compressibility effects in aerodynamics.* J. Aeronaut. Sci. **8**, 337-356 (1941) — *Proposal and tentative plans for a superspeed wind tunnel.* Repts. OSRD-13, NDRC A-9 (July 1941) — *On the propagation of plastic deformation in solids.* Repts. OSRD-365 and NDRC A-29 (Feb. 1942) — *The model supersonic wind tunnel.* Repts. OSRD-519 and NDRC A-38 (April 1942) — *Isaac NEWTON and aerodynamics.* J. Aeronaut. Sci. **9**, 521-522, 548 (1942) — *Faster than sound.* J. Wash. Acad. Sci. **35**, 144-155 (1945) — *Supersonic aerodynamics, principles and applications.* J. Aeronaut. Sci. **14**, 373-409 (1947) — *Aerothermodynamics.* Columbia University Lectures. Lectures in the Dept. of Physics prepared by William PERL (1947) — *The similarity law of transonic flow.* J. Math. Phys.

26, 182-190 (1947) — With P. DUWEZ: *On the propagation of plastic deformation in solids*. J. Appl. Phys. **21**, 987-994 (1950) — *The theory of shock waves and the second law of thermodynamics*. L'Aerotecnica **31**, 82-83 (1951) — *On the foundation of high-speed aerodynamics*. In: (E. STERNBERG, ed.) *Proc. 1st U.S. National Congress of Applied Mechanics* [Chicago, June 1951]. ASME, Ann Arbor, MI (1952), pp. 673-685 — *Aerodynamics and combustion theory*. L'Aerotecnica **33**, 80-86 (1953) — *On the foundation of high speed aerodynamics*. In: *High speed aerodynamics and jet propulsion*. Princeton University Press, Princeton, NJ; vol. VI (1954): (W.R. SEARS, ed.) *General theory of high speed aerodynamics*, pp. 3-30 — *Aerodynamics. Selected topics in the light of their historical development*. Cornell University Press, Ithaca, NJ (1954). In 1954, his book also appeared in French, German, Italian, Japanese and Spanish translations — *Höher, schneller und heißer*. Interavia **11**, 407 (1956) — *Collected works of Dr. Theodore VON KÁRMÁN (1902–1951)*. 4 vols., Butterworths Scient. Publ., London (1956) — *Aerodynamic heating, the temperature barrier in aeronautics*. In: *Proc. Symp. High Temperature – A Tool for the Future* [Berkeley, CA, June 1956]. Stanford Research Institute (SRI), Menlo Park, CA (1956), pp. 140-142 — With H.W. EMMONS, R.S. TANKIN, and G.I. TAYLOR: *Gas dynamics and detonation*. In: *High speed aerodynamics and jet propulsion*. Princeton University Press, Princeton, NJ; vol. III (1958): *Fundamentals of gas dynamics* (ed. by H.W. EMMONS), pp. 574-686 — *Some comments on applications of magnetofluidmechanics*. Proc. 3rd Biennial Gas Dynamics Symp. [Evanston, IL, Aug. 1959]. In: (A.B. CAMBEL and J.B. FENN, eds.) *Dynamics of conducting gases*. Northwestern University Press, Evanston, IL (1960); *see Introduction* — *On the existence of an exact solution of the equations of Navier Stokes*. Comm. Pure Appl. Math. **14**, 645-655 (1961) — With L. EDSON: *The wind and beyond. Theodore VON KÁRMÁN, pioneer and pathfinder in space* [with bibliography]. Little, Brown, Boston (1967) — *Collected works of Dr. Theodore VON KÁRMÁN (1952–1963)* [with bibliography]. Von Kármán Institute for Fluid Dynamics, Rhode-Saint-Genèse, Belgium (1975).

SECONDARY LITERATURE. P.A. HANLE: *Bringing aerodynamics to America*. MIT Press, Cambridge, MA (1960) — H.L. DRYDEN: *Theodore VON KÁRMÁN, Foreign Member of the Royal Society*. Nature **199**, 20-21 (July 6, 1963) — Anonymous: *Theodore VON KÁRMÁN*. Phys. Today **16**, 74 (July 1963) — H.L. DRYDEN: *Contributions of Theodore VON KÁRMÁN to applied mechanics*. Appl. Mech. Rev. **16**, 589-595 (Aug. 1963) — G.I. TAYLOR: *Memories of KÁRMÁN*. J. Fluid Mech. **16**, 478-480 (1963) — H.L. DRYDEN: *The contributions of Theodore VON KÁRMÁN: a review*. Astronaut. Aerospace Engng. **1**, No. 6, 12-17 (July 1963) — L. HOWARTH: *Theodore VON KÁRMÁN: a tribute*. In: (H.L. DRYDEN and Th. VON KÁRMÁN, eds.) *Advances in applied mechanics*. Academic Press, New York, vol. 8 (1964), pp. vii-viii — F.L. WATTENDORF and F.J. MALINA: *Theodore VON KÁRMÁN, 1881–1963*. Astronaut. Acta **10**, 81 (1964) — H.L. DRYDEN: *Theodore VON KÁRMÁN*. Biogr. Mem. Natl. Acad. Sci. (U.S.A.) **38**, 345-384 (1965) — D.S. HALACY JR.: *Father of supersonic flight: Theodore VON KÁRMÁN*. Simon & Schuster, New York (1965) — S. GOLDSTEIN: *Theodore VON KÁRMÁN*. Biogr. Mem. Fell. Roy. Soc. (Lond.) **12**, 335-365 (1966) — G.I. TAYLOR: *Memories of VON KÁRMÁN*. SIAM Rev. **15**, 447-452 (1973) — W.R. SEARS: *VON KÁRMÁN: fluid dynamics and other things*. Phys. Today **39**, 34-39 (Jan. 1986) — J.D. ANDERSON JR.: *Modern compressible flow, with historical perspective*. McGraw-Hill, New York (1990) — M.H. GORN: *The universal man: Theodore VON KÁRMÁN's life in aeronautics*. Smithsonian Institution Press, Washington, DC (1992) — A. HAMACHER: *Theodor VON KÁRMÁN und sein Einfluß auf das 20. Jahrhundert*. RWTH Themen (Berichte aus der RWTH Aachen, Jubiläums-Ausgabe) **1**, 22-23 (1995) — *Theodore VON KÁRMÁN*. In: *Encyclopædia Britannica* (2005). Encyclopædia Britannica Premium Service (19 Nov. 2005); *see* http://www.britannica.com/eb/article-9044747.

PICTURE. Courtesy Von Karman Institute (VKI) for Fluid Dynamics, Rhode-Saint-Genèse, Belgium.

# KISTIAKOWSKY, George Bogdan (1900–1982)

- Ukrainian-born U.S. chemist and high explosives specialist

George B. KISTIAKOWSKY was born in Kiev, Ukraine, a province of pre-revolutionary Russia, to Bogdan KISTIAKOWSKY, a chemistry professor, and attended private schools in Moscow and Kiev. After spending two years as a Russian soldier in the White Army, he escaped to Germany where he studied chemistry at the University of Berlin. Here he took his Ph.D. (1925) with a thesis on the photochemistry of chlorine monoxide and ozone. His thesis supervisor was the famous chemistry professor Max BODENSTEIN, an authority on chemical equilibria, catalytic reaction kinetics, and chain reactions, who was a main inspiration for him to continue research in this field. In the following year, he emigrated to the United States and resumed research at Princeton University, NJ (1926–1930), where he was promoted to the rank of assistant professor (1928). At Harvard University (1930–1970) he became professor of chemistry and performed research on the mechanism of chemical reactions, thermochemistry, and the structure of molecules. His subject compounds varied from the simplest gases to highly complex biological species such as enzymes and antibodies, and his techniques covered sound waves, ultraviolet waves, spectroscopy, shock waves, nuclear magnetism, and scanning mass spectrometry. In 1940, he was called upon by the U.S. government to serve as an explosives consultant for the National Defense Research Committee (NDRC). He became head of the committee's explosives division (1942), presiding over the preparation, testing, and manufacture of new explosives and the development of gun and rocket propellants. During the period 1943–1945, he participated in the Manhattan Project. As head of the Explosives Laboratory at Los Alamos, NM he was responsible for manufacturing the high-explosive lenses for the implosion device (the "Gadget") used in the Trinity Test of the plutonium bomb (July 16, 1945). After World War II, he was engaged in research of chemical kinetics, shock waves, and molecular spectroscopy (1946–1970).

Beginning in the early 1950s, KISTIAKOWSKY and collaborators used the shock-tube technique over a period of 20 years to study chemical reactions and relaxation processes in gases and mixtures. He developed a soft X-ray densitometer with a resolution of a few microseconds to quantitatively resolve the density in the reaction zone of gaseous detonation waves. Using data obtained from mass spectroscopic and gas conductivity probes he derived dissociation energies of $H_2$, $N_2$, CO, and $CO_2$, rate constants of free-radical reactions, and information of isotropic exchange between oxygen atoms and $CO_2$ and $SO_2$. He also studied the chemical kinetics of many high-temperature homogeneous reactions, including (1) the decomposition of $N_2O$; (2) the polymerization, decomposition, and oxidation of acetylene; and (3) chemiluminescence and chemi-ionization. Using spectroscopic methods he identified free radicals, their vibrational and rotational temperatures, and reaction rates. His biographer, the British chemist Sir Frederick DAINTON, F.R.S., appropriately wrote, "Nowhere is his experimental virtuosity and firm grasp of physico-chemical principles more evident than in a long series of papers on gaseous detonation."

KISTIAKOWSKY was a member of the Science Advisory Committee of the U.S. President (1957–1964) and Special Assistant to the President for Science and Technology (1959–1961). He received the President's Medal for Merit (1946), the Nichols Medal (1947), the Willard Gibbs Medal (1960), the President's Medal of Freedom (1961), the Bernard Lewis Medal (1962), the Parsons Medal (1963), the T.W. Richards Medal (1968), the Priestley Medal (1972), the Peter Debeye Award (1974), and a number of other medals and prizes. He was a member of the National Academy of Sciences and a foreign member of the Royal Society of London and received 13 honorary doctorates from various universities and colleges.

Since 1971, a *George B. Kistiakowsky Memorial Lecture* has been given annually at the Dept. of Chemistry and Biology of Harvard University by notable individuals in physical chemistry.

**ORIGINAL WORKS.** *Photochemical processes.* Monograph series of the American Chemical Society, vol. 43. The Chemical Catalogue Company, New York (1928) — *Report on the prediction of detonation velocities of solid explosives.* Rept. OSRD-69, NDRC Div. B (1941) — With E.B. WILSON: *The hydrodynamic theory of detonation and shock waves.* Rept. OSRD-114 (1941) — *Introduction to explosives.* Rept. OSRD-5401 (1945) — *Explosives and detonation waves. Part I: Introduction.* Rept. LA-1043, Los Alamos (1949); *Part IV: The making of explosive charges.* Repts. LA-1052 & LA-1053, Los Alamos (1949) — With D.J. BERETS and E.F. GREENE: *Gaseous detonations: (I) Stationary waves in hydrogen-oxygen mixtures.* J. Am. Chem. Soc. **72**, 1080-1086 (1950); *(II) Initiation by shock waves.* Ibid. **72**, 1086-1091 (1950) — *Density measurements in gaseous detonation waves.* J. Chem. Phys. **19**, 1611-1612 (1951) — *Initiation of detonation in gases.* Ind. Engng. Chem. **43**, 2794-2797 (1951) — *Nonstationary detonation waves in gases.* In: Conf. Chemistry and Physics of Detonation. First ONR Symp. Detonation [Washington, DC, Jan. 1951]. Rept. AD 127-20 (1951), pp. 45-51 — With H.T. KNIGHT and M.E. MALIN: *Gaseous detonations: (III) Dissociation energies of nitrogen and carbon monoxide.* J. Chem. Phys. **20**, 876-883 (1952); *(IV) The acetylene-oxygen mixture.* Ibid. **20**, 884-887 (1952); *(V) Nonsteady waves in CO-O2 mixtures.* Ibid. **20**, 994-1000 (1952) — With H.J. FISHER and D.P. MACDOUGALL: *Liquid explosives.* U.S. Patent No. 2,668,102 (1954) — With G.D. HALSEY JR., M.E. MALIN, and H.T. KNIGHT: *Detonation process of making carbon black.* U.S. Patent No. 2,690,960 (1954) — With D. GARVIN and V.P. GUINN: *The temperature pattern method in the study of fast chemical reaction.* Discuss. Faraday Soc. **17**, 32-39 (1954) — With P.H. KYDD: *The reaction zone in gaseous detonations.* J. Chem. Phys. **22**, 1940-1941 (1954); *Gaseous detonations: (VI) The rarefaction wave.* Ibid. **23**, 271-274 (1955) — With W.G. ZINMAN: *Gaseous detonations: (VII) A study of thermodynamic equilibration in acetylene-oxygen waves.* Ibid. **23**, 1889-1894 (1955) — With P.C. MANGELSDORF JR.: *Gaseous detonations: (VIII) Two-stage detonation in acetylene-oxygen mixtures.* Ibid. **25**, 516-519 (1956) — With P.H. KYDD: *Gaseous detonations: (IX) A study of the reaction zone by gas density measurement.* Ibid. **25**, 824-835 (1956) — With P.H. KYDD: *A mass spectrometer study of flash photochemical reactions.* J. Am. Chem. Soc. **79**, 4825-4830 (1957) — With J.P. CHESICK: *Gaseous detonations: (X) Study of reaction zones.* J. Chem. Phys. **28**, 956-961 (1958) — With M. CHER: *Gaseous detonations: (XI) Double waves.* Ibid. **29**, 506-511 (1958) — With F.D. TABBUTT: *Gaseous detonations: (XII) Rotational temperatures of the hydroxyl free radicals.* Ibid. **30**, 577-581 (1959) — With R.K. LYON: *Gaseous detonations: (XIII) Rotational and vibrational distributions of OH radicals.* Ibid. **34**, 995-998 (1961); *Gaseous detonations: (XIV) The CH radical in acetylene-oxygen detonations.* Ibid. **34**, 1069-1070 (1961) — With C.C. GARDINER JR.: *Density measurements in reflected shock waves.* Ibid. **34**, 1080-1081 (1961) — With H. MIYAMA and P. KYDD: *Gaseous detonations: (XV) Expansion waves in gaseous detonations.* Ibid. **34**, 2038-2045 (1961) — With N.J. BRADLEY: *Shock wave studies by mass spectrometry. (I) Thermal decomposition of nitrous oxide.* Ibid. **35**, 256-263 (1961); *(II) Polymerization and oxidation of acetylene.* Ibid. **35**, 264-270 (1961) — With L.W. RICHARDS: *Emission of vacuum ultraviolet radiation from the acetylene-oxygen and methane-oxygen reactions in shock waves.* Ibid. **36**, 1707-1714 (1962) — With C.W. HAND: *Ionization accompanying the acetylene-oxygen reaction in shock waves.* Ibid. **37**, 1239-1245 (1962) — With G.P. GLASS, J.V. MICHAEL, and H. NIKI: *Mechanism of the acetylene-oxygen reaction in shock waves.* Ibid. **42**, 608-621 (1965) — With I.D. GAY, J.V. MICHAEL, and H. NIKI: *Thermal decomposition of acetylene in shock waves.* Ibid. **43**, 1720-1726 (1965) — With I.D. GAY, G.P. GLASS, and H. NIKI: *Pyrolysis and oxidation of formaldehyde in shock waves.* Ibid. **43**, 4017-4022 (1965) — With J.B. HOMER: *Acetylene-oxygen reaction in shock waves. Origins of CH\* and CO\*.* Ibid. **45**, 1359-1360 (1966) — With I.D. GAY, R.D. KERN, and H. NIKI: *Pyrolysis of ethylene in shock waves.* Ibid. **45**, 2371-2377 (1966) — With J.B. HOMER: *Acetylene-oxygen reaction in shock waves.* Ibid. **46**, 4213-4218 (1967) — With I.D. GAY, G.P. GLASS, and R.D. KERN: *Ethylene-oxygen reaction in shock waves.* Ibid. **47**, 313-320 (1967) — With J.B. HOMER: *Oxidation and pyrolysis of ethylene in shock waves.* Ibid. **47**, 5290-5295 (1967) — With S.H. GARNETT and B.V. O'GRADY: *Isotropic exchange between oxygen and carbon monoxide in shock waves.* Ibid. **51**, 84-91 (1969) — With T.C. CLARK and S.H. GARNETT: *Reaction of carbon dioxide with atomic oxygen and the dissociation of carbon dioxide in shock waves.* Ibid. **51**, 2885-2891 (1969); *Exchange reaction of $^{18}O$ atoms with $CO_2$ and with $SO_2$ in shock waves.* Ibid. **52**, 4692-4698 (1970) — With A.M. DEAN: *Oxidation of carbon monoxide by oxygen in shock waves.* Ibid. **53**, 830-838 (1970) — With T.C. CLARK and T.P.J. IZOD: *Reaction of methyl radicals produced by the pyrolysis of azomethane or ethane in reflected shock waves.* Ibid. **54**, 1295-1303 (1971) — With T.C. CLARK, T.P.J. IZOD, and S. MATSUDA: *Oxidation of methyl radicals studied in reflected shock waves using*

*the time-of-flight mass spectrometer.* Ibid. **55**, 4644-4647 (1971); *Reactions of ethyl radicals produced in the pyrolysis of azoethane in reflected shock waves.* Ibid. **56**, 1337-1342 (1972) — *A scientist at the White House: the private diary of Pres. EISENHOWER's special assistant for Science and Technology.* Harvard University Press, Cambridge, MA (1976) — *A chemist speaks out on the neutron bomb.* Chem. Engng. News **56**, 3 (Editor's page, April 24, 1978) — *Trinity, a reminiscence.* Bull. Atom. Sci. **36**, 19-22 (June 1980).
**SECONDARY LITERATURE.** *To KISTIAKOWSKY, the 1972 Priestley Medal.* Chem. Engng. News **49**, 40-41 (1971) — S.P. PARKER (ed.) *Modern scientists and engineers.* McGraw-Hill, New York (1980), vol. 2, pp. 173-174 — J.B. WIESNER: *George KISTIAKOWSKY.* Phys. Today **36**, 70-72 (April 1983) — F. DAINTON: *George Bogdan KISTIAKOWSKY* [with a list of his 210 journal articles]. Biogr. Mem. Fell. Roy. Soc. (Lond.) **31**, 375-408 (1985) — On KISTIAKOWSKY's contributions to the Manhattan Project *see* the book by L. HODDESON, P.W. HENRIKSEN, R.A. MEADE, and C. WESTFALL: *Critical assembly: a technical history of Los Alamos during the Oppenheimer years, 1943–1945.* Cambridge University Press (1993).
**PICTURE.** Dated 25 July 1947. Courtesy Los Alamos National Laboratory Archives, Los Alamos, NM.

1910 the first shock-tube theory, which was later resumed in Berlin by Friedrich HILDEBRAND at the Knorr-Bremse AG in 1927 and Hubert SCHARDIN at the Luftkriegsakademie (Air Force Academy) in 1932.

In order to improve the level of education and to better match it to practical needs, he later reorganized the courses of instruction at the Polytechnic Vienna by asking leading authorities in industry and transportation to cooperate in the state board of examiners. He was appointed dean of the Faculty of Mechanical Engineering (1913–1915) and served as president of the Polytechnic Vienna (1919–1920). After his retirement (1935) he became a privy councilor. In 1944, he became Honorary Senator of the TU Vienna.

**ORIGINAL WORKS.** *Die Durchschlagsgeschwindigkeit bei den Luftsauge- und Druckluftbremsen.* Z. österr. Ingenieur-Architektenverein **62**, 553-579 (1910) — *Neue Versuche über die Durchschlagsgeschwindigkeit bei der Luftsauge-Schnellbremse.* Ibid. **63**, 21-22 (1911).
**SECONDARY LITERATURE.** L. SANTIFALLER (ed.) *Österreichisches Biographisches Lexikon.* H. Böhlau, Nachf., Vienna (1969).
**PICTURE.** Courtesy Archive of the Technical University of Vienna, Austria.

# KOBES, Karl (1869–1950)

- Austrian engineer and university administrator

Karl KOBES was born in Vienna and studied civil and mechanical engineering at the K.u.K. Polytechnisches Institut Wien (now TH Wien). After a 9-year period in praxis, he returned to the Polytechnic Vienna (1898) and became a university lecturer (1901), associate professor (1902), and full professor of theoretical mechanics occupying the Chair of Theoretical Mechanical Engineering (1905). In the late 1900s, he studied the problem of friction in bearings of high-speed turbines, performed power measurements on diesel engines, and determined the efficiency of belt drives. His most original contribution was certainly the investigation of the mechanism of air brakes. KOBES demonstrated that air brakes could be actuated more efficiently with a velocity exceeding the sound velocity in air, which in practice could be very important in the case of long railway trains. Based on previous works of Bernhard RIEMANN, Pierre-Henri HUGONIOT, and Győző ZEMPLÉN, he published in

# LAMB, Sir Horace (1849–1934)

- British mathematician and fluid dynamicist

Sir Horace LAMB was born in Stockport, Cheshire to John LAMB, a cotton-mill foreman. After attending the grammar school in Stockport, he won a scholarship in classics at Queen's College in Cambridge (1867) but declined the scholarship to spend a year at Owens College, a part of Victoria University (now University of Manchester), where he prepared for a mathematical scholarship. He entered Trinity College (1868), where his teachers were George STOKES and James C. MAXWELL. He graduated there as Second Wrangler (1872). In the same year, he was awarded a Smith's Prize and made a Fellow and Lecturer at Trinity College.

After a 3-year period as a lecturer at Trinity College (1872–1875), he went to Australia as the first professor of mathematics at the University of Adelaide (founded in 1874). He returned to England (1885) and became professor of pure mathematics and later of pure and applied mathematics at Owens College, where he stayed until his retirement (1885–1920). On his retirement from Manchester, LAMB returned to Cambridge, where Trinity College made him an Honorary (Rayleigh) Lecturer (1920–1934).

LAMB was an excellent teacher and noted for his writing abilities. He wrote many articles on mathematics, mostly published in the *Proceedings of the London Mathematical Society* and in his book *Elementary Course of Infinitesimal Calculus* (1897). Today he is considered one of the greatest contributors to applied mathematics. Interested in particular in fluid dynamics and the application of analysis to those physical phenomena in which wave transmission is a central feature, he worked on tidal waves, acoustics, earthquake tremors, hydrodynamics, and underwater explosions. For example, he extended Lord RAYLEIGH's work on the vibration of shells (1891) and analyzed the waves produced in an elastic solid by an impulse of short duration (1904). Here he showed that a localized impulse could separate itself out into a number of disturbances of different types that travel at different speeds.

In his book *The Dynamic Theory of Sound* (1910) he gives a comprehensive mathematical treatment of the physical aspects of sound, covering the theory of vibration, the general theory of sound, and the equations of motion of strings, bars, membranes, pipes, and resonators.

LAMB first studied the deformation of an impulsively loaded plate, which could not be determined directly using conventional bending theory (1917). To understand how earthquake tremors are transmitted around the globe, he studied the propagation of waves on the surface of an elastic solid ("Lamb waves") and investigated the seismic effects of vertical loading on the Earth's surface (1917). Other topics he worked on include electrical induction, electric waves, and the absorption of light. He also wrote textbooks entitled *Statics* (1912), *Dynamics* (1914), and *Higher Mechanics* (1920), a treatise on *The Dynamical Theory of Sound* (1910), the article *Analytical Dynamics* for the supplement to the *Encyclopaedia Britannica* (1902), and the article *Akustik* for the German *Encyclopädie der mathematischen Wissenschaften* (1906).

For the British Aeronautical Research Committee (ARC) he made studies of airflow over aircraft surfaces (1921–1927). He also treated analytically in greater detail than had hitherto been done the bubble dynamics of an underwater explosion (1923), a phenomenon that apparently was first investigated theoretically and experimentally by the German military scientist Rudolf BLOCHMANN (1898). LAMB's famous textbook *Hydrodynamics*, which originated from his courses of lectures given at Cambridge and completed in Adelaide under the title *Treatise on the Mathematical Theory of the Motion of Fluids* (1878), has meanwhile undergone eight editions of which the first six were largely revised and extended by himself over a period of 53 years. It became a standard work on fluid dynamics and wave theory on which many subsequent underwater shock physicists, aerodynamicists, and seismologists have based their work. In this book, he also treated problems of discontinuous wave motion such as tidal and aerial waves of finite amplitude; for the latter, however, he did not use the term *shock wave*.

LAMB was twice vice president of the Royal Society, which awarded him its Royal Medal (1902), and president of the British Association (1925) and the London Mathematical Society (1902–1904). He received also the De Morgan Medal (1911), the London Mathematical Society's premier award, and the Copley Medal (1923) of the Royal Society of London. He was knighted in 1931. In Australia his name is perpetuated by the *Horace Lamb Lecture Theatre* at the University of Adelaide and by the *Horace Lamb Centre for Oceanographic Research* at Flinders University, Adelaide.

A crater on the far side of the Moon is also named after him.

**ORIGINAL WORKS.** *Treatise on the mathematical theory of the motion of fluids.* Cambridge University Press (1879). The title of the second and subsequent editions changed to *Hydrodynamics* (1895, etc.) — *On the motion of a viscous fluid contained in a spherical vessel.* Proc. Lond. Math. Soc. **16** [I], 27-43 (1884/1885) — *On reciprocal theorems in dynamics.* Ibid. **19** [I], 144-151 (1888) — *On waves in a medium having a periodic discontinuity of structure.* Proc. Manch. Lit. Phil. Soc. **42**, 1-20 (1898) — *On wave propagation in two dimensions.* Proc. Lond. Math. Soc. **35** [I], 141-161 (1903) — *On the propagation of tremors over the surface of an elastic solid.* Phil. Trans. Roy. Soc. Lond. **A203**, 1-42 (1904) — *On group velocity.* Proc. Lond. Math. Soc. **1** [II], 473-479 (1904) — *On deep water waves.* Ibid. **2** [II], 371-400 (1905) — *Schwingungen elastischer Systeme, insbesondere Akustik.* In: (W.F. MEYER and H. BURKHARDT, eds.) *Enzyklopädie der Mathematischen Wissenschaften.* Teubner, Leipzig; vol. IV, Teil 4 (1907/1914): (F. KLEIN, ed.) *Mechanik*, pp. 215-310 (on collision, pp. 308-310) — *The dynamical theory of sound.* E. Arnold, London (1910) — *On the uniform motion of a sphere through a viscous fluid.* Phil. Mag. **21** [VI], 112-121 (1911) — *Dynamics.* Cambridge University Press (1914) — With L. SWAIN: *On a tidal problem.* Phil. Mag. **29** [VI], 737-744 (1911) — *On waves in an elastic plate.* Proc. Roy. Soc. Lond. **A93**, 114-128 (1917) — *On the deflection of the vertical by tidal loading of the Earth's surface.* Ibid. **A93**, 293-312 (1917) — *Higher mechanics.* Cambridge University Press, London (1920) — *The steady adiabatic flow of a gas.* H.M.S.O., London (1922) — *On the early stages of a submarine ex-*

*plosion.* Phil. Mag. **45** [VI], 257-265 (1923) — *On water waves due to a disturbance beneath the surface.* Proc. Lond. Math. Soc. **21** [II], 359-372 (1923) — *On the flow of a compressible fluid past an obstacle.* H.M.S.O., London (1928).
**SECONDARY LITERATURE.** Anonymous: *Sir Horace* LAMB [abridged from an article in the London *Times* of Dec. 4, 1935]. Science **80**, 608-609 (Dec. 28, 1934) — A.E.H. LOVE and R.T. GLAZEBROOK: *Horace* LAMB. Obit. Not. Fell. Roy. Soc. Lond. **1**, 375-392 (1935) — G.I. TAYLOR: *Sir Horace* LAMB. *F.R.S.* Nature **135**, 255-257 (1935) — A.E.H. LOVE: *Sir Horace* LAMB. J. Lond. Math. Soc. **12**, 72-80 (1937) — Anonymous: *Sir Horace* LAMB. Z. Flugwiss. **8**, 368-369 (1960) — K.E. BULLEN: *H.* LAMB. In: (C.C. GILLESPIE, ed.) *Dictionary of scientific biography.* Ch. Scribner's Sons, New York, vol. 7 (1973), pp. 594-595 — W.G.K. DUNCAN and R.A. LEONHARD: *The University of Adelaide.* Rigby, Adelaide (1973), pp. 15-17 — R. RADOK and S. RADOK: *A profile of Horace* LAMB. Rept. No. 2, Math. Dept., James Cook University of North Queensland, Townsville, Australia (1980) — J.G. JENKIN and R.W. HOME: *Horace* LAMB *and early physics teaching in Australia.* Hist. Rec. Austral. Sci. **10**, No. 4, 349-380 (1995) — A. BEN MENAHEM: *A concise history of mainstream seismology.* Bull. Seismol. Soc. Am. **85**, 1202-1225 (1995) — S. CHAPMAN and J. TOMSON: LAMB, *Sir Horace.* In: (H.C.G. MATTHEW, ed.) *Oxford dictionary of national biography.* Oxford University Press, Oxford (2004), vol. 32, pp. 271-272.
**PICTURE.** Courtesy University of Adelaide Archives, Adelaide, Australia.

# LAPORTE [or LAPORTE], Otto (1902–1971)

- German-born U.S. physicist

Otto LAPORTE was born in Mainz, Germany as the son of Wilhelm LAPORTE, a heavy-artillery officer in the Imperial German Army. He studied at the Universities of Frankfurt/Main (1920) and Munich (1921–1924). In Munich he came under the influence of Prof. Arnold SOMMERFELD, renowned for his book *Atombau und Spektrallinien* ("Atomic Structure and Spectral Lines," 1919). LAPORTE worked theoretically on the structure of atoms and analyzed measured spectra of vanadium (1922) and iron (1924), thereby discovering the fundamental principle that earned him his Ph.D. (1924). It classifies the atomic energy states into two types, today among spectroscopists known as the "Laporte rule."

With a recommendation from Prof. SOMMERFELD, he went to the United States on a postgraduate fellowship and worked for the National Bureau of Standards (NBS) in its spectroscopy section (1924–1926), coming for the first time into close contact with experimental spectroscopy. At the University of Michigan in Ann Arbor he became a faculty member and instructor of theoretical physics (1926–1927) and professor of physics (1927–1971). In addition, he lectured at NBS and was a visiting professor at Kyoto Imperial University (1928), Tokyo Imperial University (1933), and the University of Munich (1937). After the war LAPORTE worked as an intelligence analyst in the European command of the U.S. Army of Occupation at Heidelberg (1949–1950). In 1944, he entered the field of fluid dynamics and calculated the lift of an airfoil of elliptical outline. Two years later he entered shock wave physics when a member of the Michigan faculty left the university, leaving behind an unfinished project. LAPORTE, almost simultaneously with Prof. Arthur KANTROWITZ, recognized that the shock tube is a superb tool for the study of high-temperature phenomena in gases, because the gas temperature can be raised by a shock and maintained for a brief period of time. LAPORTE assumed charge of the Shock Tube Laboratory at the University of Michigan and initiated research on reflected shock waves (1951), which he first applied to produce high local temperatures. This enabled him not only to enter temperature regions hitherto inaccessible by stationary methods but also to promote high-speed spectroscopic diagnostics.

Besides his research and teaching activities, he acted temporarily as a science attaché with the American Embassy in Tokyo (1954–1956, 1961–1963). Soon speaking fluent Japanese, he became a profound connoisseur of the Japanese culture and assisted in securing the agreement between the United States and Japan on the uses of atomic energy. LAPORTE was one of the charter members of the Division of Fluid Dynamics of the American Physical Society (APS), briefly also serving as its chairman (1965). In 1971, LAPORTE was elected posthumously to the National Academy of Sciences (NAS) of the United States.

In recognition of his scholarship and early guidance the American Physical Society (APS) Division of Fluid Dynamics established the *Otto Laporte Memorial Lectureship*, to be given annually. The *Otto Laporte Award* of APS recognizes outstanding research accomplishments pertaining to the physics of fluids. In addition, at each International Symposium on Shock Waves a distinguished scientist presents the *Otto Laporte Lecture* as the final plenary lecture.

**ORIGINAL WORKS.** *Shock waves.* Scient. Am. **181**, 14-19 (Nov. 1949) — With R.N. HOLLYER, A.C. HUNTING, and E.B. TURNER: *Luminosity generated* *by shock waves.* Nature **171**, 395-396 (1953) — With R.N. HOLLYER: *Parameters characterizing a shock wave.* Am. J. Phys. **21**, 610-613 (1953) — With R.J. EMRICH: *On the interaction of a shock wave with a constriction.* Rept. LA-1740, LASL, NM (1954) — With E.B. TURNER: *On the interaction of two plane shocks facing in the same direction.* J. Appl. Phys. **25**, 678 (1954) — With H. YOSHIHARA: *A rigorous method for finding the lift of a certain class of airfoils and remarks on the meaning of SCHRENK's approximate rule.* J. Aeronaut. Sci. **22**, 787-794 (1955) — With A.L. COLE: *Construction of a section shock tube.* Progr. Rept. No. 7, Engng. Res. Inst., University of Michigan (1957) — With W.R. JOHNSON: *Interaction of cylindrical sound waves with a stationary shock wave.* Phys. Fluids **1**, 82-94 (1958) — *High temperature shock waves.* In: (A.W. MORLEY, ed.) *Proc. 3rd AGARD Colloquium on Combustion and Propulsion* [Palermo, Italy, March 1958]. Pergamon Press, Oxford (1959), pp. 499-524 — With T.D. WILKERSON: *Hydrodynamic aspects of shock tube spectroscopy.* J. Opt. Soc. Am. **50**, 1293-1299 (1960) — With E.B. TURNER: *Atomic line profiles and molecular emission spectra.* In: (A. FERRI, ed.) *Fundamental data obtained from shock-tube experiments.* NATO AGARDograph No. 41 (1961); Pergamon Press, Oxford (1961), pp. 386-404 — With T.S. CHANG: *Reflection of strong blast waves.* Phys. Fluids **7**, 1225-1232 (1964) — *Spectroscopy by means of shock waves.* Proc. Arnold Sommerfeld Centennial Memorial Meeting [Munich, Germany, Sept. 1968]. In: (F. BOPP, ed.) *Int. Symp. on the Physics of One- and Two-Electron Atoms.* North-Holland, Amsterdam (1969) — With Y.W. KIM: *Construction of a shock tube for metallic vapors.* Phys. Fluids **12**, 61-64 (1969) — With M.J. YODER: *Low temperature shock waves in molecular hydrogen.* Proc. 8th Int. Shock Tube Symp. [London, July 1971]. In: (J.L. STOLLERY, A.G. GAYDON, and P.R. OWEN, eds.) *Shock tube research.* Chapman & Hall, London (1971), Paper No. 26 — With T.S. CHANG: *Curved characteristics behind blast waves.* Phys. Fluids **15**, 502-504 (1972).

**SECONDARY LITERATURE.** *Otto LAPORTE.* Phys. Today **24**, 72-73 (June 1971) — R.G. FOWLER: *Fluid dynamics of electron gases* [adapted from the first Otto Laporte Memorial Lecture, delivered to the APS Division of Fluid Dynamics, Boulder, CO, Nov. 1972]. Phys. Today **26**, 23-29 (Nov. 1973) — A.G. GAYDON: *Light emission from shock waves, and temperature measurements* [Otto Laporte Memorial Lecture]. Proc. 9th Int. Shock Tube Symp. [Palo Alto, CA, July 1973]. In: (D. BERSHADER and W.C. GRIFFITH, eds.) *Recent developments in shock tube research: proceedings.* Stanford University Press, Stanford (1973), pp. 11-22 — H.R. CRANE and M. DENNISON: *Otto LAPORTE.* Biogr. Mem. Natl. Acad. Sci. (U.S.A.) **50**, 269-285 (1979) — H. REICHENBACH: *120 years of shock visualization, a fascinating activity* [Otto Laporte Memorial Lecture]. In: (B. STURTEVANT, J.E. SHEPHERD, and H.G. HORNUNG, eds.) *Proc. 20th Int. Symp. on Shock Waves* [Pasadena, CA, July 1995]. World Scientific, Singapore (1996), pp. 17-29.

**PICTURE.** Courtesy Dr. H. REICHENBACH, EMI, Freiburg, Germany.

# LAVAL, Carl Gustaf Patrik DE (1845–1913)

- Swedish engineer, inventor, and industrialist

C. Gustaf P. DE LAVAL was born in Orsa, a town in the Swedish province of Dalarna, and came from a Swedish noble Protestant family that had emigrated from France after the massacre of St. Bartholomew (1572). In the period 1863–1866, he studied and graduated at the newly organized Stockholm Institute of Technology, in 1877 renamed the Royal Institute of Technology (Kungliga Tekniska Högskolan, KTH). First working for the Swedish mining company Stora Kopparberg, he decided to continue his education and attended the University of Uppsala. After taking his doctorate in chemistry there (1872), he founded a sulfuric acid factory in Falun (1873) but soon went bankrupt and continued working at Stora Kopparsberg. In 1875, he worked as a metallurgical engineer in the iron works of Klosterbruck, a town in the province of Moravia, then a part of the Austro-Hungarian empire.

DE LAVAL invented various types of centrifuges; one of the best known was his cream separator (1878), which came into use in around 1880. Based on the large demand for cream separators, he founded the company AB Separator (1883). The centrifuge was driven by his first steam turbine (1883) and quickly adapted in the larger dairies for butter making. DE LAVAL's first steam turbine (Swed. *Ångturbin*) consisted of a single wheel operating at 30,000 rpm. His most renowned invention, the so-called "Laval nozzle" (1888) with its divergent exit geometry, allowed for the first time supersonic outflow velocities without detrimental choking, which immediately improved the efficiency of steam turbines. In 1893, DE LAVAL displayed his single-stage steam turbine at the World Columbian Exposition in Chicago. Today this turbine type is part of the collection of the Smithsonian Institution of Technology in Washington, DC and of the Deutsches Museum in Munich.

DE LAVAL's concept of the Laval nozzle, established by intuition rather than by scientific investigation, was later adapted by many researchers and proved to be very useful not only in supersonic aerodynamics but also in rocketry, pulsejet engines, and wind tunnel testing. DE LAVAL also performed aerodynamic drag studies of air foils and propellers using a wind tunnel; however, he did not publish these studies. His numerous notebooks and drawings are kept at the archives of the Technical Museum in Stockholm.

The extremely high revolution of the turbine wheel required special constructions of the bearings and gearing, which led to revolutionary solutions. To eliminate dangerous wobbling at high speeds, he used a thin and elastic axle for the turbine wheel, a novel concept for which he also obtained international patents (1889, 1891/1892). For purposes other than centrifuge propulsion that required a lower revolution, he designed a high-precision double-helical gear. As early as 1895 more than 37 companies and enterprises were using his patents and inventions. DE LAVAL's first factory of steam turbines, founded in 1890, was transformed three years later into a corporation, the AB de Lavals Ångturbin.

DE LAVAL also contributed with his ideas to other industrial fields, and today he is considered one of the early pioneers of Sweden's industry. In acknowledgment of his contributions to steam turbines he received the Grashof Gold Medal (1904) from the Verein Deutscher Ingenieure (VDI). In 1912, DE LAVAL became a Honorary Member of the American Society of Mechanical Engineers (ASME). In 1957, a series of *De Laval Memorial Lectures* was established by the Royal Swedish Academy of Engineering Sciences in order to commemorate DE LAVAL's contribution to the development and applications of steam turbine machinery.

**ORIGINAL WORKS.** His Swedish patents that could be connected to the Laval nozzle are: *Turbine,* No. 325 (1883) — *Method to manufacture turbines,* No. 430 (1885) — *Steam inlet channel for rotating steam engines,* No. 1,902 (1888) — *Nozzle for steam or gas turbines,* No. 6,610 (1894) — *Device at steam turbine discs,* No. 24621 (1907).
**SECONDARY LITERATURE.** Anonymous: *Carl Gustaf Patrik DE LAVAL (verstorben). Z. ges.* Turbinenwesen **10**, 113-114 (1913) — T.K.V. ALTHIN: *Gustaf DE LAVAL 1845–1913, de hoga hastghe-ternas man* [with a list of DE LAVAL's patents and extensive secondary literature]. Tekniska Museet, Stockholm (1943) — K. OSWATITSCH and W. ROTHSTEIN: *Das Strömungsfeld in einer Lavaldüse.* Jb. Dt. Akad. Luftfahrtforsch. (1942/1943), pp. 91-102 — Anonymous: *Gustaf DE LAVAL. Z.* Flugwiss. **5**, 379-380 (1957) — I. JUNG: *Dr. DE LAVAL and his early work with the steam turbine* [De Laval Memorial Lecture]. AB de Laval's Ångturbin, Stockholm (1957), pp. 7-21 — L. LEPRINCE-RINGUET (ed.) *Die berühmten Erfinder, Physiker und Ingenieure.* Aulis-Verlag Deubner, Cologne (1963), pp. 154-155 — I. JUNG: *Gustaf DE LAVAL, the flexible shaft and the gas turbine;* W. TRAUPEL: *The dynamics of the turbine rotor from Gustaf DE LAVAL to the present day* [De Laval Memorial Lecture]. The Royal Swedish Academy of Engineering Sciences (May 1973) — J. KÖRTING: *Ernst KÖRTING 1842–1921. Ein Ingenieur und Unternehmer im kaiserlichen Deutschland* [zur Vorgeschichte der Lavaldüse]. Technikgeschichte in Einzeldarstellungen, Nr. 34. Verein Deutscher Ingenieure (VDI), Düsseldorf (1975) — J.D. ANDERSON JR.: *Modern compressible flow, with historical perspective.* McGraw-Hill, New York (1990), pp. 177-179.
**PICTURE.** In the public domain, taken from Wikipedia, the free encyclopedia; http://de.wikipedia.org/wiki/bild:GustafDeLaval.jpg.
**NOTE.** The author acknowledges the assistance of Prof. Stig BORGLIN, TH Lund, Sweden, for providing worthwhile references on DE LAVAL's nozzle invention.

## LE BOULENGÉ [or LEBOULENGÉ], Paul-Emile (1832–1901)

- Belgian army-officer and inventor

Paul-Emile LE BOULENGÉ was born in Mesnil-Eglise in the province of Namur, Wallonia and educated at the Ecole Royale Militaire in Brussels (1850–1853). Thereafter, he embarked upon a military career in the Belgian artillery (1853–1859). In the early 1860s, LE BOULENGÉ invented his *clepsydre électrique,* a chronograph for measuring the velocity of artillery projectiles which in respect of accuracy and practicability surpassed Col. Auguste J.A. NAVEZ's electroballistic pendulum (1853). This instrument also allowed one to measure the trajectory time; *i.e.*, the free-flight duration that the projectile covers between leaving the muzzle and impacting the target. Later on, LE BOULENGÉ, improving the NAVEZ chronograph, used two falling rods: the first one was released electromagnetically by the projectile at the moment of leaving the muzzle, and the second rod was released after a certain time, when the projectile, interrupting in free flight the current of a second coil, released a second falling rod that marked this time instant with a sharp knife at the periphery of the first falling rod. This instrument, the so-called "Le-Boulengé chronograph," was used up to the end of World War I in most ballistic laboratories, both in Europe and overseas, particularly in Japan, Brazil, and Peru. Measured data on projectile velocities were also useful to obtain insight into drag effects as function of the projectile geometry. In the 1870s, he also invented an acoustic telemeter.

After being detached temporarily to the Ecole Pyrotechnique, Antwerp (1859–1867), he became a member of the operation staff of the artillery (1867) and was deputized to inspect military weapons.

**ORIGINAL WORKS.** *Mémoire sur un chronographe électro-balistique.* Mém. Cour. (Brux.) **32**, 1-39 (1864/1865) — *Etude de balistique expérimentale. Détermination au moyen de la clepsydre de la durée des trajectoires. Expériences exécutées avec cet instrument. Lois de la résistance de l'air sur les projectiles des canons rayés, déduite des résultats obtenus.* Acad. Roy. Sci., Lett. Beaux-Arts Belgique, Brux. **20** (1868); Hayez, Bruxelles (1868) — *Description et emploi du chronographe LE BOULENGÉ.* Muquardt, Bruxelles (1869); *Description and use of the LE BOULENGÉ's chronograph* [Engl. translation by Lieut. C. JONES]. H.M.S.O., London (1870) — *The determination of the time of flight of projectiles, etc. by means of the electric clepsydra, from researches in experimental ballistics* [trans. from French by Lieut. Commander J.D. MARVIN]. Naval Ordnance Papers, No. 4, Govt. Printing Office, Washington, DC (1873) — *Télèmètre de combat*. Liége, Bruxelles (1874) — *Description, maniement et usage des télèmètres LE BOULENGÉ*. C. Muquardt, Bruxelles (1877) — *Détermination des vitesses vélocipédiques. Vélographe*. Le Cycliste belge illustré, Bruxelles (1894).

SECONDARY LITERATURE. E. KUHN: *Über den elektroballistischen Chronographen von LE BOULENGÉ, nebst einigen Bemerkungen über die gebräuchlichen elektroballistischen Apparate*. Dingler's Polytech. J. **179**, 30-50 (1866) — Anonymous: *Nécrologie. Mort du lieutenant général LE BOULENGÉ*. La Belgique Militaire, Bruxelles (June 9, 1901), pp. 845-846 — Anonymous: *Nécrologie, LE BOULENGÉ*. Annuaire de l'Armée Belge (Bruxelles 1902), p. 521 — Anonymous: *A short description of the theory, construction and operation of the Le-Boulengé chronograph*. H.M.S.O., London (1939) — J. CORNER: *Theory of the interior ballistics of guns*. Wiley, New York (1950) — J.R. LECONTE: *BOULENGÉ (Paul-Emile LE)*. In: *Biographie Nationale*. Publ. by the Académie Royale des Sciences, des Lettres et des Beaux-Arts de Belgique, E. Bruylant, Bruxelles; vol. 34, Supplément vol. 6 (1967–1968), col. 110-114.

PICTURE. Courtesy Head Documentation Center, Royal Army and Military History Museum, Brussels, Belgium.

NOTE. LE BOULENGÉ's contribution may stand here only as an example of attempts made by many other contemporary and subsequent inventors of chronographs to improve the time resolution of diagnosing high-speed phenomena. The development of precision chronography was decisive for scientific progress in understanding the basics of interior and exterior ballistics as well as of shock and detonation wave phenomena and their technical applications.

# LE CHATELIER, Henry Louis (1850–1936)

- French chemist, metallurgist and engineer

Henry L. LE CHÂTELIER was born in Paris to Louis LE CHÂTELIER, an engineer trained at the Ecole Polytechnique and the Ecole des Mines who was responsible for building much of the French railway system. After attending the Collège Rollin in Paris, he entered the Ecole Polytechnique (1869) but shortly interrupted his study due to military service obligations during the Prussian siege of Paris (Dec. 1870 to Jan. 1871). He continued his education at the Ecole des Mines (1871) in order to be trained for the government engineering service. After graduation as a mining engineer (1873) and traveling in North Africa, he began his career in Besançon, northeastern France, as a mining engineer (1875) but soon decided dedicate his life full-time to teaching and research in chemistry. Already two years later he took up the chair of general chemistry at the Ecole des Mines in Paris (1877–1919). He also became instructor at the Ecole Polytechnique (1882) and later accepted the chair of mineral chemistry at the Collège de France (1887–1908).

When LE CHÂTELIER succeeded Ferdinand F.H. MOISSAN as professor of general chemistry at the Sorbonne, he entered the field of metallurgy and founded the journal *Revue de métallurgie* (1904). During World War I he worked for the Ministry of Armaments. Owing to a number of serious mine disasters in the 1870s in France, the Ecole des Mines was asked to investigate their cause and prevention as well (1878). Humphry DAVY had already shown that a certain temperature was needed to trigger explosions of firedamp in mines, but the propagation process of the flame was still unknown. Together with François Ernest MALLARD, a professor of metallurgy, he tackled this problem (1878–1883). They determined first the temperature of inflammation of various combustible mixtures, then measured photographically the velocity of propagation of a flame, and determined calorimetrically the specific heats of combustion up to the highest temperatures by using Robert W. BUNSEN's method of explosion in a closed vessel provided with a pressure gauge. In addition, they improved miner safety lamps and suggested to the mining industry the use of safer high explosives for mining applications.

LE CHÂTELIER developed an optical pyrometer and a platinum/rhodium thermocouple for measuring high temperatures. His early studies on the setting of cements composed of calcium silicates as well as on gases at high temperatures and later also on blast furnace reactions led to the study of chemical equilibrium and the conditions of reversible reactions. In 1888, he enunciated his *Loi de stabilité de l'équilibre chimique* (Law of Stability of Chemical Equilibrium), stating that if a system in a balanced state is disturbed, it will readjust in such a way as to tend to neutralize the disturbance and restore equilibrium – the so-called "Le Châtelier principle" (1888). His conclusion had been anticipated, independently and on a mathematical basis, in the late 1870s by the U.S. physicist J. Willard GIBBS.

He received the Davy Medal (1916) of the Royal Society of London. The Ecole des Mines still honors his memory with the Le Châtelier Prize, which is awarded for the best doctoral thesis each year.

**ORIGINAL WORKS.** With E. MALLARD: *Sur la constatation de la présence du grisou dans l'atmosphère des mines.* C. R. Acad. Sci. Paris **88**, 749-750 (1879); *Sur la température d'inflammation des mélanges gazeux.* Ibid. **91**, 825-828 (1880); *Sur les vitesses de propagation de l'inflammation dans les mélanges gazeux explosifs.* Ibid. **93**, 145-148 (1881); *Sur la vitesse de refroidissement des gaz aux températures élevées.* Ibid. **93**, 962-965 (1881); *Sur les chaleurs spécifiques des gaz aux températures élevées.* Ibid. **93**, 1014-1016 (1881); *Sur la température de combustion et sur la dissociation de l'acide carbonique et de la vapeur d'eau.* Ibid. **93**, 1076-1079 (1881); *Etudes diverses sur les lampes de sûreté.* Pièces annexes de la Commission du grisou. Dunod, Paris (1881); *Sur les procédés propres à déceler la présence du grisou dans l'atmosphère des mines.* Ann. Mines **19** [VII], 186-211 (1881); *Du rôle des poussières de houille dans les accidents des mines.* Ibid. **1** [VIII], 5-98 (1882); *Sur la nature des mouvements vibratoires qui accompagnent la propagation de la flamme dans les mélanges gazeux combustibles.* C. R. Acad. Sci. Paris **95**, 599-601 (1882); *Sur les pressions instantanées produites pendant la combustion des mélanges gazeux.* Ibid. **95**, 1352-1355 (1882); *Etude sur la combustion des mélanges gazeux explosifs.* J. Phys. Théor. Appl. **1** [II], 173-183 (1882); *Notes sur les indicateurs de grisou et les lampes de sûreté.* Ann. Mines **3** [VIII], 31-68 (1883); *Recherches expérimentales et théoriques sur la combustion des mélanges gazeux explosifs: (I) Températures d'inflammation des mélanges gazeux. (II) Sur la vitesse de propagation de la flamme dans les mélanges gazeux. (III) Sur les températures de combustion et les chaleurs spécifiques des gaz aux temperatures élevées.* Ibid. **4** [VIII], 274-568 (1883); *Sur la variation avec la pression de la température à laquelle se produit la transformation de l'iodure d'argent.* C. R. Acad. Sci. Paris **99**, 157-160 (1884); *Recherches sur la combustion des mélanges gazeux explosifs.* J. Phys. Théor. Appl. **4** [II], 59-84 (1885) — *Recherches expérimentales et théoriques sur les équilibres chimiques.* Dunod, Paris (1888); Ann. Mines **13** [VIII], 157-380 (1888) — *Sur les procédés de tirage des coups de mines dans les mines de grisou.* C. R. Acad. Sci. Paris **107**, 96-99 (1888) — *Le grisou et ses accidents.* Rev. Gén. Sci. Pures Appl. **1**, 630-635 (1890); O. Doin, Paris (1890) — *Note sur le dosage du grisou par les limites d'inflammabilité.* Ann. Mines **19** [VIII], 388-395 (1891) — *Sur le dosage du grisou.* Ibid. **2** [IX], 469-477 (1892) — *Le grisou.* In: (H. LEAUTE, ed.) *Encyclopédie scientifique des aide-mémoire.* Gauthier-Villars, Paris (1892); see Section de l'ingénieur, No. 27A — *Sur l'emploi du chlorate de potasse dans les explosifs au nitrate d'ammoniaque.* C. R. Acad. Sci. Paris **128**, 1394-1395 (1899) — With M. BERTHELOT: *Sur la vitesse de détonation dans l'acétylène.* Ibid. **129**, 427-434 (1899) — With O. BOUDONARD: *Mesure des températures élevées.* Carré, Paris (1900) — *Sur le développement et la propagation de l'onde explosive.* C. R. Acad. Sci. Paris **130**, 1755-1761 (1900) — *Sur la propagation des ondes condensées dans les gaz chauds.* Ibid. **131**, 30-33 (1900) — *Les explosifs* [Conférence fait devant la Société des Amis de l'Université]. Dunod & Pinat, Paris (1915) — *Observations sur le communication de M. E. BURLOT.* C. R. Acad. Sci. Paris **179**, 971-972 (1924).

**SECONDARY LITERATURE.** Anonymous: *Principal publications de H. LE CHATELIER.* Bull. Soc. Chim. France **4** [V], 1596-1611 (1937) — C.H. DESCH: *The Le Châtelier Memorial Lecture.* J. Chem. Soc. (Lond.) **141**, 139-150 (1938) — C.H. DESCH: *H.L. LE CHATELIER.* Obit. Not. Fell. Roy. Soc. (Lond.) **2**, 251-259 (1938) — H.M. LEICESTER: *H.L. LE CHÂTELIER.* In: (C.C. GILLESPIE, ed.) *Dictionary of scientific biography.* Scribner, New York, vol. 8 (1973), pp. 116-120.
**PICTURE.** © Collections Ecole Polytechnique, Paris, France.

# LEMAITRE, Monseigneur Georges Henri-Joseph-Édouard (1894–1966)

- Belgian civil engineer, mathematician, cosmologist and priest; cofounder of Big Bang theory

Monseigneur Georges H.J.E. LEMAÎTRE was born in Charleroi (Hainaut province, Belgium) and educated at a Jesuit school. He entered the Catholic University at Louvain (Flem. *Leuven*) to study engineering (1911) but in 1914 joined the army. After World War I, he returned to Louvain University and obtained his doctorate in applied mathematics (1920), taking up also theological studies, which led to ordination with the clerical rank of Abbé (1923). Shortly thereafter, he visited the University of Cambridge (1923), where the British astrophysicist and cosmogonist Arthur EDDINGTON initiated him into modern stellar astronomy and numerical analysis. He spent the following year for postgraduate training in the United States in Cambridge, MA at Harvard College Observatory and MIT with the U.S. astronomer Harlow SHAPLEY, an expert on nebulae. After his return to Louvain (1925) he initially worked as a part-time lecturer and on his Ph.D. thesis, which was entitled *The Gravitational Field in a Fluid Sphere of Uniform Invariant Density According to the Theory of Relativity* (1927). In the same year, he was appointed full professor of astrophysics at Louvain University. He eventually became professor of applied mathematics and was elevated to the clerical rank of Canon and later to the rank of Monseigneur. During a period of 40 years of association with Louvain University, he devoted his talents to both research and teaching.

LEMAÎTRE's contributions to astrophysics essentially covered three fields: (1) the development of a relativistic model of the Universe, his most famous work; (2) research on cosmic rays; and (3) attempts to solve the three-body problem. Independently of the Russian scientist Alexander FRIEDMANN, he published in 1927 a model of an expanding Universe that stood in contrast to the static cosmological model previously developed by Albert EINSTEIN (1917). LEMAÎTRE explained the expansion of the Universe by the explosion of a highly condensed "primeval atom" [French *atome primitif*]

– later popularly known as the "Big Bang theory" – and used Edwin P. HUBBLE's dramatic discovery of reddening in the spectra of distant nebulae as evidence for his theory.

LEMAÎTRE also pondered the origin of cosmic rays, which he considered fossils of the enormous radioactive disintegration process that must have taken place during the initial phase of the Big Bang event. Based on Carl STÖRMER's theory of charged particles in magnetic fields, LEMAÎTRE and Manuel S. VALLARTA, a Mexican-born U.S. scientist at MIT, predicted an east-west asymmetry of cosmic rays – the so-called "Störmer-Lemaître-Vallarta theory" of the trajectories of primary cosmic rays. Measurements carried out in the early 1930s by Thomas H. JOHNSON, Luis W. ALVAREZ, and Arthur H. COMPTON finally proved the charged-particle nature of cosmic rays.

LEMAÎTRE won the Prix Francqui (1934) of the Francqui Foundation at Brussels University, the highest Belgian scientific distinction, and was awarded the first Eddington Medal (1953) by the Royal Astronomical Society. In 1941, he was elected member of the Royal Academy of Sciences and Arts of Belgium. From 1960 he was president of the Pontifical Academy of Sciences in Rome.

Since 1995, the *Georges Lemaître Prize*, instituted by the Belgian *Georges Lemaître Foundation*, is awarded at least once every 2 years to a Belgian or foreign author who has made a significant contribution to increasing scientific knowledge in the fields of cosmology, astronomy, astrophysics, geophysics, and space research.

A crater on the far side of the Moon is named for him.

**ORIGINAL WORKS.** *The motion of a rigid solid according to the relativity principle.* Phil. Mag. **48** [VI], 164-176 (1924) — *Note on the theory of pulsating stars.* Circular Harvard College Observatory **282**, 1-6 (1925) — *Note on DE SITTER's Universe.* J. Math. Phys. **4**, 37-41 (1925) — *La théorie de la relativité et l'expérience.* Rev. Quest. Sci. **9** [IV], 346-374 (1926) — *The gravitational field in a fluid.* Ph.D. thesis, MIT, Cambridge, MA (1927) — *Un univers homogène de masse constante et de rayon croissant rendant compte de la vitesse radiale des nébuleuses extra-galactiques.* Ann. Soc. Sci. Brux. **47A**, 49-59 (1927); Engl. translation: *Homogeneous Universe of constant velocity and increasing radius accounting for the radial velocity of extra-galactic nebulae.* Month. Not. Roy. Astron. Soc. **91**, 483-490 (1931) — *Le mouvement varié d'un solide, d'après la théorie de la relativité.* Ann. Soc. Sci. Brux. **47A**, 103-109 (1927) — *La grandeur de l'espace.* Rev. Quest. Sci. **15** [IV], 189-216 (1929) — *The expanding Universe.* Month. Not. Roy. Astron. Soc. **91**, 490-501 (1931) — *The beginning of the world from the point of view of quantum theory.* Nature **127**, 706 (1931) — *L'univers en expansion.* Ann. Soc. Sci. Brux. **53A**, 51-85 (1933) — *Condensations sphériques dans l'univers en expansion.* C. R. Acad. Sci. Paris **196**, 903-904 (1933) — *La formation des nébuleuses dans l'univers en expansion.* Ibid. **196**, 1085-1087 (1933) — With M.S. VALLARTA: *On COMPTON's latitude effect of cosmic radiation.* Phys. Rev. **43** [II], 87-91 (1933) — *Evolution in the expanding Universe* [Note by editor on LEMAÎTRE's lecture delivered on Feb. 12, 1934 at Armstrong College, Newcastle-upon-Tyne]. Nature **133**, 654 (1934) — With M.S. VALLARTA and L. BOUCKAERT: *On the north-south asymmetry of cosmic radiation.* Phys. Rev. **47** [II], 434-436 (1935) — *On the allowed cone of cosmic radiation.* Ibid. **50**, 493-504 (1936) — *Hypothèses cosmogoniques.* Ciel et Terre (Bruxelles) **61**, 1-11 (1944/1945) — *L'hypothèse de l'atome primitif.* Verhandl. Schweiz. Naturforsch. Gesell. **125**, 77-96 (1945) — *L'hypothèse de l'atome primitif: essai de cosmogonie.* Griffon, Neuchâtel (1946); Engl. translation: *The primeval atom, an essay on cosmogony.* Van Nostrand, New York (1950) — *Cosmological application of relativity.* Rev. Mod. Phys. **21**, 357-366 (1949) — *Rayons cosmiques et cosmologie.* Nauwelaers, Louvain (1949) — *Instability in the expanding Universe and its astronomical implications.* Scripta Varia Pontificiae Acad. Sci. **16**, 475-486 (1958) — *Rencontres avec Albert EINSTEIN.* Rev. Quest. Sci. **19** [V], 129-132 (1958) — *L'œuvre scientifique d'Albert EINSTEIN.* Ibid. **19** [V], 475-487 (1958).

**SECONDARY LITERATURE.** (Ed.) Georges LEMAÎTRE [with a list of his publications and portrait]. Annuario della Pontificia Accademia delle Scienze (Città del Vaticano) **1**, 489-493 (1937) — (Ed.) Georges LEMAÎTRE [with a list of his publications and portrait]. Annuario generale, Accademia nazionale dei Quaranta, pp. 615-620 (1961) — C. MANNEBACK: *Hommage à la mémoire de Mgr. Georges LEMAÎTRE.* Rev. Quest. Sci. **27** [V], 453-461 (1966) — A.V. DOUGLAS: *Georges LEMAITRE, 1894–1966.* J. Roy. Astron. Soc. Can. **61**, 77-80 (1967) — P.A.M. DIRAC: *The scientific work of Georges LEMAÎTRE.* Pontificia Academia Scientiarvm, Vatican City (1968) — J. GIRAUDOUX: *Georges LEMAÎTRE.* Ungar, New York (1971) — C. MANNEBACK: *LEMAÎTRE.* Biogr. Natl. Acad. Roy. Belgique **38** [Suppl. 10], 453-466 (1973/1974) — A. BERGER (ed.) *The Big Bang and Georges LEMAÎTRE.* Proc. Symp. in Honor of G. LEMAÎTRE, 50 Years after his Initiation of Big-Bang Cosmology [Louvain-la-Neuve, Belgium, Oct. 1983]. Reidel, Dordrecht (1984) — *Georges LEMAITRE et l'Académie Royale de Belgique: oeuvres choisies et notice biographique.* Mém. de Cl. Sci., Acad. Roy. Belgique, Brux. (1995) — M. HELLER: *LEMAÎTRE, Big Bang, and the quantum Universe: with his original manuscript.* Pachart, Tucson, AZ (1996).
**PICTURE.** From Ann. Pont. Accad. Sci. **1**, 489 (1937); *see* above.
**NOTE.** The works of G. LEMAÎTRE and a complete list of his publications are kept at the Library of the Belgian Royal Academy in Brussels under No. 55252.

# LÉMERY [or LEMERI], Nicolas (1645–1715)

- French chemist and pharmacist

Nicolas LÉMERY was born in Rouen (Dépt. Seine-Maritime) to Julien LÉMERY, a Protestant attorney in the parliament of Normandy. After serving 6 years as an apprentice in a pharmacy of his uncle in Rouen, he went to Paris (1666), Lyon, Geneva, and Montpellier (1668–1671), where he studied pharmacy and taught chemistry. Returning to Paris, he purchased the office of "Apothecary to the King," which provided a secure financial position. Facing increasing difficulties as a Protestant, he left France for about a year and went into exile in England (1683). Losing after his return his privi-

leged apothecary, he converted to Catholicism, which allowed him to reestablish his laboratory and shop. He became an associated chemist of the Academy of Sciences (1699) and continued working, teaching, and writing in various fields of chemistry and pharmacy.

His *Cours de chymie* (1675), a textbook on chemistry that passed through numerous editions, was translated into Latin and all the major European languages. Also, his *Traité universel des drogues simples* (1698), a dictionary of various medicaments and their therapeutic action, became widely known. His chief contribution to pharmacy were his two complementary works, the *Pharmacopée universelle* (1697) and the *Traité des drogues simples* (1698). He worked out various methods of preparing antimony of mineral antimony, which he described in his book *Traité de l'antimoine* (1707), his last major work. In addition, he contributed a number of papers to the French Academy.

LÉMERY also speculated on the origin of subterranean fires, earthquakes, hurricanes, thunder, and lightning and experimented with oxyhydrogen explosions. His explosion model experiment – developed into an apparatus for easy demonstration purposes and becoming widely known as the "Volcan de Lémery" (1700) – was based on a reaction between filings of iron and sulfurous acid. It was even cited by prominent 18th-century naturalists (such as Immanuel KANT) for explaining explosive volcanic eruptions and earthquakes, although it could not provide a plausible explanation, by him or by his followers, of how nature, prior to reaction, could have provided and kept separate such huge quantities of pure iron and sulfur in the Earth's interior. Although not mentioning LÉMERY by name, Denis DIDEROT and Jean LE ROND D'ALEMBERT also referred to this curious experiment in their *Encyclopédie* [vol. 17, p. 446 (1765)]. Apparently, the British chemist Humphry DAVY first questioned his volcano model to explain volcanic explosive eruptions by subterraneous oxyhydrogen explosions (1828).

Although LÉMERY did not develop any rigorous theory of chemical reactions, he presented many attractive chemical ideas and first introduced the distinction between *chimie minérale* (inorganic chemistry) and *chimie organique* (organic chemistry).

**ORIGINAL WORKS.** *Recueil des curiosités rares et nouvelles des plus admirables effets de la nature et de l'art.* P. Vander Aa, Leiden (1684) — *Chimique des feux souterrains, des tremblements de terre, des ouragans, des éclairs et du tonnerre.* Mém. Acad. Paris (1700), pp. 140-152 — *Cours de chimie.* Chez l'auteur, Paris (1675) — *Dictionnaire, ou Traité universel des drogues simples...* J. Hofhout, Rotterdam (1727).
**SECONDARY LITERATURE.** P.A. CAP: *Nicolas LEMERY, chimiste né à Rouen le 19 Novembre 1645.* Fonderie de Fan, Paris (1839) — A.J.J. VANDEVELDE: *L'œuvre bibliographique de Nicolas LEMERY.* Bull. Soc. Chim. Belges **30**, 153-166 (1921) — O. HANNAWAY: *N. LÉMERY.* In: (C.C. GILLESPIE, ed.) *Dictionary of scientific biography.* Scribner, New York, vol. 8 (1973), pp. 171-175 — M. BOUGARD: *La chimie de Nicolas LÉMERY.* Turnhout, Brepols (1999).
**PICTURE.** Courtesy Deutsches Museum, Munich, Germany.

# LIGHTHILL, Sir Michael James (1924–1998)

- French-born British applied mathematician and fluid dynamicist; founder of aeroacoustics and biofluiddynamics

Sir M. James LIGHTHILL was born in Paris to Earnest Balzar LIGHTHILL, a mining engineer who in 1917 changed the family name from LICHTENBERG to avoid anti-German sentiment. In 1927, his family left France. Growing up in England, he was first educated by his father. Later he won a scholarship to Winchester College (1936) and was awarded a major scholarship to Trinity College, Cambridge (1939), where he studied mathematics (1941–1943) and from which he graduated (B.A., 1943). He was sent to the Aerodynamics Division of the National Physical Laboratory (NPL) at Teddington (1943–1945), where he carried out research for the Aeronautical Research Committee (ARC) and first got involved in problems of supersonic flight. He also analytically studied aerodynamic drag of fine-pointed bodies of revolution and contributed to 2-D supersonic airfoil theory.

After World War II, Sir LIGHTHILL was awarded a prized fellowship at Trinity College and became senior lecturer in mathematics at the University of Manchester (1946–1950), where he was strongly influenced by Sir Geoffrey I. TAY-

LOR, an eminent British fluid dynamicist. At the University of Manchester he succeeded Sydney GOLDSTEIN in the Beyer chair of applied mathematics (1950–1959). In 1949, he extended the classical Poincaré method of nonlinear mechanics. This so-called "Poincaré-Lighthill-Kuo (PLK) technique" of strained coordinates for obtaining uniformly vald approximations for certain classes of ordinary and partial differential equations consists in perturbing not only the unknown, but also the independent variables. In the late 1940s, the Ministry of Aviation asked him to determine if jet aircraft, originally developed for military purposes, could also be used for civilian purposes, and how jets could be made quieter and more powerful at the same time. This resulted in his renowned paper on aeroacoustics in which he formulated his "Eighth Power Law of Jet Noise," stating that the radiated acoustic power for a jet engine is proportional to the eighth power of the jet exit velocity (1952).

Sir LIGHTHILL also worked extensively on gas dynamics at very high speeds, including ionization processes during reentry, and studied diffraction effects of shock and blast waves. His generalization of the hodograph method for flow past solid boundaries allowed nonlinear phenomena to be analyzed via linear equations (1953). His book *Surveys in Mechanics*, published in 1956 to celebrate Sir G.I. TAYLOR's 70th birthday, addressed the new subject of nonlinear acoustics. In the period 1950–1966, he treated important problems of boundary layers covering the range from subsonic to supersonic velocities, as well as in fluctuating streams. He also contributed to the theory of waves in the ocean and the atmosphere. For example, he worked out the differences between nonlinear acoustics and the propagation of waves in shallow water (*e.g.*, tidal bores) – a subject of great fascination that had already attracted many early shock pioneers. When studying how animals move through air or water, he created the new discipline of biofluiddynamics, which brought him renown beyond the physics community.

In the period 1959–1964, he was director of the Royal Aircraft Establishment (RAE) at Farnborough. His research contributed to the aerodynamics of dart-shaped supersonic aircraft (1962), leading eventually to the slender delta-wing design of the SST Concorde. In the 1960s, Sir LIGHTHILL also helped NASA on its high-speed civil transport (HSCT) project, in particular on how to minimize the level of supersonic noise. However, this project was phased out in 1999 due to economic constraints.

After holding a Royal Society research professorship (1964–1969) at Imperial College, London, he returned to Cambridge as Lucasian professor of mathematics (1969–1979). Thereafter, he took on a more administrative role as provost of University College, London (1979–1989). After his retirement in 1989 he traveled and lectured worldwide, acting as chair of the International Council of Scientific Union (ICSU) Special Committee for the International Decade for Natural Disaster Reduction (IDNR).

He was one of the founding associate editors of the *Journal of Fluid Mechanics* (1956). He also served as vice president (1965–1969) of the Royal Society and was president (1984–1988) of the International Union of Theoretical and Applied Mathematics (IUTAM). Sir LIGHTHILL, who held 24 honorary doctorates, was a member of many prestigious foreign academies and was knighted (1971). He was awarded the Royal Medal (1964) and the Copley Medal (1998) of the Royal Society of London, the Cresson Medal (1975) of the Benjamin Franklin Institute, the G.I. Taylor Medal (1984) of the Society of Engineering Science, the first Theodorsen Medal (1993) of NASA Langley, and many other awards.

He made innovative contributions to such fields as applied mathematics, aerodynamics, astrophysics, and fluid mechanics. He wrote 6 books and published about 150 papers. Today he is considered one of the greatest applied mathematicians of the 20th century. To get a comprehensive survey of his enormous output, the reader is referred to his *Collected Papers* (1997). He presented numerous keynotes to various congresses, his last one, given in 1997 at the 5th International Congress on Sound and Vibration in Adelaide, Australia, was entitled "A Century of Shock Wave Dynamics." One year later, he died at age 74 while attempting to swim around Sark Island, U.K.

**ORIGINAL WORKS.** *Two-dimensional supersonic aerofoil theory.* ARC-RM 1929 Ministry of Supply, H.M.S.O., London (1944) — *The conditions behind the trailing edge of the supersonic aerofoil.* ARC-RM 1930. Ibid. (1944) — *Supersonic flow past bodies of revolution.* ARC-RM 2003, Ibid. (1945) — *A new method of two-dimensional aerodynamic design.* ARC-RM 2112, Ibid. (1945) — *The hodograph transformation in transonic flow. Parts I to IV.* Proc. Roy. Soc. Lond. **A191**, 323-369 (1947); Ibid. **A192**, 135-142 (1947) — *Supersonic flow past slender bodies of revolution at yaw.* Quart. J. Mech. Appl. Math. **1**, 76-89 (1948) — *Supersonic flow past slender bodies of revolution the slope of whose meridian section is discontinuous.* Ibid. **1**, 90-102 (1948) — *The position of the shock wave in certain aerodynamic problems.* Ibid. **1**, 309-318 (1948) — With S. GOLDSTEIN and J.W. CRAGGS: *On the hodograph transformation for high-speed flow. Part I: A flow without circulation.* Ibid. **1**, 344-357 (1948) — *Part II: A flow with circulation.* Ibid. **1**, 442-450 (1948) — *The flow behind a stationary shock.* Phil. Mag. **40** [VII], 214-220 (1949) — *A technique for rendering approximate solutions to physical problems uniformly valid.* Ibid. **40** [VII], 1179-1201 — *The shock strength in supersonic 'conical fields.'* Ibid. **40** [VII], 1202-1223 (1949) — *Methods for predicting phenomena in the high-speed flow of gases.* J. Aeronaut. Sci. **16**, 69-83 (1949) — *The diffraction of blast (I).* Proc. Roy. Soc. Lond. **A198**, 454-470 (1949) — *The diffraction of blast (II).* Ibid. **A200**, 554-565 (1950) — *Contributions to the theory of heat transfer through a laminar boundary layer.* Ibid. **A202**, 359-377 (1950) — *Reflection at a laminar boundary layer of a weak steady disturbance to a supersonic stream, neglecting viscosity and heat conduction.* Quart. J. Mech. Appl. Math. **3**, 303-325 (1950) — *The* *energy distribution behind decaying shocks. (I) Plane waves.* Phil. Mag. **41** [VII], 1101-1128 (1950) — *A new approach to thin aerofoil theory.* Aeronaut. Quart. **3**, 193-210 (1951) — *On sound generated aerodynamically. (I) General theory.* Proc. Roy. Soc. Lond. **A211**, 564-587 (1952) — *Oscillating airfoils at high Mach number.* J. Aeronaut. Sci. **20**, 402-406 (1953) — *On the energy scattered from the interaction of turbulence with sound or shock waves.* Proc. Cambr. Phil. Soc. **49**, 531-551 (1953) — *The effect of compressibility on turbulence.* In: (H.C. VAN DE HULST and J.M. BURGERS, eds.) *Proc. Symp. Gas Dynamics of Cosmic Clouds* [Cambridge, U.K., July 1953]. Int. Astronautical Union (IAU) Symp. Series, North-Holland, Amsterdam (1955), pp. 121-130 — *On sound generated aerodynamically. (II) Turbulence as a source of sound.* Proc. Roy. Soc. Lond. **A222**, 1-32 (1954) — *Higher approximations.* In: (W.R. SEARS, ed.) *General theory of high speed aerodynamics.* Princeton University Press, Princeton, NJ (1954), pp. 345-489 — With G.B. WHITHAM: *On kinematic waves. (I) Flood movement in long rivers.* Proc. Roy. Soc. Lond. **A229**, 281-316 (1955); *(II) A theory of traffic flow on long crowded roads.* Ibid. **A229**, 317-345 (1955) — *Viscosity effects in sound waves of finite amplitude.* In: (G.K. BATCHELOR and R.M. DAVIES, eds.) *Surveys in mechanics. A collection in commemoration of the 70th birthday of G.I. TAYLOR.* Cambridge University Press, Cambridge, U.K. (1956), pp. 250-351 — *The dynamics of a dissociating gas. Part I.* J. Fluid Mech. **2**, 1-32 (1957) — *Introduction to Fourier analysis and generalised functions.* Cambridge University Press, Cambridge, U.K. (1958) — *Shock waves* [Ramsden Memorial Lecture]. Mem. Manch. Lit. Phil. Soc. **101**, 1-6 (1959) — *The dynamics of a dissociating gas. Part II: Quasi-equilibrium theory.* J. Fluid Mech. **8**, 161-182 (1960) — *Higher approximations in aerodynamic theory.* Princeton University Press, Princeton, NJ (1960) — *Fluid dynamics as a branch of physics.* Phys. Today **15**, 17-20 (Feb. 1962) — *Jet noise.* AIAA J. **1**, 1507-1517 (1963) — *Physiological fluid mechanics.* Springer, Udine (1971) — *The propagation of sound through moving fluids.* J. Sound & Vibration **24**, 471-492 (1972) — *Waves in fluids.* Cambridge University Press, Cambridge, U.K. (1978) — *An informal introduction to theoretical fluid dynamics.* Clarendon Press, Oxford (1986) — *Some aspects of the aeroacoustics of high-speed jets.* J. Theor. Comp. Fluid Mech. **6**, 261-280 (1994) — (M.Y. HUSSAINI, ed.) *Collected papers of Sir James M. LIGHTHILL.* Vol. I: *Early development of supersonic hydrodynamics. More general high-speed aerodynamics. Gas dynamics interacting with gas physics;* Vol. II: *Two-dimensional potential theory. Aerodynamics with vorticity in the undisturbed flow. Contributions to boundary-layer theory. Chaotic motions;* Vol. III: *Aeroacoustics. Water waves and waterlike patterns. Wave theories of wide applicability;* Vol. IV: *External biofluiddynamics. Internal biofluiddynamics.* Oxford University Press, Oxford (1997).

**SECONDARY LITERATURE.** J. LEGRAS: *Nouvelles applications de la méthode de LIGHTHILL a l'étude des ondes de choc.* Publication ONERA, Chatillon-sous-Bagneux (1953) — D.A. SPENCE (ed.) *LIGHTHILL anniversary volume* [on the occasion of LIGHTHILL's 60th birthday on Jan. 23, 1984]. Institute of Mathematics and its Applications (IMA), University of Minnesota, Minneapolis, MN. IMA J. Appl. Math. **32**, Nos. 1-3 (1984) — Anonymous: *Sir James LIGHTHILL.* The Times. Times Newspaper, London (July 20, 1998), p. 23 — D.G. CRIGHTON: *Sir James LIGHTHILL.* The Independent, Wednesday Review. Newspaper Publ. PLC, London (July 22, 1998), p. 6 — B. BRUEN: *Sir James M. LIGHTHILL; see* http://www.coldrain.net/lucas/lighthill.html — D.G. CRIGHTON and T.J. PEDLEY: *Michael James LIGHTHILL (1924–1998).* Not. Am. Math. Soc. **46**, 1226-1229 (Nov. 1999) — T.J. PEDLEY: *LIGHTHILL, Sir (Michael) James.* In: (H.C.G. MATTHEW, ed.) *Oxford dictionary of national biography.* Oxford University Press, Oxford etc. (2004), vol. 33, pp. 762-765 — M.J. CROCKER (Dept. of Mechanical Engineering, Auburn University, AL): *Sir James LIGHTHILL and his contributions to science* [M.J. Lighthill Memorial Lecture]. 6th Int. Congress on Sound and Vibration [TU of Denmark, Lyngby, Denmark, July 1999]; http://www.iiav.org/ sirjameslighthill.pdf.

**PICTURE.** Courtesy Michael LIGHTHILL, son of Prof. LIGHTHILL, London, U.K.

# MACH, Ernst Waldfried Joseph Wenzel (1838–1916)

- Austrian physicist, psychologist, and philosopher of science; father of supersonics

Ernst W.J.W. MACH was born in Chirlitz-Turas in Moravia (now Chrlice-Tuřany, Czech Republic), at that time part of the Austro-Hungarian Empire. His father Johann MACH had an excellent classical education and was a gymnasium professor, later acting also as private tutor. Up to the age of 15 he was mostly educated by his father. After finishing the gymnasium (1855) in Kremsier (now Kroměříž, Czech Republic) in South Moravia, he studied mathematics, physics, and philosophy at the University of Vienna, where he received his Ph.D. (1860) with a thesis entitled *Über die elektrische Entladung und Induktion* ("On the Electric Discharge and Induction"). He became *Privatdozent* (university lecturer) and worked in the laboratory of Prof. Andreas VON ETTINGSHAUSEN, Christian DOPPLER's successor as chair of experimental physics. One of his tasks was the experimental verification of the Doppler effect. For this purpose he built a special machine that allowed him to demonstrate that Joseph PETZVAL's hypothesis as well as that of DOPPLER (which were rival hypotheses) were in fact correct. In 1864, he became full professor of mathematics at the University of Graz, but in 1867 accepted a professorship of experimental physics at the German Karlsuniversität (Charles University) in Prague, which provided better resources for his experimental studies (1867–1895). As Rector Magnificus (1879–1880) he fought against the introduction of Czech instead of German at Prague University.

During the 28 years he spent in that chair MACH produced most of his important work and published all of his research in gas dynamics, ballistics, and high-speed instrumentation. In 1895, he became professor of philosophy at the University of Vienna, holding there the chair of history and theory of the inductive sciences. After suffering a stroke in 1897, which left the right side of his body paralyzed, he only partly recuperated and retired prematurely in 1901. In the same year, MACH was made a member of the Austrian House of

Peers. In 1913, at the age of 75, he moved to the country home of his son Ludwig in Vaterstetten, a small town in the southeast of Munich, where he died on February 19, 1916, one day after his 78th birthday. Most of his life MACH dedicated to the philosophy of science and to problems in physiology and psychology.

In his famous book *Die Mechanik in ihrer Entwicklung historisch-kritisch dargestellt* ("The Science of Mechanics: a Critical and Historical Account of its Development," 1883), he critically discusses Sir Isaac NEWTON's mechanical views and suggests the elimination of all proportions from which observables cannot be deduced. In his later book *Beiträge zur Analyse der Empfindungen* (1886, translated in 1897 as "Contributions to the Analysis of the Sensations"), which became a classic in the physiology and psychology of sensations, MACH elaborated a new scientific positivism, exerting a powerful influence on those searching for a formula by means of which psychology might be included among the natural sciences. He argued that any physical theory that refers to objects not reducible to sensory experiences must be rejected as metaphysical. All experimental data are neutral by themselves and should merely serve to derive scientific concepts, theories, or laws in order to obtain cognition. His positivistic criteria of verifiability led him to reject the introduction of (invisibly small) atoms and molecules into physical theory. However, his rigorous concept of *"Sehen heißt Verstehen"* ("Seeing is understanding") promoted insight into high-speed phenomena that cannot be resolved by the naked eye and rely heavily upon high-speed visualization and recording methods. This greatly stimulated new physical disciplines such as shock waves, detonics, and supersonic ballistics.

Ernst MACH, who had planned supersonic ballistic experiments several years before Prof. Peter SALCHER, an Austrian physicist at the Imperial Navy, eventually succeeded in photographing them (1886). MACH immediately gave a correct interpretation of the head wave and the lines emanating from projectile surfaces. These supersonic flow phenomena were later connected with his name, such as the *Mach angle, Mach cone, Mach head wave, Mach line,* and *Mach wave.* Furthermore, MACH also studied the oblique interaction of shock waves, thereby discovering irregular reflection (*Mach reflection effect*) and the origin of a new shock wave (*Mach disk, Mach front, Mach stem*), which since the 1940s has stimulated worldwide research activities to better understand this puzzling "Mach effect." He discovered *Mach bands*, an optical illusion which is also known as the "Mach effect" in the physiology community.

MACH's postulate that the local behavior of matter is influenced by the global properties of the Universe (1893) was resumed by Albert EINSTEIN and called by him "MACH's principle." EINSTEIN (1912) wrote, "the entire inertia of a point mass is the effect of the presence of all other masses, deriving from a kind of interaction with the latter ... This is exact the point of view which Ernst MACH urged in his acute investigation on the subject."

In 1895, MACH moved with his wife to the country home of his son Ludwig in Vaterstetten, Bavaria, where he died on February 19, 1916 – one day after his 78th birthday. In his obituary EINSTEIN wrote in April 1916, "He succeeded in taking photographs of the density distribution of air in the environment of a projectile flying with supersonic speed and thus shed light on a genre of acoustic processes about which nothing was known before him." MACH's numerous pioneering discoveries in supersonics and aeroballistics were of immediate military importance and quickly brought him international fame. The Swiss aerodynamicist Jakob ACKERET (1929), honoring his contributions to gas dynamics, proposed naming the ratio of supersonic to sound velocity the "Mach number," which today is generally applied.

Since 1991 the *Ernst Mach Memorial Lecture* is presented at the International Symposium on Shock Waves by a distinguished scientist to commemorate MACH's contributions to supersonics and shock waves. Since 1995 the Academy of Sciences of the Czech Republic in Prague has presented the *Ernst Mach Honorary Medal* to recognize outstanding scientific results achieved in the field of physics (http://www.mpq.mpg.de/mpq-awards/mach-medal.html).

A crater on the far side of the Moon and a minor planet (asteroid 3949 MACH) are also named for him.

**ORIGINAL WORKS.** *Über die Änderung des Tones und der Farbe durch Bewegung.* Sitzungsber. Akad. Wiss. Wien **41** (Abth. II), 543-560 (1860) — *Über die Controverse zwischen DOPPLER und PETZVAL bezüglich der Änderung des Tones und der Farbe durch Bewegung.* Z. Math. Phys. **6**, 120-126 (1861) — *Über die Änderung von Ton und Farbe durch Bewegung.* Ann. Phys. **116** [II], 333-338 (1862) — With J. WOSYKA: *Über einige mechanische Wirkungen des elektrischen Funkens.* Sitzungsber. Akad. Wiss. Wien **72** (Abth. II), 44-52 (1875) — With J. SOMMER: *Über die Fortpflanzungsgeschwindigkeit von Explosionsschallwellen.* Ibid. **75** (Abth. II), 101-130 (1877) — With O. TUMLIRZ and C. KÖGLER: *Über die Fortpflanzungsgeschwindigkeit der Funkenwellen.* Ibid. **77** (Abth. II), 7-32 (1878) — *Neue Versuche zur Prüfung der DOPPLER'schen Theorie der Ton- und Farbänderung durch Bewegung.* Ibid. **77** (Abth. II), 299-310 (1878) — *Über den Verlauf der Funkenwellen in der Ebene und im Raum.* Ibid. **77** (Abth. II), 819-838 (1878) — With G. GRUSS: *Optische Untersuchungen der Funkenwellen.* Ibid. **78** (Abth. II), 467-480 (1878) — With J. VON WELTRUBSKY: *Über die Formen der Funkenwellen.* Ibid. **78** (Abth. II), 551-560 (1878) — With J. SIMONIDIS: *Weitere Untersuchungen der Funkenwellen.* Ibid. **80** (Abth. II), 476-486 (1879) — *Über Herrn A. GUÉBHARD's Darstellung der Äquipotentialcurven.* Ibid. **86** (Abth. II), 8-14 (1882) — *Die Mechanik in ihrer Entwicklung historisch-kritisch dargestellt.* Brockhaus, Leipzig (1883, 1888, 1897, 1901, 1904, 1908, and 1912); Engl. translation by T.J. MCCORMACK: *The science of mechanics: a critical and historical exposition of its principles.* Open Court, Chicago (1893, 1902, 1915, 1919, 1942,

and 1960) — *[Mittheilung betreffs Fixierung einer sehr flüchtigen Erscheinung durch ein photographisches Momentbild.]* Anzeiger (Akad. Wiss. Wien) **21**, 121-122 (1884) — *Photographie einer abgeschossenen Flintenkugel und andere flüchtige Erscheinungen.* Photographische Correspondenz (Vienna) **21**, 287-289 (1884) — With J. WENTZEL: *Ein Beitrag zur Mechanik der Explosionen.* Sitzungsber. Akad. Wiss. Wien **92** (Abth. II), 625-638 (1885) — With P. SALCHER: *Photographische Fixierung der durch Projectile in der Luft eingeleiteten Vorgänge.* Ibid. **95** (Abth. II), 764-780 (1887); Ann. Phys. **32** [III], 277-291 (1887) — *Über die Fortpflanzungsgeschwindigkeit des durch scharfe Schüsse erregten Schalles.* Sitzungsber. Akad. Wiss. Wien **97** (Abth. IIa), 1045-1052 (1888) — With P. SALCHER: *Über die in Pola und Meppen angestellten ballistisch-photographischen Versuche.* Ibid. **98** (Abth. IIa), 41-50 (1889) — *Über die Schallgeschwindigkeit beim scharfen Schuß nach von dem KRUPP'schen Etablissement angestellten Versuchen.* Ibid. **98** (Abth. IIa), 1257-1276 (1889) — With P. SALCHER: *Optische Untersuchungen der Luftstrahlen.* Ibid. **98** (Abth. IIa), 1303-1309 (1889) — With L. MACH: *Weitere ballistisch-photographische Versuche.* Ibid. **98** (Abth. IIa), 1310-1326 (1889) — With L. MACH: *Über longitudinale fortschreitende Wellen im Glase.* Ibid. **98** (Abth. IIa), 1327-1332 (1889) — With L. MACH: *Über die Interferenz von Schallwellen von großer Excursion.* Ibid. **98** (Abth. IIa), 1333-1336 (1889) — *Ergänzungen zu den Mitteilungen über Projektile.* Ibid. **101** (Abth. IIa), 977-983 (1892) — With B. DOSS: *Bemerkungen zu den Theorien der Schallphänomene bei Meteoritenfällen.* Ibid. **102** (Abth. IIa), 248-252 (1893) — *Die Ähnlichkeit und die Analogie als Leitmotiv der Forschung.* Ann. Naturphil. **1**, 5-14 (1902) — *Über Erscheinungen an fliegenden Projektilen.* In: E. MACH: *Populär-wissenschaftliche Vorlesungen.* J. Barth, Leipzig (1903), pp. 351-377.
SECONDARY LITERATURE. A. EINSTEIN: *Ernst MACH.* Physik. Z. **17**, 101-104 (1916) — F. AUERBACH: *MACH's Lebenswerk.* Die Naturwissenschaften **4**, 16-23 (1916) — A. SOMMERFELD: *Nekrolog auf Ernst MACH.* Jb. Bay. Akad. Wiss. (1917), pp. 51-67 — A. LAMPA: *Ernst MACH.* In: (A. BETTELHEIM, ed.) *Neue österreichische Biographie: 1815–1948.* Wiener Drucke, Vienna, Bd. 1 (1923), pp. 93-102 — J. BLACK: *Ernst MACH. Pioneer of supersonics.* J. Roy. Aeronaut. Soc. **54**, 371-377 (1950) — F. HERNECK: *Ernst MACH, eine bisher unveröffentlichte Autobiographie.* Physik. Blätter **14**, 385-390 (1958) — K. MENGER: *Introduction to the 6th American edition, 1960* [Foreword to E. MACH's book *Die Mechanik in ihrer Entwicklung*]. In: E. MACH: *The science of mechanics.* Open Court, La Salle, IL (1960), pp. v-xxi — K.D. HELLER: *Ernst MACH, Wegbereiter der modernen Physik.* Springer, Vienna (1964) — J. ACKERET: *Ernst MACH zum 50sten Todestag.* Schweiz. Bauz. **84**, 140-141 (1966) — W.F. MERZKIRCH: *MACH's contribution to the development of gas dynamics;* R.J. SEEGER: *On MACH's curiosity about shock waves.* In: (R.S. COHEN and R.J. SEEGER, eds.) *Ernst MACH, physicist and philosopher.* Boston Studies in the Philosophy of Science **6**, 42-67 (1970) — J.T. BLACKMORE: *Ernst MACH: his work, life, and influence.* University of California Press, Berkeley (1972) — E.N. HIEBERT: *Ernst MACH.* In: (C.C. GILLESPIE, ed.) *Dictionary of scientific biography.* Scribner, New York, vol. 8 (1973), pp. 595-607 — H. HÖNL: *Albert EINSTEIN und Ernst MACH. Das MACHsche Prinzip und die Krise des logischen Positivismus.* Phys. Blätter **35**, 485-494 (1979) — H. REICHENBACH: *Contributions of Ernst MACH to fluid mechanics.* Annu. Rev. Fluid. Mech. **15**, 1-28 (1983) — L.J. WEIGERT: *Ernst MACH (1838–1916).* Verlagshaus Sudetenland, München (1989) — P. KREHL and M. VAN DER GEEST: *The discovery of the Mach reflection effect and its demonstration in an auditorium.* Shock Waves **1**, 3-15 (1991) — D. GRIESER: *Ernst MACH, Vordenker der Moderne.* In: *Köpfe. Portraits der Wissenschaft.* ÖBV, Vienna (1991), pp. 93-114 — J.T. BLACKMORE: *Ernst MACH, a deeper look.* Kluwer, Dordrecht (1992) — C. HOFFMANN: *Mach-Werke* [his ballistic experiments from the epistemological point of view]. Fotogeschichte **16**, 3-18 (1996) — D. HOFFMANN: *Ernst MACH (1838–1916).* In: (K. VON MEŸENN, ed.) *Die großen Physiker.* C.H. Beck'sche Verlagsbuchhandl., München (1997), vol. 2, pp. 24-36 — P. BERZ and C. HOFFMANN: *Sichtbare Grenzen: Ernst MACHs Notizen zu den ballistisch-fotografischen Versuchen 1886/87* [Faksimile der Seiten 90-129 aus Ernst MACHs Notizbuch Nr. 25 nebst Einleitung und Kommentar]. Max-Planck-Institut (MPI) für Wissenschaftsgeschichte, Berlin (2000) — C. HOFFMANN: *Über Schall: Ernst MACHs und Peter SALCHERs Geschoßfotografien.* Wallstein, Göttingen (2001) — W. FÜSSL and M. PRUSSAT: *Der wissenschaftliche Nachlass von Ernst MACH.* Veröffentlichungen aus dem Archiv des Deutschen Museums, Bd. 4 [ISBN 3-924183-76-7]. München (2001) — W.G. POHL: *Peter SALCHER und Ernst MACH: Schlierenphotografie von Überschall-Projektilen.* PLUS LUCIS (Z. des Vereines zur Förderung des chemischen und physikalischen Unterrichts, Wien) Hefte 1/2 (2002/2003), pp. 22-26 — W.G. POHL and G. SALCHER: *Fotografien fliegender Projektile.* Mitteil. Österreich. Gesell. Wissenschaftsgeschichte (Erasmus, Vienna) **21**, 125-154 (2003) — J.T. BLACKMORE, R. ITAGAKI, and S. TANAKA (eds.): *Ernst MACH's science. Its character and influence on EINSTEIN and others.* Tokai University Press, Tokyo (2006) — *Hall of pioneers: history of heat transfer.* UCLA Dept. of Materials Science & Engineering; http://www.seas.ucla.edu/jht/pioneers/pioneers.html — *Ernst Mach Honorary Memorial Medal* for merit in the physical sciences, Academy of Sciences of the Czech Republic, Prague; http://www.mpq.mpg.de/mpq-awards/mach-medal.html.
PICTURE. Archives of Ernst-Mach-Institut (EMI), Freiburg, Germany.
NOTE. Ernst MACH's scientific legacy comprises (1) his scientific notebooks; (2) a large collection of shock wave photos recorded on silver bromide gelatin dry plates, including the famous first photographs of high-speed projectiles and shock waves generated by electric spark discharges; (3) his correspondence with famous scientists such as BOLTZMANN, BOYS, DUHEM, DVORAK, EINSTEIN, FEDDERSEN, VON HELMHOLTZ, HERTZ, KELVIN, MELSENS, PLANCK, SALCHER, TOEPLER, etc.; (4) a collection of his awards and medals of honor; and (5) a part of his personal library. The major part of his legacy was donated in 1959 by Mrs. Karma MACH, Ludwig MACH's wife, to the Ernst-Mach-Institut (EMI), Freiburg. EMI's total collection was transferred in 1998 to the Deutsches Museum at Munich; 942 photographs taken by Ernst MACH and collaborators can now be seen in the Internet, see http://www.deutsches-museum.de\bib/archiv/mach/index.htm. A part of his personal library was purchased around 1960 from L. MACH's heirs by the Institut für Aerodynamik (IfA) of the ETH Zurich (priv. comm. by Prof. Herbert SPRENGER, IfA, ETHZ).

# MACH, Ludwig (1868–1951)

- Austrian physician, physicist, and inventor

Ludwig MACH was born in Prague and was the eldest son of Ernst MACH, an eminent Austrian professor of natural philosophy. In 1887/1888, he began to study medicine at the German Charles University in Prague. Beginning in the late 1880s, however, he devoted more time to improving his father's laboratory equipment than to his own medical studies.

Ludwig MACH was a skillful experimenter. Together with his father he performed further important supersonic ballistic experiments and modified the Jamin interferometer that his father had used to obtain data on the strength of shock waves. The new instrument, put in operation in 1891, was called the "Mach-Zehnder interferometer," because its principle was invented independently in the same year by Ludwig ZEHNDER, a Ph.D. candidate of Prof. Wilhelm C. RÖNTGEN. This interferometer, which makes the distance between the measuring and reference light beams greater than in the Jamin interferometer, is very useful to visualize the density profiles in shock waves and ballistic head waves.

After his doctorate (*medicinae universae doctor*, July 1895) he did not enter medical practice but rather went to the Optischen Werke Zeiss at Jena. In 1896, Ludwig MACH was the first to introduce *particle tracer photogrammetry* into high-speed flow diagnostics: he photographed the flow of air in a wind tunnel using silk threads, cigarette smoke, and glowing particles of iron.

Because of serious health problems his father was forced to give up his chair at the University of Vienna in 1913 and moved to his son's country house outside of Munich in Vaterstetten in an isolated forest. Here Ludwig MACH assisted his father in finishing his publications, planned together with him experiments on the speed of light to discourage relativistic speculations, and promised him to complete and publish his remaining manuscripts, including also part II of *The Principles of Physical Optics*. However, his father specified that in case of a failure of the light experiments, this manuscript should be destroyed.

After his father's death (1916), he began to publish manuscripts from his father under his supervision and changed from a supporter to an opponent of Albert EINSTEIN's ideas of relativity. In the 1933 edition of his father's monograph *Science of Mechanics*, he removed the pro-Einstein afterword of his father and inserted his own anti-Einstein foreword, which brought him into controversy with some contemporaries. Since his light experiments could not be completed in Vaterstetten due to various circumstances, he destroyed all the remaining unpublished manuscripts of his father.

Ludwig MACH had dedicated himself to aiding his father in any and every possible manner. Frustrated in his efforts at the end of his life, he described himself as someone who "fought for a dead man whose shadow I always was" [Germ. *"Ich kämpfte für einen Toten, dessen Schatten ich immer war."*].

ORIGINAL WORKS. With E. MACH: *Weitere ballistisch-photographische Versuche*. Sitzungsber. Akad. Wiss. Wien **98** (Abth. IIa), 1310-1326 (1889); *Über longitudinale fortschreitende Wellen im Glase*. Ibid. **98** (Abth. IIa), 1327-1332 (1889); *Interferenz der Schallwellen von großer Excursion*. Ibid. **98** (Abth. IIa), 1333-1336 (1889) — *Über ein Interferenzrefraktor*. Ibid. **101** (Abth. IIa), 5-10 (1892); Ibid. **102** (Abth. IIa), 1035-1056 (1893); Z. Instrumentenkunde **12**, 89-93 (1892); Ibid. **14**, 279-283 (1893) — *Über die Dauer verschiedener Momentbeleuchtungen*. In: (J.M. EDER, ed.) Jb. für Photographie & Reproductionstechnik. W. Knapp, Halle a. S. (1893), pp. 195-201 — *Über die Herstellung von Rotationsflächen zweiten Grades auf der Drehbank*. Z. Instrumentenkunde **13**, 82-87 (1893) — *Das Prinzip der Zeitverkürzung in der Serienphotographie*. Photogr. Rundschau (Vienna) **7**, 121-127 (1893) — *Weitere Versuche über Projektile*. Sitzungsber. Akad. Wiss. Wien **105** (Abth. IIa), 605-633 (1896) — *Sichtbarmachung von Luftstromlinien*. Z. Luftschiffahrt Phys. Atmosphäre **15**, Heft 6, 129-139 (1896) — *Optische Untersuchung der Luftstrahlen*. Sitzungsber. Akad. Wiss. Wien **106** (Abth. IIa), 1025-1074 (1897) — *Über einige Verbesserungen an Interferenzapparaten*. Ibid. **107** (Abth. IIa), 851-859 (1898) — *Über die Herstellung schlieren- und blasenfreier Glasflüsse im Siemens'schen Ofen*. Anzeiger (Akad. Wiss. Wien) **37**, 125-127 (1900) — *Aiming device for guns*. U.S. Patent No. 1,060,469 (April 29, 1913) — With G. ULSENHEIMER: *Poliermittel, insbesondere für Glasoberflächen*. Germ. Patent Nr. 932,381 (Aug. 1955).

SECONDARY LITERATURE. L. ZEHNDER: *Ein neuer Interferenzrefraktor*. Z. Instrumentenkunde **11**, 275-285 (1891) — *Poggendorff's Bibliographisch-literarisches Handwörterbuch*. A. Barth, Leipzig (1904), vol. IV (1883–1904) — J.T. BLACKMORE: *Ernst MACH: his work, life, and influence*. University of California Press, Berkeley (1972) — G. WOLTERS: *MACH I, MACH II, EINSTEIN und die Relativitätstheorie: eine Fälschung und ihre Folgen*. De Gruyter, Berlin (1987).

PICTURE. Archives of Ernst-Mach-Institut, Freiburg, Germany.

# MAIYEVSKY [Russ. *МАИЕВСКИЙ* or *МАЙЕВСКИЙ*], Nikolai Vladimirovich (1823–1892)

- Russian physicist and ballistician

Nikolai V. MAIYEVSKY was born in the country seat of his parents' home at Pervino near Torzhok, administrative district of Tver. After studying physics and mathematics at the University of Moscow and taking his Ph.D. (1839–1843), he obtained his officer's training at the artillery school of Mikhailovskoe. Serving first in a mounted artillery brigade (1846–1850), General E.C. VESSEL called on him in the Artillery Division of the Military Academy Commission. Quickly working successfully in the field of exterior ballistics under Aleksei Vasplevich DYADIN, a general lieutenant of the artillery and a renowned Russian ballistician, he was appointed member of the commission and professor of the military academy.

His ballistic courses, which were based upon the results of his latest scientific studies but also matched practical needs, were internationally acknowledged and recorded in his two textbooks, which are entitled, in translation, *Courses on Exterior Ballistics* (1870) and *The Method of the Smallest Squares, Particularly Its Application to the Analysis of Ballistic Data* (1881), the latter introducing the theory of probabilities into practical training on artillery guns.

In exterior ballistics he investigated aeroballistic drag of spherical projectiles up to $M = 2$, thereby recognizing that the ratio of projectile velocity to sound velocity – later termed the *Mach number* – is an important parameter governing aerodynamic drag at high speeds. In the early 1860s, he studied ballistic problems arising from shooting long projectiles up to $M = 2$ from rifled barrels, such as the influence of rotation along the trajectory. Shortly after publication, this subject was resumed by the Irish engineer Robert MALLET, who proved that no rifled bullet could take anything but a curved course through tissues of the human body, an important finding for military surgeons. MAIYEVSKY also corrected existing firing tables. In interior ballistics he contributed to the design of durable and efficient gun barrels, collaborating in their fabrication with the German Krupp-Werke at Essen. He measured the pressure of explosion gases in gun barrels and, studying the choice of appropriate gunpowders for large guns, proved the superiority of prismatic powders.

MAIYEVSKY's memoir on impact phenomena of rotating projectiles, which was also translated into French, earned him wide acclaim as an international expert on ballistics and the Great Order of Mikhailovskoe (1866). In recognition of his scientific achievements the University of Moscow honored him with a doctorate in applied mathematics (1870). He was a corresponding member of the Academy of Sciences (since 1878) and an honorary member of the University of Moscow (from 1890).

**ORIGINAL WORKS.** *The pressure of the explosion gases on the walls of gun barrels and the application of results based on tests performed in Prussia in order to determine the wall thickness of guns* [in Russ.]. Artill. Zh. (St. Petersburg) **22**, No. 1, 1-75 (1856) — *Über den Druck auf die Seelenwände, und über die Anwendung der Resultate der darüber in Preußen gemachten Versuche auf die Bestimmung der Metallstärken von Geschützröhren.* Arch. Artill.-Ing.-Off. **41**, 57-91, 163-202 (1857) — *De la pression des gaz de la poudre contre les parois des bouches à feu, et de l'application des résultants des expériences faites à ce sujet en Prusse à la détermination des épaisseurs de métal des bouches à feu.* Rev. Technol. Art Milit. **2**, 173-245 (1857) — *Sur l'expression de la résistance de l'air à mouvement des projectiles sphériques.* Bull. Acad. Imp. Sci. St. Pétersbourg **1** [Cl. Phys. Math.], 337-349 (1860) — *The most important artillery systems with rifled barrels* [in Russ.]. Artill. Zh. **28**, No. 7, 621-732 (1862) — *The influence of rotation on the trajectory of long projectiles.* Ibid. **31**, No. 3, 1-191 (1865); *De l'influence du mouvement de rotation sur la trajectoire des projectiles oblongs dans l'air.* Rev. Technol. Art Milit. **5**, 1-188 (1865) — *Sur le mouvement des projectiles oblongs tirés des bouches à feu rayées.* Bull. Acad. Imp. Sci. St. Pétersbourg **8**, 181-185 (1865) — *The influence of rotation of long projectiles on their penetration in solid matter* [in Russ.]. Artill. Zh. **32**, No. 5, 1-100 (1866) — *Experiments to determine the pressure of explosion gases in the interior of guns performed in November 1867 in the Krupp steel works* [in Russ.]. Ibid. **35**, No. 5, 869-904 (1869) — *Mémoire sur les expériences faites à l'établissement de M. KRUPP, à Essen, au mois de Novembre 1867, pour déterminer les pressions des gaz de la poudre dans l'âme des bouches à feu.* Mém. Cour. Autres Mém. (Bruxelles) **XXI**, No. 4, 1-24 (1870) — *Courses in ballistics* [in Russ.]. AN, St. Petersburg (1870); *Traité de balistique extérieure.* Gauthier-Villars, Paris (1872) — *Sur les résultats des expériences concernant la résistance de l'air et leur application à la solution des problèmes du tir.* Bull. Acad. Imp. Sci. St. Pétersbourg **27**, 1-14 (1881) — *The method of the smallest squares, particularly its application on the analysis of shooting data* [in Russ.]. AN, St. Petersburg (1881) — *The solution of problems arising by aimed shooting at steep trajectory* [in Russ.]. Artill. Zh. **48**, No. 9, 1-28; No. 11, 1-2, 29-91 (1882); Germ. translation by H. KLUSSMANN: *Über die Lösung der Probleme des direkten und indirekten Schiessens.* Mittler, Berlin (1886) — *The method of interpolation* [in Russ.]. AN, St. Petersburg (1883) — *The probability of deviation from the group center of projectile impact on the target* [in Russ.]. Ibid. (1885).

**SECONDARY LITERATURE.** N. ZABUDSKII: *General of the artillery Nikolai Vladimir MAIYEVSKII* [in Russ.]. Artill. Zh. **58**, No. 4, 1-27 (1892) — F.I. FRANKL: *On the priority of EULER in the discovery of the similarity law for the resistance of air to the motion of bodies at high speeds* [in Russ.]. Dokl. AN (SSSR) **70**, No. 1, 39-42 (1950) — A.P. MANDRUIKA: *Nikolai Vladimir MAIYEVSKII* [in Russ.]. Moscow (1954) — Anonymous: *N.V. MAIYEVSKII.* In: (I.V. KUZNETSOVA, ed.) *Lyudi russkoi nauki* ("Russian men of science"). Nauka, Moscow, vol. IV (1965): *Tekhnika*, pp. 170-177.

**PICTURE.** From *Russian men of science.* Nauka, Moscow (1965), vol. I, p. 170.

**NOTE.** His name has also been transliterated as MAIEVSKIJ.

# MALLARD, François Ernest (1833–1894)

- French mineralogist, metallurgist and mining engineer

F. Ernest MALLARD was born in Châteauneuf-sur-Cher (Dépt. Cher). His father was a lawyer. After studying at the Collège de Bourges, the Ecole Polytechnique, and the Ecole des Mines in Paris, he graduated as Engineer of Mines (1853). Beginning as a geologist at the Corps des Mines, he was nominated professor of geology, mineralogy, and physics at the Ecole des Mines in Saint-Etienne (1859). His studies in crystallography began when he was chosen to fill the

vacant chair of mineralogy at the Ecole des Mines at Paris (1872), where he stayed until his death. He became Inspecteur Général des Mines (1886), was elected the second president of the Mineralogical Society of France, and was a member of the French Academy of Sciences (1890). He received the Croix de chevalier (1869) and became an officer of the Légion d'Honneur (1888). The University of Bordeaux awarded him an honorary doctorate (1888).

MALLARD's scientific studies as a mining engineer began in Saint-Etienne (Dépt. Loire) on studying the use of the safety lamp in coal mines (1868). He uncovered the dangers arising by the use of the Davy lamp and proposed various improvements. Ten years later, he became a member of the Commission du grisou, the French Firedamp Commission that was established to prevent methane explosions in mines.

In a very fruitful cooperation with Henry L. LE CHÂTELIER, then professor of general chemistry at the Ecole des Mines, he started a thorough investigation into the ignition temperature of gaseous explosions the results of which had a significant impact on practical mining. In addition, his studies on the specific heat and dissociation temperatures were of great scientific value for the understanding of explosion processes. Their joint studies revealed the importance of mixtures of coal dust and instituted the use of ammonium nitrate as a "safe explosive" in the mining industry. Its detonation temperature amounts to only 1,100 °C, which is low compared to about 2,500 °C for common explosives. Ammonium nitrate is the preferred explosive for shooting purposes to this day.

To the physics community MALLARD is probably better known for his contributions to crystallography and for his classic two-volume textbook *Traité de Cristallographie géométrique et physique* ("Treatise on Geometric and Physical Crystallography," 1879, 1884) rather than for his explosion studies. Stimulated by Auguste BRAVAIS' book *Etudes cristallographiques* ("Crystallographic Studies," 1866), he applied this theory to an understanding of the wide range of physical properties of crystals. From 1879 he published memoirs on optical anomalies, on the quasi-cubic form of all crystallized bodies, on the transformations of the polymorphous substances, and on isomorphous mixtures. He also refined the Wollaston goniometer, which resulted in his so-called "Mallard goniometer." It allows the precise measurement of the angles of crystallographic axes and indexes of refraction in glasses and crystals.

**ORIGINAL WORKS.** *De la définition de la température dans la théorie mécanique de la chaleur et de l'interprétation physique du second principe fondamental de cette théorie.* C. R. Acad. Sci. Paris **75**, 1479-1484 (1872) — *De la vitesse avec laquelle se propage l'inflammation dans un mélange d'air et de grisou, et de la théorie des lampes de sûreté.* Ann. Mines **7** [VII], 355-381 (1875) — With H.L. LE CHATELIER: *Sur la constatation de la présence du grisou dans l'atmosphère des mines.* C. R. Acad. Sci. Paris **88**, 749-750 (1879); *Sur la température d'inflammation des mélanges gazeux.* Ibid. **91**, 825-828 (1880); *Sur les vitesses de propagation de l'inflammation dans les mélanges gazeux explosifs.* Ibid. **93**, 145-148 (1881); *Sur la vitesse de refroidissement des gaz aux températures élevées.* Ibid. **93**, 962-965 (1881); *Sur les chaleurs spécifiques des gaz aux températures élevées.* Ibid. **93**, 1014-1016 (1881); *Sur la température de combustion et sur la dissociation de l'acide carbonique et de la vapeur d'eau.* Ibid. **93**, 1076-1079 (1881); *Etudes diverses sur les lampes de sûreté.* Pièces annexes de la Commission du grisou. Dunod, Paris (1881); *Sur les procédés propres à déceler la présence du grisou dans l'atmosphère des mines.* Ann. Mines **19** [VII], 186-211 (1881); *Du rôle des poussières de houille dans les accidents des mines.* Ibid. **1** [VIII], 5-98 (1882); *Sur la nature des mouvements vibratoires qui accompagnent la propagation de la flamme dans les mélanges gazeux combustibles.* C. R. Acad. Sci. Paris **95**, 599-601 (1882); *Sur les pressions instantanées produites pendant la combustion des mélanges gazeux.* Ibid. **95**, 1352-1355 (1882); *Etude sur la combustion des mélanges gazeux explosifs.* J. Phys. Théor. Appl. **1** [II], 173-183 (1882) — *Conclusions tirées des expériences de Mr. Lindsay WOOD sur la pression du grisou dans la houille.* Ann. Mines **1** [VIII], 538-551 (1882) — With H. LE CHATELIER: *Notes sur les indicateurs de grisou et les lampes de sûreté.* Ann. Mines **3** [VIII], 31-68 (1883); *Recherches expérimentales et théoriques sur la combustion des mélanges gazeux explosifs: (I) Températures d'inflammation des mélanges gazeux. (II) Sur la vitesse de propagation de la flamme dans les mélanges gazeux. (III) Sur les températures de combustion et les chaleurs spécifiques des gaz aux températures élevées.* Ibid. **4** [VIII], 274-568 (1883); *Sur la variation, avec la pression de la température à laquelle se produit la transformation de l'iodure d'argent.* C. R. Acad. Sci. Paris **99**, 157-160 (1884); *Recherches sur la combustion des mélanges gazeux explosifs.* J. Phys. Théor. Appl. **4** [II], 59-84 (1885) — *Recherches expérimentales et théoriques sur les équilibres chimiques.* Dunod, Paris (1888); Ann. Mines **13** [VIII], 157-380 (1888); *Sur les procédés de tirage des coups de mines dans les mines de grisou.* C. R. Acad. Sci. Paris **107**, 96-99 (1888) — *Etude des questions relatives à l'emploi des explosifs en présence du grisou* [Premier Rapport présenté à la Commission des substances explosives]. Ann. Mines **14** [VIII], 197-318 (1888); *Rapport supplémentaire.* Ibid. **14** [VIII], 319-376 (1888) — *Emploi des explosifs dans les mines à grisou.* In: *Congrès international des mines et de la métallurgie sur l'emploi des explosifs dans les mines à grisou.* Bull. Soc. Ind. Minér. **3** [III], 659-709 (1889) — *Le grisou et ses accidents.* Rev. Gén. Sci. Pures Appl. **1**, 630-635 (1890) — *Expériences sur les lampes de sûreté* [Rapport présenté à la Commission du grisou]. Dunod, Paris (1892).

**SECONDARY LITERATURE.** M.G. WYROUBOFF: *Ernest MALLARD.* Nature **50**, 428 (1894) — *Communications nécrologiques.* C. R. Acad. Sci. **119**, 1042-1044 (1894) — *Discours prononcés aux funérailles de M. Ernest MALLARD.* Ann Mines **6** [IX] (1894), par M. DAUBRÉE pp. 303-307; par M. LINDNER pp. 308-313; par J.N. HATON DE LA GOUPILLIERE pp. 313-316; et par A. MICHEL-LEVY pp. 316-318 — A. MICHEL-LEVY: [*Obituary notice*]. Bull. Soc. Franc. Minéral. **17**, 137-139 (1894); *François-Ernest MALLARD* [with a list of his publications]. Ibid. **17**, 241-266 (1894) — Anonymous: [*Obituary notice*]. Bull. Acad. Imp. Sci. St. Pétersbourg **2** [V], xi-xiii (1894) — A. DE

LAPPARENT: *Notice nécrologique sur Ernest MALLARD. Membre de l'Institut, inspecteur général des mines* [with a list of his publications]. Ann Mines **7** [IX], 267-303 (1895); MALLARD. In: *Livre du centenaire 1794–1894*. Tome 1: *L'Ecole et la science*. Gauthier-Villars, Paris (1895) — P. TERNIER: *Eloge d'Ernest MALLARD*. Bull. Soc. Géol. France **23** [III], 179-191 (1895) — W.T. HOLSER: *MALLARD, (François) Ernest*. In: (C.C. GILLESPIE, ed.) *Dictionary of scientific biography*. Scribner, New York; vol. 9 (1974), pp. 58-60 — Anonymous: *E.F. MALLARD* [biographie et son goniomètre]. In: (M.C. THOORIS ET AL., eds.) *Les objets scientifiques: un siècle d'enseignement et de recherche à l'Ecole polytechnique*. Ecole Polytechnique, Palaiseau (Sept. 1997).
**PICTURE.** Courtesy Bibliothèque de l'Ecole des Mines, Paris, France.

## MALLET, Robert (1810–1881)

- Irish civil engineer and seismologist; father of seismology

Robert MALLET was born in Dublin to John MALLET, an iron founder. He entered Trinity College at Dublin (1826), where he studied mathematics and science, graduating with a B.A. (1830). In the following year, he became partner in his father's works, an iron and copper foundry in the city of Dublin that ultimately became the largest works in Ireland.

His contributions to civil engineering are manifold: in 1837, he turned his attention to the hydraulic ram and produced a form of that pump that was used on the Dublin and Kingstown railway for forcing water to tanks for the engines. He built a number of swivel bridges over the River Shannon and the viaduct over the Nore, erected many terminal railway stations, built the famous Fastnet Rock lighthouse (southwest of Cape Clear, Ireland) in the period 1848–1849, and obtained a number of patents. In 1850, MALLET, among others, sought means of reinforcing the wrought-iron gun tube such as by winding sheet iron around the tube or using hoops shrunk together. He invented the buckled plate, which was used widely in structures, particularly for flooring, where it combined maximum strength with minimum depth and weight (patented in 1852).

In 1861, he gave up his father's Victoria foundry, which he had expanded into the dominant foundry in Ireland, and took the M.A. (1862) at the University of Dublin. Two years later he received a honorary LL.D. from Dublin University.

As early as the mid-1830s MALLET started his studies in physical geology, which were directed toward four main areas: glacial flowage (1837–1845), geological dynamics (from 1835), seismology (from 1845), and volcanology (from 1862). In 1846, he delivered before the Royal Irish Academy a remarkable paper on the *Dynamics of Earthquakes*, thereby addressing the vertical motion believed or supposed to accompany earthquake shocks. He was also one of the first to estimate the depth of an earthquake underground and, based upon his conclusions on the origin of earthquakes, he can be regarded as an important precursor of plate tectonics.

It was the commencement of a long series of contributions to a branch of physical geology that he called "seismology" and with which his name is associated until now. His classic paper is today regarded as one of the foundations of modern seismology. Other contributions to this subject included (1) the *Catalogue of the World's Earthquakes* (1852–1854), which he jointly compiled with his son John W. MALLET and forms his third report on earthquake phenomena and occupies nearly 600 pages in the Reports of the British Association; (2) the two-volume book *The Great Neapolitan Earthquake of 1857* (1862); (3) the *Seismographic Map of the World* (1857) published by the British Association; (4) the article *First Principles of Observational Seismology* (1862), an extension of his first article *Earthquake Phenomena* (1847) published as a three-part memoir (1850–1852); and (5) an elaborate contribution to the literature of volcanic geology, entitled *Volcanic Energy* (1872).

MALLET's idea was to look for variations in seismic velocity that would indicate variations in the properties of the earth. Using gunpowder he carried out the first seismic measurements of the velocity of earth waves. The first tests using this new method began in the wet sand of Killiney Bay and the granite of Dalkey Island (1851) and later during the progress of extensive quarrying for materials for the construction of Holyhead Harbor, North Wales, in different lengths of quartz rock, slate, and schist, thereby using for the first time the enormous blast from a large quantity of fired gunpowder – thus introducing *explosion seismology* into geological research (1860). His method is still used today, for example in oil field exploration.

MALLET was also the first to investigate the physical conditions involved in the construction of large guns, particularly of ringed ordnance. In view of the Crimean War

(1854) he designed two monster mortars for throwing 36-in. (91.44-cm) shells, but they were ultimately not used owing to the peace agreement with Russia of 1856. MALLET delivered before the Royal Irish Academy a paper on the construction of large-caliber guns and hitherto unexplained causes of destruction (1856). He also published a paper in which he analytically treated the trajectory of a rifled projectile, either flat-faced or ogival-pointed, when it travels through air or penetrates a homogeneous denser or solid-resisting medium such as sand (1867). He found that in oblique fire the ogival-pointed shot must pass at an angle significantly less than that of incidence. In consequence of his contributions to ballistics he was later elected a special honorary member of the Royal Artillery Institution, Woolwich, south-east London (1867).

When MALLET gave up his ironworks at Dublin (1861), he moved to London, where he served as a consulting engineer. MALLET was also general editor of the *Practical Mechanic's Journal* (London). As a contribution to Prof. Luigi PALMIERI's book *The Eruption of Vesuvius in 1872* (1873), MALLET wrote an introductory sketch of the present state of knowledge of terrestrial volcanicity and the cosmic nature and relations of volcanoes and earthquakes. He also urged the establishment of an international chain of seismological observatories to study the velocity of earthquake waves and to use this information to elucidate the structure of the ocean floors about which very little was then known.

In the last seven years of his life he was nearly blind, and his papers were written by the hand of an amanuensis. MALLET was a member of the Royal Irish Academy (1832), elected an honorary member of the Society of Scotland (1840), a Fellow of the Royal Society (1854) and the Geological Society of London (1859), and a corresponding member of the Physical Class of the Royal Philosophical Society of Göttingen (1869). MALLET received the Telford Medal of the Institution of Civil Engineers (1859) and the Cunningham Medal (1862) of the Royal Irish Academy. In 1877, he received the Wollaston Medal, the premier award of the Geological Society of London, "in recognition of the results of at least forty years of sedulous labor in some of the most important and difficult problems in geology."

Similar to MALLET, John MILNE, forty years younger than MALLET, made also fundamental contributions to seismology and earthquake engineering. The prestigious biennial *Mallet-Milne Lecture* is sponsored by the Society for Earthquake and Civil Engineering Dynamics (SECED) in London.

Astronomers named a crater on the near side of the Moon after him.

**ORIGINAL WORKS.** *Explanation of the vorticose movement assumed to accompany earthquakes.* Phil. Mag. **28** [III], 537-544 (1846) — *On the dynamics of earthquakes; being an attempt to reduce their observed phenomena to the known laws of wave motion in solids and fluids* [read 1846]. Trans. Roy. Irish Acad. **21**, 50-106 (1848) — *On the objects, construction, and use of certain new instruments for self-registration of earthquake shocks* [read 1846]. Ibid. **21**, 107-113 (1848) — *On the facts of earthquake phenomena.* Rept. Meet. Brit. Assoc. **17** (Part 2), 30 (1847) — *On the relation of molecular forces to geology.* J. Dublin Geol. Soc. **3**, 23-49 (1849) — *On some secular and diurnal motions of the Earth's crust.* Ibid. **3**, 180-186 (1849) — *Proposal for the general adoption of a new and uniform principle for laying down geological sections.* Ibid. **4**, 21-29 (1850) — *On the experimental determination of the limits of the transit rate of the propagation of waves or pulses, analogous to those of earthquakes through solid materials.* Proc. Roy. Irish Acad. **5** [I], 143-144 (1850–1853) — *First report on the facts of earthquake phenomena.* Rept. Meet. Brit. Assoc. **20**, 1-87 (1850) — *Second report on the facts of earthquake phenomena.* Ibid. **21**, 272-320 (1851) — *Third report on the facts of earthquake phenomena.* Ibid. **22**, 1-176 (1852); **23**, 117-212 (1853); **24**, 1-326 (1854) — *Notice of the British earthquake of 9th November 1852* [read 1854]. Trans. Roy. Irish Acad. **22**, 397-410 (1855) — *On the physical conditions involved in the construction of artillery of large caliber, and on some hitherto unexplained causes of the destruction of cannon in service* [read 1855]. Ibid. **23**, 141-436 (1856) — *Fourth report upon the facts and theory of earthquake phenomena.* Rept. Meet. Brit. Assoc. **28**, 1-136 (1858) — *On the military and naval uses of very large shells and the comparative powers of shells in relation to diameter.* United Service J. **II**, 407-451 (1859) — *Account of experiments made at Holyhead (North Wales) to ascertain the transit-velocity of waves, analogous to earthquake waves, through the local rock formations.* Phil. Trans. Roy. Soc. Lond. **151**, 655-679 (1861); Ibid. **152**, 663-676 (1862) — *Great Neapolitan Earthquake of 1857: the first principles of observational seismology.* Chapman & Hall, London (1862) — *Proposed measurement of the temperatures of active volcanic foci to the greatest attainable depth, and of the temperature, state of saturation, and velocity of issue of the steam and vapors evolved.* Rept. Meet. Brit. Assoc. **32** (Part 2), 33 (1862) — *To the late earthquake and earthquakes in general.* Quart. J. Sci. **1**, 53-69 (1864) — *On the trajectories of elongated rifled projectiles on striking and in penetrating solid resisting media.* The Engineer (Lond.) **23**, 1, 39, 52, 73 (1867) — *Volcanic energy: an attempt to develop its true origin and cosmical relations.* Proc. Roy. Soc. Lond. **20**, 438-441 (1872); Phil. Trans. Roy. Soc. Lond. **163**, 147-227 (1873); Ibid. **165**, 205-213 (1875) — *Note on the history of certain recent views in dynamical geology.* Am. J. Sci. **5**, 302-303 (1873).

**SECONDARY LITERATURE.** Anonymous: *Robert MALLET.* The Engineer (Lond.) **52**, 352-353, 371-372, 389-390 (1881) — S.H. (anonymous): Proc. Roy. Soc. Lond. **33**, xix-xx (1882) — Anonymous: Quart. J. Geol. Soc. Lond. **38**, 54-56 (1882) — Anonymous: *Memoir on R. MALLET.* Proc. Inst. Civ. Eng. **68**, 297-304 (1881/1882) — R.C. COX (ed.) *Robert MALLET 1810–1881.* Centenary Seminar Papers. The Institution of Engineers of Ireland and the Royal Irish Academy (April 1982); see G.L. HERRIES DAVIES: *Robert MALLET: Earth scientist*, pp. 35-52 — G.C.B. (anonymous): *MALLET, Robert (1810–1881). Dictionary of National Biography.* Oxford University Press, Oxford, vol. 35 (1893), pp. 429-430 — C. DAVISON: *The founders of seismology.* The University Press, Cambridge (1927), pp. 65-81 — J. WARTNABY: *Seismological investigations in the nineteenth century, with special reference to the work of John MILNE and Robert MALLET.* Ph.D. thesis, University of London (1972) — W. FISCHER: *MALLET, Robert.* In: (C.C. GILLESPIE, ed.) *Dictionary of scientific biography.* Scribner, New York, vol. 9 (1974), pp. 60-61 — R.C. COX (ed.) *Robert MALLET, F.R.S., 1810–1881: papers presented at a centenary seminar at 22 Clyde Road, Dublin, September 17, 1981.* Institution of Engineers of Ireland, Dublin (1982) — D.R. DEAN: *Robert MALLET and the founding of seismology.* Annals of Science **48**, No. 1,

39-67 (1991) — G.C. BOASE and R.C. COX: *MALLET, Robert*. In: (H.C.G. MATTHEW, ed.) *Oxford dictionary of national biography*. Oxford University Press, Oxford (2004), vol. 36, pp. 335-336.
**PICTURE.** Taken in 1865. Courtesy Grace FITZGERALD, The Institution of Engineers in Ireland, Dublin.

## MARCI VON KRONLAND, Johann [Ioannes] Marcus (1595–1667)

- Bohemian physician (the "Bohemian Plato"), physicist and mathematician; early pioneer of percussion mechanics

J. Marcus MARCI VON KRONLAND was born in Landskron, Bohemia (now Lanškroun, Czech Republic), the son of a clerk to an aristocrat. He received his basic education in a Jesuit college. He studied philosophy and theology at the University of Olomouc (now in the Czech Republic) with the intention of becoming a priest. However, for some reason he changed his mind and in 1618 commenced his studies in medicine at Karlsuniversität (German Charles University) in Prague, then a significant center of early natural sciences. There he took his M.D. (1625) and, beginning in the same year as a lecturer, soon became professor of medicine, a position he held for 40 years. His reputation as a very successful physician must have been legendary because contemporaries called him the "Hippocrates of Prague." He became physician in ordinary to Emperor Ferdinand III (1658). Besides his interest in philosophy – a kind of Platonism pointing in the direction of Johan B. VAN HELMONT and Philippus A. PARACELSUS, which earned him the nickname "Bohemian Plato" – he showed a keen interest in oriental languages and natural sciences, particularly in physics and mathematics.

It is not quite clear who stimulated and influenced his interest in tackling the problem of percussion. When MARCI published his book *De proportione motus...* ("Of Proportion in Motion...," 1639) on this subject, Galileo GALILEI (1564–1642) had published a year before his *Discorsi e Dimostrazioni Matematiche* – a treatise on mechanics, in addition to free fall and projectile motion, also including percussion (*see* his "Sixth Day"), which MARCI might have known before he published his own book on percussion. GALILEI's treatise, however, smuggled out of Italy and printed in Leiden in 1638 under the title *Discorsi e dimostrazioni matematiche intorno a due nuove scienze* ("Discourses and Mathematical Demonstrations Relating to Two New Sciences"), does not contain the passage on percussion. MARCI's contribution to the laws of percussion (or impact), although of qualitative rather than of quantitative nature, appears to be based on his own studies and experiments. He classified collisions into those between hard, soft, and fragile bodies. Recognizing the dependence of impulse on mass, he attempted to characterize impulse, which he regarded as a resistance to motion, in terms of static weight. Although this approach could not render successful results in mathematical terms because weight and impulse have different dimensions, he arrived by experiments at numerous correct conclusions. His significant results on collision phenomena, today almost forgotten, put him in first place in a line of early pioneers investigating the phenomena of percussion, which shortly afterward were taken up by renowned researchers, such as René DESCARTES, Christiaan HUYGENS, John WALLIS, and Christopher WREN.

Besides percussion, MARCI was also interested in other branches of mechanics, such as the free fall and the oscillation of a pendulum. The pendulum was used by early naturalists as a simple device for measuring an elapsed period of time. For example, the pendulum was used by GALILEI to take the pulse of a patient and by Marin MERSENNE to estimate the velocity of projectiles in relation to the sound velocity in air (1644). MARCI proposed a small pendulum for the measurement of time durations below one second (*De proportione motus*, Propositio XXXXI, Problema II).

MARCI published his optical studies in his monograph *Thaumantias. Liber de arcu coelesti...* ("The Rainbow...," 1648), in which he tried to explain the puzzle of the rainbow, his chief interest as indicated by the title of his work. He experimented with prisms to decompose white light and observed that each color corresponds to a specific refraction angle, thus anticipating several of Sir NEWTON's prismatic discoveries. To explain the change in direction in the reflection and refraction of light, he assumed that each point of a luminous source emitted rectangular rays in all directions, in a homogeneous medium bounded by a sphere, and that the points of this surface become the centers of new spheres of propagation, thus anticipating the Huygens principle.

MARCI was knighted for his achievements (1654) and granted the title Count Palatine de Kronland. Shortly after entering the Jesuit Order, he died in Prague. He was one of the great scientists of the 17th century. Since 1977, the *I.M. Marci of Kronland Medal* for outstanding achievements in the field of spectroscopy is awarded annually by the Czech Spectroscopic Society of the Czech Academy of Sciences.

A crater on the far side of the Moon is named for him.

**ORIGINAL WORKS.** *Idearum operatricium idea.* Gross, Leipzig (1635); Typis Seminarii Archiepiscopalis, Pragae (1635) — *De proportione motus seu regula sphygmica ad celeritatem et tarditatem pulsuum ex illius motu ponderibus geometricis liberato absque errore mentiendam.* Typis Ioannis Bilinae, Pragae (1639) — *De caussis naturalibus pluviae purpureae Bruxelensis...* Typis academicis, Pragae (1647) — *Thaumantias. Liber de arcu coelesti deque colorum apparentium natura, ortu et causis, in quo pellucidi opticae fontes a sua scaturigine, ab his vero colorigenii rivi dervantur.* Typis academicis, Pragae, (1648) — *De proportione motus figurarum rectilinearum et circuli quadratura ex motu.* Ex typographia academia, Pragae (1648) — *De longitudine seu differentia inter duos meridianos una cum motu vero Lunae inveniendo ad tempus datae observationis.* Typis G. Schyparz, Pragae (1650) — *Philosophia vetus restituta.* Typis academicis, Pragae (1662).

**SECONDARY LITERATURE.** M.M. GUHRAUER: *Marcus MARCI und seine philosophischen Schriften.* Z. Phil. phil. Kritik **21**, 241-259 (1852) — F.J. STUDNIČKA: *Ioannes Marcus MARCI von Cronland, sein Leben und gelehrtes Wirken.* Jahresversammlung der Königl. Böhm. Gesell. Wiss. in Prag (31. Jan. 1891) — E. HOPPE: *Marcus MARCI, ein vergessener Physiker des 17. Jahrhunderts.* Arch. Math. Naturwiss. Tech. **10**, 282-290 (1927) — J. SMOLKA: *Joannes Marcus MARCI: his times, life and work* [Lecture given on the occasion of the Int. Symposium at Prague in 1967 commemorating the tercentenary of MARCI's death]. Acta Historiae Rerum Naturalium necnon Technicarum **3** (Spec. issue), 9-50 (1967); D. LEDREROVA: *Bibliographie de Johannes Marcus MARCI.* Ibid. — J. MAREK: *Un physicien tchèque du XVIIe siècle: Ioannus Marcus MARCI de Kronland* [with bibliography]. Rev. Hist. Sci. **21**, 109-130 (1968) — E.J. AITON: *Ioannes Marcus MARCI (1595–1667).* Ann. Sci. **26**, 153-164 (1970) — L. NOVÝ: *Johannes Marcus MARCI of Kronland.* In: (C.C. GILLESPIE, ed.) *Dictionary of scientific biography.* Scribner, New York, vol. 9 (1974), pp. 96-98 — I. SZABÓ: *Geschichte der mechanischen Prinzipien.* Birkhäuser, Basel *etc.* (1977), pp. 429-436, 457-459 — D. GARBER: *On the frontlines of the scientific revolution: how MERSENNE learned to love GALILEI.* Perspectives on Science (MIT Press) **12**, No. 2, 135-163 (Summer 2004).

**PICTURE.** From M. MARCI: *De proportione motus* (see above).

# MARIOTTE [Lat. *MARIOTTUS*], Edmé (*c.*1620–1684)

- French physicist and plant physiologist; early pioneer of percussion mechanics and father of French hydraulics

Little is known with certainty of Edmé MARIOTTE's biography, such as the origin of his family, the place of his birth, his motivation to devote himself to science, his education and career before entering the French Royal Academy of Sciences, and even of his private life as an Academician.

He was born around 1620 in Dijon (Dépt. Côte d'Or) and entered the academy in 1666, some months after its official foundation. Some modern biographers state that he also served as a prior to St. Martin-sous-Beaune, but contemporary sources do not attribute to him a clerical title. All of his 9 published treatises and about 20 unpublished papers he prepared during his active period as an Academician. MARIOTTE began his scientific studies with physiological research on the vegetation of plants (delivered 1667) and the mechanism of seeing in the human eye (delivered 1668), thereby discovering the blind spot. After briefly treating an engineering problem, an improvement in the accuracy of a level instrument (1672), he soon turned to physics. His most important contribution are as follows: (1) He treated elastic and inelastic collisions of solid bodies and their deformation during impact (read 1671, publ. 1673, 1676, and 1684). His treatise soon became a standard work on collision phenomena. (2) His studies on the nature of air under isothermal conditions (1679), carried out independently of Robert BOYLE's previous findings, essentially confirmed the relationship $pV = $ const – so-called "isothermal gas law" or "Boyle law" (1660), later also known as the "Boyle-Mariotte law" which describes the isothermal behavior of an enclosed mass of air – a basic tenet of physics and chemistry. MARIOTTE even went further than BOYLE by stating that this relation only holds if there is no change in temperature. Therefore, in France his law is called the "Mariotte law." (3) He first applied his observed volume-pressure dependency to the Earth's atmosphere and showed that the pressure decreases with altitude. Coining the word "barometer," he also discussed the relation between barometric pressure and weather. (4) His investigation on the freezing of liquids (1672) showed that ice has a smaller density than water. (5) In his reports to the Paris Academy he discussed the rainbow, the refraction of light, and the nature of colors (1681). (6) In his studies on the movement of fluids (publ. posthumously in 1686), which deal with the basic properties of air and water, the balance forces of fluids due to weight, and elasticity and impact, he stated that water is practically incompressible and hence has no elastic force. (7) The first volume of the *Histoire et Mémoires de l'Académie* (1733) also contains a paper on the recoil of guns. (8) He also conducted tests on the deformation and burst pressure of cylindrical vessels hold under high pressure of water. He observed that the vessels burst when the circumferential elongation increased by a certain fraction and noted the direct proportionality between pressure and circumferential stretch. His work, published posthumously in his book *Traité du mouvement des eaux et des autres corps fluides* (1686), marked one of the first efforts to relate strength to strain (or stress).

MARIOTTE corresponded with a number of eminent scientists of his time such as Philippe DE LA HIRE, Christiaan HUYGENS and Gottfried W. LEIBNIZ. His reputation was very high not only among French colleagues but also in England. In the second edition of his *Philosophia naturais principia mathematica* (1713), Sir Isaac NEWTON acknowledged MARIOTTE's pendulum experiments as an important contributions to the laws of impact, calling him "the most illustrious MARIOTTE" [Lat. *"Clarissimus Mariottus"*]. NEWTON was the most prominent user of his two-pendulum percussion apparatus, which he applied to verify the equivalence of action and reaction, later known as "NEWTON's Third Law of Motion" {Sir NEWTON ⇨ 1687}.

A crater on the far side of the Moon is also named for him.

**ORIGINAL WORKS.** (Attribution to E. MARIOTTE by L. LAUDAN) *A discourse of local motion: undertaking to demonstrate the Laws of Motion, and with all to prove, that of the seven rules delivered by M. DES-CARTES on this subject, he hath mistaken six.* W.G, London (1670) — With C. PERRAULT: *Nouvelle découverte touchant la veüe.* F. Leonard, Paris (1671) — *Traité du nivellement, avec la description de quelques niveaux nouvellement inventez.* Cusson, Paris (1672) — *Traité de la percussion ou choq des corps, dans lequel les principales règles du mouvement, contraires à celles que Mr. DESCARTES et quelques autres modernes ont voulu établir, sont expliquées et démontrées par leurs véritables causes.* Michallet, Paris (1673) — *Discours de la nature de l'air, de la végétation des plantes. Nouvelle découverte touchant la vue.* Michallet, Paris (1676); Gauthier-Villars, Paris (1923) — *Essay de logique, contenant les principes des sciences et la manière de s'en servir pour faire de bons raisonnements.* Michallet, Paris (1678) — *De la végétation des plantes.* Michallet, Paris (1679) — *Du chaud et du froid.* Michallet, Paris (1679) — *De la nature des couleurs.* Michallet, Paris (1681) — *Traité du mouvement des eaux et des autres corps fluides.* Michallet, Paris (1686); Engl. translation by J.T. DESAGULIERS: *The motion of water and other fluids: being a treatise of hydrostaticks.* J. Senex, London (1718); Arno Press, New York (1978); Germ. translation by J.C. MEINIG: *Des Weyland vortrefflichen Herrn MARIOTTE ... Grundlehren der Hydrostatick und Hydraulick ...* J.F. Brauns, Leipzig (1723) — *Règles pour les jets d'eau.* Michallet, Paris (1693) — (P. VANDER AA, ed.) *Œuvres de Mr. MARIOTTE, de l'Académie Royale des Sciences.* 2 vols., Chez P. Vander Aa, Leiden (1717); 2 vols., Neaulme, La Haye (1740).

**SECONDARY LITERATURE.** C. MUTEAU and J. GARNIER: *Galerie bourguignonne.* 3 vols., J. Picard, Dijon (1858–1860), vol. II, p. 219 — P.G. TAIT: *Note on a singular passage in the 'Principia'* [discusses WREN's and MARIOTTE's contributions to the Laws of Motion and questions of priority]. Proc. Roy. Soc. Edinb. **13**, 72-78 (1886) — D. MCKIE: *BOYLE's law.* Endeavour **7**, 148-151 (1948) — J. PELSENEER: *Petite contribution à la connaissance de MARIOTTE.* Isis (Chicago) **42**, 299-301 (1951) — A. HELLER: *Geschichte der Physik.* Sändig, Wiesbaden (1965), vol. II, pp. 174-178 — B.

DAVIES: *Edme MARIOTTE.* Phys. Educ. **9**, 275-278 (1974) — M.S. MAHONEY: *Edme MARIOTTE.* In: (C.C. GILLESPIE, ed.) *Dictionary of scientific biography.* Scribner, NY, vol. 9 (1974), pp. 114-122 — I. SZABÓ: *Geschichte der mechanischen Prinzipien.* Birkhäuser, Basel (1977), p. 435 — F. KRAFFT (ed.) *Große Naturwissenschaftler.* VDI-Verlag, Düsseldorf (1986) — P. COSTABEL: *MARIOTTE et les règles du mouvement.* In: (P. COSTABEL, ed.) *MARIOTTE, savant et philosophe (†1684). Analyse d'une renommée.* Librairie Philosophique J. Vrin, Paris (1986), pp. 75-89 — G. PICOLET: *Etat des connaissances actuelles sur la biographie de MARIOTTE et premiers résultats d'une enquête nouvelle.* Ibid. pp. 245-276 — E. SURGOT: *Bibliographie des œuvres de MARIOTTE.* Ibid. pp. 309-320 — R.S. WESTFALL: *MARIOTTE, Edme.* The Galileo Project (1995);
http://galileo.rice.edu/Catalog/NewFiles/mariotte.html.
**PICTURE.** Courtesy Bildarchiv Preußischer Kulturbesitz (bpk), Berlin (image No. 51.295).
**NOTE.** There exists no official portrait of MARIOTTE. However, there exists a painting made by the French artist Charles LE BRUN on the occasion of the establishment of the French Academy and the foundation of the Paris Observatory on Dec. 22, 1666. This painting which is kept at the Musée National du Château et des Trianons in Versailles shows king Louis XIV, J.B. COLBERT and the 21 founding members, MARIOTTE being one of them. The enclosed picture is a detail of this painting, probably showing (from right to left) E. MARIOTTE, C. HUYGENS, and J.D. CASSINI. For further information *see* C.J. VERDUIN: *A portrait of Christiaan HUYGENS?* in Etablissement de l'Académie des Sciences et fondation de l'Observatoire. 1666. University of Leiden, The Netherlands;
http://www.leidenuniv.nl/fsw/verduin/stathist/huygens/acad1666/index.html.

# MELSENS, Louis Henri Frédéric (1814–1886)

- Belgian chemist, physicist, and physician; founder of wound ballistics

Louis H.F. MELSENS was born in Louvain (Flem. *Leuven*) in the Province of Brabant. After attending a local gymnasium in his native town, he first worked in the office of a business at Anvers, where he discovered "his incapability for commercial life." He studied organic chemistry in Paris, working in the laboratory of Jean B.A. DUMAS at the Sorbonne. To increase his knowledge of chemistry he visited Germany and came into contact with Justus VON LIEBIG at the University of Giessen, an international authority on chemistry, where he took his Ph.D. in the natural sciences. After his return to Belgium MELSENS was appointed professor of chemistry and physics at the Ecole de Médecine Vétérinaire in Brussels. Later in Brussels he became an examiner at the Ecole Royale Militaire and a regular member of the Belgian Royal Academy of Sciences (1851).

Besides his numerous contributions to chemistry (*e.g.*, hydrous saponification of fats, determination of the empiric formula of nicotine) and pharmacy (*e.g.*, treatment of lead and mercury poisonings), he applied Benjamin FRANKLIN's principle that "all electricity goes up to the free surface of the bodies without diffusing in their interior substance" (the principle of the Faraday cage) to lightning conductors. The invention of the "Melsens lightning-conductor system" perfected the lightning rod: he multiplied the terminals, the conductors, and the earth-connections, which assumed the form of an aigrette or brush with five or seven points, the central point being a little higher than the rest.

He also performed ballistic experiments, thereby studying various kinds of impact phenomena that we would classify today as terminal ballistics. He also extended this research on living tissue and investigated gunshot wounds of humans and horses, today called "wound ballistics." In 1872, MELSENS speculated that the disrupting effect during impact of a projectile is caused by compressed air that is carried along with the projectile. His hypothesis, which was only based on the analysis of target fragments and not on any high-speed visualization and recording techniques during impact, was vague. In a lecture given in 1881, MELSENS presented his theory and stimulated Ernst MACH to investigate this challenging problem in more detail, which eventually resulted in the discovery of the head wave (1886) surrounding any object flying in a fluid at supersonic velocity.

MELSENS was the first president of the Belgian Electrotechnical Committee. The *Louis Melsens Prize*, created by the Division of Sciences of the Académie Royale de Belgique in 1900, is awarded to a Belgian or naturalized Belgian author of the most outstanding work, in a given year, on "applied chemistry or physics."

**ORIGINAL WORKS.** *Note sur les poudres de guerre, de mine, et de chasse*. Bull. Acad. Roy. Sci. Brux. **11** [II], 13-42 (1861) — *Rapport sur le chronographe électro-balistique de M. P.* LE BOULENGÉ. Ibid. **17** [II], 92-123 (1864) — *Sur les paratonnerres et sur quelques expériences faites avec l'étincelle d'induction et les batteries de Leyde*. Ibid. **20** [II], 15-34 (1865) — *Sur les paratonnerres à conducteurs multiples*. C. R. Acad. Sci. Paris **61**, 84-87 (1865) — *Sur le passage des projectiles à travers les milieux résistante*. Ibid. **65**, 564-568 (1867) — *Note sur les explosions des chaudières à vapeur*. Bull. Acad. Roy. Sci. Brux. **31** [II], 123-125 (1871) — *Sur quelques effets de la pénétration des projectiles dans divers milieux, et sur l'impossibilité de la fusion des balles de plomb dans les plaies produites par les armes à feu*. C. R. Acad. Sci. Paris **74**, 1192-1195 (1872) — *Note sur les plaies des armes à feu*. J. Méd. Chir. Pharm. [Soc. Roy. Sci. Med. Natl. Bruxelles] **54**, 421-437, 513-527 (1872); Ibid. **55**, 21-29, 121-126, 217-222, 293-303 (1872) — *Notes sur les paratonnerres*. Bull. Acad. Roy. Sci. Brux. **38** [II], 320-348, 423-441 (1874); **39** [II], 831-853 (1875) — *Note sur les mines de houille dans lesquelles on constate la présence du grisou*. Ibid. **46** [II], 43-57, 382-387 (1878) — *Rapport sur un mémoire de M. G.-A.* HIRN *concernant la relation qui existe entre la résistance de l'air et sa température*. Ibid. **2** [III], 233-241 (1881) — *Sur le passage des projectiles à travers les milieux résistants, sur l'écoulement des solides et sur la résistance de l'air au mouvement des projectiles*. C. R. Acad. Sci. Paris **93**, 485-489 (1881) — *Sur les paratonnerres*. Ibid. **95**, 128-129 (1882) — With D. COLLADON: *Résistance de l'air dans les canons de fusil*. Bull. Acad. Roy. Sci. Brux. **3** [III], 721-726 (1882) — *Rapporte sur le mémoire de M. G.-A.* HIRN: *Recherches expérimentales et analytiques sur les lois de l'écoulement et du choc des gaz en fonction de sa température*. Ibid. **9** [III], 49-71 (1885) — *La balistique expérimentale*. Ibid. **11** [III], 149-156 (1886).

**SECONDARY LITERATURE.** [Brief obituary notice by editor] Leopoldina (Halle/Saale) **22**, 113 (1886) — E. MAILLY: *Discours prononcé aux funérailles de M. MELSENS*. Bull. Acad. Roy. Sci. Brux. **11** [III], 333-335 (1886) — (Ed.) *Necrologia: L.H.F. MELSENS*. Annali Chim. Farmacol. (Milano) **3** [IV], 374 (1886) — [Brief obituary notice by editor]. Ciel et Terre (Bruxelles) **7**, 117-118 (1886/1887) — P. DE HEEN: *Notice sur L.-H.-F. MELSENS*. Ann. Acad. Roy. Sci., Lett. Beaux-Arts Belgique **59**, 483-506 (1893) — J. VAN LENNEP (ed.) *Les Bustes de l'Académie Royale Belgique*. Mém. Classe Beaux-Arts, collection in-8 [III], **VI** (1993), pp. 204, 244-245.

**PICTURE.** Courtesy L'Académie Royale des Sciences, des Lettres et des Beaux-Arts de Belgique, Bruxelles, Belgium.

# MERSENNE [Lat. *MERSENIUS*], Marin (1588–1648)

• French natural philosopher, mathematician, and theologian; promoter of French science

The son of a laborer, Marin MERSENNE was born near Oizé (Dépt. Maine), France, and began his grammar studies at the College of Mans in Paris. After spending 5 years at the new Jesuit College of La Flèche in Anjou (1604–1609) he studied theology at the Sorbonne (1609–1611). Thereafter he joined the Roman-Catholic mendicant Order of Minims (1611) and entered the Minim Convent de L'Annonciade in Paris. He temporarily taught philosophy at the Minim convents at Nevers and Paris, and beginning in 1620 he traveled extensively throughout western Europe.

MERSENNE, an opponent of any mystical doctrines of alchemy, astrology, and related arcane arts, tried to approach nature by scientific methods. He devoted himself to research in mathematics, physics, and astronomy rather than adopting speculations from others. His cell became a meeting place for a number of eminent natural philosophers throughout Europe, such as Girard DESARGUES, René DESCARTES, Pierre DE FERMAT, Blaise PASCAL, Pierre GASSENDI, Gilles Personne DE ROBERVAL, Jean BEAUGRAND, and others who later formed the core of the *Academia Parisiensis* (Parisian Academy of Sciences), which he organized in 1635. Corresponding with other eminent scientists in Europe such as Galileo GALILEI, Isaac BEECKMAN, Jan Baptista VAN HELMONT, Thomas HOBBES, Christiaan HUYGENS, Nicolas-Claude Fabri DE PEIRESC, and Evangelista TORRICELLI, he played a major role in communicating knowledge and ideas throughout European mathematical circles at a time when there were not yet special scientific periodicals. (For example, the first volume of the famous journal *Philosophical Transactions of the Royal Society* was not published until 1666). It was said that "To inform MERSENNE of a discovery, meant to publish it throughout the whole of Europe." He also defended DESCARTES and GALILEI against theological criticism.

In 1634, MERSENNE translated parts of GALILEI's *Dialogo sopra i due massimi sistemi del mondo* ("Dialogue Concerning the Two Chief Systems of the World," 1632) and in 1639 his *Discorsi e dimostrazioni matematiche intorno a due nuove scienze* ("Discourses and Mathematical Demonstrations Concerning the Two New Sciences," 1638) into French. Like his friend DESCARTES, he tried to explain that the world is based on a universal mechanism and that the actions of nature are limited by quantitative laws (*Harmonie universelle*, 1636). According to his understanding of science, the scholar has at his disposal three means: (1) the use of experiments; (2) the light of reason; and (3) the use of physico-mathematical processes together with the analogies in one field of physics with another. Physics – or any other natural science – is simply the science of phenomena, where experiments substitute for syllogistic argumentation and mathematical formulation takes the place of philosophical principles.

MERSENNE, however, was not only interested in natural philosophy on a general level but contributed also to specific fields, such as mathematics and experimental physics. In mathematics, MERSENNE worked on the theory of numbers. Investigating prime numbers, he tried to find a formula that would represent all primes and wrote a synopsis of mathematics that was printed in 1664. His other fields of study were mechanics, acoustics, hydrostatics, hydraulics, and ballistics. Probably stimulated by GALILEI's pendulum experiments, he first used a pendulum to measure gravity and suggested to the Dutch physicist Christiaan HUYGENS the use of the pendulum as a timing device. He also proposed a hygrometer and a telescope with parabolic mirrors.

Today MERSENNE is mostly known for his acoustic studies. He worked out a theory of music, studied the sound generated by pipes and strings, and performed the first measurements of the speed of sound in air (1636), which later inspired his closest friend GASSENDI to resume his studies. MERSENNE also showed that the velocity of sound is independent of pitch and loudness, and that the intensity of sound, like that of light, is inversely proportional to the distance from its source. When performing ballistic studies, he noticed that the impact of a musket ball against a wall is heard by a person, stationed on the far side and close to the point of impact, at the same instant as the report of the gun. From this observation he drew the important conclusion that the velocity of a musket ball must be of the same order as the velocity of sound. Obviously, he was the first to make a correct estimation of ballistic velocities that had hitherto been based solely on crude guesses without experimental support.

A crater (*Crater Mersenius*) on the far side of the Moon and some long, narrow valleys (*Rimae Mersenius*) on its surface are named after him.

**ORIGINAL WORKS.** *La vérité des sciences, contre les sceptiques ou Pyrrhoniens.* T. Du Bray, Paris (1625) — *Les nouvelles pensées de GALILEE...* [traduit d'italien en français]. P. Ricolet, Paris (1630) — *Questions théologiques, physiques, morales et mathématiques.* H. Guenon, Paris (1634) — *Les mécaniques de GALILEE, mathematicien et ingénieur du Duc de Florence...* Guenon, Paris (1634) — *Traité de mécanique.* In MERSENNE's memoir *Harmonie universelle.* Cramoisy, Paris (1636) — *Cogitata physico-mathematica, in quibus tam naturae quam artis effectus admirandi certissimis demonstrationibus explicantur. [De mensuris, ponderibus et nummis. De arte nautica, seu histiodromia et hydrostatica. De musica teorica et pratica. De mechanicis phaenomenis. De ballisticis, seu acontismologicis phaenomenis ]* A. Bertier, Paris (1644) — *Universae geometriae mixtaeque mathematicae synopsis, et bini refractionum demonstaratum tractatus.* A. Bertier, Paris (1644) — *Novarum observationum physico-mathematicarum F. Marini MERSENNI, tomus III.* A. Bertier, Paris (1647).

**SECONDARY LITERATURE.** H. DE COSTE: *La vie du Révérend Père Marin MERSENNE, théologien, philosophe et mathématicien de l'Ordre des Pères Minim.* Cramoisy, Paris (1649) — T. VENN: *The compleat gunner ... To-*

gether with some excellent observations out of MERSENNE. In: *Military and maritime discipline.* 3 vols., 4 parts. Tyler, London (1672), see part 3 — J.C.F. HOEFER (ed.) *Nouvelle biographie générale.* Firmin-Didot Frères, Paris (1852), vol. 35 — R. LENOBLE: *MERSENNE, ou la naissance du mécanisme.* J. Vrin, Paris (1943) — P. SERGESCU: *Troisième centenaire de MERSENNE: MERSENNE l'animateur.* Rev. Hist. Sci. Appl. **2**, 5-12 (1948) — J.M.A. LENIHAN: *MERSENNE and GASSENDI, an early chapter in the history of sound.* Acustica **1**, 96-99 (1951) — H. ROUSE and S. INCE: *History of hydraulics.* Dover Publ., New York (1957), p. 74 — A.C. CROMBIE: *M. MERSENNE.* In: (C.C. GILLESPIE, ed.) *Dictionary of scientific biography.* Scribner, New York, vol. 9 (1974), pp. 316-322 — P. DEAR: *MERSENNE and the learning of the schools.* Cornell University Press, Ithaca, NY (1988) — V. BORIA: *Marin MERSENNE: educator of scientists.* Ph.D. thesis, The American University, Ann Arbor, MI (1989) — J.M. CONSTANT and A. FILLON: *1588–1988, quatrième centenaire de la naissance de Marin MERSENNE: colloque scientifique international et célébration nationale.* Faculté des Lettres, Université de Maine, Le Mans, France (1994) — A. BEAULIEU: *MERSENNE, le grand minime.* Fondation Nicolas-Claude Fabri de Peiresc, Paris (1995).

**PICTURE.** Copper engraving by Claude DUFLOS, early 17th century, Musée National des Châteaux de Versailles et de Trianon. Courtesy Gérard BLOT, Réunion des Musées Nationaux (RMN), Paris, France.

# MIKHEL'SON [Russ. *МИХЕЛЬСОН*], Vladimir Aleksandrovich (1860–1927)

- Russian physicist and combustion researcher

Vladimir A. MIKHEL'SON was born in Tul'chin (now Vinnitsa Oblast, Ukraine) to Alexander Mikhaelovich MIKHEL'SON. After his education and graduation at M.V. Lomonosov Moscow State University (1883) he worked as a secretary of the Physics Section of the Society of Amateur Naturalists (1884–1887). Beginning in 1887 as an associate professor at Moscow State University [МГУ], he established the basic law governing the relationship between the composition of a burning gas mixture and the motion of the combustion front (1890–1893), thus laying the foundations of the theory of explosive combustion six years prior to the British chemist David L. CHAPMAN. Prior to Lord RAYLEIGH he derived an expression of the conservation of mass and momentum across the detonation front which corresponds to the straight Rayleigh line in the $(p, v)$-diagram – later known as the "Mikhel'son line." With these investigations he took his Ph.D. in physics (1893).

In 1894, he became professor of physics and meteorology in Moscow at the Petrov-Razumovsky Agricultural Academy (now Timiryazev Agricultural Academy), a position he held until his death. He also calculated the Doppler shift of light passing through a medium with a varying refraction index, directed an observatory of the U.S.S.R. Academy of Sciences, and later did research on actinometry and meteorology. Together with Nikolai E. ZHUKOVSKY he can be regarded as the most prominent 19th-century Russian shock and detonation physicist.

**ORIGINAL WORKS.** *Déduction simple de la seconde loi de la thermodynamique fondés sur les principes de la mécanique analytique* [in Russ.]. Recueil Mathématique [Matematiceskij sbornik, Moskva] **13**, 229-244 (1886) — *Essai théorique sur la distribution de l'énergie dans les spectres des solides.* Russ. Phys.-Chem. Soc. J. **19** (Sect. Phys.), 79-99 (1887) — *Über die normale Entzündungsgeschwindigkeit explosiver Gasgemische.* Ann. Phys. **37** [III], 1-24 (1889) — *Sur la multiplicité des théories mécaniques des phénomènes physiques.* Russ. Phys.-Chem. Soc. J. **23**, 415-426 (1891) — *On the normal ignition velocity of explosive gaseous mixtures* [in Russ.]. Ph.D. thesis. Moscow University Printing Service, Moscow (1890). Reprinted in: Scientific Papers of the Moscow Imperial University on Mathematics and Physics **10**, 1-93 (1893) — *On the normal velocity of combustion of mixtures of detonating gas.* Moscow University Mem. (Sect. Phys./Math.) No. 10 (1893), 92 pp. — *On the subject of one of Prof. E. BOUTT's observations. Explosion of gases.* Russ. Phys.-Chem. Soc. J. **26**, 287-289 (1894) — *Contribution to the question of the correct applications of DOPPLER's principle.* Ibid. **31**, 119-125 (1899) — *Sobranie sochinenii* [Collected works]. Izd. Novyi Agronom, Moscow (1930).

**SECONDARY LITERATURE.** *V.A. MIKHELSON.* In: (I.V. KUZNETSOVA, ed.) *Lyudi russkoi nauki* ("Russian men of science"). Nauka, Moscow (1961), vol. I, pp. 223-243 — K.I. SHCHELKIN and Y.K. TROSHIN: *Gas dynamics of combustion.* Mono, Baltimore (1965), chap. 1 — H.E. SCHULZ (ed.) *Who was who in the U.S.S.R.* Scarecrow, Metuchen, NJ (1972).

**PICTURE.** From *Russian men of science.* Nauka, Moscow (1961), vol. I, p. 223.

**NOTE.** His name has also been transliterated as MICHELSEN and MICHEL'SON.

# MINTROP, Ludger (1880–1956)

- German geophysicist and seismologist; father of seismic refraction method of exploration

Ludger MINTROP was born the son of a farmer at Werden/Heidhausen, a village near Essen, capital of the Ruhr

district. After attending the Rektoratsschule in Werden and the Realgymnasium in Essen, he studied surveying at the Berliner Bergakademie (1902–1903) and the TH Aachen (1903–1905). In addition, he worked as an assistant to Prof. Karl HAUSSMANN, a renowned geophysicist, at the Lehrstuhl für Markscheidekunde (Chair of Surveying) of TH Aachen, besides serving as consultant to the Nordstern Colliery. In 1907, he went to the University of Göttingen and joined the staff at the newly established Institut für Geophysik, which was founded and headed by the famous geophysicist Prof. Emil WIECHERT, then already widely known for his experiments on artificial earthquakes and his invention of the seismograph. MINTROP studied the seismic waves resulting from the concussive force of a falling body – the so-called "Mintrop ball," a 4,000-kg iron ball – which was dropped from a 14-m-high tower. With his first portable highly sensitive seismographs he obtained records that clearly showed the P- and S-waves. In 1908, he was called by the Westfälische Berggewerkschaftskasse and became a professor of surveying at the Bergschule Bochum and director of the surveying division and the newly established earthquake and geomagnetic observatory (1909–1921). He took his Ph.D. (1911) at Göttingen University with a thesis on the propagation of ground motions produced by a large gas turbine that had been installed at the Göttingen electric power station.

During World War I he developed a seismic method to locate Allied artillery firing positions using his portable seismograph. After the war he reversed the process, and by measuring the distance from an explosion to the seismograph he found that he could estimate subsurface geological formations. He discovered the *seismic head wave* – also known as the "Mintrop wave" – a wave that travels along the interface of two media having different velocities. With his patented *Seismic Exploration Refraction Method* (1919) he could estimate subsurface formations such as the existence of salt domes in Texas where oil was found (1923–1925). In 1921, MINTROP established the Seismos GmbH in Hannover, the first geophysical company using artificial earthquakes caused by the detonation of small charges of dynamite in a shallow bore hole for the exploration of valuable oil, coal, and ore deposits. At Seismos he held a leading position until 1933. He accepted a call to teach surveying and geophysics at the University of Breslau (1928–1945). After World War II, he escaped to western Germany, where he received at the TH Aachen a professorship in surveying and geophysics.

MINTROP, one of the "Grand Old Men" of geophysics, was a founding member of the German Seismological Society (1922), an honorary member of the Deutscher Markscheider Verein (German Surveying Association), the Society of Exploration Geophysicists (1930), and the Deutsche Geophysikalische Gesellschaft (German Geophysical Society) (1950). He received the Karl-Engler Medal (1953) and the Grand Cross of the Order of Merit of the Federal Republic of Germany (1955).

Since 1980 the annual *Mintrop Seminar*, organized by the Institut für Geophysik of the Ruhr University Bochum (RUB), is dedicated to special topics of applied geophysics, particularly to seismology. The *Ludger Mintrop Award* is annually presented to the author(s) of the best paper published in the international journal *Near Surface Geophysics* (Tulsa, OK).

**ORIGINAL WORKS.** *Die Erdbebenstation der Westfälischen Berggewerkschaftskasse zu Bochum.* Glückauf **45**, 357-366, 393-403 (1909) — *Über künstliche Erdbeben.* Int. Kongress für Bergbau, Hüttenwesen, Angewandte Mechanik und Praktische Geologie [Düsseldorf, Germany, Juni 1910]. In: *Berichte der Abt. für praktische Geologie.* Selbstverlag des Arbeitsausschusses des Kongresses, Düsseldorf (1910), pp. 98-112 — *Über die Ausbreitung der von den Massendrucken einer Großgasmaschine erzeugten Bodenschwingungen.* Ph.D. thesis, University of Göttingen. W. Girardet, Essen (1911) — *Erdbeben, Schlagwetterexplosionen und Stein- und Kohlenfall.* Glückauf **50**, 330-339 (1914) — *Verfahren zur Ermittlung des Ortes künstlicher Erschütterungen.* Germ. Patent No. 304,317 (17. Mai 1917) — *Erforschung von Gebirgsschichten mittels künstlicher Erdbebenwellen.* Germ. Patent No. 371,963 (7. Dez. 1919) — *Die Ermittlung des Aufbaus von Gebirgsschichten aus seismischen Beobachtungen.* Monatsber. Dt. Geol. Gesell. **72B**, Heft 3-4, 369-370 (1920) — *Erforschung von Gebirgsschichten und nutzbaren Lagerstätten nach dem seismischen Verfahren.* I. Mitteil. Seismos-Gesell., Hannover (1920) — *Zur Geschichte des seismischen Verfahrens zur Erforschung von Gebirgsschichten und nutzbaren Lagerstätten.* II. Mitteil. Seismos-Gesell., Hannover (1930) — *Geophysikalische Verfahren zur Erforschung von Gebirgsschichten und Lagerstätten.* In: (F.W. WEDDING and R. WÜSTER, eds.) *Der Deutsche Steinkohlenbergbau: Technisches Sammelwerk* [Bergbauverein Essen]. Glückauf-Verlag, Essen, vol. 1 (1942), pp. 455-538 — *Über die Ausbreitung an der Erdoberfläche erzeugter periodischer Bodenschwingungen in die Tiefe.* Z. Geophys. **18**, Nr. 3/4, 140-149 (1943/1944) — *Hundert Jahre physikalische Erdbebenforschung und Sprengseismik.* Die Naturwissenschaften **34**, 257-262, 289-295 (1947) — *On the stratification of the Earth's crust according to seismic studies of a large explosion and of great earthquakes.* Geophysics **14**, 321-336 (1949) — *Die Entwicklung der Sprengseismik.* Z. Geophys. **19** (Sonderband), 101-122 (1953) — *Die Hypothese von AIRY verträgt sich nicht mit seismischen Beobachtungen.* Bull. Inform. Géodés. Géophy. Int. **2**(A), No. 2, 225-228 (April 1953) — *Erdölsuche mit angewandter Geophysik.* Z. Erdöl u. Kohle **8**, 677-681 (1955).

**SECONDARY LITERATURE.** H. HAALCK: *Lehrbuch der angewandten Geophysik*. Bornträger, Berlin (1934) — Anonymous: *Ludger MINTROP 70 Jahre*. Erdöl u. Kohle **3**, 413-414 (1950); *Ludger MINTROP 75 Jahre*. Ibid. **8**, 605-606 (1955) — H.A. RÜHMKORF: *Prof. Dr. Dr. h.c. Ludger MINTROP zum Gedenken*. Ibid. **9**, 48 (1956) — K. LEHMANN: *Ludger MINTROP, der große Markscheider und Geophysiker: ein Lebensbild* [with a list of his publications]. Kartenberg, Herne (1957) — E.A. ECKARDT: *Memorial: Ludger MINTROP*. Geophysics **21**, 876-877 (1956) — O.G. (anonymous): *Obituary: Prof. Dr. phil. Dr. rer. mont. hc Ludger MINTROP*. Geophysical Prospecting (The Hague) **4**, i-ii (1956) — O. RELLENSMANN: *Nachruf für Prof. Dr. phil. Dr. rer. mont. hc L. MINTROP*. Z. Vermessungswesen **81**, 76-77 (1956) — A. SCHLEUSENER: *In Memoriam Prof. Dr. Dr. hc Ludger MINTROP*. Z. Geophys. **22**, 58-61 (1956) — W. BUCHHEIM: *Ludger MINTROP*. Gerlands Beitr. Geophys. **66**, 1-3 (1957) — H. BIRETT ET AL. (eds.): *Zur Geschichte der Geophysik*. Springer, Berlin & New York (1974) — W. KERTZ: *Ludger MINTROP, der die angewandte Geophysik zum Erfolg brachte*. Mitteil. Dt. Geophys. Gesell. **3**, 2-16 (1991) — W. KERTZ: *Geschichte der Geophysik*. G. Olms, Hildesheim etc. (1999), pp. 323-328 — A. BARTH: *Ludger MINTROP*. In: (J. RITTER, ed.) *History of seismology*. Institut für Geophysik der Universität Göttingen; http://www.uni-geophys.gwdg.de/~eifel/Seismo-HTML/history.html.
**PICTURE.** From K. LEHMANN: *Ludger MINTROP, der große Markscheider und Geophysiker* (see above).

# MUNROE, Charles Edward (1849–1938)

- U.S. chemist and high explosives specialist; father of explosive engraving and explosive forming

Charles E. MUNROE was born in East Cambridge, MA, to Enoch MUNROE, an instrument maker. He studied at the Scientific Department of Harvard University, where he graduated in 1871. After having assisted Prof. Willard J. GIBBS, he remained there as an instructor in chemistry (1871–1874). In 1872, he conducted a summer course of instruction in chemistry for teachers in Cambridge, MA, which was the first of its kind. In 1874, he was called to the chair of chemistry at the U.S. Naval Academy (Annapolis, MD), where he remained for 12 years, in addition to lecturing at St. John's College (1883–1884) in Annapolis. He then accepted the appointment of chemist to the U.S. Torpedo Station and War College (Newport, RI), where he made practical demonstrations in the manufacture, testing, and use of high explosives. He was also frequently called by national authorities to conduct special investigations on explosives. Subsequently, he took the chair of chemistry at Columbian College in Washington, DC (1892) and became dean of the School of Graduate Studies connected with that institution. He received his Ph.D. from Columbia University, New York City in 1894. Later he became a consultant to numerous governmental agencies and acted as chairman of various commissions.

Because of his employment in the Navy his work was closely affiliated with military explosives. For example, he invented a stable and effective naval smokeless cannon powder, so-called "indurite" (1891), which was especially commended by U.S. President Benjamin HARRISON in an annual address to Congress.

MUNROE accidentally observed how to shape explosives to concentrate energy – later called the "Munroe effect" (1888) – thus rediscovering the "unlined shaped-charge effect," which had already been discovered twice before in Germany (by Joseph VON BAADER in 1792 and by Max VON FÖRSTER in 1883) and thereafter fallen into oblivion. MUNROE noticed that when a block of guncotton with the manufacturer's name stamped into it was detonated next to a metal plate, the lettering was cut into the plate; if letters were raised in relief above the rest of the guncotton then the letters on the plate would also be raised above its surface. MUNROE used his discovery to imprint designs on iron plates by interposing a stencil between the explosive and plates of iron, thereby laying the foundation for a new technique – so-called "explosive engraving." The adjacent surfaces of a cavity in the explosive collide, as a result of which the so-called "Munroe jet" is generated. By the turn of the century, he had recognized the possibility of forming metals using shaped charges of explosives, a concept that also formed the basis for shaped charged used so efficiently during World War II.

In the late 1890s, he participated at the Naval Proving Ground in the perfection of an armor-piercing high explosive shell that would detonate within a battleship through the agency of a delayed action fuse. He was able to prove that a loaded shell could be made to stand such a shock of impact and penetration and then to explode on the inner side of a 14.5-in. (37-cm) plate of Harveyized armor (a case-hardened plate armor based upon an invention made by the American Hayward A. HARVEY in order to strengthen armor plating for warships).

Besides publishing several books and over 100 scientific papers on chemistry and explosives, he also did a great deal of bibliographic work and compiled an *Index to the Literature of Explosives* (1886, 1893). MUNROE was elected vice president of the Chemical Section of the American Association for the Advancement of Science (1887) and president of the Chemical Society of Washington (1895) and the American Chemical Society (1898). He received many honors and was a member of the American Academy of Arts and Science, the American Philosophical Society, and the Washington Academy of Sciences. He was also a foreign member of the Chemical Society London and the Deutsche Bunsen Gesellschaft (German Bunsen Society). In 1900, he was appointed by the Royal Swedish Academy of Sciences to nominate the candidate for the Nobel Prize in chemistry.

**ORIGINAL WORKS.** *An experimental lecture upon the causes and conditions which promote explosions.* Proc. U.S. Nav. Inst. **4**, 21-35 (1878) — *Index to the literature of explosives. Part I.* Friedenwald Press, Baltimore (1886) — *Modern explosives.* Scribner's Mag. **3**, 563-576 (1888) — *Wave-like effects produced by the detonation of gun-cotton.* Am. J. Sci. **36**, 48-50 (1888) — *On certain phenomena produced by the detonation of gun-cotton.* Proc. Newport Natl. Hist. Soc. **6**, 18-23 (1888) — *The effects of explosives on civilization.* The Chautauquan (New York) **9**, 203-205 (1889) — *Experiments for demonstrating that the force of a detonating explosive is exerted in all directions about the explosive center.* Proc. Am. Assoc. Adv. Sci. **38**, 131 (1889) — *The explosiveness of celluloids.* Ibid. **38**, 177 (1889) — *The direction taken by explosives.* The Illustrated American (New York) **3**, 286-288 (1890) — *Determination of the firing points of various explosives.* J. Am. Chem. Soc. **12**, 57-64 (1890) — *Indurite.* U.S. Patent No. 489,684 (1891) — *Index to the literature of explosives. Part II.* Deutsch Lithographic Printing, Baltimore (1893) — *Determination of the relative sensitiveness of explosive substances through explosions by influence.* J. Am. Chem. Soc. **15**, 10-18 (1893) — *The applications of explosive substances.* The Polytechnic **10**, 125-133 (1894) — *Inspection of cotton for use in the manufacture of gun-cotton.* J. Am. Chem. Soc. **17**, 783-789 (1895) — *The development of smokeless powder.* Ibid. **18**, 819-846 (1896) — *Report on the explosion at the U.S. Capitol.* Annu. Rept. of the Architects of the U.S. Capitol for 1899, Washington, DC (1899), pp. 29-44 — *The applications of explosives.* Pop. Sci. Month. **56**, 300-312, 444-455 (1900) — *Development in the explosives art in the United States during the last five years (1900-1905).* In: (E. PATERNO and V. VILLAVECCHIA, eds.) *Proc. 6th Int. Congr. Applied Chemistry* [Rome, Italy, April 1906]. Tipografia nazionale di G. Bertero, Rome (1907), vol. 2, pp. 667-672 — *The detonation of gun-cotton.* Proc. Am. Phil. Soc. **48**, 69-71 (1909) — *Some modern developments in methods of testing explosives.* Proc. U.S. Naval Inst. **36**, 805-823 (1910) — *The investigation of explosives at the Pittsburgh Testing Station.* Science **33**, 470 (1911) — *Production of mercury fulminate.* J. Ind. Engng. Chem. **4**, 152-153 (1912) — *Explosives as an aid to engineering.* J. Am. Soc. Mech. Eng. **37**, 705-707 (1915) — *Explosives for use in industrial and commercial developments.* In: (G.L. SWIGETT, ed.) *Proc. 2nd Pan-American Scientific Congress* [Washington, DC, Dec. 1915/Jan. 1916]. Washington Govt. Printing Office, Washington, DC (1917), vol. 8, pp. 690-694 — *Zones of silence in sound areas from explosions.* Mem. Corps Mar. **10**, 253-260 (1918) — With S.P. HOWELL: *Products of detonation of TNT.* Proc. Am. Phil. Soc. **59**, 194-223 (1920) — *The explosivity of ammonium nitrate.* Chem. Met. Engng. **26**, 535-542 (1922) — *The literature of blasting.* Explosives Engineer **1**, 43-44 (1923) — *"Pre-blasting."* Ibid. **1**, 75-76 (1923) — *The history of gunpowder.* Ibid. **1**, 119-120 (1923) — With C.A. TAYLOR: *Methods of testing detonators.* Rept. of Investigations No. 2,558, U.S. Bureau of Mines (1923) — *Destroying a navigation menace by blasting.* Explosives Engineer **2**, 54-55 (1924) — *The father of the blasting machine.* Ibid. **3**, 383-384 (1925) — *Distribution des ondes de l'explosion de l'"Alum Chine."* Mém. Artill. Franç. **4**, 545-552 (1925).

**SECONDARY LITERATURE.** Anonymous: MUNROE, *Charles Edward 1849–1938.* In: (J.L. CHAMBERLAIN, ed.) *Universities and their sons.* 5 vols., Hendon, Boston (1899) — C.A. BROWNE: *Charles Edward MUNROE 1849–1938.* J. Am. Chem. Soc. **61**, 1301-1316 (1939) — T.L. DAVIS: *The chemistry of powder and explosives.* Angriff Press, Las Vegas (1972) — L.E. MURR: *Shock waves for industrial applications.* Noyes, Park Ridge, NJ (1988).

**PICTURE.** Courtesy Deutsches Museum, Munich, Germany.

# NEUMANN, John [born Johann] Louis VON (1903–1957)

• Hungarian-born U.S. mathematician and theoretical physicist; father of three-shock theory and cofounder of modern detonation theory

John L. VON NEUMANN was born in Budapest to Max VON NEUMANN, a wealthy Jewish banker. He studied first chemistry and then mathematics at the University of Berlin. In 1926, he received his diploma in chemical engineering from the Eidgenössische Technische Hochschule Zürich (ETHZ) and in the same year his doctorate in mathematics from the Pázmány Péter Tudományegyetem (Péter Pázmány University) in Budapest. After a period of lecturing as *Privatdozent* (university lecturer) at the Universities of Berlin (1927–1929) and Hamburg (1929–1930), he became a visiting lecturer (1930) and full professor (1931–1933) at Princeton University. Invited to join the newly founded Institute for Advanced Study (IAS), he became the youngest member of its permanent faculty and professor of mathematics – a position at Princeton that he kept for the remainder of his life.

VON NEUMANN's contributions cover a wide spectrum of contemporary scientific thought, such as quantum mechanics, theory of games and economics, computer science, numerical methods for solving nonlinear partial differential equations, and hydrodynamics. In 1932, he wrote a major work on the mathematical foundations of quantum mechanics. From the mid-1930s he was interested in hydrodynamic turbulence. In 1937, VON NEUMANN had long discussions with Stanislaw ULAM, a mathematical physicist and spiritual father of the fusion bomb, on the phenomenon of turbulence and the possibility of a statistical treatment of the Navier-Stokes equations. VON NEUMANN proposed to analyze hydrodynamical problems through replacement of the partial differential equations by a system of infinitely many total differential equations, satisfied by the Fourier coefficients in the development of the Lagrangian functions in a Fourier series. During World War II, he was invited by Robert OPPENHEIMER to bring to the implosion program his expertise on theoretical hydrodynamics and ability to solve sets of nonlinear partial differential equations numerically.

Independently of Yakov B. ZEL'DOVICH in the Soviet Union and Werner DÖRING in Germany, VON NEUMANN became engaged in the theoretical treatment of the detonation process. He improved the Chapman-Jouguet (CJ) theory by introducing a reaction zone – the so-called "Zel'dovich-von Neumann-Döring (ZND) theory." On a theoretical basis he first treated the propagation and oblique interaction of shock waves, which he classified into "regular reflection" and "Mach reflection." He performed the task of selecting the most effective Height of Burst (HOB) for the atomic bombs on the Japanese targets, made valuable proposals to the implosion method for bringing nuclear fuel to explosion, and also participated in the development of the hydrogen bomb.

After the war his interest in hydrodynamics continued, and he established a meteorological research group to tackle the mathematical problems of solving the hydrodynamic equations of the motions of the Earth's atmosphere for numerical weather prediction. Together with Stanislaw ULAM he invented the Monte Carlo method (1946), a mathematical procedure to model a complex problem stochastically, and together with Arthur W. BURKS and Herman H. GOLDSTINE he developed the idea of a parallel, stored-program electronic computer (1946–1948), a landmark in the history of computer science that was taken up and applied by most subsequent digital computer designers. At IAS he developed the so-called "IAS Computer," the first computer designed as a general purpose system with stored instructions which was optimized for scientific calculation and became operational in 1952.

From 1937, VON NEUMANN was a member of the National Academy of Sciences (NAS). In 1955, U.S. President EISENHOWER appointed him to the Atomic Energy Commission (AEC). In the following year, he received the Medal of Freedom and the AEC Enrico Fermi Award. Shortly thereafter, he died of cancer in a hospital in Washington, DC at the age of 53.

His pioneering contribution to various fields of science and technology have been widely acknowledged, and a number of terms are named for him: in detonation physics the *von Neumann (vN) spike* and the *vN state* of a detonating high explosive; in shock wave physics the *vN reflection*, the *vN criterion* and the *vN paradox*; in numerical analysis the *vN analysis*; and in computer technology the *vN architecture* and the *vN machine*.

The *John von Neumann Lecture*, a prize established in 1959 by the Society for Industrial and Applied Mathematics (SIAM), is awarded "for outstanding and distinguished contributions to the field of applied mathematical sciences and for the effective communication of these ideas to the community."

A crater on the far side of the Moon is named for him.

**ORIGINAL WORKS.** *Mathematische Grundlagen der Quantenmechanik.* Springer, Berlin (1932); Dover Publ., New York (1943) — *Shock waves started by an infinitesimally short detonation of given (positive and finite) energy.* Rept. AM-9, Informal U.S. Govt. Doc., NDRC Div. B (June 1941) — *Theory of stationary detonation waves.* Progr. Rept. OSRD-549 (April 1942) — *Theory of detonation waves.* In: (A.H. TAUB, ed.) *John VON NEUMANN. Collected works.* Pergamon Press, Oxford, vol. VI (1963), pp. 203-218 — *Theory of shock waves.* Progr. Rept. OSRD-1140 (Jan. 1943) — With R.J. SEEGER: *On oblique reflection and collision of shock waves.* U.S. Govt. Doc. PB31918 (Sept. 1943) — *Shadowgraph determination of shock-wave strength.* DC Explosive Res. Rept. No. 11, Navy Dept., Bureau of Ordnance, Washington, DC (Oct. 1943); *Oblique reflection of shocks.* Ibid. No. 12 (Oct. 1943) — *Proposal and analysis of a new numerical method for the treatment of hydrodynamical shock problems.* Applied Mathematics Group, Institute for Advanced Study, Rept. OSRD-3617 (1944) and Rept. NDRC-108 (1944) — *Introductory remarks* [Sect. I]; *Theory of the spinning detonation* [Sect. XII]; *Theory of the intermediate product* [Sect. XIII]. In: *Report of informal technical conference on the mechanism of detonation.* U.S. Govt. Doc. AM-570 (April 1944) — *Riemann method: shock waves and discontinuities (one dimensional), two-dimensional hydrodynamics.* Lectures in shock hydrodynamics and blast waves by H.A. BETHE, K. FUCHS, J. VON NEUMANN, R. PEIERLS, and W.G. PENNEY [with notes by J.O. HIRSCHFELDER]. U.S. Govt. Doc. AECD-2860 (Oct. 1944) — *Surface water waves excited by an underwater explosion.* Memo to J.R. OPPENHEIMER. Rept. LAMS-128, LASL (Aug. 1944) — *Remarks on report of R.R. HALVERSON, "The effect of air burst on the blast from bombs and small charges, Part II."* Memo to J.R. OPPENHEIMER. U.S. Govt. Doc. AM-863 (Oct. 1944) — *Digest of J. VON NEUMANN's lecture at Meeting on Optimum Heights* (on Sept. 22, 1944). Memo (secret) to N.F. RAMSEY from B. WALDMAN. LASL, Rept. LAMD-46 (Dec. 1944) — *Refraction, intersection and reflection of shock waves.* In: *Proc. Conf. on Shock Waves and Supersonic Flow* [Princeton, NJ]. U.S. Govt. Doc. AM-1663 (July 1945); NAVORD Rept. 203-45, Navy Dept., Bureau of Ordnance, Washington, DC (1943) — With D.P. MACDOUGALL and G.H. MESSERLY: *Some considerations on shaped charge assembly.* Outcome of discussions at BRL, Bruceton. LASL, Rept. LAMS-196 (Jan.

1945) — With J.W. CALKIN and R. PEIERLS: *The similarity solution for a collapsing spherical cavity near zero radius*. Rept. LA-210, LASL (Jan. 1945) — With A.H. TAUB: *Flying wind tunnel experiments*. U.S. Govt. Doc. PB33263 (Nov. 1945) — *First draft of a report on the EDVAC*. Rept. prepared for the U.S. Army Ordnance Dept. under contract W-670-ORD-4926 (1945). Reprinted in N. STERN: *From ENIAC to UNIVAC*. Digital Press, Bedford, MA (1981), pp. 177-246 — With M.M. SHAPIRO: *Underwater explosion of a nuclear bomb*. Rept. LA-545, LASL (April 1946) — *The point source solution*. In: (K. FUCHS, J.O. HIRSCHFELDER, J.L. MAGEE, R. PEIERLS, and J. VON NEUMANN, eds.) *Blast wave*. Rept. LA-2000, LASL (Aug. 1947), chpt. 2, pp. 27-55; With F. REINES: *The Mach effect and the height of burst*. Ibid, chap. 10, pp. 11-84 — With A.W. BURKS and H.H. GOLDSTINE: *(I) Preliminary discussion of the logical design of an electronic computing instrument*. Rept. prepared for the U.S. Army Ordnance Dept. under contract W-36-034-ORD-7481 (1946/1947); *(II) Planning and coding of problems for an electronic computing instrument*. Ibid. (1947/1948) — *On the theory of stationary detonation waves*. File No. X122. BRL Aberdeen Proving Ground, MD (Sept. 1948) — *Discussions of the existence and uniqueness or multiplicity of solutions of the aerodynamic equations*. In: *Proc. 1st Symp. on the Motion of Gaseous Masses of Cosmical Dimensions* [Paris, Aug. 1949]. USAF Central Air Documents 254 Office, Dayton, OH (1951) — With R.D. RICHTMYER: *A method for the numerical calculation of hydrodynamic shocks*. J. Appl. Phys. **21**, 232-237 (1950) — *Problems of cosmic aerodynamics*. Rept. prepared for the Central Air Documents Office, Dayton, OH (1951), chap. 10, pp. 75-84 — With E. FERMI: *Taylor instability at the boundary of two incompressible liquids*. Rept. AECU-2979 (1953), part II, pp. 7 13 — *Applications of the hot sphere generated by an A-bomb*. Rept. UCRL-4412, University of California Radiation Laboratory (Nov. 1954) — With H.H. GOLDSTINE: *Blast wave calculation*. Comm. Pure Appl. Math. **8**, 327-353 (1955) — *The computer and the brain* [Silliman Lecture]. Yale University Press, New Haven, CT (1958) — (A.H. TAUB, ed.) *John VON NEUMANN, Collected works*. Vol. I (1961): *Logic, theory of sets, quantum mechanics;* Vol. II (1961): *Operators, ergodic theory and almost periodic functions in a group;* Vol. III (1961): *Rings of operators;* Vol. IV (1962): *Continuous geometry and other topics;* Vol. V (1963): *Design of computers, theory of automata, and numerical analysis;* and Vol. VI (1963): *Theory of games, astrophysics, hydrodynamics and meteorology*. Pergamon Press, Oxford.

SECONDARY LITERATURE. Special issue dedicated to J. VON NEUMANN. Bull. Am. Math. Soc. **64**, No. 3, Part 2 (May 1958): S.M. ULAM: *John VON NEUMANN 1903–1957*, pp. 1-49; G. BIRKHOFF: *Von NEUMANN and lattice theory*, pp. 50-56; L. VAN HOVE: *VON NEUMANN's contribution to quantum theory*, pp. 95-99; and C.E. SHANNON: *VON NEUMANN's contribution to automata theory*, pp. 123-129 — S. BOCHNER: *John VON NEUMANN*. Biogr. Mem. Natl. Acad. Sci. (U.S.A.) **32**, 438-457 (1958) — H. GOLDSTINE: *The computer from PASCAL to VON NEUMANN*. Princeton University Press, Princeton, NJ (1972) — J. DIEUDONNÉ: *VON NEUMANN, Johann*. In: (C.C. GILLESPIE, ed.) *Dictionary of scientific biography*. Scribner, New York; vol. 10 (1974), pp. 88-92 — S.M. ULAM: *VON NEUMANN: the interaction of mathematics and computing*. In: (N. METROPOLIS, J. HOWLETT, and G.C. ROTA, eds.) *A history of computing in the twentieth century*. Academic Press, New York (1980), pp. 93-99 — S.J. HEIMS: *John VON NEUMANN and Norbert WIENER*. MIT Press, Cambridge, MA (1981) — W. ASPRAY: *John VON NEUMANN and the origins of modern computing* [with a list of VON NEUMANN's writings]. MIT Press, Cambridge, MA (1992) — W. POUNDSTONE: *Prisoner's dilemma: John VON NEUMANN, game theory and the puzzle of the bomb*. Oxford University Press, Oxford (1993) — On VON NEUMANN's contribution to the Manhattan Project see the book by L. HODDESON, P.W. HENRIKSEN, R.A. MEADE, and C. WESTFALL: *Critical assembly: a technical history of Los Alamos during the Oppenheimer years, 1943–1945*. Cambridge University Press, Cambridge, U.K. (1993) — N. MACRAE: *John VON NEUMANN*. Pantheon, New York (1996).

PICTURE. Taken in the early 1950s by Alan RICHARDS, showing VON NEUMANN in front of the IAS computer which was built under his direction. Courtesy Institute for Advanced Study (IAS), Princeton, NJ.

# NOBEL, Alfred Bernhard (1833–1896)

- Swedish chemist, inventor, and industrialist; founder of the Nobel Prizes

Alfred B. NOBEL was born the son of Immanuel NOBEL, a financially successful inventor and businessman of war materiel, and was destined to follow in his energetic father's footsteps in many ways. When the family moved to St. Petersburg, Russia for business purposes, he and his two brothers were tutored privately (1843–1850). After studying engineering at St. Petersburg and chemistry in other European countries and the United States, he returned to St. Petersburg. In 1859, his father's company had to declare bankruptcy, because the Russian government cancelled all delivery agreements after it lost the Crimean War (1853–1856). His father began to experiment with nitroglycerin, which had been invented by the Italian chemist Ascanio SOBRERO (1846), but hitherto could not be applied in practice because of its extreme sensitivity to shock and heat.

His father developed a method of producing nitroglycerin on a factory scale, and in Heleneborg, an isolated area outside of Stockholm, he put the world's first factory, Nitroglycerin Ltd Stockholm, into operation (1865). Already in 1863 NOBEL invented the *detonator cap* (charged with mercury fulminate) to detonate a liquid nitrogen explosive charge by a strong shock wave rather than by heating. Since the time instant of ignition could be determined with great precision, his invention became very important for all future applications of multiple-charge arrangements, particularly for all shock techniques based on shock wave interactions and implosion. After a long period of experimentation he patented *dynamite* (1867), initially a mixture of about 75%

nitroglycerin and 25% kieselguhr, the latter strongly absorbing the nitroglycerin, thus allowing one to handle dynamite more safely without substantially reducing the explosion efficiency. In 1875, he patented the use of active ingredients in dynamite and called the material "blasting gelatin" – also known as "oil well explosive" – a rubber-textured, water-resistant explosive made by adding nitrocellulose (guncotton) to nitroglycerin, which has a very high detonation velocity and, therefore, a greater blasting action power than dynamite. His fourth great invention was *ballistite* (1888), a smokeless blasting powder and excellent propellant that essentially contains nitroglycerin and nitrocellulose. It was used for over 75 years.

NOBEL settled in Paris (1873) and made many other inventions such as various improvements in gas burners, an automatic brake, a system of non-explodable boilers, a refrigerating apparatus, and methods of vaporizing liquids and purifying cast iron. In 1883, he obtained permission from the French government to place a cannon and establish a small shooting range in an abandoned fort close to his laboratory at Sevran, 16 km northeast of Paris – a village that had become the late-19th-century French capital of gunpowder and explosives. At the end of his life he held more than 350 patents in different countries, covering detonics, electrochemistry, metallurgy, biology, optics, and physiology. NOBEL became a wealthy man and left his estate to establish annual prizes, the so-called "Nobel Prizes," for outstanding achievements (since 1901 in physics, chemistry, medicine and physiology, peace, and literature, and since 1969 in economics). The Norwegian Nobel Institute assists the Nobel Committee in the task of selecting the recipient of the Nobel Peace Prize and organizes the annual Nobel events in Oslo.

The radioactive element "nobelium" (No; atomic number 102, mass number 253) was named in his honor. Not occurring in nature but first suggested in 1957 by a team of scientists at the Nobel Institute for Physics in Stockholm, it was eventually identified in 1958 at the University of California at Berkeley. The team produced it by bombarding curium with high-energy carbon nuclei.

A crater on the far side of the Moon and a minor planet (asteroid 6032 NOBEL) are also named for him.

**PUBLICATIONS.** *Results of blasting experiments made with nitroglycerine at Vieille-Montagne mine.* Phil. Mag. **30** [IV], 236-238 (1865) — *Expériences de sautage faites avec la nitro-glycérine, à la mine de la Vieille-Montagne.* C. R. Acad. Sci. Paris **61**, 122-124 (1865) — *On modern blasting agents.* Am. Chemist **6**, 60-68 (1876); Rev. Artill. (Paris) **7**, 262-273 (1876); Moniteur Scientifique **18**, 248-261 (1876) — *Mode d'emploi de la dynamite.* Société général pour la fabrication de la dynamite. Précédés brevetés de A. Nobel. Lahure, Paris (1878) — [*On progressive smokeless powder.*] Swed. Patent No. 7,552 (1896).

**SELECTION OF HIS BRITISH PATENTS ON EXPLOSIVES.** *Improvements in the manufacture of gunpowder and powder for blasting purposes.* No. 2,359 (1863) — *Nitroglycerine* [on manufacturing and fining]. No. 1,813 (1864) — *Explosive compounds* [dynamite]. No. 1,345 (1867) — *Explosive compounds* [dynamite with an active base]. No. 442 (1869) — *Improvements in the manufacture of explosive compounds.* No. 1,570 (1873) — *Exploding compounds* [blasting gelatin]. No. 4,179 (1875) — *Manufacture of explosive compounds* [detonators]. No. 2,399 (1879) — *Improvements in explosive projectiles.* No. 5,840 (1887) — *Improvements in regulating the pressure in guns* [method by lessening the recoil of guns]. Nos. 371 and 3,674 (1887) — *Improvements in detonators.* No. 16,919 (1887) — *Improvements in the manufacture of explosives* [ballistite]. No. 1,471 (1888) — *Improvements in the preparation of explosive compounds* [dynamite rubber]. No. 9,361 (1889) — *Improvements in fuses for mining and similar purposes.* No. 20,467 (1893).

**SECONDARY LITERATURE.** L. ROUX and E. SARRAU: *A. NOBEL: Les explosifs modernes (mémoires par A. NOBEL).* Lahure, Paris (1876) — H. DE MOSENTHAL: *The life-work of Alfred NOBEL* [with a list of his British patents including patents which received provisional protection only]. J. Soc. Chem. Ind. **18**, 443-451 (1899); *Die Tätigkeit Alfred NOBEL's.* Z. Angew. Chem. **12**, 753-757, 782-787 (1899) — C.E. MUNROE: *Alfred NOBEL and the Nobel Prize* [presented at the Washington meeting on Nov. 14, 1902]. Proc. Am. Chem. Soc (1902), p. 32 — C. HAEUSSERMANN: *Alfred NOBEL und die Erfindung der Nitroglyzerinpulver.* Union Dt. Verlagsgesell., Stuttgart (1904) — A. NOBLE: *Artillery and explosives: essays and lectures written and delivered at various times.* J. Murray, London (1906) — H. SCHÜCK and R. SOHLMANN: *The life of Alfred NOBEL.* Heinemann, London (1929) — *The history of NOBEL's Explosives Co. Limited and Nobel Industries Limited, 1871–1926.* Imperial Chemical Industries (ICI): *Imperial Chemical Industries Limited, and its founding companies.* Kynoch, Birmingham, vol. 1 (1938) — F.D. MILES: *A history of research in Nobel Division of the Imperial Chemical Industry.* ICI Ltd., Nobel Division, Ardeer, U.K. (1955) — E. BERGENGREN: *Alfred NOBEL, the man and his work.* Nelson, London (1962) — T.L. DAVIS: *The chemistry of powder and explosives.* Angriff Press, Las Vegas (1972) — T. ALTHIN: *A.B. NOBEL.* In: (C.C. GILLESPIE, ed.) *Dictionary of scientific biography.* Scribner, New York, vol. 10 (1974), pp. 132-133 — R.L. WEBER: *Pioneers of science. Nobel Prize winners in physics.* Institute of Physics, Bristol (1980), pp. 1-2 — K. FANT: *Alfred NOBEL: Idealist zwischen Wissenschaft und Wirtschaft.* Insel, Frankfurt/M. (1997).

**PICTURE.** Courtesy Deutsches Museum, Munich, Germany.

# NOBLE, Sir Andrew (1831–1915); from 1902 first Baronet

- British physicist, gunnery expert, and industrialist

Sir Andrew NOBLE was born in Greenock, Scotland. He was the third son of George NOBLE, a retired naval captain. He was first educated in Greenock and later at Edinburgh Academy and the Royal Military Academy at Woolwich, southeast London. He entered the Royal Artillery (1849), rose to be Captain (1855), and was appointed secretary to the Select

Committee on the Relative Merits of Smoothbore and Rifled Cannon and assistant inspector of artillery (1858). In the same year, the new system of breech-loaded artillery – superior to smooth-bore guns as proved by NOBLE in numerous scientific experiments and accurate observation – was adopted officially by the armed services. He also became a member of the Committee of Explosives.

After serving 12 years with his regiment in Canada and South Africa, he resigned his commission (1860) and joined Sir William George ARMSTRONG, a mechanical engineer and inventor who, emphasizing rifled bores and elongated projectiles, had just established ordnance works at Elswick, from 1882 named W.G. Armstrong & Co., which in 1897 merged with Whitworth & Co. In 1861, he became ARMSTRONG's partner and operational head of the company. After ARMSTRONG's death he rose to the position of chairman (1900).

In his interior ballistic experiments NOBLE followed earlier investigators such as Thomas J. RODMAN and Sir Benjamin THOMSON, Count VON RUMFORD; however, he carried the examination of fired gunpowder further than any of his predecessors. In order to measure the velocity of a projectile in its passage through the bore, he invented a special chronoscope for use in ballistics – the so-called "Noble chronograph" (1862) – a device for measuring very small time intervals. His instrument was essentially an improvement over the Navez chronograph, by which the velocity of a shot at any point in traveling down the bore of a gun could be ascertained. Using Sir Isaac NEWTON's Second Law of Motion, this technique gave him the means for calculating the effective mean pressure on the base of a projectile and of coupling this with the indications of his "crusher gauges," certainly his most renowned invention that, being a derivative of RODMAN's indentation gauge but more accurate, was soon used worldwide. The results of his research activities on the nature and products of explosion and other particulars in connection with this little known branch of science and guns that he jointly performed with Sir Frederick ABEL were later published in a two-part monograph under the title *Researches on Explosives* (1875, 1880). His thorough studies, which put this subject upon a scientific basis and established the science of ballistics, did much to advance the manufacture of guns and gunpowder. This resulted in longer guns with larger chambers and slower burning powders, thus superseding the use of hand-rammed black powder.

NOBLE was elected F.R.S. (1870), and awarded the Royal Medal (1880) and the Albert Medal of the Royal Society of Arts (1909). He received the honorary degree of Doctor of Science from Oxford University and was created a baronet (1902). He was a member of various orders of knighthood.

**ORIGINAL WORKS.** *Report on experiments with Major NAVEZ's electro-ballistic apparatus.* Proc. Roy. Artill. Inst. (Woolwich) **3**, 117-150 (1863) — *Note on the ratio between the forces tending to produce translation and rotation in the bores of rifled guns.* Phil. Mag. **26** [IV], 195-205 (1863) — *On the tension of fired gunpowder.* Proc. Roy. Inst. **6**, 274-283 (1872) — *Sur la force explosive de la poudre à canon.* Rev. Sci. France Etrang. **48** [II], 1125-1141 (1872) — *On the pressure required to give rotation to rifled projectiles.* Phil. Mag. **45** [IV], 204-215 (1873) — *Recherches sur les corps explosibles. Explosion de la poudre.* Ann. Chim. Phys. **3** [V], 268-288 (1874); C. R. Acad. Sci. (Paris) **79**, 204-208, 294-295, 360-364 (1874) — With F.A. ABEL: *Researches on explosives. Fired gunpowder (Part I).* Phil. Trans. Roy. Soc. Lond. **165**, 49-155 (1875); *(Part II).* Ibid. **171**, 203-279 (1880) — *Fired gunpowder. Note on the existence of potassium hyposulphite in the solid residue of fired gunpowder.* Proc. Roy. Soc. Lond. **30**, 198-208 (1880) — *Note on the energy absorbed by friction in the bores of rifled guns.* Ibid. **50**, 409-421 (1892) - *Preliminary note on the pressure developed by some new explosives.* Ibid. **52**, 123-129 (1893) — *On methods that have been adopted for measuring pressures in the bores of guns.* Rept. Meet. Brit. Assoc. **64**, 523-540 (1894) — *Researches on explosives. Preliminary note.* Proc. Roy. Soc. Lond. **56**, 205-221 (1894) — *Some modern explosives.* Nature **62**, 86-90, 111-115 (1900); Not. Proc. Meet. Memb. Roy. Inst. Lond. **16**, 329-345 (1902) — *Artillery and explosives: essays and lectures written and delivered at various times.* John Murray, London (1906) — With F.A. ABEL: *Die ältesten Untersuchungen über die Verbrennung des Schwarzpulvers* [translated by B. PLEUS]. Z. ges. Schieß- u. Sprengstoffwesen **5**, 84-87, 104-107 (1910).
**SECONDARY LITERATURE.** P.W. (anonymous): *Sir Andrew NOBLE, 1831-1915.* Proc. Roy. Soc. Lond. **A94**, i-xvi (1918) — Anonymous: *A great Victorian: Sir Andrew NOBLE...* Armstrong Whitworth Record (Spring 1932), pp. 3-13 — S.M. LINSLEY: *NOBLE, Sir Andrew.* In: (H.C.G. MATTHEW, ed.) *Oxford dictionary of national biography.* Oxford University Press, Oxford (2004), vol. 40, pp. 948-949.
**PICTURE.** Reprinted with permission from Proc. Roy. Soc. Lond. **A94**, i-xvi (1918). © The Royal Society of London, London, U.K.

# OPPENHEIM, Antoni ("Tony") Kazimierz (1915–)

- Polish-born U.S. aeronautical engineer, gas dynamicist, and combustion specialist

Antoni K. OPPENHEIM was born in Warsaw. His father, Tadeusz OPPENHEIM, was an industrialist and ceramic engi-

neer, and his mother came from an industrial family. His ambition was to become an engineer, and after graduation at the Gymnazjum of Wojciech Gorski (1933) he entered the Politechnika Warszawska (Warsaw University of Technology). With the collapse of Polish resistance after the outbreak of World War II, OPPENHEIM began his odyssey from Poland to England, where he arrived in June 1940 after many adventures. On leave from the artillery of the Polish army in Scotland (1942), he enrolled as a student at the City and Guilds College, the engineering branch of the Imperial College of Science and Technology, University of London, to complete requirements for his degree from Warsaw University of Technology. In 1945, he passed the final examination for the degree of *Dipl.-Inz.* before an Anglo-Polish committee appointed by the Ministry of Education of the Polish government in exile and completed his studies of high-speed flow of gases in channels for a Ph.D. at the University of London, in conjunction with a DIC (Diploma of Imperial College). There he became a lecturer, teaching heat transfer and gas dynamics, and, with his postgraduate students, built the first supersonic wind tunnel.

His career as a scientist was launched by Sir Owen SAUNDERS, a reader in the Mechanical Engineering Department at the City and Guilds College who received a proposal from the Napier Engine Co. to investigate the possibility of improving the performance of its piston engines used in the British fighter planes Spitfire and Hurricane. He assigned the proposal to OPPENHEIM, who worked out an analytical solution demonstrating how the total thrust (propeller plus exhaust jet) could be augmented. OPPENHEIM, having been granted an indefinite leave of absence from the Polish army, was employed by Power Jets (Res. & Dev.) at Whetstone, Leicester, U.K., which was formed in 1936 under the directorship of Sir Roxby COX, a British aeronautical engineer, to develop the gas turbine jet propulsion engine of Frank WHITTLE. Progress made on the thrust augmentation project was swift: in a few months, all the exhaust manifolds of fighter engines were replaced by individual nozzles and the valve timing adjusted for maximum total thrust. Top aircraft speed was thereby increased, giving British pilots an advantage over their German adversaries. OPPENHEIM conducted a study of secondary air mixing in the turbine combustion chamber that led to the development of the canister type, the universally accepted standard. He demonstrated that nozzles are redundant for this purpose; holes are quite sufficient.

Analyzing the operational mechanism of the pulsed jet engine that powered the German V1 (or "Flying Bomb"), he developed a systematic approach to 1-D gas dynamics that led to a paper with Joseph KESTIN on the generalized entropy chart. Published by the Institution of Mechanical Engineers, the paper won a prize for best technical article (1948). After the war, he was sent to Germany as a British intelligence objectives subcommittee officer to solicit reports from the principal scientists and engineers involved in the development of the pulsed jet engine. To obtain a value for the speed of the exothermic front that was required for the gas dynamic wave interaction analysis of this engine type, OPPENHEIM became involved in detonation phenomena – an effort that led him to the development of the theory of a double discontinuity system and its Q-curve and the locus of states immediately behind it.

After an appointment as assistant professor (1948–1950) at Stanford University, he was appointed assistant professor (1950–1954) at the University of California at Berkeley, where he was promoted to associate professor (1954–1958) and professor of mechanical engineering (1958–1986). In addition, as a staff consultant at Shell Development Co. (Emeryville, CA), he gained an impressive amount of knowledge in physical chemistry. At UC Berkeley his studies included the development of the radiation network method, the theory of heat transfer in free molecular flow, the use of vector polar methods for the analysis of interactions and intersections between gas dynamic wave fronts, the development and structure of detonation fronts, blast wave theory, turbulent combustion, plasma jets, turbulent jet plumes, and controlled combustion in engines. The common denominator of all studies throughout his career was his fascination with the combustion dynamics of exothermic systems in terms of the fundamental components: thermodynamics, thermochemistry, and aerodynamics.

OPPENHEIM was an active member of various professional and honorary societies. He organized the Northern California Section of the American Rocket Society (ARS), served as its president (1957), and was elected member of the ARS National Board of Directors. He also served on the NASA Research Advisory Committee on Fluid Mechanics (1963–1968). He was deputy editor of Combustion and Flame (1972–1973), associate editor of Astronautica Acta (1973), and editor-in-chief of Acta Astronautica (1974–1978). He was executive cochairman of the International Committee on Dynamics of Explosions and Reactive Systems (1966–1983) and organized the International Colloquium on Dynamics of Explosions and Reactive Systems (ICDERS). From 1952 was an active participant in the Symposium (International) on Combustion.

He took several sabbatical leaves, which he spent in Poitiers at the Ecole National de Mécanique et Aérothermodynamique (Prof. Numa MANSON), at the University of Marseilles (Prof. Paul CLAVIN), and in Göttingen at the MPI für Physikalische Chemie (Prof. Heinz G. WAGNER). During his tenure as a professor of engineering at UC Berkeley, he was also a visiting professor at the Sorbonne (1960–1961), a Miller Professor at UC Berkeley (1961–1962), and a Professeur Associé at the University of Poitiers (1973 and 1980). OPPENHEIM was author or coauthor of over 300 papers and supervised 21 M.S. and 34 Ph.D. students. He received numerous awards and honors from national and international societies and universities.

In 1989, the Institute for Dynamics of Explosions and Reactive Systems (IDERS) introduced the *Oppenheim Prize* to be awarded for "brilliant contributions to the theoretical or interpretive aspects of the dynamics of explosions and reactive systems."

**ORIGINAL WORKS.** *Investigation of high speed flow of gases in channels.* Ph.D. thesis, Imperial College of Science and Technology, University of London (Sept. 1944) — *Gasdynamic analysis of the development of gaseous detonation and its hydraulic analogy.* In: *4th Symp. (Int.) on Combustion.* Williams & Wilkins, Baltimore (1953), pp. 471-480 — *The thermodynamic method of attack.* J. Engng. Educ. **46**, No. 4, 332-337 (Dec. 1955) — *Radiation analysis by the network method.* Trans. ASME **27**, No. 4, 725-736 (May 1956), reprinted in (J.P. HARTNETT, ed.) *Recent advances in heat and mass transfer.* McGraw-Hill, New York (1961), pp. 372-396 — With R.A. STERN and P.A. URTIEW: *On the development of detonation with pre-ignition.* Combust. Flame **4**, No. 4, 335-341 (Dec. 1960) — With A.J. LADERMAN: *The use of schlieren methods to observe moving waves in reactive gases.* In: *Symp. on Flow Visualization* [New York, Nov. 1960]. ASME, New York (1960), pp. 11:1-8 — *Development and structure of plane detonation waves.* In: *4th AGARD Combustion and Propulsion Colloquium* [Milan, Italy, April 1960]. Pergamon Press, London (1961), pp. 186-258 — With R.A. STERN: *Les caractéristiques fondamentales de la propagation des ondes dans les milieux réactifs.* C. R. Acad. Sci. Paris **253**, 64-66 (July 3, 1961); *On the polar method in gas wave dynamics.* In: *Proc. 10th Int. Congr. on Applied Mechanics* [Stresa, Italy, Aug./Sept. 1960]; Elsevier, Amsterdam (1962), pp. 226-228 — With A.J. LADERMAN and P.A. URTIEW: *Measurement of pressure fields generated at the initiation of explosion.* In: *Proc. ASME Symp. Measurements in Unsteady Flow* [Worcester, MA, May 1962]. ASME, New York (1962), pp. 32-35 — With A.J. LADERMAN: *Initial flame acceleration in an explosive gas.* Proc. Roy. Soc. Lond. **A268**, 153-180 (July 1962) — With A.J. LADERMAN and P.A. URTIEW: *The onset of retonation.* Combust. Flame **6**, No. 3, 193-197 (1962) — With N. MANSON and H.G. WAGNER: *Recent progress in detonation research.* AIAA J. **1**, No. 10, 2243-2252 (1963) — *On the dynamics of the development of detonation in a gaseous medium.* Archiwum Mechaniki Stosowanej **16**, No. 2, 403-424 (1964) — With P.A. URTIEW

and A.J. LADERMAN: *Dynamics of the generation of pressure waves by accelerating flames.* In: *10th Symp. (Int.) on Combustion.* The Combustion Institute, Pittsburgh, PA (1965), pp. 797-804 — With P.A. URTIEW: *Experimental observations of the transition to detonation in an explosive gas.* Proc. Roy. Soc. Lond. **A295**, 13-28 (1966) — *Gasdynamics of explosions and its relevance to propulsion.* In: *Papers presented at 2nd OAR Research Applications Conference,* Washington, DC (March 1967), vol. 1, pp. 39-56 — With J.J. SMOLEN and L.J. ZAJAC: *Vector polar method for the analysis of wave intersections.* Combust. Flame **12**, No. 1, 63-76 (Feb. 1968) — With P.A. URTIEW: *Transverse flame-shock interactions in an explosive gas.* Proc. Roy. Soc. Lond. **A304**, 379-385 (1968) — *Gasdynamics of explosions.* In: (W.H.T. LOH, ed.) *Propulsion: theory and design.* Springer, New York (1968), chap. 13, pp. 508-560 — With E.A. LUNDSTROM: *On the influence of non-steadiness on the thickness of the detonation wave.* Proc. Roy. Soc. Lond. **A310**, 463-478 (1969) — With J.W. MEYER: *On the shock-induced ignition of explosive gases.* In: *13th Symp. (Int.) on Combustion.* The Combustion Institute, Pittsburgh, PA (1971), pp. 1153-1164 — E.A. LUNDSTROM, A.L. KUHL, and M.M. KAMEL: *A systematic exposition of the conservation equations for blast waves.* J. Appl. Mech. **38**, 783-794 (1971) — *Introduction to gasdynamics of explosions.* Springer, Vienna (1971, 1972) — With A.L. KUHL, E.A. LUNDSTROM, and M.M. KAMEL: *A parametric study of self-similar blast waves.* J. Fluid Mech. **52**, Part 4, 657-682 (1972) — With A.L. KUHL and M.M. KAMEL: *On self-similar blast waves headed by the Chapman-Jouguet detonation.* Ibid. **55**, Part 2, 257-270 (1972) — With R.I. SOLOUKHIN: *Experiments in gas dynamics of explosions.* Annu. Rev. Fluid Mech. **5**, 31-58 (1973) — With M.M. KAMEL: *Laser cinematography of explosions.* Springer, Vienna (1972) — With G.G. BACH and A.L. KUHL: *On blast waves in exponential atmospheres.* J. Fluid Mech. **71**, Part 1, 105-122 (1975) — With G.I. BARENBLATT, R.H. GUIRGUIS, M.M. KAMEL, A.L. KUHL, and Y.B. ZEL'DOVICH: *Self-similar explosion waves of variable energy at the front.* J. Fluid Mech. **99**, Part 4, 841-858 (1980) — *Dynamic features of combustion.* Proc. Roy. Soc. Lond. **A 315**, 471-508 (1985) — *Pulsed jet combustion – its past, present, and future.* Archivum Combustionis **11**, No. 1/2, 3-18 (1991) — *Turbulent combustion in contrast to flames.* In: (F.A. WILLIAMS, ed.) *Modern developments in energy, combustion and spectroscopy.* Pergamon Press, Oxford (1993), pp. 1-13 — *Aerodynamic control of combustion.* J. Fluids Engng. **115**, No. 4, 561-567 (1993) — With J.A. MAXSON: *A thermochemical phase space for combustion in engines.* In: *25th Symp. (Int.) on Combustion.* The Combustion Institute, Pittsburgh, PA (1994), pp. 157-165 — With A.L. KUHL, R.E. FERGUSON, K.Y. CHIEN, and J.P. COLLINS: *Gasdynamic model of turbulent combustion in an explosion.* Zel'dovich Memorial [Int. Conference on Combustion, Moscow, Sept. 1994]. In: *Combustion, detonation, shock waves.* ENAS, Moscow (1995), vol. 1, pp. 181-189 — With A.L. KUHL and R.E. FERGUSON: *Gasdynamic model of turbulent exothermic fields in explosion.* Progr. Astronaut. & Aeronaut. **173**, 251-261 and Plates 15.1-8 (1997) — With A.L. KUHL: *Aerothermodynamics of closed combustion systems.* Archivum Combustionis **19**, No. 1-4, 15-65 (1999) — With A.L. KUHL and R.E. FERGUSON: *Gasdynamics of combustion of TNT products in air.* Ibid. **19**, No. 1-4, 67-89 (1999) — With A.L. KUHL: *Dynamic features of closed combustion systems.* In: *Progress in energy and combustion science.* Pergamon Press, Oxford, vol. 26 (2000), pp. 533-564 — *Combustion in piston engines: technology evolution, diagnosis and control.* Springer, Berlin (2004).

**SECONDARY LITERATURE.** J. Ray BOWEN (ed.) *Dynamics of exothermicity: in honor of Antoni Kazimierz OPPENHEIM.* Gordon & Breach, Amsterdam (1996) — Dept. of Mechanical Engineering, UC Berkeley: *Antoni K. OPPENHEIM, professor of aeronautical sciences in graduate school* [contains the picture shown above, a biographical résumé and a list of his 320 publications]; http://www.me.berkeley.edu/faculty/oppenheim/ — A.L. KUHL: *OPPENHEIM Antoni ("Toni") Kazimierz (1915\*). On the occasion of his 90th birthday.* Shock Waves **15**, 69-71 (2006).

**PICTURE.** Courtesy Dept. of Mechanical Engineering, UC Berkeley, CA.
**NOTE.** The short biography was kindly provided by Dr. Allen L. KUHL of LLNL, Livermore, CA.

## OSWATITSCH, Klaus (1910–1993)

- Austrian physicist and gas dynamicist

Klaus OSWATITSCH was born in Marburg on the Drau, at that time part of the Austro-Hungarian Empire (now Maribor, Slovenia). His father was a judge of high rank. After studying physics and mathematics at the University of Graz and graduating (Dr. phil. nat. 1935), he also took the examinations for a grammar-school teacher (1938) to better cope with the bad economic situation of that time. Provided with a fellowship from the Deutsche Forschungsgemeinschaft (German Research Community), he joined Prof. Ludwig PRANDTL at the Kaiser-Wilhelm-Institut für Strömungsforschung in Göttingen and became *Privatdozent* (university lecturer) at the University of Göttingen (1942–1946), in addition to also carrying out research for the Heereswaffenamt (Army Ordnance Office). After the war he worked briefly for the Royal Aircraft Establishment at Farnborough (1946) and for the French military in the Bureau d'Etudes at Emmendingen, Baden (1947). In the following years, he lectured in Germany (1948–1949) at the University of Freiburg and in Sweden (1949–1956) at the Kungl. Tekniska Högskolan (KTH), the Stockholm Polytechnic. In 1956, he established and headed at Aachen the Institut für Theoretische Gasdynamik of the Deutsche Versuchsanstalt für Luft- und Raumfahrt (DVLR). In 1960, he was appointed full professor at Vienna Technical University, where he worked until his retirement (1980).

His research covers a broad spectrum of gas dynamics and treats many problems relating to supersonic flow and shock waves, such as (1) condensation phenomena in supersonic nozzles, (2) supersonic flow around delta wings and in cascades of turbines, (3) hypersonic flow, (4) the development of effective shock diffusers, (5) 3-D flow around slender bodies at all Mach numbers, (6) supersonic intake design of high-speed aircraft engines using oblique shocks to give high efficiency, (7) analysis of gas dynamic phenomena in transitional ballistics, and (8) sonic boom and other acoustic phenomena associated with the muzzle blast and explosions. In particular, OSWATITSCH dedicated himself to understanding transonic flow phenomena, which became important for the design of supersonic aircraft.

He published more than 120 papers, 3 handbook articles, and 7 books. He belonged to the board of editors of various professional journals and was chairman of the first and second Symposium Transsonicum (1962, 1975) and editor of the proceedings. His famous textbook *Gasdynamik* (1952), which was also translated into English and Chinese, is certainly his best-known work. OSWATITSCH edited volume 7 (*Two-phase flows*, 1977) of the *Vieweg Tracts in Pure and Applied Physics*. Together with Karl WIEGHARDT he edited the 8th and 9th editions of Ludwig PRANDTL's *Führer durch die Strömungslehre* (1984, 1990).

He received many honors and became an honorary member of many scientific societies. He received honorary doctorates from the Universität Karlsruhe, KTH, and ETHZ.

**ORIGINAL WORKS.** *Der Verdichtungsstoß bei der stationären Umströmung flacher Profile.* ZAMM **17**, 1-7 (1937) — *Die Nebelbildung in Windkanälen und ihr Einfluß auf Modellversuche.* Jb. Dt. Akad. Luftfahrtforsch. (1941/1942), pp. 692-703 — *Kondensationserscheinungen in Überschalldüsen.* ZAMM **22**, 1-14 (1942) — With W. ROTHSTEIN: *Das Strömungsfeld in einer Lavaldüse.* Jb. Dt. Akad. Luftfahrtforsch. (1942/1943), pp. 91-102 — *Kondensationsstöße in Laval-Düsen.* VDI-Z. **86**, 702 (1942) — *Kondensationserscheinungen in Überschalldüsen.* ZAMM **22**, 1-14 (1942) — *Zur Ableitung des CROCCO'schen Wirbelsatzes.* Luftfahrtforsch. **20**, 260 (1943) — *Zur Abschätzung der kritischen Mach-Zahl.* Tech. Bericht **10**/5, Dt. Versuchsanstalt für Luftfahrt (DVL), Berlin-Adlershof (1943) — With H. BOHM: *Luftkräfte und Strömungsvorgänge bei angetriebenen Geschossen.* Berichtswesen über Luftfahrtforsch., Berlin, HWA Bericht 1010 (1944) — *Flow research to improve the efficiency of muzzle brakes* [in German]. *Part I: Tests on baffle surfaces with one-dimensional flow.* Mitteil. Dt. Luftfahrtforsch. Nr. 6601 (1943); *Abth. II: Efficiency factor, momentum relation, two-dimensional flow, and flow with covering.* Ber. Nr. 1001, Forschung & Entwicklung, Heereswaffenamt (HWA), Berlin (1944) — With W. ROTHSTEIN: *Abth. III: The axial symmetric flow problem, a comparison of tests of muzzle brakes free of reaction.* Ber. Kaiser-Wilhelm-Institut (KWI) für Strömungsforschung, Göttingen (1945) — *Über die Grenzen der Fluggeschwindigkeit.* In: (L. PRANDTL, ed.) *Festschrift Albert BETZ zum 60. Geburtstag.* Max-Planck-Institut (MPI) für Strömungsforschung, Göttingen (1945), pp. 128-132 — *Der Luftwiderstand als Integral des Entropiestromes.* Nachr. Akad. Wiss. Göttingen [Math.-Phys. Kl.]. Vandenhoeck & Ruprecht, Göttingen (1945), pp. 88-90 — With R. SEIFERTH: *Die Nebelbildung in Hochgeschwindigkeits- und Überschallkanälen.* Aerodynamische Versuchsanstalt (AVA), Göttingen (1946) — *Über die Charakteristikenverfahren der Hydrodynamik.* ZAMM **25/27**, 195-208, 264-270 (1947) — *Gesetzmäßigkeiten der schallnahen Strömung.* Ibid. **29**, 1-2 (1949) — *Der Verdichtungsstoß bei der stationären Umströmung flacher Profile.* Ibid. **29**, 129-141 (1949) — With S.B. BERNDT: *Aerodynamic similarity at axisymmetric transonic flow around slen-*

*der bodies.* Rept. AERO-TN12, KTH Stockholm (1950) — *Die Geschwindigkeitsverteilung bei lokalen Überschallgebieten an flachen Profilen.* ZAMM **30**, 17-24 (1950) — *Die Geschwindigkeitsverteilung an symmetrischen Profilen beim Auftreten lokaler Überschallgebiete.* Acta Physica Austriaca **4**, 230-271 (1950) — *Similarity laws for hypersonic flow.* Rept. AERO-TN16, KTH Stockholm (1950) — With H. BEHRBOHM: *Flache kegelige Körper in Überschallströmung.* Ing. Arch. **17**, 370-377 (1950) — *Ähnlichkeitsgesetze für Hyperschallströmung.* ZAMP **2**, 249-264 (1951) — *Gasdynamik.* Springer, Vienna (1952); *Gas dynamics.* Academic Press, New York (1956) — With F. KEUNE: *Nicht angestellte Körper kleiner Spannweite in Unter- und Überschallströmung.* Z. Flugwiss. **1**, 137-145 (1953) — *Gasdynamik als Hochschulfach.* Physik. Blätter **9**, 271-272 (1953) — *The drag increase at high subsonic speeds.* Rept. ARC-R&M 2716, H.M.S.O., London (1954) — *Die Berechnung wirbelfreier achsensymmetrischer Überschallfelder.* Österr. Ing. Archiv **10**, 359-382 (1956) — With G. KUERTI: *Gas dynamics.* Academic Press, New York (1956) — *Der Druckwiedergewinn bei Geschossen mit Rückstoßantrieb bei hohen Überschallgeschwindigkeiten. Der Wirkungsgrad von Stoßdiffusoren.* Westdt. Verlag, Köln/Opladen (1957) — *Gelöste und ungelöste Probleme der Gasdynamik.* In: (W. GEORGII, ed.) *Aerophysikalische Flugforschung.* Westdt. Verlag, Köln/Opladen (1957) — *Was ist Gasdynamik?* Physik. Blätter **14**, 108-116 (1958) — With I. TEIPEL: *Die Pulsationen von Stoßdiffusoren* In: *Festschrift Jakob ACKERET zum 60. Geburtstag am 17. März 1958.* ZAMP **IX b**, 462-478. (1958) — *Extreme speeds and thermodynamic states in supersonic flight.* NACA-TM 1434, Washington, DC (1958) — *Antriebe mit Heizung bei Überschallgeschwindigkeiten.* Westdt. Verlag, Köln/Opladen (1959) — With J. ZIEREP: *Das Problem des senkrechten Stoßes an einer gekrümmten Wand.* ZAMM **40**, 143-144 (1960) — *Similarity and equivalence in compressible flow.* In: (H.L. DRYDEN and Th. VON KÁRMÁN, eds.) *Advances in applied mechanics.* Academic Press, New York, vol. 6 (1960), pp. 153-271 — With D. RUES: *Eine exakte Lösung der schallnahen gasdynamischen Gleichung.* Z. Flugwiss. **9**, 125-129 (1961) — *Das Ausbreiten von Wellen endlicher Amplitude.* Ibid. **10**, 130-138 (1962) — *Die Wellenausbreitung in der Ebene bei kleinen Störungen.* Archiwum Mechaniki Stosowanej (Warszawa) **14**, 621-637 (1962) — With R. SCHWARZENBERGER: *Übungen zur Gasdynamik.* Springer, Vienna (1963) — *Grundzüge der Hyperschallströmung.* ZAMM **43**, T152 (1963) — *Quellen in schallnaher Strömung.* In: (K. OSWATITSCH, ed.) *Symp. Transsonicum* [Aachen, Germany, Sept. 1962]. Springer, Berlin (1964), pp. 402-413 — *Zwischenballistik.* DLR-Forschungsbericht Nr. 64-37 (1964) — *Analytische Berechnung von Charakteristikenflächen bei Strömungsvorgängen.* DLR-Forschungsbericht 65-62 (1965) — With Y.C. SUN: *The wave formation and sonic boom due to a delta wing.* Aeronaut. Quart. **23**, 87-108 (1972) — *Spezialgebiete der Gasdynamik.* Springer, Vienna (1977) — *Wellenausbreitung in Flügeln.* Z. Flugwiss. u. Weltraumforsch. **3**, 149-156 (1979).

**SECONDARY LITERATURE.** Klaus OSWATITSCH. In: *150 Jahre Technische Hochschule Wien.* TU Vienna (1965), vol. 2, p. 521 — W. SCHNEIDER: *Klaus OSWATITSCH 70 Jahre.* ZAMM **60**, 220 (1980) — E. LEITER and J. ZIEREP (eds.) *Übersichtsbeiträge zur Gasdynamik.* In: *Symp. über Gasdynamik anlässlich des 60. Geburtstags von Prof. K. OSWATITSCH* [Vienna, Austria, March 1970]. Springer, Vienna (1971) — W. SCHNEIDER: *Klaus OSWATITSCH: His scientific career and work.* In: (W. SCHNEIDER and M. PLATZER, eds.) *K. OSWATITSCH: contributions to the development of gas dynamics* [Selected papers, translated on the occasion of K. OSWATITSCH's 70th birthday]. Vieweg, Braunschweig (1980) — *Prof. Dr. Dr. e.h. Klaus OSWATITSCH zum 75. Geburtstag.* Institut für Theoretische Strömungsmechanik, DFVLR-AVA Göttingen (1985) — W. SCHNEIDER: *Obituary for Professor Klaus OSWATITSCH.* Fluid Dyn. Res. **13**, 65-66 (1993) — A. KLUWICK: *Nachruf für OSWATITSCH.* ZAMM **74**, 223-224 (1994); *Klaus OSWATITSCH 1910–1993.* Mem. Trib. Nat. Acad. Engng. **7**, 171-174 (1994).

**PICTURE.** Courtesy Archives of TU Vienna, Austria.

# PAPIN, Denis [Lat. *Dionysius*] (1647–1712?)

- French-born British natural philosopher and inventor; father of the steam and piston engine

Denis PAPIN was born in Blois (Dépt. Loir-et-Cher) the son of Denis PAPIN, a royal counselor and district revenue collector. After attending a local Jesuit school, he graduated (M.D.) from the medical faculty of the University of Angers (1669). However, shortly afterwards he embarked upon a scientific career and worked closely with renowned scientists of his time. In Paris he became an assistant (1671–1674) to the Dutch natural philosopher Christiaan HUYGENS, who had been invited by the Paris Academy to develop his many scientific interests under French protection. In London he worked with Robert BOYLE on experiments connected with respiration, magnetism, air, and the chemistry of blood and various medicaments (1675–1679). At the Royal Society of London PAPIN became assistant (1679–1680) to Robert HOOKE. After briefly working again with HUYGENS in Paris (1680), he was appointed director at Ambrose SAROTTI's Accademia Publicca di Scienze (Public Academy of Science) in Venice (1681–1684). Since the Edict of Nantes was revoked in 1685 and PAPIN belonged to a Huguenot family, he did not return to France, but rather became an exile. After working again for the Royal Society in London (1684–1687), PAPIN went to Germany. He became professor of mathematics at the University of Marburg (1687–1696), and in Kassel he served as counselor to the landgrave Karl VON HESSEN-KASSEL (1695–1707). After 20 years he eventually returned to London (1707) but did not receive an appointment from the Royal Society and died in London in poverty (around 1712).

PAPIN was a skillful experimenter and invented various machines; the best known are an air pump (1675), his "steam digester" with the first "safety valve," which he demonstrated in London to the Royal Society (1679), and an atmospheric-pressure piston steam engine for pumping water (1690), a forerunner of the low-pressure steam engine as patented by Thomas SAVERY (1698) and developed by Thomas NEWCOMEN (1712). PAPIN's safety valve was also used in many steam engines.

Prior to his steam experiments PAPIN improved and realized HUYGENS' idea of a *pompe balistique* (piston ballistic pump) which, by applying gunpowder to generate a vacuum in a combustion cylinder, used the atmospheric pressure to perform work – particularly to operate a pump to raise water from the Seine river to the palace of Versailles. PAPIN's "gunpowder machine," which he described in his first treatise *Nouvelles expériences du vuide* ("A New Vacuum Experiment," 1674) and the principle of which he had successfully demonstrated on a model, was an early example that inspired subsequent inventors to try to use the chemical energy of explosives for periodically driving machines. The abolition of a bulky steam generator appeared very attractive to reduce weight, but the principle was still too ambitious of that time. In addition, other problems, such as achieving controlled ignition, coping with cylinder erosion, and managing detrimental shock and vibration effects in such a machine, were not even touched.

In 1839, the Syndicat Général des Industries Mécaniques et Transformatrices des Métaux de France – choosing PAPIN as their "mentor," considering that mechanical engineering would not exist without the engine – created the *Papin medal*.

**ORIGINAL WORKS.** *Nouvelles expériences du vuide, avec la description des machines servant à les faire.* Cusson, Paris (1674) — *A new digester or engine for softening bones: containing the description of its make and use in these particulars: viz. cookery, voyages at sea, confectionary, making of drinks, chymistry, and dying; with an account of the price a good big engine will cost, and of the profit it will afford.* Bonwicke, London (1681) — *An account of an experiment shown before the Royal Society, of shooting by the rarefaction of the air.* Phil. Trans. Roy. Soc. Lond. **16**, 21-22 (1686) — *A continuation of the new digester: it's improvements and new uses it hath been applyed to, both for sea and land; together with some improvements and new uses of the air-pump, tryed both in England and in Italy.* Streeter, London (1687) — *La manière d'amolir les os et de faire cuire toutes sortes de viandes en fort peu de temps, & à peu de frais: avec une description de la machine dont il se faut servir pour cet effet, ses propriétés & ses usages, confirmez par plusieurs expériences.* Desbordes, Amsterdam (1688) — *De novo pulveris pyrii usu* [describes technical improvements of HUYGENS' ballistic pump]. Actis Eruditorum (Sept. 1688), p. 497 — *Nova methodus ad vires motrices validissimas levi pretio comparandas* [contains his invention of the steam engine by replacing gunpowder in the ballistic pump by steam]. Ibid. (Aug. 1690), p. 410 — For PAPIN's letter to LEIBNIZ (dated March 6, 1704) see E. GERLAND's book below — *Ars nova ad aquam ignis adminiculo efficacissime elevandam* ("The New Art of Pumping Water by Using Steam"). Francofurti 1707); *Nouvelle manière pour lever l'eau par la force du feu.* J. Estienne, Cassell (1707).

**SECONDARY LITERATURE.** L. DE SAUSSAYE and A. PEAN: *La vie et les ouvrages de Denis PAPIN.* Franck, Paris (1869) — D. ERNOUF: *Denis PAPIN, sa vie et son œuvre.* Coulommiers, Paris (1874) — E. GERLAND: *LEIBNIZens und HUYGENS' Briefwechsel mit PAPIN, nebst der Biographie PAPIN's.* Monograph publ. by the Königl. Akad. Wiss. Berlin (1881) — C. CABANES: *Denys PAPIN, inventeur et philosophe cosmopolite.* Soc. Franç. Ed. Litt. & Tech., Paris (1935) — R. JENKINS: *The heat engine, a contribution to the history of the steam engine.* Trans. Newcomen Soc. **17**, 1-11 (1936/1937) — H.W. DICKINSON: *Tercentenary of Denis PAPIN.* Nature **160**, 422-423 (1947) — H.W. ROBINSON: *Denis PAPIN (1647–1712).* Not. Rec. Roy. Soc. **5**, 48-50 (1947) — L. LEPRINCE-RINGUET (ed.) *Die berühmten Erfinder, Physiker und Ingenieure.* Aulis-Verlag Deubner, Köln (1963) — P.P. MACLACHLAN: *D. PAPIN.* In: (C.C. GILLESPIE, ed.) *Dictionary of scientific biography.* Scribner, New York, vol. 10 (1974), pp. 292-293 — J.L. HEILBRON: *Physics at the Royal Society during NEWTON's presidency.* William Andrews Clark Memorial Library, UCLA, Los Angeles (1983) — C.A. KLEIN: *Denis PAPIN: illustre savant blaisois.* Armand, Chambray-les-Tours (1987) — (G. SCHNEIDER and H. ACKERMANN, eds.) *300 Jahre Denis PAPIN: Naturforscher und Erfinder in Hessen* [Denis Papin Ausstellung, Marburg (Aug. 1987)]. ISBN: 3-8185-0013-4, Universitätsbibliothek Marburg (1987) — A.G. RANEA: *Denis PAPIN (1647–1712?): National hero, servile technician, or natural philosopher.* History of Science Society (HSS) Annual Meeting, Halifax, Nova Scotia (1998) — A. MCCONNELL: *PAPIN, Denis.* In: (H.C.G. MATTHEW, ed.) *Oxford dictionary of national biography.* Oxford University Press, Oxford (2004), vol. 42, pp. 597-599.
**PICTURE.** Courtesy Bildarchiv Foto Marburg, Germany; image No. 206282.
**NOTE.** A computer animation showing the operation cycles of PAPIN's ballistic pump, the basic design for early steam engines, can be watched in the Internet; see http://www.geocities.com/Athens/Acropolis/6914/pappe.htm.

# PAYMAN, William Henry (1896–1946)

▪ British physical chemist and combustion specialist

William H. PAYMAN was educated at the Universities of Manchester and Sheffield. After graduating at the Manchester College of Technology (1915), he began a postgraduate study of the inflammation of gas mixtures under the guidance of Dr. Hubert Frank COWARD, with whom he was again to become associated in later years. He became a demonstrator in chemistry and later joined the staff of the Home Office Experimental Station at Eskmeals, Cumberland (1917), an organization directed by Prof. Richard V. WHEELER and devoted to the examination of explosion hazards in coal mines which later expanded into the Safety in Mines Research Board (SMRB), with stations at Buxton and Sheffield. In the period 1919–1922, he formulated his "law of flame speed"

in mixed gases, which brought him into a sharp controversy with Prof. William A. BONE, head of the chemistry department at the Imperial College and a renowned fuel technologist. Later he became principal officer of the SMRB (1926) and transferred his attention to the safe use of explosives in coal mines.

Little was known about the hazard of explosives when fired in contact with firedamp and which mechanism for triggering firedamp explosion would be dominant, the contact with flame or by adiabatic compression of the shock wave. To investigate these problems experimentally in greater detail, PAYMAN's team developed the "flame speed camera," an instrument that was used chiefly for photographing slowly moving and feebly actinic flames, and his "wave-speed camera," which allowed photography of all kinds of flames and shock waves. His original contributions to this new field of research were recognized by the award of the D.Sc. degree by the University of Manchester (1929). In his address to the participants of the First Conference on Safety in Mines (1931), which took place at Buxton Research Station, the seat of his laboratory, he summarized the present state of his applied methods and the experimental results thus obtained. Topics of practical importance were the two questions (1) whether a shock wave alone could start an explosion in a firedamp/air mixture without the aid of the flame or hot gases from an explosion and (2) why an explosive sometimes does and sometimes does not ignite firedamp. Later he supervised research on mining explosives and during most of the war conducted a large program of work on explosives that the SMRB carried out for the Ministry of Supply. In 1940, PAYMAN and coworkers performed a number of interesting shock-tube studies: they measured the shock front velocities for various combinations of driver gas and operation gas and found that stronger shocks could be achieved with a combination of hydrogen/air rather than with one with air/air. They were the first to observe a vortex ring generated by diffraction of a shock wave from the open end of a shock tube (not published until 1946).

PAYMAN's contributions are recorded in numerous papers in the Journal of the Chemical Society. He was also coauthor with Prof. I.C. Frank STATHAM of a monograph on *Mine Atmospheres* (1930), which addresses the hazards of firedamp explosions and mine fires originating from methane and carbon monoxide, and the role of mine dust and moisture. He achieved international recognition in 1938 when he was appointed president of the Explosives Section of the Congress of Applied Chemistry held at Nancy, France. He was also for some years secretary of the informal Explosives in Mines Research Committee of the SMRB. He died at the age of 50 after a short illness.

**ORIGINAL WORKS.** With R.V. WHEELER: *The propagation of flames through tubes of small diameter.* J. Chem. Soc. (Lond.) **113**, 656-666 (1918); Ibid. **115**, 36-45 (1919) — With H.F. COWARD and C.W. CARPENTER: *The dilution limits of inflammability of gaseous mixtures. Part III. The lower limits of some mixed inflammable gases with air. Part IV. The upper limits of some gases, singly and mixed in air.* Ibid. **115**, 27-36 (1919) — With R.V. WHEELER: *The combustion of complex mixtures. Part I.* Ibid. **121**, 363-379 (1922); *Part II. Mixtures of carbon monoxide and hydrogen with air.* Ibid. **123**, 1251-1259 (1923) — With W.C.F. SHEPHERD: *The pressure waves sent out by an explosive. Part I.* SMRB Paper No. 18 (1923); *Part II.* Ibid. No. 28 (1926) — *The wave-speed camera.* Ibid. No. 29 (1926) — With G.B. MAXWELL and R.V. WHEELER: *Part III. The inflammation of mixtures of carbon monoxide and hydrogen with air in a closed vessel.* J. Chem. Soc. (Lond.) **130**, 297-310 (1927) — *The study of moving flame.* Ibid. **131**, 1738-1740 (1928) — *The detonation wave in gaseous mixtures and the pre-detonation period.* Proc. Roy. Soc. Lond. **A120**, 90-109 (1928) — With I.C.F. STATHAM: *Mine atmospheres.* Methuen, London (1930) — *Firedamp explosions and their prevention.* Oxford University Press, Oxford (1931) — With D.W. WOODHEAD: *Explosion waves and shock waves. Part I: The wave-speed camera and its application to the photography of bullets in flight.* Proc. Roy. Soc. Lond. **A132**, 200-213 (1931) — *The application of schlieren photography in researches on explosives.* In: Address to the 1st Int. Conf. on Safety in Mines [Buxton, U.K., July 1931] — With H.C. GRIMSHAW: *The ignition of firedamp by coal-mining explosives 1-gallery experiments.* SMRM Paper No. 69 (1931) — With R.V. WHEELER: *Flame speeds during the inflammation of moist carbonic oxide-oxygen mixtures.* J. Chem. Soc. (Lond.) **135**, 1835-38 (1932) — With D.W. WOODHEAD: *Photographic methods for measuring velocities of explosion waves and shock waves.* Fuel Sci. Pract. (Lond.) **11**, 435-440 (1932) — With D.W. WOODHEAD: *The pressure wave sent out by an explosive.* SMRB Paper No. 88 (1934) — With R.V. WHEELER: *Speed of uniform movement of flame in mixtures of carbon monoxide and oxygen.* Nature **133**, 257 (1934) — With D.W. WOODHEAD and H. TITMAN: *Explosion waves and shock waves. Part II: The shock wave and explosion products sent out by blasting detonators.* Proc. Roy. Soc. Lond. **A148**, 604-622 (1935) — With H. TITMAN: *Explosion waves and shock waves. Part III: The initiation of detonation in mixtures of ethylene and oxygen and of carbon monoxide and oxygen.* Ibid. **A152**, 418-445 (1935) — With R.V. WHEELER: *Flame speeds during the inflammation of moist carbonic oxide – oxygen mixtures.* Nature **136**, 1028 (1935) — With C.A. NAYLOR and R.V. WHEELER: *The ignition of firedamp by coal mining explosives.* SMRB Paper No. 90 (1935), *Part II: Sheathed explosives* — With H. TITMAN: *Limits of inflammability of hydrogen and deuterium in oxygen and in air.* Nature **137**, 190 (1936) — *The properties and uses of black powder.* Mine & Quarry Engng. (Lond.) **1**, 249-254 (1936) — With W.C.F. SHEPHERD: *Explosion waves and shock waves. Part IV: Quasi-detonation in mixtures of methane and air.* Proc. Roy. Soc. Lond. **A158**, 348-367 (1937) — With D.W. WOODHEAD: *Explosion waves and shock waves. Part V: The shock wave and explosion products from detonating solid explosives.* Ibid. **A163**, 575-592 (1937) — With W.C.F. SHEPHERD and D.W. WOODHEAD: *High speed cameras for measuring the rate of detonation in solid explosives.* SMRB Paper No. 99 (1937) — *The classification of permitted explosives.* Colliery Engng. (Lond.) **14**, 21-23 (1937) — With H.F. COWARD: *Problems in flame propagation.* Chem. Rev. **21**, 359-366 (1937) — With R.V. WHEELER: *The ignition of firedamp by coal-mining explosives.* Trans. Instit. Min. Eng. (Lond.) **95**, 13-47 (1937/1938) — *The safe use of explosives.* Colliery Engng. (Lond.) **15**, 121-123 (1938) — *Les explosives de mines et leur mécanisme de détonation.* Chimie Industrie Génie Chimique (Paris) **40**, 638-648 (1938) — With R.V. WHEELER: *The ignition of firedamp by coal-mining explosives.* Colliery Guardian and J. Coal Iron Trades (Lond.) **156**, 201-204, 252-255

(1938) — *Mining explosives and their behavior on detonation.* **16**, 52-95 (1939) — *The testing of permitted explosives. Basis of the official test.* Iron Coal Trades Rev. (Lond.) **142**, 157-158 (1941) — With W.C.F. SHEPHERD: *Explosion waves and shock waves. Part VI: The disturbance produced by bursting diaphragms with compressed air.* Proc. Roy. Soc. Lond. **A186**, 293-321 (1946) — With W.B. CYBULSKI and D.W. WOODHEAD: *Explosion waves and shock waves. Part VII: The velocity of detonation in cast TNT.* Ibid. **A197**, 51-72 (1949).

**SECONDARY LITERATURE.** F.V. TIDESWELL: *Dr. William PAYMAN.* Nature **158**, 441-442 (1946).

**PICTURE.** Courtesy Mr. Mike EGGENTON, Health & Safety Laboratory, Sheffield, U.K. The photo shown is an enlargement of a group photo taken at the 1st Int. Conf. on Safety in Mines at Buxton, U.K. (July 1931).

# POISSON, Siméon-Denis (1781–1840)

- French mathematician, physicist, and astronomer; early pioneer of nonlinear acoustics

Siméon-Denis POISSON was born in Pithiviers, a small town near Orléans (Dépt. Loiret). His father, Siméon POISSON, had first chosen a military career and later became a notary. Brought up in a family of modest means, he was originally supposed to apprentice in surgery but showed little interest in this profession. At the Ecole Centrale de Fontainebleau his teacher discovered his interest in mathematics. He entered the prestigious Ecole Polytechnique, Paris (1798) and his teachers, Pierre-Simon DE LAPLACE and Joseph L. DE LAGRANGE, were impressed by his abilities and became his lifelong friends. After graduation (1800) he taught at the Ecole Polytechnique, and, drawn to the problems of integrating differential and partial differential equations, he began to look for applications in physics.

It appears that POISSON, stimulated and supported by DE LAPLACE, was the first to correctly recognize that sound is an adiabatic process, and that the adiabatic equation of state for a gas is given by $pV^\gamma = $ const, the so-called "Poisson law" or "Poisson isentrope." Here $\gamma$ is the ratio of specific heats at constant pressure and volume. Using this adiabatic law he worked out the first theory of sound of finite amplitude (1808), thus laying the theoretical foundations of nonlinear acoustics and paving the way for James CHALLIS, George STOKES, Samuel EARNSHAW and Bernhard RIEMANN to mathematically investigate waves of finite amplitude (*i.e.,* shock waves) in more detail. Shortly afterwards, he acted as an astronomer at the Bureau des Longitudes and was appointed chair of pure mathematics in the newly opened Faculté des Sciences (1809). As a member of the Conseil Royal de l'Université (1820–1840) he was intimately familiar with actual problems of national education at the highest administrative level.

He contributed significantly to the theory of differential equations and partial differential equations, game theory, and probability and today is regarded as one of the most eminent founders of modern mathematical physics. Being mostly concerned with applications of mathematics to physics, his numerous investigations advanced the progress of physics of his time, particularly in mechanics (elasticity), hydrodynamics (waves in deep water), heat conduction, electricity, celestial mechanics, gravitation, and cosmology. He also originated one of the first mathematical theories of electrostatics and magnetism. In his *Traité de mécanique* (1811, 1833), which became the standard work in mechanics for many years, he treated billiards, percussion of bodies of arbitrary geometry, and longitudinal percussion of elastic bars (*see* vol. II, chaps. 7 and 8 of his book). Stimulated by the discovery of the Coriolis effect (1835), POISSON made an analysis on the deflection of artillery shells. He ruled out any effect on a swinging pendulum which, however, was refuted by FOUCAULT's historical pendulum experiment (1851).

POISSON published around 350 papers and received many prestigious awards such as the Copley Medal (1832). He was a Fellow of the Royal Societies of London (1818) and Edinburgh (1820). He was a member of the French Academy of Science and all the scientific societies in Europe and America and an honorary member of the St. Petersburg Academy of Sciences.

In the theory of elasticity, the *Poisson ratio* is the ratio of transverse contraction strain to longitudinal extension strain in the direction of stretching force. In applied mathematics the *Poisson integral* is used in boundary-value problems. In electricity the *Poisson equation*, a partial differential equation, relates the potential to the distribution of charges; for many charges distributed randomly it allows one to calculate the dependency of the potential on the coordinates. The Poisson eqution is also used in mechanical engineering and theoretical physics. In probability theory he derived a formula known as the "Poisson distribution." A *Poisson process* is a stochastic process which is defined in terms of the occurrence of events. The *Poisson regression model* is often used to analyse count data and is fitted as a log-linear regression.

Astronomers named a crater on the near side of the Moon and a minor planet (asteroid 12874 POISSON) after him.

**ORIGINAL WORKS.** *Mémoire sur les solutions particulières des équations différentielles et les équations aux différences.* J. Ecole Polytech. **7**, No. 13, 60-125 (1806) — *Mémoire sur la théorie du son.* Ibid. **7**, No. 14, 319-392 (1808) — *Traité de mécanique.* 2 vols., Bachelier, Paris (1811, 1833) — *Mémoire sur la théorie des ondes.* Mém. Acad. Sci. Inst. France **1**, 71-186 (1816) — *Sur le mouvement des fluides élastiques dans les tuyaux cylindriques, et sur les théories des instruments à vent.* Ibid. **2**, 305-402 (1817) — *Sur les rapports qui existent entre la propagation des ondes à la surface de l'eau et leur propagation dans une plaque élastique.* Bull. Soc. Philom. Paris (1818), pp. 97-99 — *Mémoire sur la théorie des ondes* (1815). 4 vols., F. Didot, Paris (1818) — *Mémoire sur la propagation du mouvement dans les fluides élastiques.* Ann. Chim. Phys. **22**, 246-270 (1823) — *Sur la vitesse du son.* Ibid. **23**, 5-15 (1823) — *Sur la chaleur des gaz et des vapeurs.* Ibid. **23**, 337-353 (1823) — *Effets du tir d'un canon sur son affut; force et durée du recul.* Bull. Soc. Philom. Paris (1826), pp. 4-6 — *Mémoire sur l'équilibre et le mouvement des corps élastiques.* Ann. Chim. Phys. **37**, 337-355 (1828); Ibid. **42**, 145-171 (1829) — *Note sur le problème des ondes.* Mém. Acad. Sci. Inst. France **8**, 571-580 (1829) — *Mémoire sur l'équilibre des fluides.* Ibid. **9**, 1-87 (1830) — *Sur les équations générales de l'équilibre et du mouvement des corps solides élastiques et des fluides.* J. Ecole Polytech. **13**, No. 20, 1-174 (1831) — *Mémoire sur la propagation du mouvement dans les milieux élastiques.* Mém. Acad. Sci. Inst. France **10**, 549-606 (1831) — *Formules relatives au mouvement du boulet dans l'intérieur du canon, extraites des manuscrites de LAGRANGE.* J. Ecole Polytech. **13**, No. 21, 187-204 (1832) — *Théorie mathématique de la chaleur.* Ann. Chim. Phys. **59**, 71-103 (1835) — *Mémoire sur le mouvement des projectiles dans l'air en ayant égard à la rotation de la terre.* J. Ecole Polytech. **16**, No. 26, 1-69 (1838) — *Deux mémoires sur le mouvement des projectiles dans l'air en ayant égard à leur rotation.* Ibid. **16**, No. 26, 69-176 (1838); No. 27, 1-50 (1838).

**SECONDARY LITERATURE.** Anonymous: *On POISSON's investigation on the theory of sound.* Phil. Mag. **1** [IV], 410-411 (1851) — F. ARAGO: *POISSON* [Biographie liée par extraits en séance publique de l'Académie des Sciences, le 16 Déc. 1850]. In: (J.A. BARRAL, ed.) *Œuvres complètes de François ARAGO.* Gide & Baudry, Paris, vol. II (1854): *Notices biographiques*, pp. 593-698, 672-689 — H. BATEMAN: *A partial differential equation associated with POISSON's work on the theory of sound.* Am. J. Math. **60**, No. 2, 293-296 (April 1938) — I. SZABÓ: *Geschichte der mechanischen Prinzipien und ihre wichtigsten Anwendungen.* Birkhäuser, Basel (1977), pp. 460-462 — C.B. BOYER: *A history of mathematicians.* Wiley, New York (1968) — P. COSTABEL: *S.D. POISSON.* In: (C.C. GILLESPIE, ed.) *Dictionary of scientific biography.* Scribner, New York, vol. 15 (1978), pp. 480-490 — N. ROTT: *Nichtlineare Akustik: Rückblick und Ausblick.* Z. Flugwiss. u. Weltraumforsch. **4**, No. 4, 185-193 (1980) — D.H. ARNOLD: *POISSON and mechanics.* In: (M. METIVIER and P. COSTABEL, eds.) *Siméon-Denis POISSON et la science de son temps.* L'Ecole Polytechnique, Palaiseau (1981) — D.H. ARNOLD: *The mécanique physique de Denis POISSON: the evolution and isolation in France of his approach to physical theory (1800-1840).* Arch. Hist. Exact. Sci. **28**, 243-367 (1983); Ibid. **29**, 37-94, 287-307 (1984) — B. GELLER and Y. BRUK: *A portrait of POISSON.* Quantum (March/April 1991), pp. 21-25.
**PICTURE.** © Collections Ecole Polytechnique, Paris, France.

# POULTER, Thomas Charles (1897–1978)

- U.S. physicist, engineer, biosonic researcher, and inventor

Thomas Ch. POULTER was born and raised 8 miles south of Mt. Pleasant on his father's farm near Salem in the State of Iowa. His father, Micajah Louis POULTER, was a farmer and mechanic who operated a blacksmith shop part time in Salem. He graduated from Iowa Wesleyan College (1923) and was appointed professor of physics at that college (1923–1933), where he served as head of the chemistry and physics departments and the physical sciences, mathematics, and astronomy division. His most prominent student was James A. VAN ALLEN, the renowned U.S. astrophysicist, who worked as a part-time student assistant in his high-pressure research laboratory. He completed graduate work at the University of Chicago, receiving his Ph.D. in chemistry (1933). Sound and vibrations, and high pressures, both static and dynamic, were the constant unifying motifs of his scientific curiosity throughout his life. His other interests included organic chemistry, terrestrial magnetism, and Antarctic meteorological and auroral phenomena.

In the fall of 1932, POULTER took a group to the southwest United States to explore the activity of meteors. During Admiral Richard BYRD's Antarctic Expedition over the South Pole (1933–1935) he participated as second in command and chief scientist. He performed explosive seismic surveying, and using sky "reticles," which he devised and built from welding rods, he obtained the world's most comprehensive sets of observations of meteor trails at that time. He became first scientific director of the Armour Research Foundation of the Illinois Institute of Technology (IIT) in Chicago (1936–1948), where he designed the Antarctic snow cruiser used on the U.S. Antarctic Service Expedition, on which he served as scientific advisor (1939–1940). He made a total of 15 trips to the Artic and 3 to the Antarctic. As a result of techniques discovered in the Antarctic seismic studies he invented a seismic method to map underground strata by analyzing reverberations from surface explosions, so-called "air shooting" or "Poulter seismic method." It uses explosive charges detonated in the air or on poles above the ground as the source. Initially applied to measuring ice thickness on Taku Glacier, Alaska (1949), it was later also applied in other geophysical explorations.

After World War II, he conducted considerable research on detonation (such as the construction and use of shaped charges for oil well completion) and on shock pulse phenomena. His expertise in detonating explosives led to military research on shock wave and detonation phenomena related to the development of new types of nuclear weapons. He joined the Stanford Research Institute (SRI) at Menlo Park, CA as an associate director (1948) and established the Explosives and Extreme Pressure Laboratory of SRI (1953), which in 1956 was named after him ("Poulter Laboratory") for his contributions in the fields of detonation and shock pulse phenomena. He also established the Calaveras Test Site, an explosives test facility in the mountains in Calaveras County, CA. POULTER became scientific director and general manager of SRI's physical and life sciences divisions (1960). After his retirement he developed the Biological Sonar Laboratory, a private research center in Fremont Hills, CA, where he studied how marine mammals – such as seals and sea lions – hear and how they use acoustic information (1962–1973). Subsequently, he cooperated with the UC Medical School in San Francisco in a project to restore hearing in totally deaf people.

POULTER held more than 75 patents on diverse inventions. He authored more than 100 articles, books, and publications and received the Geographic Medal (1937) of the National Geographic Society and two Congressional Medals of Honor for polar exploration. From 1973 he was an honorary member of the American Polar Society.

**ORIGINAL WORKS.** *Apparatus for optical studies at high pressure.* Phys. Rev. **40** [II], 860-871 (1932) — With R.O. WILSON: *Permeability of glass and fused quartz to ether, alcohol, and water at high pressure.* Ibid. **40**, 877-880 (1932) — With F. BUCKLEY: *Diamond windows for withstanding very high pressures.* Ibid. **41**, 364-365 (1932) — With L. UFFELMAN: *The penetration of hydrogen through steel at four thousand atmospheres.* Physics (Minneapolis) **3**, 147-148 (1932) — With G.E. FRAZER: *A study of the action of acids on zinc at pressures of from one to thirty thousand atmospheres.* J. Phys. Chem. **38**, 1131-1140 (1934) — *The application of seismic methods in the discovery of new lands in the Antarctic.* Armour Engineer & Alumnus (March 1937) — *Use of extreme pressure in the investigation of lubrication.* Oil Gas J. (Dec. 1937), pp. 46, 51 — *The study of extreme pressures and their importance in the investigation of engineering problems.* J. Appl. Phys. **9**, 307-311 (1938) — *Meteor observations in the Antarctic.* Griffith Observer **3**, No. 11 (1939) — *Research on high pressures.* Scient. Am. **161**, 202-203 (Oct. 1939) — *The mechanism of cavitation erosion.* J. Appl. Mech. **9**, A31-37 (1942) — *High pressure apparatus.* U.S. Patent No. 2,554,499 (May 1947) — *Seismic exploration.* U.S. Patent No. 2,615,523 (July 1949) — *Bomb cluster.* U.S. Patent No. 2,972,946 (July 1950) — *The polar seismic methods. Oilfield techniques in the study of polar ice caps, and the contribution of glacier techniques in the discovery of oil and minerals.* Ann. Gen. Meeting, Toronto, Ont. (April 1950). Trans. LIII (1950). Can. Min. Metallurg. Bull. (1950) — *The Poulter seismic method of geophysical exploration.* Geophys. **15**, 181-207 (1950) — *Geophysical studies in the Antarctic.* SRI, Menlo Park, CA (March 1952) — With L.V. LOMBARDI: *Multiple reflections on the Edwards Plateau.* Geophys. **17**, No. 1, 107-115 (1952) — *The transmission of shock pulses through homogeneous and non-homogeneous air and possible damage to building structures from moderately small explosive charges.* SRI, Menlo Park, CA (Sept. 9, 1955) — *Meteor observations in the Antarctic.* Rept. Byrd Antarctic Expedition II, 1933–1935. Menlo Park, CA (1955) — *Controlled fracturing of solids by explosion.* U.S. Patent No. 3,076,408 (Jan. 1955) — *Thermal fusion by opposing Mach 10 detonation fronts.* Tech. Rept. GU-960 (1958) — *The development of shaped charges for oil well completion.* Petroleum Trans. **210**, 11-18 (1957) — *Multiple-jet shaped explosive charge perforating device.* U.S. Patent No. 3,013,491 (Oct. 1957) — *Systems of echolocation.* In: (R.G. BUSNEL, ed.) *Les systèmes sonars animaux. Biologie et bionique.* Frascati (Sept. 1966) — *Biosonar.* In: *McGraw-Hill yearbook of science and technology.* McGraw-Hill, New York, vol. 6 (1966), pp. 11-19 — *Over the years* [autobiography and memoirs]. SRI, Menlo Park, CA (1978).
**SECONDARY LITERATURE.** R.E. BYRD: *Alone.* Putnam, New York (1938) — Anonymous: *POULTER retires.* Research for Industries (SRI, Menlo Park, CA) **14**, No 3 (May/June 1962) — J. GEORGI: *Erinnerung an einen verdienten Antarktis-Geophysiker* [Dr. Thomas POULTER]. Polarforsch. **V**, 159-160 (Oct. 1963) — *SRI staff members honor POULTER.* Intercom (Office of Public Relations, SRI) No. 24 (Oct. 26, 1973), p. 3 — E. BLAU: *Thomas POULTER, 81, polar explorer.* The New York Times (June 17, 1978) — Anonymous: *Thomas C. POULTER, commemoration and recognition service (July 20, 1978).* SRI, Menlo Park, CA (1978) — L.A. HASELMAYER: *Thomas C. POULTER: a Renaissance man (1897–1978). A tribute.* Iowa Wesleyan College 1978 Opening Convocation. Chapel Auditorium, Mount Pleasant, IA (Sept. 7, 1978).
**PICTURE.** From SRI-Journal *Spectrum*, SRI International, Menlo Park, CA (July/Aug. 1978).

# PRANDTL, Ludwig (1875–1953)

• German engineer and physicist; father of modern fluid dynamics and aerodynamic theory

Ludwig PRANDTL was born in Freising, Upper Bavaria. His father, Alexander PRANDTL, was a professor of surveying and engineering at the Weihenstephan Agricultural College. He showed an early interest in technical matters. After studying mechanical engineering at the Technische Hochschule (TH) Munich (1894–1898), he took his Ph.D. (1900) at the University of Munich with a thesis on elastic stability, entitled *Kipperscheinungen, ein Fall von instabilem elastischen Gleichgewicht* ("Tipping Phenomena, a Case of Instable Elastic Equilibrium"). His thesis supervisor was August FÖPPL, an internationally renowned expert in mechanics who later became his father-in-law. As a scientific employee

(1900–1901) at the Maschinenfabrik Augsburg-Nürnberg (MAN) he solved practical problems such as how to remove shavings from machine tools by suction, and thereby had his first exposure to fluid mechanics, particularly aerodynamics – a field to which he dedicated the rest of his life.

After successfully solving the problems at MAN he became shortly thereafter professor at the TH Hannover (1901). Later, in pursuit of Felix KLEIN, he became *Extraordinarius* (professor without chair) of technical physics (1904) and *Ordinarius* (professor with chair) of applied mechanics (1907–1947) at the University of Göttingen, where he continued working as professor emeritus until his death. In 1907, he initiated the foundation of the Aerodynamische Versuchsanstalt (AVA) Göttingen. In 1925, PRANDTL was appointed director of the newly established Kaiser-Wilhelm-Institut (KWI) für Strömungsforschung. To scientifically support the German aircraft industry he was cofounder of the Wissenschaftliche Gesellschaft für Luftfahrt Berlin (1912) and became member of its executive board. PRANDTL presented in 1904 a paper *Über Flüssigkeitsbewegung bei sehr kleiner Reibung* ("On the Motion of a Fluid with a Very Small Viscosity"), which marked an epoch in the history of fluid mechanics: his discovery of the "boundary layer" that adjoins the surface of a body moving in a fluid led to an understanding of skin friction drag and of the way in which streamlining reduces the drag of airplane wings and other moving bodies. He also supervised the installation of Germany's first wind tunnel (1907–1909), a closed-circuit facility intended for aerodynamic testing of Zeppelin models that had a cross section of $2 \times 2$ m$^2$, a maximum speed of 36 km/h (10 m/s), and was operated at a power of 35 hp. After development of a more powerful, 300-hp wind tunnel (1915–1917), a full-scale device with a maximum speed of 50 m/s was eventually completed for propeller and air wing research (1933).

His numerous publications cover a wide range of research in fluid mechanics: he contributed to low-speed airfoil theory as well as to the theory of supersonic flow, first visualized oblique shock and expansion waves in Laval nozzles, investigated induced wing drag and the role of boundary vortices, studied compressibility effects at high-speed flight,

and invented the *Prandtl tube* (1913), which allows one to determine the dynamic pressure. He introduced into fluid dynamics the *Prandtl number* (1910), the ratio of kinematic viscosity and thermal diffusivity, a dimensionless quantity which became widely used in momentum and heat-transfer calculations. To explain turbulent fluxes he devised the *Prandtl mixing length* (1925), an average distance of air parcel turbulent movement toward a reference height. Together with the British aeronautical engineer Hermann GLAUERT he worked out the *Prandtl-Glauert rule* (1928) for subsonic airflow to describe compressibility effects of air at high speeds. The principle of his *Prandtl wind tunnel* or Göttingen-type wind tunnel (1908), a continuous-flow closed-circuit wind tunnel, was taken up by Jakob ACKERET for the construction of the first Swiss wind tunnel at ETHZ. In addition, he devised the soap-film analogy for the torsion of noncircular sections and wrote on the theory of plasticity and of meteorology.

PRANDTL received many honors and medals, such as the Daniel Guggenheim Gold Medal of the Daniel Guggenheim Fund for the Promotion of Aeronautics, the German National Prize for Art and Science, the Golden Medal of the Royal Aeronautical Society, the Lilienthal Medal of the Wissenschaftliche Gesellschaft für Luftfahrt (WGL), and the Grashof Medal of the Verein deutscher Ingenieure (VDI). His most prominent students included Jakob ACKERET, Albert BETZ, Adolf BUSEMANN, and Theodore VON KÁRMÁN. Numerous national and foreign universities honored him with a honorary doctorate. PRANDTL was a member or honorary member of a number of national and foreign societies devoted to science, engineering, and culture.

Together with Richard VON MISES, an Austrian applied mathematician, he founded in 1922 GAMM, the Gesellschaft für angewandte Mathematik und Mechanik (Society for Applied Mathematics and Mechanics). The annual *Ludwig-Prandtl-Gedächtnis-Vorlesung* (Ludwig Prandtl Memorial Lecture) of the Deutsche Gesellschaft für Luft- und Raumfahrt (German Society for Air and Space Travel) is hosted by GAMM. The *Ludwig-Prandtl-Ring* is the highest distinction awarded by the Deutsche Gesellschaft für Luft- und Raumfahrt Lilienthal-Oberth e.V. for outstanding achievements in aeronautical science and all of its disciplines.

A crater on the far side of the Moon is named for him.

**ORIGINAL WORKS.** With A. PRÖLL: *Beiträge zur Theorie der Dampfströmung durch Düsen.* VDI-Z. **48**, 348-350 (1904) — *Über stationäre Wellen in einem Gasstrahl.* Physik. Z. **5**, 599-601 (1904) — *Über Flüssigkeitsbewegung bei sehr kleiner Reibung.* In: (A. KRAZER, ed.) *Verhandl. III. Int. Mathematiker-Kongress* [Heidelberg, Germany, Aug. 1904]. Teubner, Leipzig (1905), pp. 484-491 — *Strömende Bewegung der Gase und Dämpfe.* In: *Encyclopädie der math. Wissenschaften.* Teubner, Leipzig (1903–1921), vol. V,

1 (1903), pp. 287-319 — *Zur Theorie des Verdichtungsstoßes.* Z. ges. Turbinenwesens **3**, 241-245 (1906) — *Neue Untersuchungen über die strömende Bewegung der Gase und Dämpfe.* Physik. Z. **8**, 23-30 (1907) — *Eine Beziehung zwischen Wärmeaustausch und Strömungswiderstand der Flüssigkeiten.* Ibid. **11**, 1072-1078 (1910) — *Gasbewegung.* In: (E. KORSCHELT, ed.) *Handwörterbuch der Naturwissenschaften.* Fischer, Jena, vol. 4 (1913), pp. 544-560 — *Der Luftwiderstand von Kugeln.* Nachr. Akad. Wiss. Göttingen [Math.-Phys. Kl.]. Vandenhoeck & Ruprecht, Göttingen (1914), pp. 177-190 — *Bemerkungen über den Flugzeugschall.* Z. Tech. Phys. **2**, 244-245 (1921) — *Die Aerodynamische Versuchsanstalt der Kaiser-Wilhelm-Gesellschaft und ihre Bedeutung für die Technik.* Die Naturwissenschaften **8**, 169-176 (1922) — *Theory of discontinuous fluid motion and facts.* Physik. Z. **29**, 118-119 (1928) — With A. BUSEMANN: *Näherungsverfahren zur zeichnerischen Ermittlung von ebenen Strömungen mit Überschallgeschwindigkeit* [Festschrift zum 70 Geburtstag von Prof. A. STODOLA]. Füßli, Zürich (1929), pp. 499-509 — *Über Strömungen, deren Geschwindigkeiten mit der Schallgeschwindigkeit vergleichbar sind.* J. Aeronaut. Res. Inst. (Tokyo Imperial University) **5**, 25-34 (1930) — *Über die Entstehung der Turbulenz.* ZAMM **11**, 407-409 (1931) — *Über Tragflügel kleinsten induzierten Widerstandes.* Z. Flugtech. u. Motor-Luftschiffahrt **24**, 305-306 (1933) — With O.G. TIETJENS: *Fundamentals of hydro- and aeromechanics. Applied hydro- and aeromechanics* (Based on lectures of L. PRANDTL). McGraw-Hill, New York (1934) — *Description of wind tunnels constructed since 1871.* In: (N.H. RANDERS-PEHRSON, ed.) *Pioneer wind tunnels.* Smithsonian Inst. Miscellaneous Coll. **93**, No. 4 (1935) — *Short autobiography.* In: *V Convegno Volta su "Le alte velocità in aviazione"* [Rome, Sept./Oct. 1935]. Reale Accademia d'Italia, Roma (1936), p. 679 — *Allgemeine Betrachtungen über die Strömung zusammendrückbarer Flüssigkeiten.* ZAMM **16**, 129-142 (1936) — *Theorie des Flugzeugtragflügels im zusammendrückbaren Medium.* Luftfahrtforsch. **13**, 313-319 (1936) — *Die Rolle der Zusammendrückbarkeit bei der strömenden Bewegung der Luft.* Schriften Dt. Akad. Luftfahrtforsch. Nr. 30, 1-16 (1937) — *Zur Berechnung der Grenzschichten.* ZAMM **18**, 77-82 (1938) — *Über Schallausbreitung bei rasch bewegten Körpern.* Schriften Dt. Akad. Luftfahrtforsch. Nr. 7, 1-14 (1939) — *Bemerkungen zur Theorie der freien Turbulenz.* ZAMM **22**, 241-243 (1942) — *Mein Weg zu hydrodynamischen Theorien.* Physik. Blätter **4**, 89-92 (1948) — *Essentials of fluid dynamics. With applications to hydraulics, aeronautics, meteorology and other subjects.* Blackie, London (1952) — (W. TOLLMIEN, H. SCHLICHTING, and H. GÖRTLER, eds.) *L. PRANDTL. Gesammelte Abhandlungen zur angewandten Mechanik, Hydro- und Aeromechanik.* 3 vols., Springer, Berlin (1961).
**SECONDARY LITERATURE.** W. HOFF: *Ludwig PRANDTL.* Luftfahrtforsch. **12**, 1-3 (1935) — C. WIESELSBERGER: *L. PRANDTL.* Z. Tech. Phys. **16**, 25-27 (1935) — A. SOMMERFELD: *Zu L. PRANDTL's 60. Geburtstag.* ZAMM **15**, 1-2 (1935) — Anonymous: *Ludwig PRANDTL.* Jb. Dt. Akad. Luftfahrtforsch. (1939/1940), pp. 118-120 — Anonymous: *Ludwig PRANDTL.* Phys. Today **6**, 24 (Oct. 1953) — A. BETZ: *Ludwig PRANDTL zum 70. Geburtstag.* Forschungen u. Fortschritte **21/23**, 31 (April 1947) — W. TOLLMIEN: *Ludwig PRANDTL zum Gedächtnis* [Nachruf]. Jb. Wiss. Gesell. für Luft- u. Raumfahrt, pp. 22-27 (1953) — A. BUSEMANN: *Ludwig PRANDTL.* Biogr. Mem. Fell. Roy. Soc. (Lond.) **5**, 193-205 (1959) — I. FLÜGGE-LOTZ and W. FLÜGGE: *Ludwig PRANDTL in the nineteen-thirties: reminiscences.* Annu. Rev. Fluid Mech. **5**, 1-8 (1973) — J.H. LIENHARD: *L. PRANDTL.* In: (C.C. GILLISPIE, ed.) *Dictionary of scientific biography.* Scribner, New York, vol. 11 (1975), pp. 123-125 — K. KRAEMER: *Geschichte der Gründung des Max-Planck-Instituts für Strömungsforschung in Göttingen.* In: (G. GRABITZ and H.O. VOGEL, eds.) *Max-Planck-Institut für Strömungsforschung Göttingen 1925–1975.* Festschrift zum 50jährigen Bestehen des Instituts, Göttingen (1975) — *Gedächtnisveranstaltung für Ludwig PRANDTL aus Anlaß seines 100. Geburtstag, verbunden mit der 18. Ludwig-Prandtl-Gedächtnis-Vorlesung.* Z. Flugwiss. **23**, 149-152 (1975); H. GÖRTLER: *Ludwig PRANDTL, Persönlichkeit und Wirken.* Ibid. pp. 153-162; H. SCHLICHTING: *Ludwig PRANDTL und die Aerodynamische Versuchsanstalt (AVA).* Ibid., pp. 162-167; T. BENECKE: *Einführung zur 18. Ludwig-Prandtl-Gedächtnis-Vorlesung.* Ibid., p. 174 — J.C. ROTTA: *Die Aerodynamische Versuchsanstalt in Göttingen, ein Werk Ludwig PRANDTLS.* Vandenhoeck & Ruprecht, Göttingen (1990) — J.D. ANDERSON JR.: *Modern compressible flow, with historical perspective.* McGraw-Hill, New York (1990), pp. 140-143, 286-289 — J. VOGEL-PRANDTL: *Ludwig PRANDTL: Ein Lebensbild, Erinnerungen, Dokumente.* MPI für Strömungsmechanik, Göttingen (1993) — G.E.A. MEIER (ed.) *Ludwig PRANDTL, ein Führer in der Strömungslehre* [Biographische Artikel zum Werk Ludwig PRANDTLs von H. FÖRSCHING, P. GERMAIN, K. GERSTEN, J.L. VAN INGEN, K. JACOBS, U. KALKMANN, A. KLUWICK, E. KRAUSE, K. MAGNUS, G.E.A. MEIER, J.C. ROTTA, W. SCHNEIDER, J.H. SPURK, W. WUEST, and J. ZIEREP]. Vieweg, Braunschweig (2000) — *Hall of pioneers: history of heat transfer.* UCLA, Dept. of Materials Science & Engineering, Los Angeles, CA; http://www.seas.ucla.edu/jht/pioneers/pioneers.html.
**PICTURE.** Courtesy Archiv der Max-Planck-Gesellschaft (MPG), Berlin-Dahlem, Germany.
**NOTE.** PRANDTL's written legacy, such as his voluminous correspondence, are kept at the Archiv der MPG; http://www.archiv-berlin.mpg.de.

# Rankine, William John Macquorn (1820–1872)

- Scottish civil engineer and physicist; copromoter of scientific engineering and cofounder of theoretical thermodynamics and modern shock wave physics

William J.M. RANKINE was born in Edinburgh, the second son of David RANKINE, a rifle brigade lieutenant and civil engineer. Because of his poor health, he was taught mostly privately. He entered the University of Edinburgh (1836), where he studied natural philosophy under James David FORBES, who conducted important research on heat and the movement of glaciers. RANKINE received a gold medal for an essay on the *Undulation Theory of Light* (1836) and a prize for his essay on *Methods of Physical Investigation* (1838). After being introduced to railroad engineering by his father, who had become a superintendent for the Edinburgh and Dalkeith Railway, he left the university without a degree (1838). He went to Ireland, working there mainly on surveys, harbors, and railroads (1839–1841). He was trained as an engineer under Sir John Benjamin MACNEILL, an engineer for the Dublin & Drogheda Railway. After finishing his apprenticeship he returned to Edinburgh, practiced there civil engineering, and made important contributions to the science of railway locomotion. For example, he delivered a paper to the Institution of Civil Engineers (ICE) in London on the fracture of axes by referring to molecular structure (1843). While engaged on railway problems, he devised a method of setting out curves "by chaining and angles at circumferences combined," which has since been known as the "Rankine method."

In 1848, RANKINE began a series of researches on molecular physics. In his theory of matter, which he called "the hypothesis of molecular vortices," matter was composed of atoms, each comprising an atmosphere consisting of innumerable vortices surrounding a comparatively small nucleus: the absolute temperature of an atom being proportional to the square of the vortical velocity and the quantity of heat in a body being the energy of the molecular vortices. Light and heat were regarded by him as the result of vibrations of the nuclei. He applied his hypothesis of molecular vortices to the theory of heat (1850) and attempted to explain the phenomena of double refraction and elasticity of solid bodies in a similar way.

RANKINE is best known for describing the operational thermodynamic cycle of an ideal engine using steam or another vapor. His *Rankine cycle* is less efficient than the Carnot cycle, but has less practical difficulties and is more economic. After his appointment to chair of civil engineering and mechanics at the University of Glasgow (1855–1872), he spent much of his time on educational activities, developing new analytical techniques such as his "reciprocal diagrams" of frames and forces (1856), which allowed an engineer designer greater scope in studying the stresses in structures. He coined new, enduring terms such as "stress," "strain," and "adiabatic." He published most of his original work on the strength of materials and the theory of structures in his two books *A Manual of Applied Mechanics* (1858) and *A Manual of Civil Engineering* (1862).

After settling permanently in Glasgow, RANKINE became increasingly interested in naval architecture. Together with the shipbuilders James Robert NAPIER, Isaac WATTS, and Frederick K. BARNES he wrote *Shipbuilding, Theoretical and Practical* (1866) with the objective of bringing precision and theory to the British naval industry, which was largely empirical. Independently of William FROUDE, he worked out a possible theory of sea waves of finite height (or displacement) on the surface of deep water (1862). Recognizing that the action of a ship propeller is based on the acceleration of water masses swept by the propeller blades, he derived a momentum theory of the propeller (1865).

In August 1869, three years before his death, he first reported on the thermodynamic state of a "wave of finite longitudinal disturbance," assuming that the abrupt thermodynamic change from upstream to downstream – the term *shock front* had not yet been coined – is not isentropic, but rather a region of dissipation; *i.e.*, the fluid is thermally conductive but nonviscous. He derived the equations for continuity, momentum, and energy. Eight years later, the French physicist Pierre-Henri HUGONIOT, apparently unaware of RANKINE's work, derived the same equations, today termed the *Rankine-Hugoniot equations*. RANKINE also worked as a consultant engineer, and together with Stevenson MACADAM, a lecturer in chemistry at the University of Edinburgh, he investigated the accident at the Tradeston Flour-Mills in Glasgow, reflecting on possible causes of flour-dust explosions (1872).

RANKINE published more than 150 papers and wrote 4 books. He was a Fellow of the Royal Society of Edinburgh (from 1849) and the Royal Society of London (from 1853), and was president of the Scottish Institution of Engineers (1858). He received many honors and was awarded the Gold Medal of the Institution of Engineers in Scotland for his contributions to thermodynamics and engineering. Trinity College, Dublin, conferred on him the degree of LL.D. (1857). James C. MAXWELL, promoting substantially the evolution of the kinetic theory of gases, placed RANKINE alongside William THOMSON (from 1892 Lord KELVIN) and Rudolf J.E. CLAUSIUS as one of the main founders of theoretical thermodynamics.

RANKINE's scientific findings also form the foundation of modern soil mechanics. The prestigious *Rankine Lectures*, established by the British Geotechnical Association, have been given annually since 1961 at Imperial College, London, by eminent contributors to geotechnical engineering. They are subsequently published as papers by The Institution of Civil Engineers in the British journal *Géotechnique*.

In fluid dynamics the *Rankine vortex* is an idealized vortex in unbounded fluid with uniform vorticity inside a circular patch and zero vorticity outside. The *Rankine body* consists of a source and a sink. RANKINE is best known for describing the operational thermodynamic cycle of an ideal engine using steam or another vapor. This *Rankine cycle* is less efficient than the Carnot cycle, but has less practical difficulties and is more economic.

Astronomers named a crater on the near side of the Moon after him.

**ORIGINAL WORKS.** *Über die mechanische Theorie der Wärme.* Ann. Phys. **81** [II], 172-176 (1850) — *On LAPLACE's theory of sound.* Phil. Mag. **1** [IV], 225-227 (1851) — *On POISSON's investigation on the theory of sound.* Ibid. **1** [IV], 410-411 (1851) — *On the law of the compressibility of water at different temperatures.* Ibid. **1** [IV], 548-549 (1851) — *Mechanical theory of heat. (I) The specific heat of air.* Ibid. **5** [IV], 437-439 (1853); *(II) Velocity of sound in gases.* Ibid. **5** [IV], 483-486 (1853) — *On the mechanical action of heat.* Ibid. **7** [IV], 1-21, 111-122, 172-185, 239-254 (1854) — *On the compressibility of water.* Proc. Roy. Soc. Edinb. **3**, 58-59 (1857) — *A manual of applied mechanics.* Griffin, London (1858). Reprinted in: *Encyclopaedia Metropolitana.* 40 vols., Griffin, London (1849–1858), vol. 39: *On flow of fluids of varying density,* §§ 416-420, 646-647 — *On the conservation of energy.* Phil. Mag. **17** [IV], 250-253, 347-348 (1859) — *A manual of civil engineering.* Griffin, London (1862) — *On the exact form and motion of waves at and near the surface of deep water.* Phil. Mag. **24** [IV], 420-422 (1862) — *On the dynamical theory of heat.* Ibid. **27** [IV], 194-196 (1864) — *On the mechanical principles of the action of propellers.* Trans. Instit. Naval Architects **6**, 13-30 (1865) — *On the theory of explosive gas-engines.* The Engineer (Lond.) **22** (July 27, 1866), pp. 55-56 — *On waves in liquids.* Proc. Roy. Soc. Lond. **16**, 344-347 (1868) — *Summary of the thermodynamic theory of waves of finite longitudinal disturbance.* Rept. Meet. Brit. Assoc. **39**, 14-15 (1869) — *On the explosive energy of heated liquids.* The Engineer (Lond.) **29**, 323 (Nov. 11, 1870) — *On the thermodynamic theory of waves of finite longitudinal disturbance.* Proc. Roy. Soc. Lond. **18**, 80-83 (1870); Phil. Mag. **39** [IV], 306-309 (1870); Phil. Trans. Roy. Soc. Lond. **160**, 277-286 (1870); Supplement. Ibid. **160**, 287-288 (1870) — *Theorie des hydraulischen Widders.* Der Civilingenieur **18**, 297-302 (1872) — *Sur la force explosive des liquides chauffés.* Rev. University Mines, Métall. Méc. Trav. Publ. Sci. Arts Appl. Ind. (Liège) **31** [I], 65-71 (1872) — *On sea-waves.* Proc. Roy. Inst. **6**, 355-356 (1872) — (W.J. MILLAR, ed.) *Miscellaneous scientific papers by W.J.M. RANKINE.* Griffin, London (1881).

**SECONDARY LITERATURE.** P.G. TAIT: *Notice of Professor Macquorn RANKINE.* Glasgow Herald (Dec. 28, 1872) — S. MACADAM: *On flour-mill fire-explosions.* Proc. Roy. Phil. Soc. Glasgow **8**, 280-288 (1873) — Anonymous: *William John Macquorn RANKINE.* Proc. Roy. Soc. Lond. **21**, i-iv (1873); Proc. Am. Acad. Arts Sci. **9**, 276-278 (1874) — P.G. TAIT: *Memoir on W.J.M. RANKINE.* In: (W.J. MILLAR, ed.) *Miscellaneous scientific papers by W.J. Macquorn RANKINE.* Griffin, London (1881), pp. xix-xxxvi — L.D.B. GORDON: *Obituary notice of Professor RANKINE.* Proc. Roy. Soc. Edinb. **8**, 296-306 (1875) — A. BARR: *W.J. Macquorn RANKINE, a centenary address.* Proc. Roy. Phil. Soc. Glasgow **51**, 167-187 (1923) — R. SOUTHWELL: *W.J.M. RANKINE: a commemorative lecture delivered on 12 December, 1955, in Glasgow.* Proc. Inst. Civ. Eng. (Lond.) **5**, pt. 1, 177-193 (1956) — E.E. DAUB: *Atomism and thermodynamics.* Isis **58**, 293-303 (1967) — H.B. SUTHERLAND: *RANKINE, his life and times.* Institution of Civil Engineers, London (1973) — V.V. RAMAN: *William John Macquorn RANKINE: 1820–1872.* J. Chem. Educ. **50**, 274-276 (April 1973) — E.M. PARKINSON: *W.J.M. RANKINE.* In: (C.C. GILLESPIE, ed.) *Dictionary of scientific biography.* Scribner, New York, vol. 11 (1975), pp. 291-295 — K. HUTCHISON: *W.J.M. RANKINE and the rise of thermodynamics.* Brit. J. Hist. Sci. **14**, 1-26 (1981) — D.F CHANNELL: *RANKINE, William John Macquorn, 1820–1872.* Scotland's Cultural Heritage, Edinburgh (1986) — B. MARSDEN: *Engineering science in Glasgow: economy, efficiency and measurement as prime movers in the differentiation of an academic discipline.* Brit. J. Hist. Sci. **25**, 319-346 (1992) — B. MARSDEN: *RANKINE, (William John) Macquorn.* In: (H.C.G. MATTHEW, ed.) *Oxford dictionary of national biography.* Oxford University Press, Oxford (2004), vol. 46, pp. 38-42.

**PICTURE.** Courtesy Deutsches Museum, Munich, Germany.

**NOTE.** For collection and biographical history of W.J.M. RANKINE see Glasgow University Archive Services; NAHSTE Project; http://www.nahste.ac.uk/cgi-bin/view_isad.pl?id=GB-0248-DC-320&view=basic.

# RAYLEIGH J.W. STRUTT, Lord (1842–1919) →STRUTT, John William

# REGNAULT, Henri Victor (1810–1878)

▪ French chemist and physicist

Henri V. REGNAULT was born in Aix-la-Chapelle (now Aachen, Germany) to André-Privat REGNAULT, a captain in the Corps des Ingénieurs Géographes Militaires of Napoleon's Army. He lost both his parents when he was only 8 years old. Working first in a textile establishment in Paris, he graduated from the Ecole Polytechnique (1830–1832) and continued his studies in Paris at the Ecole des Mines (1832–1834). After short periods of research under Justus VON LIEBIG at Gießen and Jean-Baptiste BOUSSINGAULT at Lyon, REGNAULT returned to the Ecole Polytechnique. Working there first as an assistant (1836–1840) to Joseph-Louis GAY-LUSSAC, he succeeded him later as chair of chemistry (1840). During this period he conducted most of his famous experimental work in organic chemistry, such as on the halogen and other derivatives of unsaturated hydrocarbons, and in the process discovered vinyl chloride, dichlorethylene, trichlorethylene, and carbon tetrachloride. Shortly afterwards, he succeeded Pierre L. DULONG, a French chemist and physicist, as chair of physics at the Collège de France (1841) in Paris, a position he held until his retirement (1872). His renowned textbook *Cours élémentaire de chimie*

(1847) soon became a classic and was translated into various languages.

Appointed by the minister of public works to redetermine all the physical constants involved in the design and operation of steam engines, REGNAULT produced the first classic work on the properties of steam, which was subsidized by the French government, and the findings were published in 1847, 1862, and 1870. His work was the standard up to the early 20th century.

REGNAULT performed a long series of very careful measurements of the specific heats of many gases, liquids, and solids and also investigated the expansion of gases and devised new instruments. When he became director of the famous porcelain manufactory at Sèvres, he lived and continued his research there in his private laboratory, but all his results and instruments were destroyed during the Franco-German War (1870–1871) in which his son Henri was also killed. REGNAULT never recovered from the double blow, and although he lived until 1878, his scientific labors ended in 1872.

REGNAULT's contributions to shock waves were barely recognized by either his contemporaries or modern shock physicists. Almost 10 years prior to Ernst MACH and Jan SOMMER at Charles University in Prague he first proved by accurate measurements in long pipeline systems that the velocity of sound – in his experiments a blast wave originated by a pistol or the sudden opening of a high-pressure gas reservoir – decreases with the diameter and, therefore, with the intensity, tending to a limit for very feeble sounds. His motivation for turning to acoustics was to precisely determine the mechanical equivalent of heat using the ratio of specific heats, $\gamma = c_p/c_v$. To determine $\gamma$ he applied POISSON's formula $\gamma = a^2 \rho/p$ and measured the sound velocity $a$ in a test substance at given pressure $p$ and density $\rho$.

The Royal Society of London honored his contributions to organic chemistry and thermodynamics by awarding him the Rumford Medal (1848) and the Copley Medal (1869), respectively. He was elected member of the chemical section of the Berlin and St. Petersburg Academies and member of the French Academy of Sciences (1840) and was a foreign member of the Royal Society of London. In 1863, he was made Commander of the Legion of Honor.

Astronomers named a crater on the near side of the Moon after him.

**ORIGINAL WORKS.** *Recherches sur la chaleur spécifique des corps simples et composés. 1er Mémoire.* Ann. Chim. Phys. **73** [II], 5-72 (1840); *2e Mémoire.* Ibid. **1** [III], 129-207 (1841); *3e Mémoire.* Ibid. **9** [III], 322-349 (1843) — *Sur le coefficient de dilatation des gaz.* C. R. Acad. Sci. Paris **13**, 1077-1079 (1841) — *Mémoire sur les forces élastiques de la vapeur d'eau.* Ann. Chim. Phys. **11** [III], 273-335 (1844) — *Sur le détermination de la densité des gaz.* Ann. Chim. Phys. **14** [III], 211-239 (1845) — *Sur la loi de la compressibilité des fluides élastiques.* C. R. Acad. Sci. Paris **23**, 787-798 (1846) — *Relation des expériences entreprises, pour déterminer les principales lois et les données numériques qui entrent dans le calcul des machines à vapeur.* Mém. Acad. Sci. Inst. France **21**, 1-767 (1847) — *Cours élémentaire de chimie.* 3 vols. Langlois, Leclercq & Masson, Paris (1847–1849) — *Sur la vitesse de propagation des ondes dans les milieux gazeux.* C. R. Acad. Sci. Paris **66**, 209-220 (1868) — *Mémoire sur la vitesse de propagation des ondes dans les milieux gazeux.* Mém. Acad. Sci. Inst. France **37**, 3-575 (1868) — *On the velocity of the propagation of waves in gaseous media.* Phil. Mag. **35** [IV], 161-171 (1868) — *Recherches sur la vitesse de propagation des ondes dans les milieux élastiques.* Arch. Sci. Phys. Natl. (Genève) **42**, 147-163 (1871).

**SECONDARY LITERATURE.** G.A. DAUBREE, J.H. DEBRAY, J.C. JAMIN, and C. LABOULAYE: *Discours prononcés par des membres de l'académie aux funérailles de M. REGNAULT.* C. R. Acad. Sci. Paris **86**, 131-143 (1878) — T.H. NORTON: *Henri Victor REGNAULT.* Nature **17**, 263-264 (Jan. 1878) — J.H. GLADSTONE: *Henri Victor REGNAULT.* Chem. Soc. J. **33**, 235-239 (1878) — A. DUANE: *The scientific life and work of H.V. REGNAULT.* Trans. Albany Inst. **9**, 270-284 (1879) — J.B. DUMAS: *Eloge historique de H.V. REGNAULT.* Dunod, Paris (1881); Ann. Mines **19** [VII], 212-244 (1881); Rev. Sci. **1**, 354-363 (1881); Mém. Acad. Sci. Inst. France **42**, xxxvii-lxxv (1883) — M. BERTHELOT: *Science et philosophie.* Calman-Lévy, Paris (1886), pp. 218-235 — C. KELLER: *Henri Victor REGNAULT.* In: (C. MATSCHOSS, ed.) *Beiträge zur Geschichte der Technik und Industrie.* Jb. VDI, Springer, Berlin (1910), vol. 2, pp. 58-63 — R. FOX: *The caloric theory of gases from LAVOISIER to REGNAULT.* Clarendon Press, Oxford (1971); *Henri Victor REGNAULT.* In: (C.C. GILLESPIE, ed.) *Dictionary of scientific biography.* Scribner, New York, vol. 11 (1975), pp. 352-354.

**PICTURE.** Courtesy Collections Ecole Polytechnique, Paris, France.

# RIEMANN, Georg Friedrich Bernhard (1826–1866)

- German mathematician and physicist

G.F. Bernhard RIEMANN was born in Breselenz, a village approx. 45 km west of Wittenberge, Lower Saxony. His father was Friedrich Bernhard RIEMANN, a Protestant minister. Before entering secondary school he was mostly educated by his father and enjoyed learning since early childhood. After attending the lyceum in Hannover and the Johanneum in Lüneburg (1840–1846) he began to study theology and philology at Göttingen University, but the lectures of the famous mathematician Carl Friedrich GAUSS influenced him to dedicate himself wholly to mathematics (1846–1847). He continued his education at the University of Berlin under the German mathematician Carl G.J. JACOBI (1847–1849), a codiscoverer of elliptic functions and important contributor to the theory of partial differential equations of the first order with applications to dynamic problems. Upon his return to Göttingen he attended courses in physics given by the famous experimental physicist Wilhelm E. WEBER. In his Ph.D. thesis (1851), supervised by GAUSS, he investigated the geometry of "Riemann surfaces." After preparing his *Habilitationsschrift* (habilitation thesis) on trigonometric series expansion of an arbitrary function, in which he also gave a historical review of this problem, he became in 1853 *Privatdozent* (university lecturer). In addition, he worked as an assistant, probably unpaid, to the mathematician Heinrich M. WEBER, who doubtless inspired him to tackle physical problems mathematically. In 1857, RIEMANN became assistant professor and in 1859 was appointed chair of mathematics at Göttingen University. A few days later he was elected to the Berlin Academy of Sciences.

He gave his first course on partial differential equations with applications to physics entitled *Die Theorie der Integration der partiellen Differentialgleichungen nebst Anwendungen derselben auf verschiedene Probleme der Physik* (1854–1855), a subject on which he deepened his understanding in the following years (1860–1864). In 1859, he published his famous paper on plane sound waves of finite amplitude and purely mathematically discovered the phenomenon of a discontinuous wave structure, which he termed *Verdichtungsstoß* ("condensation shock"). RIEMANN's main motivation to treat this subject was without question the application of linear partial differential equations to problems of applied acoustics. In his introduction he mentioned that Hermann VON HELMHOLTZ, addressing the problem of the origin of combination tones at high sound levels, recommended integrating the differential equations by taking into account also pressure terms of higher order. It appears that RIEMANN's interest in acoustics continued in the following years, and in the last months of his life he wrote a long article on the mechanism of hearing that, being in opposition to VON HELMHOLTZ's theory, was published posthumously (1867). It is interesting here to note that this article contained no mathematics and is a pure physiological-physical description of hearing.

Also in 1859, he reported to the Berlin Academy of Sciences on his most recent research. In his report entitled (in translation) *On the Number of Primes Less Than a Given Magnitude*, he extended the zeta function to complex values and made the famous conjecture about how prime numbers were distributed among other numbers, now known as the "Riemann hypothesis," which is of fundamental importance in number theory. It remains today one of the most important of the unsolved problems of mathematics. In 2001, the Clay Mathematics Institute (CMI) in Cambridge, MA offered a $1 million prize to the first person to prove or disprove the Riemann hypothesis.

In the period 1862–1866, he suffered several attacks of pleurisy, and, despite attempts to recover in Italy, where he spent two winters in Pisa, he died in 1866 at Selasca, a town in Piedmont near Intra on the Lago Maggiore, at the early age of 39. Despite his short life, RIEMANN's contributions to mathematical physics are huge and encompass mechanics, gravity, acoustics, electricity, and magnetism. He was a corresponding member of the Berlin Academy of Sciences (from 1859) and a member of the Göttingen Academy of Sciences (from 1860).

Karl HATTENDORFF, one of RIEMANN's former auditors, edited RIEMANN's lecturers after his death in a book entitled *Partielle Differentialgleichungen und deren Anwendung auf physikalische Fragen: Vorlesungen von Bernhard RIEMANN* (1869). Seven years later, Heinrich M. WEBER, a mathematics professor at the University of Königsberg who was interested in the same subject, published the textbook *Die partiellen Differentialgleichungen der mathematischen Physik, nach RIEMANN's Vorlesungen* (1876), in which he followed RIEMANN's concept. WEBER's book, in which he also treated shock waves, went through six editions and became a classic of mathematics.

Among mathematicians RIEMANN became widely known for devising *Riemann surfaces* and *Riemannian geometry*. Among fluid dynamicists his *Riemann method* for solving linear hyperbolic second-order partial differential equations using the *Riemann function* proved most useful. The *Riemann invariants* are very helpful for analyzing linear solutions of the wave equation by determining "simple waves" as building blocks for constructing more complex solutions.

Astronomers named a crater on the near side of the Moon and a minor planet (asteroid 4167 RIEMANN) after him.

**ORIGINAL WORKS.** *Grundlagen für eine allgemeine Theorie der Functionen einer veränderlichen complexen Größe.* Ph.D. thesis, University of Göttingen (Nov. 1851) — *Über die Darstellbarkeit einer Funktion durch eine trigonometrische Reihe.* Habilitationsschrift, Universität Göttingen (1854). Publ. posthumously in: Abhandl. Königl. Gesell. Wiss. Gött. **XIII** [Math.-Physik. Kl.], 87-132 (1866/1867) — *Über die Fortpflanzung ebener Luftwellen von endlicher Schwingungsweite* [vorgelegt der königl. Societät am 22. Nov. 1859]. Abhandl. Königl. Gesell. Wiss. Gött. **VIII** [Math.-Physik. Kl.], 43-65 (1858/1859) — Review in: Fortschr. Phys. (Berlin) **15**, 123-130

(1859) — Selbstanzeige der vorstehenden Abhandlung in: Nachr. Georg-August-University & Königl. Gesellsch. Wiss. Göttingen Nr. 19, 192-196 (5. Dez. 1859) — *Mechanik des Ohres* [publ. posthumously by Profs. J. HENLE and E. SCHERING]. Z. Rat. Medicin **29** [III], 129-143 (1867) — *Über die Fläche von kleinstem Inhalt bei gegebener Begrenzung* [publ. posthumously by K. HATTENDORFF]. Abhandl. Königl. Gesell. Wiss. Gött. **XIII**, 2-52 (1866/1867) — (K. HATTENDORFF, ed.) *Partielle Differentialgleichungen und deren Anwendung auf physikalische Fragen. Vorlesungen von B. RIEMANN*. Vieweg, Braunschweig (1869, 1876, 1882) — (H. WEBER and R. DEDEKIND, eds.) *Bernhard RIEMANN's gesammelte mathematische Werke und wissenschaftlicher Nachlass von Bernhard RIEMANN*. Teubner, Leipzig (1876, 1892); (M. NOETHER and W. WIRTINGER, eds.) *Nachträge*. Teubner, Leipzig (1902) — (H. WEBER, ed.) *Collected works of Bernhard RIEMANN*. Dover Publ., New York (1953).

**SECONDARY LITERATURE.** E. SCHERING: *Bernhard RIEMANN zum Gedächtnis*. Nachr. Königl. Gesell. Wiss. Gött. Nr. 15 (Juni 1867), pp. 305-314 (1867) — Anonymous: *Georg Friedrich Bernhard RIEMANN*. Proc. Roy. Soc. Lond. **16**, lxix-lxx (1868) — Anonymous: *Biography of Bernhard RIEMANN* [in Russ.]. Recueil Math. (Moscow) **3**, 153-158 (1868) — R. DEDEKIND: *B. RIEMANN's Lebenslauf*. In: (H. WEBER and R. DEDEKIND, eds.) *Bernhard RIEMANN's gesammelte Werke und wissenschaftlicher Nachlass*. Teubner, Leipzig (1876), pp. 541-558 — F. KLEIN: *RIEMANN und seine Bedeutung für die Entwicklung der modernen Mathematik*. Jber. Dt. Math.-Vereinigung **4**, 71-87 (1897) — H.M. WEBER: *Fortpflanzung einer Unstetigkeit*. In: H. WEBER: *Die partiellen Differentialgleichungen der mathematischen Physik, nach RIEMANN's Vorlesungen*. Vieweg, Braunschweig, vol. 1 (1900), pp. 497-506; *Fortpflanzung von Stoßen in einem Gas*. Ibid. vol. 2 (1901), pp. 469-498; *Luftschwingungen von endlicher Amplitude*, Ibid. vol. 2 (1901), pp. 499-518 — R. COURANT: *Bernhard RIEMANN und die Mathematik der letzten hundert Jahre*. Die Naturwissenschaften **14**, 813-818 (1926) — I. GRATTAN-GUINNESS: *The development of the foundations of mathematical analysis from EULER to RIEMANN*. MIT Press, Cambridge, MA (1970) — H. FREUDENTHAL: *B. RIEMANN*. In: (C.C. GILLESPIE, ed.) *Dictionary of scientific biography*. Scribner, New York, vol. 11 (1975), pp. 447-456 — L. WEICHSEL (ed.): *J.C. Poggendorf's Biogr.-Lit. Handwörterbuch (1932–1962)*. Akademischer Verlag, Berlin (1985), vol. VIIb, pp. 4781-4789.

**PICTURE.** After a copper engraving by A. WEGER. From (R. NARASIMHAN, ed.) *Bernhard RIEMANN's gesammelte mathematische Werke, wissenschaftlicher Nachlaß und Nachträge*. Springer, Berlin (1990).

# ROBINS, Benjamin (1707–1751)

- British mathematician and military engineer; founder of scientific ballistics

Benjamin ROBINS was born in Bath, Somerset. His father, John ROBINS, was a tailor, and his parents were Quakers. Originally trained as a teacher, he soon left that profession and started as a mathematician but only gave private lessons. He was elected to the Royal Society of London (1727) and got involved in mechanics and various projects of civil engineering. In addition, he began to study gunnery and fortification. With his book *New Principles of Gunnery* (1742) he became an internationally reputed ballistician. Translated into German and comprehensively annotated by Leonard EULER (1745), the book laid the groundwork for modern gun theory and practical ordnance (field artillery). At the request of the British Admiralty, EULER's work was translated into English by the English mathematician Hugh BROWN, who supplemented it with his own annotations (1777). The French Charles LE ROY translated ROBINS' book for the Paris Académie des Sciences (1751).

ROBINS' theory of fired gunpowder assumed that all the powder of which a charge consists is not only set on fire, but that it is actually consumed and "converted into an *elastic fluid* before the bullet is sensibly moved from its place." To measure the muzzle velocity of projectiles, he first applied the ballistic pendulum (1740), which was already suggested earlier by Jacques CASSINI (1707). His pendulum was suspended from a tripod, and the bullet was shot into a wooden block screwed to the pendulum. It enabled ballisticians for the first time to also measure the influence of drag on projectile velocity and geometry by positioning the ballistic pendulum at different distances from the muzzle. He discovered that the force of aerodynamic drag could be as high as 120 times the projectile's weight. Since ROBINS already noticed a significant increase in drag upon approaching the sound barrier and even studied supersonic projectile velocities up to $M = 1.5$, his investigations can be regarded as the birth of scientific supersonic aerodynamics.

ROBINS, also studying the lateral deflection of high-speed projectiles by setting up a series of evenly spaced paper curtains, observed that a musket ball in flight is enormously deflected and identified the spin of the ball as the cause of this deflection. He also investigated pressures on projectiles inside a gun barrel and the shape of actual – as opposed to ideal – trajectories (1742). He predicted the superiority of rifled cannon; however, the first comprehensive experiments on realizing this technically demanding concept eventually changing warfare did not start until in the 1830s in Europe and the United States. In the famous *vis visa* controversy, initiated in 1686 by the German natural philosopher Gottfried Wilhelm LEIBNIZ, ROBINS took part in a polemic manner attacking Sir NEWTON's enemies who, besides LEIBNIZ, included the Italian mathematician Giovanni POLENI, the Swiss mathematicians Daniel and Johann BERNOULLI, and others.

ROBINS' last work consisted of investigations on rockets for the purpose of military signaling. Appointed Engineer General by London's East India Company he went to Madras, India to repair and improve the forts of the Company.

Starting at Fort St. David in Cuddalore, India in 1750, he died there of a fever only a year later.

From the Royal Society of London ROBINS received the Copley Medal (1746) for his contributions to ballistics, particularly for his studies on aerodynamic drag at high speeds. Many of his writings were published posthumously by his friend James WILSON in the *Mathematical Tracts* (1761). The numerous secondary literature recently published on ROBINS' life and work well illustrates the curiosity of modern ballisticians about his unique contributions. In the 1830s, ROBINS' ballistic pendulum was significantly improved by the Commission des principes du tir, a French ballistic commission that consisted of the French mathematicians Isidore DIDION, Arthur MORIN, and Guillaume PIOBERT.

**ORIGINAL WORKS.** *Remarks on Mr. EULER's treatise of motion.* Nourse, London (1739) — *New principles of gunnery.* J. Nourse, London (1742); Translated into French by M. DUPUTY fils: *Nouveaux principes d'artillerie.* Durand, Grenoble & Jombert, Paris (1771); translated into German and supplemented with comments by L. EULER: *Neue Grundsätze der Artillerie, enthaltend die Bestimmung der Gewalt des Pulvers nebst einer Untersuchung über den Unterschied des Widerstands der Luft in schnellen und langsamen Bewegungen.* A. Haude, Berlin (1745). Also in: (F.R. SCHERRER, ed.) *Leonardi EULERI opera omnia.* Teubner, Leipzig (1922), vol. 14 [II]; retranslated into English by H. BROWN: *The true principles of gunnery.* J. Nourse, London (1777); retranslated into French by J.L. LOMBARD: *Nouveaux principes d'artillerie, commentés par M. Léonard EULER.* Frantin, Dijon (1783) — *On the height to which rockets will ascend.* Phil. Trans. Roy. Soc. Lond. **46**, No. 492, 131-133 (1749) — With J. ELLICOTT: *An account of some experiments in order to discover the height to which rockets may be made to ascend, and to what distance their height may be seen.* Ibid. **46**, No. 496, 578-584 (1750) — *On the resistance of the air* [1746]. In: (J. WILSON, ed.) *Mathematical tracts of the late Benjamin ROBINS, Esq.* 2 vols., Nourse, London (1761), vol. I, pp. 175-178 — *On the resistance of the air; together with the method of computing the motions of bodies projected in that medium* [1746]. Ibid. vol. I, 179-199 — *An account of the experiments relative to the resistance of the air* [1747]. Ibid. vol. I, 200-217 — *On the force of gunpowder, together with the computation of the velocities; thereby communicated to military projectiles.* [1747]. Ibid. vol. I, 218-229 — *A comparison of the experimental ranges of cannon and mortars, with the theory contained in the preceding papers.* Ibid. vol. I, 230-244 — *On pointing, or directing of cannon to strike distant objects.* Ibid. vol. I, 312-316 — *Remarks on a treatise lately printed in Paris and entitled 'Discours sur les Loix de la Communication du Mouvement,' par Mons. BERNOULLI.* Ibid. vol. II, 174-188.

**SECONDARY LITERATURE.** O. FLACHSBART: *Geschichte der experimentellen Hydro- und Aeromechanik, insbesondere der Widerstandsforschung.* In: (L. SCHILLER, ed.) *Handbuch der Experimentalphysik.* Akad. Verlagsgesell., Leipzig, vol. IV, pt. 2 (1932), pp. 1-61 — I. DIDION: *Traité de balistique.* A. Leneveu, Paris (1848) — R.S. HARTENBERG: *Benjamin ROBINS, ein Aerodynamiker.* Z. Flugwiss. **4**, 213-217 (1956) — S.F. HOERNER: *Aerodynamic drag.* Dayton, OH (1951). Extended edition: *Fluid-dynamic drag.* Publ. by the author, Midland Park, NJ (1958) — H.M. BARKLA: *Benjamin ROBINS and the resistance of air.* Ann. Sci. **30**, 107-122 (1973) — J.M. BRIGGS JR.: *B. ROBINS.* In: (C.C. GILLESPIE, ed.) *Dictionary of scientific biography.* Scribner, New York, vol. 11 (1975), pp. 493-495 — E.W.E. ROGERS: *Aerodynamics, retrospect and prospect.* Aeronaut. J. **86**, 43-67 (1982) — W. JOHNSON: *Benjamin ROBINS' New Principles of Gunnery.* Int. J. Impact Engng. **4**, 205-219 (1986); *The origin of the ballistic pendulum: the claims of Jacques CASSINI and Benjamin ROBINS.* Int. J. Mech. Sci. **32**, 345-374 (1990); *Benjamin ROBINS (1707–1751): opting not to be a commissary for Arcadia but a fortification engineer in East India.* Int. J. Impact Engng. **11**, 547-71 (1991); *Encounters between ROBINS, and EULER and the BERNOULLIs; Benjamin ROBINS (18th century founder of scientific ballistics): some European dimension, and past and future perceptions.* Ibid. **12**, 293-323 (1992); *Artillery and related subjects.* Int. J. Mech. Sci. **34**, 651-679 (1992) — B.D. STEELE: *The ballistic revolution: military and scientific change from ROBINS to NAPOLÉON.* Ph.D. thesis, University of Minnesota (1994); *Muskets and pendulums. B. ROBINS and L. EULER, and the ballistic revolution.* Technol. Culture **35**, 348-382 (1994) — W. JOHNSON: *Benjamin ROBINS during 1739–1742.* Notes Rec. Roy. Soc. Lond. **48**, 31-42 (1994); *Executors of the published work of Benjamin ROBINS, James WILSON and Martin FOLKES.* Int. J. Impact Engng. **21**, 694 (1998); *Collected works on Benjamin ROBINS & Charles HUTTON.* Phoenix, New Delhi (2001) — B.D. STEELE: *ROBINS, Benjamin.* In: (H.C.G. MATTHEW, ed.) *Oxford dictionary of national biography.* Oxford University Press, Oxford (2004), vol. 47, pp. 292-294.

**NOTE.** There exists no picture of ROBINS; private communication from Prof. emer. William JOHNSON, F.R.S., Cambridge, U.K.

# RODMAN, Thomas Jackson (1815–1871)

- U.S. brigadier-general, military engineer, and inventor

Thomas J. RODMAN was born in Salem, IN. His father was a farmer. He graduated from the ordnance department of the U.S. Military Academy at West Point, NY (1841) and was the 1065th graduate of the Academy. Assigned to the U.S. Ordnance Department, he was assigned to various arsenals. Serving at Alleghany Arsenal, PA, until 1848, he carried out experiments on war materiel and summarized the results in numerous Alleghany Arsenal reports (1857–1858). His fifth report contains the first description of his cutter gauge, the so-called "Rodman gauge" (1857). His indenting apparatus is a pressure gauge consisting of a piston working in a hole bored into the wall of a gun and acting on an indenting tool. Measurement of the depth of the indentations indicate the relative pressure along the tube. In a simple manner it allows one to measure the absolute pressure of fired gunpowder or gas in the bore of a gun. RODMAN's invention was immediately resumed by various European countries. However, in 1894 the British ballistician and gun expert Sir Andrew NOBLE first demonstrated that the maximum pressures in gun barrels derived by RODMAN from his indentation gauge were far too high (1894).

RODMAN served as a superintendent of Watertown Arsenal, MA, where he spent the Civil War producing cannon for the Union. In particular, he devised the method of casting guns on a hollow core, the metal being cooled by a stream of water running through the inside – a technique that showed greater power of resistance than those that involved casting in the usual way; *i.e.*, in one piece, and afterwards bored out. The first of his famous so-called "15-in. smooth-bore Rodman gun," a new form of columbiad intended primarily for sea-coast defense, was successfully tested in March 1861. Three years later he even increased the caliber to 20-in., which required a four-piece mold taking 160,000 lb. of molten iron. His new 20-in. smooth-bore cannon, then the largest in the world, brought him into prominence; it was capable of firing a 1,080-lb. projectile at a maximum range of 4.5 miles. However, it was too big and because of technical problems never used in practice.

For proper operation of these huge guns he proposed his "mammoth powder," later called "Rodman powder" (1860), a large-grained, progressively-burning powder that became a useful propellant because it burned slowly. Thus, a high velocity would be given to a shot without subjecting the gun to excessive strain. His mammoth powder was adopted by the U.S. government, and soon after by Russia, England, and Prussia.

In the Civil War he was appointed brigadier-general (1861). Four years before his death, he was promoted to the rank of lieutenant-colonel (1867). Worn down by his many activities and responsibilities, he died prematurely at the age of 56 at Rock Island Arsenal, IL. Besides Charles E. MUNROE, he is certainly the most prominent figure of 19th-century American research on explosives and propellants and their use for military purposes.

**ORIGINAL WORKS.** *Reports of experiments on the properties of metals for cannon, and the qualities of cannon powder; with an account of the fabrication and trial of a 15-inch gun (Jan. 1857 – May 1860).* Crosby, Boston (1861).
**SECONDARY LITERATURE.** Anonymous: *The biggest gun in the world.* Harper's Weekly: J. Civilization (New York) **5**, No. 222 (March 30, 1861) — Editor and A.B. DYER: *General RODMAN.* U.S. Army and Navy J. **10**, 702-703, 817 (June 17 and Aug. 5, 1871) — *The national cyclopaedia of American biography.* White, New York (1898–1984), vol. 1-1 (1898) — G.W. CULLUM: *Biographical register of the officers and graduates of the U.S. Military Academy at West Point from 1802–1890.* Houghton Mifflin, Boston (1891) — C.D. RHODES: *T.J. RODMAN.* In: (D. MALONE, ed.) *Dictionary of scientific biography.* Scribner, New York, vol. 8 (1935), pp. 80-81 — D.B. WEBSTER JR.: *RODMAN's great guns.* Ordnance **47**, 60-62 (July/Aug. 1962); see http://www.cwartillery.-org/ve/tjrodman.html — E. LAYTON: *Mirror-image twins: the communities of science and technology in 19th-century America.* Technology & Culture **12**, No. 4, 562-580 (Oct. 1971) — W. JOHNSON: *T.J. RODMAN, mid-19th century gun barrel research and design for the U.S. Army.* Int. J. Impact Engng. **9**, 127-159 (1990) — E. OLMSTEAD, W.E. STARK, and S.C. TUCKER: *The big guns: civil war siege, sea-coast and naval cannon.* Museum Restoration Service, Bloomfield (1997) — S.C. MOWBRAY and J. HEROUX (eds.): *Civil war arms makers.* A. Mowbray, Lincoln, RI (1998); see Case No. 103: *Captain T.J. RODMAN and Charles KNAPP*, pp. 549-570 — *Civil War artillery – famous men.* Civil War Virtual Archives; see http://www.cwartillery.org/afammen.html
**PICTURE.** Courtesy John L. CARNPROBST, Pittsburgh, PA.

## RUSSELL, John Scotch (1808–1882)

- Scottish engineer and naval architect; father of the solitary wave (soliton) and irregular wave reflection

John S. RUSSELL was born in Parkhead near Glasgow. He was the eldest son of Rev. David RUSSELL, a Scottish minister. Destined by his father for the Church, he showed a great love for mechanics at a very early age and eventually prevailed upon his father to allow him to study science and practical mechanics. He studied at the University of Glasgow, where he also graduated with an M.A. (1825). He moved to Edinburgh, where the University of Edinburgh offered him a temporary appointment as professor of natural philosophy (1832–1833), which became vacant by the death of Sir John LESLIE, a Scottish physicist and mathematician.

In the following years, he carried out investigations for the Scotch Canal Company on the practicability of steam navigation on inland waterways. These studies sparked his interest in the effects of water waves on hulls. About this time, he commenced his well-known researches on the nature of waves and the resistance of fluids to the motion of floating bodies (1834). RUSSELL discovered and studied experimentally the "wave of translation," which he also called "great solitary wave" – a wave consisting of a single hump of constant shape and constant speed. His first paper, submitted to the British Association for the Advancement of Science (BAAS), showed how the wave of translation could be used to reduce the resistance of barges moving fast in a restricted waterway (1835). He spent a major portion of his professional life carrying out experiments to determine the properties of the solitary wave. While investigating unsteady, non-uniform, open-channel flow, he discovered a new type of irregular reflection, in appearance very similar to the Mach effect in gas dynamics, and presented his numerous experimental results to the BAAS (1834–1835, 1837). For these wave studies the Royal Society of Edinburgh awarded him the Gold Medal.

As an outcome of his study on water waves he proposed his "wave-line design," a reverse-curve form of the bow in order to produce a solitary wave of the smallest possible amplitude – thus reducing the water resistance of the hull. The first vessel based on this wave line system, the 18.3-m-long SS *Wave*, was built in 1835, followed by three longer experimental vessels of differing forms to test his theory. Having moved to London (1844), he became a well-known shipbuilder on the Thames and constructed a number of ships on his principle such as the HMS *Warrior* (1860), the world's first wholly ironclad battleship. He believed that his wave-line theory could also be extended to reduce the resistance of ships in the open sea based upon the idea of pushing the water aside with minimum loss of energy. Although ships of his design were faster and could be operated more economically, he was unable to derive a sound mathematical theory of ship design. Together with the British naval engineer Isambard K. BRUNEL he cooperated on the design of big steamers, one of which was the enormous 211-m, 18,914-ton SS *Great Eastern*, RUSSELL being responsible for designing the hull form and paddle engines as well as for the actual building (1851–1858). It was the first ship to be built with a double iron hull and had sufficient tonnage to store the first transatlantic cable (1866).

Besides naval construction, RUSSELL practiced in other fields of engineering: for example, he wrote an article on steam engines in the *Encyclopædia Britannica* (7th edn., 1842), constructed a steam-ferry for carrying railways across the Lake of Constance (1868), built a steam coach for common roads, and designed the great dome of the Great Exhibition of Vienna (1873), whose clear span was about 110 m.

In addition, he took an active part in the foundation of the Royal Institution of Naval Architects and in the management of the Royal Society for the Encouragement of Arts, Manufactures and Commerce (RSA). He wrote many professional and scientific papers, which he read before various institutions and societies.

**ORIGINAL WORKS.** With J. ROBISON: *Report of the Committee on Waves.* Rept. Meet. Brit. Assoc. **7**, 417-496 (1838) — *Supplementary report of the Committee on Waves.* Ibid. **12**, 19-21 (1842) — *On the abnormal tides of the Firth of Forth.* Ibid. **12**, 115-116 (1842) — *Report of a series of observations on the tides of the Firth of Forth and the east coast of Scotland.* Ibid. **13**, 110-112 (1843) — *On the tides of the east coast of Scotland.* Ibid. **14**, 6 (1844) — *On the nature of the sound-wave.* Ibid. **14**, 11 (1844) — *Report on waves. Made to the Meeting of the British Association for the Advancement of Science (1834–1843).* Ibid. **14**, 311-390 (1844); later publ. in: *The wave of translation in the oceans of water, air, and ether.* Trübner, London (1885) — *Account of a cheap and portable self-registering tide-gauge invented by John WOOD.* Edinb. New Phil. J. **38**, 179-182 (1845) — *On the terrestrial mechanism of the tides.* Proc. Roy. Soc. Edinb. **1**, 179-182 (1845) — *On certain effects produced on sound by the rapid motion of the observer.* Rept. Meet. Brit. Assoc. **18**, 37-38 (1848) — *On gun-cotton.* Quart. J. Sci. **1**, 401-412 (1864) — *On the mechanical nature and uses of gun-cotton.* Proc. Roy. Inst. **4**, 292-299 (1866) — *On the true nature of the wave of translation and the part it plays in removing the water out of the way of a ship with least resistance.* Trans. Roy. Inst. Nav. Arch. **20**, 59-84 (1879) — *On the true nature of the resistance of armor to shot.* Ibid. **21**, 69-92 (1880) — *The wave of translation and the work it does as the carrier wave of sound.* Proc. Roy. Soc. Lond. **32**, 382-383 (1881).
**SECONDARY LITERATURE.** J.W. STRUTT (Lord RAYLEIGH): *On waves: the solitary wave. Periodic waves in deep water.* Phil. Mag. **1** [V], 257-279 (1876) — Anonymous: *Memoir of Mr. John RUSSELL, F.R.S.* Trans. Roy. Inst. Nav. Arch. **23**, 258-261 (1882) — W.H.B. (anonymous): *Obituary note of J.S. RUSSELL.* Proc. Roy. Soc. Lond. **34**, xv-xvi (1883) — H. ROUSE and S. INCE: *History of hydraulics.* Dover Publ., New York (1957) — G.S. EMMERSON: *John Scott RUSSELL. A great Victorian engineer and naval architect.* Murray, London (1977) — A.C. NEWELL: *Solitons in mathematics and physics.* Society for Industrial and Applied Mathematics (SIAM), Philadelphia, PA (1985); *see The history of the soliton*, pp. 1-21 — D.K. BROWN: *RUSSELL, John Scott.* In: (H.C.G. MATTHEW, ed.) *Oxford dictionary of national biography.* Oxford University Press, Oxford (2004), vol. 48, pp. 312-314.
**PICTURE.** Courtesy Science Museum Library, London, U.K.

# SAINT-VENANT, Adhémar Jean Claude, Barré DE SAINT-VENANT (1797–1886)

▪ French mathematician and civil engineer

Adhémar J.C. SAINT-VENANT was born at the castle of Fortoiseau (Dépt. Seine-et-Marne). His father was a distinguished agronomist and a former officer. He entered the Ecole Polytechnique in 1813. However, because of his political activities he was not permitted to continue his study.

Working already early in his career very successfully as an assistant in the Service des Poudres et Salpêtres (1814–1823), he was soon permitted by the government to enter the Service des Ponts et Chaussées without a final examination. There he finished his study at the school of this organization (1825). Already in 1834 he presented two remarkable papers to the Paris Academy on theoretical mechanics and fluid dynamics, which were published posthumously (1888). After working for some time on various French channels, he devoted himself to teaching on the subject of the strength of materials at the Ecole des Ponts et Chaussées (1837–1842). Together with Pierre L. WANTZEL (1814–1848), a French road engineer, he studied the exhaust of gases from nozzles and provided fundamental formula relating pressure and speed in compressible flow (1839). His *Mémoires d'hydraulique agricole* earned him a medal of the Society of Agriculture (1849) and a position as a professor at the newly founded Institute of Agriculture at Versailles (1850).

In 1852, he retired from his duties as Chief Engineer of the Paris constructions authority in order to fully devote himself to science, in particular to mechanics with an emphasis on the elasticity and rigidity of solids. Investigating the bending of rectangular beams both experimentally and theoretically, he stated that (1) the cross sections of a beam remain plane during the deformation and (2) the longitudinal fibers of a beam do not press upon but rather glide along each other [French *glissement*], being in a state of tension or compression. In 1855, he published a paper titled *Mémoire sur la torsion des prismes, avec des considérations sur leur flexion, ainsi que sur l'équilibre intérieur des solides élastiques en général* ("Memoir on the torsion of prisms, with considerations on their deflection, as well as on the internal equilibrium of elastic solids in general"), where he presented a new engineering method of calculating stresses in structures, which he called "la méthode mixte" and which others (*e.g.*, CLÉBSCH, TAIT, KELVIN, KIRCHHOFF) called the "de Saint-Venant problem" – today better known as the "Saint-Venant principle." This memoir, together with an article on the bending of bars, published in the following year in the *Journal de mathématiques pures et appliquées*, constituted the most complete and exact solution to the problem of deformation hitherto known. Treating the head-on collision of long, cylindrical bars (1862–1867), which was later applied in the Hopkinson pressure bar and derived devices for measuring mechanical properties of materials, he pointed out that elastic waves were generated during collisions that, even in the case of an ideal-elastic material, absorb kinetic energy, a fact that was later confirmed experimentally by Carl W. RAMSAUER (1909). In connection with the application of nonuniform load distribution at the bar end of Hopkinson devices and their proper measurement using strain gauges bonded to the bar outside, an investigation was made in the 1950s into whether the Saint-Venant principle could also be extended to dynamic "nonequilibrium" loading problems.

DE SAINT-VENANT published about 160 papers, notes, and memoirs covering geometry, mathematics, mechanics, hydrostatics, hydrodynamics, and agriculture. His significant contributions to the mechanics of solid bodies made him soon internationally renowned, and after the death of Gaspar G. DE CORIOLIS he was elected head of the mechanics section of the Paris Academy of Sciences (1843). He was also a member of the Institut de France, of the Royal Society of Göttingen, and of the Academies of Manchester, Louvain, and Brussels.

**ORIGINAL WORKS.** With P.L. WANTZEL: *Mémoire et expériences sur l'écoulement de l'air.* J. Ecole Polytech. (Paris) **16**, 85-122 (1839); *Mémoire et expériences sur l'écoulement de l'air, déterminé par des différences de pression considérables.* C. R. Acad. Sci. Paris **8**, 294-298 (1839); *Nouvelles expériences sur l'écoulement de l'air déterminé par des différences de pression considérables.* Ibid. **17**, 1140-1142 (1843) — *Note à joindre au mémoire sur la dynamique des fluides.* Ibid. **17**, 1240-1243 (1843) — *Sur la question 'Si la matière est continue ou discontinué.'* Proc. Verb. Soc. Philom. Paris (1844), pp. 3-16 — With P.L. WANTZEL: *Note sur l'écoulement de l'air.* C. R. Acad. Sci. Paris **21**, 366-369 (1845) — *Mémoire sur la perte de force vive d'un fluide, aux endroits où sa section d'écoulement augmente brusquement ou rapidement.* Ibid. **23**, 147-149 (1846) — *Solutions des problèmes du choc transversal et de la résistance vive des barres élastiques appuyées aux extrémités.* J. de l'Institut (Société Philomathique) **22**, 61-63 (1854) — *Mémoire sur la torsion des prismes, avec des considérations sur leur flexion, etc.* Mém. Savants Étrangers (Paris) **14**, 233-560 (1856) — *Mémoire sur la flexion des prismes élastiques.* J. Math. Pures Appl. **1** [II], 89-189 (1856) — *Sur la vitesse du son.* J. de l'Institut (Société Philomathique) **24**, 212-216 (1856) — *Mémoire sur l'impulsion transversale et la résistance vive des barres élastiques appuyées aux extrémités.* C. R. Acad. Sci. Paris **45**, 204-208 (1857) — *Mécanique appliquée de NAVIER, annotée par SAINT-VENANT.* Dunod, Paris (1858) — *Mémoire sur l'impulsion, la résistance vive et les vibrations des pièces solides etc.* C. R. Acad. Sci. **62**, 130-134 (1866) — *Note sur les pertes apparentes de force vive dans le choc des pièces extensibles et flexibles.* Ibid. **62**, 1195-1199 (1866) — *Sur le choc longitudinal de deux barres élastiques de grosseurs et de matières semblables ou différentes et sur la proportion de leur force vive, qui est perdue pour la translation ultérieure. Et généralement sur le mouvement longitudinal d'un système de deux ou plusieurs prismes élastiques.* Ibid. **63**, 1108-1111 (1866); J. Math. Pures Appl. **12** [II], 237-376

(1867) — *Mémoire sur le choc longitudinal de deux barres, etc.* C. R. Acad. Sci. Paris **64**, 1009-1013 (1867) — *Démonstration élémentaire: 1e de l'expression de la vitesse de propagation du son dans une barre élastique; 2e des formules données pour le choc longitudinal de deux barres.* Ibid. **64**, 1192-1195 (1867) — *Choc longitudinal de deux barres élastiques, dont l'une est extrêmement courte ou extrêmement vide par rapport à l'autre.* Ibid. **66**, 650-653 (1868) — *Solution, en termes finis, du problème du choc longitudinal de deux barres élastiques en forme de tronc de cône ou de pyramide.* Ibid. **66**, 877-881 (1868) — *Théorie du mouvement non permanent des eaux.* Ibid. **73**, 147-154, 237-240 (1871) — *Du choc longitudinal d'une barre élastique libre contre une barre élastique d'autre matière ou d'autre grosseur fixée au bout non heurté, etc.* Ibid. **95**, 359-365 (1882) — With M. FLAMANT: *Résistance vive ou dynamique des solides. Représentation graphique des lois du choc longitudinal, subi à une de ses extrémités par une tige ou barre prismatique assujettie à l'extrémité opposée.* Ibid. **97**, 444-447 (1883) — *Mouvement des molécules de l'onde dite solitaire, propagée à la surface de l'eau d'un canal.* Ibid. **101**, 1101-1105, 1215-1218 (1885) — *Résistance des fluides.* Mém. Acad. Sci. Inst. France **44**, Mémoire No. 5, 1-280 (1888).
SECONDARY LITERATURE. J.V. BOUSSINESQ and M. FLAMANT: *Notice nécrologique sur la vie et les travaux de M. DE SAINT-VENANT.* Ann. Ponts Chaussées **12** [VI], 557-595 (1886) — E. PHILLIPS: *Notice sur M. de SAINT-VENANT et sur ses travaux.* C. R. Acad. Sci. Paris **102**, 141-147 (1886) — J. BOUSSINESQ: *Quelques mots sur la vie et l'œuvre de M. DE SAINT-VENANT.* Mém. Soc. Sci. (Lille) **15** [IV], 147-151 (1888) — S.P. TIMOSHENKO: *History of strength of materials.* McGraw-Hill, New York (1953), pp. 139-140, 229-233 — B.A. BOLEY: *Application of SAINT-VENANT's principle in dynamical problems.* J. Appl. Mech. **22**, 204-206 (1955) — J. ITARD: *A.J.C. DE SAINT-VENANT.* In: (C.C. GILLESPIE, ed.) *Dictionary of scientific biography.* Scribner, New York, vol. 12 (1975), pp. 73-74.
PICTURE. Courtesy J. BARANDE, Bibliothèque de l'Ecole Polytechnique, Paris, France.

# SALCHER, Peter (1848–1928)

- Austrian physicist

Peter SALCHER was born in Kreuzen-Ebene, a small village in Carintia, Austro-Hungarian Empire (now southern Austria). His father, Peter SALCHER III, was a school teacher, mayor, farmer, and innkeeper. He attended the gymnasium in Klagenfurt (1861–1868) and studied natural philosophy at the University of Graz, where he graduated with a *Dr. phil.* (1872). After a short period of teaching at various gymnasia in Graz and Triest (1872–1875), he was appointed full professor of physics and mechanics (1875–1909) at the Austro-Hungarian Imperial Royal Naval Academy in Fiume (now Rijeka, Croatia), where he managed from 1880 the local meteorological station. Being a skillful experimenter and inventor of scientific instruments, SALCHER became internationally famous for his supersonic ballistic experiments, which he

performed at the Naval Academy in Fiume (1886–1889). He corresponded with the Austrian physicist Ernst MACH, spiritual father and originator of this ballistic study, on the progress of his work, also contributing many ideas of his own. Together with John WHITEHEAD, son of the British torpedo expert and manufacturer Robert WHITEHEAD, he visualized free air jets generated by discharging a high-pressure reservoir through a small opening. They first noticed a characteristic oblique pattern of stationary waves in the jet, now known as "shock diamonds" – a curious irregular interaction phenomenon of shock waves which creates a sequence of Mach disks in the jet and occurs anytime a flow exits a nozzle at supersonic speeds. These experiments, no less important than the discovery of the head wave phenomenon, were later resumed in Göttingen by Ludwig PRANDTL and Theodor MEYER, stimulating the understanding of supersonic flow both inside and outside of Laval nozzles.

SALCHER published numerous papers dealing with physical, meteorological, and fluid dynamic observations and wrote several books on mechanics, physics, oceanography, meteorology, and the history of the Austro-Hungarian Navy. These papers were primarily intended to serve the education of K.u.K. naval cadets. He received various high-ranking Austrian awards, such as the Knight's Cross of the Franz Joseph Order and the *Signum laudis* (an Austrian medal of honor). He was a corresponding member of the Austrian and Australian Academies of Sciences, and a member of the French Physical Society.

ORIGINAL WORKS. *Elemente der theoretischen Mechanik.* Gerold, Wien (1881) — With E. MACH: *Photographische Fixierung der durch Projectile in der Luft eingeleiteten Vorgänge.* Sitzungsber. Akad. Wiss. Wien **95** (Abth. IIa), 764-778 (1887); Ann. Phys. **32** [III], 277-291 (1887) — *Über die Fortpflanzungsgeschwindigkeit des durch scharfe Schüsse erregten Schalls.* Sitzungsber. Akad. Wiss. Wien **97** (Abth. IIa), 1045-1052 (1888) — With E. MACH: *Über die in Pola und Meppen angestellten ballistisch-photographischen Versuche.* Ibid. **98** (Abth. IIa), 41-50 (1889) — With J. WHITEHEAD: *Über den Ausfluß stark verdichteter Luft.* Ibid. **98** (Abth. IIa), 267-287 (1889) — With E. MACH: *Optische Untersuchungen der Luftstrahlen.* Ibid. **98** (Abth. IIa), 1303-1309 (1889) — *Optische Untersuchungen der Luftstrahlen.* Ann. Phys. **41** [III], 144-150 (1890) — *Physik und Mechanik.* 2 vols., Hof- und Staatsdruckerei, Wien (1891, 1895) — *Die Wasser-Spiegelbilder, Angaben für Zeichner, Maler und Photographen.* V. Knapp, Halle a. S. (1903).

SECONDARY LITERATURE. H. BAYER VON BAYERSBERG: *Österreichs Admirale und bedeutende Persönlichkeiten der k. und k. Kriegsmarine 1867–1918.* Bergland, Wien (1962), pp. 158-160 — P. BERZ and C. HOFFMANN: *Sichtbare Grenzen: Ernst MACHs Notizen zu den ballistisch-fotografischen Versuchen 1886/87.* Max-Planck-Institut für Wissenschaftsgeschichte, Berlin (2000) — C. HOFFMANN: *Über Schall: Ernst MACHs und Peter SALCHERs Geschoßfotografien.* Wallstein, Göttingen (2001) — W.G. POHL: *Peter SALCHER und Ernst MACH – Schierenphotografie von Überschall-Projektilen.* Plus Lucis (Wien), Hefte 1/2 (2002/2003), pp. 22-26 — W.G. POHL and G. SALCHER: *Fotografien fliegender Projektile.* Mitteil. Österreich. Gesell. Wissenschaftsgeschichte (Erasmus, Wien) **21**, 125-154 (2003).

PICTURE. Courtesy Dr. Günther SALCHER, Hermagor, Austria.

NOTE. While photographing supersonic projectiles (1886–1889), SALCHER continuously reported to E. MACH on the progress of his investigations in 140 letters. MACH's response letters, in 1993 rediscovered by the author, are kept by G. SALCHER at Hermagor. The Max-Planck-Institut (MPI) für Wissenschaftsgeschichte at Berlin plans to edit and annotate this correspondence.

## SCHALL, Rudi Joachim (1913–2002)

- German detonation and shock wave physicist

Rudi J. SCHALL was born in Berlin-Zehlendorf to Wilhelm SCHALL, a high-school teacher. After attending the secondary school in Zehlendorf, he entered the Friedrich-Wilhelms-Universität (since 1945 Humboldt-Universität) in Berlin and began to study mathematics, physics, and chemistry (1931). After his diploma he took the degree of *Dr. phil.* (1937) with a work on acoustics entitled *Symmetrisch geschaltete kapazitive Mikrophone* ("Symmetrically Connected Capacitive Microphones"); his thesis adviser was Prof. Arthur Rudolph WEHNELT, renowned for his invention of the "Wehnelt cylinder." SCHALL continued to work at the Friedrich-Wilhelms-Universität as an assistant to Prof. Erich SCHUMANN, an acoustician and military physicist who also headed the research division of the Heereswaffenamt (Army Ordnance Office) and was Deputy of Detonation Physics in the Reichswehr Ministry (1942–1945). Asked by SCHUMANN to explore the actual state of the art of nonlinear acoustics, SCHALL got involved in explosives and detonation physics. He performed military research for the army and became an expert on explosives and detonation.

In 1946, SCHALL joined Prof. Hubert SCHARDIN at the newly founded Laboratoire de Recherches Techniques de Saint-Louis (LRSL), which, by a German-French treaty signed in 1959, was transformed into a binational research institute, named the *Institut Franco-Allemand de Recherches de Saint-Louis* or *Deutsch-Französisches Forschungsinstitut Saint-Louis* (ISL). There he continued his research in detonation physics and did pioneering work on hollow charges, shock waves, detonation waves, and high-speed diagnostics. SCHALL's contributions are manifold. (1) In 1950, he was the first to obtain a flash X-ray diffraction pattern of a shock-loaded crystal within ultrashort times, a subject that was taken up again in the United States at LLNL not until the late 1960s. (2) In 1950, he also determined the Hugoniot curves of some liquids and solids up to very high pressures using an explosive compression technique and flash radiography. With this combined method he measured the detonation pressure of explosives. (3) Based upon the first systematic series of flash radiographs of jet formation in shaped charges, which he obtained in cooperation with Gustav THOMER, a German physicist at ISL, he derived a theory of twist-stabilized hollow charges. Later, in cooperation with the Schlumberger Company, he also explored civil applications of short hollow charges for use in the petroleum industry. (4) Based on the hydrodynamic theory of detonation and on the assumption of an incomplete decomposition of solid explosives, he concluded that it should be possible to produce stable low-velocity detonations over a fairly wide range of velocities. (5) He studied the initiation process of secondary explosives by a spark discharge, which would allow a safer handling of explosives, particularly in mining applications. (6) He also took a continuous interest in the progress of modern high-speed diagnostic methods, initiated their further development at ISL, and reviewed modern achievements at international congresses.

In the period 1962–1969, SCHALL worked for NATO in Paris and Brussels, serving there as deputy general secretary of the Division of Scientific Affairs. After his return to ISL, he was appointed German Director, a position which he held until his retirement (1969–1979). Thereafter, he directed the Carl-Cranz-Gesellschaft (CCG) in Weil am Rhein, a society devoted to post-academic training in engineering sciences. He published more than 70 papers in the open literature, which brought him renown beyond the military research community. At the International Congresses on High-Speed Photography he was National Delegate of the Federal Republic of Germany (1968–1980). In 1980, he received the Photo-Sonics Award of the American Society of Motion Pictures and Television Engineers (SMPTE).

**ORIGINAL WORKS.** With E. SCHUMANN and G. HINRICHS: *Röntgenblitzuntersuchungen an Hohlkörpern.* 1. Bericht des Bevollmächtigten für Sprengstoffphysik (Reichswehrministerium, Berlin) (1943) — *Detonationswelle und Schwadenströmung.* Ibid. 4. Bericht (1943) — *Röntgenblitzuntersuchungen an Detonationsvorgängen.* Ibid. 10. Bericht (1943) — *Feinstrukturaufnahmen in ultrakurzen Zeiten mit dem Röntgenblitzrohr.* Z. Angew. Phys. **2**, 83-88 (1950) — *Die Zustandsgleichung des Wassers bei hohen Drucken nach Röntgenblitzaufnahmen intensiver Stoßwellen.* Ibid. **2**, 252-254 (1950) — With G. THOMER: *Röntgenblitzaufnahmen von Stoßwellen in festen, flüssigen und gasförmigen Medien.* Ibid. **3**, 41-44 (1951) — *Röntgenblitzuntersuchungen bei nichtidealen Detonationswellen.* Ibid. **4**, 291-293 (1952) — *Röntgenröhre in Betrieb und Anwendung.* Archiv Tech. Messen (ATM) Lfg. **74-13**, pp. 117-120 (Mai 1953) — *Die Initiierungsempfindlichkeit von Sekundärsprengstoffen.* Z. Naturforsch. **8a**, 676 (1953); Nobel-Hefte **20**, 75-79 (1954) — *Etudes par radiographie-éclair des phénomènes de charges creuses.* In: (P. NASLIN and J. VIVIE, eds.) *Actes 2e Congrès Int. de Photographie et Cinématographie Ultra-Rapide* [Paris, France, Sept. 1954]. Dunod, Paris (1956), pp. 261-266 — *Die Stabilität langsamer Detonationen.* Z. Angew. Phys. **6**, 470-475 (1954) — *Zur Frage der Röntgenabsorption in komprimierten Flüssigkeiten.* Z. Phys. **142**, 637-641 (1955) — *Röntgenblitz-Untersuchungen bei Sicherheitssprengstoffen.* Nobel-Hefte **21**, 1-10 (1955) — *X-ray flash measurements on shock waves in solids.* In: (R.B. COLLINS, ed.) *Proc. 3rd Int. Congress on High-Speed Photography* [London, U.K., Sept. 1956]. Butterworths, London (1959), pp. 228-237 — *Methoden und Ergebnisse der Detonationsdruckbestimmung bei festen Sprengstoffen.* Z. Elektrochem. **61**, 629-635 (1957) — *Fortschritte der militärischen Sprengstoff-Forschung.* Wehrtech. Monatshefte **54**, 386-394 (1957) — *Untersuchung an Detonationsstoßwellen in Leichtmetallen zur Bestimmung der Zustandsgleichung der Metalle.* Explosivstoffe **6**, 120-124 (1958) — With K. VOLLRATH: *Zum Verhalten ferroelektrischer Keramik bei intensiven Stoßwellen.* Proc. 4th Int. Congress on High-Speed Photography [Cologne, Germany, Sept. 1958]. In: (O. HELWICH and H. SCHARDIN, eds.) *Kurzzeitphotographie.* Helwich, Darmstadt (1959), pp. 329-334 — With J. DELACOUR: *Le développement de revêtements spéciaux pour les charges creuses appliqués à la perforation des sondages pétrolifères.* Rev. Inst. Franç. Pétrole **14**, 1423-1467 (1959) — *Über Druckwellen bei Sprengungen in Sand und sandigen Böden.* Nobel-Hefte **26**, 76-90 (1960) — *High-speed measurement of shock compressibility of solids in the 1-Mb range.* In: (J.S. COURTNEY-PRATT, ed.) *Proc. 5th Int. Congr. on High-Speed Photography* [Washington, DC, Oct. 1960]. SMPTE, New York (1962), pp. 184-187 — *Der Meteoriteneinschlag, eine Aufgabe der Raumfahrt an die Ballistik.* Physik. Blätter **18**, 308-313 (1962) — *Präzisionsmessungen der Detonationseigenschaften gegossener TNT-Hexogen-Gemische.* Nobel-Hefte **28**, 133-143 (1962) — With K. VOLLRATH: *Sur la conductibilité électrique provoquée par les ondes de détonation dans les explosives solides.* In: *Les Ondes de Détonation* [Int. Conf., Gif-sur-Yvette, France, Aug./Sept. 1961]. Editions du CNRS, No. 109, Paris (1962); Rapport 2/61, Institut Franco-Allemand de Recherches de Saint-Louis, France (1961) — With K. VOLLRATH: *Piezoelektrische Funkengeneratoren.* In: (J.G.A. DE GRAAF and P. TEGELAAR, eds.) *Proc. 6th Int. Congr. on High-Speed Photography* [Scheveningen, The Netherlands, Sept. 1962]. Willink & Zoon, Haarlem (1963), pp. 403-408 — With E. DAVID and H. SCHARDIN: *Visualization of wave propagation in impulse-loaded bars.* IUTAM Symp. on Stress Waves in Anelastic Solids [Brown University, Providence, RI, April 1963]. In: (H. KOLSKY and W. PRAGER, eds.) *Stress waves in anelastic solids.* Springer, Berlin (1964), pp. 183-192 — With H. SCHARDIN: *Die Sprengformung.* VDI-Nachrichten (1. April 1964) — With W. KEGLER: *Mechanical and detonation properties of rubber bonded sheet explosives.* Rept. T36/65, ISL, Saint-Louis, France (1965); In: (S.J. JACOBS, ed.) *Proc. 4th (Int.) Symp. on Detonation* [White Oak, MD, Oct. 1965]. NTIS, Springfield, VA (1965), pp. 496-501 — With C.L. LECOMTE: *Etude d'impacts à grande vitesse à l'aide d'un canon à gaz léger.* In: (B.H. GOETHERT and H.H. KURZWEG, eds.) *The fluid dynamic aspects of space flight* [AGARD-NATO Specialists' Meeting, Marseille, France, April 1964]. AGARDograph No. 87, Gordon & Breach, New York (1966); vol. 1, pp. 315-330 — With H.D. VON STEIN and B. KOCH: *Eine Mikrowellen-Meßmethode zum Nachweis intensiver Stoßwellen in nichtleitenden Festkörpern.* Z. Naturforsch. **20a**, 157-158 (1965) — With H.D. VON STEIN, B. KOCH, and W. KEGLER: *Messung von Detonationsgeschwindigkeiten in gegossenen Sprengstoffen mit Hilfe von Mikrowellen.* Explosivstoffe **14**, 145-154 (1966) — *Dynamische Hochdruckphysik.* In: *Beiträge zur Ballistik und Technischen Physik.* Wehrtech. Monatshefte **64**, Beiheft Nr. 7, 128-143 (1967) — *Detonationsphysik.* In: (K. VOLLRATH and G. THOMER, eds.) *Kurzzeitphysik.* Springer, Wien (1967), pp. 849-907 — *Detonation physics.* In: (P. CALDIROLA and H. KNOEPFEL, eds.) *Physics of high energy density.* Academic Press, New York (1971), pp. 230-244 — *High-speed photomicrography.* HSSP Newsletter **1**, No. 1 (Winter 1981) — *On the hydrodynamic theory of detonation.* Propellants, Explosives, Pyrotechnics **14**, 133-139 (1989) — With M. HELD: *Geformte Ladung.* Germ. Patent No. 3,808,110 (1994) — *Schwerpunkte der ballistischen Forschung des ISL in den 80er Jahren.* In: (G. WEIHRAUCH, ed.) *Ballistische Forschung im ISL 1945–1994.* Festschrift zu Ehren von Prof. Dr. R.E. KUTTERER anlässlich seines 90. Geburtstages. ISL, Saint-Louis, France (1994).

**SECONDARY LITERATURE.** K. VOLLRATH: *Beiträge zur Ballistik, Detonations- und Kurzzeitphysik. Herrn Dr. phil. Rudi SCHALL zum 65. Geburtstag.* Bundesministerium der Verteidigung, Bericht Nr. BMVg-FBWT 79-1 (1979) — M. HUGENSCHMIDT: *Dr. Rudi SCHALL feiert seinen 75. Geburtstag.* Badische Z. (Freiburg, Germany, 29. Nov. 1988).

**PICTURE.** Courtesy ISL, Saint-Louis, France.

# SCHARDIN, Hubert (1902–1965)

- German physicist, gas dynamicist and ballistician, father of numerous color schlieren methods

Hubert SCHARDIN was born in Plassow, Pomerania (now Plaszewo, Poland). He was the eldest son of Reinhold SCHARDIN, a school teacher. He studied physics at the Technische Hochschule Berlin-Charlottenburg (TH Berlin) and the University of Munich (1922–1928). After graduating in technical physics (1926), he became assistant of Geheimrat Prof. Carl

CRANZ at the Institut für Technische Physik of the TH Berlin, which was essentially the ballistic laboratory of the former Military Academy. Under CRANZ's supervision he worked on the theoretical foundations of the Toepler schlieren method and its applications, which earned him the Ph.D. (1934). CRANZ and SCHARDIN went together to Nanking to establish the first ballistic research institute in China (1934–1936). Thereafter, SCHARDIN became director (1936) of the Department of Technical Physics and Ballistics at the Technische Akademie der Luftwaffe (TAL) (Air Force Technical Academy) in Berlin-Gatow. In addition, he served as assistant professor (1936) and full professor (1941) at the TH Berlin.

At the end of World War II, his ballistic department at TAL was evacuated to southern Germany and after the war transferred by the French military to the Alsatian town of Saint-Louis. He became scientific director of the newly founded Laboratoire de Recherches Techniques de Saint-Louis. From 1945 SCHARDIN and the French Ingénieur-Général Robert CASSAGNOU expanded military research and in 1959 established the binational Institut Franco-Allemand de Recherches de Saint-Louis or Deutsch-Französisches Forschungsinstitut Saint Louis (ISL). SCHARDIN, who became honorary professor at the German Universities of Freiburg (1947) and Cologne (1965), also established two departments of applied physics in Weil am Rhein and at the University of Freiburg (1949), which, ten years later, merged into the Ernst-Mach-Institut (EMI), a subsidiary research institute of the Fraunhofer-Gesellschaft (FhG) with headquarters in Munich.

In the very fruitful period of cooperation with CRANZ (1926–1936), SCHARDIN tackled many technical problems of his time, which resulted in numerous milestone achievements. (1) He provided a solid theoretical background for schlieren imaging and applied it to a broad range of scientific pursuits. (2) In high-speed photography, he applied new spark light sources and the Kerr-effect for chronography. His ultrahigh-speed framing camera, which he developed in cooperation with CRANZ – the so-called "Cranz-Schardin multiple spark camera" [Germ. *Cranz-Schardin Mehrfach-Funkenkamera*] – earned him a Gold Medal at the Paris World Exhibition (1937). (3) He refined some main optical visualization techniques in gas dynamics and extended the theory of Mach-Zehnder interferometry and its applications. (4) He studied the ballistic drag of projectiles, both inside the barrel and during flight. (5) He extended the shock-tube theory and applied it to a number of practical examples. (6) He also worked on the fundamentals and applications of detonation physics, particularly on hollow charges. (7) After the war he mainly worked on problems of nonstationary gas dynamics. He theoretically and experimentally studied the nature of spherical blast waves generated in chemical explosions and blast effects on architectural structures. (8) Phenomena observed during interactions of blast waves with window panes also stimulated his interest in the nature of fracture mechanics. He was the first to use high-speed cinematography to resolve the fracture mechanism in brittle materials such as glass.

SCHARDIN presided over the Schutzkommission, a committee for the protection of civil buildings established by the German Ministry of the Interior, was cofounder and member of the executive board of the Deutsche Gesellschaft für Wehrtechnik (German Society for Military Technology), and initiator of the Carl-Cranz-Gesellschaft (CCG), then primarily an institution for scientific training of young military talents. He was one of the first and most enthusiastic promoters of founding an International Congress on High-Speed Photography; the first one was held in Washington, DC (1952).

The *Hubert Schardin Medal* award, instituted in his honor by the 8th International Congress on High-Speed Photography (1968), recognize achievements in high-speed diagnostics and in their many applications, both in fundamental research and in technical engineering. This award is particularly aimed at encouraging young scientists for performing outstanding work in this field.

**ORIGINAL WORKS.** With C. CRANZ: *Kinematographie auf ruhendem Film und mit extrem hoher Bildfrequenz.* Z. Phys. **56**, 147-183 (1929) — *Bemerkungen zum Druckausgleichsvorgang in einer Rohrleitung.* Physik. Z. **33**, 60-64 (1932) — With C. CRANZ: *Eine neue Methode zur Messung des Geschosswiderstandes im Rohr.* Z. Tech. Phys. **13**, 124-132 (1932) — With C. CRANZ and R.E. KUTTERER: *Der Kerreffekt-Chronograph, ein neuer Geschossgeschwindigkeitsmesser.* Wehr u. Waffen **10**, Nr. 9, 385-394 (1932) — *Einige interessante Einzelheiten und Erscheinungen beim Abfeuern eines Geschosses.* Dt. Jägerzeitung (Abt. Schießwesen) **100**, Nr. 22, 1-6 (Juni 1933) — *Theorie und Anwendungen des Mach-Zehnder'schen Interferenz-Refraktometers.* Z. Instrumentenkunde **53**, 396-403, 424-436 (1933) — *Das TOEPLER'sche Schlierenverfahren: Grundlagen für seine Anwendung und quantitative Auswertung.* Dissertation, TH Berlin (1934); VDI-Forschungsheft Nr. 367, 1-32 (1934) — *Die Luftströmung in der Umgebung eines fliegenden Geschosses als Ursache des Flugwiderstandes.* Dt. Jagd-Ausgabe **A35**, 701-702; Ibid. **A37**, 741-743 (1934) — *Neue Ergebnisse der Funkenkinematographie.* Z. Tech. Phys. **18**, 474-477 (1937) — *Beiträge zur Ballistik und Technischen Physik.* Barth, Leipzig (1938) — *Experimentelle Arbeiten zur Klärung sprengstoffphysikalischer Fragen.* In: *Probleme der Detonation.* Schriften Dt. Akad. Luftfahrtforsch. Nr. 1023/41G, 314-334 (1940/1941) — With G. THOMER: *Untersuchung des Hohlladungsproblems mit Hilfe der Röntgenblitz-Methode.* Prüfbericht Nr. 9, Ballist. Inst., Tech. Akad. Luftwaffe (TAL), Berlin-Gatow (Nov. 1941) — *Vorgänge bei hohen Belastungen und Belastungsgeschwindigkeiten.* Schriften Dt. Akad. Luftfahrtforsch. Nr. 40, 21-91 (1941) — *Gerät zur kinematographischen Aufnahme einer Bilderreihe mittels einer Reihe nacheinander ausgelöster Beleuchtungsfunkenstrecken.* Germ. Patent No. 703,039 (1941) — *Möglichkeiten der experimentellen Untersuchung am fliegenden Geschoss.* Berichte Lilienthal-Gesell. Luftfahrtforsch. (Berlin), Nr. 139, 38-51 (1941) — *Die Schlierenverfahren und ihre Anwendungen.* Ergebn. exakt.

Naturwiss. **20**, 303-349 (1942) — *Die physikalischen Grundlagen der Wirkungen der Detonation.* Jb. Dt. Akad. Luftfahrtforsch. (1943/1944), pp. 56-73 — *Bezüglich der abnehmenden Druckwellenwirkung von Explosionen in großer Höhe.* Bericht 9/43, Ballist. Inst., TAL, Berlin-Gatow (1943) — *Ergebnisse der kinematographischen Untersuchung des Glasbruchvorganges.* Glastech. Berichte **23**, 1-10, 67-79, 325-336 (1950) — *Über das Bruchkriterium bei kurzzeitiger Belastung.* Ibid. **23**, 189-193 (1950) — *Die Mehrfachfunkenkamera und ihre Anwendung in der technischen Physik.* Z. Angew. Phys. **5**, 19-24 (1953) — With W. DÖRING: *Detonationen.* In: *Naturforschung und Medizin in Deutschland 1939–1946* [FIAT Review of German Science]. Dieterich, Wiesbaden.; Bd. 11 (1953): (A. BETZ, ed.) *Hydro- und Aerodynamik,* pp. 97-125 — *Measurement of spherical shock waves.* Comm. Pure Appl. Math. **7**, 223-243 (1954) — *Über die Entwicklung der Hohlladung.* Wehrtech. Monatshefte **51**, Nr. 4, 97-120 (1954) — *Application de la cinématographie par étincelles à l'examen des phénomènes de rupture.* 2e Congrès International de Photographie et Cinématographie Ultra-Rapides [Paris, France, Sept. 1954]. In: (P. NASLIN and J. VIVIE, eds.) *Photographie et Cinématographie Ultra-Rapides.* Dunod, Paris (1956), pp. 301-314 — With H. MORITZ and G. SCHÖNER: *Wirkungen von Spreng- und Atombomben auf Bauwerke.* Ziv. Luftschutz **18**, 283-291 (1954) — *Entstehung und Wirkung von Stoßwellen bei Explosionsvorgängen.* Kölner Tech. Mitteil. KTM/VDI Nr.3, 3-8 (1955) — *Die untere Grenze der Zeitmessung.* 5e Congrès Int. de Chronométrie [Paris, Oct. 1954]. Procès Verb. Mém. (Observatoire National de Besançon) **3**, 729-738 (1956) — *Ausbreitung von elastischen Oberflächendeformationen bei Stoßbelastung fester Körper.* VDI-Berichte **8**, 124-129 (1956) — *High-frequency cinematography in the shock tube.* In: (R.B. COLLINS, ed.) *Proc. 3rd Int. Congr. on High-Speed Photography* [London, U.K., Sept. 1956]. Butterworths, London (1957), pp. 365-369 — *Naturwissenschaft und Wehrforschung in gegenseitiger Beeinflussung.* Doube, Essen (1957) — *Die Kurzzeitphotographie in der Ballistik.* Wehrtech. Monatshefte **54**, Heft 8/9, 292-314 (1957) — *Ein Beispiel zur Verwendung des Stoßwellenrohres für Probleme der instationären Gasdynamik* [Festschrift J. ACKERET zum 60. Geburtstag]. ZAMP **9b**, 606-621 (1958) — *C. CRANZ als Mitbegründer der Kurzzeitphotographie.* Proc. 4th Int. Congr. on High-Speed Photography [Cologne, Germany, Sept. 1958]. In: (O. HELWICH and H. SCHARDIN, eds.) *Kurzzeitphotographie.* Helwich, Darmstadt (1959), pp. 1-8 — *Filmaufnahmen von Vorgängen im Stoßwellenrohr.* Ibid., pp. 139-141 — *Modellversuche für Explosionsvorgänge.* Ibid. pp. 142-150 — *Essais sur les effets d'ondes de choc.* Rept. 6/62, ISL, Saint-Louis, France (1962) — With E. DAVID and R. SCHALL: *Visualization of wave propagation in impulse-loaded bars.* IUTAM Symp. on Stress Waves in Anelastic Solids [Providence, RI, April 1963]. In: (H. KOLSKY and W. PRAGER, eds.) *Stress waves in anelastic solids.* Springer, Berlin (1964), pp. 183-192 — (H. SCHARDIN, ed.) Symp. über wissenschaftliche Grundlagen des Schutzbaues [Freiburg, Germany, Sept. 1960]. Bericht Nr. 13/64, EMI, Freiburg (1964) — With H. REICHENBACH: *The behavior of shock waves in ducts and when entering entrance structures.* [presened at a symposium held in Washington, DC, 1965]. In: *Protective structures for civilian populations.* Natl. Acad. Sci., Washington, DC (1965); *Verhalten von Stoßwellen in Kanälen und beim Eindringen in Eingangsbauwerke.* Bericht 8/65, EMI, Freiburg (1965) — *Untersuchung instationärer gasdynamischer Vorgänge als Beispiel für den zweckmäßigen Einsatz der Hochgeschwindigkeitskinematographie.* Proc. 7th Int. Congr. on High-Speed Photography [Zurich, Sept. 1965]. In: (O. HELWICH, ed.) *Kurzzeitphotographie.* Helwich, Darmstadt (1967), pp. 17-23. — With H. REICHENBACH: *Stosswellenrohr zur Erzeugung von Gleichdruckstoßwellen.* Germ. Patent No. 1,273,850 (1968).

**SECONDARY LITERATURE.** (Ed.) *Hubert SCHARDIN, 1902–1965.* Astronaut. Acta **12**, 192 (1966) — K. VOLLRATH: *Hubert SCHARDIN.* Proc. 7th Int. Congress on High-Speed Photography [Zurich, Sept. 1965]. In: (O. HELWICH, ed.) *Kurzzeitphotographie.* Helwich, Darmstadt (1967), pp. vi-vii — *Various authors: Beiträge zur Ballistik und Technischen Physik, Gedenkschrift für Hubert SCHARDIN.* Beiheft Nr. 7 der Wehrtech. Monatshefte. Mittler, Frankfurt/M. (1967) — R. SCHALL: *Hubert SCHARDIN in memoriam. List of publications by Prof. Dr.-Ing. Hubert SCHARDIN.* In: (N.R. NILSSON and L. HÖGBERG, eds.) *Proc. 8th Int. Congr. on High-Speed Photography* [Stockholm, Sweden, June 1968]. Almquist & Wiksell, Stockholm (1968), pp. 386-388, 497-500 — H. REICHENBACH: *Contributions of H. SCHARDIN to the theory and applications of the shock tube.* Phys. Fluids **5**, 1-3 (1969) — *Kolloquium zum Andenken an Prof. Dr.-Ing. Hubert SCHARDIN. Festreden* [with a detailed list of his publications]. ISL, Saint-Louis, France (June 1982) — K. VOLLRATH: *Hubert SCHARDIN. His work and his influence on modern high-speed physics and photonics.* In: (L.L. ENDELMAN, ed.) *Proc. 15th Int. Congr. on High-Speed Photography and Photonics* [San Diego, CA, Aug. 1982]. Proc. SPIE **348**, 1008-1011 (1982) — H. REICHENBACH: *Hubert SCHARDIN (1902–1965). His life and work.* In: (J.M. DEWEY, ed.) *Proc. 20th Int. Congr. on High-Speed Photography and Photonics* [Victoria, B.C., Canada, Sept. 1992]. Proc. SPIE **1801**, 2-9 (1993).

**PICTURE.** Courtesy Archives of Ernst-Mach-Institut (EMI), Freiburg, Germany.

## SCHMIDT, Oswald VON (1889–1945)

- Estonian-born German physicist

There is little information about Oswald VON SCHMIDT's life except what can be found in his autobiography attached to his Ph.D. thesis. He was born in Dorpat (at that time a part of Russia, now Tartu, Estonia) to Arved VON SCHMIDT, an attorney. After attending a private gymnasium and the Dorpat State Gymnasium, he studied chemistry and physics (1909–1912) at the University of Dorpat. He graduated from the University of Göttingen in physical chemistry and metallurgy and began to prepare his Ph.D. thesis under the guidance of Prof. Gustav TAMMANN and Prof. Otto WALLACH but was drafted at the beginning of World War I. Thereafter, he continued his studies at the Friedrich-Wilhelms-Universität (since 1945 Humboldt-Universität) in Berlin and took his Ph.D. (1919–1921) under the guidance of Prof. Walther Hermann NERNST, an eminent physical chemist.

Later he investigated the propagation and reflection phenomena of waves at boundaries of different states of matter. Theoretical foundations were already laid by him in the late 1920s and later continued at the Technische Hochschule Berlin by a study group that was financially supported by the Notgemeinschaft der Deutschen Wissenschaft. While employed at the Ballistic Institute of the Technische Akademie der Luftwaffe (TAL), a recently established technical academy of the German Air Force in Berlin-Gatow (1938), he performed his famous photography of head wave phenomena at liquid and solid boundaries, the so-called "Schmidt head wave" [Germ. *VON SCHMIDTsche Kopfwelle*]. This pseudo-supersonic wave effect which can be observed both in acoustics and shock wave physics is of great importance

VON SCHMIDT died in Berlin at the end of World War II.

**ORIGINAL WORKS.** *Die Dissoziationswärme des Wasserstoffes und des Chlors.* Dissertation, Universität Berlin (1921) — *Angewandte Seismik.* Z. Geophys. **4**, 134-146 (1928) — *Theorie der 3-Schichten-Seismik.* Ibid. **7**, 37-55 (1931) — *Brechungsgesetz oder senkrechter Strahl?* Ibid. **8**, 376-396 (1932) — *Über den Energietransport bei der Sprengseismik.* Ibid. **10**, 378-385 (1934) — *Über die Totalreflexion in der Akustik und Optik (Auf Grund experimenteller Ergebnisse der Sprengseismik).* Ann Phys. **19** [V], 891-912 (1934) — *Sprengseismische Untersuchungen.* Z. Geophys. **11**, 83-89 (1935) — *Zur Theorie der Erdbebenwellen. Die "wachsende" Reflexion der Seismik als Analogon zur "Kopfwelle" der Ballistik.* Ibid. **12**, 199-205 (1936) — *Neue Erklärung des Kurzwellenumlaufs um die Erde.* Z. Tech. Phys. **17**, 443-446 (1936) — *Über Knallwellenausbreitung in Flüssigkeiten und festen Körpern.* Ibid. **19**, 554-561 (1938); Physik. Z. **39**, 868-875 (1938) — *Wissenschaftliche Auswertung der von F. THOMANEK mit Hilfe der Hochfrequenz-Funkenkinematographie durchgeführten Aufnahmen zur Richtungsdetonation.* Bericht 5/40, Ballist. Inst., Tech. Akad. Luftwaffe (TAL), Berlin-Gatow (1940).

**SECONDARY LITERATURE.** G. JOOS and J. TELTOW: *Zur Deutung der Knallwellenausbreitung an der Trennschicht zweier Medien.* Physik. Z. **40**, 289-293 (1939) — A. SCHOCH: *Schallreflexion, Schallbrechung und Schallbeugung.* Ergebn. exakt. Naturwiss. **23**, 127-234 (1950) — H. BERCKHEMER and J. ANSORGE: *Wave front investigations in model seismology.* Geophys. Prospecting **11**, No. 4, 459-470 (Dec. 1963) — P. KREHL: *Röntgenblitzaufnahmen von Kopfwellen, die sich beim dielektrischen Durchschlag an den Grenzen flüssiger und fester Stoffe ausbilden.* In: (N.R. NILSSON and L. HÖGBERG, eds.) *Proc. 8th Int. Congress on High-Speed Photography* [Stockholm, Sweden, June 1968]. Almquist & Wiksell, Stockholm (1968), pp. 267-271 — H.P. ROSSMANITH and W.L. FOURNEY: *Fracture initiation and stress wave diffraction at cracked interfaces in layered media. Part I: brittle/brittle transition.* Rock Mechanics & Rock Engineering **14**, No. 4, 209-233 (1982) — H.P. ROSSMANITH: *Elastic wave interaction with a cracked quarter-plane* [explosively generated body and surface wave]. Meccanica **20**, No. 2, 127-135 (1985).

# SCHMIDT, Paul (1898–1976)

- German mechanical engineer and inventor

Paul SCHMIDT was born in Hagen, a town in the Ruhr Area, North Rhine-Westphalia. After fighting in World War I, he studied mechanical and electrical engineering at the Universities of Münster, Hannover, and Munich (1919–1924). He became employee (1924) and later partner (1928–1934) of a Munich engineering office that performed consulting on the design of centrifugal pumps. In this period he worked out

his idea of a recurrently operating pulsejet engine for aircraft. He began experiments that were focused on the ignition of combustible mixtures by shock waves and supported by the German Ministry of Transportation. In 1934, he suggested to the Reichsluftfahrtministerium (RML) to apply his invention in the building of a flying bomb.

Based on his favorable results, he received a grant from the RML that allowed him to continue his research on a larger scale. At the outbreak of World War II, his experiments resulted in in the serviceable prototype of a periodically operating pulsejet engine – the so-called "Schmidt tube" [Germ. *Schmidtrohr*]. Erhard MILCH, Hermann GÖRING's deputy and director of air armament, being aware of the potentials of the Schmidt tube engine, approached the engineer Robert LUSSER at the Gerhard Fieseler Werke in Kassel to draw up plans for what later became known as "Vengeance Weapon One" [Germ. "*Vergeltungswaffe Eins*"] or "V1". This induced the RML to put Argus GmbH in Berlin in charge of the final development – the so-called "Argus-Schmidt tube" [Germ. *Argus-Schmidtrohr*]; the project leader was Fritz GOSSLAU. Development of such engines also began in mid-1942 at the secret research facility at Peenemünde-West on the Baltic Sea. The first flights were successfully performed in 1941 with a carrier aircraft and continued in 1943 with a Messerschmitt Me-328, a fighter that was propelled by a twin Argus-Schmidt tube engine. However, the extreme vibrations created by the pulsejets proved highly detrimental to the light airframe.

The pulsejet engine was analyzed in more detail at the Deutsche Versuchsanstalt für Luftfahrtforschung (German Aviation Research Facility) by Adolf BUSEMANN (1936) and at the TH Aachen by Fritz SCHULTZ-GRUNOW (1943). In 1944, its first mass production and application as a flying bomb began. It was also used to warm up jet aircraft engines by directing the jet into the turbine blades. The V1 so impressed the U.S. Army that it planned to build large numbers of first-generation cruise missiles, designated the JB-2 "Loon." It was based on the V1 but incorporated improved guidance systems using airborne and shipborne radars, radio control, and human operators, giving them much greater accuracy. About 1,000 JB-2s had been delivered to the U.S. Army and Navy by the end of World War II.

After the war, the Schmidt tube was applied by the American Helicopter Corporation to drive the rotor blades directly with jets at the blade tips, which eliminated the heavy engine and its gearing to the blades. In France it was used under the name *Escopette* to assist gliders in takeoff. In the 1950s, SCHMIDT came up with the idea of using a multitude of pulsejets in a large spherical arrangement in order to focus the generated pressure waves in the center and to possibly initiate a thermonuclear reaction (Germ. Patent No. 1,016,376).

**ORIGINAL WORKS.** *The Schmidt tube.* Zentrale für wiss. Berichtswesen (ZWB) der Luftfahrtforschung des Generalluftzeugmeisters, THM/Schmidt 44 (Juni 1944). Engl. translation by the University of Michigan Bureau of Aeronautics, Dept. of the Navy — *Die Geschichte der V1.* Motor-Rundschau **2**, Heft 7/8 (1948) — *Die Entwicklung der Zündung periodisch arbeitender Strahlgeräte.* VDI-Z. **92**, 393-399 (1950) — *On the history of the development of the Schmidtrohr.* In: (T. BENECKE and A.W. QUICK, eds.) *History of German guided missiles.* AGARDograph No. 20, Appelhaus, Braunschweig (1957), pp. 375-399 — *Instationäre gasdynamische Vorgänge als Grundlage technischer Verfahrensweisen.* VDI-Z. **106**, 435-440, 559-564 (1964).
**GERMAN PATENTS.** *Verfahren zum Erzeugen von Antriebskräften (Reaktionskräften) an Luftfahrzeugen.* No. 523,655 (9. April 1931) — *Einrichtung zur Erzeugung von Reaktionskräften an Luftfahrzeugen.* No. 567,042 (15. Dez. 1932) — *Zündung von Kraftstoff-Luftgemischen durch eine Stoßwelle.* Sch 980,44 l/46g (geheim, 11. Juni 1932) — *Wurfgerät mit rotierender Bewegung beim Flug.* No. 836,448 (1952) — *Gasturbine mit absatzweiser wiederholter, selbsttätiger Zündung durch Stoßwelle.* No. 926,396 (1955) — *Einrichtung zum Erzeugen von Stoßwellen in schneller Folge, insbesondere für einen thermonuklearen Reaktor.* No. 1,016,376 (1957) — *Flugzeug mit einem periodisch und mit Stoßwellenzündung arbeitenden rohrförmigen Rückstossantrieb.* No. 1,022,912 (1958) — *Vorrichtung zur Erzeugung eines Plasmastrahles.* No. 1,212,228 (1966) — *Verfahren zum Behandeln von in den inneren Bereich eines Stosswellenraums eingeführten Stoff, insbesondere zum Überführen des Stoffes in den Plasmazustand.* No. 1,212,229 (1966) — *Vorrichtung zur Erzeugung hoher Temperaturen und hoher Drücke mit einer für Stoßwellen reflexionsfähigen Innenfläche eines einen Hohlraum begrenzenden Körpers.* No. 1,233,508 (1967).
**BRITISH PATENTS.** *Improved method of producing motive forces for the propulsion of vehicles or aircraft.* No. 368,564 (1931) — *Process for carrying out chemical reactions in gases and aerosols.* No. 737,555 (1950).
**SECONDARY LITERATURE.** A. BUSEMANN: *Bericht über den Paul SCHMIDTschen Strahlrohrantrieb.* Bericht FB530, Deutsche Versuchsanstalt für Luftfahrtforschung (DVL), Berlin-Adlershof (1936) — F. SCHULTZ-GRUNOW: *Gasdynamische Untersuchungen am Verpuffungsstrahlrohr.* Institut für Mechanik, TH Aachen. Interne Berichte FB 2015/1 (1943) und FB 2015/2 (1944) — L.B. EDELMANN: *The pulsating jet engine, its evolution and future prospects.* SAE Quart. Trans. **1**, 205-216 (1947) — F. SCHULTZ-GRUNOW: *Wirkungsweise des Paul-Schmidt-Verpuffungsstrahlrohres. V1 Propulsion.* Flugwehr u. Flugtechnik (Zurich) **19**, 141-143 (1948) — H. LEMBKE: *Das Schmidtrohr.* VDI-Z. **94**, 1005-1008 (1952) — J. BERTIN, F. PARIS, and J. LE FOLL: *Das Pulso-Düsentriebwerk SNECMA "Escopette."* Interavia **8**, Nr. 6, 343-347 (1953) — F. STAAB: *Über Strahltriebwerke auf der Grundlage des Schmidtrohres.* Z. Flugwiss. **2**, 129-141 (1954) — H.G. KLIEMANN and S.S. TAYLER (eds.): *Who's Who in Germany.* Oldenbourg, Munich (1956), p. 1027 — F. GOSSLAU: *Development of the V1 pulse jet.* In: (T. BENECKE and A.W. QUICK, eds.) *History of German guided missiles.* AGARDograph No. 20, Appelhaus, Braunschweig (1957), pp. 400-418 — W. KAMM: *Zur Geschichte des Schmidtrohres.* Z. Flugwiss. **11**, 120-126 (1963); *Paul SCHMIDT 65 Jahre.* Ibid., p. 127 — T. BENECKE, K.H. HEDWIG, and J. HERMANN: *Flugkörper und Lenkraketen.* Bernard & Graefe, Koblenz (1987) — W. HELLMOND: *Die V1. Eine Dokumentation.* Bechtle, Esslingen (1991) — *Schmidtrohr. Zeittafel.* Deutsches Museum (1999), priv. communication — *JB-2 Loon (V1 buzz bomb).* Natl. Museum of the USAF;
http://www.nationalmuseum.af.mil/factsheets/factsheet.asp?id=510.
**PICTURE.** Courtesy Mrs. Grete KLIMM, Paul SCHMIDT's daughter, Rummelsberg, Bavaria, Germany.

## SCHULTZ-GRUNOW, Fritz Claus (1906–1987)

▪ German mechanical engineer and fluid dynamicist, founder of the Aachen Shock Wave Laboratory

Fritz C. SCHULTZ-GRUNOW was born in Munich. After taking his Abitur at Konstanz, he studied mechanical engineering at the ETH Zurich. He began working as a constructional and experimental engineer at the Escher-Wyss AG in Zurich and the Henschel Co. in Kassel, but then decided to enter research. He went to the Kaiser-Wilhelm-Institut (KWI) für Strömungsforschung in Göttingen and became one of Ludwig PRANDTL's assistants (1935). There he habilitated with a study on the after-effects of turbulence in the cases of locally and temporarily delayed boundary layer flows (1938). He was appointed lecturer in applied mechanics at the University of Göttingen (1939–1941) and subsequently took the prestigious chair of mechanics at the RWTH Aachen, which had been offered to him; in this position he also headed the affiliated Institute of Mechanics (1941–1975). During his directorate he considerably extended the institute's scope of research, including also a newly founded Shock Wave Laboratory, the *Stoßwellenlabor RWTH Aachen* (1956). In addition to his duties as a university teacher he also temporarily directed the Ernst-Mach-Institut at Freiburg (1967–1972).

During this period he worked out a gasdynamic theory of spinning detonation (1973).

In his Göttingen period he worked initially on classical problems of mechanics (theory of shells and fundamentals of modern construction engineering) and on fluid mechanics (turbulence, boundary layer theory, and rheology of lubricating oils). In his Aachen period he dedicated himself to the investigation of specific problems in aeronautics and astronautics, high-temperature engineering, and energy technology (flow, transport and reaction phenomena in hot gases). His Shock Wave Laboratory, the *Stoßwellenlabor der RWTH Aachen*, became widely known for its hypersonic studies on the mechanical and thermal loading of model orbital space vehicles and the Space Shuttle, and the development of various high-speed diagnostic techniques.

Prof. SCHULTZ-GRUNOW supervised more than 50 Ph.D. theses and 5 habilitation theses. His numerous contributions to research and education were acknowledged by the awarding of the honorary doctorate of the University of Stuttgart (1977), the Ludwig-Prandtl-Ring (1979), and the Grand Cross of the Order of Merit of the Federal Republic of Germany (1987). He was a Fellow of the AIAA (from 1957) and a corresponding member of the Institute of Aeronautical Sciences, New York. He edited the proceedings of the GAMM conference *Elektro- und Magnethydrodynamik* (Aachen, Oct. 1967).

**ORIGINAL WORKS.** *Modellversuche über den instationären, ebenen Spülvorgang im Zweitaktmotor.* Forsch. Ing.-Wes. **9**, 235-248 (1938) — *Anwendung gasdynamischer Methoden auf Wasserströmungen mit freier Oberfläche.* Ibid. **10**, 301-302 (1939) — *Zur Vorausbestimmung der turbulenten Ablösestelle.* Luftfahrtforsch. **16**, 425-428 (1939) — With K. WIEGHARDT: *Der Spülvorgang auf Grund einer neuen Auffassung der Expansionsströmung.* VDI, Berlin (1941) — *Pulsierender Durchfluß durch Rohre.* VDI, Berlin (1941) — *Nichtstationäre eindimensionale Gasbewegung.* Forsch. Ing.-Wesen **13**, 125-134 (1942) — *Nichtstationäre, kugelsymmetrische Gasbewegung und nichtstationäre Gasströmung in Düsen und Diffusoren.* Ing.-Archiv **14**, 21-29 (1943) — *Gasdynamische Untersuchungen am Verpuffungsstrahlrohr.* Zwischenberichte FB 2015/1 (1943) und FB 2015/2 (1944), Institut für Mechanik, TH Aachen; Engl. translation: *Gas-dynamic investigations of the pulse-jet tube. Parts I and II.* NACA-TM 1131 (1947) — *Zur Behandlung nichtstationärer Verdichtungsstöße und Detonationswellen.* ZAMM **24**, 284-288 (1944) — *Theoretisch und experimentell ermittelter Durchfluss einer nichtstationären Gasströmung.* Ibid. **25/27**, 155-156 (1947) — *Über die MACH'sche V-Ausbreitung.* ZAMM **28**, 30-31 (1948) — *Wirkungsweise des Paul-Schmidt-Verpuffungsstrahlrohres. V1 Propulsion.* Flugwehr u. Flugtech. **19**, 141-143 (1948) — *Der CARNOT'sche Stoßverlust in nichtstationärer Gasströmung.* ZAMM **29**, 257-267 (1949) — *Einführung in die Festigkeitslehre.* Werner, Düsseldorf (1949) — *Similarity laws of deflagration.* In: (B. LEWIS and H.C. HOTTEL, eds.) *4th Symp. (Int.) on Combustion* [Cambridge, MA, Sept. 1952]. Williams & Wilkins, Cambridge (1953), pp. 439-443 — *Gesetzmäßigkeiten der laminaren Flammenfortpflanzung und ihre Grenzen.* Z. Physik. Chem. **5**, 204-231 (1955) — *Das Institut für Mechanik an der RWTH Aachen.* Achema-Jb., Achema, Frankfurt/M. (1956–1958) — *Strömungsversuche und Widerstandsmessungen an Eisenbahnwaggonformen zur Verminderung des Querwiderstandes.* Westdt. Verlag, Köln/Opladen (1957) — *Theoretische und experimentelle Beiträge zur Grenzschichtströmung.* Ibid. (1959) — *Neues zur Physik der Stoßwellen.* Jb. Wiss. Gesell. Luft- u. Raumfahrt e.V., Vieweg, Braunschweig (1964), pp. 76-84 — With A. FROHN: *Density distribution in shock waves traveling in rarefied monatomic gases.* In: (J.H. DE LEEUW, ed.) *Proc. 4th Int. Symp. on Rarefied Gas Dynamics* [Toronto, Canada, July 1964]. Academic Press, London (1965), pp. 250-264 — With H. ZEIBIG: *Isotopentrennung von Gasen durch Thermodiffusion mit einer in einem geschlossenen Gehäuse rotierenden Scheibe.* Westdt. Verlag, Köln/Opladen (1969) — With C.M. CHANG: *Zündung kondensierter Oberflächen.* Ibid. (1969) — With R. EWALD and M. GRÖNIG: *Erforschung der Wärmeleitfähigkeit von Gasen bei extrem hohen Temperaturen.* Westdt. Verlag, Köln/Opladen (1970) — With W.G. STRUCK: *Zellenstruktur von Detonationswellen.* Bericht Nr. 5/69, EMI, Freiburg (1969); ZAMM **51**, T95 (1971) — With C.M. CHANG: *Zündung kondensierter Oberflächen.* Ibid. (1969) — *Mikrostruktur der Detonationswellen in Gasen.* Bericht Nr. 6/71, EMI, Freiburg (1971) — *Diffuse Reflexion einer Stoßwelle.* Bericht Nr. 7/72, Ibid. (1972) — With R. EWALD, E. SCHMOLINSKE, and W.G. STRUCK: *Mach-stem-induced detonation and its relation to the structure of detonation waves.* Astronaut. Acta **17**, 467-473 (1972) — *Dreidimensionale Grenzschicht-Instabilität.* ZAMM **52**, T113-114 (1972) — *Gasdynamic theory of spinning detonation.* Bericht Nr. 4/73, EMI, Freiburg (1973) — *Structure of shock waves including dissociation and ionization.* Z. Flugwiss. **23**, 51-57 (1975) — *Exakte Zugänge zu hydrodynamischen Problemen* [18. Ludwig-Prandtl-Gedächtnis-Vorlesung (1975)]. Ibid. **23**, pp. 175-183 (1975) — *The mechanism of spinning detonation.* Abhandl. Aerodyn. Inst. RWTH Aachen, Sonderband Heft 22 (1975) — *Production and inhibition of Mach stems.* 10th Int. Shock Tube Symp. [Kyoto, Japan, July 1975]. In: (G. KAMIMOTO, ed.) *Modern developments in shock tube research.* Shock Tube Research Society, Japan (1975) — *Generation of the patterns in gaseous detonations.* In: (C.E. TREANOR and J.G. HALL, eds.) *Proc. 13th Int. Symp. on Shock Tubes and Waves* [Niagara Falls, NY, July 1981]. State University of New York Press, Albany, NY (1982), pp. 352-357.

**SECONDARY LITERATURE.** *Festschrift Herrn Prof. F. SCHULTZ-GRUNOW zum 65. Geburtstag.* Institut für Allgemeine Mechanik, RWTH Aachen (Okt. 1971) — G. ADOMEIT and H.J. FRIESKE (eds.) *Neue Wege in der Mechanik. Festschrift zum 75. Geburtstag von Prof. Dr. sc. tech. Dr.-Ing. E. h. F. SCHULTZ-GRUNOW* [with a complete list of his publications]. VDI-Verlag, Düsseldorf (1981) — *Nachruf Prof. Dr. SCHULTZ-GRUNOW.* VDI-Nachrichten (11. Dez. 1987) — Anonymous: *Ein bedeutender Strömungsforscher. Ehrendoktor Prof. Dr. F. SCHULTZ-GRUNOW in Aachen gestorben.* Stuttgarter Uni-Kurier **34**, 8 (März 1988).

**PICTURE.** Courtesy Archives of the Ernst-Mach-Institut, Freiburg, Germany.

# SEDOV [Russ. *СЕДОВ*], Leonid Ivanovich (1907–1999)

- Soviet mathematical physicist and hydrodynamicist

Leonid I. SEDOV was born in Rostov-on-Don, south-west Russia. His father was Ivan Grigorevich SEDOV, a mining

engineer. In 1931, he graduated from the Physical-Mathematical Faculty of Moscow State University [Russ. *МГУ*]. At the Central Aero-Hydrodynamic Institute [Russ. *ЦАГИ*], then the leading Russian institution in fluid dynamics and directed by Prof. Sergey A. CHAPLYGIN, he specialized in mechanics and hydrodynamics and became a staff member (1931–1947). After taking his Ph.D. he was given the title of professor of dynamics at Moscow State University (1937). Concurrently he worked at the Institute of Mathematics of the U.S.S.R. Academy of Sciences (1945) and at the Central Institute of Aircraft Engine Construction [Russ. *ЦИАМ*] (1947–1956).

His numerous contributions to theoretical fluid dynamics and mechanics comprise a wide field, such as (1) a theory of ideal fluidity (1937) that was stimulated by CHAPLYGIN's early research; (2) a thorough treatment of percussion and gliding phenomena, so-called "ricocheting" (1941–1942), then of great practical interest for designers of hydroplanes and in the 1930s also treated by German researchers; (3) an aerodynamic theory of flat airfoils; (4) a method of using ordinary differential equations in order to obtain accurate solutions for gasdynamic equations; (5) a theory of strong explosions, resulting in the so-called "Sedov equation" (1946), which was later confirmed by the first Soviet nuclear weapons test (1949); (6) a phenomenological theory for the construction of a model of continuous media on the basis of variational principles (1965); and (7) a theory of unstabilized gas motion and dispersion of strong shock waves, derived from the theory of similarity and dimensionality. He published about 200 articles, and most of his monographs were also translated into English. Among shock physicists the most renowned books are (in translation) *Similarity and Dimensional Methods in Mechanics* (1951), *Two Dimensional Problems in Hydrodynamics and Aerodynamics* (1965), and *Unsteady Motion of Compressible Media with Blast Waves* (1967).

SEDOV was elected corresponding member of the U.S.S.R. Academy of Sciences (1946) and became Academician (1953). He served as chairman of the Interdepartmental Commission on Interplanetary Communications of the U.S.S.R. Academy of Sciences "to coordinate and direct all work concerned with solving the problem of mastering cosmic space." In the period 1959–1961, he was president of the International Astronautical Federation (IAF). He was awarded the Chaplygin Medal (1946) of the U.S.S.R. Academy of Sciences, won a Stalin Prize (1952), and received the Order of the French Legion of Honor (1952). He also received the first Lomonosov Prize (1954) for his contribution of applying methods of gas dynamics to astrophysical problems and was Hero of Socialist Labor (1967). In 1981, he received, together with the Austrian politician Peter JANKOWITSCH, the Allen D. Emil Memorial Award of the International Astronautical Federation (IAF). In 1993, he received the Nikolai Zhukovsky Prize and the Academy Medal for his outstanding contribution to the development of mechanics.

The second phase (or adiabatic phase) in the lifetime of a supernova remnant, in which the material begins to decelerate and the ejecta of the supernova remnant mix up with the gas that was just shocked by the initial shock wave, is named after him (*Sedov phase*).

Astronomers named a minor planet (asteroid 2785 SEDOV) after him.

**ORIGINAL WORKS.** *On the theory of unsteady planning and the motion of a wing with vortex separation* [in Russ.]. Rept. No. 252, Trudy CAGI [Central Aerohydrodynamics Institute, Moscow] (1936); Engl. translation in NACA-TM 942 (May 1940) — With M.V. KELDISCHEM: *Sur la solution effective de quelques problèmes limites pour les fonctions harmoniques.* Dokl. AN (SSSR) **16**, No. 1, 7-10 (1937) — *On scale effect and optimum proportions for gliding* [in Russ.]. Sudostroenie (Leningrad) **10**, No. 3, 123-135 (1940) — With A.N. WLADIMIROW: *Gleiten einer flach-kielartigen Platte.* Dokl. AN (SSSR) **33**, No. 3, 116-119 (1941); *Die Stabilität einer flach-kielförmigen Platte.* Ibid. **33**, No. 3, 194-197 (1941) — *Water ricochets.* Ibid. **37**, No. 9, 254-257 (1942) — With A.N. WLADIMIROV: *Einfluß der mechanischen Parameter auf das Gleiten einer flach-kielartigen Platte* [in Russ.]. Bull. Acad. Sci. URSS (Otdel. Tekh. Nauk) **1** [Classe Sci. Tech.], 44-66 (1943) — *Decay of isotropic turbulent motion of incompressible fluid* [in Russ.]. Dokl. AN (SSSR) **42**, No. 3, 116-119 (1944) — *On unsteady motion of a compressible fluid* [in Russ.]. Ibid. **47**, No. 2, 91-93 (1945) — *On unsteady motions of compressible fluids* [in Russ.]. Prikl. Mat. Mekh. (SSSR) **9**, No. 4, 293-311 (1945) — *Propagation of strong explosive waves* [in Russ.]. Ibid. **10**, No. 2, 241-250 (1946) — *Le mouvement d'air en cas d'une forte explosion.* Dokl. AN (SSSR) **52**, No. 1, 17-20 (1946) — *On the theory of the unsteady motion of an airfoil.* NACA-TM 1156 (July 1947) — *Scale effect and optimum relations for sea surface planning.* NACA-TM 1097 (Feb. 1947) — *On the general form of the kinetic equations of chemical reactions in gases* [in Russ.]. Dokl. AN (SSSR) **60**, No. 1, 73-76 (1948) — *Hydro-aerodynamic forces in flows of compressible fluids about airfoil profiles* [in Russ.]. Ibid. **63**, No. 6, 627-628 (1948) — *Plane problems of hydrodynamics and aerodynamics* [in Russ.]. Gostekhizdat, Moscow (1950) — With M. KELDYSCH: *Die Entwicklung der Mechanik in Rußland.* In: (S.I. WAWILOW ET AL., eds.) *Enzyklopädie der Union der Sozialistischen Sowjetrepubliken.* Kultur & Fortschritt, Berlin (1950), vol. II, pp. 1385-1393 — *Metody podobija i razmernosti v mechanike.* Nauka, Moscow (1951, 1957, 1959, 1967, 1972,

1977, 1987); Engl. translation by M. HOLT (ed.): *Similarity and dimensional methods in mechanics*. Academic Press, New York (1959, 1961); CRC, Boca Raton, FL (1993) — *General theory on the steady movement of gases* [in Russ.]. Dokl. AN (SSSR) **85**, No. 4, 723-726 (1952) — *The integration of the equations for one-dimensional motion of a gas* [in Russ.]. Ibid. **90**, No. 5, 735 (1953) — *Formula on the laws of stars governing "luminous power-mass" and "radius-mass"* [in Russ.]. Ibid. **94**, No. 4, 643-646 (1954) — *O poletakh v mirovoe prostranstvo* ("On flights into space"). Pravda, Moscow (Sept. 26, 1955). Engl. translation in F.J. KRIEGER's book *Behind the Sputniks: a survey of Soviet space science*. Public Affairs, Washington, DC (1958), pp. 112-115 — *On the motion of a gas in stellar bursts* [in Russ.]. Dokl. AN (SSSR) **111**, No. 4, 780-782 (1956) — *On the dynamical explosion equilibrium* [in Russ.]. Ibid. **112**, No. 2, 211-212 (1957) — *Examples of gas motion and certain hypotheses on the mechanism of stellar outbursts*. Rev. Mod. Phys. **30**, 1077-1079 (1958) — *The concept of different rates of change in a tensor* [in Russ.]. Prikl. Mat. Mekh. (SSSR) **24**, 393-398 (1960) — *Concepts of simple leading and possible paths of deformation* [in Russ.]. Ibid. **24**, 400-402 (1960) — With M.E. EGLIT: *Construction of non-holonomous models of continuous media with allowance for the finite nature of deformation and certain physical-chemical effects* [in Russ.]. Dokl. AN (SSSR) **142**, No. 1, 54-57 (1962) — *Two-dimensional problems in hydrodynamics and aerodynamics*. Interscience, New York (1965) — *Introduction to the mechanics of a continuous medium*. Addison-Wesley, Reading, MA (1965) — *Foundations of the non-linear mechanics of continua*. Pergamon Press, Oxford (1966) — *Unsteady motion of compressible media with blast waves*. Proc. Steklov Inst. Math. (Birmingham, AL) No. 87 (1966). Am. Math. Soc., Providence, RI (1967) — (J.R.M. RADOK, ed.) *A course in continuum mechanics*. Wolters-Noordhoff, Groningen (1971/1972). Vol. 1: *Basic equations and analytical techniques*; Vol. 2: *Physical foundations and formulations of problems*; Vol. 3: *Fluids, gases and the generation of thrust*; Vol. 4: *Elastic and plastic solids and the formation of cracks: macroscopic theories of matter and fields. A thermodynamic approach*. In: *Advances in science and technology in the U.S.S.R.* Mir, Moscow (1983) — (L.S. SEDOV and V.V. GOGOSOV, eds.) *Sovremennye problemy elektrogidrodinamiki*. Izd. University, Moscow (1984).

**SECONDARY LITERATURE.** N.D. MOISEEV: *General outline of the development of mechanics in Russia and in the U.S.S.R.* [in Russ.]. In: (V.Z. VLASOV, V.V. GOLUBEV, and N.D. MOISEEV, eds.) *Mekhanika v SSSR za tridtsat' let 1917–1947* ("Mechanics in the U.S.S.R. for 30 years"). Nauka, Moscow (1950), pp. 11-57 — E.A. KRASIL'SHCHIKOEA and G.V. RUDNEV: *Stalin Prize: L.I. SEDOV*. Priroda (AN SSSR) **41**, No. 9, 57-59 (1952) — G. WEDENSKY: *L.I. SEDOV*. In: *Porträts der UdSSR-Prominenz*. Institut zur Erforschung der UdSSR e.V., München (1960) — J. TURKEVICH (ed.): *Soviet men of science*. Van Nostrand, Princeton, NJ (1963), pp. 329-330 — F. BISSHOPP (ed.) *Problems of hydrodynamics and continuum mechanics* [contributions in honor of the 60th birthday of Academician L.I. SEDOV, Nov. 14, 1967]. SIAM, Philadelphia (1969) — A.A. BARNIM: *Current mathematical problems of mechanics and their applications* [dedicated to Academician L.I. SEDOV on his 80th birthday]. Am. Math. Soc., Providence, RI (1989) — *Leonid Ivanovich SEDOV on his 90th birthday*. Cosmic Res. **35**, No. 6, 525 (1997) — (G.G. CHERNYI and E.F. MISHCHENKO, eds.) *Modern methods in continuum mechanics* [collected papers dedicated to the 90th birthday of Academician Leonid Ivanovich SEDOV]. Am. Math. Soc., Providence, RI (1998) — *Leonid Ivanovich SEDOV (1907–1999): obituary*. Cosmic Res. **37**, No. 6, 529 (Nov. 1999) — A.G. KULIKOVSKII and G.A. LYUBIMOV: *Leonid Ivanovich SEDOV* (obituary). Russ. Math. Surv. **55**, No. 2, 323-326 (2000) — A.A. SIDDIQI: *The Soviet space race with Apollo*. University Press of Florida, Gainesville, FL (2003).

**PICTURE.** Taken by Howard SOCHUREK in October 1957. Courtesy GettyImages / Time & Life Pictures.

# SEMENOV [Russ. *СЕМЕНОВ*], Nikolai Nikolaevich (1896–1986)

▪ Soviet physical chemist; discoverer of branched chain reactions in chemistry

Nikolai N. SEMENOV was born in Saratov, a town in south-central Russia. His father was Nikolai Aleksandrovich SEMENOV, a professional infantry soldier. Educated at the University of Petrograd (now St. Petersburg), he graduated with high honors in physics in 1917, the year of the Bolshevik Revolution. Invited by the University of Tomsk (western Siberia), he accepted an assistant professorship and in addition to lecturing worked at the Institute of Technology (1918–1920). Recalled to Petrograd by his former tutor Abram F. IOFFE, he worked mainly on electrical discharges in gases and electron impact phenomena at the Laboratory of Electronic Chemistry of the Leningrad Physico-Technical Institute (1927–1931), which was incorporated into the newly founded Institute of Chemical Physics of the Russian Academy of Sciences (1931) and later moved to Moscow (1943).

The Institute of Chemical Physics was created with the aim of "introducing physical theories and methods into chemistry, chemical industry, and other branches of economics." SEMENOV defined chemical physics as a "science describing the fundamentals of chemical transformations and the associated problems of substance structure." He was appointed professor and head of the Department of Combustion and Explosion Research, where he worked out his theory of chain reactions in combustion processes, self-ignition phenomena, and kinetics of chemical reactions (1931–1939). At the Institute of Chemical Physics other renowned Soviet researchers also worked temporarily on chemical kinetics such as A.F. BELAJEV (1938–1940), D.A. FRANK-KAMENETSKY (1938), Y.B. KHARITON (1926–1936), O.M. TODES (1936), and Y.B. ZEL'DOVICH (1938–1946), who developed their famous theories of combustion and detonations as well as the fundamentals of the thermal decomposition of explosives.

Prior to or simultaneously with SEMENOV's investigations, the conception of chain reactions to thermochemical reaction was advanced also independently by various other eminent researchers, such as in Germany by M. BODENSTEIN (1913), E. CREMER (1927), F. HABER (1931), and W. NERNST (1916), in the United States by H.L.J. BÄCKSTRÖM (1927), in Denmark by J.A. CHRISTIANSEN and H.A. KRAMERS (1923), and in England by C.N. HINSHELWOOD, C.H. GIBSON, and H.W. THOMPSON (1928–1929).

SEMENOV investigated the question why chemical chain reactions occur and revealed their importance in connection with the phenomenon of explosion. In his monograph *Chain Reactions* (1934) he developed the theory of nonbranching reactions and showed the wide distribution of chain reactions in chemistry. He and his associates investigated experimentally the breaking of reaction chains on walls and in the volume of a container, the degeneration of chain branching, positive and negative interaction of chains, and the mechanism and role of free atoms and radicals in series chain processes. SEMENOV and the British chemist Sir Cyril HINSHELWOOD received together the 1956 Nobel Prize for Chemistry "for their researches into the mechanism of chemical reactions." HINSHELWOOD had worked on reaction rates and reaction mechanisms, particularly that of the combination of hydrogen and oxygen to form water.

SEMENOV was Secretary of the Department of Chemical Sciences of the U.S.S.R. Academy of Sciences (1961) and honorary doctor of the Universities of Oxford (1960), Brussels (1962), Milan (1964), Budapest, Prague, and Berlin (1965). He also obtained many national awards for his numerous contributions to chemical physics. Since 1990 the Institute of Chemical Physics at Moscow has carried his name, the *N.N. Semenov Institute of Chemical Physics* of the Russian Academy of Sciences. On his initiative, a number of scientifiia institutes, dedicated to research on combustion and chemical kinetics, were established in the former Soviet Union such as at Novosibirsk, Tomsk, and Yerevan. SEMENOV was also cofounder of the Moscow Institute of Physics and Technology (MIPT), which was established in 1946. The *N.N. Semenov Gold Medal* is awarded by the Russian Academy of Sciences to Russian and foreign scientists for outstanding contributions in Chemical Sciences.

Astronomers named a minor planet (asteroid 2475 SEMENOV) after him.

**ORIGINAL WORKS.** *Die Oxidation des Phosphordampfes bei niedrigen Drucken.* Z. Phys. **46**, 109-131 (1928) — *Theorie des Verbrennungsprozesses.* Ibid. **48**, 571-582 (1928) — With J.N. RJABININ: *Die physikalischen Grundlagen der elektrischen Festigkeitslehre.* Springer, Berlin (1928); *Die Oxidation des Schwefeldampfes bei niedrigen Drucken.* Z. Physik. Chem. **B1**, 192-204 (1928) — *Zur Theorie der chemischen Reaktionsgeschwindigkeit.* Ibid. **B2**, 161-168 (1929) — *Die Kinetik der Vereinigung von Wasserstoff und Sauerstoff.* Ibid. **B2**, 169-180 (1929) — *Entartete Explosionen und Induktionsperiode.* Ibid. **B11**, 464-469 (1931) — *Upper pressure limit of ignition.* Nature **132**, 566-567 (1933) — *The evolution of the theory of gas explosions in the Soviet Union* [in Russ.]. Socialisticeskaja Rekonstr. i Nauka No. 3, 54-67 (1933) — *Chemical kinetics and chain reactions.* Goschimizdat, Leningrad (1934); Clarendon Press, Oxford (1935) — *On the kinetics of complex reactions.* J. Chem. Phys. **7**, 683-699 (1939) — *Thermal theory of burning and explosions* [in Russ.]. Uspekhi Fiz. Nauk **23**, 251-292 (1940) — *Theory of normal flame propagation* [in Russ.]. Ibid. **24**, 433-486 (1940) — *Advances of chemical kinetics in the Soviet Union.* Nature **151**, 185-187 (1943) — *On types of kinetic curves in chain reactions. I. Laws of the autocatalytic type* [in Russ.]. Dokl. AN (SSSR) **42**, No. 8, 342-348 (1944) — *II. Consideration of the interaction of active particles* [in Russ.]. Ibid. **44**, No. 2, 62-66 (1944) — *On the constants of the reaction $H + O_2 = OH + O$ and $H_2 + O_2 = 2 OH$.* Acta Physicochimica URSS **20**, 291-303 (1945) — *Einige Probleme der Kettenreaktionen und der Verbrennungstheorie* [Nobel-Vortrag gehalten am 11. Dez. 1956]. Angew. Chem. **69**, 767-777 (1957) — *Some problems in chemical kinetics and reactivity.* 2 vols., Princeton University Press, Princeton, NJ (1958/1959) — *Certain chemical reactions at reduced temperatures and related problems of energy transfer.* Pure Appl. Chem. **5**, 353-376 (1962) — Contribution to a paper written by G. MENDELEVICH: *What Marcellin BERTHELOT could not foresee* [in Russ.]. Technika Molodeži **32**, No. 4, 16-17 (1964).

**SECONDARY LITERATURE.** Y.B. KHARITON: *Creator of the theory of chemical chain reactions* [in Russ.]. Sovetskaja Nauka No. 4, 16-21 (1941) — J.G. GROWTHER: *Science in the U.S.S.R.* Endeavour **1**, No. 4, 21-25 (1942) — N.M. EMANUEL: *Academician Nikolai Nikolaevich SEMENOV* [on occasion of his 50th birthday, in Russ.]. Bull. Acad. Sci. URSS (Série Sciences Chimique) No. 4, pp. 337-343 (1946) — Anonymous: *Nikolai Nikolaevich SEMENOV. On the 60th anniversary since the date of birth* [in Russ.]. Zh. Eksp. Teor. Fiz. (SSSR) **30**, No. 4, 625-627 (1956) — C.N. HINSHELWOOD: *Reaktionskinetik in den letzten Jahrzehnten* [Nobel-Vortrag gehalten im Dez. 1956]. Angew. Chem. **69**, 445-449 (1957) — N.M. EMANUEL: *An important contribution to the world of science* [on Nobel Prize award to Academician N.N. SEMENOV, member of Academy, in Russ.]. Priroda (AN SSSR) **46**, No. 2, 43-48 (1957) — J. TURKEVICH (ed.) *Soviet men of science.* Van Nostrand, Princeton, NJ (1963), pp. 330-332 — B.N. TARUSOV: *On the 70th anniversary of the Laureate of the Nobel Prize of Academician Nikolai Nikolaevich SEMENOV. The influence of N.N. SEMENOV and his school on the development of radiation biophysics* [in Russ.]. Radiobiologiia **6**, No. 2, 161-165 (1966) — N.M. EMANUEL: *The founder of the Soviet school of chemical physics* [in Russ.]. Priroda (AN SSSR) **59**, No. 3, 76-77 (1970) — *Nikolai Nikolaevich SEMENOV* [biography with emphasis on his work in physics and chemistry, in Russ.]. Zh. Vses. Khim. **20**, No. 6, 674-675 (1975) — *A short account of scientific trends developed at the Chair of Chemical Kinetics, Moscow State University, in the period from 1944–1974* [in Russ.]. Vestn. Mosk. University Khim. **16**, No. 6, 643-658 (1975) — V.N. KONDRATE'EV: *To the theory of chain reactions by N.N. SEMENOV* [dedicated to his 80th birthday, in Russ.]. Zh. Fiz. Khim. **50**, 825-829 (1976) — (S. NEUFELDT, ed.): *CREMER, SEMENOV, HINSHELWOOD.* In: *Chronologie Chemie: 1800–1980.* VCH, Weinheim/Bergstr. (1987) — Lord F. DAINTON: *Nikolai Nikolaevich SEMENOV.* Biogr. Mem. Fell. Roy. Soc. (Lond.) **36**, 527-548 (1990) — A.E. SHILOV: *N.N. SEMENOV and the chemistry of the 20th century* [to the 100th anniversary of his birth]. Pure & Appl. Chem. **69**, 857-863 (1997).

**PICTURE.** © The Nobel Foundation, Stockholm, Sweden.

**NOTE.** His name has also been transliterated as SEMENOFF, SEMYENOV, and SEMYONOV.

## SHOEMAKER, Eugene Merle (1928–1997)

- U.S. geologist and planetary scientist; founder of the scientific study of impact cratering and father of astrogeology

Eugene M. SHOEMAKER was born in Los Angeles, CA. His father was George Estel SHOEMAKER, a teacher, farmer, trucker, and studio grip. After schooling in New York and graduation from high school in Los Angeles he entered CalTech in Pasadena at the age of 16 (1944). There he graduated and got a master's degree (1948). SHOEMAKER, who was interested in rocks and minerals in his childhood, followed this inclination all his life. His first work was for the USGS in the uranium exploration program in Grand Junction, CO. Assigned to map craters formed by nuclear explosions at the Nevada test site, he discovered that both nuclear craters and the Meteor Crater in Arizona had a very similar crater ejecta stratigraphy. Continuing his education at Princeton University and becoming interested in terrestrial and lunar craters and the possible role of asteroids, he received his Ph.D. with a thesis entitled *Penetration Mechanics of High Velocity Meteorites, Illustrated by Meteor Crater, Arizona* (1960). Together with the U.S. astronomer Edward C.T. CHAO he discovered in 1960 the natural occurrence of coesite – a shock-induced high-pressure polymorph of silica – at Meteor Crater and shortly after also in the Ries Basin, Bavaria, thus not only solving an old geologic puzzle that these crater structures were indeed formed by meteorite impact but also introducing coesite as a diagnostic tool in distinguishing impact in other geologic structures. Several months later, he discovered at Meteor Crater another, even higher-temperature, higher-pressure polymorph of quartz, which had been produced shortly before by the Soviet mineralogist Sergei M. STISHOV using static high pressures and which SHOEMAKER called "stishovite."

SHOEMAKER is credited with creating the Astrogeology branch of the USGS: the *Astrogeology Research Program*, dedicated to the study of the geology of extraterrestrial solid objects, was founded in 1961 as a subgroup of the USGS, with SHOEMAKER serving as its first director. In addition, he established the Flagstaff Field Center in 1963. By mapping and analyzing craters on the Moon, both from telescopes and from a series of lunar spacecraft, such as Ranger, Surveyor, Lunar Orbiter, and Apollo, and with Voyager the cratered moons of some outer planets, he was among the first to recognize that by measuring the relationship between crater density and diameter the age of a planetary surface could be inferred. To confirm the impact of craters, he also demonstrated that the size distribution and flux of existing objects that could hit the Earth is closely correlated with the number, size, and age distribution of the relatively recent craters found on the Earth and Moon. Together with Robert J. HACKMAN, a USGS geologist, he suggested in 1962 that radial streaks extending from some lunar impact craters for several multiples of the crater's diameter – so-called "lunar rays" – were the result of fragmented ejecta material superposed on all other surrounding terrains, which places rayed craters in the youngest (Copernican) system of their defined time-stratigraphic classification.

Together with his wife Carolyn and with David H. LEVY, both astronomers, he made the famous discovery of the comet SHOEMAKER-LEVY 9 (or SL9), a fragmented comet that looked like "a string of pearls" whose spectacular impact on Jupiter could be observed from telescopes of the Earth (1994). Ironically, he died "on impact" in a car accident in Central Australia on his way to doing fieldwork on some impact craters.

His scientific output includes 195 publications and more than 200 abstracts. He received many honors and awards, such as the Wetherill Medal (1965) of the Franklin Institute, the NASA Medal (1967), the Day Medal (1982) and the Gilbert Award (1983) of the Geological Society of America, the first Barringer Award (1984) and the Leonard Medal (1985) of the American Meteoritical Society, the Kuiper Prize (1984) of the American Astronomical Society, the National Medal of Science (1992), and the Bowie Medal (1996) of the American Geological Union. From 1980 he was also a member of the National Academy of Sciences. The formerly Teague ring structure in western Australia has been named *Shoemaker Impact Structure* (1998). A minor planet (asteroid 2074 SHOEMAKER) is also named for him.

His greatest honors came posthumously: a small portion of his ashes was sent by NASA to the Moon aboard Lunar Prospector (1999), and in March 2000 NASA's NEAR (Near-Earth Asteroid Rendezvous) spacecraft was renamed *NEAR Shoemaker*. In 2001, the *Shoemaker Center for Astrogeology Building* on the USGS Flagstaff Science Center campus was named in memory of him.

ORIGINAL WORKS. *Penetration mechanics of high velocity meteorites, illustrated by Meteor Crater, Arizona.* In: (T. SORGENFREI, ed.) *Proc. 21st Int. Geological Congr.* [Copenhagen, 1960]. Det Berlingske Bogtrykkeri, Copenhagen, Denmark (1960), part XVIII, pp. 418-434 — With E.C.T. CHAO and B.M. MADSEN: *First natural occurrence of coesite from Meteor Crater, Arizona.* Science **132**, 220-222 (July 22, 1960) — With E.C.T. CHAO: *New evidence for the impact of the Ries Basin, Bavaria, Germany.* J. Geophys. Res. **66**, 3371-3378 (1961) — With D.E. GAULT and R.V. LUGN: *Shatter cones formed by high-speed impact in Dolomite.* USGS Prof. Paper No. 424-D (1961), pp. 365-368 — With R.J. HACKMAN: *Stratigraphic basis for a lunar time scale.* In: (Z. KOPAL and Z.K. MIKHAILOV, eds.) *The Moon – Symp. No. 14 of the International Astronomical Union (IAU).* Academic Press, London (1962), pp. 289-300 — *Geologic interpretation of lunar craters.* In: (Z. KOPAL, ed.) *Physics and astronomy of the Moon.* Academic Press, New York (1962), pp. 283-359 — *Impact mechanics at Meteor Crater, Arizona.* In: (B.M. MIDDLEHURST and G.P. KUIPER, eds.) *The Solar System.* Vol. IV: *The Moon, meteorites, and comets.* University of Chicago Press, Chicago (1963), pp. 301-336 — *The geology of the Moon.* Scient. Am. **211**, 38-47 (Dec. 1964) — *When the irresistible force meets the immovable object.* Engng. & Sci. (Pasadena) **29**, No. 5, 11-15 (1966) — With E.F. HELIN: *Populations of planet-crossing asteroids and the relationship of Apollo objects to main-belt asteroids and comets.* In: (A.H. DELSEMME, ed.) *Comets, asteroids, meteorites: interrelations, evolution, and origins.* University of Toledo, Toledo, OH (1977), pp. 297-300 — *Why study impact craters?* In: (D.J. RODDY, R.O. PEPIN, and R.B. MERILL, eds.) *Impact and explosion cratering: planetary and terrestrial implications.* Pergamon Press, New York (1977), pp. 1-10; *Astronomically observable crater-forming projectiles.* Ibid., pp. 617-628 — With J.G. WILLIAMS, E.F. HELIN, and R.F. WOLFE: *Earth-crossing asteroids: orbital classes, collision rates with Earth and origin.* In: (T. GEHRELS, ed.) *Asteroids.* University of Arizona Press, Tucson, AZ (1979), pp. 253-282 — *The new Solar System.* The Press Syndicate of the University of Cambridge (1981) & Sky Publ. Corp., Cambridge, MA (1982); Germ. translation: *Kollision fester Körper.* In: (J.K. BEATTY and B.O.A. CHAIKIN, eds.) *Die Sonne und ihre Planeten.* Physik-Verlag GmbH, Weinheim/Bergstr. (1983), pp. 31-42 — *Asteroid and comet bombardment of the Earth.* Annu. Rev. Earth & Planetary Sci. **11**, 461-494 (1983) — (Ed.) *The first presentation of the Barringer Award* [to E.M. SHOEMAKER]. Meteoritics **19**, 177 (1984) — With P. HUT, W. ALVAREZ, W.P. ELDER, T.A. HANSON, E.G. KAUFFMAN, G. KELLER, and P.R. WEISSMANN: *Comet showers as a cause of mass extinction.* Nature **329**, 118-126 (1987) — With R.A.F. GRIEVE: *The record of past impacts on Earth.* In: (T. GEHRELS, ed.) *Hazards due to comets and asteroids.* University of Arizona Press, Tucson (1994), pp. 417-462 — With C.S. SHOEMAKER and D.H. LEVY: *Discovering comet Shoemaker-Levy 9.* In: *Once in a thousand lifetimes: a guide to the collision of comet Shoemaker-Levy 9 with Jupiter.* Planetary Society, Pasadena, CA (1994), pp. 2-3 — With D.H. LEVY and C.S. SHOEMAKER: *Comet Shoemaker-Levy 9 meets Jupiter.* Scient. Am. **273**, 69-75 (Aug. 1995) — With C.S. SHOEMAKER: *The proterozoic impact record of Australia.* AGSO J. Austral. Geol. Geophys. **16**, 379-398 (1996) — With J.C. WYNN: *The day the sands caught fire. A desert impact site demonstrates the wrath of rocks from space.* Scient. Am. **279**, 36-43 (Nov. 1998) — *Long-term variations in the impact cratering rate on Earth.* In: (M.M. GRADY, ed.) *Meteorites: flux with time and impact effects.* Special Publ. No. 140, Geological Society, London (1998) — With C.S. SHOEMAKER: *The role of collisions.* In: (J.K. BEATTY, C.C. PETERSEN, and A. CHAIKIN, eds.) *The new Solar System.* Sky, Cambridge, MA (1998).

SECONDARY LITERATURE. J.N. BELL: *Eugene* SHOEMAKER. In: *Science Year, the World Book Science Annual 1972.* Pasadena, CA, pp. 398-413 — S.W. KIEFFER: *Eugene M. SHOEMAKER (1928–1997).* Science **277**, 776-777 (Aug. 8, 1997) — T. AHRENS: *Eugene SHOEMAKER (1928–1997). Founder of the scientific study of impact cratering.* Nature **389**, 132 (Sept. 11, 1997) — B.M. FRENCH: *Eugene M. SHOEMAKER (1928–1997).* Meteorit. Planet. Sci. **32**, 985-986 (1997) — A. BRAKEL: *Eugene SHOEMAKER 1928–1997: An appreciation.* Southern Cross. J. Canberra Astron. Soc. (Aug. 1997); see http://msowww.anu. edu.au/cas/sc_0897.html — F. PIRAJNO and A.Y. GLIKSON: *Shoemaker impact structure (formerly Teague ring structure), western Australia.* Austral. Geol. **106**, 16-18 (1998) — A.Y. GLIKSON: *The astronomical connection of terrestrial evolution.* Ibid. **108**, 34-37 (1998); *Eugene SHOEMAKER and the impact paradigm in Earth and planetary science* [with a bibliography of SHOEMAKER's impact papers]. In: (S. YABUSHITA and J. HENRARD, eds.) *Workshop on Dynamics of Comets and Asteroids and their Role in Earth History* [Ten-Kyu-Kan, Japan, Aug. 1997]. Kluwer, Dordrecht (1998); Cel. Mech. Dyn. Astron. **68**, 9-24 (1998) — C.S. SHOEMAKER: *SHOEMAKER, Gene.* Am. Natl. Biogr. Online (June 2000), Oxford University Press, Oxford; http://www.anb. org/articles/13/13-02618.html — D. WHITEHOUSE: *Moon burial for geologist.* BBC News Online Network; http://news.bbc.co.uk/hi/english/sci/tech/newsid_405000/405944.stm — D.H. LEVY: *SHOEMAKER by LEVY: the man who made an impact* [with selected bibliography]. Princeton University Press, Princeton, NJ (2000) — A.Y. GLIKSON: *The world's largest impact and the tale of two craters.* Meteorite **6**, 18-20 (2000) — Astrogeology Team, USGS at Flagstaff, AZ: *E. SHOEMAKER;* http://wwwflag.wr.usgs.gov/USGSFlag/Space/Shoemaker/GeneObit.html — D.R. WILLIAMS: *Comet P/Shoemaker-Levy 9.* NASA-GSFC, Greenbelt, MD; http://nssdc.gsfc.nasa.gov/planetary/comet_body.html.

PICTURE. From the USGS Photographic Library, Portraits Collection; http://libraryphoto.cr.usgs.gov/ports.htm.

NOTE. SHOEMAKER was author or coauthor of 202 publications, see list http://astrogeology.usgs.gov/About/People/GeneShoemaker/00gene_bib.txt.

# SOLOUKHIN [Russ. *СОЛОУХИН*], Rem Ivanovich (1930–1988)

- Soviet combustion physicist and gas dynamicist

Rem I. SOLOUKHIN was born in Gus-Khrustalny, a village near Vladimir in Central Russia. After graduating from Lomonosov University in Moscow, he received a diploma from the Faculty of Thermal and Molecular Physics (1953), which involved first pioneering shock-tube experiments in the U.S.S.R. and the development of gauges for recording detonation pressure. After working at the Power Engineering Institute of the U.S.S.R. Academy of Sciences (1953–1958) and the Moscow Physico-Technical Institute (1958–1959), he became head of the laboratory of the AN SSSR Siberian Division's Institute of Hydrodynamics (1959–1967). SOLOUKHIN received the first Ph.D. in physics and mathematics of the newly founded Novosibirsk State University (1962) to which he later served as the first Dean of Physics and Pro-Rector for Education and Research. He was appointed professor at the University of Novosibirsk (1965), became deputy director of the AN SSSR Siberian Division's Institute of Nuclear Physics (1967–1971), director of the AN SSSR Siberian Division's Institute of Pure and Applied Mechanics at Academgorod, and eventually director of the Heat and Mass Transfer Institute of the Belarus Academy at Minsk (1976–1988).

His research was mainly devoted to the combustion of gases and the kinetics of high-temperature chemical reactions in shock waves, in particular to the structure of detonation waves in gases and the processes by which combustion is transformed into detonation. He also studied the propagation of shock waves in inhomogeneous media and the problem of shielding shock waves, developing new techniques for the diagnosis of the main shock wave parameters, such as pressure, density, and temperature. In cooperation with Nikita A. FOMIN, N.N. KUDRIAVTSEV, and S.S. NOVIKOV he conducted research on the diagnostics of molecular levels in nonequilibrium flows of gasdynamic lasers.

The results of his investigations were published in several books and over 300 research articles. Together with Bogdan V. VOITSEKHOVSKY and Yakov K. TROSHIN he won the reputed Lenin Prize (1965) for fundamental investigation of detonation. He also obtained a number of other medals. SOLOUKHIN was a member of the Belarus Academy of Sciences and a corresponding member of the U.S.S.R. Academy of Sciences. He also served as a member of the editorial board of the international journal *Experiments in Fluids*.

The *Soloukhin Medal*, instituted by the 12th International Colloquium on Gas Dynamics of Explosions and Reactive Systems (1989), is awarded to scientists for outstanding experimental studies in the field of gas dynamics.

Astronomers have named a minor planet (asteroid 9741 SOLOUKHIN) after him.

**ORIGINAL WORKS.** *Shock wave studies of the physical properties of gases.* Sov. Phys. Uspekhi **2**, 547-556 (1959) — *Detonation waves formed by an electric discharge in $H_2O$* [in Russ.]. Energet. Inst. Im. G.M. Krzhizhanovskogo (1959), pp. 143-145 — *Gas flow behind the shock wave during ignition* [in Russ.]. In: (G.M. KRZIZANOVSKOMU, ed.) Problemy Energetiki. Izd. Akad. Nauk SSSR (1959), pp. 735-744 — *Physical investigations of gases by means of shock waves.* Joint Publications Research Service, Arlington, VA (1960) — *Application of shock waves for study of gas ignition* [in Russ.]. Zh. Prikl. Mekh. Tekh. Fiz., No. 2, 90-92 (1960) — *Transition from burning to detonation* [in Russ.]. Ibid. No. 4, 128-132 (1961) — With T.V. BAZHENOVA: *Pressure field occurring in water during an electrical discharge.* In: (A.S. PREDVODITELEV, ed.) *Physical gas dynamics.* Pergamon Press, Oxford (1961), pp. 144-153 — *Shock waves forming during an electrical discharge in water.* Ibid., pp. 154-157 — *The effect of bubbles in shock detonation in a liquid.* Sov. Phys. Dokl. **6**, No. 1, 16-17 (1961) — With T.A. SHARAPOVA: *Spectroscopic investigation of the state of a gas behind the front of a detonation.* Ibid. **7**, No. 2, 37-41 (1962) — *The non-equilibrium state of carbon dioxide behind the shock wave* [in Russ.]. Zh. Prikl. Mekh. Tekh. Fiz No. 6, 138-140 (1963) — *Shock waves and detonations in gases* [in Russ.]. Gos. Izd. Fiz. Mat. Literatury, Moscow (1963); Engl. translation by B.W. KUVSHINOFF. Mono Book Corp., Baltimore (1966) — *Detonation waves in gases* [a review]. Sov. Phys. Uspekhi **6**, 523-541 (1963) — With V.V. VOEVODSKII: *Mechanism and limits of chain self-ignition of hydrogen with oxygen in shock waves.* Dokl. AN (SSSR) **154**, No. 6, 1425-1428 (1964) — With V.V. MITROFANOV: *The diffraction of multifront detonation waves.* Sov. Phys. Dokl. **9**, No. 12, 1055-1058 (1964) — With V.V. VOEVODSKII: *On the mechanism and explosion limit of hydrogen-oxygen chain self-ignition in shock waves.* In: *10th Symp. (Int.) on Combustion.* The Combustion Institute, Pittsburgh, PA (1965), pp. 279-283 — With M.I. VOROTNIKOVA and V.K. KEDRINSKII: *Shock tube for studying one-dimensional waves in liquid* [in Russ.]. Fiz. Gor. Vzryva (Novosibirsk) **1**, 5-14 (1965) — *Multi-headed structure of gaseous detonation.* Combust. Flame **9**, 51-58 (1965) — *Direct shock tube measurements of oxygen from recombination rates.* Ibid. **11**, 489-495 (1967) — With Y.E. NESTERIKHIN: *Methods of high-speed measurements in gas dynamics and plasma physics.* Nauka, Moscow (1967) — *Quasi-stationary reaction zone in gaseous detonation.* In: *Proc. 11th Symp. (Int.) on Combustion* [Berkeley, CA, Aug. 1966]. The Combustion Institute, Pittsburgh, PA (1967), pp. 671-676 — With V.K. KEDRINSKII ET AL.: *The study of high-rate reactions in solutions behind a powerful shock front.* Dokl. AN (SSSR) **187**, 130-133 (1969) — With J.H. LEE and A.K. OPPENHEIM: *Current views on gaseous detonations.* Astronaut. Acta **14**, 565-584 (1969) — With K.W. RAGLAND: *Ignition processes in expanding detonations.* Combust. Flame **13**, 295-302 (1969) — With C. BROCHET: *The development of instabilities in a shocked exothermic gas flow.* Ibid. **18**, 59-64 (1972) — With V.F. KLIMKIN and P. WOLANSKY: *Initial stages of a spherical detonation directly initiated by a laser spark.* Ibid. **21**, 111-117 (1973) — With A.K. OPPENHEIM: *Experiments in gas dynamics of explosions.* Annu. Rev. Fluid Mech. **5**, 31-58 (1973) — With YU.A. YACOBI and V.I. YACOVLEV: *Studying ionizing shock wave by IR diagnostic techniques.* Arch. Mech. (Warszawa) **26**, 637-646 (1974) — *Shock tubes in laser research; modeling and application* [Otto-Laporte Lecture]. Proc. 11th Int. Symp. on Shock Tubes and Waves [Seattle, WA, July 1977]. In: (B. AHLBORN, A. HERTZBERG, and D. RUSSELL, eds.) *Shock tube and shock wave research* [University of Washington Press, Seattle (1978), pp. 629-644 — With V.F. KLIMKIN and A.N. PAPIRIN: *Optical recording methods of high-speed processes* [in Russ.]. Nauka, Novosibirsk (1980) — With N.A. FOMIN: *Mixing gasdynamic lasers* [in Russ.]. Nauka & Technika, Minsk (1984) — (R.I. SOLOUKHIN, ed.): *Measurement techniques in heat and mass transfer.* Proc. Int. Center for Heat and Mass Transfer. Hemisphere Publ., Washington, DC (1985) — (R.I. SOLOUKHIN, ed.): *Handbook of radiative heat transfer in high-temperature gases.* Hemisphere, Washington, DC (1987).

**SECONDARY LITERATURE.** R.J. EMRICH and N.A. FOMIN: *In memoriam, Rem Ivanovich* SOLOUKHIN, *1930–1988.* Exp. Fluids **7**, 433-434 (1989); http://www.itmo.by/calendar/rem/rem.html.
**PICTURE.** Courtesy Prof. Nikita A. FOMIN, Heat & Mass Transfer Institute, Minsk, Russia.
**NOTE.** His name has also been transliterated as SOLOUCHIN.

## STANTON, Sir Thomas Edward (1865–1931)

- British engineer and physicist; pioneer of supersonic wind tunnel testing

Sir Thomas E. STANTON was born in Atherstone in Warwickshire (central England), the son of the landowner Thomas STANTON. After attending the Atherstone Grammar School, he became an apprentice at a company in Leicester, being trained there in general engineering and millwrights (1884–1887). In 1888, he entered Owens College at Manchester and followed the engineering course given at the Whitworth Laboratory under the engineer and physicist Osborne REYNOLDS. He took his B.Sc. at Victoria University, Manchester, and continued working in REYNOLDS' laboratory as a Junior and later as Senior demonstrator (1891–1896), and later as resident tutor in mathematics and engineering at the Hulme Hall of Residence, Manchester (1892–1896). After working as a senior assistant lecturer in engineering at the University College, Liverpool (1896–1900), and shortly serving as professor of engineering at Bristol University College (1899–1900), he was appointed Superintendent of the Engineering Department of the National Physical Laboratory (NPL) in Bristol, a position he held from 1901 until his retirement in 1930.

STANTON's main field of interest was fluid flow and friction and the related problem of heat transmission. At NPL he began to tackle various problems of public interest, such as the study of wind forces on bridges, roofs, and other structures. He performed model experiments at reduced scale and, using a small vertical wind tunnel of circular section 2 ft in diameter, exposed his model structures to an artificial current of air (1902–1907). After verifying his model analysis by full-scale experiments and realizing that invaluable information could be obtained by applying this concept, he extended this method to various problems of airplane and airship design. After 1908 he devoted himself to problems of airplane and aircraft design and the dissipation of heat from air-cooled engines.

In the early 1920s STANTON conducted research into the subject of aerodynamic drag to motion at very high speeds, which became important for the motion of fast projectiles and propeller blades with high tip speeds. For this purpose he designed at NPL the first British supersonic wind tunnel which he called "wind channel." It had a diameter of 0.8 in. (20.3 mm), which only allowed the testing of tiny-scale model projectiles 0.09 in. (2.3 mm) in diameter up to approx. $M = 2$. However, he was able to correlate his results obtained from testing tiny model projectiles with those from firing tests on artillery ranges. Early tests were also made on airscrew blade sections and bullet shapes. After provision of a more powerful air-compression capability, a larger and continuously operated wind tunnel, 3.07 in. (78 mm) in diameter, $M = 3.25$, was put into operation (1922). He tested airfoils in this tunnel, even to the extent of determining the pressure distribution over model airfoils of only half-an-inch (12.7-mm) chord.

STANTON also contributed to other fields of aircraft engineering, such as film and boundary lubrication in engines, and the understanding and testing of metal fatigue.

**ORIGINAL WORKS.** *Alternating stress testing-machine at the National Physical Laboratory.* The Engineer (Lond.) **99**, 201-203 (1905) — *The mechanical viscosity of fluids.* Proc. Roy. Soc. Lond. **A85**, 366-376 (1911) — With D. MARSHALL and C.N. BRYANT: *On the conditions at the boundary of a fluid in a turbulent motion.* Ibid. **A97**, 413-434 (1920) — *On the flow of gases at high speeds.* Ibid. **A111**, 306-339 (1926) — *A high-speed wind channel for tests on aerofoils.* ARC-R&M 1130, Aeronautical Research Committee (1928) — With D. MARSHALL: *On the eddy system in the wake of flat circular plates in three-dimensional flow.* Proc. Roy. Soc. Lond. **A130**, 295-301 (1931) — *The development of a high-speed wind tunnel for research in external ballistics.* Ibid. **A131**, 122-132 (1931).
**SECONDARY LITERATURE.** J.E.P. and F.J.S (anonymous): *Sir Thomas Edward STANTON, 1865–1931.* Proc. Roy. Soc. Lond. **A135**, ix-xv (1932) — U. GRIGULL, H. SANDNER, J. STRAUB, and H. WINKLER: *Origins of dimensionless groups of heat and mass transfer.* In: (U. GRIGULL ET AL., eds.) *Proc. 7th Int. Heat Transfer Conference* [Munich, Germany, Sept. 1982]. Hemisphere Publ. Corp., Washington, DC (1982) — *Sir Thomas STANTON 1865–1931.* Dept. of Mechanical Engineering, University of Texas; http://www.me.utexas.edu/~me339\bios/stanton.html — *Famous names at NPL: Sir Thomas STANTON (1865–1931).* NPL, Teddington, U.K.; http://www.npl.co.uk/about/famous_names/sir_thomas_stanton.html — *Hall of pioneers: history of heat transfer.* UCLA Dept. of Materials Science & Engineering; http://www.seas.ucla.edu/jht/pioneers/pioneers.html.
**PICTURE.** Courtesy NPL Crown Copyright 1950. Reproduced with the permission of the Controller of H.M.S.O. and the Queen's Printer for Scotland.

## STANYUKOVICH [Russ. Станюкович], Kirill Petrovich (1916–1989)

- Soviet theoretical physicist and eminent gas dynamicist

Kirill P. STANYUKOVICH studied at the Mechanical-Technical Faculty of Moscow State University and graduated with a

specialization in astronomy (1939). While still a student he already showed a keen interest in meteorite impact and published a theory on the nature of craters on the Moon, expounding the idea that the craters are the result of bombardment of the Moon over millions of years (1937). In 1942, he entered the fields of shock wave and detonation physics and worked at the Soviet Artillery Academy under the supervision of Prof. Lew D. LANDAU, then director of the section of theoretical physics at the Institute of Physical Problems of the U.S.S.R. Academy of Sciences. Together with LANDAU he developed a concept of a barotropic equation of state for detonation products that formed the basis of modern explosion physics, and he analytically treated convergent spherical shock waves (1944, published in 1955). In 1947, he received his Ph.D. in technical sciences. As a result of his courses on gas dynamics he published his first monograph on the theory of unsteady flows (1948), which, later translated into English under the title *Unsteady Motion of Continuous Media* (1960), quickly became a classic textbook on the field of shock waves and detonation also in the Western World. In 1954, he first showed that a power law entropy profile (with index $b = 3\gamma - 1$) is mathematically equivalent to the case of an isentropic gas ($b = 0$); *i.e.*, presenting Riemann invariants and thus resulting in an essential simplification of the system of Euler equations which become linear.

From 1950 he worked at the N.E. Baumann Institute of Moscow State Technical University (BMSTU) and became professor in 1952. Together with various other Soviet professors he laid the scientific foundations of the M4 Department Gas-Dynamic Impulse Devices at BMSTU. In an article entitled *Problems of Interplanetary Flights* (1954), he treated the subject of nuclear-powered rockets. Prior to Sputnik's flight (1957), he wrote a paper on the problems associated with artificial Earth satellites (1955). As a coauthor he contributed to Filipp A. BAUM's monograph *Physics of Explosions* (1959, 1975). He also wrote two popular-science books on cosmology that were translated into English: *Matter and Man* (1962) and *There Are Seven Elements in the World* (1963).

STANYUKOVICH was a member of the Interdepartmental Commission on Interplanetary Communications (ICIC), an organization established by the Soviet Presidium (1957), and from 1954 a charter member of the V.P. Chkalov Central Aero Club of the Soviet Union.

**ORIGINAL WORKS.** With G.I. POKROVSKY: *The problem of directed explosions* [in Russ.]. Izvestiia Akad. Nauk Ser. Fiz. **8**, 214-223 (1944) — With L.D. LANDAU: *On a study of detonation of condensed explosives* [in Russ.]. Dokl. AN (SSSR) **46**, No. 9, 362-364 (1945); *Determination of the flow velocity of the detonation products of condensed explosives* [in Russ.]. Ibid. **47**, No. 4, 271-274 (1945); *The determination of the flow velocity of the detonation products of some gaseous mixtures* [in Russ.]. Ibid. **47**, No. 8, 199-201 (1945); *On self-similar solutions of the hydrodynamic equations with central symmetry* [in Russ.]. Ibid. **48**, No. 5, 310-312 (1945) — *Application of particular solutions of gas dynamics equations to the study of detonation and shock waves*. Ibid. **52**, No. 7, 589-592 (1946) — *On the reflection of a detonation wave front*. Ibid. **52**, No. 9, 771-772 (1946) — *The flow of the detonation products in the case of an oblique detonation wave*. Ibid. **55**, No. 4, 311-314 (1947) — With Y.B. ZEL'DOVICH: *On the reflection of a plane detonation wave*. Ibid. **55**, No. 7, 587-590 (1947) — With V.V. FEDYNSKY: *On the destructive effect of meteorite impacts* [in Russ.]. Ibid. **57**, No. 2, 129-132 (1947) — *Two-sided escape of gas from cylindrical vessel into tube* [in Russ.]. Ibid. **58**, No. 2, 201-204 (1947) — *The particle motion of the detonation products from a long charge* [in Russ.]. Ibid. **58**, No. 5, 763-766 (1947) — *Theory of irregular gas movements* [in Russ.]. Izd. Byro novoi tekhniki, Moscow (1948) — *Some exact solutions of the gasdynamic equations for central-symmetric motions* [in Russ.]. Dokl. AN (SSSR) **60**, No. 7, 1141-1144 (1948) — *Reflection of a plane detonation wave* [in Russ.] Ibid. **61**, No. 2, 227-230 (1948) — *Self-similar plane and axial-symmetric stationary motions of a gas* [in Russ.]. Ibid. **64**, No. 1, 29-32 (1949) — *Self-similar plane and axial-symmetric nonstationary gas flows*. Ibid. **64**, No. 2, 179-181 (1949) — *Self-similar motion of a gas in its center of gravity* [in Russ.]. Ibid. **64**, No. 4, 467-470 (1949) — *Elements of a physical theory of meteors and crater formation by meteorites* [in Russ.]. Meteoritika **7**, 39-62 (1950) — *Elements of applied theory of unsteady motion of a gas* [in Russ.]. Oborongiz, Moscow (1953) — *A new method of approximation for integrating equations of the hyperbolic type* [in Russ.]. Dokl. AN (SSSR) **93**, 979-982 (1953) — *Trip to the Moon: Fantasy and reality*. News: A Soviet Review of World Events (June 1, 1954) — *Problems of interplanetary flights*. Krasnaya Zvezda ("Red Star") (Aug. 10, 1954). Engl. translation and commentary by F.J. KRIEGER. Santa Monica Corp. (1954) — *General solutions of the gasdynamic equations for one-dimensional motions for a given equation of state or process* [in Russ.]. Dokl. AN (SSSR) **96**, 441-444 (1954) — *Neustanovivsjesja dvizenija splosnoj sredy*. Gostekhizdat, Moscow (1955); (M. HOLT, ed.) *Unsteady motion of continuous media* (Engl. translation by J.G. ADASHKO). Pergamon Press, London (1960) — *Elements of relativistic magnetogasdynamics*. Bull. Acad. Sci. U.S.S.R. (Phys. Series) **6**, 578-589 (1955) — *Shock wave configuration during flight and explosions of meteorites* [in Russ.]. Meteoritika **14**, 62-69 (1956) — *On space flight* [in Russ.]. Molodaya Gvardiia, Moscow (1956) — *On the impact of solids at high velocities*. Zh. Eksp. Teor. Fiz. (SSSR) **36**, 1605-1606 (1956) — *Gasdynamic foundations of inner ballistics* [in Russ.]. Gostekhizdat, Moscow (1957) — *Some unsteady two- and three-dimensional gas flows*. Dokl. AN (SSSR) **112**, 595-598 (1957) — With B.I. SHEKHTER: *Fizika vzryva*. Fizmatgiz, Moskva (1959). With F.A. BAUM and B.I. SHEKHTER: *Physics of an explosion*. Engl. translation for U.S. Army Engineer Research and Development Laboratories, Information Resources Branch, Technical Information Section by Research Information Service, Fort Belvoir, VA (1962) — *Elements of the theory of solids with great (cosmic) velocities* [in Russ.]. Iskusstvennye sputniki zemli ("Artificial Earth Satellites") No. 4, 86-117 (1960) — With L.P. ORLENKO: *Shock waves in solids*. U.S. Joint Publ. Res. Service, New York (1960) — With V.V. FEDYNSKY: *In the wake of the fiery stone* [book review]. Sov. Astron. **3**, 374-375 (1960) — *On an effect in the area of the aerodynamics of meteors* [in Russ.].

Izv. AN (SSSR), Otd. Tekh. Nauk Mekh. Mashinostr. (1960), No. 5, pp. 3-8 — With V.A. BRONSHTEN: *Velocity and energy of the Tunguska meteorite.* Dokl. AN (SSSR) **140**, 1053-1055 (1961) — With M.V. VASIL'EV: *Matter and man* [in Russ.]. Peace, Moscow (1962); Engl. translation by V. TALMY. University Press of the Pacific, Stockton, CA (2000) — *The Golub-Yavan.* In: (A.F. BELAYEV ET AL.) *Destination: Amaltheia.* Engl. translation by L. KOLESNIKOV. Moscow Foreign Languages Publishing House, Moscow (1962) — *There are seven elements in the world.* Foreign Tech. Div., Air Force Systems Command, Wright-Patterson Air Force Base, OH, (1963) — *Gravitational field and elementary particles* [in Russ.]. Nauka, Moscow (1965).

**SECONDARY LITERATURE.** V.A. BRONSHTEN: *Kirill Petrovich STANYUKOVICH (1916–1989)* [in Russ.]. Nauka, Moskva (2005).

**PICTURE.** Courtesy Prof. Gennadij I. KANEL', Institute for High Energy Densities of the Russian Academy of Sciences, Moscow, Russia.

**NOTE.** His name has also been transliterated as STANJUKOVIČ.

## STODOLA, Aurel Boleslav (1859–1942)

• Hungarian-born Swiss mechanical engineer and fluid dynamicist

Aurel B. STODOLA was born in the Hungarian town Liptovs-ký Mikuláš (now Liptószentmiklós, eastern Czechia). He was the second son of Andreas STODOLA, a leather-belt manufacturer. He studied at the Polytechnic School of Budapest (1876–1877), the University of Zurich (1877–1878), and finally at the Eidgenössische Technische Hochschule Zürich (ETHZ, 1878–1880), from which he graduated as a mechanical engineer with the highest distinction (1881). He worked briefly in the workshops of the Hungarian State Railways in Budapest and continued his studies in Berlin, where he attended lectures of the German physicist Hermann VON HELMHOLTZ and the German mathematician Paul DU BOIS-REYMOND. After gaining further practical experiences in Paris, he worked as a designer in the Prague steam engine factory Ruston & Co. In 1892, he was offered the chair of thermal machinery at ETHZ, where he remained until his retirement (1892–1929).

His main subject of research was the theoretical treatment of steam and gas turbines and centrifugal compressors. Most of his results were published in his internationally widely acknowledged book *Die Dampfturbinen und die Aussichten der Wärmekraftmaschinen* (1903), which, translated into English and French, saw six editions and substantive revisions. STODOLA investigated in great detail the principal parts of the steam turbine, namely, the nozzles and blades. When STODOLA began his studies, only a rudimentary theory of supersonic flow in Laval nozzles was available. He was the first to perform systematic measurements of the axial pressure distribution at different back pressures and graphically demonstrated the conditions under which a supersonic flow in Laval nozzles occurs. When driving Laval nozzles at high back pressures he noticed a dramatic increase in pressure and correctly related his observation to a *Verdichtungsstoß* (condensation shock); *i.e.*, a shock wave, as was previously derived mathematically by Bernhard RIEMANN (1859). Since a profound knowledge of the proper operation characteristics of a Laval nozzle is of greatest practical importance for the efficiency of steam turbines, STODOLA verified his theoretical results by carrying out many experiments and also established the various sources of losses and their magnitude. Typically for him, he managed to express complicated fluid dynamic conditions in simple formulae ready for practical use in engineering, for example steam flow through clearances and labyrinths, and friction effects of disks. He eventually became the leading expert in Europe on steam turbines.

Besides the thermodynamic and fluid dynamic treatment of steam, STODOLA investigated mechanical and thermal stresses in bodies of basic geometry, such as plate shells and rotating disks as well as of real turbine parts in order to estimate the forces due to the steam pressure, steam temperature, and the motion itself. He also had a strong sense of social responsibility, as evidenced in his later work *Gedanken zu einer Weltanschauung vom Standpunkt des Ingenieurs* ("Thoughts on a Worldview from the Standpoint of an Engineer," 1931). STODOLA's assistants Jakob ACKERET and Gino FANNO became famous gas dynamicists.

The *Stodola Lecture*, established by ETHZ in 2004, is given by an internationally recognized expert in the field of mechanical and process engineering, which is selected by the department faculty and is awarded the *Stodola Medal*.

**ORIGINAL WORKS.** *Die Dampfmotoren auf der Weltausstellung in Paris 1900.* Rascher, Zürich (1901) — *Die Dampfturbinen und die Aussichten der Wärmekraftmaschinen. Versuche und Studien.* Springer, Berlin (1903); *Steam turbines: with an appendix on gas turbines and the future of heat engines* (Engl. translation by L.C. LOEWENSTEIN). Van Nostrand, New York (1905). Reprint by P. Smith, New York (1945) — *Strömungen von Gasen und Dämpfen durch Rohre mit veränderlichem Querschnitt.* VDI-Z. **49**, 1787-

1788 (1903) — *Improvements in explosion gas turbines*. Brit. Patent No. 190,513,473A (1905) — *Wirkungsgrad der Explosionsturbine*. VDI-Z. **56**, 1005-1009 (1912) — *On undercooling of steam in nozzles*. Engineering (Lond.) **99**, 81-82, 643-644, 685-686 (1915) — *The steam flow in a compound steam turbine*. Ibid. **102**, 2-3 (1916) — *Neue kritische Wellengeschwindigkeit*. Dingler's Polytech. J. **99**, 117-120 (1918) — *Strömung in Düsen und Strahlvorrichtungen, mehrdimensional betrachtet*. VDI-Z. **63**, 31-36, 96-100 (1919) — *Turbines à vapeur et à gaz; ouvrage suivi de considérations sur les machines thermiques et leur avenir* (French translation by E. HAHN). Dunod, Paris (1925) — *Zur Theorie des Wärmeübergangs von Flüssigkeiten oder Gasen an feste Wände*. Schweiz. Bauz. **88**, 243-244 (1926) — *Wärmeübergang in Grenzschichten bei stark veränderlicher Grundströmung*. Ibid. **89**, 193-196 (1928) — *Wärmeübergang in Grenzschichten bei großen Temperaturunterschieden zwischen Wand und Flüssigkeit*. Ibid. **89**, 261-262 (1928) — *Über die Strömungsverhältnisse an der Austrittskante eines vereinfachten Dampfturbinen-Leitschaufel-Modells*. Ibid. **95**, 309-310 (1930) — *Gedanken zu einer Weltanschauung vom Standpunkt des Ingenieurs*. Springer, Berlin (1931) — *Die geheimnisvolle Natur: Weltanschauliche Betrachtungen* [gekürzte Umarbeitung]. Orell Füßli, Zurich (1940) — *Sulzer single-tube steam generator*. The Engineer (Lond.) **156**, 496-500, 525-527 (1933).

**SECONDARY LITERATURE.** E. HONEGGER (ed.): *(1) Festschrift Prof. Dr. A. STODOLA zum 70. Geburtstag. (2) Zusammenstellung der literarischen Arbeiten von Prof. Dr. A. STODOLA*. Orell Füssli, Zürich & Leipzig (1929) — (Ed.) *Dr. STODOLA and the James WATT International Medal*. The Engineer (Lond.) **171**, 78 (1941) — G. EICHELBERG: † *Aurel STODOLA*. Schweiz. Bauz. **121**, 73-74 (1943) — H. QUIBY: *Aurel STODOLA, sein wissenschaftliches Werk*. Ibid., 74-77 (1943) — E. SÖRENSEN: *Aurel STODOLA*. VDI-Z. **87**, 169-170 (1943) — *Obituary notice of Dr. STODOLA* [with portrait]. Engineering **153**, 334 (1943) — B. POCHOBRADSKY: *Aurel STODOLA: the teacher, the scientist, the man*. Ibid. **156**, 454-455 (1943) — A. SONNTAG: *Aurel STODOLA 1859–1942*. Brennstoff-Wärme-Kraft **11**, 211-212 (1959) — A. TURECKÝ and J. VODA (eds.): *Aurel STODOLA 1859–1942; Pamiatke storocnice narodenia v pamjat' stoletnej godovsciny* [Memoir on his 100th birthday on May 10, 1959]. Vydavatel'stvo Slovenskej akadémie vied, Bratislava (1959) — O. MAYR: *A.B. STODOLA*. In: (C.C. GILLESPIE, ed.) *Dictionary of scientific biography*. Scribner, New York, vol. 13 (1976), pp. 72-74 — J.D. ANDERSON JR.: *Modern compressible flow, with historical perspective*. McGraw-Hill, New York (1990), pp. 179-181 — N. LANG: *Aurel STODOLA (1859–1942): Wegbereiter der Dampf- und Gasturbine*. Serie Schweizer Pioniere der Wirtschaft und Technik, Nr. 75. Verein für Wirtschaftshistorische Studien, Meilen, Switzerland (2003).

**PICTURE.** Courtesy Bildarchiv ETH-Bibliothek, Zurich, Switzerland.

**NOTE.** See also 23 pictures of Aurel STODOLA, "bildarchivonline" der ETH Bibliothek, Zürich; http:/\ba.e-pics.ethz.ch/ETH_Bibliothek/Standard/.

# STOKES, Sir George Gabriel (1819–1903); from 1889 first Baronet

- Irish mathematical physicist; founder of geodesy

Sir George G. STOKES was born in Skreen (County Sligo, Ireland) into an Anglo-Irish family of clerical tradition. His father Gabriel STOKES was rector of Skreen and vicar-general of Killala (County Mayo). After receiving an early education in mathematics and geometry in Skreen, Sligo and Dublin (1832–1835), he went to Bristol College to study the traditional curriculum, but with an emphasis on mathematics, under Francis NEWMAN (1835–1837). At his final examination (1837) he won a prize "for eminent proficiency in mathematics" and continued his study of mathematics at Pembroke College, Cambridge (1837–1841). After graduation he was immediately elected a Fellow of the college. Following the suggestion of William HOPKINS, one of his teachers, he took up the subjects of hydrodynamics and optics in addition to mathematics.

STOKES' earliest papers were on fluid motion (1842–1846): he worked out (1) a mathematical solution to the problem of finding the steady motion of an incompressible fluid in the interior of a rectangular box that is given any motion whatever, starting from rest with the contained liquid at rest (1843, 1846); (2) a theory of the viscosity of fluids and its influence in fluid motion, which constitute the complete foundation of hydrokinetics and of the equilibrium and motion of elastic solids (1845); (3) a classical theory on a discrepancy concerning some solutions of equations of compressible gas dynamics, thereby correctly deriving the jump conditions that discontinuous solutions must satisfy (1848); (4) a theory of oscillatory waves (1849), thereby also addressing the determination of the motion of steep deep-sea waves and showing that the difference in level between crest and hollow is $7/40$ of the wavelength; and (5) a solution to the difficult problem of internal friction effects of fluids on the motion of pendulums (1850).

Already in 1846 regarded as an authority on hydrodynamics, STOKES was asked by the British Association for the Advancement of Science (BAAS) to give a report on its current state, which he did in a concise, masterly manner. With this profound knowledge of fluid dynamics he began 2 years later to tackle the difficult problem of waves of finite amplitude, which he solved by introducing surfaces of discontinuity in the velocity and density of the medium, thus pioneering the modern concept of a shock wave profile. In the period 1847–1849, he collaborated with the British engineer William THOMSON (the later Lord KELVIN), his close friend, in a series of articles on hydrodynamic principles, which THOMSON applied to electrical and atomic theory.

In 1849, STOKES was elected to the Lucasian professorship of mathematics at Cambridge, which came with the direction of the observatory. Since the previous holders of that office, James CHALLIS and George B. AIRY, used to give lectures not only on hydrodynamics but also on acoustics, he continued this tradition and entered a branch of physics to which he made many outstanding contributions. STOKES' proposed theory of the propagation of a wave of finite amplitude (1848) was in contradiction to CHALLIS' proposed theory. But later on, due to arguments of Lord KELVIN and Lord RAYLEIGH, his former student, that his proposed motion would violate the conservation of energy, he retracted the idea of such motion, and when he edited his complete works in 1880, he did not reproduce his (correct) 1848 proof of the jump conditions. His treatise on the shock wave problem only covered a very small part of his activities; his main research topics were investigations of the internal friction (or viscosity) in fluids, *fluorescence* (a term he invented in 1852), geodesy, optical spectroscopy, the wave theory of light, and pure mathematics.

STOKES became a Fellow (1851) of the Royal Society and was the first man since Sir Isaac NEWTON to hold the three positions of Lucasian professor of mathematics (1849–1903), secretary (1854–1885), and president (1885–1890) of the Society. He was also president of the Cambridge Philosophical Society (1859–1861) and the BAAS (1869). He received honorary degrees from several universities and awards from numerous scientific societies, such as the Rumford Medal (1852) and the Copley Medal (1893) of the Royal Society of London, and was made a baronet (1889). When he died at the age of 84, the world lost not only one of its greatest promoters of general fluid dynamics, but also a pioneer of shock wave physics. Like Lord KELVIN and James C. MAXWELL he was a member of the "Cambridge School of Natural Philosophers," which helped to revolutionize science in Victorian Britain.

In radiation physics the *Stokes parameters* are a set of four values that describe completely the state of polarization of a beam of electromagnetic radiation. In fluid dynamics his name is connected with the terms *Navier-Stokes equations* and *Stokes law*, and in optics with *Stokes shift* and *Stokes lines*. The *Stokes Medal*, established in 1999 by the Royal Society of Chemistry (RSC), London is awarded biennially for outstanding and sustained contributions to analytical science by someone working in a complementary field, which has led to developments of seminal importance to chemical analysis.

Astronomers named a crater on the near side of the Moon, a crater on Mars, and a minor planet (asteroid 30566 STOKES) after him.

ORIGINAL WORKS. *On the steady motion of incompressible fluids.* Trans. Cambr. Phil. Soc. **7**, 439-454, 465 (1842) — *On the motion of a piston and of the air in a cylinder.* Cambr. Math. J. **4**, 28-33 (1845) — *Report on recent researches in hydrodynamics.* Rept. Meet. Brit. Assoc. **16**, 1-20 (1846) — *On a difficulty in the theory of sound.* Phil. Mag. **33** [III], 349-356 (1848) — *Notes on hydrodynamics. (I) On the dynamical equations.* Cambr. Dublin Math. J. **3**, 121-127 (1848) — *(II) Demonstration of a fundamental theorem.* Ibid. **3**, 209-219 (1848) — *(III) On waves.* Ibid. **4**, 219-240 (1849) — *On some points in the received theory of sound.* Phil. Mag. **34** [III], 52-60 (1849) — *On the theory of sound, in reply to Prof. CHALLIS.* Ibid. **34** [III], 203-204, 348-350, 501-502 (1849) — *On some cases of fluid motion.* Trans. Cambr. Phil. Soc. **8**, 105-137, 409-414 (1849) — *On the theories of the internal friction of fluids in motion, and of the equilibrium and motion of elastic solids.* Ibid. **8**, 287-319 (1849) — *On the theory of oscillatory waves.* Ibid. **8**, 441-455 (1849) — *An examination of the possible effect of the radiation of heat on the propagation of sound.* Phil. Mag. **1** [IV], 305-317 (1851) — *On the effect of wind on the intensity of sounds.* Rept. Meet. Brit. Assoc. **27** (Part 2), 22-23 (1857) — *Report on steam-boiler explosions.* Ibid. **33**, 686-688 (1863) — *Note on the theory of the solitary wave.* Phil. Mag. **32** [V], 31-316 (1891) — *Mathematical and physical papers by the late Sir George Gabriel STOKES.* 4 vols., Cambridge University Press, Cambridge, UK (1880–1905). Reprint with a Preface by C.A. TRUESDELL. Johnson Reprint, New York (1966).

SECONDARY LITERATURE. P.G. TAIT: *George Gabriel STOKES.* Nature **12**, 201-203 (1875); *Professor STOKES works.* Ibid. **29**, 145-146 (Dec. 13, 1883) — H.P. STOKES: *Reminiscence of Sir George STOKES.* Cambridge Chronicle (Feb. 13, 1903) — W. VOIGT: *G.G. STOKES.* Nachr. Königl. Gesell. Wiss. Gött. [Math.-Phys. Kl.], 70-80 (1903) — Lord KELVIN: *The scientific work of Sir G.G. STOKES.* Nature **67**, 337-338 (1903) — E.W. BROWN: *Note on George Gabriel STOKES.* Phys. Rev. **18** [I], 58-62 (1904) — J.W. STRUTT (Lord RAYLEIGH): *George Gabriel STOKES 1819–1903.* Proc. Roy. Soc. Lond. **75**, 199-216 (1905) — J. LARMOR (ed.): *Memoir and scientific correspondence of the late Sir George Gabriel STOKES.* 2 vols., Cambridge University Press, Cambridge (1907) — P.A. THOMPSON: *Compressible-fluid dynamics.* McGraw-Hill, New York (1972), pp. 311-313 — E.M. PARKINSON: *G.G. STOKES.* In: (C.C. GILLESPIE, ed.) *Dictionary of scientific biography.* Scribner, New York, vol. 13 (1976), pp. 74-79 — F.A.J.L. JAMES: *The conservation of energy, theories of absorption, and resonating molecules, 1851–1854: G.G. STOKES, A.J. ANGSTROM and W. THOMSON.* Notes Rec. Roy. Soc. **38**, 79-107 (1983/1984) — D.B. WILSON: *KELVIN and STOKES: a comparative study in Victorian physics.* Hilger, Bristol (1987) — *The correspondence between Sir George Gabriel STOKES and Sir William THOMSON, Baron KELVIN of Largs.* 2 vols., Cambridge University Press, Cambridge & New York (1990) — A. WOOD: *George Gabriel STOKES 1819–1903, an Irish mathematical physicist.* Bull. Irish Math. Soc. **35**, 49-58 (1995) — R. PARIS: *The mathematical work of G.G. STOKES.* Math. Today **32**, No. 3/4, 43-46 (1996) — D.B. WILSON: *STOKES, Sir George Gabriel.* In: (H.C.G. MATTHEW, ed.) *Oxford dictionary of national biography.* Oxford University Press, Oxford (2004), vol. 52, pp. 860-865.

PICTURE. Reprinted with permission from Proc. Roy. Soc. Lond. **75**, 199 (1905). © The Royal Society of London, London, U.K.

## STRUTT, John William (1842–1919); from 1873 third Baron RAYLEIGH

- British experimental and mathematical physicist

John W. STRUTT was born at Langford Grove, Maldon, in the County of Essex. He was the eldest son of John James STRUTT, second Baron RAYLEIGH. After schooling at Eton and Harrow and private tutoring, he entered Trinity College, Cambridge (1861), graduated from there with a B.A. in mathematics (1865), and became a Fellow of Trinity College (1866). This was followed by a period of touring the Continent and visiting the United States. Upon returning to England (1868) he decided to fully dedicate himself to science and began to install his private laboratory at the family seat in Terling Place at Witham, Essex. After the death of his father (1873), he succeeded to the title of third Baron RAYLEIGH. After the death of James C. MAXWELL he took over as chair at Cambridge University and became second Cavendish professor of experimental physics (1879–1884). After his retirement (1885) he continued his research in his private laboratory at Terling Place. At the Royal Institution of Great Britain he served as professor of natural philosophy (1887–1896).

He also gave freely of his time and energy to matters of public interest and scientific committees of government and professional organizations. RAYLEIGH served as secretary (1884–1895) and president (1905–1908) of the Royal Society, chancellor of Cambridge University (1908–1919), president of the special government advisory committee on aeronautics (established in 1909), chairman of the Explosives Committee of the War Office, president of the British Association for the Advancement of Science (BAAS), and president of the first executive committee of the National Physical Laboratory (NPL) at Teddington.

RAYLEIGH's research covered almost the entire field of physics, but nearly three quarters of his papers deal with problems in acoustics and optics. In acoustics, starting from physiological research, he worked out a theory of resonance that established his reputation as a leading authority on sound. On a trip to Egypt (1871), which he made for health reasons, he started working on his two-volume book *The Theory of Sound* (1877–1878). Addressing questions of vibrations and the resonance of elastic solids and gases, it has remained a standard work to this day and has undergone several reprintings. In 1877, he had a dispute with George G. STOKES on his early papers (1848–1849) in which STOKES had derived a theory of sounds of finite amplitude; *i.e.*, of shock waves. He refuted STOKES' and RIEMANN's hypothesis of a discontinuous change in the thermodynamic quantities and maintained this view also in the second edition of his *Theory of Sound*, although August TOEPLER (1864) and Ernst MACH (1885, 1887, 1889) had already visualized the discontinuous nature of a shock front. Ten years before his death, however, RAYLEIGH took up this problem again and extended earlier investigations of RANKINE and HUGONIOT in his famous paper on *Aerial Waves of Finite Amplitude* (1910), a masterpiece on early shock wave research. In his later years, he also returned to his early physiological research and studied the binaural effect of hearing (1905). Today RAYLEIGH is considered one of the leading pioneers in fluid dynamics and acoustics, his studies ranging from the measurement of the minimum audible intensity of sound to the analytical treatment of shock waves, including also the first molecular-acoustic studies.

RAYLEIGH contributed much to fluid dynamics. Related to problems in acoustics, RAYLEIGH (in order to model the oscillations of a clarinet reed) described a typical nonlinear system with one degree of freedom which admits auto-oscillations, so-called "Rayleigh equation" (1877/1878). RAYLEIGH also modeled the instability of a column of liquid and predicted its decay into a chain of droplets, so-called "Rayleigh instability" (1879). He derived the growth equation for an immobile spherical gas bubble in a finite spherical volume of liquid, so-called "Rayleigh vapor bubble model" (1917). Apparently, he never studied shock waves experimentally. However, during his period as chief scientific advisor (1896–1911) to Trinity House, which maintained the fog warnings and lighthouses around the English coast, he thought about how to spread the range of fog horns within which their sound is heard as widely as possible. This requires not only sound generators of very high intensity, for example the use of trumpets and an improvement in their beam characteristic by choosing the opening of the trumpet to be small in the horizontal and large in the vertical direction. However, he estimated that even by choosing trumpets emitting a high tone (*i.e.*, a short wavelength) this would result in a very large structure, the vertical dimension required for waves about 1.2 m long being of the order of 6 m (Scient. Papers, vol. 5, p. 133). In shock wave and detonation physics a straight line on a Hugoniot plot connecting the initial and final states is called a "Rayleigh line" (1910).

Rayleigh also contributed to wave theory, optics, radiation, electromagnetism, spectroscopy, and to the redetermination of electrical units in absolute measure. His theory of the scattering of sunlight by small particles in the atmosphere, evolved in the so-called "Rayleigh scattering law" (1871), provided the first explanation of why the sky is blue. He investigated elastic surface waves that are generated at a boundary surface and showed that their effect decreases rapidly with depth and that their velocity of propagation is smaller than that of body waves. These so-called "Rayleigh waves" (1887) are of greatest importance for the interpretation of seismograms. In probability theory and statistics, the "Rayleigh distribution" is a continuous probability distribution. Together with the British chemist Sir William Ramsay he discovered the inert gas argon (1895).

Rayleigh wrote one book and about 450 scientific articles. He was awarded the Royal Medal (1882), the Copley Medal (1899) and the Rumford Medal (1914) of the Royal Society of London, and the De Morgan Medal (1890), the London Mathematical Society's premier award. He also received special awards from over 50 learned societies and was the recipient of 13 honorary degrees. In 1904, Rayleigh earned the Nobel Prize for Physics and Ramsay the Nobel Prize for Chemistry "for their investigation of the densities of the most important gases, and for their discovery of argon in connection with these studies." He died on June 30, 1919 at Terling Place, where he carried out practically all of his scientific investigations.

The *Rayleigh Medal* is the premier award of the British Institute of Acoustics (IOA) and awarded to persons renowned for outstanding contributions to acoustics. The *Rayleigh Lectures Series* is given at the International Mechanical Engineering Congress & Exhibition (IMECE) by researchers who have made pioneering contributions to the sciences and applications of noise control and acoustics.

Astronomers named a crater on the near side of the Moon and a crater on Mars after him.

**ORIGINAL WORKS.** *On waves.* Phil. Mag. **1** [V], 257-279 (1876) — *Notes on hydrodynamics.* Ibid. **2** [V], 441-447 (1876) — *The theory of sound.* 2 vols., Macmillan, London (1877/1878); *Die Theorie des Schalles.* 2 vols., Vieweg, Braunschweig (1879/1880) — *On the instability of jets.* Proc. Lond. Math. Soc. **10**, 4-13 (1878/1879) — *On the stability or instability of certain fluid motions (Part I).* Ibid. **11**, 57-70 (1879/1880) — *Investigation of the character of the equilibrium of an incompressible heavy fluid of variable density.* Ibid. **14**, 170-177 (1882/1883) — *On waves propagated along the plane surface of an elastic solid.* Ibid. **17**, 4-11 (1885/1886) — *On the stability or instability of certain fluid motions (Part II).* Ibid. **19**, 67-74 (1887/1888) — *Instantaneous photographs of water jets.* Rept. Meet. Brit. Assoc. **60**, 752 (1890) — *On the virial of a system of colliding bodies.* Nature **45**, 80-82 (1892) — *Experiments in aerodynamics.* Ibid. **45**, 108-109 (1892) — *On the amplitude of aerial waves which are but just audible.* Phil. Mag. **38** [V], 365-370 (1894) — *On the stability or instability of certain fluid motions (Part III).* Proc. Lond. Math. Soc. **27**, 5-12 (1896) — *On the propagation of waves upon the plane surface separating two portions of fluid of different vorticities.* Ibid. **27**, 13-18 (1896) — *On the cooling of air by radiation and conduction, and on the propagation of sound.* Phil. Mag. **47** [V], 308-314 (1899) — *On the production of vibrations by forces of relatively long duration, with application to the theory of collisions.* Ibid. **11** [VI], 283-291 (1906) — *On the instantaneous propagation of disturbance in a dispersive medium, exemplified by waves on water deep and shallow.* Ibid. **18** [VI], 1-6 (1909) — *Note on tidal bores.* Proc. Roy. Soc. Lond. **A81**, 448-449 (1908) — *Aerial plane waves of finite amplitude.* Ibid. **A84**, 247-284 (1910) — *On the propagation of waves through a stratified medium, with special reference to the question of reflection.* Ibid. **A86**, 207-266 (1912) — *On the theory of long waves and bores.* Ibid. **A90**, 324-328 (1914) — *Deep water waves, progressive or stationary, to the third order of approximation.* Ibid. **A91**, 345-353 (1915) — *On the principle of similitude.* Nature **95**, 66-68 (March 1915) — *On the flow of compressible fluid past an obstacle.* Phil. Mag. **32** [VI], 1-6 (1916) — *On the discharge of gases under high pressure.* Ibid. **32** [VI], 177-187 (1916) — *On periodic irrational waves at the surface of deep water.* Ibid. **33** [VI], 381-389 (1917) — *On the pressure developed in a liquid during the collapse of a spherical cavity.* Ibid. **34** [VI], 94-98 (1917) — *On the dynamics of revolving fluids.* Proc. Roy. Soc. Lond. **A93**, 148-154 (1917) — *Scientific papers by John William Strutt, Baron Rayleigh* 6 vols., Cambridge University Press (1899–1920); Vol. 1: *1869–1881* (1899), vol. 2: *1881–1887* (1900), vol. 3: *1887–1892* (1902), vol. 4: *1892–1901* (1903), vol. 5: *1902–1910* (1912), vol. 6: *1911–1919* (1920); (J.N. Howard, ed.) 3 vols., Dover Publ., New York (1964).

**SECONDARY LITERATURE.** A. Schuster: *John William Strutt, Baron Rayleigh, 1842–1919.* Proc. Roy. Soc. Lond. **A98**, pp. i-l (1922) — R.J. Strutt (4th Baron Rayleigh): *Life of John William Strutt, third Baron Rayleigh.* Arnold, London (1924); University of Wisconsin Press, Madison (1968) — J.N. Howard: *John William Strutt, third Baron Rayleigh.* Appl. Optics **3**, 1091-1101 (1964) — R.B. Lindsay: *Men of physics: Lord Rayleigh, the man and his works.* Oxford Press, London (1970) — P.A. Thompson: *Compressible-fluid dynamics.* McGraw-Hill, New York (1972), pp. 311-313 — R.B. Lindsay: *Rayleigh, John William Strutt, Lord.* In: *Encyclopædia Britannica, Macropædia.* Benton & Hemingway, Chicago (1974), vol. 15, 538-539 — E.M. Parkinson: *J.W. Strutt, third Baron Rayleigh.* In: (C.C. Gillespie, ed.) *Dictionary of scientific biography.* Scribner, New York, vol. 13 (1976), pp. 100-107 — N. Rott: *Lord Rayleigh and hydrodynamic similarity.* Phys. Fluids **A12**, 2595-2600 (1992) — A.T. Humphrey: *Lord Rayleigh, the last of the great Victorian polymaths.* Bull. Inst. Math. Appl. **31**, 113-120 (1995) — K. Gavroglu: *Strutt, John William, third Baron Rayleigh.* In: (H.C.G. Matthew, ed.) *Oxford dictionary of national biography.* Oxford University Press, Oxford (2004), vol. 53, pp. 116-120 — *Hall of pioneers: history of heat transfer.* UCLA Dept. of Materials Science & Engineering; http://www.seas.ucla.edu/jht/pioneers/pioneers.html.

**PICTURE.** Reprinted with permission from Proc. Roy. Soc. Lond. **A98**, i (1922). © The Royal Society of London, U.K.

# Sturtevant, Bradford ("Brad") (1933–2000)

- U.S. aeronautical engineer and fluid dynamicist

After receiving his B.S. (1955) in mechanical engineering from Yale University, Bradford Sturtevant joined Cal-

Tech's Graduate Aeronautical Laboratory (GALCIT), where he stayed for the rest of his professional career, receiving an M.S. (1956) in aeronautics and a Ph.D. (1960) in fluid mechanics. He joined the faculty in 1960 and became associate professor (1966–1971) and professor of aeronautics (1971–1995). In 1995, he was appointed Hans W. Liepmann Professor of Aeronautics.

STURTEVANT was best known in the fluid dynamics community for his research on shock waves and nonsteady gas dynamics. He interpreted nonsteady gas dynamics in the most imaginative way possible with applications ranging from noise control in motorcycle exhausts to volcanic eruptions and treatment of kidney stones with shock waves. His projects included (1) experimental and theoretical investigations of the propagation of shock waves through inhomogeneous media, including shock-excited Rayleigh-Taylor instability; (2) hydrodynamic sources of earthquakes and harmonic tremor; (3) sonic boom; (4) the effects of dissociation relaxation in hypervelocity flow; (5) shock wave physics of extracorporeal shock wave lithotripsy, including the focusing of weak shock waves; and (6) the fluid mechanics of explosive volcanic eruptions, including the explosive evolution of dissolved gas from rapidly depressurized liquids.

STURTEVANT was a dedicated member of the CalTech community, a vigorous athlete, and a proponent of fluid mechanics as a rigorous intellectual activity that spanned across scientific disciplines from medicine to geology. He held a number of administrative positions such as Executive Officer for Aeronautics and Secretary-Treasurer for Aeronautics of CalTech (1971–1976), and as an active sportsman he served terms as chairman and secretary-treasurer of the Southern California Intercollegiate Athletic Conference (1980–1986). He was a co-organizer of the 20th International Symposium on Shock Waves, held at CalTech in Pasadena (1995), and editor of the proceedings.

**ORIGINAL WORKS.** With H.W. LIEPMANN, A. ROSHKO, and D. COLES: *A 17-inch diameter shock tube for studies in rarefied gas dynamics.* Rev. Scient. Instrum. **33**, 625-631 (1962) — With E. SLACHMUYLDERS: *End-wall heat-transfer effects on the trajectory of a reflected shock wave.* Phys. Fluids **7**, 1201-1207 (1964) — *Implications of experiments on the weak undular bore.* Ibid. **8**, 1052-1055 (1965) — *Application of a magnetic mass spectrometer to ionization studies of impure shock-heated argon.* J. Fluid Mech. **25**, 641-656 (1966) — *Optical depth gauge for laboratory studies of water waves.* Rev. Scient. Instrum. **37**, 1460-1463 (1966) — With T.T. OKAMURA: *Dependence of shock tube boundary layers on shock strength.* Phys. Fluids **12**, 1723-1725 (1969) — With R.E. SETCHELL and E. STORM: *An investigation of shock strengthening in a conical convergent channel.* J. Fluid Mech. **56**, 505-522 (1972) — *Gas dynamics and shock dynamics of weak shock waves.* Fluid Dyn. Trans. **8**, 219-254 (1976) — With V.A. KULKARNY: *The focusing of weak shock waves.* J. Fluid Mech. **73**, 651-671 (1976) — *Shock waves in non-uniform media: real-life gas dynamics* [5th Paul Vieille Memorial Lecture]. Proc. 11th Int. Symp. on Shock Tubes and Waves [Seattle, WA, July 1977]. In: (B. AHLBORN, A. HERTZBERG, and D. RUSSELL, eds.) *Shock tube and shock wave research.* University of Washington Press, Seattle (1978), pp. 12-23 — With C.J. CATHERASOO: *Shock dynamics in non-uniform media.* J. Fluid Mech. **127**, 539-561 (1983) — With S.W. KIEFFER: *Laboratory studies of volcanic jets.* J. Geophys. Res. **89**, 8253-8268 (1984) — With D.M. MOODY: *Shock waves in superfluid helium.* Phys. Fluids **27**, 1125-1137 (1984) — With J.F. HAAS: *Interaction of weak shock waves with cylindrical and spherical gas inhomogeneities.* J. Fluid Mech. **181**, 41-76 (1987) — With S.W. KIEFFER: *Erosional furrows formed during the lateral blast at Mount St. Helens, May 18, 1980.* J. Geophys. Res. **93**, 14793-14816 (1988) — With L. HESSELINK: *Propagation of weak shocks through a random medium.* J. Fluid Mech. **196**, 513-553 (1988) — *The physics of shock focusing in the context of extracorporeal shock wave lithotripsy.* In: (K. TAKAYAMA, ed.) *Proc. Int. Workshop on Shock Focusing* [Sendai, Japan, March 1989]. Shock Wave Research Center, Institute of Fluid Science, Tohoku University (1990), pp. 39-64 — With B. HARTENBAUM: *Performance of an electrohydraulic shock wave lithotripter measured with thin film PVDF transducers.* JASA **90**, 2339 (1991) — With H. GLICKEN, L. HILL, and A.V. ANILKUMAR: *Explosive volcanism in Japan and the United States: gaining an understanding by shock tube experiments.* Proc. 18th Int. Symp. on Shock Waves [Sendai, Japan, July 1991]. In: (K. TAKAYAMA, ed.) *Shock Waves. Proceedings, Sendai, Japan.* Springer, Berlin (1992), vol. 1, pp. 129-140 — With M. BROUILLETTE: *Experiments on the Richtmyer-Meshkov instability: small-scale perturbations on a planar interface.* Phys. Fluids **A5**, 916-930 (1993) — With R. BONAZZA: *X-ray measurements of growth rate at a gas interface accelerated by shock waves.* Ibid. **A8**, 2496-2512 (1996) — With H.M. MADER, E.E. BRODSKY, and D. HOWARD: *Laboratory simulations of sustained volcanic eruptions.* Nature **388**, 462-464 (1997) — With J. CATES: *Shock wave focusing using geometrical shock dynamics.* Phys. Fluids **A9**, 3058-3068 (1997) — With D.D. HOWARD: *In vitro study of the mechanical effects of shock wave lithotripsy.* Ultrasound Med. Biol. **23**, 1107-1122 (1997) — With E.E. BRODSKY and H. KANAMORI: *A seismically constrained mass discharge rate for the initiation of the May 18, 1980 Mount St. Helens eruption.* J. Geophys. Res. **104**, 29387-29400 (1999).

**SECONDARY LITERATURE.** Anonymous: *Obituaries: Brad STURTEVANT 1933–2000.* Engng. & Sci. (Pasadena) **64**, No. 4, 39-41 (2000) — GALCIT: *Curriculum vitae: Bradford STURTEVANT* [with a list of his publications]; http://www.galcit.caltech.edu/~jeshep\brad/cv_bbs.pdf (Feb. 2001) — Y. ZHANG, H.M. MADER, E. BRODSKY, H. KANAMORI, S. KIEFFER, J. PHILLIPS, J. SHEPHERD, and S. SPARKS: *Brad STURTEVANT, 1933–2000.* Bull. Volcanology **63**, 569-571 (2002) — *In memoriam B. STURTEVANT & curriculum vitae*; see http://www.galcit.caltech.edu/~jeshep\brad/.

**PICTURE.** Courtesy Prof. Joseph E. SHEPHERD, Graduate Aeronautical Laboratories at CalTech, Pasadena, CA.

## TAIT, Peter Guthrie (1831–1901)

- Scottish physicist and mathematician; father of science and golf

Peter G. TAIT was born in Dalkeith (a town 10 km south of Edinburgh), the son of John TAIT, a secretary to the Duke of Buccleuch. After schooling, partly together with James C. MAXWELL, he entered the University of Edinburgh to study mathematics (1847) and continued at Peterhouse College in Cambridge (1848–1852), where the mathematician and geologist William HOPKINS was his tutor. After graduating from Cambridge, TAIT became professor of mathematics at Queen's College in Belfast (1854–1860) and thereafter took the chair of natural philosophy at the University of Edinburgh (1860–1901), which he held until shortly before his death.

In the 1860s, he became interested in thermodynamics, working particularly on thermoelectricity and thermal conductivity. He collected deep-sea temperatures during the HMS *Challenger* Expedition, a prolonged oceanographic exploration cruise (1872–1876) carried out through cooperation of the British Admiralty and the Royal Society. Returning to England, TAIT began to work out corrections of his measured temperature data, which were required because of the great pressures to which his thermometers had been subjected. He investigated the compressibility of water, sea water, glass, and mercury. His isothermal equation of state (1888), the so-called "Tait equation," was later modified by others to also match compressibility data of statically compressed organic liquids (HIRSCHFELDER ET AL. 1964; ATANOV 1966) as well as for use in underwater explosions to shock pressures up to 90 kbar (KIRKWOOD 1942; COLE 1948).

TAIT also studied percussion phenomena both theoretically and experimentally by a special guillotine-like percussion machine that was capable of recording contact times. He collaborated with William THOMSON, (from 1892 Lord KELVIN), on the book *Treatise on Natural Philosophy* (1888). Here they treated the collision of spherical bodies and introduced a "restitution coefficient," which they defined as the quotient of velocities after and before impact. To study the impact of a golf ball, TAIT used a ballistic pendulum faced with clay into which the ball was driven. For the contact phase between club and ball he obtained a time duration as short as about 5 ms and estimated the impact force on the order of tons. TAIT was apparently the first to study the lift on a spinning ball in the game of golf. His golf-ball studies, published in the British journal *Nature* in a series of articles, revealed that underspin provided the great secret of long driving (1890–1891). However, the markings on balls went through extensive development before the present dimpled surface was considered to be near the optimum design. TAIT's third son was Frederick Guthrie TAIT, who became the leading amateur golfer (1893) and won two Open golf championships (1896, 1898).

In 1853, he became interested in quaternions, a new advanced algebra of complex numbers in more than two dimensions invented in 1843 by the Irish mathematician William Rowan HAMILTON that gave rise to vector analysis and was instrumental in the development of modern mathematical physics. He also carried out pioneering studies on the topology of knots, which he published in his book *An Elementary Treatise on Quaternions* (1873; 3rd edn. 1890). In mathematics knots are defined as closed, non-self-intersecting curves that are embedded in three dimensions and cannot be untangled to produce simple loops. In collaboration with William John STEELE, he published *A Treatise on Dynamics of a Particle* (1882) and contributed to the kinetic theory of gases (1886–1892). He published about 360 articles and reviews.

TAIT was a Fellow of the Royal Society of Edinburgh (from 1860) and acted as its general secretary (1879–1901). For his various mathematical and physical researches TAIT was awarded the Royal Medal (1886) of the Royal Society of London. He was also awarded other prizes in England and Scotland. He was a fellow or member of the Danish, Dutch, Swedish, and Irish scientific academies.

The *Tait Professorship of Mathematical Physics* is an established chair within the University of Edinburgh.

**ORIGINAL WORKS.** *Sketch of thermodynamics.* Edmonston & Douglas, Edinburgh (1868) — *Note on a mode of producing sounds of very great intensity.* Proc. Roy. Soc. Edinb. **9**, 737-738 (1878) — *On comets.* Ibid. **10**, 367-370 (1880) — *On the accurate measurement of high pressures.* Ibid. **10**, 572-576 (1880) — *Thunderstorms* [Lecture]. Nature **22**, 339-341, 364-366, 408-410, 436-438 (1880) — *The pressure errors of the »Challenger« thermometers.* Ibid. **25**, 90-93, 127-130 (1882) — *Note on the temperature changes due to compression.* Proc. Roy. Soc. Edinb. **11**, 51-52, 217-219 (1882) — *On the crushing of glass by pressure.* Ibid. **11**, 204-206 (1882) — *On the laws of motion. Part I.* Ibid. **12**, 8-18 (1884) — *Note on the compressibility of water.* Ibid. **12**, 45-46 (1884) — *Note on the com-*

*pressibility of water, sea-water, and alcohol, at high pressures.* Ibid. **12**, 223-224 (1884) — *Further note on the compressibility of water.* Ibid. **12**, 757-758 (1884) — *Properties of matter.* Black, Edinburgh (1885); *Die Eigenschaften der Materie.* Pichler, Wien (1888) — *On an improved method of measuring compressibility.* Proc. Roy. Soc. Edinb. **13**, 2-4 (1886) — *Note on a singular passage in the 'Principia.'* Ibid. **13**, 72-78 (1886) — *On the partition of energy among groups of colliding spheres.* Ibid. **13**, 537-539 (1886) — *Note on the effects of explosives.* Ibid. **14**, 110-111 (1888) — *On the foundations of the kinetic theory of gases. Part I.* Trans. Roy. Soc. Edinb. **33**, 65-95 (1888); *Part II.* Ibid. **33**, 251-277 (1888) — With W. THOMSON (Lord KELVIN): *Treatise on natural philosophy.* Cambridge University Press, Cambridge, U.K. (1888) — *Voyage of HMS »Challenger«.* Rept. H.M.S.O., London (1888), vol. II, Part IV: *Physics and chemistry* — *Preliminary note on the duration of impact.* Proc. Roy. Soc. Edinb. **15**, 159 (1889) — *On the mean free path, and the average number of collisions per particle per second in a group of equal spheres.* Ibid. **15**, 225-226 (1889) — *On the foundations of the kinetic theory of gases. Part III.* Trans. Roy. Soc. Edinb. **35**, 1029-1041 (1890) — *On the foundations of the kinetic theory of gases. Part IV.* Proc. Roy. Soc. Edinb. **16**, 65-72 (1890) — *Some points in the physics of golf.* Nature **42**, 420-423 (1890); Ibid. **44**, 497-498 (1891) — *Sur la durée du choc.* Rev. Gén. Sci. Pures Appl. **3**, 777-781 (1892) — *Heat.* Macmillan, London (1892) — *On impact* [1891-1892]. Trans. Roy. Soc. Edinb. **36**, 225-252 (1892) — *On the foundations of the kinetic theory of gases. Part V.* Proc. Roy. Soc. Edinb. **19**, 32-35 (1893) — *Note on the thermal effect of pressure on water.* Ibid. **19**, 133-135 (1893) — *On the compressibility of liquids in connection with their molecular pressure.* Ibid. **20**, 63-68 (1895) — *On the compressibility of fluids.* Ibid. **20**, 245-251 (1895) — *On the application of VAN DER WAAL's equation to the compression of ordinary liquids.* Ibid. **20**, 285-289 (1895) — *Dynamics.* Black, London (1895) — *On impact.* Trans. Roy. Soc. Edinb. **37**, 381-397 (1895) — *On the path of a rotating spherical projectile.* Ibid. **37**, 427-440 (1895); Ibid. **39**, 491-506 (1900) — *Scientific papers.* Cambridge University Press, Cambridge, U.K.; vol. 1. (1898), vol. 2 (1900).

**SECONDARY LITERATURE.** G. CHRYSTAL: *Professor TAIT.* Nature **64**, 305-307 (July 25, 1901) — A. MACFARLANE: *Peter Guthrie TAIT.* Phys. Rev. **15** [I], 51-64 (1902) — *Peter Guthrie TAIT, his life and works.* Bibliotheca Mathematica **4** [III], 185-200 (1903) — C.G. KNOTT: *Life and scientific work of Peter Guthrie TAIT.* Cambridge University Press (1911) — W. THOMSON (Lord KELVIN): *Obituary notice of Professor TAIT.* In: W. THOMSON: *Mathematical and physical papers.* Cambridge University Press, Cambridge, U.K. (1911), vol. VI — J.D. NORTH: *P.G. TAIT.* In: (C.C. GILLESPIE, ed.) *Dictionary of scientific biography.* Scribner, New York, vol. 13 (1976), pp. 236-237 — C.W. SMITH and M.N. WISE. *Energy and empire: a biographical study of Lord KELVIN.* Cambridge University Press, Cambridge, U.K. (1989) — D.B. WILSON: *P.G. TAIT and Edinburgh natural philosophy, 1860-1901.* Ann. Sci. **48**, 267-287 (1991) — C. DENLEY and C. PRITCHARD: *The golf ball aerodynamics of Peter Guthrie TAIT.* Math. Gazette **77**, 298-313 (1993) — C.W. SMITH: *TAIT, Peter Guthrie.* In: (H.C.G. MATTHEW, ed.) *Oxford dictionary of national biography.* Oxford University Press, Oxford (2004), vol. 53, pp 665-668.

**PICTURE.** From C.G. KNOTT: *Life and scientific work of Peter Guthrie TAIT* (see above). Reprinted with permission of Cambridge University Press, Cambridge, U.K.

**NOTE.** A comprehensive *Provisional bibliography of Peter Guthrie TAIT* was compiled by Chris PRITCHARD (British Society for the History of Mathematics) for the *Peter Guthrie TAIT (1831-1901): Centenary Meeting.* Roy. Soc. Edinburgh (July 2001); see
http://www.maths.ed.ac.uk/~aar/knots/taitbib.htm.

# TAUB, Abraham Haskel (1911–1999)

- U.S. mathematician and theoretical fluid dynamicist; founder of the theory of relativistic simple waves and shocks

Abraham H. TAUB was born in Chicago, where he studied mathematics at the University of Chicago and earned his B.S. (1931). After working as an instructor in mathematics and earning a Ph.D. (1935) in mathematical physics at Princeton University, he became at Princeton's Institute for Advanced Study (IAS) assistant (1935–1936), member (1940–1941), theoretical physicist (1942–1946), and Guggenheim Fellow (1947–1948).

Closely working there together with John VON NEUMANN, he got involved in digital computers, numerical analysis, and problems in shock physics theory, particularly in shock interactions. In 1942, he collaborated with Prof. Walker BLEAKNEY, then head of the Palmer Physical Laboratory at Princeton University, and developed a theory of the shock tube, independently of previous existing theories in Germany (KOBES 1910; HILDEBRAND 1927; SCHARDIN 1932). Together they investigated shock wave interactions in gases, from both the experimental and theoretical points of view. This work was for a long time the basis for many subsequent studies on this subject. During the war he also collaborated in a number of projects on dynamic effects in structures caused by blast waves and the detonation of high explosives.

After World War II, he was called by the University of Illinois in Urbana and became professor of mathematics (1948–1964) and head of the Digital Computer Laboratory (1961–1964). He built a computer that, based on John VON NEUMANN's plans and called "ORDVAC" (Ordnance Variable Automated Computer), was completed in 1952 and delivered to the Aberdeen Proving Ground. A subsequently built computer, named ILLIAC (Illinois Automated Computer), remained at Illinois and was the prototype for several other computers.

He became professor of mathematics at UC Berkeley in 1968 and remained there until his retirement (1978). In the late 1940s, TAUB began working on 1-D shock wave propagation in relativistic hydrodynamics, which is important in the numerical simulation of supernova explosions, and on impulsive gravitational waves. He derived the relativistic Rankine-Hugoniot equations (1948). The relativistic version of the Hugoniot adiabat is called the "Taub adiabat." The relativistic hydrodynamic equations introduced by TAUB are called the "Taub equations." A peculiar unstable solution of EINSTEIN's equations of general relativity is called the "Taub Universe."

TAUB volunteered to undertake the labor of assembling and editing the manuscripts of his friend John VON NEUMANN, which made possible the publication of his six-volume *Collected Works John von Neumann* (1963). He also edited the book *Studies in Applied Mathematics* (1971) and, with Sidney FERNBACH, coedited the book *Computers and Their Role in the Physical Sciences* (1970). Besides his numerous engagements in research and teaching, he was also interested in foreign research activities and worked as a visiting scholar at the ETH Zurich (1954) and the Universities of Cambridge (1969–1970), Chile (1972), and Oxford and Paris (1975).

ORIGINAL WORKS. With O. VEBLEN and J. VON NEUMANN: *The Dirac equation in projective relativity*. Proc. Natl. Acad. Sci. **20**, 383-388 (1934) — *Peak pressure dependence on height of detonation*. Rept. OSRD-4078a (Aug. 1944) — With W. BLEAKNEY: *Remarks on fortification design*. Interim Memorandum M-10, CFD Natl. Res. Council (Nov. 1944) — With D. MONTGOMERY: *Reactions of simple systems under blast loading*. Rept. OSRD-5303a, AES-12a (July 1945); *Remarks on reactions under blast loading*. Rept. OSRD-6007a (Sept. 1945) — With J.A. WISE: *Effects of confined blast on brick curtain walls*. Rept. OSRD-6007d, NDRC AES-14d (Sept. 1945) — With J. VON NEUMANN: *Flying wind tunnel experiments*. U.S. Govt. Doc. PB33263 (Nov. 1945) — With I.M. FREEMAN and D.G. KRETSINGER: *Contact explosions against concrete*. Rept. OSRD-6319, NDRC A-354 (Dec. 1945) — With L.G. SMITH: *Theory of reflection of shock waves*. Phys. Rev. **69** [II], 678 (1946) — *Interaction of progressive rarefaction waves*. Annals of Math. **47**, 811-828 (1946) — *Refraction of plane shock waves*. Phys. Rev. **72** [II], 51-60 (1947) — *Relativistic Rankine-Hugoniot equations*. Ibid. **74** [II], 328-334 (1948) — With W. BLEAKNEY: *Interaction of shock waves*. Rev. Mod. Phys. **21**, 583-605 (1949) — *On HAMILTON's principle for perfect compressible fluids*. In: (E. REISSNER, ed.) *Proc. 1st Symp. on Applied Mathematics of the Mathematical Society* [Providence, RI, Aug. 1947]. American Mathematical Society, New York (1949) — *A sampling method for solving the equation of compressible flow in a permeable medium*. In: *Proc. 1st Midwestern Conf. on Fluid Dynamics* [Ann Arbor, MI, May 1950]. J.W. Edwards, Ann Arbor (1951) — With W. BLEAKNEY and C.H. FLETCHER: *Reflection of shock waves at nearly glancing incidence*. Rev. Mod. Phys. **23**, 271-286 (1951) — *Refraction of plane shock waves*. Phys. Rev. **84** [II], 922-929 (1951) — With C.H. FLETCHER: *Reflection of shock waves as a pseudo-stationary phenomenon*. Ibid. **87**, 912-913 (1952) — *Curved shocks in pseudo-stationary flows*. Ann. Math. **58** [II], 501-527 (1953) — *Singularities on shocks*. Ibid. **61** [II], 11-12 (1955) — *Determination of flows behind stationary and pseudo-stationary shocks*. Ibid. **62** [II], 300-325 (1955) — With D.R. CLUTTERHAM: *Numerical results on the shock configuration in Mach reflection*. In: (J.H. CURTISS, ed.) *Proc. 6th Symp. on Applied Mathematics of the American Mathematical Society* [Santa Monica, CA, Aug. 1955]. McGraw-Hill, New York (1956); *see* vol. IV: *Numerical analysis*, pp. 45-72 — *Wave propagation in fluids*. In: (E.U. CONDON and H. ODISHAW, eds.) *Handbook of physics*. McGraw-Hill, New York (1958), chap. 4, part III — *On circulation in relativistic hydrodynamics*. Arch. Rat. Mech. Anal. **3**, 312-324 (1959) — *Relativistic hydrodynamics*. Rept. A840693, University of Illinois (March 1959) — *Stability of general relativistic gaseous masses and variational principles*. Comm. Math. Phys. **15**, 235-254 (1969) — With S. FERNBACH: *Computers and their role in the physical sciences*. Gordon & Breach, New York (1970) — *General relativistic shock waves in fluids for which pressure equals energy density*. Comm. Math. Phys. **29**, 79-88 (1973) — *Relativistic fluid mechanics*. Annu. Rev. Fluid Mech. **10**, 301-332 (1978) — *Singular shocks*. Math. Today **1**, 25-42 (1983) — *On the collision of planar impulsive gravitational waves*. J. Math. Phys. **29**, 690-695 (1988) — *Collision of impulsive gravitational waves followed by dust clouds*. Ibid. **29**, 2622-2627 (1988) — *Interaction of null dust clouds fronted by impulsive plane waves. Part I*. Ibid. **31**, 664-668 (1990); *Part II*. Ibid. **32**, 1322-1327 (1991).

SECONDARY LITERATURE. F.J. TIPLER (ed.) *Essays in general relativity: a Festschrift for Abraham TAUB*. Academic Press, New York (1980) — B. MASHHOON: *In memoriam: Abraham Haskel TAUB*. SIAM News **34**, No. 7 (Sept. 2001); http://www.siam.org/siamnews/09-01/taub.htm.
PICTURE. Courtesy Prof. Calvin C. MOORE, Dept. of Mathematics, University of Berkeley, CA. The picture shows Dr. TAUB at the blackboard treating Mach reflection.

# TAYLOR, Sir Geoffrey Ingram (1886–1975)

• British physicist, applied mathematician, and engineer

Sir Geoffrey I. TAYLOR was born in London. He was the elder son of the artist Edward Ingram TAYLOR. He was educated in mathematics and natural sciences at University College School, London and Trinity College, Cambridge. After earning a prize fellowship (1910) he carried out research at Cavendish Laboratory in Cambridge on the mechanics of fluids and solid materials and their applications in geophysics and engineering. Already in his second paper, published in the same issue of the *Proceedings of the Royal Society* as Lord RAYLEIGH's famous paper on shock waves (1910), TAYLOR made independently a first estimate of the thickness of a shock front, which essentially supported RAYLEIGH's conclusions on the structure of shock waves. In the same year, he was elected Fellow at Trinity College, which provided support and freedom to pursue his research for up to 6 years, and became Schuster Reader at

Cambridge in dynamical meteorology (1911). Following the HMS *Titanic* disaster (April 15, 1912), he was invited as a meteorologist on an expedition to observe the paths of icebergs in the North Atlantic. During the expedition of the HMS *Scotia* to the North Atlantic (1913) he measured the vertical distributions of wind strength and direction, temperature, and humidity to a height of about 2,500 m using instrument-carrying kites. These studies of vertical transfer of heat and momentum and water vapor in the friction layer of the atmosphere stimulated other fields of his research in fluid dynamics. TAYLOR's work on turbulent motion in fluids, leading to his publication *Eddy Motion in the Atmosphere* (1915), won him the prestigious Adams Prize (1915) of the Faculty of Mathematics at the University of Cambridge. His investigations of the turbulence transfer process in the friction layer of the Earth's atmosphere led to his paper *Skin Friction of the Wind on the Earth's Surface* (1916).

During World War I, he worked at the Royal Aircraft Factory in Farnborough on the design and operation of airplanes. Stimulated by aeronautical problems, such as induced stress in propeller shafts under torsion, he got involved with studying the physical processes that limit the strength of solid materials. Together with his student Alan Arnold GRIFFITH he set up a method for detecting stress concentrations in shafts under torsion by observing the displacement of a soap film (1917). After various teaching activities at Trinity College, TAYLOR continued his research at Cavendish Laboratory as Yarrow Research Professor (1923–1952). In this period, he studied the mechanism of plastic deformation of metal crystals and published his famous "dislocation theory" (1934), which assumes that during work hardening the flow stress $S$ is proportional to the square root of the strain $s$. His experimental observations on the deformation of crystals (1938) indicated that the stress-strain curve for many cubic crystals is parabolic (*i.e.*, $S \propto s^{1/2}$), the constant being temperature dependent, and that the crystals deform by the motion of dislocation on specific crystallographic planes in specific directions.

TAYLOR also contributed much to the understanding of the turbulent motion of fluids and developed a theory describing velocity fluctuations by a statistical method (1935–1939). The development of his ideas about turbulence was aided by his previous experience in meteorology and his increasing acquaintance in the 1930s with turbulent flow systems relative to aeronautics and supersonic wind tunnel measurements, which he carried out together with John W. MACCOLL.

In the period 1939–1945, he was a consultant to various civil and military authorities on high explosives. For example, he reported to the Civil Defense Research Committee (CDRC) of the Ministry of Home Security and was a member of the Physics of Explosives Committee (Physex) of the Ministry of Supply. His numerous studies cover a wide field on detonation, blast waves, underwater explosions, and associated mechanical damage effects. He participated also in the Manhattan Project (1942–1945) at Los Alamos and analytically treated blast effects of the first nuclear explosion, which earned him the U.S. Medal for Merit (1946). After World War II, he resumed his work at the Cavendish Laboratory for another 20 years. Together with A.C. WHIFFIN he worked out a test method to determine the strength of shock-loaded samples at high strain rates, the so-called "Taylor test" (1948). He theoretically demonstrated that the high-pressure wave of a planar detonation passing through a high explosive is immediately reduced, leading to a pressure-relief wave, a rarefaction, the so-called "Taylor wave" (1949). He also showed that the interface of two fluids becomes unstable when the fluid of higher density is forced into another one of lower density, the so-called "Rayleigh-Taylor instability" (1950).

TAYLOR wrote over 200 scientific papers and articles, nearly all of which were later republished by Cambridge University Press in four volumes (1958–1971). He was knighted (1944) and elected to honorary membership or fellowship by many international societies. He received many honorary degrees and awards such as the Copley Medal (1944) of the Royal Society of London, the Kelvin Medal (1959) of the Institute of Physics, the James Watt Medal (1965) of the Institution of Mechanical Engineers, and the von Kármán Medal (1969) of the American Institute of Aeronautics and Astronautics (AIAA).

Since 1976, the *G.I. Taylor Memorial Lecture* is given annually by renowned fluid dynamicists at the University of Florida. Since 2003, the *G. I. Taylor Medal* is awarded by the Society of Engineering Science, Inc. (USA) for outstanding research contributions in either theoretical or experimental fluid mechanics or both.

**ORIGINAL WORKS.** *The conditions necessary for discontinuous motion in gases.* Proc. Roy. Soc. Lond. **A84**, 371-377 (1910) — *Skin friction of the wind on the Earth's surface.* Ibid. **A92**, 196-199 (1916) — *Diffusion of continuous movements.* Proc. Lond. Math. Soc. **20** [II], 196-212 (1921) — *Tides in the Bristol Channel.* Proc. Cambr. Phil. Soc. **20**, 320-325 (1921) — *A relation between* BERTRAND's *and* KELVIN's *theorems on impulses.* Proc. Lond. Math. Soc. **21** [II], 413-414 (1922) — *Stability of a viscous liquid contained between two rotating cylinders.* Phil. Trans. Roy. Soc. Lond. **A223**, 289-343 (1923) — With C.F. SHARMAN: *A mechanical method for solving problems of flow in compressible fluids.* Proc. Roy. Soc. Lond. **A121**, 194-217 (1928) — *The force acting on a body placed in a curved and converging stream of fluid.* Rept. ARC-R&M No. 1166 (1928) — *The air wave from the great explosion at Krakatao.* In: *Proc. 4th Pan-Pacific Science Congr.* [Djakarta/Bandung, Indonesia, 1929]. Pacific Science Assoc., Melbourne

(1929), vol. IIB, pp. 645-655 — *The flow of air at high speeds past curved surfaces*. Rept. ARC-R&M No. 1381 (1930) — *Recent work on the flow of compressible fluids*. J. Lond. Math. Soc. **5**, 224-240 (1930) — *Strömung um einen Körper in einer kompressiblen Flüssigkeit*. ZAMM **10**, 334-345 (1930) — *Applications to aeronautics of ACKERET's theory of airfoils moving at speeds greater than that of sound*. Rept. ARC-R&M No. 1467, WA-4218-5a (1932) — With J.W. MACCOLL: *The air pressure on a cone moving with high speeds*. Proc. Roy. Soc. Lond. **A139**, 278-311 (1933); *L'onde balistique d'un projectile à tête conique*. Mém. Artill. Franç. **12**, 651-683 (1933) — *The mechanism of plastic deformation of crystals. (I) Theoretical*. Proc. Roy. Soc. Lond. **A145**, 362-387 (1934); *(II) Comparison with observations*. Ibid. **A145**, 405-415 (1934) — *A theory of the plasticity of crystals*. Z. Kristall. Mineral. Petrogr. **A89**, 375-385 (1934) — With J.W. MACCOLL: *The mechanics of compressible fluids*. In: (W.F. DURAND, ed.) *Aerodynamic theory*. Springer, Berlin (1935), vol. III, pp. 209-250 — *Statistical theory of turbulence. Parts I-IV*. Proc. Roy. Soc. Lond. **A151**, 421-478 (1935) — Short autobiography in: *V Convegno Volta su "Le alte velocità in aviazione"* [Rome, Sept./Oct. 1935]. Reale Accademia d'Italia, Roma (1936), p. 683 — *Statistical theory of turbulence. Part V: Effect of turbulence on boundary layer*. Proc. Roy. Soc. Lond. **A156**, 307-317 (1936) — *The propagation and decay of blast waves*. Paper for Civil Defense Research Commission, London (1939); *Propagation of earth waves from an explosion*. Ibid. (1940); *Notes on the dynamics of shock waves from bare explosive charges*. Ibid. (1940); *Pressures on solid bodies near an explosion*. Ibid. (1940); *The propagation of blast waves over the ground*. Ibid. (1941); *Analysis of the explosion of a long cylindrical bomb detonated at one end*. Ibid. (1941); *The plastic wave in a wire extended by an impact load*. Ibid. (1942) — *Detonation waves*. Minist. of Supply, Adv. Council on Sci. Res. Tech. Development, Explosion Res. Comm., Paper AC-639, Res. Comm. 178, W-12-144 (1941) — *Underwater explosion research*. David Taylor Model Basin Rept. TMB-510, Brit. Ministry of Supply — *The pressure and impulse of submarine explosion waves on plates*. Rept. RC-235 (1941) — *The formation of a blast wave by a very intense explosion*. Rept. RC-210, II-5-153, Res. Comm., Ministry of Home Security (1941) — With R.M. DAVIES: *The effect of the method of support in tests of damage to thin-walled structures by underwater explosions*. Paper for Coord. Comm. "Shock Waves" (1942) — *The plastic wave in a wire extended by an impact load*. Rept. RC-320 (1942) — *The propagation and decay of blast waves*. Brit. Civ. Def. Res. Comm. (1944) — With R.M. DAVIES: *A measurement of the pressure close to an explosive under water*. UNDEX Rept. (1944) — *The air wave surrounding an expanding sphere*. Proc. Roy. Soc. Lond. **A186**, 273-292 (1946) — *The testing of materials at high rates of loading*. J. Inst. Civ. Eng. **26**, 486-518 (1946) — *The use of flat-ended projectiles for determining dynamic yield stress. (I) Theoretical considerations*. Proc. Roy. Soc. Lond. **A194**, 289-299 (1948); see also A.C. WHIFFIN: *The use of flat-ended projectiles for determining dynamic yield stress. (II) Tests on various metallic materials*. Ibid. **A194**, 300-322 (1948) — With G. BIRKHOFF, D.P. MACDOUGALL, and E.M. PUGH: *Explosives with lined cavities*. J. Appl. Phys. **19**, 563-582 (1948) — *The dynamics of the combustion products behind plane and spherical detonation fronts in explosives*. Proc. Roy. Soc. Lond. **A200**, 235-247 (1950) — *The formation of a blast wave by a very intense explosion. (I) Theoretical discussion*. Ibid. **A201**, 159-174 (1950); *(II) The atomic explosion of 1945*. Ibid. **A201**, 175-186 (1950) — *The instability of liquid surfaces when accelerated in a direction perpendicular to their planes. (I)*. Ibid. **A201**, 192-196 (1950) — *Similarity solutions to problems involving gas flow and shock waves*. Ibid. **A204**, 8-9 (1950) — *Distribution of stress when a spherical compression pulse is reflected at a free surface*. Research (Lond.) **5**, 508-509 (1952) — *Detonation in condensed explosives*. Clarendon Press, London (1952) — With R.S. TANKIN: *Gas dynamical aspects of detonation*. In: *High speed aerodynamics and jet propulsion*. Princeton University Press, Princeton, NJ; vol. III (1958): (H.W. EMMONS, ed.) *Fundamentals of gas dynamics*, pp. 622-686 — *Oblique impact of a jet on a plane surface*. Phil. Trans. Roy. Soc. Lond. **A260**, 95-100 (1966) — (G.K. BATCHELOR, ed.) *The scientific papers of Sir G.I. TAYLOR*. Cambridge University Press, Cambridge, MA. Vol. 1 (1958): *Mechanics of solids*; vol. 2 (1960): *Meteorology, oceanography and turbulent flow*; vol. 3 (1963): *Aerodynamics and the mechanics of projectiles and explosions*; vol. 4 (1971): *Mechanics of fluids. Miscellaneous papers*.

**SECONDARY LITERATURE.** R.V. SOUTHWELL: *G.I. TAYLOR: a biographical note*. In: (G.K. BATCHELOR and R.M. DAVIES, eds.) *Surveys in mechanics. A collection of surveys of the present position of research in some branches of mechanics, written in commemoration of the 70th birthday of G.I. TAYLOR*. Cambridge University Press, Cambridge, U.K. (1956), pp. 1-6 — N. ROY: *Notice nécrologique sur Sir Geoffrey Ingram TAYLOR, correspondant pour la section de mécanique*. Vie académique (Acad. Sci. Paris) **281**, 80-82 (1975) — B.A. PIPPARD: *Sir Geoffrey TAYLOR*. Phys. Today **28**, 67 (Sept. 1975) — G.K. BATCHELOR: *G.I. TAYLOR*. Biogr. Mem. Fell. Roy. Soc. (Lond.) **22**, 565-633 (1976); *G.I. TAYLOR as I knew him*. Adv. Appl. Mech. **16**, 1-8 (1976); *Geoffrey Ingram TAYLOR, 7 March 1886 – 27 June 1975*. J. Fluid Mech. **173**, 1-14 (1986) — J.K. BELL: *A retrospect on the contributions of G.I. TAYLOR to the continuum physics of solids*. Exp. Mech. **35**, No. 1, 1-10 (1995) — G.K. BATCHELOR: *The life and legacy of G.I. TAYLOR*. Cambridge University Press, Cambridge (1996) — J.S. TURNER: *G.I. TAYLOR (1886–1975) in his later years*. Annu. Rev. Fluid Mech. **29**, 1-25 (1997) — M.P. BRENNER and H.A. STONE: *Modern classical physics through the work of G.I. TAYLOR*. Phys. Today **53**, No. 5, 30-35 (May 2000); http://www.deas.harvard.edu/brenner/taylor/physic_today/taylor.htm — G.K. BATCHELOR: *TAYLOR, Sir Geoffrey Ingram*. In: (H.C.G. MATTHEW, ed.) *Oxford dictionary of national biography*. Oxford University Press, Oxford (2004), vol. 53, pp. 885-886.

**PICTURE.** Reprinted with permission from Biogr. Mem. Fell. Roy. Soc. (Lond.) **22**, 565 (1976). © The Godfrey Argent Studio, London, U.K.

**NOTE.** For biographical information on G.I. TAYLOR see also the Navigational Aids for the History of Science, Technology & the Environment (NAHSTE) Project;

http://www.nahste.ac.uk/pers/t/GB_0237_NAHSTE_P1870/.

# TOEPLER [or TÖPLER], August Joseph Ignatz (1836–1912)

- German experimental physicist and inventor; the first to visualize shock waves

August J.I. TOEPLER was born in Brühl (near Cologne), the son of Michael TOEPLER, an instructor of the Catholic school teacher seminar and royal music director. Although very talented in music and painting, he decided to dedicate his life to natural philosophy. He studied physics, chemistry, and mathematics at the Royal Technical Institute in Berlin (1854–1858) and graduated as a chemist. After completing his military service, he entered the Agricultural College at Poppelsdorf, a small town near Bonn. Initially working as an experimental chemist (1859–1862), he soon became a lec-

turer (1862–1864) on physics and chemistry and took his Ph.D. (1860) at the University of Jena.

TOEPLER invented various scientific instruments and contributed to the measurements of acoustic, optical, and magnetic parameters. At Poppelsdorf he also invented his famous "schlieren method" (1864). Initially he intended to apply this high-sensitivity optical method to the visualization of sound waves; however, when experiments with organ pipes failed, he turned to strong acoustic waves, which he generated by electric sparks. Since high-sensitivity films were not yet available to him, he observed the propagation and reflection wave phenomena subjectively using a stroboscopic method. He modified the Knochenhauer circuit (1858), a loose coupling of two electric discharge circuits, and applied the primary spark to generate the shock wave and the second spark, which, being fired shortly thereafter, acted as a light source. This allowed him to "freeze" the supersonic wave motion repetitively at about 20 Hz. TOEPLER made correct pen-and-ink drawings of the observed shock fronts and even from the pinched discharge channel, which looked so realistic that often modern shock physicists regarded them as photographs. He first noticed the density jump at the shock front, but he did not attempt to investigate shock waves in more detail. In order to generate stronger spark waves he improved his high-voltage discharge circuit and devised a 2-disc electrostatic influence generator. Manufactured in Berlin by the German Wilhelm HOLTZ, so-called "Toepler-Holtz machine" (1865), it was capable of quickly charging the Leiden jars to a higher voltage.

After holding the chair of chemistry and chemical technology at the Polytechnikum Riga (1864–1868), TOEPLER was appointed professor of physics at the prestigious University of Graz (1868–1876) and the Polytechnikum Dresden (1876–1900). In Graz, TOEPLER, together with his friend Ludwig BOLTZMANN, tackled again the problem of visualizing sound waves and measuring the threshold of hearing. Using a Jamin-type interferometer, they obtained for the eardrum a displacement of about 50 nm (1870). Modern measurements, however, have shown that the displacement at the threshold of hearing is even much smaller and only on the order of 0.01 nm.

TOEPLER became a corresponding member of the Academies of Vienna (1874), Berlin (1879), and Munich (1896) and a full member of the Royal Saxon Academy (1885) and the German Academy of Sciences Leopoldina Carolina (1879). He was awarded an honorary doctorate from the Universities of Heidelberg (1886), Dresden (1895), and Riga (1906). To commemorate his greatest scientific achievement, in 1912 his wife inscribed on his tombstone the epitaph *August Toepler – Er sah als Erster den Schall* ("August Toepler – The first to see sound"). Obviously, she was not aware that he visualized with his invented schlieren method spark waves (*i.e.*, weak shock waves) rather than sound waves.

**ORIGINAL WORKS.** *Beobachtungen nach einer neuen optischen Methode.* Cohen, Bonn (1864) — *Über die Erzeugung einer eigenthümlichen Art von intensiven electrischen Strömen vermittels eines Influenz-Electrometers.* Ann. Phys. **125** [II], 469-496 (1865) — *Neue optische Methode.* Riga Corresp. Blatt **15**, 44-46 (1866) — *Vergleichende Versuche über die Leistungen der Influenzmaschine mit und ohne Metall-Belegung.* Ann. Phys. **127** [II], 177-198 (1866) — *Über die Methode der Schlierenbeobachtung als mikroskopisches Hilfsmittel, nebst Bemerkungen zur Theorie der schiefen Beleuchtung.* Ibid. **127** [II], 556-580 (1866) — *Das Princip der stroboscopischen Scheiben als vortheilhaftes Hülfsmittel zur optischen Analyse tönender Körper.* Ibid. **128** [II], 108-125 (1866) — *Vibroscopische Beobachtungen über die Schwingungsphasen singender Flammen mit Benutzung des Schlierenapparates.* Ibid. **128** [II], 126-139 (1866) — *New optical method.* Phil. Mag. **33** [IV], 75-76 (1867) — *II. Optische Studien nach der Methode der Schlierenbeobachtung.* Ann. Phys. **131** [II], 33-55, 180-215 (1867); Ibid. **134** [II], 194-217 (1868) — With L. BOLTZMANN: *Über eine neue optische Methode, die Schwingungen tönender Luftsäulen zu analysieren.* Anzeiger Wien. Akad. Ber. No. IX, 73-75 (1870); Ann. Phys. **141** [II], 321-352 (1870).
**SECONDARY LITERATURE.** W. HALLWACHS: *Nachruf für August TOEPLER.* Ber. Verhandl. Königl.-Sächs. Gesell. Wiss. Leipzig **64** [Math.-Phys. Kl.], 479-497 (1912) — M. TOEPLER: *Zu August TOEPLERs 100. Geburtstag.* Focken & Oltmann, Dresden (1936) — P. KREHL and S. ENGEMANN: *August TOEPLER: the first who visualized shock waves.* Shock Waves **5**, 1-18 (1995).
**PICTURE.** Courtesy Archives of TU Dresden, Photo Collection, Dresden, Gemany.
**NOTE.** TOEPLER spelled his name TOEPLER. However, in most of his publications his name was printed as TÖPLER.

## TOEPLER [or TÖPLER], Maximilian August (1870–1960)

▪ German theoretical physicist; father of electric spark research

The eldest son of August TOEPLER, Maximilian A. TOEPLER was born in Graz, Austro-Hungarian Empire, and followed in his father's footsteps. He studied physics, chemistry, and

mathematics at the Polytechnikum Dresden (1890–1891) and at the Universities of Leipzig (1891–1894) and Göttingen (1894–1895). Already in his habilitation thesis he treated problems of electric discharges, a subject of research with which he was associated until his retirement. In 1900, he was appointed *Privatdozent* (university lecturer) and became associate professor of theoretical physics at the Technische Hochschule (TH) Dresden (1903) and professor and director of the newly founded chair of theoretical physics (1926–1935).

In the period 1907–1908, M.A. TOEPLER repeated the shock visualization experiments of his father (1864), but recorded the propagation and reflection phenomena on film rather than visualizing them subjectively. His technique was similar to the setup used by the U.S. physicist Robert W. WOOD, who had photographed spark waves in the late 1890s. Stimulated by previous observations of his father, who was the first to study the geometry of expanding spark channels – the actual driving force of the *Funkenwelle* (spark wave), a weak shock wave – he focused his interest on the spark channel itself, and investigated the nature of all kinds of spark discharges ranging from point and gliding discharges to ball-lightning. His numerous investigations on spark discharges established his reputation as a leading authority on this subject. He worked out a relationship for the nonlinear, time-dependent impedance of a spark gap, the so-called "Toepler law."

After his retirement (1935), TOEPLER worked temporarily for the German Army as a consultant on schlieren instrumentation for the Heeresversuchsanstalt (HVA) Peenemünde-Ost, which had installed a huge supersonic wind tunnel for aerodynamic rocket model testing. When the TH Dresden, which was heavily damaged in World War II, was reopened, TOEPLER, then already at the age of 78, gave lectures on theoretical physics (1948–1951). In honor of his numerous contributions to research and education, a building of the Faculty of Electrical Engineering on the campus of the TU Dresden, the *Töpler Bau*, was named after him.

ORIGINAL WORKS. *Objektive Sichtbarmachung von Funkenschallwellen nach der Schlierenmethode mit Hilfe von Gleitfunken.* Ann. Phys. **14** [IV], 838-842 (1904) — *Über gleitende Entladung.* Physik. Z. **8**, 743-748 (1907) — *Neue, einfache Versuchsanordnung zur bequemen subjektiven Sichtbarmachung von Funkenschallwellen nach der Schlierenmethode.* Ann. Phys. **27** [IV], 1043-1050 (1908) — *Zur Kenntnis der Funkenschallwellen elektrischer Oszillationen.* Ibid. **27**, 1051-1058 (1908) — *Schlierenmethode.* In: (R. DITTLER and G. JOOS, eds.) *Handwörterbuch der Naturwissenschaft.* G. Fischer, Jena (1931–1935), vol. VIII, pp. 924-929.

SECONDARY LITERATURE. H. FALKENHAGEN: *M. TOEPLER zum 85. Geburtstag.* Forsch. Fortschr. **29**, 222-223 (1955) — E. BRÜCHE: *Maximilian TOEPLER †.* Phys. Blätter **16**, 334 (1960) — G. MIERDEL and G. LEHMANN: *In memoriam Prof. Dr. phil. habil. Maximilian TOEPLER.* Wiss. Z. TU Dresden **11**, 103-110 (1962).

PICTURE. Courtesy Archives of TU Dresden, Photo Collection, Dresden, Germany.

# TUPOLEV [Russ. *ТУПОЛЕВ*], Andrei Nikolaevich (1888–1972)

▪ Soviet pioneer in the development of wind tunnels, all-metal airplanes, and supersonic aircraft; father of Russian aviation

Andrei N. TUPOLEV was born in the village of Pustomazovo (near the town of Kimry, in northwestern Russia). His father, Nikolai Ivanovich TUPOLEV, was an educated man who worked first as a mathematics teacher, then as a notary, and eventually became a farmer. In 1909, he entered the Moscow Higher Technical School (MWTU) and soon joined Prof. Nikolai E. ZHUKOVSKY's aeronautical course, taking part also in early gliding experiments and in designing Russia's first wind tunnel. After graduation from MWTU (1918) he assisted ZHUKOVSKY in organizing the Central Aerohydrodynamics Institute [ЦАГИ] at Moscow State University [МГУ], of which TUPOLEV became director (1918–1935). TUPOLEV was put in charge of the aircraft design department (1920) and appointed head of the Institute's design bureau with the goal of building military and civilian all-metal aircraft (1922). The first Tupolev designs were built at the German Junkers factory in Moscow, after being taken over by Soviet authorities (1926). His eight-engine ANT-20, the unique "Maxim Gorky" (1934), was the largest aircraft flying anywhere in the world, and his ANT-25, a long-range monoplane, flew across the Arctic to America (1937). After returning from a trip to Germany and the United States (1936), TUPOLEV was unjustly accused of selling secrets to Germany. He was arrested and sent to the Gulag but later placed in charge of a team that was to design military aircraft (1936–1943). In 1944, he was released, received the Stalin Prize for designing the Tu-2 medium-range bomber (1941), and was given a job copying the U.S. B-29 "Superfortress" Heavy Strategic Bomber – three of this type had force-landed in the Soviet Far East – and supervising the building of about 2,000 copies, designated as Tu-4s.

TUPOLEV designed more than 130 types of aircraft. His Tu-104 (870 km/h), a twin-engine turbojet airliner, was introduced in 1955. The turboprop heavy bomber Tu-20 (950 km/h) and the airliner Tu-114 (800 km/h) were the world's fastest propeller-driven aircraft that used the same wings and tail unit. Early in 1963, he set about designing a Supersonic Transport (SST) and appointed his son Alexei Andreevich as chief designer of the Experimental Construction Bureau, and in 1972 his son became general designer after his father's death. The SST Tu-144 ($M = 2.65$) was the world's first supersonic transporter to break the sound barrier during a test flight on December 31, 1968 – only a few months before the British-French "Concorde," the only rival, passed this crucial test. However, plans to also use the Tu-144 as a passenger jet were eventually scrapped in the 1970s due to mismanagement and design problems. One of the four Tu-144 that remained in open storage in the Moscow Zhukovsky Test Base was reactivated in 1995 and used for a series of supersonic test flights operated jointly by Russia and the United States. Besides designing airplanes, he also constructed various types of naval torpedo boats.

TUPOLEV was one of the most decorated Soviet citizens. He was made Hero of Socialist Labor (1945), received many honors and governmental awards such as the Lenin Prize (1957) and three Stalin Prizes for his aircraft design, was elected corresponding member (1933) and Academician (1953) of the U.S.S.R. Academy of Sciences, and acted temporarily as a member of the Commission for Foreign Affairs (1958) and Deputy of the Supreme Soviet. Seventy-three Soviet dignitaries signed his obituary, and tributes to his work were published all over the world.

In 1973, TUPOLEV was honored posthumously when the former Kazan State Technical University (KSTU) – established in 1932 in Kazan, Republic of Tatarstan, Russian Federation, hitherto better known as the "Kazan Aeronautical Institute" – was renamed *A.N. Tupolev Kazan State Technical University*. The *Andrei Tupolev Medal*, awarded by the Aeromodelling Commission of the Fédération Aéronautique Internationale (FAI), was established in 1989 and is donated annually by the National Airsport Control (NAC) of Russia.

**PUBLICATIONS.** With D. PIPKO: *Ekranoplany – winged craft of the future (by A.N. TUPOLEV). An air sled-amphibian from the Design Bureau (by D. PIPKO)* [in Russ.]. Nauka i Zhizn' ("Science and Life") **1**, No. 1, 32-41 (1966). Engl. translation by the Foreign Science and Technology Center (FSTC), U.S. Army, Charlottesville, VA (1969) — *Into the supersonic era.* Science J. **4**, No. 9, 3 (1968) — With M.S. LISTOV: *Grani derznovennogo tvorchestva* ("Aspects of technical frontiers of TUPOLEV"). Selected Works, Nauka, Moskva (1988) — *Zhizn I deiatelnost* ("Life and creative activities"). TsAGI Publ., Moscow 1989).
**SECONDARY LITERATURE.** M. GLADYCH: *Andrei TUPOLEV, Soviet airpower's.* Air Force Mag. (U.S.A.) **41**, 40-42 (July 1958) — G. WEDENSKY: *A.N. TUPOLEV.* In: *Porträts der UdSSR-Prominenz.* Institut zur Erforschung der UdSSR e.V., Munich (1960) — J. TURKEVICH (ed.) *Soviet men of science.* Van Nostrand, Princeton, NJ (1963) — G. BERKENKOPF: *Die Wissenschaft im "wissenschaftlichen Sozialismus."* Die Neue Ordnung in Kirche, Staat, Gesellschaft, Kultur (Paderborn) **26**, 125-135 (1972) — Anonymous: *Andrei N. TUPOLEV †.* Der Flieger **53**, 54-55 (1973) — J. ALEXANDER (ed.) *Encyclopaedia of aviation.* Scribner, New York (1977) — *60 let OKA A.N. Tupoleva* ("60 years of the A.N. Tupolev Design Bureau"). TsAGI, Moscow (1982) — Akad. Nauk SSSR (ed.): *Andrej Nikolaevic TUPOLEV, grani derznovennogo tvorcestva; k 100-letiju so dnja rozdenija.* Nauka, Moscow (1988) — H. MOON: *Soviet SST. The technopolitics of the Tupolev-144.* Orion Books, New York (1989) — B. GUNSTON: *TUPOLEV aircraft since 1922.* Naval Institute Press, Annapolis, MD (1995), pp. 7-14 [biography], pp. 202-207 [about the Tu-144] — P. DUFFY and A.N. KANDALOV: *TUPOLEV: the man and his aircraft.* Airlife, Shrewsbury (1996), pp. 9-18 [biography], pp. 153-157 [about the Tu-144] — L.L. KERBER (ed.): *STALIN's aviation gulag. A memoir of Andrei TUPOLEV.* Smithsonian Institution Press, Washington, DC (1996) — L.L. KERBER: *TUPOLEV* [in Russ.]. Politekhnika, St. Petersburg (1999) — P. DUFFY and A.N. KANDALOV: *A.N. TUPOLEV: Mu a jeo letadla.* Vaut, Prague (1999); *A.N. TUPOLEV: chelovek i ego samolety.* Moskovskii rabochii, Moscow (1999) — *General designers: A.N. and A.A. TUPOLEV.* Tupolev Public-Stock (TPS) Company, Moscow, Russia; http://www.tupolev.ru/English/Show.asp?SectionID=47.
**PICTURE.** Courtesy Bildarchiv Preussischer Kulturbesitz (bpk), Berlin, Germany.

# VIEILLE, Paul Marie Eugène (1854–1934)

▪ French experimental physicist, explosives specialist and inventor

Paul M.E. VIEILLE was born in Paris, where his father, Jules-Marie-Louis VIEILLE, taught mathematics at the Ecole Nor-

male Supérieure. After schooling in Paris and Aix-en-Provence, he decided to study science rather than literature. He began studying mathematics, physics, and chemistry at the secondary school in Marseilles and later at the Ecole Polytechnique (1873–1875). After graduation VIEILLE entered the Service des Poudres et Salpêtres, a governmental agency in Paris, and was soon entrusted with the office of *Ingénieur du Corps des Poudres*. He was appointed assistant director (1879) of the Laboratoire Central des Poudres et Salpêtres in Paris, then headed by Prof. Emile SARRAU, and later director (1897–1918). He became a member and secretary of the newly founded Commission des Substances Explosives (1878), which was headed by P.E. Marcellin BERTHELOT and to which Henry L. LE CHÂTELIER also belonged. In addition to his research activities at the Laboratoire Central he took up the respected position of *Répétiteur* at the Ecole Polytechnique (1882–1913).

VIEILLE improved the bomb calorimeter, which was used in France since 1870 in the study of combustion processes at constant volume, particularly for the measurement of heat generated in an explosion. Together with BERTHELOT he applied it to thermochemical studies of mercury fulminate and other explosives (1880), performed studies on flame propagation in explosives, and determined the specific heat of gaseous detonation products up to 2,000 °C (1883–1884). Together with Prof. SARRAU he investigated aerodynamic drag of projectiles and invented a crusher gauge to measure the internal pressure of guns (1882). Already in 1884 VIEILLE had the idea of using a tube of great length, which, eventually leading to the invention of the *shock tube* (1899), is regarded today as his greatest contribution to shock physics. With this technique he studied flame propagation in reactive gases and discovered two basic modes of flame propagation in combustible mixtures: deflagration (slow) and detonation (fast). VIEILLE's study of colloidal explosives (such as gelatinized nitrocellulose, which possess a nearly homogeneous character) led to the invention of the smokeless powder (1884), known as *"Vieille poudre"* or *"Poudre B,"* a straight nitrocellulose of white color [French *poudre blanche*], which was soon after introduced into the French military. It revoluzionized the effectiveness of small guns and rifles, and and later earned him a prize of 50,000 francs set by the French Academy of Sciences (1889). The "Vieille test" (1896) is used in the thermochemistry of propellants to test their thermal stability and compatibility. After 1900, he spent most of his time solving practical problems related to his Poudre B. He also studied the important problem of erosion in gun barrels and found that the gases from the explosion of nitroguanidine were much less erosive than those from other explosives of comparable force (1901).

VIEILLE became officer of the Légion d'honneur (1890) and member of the Paris Academy of Sciences (1904). In 1952, a new hall at the Laboratoire Central des Poudres in Paris, the *Salle Paul Vieille*, was named after him. Since 1975 the *Paul Vieille Memorial Lecture* has been presented by a distinguished scientist at the biennial International Symposium on Shock Waves in memory of VIEILLE's numerous contributions to shock waves, ballistics, and detonics. From 1989, the *Prix Pal Vieille* is awarded by the Association Française de Pyrotechnie (AFP) for the promotion of the technical and scientific training in the field of pyrotechnics or related disciplines. The Association des Amis du Patrimoine Poudrier et Pyrotechnique (Association of the Friends of the Heritage in Gun Powders and Pyrotechnics), or Association 3P (A3P), at Paris organizes international meetings called "Journées Scientifiques Paul Vieille," as a tribute to VIEILLE's famous research in the field of energetic materials during the 19th century.

ORIGINAL WORKS. With E. SARRAU: *Recherches expérimentales sur la décomposition du coton-poudre en vase clos*. C. R. Acad. Sci. Paris **89**, 165-167 (1879) — With M. BERTHELOT: *Etudes des propriétés explosives du fulminate de mercure*. Ibid. **90**, 946-952 (1880) — With M. BERTHELOT: *Recherches sur le sulfure d'azote*. Ibid. **92**, 1307-1309 (1881) — With E. SARRAU: *Sur la chaleur de formation des explosifs*. Ibid. **93**, 213-215, 268-271 (1881) — With M. BERTHELOT: *Sur la vitesse de propagation des phénomènes explosifs dans les gaz*. Ibid. **94**, 101-108, 822-823 (1882) — With E. SARRAU: *Etude sur l'emploi des manomètres à écrasement pour la mesure des pressions développées par les substances explosives*. Mém. Poudres Salpêtres **1**, 356-431 (1882); *Recherches sur l'emploi de manomètres à écrasement pour la mesure des pressions développées par les substances explosives*. C. R. Acad. Sci. Paris **95**, 26-29, 130-132, 180-182 (1882) — With M. BERTHELOT: *Nouvelles recherches sur la propagation des phénomènes explosifs dans les gaz*. Ibid. **95**, 151-157 (1882); *Sur la période d'état variable qui précède le régime de détonation et sur les conditions d'établissement de l'onde explosive*. Ibid. **95**, 199-205 (1882) — *Sur la mesure des pressions développées en vase close par les mélanges gazeux explosifs*. Ibid. **95**, 1280-1282 (1882) — *De l'influence du refroidissement sur la valeur des pressions maxima développées en vase clos par les gaz tonnants*. Ibid. **96**, 116-118 (1883) — *Sur les chaleurs de quelques gaz aux températures élevées*. Ibid. **96**, 1358-1361 (1883) — With M. BERTHELOT: *L'onde explosive*. Ann. Chim. Phys. **28** [V], 289-332 (1883) — With E. SARRAU: *Etude sur le mode de décomposition de quelques explosifs*. Mém. Poudres Salpêtres **2**, 126-167 (1885); *Sur l'emploi des manomètres à écrasement pour la mesure des pressions développées par les substances explosives*. C. R. Acad. Sci. Paris **102**, 1054-1059 (1886) — *Etude des pressions ondulatoires produites en vase clos par les explosifs*. Mém. Poudres Salpêtres **3**, 177-236 (1890) — *Méthodes appliquées à l'étude des matières explosives*. Lecture presented on March 1, 1890 at the "Association française pour le développement des sciences" and published as a brochure in the same year — *Influence du covolume des gaz sur la vitesse de propagation des phénomènes explosifs.* Mém. Poudres Salpêtres **4**, 20-22 (1891); C. R. Acad. Sci. Paris **112**, 43-45 (1891) — *Etude sur le mode de combustion des matières explosives.* Mém. Poudres Salpêtres **6**, 256-391 (1894) — *Sur la vitesse de propagation d'un mouvement dans un milieu en repos.* C. R. Acad. Sci. Paris **126**, 31-33 (1898) — *Vitesse de propagation des discontinuités dans les milieux en repos.* Ibid. **127**, 41-43 (1898) — *Déformation des ondes au cours de leur propagation.* Ibid. **128**, 1437-1440 (1899) — *Sur les discontinuités produites par la détente brusque des gaz comprimés.* Ibid. **129**, 1228-30 (1899) — *Sur la loi de résistance de l'air au mouvement des projectiles.* Ibid. **130**, 235-238 (1900) — *Rôle des discontinuités dans la propagation des phénomènes explosifs.* Ibid. **131**, 413-416 (1900) — *Etude sur le rôle des discontinuités dans les phénomènes de propagation.* J. Phys. Théor. Appl. **9** [III], 621-644 (1900); Mém. Poudres Salpêtres **10**, 177-260 (1900) — *Etude sur les phénomènes d'érosion.* Ibid. **11**, 157-210 (1901).

**SECONDARY LITERATURE.** J. CHALLEAT: *L'artillerie de terre en France pendant un siècle (1816–1919).* Lavauzelle, Paris (1935) — R.J. EMRICH: *Early development of the shock tube and its role in current research.* In: (Z.I. SLAWSKY, J.F. MOULTON JR., and W.S. FILLER, eds.) *Proc. 5th Int. Shock Tube Symp.* [Silver Spring, MD, April 1965]. Fluid Dynamics Div., APS (1965), pp. 1-10 — J. VALENSI: *Paul VIEILLE.* In: (G. KAMIMOTO, ed.) *Proc. 10th Int. Shock Tube Symp.* [Kyoto, Japan, July 1975]. Shock Tube Research Society, Japan (1975), p. 2 — *Journées scientifiques Paul VIEILLE: Colloque scientifique international organisé à l'occasion du centenaire de l'invention de la poudre* [Vert-le-Petit, France, Sept. 1984]. Centre de Recherches du Bouchet, Société Nationale des Poudres et Explosifs (SNPE), Délégation Générale pour L'Armement, Paris (1986) — L. MEDARD: *Paul VIEILLE et son œuvre.* Sci. Tech. l'Armement. Mém. Artill. Franç. **60**, No. 2, 11-23 (1986) — N. MANSON: *Contribution de Paul VIEILLE à la connaissance des détonations et des ondes de choc.* Ibid. **60**, No. 2, 26-59 (1986) — L. MEDARD: *L'œuvre scientifique de Paul VIEILLE (1854–1934).* Rev. Hist. Sci. **47**, 381-404 (1994).
**PICTURE.** Courtesy Collections Ecole Polytechnique, Paris, France.
**NOTE.** For a list of recipients of *Le Prix Paul Vieille* see http://www.afpyro.org/pvieille. Hitherto five meetings, entitled *Journées Scientifiques Paul Vieille* and organized by the Association 3P, were held in France with the following topics: (1) *100th anniversary of the invention of solvent-less gun propellants* [Vert-le-Petit, Sept. 1984]; (2) *Influence of explosives materials in reducing the vulnerability of ammunitions* [Brest, Oct. 1991]; (3) *Instrumentation, experimentation and expertise of energetic materials since the XVIth century* [Paris, Oct. 2000]; (4) *History of solid propellants during the XXth century* [Paris, Oct. 2003]; and (5) *1945-2005: sixty years of modern pyrotechnics* [Paris, Nov. 2006]; see http://association.a3p.free.fr/anglais/colloques_a.htm.

# WALLIS, John [Lat. *Johannus*] (1616–1703)

- English mathematician and cryptographer; early pioneer of percussion mechanics

John WALLIS was born in Ashford, a town in Kent. His father was the Rev. John WALLIS. He began his education in Latin at a private school in Tenterden, Kent (1625), and continued in Greek, Hebrew, and logic at Felsted School, Essex (1630). At Emmanuel College in Cambridge he studied natural philosophy, ethics, metaphysics, theology, anatomy, and medicine and graduated with a B.A. (1637) and an M.A. (1640). After ordination (1640) he became domestic chaplain, first to Sir Richard DARLEY and later to Baroness Mary VERE in London. Because of his particular ability to decipher letters in code, he was employed by the parliament as a cryptographer to decipher intercepted dispatches (1642–1645), an activity which he later resumed for king WILLIAM III (1690). After inheriting a considerable estate (1643), he dedicated himself entirely to science and became a Fellow of Queen's College in Cambridge (1644). WALLIS was appointed to the Savilian Chair of geometry (1649–1703) at Oxford University by Sir Oliver CROMWELL and later became keeper of the University Archives (1658–1703). He was made a doctor of divinity (1654) and became a royal chaplain (1660) and F.R.S. (1663). In addition, he acted temporarily as secretary to the assembly of divines at Westminster in London. During his period in London (beginning in 1645) he made the acquaintance of numerous scientists who met regularly at the lodgings of the physician Dr. Jonathan GODDARD in Woodstreet, London. These meetings formed the nucleus of the later Royal Society of London (founded in 1660). WALLIS was one of the founding members.

During his Savilian professorship he published his famous book *Arithmetica infinitorum* ("The Arithmetic of Infinitesimals," 1655), which contains the germs of the differential calculus, and extended higher algebra. With his method of interpolation he introduced the principles of analogy and continuity into mathematical sciences and invented the symbol $\infty$ for infinity using $1/\infty$ to represent an infinitesimal height. Under the heading *Johannis Wallisii ... operum mathematicorum* (*Pars altera* 1656, *Pars prima* 1657) he also published several mathematical treatises that were partly an outcome of his university lectures. His last great mathematical work was *A Treatise of Algebra, both Historical and Practical* (1685).

WALLIS, like Christiaan HUYGENS and Christopher WREN, also contributed considerably to the problem of percussion and impact, which in the 1660s was a much discussed subject at the Royal Society and the French Academy. In a paper submitted in 1668 to the Royal Society and published in the following year in the journal *Philosophical Transactions*, he treated inelastic percussion, equating elasticity with less-than-perfect hardness. These studies were extended to elastic percussion in the third part of his voluminous book *Mechanica, sive de moti tractatus geometricus* ("Mechanics, or Geometrical Tracts on Motion"), published in the period 1670–1671, then the most thorough study of mechanics and motion prior to Sir Isaac NEWTON's *Principia* (1687). Refuting many of the errors regarding motion that had persisted since the time of ARCHIMEDES, he gave a more rigorous meaning to such terms as force and momentum and widened the scope of treating mechanical problems analytically. The first part of his *Mechanica*... deals with various forms of motion in a strictly "geometrical" (*i.e.*, Euclidean) form, introducing the idea of moment, in modern terms the moment of inertia, which is essential for inquiries into the center of gravity and the center of percussion. The second – major – part deals with the calculation of the center of gravity. In the third part he discusses, in addition to problems on percussion, elementary machines that, according to ancient tradition, encompassed the lever, wedge, wheel and axle, pulley, and screw.

Later, in the first volume of his three-volume *Opera mathematica* (Oxford 1695) in which he edited his own mathematical works, he resumed the subject of percussion: he discovered in bodies that are freely rotating around a fixed axis and eccentrically struck by an impulsive force particular properties of a so-called "center of percussion" [Lat. *centrum percussionis*]. WALLIS first noticed that in a compound pendulum the center of percussion is identical to the "center of oscillation" as described by HUYGENS in his *Horologium oscillatorium* (Paris 1673).

WALLIS also published sermons and collections of his own theological tracts (1691) and an English grammar (1652). He edited classical mathematical authors (1676–1688) and published more than 60 papers and book reviews in the *Philosophical Transactions* (1666–1702).

**ORIGINAL WORKS.** *Arithmetica infinitorum.* Paris (1655), Oxoniae (1656); Engl. translation by J.A. STEDALL: *The arithmetic of infinitesimals.* Springer, New York (2003) — *Johannis VVallisii, S.S. Th.D., Geometriæ Professoris Saviliani in celeberrimâ Academia Oxoniensi; Operum mathematicorum.* Typis Leon. Lichfield Academiæ typographi, Oxonii. Pars altera (1656), pars prima (1657) — *On the mechanical theories of tides.* Phil. Trans. Roy. Soc. Lond. **1**, No. 16, 263-294 (1666) — *Mechanica: sive de motu tractatus geometricus.* G. Godbid, London (1670/1671); *see De percussione*, pp. 660-682 — *A summary account of the general laws of motion, by way of letter written by him to the Publisher, and communicated to the Roy. Society (Nov. 26, 1668)* [in Latin]. Phil. Trans. Roy. Soc. Lond. **3**, No. 43, 864-866 (Jan. 11, 1669) — *A discourse of gravity and gravitation.* J. Martyn, London (1675) — *Opera mathematica* [in Latin]. 3 vols., E theatro Sheldoniano, Oxoniae (1693–1699). 3 vols., Olms, Hildesheim (1972); *see* vol. 1 (1695): *De percussione*, pp. 1002-1015 — *An essay on the art of deciphering.* Gilliver & Clarke, London (1737).

**SECONDARY LITERATURE.** J.F. SCOTT: *The mathematical work of J. WALLIS, D.D., F.R.S. (1616–1703).* Taylor & Francis, London (1938) — G.U. YULE: *John WALLIS, D.D., F.R.S. 1616–1703.* Notes Rec. Roy. Soc. Lond. **2**, 74-82 (1939) — J.F. SCOTT: *The Reverend John WALLIS, F.R.S. (1616–1703).* Ibid. **15**, 57-67 (1960) — H. HARTLEY (ed.) *The Royal Society: its origin and founders.* The Royal Society of London, London (1960) — C.J. SCRIBA: *WALLIS and HARRIOT.* Centaurus (Copenhagen) **10**, 248-257 (1964) — A.R. HALL and M.B. HALL (eds.): *The correspondence of Henry OLDENBURG.* University of Wisconsin Press, Madison; *see* vol. 5 (1968): *1668–1669*, pp. 167-170, 192-194, 220-222 — A.R. HALL: *Mechanics and the Royal Society, 1668–1670.* Brit. J. Hist. Sci. **3**, 24-38 (1966/1967) — C.J. SCRIBA: *The autobiography of John WALLIS.* Notes Rec. Roy. Soc. **25**, 17-46 (1970); *J. WALLIS.* In: (C.C. GILLESPIE, ed.) *Dictionary of scientific biography.* Scribner, New York, vol. 14 (1976), pp. 146-155 — I. SZABÓ: *Geschichte der mechanischen Prinzipien und ihre wichtigsten Anwendungen.* Birkhäuser, Basel (1977), pp. 439-445 — R.S. WESTFALL: *Force in NEWTON's physics: the science of dynamics in the seventeenth century.* Macdonald, London (1979), pp. 231-244 — D.B. MELI: *WALLIS, John.* In: (H.C.G. MATTHEW, ed.) *Oxford dictionary of national biography.* Oxford University Press, Oxford (2004), vol. 57, pp. 15-18.
**PICTURE.** From his book *Opera mathematica* (1695), frontispiece of vol. 1.

# WEBER, Heinrich Martin (1842–1913)

▪ German theoretical and applied mathematician

Heinrich M. WEBER was born in Heidelberg to the noted historian Georg WEBER and studied mathematics and physics at the Universities of Heidelberg (1860–1861) and Leipzig (1860–1862). At Heidelberg University he earned his Ph.D. (1863). In 1866, he prepared for his habilitation as *Privatdozent* (university lecturer) and 3 years later became extraordinary professor. After teaching mathematics at the ETH Zürich (1869–1875) and the University of Königsberg (1875–1883), he was appointed professor of mathematics at the Technische Hochschule Berlin-Charlottenburg (1883) and subsequently taught in this discipline at the Universities of Marburg (1884–1892), Göttingen (1892–1895), and Strassburg (1895–1913). The direction of his work was decisively influenced by his stay at Königsberg, where in the period 1827–1842 the mathematician and greatest algorist of all time Carl G.J. JACOBI had worked on elliptic functions and made important contributions to the theory of determinants, so-called "Jacobian mathematics." At Königsberg WEBER was encouraged by Franz E. NEUMANN to investigate physical problems and by Friedrich Julius RICHELOT to study algebraic functions.

In 1868, WEBER published a paper entitled *Über eine Transformation der hydrodynamischen Gleichungen* ("On a Transformation of the Hydrodynamic Equations") in which he shows that in the case of using *Lagrangian* coordinates (in opposition to *Eulerian* coordinates) the system of differential equations of the second order can be transformed by a partial integration into a system of first-order differential equations, which becomes particularly simple in the case of using a velocity potential.

Together with J.W. Richard DEDEKIND, a mathematician and his closest friend, he edited G.F. Bernhard RIEMANN's collected works (1876). Based upon the lecture notes of Karl HATTENDORFF – one of RIEMANN's former students at Göttingen who had published in 1869 the book *Partielle Differentialgleichungen und deren Anwendung auf physikalische Fragen: Vorlesungen von Bernhard RIEMANN* ("Partial Differential Equations and Their Application on Physical Problems. Lectures of Bernhard RIEMANN") – WEBER reworked HATTENDORFF's book and in 1900/1901 edited Bernhard RIEMANN's posthumously published lectures on the theory and application of partial differential equations in the two-volume book *Die partiellen Differentialgleichungen der mathematischen Physik* ("Partial Differential Equations of Mathematical Physics"). In this book, which went through three editions, WEBER also addressed the fundamentals of thermodynamics and shock waves in gases with numerous examples.

In 1871, WEBER edited Charles BRIOT's *Lehrbuch der mechanischen Wärmetheorie* ("Textbook on the Mechanical Theory of Heat") and in 1907, together with Josef WELLSTEIN, the three-volume *Enzyklopädie der Elementar-Mathematik* ("Encyclopedia of Elementary Mathematics"). He also published the three-volume *Lehrbuch der Algebra* (1895–1896), an important textbook on algebra that was widely used in teaching and research for decades.

WEBER was a cofounder of the Deutsche Mathematiker-Vereinigung (1890) and a member of the editorial board of the journal *Mathematische Annalen* (1893–1913). His most prominent students were Hermann MINKOWSKI, David HILBERT, Adolf KNESER, and Robert FRICKE.

**ORIGINAL WORKS.** *Zur Theorie der singulären Lösungen partieller Differentialgleichungen erster Ordnung.* Engelmann, Leipzig (1866) — *Über eine Transformation der hydrodynamischen Gleichungen.* J. Math. **68**, 286-292 (1868) — *Neuer Beweis des ABEL'schen Theorems.* Math. Ann. **8**, 49-53 (1874) — With R. DEDEKIND: *Gesammelte mathematische Werke und wissenschaftlicher Nachlaß von Bernhard RIEMANN.* Teubner, Leipzig (1876) — With R. DEDEKIND: *Theorie der algebraischen Funktionen einer Veränderlichen.* J. Reine u. Angew. Math. **92**, 181-290 (1882) — *Theorie der ABEL'schen Zahlkörper. I, II, and III.* Acta Mathematica **8**, 193-263 (1886), ibid. **9**, 105-130 (1886) — *Die partiellen Differentialgleichungen der mathematischen Physik. Nach RIEMANN's Vorlesungen bearbeitet von Heinrich WEBER.* Vieweg, Braunschweig. Vol. 1 (1900), vol. 2 (1901) — *Lehrbuch der Algebra.* Vieweg, Braunschweig. Vol. 1: 1895, 1898; vol. 2: 1896, 1899; vol. 3: *Elliptische Funktionen und algebraische Zahlen* (1891, 1908) — With J. WELLSTEIN ET AL.: *Enzyklopädie der Elementar-Mathematik: ein Handbuch für Lehrer und Studierende.* 3 vols., Teubner, Leipzig (1903–1907) — *Mathematischer Teil.* In: (P. FRANK and R. VON MISES, eds.) *Die Differential- und Integralgleichungen der Mechanik und Physik.* 7. Auflage von RIEMANN-WEBER's *Partielle Differentialgleichungen der mathematischen Physik.* Vieweg, Braunschweig. Vol. 1: *Mathematischer Teil* (1925); vol. 2: *Physikalischer Teil* (1927).

**SECONDARY LITERATURE.** (Various authors): *Festschrift Heinrich WEBER zu seinem siebzigsten Geburtstag am 5. März 1912.* Teubner, Leipzig (1912); reprinted by ASM Chelsea, New York (1971) — R.C. ARCHIBALD: *Heinrich Martin WEBER* [Reprinted from the Bulletin of the American Mathematical Society (AMS)]. AMS, New York (1913) — A. VOSS: *Heinrich WEBER.* Jb. Dt. Mathematiker-Vereinigung **23**, 431-444 (1914) — B. SCHOENBERG: *H. WEBER.* In: (C.C. GILLESPIE, ed.) *Dictionary of scientific biography.* Scribner, New York, vol. 14 (1976), pp. 202-203 — G. FREI: *Heinrich WEBER and the emergence of classic field theory.* Hist. Mod. Math. **1**, 425-450 (1989).

**PICTURE.** From the frontispiece of the *Festschrift H. WEBER* (1912).

# WIESELSBERGER, Carl (1887–1941)

- German aerodynamicist

Carl WIESELSBERGER was born in Eberstahl, Lower Bavaria, and took his Ph.D. in general mechanical engineering at the TH Munich (1913). From 1913 to 1922 he was assistant and department leader at the Aerodynamische Versuchsanstalt (AVA), an aerodynamic testing facility in Göttingen. He made important contributions to the design of the Gottingen wind tunnel (1917) and carried out numerous investigations on aerodynamic drag and its dependency on surface friction and roughness. Invited by the Japanese government, he stayed at the Imperial University in Tokyo, where he significantly contributed to the design of Japanese wind tunnels

(1922–1930). After his return to Germany he was appointed chair of aeronautics at the TH Aachen and became director of the Aerodynamics Institute (1930–1941).

During this period WIESELSBERGER got involved in problems of high-speed aviation, both theoretically and experimentally. At the 5th Volta Conference in Rome (1935) he gave the first correct explanation of the puzzling phenomenon of the "X-shock," hitherto observed in high-speed wind tunnels. Since it is caused by the humidity of the air, it is therefore also called "condensation shock." He also suggested a specific wind tunnel configuration with 46% of the perimeter open via two wide longitudinal slots to reduce the blockage effect (1942). When, after the war, Ray H. WRIGHT and John STACK took up again the wind tunnel technique that used longitudinal slots (1947), WIESELSBERGER's concept resulted in NACA's many-slotted wind tunnel (1947). WIESELSBERGER developed various measurement techniques and instruments related to fluid dynamics and practical aviation, such as the wind tunnel balance (1934). Under his direction, Rudolf HERMANN, one his assistants, designed, built, and operated the Aachen supersonic wind tunnel, a little blow-down-type facility with a $10 \times 10$ cm$^2$ test section, operable up to $M = 3.3$ and equipped with a three-component balance (1931–1933). This very efficient concept of the Aachen supersonic wind tunnel was later closely followed up in the design of the larger Peenemünde wind tunnels (1937–1945).

ORIGINAL WORKS. *Luftwiderstand von Kugeln.* Z. Flugtech. u. Motor-Luftschiffahrt 5, 140-145 (1914) — *Feststellungen über die Gesetze des Flüssigkeits- und Luftwiderstandes.* Phys. Z. 22, 321-328 (1921); Ibid. 23, 219-224 (1922) — *Ein Manometer zur Aufzeichnung von Fluggeschwindigkeiten.* Z. Flugtech. u. Motor-Luftschiffahrt 12, 4-6 (1921) — *Einfluß der Modellaufhängung auf die Messergebnisse.* Ibid. 13, 188-191 (1922) — With A. BETZ: *Aerodynamik.* In: (R. SÜRING and K. WEGENER, eds.) *Moedebecks Taschenbuch für Flugtechniker und Luftschiffer.* Krayn, Berlin (1923), ch. 5 — *Hydrodynamische Versuche von JOUKOWSKY.* ZAMM 4, 184-186 (1924) — *Die wichtigsten Ergebnisse der Tragflügeltheorie und ihre Prüfung durch die Versuche.* In: (Th. VON KÁRMÁN and T. LEVI-CIVITA, eds.) *Vorträge aus dem Gebiet der Hydro- und Aerodynamik* [Innsbruck, Austria, 1922]. Springer, Berlin (1924), pp. 47-58 — *Einfluß von eingebauten Motorgondeln auf die Luftkräfte eines Tragflügels.* In: (World Engng. Congr., Publ. Committee, ed.) *Industrial Japan.* Kokusai Shuppan Insatsusha, Tokyo (1929), Nr. 203 — *Über den Luftwiderstand bei gleichzeitiger Rotation des Versuchskörpers.* Phys. Z. 28, 84-88 (1927) — *Zur theoretischen Behandlung der gegenseitigen Beeinflussung.* Z. Flugtech. u. Motor-Luftschiffahrt 23, 533-535 (1932) — With H. LORENZ: *Beitrag zur gegenseitigen Beeinflussung von Flügel und Luftschraube.* Springer, Berlin (1933) — *Ein Apparat zur Demonstration überlagerter Strömungen.* Phys. Z. 34, 46-47 (1933) — *Die aerodynamische Waage des Aachener Windkanals.* Abhandl. Aerodyn. Inst. Aachen (AIA), TH Aachen (AIA) 14, 24-27 (1934); *The Aachen wind-tunnel balance.* NACA-TM 757 (1934) — *Airplane body (nonlifting system), drag and influence on lifting system.* In: (W.F. DURAND, ed.) *Aerodynamic theory.* Springer, Berlin (1935), vol. 4, Div. K, pp. 130-168 — *Strömung in räumlich gekrümmten Rohren.* ZAMM 15, 109-111 (1935) — *Discussion note on the condensation effect.* In: V Convegno Volta su "Le alte velocità in aviazione" [Rome, Sept./Oct. 1935]. Reale Accademia d'Italia, Roma (1936), p. 558; Short autobiography. Ibid., p. 686 — *Über die Verteilung des Auftriebs längs der Spannweite bei hohen Anstellwinkeln.* Forschungsbericht NRW Nr. 493 (1935) — *On the distribution of lift across the span near and beyond the stall.* J. Aeronaut. Sci. 4, 363-365 (1937) — *Elektrische Anzeige von Kräften durch Änderung einer Induktivität.* Jb. Dt. Akad. Luftfahrtforsch. (1937/1938), p. 592 — *Die Überschallanlage des Aerodynamischen Instituts der Technischen Hochschule Aachen.* Luftwissen 4, 301-303 (1937); VDI-Z. 82, 1230 (1938) — *Arbeiten des Aerodynamischen Instituts der Technischen Hochschule Aachen.* Forsch. Gebiet Ing. Wes. 10, 55-56 (1939) — *Windkanalkorrekturen bei kompressibler Strömung.* Bericht Nr. 127, Lilienthal-Gesell., Berlin (1940) — *Über den Einfluß der Windkanalbegrenzung auf den Widerstand, insbesondere im Bereiche der kompressiblen Strömung.* Luftfahrtforsch. 19, 124-128 (1942).

SECONDARY LITERATURE. Anonymous: Jb. Dt. Akad. Luftfahrtforsch. (1937/1938), p. 99 — B. DIRKSEN: *Carl WIESELSBERGER* [with list of his publications]. Luftfahrt-Forsch. 19, 121-123 (1942) — A. BETZ: *Nachruf Carl WIESELSBERGER.* Jb. Dt. Akad. Luftfahrtforsch. (1942), p. 305 — Anonymous: *Carl WIESELSBERGER.* Z. Flugwiss. 2, 343 (1954) — A. NAUMANN (ed.): *Das Institut unter der Leitung von C. WIESELSBERGER.* In: *50 Jahre Aerodynamisches Institut der RWTH Aachen.* Abhandl. Aerodyn. Inst. RWTH Aachen, Heft 17 (1963), pp. 20-24 — P.P. WEGENER: *The Peenemünde wind tunnels.* Yale University Press, New Haven, CT (1996), pp. 23, 25, 161, 166 — U. KALKMANN: *Die Technische Hochschule Aachen im dritten Reich (1933–1945).* Mainz, Aachen (2003).

PICTURE. From the commemorative publication *50 Jahre Aerodynamisches Institut der RWTH Aachen 1913–1963,* courtesy Universitätsarchiv der RWTH Aachen, Germany.

## WOOD, Robert Williams (1868–1955)

▪ U.S. experimental physicist, father of infrared and ultraviolet photography

Robert W. WOOD was born in Concord, MA, the son of Robert Williams WOOD, a physician and pioneer in the sugar industry. After graduating in chemistry from Harvard University (1891), he entered Johns Hopkins University (1892) and became an Honorary Fellow of the University of Chicago (1892–1894). In order to complete his studies in chemistry, WOOD decided to continue his studies at the University of Berlin (1894–1896), but, coming there under the influ-

ence of Heinrich RUBENS, a professor of experimental physics, he became more interested in physics. Two years later, after his return to the United States, he spent a short time at MIT, where he was appointed instructor (1897). Thereafter, he became assistant professor (1899) of physics at the University of Wisconsin (1899).

WOOD's main interest was physical optics, but in order to demonstrate to his students the wave properties of light, he used wave fronts of sound waves, which he generated by electric spark discharges. Like August TOEPLER, he studied the propagation and reflection of spark waves (*i.e.*, weak shock waves) and, with the support of Sir Charles V. BOYS, was the first to photograph and present this subject to a British public. In the period 1901–1938 he again joined Johns Hopkins University, where he served as professor of experimental physics until his retirement, achieving there great popularity as a teacher of physics and as an outstanding demonstration lecturer. WOOD contributed significantly to spectroscopy by developing the technique of Raman spectroscopy, helped to develop color photography, conducted significant research on the physical properties and biological effects of ultrasonic waves, and devised practical uses for infrared and ultraviolet light. He invented a filter for ultraviolet transmission, which excluded all visible light, and was the first to publish infrared photographs (landscapes) taken on experimental film (1910).

After his retirement WOOD worked as research professor (1938–1953), serving also as a consultant to the Manhattan Project and to Aberdeen Proving Ground. He was the first to propose the use of tear gas and the use of air space around warships to dissipate the destructive power of torpedoes when the ship is attacked. Towards the end of World War II, the irregular reflection of shock waves (Mach reflection) became of great interest to optimize the Height of Burst (HOB) of the first atomic bomb to be dropped on Japan in regard of maximizing the overpressure (and supposedly also the destruction) on the ground. Renowned for his early sound wave reflection studies (1899), WOOD entered the field of shock waves again in 1943, when – then at the age of 75 – he was enlisted by John von NEUMANN at Princeton's Institute of Advanced Study (IAS) to repeat Ernst MACH and Jaromir WOSYKA's soot experiments on the oblique interaction of shock waves, which they had performed in 1875 at the German Charles University in Prague. WOOD's results, fully confirming the existence of Mach reflection, were an important basis for subsequent refined shock-tube measurements carried out by Walter BLEAKNEY's group at Princeton.

WOOD wrote over 200 scientific articles and two books. His textbook *Physical Optics* (1905) became a classic in the art of experimental optics, of which he finished a revised 4th edition shortly before he died in 1955. After 1920, WOOD made important contributions to the fields of ultrasound and biophysics by studying the biological and physiological effects of high-frequency sound waves. He wrote the book *Supersonics, the Science of Inaudible Sounds* (1939); at that time the term "supersonics" (dealing with phenomena arising when the velocity of a solid body exceeds the speed of sound) was formerly applied to "ultrasonics" (dealing with high-frequency sound waves).

WOOD's numerous contribution to physics, particularly to physical optics, were honored by medals from the Optical Society of America (OSA), the American Academy of Arts and Sciences (AAAS), the National Academy of Sciences (NAS), the Royal Society of London, the London Society of Arts, and the Franklin Institute. He was a Fellow of the American Physical Society and an honorary member of OSA. The *R.W. Wood Prize*, established by OSA in 1975 to honor the many contributions that WOOD made to optics, recognizes an outstanding discovery, scientific or technical achievement, or invention in the field of optics.

A crater on the far side of the Moon is named for him.

**ORIGINAL WORKS.** *A brilliant meteor.* Science No. 9, 13 (July 1, 1887) — *Effects of pressure on ice.* Am. J. Sci. **41**, 30-33 (1891) — *Combustion of gas jets under pressure.* Ibid. **41**, 477-482 (1891) — *Demonstration of caustics.* Ibid. **50**, 301-302 (1895) — *The duration of the flash of exploding oxyhydrogen.* Phil. Mag. **41** [V], 120-122 (1896) — *Photography of sound-waves by the "schlierenmethode."* Ibid. **48** [V], 218-227 (1899) — *Dark lightning.* Nature **60**, 460-461 (1899) — *Photography of sound waves.* The Photographic J. **24**, 250-256 (1900) — *The photography of sound-waves and the demonstration of the evolutions of reflected wavefronts with the cinematograph.* Nature **62**, 342-349 (1900) — *Photography of sound-waves and the cinematograph demonstration of the evolution of reflected wave fronts.* Proc. Roy. Soc. Lond. **66**, 283-290 (1900); Phil. Mag. **50** [V], 148-156 (1900) — *The kinetic theory of the expansion of compressed gas into a vacuum.* Science **16**, 908-909 (Dec. 5, 1902) — *On screens transparent only to ultraviolet light and their use in spectrum photography.* Phil. Mag. **5** [VI], 257-263 (1903) — *Apparatus to illustrate the pressure of sound waves.* Phys. Rev. **20** [I], 113-114 (1905) — *Physical optics.* Macmillan, New York (1905); Optical Society of America (OSA), Washington, DC (1988) — *Photography by invisible rays.* The Photographic J. **50**, 329-338 (Oct. 1910) — With A.L. LOOMIS: *The physical and biological effects of high-frequency sound waves of great intensity.* Phil. Mag. **4** [VII], 417-436 (1927) — *Optical and physical effects of high explosives.* Proc. Roy. Soc.

Lond. **A157**, 249-261 (1936) — *Supersonics, the science of inaudible sound.* Brown University Press, Providence, RI (1939) — *On the interaction of shock waves.* Progr. Rept. OSRD-1996 (1943).

**SECONDARY LITERATURE.** Biographical note in Proc. Inst. Mech. Engng. (1895), p. 313 — J. MCKEEN CATELL (ed.): *Leaders in education, a biographical dictionary.* Science, New York (1932) — W. SEABROOK: *Doctor WOOD, modern wizard of the laboratory.* Harcourt & Brace, New York; University of Michigan Press, Ann Arbor (1941) — R.J. STRUTT: *R.W. WOOD of Baltimore.* Nature **149**, 650-651 (1942) — Anonymous: *R.W. WOOD.* Phys. Today **8**, 33 (Nov. 1955) — G.H. DIEKE: *Robert Williams WOOD.* Biogr. Mem. Fell. Roy. Soc. (Lond.) **2**, 327-345 (1956) — R.J. SEEGER: *On MACH's curiosity about shock waves.* In: (R.S. COHEN and R.J. SEEGER, eds.) *Ernst MACH, physicist and philosopher.* Boston Studies Phil. Sci. **6**, 42-67 (1970) — R.B. LINDSAY: *R.W. WOOD.* In: (C.C. GILLESPIE, ed.) *Dictionary of scientific biography.* Scribner, New York, vol. 14 (1976), pp. 497-499.

**PICTURE.** Courtesy The Ferdinand Hamburger Archives of The Johns Hopkins University, Baltimore, MD.

**NOTE.** WOOD's manuscripts are kept at the Milton S. Eisenhower Library, Johns Hopkins University.

## WREN, Sir Christopher (1632–1723)

• English architect, designer, geometrician, and astronomer; early pioneer of percussion mechanics

Sir Christopher WREN was born in East Knoyle in the County of Wiltshire as the son of Christopher WREN, then serving as rector of East Knoyle and chaplain to King Charles I, king of Great Britain and Ireland. He began his studies of the natural sciences at Westminster School in London and graduated (M.A.) from Wadham College in Oxford (1653). He became Fellow of All Souls College (1653–1661) in Oxford and professor of astronomy at Gresham College in London (1657–1661) and at Oxford University (1661–1673). His early life was marked by an interest in mathematics and natural science, particular in mechanics, but at the age of 30 he turned to architecture and made his first architectural designs. He probably applied himself to architecture and visited France (1665), where he studied the buildings erected in the period of LOUIS XIV, king of France (1643–1715). After his return to England, he prepared a scheme for rebuilding London after the Great Fire (1666) and was appointed Surveyor-General and Principal Architect for rebuilding the entire city (1668).

In the same year, Sir WREN submitted a paper on percussion to the Royal Society, which he had already written several years earlier – thus fulfilling an urgent request by Henry OLDENBURG, secretary of the Society who had asked several leading authorities in Europe to treat this subject of general interest. Together with DESCARTES, GALILEI, HUYGENS, MARCI, MARIOTTE, Sir NEWTON, and WALLIS, he belongs to the early-17th-century pioneers who investigated percussion phenomena scientifically, their results forming a significant base for subsequent researchers on this subject.

Sir WREN's short paper on the Laws of Motion in the case of two bodies impacting head-on was published in the journal *Philosophical Transactions*, immediately following John WALLIS' paper treating the same subject. Sir WREN limited his study to the case of a perfectly elastic collision, so that no motion is lost and the sum of the two products of mass and velocity is constant; *i.e.*, using correctly the Law of Conservation of Momentum. Attempting to solve a dynamic problem by a static approach, he plotted the velocities before and after percussion along a velocity axis forming a balance swinging about its center of gravity. Unfortunately, his geometrical solution – somewhat obscure but providing correct results and proved by him experimentally – was given by him without proof. Generally, the solution to this two-body percussion problem requires the application of both the Law of Conservation of Momentum and the Law of Conservation of Energy. At that time, however, the latter fundamental principle of mechanics was not yet recognized, and this fact even underlines the ingenuity of his solution. Very recently, Kerry DOWNES, a history of art professor at Reading University and WREN biographer (*see below*), appropriately wrote: "Contemporaries complained that he [WREN] valued the neatness of a solution above the presentation of proofs, claiming that the truth, once stated, was self-evident… As in the case of PASCAL, he preferred geometrical, visual, and intuitive solutions for mathematical problems."

Sir WREN was a founding member of the Royal Society of London (founded in 1660) and served as its third president (1680–1682). He was knighted in 1673. His scientific work was highly regarded by Sir Isaac NEWTON, who specifically recognized his experiments with suspended balls as confirming one of his three Laws of Motion. Today Sir WREN is more renowned for his contributions to British architecture rather than to science. He designed and rebuilt 52 city churches in London; his best-known work is certainly St. Paul's Cathedral (1675–1710), the world's third largest church and his burial place. Other famous buildings he designed are the Sheldon Theatre in Oxford and the Pembroke College and the Library of Trinity College in Cambridge. Together with his colleague and friend Robert HOOKE, WREN also designed the monument on Fish Street, which commemorates the Great Fire.

A crater on Mercury is named for him.

**ORIGINAL WORKS.** *Lex Naturae de collisione corporum* [read on Dec. 17, 1668]. Phil. Trans. Roy. Soc. Lond. **3** (No. 43), 867-868 (March 1669) — *Life and works of Sir Christopher WREN. From the parentalia or memoirs by his son Christopher.* Essex House, Campden (1903).
**SECONDARY LITERATURE.** J. WARD: *The lives of the Professors of Gresham College.* J. Moore, London (1740) — L. MILMAN: *Sir Christopher WREN.* Duckworth, London (1908) — L. WEAVER: *Sir Christopher WREN: scientist, scholar and architect.* G. Newnes, London (1923) — J.N. SUMMERSON: *Sir Christopher WREN.* Collins, London (1953) — E.F. SEKLER: *WREN and his place in European architecture.* Faber & Faber, London (1956) — H. HARTLEY (ed.): *The Royal Society: its origin and founders.* Royal Society, London (1960) — A.R. HALL: *Mechanics and the Royal Society, 1668–1670.* Brit. J. Hist. Sci. **3**, 24-38 (1966/1967) — A.R. HALL and M.B. HALL (eds.): *The correspondence of Henry OLDENBURG.* University of Wisconsin Press, Madison, WI, vol. 5 (1968): *1668–1669*, pp. 117-118, 134-135, 193, 263, 265, 320-322, 582 — R.S. WESTFALL: *Force in NEWTON's physics: the science of dynamics in the seventeenth century.* Macdonald, London (1971), pp. 203-206 — B. LITTLE: *Sir Christopher WREN: a historical biography.* Hale, London (1975) — J.F. SCOTT: *Christopher WREN.* In: (C.C. GILLESPIE, ed.) *Dictionary of scientific biography.* Scribner, New York, vol. 14 (1976), pp. 509-511 — I. SZABÓ: *Geschichte der mechanischen Prinzipien und ihre wichtigsten Anwendungen.* Birkhäuser, Basel (1977), pp. 439-445 — J.A. BENNETT: *The mathematical science of Christopher WREN.* Cambridge University Press (1982) — K. DOWNES: *WREN, Sir Christopher.* In: (H.C.G. MATTHEW, ed.) *Oxford dictionary of national biography.* Oxford University Press, Oxford (2004), vol. 60, pp. 406-419.
**PICTURE.** Courtesy Deutsches Museum, Munich, Germany.

# YEAGER, Charles ("Chuck") Elwood (1923–)

▪ U.S. aviator and test pilot, the "Supersonic Man"

Charles E. YEAGER was born in Myra, WV, the son of Albert YEAGER, a driller for natural gas in the West Virginia coal fields. Enlisting in the U.S. Army (1941), he was commissioned a reserve flight officer (1943) and became a pilot in the fighter command of the 8th Air Force stationed in England. He flew 64 missions over Europe during World War II and

was shot down by German aircraft. After the war he became a flight instructor and then a test pilot. He was chosen from several volunteers to test-fly the X-1, a secret experimental aircraft built by the Bell Aircraft Company to test the capabilities of human pilots and fixed-wing aircraft against the severe aerodynamic stresses of sonic flight. On October 14, 1947 – only days after cracking several ribs in a horseback riding accident but not wishing to disrupt the testing schedule – YEAGER broke the sound barrier over Rogers Dry Lake in southern California. His X-1, attached to a B-29 mother ship and carried to an altitude of 7,600 m, then rocketed separately to 12,000 m, reaching in level flight a supersonic velocity of 1,066 km/h (296 m/s). The feat was not announced publicly until June 1948. He continued to make test flights, and on December 12, 1953, flying the rocket plane Bell X-1A, he became the first person to exceed Mach 2, thereby establishing a world speed record of 2,660 km/h. Today, the orange rocket plane is part of the collection of the Smithsonian Institution of Technology in Washington, DC.

YEAGER, renowned as the "Supersonic Man," held various air force command assignments between 1954–1962, was vice-commander of the Ramstein Air Force Base in Germany (1968–1969), and retired from active duty in the Air Force with the rank of brigadier general (1975). For his numerous contributions to the advance of aeroscience he was honored with the Fédération Aéronautique International Gold Medal (1947), the Collier Trophy (1948), the Harmon International Trophy (1953), and the Presidential Medal of Freedom (1985). In 1966, he was inducted into the International Aerospace Hall of Fame. YEAGER's autobiography was published in 1985. He appears as the main character in Tom WOLFE's book *The Right Stuff* (1979), and as the epitome of that virtue he appeared in numerous commercial endorsements.

**PUBLICATIONS.** With L. JANOS: *YEAGER: an autobiography,* Bantam, Toronto (1985); Hall, Boston (1986) — *Breaking the sound barrier.* Popular Mechanics **164**, 91-92, 146-148 (Nov. 1987); article reprinted in the Oct. 1997 issue — With C. LEERHSEN: *Press on! Further adventures in the good life.* Bantam, Toronto (1988) — With J. ETHELL (ed.) *There once was a war.* Viking Studio Books, New York (1995).
**SECONDARY LITERATURE.** W.R. LUNDGREN: *Across the high frontier: the story of a test pilot-major Charles E. YEAGER, USAF.* Morrow, New York

(1955) — R.P. HALLION: *Supersonic flight: the story of the Bell X-1 and Douglas D-558.* Macmillan, New York (1972) — T. WOLFE: *The right stuff.* Farrar, Straus & Giroux, New York (1979) — *Interview: Chuck YEAGER.* Omni Magazine (Aug. 1986) — J.D. ANDERSON JR.: *Modern compressible flow, with historical perspective.* McGraw-Hill, New York (1990) — *Charles E. ("Chuck") YEAGER.* NASA History Division, Washington, DC (Sept. 1997); http://www.hq.nasa.gov/office/pao/History/x1/chuck.html — *Brig. Gen. Charles E. "Chuck" YEAGER.* History of Edwards AFB, CA (Feb. 2005); http://www.edwards.af.mil/history/docs_html/people/yeager_biography.html — *Brigadier General Charles E. "Chuck" YEAGER.* USAF Military Biographies; http://www.findarticles.com/p/articles/mi_m0RBE/is_2004_Annual/ai_n8566107.
**PICTURE.** Courtesy Deutsches Museum, Munich, Germany.

## ZEL'DOVICH [Russ. ЗЕЛЬДОВИЧ], Yakov Borisovich (1914–1987)

- Soviet chemical physicist, nuclear physicist, astrophysicist, and shock wave and detonation physicist; cofounder of modern detonation theory

Born in Minsk, Belorussia (Belarus) the son of an attorney, Yakov B. ZEL'DOVICH began his scientific career as a laboratory assistant in the department of chemical physics at the Leningrad Institute of Chemical Physics of the U.S.S.R. Academy of Sciences (1931). The Institute, then directed by Prof. Nikolai N. SEMENOV, was also involved in the crystallization of nitroglycerin in two modifications. Very soon his talent for theory was discovered, and he was transferred to the theoretical department. Here mathematicians and physicists provided him with continuous instruction on the foundations of theoretical physics, and with their help he received a thorough education. He never graduated from a university, although he was the author of over 20 books and 500 scientific articles on subjects ranging from chemical catalysis to large-scale cosmic structures, with major contributions to the theory of combustion, detonation, elementary particles, and astrophysics. However, his exceptional talent enabled him to embark on a postgraduate career, which he successfully completed with a Ph.D. thesis entitled *Oxidation of Nitrogen in Combustion and Explosions* (1939), which, containing a multitude of experimental and theoretical data, proved that the oxidation of nitrogen is an unbranched chain reaction.

ZEL'DOVICH and his colleague David A. FRANK-KAMENETSKII, an outstanding theorist, worked out a theory of flame propagation and proposed a mechanism for chemical reactions in a shock wave (1938–1943). An understanding of the features of the combustion of powders served as a basis for creating the internal ballistics of solid-fuel rockets: research carried out during World War II was orientated toward the *Katyusha* ("Little Katie," the title of a famous Russian song). This rocket weapon consists of a launcher capable of being fired in a salvo, popularly known as the "Stalin organ." Motivated by his patriotism, he also contributed to the development of the first Soviet atomic bomb and together with Yulii B. KHARITON did important theoretical research on nuclear chain reactions (1939–1941).

In the shock physics community his reputation rests primarily on his theoretical contributions to shock waves and detonation. In 1940, he completed his 1-D theory of detonation, 2 years prior to John VON NEUMANN. His work on the theory of shock waves fell into the period 1946–1969 and encompassed (1) the systematic use of physical concepts in gas dynamics; (2) experimental studies of physicochemical kinetics in gases at high temperatures using shock tubes; (3) investigations on the structure of shock waves, particularly the possibility of rarefaction waves with discontinuities; (4) a quantitative theory of observed front patterns of shock-induced luminescence; and (5) experimental methods to study the dependence of the refraction and polarization coefficients of optically transparent substances during strong shocks.

He also made significant contributions to the theory of combustion and detonation by bringing together gas dynamics, gas-kinetic theory, molecular transport effects, and the kinetics of high-temperature chemical reactions. In 1940, he developed a steady-state detonation model that was independently proposed in the United States by John VON NEUMANN (1942) and in Germany by Werner DÖRING (1941) and that is known today as the "Zel'dovich-von Neumann-Döring (ZND) theory." In the same year, ZEL'DOVICH also considered the application of detonations to propulsion and power engineering.

Together with Yuri P. RAIZER he published in 1963 the Russian textbook *Fizika udarnych voln i vysokotemperaturnych dinamiceskich javlenij* ("Physics of Shock Waves and High-Temperature Hydrodynamic Phenomena"). It was translated into English (1966–1967) and recently reprinted (2002). ZEL'DOVICH's contribution to the theory of shock waves has been reviewed by Y.P. RAIZER, one of his former coworkers (see also his introduction in the Selected Works of Yakov B. ZEL'DOVICH). Beginning in the 1960s, ZEL'DOVICH's interest in basic research shifted increasingly to cosmology and cosmogony, particularly to the origin of the Universe.

In the 1960s, ZEL'DOVICH worked on astrophysics and cosmology. First suggesting a "cold model variant" to the origin of the Universe (1962), he immediately became an enthusiastic advocate of the "Hot Big Bang model" when the phenomenon of a residual microwave background radiation at about 3 K was discovered by Arno A. PENZIAS and Robert W. WILSON (1965). Together with the Russian theoretical astrophysicist Rashid A. SUNYAEV he predicted in 1969 the so-called "Sunyaev-Zel'dovich effect." Measurements of this effect provide distinctly different information about cluster properties than X-ray imaging data, while combining X-ray and Sunyaev-Zel'dovich effect data leads to new insights into the structures of cluster atmospheres and the structure of the Universe on the largest scales.

He also worked on the dynamics of neutron emission during the formation of black holes, the formation of galaxies and clusters, and the large structure of the Universe. Together with his followers he made his Moscow Sternberg Astronomical Institute a stronghold of relativistic astrophysics.

He received many national and international awards and medals. He was elected corresponding member (1946) and Academician (1958) of the U.S.S.R. Academy of Sciences and foreign member of various other highly regarded academies. Cambridge University awarded him an honorary doctorate. In his later years he tried to emigrate, but this was twice prevented. Recently, Prof. Lev V. AL'TSHULER, the patriarch of Russian shock wave research who closely collaborated with ZEL'DOVICH for a long time, retrospectively summarized his scientific biography: "YA. B. ZEL'DOVICH lived through a whole number of careers, each devoted to explosions of one kind or another, whose power progressively increased as his interest shifted from the detonations and explosions of chemical substances, through increasingly powerful chain reactions and nuclear explosions, and finally to the Big Bang, from which the Universe emerged 15 billion years ago. Combustion and detonation were ZEL'DOVICH's first and lifelong love, a passion to which he remained faithful until his last days."

The 1994 and 2004 International Conferences on Combustion Detonation and Shock Waves, organized by the Scientific Council of the Russian Academy of Sciences and held in Moscow, were dedicated to his 80th and 90th birthday anniversaries and renamed the Zel'dovich Memorial I and Zel'dovich Memorial II, respectively. The Ya.B. Zel'dovich Gold Medal, established in 1990 by The Combustion Institute, is awarded "for outstanding contribution to the theory of combustion or detonation."

A minor planet (asteroid 11,438 ZEL'DOVICH) was named after him.

**ORIGINAL WORKS.** With D.A. FRANK-KAMENETSKII: *A theory of thermal flame propagation* [in Russ.]. Zh. Fiz. Khimii **12**, No. 1, 100-105 (1938) — With Y.B. KHARITON: *On the problem of the chain decay of the main isotope of uranium* [in Russ.]. Zh. Eksp. Teor. Fiz. **9**, 1425-1427 (1939) — *Theory of combustion and explosion* [in Russ.]. Nauka, Moskva (1940) — *To the question of energy use of detonation combustion* [in Russ.]. Zh. Tekh. Fiz. **10**, 1453-1461 (1940); Engl. translation and comments by E. WINTENBERGER and J.E. SHEPHERD in J. Propulsion & Power **22**, 586-587 (2006) — *On the theory of the propagation of detonation in gaseous systems* [in Russ.]. Ibid. **10**, 542-568 (1940); Engl. translation: NACA-TM 1261 (Nov. 1950) — *The theory of the limit of propagation of a slow flame* [in Russ.]. Zh. Eksp. Teor. Fiz. **11**, 159-169 (1941) — *Pressure and velocity distributions in the detonation products of a divergent, spherical wave* [in Russ.]. Ibid. **12**, 389-406 (1942) — *Theory of the formation of a new phase. Cavitation* [in Russ.]. Ibid. **12**, 525-538 (1942) — *Theory of combustion and detonation of gases* [in Russ.]. AN, Moscow & Leningrad (1944); Engl. translation by Air Materiel Command, Wright-Patterson Air Force Base, Dayton, OH (1949) — *On the possibility of rarefaction shock waves*. Zh. Eksp. Teor. Fiz. **16**, 363-364 (1946) — *On the propagation of shock waves in a gas with reversible chemical reactions* [in Russ.]. Ibid. **16**, 365-368 (1946) — *Theory of shock waves and introduction to gas dynamics* [in Russ.]. Izdat. AN, Moscow (1946) — *Theory of detonation spin*. Dokl. AN **52**, No. 2, 147-150 (1946) — With A.I. ROZLOVSKII: *Transition from a spherical into a plane detonation* [in Russ.]. Ibid. **57**, No. 4, 365-368 (1947) — *Theory of the detonation onset in gases* [in Russ.]. Zh. Tekh. Fiz. **17**, 3-26 (1947) — *Theory of combustion of initially unmixed gases* [in Russ.]. Ibid. **19**, No. 10, 1199-1210 (1949) — With I.A. SHLYAPINTOKH: *Ignition of explosive gas mixture in shock waves*. Dokl. AN **65**, 871-876 (1949) — With YA.T. GERSHANIK and A.I. ROZLOVSKII: *Adiabatic combustion processes of inflammable gas mixtures* [in Russ.]. Zh. Fiz. Khimii **24**, 85-95 (1950) — *The theory of the propagation of detonation in gaseous systems*. NACA-TM 1261 (1950) — With A.S. KOMPANEETS: *Teoriya detonatsii*. Gostekhizdat, Moskva (1955); *The theory of detonation*. Academic Press, New York (1960) — With S.M. KOGARKO and N.N. SIMONOV: *Experimental investigation of spherical gaseous detonation*. Sov. Phys. Tech. Phys. **1**, 1689-1713 (1957) — With Y.P. RAIZER: *Shock waves of large amplitude in air*. Sov. Phys. JETP **5**, 919-927 (1957) — With S.B. KORMER, M.V. SINITSYN, and A.I. KURYAPIN: *Temperature and specific heat of plexiglas under shock wave compression*. Sov. Phys. Dokl. **3**, 938-939 (1958) — With Y.P. RAIZER: *Physical phenomena that occur when bodies compressed by strong shock waves expand in vacuo*. Sov. Phys. JETP **8**, 980-982 (1959) — With G.I. BARENBLATT: *Theory of flame propagation*. Combust. Flame **3**, 61-74 (1959) — *Converging cylindrical detonation wave*. Sov. Phys. JETP **9**, 550-557 (1959) — With S.B. KORMER, M.V. SINITSYN, and K.B. YUSHKO: *A study of the optical properties of transparent materials under high pressure*. Sov. Phys. Dokl. **6**, No. 6, 494-496 (1961) — *Prestellar state of matter*. Zh. Eksp. Teor. Fiz **43**, 1561-1562 (1962) — *The theory of the expanding Universe as originated by A.A. FRIEDMANN*. Sov. Phys. Uspekhi **6**, 475-494 (1963) — With R.M. ZAIDEL': *One-dimensional instability and at-*

tenuation of detonation [in Russ.]. Zh. Prikl. Mekh. Tekh. Fiz. No. 6, 59-65 (1963) — With S.B. KORMER, G.V. KRISHKEVICH, and K.B. YUSHKO: *Investigation of the smoothness of the detonation front in a liquid explosive.* Sov. Phys. Dokl. **9**, 851-853 (1964) — With Y.P. RAIZER: *Physics of shock waves and high-temperature hydrodynamics phenomena.* 2 vols., Academic Press, New York (1966/1967); Dover Publ., Mineola, NY (2002) — *EMF produced by a shock wave moving in a dielectric.* Sov. Phys. JETP **26**, 159-162 (1968) — With Y.P. RAIZER: *Elements of gas dynamics and the classical theory of shock waves.* Academic Press, New York (1968) — With R.A. SUNYAEV: *The interaction of matter and radiation in a hot-model Universe.* Astrophys. Space Sci. **4**, 301-316 (1969) — With S.B. KORMER and V.D. URLIN: *Non-equilibrium radiation from shock compressed ionic crystals at temperatures above 1 eV.* Sov. Phys. JETP **28**, 855-859 (1969) — With I.D. NOVIKOV: *Relativistic astrophysics.* Vol. 1: *Stars and relativity*; vol. 2: *The structure and evolution of the Universe.* University of Chicago Press, Chicago (1971) — With R.A. SUNYAEV: *Shock wave structure of the radiation spectrum during bose condensation of photons.* Zh. Eksp. Teor. Fiz. **62**, No. 1, 153-160 (1972) — *Hydrodynamics of the Universe.* Annu. Rev. Fluid Mech. **9**, 215-228 (1977) — With G.I. BARENBLATT, V.B. LIBROVICH, and G.M. MAKHVILADZE: *Mathematical theory of combustion and explosion* [in Russ.]. Nauka, Moskva (1980); Engl. translation by D.H. MCNEIL: Consultants Bureau, New York, NY (1985) — *Cosmology from ROBERTSON to today* [H.P. Robertson Memorial Lecture, Washington, DC, April 1987]. Phys. Today **41**, 27-29 (March 1988) — (J.P. OSTRIKER, G.I. BARENBLATT, and R.A. SUNYAEV, eds.) *Selected works of Yakov Borisovich ZEL'DOVICH.* Princeton University Press, Princeton, NJ. Vol. 1 (1992): *Chemical physics and hydrodynamics*; vol. 2 (1993): *Particles, nuclei, and the Universe — An autobiographical afterword.* In: *Selected works*, vol. 2 (1993), pp. 632-644.

**SECONDARY LITERATURE.** J. TURKEVICH (ed.): *Soviet men of science.* Van Nostrand, Princeton, NJ (1963), pp. 435-436 — J.A. CHRAMOV: *ZEL'DOVICH Yakov Borisovich* [in Russ.]. *Fiziki: biograficeskij spravocnik.* Naukova Duma, Kiev (1977), pp. 140-141 — W. PRIESTER: *Y.B. ZEL'DOVICH (1914–1987)* [in Germ.]. Physik. Blätter **44**, 45-46 (1988) — *Yakov Borisovich ZEL'DOVICH.* Combustion, Explosion & Shock Waves **24**, No. 3, 378-380 (May 1988) — V.I. GOLDANSKII: *Ya. B. ZEL'DOVICH.* Phys. Today **41**, 98-102 (Dec. 1988) — A. SAKHAROV: *A man of universal interests.* Nature **331**, 671-672 (1988) — Anonymous: *Yakov Borisovich ZEL'DOVICH, 1914–1987.* Sov. Astron. Lett. **14**, 121-122 (1988) — J. VRONSKAYA (ed.): *A biographical dictionary of the Soviet Union 1917–1988.* K.G. Saur, London (1989) — (Eds.) *Yacov Borisov ZEL'DOVICH.* High Pressure Res. **1**, 157-159 (1989) — H. FRIEDMAN: *Yakov Borisovich ZEL'DOVICH (March 8, 1914 – December 2, 1987).* Proc. Am. Phil. Soc. **135**, No. 2, 322-326 (June 1991) — *The scientific and creative career of Yakov Borisovich ZEL'DOVICH.* In: (J.P. OSTRIKER, G.I. BARENBLATT, and R.A. SUNYAEV, eds.) *Selected works of Yakov Borisovich ZEL'DOVICH.* Princeton University Press, Princeton, NJ, vol. 1 (1992), pp. 3-56 — V.L. GINZBURG: *Yakov Borisovitch ZEL'DOVICH.* Biogr. Mem. Fell. Roy. Soc. (Lond.) **40**, 429-441 (1994) — D. HOLLOWAY: *STALIN and the bomb.* Yale University Press, New Haven, CT (1994) — R.N. KEELER: *Some thoughts of an American scientist on the dynamic high-pressure work of Academician Y.B. ZEL'DOVICH.* Sov. Phys. Uspekhi **38**, 559-563 (1995) — (S.M. FROLOV and A.G. MESZHANOV, eds.) *Proc. Zel'dovich Memorial Int. Conf. on Combustion* [Moscow, Russia, Sept. 1994]; *The International Conference on Combustion devoted to the 80-th birthday of Y.B. ZEL'DOVICH.* Russ. Chem. Bull. **44**, No. 2, 392-394 (Feb. 1995) — L.V. AL'TSHULER, R.F. TRUNIN, V.D. URLIN, V.E. FORTOV, and A.I. FUNTIKOV: *Development of dynamic high-pressure techniques in Russia.* Phys. Uspekhi **42**, 261-280 (1999) — R.A. SUNYAEV (ed.): *ZEL'DOVICH reminescences.* Boca Raton, London (2004) — A.A. BORISOV (ed.): *Progress in combustion and detonation* [Zel'dovich Memorial]. Torus Press, Moscow (2004).

**PICTURE.** Courtesy Deutsches Museum, Munich, Germany. The photo shows Y.B. ZEL'DOVICH (*right*) together with Y.B. KHARITON.
**NOTE:** His name has also been transliterated as ZEL'DOVIĆ.

## ZEMPLÉN, Gyözö Victor (1879–1916)

• Hungarian physicist

Gyözö V. ZEMPLÉN was born in Nagy-Kanizsám, then a town in the Hungarian part of the Austro-Hungarian Empire, and attended the gymnasium in Fiume (now Rijeka, Croatia). Already very early interested in natural sciences, he won at the age of 19 the Johann Pasquich Prize with a work on the inner friction of gases. After studying mathematics and physics at the Múegyetem (Royal Josephs Polytechnic), he took his *Magister* degree and became assistant at the institute of experimental physics of Prof. Roländ VON EÖTVÖS, the most significant Hungarian physicist of that time.

ZEMPLÉN critically reviewed contemporary hypotheses on the kinetic theory of gases (1900) and proposed a new method of measuring the viscosity of gases and liquids using a torsion pendulum, which earned him the Ph.D. (*sub auspiciis regis*) and the Than Károly Prize (1901). Thereafter, he won a research fellowship (1904–1906) and visited the Universities of Göttingen and Paris. At Göttingen he became interested in the mathematical treatment of hydrodynamic nonlinear differential equations and the physical interpretation of their solutions (1905).

ZEMPLÉN also propounded the idea that shock waves can only be compression waves, meaning that rarefaction shocks (*i.e.*, negative shocks) cannot exist, the so-called "Zemplén theorem." He presented his results in Göttingen at Felix KLEIN's seminar, then an international authority on mathematics, and in Paris at the French Academy, which resulted in fruitful discussions with Jacques HADAMARD and Pierre DUHEM, two experts in the young discipline of shock waves. In the following year, he published his study in the prestigious French journal *Comptes rendus de l'Académie des sciences* and the German *Enzyklopädie der Mathematischen Wissenschaften*.

ZEMPLÉN's personal contacts to leading French shock scientists of his time, his great expertise, and his publications in Hungarian, French, and German certainly contributed to a quick spreading of the new term *shock wave*, which, used first by E. MACH and J. WENTZEL [Germ. *Stoßwelle*, 1885] and later by HADAMARD [French *onde de choc*, 1904], was picked up immediately by ZEMPLÉN (1905).

Returning to Budapest, ZEMPLÉN habilitated (1907), worked as a *Studienrat* (Assistant Master) at the University of Budapest, and eventually became full professor and director of the newly founded Chair and Institute of Theoretical Physics. Besides his contributions to gas dynamics and hydrodynamics, ZEMPLÉN began working in mechanics, electrodynamics, and X-rays. However, at the beginning of World War I he was called up and killed in the battle at Monte Dolore (1916), northern Italy, at the early age of 37.

**ORIGINAL WORKS.** *Über die Grundhypothesen der kinetischen Gastheorie.* Ann. Phys. **2** [IV], 404-413 (1900); Ibid. **3** [IV], 761-763 (1900) — *Über den Energieumsatz in der Mechanik*. Ibid. **10** [IV], 419-428 (1903) — *Über die Anwendung der mechanischen Prinzipien auf reibende Bewegungen*. Ibid. **12** [IV], 356-372 (1904), Ibid. **13** [IV], 216 (1904) — *Kriterien für die physikalische Bedeutung der unstetigen Lösungen der hydrodynamischen Bewegungsgleichungen*. Math. Ann. **61**, 437-449 (1905) — *Sur l'impossibilité d'ondes de choc négatives dans les gaz*. C. R. Acad. Sci. Paris **141**, 710-712 (1905); Ibid. **142**, 142-143 (1906) — *Bestimmung des Koeffizienten der inneren Reibung der Gase nach einer neuen experimentellen Methode*. Ann. Phys. **19** [IV], 784-806 (1906) — *Besondere Ausführungen über unstetige Bewegungen in Flüssigkeiten*. Enzyklopädie der math. Wiss. **IV** [3], Teubner, Leipzig (1906), pp. 281-323 — *Über die Kompatibilitätsbedingungen bei Unstetigkeiten in der Elektrodynamik*. Math. Ann. **62**, 568-581 (1907); Ibid. **63**, 144 (1907) — *(I) Untersuchungen über die innere Reibung der Gase*. Ann. Phys. **29** [IV], 860-908 (1909); *(II) Versuche bei sehr kleinen und sehr großen Geschwindigkeiten*. Ibid. **38** [IV], 71-125 (1912) — *Über die Theorie der Stoßwellen*. Physik. Z. **13**, 498-501 (1912) — *Untersuchungen über die innere Reibung von Flüssigkeiten (innere Reibung und Gleitung tropfbarer Flüssigkeiten)*. Ann. Phys. **49** [IV], 30-70 (1916).

**SECONDARY LITERATURE.** J.S. HADAMARD: *Remarque au sujet de la note de M. Győző ZEMPLEN.* C. R. Acad. Sci. Paris **141**, 713 (1905) — *ZEMPLEN Győző* [in Hung.]. In: *Révai Nagy Lexikona*. Révai, Budapest (1926), vol. XIX, p. 658 — *ZEMPLÉN, Győző.* In: *Poggendorffs biographisch-literarisches Handwörterbuch*. Verlag Chemie, Leipzig. Bd. V: *1904–1922*, p. 1408 (1926) — I. ABONYI: *ZEMPLÉN Győző* [in Hung.]. In: *Műszaki nagyjaink*. 4. köt./szerk. Pénzes István Bp. (1981). GTE., pp. 308-323 — A. IVÁN: *ZEMPLÉN Győző* [in Hung.]. Fizikai Szemle (Budapest) **16**, No. 10, 289 (1966) — L. KOVÁCS: *ZEMPLÉN Győző élete és munkássága*. Városi Tanács Hiradója, Nagykanizsa (1974) — G. GENDRON: *English translation of three milestone papers* [written by DUHEM (1909), JOUGUET (1901), and ZEMPLÉN (1905)] *on the existence of shock waves*. Rept. VPI-E-89-12, Virginia Polytechnic Institute, Blacksburg, VA (1989) — L. KOVÁCS (ed.): *ZEMPLÉN – the scientist and the teacher*. In: *Studia Physica Savariensia (SPS)*, vol. XI. Berzsenyi College Physics Dept., Szombathely, Hungary (2004); http://sci-ed.org/Conference-2004/Proceedings/anett-zempl.pdf.

**PICTURE.** Courtesy Library & Information Center, TU Budapest, Hungary.

## ZHUKOVSKY [Russ. ЖУКОВСКИЙ], Nikolai Egorovich (1847–1921)

- Russian physicist and mathematician; father of Russian aviation and founder of Russian hydro- and aeromechanics

Nikolai E. ZHUKOVSKY was born in Orekhovo, Vladimir Oblast (western Russia). His father was a railway engineer. After attending the 4th Gymnasium for Men in Moscow (1864), he graduated in physics and mathematics at the Moscow State University [*МГУ*] (1868) and was appointed professor of mechanics at the Moscow Technical School (1872). After earning his Ph.D. (1882) from Moscow State University for a thesis on the stability of motion, he became professor of theoretical mechanics and head of the department of mechanics (1886). He built the first Russian wind tunnel to study various profile forms (1891). At the Moscow Higher Technical School he gave the world's first systematic lectures on the theoretical foundations of aeronautics (1911–1912). During World War I he taught a special course for pilots at Moscow State University and organized an association of aeronautics that was later transformed into the famous Central Aerohydrodynamics Institute [CAGI, Russ. *ЦАГИ*], which soon became the leading Russian school of hydrodynamics and aerodynamics (1918). In the 1920s, he also organized the Aviation Engineering Academy. At the turn of the century ZHUKOVSKY and collaborators investigated the generation and propagation of shock waves in water supply lines. They studied the dreaded hydraulic impact, which they called the "water hammer effect," and possible measures of its prevention.

ZHUKOVSKY was a very versatile researcher: he invented a device for testing airscrews, wrote on the theory of ships, investigated propeller thrust and the most efficient angle of attack of a wing, the resistance of moving bodies in a stationary liquid, and the generation of turbulence, established the kinematic laws of particles in a current, and was the first to apply conformal transformation to convert a circle into an airfoil profile, the so-called "Zhukovsky profile" (1910). Independently of the German mathematician Wilhelm KUTTA, ZHUKOVSKY discovered the condition of smooth flow at the trailing edge of a 2-D wing, the so-called "Kutta-Zhukovsky equation" (1912).

Known already during his lifetime as one of the foremost authorities in the world of aeronautics, ballistics, and hydromechanics, a governmental decree honored him with the title "Father of Russian Aviation" (1920). He published about 200 papers on mechanics, which mostly treated problems of aerodynamics and hydrodynamics, and laid the basis of Soviet aviation and industry. His collected works were published in seven volumes from 1948 to 1950.

He was a corresponding member of the Russian Academy of Sciences, and a member of the Moscow Mathematical Society, serving also as its president (1905–1921). In 1922, the Russian Academy of Science established the *Zhukovsky Prize* for research in the area of aerodynamics of aircraft. Today ZHUKOVSKY is considered the founder of Russian hydromechanics and aeromechanics. In commemoration of his services the School of Aviation, which evolved from *ЦАГИ*, was named *N.E. Zhukovsky Academy of Military and Aeronautical Engineering* (1922). On the occasion of the 100th anniversary of his birth, the city of Stakhanovo, located on the Moskva River about 20 km southeast from Moscow and the home of *ЦАГИ*, was renamed in 1947 "Zhukovsky (City)" [*Жуко́вский*].

A crater on the far side of the Moon is also named for him.

**ORIGINAL WORKS.** *Sur la percussion des corps.* J. Math. Pures Appl. **4** [III], 417-424 (1878) — *On the stability of motion* [in Russ.]. Mem. Moscow Univ. **4**, 1-104 (1882) — *Über den Stoß absolut fester Körper* [in Russ.]. J. Russ. Phys.-Chem. Soc. **16** [Phys.], 388-399 (1884); Ibid. **17**, 47-51 (1885); Abstract in: Fortschr. Phys. **40** [Abth. 1], 271 (1884); Ibid. **41** [Abth. 1], 278-279 (1885) — *Über den Stoß zweier Kugeln, wenn eine von ihnen in einer Flüssigkeit schwimmt* [in Russ.]. Natl. Mem. New Russ. Soc. **5** [Math. Sect.], 43-48 (1884); Abstract in: Fortschr. Phys. **40** [Abth. 1], 323 (1884) — *Lectures on hydrodynamics* [in Russ.]. Mem. Moscow Univ. (Phys. Math.) **7** (1887), 178 pp. — *Über das Geschoss von* CHAPELLE [in Russ.]. Bull. Moscow Soc. Sci. **65**, No. 1, 55-56 (1890); Abstract in Fortschr. Phys. **46** [Abth. 1], 325 (1890) — *Zur Theorie des Fluges* [in Russ.]. J. Russ. Phys.-Chem. Soc. **22** [Phys. Sect.], 1-10, 120-125 (1890); Abstract in: Fortschr. Phys. **46** [Abth. 1], 377-378 (1890) — *Über den Einfluß des Druckes auf den mit Wasser gesättigten Sand* [in Russ.]. Bull. Moscow Soc. Sci. **65**, No. 1, 52-54 (1890); Abstract in: Fortschr. Phys. **46** [Abth. 1], 325 (1890) — *Über den Vogelflug* [in Russ.]. Bull. Moscow Soc. Sci. **73**, No. 2, 29-43 (1891); Abstract in: Fortschr. Phys. **47** [Abth. 1], 283-284 (1891) — *Sur un appareil nouveau pour la détermination des moments de l'inertie des corps.* Bull. Moscou Soc. **5**, 415-416 (1892) — *On the importance of geometrical interpretation in theoretical mechanics* [in Russ.]. Rec. Math. (Moscow) **18**, xxxvii-xlii (1896) — *On aeronautics* [in Russ.]. Bull. Moscow Soc. Sci. **93**, No. 2, 23-29 (1898) — *Analogy between two problems of mechanics* [in Russ.]. Rec. Math. (Moscow) **21**, 542-551 (1901); Abstract in: Jb. Fortschr. Math. **31**, 703 (1900) — *Über den hydraulischen Stoß in Wasserleitungsröhren.* Mém. Acad. Sci. Imp. St. Pétersbourg **9** [VIII], No. 5, 1-72 (1900) — *On the adjunct vortices* [in Russ.]. Trans. Phys. Sec. Imp. Soc. Friends Natl. Sci. Moscow **13**, No. 2, 12-25 (1907) — *Über die Konturen der Tragflächen der Drachenflieger.* Z. Flugtech. u. Motor-Luftschiffahrt **1**, 281-284 (1910); Ibid. **3**, 81-86 (1912) — *De la chute dans l'air de corps légers de forme allongée, animés d'un mouvement rotatoire.* Bull. Inst. Aérodyn. Koutchino (Moscou) **1**, 51-65 (1912) — *Aérodynamique: bases théoriques de l'aéronautique.* Cours professé à l'Ecole Impériale Technique de Moscou. Gauthier-Villars, Paris (1916) — *The theoretical foundations of aeronautics* [in Russ.]. GTI, Moscow (1925) — *Théorie tourbillonnaire de l'hélice propulsive.* Gauthier-Villars, Paris (1929) — (L.S. LEIBENZON ET AL, eds.) *Sobranie sochinenii* ("Collected works"). Gostekhizdat, Moskva (1948–1950). Vol. I: *Obshaia mekhanika, matematika i astronomiia* ("General mechanics, mathematics and astronomy"); vol. II: *Gidrodinamika* ("Lectures on hydrodynamics"); vol. III: *Gidravlika, prikladnaia mekhanika* ("Hydrodynamics, applied mechanics"); vol. IV: *Aerodinamika* ("Aerodynamics"); vol. V: *Teoreticheskaia mekhanika* ("Theoretical mechanics"); vol. VI: *Teoreticheskie osnovy vozdukhoplavaniia* ("Theoretical foundations of aeronautics"); vol. VII: *Rechi i doklady. Kharakteristiki i otzyvy. Raboty, publikuemye vpervye* ("Speeches and lectures. Biography and laudatory. Original works").

**SECONDARY LITERATURE.** O. SIMIN: *Water hammer, with special reference to the researches of Prof. N. JOUKOVSKY.* Reprinted from the Proc. Am. Water Work Assoc. (1904) — V.A. KOSTICIN: *Professor N.E. JOUKOWSKI (1847–1921)* [in Russ.]. Recueil Math. [Matematiceskij sbornik, Moskva] **31**, 5-6 (1922–1924) — C. WIESELSBERGER: *Hydrodynamische Versuche von* JOUKOWSKY. ZAMM **4**, 184-186 (1924) — V.V. GOLUBEV: *Nikolay Egorovich* ZHUKOVSKY. Central Aerohydrodynamic Institute, Moscow (1941) — Y.L. GERONIMUS: *Nikolai Egorovich* ZHUKOVSKI, *ein Pionier der russischen Luftfahrtwissenschaft.* Verlag Technik, Berlin (1954) — Anonymous: *Nikolai E.* JOUKOWSKI. Z. Flugwiss. **2**, 343 (1954) — S.I. STRIZHEVSKII: *Nikolai* ZHUKOVSKII, *founder of aeronautics.* In series: *Men of Russian science.* Foreign Languages, Moscow (1957) — H. ROUSE and S. INCE: *History of hydraulics.* Dover Publ., New York (1957), pp. 213-215 — N.S. ARSHANIKOW and W.N. MALZEW: *Die Rolle* SHUKOWSKIS *und* TSCHAPLYGINS *in der Entwicklung der modernen Aerodynamik.* In: (N.S. ARSHANIKOW and W.N. MALZEW, eds.) *Aerodynamik.* VEB Verlag Technik, Berlin (1959), pp. 24-29 — Anonymous: *N.E.* ZHUKOVSKY. In: (I.V. KUZNETSOVA, ed.) *Lyudi russkoi nauki* ("Russian men of science") [in Russ.]. Nauka, Moscow (1961), vol. I, pp. 169-177 — M.S. ARLAZAAROV: *On N.E.* ZHUKOVSKY. In: *A guide through the museums and expositions of Moscow and environs.* Museum of the History of Sciences, 17 Radiostreet, Moscow (1963) — (N.M. SEMENOVA ET AL., eds.) *N.E.* ZUKOVSKIJ *– bibliografia pecatnyh trudov.* Izd. Otdel Cagi, Moskva (1968) — H.E. SCHULZ (ed.): *Who was who in the U.S.S.R.* Scarecrow, Metuchen, NJ (1972) — A.T. GRIGOR'YAN: *N.E.* ZHUKOVSKY. In: (C.C. GILLESPIE, ed.) *Dictionary of scientific biography.* Scribner, New York, vol. 14 (1976), pp. 619-622 — A.T. GRIGOR'YAN: *Development of the theoretical foundations of aviation in the work of N.E.* ZHUKOVSKY *and S.A.* CHAPLYGIN [in Russ.]. In: (A.T. GRIGOR'YAN, ed.) *Issledovanija po istorii mechaniki* ("Investigations in the history of mechanics"). Nauka, Moscow (1983), vol. 2, pp. 183-192; N.M. MERKULOVA: *N.E.* ZHUKOVSKII *– founder of the scientific school of aerodynamics* [in Russ.]. Ibid., 268-292 — ZHUKOVSKY *(Nikolay Egorovich).* In: *Enciklopediceskij slovar' Brokgauz I Efron: biografi, Rossija.* 2 CR-ROMs, ElectroTECH Multimedia, Moscow (1997).

**PICTURE.** From *The MacTutor History of Mathematics Archive,* School of Mathematics and Statistics at the University of St Andrews, Scotland; http://www-groups.dcs.st-and.ac.uk/~history//PictDisplay/Zhukovsky.html.

**NOTE.** His name has also been transliterated as JOUKOVSKY, JOUKOWSKI, SCHUKOWSKII, SHUKOWSKI, ZHUKOVSKII, and ŽUKOVSKIJ.

# 6

## SUPPLEMENTARY REFERENCES

# 6 SUPPLEMENTARY REFERENCES

## 6.1 GENERAL ENCYCLOPEDIAS

*Bilder-Konversationslexikon.* 4 vols., Brockhaus, Leipzig (1834).
*Brockhaus Enzyklopädie.* 20 vols., Brockhaus, Wiesbaden (1966–1974).
*Brockhaus-Konversationslexikon.* 16 vols., Brockhaus, Leipzig (1892–1897).
*Chambers's encyclopaedia* (ed. by M.D. LAW). 15 vols., International Learning Systems Corporation Ltd., London (1963).
*Columbia encyclopedia.* Columbia University Press, New York (6th edn., 2001–2005); http://www.bartleby.com/65/.
*Collier's encyclopedia.* 24 vols., Macmillan Education Co, New York (1987).
*Der Große Herder.* 13 vols., Herder, Freiburg (1932–1935).
*Encyclopaedia Britannica.* 29 vols. (11th edn., 1911). LoveToKnow™ free online Encyclopedia; http://www.1911encyclopedia.org/.
*Encyclopaedia Britannica.* 24 vols. (1875–1889), 24 vols. (1929), 24 vols. (1959); 30 vols. (1974–1984); 32 vols. (1985–2002)
*Encyclopaedia Metropolitana* Various publishers, London (1822–1858).
*Encyclopaedia Universalis Larousse.* 24 vols., Larousse, Paris (1985).
*Encyclopedia Americana.* 30 vols., Grolier, Danbury, CT (1997).
*Encyclopédie ou dictionnaire raisonné des sciences, des arts et des métiers* (ed. by D. DIDEROT and J. LE ROND D'ALEM-BERT). 35 vols., S. Faulche, Neufchatel (1751–1780).
*Grand Larousse encyclopédique.* 10 vols., Librairie Larousse, Paris (1963).
*Great Soviet encyclopedia* (ed. by A.M. PROKHOROV and J. PARADISE). 31 vols., Macmillan Inc., New York; Collier Macmillan Publ., London (1973–1982).
*Großes vollständiges Universal-Lexikon aller Wissenschaften u. Künste.* 68 vols., Zedler, Halle *etc.* (1732–1754); Akad. Druck- und Verlagsanstalt, Graz (1961–1964).
*Lessico Universale Italiano di lingua, lettere, arti, scienze e tecnica.* 27 vols., Istituto della Enciclopedia Italiana, Roma (1968–1986).
*Meyers Konversations-Lexikon.* 18 vols., Bibliographisches Institut, Leipzig *etc.* (1894–1898).
*Microsoft® Encarta® Online Encyclopedia*, Microsoft Corporation (2005); http://encarta.msn.com/artcenter_/browse.html.
*Révai Nagy Lexikona.* 28 vols., Révai Testvérek irodalmi intézet részvénytársaság, Budapest (1926).
*Wikipedia, the free encyclopedia*; http://en.wikipedia.org/wiki/Main_Page.

## 6.2 SPECIAL ENCYCLOPEDIAS, DICTIONARIES & GLOSSARIES

*A mathematical and philosophical dictionary* (by Ch. HUTTON). 2 vols., Johnson & Robinson, London (1795–1796); reprinted by Olms Verlag, Hildesheim (1973).
*Aerospace science and technology dictionary.* NASA Headquarters, Washington, DC; http://www.hq.nasa.gov/office/hqlibrary/aerospacedictionary/.
*Artillerie und Ballistik in Stichworten* (ed. by H.H. KRITZINGER and F. STUHLMANN). Springer, Berlin (1939).
*Astrophysics glossary.* Interactions.org, Amsterdam, The Netherlands (2006); http://www.interactions.org/cms/?pid=1003020.
*Chambers science and technology dictionary* (ed. by P.M.B. WALKER). Chambers, Cambridge *etc.* (1988).
*Chronologie Chemie 1800–1970* (ed. by S. NEUFELDT). Verlag Chemie, Weinheim/Bergstr. (1977).
*Compressible aerodynamics index.* NASA Glenn Research Center, Cleveland, OH; http://www.grc.nasa.gov/WWW/K-12/airplane/short.html.
*Cosmos: the Swinburne astronomy online encyclopedia.* Swinburne University of Technology, Hawthorn, Australia; http://www.cosmos.swin.edu.au/.
*Cosmology glossary.* Western Washington Univ. Planetarium, Bellingham, WA; http://www.wwu.edu/depts/skywise/a101_cosmologyglossary.html.
*CXC Glossary of astrophysical terms.* Chandra X-ray Center (CXC), operated for NASA by Harvard-Smithsonian Center for Astrophysics, Cambridge, MA; http://chandra.harvard.edu/resources/glossaryA.html.
*Dictionnaire des sciences naturelles* (ed. by F.G. CUVIER). Levrault, Strasbourg (1816–1826).
*Dictionary of medieval Latin from British sources* (ed. by R.E. LATHAM and D.R. HOWLETT). Oxford University Press, London; vol. 1 (1975) to vol. 6 (2003).
*Dictionary of mining, mineral, and related terms* [compiled and edited by the U.S. Bureau of Mines, U.S. Dept. of the Interior]. Am. Geol. Inst., Alexandria, VA (1997); http://www.maden.hacettepe.edu.tr/dmmrt/index.html.
*Dictionary of SDI* (ed. by H. WALDMAN). Scholarly Resources Imprint, Wilmington, DE (1988).
*Dictionary of scientific and technical terms.* McGraw-Hill, New York (1989).
*Dictionary of technical terms for aerospace use* (ed. by D.R. GLOVER JR.). NASA-LRC, Cleveland, OH (May 2003); http://roland.lerc.nasa.gov/~dglover/dictionary/content.html.
*Encyclopaedic history of science and technology* (ed. by G.R. CHATWAL). 12 vols., Anmol Publ., New Delhi (1994).
*Encyclopedia astronautica* [compiled by M. WADE]; http://www.astronautix.com/.
*Encyclopedia of acoustics* (ed. by M.J. CROCKER). 4 vols., J. Wiley, New York *etc.* (1997).
*Encyclopedia of applied physics* (ed. by G.L. TRIGG). 23 vols., VCH Publ., Inc., New York *etc.* (1991–1998).
*Encyclopedia of astronomy* (ed. by C.A. RONAN), Hamlyn, London (1979).
*Encyclopedia of astronomy and astrophysics* (ed. by P. MURDIN). 4 vols., Institute of Physics Publ., London *etc.* (2001).
*Encyclopedia of aviation* (ed. by J. ALEXANDER). Ch. Scribner's Sons, New York (1977).
*Encyclopedia of chemical technology* (ed. by M. HOWE-GRANT). 27 vols., Wiley, New York *etc.* (1991–1998).
*Encyclopedia of computational mechanics* (ed. by E. STEIN, R. DE BORST, and T.J.R. HUGHES). 3 vols., Wiley, Chichester *etc.* (2004).
*Encyclopedia of condensed matter physics* (ed. by G.F. BASSANI, G. LIEDL, and P. WYDER). Elsevier/Academic Press, Oxford (2005).
*Encyclopedia of cosmology. Historical, philosophical and scientific foundations of modern cosmology* (ed. by N.S. HETHERINGTON). Garland Publ., New York (1993).
*Encyclopedia of earthquakes and volcanoes* (ed. by D. RITCHIE and A.E. GATES). Checkmark Books, New York (2001).
*Encyclopedia of Earth system science* (ed. by W.A. NIERENBERG). 4 vols., Academic Press, San Diego, CA (1992).
*Encyclopedia of explosives and related items* (ed. by B.T. FEDOROFF and S.M. KAYE). 10 vols., Picatinny Arsenal, Large Caliber Weapons Systems Laboratory, Dover Publ., NJ (1960–1983).
*Encyclopedia of fluid mechanics* (ed. by N.P. CHEREMISINOFF). 10 vols., Gulf Publ., Book Div., Houston, TX (1986–1990).
*Encyclopedia of human evolution and prehistory* (ed. by I. TATTERSALL ET AL.). Garland Publ., New York & London (1988).
*Encyclopedia of natural disasters* (ed. by L. DAVIS). Headline, London (1993).
*Encyclopedia of nonlinear science* (ed. by A. SCOTT). Routledge, New York (2005).
*Encyclopedia of physical science and technology* (ed. by R.A. MEYERS). 15 vols., Academic Press, Orlando *etc.* (1992); 18 vols., Academic Press, San Diego, CA *etc.* (2002).
*Encyclopedia of physics* (ed. by R.G. LERNER and G.L. TRIGG). VCH Publ., New York *etc.* (1991).

*Encyclopedia of planetary sciences* (ed. by J.H. SHIRLEY and R.W. FAIRBRIDGE). Chapman & Hall, London etc. (1997).

*Encyclopedia of space science and technology* (ed. by H. MARK). Wiley-Interscience, Hoboken, NJ (2003).

*Encyclopedia of strange and unexplained physical phenomena* (ed. by J. CLARK). Gale Research Inc., Detroit, MI (1993).

*Encyclopedia of the Solar System* (ed. by P.R. WEISSMAN, L.A. MCFADDEN, and T.V. JOHNSON). Academic Press, San Diego, (1999).

*Encyclopedia of the Universe* (ed. by A. ROY). University Press, Oxford (1992).

*Encyclopedia of volcanoes* (ed. by H. SIGURDSSON). Academic Press, San Diego, CA (2000).

*Encyclopedia of world mythology* (ed. by R. WARNER). Peerage Books, London (1975).

*Encyclopedic dictionary of physics* (ed. by J. THEWLIS). Pergamon Press, Oxford etc. (1962).

*Enzyklopädie der mathematischen Wissenschaften* (ed. by M. DEURING). Teubner, Stuttgart (1953–1972).

*Enzyklopädie der Union der sozialistischen Sowjetrepubliken.* (ed. by S.I. WAWILOW ET AL.). 2 vols., Verlag Kultur und Fortschritt, Berlin (1950).

*Fact Guru Astronomy knowledge base* [glossary]. University of Ottawa, Canada; http://www.site.uottawa.ca:4321/astronomy/index.html.

*General terminology.* Int. Hydro Cut Technologies Corp., North Vancouver, B.C., Canada; http://www.hydrocut.com/Terminology.html.

*Glossary of Eric WEISSTEIN's World of Physics.* Wolfram Research; http://scienceworld.wolfram.com/physics/letters/.

*Glossary of astronomy and astrophysics* (J. HOPKINS, ed.). University of Chicago Press, Chicago (1980).

*Glossary of astrophysics terms.* Lawrence Berkeley National Laboratory, Berkeley, CA (1999); http://ie.lbl.gov/education/glossary/glossaryfa.htm.

*Glossary of coined names & terms used in science* (by J. ANDRAOS). Dept. of Chemistry, York University, Toronto, Ontario, Canada; http://careerchem.com/NAMED/Glossary-Coined-Terms.pdf.

*Glossary of earthquake terms.* USGS Earthquake Hazards Program (May 2002); http://earthquake.usgs.gov/image_glossary/.

*Glossary of explosives and propellants.* In: S. FORDHAM: *High explosives and propellants.* Pergamon Press, Oxford etc. (1980); http://www.chimicando.it/e-book/%5Bebook%5DHigh%20Explosives%-20And%20Propellants.pdf.

*Glossary of geology* (ed. by R.L. BATES). Am. Geol. Inst., Falls Church, VA (1987); (ed. by J.A. JACKSON). Am. Geol. Inst., Alexandria, VA (1997).

*Glossary of meteorites and asteroids* (R. KALLENBACH). Space Science Rev. 92, 415-419 (2000).

*Glossary of meteorology.* American Meteorological Society (AMS), Boston, MA; http://amsglossary.allenpress.com/glossary/browse?s=A&p=1.

*Glossary of oceanography and the related geosciences with references* (compiled by S.K. BAUM). Texas Center for Climate Studies, Texas A&M University, College Station, TX (May 2004); http://stommel.tamu.edu/~baum/paleo/paleogloss/paleogloss.html.

*Glossary of seismological terms.* National resources Canada, Earthquakes Canada; http://earthquakescanada.nrcan.gc.ca/gen_info/glossa_e.php.

*Glossary of solar systems dynamics.* NASA-JPL, CalTech, Pasadena, CA (Jan. 2007); http://ssd.jpl.nasa.gov/?glossary&term=H.

*Glossary of terms associated with explosive devices.* The Metalith, a division of Infrastructure Defense Technologies. Belvidere, IL; http://www.themetalith.com/glossary.html#numeric.

*Glossary of terrestrial impact craters.* Lunar and Planetary Institute, Houston, TX (2007); http://www.lpi.usra.edu/publications/slidesets/craters/glossary.shtml.

*Glossary of the construction, decoration and use of arms and armor...* (by G.C. STONE). Southworth Press, Portland, ME (1934); reprinted by J. Brussel, New York (1961).

*Glossary of volcanic and geologic terms.* Volcanoworld. Dept. of Geosciences, Oregon State University, Corvallis, OR; http://volcano.und.edu/vwdocs/glossary.html.

*Glossary on explosion dynamics* (J. SHEPHERD ET AL.). Explosion Dynamics Laboratory, GALCIT, Pasadena, CA; http://www.galcit.caltech.edu/EDL/projects/JetA/Glossary.html.

*Handwörterbuch der Naturwissenschaften* (ed. by R. DITTLER). 11 vols., Fischer, Jena (1931–1935).

*High-energy astrophysics dictionary.* SkyView NASA-GSFC (June 2005); http://skyview.gsfc.nasa.gov/help/dictionary.html.

*Hypersonic Index.* NASA-GRC, Cleveland, OH; http://www.grc.nasa.gov/WWW/BGH/shorth.html.

*Image the Universe! Dictionary.* NASA-GSFC; http://imagine.gsfc.nasa.gov/docs/dictionary.html.

*International dictionary of geophysics: seismology, geomagnetism, aeronomy, oceanography, geodesy, gravity, marine geophysics, meteorology, the Earth as a planet and its evolution* (S.K. RUNCORN, ed.). 7 vols., Pergamon Press, Oxford (1967).

*Lexikon der Astronomie* (ed. by H. ELSÄSSER). 2 vols., Herder, Freiburg etc. (1989).

*Lexikon der gesamten Technik und ihrer Hilfswissenschaften* (ed. by O. LUEGER). 10 vols., Dt. Verlagsanstalt, Stuttgart (1905–1920).

*Lightning: physics and effects. Encyclopedia of lightning* (by V.A. RAKOV and M.A. UMAN), Cambridge University Press, Cambridge (2003).

*Mathematical and philosophical dictionary* (ed. by C. HUTTON). 2 vols., Johnson, London (1795–1796); Reprinted by Olms, Hildesheim & New York (1973).

*McGraw-Hill concise encyclopedia of science and technology* (ed. by S.P. PARKER). McGraw-Hill, New York (1989).

*McGraw-Hill dictionary of scientific and technical terms* (S.P. PARKER, ed.). McGraw-Hill, New York (1984, 1989, 1994).

*McGraw-Hill encyclopedia of the geological sciences* (ed. by D.N. LAPEDES). McGraw-Hill, New York etc. (1977).

*NASA's Imagine the Universe! Dictionary.* NASA-GSFC; http://imagine.gsfc.nasa.gov/docs/dictionary.html.

*National Geophysical Data Center (NGDC), Boulder, CO. Hazards.* (I): *General earthquake topics*; (II): *Earthquake events*; (III): *Landslides, tsunamis and volcanoes*; http://www.ngdc.noaa.gov/seg/mainmeta.shtml.

*New Larousse encyclopedia of mythology* (ed. by R. GRAVES). Hamlyn, London (1959).

*Österreich-Lexikon*, Projekt AEIOU des BM:BWK; http://www.aeiou.at/.

*Oxford illustrated encyclopedia.* Oxford University Press, Oxford, vol. 8 (1992): *The Universe* (ed. by A. ROY).

*Perspective of aeronautical & astronautical research* [with index]; http://pw1.netcom.com/~indexer/frorders.htm.

*Photo glossary of volcano terms.* USGS Hazards Program; http://volcanoes.usgs.gov/Products/Pglossary/pglossary.html.

*Porträts der UdSSR-Prominenz* (ed. by H.E. SCHULZ). Institut zur Erforschung der UdSSR e.V., München (1960).

*Propyläen Technikgeschichte* (ed. by W. KÖNIG). 5 vols., Ullstein, Berlin (1990–1992).

*Reallexikon der Vorgeschichte* (ed. by M. EBERT). 15 vols., De Gruyter, Berlin (1924–1932).

*Terrestrial impact craters, glossary.* Lunar and Planetary Institute, Houston, TX; http://www.lpi.usra.edu/publications/slidesets/craters/glossary.shtml.

*The astronomy and astrophysics encyclopedia* (ed. by S.P. MARAN). Van Nostrand & Reinhold, New York (1992).

*The Cambridge encyclopedia of Earth sciences* (ed. by D.G. SMITH). Cambridge University Press, Cambridge (1981).

*The Cambridge encyclopedia of human evolution* (ed. by S. JONES, R. MARTIN, and D. PILBEAM). Cambridge University Press, Cambridge (1992).

*The Cambridge encyclopedia of stars* (ed. by J. KALER). Cambridge University Press, Cambridge (2001).

*The Cambridge encyclopedia of the Sun* (ed. by K.R. LANG). Cambridge University Press, Cambridge (2001).

*The encyclopedia of the solid Earth sciences* (ed. by P. KEAREY). Blackwell Scientific Publ., Oxford *etc.* (1993).

*The illustrated encyclopedia of artillery* (ed. by I.V. HOGG). Stanley Paul, London (1987).

*The international encyclopedia of physical chemistry and chemical physics.* Topic 19: *Gas kinetics* (ed. by E.A. GUGGENHEIM). Pergamon Press, Oxford (1964–1966).

*The Internet encyclopedia of science*; The worlds of David DARLING; http://www.daviddarling.info/encyclopedia/ETEmain.html.

*The Oxford companion to the Earth* (ed. by P.L. HANCOCK and B.J. SKINNER). Oxford University Press, Oxford (2000).

*The Oxford English dictionary* (ed. by J.A. SIMPSON and E.S.C. WEINER). Clarendon Press, Oxford (1989).

*The Oxford Latin dictionary* (ed. by P.G.W. GLARE). Clarendon Press, Oxford (1968).

*Tsunami glossary.* International Tsunami Information Center (ITIC), Honolulu, HI; http://ioc3.unesco.org/itic/files/tsunami_glossary.pdf.

*Tsunami glossary.* Pacific Tsunami Museum, Hilo, HI; http://www.tsunami.org/definitions.htm.

*Tsunami glossary*. UNESCO-IOC Information document No. 1221, Paris (2006); http://ioc3.unesco.org/itic/files/tsunami_glossary_en_small.pdf.

*Ullmann's encyclopedia of industrial chemistry* (ed. by L. KAUDY, J.F. ROUNSAVILLE, and G. SCHULZ). VCH, Weinheim/Bergstr. (1987).

*Van Nostrand's scientific encyclopaedia.* (D.M. CONSIDINE, ed) Van Nostrand & Reinhold, New York (1989).

## 6.3 HANDBOOKS

*A ballistic handbook* (by G. KOLBE). Pisces Press, Newcastleton, Scotland (20000).

*A destruction handbook: small arms, light weapons, ammunitions and explosives*. Dept. or Disarmament Affairs, United Nations, New York (2001).

*Aerospace ordnance handbook* (ed. by F.B. POLLAD and J.A. ARNOLD). Prentice-Hall, Englewood Cliffs, NJ (1966).

*American Institute of Physics handbook* (ed. by D.E. GRAY). McGraw-Hill Book Co., New York *etc.* (1972).

*Blasters' handbook*. E.I. du Pont de Nemours & Co., Wilmington, DE (1989).

*Feuerwaffen. Ein waffenhistorisches Handbuch* (by A. HOFF). Klinkhardt & Biermann, Braunschweig (1969).

*Gas explosion handbook* (ed. by D. BJERKETVEDT ET AL.). Rept. CMR-93-A25034, Christian Michelsen Research AS, Bergen, Norway (1993); http://www.gexcon.com/index.php?src=/handbook/GEXHBcontents.htm.

*Handbook of generalized gas dynamics* (ed. by R.P. BENEDICT). Plenum Press, New York (1966).

*Handbook of physics* (ed. by E.U. CONDON and H. ODISHAW). McGraw-Hill Book Co., New York *etc.* (1967).

*Handbook of problems in exterior ballistics* (ed. by J.M. INGALLS). Artillery School Press, Virginia (1890).

*Handbook of shock waves* (ed. by G. BEN-DOR ET AL.). 3 vols., Academic Press, New York (2001).

*Handbook of supersonic aerodynamics*, NavOrd Rept. No. 1488, publ. by the Bureau of Ordnance and Hydrography, Washington, DC. U.S. Govt. Printing Office, Washington, DC (1950–1959).

*Handbook on modern explosives, being a practical treatise on the manufacture and application of dynamite, gun-cotton, nitro-glycerine, and other explosive compounds, including the manufacture of collodion-cotton* (ed. by M. EISSLER). C. Lockwood, London (1890).

*Handbuch der Experimentalphysik*. Akad. Verlag, Leipzig Vol. IV (1931): *Hydro- und Aerodynamik.*

*Handbuch der Ozeanographie* (ed. by G.G. VON BOGUSLAWSKI and O. KRÜMMEL). 2 vols., Engelhorn, Stuttgart (1884, 1887).

*Handbuch der Physik* (ed. by A. WINKELMANN). J.A. Barth, Leipzig. Vol. I (1908): *Allgemeine Physik*; vol. II (1909): *Akustik*; vol. VIII (1927): *Akustik*; vol. X (1926): *Thermische Eigenschaften der Stoffe.*

*Handbuch der Raumexplosionen* (ed. by H.H. FREYTAG). Verlag Chemie GmbH, Weinheim/Bergstr. (1965).

*Handbuch der Seenkunde: allgemeine Limnologie* (ed. by F.A. FOREL). Engelhorn, Stuttgart (1901).

*Handbuch der Sprengarbeit* (ed. by O. GUTTMANN). Vieweg, Braunschweig (1892).

*Handbuch der theoretischen Physik* (ed. by H. HELMHOLTZ and G. WERTHEIM). 2 vols., Vieweg, Braunschweig (1871, 1874).

*Handbuch zur Geschichte der Naturwissenschaften und der Technik* (ed. by L. DARMSTAEDTERS). Springer, Berlin (1908).

*LLNL explosives handbook: properties of chemical explosives and explosive simulants* (ed. by B.M. DOBRATZ). NTIS, Springfield, VA (1981).

*Shock and vibration handbook* (ed. by C.M. HARRIS). McGraw-Hill, New York *etc.* (1961).

*SME mining engineering handbook*. Society for Mining, Metallurgy, and Exploration, Inc., Littleton, CA (1992).

*Traces of catastrophe. A handbook of shock-metamorphic effects in terrestrial meteorite impact structures* (ed. by B.M. FRENCH). LPI Contribution No. 954, Lunar and Planetary Institute (LPI), Houston, TX (1998); http://www.lpi.usra.edu/publications/books/CB-954/CB-954.intro.html.

## 6.4 BIOGRAPHICAL CATALOGUES, DICTIONARIES, ENCYCLOPEDIAS & PERIODICALS

*A biographical dictionary of the Soviet Union 1917–1988* (ed. by J. VRONSKAYA and V. CHUGUEV). K.G. Saur, London *etc.* (1989, 1992).

*A chronology of American history of science 1450–1900* (ed. by R.M. GASCOIGNE). Garland Publ., New York (1987).

*A dictionary of science, comprising astronomy, chemistry, dynamics, electricity, heat, hydrodynamics, hydrostatics, light, magnetism, mechanics, meteorology, pneumatics, sound, and statics; preceded by an essay on the history of the physical sciences* (ed. by G.F. RODWELL). E. Moxon, London (1871).

*American biographical index* (ed. by L. BAILLIE). Microfiches, K.G. Saur, Munich (1998).

*American men & women of science* (ed. by S.L. TORPHIE). Bowker, New Providence, NJ (1992–1993).

*Biografisch archief van de Benelux* (ed. by W. GORZNY). Microfiches, K.G. Saur, Munich (1992–1994).

*Biographical database of the British chemical community, 1880–1970.* The Open University; http://www5.open.ac.uk/Arts/chemists/index.htm.

*Biographical dictionary of American science: the 17th through the 19th centuries* (ed. by C.A. ELLIOTT). Greenwood Press, Westport, CT & London (1979).

*Biographical memoirs*. National Academy of Sciences of the U.S.A., Columbia University Press, New York; vol. (1877) – vol. 87 (2005).

*Biographic dictionary of the U.S.S.R.* (ed. by W.S. MERZALOW). Scarecrow Press, New York (1958).

*Biographical memoirs of Fellows of the Royal Society*. 26 vols., Royal Society of London (1955–1980).

*Biographie-Links: Biographien und Homepages von bekannten Persönlichkeiten*; http://www.polarluft.de/suche_im_web/sp/a/personen_a.html.

*Biographien bedeutender Physiker: eine Sammlung von Biographien* (ed. by W. SCHREIER). Volk und Wissen, Berlin (1984).

*Biographie nationale.* Publiée par l'Académie royale des sciences, des lettres et des beaux arts de Belgique. 44 vols., Thiry-van Buggenhoudt, Bruylant-Christophe & Emile Bruylant, Bruxelles (1866–1986).

*Biographies de grands chimistes*;
http://isimabomba.free.fr/biographies/liste_biographie.htm.

*Biographies of astronomers. List of obituary notes of astronomers*;
http://www.astro.uni-bonn.de/~pbrosche/hist_astr/ha_pers.htm.

*Biographies of British civil engineers, architects, etc.* Steamindex;
http://www.steamindex.com/people/civils.htm.

*Biography and genealogy master index: a consolidated index to ... biographical sketches in current and retrospective biographical dictionaries* [Periodical]. Gale Research Co., Detroit, MI (1980–).

*Biography.com.* The Web's Best Bios, A&E Television Networks;
http://www.biography.com/index.html.

*British biographical index* (ed. by D. BANK), Microfiches, K.G. Saur, Munich (1990).

*Cesky biograficky archiv a slovensky biograficky archiv (CSBA)* (ed. by U. KRAMME). Microfiches, K.G. Saur, Munich (1993).

*Chambers encyclopaedia of scientists* (ed. by J. DAINTITH, S. MITCHELL, and E. TOOTILL). Chambers, Edinburgh (1983).

*Chemistry, biographies articles*;
http://reference.allrefer.com/encyclopedia/categories/chembio.html.

*Appleton's Cyclopedia of American Biography* (ed. by J.G. WILSON and J. FISKE). 6 vols., D. Appleton & Co., New York (1887–1889).

*Deutsche Biographische Enzyklopädie*. K.G. Saur, Munich (1995).

*Deutscher Biographischer Index* (ed. by A. FREY). Microfiches, K.G. Saur, Munich (1998).

*Dictionnaire de biographie française* (ed. by M. PREVOST and J. BALTEAU). 14 vols., Letouzey & Ané, Paris (1929–1979).

*Dictionary of national biography* (ed. by L. STEPHEN and S. LEE). Smith, Elder & Co, London; vol. 1 (1885) – vol. 63 (1900).

*Dictionary of scientific biography* (ed. by C.C. GILLESPIE). 14 vols., Ch. Scribner's Sons, New York (1970–1980).

*Die berühmten Erfinder, Physiker und Ingenieure* (ed. by L. LEPRINCE-RINGUET). Aulis-Verlag Deubner & Co, Köln (1963).

*Die Biographie in der Physikgeschichte.* In: (K. VON MEŸENN, ed.) *Die großen Physiker.* C.H. Beck'sche Verlagsbuchhandl., München (1997), vol. 1, pp. 7-25.

*Die Porträtsammlung der Herzog-August-Bibliothek Wolfenbüttel* (bearbeitet von P. MORTZFELD). K.G. Saur, München etc. (1993).

*Eric's treasure trove of scientific biography*;
http://www.astro.virginia.edu/~eww6n/bios/.

*Hall of heat transfer pioneers. Journal of Heat Transfer, Dept. of Mech. Engng., University of Texas*;
http://www.me.utexas.edu/~me339/history.html.

*History of astronomy: persons.* University of Bonn, Germany;
http://www.astro.uni-bonn.de/~pbrosche/hist_astr/ha_pers.html.

*Index biographique français* (ed. by H. DWYER and B. DWYER). Microfiches, K.G. Saur, Munich (1993).

*Indice biografico italiano* (ed. by T. NAPPO). K.G. Saur, Munich (1987–1992).

*Internationaler biographischer Index der Naturwissenschaften* (ed. by B. KOPP). K.G. Saur, Munich (1998).

*Internationaler biographischer Index der Technik.* K.G. Saur, Munich (1998).

*Kürschners deutscher Gelehrtenkalender* (ed. by H. STROBEL ET AL.). De Gruyter, Berlin (1925–present).

*Lexikon der Naturwissenschaftler* (ed. by K.G. COLLATZ and D. FREUDIG). Spektrum Akad. Verlag, Heidelberg etc. (1996).

*Lyudi russkoi nauki ("Russian men of science")*, ed. by I.V. KUZNETSOVA. Nauka, Moskva. Vol. 1 (1961): *Matematika, mekhanika, astronomiya, fizika, khimiya*; vol. 2 (1962): *Teologiya, geografia*; vol. 3 (1963): *Biologiya, meditsina, sel'skokhozyaistvennye nauki*; vol. 4 (1965): *Tekhnika*.

*McGraw-Hill modern scientists and engineers* (ed. by S.P. PARKER). 3 vols., McGraw-Hill, New York (1980).

*MacTutor history of mathematics archive*;
http://www-history.mcs.st-andrews.ac.uk/history/.

*Magyar életrajzi archivum* (ed. by U. KRAMME). Microfiches, K.G. Saur, Munich (1995).

*Matematika v SSSR za sorok let.* Gos. Tom vtoroj: Biobibliografia Izd. Fiz.-mat. Lit., Moskva (1959).

*Mechanical engineering biographies index.* History and Heritage Center, ASME, New York, NY; http://www.asme.org/history/biog_dex.html#J.

*Modern English biography* (ed. by F. BOASE). 6 vols., Cass, London (1892–1921).

*NAHSTE (Navigational Aids for the History of Science, Technology & the Environment): People Index*; http://www.nahste.ac.uk/pers/a/.

*Neue Deutsche Biographie* (Hist. Komm. Bay. Akad. Wiss., Hrsg.). 16 vols., Duncker & Humblot, Berlin (1953–1997).

*Nobel e-Museum*, the Official Web Site of The Nobel Foundation, Stockholm; http://www.nobel.se/index.html.

*Obituary notes of astronomers (1900–1997)* (by H.W. DUERBECK and B. OTT). Argelander-Institut für Astronomie der Universität Bonn (May 2000); http://www.astro.uni-bonn.de/~pbrosche/persons/obit/index.html.

*Obituary notices of Fellows of the Royal Society.* 9 vols., Royal Society of London (1932–1954).

*Österreichisches Biographisches Lexikon 1815–1950* (ed. by L. SANTIFALLER). Verlag der Österreichischen Akademie der Wissenschaften, Wien (1969); http://hw.oeaw.ac.at/oebl?frames=yes.

*Oxford dictionary of national biography: from the earliest times to the year 2000* (ed. by H.C.G. MATTHEW and B. HARRISON). 61 vols., Oxford University Press, Oxford (2004).

*Physics, Biographies articles*;
http://reference.allrefer.com/encyclopedia/categories/physicsbio.html.

*Portraits collection of U.S. geologists.* USGS Photographic Library; http://libraryphoto.cr.usgs.gov/ports.htm.

*Prominent personalities in the U.S.S.R.* (ed. by E.L. CROWLEY). Scarecrow Press, Metuchen, NJ (1968).

*Poggendorff biographisch-literarisches Handwörterbuch.* Verlag Chemie GmbH, Leipzig etc. (1926–present).

*Scientists & inventors* (by A. FELDMANN and P. FORD). Aldus Books, London (1979).

*Smith Collection: portraits of scientists and philosophers.* University of Pennsylvania Library, Center for Electronic Text and Image; http://www.library.upenn.edu/etext/smith/c/.

*Soviet men of science. Academicians and corresponding members of the Academy of Sciences of the U.S.S.R.* (ed. by J. TURKEVICH). Van Nostrand, Princeton, NJ etc. (1963).

*Sowjetische Kurzbiographien* (ed. by B. LEWYTZKYJ and K. MÜLLER). Verlag für Literatur- und Zeitgeschichte, Hannover (1964).

*Svenskt biografiskt lexikon* (ed. by B. BOËTHIUS). 21 vols., Bonnier, Stockholm (1918–1977).

*The biographical encyclopedia of astronomers* (ed. by T. HOCKEY). Springer, New York (2006).

*The Cambridge biographical encyclopedia* (ed. by D. CRYSTAL). Cambridge University Press, Cambridge (1998).

*The Cambridge dictionary of American biography* (ed. by J.S. BOWMAN). Cambridge University Press, Cambridge (1995).

*The concise dictionary of national biography. From earliest time to 1985.* Oxford University Press, Oxford etc. (1992).

*The dictionary of national biography: from the beginnings to 1911* (ed. by S. LEE). 63 vols., Oxford University Press, London (1917).

*The general biographical dictionary: containing an historical and critical account of the lives and writings of the most eminent persons in every nation* (ed. by A. CHALMERS). 32 vols., Nichols & Son et al., London (1812–1817).

*The Hutchinson dictionary of scientific biography* (ed. by R. PORTER and M. OGILVIE). Helicon, Oxford. 6 vols. (1983–1985); 1 vol. (1994); 2 vols. (2000).

*The Mathematics Genealogy Project* [Database of international Ph.D. theses]. Dept. of Mathematics, North Dakota State University, Fargo, ND; http://genealogy.math.uni-bielefeld.de/html/search.phtml.

*The Royal Society, London:* for biographies, memoirs and obituaries of past fellows see http://www.royalsoc.ac.uk/page.asp?id=1728; for lists of Royal Society Fellows 1660–2004 see http://www.royalsoc.ac.uk/page.asp?id=1727.

*U.S. Air Force military biographies* [biographical information on Air Force leaders]; http://www.af.mil/bios/alpha.asp?alpha=Y.

*Virtual American biographies.* NASA Center for Educational Technology; http://famousamericans.net/famousamericans-aar-ada/.

*Who is Who in America.* Marquis-Who's Who Inc., New Providence, NJ (1998).

*Who is who of British scientists 1971–1972.* Longmans, London (1972).

*Who's who in central and east Europe* (ed. by S. TAYLOR). Central European Times Publ. Co., Zurich (1937).

*Who's who in Nazi Germany* (ed. by R.S. WISTRICH). Weidenfeld & Nicolson, London (1982).

*Who's who in the Soviet Union* (ed. by B. LEWYTZKYJ). K.G. Saur, Munich (1984).

*Who was who in the U.S.S.R.* (ed. by H.E. SCHULZ). Scarecrow Press, Metuchen, NJ (1972).

*World Who's Who in science.* Marquis-Who's Who Inc., Chicago, IL (1968).

*5000 Sovietköpfe* (ed. by H. KOCH). Dt. Industrie-Verlag, Köln (1959).

## 6.5 BIBLIOGRAPHICAL CATALOGUES

*Poggendorffs biographisch-literarisches Handwörterbuch.* Verlag Chemie GmbH, Leipzig *etc.* (1926); vols. 1-2: Pre–1863; vol. 3: 1863–1898; vol. 4: 1898–1904; vol. 5: 1904–1922; vol. 6: 1923–1931; vol. 7a: 1932–1953; vol. 7b: 1932–1962.

*Catalogue of scientific papers.* Vols. I-VI: *1800–1863*; vols. VII-VIII: *1864–1873*; vols. IX-XI: *1874–1883*; vol. XII: *1800–1883*, vols. XIII-XIX: *1884–1900*. Compiled and published by the Royal Society of London, Murray, London (1800–1900). *See* the catalogue *Gallica, la bibliothèque numérique.* Bibliothèque nationale de France; http://gallica.bnf.fr/.

*The National Union Catalogue of manuscript collections.* Publ. by the Library of Congress, Washington, DC: *Period Pre–1956,* 754 vols., Mansell, London (1980); *Period 1958–1962,* 50 vols. Rowman & Littlefield, New York (1963); *Period 1963–1967,* 59 vols., Edwards Publ., Ann Arbor, MI (1969); *Period 1968–1972,* 104 vols., Edwards Publ., Ann Arbor, MI (1972); *Period 1973–1977,* 135 vols., Rowman & Littlefield, Totowa, NY (1978). Catalogues covering the periods 1978–1982 and 1983–2002 are only available as microfiches.

*General catalogue of printed books.* The Trustees of the British Museum, London (1964).

## 6.6 BOOKS & REVIEW ARTICLES WITH HISTORICAL PERSPECTIVE

### ACOUSTICS & NONLINEAR ACOUSTICS

R.T. BEYER: *Nonlinear acoustics.* Acoustical Society of America. Woodbury, N.Y. (1997).

R.G. BUSNEL (ed.): *Acoustic behavior of animals.* Elsevier, Amsterdam *etc.* (1963).

J.M.A. LENIHAN: MERSENNE *and* GASSENDI. *An early chapter in the history of sound.* Acustica 1, 96-99 (1951).

E. MACH: *Zur Geschichte der Akustik.* Mittheil. Dt. Gesell. zu Prag, Prag, Wien & Leipzig (1892).

N. ROTT: *Nichtlineare Akustik – Rückblick und Ausblick.* Z. Flugwiss. & Weltraumforsch. **4**, 185-193 (1980).

### AERODYNAMICS, AERONAUTICS & ASTRONAUTICS

*Aerospaceweb: Historical background of hypersonic waveriders*; http://www.aerospaceweb.org/design/waverider/history.shtml.

J.D. ANDERSON JR.: *Introduction to flight. Its engineering and history.* McGraw-Hill, New York (1978).

J.D. ANDERSON JR.: *A history of aerodynamics and its impact on flying machines.* Cambridge University Press, Cambridge (1997).

J.D. ANDERSON JR.: *History of high-speed flight and its technical development.* AIAA J. **39**, 761-771 (2001).

J.D. ANDERSON JR.: *Research in supersonic flight and the breaking of the sound barrier.* In: (P.E. MACK, ed.) *From engineering science to big science.* The NASA History Series, NASA SP-4219 (2001), chap. 3; http://history.nasa.gov/SP-4219/Cover4219.htm.

N.S. ARSHANIKOW and W.N. MALZEW: *Aerodynamik.* VEB Verlag Technik, Berlin (1959).

D. BAKER: *The history of manned space flight.* Crown Publ., New York (1982).

T. BENECKE and A.W. QUICK: *History of German guided missiles development.* Appelhans, Brunswick (1957).

A. BETZ: *Die Entwicklung der Fluggeschwindigkeit.* Die Naturwissenschaften **41**, 101-107 (1954).

W. DORNBERGER: *Peenemünde: die Geschichte der V-Waffen.* Bechtle, Esslingen (1981).

H. DRYDEN: *Supersonic travel within the last two hundred years.* Scient. Month. **78**, 289-295 (May 1954).

W.F. DURAND: *Historical sketch of the development of aerodynamic theory.* In: *Int. Civil Aeronaut. Conference* [Washington, DC, Dec. 1928].

W.F. DURAND (ed.): *Aerodynamic theory.* Dover Publ., New York (1963).

E.E. EMME: *Aeronautics and astronautics: an American chronology of science and technology in the exploration of space, 1915–1960.* NASA, Washington, DC (1961). See *NASA history timelines – NASA chronology of aeronautics and astronautics*; http://www.hq.nasa.gov/office/pao/History/timeline.html.

*ESA Space Science: History of space science*; http://www.esa.int/esaSC/SEMXSX57ESD_index_0.html

O. FLACHSBART: *Geschichte der experimentellen Hydro- und Aeromechanik, insbesondere der Widerstandsforschung.* In: (L. SCHILLER, ed.) *Handbuch der Experimentalphysik.* Akad. Verlagsgesell., Leipzig; Bd. 4 (1932): *Hydro- und Aerodynamik,* 2. Teil: *Widerstand und Auftrieb,* pp. 1-61.

R. GIACOMELLI and E. PISTOLESI: *Historical sketch* [of the development of aerodynamics]. In: (W.F. DURAND, ed.) *Aerodynamic theory.* Dover Publ., New York (1934), vol. 1, pp. 305-394.

K.L. GOIN: *The history, evolution and use of wind tunnels.* AIAA Student J. **9**, 3-13 (Feb. 1971).

M.H. GORN: *The universal man: Theodore* VON KÁRMÁN's *life in aeronautics.* Smithsonian Institution Press, Washington, DC (1992).

J.J. GREEN ET AL.: *Die deutsche Luftfahrtforschung im zweiten Weltkrieg.* Bericht einer kanadischen Kommission aus dem Jahre 1945. Historisches Archiv der DFVLR, Göttingen (1983).

W.F. HILTON: *British aeronautical research facilities* [covering the period 1871–1950s]. J. Roy. Aeronaut. Soc. **70**, 103-107 (Jan. 1966).

D.R. JENKINS: *Hypersonics before the shuttle: a concise history of the X-15 research airplane.* NASA Rept. SP-2000-4518 (2000).

T. VON KÁRMÁN: *Aerodynamics. Selected topics in the light of their historical development.* Cornell University Press, Ithaca, NJ (1954).

H. KURZWEG: *The aerodynamic development of the V-2.* In: (T. BENECKE and A.W. QUICK, eds.) *History of German guided missiles development.* AGARD 1st Guided Missiles Seminar [Munich, April 1956]. Appelhaus, Brunswick (1957), pp. 50-69.

*Land speed record history.* Bluebird Electric Racing Ltd., Oakfields, U.K.; http://www.bluebird-electric.net/land_speed_record_history.htm.

F.K. LU and D.E. MARREN (eds.): *Advanced hypersonic test facilities* [covers mostly American hypersonic shock tunnels; long-duration hypersonic facilities; ballistic ranges, sleds, and tracks; and advanced technologies for next generation facilities]. Progress in Astronautics and Aeronautics Series, AIAA (2002).

J. LUKASIEWICZ: *Experimental methods of hypersonics.* Dekker, New York (1973); chap. 3: *Historical perspective* [on hypersonic wind tunnel design], pp. 25-51.

F.A. MAGOUN and E. HODGINS: *A history of aircraft.* Arno Press, New York (1972).

W.A. MAIR (ed.): *Research on high-speed aerodynamics at the Royal Aircraft Establishment from 1942 to 1945.* H.M.S.O., London (1950).

F.J. MALINA, R.C. TRUAX, and A.D. BAXTER: *Historical development of jet propulsion.* In: (O.E. LANCASTER, ed.) *Jet propulsion engines.* Princeton University Press, Princeton (1959).

NASA History Office: *Rocketry through the ages: a timeline of rocket history.* NASA-MSFC, Huntsville, AL; http://history.msfc.nasa.gov/rocketry/index.html.

N.H. RANDERS-PEHRSON: *Pioneer wind tunnels.* Smithsonian Inst. Misc. Coll. **93**, No. 4 (Jan. 19, 1935), 20 pages.

G.H.R. REISIG: *Raketenforschung in Deutschland.* Wissenschaft u. Technik Verlag, Berlin (1999).

E.W.F. ROGERS: *Aerodynamics – retrospect and prospect.* Aeronaut. J. **86**, 43-67 (1982).

K.W. STREIT and J.W.R. TAYLOR: *History of aviation.* New English Library, London (1972).

P.P. WEGENER: *The Peenemünde wind tunnels: a memoir.* Yale University Press, New Haven *etc.* (1996).

## ASTRONOMY & ASTROPHYSICS

A.G.W. CAMERON: *The origin and evolution of the Solar System.* Scient. Am. **233**, 66-75 (Sept. 1975).

D.H. CLARK and F.R. STEPHENSON: *The historical supernovae.* Pergamon Press, New York (1976).

D.H. CLARK: *Supernovae, historical.* In: (S.P. MARAN, ed.) *The astronomy and astrophysics encyclopedia.* Van Nostrand & Reinhold, New York (1992), pp. 886-887.

W.R. DICK: *History of astronomy* [with a topic and name index]. Int. Earth Rotation & Reference Systems Service (IERS) Central Bureau, Frankfurt/M., Germany; http://www.astro.uni-bonn.de/~pbrosche/astoria.html.

P. DOIG: *A concise history of astronomy.* Chapman & Hall Ltd., London (1950).

V.C.A. FERRARO: *The birth of a theory* [about the beginnings of the Chapman-Ferraro theory]. In: (S.I. AKASOFU, B. FOGLE, and B. HAURWITZ, eds.) *Sidney CHAPMAN, eighty: from his friends.* University of Colorado Press, Boulder (1969), pp. 14-18.

D.A. GREEN and F.R. STEPHENSON: *The historical supernovae.* In: (K.W. WEILER, ed.) *Supernovae and gamma ray bursters.* Lecture notes in physics. Springer, Berlin *etc.* (2003).

W. ISRAEL: *Dark stars: the evolution of an idea.* In: (S.W. HAWKING and W. ISRAEL, eds.) *300 years of gravitation.* Cambridge University Press, Cambridge (1987), pp. 199-276.

W.M. MITCHELL: *The history of the discovery of solar spots.* Popular Astronomy **24**, 22, 82, 149, 206, 290, 341, 428, 488, 562 (1916).

J. NORTH: *The Fontana history of astronomy and cosmology.* Fontana Press, London (1994). Germ. translation: *Viewegs Geschichte der Astronomie und Kosmologie.* Vieweg, Braunschweig & Wiesbaden (1997).

L. QIBIN: *A recent study on the historical novae and supernovae.* Proc. 2nd Workshop on High Energy Astrophysics. [Schloss Ringberg, Germany, July 1987]. In: (Max-Planck-Gesellschaft & Academia Sinica, eds.) *High energy astrophysics.* Springer, Berlin *etc.* (1988), pp. 2-25.

C.A. RONAN: *The natural history of the Universe.* Doubleday, London (1991).

F.R. STEPHENSON: *Historical supernovae and their remnants.* Oxford University Press, Oxford (2002).

D.P. STERN and M. PEREDO: *Chronology of magnetospheric exploration (1000–1991).* NASA-GSFC, Greenbelt, MD (Nov. 10, 2003); http://www-spof.gsfc.nasa.gov/Education/whchron.html.
*The solar wind*; http://www-spof.gsfc.nasa.gov/Education/wsolwind.html.
*The magnetopause – history*; http://www-spof.gsfc.nasa.gov/Education/whmpause.html.

R. VILLARD: *A brief history of milestones astronomy.* Space Telescope Science Institute (STScI), Baltimore, MD; http://rayvillard.com/history_of_ASTRONOMY.htm.

D.H. DE VORKIN: *The history of modern astronomy and astrophysics: a selected, annotated bibliography.* Garland, New York (1986).

B. WARNER: *Cataclysmic variable stars* [chap. 1 gives a historical perspective on the subject of cataclysmic variable stars]. Cambridge University Press (1995); http://www.amazon.com/gp/reader/052154209X/ref=sib_rdr_zmout/104-7305530-1966321?%5Fencoding=UTF8&p=S0GM#reader-page.

## BALLISTICS

G. BARBER, J.G. SCHMIDT, and H.L. REED (eds.): *Ballisticians in war and peace: a history of the United States Army Ballistic Research Laboratories.* NTIS, Springfield, VA; vol. 1: *1914–1956*, vol. 2: *1957–1976*, vol. 3: *1977–1992*.

W.Y. CARMAN: *A history of firearms, from earliest times to 1914.* Routledge & Kegan, London (1956).

M.P. CHARBONNIER: *Essais sur l'histoire de la balistique.* Mém. Artill. Franç. **6**, 955-1251 (1927).

J. CORNER: *Theory of the interior ballistics of guns.* Wiley, New York and Chapman & Hall, London (1950).

C. CRANZ: *Lehrbuch der Ballistik.* Springer, Berlin (1925–1927); vol. I: *Äußere Ballistik*; vol. II: *Innere Ballistik*; vol. III: *Experimentelle Ballistik*.

J.B. GOOD: *Forty years of British internal ballistic research.* In: (W.C. NELSON, ed.) *Selected topics on ballistics.* Pergamon Press, London *etc.* (1959), pp. 213-223.

A.R. HALL: *Ballistics in the seventeenth century.* Cambridge University Press, New York (1952).

E.W. HAMMER: *Muzzle brakes, volume I: History and design.* The Franklin Institute Research Laboratories, Philadelphia, PA (June 1949).

C. HOFFMANN and P. BERZ: *Über Schall. Ernst MACHs und Peter SALCHERs Geschoßfotografien.* Wallstein, Göttingen (2001).

R.S. HAWKE: *A decade of railgun development for high pressure research.* In: (Y.M. GUPTA, ed.) *Shock wave compression of condensed matter.* Proc. Symposium in honor of G.E. DUVALL. Rept. 99164-2814, Shock Dynamics Laboratory, Washington State University, Pullman, WA (1988), pp. 53-57.

H. KLEIN: *Vom Geschoss zum Feuerpfeil. Der große Umbruch der Waffentechnik in Deutschland 1900–1970.* Vowinckel, Neckargemünd (1977).

E.J. MCSHANE, J.L. KELLEY, and F.V. RENO: *Exterior ballistics.* University of Denver Press, Denver, CO (1953).

J.D. NICOLAIDES: *A history of ordnance flight dynamics.* In: *Proc. 1st AIAA Atmospheric Flight Mechanics Conference* [Tullahoma, TN, May 1970]. AIAA, Arlington, TX, vol. 1 (1970), Paper AIAA-1970-533.

F. PFLUG: *Nikolaus DREYSE und die Geschichte des preußischen Zündnadelgewehrs.* Haude & Spener, Berlin (1866).

R. VINCENT (ed.): *Textbook of ballistics and gunnery.* 2 vols., H.M.S.O., London (1984).

L.A. WATERMEIER and J.M. HURBAN: *A historical perspective on gun ballistics at the U.S. Army Ballistic Research Laboratory.* In: (H. KRIER and M. SUMMERFIELD, eds.) *Interior ballistics of guns.* Progr. Astronaut. & Aeronaut. **66**, 349-384 (1979).

G. WEIHRAUCH (ed.): *Ballistische Forschung im ISL. Von 1945 bis 1994.* ISL, Saint-Louis, France (1994).

W. WITT and M. LÖFFLER: *Entwicklungsgeschichte der elektromagnetischen Beschleunigung: Die elektrische Kanone auf dem Weg zum Waffensystem.* Rheinmetall GmbH, Düsseldorf; Technologie Zentrum Nord, Unterlüß. In: *Elektromagnetische Kanone.* Lehrgang der Carl-Cranz-Gesellschaft (CCG), Weil am Rhein (Dez. 1987).

## BORES, HYDRAULIC JUMPS, SOLITARY WAVES, TIDES & TSUNAMIS

G.B. AIRY: *Tides and waves.* In: *Encyclopaedia Metropolitana* (ed. by E. SMEDLEY ET AL.). Fellowes, London, vol. 5 (1845), pp. 241-396.

J.E. ALLEN: *The early history of solitons (solitary waves).* Physica Scripta **57**, 436-441 (1998).

A.D.D. CRAIK: *The origins of water wave theory.* Annu. Rev. Fluid Mech. **36**, 1-28 (2004).

B.W. LEVIN: *Tsunamis: causes, consequences, prediction, and response.* In: *Natural disasters, Encyclopedia of Life Support Systems (EOLSS).* Developed under the Auspices of the UNESCO, EOLSS Publ., Oxford, U.K. (July 9, 2005); http://www.undl.org/unlcorpus/eolss/tsunamis.doc.

D.K. LYNCH: *Tidal bores.* Scient. Am. **247**, No. 4, 134-144 (1982).

J.W. MILES: *The Korteweg-de Vries equation: a historical essay.* J. Fluid Mech. **106**, 131-147 (1981).

A.C. NEWELL: *Solitons in mathematics and physics.* Society for Industrial and Applied Mathematics (SIAM), Philadelphia, PA (1985); see *The history of the soliton*, pp. 1-21.

S.L. SOLOVIEV ET AL.: *Tsunamis in the Mediterranean Sea. 2000 B.C. – 1991 A.D.* Kluwer/Academic Press, Dordrecht (2000).

J.J. STOKER: *Water waves.* Interscience Publ., New York (1957).

S.O. WIGEN: *Historical study of tsunamis.* Institute of Ocean Sciences, Sidney, B.C. (1978).

*Historical tsunami database for the Pacific 47 B.C. to present.* Intergovernmental Commission, Russian Foundation for Basic Research (Sib. Division Russ. Acad. Sci.), Institute of Computational Mathematics and Mathematical Geophysics (Feb. 21, 2005); http://tsun.sscc.ru/htdbpac.

## CHEMISTRY

P.E.M. BERTHELOT: *Sur la force des matières explosives d'après la thermochimie.* Gauthier-Villars, Paris (1883); *Explosives and their power.* J. Murray, London (1892).

J.N. BRADLEY: *Shock waves in chemistry and physics.* Methuen, London; Wiley, New York (1962).

W.H. BROCK: *The Fontana history of chemistry.* Fontana Press, London (1992).

G. BUGGE (ed.): *Das Buch der großen Chemiker.* 2 vols., Verlag Chemie, Weinheim/Bergstr. (1955).

T.L. DAVIS: *The chemistry of powder and explosives.* Angriff Press, Las Vegas (1972).

E.F. GREENE and J.P. TOENNIES: *Chemical reactions in shock waves.* Arnold Publ., London etc. (1964).

F. HABASHI: *From alchemy to atomic bombs: history of chemistry, metallurgy, and civilization.* Metallurgy Extractive Québec, Sainte-Foy (2002).

E. HJELT: *Geschichte der organischen Chemie von ältester Zeit bis in die Gegenwart.* Vieweg, Braunschweig (1916).

J. HUDSON: *The history of chemistry.* Chapman & Hall, New York (1992).

H. LANGHANS: *Geschichtstafel zu Gebieten der Wehrchemie.* Z. ges. Schieß- u. Sprengstoffwesen **33**, Beilage (1938).

L.A. MÉDARD and H. TACHOIRE: *Histoire de la thermochimie.* Publication de l'Université de Provence, Aix-en-Provence (1994).

S. NEUFELDT (ed.): *Chronologie Chemie: 1800–1980.* VCH, Weinheim/Bergstr. etc. (1987).

G.M. SCHWAB: *Zur Frühgeschichte der chemischen Kinetik.* In: *Rete, Strukturgeschichte der Naturwissenschaften.* Gerstenberg, Hildesheim (1972), vol. 1, pp. 125-134.

J. VAN HOUTEN: *A century of chemical dynamics traced through the Nobel Prizes.* Chemical Education Today **79**, 667–669 (June 2002).

## COLLISION, PERCUSSION & IMPACT

F. BERGER: *Das Gesetz des Kraftverlaufes beim Stoß.* Vieweg, Braunschweig (1924).

H. FAIR: *Hypervelocity – then and now.* Proc. 1st High Velocity Impact Symposium [San Antonio, TX, 1986]. Int. J. Impact Engng. **5**, 1-11 (1987).

J.C. FISCHER (ed.): *Geschichte der Physik.* Röwer, Göttingen; *Stoß der Körper,* vol. I (1801), pp. 353-382.

L. FÖPPL: *Der Stoß.* In: *Handbuch der Experimentalphysik; Mechanik.* 2. Teil: L. FÖPPL: *Technische Mechanik.* Akad. Verlagsgesell., Leipzig (1926), chap. 6, pp. 151-170.

D. FRANÇOIS and A. PINEAU: *From CHARPY to present impact testing.* In: *Charpy Centenary Conference* [Poitiers, France, Oct. 2001]. Elsevier, Amsterdam (2002).

L. PFEIFFER: *Die steinzeitliche Technik und ihre Beziehungen zur Gegenwart.* Fischer, Jena (1912).

T. PÖSCHL: *Der Stoß* [with a review of the literature on impact]. In: *Handbuch der Physik.* Springer, Berlin, vol. 6 (1928): (R. GRAMMEL and G. ANGENHEISTER, eds.) *Mechanik der elastischen Körper*, chap. 7, pp. 501-565.

T.A. SIEWERT ET AL.: *The history and importance of impact testing.* In: (T.A. SIEWERT and M.P. MANAHAN, eds.) *Pendulum impact testing: a century of progress.* ASTM, West Conshohocken, PA (1999).

A.J. STILP and V. HOHLER: *Aeroballistic and impact physics research at EMI: an historical overview.* Int. J. Impact Engng. **17**, 785–805 (1995).

I. SZABÓ: *Geschichte der Theorie des Stoßes.* Humanismus u. Technik **17**, 14-44, 128-144 (1973).

## COMBUSTION RESEARCH

A. FERRI: *Review of problems in application of supersonic combustion.* J. Roy. Aeronaut. Soc. **68**, 575-597 (1964).

W.C. GARDINER: *Shock tube studies of combustion chemistry* [see chap. *Early history of shock-tube studies of combustion*]. In: (K. TAKAYAMA, ed.) Proc. 18th Int. Symposium on Shock Waves [Sendai, Japan, July 1991]. Springer, Berlin etc. (1992); vol. 1, pp. 49-60.

P.A. LIBBY: *Observations concerning supersonic combustion* [with a review of the early history of supersonic combustion]. In: (M. CHAMPION and B. DESHAIES, eds.) *IUTAM Symposium on Combustion in Supersonic Flows.* Kluwer, Dordrecht (1997), pp. 1-11.

T.A. SIEWERT ET AL.: *The history and importance of impact testing.* In: (T.A. SIEWERT and M.P. MANAHAN, eds.) *Pendulum impact testing: a century of progress.* ASTM, West Conshohocken, PA (1999).

F.J. WEINBERG: *The first half-million years of combustion research and today's burning problems.* In: Proc. 15th Symp. (Int.) on Combustion [Tokyo, Japan, Aug. 1974]. The Combustion Institute, Pittsburgh, PA (1974), pp. 1-17.

## COMPUTER, EARLY DEVELOPMENTS & APPLICATIONS

T.J. BERGIN (ed.): *50 years of army computing. From ENIAC to MSRC* [A record of a symposium and celebration]. ARL, Aberdeen Proving Ground, MD (Nov. 1996). Rept. ARL-SR-93 (Sept. 2000); http://www.arl.mil/main/Main/eniac.pdf.

U. BERNHARDT and I. RUHMANN: *Computer im Krieg: die elektronische Potenzmaschine*. In: (N. BOLZ, F.A. KITTLER, and C. THOLERN, eds.) *Computer als Medium*. Fink, München (1992), pp. 183-207.

H.H. GOLDSTINE: *A history of numerical analysis. From the 16th through the 19th century*. Springer, New York etc. (1977).

H.H. GOLDSTINE: *The computer from PASCAL to VON NEUMANN*. Princeton University Press, Princeton. NJ (1972).

J.A.N. LEE, S. WINKLER, and M. SMITH: *Key events in the history of computing* [prepared in 1996 for the IEEE Computer Society, Washington, DC]; http://ei.cs.vt.edu/~history/50th/30.minute.show.html.

P.J. ROACHE: *Computational fluid dynamics*. Hermosa Publ., Albuquerque, NM (1972); see *Historical outline of computational fluid dynamics* [1910–1969], pp. 2-6.

## DIAGNOSTICS, HIGH-SPEED PHOTOGRAPHY & SCIENTIFIC INSTRUMENTS

W. BAIER: *Geschichte der Fotografie*. Schirmer & Moser, München (1977).

M. BERTOLOTTI: *Masers and lasers: an historical approach*. A. Hilger, Bristol (1983, 1987).

C. CRANZ: *Lehrbuch der Ballistik*. Springer, Berlin; vol. III (1927): *Experimentelle Ballistik*.

A. DUBOVIK: *The photographic recording of high-speed processes*. Wiley, New York etc. (1981).

H.E. EDGERTON: *Electronic flash, strobe*. MIT Press, Cambridge, MA (1979).

S.M. GORLIN and I.I. SLEZINGER: *Wind tunnels and their instrumentation*. S. Monson, Jerusalem (1966).

R.A. GRAHAM and J.R. ASAY: *Measurement of wave profiles in shock-loaded solids*. High Temperatures, High Pressures **10**, 355-390 (1978).

W.G. HYZER: *Engineering and scientific high-speed photography*. Macmillan, New York (1962).

F. JAMET and G. THOMER: *Flash radiography*. Elsevier, Amsterdam (1976).

W. MERZKIRCH: *Flow visualization*. Academic Press, Orlando, FL (1987).

A. POPE and K.L. GOIN: *High-speed wind tunnel testing*. Wiley, New York etc. (1965).

S.F. RAY: *High-speed photography and photonics*. Focal Press, Oxford (1997).

G. SETTLES: *Schlieren and shadowgraph techniques*. Springer, Berlin etc. (2001).

SPIE (The International Society for Optical Engineering, Bellingham, WA): *Ultrahigh- and high-speed photography, videography, and photonics* [San Diego, CA]. Proceedings of SPIE **1346** (1990), **1539** (1991), **1757** (1992), **2002** (1993), **2273** (1994), **2549** (1995); **5210** (2003).

G. THOMER: *History of flash radiography*. 36th National Fall Conference of the American Society for Nondestructive Testing (ASNT) [Houston, TX, Sept. 1976]. In: (L.E. BRYANT JR., ed.) *Flash Radiography Symposium*. ASNT, Columbus, OH (1977), pp. 1-14.

G.L.E. TURNER: *Nineteenth-century scientific instruments*. Sotheby Publ., University of California Press, Berkeley (1983).

K. VOLLRATH and G. THOMER (eds.): *Kurzzeitphysik*. Springer, Wien etc. (1967).

## EARTHQUAKES & SEAQUAKES

L. ALBERDI: *POMBAL y el terremoto de Lisboa de 1755*. Historia y Vida (Barcelona) **21** (247), 116-125 (1988).

J. BEVIS: *The history and philosophy of earthquakes, from the remotest to the present times*. J. Nourse, London (1757).

W. BREIDERT: *Die Erschütterung der vollkommenen Welt*. Wiss. Buchgesell., Darmstadt (1994).

W. BREIDERT: *Das Erdbeben von Lissabon und die Erschütterung seiner Zeitgenossen*. Lichtenberg Jb. SDV, Saarbrücken (1994).

E. OESER: *Historical earthquake theories from ARISTOTLE to KANT*. Abhandl. Geolog. Bundesanstalt Wien **48**, 11-31 (1992); *Heat* [discusses early mechanical, chemical and steam pressure theories on earthquake generation such as proposed by MAYER, LEMERY, KRÜGER, and JACOBI]; http://www.univie.ac.at/Wissenschaftstheorie/heat/heatcont.htm.

The Jan T. Kozak Collection: *Images of historical earthquakes*. National Information Service for Earthquake Engineering (NISEE), UC Berkeley; http://nisee.berkeley.edu/kozak/index.html.

B. WALKER: *Earthquake*. Time-Life Books Inc., Alexandria, VA (1982).

D. WILLIAMS: *Historical observations of seaquakes*; http://www.deafwhale.com/stranding/observations.html.

## EXPLOSIONS & DETONATIONS

F.A. ABEL: *Contributions to the history of explosive agents*. Phil. Trans. Roy. Soc. Lond. **159**, 489-516 (1869); Ibid. **164**, 337-395 (1874).

Anonymous: *Joshua SHAW, artist and inventor. The early history of the copper percussion cap*. Scient. Am. **21**, 90 (Aug. 7, 1869).

R. ASSEHTON: *History of explosions on which the American Table of Distances was based, included other explosions of large quantities of explosives*. Rept. AD 493246, The Institute of Makers of Explosives, New York (1930); *History of explosives*. Ibid. (1940).

P.A. BAUER, E.K. DABORA, and N. MANSON: *Chronology of early research on detonation wave*. In: (A.L. KUHL, J.C. LEYER, A.A. BORISOV, and W.A. SIRIGNANO, eds.) *Dynamics of detonations and explosions: detonations*. Progr. Astro- & Aeronautics **133**, 3-18 (1991).

R. BECKER: *Stoßwelle und Detonation*. Z. Phys. **8**, 321-362 (1922); Engl. translation: *Impact, waves and detonation*. NACA-TM 505 (1929) and NACA-TM 506 (1929).

F. BÖCKMANN: *Die explosiven Stoffe, ihre Geschichte, Fabrikation, Eigenschaften, Prüfung und praktische Anwendung in der Sprengtechnik*. Hartleben, Wien (1895).

G. BUGGE: *Schieß- und Sprengstoffe und die Männer, die sie schufen*. Franckh, Stuttgart (1942).

W.G. CHACE: *A survey of exploding wire progress*. In: (W.G. CHACE and H.K. MOORE, eds.) *Exploding wires III*. Plenum Press, New York (1959), pp. 7-14.

R. CHÉRET: *Detonation of condensed explosives*. Springer, New York etc. (1993).

W.M. CONN: *Studien zum Mechanismus von elektrischen Drahtexplosionen (Metallniederschläge und Stoßwellen)*. Z. Angew. Phys. **7**, 539-554 (1955).

J. CROSS and D. FARRER: *Dust explosions*. Plenum Press, New York etc. (1982).

T.L. DAVIS: *The chemistry of powder and explosives*; J. Wiley & Sons, New York (1943); Angriff Press, Las Vegas, NV (1972); http://www.sciencemadness.org/library/books/the_chemistry_of_powder_and_explosives.pdf.

DUPONT's *Blasters' handbook*. E.I. du Pont de Nemours & Co., Wilmington, DE (1989).

R. ENGELKE and S.A. SHEFFIELD: *History of condensed-phase explosives*. Chap. 1 in: *Explosives*. In: (G.L. TRIGG, ed.) *Encyclopedia of applied physics*. VCH Publ., New York etc. (1991), vol. 6, pp. 327-357 and vol. "Update 1," pp. 371-379.

P. FIELD: *Dust explosions*. Elsevier Scient. Publ., Amsterdam etc. (1982).

W. FICKETT and W.C. DAVIS: *Detonation*. University of California Press, Berkeley etc. (1979). See chap. *History* [of detonation research], pp. 2-11.

S. FORDHAM: *High explosives and propellants*. Pergamon Press, Oxford etc. (1966, 1980); http://www.chimicando.it/e-book/%5Bebook%5DHigh%20Explosives%20And%20Propellants.pdf.

A.P. VAN GELDER and H. SCHLATTER: *History of the explosives industry in America*. Institute of Makers of Explosives, New York (1927).

G.J. HOLLISTER-SHORT: *Gunpowder and mining in sixteenth- and seventeenth-century Europe*. History of Technology (London & New York) **10**, 31-66 (1985).

E. JOUGUET: *Résumé des théories sur la propagation des explosions*. La Science Aérienne **3**, 138-155 (1934).

D.R. KENNEDY: *History of the shaped charge effect. The first 100 years*. Company brochure prepared by D.R. Kennedy & Associates, Inc., Mountain View, CA (1983).

V. LINDNER: *Explosives and propellants*. In: (M. HOWE-GRANT, ed.) *Encyclopedia of chemical technology*. 27 vols., Wiley, New York etc. (1991–1998); vol. 10 (1993), pp. 1-125.

L.A. MÉDARD: *Accidental explosions*. 2 vols., Ellis Horwood, Chichester (1989).

N. MANSON: *Historique de la découverte de l'onde de détonation*. Colloque C4. J. de Physique **48**, 7-37 (1987).

N. MANSON and E.K. DABORA: *Chronology of research on detonation waves: 1920–1950*. In: (A.L. KUHL, J.C. LEYER, A.A. BORISOV, and W.A. SIRIGNANO, eds.) *Dynamic aspects of detonations*. Progr. Astro- & Aeronautics **153**, 3-39 (1993).

C.E. MUNROE: *The effects of explosives on civilization*. The Chautauquan (New York) **9**, 203-205 (1889).

C.E. MUNROE: *The history of gunpowder*. J. Am. Chem. Soc. **1**, 119-120 (1923).

V. MUTHESIUS: *Zur Geschichte der Sprengstoffe und des Pulvers*. Im Auftrag der Westfälischen Anhaltischen Sprengstoff-Actien-Gesellschaft, Berlin (1941).

A.K. OPPENHEIM: *Introduction – the past, present and future of discussions on gaseous detonation phenomena*. Proc. 12th Symp. (Int.) on Combustion. The Combustion Institute, Pittsburgh, PA (1969), pp. 795-797.

J.R. PARTINGTON: *A history of Greek fire and gunpowder*. Heffer, Cambridge (1960).

R.A. PRÜMMER: *Explosivverdichtung pulvriger Substanzen*. Springer, Berlin etc. (1987).

S.J. VON ROMOCKI: *Geschichte der Explosivstoffe*. Oppenheim, Berlin. Vol. I (1895): *Geschichte der Sprengstoffchemie, der Sprengtechnik und des Torpedowesens*; vol. II (1896): *Die rauchschwachen Pulver in ihrer Entwicklung*.

F. SASS: *Geschichte des Deutschen Verbrennungsmotorenbaues*. Springer, Berlin etc. (1962).

W.P. WALTERS and J.A. ZUKAS: *Fundamentals of shaped charges*. Wiley & Sons, New York etc. (1989).

N.B. WILKINSON: *Explosives in history: the story of black powder*. Rand McNally, Chicago (1966).

# FLUID DYNAMICS & GAS DYNAMICS

J.D. ANDERSON JR.: *Modern compressible flow, with historical perspective*. McGraw-Hill, New York (1990).

A. BUSEMANN: *Compressible flow in the thirties*. Annu. Rev. Fluid Mech. **3**, 1-12 (1971).

Cambridge University Press: *Fluid Mechanics. History of science sites* [covering aerodynamics, fluid dynamics, heat transfer, hydraulics, mathematics etc.]; http://www.fluidmech.net/msc/hist.htm.

D.A. CAUGHEY and A. JAMESON: *Development of computational techniques for transonic flows: an historical perspective*. IUTAM Symposium Transsonicum IV [Göttingen, Sept. 2002], Kluwer Academic, Dordrecht (2003), pp. 183-194.

A. KANTROWITZ: *One-dimensional treatment of non-steady gas dynamics*. In: *High-speed aerodynamics and jet propulsion*. In: (H.W. EMMONS, ed.) *Fundamentals of gas dynamics*. Princeton University Press, Princeton, NJ, vol. III (1958), pp. 350-415.

L. PRANDTL: *Strömende Bewegung der Gase und Dämpfe*. Encyclopädie der math. Wissenschaften. Teubner, Leipzig (1903–1921), vol. V (1), pp. 287-319.

R. SMELT: *A critical review of German research on high-speed airflow*. J. Roy. Aeronaut. Soc. **50**, 899-934 (1946).

P.A. THOMPSON: *Compressible-fluid dynamics*. McGraw-Hill, New York (1972).

C.A. TRUESDELL: *The mechanical foundations of elasticity and fluid dynamics*. J. Ration. Mech. Anal. **1**, 125-171, 173-300 (1952).

C.A. TRUESDELL: *Rational fluid mechanics*. In: (C.A. TRUESDELL, ed.) *Leonardi EULERI opera omnia* XII [II]. Teubner, Leipzig etc. (1954); see editor's introduction, pp. I-CXXV.

# GENERAL REVIEWS

R. TATON (ed.): *Histoire générale des sciences*. Presses Universitaires de France, Paris (1969), tome II: *La science moderne (de 1450 à 1800)*.

B. VODAR and J. KIEFFER: *Historical introduction* [on static and dynamic high pressure research in liquids and solids]. In: (H.L.D. PUGH, ed.) *Mechanical behavior of materials under pressure*. Applied Science Publ., London (1971), pp. 1-53.

L.P. WILLIAMS and H.J. STEFFENS: *The history of science in western civilization*. University Press of America, Washington, DC; vol. 1 (1977): *Antiquity and Middle Ages*; vol. 2 (1978): *The scientific revolution*; vol. 3 (1978): *Modern science, 1700–1900*; http://www.worldwideschool.org/library/books/sci/history/AHistoryofScienceVolumeIII/toc.html.

# GEOLOGY & GEOPHYSICS

H. BIRETT, K. HELBIG, W. KERTZ, and U. SCHMUCKER: *Zur Geschichte der Geophysik. Festschrift zur 50jährigen Wiederkehr der Gründung der Deutschen Geophysikalischen Gesellschaft*. Springer, Berlin etc. (1974).

S.G. BRUSH and H.E. LANDSBERG: *History of geophysics and meteorology: an annotated bibliography*. Garland, New York (1985).

C.L.O. BUCHNER: *Die Meteoriten in Sammlungen: ihre Geschichte, mineralogische und chemische Beschaffenheit*. Engelmann, Leipzig (1863).

F. ELLENBERGER: *History of geology*. 2 vols., Balkema, Rotterdam etc. (1996).

W. KERTZ: *Geschichte der Geophysik*. Olms, Hildesheim etc. (1999).

L.C. LAWYER ET AL.: *Geophysics in the affairs of mankind: a personalized history of exploration geophysics*. Society of Exploration Geophysicists, Tulsa, OK (2002).

W.A.S. SARJEANT: *Geologists and the history of geology: an international bibliography from the origins to 1978*. 2 vols., Krieger, Malabar, FL (1987).

W. SCHRÖDER: *Geophysics: past achievements and future challenges*. Science Ed. Schröder, Bremen-Rönnebeck (1994).

W. SCHRÖDER (ed.): *Geschichte und Philosophie der Geophysik*. Arbeitskreis Geschichte der Geophysik und Kosmischen Physik, Bremen (2000).

## HYDRAULICS & HYDRODYNAMICS

O. DARRIGOL: *Worlds of flow: a history of hydrodynamics from BERNOULLI to PRANDTL*. Oxford University Press, Oxford (2005).

O. FLACHSBART: *Geschichte der experimentellen Hydro- und Aeromechanik, insbesondere der Widerstandsforschung*. In: (L. SCHILLER, ed.) *Handbuch der Experimentalphysik*. Akad. Verlagsgesell., Leipzig; Bd. 4 (1932): *Hydro- und Aerodynamik*, 2. Teil: *Widerstand und Auftrieb*, pp. 1-61.

H. ROUSE and S. INCE: *History of hydraulics*. Dover Publ., New York (1957).

## MATHEMATICS

C.B. BOYER: *A history of mathematicians*. Wiley, New York (1968).

J.W. DAUBEN and C.J. SCRIBA (eds.): *Writing the history of mathematics: its historical development*. Birkhäuser, Basel *etc.* (2002).

J.A. DIEUDONNÉ: *Geschichte der Mathematik 1700–1900*. Vieweg, Braunschweig (1985).

W.W. ROUSE BALL: *A short account of the history of mathematics*. Dover Publ., New York (1960).

J. STILLWELL: *Mathematics and its history*. Springer, New York *etc.* (2004).

J. SUZUKI: *A history of mathematics*. Prentice Hall, Upper Saddle River, NJ (2002).

## MECHANICS

R. DUGAS: *Histoire de la mécanique*. Griffon, Neuchâtel (1950); Engl. translation by J.R. MADDOX: *A history of mechanics*. Dover Publ., New York (1988).

L.G. LOITSYANSKII: *Mechanics of liquids and gases*. Pergamon Press, Oxford (1966).

E. MACH: *Die Mechanik in ihrer Entwicklung, historisch-kritisch dargestellt*. Brockhaus, Leipzig (1883); *The laws of collision*, pp. 310-331. Engl. translation by T.J. MCCORMACK: *The science of mechanics, a critical and historical account of its development*. Open Court Publ., La Salle, IL & London (1942).

R. MEWES: *Geschichtliche Entwicklung der Prinzipien der Mechanik und Physik, Grundgesetze der Thermodynamik*. Degener, Leipzig (1910).

I. SZABÓ: *Einführung in die Technische Mechanik*. Springer, Berlin *etc.* (1966).

I. SZABÓ: *Geschichte der mechanischen Prinzipien und ihre wichtigsten Anwendungen*. Birkhäuser, Basel *etc.* (1977).

S.P. TIMOSHENKO: *History of strength of materials*. McGraw-Hill, New York *etc.* (1953).

I. TODHUNTER and K. PEARSON: *A history of the theory of elasticity and of the strength of materials from GALILEI to the present time*. Cambridge University Press, Cambridge; vol. 1 (1886): *GALILEI to SAINT-VENANT 1639–1850*, vol. 2 (1893): *SAINT-VENANT to Lord KELVIN*.

C.A. TRUESDELL: *The mechanical foundations of elasticity and fluid dynamics*. J. Ration. Mech. Anal. **1**, 125-171, 173-300 (1952).

## MINING INDUSTRY

*A history of mine safety research in Great Britain*; http://www.users.zetnet.co.uk/mmartin/fifepits/starter/safe-2.htm.

M. FUNK and H. KREBS: *Die Entwicklung der Sprengtechnik im Bergbau, dargestellt am Beispiel des Freiberger Bergbaus*. Sprengstoffe, Pyrotechnik [Schönebeck/Elbe] **24**, Nr. 2, 3-10; Nr. 3, 3-15; Nr. 4, 3-16 (1987).

M. FUNK and H. KREBS: *Entwicklung der Zündtechnik im Bergbau*. 9. Int. Fachtagung Sprengwesen [Gera, Germany, April 1988]; Arbeitsgruppe III „Sprengstoffe, Zündmittel, Sprengzubehör." Publ. in: Sprengstoffe, Pyrotechnik (Schönebeck/Elbe) **26**, Nr. 3, 3-16; Nr. 4/5, 3-15 (1989); Ibid. **27**, Nr. 1, 4-17 (1990).

G.J. HOLLISTER-SHORT: *Gunpowder and mining in sixteenth and seventeenth century Europe*. History of Technology **10**, 31-66 (1985).

F. VON RŽIHA: *Lehrbuch der gesamten Tunnelbaukunst*. Ernst & Korn, Berlin (1867); *see Geschichte der bergmännischen Sprengarbeit* [on the history of blasting in the mining industry], vol. I, pp. 37-65.

## NUCLEAR WEAPONS DEVELOPMENT

H.A. BETHE: *Comments on the history of the H-bomb*. Los Alamos Science **3**, 42-53 (Fall 1982).

H.A. BETHE: *Bombs after Hiroshima* [Book review of R. RHODES: *Dark Sun: the making of the Hiroshima bomb*. Simon & Schuster, New York (1995)] Science **269**, 1455–1457 (Sept. 8, 1995).

H.A. BETHE's letter to U.S. President CLINTON, sent on April 25, 1997; http://www.fas.org/bethepr.htm.

H.A. BETHE: *The German uranium project*. Phys. Today **53**, 34-36 (July 2000).

R. DEL TREDICI: *At work in the fields of the bomb*. Harrap, London (1987).

S. GOLDBERG: *GROVES and the scientists: compartmentalization and the building of the bomb*. Phys. Today **48**, 38-43 (Aug. 1995).

G.A. GONCHAROV: *American and Soviet H-bomb development programmes: historical background*. Phys. Uspekhi **39**, 1033–1044 (1996).

C. HARDY: *Atomic rise and fall, the Australian Atomic Energy Commission 1953–1987*. Glen Haven, Peakhurst, N.S.W. (1999). For atomic weapons development in the period 1939–1945 *see* chap. 1.

W. HEISENBERG: *The Third Reich and the atomic bomb*. Bull. Atom. Sci. **24**, 34-35 (June 1968).

L. HODDESON, P.W. HENRIKSEN, R.A. MEADE, and C. WESTFALL: *Critical assembly: a technical history of Los Alamos during the Oppenheimer years, 1943–1945*. Cambridge University Press, Cambridge (1993).

R. KARLSCH: *HITLER's Bombe*. Dt. Verlagsanstalt, München (2005).

G.B. KISTIAKOWSKY: *Trinity – a reminiscence*. Bull. Atom. Sci. **36**, 19-22 (June 1980).

J.W. KUNETKA: *City of fire, Los Alamos and the atomic age, 1943–1945*. Prentice-Hall, Englewood Cliffs, NJ (1978).

R. RHODES: *The making of the atomic bomb*. Simon & Schuster, New York (1986).

R. RHODES: *Dark Sun: the making of the Hiroshima bomb*. Simon & Schuster, New York (1995).

F.H. SHELTON: *Reflections of a nuclear weaponeer*. Shelton Enterprise Inc., Colorado Springs, CO (1988); *Reflections on the big red bombs*. Ibid. (1995).

N. SKENTELBERY: *Arrows to atom bombs. A history of the Ordnance Board*. H.M.S.O., London (1975).

V.A. TSUKERMAN and Z.M. AZARKH: *Arzamas-16: Soviet scientists in the nuclear age: a memoir*. Bramcote Press, Nottingham (1999).

*Los Alamos, NM (Project Y)*. Manhattan Project Heritage Preservation Association, Inc., New York; http://www.childrenofthemanhattanproject.org/LA/Photo-Pages-2/LAPG_13.htm.

*United States nuclear tests July 1945 through September 1992*. Rept. DOE/NV-209-REV 15, U.S. Dept. of Energy, Nevada Operations Office, Las Vega, NV (Dec. 2000); http://www.nv.doe.gov/news&pubs/publications-/historyreports/pdfs/DOENV209_REV15.pdf.

## PHYSICS

J.C. FISCHER (ed.): *Geschichte der Physik*. 2 vols., Röwer, Göttingen (1801, 1802).

A. HELLER: *Geschichte der Physik von ARISTOTELES bis auf die neueste Zeit*. 2 vols., Enke, Stuttgart (1882, 1884).

P. LA COUR and J. APPEL: *Die Physik aufgrund ihrer geschichtlichen Entwicklung.* Vieweg, Braunschweig (1905).

R. MEWES: *Geschichtliche Entwicklung der Prinzipien der Mechanik und Physik, Grundgesetze der Thermodynamik.* Degener, Leipzig (1910).

K. VON MEŸENN (ed.): *Die großen Physiker.* 2 vols., C.H. Beck'sche Verlagsbuchhandl., München (1997).

## SEISMOLOGY & GEOPHYSICAL PROSPECTING

A. BEN-MENAHEM: *A concise history of mainstream seismology. Origins, legacy and perspectives.* Bull. Seismol. Soc. Am. **85**, No. 4, 1202-1225 (1995).

J. DEWEY and P. BYERLY: *The early history of seismometry (to 1900).* Bull. Seismol. Soc. Am. **59**, No. 1, 183-227 (1969).

R. EHLERT: *Zusammenstellung, Erläuterung und kritische Beurteilung der wichtigsten Seismometer mit besonderer Berücksichtigung ihrer praktischen Verwendbarkeit.* Beiträge zur Geophysik **3**, 350-475 (1898).

*Geschichte der Seismologie, Seismik und Erdbebenforschung.* Tagung in Eisenach, Germany (Dez. 1979). Veröffentlichung des Zentralinstituts für Physik der Erde (Potsdam, Germany) Nr. 64, 17-24 (1981).

L. MINTROP: *Zur Geschichte des seismischen Verfahrens zur Erforschung von Gebirgsschichten und nutzbaren Lagerstätten.* II. Mitteil. der Seismos-Gesell., Hannover (1930).

L. MINTROP: *100 Jahre physikalische Erdbebenforschung und Sprengseismik.* Die Naturwissenschaften **9/10**, 258-262, 289-295 (1947).

J. RITTER: *History of seismology in Göttingen*; http://www.uni-geophys.gwdg.de/~eifel/Seismo_HTML/history.html.

J.P. ROTHÉ: *Fifty years of history of the International Association of Seismology (1901-1951).* Bull. Seismol. Soc. Am. **71**, 905-923 (1981).

W. SCHRIEVER: *Reflection seismograph prospecting – how it started.* Geophys. **17**, 936-942 (1952).

P.M. SHEARER: *Introduction to seismology.* Cambridge University Press, Cambridge & New York (1999).

G.E. SWEET: *The history of geophysical prospecting.* Science Press, Los Angeles, CA (1978).

T. TAYMAZ: *History of seismology.* In: *History of geophysics* (1994); http://www.geocities.com/nmadentr/history.htm.

USGS: *The men and women of seismology.* National Earthquake Information Center (NEIC), World Data Center for Seismology, Denver, CO; http://wwwneic.cr.usgs.gov/neis-/seismology/people/people.html.

## SHOCK WAVES & BLAST WAVES

M.B. ABBOTT: *The method of characteristics.* American Elsevier, New York (1966).

L.V. AL'TSHULER and V.A. SIMONENKO: *History and prospects of shock wave physics.* High Pressure Research **5**, 813-815 (1990).

L.V. AL'TSHULER, R.F. TRUNIN, K.K. KRUPNIKOV, and N.V. PANOV: *Explosive laboratory devices for shock wave compression* [historic Arzamas-16 explosive laboratory devices]. Phys. Uspekhi **39**, No. 5, 539-544 (1996).

S.S. BATSANOV: *On the kinetics of chemical reactions in solids under shock compression* [a review]. Propellants, Explosives, Pyrotechnics **18**, 100-105 (1993).

G. BEN-DOR and K. TAKAYAMA: *The phenomena of shock wave reflection, a review of unsolved problems and future research needs.* Shock Waves **2**, 211-223 (1992).

J.N. BRADLEY: *Shock waves in chemistry and physics.* Methuen, London; Wiley, New York (1962).

S. CHAPMAN and T.G. COWLING: *The mathematical theory of non-uniform gases.* Cambridge University Press, Cambridge (1960).

R. COURANT and K.O. FRIEDRICHS: *Supersonic flow and shock waves.* Interscience Publ., New York (1948).

G.E. DUVALL: *Problems in shock wave research* [invited paper]. In: (R.W. ROHDE, B.M. BUTCHER, J.R. HOLLAND, and C.H. KARNES, eds.) *Conference on Metallurgical Effects at High Strain Rates* [Sandia Laboratories, Albuquerque, NM, Nov. 1973]. Plenum Press, New York *etc.* (1973), pp. 1-13.

G.E. DUVALL: *Shock wave research: yesterday, today and tomorrow.* In: (Y.M. GUPTA, ed.) *Proc. 4th APS Topical Conference on Shock Waves in Condensed Matter* [Spokane, WA, July 1985]. Plenum Press, New York (1986), pp. 1-12.

R.J. EMRICH: *Early development of the shock tube and its role in current research.* In: (Z.I. SLAWSKY, J.F. MOULTON JR., and W.S. FILLER, eds.) *Proc. 5th Int. Shock Tube Symposium* [White Oak, Silver Spring, MA, April 1965], pp. 1-10.

A.G. GAYDON and I.R. HURLE: *The shock tube in high-temperature chemical physics.* Chapman & Hall, London (1963).

R.A. GRAHAM: *Solids under high-pressure shock compression.* Springer, New York *etc.* (1993); see *Introduction*, pp. 3-12.

E.F. GREENE and J.P. TOENNIES: *Chemical reactions in shock waves.* Arnold Publ., London *etc.* (1964).

W.C. GRIFFITH: *Shock waves.* J. Fluid Mech. **106**, 81–101 (1981).

H. GRÖNIG: *Past, present and future of shock focusing research.* In: (K. TAKAYAMA, ed.) *Proc. Int. Workshop on Shock Focusing* [Sendai, Japan, March 1989]. Shock Wave Research Center, Institute of Fluid Science, Tohoku University (1990), pp. 1-37.

A. HERTZBERG: *Shock tube research, past, present and future.* In: (I.I. GLASS, ed.) *Proc. 7th Int. Shock Tube Symposium* [Toronto, Canada, June 1969]. University of Toronto Press, Toronto (1970), pp. 3-5.

J.N. JOHNSON and R. CHÉRET (eds.): *Classic papers in shock compression.* Springer, New York *etc.* (1998).

J.N. JOHNSON and R. CHÉRET: *Shock waves in solids: an evolutionary perspective.* Shock Waves **9**, 193-200 (1999).

C.F. KENNEL, J.P. EDMISTON, and T. HADA: *A quarter century of collisionless shock research.* In: (R.G. STONE and B.T. TSURUTANI, eds.) *Shocks in the heliosphere.* Geophysical monograph series, AGU, Washington, DC, vol. 35 (1985), pp. 1-36.

J. LUKASIEWICZ: *Experimental methods of hypersonics.* Dekker, New York (1973).

R. VON MISES: *Mathematical theory of compressible fluid flow.* Academic Press, New York (1958).

H. OERTEL: *Stoßrohre. Mit einer Einführung in die Physik der Gase.* Springer, Vienna *etc.* (1966).

J.S. RINEHART: *Historical perspective* [on dynamic deformation]. In: (M.A. MEYERS, ed.) *Shock waves and high-strain-rate phenomena.* Plenum Press, New York (1981), pp. 3-19.

R.J. SEEGER: *On MACH's curiosity about shock waves.* In: (R.S. COHEN and R.J. SEEGER, eds.) *Ernst MACH, physicist and philosopher.* Boston Studies Phil. Sci. **6**, 42-67 (1970).

I. SZABÓ: *Geschichte der mechanischen Prinzipien und ihre wichtigsten Anwendungen.* Birkhäuser, Basel *etc.* (1977); see *Geschichte der Stoßwellen*, pp. 281-314.

V.A. TSUKERMAN and Z.M. AZARKH: *Arzamas-16: Soviet scientists in the nuclear age: a memoir.* Bramcote Press, Nottingham (1999).

## SHOCK WAVES IN BIOLOGY & MEDICINE

L.W. ALVAREZ, W. ALVAREZ, F. ASARO, and H.V. MICHEL. *Extraterrestrial cause for the cretaceous-tertiary extinction.* Science **208**, 1095-1108 (June 6, 1980).

A. BAR-NUN: *Shock waves and the origin of life.* UTIAS Rev. No. 41. Institute for Aerospace Studies, University of Toronto (1976).

M. DELIUS and W. BRENDEL: *Historical roots of lithotripsy.* J. Lithotripsy & Stone Disease **2**, 161-163 (1990).

R.M. ENGEL: *De historia lithotomiae.* W.P. Didusch Museum of the AUA; http://www.urolog.nl/urolog/php/content.php?doc=lithotomia&profmenu=yes.

B. FORSSMANN: *25 years in ESWL – from the past to the future.* In: (A.A. ATCHLEY, V.W. SPARROW, and R.M. KEOLIAN, eds.) *Innovations in nonlinear acoustics: 17th Int. Symp. on Nonlinear Acoustics* [State College, PA, July 2005]. American Institute of Physics (AIP), Melville, New York (2006).

V. GOLDBERG: *Geschichte der Urologie: eine neue Methode der Harnsteinzertrümmerung – elektrohydraulische Lithotripsie.* Der Urologe **B 19**, 23-27 (1979).

D. JOCHAM: *Historical development of ESWL.* In: (R.A. RIEHLE and R.C. NEWMAN, eds.) *Principles of extracorporeal shock wave lithotripsy.* Churchill Livingstone, New York (1987), pp. 1-11.

S.L. MILLER and H.C. UREY: *Organic compound synthesis on the primitive Earth.* Science **130**, 245-251 (July 31, 1959).

S.L. MILLER and L. ORGEL: *Origins of life on the Earth.* Prentice Hall, Englewood, NJ (1974).

S.L. MILLER: *The prebiotic synthesis of organic compounds as a step toward the origin of life.* In: (J.W. SCHOPF, ed.) *Major events in the history of life.* Jones & Bartlett, Boston (1992), pp. 1-28.

K.O. POPE ET AL.: *Meteorite impact and the mass extinction of species at the Cretaceous Tertiary boundary.* Proc. Natl. Acad. Sci. **95**, 11028-11029 (Sept. 15, 1998); http://www.pnas.org/cgi/content/full/95/19/11028.

K. TAKAYAMA: *Application of shock wave research to medicine* (3rd Plenary Lecture). In: (G.J. BALL, R. HILLIER, and G.T. ROBERTS, eds.) *Proc. 22nd Int. Symposium on Shock Waves* [Imperial College, London, July 1999]. University of Southampton, U.K. (2000).

M. THIEL, M. NIESWAND, and M. DÖRFFEL: *Shock wave application in medicine, a tool of modern operating theatre – An overview of basic physical principles, history and basic research.* HMT High Medical Technologies AG, Lengwil, Switzerland. *See* chap. 4: *Historical overview*; http://www.ismst.com/history.htm.

## STEAM BOILER EXPLOSIONS

R. ARMSTRONG and J. BOURNE: *The modern practice of boiler engineering, containing observations on the construction of steam boilers.* Spon, London (1856).

F. FISCHER: *Zur Geschichte der Dampfkesselexplosionen.* Dingler's Polytech. J. **213**, 296-308 (1874).

H.H.P. POWLES: *Steam boilers, their history and development.* A. Constable, London (1905).

*Steam boiler explosions in England.* The Manufacturer & Builder **3**, 176-177 (Aug. 1871); Ibid. **5**, 111-112 (May 1873); Ibid. **7**, 130 (June 1875); Ibid. **10**, 266 (Dec. 1878).

## THERMODYNAMICS

S.G. BRUSH: *The kind of motion we call heat: a history of the kinetic theory of gases in the 19th century.* North-Holland Publ., Amsterdam (1976).

N.S. HALL (ed.): *The kinetic theory of gases: an anthology of classic papers with historical commentary by S.G. BRUSH.* Imperial College Press, London (c.2003).

D. LEVERMORE: *History of kinetic theory.* Dept. of Mathematics, University of Maryland; http://www.math.umd.edu/~lvrmr/History/index.shtml.

R. MEWES: *Geschichtliche Entwicklung der Prinzipien der Mechanik und Physik, Grundsetze der Thermodynamik.* Degener, Leipzig (1910).

C.A. TRUESDELL: *The tragicomical history of thermodynamics, 1822–1854.* Springer, New York (1980).

## THUNDER

D.F.J. ARAGO: *On thunder and lightning* [a historical sketch]. Edinb. New Phil. J. **26**, 81-144, 275-291 (1839).

A. BENNETT: *New experiments on electricity, wherein the causes of thunder and lightning as well as the constant state of positive or negative electricity in the air or clouds, are explained...* J. Drewry, Derby (1789).

V.A. RAKOV and M.A. UMAN: *Lightning: physics and effects* (Encyclopedia of lightning). Cambridge University Press, Cambridge (2003).

M.A. UMAN: *Lightning.* Dover Publ., New York (1984), chap. 6.

## VOLCANOES

F.M. BULLARD: *Volcanoes in history, in theory, in eruptions.* University of Texas Press, Austin, TX (1963).

N. MORELLO (ed.): *Volcanoes and history.* In: *Proc. 20th INHIGEO (International Commission on History of Geological Sciences) Symp.* [Neaples, Italy, Sept. 1995].

H. SIGURDSSON: *The history of volcanology.* In: (H. SIGURDSSON, ed.) *Encyclopedia of volcanoes.* Academic Press, San Diego, CA *etc.* (2000), pp. 15-37.

## 6.7 INTERNET HOMEPAGES

### 6.7.1 RESEARCH INSTITUTIONS

A.F. IOFFE PHYSICO-TECHNICAL INSTITUTE, Russian Academy of Sciences, St. Petersburg, Russia; http://www.ioffe.rssi.ru/.

ARMY RESEARCH LABORATORY ⇨ U.S. ARMY RESEARCH LABORATORY.

ATOMIC WEAPONS ESTABLISHMENT (AWE), Aldermaston, U. K.; http://www.awe.co.uk/main_site/scientific_and_technical/featured_areas/hydrodynamics_contents/esr-m1/.

CAVENDISH LABORATORY, Dept. of Physics, University of Cambridge, U.K.; http://www.phy.cam.ac.uk/. History of the Laboratory; http://www.phy.cam.ac.uk/cavendish/history/.

COURANT INSTITUTE OF MATHEMATICS, New York University, NY, U.S.A.; http://www.cims.nyu.edu/. *See* C. REID: *COURANT in Göttingen and New York.* Springer, New York (1986), and History of foundation; http://www.cims.nyu.edu/information/brochure/history.html.

DAVID TAYLOR RESEARCH CENTER, Carderock & Annapolis, MA, U.S.A.; http://www.aacounty.org/hot_new_davidtaylor.htm. *See* R.P. CARLISLE: *Where the fleet begins. A history of the David Taylor Research Center (1898–1998).* Naval Historical Center, Dept. of the Navy, Washington, DC (1998).

DEFENCE RESEARCH & DEVELOPMENT CANADA (DRDC) – SUFFIELD, Suffield, Alberta, Canada; http://www.suffield.drdc-rddc.gc.ca/Home/index_e.html.

DEUTSCHES ZENTRUM FÜR LUFT- UND RAUMFAHRT (DLR), Germany; http://www.dlr.de/DLR-Homepage.

ECOLE POLYTECHNIQUE, Palaiseau (Paris), France; http://www.polytechnique.fr/. For its history *see* http://www.polytechnique.fr/institution/historique.php.

ERNST-MACH-INSTITUT (EMI), Fraunhofer-Institut für Kurzzeitdynamik, Freiburg, Germany; http://www.emi.fhg.de/.

ESA, EUROPEAN SPACE AGENCY, Headquarters in Paris, France; http://www.esa.int/esaCP/index.html.

EUROPEAN SPACE OPERATIONS CENTRE (ESOC), Darmstadt, Germany; http://www.esa.int/esoc.

EUROPEAN SPACE RESEARCH & TECHNOLOGY CENTRE (ESTEC), Noordwijk, The Netherlands; http://www.esa.int/estec.

GAS DYNAMICS LABORATORY (GDL), Princeton University, Princeton, NJ, U.S.A.; http://www.princeton.edu/~gasdyn/.

IMPACT AND CRASHWORTHINESS LABORATORY (ICL), MIT, Cambridge, PA, U.S.A.; http://web.mit.edu/icl/.
IMPACT RESEARCH CENTRE (IRC), University of Liverpool, U.K.; http://dbweb.liv.ac.uk/engdept/content/centres/irc/default.htm.
INSTITUT CHEMISCHE TECHNOLOGIE (ICT), Fraunhofer-Institut, Pfinztal, Germany; http://www.ict.fhg.de.
INSTITUT SAINT-LOUIS (ISL), Saint-Louis, Alsace, France; http://www.isl.tm.fr/fr/accueil_f.html.
INSTITUTE FOR HIGH TEMPERATURES (IVTAN), Russian Academy of Sciences, Moscow; Russia; http://www.tech-db.ru/istc/db/inst.nsf/wu/1259.
INSTITUTE FOR SHOCK PHYSICS (ISP), Washington State University (WSU), Pullman, WA, U.S.A.; http://www.shock.wsu.edu/contactus.html.
INSTITUT FÜR FLUIDDYNAMIK (IFD), ETH Zurich, Switzerland; http://www.ifd.mavt.ethz.ch/.
KUNGLIGA TEKNISKA HÖGSKOLAN (KTH, Royal Institute of Technology), Stockholm, Sweden; http://www.kth.se/eng/about/index.html.
LABORATOIRE DE COMBUSTION ET DE DÉTONIQUE (LCD), Futuroscope Chasseneuil du Poitou, France; http://www.lcd.ensma.fr/bienvenue.html.
LABORATORY FOR COMPUTATIONAL SCIENCE AND ENGINEERING (LCSE), University of Minnesota, Minneapolis, MN, U.S.A.; http://www.lcse.umn.edu/index.php?c=home.
LAVRENTYEV INSTITUTE OF HYDRODYNAMICS of the Siberian Branch of the Russian Academy of Sciences, Novosibirsk, Russia; http://www-sbras.nsc.ru/sbras/db/dep.phtml?3+31+eng.
LAWRENCE LIVERMORE NATIONAL LABORATORY (LLNL), Livermore, CA, U.S.A.; http://www.llnl.gov/. *History of LLNL* [from the 1950s to the 2000s]; http://www.llnl.gov/timeline/60s.html#.
LOS ALAMOS NATIONAL LABORATORY (LANL) Los Alamos, NM, U.S.A.; http://www.lanl.gov.
MAX-PLANCK INSTITUTES (MPIs), Germany:
– MPI FÜR ASTRONOMIE, Heidelberg; http://www.mpia.de/Public/menu.php.
– MPI FÜR ASTROPHYSIK, Garching; http://www.mpa-garching.mpg.de/.
– MPI FÜR CHEMIE [Otto-Hahn-Institut], Mainz; http://www.mpch-mainz.mpg.de/.
– MPI FÜR DYNAMIK UND SELBSTORGANISATION [previously MPI für Strömungsforschung], Göttingen, Germany; http://www.mpisf.mpg.de/index.html.
– MPI FÜR FESTKÖRPERFORSCHUNG, Stuttgart, Germany; http://www.fkf.mpg.de/start.html.
– MPI FÜR METALLFORSCHUNG, Stuttgart, Germany; http://www.mf.mpg.de/de/.
– MPI FÜR PHYSIK [Werner-Heisenberg-Institut], München, Germany; http://www.mppmu.mpg.de/.
– MPI FÜR RADIOASTRONOMIE, Bonn, Germany; http://www.mpifr-bonn.mpg.de/.
– MPI FÜR SONNENSYSTEMFORSCHUNG, Katlenburg-Lindau, Germany; http://www.mps.mpg.de/.
– MPI FÜR WISSENSCHAFTSGESCHICHTE, Berlin, Germany; http://www.mpiwg-berlin.mpg.de.
NASA, NATIONAL AERONAUTICS & SPACE ADMINISTRATION, Washington, DC, U.S.A.; http://www.nasa.gov/home/index.html?skipIntro=1.
– AMES RESEARCH CENTER (ARC), Moffet Field, CA; http://www.nasa.gov/centers/ames/home/index.html.
– DRYDEN FLIGHT RESEARCH CENTER (DFRC), Edwards, CA; http://www.nasa.gov/centers/dryden/home/index.html.
– GODDARD SPACE FLIGHT CENTER (GSFC), Greenbelt, MD; http://www.nasa.gov/centers/goddard/home/index.html.
– HISTORY OFFICE, Washington, DC: http://history.nasa.gov/; NASA History Series; http://www.hq.nasa.gov/office/pao/History/series95.html.
– JET PROPULSION LABORATORY (JPL), CalTech, Pasadena, CA: http://www.jpl.nasa.gov/.
– JOHNSON SPACE CENTER (JSC); Houston, TX; http://www.nasa.gov/centers/johnson/home/index.html.
– KENNEDY SPACE CENTER (KSC), FL; http://www.nasa.gov/centers/kennedy/home/index.html.
– LANGLEY RESEARCH CENTER (LaRC), Hampton, VA; http://www.nasa.gov/centers/langley/home/index.html.
– LEWIS RESEARCH CENTER (LeRC), Cleveland, OH; http://www.lerc.nasa.gov
– MARSHALL SPACE FLIGHT CENTER (MSFC), Huntsville, AL; http://www.nasa.gov/centers/marshall/home/index.html.
NATIONAL AEROSPACE LABORATORY OF JAPAN: KAKUDA SPACE PROPULSION LABORATORY, Miyagi, Japan; http://www.nal.go.jp/krc/eng/index.html.
National Physics Laboratory (NPL), Teddington, U.K.; http://www.npl.co.uk/.
N.N. SEMENOV INSTITUTE OF CHEMICAL PHYSICS, Russian Academy of Sciences, Moscow, Russia; http://www.chph.ras.ru/.
OFFICE NATIONAL D'ÉTUDES ET DE RECHERCHES AÉROSPATIALES (ONERA), France; http://www.onera.fr/.
PEARLSTONE CENTER FOR AERONAUTICAL ENGINEERING STUDIES, Ben-Gurion University, Israel; http://profiler.bgu.ac.il/site/public_site/Show_Unit.cfm?unit_id=94.
PENN STATE GAS DYNAMICS LABORATORY (PSGDL), Pennsylvania State University, University Park, PA; http://www.me.psu.edu/psgdl/.
POULTER LABORATORY, SRI International, Menlo Park, CA; http://www.sri.com/poulter/.
PRINS MAURITS LABORATORY (TNO-PML), Rijswijk, The Netherlands; http://www.pml.tno.nl/homepage.html.
SANDIA NATIONAL LABORATORIES (SNL), Albuquerque, NM & Livermore; http://www.sandia.gov.
SHOCK WAVE RESEARCH CENTER (SWRC), Tohoku University, Sendai, Japan; http://ceres.ifs.tohoku.ac.jp/~swrc/.
SOUTHWEST RESEARCH INSTITUTE (SwRI), San Antonio, TX; http://www.swri.edu/default.htm.
STOSSWELLENLABOR (SWL), RWTH Aachen, Germany; http://www.swl.rwth-aachen.de/.
THE INSTITUTE FOR HIGH PRESSURE PHYSICS, Russian Academy of Sciences (IFVD RAN), Troitsk, Russia; http://www.hppi.troitsk.ru/history.htm.
THE RUSSIAN FEDERAL NUCLEAR CENTER – ALL-RUSSIA SCIENTIFIC RESEARCH INSTITUTE OF EXPERIMENTAL PHYSICS (VNIIEF), Sarov, N. Novgorod, Russia; http://www.vniief.ru/english/index.html.
TOULOUSE RESEARCH CENTER, Toulouse, France; http://www.cert.fr/index.a.html.
UNIVERSITY OF TORONTO INSTITUTE OF AEROSPACE STUDIES (UTIAS), Toronto, Canada; http://www.utias.utoronto.ca/.
UNIVERSITY OF WASHINGTON AERONAUTICAL LABORATORIES (UWAL), Seattle, WA; http://www.aa.washington.edu/index.html. For the history of UWAL *see* http://www.aa.washington.edu/about/.
U.S. ARMY RESEARCH LABORATORY (ARL), Aberdeen, MD; http://www.arl.army.mil/main/Main/default.cfm.
VON KARMAN INSTITUTE FOR FLUID DYNAMICS (VKI), Rhode-Saint-Genèse, Belgique; http://www.vki.ac.be/index.html.
WISCONSIN SHOCK TUBE LABORATORY (WiSTL), University of Wisconsin, Madison, WI; http://silver.neep.wisc.edu/~shock/.

## 6.7.2 TECHNICAL MUSEUMS

AMERICAN MUSEUM OF NATURAL HISTORY, New York City, NY; http://www.amnh.org/rose/meteorite.html.
ATOMIC TESTING MUSEUM (an Affiliate of the Smithsonian Institution), operated by the Nevada Test Site Historical Foundation (NTSHF), Las Vegas, NV; http://www.nv.doe.gov/nts/museum.htm.

AUTO & TECHNIK MUSEUM SINSHEIM, Sinsheim, Germany; http://www.technik-museum.de/uk/sinsheim/.
BRADBURY SCIENCE MUSEUM, Los Alamos, NM; http://www.lanl.gov/museum/index.shtml.
DEUTSCHES MUSEUM, Munich, Germany; http://www.deutsches-museum.de/e_index.htm.
HISTORISCH-TECHNISCHES INFORMATIONSZENTRUM PEENEMÜNDE, Peenemünde, Germany; http://www.peenemuende.de/flash/museum.html.
LOS ALAMOS HISTORICAL MUSEUM, Los Alamos, NM; http://www.losalamos.com/Historicalsociety/museum.asp.
METEOR CRATER MUSEUM, Sedona, AZ; http://www.meteorcrater.com/Meteorpress.htm.
MUSEUM "LA RÉGENCE," Ensisheim, Alsace, France; http://www.ville-ensisheim.fr/page_05d.htm#reg.
NATIONAL AIR & SPACE MUSEUM (NASM), Smithsonian Institution, Washington, DC; http://www.nasm.si.edu/.
PACIFIC TSUNAMI MUSEUM, Hilo, HI, U.S.A.; http://www.tsunami.org/.
RIES CRATER MUSEUM, Nördlingen, Bavaria, Germany; http://www.noerdlingen.de/englisch/text_rieskratermuseum.htm.
SCIENCE MUSEUM, South Kensington, London, U.K.; http://www.sciencemuseum.org.uk.
TECHNISCHES MUSEUM, Vienna, Austria; http://www.tmw.ac.at/.
THE BRITISH MUSEUM, London, U.K.; http://www.thebritishmuseum.ac.uk/.
U.S. AIR FORCE MUSEUM, Wright-Patterson Air Force Base, Dayton, OH; http://www.atomictourist.com/afm.htm.
VNIIEF, NUCLEAR WEAPONS MUSEUM, Sarov, Russia; http://www.vniief.ru/museum/museum_e.html.

# Name Index

# NAME INDEX

Note: For persons with names in bold print short biographies are given in the BIOGRAPHIES INDEX.

## A

ABADIE, M. 814
ABBE, C. 411
ABBOT, H.L. 91, 349, 354, 359-360, 371, 384, 979
ABBOTT, I.H. 1071
ABBOTT, M.B. 256, 1213
ABEL, J.L. 1037
**ABEL, Sir F.A.** 23, 107-108, 110, 133, 284, 307-308, 348, 354, 358-360, 372, 393, 998, **1037 1038**, 1137, 1210
ABELL, G.O. 795
ABONYI, I. 1199
ABOU-SAYED, A.S. 737
ABRAHAM, F.F. 809, 821, 838
ABRAHAMSON, J. 811
ABRAMOVIC, G.N. 568
ABRAMOWICZ, W. 444
ACHENBACH, J.D. 98
**ACKERET, J.** 6, 36, 80, 101, 442-443, 473, 476, 480, 496-497, 503, 512, 554, 595, 637, 940, 969, **1038-1039**, 1054, 1118-1119, 1147, 1162, 1174
ACKERMANN, H. 1142
ACRIVOS, A. 780
ADADUROV, G.A. 674, 680
ADAM, C. 204
ADAMAR, Z. 1095, 1098
ADAMS, H.S. 617
ADAMS, J.C. 1058
ADAMS, M.C. 1048
ADAMS, W. 340
ADAMS, W.M. 578, 761
ADAMS, W.S. 1094
ADAMSKII, K. 96, 542
ADAMSKII, V.B. 599
ADASHKO, J.G. 1173
ADOMEIT, G. 1165
ADRIAN, R.J. 833
AGUI, J.H 828
AHARONIAN, F.A. 835
A'HEARN, M.F. 815
AHLBORN, B. 744, 1171, 1179

AHRENS, T.J. 658, 721, 772, 774, 1074, 1170
AIELLO, G. 735
AIKEN, H.H. 141
**AIRY, Sir G.B.** 29, 76-78, 306, 310-311, 315, 334, 349, **1039-1041**, 1057-1058, 1176, 1209
AIRY, W. 1039-1040
AITON, E.J. 1126
AIZIK, F. 827
AKASOFU, S.I. 492, 1208
AKIRA, M. 333
AL HASASINI, S.T.S. 1100
ALBERDI, L. 1210
ALBRITTON, C.C. 510
ALCON, R.R. 783, 806
ALDER, B.J. 392, 615
ALDRIN, E.E. 704-705
ALEKSANDROV, V.A. 630
ALEMBERT, J. LE ROND D' v, 43-45, 48, 139, 223, 234, 237, 239, 243, 246, 249, 262, 274, 276, 385, 1115, 1203
ALEXANDER, C.C. 564, 651
ALEXANDER, J. 1187, 1203
ALEXANDER, J.M. 444
ALEXANDER, R.M. 780
ALEXANDER, W. 1057
ALFTHAN, O. 638
ALFVÉN, H.O.G. 574, 632, 642
AL-ĞABARTĪ, A. 270
ALLAMAND, J.N.S. 233
ALLEN, A.O. 504
ALLEN, H.J. 595, 925
ALLEN, J.A. van 1145
ALLEN, J.E. 297, 1209
ALLEN, S. 793
ALLIÉVI, L. 419
ALLISON, H.C. 1048
ALPERS, W. 294
ALPHER, R.A. 68, 568-569, 1083
ALTEVOGT, R. 92
ALTHIN, T.K.V. 1111, 1136
AL'TSHULER, B.L. 1042
**AL'TSHULER, L.V.** 45, 93, 96, 116, 145, 542, 571, 575, 630, 647, 661,

745, 774, 788, 799-800, 907, 976, **1041-1042**, 1197-1198, 1213
ALVAREZ, L.W. 542, 747, 772, 1114, 1213
ALVAREZ, W. 747, 772, 1170, 1213
AMANN, H.O. 713
AMBROSE, S.H. 182
AMELSFORT, R.J.M. van 238
AMES, J.S. 457, 515, 1070
ANDERHOLM, N.C. 953
ANDERS, E. 646, 673, 774
ANDERSON JR., C.E. 772
ANDERSON JR., J.D. 30, 462, 692, 705, 1055, 1085, 1105, 1111, 1148, 1175, 1196, 1207, 1211
ANDERSON JR., O.E. 523
ANDERSON, B.W. 765
ANDERSON, C.D. 116, 483
ANDERSON, C.E. 145
ANDERSON, D.L. 736
ANDERSON, G.D. 1083
ANDERSON, G.Y. 639
ANDERSON, J.A. 465
ANDERSON, J.E. 584
ANDERSON, R.E. 1096
ANDERSON, R.R. 56
ANDERSON, S.E. 807
ANDERSON, T.D. 425-426
ANDRAOS, J. 1204
ANESHANSLEY, D.J. 713
ANGELES, J. 785
ANGENHEISTER, G. 1209
ANGSTROM, A.J. 1176
ANILKUMAR, A.V. 786, 1179
ANKENEY, D.P. 743-744
ANKETELL-JONES, S. 377
ANKLITZEN, B. 195
ANSORGE, J. 1163
**ANTOLIK, K.** 130, 161, 355-357, 443, 911-912, 995, **1042-1043**
ANTONIADI, E.M. 426
ANTOUN, T. 98, 837
AOKI, T. 759
APPEL, J. 1213
APPLEBEY, M.P. 1059
APPLEMAN, D.E. 674, 758

APPLETON, P.N. 177
APROINO 889
ARAGO, D.F.J. 264, 266, 288, 291, 300, 334, 1145, 1214
ARAGO, F.J.D. 284
ARBOGAST, L.F.A. 244
ARCHER, R.D. 544, 547, 765
ARCHIBALD, R.C. 1191
ARCHIMEDES, Greek philosopher 38, 1190
ARCHULETA, R. 817
ARDAVAN, A. 781
ARDAVAN, H. 781
ARGAND, A. 89, 268
ARISTOTLE, Greek philosopher 11, 22-23, 186-187, 195, 202, 244, 889
ARLAZAAROV, M.S. 1200
ARMSTRONG, J.G. 459
ARMSTRONG, N.A. 704-705
ARMSTRONG, R. 334, 1214
ARMSTRONG, Sir G.E. 349
ARMSTRONG, Sir W.G. 366, 1137
ARNETT, W.D. 712
ARNOLD, D.H. 1145
ARNOLD, G. 621
ARNOLD, H. 582, 991
ARNOLD, H.H. 457
ARNOLD, J.A. 514, 1205
ARNOULD, C. 374
ARONS, A.B. 532, 1049
AROUET, F.-M. ⇨ VOLTAIRE
ARRHENIUS, S.A. 398
ARSHANIKOW, N.S. 1200, 1207
ARTSIMOVICH, L.A. 630
ASANUMA, T. 742
ASARO, F. 747, 1213
ASAY, J.R. 93, 97, 134, 542, 664, 738, 775, 791, 901, 1210
ASCHENBACH, B. 773, 798, 835, 873
ASHCROFT, N.W. 741, 802
ASHLEY, S. 92, 105, 825
ASIMOW, P.D. 658
ASKAR'YAN, G.A. 661
ASPDIN, J. 71
ASPRAY, W. 144-145, 1135
ASSEHTON, R. 1210
ASTON, F.W. 505
ATANOV, Y.A. 398, 1180
ATCHLEY, A.A. 1214
ATEN, C.F. 373
ATHANASSOULIS, G. 821
ATWOOD, CAPT. 333
ATWOOD, T. 525
AUBREY, J. 200
AUDIC, J.M.S. 349
AUENBRUGGER, L. VON 247
AUERBACH, F. 80, 443, 1119
AUTERY, J. 859

AVDEENKO, A.I. 717
AVERY, W.H. 444
AXFORD, W.I. 656, 662, 746, 840, 872
AYERS, R. 1027
AZARKH, Z.A. 514
AZARKH, Z.M. 96, 1212-1213
AZEVEDO, R. 769
AZIZ KHAN, A. 1037

# B

BAADE, W.H.W. 491, 499
BAADER, F.X. VON 264-265
BAADER, J. VON 1132
BAALS, D.D. 84, 388, 561, 677
BABBAGE, C. 139
BACH, G.G. 1139
BACHER, R.F. 1048
BACKAS, R. 202
BACKOFEN JR., J.E. 760
BÄCKSTRÖM, H.L.J. 1168
BACON, R. 23, 106, 193-194
BACON, Sir F. 122, 210
BADASH, L. 317
BADER, F. 726
BAHCALL, J.N. 712
BAHNSEN, U. 803
BAIER, W. 351, 1210
BAILEY, A. 485
BAILEY, A.B. 338
BAILEY, C. 187
BAILEY, N.P. 482
BAILLIE, L. 1205
BAIRD, E.G. 559
BAIRSTOW, Sir L. 463, 465, 482, 596, 924
BAKANOVA, A.A. 630, 650, 661, 1041-1042
BAKER, B.B. 1098
BAKER, D. 1207
BAKER, H. 223
BAKER, H.B. 1068
BAKER, J.G. 825
BAKER, W.E. 729, 740
BAKIROV, I.T. 738
BAKKE, J.R. 791
BALARD, A.J. 1045
BALCHAN, A.S. 616
BALDWIN, R.B. 410, 576
BALFOUR, H. 273
BALIBAR, S. 313
BALL, G.J. 777, 1214
BALL, W.W.R. 1212
BALLY, J. 873
BALOGH, A. 807
BALTEAU, J. 1206
BALTRUSKA, A. 685
BAMES, E. 1063
BANCROFT, D. 548, 610, 616

BANK, D. 1206
BANKS, Sir J. 279, 289, 1096
BARANDE, J. 1158
BARANOWSKI, B. 741
BARBEE, T.W. 98
BARBER, G. 1208
BARBER, M.S. 207, 726, 1074
BARBOUR, J.P. 977
BARENBLATT, G.I. 1139, 1197-1198
BARÉNYI, B. 587
BARGMANN, H.W. 31, 300
BARKER, L.M. 681, 696, 775, 956
BARKLA, H.M. 242, 1154
BARLOW, J. 100
BARNABY, F. 118
BARNABY, S.W. 414
BARNES, E. 789
BARNES, F.K. 1149
BARNES, H.T. 476-477
BARNES, J. 186
BARNES, J.E. 819
BARNES, M.A. 697, 721
BARNES, V.E. 618, 697, 721
BARNETT, J.D. 730
BARNIM, A.A. 1167
BAR-NUN, A. 714, 966, 1213
BAR-NUN, N. 714
BARON, M.E. 1097
BARR, A. 1150
BARR, R.A. 765
BARRAL, J.A. 1145
BARRIE, D.S. 282, 1029
BARRINGER, D.M. 436, 848
BARROW, I. 1039
BARTEL, N. 789
BARTH, A. 1132
BARTHEL, P.D. 770
BASHE, C.J. 141
BASHFORTH, F.A. 124, 337-338, 973
BASOV, N.G. 662
BASS, H.E. 54-55, 754-755
BASSANI, G.F. 1203
BASSETT, W.A. 741
BASTEA, M. 802
BASTING, D. 736
BATCHELOR, G.K. 485, 615, 617, 1117, 1184
BATEMAN, H. 1145
BATEMAN, N.H. 1071
BATES, R.L. 32, 363, 1204
BATSANOV, S.S. 755, 1213
BAUER, F. 710, 726, 729
BAUER, P.A. 1210
BAUER, S.H. 714
BAUM, F.A. 542, 1173
BAUM, J.D. 1032
BAUM, S.K. 1204
BAUMANN, W. 381
BAUMUNG, K. 768, 843

NAME INDEX

BAUSCHINGER, J. 614
BAXTER, A.D. 1208
BAXTER, H.W. 582, 991
BAYER VON BAYERSBERG, H. 1159
BAŽANT, Z.P. 825, 1033
BAZHENOVA, T.V. 1171
BEALES, F.W. 176
BEATTY, J.K. 1170
BEATY, W. 909
BEAUDOUIN, L. 510
BEAUGRAND, J. 1129
BEAULIEU, A. 1130
BEAVERS, G.S. 616
BECCARIA, G.B. 244
BECHANAN, H.G. 593
BECHER, F. 599, 874
BECKER, E. 567
BECKER, H. 476
BECKER, J.V. 456, 499, 533, 541, 561, 638, 688, 943
BECKER, K. 125, 243, 429, 451, 454, 1016, 1063
**BECKER, R.A.** 284, 353, 466-468, 525, 544, 697, 907, **1043-1044**, 1069-1070, 1210
BECQUEREL, A.H. 116
BEDDOES, T. 1065
BEECHEY, F.W. 57
BEECKMAN, I. 1129
BEGELMAN, M.C. 770
BEGÉT, J.E. 27, 866
BEHRBOHM, H. 1141
BEIGLBOECK, W. 803
BEKHMETIEV, E.F. 1041
BELAJEV, A.F. 510, 1167
BELAK, A.A. 657
BELANGER, J. 616
BÉLANGER, J.B. 78, 266, 290
BELAYEV, A.F. 1174
BELGER, M. 222
BELL, A.E. 1098
BELL, A.R. 746
BELL, C.E. 696
BELL, J.K. 1184
BELL, J.N. 1170
BELL, L.D. 450, 561
BELL, P.M. 730
BELL, S.J. 694
BELL, W.W. 765
BELLA, S. DELLA 889
BELLAMY, P.M. 1074
BELLANI, A. 194
BELL-BURNELL, S.J. 694
BELLEUDY, P. 861
BELLO, F. 670
BELOTSERKOVSKY, S.M. 128
BELUGIN, V.A. 550
BEN CLYMER, A. 140
BEN MENAHEM, A. 1109

BEN-DOR, G. v, 79, 82, 145, 335, 446, 555, 583, 610, 621, 639, 673, 708, 716, 744, 746, 758, 827, 1084, 1205, 1213
BENECKE, T. 502, 517, 1088, 1148, 1164, 1207-1208
BENEDICK, W.B. 664
BENEDICT, R.P. 690, 1205
BENFORD, J. 648
BEN-MENAHEM, A. 644, 1213
BENNET, A. 263
BENNETT, A. 1214
BENNETT, B.I. 747
BENNETT, C. 351
BENNETT, C.L. 833
BENNETT, F.D. 991
BENNETT, J.A. 1195
BENNETT, W.H. 442
BENNETT, W.R. 649
BENSBERG, A. 1063
BENSON JR., O.O. 564
BENZ, C. 114
BENZ, W. 175
BERCKHEIMER, H. 1163
BEREST, Y.Y. 1098
BERETS, D.J. 1106
BEREZHKO, E.G. 746
BERG, U.I. 649
BERGE, D. 835
BERGENGREN, E. 336, 1136
BERGER, A. 1114
BERGER, D. 602
BERGER, F. 416, 471, 898, 1209
BERGER, H. 493
BERGER, J. 649
BERGER, S.A. 605
BERGERON, L.J.B. 508
BERGGREN, R.E. 552
BERGHAUS, H. 307, 857
BERGIN, T.J. 143, 1210
BERGIUS, F. 1043
BERJOSOWSKY, N. 418
BERKENKOPF, G. 1187
BERLINER, A. 473, 1063
BERNARD, F.M. Comte DE MONTESSUS DE BALLORE 441, 516
BERNDT, S.B. 1140
BERNEGGER, M. 205
BERNHARDT, U. 1210
**BERNOULLI, DANIEL** 38, 45, 77, 223, 234-237, 249, 256, 262, 269, 289, 301, 320, 358, 401, 485, **1044-1045**, 1153
BERNOULLI, JAKOB 38, 234, 1044
BERNOULLI, JOHANN 38-39, 43, 223, 233-234, 236, 246, 249, 358, 1044, 1075-1076, 1153
BERNOULLI, NIKOLAUS 1044
BERNSTEIN, B. 455, 630, 914

BERNSTEIN, D. 673, 696
BERNSTEIN, J. 517, 520, 1048
BEROALD, F. 199
BEROZA, G.C. 438
BERRY, A.J. 1057
BERSHADER, D. 129, 1110
BERTELLI, T. 349
BERTHELOT, J.M. 1045
**BERTHELOT, P.E.M.** 23, 110-111, 313, 323, 344, 350, 359-360, 369, 372-375, 383, 409, 460, 619, **1045-1047**, 1072, 1113, 1151, 1188, 1209
BERTHOLLET, C.L. 45, 201, 263, 271, 275-278, 316, 362, 1067
BERTIN, J. 1164
BERTOLOTTI, M. 634, 1210
BERTOUILLE, H. 711
BERTRAND, E. 253
BERTRAND, G. 1101
BERTSCH, W. 689
BERZ, P. 1063, 1119, 1159, 1208
BERZELIUS, J.J. 40, 303, 1046
BESANT, W.H. 81, 277, 329, 392, 461
BESSEL, F.W. 333
BESSEMER, H. 286
BESSON, J. 199
**BETHE, H.A.** 115-116, 142, 317, 487, 517, 524, 530-531, 569, 594, **1047-1048**, 1083, 1134, 1212
BETTELHEIM, A. 1069, 1119
BETTS, J. 287
BETZ, A. 80, 101, 442-443, 499, 516, 526, 587, 1039, 1070, 1147-1148, 1162, 1192, 1207
BEUDANT, F.S. 290-291, 293
BEVIS, J. 192, 253, 1210
BEWICK, T. 1096
BEYER, R.T. 30, 73, 1207
BHATTACHARJEE, A. 818
BICHEL, C.E. 417, 440, 1000
BICKFORD, W. 109, 291-292
BIELINSKI, Y. 1095
BIERMANN, L.F. 588
BIEZENO, C.B. 1080
BIGELOW, P. 767
BIGGER, R. 70
BILLIG, F. 639
BILSTEIN, R.E. 457, 628
BINA, A. 190, 244
BINET, J. 1061
BINNEY, J. 17
BINZEL, R.P. 813
BIOT, J.B. 81, 266, 271-273, 276-277, 331, 1056
BIRCH, F. 741, 1052
BIRCH, T. 212
BIRD, R. 398
BIRETT, H. 1132, 1211

BIRKELAND, K. 426
BIRKHOFF, G.D. 330, 565, 582, 1135, 1184
BISHOP, G. 318
BISNOVATYI-KOGAN, G.S. 666
BISPLINGHOFF, R.L. 104, 1071
BISSHOPP, F. 1167
BITTER, F. 714
BITTERLY, J.G. 559
BIZONY, P. 1082
BJERKETVEDT, D. 791, 1205
BJOERNOE, L. 414, 728
BLACK, J. 482, 1119
BLACKBURN, J.H. 1006
BLACKMORE, J.T. 1119-1120
BLACKSTOCK, D.T. 325
BLACKWELL, D.E. 657
BLAIR, N. 706
BLANC, C. 1076
BLANDFORD, R.D. 545, 655, 746
BLAU, E. 1146
BLAYLOCK, N.W. 772
**BLEAKNEY, W.** 81, 85, 484, 495-496, 536, 551, 577, 591, 595, 945, **1048-1049**, 1181-1182, 1193
BLECHAR, T. 610
BLESS, S.J. 90, 714
BLIZNIUK, V. 699
BLOCHMANN, G.F.R. 91, 349, 417, 985
BLOCHMANN, R. 417, 1108
BLOCK, A. 384
BLOCK, S. 730
BLONDIN, J.M. 66
BLOOM, J.S. 813
BLOOM, M.H. 1078
BLOT, G. 1130
BLOUKE, M.M. 707
BLOXAM, C.L. 110, 358, 1037
BLOXSOM JR., D.E. 990
BLUHM, H. 768
BLUMENSCHINE, R.J. 48
BLYTH, W.J. 786
BOAS, M. 202
BOASE, F. 1075, 1206
BOASE, G.C. 1125
BOAS-HALL, M. 208
BOBBITT, P.J. 1078
BOBRIK, E. 885
BOCHNER, S. 1135
BÖCKMANN, F. 1210
BODENSTEIN, M.E.A. 113, 216, 452, 1053, 1059, 1105, 1168
BOER, J.Z. DE 280
BOËTHIUS, B. 1206
BOGDANOFF, D.W. 772, 1091
BOGDONOFF, S.M. 1065, 1078
BOGGS, E.M. 537
BOGUSLAWSKI, G.G. VON 361, 1205

BOHLEN UND HALBACH, B. VON 446, 887
BOHM, H. 1140
BOHOR, B.F. 767
BOHR, H.D. 39
BOHR, N. 117, 514
BOILEAU, J. 109
BOIS DUDDELL, W. DU 471
BOLEY, B.A. 319, 1158
BOLKHOVITINOV, L.G. 341
BOLSTER, C.M. 587
BOLTZMANN, L. 86, 227, 320, 352, 373, 1010, 1119, 1185
BOLTZMANN, L.E. 320, 352, 373
BOLZ, N. 1210
BOLZANO, B. 1069
BONANNO, A. 174
BONAPARTE, N. 265, 270, 275-276, 1066, 1150
BONAZZA, R. 1179
BOND JR., J.W. 681
BONDALETOV, V.N. 709
BONDI, H. 68, 566
BONDORF, J.P. 1048
BONE, W.A. 485, 994, 1067-1068, 1143
BONNELL, J. 183
BONNET-BIDAUD, J.M. 1048
BONNEY, W.T. 561
BONWETCH, A. 475
BOOK, D.L. 761
BOOKER, J.D. 680
BOON, J.D. 510
BOOTH, E.T. 117
BOPP, F. 1089, 1110
BORELLI, G.A. 215
BORGLIN, S. 1111
BORIA, V. 1130
BORIS, J.P. 761
BORISOV, A.A. 748, 775, 1198, 1210-1211
BORLINETTO, L. 256
BORMANN, P. 1103
BORN, M. 86, 128, 1103
BORNE, G. VON DEM 447
BORNHARDT, A. 366
BORST, R. DE 1203
BORZESZKOWSKI, H.H. VON 1077
BOS, H.J.M. 209, 1098
BOSCOVICH, R.J. [R.G.] 225, 1056
BOSLOUGH, M. 794
BOSSUT, C. 262
BOSTOCK, J. 188
BOUCHER DE PERTHES, J. 47, 299
BOUCHER, G. 707
BOUCHON, M. 817, 826
BOUCKAERT, L. 1114
BOUDONARD, O. 1113
BOUGARD, M. 1115

BOURCIER DE CARBON, CH. 599
BOURGEOIS, J. 178
BOURNE, J.F. 334, 1214
BOURNE, N.K. 696
BOURNE, W. 237
BOUSSINESQ, J.V 297, 1158
BOUSSINGAULT, J.P. 1150
BOUTARIC, A. 1046
BOUTRON-CHARLAND, A. 294
BOUTT, E. 1130
BOVY, P.H.L. 613
BOWDEN, F.P. 680-682, 936
BOWEN, E.J. 1059
BOWEN, I.S. 1094
BOWEN, J.R. 1139
BOWER, J. 430
BOWER, L. 1067
BOWERS, A.H. 790
BOWERS, B. 280, 1066
BOWMAN, J.S. 1206
BOWNE, J. 834
BOWSER, M.L. 622
BOWSHER, G. 1027
BOYD JR., T.J. 686
BOYD, F.R. 686, 848
BOYER, C.B. 1145, 1212
BOYER, D.W. 623, 986
BOYLE, E. 273
BOYLE, R. 23, 44, 80, 128, 196, 200, 208-211, 221-222, 226-227, 1126, 1141
BOYLE, W.S. 707
BOYS, C. 1049
**BOYS, Sir C.V.** xxi, 128, 401, 448, 922, **1049-1051**, 1119, 1193
BRADBURY, N. 987
BRADFIELD, W.S. 341
BRADLEY, B. 47
BRADLEY, J.N. 653, 1106, 1209, 1213
BRADLEY, R.S. 96, 691, 909, 1052, 1074
BRADSHAW, L. 1067
BRAGG, W.H. 133
BRAHE, T. 66-67, 199-200
BRAKEL, A. 1170
BRAMAH, J. 90, 267
BRANCA, G. 926
BRANCH, D. 469
BRANCO, W. 844
BRANT, S. 850
BRAUER, F.E. 702
BRAUN, F. 125
BRAUN, W. VON 102, 490, 507, 529-530, 608, 631, 1054, 1088
BRAVAIS, A. 1122
BRAZHNIK, M.I. 647, 650, 661, 1042
BREDEHOEFT, J.D. 659
BREDIG, G. 1046
BREDT, I. 91

# NAME INDEX

BREIDENTHAL, R.E. 689, 1091
BREIDERT, W. 247, 1103, 1210
BRENDEL, W. 715, 1213
BRENNER, A. 1073
BRENNER, M.P. 1184
BRESCIANI, B. 205
BREUSOV, O.N. 674
BRIDEL, G. 1039
BRIDGMAN, R.L. 1051
**BRIDGMAN, Sir P.W.** 6, 87, 94-97, 135, 210, 407, 449, 473, 491, 509, 521, 523, 542, 566, 610, 615-616, 626, 667-668, 740, **1051-1052**
BRIERS, F. 1059
BRIGGS JR., J.M. 48, 1154
BRIGGS, L.J. 463, 469, 1071
BRIOT, C. 1191
BRISH, A.A. 1042
BRITAN, A. 827
BRIXNER, B. 545, 556, 593, 988, 1006, 1018
BROAD, W.J. 1048
BROCHET, C. 1171
BROCK, N. 126
BROCK, W.H. 1209
BRODE, H.L. 20, 585, 605, 623, 639, 1084
BRODSKY, E.E. 1179
BRODY, H. 750-751, 904
BROGLIATO, B. 377, 814
BROGLIE, L. DE 1072
BRONSHTEN, V.A. 1174
BROOKS, R.E. 562, 690
BROSCH, G. 831
BROSE, (?) 252
BROUGHAM, H. 1057
BROUILLETTE, M. 712, 837, 1179
BROUNCKER, V. 210
BROUZENG, P. 1073
BROWN, D.K. 1156
BROWN, E.O. 354
BROWN, E.W. 1176
BROWN, F.N.M. 641
BROWN, G.E. 1048
BROWN, G.G. 480, 619
BROWN, H.N. 621, 1076, 1153-1154
BROWN, J.A. 792
BROWN, K. 822
BROWN, L.M. 5
BROWN, P.J. 807
BROWN, R. 45, 289, 906
BROWNE, C.A. 407, 1133
BROWNE, J. 202
BROWNING, J. 352
BROWNLEE, D. 813
BRUCE, J. 1096
BRÜCHE, E. 1186
BRUCKNER, A. 589
BRUCKNER, A.P. 689, 772-773, 1091

BRUEN, B. 1117
BRUGGERT, G.W. 826
BRUGMANS, H.L. 1098
BRUGNATELLI, L.V. 263
BRUK, Y. 1145
BRUNEL, I.K. 1156
BRUNO, C. 1065
BRUNSWICK, H. 429
BRUNTON, J.H. 682, 715, 936
BRUSH, S.G. 1211, 1214
BRYAN, W.F. 874
BRYANT JR., L.E. 510, 1210
BRYANT, C.N. 1172
BRYANT, E.J. 604, 607, 732
BRYANT, L.E. 132
BRYNJOLFSSON, I. 859
BUCHANAN, D.J. 728
BUCHER, W.H. 497
BUCHHEIM, W. 1132
BUCHNER, C.L.O. 347, 1211
BUCKINGHAM, E. 729
BUCKLAND, G. 316
BUCKLEY, F. 1146
BUCKMASTER, J.D. 684, 1091
BUDDE, E. 1040
BUEHLER, M.J. 838
BÜRGER, W. 213
BUFFON, Comte DE ⇒LECLERC, G.L.
BUGGE, G. 108, 1047, 1209-1210
BUGGE, T. 279
BULL, G.V. 224, 588, 653
BULL, W.H. 1074
BULLARD, F.M. 1214
BULLEN, K.E. 122, 1109
BULSON, P.S. 780
BUNDY, F.P. 607, 741
BUNDY, M.L. 654
BUNNEY, S. 47
BUNSEN, C. 1052
**BUNSEN, R.W.** 107, 111, 268, 303, 316, 322-323, 338, 343-344, 350, 362, 370, 397, 409, 999, 1042, **1052-1054**, 1066-1067, 1112
BUNTZEN, R.R. 663
BUONAVENTURI, T. 205
BURBACH, T. 1088
BURCHETT, W. 1081
BURGER, W. 860
BURGERS, J.M. 499, 574, 601, 1080, 1104, 1117
BURGESS, C.H. 1059
BURGESS, D. 65, 955
BURGESS, T.J. 207
BURIDAN, J. 43, 195
BURKE, R.B. 193
BURKHARDT, G. 523, 534-535, 1070
BURKHARDT, H. 1108
BURKS, A.W. 138, 141, 556, 1134-1135

BURLAGA, L.F. 840
BURLOT, M.E. 1113
BURNETT, D. 353
BURNS, K. 589, 1091
BUROV, M. 1080
BURRILL, L.C. 384
BURROWS, A. 773
BURSILL-HALL, P. 225
BURTON, B.L. 610
BURTON, C.V. 145, 393, 408, 448, 907
BURTSCHELL, Y. 838
BURY, R.G. 185
**BUSEMANN, A.** 80, 101, 140, 226, 443, 473, 488, 490, 498, 503, 516, 525, 718, 1022, **1054-1055**, 1078, 1147-1148, 1163-1164, 1211
BUSH, G. 742
BUSH, V. 139-140, 519
BUSNEL, R.G. 304, 1146, 1207
BUSSE, A. 698
BUTCHER, B.M. 708, 1074, 1213
BUTLER, C.P. 969
BUTLER, D.S. 531
BUTOWSKY, H.A. 591
BUTTER Bros. 982
BUTTER, K. 504
BUTTER, O. 504
BUTTERWORTH, S. 470
BÜTTNER, R. 20
BUXTON, L.D. 724
BUYS-BALLOT, C.H.D. 302, 1068
BUZZARD, R.D. 700
BYCHKOV, A.V. 811
BYCHKOV, V.L. 811
BYERLY, P. 61, 190, 1213
BYKOV, A.I. 648
BYKOV, A.M. 1083
BYRD, R.E. 121, 1145-1146

# C

CABANES, C. 1142
CACCIATORE, G 896
CAESAR, G.J., Roman dictator 56
CAIN, J.C. 1067
CAIRD, R.S. 207, 647
CAJORI, F. 223-224, 226
CALDER, R. 455
CALDIROLA, P. 1160
CALDWELL, F.W. 462
CALKIN, J.W. 1135
CALZIA, J. 430
CAMBEL, A.B. 37, 612, 642, 1105
CAMERON, A.G.W. 174-175, 267, 699, 1208
CAMERON, S. 286
CAMPBELL, A.W. 654
CAMPBELL, C. 475-476, 994, 1067
CAMPBELL, D.C. 526, 537

CAMPBELL, R.G. 647
CAMUS, J.J. 223, 715
CANNING, T.N. 14, 520, 552, 977
CANT, J.G. 1102
CANTON, J. 86, 254
CANUP, R.M. 175
CAP, P.A. 1115
CARABIN, H. 430
CARDOSO, L. 247
CARILLI, C.L. 770
CARL, L.R. 541
CARLISLE, R.P. 564, 627, 1214
CARLOMAGNO, G.M. 567
CARLSON, H.W. 586-587, 923
CARMAN, W.Y. 1208
CARMODY, T. 236, 1045
CARNOT, H. 277
CARNOT, L.N.M. 259-260, 265, 275, 277, 1060
CARNOT, S. 45, 259, 266, 276-277
CARNPROBST, J.L. 1155
CARPENTER, C.W. 1143
CARPENTER, R.C. 419
CARPENTER, S.H. 541
CARR, J. 49
CARR, M.J. 668
**CARRÉ, L.** 89, 231-232, 251, **1055**
CARRIÈRE, Z. 477, 914
CARRINGTON, R.C. 326-327, 870
CARRINGTON, T. 600
CARROLL, (?) 747
CARSON, M. 117
CARTER, D. 794
CARTER, N.L. 673
CARTLIDGE, E. 781
CARTWRIGHT, M.L. 1086
CARVALHO E MELLO, SEBASTIÃO JOSÉ DE ⇨ POMBAL, MARQUÉS DE
CASALEGNO, R. 873
CASCI, C. 1065
CASSAGNOU, R. 547, 1161
CASSEN, P. 749
CASSIDY, D. 517
CASSINI DE THURY, C.F. 236
CASSINI JR., J. 156, 231-232, 236
CASSINI, J. 231, 1055, 1099, 1153
CASSINI, J.D. 227, 737, 1127
CAST, M. 798
CASTELFRANCO, J. 1064
CATES, J. 1179
CATHERASOO, C.J. 1179
CAUBLE, R. 715
CAUCHY, A.L. 60, 77, 221, 266, 282, 284, 288, 318, 1060
CAUGHEY, D.A. 761, 1211
CAVALLERI, R.P.A. 77
CAVALLI, A. 229
CAVENDISH, C. 1056
CAVENDISH, F. 1056

**CAVENDISH, H.** 253, 258, **1056-1057**
CAVENDISH, S.C. 1057
CAVENDISH, W. 1056
CAYNE, B.S. 241, 352, 524
CEANGA, V. 785, 814
ČERENKOV, P.A. 500, 773
ČERKES, H.K. 271
CERNAN, E.A. 724
CHACE, W.G. 582, 624, 632, 637, 663-664, 674, 689-690, 713, 716, 991, 1006, 1210
CHADWICK, J. 493
CHAHINE, G.L. 15, 985
CHAIKEN, R.F. 631
CHAIKIN, A. 1170
CHAIKIN, B.O.A. 1170
CHALLEAT, J. 1189
**CHALLIS, J.** 77-78, 165, 250, 276, 310-311, 316, 1040, **1057-1058**, 1144, 1176
CHALMERS, A. 1055, 1206
CHAMBERLAIN, J.L. 1133
CHAMBERLAIN, O. 609
CHAMBRÉ, P.L. 235
CHAMPION, M. 723, 773, 796, 1078, 1209
CHAMPION, P. 110
CHAN, S.K.L. 1084
CHANCE, B. 127
CHANDLER, D.L. 333
CHANDRASEKHAR, S. 212, 333, 491-492, 602, 819, 1100
CHANETZ, B. 827
CHANG, C.M. 1165
CHANG, J. 764
CHANG, S. 706
CHANG, T.S. 1110
CHANG HÊNG 190
CHANNELL, D.F. 1150
CHAO, E.C.T. 618, 645-646, 658, 663, 689, 848, 1169-1170
CHAPELLE, (?) 1200
CHAPLYGIN, S.A. 1166, 1200
CHAPMAN, A. 1041
**CHAPMAN, D.L.** 23, 39, 112-113, 409, 420, 425, 434-435, 460, 490, 993, **1058-1059**, 1067, 1101, 1130
CHAPMAN, D.R. 305, 697, 721, 749
CHAPMAN, J. 16
CHAPMAN, S. 353, 492, 872, 1109, 1213
CHAPMAN, S.C. 451
CHAPRON, P. 738
CHAPTAL, J.A. DE 272
CHARBONNIER, M.P. 1208
CHAREIX, F. 207
CHARLES I, king of Great Britain and Ireland 1194

CHARLES II, king of Great Britain and Ireland 212, 218
CHARLTON, T.M. 1093
CHARPY, G. 402, 427, 437
CHARTAGNAC, P. 573
CHARTERS, A.C. 534, 552
CHASSELOUP-LAUBAT, F. DE 286
CHAU, H. 756, 785
CHAUDRI, M.M. 212
CHAUSSY, C. 715, 738
CHECHEROV, K.P. 771
CHEN, B. 785
CHEN, C.H.R. 816
CHEN, Y. 835
CHENG, S. 1065
CHER, M. 1106
CHEREMISINOFF, N.P. 1203
CHEREPANOV, G.P. 377
CHEREPASHCHUK, A.M. 737
CHÉRET, R. 94, 1059, 1095, 1101, 1210, 1213
CHERKASHOV, Y.M. 771, 1031
CHERNIN, A.D. 1080
CHERNYI, G.G. 1167
CHESICK, J.P. 1106
CHESTER, D.K. 248
CHEVREUL, M.E. 107, 287
CHHABILDAS, L.C. 791, 812
CHHATWAL, G.R. 1203
CHIBA, A. 816
CHICK, M.C. 430
CHIDAMBARAM, R. 96
CHIEN, K.Y. 1139
CHILDERS, R.L. 46
CHILOWSKY, C. 84, 462
CHINCARINI, G.L. 795
CHIN-SHU 191
CHISNELL, R.F. 621
CHISOLM, E.D. 86
CHIU, H.Y. 670
CHLADNI, E.F.F. 63, 266, 273, 1042
CHOCHOL, D. 750
CHOU, P.C. 1074
CHPOUN, A. 827
CHRAMOV, J.A. 1198
CHRISTENSEN, C. 705
CHRISTIAN, R.H. 615, 677, 978
CHRISTIANSEN, J.A. 1168
CHRISTOFFEL, E.B. 362, 542
CHRYSTAL, G. 1181
CHUGUEV, V. 1205
CH'UNG, W. 189
CHURCH, I.P. 419
CHURCH, P.D. 573
CHUSHKIN, P.I. 443
CHYBA, C.F. 443
CIOFFI, D.F. 770
CIOLKOSZ, T. 709
CIOLKOWSKY, S. 1081

NAME INDEX

CIPOLLA, J. 581
CLAEYS, P. 178
CLAIRAUT, A.C. 249
CLANNY, W.R. 281, 1029
CLARK, (?) 430
CLARK, A. 333
CLARK, D.H. 192, 1208
CLARK, D.S. 223, 577
CLARK, F.G. 704
CLARK, J. 1204
CLARK, P.W. 515
CLARK, R 1033
CLARK, T.C. 1106
CLARKE JR., R.S. 674, 758
CLARKE, J.H. 1078
CLAUS, J.R. 829
CLAUSEN, H.B. 184
CLAUSIUS, R.J.E. 45, 313, 320, 323-324, 341, 500, 1089, 1149
CLAVIN, P. 723, 1139
CLAY, C.S. 288
CLÉBSCH, A. 1157
CLEMENCEAU, P. 506
CLEMENTS, R.E. 778
CLERK, Sir D. 1092
CLERKE, A,M, 1058
CLIFTON, R.J. 579, 737-738
CLINTON, B. 809
CLINTON, B., U.S. President 115, 1048, 1212
CLIVER, E. 327
CLOTFELTER, B.E. 617
CLUTTERHAM, D.R. 1182
COCHRAN, A.J. 408
COCKROFT, J.D. 116
COES, L. 600
COHEN, C.B. 1065
COHEN, E.R. 553
COHEN, R.S. 532, 534, 537, 1119, 1194, 1213
COHEN, S. 741
COIFMAN, B. 806
COKER, D. 817
COLBERT, J.B. 213, 221, 1127
COLE JR., T.W. 823
COLE, A.L. 647, 1110
COLE, J.D. 724
COLE, R.H. 91, 150, 354, 371, 526, 535, 537, 566, 729, 962, 967, 978, 1180
COLEBURN, N.L. 700, 962
COLELLA, P. 548, 582
COLEMAN, A.J. 710
COLES, D. 1179
COLEY, G.D. 716
COLGATE, S.A. 643, 696-697, 728
COLLADON, D. 288, 291, 1128
COLLAR, A.R. 558
COLLATZ, K.G. 1206

COLLENS, H. 1050
COLLIER, R.J. 449
COLLINS, H.E. 463
COLLINS, J.P. 1139
COLLINS, M. 704-705
COLLINS, R. 700
COLLINS, R.A. 694
COLLINS, R.B. 1160, 1162
COLLINSON, P. 241, 244-245
COLTON, J.D. 682
COMPTON, A.H. 1114
COMRIE, L.J. 139
CONDON, E.U. 254, 1182, 1205
CONN, W.M. 582, 991, 1210
CONNOR, J.G. 723
CONSELICE, C. 873
CONSIDINE, D.M. 1205
CONSTANT, J.M. 1130
CONWAY, E.M 84
COOK, J.H. 110, 567, 950
COOK, M.A. 24, 484, 616, 663
COOK, N.G.W. 40
COOK, S.S. 457, 461, 482, 967
COOKSEY, M. 773
COOKSON, A.H. 719
COOMBS, A. 620
COOPER, A.W. 735
COOPER, G. 706
COOPER, G.E. 567
COOPER, G.R. 654
COOPER, P.W. 435
COOPERSTEIN, J. 1048
COPERNICUS, N. 889, 1079
COPSON, E.T. 1098
CORCORAN, T.H. 188
CORDS JR., P.H. 596
**CORIOLIS, G.G. DE** 49, 240, 266, 288, 295, 298, 895, **1059-1061**, 1157
CORIOLIS, J.B.E. DE 1059
CORLISS, W.R. 84, 388, 561, 677
CORNER, J. 552, 1112, 1208
CORNISH, V. 82, 447, 861
COSTA, A.B. 1068
COSTABEL, P. 208, 1061, 1127, 1145
COSTE, H. DE 1129
COUGHLIN, W.J. 1071
COULOMB, C.A. 262, 295
COULTER, W.A. 853
**COURANT, R.** 15, 23, 81, 256, 389, 392, 482, 535, 567, 614, **1061-1062**, 1080-1081, 1153, 1213
COURANT, S. 1061
COURTÈS, G. 655
COURTNEY-PRATT, J.S. 805, 1160
COWAN, G.R. 583, 954
COWARD, H.F. 1142-1143
COWLING, T.G. 353, 1213
COX, D.C. 57, 420, 559, 651
COX, H. 295

COX, R.C. 1124-1125
COX, R.N. 925
COX, Sir R. 1138
CRABTREE, L.F. 925
CRAGGS, J.W. 1116
CRAIK, A.D.D. 1209
CRANDELL, D.R. 35
CRANE, H.R. 1110
**CRANZ, C.J.** 35, 72, 125, 204, 213, 221, 231, 237, 243, 292, 324, 355-356, 358, 387, 429, 432, 451, 454, 473, 477, 486-487, 495, 512, 658, 915, 959, 973, 1000, 1014, 1016, 1018, **1062-1063**, 1161, 1208, 1210
CRARY, A. 526
CRAWFORD, D. 794
CRAWFORD, F.C. 457
CRAWFORD, J. 120
CRAWFORD, P.C. 759
CREDE, C.E. 653
CREHORE, A.C. 416
CREMER, E. 1168
CREMER, L. 640
CREMERS, D.A. 1074
CREW, H. 204
CREWDSON, R.C. 98, 733
CRIGHTON, D.G. 73, 1117
CRIST, S. 301
CRITCHFIELD, C.L. 1047
CROCCO, G.A. 321, 492-493, 502, 505, 558, 1064
**CROCCO, L.** 31, 503, 505, 508, 631, **1064-1065**
CROCKER, M.J. 30, 73, 1117, 1203
CROFT, H.O. 535
CROLL, O. 108, 200
CROMBIE, A.C. 1130
CROMWELL, O. 1189
CRONIN, J.R. 706
CRONIN, J.W. 451, 500
CROSLAND, M.P. 1046
CROSLEY JR., H.E. 259
CROSS, J. 318, 743, 1210
CROSS, P.E. 566
CROSS, R. 809
CROSSE, A. 283, 990
CROSSFIELD, A.S. 450
CROSSLAND, Sir B. 541
CROSSLEY JR., H.E. 559, 583, 694, 957-958
CROW, A D. 494
CROW, H.E. 483
CROWDER, J.P. 765
CROWLEY, E.L. 1206
CROZIER, W.D. 551, 951
CRUM, L.A. 777
CRUSSARD, L. 390, 441, 460, 1101
CRUVELLIER, P. 655

CRYSTAL, D. 1206
CULBERTSON, D.W. 708, 945
CULLUM, G.W. 1155
CULP, R.D. 1055
CUMMINS, H.Z. 303
CUNAEUS, A. 241
CUNDILL, J.P. 398
CURIE Bros. 135, 370
CURIE, J. 370
CURIE, P. 370
CURIE, P.J. 70, 781
CURRAN, D.R. vii, 160, 655, 720, 733, 775, 817, 837
CURRIE, P. 1086
CURRIE, P.J. 178
CURTIS, G.H. 449
CURTIS, H.D. 462, 491
CURTISS, C. 398
CURTISS, J.H. 1182
CURTIUS, T. 402
CURY, J.E. 570
CUSHMAN, R.H. 1078
CUSTANCE, A.C. 692
CUTLER, J. 221
CUVIER, F.G. 287, 1203
CYBULSKI, W.B. 1144

# D

D'ALEMBERT, J. LE ROND ⇨ ALEMBERT, J. LE ROND D'
D'ALMEIDA, T. 812
D'ELIA, A. 1098
DABORA, E.K. 1210-1211
DAGUERRE, L.J.M. 300
DAHAN, M. 785
DAHENG, W. 778
DAHLGREN, J.A.B. 1100
DAHMS, H.J. 476
DAI, Z.R. 778
DAIBER, J.W. 690, 1091
DAIMLER, G. 114
DAINTITH, J. 1206
DAINTON, Sir F. 1106-1107, 1168
DALYMPLE, D.G. 1100
DANA, J.D. 64
DANIELS, W.B. 97
DANILENKO, V.V. 778
DANILOVSKAYA, V.Y. 300, 583
DANNESKIOLD, J. 687
DARBES, J.F.A. 1077
DARDEN, C.M. 554
DARLEY, Sir R. 1189
DARLING, D. 1205
DARMSTAEDTERS, L. 1205
DARRIGOL, O. 1212
DARTON, N.H. 436
DARWIN, C.R. 40, 293, 304, 317, 350, 413, 418, 692

DARWIN, Sir G.H. 413, 418
DASKIN, W. 622, 1078
DAUB, E.E. 1150
DAUBEN, J.W. 1212
DAUBENTON, L.J.M. 245, 257
DAUBREE, G.A. 1122, 1151
DAUTRICHE, H.J. 440, 460, 1002
DAVEY, J. 1027
DAVID, E. 1160, 1162
DAVID, H.G. 625
DAVID, L. 602
DAVIDSON, D.F. 827
DAVIDSON, L.W. 1095
DAVIDSON, N. 600
DAVIES, B. 1127
DAVIES, P.C.W. 795
DAVIES, R.E. 598
DAVIES, R.M. 568, 617, 900, 1117, 1184
DAVIES, T.L. 203
DAVIS JR., L. 633
DAVIS, D.R. 174
DAVIS, K. 426
DAVIS, L. 1203
DAVIS, L.K. 658
DAVIS, T.L. 107, 1133, 1136, 1209-1210
DAVIS, W.C. 113, 523, 654, 697, 723, 751, 1211
DAVISON, C. 32, 307, 857, 1124
DAVISON, G. 302
DAVISON, L. 96, 791
DAVISON, L.W. 784
DAVY, J. 1066
DAVY, R. 1065
**DAVY, Sir H.** 111, 228, 274, 279-281, 374, 1029, **1065-1066**, 1112, 1115
DAWKINS, R. 47
DAWSON, J.M. 1091
DAY, C.A. 739
DEAL, W.E. 587, 601, 604, 621
DEAN, A.M. 1106
DEAN, D.R. 1124
DEAN, J. 713
DEAN, R.G. 28
DEAR, J.P. 92, 777
DEAR, P. 1130
DEBRAY, J.H. 1151
DEBUS, H. 107, 375
DEBYE, P.J.W. 133
DECARLI, P.S. 65, 656, 658
DEDEKIND, J.W.R. 1191
DEDEKIND, R. 1153
DEEDS, E. 553
DEFFET, L. 695
DEFOE, D. 32
DEITZ, R. 60
DELACOUR, J. 1160
DELASSUS, E. 472

DÉLERY, J.M. 827
DELIUS, M. 777, 1213
DELSEMME, A.H. 1170
DEMING, D. 15
DEMOCRITUS, Greek philosopher 44
DEMPSTER, A.J. 505
DENARDO, B.P. 14
DENCE, M.R. 689
DENDY, R.O. 451
DENGLER, R.P. 744
DENISOV, Y. 640
DENISOV, Y.N. 640, 995
DENLEY, C. 1181
DENNISON, M. 1110
DEPARCIEUX ⇨ PARCIEUX, A. DE
DEPREZ, M. 124, 355
DERIBAS, A. 784
DES MARAIS, D.J. 706
DESAGULIERS, J.T. 223, 233, 883, 892, 1127
DESARGUES, G. 1129
DESCARTES, R. 43-44, 67, 151, 204, 207-209, 215, 217, 220, 222-223, 229, 1097-1098, 1125, 1127, 1129, 1194
DESCH, C.H. 1113
DESCHOUX, S. vii
DESHAIES, B. 723, 773, 796, 1078, 1209
DESMAREST, N. 250-251
DÉSORTAUX, E. 374
DETTLEFF, G. 751
DEUBNER, F.L. 810
DEURING, M. 1204
DEUTSCH, F.W. 356
DEVIENNE, F.M. 626
DEWAR, Sir J. 393, 430, 1037
DEWEY, J.M. 61, 79, 134, 190, 555, 593, 608, 639, 672-673, 753, 757, 827, 992, 1162, 1213
DEWITT, C. 662
DEXTER, R.R. 688
DHOMBRES, J. 1047
DÍAZ, F.G. 789
DICK, R.D. 708
DICK, S.J. 787
DICK, W.R. 1208
DICKINSON, H.W. 1142
DIDEROT, D. v, 246, 1115, 1203
DIDION, I. 1154
DIECKMANN, M.E. 451
DIEKE, G.H. 1194
DIESEL, E. 408
DIESEL, R. 273, 408
DIETZ, R.S. 64-65, 552, 646, 844
DIEUDONNÉ, J.A. 1135, 1212
DIJKSTERHUIS, E.J. 1098
DINNIK, A.N. 376
DIRAC, P.A.M. 483, 1114

DIRKSEN, B. 1192
DITMIRE, T. 604, 662
DITTBENNER, G. 756
DITTLER, R. 434, 1186, 1204
DITTMAR, H. 525, 1022
DITTMAR, P. 683
DIVEN, B.C. 745
DIVNOV, I.I. 1042
**DIXON, H.B.** 23-24, 316, 362-363, 370, 409, 420-421, 424, 430, 460, 476, 1003, 1046, 1053, 1058-1059, **1066-1068**
DIXON, W.H. 1066
DLOTT, D.D. 31, 822
DÖBLER, G. 1103
DOBRATZ, B. 759, 1205
DODD, R.J. 180
DODGE, D.A. 823
DODGE, F.T. 729
DODSON, B.W. 755
DODSON, R.W. 514
DOERFFER, R. 919
DOIG, P. 1208
DOLAN, D.H. 842
DOLAN, W.W. 518
DOLD, B. 805
DOLOTENKO, M.I. 648
DOMENICO, N. 445
DOMICH, E.G. 616-617
DONAHUE, L. 570
DONAR, TEUTONIC DEITY 859
DONCEL, M.G. 1089
DONGEN, M. van 751, 803
DONGEN, M.E.H. van 827
DONNELL, L.H. 489
DOPPELMAYR, J.G. 197
**DOPPLER, C.A.** 82, 100, 294, 302, 309-310, 387-388, 395, 433, 437, 463, 481, 918-919, **1068-1069**, 1098, 1117
DOPPLER, J.C. 1069
DOPPLER, J.E. 1068
DORAN, J. 1082
DÖRFFEL, M. 1214
DÖRING, G. 1069
**DÖRING, W.S.** 112, 522-523, 525-526, 532, 534-535, 993, 1044, **1069-1070**, 1134, 1162, 1196
DORMEVAL, R. 573
DORNBERGER, W. 101-102, 529-530, 1054, 1207
DORSCHNER, J. 1069
DORSETT JR., H.G. 646
DOSS, C.B. 54, 411, 1119
DOTT JR., R.H. 1066
DOTTY, B. 120
DOUGHERTY, S.M. 737
DOUGLAS, A.V. 479, 1114
DOUGLAS, G.P. 469

DOUGLAS, Sir H. 314
DOUKAS, A.G. 817
DOVE, J.E. 716
DOWD, R.S. 731
DOWNES, K. 1194-1195
DRAGOSET, W. 51
DRAINE, B.T. 34, 791
DRAKE, E.T. 212
DRAKIN, V.P. 1042
DRAPER, J.W. 304
DRAPER, S. 695
DREMIN, A.N. 476, 674
DREMIN, I.M. 93, 575, 1042
DRESSLER, B.O. 788
DREYSE, N. 289, 1209
DRICKAMER, H.G. 616, 740
DRINKER, H.S. 352
DRON, A. 1080
DROSDOWSKI, G. 348
DROUIN, G. 772
DRUMMOND, W.E. 433, 621
DRURY, L.O'C. 451
**DRYDEN, H.L.** 104, 433, 463, 469, 521, 541, 575, 609, 652, **1070-1071**, 1105, 1141, 1207
DRYDEN, S.I. 1070
DRYFHOUT, A.G. 60
DRYFHOUT, J.F. 254
DU MONTIER, (?) 273
DU PONT, L. 106
DUANE, A. 1151
DUBERG, J.E. 609
DUBINSKI, J. 179, 821
DUBOIS-REYMOND, E. 1062
DUBOIS-REYMOND, P. 1062, 1174
DUBOVIK, A. 1210
DUBS, R. 419
DUDOLADOV, I.P. 1042
DUERBECK, H.W. 1206
DUERRE, D.E. 648
DUFF, (?) 532
DUFF, R.E. 600
DUFFY, P. 591, 1187
DUFLOS, C. 1130
DUFOUR, A.E. 125-126
DUFTSCHMID, C.E. 713
DUGAS, R. 1061, 1095, 1212
DUHAMEL, J.M.C. 220, 266, 299
DUHEM, P.J. 1072
**DUHEM, P.M.M.** 29, 72, 425, 427, 440, 446, 460, 542, **1072**-1073, 1086, 1095, 1100-1101, 1119, 1198-1199
DUKE JR., C.M. 595, 724, 1049
DUKE, M.B. 427
DULLER, C.E. 178
DULONG, P.L. 278-279, 1060, 1150
DUMAS, J.B.A. 40, 256, 1127, 1151
DUMOND, J.W.M. 553

DUNCAN, J.E. 465
DUNCAN, J.L. 1100
DUNCAN, W.G.K. 1109
DUNGEY, J.W. 632, 662
DUNLEAVY, J.G. 584
DUNN, B.W. 416
DUNNING, J.R. 117
DUNN-RANKIN, D. 738
DUPUTY, M. 1154
DURAND, W.F. 72, 80, 198, 467, 499, 1104, 1184, 1192, 1207
**DUVALL, G.E.** 96, 146, 160, 610, 616, 663, 682, 699, 731, 744, 749-750, 774, 909, **1073-1074**, 1208, 1213
DUVALL, G.W. 1073
DUWEZ, P. 1105
DVOŘÁK, C.[V.] 128, 370, 1009, 1050, 1119
DWYER, B. 1206
DWYER, H. 1206
DYADIN, A.V. 1121
DYER, A.B. 1155
DYKE, W.P. 518
DYRENFORTH, R. 403
DYSON, F. 1048
DYSTHE, K.B. 821
DZIEWONSKI, A.M. 122

# E

EAMES, J. 223, 234
EARLE, P.S. 823
EARNSHAW, J. 1074
**EARNSHAW, S.** 26, 29, 54, 56, 78, 81, 156-157, 255, 287, 315, 324-325, 328, 330-331, 339, 349, 1057, **1074-1075**, 1144
EAST, R.A. 1011
EASTMAN, R.G. 834
EATON, D. 700
EATON, P.E. 823
EBDING, T. 490
EBERHARD, O. VON 473, 1063
EBERHARDT, D.S. 689, 1091
EBERT, H. 1063
EBERT, M. 1204
ECCLES, W.H. 464
ECKARDT, E.A. 1132
ECKERT, J.P. 138, 142-143
ECKERT, W.J. 138-139
EDDINGTON, Sir A. 469, 1113
EDELINCK, G. 1099
EDELMANN, L.B. 1164
EDELMANN, M.T. 1014
EDEN, A. 1069
EDEN, G. 698
EDER, J.M. 351, 1120
EDGELL, B. 302

EDGERTON, H.E. xxi, 413, 477, 526, 632, 640, 713, 902, 904, 1009, 1210
EDISON, T.A. 485
EDMISTON, J.P. 1213
EDSER, E. 79
EDSON, L. 103, 529, 1055, 1105
EDWARDS, D.G. 760
EGELAND, A. 427
EGGEBRECHT, A. 193
EGGEN, O.J. 1041, 1058
EGGENTON, M. 1144
EGGER, R. 798
EGGERS, A.J. 595
EGLIT, M.E. 1167
EGOROV, L.A. 726
EGOROV, N.I. 648
EGUCHI, T. 754
EHLERT, R. 1213
EHRENFREUND, P. 844
EICHELBERG, G. 1175
EICHELBERGER, R.J. 656-657
EICHLER, D. 836
EIFFEL, G. 385
EINSTEIN, A. 40, 67-68, 268, 289, 381, 453, 459-460, 469, 479, 514-515, 559, 812, 825, 1079, 1113, 1118-1120, 1182
EINSTEIN, H.A. 559
EISENBERGER, F. 715, 738, 776
EISENHOWER, D.D., U.S. President 1107, 1134
EISENMENGER, W. 640, 674, 779, 950
EISNER, T. 713
EISSLER, M. 398, 1205
ELDER, W.P. 1170
ELDREDGE, D.H. 610
ELDRIDGE, J.E. 554, 967
ELIA, P. 738
ELIOT, C.W. 329
ELLENBERGER, F. 1211
ELLICOTT, J. 1154
ELLINGTON, J.W. 250, 1103
ELLIOTT, C.A. 1205
ELLIOTT, R.F. 809
ELLIS, A.T. 461
ELPERIN, T. 827
ELPHIC, R.C. 752
ELSÄSSER, H. 1204
ELSEVIER, L. 205
EMANUEL, G. 827
EMANUEL, N.M. 1168
EMDEN Bros. 932
EMDEN, P. 396, 430-431, 932
EMDEN, R. 430, 932
EMME, E.E. 1207
EMME, E.M. 1071
EMMERSON, G.S. 1156
EMMONS, H.W. 631, 1065, 1105, 1184, 1211

EMRICH, R.J. 486, 496, 537, 577, 1049, 1110, 1171, 1189, 1213
EMRICK, R.M. 1074
ENCRENAZ, T. 794
ENDELMAN, L.L. 1162
ENGEL, C.D. 640
ENGEL, M.H. 706
ENGEL, O.G. 616
ENGEL, R.M. 638, 1085, 1214
ENGELHARDT, W. VON 689
ENGELKE, R. 532, 654, 783, 1211
ENGEMANN, S. vii, 148, 399, 796, 1185
ENIG, J.W. 730
ENSKOG, D 353
ENBLIN, T.A. 66
EÖTVÖS, R. VON 1043, 1198
EPICURUS, Greek philosopher 44, 187
EPSTEIN, H.T. 570
ERB, W.H. 302
ERDMANN, S.F. 538, 943
EREMETS, M.I. 843
ERGGELET, H. 578
ERLENMEYER, E. 1087
ERNOUF, D. 1142
ERNST AUGUST, Duke of Lüneburg-Braunschweig 222
ERSCH, I.S. 275
ESASHI, H. 765
ESCALES, R. 440
ESCHARD, G. 489
ESCLANGON, E. 474
ESPINOSA, H.D. 579
ESSER, D. 734
ETHELL, J. 1195
ETHRIDGE, N.H. 607
ETTINGSHAUSEN, A. VON 1117
EULER, J.A. 298
**EULER, L.** 29, 77-78, 138-139, 234-235, 239, 241, 243-244, 246, 249, 253, 255, 285, 298, 374, 392, 401, 728, 1044-1045, **1075-1077**, 1096, 1153-1154
EULER, P. 1075
EVANS, L. 717
EVANS, Sir J. 47, 876
EVANS, W.H. 571
EVERS, L.G. 815
EWALD, A.H. 625
EWALD, P.P. 1048
EWALD, R. 713, 1165
EWEN, H.I. 589
EWING, M. 526
EWING, Sir J.A. 382, 450, 897, 1093
EXNER, F. 132

# F

FABRI, H. 215, 1097
FABRICIUS, D. 199, 201

FABRICIUS, J. 201
FABRY, C. 421
FAGE, A. 463
FAHEY, J.J. 663
FAIR, H. 14, 609, 1209
FAIRBRIDGE, R.W. 30, 174, 192, 794, 1204
FALCONER, H. 47, 876
FALCOVITZ, J. 537
FALEMPIN, F. 444
FALES, E.N. 462
FALKENHAGEN, H. 1186
FALLOW, A. 859
FANNO, G. 1174
FANT, K. 1136
FARADAY, M. 111, 280, 302, 307, 329, 394, 1015, 1037, 1065
FÄRBER, E. 1047
FARBER, J. 709
FARLEY, D.R. 838
FARNSWORTH, P.T. 489
FARREN, W.S. 1085
FARRER, D. 318, 743, 1210
FASTL, H. 53
FAUCHON-VILLEPLÉE, A. 459, 951
FAUGUIGNON, C. 513
FAUQUIGNON, C. 109, 649
FAUST, J.L. 15
FAVARGER, A. 355
FAVARO, A. 204
FAY, J.A. 590
FAYE, M. 1040
FECHNER, G.T. 293-294
FEDDERSEN, B.W. 297, 1119
FEDOROFF, B.T. 645, 1203
FEDYNSKY, V.V. 1173
FEIGEN, M. 595, 1049
FEIKEMA, D. 804
FEILITZSCH, F.C.O. VON 329
FEIT, D. 16
FELD, M.S. 1091
FELDHAUS, F.M. 195
FELDMAN, L. 622, 1078
FELDMANN, A. 1206
FELDMANN, F. 554, 637, 969, 1039
FELDMANN, J. 69
FELTHUN, L.T. 668
FELTON, J.E. 426
FENG, R. 756
FENN, J.B. 642, 1105
FENTON, J.D. 297
FEOKTISTOVA, E.A. 647, 964
FERDINAND III, Holy Roman emperor 1125
FERGUSON, R.E. 1139
FERMAT, P. DE 222, 1129
FERMI, E. 116-117, 146, 514, 524, 545, 1135
FERMI, R. 116

NAME INDEX

FERNBACH, S. 142, 392, 1048, 1182
FERRARO, V.C.A. 492, 872, 1208
**FERRI, A.** 141, 256, 562, 622, 641, 658, 1055, **1077-1078**, 1110, 1209
FERRILL, D.A. 843
FESSENDEN, R.A. 445
FEW JR., A.A. 54, 719
FEYNMAN, R.P. 577
FFOWCS WILLIAMS, J.E. 781
FICKEN JR., G.W. 219
FICKEN, G.W. 219
FICKER, H. VON 1080
FICKETT, W. 113, 523, 697, 723, 751, 1211
FIELD, J.E. 92, 573, 682, 715-716, 764, 777
FIELD, P. 1211
FIERZ, M. 1098
FILENKO, Y.I. 690
FILLER, W.S. 486, 647, 686, 708, 1189, 1213
FILLON, A. 1130
FINCH-CRISP, W. 283
FINEBERG, J. 844
FINKELSTEIN, R.J. 436, 1083
FINLAYSON, B.A. 761
FINN, B.S. 283
FISCHER, F. 334, 1214
FISCHER, J.C. 1209, 1212
FISCHER, M. 579
FISCHER, P. 816
FISCHER, W. 1124
FISHBINE, J.M. 712
FISHER, D.F. 567
FISHER, H.J. 1106
FISHER, R.V. 550
FISHMAN, G.J. 804
FISK, L.A. 147
FISKE, J. 1206
FISKE, R.S. 379, 854
FITZGERALD, G. 1125
FITZPATRICK, A. 143
FLACHSBART, O. 72, 1154, 1207, 1212
FLAGG, R.F. 1084
FLAMANT, M. 1158
FLAMMARION, C. 426
FLANDERN, T. VAN 180, 727, 792
FLANDERS, D.A. 1081
FLEMING, D.K. 612
FLEMING, K.J. 681
FLEROV, G.N. 117
FLETCHER, C.H. 485, 551, 577, 1049, 1182
FLIEGNER, A. 334-335, 431, 439
FLIERL, K. 1045
FLOTTE, T. 817
FLÜGGE, S. 511, 1073, 1086
FLÜGGE, W. 1148
FLÜGGE-LOTZ, I. 1148

FLURANCE-RIVAULT, D. DE 197
FOA, J.V. 491
FOALE, M. 802
FOGLE, B. 492, 1208
FOLBERTH, W. 710
FOLDVARY, F.E. 210
FOLEY, A.L. 365
FOLK, R.L. 703-704
FOLKES, M. 71, 1099
FOLTA, J. 1043
FOMICHEV, A.G. 689
FOMIN, N.A. 1171
FONO, A. 483
FONSECA, J.F.B.D. 248
FONTENELLE, B.L.B. 1055
FOORD, E.E. 767
FOOTE, A.E. 403
FÖPPL, A. 376, 1146
FÖPPL, L. 1209
FÖPPL, O. 1054
FORBES, J.B. 1074
FORBES, J.D. 1148
FORBES, J.W. 750
FORD, B.J. 289, 906
FORD, P. 1206
FORDHAM, S. 691, 729, 1204, 1211
FOREL, F.A. 27, 32, 251, 506, 1102, 1205
FOREST, C. 665
FORKE, A. 189
FORRESTER, J. 143
FÖRSCHING, H. 1148
FORSSMANN, B. 738, 981, 1214
FÖRSTER, K. 277
FÖRSTER, M. VON 380, 984, 1132
FORSYTH, A.J. 51, 274-275, 280
FORTOV, V.E. 542, 741, 775, 791, 843, 1042, 1198
FOSTER JR., J.S. 524
FOSTER, H. 53, 81, 156, 286-287
FÖTTINGER, H. 457
FOUCAULT, J.B.L. 128, 297, 340, 1144
FOURIER, J.B.J. 39, 244, 284, 318, 458
FOURNEY, W.L. 1163
FOURNIER, E. 105, 202
FOWLER, C.M. 207, 647
FOWLER, G.J. 1068
FOWLER, R.D. 514
FOWLER, R.G. 617, 1110
FOWLER, R.H. 465, 924
FOWLES, G.R. 95-96, 160, 615, 682, 909, 964, 1074
FOX, R. 273, 276-277, 1151
FOXALL, H. 100
FRAAS, E. 844
FRANCHI, J. 104
FRANCK, E.U. 741
FRANÇOIS, D. 427, 1209
FRANK JR., R.G. 200

FRANK, A.M. 778
FRANK, I.M. 500
FRANK, P. 1191
FRANKE, H. 43
FRANK-KAMENETSKII, D.A. 1167, 1196-1197
FRANKL, F.I. 241, 374, 464, 473, 1076, 1121
FRANKLIN, B. 241, 244-246, 252, 255, 257, 1057, 1128
FRANKLIN, W.T. 1057
FRASER, R.P. 485, 994
FRAUENHOLZ, E. VON 1063
FRAZER, G.E. 1146
FREDERICK II, king of Prussia 240, 315, 1076
FREDRICKS, R.W. 33
FREEDMAN, W. 487
FREEMAN, I.M. 1182
FREER, S. 40
FREI, G. 1191
FREIMAN, L.S. 1061
FREITAG, M. 18
FREIWALD, H. 380, 683
FRENCH, B.M. 179, 688-689, 809, 1074, 1170, 1205
FRENKEL, V.Y. 1080
FRENKIEL, F.N. 629
FRESNEL, A. 222, 1098
FREUDENBERG, K. 1054
FREUDENTHAL, H. 1153
FREUDIG, D. 1206
FREY, A. 1206
FREYTAG LÖRINGHOFF, B. VON 139
FREYTAG, H.H. 683, 1205
FRIAUF, J.B. 435, 490
FRICKE, R. 1191
FRIEDEL, C. 406
FRIEDLÄNDER, F.G. 554
FRIEDMAN, G. 644
FRIEDMAN, H. 1198
**FRIEDMANN, A.A.** 68, 460, 469, 478-479, **1078-1080**, 1082, 1094, 1113
FRIEDRICH, W.L. 184
**FRIEDRICHS, K.O.** 15, 23, 81, 256, 389, 392, 482, 535, 553, 567, 605, 614, 1049, 1061-1062, **1080-1081**, 1213
FRIESKE, H.J. 1165
FRISCH, O.R. 117, 514, 519, 540-541
FRISCH, P. 66
FROBÖSE, M. 555, 947
FROHN, A. 1165
FROLOV, S.M. 1198
FROMAN, D. 541
FROUDE, W. 1149
FRÜNGEL, F.B.A. 568, 683
FUBINI-GHIRON, E. 325

FUCHS, K. 529, 1048, 1134-1135
FUETER, E. 1045
FUJITA, T.T. 722
FUJIWARA, T. 781
FULLENIUS, B. 229, 1098
FULLER, P.J.A. 664
FULLER, P.W.W. xxii, 297, 1063
FULLMER, J.Z. 1066
FUNAKOSHI, K. 88
FÜNFER, E. 399, 521
FUNG, W.K. 744
FUNK, M. 203, 1212
FUNTIKOV, A.I. 542, 647, 661, 1042, 1198
FURNA 1091
FURNISH, M.D. 113, 696, 745, 806, 812
FÜRTH, R.H. 566
FURTTENBACH, J. 237
FURUMOTO, A.S. 761
FUSCH, Y. 396
FÜSSL, W. 1119
FYALL, A.A. 548, 680, 715
FYE, P.M. 554

# G

GABEAUD, M.L. 482
GABOR, D 562, 1011
GAFFET, B. 328
GAFFNEY, E.S. 792
**GAGARIN, Y.A.** 103, 651, 1025, **1081-1082**
GAGE, W.R. 665
GAISSER, T.K. 451, 500
GAITAN, D.F. 93
GALBRAITH, W. 78, 156, 287
GALILEI, G. 3, 11, 38, 41, 64, 67, 151, 155, 175, 200-202, 204-205, 207, 213, 215, 222, 240, 410, 416, 737, 847, 889, 891, 1125, 1129, 1194
GALITZIN [GOLITZIN], B.B., Prince 386
GALLE, J.G. 1058
GALLI, G. 774
GALLOIS, M. 220
GALLOWAY, Sir W. 111, 354, 356
GALTON, F. 329
GAMOW, A. 1082
**GAMOW, G.A.** 68, 436, 568-569, 609, 1079, **1082-1083**, 1094
GAO, H. 821, 838
GARABEDIAN, P. 726, 729
GARBER, D. 1126
GARCET, H. 100
GARDINER JR., C.C. 1106
GARDINER, W.C. 635, 1209
GARN, W.R.B. 647
GARNERO, E.J. 679

GARNETT, S.H. 1106
GARNIER, J. 1127
GARNOV, V.V. 689
GARRY, K.P. 760
GARTMANN, H. 608
GARVIN, D. 1106
GASCOIGNE, R.M. 1205
GASSENDI, P. 44, 203, 1129
GATES, A.E. 1203
GATES, J.W.C. 690
GATHERS, R. 785
GATTO, J.A. 765
GÄTZSCHMANN, M.F. 316
GAULD, C. 224
GAULT, D.E. 657, 1170
GAUSS, C.F. 255, 1152
GAVRILOV, A.F. 1080
GAVROGLU, K. 1178
GAY, I.D. 1106
GAY, J. 968
GAYDON, A.G. 422, 1110, 1213
GAY-LUSSAC, J.L. 113, 266, 278, 284, 1150
GAZIS, P.R. 808
GEDALIN, M. 746, 827
GEEST, M. van DER 559, 1119
GEHLHOFF, G. 1044
GEHRELS, T. 1170
GEHRI, W. 913
GEHRING, J.W. 552
GEIGER, F.W. 570
GEIGER, H. 475-476, 1039, 1048, 1054
GEIGER, L. 452
GEINITZ, F.E. 307, 361, 392, 403, 864
GEIRINGER, H. 543
GELDER, A.P. van 1211
GELDORF, A.A. 777
GELFAND, B. 808
GELLER, B. 1145
GENDLER, R. 655
GENDRON, G. 427, 1073, 1101, 1199
GENTNER, W. 231
GEORGE III, king of Great Britain and Ireland 1057
GEORGI, J. 1146
GEORGII, W. 1141
GEPPERT, S.T. 269
GERDESMEYER, L. 797
GERLACH, W. 1089
GERLAND, E. 221, 231, 882, 1142
GERMAIN, P. 1148
GERMAN, V.N. 1042
GERNET, A. VON 29, 397, 1017, 1067
GERONIMUS, Y.L. 1200
GERSHANIK, Y.T. 535, 1197
GERSTEN, K. 1148
GERSTNER, F.J. VON 273
GERTHSEN, C. 709
GIACCONI, R. 703

GIACOBINI, M. 743
GIACOMELLI, R. 72, 198, 262, 1207
GIBBS, F.W. 1066
GIBBS, J.W. 348, 1072, 1112, 1132
GIBBS, T. 756
GIBSON, A. 704
GIBSON, C.H. 1168
GIBSON, F.C. 622-623, 666
GIBSON, R. 734
GIBSON, R.L. 176
GIERKE, H.E. VON 34
GIFFORD, A.C. 472
GILARDI, R. 823
GILBERT, F. 122
GILBERT, G.K. 13, 64, 410, 472, 496, 597, 847
GILBERT, L.W. 283
GILDEMEISTER, H.H. 50
GILES, B.F. 197
GILHEANY, J.J. 629
GILKE, S.A. 524
GILLES, A. 1055
GILLESPIE, C.C. 40, 48, 209, 250, 255, 259, 263, 275, 381, 1041, 1045-1046, 1051-1052, 1054, 1057-1059, 1061, 1066, 1068-1069, 1073, 1076, 1080, 1083, 1086, 1089, 1094-1095, 1097-1098, 1103, 1109, 1113, 1115, 1119, 1123-1124, 1126-1127, 1130, 1135-1136, 1142, 1145, 1148, 1150-1151, 1153-1154, 1158, 1175-1176, 1178, 1181, 1190-1191, 1194-1195, 1200, 1206
GILMORE, F.R. 259, 583
GINZBURG, V.L. 608, 656, 1042, 1198
GIRARD, P.H. DE 286
GIRAUDOUX, J. 1114
GITTINGS, E.F. 610
GIUA, M. 256
GLADECK, F.R. 593
GLADSTONE, J.H. 1151
GLADYCH, M. 1187
GLAISHER, J.W.L. 1058
GLARE, P.G.W. 19, 1205
GLASS, D.R. 301, 758
GLASS, G.P. 1106
**GLASS, I.I.** 423, 574, 623, 708, 759-760, 785, 1049, **1083-1084**, 1091, 1213
GLASSER, O. 132
GLASSMAN, I. 1065
GLASSTONE, S. 19, 29, 404, 536, 623
GLATZEL, B. 326, 1063
GLAUBER, J.R. 209
**GLAUERT, H.** 463, 483, **1084-1085**, 1147
GLAUERT, L. 1084

GLAZEBROOK, R.T. 1109
GLENIE, J. 1099
GLENNAN, T.K. 628
GLICK, H.S. 610, 946, 1091
GLICKEN, H. 786, 1179
GLIKSON, A.Y. 1170
GLOVER JR., D.R. 1203
GMELIN, L. 1053
GNOS, E. 410
GOBRECHT, H. 310, 320, 377
GOBUSH, W. 408
GODDARD, F.E. 496, 1049
GODDARD, J. 1189
GODDARD, R.H. 238, 474
GODDARD, V.P. 129, 641
GODFREY, C. 696
GODWAL, B.K. 96
GOETHERT, B.H. 552, 1160
GOGOSOV, V.V. 1167
GOIN, K.L. 85, 588, 642, 1207, 1210
GOLD, T. 68, 566, 601-602
GOLDANSKII, V.I. 1198
GOLDBERG, S. 1212
**GOLDBERG, V.** 93, 319, 584, 637-638, 981-982, **1085**, 1214
GOLDMAN, D.E. 34
GOLDSMITH, W. 14, 314, 390, 648
GOLDSTEIN, J.S. 617
GOLDSTEIN, S. 1105, 1116
GOLDSTINE, H.H. 138, 142, 144-145, 486, 556, 1134-1135, 1210
GOLOVANOV, V.J. 771
GOLUBEV, V.V. 1167, 1200
GOMBERG, M. 295
GONCHAROV, G.A. 524, 575, 608, 1212
GOOD, J.B. 1208
GOODAY, G.J.N. 1051
GOODIER, J.N. 319
GOODRICKE, J. 260-261
GOODWIN, I. 1048
GOOSMAN, D.R. 421
GORANSON, R.W. 95, 542, 548, 600, 610, 621, 951
GORDON, K.D. 844
GORDON, L.D.B. 1150
GOREE, J. 818
GOREE, J.A. 818
GORIELY, A. 631, 796
GÖRING, H. 1163
GORLIN, S.M. 1210
GORN, M.H. 1071, 1105, 1207
GÖRTLER, H. 385, 520, 1148
GORZNY, W. 1205
GOSSLAU, F. 1163-1164
GOSSOT, F.F. 404
GÖTHERT, B. 488
GOTO, A. 754
GOTTFRIED, K. 1048

GOTTLIEB, J.J. 1084
GOTTWALD, M. 849
GÖTZMANN, E. vii
GOUGH, D.O. 119, 555
GOULARD, R.J. 693
GOULD, G. 623-624
GOURLEY, L.E. 552
GOURLEY, M.F. 552
GOWING, M. 519
GOWLETT, J.A.J. 47
GOYER, G.G. 55-56, 677
GRAAF, J.G.A. DE 666, 1160
GRABITZ, G. 476, 1148
GRADY, D.E. 791, 1074
GRADY, M.M. 1170
GRAEBE, C. 1046
GRÆCUS, M. 193
GRAFF, J 797
GRAHAM, E. 1067
GRAHAM, R.A. 14, 93, 95-96, 134, 148, 542, 616, 664, 668, 674, 705, 710, 744, 755, 774, 792, 1074, 1210, 1213
GRAHAM-ROWE, D. 805
GRAMMEL, R. 525, 578, 1209
GRANDI, G. 205
GRANEAU, N. 675, 768-769, 782
GRANEAU, P. 54, 675, 768-769, 782
GRANHOLM, R.H. 1074
GRANT, I. 567
GRATTAN-GUINNESS, I. 244, 1076, 1153
GRAVES, A. 541
GRAVES, R. 1204
GRAVESANDE, W.J.'S 44-45, 223, 233, 891-892, 1098
GRAY, D.E. 284, 312, 1205
GRAY, S. 232
GRAY, T. 381-382, 896
GREEN JR., R.E. 732
GREEN, A. 104, 803, 923, 1027
GREEN, D.A. 1208
GREEN, J.J. 629, 1207
GREENE, E.F. 373, 674, 1106, 1209, 1213
GREENE, J.O. 701
GREENER, W. 13
GREENHILL, Sir A.G. 367, 402
GREENSFELDER, N.S. 469
GREENSLAD JR., T.B. 257
GREENSTADT, E.W. 33, 675
GREENSTEIN, J.L. 670
GREGG, D.W. 691, 735
GREGORY, O. 1097
GREIFER, B. 666
GREIG, E.F. 476
GREINER, N.R. 778
GREISEN, K. 545
GREN, F.A.C. 267

GRIESER, D. 1119
GRIEVE, R. 176, 179
GRIEVE, R.A.F. 1170
GRIFFIN, D. R. 874
GRIFFITH, A.A. 1183
GRIFFITH, W.C. 434, 551, 595-596, 618, 759, 1049, 1110, 1213
GRIGG, P.P. 1059
GRIGGS, D.T. 659
GRIGOR'YAN, A.T. 1045, 1080, 1200
GRIGULL, U. 1172
GRIMSHAW, H.C. 494, 1143
GRIMSHAW, W.E. 494
GRIMWOOD, J.M. 564, 651
GROBERT, J.F. 273
GROLIER, J. 232
GROLLIER DE SERVIÈRE, G. 232
GROLLIER DE SERVIÈRE, N. 232
GRÖNIG, H. 85, 615, 650, 772, 782, 803, 808, 995, 1026, 1165, 1213
GROOT, P.J. 318
GROSE, F. 879
GROSE, V.L. 743
GROSS, R.A. 617
GRÖSSING, H. 1069
GROVES, L.R. 546
GROWTHER, J.G. 1168
GRUBER, H.E. 40
GRUBER, J.G. 275
GRÜNEISEN, E. 94, 475, 736
GRÜNEWALD, B. vii, 850
GRUSS, G. 126, 365, 1118
GRZIMEK, B. 92
GUBKIN, A.A. 1042
GUBKIN, K.E. 30
GUDERLEY, K.G. 531, 718, 1055
GUDMESTAD, O.T. 28, 869
GUÉBHARD, P.E.A. 389, 485, 1118
GUENTHER, A.H. 664
GUERICKE, O. VON 22, 210
GUGGENHEIM, E.A. 1205
GUGGENHEIM, H.F. 457, 1064, 1103
GUHRAUER, M.M. 1126
GUICCIARDINI, N. 1097
GUIDONI, A. 497
GUIEYSSE, P. 1040
GUINN, V.P. 1106
GUIRGUIS, R.H. 1139
GUM, C. 183
GUMBSCH, P. 830
GUMMERT, P. 154
GUNSTON, B. 1187
GUNSTON, W.T. 699
GÜNTHER, P.A. 213, 231, 1063
GUPTA, V. 662
GUPTA, Y.M. xx, 96, 146, 656, 726, 730, 749, 755-756, 775, 812, 842, 1074, 1208, 1213
GURNETT, D.A. 56

GUSTAVSEN, R.L. 806
GUTENBERG, B. 26, 122, 452, 475, 505, 516, 611
GUTER, (?) 196
GÜTH, W. 461, 634, 967
GUTSCHER, M.-A. 248
GUTTMANN, O. 203, 407, 1205
GYUNTER, N.M. 1080

# H

HAACK, W. 243, 526, 557
HAAGE, K. 475
HAAK, H.W. 815
HAALCK, H. 1132
HAAS, J.F. 1179
HABASHI, F. 1209
HABER, F. 107, 1168
HACHETTE, J.N.P. 290, 293
HACKMAN, R.J. 1169-1170
HADA, T. 1213
HADAMARD, A. 1086
**HADAMARD, J.S.C.** 29, 72, 265-266, 431, 542, 1072-1073, **1086-1087**, 1101, 1198-1199
HADDOW, J.B. 1099-1100
HAENSEL, H. 607
HAERING JR., E.A. 790
**HAEUSSERMANN, C.** 308, 404, **1087**, 1136
HAFEZ, M.M. 761
HAGELWEIDE, J.B. 543
HAHN, E. 1175
HAHN, O. 511
HAID, A. 270
HAJEK, H.V. 611
HALACY JR., D.S. 1105
HALBHUBER, K.J. 816
HALE, E.E. 350, 620
HALE, G.E. 458, 1093
HALL, A.R. 72, 198, 203, 214, 216-218, 692, 1098, 1190, 1195, 1208
HALL, D.A. 630
HALL, G. 1084
HALL, G.F. 469
HALL, H.T. 607
HALL, J.G. 595, 623, 642, 760, 1084, 1091, 1165
HALL, M.B. 214, 216-217, 1098, 1190, 1195
HALL, N.S. 1214
HALL, R. 214, 1082
HALL, R.G.N. 690
HALLER, A.R. 155
HALLER, P. DE 265, 1039
HALLEY, E. 239
HALLIDAY, D. 216
HALLION, R.P. 561, 638, 1196

HALLWACHS, W. 1185
HALPERIN, B. 688
HALSEY JR., G.D. 1106
HALVERSON, R.R. 1134
HAMACHER, A. 1105
HAMANN, S.D. 96, 691
HAMBIR, S.A. 822
HAMBURGER, M. 251, 384
HAMEL, G.K.W. 478
HAMILL, J. 52
HAMILTON, C.J. 175, 749, 754, 830
HAMILTON, W.R. 1180
HAMMACK, J.B. 598
HAMMELMANN, F. 51, 685
HAMMER, C.U. 184
HAMMER, E.W. 1208
HAMMICK, D.L. 1059
HAMPTON, L.D. 600
HAN, Z.Y. 827
HANCOCK, P.L. 18, 1205
HAND, C.W. 1106
HANKEL, H. 392
HANKS, T.C. 752
HANLE, P.A. 1105
HANNAWAY, O. 1115
HANSEN, C.J. 1048
HANSEN, G. 405
HANSEN, J.R. 456, 1078
HANSEN, T.A. 178, 1170
HANSJAKOB, H. 195
HÄNSLER, J. 544, 951
HANSON, A.R. 616-617
HANSON, D.J. 108
HANSON, R.K. 827
HARBECK, M. 726
HARCOURT, A.V. 370, 1067
HARDING, J. 579, 731
HARDING, K. 271
HARDY, C. 1212
HARGATHER, M. 70
HÄRING, M. 659
HARITOS, N. 59
HARLEMAN, D.R.F. 495, 583
HARLOW, F. 137
HARLOW, F.H. 392
HARO, G. 580
HARPER, E. 1083
HARRINGTON, J. 15
HARRIOT, T. 201, 1190
HARRIS, C.M. 34, 653, 1205
HARRIS, D.L. 602
HARRISON, B., U.S. President 407, 1132, 1206
HARRISON, E.R. 795, 839
HARRISON, F.B. 1049
HARRISON, M. 461, 634
HART, G.A. 736
HARTE, H.H. 278
HARTENBAUM, B. 1179

HARTENBERG, R.S. 42, 181, 1154
HARTER, W.G. 702
HARTLEY, H. 1066, 1190, 1195
HARTLIB, S. 208
HARTMAN, E.P. 516
HARTMAN, H.L. 51, 685
HARTMANN, D.H. 804
HARTMANN, G.K. 470, 581
HARTMANN, W.K. 174
HARTNETT, J.P. 1139
HARTREE, D.R. 140, 465, 924
HARVEY, E.N. 231, 952
HARVEY, R.P. 704
HASELMAYER, L.A. 1146
HASINGER, G. 773
HASSALL, J.R. 103
HASSLACHER, A. 360
HATCH, R.A. 213
HATHAWAY, D.H. 327
HATHAWAY, G. 782
HATON DE LA GOUPILLIERE, J.N. 360, 1122
HATT, W.K. 416, 430
HATTENDORFF, K. 427, 1152-1153, 1191
HATZE, H. 809, 904
HAUGHTON, S. 368
HAUPT, G. 797
HAURWITZ, B. 492, 1208
HAUSDORFF, F. 229, 231
HÄUSLER, E. 722-723
HAUSSMANN, K. 1131
HAUTE-FEUILLE, J. DE 221, 229, 896
HAUVER, G.E. 135, 649, 656-657
HAVER, S. 869
HAVNES, O. 818
HAWKE, R.S. 207, 648, 749, 1208
HAWKING, S.W. 1208
HAWKYARD, J.B. 700, 1100
HAYAKAWA, H. 369
HAYASHI, A.K. 738
HAYES, D.B. 731
HAYES, M. 1086
HAYES, W.D. 224, 226
HAYWARD, G. 640
HAZEN, A. 88
HEALY, J.H. 659
HEARN, L. 26
HECHT, F. 608
HECKER, O. 386
HECKSCHER, A. 1098
HEDWIG, K.H. 517, 1164
HEEN, P. DE 1128
HEFLINGER, L.O. 562, 690
HEGLAND, J.E. 1074
HEIDER, N. 934
HEIL, P. 454
HEILBRON, J.L. 240, 1142

# NAME INDEX

HEILBROWN, H. 1086
HEILIG, W. 621, 744, 827, 935, 961
HEIMERL, J.M. 557
HEIMS, S.J. 1135
HEINKEL, E. 518
HEISE, F. 366, 424, 893, 1001
HEISENBERG, W. 79, 515, 517, 574, 611, 839, 1212
HEISKELL, R.H. 635
HELBIG, K. 1211
HELD, M. 430, 717, 725, 732, 1160
HELFAND, D. 773
HELFER, H.L. 602
HÉLIE, F. 1094-1095
HELIN, E.F. 1170
HELIOS, GREEK DEITY 100, 185, 1020
HELL, C. 1087
HELLER, A. 1127, 1212
HELLER, K.D. 1119
HELLER, M. 1114
HELLMOND, W. 490, 1164
HELLWARTH, R.W. 665
HELM, J. 552
HELMHOLTZ, H. VON 79, 110, 216, 311, 317, 335, 368, 1042, 1062, 1072, 1089, 1119, 1152, 1174, 1205
HELMONT, J.B. van 202, 1125, 1129
HELVIG, C.G. VON 267, 282
HELWICH, O. 356, 477, 633, 684, 1063, 1160, 1162
HEMSING, W.F. 681
HENDERSON, L.F. 536, 548, 582, 827
HENDRICKS, S. 840
HENLE, J. 1153
HENLEY, D. 755
HENNECKE, D.K. 639
HENNING, G.F. 421
HENNING, T. 844
HENRARD, J. 1170
HENRIKSEN, P.W. 529, 541-542, 547, 578, 1107, 1135, 1212
HENSEL, G. 525
HENTSCHEL, K. 1054
HEPP, P. 404
HEPP, W. 738
HERBIG, G.H. 580
HERDER, J.G. VON 39
HERING, C. 442, 509
HERIVEL, J. 1098
HERLOFSON, A.N. 632
HERMAN, L. 737
HERMAN, R. 68, 569, 1083
HERMANN, A. 1072
HERMANN, J. 139, 517, 1164
**HERMANN, R.** 502, 504, 507-508, 517, 939, 941, **1087-1089**, 1192
HERNECK, F. 1119

HERO [HERON] of Alexandria, Greek physicist 188, 926
HEROUX, J. 1155
HERRIES DAVIES, G.L. 1124
HERRIOTT, D.R. 649
HERRMANN, (?) 590
HERRMANN, F. 759
HERRMANN, K.H. 632, 912, 1009
HERSCHEL, Sir F.W. 64, 178, 271, 274, 754
HERSEY, J.B. 640
HERTZ, F.C. 377
HERTZ, G. 1089
**HERTZ, H.R.** 46, 110, 360, 376-377, 384, 406, **1089**, 1119
**HERTZBERG, A.** 423, 589, 610, 642, 684, 686, 689-690, 744, 766, 772, 949, 975, 1049, 1084, **1090-1091**, 1171, 1179, 1213
HERTZMANN, C.E. 334
HERTZSPRUNG, E. 454
HERZ, E. VON 421
HESS, H. 60
HESS, P. 424
HESS, V.F. 451, 835
HESSELINK, L. 1179
HETHERINGTON, N.S. 1203
HEUCKROTH, L.E. 1084
HEWISH, A. 694
HEWLETT, R.G. 523
HEYDT, A. VON DER 822
HEYMAN, J. 1093
HEYMANN, F.J. 709
HICKERSON, N.L. 386
HICKMAN, R. 1052
HICKS, D.L. 543
HIEBERT, E.N. 381, 1119
HIEBLOT, J. 662
HIEMENZ, R.J. 636
HIGASHINO, F. 827
HIGAWARA, T. 524
HIGGINS, A.J. 738
HILBERT, D. 353, 1062, 1103, 1191
HILDEBRAND, A.R. 177-178
HILDEBRAND, F. 85, 478, 1107, 1181
HILDEBRAND, W. 478
HILL, A. 1041
HILL, A.V. 1093
HILL, E.G. 470, 581
HILL, L. 786, 1179
HILL, R. 563
HILLIAR, H.W. 458
HILLIER, R. 777, 1051, 1214
HILLS, J.G. 739
HILTON, W.F. 383, 468, 482, 503, 505, 511, 522, 596, 1207
HIND, J.R. 318
HINDE, R.A. 3
HINRICHS, G. 1160

HINSCH, K. 726
HINSHELWOOD, Sir C.N. 475, 619, 1168
HIPP, M. 123, 302, 1012
HIRAHARA, H. 129
HIRE, P. DE LA 259
HIRN, G.A. 330, 341
HIROSHI, Y. 333
HIRSCH, F.G. 701
HIRSCHFELDER, J.O. 398, 529, 1048, 1134-1135, 1180
HIRTH, A. 690, 922
HITLER, A. 530, 611, 1212
HIXON, R.S. 1074
HIXSON, R.S. 812
HJELT, E. 1209
HJORTH, J. 834
HOBBES, T. 1129
HOBLITT, R.P. 35, 62, 189, 753
HOCHSTETTER, F. VON 307, 346-347, 361, 392, 403, 864
HOCKEY, T. 1206
HODDESON, L. 5, 116, 529, 541-542, 547, 578, 1048, 1107, 1135, 1212
HODES, I. 1064
HODGINS, E. 1208
HODGKINSON, E. 295
HODGSON, R. 327, 870
HOEFER, J.C.F. 1130
HOELZER, H. 140
HOERLIN, H. 660
HOERNER, S.F. 49, 242, 683, 1154
HÖFER VON HEIMHALT, H. 370-371
HOFER, E. 148
HOFF, A. 197, 1205
HOFF, G. 738
HOFF, J.H. VAN'T 398
HOFF, K.E.A. VON 307
HOFF, N.J. 1055, 1077
HOFF, W. 1148
HOFFMAN, F. DE 584
HOFFMANN, C. 1119, 1159, 1208
HOFFMANN, D.M. 775, 1119
HOFFMEISTER, C. 588
HOFMANN, J.E. 234, 1076
HÖGBERG, L. 128, 621, 625, 689, 700, 702, 1162-1163
HOGG, I.V. 1205
HOGGATT, C.R. 709
HOHLER, V. 1209
HOIGUCHI, Y. 668
HOJEL, W.C. 973
HOLBERG, J. 469
HOLDER, D.W. 370, 596, 1008
HOLL, H. 1085
HOLLAND, J.R. 1074, 1213
HOLLENBACH, D.J. 29, 756
HOLLENBACH, R.E. 681
HOLLENBERG, K. 962

HOLLERITH, H. 139
HOLLISTER-SHORT, G.J. 199, 1211-1212
HOLLMANN, S.C. 245
HOLLOWAY, D. 116, 144, 575, 598, 608, 1198
HOLLYER, R.N. 603, 1109-1110
HOLMAN, G.D. 327
HOLMBERG, E. 820
HOLMES, A. 60
HOLMES, N.C. 148, 674
HOLMES, S. 953
HOLMYARD, E.J. 692
HOLSER, W.T. 1123
HOLSTEN, (?) 455
HOLT, M. 567, 605, 614, 635, 700, 744, 1167, 1173
HOLTEN, A.J. 90
HOLTON, G. 1052
HOLTON, W.C. 136, 663
HOLTZ, W. 211, 1185
HOLZER, R.E. 662
HOME, R.W. 1109
HOMER, Greek poet 64, 185
HOMER, J.B. 1106
HONEGGER, E. 1175
HÖNIGER, W. 416, 452
HÖNL, H. 1119
HONMA, H. 536
HONORÉ, P. 876
HONOUR, J. 791
HOOGENDOORN, S.P. 613
HOOKE, R. xix, 22, 44, 46, 64, 128, 201, 210-212, 214, 220-221, 226, 340, 370, 597, 1060, 1141, 1195
HOOVER, A. 806
HOOVER, R.B. 706, 850
HOPF, L. 1055
HOPKINS, A.K. 1074
HOPKINS, J. 1204
HOPKINS, W. 1175, 1180
**HOPKINSON, B.** xx, 94, 436, 450, 455-458, 900, 976, **1092-1093**
HOPKINSON, J. 94, 353, 436, 450, 1092
HOPKIRK, R. 659
HOPPE, E. 1126
HOREAU, A. 1047
HORGAN, C.O. 319
HORIE, Y. 30, 113, 791, 803, 1074
HORIGUCHI, Y. 668
HORN, P.D. 730
HORNEMANN, U. 744-745, 935
HORNIG, D.F. 583, 954
HORNIG, H.C. 570
HORNUNG, H.G. 536, 1110
HOROCK, C.L. 599
HÖRZ, F. 689, 706
HOTTEL, H.C. 1165

HOTTNER, T. 613
HOTZ, R. 104
HOUAS, L. 838
HOUSTON, E.E. 532, 600, 610
HOUTEN, J. van 1209
HOVE, L. van 1135
HOWARD, D.D. 1179
HOWARD, E.C. 201, 270, 279
HOWARD, J.N. 1178
HOWARTH, L. 1086, 1105
HOWE-GRANT, M. 109, 1203, 1211
HOWELL, S.P. 1133
HOWLETT, D.R. 1203
HOWLETT, J. 138, 1135
HOWSON, A.G. 1097
HOYLE, Sir F. 68, 555, 566, 574, 694
HRUBES, J.D. 968
HU, T.C.J. 759
HUANG, Y. 822
HUBBARD, H.W. 641
**HUBBLE, E.P.** 487, 491, 540, 670, 703, 1068, **1093-1094**, 1102, 1114
HUBBLE, J.P. 1093
HUBLER, G.K. 811
HUCHRA, J.P. 569
HUCKELL, B. 47
HUDSON, J. 1209
HUEBEL, J.G. 648
HUEBNER, W.F. 799
HUGENSCHMIDT, M. 1160
HUGGINS, Sir W. 303, 338, 1068
HUGHES, D.S. 552
HUGHES, T.J.R. 1203
**HUGONIOT, P.H.** 29, 76, 79, 150, 166-167, 265-266, 290, 311, 328, 342, 346, 348, 383, 389-390, 409, 424, 431, 439-440, 542, 602, 614, 1072, 1086, **1094-1095**, 1107, 1149, 1177
HUGUENARD, E. 84, 401, 462, 938
HULL, G.F. 1071
HULST, H.C. van DE 574, 601, 1117
HUMASON, M.L. 487, 1093-1094
HUMBOLDT, A. VON 300, 327
HUME, W. 551, 951
HUMPHREY, A.T. 1178
HUNG, C.M. 761
HUNSAKER, J.C. 1071
HUNTING, A.C. 603, 1109
HUNTINGTON, H.B. 507, 802
HURBAN, J.M. 1209
HURLE, I.R. 422, 684, 975, 1091, 1213
HURMUZLU, Y. 785, 814
HUSEBYE, E.S. 679
HUSS, G.R. 774
HUSSAINI, M.Y. 1117
HUSTON, A.E. 684, 1019

HUT, P. 1170
HUTCHISON, K. 1150
HUTTMANN, A. 637, 1085
**HUTTON, C.** 71, 100, 123, 156, 223, 237, 239, 242, 260-261, 279, 301, 358, 702, 893, 1076, **1095-1097**, 1099, 1203-1204
HUTTON, H. 1095
**HUYGENS, C.** 42-43, 48, 151, 153-154, 164, 207, 209, 211, 213-222, 224-225, 229-231, 240, 387, 737, 882, 890, 918, 963, 974, 1068, **1097-1099**, 1125, 1127, 1129, 1141-1142, 1189-1190, 1194
HYERS, D.H. 1081
HYZER, W.G. 543, 632, 689-690, 713, 716, 1007, 1210

# I

IGRA, O. 537, 621, 827, 1084
IIDA, K. 57, 578
IKUI, T. 759
IL'KAEV, R.I. 1042
ILLGEN, H. 494, 1013
ILLGEN, H.J. 494
IMAIDE, Y. 760
IMAMURA, A. 28, 264, 578, 801
IMAMURA, F. 578
IMBODEN, O. 184
INABA, K. 130
INCE, S. 1045, 1130, 1156, 1200, 1212
INGALLS, J.M. 386-387, 1205
INGEN, J.L. van 1148
INGHAM, M.F. 657
INGRASSIA, M. 1078
INOGAMOV, N.A. 712
IOFFE, A.F. 1167
IPPEN, A.T. 495, 583
IRION, R. 67, 824
IRO, H. 716
IRVING, D. 490
ISAACSON, E. 1062
ISBELL, W.M. 95, 682, 691, 964
ISMAIL, A. 1078
ISNARD, (?) 244
ISRAEL, W. 1208
ITAGAKI, R. 1119
ITARD, J. 1158
ITOH, K. 686
ITOH, S. 583, 827, 839
IVÁN, A. 1199
IVANENKO, D.D. 1082
IVANOV, A.P 814
IVANOV, E.N. 709
IVES, R.L. 56, 526
IZOD, E.A. 436-437
IZOD, T.P.J. 1106

NAME INDEX

## J

JACKSON, J.A. 13, 1204
JACKSON, J.D. 386
JACKSON, J.P. 238
JACKSON, J.W. 579
JACKSON, W.F. 675
JACOBI, C.G.J. 1152, 1191
JACOBI, J.F. 250, 253, 1210
JACOBS, E.N. 101, 499, 503, 972
JACOBS, K. 1148
JACOBS, S.J. 24, 686-687, 1074, 1160
JACQUES, W.W. 367
JAEGGY, B.C. 925
JAGADEESH, G. 832
JÄGER, G. 1069
JAGGAR, T.A. 549
JÄHN, F. 710
JAKI, S.L. 1073
JAMES I, king of England 212
JAMES, C.S. 14, 552
JAMES, F.A.J.L. 1176
JAMES, H.J. 571
JAMES, R. 217
JAMESON, A. 724, 726, 1211
JAMESON, R.L. 1007
JAMET, F. 414, 625, 725-726, 732, 934, 991, 1210
JAMIESON, J.C. 656
JAMIN, J.C. 320, 1010, 1151
JANKOWITSCH, P. 1166
JANOS, L. 1195
JANSSEN, M. 97
JANZON, B. 731
JAROSEWICH, E. 199, 273, 705
JAUMOTTE, A.L. 695, 1078
JAVAN, A. 649, 1091
JAVAULT, B. 1047
JAVOR, G.T. 55
JEANS, J. 487-488
JEDLIK, A. 1042
JEFFERSON, T., U.S. President 64
JEFFREYS, Sir H. 175, 379
JELINEK, A.J. 47
JENKIN, J.G. 1109
JENKINS, D.C. 680
JENKINS, D.R. 638, 1207
JENKINS, R. 1142
JENNINGS, B.H. 612
JENNISON, R.C 811
JENTSCHKE, W. 514
JERVIS-SMITH, F.J. 125, 398-399
JIMBO, M. 228
JING, D. 430
JOACHIM, H. 494, 1013
JOCHAM, D. 716, 1214
JOHANSEN, L. 185
JOHANSSON, C.H. 24, 669, 717, 725
JOHANSSON, L. 442

JOHN, archduke of Austria 281
JOHNSON, B. 525
JOHNSON, J. 1099
JOHNSON, J.B. 792
JOHNSON, J.D. 747, 778
JOHNSON, J.N. 94, 784, 842, 956, 1095, 1213
JOHNSON, K.L. 377
JOHNSON, L. 753
JOHNSON, L.R. 141
JOHNSON, M.H. 641
JOHNSON, N.G. 291, 309, 385
JOHNSON, N.L. 145
JOHNSON, Q. 717, 726, 955
JOHNSON, T.H. 1114
JOHNSON, T.V. 1204
**JOHNSON, W.** vii, 196-197, 213, 232, 237, 300, 315, 444, 700, 747, 753, 764, 1045, 1077, 1097, **1099-1100**, 1154-1155
JOHNSON, W.E. 145
JOHNSON, W.R. 1110
JOHNSTON, A.C. 248
JOHNSTON, A.K. 857
JOHNSTON, L.H. 109, 542
JOHNSTON, R.C. 197
JOHNSTONE, W.E. 650
JOIGNEAU, S. 649
JOKIPII, J.R. 33, 836
JOLIOT-CURIE, I. 117
JONES, A.E. 707
JONES, A.H. 160, 691
JONES, B.M. 499
JONES, C. 1112
JONES, E.R. 46
JONES, H. 430, 569
JONES, I. 811
JONES, N. 444, 763, 768, 783
JONES, O.E. 664
JONES, R. 138
JONES, R.H. 1067
JONES, R.T. 503, 1055
JONES, R.V. 1051
JONES, S. 3, 47, 1204
JONGENEELEN, P. 1024
JONQUIÈRES, E. DE 314
JOOS, G. 434, 1163, 1186
JORDAN, C. 1086
JORDAN, F.W. 464
JOSEPH, D.D. 616
JOSEPH, KING 852
JOSHUA, Israelite commander 70
JOST, W. 532, 535, 683
**JOUGUET, J.C.E.** 23, 29, 76-78, 112, 226, 290, 421, 427, 431, 433-435, 438, 446, 460, 466-467, 490, 528, 542, 993, 1059, 1072-1073, **1100-1101**, 1199, 1211
JOUKOVSKY ⇨ ZHUKOVSKY, N.E.

JOULE, J.P. 100, 314, 317, 320-321, 342
JOURDON, G. 838
JOURNÉE, F.A. 395
JUAN, D.G. 13
JUNG, I. 378, 928, 1111
JUNGFLEISCH, E. 1046
JUNGNICKEL, C. 1057, 1066
JURIN, J. 223, 234
JUST, T. 532
JUTKIN, L.A. 283, 584, 638, 982

## K

KADLETZ, K. 1069
KALER, J. 1204
KALITKIN, N.N. 1042
KALKMANN, U. 1148, 1192
KALLENBACH, R. 1204
KALUMUCK, K.M. 15
KAMEL, M.M. 1139
KAMIMOTO, G. 423, 1165, 1189
KAMM, W. 1164
KANAMORI, H. 752, 817, 1179
KANDALOV, A.N. 591, 1187
KANE, E.J. 481
KANEL', G.I. 745, 768, 837, 843, 1042, 1174
**KANT, I.** 11, 27, 60, 67, 250-252, 317, 862, **1102-1103**, 1115
KANTER, E.P. 784
KANTROWITZ, A. 590, 601, 692, 803, 971, 1090-1091, 1109, 1211
KAPITZA, P.L. 685
KAPLITA, T. 1078
KAPOOR, S.K. 263
KARASIK, B.S. 605
KARAVODINE, V. 490
KARCHER, J.C. 445
KARELJSKICH, K. 418
KARL, Landgraf von Hessen-Kassel 1141
KARLSCH, R. 611, 1212
KÁRMÁN, M. 1103
**KÁRMÁN, T. VON** 30, 37, 72, 91, 94, 101, 103-104, 226, 428, 431, 433, 436, 478, 481, 492, 494, 497, 499, 503, 505, 507, 521, 529, 540, 543, 574-575, 592, 611, 615, 642, 645, 718, 1021, 1055, 1071, 1080, **1103-1105**, 1141, 1147, 1192, 1208
KARNES, C.H. 708, 1074, 1213
KARPOVICH, E.A. 473
KAST, H. 270
KASTLER, A. 1049
KATAYAMA, Y. 88
KATZ, B.S. 723
KAUDY, L. 109, 1205
KAUFFMAN, E.G. 178, 1170

KAUSEL, E. 825
KAWAHASHI, M. 129
KAWAI, H. 710
KAWALKI, K.H. 488
KAYAFAS, G. 608
KAYE, S.M. 1203
KAYSER, E. 478
KEANE, B.T. 765
KEAREY, P. 1205
KEDRINSKII, V. 827
KEDRINSKII, V.K. 684, 1171
KEEFER, J.H. 607
KEELE, K.D. 198
KEELER, R.N. 717, 1198
KEGLER, W. 1160
KEIHN, G. 786
KELDISCHEM, M.V. 1166
KELDYSCH, M. 1166
KELLER, C. 1151
KELLER, D.V. 664
KELLER, G. 1170
KELLER, H.B. 636
KELLER, J.B. 605, 1080-1081
KELLER, L. 1080
KELLER, R. 465
KELLERMANN, K. 670
KELLEY, J.L. 1208
KELLOGG, P.J. 662, 872
KELLY, H.C. 118
KELVIN, Lord (or W. THOMSON)
    4, 22, 42, 100, 135, 139, 148,
    225, 294, 302, 311, 317, 320,
    349, 364, 368, 394, 406, 422,
    448, 890, 919, 1119, 1149, 1157,
    1176, 1180-1181
KEMBLE, E.C. 1052
KEMPF, K. 142
KENDALL, M.A.F. 817
KENNEDY, D.R. 265, 380, 513, 1211
KENNEDY, G.C. 673
KENNEDY, G.P. 1082
KENNEDY, J.F., U.S. President
    651, 1104
KENNEDY, J.W. 523
KENNEDY, W.D. 555
KENNEL, C.F. 675, 1213
KENNY, A. 618
KENT, R.H. 265
KEOLIAN, R.M. 1214
KEOUGH, D.D. 673, 755
KEPLER, J. 64, 66-67, 138-139,
    200, 588
KERBER, L.L. 1187
KERKHOF, F. 710, 876
KERLEY, G.I. 97, 747
KERN, R.D. 1106
KERR, J. 356
KERR, R.A. 766-767
KERSTEN, M. 1044

KERSTEN, R.J.A. 560
KERTZ, W. 464, 560, 1132, 1211
KESTIN, J. 1138
KEUNE, F. 1141
KEYES, R.T. 663
KEYS, D.A. 126, 467, 1013
KHARITON, Y.B. 96, 116-117, 511,
    513, 542, 598-599, 799, 1041,
    1167-1168, 1196-1198
KHVOSTOV Y.B. 341
KIEFER, W. 722-723
KIEFFER, J. 1211
KIEFFER, S.W. 757, 767, 1170, 1179
KIKUCHI, Y. 393-394
KILLIAN JR., J.R. 413
KIM, H. 822
KIM, K.S. 738
KIM, Y.W. 711, 948, 1110
KIMURA, Y. 668
KIND, K.G. 50, 295-296, 885
KINDLEBERGER, J.H. 450
KING, R.B. 680, 715
KINGDON, K.H. 132, 510
KINLEY JR., K. 453, 983
KINLEY, J.D. 453
KINLEY, K. 983
KINLEY, M.M. 453
KINSEY, J.L. 635
KINSLOW, R. 386, 718
KIP, A.F. 241
KIRCHHOFF, G.R. 222, 335, 338,
    347, 418, 1053, 1062, 1066, 1089,
    1098, 1157
KIRILLOV, G.A. 684
KIRK, J.F. 1075
KIRK, R.E. 209, 406
KIRKWOOD, J.G. 398, 531,
    1047-1048, 1180
KIRSCHNER, R.P. 712
KISSLINGER, C. 119
**KISTIAKOWSKY, G.B.** 86, 430, 521,
    527, 653, 741, **1105-1107**, 1212
KISTLER, M. 825, 1090
KITAMURA. K. 760
KITTINGER, J.W. 105
KITTLER, F.A. 1210
KIUCHI, T. 333
KLAPPER, H. 648
KLAUS-GOLDBERG, M. 1085
KLEBESADEL, R.W. 643
KLEIN, A. 696
KLEIN, C.A. 1142
KLEIN, C.F. 1061
KLEIN, E. 562
KLEIN, E.J. 701
KLEIN, F. 72, 332, 1063, 1103, 1108,
    1147, 1153, 1198
KLEIN, H. 1208
KLEIN, L. 396

KLEINE, H. 827, 916, 969
KLEISER, L. 940, 1039
KLEIST, G.E. VON 241
KLEIST, G.G. VON 474
KLEMM, R. 878
KLEPACZKO, J.R. 579
KLIEMANN, H.G. 1164
KLIMKIN, V.F. 1171
KLIMM, G. 1164
KLINGENBERG, G. 557
KLOCEK, P. 682
KLUSSMANN, H. 1121
KLUWICK, A. 446, 827, 1141, 1148
KNAPP, C. 1155
KNESER, A. 1191
KNIGHT, D. 1066
KNIGHT, D.M. 1066
KNIGHT, H.T. 1106
KNOCHENHAUER, K.W. 325, 1014
KNOEPFEL, H. 1160
KNOKE, G.S. 613
KNOTT, C.G. 422, 1181
KNOWLEN, C. 772, 1091
KNUDSEN, M. 500-501
KNYSTAUTAS, R. 711
**KOBES, K.** 85, 447, 478, **1107**, 1181
KOBUS, H. 236, 1045
KOCH, B. 597, 1160
KOCH, H. 1207
KOCH, K.R. 1063
KODAMA, T. 817
KOEBERL, C. 65, 722
KOENE, E. 826
KOGA, T. 320
KOGARKO, S.M. 1197
KÖGLER, C. 364, 1118
KOHKHLOV, A.M. 24
KÖHLER, G. 710
KOLB, A.C. 617, 947
KOLBE, G. 1205
KOLESNIKOV, L. 1174
KOLESNIKOVA, A.N. 661
KOLKERT, W.J. 238
KOLLIAS, N. 817
KOLM, H.H. 427
KOLSKY, H. xx, 579, 605, 711, 900,
    1074, 1092-1093, 1160, 1162
KOMAR, P.D. 212
KOMEL'KOV, V.S. 575, 598
KOMOSSA, S. 825
KOMPANEETS, A.S. 412, 529, 1197
KONDO, K. 726
KONDOH, N. 759
KONDRATE'EV, V.N. 1168
KÖNIG, K. 816
KÖNIG, W. 1204
KONYA, C.J. 732
KONYUKHOV, V.K. 663
KOPAL, Z. 629, 1170

KOPP, B. 1206
KOPP, C. 648
KORMER, S.B. 542, 623, 630, 636, 647, 661, 684, 1042, 1197-1198
KORN, D. 726, 729
KORNFELD, M.I. 543, 589, 903
KOROBEINIKOV, V.P. 443
KORSCHELT, E. 226, 443, 454, 1055, 1104, 1148
KORTEWEG, D.J. 296, 418
KÖRTING Bros. 396, 927
KÖRTING, B. 396
KÖRTING, E. 396
KÖRTING, J. 396, 1111
KORYCANSKY, D.G. 794
KOSOVICHEV, A.G. 810, 871
KOSTICIN, V.A. 1200
KOUVELIOTOU, C. 643
KOVACH, R.L. 721
KOVÁCS, L. 1199
KOVALEV, B.D. 1045
KÖVECSES, J. 785
KOYAMA, K. 835
KOZYREV, A.S. 630
KOZYREV, N.A. 629
KOZYREV, S.P. 718, 903
KRAEMER, K. 476, 1148
KRAFFT, F. 1127
KRAFFT, G.W. 262
KRAFT, R. 517
KRAMER, R.L. 238
KRAMER, W.G. 195
KRAMERS, H.A. 1168
KRAMME, U. 1206
KRANZ, W. 434
KRASIL'SHCHIKOEA, E.A. 1167
KRASINSKI, J.S. 793
KRÄTZ, O. 323
KRAUS, W. 517, 1088
KRAUSE, E. 1148
KRAUSS, F. 304, 413-414
KRAUSS, L.M. 483
KRAZER, A. 433, 1147
KREBS, H. 203, 1212
KREHL, P. 148, 357, 399, 508, 543, 552, 559, 571-572, 715, 735, 744, 796, 827, 910, 912, 914, 935, 954, 963, 976, 996, 1119, 1163, 1185
KRETSINGER, D.G. 1182
KRIDER, E.P. 54
KRIEGER, F.J. 620, 1167, 1173
KRIER, H. 1209
KRIGE, J. 734
KRIMIGIS, S.M. 836
KRINOV, E.L. 443
KRISHKEVICH, G.V. 1198
KRITZINGER, H.H. 1203
KROGSTAD, H.E. 821
KRÖNIG, A.K 45, 320, 323

KROPATSCHECK, W. 1020
KRUEGER, R.R. 793
KRÜGER, J.G. 252, 1210
KRÜMMEL, O. 361, 392, 1205
KRUPNIKOV, K.K. 542, 650, 661, 800, 1041-1042, 1213
KRUPP, M. 1121
KRUSKAL, M.D. 26
KRYMSKY, G.F. 746
KRYUKOV, P.V. 814
KRZIZANOVSKOMU, G.M. 1171
KUCHARSKI, W. 527
KÜCHEMANN, D. 1077
KUDO, H. 1099
KUDRIATSEV, N.N. 1171
KUENTZMANN, P. 444
KUERTI, G. 1141
KUGLER, F.X. 185
KUHL, A.L. 1139-1140, 1210-1211
KÜHN, A. 701
KUHN, D. 860
KUHN, E. 1112
KUIPER, G.P. 1170
KUKARKIN, B.V. 564
KUKLJA, M.M. 657
KULESHOVA, L.V. 1042
KULIK, L. 63
KULIKOVSKII, A.G. 1167
KULKARNI, S.R. 804
KULKARNY, V.A. 1179
KÜLP, F. 213, 231, 1063
KUMAGAI, T. 540
KUMAR, P. 738
KUNDT, A. 53
KUNETKA, J.W. 1212
KUNINAKA, H. 369
KUNZE, M. 814
KUO, A.C. 495
KUPSCHUS, P. 552
KURBJUN, M.C. 598
KURCHATOV, I.V. 116
KURNIT, N.A. 1091
KURTH, W.S. 56, 662
KURTZ, R.L. 690
KURYAPIN, A.I. 623, 636, 1197
KURZER, F. 201
KURZWEG, H.H. 502, 552, 1088, 1160, 1208
KUSUBOV, A.S. 745
KUTATELADZE, S.S. 31, 748, 775
KUTLER, P. 735
KUTTA, W. 1084, 1199
KUTTER, A. 367
KUTTER, H.K. 584
KUTTERER, R.E. 356, 1063, 1160-1161
KUVSHINOFF, B.W. 1171
KUZNETSOV, N.M. 827
KUZNETSOVA, I.V. 1121, 1130, 1200, 1206

KVENOLDEN, K.A. 706
KYDD, P.H. 1106

# L

LA CAILLE, N.L. DE 236
LA COUR, P. 1213
LA HIRE, P. DE 1127
LABOULAYE, C. 342, 1151
LABOURET, C.L.E. DE 395
LACROIX, A. 428
LADENBURG, R. 129, 405, 1049
LADENBURG, R.W. 129, 235, 495
LADERMAN, A.J. 666, 1139
LAFFITTE, P.F. 528, 1101
LAGRANGE, J.L. DE 41, 77-78, 139, 145, 243, 253, 255, 258-259, 262, 265-266, 271, 276, 279, 285, 1076, 1144
LAITHEWAITE, E. 1100
LAM, S.H. 1065
LAMB, J. 1107
**LAMB, Sir H.** 77, 282, 412, 433, 470, **1107-1109**
LAMBERT, D. 833
LAMBOURN, B.D. 698
LAMOR, J. 450
LAMPA, A. 1119
LANCASTER, O.E. 491, 1208
LANCESTER, F. 1084
LANCHESTER, F.W. 441
LAND, E.H. 563
LANDAU, L.D. 493, 548, 553, 1173
LANDEEN, S.A. 610
LANDON, J.W. 1093
LANDSBERG, H.E. 1211
LANDSHOFF, R.K.M. 617
LANDT, J.A. 696
LANE, J.C. 748
LANG, C.J. 303, 437
LANG, K.R. 1205
LANG, N. 1175
LANGENHORST, F. 689
LANGEVIN, P. 84, 462
LANGHANS, H. 1209
LANGMUIR, I. 470, 480
LANGSTON, C.A. 386
LANGWEILER, H. 512
LANIER, N.E. 712
LANLEY, C.B. 810
LANZEROTTI, L.J. 769
LAPEDES, D.N. 645, 1204
LAPLACE, P.S. DE 67, 77-78, 139, 155, 227, 239, 261, 266-268, 271, 276, 283, 285, 287, 306, 315-316, 318, 336, 339, 1056, 1096, 1102, 1144
LAPORTE DU THEIL, F. J. G. DE 193
**LAPORTE, O.** 86, 603, 647, 711, 948, **1109-1110**

LAPORTE, W. 1109
LAPPARENT, A. DE 1061, 1123
LAPWORTH, A. 1068
LARMOR, J. 363, 1176
LAROCCA, E.W. 633
LARSEN, A. 417
LARSON, H.K. 305
LARSON, R.J. 466
LATHAM, R.E. 1203
LATOURTE, M. 1101
LAUDAN, L. 1127
LAUDONE, V.P. 777
LAUE, M. VON 133
LAURENT, A. 331
LAUTERBORN, W. 414, 462, 690, 726, 798
**LAVAL, C.G.P. DE** 83, 378, 384, 395-396, 410, 926-928, **1110-1111**
LAVIRON, E. 572, 690, 725, 805
LAVOISIER, A.L. 202, 258, 1056
LAVRENTYEV, M.A. 680
LAW, C. 668
LAW, C.K. 1084
LAW, M.D. 273, 1203
LAWICK, H. van 692, 874
LAWICK-GOODALL, J. van 3, 692, 874
LAWN, B.R. 377
LAWRENCE, E. 116
LAWRENCE, E.O. 592
LAWRENCE, R.J. 15
LAWSON, M. 70
LAWYER, L.C. 1211
LAX, P.D. 1062
LAYTON, E. 1155
LAZUTKIN, A. 802
**LE BOULENGÉ, P.E.** 123, 329, 1012, **1111-1112**
LE BRUN, C. 1127
**LE CHÂTELIER, H.L.** 111, 266, 360, 372, 374, 381, 402, 460, 999, 1046, 1067, **1112-1113**, 1122, 1188
LE CONTE, J. 78, 338-339, 1112
LE FOLL, J. 1164
LE MÉHAUTÉ, B. 739
LE.CONTE, J. 338
LEACH, S.J. 51, 684
LEAKE, J. 813
LEAKEY, L.S.B. 180-181, 501, 877
LEAKEY, M.D. 180-181, 501, 877
LEAN, B. 1067
LEAN, D. 553
LEAUTE, H. 1113
LEBEAU, A. 662
LEBEDEV, S.A. 144
LEBRUN, P.-D. E. 252
LECLERC, G.L., Comte DE BUFFON 245, 257

LECOMTE, C.L. 1160
LEDENEV, B.N. 1042
LEDERBERG, J. 707
LEDREROVA, D. 1126
LEE, E.H. 589
LEE, E.L. 570
LEE, J. 689, 1091
LEE, J.A.N. 1210
LEE, J.H. 711, 759, 1084, 1171
LEE, J.H.S. 827
LEE, J.J. 24
LEE, K. 777
LEE, M. 120
LEE, M.A. 602
LEE, N. 793
LEE, R. 785
LEE, S. 817, 1206
LEER, E 746
LEERHSEN, C. 1195
LEES, L. 1065
LEEUW, J.H. DE 1165
LEGALLEY, D.P. 552
LEGENDRE, A.M. 318
LEGNER, H.H. 794
LEGRAS, J. 1117
LEHMAN, D.S. 3
LEHMANN, G. 1186
LEHMANN, I. 122
LEHMANN, K. 1132
LEHMANN, P. 934
LEHR, H.F. 723
LEIBENZON, L.S. 1200
LEIBER, V. 445
LEIBFRIED, G. 1044
LEIBNIZ, G.W. 39, 41, 43, 139, 207, 222-223, 231, 234, 246, 248, 250, 274, 1097-1098, 1127, 1142, 1153
LEICESTER, H.M. 1113
LEIDLER, K.J. 1059
LEIGHTON, R. 810
LEITER, E. 1141
LEKAWA, E. 590
**LEMAÎTRE, G.E.** 68, 139, 460, 478-479, 685, 1079, 1082, 1094, **1113-1114**
LEMBCKE, B. 668, 948, 975
LEMBKE, H. 1164
LÉMERY, J. 1114
**LÉMERY, N.** 60, 228, 252, 1102, **1114-1115**, 1210
LENARD, (?) 446
LENIHAN, J.M.A. 203, 1130, 1207
LENK VON WOLFSBERG, W. 308
LENNEP, J. van 1128
LENOBLE, R. 1130
LEONHARD, R.A. 1109
LEPPING, R.P. 779
LEPRINCE-RINGUET, L. 1111, 1142, 1206

LEPSIUS, B. 1087
LERNER, R.G. 30, 54, 1203
LE ROY, C. 1153
LESAGE, G.L. 262
LESEVICH, V.V. 363
LESLIE, F. 1028
LESLIE, Sir J. 1155
LESSER, M.B. 715, 764
LEVASHOV, P.R. 745
LEVEQUE, R.J. 145-146
LEVERMORE, D. 1214
LEVERRIER, U.J. 1058
LEVI, M.A. 118
LEVI-CIVITA, T. 468, 1192
LEVI-HEVRONI, D. 827
LEVIN, B.V. 1042
LEVIN, B.W. 1209
LEVINE, D.A. 636
LEVINE, M.A. 637
LEVY, A. 827
LEVY, D.H. 794, 1169-1170
LEVY, P. 1086
LEWIS, B. 129, 435, 490, 572, 1049, 1165
LEWIS, D.J. 539
LEWIS, J.A. 801
LEWIS, R.S. 774
LÉWY, H. 482, 1061-1062, 1080
LEWYTZKYJ, B. 1206-1207
LEY, W. 1103
LEYER, J.C. 1210-1211
LI, Y.H. 398
LIBBY, P.A. 579, 622, 1055, 1077-1078, 1209
LIBDURY, F.A. 1059
LIBESSART, P. 535, 543
LIBRESCU, L. 579
LIBROVICH, V.B. 1198
LICHTENBERG, A. VON 1042, 1085
LICHTENBERG, G. 266
LICHTENBERGER, A. 456, 573
LIEBIG, J. VON 40, 201, 294, 303, 309, 1127, 1150
LIEDL, G. 1203
LIENARD, M. 1101
LIENHARD, J.H. 1148
LIEPMANN, H.W 1179
LIFSHITZ, A. 610, 714
LIGHTHILL, E.B. 1115
**LIGHTHILL, Sir M.J.** 69, 101, 580, 612, 617, 748, 907, **1115-1117**
LILIENTHAL, O. VON 405
LIN, C.C. 1071
LIN, C.P. 776
LIN, S.C. 30, 606
LINCOLN, A., U.S. President 286
LINDBERGH, C.A. 457
LINDNER, M. 1122
LINDNER, N.J. 559

# NAME INDEX

LINDNER, V. 109, 1211
LINDROOS, M. 1026
LINDROTH, S. 40
LINDSAY, R.B. 1178, 1194
LINNÉ, C. VON 40, 713
LINSLEY, S.M. 1137
LIOUVILLE, R. 389, 1095
LIPELES, J. 831
LIPPISCH, A. 525, 1022
LIPPMANN, H. 1100
LIPSCHUTZ, M.E. 646, 673
LISTOV, M.S. 1187
LITTLE, B. 1195
LITTLER, J.J. 663
LIU, C. 785
LIU, P.L.F. 842
LIU, W.S. 1084
LIU, Y. 817
LIVEING, G.D. 430
LIVIO, M. 739
LOBB, R.K. 588
LOBSINGER, H. 197
LOCHTE-HOLTGREVEN, W. 674
LOCKE, J. 39
LODGE JR., H.C. 1071
LOEWENSTEIN, L.C. 1174
LÖFFLER, M. 748, 1209
LÖFFLER, R. 434
LOGAN, J. 517
LOGAN, J.G. 1091
LOGORY, L.M. 838
LOH, H.Y. 690
LOH, W.H.T. 37, 693, 1139
LOHSE, D. 822, 826
LOITSYANSKII, L.G. 1212
LOMBARD, J.L. 1154
LOMBARDI, L.V. 1146
LOMMEL, E. 1098
LONG, E.F. 1025
LONG, G. 624
LONGDON, L.W. 71, 718
LONGRIDGE, J.A. 387
LONSDALE, D.K. 403
LONSDALE, J. 1066
LOOMIS, A.L. 1050, 1193
LORENTZ, H.A. 459
LORENTZ, R. 273
LORENZ, H. 431, 452, 1192
LORIA, G. 1098
LORIN, R. 443-444
LORIS-MEKHOV, M.A. 1080
LOSKE, A.M. 827
LOSKE-MEHLING, A.M. 716
LOUIS XI, king of France 48
LOUIS XIV, king of France 48, 888, 1127, 1194
LOVE, A.E.H. 121, 265, 1109
LOVE, H.M. 1067
LOVE, S.G. 831

LOVELL, C.A. 140
LOVELL, M. 359
LOVETT, D.R. 377
LOVETT, E.O. 139
LOW, B.C. 829
LOWMAN, P.D. 740
LOWNDES, A.G. 1050
LOWTHER, W. 224
LÖWY, R. 508
LOYTSYANSKY, L.G. 1080
LOZEJ, G.P. 176
LOZZI, A. 536
LU, E.T. 831
LU, F.K. 838, 1208
LUCRETIUS CARUS, T. 44, 187
LUDFORD, G.S.S. 328
LUDOLFF JR., C.F. 240
LUDWIEG, H. 613, 942
LUDWIG, (?) 430
LUEGER, O. 1204
LUESSEN, L.H. 726, 764, 805
LUGN, R.V. 657, 1170
LUKASIEWICZ, J. 85, 1208, 1213
LUKYANOV, A. 675
LUMMER, O.R. 69, 437
LUNDGREN, W.R. 1195
LUNDMARK, K.E. 193, 491
LUNDSTROM, E.A. 744, 1139
LUN-HÊNG 189
LUO, H. 96
LUO, S.N. 658
LURIE, A.I. 1080
LUSSER, R. 1163
LUSTWERK, F 974
LUYTEN, W.J. 469
LYAMKIN, A.E. 778
LYDALL, F.R.S. 200
LYELL, C. 307-308
LYNAM, E.J.H. 463
LYNCH, D.K. 1209
LYNCH, F.L. 703
LYNDS, R. 176, 820
LYON, R.K. 1106
LYUBIMOV, G.A. 1167

# M

MA, Z.W. 818
MACADAM, S. 111, 1149-1150
MACAULAY, M. 13, 444
MACBETH, C. 621
MACCHETTO, F. 795
MACCOLL, J.W. 467, 485, 494
MACCOLL, J.W. 80, 1183-1184
MACCORMACK, R.W. 761
MACDOUGALL, D.P. 537, 565, 1106, 1134, 1184
MACFADYEN, A. 643, 834
MACFARLANE, A. 1181

**MACH, E.** ii, xxi, 12, 29, 38, 54, 64, 80-82, 122, 126-130, 136, 161, 198, 216, 226-227, 302, 310, 326, 336, 348, 353, 356-357, 360, 362-365, 370, 380-381, 383, 385, 387-389, 393, 395, 399-401, 411, 420, 423, 440, 454, 473, 480-481, 534, 537-538, 641, 671, 907, 910-912, 918, 920-922, 931, 954, 958-959, 993, 1002, 1005, 1010, 1014-1015, 1043, 1050, 1052, 1068, 1072, 1098, **1117**-1120, 1128, 1151, 1158-1159, 1177, 1193, 1199, 1207, 1212
MACH, J. 1117
MACH, K. 383, 1119
**MACH, L.** ii, 29, 81, 127, 129, 136, 198, 399-400, 405, 415, 417, 641, 910, 912, 924, 931, 959, 1010, 1014, **1119-1120**
MACK, P.E. 462, 1207
MACKO, S.A. 706
MACLACHLAN, P.P. 1142
MACLEOD, K.G. 65
MACMAHON, P.S. 1059
MACNEILL, Sir J.B. 1149
MACNEVIN, W.M. 603
MACRAE, N 1135
MADDEN, R. 256
MADDOX, J.R. 1212
MADDOX, R.L. 351
MADER, C.L. 60, 248, 430, 527, 627, 654, 664-665, 669, 687
MADER, H.M. 1179
MADLER, J.H. 410
MADSEN, B.M. 618, 645, 1170
MAEKAWA, I. 788
MAGEE, J.L. 529, 1048, 1135
MAGLIERI, D.J. 1078
MAGNUS, H. 344
MAGNUS, K. 1148
MAGNUS, R.J. 718
MAGOUN, F.A. 1208
MAHONEY, M.S. 219, 1127
MAIA, M. DA 852
MAIDEN, C.J. 552, 691
MAIENSCHEIN, J.L. 681
MAILLY, E. 1128
MAIMAN, T.H. 634, 644
MAIR, W.A. 1208
MAIRAN, M. DE 223
**MAIYEVSKII, N.V.** 374, 973, **1120-1121**
MAK, W.A. 560
MAKHVILADZE, G.M. 1198
MALEBRANCHE, N. 231, 1055
MALGRANGE, B. 1086
MALIK, J. 547
MALIN, M.E. 1106
MALINA, F.J. 1105, 1208

**MALLARD, E.** 360, 372, 374, 381, 460, 999, 1067, 1112-1113, **1121-1123**
MALLARD, F.E. 111
MALLET, J. 1123
MALLET, J.W. 1123
**MALLET, R.** 61, 119, 196, 254, 298-299, 307-309, 311, 319-322, 332, 359-360, 420, 857, 862, 979, 1121, **1123-1125**
MALLIAVIN, P. 1086
MALLOCK, A. 1050
MALLOCK, H.R.A. 444
MALLORY, H.D. 613
MALONE, D. 1155
MALYUSHEVSKY, P.P. 584
MALZEW, W.N. 1200, 1207
MAMALIS, A.G. 747, 1099-1100
MAMEN, J. 738
MAMUM, A.A. 818
MANAHAN, M.P. 437, 1209
MANAKOVA, M.A. 625
MANCIU, F.S. 828
MANCIU, M.M. 828
MANDELBROJT, S. 1086
MANDRUIKA, A.P. 1121
MANGELSDORF JR., P.C. 1106
MANNEBACK, C. 1114
MANNHEIM, A. 139
MANSON, N. 695, 1047, 1139, 1189, 1210-1211
MANVILLE, O. 1073
MAO, H.K. 730, 741
MARALDI, J.D. 236, 239
MARAN, S.P. 29, 492, 560, 602, 699, 1204, 1208
MARCAIDE, J.M. 789
MARCELLINUS, A. 190
**MARCI VON KRONLAND, I.M.** 38, 45, 48, 91, 205-206, 215, 220, 246, 497, 888, 890, 895, **1125-1126**, 1194
MARCONNET, G 490
MARDEL, C. 852
MAREK, J. 1126
MARINATOS, S. 184
MARINESCO, N. 93
**MARIOTTE, E.** 43-45, 86, 213, 220-221, 225-226, 240, 246, 336, 890, **1126-1127**, 1194
MARIS, H. 313
MARIUS, S. 175
MARK, H. 633, 969, 1204
MARK, J.C. 660
MARKOV, M.A. 1080
MARKS, P.H. 1065
MARLOW, W.R. 430
MARREN, D.E. 1208
MARRION, M.C. 817
MARRISON, W.A. 464
MARSDEN, B. 1150

MARSDEN, B.G. 333
MARSH, S.P. 649, 756, 951
MARSHAK, R.E. 1048
MARSHALL, A. 397
MARSHALL, D. 1172
MARSHALL, R.A. 748-749, 951, 969
MARTEAU, P. 756
MARTEN, E.B. 1028
MARTIN, M.H. 328
MARTIN, R. 1204
MARTIN, R.N. 1073
MARTIN, W.A. 1084
MARTINENGO, G.B. 199
MARTINEZ, T. 687
MARTINI, W. 878
MARTON, L. 518
MARTSON, P.L. 1074
MARUM, M. van 270
MARVIN, J.D. 1112
MARVIN, U.B. 198
MASAITIS, V.L. 179
MASAJI, G. 333
MASHHOON, B. 1182
MASKELYNE, N. 1096
MASON, C.M. 622, 666
MASON, H.B. 1038
MASON, W. 749
MATHER, J.C. 787
MATSCHOSS, C. 1151
MATSUDA, S. 1106
MATSUO, A. 130
MATSUO, K. 759
MATTHEW, H.C.G. 1038, 1041, 1051, 1057-1059, 1066, 1093, 1097, 1109, 1117, 1125, 1137, 1142, 1150, 1154, 1156, 1176, 1178, 1181, 1184, 1190, 1195, 1206
MATTHEWS, T.A. 649
MATTICK, A.T. 1091
MATTMÜLLER, C.R. 180
MATURA, E. 710
MAUCHLY, J.W. 138, 142-143
MAUPERTUIS, P.L.M. DE 239, 267
MAURER, R.E. 640
MAURIN, J. 563
MAUTZ, C.W. 401, 570
MAXIM, H.P. 437, 978
MAXIM, Sir H.S. 437, 978
MAXSON, J.A. 1139
MAXWELL, G.B. 1143
MAXWELL, J.C. 45, 262, 284, 314, 320, 324, 344, 365, 414, 500, 1040, 1107, 1149, 1176-1177, 1180
MAXWELL, R. 1099
MAY, R.P. 740
MAYALL, N.U. 1094
MAYER, H. 530
MAYER, J.R. 216, 311
MAYER, T. 252, 1210

MAYERS, R.A. 569
MAYR, O. 1175
MAZ'YA, V. 1087
MCAULIFFE, D.J. 817
MCCHESNEY, M. 45
MCCLEMENTS, K.G. 451
MCCLINTON, C.R. 808
MCCLUNG, F.J. 665, 685
MCCOLLUM, A. 806-807
MCCONNELL, A. 1142
MCCORMACK, T.J. ii, 12, 380, 1118, 1212
MCCORMMACH, R. 1057, 1066, 1089
MCCREERY, C. 549
MCCURDY, D.A. 790
MCDONALD, F.B. 836
MCDONALD, K.T. 781
MCFADDEN, L.A. 1204
MCFALL, K.A. 1091
MCKAY, D.S. 427, 766
MCKAY, M.F. 427
MCKEE, C.F. 22, 29, 666, 756, 791
MCKEE, J.W. 552
MCKEEN CATELL, J. 1194
MCKIE, D. 1127
MCKINNON, W.B. 175
MCLAUGHLIN, J.A. 641
MCLAURIN, C. 43, 77, 223, 233-234
MCLEAN, F.B. 724
MCLEMORE, R.H. 556
MCMILLEN, J.H. 231, 952
MCMILLEN, T. 631, 796
MCMULLIN, E. 227
MCNAB, I. 749
MCNEIL, D.H. 1198
MCONIE, M.P. 681, 936
MCQUEEN, R.G. 625, 634, 649, 774, 951
MCSHANE, E.J. 1208
MEAD, F. 803
MEADE, R.A. 529, 541-542, 547, 578, 1107, 1135, 1212
MÉDARD, L.A. 563, 1047, 1189, 1209, 1211
MEDWIN, H. 288
MEER, D. van der 826
MEHMED II, Ottoman Sultan 196-197
MEIER, G.E.A. 483, 751, 778, 1148
MEINIG, J.C. 1127
MEISSNER, E. 1101
MEITNER, L. 514
MELI, D.B. 1190
MELLEN, W.R. 701
MELLON, W.R. 701
MELLOR, P.B. 1099
MELO, F. 828
MELOSH, H.J. 15, 174-175, 766, 783
**MELSENS, L.H.F.** 353, 924, 1119, **1127-1128**

# NAME INDEX

MENDELEVICH, G. 1046, 1168
MENGER, K. 1119
MENZEL, E.R. 730
MENZEL, P. 859
MERCER, C. 790
MEREDITH, D. 444
MEREDITH-HARDY, R. 923
MERILL, R.B. 1170
MERKOULOVA, N.M. 1095
MERKULOV, M. 805
MERKULOVA, N.M. 1200
MERLET, J.P. 785
MERLINO, R.L. 818
MERRILL, P.W. 815
MERRILL, R.B. 658
**MERSENNE, M.** 53, 155, 203, 208, 219-220, 1097-1098, 1125, **1128-1130**
MERSTALLINGER, D. 1069
MERTENS, K.H. 368
MERZALOW, W.S. 1205
MERZHIEVSKII, L.A. 775
MERZKIRCH, W.F. 764, 1119, 1210
MESHCHERSKY, I.V. 1080
MESHKOV, Y.Y. 711-712
MESLER, R.B. 702
MESSERLY, G.H. 537, 1134
MESSERSCHMIDT, W. 518
MESSIER, C. 795
MÉSZÁROS, P. 697
MESZHANOV, A.G. 1198
METIVIER, M. 1145
METROPOLIS, N. 137
METROPOLIS, N.C. 138, 141, 143, 577, 1048, 1135
MEWES, R. 1212-1214
MEWES, S.A. 884
MEŸENN, K. VON 1054, 1077, 1089, 1099, 1119, 1206, 1213
MEYER, C. 108, 368
MEYER, E. 513, 974
MEYER, J.W. 1139
MEYER, L. 1039
MEYER, O.E. 384
MEYER, R. 1087
MEYER, R.A. 30
MEYER, TH. 101, 445, 473, 933, 1158
MEYER, V. 368
MEYER, W.B. 403
MEYER, W.F. 1108
MEYERAND JR., R.G. 952
MEYER-LARSEN, W. 733
MEYERS, M.A. 606, 1213
MEYERS, R.A. 634, 1203
MEZIERE, Y. 696
MEZZACAPPA, A. 66, 824
MICHAEL, E. 819
MICHAEL, J.V. 1106
MICHAELIS, A. 694

MICHEL, H.V. 747, 1213
MICHELL, J. 60, 228, 253-254, 261, 267, 862, 1056
MICHEL-LÉVY, A. 501, 527, 1122
MICHELMORE, P. 479
MICHELOTTI, P. 1044
MICHELS, D.J. 750
MICHLER, W. 108, 368
MIDDLEHURST, B.M. 1170
MIE, G. 94, 475, 736
MIELE, A. 224, 226, 526
MIERDEL, G. 1186
MIKHAILOV, M.V. 179
MIKHAILOV, Z.K. 629, 1170
MIKHAYLOV, A. 1042
MIKHEL'SON, A.M. 1130
**MIKHEL'SON, V.A.** 411-412, 993, **1130**
MIKKELSEN, R. 826
MILBANKE, Sir R. 111, 279
MILCH, E. 1163
MILES, F.D. 1038, 1136
MILES, J.W. 297, 1209
MILES, P. 286
MILIMETE, W. DE 194, 880
MILLAR, W.J. 1150
MILLARD, M.A. 399
MILLER, A.R. 569
MILLER, C.D. 62, 189
MILLER, CEARCY D. 556, 1018
MILLER, D.C. 424
MILLER, D.G. 338, 367, 1073
MILLER, D.J. 627
MILLER, D.P. 1097
MILLER, H.W. 459
MILLER, J.D. 917
MILLER, J.R. 612
MILLER, R. VON 927
MILLER, R.P. 207
MILLER, S.L. 55, 603, 714, 966, 1214
MILLER, V.R. 207
MILLET, C. 769
MILLET, J.C.F. 696
MILLIKAN, R.A. 1103
MILLMAN, P.M. 560
MILMAN, L. 1195
MILNE, E.A. 1094
MILNE, J. 190, 378, 381-382, 412, 419, 516, 896, 1124
MILTON, B.E. 544, 547, 765
MILTON, D.J. 658, 663
MILTON, K. 3
MINEEV, V.N. 667
MING, T. 774
MINGAUD, F. 49, 298, 1060
MINKOWSKI, H. 1191
MINOMURA, S. 741
MINSHALL, S. 613, 616
MINTO, D. 834

**MINTROP, L.** 120, 382, 445, 450, 456, 464-465, 513, 902, 980, **1130-1132**, 1213
MINUTAGLIO, B. 560
MIRELS, H. 596
MIRKIN, L.I. 1042
MISES, R. VON 433, 543, 718, 1071, 1081, 1147, 1191, 1213
MISHCHENKO, E.F. 1167
MITCHELL, A.C. 717, 726, 745, 774, 802
MITCHELL, C.E. 766
MITCHELL, E.D. 720
MITCHELL, S. 1206
MITCHELL, W.M. 201, 1208
MITROFANOV, V.V. 738, 1171
MITTER, H. 492
MIURA, Y. 705
MIX, L.P. 738
MIYAMA, H. 1106
MIZUTANI, T. 88
MKHITARYAN, A.M. 568
MOCHIZUKI, S. 790, 829
MÖCKEL, K. 323
MODARRES NAJAFABADI, S.A. 785
MODRESKI, P.J. 767
MOHAUPT, H. 514
MOISEEV, B.N. 745, 1042
MOISEEV, N.D. 1167
MOISSAN, F.F.H. 94, 406-407, 1112
MOISSON, A. 349, 382, 482
MOLLET, J. 273
MOLOKOV, S. 675
MONGE, G. 138, 255, 265-266, 294
MONIER, J. 71
MONKHOVEN, D. VAN 351
MONS, J.B. VON 267
MONTEIRO-MARQUES, M.D.P. 814
MONTGOLFIER, J.M. 89, 268, 882
MONTGOMERY, D. 1182
MONTGOMERY, W.T. 650
MONTIGNY, C.M.V. 330-331, 339
MONTROLL, E. 1048
MOODY, D.M. 1179
MOODY, G.L. 775
MOON, H. 493, 1187
MOON, P. 541
MOORE, C.C. 1182
MOORE, D.W. 1061
MOORE, F.E. 593
MOORE, H.K. 582, 624, 637, 663-664, 674, 991, 1006, 1210
MOORE, J.G. 678, 728
MOORE, N.B. 30, 101, 494, 1103-1104
MOORE, W.U. 189
MORABITO, L. 749
MORAWETZ, C.S. 1081
MORAY, Sir R. 210
MORDECAI, A. 1099

MORDUCHOW, M. 579
MOREAU, J.J. 814
MORELAND, J.B. 719
MORELLO, N. 1214
MORFILL, G.E. 818
MORGAN, J.V. 178
MORGAN, R. 827
MORIN, A. 1154
MORIN, A.J. 266, 295
MORITA, R.Y. 704
MORITZ, H. 1162
MORLEY, A.W. 1110
MOROSINI, F. 227
MOROVICS, M.T. 1043
MOROZ, E.M. 661
MOROZZO, Count 262
MORRIS, C. 24, 36, 137
MORRIS, C.E. 95, 240, 542, 642, 761
MORRIS, J. 191
MORRIS, O.A. 587, 923
MORRIS, P.J. 723
MORROCCO, J.D. 90
MORSE, C.R. 744
MORSE, R.L. 702
MORTON-BRIGGS, J. 48
MORTZFELD, P. 1206
MOSENFELDER, J.L. 658
MOSENTHAL, H. DE 345, 374, 1136
MOSES, Hebrew prophet 70
MOTTE, A. 223, 226
MOTT-SMITH, H.M. 353, 470
MOULDING, K.M. 377
MOULTON JR., J.F. 486, 686, 1189, 1213
MOWBRAY, G. 309
MOWBRAY, S.C. 1155
MOYE, W.T. 142
MOYER, A.E. 221
MUDFORD, N. 785
MUELLER, D. 775
MUELLER, T.J. 641
MUHAMMED II, Arab ruler 1037
MÜHLENPFORDT, J. 518
MULLEN, L. 178
MÜLLER, C. 72, 1063
MULLER, C.A. 589
MÜLLER, E. 146
MÜLLER, F. 718, 726, 962
MÜLLER, H. 197, 257, 880
MÜLLER, K. 1206
MÜLLER, M. 710
MÜLLER, W. 356, 624
MÜLLER-BECK, H.J. 710, 876
MULLIGAN, J.F. 1089
MUNROE, C. 984
**MUNROE, C.E.** 209, 318, 396-398, 405, 407, 453, **1132-1133**, 1136, 1155, 1211
MUNROE, E. 1132

MUNSON, D.E. 740
MURAOUR, H. 501, 527, 969
MURDIN, P. 40, 59, 65, 327, 655, 694, 794, 829, 1094, 1203
MURMAN, E.M. 724
MURNAGHAN, F.D. 509, 1071
MURPHY, C.H. 653
MURPHY, G. 729
MURPHY, L. 652
MURR, L.E. 754, 779, 1133
MURRAY, E.W. 329, 805
MURRI, W.J. 96, 134, 160, 733
MURTY, T.S. 59
MUSSCHENBROEK, P. van 197, 223, 241
MUTEAU, C. 1127
MUTHESIUS, V. 1211
MUTKE, H.G. 524
MYLON, C. 153, 209, 231
MYRABO, L. 803, 953
MYRHVOLD, N.P. 70, 781

# N

NAAMAN, R. 784
NADLER, D. 1060
NAGAYAMA, K. 827
NAGAYASU, N. 386
NAGLE, D. 830
NAGY, B. 706
NAIRNE, E. 256
NAKAGAWA, M. 828
NAKANO, M. 593
NAKORYAKOV, V.E. 748, 775
NAPIER SHAW, Sir ⇨ SHAW, W.N.
NAPIER, J. 139
NAPIER, J.R. 1149
NAPIER, R.D. 335, 343
NAPIER, W.MCD. 180
NAPOLITANO, L.G. 508, 1078, 1088
NAPOLY, C. 109
NAPPO, T. 1206
NARASIMHAN, R. 1153
NASER, G. 640
NASILOWSKI, J. 674
NASLIN, P. 556, 607, 1160, 1162
NASMYTH, J. 300
NATEGHI-A, F. 722
NAUDÉ, C.F. 461
NAUMANN, A. 502, 507, 1192
NAUMANN, J.F. 874
NAVEZ, A.J.A. 123, 1111
NAVIER, C.L.M.H. 45, 221, 266, 284-285, 288, 1060
NAYLOR, C.A. 1143
NAZAROV, V.E. 73
NAZE, Y. 737
NEAL, T.R. 687, 736, 965
NEDDERMEYER, S. 110, 117, 142, 524

NEHER, F.L. 275
NEILSON, F.W. 664, 666
NEKRASOV, A.I. 467-468
NELIS, M. DE 283
NELLIS, W.J. 741, 774, 802, 1042, 1074
NELSON, W.C. 243, 557, 635, 1063, 1208
NEMAT-NASSER, S. 98
NEMIROFF, R. 183
NENNIUS 191
NERNST, W.H. 113, 438, 453, 472, 1059, 1162, 1168
NERO, Roman emperor 188
NESS, N.F. 33, 675, 1078
NESTERENKO, V.F. 759, 827-828
NESTERIKHIN, Y.E. 1171
NESVETAILOV, G. 584
NEUBAUER, E. 283
NEUBERT, U. 340
NEUENDORF, K.K.E. 47
NEUFELD, M.J. 507, 930
NEUFELDT, S. 1168, 1203, 1209
NEUGEBAUER, M. 665
NEUMANN, E.P. 974
NEUMANN, F.E. 46, 288, 385, 1191
**NEUMANN, J. (L.) VON** 81, 112, 117-118, 137-138, 142-145, 306, 437, 487, 522, 524, 529, 532, 534-536, 543-544, 548-549, 556, 574, 582, 584-585, 611, 911, 962, 965, 993, 1048, 1061, 1069, **1133-1135**, 1181-1182, 1193, 1196
NEUMANN, M. VON 1133
NEUMAYER, G.B. 379
NEUWALD, P. vii, 155, 916, 993
NEWCOMEN, T. 228, 882, 1141
NEWELL, A.C. 296, 1156, 1209
NEWHALL, C. 762
NEWITT, D.M. 1052
NEWMAN, A. 639
NEWMAN, F. 1175
NEWMAN, R.C. 716, 1214
NEWTH, G.S. 323
NEWTON, J. 359-360, 384
NEWTON, Sir I. xix, 11, 43-45, 67, 71, 77-78, 83, 122, 124, 129, 155, 195, 198, 204, 207, 211, 218, 220, 223-227, 234, 250, 252, 267, 274, 283, 285, 287, 297, 306, 315, 346, 368, 759, 785, 890, 906, 926, 1039, 1056, 1060, 1102, 1118, 1125, 1127, 1137, 1153, 1176, 1190, 1194
NICHOLSON, W. 273
NICK, R.D. 687
NICOLAIDES, J.D. 1208
NIELSEN, A.T. 775
NIEMANN, H. 587
NIÉPCE Bros. 275

# NAME INDEX

NIÉPCE, C. 275
NIÉPCE, I. 300
NIÉPCE, J.N. 275, 283, 300
NIER, A.O.C. 505
NIERENBERG, W.A. 1203
NIESIOLOWSKI-GAWIN, V. VON 1063
NIESWAND, M. 1214
NIGGLI, P. 898
NIKI, H. 1106
NIKOWITSCH, P. 430
NILSON, C.G. 396
NILSSON, H.R. 621, 625, 689, 700, 702, 1162-1163
NILSSON, N.R. 128
NININGER, H.H. 64, 618, 645, 848
NISHIDA, M. 827
NITOCHKINA, E.V. 726
**NOBEL, A.B.** 106, 108-109, 270, 309, 333-334, 336, 342, 345, 357, 374, 392-393, **1135-1136**
NOBEL, I. 309, 1135
NOBILI, L. 389, 485
NOBLE, G. 1136
NOBLE, M.M. 1046
NOBLE, R. 803, 1027
**NOBLE, Sir A.** 105, 107, 124, 133, 284, 353, 358-359, 366, 413, 998, 1003-1004, 1037, 1136-**1137**, 1155
NOETHER, M. 1153
NOLLET, J.A. 240-241
NOMURA, Y. 668
NONWEILER, T.R.F. 669
NORDWALL, B.D. 790
NORMAN, M.L. 770, 937
NORRBIN, J.H. 345
NORRIS, W.G. 1038
NORTH, J. 1208
NORTH, J.D. 1181
NORTH, R.J. 370, 596, 1008
NORTHRUP, E.F. 341, 441
NORTON, T.H. 1151
NOVIKOV, I.D. 1198
NOVIKOV, S.A. 1042
NOVIKOV, S.S. 1171
NOVÝ, L. 1126
NOYES, R. 810
NUCCI, L.N. 1078
NYE, N 237

# O

O'BRYAN, T.C. 598
O'CONNOR, J.J. 1080
O'GRADY, B.V. 1106
O'KEEFE, J.A. 304, 697, 740
O'KEEFE, J.D. 772
OAKLEY, K.P. 692
OCAMPO, A.C. 178
OCCHIALINI, G. 804
ODISHAW, H. 254, 1182, 1205
ODOM, W. 1075
OERTEL, H. 536, 547, 597, 693, 778, 1213
OESER, E. 186, 244, 248, 1103, 1210
OETTINGEN, A. VON 3, 29, 204, 397, 1017, 1067, 1098
OEYNHAUSEN, E.H.K. VON 295
OGGIONI, S. 353
OGILVIE, K. 1074
OGILVIE, M. 1207
OHL, C.D. 798, 967
OHLSSON, J. 345
OHYAMA, N. 657
OKAMURA, T.T. 1179
OLAGNON, M. 821
OLAH, G.A. 775
OLBERS, H.W.M. 271, 274, 279
OLDENBOURG, ? 1098
OLDENBURG, H. 48, 214, 216, 1097, 1194
OLDENBURGER, W. 418
OLDHAM, E.U. 606
OLDROYD, D.R. 1103
OLEARY, A. 706
OLEĬNIK, A.G. 667
OLIVER, J. 652
OLIVIER, C. 180
OLIVIER, H. 615
OL'KHOVATOV, A.Y. 443
OLMERT, E. 575
OLMSTEAD, E. 1155
OLSON, R.A. 643
OLSON, W. 466
ONO, Y. 657
OOM, F.A. 329
OORT, J.H. 589
ÖPIK, E.J. 727
**OPPENHEIM, A.K.** 381, 609, 666, 671, 695, 769, 995, 1084, **1137-1140**, 1171, 1211
OPPENHEIM, T. 1137
OPPENHEIMER, D. 549
OPPENHEIMER, J.R. 529-530, 547, 1051, 1134
OPPENHEIMER, R. 7
ORAEVSKII, A.N. 662
ORDWAY, F.I. 523, 574
OREKIN, Y.K. 726
ORGEL, L.E. 1214
ORLENKO, L.P. 1173
ORLIN, W.J. 559
ORNELLAS, D.L. 775
ÖRSTED, H.C. 254
ORT, M. 180
ORVILLE, R.E. 719
OSEEN, C.W. 1055, 1071
OSHER, J.E. 785
OSTRIKER, J.P. 481, 666, 746, 1198
OSTROGRADSKY, M.V. 318
OSTWALD, W. 398, 1053-1054
OSWALD JR., R.B. 724
**OSWATITSCH, K.** 36, 256, 277, 304, 387, 476, 504, 544, 563, 567, 660, 1055, 1111, **1140-1141**
OTHMER, D.F. 209, 406
OTT, B. 1206
OTTENS, F. 1099
OTTO, N.A. 114
OUGHTRED, W. 139
OVENDEN, M.W. 727
OWEN, P.L. 570
OWEN, P.R. 1110

# P

PAAR, F.W. 192
PABST VON OHAIN, H.J. 515
PACK, D.C. 95, 563, 571, 748
PACZYŃSKI, B. 67, 804
PAGET, R.A.S. 1051
PAI, S.I. 693
PAISLEY, D.L. 834, 1063
PALMER, J.H. 141
PALMER, S.B. 191
PALMER, T. 839
PALMERSTON, Lord 196
PALMIERI, L. 229, 1124
PANDYA, B.H. 917
PANOFSKY, W.K.H. 553
PANOV, N.V. 800, 1042, 1213
PAO, Y.H. 377
PAOLI, R.M. 702
PAPADOPOULOS, G.A. 578
PAPAGEORGIOU, S. 191
**PAPIN, D.** 83, 221-223, 228, 231, 251, 291, 882, 926, 1097, **1141-1142**
PAPINEAU, D. 225
PAPIRIN, A.N. 1171
PARACELSUS, P.A. 202, 1125
PARADIJS, J. VAN 643
PARADISE, J. 30, 1203
PARARAS-CARAYANNIS, G. 627, 672, 761
PARCIEUX, A. DE 267, 282
PARENAGO, P.P. 564
PARIS, F. 1164
PARIS, J.A. 1066
PARIS, R. 1176
PARKE, W.C. 1083
PARKER, A. 1067
PARKER, E.N. 33, 65, 633, 657
PARKER, S.P. 16, 30, 96, 1048, 1107, 1204, 1206
PARKER, V. 297
PARKINSON, D.B. 140
PARKINSON, E.M. 1150, 1176, 1178
PARNELL, W.J. 583

PARR, W. 1005
PARRY, Sir W.E. 53, 78, 81, 156, 286-287, 325, 339
PARSON, C. 926
PARSON, W., 3rd Earl of Rosse 192
PARSONS, P.W. 192
PARSONS, Sir C.A. 94, 384, 407, 417, 457, 461
PARSONS, W. ('DEKE') 533
PARTINGTON, J.R. 1211
PASCAL, B. 11, 138-139, 206, 267, 645, 1129, 1194
PASTA, J. 146
PASTOR, J. 797
PATANELLA, A. 579
PATERN, L. 174
PATERNO, E. 1133
PATERSON, W.F. 194
PATTERSON, A.M. 695
PATTERSON, G.N. 574, 1084
PAUL, W. 667, 1052
PAULA GRUITHUISEN, F. VON 64
PAULL, A. 785
PAVLOVSKII, A.I. 648
PAVLOVSKII, M.N 1041-1042
PAWLAS, K.R. 437, 978
**PAYMAN, W.H.** 109, 131, 369, 484, 493, 509, 522, 996, 1017, **1142-1144**
PEALE, S.J. 749
PEAN, A. 1142
PEARCE, J.M.S. 483
PEARSALL, I.S. 633-634
PEARSON, J. 29, 606, 633, 669, 749, 982
PEARSON, K. 1212
PEASE, R.N. 129, 1049
PECK, H.A. 362
PEDLEY, T.J. 1117
PEIERLS, R.E. 117, 514, 519, 529, 540-541, 1047-1048, 1134-1135
PEIRESC, N.C.F. DE 1129
PELÉ, P.M. 850
PELLET, H.J.B. 110
PELLEW, F.H. 350
PELOUZE, T.J. 309
PELSENEER, J. 1127
PEMBERTON, H. 223
PENG-YOKE, H. 192
PENNEY, Sir W.G. 354, 547, 1048, 1134
PENNICOTT, K. 828
PENNIMAN, R.S. 385
PENNING JR., J.R. 664
PENNING, J.R. 664
PENZIAS, A.A. 685, 1082, 1197
PEPIN, R.O. 658, 1170
PEREDO, M. 872, 1208
PEREIRA, G. 247

PÉRÈS, J.J.C. 472
PERESSINI, E.R. 665
PÉRIER, F. 206
PERKINS JR., B. 675
PERKINS, J. 254, 285-286
PERL, W. 1104
PÉROT, A. 421
PERRAULT, C. 1127
PERRIN, F. 514
PERRIN, G. 200
PERRIN, J.B. 45, 289, 906
PERRY, F.C. 738
PERRY, R.W. 590, 971
PERSSON, A. 666
PERSSON, P.A. 24, 669, 717, 725
PESSES, M.E. 836
PETAVEL, J.E. 135
PETERMANN, A. 361, 864
PETERSEN, C.C. 1170
PETERSEN, F. 450
PETERSON, C.F. 733
PETERSON, E.L. 616
PETRARCA, F. 194
PETRUNIN, A.P. 1042
PETRZHAK, K.A. 117
PETZOLD, H. 140
PETZVAL, J. 1117
PEYRET, R. 800
PFEFFERMANN, E. 835
PFEIFER, W. 348
PFEIFFER, L. 876, 1209
PFLUG, F. 289, 1209
PFLÜGER, A. 475
PHAETON, son of HELIOS 185
PHAN, K.C. 544, 760
PHILBY, H.ST.J.B. 493
PHILIBERT, J. 767
PHILIPP, A. 462, 798
PHILIPPS, H. 927
PHILLIPS, D.S. 778
PHILLIPS, E. 312, 1158
PHILLIPS, H.F. 385
PHILLIPS, J. 1179
PHILOPONUS, J. 195
PHIPPS JR, T. 675
PIAZZI, G. 271
PICARD, E. 1073
PICKLES, C.S.J. 682
PICOLET, G. 1127
PICTET, M.A. 276
PIDDUCK, F.B. 265
PIERMARINI, G.J. 730
PIKELNER, S.B. 574, 606
PILBEAM, D. 1204
PILKEY, W.D. 790
PILKINGTON, J.D.H. 694
PINEAU, A. 427, 1209
PINET, G. 266
PINTO, F. 302

PIOBERT, G. 292, 358, 1154
PIPICH, P.V. 380
PIPKO, D. 1187
PIPPARD, B.A. 1184
PIPPERT, K. 736
PIRAJNO, F. 1170
PISKO, F.J. 346
PISTOLESI, E. 72, 198, 1207
PITOT, H. 234
PIZZARELLO, S. 706
PLACH, F. 349
PLANCK, M. 11, 38-39, 69, 1089, 1119
PLARR, V.G. 1038
PLASKOWSKI, Z. 1039
PLATO, Greek philosopher 22, 184-185, 1072
PLATZER, M. 1141
PLESSET, M.S. 259, 583
PLEUS, B. 358, 1137
PLINY, the Elder 188
PLINY, the Younger 188-189
PLUNKETT, J. 648-649
POCHOBRADSKY, B. 1175
POCKELS, F.C.A. 412
PODURETS, A.M. 600, 1041
POGGENDORFF, J.C. 241
POGSON, N.R. 260, 425
POHL, W.G. 1119, 1159
POHLHAUSEN, E. 468
POHLHAUSEN, K. 468
POINSOT, L. 324
POISON, S.D. 77
POISSON, S. 1144
**POISSON, S.D.** 5, 29, 54, 60, 77-78, 154, 162, 165, 265-266, 278, 282, 284-285, 291-292, 295, 298, 310, 312, 318, 324-325, 331, 349, 390, 412, 431, 1094, **1144-1145**, 1151
POKROVSKY, G.I. 1173
POLACHEK, H. 590
POLENI, G. 223, 1153
POLLAD, F.B. 514, 1205
POLO, M. 106
POLVANI, L.M. 495
POMBAL, Marqués DE 60, 247, 852, 1210
POMYKAL, G. 785
PONCELET, J.V. 266, 294, 300, 1060
PONT DE NEMOURS, L. DU 326, 377, 384, 675
POOLE, C.J. 262
POPE, A. 85, 248, 588, 642, 1210
POPE, K.O. 178, 1214
POPOLATO, A. 756
POPOV, L.V. 745, 1042
POPOV, N.A. 630
POPOVA, S.V. 658
POREMBKA, S.W. 650

# NAME INDEX

PORTER, D.H. 807
PORTER, R. 1207
PÖSCHL, T. 1209
POSEIDON, GREEK DEITY 185
PÖSGES, G. 906
POSPICHAL, J.J. 178
POST, A. 856
POUILLET, C.S.M. 125, 266, 305, 894
POULTER, M.L. 1145
**POULTER, T.C.** 96, 121, 556, 571, 682, 1073, **1145-1146**
POUNDSTONE, W. 1135
POWERS, E. 403
POWLES, H.H.P. 1214
PRAGER, W. 1074, 1160, 1162
PRANDTL, A. 1146
**PRANDTL, L.** 80, 83, 101, 310, 335, 386, 388, 396, 431-433, 439-443, 445, 448, 454, 463, 468, 473, 483, 488, 498, 504, 933, 968, 1038, 1054-1055, 1084-1085, 1103, 1140, **1146-1148**, 1158, 1164, 1211
PRANDTL, W. 1066
PRANKL, F. 514
PRATT, A.T.C. 1038
PRAVICA, M.G. 507
PREDVODITELEV, A.S. 568, 1086, 1171
PREISWERK, E. 512, 861
PRESS, F. 644
PREVOST, M. 1206
PRÉVOST, P. 262
PRICE, J.H. 664
PRICE, R.S. 537, 962
PRIESTER, W. 1198
PRILUTSKII, O.F. 737
PRINGLE, J.E. 598
PRITCHARD, C. 1181
PROBSTEIN, R.F. 224, 1065
PROCTOR, P. 24
PROCTOR, R. 64, 354
PROKHOROV, A.M. 30, 574, 663, 1203
PRÖLL, A. 1147
PROPPER, R. 438
PROSSER, V. 1043
PRÜMMER, R.A. 776, 1211
PRUSSAT, M. 1119
PTOLEMAEUS, C. 889
PUCKETT, A.E. 540
PUGH, E.M. 565, 580, 1184
PUGH, E.W. 141
PUGH, H.L.D. 1211
PUGSLEY, A. 444
PULIAFITO, C.A. 776, 997
PULVERMACHER, G. 429
PURCELL, E.M. 589
PURDY, A. 1081
PYRAK-NOLTE, L.J. 40

# Q

QIBIN, L. 192, 1208
QU, Q.-Y 835
QUIBY, H. 1175
QUICK, A.W. 502, 1088, 1164, 1207-1208
QUINCKE, G.H. 1043
QUINNEY, H. 1093

# R

RAAB, J.L. 1103
RABEL, G. 1103
RADICE, B. 188
RADOK, J.R.M. 1167
RADOK, R. 1109
RADOK, S. 1109
RADZEWICZ, C. 793
RAE, J. 692
RAE, W.J. 1091
RAGAN III, C.E. 745
RAGLAND, K.W. 1171
RAILLARD, F. 331
RAIZER, Y.P. 398, 568, 666, 672, 676, 1196-1198
RAJAN, R. 640
RAKHMATULIN, K.A. 94, 436
RAKOV, V.A. 720, 1204, 1214
RALEIGH, C.B. 659
RAMAN, V.V. 1150
RAMPINO, M.R. 182
RAMSAUER, C.W. 305, 416, 446, 470, 498, 894, 899, 985, 1063, 1157
RAMSAY, J.B. 587, 654
RAMSAY, Sir W. 457, 1056, 1178
RAMSBOTTOM, J.E. 1059
RAMSDEN, S.A. 675
RAMSES II, king of Egypt 352
RAMSEY, N.F. 1134
RAÑADA, A.F. 811
RAND, J.L. 579
RANDAZZO, R.F. 777
RANDERS-PEHRSON, N.H. 1148, 1208
RANEA, A.G. 1142
RANI, S.L. 694
RANKINE, D. 1148
**RANKINE, W.J.M.** 29, 73, 78-79, 94, 273, 276, 311, 315, 328-329, 348-349, 389-390, 412, 602, **1148-1150**, 1177
RANYARD, C.A. 330
RAS, M. 1039
RASHLEIGH, S.C. 748, 951, 969
RASPE, R.E. 39
RASPET, R. 30
RATEAU, A. 1101
RATHERT, G.A. 567
RATHKE, R. 1053

RATY, J.Y. 774
RAUSENBERGER, F. 445, 459
RAVANI, B. 217
RAVEAU, C. 277
RAY, S.F. 297, 732, 805, 1210
**RAYLEIGH, Lord** (OR J.W. STRUTT) 29, 60, 77-79, 313, 335, 363-365, 386, 390, 408-410, 412, 417, 428, 433, 448-450, 458, 461, 539, 957, 1057, 1093, 1108, 1130, **1150**, 1156, 1176-**1178**, 1182
RAYMOND, J.C. 580
RAYMOND, S.O. 640
RAZORENOV, S.V. 745, 837, 843
READY, J.F. 669
REAGAN, R., U.S. President 741, 748, 763, 770
REBEUR-PASCHWITZ, E. VON 400
RECHES, Z. 844
RECHT, R.F. 709
RECKLING, K.A. 154
REDDY, K.V. 781
REDMAN, R.O. 1048
REED, E.J. 835
REED, H.L. 1208
REED, S.A. 463
REES, M. 1062
REES, M.J. 697
REGNAULT, A.P. 1150
**REGNAULT, H.V.** 81-82, 157, 266, 335-336, 363, **1150-1151**
REICHENBACH, H. 231, 383, 388, 548, 621, 637, 686, 913, 946, 960, 996, 1032, 1110, 1119, 1162
REICHENBERGER, H. 640
REID, C. 1062, 1081, 1214
REID, H.F. 439
REID, S.R. 765, 1100
REIMOLD, W.U. 176
REIN, J. 415
REINECKE, W.G. 794
REINES, F. 487, 536, 1083, 1135
REINHARDT, O. 1103
REISIG, G.H.R. 1208
REISSNER, E. 1182
REITHEL, R.J. 1006
REKHMIRA 878
RELF, E.F. 103, 522
RELLENSMANN, O. 1132
REMBE, C. 148
RENARD, N.A. 1061
RENDELL, J.T. 297
RENO, F.V. 1208
RÉSAL, H.A. 1060-1061
RESNICK, R. 216
RETALLACK, G.J. 178
RETHORST, S. 554
REUSCH, F.E. 346, 898, 1062
REUTER, H.J. 638, 1085

REUTER, M.A. 638, 1085
REYNOLDS, G. 626
REYNOLDS, G.T. 238, 536, 551
REYNOLDS, O. 91, 198, 343, 382-383, 386, 439, 468, 1172
REYNOLDS, P.W. 1059
REYNOLDS, R.M. 552
REYNOLDS, R.T. 749
RHEINBERG, J.H. 128
RHODES, C.D. 1155
RHODES, R. 116, 505, 533, 545, 987, 1048, 1212
RIABOUCHINSKY, D. 77, 466, 495
RIBNER, H.S. 1084
RICARDO, H. 1092
RICE JR., J.W. 48
RICE, J.R. 817
RICE, M.H. 96, 625, 634, 774
RICE, O.K. 504
RICHARDS, A. 1135
RICHARDS, L.W. 1106
RICHARDS, P.I. 612
RICHARDSON, J.M. 398
RICHARDSON, W. 205
RICHELOT, F.J. 1191
RICHET, P. 1053
RICHTER, C.F. 505-506, 659
RICHTER, K.D. 797
RICHTMYER, R.D. 144-145, 584-585, 650, 1135
RIDDELL, F.R. 642
RIDGWAY, S.T. 200
RIEBER, F. 360, 588
RIEHLE, R.A. 716, 1214
RIEMANN, F.B. 1151
**RIEMANN, G.F.B.** 29, 76, 79, 255, 328, 362-364, 385, 393, 408-409, 427-428, 431-433, 440, 448, 543, 929, 1059, 1107, 1144, **1151-1153**, 1174, 1177, 1191
RIEMANN, I. 816
RIENITZ, R. 220
RIGG, P.A. 726, 812
RILEY, D. 786
RILEY, H.T. 188
RINEHART, J.S. 29, 590, 597, 606, 669, 982, 1213
RINK, H.J. 336
RINNERT, K. 769
RIPLEY, R.C. 570
RIQUET, P.P. 222
RITCHIE, D. 1203
RITTER, A. 396
RITTER, J. 420, 1132, 1213
RJABININ, J.N. 1168
ROACH, A.M. 587
ROACHE, P.J. 1210
ROBBINS, H. 1062
ROBERTS JR., H. 469
ROBERTS, C. 297
ROBERTS, E.A.L. 339-340
ROBERTS, G. 27
ROBERTS, G.T. 777, 1214
ROBERTS, P.H. 31
ROBERTSON, E.F. 1080
ROBERTSON, H.P. 1094
ROBERTSON, J. 1057
ROBERTSON, P. 670
ROBERTSON, P.B. 689
ROBERVAL, G.P. DE 1129
ROBEY, R. 335, 827
**ROBINS, B.** 49, 53, 71, 83, 100, 123, 156, 204, 223, 237, 239, 241-242, 268, 358, 893, 1076, 1096, 1099-1100, **1153-1154**
ROBINS, J. 1153
ROBINSON, A. 794
ROBINSON, G. 318-319
ROBINSON, H.W. 340, 1142
ROBINSON, T.R. 332, 979
ROBIQUET, P. 294
ROBISON, J. 296, 1156
RODDY, D.J. 658, 848, 1170
RÖDL, E. 408
**RODMAN, T.J.** 100, 324, 331, 336-337, 353, 358, 413, 886, 1003, 1100, 1137, **1154-1155**
RODWELL, G.F. 1205
ROE, P. 827
ROE, P.L. 145
ROEBUCK, J. 243
ROESKY, H.W. 323
ROGERS, E.W.E. 242, 1154
ROGERS, E.W.F. 1208
ROGERS, P.H. 532
ROGIER, H. 506
ROHDE, R.W. 1074, 1213
RÖHR, C. 848
ROHWEDDER, A. 710
ROISMAN, I.V. 826
ROLFE, J.C. 190
ROLL, P.G. 1083
ROMBURGH, P. van 368
ROMISHEVSKII, E.A. 662
ROMOCKI, S.J. VON 200, 1211
RONAN, C.A. 1203, 1208
RÖNTGEN, W.C. 132, 405, 414, 931, 1120
ROOD, G.T. 747
ROOK, L. 216-218
ROOSEVELT, F.D., U.S. President 514-515
ROOSEVELT, TH., U.S. President 117
ROSAKIS, A.J. 227, 817, 821-822
ROSALES, R.R. 583
ROSCOE, D. 675
ROSCOE, Sir H.E. 322-323, 1053
ROSE, H.J. 1040
ROSE, HENRY J. 76
ROSE, HUGH J. 76
ROSE, L.W. 638
ROSE, M.F. 768
ROSEN, R. 699
ROSENBERG, G. 1057, 1074
ROSHKO, A. 1179
ROSICKÝ, W. 360, 912
ROSLUND, L.A. 700, 962
ROSS, D.R. 674, 758
ROSS, I.N. 690
ROSS, M. 30
ROSS, Sir J.C. 53, 81, 156, 287, 325
ROSSI, C. 923
ROSSMANITH, H.P. 1163
ROSSMANN, TH. 387, 525
ROST, S. 679
ROSTOKER, W. 238
ROTA, G.C. 138, 1081, 1135
ROTENBERG, M. 392
ROTHÉ, J.P. 426, 1213
ROTHE, R. 1063
ROTHSTEIN, W. 1111, 1140
ROTT, N. 382, 482, 637, 1039, 1145, 1178, 1207
ROTTA, J.C. 442, 1148
ROTTOK, D. 403
ROUILLE, M. 234
ROUNSAVILLE, J.F. 109, 1205
ROUSE, H. 1045, 1130, 1156, 1200, 1212
ROUSSEAU, J. 248, 252, 1102
ROUTH, E.J. 12, 260, 295, 332, 1041
ROUX, L. 107, 322, 354-355, 359, 1136
ROWBOTHAM, F. 57, 908
ROWLAND, I.D. 198
ROWSEY, J.J. 793
ROY, A. 1204
ROY, M. 441, 444, 579, 595, 681, 696-698, 1074, 1078, 1101
ROY, N. 1184
ROYCE, E.B. 1074
ROYER, H. 1011
ROZLOVSKII, A.I. 535, 1197
RUBACH, H. 1104
RUBENS, H. 1193
RUBEY, W.W. 659
RUBINCAM, D.P. 740
RÜCKER, A.W. 79
RUDIO, F. 1076
RUDNEV, G.V. 1167
RUDOLF II, Holy Roman emperor 200
RUDOLPH, E. 60, 371, 393, 403
RUES, D. 1141
RUHMANN, I. 1210
RÜHMKORF, H.A. 1132
RÜHMKORFF, H.D. 316
RUMFORD, Count VON 45, 268-269, 277, 358, 998, 1137

RUMYANTSEV, A.Y. 1042
RUNCORN, S.K. 1204
RUOFF, A.L. 96-97, 741
RUPERT of the Palatinate, Prince
    19, 203, 212, 1100
RUPIEPER, H.J. 1063
RUPPE, H.O. 1089
RÜRUP, R. 1063
RUSSELL, C.T. 662, 752
RUSSELL, D. 744, 1091, 1155,
    1171, 1179
RUSSELL, E.J. 1067
RUSSELL, H.N. 454
RUSSELL, J.B. 341
**RUSSELL, J.S.** 78, 82, 296, 306-307,
    958, **1155-1156**
RUSSELL, P. 359
RUSSO, A. 734, 1091
RUTHERFORD, E. 16
RUTISHAUSER, H. 141
RYABEV, L.D. 575, 1042
RYABININ, Y.N. 618
RYALL, A. 707
RYDER, G. 178
RYLE, Sir M. 694
RYNIN, N.A. 494
RZIHA, F. VON 355, 429, 1212

# S

SABINE, R. 48, 125, 305, 360, 894
SABINE, R.A. 300
SABINE, Sir E. 297
SACHS, R.G. 544
SAFFER, R.A. 739
SAFFIR, H. 722
SAGAN, C. 714, 750
SAGDEEV, R.Z. 675
SAGREDO 889
SAGY, A. 844
SAINTE-LAGUË, J.A. 84, 401, 463, 938
**SAINT-VENANT, A.J.C. DE** 46, 266,
    284, 288, 301, 319, 346, 899,
    **1156-1158**
SAITO, S. 726
SAITO, T. 62, 754, 761
SAKASHITA, S. 657
SAKELL, L. 735
SAKHAROV, A.D. 116, 575, 590, 594,
    598, 608, 667, 1041, 1198
SALCHER III, P. 1158
SALCHER, G. 387, 920, 1119, 1159
**SALCHER, P.** 993
SALCHER, P. 29, 64, 82-83, 226,
    387-388, 395, 399-401, 454, 469,
    473, 918, 920-922, 931, 1015, 1050,
    1098, 1118-1119, **1158-1159**
SALLO, D. DE 211
SALPETER, E.E. 1048

SALVIATI, GALILEI's fictional advocate
    889
SALVIO, A. DE 204
SAMRA, E. 116
SAMSONOV, D. 818
SAMUDRALA, O. 817, 822
SAMUELS, D.E.J. 547
SANADA, N. 765
SANDAGE, A. 1094
SANDAGE, A.R. 649
SANDEMAN, R.J. 536
SANDERS, D.T. 280
SANDNER, H. 1172
SANFORD, P.G. 415
SÄNGER, E. 91, 498
SANTIFALLER, L. 1107, 1206
SANTOS, E. DOS 852
SARGENT, R.M. 227
SARGENT, W.L.W. 555
SARJEANT, W.A.S. 178, 1211
SAROTTI, A. 1141
SARRAU, E. 107, 266, 322, 354-355,
    359, 374, 1136, 1188
SASS, F. 1211
SATAKE, K. 59
SAUER, A. 434
SAUER, F. 525
SAUER, R. 80, 256, 567-568
SAUERBRUCH, T. 716
SAUNDERS, I.E. 710
SAUNDERS, R. ⇨ FRANKLIN, B.
SAUNDERS, Sir O. 1138
SAUSSAYE, L. DE 1142
SAVART, F. 290
SAVERY, T. 228, 1141
SAVIC, P. 675
SAVINOV, E.V. 667
SAWAOKA, A. 30, 726
SAXE, R.F. 1019
SCALLION, W. 794
SCAVUZZO, R. 581
SCEARCE, C.S. 675
SCHAAF, S.A. 37, 235
SCHAAFFS, W. 510, 558, 571-572, 976
SCHACHER, S.G. 1054
SCHÄFER, E. vii
SCHÄFER, V.J. 677
SCHAFFER, S. 261, 1057
**SCHALL, R.J.** 435, 585-586, 590, 614,
    666, 697, 717, 978, **1159-1160**, 1162
SCHALL, W. 1159
SCHALLHORN, D.R. 724
**SCHARDIN, H.** 85, 88, 128, 132, 243,
    356, 380, 387, 477, 486-487, 495,
    510, 513, 523, 525-526, 533, 547,
    557, 607, 633-634, 637, 639, 684,
    686, 916-917, 922, 934, 946-947,
    959, 1014, 1018, 1062-1063, 1070,
    1107, 1159-**1162**, 1181

SCHARDIN, R. 1160
SCHÄTZLE, W. 913
SCHATZMAN, E. 574
SCHAWLOW, A.L. 624, 634
SCHEEL, K. 473, 475-477, 1039,
    1048, 1063
SCHEIE, C.E. 408
SCHEINER, J. 1069
SCHERING, E. 1153
SCHERRER, F.R. 1076, 1154
SCHERRER, P. 133
SCHICHKOFF, L. 1053
SCHICKARD, W. 138
SCHILDKNECHT, H. 713
SCHILLER, L. 72, 443, 1039, 1055,
    1154, 1207, 1212
SCHILPP, P.A. 381
SCHIMMING, R. 222
SCHIRMER, H. 1029
SCHISCHKOFF, L. 107, 322-323, 1053
SCHLATTER, H. 1211
SCHLATTL, H. 174
SCHLEGEL, F. VON 39
SCHLEUSENER, A. 1132
SCHLICHTING, H. 385, 520,
    1070-1071, 1148
SCHMÄLZLE, P. 759
SCHMETTAU, (?) VON 197
SCHMIDT, A. VON 1162
SCHMIDT, B.P. 834
SCHMIDT, E.M. 544, 557, 654
SCHMIDT, H. 1070
SCHMIDT, I.E.E. 278
SCHMIDT, J.F.J. 32
SCHMIDT, J.G. 1208
SCHMIDT, M. 52, 670
**SCHMIDT, O. VON** 120, 512-513, 963,
    974, **1162-1163**
**SCHMIDT, P.** 140, 490, 1103,
    **1163-1164**
SCHMIDT, S.C. 148, 674, 784, 1042,
    1095
SCHMIEDT, E. 715, 764, 1085
SCHMITT, H.H. 724
SCHMITT, O.H. 127
SCHMITZ, B. 822, 875
SCHMITZ, H. 451
SCHMOLINSKE, E. 713, 1165
SCHMUCKER, U. 1211
SCHMUNDT, W. 1063
SCHNAITER, M. 844
SCHNEIDER, E. 657
SCHNEIDER, G. 1142
SCHNEIDER, H. 745
SCHNEIDER, W. 525, 1141, 1148
SCHNEIDER, W.G. 566
SCHNEPF, E. 713
SCHNYDER, O. 508
SCHOCH, A. 1163

SCHOENBERG, B. 1191
SCHOENBERG, M. 40, 1083
SCHÖLER, H. 701
SCHÖN, G. 1070
SCHÖNBEIN, C.F. 308
SCHÖNER, G. 1162
SCHONLAND, B.F.J. 1050
SCHOPF, J.W. 603, 1214
SCHOTT, G.L. 635
SCHREFFLER, R.G. 677
SCHREIER, W. 1205
SCHRENK, O. 1110
SCHRIEVER, B. 609
SCHRIEVER, W. 1213
SCHRÖDER van der KOLK, H.W. 342
SCHRÖDER, W. 1211
SCHRÖTER, J.H. 64, 272, 472
SCHRÖTTER, A. 1069
SCHUBAUER, G.B. 1070-1071
SCHUBERT, H. 737
SCHÜCK, H. 1136
SCHULLER, M. 357
SCHÜLLER, V. 1099
SCHULTE, E. 718, 726
**SCHULTZ-GRUNOW, F.** 140, 356, 473, 490, 537, 615, 713, 1039, 1163-**1165**
SCHULZ, G. 109, 1205
SCHULZ, H.E. 1130, 1200, 1204, 1207
SCHULZ, M. 524
SCHUMANN, E. 611, 1159-1160
SCHUSTER, A. 409-410, 425, 460, 1059, 1178
SCHUSTER, P.M. 1069
SCHWAB, G.M. 1209
SCHWAB, I.R. 599
SCHWARTZ, (?) 430
SCHWARTZ, L.W. 297
SCHWARTZ, R.D. 580
SCHWARZ, B. 195
SCHWARZENBERGER, R. 1141
SCHWARZSCHILD, K. 268
SCHWEBER, S.S. 594, 1048
SCHWEIKERT, G. 1044
SCHWEITZER, A. 1087
SCHWENKEL, D. 796
SCHWIRZKE, F. 735
SCORGIE, G.C. 547
SCOTT, A. 1203
SCOTT, F.H. 622
SCOTT, G.B. 413
SCOTT, J.F. 1190, 1195
SCOTT, P.F. 694
SCOTT, R.H. 354
SCOTT, W.E. 62, 189
SCRIBA, C.J. 1190, 1212
SCRIBNER, K.J. 775
SEABORG, G.T. 523

SEABROOK, W. 1194
SEAL, M. 407
SEAMAN, L. 98, 775, 837, 1074
SEAMANS JR., R.C. 1071
SEARLES, S.K. 736
SEARS, W.R. 256, 481, 526, 667, 1078, 1081, 1104-1105, 1117
SÉBERT, H. 124-125, 346, 360, 383, 395, 1094-1095
SEBESTA, L. 734
SECK, F. 138
SEDOV, I.G. 1165
**SEDOV, L.I.** 91, 497, 528, 545, 620, 642, 666, **1165**-1167
SEDOV, L.S. 1167
SEEGER, R.J. 532, 534, 537, 574, 590, 1119, 1134, 1194, 1213
SEEK, J.B. 675
SEGALL, P. 438
SEGNER, A. 926
SEGRÉ, E.G. 523, 609
SEHGAL, C.M. 93
SEHMER, A. 1027
SEIBEL, B. 875
SEIFERTH, R. 1140
SEIFF, A. 14, 552
SEITZ, F. 258, 634, 978, 1057
SEKHTER, B.I. 542
SEKINE, T. 791
SEKIYA, S. 393
SEKLER, E.F. 1195
SELF, S. 182, 762
SELIVANOVSKAYA, T.V. 179
SELVAGGIO, M.M. 48
SEMAN, G.W. 709
SEMENOV, N.A. 1167
**SEMENOV, N.N.** 484, 502, 504, 619, **1167-1168**, 1196
SEMENOVA, N.M. 1200
SEMIKHATOV, A.M. 93, 575, 1042
SEN, S. 828
SENECA, L.A. 104, 188
SENO, T. 549
SERBER, R. 524, 529, 533, 987
SEREBRIAKOV, E. 584
SERGESCU, P. 1130
SERRES, M. 267
SETCHELL, R.E. 971, 1179
SETTLES, G. 70
SETTLES, G.S. vii, 70, 269, 341, 597, 632, 765, 790, 828-829, 915, 917, 1210
SEWARD, C.R. 682
SEWARD, T.M. 1053
SEYFERT, C.K. 538
SHAHINPOOR, M. 791
SHAKESPEARE, W. 48
SHAL'NEV, K.K. 718, 903
SHANER, J. 730

SHANNON, C.E. 1135
SHAPIRO, A.H. 495
SHAPIRO, M.M. 656, 1135
SHAPLEY, H. 1113
SHAPOSHNIKOVA, T. 1087
SHARA, M.M. 739, 819
SHARAPOVA, T.A. 1171
SHARIPDZHANOV, I.I. 1042
SHARMA, S.P. 1084
SHARMAN, C.F. 485, 1183
SHARP, D.H. 540
SHARPTON, V.L. 777
SHAW, B.D. 766
SHAW, E. 3
SHAW, J. 279-280
SHAW, L.L. 778
SHAW, R.R. 56
SHAW, W.N. 446
SHAYLER, D.J. 1082
SHCHELKIN, K.I. 486, 522, 640, 671, 993, 1130
SHEAR, D.D. 991
SHEARER, P.M. 1213
SHEELEY JR., N.R. 750
SHEETS, B. 670
SHEFFIELD, S.A. 532, 654, 684, 783, 806, 1074, 1211
SHEKHTER, B.I. 1173
SHELLARD, P. 69, 174
SHELTON, F.H. 116, 354, 529, 539, 550, 581, 593, 603-604, 607, 659, 987-988, 1212
SHELTON, I. 773
SHEM ED DIN MOHAMED 194
SHEPARD JR., A.B. 651, 720
SHEPHERD, J. 1204
SHEPHERD, J.E. 130, 778, 1110, 1179, 1197
SHEPHERD, W.C.F. 509, 522, 572, 1143-1144
SHEPPARD JR., J.J. 341
SHERMAN, F.S. 567
SHERMAN, P.M. 301
SHETSOV, G.A. 648
SHEWMON, P.G. 644, 1074
SHI, H.H. 386
SHIBATA, K. 59
SHILOV, A.E. 1168
SHIMMIN, W. 696
SHIMOMURA, O. 88
SHINN, R. 613
SHIRLEY, J.H. 30, 174, 192, 794, 1204
SHIROUZU, M. 1084
SHIVA, HINDU DEITY 859
SHKLOVSKII, I.S. 666
SHLYAPINTOKH, I.A. 1197
SHOCKEY, D.A. 720, 775, 817
SHOEMAKER, C.S. 794, 1169-1170

# NAME INDEX

**SHOEMAKER, E.M.** 363, 618, 645-646, 657-658, 794, 799, **1169-1170**
SHOEMAKER, G.E. 1169
SHONK, C.R. 702
SHORT, J.L. 681
SHORT, N.M. 688, 705, 1074
SHRAPNEL, H. 262
SHREFFLER, R.G. 604
SHU, F.H. 491
SHUKLA, P.K. 818
SHURSHALOV, L.V. 443
SHUTO, N. 59, 644
SHVARTS, D. 827
SIACCI, F. 973
SIBERT, H.W. 15, 572
SIDDIQI, A.A. 1167
SIEBERG, A. 578
SIEBERT, L. 762
SIEGFRIED, R. 1066
SIEMENS, P.J. 1048
SIEMENS, W. VON 124, 366
SIERSCH, A. 415, 501, 1001
SIEWERT, T.A. 437, 1209
SIGURDSSON, H. 27, 61, 1204, 1214
SIGURDSSON, S. 175
SIGURGEIRSSON, T. 728
SIKKA, S.K. 96
SILBERRAD, D. 457
SILBERT, M.G. 745
SILFER, J.A. 706
SILFVAST, W.T. 634
SILICH, S.A. 666
SILVERA, I.F. 507
SIMAKOV, G.V. 745, 1042
SIMIN, O. 1200
SIMKIN, T. 379, 762, 854
SIMON, G. 810, 1070
SIMON, I. 1052
SIMONENKO, V.A. 1042, 1213
SIMONIDIS, J. 357, 1118
SIMONOV, N.N. 1197
SIMONS, C.C. 650
SIMPSON, J.A. 12-13, 16, 22-23, 26, 89, 138, 1205
SIMPSON, J.E. 57
SIMPSON, R. 722
SIMPSON, R.L. 775
SINGER, C. 692
SINGER, G.J. 283, 990, 1080
SINITSYN, M.V. 623, 630, 636, 684, 1197
SIRIGNANO, W.A. 1065, 1210-1211
SISLIAN, J.P. 623, 1083-1084
SITTER, W. DE 68, 460, 1114
SKADRON, G. 746
SKALAK, R. 16
SKENTELBERY, N. 1212
SKEWS, B.W. 668, 708, 947

SKEWS, S. 827
SKINNER, B.J. 18, 1205
SKRAMSTAD, H.K. 1070
SKRYL, Y. 657
SKUMBURDIS, K. 1085
SLACHMUYLDERS, E. 1179
SLACK, C.M. 902
SLADE JR., J.J. 557
SLATER, R.A.C. 1100
SLATER, W. 1067
SLATTERY, W.L. 175
SLAVIN, J.A. 662, 752
SLAWSKY, Z.I. 486, 686, 1189, 1213
SLEZINGER, I.I. 1210
SLOAN, E.D. 58
SLUTSKY, S. 1078
SMARR, L. 770, 937
SMEDLEY, E. 76, 1040, 1209
SMEETS, G. 773, 949
SMELT, R. 516, 1211
SMIGIELKI, P. 690
SMIGIELSKI, P. 690, 922, 1011
SMIRNOV, Y. 96, 511, 542, 598-599
SMITH, C. 1181
SMITH, D.G. 1204
SMITH, E. 793
SMITH, F. 557
SMITH, G.E. 223, 227, 707
SMITH, H.J. 365
SMITH, H.M. 597
SMITH, H.W. 1067
SMITH, L.G. 536, 548, 551, 1182
SMITH, M. 1210
SMITH, M.D. 770, 937
SMITH, R.K. 1071
SMITH, R.W. 1094
SMITH, S. 1181
SMITH, W.E. 1091
SMOLEN, J.J. 1139
SMOLKA, J. 1126
SMOLUCHOWSKI, M. 80, 476
SMOLUCHOWSKI, M.M. 432
SMOOKE, M. 805
SMOOT, G. 787
SMYTH, A.H. 245
SNAY, H.G. 567, 635
SNOW, W.R. 427
SNYDER, C.W. 665
SOAPES, T.D. 664
SOBEL, M.I. 1048
SOBRERO, A. 309, 1135
SOCHUREK, H. 1167
SOD, G.A. 496
SODDY, F. 455
SODEN, P.D. 1100
SÖDERBERG, O. 395
SOFFEN, G.A. 736
SOHLMANN, R. 1136
SOKOLOWSKI, T.J. 672

SOLER, M. 811
SOLIE, D.J. 792
SOLIMAN, A.S. 1100
SOLOMON, M.B. 829
SOLON, Greek philosopher 185
**SOLOUKHIN, R.I.** 130, 640, 684, 695, 1139, **1170-1171**
SOLOVIEV, S.L. 1209
SOMMER, J. 82, 336, 363, 1118, 1151
SOMMERFELD, A. 260, 298, 1043, 1047-1048, 1109, 1119, 1148
SONDIN-KLAUSNER, G 877
SONETT, S.P. 30
SONG, S.G. 438
SONNENBERG, K. 552
SONNTAG, A. 1175
SÖRENSEN, E. 1175
SORENSON, H. 1078
SORGENFREI, T. 1170
SOULARD, R. 275
SOUTHWELL, Sir R.V. 73, 1150, 1184
SOWERBY, R. 1099
SPALLANZANI, L. 22, 269-270
SPARACO, P. 90
SPARKS, S. 1179
SPARRE, M. DE 1101
SPARROW, V.W. 1214
SPECKMANN, H.D. 751
SPEISER, A. 237, 243, 1076
SPEISER, D. 1045
SPENCE, D.A. 1117
SPENCE, P. 16
SPENCER, L.J. 410, 496, 618
SPERANSKAYA, M.P. 647, 1042
SPEYER, B. 1019
SPIEGEL, E.A. 176
SPIESS, O. 1045, 1076
SPIGA, G. 353
SPILLER, J. 1038
SPIROV, G.M. 648
SPITZER, R.W. 537, 554, 574, 962
SPREITER, J.R. 662
SPRENGEL, H. 351
SPRENGER, H. 1039, 1119
SPURK, J.H. 1148
SPURR, J.E. 579
SQUIER, B. 688
SQUIER, G.O. 416
SQUIRE, D.R. 775
SQUIRE, W. 610, 1091
SRINIVASAN, R. 776
ST. LAZARE, B. DE 244
STAAB, F. 1164
STABENOW, E. 849
STACK, J. 101, 450, 499, 533, 541, 561, 591, 1192
STACY, K. 790
STADLER, J. 355
STAHARA, S.S. 662

STALIN, J. 1187, 1198
STALKER, R.J. 686, 949
STAMM, U. 736
**STANTON, Sir T.E.** 84, 101, 468, 484, 496, 938, **1172**
STANTON, T. 1172
STANTON, T.E. 462, 468, 484
**STANYUKOVICH, K.P.** 36, 265, 328, 542, 548, 568, 614, **1172-1174**
STAPERT, D. 185
STARK, W.E. 1155
STARKENBERG, J. 654
STATHAM, I.C.F. 1143
STAUDENRAUS, J. 779
STEDALL, J.A. 1190
STEEL, E. 288, 774
STEELE, B.D. 1154
STEELE, R. 23, 1038
STEELE, W.J. 1180
STEENBECK, M. 132, 509-510
STEFAN, J. 227
STEFFENS, H.J. 1211
STEIN, E. 1203
STEIN, H.D. VON 1160
STEINBERG, D.J. 648, 730, 756
STEINER, O. 146
STEKLOV, V.A. 1079
STELTZ, W.G. 690
STENGER, R. 830
STENHOFF, M. 395
STENZEL, H.B. 618
STEPANOV, B.M. 690
STEPHEN, L. 1038, 1206
STEPHENSON, F.R. 192, 290, 1208
STEPHENSON, G. 281, 307, 1029
STEPKA, F.S. 744
STERN, D. 776
STERN, D.P. 174, 238, 872, 1208
STERN, N. 1135
STERN, R.A. 1139
STERNBERG, E. 1105
STERNBERG, H.M. 687-688
STERNE, L.H.G. 1077
STEWART, D.S. 832-833
STEWART, G.A. 280
STIBITZ, G.R. 138, 141
STIGER, A. 403
STILLWELL, J. 1212
STILLWELL, W.H. 1025
STILP, A.J. 901, 1209
STINGLEY, N. 676
STINNER, A. 317
STIROS, S.C. 191
STISHOV, S.M. 658, 741, 1169
STOCK, J.T. 1051
STODDARD, F.J. 1091
STODOLA, A. 1174
**STODOLA, A.B.** 80, 432, 439, 488, 929, 1038, 1055, 1148, **1174-1175**

STOFF, M.B. 529
STÖFFLER, D. 179, 689
STOKER, J.J. 25, 77, 572, 624, 957, 1061, 1080-1081, 1209
STOKES, G. 1175
STOKES, H.P. 1176
**STOKES, Sir G.G.** 29, 60, 77-78, 165, 276, 282, 284, 311-313, 316, 349, 363-364, 448, 1040, 1057, 1107, 1144, **1175**-1177, 1179
STOLL, I. 1069
STOLLERY, J.L. 544, 760, 1084, 1110
STOMMEL, H.M. 1061
STONE, G.C. 237, 1204
STONE, H.A. 1184
STONE, R.G. 662, 1213
STONECIPHER, K.G. 793
STONER, R.G. 484, 551, 1049
STORM, E. 971, 1179
STÖRMER, C. 1114
STOTHERS, R. 191
STRACHEY, Sir R. 379
STRANGE, E.H. 1067
STRASSMANN, F. 511
STRAUB, G.K. 93, 542
STRAUB, H. 1045
STRAUB, J. 1172
STRAUMANN, N. 492
STREHLOW, R.A. 640, 740
STREIT, K.W. 1208
STRIZHEVSKII, S.I. 1200
STROBEL, H. 1206
STRÖDEL 429
STROGATZ, S. 146
STROMEYER, F. 1052
STRONG, H.M. 607
STRONG, I.B. 643
STRONGE, W.J. 217, 377, 785, 814, 819
STRUCK, C.J. 18, 820
STRUCK, W.G. 713, 1165
STRUTT, J.J. 1177
**STRUTT, J.W.** ⇨**RAYLEIGH, Lord**
STRUTT, R.J. 1051, 1178, 1194
STUDNIČKA, F.J. 1126
STUEWER, R.H. 1083
STUHLINGER, E. 102
STÜHLINGER, E. 523, 574
STUHLMANN, F. 1203
STUKELEY, W. 244
STULL, F.D. 639
STURM, C.F. 291
**STURTEVANT, B.** 62, 393, 693, 757, 767, 786, 957, 971, 1110, **1178-1179**
STUTZER, K.O. 508
STÜWE, B. 518, 525
STYAZHKIN, Y.M. 799, 1041-1042
SUESS, F.E. 305

SULLIVAN, D.A. 751
SULLIVAN, W. 685
SULTANOFF, M. 1007
SUMMERFIELD, M. 667, 1209
SUMMERS, C.R. 622
SUMMERS, D.A. 51, 685
SUMMERS, J.L. 14
SUMMERSON, J.N. 1195
SUN, Y.C. 1141
SUNG, L. 860
SUNYAEV, R.A. 1197-1198
SUPAN, A.G. 307, 361
SURGOT, E. 1127
SÜRING, R. 1192
SUSSKIND, C. 1089
SUTHERLAND, H.B. 1150
SUTIN, A.M. 73
SUTO, Y. 481
SUVOROV, L. 543
SUZUKI, J. 1212
SUZUKI-KAMATA, K. 754
SWAIN, L. 1108
SWEDENBORG, E. 250
SWEET, G.E. 1213
SWEGLE, J. 648
SWENSON JR., L.S. 564, 651
SWESTY, F.D. 812
SWIFT, L.A. 333
SWIGETT, G.L. 1133
SWISHER, C.C. 177
SWORDY, S.P. 17, 451, 500
SYMES, W.L. 302
SYMONS, G.J. 379, 854-855
SYMONS, L. 280, 1066
SYROVATSKII, S.I. 656
SZABO, A. 779
SZABÓ, I. 42-43, 152, 207, 218, 256, 369, 889, 1045, 1077, 1089, 1098, 1126-1127, 1145, 1190, 1195, 1209, 1212-1213
SZILARD, L. 116, 455, 514-515

# T

TABAK, E.G. 583
TABBUTT, F.D. 1106
TABE, I. 228
TACHOIRE, H. 1047, 1209
TAFFANEL, J. 460
TAIT, F.G. 1180
TAIT, J. 1180
**TAIT, P.G.** 4, 48, 225, 352, 360, 368, 397, 406-408, 890, 894, 1098, 1127, 1150, 1157, 1176, **1180-1181**
TAKALA, B.E. 587
TAKANO, A. 1084
TAKASHI, A. 333
TAKASHIMA, K. 784, 982

NAME INDEX

TAKAYAMA, K. 62, 85, 146, 357, 386, 536, 708, 715, 738, 744, 754, 758, 760-761, 765, 777, 780, 782, 786, 803, 832, 840, 1091, 1179, 1209, 1213-1214
TALBOT, W.H.F. 130-131, 316, 1015
TALEYARKHAN, R. 624
TALMY, V. 1174
TAMAKI, T. 631
TAMARKIN, J.D. 1079-1080
TAMM, I.Y. 500
TAMMANN, G. 467, 1162
TAN, Z. 793
TANAKA, S. 1119
TANG, F.C. 1084
TANGUAY, M. 716
TANGUAY, V. 738
TANIGUCHI, H. 754
TANIMORI, T. 192
TANIOKA, Y. 549
TANIS, H.E. 132, 510
TANK, F. 1039
TANKIN, R.S. 1105, 1184
TANNER, R.I. 1100
TAO, W.C. 778
TARASOV, D.M. 1042
TARLÉ, G. 17
TARTER, J.C. 469
TARUSOV, B.N. 1168
TARVER, C.M. 113, 762-763
TATON, R.A. 236, 255, 1211
TATTERSALL, I. 182, 877, 1203
**TAUB, A.H.** 142-143, 487, 495, 529, 534, 536, 549, 551, 572-573, 577, 1048-1049, 1134-1135, **1181-1182**
TAUBER, H. 184
TAUBER, M.E. 520
TAYLER, S.S. 1164
TAYLOR, C.A. 1133
TAYLOR, D.W. 457, 526
TAYLOR, E.I. 1182
TAYLOR, H.S. 129, 575, 1049
TAYLOR, J. 238, 302, 308, 650
TAYLOR, J.W.R. 93, 95, 542, 1208
TAYLOR, S. 1085, 1207
**TAYLOR, Sir G.I.** 80, 94, 436, 449, 467, 473, 481, 485, 488, 494, 496, 520, 527-528, 539, 554, 565, 568, 573, 580, 666, 901, 977, 1038-1039, 1105, 1109, 1116, **1182-1184**
TAYMAZ, T. 1213
TEGELAAR, P. 666, 1160
TEICHMANN, J. 317
TEIPEL, I. 1141
TELLER, E. 117, 143, 521, 524, 575, 577, 584, 592, 594, 752, 1047-1048, 1082
TELTOW, J. 1163
TEMPEL, E.W. 329

TEMPLE, G. 482
TENG, X. 825
TENNANT, S. 406
TER HAAR, D. 267
TERLETSKII, Y.P. 625
TERMAN, F.E. 499
TERNIER, P. 1123
TERRALL, M. 43
THADANI, N.N. 791
THADHANI, N.N. 113
THAM, R. 901, 996
THÉNARD, L.J. 113, 278
THEODORSEN, T. 499
THERRIAULT, A. 176
THEWLIS, J. 1204
THEYS, J.C. 176
THIEL, M. 745, 1214
THIFFEAULT, J.L. 33
THOLERN, C. 1210
THOLOZÉ, H.A. DE 49
THOMA, K. vii, 830
THOMANEK, F. 380, 510, 513
THOMAS, C. 972
THOMAS, H. 583, 650
THOMAS, H.M. 818
THOMAS, L.H. 544, 1044
THOMAS, P.J. 443
THOMAS, S. 1071
THOMAS, S.J. 691, 735
THOMER, G. 147, 314, 356, 414, 510, 513, 590, 597, 625, 725-726, 732, 934, 978, 991, 1159-1161, 1210
THOMPSON, H.W. 1168
THOMPSON, J.E. 726, 764, 805
THOMPSON, P.A. 751, 909, 1176, 1178, 1211
THOMPSON, S.P. 222
THOMPSON, Sir B. ⇨ RUMFORD, Count VON
THOMSON, E. 709
THOMSON, J.J. 126, 139, 467, 471, 1013
THOMSON, R.M. 1074
THOMSON, Sir W. ⇨ KELVIN, Lord
THOORIS, M.C. 1123
THOR, NORSE DEITY 859
THORNE, K.S. 545
THORNHILL, C.K. 570, 620
THORNTON, E.A. 705
THORNYCROFT, J.I. 414
THORPE, E. 253, 1057
THORPE, J.F. 1068
THORSETT, S.E. 694
THUTMOSE III, Egyptian pharaoh 878
TIBENSKY, J. 1043
TIDESWELL, F.V. 1144
TIDMAN, D.A. 654
TIETJENS, O.G. 1148
TIFFORD, A. 1088

TIFONOWA, G. 1085
TILGHMAN, B.C. 436
TILLMANN, W. 1055
TIMMERHAUS, K.D. 207, 726, 1074
TIMOSHENKO, S.P. 221, 319, 614, 1089, 1158, 1212
TING, L. 1078
TINGLE, T.N. 15
TINKLER, J. 558
TIPLER, F.J. 1182
TISSOT, J. 202
TITMAN, H. 1143
TITOV, V.M. 648, 775, 1042
TITTERTON, E. 541
TIZARD, H.T. 1085
TOBAK, M. 520
TOBIAS, S.A. 1100
TODD, R. 459
TODES, O.M. 1167
TODHUNTER, I. 1099, 1212
TOENNIES, J.P. 674, 1209, 1213
**TOEPLER, A.** 29, 123, 126-128, 131, 147, 211, 220, 320, 325, 340-341, 344, 346-347, 375, 399, 424, 434, 910, 921, 970, 1008, 1010, 1014, 1050, 1119, 1177, **1184-1185**, 1193
**TOEPLER, M.** 434, 1184, **1185-1186**
TOKSOZ, M.N. 644
TOLLENS, B.C.G. 405
TOLLMIEN, C. 476
TOLLMIEN, W. 385, 1070, 1148
TOLMAN, R.C. 1094
TOMITA, Y. 631
TOMIZUKA, C.T. 1074
TOMSON, J. 1109
TONE, D.R. 352
TONKS, L. 442, 509
TOOMRE, A. 176
TOOTILL, E. 1206
TOPINKA, L. 854
TORPHIE, S.L. 1205
TORRICELLI 206
TORRICELLI, E. 205, 1129
TØRUM, A. 28
TOUPIN, R. 1073, 1086
TOWNEND, L.H. 760
TOWNES, C.H. 623-624, 634
TOWNLEY, R. 210
TOWNSEND, C.H.T. 479-480
TOWNSLEY, R. 573
TRAPPENIERS, N.J. 741
TRAUBE, I. 1085
TRAUPEL, W. 396, 1111
TRAUZL, I. 366, 429, 1001
TRAVIS, F.W. 1100
TRAVIS, J.R. 654
TREANOR, C.E. 760, 1165
TREDICI, R. DEL 1212
TREMAINE, S. 17

TREMAYNE, D. 803
TRENDELENBURG, F. 571
TRENT, H.M. 630
TRESCA, H. 1100
TRÉVILLE DE BEAULIEU, C. 303
TRIBBECK, T.D. 716
TRICKER, R.A.R. 25, 57, 83
TRIGG, G.L. 30-31, 54, 93, 97, 119, 126, 532, 654, 707, 754, 1203, 1211
TRILLAT, J.J. 93
TRIPLEHORN, D.M. 767
TROKEL, S.L. 776
TROPEA, C. 826
TROPP, E.A. 1079-1080
TROSHIN, Y.K. 640, 671, 993, 995, 1130, 1171
TROTMAN, C.G. 1059
TROUVAIN, A. 888
TROUVELOT, E.L. 870
TROWER, W.P. 757
TRUAX, R.C. 1208
TRUBY, J.D. 379
TRUCANO, T. 794
TRUEBA, J.L 811
TRUELOVE, J.K. 666
TRUEMAN, E.R. 1100
TRUESDELL, C. 1073, 1086
TRUESDELL, C.A. 237, 243, 249, 253, 1045, 1176, 1211-1212, 1214
TRULSEN, K. 821, 869
TRUMAN, H.S., U.S. President 581
TRÜMPER, J. 773, 798
TRUMPLER, R.J. 338
TRUNIN, R.F. 542, 630, 650, 661, 745, 788, 800, 952, 1041-1042, 1198, 1213
TRUONG, N.K. 817
TRYON, E.P. 569
TSANG, W. 610
TSIBLIEV, V. 802
TSIEN, H.S. 37, 558, 1104
TSIOLKOVSKY, K. 432, 523, 930
TSUKERMAN, V.A. 96, 513-514, 542, 550, 625, 717, 1041, 1212-1213
TSURUTANI, B.T. 662, 1213
TSYMBAL, N. 1082
TUCK, J. 509, 544
TUCKER, M.J. 693
TUCKER, S.C. 1155
TULIN, M. 627
TUMLIRZ, O. 364, 393, 1118
TUPOLEV, A.A. 699, 1187
**TUPOLEV, A.N.** 462, 591, 699, **1186-1187**
TUPOLEV, N.I. 1186
TURCHI, P.J. 768
TURECKÝ, A. 1175
TURETSCHEK, G. 525

TURKEVICH, J. 1167-1168, 1187, 1198, 1206
TURNBULL, D. 634, 978
TURNER, E.B. 603, 1109-1110
TURNER, G.L. 42
TURNER, G.L.E 990, 1210
TURNER, H.H. 1041
TURNER, J.S. 1184
TURPIN, E. 256
TURPIN, G.S. 1067
TURQUET DE MAYERNE, T. 202
TUTTLE, H.P. 333
TWAIN, M. 20
TYLOR, E.B. 273
TYNDALL, J. 13, 1015
TYTLER, D. 69

# U

UCHIDA, M. 760
UFFELMAN, L. 1146
ULAM, S.M. 138, 145-146, 575, 594, 1082, 1134-1135
ULLOA, D.A. DE 13
ULSENHEIMER, G. 1120
UMAN, M.A. 54-55, 719-720, 1204, 1214
UNDERHILL, L.K. 1059
UNOKI, S. 593
URBANOWICZ, J.T. 760, 961
URE, A. 266
UREY, H. 618, 629, 645
UREY, H.C. 55, 116, 603, 714, 966, 1214
URLIN, V.D. 542, 661, 684, 1042, 1198
URSENBACH, W.O. 663
URTIEW, P.A. 381, 666, 775, 1139
USOV, V.V. 737
USSHER, J. 47, 299
UTKIN, A.V. 843
UTSUMI, W. 88

# V

VAGER, Z. 784
VALENSI, J. 1189
VALENTINIAN, F., Roman emperor 190
VALENTINUS, B. 200
VALLARTA, M.S. 139, 1114
VALLÉE, M. 826
VANDER, P. 1127
VANDEVELDE, A.J.J. 1115
VANGENGEIM, A.F. 1080
VARGHESE, P.L. 793
VARLEY, E. 1074
VASIL'EV, M.V. 1174
VASSY, E. 527
VAUBAN, S. LE PRESTRE DE 91, 315, 885

VAULOUE, J. 883
VAZSONYI, A. 508
VEBLEN, O. 1182
VEILLET, C. 836
VELARDE, M.G. 82
VELIKOVSKY, I. 586
VELLUZ, L. 1046
VELOSO, M.B. 247
VENABLE, D. 686-687
VENN, J.A. 1075
VENN, T. 1129
VENTER, R.D. 1099
VENTURI, G.B. 268-269, 927
VERBEEK, R.D.M. 379, 854
VERDÚ, G. 316
VERDUIN, C.J. 1127
VERE, M. 1189
VERHOEVEN, K. 50
VERNE, J. 100, 337, 342-343, 432, 886
VERPILLEUX, C. 307
VERSCHUUR, G.L. 801
VERSLUIS, M. 92, 822, 826, 875
VESELOV, A.P. 1099
VESSEL, E.C. 1120
VICKERY, A.M. 15, 766
VIDALE, J.E. 823
VIEILLE, J.M.L. 1187
**VIEILLE, P.M.E.** 29, 85, 111, 266, 292, 342, 360, 369, 374-375, 378, 383, 420, 422, 424-425, 460, 472, 551, 944-945, 1046, **1187-1189**
VIEIRA, G.J. 790
VIGNES, D. 678
VIJH, U.P. 844
VILLARD, R. 1208
VILLAVECCHIA, V. 1133
VINCENT, R. 71, 718, 1209
VINCENT, W.T. 238
VINCI, L. DA 41, 83, 197-198, 208, 285, 883
VINH, N.X. 1055
VIRDEN, R. 524
VIRGONA, R.J. 536
VIRTANEN, R. 1046
VITAL, P. 307
VIVIANI, V. 205
VIVIE, J. 556, 607, 1160, 1162
VLADIMIROV, L.A. 647, 1042
VLASOV, V.Z. 1167
VLECK, J.H. van 1049
VODA, J. 1175
VODAR, B. 679, 741, 756, 1211
VOELKEL, J.R. 227
VOEVODSKII, V.V. 1171
VOGEL-PRANDTL, J. 1148
VOGEL, H.O. 476, 1148
VOGEL, H.U. 193
VOGEL, H.W. 351
VOHRA, Y.K. 96

# NAME INDEX

VOIGT, W. 1176
VOITENKO, A.E. 676-677
VOITSEKHOVSKY, B.V. 717
VOITSEKOVSKY, B.V. 1171
VOLDER, B. DE 229, 1098
VOLK, F. 778
VOLLGRAFF, J.A. 1098
VOLLRATH, K. 147, 314, 356, 597, 666, 1160, 1162, 1210
VOLSKY, L.H. 1048
VOLTA, A. 256-257
VOLTAIRE 252, 851, 1099, 1102
VOORHIS, C.C. van 405
VORKIN, D.H. DE 1208
VOROTNIKOVA, M.I. 684, 1171
VORTHMAN, J.E. 1074
VOS, M.A. 689
VOSS, A. 1191
VREELAND JR., T. 577
VRIES, G. DE 296
VRONSKAYA, J. 1198, 1205
VULCAN, P. 748

# W

WAALS, J.D. van der 355, 1181
WACKER, J.F. 774
WACKERLE, J.D. 666
WADDELL, J.H. 594
WADE, M. 574, 1203
WAGNER, G.A. 657
WAGNER, H. 497
WAGNER, H.G. 532, 658, 683, 1139
WAGNER, H.J. 584
WAGNER, L. 1043
WAHA, A. DE 357
WAHL, A.C. 523
WAHSNER, R. 1077
WALCHNER, O. 498, 1055
WALDMAN, B. 1134
WALDMAN, H. 1203
WALKER, B. 51, 121, 190, 228, 254, 1203, 1210
WALKER, G.L. 684
WALKER, J.A. 450
WALKER, J.D. 799
WALKER, W.A. 687-688
WALL, (?) 232
WALLACE, D.C. 86, 763
WALLACE, R.E. 858
WALLACE, W.A. 195
WALLACH, O. 1162
WALLENBERG, C.F. VON 329
WALLER, R. 214
WALLEY, S.M. 573
**WALLIS, J.** 42-43, 154, 214-219, 224-225, 890, 1097-1098, 1125, **1189-1190**, 1194
WALLS, N.S. 1067

WALLWORK, A. 696
WALSH, J.M. 604, 615, 625, 634, 774, 978
WALTER, D.F. 755
WALTER, H. 518
WALTER, K. 800
WALTER, W.R. 905
WALTERS, E. 1059
WALTERS, W.P. 1211
WALTHER, M.F. 555
WALTON, E.T.S. 116
WALTRUP, P.J. 639
WANG, H.C. 444, 1078
WANG, L. 537
WANG, Q.D. 815
WANG, S. 739
WANG, X. 842
WANG, Z.R. 191, 835
WANHILL, R.J.H. 97
WAN-HOO 199, 1020
WANNER, K. 738
WANTZEL, P.L. 266, 301, 1157
WARD, D.J. 838
WARD, G.N. 580
WARD, J. 1195
WARD, P.D. 777
WARD, V.G. 591
WARD, W.R. 174
WARDELL, W.H. 110
WARK, J.S. 726, 786, 956
WARKEN, D. 935
WARLTIRE, J. 258, 1056
WARNER, B. 1208
WARNER, R. 1204
WARSCHAUER, D.M. 667, 1052
WARSITZ, E. 518
WARTNABY, J. 1124
WASLEY, R.J. 99
WATANABE, H. 760
WATANABE, J. 228
WATERMEIER, L.A. 1209
WATERSTON, J.J. 311
WATKINS, J.S. 1059
WATSON, G.M. 766
WATSON, K.D. 1038
WATSON, K.M. 681
WATSON, R. 13, 1000
WATSON, R.W. 697
WATSON, W. 242, 1057
WATT, J. 264, 286, 412
WATTENDORF, F.L. 91, 497, 1104-1105
WATTS, I. 1149
WAWILOW, S.I. 1166, 1204
WAYNE, N. 702
WEATHERSTON, R.C. 1091
WEAVER, L. 1195
WEBB, F.H. 637
WEBB, J.E. 1071

WEBB, K.R. 1038
WEBER, E. 1152
WEBER, E.H. 287, 559
WEBER, G. 1190
WEBER, H. 448
**WEBER, H.M.** 79, 392, 427-428, 448, 1152-1153, **1190-1191**
WEBER, M. 847
WEBER, R. 281, 366, 1002
WEBER, R.L. 1048, 1052, 1136
WEBER, W. 287, 559
WEBSTER JR., D.B. 336, 1155
WEBSTER, A.G. 108
WECKEN, F. 555, 658
WEDDING, F.W. 1131
WEDENSKY, G. 1167, 1187
WEELE, K. van der 826
WEGELER, C. 476
WEGENER, A. 60, 410, 466, 597
WEGENER, K. 1192
WEGENER, P.P. 517, 538, 562, 588, 941, 943, 1021, 1089, 1192, 1208
WEGENER, W. 683
WEGER, A. 1153
WEHNELT, A.R. 1159
WEICHSEL, L. 1153
WEIGERT, L.J. 1119
WEIHRAUCH, G. 547, 1160, 1209
WEIK, M.H. 143
WEILER, K.W. 1208
WEIMER, D.K. 551, 577, 1049
WEINBERG, F.J. 1209
WEINBERG, S. 746
WEINDL, C. 203
WEINER, D. 685
WEINER, E.S.C. 12-13, 16, 22-23, 26, 89, 138, 1205
WEINERT, (?) 526
WEINGART, R. 756, 785
WEINSTEIN, L.M. 790, 834, 923, 1027
WEIR, S.T. 802
WEISSKOPF, V.F. 545, 1048
WEISSMAN, P.R. 1204
WEISSMANN, P.R. 1170
WEIZSÄCKER, C.F. VON 250, 317, 515, 520, 574, 1047
WELCH JR., J.A. 681
WELLS, H.G. 455
WELLSTEIN, J. 1191
WELTRUBSKY, J. VON 364, 954, 1010, 1118
WENDEL, F. 529
WENDLANDT, R. 472
WENHAM, F.H. 352
WENNEIS, W. 713
WENTORF JR., R.H. 607, 733
WENTZEL, J. 383, 385, 1002, 1119, 1199
WERNER, E. 434

WERTHEIM, G. 368, 1205
WESTFALL, C. 529, 541-542, 547, 578, 1107, 1135, 1212
WESTFALL, R.S. 218, 1098, 1127, 1190, 1195
WESTINE, P.S. 729
WEYL, H. 544
WEYMANN, H.D. 650
WHEATSTONE, Sir C 131, 297, 301-302, 332, 979, 1012, 1015, 1037
WHEELER, J.A. 117, 261, 268, 469, 703
WHEELER, J.C. 492, 823
WHEELER, R.V. 1142-1143
WHEELER, W.H 485
WHIFFIN, A.C. 573, 901, 1183-1184
WHIPPLE, F. 609
WHIPPLE, F.J.W. 446
WHIPPLE, F.L. 488, 563-564, 743, 977
WHITAKER, E.A. 201
WHITCOMB, R.T. 450, 591, 688, 973
WHITE, A. 831
WHITE, C.S. 564
WHITE, D.R. 81, 591, 595, 640, 960, 1049
WHITE, E.P. 95
WHITE, M.P. 551, 566
WHITE, R.M. 103, 450, 652, 671
WHITE, W.C. 597
WHITE, W.T. 555
WHITEHEAD, J. 400, 1158
WHITEHEAD, R. 400, 1158
WHITEHOUSE, D. 825, 1170
WHITELAW, J.H. 764
WHITHAM, G.B. 586-587, 612-613, 619, 636, 733, 1117
WHITLOCK, R.R. 786
WHITROW, G.J. 1094
WHITTLE, F. 515, 1138
WIBERG, E. 243
WIBERT, P.L. 178
WICKHAM, J.E.A. 766, 776
WICKMAN, R. 629
WIDDER, F. 492
WIECHERT, E. 420, 450, 452, 1131
WIEGAND, P. 405
WIEGEL, R.L. 559, 693
WIEGHARDT, K. 1140, 1165
WIENER, N. 1063, 1135
WIENER, P.P. 195
WIERZBICKI, T. 825
**WIESELSBERGER, C.** 442, 502, 504, 507, 517, 533, 939, 968, 1088, 1148, **1191-1192**, 1200
WIESNER, J.B. 1107
WIGEN, S.O. 1209
WIGGINS JR., J.H. 554
WIGNER, E.P. 507, 802
WIINTENBERGER, E. 1197

WIJERS, R.A.M.J. 643
WILBECK, J.S. 772
WILBRAND, J. 404
WILBUR, P.J. 766
WILCKE, J.C. 252
WILD, P. 813
WILKENS, M.E. 520
WILKERSON, T.D. 1110
WILKES, M.V. 138, 143
WILKINS, J. 210
WILKINS, M.L. 570, 688
WILKINSON, D.T. 824, 1083
WILKINSON, N.B. 1211
WILLIAM III 1189
WILLIAMS JR., R.S. 728
WILLIAMS, D. 1210
WILLIAMS, D.R. 1170
WILLIAMS, F.A. 1139
WILLIAMS, H. 773
WILLIAMS, J.G. 1170
WILLIAMS, L.P. 1211
WILLIAMS, M.R. 138
WILLIAMS, R.F. 90
WILLIG, F.J. 604
WILLIS, T. 201
WILLMANN, U. 803
WILLSON, A.J. 1086
WILSON JR., E.B. 527
WILSON JR., H.A. 481
WILSON, C.R. 1074
WILSON, C.T.R. 1051
WILSON, D. 793
WILSON, D.B. 1058, 1176, 1181
WILSON, D.M. 723
WILSON, E.B. 527, 566, 1106
WILSON, G. 1066
WILSON, J. 71, 242, 1154
WILSON, J.G. 1206
WILSON, J.R. 1048
WILSON, M.R. 636, 642
WILSON, R. 300
WILSON, R.O. 1146
WILSON, R.W. 685, 1082, 1197
WILSON, W., U.S. President 456, 458
WILSONS, G.T. 1092
WIMSHURST, J. 211
WINCKLER, J. 405
WINGERDEN, K. van 791
WINKELMANN, A. 443, 1205
WINKHAUS, F. 429
WINKLER, H. 1172
WINKLER, K.H.A. 770, 937
WINKLER, S. 720, 817, 1210
WINTERBERG, F. 839
WINTERS, L.M. 793
WINTHROP, J. 248-249
WINTRICH, M.A. 302
WIRTINGER, W. 1153
WISE, J.A. 1182

WISE, M.N. 1181
WISSHOFER, F.A. 257
WISTRICH, R.S. 1207
WITCOFSKI, R. 794
WITHERING, W. 264
WITOSZY'NSKI, C. 499
WITT, A.N. 844
WITT, C.G. 798
WITT, O.N. 429
WITT, W. 748, 1209
WITTLIFF, C.E. 642, 690, 1091
WITTMANN, R.H. 541
WIVELESLIE ABNEY, Sir W. DE 351
WLADIMIROW, A.N. 91, 497, 1166
WÖHLER, F. 40, 294, 303, 1046, 1053
WOLANSKY, P. 1171
WOLF, E. 128
WOLF, H. 589
WOLFE, R.F. 1170
WOLFE, T. 1195-1196
WOLFF, C.L. 727
WOLFF, E.G. 210
WOLFF, W. 423-424, 1005
WOLFIUS, J.C. 223
WOLLASTON, W.H. 89, 195, 223, 274, 1066
WOLTERS, G. 1120
WONG, J. 821
WONG, M.K.W. 756
WOOD, (?) 535
WOOD, A. 1176
WOOD, A.B. 126, 471, 1013
WOOD, D.S. 577
WOOD, R. 1085
WOOD, R.M. 469
**WOOD, R.W.** 424-425, 537-538, 970, 1186, **1192-1194**
WOOD, S.A. 485
WOODBRIDGE, W.E. 402
WOODHEAD, A.W. 489
WOODHEAD, D.W. 475-476, 493, 994, 1143-1144
WOODRUFF, A.E. 1069
WOODWARD, P.R. 807
WOOLRIDGE, M.S. 694
WOOLNOUGH, P. 831
WOOLSEY, N.C. 786
WOOSLEY, S.E. 712, 834
WORTHINGTON, A.M. 413
WOSYKA MANDANSKY, J. 538
WOSYKA, J. 130, 356-357, 911, 958, 1118, 1193
WOULFE, P. 223, 256
**WREN, Sir C.** 43, 152, 210, 214-218, 224-225, 890, 1125, 1127, 1189, **1194-1195**
WRIGGERS, P. 377

# NAME INDEX

WRIGHT Bros. 83, 428
WRIGHT, O. 428, 457
WRIGHT, P.W. 698
WRIGHT, R.H. 533, 591, 1192
WRIGHT, T. 67
WRIGHT, W 428
WU, B.J.C. 562
WU, D. 828
WUERKER, R.F. 562, 690
WUEST, W. 1148
WUNSCH, D.C. 664
WÜNSCH, V. 222
WÜSTER, R. 1131
WYDER, P. 1203
WYNN, J.C. 1170
WYNN-WILLIAMS, C.E. 125
WYROUBOFF, M.G. 1122

# X

XIA, H. 96-97
XIA, K. 817
XIA, Q. 97
XU, D.Q. 536

# Y

YABUSHITA, S. 1170
YACOBI, Y.A. 1171
YACOVLEV, V.I. 1171
YAGI, T. 741
YAMAKATA, M. 88
YAMALIDOU, M. 1058
YAMAMOTO, M. 130
YAMANE, R. 631
YANG, Y. 822
YAO, W. 785
YARGER, F.L. 625
YASKO, M. 297
YCAS, M. 1083
YEAGER, A. 1195
**YEAGER, C.E.** 103-104, 450, 561, 734, 803, 1023, **1195-1196**
YEH, Y. 303
YELLA-REDDY, T. 1100

YELTSIN, B.N. 550
YIN, X.Z. 827
YODER, M.J. 1110
YONEDA, K. 760
YORK, H.F. 592
YOSHIHARA, H. 718, 1110
YOUNG, G. 541
YOUNG, J.W. 724
YOUNG, T. 249
YOUSCHKEVITCH, A.P. 244, 1076
YU, H. 708
YUEN, G. 706
YULE, G.U. 1190
YURKOVICH, R. 744
YUSHKO, K.B. 630, 1197-1198
YUTKIN, L.A. ⇨ JUTKIN, L.A.

# Z

ZABABAKHIN, E.I. 575
ZABUDSKIJ, N. 1121
ZABUSKY, N.J. 26
ZACKAY, V.F. 644, 1074
ZAHNLE, K.J. 443
ZAÏDEL', R.M. 667, 1197
ZAJAC, L.J. 1139
ZAK, A. 1025
ZAKKAY, V. 1078
ZAVERI, K. 103
ZECH, P. VON 1063
ZEDLER, J.H. 22, 197
ZEHLER, V. 1070
ZEHNDER, L. 129, 405, 931, 1010, 1120
ZEHR, J. 683
ZEIBIG, H. 1165
ZEITOUN, D.E. 838
**ZEL'DOVICH, Y.B.** 31, 96, 112, 116-117, 398, 412, 433, 486, 511, 522, 528, 532, 534-535, 558, 568, 575, 594, 598, 623, 630, 636, 666, 672, 748, 799, 993, 1041-1042, 1069, 1134, 1139, 1167, 1173, **1196-1198**
ZELIKOFF, M. 1091

ZELLER, K. 848
ZEMAN, J.L. 300
**ZEMPLÉN, G.V.** 72, 427, 438, 1073, 1101, 1107, **1198-1199**
ZEUS, GREEK DEITY 185, 859
ZHANG HENG ⇨ CHANG HÊNG
ZHANG, B. 175
ZHANG, F. 995
ZHANG, M.X. 823
ZHANG, T.L. 752
ZHANG, X. 838
ZHANG, Y. 1179
ZHARKOVA, V.V. 810
ZHIGULEV, V.N. 662
ZHUCHIKHIN, V.I. 1042
**ZHUKOVSKY, N.E.** 418, 462, 709, 1130, 1186, **1199-1200**
ZIEGLER, F. 300
ZIEREP, J. 467, 483, 567-568, 637, 778, 1141, 1148
ZIMANOWSKI, B. 20
ZIMMERMANN, E. 1012
ZIMMERMANN, H.U. 835
ZINAIDA, A. 550
ZINMAN, W.G. 1106
ZINNER, E. 743
ZIOLKOWSKI, K. ⇨ TSIOLKOVSKY, K.
ZOEPPRITZ, K. 452
ZOLLER, K. 578
ZÖLLNER, J.K.F. 343
ZORNIG, H.H. 486, 504
ZORZI, F. DE 199
ZUIDEMA, J. 97
ZUKAS, J.A. 1211
ZUSE, K. 138, 141, 143
ZWERGER, M. 401, 1098
ZWICKER, E. 53
ZWICKY, F. 491, 499, 518, 820
ZWOLINSKI, B. 610, 1074

# Subject Index

# SUBJECT INDEX

Acronyms as well as names of scientists following main entries appear in capital letters; years italicized and in parentheses.

## A

A-3 rocket, liquid fuel 507
   aerodynamic optimization 517
   supersonic 529
A-4b rocket 498
A-5 rocket 517
A-9/A-10/A-11, multistage rockets 523
Abel equation (Abel law) 358
Abel heat test (explosive stability test) 1037
Aberdeen Proving Ground 594
Ablation plume 20
Ablative photodecomposition 776, 997
Abrasion 680
Acceleration 224, 230, 236, 301, 483, 576
   charged particles 576, 835
   cosmic rays 656, 746
   diffusive 746
   interstellar particles 836
   shock drift 746
   ultrahigh 709
Acceleration waves 431, 1086
Accelerators, blast wave 35, 793
   flat plate accelerator 97, 114, 604, 649, 951
   high-velocity, gun-type 951
   multiple-collision 697
   ram accelerator (RAMAC) 772, 773, 786, 1090
Accretion disk 318
Accretion shock waves 66
Ackeret theory 1038
Acoustic studies 1129
Acoustical Society of America (ASA) 485
Acoustics, nonlinear 72, 1075, 1116, 1144, 1159
Action turbine 378, 384
Adiabatic, RANKINE 329, 348
Adiabatic blast wave theory, supernova remnant theory 666

Advisory Committee on Uranium, ROOSEVELT 515
Advisory Group for Aerospace Research & Development (AGARD) 592
Aeolipile 188
Aerial waves 76, 446
   finite amplitude 29, 363
Aeriform bodies, VAN HELMONT 202
Aeroacoustics, LIGHTHILL 1115
Aeroballistic drag 1121, 1153–1154, 1191
Aeroballistic Range Association (ARA) 652
Aeroballistic studies 938
Aeroballistics 49, 102, 187, 239
Aerodynamic drag 83, 156, 452, 683, 973, 1016, 1096, 1110, 1115, 1121, 1153–1154, 1172, 1188, 1191
   projectiles 374, 382
Aerodynamic heating 100, 203
   supersonic speed 609
Aerodynamic lift of rotating cylinders 1038
Aerodynamic shock heating 969
Aerodynamic theory 499
   PRANDTL 1146
Aerodynamics 80, 441
   airfoils 483
   compressible flows 432
   cosmical 574, 589
   high-speed 572, 596
   supersonic 581, 1088, 1090, 1104, 1110, 1153
Aeromechanics, ZHUKOVSKY 1199–1200
Aeronautics (aeronautical engineering) 73, 930, 1064, 1070–1071, 1077, 1083–1084, 1090–1092, 1103–1104, 1137, 1147, 1165, 1177–1179, 1183, 1199–1200
   high-speed 1022, 1054
Aeronautics Publication Program 575
Aerospace ballistics 187

Aerospace Research Laboratory (ARL) 689
Aerothermochemistry 611
Aerothermodynamics 321, 492, 611
Aftershocks (earthquake) 247–248, 672, 853, 868
Age of the Earth Debate 317
Air, dynamic elasticity 210
   resistance 241
   sound propagation, wave equation 243
Air/methane, explosivity 381
Air blast focusing, prediction 675
Air blasts, elastic structures 621
Air-breathing jet engine 443
Air compressibility, drag/lift 483
Air condensation 517
Air gun (wind gun) 80, 196, 880
Air jet engine 483
Air-methane mixtures 509
Air percussion wave 340
Air shocks, discharge of Leiden jars 241
Air shooting 1145
Air waves, intense 81, 287
   Krakatau 379
Airfoils (aerofoils) 724, 726, 1038, 1077, 1104, 1109, 1115, 1117, 1147, 1166, 1172, 1199
   biconvex 1077
   high speed, schlieren photography 554
   high-aspect-ratio 352
   shock-free transonic 726
   supercritical wings 688
   supersonic 484, 496, 1038, 1077, 1115
   theory 496, 1084, 1115, 1147
Airplane, area rule 591
Airscrew tip phenomena, high speeds 463
Airscrews, speed of sound, lift/drag 470
Airy disc/spiral 1040
Alchemy 1046, 1129
Aleutian earthquake (*1946*) 549

Algol (Beta Persei), period 260
Alkali halide, electrical
    conductivity 615
Allan Hills meteorite 766–767
Allende meteorite 703
Alpha-beta-gamma theory 569, 1082
Alum Chine, dynamite explosion 453
Aluminum/water steam explosions 624
Amatol 404
American Institute of Aeronautics
    and Astronautics (AIAA) 667
Ames Aeronautical Laboratory
    (ARL) 515
Amino acid generation, Earth's
    primitive atmosphere 714, 966
Ammonia dynamites 345, 385
Ammonium nitrate (AN) 209, 385
    fertilizer-grade (FGAN),
        explosion 560
Ammonium nitrate/aluminum
    powder/water 616
Ammonium sulfate-nitrate double salt
    explosion (Oppau, Germany) 209
Anak Krakatau 378–379
Analytical engine 139
Andalusite 745
Andromeda (galaxy) 820
ANFO (ammonium nitrate/
    fuel oil) 210
Annihilation 17, 609
Antennae Galaxies 178
Anti-aircraft supersonic missile,
    guided 517
Antibang 795
Antimony sulfide/potassium
    chlorate 316
Antiparallel method
    (detonation velocity) 1002
Antipersonnel projectile, shrapnel 262
Antiprotons, highenergy collisions 610
Antolik's soot figures 130, 355–356,
    360, 385, 443
    irregular interactions of shock
        waves in air 355–356
*Apatosaurus* (dinosaur) 780–781
Apollo 11, Moon landing 704–705
Apollo 14 720–721
Apollo 16/17 724
Aqua fortis 200
Aquarium technique 663
Ar-40/Ar-39 isotope dating 177
Arbitrary fluid, stability of shock
    profiles 544
Arbitrary Lagrangian Eulerian
    (ALE) techniques 392
Area rule, Whitcomb's 973
Argus-Schmidt tube 1163
Arica Earthquake
    (Peru/Chile) 346

Arizona meteorite (Canyon Diablo
    meteorite), meteoritic diamonds
    403, 406
Arizona's Meteor Crater (Barringer
    Crater) 64, 182, 436, 465, 508, 618,
    645, 658, 663
Array seismology 679
Arrhenius equation
    (Arrhenius rate law) 398
Arzamas-16, Soviet nuclear weapons
    research 550
Asteroid belt 706
Asteroid impacts 747, 849
    organic compounds 706
Asteroid seismology 119
Asteroids 271, 274, 798–799, 801,
    830, 1058, 1169
    collision with asteroids 354
    near-Earth (NEAs) 814
Asteroseismology 119
Astigmatism 1040
Astrobiology 706–707
AstroBlaster, multiple-collision
    accelerator 697
Astroblemes 65, 497, 646
Astrogeology, SHOEMAKER 1169
Astronautics 75, 104, 645, 1090
    birth of 432
Astronomical imaging 1094
Astronomy, extragalactic,
    HUBBLE 1093
Astrophysical explosions 1094
Astrophysical jets 6, 462, 655, 937
Astrophysics, nuclear,
    BETHE 1047
Asymptotic law of shock wave
    attenuation 642
Athodyd 444
Atlantic telegraph cable 1040
Atmospheric friction 1025
Atmospheric steam engine 228
Atomic bombs 117, 404, 455, 505,
    511, 514, 517, 528, 530, 533, 651,
    987, 1134, 1193, 1196
    China (*1964*) 673
    critical mass 455
    GADGET 987, 1105
    HEISENBERG 517
    Hiroshima 546
    Nagasaki 546
    testing 75
Atomic explosions 988
    Trinity Test (*1945*) 988,
        1019, 1105
Atomic power/atomic bombs,
    Herbert G. WELLS 455
Audibility, zone of normal 447
Aurum fulminans
    (fulminating gold) 200

Australites 304
Autoclave, forerunner 222
Automatic digital computer 141
Automatic tabulating machine 139
Aviation, high velocity 502
    Russian,
        TUPOLEV/ZHUKOVSKY
        1186, 1199–1200
Azide salts 402, 605
Azomethane 504

# B

Baade-Zwicky hypothesis 500
Babylon Gun (Iraq Gun) 224
Babylonian eggs 321
Back-spalling 450
Balance balls 46, 906
Ball lightning 300, 302, 811,
    860, 1186
Ball resilience 676
Ballista 879
Ballistic coefficient 338
Ballistic computer 140
Ballistic crack 103
Ballistic drag 242, 938, 973
Ballistic galvanometer 125, 305, 894
Ballistic head wave, single shock 404
Ballistic missile reentry shapes,
    blunt-body design 595
Ballistic mortars 237
Ballistic pendulum 893, 900, 1055,
    1092, 1096, 1100, 1111, 1153, 1180
    CASSINI Jr. 49, 156, 455
    double 238
    momentum of giant laser pulse 691
    small shot 237
Ballistic problem, interior/exterior 474
Ballistic pump 882
Ballistic resistance 374
Ballistic studies, exterior/interior
    1003–1004, 1062, 1093,
    1121, 1137, 1196
Ballistics 42, 72, 187
    CRANZ 473
    exterior 387
    gas dynamics 557
    geoballistics 387
    interior (internal ballistics) 386
    intermediate 915
    penetration (terminal ballistics) 387
    reentry vehicles 653
    scientific, ROBINS 1153
    supersonic 918, 993, 1118,
        1120, 1158
    ROBINS 242
    terminal 387, 718
    wound, MELSENS 1127–1128
Ballistite 357, 392, 1136

# SUBJECT INDEX

Barisal guns (fog guns) 350, 413
Barometer 1126
Barrel rupture, fatal "banana split"-type 506
Barrier line 432
Barringer Crater (Meteor Crater) 64, 182, 436, 465, 508, 618, 645, 658, 663, 848, 1169
Base surges 26, 550
   water/magma interactions 678
Baseball bat investigations 904
Bashforth chronograph 124, 337
Battering ram 879
Bazooka 115
Beam, coherent visible light 623
   electrons 671, 724, 736
   laser 675, 696, 763
Beam cap 937
Beam splitter 274, 405, 791
Becker equation of state 468
Becker piston model 466
Becker-Kistiakowsky-Wilson (BKW) equation of state 527, 669
Beelzebub 337
Bélier hydraulique (hydraulic ram) 268
Bell X-1 103
Belt-snap phenomenon 70
Bending theory 1108
Benign shock concept 674
Bent distance (explosion) 980
Bergeron method 508
Bernoulli equation 1044
Bernoulli principle (Bernoulli theorem) 236
Berthelot wave 397
Berthollet's Law of Mass Action 362
BESM (Bolshaja Elektronno-Schetnaja Mashina) 144
Beta Lyrae, variability 261
Bethe-Gamow theory 569
Bethe-Weizsäcker cycle 317, 1047
Bethlehem Zinc Works, metal-dust explosion 318
Bhangmeters 643
Bickford fuse (safety fuse) 106, 127, 291
Big Bang 21–22, 67, 74, 174, 479, 566, 824, 833, 872, 1079, 1082, 1094, 1113–1114, 1197
   FRIEDMANN/LEMAÎTRE 1078–1079, 1113
   hot 569, 685, 1079, 1082, 1197
   nucleosynthesis 68
Big Bertha 887
Big Crunch 6, 22, 795, 833
Big guns 402
Big Jump 105

Billiard balls, mathematical studies 895, 1060, 1144
Billiards, collision 298
   percussion research 48
Binary collisions 4
Binary digital computer 141
Binary star system (WR+OB) 737
Biofluid dynamics, LIGHTHILL 1115–1116
Biomechanics 52
Birefringence, strong electric fields 356
Bistable multivibrator (flip-flop) 464
Black dwarf 469, 491
Black holes 268, 670, 703
   supermassive 795, 825
Black powder 106, 111, 984
   explosions 193
   first demonstration 203
   mining industry 199
Black stars 261
Blackbody 787–788
Blackbody radiation 569
Bladder calculus, disintegration in vivo, shock waves 637
Blast 35, 62, 282
   energy 264
   largest non-nuclear, Ripple Rock 626
Blast damage 36
Blast ditching 989
Blast effects, on man 610
Blast suppression technique 760
Blast trauma, otic (explosion) 701
Blast tube 335
Blast wave accelerator 35, 793
Blast wave noise, shoulder-launched weapon 760
Blast wave shock 35, 606
Blast wave simulator, equal pressure shock tube 686
Blast waves 29, 34–35, 385, 723, 760, 819, 856, 917, 945, 952, 954, 975, 992–993, 1005, 1032, 1138, 1151, 1161, 1181, 1183
   diffraction 580
   Friedlander type 73
   propagation 917, 1032
   recording 1005
   velocity, in a tube 157
   velocity-distance profile 363
Blasting 36, 115, 211
   Bickford fuse 291
Blasting agents 345, 357
   ammonium nitrate (AN)/aluminum powder/water 616
   black powder 199
   nitroglycerin 345
Blasting cap 109, 201, 336, 342

Blasting gelatin 357, 1136
Blasting powder 199
   A 326
   B 106, 326
Blind rivet technique 504
Blister 1092
Blow, hydraulic 91
Blowdown wind tunnel 517, 561, 613, 939
Blowguns 210, 273–274, 880
Blue straggler 739
Blunt-body design 595
Blunt-body theory 925
Body waves 60, 292
Boiler explosions, steam engines 291, 359
   synchronous 288
Boiling, delayed 334
Bolides 266
Boltzmann equation 352
Bomb(s), fusion 117, 144, 593, 598, 651
   nuclear/atomic 117, 517, 530, 608, 651
   thermobaric 71
   thermonuclear (H-bomb) 117, 144, 593
   Eluklab Island 593
   volcanic 62, 304, 489
Bomb calorimeter 383
Bombardier beetle (*Brachinus crepitans*) 713
Bores 25, 56
Boron nitrite 726
Bosumtwi Crater (Ghana) 63, 65
Botfly, supersonic 171
Boundary friction 259
Boundary layer 433, 943, 947, 969, 974, 977, 1038, 1064, 1070, 1077, 1104, 1116, 1147, 1164–1165
   effect 947
   equation 1064
   flow visualization 701
   theory 520, 1104, 1165
Boundary wave 464, 980
Bourdon gauge 999
Bow shock 65, 147, 662
   cometary 743
   cosmic 770, 797, 873, 937
   Earth 752, 872
   Neptun 779
   planetary 6, 662, 752
   quasi-parallel 955
   quasi-perpendicular 955
   Saturn 840
Bow waves 82, 919
Boyle law 210, 221
Boyle-Mariotte law (Boyle law) 1126
Boys camera 1050

Bramah press 267
Branched-chain explosions 295
Braun tube (oscillograph) 1013
Bridgman opposed-anvil
  apparatus 1051
Bridgman unsupported area seal 1051
Brisance gauge 1000
Brisance test 1000–1001
Brittle fracture 903
Broadcasting, radio waves 1089
Brontidi (baturlio marinas) 350
Bronze Age 184
Brown dwarf 469
Brownian motion/movement
  288, 906
Bubble(s) (cavities) 91, 526
  cavitation bubbles 414, 417, 457,
    461, 543, 627, 633, 726, 798
  collapse 461
  gas bubbles, underwater
    explosions 470
  gravity migration 458
  HOOKE gas-bubble theory 212
  jet formation 458, 461
  pulsations, underwater explosion
    458, 985
Bubble oscillation 417–418
Bubble pulses 372
Buffeting 525
Buffon cometary collision
  hypothesis 257
BUGGY Test 698
Bulb of percussion 46
Bull whip, cracking sound 630
Bullets, penetration 450
Bumper-WAC 574
Bunsen burner 1053
Bunsen waves 397
Bunsen's exploding vessel 999
Bunsenbatterie (carbon-zinc
  battery) 1053
Burning harmonica hypothesis,
  Faraday 1015
Bursting diaphragm shock tube 422
Busemann biplane 1054
Buxton test 369
Buzz-bomb 490

## C

Cacodyl cyanide 303
Calculating machine
  (computing machine) 138
Callisto, Valhalla crater 175
Caloric 269
Caloric theory, Lavoisier 1056
Camera,
  BOYS 1050
  CCD 708, 791

Cranz-Schardin multiple-spark 131,
  486, 625, 913, 1018
curved MCP 764
flame speed 1143
high-speed 556
high-speed rotating drum 485, 636
high-speed sweeping image 1006
mechanical framing 131
multi-CCD 708
NACA 556
Polaroid Land 563
schlieren 624, 790
streak 493, 784, 1017, 1050
tube camera 684
ultrafast multiple digital
  framing 131
ultrahigh-speed 297
wave speed 131, 493, 1017, 1143
Cannon 880
Cannon shot 194
Cannonball, velocity 208, 1096
Canyon Diablo meteorite
  (Arizona meteorite) 182
Capacitor (Kleist's jar) 241
Carbon gauges 136
Carbon-nitrogen cycle
  (astrophysics) 1047
Carbon-nitrogen-oxygen (CNO) cycle
  (Bethe-Weizsäcker cycle) 317
Carbon-zinc battery 1053
Carnot cycle 1149
Carnot shock 277
Carnot theorem 260
Carrington's solar flare 326
Cartwheel Galaxy 176
Cassini-Huygens 735
Catalina Sky Survey (CSS) 65
Catalysis, heterogeneous 281
Catalytic reaction kinetics 1105
Catapult (ballista) 187, 879
Catastrophic instabilities, partial
  differential equations 482
Catastrophic shock concept 674
Cathode rays 414
Cathode-ray oscillographs/
  oscilloscopes 125, 1013
  high-speed single-shot 471
Cauchy-Poisson waves 77, 282, 312
  submarine eruption 593
Cavendish experiment 1057
Cavitation 91, 414, 726, 798, 816,
  967, 1038
  transient 461
  water-hammer of collapsing
    vortices 417
Cavitation bubbles 414, 417, 457,
  461, 543, 627, 633, 726, 798
  collapse 461
  high-speed photography 633

Cavitation erosion 461
Cavitators 92
Cavity (cavities), collapsing,
  hydrodynamic properties 482
Cavity collapse 461, 584, 777
Cavity collapse phase 90
Cavity effect, von BAADER 265, 984
Cavity liner 513
CCD (charge coupled device) 707
CCD camera 708, 791
  intensified (ICCDs) 489, 791
Cellulose nitrate 107, 308
Center of gravity 184, 218, 401,
  1190, 1194
Center of mass (CM) 904
Center of oscillation 1097, 1190
  compound pendulum 154, 219
Center of percussion (COP) 42, 154,
  219, 750, 809, 904, 1097, 1190
Center punch test 346
Central Aerohydrodynamic Institute,
  TUPOLEV 462
Central percussion 12
Central Processing Unit
  (CPU) 143
CERES 274
Chain reactions 7, 278
  branched, chemistry 1167
  combustion processes 1167
  explosion 1168
  initiation/propagation/
    termination 453
  nonbranching 502
  nonexplosive 799
  nuclear 987, 1196
Challis paradox 310–311
Chamber pressure, detonating
  explosives 1000
Chandrasekhar limit 491
Channel effect 738
Chaos (disorder) 202
Chapman equation 1059
Chapman-Ferraro cavity 492
Chapman-Jouguet condition 435
Chapman-Jouguet model (detonation)
  435, 1059, 1101, 1134
Chapman-Jouguet plane 522
Chapman-Jouguet point 435
Chapman-Jouguet pressure, high
  explosives 600, 621
Chapman-Jouguet relation,
  supplementary 441, 460
Chapman-Jouguet state 435
Chapman-Jouguet theory/hypothesis
  112, 435, 490, 1059, 1101, 1134
Chapman-Jouguet zone
  (detonation) 1101
Characteristic cone
  (MONGE's cone) 255

# SUBJECT INDEX

Characteristic variable 255
Charge coupled device (CCD) 707, 791
Charpy impact test 97, 427
Chemical changes of explosives 1037
Chemical elements, primeval thermonuclear explosion 568
Chemical evolution 55
Chemical harmonica hypothesis, Faraday 1015
Chemical kinetics 113, 619
Chemical reactions, kinetics 148
   shock waves 674
Chemical theory of gunpowder 375
Cherenkov effect/radiation 500
Chernobyl 771, 1031
Chester-Chisnell-Whitham theory 971
Chicxulub Crater (Yucatan Crater) 63, 177
Chile saltpeter ($NaNO_3$) 326
Chilean tsunami (1960) 644
Chisel 878
Chlorine detonating gas 278, 304
Chlorine-hydrogen explosion 453
   chain reaction 113
Choking, air flow through well-rounded mouthpieces 334
Choking effect 343, 462, 942
Christmas Meteor (Washington Meteorite) 361
Chronographs 123, 1012
   Bashforth 337
   Cranz-Becker 125, 451
   Deprez 355
   electric tram 125, 399
   Le-Boulengé 329
   multichannel 124
   Navez 123
   Noble 124, 354, 359, 1137
   photochronograph 416
   rotating-drum 337
   Sébert 125
   Siemens 124
Chronoscopes 123, 1012
   Pouillet 305
   Sabine 125, 360
   Wheatstone 123, 301, 1012
Cinematography 913, 992, 997, 1007
   high-speed 131, 708, 915, 967, 971, 1018–1019
Cineradiography 625
Circum-Pacific seismic zone 865
CJ hypothesis (see also Chapman-Jouguet) 435
CL-20 (super explosive) 109, 775
Clapping of hands 69
Clinical extracorporeal shock wave lithotripsy 640
Cnoidal waves 296

Coal dust explosions 21, 111, 193, 307, 356, 369, 372, 402, 408, 441, 447, 647
   mine galleries 372
Coal liquefaction 1043
Coal mines,
   firedamp explosions 202, 279–281, 402, 415, 441, 456
   methanometers 281
   permitted explosives 369
Coefficient of restitution 225, 368, 890
Coesite 64, 497, 508, 600, 618, 645, 658, 663, 848–849, 1169
   high-pressure silica polymorph, synthesis 600
   meteorite impact site 618
Coil gun 426
Collapsar model 67, 834
Collapse 825, 853, 875, 934, 967, 982, 987, 1032
Collapsing cavities, hydrodynamic properties 482
Colliding bodies, force-time profiles 897
Colliery explosions, weather 354
Collision(s) 16, 18, 80, 151–152, 586
   asymmetric 235
   billiard 888, 1060, 1098
   binary 4
   charged particles 1047
   compact bodies 775
   continental 18
   critical collision angle 604
   direct central 151
   elastic 471, 701, 888, 890, 899, 1060, 1089, 1126, 1194
   galaxies 820, 825
   inelastic 820, 1126
   oblique central 151
   oblique eccentric 151
   parabolic/hyperbolic 18
   perfectly elastic 150, 153
   reactive 17
   rear-end 17
   solid bodies 376
   spacecrafts 802
   spherical shock waves 700
   stars 739, 812, 819
   sungrazing comet 750
Collision-free hydromagnetic shock theory 602
Collision phenomena 1125–1126
Collision theory 426
Collisional accretion 17
Collisional orogeny 18
Collisional shocks 33
Collodion cotton,
   dissolved in nitroglycerin 357

Color schlieren methods 128
   photographs, source-filter technique 597
Columbia Space Shuttle 628
Columbiads 100, 194, 336
Combustion 114, 480, 604
   diamond 1065
   gases 1053, 1130, 1142, 1171
   transition to detonation 522
Combustion dynamics 769
Combustion engines 1092, 1138
   supersonic 831
Combustion laws, propellants 1064
Combustion phenomena 769
Combustion processes, chain reactions 1167
Combustion ramjet, supersonic 639
Combustion research 1067, 1112, 1188
Combustion theory 114, 805, 1167, 1196
Combustion waves 22
Cometary collision hypothesis, BUFFON 257
Comets 257, 271, 279, 588, 792, 794, 801, 815, 1058
   collision with Earth 239
   exploded planet hypothesis 271
   SWIFT-TUTTLE 333
Composition B/Comp-B (60/40 RDX/TNT) 421, 604, 621
   CJ pressure 601
Compound pendulum 1097, 1190
   center of oscillation 401
Comprehensive Test Ban Treaty (CTBT) 629, 799
Compressibility 291
   laminar boundary layers 1064
   stall 522
   uranium/plutonium 1051
Compressible flow 37
Compressible fluids, high-speed flows 599
Compression 208, 336
   isotropic 717, 812
Compression force 260
Compression lighter 881
Compression period 48
Compression ratio 87
Compression ring 571
Compression shock 432, 467
Compression strength, dynamic 901
Compressional waves/longitudinal waves (P-waves/primary waves) 121
Compressive stresses, lateral 755
Computational analysis 137
Computational fluid dynamics (CFD) 145

Computer(s) 137
  digital computer, early 138, 142, 1134, 1181
  experiment 146
  stored-program 143, 556
Conchoidal fractures 47, 183
Concorde, supersonic 704, 733
Concussion 12, 201, 313, 599
Concussion fracture 13
Concussion fuse 13
Concussion grenade 13
Condensation phenomena 1140
Condensation shock 29, 328, 385, 504, 751, 929, 968, 1152, 1174, 1192
Condensed detonation 23
Condensed matter 750, 774–775, 791, 803, 816, 843, 1042, 1073
Conflagrations 22
Conservation of motion, LEIBNIZ 222
Contact discontinuities 5
Contact discontinuity lines (slipstreams) 534
Contact mechanics 377
Contact time 48, 150
Continental collision 18
Continuity 244
Cookie cutter, shock tube technique 577
Copper, cold hammering 184
Copper detonator 345
Cordite (explosive) 1037
Coriolis force/effect 1060, 1144
Corona, magnetohydrodynamic waves 327
Coronal mass ejections (CME, Sun) 21, 829, 871
Corotating interaction regions (CIRs) 807–808
Corpuscular fluid model (NEWTON) 226
Corpuscular models 44
  shock wave 906
Corpuscular radiation, comet tail plasma deflection 588
Cosmic Background Explorer (COBE) 787
Cosmic background radiation 174, 787–788, 833, 872
Cosmic detonations 53
Cosmic electrodynamics 632
Cosmic explosion phenomena 66
Cosmic gas dynamics 574, 589
Cosmic jets 873
Cosmic magnetohydrodynamics 574
Cosmic rays (cosmic radiation) 451, 1113
  acceleration in supernova remnants 656
Cosmic shock waves 65–66, 873

Cosmic tornadoes 580, 873
Cosmic velocities 620
  first 1025
Cosmogony 67
Cosmology 487, 566, 569, 820, 1082, 1102, 1113–1114, 1144, 1173, 1197
  steady-state 566
Coulomb collisions 16, 262
Coulomb explosion (electrons) 784
Coulomb lattice 818
Coupled Euler-Lagrange (CEL) 392
Courant criterion 144, 482, 1061
Courant number 1061
Courant-Friedrichs-Léwy condition 1061
Cow-dung bombs 62
Crab Nebula, SN 1054 192
Crack propagation 98
  computer simulation 821
  supersonic 720, 809, 817, 838
Cranz-Becker chronograph 125, 451, 1016
Cranz-Schardin arrangement 625
Cranz-Schardin multiple-spark camera 131, 486, 913, 1018
Crashworthiness 747–748, 763
Craters, extraterrestrial 783
  geometry, high-speed pellets, plaster of Paris 597
  lunar 847, 1169, 1173
  terrestrial 783, 847–848, 1169
Creation of the Sun 174
Creation of the Universe 174
Critical collision angle 604
Critical Mach number 462
Critical mass, nuclear explosion 455, 514
Critical speed 396
Criticality 524
Crocco energy integral 1064
Crocco equation (Crocco vorticity law) 508, 1064
Crocco-Vazsonyi equation 508
Crocco vorticity law 1064
Crown-and-jet phenomenon 413
Crumple zones, passive safety 587
Crusher gauge, Andrew NOBLE 135, 353
Crushing 50
Cryogenic engines 1090
Cryptography 1189
Cryptovolcanic structure (cryptoexplosion structure) 497
Current-carrying conductors, explosive deformation 590
Current pulse experiment, in water 768
Curvature of space, FRIEDMANN 469
Cutting-wire gauge (gun) 1004

Cyanide gas explosion, WTC (*1993*) 789
Cyclonite 421
Cyclotetramethylenetetranitramine (HMX) 108
Cyclotol (60/40 RDX/TNT) 404, 421
Cyclotrimethylenetrinitramine (RDX) 421
Cygnus (galaxy) 770, 787
Cygnus Loop, supernova explosion, expanding blast waves 183, 606

# D

d'Alembert principle 240
DARHT (Dual Axis Radiographic Hydrodynamic Test) 687
Dark matter 68, 792
DART (Deep-Ocean Assessment and Reporting of Tsunamis) 679
David Taylor Model Basin (DTMB), hypersonic wind tunnel 645
Davies bar 456, 568
Davy lamp 111, 281, 1029, 1066, 1122
de Hoffman-Teller shock-wave equations 602
de Sitter cosmological model 460
Death of Gun Barrels on the Battlefield, ROGIER 506
Deep fueling 552
Deep Impact (discovery mission) 815
Deep-sea temperatures 352, 368
Deer botfly ("supersonic"), wing structure 479
Deflagration 22, 343, 354, 381, 1080, 1101, 1188
Deflagration to detonation transition (DDT) 24
Deformation, elastic 812, 902, 904
  high strain rates 682
  inelastic, shocked crystals 1073
  plastic, liquids 903
  metal crystals 1183
  tensile, rubber 711
Delayed boiling 334
Delta Cephei, variability 261
Delta wings 973, 1024, 1116, 1140
Density jump 1009
Density variations, optical method, HOOKE 220
Deprez chronograph 124, 355
Detonating gas 316
Detonation
  (explosion) 22–23, 348, 354, 360, 460, 587, 751, 781, 953, 984–1007, 1037, 1044, 1058, 1067, 1069, 1073, 1101, 1134–1138, 1146, 1159,

# SUBJECT INDEX

1161, 1167, 1170–1173, 1181,
    1183, 1188, 1196–1197
  CO/O$_2$ 472, 475
  condensed 23
  cosmic 53
  definition 769
  electrical disturbance 605
  explosives, electrical 1037
  firedamp 999, 1065
  galloping 24
  gaseous 658, 995, 999, 1017,
    1067, 1106, 1188
  H$_2$/O$_2$ 472
  in air, shock wave 484
  initiation 641
  large yield surface 992
  laser-supported (LSD) 953
  nuclear 988, 1047
  overdriven 24
  reverse (retonation) 24, 430
  self-sustaining 781
  shock initiation 664
  spin phenomenon 475
  spinning 993–994, 1067, 1165
  steady detonation model 522
  stuttering 24
  superdetonation 24, 631
  theory 411, 434
  time delay/induction time 641
Detonation caps 211, 280, 1135
Detonation diagnostics 998–1007
Detonation front 420, 993, 1017,
    1130, 1138
  edge effects 110
  propagation,
    streak record 374
Detonation initiation, nonelectric 729
Detonation physics 105, 525
Detonation pressure 1159, 1170
Detonation process, 1-D model 534
Detonation spin 476
Detonation studies 1073, 1173, 1188
Detonation theory 1044, 1067, 1069,
    1133, 1196
  first, CHAPMAN/JOUGUET 1058,
    1067, 1100–1101
  modern, DÖRING/von
    NEUMANN/ZEL'DOVICH
    1069, 1133, 1196
Detonation velocity 440, 1002–1003,
    1037, 1137
  gaseous explosion 1003
  high explosives 621, 1002
  mercury fulminate 468
  nitroglycerin 468
Detonation waves 111–112, 381,
    949–950, 964, 982, 994–995, 1017,
    1041, 1059, 1069, 1082, 1101,
    1159, 1171

  aquarium technique 663
  multifront/"cellular" structure 640
  periodic cell structure 435
  shock wave 472
  spherical 528
  spinning structure 435
  three dimensions 527
  velocity of sound 409
Detonator caps (blasting cap)
    280, 1135
  electric, ignition 363
Detonics, high explosives 717
  micro- 832–833
Deviatoric stress gauge 696
Devices, periodically operating 882
  rapid compression 881
  rapid expansion 880–881
Diamonds, artificial 406
  extra-solar origin
    (or interstellar) 774
  lonsdaleite 403
  meteorite origin 758
  meteoritic
    (Arizona meteorite/
    Canyon Diablo meteorite)
    403, 406
  nanodiamonds 778
  nitrogen 843
  preparation by explosive shock 778
  shock diamonds 417, 433,
    673, 1158
  shock-induced transformation of
    graphite 656
Dicke Bertha
    (Big Bertha) 196, 446, 887, 1063
Diesel engines 408, 1107
Differential analyzer 139, 140
Differential geometry, MONGE 255
Diffusive acceleration of particles 746
Digital computer, early 138, 142,
    1134, 1181
Digital storage oscilloscopes 126
Dinosaurs 780
Diode 464
Direct central collision
    (collinear collision) 151
Dirty bomb (radiological dispersal
    device RDD) 118
  explosion 118
Disasters, man-made 1028–33
Discontiguity 244
Discontinuities 244, 431
  fast-propagating 420
  lines, opposite V gliding spark 534
  reactive (detonation waves) 427
Discontinuous flow 1094, 1195
Discontinuous function,
    EULER 243
Discontinuous phenomena 5

Dislocation theory 1183
Displacement discontinuity model 40
Ditch blasting 698
Divergent nozzle 929
DNT (2,4-dinitrotoluene) 345, 404
Doppler effect 302, 309, 1068
Doppler radar technique, high-speed
    diagnostics 303
Doppler shift, reflected laser light 681
Dora 887, 1063
Döring's detonation model 993
Double ballistic pendulum 238
Double exposure holographic
    interferometry 690
Double Mach reflection
    (DMR) 130, 357
Double-acting hammer 300
Double-driver shock tube 668
Double-piston press,
    explosively activated 633
Double sonic boom 595
Drag, projectiles 261
  second power
    of the flow velocity 221
Drag coefficient 1016
Driving piston 725
Drop forging 50
Dropping weight timing system,
    electrically triggerable 329
Drug delivery,
    needle-free 817
Dufour-type oscillograph 125
Dulong's explosive oil 278
Dust clouds,
    ignition temperature 646
Dust explosions 21, 364, 1002, 1030
  tester 1002
Dwarfs 426, 454
  black 469, 491
  brown 469
  white 333, 469, 491, 787
Dynamic adiabat/shock adiabat
    (Hugoniot curve) 167, 390
Dynamic elastic limit 614
Dynamic fracture 97, 775,
    837–838, 844
Dynamic materials testing 607
Dynamic photoelasticity 129
Dynamic theories 44
  gases 323
Dynamics 41, 240, 246
Dynamite
  (Guhr dynamite) 100, 108,
    345, 1135–1136
  addition of ammonium nitrate 345
  East River (New York) 359
  gelatin 357
Dynamo exploders 364
Dynamometer 417, 985

## E

E (electromagnetic) bomb 648
Ear, shock/blast injury 34
Earnshaw paradox 330
Earth-Moon system 174
Earth waves, elastic 122
   elastic compression 311
Earth's bow shock, fast shock 602
Earth's density 1096
Earth's fluid core 452
Earth's magnetosphere 601
Earth's mantle, Gutenberg
   discontinuity 452
Earthquake(s) 60, 171, 190, 795, 826,
   851–853, 857, 858, 862–868,
   896–897, 905, 979–980, 1102,
   1108, 1115, 1123–1124, 1131, 1179
   angle of emergence 321
   artificial 309, 381, 450, 464
   controllable 659
   firedamp explosions in coal
      mines 456
   intensity vs. magnitude 505
   Iquique 361
   isoseismal lines 321
   largest (*1960*) 644
   liquid waste deep well 658
   Lisbon (*1755*) 247–252
   Lituya Bay 627
   magnitude scale 505
   megathrust 60
   meizoseismal area 321
   meizoseismic line 321
   moment magnitude 506
   moment magnitude scale 752
   Port Royal (*1692*) 228
   progressive motion 269
   propagating wave 249
   Rayleigh wave 386
   Richter scale 62, 506, 752
   rupture, shear slip on faults 611
   seismic focus 321
   seismic vertical 321
   sequence of waves 249
   shear slip on faults 611
   undulation 249
   visible waves 228
   waves of elastic compression 308
Eccles-Jordan circuit 464
Echo 1073
EDVAC 143
EIC (explosive-in-cell) code 392
Einstein field equations 460
Elastic bodies 888, 890, 894, 1089
   totally 217
Elastic collision 701, 888, 890, 899,
   1060, 1194
Elastic deformation 812, 902, 904

Elastic fluids 906, 998, 1044, 1153
Elastic percussion 154, 1097, 1190
Elastic-plastic impact, extended
   HERTZ's theory 377
Elastic rebound theory 439
Elastic solids 1086, 1108, 1149, 1157,
   1175, 1177
Elastic vibrations 1045
Elastic waves 900, 979, 1157, 1178
   high-frequency 671
Elasticity 1080, 1103–1104, 1126,
   1144, 1190
   nonlinear 838
Electric guns (WISSHOFER) 257,
   756, 990
   ultrahigh-velocity 785
Electric pin method, implosion
   timing 541
Electric shock 28, 241
Electric shock tube 617
Electric spark(s) 6, 70
   source of sound 344
Electric spark photochronography
   125, 429
Electric spark research, TOEPLER
   1185
Electric tram chronograph 125, 399
Electric wire explosions 270, 283,
   318, 465, 582, 637
Electrical action, animals 1065
Electrical inertia 1089
Electrical pin method 951
Electrical system failure,
   EMP-induced 659
Electrical triggering 124
Electricity, FRANKLIN 244–245
Electricity and magnetism,
   COULOMB 262
Electrodynamic forces 782
Electrohydraulic effect,
   YUTKIN 93, 584, 982
Electrohydraulic lithotriptor
   (EHL) 638
Electrohydraulic shock lithotripsy
   584, 588
Electrohydraulic shock-wave
   generator 588
Electro-lithotripsy 638
Electromagentism 1072
Electromagnetic launcher 426
Electromagnetic shock tube 617
Electron beam 671, 724, 736
   gun 736
Electron impact phenomena 1167
Electron multiplier,
   continuous dynode 489
Electroshock guns
   (stun guns) 242
Electrothermal plasma gun 990

Eluklab Island, thermonuclear
   (fusion) bomb 593
EMP (ElectroMagnetic Pulse)
   devices 648
Enantiomeric excesses, extraterrestrial
   molecules 706
Energetic coefficient of restitution 785
Energy leaps/energy jumps 39
Energy of motion, gas molecules 482
Enewetak Atoll, plutonium bomb 593
Enhanced radiation warhead
   (ERW) 741
ENIAC 142
Enrichment, U-235 517
Ensisheim meteorite 198
Entropy, CLAUSIUS 313, 328, 341
   small-amplitude shock wave 433
Éprouvette 237
Equations of state (EOS) 98
   elements 566
   of matter, very high densities 577
   water/lucite 663
Equation of state data 745, 747, 756
   shock-compressed materials 542
Ernst-Mach-Institut (EMI) 637
Erosion 679
Erosion pattern, banana-split-type 996
Escape velocity 342, 620
Eudiometric measurements 1056
Euler equations 1076, 1173
Eulerian coordinates 391
European Space Agency (ESA) 734
Evolution, Universe 67
Ewing seismometer 382
Expanding world models 1079, 1093
Expansion, Universe 1079, 1082,
   1094, 1113
Expansion nozzle 928
Expansion waves (rarefaction waves)
   473, 1090
Exploding Bridge-Wire
   (EBW) detonator 542
Exploding foils, shock waves
   in solids 664
Exploding vessel, Bunsen 999
Exploding wires 270, 637,
   990–991, 1006
   detonation of nitrogen
      chloride/nitroglycerin 510
   electric fuses 582
   electrical 256
   fragmentation, striations 582
   schlieren camera 624
   underwater 662
   unduloids 674
Explorer Missions 628
Explosifs S.G.P. (sécurité, grisou,
   poussière) 345
Explosion (definition) 769

Explosion(s)  18, 22, 984–1007,
    1043, 1046, 1053, 1056, 1092,
    1112, 1138, 1142, 1188
  1-D/2-D/3-D  19
  accelerating force  378
  astrophysical  1094
  atomic  988–989
  branched-chain  114, 295
  chemical  20
  coal mine  1065, 1112, 1122,
    1142–1143
  dirty nuclear  118
  dust  743, 746–747, 1002,
    1030, 1037, 1149
  effects  551
  electrodynamic  769
  finite source  20
  fireball  705
  firedamp  1029, 1037, 1065,
    1112, 1143
  first kind (detonation),
    ROUX/SARRAU  354
  gasdynamics  695
  gaseous  711, 986, 1056, 1059,
    1066, 1167, 1121
  heavy steam (volcanic)  786
  in air  729, 740
  in air, large-scale  423
  intense point-source  700
  liquid  716
  micro-  713
  molten metals by water  702
  near-field  996
  nuclear  115, 511, 707, 721,
    745, 771, 788, 905, 988–
    989, 1031, 1169, 1183
  oxyhydrogen  1115
  period of induction  381
  second kind (deflagration),
    ROUX/SARRAU  354
  spherical  20
  spherically symmetric,
    blast scaling laws  658
  steam-boiler  1028, 1040
  stellar  823
  subterraneous  1115
  supernova  712
  sympathetic  110
  thermal  21, 114
  thermohydraulic  20
  thermonuclear  115
  underwater  90, 739, 744, 798, 801,
    985, 1013, 1108, 1180, 1183
Explosion diagnostics  998–1007
Explosion hazards,
  moderate/strong/severe  646
Explosion of the first kind
  (detonation),
    ROUX/SARRAU  354

Explosion of the second kind
  (deflagration),
    ROUX/SARRAU  354
Explosion phenomena, cosmic  66
Explosion ram  475, 884
Explosion seismology  115, 119,
    309, 333
Explosion tamper  884
Explosion theory  426
Explosion wave  385
  layer to layer  373
  velocity  375
Explosion welding
  (explosive welding)  541
Explosive(s)  106, 398, 759, 776, 1043,
    1105, 1132, 1146, 1159, 1187–1188
  colloidal  1188
  fuel-air  830
  heterogeneous  109
  high explosives  106, 108–109,
    114, 936, 945, 959, 962, 964,
    979, 982, 987, 1001–1002,
    1007, 1082, 1092, 1105, 1112,
    1132, 1134, 1181, 1183
  high explosives,
    shock-insensitive  823
  high mechanical stress  521
  HMX-based  806
  homogeneous  109
  low explosives (black powder)
    106, 109
  mining  1001
  oil well  1136
  permissible  115, 345, 404
  permitted  345, 369, 404
  plastic (BICHEL)  440
  reaction zone  435
  safe  1122
  secondary  405
  slurry explosives  616
  super (CL-20)  775
Explosive ablation,
  biological tissue  997
Explosive bonding (explosive
  cladding)  541, 982
Explosive carbon burning  55
Explosive combustion theory  1130
Explosive engraving  265, 396, 1132
Explosive flash charge  554, 978
Explosive force  1046
Explosive-formed projectiles
  (EFPs)  110
Explosive forming, MUNROE  1132
Explosive liquids,
  detonation initiation  681
Explosive mixtures  1017, 1046
Explosive oxygen burning  555
Explosive pressing, hot  776
Explosive Property Data, LASL  756

Explosive-pumped laser  752
Explosive Reactive Armor (ERA)  717
Explosive rivet  504, 982
Explosive studies  1043, 1046, 1059
Explosive variables (catastrophic
    variables)  787, 815
Explosive volcanic eruptions  61
Explosive wave  112, 420
Explosive welding (metal working
    technique)  982
Explosive working  115, 669
Explosives antigrisouteux  345
Exterior ballistics  387
Extragalactic astronomy  1093
Extragalactic jets  462, 937
Extragalactic nebulae  1093
Extraterrestrial life  704, 706, 767, 813
Extraterrestrial solid objects  1169

# F

Fabry-Pérot interferometer  421
Falling drops  413
Faraday cage  1128
Faraday's chemical harmonica
    hypothesis  1015
Faulting, Mars  843
Faults, earthquakes/plate tectonics
    439, 858
Ferri sled  622, 1077
Ferroelectrics, shock compression  666
Ferromagnetism  1044, 1069, 1170
Feynman-Metropolis-Teller (FMT)
    equation of state  577–578
Filament eruption (Sun)  871
Filler tube  647, 945
Finite element analysis (FEA)  1061
Finite source explosion  20
Fire pump  881
Firearms  51, 194
    propellants, maximum velocity  512
Fireball blackout  660
Fireballs  266
Fire-corpuscles, GALILEI  202
Firedamp  280
Firedamp/air, shock wave  509
Firedamp explosions  111, 131,
    193, 402, 441, 1029
    proper ventilation  307
Fireworks (black powder)  193,
    263, 357
Fireworks theory  479
Firing devices  109
First cosmic velocity
    (circular velocity)  620
First Law of Motion  224
Fish Canyon Tuff Eruption  180
Fissile materials  524, 539, 1041
Fission bomb  117, 545, 575

Fission-fusion-fission 651
Five-gallon-can blast pressure gauges 354
Fixed air (carbon dioxide) 1056
Flakes 876
   consistent fracture 180
Flame front, propagation 374, 397
Flame propagation 1188
   speeds 373
Flame speed camera 1143
Flame velocities 373
Flash gap technique 615, 978
Flash interferogram 924, 931
Flash neutron radiography 493
Flash radiography 132, 509–510, 934–935, 962, 996, 1009, 1063
   detonation pressure, high explosive 614
Flash soft radiography 132
Flash-to-bang rule/method 53, 155, 208
Flash X-ray diffraction analysis 133
Flash X-ray techniques 132
Flash X-ray tubes, open-vacuum discharge 518
Fliegner formula 335
Flint tools 299
Flint, Stone Age man/natural agencies, LEAKEY 501
Flour-dust explosions 262, 364, 1002, 1030, 1149
Flow phenomena, nonequilibrium 1090, 1171
Flow visualization, optical 370, 742, 1008–1011
Fluid dynamic drag 683
Fluid-dynamic trigger device 1015
Fluid dynamics, PRANDTL 1146
   theoretical, BERNOULLI/EULER 1044, 1075
Fluid jet impact 50
Fluid phenomena 128
Fluid viscosity 1175–1176
Fluids, chemical equilibrium (Crocco equation) 508
   compressible 86
   convection regime 448
   elasticity/strength 589
   motion 258
   shock-compressed 87
Fluorescence 1176
Flux-corrected transport (FCT) 761, 781
Flyer plate 14, 160
Flyer plate method (flat plate accelerator) 97, 114, 604, 649, 951
Flying Bomb 1038
Foot/ground impact shock 52
Force, moving body (mass×velocity) 207

Force of percussion 43, 153–154, 205, 452
Force-time profiles, colliding bodies 897
Foreshock/aftershock 32, 65
Forge steam ram 884
Forward shock 22
Fossil records of human stone tools 877, 878
Foucault pendulum 1060, 1144
Fourier series 1076, 1134
Fracture mechanics 97
Fracture phenomena 450
Fracturing 97
Fragmentation 50
Framing cameras, mechanical 131
Fraser high-speed photographic machine 485
Freak waves 869
Free air jet 931
   discharge parameters 400
Free fall 1012, 1033
Free-flight duration 1111
Free molecular flow 1138
Free radicals 40, 114
Free surface phase 90
Freezing, shock-induced 87
Fresnel principle 924
Friction, atmospheric 1025
   boundary 259
   clouds, lightning 188
   fault friction 752
   kinetic friction coefficient 579
   shock friction (impulsive friction) 295
   surface sliding 681
   tidal 1040
Friedlander function, diffraction effects of planar sound pulses 554
Friedmann closed Universe 1079
Frog tamper 884
Fuel-coolant interaction (volcanism) 20, 728
Fulgurites (lightning tubes) 264, 290, 806, 859
Fulminates 201, 415
Fulminating cap 345
Fulminating gold 21, 106, 200
Fulminating silver 263, 279
Fulmination of water, LEIBNIZ 231
Fulminations (chemical explosions) 246
Function, continuous/discontinuous 243
Fundamental vibration node 809
Furrows, volcanic 753, 767
   volcanoe cones, Taylor-Görtler vortices 520

Fused quartz, stressvolume relation 666
Fuses, exploding wire 582
Fusion bomb 117, 593, 598, 608, 651, 703
   air-dropped 608
   ULAM 1134

# G

Galaxies 770, 820
   M87 795
   Seyfert 538
   velocities 487
   violent explosions 538
Galaxy-galaxy collisions 18
Galilei's vacuum theory (parabolic trajectory theory) 204
Galloping detonation 24
Gamma-ray bursts (GRBs) 21, 67, 643, 696, 804–805, 816, 823, 834
Gamow-Teller rules 1082
Gas bubble boundary, electrolytic probe 471
Gas calorimeter 1050
Gas density measurement 700
Gas-discharge (helium-neon) laser 649
Gas dynamics 36, 72, 74, 80, 432, 476, 612, 614, 690
   atomic theory 681
   ballistics 557
   combustion 671
   cosmic 574, 589, 632
   interstellar 38
   of explosions 695
   radiation 693
   rarefied 501, 558, 626
   thin bodies 101, 473
Gas explosions 20
Gas flow, one-dimensional discontinuous 383
   two-dimensional compressible 512
Gas globe oscillation 349
Gas laws for adiabatic compression, POISSON 285
Gas turbines 1038, 1131, 1138, 1174
Gasdynamic laser 684, 692, 975
Gasdynamic theory, spinning detonation 1165
Gaseous detonations, spin 485
Gaseous explosion, photography 430
Gases, adiabatic conditions 210
   intense shock waves, luminous/spectroscopic phenomena 501
   isothermal conditions 210
   kinetic theory 500, 1044, 1180, 1198
   turbulization 522

Gateshead pit accident 202
Gauges, piezoelectric 135
Geissler-discharge tube 931
Gelatin dry plate 351
Gelatin dynamite 357
General Catalogue of Variable Stars (GCVS) 564
Geoballistics 387
Geodesy, STOKES 1175–1176
Geomagnetic storms 327, 601
Geometrical theory of diffraction, KELLER 605
Geophones 120
Geothermal power plant, earthquake by drilling 659
German Uranprojekt 520
Gerstner wave 273
Giant impact theory/hypothesis (Big Splash/Big Whack) 174
Giant laser pulses 665
Giants 454
Gilbert's impact theory 496
Glass bombs, imploding 282
Gliding sparks (guided sparks) 399, 1042, 1143
Global catastrophes 757
Gold fulminate 21, 106, 200
Golf-ball studies 1180
Göttingen wind tunnel (Prandtl wind tunnel) 442, 1147
Grain dust explosions 1002, 1030
Grain elevator explosions 743, 747
Granular jet 826–827
Graphite, into diamond, shock-induced transformation 656
  transformation to diamond 607
Gravitational constant 1050, 1057
Gravitational implosion model, Sun 317
Gravity, kinetic theory 262
Gravity waves, long atmospheric 488
Gray seismograph 121
Gray-Ewing seismograph 382
Great Alaskan earthquake (*1964*) 672
Great Chilean earthquake (Valdivia earthquake) (*1960*) 644
Greenhill rule 367
Greenwich meridian 1039
Ground shock 32, 545
Guided anti-aircraft supersonic missile 517
Guided bullet 831
Gun, electric 756, 785, 990
  long-range 459
  multiple-baffle system 978
  Noble gauge 1003
Gun-assembly device 117
Gun barrel bursts 1029

Gun blasts 557
Gun carriage ballistics 387
Gun fire, protection 717
Gun muzzle blasts, projectiles 474
Gun pendulum 893
Gun shots, drag studies 73
Gun tunnel (Stalker tube) 85, 686
Guncotton, cellulose nitrate (high nitrogen content) 308
Gunnary 1153
Gunpowder 195, 203
  artificial quakes 308
  brisance test 1000
  chemical theory 375
  decomposition 284, 287
  electric spark 245, 252, 316
  exploding 1053
  explosion engine 221
  explosive effects in guns 358
  firearms/guns 266, 358
  fired 998, 1003, 1037, 1096, 1137, 1153, 1155
  firing by electric flame 245
  ignition, safe device 291
  metamorphosis 107
  products 322
Gunpowder machine 1142
Gutenberg discontinuity 452
Gutenberg-Richter magnitude-energy relation 506

# H

H-bomb *see* Hydrogen bomb
$H_2/O_2/Ar$, ignition delay times 635
Haack-Sears body 243, 526
Haber process 423
Hadamard material 1086
Hadamard matrix 1086
Hadamard theorem 431, 1086
Halifax explosion 459
Hammer, and nail 404
  double-acting 300
  drilling 296
  meat hammer 829
  percussion 302
  steam 300
  Taylor hammer (tomahawk reflex hammer) 302
  water hammer 419, 482, 508, 709
Hammering mechanism, woodpeckers 599
Hand calculators 139
Handaxe 878
Harmony of continuity 38
Hawaiian-type eruptions 638
Head-on collision 17

Head waves 100, 106, 310 388, 454, 915, 918–925, 963, 974, 993, 1014, 1027, 1050, 1068, 1098, 1120, 1128, 1158
  elastic 464
  explosion-like sound effects 411
  micro- 608
  single shock 404
Heat transfer measurements 701
Heavy Gustav 887
Heeresversuchsanstalt (HVA) Peenemünde 508
Height of Burst (HOB) 1134, 1193
Heinkel He176 518
Helgoland blast, largest non-nuclear man-made explosion 560
Helioseismology 119, 810
Heliospheric termination shock 75, 147, 602, 836, 840
Hell Creek Formation 178
Helmholtz pendulum 127, 1014
Helmholtz resonator 110
Helmholtz-Kelvin contraction theory 317
Herbig-Haro (HH) objects 580
Hermes space plane 1026
Herschel Crater (Saturn) 754
Hertz contact model/law of contact 376
Hertz theory of impact 376, 384
Hertzian cone 45, 183, 377, 876, 1089
Hertzian cracks 183, 378, 1089
Hertzian fracture 377, 710, 1089
Hertzian waves 1089
Heterogeneous catalysis 281
Hexogen/T4/cyclonite 421
High-aspect-ratio airfoils 352
High Enthalpy Shock Tunnel (HIEST) 686
High explosives (HE) 108, 200, 237, 263, 270, 279, 345, 368, 405, 429, 502, 527, 570, 600, 622, 647, 687, 691, 717, 728, 759, 823
  brisance test 1001
  detonation 23
  detonation velocity 1002, 1082
High pressures 491, 1051, 1145
  dynamic 571
  shock experiments 768
  techniques, Old Stone Age 877
High-rate chemical processes 148
High-rate physical processes 148
High speed aerodynamics 1064, 1153–1154
High-speed aviation 1192
High-speed camera 556
High-speed cinematography 131, 765, 915, 967, 971, 1018–1019, 1161
High-speed civil transport (HSCT) 1116

High-speed diagnostics 122, 805, 1012–1019, 1050, 1062–1063, 1068, 1109, 1120, 1159, 1165
High-speed fatigue tester 1092
High-speed flight 104, 1038, 1054, 1085, 1147
High-speed flow 1008, 1070, 1077, 1138
High-speed framing cameras 1018–1019, 1161
High-speed impact 609
High-speed multiframe image tube camera 684
High-speed phenomena 1118
High-speed photography 6, 130
High-speed rotating-lens camera 1050
High-speed rotating-mirror streak camera 1050
High-speed rotating-prism camera 1019
High-speed sweeping image camera 1006
High-speed turbines 1107
High-speed vehicles 1020–1027, 1054
  mythology 1020
High-speed videography 875, 914, 1019
High-speed visualization 123, 148, 1118, 1128, 1161
High-strain rate phenomena 754
High-velocity accelerators, gun-type 951
High-velocity flight 6
High-velocity impact phenomena 718
High-voltage electric friction machines 211
High-voltage generator, VON GUERICKE 210
Hildebrand-Knorr pressure brake 478
Hipp-Wheatstone chronometer/clock 302, 979, 1012
Hiroshima nuclear bomb, LITTLE BOY 546
HMX (high melting explosive)/ cyclotetramethylenetetranitramine 108
HOB (height of burst) 536
Hollerith tabulator 139
Holloman High-Speed Test Track (HHSTT) 1027
Hollow cavity effect 380
Hollow charge effect (shaped charge unlined cavity effect) 380, 984
Hollow charges 1159, 1161
Holography (lensless photography) 562, 1011
  high-speed 726
  pulsed laser 700
Homocyclonite 108

Hooke gas-bubble theory 212
Hooke law (Hooke's law) 221, 344
Hopkinson effect (back-spalling/spallation) 450
Hopkinson pressure bar 97, 451, 455, 568, 1092, 1157
  split 579
Hopkinson scaling, blast measurements 555
Hopkinson scaling law (Hopkinson rule) 457, 544, 1092
Horizontal blast (volcanic) 753
Horror vacui 206
Hot Big Bang theory, GAMOW 6, 68, 174, 1079, 1082, 1197
Hot hurricanes 35
Hot ice 1051
Hubble constant (Hubble parameter) 487, 1093
Hubble law 303, 487, 1093
Hubble relation 1093
Hubble Space Telescope (ESA) 734, 797
Huckleberry Ridge Tuff 181
Hugoniot curve (dynamic adiabat) 150, 160, 167, 285, 290, 342, 390, 951, 1095, 1159
  high temperatures/pressures 530
  ideal gas 431
  principal 615, 698
  shock adiabat, water 585
Hugoniot data 708, 951, 978
Hugoniot data banks 99, 745
Hugoniot elastic limit (HEL) 614, 812
Hugoniot equation (Hugoniot relation) 391, 964
Hugoniot relation 391, 424
Hugoniot waves 1072
Human space flight (Mercury/Gemini/Apollo) 628
Hurricane scale, SAFIR-SIMPSON 722
Hurricane storm tides 59
Hurricanes 722, 1115
  superheated 62
Huygens principle 81, 222, 918, 963, 974, 1068, 1086, 1098, 1125
  waves of finite amplitude 532
Hybrid numerical methods 392
Hydraulic blows 91, 457
Hydraulic jumps (shooting flow) 25, 259, 290, 583, 748, 760, 860, 908, 957–958, 1101
  models 908
  shallow water 447
Hydraulic press, BRAMAH 267, 283
Hydraulic ram (hydrodynamic ram) 89, 231, 268, 275, 743, 882, 1055, 1123

Hydraulics, French, MARIOTTE 1126
Hydroballistics 187
Hydrocodes 145
Hydrodynamic attenuation 1073
Hydrodynamic blast wave 661
  solar flares 657
Hydrodynamic equations (shock wave equations) 75
Hydrodynamic evacuator (urology) 1085
Hydrodynamic ram (hydraulic ram) 89, 231, 268, 275, 743, 882, 1055, 1123
Hydrodynamic resistance 683
Hydrodynamic shock tube, two-diaphragm 684
Hydrodynamic tests 687
Hydrodynamics 236, 262, 563, 582
  general Euler equations 249
Hydroexplosions 253
Hydrofoils 91
Hydrogen 258, 507
  metallic 802
Hydrogen bomb (H-bomb) 95, 117, 524, 575, 581, 593, 1082, 1134
  British 593
  Soviet 598
  true (staged), Soviet 608
Hydrogen chloride 278
Hydrogen nuclei, fusion 1047
  thermonuclear reaction 524
Hydrogen-oxygen reactions 716, 723, 994–995, 1053, 1056, 1168
Hydromechanics, ZHUKOVSKY 1199–1200
Hydrometeors 56, 677
Hydronuclear tests 799
Hydrophone, fiber optic probe 779
Hydrospark forming of metals 982
Hydroxyl radical 635
Hyperelasticity 838
Hypernovae 67, 804, 816, 824, 834
  explosion 815
Hypersonic 558
Hypersonic aerodynamics 1090
Hypersonic aircraft 1025, 1071
Hypersonic boundary layers 411
Hypersonic flight 424, 628, 808, 925, 1021, 1081
Hypersonic flow 641, 701, 749, 943, 1011, 1026, 1083, 1140
Hypersonic launchers 1083
Hypersonic rocket plane 638
Hypersonic shock tubes 1077
Hypersonic sled test 834
Hypersonic tests, sled vehicles, single-/ multiple-stage rockets 608
Hypersonic velocity 1081

SUBJECT INDEX

Hypersonic wind tunnel 942–943, 1077, 1088
Hypervelocities 14, 620
Hypervelocity facilities 839, 943
Hypervelocity flow 1179
Hypervelocity impact 609, 706, 708, 772, 902
Hypervelocity impact physics 14
Hypervelocity launchers 772–773, 1090
Hypervelocity tunnels 85
Hyper-X (X-30, hypersonic plane) 770, 808, 839

# I

IAS Computer 1134
ICCDs 489
Ice I, high-pressure experiments 523
Ice VII (cubic phase) 626
Iceberg, thermite reaction 476
Ideal fluidity theory 1166
Ignition (combustion) 769
    shock waves 572
Ignition temperature of dust clouds 646
Image converter 1019
Imamura-Iida scale 578
Impact 13, 17, 50
    effects 551
    GOLDSMITH 648
    liquid 764
    liquid-solid 709, 715, 718
    plastic 154
    symmetric 14
Impact breccia, Suevit 434
Impact chemistry 15
Impact cratering 783, 1169
Impact craters 177, 496
Impact engineering 13, 788
Impact erosion 15
Impact-explosion analogy 472
Impact flash 14
Impact fusion 839
Impact geology 15
Impact-induced wave profiles 899
Impact loading 13
Impact mechanics 13
Impact melt 15
Impact metamorphism 15, 688
Impact physics 13
Impact plume 15
Impact pressure 15
Impact shock, foot/ground 52
Impact structures 15
Impact theory 496
    lunar craters, BALDWIN 576
Impact tube 234
Impact velocity 903

Impact welding 13, 1073
Impact winter 16
    Chicxulub comet/asteroid 178
Impact zone, shatter cones 498
Impactites (impact glasses) 64, 618
Impactor 15
Impedance match method 634
Impetus (impressed force) 43, 195, 217, 274
Implosion 22, 984–997
    gas 986
    nuclear 987
    sharp report 486
Implosion-assembly device 117
Implosion bomb 987
Impulse 205
Impulse propagation, in elastic bead chain 828
Impulse steam turbine 396
Impulsive force 154
    photo-chronography 416
Impulsive loading 13, 29, 606
Incompressible flow 1084
Indentation gauge 324
Index fossils 497, 646
Indirect percussion technique (IV) (punch technique) 183
Indurite 407, 1132
Inelastic bodies, totally 217
Infinity symbol ($\infty$), WALLIS's 1189
Inflammable air (hydrogen) 258, 1056
Inflammation (combustion) 769
Infrared photography, WOOD 1192–1193
Initiation experiments, detonation 783–784
Instability, RAYLEIGH-TAYLOR 712, 1179, 1183
    RICHTMYER-MESHKOV (RMI) 711–712
Institute for Shock Physics (ISP) 699
Insulator, electrical conductivity 615
Integraph 1050
Intensified CCD (ICCD) 131, 489
Interference heating 701
Interferometers 1010, 1014
    FABRY-PÉROT 421
    JAMIN 129, 136, 320, 363, 1010, 1120, 1185
    MACH-ZEHNDER 129, 405, 415, 924, 931, 940, 960, 1010, 1049, 1120, 1161
    optical 320
    radiofrequency 597
Interferometry 128, 1010–1011
    double exposure holographic 690
    holographic 922
    laser 956

Intergalactic shock wave, Stephan's Quintet 177
Interior ballistics (internal ballistics) 386
Intermediate ballistics (transitional ballistics) 387
International Academy of Astronautics (IAA) 645
International Sun-Earth Explorer (ISEE) Program 742
International Tsunami Information Center (ITIC) 679
Interplanetary flights 1173
Interplanetary shock 601
Intersonic rupture 821
Interstellar medium 338
Interstellar space 873
Io (Jupiter's satellites) 749
Ionization processes 1048–1049, 1116
Iquique earthquake 361
Iridium signature 177
Iron,
    second-order phase transitions 655
    shock-induced, high-pressure polymorphic transition 616
Iron pyrite, striker 185
Iron-sulfur hypothesis, LÉMERY 228
Irregular reflection (Mach reflection), three-shock theory 535
Irregular wave reflection, RUSSELL 759, 916, 935, 958, 1118, 1155–1156, 1193
ISEE (International Sun-Earth Explorer) 734
Isentrope (static adiabat) 389
Isentropic-compression guns 552
Isothermal gas law 1026
Isothermal theory of sound, NEWTON 226
IUGG Tsunami Committee 644
Izod impact, definition 436
Izod impact test 97, 436–437

# J

Jacobian mathematics 1191
Jamaica Earthquake 228
Jamin interferometer 129, 136, 363
Jet-assisted takeoff (JATO) rockets 507
Jet engines (turbojet, fanjet, turboprop, ramjet, scramjet) 515
    compressorless (propulseur à réaction) 443
Jet fighters 516, 529, 591
    coke-bottle (wasp-waisted) 591
    delta-winged 591

Jet propulsion 515, 576
Jet Propulsion Laboratory (JPL) 507, 628, 665, 678, 729, 815, 836, 840
Jet shock 33, 937
  terminal 770
Jones-Miller equation of state (JM-EOS) 570
Jones-Wilkins-Lee equation of state (JWL-EOS) 570
Joule-Thomson effect 320
Jupiter, impact spot (CASSINI) 227
Jupiter's atmosphere 794

## K

K/T boundary 177
Kaiser-Wilhelm-Geschütz 887
Kakodyl (cacodyl) 303
Kamaishi tsunami 414
Katyusha (rocket weapon) 1196
KB-11 (Design Bureau 11) 550
Kelvin arms 294
Kelvin envelope 11, 294, 919
Kelvin wake 11, 294, 919
Kepler's nova 200
Kerr cell 356, 521, 597, 624, 665
  ultrafast shutter 624
Kerr effects 412
Kerr electro-optic effect 356, 412
Kidney stone destruction 722, 738, 760–761, 765–766, 981
Kieselguhr 345
Killer waves see Tsunamis
Kinetic energy (KE) projectiles 44
Kinetic theory of gases, HOOKE 211, 320, 500
Kinetic theory of gravity 262
Kirkwood-Bethe propagation theory 531, 1047
Kistler pressure gauge 1032
Kites 352
Klick-Klack 46, 213
Knochenhauer circuit 126, 325, 340, 910, 1014
Knocking, internal combustion engines 114, 438
Knudsen number 501
Knudsen region 501
Kolsky bar (split Hopkinson pressure bar) 456, 579
Komet (transonic rocket plane) 1022
Korn airfoil 726
Korn equation 726
Korteweg-de Vries equation 296
Krähenberg meteorite (*1869*) 347
Krakatau 378, 854–855, 865
Kryophor 89

Kucharski effect 527
  whiplash dynamics 578
Kutta-Zhukovsky equation 1199

## L

L'Aigle meteorite fall 272
L'atome primitif (primeval atom) 479
L'oeuf cosmique (cosmic egg) 479
La Garita Caldera 180
Lagrange problem 140, 265
Lagrangian coordinates 144, 324, 390
Lake Toba 182
Lamb waves 433, 1108
Lambda shocks 631
Laminate 97
Landau-Stanyukovich polytrope 548
Laplace-Poisson law (adiabatic gas law) 272
Laporte rule 1109
Large Aperture Seismic Array (LASA) 678
Laser, first operational 634
  gas-discharge (helium-neon) 649
  gasdynamic 684, 692, 975
  prototype 624
  ruby laser 644
  strong shock waves in water 696
Laser ablation, corneal 776, 793, 997
Laser Doppler anemometry 303
Laser Doppler velocimetry 303
Laser-induced fusion systems 1090
Laser-induced spark 952
  detonation model 675
Laser-induced stress wave 735
Laser-irradiated targets, shock effects 652
Laser intensity, optimum 691
Laser interferometry 956
Laser light, reflected, Doppler shift 681
Laser oscillations, mode locking 685
Laser percussion drilling 50
Laser-propelled "lightcraft" 953
Laser pulses, giant 665
  momentum 691
Laser Q-switching, Kerr cell 665
Laser-shocked crystals 786–787
Laser sparks 711
Laser speckle photography 129
Laser-supported detonation (LSD) 676, 953
Laser-supported shock wave 21
Laser weapons technology 669
LASL PHERMEX Data 687
Late compressional wave (primary wave/P-wave) 452
Lateral blasts (volcanic) 35, 754, 767
Lattice 93, 725, 818

Lattice compression 717
Lattice vibration theory 475
Laue diffraction pattern 717
Launcher, electromagnetic 426
Lava fountains, Mt. Kilauea-Iki 638
Laval nozzle 74, 83, 395, 926–930, 932–933, 938–941, 968, 1110, 1147, 1158, 1174
Lavoisier's caloric theory 1056
Law, cube-root scaling 1092
Law of center of mass 218
Law of comparison 1092
Law of conservation of kinetic energy 151, 215, 231, 1097
Law of conservation of mass 165
Law of conservation of momentum 152, 165, 215, 218, 230, 906, 1194
Law of conservation of vis viva (living force) 151, 240
Law of elasticity (law of deformation) 4, 221
Law of equivalents 453
Law of flame speed 1142
Law of gravitation, NEWTON 224
Law of jet noise, eighth power 1116
Law of mass action for chemical explosion 1067
Law of motion, third 218, 224, 240, 274
Law of nature in the collision of bodies of motion, WREN 152
Law of reflection 222, 888, 895
Law of similarity 1062, 1101
Law of stability of chemical equilibrium 1112
Laws of conservation of momentum and energy 906, 1194
Laws of impact 380, 407
Laws of motion, NEWTON 223, 906, 1127, 1137, 1194
  WREN 1194
Laws of motion on the impact of bodies 214
Laws of percussion, HUYGENS 211, 380, 1125
Laws of reflection and refraction 222
Lawrence Livermore National Laboratory (LLNL) 592
Le Châtelier principle 1112
Lead azide 212, 402, 605, 635, 738, 760, 765, 832
Lead-chamber process, sulfuric acid 243
Le-Boulengé chronograph 329, 1012, 1111
Leiden jar 990, 1002, 1015–1017, 1185
Leidenfrost layer 334
Libessart spark, open/confined 543

Lift coefficient, critical speed 462
Light booms 781
Light-gas gun 551, 740, 951
   equation of state studies 691
Lightcraft 105, 803, 953
Lighthill-Whitham model 612
Lighthill-Whitham-Richards
   model 612
Light-initiated explosion 323
Lightning 204, 677, 754, 859,
   966, 1057, 1115, 1128
   ball lightning 811, 860, 1186
   extraterrestrial 769
Lightning conductor, FRANKLIN 244
Lightning flash 55, 157
Lightning rod, FRANKLIN 257
Lightning simulation 719–720
Lightning strikes 55, 157
Lightning tubes 859
Limit of elasticity 422
Limited Test Ban Treaty (LTBT) 667
Line source explosion 55
Lines of discontinuity, opposite V
   gliding spark 534
Liquefaction 88
Liquefaction shock wave 88, 751
Liquid crystal surface
   thermography 701
Liquid impact 16
Liquid jet 718, 798, 935
Liquids, cavities
   (cavitation bubble) 414, 680
   compressibility 240, 254, 291, 491
   shock waves 86, 467, 523, 640,
      684, 744
Lisbon earthquake (1755) 247–252
Litholapaxie (urology) 1085
Lithotripsy 360, 584, 588, 638, 710,
   715, 739, 760, 764–766, 776, 783,
   950, 981, 1085, 1179
   clinical extracorporeal
      shock wave 640
   extracorporeal shock wave
      (ESWL) 715–716, 765, 1179
   laser-induced shock wave 766
   microexplosive 760, 765
Lituya Bay, mega-tsunami 627
Local tsunami (near-field tsunami) 361
Lonsdaleite 403
Los Alamos Scientific Laboratory
   (LASL) 530
Love surface waves 121
Low-sonic-boom technology 1077
Löwy-Schnyder method 141, 508
Ludwieg tube 613
   facility 942
Lunar craters 201, 212, 272, 343,
   466, 472, 552, 576, 629, 638
   Alphonsus 629

   explosions 472
   impacts 552
   meteorite impacts 354, 466
   volcanic collapse structures
      (calderas) 579
Lunar rays 1169
Lycopodium dust explosions 275
Lyddite, picric acid 256

# M

MacCormack FCT
   (flux correction transport) 761, 781
Mach angle 310, 388, 433, 454,
   817, 1118
Mach bands 1118
Mach bridge 959, 962
Mach circuit 127, 326, 1014
Mach cones 11, 310, 388, 454, 773,
   798, 817–818, 821, 838, 873, 918,
   921–922, 1021, 1068, 1118
   cosmic 797
   planetary 662, 773
Mach cutoff 554
Mach disks (Mach bridges) 79, 356,
   365, 537, 682, 770, 911, 916, 931,
   958–959, 962–965, 1118, 1158
   micro 608
   schlieren photos 400
Mach effect (Mach reflection effect)
   78–79, 81, 357, 958–965,
   1118, 1156
Mach equation 388
Mach front 992, 1118
Mach funnels 356
Mach head wave 294, 310, 388, 968,
   1068, 1118
Mach indicator 1022
Mach lines 388, 440, 920–921,
   933, 1118
Mach number 74, 241, 374, 480,
   924–925, 1038, 1118, 1121
   cosmic 481
   critical 462, 972
   cutoff 481
   local 481
   magnetic
      (Alfvén Mach number) 481
   shock 480
   stream 481
Mach reflection 81, 130, 161, 306,
   365, 447, 537, 730, 736, 757–760,
   861, 908, 912–913, 916, 932, 935,
   958–965, 971, 975, 995, 1049,
   1118, 1134, 1193
   complex 548
   different kinds of 758
   double 591
   inverse 736, 744, 758, 935

   progressive 744, 758
   regressive 744, 758
   steady 647
   three-shock theory 535
   transitional 548
   water 535, 537
Mach shock 79, 582
Mach stem 79, 81, 365, 549, 671
Mach warning (Mach indicator, in
   airplanes) 1022
Mach waves 440, 700, 713, 736
Mach's principle 1118
Mach-Zehnder interferometer 129,
   405, 415
Machmeter 1024
Magellanic cloud 773
Magnetic disturbance, sudden
   commencement 601
Magnetic fields, compression 114, 625
Magnetic flux compression 647
Magnetic pinch compression 714
Magnetic pressures 207
Magnetic storms 492
Magnetism 262, 1069, 1072,
   1141, 1144–1145, 1178
   micromagnetism 1069
Magnetoaerodynamics (in air) 642
Magnetofluidmechanics 642
Magnetogasdynamics 642
Magnetohydrodynamics 642
   cosmic 574
Magneto-induction exploder 363
Magnetopause, collision-free bow
   shock 675
Magnetosphere 601, 662
   Earth's 734, 742, 872
Magnetothermoelastic model 675
Magnus effect 1038
Mallard goniometer 1122
Mallet 878
Mammoth powder 100, 324, 331
Man in space, first (GAGARIN) 1081
Man-made chemical explosions 193
Man-made tools 501
Man on the Moon,
   ARMSTRONG 704
Manganin 449
   change in resistance 673
   piezoresistive gauge 664
   wire gauge, deviatoric stress 696
Manhattan Project (Project Y) 529
Manicouagan Crater 63
Manned orbital flight, GAGARIN 651
Mare Imbrium, meteorite crater 410
Marine propulsion 417
Mariner 10 (space launcher) 729
Mars Express/Beagle (ESA) 735
Martens precision pendulum 452
Martian meteorite 766–767

Maser 623, 634, 649
Mass extinction 757
Mass surge 26
Mass velocity 907
Material strength effects 1073
Materials, dynamic loading 767
Materials research 982
Materials testing, high-rate 97
Mathematical physics,
 BERNOULLI 1045
 modern, POISSON 1144
Mathematical theory of sound,
 EARNSHAW 324
Maxim Gorky
 (Tupolev's airplain) 1187
Maxim silencer 437
Mayer-Waterston solar meteoric
 hypothesis 317
Mean free path 324
Meat hammer 829
Mechanical framing cameras 131
Mechanical momentum 1056
Megathrust earthquakes 60
Mega-tsunamis 27
Melinite, picric acid 256
Melsens lightning-conductor
 system 1128
Mercury 729
Mercury azide 402
Mercury fulminate (detonating
 mercury) 108, 201, 270
 detonation velocity 468
 explosive properties 369
Mercury seismoscope 229, 308
Mercury spacecraft, SHEPARD 651
Mesomodel (meso-scale model) 97
Messerschmitt Me 163
 (Komet) interceptor,
 rocket-powered 525
Messerschmitt Me 262,
 twin-jet 529
Metal powder, compacting 650
Metal-dust explosions 21
 Bethlehem Zinc Works 318
Metallic surface, high-power laser
 pulse, luminous plume 669
Metallurgy 754, 776, 1049,
 1073, 1094, 1110, 1112,
 1121, 1136, 1162,
Metal-metal impact 613
Metals,
 compressible perfect liquids 565
 explosive system 1041
 high velocity deformation 644
 impulsive loads 606, 670
 multiple shocks 621
 ultrahigh pressures 661
Metamorphism, shock-induced
 ultrahigh-pressure 600

Meteor (shooting star) 63
Meteor bumper 564
Meteor Crater (Barringer Crater)
 64, 182, 436, 465, 508, 618,
 645, 658, 663, 848, 1169
Meteor explosion 815
Meteoric shower 63
Meteorite approach, supersonic,
 ballistic head wave 411
Meteorite explosions, Earth's
 atmosphere 703
Meteorite fall, L'Aigle 272
Meteorite impacts 63, 788
 astroblemes 646
 hypothesis 410
 in space 775
 lunar surface 847, 1173
 shatter cones 510
 terrestrial 703–706, 772, 809, 848
Meteorites 63, 266, 347,
 618, 703–706, 758, 766,
 774, 850, 969, 1046
 Allan Hills 766–767
 Allende 703
 hypervelocity impact 689
 iron 616, 673–674, 758
 Martian 766–767
Meteoritic diamonds,
 Arizona meteorite
 (Canyon Diablo meteorite) 403
Meteoroid impact 510, 609, 772, 775
Meteoroids 63, 703, 744, 808
 debris from comet
 Tempel-Tuttle 808
 micrometeoroids 645, 785
 potential hazard to spacecraft 744
 Whipple shield 563
Methanometers, coal mines 281
Method of characteristics 140
Method of finite differences 137
Method of operator differences 1086
Method of parallel sections 258
Method of smearing out the shock 585
Meyer-Prandtl expansion fan 74
MHD shocks,
 magnetohydrodynamics 31
Micro head waves 608
Micro steam explosions 20
Microactuator 148
Microchannel plates (MCPs) 131,
 489, 764
Microdetonics 25
 facility LLNL 778
Microexplosions 21
Microexplosives, laser-ignited 765
Microfossils 850
Microjets (tiny Munroe jet) 681, 936
Micromagnetism 1069
Micrometeoroids 645, 785

Microphone, supersonic projectile
 velocities 423
Microphotography 778
Microschlieren 341
Microsecond snapshot photo 316
Microseismic shocks 33, 456
Micro-shock waves 31
Microsystems technology 148
Microtektites 305
Microwave background radiation,
 residual cosmic 685
Mie-Grüneisen equation 98, 475
Mikhel'son line (Rayleigh line) 112,
 363, 391, 412, 1130
Military gunpowder 292
Milk-Drop Coronet 413
Milky Way 1093, 1102
Miller-Urey experiment
 (Miller experiment) 55, 603
Miller-Urey spark-discharge
 apparatus 966
Milne seismograph 412
Mine explosions 345, 1065, 1112,
 1122, 1142–1143
Miniature shock devices 837–838
Minimally invasive therapy/surgery
 776–777
Minimum explosive concentration 646
Mining explosives 351
Minoan civilization, Thera event 184
Minor planets
 (planetoids /asteroids) 271, 354
Mintrop ball 902, 1131
Mintrop seismograph 445, 464
Mintrop wave 120, 464, 980, 1131
Mintrop's refraction method 980, 1131
Missile weapons 4
Missing planet 727, 792
Missing shear wave
 (secondary wave/S-wave) 452
Mist pouffers 350
Mode locking, laser oscillations 685
Model volcano experiment 1102
Mohorovičić discontinuity 40
Moiré techniques 129
Moisson number 482
Moldable plastic explosives 108
Molecular chaos,
 BOLTZMANN 352
Molecular dynamics 98
Molecular vortices hypothesis 1149
Mollet pump 273
Molten fuel-coolant interaction 20
Moment magnitude 506
 scale, earthquakes 752
Moment of inertia 42
Momentum (quantitas motus) 274
Monads 41
Monge's curve 255

SUBJECT INDEX

Monostable multivibrator
  (monoflop) 127
Monotube high-pressure gas shock
  absorber 599
Monster blast (solar flare) 837
Monster gun (Dulle Griete –
  "Mad Margaret") 196, 886, 1100
Monster wave 28
Monte Carlo method 1134
Moon 172, 174, 189, 212, 337,
  342, 576, 579, 629, 830, 835
  Earth-Moon system 174
  meteor strikes 410, 472
  tektites 697, 740
  wicker chair, 47 large rockets 199
Moon craters 201, 410, 552
Moon landing 651, 704–705, 720, 724
  Apollo 704, 720, 736
  VERNE 343
Moonquakes 629
Moteur à explosion
  (explosion engine) 221
Motion of fluids 258
Motions of molecules 4
Mount Kilauea-Iki, lava fountains 638
Mount Pelée (fiery cloud/
  nuée ardente) 428
Mount Pinatubo 786
Mount St. Helens 63, 752–754,
  757, 767, 856
Mount Tambora 280
Mount Toba eruption 182
Mount Vesuvius 188, 539
  explosive eruptions (1944) 539
Moving flame, photographic
  records 430
Moving shock wave, boundary layer,
  schlieren technique 596
Multi-frame cameras 1019
Multichannel chronograph 124
Multiple-baffle system (gun) 978
Multiple-ball percussion experiment,
  NEWTON 759
Multiple percussion pendulum 46
Multiple-spark camera 913, 959,
  1014, 1018, 1062, 1161
Multishot flash X-ray system 625
Multivibrator (flip-flop) 125
  bistable (flip-flop) 464
  monostable (monoflop/one-shot
    multivibrator/Eccles-Jordan
    circuit) 127, 464
Munroe effect 396, 984, 1132
Munroe jet 396, 716, 936, 1132
Murchison meteorite 705, 706
Murman-Cole procedure 724
Murnaghan equation of state 98, 509
Musconetcong Tunnel,
  use of dynamite 352

Mushroom cloud 539
Musical instruments 70
Musical sounds, velocity 277, 329
Musket shots drag studies 73
Muzzle blast 35, 106, 915
Muzzle brake 303
  (recoil) efficiency 544
Muzzle velocity 123, 478

# N

N-wave, sonic boom 553, 586
NACA camera 556
Nagasaki, FAT MAN 546
Nanobacterial cells,
  Allende meteorite 704
Nanodiamonds 778
Nanoscalpel 148, 816
Nanoshocks 31, 148, 822
  wave physics 149
Napier bones 139
NASA (1958) 628
  Project Apollo 651
National Defense Research Committee
  (NDRC) 519
Navez chronograph 123
Navier-Stokes equations 284, 954,
  1076, 1176
NEAR
  (Near-Earth Asteroid Rendezvous)
  spacecraft 798–799
Near-Earth objects
  (NEOs) 65, 813
Near-point high-intensity spark light
  source 543
Nebulae 814, 844, 1093, 1102,
  1113–1114
  Herbig-Haro (HH) objects 580
Nebulous stars 1093
Negative shocks 543
Neptun 779–780
Network analyzer 139
Neutron, discovery 493
Neutron bomb 741
Neutron multiplication,
  critical mass 540
Neutron stars 493
  supersonic propeller 598
Nevada Proving Ground 581
Newton rings 129, 1040
Newton-Busemann pressure law 226
Newton's cradle 46, 213, 759,
  827, 906
Newton's demonstrator 46, 213
Newtonian fluids, high subsonic and
  supersonic velocities, break-up 616
Newtonian pressure law 226
Newtonian principles 1002
Nitrates, decomposition 287

Nitroexplosives 415, 716,
  783–784, 935, 993, 1005,
  1037, 1087, 1188, 1196
Nitrogen diamond 843
Nitrogen fertilizer,
  ammonium nitrate 210
Nitrogen iodine ($NI_3 \cdot 2\,NH_3$) 278
Nitrogen trichloride ($NCl_3$) 278–279
Nitroglycerin
  (glycerol trinitrate)
  ($C_3H_5O_3(NO_2)_3$) 100, 107,
  309, 333, 336, 936, 983, 1001,
  1037, 1046, 1135–1136, 1196
  detonation velocity 468
  mining industry 342
Nitrolite
  (76/12/5 AN/TNT/
  nitroglycerin) 404
Nitromethane,
  initiation of detonation 631
  shock initiation of detonation 664
Nitropenta 405
Nitrum flammans 209
Nobel Prize, creation of 1135–1136
Nobel's blasting oil 333
Nobel's igniter 336
Nobelium 1136
Nobili rings 389
Noble chronograph 124, 354,
  359, 1137
Noble gauge (gun) 1003
Nonlinear acoustics 72
Non-Newtonian fluids 1088
Non-nuclear blast, largest,
  Ripple Rock 626
Nonstationary flow 1083
Nova PERSEI 1901 425
Nova stella 66
Novae (see also Supernovae) 192,
  491, 787
Nozzle, convergent-divergent 396, 411
  critical speed 301
Nozzle outflow 83, 926
Nozzle studies 926–930
Nuclear bomb(s)
  (see also Atomic bombs) 117, 517,
  530, 651
  HEISENBERG 517
Nuclear bomb explosion, optimum
  HOB (Height of Burst) 536
Nuclear bomb physics,
  superbomb 519
Nuclear chain reaction,
  critical mass 514
  danger, EINSTEIN 514
Nuclear detonations, gamma-ray
  signatures 643
Nuclear excavation experiments
  659, 989

Nuclear explosion(s) 21, 118,
    707, 721, 771, 905, 988–989,
    1031, 1169, 1183
  destructive force estimation 511
  dirty 118
  largest (Tsar Bomba) 651
  underground 626, 707, 745, 788
  underwater 988
Nuclear explosion test, CHAGAN 677
Nuclear fireball 1019
Nuclear fission 505, 511, 514
  artificial, Pu-239 520, 523, 575
  artificial, U-235 511, 524
  artificial, U-238 520
Nuclear fusion 787, 839
Nuclear implosion 987
Nuclear Non-Proliferation Treaty
    (NNPT) 699
Nuclear physics, milestones 116
Nuclear power plant (nuclear reactor)
    831–832, 1031
Nuclear reactor disaster,
    Chernobyl 771
Nuclear reactor explosion 1031
Nuclear stimulation, recovery of
    natural gas 695
Nuclear Test Ban Treaty 667
Nuclear war 741–742
Nuclear weapons 116, 741–742, 799,
    1047–1048, 1146, 1196
  development, Feynman-Metropolis-
    Teller (FMT) equation of state
    577–578
  effects 623
  research 96
  technology 75
  tests 626, 660, 800, 808
  atmospheric, U.S. 581, 607
  ban 799
  Britain 593
  final atmospheric, U.S. 660, 667
  France 643
Nuclear winter 16
Nucleosynthesis, explosive,
    supernovae 555
Nucleus formation 1044
Nuée ardente 428

# O

Oblique central collision 151–152
Oblique collision 744, 976
Oblique eccentric collision 151
Oblique impact 737
Oblique percussion 12
Ocean tides, theory 77
Octanitrocubane 823
Octogen 108
Off-Hugoniot states 99

Oil refinery 1087
Oil well(s) 299, 339, 340, 689
  Kuwait 689
Oil well fire, blast wave to shoot out
    flame 115, 453
Oil well flow, shaped charges 556
Oil well shooting 115, 339
Olduvan (Oldowan) pebble tools 181
Olduvan-type choppers 181
Operation ARGUS 660
Operation CROSSROADS 550
Operation IVY 593
Operation TEAPOT 607
Opposite V gliding spark, lines of
    discontinuity 534
Optical depth gauge, water waves 693
Optical flow visualization 1008–1011
Optical interferometer 320
Optical masers 624, 649
Optical reflection techniques 129
Optical reflectivity, shock front
    in a gas 583
Optical streak recording 732
Orbital motions (planets/comets) 1039
ORDVAC (Ordnance Variable
    Automated Computer) 1181
Organic materials in space 706,
    813, 844
Origin of life 149
Oscillation 750, 1044–
    1045, 1097, 1125
Oscillatory galloping detonation 723
Oscillatory waves 1175
Oscillographs 125, 1013
  Dufour-type 125
  hybrid-type 126
  Wood 1013
Oscilloscopes 125
  cathode-ray 1013
  digital storage 126
  ultrafast traveling-wave 652
Overdriven detonation 24
Oxygen, metallic 802
Oxyhydrogen explosions 202, 258
  Volcan de Lémery 228
Ozone 1059, 1105

# P

PALLAS 274
Palmieri seismograph 121
Pancake bombs 62
Panspermia hypothesis 706
Panzerfaust ("tank fist") 514
Papin's gunpowder machine 1142
Papin's steam digester 926, 1141
Paris gun 459, 512, 887
Partial Test Ban Treaty
    (PTBT) 667

Particle-in-cell (PIC) method 392
Particle tracer analysis 84, 129
Particle tracer photogrammetry 672
Partition wave 29
Pascal law 206
Patellar reflex (knee-jerk),
    percussion hammer 302
PBX 9404 687
Pebble powder 331
Pebble tools 180
Peléan-type eruption 428
Pendulum, ballistic 49
  compound,
    center of oscillation 154
  precision, Martens 452
Pendulum clock 1097
Pendulum experiments
    1125–1126, 1129
Pentaerythritol 405
Pentaerythritol tetranitrate
    (PETN/nitropenta) 405
Pentolite (50/50 PETN/TNT) 404,
    406, 688, 730
Perbuatan Volcano (Krakatau) 378
Percussion 12, 17, 246
  animal 874–875
  ball 902
  bulb of 46
  center of 42, 154, 219
  central 891
  classical theory of 1097
  discontinuity problem 150
  elastic 41, 154, 209, 446, 888,
    890–891, 906, 1097
  elastic bodies 1097, 1190
  force of 43, 48, 153–154, 233, 241
  gain in 153
  heating effects
    (BERTHOLLET) 276
  immediate 51
  inelastic 4, 41, 902
  mechanical 3
  medical diagnostics 51, 247
  oblique 217, 892
  perfectly elastic
    (HUYGENS) 229
  perfectly elastic central and lateral
    (MARCI DE KRONLAND) 205
  phenomena
    (GALILEI) 204
  theory 300
  three-body 153
  tool-using California sea otter
    171, 692
  tool-using Egyptian vulture
    171, 692
Percussion boring 295, 885
Percussion bulb 876
Percussion bullet (explosive bullet) 51

Percussion cap (percussion primer), SHAW  51, 279, 290
Percussion drilling  50, 885
Percussion figures  47, 346, 898
Percussion force  48
Percussion grinder  50
Percussion hammer  302
Percussion ignition  274
Percussion lock  274
Percussion machines  212, 890–892
Percussion marks  47
Percussion match  51
Percussion models  906–909
Percussion number (NEWTON)  890
Percussion pendulum  890
Percussion powder  51
Percussion pressure  133
Percussion studies, 17–20th centuries  888–905
Percussion techniques, stone tools  877, 878
Periodic cell structure  113, 671, 995
Permissible explosives  345
PETN (nitropenta)  404–405, 542, 605, 623, 664
   shock initiation of detonation  664
Photochronography  416
   electric spark  429, 451
Photodisruptive effect  92, 816
Photoelastic stress analysis (MAXWELL)  314
Photo-gelatin dry plate  351
Photographic image, first  283
Photography  300
   single-shot/high-speed  130–131, 399
   spark waves (weak shock waves)  425
Photonic crystal  835
Picric acid  108, 256
Piezoelectric crystals, electrostatic fields  412
Piezoelectric effect  710
   direct  135
   indirect/inverse  135
Piezoelectric gauges  954
Piezoelectric pressure gauges  135
Piezoelectric transducers  1073
Piezoelectricity  370
Piezo-indicator  494
Piezoresistive effect  135, 473
Piezoresistive gauges  135, 755–756
   shock pressure, manganin  664
Piezoresistivity  135, 473
Pile driver (pile engine)  232, 279, 883
Pile driving  50, 198, 295, 300
   Neolithic  183
Pile-engines, efficiency  279

Pinch effect/phenomenon  341, 442, 509
Piston ballistic pump  1142
Piston engine, PAPIN  1141
Pitot pressure  15
Pitot tube  235
Planar impact test  97
Plane wave generator  567, 908, 950–951
Planet(s)  727
   motions, NEWTON  227
Planetary evolution  788
Planetary explosion theory  271, 274
Planetary nebulae  814, 1093
Planimeter  139
Plasma (quasi-neutral system of ionized gas)  470
   dusty  818
Plasma fusion  690
Plastic explosive, BICHEL  440
Plastic flow  956, 1051
Plastic impact  154
Plasticity  1044
Plasticity theory  1099, 1147
Plate convergence  18
Plate tectonics  60, 857, 1123
Plinian eruptions  189
Plowshare Program  117, 626, 698, 989
   largest cratering test  659
Pluto-Charon binary  175
Plutonium (Pu-239)  519–524, 660, 987, 1051
   implosion assembly  524
   reactor-grade  660
   underground nuclear test  660
Plutonium bomb  546, 593, 988, 1105
   von WEIZSÄCKER  520
Pneumatic drilling hammers  296
Pneumatic gun  881
Pneumatic lighter  881
   fire pump (tachopyrion)  210, 273
Pneumatics  1057
Pockels effect  412
Poincaré-Lighthill-Kuo (PLK) technique  1116
Point explosions  19, 606
   high-intensity, similarity solution  528
Point source  19
   explosion  55
Poisson distribution  1145
Poisson isentrope (Poisson adiabatic law/static adiabat)  285, 1144
Poisson process  1145
Poisson regression model  1145
Polarization of light  1040
Polaroid film  126
Polaroid Land camera  563

Polycyclic aromatic hydrocarbons (PAHs)  149, 844
Polymerization, shock-induced, organic monomers  680
Polytropic gas  389
Pompe balistique (ballistic pump)  221, 228
Popigai Crater  179
Porous material studies  1073
Positrons  483
Potassium chlorate ($KClO_3$)  263
Potassium nitrate  106
Poudre blanche (white powder)  383
Pouillet chronoscope  305
Poulter seismic method  1145
Powders, to solid density  633
Prandtl airfoil and wing theory  1084
Prandtl mixing length  1147
Prandtl number  74, 448, 1147
Prandtl rule  483
Prandtl tube  1147
Prandtl wind tunnel  1147
Prandtl-Busemann method of characteristics  488, 1054
Prandtl-Glauert rule/analogy  483, 1085, 1147
Prandtl-Meyer angle  445
Prandtl-Meyer expansion fan  445
Prandtl-Meyer function  445
Prandtl-Meyer nozzle  933
Prandtl-Meyer theory of expansion  101, 445, 473
Prebiological organic synthesis  714
Prebiotic substances, shock-induced creation  966
Precision pendulum, Martens  452
Precursor detonation  993
Precursor shock wave  604
Prerigger framing photography  127
Pressure, Chapman-Jouguet, high explosives  600, 621
   detonation pressure  1159, 1170
   diaphragm sensors  134
   high  491, 1051, 1145
   impact pressure  15
   magnetic  207
   mechanical gauges  134
   of a blow, HOPKINSON  450
   piezoelectric gauges  135
   stagnation overpressure  133
Pressure bar, DAVIES  900
   devices  900
   HOPKINSON  900, 1092, 1157
   KOLSKY  900
Pressure brake, Hildebrand-Knorr  478
Pressure cooker ("digester"), forerunner of autoclave  222
Pressure field phase (drag phase)  90
Pressure figure  346

Pressure flaking technique (V) 183
Pressure gauge, Kistler 1032
 piezoresistive 135
Pressure law,
 NEWTON 226
 NEWTON-BUSEMANN 226
Pressure-time profile 133
Pressure wave 484
Primacord 406
Primary wave (P-wave) 452, 721
Prime numbers 1086, 1129
Primeval atom 1113
Prince Rupert drops 212, 329, 996
Princeton shock tube technique 595
Principal Hugoniots (curves) 99, 615
Principle of mechanical percussion 3
Probability theory 1086
Profiles, force-time 897
 free-surface velocity, metals 745
 pressure-time 133
 shock waves, spatial density 632
 Zhukovsky 1199
Projectile, flat-ended 700
 hypervelocity 740
Projectile head,
 pressure distribution 494
Projectile trajectories, analysis
 (EULER) 246
 GALILEI 204
Projectile velocities, Wheatstone
 chronoscope 302
Prominences 870
Propagation, V-shaped 818, 911, 1043
Propellant(s) 691
 double-base powder 1064
 liquid 1064
 rockets 474
Propellant gases,
 equations of state 494
Propeller(s), supercavitating 627
 supersonic 598
Propeller blades 428
 erosion 457, 482
 shock waves 511
Propeller tests, high-speed 469
Propulsive duct 443
Protection against gun fire 717
Proton(s) 14, 609, 806, 835
 antiprotons 609
Proton beam 650, 768
Proton density 657
Proton-proton chain
 (astrophysics) 1047
Proton radiography 806
Pseudo-shocks 31, 631, 974
 λ-type 31
Pseudostationary reflection 759
Pseudo-supersonic wave effect 974, 980, 1163

Pu (plutonium) 519–524, 660, 987, 1051
Pu-238/239, nonfissile/fissile 523
Pulsar (pulsating radio star) 67, 500, 694, 699
Pulsating variable stars 261
 Mira Ceti 200
Pulse generator, electromagnetic, weak planar shock waves in liquids 640
Pulse technology 683
Pulse thickness 554
Pulsed-laser holography 129
Pulsejet, bombardier beetle 171
Pulsejet engines 1110, 1138, 1163
 deflagration 23
Punched-card machines 139
PVDF (polyvinylidene fluoride) 710
Pyréolophore (explosion-driven piston engine) 275
Pyroclastic flows 539, 754, 786, 856
Pyroclastic surges 26, 428
Pyroglycerin 309, 333

## Q

QSO (quasi-stellar object) 670
Quantity of motion, DESCARTES 151, 215
Quantum mechanics 1048, 1062, 1070, 1133–1134
Quantum of time (Planck time) 69
Quasars (quasi-stellar radio source) 67, 649
 3C-9, farthest known object 670
 extreme redshift 670
Quasi-shock waves 440, 1072
Quasi-stellar object (QSO) 670

## R

Radiation, blackbody 569
 bomb related 547
 Cherenkov 500
 corpuscular 588
 cosmic 451
 cosmic background 174, 787–788, 833, 872
 γ 451
 microwave background 685
 solar 311
 Stokes parameters 1176
 thermal 21, 603
 ultra γ 451
 ultraviolet 776
 X-rays 414
Radiation belt, STARFISH 660
Radiation gasdynamics 37, 693
Radiation implosion 608
Radiation mean free path 693

Radiation-supported shock wave 676
Radiative-precursor shock wave 604
Radiative shock 32
Radical (free radical/uncombined radical) 294
Radio micrometer 1050
Radio waves 1089
Radioactivity 40
Radiofrequency interferometer 597
Radiography 132
 flash 934–935, 962, 996, 1009, 1063
 pseudocine 934
 pulsed 1041
Rail-guided rocket sled test facility 608
Railgun 748, 951
 electromagnetic 459, 544
Railway airbrake systems, nonstationary flow 478
Railway air-suction brakes 447
Railway torpedo 109
Rain, artificial, disrupting air currents 403
Rain erosion 547, 679–680
Rainbow 1125
Raindrops, disintegration by shock waves 680
Ram accelerator (RAMAC) 772–773, 786, 1090
Ramjet engine 443, 766, 808
Random Access Memory (RAM) 143
Rankine body 1149
Rankine cycle 1149
Rankine method 1149
Rankine-Hugoniot curve 390–391
Rankine-Hugoniot equations (Rankine-Hugoniot shock relations) 348, 391, 1149, 1182
 reactive fluid 441
 reactive waves 460
 relativistic 572
Rarefaction shock 31, 433, 438, 446, 558, 622
 overtaking second compressive shock 622
Rarefaction (shock) wave 328, 349, 471, 473, 622, 700, 748, 771, 775
 implosion 486
Rarefied gas dynamics, MAXWELL 37, 363, 501
Ray method (shock wave studies) 970
Rayleigh distribution 1178
Rayleigh equation 1177
Rayleigh instability 712, 1177
Rayleigh line 112, 363, 391, 412, 1177
Rayleigh scattering law 1178
Rayleigh surface waves 121, 386, 1178

Rayleigh vapor bubble model 1177
Rayleigh waves 121, 386, 1178
Rayleigh-Taylor (R-T) instability 539
RDX 108, 270, 421, 510, 601
Reaction turbine 384
Reaction zone parameters 806
Reactive discontinuities
 (detonation waves) 427
Reactive shock wave 30, 112
Rear-end collision 17
Rebounding (skipping) 91, 151,
 314, 497
Recurrent novae 787
Red giant stars 455, 674, 774, 807,
 814, 844
 simulation 807
Reentry capsules 1025
Reentry vehicles, ballistics 653
Reflected-light photography 127
 detonics 1007
Reflection, irregular 74, 161
 Mach reflection (irregular
  reflection), three-shock theory
  81, 130, 161, 535
 regular, two-shock theory 161, 535
Reflection method 121
 seismic survey technique 980
Reflection wave 1067, 1185–1186
Reflex hammer 302
Refraction method 121
Refraction wave dynamics 1083
Relativistic cosmological model,
 EINSTEIN 459
Relaxation time 344
Relay ballistic computer 141
Relay calculator 141
Relay interpolator 141
Resilience, compressed synthetic
 rubber 676
Resonance 1177
Restitution coefficient 4, 1180
Restitution force 260
Restitution period 48
Retonation 24
Retonation wave 430, 666, 1067
Retroreflective foils 765
Retroreflective screens 632, 1009
Reverse detonation (retonation)
 24, 430
Reverse shock 22
Revolving mirror, high-speed
 phenomena 297
Reynolds number 74, 382
Richter scale 62, 506, 752
Richtmyer-Meshkov instabilities 423
Ricochet firing 91
Ricocheting (skipping) 91, 314, 497,
 885, 1055, 1166
Riemann function 1152

Riemann geometry 1152
Riemann hypothesis
 (number theory) 1152
Riemann invariants 328, 1152
Riemann method 1152
Riemann problem 329, 496
Riemann surfaces 1152
Riemann waves 911
Ries Basin, impact crater (Germany)
 63, 180, 434, 508, 849, 1169
Ring expansion experiment 710
Ripple Rock 626
Ripples 25, 77, 294
Rivet, explosive 504
Robert J. Collier Trophy 449
Roberts torpedo 339
Rock cutting, high-speed
 (subsonic) water jets 684
Rocket(s) 930, 941, 943, 1020–1023,
 1027, 1088, 1153, 1186, 1196
 black powder 193
 liquid-fueled, first flight 474
 multi-stage 432
Rocket engines, liquid-propellant 507
Rocket motors 1064,
Rocket plane, hypersonic 638
Rocket sleds 1026
Rocketry 930, 1063–1064, 1110
Rodman gauge 135, 324, 1003, 1154
Rodman guns 336, 886, 1155
Rodman powder
 (mammoth powder) 331, 1155
Roentgen flash technique 509
Rogue (freak) waves 25, 28, 869
Rohrwindkanal
 (tube wind tunnel) 613
Roller pendulum, ballistic 1005
Ropes, folded stretch-free,
 kinetics 527
Rotating mirror 131, 1015, 1017–1018
Rotating mirror streak camera
 1017, 1050
Rotating stylus 671
Rotating-drum chronograph 337
Rotating-prism camera,
 high-speed 1019
Rotation, Earth's inner core 823
R-T instabilities, mushroom cloud 539
Ruby fluorescence method 730
Ruby laser 644
Ruby scale 97
Ruby standard 730
Rutschschere (gliding jar) 295

## S

S-bullet (*Spitzgeschoss*) 101, 429, 1016
S-wave (secondary wave) 292,
 450, 452

Sabine chronoscope 125, 360
Sachs scaling law 544
Safe lamp, coal mines 279
Safety detonator 109, 336, 542
Safety fuse 106, 109, 291
Safety in Mines Research Board
 (SMRB) 466
Safety lamp, Davy-type 281
Safety valve 222, 254, 288, 359,
 926, 1141
 failure 359
 pressure cooker 222
Sagami Bay, earthquake/tsunami 229
Saint-Venant principle
 (de Saint-Venant problem)
 319, 1157
Saltpeter (KNO3) 106, 194, 263
San Andreas Fault 438
San Francisco earthquake (*1906*) 438
Sandia quartz gauge 664
Santorini volcano 184
Satellites 350, 353, 387, 501, 564,
 628, 643
 COS-B gamma-ray 734
 ESA 735, 833
 first artificial (*1957*) 620
 geodetic 679
 ROSAT 183, 773, 794, 798,
  815, 835
 Vela 643
Saturn rings, discovery of 1097, 1098
Scabbing 450
Schießbaumwolle (guncotton) 308
*Schlagwettersichere Sprengstoffe*
 (*Wetter-Sprengstoffe*) 345
Schlieren, microschlieren 341
Schlieren experiments, HOOKE 212
Schlieren methods 128, 220, 340, 533,
 765, 828–829, 910, 912, 921, 1008,
 1160–1161, 1185
 color, SCHARDIN 1160
Schlieren photography 790, 914,
 916–917, 969, 995, 1008
Schlieren system, 3-D 700
 color, constant-dispersion
  prism 596
 focusing 790
Schlieren visualization,
 spark waves 347
Schmidt head waves (SHW) 120, 310,
 464, 512, 963, 974, 1162
 pseudo-supersonic 513
Schmidt tube 140, 1163, 1164
 pulsejet engine 490
Schmitt trigger circuit 127
Schröter rule (lunar craters) 272
*Schwarzpulver* ("black powder") 195
Schwerer Gustav 887
Scramjet 639, 808

Scramjet engine, hydrogen-fueled 770, 831
Sea mine explosions 349
Sea shock 28, 333
Sea waves 190, 247, 311, 361, 864
 giant
 (*see also* Tsunamis) 190, 308, 346, 361, 378, 419, 651, 821
Seaquakes 60, 254, 308, 392–393, 789, 841
 seismic sea waves 308
 sound, sea surface and seismic shock phenomena 403
 tsunamis 392, 414, 789, 841
Seawater, static compressibility 368
Sébert chronograph 125
Second cosmic velocity (escape velocity) 620
Secondary blast-wave shock 22
Secondary high explosives (TNT, RDX, PETN) 109
Secondary wave (S-wave) 60–61, 292, 450, 452
Sectional density 338
Sedov equation 1166
Sedov phase (supernova) 1166
Seiches 27, 251
Seismic air-guns 51
Seismic bands 307, 857
Seismic discontinuities 5
Seismic experiments,
 active/passive 119
 lunar 720–721, 724–725
Seismic focus 321
Seismic line 120
Seismic prospecting 121
Seismic reflection method (MINTROP) 445, 464–465, 1130–1131
Seismic sea waves 864
 arrival times 361
 propagation 347
Seismic shocks 32, 119, 292, 853, 858, 896, 902
Seismic studies 1145
Seismic waves 727, 862, 905, 979–980, 992, 1123, 1131
 velocity 979, 1123
Seismograms 121
 rupture propagation 611
Seismographs/seismography 61, 121, 896–897, 902, 1131
 Milne 412
 Mintrop 445, 464
 rolling-sphere 896
 teleseismic event 400
 Wiechert 420
Seismography,
 19th century 896–897

Seismology 60, 119, 370, 378, 1002, 1123, 1130–1131
 explosion 980, 1123
 MALLET 1123
 nuclear explosions 707, 905
 three-dimensional problem of shock wave 664
Seismometers/seismometry 61, 120, 332, 359, 420, 678
 Ewing 382
 mercury 312, 896, 979
 miniature (geophones) 720
 Wood-Anderson 506
Seismoscopes 120, 896
 mercury 229
Self-propagation high-temperature synthesis 776
Self-similar solution 666
Severn bore 25, 57
Severn's tidal bore 191
Seyfert galaxies 538
Shadow zones 454
Shadowgraphy 128, 370, 790, 793, 828–829, 963, 1009, 1011, 1049–1050
 air flow around bullets 401
 steam jets from orifices 430
Shaken baby syndrome 52, 649
Shallow-water equations, gas dynamics 495
Shallow-water waves 306, 863
Shaped charge(s) 732
 petroleum industry 983
Shaped charge cavity effect 115
Shaped charge lined cavity effect 115
Shaped charge perforating guns (petroleum industry) 983
Shatter cones 64, 497, 552, 844
 artificial 657
 Kentland type 510
Shear ruptures 863
Shear waves (S-waves) 60–61, 121, 292, 450, 452
 velocity 838, 863
Shimose, picric acid 256
Shiva Crater 178
Shkval 105
Shock 28, 653
 astrophysical 746
 bow shock 33
 foreshock/aftershock 32, 65
 interstellar 756, 791
 $\lambda$-shock 554, 969
 longitudinal 584
 main 433
 multiple 698
 negative 543
 sudden/violent change in the state of motion 562

 thermal precursor 32
 transverse 584
 vibration 34
Shock absorber 198
 monotube high-pressure gas 599
 tapered chain 828
Shock adiabat (Hugoniot curve), water 585
Shock bottle (barrel shock pattern) 35, 241, 474, 557
Shock compressibility, ultrahigh pressures 661
Shock compression 1041–1042
 hydrodynamic theory 610
 liquids 625
 materials 725, 745, 1041
 plasma 703
 solids, high-pressure 791–792
 temperature measurements 636
Shock-compression guns 552
Shock compression ratio, perfect gas 390
Shock concept, benign 674
 catastrophic 674
Shock deceleration, circumstellar material 834
Shock diamonds 29, 74, 417, 433, 1158
Shock drift acceleration of particles 746
Shock duration measurement 894
Shock Dynamics Center (SDC) 699
Shock Dynamics Laboratory (SDL) 699
Shock friction (impulsive friction) 295, 332
Shock front 290, 311, 340, 348, 364, 431, 531, 1149, 1182
 entropy increase 610
 in air, thickness 466
 local density profile 363
 luminous phenomena 527
 phase transition of solid 726
 Sun's collapse 174
 thermodynamic conditions 449
 theory 544
 tornado 581
Shock front analysis 954–957
 gaseous matter 954
 solid matter 955–956
 space 955
Shock front thickness 441, 448, 467, 583, 674
Shock front velocity 160, 424, 481, 696
Shock-generated mercury plasma 711
Shock-heated plasma 737
Shock-heated substances, MS analysis 653

Shock heating, aerodynamic 969
Shock implosion techniques 1083
Shock-induced
   cavity collapse 777
   chemical kinetics 1083
   chemical reactions 618, 755, 1073
   combustion 723
   freezing 87, 523
   lattice compression 717
   liquefaction 751
   seismic 795
   luminescence 1196
   phase transitions 630, 708,
      726, 731, 744, 1073
   polymerization,
      organic monomers 680
   solidification 976
   water freezing 842
Shock interactions, gases
   959–961, 1083
   liquids 962–963
Shock lamellae 767
Shock line 33, 371
Shock liquefaction 88
Shock lithotripsy 950, 981, 1085, 1179
   electrohydraulic 584, 588
Shock loading 738, 744, 792
   conditions 97
Shock metamorphism
   (impact metamorphism) 646,
   688, 848
   lunar 705
Shock-metamorphosed quartz
   grains 767
Shock-on-shock (SOS) interaction
   668, 735, 975
Shock precursor phenomenon 603
Shock pressure 133
   ultrahigh 745, 756, 788, 800
Shock propagation 158
   layer to layer 350
Shock pulse phenomena 1146
Shock reflection 1049
   irregular 486
Shock stall 521, 972
Shock synthesis 7, 115
   inorganic materials 668
Shock theory 437
   collision-free hydromagnetic 602
Shock tube(s) 85, 422, 496, 509,
   551, 693, 944–949, 966, 1048–
   1049, 1077, 1083, 1106–1107,
   1109, 1143, 1188, 1196
   bifurcated 947
   Bleakney's triggerable 945
   chemical 610
   double shock 948
   drug delivery 817
   electric 617

   electromagnetic 617
   equal-pressure 946
   heated 711
   hypersonic 589
   mercury 948
   nitrogen tetroxide 600
   single-pulse 610, 714, 946, 966
   spherical 605
   supersonic flow patterns 570
   vertical 786
   Vieille's 944
Shock tube equations 495
Shock tube nuclear air blast
   simulator 708
Shock tube problem
   (Sod problem) 496
Shock tube reflection nozzle 713
Shock tube studies 784–785, 913,
   1048–1049, 1090, 1143, 1170, 1193
Shock tube technique, chemical
   reaction studies 1106
   Princeton 595
   reflection of shock waves 577
Shock tube theory 447, 1107,
   1161, 1181
Shock tunnel 85, 642
   facilities 1090
   free-piston-driven 686, 949
Shock wave(s) 5, 29, 38, 74, 159, 241,
   340, 397, 431, 484
   animal 874–875
   boundary layer interaction 31
   channel 621
   chemical reactions 674
   collisionless 33
   converging 531
   cosmic 66
   cylindrical 963, 971
   density-time profiles 597
   diagnostics 594
   discontinuity problem 150
   disintegrating bladder calculus in
      vivo 637
   disintegration of raindrops 680
   dispersed 617
   effects 966–977
   electromagnetic 500
   evolution of knowledge 159
   first and second 639
   flash soft radiography, low-density
      materials 590
   flyer plate 160
   focusing 780, 782–783, 808,
      970–971
   ignition 572
   in air, schlieren photo 400
   in condensed matter 750, 774–775,
      791, 803, 816, 843, 1042, 1073
   in dense fluids 763

   in gases 80, 438
   in liquids 86, 466
   in solids 390
   exploding foils 664
   industrial applications 779
   intense 691
   intensity 30
   interstellar 34
   isentropic process 328
   large-amplitude planar 610
   laser-supported 21
   liquefaction 751
   lunar 704
   melting 976
   micro- 31, 832
   nanosecond duration 709
   oblique interactions 114, 911,
      933, 958–959, 961, 965, 969,
      975, 1083, 1147, 1158, 1193
   piezoelectricity 649
   plane 913, 916, 947, 951, 971,
      981, 1086
   pressure, hydrodynamic 829
   propagation 427
   pulsed laser hologram 690
   quasi 440
   radiation supported 676
   radiative-precursor 604
   reactive 112
   reverse 819
   shielding 1171
   solar 790
   spatial density profile 632
   spherical 700, 1173
   spiral 511
   splitting 87
   stability 530
   stationary/nonstationary 29
   steady speed 528
   supercooled water droplets 677
   superposition 79
   supersonic aircraft 586
   supersonic flow 567
   surface 958
   target plate 160
   threshold overpressure 563
   transmission,
      longitudinal,
      solid bodies 489
   underwater, snapping shrimp
      92, 171
   velocity 363
   visualization 910–917, 1184–1186
   water-like substances 590
   Whitham rule 636
Shock wave applications 978–983
   medical 981
   metal industry 982
   oil production 983

Shock wave attenuation,
  asymptotic law 642
Shock Wave Database,
  Russian 745
  living tissue 777
Shock-wave equations,
  de Hoffman-Teller 602
Shock wave front 718
Shock wave generation 950
Shock wave generator,
  electrohydraulic 588
  planar, COOK-type 955
Shock wave-induced polarization 656
Shock wave interactions,
  in gases 1181
  in metals 976
  in solids 964–965, 1101
  oblique 911, 933, 958–961, 965,
    969, 975, 1083, 1118, 1134,
    1140, 1147, 1158, 1193
Shock wave lithotripsy, clinical
  extracorporeal 640
Shock wave models 906–909
  piston 907
Shock wave particle velocity,
  impedance match method 634
Shock wave phase 90
Shock wave photography, gases 912
Shock wave physics,
  classification 158
  development 72
  RANKINE 1148
  solid-state 93, 542
Shock wave precursor 607
Shock wave-proof spacesuit 105
Shock wave propagation 910,
  1047, 1062, 1073, 1075, 1101,
  1108, 1162, 1181, 1186, 1193
  collisionless plasma 702–703
  inhomogeneous media 1171, 1179
  liquid 777
  porous medium 708
Shock wave superposition 1049
Shock wave theory 1196, 1197
  HUGONIOT 1094
Shock wave therapy,
  extracorporeal 797
Shock weapons 4
Shooting channel flow,
  supersonic compressible flow 466
Shot, spin rate 401
Shot firing 36
Shrapnel 262, 521
Shrimpoluminescence 93, 822
Shroud technique 622, 1077
SI units
  (Système International) 645
Siemens chronograph 124
Sikhote Alin meteorite 63, 560

Silencer,
  firearms 978
  small firearms 437
Silica glass, meteorite impact 493, 496
Silicon carbide, interstellar 774
Silicon chip, light-sensitive 707
Silver azide 402
Silver bromide 351
Silver fulminate (AgONC) 201,
  263, 385
Single-blow impact test (Izod impact
  test) 436
Single crystals, shock-compressed
  ionic 684
Single Mach reflection (SMR) 130
Single-pulse shock tube 610
Single-shot photography 131, 1015
Sirius A/B (white dwarf) 333
Skipping effect 91
Skull injuries 52
Sled vehicles, hypersonic test
  834, 1026
Slide rule 139
Slipstreams (contact discontinuity
  lines) 534
  second 591
Slurry explosives 616
Smoke-coated surfaces, periodic cell
  structure 671
Smoke flow visualization 129, 641
Smoked-foil technique 995
Smokeless powder 383
  nitrocellulose/nitrobenzene 405
Smooth-bore guns 886, 1137, 1155
Smoothed particle hydrodynamics
  (SPH) 175, 392
Snapping belts 70, 950
Snapping shrimp 92, 171, 304, 822
Snapping towel 70
Snapshot photography,
  microsecond 316
  triggered 1015
  underwater,
    explosive flash charge 554
Snow compaction, shock loading 792
Snowplow concept 509
Sodium nitrate/Chile saltpeter
  ($NaNO_3$) 326
Soft coal 1043
Solar flares 6, 21, 727, 810, 837,
  870, 871
  CARRINGTON 326
  hydrodynamic blast wave 657
Solar quakes 810, 871
Solar System 727, 780
  origin 257
Solar wind 6, 65, 588, 734,
  742, 808, 872, 955
  collision 17

  Earth's shock 675
  supersonic 665
  termination shock 147, 836
  theory 633
Solid-state shock wave physics 542
Solids deformation, high strain rates
  682
Solitary wave (soliton) 26, 296–297,
  827, 1156
Solitary wave of translation 296
Solitary wave reflection,
  RUSSELL 1155
Soliton (solitary wave) 26, 296–297,
  827, 1156
Sonic booms (sonic bangs) 64, 102,
  454, 474, 481, 500, 553, 753, 781,
  1077, 1083, 1179
  double 595
  effects, projectiles 474
  N-/U-wave 73, 553
Sonic wall 102
Sonoluminescence 92, 777
Soot funnel 911
Soot method, ANTOLIK 130,
  355–356, 360, 385, 443, 537, 713,
  911–912, 1042
Sound, adiabatic, DE LAPLACE 155
  finite amplitude 244, 1075,
    1144, 1177
  intense 342
  triplicity 330
  underwater 1073
Sound barrier (sonic barrier) 505,
  530, 561
Sound generation, intense 51
Sound propagation, in air, wave
  equation 243
  zone of normal/abnormal
    audibility 447
  zone of silence 447
Sound waves 340, 344, 375
  finite amplitude 39
Soviet atomic bomb (*1949*) 575
Soviet fusion bomb (*1953*) 598
Space debris,
  protecting spacecraft 563
Space exploration, multistage
  (or multistep) rockets 523
Space flight 342, 1081
  first man, GAGARIN 651, 1081
  human 628
  Mercury, Gemini, and Apollo 628
Space protection of the Earth 795
Space shuttles 613, 628, 652, 1026
Spacecraft, lunar 704, 1169
  manned 651, 1025–1026
  reentry shapes,
    blunt-body design 595
Spacesuit, shock wave-proof 105

Spall fracture 837
Spallation 94, 172, 450, 662
Spark(s), crackling sound 232
　electric 70
Spark chronography 1016
Spark discharge 1042–1043,
　1186, 1193
Spark eudiometer (VOLTA) 256
Spark gap 256, 486, 632, 638,
　921, 1186
Spark photochronography, electric 429
Spark waves 29, 385, 1186, 1193
　propagation 346, 424
　schlieren photographs 434
Sphere drag, high speeds 242
Spherical spark waves 959
Spin phenomenon, detonations 475
Spinning detonation (detonation spin)
　476, 993–994, 1067, 1165
　gasdynamic theory 486, 1165
Spiral galaxy, optical jet 655
Spiral shock wave 511
Split Hopkinson pressure bar
　(Kolsky bar) 97
Sprengel explosives 351
Sputnik 1, first artificial Earth
　satellite 620
Sputnik shock 620
St. Petersburg Declaration, ban of
　explosive bullets 347, 353
Stagnation overpressure 133
Stalker tube, gun tunnel 686, 949
Stall 522
Stand-off shock 770
Stardust (spacecraft) 813
STARFISH radiation belt 660
Stars, black 261
　collapsed 172
　collision 739, 812, 819
　core collapse supernova 823
　life cycle 807
　nebulous 1093
　neutron stars 499, 781, 812
　pulsating variable 261
　red giants 455, 674, 774, 807,
　　814, 844
　red supergiants 807
　symbiotic 815
Starspot 343
Static compression 1051
Static high pressure, effects of 1051
　physics 667
Stationary discontinuities 5
Statoréacteur 444
Steady-state cosmology 566
Steam ball 926
Steam-boiler explosions 20, 291,
　334, 1028, 1040
Steam digester 83, 926, 1141

Steam engine 228, 329, 396, 884, 926,
　1028, 1141, 1151, 1156, 1174
　atmospheric 228
　boiler explosions 291, 334
　PAPIN 228, 1141
　rotating 395
Steam engine indicator 417
Steam explosion, greatest in historic
　times (Krakatau) 378
Steam gun, PERKINS 285
Steam hammer 50, 300
Steam jets 378, 396, 430, 929, 931
　from orifices, shadowgraphy 430
Steam nozzle 396, 927
Steam ram 300
Steam turbines 378, 432, 926, 928,
　1038, 1101, 1110–1111, 1174
　impulse 396
　multistage 384
Steam wheel 926
Steel wires, dynamic strength 436
Steinbüchse (stone gun) 195
Steinheim Basin 63, 180, 849
Stellar collisions 18, 488, 739, 819
Stellar nuclear energy 1047
Stephan's Quintet,
　intergalactic shock wave 177
Stereomechanics 648
Stereoscopic camera, high-speed 521
Sternberg-Walker (SW) equation
　of state 688
Stishovite 64, 497, 600, 646, 658, 663
　statical synthesis 658
Stokes law (fluid dynamics) 1176
Stokes lines (optics) 1176
Stokes parameters (radiation) 1176
Stokes shift (optics) 1176
Stone tools 299
　Great Rift Valley 180
　human 710, 876–878
Stony Tunguska 63, 171, 442,
　446, 830
Stored-program computer 143, 556
Stored-program serial computer 143
Storm, geomagnetic 327, 601, 837
　magnetic 442
Storm surge 59, 418
Storm tides 59
Störmer-Lemaître-Vallarta
　theory 1114
Strain rate 99
Strategic Defence Initiative (SDI) 763
Streak camera, rotating mirror
　1017, 1050
Streak photography, detonics 1006
Streak record 493
Streak technique, explosion/
　shock wave phenomena,
　straight tubes 493

Streaming flow 259
STRELA 144
Stress-strain data,
　uniaxial tensile 709 ?
Stroboscopy 970, 1010, 1053
　laser 477, 796, 914, 995
　EDGERTON 526
　TOEPLER 910, 1185
Strombolian-type eruptions
　(Strombolian bursts) 489
Structural impact 13, 783
　engineering 13
Stuttering detonation 24
Subatomic particle behavior 1082
Subduction (plate tectonics) 841, 863,
　868, 789–790
　tsunamis 58, 60
Submarine landslides 864
Submarine ridges 864
Suboceanic landslides 419
　tsunamis 320
Subsonic airflow 1085
Subsonic flow, schlieren photos of
　shock waves 503
Subterranean fires 1115
Suction phase, blast wave 423
Sudbury Basin 176
Suevite 434, 646
Sulfuric acid,
　lead-chamber process 243
Sun, axial rotation 201
　creation of 174
　energy 311
　filament eruption 871
　gravitational implosion model 317
　heat, meteoric hypothesis 311
　solar system 257, 727, 780
　solar wind 6, 65, 588, 734, 742,
　　808, 872, 955
Sunquake 174, 810
Sunspots 173, 201, 870, 1039
Sunyaev-Zel'dovich effect 1197
Super Jump 105
Super penetrators 627
Super-relativistic blast waves 451
Super shake 61
Superaerodynamics 558
Superaviation 493, 558
Superballs 701
Superbomb 143, 519, 594
　true (staged) H-bomb, Soviet 608
Supercavitating propeller 627
Supercavitation 92, 627, 805, 968
Supercollision 17
Supercomputers 146, 808–809
Superconductive material,
　high-density 982
Supercooled fog droplets, "flash"
　into ice 526

Supercriticality (nuclear) 987
Superdetonation 24, 631, 654, 784
Superdynamics 147
Superguns 196, 224, 887
   140-ton 459
Superheated hurricane 62
Supernova(e) 21–22, 34, 66, 192, 491, 773, 789, 798, 804, 823
   catalogue 564
   core collapse 823–824
   explosions 451, 499, 712, 819, 1047
   neutron stars 499
   explosive nucleosynthesis 555
   shock waves 338
   supergiant 492
   visible 773
   Vela 798
   white dwarfs 333, 469, 491, 787
Supernova progenitor, photosphere, breakout of a shock 696
Supernova remnant theory, adiabatic blast wave theory 666
Supernova remnants (SNR) 66, 540, 573, 656, 666
   acceleration of cosmic rays 656
   interstellar shocks 756
Supershear 438
Supershear fault rupture 61
Supersonic aircraft 1077, 1116, 1140, 1186–1187
   no approach noise 454
Supersonic airfoil theory 1115
Supersonic airframe design 1054
Supersonic airliners 75, 1024
Supersonic ballistics (ROBINS) 242
   experiments 396
Supersonic botfly 171
Supersonic bullets 1050, 1092
   head wave 513
Supersonic car (SSC) 803, 1027
Supersonic combustion ramjet 639
Supersonic facility, continuous-flow
   closed-circuit 940
   prewar 939
   slotted throat 942
Supersonic flight 104, 763, 1021–1027, 1038, 1103–1104, 1115, 1128, 1187, 1195
   von KÁRMÁN 1103
Supersonic flow 37, 749, 757, 786, 796, 873, 938, 941, 949, 1054, 1062, 1080, 1083, 1103, 1118, 1140, 1147, 1158, 1174
   shock waves 567
Supersonic flying laboratory 800
Supersonic intermittent indraft facility 941

Supersonic jet 931–937
   astrophysical 770
Supersonic Man (YEAGER) 1195
Supersonic motion 1103, 1185
Supersonic phenomena 52
Supersonic plasma flow 1083
Supersonic propeller 598, 1023
Supersonic ricochet 314
Supersonic rocket 941, 1021, 1023, 1027
Supersonic rocket plane 1023
Supersonic speed, aerodynamic heating 609
Supersonic tail cracking, giant sauropod dinosaurs 171
Supersonic testing 1077
Supersonic transport (SST) 704, 1024, 1054, 1187
Supersonic Tunnel Association 604
Supersonic velocities 1068, 1128, 1195
   rail-guided rocket sled test facility 608
Supersonic wave model 1068
Supersonic wind tunnels 938–942, 1038, 1064, 1077, 1088, 1138, 1172, 1183, 1186, 1192
   testing, pioneer (STANTON) 1172
Supersonics 6, 53, 82, 1117–1118, 1193
   birth 71
   MACH 1117
   outflow into vacuum 301
Supervolcano 762
Supra conductivity 1044
Surface boundary layer 433
Surface detonations, large yield 992
Surface jetting effect 982
Surface of discontinuity 29
Surface shock unloading 471
Surface thermography 129
   liquid crystal 1011
Surface waves 60–61, 292
Surge of the trades 59
Surges 26, 59, 550, 678
Sweepback jet fighter 529
Sweepback wings 503, 1054
Sweet spots, sport equipment 751, 904
Swept-back wings, high-speed aviation 516
Swept wings 587, 1054
Swing wing 1024
Swiss EDISON 302
Symbiotic stars 814–815
Synchrotron shock model 697
Synergetic effects 723
Système International (SI) 645

# T

Taal Volcano, phreatomagmatic eruptions 678
Tail wave 388
Taisho Kanto Earthquake 229
Tait equation 98, 397, 509, 1180
Tamper, reflection of neutrons 524
Tangency solution 420
Target plate 160
Taub adiabat 1182
Taub equations 1182
Taub Universe 1182
Taylor expansion waves (Taylor waves) 512, 580
Taylor hammer (tomahawk reflex hammer) 302
Taylor instability 520, 712, 977, 1183
Taylor series (mathematics) 1086
Taylor test 97, 573, 700, 901, 1183
Taylor wave 512, 580, 1183
Taylor-Görtler vortices 520
Taylor-Sedov phase 666
Tear gas 1193
Tektites 172, 305, 618, 697, 721, 740
   lunar origin theory 721, 740
   volcanic ejecta from the Moon (?) 697
Tektronix model 519, first ultrafast traveling-wave oscilloscope 652
Telescope 789, 808, 1093, 1097, 1129
Telescope shock absorber 198
Teleseismic event 400
Teletsunamis (far-field tsunami) 27, 361, 865
Teller-Ulam design 594, 608
Tennis racket investigations 750, 904
Tensile strength 313
Terminal jet shock 33, 937
Termination shock 17, 33, 66, 861
   heliospheric 75
Terrestrial lifeforms, first 706
Terrorist attacks 789, 796, 825–826, 1032–1033
   WTC (2001) 825–826, 1032–1033
Tetryl 368, 623, 672
Tetrytol 368
Theia, protoplanet impactor 174
Theory of detonation, theory of shock waves 411
Theory of elasticity 314
Theory of explosive combustion 1130
Theory of germ formation in supersaturated phases 1070
Theory of heat (CLAUSIUS) 313
Theory of nucleosynthesis (CAMERON) 267
Theory of oscillatory waves 312
Theory of percussion 300

# SUBJECT INDEX

Theory of relativistic simple waves and shocks, founder (TAUB) 1181
Theory of relativity 67, 460
Theory of resonance, HELMHOLTZ 110
Theory of shock waves, electrically conducting fluids 584
Theory of sound, BIOT 271
  RAYLEIGH 363
  perfect fluids, POISSON 276
Theory of spinning detonation 486
Theory of surfaces, MONGE 255
Theory of waves of finite amplitude, RIEMANN 328
Thermal explosions 21, 114
  theories 484, 504
Thermal precursor shock 32
Thermal radiation 21, 603, 607
Thermal shock(ing) 31
  iceberg 476
Thermobaric bombs 830
Thermochemistry, BERTHELOT 1045
Thermodynamics 1101, 1148–1149, 1180, 1191
  theoretical, RANKINE 1148–1149
Thermoelastic response 724
Thermoelastic theory, DUHAMEL 299
Thermography 84
  liquid crystal surface 701
  surface 1011
Thermohydraulic explosions 20
Thermonuclear bomb (fusion bomb/H-bomb) 117, 524, 593, 608, 651, 659
  Eluklab Island 593
Thermonuclear explosion 426, 491, 598
  primeval 568
Thermonuclear fusion 617, 839
Thermosphere, first entry 530
Third cosmic velocity 620
Three-body problem 1060, 1113
Three-shock theory, von NEUMANN 965, 1133
Thrust-faulting 868
ThrustSSC (supersonic car) 803, 1027
Thumper 720–721
Thunder 53–54, 186, 204, 300, 754–755, 769, 782, 859, 966, 1075, 1115
  propagation velocity 331
  touching of clouds, FRANKLIN 244
  unusual sound phenomena 315
Thunderhouse, FRANKLIN 257
Tidal bores 25, 56, 77, 189–191, 418, 860, 1116

Tidal friction 1040
Tidal interaction 18
Tidal waves 25–26, 651, 1108
  velocity 347
Tides 306, 1040
  tidal bores, lunar cycle 189
TIGHTROPE, final atmospheric U.S. nuclear weapons test 660
Time, quantum of time (Planck time) 69
Time of flight, projectile 123
Time-delay generators 1014
Timelapse video recording 127
TNT 404, 794, 1087
  shock initiation of detonation 664
TNT equivalent energy 404
Toepler law 1186
Toepler schlieren method 128
Toepler-Holtz machine 1185
Tool-use 3
Torino Impact Hazard Scale 813
Tornado Scale, Fujita 722
Tornadoes, nuclear explosions 602
  shock front 581
Torpedo 340, 349, 400
Torsion pendulum 1198
Townley-Boyle law
Traffic shocks 909
  Lighthill-Whitham model 612
  Lighthill-Whitham-Richards model 612
Trajectory of characteristics 255
Transatlantic cable 1156
Transient cavitation 461
Transient discontinuities 5
Transient recorders (transient digitizers) 126
Transition, α to ε 616
Transonic flight 779, 973
Transonic flow 749, 942
Transonic phenomena, unsteady 778
Transonic rocket plane 1022
Transonic shock 972
Transonic small disturbance equation 724
Transonic speeds 521
Transonic tunnel, many-slotted 533
Transparent solids, high-dynamic fracture 607
Transsonicum 660
Transurethral shock lithotripsy 638
Trauzl lead block test (Trauzl Test) 429, 1001
Tremor, minor 433
Triggered snapshot photography 1015
Triggering 126
Trinitrotoluol/trinitrotoluene (TNT) 404, 794, 1087

Trinity Test (atomic explosion, 1945) 988, 1019, 1105
  ignition of first nuclear fission bomb ("The Gadget") 545
Triple point trajectories 130
Triple shock paradox 582
Triplicity of sound 330
Tritonal (80/20 TNT/aluminum) 404
Trumpets of Jericho 70
Tsar Bomba 651
Tsunamis (seismic sea waves) 25, 26, 57, 60, 77, 251, 306, 346, 378, 414, 592, 651, 679, 761, 841, 842, 851, 855, 862–868, 957, 1102
  animation 59
  earthquakes 516, 549
  effects 867, 868
  forerunners 59
  largest trans-Pacific 549
  Lituya Bay 627
  local (near-field tsunami) 361
  magnitude 27
  maximum wave height at the coast 578
  mega-tsunami 627
  propagation velocity 307
  research 761, 864, 865
  runup height 27
  slides 441
  suboceanic landslides 320, 419
  teletsunamis (far-field tsunamis) 361, 865
  volcanic 866
Tu-144, supersonic (140 passengers) 699
Tube camera, high-speed multiframe image 684
Tube wind tunnel 613
Tunguska (meteorite) 63, 171, 442, 446, 830
Tupolev's airplanes 800, 1187
Turbojet-propelled aircraft 515
Turbulance 839, 1070, 1104, 1134, 1138, 1164, 1183
Turbulent flow 382
Turbulent wake (vortex wake) 294
Two-digit binary adder 141
Two-dimensional (2-D) linearized wing theory 473
Two-jet process 744
Tycho's nova 199

# U

U Gem (Geminorum), accretion disk 318
U-wave (focused boom), sonic boom 553
U.S. Bureau of Mines (USBM) 447

U.S. National Research Council
    (NRC)  458
U-235, enrichment  517
UD2T target, neutrons  630
Ultra γ-radiation  451
Ultrafast image tube camera
    technology  131
Ultrafast multiple digital framing
    cameras  131
Ultramundane corpuscles, gravity  262
Ultra-relativistic matter  805
Ultrasonic waves,
    biological effects  1193
Ultrasonics  1193
Ultraviolet  603, 635, 649, 736, 776
Ultraviolet photography  1192–1193
    WOOD  1192
Uncorking (volcanic)  753
Uncorking effect  106
Underground nuclear explosions/
    tests  652
    reactor-grade plutonium  660
Underwater detonations,
    interference effects of spherical
    shock waves  537
Underwater explosions  90, 349, 371,
    470, 581
    bubble pulsations  985
    COLE  566
    gas bubble expansion  470
    gravity migration  458
    jet formation  458
    oscillation of the gas globe
      ("bubble")  526
    pressure-time profiles  471
    theory on the shock wave  531
    von Neumann reflections  583
Underwater explosive gun  620
Underwater nuclear weapons tests  550
Underwater shock waves,
    hydrodynamic conical
    shock tube  647
UNDEX (Underwater Explosions)
    Reports  526, 581
Undulation theory  1039, 1148
Unduloid  582
Universal fireball  174
Universe, age of  833
    creation  174
    curvature, EINSTEIN  460
    dynamic cosmological model  478
    expanding  833, 1079, 1082,
      1094, 1113
    finite spherical, EINSTEIN  459
    Friedmann closed  1079
    origin of  746, 783, 795, 1082,
      1102, 1113, 1197
    primordial fireball, relic radiation
      field  685

Universe model  469
University of Toronto Institute
    of Aerophysics (UTIA)  574
Unlined shaped-charge effect  1132
Unsteady compressible viscous
    flow  761
Unsteady waves  733
Unyawed circular cone, pressure  494
Unzen-dake, tsunami  264
Upper atmosphere, temperature/sound
    velocity  475
URAL  144
Uranium, U-235/U-238  505
    exploration program  1169
Urea nitrate  789
Ureteral calculus, disintegration in
    vivo, shock waves  638

# V

V-propagation  486, 818, 911, 1043
V1 (Vengeance Weapon 1)  490, 1163
V2 (Vengeance Weapon 2)  930
Vacuum theory
    (parabolic trajectory theory),
    GALILEI  204
van der Waals equation  355
Variable stars, brightness  642
    eclipsing binary  565
    eruptive (cataclysmic)  564
    explosive (catastrophic)  318, 565
    extrinsic  564
    General Catalogue of Variable Stars
      (GCVS)  564
    intrinsic  564
    pulsating  564
    rotating  565
Variable swept wing  587
Vela, supernova  183
Velikovsky Affair  586
Velocity of sound  285, 309
    in air  155–157, 203, 236, 330, 335
    in water  288, 291
Vengeance Weapon 1 (V1)  490, 1163
Vengeance Weapon 2 (V2)  930
Venturi tube  269, 927
Vesuvius, explosive eruptions  228
Vibrations  240, 348, 562, 653, 707,
    750, 790, 809
    periodic motion  562
Vibroseis  121
Vieille poudre (Poudre B)  1188
Vieille test  1188
Viking 2 (NASA)  736
VIRCATOR
    (Virtual Cathode Oscillator)  648
Vis mortua  223
Vis viva  195, 222, 231, 234, 240, 260
    conservation of energy  274

VISAR (Velocity Interferometer
    System for Any Reflector)  681, 696
Viscoelasticity  344
Viscosity, shock-wave front  667
Visualization  127
Voitenko compressor  676
    shock tube  677
Volcan de Lémery (explosion model
    experiment)  1115
Volcanic blasts  35, 62
Volcanic bombs  62, 304, 489
Volcanic eruption  728, 752–754, 786,
    854–858, 866, 1115, 1179
    extraterrestrial  749
    lateral blast  62
    shock wave dynamics  62
    submarine  862
Volcanic explosivity index
    (VEI)  62, 762
Volcanic tsunamis  27
Volcanic winter  16, 182
Volcanism  777, 1065, 1123–1124
Volcano experiment  1102
Volcano zones  856
Volcanoes, Mars  829–830
    Mount Kilauea-Iki, lava fountains
      (1959)  638
    Mount Pelée (fiery cloud/
      nuée ardente)  428
    Mount Pinatubo  786
    Mount St. Helens  63, 752–754,
      757, 767, 856
    Mount Vesuvius  188, 539
    Mt. Bandai-San  393
    Mt. Tambora  280
    Mt. Toba  182
    Mt. Vulcano  394
    Myojin-sho Reef (1952)  592
    Perbuatan (Krakatau)  378
    Santorini  184
    Stromboli  489
    supervolcano  762
    Taal,
      phreatomagmatic eruptions  678
    Unzen-dake, tsunami  264
Volta cannon  257
Volta cell  1065
Volta pistol  257
von Foerster (Förster) effect  380
von Kármán Institute for Fluid
    Dynamics (VKI)  615
von Kármán ogive  526
von Kármán vortex street  1104
von Neumann architecture  143
von Neumann criterion
    (detachment criterion)  535
von Neumann paradox
    (Birkhoff triple shock paradox)
    536, 548, 582

von Neumann reflection (von Neumann-Mach reflection) 582
von Neumann spike 532, 601
von Neumann state 532
Vostok 1 (spacecraft) 651, 1025, 1081
   first human to travel at hypersonic speeds 103, 651
Voyager 1 836, 840
Voyager 2 779–780, 836
Vredefort Basin 171, 176
Vredefort Dome 176
Vredefort Ring 65, 646

# W

WAC-Corporal 574
Wake pattern 294
War machines, early 879
War of Engineers 196
Washington Meteorite (*1873*) 361
Water arc explosions 769
Water hammer (water ram) 89
   hydraulic shocks 508
Water hammer effect 89, 141, 1199
Water-hammer pressure 418
Water-hammer waves, plumbing systems 418
Water impact waves 952
Water pipes, propagation of hydraulic shocks 418
Water ricochets 91
Water table studies (hydraulic jumps) 957
Water waves 77, 624
   CAUCHY 282
   reflection 559
   shallow 306
   velocity of propagation 296
Wave(s), aerial 76
   body waves 60–61
   cnoidal 296
   elastic 900, 979, 1157, 1178
   elastic-plastic 571
   finite amplitude 381, 728–729, 1040, 1057, 1075, 1086, 1108, 1144, 1152, 1175–1177
   finite height 1149
   finite longitudinal disturbance 1149 ?
   generation 77
   great solitary wave (wave of translation) 78
   longitudinal 412
   N-waves 553
   primary (P-waves) 60
   secondary (shear waves) (S-waves) 60–61, 292, 450, 452

   seismic 727, 862, 905, 979–980, 992, 1123, 1131
   shallow-water 306, 863
   solitary 296
   surface waves 60–61
   tidal bores 25
   transversal 412
   unsteady 733
Wave drag 72, 494, 973
Wave-line design/theory 1156
Wave of translation (great solitary wave/solitary wave of translation) 296, 306–307, 1156
Wave reflection, irregular 306
   regular 730, 1134
   studies 1193
Wave speed camera 131, 493, 1017, 1143
Wave theory, elastic percussion 256
   of light 1097, 1176, 1193
Wavefront method 970
Wavefront reconstruction (lensless photography) 562
Wavefronts 222
   overtaking effect 315
Wavelets (outgoing spherical waves) 222
Waverider, lifting system 669
Weak shock theory 55, 619
Weak shock waves 69, 346, 424–425, 566
Weak-shock solution 532
Weapons 879–882, 886–887
Wehnelt cylinder 1159
Weight timing system, dropping, electrically triggerable 329
Well fracturing 298
Wells, oil 299, 339–340, 689
Wheatstone bridge 131, 954
Wheatstone chronoscope 123, 301, 1012
Whip-cracking 69, 477, 796, 914, 950
   Kucharski effect 527
   supersonic 437
Whiplash 52, 483
Whipple shield 977
   spacecraft protection 563
Whiskey/black powder, 100-proof 108
White dwarfs 333, 455, 469, 491
Whitham equation 586
Whitham rule 636
Wiechert seismograph 121, 420
Wind channels 1172
Wind-gun 196, 208, 210, 880
Wind tunnels 83, 352, 927, 938–943, 1038, 1064, 1070, 1077, 1088, 1103, 1110, 1138, 1147, 1172, 1183, 1186, 1191–1192, 1199

   choking 559
   closed-circuit 442
   continuous-flow closed-loop supersonic 497
   Göttingen-type 1147
   high-enthalpy hypersonic 558
   high-speed 462
   hypersonic 84, 645
   tailored interface 642
   hypersonic-flow 561–562, 588
   hypervelocity 85
   intermittent supersonic 588
   Japanese 478
   Prandtl 1147
   slotted throat 591
   Stalker tube 686
   subsonic 83
   supersonic 83, 462, 468, 498, 502–503, 505, 507–508, 516, 540
   tests, hypersonic, Mach 9 538
   transonic 83
   tube 613
Windrohr/Windbüchse ("wind-gun"/"air-gun") 196
Wing(s), supercritical 688
   swept-back, high-speed aviation 516
Wing sections, sonic/supersonic speeds 469
Wing theory, two-dimensional (2-D) linearized 473
Wire gun 402
Wire(s), exploding, detonation of nitrogen chloride/nitroglycerin 510
   exploding, electric fuses 582
   electrical 256
   fragmentation, striations 582
   schlieren camera 624
   underwater 662
   unduloids 674
   explosions 270, 283, 318, 465, 582, 637
   electric fuses 582
   high pulsed currents 675
Wollaston prism 274
Wood oscillograph 1013
Wood-type oscillograph 126
Woodpeckers, hammering mechanism 52, 599
World Trade Center (WTC) 789, 825–826, 1032, 1033
World-Wide Standardized Seismograph Network (WWSSN) 652
Worthington crown 413
Worthington jet 413, 827
Wound ballistics 387
Wrecking ball 884

## X

X-15 rocket plane, hypersonic 652
X-ray diffraction analysis 133
    flash 725–726, 732, 744, 764, 786–787, 812, 955, 1159
X-ray pulses, high-intensity 509
X-ray systems, multiple flash 625
X-ray tube, flash 962, 996, 1009, 1063
X-rays,
    RÖNTGEN 414
    MCPs 489
X-shock 968, 1192

## Y

Year of Earthquakes, England (*1750*) 245
Yellowstone Caldera 181
Yield stress 98

## Z

Z-pinch, shock compression 509
Zel'dovich-von Neumann-Döring (ZND) theory 112, 522, 532, 534, 716, 1044, 1134, 1196
    nonequilibrium (NE-ZND) 762
Zemplén theorem 438, 1198
Zhukovsky profile 1199
ZND model, one-dimensional nonequilibrium 113
Zone of normal/abnormal audibility 447
Zone of action 454
Zones of silence 447, 454, 922
Zündnadelgewehr (needle gun) 288